THE ENCYCLOPEDIA OF

Chemistry

THE ENCYCLOPEDIA OF

Chemistry

THIRD EDITION

Edited by

CLIFFORD A. HAMPEL

Consulting Chemical Engineer
Skokie, Illinois

GESSNER G. HAWLEY

Formerly Executive Editor
Reinhold Publishing Corporation

Editor Condensed Chemical Dictionary
Eighth Edition

VNR VAN NOSTRAND REINHOLD COMPANY

NEW YORK   CINCINNATI   ATLANTA   DALLAS   SAN FRANCISCO
LONDON   TORONTO   MELBOURNE

Van Nostrand Reinhold Company Regional Offices:
New York   Cincinnati   Chicago   Millbrae   Dallas

Van Nostrand Reinhold Company International Offices:
London   Toronto   Melbourne

Manufactured in the United States of America

Published by Van Nostrand Reinhold Company
135 West 50th Street, New York, N.Y. 10020

Published simultaneously in Canada by Van Nostrand Reinhold Ltd.

15   14   13   12   11   10   9   8   7   6   5   4

Library of Congress Cataloging in Publication Data
Main entry under title:

The Encyclopedia of chemistry.

   Second ed., 1966, edited by G. L. Clark and G. G. Hawley.
   1. Chemistry—Dictionaries. I. Hampel, Clifford A., ed. II. Hawley, Gessner Goodrich, 1905-   ed.
QD5.E58 1973   540'.3   73-244
ISBN 0-442-23095-8

CONTRIBUTORS

ABELL, JERROLD J., Rogers Corporation, Rogers, Conn., *Poromeric Materials*.

ABRAHAM, E. P., Oxford University, Oxford, England, *Penicillin*.

ADAMS, ROY M., Geneva College, Beaver Falls, Pa. *Boron*.

ADLER, GEORGE, Dept. of Applied Science, Brookhaven National Laboratory, Upton, N.Y. *Crosslinking of Polymers*.

AKIN, RUSSELL B., Plastics Dept., E. I. DuPont de Nemours & Co., Wilmington, Dela. *Acetal Resins*.

ALBRIGHT, PENROSE S., Wichita, Kans., *Electroplating*.

ALLEN, DOUGLAS L., U.S. Industrial Chemicals Co. Tuscola, Ill., *Petrochemical Feedstocks; Vapor Pressure*.

ALLEN, M. J., Shell Research Ltd., Braun Microbiological Lab., Sittingbourne, Kent, England, *Electrodes, Organic Electrochemical*.

ALLYN, GEROULD, Rohm and Haas Co., Philadelphia, Pa., *Acrylic Resins*.

ALPER, CARL, Bio-Science Laboratories, Philadelphia, Pa., *Blood*.

ALSTON, R. H. (Deceased), *Chemotaxonomy*.

AMIS, EDWARD S., Dept. of Chemistry, University of Arkansas, Fayetteville, Ark., *Dissociation; Transference Numbers; Activity Coefficients*.

AMPHLETT, C. B., Atomic Energy Research Establishment, Harwell, England, *Ion Exchange in Inorganic Materials*.

ANDERSON, DUWAYNE, M., U.S. Army Cold Regions Lab., Hanover, N. H., *Thixotropy*.

APPELMAN, E. H., Argonne National Laboratory, Argonne, Ill., *Astatine*.

ASIMOV, ISAAC, New York, N.Y., *Genetic Code; Nucleic Acids*

BAAB, KENNETH A., Homer Research Laboratories, Bethlehem Steel Corp., Bethlehem, Pa., *Refractories*.

BABCOCK, C. S., National Fire Protection Assn., Boston, Mass., *Extinguishing Agents*.

BACK, R. A., National Research Council of Canada, Ottawa, Ontario, *Free Radicals*.

BACKMAN, JULES, Dept. of Economics, New York University, New York, N.Y., *Economics of the Chemical Industry*.

BACON, F. E., Mining and Metals Div., Union Carbide Corp., Niagara Falls, N.Y., *Chromium*.

BAILAR, JOHN C., JR., Dept of Chemistry, University of Illinois, Urbana, Ill., *Chelation; Sequestering Agents*.

BARKSDALE, JELKS, Auburn, Ala., *Titanium*.

BARNARD, A. J., JR., J. T. Baker Chemical Co., Phillipsburg, N.J., *Complexing Agents in Chemical Analysis*.

BARNES, ROBERT K., Union Carbide Corp., So. Charleston, West Va., *Glycols*.

BARON, SAMUEL, Laboratory of Viral Diseases, National Institutes of Health, Bethesda, Md., *Interferon*.

BARR, ROBERT Q., Climax Molybdenum Co., New York, N.Y., *Molybdenum*.

BARTHAUER, G. L., Consolidation Coal Co., Library, Pa., *Gas Laws*.

BARTLETT, EDWIN S., Battelle-Columbus Laboratories, Columbus, Ohio, *Niobium*.

BEDOUKIAN, PAUL Z., Hastings, New York, *Perfumes*.

BEECHAM, CHARLES R., Homer Research Laboratories, Bethlehem Steel Corp., Bethlehem, Pa., *Refractories*.

BEMENT, ARDEN L., JR., Professor of Nuclear Materials, Mass. Institute of Technology, Cambridge, Mass., *Biomaterials*.

BENT, RICHARD L., Eastman Kodak Co., Rochester, New York, *Steric Hindrance*.

BERG, DANIEL, Westinghouse Research Laboratories, Pittsburgh, Pa., *Ammonia*.

BERGMAN, PAUL, General Electric Co., West Lynn, Mass., *Steel*.

BERL, WALTER G., Johns Hopkins University, Silver Spring, Md., *Combustion*.

BERNIER, CHARLES L., School of Information and Library Studies, State Univ. of New York at Buffalo, Buffalo, N.Y. *Chemical Literature*.

BERTIN, EUGENE, RCA, Princeton, N.J., *Getters*.

BESANCON, R. M., Physical Sciences Administrator, U.S. Air Force Materials Laboratory, Wright-Patterson Air Force Base, Dayton, Ohio, *Atoms*.

BESTE, LAWRENCE F., Rapid City, S.D., *Gibbs-Duhem Equation*.

BEVINGTON, J. C., Dept. of Chemistry, University of Lancaster, Lancaster, England, *Radical Polymerization*.

BEYER, GEORGE L., Webster, New York, *Molecular Weight*.

BIKERMAN, J. J., Shaker Heights, Ohio, *Diffusion*.

BJORKSTEN, JOHAN, Madison, Wisconsin, *Polyester Resins*.

BLANDER, MILTON, North American Rockwell Co., Thousand Oaks, Cal., *Molten Salt Chemistry*.

BONDI, A. A., Shell Development Co., Bellaire Research Center, Houston Tex., *Lubrication*.

BONER, C. J., Southwest Grease & Oil Co., Inc., Kansas City, Mo., *Lubricating Greases*.

BONNEY, DONALD T., (Address unknown), *Gas Analysis*.

BOOTS, MARVIN R., Associate Professor, Dept. of Chemistry & Pharmaceutical Chemistry, Virginia Commonwealth University, Richmond, Virginia, *Esters; Thiols.*

BOOTS, SHARON G., Research Fellow, Dept. of Chemistry and Pharmaceutical Chemistry, Virginia Commonwealth University, Richmond, Va., *Esters; Thiols.*

BORASKY, RUBIN, Silver Spring, Md. *Collagen.*

BRADLEY, W. W., Alcoa Professor, Institute of Colloid and Surface Science and Department of Civil and Environmental Engineering, Clarkson College of Technology, Potsdam, New York, *Cementation.*

BRANDT, WARREN W., Virginia Commonwealth University, Richmond, Va., *Amperometric Titration.*

BREWER, JEROME (Address unknown), *Tritium.*

BREWER, LEO, University of California, Lawrence Berkeley Laboratory, Berkeley, Cal., *Nonstoichiometric Compounds.*

BREWSTER, JAMES H., Purdue University, W. Lafayette, Ind., *Infrared Absorption Spectroscopy.*

BRIDGERS, HENRY E., Bell Telephone Laboratories, Allentown, Pa., *Germanium; Semiconductors.*

BRIGGS, J. Z., Climax Molybdenum Co., New York, N.Y. *Molybdenum.*

BRODIE, ARNOLD F., University of Southern California, School of Medicine, Los Angeles, Cal., *Venoms.*

BROOKER, L. G. S., Rochester, N.Y., *Sensitizing Dyes.*

BROWN, CHRISTINE F., University of Tennessee Medical School, Memphis, Tenn., *Hallucinogenic Drugs.*

BROWN, THEODORE M., Dept. of Chemistry, Arizona State University, Tempe, Ariz., *Carbonyls, Metal.*

BROWNING, B. L., Appleton, Wis., *Moisture Content.*

BRUNAUER, STEPHEN, Clarkson Institute of Technology, Postdam, N.Y., *BET Theory.*

BRUNER, WALTER M., Plastics Dept., E. I. DuPont de Nemours & Co., Wilmington, Del., *Acetal Resins.*

BURCHFIELD, H. P., Gulf South Research Institute, New Iberia, La., *Hydroquinones; Quinones.*

BURLANT, W. J., Ford Motor Co., Dearborn, Mich., *Block and Graft Polymers; Ladder Polymers.*

BURNS, ROGER G., University of California, Berkeley, Cal., *Crystal Field Theory.*

BURTNER, R. R. (Retired), G. D. Searle & Co., Chicago, Ill., *Inhibitors, Gastric.*

BUSH, J. D., (Address unknown), *Molality.*

BUTTS, ALLISON, Dept. of Metallurgy and Materials Science, Lehigh University, Bethlehem, Pa., *Copper; Silver.*

CADLE, RICHARD D., Head, Chemistry Dept., National Center for Atmospheric Research, Boulder, Colo., *Aerosols; Particle Size.*

CAGLE, F. WILLIAM, JR., Dept. of Chemistry, University of Utah, Salt Lake City, Utah, *Reaction Rates.*

CALEY, EARLE R., Dept. of Chemistry, Ohio State University, Columbus, Ohio, *Archeological Chemistry; Chemical Dating; Reagents.*

CALVERT, ROBERT (Deceased), *Legal Chemistry; Patents*

CAMP, THOMAS R. (Deceased), Camp, Dresser & McKee, *Water.*

CAMPAIGNE, E., Dept. of Chemistry, University of Indiana, Bloomington, Ind., *Antihistamines.*

CANNON, MARTIN, Product Development Division, Procter & Gamble Co., Cincinnati, Ohio, *Detergents, Synthetic.*

CARAPELLA, S. C., JR., American Smelting and Refining Co. Central Research Laboratories, South Plainfield, N.J., *Antimony; Arsenic; Bismuth; Thallium.*

CARPENTER, FRANK J., Cane Sugar Refining Research Project, Inc., New Orleans, La., *Sampling.*

CARPENTER, RICHARD A., Chief, Environmental Policy Div., Library of Congress, Washington, D.C., *Syneresis; Synergism.*

CARTER, FRED E. (Retired), Maplewood, N.J., *Gold.*

CASIDA, JOHN E., Professor of Entomology, University of California, Berkeley, Cal., *Insecticides.*

CAUGHEY, W. S., University of Southern Florida, Tampa, Fla., *Porphyrins.*

CHAGNON, CHARLES W., Air Force Cambridge Research Lab., Bedford, Mass., *Absorbents.*

CHAPPELOW, C. C., Midwest Research Institute, Kansas City, Mo., *Acid Number.*

CHRISTMANN, L. J. (Deceased), *Technology of Nitrogen Compounds.*

CLAPP, LEALLYN B., Dept. of Chemistry, Brown University, Providence, R. I., *Indicators; Nitrogen and Compounds; Radicals.*

CLARK, F. M. (Deceased), *Dielectric Materials.*

CLARK, GEORGE L. (Deceased), *Conjugation; Radiation; Particles and Antiparticles; Radioactivity; Wittig Reaction.*

CLAUSER, HENRY R., Consultant, Armonk, N.Y., *Cermets; Composites.*

CLEVENGER, SARAH, Bloomington, Ind., *Flower Pigments.*

COHN, ERNST M., Manager, Solar and Chemical Power, NASA, Code RPP. Washington, D.C., *Fischer-Tropsch Process; Fuel.*

COLBURN, CHARLES B., Auburn, Ala., *Nitrogen-Fluorine Chemistry.*

COLL, HANS, Shell Development Co., Emeryville, Cal., *Osmosis.*

COOK, GERHARD A., Dept. of Chemistry, State University of New York, Buffalo, N.Y., *Oxygen.*

COOK, M. A., Salt Lake City, Utah, *Azides; Chlorates and Perchlorates; Explosives; Nitrates.*

COOK, RALPH L., Dept. of Ceramic Engineering, University of Illinois, Urbana, Ill., *Ceramics; Glazes; Porcelain Enamels.*

COOK, WARREN A., Professor Emeritus of Industrial Health, University of Michigan and Adjunct Professor of Industrial Health, University of North Carolina, Chapel Hill, N.C., *Noxious Gases.*

COOPER, W. CHARLES, United Nations Development Program, Santiago, Chile, *Tellurium.*

COPELAND, L. E., Portland Cement Association, Skokie, Ill., *Cement, Portland.*

CORWIN, ALSOPH H., Dept. of Chemistry, Johns Hopkins University, Baltimore, Md., *Chlorophylls; Pyrroles.*

CRAIG, LYMAN C., Rockefeller University, New York, N.Y., *Impurities.*

CRATTY, L. E., JR., Dept. of Chemistry, Hamilton College, Clinton, N.Y., *Chemisorption.*

CROVETTI, A. J., Manager, Agricultural Chemicals Research, Abbott Laboratories, North Chicago, Ill., *Repellents, Insect.*

CURTIN, D. Y., University of Illinois, Urbana Ill., *Tautomerism.*

DAANE, ADRIAN H., Dean, College of Arts & Sciences, University of Missouri-Rolla, Rolla, Mo., *Scandium; Yttrium.*

DANIELS, RALPH, University of Illinois Medical Center, College of Pharmacy, Chicago, Ill., *Quinoline.*

DAVIDSON, ROBERT L., Princeton, N.J., *Synthetic Water-soluble Resins.*

DAVIS, DALE S., Bailey Island, Me., *Nomography.*

DEATON, WILLIAM D., Amarillo, Tex., *Helium.*

DeBETHUNE, ANDRE J., Dept. of Chemistry, Boston College, Chestnut Hill, Mass., *Electrode Potentials; Temperature Scales; Frost Diagrams.*

DEHNE, G. CLARK, Dept. of Chemistry, Capital University, Columbus, Ohio, *Spectroscopy, Absorption.*

DEITZ, VICTOR R., Naval Research Laboratory, Washington, D.C., *Sampling.*

DEICHMANN, WILLIAM B., University of Miami School of Medicine, Coral Gables, Fla., *Toxicology.*

DELMONTE, JOHN, President, Furane Plastics, Inc., Los Angeles, Cal., *Furane Resins.*

DEMONSABERT, WINSTON R., Scientist Administrator, Dept. of Health, Education & Welfare, Health Services & Mental Health Administration, Rockville, Md., *Zirconium.*

DENNEY, D. B., Rutgers University, New Brunswick, N.J., *Organophosphorus Compounds.*

DENO, NORMAN C., Pennsylvania State Univ., University Park, Pa., *Carbonium Ions and Carbanions; Nitration.*

DE RADZITSKY, P., Research Manager, Labofina, Brussels, Belgium, *Clathrate Compounds.*

DEXTER, T. H., Hooker Research Center, Niagara Falls, N.Y., *Phosphorus Sulfides.*

DiCARLO, FREDERICK J., Warner-Lambert Research Institute, Morris Plains, N.J., *Indoles.*

DIEHL, HARVEY, Iowa State University, Ames, Iowa, *Standards.*

DIETZ, ALBERT A., Veterans Administration Hospital, Hines, Ill., *Cereal Chemistry.*

DILLON, J. H., Seneca, S.C., *Fibers, Synthetic.*

DiMASCIO, ALBERTO, Mass. Dept. of Mental Health, Boston, Mass., *Psychotropic Drugs.*

DOBRY, ALAN, American Oil Co., Whiting, Ind., *Antioxidants for Lubricants.*

DONDES, SEYMOUR, Dept. of Chemistry, Rensselaer Polytechnic Institute, Troy, N.Y., *Nitrogen Fixation.*

DORFMAN, RALPH I., Syntex Research, Palo Alto, Cal., *Hormones, Steroid.*

DRUDING, LEONARD F., Dept. of Chemistry, Newark College of Arts and Sciences, Newark, N.J., *Metal-Metal Salt Solutions.*

DuBOIS, KENNETH P., Director, Toxicity Laboratory, University of Chicago, Chicago, Ill., *Cholinesterase Inhibitors.*

DUECKER, W. W., (Formerly with Texas Gulf Sulphur Co.), *Sulfur and Compounds.*

DUKE, F. R., University of Iowa, Ames, Iowa, *Conductometric Titrations.*

DUNLOP, A. P., Director of Chemical Research & Development, Quaker Oats Co., Barrington, Ill., *Furans.*

DUNSTON, JOYCE M., National Research Council of Canada, Ottawa, Ontario, *Fluorescence.*

DUX, J. P., FMC Corporation, Marcus Hook, Pa., *Acetates.*

EDELSON, D., Bell Telephone Laboratories, Murray Hill, N.J., *Polar Molecules.*

EDGELL, WALTER F., Purdue University, W. Lafayette, Ind., *Infrared Absorption Spectroscopy.*

EDSALL, JOHN T., Biological Laboratory, Harvard University, Cambridge, Mass., *Proteins.*

EDWARDS, M. B., Longview, Tex., *Polyallomer Resins.*

EGGERT, E. F., (Address unknown), *Fluorocarbon Resins.*

EHMAN, P. J., Ansul Company Research Center, Madison, Wis., *Chlorinated Hydrocarbons.*

EHMANN, WILLIAM D., Dept. of Chemistry, University of Kentucky, Lexington, Ky., *Elements, Abundance of.*

EISCH, J. J., Department of Chemistry, State University of New York at Binghampton, Binghamton, N.Y., *Lithium in Organic Chemistry; Organometallic Compounds.*

ELIEL, E. L., Kenan Professor of Chemistry, Department of Chemistry, University of North Carolina, Chapel Hill, N.C., *Conformational Analysis.*

ELLIOTT, STANLEY B., Bedford, Ohio, *Driers; Metallic Soaps; Saponification.*

EMMETT, PAUL H., Milwaukie, Ore., *Adsorption; Catalysis.*

ETHERINGTON, T. L., Lighting Systems Dept., General Electric Co., Hendersonville, N.C., *Refrigerants.*

EWING, W. W. (Deceased), *Conductance.*

EYRING, HENRY, Distinguished Professor, University of Utah, Salt Lake City, Utah, *Reaction Rates.*

FAIRBANK, HENRY A., Dept. of Physics, Duke University, Durham, N.C., *Argon; Neon; Xenon.*

FALLER, LARRY, Dept. of Chemistry, Wesleyan University, Middletown, Conn., *Enzymes.*

FARR, WANDA, Nanuet, N.Y., *Cytochemistry; Histochemistry.*

FARRIS, RUSSELL E., GAF Corp., Rennselaer, N.Y., *Diazotization.*

FAUST, CHARLES L., Battelle-Columbus Institute, Columbus, Ohio, *Pickling (Metals).*

FAUST, GEORGE T., Silver Spring, Md., *Mineralogy, Chemical.*

FERGUSON, HUGH E., Mobil Research and Development Corp., Princeton, N.J., *Fuel Gases.*

FERINGTON, T. E., Washington Research Center, W. R. Grace & Co., Clarksville, Md., *Polymers, Inorganic and Organic.*

FIESER, L. F. and FIESER, MARY, Dept. of Chemistry, Harvard University, Cambridge, Mass., *Gatterman Synthesis; Clemmenson Reaction; Knoevenagel Reaction; Reformatsky Reaction; Willgerodt Reaction; Wolff-Kishner Reaction; Wurtz Reaction.*

FILLER, ROBERT, Dept. of Chemistry, Illinois Institute of Technology, Chicago Ill., *Parachor.*

FISCHER, ALBRECHT G., Pittsburgh, Pa., *Electroluminescence.*

FISCHER, ROBERT B., California State College, Dominguez Hills, Cal., *Spectroscopy, Microwave.*

FOLEY, DENNIS D., Cheshire, Conn., *Solvent Extraction.*

FOLKINS, H. O., Research Dept., Union Oil Co. of California, Brea, Cal., *Alkylation.*

FONDA, G. R., (Address unknown), *Luminescence.*

FOSTER, J. F., Battelle-Columbus Laboratories, Columbus, Ohio, *Films, Surface.*

FOSTER, L. M., IBM Research Center, Yorktown, Heights, N.Y., *Gallium.*

FOWLER, FRANK C., Kansas City, Mo., *Partition.*

FRANKLIN, J. L., Rice University, Houston, Tex., *Ions, Gaseous.*

FRANTA, WILLIAM, A., Director, R & D. Division, Plastics Dept., E. I. DuPont de Nemours & Co., Wilmington, Del., *Industrial Research.*

FRAZIER, JOHN C., Kansas Agricultural Experiment Station, Manhattan, Kan., *Seeds.*

FRIEDMAN, A. M., Argonne National Laboratory, Argonne, Ill., *Radium.*

FREEDMAN, M. L., Great Lakes Research Corp., Elizabethton, Tenn., *Refractory Carbides.*

FRISTROM, R. M., Applied Physics Laboratory, John Hopkins University, Silver Spring, Md., *Flame Chemistry.*

FRITZ, JAMES S., Iowa State University, Ames, Iowa, *Nonaqueous Titrations.*

FYFE, WILLIAM S., Manchester University, Manchester, England, *Crystal Field Theory.*

GADBERRY, HOWARD M., Midwest Research Institute, Kansas City, Mo., *Dispersing Agents; Fumigants.*

GAFFRON, HANS, Institute of Molecular Biophysics, Florida State University, Tallahassee, Fla., *Carbon Cycle; Photosynthesis.*

GALLO, DUANE G., Pharmacology Dept., Mead Johnson Research Center, Evansville, Ind., *Amino Acids.*

GALLONE, PATRIZIO, Professor of Electrochemistry, University of Genoa, Genoa, Italy, *Chlorine.*

GANS, DAVID M., Coatings Research Group, Inc., Shaker Heights, Ohio, *Binders.*

GARN, PAUL D., University of Akron, Akron, Ohio, *Electrolysis.*

GEARIEN, JAMES E., University of Illinois Medical Center, Chicago, Ill., *Antiseptics; Disinfectants.*

GIARDINI, A. A., Dept. of Geology, University of Georgia, Athens, Ga., *Diamond Synthesis.*

GILL, S. J., University of Colorado, Boulder, Colo., *Chemical Potential.*

GILMAN, HENRY, Dept. of Chemistry, Iowa State University, Ames, Iowa, *Lithium in Organic Chemistry; Organometallic Compounds.*

GILMONT, ROGER, Roger Gilmont Instruments, Inc., Great Neck, N.Y., *Kinetics, Chemical.*

GODDARD, E. D., Research and Development Division, Lever Brothers Co., Edgewater, N.J., *Soaps.*

GOLDBERG, RICHARD J., Chairman of the Board, Dymat International Corp., Sherman Oaks, Cal., *Antibodies and Antigens.*

GOLDMAN, LEON, Children's Hospital Research Foundation, Cincinnati, Ohio, *Lasers in Chemistry.*

GOLUB, MORTON A., Ames Research Center, Moffett Field, Cal., *Cyclized Rubber.*

GOOD, R. H., JR., Department of Physics, Pennsylvania State University, University Park, Pa., *Photon.*

GOODSON, LOUIS H., Midwest Research Institute, Kansas City, Mo., *Racemization.*

GOULD, DAVID H., Merck Sharp & Dohme International, Rahway, N.J., *Ethers.*

GRAY, G. W., Dept. of Chemistry, University of Hull, Hull, England, *Liquid Crystals.*

GREENE, C. H., College of Ceramics, State University of N.Y., Alfred, N.Y., *Glass.*

GREENE, JANICE L., Standard Oil Co., Cleveland, Ohio, *Nitriles.*

GRIER, NATHANIEL, Merck Sharp & Dohme Research Laboratories, Rahway, N.J., *Mercury and Compounds.*

GRIFFIN, D. J., Dow Chemical Co., Midland, Mich., *Polystyrene.*

GRIFFITTS, F. A., Huntingdon College, Montgomery, Ala., *Chemical Nomenclature; Poison, Catalyst.*

GRINSTEAD, R. R., Dow Chemical Co., Walnut Creek, Cal., *Copper, Biochemical Behavior.*

GROSS, WILLIAM H. (Retired), Dow Chemical Co., Midland, Mich., *Magnesium.*

GROSSWEINER, LEONARD I., Dept. of Physics, Illinois Inst. of Technology, Chicago, Ill., *Flash Photolysis; Photosensitization in Solids.*

GROTZ, LEONARD C., University of Wisconsin, Waukesha, Wis., *Heterogeneous Equilibria.*

GUENTHER, ERNEST, Fritzsche Bros., New York, N.Y., *Essential Oils.*

GUINN, VINCENT P., Dept. of Chemistry, University of California, Irvine, Cal., *Neutron Activation Analysis.*

GUNDERSON, REIGH C., Dow Chemical Co., Midland, Mich., *Lubricants, Synthetic.*

GURLEY, MARTIN H., JR., Lexington, Va., *Textile Processing.*

GUTSCHE, C. DAVID, Dept. of Chemistry, Washington University, St. Louis, Mo., *Diazoalkanes.*

HADLEY, ELBERT, H., Southern Illinois University, Carbondale, Ill., *Aldehydes; Organic Chemistry.*

HAGAN, M. A., Hagan Associates, Palos Verdes, Cal., *Fluxes.*

HAGEMEYER, H. J., Texas Eastman Co., Longview, Tex., *Polyallomer Resins.*

HALL, CARL W., Dean, College of Engineering, Washington State University, Pullman, Wash., *Pasteurization.*

HALLETT, JOHN V., Imperial Color Division, Hercules Co., Glens Falls, N.Y., *Lakes.*

HAMPEL, CLIFFORD A., Skokie, Ill., *Barium; Hafnium; Acetylene; Iron; Manganese; Cesium; Chromium; Powdered Metals; Rubidium; Strontium; Rare Metals; Flame Retardants; Natural Gas; Organic Carbonates.*

HANSEN, ROBERT S., Director, Ames Laboratory USAEC, Iowa State University, Ames, Iowa, *Colloid Chemistry.*

HARDING-BARLOW, I., Palo Alto, Cal., *Ultramicroanalysis with Laser.*

HARDY, W. B., American Cyanamid Co., Bound Brook, N.J., *Ultraviolet Stabilizers.*

HARKIN, JOHN M., Forest Products Labortory, Madison, Wis., *Lignin, Biosynethetic.*

HARLEY, JOHN H., Director Health and Safety Lab., U.S. Atomic Energy Commission, New York, N.Y., *Dosimetry; Analytical Chemistry of Radioactive Elements; Waste Treatment, Radioactive.*

HART, EDWIN J., Argonne National Laboratory, Argonne, Ill., *Hydrated Electron.*

HARTECK, PAUL, Dept. of Chemistry, Rennsselaer Polytechnic Institute, Troy, N.Y., *Nitrogen Fixation.*

HARWOOD, H. JAMES, Institute of Polymer Science, University of Akron, Akron, Ohio, *Copolymers.*

HATFIELD, W. D., Decatur, Ill., *Activated Sludge, Waste Treatment.*

HAUSLER, R. H., Universal Oil Products Co., Des Plaines, Ill., *Free Energy.*

HAWKINS, W. LINCOLN, Bell Telephone Laboratories, Murray Hill, N.J., *Autoxidation.*

HAWLEY, G. G., Newton Center, Mass., *Combustible and Flammable Materials.*

HAY, A. S., Chemical Laboratory, General Electric Co., Schenectady, N.Y., *Polymerization by Oxidative Coupling.*

HEFTMANN, ERICH, Orinda, Cal., *Bile Acids; Chromatography; Cholesterol; Glycosides, Steroid.*

HEINE, H. W., Bucknell University, Lewisburg, Pa., *Chlorohydrins.*

HELBIG, W. A., Wilmington, Del., *Carbon, Activated.*

HEMEON, W. C. L., Hemeon Associates, Pittsburgh, Pa., *Air Pollution; Dusts, Industrial.*

HENN, R. W., Research Laboratories, Eastman Kodak Co., Rochester, N.Y., *Developers.*

HENTZ, ROBERT R., Radiation Laboratory, University of Notre Dame, Notre Dame, Ind., *Radiation Chemistry.*

HERSH, CHARLES K., Research Director, Sonneborn Division, Witco Chemical Co., New York, N.Y., *Molecular Sieves; Ozone; Drying; Ozonization.*

HICKSON, JOHN L., International Sugar Research Fdn. Inc., Bethesda, Md., *Sugars.*

HILDEBRAND, JOEL H., Dept. of Chemistry, University of California, Berkeley, Cal., *Solutions.*

HILL, ROY O., JR., Eastman Chemical Products, Inc., Kingsport, Tenn., *Cellulose Ester Plastics.*

HILLYER, J. C. (Retired), Research and Development Dept., Phillips Petroleum Co., Bartlesville, Okla., *Dienes.*

HINE, JACK, Dept. of Chemistry, Ohio State University, Columbus, Ohio, *Methylenes.*

HIRSCH, H. E., Cominco Ltd., Trail, B.C. Canada, *Indium.*

HIRSCHLER, D. A., Ethyl Corp., Ferndale, Mich., *Antiknock Agents.*

HITCHCOCK, DAVID I., New Haven, Conn., *Buffers; pH; Membrane Equilibrium.*

HOFER, L. J. E., Mellon Institute, Pittsburgh, Pa., *Carbides.*

HOLAPPA, H. S., President, H. S. Holappa & Associates, Lynnfield, Mass., *Antistatic Resins.*

HOOD, C. B., Mobil Research and Development Corp., Paulsboro, N.J., *Gasoline.*

HOOGENBOOM, B. E., Gustavus Adolphus College, St. Peter, Minn., *Organic Chemistry, Reactions in.*

HOWE, HERBERT E., American Smelting and Refining Co., South Plainfield, N.J., *Bismuth; Thallium.*

HUDGIN, D. E., Princeton Chemical Research, Summit, N.J., *Ketones.*

HUGGINS, MAURICE L., Woodside, Cal., *Valence.*

HUMM, HAROLD J., Marine Science Institute, University of South Florida, St. Petersburg, Fla., *Phycocolloids; Oil on the Sea.*

HULTGREN, R. R., University of California, Berkeley, Cal., *Metals, Classification of.*

HURD, CHARLES D., Northwestern University, Evanston, Ill., *Ketenes; Pyrolysis.*

HURD, DALLAS T., Lamp Metals and Components Dept., General Electric Co., Cleveland, Ohio, *Carbonyl Compounds.*

HYDE, EARL K., Lawrence Radiation Laboratory, University of California, Berkeley, Cal., *Francium.*

HYMAN, HERBERT H., Argonne National Laboratory, Argonne, Ill., *Hexafluoride Molecules; Noble Gas Compounds.*

IFFLAND, DON C., Western Michigan University, Kalamazoo, Mich., *Oximes.*

IHDE, AARON J., Dept. of Chemistry, University of Wisconsin, Madison, Wis., *Isomers and Isomerization.*

INMAN, CHARLES G., Hercules Incorporated, Coatings & Specialty Products Dept., Glens Falls, N.Y., *Pigments.*

INNES, W. B., Upland, Cal., *Promoters.*

INTERNATIONAL NICKEL CO., INC., Product Research and Development Div., New York, N.Y., *Nickel and Compounds.*

IRELAND, LOUISE, Lake Bluff, Ill., *Nutrition.*

ISBIN, H. S., Dept. of Chemical Engineering, University of Minnesota, Minneapolis, Minn., *Heat-transfer Agents; Nuclear Reactors.*

ISSEROW, SAUL, Department of the Army, Army Materials and Mechanics Research Center, Watertown, Mass., *Allotropy; Alloys.*

JACOBSON, MARTIN, Silver Spring, Md., *Repellents, Animal.*

JACOBY, H., (Address unknown), *Polycarbonate Resins.*

JACKSON, M. L., University of Wisconsin, Madison, Wis., *Soil Chemistry.*

JAFFÉ, H. H., Dept. of Chemistry, University of Cincinnati, Cincinnati, Ohio, *Resonance.*

JAFFE, PHILIP M., Department of Chemistry, Oakton Community College, Morton Grove, Ill., *Phosphors.*

JAHN, EDWIN C., N.Y. State College of Forestry, Syracuse, N.Y., *Hemicellulose; Hydrolysis; Wood.*

JOHNSON, C. A. (Retired), Oak Park, Ill., *Iron, Biochemical and Biological Aspects.*

JOHNSON, ERNEST F., Dept. of Chemical Engineering, Princeton University, Princeton, N.J., *Chemical Engineering.*

JOHNSON, V. A., Dept. of Physics, Purdue University, Lafayette, Ind., *Solid State.*

JONES, MARK M., Vanderbilt University, Nashville, Tenn., *Activation, Molecular.*

JONES, R. NORMAN, National Research Council of Canada, Ottawa, Ontario, *Fluorescence.*

JUEL, LESTER H., Great Lakes Carbon Co., Niagara Falls, N.Y., *Carbon.*

JUSTER, NORMAN J., Pasedena City College, Pasedena, Cal., *Organic Semiconductors.*

JUVET, RICHARD S., JR., Dept. of Chemistry, Arizona State University, Tempe, Ariz., *Gas Chromatography.*

KASWELL, ERNEST R., President, Fabric Research Laboratories, Dedham, Mass., *Fibers, Natural.*

KAUFMAN, MURRAY, Thomson Laboratory, General Electric Co., West Lynn, Mass., *Alloys, Refractory.*

KAUFMAN, SAMUEL, Surface Chemistry Branch, Naval Research Laboratory, Washington, D.C., *Solubilization.*

KEISCH, BERNARD, (Address unknown), *Iodine as Radioactive Tracer.*

KELLETT, J. C., JR., National Science Foundation, Washington, D.C., *Diels-Alder Reaction.*

KERFMAN, H. D., General American Transportation Corp., Niles, Ill., *Storage of Chemicals; Tank Cars.*

KERN, JOYCE C., Glycerine Producers' Association, New York, N.Y., *Glycerol.*

KERR, D. L. CAMPBELL, General Traffic Manager, Freeport Sulphur Co., New York, N.Y., *Transportation of Chemicals.*

KERTESZ, ZOLTON I. (Deceased), *Pectins.*

KIEFFER, WILLIAM F., College of Wooster, Wooster, Ohio, *Mole Concept.*

KING, CECIL V., President, Electrochemical Society, 1971, American Gas & Chemicals, Inc., New York, N.Y., *Electrochemistry.*

KING, EDWARD J., Barnard College, Columbia University, New York, N.Y., *Activity.*

KING, W. R., Kaiser Aluminum & Chemical Corp., Raw Materials Research, Center for Technology, Pleasanton, Cal., *Aluminum.*

KIPPUR, PERRY R., (Address unknown), *Hydrazine.*

KIRCH, ERNEST R., Mountain Home, Ark., *Phosphorus.*

KITSON, R. E., Textile Research Laboratory, E. I. DuPont de Nemours & Co., Wilmington, Del., *Optical Spectroscopy; Mass Spectrometry.*

KLIMSTRA, PAUL D., Director of Chemical Research, G. D. Searle & Co., Chicago, Ill., *Repellents, Bird; Steroids.*

KNIGHT, C. A., Dept. of Molecular Biology, University of California, Berkeley, Cal., *Viruses.*

KNOX, BRUCE E., Materials Research Laboratory, Penn State University, University Park, Pa., *Bond Energies.*

KNUTH, CHARLES J., Pfizer, Inc., New York, N.Y., *Carboxylic Acids; Plasticizers.*

KOHAN, M. I., Plastics Dept., E. I. DuPont de Nemours, Inc., Wilmington, Del., *Polyamide Resins.*

KOPPLE, KENNETH D., Dept. of Chemistry, Illinois Institute of Technology, Chicago, Ill., *Polymers, Electroconductive.*

KRAUS, GERARD, Phillips Petroleum Co., Bartlesville, Okla., *Reinforcing Agents.*

KREMERS, HOWARD E., Kerr-McGee Chemical Corp., Oklahoma City, Okla., *Rare Earths.*

KUHLMANN-WILSDORF, DORIS, University of Virginia, Charlottesville, Va., *Dislocations, Crystals.*

KUNIN, ROBERT, Rohm & Haas Co., Philadelphia, Pa., *Ions.*

KUTSCHKE, K. O., National Research Council of Canada, Ottawa, Ontario, *Photochemistry.*

LABBAUF, A., Dept. of Chemistry, Pennsylvania State University, McKeesport, Pa., *Carbon-12 Scale.*

LAFFERTY, J. M., General Physics Laboratory, General Electric Co., Schenectady, N.Y., *Borides.*

LAFLEUR, PHILIP D., Analytical Chemistry Division, National Bureau of Standards, Washington, D.C., *Spectroscopy, Gamma Ray.*

LAITINEN, HERBERT A., Dept. of Chemistry, University of Illinois, Urbana, Ill., *Polarography.*

LALONDE, ROBERT T., State University of New York, College of Environmental Science and Forestry, Syracuse, N.Y., *Terpenes-Terpenoids.*

LAMBERT, JOSEPH B., Dept. of Chemistry, Northwestern University, Evanston, Ill., *Stereochemistry.*

LANDOLT, PERCY (Deceased), *Lithium.*

LANDS, WILLIAM E. M., Department of Biological Chemistry, University of Michigan, Ann Arbor, Mich., *Phosphatides and Phosphatidic Acids.*

LANNING, F. C., Dept. of Chemistry, Kansas State University, Manhattan, Kan., *Silicon.*

LAPPLE, WALTER C., Alliance, Ohio, *Stoichiometry.*

LAQUE, FRANK L., President, American National Standards Institute, Inc., New York, N.Y., *Antifouling Agents; Corrosion.*

LARSEN, DELMAR H., Los Angeles, Cal., *Drilling Fluids.*

LAUFFER, PAUL G., Hastings-on-Hudson, N.Y., *Cosmetics; Odor.*

LEDERLE, HENRY F., Olin Corp., Research Center, New Haven, Conn., *Organic Synthesis.*

LEDNICER, DANIEL, Upjohn Co., Kalamazoo, Mich., *Antifertility Agents.*

LEE, HENRY, Lee Pharmaceuticals, South El Monte, Cal., *Epoxy Resins.*

LESHIN, RICHARD, Section Head, Research Div., Goodyear Tire & Rubber Co., Akron, Ohio, *Accelerators.*

LIETZKE, M. H., Oak Ridge National Laboratory, Oak Ridge, Tenn., *Elements; Electrons; Isotopes; Radioisotopes; Neutrons; Polonium; Protons; X-rays.*

LINDSAY, KENNETH L., Ethyl Corporation, Baton Rouge, La., *Sodium.*

LITTLE, WILLIAM F., Dept. of Chemistry, University of North Carolina, Chapel Hill, N.C., *Metallocenes.*

LOONAM, A. C. (Deceased), *Iodates.*

LOSHBAUGH, ROYCE E., Mobil Research and Development Corp., Princeton, N.J., *Fuel Gases.*

LOVERING, D. W., Arthur D. Little Co., Cambridge, Mass., *Fillers and Extenders.*

LUNDBERG, C. V., Bell Laboratories, Murray Hill, N.J., *Antiozonants.*

LUNDGREN, HAROLD P., U.S. Dept. Agriculture, Agricultural Research Service, Western Regional Research Laboratory, Berkeley, Cal., *Keratins.*

LUPINSKI, JOHN H., General Electric Research and Development Center, Schenectady, N.Y., *Polymers, Electroconductive.*

MABRY, T. J., University of Texas, Austin, Tex., *Chemotaxonomy (Biochemical Systematics).*

MacDonald, Roderick P., Director of Clinical Chemistry, Harper Hospital, Detroit, Mich., *Clinical Chemistry.*

MacDonald, W. E., University of Miami, Coral Gables, Fla., *Toxicology.*

Machurek, Joseph E., U.S. Atomic Energy Commission, Washington, D.C., *Nucleonics.*

MacIntosh, Robert M., Tin Research Institute, Inc., Columbus, Ohio, *Tin.*

Mack, G. P., Jackson Heights, N.Y., *Stabilizers (Plastics).*

Mackenzie, Fred T., Dept. of Geological Sciences, Northwestern University, Evanston, Ill., *Ocean Water Chemistry.*

MacKinnon, C. E., International Salt Co., Clarks Summit, Pa., *Sodium Chloride.*

Maddock, A. G., University Chemical Labortory, Cambridge, England, *Protactinium.*

Mahncke, Henry E., SKF Industries Research Division, King of Prussia, Pa., *Lubricating Oils.*

Malmstadt, Howard V., Dept. of Chemistry, University of Illinois, Urbana, Ill., *Titration (Automatic, Spectrophotometric).*

Mantell, Charles L., Manhasset, N.Y., *Calcium.*

Marberg, Carl M., Pittsburgh, Pa., *Lactones.*

Marchi, R. P., Fort Ord, Cal., *Liquid Structures.*

Margenau, Henry, Dept. of Physics, Yale University, New Haven, Conn., *Van der Waals Forces.*

Margrave, John L., Rice University, Houston, Tex., *High-temperature Chemistry.*

Marinetti, G. V., School of Medicine and Dentistry, University of Rochester, Rochester, N.Y., *Lipides.*

Mark, Herman F., Brooklyn Polytechnic Institute, Brooklyn, N.Y., *Monomers; Polymerization.*

Markley, Klare S., Rio de Janeiro, Brazil, *Fats; Fatty Acids; Laurates; Oleates.*

Marsh, Walton H., State University of New York, Brooklyn, N.Y., *Automatic Clinical Chemical Analysis.*

Martens, Charles R., Parma Heights, Ohio, *Alkyd Resins; Paints.*

Martin, Robert L., Battelle-Columbus Laboratories, Columbus, Ohio, *Thorium.*

Masci, J. N., Johnson & Johnson Laboratories, New Brunswick, N.J., *Sterilization.*

Massey, A. G., University of London, Queen Mary College, London, England, *Boron-10.*

Mausteller, J. W., MSA Research Corp., Evans City, Pa., *Liquid Metals.*

Mayer, J. E., Revelle College, University of California, LaJolla, Cal., *Liquid State.*

McBryde, W. A. E., Department of Chemistry, University of Waterloo, Waterloo, Ontario, *Iridium; Platinum; Palladium; Osmium; Rhodium; Ruthenium.*

McConnell, Duncan, College of Dentistry, Ohio State University, Columbus, Ohio, *Crystal Optics; Isomorphism.*

McDonald, Hugh J., Dept. of Biochemistry, Loyola University, Maywood, Ill., *Electrophoresis.*

McMahon, Howard O., President, Arthur D. Little Co., West Cambridge, Mass., *Chemical Research.*

McNally, J. Rand, Jr., Thermonuclear Division, Oak Ridge, National Laboratory, Oak Ridge, Tenn., *Fusion, Nuclear.*

McNew, George L., Boyce Thompson Laboratories, Yonkers, N.Y., *Fungicides; Quinones.*

Meek, Devon W., Dept. of Chemistry, Ohio State University, Columbus, Ohio, *Ligand Field Theory.*

Melavan, A. D., University of Tennessee, Knoxville, Tenn., *Rhenium.*

Mennear, John H., Professor of Toxicology, Purdue University, Lafayette, Ind., *Antidotes.*

Merrill, H. B., Long Beach, Cal., *Leather.*

Merritt, Lynne L., Jr., Dept. of Chemistry, Indiana University, Bloomington, Ind., *Analytical Chemistry.*

Miller, Lawrence P., Boyce Thompson Institute, Yonkers, N.Y., *Phytochemistry.*

Miller, W. R., U.S. Dept. of Agriculture, Peoria, Ill., *Hydrolysis Polymerization.*

Mills, Jack F., Dow Chemical Co., Midland, Mich., *Iodine and Compounds.*

Minckler, L. S., Jr., Watchung, N.J., *Elastomers, Ozone-resistant.*

Miron, Jerry, Skeist Laboratories, Inc., Livingston, N.J., *Adhesives.*

Moe, Owen A., (Address unknown), *Mannans.*

Moeller, Therald, Dept. of Chemistry, Arizona State University, Tempe, Ariz., *Lanthanide Elements.*

Mohr, J. Gilbert, Johns-Manville Fiber Glass, Inc., Waterville, Ohio, *Reinforced Plastics.*

Monick, John A., Colgate-Palmolive Research Center, Piscataway, N.J., *Alcohols.*

Montgomery, Rex, University of Iowa, Iowa City, Iowa, *Gums and Mucilages.*

Morin, Richard D., University of Alabama Medical Center, Birmingham, Ala., *Flavones.*

Morris, H. H., Freeport Kaolin Co., Gordon, Ga., *Clays.*

Morris, Harold P., Dept. of Chemistry, Howard University, Washington, D.C., *Iodine (Physiological Effects).*

Morris, Ralph W., College of Pharmacy, Univ. of Illinois Medical Center, Chicago, Ill., *Pharmacology.*

Morton, Maurice, University of Akron, Akron, Ohio, *Elastomers; Vulcanization; Latex.*

Moskowitz, Albert, (Address unknown), *Optical Rotatory Dispersion.*

Moser, F. H., Holland, Mich., *Phthalocyanine Compounds.*

Mueller, Richard A., G. D. Searle & Co., Chicago, Ill., *Prostaglandins.*

Mueller, Robert F., Goddard Space Flight Center, Greenbelt, Md., *Chemical Planetology.*

Mulay, L. N., Materials Research Laboratory, Penn. State University, University Park, Pa., *Magnetic Phenomena.*

Murmann, R. Kent, Dept. of Chemistry, University of Missouri, Columbia, Mo., *Coordination Compounds.*

Musgrave, John R., Eagle-Picher Co., Miami, Okla., *Cadmium.*

Nadeau, Gale F., Eastman Kodak Co., Rochester, N.Y., *Rheology.*

Nahin, Paul G., Union Oil of California, Brea, Cal., *Polyorganosilicate Graft Polymers.*

Nairn, R. C., Dept. of Pathology, Monash University, Victoria, Australia, *Fluorescent Protein Tracing.*

NASSET, E. S., Children's Hospital Medical Center, Oakland, Cal., *Digestion, Physiological.*

NEARN, WILLIAM T., Weyerhaeuser Company, Seattle, Wash., *Wood Preservatives.*

NELSON, L. S., (Address unknown), *Olefin Compounds.*

NEUMEYER, JOHN L., Dept. of Medicinal Chemistry, Northeastern University, Boston, Mass., *Herbicides.*

NEVILLE, KRIS, Epoxylite Corp., South El Monte, Cal., *Epoxy Resins.*

NORDELL, ESKEL, Edison, N.J., *Water Conditioning.*

OGRYZLO, E. A., University of British Columbia, Vancouver, B.C., *Orbitals.*

O'NEILL, M. J., Perkin-Elmer Corp., Norwalk, Conn., *Calorimetry.*

ORCHIN, MILTON, Dept. of Chemistry, University of Cincinnati, Cincinnati, Ohio, *Oxo Process.*

ORR, CLYDE, School of Chemical Engineering, Georgia Institute of Technology, Atlanta, Ga., *Aerools, Solid/Gas.*

OSIPOW, LLOYD, New York, N.Y., *Surface Chemistry.*

PARIS, J. P., (Address unknown), *Chemiluminescence.*

PARRAVANO, G., Dept. of Chemical Engineering, University of Michigan, Ann Arbor, Mich., *Oxides.*

PARSI, E. J., Ionics, Inc., Watertown, Mass., *Desalination.*

PATTERSON, JOHN A., Chempro, Inc., Morristown, N.J., *Fluidization.*

PAUL, FRED W., Goddard Space Flight Center, Greenbelt, Md., *Phosphorescence.*

PEARL, IRWIN A., Institute of Paper Chemistry, Appleton, Wis., *Waste Treatment, Spent Sulfite Liquor.*

PEDERSON, CARL S., N.Y. State Agrictultural Exp. Station, Geneva, N.Y., *Pickling (Foods).*

PELLETIER, S. W., University of Georgia, Athens, Ga., *Alkaloids.*

PENCE, ROY J., Dept. of Agricultural Sciences, University of California, Los Angeles, Cal., *Pest Control by Antimetabolites.*

PENNINGTON, WILLIAM A. (Deceased), *Halocarbon Compounds.*

PEPPLER, H. J., Universal Foods Corp., Milwaukee, Wis., *Fermentation by Yeast.*

PETERS, F. N. (Retired—address unknown), *Furans.*

PETERSON, MARTIN S., U.S. Quartermaster Laboratory, Natick, Mass., *Foods.*

PETICOLAS, W. L., University of Oregon, Eugene, Ore., *Order-Disorder Theory.*

PETROCELLI, A. W., Westerley, R.I., *Superoxides.*

PIERCE, OGDEN R., Midland, Mich., *Fluorine and Compounds.*

PLUMMER, ALBERT J., Pharmaceutical Division, Ciba-Geigy Corp., Summit, N.J., *Antihypertensive Agents.*

POHLAND, HERMANN W., E. I. DuPont de Nemours & Co., Organic Chemicals Dept., R & D Division, Jackson Laboratory, Wilmington, Del., *Dyes; Dyeing.*

POSNER, AARON S., Cornell University Medical College, New York, N.Y., *Fluoridation.*

POST, HOWARD W., Williamsville, N.Y., *Silicone Resins.*

POTTER, J. F., (Address unknown), *Porosity.*

POTTER, WENDELL J., U.S. Industrial Chemical Co., New York, N.Y., *Polyethylene, Cross-linked.*

POWELL, ALFRED R., Pittsburgh, Pa., *Coal and Coke; Coal Tar.*

PUNDSACK, FRED L., Johns Manville Research Center, Manville, N.J., *Geochemistry.*

RAACKE, I. D., Dept. of Biology, Boston University, Boston, Mass., *Nucleoprotein.*

RATHMANN, H. W., Consultant, Foote Mineral Co., Exton, Pa., *Vanadium.*

READ, HAROLD J., State College Pa., *Electrodeposition; Galvanizing; Passivity.*

RECHNITZ, G. A., Dept. of Chemistry, State Univ. of New York at Buffalo, Buffalo, N.Y., *Kinetics in Analytical Chemistry; Glass Electrode.*

REED, ROBERT M., C & I/Girdler, Inc., Louisville, Ky., *Carbon Dioxide; Hydrogen.*

REICHENBERG, D., National Physical Laboratory, Teddington, England, *Ion Exchangers.*

REINMUTH, OTTO, (Address unknown), *Grignard Reactions.*

REISS, HOWARD, Dept. of Chemistry, University of California, Los Angeles, Cal., *Thermodynamics.*

RICE, ROBB V., Gane's Chemical Works, Carlstadt, N.J., *Barbiturates.*

RICHARDSON, A. S., Proctor & Gamble Co., Cincinnati, Ohio, *Glycerides; Iodine Value; Sulfonation.*

RIDDICK, JOHN A. (Retired from Commercial Solvents), Baton Rouge, La., *Assay; Association; Nitroparaffins; Solvents.*

RIEMAN, WILLIAM III (Deceased), School of Chemistry, Rutgers University, New Brunswick, N.J., *Acidimetry and Alkalimetry.*

RIES, HERMAN E., JR. (Retired), American Oil Co., Whiting, Ind., *Monomolecular Films.*

RIESSER, G. H., Shell Chemical Co., Houston, Tex., *Glycerol, Synthetic.*

RIPPLE, M. CONSTANCE, Librarian, Carborundum Co., Niagara Falls, N.Y., *Abrasives.*

RODDY, WILLIAM T., Tanners' Council Laboratory, University of Cincinnati, Cincinnati, Ohio, *Leather.*

RODERICK, WILLIAM R., Roosevelt University, Chicago, Ill., *Natural Products.*

ROSE, ARTHUR, President, Applied Science Labs., State College, Pa., *Distillation; Fractionation.*

ROSENWALD, R. H., Universal Oil Products Co., Des Plaines, Ill., *Inhibitors.*

ROSS, R. D., Oreland, Pa., *Solid Waste.*

ROSS, SYDNEY, Rensselaer Polytechnic Institute, Troy, N.Y., *Emulsions; Foams; Interfaces; Micelles.*

ROSSINI, FREDERICK D., President, World Petroleum Congresses, Chemistry, Rice University, Houston, Tex., *Petroleum.*

ROTHEMUND, PAUL (Deceased), *Microchemistry; Purines.*

ROWLAND, ALEX T., Dept. of Chemistry, Gettysburg College, Gettysburg, Pa., *Stereoisomerism.*

RUDMAN, P. S., Dept. of Physics, Technion-Israel Institute, Haifa, Israel, *Metallic Solid Solutions*.

RUSHTON, J. H., School of Chemical Engineering, Purdue University, Lafayette, Ind., *Mixing*.

RUSHTON, W. A. H., University of Cambridge, Cambridge, England, *Color Vision Chemistry*.

RUSSELL, FINDLAY E., Univ. of Southern California, School of Medicine, Los Angeles, Cal., *Venoms*.

RUTLEDGE, T. F., ICI America, Inc., Wilmington, Del., *Acetylene Compounds*.

RYGE, GUNNAR, Director, Dental Health Center, San Francisco, Cal., *Amalgams; Amalgams, Dental*.

SANDERSON, R. T., Dept. of Chemistry, Arizona State University, Tempe, Ariz., *Bonding, Chemical; Electronic Configuration; Electronegativity; Periodic Law; Oxide Chemistry; Halogen Chemistry; Hydrogen Chemistry; Nonmetals; Transition Elements*.

SARVETNICK, HAROLD A., Westfield, N.J., *Polyvinyl Chloride*.

SATTERFIELD, CHARLES N., South Lincoln, Mass., *Oxidation; Peroxides*.

SCHAEFFER, WILLIAM D., Graphic Arts Technical Foundation, Pittsburgh, Pa., *Printing Inks*.

SCHAFFERT, R. M., Saratoga, Cal., *Xerography*.

SCHAPPEL, J. W., FMC Corporation, American Viscose Division, Marcus Hook, Pa., *Viscose*.

SCHIEBER, MICHAEL M., Hebrew University of Jerusalem, Jerusalem, Israel, *Magnetochemistry*.

SCHILDKNECHT, CALVIN E., Consultant and Professor, Gettysburg College, Gettysburg, Pa., *Stereoregular Polymers*.

SCHMERLING, LOUIS, Universal Oil Products Co., Des Plaines, Ill., *Paraffins*.

SCHOELD, E. A., Carlsbad, N. M., *Potassium*.

SCHREINER, FELIX, Argonne National Laboratory, Argonne, Ill., *Krypton*.

SCHUBERT, A. E., Los Gatos, Cal., *Refrigerants*.

SCHUERCH, CONRAD, State College of Forestry, Syracuse, N.Y., *Lignin*.

SCOTT, ALLEN B., Dept. of Chemistry, Oregon State University, Corvallis, Ore., *Electrochemistry in Solid State*.

SEABORG, GLENN T., Lawrence Berkeley Laboratory, University of California, Berkeley, Cal., *Transuranium Elements; Uranium*.

SEARLES, SCOTT, JR., Columbia, Mo., *Alicyclic and Cyclic Compounds*.

SEELEY, SHERWOOD B., Marco Island, Fla., *Graphite*.

SELBIN, JOEL, Dept. of Chemistry, Louisiana State University, Baton Rouge, La., *Oxocations*.

SEYBOLD, A. M., (Address unknown), *Getters*.

SEYMOUR, RAYMOND B., Dept. of Chemistry, University of Houston, Houston, Tex., *Allyl Resins; Polyethylene; Cellular Plastics; Emulsion Polymerization; Peroxides as Chain Initiators; Polyimides; Polypropylene; Polyurethanes; Poly(vinylidene Chloride); Poly(vinylidene Fluoride); Vinyl Resins; Polyvinyl Alkyl Ethers*.

SHAPIRO, S. H., McCook Research Laboratories, Armak Company, McCook, Ill., *Cationic Surfactants*.

SHEPPARD, W. J., Battelle-Columbus Laboratories, Columbus, Ohio, *Fuel Oil*.

SHERWIN, E. R., Senior Chemist, DPI Division, Eastman Chemical Products, Inc., Kingsport, Tenn., *Antioxidants, Food*.

SHIVE, WILLIAM, University of Texas, Austin, Tex., *Structural Antagonism*.

SHREVE, R. NORRIS, Dean of Engineering, Purdue University, Lafayette, Ind., *Intermediates*.

SHRINER, RALPH L., Dept. of Chemistry, Southern Methodist University, Dallas, Tex., *Carcinogenic Substances*.

SIEGEL, BERNARD, Palos Verdes Estates, Cal., *Gases, Homonuclear; Hydrides*.

SILVERBERG, JULIUS, Tennessee Valley Authority, Muscle Shoals, Ala., *Fertilizers*.

SILVESTRI, A. J., Mobil Research & Development Corp., Princeton, N.J., *Reforming*.

SIMPSON, KENNETH L., Agricultural Experiment Station, University of Rhode Island, Kingston, R.I., *Carotenoids*.

SIMPSON, W. C., Shell Development Co., Emeryville, Cal., *Asphalt*.

SISLER, HARRY H., University of Florida, Gainesville, Fla., *Chlorides; Inorganic Chemistry; Halogen-Nitrogen Compounds*.

SKAU, E. L., New Orleans, La., *Melting Point*.

SMITH, GAIL P., Corning Glass Co., Corning, N.Y., *Photochromic Glass*.

SKEIST, IRVING, Skeist Laboratories, Inc., Livingston, N.J., *Adhesives*.

SMITH, WILLIAM B., Dept. of Chemistry, Texas Christian University, Fort Worth, Tex., *NMR, Quantitative Analysis by*.

SMYTH, CHARLES P., Dept. of Chemistry, Princeton University, Princeton, N.J., *Dipole Moment*.

SMYTH, HENRY F., JR., Mellon Institute, Pittsburgh, Pa., *Alcohols, Physiological Aspects*.

SNELL, CORNELIA T., New York, N.Y., *Colorimetry*.

SNETSINGER, K. G., NASA-Ames Research Center, Moffet Field, Cal., *Ultramicroanalysis with Laser*.

SNYDER, C. F., (Address unknown), *Polarimetry*.

SPACHT, R. B., Goodyear Tire & Rubber Co., Akron, Ohio, *Antioxidants, Rubber*.

STANLEY, L. N., GAF Corp., Rensselaer, N.Y., *Diazotization*.

STANLEY, WENDELL M., (Deceased), *Viruses*.

STANNET, V., No. Carolina State College, Raleigh, N.C., *Hydrolysis*.

STANSBY, MAURICE E., Director, Pioneer Research Laboratory, Seattle, Wash., *Fish, Chemistry and Preservation*.

STAPLEY, EDWARD O., Merck Institute for Therapeutic Research, Rahway, N.J., *Fermentation, Antibiotics by*.

STEIN, LAWRENCE, Argonne National Laboratory, Argonne, Ill., *Radon*.

STEINBERG, ELLIS P., Argonne National Laboratory, Argonne, Ill., *Fission*.

STENGER, V. A., Analytical Laboratory, Dow Chemical Co., Midland, Mich., *Bromine*.

STEPHENSON, RICHARD, Dept. of Chemical Engineering, University of Connecticut, Storrs, Conn., *Separation Methods; Solvay Process*.

STESLOW, FRANK J., Sherwin-Williams Co., Chicago, Ill., *Lacquers*.

STEWART, BURCH B., Allied Chemical Co., Morristown, N.J., *Nuclear Magnetic Resonance.*

STEWART, JAMES E., Durrum Instrument Corp., Palo Alto, Cal., *Infrared Emission Spectroscopy.*

STOKES, CHARLES A., Princeton, N.J., *Carbon Black; Carbon, Industrial.*

STONE, JOHN R., American Smelting & Refining Co., Perth Amboy, N.J., *Selenium.*

STOUGHTON, R. W., Oak Ridge National Laboratory, Oak Ridge, Tenn., *Elements; Electrons; Isotopes; Radioisotopes; Neutrons; Polonium; Protons; X-rays.*

STRAUMANIS, M. E., Materials Research Center, University of Missouri, Rolla, Mo., *Crystal Defects.*

STRAUSS, ULRICH P., School of Chemistry, Rutgers, University, New Brunswick, N.J., *Polyelectrolytes.*

STURTEVANT, JULIAN M., Sterling Chemistry Laboratory, New Haven, Conn., *Thermochemistry.*

SULLIVAN, ROYAL A., Binghamton, N.Y., *Aggregation.*

SVETLIK, J. F., (Address unknown), *Oil Resistance.*

SWANN, SHERLOCK, JR., University of Illinois, Urbana, Ill., *Electroorganic Chemistry.*

SWEENEY, ROBERT A., Great Lakes Laboratory, State University College at Buffalo, Buffalo, N.Y., *Environmental Chemistry.*

SWEENEY, T. R., Division of Medicinal Chemistry, Walter Reed Army Institute of Research, Washington, D.C., *Antimalarials.*

TATLOW, J. C., Dept. of Chemistry, University of Birmingham, Birmingham, England, *Fluorocarbons, Aromatic.*

TAYLOR, DONALD F., Waukegan, Ill., *Tantalum; Tungsten.*

TERRY, PAUL H., Agricultural Research Service, Beltsville, Md., *Insect Chemosterilants.*

THOMAS, CHARLES ALLEN, St. Louis, Mo., *Friedel-Crafts Reaction.*

THOMPSON, A. PAUL, Eagle-Picher Industries, Inc., Joplin, Mo., *Lead; Zinc.*

THOMPSON, BARBARA A., Analytical Chemistry Division, National Bureau of Standards, Washington, D.C., *Spectroscopy, Gamma Ray.*

THOMPSON, NORMAN S., Institute of Paper Chemistry, Appleton, Wis., *Paper.*

TOME, J., (Address unknown), *Polycarbonate Resins.*

TRESSLER, DONALD K., AVI Publishing Co., Westport, Conn., *Food Preservation Methods.*

TRURAN, J. W., Belfer Graduate School of Science, Yeshiva University, New York, N.Y., *Carbon Cycle (Stellar).*

TURNER, B. L., U. of Texas, *Chemotaxonomy.*

UPDEGRAFF, NORMAN C., President, Louisville Fire Brick Works, Louisville, Ky., *Carbon Dioxide; Hydrogen.*

URANECK, C. A., Bartlesville, Okla., *Rubber.*

VALLE-RIESTRA, J. FRANK, Dow Chemical Co., Walnut Creek, Cal., *Recycling.*

VAN ATTA, F. A., Washington, D.C., *Safety Practice.*

VAN ATTA, R. E., Dept. of Chemistry, Ball State University, Muncie, Ind., *Coulometric Titrations.*

VAN HOOK, H. J., Research Division, Raytheon Co., Waltham, Mass., *Ferrites.*

VAN LENTE, K. A., Southern Illinois University, Carbondale, Ill., *Coulometric Titrations.*

VAN UITERT L. G., Bell Telephone Laboratories, Murray Hill, N.J., *Equilibrium.*

VANWAZER, JOHN R., (Address unknown), *Phosphates.*

VITTUM, PAUL, Eastman Kodak Co., Rochester, N.Y., *Photography, Color.*

WAACK, RICHARD, Dow Chemical Co., Midland, Mich. *Aromaticity.*

WAGAR, N. WILLIAM, Hercules Incorporated, Coatings & Specialty Products Dept., Glens Falls, N.Y., *Pigments.*

WAGNER, WILLIAM F., Dept. of Chemistry, University of Kentucky, Lexington, Ky., *Bases.*

WAITKUS, PHILLIP A., Plastics Engineering Co., Sheboygan, Wis., *Aminoplast Resins; Phenolic Resins.*

WAKSMAN, SELMAN A., New Haven, Conn., *Antibiotics.*

WALKER, J. FREDERIC (Deceased), *Formaldehyde.*

WALLES, W. E., Plastics Laboratory, Dow Chemical Co., Midland, Mich., *Autocatalysis.*

WALSH, KENNETH A., Brush Wellaman, Inc., Elmore, Ohio, *Beryllium.*

WANG, JUI H., Dept. of Chemistry, Yale University, New Haven, Conn., *Debye-Huckel Theory.*

WARREN, FRANCIS A., Waco, Tex., *Propellants.*

WARSCHAUER, D., Department of the Navy, Naval Weapons Center, China Lake, Cal., *Hall Effect; Phosphorus, Black.*

WARTH, ALBIN H., Cape May, N.J., *Waxes.*

WATREL, WARREN G., Somerville, N.J., *Gel Filtration Chromatography.*

WATROUS, G. H., JR., College of Agriculture, Pennsylvania State University, University Park, Pa., *Homogenization.*

WATSON, W. F., Rubber and Plastics Research Assn., Shrewsbury, England, *Mechanochemistry.*

WEAVER, ELBERT C., Madison, Conn., *Acids; Air; Aliphatic Compounds; Amides; Amines; Anhydrides; Aromatic Compounds; Asymmetry; Combining Number; Combinng Weight; Deliquescence; Formulas; Heterocyclic Compounds; Hydration; Hydrochloric Acid; Iodides; Molecular Weight; Molecules; Nitric Acid; Normality; Phosphoric Acid; Precipitation; Reaction Types; Rearrangement Reactions; Replacement Reactions; Salts; Saturation; Specific Gravity; Sublimation; Sulfuric Acid.*

WEBB, BYRON H., Dairy Products Laboratory, U.S. Dept. of Agriculture, Washington, D.C., *Milk and Milk Products.*

WEBSTER, G. L., Wilmette, Ill., *Anesthetics.*

WEINSTOCK, B., Ford Motor Co., Dearborn, Mich., *Hexafluoride Molecules.*

WEISZ, P. B., Mobil R & D Corp., Central Research Division, Princeton, N.J., *Cracking; Shape-selective Catalysis; Reforming.*

WELLER, SOL W., Dept. of Chemical Engineering, State University of New York, Buffalo, N.Y., *Hydrogenation.*

WENDER, IRVING, Research Director, Bureau of Mines, U.S. Dept. of the Interior, Pittsburgh, Pa., *Carbon Monoxide.*

WERNICK, J. H., Physical Metallurgical Research Dept., Bell Telephone Laboratories, Murray Hill, N.J., *Carbonates.*

WERTZ, JOHN E., Dept. of Chemistry, Univ. of Minnesota, Minneapolis, Minn., *Crystals, Color Centers.*

WEST, JAMES R., Texas Gulf Sulphur Co., New York, N.Y., *Sulfur and Compounds.*

WEST, JOSEPH, St. Xavier College, Chicago, Ill., *Enamines.*

WEST, W., Rochester, N.Y., *Photography.*

WHALEY, THOMAS P., International Minerals & Chemical Corp., Libertyville, Ill., *Phosphates.*

WHISTLER, ROY L., Dept. of Biochemistry, Purdue University, Lafayette, Ind., *Carbohydrates; Starches.*

WHITE, A. G., Cominco Ltd., Trail, B.C. Canada, *Indium.*

WHITE, CHESTER M., Rochester, N.Y., *Antifreeze Agents; Hydraulic Fluids.*

WHITE, J. CRAIG, Hydrocarbons & Polymers Div., Monsanto Co., Indian Orchard, Mass., *ABS Resins.*

WHITE, J. C., Naval Research Laboratory, Washington, D.C., *Batteries.*

WHITTEMORE, C.R., Belleville, Ontario, Canada, *Cobalt.*

WILBER, CHARLES G., Dept. of Zoology and Entymology, Colorado State University, Fort Collins, Colo., *Carbon Dioxide, Physiological; Carbon Monoxide, Physiological; Nerve Gases.*

WILLAMAN, J. J., Morris Arboratum, University of Pennsylvania, Philadelphia, Pa., *Agricultural Chemistry.*

WILLARD, JOHN J., J. P. Stevens & Co., Garfield, N.J., *Cellulose.*

WILLIAMS, FRED, Dept. of Physics, University of Delaware, Newark, Del., *Luminescent Inorganic Crystals.*

WILLIAMS, JONATHON, Wilmington, Del., *Chemotherapy.*

WILLIAMS, ROBERT E., Chemical Systems, Inc., Santa Ana, Cal., *Carboranes.*

WILLIAMS, ROGER J., University of Texas, Austin, Tex., *Biochemistry; Vitamins.*

WILLIAMS, THEODORE J., School of Engineering, Purdue University, Lafayette, Ind., *Computer Simulation of Chemical Reactions.*

WILLIHNGANZ, E. A., C & D Batteries, Division of Eltra Corp., Plymouth Meeting, Pa., *Storage Batteries.*

WILLS, JOHN H., Cheney, Pa., *Clays; Phase Rule; Silicates.*

WILSON, J. N., Shell Development Co., Emeryville, Cal., *Surface Tension.*

WINCHELL, HORACE, Dept. of Geology and Geophysics, Yale University, New Haven, Conn., *Refractive Index.*

WINCHESTER, JOHN W., Dept. of Oceanography, Florida State University, Tallahassee, Fla., *Trace Elements Analysis.*

WINDSOR, M. W., Elyria, Ohio, *Photochromism.*

WINDUS, WALLACE, USDA, Agricultural Research Service, Philadelphia, Pa., *Tanning.*

WINGER, ALVIN, Rohm & Haas Co., Philadelphia, Pa., *Ions.*

WINITZ, MILTON, Vivonex Corp., Mountain View, Cal., *Polypeptides.*

WINSLOW, N. M., Cleveland, Ohio, *Electroorganic Chemistry.*

WRIGHT, JACK P., Bell Telephone Laboratories, Murray Hill, N.J., *Photometric Analysis.*

WRIGHT, WALTER E., Lilly Research Laboratories, Indianapolis, Ind., *Pharmaceuticals.*

WUNDERLICH, BERNHARD, Rensselaer Polytechnic Institute, Troy, N.Y., *Polyethylene, Crystallization of.*

YAMAZAKI, W. T., U.S. Dept. of Agriculture, Wooster, Ohio, *Baking Chemistry.*

YORK, J. LYNDAL, School of Medical Science, University of Tennessee, Memphis Tenn., *Porphyrins.*

YOUNG, ARCHIE R., Montclair, N.J., *Dioxygenyl Salts.*

YOUNG, HARLAND H., Western Springs, Ill., *Gelatin.*

ZAHN, CHARLES, Bureau of Mines, U.S. Dept. of the Interior, Pittsburgh, Pa., *Carbon Monoxide.*

ZULESKI, F. R., Warner-Lambert Research Institute, Morris Plains, N.J., *Indoles.*

PREFACE

The idea of a one-volume encyclopedia devoted to a single major scientific discipline was originated about twenty years ago by the editorial staff of the Reinhold Publishing Corporation, now the Van Nostrand Reinhold Company. Conceived as an experiment in technical publishing, it was brilliantly executed under the skilled leadership of the late Dr. George L. Clark. Both the validity of this concept and the noteworthy success of the endeavor as a publishing enterprise are attested by the fact that over twenty similar volumes have been organized and published in a wide range of fields since the appearance of the first edition of the *Encyclopedia of Chemistry* in 1956. The present editors have endeavored to maintain the original purpose and high quality of the 800-odd articles prepared by almost 600 leading scientists, both in the United States and abroad.

No better summary of the philosophy and methodology of this work can be presented than that stated by Dr. Clark: "The primary function of a one-volume encyclopedia of chemistry is to introduce to this vast subject a factor of convergence instead of divergence, of focal condensation instead of scattering, of unity instead of multiplicity—in a word, *chemistry* instead of a multitude of qualifying terms.

"Despite its necessary brevity of treatment, it must somehow communicate, in an irreducible minimum of words, the living, ever-changing habiliments of one of the greatest areas of organized knowledge. An encyclopedia prepared by skilled authorities in so many fields can have simultaneously the qualities of timeliness—an up-to-date representation of the chemistry of its era—and time*less*ness. By this is meant a preparation so sound in terms of fundamentals that there will be no feeling of obsolescence when these articles are read in later years. Expansion, changes, and new discoveries there will inevitably be; but these should constitute merely superstructures on an established and enduring foundation. . . . The plan of this encyclopedia is to meet the needs of students, intelligent laymen, and experts outside their own fields of specialization for a concise, well-presented, completely modern cross-sectional view of an intricate science that is constantly in progressive forward motion."

The preparation of the third edition has been affected by a number of factors. First, the death on January 8, 1969, of Dr. George L. Clark, Emeritus Professor of Chemistry at the University of Illinois, required the selection of a replacement. The new senior editor has had previous experience in the editing of one-volume reference works of this sort, namely, *Rare Metals Handbook* (1954 and 1961), *Encyclopedia of Electrochemistry* (1964), and *Encyclopedia of the Chemical Elements* (1968). Since the coeditor worked with Dr. Clark on previous editions, maintaining the vital continuity of this original purpose of the book was not as difficult as it might have been. As a result, the transition from Clark and Hawley to Hampel and Hawley as editors has been smooth and rewarding.

Second, the attrition of time has removed many authors from the roster of original contributors because of death, incapacity, retirement, change of fields of interest and expertise, etc. This necessitated a search for and selection of new authors for a large number of entries. Rather surprisingly, it was not possible to locate many previous contributors because their employers (academic, foundation, and industrial) declined to forward letters sent to the contributors at the addresses of their employers. The editors were thus unable to communicate with many previous authors, even though such data sources as *American Men of Science* were searched.

Third, considerable changes have occurred in the emphasis and degree of attention devoted to the several branches of chemistry during the last few years. For example, increasing importance has been ascribed to environmental chemistry and to the chemistry of life processes. The editors have endeavored to recognize these new developments by including relevant entries. Among these are articles on Aerosols, Biomaterials, Blood, Carbon Cycle, Cholinesterase Inhibitors, Desalination, Chemical Industry Economics, Environmental Chemistry, Eutrophication, Food Additives, Genetic Code, Herbicides, Hallucinogens, Insecticides, Prostaglandins, Psychotropic Drugs, Repellents (Animal, Insect, and Bird), Sea Water Chemistry, Transportation, and Venoms.

Providing space for these and other new entries while keeping the volume to reasonable limits of size required elimination of much material that appeared in the previous editions. Most of this dealt with subjects related to nonchemical fields; also, duplication of topics listed under two or more separate headings has been avoided by appropriate combination, as for example Freezing Point and Melting Point, and Collagen and Gelatin.

Another new and valuable feature is the inclusion of a few critical references to additional literature on the specific subject.

The success of the volume has been vitally dependent on the effective cooperation of almost 600 contributors, all of whom are busy with a host of other commitments. To each of these the editors express their sincere appreciation for the effort and time so thoughtfully expended without commensurate reward. Special gratitude is due to those who were so kind as to prepare several entries. Besides the scores who wrote two or three articles, the following contributed five or more: Professor Raymond B. Seymour of the University of Houston; Professor Elbert C. Weaver (retired), Philips Andover Academy; Professor R. T. Sanderson of Arizona State University; Drs. M. H. Lietzke and R. W. Stoughton of Oak Ridge National Laboratory; and Professor W. A. E. McBryde of University of Waterloo, Ontario.

The efficient preparation of the project has been enhanced by the expert processing of manuscripts and proofs by Alberta Gordon, Managing Editor, Van Nostrand Reinhold Company, and her staff, whom the editors thank for their conscientious work.

Clifford A. Hampel
Gessner G. Hawley

A

ABRASIVES

An abrasive is a substance which will wear away another substance faster than it is worn away itself when the two are rubbed together. This offers a means of altering the shape or appearance of a material to produce a needed product. Today all industry depends on abrasives for making the tools required as well as many final products.

Hard, rough abrasives are used for grinding, and softer, finer abrasives for polishing and buffing. In grinding, the sharp edges of the hard abrasive grain dig into the body of the material to be shaped, each edge removing a tiny chip of the workpiece. Chip size depends on the size of the abrasive grain and the pressure applied. Since abasion in a reciprocal process, the abrasive grain also wears away in use and its edges become dull. For this reason the abrasive should be somewhat brittle so that, when dull, the grain will break down under the grinding pressure and heat of friction and present a new sharp edge to the work; however, it should not be so brittle as to break down too fast.

In polishing metal or mineral surfaces, much finer and generally softer abrasives are used. A factor here is the high melting point of the abrasive, higher than that of the work. Little material is removed from the surface, but the heat of friction causes incipient melting of the work surface. There is formed an extremely thin film, called the Beilby layer, which spreads over the work surface and masks surface scratches or chatter marks, thus producing the high finish. Buffing and lapping are polishing operations.

The high heat and pressure at the contact points in both grinding and polishing can also make possible chemical reactions between the abrasive and the work material. This can be either helpful or detrimental, and is another factor entering into the choice of an abrasive.

Natural Abrasives

Diamond, the hardest of all materials, is pure carbon. Those used for abrasive purposes are commonly called boart; they are the crystals not suitable for gem stones because of poor color and shape. Corundum is aluminum oxide, the highly crystalline form of which is sapphire. Emery is an impure corundum and garnet a crystalline form of mixed silicates and oxides. The soft abrasives are more or less impure silicates and oxides: feldspar, a mixed aluminum silicate of potassium; rottenstone, the siliceous material formed by the weathering of a clayey limestone; Vienna lime, a mixed calcium oxide produced from dolomite by calcination. Rouge and crocus are iron oxides and putty powder is tin oxide.

Hard	Siliceous	Soft
diamond	quartz and flint	feldspar
corundum	sandstone and sand	rottenstone
emery	pumice	vienna lime
garnet	diatomite	rouge
	tripoli	crocus
	siliceous clay	putty powder

The earliest grinding wheels were cut from deposits of natural sandstone. Coated abrasives were made as early as 1700 by adhering sand or other grit with glue to a paper backing. Loose natural abrasive grains and powders have been used for polishing from early times.

Synthetic Abrasives.

Synthetic abrasives are produced by high-temperature reactions. Many such compounds were prepared by Henri Moissan in the 1880's in his experiments with the electric furnace, but commercial development awaited further refinements of furnacing. In 1891 Edward G. Acheson first produced silicon carbide in a small electric furnace and recognized its potential. He developed a large trough-shaped electric resistance furnace filled with a charge of sand and petroleum coke through which a current was passed. Temperature rises to 1900–2595°C after a run of 18 hours and then falls to 2040°C until completion of the run. After cooling, the charge is found to have converted to large, well-formed crystals around a central core, smaller

Diamond	10	Mohs Hardness
Boron carbide	$9\frac{3}{4}$	" "
Silicon carbide	$9\frac{1}{2}$	" "
Aluminum oxide	9	" "

crystals further away, and semiconverted and unconverted mix around that. The latter, which acts as self-insulation for the firing operation, can be refurnaced in the next run. The over-all chemical reaction is quite simple:

$$SiO_2 + 3C \rightarrow SiC + 2CO\uparrow$$

Abrasive aluminum oxide or fused alumina was first produced commercially in 1897 by Charles B. Jacobs by fusing bauxite in an electric arc furnace. On melting, the iron silicates in the bauxite collect as ferrosilicon in the bottom of the large pig which, after cooling, is crushed for further processing.

Boron carbide, another electric furnace product was also made experimentally by Moissan. It has

1

been produced commercially since 1934 in a graphite resistance furnace by reacting boric oxide and petroleum coke at 2500°C in a reducing atmosphere.

$$2B_2O_3 + 7C \rightarrow B_4C + 6CO\uparrow$$

Cerium oxide or ceria, CeO_2, is a later addition to the list of synthetic abrasives. It was introduced in Europe in 1941, for glass polishing as an alternative to rouge. Crude cerium hydroxide is precipitated as ceric phosphate in nitric acid solution to separate out the other unwanted rare earths. The phosphate is reacted with oxalic acid and converted to ceric oxalate which is heated first at 475°–550°C and then to 1050°–1170°C in a gas furnace. The ignited material is passed through a 150 mesh screen to produce a fine powder. Not as hard as the other synthetics, its hardness is a function of firing temperature, the soft end of the range being suitable for polishing the softer glasses and the harder, higher fired product being used for crown and optical glass.

Most recently diamond itself has been synthesized. Ever since Lavoisier first proved that diamond is simply a crystalline form of carbon efforts have been made to synthesize it. J. B. Hannay in 1880 and Henri Moissan in 1890 thought they had produced diamonds but subsequent investigators have been unable to reproduce their results from the methods described. Extremely high pressures at high temperatures are required, above 2500°K and 30 k bar, as reported by P. W. Bridgman in 1947. It remained until 1955 for a team from General Electric Company, F. P. Bundy, H. T. Hall, H. M. Strong and R. H. Wentorf, Jr. to design an apparatus, called a belt press, which is capable to 2000°K and 70 k bar. Using this press and a catalyst in the mix, synthetic diamonds are now available commercially and represent a great advance in abrasive technology.

Abrasive Products

After the crude abrasive is made it is crushed into granules which are then classified into size grades by passing through screens or by sedimentation methods for sub screen sizes. The grain may be further treated by mulling or by chemical wash to improve its shape or surface characteristics such as affinity for the bond.

Grinding wheels, sharpening stones, hones and rubs are made by mixing the selected grain with bond and shaping in a mold. The bond may be ceramic, rubber, resin, oxychloride, or shellac. If ceramic, the bond will be clay or frit, which is a powdered glassy material. In this case, a temporary binder is added to hold the shape until the bond is cured by firing in a kiln when the binder burns off. Other bonded shapes are heat cured in an oven at lower temperature than that in a kiln. Metal bonds are also used for many diamond wheels. These wheels are generally cured by sintering, but may also be produced by electroplating.

Coated abrasives are made by applying glue or resin adhesive to a paper or cloth backing, projecting the grain electrostatically against the wet adhesive and curing the adhesive by drying and heating. The cured sheet, usually in jumbo rolls is then cut to produce the final product, ream goods, belts, discs or sleeves.

Abrasive compounds include pastes, the abrasive grain in a grease carrier, and sticks using a wax carrier. Loose grain is available in closely graded powders and coarser tumbling nuggets or blasting grits.

Selection

With this variety of abrasives, products and processes available, the abrasive engineer must select what is best suited to the particular operation considering:

1. Properties of the abrasive relative to the material to be worked: hardness, toughness, brittleness, grain size, chemical properties, melting point. In general the harder, more brittle silicon carbide is used on low tensile strength materials, most non-ferrous metals, cast iron, stone, marble, and wood. The tougher fused alumina is used on alloy steels.

2. Abrasive process and machine to be used: *grinding* operations include surface grinding, internal grinding, cylindrical grinding, cutting-off, and honing; *polishing* is done by buffing or lapping; *tumbling* and *blasting* are also abrasive operations.

3. Abrasive product: wheel shape or type; coated abrasive product type, regular or waterproof; type of compound or loose abrasive.

The choice must optimize for best results achieved most economically to hold the final tolerances and the finish desired.

(MRS.) M. CONSTANCE RIPPLE

Cross-References: *Diamond Synthesis.*

ABS RESINS

ABS resins are a family of tough, rigid thermoplastics deriving their name from the three letters of the monomers which produce them; *A*crylonitrile-*B*utadiene-*S*tyrene. Each of the monomers contributes to the final properties of the ABS resin: acrylonitrile gives chemical resistance and rigidity; butadiene gives toughness and impact resistance, and styrene gives ease of processing and rigidity.

Most contemporary ABS resins are true graft polymers consisting of an elastomeric polybutadiene or rubber phase, grafted with styrene and acrylonitrile monomers for compatibility, dispersed in a rigid styrene-acrylonitrile (SAN) matrix. Mechanical polyblends of elastomeric and rigid copolymers, e.g., butadiene-acrylonitrile rubber and SAN, historically the first ABS resins, are also marketed.

Varying the composition of the polymer through changing the ratios of the three monomers and through use of other comonomers and additives results in ABS resins with a wide variation in properties, a family of resins. Alloys of ABS with other polymers further extend its properties.

The natural resin is an opaque, ivory color. Its physical form is usually a small pellet although a fine powder and a flat sheet form are also sold. The pelleted form is easily fabricated into simple or complex shapes via injection or blow molding, extrusion-vacuum forming, calendering or other processes combining the elements of heat and pressure. The powder form of the resin is primarily used for enhancing the properties of other resins via polyblending, e.g., ABS/polyvinylchloride resins. The ABS sheet stock may be hot or cold formed, machined, solvent or heat welded, or otherwise shaped into finished articles.

ABS resins are classified as engineering plastics (can be fabricated into structural and load-bearing parts) because of their combination of impact resistance, high mechanical strength, and resistance to creep under load. This desirable combination of properties is retained over a temperature range of −40 to 140°F with little change. The distortion temperature of most grades of ABS resins ranges from 185–200°F under a moderate load (264 psi). Higher heat-resistant grades are available via a comonomer addition.

ABS resins exhibit resistance to chemical attack by water, aqueous salt solutions, alkalies, nonoxidizing inorganic acids, many foodstuffs and household cleaners and oils. They are not resistant to oxidizing acids, aromatic and chlorinated hydrocarbons, aldehydes, ketones and esters.

ABS resins in their natural color are embrittled by sunlight (ultraviolet light); heavily pigmented and black formulations improve the resistance but continuous, unprotected outdoor exposure is not recommended. ABS resins are classified as slow burning but they can be rendered self-extinguishing via coating, additives, or polyblend techniques. They are nonconductors of electricity.

Specialty grades of ABS resins currently available include: electroplating, foam, self-extinguishing, low gloss, cold forming, transparent, high-heat resistant, antistatic, improved weathering and other polymer alloys.

The three major suppliers of ABS resins in this country are Monsanto (Lustran ABS), Marbon (Cycolac), and Uniroyal (Kralastic). Principal markets include automotive, appliances, pipe and fittings, and recreation. Sales of 510 million pounds were recorded in 1970. By 1975 it is estimated that ABS resin sales will be near one billion pounds.

J. CRAIG WHITE

References

1. Basdekis, C. H., "ABS Plastics," New York, Van Nostrand Reinhold, 1964.
2. Rathbone, W. J., Jr., "ABS," Modern Plastics Encyclopedia Issue, 1971–72, Vol. 48, No. 10A, pp. 10 and 11.

ABSORBENTS

Absorbents are substances having the capacity to take up other substances by the process of absorption, i.e., the penetration of one substance (the absorbate) into the inner structure of another (the absorbent), with a resulting loss of identity of the original substance. Either the absorbate merely dissolves in the absorbent, as in the solution of hydrogen chloride gas in water to form hydrochloric acid, or it reacts chemically with the absorbent, as in the absorption of carbon dioxide by sodium hydroxide solution, yielding sodium carbonate. By the reverse process of desorption, or stripping, the absorbed materials may sometimes be recovered in their original form. For instance, passing air through a solution of hydrochloric acid liberates hydrogen chloride, and the treatment of sodium carbonate with an acid regenerates carbon dioxide.

Absorbents are widely used in both laboratory and large-scale operations for the concentration and recovery of desirable materials, or for the elimination of interfering, obnoxious, or otherwise objectionable substances. Very often they are selective, so that only a specific constituent is removed from the gas or liquid stream for subsequent regeneration or discard.

Types of Absorbents. The effectiveness of the *nonreactive* absorbents depends on the relative resistances of the fluid films adjoining the interface between the absorbate and the absorbing medium, the relative motion of the two fluids, and the solubility of the solute in the absorbent. It is only with proper engineering of the absorption apparatus to bring the separate phases into intimate contact that maximum efficiency can be achieved. The usual effect of increasing temperature is to diminish the solubility of a gaseous component. Solubility is further reduced by increasing the concentration of solute in the absorbent. Relatively large quantities of the nonreactive absorbents and well-designed equipment are required. Water is generally used in this type of absorption because it is the most available and the cheapest solvent. If the material to be absorbed is not soluble in water, other solvents must, of course, be used. With nonreactive absorbents, the absorbate is rather easily stripped or desorbed by such physical processes as heating, flushing with an inert gas, or fractional distillation.

Absorption by chemical reaction is often the most economical method, since virtually complete absorption is possible with minimum amounts of the *reactive* type of absorbent. Although intimate contact between phases is necessary, the limiting factor is only the rate of reaction, which is controlling if the reaction takes place slowly. The rate of absorption is then equal to the rate of reaction. If the reaction is rapid, as in most instances, the chemical resistance is small, and the diffusional resistances are controlling, just as though no chemical reaction were involved. A second chemical reaction may be required, to liberate the absorbed material, but this is not usually objectionable. In some processes, it is advantageous to use as an absorbent a liquid compound or a solution which reacts with the gas to be absorbed, forming a loose chemical compound which may readily be decomposed.

For laboratory applications, such as the analysis and purification of gases, the recovery of vapors, or the elimination of obnoxious fumes, absorption is easily accomplished by passing the stream through bubble towers, absorption tubes, contact pipettes or other apparatus containing the absorbent material. Some of these have been developed and improved for such specific purposes as direct weighing of the absorbed constituent, regeneration of a gas, or the simultaneous absorption of two or more gases.

Equipment for Absorption. Because of the wide variety of purposes and specifications of industrial absorption equipment, a large number of different types are in use. The objective of each design, however, is the promotion of intimate contact of gas and liquid over a large interphase surface, to produce a high rate of absorption with a minimum quantity of absorbent. The various types may be divided into three groups: (a) plate or bubble-cap towers, (b) packed towers, and (c) miscellaneous types, including spray towers. Equipment of the same basic design is effective also for the desorption of the non-reactive absorbates, since intimate surface contact is the prime factor. Solid absorbents, extremely useful in small-scale operations because of the high flow rates permissible, are not commonly utilized in industrial absorption processes because of the prohibitive cost of manufacture and handling. These should be distinguished from *adsorbents* which hold layers of adsorbates on surfaces rather than throughout their bulk.

CHARLES W. CHAGNON

ACCELERATORS

Accelerators are substances which increase the rate of sulfur vulcanization of rubber. In the absence of an accelerator, the rate of reaction between sulfur and rubber is very slow. Even with the inorganic bases first discovered to be accelerators by Charles Goodyear in 1839 and used until the early part of this century, it took about four hours to cure a tire. In 1906, Oenslager found that aniline, an organic base, was a far more effective accelerator. Continued research has determined that other amines, and especially many of their derivatives, are accelerators. Most accelerators are used in a system consisting of accelerator, sulfur, zinc oxide and fatty acid. They vary widely in their ability to speed vulcanization, and the choice of an accelerator depends upon the desired use. It is possible to select an accelerator system which can achieve a cure in seconds. Modern tires may be cured in less than 20 minutes.

Accelerators do much more than increase the rate of vulcanization. In the early years of the rubber industry the variation in composition of natural rubber resulted in fluctuations of the cure rate. Accelerators have the ability to level out these differences. They also reduce the sulfur requirement considerably—from as much as 10% without an accelerator to less than 3% with one. Certain sulfur-donating accelerators such as the thiuram disulfides and 2-(morpholinodithio)-benzothiazole can give practical cures without any elemental sulfur. There is a direct correlation between the amount of combined sulfur in vulcanized rubber and the rate of deterioration of the rubber in air. The use of accelerators which require less sulfur gives products with greatly improved aging properties. Shorter curing times have the additional advantage of allowing the use of organic dyes, which normally decompose under prolonged heating at elevated temperatures.

Certain accelerators have the ability to broaden safe vulcanization time before the rubber undergoes reversion. This is called the plateau effect. Normally, the physical properties such as modulus and tensile strength increase, go through a maximum, and then decrease or revert on extended cure. The plateau effect is of particular importance in the case of thick articles which do not cure at a uniform rate because of the poor heat conductance of the rubber. Proper selection of the accelerator system controls not only the cure rate, but the nature of the cure and the resultant physical properties as well. Recent work has shown the advantages of using high accelerator/low sulfur ratios. These "efficient vulcanization" systems result in combinations of properties not possible with the ratios previously standard in the rubber industry.

Classification. The first accelerators were the basic carbonates and oxides of lead, supplemented by other basic metal oxides or hydroxides such as magnesia or lime. These are no longer used by themselves to any extent. However, metal oxides are almost always used as activators of organic accelerators, the most common being zinc oxide.

Only a few organic accelerators contain no nitrogen in their composition. They include the xanthates, the dithiocarboxylic acids and their derivatives. Because of relatively poor stability and a tendency to precure or "scorch," these have been of little value as accelerators. All commercially important accelerators are either amines or amine derivatives. There are over 50 different chemicals currently being used as accelerators. Current annual production of organic accelerators is about 110 million pounds. The most important are considered below.

Aldehyde-amines. Most aldehyde-amine accelerators are condensation products of aniline with such aldehydes as formaldehyde, acetaldehyde, butyraldehyde, heptaldehyde, etc., or combinations of two or more such aldehydes. They are liquid or resinous products, often polymeric, and of uncertain composition. They are used largely as activators with thiazole-type accelerators, and give high modulus, good-aging stocks.

Guanidines. The guanidines, the most important of which are diphenylguanidine and di-*o*-tolylguanidine, are white, crystalline organic bases. They are obtained by reacting aniline or *o*-toluidine with cyanogen chloride. At present, they are chiefly used as activators with thiazole-type accelerators, particularly in the synthetic rubbers SBR and NBR.

Thiuram sulfides. The thiuram sulfides are fast-curing or "ultra" accelerators. The most important member of this class is tetramethylthiuram disulfide, a white crystalline solid prepared by oxidation of dimethyldithiocarbamic acid. It is the primary accelerator in many synthetic rubber compounds. Small amounts are useful as an activator of thiazole-type accelerators. In addition, high ratios act as a curing agent without additional sulfur to provide rubber compounds having excellent aging and heat-resistant properties. Tetramethylthiuram monosulfide is a yellow crystalline powder obtained by treating the corresponding disulfide with sodium cyanide. It does not cure without elemental sulfur.

Dithiocarbamates. A number of metal salts of various dithiocarbamic acids are used as accelerators. The most important are zinc dimethyl- and zinc diethyldithiocarbamate. The former is used largely as an activator while the latter is used extensively in latex foamed rubber. In addition to the zinc compounds, lead, copper, cadmium, bismuth, selenium and tellurium dithiocarbamates are commercial "ultra" accelerators. The zinc compounds are prepared by adding carbon disulfide to an aqueous solution of the amine and sodium hydroxide. Upon addition of a soluble zinc salt, the zinc dithiocarbamate precipitates. The dithiocarbamates and the thiuram disulfides together constitute the bulk of the acyclic accelerators. These account for about 25% of the total organic accelerators produced.

Thiazole derivatives. 2-Mercaptobenzothiazole and its derivatives have been the outstanding accelerators in the tire manufacturing field for over 30 years. They combine many of the most desirable features of an accelerator and are easily manufactured at relatively low cost. Thiazole derivatives represent over 65% of the total accelerator production.

The parent compound, 2-mercaptobenzothiazole, is prepared by the reaction of aniline, carbon disulfide and sulfur. It has a tendency to precure. This difficulty was overcome by its more generally used oxidation product, 2,2'-dithiobisbenzothiazole. With the introduction of the highly reinforcing and somewhat alkaline furnace blacks, the scorching problem became still more acute. For this purpose, the benzothiazole sulfenamide derivatives were developed and have become the most widely used class of accelerator. They are prepared by oxidative condensation of an amine and 2-mercaptobenzothiazole.

There are several sulfenamides commercially available, of which the three most important are N-cyclohexyl-2-benzothiazolesulfenamide, N-*tert* butyl-2-benzothiazolesulfenamide, and 2-(morpholinothio)-benzothiazole. Their greatest shortcoming is their limited stability during storage.

Mechanism of Action. Evidence now indicates that sulfur crosslinks are formed by a mechanism which is predominantly polar, even though free radical reactions are believed to operate to some extent. The actual contribution of each of these mechanisms to crosslink formation is undoubtedly a function of the nature of the accelerator system itself. Although the process is very complex, it may be stated simply that there are three principal stages involved: (1) there is an initial delay, or induction period, during which the accelerator, sulfur, zinc oxide and fatty acid interact to form a soluble reactive complex; (2) it is this complex which attacks the rubber molecule to form the initial polysulfidic crosslinks (the nature of the complex determines both the rate of crosslink formation and the structure of the rubber at the site of the crosslink); (3) upon continued heating, the initial crosslinks are shortened by loss of sulfur and further sulfidic crosslinks, of lower sulfur content, are also formed. Much of the current research is directed toward elucidating the exact nature of the crosslinks, as well as the ultimate fate of the accelerator, or its decomposition products, during the vulcanization process.

RICHARD LESHIN

References

Alliger, G., and Sjothun, I. J., "Vulcanization of Elastomers," Van Nostrand Reinhold, N. Y., 1964.

Bateman, L., "The Chemistry and Physics of Rubber-Like Substances," John Wiley & Sons, N. Y., 1963.

Blokh, G. A., "Organic Accelerators in the Vulcanization of Rubber," Israel Program for Scientific Translations, Jerusalem, 1968.

Hofmann, W., "Vulcanization and Vulcanizing Agents," Palmerton Publishing Co., Inc., N. Y., 1967.

Winspear, G. G., "The Vanderbilt Rubber Handbook," R. T. Vanderbilt Co., Inc., 1968.

ACETAL RESINS

Acetal resins are linear thermoplastic polymers of formaldehyde, or copolymers of formaldehyde with a small proportion of a comonomer. The structure is repeating $-O-CH_2-O-CH_2-$, and typically contains over 1500 formaldehyde units. Although acetal linkages $-O-R-O-$ exist in side chains of other plastics, such as polyvinyl butyral for laminated safety glass, these resins are not included in the commercial understanding of "acetal resins." The regular structure provides highly crystalline products, among thermoplastics the stiffest and most resistant to fatigue on repeated impact or flexing. These properties are maintained over a wide range of temperature, humidity, and exposure to organic solvents.

The first acetal polymer was made in 1859 when Butlerov, trying to make methylene glycol, actually synthesized a low molecular weight formaldehyde polymer which decomposed at about 300°F. In the 1920's, Staudinger intensively studied formaldehyde polymerization and established that the polymers comprise molecular chains of polyoxymethylene units. Staudinger increased molecular weight to produce somewhat tougher semitransparent solids. Since these resembled the "eucolloids," he called his formaldehyde polymers eu-polyoxymethylene. However, these polymers were too weak and unstable for commercial value.

Several years later conditions were uncovered for making high molecular weight polymers that were relatively tough and stable. Stability was greatly improved by end-capping of the hemiacetal ends with ester or alkyl groups and by suitable additives.

The polymerization is ionically initiated and can proceed by either anionic or cationic route. A variety of acidic and basic species are effective initiators, and include tertiary amines and quaternary ammonium salts. Molecular weight is controlled chiefly by chain transfer. Chain transfer with a water molecule occurs as shown by the following equations:

$$-CH_2OCH_2O^- + H_2O \rightarrow$$
$$-CH_2OCH_2OH + OH^- \quad (1)$$
$$HO^- + CH_2O \rightarrow HOCH_2O^- \quad (2)$$
$$HOCH_2O^- + nCH_2O \rightarrow$$
$$HOCH_2O(CH_2O)_{n-1}CH_2O^- \quad (3)$$

In Step (1) the growing chain reacts with water, releasing a hydroxyl ion. In Step (2) a new chain is initiated. In Step (3) added monomer gives a new polymer chain.

Such a process produces a high molecular weight polyoxymethylene glycol ready for end-capping. This step may be accomplished by reaction with an anhydride or other agent as follows:

$$\underset{\text{Uncapped polymer}}{HO\ (CH_2O)_nH} + 2Ac_2O \xrightarrow[\text{Catalyst}]{\text{Base}}$$

$$\underset{\text{Esterified polymer}}{AcO\ (CH_2O)_nAc + 2HOAc}$$

Instead of formaldehyde monomer, the starting material may be the cyclic trimer, trioxane ($\underline{CH_2OCH_2OCH_2O}$), with the formation of essentially equivalent polymers. Boron trifluoride or other Lewis acids are preferred initiators. If polymerization occurs in the presence of ethylene oxide or 1,3-dioxolane, the product is a copolymer containing ethylenoxy ($-OCH_2CH_2-$) units distributed throughout the chain. In lieu of ethylene oxide other comonomers such as tetrahydrofuran may be used.

In 1960, the Du Pont Co. provided commercial quantities of the homopolymer Delrin® the first acetal plastic. In 1962, Celanese Corporation announced 'Celcon' acetal copolymer. In 1971, world production facilities for these two types exceeded 200 million annual pounds, and expansions were announced in U.S., Europe, and Japan. Acetal plastics are produced in a range of molding flows, including high viscosity for extrusion. They are naturally nearly opaque white, and are available in a wide range of opaque and pastel colors, including metallics.

Mechanical properties of moldings from acetal resins are between those of a die cast metal and other thermoplastics. Rockwell hardness is M94 for the homopolymer, and M80 for the copolymer. The surface is hard, reproduces mold finish very well, and feels slippery. Creep is low and uniform for a wide range of temperature and humidity, so that design calculations, such as for metals, accurately forecast service life under physical stress. The homopolymer tensile strength is 10,000 psi at 73°F; flexural modulus is 410,000 psi; and elongation, 25% for injection grades and 75% for extrusion grade. Typical copolymer injection grade has slightly lower strength and modulus, and higher elongation. For both types, notched Izod impact strength is essentially unchanged from 150°F down to minus 40°F. The upper limit of service temperature will vary with conditions of use and design of part. Until prototype testing has indicated other temperatures, it is usual to design for 200°F for long continuous use and 260°F for intermittent use. Specific gravity is 1.4.

Fatigue resistance of homopolymer is outstanding among plastics. At 73°F fatigue endurance limit is 5000 psi, and at 150°F it is 3000 psi; these are stresses which can be repeatedly applied without failure. Among injection—moldable plastics, abrasion resistance is outstanding.

Resistance to organic solvents is high, no common solvent being known at room temperature. Moisture absorption of ⅛ inch thick homopolymer molding is only 0.25% after 24 hour immersion, 0.22% at equilibrium with 50% relative humidity, and 0.9% on prolonged immersion. Moisture reduces stiffness only slightly, and saturation at 73°F causes expansion of only 0.004 inch per inch. Copolymers and some homopolymer types have good resistance to aqueous alkali, but neither type is recommended for long service in aqueous acids or strong alkalies. Most detergents and mild oxidants are not harmful at dilutions normally encountered.

Electrical properties of acetals are good. The dielectric constant is close to 3.7 from 10^2 to 10^6 Hz, and hardly affected by moisture. Dielectric strength is 1200 volts/mil at 20 mil specimen thickness, and 600 v/m at 80 mils. Arc resistance is over 120 seconds, erosion then occurring without carbon tracking; longer exposure or confinement of fumes can cause ignition. Acetals are classed as slow-burning at 1.1 inch per minute.

Acetal resins do not present any dermatitis hazard, as demonstrated by human skin patch tests, and extensive use as wristwatch cases, bra buckles and garter clips. Strong chemical exposure or overheating can generate traces of formaldehyde; hence, each application for contact with food or medicine should be separately evaluated. Compositions approved by the National Sanitation Foundation are used in parts for water softeners and filters and have not presented taste problems.

Uses of acetal resins are largely for strength, stiffness, dimensional stability, solvent resistance, and ability to be molded into intricate shapes. Gears, bearings, and counting wheels for business machines are widely used because of a low coefficient of friction, rapid recovery from deformation, and fatigue-resistance. Plumbing valves and shower heads of acetal are common, because of dimensional stability, resistance to corrosion, and low adhesion of mineral deposits. The same reasons apply for use as pump impellers, valves and couplings for organic solvents and aqueous dispersions of insecticides. Clothing fitments are used for stiffness, colorability, and resistance to laundering and dry cleaning.

Injection molding is readily done on all standard equipment. The crystalline nature of acetals provides quick setting and short cycles. Undercuts are often molded in, and assembly is by snap-fit of acetal-to-acetal or acetal-to-metal. Hot plate, spin and ultrasonic welding are fast and provide strong assemblies because of high fluidity of melt and rapid freezing. Cold heading, riveting, and self-tapping screws are effective because acetals have high tear strength, low creep, and only low dimensional change when exposed to moisture or organic fluids.

Some properties of acetal resins are enhanced by compounding with other materials. Fibers or powder of polytetrafluoroethylene dispersed in acetal polymers provide parts with unusually low coefficient of friction and wear rates. Stiffness and dimensional stability, particularly at high use temperatures, are enhanced by reinforcing glass fibers blended into granules for injection molding. Copolymers have been devised to bond chemically to glass reinforcement, providing high strength and retention in water or solvent. Decorating acetal resins is easily done by hot-stamping; painting and adhesive assemblies may require priming or etching systems, which are commercially proved.

Russell B. Akin and
Walter M. Bruner

References

Akin, R. B., "Acetal Resins," Van Nostrand Reinhold, New York, 1962.
Annual reviews by various authors, Modern Plastics Encyclopedia, McGraw-Hill Co., New York.
"Polyaldehydes," Edited by O. Vogl, Marcel Dekker, Inc., New York.
Sittig, M., "Polyacetal Resins," Gulf Publishing Co., Houston, Texas, 1963.

ACETATES

Acetates are compounds derived from acetic acid, CH_3COOH, by replacing the acid hydrogen by another atom or group. Inorganic acetates are metallic salts of acetic acid, while organic acetates are esters, CH_3COOR.

Inorganic Acetates. Inorganic acetates are generally water-soluble and are widely used as neutralizing or buffering agents in industrial chemical reactions. Among the commercially available inorganic acetates are those of: aluminum, ammonium, barium, calcium, copper, ferric iron, sodium, lead and zinc. They are used as mordants in the dyeing industry, in photography, tanning and curing animal hide, and lead and zinc acetates have limited pharmacological applications.

Organic Acetates. Some of the commercially available acetate esters of simple alcohols are those of the following: ethyl, propyl, isopropyl, butyl, isobutyl, amyl, and benzyl. The mono, di-, and triesters of glycerol are also commercially available. These esters are all liquids, most of them have low boiling points, low toxicity, and are relatively inexpensive. Hence they find extensive application as organic solvents in the chemical processing industry and in some consumer products, e.g., lacquers, paints, degreasing solvents, etc. The lower molecular weight products have a characteristic pleasant, "fruity" odor which leads to their use in flavors, essences, and perfumes.

These esters are generally produced by direct esterification of the alcohol with acetic acid, using a strong acid, such as sulfuric, as a catalyst. Since the esterification reaction is an equilibrium, the reaction is forced by using excess alcohol, and removing the ester and the by-product water. In the case of ethyl acetate this is done by azeotropic distillation, the overhead of the esterification column being the ternary azeotrope: 82.2% ester, 8.4% alcohol and 9% water. The alcohol is then washed out of the ester with excess water, and the water removed via a rectification column, the overhead of which is a binary azeotrope.

Cellulose Acetate. This ester is the most widely used of all acetates, since from it are made fibers, films, lacquers, and molding plastics. As an example, over half a billion pounds of cellulose acetate were used in the synthetic fiber industry alone in 1970. The ester is made from cellulose and acetic anhydride.

The cellulose used is a highly purified form made from wood pulp or cotton linters. It is pretreated with acetic acid to swell the fibers and make them more accessible to the anhydride reagent. The pretreated pulp is mixed with acetic acid, excess acetic anhydride, and a small amount of sulfuric acid catalyst. The reaction proceeds for several hours during which the reacting pulp becomes more soluble in acetic acid until a clear, viscous solution of the triacetate is produced. At this point, if the triester is desired, the catalyst is neutralized and the acetate precipitated with dilute acetic acid, washed, and dried.

Most of the commercial acetate used, however, is not the triester, but a material called "secondary acetate." This is cellulose in which only 83–86% of the hydroxyls is esterfied. This compound is more soluble in common solvents, particularly acetone, than the triester. It is made by adding water in the form of dilute acetic acid to the acetylating mixture containing the fully acetylated cellulose. When hydrolysis has proceeded to the desired extent, the catalyst is neutralized with sodium acetate and the secondary acetate precipitated with dilute acetic acid. The precipitate is washed, neutralized and dried.

For spinning into fibers, the acetate is dissolved in acetone, containing a small amount of water to aid solubility, to the extent of about 25% by weight. This "spinning dope" is extruded through small holes in a metal plate ("spinneret") into a chamber containing hot air where the acetone is evaporated, leaving the acetate in the form of continuous filaments which are wound up on bobbins. Acetate fibers are prized for their luxurious "hand," drapability, and ability to accept brilliant dyes.

Cellulose acetate films are made by coating a revolving drum with the dope, evaporating the solvent, and stripping off the film. Acetate film is used for photographic purposes, recording tapes, and a wide variety of packaging applications.

Cellulose acetate for general plastic use is made by mixing the ester with plasticizing agents and pigments. The plasticizing agents are high-boiling, inert esters, which reduce molecular friction and enable the material to be molded.

In recent years, the pure cellulose acetate used for molding purposes and films has been largely superseded by mixed cellulose esters containing, in addition to acetate, esters of butyric and propionic acids. These are manufactured as above, except that the alternative acids are added to the acetylating mixture. By controlling the composition of the resulting ester, a wide range of properties can be imparted to the material, enabling it to be tailored to specific end-uses. Cellulose acetate plastics are strong and among the toughest plastic materials used. They are used in manufacturing a wide variety of articles, e.g., toys, tool handles, radio cabinets, etc.

Vinyl Acetate. Another acetate of great commercial importance is vinyl acetate:

$$CH_2{=}CH{-}O\overset{\displaystyle O}{\overset{\|}{C}}CH_3.$$

Total manufacturing capacity for vinyl acetate in the U.S. in 1967 was about 600 million pounds and was predicted to pass one billion pounds in 1972. Unlike cellulose acetate, vinyl acetate is seldom used as is, but is the raw material for the manufacture of polyvinyl acetate, polyvinyl alcohol and copolymers of vinyl acetate with other vinyl monomers.

Most of the vinyl acetate manufactured today is made by the vapor phase reaction of acetylene with acetic acid. Acetylene is bubbled through hot acetic acid and the mixed vapors heated to about 170°C and passed over a catalyst consisting of charcoal impregnated with zinc acetate. The reaction mixture is condensed and purified by distillation. A recent development is that of various techniques for manufacturing vinyl acetate from ethylene. Ethylene is considerably cheaper than acetylene and this process may eventually replace the older acetylene-acetic acid method.

Vinyl acetate is a clear, colorless, relatively non-toxic, volatile liquid. It may be polymerized by light or any of the standard catalysts for free radical polymerization, e.g., benzoyl peroxide. The most important commercial polymerization process is emulsion polymerization. In this process the resulting polyvinyl acetate is obtained as a high molecular weight polymer in an aqueous emulsion or latex. Polyvinyl acetate latexes are the base from which water-based latex paints and adhesives are made. These two end-uses are the most important applications of vinyl acetate. Polyvinyl acetate and polyvinyl alcohol produced from the acetate are also important textile sizes and binders. In all of these applications copolymers of vinyl acetate are used, as well as the homopolymer.

J. P. Dux

References

Vol. II, Part IX, Ott, E., Spurlin, H. M., and Grafflin, M. W., "Cellulose and Cellulose Derivatives," Interscience Pub. Co., New York, 1954.

Part I., Leonard, E. C., Ed., "Vinyl and Diene Monomers," Wiley-Interscience, New York, 1970.

ACETYLENE

Acetylene is a colorless gas of formula CH≡CH, molecular weight 26.04, melting point $-81.8°C$ at 890 mm, boiling point $-84°C$, and specific gravity 0.91 (air = 1). It is the first of the acetylene or triple bond hydrocarbons. Differing from other hydrocarbons, it is a highly endothermic compound and can be formed from other hydrocarbons at high temperatures (in excess of 1200°C). Further, its free energy of formation decreases with increasing temperature, following approximately the equation:

$$\Delta F \text{ (cal/g-mole)} = 54,900 - 13.6 T°K.$$

Manufacture. The oldest and still most widely used method of producing acetylene is by the reaction of water on calcium carbide: $CaC_2 + 2H_2O \rightarrow C_2H_2 + Ca(OH)_2$, and almost all the approximately 800,000 tons of calcium carbide made annually in this country is used to manufacture acetylene. This process accounts for roughly 60% of acetylene production, using generators of safe and efficient design to bring the carbide and water into contact and control the rate of generation of the acetylene.

The other 40% of acetylene is derived from thermal decomposition of other hydrocarbons by (1) partial oxidation, (2) electric arc, and (3) regenerative heating to attain the high temperature required and the rapid heating and cooling needed to minimize side reactions and decomposition of the C_2H_2 formed. The raw materials include natural gas and other gaseous and liquid hydrocarbons. *Partial oxidation* (BASF, SBA and Montecatini) processes involve the rapid oxidation by air or oxygen of a part of the preheated hydrocarbon feed. The *electric arc* (Hüls, duPont, Knapsack, etc.) processes utilize the rapid passage of the hydrocarbon through an electric arc in a furnace designed to achieve rapid heating, passage through the arc and rapid chilling of the products. The energy requirements are from 4 to 5 kwh/lb of acetylene, the higher value associated with methane

feed and the lower with higher paraffins. A typical cracked gas from the process contains 12 to 16 mole% acetylene and 7 to 10% ethylene. The *regenerative* (Wulff) process uses one of a pair of furnaces to burn the fuel to heat the furnace and another through which the feed is passed at 1000–1300°C for a contact time of about 0.1 second. The flow through the furnaces is alternated as required to maintain the desired temperature range.

The hydrocarbon processes yield by-products, e.g., carbon (usually small in quantity), hydrogen, and hydrocarbons comprising higher acetylenes, olefins, and other paraffins than those in the feed stock. These are usually removed by absorbing the acetylene from them in a solvent. The calcium carbide process produces a high-quality acetylene.

Uses. The annual United States production of acetylene is over 15,000 million cubic feet or about 1 billion pounds. The use of it to make other chemicals accounts for 85–90% of the consumption. The rest is used as a fuel, chiefly for the oxyacetylene flame employed for welding, cutting and scarfing of metals, where its 6000°F temperature, the highest achieved by combustion of any carbonaceous fuel, is of value.

In the chemical industry, about ⅓ of the acetylene used goes into vinyl chloride production, about ¼ to neoprene, about 15% to acrylonitrile, 10% to vinyl acetate and the rest into a variety of products, such as perchloroethylene, acrylates, vinyl esters and ethers, and acetaldehyde.

Clifford A. Hampel

References

Howard, Walter B., "Acetylene by Electric Discharge," pp. 1–4, in "Encyclopedia of Electrochemistry," Hampel, C. A., Editor, Van Nostrand Reinhold, New York, 1964.

Acetylene Compounds

Structure and Characteristics of the Triple Bond. Ethylene has one normal σ bond and one π bond; acetylene has two π-electron bonds along with the ordinary σ bond. The electrons in the C—H bonds and in the normal C—C bond move in overlapping hybridized sp orbitals. This stabilizes the molecule, and shortens the C-C distance to 1.2Å. It also forces all four atoms into a linear configuration. One result of the overlapping and shortening is that the C≡C bond is stronger than the C=C bond. For example, the C≡C bond exists under temperature conditions which rupture the C=C and C—C bonds. In fact, the pyrolysis of saturated and olefinic hydrocarbons is one commercial route to acetylene.

Triple bonds are generally more reactive toward nucleophilic reagents (e.g., water, alcohols, amines, etc.) than are double bonds; that is, the electrons of the triple bond are more electrophilic in character than the electrons of ethylenic bonds. The reverse is true for electrophilic reagents such as halogens, ozone, peracids, etc. This is explained by the fact that the π electrons of the acetylenic bond are concentrated more nearly in the center of the carbon-carbon bond. This arrangement of the triple bond π electrons then accounts for the much greater acidity of the ethynyl hydrogens. Because of the planar, symmetrical configuration of the acetylenic bond, the π electrons are not polarized as readily as corresponding ethylenic electrons.

The acetylenic bond resembles the benzene ring in electron withdrawing power, i.e., it is an "electron sink." Thus, α-acetylenic carboxylic acids, such as propiolic acid (acetylene monocarboxylic acid) are very powerful organic acids. The ionization of propiolic acid ($pK_a = 1.4 \times 10^{-2}$) is about the same as that of chloroacetic acid, and is 250 times greater than the ionization of the corresponding olefinic acid (acrylic acid). Another example of this effect is the acidity of acetylenic hydrocarbons:

Hydrocarbon	pK_a
Cyclopentadiene	15
Phenylacetylene	18.5
Acetylene	25
Triphenylmethane	32.5

Preparation of Acetylene Compounds by Elimination Reactions. Some acetylene compounds are most conveniently made by elimination reactions. Others such as secondary and tertiary alkynes and various acetylenic ethers cannot be made by other methods. Dehydrohalogenation is the first and still most widely used elimination. Alkali metal hydroxides are suitable bases. Acetylene compounds usually made by eliminations include: arylacetylenes, diacetylenes, various enynes, α-acetylenic ketones, α-acetylenic acids, organotin acetylenes, α-acetylenic ethers, α-acetylenic thio-, seleno-, and telluroethers, and α-alkynylamines(ynamines). Many new acetylenic compounds with very interesting properties have been made in recent years. For example, ynamines are very powerful dehydrating agents (acetic acid $\rightarrow$ acetic anhydride at room temperature).

Substitution Reactions of Terminal Acetylenes. Acetylenes containing at least one ethynyl hydrogen are a special case in which the positive hydrogen can be replaced by other elements. Alkali and alkaline earth metals react to liberate hydrogen and form the metallo-derivatives. The corresponding metal amides react similarly, liberating ammonia. In the first reaction, use of free metal reduces some of the acetylenic compound. Consequently the second method has been more widely used. Thus the classical method of preparation of sodium acetylide is passing acetylene into liquid ammonia solution of sodium amide (formed *in situ*). Calcium, lithium, potassium, and barium derivatives have also been made this way. Pure sodium acetylide can be prepared by reacting pure acetylene with sodium dispersion, with very little reduction of acetylene. Lithium acetylide-ethylenediamine complex is offered commercially as a stable dry powder. The complex reacts like pure lithium acetylide.

Salts and oxides of 1-B metals react to form acetylides. Cuprous, cupric, mercuric, silver, and gold acetylides form readily in neutral or basic media. All these compounds are potentially explosive, but can be used safely under certain conditions. Cuprous acetylide is the most important, and was used by Reppe and other workers as catalyst for ethynylation reactions. Cuprous acetylides react with aryl iodides to form aryl acetylenes in high yields. Furan iodides and allylic chlorides react similarly.

Grignard reagents are formed readily, and resemble alkali metal acetylides in many reactions.

Recently a large number of new compounds ($-C\equiv C-)_n M-$ have been made. M can be zinc, cadmium, boron, aluminum, gallium, silicon, germanium, tin, arsenic, bismuth, antimony, lead, nitrogen, or phosphorus.

Coupling of Terminal Acetylenes. One of the most useful properties of terminal acetylenes is their ability to couple under mild conditions to form conjugated polyacetylenes. Intense interest in naturally occurring polyacetylenes and in macrocyclic polyacetylenes has spurred research on coupling reactions in recent years. Linear polyacetylenes can also be made. The two most widely used systems are: (1) Oxidative coupling (Glaser coupling). Cuprous compounds catalyze the oxidative coupling to give either linear or macrocyclic polyacetylenes. (2) Chodkiewicz-Cadiot coupling. A cuprous acetylide reacts with a 1-bromoacetylene ($-C\equiv C-Cu + BrC\equiv C- \rightarrow -C\equiv C-C\equiv C-$).

Macrocyclic polyacetylenes can be isomerized to form macrocyclic polyenes ("annulenes"). They can also be isomerized to form macrocyclic allenes.

Ethynylations. Compounds containing at least one ethynyl hydrogen can add across polarized bonds to produce new acetylenic compounds. One of Reppe's contributions to acetylenic chemistry is the technique for reacting acetylene with formaldehyde to produce propargyl alcohol and butynediol. The points of greatest technical significance are that pressure acetylene reactions can be conducted if the acetylene is diluted with an inert gas, and that wet supported cuprous acetylide is safe enough for use as a catalyst.

An important ethylnylation technique uses stoichiometric or excess powdered potassium hydroxide as condensing agent. An ether such as ethyl ether, acetal, or a polyether can be used as liquid medium. If special solvents like dimethylsulfoxide are used, requirements for potassium hydroxide are lowered, and the reaction is semicatalytic. Thus, one mole of base gives more than one mole of 2-methyl-3-butyn-2-ol from acetone and acetylene. Semicatalytic reactions can also be done in liquid ammonia and even in liquid acetylene. Ethynylation and alkynylation reactions are used to prepare a great variety of α-acetylenic alcohols.

Isomerization of Acetylenic Bonds. Acetylenic compounds can be isomerized to conjugated diolefins by an alkali metal. Thus, acetylene reacts with formaldehyde and dimethylamine in the presence of cuprous chloride to form 1,4-bis-(dimethylamino)-2-butyne, which is converted by treatment with sodium metal into 1,4-bis-(dimethylamino)-1,3-butadiene. Acetylenic bonds can also be isomerized to allenic bonds (e.g., propyne-allene). The propargylic-allenic rearrangement is a practical synthetic tool. Acetylenic bonds located internally in a hydrocarbon chain can be isomerized to terminal acetylenic bonds by treatment with a base.

Additions to the Triple Bond. Many substances add across the triple bond of acetylenic compounds. Among these are hydrogen, halogens, HX, water, HCN, acrylates, and carbonate esters. The halogens and halogen acids will add in the absence of catalysts. Best reactions are obtained if a catalyst is used. The function of catalysts is to "activate" or polarize the acetylene. Among the important products obtained by such reactions are trichloroethylene, vinyl chloride, acetaldehyde, acetone, and acrylonitrile. Many unusual addition reactions are known. Nitric oxide-nitrogen dioxide mixture adds to acetylene to form 3,3'biisoxazole. Bis-diazoalkanes react with acetylene to give high yields of bis-pyrazoles. Bis-sydnones and bis-acetylenes react to form interesting new polymers. Carbenes add to acetylenic bonds to form cyclopropenes and cyclopropanes.

Hydration. Hydration of the acetylenic bond is usually carried out with a mercury salt-acid catalyst (such as $HgSO_4$-H_2SO_4). The exact nature of the intermediate acetylenic-catalyst complex is unknown. There is considerable evidence that a coordination compound is the active species. Acetylene itself forms acetaldehyde under the conditions. When a zinc oxide-vanadium pentoxide catalyst is used, acetone is the chief product. (see **Hydration**).

Hydrogenation. Stepwise hydrogenation can be accomplished by proper choice of catalysts. "Poisoned" palladium is used to convert acetylenics to olefins. Mild conditions of pressure and temperature are required. Acetylene in ethylene and propylene is selectively hydrogenated to ethylene in the important industrial processes for preparing polymerization grade olefins. Acetylene is a strong poison for many olefin polymerization catalysts. Complete hydrogenations occur when nickel, iron, platinum, or palladium catalysts are used under more drastic conditions. Hydroboration is one of the most interesting new reactions of acetylenes. The techniques can be modified to give *cis*-olefins, ketones, alcohols, aldehydes, carboxylic acids, glycols, and higher olefins. Hydroboration is a versatile synthetic method. (See **Hydrogenation**).

Diels-Alder Reactions. Diels-Alder type additions are difficult unless the acetylenic bond is activated (polarized) by an adjacent group. In compounds such as propyl ethynyl ketone and ethyl propriolate the triple bond is a good dienophile. For example, the above ketone reacts with butadiene at 120°C to form 2,5-dihydrobutyrophenone in nearly quantitative yield. In compounds which contain conjugated olefinic and acetylenic bonds, the acetylenic bond can take part in Diels-Alder reactions as part of a diene system. In such cases only one pair of electrons of the triple bond enters into the reaction. Diels-Alder reactions have been used as polymer forming reactions, e.g., bis-cyclopentadienone + diacetylene → aromatic polymer. (See **Diels-Alder Reactions**).

Radiation Chemistry. Aqueous solution of acetylene irradiated with γ-rays in absence of oxygen formed a solid polymer and acetaldehyde and propionaldehyde. In presence of oxygen, the products were glyoxal and hydrogen peroxide. G-value for formation of glyoxal was 50, showing a relatively efficient utilization of the γ-rays. Irradiation of acetylene-propylene in an atomic reactor gave 3-methyl-l-pentene by a new radical reaction which has no counterpart in thermal chemistry. Several other interesting free radical additions to acetylenes were reported in the past few years: addition of halogen, thiols, phosphines, nitrogen oxides, and carbon radicals (acyl, carboxylic acids, hydrocarbons, ethers, ketones, amines, and alcohols). De-

pending on conditions one or two radicals add. In some cases the free radicals were generated by nonradiation methods.

Vinylation. This is a special case of addition across a triple bond. The two most important vinylation reactions involve (1) organic acids and (2) alcohols, to form vinyl esters and ethers, respectively. Vinyl esters are prepared commercially on a very large scale for use as monomers. The most important is vinyl acetate. Vinyl ethers are specialty monomers. Much of the pioneer work on vinylation reactions was due to the German chemist Reppe before and during World War II.

Acetylenic bonds undergo a form of self-vinylation reaction. When acetylene is treated with a cuprous ammonium chloride catalyst, one molecule adds across another to form vinyl acetylene. The chief by-product is divinyl acetylene. Treatment of vinyl acetylene with HCl results in chloroprene. This reaction was discovered by Nieuwland about 1930 and is important because polychloroprene became neoprene, the first practical synthetic rubber. (See **Rubber**).

Another self-vinylation reaction was discovered by Reppe. When compressed acetylene is heated in the presence of a nickel catalyst, (e.g., nickel acetylacetonate) in tetrahydrofuran solvent, four molecules of acetylene react to form cycloöctatetraene in 75–85% yield. The chief difficulty with this reaction is that it is apparently very sensitive to "poisons." A similar reaction, which may also be considered as a special case of the Diels-Alder reaction, involves acetylene and butadiene to produce cycloöctatriene. Acetylene and allene are condensed by nickel complex catalysts in solvents under mild conditions to form cyclic exomethylene hydrocarbons, mainly 3,5-bis(methylene)-cyclohexene-1. A large variety of complexed metals have been used for cyclization, oligomerization, and polymerization reactions. The unusual reactions caused by metal complex catalysts may well be the basis for important developments in acetylene chemistry.

Primary amines react with acetylene in the presence of cadmium acetate or zinc acetate. Products are not the expected vinylamines, but are ethylidineimines.

Polymerization. Acetylene and monosubstituted acetylenes are polymerized by Ziegler catalysts (e.g., triethylaluminum and titanium tetrachloride) to dark powders which are linear conjugated polyolefins. The polymers are organic semiconductors. Ziegler catalysts can copolymerize acetylenes with olefins. Nickel complex catalysts, such as bisacrylonitrile nickel, are also active polymerization catalysts. Reaction conditions are quite mild.

Carbonylation. Acetylene, carbon monoxide, and alcohols in the presence of nickel carbonyl form acrylate esters. This is a commercial reaction. Acetylene reacts with carbon monoxide in presence of dicobalt octacarbonyl to form bifurandione via a series of unusual addition-cyclization steps.

T. F. RUTLEDGE

References

Johnson, A. W., "Acetylenic Compounds, Vol. I, The Acetylenic Alcohols," Edward Arnold Co., London, 1946.

Nieuwland, J. A., and Vogt, R. R., "The Chemistry of Acetylene," Reinhold, New York, 1945.

Raphael, R. A., "Acetylenic Compounds in Organic Synthesis," Butterworth, London, 1955.

Rutledge, T. F., "Acetylenic Compounds. Preparation and Substitution Reactions," Van Nostrand Reinhold, New York, 1968. "Acetylenes and Allenes. Additions, Cyclization and Polymerization Reactions," Reinhold, New York, 1969.

Viehe, H. G., Ed., "Chemistry of Acetylenes," Dekker, New York, 1970.

ACIDIMETRY AND ALKALIMETRY

Acidimetry is the determination of acids by titration with a standard solution of a base. *Alkalimetry* is the determination of bases by titration with a standard solution of an acid. (Some authors reverse these definitions.)

Sodium hydroxide is the most commonly used standard base. Hydrochloric and sulfuric acids are the most commonly used standard acids. The standard solutions are generally 0.01 *N* to 1 *N*. With weaker acids and bases and with less concentrated solutions, the end points are less distinct, and the indicator blanks are larger.

To standardize a solution of NaOH for use in acidimetry, one usually weighs accurately a suitable quantity of potassium hydrogen phthalate, $KHC_8H_4O_4$, dissolves it in a small volume of water and tritrates with the solution of NaOH. The concentration of the solution is then calculated by the equation

$$N = \frac{W}{Qv}$$

where N is the normality of the NaOH, v the volume (in milliliters) of the NaOH solution used in the titration, W the weight (in milligrams) of $KHC_8H_4O_4$ taken, and Q the equivalent weight of the $KHC_8H_4O_4$. Similarly, Na_2CO_3 is generally used as the *primary standard* in the standardization of solutions of HCl or H_2SO_4. The reaction for HCl is

$$2H^+ + 2Cl^- + 2Na^+ + CO_3^= \rightarrow 2Na^+ + 2Cl^- + H_2CO_3.$$

The principles of acidimetry are best understood by studying curves of potentiometric titrations, where pH is plotted against v, the volume of NaOH added. Fig. 1 shows such a curve for HCl. It is typical of the titration of any strong acid with any strong base. The titrations of the weak acids, acetic acid, phenol, and hydrogen peroxide are shown in Fig. 2, curves A, B, and C, respectively. The equivalence point in each case is the jump, *i.e.*, the steepest point of the curve, the steep part at the beginning being disregarded. With an acid as weak as phenol ($K = 10^{-10}$) the jump cannot be located accurately; hence an acidimetric determination is not satisfactory. Hydrogen peroxide is so weak an acid ($K = 10^{-11}$) that the jump does not appear at all.

Curve 2D indicates that a mixture of two acids of sufficiently different ionization constants will have two jumps. Thus both acids can be determined by a single titration. The volume of base used up to the first jump is a measure of the quantity of the stronger acid. The volume of base used between the jumps is a measure of the quantity of the weaker acid.

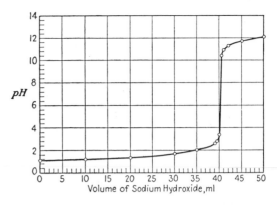

FIG. 1. Titration curve of HCl with NaOH. Approximately 8 millimoles of HCl in 60 ml of water were titrated with 0.2*N* NaOH.

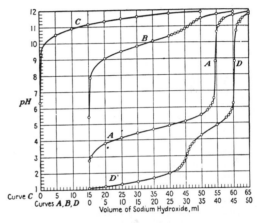

FIG. 2. Titration curves of some monoprotic acids with 0.246*N* NaOH. Initial volume = 60 ml in all cases. (A) 8.06 millimoles of acetic acid; (B) 6.75 millimoles of phenol; (C) 8.0 millimoles of hydrogen peroxide; (D) 6.26 millimoles of hydrochloric acid plus 3.02 millimoles of acetic acid.

Fig. 3 presents the titration curves of some polyprotic acids (acids with more than one ionizable hydrogen atom) and indicates that such an acid may have as many jumps as the number of ionizable hydrogen atoms or may have fewer jumps. Thus oxalic acid, $H_2C_2O_4$ (curve B), has two jumps, but tartaric acid, $H_2C_4H_4O_6$ (curve A), has only one corresponding to the formation of $Na_2C_4H_4O_6$. Phosphoric acid, H_3PO_4 (curve C), has two jumps corresponding to the formation of NaH_2PO_4 and Na_2HPO_4. Citric acid, $H_3C_6H_5O_7$ (curve D), has only one jump at the formation of $Na_3C_6H_5O_7$.

The slopes of the jumps dpH/dv at the various equivalence points are given by the approximate expressions tabulated below. L is the quantity of acid (in millimoles) being titrated. N is the normality of the standard NaOH. V is the total volume of solution (in milliliters), *i.e.*, the original volume plus that added during the titration. K is the ionization constant of a monoprotic acid. K_1, K_2, and K_3 are the first, second, and third onization constants of polyprotic acids. K_w is the ion product of water,

Acid	First		Second		Third
Monoprotic	$0.22N \sqrt{\dfrac{K}{K_w L V}}$				
Diprotic	$\dfrac{0.22N}{L} \sqrt{\dfrac{K_1}{K_2}}$		$0.22N \sqrt{\dfrac{K_2}{K_w L V}}$		
Triprotic	$\dfrac{0.22N}{L} \sqrt{\dfrac{K_1}{K_2}}$		$\dfrac{0.22N}{L} \sqrt{\dfrac{K_2}{K_3}}$		$0.22N \sqrt{\dfrac{K_3}{K_w L V}}$

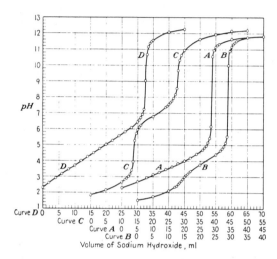

FIG. 3. Titration curves of some polyprotic acids with $0.246N$ NaOH. Initial volume = 60 ml in all cases. (A) 2.94 millimoles of tartaric acid; (B) 2.94 millimoles of oxalic acid; (C) 2.84 millimoles of phosphoric acid; (D) 2.21 millimoles of citric acid.

All figures reproduced from "Quantitative Analysis" by Rieman, Neuss and Naiman, McGraw-Hill Book Company, Inc., 1951.

about 10^{-14}. In general, the greater the slope at the equivalence point, the more accurate is the titration. With slopes of 0.5 or less, the titration is usually impracticable. Analogous expressions apply to the titration of weak bases by strong acids.

Titrations with indicators are more convenient than potentiometric titrations. The indicator for any given titration should be one which changes color at the steepest point of the titration curve. A gradual change in the indicator at the equivalence point may be due to the use of an unsuitable indicator or to insufficient steepness of the titration curve at the equivalence point. In the latter case, the accuracy can be improved by adding the indicator to a buffer solution of the pH that the titrated solution will have at the equivalence point, then titrating until the colors of the two solutions match.

WILLIAM RIEMAN III

Cross-references: *Acids, Analytical Chemistry, Bases, Buffers, Indicators, Normality, pH, Titration.*

ACID NUMBER

Acid number or acid value is a chemical term used to express the degree of acidity of a substance. It is defined as the number of milligrams of potassium hydroxide required to neutralize the acidic constituents in one gram of material. The determination is usually performed by titrating a solution of the substance with 0.1 N alkali to an end point of pH 8.7. The acid number determination is used mainly in the analysis of animal and vegetable oils, fats, and waxes and, to a lesser extent, petroleum oils and waxes. Also, products derived from these substances, such as soaps, paints, fuels, coatings, and lubricants are sometimes evaluated in terms of acid numbers.

The acidity of animal and vegetable oils, fats and waxes is due almost entirely to the presence of free fatty acids, which are formed by hydrolysis of the component glycerides as a result of chemical treatment, enzymatic action, or bacterial decomposition. Therefore, the acid number of natural oils and fats is a variable property, which is related to conditions of manufacture, age and storage. Fresh vegetable oils seem to contain small percentages of free fatty acids, whereas animal fats in the fresh state are practically devoid of them. Generally, in fats of good quality, the amount of free fatty acids is not greater than about 1%, although in fats obtained from damaged materials it may be much higher. Palm oil and inedible tallows and greases are characteristically high in free fatty acids, ranging from 3 to 30%. Thus, the importance of the acid number lies in the fact that it indicates the quality of a fatty substance. Certain arbitrary limits have been established for fats and oils in good condition. The range of values frequently reported in the literature is: butter fat, 0.45–2.0; coconut oil, 2.5–10.0; corn oil, 1.4–2.0; cottonseed oil, 0.6–0.9; lard, 0.5–0.8; menhaden oil, 5.0–8.0; mutton tallow, 1.7–14.0; olive oil, 0.3–1.0; rape oil, 0.4–1.0; and soya oil, 0.3–1.8.

The acidity of fats frequently is expressed directly as percentages of free fatty acids. In these calculations, the assumption is made that the free fatty acids in the substance have a molecular weight equal to a specific fatty acid. For example, the free fatty acids (F.F.A.) are calculated as lauric acid for coconut and palm-kernel oils; as palmitic acid for palm oil; as ricinoleic acid for castor oil; and as oleic acid for most other oils. One unit of acid number is equivalent to: 0.357 per cent F.F.A. as lauric acid; 0.456 per cent F.F.A. as palmitic acid; 0.531 per cent F.F.A. as ricinoleic acid; and 0.503 per cent F.F.A. as oleic acid.

The scope of the acid number determination, as designed for petroleum products and lubricants, has been extended to include as acidic components: organic and inorganic acids, esters, phenolic compounds, lactones, resins, salts of heavy metals, and addition agents such as inhibitors and detergents. A set of standards on petroleum products specifies maximum allowable acid numbers for the following lubricating oils: general use oil (medium grades), 0.10–0.30; steam cylinder oil, 0.80; and diesel and marine engine oils, 3.00. The acid number may be used to express changes that occur in a lubricating oil used under oxidizing conditions. However, the acid number cannot be used to predict the performance of an oil under service conditions.

C. C. CHAPPELOW, JR.

Cross-references: *Acids, Fats, Fatty Acids, Glycerides, Waxes.*

ACIDS (GENERAL)

Acids comprise a group of organic or inorganic compounds which in water solution taste sour, liberate hydrogen when they react with metals, neutralize alkaline compounds, change the color of litmus indicator from blue to red, make characteristic changes in the colors of other indicators, and have pH value less than 7. With solvents other than water and extending acid-base theory broadly, the proton-donor definition (Lowry-Brønsted) and the electron acceptor (Lewis) definition of acids are fruitful. In water solution, however, acids may be considered compounds capable of producing hydrogen or, better, hydronium (H_3O^+) ions.

Well over 100 kinds of acids are commercially available. They are prepared in several ways: (1) By the reaction of a salt with an acid of a higher boiling point than the acid sought. For example, sulfuric acid has a high boiling point and it is stable. It reacts with chlorides and liberates hydrogen chloride which has a lower boiling point than sulfuric acid. Hydrogen chloride dissolved in water forms hydrochloric acid:

$$NaCl + H_2SO_4 \rightarrow HCl + NaHSO_4$$

$$HCl + H_2O \rightarrow H_3O^+ + Cl^-.$$

(2) The addition of water to an oxide, usually of a nonmetal (and called an acid anhydride):

$$SO_3 + H_2O \rightarrow H_2SO_4,$$

(3) *Special methods:* Sulfuric acid is made by the chamber and the contact processes. Nitric acid is made by the catalytic oxidation of ammonia. Hydrogen chloride is also made as a by-product of the chlorination of hydrocarbons.

The *mineral acids* include sulfuric (H_2SO_4), hydrochloric (HCl), nitric (HNO_3), perchloric ($HClO_4$), and phosphoric (H_3PO_4). Organic acids contain carbon, and usually contain the group or radical —COOH, called the carboxyl radical. Such acids are *carboxylic acids.* Formic acid (HCOOH), acetic acid (CH_3COOH), and propionic acid (C_2H_5COOH) are in this group. Acids with a larger number of carbon atoms such as palmitic ($C_{15}H_{31}COOH$) and stearic ($C_{17}H_{35}COOH$), are acids that may be derived from fats and oils. Hence this entire group of carboxylic acids is called the *fatty acids.* Some monocarboxylic acids may be unsaturated such as acrylic (or propenoic) acid ($CH_2=CHCOOH$) and oleic acid ($C_{17}H_{31}COOH$).

Dicarboxylic acids contain two carboxyl groups. The simplest is oxalic acid (COOH·COOH), a white solid, melting point 189°C as the dihydrate [$(COOH)_2 \cdot 2H_2O$]. This compound might also be called ethanedioic acid. Propanedioic acid, better known as malonic acid ($COOHCH_2COOH$), m.p. 135.6°C, is also in this group.

Aromatic carboxylic acids include benzoic acid (C_6H_5COOH), found in cranberries. It is used as a preservative for foods.

Three phthalic acids are known, of which the ortho- is the most important. It is $C_6H_4(COOH)_2$, 1,2-benzenedicarboxylic acid, m.p. 206–208°C. Its anhydride, phthalic anhydride [$C_6H_4(CO)_2O$] polymerizes with glycerol and forms a useful resin.

Phenol, sometimes called carbolic acid, (C_6H_5-OH), is an organic acid that reacts with sodium hydroxide, a typical base, and forms sodium phenoxide (or sodium phenolate) and water: $C_6H_5OH + NaOH \rightarrow C_6H_5ONa + H_2O$. In contrast to carboxylic acids, phenols in general do not react with sodium hydrogen carbonate ($NaHCO_3$), but many carboxylic acids do.

Amino acids are a group of 21 nitrogen-containing acids that may be derived from the decomposition of complex proteins. They all contain at least one carboxyl group.

Strong acids are those that are completely dissociated into ions in water solution. Weak acids form few ions. For example, the reaction $HCl + H_2O \rightarrow H_3O^+ + Cl^-$ goes practically to completion when hydrochloric acid, a strong acid, is formed. Acetic acid solution $CH_3COOH + H_2O \rightleftharpoons CH_3COO^- + H_3O^+$ forms only a few ions, and its properties are those of a weak acid. In this sense, glacial acetic acid is even weaker than a solution of hydrogen acetate in water. Concentrated sulfuric acid is weak, and dilute sulfuric acid is strong.

The dissociation of the first hydronium ion from an acid in general proceeds more readily than a second. For example, $H_3PO_4 + H_2O \rightarrow H_3O^+ + H_2PO_4^-$ is easier than $H_2PO_4^- + H_2O \rightarrow H_3O^+ + HPO_4^{--}$, and far easier than $HPO_4^{--} + H_2O \rightarrow H_3O^+ + PO_4^{---}$.

The dissociation constant (or ionization constant) for an acid is a measure of the extent to which it forms hydronium ions (H_3O^+) in water solution. In the case of 0.1M acetic acid at 25°C,

$$K_A = \frac{[H^+] \times [CH_3COO^-]}{[CH_3COOH]} = 1.76 \times 10^{-5}$$

This dissociation constant has a larger value for strong acids than for weak ones. In cases in which an acid dissociates more than one ion, the value for K_A for the second hydronium ion is always smaller than that for the first.

ELBERT C. WEAVER

Cross-references: *Acid Number, Amino Acids, Carboxylic Acid, Anhydrides, Fatty Acids, Ions.*

ACRYLIC RESINS

Acrylic resins include a wide range of materials with a variety of properties. The items may be tough or brittle or be very hard or very soft. These materials range from hard transparent sheet to soft flexible coatings to fibers to viscous oils. The common bond linking these materials is the basic chemical family called acrylic monomers.

Acrylic Monomers. The family of acrylic monomers used in preparing acrylic resins is based on acrylic and methacrylic acids. Acrylic acid has the structure:

$$\underset{\displaystyle H_2C=C-C-OH}{\overset{\displaystyle H \quad O}{}}$$

Methacrylic acid differs by having a methyl group in place of the alpha hydrogen:

$$\underset{\displaystyle H_2C=C-C-OH}{\overset{\displaystyle CH_3 \quad O}{}}$$

These acids can react at the carboxylic functionality like other organic acids to form a variety of derivatives such as esters, salts, anhydrides, and nitriles. The most important of these derivatives are the esters.

The acids and their derivatives are known collectively as acrylic monomers. They react, particularly the methacrylic and acrylic esters, at the double bond to form long chains of repeating monomer units called acrylic polymers.

There are a number of commercial processes in use for the preparation of acrylic monomers. The accelerating demand for acrylic resins has resulted in the improvement of these processes and in reductions of the cost of the monomers.

Acrylic Polymers. The wide variety of properties possessed by acrylic resins is dependent on the particular monomer (homopolymerization) or mixture of monomers (copolymerization) used and on the technique of polymerization used.

Monomer Selection. By proper selection of acrylic monomers in the polymerization recipe, a resin with the desired properties for a particular application can be obtained. The acrylates are softer and more extensible than the corresponding methacrylates. Softness and extensibility also increase with increasing number of carbon atoms in the alcohol side chain of the acrylic esters. This allows formulation of a polymer with the desired softness without the use of plasticizers. This property is important because of the unsurpassed clarity and resistance to discoloration and degradation by heat, aging or sunlight of unmodified acrylic polymers. As an example, the polymerization of methyl methacrylate yields a hard polymer with a high tensile strength but low elongation, while butyl acrylate polymer is a very soft, low-tensile strength material with high elongation. A copolymer of these two monomers would be intermediate in these properties.

Both thermoplastic or thermosetting polymers can be produced from acrylic monomers. A thermoplastic polymer is one which will soften on heating but on cooling will revert to the starting properties. A thermosetting acrylic resin is one which permanently hardens on heating or curing. Thermosettable acrylic resins are produced if the monomers used have further reactive functionality such as the carboxyl of acrylic acid or the hydroxyl of hydroxyethyl methacrylate (HEMA). These resins are thermoset by crosslinking the linear chain of monomer units at the reactive site. The resins may be self-curing or require the addition of another polyfunctional resin. Examples are epoxy with acid and nitrogen or isocyanate with hydroxyl.

Polymerization Techniques. There are four basic techniques used in the production of acrylic resins; bulk, suspension, solution and emulsion polymerization. *Bulk polymerization* is the polymerization of monomers without dilution. This process is most commonly used to produce cast sheet, rods or tubes. *Suspension polymerization* is a special case of bulk polymerization where droplets of monomer are suspended with agitation in water, then polymerized. Small spheres of polymer are produced which can be injection-molded, extruded, or otherwise processed into usable items. *Solution polymerization* is a form of polymerization in which the reaction is carried out in a solvent in which both the monomer and the polymer are soluble. *Emulsion polymerization* is, by the generally accepted theory, accomplished when small droplets of monomers are dispersed in a water-surfactant system. The surfactant is concentrated in the aqueous phase in "micelles." The monomer diffuses through the water into a micelle where polymerization takes place. The resultant system is a dispersion of polymer solids in water.

Uses of Acrylic Resin. The uses of acrylic resins produced by these processes are found in most homes, vehicles and factories. Cast acrylic sheet such as Plexiglas is widely used as shatter-resistant substitute for glass in doors, windows and skylights. Sheet and rods are available clear or translucent, colored or colorless for a wide variety of practical and decorative uses. Backlighted acrylic sheet signs of distinctive color have become an important means of customer recognition to many businesses. The ease of fabrication allows attractive designs to be produced easily for use in modern furniture. Translucent lighting diffusers have enabled architects to design brightly lit structures which are free of glare. Valuable or attractive specimens have been embedded in clear acrylic shapes to be handled and studied from all angles without damage. The excellent optical properties of acrylic resins also allow them to be ground or molded as lenses in many applications including contact lenses.

Suspension-polymerized acrylic resins have found uses as molding powders known by the tradenames Plexiglas and Lucite. Automotive lenses produced from molding powders, for example, have clarity approaching that of cast acrylic resins. In fact, many of the uses described for cast acrylics can be filled by molding powders processed into the proper shapes. Another interesting use of suspension polymerized beads is as chromatographic adsorbents and as ion-exchange resins. Inclusion of methacrylic acid in the copolymer formulation produces an ion-exchange resin of the weak carboxylic acid type. In the future, suspension-polymerized acrylic resin may be of interest in development of 100% solids coatings. Hot-melt adhesives are another use of suspension polymerized powders.

Acrylic solution polymers are used most often in coatings applications. The coating may be used to hide, protect or decorate many types of surfaces such as metal, wood, paper, textile, and masonry. The coatings can be designed to be applied by conventional or airless spray, brush, roller, aerosol, etc. New formulations are continually being developed to meet modern applications methods. The water-soluble acrylic polymers such as sodium salts of polyacrylic acid have found use as thickeners in aqueous systems as widely diverse as latex paints and shaving cream. Aqueous solution polymers are available as low-viscosity materials which can become highly viscous on pH adjustment. Solution polymers are also used in automotive lubricants as viscosity index improvers, pour-point depressants and sludge dispersants.

Emulsion-polymerized acrylic resins offer the advantages of high molecular weight including toughness and durability in a low-viscosity system. Acrylic emulsions find use in wood and masonry coating where flat, semigloss and gloss paints are available for interior and exterior use in primers, sealers, and in white, tinted or deep-tone topcoats. Techniques are also available for using acrylic emulsions in high-speed factory coating operations. In this application, the advantages of the aqueous systems of no fire hazard or solvent air pollution are readily apparent.

Emulsions are widely used in the textile industry as laminating and flocking adhesives, for fabric backcoating and as bonding agents for nonwovens. Self-crosslinking emulsions are also available for fabric finishing to modify the "hand" or feel of the fabric. The copolymer recipe of the emulsion used imparts the desired "hand" without additional plasticization being required. These acrylic emulsions also give the fabric abrasion resistance, tear strength, soil resistance, and durability to washing and dry cleaning. In the paper industry, acrylic resins are used to obtain high gloss and smoothness, better printability, freedom from undesirable color and resistance to ultraviolet light. Artificial leather, window shades, and gaskets are also produced by saturating paper with acrylic emulsion. Good water resistance can also be obtained for paper to be used in cartons, wall paper and building materials.

Emulsions containing relatively high amounts of methacrylic acid crosslinked with zinc in the copolymer are used in detergent-resistant floor polish formulations. When the polish is to be removed, ammonia is used to replace the zinc, yielding an easily lifted water-soluble ammonium polymethacrylate copolymer.

Acrylic resins are represented in fibers where polyacrylonitrile is the basic material in the manufacture of "Orlon," "Acrilan," "Dynel" and others. Anidex is a new generic fiber classification issued by the Federal Trade Commission for a chemically crosslinked polyacrylate elastomer.

Acrylic films such as Korad are available. These films are prefabricated 100% solid acrylic coatings. They can be applied to a variety of substrates with freedom from problems of pigment settling, solvent and thickness variations. They are also used for base films for decals and labels.

Acrylic resins are used as modifiers of other plastic materials to improve their properties. A few percent of certain acrylic resins added to rigid PVC will improve fabrication and surface smoothness. Other acrylic polymers are designed to increase the impact strength of PVC.

The wide variety of acrylic resins available now show the need for materials with this desirable balance of properties and predict a continued rapid increase in use of acrylics as polymerization technology advances.

GEROULD ALLYN

References

"Preparation, Properties and Uses of Acrylic Polymers," Plastics Department, Rohm and Haas Co., Bulletin SP-229, May, 1967.

Horn, Milton B., "Acrylic Resins," Van Nostrand Reinhold, New York, 1960.

Riddle, Edward H., "Monomeric Acrylic Esters," Reinhold Publishing Corp., New York, 1954.

Luskin, L. S., and Meyers, R. J., "Acrylic Ester Polymers," in "Encyclopedia of Polymer Science and Technology," Vol. I, John Wiley & Sons, Inc., New York, 1964.

ACTIVATED SLUDGE

Activated sludge is the biologically active sediment produced by the repeated aeration and settling of sewage and/or organic wastes. The dissolved organic matter acts as food for the growth of an aerobic flora. This flora produces a biologically active sludge which is usually brown in color and which destroys the polluting organic matter in the sewage and waste. The process is known as the activated sludge process.

The activated sludge process, with minor variations, consists of aeration through submerged porous diffusers or by mechanical surface agitation, of either raw or settled sewage for a period of 2–6 hours, followed by settling of the solids for a period of 1–2 hours. These solids, which are made up of the solids in the sewage and the biological growths which develop, are returned to the sewage flowing into the aeration tanks. As this cycle is repeated, the aerobic organisms in the sludge develop until there is 1000–3000 ppm of suspended sludge in the aeration liquor. After a while more of the active sludge is developed than is needed to purify the incoming sewage, and this excess is withdrawn from the process and either dried for fertilizer or digested anaerobically with raw sewage sludge. This anaerobic digestion produces a gas consisting of approximately 65% methane and 35% CO_2, and changes the water-binding properties so that the sludge is easier to filter or dry.

The activated sludge is made up of a mixture of zoogleal bacteria, filamentous bacteria, protozoa, rotifera, and miscellaneous higher forms of life. The types and numbers of the various organisms will vary with the types of food present and with the length of the aeration period. The settled sludge withdrawn from the process contains from 0.6 to 1.5% dry solids, although by further settling it may be concentrated to 3–6% solids. Analysis of the dried sludge for the usual fertilizer constituents show that it contains 5–6% of slowly available N and 2–3% of P. The fertilizing value appears to be greater than the analysis would indicate, thus suggesting that it contains beneficial trace elements and growth-promoting compounds. Recent developments indicate that the sludge is a source of vitamin B_{12}, and has been added to mixed foods for cattle and poultry.

The quality of excess activated sludge produced will vary with the food and the extent of oxidation to which the process is carried. In general, about 1 lb of sludge is produced for each lb of organic matter destroyed. Prolonged or over-aeration will cause the sludge to partially disperse and digest itself. The amount of air or more precisely oxygen that is necessary to keep the sludge in an active and aerobic condition depends on the oxygen demand of the sludge organisms, the quantity of active sludge, and the amount of food to be utilized. Given sufficient food and sufficient organisms to eat the food, the process seems to be limited only by the rate at which oxygen or air can be dissolved into the mixed liquor. This rate depends on the oxygen deficit, turbulence, bubble size, and temperature and at present is restricted by the physical methods of forcing the air through the diffuser tubers and/or mechanical agitation.

In practice, the excess activated sludge is conditioned with 3–6% $FeCl_3$ and filtered on vacuum filters. This reduces the moisture to about 80% and produces a filter cake which is dried in rotary or spray driers to a moisture content of less than 5%. It is bagged and sold direct as a fertilizer, or to fertilizer manufacturers who use it in mixed fertilizer.

The mechanism of purification of sewage by the activated sludge is two-fold i.e., (1) absorption of colloidal and soluble organic matter on the floc with subsequent oxidation by the organisms, and (2) chemical splitting and oxidation of the soluble carbohydrates and proteins to CO_2, H_2O, NH_3, NO_2, NO_3, SO_4, PO_4 and humus. The process of digestion proceeds by hydrolysis, decarboxylation, deaminization and splitting of S and P from the organic molecules before oxidation.

The process is applicable to the treatment of almost any type of organic waste waters which can serve as food for biological growth. It has been applied to cannery wastes, milk products wastes, corn products wastes, and even phenolic wastes. In the treatment of phenolic wastes a special flora is developed which thrives on phenol as food.

W. D. Hatfield

Cross-references: *Fertilizers.*

ACTIVATION, MOLECULAR

When a molecule which is a Lewis base forms a coordinate bond with a metal ion or with a molecular Lewis acid (such as $AlCl_3$ or BF_3), its electronic density pattern is altered, and with this, the ease with which it undergoes certain reactions. In some instances the polarization that ensues is sufficient to lead to the formation of ions by a process of the type

$$A:B: + M = A^+ + :B:M^-$$

More commonly the molecule is polarized by coordination in such a manner that A bears a partial positive charge and B a partial negative charge in a complex A:B:M. Where M is a reducible or oxidizable metal ion, an electron-transfer process may result in which a free radical is generated from A:B: or its fragments and the metal assumes a different oxidation state. In any case the resulting species is often in a state in which it undergoes one or more types of chemical reaction much more readily.

Theoretical Basis. The theoretical basis underlying these activation processes is in the description of the bonding which occurs between the ligand (Lewis base) and the coordination center. This consists of variable contributions from two types of bond. The first is from the sigma bond in which both the electrons in the bonding orbital come from the ligand. This kind of bonding occurs with ligands with available lone pairs (as NH_3, H_2O and their derivatives) and leads to a depletion of electronic charge from the substrate. The second is from the pi bond; here the electrons may come from the metal (if it is a transition metal with suitably occupied d orbitals) or from the ligand (where it has filled p orbitals or molecular orbitals of suitable symmetry). In this case the electronic shifts may partially compensate those arising from the sigma bond if the metal "back-donates" electrons to the ligand. Such shifts may also accentuate those of sigma bonding where ligand electrons are used for both bonds. It is generally found that the net drift of the electronic density is *away* from the ligand.

Activation of Electrophiles. The activation of electrophiles by coordination is a direct result of the weakening of the bond between the donor atom and the rest of the ligand molecule after the donor atom has become bonded to the coordination center. There is considerable evidence to support the claim that this bond need not be broken heterolytically prior to reaction as an electrophile. When this does not occur the literal electrophile is that portion of the ligand which bears a partial positive charge. In such a case the activation process can be more accurately represented by:

$$A:B: + M = \overset{\delta+ \ \ \delta-}{A:B:M}$$

Examples of this type of process are:

$A:B:$	$+$	M	$\rightarrow$	$\overset{\delta+ \ \ \delta-}{A:B:M}$
$Cl:Cl:$		$FeCl_3$		$Cl:Cl:FeCl_3$
$NC:Cl:$		$AlCl_3$		$NC:Cl:AlCl_3$
$RC:Cl:$ ‖ O		$AlCl_3$		$RC:Cl:AlCl_3$ ‖ O
$O_2N:Cl:$		$AlCl_3$		$O_2N:Cl:AlCl_3$
O ‖ $RS:Cl:$ ‖ O		$AlCl_3$		O ‖ $RS:Cl:AlCl_3$ ‖ O
$R:Cl:$		BF_3		$R:Cl:BF_3$

The resultant electrophiles are effective attacking species and can be used to replace an aromatic hydrogen by the group A. When coordination is used to activate an electrophile which has additional donor groups not involved in the principal reaction, and if these are sufficiently effective as donors, they will react with the coordination centers initially added and stop the activation process. In these cases a larger amount of the Lewis acid must be added so that there is more than enough to complex with all the uninvolved donor groups. The extra reagent then provides the Lewis acid needed for the activation process. This is encountered in the Fries reaction and in Friedel-Crafts reactions where the substrate has additional coordination sites. This particular procedure is also used in the "swamping catalyst" procedure for the catalytic halogenation of aromatic compounds.

The usual activation of carbon monoxide by coordination appears to involve complexes in which the carbon atom bonded to the metal is rendered slightly positive, and thus more readily attacked by electron rich species such as ethylenic or acetylenic linkages. An example is seen in the reaction of nickel carbonyl and aqueous acetylene, which results in the production of acrylic acid.

Activation of Free Radicals. The formation of free radicals results from a very similar process when the species M can be oxidized or reduced in a one-electron step with the resultant heterolytic splitting of the A:B bond. The basic reaction in an oxidation reaction of this sort is

$$A:B: + M^{+x} = A\cdot + :B:M^{+x+1}$$

where A and B may be the same or different. The most thoroughly characterized of these reactions is the one found with Fenton's reagent:

$$Fe^{2+} + HOOH = HO\cdot + Fe(OH)^{2+}.$$

The resultant hydroxyl radicals are effective in initiating many chain reactions. The number of metal ions and complexes which are capable of activating hydrogen peroxide in this manner is quite large and is determined in part by the redox potentials of the activator. Related systems in which free radicals are generated by the intervention of suitable metallic catalysts include many in which oxygen is consumed in autoxidations. Cobalt(II) compounds which act as oxygen carriers can often activate radicals in such systems by reactions of the type:

$$Co(II)L + O_2 = Co(III)L + O_2^-, \text{ etc.}$$

Processes of this sort have been used for catalytic oxidations and in cases where a complex with O_2 is formed, the reversibility of the reaction has been studied as a potential process for separating oxygen from the atmosphere.

Radical generating systems of this sort may be used for the initiation of many addition polymerization reactions including those of acrylonitrile and unsaturated hydrocarbons. The information on systems other than those derived from hydrogen peroxide is very meager.

The activation of O_2 by low oxidation states of platinum has also been demonstrated in reactions such as

$$2P(C_6H_5)_3 + O_2 \xrightarrow{Pt(P(C_6H_5)_3)_3} 2(C_6H_5)_3PO.$$

Ligands such as ethylene are Lewis bases by virtue of the availability of the electrons of their pi bonds to external reagents and the coordination of unsaturated organic compounds to species such as Cu(I), Ag(I), Pd(II), and Pt(II) is a well established phenomenon. The coordination process with such ligands usually involves a considerable element of back bonding from the filled d orbitals of the metal ion. The coordination process activates olefins towards cis-trans isomerizations and attack by reagents such as hydrogen halides. Coordination to palladium(II) facilitates attack of olefins by water via a redox process as seen in the Smid reaction:

$$PdCl_2 + H_2C{=}CH_2 + H_2O =$$
$$Pd + CH_3CHO + 2HCl$$
$$Pd + 2CuCl_2 = 2CuCl + PdCl_2$$
$$2HCl + 2CuCl + \tfrac{1}{2}O_2 = 2CuCl_2 + H_2O$$

A similar reaction also occurs for carbon monoxide:

$$Pd^{2+} + CO + H_2O = Pd + CO_2 + 2H^+.$$

Activation of nucleophiles by coordination, best exemplified by various complexes used as catalysts for hydrogenation, and coordination assistance to photochemical activation, may be similarly demonstrated.

MARK M. JONES

Cross-References: *Ligand Field Theory, Chelation, Free Radicals, Oxidation, Hydrogenation.*

ACTIVITY

The term "activity" is used in a general way to refer to the rate or extent of a change associated with some substance or system. Thus the activity of magnesium metal is greater than that of copper, for it will combine more rapidly and to a larger extent with a variety of substances and will displace copper from many of its compounds. This general meaning is also implied when activity is used in words or phrases like radioactivity, optical activity, catalytic activity and biological activity.

Activity has a more specialized meaning in chemical thermodynamics. The behavior of ideal systems at equilibrium is described by laws which are relations involving the concentrations of the species present. The substitution of activities for concentrations, as suggested by G. N. Lewis, preserves the mathematical form of the ideal law and enables it to be extended to real systems, particularly liquid solutions and alloys. For example, the ratio of the product of the activities of hydrogen and acetate ions to the activity of unionized acetic acid has a fixed value independent of the concentration of acid or the presence of salts, whereas the ideal mass action constant, which is the ratio of concentrations, is not strictly constant. The activity of a chemical species can be expressed as the product of its concentration and an activity coefficient that measures the deviation of the species from ideal behavior. The numerical value of the activity and activity coefficient will depend upon the unit of concentration employed. For electrolytes the mean activity ($a_{\pm}$) and mean activity coefficient are used because the activity and activity coefficient of a single ionic species cannot be measured and can thus be used only in a formal way. The relation between mean and ionic activities is illustrated by $a_{\pm} = (a_{K^+}a_{Cl^-})^{1/2}$ for potassium chloride and $a_{\pm} = (a_{Ba^{++}}a_{Cl^{--}})^{1/3}$ for barium chloride.

Activity coefficients and activities are most commonly obtained from measurements of vapor pressure lowering, freezing point depression, boiling point elevation, solubility, and electromotive force. In certain cases they can also be estimated theoretically. As commonly used, activity is a relative quantity having unit value in some chosen standard state. Thus the standard state of unit activity for water in aqueous solutions of potassium chloride is pure liquid water at 1 atm. pressure and the given temperature. The standard state for the activity of a solute like potassium chloride is often so defined as to make the ratio of the activity to the concentration of solute approach unity as the concentration decreases to zero.

The relative nature of activity is shown also by its formal definition as the ratio of the fugacity of the substance in the given state to its fugacity in the standard state or by its relation to the chemical potentials (μ and μ^0) of the substance in the given and standard states: $\mu = \mu^0 + RT \ln a$ (R is the gas constant and T the absolute temperature.) An absolute activity (λ) can be defined by $\mu = RT \ln \lambda$, but this has not come into general use. The variation of the activity of a species i with temperature is related to its relative partial molal heat content (L_i) by $((\partial \ln a_i/\partial T)_{P,N} = (L/RT^2)$. The influence of total pressure on activity is given by $(\partial \ln a_i/\partial P)_{T,N} = V_i/RT$ where V_i is the partial molal volume of the species.

EDWARD J. KING

Reference

"Thermodynamics" by G. N. Lewis and M. Randall, 2nd ed. revised by K. S. Pitzer and L. Brewer, McGraw-Hill Book Co., New York, N. Y. (1961).

Activity Coefficient

In general, the activity coefficient of a substance may be defined as the ratio of the effective contribution of the substance to a phenomenon to the actual contribution of the substance to the phenomenon. In the case of gases the effective pressure of a gas is represented by the fugacity f and the actual pressure of the gas by P. The activity coefficient, γ, of the gas is given by $\gamma = f/P$.

One method of calculating fugacity and hence γ is based on the measured deviation of the volume of a real gas from that of an ideal gas. Consider the case of a pure gas. The free energy F and chemical potential μ changes with pressure according to the equation

$$dF = d\mu = V \, dP. \qquad (2)$$

but by definition

$$d\mu = V \, dP = RT \, d \ln f \qquad (3)$$

If the gas is ideal, the molal volume V_i is given by

$$V_i = \frac{RT}{P} \qquad (4)$$

but for a nonideal gas this is not true. Let the molal volume of the nonideal gas be V_n and define the quantity α by the equation

$$\alpha = V_i - V_n = \frac{RT}{P} - V_n \qquad (5)$$

Then V of Eq. (2) is V_n of Eq. (5) and hence from Eq. (5)

$$V = \frac{RT}{P} - \alpha \qquad (6)$$

Therefore from Eqs. (2), (3), and (6)

$$RT \, d \ln f = dF = d\mu = RT \, d \ln P - \alpha \, dP \qquad (7)$$

and

$$RT \ln f = RT \ln P - \int_0^P \alpha \, dP \qquad (8)$$

Thus knowing PVT data for a gas it is possible to calculate f. The integral in Eq. (8) can be evaluated graphically by plotting α, the deviation of gas volume from ideality, versus P and finding the area under the curve out to the desired pressure. Also it may be found by mathematically relating α to P by an equation of state, or by using the method of least squares or other acceptable procedure the integral may be evaluated analytically for any value of P. The value of f at the desired value of P may thus be found and consequently the activity coefficient calculated. Other methods are available for the calculation of f and hence of γ, the simplest perhaps being the relationship

$$f = \frac{P^2}{P_i} \qquad (9)$$

where P_i is the ideal and P the actual pressure of the gas.

In the case of nonideal solutions, we can relate the activity α_A of any component A of the solution to the chemical potential μ_A of that component by the equation

$$\mu_A = \mu_A^{0\prime} + RT \ln a_A \qquad (10)$$

$$= \mu_A^{0\prime} + RT \ln \gamma_A X_A \qquad (11)$$

where $\mu_A^{0\prime}$ is the chemical potential in the reference state where a_i is unity and is a function of temperature and pressure only, whereas γ_A is a function of temperature pressure and concentration. It is necessary to find the conditions under which γ_A is unity in order to complete its definition. This can be done using two approaches—one using Raoult's law which for solutions composed of two liquid components is approached as $X_A \to 1$; and two using Henry's law which applies to solutions, one component of which may be a gas or a solid and which is approached at $X_A \to 0$. Here X_A represents the mole fraction of component A.

For liquid components using Raoult's law

$$\gamma_A \to 1 \text{ as } X_A \to 1 \qquad (12)$$

Since the logarithmic term is zero in Eq. (11) under this limiting condition, $\mu_A^{0\prime}$ is the chemical potential of pure component A at the temperature and pressure under consideration. For ideal solutions the activity coefficients of both components will be unity over the whole range of composition.

The convention using Henry's law is convenient to apply when it is impossible to vary the mole fraction of both components up to unity. Solvent and solute require different conventions for such solutions. As before, the activity of the solvent, usually taken as the component present in the higher concentration, is given by

$$\gamma_A \to 1 \text{ as } X_A \to 0 \qquad (14)$$

Thus, $\mu_A^{0\prime}$ for the solute in Eq. (11) is the chemical potential of the solute in a hypothetical standard state in which the solute at unit concentration has the properties which it has at infinite dilution.

γ_A is the activity coefficient of component A in

the solution and is given by the expression

$$\gamma_A = \frac{a_A}{X_A} \qquad (15)$$

In Eq. (15) a_A is the activity or in a sense the effective mole fraction of component A in the solution.

The activity a_A of a component A in solution may be found by considering component A as the solvent. Then its activity at any mole fraction is the ratio of the partial pressure of the vapor of A in the solution to the vapor pressure of pure A. If B is the solute, its standard reference state is taken as a hypothetical B with properties which it possesses at infinite dilution.

The equilibrium constant for the process

$$B \text{ (gas)} \longleftrightarrow B \text{ (solution)} \qquad (16)$$

is

$$K = \frac{a_{\text{solution}}}{a_{\text{gas}}} \qquad (17)$$

Since the gas is sufficiently ideal its activity a_{gas} is equivalent to its pressure P_2. Since the solution is far from ideal, the activity a_{solute} of the liquid B is not equal to its mole fraction N_2 in the solution. However,

$$K' = N_2/P_2 \qquad (18)$$

and extrapolating a plot of this value versus N_2 to $N_2 = 0$ one obtains the ratio where the solution is ideal. This extrapolated value of K' is the true equilibrium constant K when the activity is equal to the mole fraction

$$K = a_2/P_2 \qquad (19)$$

Thus a_2 can be found. The methods involved in Eqs. (16) through (19) arrive at the activities directly and thus obviate the determination of the activity coefficient. However, from the determined activities and known mole fractions γ can be found as indicated in Eq. (15).

In the case of ions the activities, a_+ and a_- of the positive and negative ions, respectively, are related to the activity, a, of the solute as a whole by the equation

$$a_+^p \times a_-^q \qquad (20)$$

and the activity coefficients γ_+ and γ_- of the two charge types of ions are related to the molality, m, of the electrolyte and ion activities a_+ and a_- by the equations

$$\gamma_+ = \frac{a_+}{pm} ; \quad \gamma_- = \frac{a_-}{qm} \qquad (21)$$

Also the activity coefficient of the electrolyte is given by the equation

$$\gamma = (\gamma_+^p \times \gamma_-^q)^{(1/p+q)} \qquad (22)$$

In Eqs. (20), (21) and (22) p and q are numbers of positive and negative ions, respectively, in the molecule of electrolyte. In dilute solutions it is considered that ionic activities are equal for uni-univalent electrolytes, i.e., $\gamma_+ = \gamma_-$.

Consider the case of $BaCl_2$.

$$\gamma = (\gamma_+ \times \gamma_-^2)^{(1/1+2)} = (\gamma_+ \times \gamma_-^2)^{1/3} \qquad (23)$$

or

$$\gamma^3 = \gamma_+\gamma_-^2 \qquad (24)$$

also

$$a = a_+ \times a_-^2 = (m\gamma_+)(2m\gamma_-)^2 \qquad (25)$$

$$= 4m^3\gamma_+\gamma_-^2 = 4m^3\gamma^3 \qquad (26)$$

Activity coefficients of ions are determined using electromotive force, freezing point, and solubility measurements or are calculated using the theoretical equation of Debye and Hückel.

The solubility, s, of AgCl can be determined at a given temperature and the activity coefficient γ determined at that temperature from the solubility and the solubility product constant K. Thus

$$K = a_+ a_- = \gamma_+c_+\gamma_-c_- \qquad (27)$$

where c_+ and c_- are the molar concentrations of the positive silver and negative chloride ions, respectively. The solubility s of the silver chloride is simply $s = c_+ = c_-$. The expression for K is then

$$K = \gamma^2s^2 \qquad (28)$$

and

$$\gamma = \frac{K^{1/2}}{s} \qquad (29)$$

By measuring the solubility, s, of the silver chloride in different concentration of added salt and extrapolating the solubilities to zero salt concentration, or better, to zero ionic strength, one obtains the solubility when $\gamma = 1$, and from Eq. (29) K can be found. Then γ can be calculated using this value of K and any measured solubility. Actually, this method is only applicable to sparingly soluble salts. Activity coefficients of ions and of electrolytes can be calculated from the Debye-Hückel equations. For a uni-univalent electrolyte, in water at 25°C, the equation for the activity coefficient of an electrolyte is

$$\log \gamma = -0.509z_+z_-\sqrt{\mu} \qquad (30)$$

where z_+ and z_- are the valences of the ion and μ is the ionic strength of the solution, i.e.,

$$\mu = \tfrac{1}{2} \Sigma \, c_iz_i^2 \qquad (31)$$

where c_i is the concentration and z_i the valence of the ith type of ion.

To illustrate a use of activity coefficients, consider the cell without liquid junction

$$Pt,H_2(g); HCl(m); AgCl,Ag \qquad (32)$$

for which the chemical reaction is

$$\tfrac{1}{2} H_2(g) + AgCl \text{ (solid)} = HCl \text{ (molality, } m)$$
$$+ Ag \text{ (solid.)} \quad (33)$$

The electromotive force, E, of this cell is given by the equation

$$E = E° - \frac{2.303RT}{nF} \log \frac{a_{HCl}}{P_{H_2}}$$
$$= E° - 0.05915 \log m^2\gamma^2 \qquad (34)$$

where $E°$ is the standard potential of the cell, n is the number of electrons per ion involved in the electrode reaction (here n = 1), $\mathbf{F}$ is the coulombs per faraday, a (equal to $m^2\gamma^2$) is the activity of the electrolyte HCl, P_{H_2} is the pressure (1 atm) and is equal to the activity of the hydrogen gas, and AgCl (solid) and Ag (solid) have unit activities. Transferring the exponents in front of the logarithmic term in Eq. (34), the equation can be written,

$$E = E° - 0.1183 \log m - 0.1183 \log \gamma \quad (35)$$

which by transposing the log m term to the left of the equation becomes

$$E + 0.1183 \log m = E° - 0.1183 \log \gamma \quad (36)$$

For extrapolation purposes, the extended form of the Debye-Hückel equation involving the molality of a dilute univalent electrolyte in water at 25°C is used:

$$\log \gamma = -0.509\sqrt{m} + bm \quad (37)$$

where b is an empirical constant.

Substitution of log γ from Eq. (37) into Eq. (36) gives

$$E + 0.1183 \log m - 0.0602m^{1/2}$$
$$= E° - 0.1183bm \quad (38)$$

A plot of the left hand side of Eq. (38) versus m yields a practically straight line, the extrapolation of which to $m = 0$ gives $E°$ the standard potential of the cell. This value of $E°$ together with measured values of E at specified m values can be used to calculate γ for HCl in dilute aqueous solutions at 25° for different m — values. Similar treatment can be applied to other solvents and other solutes at selected temperatures.

Activity coefficients are used in calculation of equilibrium constants, rates of reactions, electrochemical phenomena, and almost all quantities involving solutes or solvents in solution.

EDWARD S. AMIS

Cross-references: *Activity, Solutions, Debye-Hückel Theory, Electrochemistry.*

ADHESIVES

An adhesive is a substance that holds materials together by surface attachment. The materials bonded are termed *adherends* or *substrates*. There are many conflicting theories to explain the source and strength of adhesive bonds; but on one point there is agreement: the adhesive and substrate must be in intimate contact. Consequently, the adhesive must "wet" the substrate. At the time of application, therefore, the adhesive must be fluid. Subsequently, however, it must harden sufficiently to develop the high cohesive strength that is characteristic of high molecular weight thermoplastics or crosslinked thermoset materials.

Mechanisms of Setting

The *setting* or transformation from liquid to solid is accomplished in various ways:

(1) *Cooling of a thermoplastic or hot melt.* A thermoplastic is a material which becomes fluid when heated, then solidifies on cooling. Among adhesives in this category are waxes, asphalt, polyethylene, compositions comprising ethylene/vinyl acetate copolymers, and polyvinyl butyral.

Hot-melt adhesives are being used in breadwraps and other packages, for binding paperback books, for attaching soles and in some lasting operations in shoe manufacture, and as the inner layer in the manufacture of safety glass.

(2) *Release of water or organic solvent.* Most adhesives for paper are based on water solutions, dispersions, or latexes, using such materials as starch and dextrin, polyvinyl acetate, natural and synthetic rubbers, animal glue, and sodium silicate. As the water evaporates or diffuses through the porous substrate, the polymeric composition becomes a film.

Solvent cements are compositions usually based on rubbers, in which the evaporation of organic solvent results in the formation of a film. Aliphatic and aromatic hydrocarbons are the most widely used solvents for the rubbers, while ketones and esters are preferred for nitrocellulose and other polar resins. Flammability, cost and toxicity are sometimes objections to the use of solvents.

Contact cements are usually solutions of neoprene and phenolic resin which give strong bonds when they are applied to metal surfaces, allowed to stand until the solvent has evaporated and then brought into contact with other adherends similarly coated.

(3) *Polymerization in situ.* Resin intermediates capable of crosslinking are applied to the adherend as reactants or polymers of low to intermediate molecular weight. Then they are allowed to react to form crosslinked structures that are characteristically strong and resistant to heat, water and organic solvents. In this category are the phenolics, urea resins, epoxies, and other thermosetting materials.

The bonding of wood to make plywood and particle-board is accomplished principally through the use of phenol-formaldehyde and urea-formaldehyde resins, applied as intermediates (together with catalysts and other ingredients) in aqueous solution, and generally cured with the aid of heat. Pressure is required to contain the water of condensation that is formed in the final curing reaction.

A particular advantage of epoxy resins, unsaturated polyesters and isocyanates is that they are capable of crosslinking without emission of volatiles. Consequently they can be applied to impervious surfaces, such as steel, aluminum or glass, which can then be brought together and cured with little or no pressure.

Some thermoplastics can also be made by polymerization in situ. For example, methyl methacrylate monomer is utilized as the polymerizable binder for polymethylmethacrylate powder in tooth fillings.

(4) *Plastisols* are dispersions of polyvinyl chloride resin in plasticizers, stable and moderately viscous at room temperature. Upon heating, the plasticizer solvates the resin, causing it to fuse to a continuous tough film. In *organosols*, some of the nonvolatile is replaced by volatile organic solvent, resulting in lower fusion temperatures and harder products. A significant recent development in adhesive technology is a plastisol containing a polymerizable plasticizer, thus capable of giving a high strength bond.

(5) *Pressure-sensitive* tapes and labels are materials in which a paper, cloth or plastic backing is coated with an elastomeric *mass coat* that will give a bond of at least moderate strength upon application of only light pressure at room temperature.

(6) *Encapsulated* adhesives are a novel approach to prepare one-part systems. Here the hardener is surrounded by a thin membrane, which separates it from the bulk resin. The membrane is destroyed

under the influence of heat or pressure, and the adhesive will cure.

The Adhesive Materials

A great variety of polymers can be used in adhesive applications. The selection of adhesive depends upon the adherend and the end use. Consideration must be taken of such factors as the porosity or impermeability of the adherend, its polarity, and its modulus of rigidity. The adhesive film must be no stiffer than the adherend else flexing will cause stresses to concentrate in the adhesive layer and bring about failure.

Cost is a major consideration in products such as corrugated cartons, in which the adhesive contributes heavily to overall expense. For such products, starch and sodium silicate are still the preferred materials, despite poor water resistance.

Natural Adhesives. All organic adhesives are polymers, deriving their cohesive and adhesive strength from a multiplicity of contacts with each other and with the adherends.

Animal glues are derived from the bones or hides of animals, particularly cattle. Hide glues are generally higher than bone glues in molecular weight, solution viscosity, strength, and price. The animal glues are used in gummed tapes, sandpaper, bookbinding, and the joining of furniture. The supply is diminishing.

Casein, obtained from milk, is largely imported from Argentina. It, too, is becoming less available. It is used as a constituent of binders for clay coatings for paper. In large wooden arches and other structural laminates for indoor use, casein is an effective laminant. Joint cements for gypsum wallboard are fortified with casein, and ammonical casein solutions are employed in conjunction with various synthetic rubber latexes for the bonding of aluminum foil in packaging applications.

Soybean and *blood glues* are employed as adhesives extenders for plywood. Purified soy protein is also going into clay coated paper as a pigment binder.

Starches and dextrins, low in cost, nontoxic and available in large quantity, are the main adhesives for the bonding of paper products. Domestic corn is the chief source, but imported tapioca and potato starch are also used. The dextrins are derived from the starches by acid or heat treatments, which not only reduce the molecular weight but sometimes reconstitute the molecules to give them more branched structures which are easier to dissolve and more tacky. Corrugated boxboard is the principal outlet for starch adhesives. Other main uses include the manufacture of bags and sacks and the clay coating of paper. Dextrin goes into envelopes, gummed tapes and labels, case-sealing adhesives and many other paper products.

Nitrocellulose (pyroxylin, cellulose nitrate) has had more than a century of use as an adhesive, and still goes into household cements, hobby cements and heat-sealable coatings for cellophane.

Natural rubber is employed in both latex and solvent cements. As a latex, it is used for the bonding of paper and fabrics, especially in carpet backing and shoe manufacture. Solvent cements based on natural rubber are favored as pressure-sensitive adhesives, providing high "tack" (the ability to develop at least moderate adhesive strength immediately upon contact with the adherend).

Synthetic Adhesives. *Styrene-butadiene rubber* (SBR) goes into much the same applications as natural rubber. It is not as tacky but is more uniform in composition and has usually been lower in price. *Styrene-butadiene resin,* containing 50% or more styrene, is a harder material, suitable for the binding of clay and other pigments in the coating of paper and paperboard, as well as carpet backing. *Reclaimed rubber* is derived from tires and inner tubes and is thus of variable composition, but with SBR and fillers as its main components. It goes into pressure-sensitive tapes and miscellaneous and inexpensive cements.

Styrene-butadiene-styrene block copolymers display both elastomeric and thermoplastic properties, and are therefore promising for hot-melt, pressure-sensitive adhesive applications.

Neoprene is the elastomer derived from 2-chlorobutadiene. It is used in adhesives principally as a solvent cement, often compounded with phenolic resins, terpene resins, partially hydrogenated rosin esters, etc. It makes an excellent contact cement for the attaching of shoe soles, the manufacture of laminated panels, and the laying down of high pressure decorative laminates in kitchens and table tops.

Nitrile rubbers, known abroad as buna N, are copolymers of butadiene with acrylonitrile. They have excellent resistance to oil. Compounded with phenolic resins, they provide high strength bonding in brake linings, aircraft wings, and miscellaneous applications. The latexes are useful in nonwoven fabrics where high elasticity is required.

Polysulfides and *silicones* are utilized principally as elastomeric sealants in the construction of highrise buildings.

Thermosetting Resins. *Phenolic resins* made by the condensation of phenol and formaldehyde are the adhesives for exterior grade plywood. Historically, this highly water resistant plywood has been made from Douglas Fir, principally in the northwestern states. But Southern Pine is coming more and more into use for this purpose. Other uses for phenolic adhesives include high pressure laminates, for table tops, counters and electrical panels; coated and bonded abrasives; brake linings and other friction materials, thermal insulation based on glass fiber; and foundry sand binders.

Resorcinol formaldehyde is sometimes added to phenolic resins to facilitate curing at a lower temperature.

Amino resins may be either urea formaldehyde, melamine formaldehyde or combinations. Melamine resins are more expensive than urea but provide considerably better resistance to heat and water. Both types are light in color, in contrast to the dark brown of phenolics and resorcinol resins. Hardwood plywood and particleboard are principal consumers of these materials. Wet strength papers and the surface layers of high pressure laminates are additional significant outlets.

Epoxy resins have in a short time shown outstanding utility for the bonding of metals and glass. The most common types used in adhesives are liquid resins derived from bisphenol A and epichlorohydrin, and hardened by reaction with polyamines.

Isocyanates and the *polyurethanes* made from them are among the newest of the adhesive materials with promise of large scale usage. Elastomeric urethanes are showing marked growth as sealants for buildings, clay pipe, and varied other applications. In solvent based adhesives, the highly reactive isocyanate group facilitates good adhesion to fabrics and many plastics. Thermoplastic polyurethanes are high-molecular-weight polymers which do not contain free isocyanate groups. They are used in solution or as a hot melt, especially in textile bonding and laminating.

Polyamides of the type most used in adhesives are made by the condensation of dimer acids (dimerized unsaturated fatty acids) with diamines and polyamines. Neutral polyamides are utilized as hot melts, while those having an excess of amino groups are among the favored curing agents for epoxy resins where flexibility and toughness are required. Soluble grades of polyamides also go into epoxy resin adhesive formulations as toughening ingredients providing unusually high peel strength. Nylon 11 and nylon 12 are employed for hot-melt bonding when higher heat resistance and strength are required.

Acrylics are adhesives based on acrylate and methacrylate polymers. Whereas acrylic plastics are usually thermoplastic, acrylic adhesives are more often crosslinkable via amide, methylol amide, carboxyl or hydroxyl groups. With excellent lightfastness and good resistance to washing and drycleaning, they are achieving acceptance in nonwoven fabrics in flocked fabrics.

Cyanoacrylate monomers polymerize in the presence of traces of water. They are useful in the electrical and medical fields to accomplish instant room-temperature bonding.

Anaerobic adhesives are based on diacrylate esters of glycols or on the reaction products of hydroxyethyl methacrylate with diisocyanate compounds. These will cure when the oxygen is excluded and therefore provide for a one-package, room-temperature-curable system employed for metal-to-metal and metal-to-glass bonding.

Polyimides and polybenzimidazoles are examples of polymers used as heat-resistant structural adhesives. They contain heterocyclic rings in their structure which impart high-temperature properties, and are used in the aircraft-aerospace industry. Other heat-resistant adhesives developed recently include polyquinoxalines, polyimidazoquinazolines, etc.

Thermoplastics. *Polyvinyl acetate* and copolymers of vinyl acetate are the leading synthetic thermoplastic adhesive based materials. Usually polymerized and used in latex form, they are leading materials for the bonding of paper in myriad packaging applications. They are also used for the gluing of furniture, the clay coating of paper, nonwoven fabrics and many other applications.

Polyvinyl alcohol is derived from polyvinyl acetate by hydrolysis. The 88% hydrolized grades are utilized in the polymerization of vinyl acetate and other monomers. Paradoxically, the 98–99% hydrolized materials have better water resistance than the 88% hydrolized products, and are therefore favored for many paper-bonding applications. *Polyvinyl butyral,* made by reaction of polyvinyl alcohol with butyraldehyde, is the favored interliner in the manufacture of safety glass.

Inorganic Materials. The inorganic resins are at the same time the oldest and the newest of adhesive materials. *Sodium silicate,* among the cheapest of adhesive materials, has long been used for corrugated board manufacture, but has been losing out to starch compositions. In the field of very high temperature adhesives, the expensive *silicones* and more exotic and costly polymers are the subject of much current research.

IRVING SKEIST
JERRY MIRON

Reference

"Handbook of Adhesives," Irving Skeist, Editor, Reinhold Publishing Corp., New York, 1962.

ADSORPTION

When any gas is brought into contact with any solid under the right conditions of temperature and pressure it is attracted to and partially covers the surface of the solid by a process known as adsorption. A common example is the adsorption of gases by activated charcoal. The solid is called the *adsorbent;* the adsorbed substance is called the *adsorbate.* Actually, either liquids or solids can act as adsorbents. Furthermore, adsorbates can be gases, liquids, solutes, or solvents. The most studied and best known examples of adsorption are gas-solid and liquid-solid systems. This article is restricted to a discussion of the adsorption of gases on solids.

If the adsorbate molecules are held exclusively at the adsorbent interface, the process is called *adsorption;* if they penetrate into the interior of the adsorbent, the process is called *absorption.* In this connection the walls of capillaries, cracks and crevices in the solid are still considered as being part of the surface, (sometimes called the "inner surface"). Penetration into the actual lattice structure of the solid or into the interior of a liquid absorbent must take place before the process is called *absorption.*

The picking up of a gas or liquid by an adsorbent is sometimes referred to as "sorption" when one does not wish to specify, or does not know whether a given process is adsorption or absorption.

Gas-Solid Adsorption. The adsorption of gases on solids has been found to fall into convenient classifications, known respectively as *physical* and *chemical* adsorption. Physical adsorption is so named because presumably it occurs through the operation of physical or van der Waals forces between the solid adsorbent and the adsorbate molecules. As one would expect, the extent of physical adsorption by a given solid will be related to the boiling point of the adsorbates rather than to the chemical nature of either the solid or the adsorbate. Chemical adsorption differs from physical adsorption in that it depends on chemical bond formation between the adsorbent and the adsorbate. Chemical adsorption is therefore highly specific. For example, the noble gases are not chemisorbed by any known solid because they cannot form adsorbate-adsorbent chemical bonds. Similarly, O_2 or N_2 at room temperature would not be expected to chemisorb on a relatively inert solid such as silica gel, but would be expected to chemisorb on reactive materials such as iron catalysts.

Physical Adsorption. The phenomenon of physical adsorption has been recognized for many years. However, because of the absence of any method for measuring surface areas, considerable confusion exists in much of the early work as to the thickness of the adsorbed layers. At the present time, with the help of new methods (See **BET Theory**) for measuring surface areas of solids, it is possible to state with fair certainty that physical adsorption may consist of monolayers, multilayers, or condensation of the adsorbate as a liquid in tiny capillaries of the adsorbent; indeed it may involve all three of these forms at one time. In all cases, however, it is well to keep in mind that physical adsorption is closely related to condensation in that it involves physical forces operating between the adsorbate molecules and between the adsorbate and the adsorbent.

A survey of the literature a number of years ago indicated that all physical adsorption could be expressed by one of five types of adsorption isotherms or plots of the volume of gas adsorbed against pressure. These five isotherms types are shown in Figure 1. Type 1 is characteristic of physical adsorption of vapors on solids, the pores of which are so small as to prevent the building of layers thicker than a single layer. The adsorption of most vapors (particularly nonpolar) on charcoal yields isotherms of Type I. If the physical adsorption can build up indefinitely thick multilayers at sufficiently high relative pressures, isotherms of Type II are obtained. These have proved to be very useful (See **BET Theory**) for measuring surface areas of finely divided or porous solids by an adsorption method. If the heat of adsorption of a gas or vapor is less than the heat of liquefaction, isotherms of Type III are observed. The adsorption of water vapor on graphite or on de-oxygenated carbon black is of this type. If a solid has pores in a medium size range (20 to 500 Å in diameter, for example) they yield adsorption isotherms of Type IV, which presumably represent a combination of monolayer adsorption, multilayer adsorption and capillary condensation. Silica-alumina catalysts of the type used for cracking hydrocarbons to form gasoline usually yield this type of isotherm. Finally, if the heat of

adsorption is small, solids having small pores will yield isotherms of Type V. This can be illustrated by curves for the adsorption of water vapor on charcoal.

Space does not permit a detailed discussion of the thermodynamics of physical adsorption. It will perhaps suffice to point out two things. To begin with, since the adsorption of a gas on a solid involves a free energy decrease and an entropy decrease it follows that adsorption of a gas on a solid is always exothermic. Secondly, the isosteric head of "adsorption" can be calculated from adsorption data by the Clausius-Clapeyron equation

$$\ln \frac{p_2}{p_1} = \frac{\lambda}{R} \left(\frac{1}{T_1} - \frac{1}{T_2} \right)$$

where p_1 and p_2 are the adsorption pressures required to cause a given volume of gas to be adsorbed at temperatures T_1 and T_2, respectively, R is the gas constant. For Types I, II, and IV isotherms the heat of adsorption is frequently 50 to 100% higher than the heat of liquefaction; whereas, for Types III and V it is equal to or less than the heat of liquefaction of the adsorbate.

One other factor related to the thermodynamics of physical adsorption should be mentioned. In the past it has seemed reasonable to base all derivations on the assumption that the solid itself is not influenced by the physical adsorption. But recent experiments by Yates have shown that the adsorption of certain gases such as nitrogen, argon, and krypton at −195°C on a sample of porous glass causes a considerable expansion, even for a small fraction of a monolayer of adsorbed gas on the glass. Other gases cause sharp contraction with small coverages and expansion with large coverage. Clearly then, a detailed treatment of physical adsorption will have to include a consideration of changes both in the solid adsorbent and in the adsorbate.

The study of physical adsorption has received added attention since the discovery of evidence for the existence of multimolecular adsorption. S-shaped adsorption isotherms obtained, using N_2 as adsorbate and iron synthetic ammonia catalysts as adsorbents, seemed most easily interpretable as representing the building up of multilayers at the higher relative pressures. Actually, it appeared that the low pressure end of the long linear part of such an isotherm (designated as "Point B") corresponded to a statistical monolayer of adsorbed gas. This led in turn to the suggestion that one could calculate the area of a solid by merely multiplying the number of molecules adsorbed at Point B by a value for the cross-sectional area of the adsorbed nitrogen molecule (16.2 Å^2 is usually employed). A theoretical treatment of the S-shaped isotherms soon led to an equation known as the BET (Brunauer-Emmett-Teller) equation for obtaining the volume of gas in a monolayer and hence the surface area of the solid adsorbent (See **BET Theory**).

Capillary condensation, another component of physical adsorption, is of particular interest because of information it can give about pore size and pore size distribution of tiny capillaries in porous solids. The Kelvin equation

$$\ln \frac{p}{p_0} = -\frac{2\sigma V \cos \theta}{rRT}$$

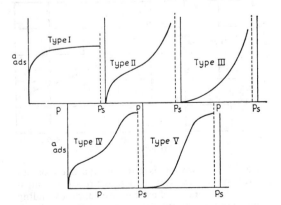

FIG. 1. Types of physical adsorption isotherms (taken from section on BET theory).

was derived in 1871 to show the relation between r, the radius of a capillary; p/p_0, the relative pressure of the vapor in the capillary; T, the temperature; V, the molal volume of the adsorbate; σ the surface tension of the adsorbate, and θ the angle of wetting of the capillary by the liquid adsorbate. It expresses the fact that the pressure p above a liquid condensed in a capillary is less than the vapor pressure p_0 of the bulk liquid. Zsigmondy was the first to apply this equation to show the relation between capillary condensation and pore size. His calculations involved a number of assumptions but were generally considered to furnish at least a semi-quantitative estimate of pore size and distribution.

Soon after the existence of multimolecular adsorption was established, Wheeler called attention to the possibility of modifying the Kelvin equation in such a way as to improve the accuracy of pore size estimates. Specifically, he recognized that the evaporation of capillarily condensed liquid from a given pore at its equilibrium pressure p would leave a multilayer adsorbed on the capillary wall. The pore size calculated from the desorption curves of an adsorption-desorption isotherm such as is shown in Fig. 2 would then give a capillary radius r smaller than the true radius r_c by the thickness of the multilayer left on the walls. Methods have been worked out for taking multilayer adsorption into consideration and calculating from desorption isotherms the pore size and pore size distribution for capillaries in porous solids.

In summary then, one can say that new uses for physical adsorption have been found in the form of a new tool for measuring both the surface area of finely divided solids and the surface area, pore size, and pore size distribution of porous solids. In addition, it finds traditional use as the means by which adsorbents such as charcoal can be used in gas masks for removing poisonous gases from a stream of air or in commercial apparatus for solvent recovery and for air conditioning. Another rapidly growing use for physical adsorption is in the field of gas chromatography that is now being employed for analyzing a variety of gases.

Chemical Adsorption. Chemical adsorption (or chemisorption) partakes of most of the properties of chemical reactions. It occurs only under conditions in which on chemical grounds incipient bond formation between adsorbate and the adsorbent would be expected. It evolves an amount of heat that is usually in the range encountered in chemical reactions (5 to 100 kcal per mole of adsorbate in most cases). Furthermore, chemical adsorption is often slow and temperature-sensitive. Energies of activation for chemical adsorption are frequently in the range 10 to 20 kcal per mole. Occasionally, however, chemical adsorption or chemisorption, as it is sometimes called, occurs instantaneously. This is true, for example, of the chemisorption of carbon monoxide on iron catalysts at $-195°C$. The heat of this adsorption is as high as 35 kcal per mole, but the energy of activation is so low as to be practically unmeasurable.

Multilayer chemisorption is, by definition, assumed not to exist. If an adsorbate is taken up chemically in quantities in excess of a monolayer, it is assumed to have reacted with at least some of the underlying layers of the solid and no longer represents chemisorption in the strictest sense of the word.

Although it is realized that an energy of activation is frequently involved in chemical adsorption, it is also recognized that extreme care must be used in calculating it. The difficulty can perhaps be best explained with the help of the adsorption isobar (a plot of the volume of gas adsorbed against temperature at constant pressure) for hydrogen on an iron catalyst promoted with Al_2O_3 and K_2O. Such a plot is shown in Fig. 3. Types A and B are two different kinds of chemical adsorption of hydrogen. It is evident that any heat values obtained by

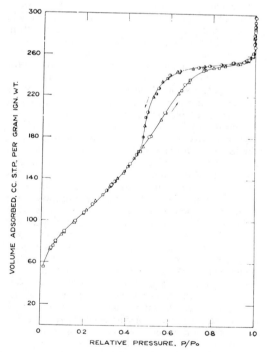

FIG. 2. Nitrogen adsorption-desorption isotherms for a silica alumina cracking catalyst as obtained by Ries and co-workers.

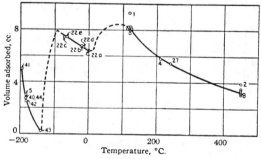

FIG. 3. Adsorption isobar for hydrogen on an Fe–K$_2$O–Al$_2$O$_3$ catalyst at a pressure of 1 atmosphere. The type of chemical adsorption extending from about $-100°$ to $0°C$ is called "type A" adsorption; that occurring at and above about $100°C$, type B adsorption.

applying the Clausius-Clapeyron equation to adsorption data taken partly from Type A and partly from Type B would be of little significance. Only when the data are taken from isotherms for a single type of adsorption can calculated energies of activation be considered as even approximately correct. Actually, there is no assurance that only two types of chemisorption can exist between an adsorbate and a solid. Indeed, on iron catalysts promoted with Al_2O_3, a third type of chemical adsorption of hydrogen (called Type C) has recently been found to exist in the temperature range -100 to $-195°C$. Furthermore, there is ample evidence that chemical adsorption behaves as though the surface were heterogeneous. Originally it was suggested by Taylor that solid catalysts functioned through a number of "active centers" and that these active centers comprised a collection of surface atoms covering a wide range of energies of activation and heats of adsorption. Modern concepts suggest that these "active centers" may not actually represent fixed positions on a surface but may be continually created and destroyed as a result of electron motion within the solid attending the chemisorption of molecules on the surface. The difficulty therefore of calculating significant energies of activation of gases on solid should be evident.

Chemical adsorption is important chiefly because it is one of the essential steps in catalytic reactions (See **Catalysis**). It is generally believed that at least one of the reactants of a catalytic reaction has to be chemically adsorbed during reaction. A second application of chemical adsorption is its use in obtaining estimates of the fraction of the surface of certain metallic catalysts composed of metal atoms. Thus, for example, iron-synthetic ammonia catalysts containing a few percent of aluminum oxide and potassium oxide as promoters appear to have about one-third of their surface covered with iron atoms and the other two-thirds with the promoter molecules. This can be established by comparing the chemisorption of carbon monoxide with the volume of nitrogen required according to the BET theory to cover the entire surface of the catalyst.

Until recently no means existed for telling in detail the properties of molecules chemisorbed on the surface of a catalyst. There are now available a variety of electronic, magnetic, and spectroscopic methods for studying both the catalyst surface itself (See **Catalysis** Chemisorption) and molecules adsorbed on the surface. These new approaches should lead to a much better understanding of adsorbed species and, in consequence, to an elucidation of the nature of the factors entering into the activation of adsorbed molecules for catalytic reactions.

PAUL H. EMMETT

References

Brunauer, S., "Physical Adsorption," Princeton Univ. Press, 1943.
Flood, E. A., "The Solid-Gas Interface," Vols. I and II, Dekker, 1966–67.
Hayward, D. O., and Trapnell, B. M. W., "Chemisorption," Butterworths, 1964.
Wheeler, A., "Catalysis," Vol. II, Chap. 2., Reinhold Publishing Corp., New York (1955).

AEROSOLS

The term "aerosol" was first proposed to mean a system of particles suspended in air and to include both the continuous (gaseous) and discontinuous (particulate) phases. It was originally intended to apply only to systems in which the particle size is sufficiently small that gravitational settling is very slow, but in recent years the term has unfortunately been applied to almost any dispersion in air such as those formed by the aerosol packages of commerce or even to the packages themselves. The original meaning will be used in this discussion unless otherwise indicated.

Particle Dynamics. A spherical particle suspended in a gas quickly attains a constant velocity. If the gas can be assumed to be continuous, and viscous drag is the only restraining force on the particle, the settling velocity (μ) can be calculated from the Stokes equation:

$$\mu = \frac{2gr^2(\rho - \rho^1)}{9\eta}$$

where g is the acceleration of gravity, η is the viscosity of the gas, r is the particle radius, and ρ and ρ^1 are the densities of the particle and gas, respectively. These assumptions cannot be made when the particles are so large that a turbulent wake forms, or so small that the diameters are approximately equal to or less than the mean free path of the gas. In the latter case, the particles "slip" between the atoms or molecules of the gas and settle at a higher velocity than predicted by the Stokes equation. The correction introduced by Cunningham is satisfactory for most purposes:

$$\mu = \frac{2gr^2(\rho - \rho^1)}{9\eta}\left(1 + \frac{Al}{r}\right)$$

where A is a constant close to unity (0.9 is often used) and l is the mean free path of the fluid. Turbulence develops behind a particle when the Reynolds number ($2\mu r/\nu$) exceeds about 1.2 (ν is the kinematic viscosity of the gas), in which case the Stokes equation no longer applies.

The aerosols in which particles are settling are usually somewhat turbulent, and a random motion is superposed on the downward movement of the particles. Such turbulence does not affect the average rate of fall but tends to maintain a uniform concentration throughout the aerosol which decreases with time.

Coagulation of aerosol particles can occur by several processes. When the particles are not all the same size (a "polydisperse" aerosol), the different settling rates can lead to collisions. This does not occur to an appreciable extent in aerosols as defined above, but is an important process in natural clouds and fogs. A much more important mechanism when the particles are small is coagulation by Brownian motion (diffusion). The Smoluchowski equation describing this coagulation for spherical particles in a monodisperse aerosol, and corrected for slippage, is:

$$\frac{1}{n} - \frac{1}{n_o} = \frac{4RT}{3\eta N}\left(1 + \frac{Al}{r}\right)t$$

where n_o and n are the particle number concentrations before and after elapsed time t, R is the gas constant, T is the absolute temperature, and N is Avogadro's number. Coagulation is faster for polydisperse aerosols and if the particles are not spheres.

Electrically charging aerosol particles may markedly affect the rate of coagulation, but even the direction of the effect is often difficult to predict. For example, the rate of coagulation in a unipolar electrically charged aerosol is not necessarily zero even if all the particles are charged, since the close approach of the particles may induce charges of opposite sign. When an aerosol is charged in a bipolar symmetrical manner, the charging may have little effect on the coagulation rate, but when the charging is unsymmetrical, there may be a large increase in the rate.

When an aerosol is accelerated (or decelerated), the inertia of the particles results in their moving relative to the suspending air. If the flow of an aerosol is diverted by an object immersed in the aerosol, particles will tend to impact on the object. If an aerosol flows around a bend in a duct, particles will tend to collect on the walls of the duct. This behavior plays a major role in the laws governing many methods for the collection of particles from aerosols, such as filtration and the impingement of aerosol jets on microscope slides. The collection efficiency for a jet of aerosol impinging on a surface or for an object such as a rod immersed in a flowing aerosol is usually expressed in terms of the ratio of the particle "stopping distance" to the width of the jet or rod. The stopping distance is defined as the distance a particle will continue to travel when the moving aerosol is stopped, and equals $2r^2p\mu/9\eta$. A convenient "rule of thumb" is that the collection efficiency usually is low when this ratio is much less than unity.

When a thermal gradient exists in an aerosol, a force (F) is exerted on the particles such that they move in the direction of decreasing temperature (thermophoresis). The theory of this behavior varies with r/l. When the value is large, the rather complicated equations developed by J. R. Brock of the University of Texas seem to be reasonably satisfactory. When this value is much less than unity, the Waldmann equation is valid:

$$F = -\frac{32}{15} r^2 \lambda_{trans} \Delta T/\bar{c}$$

where λ_{trans} is the translational part of the thermal conductivity of the gas, ΔT is the temperature gradient, and $\bar{c}$ is the mean thermal velocity of the gas molecules. No really satisfactory quantitative theory has been developed for the intermediate region where the ratio is close to unity.

A closely related phenomenon is photophoresis, the particle movement that may occur when an aerosol is irradiated with light. The motion is believed to result from the nonuniform heating of the particles when they absorb the light.

Diffusiophoresis can be defined as the motion imparted to particles suspended in a mixture of gases in which there is a concentration gradient. Several situations are possible. An example is the motion of a small nonvolatile particle in the vicinity of a drop of evaporating liquid.

Production Methods. The methods for producing aerosols are based on one of two general approaches: dispersion and condensation. However, a great variety of techniques exists, and only a few important examples can be mentioned.

One of the oldest methods of controlled aerosol production is by means of aspirators. These utilize the pressure drop produced by an airstream passing by an orifice to draw a liquid to be dispersed into the airstream where it breaks up into droplets. The droplet size distribution is very wide, but by forcing the resulting aerosol to pass around a series of baffles or through a bed of beads, the larger droplets are removed by impaction. Aerosols containing droplets all of which are smaller than one micron diameter, can be prepared in this way. If the liquid is a solution of a nonvolatile solute in a volatile solvent, an aerosol of the solute is obtained with median particle size largely controlled by the concentration of the solution. Individual particles of a powder can also be dispersed in this manner using a suspension of the powder in a volatile liquid. This suspension must be so dilute that there is only a small probability that a droplet will contain more than one particle; the suspension must also be free of dissolved solids such as dispersing agents.

A method for producing nearly monodisperse aerosols from high-boiling liquids was devised by Sinclair and LaMer and has been used in many modifications. Filtered air or other gas is passed through a condensation nuclei generator, over the heated liquid, into a "super heater" at a somewhat higher temperature than the liquid, and finally through a long, double walled condenser where the vapor condenses largely on the nuclei to form the aerosol. It is then diluted to slow coagulation. Additional filtered air may be passed over or bubbled through the heated liquid to dilute that coming from the nuclei generator. The latter may be simply a coil of electrically heated resistance wire coated with sodium chloride. Ions produced from an arc or spark can also be used. The particle size can be varied within certain limits by varying the concentration of nuclei and of vapor. Low concentrations of nuclei and high ones of vapor produce relatively large particles. Modifications of this approach have been used extensively in aerosol research.

Disk atomizers also produce nearly monodisperse aerosols. They consist of horizontal disks rotating about the vertical axis and may be compressed-air driven at very high speeds. The liquid to be dispersed flows onto the center of the disk and moves as a film to the edges where droplets break away. Very small satellite droplets are also produced and must be removed. The retained droplets are tens or hundreds of microns in diameter, much larger than those usually produced by the Sinclair-LaMer generators.

Many generators have used a vibrating capillary. One consists of a glass capillary 20 cm long and 0.6 mm diameter. With part of a hypodermic needle attached to the capillary and an electromagnet, the capillary is forced to vibrate, throwing out droplets. As with the use of the spinning disk, the droplets obtained are quite large and most of the dispersions obtained should probably not be considered to be aerosols.

Aerosols can be produced by a number of gas-phase chemical reactions, although they are usually polydisperse. A classic example is the reaction of hydrogen chloride with ammonia to form clouds of ammonium chloride. Another example is the hydrolysis of titanium tetrachloride vapor in air to produce copious white clouds. This reaction has been used to form smoke screens. Photochemical smog is a chemically-produced aerosol which we could do without. A closely related method is to produce aerosols as "smoke" from combustion, as by burning magnesium ribbon.

Various methods have been developed for the direct dispersion of powders. They usually consist of some method for feeding the powder into a jet of air. The greatest difficulties are achieving a uniform rate of feeding and achieving the dispersion of ultimate particles instead of aggregates.

The simple spraying technique of forcing a liquid through a nozzle, as with a garden hose, produces a very coarse dispersion. However, one application of this technique should be mentioned, namely the aerosol package of commerce. In fact, the term has been broadened to refer to a package which can convert its contents to foams, or which can emit, upon release of a valve, streams, pastes, or creams. The interior of the package is pressurized by a gas or a very high-vapor pressure liquid (the propellant), and this pressure acting on the contents expels part or all of them upon opening a valve. The container may consist of metal, plastic, or glass. The propellant is usually a halogenated hydrocarbon having a vapor pressure at 70°F between 15 and 100 psig. The most common gaseous propellents are carbon dioxide, nitrous oxide, nitrogen, and air.

Finally, it should be emphasized that nature is a great producer of aerosols and our atmosphere is a huge aerosol system. The importance of this can hardly be overemphasized. For example, all rainfall depends on the existence of fine particles serving as condensation or freezing nuclei.

Particle Collection and Characterization. The nature of the particles in an aerosol can be determined at least in part *in situ* and more completely by first collecting the particles. Collection by filtration is commonly used. Filters consisting of submicron diameter fibers have a very high collection efficiency even for particles much smaller than the fiber diameters and produce a low-pressure drop. They collect by at least three mechanisms. Particles of diameter about equal to or larger than the fiber diameters are collected mainly by direct interception or impaction. Smaller particles are collected mainly by Brownian diffusion. Such filters are excellent for collecting large amounts of particulate material for bulk analyses but much less satisfactory for the examination of individual particles. Membrane filters are continuous membranes penetrated by numerous small holes. Particles larger than the holes are collected on the membrane surface, while smaller particles may be collected within the holes. Many such filters can be rendered transparent or even dissolved by immersing them in appropriate liquids, so they are useful for the examination of individual particles with a microscope.

Impactors and impingers collect by sucking the aerosol at a high velocity through a slit or other orifice facing a plate such as a microscope slide or the bottom of a bubbler. Collection is by inertial impaction as described earlier. By arranging several of these in series with slits of decreasing size, some separation according to particle size can be achieved.

Collections based on thermophoresis are made by thermal precipitators. The aerosol flow rates through them are very slow, so they are usually unsatisfactory unless the particle number concentrations are very high.

Small electrostatic precipitators that charge the particles and collect them on a surface charged to the opposite sign are sometimes used for aerosol particle sampling, especially during industrial hygiene studies. They are most useful for bulk analyses.

Collection by simple sedimentation is seldom used except for particles too large to be considered aerosol particles. However, several centrifugal methods are very effective both for collecting the particles and separating them according to size. Examples are the "conifuge," the Goetz "aerosol spectrometer," and the Hochrainer and Brown "spectrometer."

Once the particles are collected they can be studied by any of a large assortment of methods such as by optical and electron microscopy, x-ray and electron diffraction, electron microprobes, and wet chemical analytical methods.

Most of the techniques for studying the particles *in situ* depend on light scattering. Some depend on determining certain properties of light scattered from a considerable volume of the aerosol but most of these have very limited application. They provide either a mean particle size or a size distribution. The most widely-used of the light-scattering methods involves passing the aerosol past a beam of light in such a manner that only one particle is illuminated at one time. Light scattered by the particle is focused onto a photocell, the output of which is fed to a pulse-height discriminator. The output is presented as a cumulative particle size distribution. Another technique occasionally used is essentially very short exposure time photomicrography, without removing the particles from the aerosol. Holography provides a modification of this approach. Aerosol concentrations are sometimes monitored simply by measuring the decrease in the intensity of a beam of light traversing the aerosol. For example, atmospheric turbidity is often measured using the sun as the light source.

RICHARD D. CADLE

References

Cadle, R. D., "Particle Size. Theory and Industrial Applications," Reinhold, (Van Nostrand Reinhold), New York, 1965.

Cadle, R. D., "Particles in the Atmosphere and Space," Reinhold, (Van Nostrand Reinhold) New York, 1966.

Davies, C. N., Ed., "Aerosol Science," Academic Press, New York, 1966.

Fuks, N. A., "The Mechanics of Aerosols," Macmillan, New York, 1964.

Green, H. L., and Lane, W. R., "Particulate Clouds: Dusts, Smokes and Mists," Van Nostrand, Princeton, New Jersey, 1964.

Solid/Gas Aerosols

A relatively stable dispersion of solid particles in air (or more generally, in a gas) is termed an aerosol in conformity with the definition of particles dispersed in a liquid (specifically, water) as a hydrosol. The terms are not entirely analogous because of the much greater instability of aerosols. Aerosols inevitably decay through coagulation and gravity settling; hydrosols may persist indefinitely. The greatest diameter for a particle to be considered as constituting a part of an aerosol is quite arbitrary. Generally accepted is the limitation that particles must have terminal settling velocities under the influence of gravity less than that of a 100 μm water droplet, approximately 25 cm/sec.

Solid/gas aerosols are commonly referred to by a number of other terms arising from the nature of the particles and their mode of generation. A *dust,* in the popular usage of the word, usually refers to solid particles that have settled on a horizontal surface. In the context here, it is understood to be an aerosol of solid particles resulting from some mechanical action. Industrially, dusts are created by crushing, grinding, drilling, polishing and blasting operations. They arise because of the movement of traffic and the interaction of wind and soil. Human beings give rise to fine dusts of all kinds, including those from cosmetics, dried ointments, bits of clothing and leather, dandruff, and even debris of the skin itself. Dusts tend to be very heterogeneous systems of poor stability and to contain lower concentrations of particles with more large particles than smokes.

Smoke particles may be solid, semisolid, or even liquid. They most often arise from a chemical reaction, especially combustion. At one time the term was restricted solely to the products of combustion and destructive distillation. Now the designation is often employed to include other gaseous suspensions not otherwise readily classified, examples being volatilization products and the products of electrical disintegration. Smoke particles generally are less than 5 μm in diameter. *Fume* is a special type of smoke generated in connection with a chemical process and resulting simultaneously in the generation of a gas or vapor. The term is applied, for example, when metallic or metallic oxide particles and sulfur dioxide are created as in a smelting operation. The word *smog,* composed from parts of the words "smoke" and "fog," designates an aerosol irritating to mucuous membranes in the eye, nose, and throat. The term was initially meant to apply to noxious airborne suspensions created from a combination of domestic and industrial emissions with natural fog. Usage now applies equally well to irritating aerosols created photochemically within the lower atmosphere from exhaust gases and vapors without any evident contribution of fog.

Dust particles tend to be of quite irregular shapes. Magnesium oxide fume, on the other hand, consists exclusively of minute cubes; zinc oxide particles are needle-like; while the particles of mercuric oxide are approximately spherical. Some smokes are a mixture of shapes. Acetanilide, for example, gives a mixture of needle-shaped and spherical particles. A few dyestuffs even result in crystalline particles with very long, hair-like tails suggestive of a miniature tadpole.

The concentration of particles in an aerosol may be high. Concentrations up to 10^5 and 10^6 particles per cubic centimeter are not unusual in urban and industrial atmospheres and are encountered in nature. A single dust particle 1 μm in diameter and having a specific gravity of 2.5 weighs 1.3×10^{-12} gram, or, stated differently, 7.6×10^{11} of them are required to make a gram. Under normal illumination, such a particle is invisible to the unaided eye. An aerosol composed of 10^4 such particles per cubic centimeter is not noticeable unless viewed against a bright light.

The property of coagulating continuously and spontaneously is one of the most striking characteristics of aerosols. The particle number concentration diminishes due to collision and cohesion as a consequence, first, of Brownian motion. The reduction rate varies with gas temperature, pressure, and viscosity, and with particle concentration, size, size distribution, and shape. Considering the particles to be spheres of one size only and that every collision results in cohesion, the change in number with time is expressed theoretically by

$$\frac{1}{n} - \frac{1}{n_0} = \frac{4RT}{3\eta A}\left(1 + \frac{0.9\lambda}{r}\right)t \qquad (1)$$

where n is the number of distinct aerosol particles remaining after the elapse of time t; n_0 the number at time zero; R and T the gas constant and absolute temperature, respectively; η the gas viscosity; A Avogadro's number; λ the mean free path length for the gas molecules; and r the radius of the individual particle. Coagulation is thus very great when the number concentration is high, but falls off rapidly with time. The half-life, or time required for reduction to half the number of particles, of aerosols due to Brownian motion alone is given in Table 1. Turbulence within the gas and settling due to gravity increase the rate of aerosol decay, but small-particle aerosols can persist for relatively long times at quite high concentrations.

TABLE 1.

HALF-LIFE OF AEROSOLS BECAUSE OF BROWNIAN COAGULATION

Particle diameter, μm	Half-life, in minutes, at aerosol concentrations, in particles per cubic centimeter, of			
	10^5	10^6	10^7	10^8
2.0	500	50	5	0.5
0.2	300	30	3	0.3
0.02	167	16.7	1.67	0.17
0.1	50	5.0	0.50	0.05

While coagulation is resulting in the formation of ever coarser aerosols, it is creating particle shapes ranging from nodules of roughly spherical character to straight and branched chains, the latter often having the skeletal appearance of a tree. The effective density of the coagulated particles depends on their form and may be as little as one-tenth of the actual material density. Nevertheless, settling out of suspension under the influence of gravity is another attribute of aerosols. The terminal velocity u_s of particles of diameter d is given by Stokes' law:

$$u_8 = \mathrm{kg}(\rho - \rho_g)d^2/\eta \qquad (2)$$

where k is a constant depending on particle shape ($= 1/18$ for spheres); g the gravitational constant; p and p_g the particle and gas density, respectively; and η, as before, the gas viscosity. Settling velocities are listed in Table 2.

TABLE 2. SETTLING VELOCITY OF UNIT DENSITY SPHERES IN THE ATMOSPHERE

Particle diameter, μm	Velocity, cm/min
10	18.11
1	0.208
0.1	0.00514
0.01	0.000403

Aerosols acquire electric charges both during the generation process and from ions in the air. Charging during the process of generation usually results in approximately equal numbers of positively and negatively charged particles, but occasionally one sign predominates. An excess of positive charges accompanies the generation of resin smokes, for example. High temperature liberates electrons and positive ions from the air; as a result, the proportion of particles having charges is initially high— of the order of 90% or more in the case of magnesium oxide aerosols produced by burning magnesium. Low-temperature smoke generation is likely to result in something like 5% of the particles being charged. Atmospheric ions—normally present in a quantity of about 1700 of both signs per cubic centimeter—rapidly discharge highly charged aerosols to a lower equilibrium level and they also charge uncharged aerosols to the equilibrium state. An aerosol that contained 5% charged particles soon after generation was found after 2 hours to be 70% charged. Highly charged aerosols tend to coagulate somewhat more rapidly than lesser charged ones. Charged particles, regardless of the origin of the charge, move in a unidirectional electric field toward one of the electrodes establishing the field until they are deposited upon it. Electrostatic precipitators, making use of this phenomenon, afford one of the primary means for collecting aerosols.

Electromagnetic radiation is scattered, absorbed, and reflected by aerosols in accordance with complex relationships that depend on particle size, shape, and refractive index, and on the radiation wavelength. Atmospheric aerosols are responsible for many of the phenomena relating to the color of the sky, rings around the sun or moon, coronas, and the like. Visibility is influenced by aerosols both because of the attenuation of light from the object being viewed and because of the scattering of light coming from other directions.

Aerosol particles exposed to a thermal field, i.e., the region between two bodies of different temperature, move in the direction of the lower temperature, depositing ultimately on the cooler of the two bodies. The phenomenon is termed thermophoresis. Brownian motion is basically responsible for this thermal migration. Gas molecules in the vicinity of the hotter body, being in more violent agitation than those nearer the cooler object, exert an excess force on the interposed particles.

Photophoresis, or motion of particles under the influence of light, is a related phenomenon. Photophoresis is evident, however, only with very small particles under greatly reduced gas pressures. Unlike unidirectional thermophoresis, photophoretic migration can be either to or away from the light, apparently depending on the properties of individual particles. The phenomenon is not completely explained.

An airborne solid-particle aerosol, like a gas, has the capability of propagating a flame and, in a confined space, producing an explosion. Aerosols having combustible particles, such as those of coal, starch, sugar, wood, sulfur, magnesium, and aluminum, constitute a great hazard when their concentration lies within certain limits. The danger of explosion always increases as particle size decreases. Ignition can be caused from accidental fires, radiant energy, and electric sparks. The latter is of special significance because frictional contact between solid aerosol particles and confining walls may be sufficient to create a static electrical discharge. Calculations show that a powder passing at a rate of 60 lb/min down a chute which, because of corrosion or improper grounding, has a high resistance to earth, specifically 10^{11} ohms, can generate a potential of 10,000 volts in 20 sec. and cause a spark of 0.005 joule, which is sufficient to initiate an explosion.

Laboratory methods of investigating aerosols utilize light transmission and scattering and individual particle photography, but usually examination is carried out after the particles are collected on filters, in electrostatic or thermal precipitators, or with impactors, the latter being devices in which an aerosol stream is deflected causing the particles to impinge on a judiciously located surface. Industrial collectors for aerosols include scrubbers in which liquid sprays impinge on the aerosol collecting the particles; filters; centrifugal separators, that remove the particles by virtue of their inertia; and electrostatic precipitators.

Solid/gas aerosols are intimately associated with air pollution. As such, they are a hazard to the extent that they penetrate the respiratory system, or irritate sensitive eye, nose, and throat regions. Particles larger than about 5 μm in diameter are generally captured and rejected in the upper respiratory passages; smaller than about 0.25 μm, they resist capture anywhere and are mostly exhaled. Intermediate sizes are partially captured in the lungs, the peak retention being at about 1 μm. Polycyclic hydrocarbons, lead, and asbestos particles in urban smog are considered particularly hazardous because of their special properties and small sizes. The respiratory disease byssinosis is produced by prolonged inhalation of cotton dusts by textile workers. Silicosis, a fibriotic pneumoconiosis caused by breathing free silica, usually in the form of quartz particles, is an ever-present danger to miners and others who work with rocks. Many other dusts irritate respiratory passages and cause bronchitis, asthma, or catarrh and some produce dermatitis. Certain metal oxide fumes and metals themselves are poisonous, among them zinc, copper, and cadmium oxides, beryllium and manganese. Natural microflora aerosols spread infectious plant diseases, and, in the case of pollen, afflict hay fever upon millions of sufferers.

Aerosols also enter into many meteorological phenomena. The energy balance of the low atmosphere is influenced by their thermodynamic characteristics since cloud droplets and raindrops each form about a particle as a nucleus. Stratospheric aerosols, such as those produced by volcanic eruptions, have a bearing on the temperature and movement of upper air masses. Radioactive aerosols are responsible for the spreading throughout a hemisphere of the discharge from a nuclear explosion. As dense smokes, aerosols can screen a military operation; confuse a radar signal; and, according to preliminary high-altitude tests, even trigger the premature explosion of a missile. As an orchard protective measure, dense smokes over a region hinder radiative cooling of the earth's surface and reduce frost formation. Dusts provide the most common means for spreading insecticides. Finally, aerosols have contributed their part to pure science. Avogadro's number was calculated in 1909 from the concentration gradient established in a stagnant aerosol. C.T.R. Wilson's cloud chamber first made visible the traces of elementary particles. And, by means of a suspended particle, A. Millikan determined the fundamental electron charge.

CLYDE ORR, JR.

AGGREGATION

This term is a general one that may be considered to include the more specific designations *agglutination, coagulation* and *flocculation*. The reverse phenomenon termed *peptization* will also be considered. All these imply some change in the state of dispersion of sols or of macromolecules in solution. Aqueous colloid systems are generally classified as hydrophilic or hydrophobic. Such terms were adopted by colloid chemists before adequate theories of the colloidal state were developed and there has been a tendency to retain the terminology in application to systems that are true solutions which are thermodynamically stable. An example would be the application of these terms to the dissolution of many biocolloids whose molecular weights were unknown at the time of the original studies. This appeared natural since previous experience had demonstrated that sols of either type might be prepared from small molecules. The situation was clarified considerably when solutions of organic monomers and polymers were studied in various organic solvents. However, solutions of polyelectrolytes in water present special problems both because of the nature of the solute and the nature of the solvent.

Many proteins undergo limited and reversible association-dissociation reactions and appear to have a forced structure so that they occupy more space than would the unfolded molecule. Studies of electrostriction, denaturation, and enzymatic hydrolysis indicate that this excess space may amount to as much as 10% of the total volume. As pointed out by Schellman, the various techniques used for studying proteins are based on characteristic models so that different authors may refer to proteins "as hydrodynamic ellipsoids of revolution, charged spheres, bundles of crystallographic rods, acid-base catalysts, adsorbing surfaces, supporters of lock and key arrangements, charge-studded regions of low dielectric constant but high refractive index, etc." Clarification of some of these terms is facilitated by giving separate consideration to the primary, secondary, and tertiary structures of proteins. Conformational changes occur with various treatments and as a result the proportion of the molecule present in helical form or as a random coil will vary. The α-helix may be stabilized by hydrogen bonds, ionic bonds, and hydrophobic bonds. Many of the properties of native proteins suggest that the majority of the nonpolar groups are found in the internal volume. Thus, changes in conformation may have a marked effect on solubility and on protein-protein interactions. Also, the structure of liquid water is an important factor, since the formation of a hydrophobic bond may be considered as a partial reversal of the solution process. In the water, hydrogen bonds are forming and breaking and in some regions there are "clusters" of highly hydrogen-bonded molecules surrounded by non-hydrogen molecules. Both the size of the clusters and their number will, of course, decrease with rising temperature. It is claimed that the existence of hydrophobic bonds may be due mainly to the entropy increase connected with changes in the water structure around the side chains.

Early attempts at formulating a theory of the stability of hydrophobic colloids were based on the observation that an electric double layer exists at surfaces separating solid and liquid phases. This would adequately account for repulsive forces. Coagulation by electrolytes was thought to be due to the adsorption of oppositely charged ions with resultant elimination of the charge on the sol. It was observed that these systems are much more sensitive to added electrolytes than hydrophilic systems. Electrophoretic measurements showed that, prior to coagulation, some reduction in zeta potential generally could be observed, but coagulation frequently took place before the potential had been reduced to zero. Furthermore, a comparison of different electrolytes of the same valency type revealed an absence of specific ion effects such as would be expected based on an adsorption mechanism. A theory of stability must include an explanation of the Schultze-Hardy rules, which state that coagulation is nearly independent of the nature of the ions of like charge but is strongly dependent on the valency of the ions of opposite charge. The concentration of electrolytes required for coagulation is, in the limiting case, inversely proportional to the sixth power of the valence.

The theory of the stability of lyophobic colloids was developed independently by Verwey and Overbeek and by Darjaguin and Landau. Starting at the surface of the particle, one may consider that there is a molecular condenser with a Helmholtz inner plane and a Helmholtz outer plane. Beyond the outer plane will be the Gouy-Chapman diffuse layer. It was soon found that this model must be corrected to include the Stern theory of the double layer in which the energy of adsorption of the cation consists of a chemical term and an electrostatic term, but the discrete nature of the ionic charge was ignored. Based on this model and the inclusion of London-van der Waals forces, a theory of stability was developed that accounted for many experimental observations including the application of the Schultze-Hardy rule to sols such as arsenic sulfide. Theoretical considerations indicate that the repulsive potential, represented by V_R, decays exponentially with distance while the attractive potential V_A varies inversely with the square of the distance. The conditions for stability or instability become apparent from plots of the total potential $(V_R + V_A)$ as a function of distance. Because of the nature of the above functions, curves may be obtained which pass through a maximum at distances that are of the order of the thickness of the double layer. Positive values denote stability while negative values denote instability. The original calculations were based on a model consisting of infinitely thick parallel plates.

However, some experimental results on silver iodide did not show the proper dependence on the valence of the cation and numerous studies on silver halide sols in *statu nascendi* by Tezak *et al* indicated that the logarithm of the concentration of coagulating ion varies linearly with its valency. Levine *et al* have shown that a new term which corrects for the discrete-ion effect and is proportional to the density of adsorbed counter-ions in the Stern layer occurs in the adsorption energy of these ions. As a consequence, the theory predicts that the electrolyte concentration required to produce coagulation increases with the potential at the colloidal surface for small potentials, but reaches a maximum and then decreases with further increase in surface potential. Using this modification, it is possible to account for the inverse sixth power rule and the logarithmic dependence on concentration. Recent developments include calculation of the double-layer interaction free energy, stabilization and the Hofmeister series, viscous interactions, and consideration of a diffuse layer in the solid phase. (See *J. Colloid and Interface Science*, **33**, 335–455, 562 (1970); **34**, 549 (1970).

Agglutination. Agglutination refers to the aggregation of particulate matter mediated by an interaction with a specific protein. Generally, and perhaps preferably, the term refers to antigen-antibody reactions characterized by a clumping together of visible cells such as bacteria or erythrocytes. The precipitin test, performed with soluble antigens, originally appeared quite distinct from agglutination. However, the same amount of antibody is found in a type I antipneumococcus serum whether determined as a precipitin or as an agglutinin. An extremely sensitive test has recently been devised which involves the preparation of protein-bacteriophage T 4 conjugates using bifunctional reagents such as gluteraldehyde. Antiprotein antibodies can then be detected at levels of 0.2-2 ng of antibody per mil of serum.

The distinguishing feature of these reactions appears to be the presence of special areas where the orientation of active groups permits specific interaction of antigen with antibody. There is evidence that the forces involved may include hydrogen bonding, electrostatic attraction, and London-van der Waals forces. The binding of hapten to antibody stabilizes the tertiary structure of the latter. Therefore, it has been suggested that the amino acid residues involved in the binding site are in nonconsecutive portions of the polypeptide chain. Further details of the sequence of the amino acids has been obtained by replacing the antigen with a very reactive labeled compound and then analyzing the points of attachment of the radioactive tag.

Agglutination reactions have been described in which the agglutinin does not exhibit the usual high degree of specificity or is present without known infection. Antisheep hemolysins are found in rabbits following invasion by certain bacteria. Isoagglutinins of the human blood groups show some distinctive properties suggesting that they may be of genetic origin rather than arising from an infection by a chemically related antigen. Although the agglutinins are generally considered to be globulins, certain viruses can also cause agglutination of erythrocytes. The determinant groups in this interaction may include accurate orientation of an enzyme and its substrate. The surface of the erythrocyte becomes altered and gradual elution of the virus takes place. The altered cell can no longer be agglutinated by fresh virus of the same strain but may be agglutinated by other viruses.

Other reactions resembling agglutination have been observed in which either a certain specificity is evident or a globulin., which may act as an antibody, is serving a less specific purpose. Plant agglutinins have been found which show a high degree of specificity for blood group antigens. The clumping of the fat globules in milk has been described as an agglutination by the proteins of the euglobulin fraction. When the fat globules are dispersed in a dilute salt solution, the addition of this protein fraction induces normal creaming. Although this action is nonspecific, the same protein fraction causes agglutination of certain bacteria when they are added to milk.

Coagulation. Coagulation of *hydrophobic* sols may be brought about by the addition of small amounts of electrolytes. It was pointed out above that the concentration of electrolyte required may show rather complex dependence on the valency and is related to the surface potential. Other methods of causing coagulation are by mechanical agitation, changes in temperature (particularly freezing), removal of stabilizing ions (as by dialysis), or addition of nonelectrolytes. The change may be observed at times by noting the Brownian movement. Coagulation may be rapid, occurring in seconds, or slow, requiring months for completion. The resultant coagula contain relatively small proportions of the dispersions medium in contrast with jellies formed from hydrophilic systems. Sometimes it is found that what at first appeared to be a homogeneous liquid becomes turbid and distinctly nonhomogeneous. Systems which are intermediate between true hydrophobic sols and hydrophilic sols are encountered quite frequently so that, in commons usage, the term "coagulation" is applied to such diverse phenomena as the clotting of blood by thrombin or the clotting of milk by rennin.

Flocculation. Flocculation of sols implies a transformation from a soluble or highly dispersed state into a condition in which loose aggregates of material appear that may resemble small tufts of wool, whence the name. Coagulation and flocculation are generally considered as synonyms, due consideration being given to the slight difference in appearance of the aggregates. To simplify the present discussion, the term flocculation will be reserved for describing *hydrophilic* systems. Based on this differentiation, it is apparent that an important factor in flocculation must be the solvation of the particles, despite the common presence of an electric charge.

Since stability appears to depend on solute-solvent interactions, solute-solute interactions, and solubility properties, flocculation can frequently be brought about by either of two pathways. The addition of salts may compress the double layer, leaving the macromolecules stabilized by a diffuse solvation shell. The addition of alcohol or acetone will dehydrate the particles leading to instability and flocculation. Alternatively, the alcohol or acetone may be added first, which will convert the particle to one of hydrophobic character stabilized largely by the electric double layer. Such a sol can, of course, be coagulated by the addition of small amounts of electrolytes.

Flocculation studies have been carried out on many other hydrophilic systems such as agar, carrageenin, the pectins, the nucleic acids, etc. Electrophoretic studies, made in the presence of various salts, have shown a marked dependence on the valence of the cation. However, specific ion effects have been observed which are related to the radius and polarizing power of the cation and to the polarizability of the negatively charged groups of the colloid. Colloidal clays may take up large amounts of water and exhibit properties intermediate between the extreme types of lyophilic and lyophobic sols. The flocculation of kaolinite has been studied extensively and a sensitivity to the valency of both cations and anions has been noted. This has been referred to as an example of a double Schultze-Hardy rule. (See **Clays**).

Peptization. The term "peptization" was proposed by Thomas Graham over 100 years ago, to describe the dispersion of gelatinous precipitates. The liquefaction of silicic acid by a trace of alkali appeared analogous to the peptic digestion of a protein gel. As colloid science developed, numerous techniques were found for preparing sols and it was discovered that the stability of hydrophobic sols was due to the presence of an electrical double layer. Since this essential charge was frequently the result of adding small amounts of electrolytes, the latter became known as peptizing agents. Non-ionogenic materials also acquire a charge in contact with water as a result of preferential adsorption of hydroxyl ions, which may consequently be considered to be peptizing agents. However, at times, the application of this term has been broadened to include the preparation of solutions of polyelectrolytes.

One of the classic examples of peptization is the preparation of sols of the silver halides. Positively or negatively charged sols may be obtained by the addition of a slight excess of $AgNO_3$ or KI. The Ag^+ or I^- ions are the peptizing ions, being strongly adsorbed because they fit into the lattice surface. Isomorphic ions may be substituted but other ions

merely influence the potential of the double layer. The double layer potential D is related to the excess concentration of the potential-determining ion. If C_{Ag_0} and C_{I_0} represent the concentrations at zero charge,

$$D = \frac{RT}{F} \ln \frac{C_{AG}}{C_{Ag_0}} = -\frac{RT}{F} \ln \frac{C_I}{C_{I_0}}$$

In some cases, the mechanism of charging has been the subject of dispute. Platinum or gold sols, prepared by the Bredig arc method, are said to be stabilized respectively by platinic acid and chloraurates. Precipitated $Fe(OH)_3$ may be peptized by acidification or by the addition of $FeCl_3$. Vanadium pentoxide sols are prepared by thorough washing of the precipitated oxide. Precipitated sulfides, such as HgS, may be peptized by thorough washing followed by the introduction of H_2S. Ferric oxide in xylene is stabilized by copper oleate.

Almost every procedure used to prepare a precipitate may, under proper circumstances, be used to prepare a sol. In fact, in many precipitation procedures, it is probable that the material passes through the sol state. Consequently, peptization is an important and disturbing phenomenon in analytical chemistry. Thorough washing of a precipitate may remove the coagulating electrolyte and the peptized sol will pass through the filter paper. Also, with readily peptized materials, quantitative results can be obtained only if great care is exercised to avoid an excess of either reagent.

ROYAL A. SULLIVAN

AGRICULTURAL CHEMISTRY

Agricultural chemistry in its broadest sense involves the application of chemistry to all phases of agriculture: the composition and modification of soils, the chemical nature of the plants grown on them, the nutritional needs of livestock, the use of pesticides to control weeds and to protect plants and animals against their enemies, and the chemical make-up of products grown for industrial use.

In the latter half of the 19th century agricultural experiment stations began to appear in European countries, then in every American state and in the United States Department of Agriculture. They established the connecting links between laboratory findings and the needs of agriculture. They have largely been responsible for the trend which has been termed "the chemicalization of agriculture." Chemical manufacturers are now tailoring their products for specific uses in agriculture, thus greatly aiding in the improvement in both quantity and quality of farm products.

Soil is not only a mechanical means of holding plants in place, but it is a reservoir of water and of chemicals from which plants can obtain their needs of inorganic elements. Chemistry has determined why some soils are "rich" and others "poor," has devised ways of correcting the chemical faults, and has thus greatly improved their crop-producing potential. By far the greatest contribution to this end has been the development of chemical fertilizers to make up the deficits in soils, and practically all soils are deficient in one way or another for maximal productivity. The tonnage in these commercial fertilizers mainly concerns nitrogen, phosphorus and potassium.

Four momentous discoveries of latent natural resources made possible these commercial fertilizers. One was finding the rather limited guano deposits on a barren island off the west coast of South America. Another was the uncovering of the vast Chilean nitrate deposits. A third was the more or less accidental discovery of potash salts in large quantities in certain parts of Western Europe; and then of finding large phosphate deposits in many parts of the world, notably North Africa, some islands in the Pacific and in Florida and Tennessee. (See Soil Chemistry).

Thus chemists had at their command reasonably adequate and economical supplies of the big three of commercial fertilizers—nitrogen, phosphorus and potassium. In addition to these natural products, synthetic nitrogen compounds have developed apace in recent decades. Nitrogen from the air is converted to solid compounds which can be bagged and shipped to the farms. These compounds are mostly ammonium salts, nitrates, calcium cyanamide and urea. Thus nitrogen, the most critical and the most expensive of the main fertilizer elements, became available in unlimited quantities, since there will always be plenty of atmospheric nitrogen.

From the soil come the plants, but only in part. About 1840 Liebig demonstrated that only the water and the ash constituents are drawn from the soil. All of the rest—the carbon compounds that constitute 90 percent of the dry matter—come from the carbon dioxide of the air. The noncarbon, or inorganic, elements are, however, of vital importance to the plant. Of the 36 or more chemical elements identified in plants, ten are required in relatively high amounts; carbon, hydrogen, oxygen, nitrogen, phosphorus, sulfur, potassium, calcium, magnesium and iron. In addition, trace amounts of boron, manganese, zinc, copper and molybdenum are required. The rest of the total 36 elements are adventitious and useless to the plant. (See Phytochemistry).

Photosynthesis, the fixation of atmospheric carbon dioxide by chlorophyll-bearing plants, is considered to be the most fundamental reaction, or series of reactions, in the living world. By it solar energy is converted to chemical energy in the immediate form of sugars. Indeed, the essence of the whole of agriculture is the trapping of radiant energy and its conversion to chemical energy for human use. With sugar as a start and with compounds of nitrogen and of inorganic elements available the plant proceeds to make the host of organic compounds which characterize its various tissues— cellulose and other carbohydrates, proteins, fats, pigments, acids, vitamins, tannins, drugs, poisons and so on almost indefinitely. (See Photosynthesis).

Chemical processes in animals are different in that the animal cannot use the radiant energy of the sun, but must eat plants (or other animals) to obtain chemical energy. It breaks down the above groups of constituents, burns a portion for fuel and out of the rest resynthesizes its own particular fats, proteins, pigments, fibers, hormones, etc. The residues are excreted in the urine and feces.

The great advances in the chemistry of human nutrition have been paralleled in the nutrition of livestock. In fact farm animals are probably better fed than humans because of easier control over the diet. Such terms as protein requirements, total calories and mineral supplements, so familiar among urban humans, are equally familiar among stockmen.

The volume of food, feed and fiber produced depends not only on the soil conditions under which the crops are grown but also on the success man has in thwarting ravages by innumerable pests, both during growth and during storage after harvest. For many years insecticides and fungicides were limited almost exclusively to inorganic compounds, such as arsenates, copper sulfate and Paris green. Nicotine sulfate, from tobacco waste, was the most widely used organic insecticide. In more recent years a whole flood of new synthetic organic pesticides has been developed, with new ones still being added with almost bewildering rapidity. These have completely revolutionized the farmer's ability to cope with the ravages caused by insects, fungi, rodents and bacteria. (See Pesticides).

While these new pesticides have greatly increased agricultural effectiveness they have also greatly increased the responsibility of the farmer. These chemicals, in one degree or another, are also toxic to humans, sometimes extremely so. The farmer must use them on food crops, but pesticide residues must not be on or in the marketed product in harmful amount. Manufacturers' directions, and information supplied by the experiment stations, must be carefully followed; and the Food and Drug Administration stands guard continuously. Operators of the spray and dust rigs and the wild life around must also be protected. The tremendous persistence built into certain pesticides as a desirable feature has led to unforeseen complications. They have been stored in plant and animal tissues in concentrations which increase upward in the food chains, with increasingly serious ecological consequences. This has led to efforts by agricultural chemists and others to seek out and apply other principles to pest control and to design rapid decomposition into all new pesticides.

Probably the most dramatic use of chemicals in agriculture is in the field of hormones. Most of these are synthetic chemicals which, in extremely minute amounts, can modify plant and animal physiology. Their effects and their uses are amazingly varied. One or another of them can hasten root formation on plant cuttings: induce seedless tomato fruits; selectively kill weeds without harm to desired plants; cause a portion of young fruit to drop off for thinning purposes; prevent premature dropping of fruit before harvest; retard the sprouting of potato tubers or hasten it; speed up weight increase in poultry and lambs; retard the growth of hedges and lawns; defoliate cotton and potato plants previous to mechanical harvest. Further bizarre and useful effects are in the making.

Thus chemical industry supplies a considerable number of chemicals, as fertilizers, pesticides and hormones, for farm use. These are generally known as "agricultural chemicals." But the reverse also holds. Chemical industry has always looked to agriculture for some of its raw materials—sugar, starch, drying oils and food oils as examples—and it is now looking more and more to this basic source of renewable supplies, in contrast to the non-renewable fossil sources—coal, petroleum and natural gas.

Since for several decades agriculture has been producing a surplus of goods for the usual channels of outlet it has become urgently important to develop further materials for industry. For this purpose the United States Department of Agriculture established four Regional Laboratories in 1941. Their objective was utilization research, an organized effort through science and technology to increase present uses of farm products and to discover and develop varied new uses for them, especially nonfood uses. This effort (also called "chemurgy") has met with outstanding success. New directions for these laboratories reflect increasing concern for environmental and nutritional problems. More attention is being given to finding ways to alleviate pollution in farm and food processing operations, and in finding better ways to preserve and deliver food, feed, and fiber products of the highest quality.

J. J. WILLAMAN

AIR

The atmosphere, or air, is the gaseous envelope that surrounds the earth and even penetrates into it to some extent. Air is a mixture because: (1) The composition of air varies slightly, more than can be accounted for by experimental error; the composition of a compound, on the other hand, is constant. (2) Air has the boiling points of its several components, and not one single boiling point at a given pressure. Compounds have one single boiling point. (3) When air containing 21 percent oxygen by volume is dissolved in water, and the air is later removed from the water, the percentage of oxygen recovered in the resulting gas is 35 percent. If air were a compound there would be no change in composition upon dissolving and recovering. (4) When air diffuses through a porous membrane or porous porcelain, the lighter nitrogen molecules pass through more readily than the heavier oxygen molecules. The resulting air changes in composition. A compound shows no change in composition after such a diffusion. These facts are all evidence that air is a mixture and not a compound.

Analysis of air is usually made after it is freed from solids such as dust, spores and bacteria, and water vapor. The following is a representative composite of analyses of dry air.

Substance	% by Weight	% by Volume
Nitrogen, N_2	75.53	78.00
Oxygen, O_2	23.16	20.95
Argon, Ar	1.27	0.93
Carbon dioxide, CO_2	0.033	0.03
Neon, Ne		0.0018
Helium, He		0.0005
Methane, CH_4		0.0002
Krypton, Kr		0.0001
Nitrous oxide, N_2O		0.000,05
Hydrogen, H_2		0.000,05
Xenon, Xe		0.000,008
Ozone, O_3		0.000,001

The composition of air varies with altitude at which the sample is taken. Most of the shorter waves of radiant energy that reach the earth from the sun are absorbed in the stratosphere, about 15 miles above the earth. The reaction $3O_2 +$ energy $\rightarrow 2O_3$ is caused by ultraviolet waves producing an ozone-rich region of the atmosphere.

The air, of course, is the source of oxygen for burning, respiration of plants and animals, decay, and industrial oxidations. Some of these processes return CO_2 to the air, in which it is present in the amount of 0.03% by volume. Thus the air is the source of the CO_2 utilized in the weathering of rocks and in photosynthesis. Photosynthesis returns oxygen to the air. The changes such as those mentioned more or less offset one another, and the winds keep the air well mixed. Variations in composition are usually caused by some local effect.

In spite of the extensive use of oxygen in making steel and in combustion of fuels, no change in the percent of oxygen in air has been detected.

Liquid air is the source of oxygen that is used extensively, and marketed in steel tanks at low temperature. When liquid air boils, nitrogen distills first at $-196°C$, and then oxygen at its boiling point, $-183°C$. When liquid air is fractionated at $-196°C$ to prepare nitrogen, the resulting gas contains about 1.25 percent noble gases. This mixture is used to fill electric light bulbs, to pack with oxidizable foods and to maintain pressure in excess of atmospheric within telephone cables.

The noble gases, except helium, are obtained from air. The fraction from liquid air that contains the noble gases consists of 60 percent argon, 30 percent oxygen, and 10 percent nitrogen. Hydrogen is added to combine with oxygen and remove it as water vapor, which can be readily separated. The commercial mixture that remains is used to fill electric light bulbs. It contains 85 percent noble gases, chiefly argon, and the balance nitrogen.

Isolation of neon, krypton, and xenon from the air is a difficult process. Fractional distillation first makes a gross separation. Oxygen in the mixture can be removed by combining it with hydrogen or by passing the gas over hot metallic copper with which the oxygen forms copper oxide. Nitrogen is then removed by passing the gaseous mixture over hot magnesium. The nitrogen combines with this metal and forms magnesium nitride (Mg_3N_2). Adsorption of the remaining noble gases by activated charcoal is at different rates, as is desorption. If charcoal at a low temperature is saturated with a mixture of noble gases, and then the temperature raised to $-80°C$, the gas that escapes is almost pure argon.

ELBERT C. WEAVER

Cross-references: *Nitrogen, Oxygen, Neon, Gas Laws, Air Pollution.*

AIR POLLUTION

Pollution of the atmosphere may be objectionable because of (a) its health aspects, (b) deleterious effects on vegetation or animals, (c) corrosion of materials, or (d) its nuisance effects. With the exception of a few isolated cases there is no concrete evidence establishing the relationship between common types of air pollution and *chronic* health effects, although there has been considerable speculation concerning a relationship between lung cancer and fuel smoke. In the case of *acute* health effects the evidence is clear-cut as in the case of an episode in the Meuse Valley (1930), several in England and the Donora smog (1948). In all these instances effects were evidently related to the ability of the pollutant to cause irritation of the respiratory tract, but the specific irritant in the atmosphere was not identified, although it is now generally believed that the irritant in these episodes was a particulate sulfate derived from oxidation of SO_2.

Many of the toxic dusts and gases that are of concern in industrial health protection are found in the outdoor atmosphere but in concentrations that are but a small fraction of those that engage the attention of workers in industrial hygiene. Typical concentrations in United States cities: total solids 100–500; lead 0.5–5; manganese 0.1–0.5; cadmium 0.01–0.1; (units in micrograms per cubic meter). Concentrations regarded as acceptable in in-plant exposures are as follows: lead 150, manganese over 6000, and cadmium (fume) 100. On the basis of present knowledge, therefore, there does not appear to be any reason to suppose that systematic toxic effects occur, as distinguished from any that may result from deposition of an irritant substance in the respiratory tract.

Damage to vegetation results from exposure to low concentrations of a number of gases, but those of primary economic significance are sulfur dioxide and hydrofluoric acid. Significant concentrations are dependent on many factors, e.g., type of plant and its age, season, humidity, light conditions, time of exposure and character of soil. In the case of sulfur dioxide in favorable conditions one-half part per million may produce significant effects in some plants. In the case of hydrofluoric acid gas concentrations one-twentieth of those or even less may produce deleterious results in sensitive plants.

Aside from its effects on vegetation, sulfur dioxide occupies a prominent place in atmospheric pollution because of its widespread occurrence due to the combustion of sulfur-containing fuels. One of its properties of economic concern is its well-known ability to aggravate the corrosion of materials. It is also suspected of contributing to an irritant quality of the atmosphere in acute smog episodes. The spontaneous oxidation of sulfur dioxide gas is assisted by actinic energy in sunlight, high humidity and by the catalytic effect of particulate matter in atmospheric suspension.

Nuisances such as those due to deposited dirt, decreased visibility and foreign odors are of *subjective* nature and it is therefore inherently impossible to fix limits of concentrations or intensity to define such conditions, in contrast to the *objective* effects cited above. In the interpretation of the measurements of air pollution nuisances, therefore, one must resort to the adduction of comparative data. The most common and widespread atmospheric problem stems from the burning of coal, both in domestic and industrial equipment, giving rise to two distinct types of air pollution: that due to smoke which is avoided by efficient combustion of volatile material of fuel and that due to fuel dust which is blown off the fuel bed or, in industrial installations, to the incombustible ash fraction from the burning of pulverized coal. Standards of good practice for the control of solids emissions from coal-burning installations having common acceptance are based on (a) the use of the Ringelmann chart for describing limits on smoke emission and (b) on the determination of weight concentration of solids in the flue gas.

Photochemical smog, first identified around 1950, results from the interaction of oxides of nitrogen with hydrocarbons and other organic compounds, catalyzed by natural ultraviolet irradiation. Products of the reaction include: ozone and other oxidants, aerosols, and numerous organic compounds. Peroxyacyl nitrites (nitrates), etc. appear importantly in the reaction.

In circumstances conducive to development of high concentrations, the perceptible effects include: visible haze, eye irritation, excessive cracking of rubber (ozone), and damage to some types of vegetation.

A major source of hydrocarbons for the reaction are the emissions from automobiles (not diesel engines); nitrogen oxides are formed from all high temperature combustion processes, including the automobile. A dominant characteristic of air pollution in the Los Angeles area, it is of lesser significance in other regions where natural meteorological ventilation is greater and where other types of air pollution dominate.

The contribution of internal combustion engines to photochemical smog is not to be confused with other emissions having a direct air pollution effect, e.g., obnoxious odors from diesel engines in busses and trucks.

Methods that are appropriate for measurement and analysis of air pollution intensity at ground level are dependent on the objective. If the purpose is to ascertain the origin of the pollution, the entire gamut of scientific techniques may be applied, as in the Los Angeles problem (1945 et. seq.). If the objective is simply to establish levels of atmospheric pollution for historical purposes, i.e., time comparisons, or for geographical comparisons, a relatively few techniques suffice.

Where, for example, it is sought to establish records of the coarse dust content in the air which deposits by gravity, dust fall observations can be made by determining the rate at which solids are deposited within an open-top vessel. When a cylindrical vessel with a continuously wet bottom is employed (U.S. practice) and the results expressed in the units, tons per square mile per month, the following figures may be taken as typical of those in the northern United States: suburban areas (summertime) 5–15, (wintertime) 10–20; in urban residential areas removed from nearby industrial sources (summertime) 10–30, (wintertime) 20–60. Analogous measurements for smoke content of the atmosphere have a relatively brief history and procedures are less standardized. One common method used in Great Britain and the United States involves measurement of the light scattering or staining potential of smoke that has been deposited by aspiration onto a filter paper. Similar data can be obtained by direct measurement of light scattering in suitable electrophotometric devices.

In the case of odors there is no objective basis for measurement, except in a few instances where the odor chemistry is well-known and it becomes necessary therefore to rely entirely on subjective impressions of humans. This circumstance proposes nearly insuperable problems in the evaluation of odor intensity in the open, but odor measurements in a stream of gas at the point of emission are readily employed in objective analyses. Such measurements are effected by instrumental arrangements in which the odor-carrying gas stream is diluted in known proportions with fresh air, and the dilution ratio value corresponding to zero odor threshold determined subjectively by a number of persons. The result can be related to estimates of dilution occurring by natural turbulence between the point of emission and some point at ground level or to the performance of a pilot odor-removal unit.

The concentration of air pollution in any locality is profoundly influenced by the natural rate of ventilation characteristic of the place and the time, as well as by the rate at which pollutants are emitted to the atmosphere. Ventilation results not only from horizontal winds, but also from vertical air currents, which are largely determined by the relative temperatures of the atmosphere at different elevations. When, due to such temperature variation, the air layers nearer the ground level are more dense than those aloft, vertical motion of the atmosphere and consequent dispersal of polluted air is prevented. This condition termed *inversion* by meteorologists occurs almost every night and not infrequently is persistent over a longer period of time. Due to meteorological and geographical peculiarities some regions are characteristically ventilated at greater rates than others. This meteorological circumstance is largely responsible for the commonly observed wide variations in air pollution intensity in consecutive periods.

The concentration of any pollutant at ground level due to emission from a single source is directly proportional to the mass rate of emission, to the wind velocity and to the square of the chimney height, as well as to a variable parameter incorporating meteorological factors that describe the prevailing atmospheric turbulence. The location of the point of maximum ground level concentration varies markedly with the meteorological conditions but falls most commonly at a distance of 5 to 25 stack lengths from the source. A theoretically derived expression for certain common meteorological conditions gives this maximum concentration (mass per unit volume) as $2Q/e\pi u h^2$, where Q is rate of emission of the contaminant, (mass per unit time), u is wind velocity and h is stack height, employing consistent units. Corresponding equations for concentrations at variable distance are complex because they are dependent upon atmospheric conditions of turbulence which are not readily defined in quantitative terms.

The abatement of air pollution when the sources are identified is almost entirely an economic problem. Coal smoke is only controllable practically by automatic arrangements for efficient combustion. This has not been found applicable to the domestic coal smoke problem; substitute fuels have reduced that problem in many areas. Dust emissions can be suppressed by application of various types of collectors listed, as follows, in the order of their

characteristic collection efficiency: bag filters, Cottrell precipitators, and the miscellaneous group that includes centrifugal precipitators, inertial separators and water scrubbers. Technologically, sulfur dioxide can be removed by conventional scrubbing techniques, but serious engineering and economic problems have appeared in the early installations in large fuel-burning power plants.

Odorous elements of organic nature can be destroyed by heating the gas stream to a temperature of approximately 1500°F and, in the presence of a catalyst, to about 500°F excepting in some particular circumstances. If the odorous substance is present in very small quantities, it is sometimes feasible to absorb it in activated carbon.

Dilution of moderate quantities of gases can often be effected by the use of tall stacks, a method that is commonly applied to the sulfur dioxide problem. It is to be noted that initial dilution in the stack itself is without effect.

W. C. L. Hemeon

Cross-references: *Air; Precipitation; Dust, Industrial.*

ALCOHOLS

Chemical Aspects. Alcohols, from a chemical viewpoint, are those classes of organic compounds where one or more hydroxyl (OH) groups are present in a hydrocarbon molecule with no more than one (OH) group attached to a single carbon atom. Three principal types of nomenclature are used;

(a) Common or radical names with the word alcohol, derived from natural sources, e.g., cetyl alcohol or from the hydrocarbon portion, e.g., ethyl alcohol from ethane.

(b) Substituted carbinol wherein this represents derivatives of methanol. Triphenylmethanol is called triphenylcarbinol, $(C_6H_5)_3COH$, and naphthalene-methanol is naphthycarbinol, $C_{10}H_7CH_2OH$.

(c) I.U.P.A.C. or Geneva name. The name of the alcohol is derived from the hydrocarbon having the longest straight chain and the alcohol function. The final hydrocarbon "e" is replaced by "-ol" (or "-diol," "-triol," etc., according to the number of hydroxyl groups). Examples of these names for the same compound are as follows: $(CH_3)_3COH$, *tert*-Butyl alcohol, Trimethylcarbinol, 2-Methyl-2-propanol.

Alcohols are classified as primary, secondary, and tertiary, depending on the number of hydrogen atoms attached to the carbon atom with the hydroxyl group. A primary alcohol would contain two or more hydrogens on the carbon atom next to the hydroxyl function, one hydrogen for a secondary alcohol, and no attached hydrogens for a tertiary alcohol.

Typical examples are:

Methyl alcohol	CH_3OH	primary	CH_3OH
Propyl alcohol	$CH_3CH_2CH_2OH$	primary	$-CH_2OH$
Isopropyl alcohol	$(CH_3)_2CHOH$	secondary	$>CHOH$
Trimethyl carbinol	$(CH_3)_3COH$	tertiary	$\geqq COH$

Alcohols are further characterized by the nature of the hydrocarbon portion of the molecule.

Class	Example
1. Simple aliphatic alcohol	Ethyl alcohol (CH_3CH_2OH)
2. Unsaturated aliphatic alcohol	Ethenol ($CH_2{=}CHOH$)
3. Substituted aliphatic alcohol	Ethanolamine ($NH_2CH_2CH_2OH$)
4. Aromatic alcohol	Benzyl alcohol ($C_6H_5CH_2OH$)
5. Alicyclic alcohol	Cyclohexanol ($C_6H_{11}OH$)
6. Heterocyclic alcohol	Furfuryl alcohol ($C_4H_3OCH_2OH$)

Alcohols may contain more than one OH group provided that they are attached to different carbon atoms. Hence, we have the description of mono-hydric, dihydric, trihydric, etc., acording to the number of hydroxyl groups present.

Description	Example
1. Monohydric	Ethyl alcohol (CH_3CH_2OH)
2. Dihydric	Ethylene glycol ($CH_2OH—CH_2OH$)
3. Trihydric	Glycerol ($CH_2OH—CHOH—CH_2OH$)
4. Hexahydric	Sorbitol [$CH_2OH(CHOH)_4CH_2OH$]

Properties. Alcohols in general are colorless liquids or solids. Normal primary alcohols from methanol to butanol are fluid liquids; from C_5 to C_{11} they have an oily consistency, and those from C_{12} and higher are solids at room temperature. All monohydroxy alcohols are soluble in organic solvents, and those with one to three carbon atoms are soluble in water. Polyhydric alcohols tend to be syrups, and alcohols with complex structures are generally solids. In general, water solubility decreases with increasing molecular weight or complexity of the formula. Conversely, boiling points rise with increases in the same factors. Additional hydroxyl groups increase water solubility and decrease solubility in ether and alcohol.

An increase in OH groups tends to enhance sweetness to taste buds; e.g., ethanol, C_2H_5OH, is not sweet; propylene glycol, $C_3H_6(OH)_2$, is slightly sweet; glycerol, $C_3H_5(OH)_3$, is more so; and mannitol, $C_6H_8(OH)_6$, is the sweetest. Lower alcohols have characteristic odors; C_8–C_{12} alcohols have rose- or lily-like odors which find use in perfumes; and the higher alcohols are practically odorless.

Alcohols resemble water in many ways and are neither acid nor alkaline in reaction. They can combine with compounds to form crystalline products such as $MgCl_2 \cdot 6CH_3OH$. Primary alcohols, on oxidation, yield aldehydes and acids; secondary alcohols go to ketones; and tertiary alcohols yield various decomposition products.

In general, the reactions of alcohols are primarily those of the hydroxyl groups (unless other active groups are present.) The hydroxyl group may be replaced by halogen or amino groups; the alcohol may be dehydrated; the hydrogen of the hydroxyl group replaced by a metal. Alcohols combine with organic acids to form organic esters and water, and can also combine with inorganic acids such as nitric and sulfuric acid.

Sources and Preparation of Alcohols. Alcohols occur widely in nature as volatile or essential oils, and as esters in volatile oils, fats and waxes. For example, n-nonyl alcohol is found in the oil of unripe orange skins, cetyl alcohol is part of the ester in spermaceti wax, and glycerol is combined with fatty acids in natural fats and oils. Alicyclic alcohols are found in plant and animal tissue as sterols, and the best known substituted alcohols are the sugars, which are aldehydo or ketopolyhydroxy compounds.

Alcohols can be synthesized by various means.

(1) *Hydration of Alkenes*

$$\text{>}C{=}C\text{<} + H—SO_4H \rightarrow \text{>}\underset{\underset{H}{|}}{C}—\underset{\underset{SO_4H}{|}}{C}\text{<} +$$

$$HOH \rightarrow \text{>}\underset{\underset{H}{|}}{C}—\underset{\underset{OH}{|}}{C}\text{<} + H_2SO_4$$

The above method produces secondary and tertiary alcohols.

(2) *Hydrolysis of Halides and Sulfates*

$$RX + HOH \xrightarrow{\text{alkali}} ROH + HX$$

This method produces primary and secondary alcohols, and also will synthesize glycerol from propylene.

(3) *Hydrolysis of Organic Esters*

$$R'COOR + HOH \rightarrow R'COOH + ROH$$

The hydrolysis is catalyzed by acids and bases, and has considerable commercial importance for producing alcohols from naturally occurring esters.

(4) *Oxidation of Hydrocarbons*

Saturated hydrocarbons are oxidized to alcohols by employing high temperature (100–300°C), pressure of 15–50 atmospheres and short contact time.

(5) *Reduction of Carbonyl Compounds*

$$\text{>}C{=}O + 2H \rightarrow \text{>}CHOH$$

This again is a high temperature and high pressure operation in the presence of complex catalysts.

(6) *Reduction of Esters*

$$RCOOR' + 4H \rightarrow RCH_2OH + R'OH$$

Used industrially to produce alcohols from higher fatty esters. Esters are reduced by alkali metal in alcohol or by high pressure hydrogenation over copper oxide-chromium oxide catalyst.

(7) *Aldol Condensation*

$$RCH_2CHO + RCH_2CHO \rightarrow$$

$$RCH_2CHOHCHRCHO$$

Aldehydes or ketones having an α-hydrogen atom are condensed to β-hydroxy aldehyde or ketone.

(8) *The Grignard Reaction*

$$\text{>C=O} + \text{RMgX} \rightarrow \text{R}-\overset{|}{\underset{|}{\text{C}}}-\text{OMgX} \xrightarrow{\text{HOH}}$$

$$\text{R}-\overset{|}{\underset{|}{\text{C}}}-\text{OH} + \text{Mg(OH)X}$$

This reaction involves addition of an alky (or aryl) magnesium halide to an aldehyde, ketone, ester, or alkene oxide.

(9) *Fermentation of Carbohydrates*

Ethanol and *n*-butanol are the most important alcohols made in this manner.

(10) *Unsaturated Alcohols*

(a) High temperature chlorination followed by hydrolysis

$$\text{CH}_2\text{=CHCH}_3 + \text{Cl}_2 \rightarrow$$
$$\text{CH}_2\text{=CHCH}_2\text{Cl} + \text{HCl}$$

$$\text{CH}_2\text{=CHCH}_2\text{Cl} + \text{NaOH} \rightarrow$$
$$\text{CH}_2\text{=CHCH}_2\text{OH} + \text{NaCl}$$

(b) Dehydrohalogenation of an α, β-dihalo alcohol

$$\text{RCHBr·CHBr·CH}_2\text{OH} + \text{2KOH} \rightarrow$$
$$\text{RC≡CCH}_2\text{OH} + \text{2KBr} + \text{2H}_2\text{O}$$

(c) Reaction of acetylene with aldehydes in in presence of copper, silver, gold or mercury catalyst under pressure.

$$\text{HC≡CH} + \text{2HCHO} \rightarrow$$
$$\text{CH}_2\text{OHC≡CCH}_2\text{OH}$$

(11) *Ziegler Synthesis*

Addition of an olefin to triethylaluminum forms trialkylaluminum, which is subsequently oxidized and hydrolyzed to straight-chain alcohols with an even number of carbon atoms (section on Higher Aliphatic Alcohols).

Aliphatic Alcohols

Methyl alcohol. Methyl alcohol (methanol, wood alcohol, carbinol) is produced by destructive distillation of hard wood, or by catalytic synthesis from hydrogen and carbon monoxide. Annual synthetic production in 1971 was 5,010 million pounds. Methanol is a solvent for fats, oils, varnishes and lacquers; is used for methylation of organic compounds; is a raw material for manufacture of formaldehyde and special resins; and finds application in special fuels and in antifreeze solutions. Recent manufacturing technology uses a copper-based catalyst and low pressure (50 atm).

Ethyl alcohol. Ethyl alcohol (ethanol, grain alcohol) is obtained by the fermentation of sugar solution and saccharified mashes of starch-containing materials, or synthetically from ethylene. Molasses is the most common raw material for the fermen-

tation process, but natural gas supplies ethane which is cracked to ethylene in order to produce the 85% of industrial ethanol which is made synthetically.

Pure ethyl alcohol is supplied as 200 proof (anhydrous) and 190 proof (95% by volume), and also comes as completely denatured alcohol, specially denatured alcohol or proprietary solvent. The uses are many: solvent action on lacquers, varnishes, stains, explosives, rubbers, antiseptics, toilet preparations; antifreeze solutions; building block for chemicals; and miscellaneous applications as a preservative and antiseptic. The total production of synthetic ethanol in 1971 was 1,960 million pounds.

Isopropyl alcohol. Isopropyl alcohol (isopropanol) is made by hydration of a propylene-sulfuric acid mixture, although smaller amounts are a by-product of special fermentations. The industrial uses are similar to ethanol, namely solvent, extractant, antifreeze and a chemical intermediate. About 1,820 million pounds were produced during the year 1971.

Normal propyl alcohol. Normal propyl alcohol comes from the Fischer-Tropsch process and is also a co-product of air oxidation of propane and butane mixtures. In addition to being a solvent for lacquers, resins, coating, films, and waxes, it is also a component of brake fluids, and is used to make propionic acid and plasticizers.

Butyl alcohol. Butyl alcohol (*n*-butanol) is produced by fermentation of starches and carbohydrates (Weizmann process), and also synthetically from ethanol or acetylene. Used as a solvent for oils, lacquers and resins; a diluent in hydraulic fluids; an extractant; and a chemical intermediate for butyl compounds. About 405 million pounds were produced during 1969.

Isobutyl alcohol. Isobutyl alcohol is mostly synthesized from carbon monoxide and hydrogen at high pressure. It is used for esterification of mono- and dibasic acids, a solvent for castor oil type of brake fluids, and a solvent for alkylated urea resins.

Secondary butyl alcohol. Secondary butyl alcohol is produced by hydration of 1-butane. Uses are a lacquer solvent, brake fluid formulations, industrial cleaning, solvent for fats and greases, and a chemical intermediate.

Tertiary butyl alcohol. Tertiary butyl alcohol is produced by hydration of isobutylene. It is used as a solvent for drugs and cleaning compounds, a chemical intermediate in the chemical and drug fields, and a denaturant for ethyl alcohol.

Amyl alcohols (pentyl alcohols) occur in 8 isomeric forms, not including optical isomers. Some can be obtained by fractional distillation of fusel oil, and produced by the chlorination and alkaline hydrolysis of pentanes. These alcohols are resin solvents, diluents in hydraulic fluids, printing inks and lacquers. They are also converted to acetates and phthalates for protective coatings, and other amyl compounds for medicinals and flotation products.

Higher Aliphatic Alcohols

The rapid rise of synthetic detergent utilization in the U.S. since 1946 has intensified the study of processes to make higher aliphatic alcohols. The earliest plants in the United States used the sodium reduction route whereby natural fats and oils such as tallow and coconut oil were reduced to straight-chain fatty alcohols with a carbon length of C-6 to about C-22. The major advantage of this process was that unsaturated fatty alcohols could be obtained.

For reasons of economics, commercial plants in recent years have swung over to the high temperature, high pressure (about 3000 psig) catalytic hydrogenation of fatty esters or fatty acids.

$$RCOOR' + 4H \rightarrow RCH_2OH + R'OH$$

Synthetic alcohols of molecular weight equivalent to those obtained from neutral fats and oils have been produced by the "oxo" and "oxyl" processes, but they tend to be branched chain alcohols. A major development since 1955 has been the manufacture of straight chain primary alcohols via the Ziegler process.

$$Al + 3/2H_2 + 2Et_3Al \rightarrow 3Et_2Al\,H$$

$$3Et_2Al\,H + CH_2{=}CH_2 \rightarrow 3Et_3Al$$

$$Et_3Al + nCH_2{=}CH_2 \rightarrow R\cdot Al{\Big\langle}{}^{R'}_{R''}$$

$$R\cdot Al{\Big\langle}{}^{R'}_{R''} + O_2 \xrightarrow{\ H_2O\ }$$

$$(C_nH_{2n+1})OH + Al(OH)_3$$

The first manufacturing plant was a 50-million lb/yr unit put up by Continental Oil Co. in 1962 to produce higher aliphatic alcohols. These are a homologous series of straight-chain (normal) primary alcohols containing an even number of carbon atoms, and ranging from C-6 to C-20.

Industrial manufacturing capacity for linear and branched alcohols was estimated at more than one billion pounds in 1970, primarily based on the Oxo process.

Unsaturated Alcohols

Allyl alcohol. An ethylenic alcohol made by chlorination of propene and used as a chemical intermediate.

Propargyl alcohol. An acetylenic alcohol synthesized from acetylene and formaldehyde; used as a chemical intermediate, rust inhibitor, and special solvent.

Pentaerythritol. Pentaerythritol is made by the condensation of acetaldehyde and formaldehyde. Major application is in synthetic resins and explosives. The annual production was 92 million pounds in the year 1969.

Sorbitol. Sorbitol comes from corn sugar by reduction with hydrogen and is a solid. It is used in foods and pharmaceuticals; as a chemical intermediate; and as a conditioning or softening agent in paper, textiles, tobacco, glue and cosmetics.

Cyclohexanol. Cyclohexanol is made by catalytic hydrogenation of phenol or catalytic air oxidation of cyclohexane. It is an intermediate for nylon via adipic acid; a stabilizer and homogenizer for soap and synthetic detergent emulsions; a solvent in the dye, textile, lacquer and resin industry, and a chemical intermediate.

Benzyl alcohol. Benzyl alcohol is usually made from benzyl chloride and sodium or potassium carbonate. It is used as a solvent in cosmetics, inks and special coatings; to make esters for soap, perfume and flavors; for suntan lotions and medicinal preparations. The annual production was 7.7 million pounds in the year 1969.

Phenyl ethyl alcohol. Phenyl ethyl alcohol is produced by a Friedel-Crafts synthesis from benzene and ethylene oxide. The principal use is in perfumes.

Furfuryl alcohol. Furfuryl alcohol is made by reduction of furfural. It is used as a solvent, in resins, as a dispersant in textile printing, and for production of tetrahydrofurfuryl alcohol.

Industrial Ethanol

The ethanol content of industrial spirits is given in terms of "proof," the proof being twice the alcohol content by volume. A wine gallon is 231 cu. in. of any proof. A proof gallon or tax gallon is a wine gallon of 100-proof spirits, or its equivalent.

The Federal Government levies a tax on alcohol for beverage purposes and certain industrial uses. The manufacture, sale and use of ethanol is therefore controlled by regulations issued by the Alcohol, Tobacco and Firearms Division of the Internal Revenue Service, U.S. Treasury Department; bond must be posted for many purposes and penalties are provided for infringements. Tax-paid industrial alcohol is pure ethyl alcohol which has been released from bond by payment of the federal tax of $10.50 per proof gallon (1960), which is $21.00 per wine gallon at 200 proof. Tax-paid alcohol may be purchased by manufacturers and others for nonbeverage use without a federal permit or bond. Pure alcohol may be tax-free when used for special purposes such as scientific research or sold to hospitals, municipalities, states, or agencies of the Federal Government.

Completely denatured alcohol contains sufficient malodorous and obnoxious constituents to prevent completely its use or recovery for beverage purposes. When the Tax-Free Alcohol Act was passed to free industrial users of ethyl alcohol from excessive taxes, denatured alcohol was put into production by the U.S. Industrial Alcohol Co. in 1906. As mentioned previously, the U.S. Government exercises a very close control of alcohol production, denaturation, and use. Federal Industrial Alcohol Regulations #3 gives all details of the requirements.

Completely denatured alcohol is sold tax-free, in reasonable quantities, and finds widespread use in antifreeze, fuels, paints, varnishes, stains, soaps and dipping fluids. Two formulas, C.D.A. 18 and C.D.A. 19 have been approved by the Department of Internal Revenue.

Specially denatured alcohol is alcohol to which certain denaturants have been added to adapt it for the more specialized uses in the arts and industries and, at the same time, to render it unfit for beverage purposes. About 50 formulas have been authorized for over 400 specific uses. Formula S.D.A. 3, "methylated spirits," is obtained by adding 5 gallons of methyl alcohol to 100 gallons of 190 or 200 proof ethyl alcohol. Some denaturants used in other formulas are acetaldehyde, ethyl propionate, and ethylamines.

Proprietary solvents are made by mixing Specially Denatured Alcohol (usually Formula No. 1) with small quantities of additional chemicals, such as ethyl acetate, methyl isobutyl ketone, methanol, rubber solvent and gasoline. Whereas the Alcohol, Tobacco and Firearms Division does not prescribe the chemicals or the quantities to be used, it is necessary for the manufacturer of Proprietary Solvent to obtain a permit to use the Specially Denatured Alcohol and to have his formulation approved. Proprietary Solvents are made from 190-proof or 200-proof alcohol, and are used widely for shellac, lacquers, varnish, printing inks, fuels, solvent cleaning and chemical processing.

In the year ending June, 1969, the total U.S. production of industrial alcohol was 338 million wine gallons of 190 proof, in addition to 141 million gallons of 190 proof beverage alcohol.

JOHN A. MONICK

References

Bent, R. L., "Alcohols," in "Encyclopedia of Chemical Technology," 2nd ed., Vol. I, p. 531, New York, Interscience Publishers, 1963.

Hatch, L. F., "Ethyl Alcohol," New York, Enjay Chemical Co., 1962.

Monick, J. A., "Alcohols," New York, Van Nostrand Reinhold, 1968.

Rudd, E. H., "Chemistry of Carbon Compounds," Vols. 1A–2D, New York, Elsevier Publishing Co., 1965–1970.

Physiological Aspects

The toxicity of the aliphatic alcohols is a narcosis (anesthesia) without cumulative features. Toxicity increases by a factor of about three times for each carbon atom added in an homologous series, due to increased lipoid-to-water distribution which favors alcohol entering the lipoids of the nervous system. However, the difference in toxicity between methyl and ethyl alcohols is less than for higher successive members of the primary series. Toxicity decreases somewhat in the series primary, secondary, tertiary alcohols. After there are some eight carbons in the molecule, low water solubility and low volatility reduce the toxic hazard of swallowing and of inhalation. The unsaturated alcohols are more toxic than the corresponding saturated ones, the difference between allyl and *n*-propyl alcohols being about thirty times. There are not great species differences among warm-blooded animals in response to the alcohols.

Irritating effects upon mucous membrane and respiratory tract increase in an homologous series until decrease in water solubility limits concentrations in body fluids. The unsaturated alcohols are much more irritating than their saturated analogs. A low concentration of allyl alcohol vapors can cause pulmonary edema and can incapacitate a man through the pain and temporary blindness of corneal injury. Permanent lost vision can result from a high concentration. The alcohols are not active skin sensitizers.

The alcohols enter the bloodstream rapidly from the stomach at rates that are proportional to their water solubility. The vapors enter the bloodstream through the lungs at rates proportional to their distribution coefficients between air and blood. They penetrate the intact skin, but only with the unsaturated alcohols is toxicity sufficient to make this route of entry a probable source of poisoning.

As soon as the primary saturated alcohols reach the liver, oxidation to aldehyde and carboxylic acid begins, and it proceeds at a constant rate. With methyl alcohol, further oxidation of formaldehyde to formic acid is very slow, accounting for the great danger of this alcohol. Accumulation of these oxidation products apparently accounts for the blindness from injuries to retina and optic nerve which is characteristic of methyl alcohol poisoning. In the presence of ethyl alcohol, methyl alcohol is not oxidized, but is excreted unchanged, preventing effects on vision. With ethyl and higher primary saturated alcohols, oxidation of the carboxylic acid to carbon dioxide and water proceeds promptly, apparently in the muscles. The secondary alcohols are oxidized only to the ketones, and the tertiary alcohols are not oxidized in the body. The manner in which the body handles unsaturated alcohols is not known.

Small amounts of the alcohols are excreted unchanged in the urine and as vapors from the lungs. A portion of the oxidation products may also appear in the urine, some being conjugated with glucuronic acid, and a portion is in the expired air. It is the experience of industrial hygienists and physicians that a workman inhaling 8 hours a day, every working day, the following concentrations of vapors (in ppm by volume) is unlikely to be affected: methyl alcohol 200, ethyl alcohol 1000, isopropyl alcohol 400, *n*-butyl alcohol 100. Slightly higher concentrates irritate eye, nose and throat but are not systemically toxic.

Like any other anesthesia, the first action of a dose of ethyl alcohol is on the higher centers of the brain, with emotions, inhibitions, judgment, and mental clarity being affected. Increasing doses increasingly affect the central nervous system, decreasing muscular coordination, perception and reaction time. Loss of muscular control follows and the terminal stage is an anesthetic death from respiratory and circulatory arrest. Diuresis results early from action upon the pituitary. If the stomach is empty of food, alcohol starts to reach the bloodstream one to two minutes after it is swallowed. Food in the stomach somewhat retards absorption. A man can oxidize about 8 grams an hour. The oxidation yields calories to the body, as with any food, somewhat more from alcohol than from carbohydrate, but less than from fat.

The severity of symptoms is proportional to the amount of unoxidized alcohol remaining in the bloodstream. A level of 0.05% ethyl alcohol in the blood is compatible with sobriety. Some people are intoxicated at 0.06% and practically all at 0.26%. The fatal level in the blood is from 0.3 to 1.3%, and 6 to 8 grams per kilogram body weight taken at one time on an empty stomach is fatal to a man. The habituation of the chronic drinker consists in an acquired ability of brain cells to function in the presence of ethyl alcohol, not in a special ability to oxidize it. The social problem of ethyl alcohol in beverages is due to psychological and nutritional factors, not to toxicity. The symptoms of the chronic alcoholic are not chronic ethyl alcohol poisoning. They are nutritional in origin, due to vitamin deficiencies caused by obtaining calories from alcohol to the neglect of vitamin-containing foods and proteins.

HENRY F. SMYTH, JR.

ALDEHYDES

Aldehydes are members of an important class of organic compounds characterized by the presence of a carbonyl group ($-\overset{\|}{\underset{O}{C}}-$). The name aldehyde is derived from the fact that members of this series of compounds can be obtained when primary ALcohols are DEHYDrogenatEd either by dehydrogenation or oxidation reactions. This indicates that aldehydes are intermediary between primary alcohols and acids, which are formed on further oxidation. Aldehydes are closely related to ketones, which also contain the carbonyl group, the difference being that aldehydes have at least one of the bonds from the carbon atom of the carbonyl group to a hydrogen atom, while with ketones each bond must be to some organic radical. Consequently the aldehyde group can occur only at the end of an aliphatic chain or attached directly to the aromatic or heterocyclic ring. As aldehydes are much more reactive than ketones, they require separate treatment. (See **Carbonyl Compounds.**)

Physical Properties. The aldehyde group is an osmorphic group; thus the odor is often characteristic. The lowest members of the series are irritating and unpleasant but a pleasing odor develops with an increasing molecular weight, so that C_8-C_{12} members are often used in synthetic perfumes. Also many of the pleasing, volatile oils of nature are higher molecular weight aldehydes.

The first member of the series of aldehydes is a gas at room temperature and pressure. The boiling points of succeeding members increase with the molecular weight. The first two members are miscible with water in all proportions, and in general solubility in water decreases with increasing molecular weight except where the presence of other groups within the molecule may influence it.

Nomenclature. Many common aldehydes are given nonsystematic names usually relating to their original source, such as vanillin, acrolein, cinnamylaldehyde, anisaldehyde, etc.; but chemically, they are usually named by one of two methods: (1) After the common name of the acid which it forms upon oxidation; the ending -ic or -oic of the acid is deleted and the word aldehyde is added, as for example in formaldehyde (which oxidizes to formic acid), acetaldehyde (which forms acetic acid), benzaldehyde (which forms benzoic acid), etc. (2) The I.U.P.A.C. system which names the aldehyde after the parent hydrocarbon with the longest continuous carbon chain containing the aldehyde group on the terminal carbon atom. The "e" of the characteristic I.U.P.A.C. hydrocarbon ending -ane is deleted and the characteristic ending -al is added. Thus for example formaldehyde becomes methanal, acetaldehyde becomes ethanal, etc.

Preparation. Aldehydes can be prepared in the laboratory by a variety of methods. Many are applicable both to the aliphatic and aromatic series, but a few can be used only for aromatic aldehydes. The industrial procedure for the preparation of aldehydes is often quite different from that used in the laboratory. The exact procedure often varies with the aldehyde being produced.

(1) Mild oxidation of primary alcohols. This is often a difficult reaction to carry out, since the aldehyde formed is often more susceptible to oxidation than the alcohol from which it was formed. Thus oxidation may continue, with the resulting formation of the corresponding acid instead. This reaction is possible only because the aldehyde has a lower boiling point than either the alcohol or the acid, and thus it can be removed during the reaction. (2) Hydrolysis of a dihalogen compound if both halogen atoms are on the terminal carbon atom. (3) Pyrolysis of a mixture of calcium formate and the calcium salt of the corresponding acid. (4) Treatment of an olefin with ozone, followed by hydrolysis with heat. (5) The Grignard reaction, using an excess of ethyl formate. (6) Treatment of the corresponding acid halide with a mild reducing agent.

The following methods are applicable only to aromatic aldehydes:

(1) Mild oxidation of methyl side-chains attached to the aromatic ring. (2) Gattermann-Koch procedure, treatment of benzene or its homologs with a mixture of carbon monoxide, hydrogen chloride, cuprous chloride and aluminum chloride. A variation of this is the Gattermann procedure, which uses sodium cyanide in place of carbon monoxide and eliminates the cuprous chloride. (3) Reimer-Tiemann procedure for production of hydroxyaldehydes, heating an aqueous, basic solution of a phenol with chloroform.

Reactions. Due to the presence of the reactive carbonyl group, aldehydes can be considered one of the more reactive series of organic compounds. The more important reactions can be conveniently grouped into six classes:

(1) *Oxidation.* Aldehydes are easily oxidized by mild oxidizing agents, and sometimes even by oxygen of the air alone, to form the corresponding acid. This ease of oxidation is often used as a means of detecting their presence, and differentiating them from the related ketones.

(2) *Reduction.* Aldehydes can be reduced catalytically or by any mild reducing agent to form the corresponding primary alcohol. The Cannizzaro reaction is an intermolecular oxidation-reduction limited to aldehydes with no hydrogen on the carbon atom adjacent to the carbonyl group. Two molecules of the aldehyde under the influence of sodium hydroxide form one molecule each of the corresponding alcohol and acid.

(3) *Addition to the carbonyl oxygen.* (a) With water to form the 1,1-dihydroxy derivative. The product is stable only in water solutions or with certain substituted aldehydes such as trichloroacetaldehyde to form the product known as chloral hydrate. (b) With ammonia to form the aldehyde-ammonia complex. The simple complex is unstable but readily polymerizes to form a stable, crystalline compound. (c) With alkali bisulfite to

form a crystalline bisulfite addition product. (d) With hydrogen cyanide to form the important cyanohydrins. (e) With alcohols to form first the hemiacetal and further to form the acetals. These are stable compounds derived from the unstable 1,1-dihydric alcohols. Aldehydes also react with the poly-alcohols such as polyvinyl alcohol with the elimination of a molecule of water to form useful resins.

(4) *Reaction with the carbonyl oxygen with loss of water.* (a) With hydroxyl amine to form the oxime. (b) With phenyl hydrazine and substituted derivatives to form the hydrazones. This reaction is used particularly in the identification of carbohydrates. (c) With semicarbazide to form the semicarbazone.

The above reactions are often used for the purification and/or identification of aldehydes, since most derivatives are crystalline products, easily purified, easily regenerate the aldehyde and have definite melting points.

(5) *Polymerization.* (a) Aldehydes with an active hydrogen on the carbon adjacent to the carbonyl group undergo "adolization" to form hydroxy aldehydes. With formaldehyde some of the hexose sugars have been isolated. (b) Many of the lower molecular weight aldehydes form simple linear and cyclic polymers from which the monomeric aldehyde can be easily regenerated (see **Formaldehyde** in separate article and **Acetaldehyde** below). (c) Many higher molecular weight aldehydes form polymers of unknown structures (resins) under the influence of acids. This procedure can sometimes be used to differentiate aldehydes and ketones since the latter will not resinify readily. (d) Many aldehydes form useful resins when copolymerized with other molecules. See phenol- and urea-formaldehyde resins under **Phenolic Resins** and **Aminoplast Resins** respectively.

(6) *Miscellaneous reactions.* (a) With the Grignard reagent (RMgX) formaldehyde forms primary alcohols and all other aldehydes form secondary alcohols. (b) With phosphorus pentachloride to form the corresponding 1,1-dichloro derivatives by replacement of the carbonyl oxygen atom. (c) There are many other isolated reactions of varying importance for example the Benzoin condensation, Perkin condensation, other condensations and other elimination reactions.

Acetaldehyde (Ethanol). The second most important aldehyde from the commercial standpoint is acetaldehyde. Exact production figures are impossible to obtain since most is used by the producer to prepare other compounds. However, the best estimate of production for 1970 was 1.7×10^9 pounds. Acetaldehyde is an irritating liquid that boils at 20°C and is thus difficult to handle without refrigeration. The chief method of production is the oxidation of ethylene, replacing the older method involving the catalytic hydration of acetylene. Acetaldehyde can also be produced by the mild oxidation of ethyl alcohol and is often recovered as a by-product in the first distillation in ethyl alcohol refineries.

Acetaldehyde, like formaldehyde, polymerizes readily. In the presence of sulfuric acid a liquid, cyclic trimer commonly known as paraldehyde is rapidly formed. When cold, an unstable solid poly-mer known as metaldehyde is formed, but at room temperature this slowly decomposes to paraldehyde and acetaldehyde. From either polymer pure acetaldehyde is easily regenerated and is thus used as a convenient source of the pure monomer.

The chief use of acetaldehyde is as an intermediate in the production of other organic compounds, particularly in the catalytic conversion to glacial acetic acid, acetic anhydride, ethyl acetate, and vinyl acetate. Another important use is in the production of the compound known as aldol (butanol-3-al-1). This compound is easily dehydrated to crotonaldehyde (butene-2-al-1), which can then be hydrogenated to either butyraldehyde (butanal-1) or normal butyl alcohol (butanol-1). Other important uses are in (1) preservatives for fruit, (2) hardening of leather and other protein products, (3) organic syntheses, (4) in the production of resins and (5) dyes.

ELBERT H. HADLEY

Cross-reference: Formaldehyde.

Formaldehyde

Formaldehyde (CH_2O) is the unique first member of the series of aliphatic aldehydes. In the pure state, it is a colorless, highly toxic gas of strong, disagreeable odor; it condenses to a liquid at low temperatures (b.p. −19°C; f.p. −118°C). Both liquid and gas polymerize readily and can be kept in the monomeric state for only a limited time. It is readily soluble in water and is marketed chiefly as an aqueous solution although solutions in methanol, propanol and butanol are commercially available. It is also sold in the form of its hydrated, linear polymer, paraformaldehyde, $HO \cdot (CH_2O)_n \cdot H$, and the cyclic trimer, trioxane, $(CH_2O)_3$ (m.p. 62−4°C; b.p. 115°C). High molecular weight polymers of formaldehyde are characterized by their plastic properties. The commercial polyoxymethylene acetal resins of polyformaldehydes are stable high molecular weight formaldehyde polymers or copolymers modified with minor amounts of other materials.

As a result of its chemical reactivity, high purity and low cost, formaldehyde has become an industrial product of outstanding commercial value. Production in terms of the 37% aqueous solution reached 2.75 billion pounds in the United States alone in 1964. Total U.S. production capacity was estimated as over 5 billion pounds in 1971.

Formaldehyde is produced principally from methanol, but appreciable quantities are also derived from the direct oxidation of hydrocarbon gases.

Two methanol processes are in use. In general, these procedures involve passing a mixture of methanol vapor and air over a stationary catalyst at approximately atmospheric pressure and scrubbing the offgases with water to obtain aqueous formaldehyde. The first or classical procedure makes use of a silver catalyst and employs a rich mixture of methanol with air. The second method makes use of an oxide catalyst, e.g., iron-molybdenum oxide, employs a lean mixture of methanol with air and produces a formaldehyde solution which is substantially free of unreacted methanol.

Hydrocarbon oxidation processes for formaldehyde manufacture involve a controlled gas-phase oxidation of 2- to 4-carbon paraffin gases followed by abrupt cooling and condensation of products. This is usually accomplished by scrubbing with water to yield a crude solution which must be refined to separate formaldehyde from the other products. Formaldehyde is not the major product of these petrochemical processes. Other products include acetaldehyde, acetic acid, methanol, acetone, etc.

Formaldehyde is principally marketed in the form of 37 to 50% by weight solutions, which contain 1% or less methanol and the 37% U.S.P. grade containing sufficient methanol (6 to 15%) to prevent precipitation of polymer under ordinary conditions of shipping and storage. The low-methanol solutions must be kept warm to prevent polymer formation. However, solutions of this type containing trace amounts of catalytic inhibitors or stabilizers are available. These solutions do not show the temperature sensivity of the uninhibited low-methanol variety.

Formaldehyde is hydrated by water and the aqueous solutions contain an equilibrium mixture of the monohydrate methylene glycol, $CH_2(OH)_2$, and polyoxymethylene glycols, such as $HOCH_2$-OCH_2OH and $HOCH_2OCH_2OCH_2OH$. A small concentration of monomeric formaldehyde is also present, but this is well under 0.1%. Solution composition plays an important role in the kinetics involved in the handling and use of formaldehyde solutions.

Formaldehyde has a pungent odor and its solutions and vapors are highly irritating to the mucous membranes of the eyes, nose and upper respiratory tract. The threshold value for these effects is as low as 5 ppm by volume in air. Contact of solutions with the skin should be avoided since it causes hardening or tanning, diminished secretion and, in some cases, leads to dermatitis.

Paraformaldehyde, $HO \cdot (CH_2O)_n \cdot H$, is usually produced by vacuum concentration of the aqueous solution until a solid containing 91% or more formaldehyde is obtained. It evolves formaldehyde gas and should be handled with essentially the same precautions as formaldehyde solutions. The high molecular weight polyoxymethylenes are obtained by the catalytic polymerization of pure monomeric formaldehyde, trioxane and formaldehyde solutions. Modification of these polymers to obtain acetal resins may include esterification of hydroxyl end-groups, copolymerization with small amounts of monomers, such as ethylene oxide, as well as addition of antioxidants and other additives to improve the chemical stability and physical properties of the resultant plastic.

Pure formaldehyde has a good degree of chemical stability and shows no appreciable decomposition at temperatures below 300°C. At higher temperatures, it decomposes almost exclusively to carbon monoxide and hydrogen. In aqueous solutions, the Cannizzaro reaction yields formic acid and methanol in equimolar proportions. This reaction accounts for the gradual increase in acidity of commercial solutions on storage. Formaldehyde also undergoes an aldol-type condensation yielding hydroxy aldehydes, hydroxy ketones and hexose sugars.

Formaldehyde has a high degree of chemical reactivity and will combine with almost every type of chemical compound. These reactions give it unique value as a synthetic agent. Its primary reactions result in the formation of methylol (hydroxymethyl),—CH_2OH, or methylene ($=CH_2$) derivatives. Resins are produced by polycondensations in which reactant molecules are linked by methylene groups.

Alcohols and polyhydroxy compounds combine with formaldehyde under acidic conditions to give the formaldehyde acetals or formals. Relatively unstable hemiformals, $HOCH_2OR$, are produced under neutral or alkaline circumstances.

Aldehydes and ketones containing alpha hydrogen atoms give mono- and polymethylol derivatives under alkaline catalysts followed by reduction of the carbonyl group in some instances. This is exemplified by the production of pentaerythritol, $C(CH_2OH)_4$, from acetaldehyde. Phenols form nuclear methylol derivatives which undergo polycondensation to give the methylene-linked phenolformaldehyde resins.

Of outstanding importance in organic syntheses are the reactions by which formaldehyde can be used to introduce substituted methyl groups in organic molecules. These include the halomethylation reactions, aminomethylations including the Mannich reaction, cyanomethylation, alkoxymethylation, sulfomethylation, thiocyanomethylation, etc.

The estimated production of formaldehyde in 1970 was 4.75 billion lbs of 37% solution. Its applications by type were about as follows:

Use	Million lbs, 37% Solution
Urea formaldehyde resins	1030
Melamine resins	387
Phenolic resins	1030
Ethylene glycol	172
Pentaerythritol	300
Hexamethylenetetramine	475
Fertilizer	172
Acetal resins	430
Other	300

Hexamethylenetetramine is employed commercially as a special form of formaldehyde in the production of synthetic resins, the hardening of proteins, organic syntheses, etc.

The phenolic, urea and melamine resins find use in adhesives and protective coatings as well as in molded products. Pentaerythritol is chiefly utilized in alkyd resin manufacture. Urea-formaldehyde concentrates find industrial application in bonding resins, coating materials, textile modifiers and compositions for treatment of paper, paperboard, etc. Agriculturally, they are of value in slow nitrogen release fertilizers. Formaldehyde is of value alone or, more commonly, in combination with various nitrogen compounds in the production of "washwear" and "permanent press" textiles.

The polyoxymethylene acetal resins have outstanding mechanical properties which have made it possible for them to replace metal, wood, glass, rubber and other plastics in many instances since their production was initiated in 1959.

Formaldehyde is used in the manufacture of ethylene glycol by a process involving hydrogenation of glycolates ex formaldehyde and carbon monoxide. Sequestering agents are made by the cyanomethylation of ammonia and amines followed by hydrolysis; ethylenediaminetetraacetic acid (EDTA) is synthesized from ethylenediamine in this manner.

Numerous though important minor applications of formaldehyde are found in the synthesis of drugs and other specialty chemicals. Of particular interest is its utility for the deactivation of viruses in vaccine preparation. The bactericidal, fungicidal and preservative action of formaldehyde are of value for disinfection, control of plant diseases in agriculture, preservation of biological specimens and embalming.

J. Frederic Walker*

Cross-references: *Aldehydes; Phenolic Resins.*
* Deceased.

ALICYCLIC AND CYCLIC COMPOUNDS

Compounds having three or more atoms joined to form a closed ring are known as cyclic compounds. Cyclic compounds with properties resembling those of aliphatic compounds of the same functional type, rather than those of aromatic compounds, are termed alicyclic compounds. This term strictly applies to both the carbocyclic type, in which all the ring atoms are carbon, and to the heterocyclic type, having other than carbon atoms in the ring. In practice, however, the term is usually limited to carbocyclic compounds. These compounds have received a great deal of attention because of their widespread occurrence in nature and their commercial and theoretical importance.

The simplest alicyclic, cyclopropane, is prepared by the reaction of zinc dust with trimethylene chloride. It is frequently used as an anesthetic in surgery. Cyclohexane is prepared by catalytic hydrogenation of benzene and is a commercial solvent. Cyclohexanol, obtained by hydrogenation of phenol, is an intermediate in the synthesis of nylon. Addition of chlorine to benzene gives a mixture of the isomers of 1,2,3,4,5,6-hexachlorocyclohexane, which is the valuable insecticide, lindane or gammaexane. Ethylene oxide, an alicyclic with a hetero atom, is a very reactive compound, useful in many synthetic processes.

Cyclopentane, cyclohexane and their methyl derivatives occur in petroleum. The potent insecticide pyrethrin I occurs in pyrethrum flowers; muscone is the odoriferous principle of musk and important in perfumery. α-Pinene is the chief constituent of turpentine oil. Another terpene, camphor, has long been valued in medicine and has found much use in the manufacture of "Celluloid." Naturally occurring polycyclic representatives include triterpenes such as abietic acid (pine rosin) and steroids such as cholesterol (widely distributed in the animal body).

The chemical properties of alicyclic compounds are affected so much by ring size that it has been found convenient to classify them in this manner: (a) Small rings of three and four members, which are relatively difficult to form and are most reactive; (b) common rings of five to seven members, which are usually readily formed and are very stable, like open-chain aliphatic chains; (c) medium rings of eight to twelve members, which are difficult to synthesize by most methods but are generally stable, and in which class ordinary reactions may occur in an abnormal "transannular" manner; and (d) large rings (or macrocyclics), having more than 12 members, which also are difficult to synthesize but once formed are very stable. The chemical properties of functional groups on alicyclic rings are also affected by the ring size; for instance, small rings seem abnormally electronegative.

An ingenious explanation for the effect of ring size on the reactivity of alicyclic rings was proposed by Baeyer in 1885 and is famous as the "strain theory." It is based on two assumptions: (1) that the carbon atoms in any alicyclic ring are coplanar, and (2) that the bonds of a tetra-substituted carbon atom are normally at the tetrahedral angle of 109°28, but that this angle can be altered. Any deviation of this angle, however, is considered to result in a condition of internal strain in the molecule, decreasing its stability and causing it to be more difficult to form. The strain was measured as one-half of the difference between the tetrahedral angle and the actual angle.

For 3-, 4- and 5-membered rings, the size of the bond angle distortion does indeed parallel the reactivity of the ring system and the theory can be extended successfully to ethylene, if it is considered to have a 2-membered ring. Thus hydrogen adds to ethylene with a nickel catalyst at 40°; under such conditions 100° is required for hydrogenolysis of cyclopropane, 180° for cyclobutane and 300° for cyclopentane. Cyclopropane adds bromine and iodine to produce 1,3-dihalopropanes, and hydrogen bromide and hydrogen iodide to form the 1-halopropanes, while the larger rings are inert toward such reagents under usual conditions. Cyclopropyl ketones, acids and esters are capable of undergoing conjugate addition with active methylene compounds, and their ultraviolet spectra indicates that the cyclopropyl group can conjugate with a carbonyl group, like an olefinic double bond. Small ring compounds containing a hetero atom in the ring also show an unusually high reactivity. The 3-membered rings are more reactive than the 4-membered, and some, like ethylene oxide, are very important reagents in organic synthesis.

The latest modification of the "strain theory" for small rings is that the direction of bonding of the carbon atoms remains at the tetrahedral angles, but that the bonds are bent. This results in less than the maximum overlap of the bond orbitals in the ring and therefore weaker bonds than normal.

Baeyer's prediction that cyclopentane would be easily formed and stable was quickly confirmed by its synthesis and the determination of its properties. This work also showed, however, that cyclopentane and cyclohexane are of nearly equal stability. Precise work done later on heats of combustion showed that the cyclohexane ring is actually slightly more stable than the cyclopentane ring, and the larger rings are also approximately strainless.

Since Baeyer's strain theory pictured cyclohexane as more strained than cyclopentane, and greater strain was associated with the larger rings, due to the assumption that all rings were coplanar, this part of the theory had to be modified. It was recognized that rings of 6 members and higher were not planar but puckered, to allow the normal tetrahedral angle between the carbon atoms. Cyclohexane thus can exist in two modifications, a "chair" form and a "boat" form (referring to the shape of the structural formulas). There is apparently an equilibrium between them, the former predominating. As a consequence of this, there are two kinds of hydrogens in cyclohexane, the ones above and below the carbon skeleton being known as axial (a) and those around its periphery as equatorial (e). Substituents also can occupy either axial or equatorial positions, with the result that different conformations of the same "isomer" are possible; of course, the conformation with minimum internal repulsive forces is favored in such cases.

Recent spectral studies have shown that cyclopentane and even cyclobutane are also somewhat puckered. Apparently repulsive forces between hydrogen atoms causes the carbon atoms to twist so as to stagger the hydrogen atoms on adjacent methylene groups and thus increase H—H distances. In rings of 8 to about 12 members, the puckering causes some atoms which are not bonded to each other to be in very close proximity, resulting in some very interesting "transannular" reactions. Thus hydrolysis of 1,2-epoxycyclooctane gives rise to 1,4-cyclooctanediol as well as the expected 1,2 isomer, apparently due to an intramolecular hydride ion shift. Likewise oxidation of cyclononene gives 1,5-cyclononanediol and that of cis-cyclodecene gives 1,6-cyclodecanediol. These medium rings appear to be slightly strained because of the closeness of some of the nonbonded atoms, but when the ring size is increased further, the rings become strainless, loose structures which tend to "collapse." X-ray spectral data indicate, for example, that a ring ketone of more than 16 carbon atoms resembles a double paraffin chain joined to a carbonyl group.

There is a marked effect of ring size on ease of formation of alicyclic compounds, although it is not parallel to ring stability. Generally 5- and 6-membered rings are most easily formed, and large rings of 10 to 12 members are the hardest to form. At this size, the two ends of the coiled carbon chain apparently are least likely to collide, because the other atoms in the chain get in the way. It is somewhat easier to close a 3-membered ring than a 4-membered one, proximity thus appearing more important than deformation of bond angles.

Some general methods of synthesis of alicyclic compounds are:

(1) Freund synthesis from dihalides: $(CH_2)_n X_2 + Zn \rightarrow (CH_2)_n + ZnX_2$

(2) Malonic and acetoacetic ester methods, for example:

$$(CH_2)_n X_2 + CH_2(COOEt)_2 \xrightarrow{NaOEt}$$

$$(CH_2)_n C(COOEt)_2 \xrightarrow[2.\ HCl]{1.\ NaOH}$$

$$(CH_2)_n CHCOOH$$

(3) Thermal decarboxylation of dicarboxylic acids or their salts, giving cyclic ketones:

$$(CH_2)_n(COOH)_2 \rightarrow (CH_2)_n\ C{=}O$$

(4) Dieckmann reaction (mainly for 5- and 6-membered rings)

$$(CH_2)_n(CO_2Et)_2 \xrightarrow{NaOEt\ +\ Na}$$

$$(CH_2)_{n-1} \begin{array}{l} {-}CH{-}COOEt \\ | \\ {-}CO \end{array} + EtOH$$

(5) Acyloin condensation (best for large rings)

$$(CH_2)_n(COOEt)_2 \xrightarrow{Na} (CH_2)_n \begin{array}{l} {-}C{=}O \\ | \\ {-}CHOH \end{array}$$

(6) Thorpe reaction (good with large rings)

$$(CH_2)_n(CN)_2 \xrightarrow[or\ LiNR_2]{NaOEt}$$

$$(CH_2)_{n-1} \begin{array}{l} {-}CH{-}C{\equiv}N \\ | \\ {-}C{=}NH \end{array}$$

(7) Diene Synthesis (particularly valuable in the synthesis of 6-membered rings):

(8) Hydrogenation of aromatic compounds.

Methods for going from one ring size to another are of importance, such as:

(1) Ring enlargement of cyclic ketones with diazomethane:

$$(CH_2)_5CO + CH_2N_2 \rightarrow (CH_2)_6CO + N_2$$

(2) Demjanov ring expansions and contractions:

$$\begin{array}{l} CH_2{-}CH{-}CH_2NH_2 \\ |\quad\quad\ | \\ CH_2{-}CH_2 \end{array} + HNO_2 \rightarrow$$

(and other products).

(3) Pinacol-pinacolone rearrangements:

SCOTT SEARLES, JR.

Cross-references: *Aliphatic Compounds, Aromatic Compounds, Heterocyclic Compounds.*

ALIPHATIC COMPOUNDS

Organic compounds characterized by open-chain structures are called aliphatic; contrasted with these are aromatic compounds, which have ring structures. Aliphatic compounds as a class are named from the Greek word *aleiphar*, which means fat or oil. While it is true that fats and oils are included among the aliphatic compounds, the group is far larger than fats and oils alone. All open-chain hydrocarbons, both saturated and unsaturated, are aliphatic compounds. Also all esters, ethers, ketones, aldehydes, amines, amides, alcohols, carbohydrates, and substitution products of all the groups mentioned are aliphatic compounds, provided that they have no aromatic-type ring structure.

Hydrocarbons contain carbon and hydrogen only; those that contain the maximum possible quantity of hydrogen are called *saturated* hydrocarbons. Saturated hydrocarbons are thought to contain bonds between adjacent carbon atoms that consist of a single shared electron pair (covalent bond). Saturated hydrocarbons all burn in air or oxygen. When the supply of air is adequate, the products of burning a hydrocarbon are carbon dioxide and steam (water). Saturated hydrocarbons do not combine readily with bromine or with halogen acids. Their chemical activity, aside from the ability to burn, is in general small.

Methane (CH_4) is the simplest and most abundant hydrocarbon. It is the chief constituent of natural gas. Methane (b.p.—161.5) is thought to have a tetragonal molecule in which all the hydrogen atoms are spaced an equal distance from one another.

The methyl radical (CH_3—) H—$\overset{\overset{\displaystyle H}{|}}{\underset{|}{C}}$— is related to

methane (CH_4) H—$\overset{\overset{\displaystyle H}{|}}{\underset{\underset{\displaystyle H}{|}}{C}}$—$H$. A compound in which

one hydrogen of methane is replaced by the methyl

radical is called ethane (C_2H_6) H—$\overset{\overset{\displaystyle H}{|}}{\underset{\underset{\displaystyle H}{|}}{C}}$—$\overset{\overset{\displaystyle H}{|}}{\underset{\underset{\displaystyle H}{|}}{C}}$—$H$. The

radical related to ethane is called the ethyl radical

(C_2H_5—) H—$\overset{\overset{\displaystyle H}{|}}{\underset{\underset{\displaystyle H}{|}}{C}}$—$\overset{\overset{\displaystyle H}{|}}{\underset{\underset{\displaystyle H}{|}}{C}}$—. In a similar manner, the

propyl radical (C_3H_7—) is related to propane (C_3H_8) and the butyl radical (C_4H_9—) to butane (C_4H_{10}).

As the number of carbon atoms increases, the molecular weight of the compounds in this the paraffin series increases, the boiling point rises, the specific gravity increases, but the general chemical properties are substantially the same. Pentane (C_5H_{12}), hexane (C_6H_{14}), heptane (C_7H_{16}), octane (C_8H_{18}), nonane (C_9H_{20}), and decane ($C_{10}H_{22}$) all correspond to the general formula C_nH_{2n+2}.

Two formulas for butane are possible. One has the carbon atoms arranged in a straight chain

$\overset{\overset{\displaystyle H}{|}}{\underset{\underset{\displaystyle H}{|}}{H-C}}$—$\overset{\overset{\displaystyle H}{|}}{\underset{\underset{\displaystyle H}{|}}{C}}$—$\overset{\overset{\displaystyle H}{|}}{\underset{\underset{\displaystyle H}{|}}{C}}$—$\overset{\overset{\displaystyle H}{|}}{\underset{\underset{\displaystyle H}{|}}{C}}$—H and the other has a branched

chain, H—$\overset{\overset{\displaystyle H}{|}}{\underset{\underset{\displaystyle H-\overset{\overset{\displaystyle H}{|}}{\underset{\underset{\displaystyle H}{|}}{C}}-H}{|}}{C}}$—$\overset{\overset{\displaystyle H}{|}}{C}$—$\overset{\overset{\displaystyle H}{|}}{\underset{\underset{\displaystyle H}{|}}{C}}$—H. Both arrange-

ments correspond to the formula C_4H_{10}. Corresponding to these two formulas, two different butanes, and only two, are known. The straight-chain butane (normal butane) boils at —0.6°, and the branched chain compound (isobutane) boils at —10.2°. These compounds are isomers of one another, that is, they differ in structure, but not in composition. Their properties differ slightly. As the number of carbon atoms increases, the number of isomers increases greatly. For example, nonane has 35 isomers.

Unsaturated hydrocarbons are called alkenes (or olefins) if they have one double bond, alkadienes (or diolefins) if they contain two double bonds. Those having a triple bond are called alkynes (or acetylenes). Saturated hydrocarbons are named alkanes (or paraffins). A double bond represents the bonding force between two adjacent carbon atoms as being attributed to two covalent electron pairs; a triple bond represents three covalent electron pairs between adjacent carbon atoms. Double and triple bonds often represent points for chemical attack on a molecule. For example, one mole of a double-bonded hydrocarbon reacts with one mole of bromine. One mole of a triple-bonded hydrocarbon reacts with two moles of bromine. A triple band can exist under some temperature conditions that rupture single and double bonds.

Propane (C_3H_8) is a typical alkane. It is found in petroleum and in natural gas. It is also a by-product of the refining of petroleum. It boils at —42°C, and is a gas at room temperature. Huge quantities of this gas are used for rural domestic fuel gas.

Ethylene (ethene) (C_2H_4) is the simplest alkene. It is found in natural gas, and is made to some extent by the destructive distillation of coal or wood; it is also formed by cracking petroleum. It may be made by dehydration of ethanol and by several other methods. Like all hydrocarbons, ethylene burns. It also undergoes a number of reactions that depend upon addition to the double bond. Chlorine, bromine, halogen acids, hypochlorous acid, hydrogen, and ozone all add characteristically at the double bond. Ethylene also polymerizes and forms the well-known plastic material, polyethylene.

Butadiene is a well-known example of an alkadiene. 1,3-Butadiene (CH_2=CH—CH=CH_2) contains alternate double and single bonds, an arrangement called a conjugated system. Isoprene [CH_2=C(CH_3)—CH=CH_2] is closely related to butadiene. Butadiene may be obtained by distilling natural rubber. Isoprene is used in the synthesis of synthetic elastomers. (See **Butadiene**).

Acetylene (C_2H_2) is the simplest and best-known

member of the alkyne series. It contains a triple bond ($HC{\equiv}CH$), and it is correspondingly more reactive than ethylene. Acetylene is made from the reaction of calcium carbide with water: $CaC_2 + 2H_2O \rightarrow C_2H_2\uparrow + Ca(OH)_2$ and from natural gas. Acetylene is stored in tanks that contain porous material and acetone. Acetylene burns with an exceedingly hot flame which has the combined effect of the heat of decomposition and the heat of combustion. The well-known oxyacetylene torch uses this gas for fuel. Acetylene is also an important raw material for several syntheses. (See **Acetylene**).

Alcohols are related to hydrocarbons in that one or more of the hydrogen atoms is replaced by an OH group. Ethanol or grain alcohol is manufactured by the fermentation of grain. More than one-half of the supply of ethanol (C_2H_5OH) is made by processes that consists essentially of adding the elements of water to the double bond of ethene: $C_2H_4 + HOH \rightarrow C_2H_5OH$. (See **Alcohols**).

Ethers are organic oxides. If R represents a hydrocarbon radical such as methyl ($CH_3{-}$), ethyl ($C_2H_5{-}$) or vinyl ($CH{=}CH{-}$), then ROH is an alcohol and R—O—R′ is an ether. Diethyl ether is an ether frequently used. It is made by dehydration of alcohol. It is well-known as a solvent, fuel, anesthetic, and medium for organic reactions. Ethers that have two different substituents for R are called mixed ethers. Diethyl ether ($C_2H_5{-}O{-}C_2H_5$) is not a mixed ether, but methyl ethyl ether ($CH_3{-}O{-}C_2H_5$) is called a mixed ether. (See **Ethers**).

Aldehydes have the group $-\overset{\overset{\text{O}}{\|}}{\text{C}}-\text{H}$. The general formula for an aldehyde is RCHO, and the simplest member of the aldehyde class is formaldehyde (HCHO). Like many other compounds that have hydrogen instead of a hydrocarbon radical for R, formaldehyde is not typical of the entire group of aldehydes. Acetaldehyde (CH_3CHO) is more closely a typical compound. Aldehydes as a group are midway in oxidation condition between an alcohol and an organic acid. That is, oxidation of an aldehyde produces an acid; reduction of an aldehyde produces an alcohol. (See **Aldehydes**).

Ketones are R—CO—R′. Again both mixed and simple ketones are known, depending upon whether the substituent for R is different or the same. The R substituent may be a saturated or an unsaturated hydrocarbon radical. $CH_3{-}CO{-}CH_3$ is dimethyl ketone, more commonly called acetone. Acetone boils at 56.5°C. It is a colorless liquid that has a characteristic pungent penetrating somewhat sweet odor. It mixes completely with water and with gasoline. It is used as a solvent, and as a starting material in chemical syntheses. (See **Ketones**).

Amines (RNH_2, RR′NH, RR′R″N) and amides ($RCONH_2$, R′CO RCONH, R″CO R′CO RCON) are derivatives of ammonia. (See **Amines**).

Organic acids (RCOOH) are characterized by the carboxyl group. Formic acid (HCOOH), a liquid that boils at 100.7°C, is slightly atypical. Acetic acid (CH_3COOH) is a more typical member of the so-called fatty-acid series. The members of this series with larger numbers of carbon atoms, such as stearic acid ($C_{17}H_{35}COOH$), are found in natural fats and oils in esters. Acetic acid is found

in vinegar. It is made by fermentation of sugar followed by oxidation of alcohol, by destructive distillation of wood, and synthetically from ethanol or acetylene. Acetic acid is a weak acid, but it forms many well-known salts such as those of sodium, calcium, copper (the basic acetate), and lead. Acetic acid finds many uses in the arts such as in dyeing and in photographic work. It is used to make cellulose acetate, widely employed as a base for photographic film and for other purposes. One type of rayon (Celanese) uses acetic acid in its manufacture. (See **Acids, Acetates**).

When an acid and an alcohol react in the presence of a catalyst, a compound called an *ester* may be formed as well as water. For example, $CH_3COOH + C_2H_5OH \rightarrow CH_3COOC_2H_5$ (ethyl acetate) + H_2O. Esters have the general formula RCOOR′, and there are many possibilities. Theoretically, every alcohol and every acid should form an ester. Since mono-, di-, and tri-hydroxy alcohols are known, and since mono-, di-, and tricarboxy acids are known, the possible combinations are numerous. Esters often have a fragrant odor. Many of them are found in the flavors of fruits and flowers. Ethyl acetate, an ester, boils at 77°C. It is a colorless liquid, slightly soluble in water, but miscible in all proportions with alcohol. It is a solvent, but somewhat toxic.

In many cases a hydrogen atom can be replaced by a halogen atom, the nitro group, or a similar element or radical. The resulting compound is still open-chain, hence aliphatic. Familiar examples are trichloroacetic acid (CCl_3COOH), chloroform ($CHCl_3$), carbon tetrachloride (CCl_4), and nitromethane (CH_3NO_2).

Still another large group of aliphatic compounds is the *carbohydrates*. These include starches and sugars. The general formula is $C_xH_{2n}O_n$. Common table sugar (sucrose, $C_{12}H_{22}O_{11}$) is the most familiar example of a carbohydrate. Glucose, corn sirup ($C_6H_{12}O_6$), and lactose in mammalian milk ($C_{12}H_{22}O_{11}$) are other examples of carbohydrates.

ELBERT C. WEAVER

Cross-references: *Hydrocarbons, Alcohols, Paraffins, Ethers, Aldehydes, Ketones, Amides, Amines, Fatty Acids, Esters, Carbohydrates, Fischer-Tropsch Process, Acetates.*

ALKALOIDS

Introduction and Definition. From ancient times man has utilized alkaloids as medicines, poisons and magical potions. Only recently has he gained precise knowledge about the chemical structures of many of these interesting compounds. The term *alkaloid,* or "alkali-like," was first proposed by the pharmacist, W. Meissner, in 1819. It is usually applied to basic, nitrogen-containing compounds of plant origin. Two further qualifications are usually added to this definition—the compounds have complex molecular structures and manifest significant pharmacological activity. Such compounds occur only in certain genera and families, rarely being universally distributed in larger groups of plants. Chemical, pharmacological and botanical properties must all be considered when classifying a compound as an alkaloid. Examples of well-known alkaloids are *morphine* (opium poppy), *nicotine*

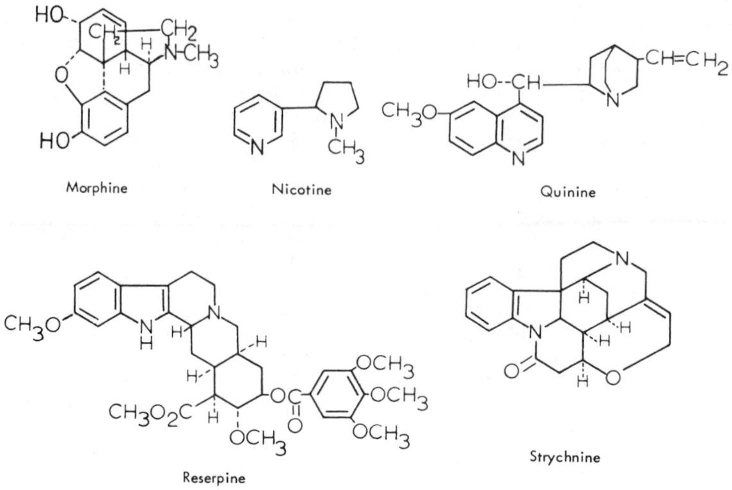

Morphine Nicotine Quinine

Reserpine Strychnine

FIG. 1.

(tobacco), *quinine* (cinchona bark), *reserpine* (rauwolfia) and *strychnine* (strychnos nux-vomica) (Fig. 1).

It should be emphasized that many widely distributed bases of plant origin, such as methyl, trimethyl and other open-chain simple alkyl amines, the cholines and the phenylalkylamines, are not classed as alkaloids. These are designated by some authorities as "biological amines" or "protoalkaloids." In certain cases, the distinction drawn between the "biological amines" and alkaloids is rather arbitrary. Alkaloids usually have a rather complex structure with the nitrogen atom involved in a heterocyclic ring. Yet, THIAMINE, a heterocyclic nitrogenous base, is not regarded as an alkaloid mainly because of its almost universal distribution in living matter. Other heterocyclic nitrogenous bases not classed as alkaloids by some authorities include the purines, of which caffeine, xanthine, and theobromine are examples. Interestingly, *colchicine* is classed as an alkaloid even though it is not basic and its nitrogen atom is not incorporated into a heterocyclic ring, because of its particular pharmacological activity and limited distribution in the plant world.

Occurrence and Distribution. Today a little over two thousand alkaloids are known, and it is estimated that they are present in only 10–15% of all vascular plants. They are rarely found in cryptogamia (exception—ergot alkaloids), gymnosperms or monocotyledons. They occur abundantly in certain dicotyledons and particularly in the families: **Apocynaceae** (dogbane, quebracho, pereiro bark), **Papaveraceae** (poppies, chelidonium), **Papilionaceae** (lupins, butterfly-shaped flowers), **Ranunculaceae** (aconitum, delphinium), **Rubiaceae** (cinchona bark, ipecacuanha), **Rutaceae** (citrus, fagara), and **Solanaceae** (tobacco, deadly nightshade, tomato, potato, thorn apple). Well characterized alkaloids have been isolated from the roots, seeds, leaves or bark of some forty plant families, and it is probable that the remaining families will provide only an occasional alkaloid-bearing plant. **Papaveraceae** is an unusual family in that all of its species contain alkaloids. The majority of plant families occupy an intermediate position in which most species within a genus or closely related genera either do or do not contain alkaloids. Thus all *Aconitum* and *Delphinium* species elaborate alkaloids whereas most of the other genera (*Anemone, Ranunculus, Trocleus*) in the family **Ranunculaceae** do not. It is generally true that a given genus or related genera yield the same or structurally related alkaloids; for example, seven different genera of **Solanaceae** contain hyoscyamine. It is also true that simple alkaloids often occur in numerous, botanically unrelated plants while the more complicated ones such as quinine, nicotine and colchicine are usually limited to one specie or genus of plant and form a distinguishing characteristic of it.

Nomenclature and Classification. The nomenclature of alkaloids has not been systematized—both because of the complexity of the compounds involved and for historical reasons. The two commonly used systems classify alkaloids either according to the plant genera in which they occur or on the basis of similarity of molecular structure. Important classes of alkaloids containing generically related members are the aconitum, cinchona, ephedra, lupin, opium, rauwolfia, senecio, solanum and strychnos alkaloids. Chemically derived alkaloid names are based on the skeletal feature which members of a group possess in common. Thus indole alkaloids (*e.g.* psilocybin, the active principle of Mexican hallucinogenic mushrooms) contain an indole or modified indole nucleus and pyrrolidine alkaloids (*e.g.* hygrine) contain the pyrrolidine ring system. Other examples (Fig. 2) of this classification include the pyridine, quinoline, (cf. quinine) isoquinoline, imidazole, pyridine-pyrrolidine, (cf. nicotine) and piperidine-pyrrolidine type alkaloids. A large number of important alkaloids have received names derived directly from those of plants —as papaverine, hydrastine, berberine. A few have been named for their physiological action, e.g. morphine, (Ger. *morphin,* God of dreams), narcotin (Gk. *narkoō,* to benumb), and emetine (Gk. *emetikos* to vomit). Only one, the pelletierine

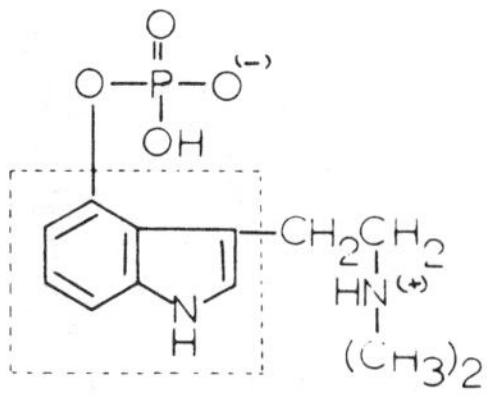

Psilocybin (indole type) 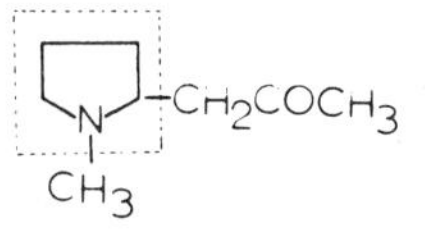   Hygrine (pyrrolidine type) 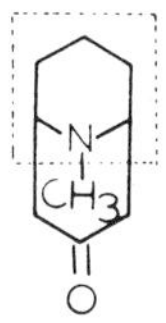   Pseudopelletierine
                                                                      (pyridine type)

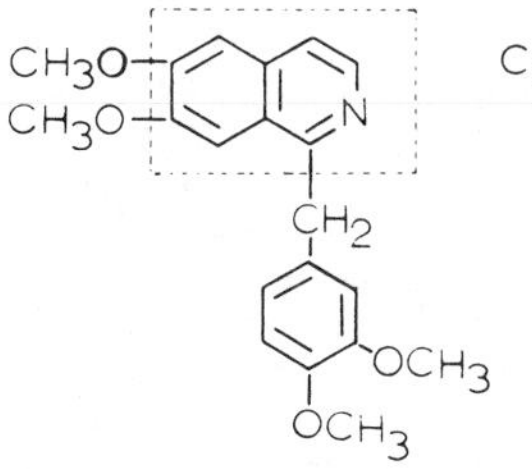

Papaverine   Pilocarpine   Atropine, Hyoscyamine
(isoquinoline type)   (imidazole type)   (piperidine-pyrrolidine type)

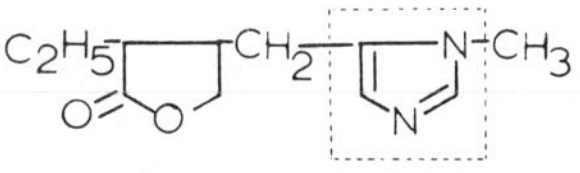

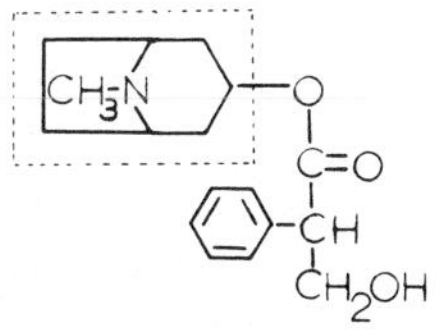

Note: The dashed lines enclose the parent chemical nucleus upon which the type name is based.

FIG. 2.

group, has been named after a chemist—Pierre Joseph Pelletier, the discoverer of emetine (1817), colchicine (1819), quinine (1820), cinchonine (1820), strychnine (1820), brucine (1820), caffeine (1820), piperine (1821), thebaine (1835) and incidentally also chlorophyll.

History. The beginning of alkaloid chemistry is usually dated back about 170 years when F. W. Sertürner announced the isolation of morphine in 1805. He prepared several salts of morphine and demonstrated that it was the principle responsible for the physiological effect of opium. Later (1810) Gomes treated an alcoholic extract of cinchona bark with alkali and obtained a crystalline precipitate which he named "'cinchonino." Subsequent studies (1820) by P. J. Pelletier and J. B. Caventou of the Faculty of Pharmacy, Paris, showed that "cinchonino" was a mixture which they separated into two new alkaloids named quinine and cinchonine. Subsequently various investigators isolated more than two dozen additional bases from species of *Cinchona* and *Remijia*. Between 1820–1850 investigations were intensified and a large number of alkaloids of new and varied types were isolated and characterized. Among the important representatives discovered during this period were aconitine, one of the most toxic materials of plant origin known to man; atropine, a powerful mydriatic agent (4.3×10^{-6} g. will cause dilation of the pupil of the eye); colchicine, the alkaloid of the meadow saffron, which is used extensively in the treatment of gout; coniine, the principle responsible for the death of Socrates when he drank the cup of poison hemlock; codeine, a close relative of morphine, and a valuable pain killer and cough repressant; hyoscyamine, the optically active form of atropine; piperine, an alkaloid of pepper plants; berberine of barberry root; strychnine, a highly poisonous alkaloid used in certain cardiac disorders and for exterminating rodents; and emetine, a powerful emetic sometimes used for treating amoebic dysentery.

Determination of Molecular Structure. Because of the complex molecular architecture of these compounds, with few exceptions, little progress was made in the elucidation of their structures during the 19th century. Only within the past few years have certain of the more complex alkaloids yielded the secrets of their molecular structure to the chemist.

The first step in the determination of the structure of a pure alkaloid consists in ascertaining the molecular formula. This information can be obtained by a combustion analysis for the elements present (always carbon, hydrogen and nitrogen and often oxygen) together with a molecular weight determination, or where high resolution mass spectrometry is available, by this modern technique. The chemist next proceeds to ascertain the function of oxygen and nitrogen atoms in the molecule. Oxygen is most frequently present in the form of hydroxyl or phenol (—OH), methoxyl (—OCH₃), acetoxyl (—OCOCH₃), benzoxyl (—OCOC₆H₅), carboxyl (—CO₂H) or ketone (>C=O) groups. Oxygen in the form of a phenolic group can be recognized by alkali solubility, a color reaction with ferric chloride, acylation to an ester or alkylation to an ether; in the form of an alcohol by alkylation, dehydration or oxidation and by characteristic absorption in the 3.0μ region of the infrared spectrum. Carboxyl groups are suggested by solubility in weak base, by esterification and by specific absorption in the infrared. Methoxyl groups can be determined quantitatively by the method of Zeisel which involves boiling the alkaloid with hydriodic acid and determining the quantity of methyl iodide formed. The estimation of the methylenedioxyl group is accomplished by reactions in which formaldehyde is split out by means of sulfuric acid. Carbonyl groups such as ketones or aldehydes may be detected by spectroscopic means (infrared, ultraviolet, nuclear magnetic resonance) as well as by standard chemical tests.

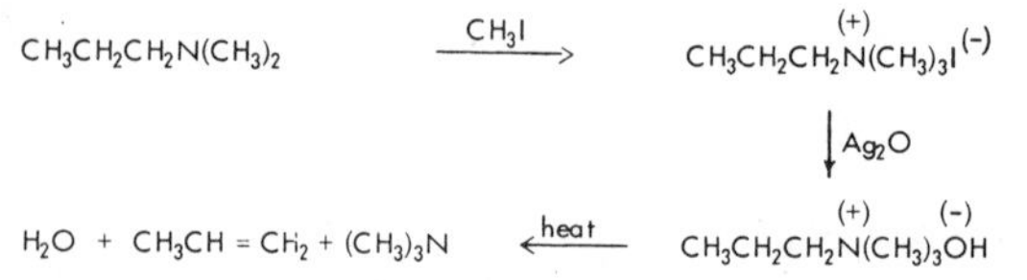

FIG. 3.

The determination of alkyl groups on nitrogen is carried out by heating the alkaloid hydriodide salt at 200–300° and measuring the amount of alkyl iodide produced. Thus an alkaloid bearing an N—CH_3 group (*e.g.* nicotine) will produce methyl iodide and one containing an N—CH_2CH_3 group (*e.g.* aconitine) will liberate ethyl iodide. In most cases the nitrogen in an alkaloid is involved in a ring structure and will usually be secondary or tertiary. It is sometimes difficult to distinguish between these forms though several tests are available. The most widely used method for ascertaining the environment about the nitrogen is exhaustive methylation, also known as the Hofmann degradation. This method is based on the property of quaternary ammonium hydroxides when heated of decomposing with loss of water and cleavage of a carbon-nitrogen linkage to give an olefin. Thus with the acyclic tertiary amine (Fig. 3), a single methylation and decomposition eliminates the nitrogen as trimethyl amine. If the nitrogen atom is involved in a cyclic structure, two or three such cycles of methylation and decomposition will be necessary to liberate the nitrogen and expose the carbon skeleton. Thus in the classic degradation of pseudopelletierine by Willstätter, the alkaloid

was converted to the methohydroxide via methyl granatinine and this decomposed to yield the mono-olefin, (Fig. 4). Repetition of the methylation and decomposition yielded 1,5-cycloöctadiene. Where nitrogen is linked in ring structures through three valancies, three methylations and decompositions are necessary to eliminate the nitrogen.

Numerous other degradative methods are available which transform the alkaloid into stable, easily recognized compounds. Thus distillation over hot zinc dust will sometimes degrade an alkaloid to a stable aromatic derivative, *e.g.* morphine gives phenanthrene, and cinchonine yields quinoline and picoline. Another valuable method involves dehydrogenating the alkaloid with a catalyst such as sulfur, selenium or palladium. During dehydrogenation peripheral groups such as hydroxyls and C-methyls are eliminated. From the dehydrogenation reaction relatively simple, easily recognized products can sometimes be isolated. These may provide a ready clue to the gross skeleton of the alkaloid. A good example of the value of this method is seen in the dehydrogenation of atisine, an alkaloid of *Acontium heterophyllum*. The products (Fig. 5), 1-methyl-6-ethylphenanthrene and 1-methyl-6-ethyl-3-azaphenanthrene, provided early information as to the gross skeleton of atisine and thereby saved much valuable time in the elucidation of the structure of this alkaloid. Another interesting example of the use of this technique is the dehydrogenation of isorubijervine, an alkaloid found in a number of *Veratrum* species. A crystalline hydrocarbon, cyclopentanophenanthrene, and a liquid aromatic amine, 2-ethyl-5-methylpyridine are the main products of the reaction and provide immediate insight into the gross structure of isorubijervine. (Fig. 6).

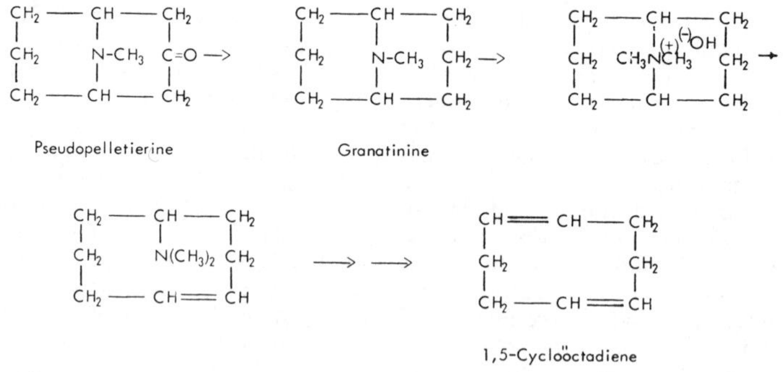

Pseudopelletierine   Granatinine

1,5-Cycloöctadiene

FIG. 4.

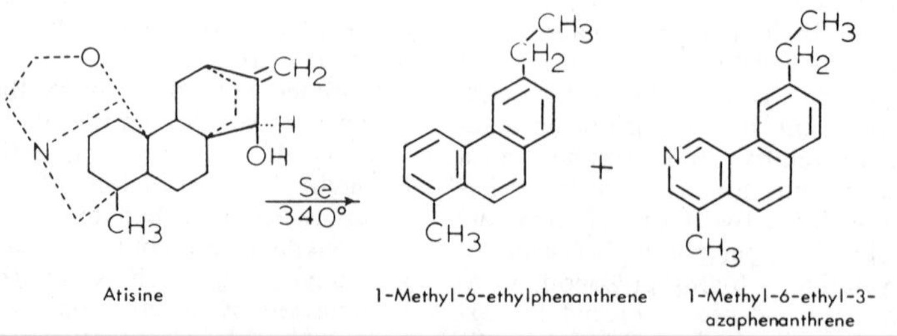

Atisine   1-Methyl-6-ethylphenanthrene   1-Methyl-6-ethyl-3-azaphenanthrene

FIG. 5.

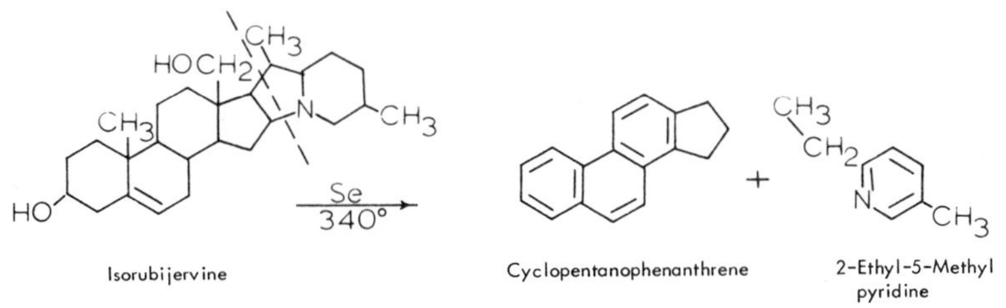

Isorubijervine   Cyclopentanophenanthrene   2-Ethyl-5-Methyl pyridine

FIG. 6

Other methods of degradation are available which allow the removal of particular groups or the cleavage of the molecule at specific points. The aim is to break the molecule into smaller fragments which can be identified and furnish clues which will allow the chemist to visualize the complete structure. The final part of the elucidation of an unknown alkaloid is usually concerned with the stereochemistry of the molecule, viz. the determination of the complete three dimensional representation of all the atoms in the molecule. In the case of complex alkaloids such degradation and detective work can often take years to complete. Fortunately, in recent times the techniques of single crystal x-ray crystallography have been refined to such a point that where a suitable heavy-atom crystalline derivative of an unknown alkaloid is available, the complete structure and stereochemistry may often be elucidated in a matter of a few weeks' time, particularly if large electronic computers are available to solve the required Fourier equations. However, even in situations where the structure of an alkaloid has been deduced by physical methods such as spectroscopy, mass spectrometry and/or x-ray crystallography, an organic chemist may still wish to explore the detailed chemistry of the compound by degradative procedures and even to develop a total synthesis of the molecule. Much of our current knowledge of organic chemical reactions has been derived from detailed degradative studies of complex natural products such as alkaloids.

Biogenesis of Alkaloids. One of the most exciting and fascinating subjects pertaining to alkaloids is the mode of their synthesis in plants. Over the past few decades chemists have proposed many biogenetic schemes for the synthesis of individual alkaloids. Most of these schemes have been based upon the idea that alkaloids are derived from relatively simple precursors such as phenylalanine, tryptophan, "acetate" units, terpene units, methionine and a few other amino acids such as ornithine. On paper at least, the structures of most alkaloids can be derived from such simple precursors using a few well-known chemical reactions. As a matter of fact several simple alkaloids have been synthesized in the laboratory from amino acid derivatives under physiological conditions using just such biogenetic concepts. With the advent of isotopically labeled compounds these theories have been subjected to experimental test.

The modern approach to biosynthetic studies of alkaloids involves administration of labeled precursors to selected plants and after a suitable period of growth, isolation of the alkaloids. These are then degraded in a systematic fashion to determine the position of the labeled atoms. Using this technique it has been demonstrated that morphine and its companions are produced in the plant from TYROSINE. Thus when tyrosine-2C^{14} was fed to *Papaver somniferum* plants it was incorporated into morphine, codeine and thebaine. Degradation of the morphine revealed a distribution of radioactivity in the molecule which can be accounted for by the construction of morphine in the plant from two molecules of tyrosine (or a close biological equivalent) by way of norlaudanosoline. A study of the comparative rates of $C^{14}O_2$ and tyrosine-2C^{14} incorporation indicates that thebaine is the first alkaloid synthesized and that it is converted by demethylation successively into codeine and morphine. By similar experiments many other alkaloids (e.g. nicotine, hyoscyamine, pellotine, papaverine, colchicine, gramine) have been shown to be synthesized from amino acids. The present state of research on alkaloid biogenesis represents a dramatic breakthrough. Not only are plants known to incorporate amino acids, acetate and mevalonolactone, but in specific cases large intermediates have been successfully introduced into the plant's biosynthetic system.

Function of Alkaloids. Unlike biogenesis, the function of alkaloids in the plants remain a subject for speculation. Many authorities regard them as by-products of plant metabolism. Still others conceive of alkaloids as reservoirs for protein synthesis; as protective materials discouraging animal or insect attacks; as plant stimulants or regulators in such activities as growth, metabolism and reproduction; as detoxicating agents, which render harmless, by processes such as methylation, condensation, and ring closure, substances whose accumulation might otherwise cause damage to the plant. While a particular explanation may have application to a given plant, it should be remembered that 85–90% of all plants manage very well without elaborating any alkaloids. In conclusion one might say that while much has been learned about the biogenesis and metabolism of alkaloids, their functions in the plant, if any, are still largely unknown.

S. W. PELLETIER

References

Pelletier, S. W., Ed., "Chemistry of the Alkaloids," Van Nostrand Reinhold, New York, 1970.

Manske, R. H. F., and Holmes, H. L., *Alkaloids,* 1–5 (1950–1955).

Manske, R. H. F., *Alkaloids,* 6–12 (1960–1969).

Henry, T. A., "The Plant Alkaloids," 4th ed., Blakeston, Philadelphia, Pa. 1949.

Cross-references: *Phytochemistry, Pharmacology.*

ALKYD RESINS

Alkyd is a term which is applied to a group of synthetic resins best described as oil-modified polyester resins. This group of materials comprises the reaction products derived from polyhydric alcohols, polybasic acids and fatty monobasic acids. The term "alkyd" was coined from the "al" of alcohol and the "kyd" being representative of the last syllable of acid.

These resins are used in every type of organic coating; i.e., paints, varnishes and lacquers. They impart high gloss, good adhesion, good weathering, and long life at reasonable cost.

The first alkyd resin was prepared by Berzelius in 1847 from glycerol and tartaric acid. In 1851 Van Bemmelen obtained a resin from glycerol and adipic acid. The first alkyd resins to become industrially important were glyceryl phthalate resins prepared by the polycondensation of phthalic anhydride and glycerine. The glyceryl phthalate resins were introduced in 1901 by W. Smith but did not gain wide application until the development of the phthalic anhydride industry after World War I.

Phthalic anhydride is the principal dibasic acid used in alkyd resins although isophthalic, adipic, maleic, sebacic, benzoic and other acids are used to some extent.

Glycerol is the principal polyhydric alcohol used in alkyd resins. Others are pentaerythritol, trimethylol ethane, trimethylol propane, ethylene glycol, propylene glycol, etc.

Long-chain fatty acids are incorporated in the alkyd molecule to import flexibility. Both drying and nondrying oils are used. With drying oil, the alkyds are converted by the oxygen in the air. Nondrying alkyds are used to plasticize other film formers such as cellulose nitrate, etc. The drying oils and semidrying oils used include linseed, soybean, tall, tung and dehydrated castor oil. Cottonseed, castor and coconut oils are the commonly used nondrying oils.

The higher the oil content the more flexible is the resin. Alkyds are usually classified as short oil alkyds, less than 45% oil; medium alkyd resins, 45 to 55% oil, and long oil alkyd 55–75% oil. Long oil alkyds are soluble in aliphatic solvents and are used mainly for air-drying finishes, such as architectural finishes. Medium oil alkyds are soluble in aliphatic and aromatic solvents and are used for both air drying and baking finishes. Short oil alkyds are soluble in aromatic solvents and sometimes require some alcohol solvent.

Alkyd resins are manufactured by the direct fusion of the raw materials with an inert gas atmosphere or by solvent processing. Stainless steel equipment is used and the reaction temperature ranges from 350–470°F. Water is given off in the reaction and when the polymer has attained the desired molecular weight it is dropped into solvent.

Alkyds are used alone and in combination with other resins. Styrenated alkyd resins have very fast drying characteristics. Phenolic modified alkyds have improved water and alkali resistance. Amino-alkyd resins are used in baking applications for applications requiring good color and hardness.

The addition of silicone resins imparts heat resistance and improved weathering to the coating.

C. R. Martens

References

Martens, C. R., "Alkyd Resins," Reinhold Publishing, New York, 1961.
Patton, T. C., "Alkyd Resin Technology," Interscience Publishers, New York, 1962.

ALKYLATION

Alkylation, in broad terms, is the process by which an alkyl radical is introduced, by addition or substitution, into an organic compound. Alkylation may be classified according to the type of chemical bonding formed with the alkyl group. The process includes reactions whereby alkyl groups are bonded with carbon, nitrogen, oxygen, sulfur or metals to form compounds such as isoparaffins, alkyl amines, ethers, mercaptans, metal alkyls, etc. Commonly used alkylating agents are olefins, alkyl halides and alcohols.

In the petroleum industry alkylation generally refers to the reaction of an olefin with a paraffin, aromatic or other cyclic hydrocarbon to form a hydrocarbon equal in molecular weight to that of the combined reactants. Alkylation reactions assumed great importance in World War II in the production of high-octane isoparaffins, cumene and tetraethyllead for use in aviation gasoline as well as ethylbenzene for the production of styrene, used in the preparation of synthetic rubber and plastics.

Alkylation of paraffins, particularly isobutane, with low molecular weight olefins to form high-octane isoparaffins in the gasoline boiling range, was first described in the 1930s. The reaction of isobutane with butenes is:

$$(CH_3)_3CH + C_4H_8 \rightarrow$$

$$C_8H_{18}(2,2,4\text{-trimethylpentane, etc.})$$

Paraffin alkylation proceeds thermally at high temperatures and pressures and at low temperatures and pressures in the presence of certain acid catalysts. A high paraffin-olefin ratio is required to prevent polymerization. The thermal reaction is carried out at about 900°F and at 3000–8000 psi. Ethylene, in contrast to its low activity in most catalytic processes, is more reactive than higher molecular weight olefins. The reaction of isobutane with ethylene yields neohexane (2,2-dimethyl butane) as the principal product. Under thermal operation both normal and isoparaffins are alkylated. Yields per pass range from 11 to 35 percent.

Alkylation of isoparaffins with olefins occurs in the liquid phase at low temperatures with catalysts such as sulfuric acid, hydrogen fluoride, aluminum chloride, boron trifluoride, etc. Under catalytic conditions only isoparaffins are alkylated. Alkylation of isobutane with ethylene in the presence of aluminum chloride yields biisopropyl (2,3-dimethylbutane). Commercially, alkylate is produced by processes employing either sulfuric acid or hydrogen fluoride as catalysts.

In the sulfuric acid process, olefins such as propylene, butenes and pentenes react with isobutane in the presence of 98 percent acid at 30–70°F and at pressures of 100–150 psi. About 3 to 8 parts of isobutane per part of olefin are charged to prevent polymerization. Practically all the olefin reacts and the excess isobutane is recycled. Consumption of acid is 0.5 to 1.0 pound per gallon of alkylate produced.

The hydrogen fluoride process is somewhat similar to the sulfuric acid operation. Temperatures of 70–115°F and pressures of 125–175 psi are employed. Consumption of acid is lower than in the sulfuric acid process thus making hydrogen fluoride competitive in spite of its higher cost.

Alkylate composion depends upon the isoparaffin-olefin ratio and upon the reactants used. Temperature, acid strength, and degree of mixing also affect composition. Isobutylene is the most reactive olefin, whereas ethylene is not reacted by these acids. Aluminum chloride is used as a catalyst to produce diisopropyl from isobutane and ethylene.

In the alkylation of aromatics the production of ethyl benzene and cumene are important industrially. Ethyl benzene results from the reaction of benzene with ethylene at 190°F over aluminum chloride as catalyst. Cumene, a component of aviation gasoline and a precursor for the preparation of phenol, is produced by vapor phase reaction of benzene and propylene at 400–500°F over a phosphoric acid catalyst.

Although recent emphasis has been placed upon hydrocarbon alkylation, other types of alkylation have received a great deal of attention and numerous chemicals are derived by reactions of this nature. The manufacture of tetraethyllead proceeds according to the reaction:

$$4PbNa + 4C_2H_5Cl \rightarrow Pb(C_2H_5)_4 + 3Pb + 4NaCl.$$

Consumption of this chemical, as a fuel antiknock, amounted to 720,000,000 pounds in 1971. Among the many other chemicals produced commercially by alkylation processes a few may be mentioned: aliphatic amines result from the reaction of alkyl halides with ammonia in alcohol solution; ethers such as "Cellosolve" and "Carbitol," used as solvents, are products in which ethanol acts as the alkylating agent; ethyl cellulose, for plastics and protective coatings, is prepared by the action of alkyl halides on alkali cellulose. Other alkylation products include mercaptans from alcohols and hydrogen sulfide and silicon organics which show promise of application in lacquers, resins and lubricants.

H. O. FOLKINS

Cross-references: *Gasoline, Antiknock Agents, Petroleum.*

ALLOTROPY

Allotropy is the existence of a substance in two or more different forms. Polymorphism is a more suitable term, especially for different crystalline phases, but the term allotropy has been generally adopted for the elements. (The previously recognized allotropy of silver iodide helped lead to the recognition of allotropy in steels). The term allotropy is applied even when the forms are not crystalline, such as the different forms of liquid sulfur or helium or the O_2-O_3 (ozone) forms of molecular oxygen. The different forms, designated as allotropes or polymorphs, and often denoted by different Greek letters, are thermodynamically different and therefore manifest significant differences in various properties, both physical and chemical. Thus, allotropes can differ in atomic configuration and dimensions (hence in density), melting and boiling points, color, and thermal conductivity. In addition, allotropes show marked differences in reactivity with various reagents and in solubility in different solvents. The conditions for the preparation of an element can determine which allotrope is obtained.

The transformation between the allotropes is unidirectional if one of the forms always has a higher free energy, being metastable and requiring an energy input for its initial formation. Each phase is stable within a different range of temperature and pressure, as represented in a phase diagram. In most cases, the transformations between phases (represented by boundaries between the regions of stability for each phase) are reversible and are governed by the same physicochemical considerations that govern other phase equilibria such as melting and vaporization. Thus, a single-component two-phase system is univariant, i.e., according to the phase rule either the temperature or the pressure, but not both, may be independently specified for the coexistence of two phases. The relation between the equilibrium temperature and pressure is determined, according to the Clapeyron equation, by the enthalpy and volume changes associated with the phase transformation ($dp/dT = \Delta H/T\Delta V$). Both the magnitude and the sign of the volume change of the transformation are of practical importance. Thus, the design of metal components has to allow for such volume changes which (like the heat of transformation) are much more consequential than changes associated solely with temperature.

From a practical point of view, it is very important that the temperature for transformation between two phases of an element (usually specified at atmospheric pressure) can be altered by the presence of other elements. If these solute elements dissolve to different extents in the allotropes of the base element, the dependence of the free energies of the individual allotropes on temperature is so modified as to shift the equilibrium temperature, i.e., the temperature at which the free energies are equal. The allotrope in which solubility is greater is stabilized, since the temperature at which it prevails has been lowered or raised, depending on whether it is the high- or low-temperature phase. (The situation is analogous to depression of the melting point by a solute that is more soluble in the liquid than in the solid.) The dependence of the transformation temperature on solute concentration is of course an essential aspect of the phase diagram for the base element and the solute. Such phase diagrams thus serve as guides in the formulation of alloy compositions.

Another important consideration in the relation between the allotropes is the delay in reaching equilibrium, especially in the presence of solutes. A metastable phase may persist, a situation that is not necessarily undesirable and is in fact often exploited to retain a desirable allotrope. This situation may be encouraged by incorporation of solute elements that, even if they do not effect retention of the desirable allotrope under equilibrium conditions, are so effective in retarding attainment of equilibrium that, for all practical purposes, they have effected retention of a metastable phase.

Examples of allotropy are abundant; indeed it is the rule rather than the exception, especially when one considers the effect of pressure as well as temperature on phase stability. Even at atmospheric pressure, some elements exhibit many crystalline forms: uranium, 3; manganese, 4; plutonium, 6. More extended information can be found in the entries on these elements and also others such as helium, hydrogen, carbon, phosphorus and sùlfur, in which allotropy plays an important part. The following two paragraphs respectively cite examples of metallic and non-metallic elements in which allotropy has to be taken into account.

The most common example of the manipulation of allotropy is found in the heat treatment of steel to obtain and retain phases with the desired metallurgical properties, achieving maximum benefit from the alloying elements. Carbon is the most prominent of the alloying elements introduced into iron for the effect on the relative stability of the phases, i.e., body-centered cubic ferrite (α) and face-centered cubic austenite (γ). In other metallic systems, consideration should be given to the differences between properties of the allotropes. Thus, it may be essential to avoid transformations than can lead to dimensional changes or deterioration of mechanical properties (grey tin). In alloy systems, attention must be paid to the partitioning of solute elements between phases; such partitioning can cause serious inhomogeneities (the hexagonal alpha phase of titanium or zirconium versus the body-centered cubic beta). Transformations may be used to wipe out preferred orientation imparted by working (beta phase treatment of uranium worked in the alpha phase). Metals difficult to work in their common structure may become workable in a structure that is stable under different conditions. These metals may then be heat-treated to obtain a structure that is more desirable, for instance, for greater strength.

Nonmetallic elements provide some of the classical examples of allotropy. White phosphorus is obtained by condensation from the vapors in the usual reduction. The more stable red allotrope can be obtained by heating at 240°–250°C in the presence of a catalyst such as iodine. It differs from white phosphorus in its substantially lower reactivity and in its solubility in a variety of solvents. A black allotrope can be obtained under pressure. Another white allotrope is stable below −70°C. Sulfur has a variety of allotropic forms in both the solid and the liquid phases. The well-known allotropy of graphite and diamond has provided the basis for the goal of preparing the more valuable form, an accomplishment requiring the extreme conditions of temperature and pressure where it is the stable allotrope.

SAUL ISSEROW

Reference

Klement, W., and Jayaraman, A., "Phase Relations and Structures of Solids at High Pressures," pp. 289–376 in, "Progress in Solid State Chemistry," Vol. 3, H. Reiss, editor. Pergamon Press, 1967.

ALLOYS

Alloys are metallic substances composed of two or more chemical elements, at least one of which is a metal. This definition is deliberately made so broad as to encompass every form of combination or interaction of a metallic element with other elements. At one extreme, the elements may be blended on the atomic level as a homogeneous solid solution. At the other extreme, one or more of the alloying elements or components may be in the form of discrete particles, fibers, or even continuous filaments. Alloying has come to represent such a wide range of possibilities in materials blending that the term has been extended to other classes of materials such as ceramics or polymers. Alloying has been most extensively applied in metal systems and the balance of this article is devoted exclusively to metal alloys.

Metals now recognized as alloys were used from the dawn of metallurgy—long before the recognition of the chemical elements—since many ores contain more than one metallic element, so that the melting or reduction processes yielded metals now recognized as alloys. In due time, it was found beneficial to mix different ores to control the properties of the product; this was the beginning of physical metallurgy, the application of guidelines for the achievement of metallic materials with properties controlled or modified for certain applications. These guidelines have evolved from empirical observations to principles related to modern concepts of the electronic structures of atoms and their combinations. Thus, it is increasingly possible to predict the stability of various phases, their ability to accommodate (dissolve) other elements, and the tendency of elements to form new phases. The presence, distribution and interactions of the various phases determine the properties of the metal.

Any general discussion of alloys must make special mention of steels. Alloys based on the iron-carbon system have been used since ancient times and are still produced and used in largest quantity. The study of such alloys has been dominant in the establishment of metallurgical principles applicable to all metal systems. Hence, iron-base alloys often serve as prototypes for alloys in other systems and approaches developed for the ferrous alloys are extended to others. Thus, martensitic transformations are investigated in non-ferrous alloys; in a newly evolving alloy system, an element is sought for a role analogous to that of carbon in iron.

The composition of a metal is modified, i.e., the metal is alloyed, with the intention of achieving a more desirable combination of properties. Service requirements dictate the selection of the heat treatment of the alloy, since the properties of a metal (particularly an alloy) are not a unique function of the composition alone, but can be very strongly influenced by its history, notably the conditions of exposure to heat and deformation. Also important are the conditions of cooling following such exposures, since the conditions determine the extent of retention of any metastable phases. Such phases can play a crucial role in determining the properties of an alloy; hence their generation and preservation is essential for the achievement of the desired properties. Consequently, it is often necessary to specify the thermal history of an alloy as well as its composition for a specific application and the properties desired therefor.

The following are some of the changes that are being sought and achieved through alloying: greater strength; improved toughness; reduced sensitivity to attack by hostile media; responsiveness to heat treatment (an additional element may be incorporated to increase the responsiveness of an alloy to a heat treatment involving other elements); improved fabricability or formability (a more specific expression of the broadly used term "ductility"); more effective castability; easier machinability; amenability to joining processes such as welding; more suitable values of physical properties such as electrical conductivity, magnetic permeability, or coefficient of expansion; lower density. Different compositional modifications and different alloying approaches are available to achieve these goals. Often, alloying to improve one property impairs another property and the alloy developer or selector (hopefully in conjunction with the designer) makes a decision representing the most effective tradeoff for the particular service requirement. In some cases, particularly for electrical or electronic requirements, the best alloy is the least alloyed, i.e., a pure element. Yet a concurrent requirement such as strength may necessitate some alloying. An alloying approach is then selected that is least compromising of the desired property. Thus, to maintain the electrical and thermal conductivity of a pure metal such as copper, it is better to strengthen by a dispersed second phase rather than by formation of a solid solution.

The effects of alloying elements may be divided on the basis of their effects on the phases present. (This classification assumes some interaction of the alloying element as occurs in the classical cases derived from dissolution in a melt). An alloying element may dissolve interstitially or substitutionally in the base element, so that only one phase is present, its properties modified by the dissolved element(s). Alternatively, an alloying element contributes to the formation of a new phase. The alloying element may be a predominant component in the new phase, for instance an intermetallic compound, or it may dissolve and affect the phase relationships in the base compositions, so that two phases now coexist. Other effects of alloying relate to the above but are not adequately described above. Thus, an alloying element may give a solid solution (single phase) which is so affected by the solute that a different phase of the base metal solvent is now stable. Or the alloying element may so affect the thermodynamic relationships in the solvent as to modify the dissolution or precipitation of other phases. In summary, the various alloying elements differ in their influences on how many and which phases are present, and also in their influences on the phases in which they are present. The skillful application of metallurgy then presuppose defining the effects of other elements on a base composition. In various alloy families general principles regarding effects of individual elements have been established. In an analogous manner the alloys based on a new element evolve from an understanding of the effects of individual alloying elements. With the help of generalizations regarding the alloying elements, it is possible to see (1) the evolution of families of alloys (such as the type 300 stainless steels) by incremental additions or removals of individual elements and (2) the introduction of a new type of alloy through a new alloying addition (rare earths or thorium to magnesium alloys).

The alloy composition and method of preparation are closely related—so much so that, even with newer methods of alloying (see below), the usefulness of an alloy system has been determined by its amenability to preparation by conventional methods, starting with a casting derived from a homogeneous melt. The distribution of phases in the solid is then determined by the method of cooling this casting since the alloying elements are likely to be less soluble in the solid and hence form additional phases as the casting freezes and cools. The distribution of phases and other microstructural features such as grain size are subsequently modified by working and heat treating. Current metallurgical practice, notably powder metallurgy, extends alloying to elements that are completely insoluble in the solid and even to combinations that do not dissolve in the liquid or whose heating above the melting point is not desirable or warranted. This view of a predominantly mechanical alloying can be broadened to include fiber-reinforced composites, where the interaction between phases is primarily mechanical and limited to the interface.

Alloy groups are often designated in ways depicting either the phase (austenitic), the type of strengthening (age hardening, maraging) or the application (tool steels). More specifically, codes have been developed to designate alloy compositions by numerals and/or letters. Such codes have been established for various alloy families by technical societies or trade associations representing commercial organizations such as the producers or users. These organizations also establish specifications for standard compositions within a country such as the United States. Various government specifications are also available.

Each alloy code seeks rational guides for associating letters or numbers with the nature and content of the various alloying elements. Fairly systematic codes are established for various families of steels (structural, stainless, tool, high temperature), aluminum and magnesium. For copper alloys, historical terms persist, but arbitrary numerical designations have been established by the Copper Development Association (CDA). The special nickel and cobalt alloys are still largely proprietary and are labeled by designations, often trade marked, which are more informative about their developers and owners than about their chemistries. Titanium alloys are usually designated by a concise representation of the composition. Refractory alloys have been labeled like the nickel alloys. A possible exception is molybdenum where the few alloys carry designations with some clue as to composition.

For more detailed information regarding the import of alloy designations, the reader is referred to compilations in the handbooks, including those available from suppliers as well as less commercially oriented publications. Treatises on individual elements include both tabulation and systematization of codes for alloy designations. Compilations combining the more common compositions in diverse families are available from some of the producers or are published at various intervals in the trade journals. The user of an alloy family finds that before long he becomes conversant with that family and readily associates designations, compositions, and unique properties.

SAUL ISSEROW

References

American Society for Metals, "Metals Handbook," Vol. 1, 8th edition, Metals Parks, Ohio, 1961.

Cahn, R. W. ed., "Physical Metallurgy," 2nd rev. ed., North-Holland Publishing Co., Amsterdam, 1971.

Guy, A. G. "Elements of Physical Metallurgy," 2nd edition, Addison-Wesley Press, Inc., Reading, Mass., 1959.

Reed-Hill, R. E. "Physical Metallurgy Principles," D. Van Nostrand Co., Inc., Princeton, N.J., 1964.

Woldman, N. E. "Engineering Alloys", 4th edition, Reinhold Publishing Corp., New York, N. Y., 1962.

"Properties of Some Metals and Alloys," International Nickel Co., New York, N. Y., 1968.

Refractory Alloys

The refractory alloys are alloys based on the relatively common metals having melting points significantly higher than iron, nickel or cobalt, namely tungsten, tantalum, molybdenum and columbium, although chromium and vanadium are frequently included.

The nuclear, aerospace, chemical processing, etc., fields have many applications which require metallic-type properties at temperatures over 2200°F (1205°C). None of the more well-known alloy systems, including the superalloys, can be used. A glance at the melting points of the refractory metals (Table 1) compared to iron, nickel and cobalt shows the great increase in temperature capability possible. The refractory metals, unalloyed, have

TABLE 1. MELTING POINTS OF REFRACTORY METALS

Metal	Melting Temperature	
	°F	°C
Tungsten (W)	6170	3410
Tantalum (Ta)	5425	2996
Molybdenum (Mo)	4730	2610
Columbium (Cb)	4474	2468
Vanadium (V)	3560	1919
Chromium (Cr)	3407	1875
Iron (Fe)	2795	1535
Cobalt (Co)	2719	1493
Nickel (Ni)	2646	1452

Data taken from "Rare Metals Handbook," Clifford A. Hampel, Editor, 2nd Ed., Reinhold Publ. Corp., 1961.

such a great advantage over other metals that the impetus for development of stronger refractory alloys has not been as great as in some other alloy systems, i.e., superalloys.

Two major problems have not yet been solved completely by alloying: (1) low-temperature brittleness of tungsten, molybdenum and chromium (below a "transition temperature"); (2) poor oxidation resistance or tendency for oxygen or nitrogen pickup of all the alloys.

The latter problem virtually prevents use in air at elevated temperatures without protective coatings. None of the coatings are really satisfactory for long exposures, especially where rapid thermal cycling occurs. Basic oxidation resistance can be improved somewhat by alloying, i.e., titanium in columbium-base alloys, and nitrification of chromium can be improved by addition of yttrium, but long-time maintenance of resistance has not been accomplished. The underlying cause of the problem is that the surface oxides that form either vaporize or are not tightly adherent and diffusion-resistant.

The existence of a ductile-to-brittle "transition temperature" is not unexpected in the body-centered cubic structure of the refractory metals. While columbium and tantalum alloys generally are ductile to well below room temperature, molybdenum and tungsten alloys have transition temperatures occasionally over room temperature, as do chromium-base alloys. The transition temperatures are dependent on the alloy composition, the working history and the heat treatments used. Most dissolved elements, particularly interstitial elements such as oxygen, nitrogen, etc., act to raise the transition temperature. One alloying element, rhenium, has shown a marked effect in lowering transition temperatures of tungsten, molybdenum and chromium.

Strengthening can be obtained by solution hardening or by particle dispersion, or both. Since W, Ta, Mo and Cb are soluble in one another and do not affect melting points greatly, they are widely used for mutual solution-hardening. Hafnium and vanadium are fairly soluble, and are used as solution hardeners in tantalum and columbium. Dispersed particle hardening has been virtually limited to carbide dispersions in molybdenum and columbium and thoria dispersion in tungsten. The presence of as little as 0.01% carbon by weight will lead to carbide formation in any of the refractory alloys. The monocarbides TiC and ZrC, with substitutions of Cb, Mo, etc., for some of the Ti or Zr, are mainly responsible for the strengthening. Other carbides, such as Mo_2C, Cb_2C, etc., can form, and these have regions of stability. In some cases, where the carbides can be dissolved at an elevated temperature and precipitated at a lower temperature, true precipitation hardening can be obtained. In other cases, the carbides are formed during solidification and their dispersion remains fairly constant during use of the alloy. Oxides and nitrides are also found in small amounts.

Many of the mechanical properties depend on the condition of the alloy with respect to prior working and heat treatments. Precipitation of carbides may be induced or accelerated by the presence of strain. Grain size affects both tensile and stress-rupture properties. The presence and location of all phases (intentional or impurity) affect properties. Heat treatments, recrystallization in particular, change the structures produced by prior deformation and, therefore, the properties. At low temperatures, recrystallization results in the lowest properties. At elevated temperatures, the differences tend to diminish. Existence of stable carbide or oxide particles can inhibit recrystallization, and is responsible for the relatively high recrystallization temperatures in the carbide-hardened alloys TZC molybdenum and F48 columbium, which in turn is responsible in part for the retention of strength of these alloys.

Production of the refractory alloys is understandably more difficult than of the lower melting metals. The high melting temperatures, high working temperatures, poor oxidation resistance and low temperature brittleness (in some cases) of the refractory alloys all contribute to the difficulty. Two general methods are in use for production: consolidation of powder and vacuum melting (arc or electron beam). The latter method has been limited mainly to molybdenum and lower melting temperature systems. Close control of impurities must be maintained. After melting, the ingot formed is usually "canned" (clad in an oxidation-resistant metal container) and either extruded, forged or rolled into shape. The powder metal method involves compaction and sintering at high temperature in a protective atmosphere, followed by "canning" and extrusion or forging. Temperatures over recrystallization must be used to accomplish "hot working." Since these temperatures are so high, the alloys are frequently "warm worked" (worked at an elevated temperature below the recrystallization temperature), with intermediate annealing. The amount of deformation and the working temperatures must be controlled accurately since they have such an important effect on properties.

Fabrication of refractory alloy parts can be accomplished by standard processes, with some precautions. Those alloys having transition temperatures at room temperature or higher should be handled carefully because of their low ductility and bending or forming must be done above the transition temperature. If the "warm" working temperature is high enough, protection from air may be necessary. Similarly, refractory alloys can be welded or brazed by standard methods which provide oxidation protection, such as inert arc welding, electron beam welding, inert atmosphere or vacuum brazing, etc. The high melting temperatures coupled with relatively good thermal conductivity make welding somewhat more difficult than conventional materials. For molybdenum and tungsten, the large temperature differences between melting and room conditions plus the high modulus of elasticity can set up large thermal stresses which can cause cracking, especially in high transition temperature alloys. Preheating the material, localizing heat input, etc., aids in reducing the stresses.

Superalloys

Many applications in process equipment, steam and gas turbines, etc., require materials that will pro-

vide strength and oxidation-corrosion resistance for long periods of time at elevated temperatures. The common steels, aluminum, copper and titanium-base alloys generally are limited to about 1000°F (540°C). Superalloys are the class of iron-nickel, nickel, and cobalt-base materials whose usefulness starts here and extends, in some cases, up to about 2200°F (1200°C).

Oxidation and corrosion resistance are imparted by the formation of tightly adherent surface oxide films. The major contributors to the protective films are chromium, aluminum, nickel and cobalt. Typical oxides formed are NiO, Cr_2O_3, Al_2O_3, various spinels, etc. Under certain corrosive conditions at 1400–2000°F (presence of some salts plus sulfur) these oxides may be attacked and rapid "hot corrosion" can take place. Protection by coatings generally containing high aluminum, chromium or titanium may be required.

The strengths of superalloys depend on one or both of the conventional alloying means: solution hardening and dispersed particle hardening. Any element dissolving into the base metal will act to strengthen the alloy. The most common elements used for this purpose include chromium, molybdenum and tungsten. Dispersed particle hardening may be produced either by a solid state precipitation of a phase or phases during a heat treating procedure or by the presence of a relatively insoluble phase during solidification. The behavior of these phases is so important that a more detailed description is necessary. Three broad groups of phases should be distinguished: carbides-oxides-nitrides-borides, precipitation-hardening intermetallics and complex intermetallics.

The carbides, etc., group can be subdivided into four types:

(1) MC, MN or M(C, N), where M is usually Ti, Cb, Ta, W or combinations of these. All have high solution temperatures of over 2250°F (1230°C), and are usually formed during solidification as coarse, dispersed particles. They have little effect on low temperature strength, but they do aid high temperature creep and rupture resistance. If the carbo-nitrides form as clusters or stringers, ductility and fabrication problems may exist. Vacuum-melting will eliminate nitrogen as a constituent. (Most of the nickel-base, and some of the others, are normally vacuum melted.) Tantalum and columbium will displace some of the titanium and tungsten when they are present.

(2) Cr_7C_3, $M_{23}C_6$, M_6C, etc., where M is usually Cr, Mo, W or combinations. The complex carbides have lower solution temperatures, ranging from 1500–2150°F (815–1180°C), and are therefore subject to solution and precipitation during heat treatment or operation. Greater molybdenum or tungsten content tends to favor production of the M_6C carbide in preference to $M_{23}C_6$ or Cr_7C_3, with a consequent higher solution temperature. Both $M_{23}C_6$ and M_6C will form at the expense of initially present MC over long periods of time at elevated temperatures. The complex carbides are less effective than the mono-carbo-nitrides in

strengthening at high temperatures due partially to their lower solution temperatures. In addition, the complex carbides, particularly $M_{23}C_6$, often concentrate at grain boundaries as brittle films and cause lower ductility and impact resistance.

(3) *Oxides.* Occasionally, due to imperfect melting or fabricating, oxides of Cr, Ni, Co, etc., are found in the alloys, and are considered detrimental to mechanical properties. More recently, intentional dispersions of high melting oxides have been used to cause an increase in high temperature strength. The only commercial superalloy of this type is TD nickel, which contains a dispersion of ThO_2 in a nickel base and provides excellent retention of strength over 2000°F.

(4) M_3B_2 where M may be Mo or W and Ni, Co, etc., and other borides. The borides are high solution temperature phases like the mono-carbo-nitrides and are mostly formed in grain boundaries. It is believed that the borides aid in creep strength while inhibiting the precipitation of embrittling carbides in the grain boundaries. However, they also lower the melting temperature of the alloy, necessitating a careful balance in boron content.

The major intermetallic precipitation phase system in the superalloys is γ' or $Ni_3(Al, Ti, Cb, Ta)$. Initial precipitation forms a fine dispersion of a face-centered cubic structure which has a lattice parameter close to that of the matrix and causes an increase in strength. Ni_3Al is normally face-centered, and can dissolve into its structure considerable amounts of titanium, columbium and other elements. Its solution temperature varies from 1800°F (980°C) to 2200°F (1200°C) depending on the dissolved elements and the matrix. Precipitation is extremely rapid at temperatures over 1400°F (760°C). Apparently, a stable precipitate particle size is associated with each temperature, and is quickly reached on heating or cooling. This peculiarity leads to quick recovery of mechanical properties after exposures to higher-than-normal temperatures.

The equilibrium structures of Ni_3Ti and Ni_3Cb are hexagonal. Alloys containing titanium or columbium with exposure to temperatures of about 1300°F (700°C) and higher, and in the absence of a sufficient amount of aluminum to stabilize the initially formed face-centered cubic structure, will produce the hexagonal structure. Coherency with the matrix is lost, and strength decreases. Occasionally, the Ni_3Ti or Ni_3Cb will become acicular, which can cause a loss in ductility as well. Unlike γ', these phases require a re-solution and precipitation treatment before strength can be regained. Solution temperatures for Ni_3Ti and Ni_3Cb vary from 1650°F (900°C) to 2050°F (1120°C), depending mainly on the matrix composition. Higher nickel and lower iron alloys permit the higher solution temperatures. The major advantages of the high titanium and columbium-containing alloys are higher strength at lower temperatures and slower aging rates compared to high aluminum alloys. In alloys containing very high aluminum in relation to nickel, the phase NiAl can form. It

is deleterious to high temperature properties and is avoided in the commercial alloys.

The complex intermetallics include phases such as σ, Laves, μ, etc. These form when the residual matrix (after precipitation of γ' and carbides) is too rich in chromium and the refractory metals. They are generally detrimental to long life and ductility and should be avoided by proper alloy design and heat treatment.

Based on the major strengthening mechanisms, the superalloys may be divided into three categories: solution strengthened (N155, "Hastelloy" X, L605, for example); dispersed carbide or oxide strengthened (X40, Mar M 302, TD nickel); and precipitation-hardened with $Ni_3(Al, Ti, Cb)$ (A286, Inconel 718, SEL, J1570, etc.). Of course, combinations of several mechanisms are often present in a single alloy. Each class has its own area of usefulness. The precipitation-hardened alloys have the highest strengths at low and intermediate temperatures and are the most difficult to fabricate and join. The solution strengthened alloys are weakest at low temperatures, but at higher temperatures, where the precipitates start to dissolve, their strengths become much closer. Fabrication and joining of the solution strengthened alloys is the easiest of the three classes. The carbide-oxide strengthened alloys fall in between the other two types in both strength at low temperature and fabricability. At high temperatures (over the solution range for $Ni_3(Al, Ti, Cb)$), due to the relative stability of the carbides and oxides, these types of alloys have the highest strengths.

MURRAY KAUFMAN

ALLYL RESINS

Allyl resins and monomers—of which diallyl phthalate is commercially the best known—are premium plastic materials noted for their unusual electrical, chemical, and water-resistant characteristics and ability to maintain these properties under severe environmental conditions.

Compounds based on diallyl phthalate and diallyl isophthalate resins have been used for some time as molding materials for rocket and missile components that must withstand extraordinary conditions. They are finding growing acceptance for such uses not only in the military and aerospace fields but also in civilian applications.

These materials are also widely used to impregnate glass cloth and roving (called prepregs) in reinforced plastics. In addition, they are employed in insulating varnishes, decorative laminates, and as additives for other resin systems.

Distinguishing properties of the allyls include exceptional insulation resistance, heat resistance up to 450°F, unusual dimensional stability, moisture resistance, exceptional shelf stability, controllable flow characteristics, low mold shrinkage, and absence of post mold deformation. This combination has helped make the use of allylic systems commercially practical and competitive—in terms of costs per unit finished part—in many applications.

Because of degradative chain transfer in which a hydrogen atom is abstracted by a macroradical to yield a dead polymer and a stable allyl radical, it is not feasible to produce high molecular weight polymers from monoallyl compounds. As shown in the following equation, the allyl radical produced by this chain transfer reaction is resonance-stabilized and has little tendency to initiate a reaction to form new polymer chains.

$$R\cdot \;+\; H_2C\!\!=\!\!CHCH_2X \rightarrow$$
macroradical   allyl compound

$$RH \;+\; H_2C\!\!=\!\!CH\!-\!\overset{\bullet}{C}HX$$
dead polymer $\updownarrow$

$$H_2\overset{\bullet}{C}\!-\!CH\!\!=\!\!CHX$$
allyl radical

Poly(allyl alcohol) has been produced indirectly by the lithium aluminum hydride reduction of poly(methyl acrylate) as shown by the following equation:

$$\left[\begin{array}{cc} H & H \\ | & | \\ -C\!-\!\!&\!\!C- \\ | & | \\ H & O\!\!=\!\!C\!-\!OCH_3 \end{array}\right]_n \xrightarrow{LiAlH_4} \left[\begin{array}{cc} H & H \\ | & | \\ -C\!-\!\!&\!\!C- \\ | & | \\ H & H_2\!-\!C\!-\!OH \end{array}\right]_n$$

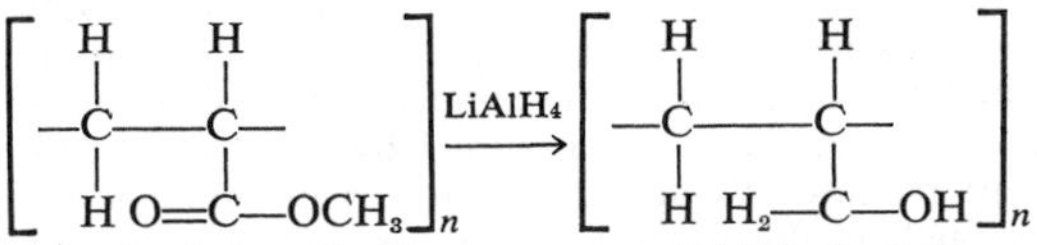

poly(methyl acrylate)   poly(allyl alcohol)

However, since the properties of this polymer are similar to those of the less expensive poly(vinyl alcohol), poly(allyl alcohol) is not produced commercially. Polysulfones may be produced by the reaction of sulfur dioxide and a monoallyl compound.

In spite of the difficulties associated with the polymerization of monoallyl alcohol, halides, ethers and esters, useful cross-linked copolymers of vinyl acetate and diallyl succinate have been known since 1940. Polymers of diallyl esters and diallyl ethers are used to a moderate extent as thermosetting resins which cure with a low exotherm and do not release volatile byproducts during the curing cycle.

Allyl ethers of sucrose and starch may be used as air-cured coatings. The curing rate of these allyl ethers is accelerated in the presence of cobalt naphthenate and at elevated temperatures. As shown by the following equation, diethylene bis(allyl carbonate) is produced by the reaction of phosgene, ethylene glycol and allyl alcohol.

$$HO\!-\!CH_2CH_2\!-\!OH \;+\; Cl\!-\!\underset{\underset{O}{\|}}{C}\!-\!Cl \rightarrow$$
ethylene glycol   phosgene

$$Cl\!-\!\underset{\underset{O}{\|}}{C}\!-\!OCH_2CH_2\!-\!O\!-\!\underset{\underset{O}{\|}}{C}\!-\!Cl$$
intermediate

$$Cl\!-\!\underset{\underset{O}{\|}}{C}\!-\!OCH_2CH_2\!-\!O\!-\!\underset{\underset{O}{\|}}{C}\!-\!Cl \;+$$
intermediate

$$2HOCH_2\!-\!CH\!\!=\!\!CH_2 \xrightarrow[\substack{-2NaCl \\ -2H_2O}]{2NaOH}$$
allyl alcohol

$$H_2C\!\!=\!\!CH\!-\!CH_2O\!-\!\underset{\underset{O}{\|}}{C}\!-\!OCH_2CH_2\!-$$

$$O\!-\!\underset{\underset{O}{\|}}{C}\!-\!O\!-\!CH_2\!-\!CH\!\!=\!\!CH_2$$
diethylene bis(allyl carbonate)

This diallyl ether may be cured with peroxydicarbonate in appropriate molds in the absence of oxygen to produce a colorless, scratch resistant, optically pure polymer called Allymer CR-39. This polymer, which is used for casting contact lenses and prescription sun glasses, transmits 89–92 per cent of ordinary light and filters out over 90 per cent of the ultraviolet radiation present in this light. The monomer may be used as a crosslinking agent with methyl methacrylate.

Allyl ethers with residual hydroxyl groups have been produced by the partial etherification of pentaerythritol with allyl alcohol. Polymers from these hydroxyethers may be reacted further with alkyl or aryl diisocyanates or with the free carboxyl groups in alkyd resins. However, the only polyallyl ether other than the Allymer CR-39 monomer used commercially is triallyl cyanurate. As shown by the following equations, the latter may be produced by (1) the nucleophilic displacement of the chloride in cyanuric chloride by allyl alcohol or by (2) the tranesterification of trimethyl cyanurate with allyl alcohol.

(1)

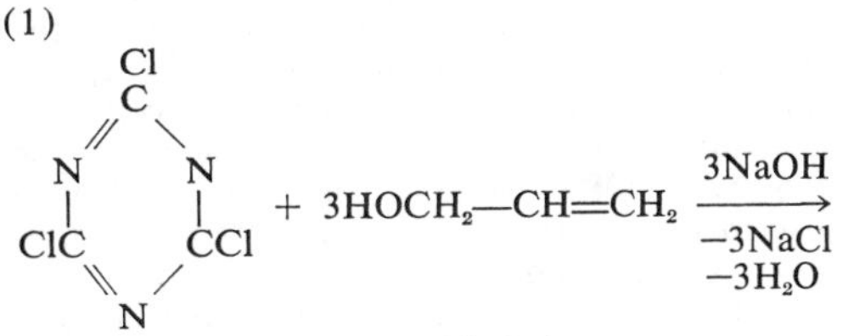

$$\text{Cyanuric chloride} \;+\; 3HOCH_2\!-\!CH\!\!=\!\!CH_2 \xrightarrow[\substack{-3NaCl \\ -3H_2O}]{3NaOH}$$

Cyanuric chloride   allyl alcohol

or

(2)

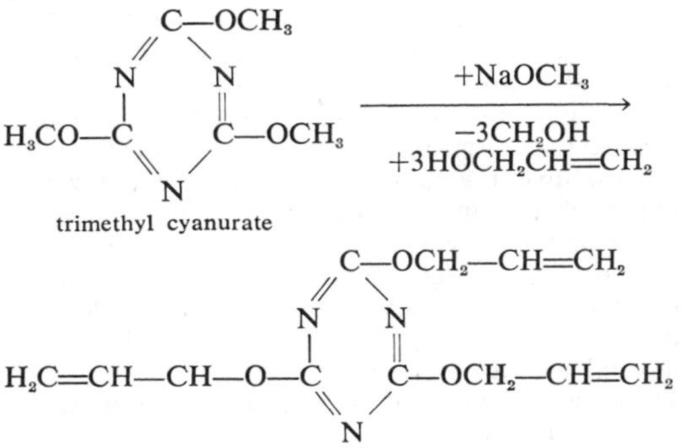

trimethyl cyanurate

$$\xrightarrow[\substack{-3CH_2OH \\ +3HOCH_2CH=CH_2}]{+NaOCH_3}$$

triallyl cyanurate

Triallyl cyanurate (TAC) and triallyl isocyanurate readily polymerize in the presence of peroxide initiators to produce heat resistant polymers. These monomers are mixed with unsaturated polyesters to produce heat resistant polyester alkyds such as Laminac 4232 and 4233 and Vibrin 135 and 136A.

The largest single use of allyl alcohol is for the production of diallyl esters which are used for the production of thermosetting polymers with excellent electrical properties, good resistance to high temperatures and fair resistance to corrosive environments. Diallyl maleate (DAM) is a very reactive trifunctional monomer which has limited use. Polymers obtained from diallyl chlorendate (DAC) which contains 45 per cent chlorine and allyl phosphates such as diallylphenyl phosphate have outstanding flame resistance.

The most widely used diallyl esters are diallyl phthalate (DAP) and diallyl isophthalate (DAIP). These are produced by the conventional esterification of phthalic anhydride and isophthalic acid respectively. It is customary to use a solution of a diallyl ester prepolymer in the liquid diallyl ester monomer. Inhibitors such as alkylphenols or phenol ethers may be added to assure long shelf life. Tert. butyl perbenzoate and/or benzoyl peroxide may be used as curing agents.

The diallyl ester prepolymer is a free flowing thermoplastic polymeric powder which is obtained by the partial polymerization of the monomer to the B stage. The prepolymer of DAP or DAIP contracts less than 1 per cent when cured to an infusible thermosetting resin. The contraction when the monomers are cured to crosslinked resins is about 12 volume per cent and is much less than other monomers such as styrene. The prepolymers are stable at room temperature but polymerize in less than 30 seconds at 370°C.

The prepolymer may be used to provide a durable surface for plywood, and for encapsulation and the sealing of porous metal castings and for low cost decorative laminates. Solutions of the prepolymer are used for the production of prepregs by the impregnation of fibrous glass sheet or paper. The diallyl ester monomers or solutions of the prepolymer in the monomer may be used as a replacement for styrene in reinforced polyester plastics. The prepolymers from diallyl phthalate and diallyl isophthalate are called Dapon 35 and Dapon M respectively. The latter provides polymers with superior heat resistance.

Molding compounds may be prepared by advancing a solution of prepolymer and monomer in the presence of initiator, pigment and filler. The filler may be chopped strands of fibrous glass, acrylic fibers or asbestos. These molding compounds may be compression or transfer molded to produce electrical or electronic parts or appliance fixtures. For example, the basic components of coffee vending machines are molded from Dapon molding compounds. A typical glass fiber filled molding of Dapon may be used continuously at temperatures up to 400°C. In addition to a high heat deflection temperature, the allyl molded product will have a flexural strength of 18,000 psi, a tensile strength of 7500 psi and an izod impact resistance of 6 ft. lbs/in of notch.

RAYMOND B. SEYMOUR

ALUMINUM

The chemical element, Al, atomic number 13, is the second member in the third group of the periodic system of elements. Its atomic mass is 26.9815. Aluminum is a low-density metal of very high electromagnetic reflectivity, silver-white when pure, but it usually has a faint blue tinge. It forms a thin transparent oxide coating in air which stabilizes it against further corrosion. This makes aluminum a highly useful material, as it is strong, light, ductile, and malleable.

Aluminum is an important commercial commodity. The world production in 1970 was over 10 million metric tons. Consumption and distribution among industries vary greatly throughout the world. In the United States aluminum was consumed in 1969 and 1970 at an average rate of about 4.3 million metric tons per year approximately as follows: building and construction, 25%; transportation equipment including automobiles, 20%; electrical products, 15%; consumer durables, 10%; containers and packaging, 12%; machinery and equipment, 7%; and other uses, 11%.

History. Aluminum derives its name from *alumen*, the Latin name for alum, known from very early times. Although aluminum bronzes were produced in China 500 years ago, probably by carbothermic reduction, the relatively pure metal was first made by Oersted in 1824. He did this by heating anhydrous aluminum chloride with potassium amalgam and distilling off the mercury. In 1854 Deville produced commercial quantities of aluminum by reacting sodium metal with sodium aluminum chloride. In 1886, Hall and Héroult independently invented the process in which aluminum oxide (alumina) is dissolved in molten cryolite, Na_3AlF_6, and decomposed by an electric current passing through the melt. Aluminum is deposited at the cathode. This is still the present commercial process.

Occurrence. Aluminum is the third most abundant (8.13% by weight) element in the 16 kilometer thick earth's crust, after oxygen and silicon. This makes it the most abundant crustal metallic element, occurring in almost all silicate rocks. Natural leaching of rocks has led to the geologic formation of large deposits of clays, aluminum silicates, and bauxite, a generic term for impure hydrated alumina. Bauxites contain variable quantities of iron oxide, silica and titanium dioxide, and traces of other insoluble oxides. The alumina in bauxite is mostly hydrated, occurring variably as the trihydrate gibbsite, the monohydrate boehmite, or the monohydrate diaspore. The usual range of alumina content in bauxites is 40–60%. Bauxite is the major source of aluminum.

Production. The alumina in bauxite is extracted and purified by the Bayer process in which the bauxite is first digested hydrothermally in a strong sodium hydroxide solution, the hydrated alumina being dissolved and held in solution as the aluminate ion, AlO_2^-. This reaction may be represented as $Al_2O_3 \cdot 3H_2O + 2NaOH = 2NaAlO_2 + 4H_2O$. Silica dissolved initially during the digestion is reprecipitated as a sodium aluminum silicate and remains with the other insoluble oxides which are separated after cooling. The clarified solution, supersaturated with alumina, is seeded with alumina trihydrate from previous cycles and agitated; this reverses the dissolution reaction and precipitates alumina trihydrate. The product, an artificial gibbsite, is recovered by filtration, washed, and finally calcined (dehydrated) above 1000°C to produce a very pure alumina.

The metal is won from the alumina by the Hall-Héroult process outlined before. Carbon anodes used in this process are consumed during the electrolysis so that the net reaction may be represented as $2Al_2O_3 + 3C = 4Al + 3CO_2$.

Properties and Chemistry. Aluminum has only one stable isotope of mass number 27. Several artificial radioisotopes are known. The neutral aluminum atom has a radius of 1.43 Å; its electronic configuration in the ground state is $1s^2 2s^2 2p^6$-$3s^2 3p^1$, term symbol $^2P_{1/2}$. The crystalline form of aluminum has a face-centered cubic lattice with a unit-cell dimension of 4.04958 ± 0.000025 Å (25°C). The distance between an atom and any of its 12 nearest neighbors is 2.86 Å. Aluminum's density is 2.6989 g/cc at 20°C, and it melts at 660.24°C. Its normal boiling point is 2467°C. The heats of fusion and vaporization are 2.55 and 70.7 kcal/mole, respectively.

The electrical conductivity of 99.999% aluminum at 20°C is 65.45% of that of copper, the value being 38.02×10^4 mho/cm. The thermal conductivity at 200°C is 0.82 cal/cm/sec/°C. Aluminum is weakly paramagnetic with a susceptibility of 0.60×10^{-6} c.g.s. units.

Pure aluminum has good working and forming properties and high ductility but low mechanical strength. The strength properties may be improved by alloying, strain-hardening (cold-working), solution heat treatment, and aging.

Aluminum is ordinarily trivalent (Al^{+3}) in its compounds, but the monovalent (Al^+) and divalent (Al^{+2}) forms are known in high temperature compounds, such as AlCl and AlS. Aluminum is of potentially high chemical reactivity as is seen in the very high energies of formation of its compounds with oxygen, halogens, carbon, and sulfur. The standard oxidation reduction potential is 1.67 volts for the couple $Al \rightarrow Al^{+3} + 3e^-$.

Finely divided aluminum finds application in high explosives and solid rocket propellant mixtures by virtue of its very rapid oxidation at high temperatures. Aluminum is only superficially attacked by pure water at ordinary temperature but is rapidly oxidized at 180°C. Dilute or cold concentrated sulfuric acid or concentrated nitric acid have little effect on pure aluminum. Hydrochloric acid solutions attack aluminum with hydrogen evolution. Strong alkalis attack aluminum violently, also with the formation of hydrogen.

Aluminum forms many compounds the most useful of which is possibly its oxide, alumina. Alumina and its hydrates are starting materials for such products as refractories, abrasives, instrument and watch bearings, artificial rubies and sapphires, cement, ceramics, pigments, desiccants, aluminum salts, antacids, antiperspirants, and catalysts.

Lithium aluminum hydride, aluminum alkyls and alkoxides, and aluminum chloride are very useful compounds in organic synthesis.

W. R. KING

References

Gerard, Gary, and Stroup, P. T., Editors, "Extractive Metallurgy of Aluminum," Vol. 1, New York, Interscience Publishers, 1963.

Ginsberg, Hans, and Wrigge, F. W., "Tonerde und Aluminium," Vol. 1, Berlin, Walter de Gruyter, 1964.

Hampel, C. A., Editor, "The Encyclopedia of Chemical Elements," New York, Van Nostrand Reinhold, 1968.

AMALGAMS

An amalgam is commonly defined as "an alloy of mercury with some other metal or metals," or as "an alloy in which mercury is an important component." Mercury forms amalgams not only with metals but also with tellurium, ammonium and other constituents not always considered metals.

Amalgams may be liquid or solid, depending on the temperature, the component or components associated with the mercury, the proportions of the components, and nature of their association. The natures of the associations include; solutions in mercury of individual atoms or associated atoms; suspensions in mercury of particles of colloidal size or larger; compounds in solution in mercury, or in mixtures, or nearly pure; solid solutions; and solid mixtures. Some examples of these states and conditions follow.

A dilute solution of cesium in mercury (m., −38.87°C) melts at −46.6°C. This and many other dilute solutions of metals show a lowering of freezing point corresponding to monatomic particles, and these particles obey the gas laws. Other amalgans have melting points which would indicate diatomic or polyatomic associated-atom solutes. Only one part of iron by weight is soluble in 10^{17} parts of mercury. But an amalgam containing 1% of iron can be made. It has a very large magnetic susceptibility which decreases on standing. When the mercury is evaporated at low temperatures a highly magnetic pyrophoric powder remains. The iron is supposed to form in single magnetic units which combine into much larger particles and partly neutralize each other. Mercury (m., −38.87°C) mixes with: lithium (m., 179°C); sodium (m., 97.0°C); potassium (m., 63.4°C); and cesium (m., 28.5°C), to form four continuous series of amalgams.

If the percentage of each alkali metal is plotted against the melting point of the alloy it forms, the curve for each metal is a succession of rounded maxima interspersed with sharp minima. The highest melting point on each curve preceded by a formula which represents the composition of that amalgam is, respectively: LiHg(m., 600.5°) NaHg$_2$(m., 360°C); KHg$_2$(m., 279°C); CsHg$_2$(m., 208.2°C). Such a series of maxima and minima suggest that a rather pure compound is represented by each maximum, and each minimum represents the melting point of a mixture, a eutectic. The maximum melting points, some of them hundreds of degrees higher than the melting point of either component, cannot be explained on any other basis, than formation of compounds.

Amalgams may be prepared as follows: (1) by simple contact between mercury and any of many metals at low temperatures; higher temperatures and/or the presence of a dilute acid either increase the rate of amalgamation or increase the number of metals which can be amalgamated by this method; (2) by contact between mercury and an aqueous solution of a salt of a noble metal; (3) by contact between an aqueous solution of a salt of mercury and an active metal; (4) by contact between an aqueous solution of a salt of the appropriate metal and an amalgam of a more active metal; (5) by electrolysis of an aqueous solution of a salt of the appropriate metal using a mercury cathode; and (6) by the electrolysis of an aqueous solution of a mercury salt using the appropriate metal as a cathode.

Amalgams may be used: (1) to deposit metals in thin layers as in silvering mirrors; (2) to plasticize metals or alloys so that irregular cavities may be fitted, as in the use of dental amalgams; (3) to control by dilution the reaction rates of, or facilitate the application of active metals such as sodium, aluminum and zinc when used in the preparation of titanium and other metals, or in the reduction of a great variety of organic compounds; (4) to separate such metals as iron and uranium in analytical chemistry by use of alloys of bismuth and zinc; and (5) as catalysts.

In the preparation of pure sodium hydroxide by the electrolysis of brine, the formation of sodium amalgam is a step in one of the processes. Standard cells such as the Weston cell and the Clark cell are based upon the use of two-phase amalgams of cadmium and zinc, respectively, to give constant electromotive forces.

Discharge of mercury into rivers and lakes from producers of caustic soda and chlorine by means of mercury cells appears to be partly responsible for high levels of organic mercury compounds (mainly methyl mercury) in some species of fresh water fish. In December of 1970 the Food and Drug Administration announced withdrawal of an estimated one million cans of tuna fish, and banned some species of fresh and frozen fish. However, mercury concentrations in preserved specimens of tuna and cod caught about 100 years ago are about the same as those in fish caught recently. This indicates that man's discharge of mercury into the oceans is not the sole cause of mercury found in ocean fish. Efforts to control pollution of waterways may require extensive and expensive changes in production methods or waste disposal methods for industries that rely on the use of amalgams.

F. E. BROWN

Dental Amalgams

Dental amalgam is used extensively as a restorative dental material. When mercury is mixed ("triturated") with small particles (lathe cut "filings" or spherical particles produced by aerosolization) of an alloy containing silver and tin with small amounts of copper (maximum 6%) and zinc (maximum 2%), a plastic mass is obtained. While plastic, the amalgam is placed into properly prepared cavities in the teeth after which it solidifies to form "silver fillings."

Ag_3Sn is the main component of the alloy triturated with mercury to form dental amalgam. The setting (hardening) mechanism is thought to be one of partial solution and precipitation.

Normally, a considerable amount of the original alloy remains in a matrix of silver-mercury (γ_1) phase and tin-mercury (γ_2) phase. The solution stage is accompanied by contraction, whereas the precipitation results in some expansion. The (γ_2) phase has been identified as the least desirable component of the hardened dental amalgam.

The addition of 10% gold to spherical dental amalgam alloy (substitution of 10% Au for Ag in Ag_3Sn) has been shown to reduce the (γ_2) phase, as has dispersion of copper in the alloy. Laboratory data indicate higher corrosion resistance (lower potential) and/or better mechanical properties, but

controlled clinical studies are needed before the significance of the laboratory findings can be established. Variables such as particle size of the alloy, trituration time, mercury content, packing ("condensation") pressure and other manipulative variables influence the final strength and dimensional behavior of the resulting amalgam restoration.

Polishing of the hardened restoration provides some protection against chemical attack by saliva and foodstuff. Still, amalgam restorations are subject to some tarnish and corrosion which, incidentally, tend to improve the marginal adaptation of the restoration thus minimizing microleakage around the periphery.

Gallium has been considered as a component of dental restorations to replace mercury but so far without success because of the highly exothermic reactions resulting from mixtures of gallium with most dentally suitable metals.

Preamalgamated copper amalgam pellets have found some limited use as dental restorative materials because of their alleged cariostatic effect. Copper amalgam is plasticized by slight heating after which it is triturated in a mortar to a suitable consistency for dental application. Copper amalgam restorations discolor (blacken) in service and their use is decreasing.

GUNNAR RYGE

Editor's Note: In view of the concern about mercury hazards, the recent comment of Nelson W. Rupp, D.D.S., Research Associate, American Dental Association, National Bureau of Standards, on dental amalgam fillings is pertinent. Citing studies on the subject, he says, "Currently it is felt that there is no danger to the patient, the environment, or dental office personnel when good mercury hygiene is practiced." (*Chem. Eng. News*, p. 34, Aug. 2, 1971.)

AMIDES

Amides are nitrogen-containing compounds which may be divided into metallic amides and organic amides. Metallic amides are chiefly those of groups I and II metals, including Zn, and they have the general formula $Me(NH_2)_x$. They are strong bases in a system in which ammonia is the solvent, and correspond to alkali metallic hydroxides in water systems.

Organic amides contain a —$CONH_2$ group and are closely related in structure to organic acids that contain a carboxyl (—COOH) group.

Sodium amide (sodamide) ($NaNH_2$, m.p. 210°C) is made by the reaction of metallic sodium with ammonia. If sodium amide is left in contact with air, sodium nitrite forms, and the resulting mixture is explosive. Sodium amide reacts with (1) water and forms sodium hydroxide and ammonia; (2) hydrogen forming sodium hydride and ammonia; (3) carbon monoxide forming sodium cyanide, sodium hydroxide, and ammonia; (4) carbon dioxide forming cyanamide (NH_2CN) and sodium hydroxide; (5) carbon disulfide forming sodium thiocyanate and hydrogen sulfide; (6) magnesium forming magnesium nitride; (7) aluminum forming aluminum amide [$Al(NH_2)_3$]; (8) alkynes in ether ($RC{\equiv}CH + NaNH_2 \rightarrow RC{\equiv}CNa + NH_3$); (9) alkyl halides forming an alkene ($C_2H_5I + NaNH_2 \rightarrow C_2H_4\uparrow + NaI + NH_3\uparrow$). When heated above 330°C, sodium amide decomposes into its elements.

Other alkali amides are known. *Lithium amide* is less reactive than sodium amide. *Potassium amide* is quite similar to sodium amide, but it is much more soluble in liquid ammonia.

Organic amides may be considered to be derivatives of ammonia in which a hydrogen atom has been replaced by an acyl group (RCO—). The simplest amide is *formamide* ($HCONH_2$, b.p. 210°C). Amides, like amines, are primary, secondary, or tertiary depending upon the number of hydrogen atoms of ammonia that has been replaced.

Amides may be prepared by (1) rearrangement caused by heating the corresponding ammonium salt ($CH_3COONH_4 \rightleftharpoons CH_3CONH_2 + H_2O$); (2) the reaction between an ester and concentrated ammonia water ($CH_3COOC_2H_5 + NH_3 \rightleftharpoons CH_3CONH_2 + C_2H_5OH$); (3) ammonolysis of an acid halide or an acid anhydride by ammonia or by a primary or a secondary amine

$$(CH_3COCl + 2NH_3 \rightarrow CH_3CONH_2 + NH_4Cl);$$

(4) partial hydrolysis of a nitrile ($C_2H_5CN + H_2O \rightarrow C_2H_5CONH_2$).

Amides undergo many reactions. Among them are (1) the formation of salts with strong acids (amides, however, are relatively weak as proton-acceptors); (2) hydrolysis in either acid or alkaline solution. An example of each follows: (1) $CH_3CONH_2 + H_3O^+ \rightarrow CH_3COOH + NH_4^+$; (2) $CH_3CONH_2 + OH^- \rightarrow CH_3COO^- + NH_3 \uparrow$; (3) reduction essentially by hydrogen forming a primary amine; (4) reaction with nitrous acid forming the corresponding acid, nitrogen, and water; (5) the reaction with an alcohol in the presence of boron trifluoride or hydrochloric acid yields an ester as the chief product; (6) when treated with a strong dehydrating agent such as P_2O_5 or $SOCl_2$, water is lost and a nitrile is formed that contains the same number of carbon atoms as the original amide; (7) conversion to a primary amine that contains one fewer carbon atoms than the original amide is accomplished by the Hofmann reaction in which alkaline hypohalite is used.

Acetamide (ethanamide, CH_3CONH_2, m.p. 81°C) is probably the best known amide. Its colorless crystals are deliquescent. The compound dissolves well in water, glycerol, and ethanol. It lacks odor, and it boils at 222°C. Acetamide is used as a solvent, and in organic syntheses. It is used to make lacquers, explosives, and surface-active agents.

Urea (carbamide, $CO(NH_2)_2$, m.p. 132.7°C) is an example of a diamide. Many derivatives of this compound are known. Urea is soluble in water, but it is not generally soluble in most organic solvents. Urea is formed as the end product from the metabolism of protein food. It is excreted in the urine of mammals. Synthetic urea is now manufactured in large quantities by a process in which ammonia and carbon dioxide are carefully heated together under controlled pressure conditions. Synthetic urea is used as an animal-food additive, as a fertilizer, and for the manufacture of resins and plastic materials.

ELBERT C. WEAVER

Cross-references: *Amines, Acids, Ammonia.*

AMINES

Amines are nitrogen-containing organic compounds derived from ammonia. Amines are used extensively for a number of purposes: they are solvents, surface-active agents, rust inhibitors in anti-freeze mixtures, and are used to make special soaps employed in the cosmetic and dry-cleaning industries. Amines are intermediates in the synthesis of many useful compounds, some of which become dyes or medicines. Amines also yield salts such as methyl ammonium chloride (CH_3NH_3Cl). On extreme alkylation, tertiary amines form quaternary ammonium salts, for example, tetramethyl ammonium chloride [$(CH_3)_4NCl$].

Primary amines (RNH_2) such as methyl amine (CH_3NH_2, b.p. −7°C) have one hydrogen atom of ammonia replaced by an organic radical; secondary amines ($RR'NH$) such as dimethyl amine [$(CH_3)_2$ NH, b.p. 7°C] have two hydrogen atoms of ammonia replaced; and tertiary amines [$(CH_3)_3N$, b.p. 4°C] have no hydrogen attached to nitrogen. These are all alkyl amines. The same three sorts of amines are known with aryl substituents. The simplest primary aryl amine is *aniline* ($C_6H_5NH_2$) or phenyl amine. Mixed amines are also known.

Like ammonia, amines tend to ionize in water and form alkaline solutions. Some of them are even more active than ammonia in this respect. If K is the dissociation constant,

$$NH_3 + H_2O \rightleftharpoons NH_4^+ + OH^-$$
$$K = 1.7 \times 10^{-5}$$

$$CH_3NH_2 + H_2O \rightleftharpoons CH_3NH_3^+ + OH^-$$
$$5 \times 10^{-4}$$

$$(CH_3)_2NH + H_2O \rightleftharpoons (CH_3)_2NH_2^+ + OH^-$$
$$5.4 \times 10^{-4}$$

$$(CH_3)_3N + H_2O \rightleftharpoons (CH_3)_3NH^+ + OH^-$$
$$5.9 \times 10^{-5}$$

Primary amines may be prepared by (1) reduction of nitro compounds by hydrogen; (2) alkaline hydrolysis of isocyanates or isocyanides; (3) the Hofmann reaction of sodium hypobromite on an amide; (4) action of strong alkali on an amine salt; (5) reaction of an alkyl halide with ammonia; (6) reduction of a nitrile by sodium and alcohol. A special case is the reduction of adipic acid by ammonia, followed by dehydration and then hydrogenation (1,6-hexanediamine [$CH_2(CH_2)_4CH_2NH_2$] forms a compound used in the manufacture of nylon); (7) sodium phthalimide and an alkyl halide followed by alkaline hydrolysis (Gabriel synthesis); (8) reduction of oximes by sodium and alcohol; (9) alcohol vapor and ammonia heated with thorium oxide as a catalyst at 360°C; (10) from amino acids, as accomplished in the decomposition of fish.

Secondary amines are prepared by (1) alkyl halides reacting with primary amines; (2) the reaction of alkyl halides with sodium cyanamide (Na_2NCN); (3) decomposition of amino acids.

Tertiary amines may be prepared by treating ammonia with an excess of alkyl halide.

Primary amines react (1) with water and form substituted ammonium hydroxide; (2) with acids and form addition compounds; (3) with alkyl halides and form the halides of the corresponding secondary amine; (4) with acid chlorides and form substituted amides (for example, acetyl chloride and ethylamine form N ethylacetamide and ethylamine hydrochloride); (5) with nitrous acid and form a primary alcohol; (6) with chloroform and form the corresponding isocyanide; (7) with Grignard reagents.

Secondary and tertiary amines also react with many of these substances, but the products differ.

Mono-, di-, and trimethyl amines are low-cost sources of basic organic nitrogen.

Dimethylamine [$(CH_3)_2NH$, sp. gr. 0.68 at 0°C] may be made by the reaction of methanol vapor and ammonia with a catalyst at a high temperature. It is used to unhair hides, to absorb acid gases, as a flotation agent, as a gasoline stabilizer, and as an intermediate in the preparation of local anesthetics and antihistamine, as a rubber-curing accelerator, and in electroplating. *Trimethylamine* [$(CH_3)_3N$, sp. gr. 0.662 at −5°C] is used to make choline chloride [$(CH_2OHCH_2N(CH_3)_3)Cl$], a poultry feed additive.

If a 12- to 24-carbon atom chain is attached to the carbon atom adjacent to the nitrogen atom in a primary amine, the result is a stable liquid that resists oxidation, and which can be used as an oil additive.

A more branched-chain amine such as tertiary butyl amine [CH_3—$C(CH_2)_2$—NH_2] reacts with aldehydes, cyanogen chloride, and alkyl halides. Uses for the products of these reactions include intermediates for rubber-processing chemicals, insecticides, oil additives, photographic chemicals, dyestuffs, pharmaceuticals, surface-active agents, and corrosion inhibitors.

Hexamethylenetetramine [$(CH_2)_6N_4$ m.p. 263°C] is made by the combination of formaldehyde and ammonia. It is a sweet-tasting solid, irritating to the skin, and soluble in water. Under the name "Urotropin" or Methenamine, it is used as an urinary antiseptic. It is also used as an accelerator in the curing of rubber, and as a raw material in the manufacture of some plastics. It has a slightly irritating effect on the skin if used regularly.

Other interesting amines include choline, discovered in hog bile and also found in beer; acetylcholine, a medicinal; ethylenediamine, a rubber latex stabilizer and a corrosion inhibitor; and numerous compounds from the decomposition of proteins such as cadaverine, peutrescin, and neurine.

ELBERT C. WEAVER

Cross-references: *Nitrogen and Organic Compounds.*

AMINO ACIDS

Amino acids, by definition, are organic compounds having both amino and carboxylic functional groups. They occur in nature in both free and combined states. Most of the more than 200 natural amino acids which have been characterized are found only in the free state where, with a few exceptions, their function is not clearly known. In the combined state, amino acids serve as the monomers which form the carbon skeleton of the polymers, proteins. Only about 35 naturally occurring amino acids have been clearly established as pro-

tein constituents, and only the most abundant of these will be discussed in detail.

General Considerations. Most of the amino acids from natural sources may be represented by the general structure, RCHCOOH, in which R represents an aliphatic, aromatic or heterocyclic grouping. The more common exceptions to this generalization include amino acids having a second amino or carboxyl group, and those having in addition an amide or guanidino group. All the more common natural amino acids have a free amino group α^- to the carboxyl; exceptions include proline and hydroxyproline, in which the nitrogen, while α^- to a carboxyl, is part of a pyrrolidine ring. In general, the natural amino acids possess one or more asymmetric centers and therefore may exist in at least two optically active forms. Most of the natural amino acids are of the L-configuration; however, a number of D-amino acids occur, most commonly in bacterial cell wall hydrolyzates or in many of the fungal antibiotics.

Because they contain one or more basic amino and acidic carboxyl groups, amino acids are amphoteric compounds and exist in aqueous solution as dipolar ions (zwitterions); *e.g.*, glycine may be represented as, $^+H_3NCH_2COO^-$. The covalent bond formed when an amino acid reacts with another amino acid, is called a peptide bond. The product of the reaction between two amino acids would have one such peptide bond and is termed a dipeptide. Similarly, a tripeptide would contain three amino acid residues and two peptide bonds. Polypeptides contain several, but an unspecified number of amino acid residues. Proteins, by definition, are polypeptide substances having a molecular weight greater than about 5000. The critical importance of polypeptides and proteins to life cannot be overemphasized. Proteins are a basic structural component of all living tissues; as enzymes they catalyze the chemical reactions necessary for life; and the polypeptide, *e.g.*, insulin, secretin, gastrin, and the protein, *e.g.*, adrenocorticotropic, thyroid stimulating, gonadotropic, hormones regulate the functioning of many vital organs.

Determination of Amino Acids. A variety of methods are used for the quantitative determination of amino acids. Before they may be measured, however, the amino acids of proteins must be freed from the peptide bond. This may be accomplished by hydrolysis with acid, base or proteolytic enzymes. Each method has advantages and disadvantages, and the procedure to be used must be carefully chosen.

General methods for amino acid determinations include the ninhydrin and nitrous acid reactions. A number of procedures have been developed which are specific for certain amino acids. These include (1) specific enzymatic decarboxylation, (2) the oxidation of hydroxy amino acids by periodate, applicable to serine and threonine, (3) the Sakaguchi reaction for arginine, (4) the Sullivan reaction for cysteine and (5) isotope dilution procedures.

The most generally applicable procedure for amino acid analysis is based upon a preliminary separation of the amino acids, followed by their quantitative determination with ninhydrin. The separations are accomplished chromatographically, and paper, thin layer, column and gas phase techniques have all been used. The column method has the advantage of complete automation while thin layer and gas chromatography are more rapid.

Amino Acids in Human Nutrition. The need for protein-containing foods has long been recognized, as has the fact that some proteins support growth better than do others. Since proteins are hydrolyzed to their constituent amino acids during digestion, and before absorption, it is quite clear that the nutritional requirement is not for proteins *per se*, but rather for amino acids. The amount of a given amino acid which must be ingested each day has been determined in humans by use of the nitrogen balance technique. This technique is based on the experimental observation that animals and man quickly achieve nitrogen balance—between the amount of nitrogen consumed and that excreted—under constant dietary conditions. Under these conditions, the addition or removal of a single amino acid from a purified diet can alter the nitrogen balance in a profound manner. From such studies it has been determined that some amino acids can be omitted from the diet without affecting nitrogen balance, while the omission of others has a marked effect on the utilization of all other amino acids. Out of this work has evolved the concept of "essential" and "non-essential" amino acids. The "essential" amino acids are those which must be present in the diet, and which therefore cannot be synthesized by the body, at least in sufficient quantity. The "non-essential" amino acids need not be present in the diet. The requirement for each of the "essential" amino acids by adult humans was studied by Rose and his coworkers. On the basis of their results daily minimal and safe dietary levels were estimated and are given in Table 1. In addition it appears that L-histidine, and possibly L-arginine, are required for normal growth by the human infant. (See **Nutrition**).

Alanine (α-Aminopropionic acid). [$CH_3CH(NH_2)$ COOH]. Mol. wt. 89.10. $[M]_D$ of L-antipode, +1.6° in H_2O, +13.0° in 5N HCl, +29.4° in HAc, and +2.7° in N NaOH;c = 2,T = 25°. Two amino

TABLE 1. COMMON AMINO ACIDS AND THOSE ESSENTIAL IN HUMAN NUTRITION

Essential Amino Acids			Non-Essential Amino Acids
Name	Minimal Daily Level	Safe Daily Level	
	gm	gm	
L-Isoleucine	0.70	1.4	Glycine
L-Leucine	1.10	2.2	Alanine
L-Lysine	0.80	1.6	Serine
L-Methionine	1.10	2.2	Cystine
L-Phenylalanine	1.10	2.2	Cysteine
L-Threonine	0.50	1.0	Tyrosine
L-Tryptophan	0.25	0.5	Glutamic Acid
L-Valine	0.80	1.6	Aspartic Acid
			Proline
			Hydroxyproline
			Histidine
			Arginine

acids were prepared by chemical synthesis prior to their being identified as constituents of protein hydrolyzates. The first of these was alanine, the second proline. While searching for a preparative method for lactic acid, Strecker in 1850 treated a mixture of acetaldehyde and ammonia with hydrocyanic acid, followed by hydrolysis with excess HCl. The amino acid he isolated was named alanine, denoting its origin from aldehyde. His goal of producing lactic acid was achieved when he noted that the hydroxy acid was formed by the reaction of nitrous acid with alanine. This reaction also served to confirm the structure of the amino acid. Thirty eight years later Weyl isolated alanine in crystalline form from silk fibroin, which contains about 25 per cent alanine, and established its identity as a constituent of protein hydrolyzates.

Alanine is a non-essential amino acid and can therefore be synthesized in adequate quantities by the body. This is accomplished in several ways: (1) by β-decarboxylation of aspartic acid. (2) from kynurenine by the action of the enzyme kynureninase, and (3) by the enzymatic transamination of pyruvic acid, in which an α-amino group and an α-keto group are interchanged, as follows:

$$CH_3\overset{O}{\overset{\|}{C}}COOH + R\overset{|}{\underset{NH_2}{C}}HCOOH \leftrightarrow$$

$$CH_3\overset{|}{\underset{NH_2}{C}}HCOOH + R\overset{O}{\overset{\|}{C}}COOH$$

Arginine (1-Amino-4-guanidinovaleric acid), $H_2N\overset{\|}{\underset{NH}{C}}NHCH_2CH_2CH_2\overset{|}{\underset{NH_2}{C}}HCOOH$. Mol. wt. 174.21

$[M]_D$ of L-antipode, +21.8° in H_2O, +48.1° in 5N HCl and +51.3° in HAc, c= 2, T = 25°. Schulze and Steiger isolated arginine from aqueous extracts of lupine seedlings in 1886. Its presence in casein hydrolyzates was noted by Drechsel in 1890, although he did not recognize its identity with the material isolated by Schulze. Hedin, in 1895, isolated the silver nitrate salt of arginine from horn hydrolyzates, thus confirming its presence in protein hydrolyzates, and Sörenson synthesized it in 1910 from benzoylornithine. Arginine may be isolated from acid hydrolyzates of proteins such as hair or gelatin; however, the richest natural source is the basic protein salmine, which contains about 85 per cent of this amino acid. Salmine belongs to a class of basic nuclear proteins termed histones. The histones are basic because of their high content of the basic amino acids arginine and lysine. It is believed that the histones regulate messenger RNA synthesis by reversible combination with template DNA.

Arginine is not required in human nutrition. The growth of rats is stimulated by dietary arginine, while in chicks this amino acid is considered essential. The major end-product of protein nitrogen metabolism in man is urea, which is formed via the so-called Krebs-Henseleit Urea Cycle. Arginine forms a part of this cyclic series of enzyme catalyzed reactions, which function to convert waste

CO_2 and NH_3 to urea. The essentials of this cycle may be outlined as follows:

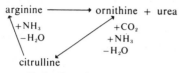

Krebs-Henseleit Urea Cycle

Aspartic acid (Aminosuccinic acid), [HOOCCH$_2$CH(NH$_2$)COOH]. Mol. wt. 133.11 [M]D of L-antipode +6.7° in H_2O, +33.8° in $5N$ HCl and −2.3° in N NaOH, c = 2, T = 25°.

Asparagine (Aminosuccinamic acid), [H$_2$NCOCH$_2$CH(NH$_2$)COOH]. Mol. wt. 132.12, [M]D of L-antipode, −7.4° in H_2O, +37.8° in N HCl, and −12.4° in N NaOH; c = 2, T = 25°.

Because aspartic acid and asparagine are interconvertible in the body they will be considered together. Asparagine, the first amino acid to be isolated from natural sources, was crystallized from the juice of the asparagus plant by Vauquelin and Robiquet in 1806. Although the formation of ammonia during the hydrolysis of proteins suggested the presence of amide groups, it was not until 1932 that Damodaran reported the isolation of asparagine from enzymatic hydrolyzates of edestin. Asparagine is widely distributed in both plant and animal materials and occurs both free, as well as combined in proteins. Aspartic acid was first discovered as a product of asparagine hydrolysis; Ritthausen later isolated and established its presence in protein hydrolyzates.

L-Aspartic acid is not required in human or animal nutrition. It is of considerable interest in metabolism, however, and serves an important role in transamination. The α-keto acid corresponding to aspartic acid, oxaloacetic acid, forms a part of the tricarboxylic acid energy cycle. Thus, aspartic acid is reversibly linked to carbohydrate and energy metabolism and, through transamination, to amino acid synthesis. Aspartic acid is also involved in nucleic acid metabolism via orotic acid, a pyrimidine precursor.

Cysteine (2-Amino-3-mercaptopropanoic acid), [HSCH$_2$CH(NH$_2$)COOH]. Mol. wt. 121.16, [M]D of L-antipode, −20.0° in H_2O, +7.9° in $5N$ HCl, and +15.7° in HAc; c= 2, T = 25°.

Cystine [3,3′-Dithiobis(2-aminopropanoic acid)], [S-CH$_2$CH(NH$_2$)COOH]$_2$. Mol. wt. 240.31([M]D of L-antipode, −509.2° in N HCl, and −168.1° in N NaOH; c = 1, T = 25°.

Again, cysteine and cystine are readily interconvertible in the body and will be considered together. Wollaston isolated cystine from urinary calculi in 1810. Its presence in proteins was not established until nearly 90 years later when Mörner, closely followed by Embden, isolated it from a horn hydrolyzate. Erlenmeyer synthesized both cystine and cysteine in 1903. Cystine is present in many proteins and is a major constituent of keratin, the principal protein of hair, skin and wool. There is good evidence for the presence of cysteine in proteins; however, acid hydrolyzates of proteins usually contain only cystine, the oxidation product of cysteine. The disulfide bonds of cystine impart a configurational rigidity to protein structure; the free sulfhydryl of cysteine plays an important role in the action of many enzymes.

L-Cystine, or L-cysteine, is a non-essential amino acid, but can serve to spare methionine. In addition to its participation in protein structure and enzyme action, noted above, cystine is a precursor of taurine and of sulfate, which are utilized by the body for the formation of soluble derivatives of toxic substances (detoxication). The reversible oxidation of cysteine to cystine also is important in the regulation of the oxidation-reduction potential of tissue fluids.

Glutamic Acid (2-Aminopentanedioic acid), [HOOCCH$_2$CH$_2$CH(NH)$_2$COOH]. Mol. wt. 147.14, [M]D of L-antipode, +17.7° in H_2O and +46.8° in $5N$ HCl; c = 2, T = 25°.

Glutamine (2-Amino-4-carbamoylbutanoic acid), [H$_2$NCOCH$_2$CH$_2$CH(NH$_2$)COOH]. Mol. wt. 146.15, [M]D of L-antipode, +9.2° in H_2O and +46.5° in N HCl; c = 2, T = 25°. These two amino acids bear the same relationships to each other as do aspartic acid and asparagine. Glutamic acid was isolated by Ritthausen in 1866 from acid hydrolyzates of wheat gluten; hence, the name glutamic acid. Wolff synthesized glutamic acid in 1890. When heated in aqueous solution glutamic acid is converted to the much more soluble pyrrolidone carboxylic acid, which has the property of reducing blood glucose levels in adrenalectomized or alloxan diabetic rats. Interestingly, pyroglutamic acid has recently been found to form the C-terminus amino acid of the polypeptide hormones gastrin and cholecystokinin, and of the hypothalamic tripeptide hormone, thyrotropin-releasing hormone.

Although the existence of glutamine in natural products was indicated from numerous studies, it was not until 1883 that Schulz and Bosshard isolated pure glutamine from beet root extracts. Like asparagine, the presence of glutamine in proteins was suspected, but not proved until 1932 when it was isolated from enzymic digests of gliadin (gluten).

L-Glutamic acid is not essential in human nutrition, but is required by the chicken. Like aspartic acid, glutamate is important in nitrogen metabolism via transamination and enters the tricarboxylic acid energy cycle via α-ketoglutarate. In this manner it is linked to carbohydrate metabolism. Glutamate also serves in the formation of numerous biologically important substances, which may be summarized as follows:

Carbohydrate ↔ α-Ketoglutarate → Energy

Glutamine ↔ Glutamate → Ornithine → Urea Cycle
Purine   Proline → Hydroxyproline
                     Glutathione

Of commercial importance as a condiment is the monosodium salt of glutamate, which possesses the property of enhancing the flavors of meats.

Glycine (Aminoacetic acid). (H$_2$NCH$_2$COOH). Mol. wt. 75.07. Glycine is the only amino acid which does not possess an asymmetric center. It was isolated from an acid hydrolyzate of gelatin in

1820 by Braconnot, who also noted its sweet taste. First named glycocoll, Berzelius later synthesized glycine and suggested the name. Glycine is common in most proteins and is particularly abundant in gelatin (28%) and silk fibroin (43%).

Glycine is not essential in human nutrition but is required by the chicken. Its importance in metabolism is principally related to its conversion to many essential metabolites, and to its use in the detoxication of toxic aromatic compounds.

$$\begin{array}{c}
\text{Carbohydrate} \\
\updownarrow \\
\text{Purines} \leftarrow \quad \nearrow \text{Pyruvate} \rightarrow \text{Energy} \\
\text{Glycine} \leftrightarrow \text{Serine} \\
\text{Porphyrins} \nearrow \quad \searrow \text{Ethanolamine} \rightarrow \text{Choline} \\
\text{Phospholipids} \\
\text{Creatinine} \leftarrow \text{Creatine} \\
\text{Detoxication}
\end{array}$$

Histidine [α-amino-4-(or 5)-imidazolepropionic acid]. Mol. wt. 155.16, [M]$_D$ of L-form, $-59.8°$ in

$$H-C=\!=\!=CCH_2CH(NH_2)COOH$$
$$\underset{\underset{CH}{\diagdown\!\diagup}}{N\qquad NH}$$

H_2O, $+18.3°$ in 5N HCl, and $+11.6°$ in HAc; c = 2, T = 25°. Kossel isolated and characterized an amino acid from an acid hydrolyzate of sturgeon sperm protamine in 1896, and named it histidine. Hedin simultaneously reported the isolation of the same amino acid from a number of protein hydrolyzates. Pyman synthesized histidine in 1911.

L-Histidine is not an essential amino acid in adult man; however, several lower species of animals including the rat, dog and chicken do require it. Histidine is a constituent of most proteins. It plays a role in the binding of substrate molecules by enzymes, and in the maintenance of protein configuration by hydrogen bonding. Histidine participates in transamination as well as other metabolic processes, including the formyl or one-carbon transformations involving various folic acid derivatives. Histidine may be decarboxylated by various tissues, as well as by the bacterial flora of the intestine, to give the corresponding amino compound, histamine. This amine has profound pharmacological activities, including bronchiolar constriction, capillary and arteriolar dilation and stimulation of gastric secretion. Histamine injection causes a marked decrease in blood pressure and may be responsible, at least in part, for the physiological alterations seen in traumatic and anaphylactic shock and the local inflammatory response. These effects of histamine form the basis for the clinical use of the antihistamine drugs.

Hyroxyproline (4-Hydroxy-2-pyrrolidine carboxylic acid). Mol. wt. 134.14, [M]$_D$ of L-antipode,

$$\begin{array}{c}
\text{HO—CH——CH}_2 \\
\text{H}_2\text{C} \qquad \text{CH—COOH} \\
\diagdown\text{NH}\diagup
\end{array}$$

$-99.6°$ in H_2O, and $-66.2°$ in 5N HCl; c = 2, T = 25°. This amino acid was isolated by Fischer from acid hydrolyzates of gelatin in 1902. The presence

of two asymmetric centers complicated the problem of identifying the configuration of the natural isomer, which was finally accomplished by Leuchs in 1913. Hydroxyproline has been found in only a limited number of proteins, mainly collagen and elastin which form the principal proteins of the connective tissues of animals.

L-Hydroxyproline is not essential in human or animal nutrition. It is biosynthesized from proline but, rather unexpectedly, not until after the proline has been incorporated into a collagen precursor called protocollagen. Proline in turn is derived from glutamate or arginine via ornithine.

Isoleucine (2-Amino-3methylpentanoic acid).

$$\begin{array}{c}
\text{CH}_3 \\
\diagdown\text{CHCH(NH}_2\text{)COOH} \\
\text{CH}_3\text{CH}_2\diagup
\end{array}$$

Mol. wt. 131.18, [M]$_D$ of L-antipode, $+16.3°$ in H_2O, $+51.8°$ in 5N HCl, and $+64.2°$ in HAc; c = 1, T = 25°. Ehrlich, who first isolated isoleucine from beet sugar molasses in 1904, later also found this amino acid in fibrin, gluten and other proteins. Isoleucine was synthesized in 1906 by Bouveault and Locquin.

A dietary source of L-isoleucine is essential in the nutrition of both man and other animal species. In its metabolism, isoleucine is transaminated to give α-keto-β-methylvaleric acid, which is then further degraded to eventually give two fragments, acetylated coenzyme A and pyruvate, which may be utilized for energy or for the synthesis of fats, carbohydrates or amino acids.

Leucine (2-Amino-4-methylpentanoic acid), [(CH$_3$)$_2$CHCH$_2$CH(NH$_2$)COOH]. Mol. wt. 131.18, [M]$_D$ of L-antipode, $-14.4°$ in H_2O, $+21.0°$ in 5N HCl, and $+29.5°$ in HAc: c = 2, T = 25°. Proust isolated leucine in crude form from cheese byproducts in 1819; in 1820 Braconnot obtained the amino acid in crystalline form from acid hydrolyzates of wool and muscle proteins and named it leucine. Its synthesis was accomplished from isovaleraldehyde by Schulze and Likiernik by means of the Strecker reaction.

L-Leucine is an essential amino acid in human and animal nutrition. Its metabolism involves transamination to α-ketoisicaproate, which is cleaved to acetylated coenzyme A derivatives. These are utilized for energy production and for various synthesis processes.

Lysine (2,6-Diaminohexanoic acid) [H$_2$NCH$_2$(CH$_2$)$_3$CH(NH$_2$)COOH]. Mol. wt. 146.19, [M]$_D$ of L-antipode, $+19.7°$ in H_2O and $+37.9°$ in 5N HCl; c = 2, T = 25°. Lysine was isolated in 1889 from casein hydrolyzate by Drechsel. Its structure was established in 1902 by Fischer and Weigert. The 5-hydroxy derivative of lysine, hydroxylysine, is a minor component of some proteins. Lysine is the only known amino acid constituent of mammalian proteins which contains a terminal amino group in addition to the α-amino. As mentioned previously lysine, and arginine, are responsible for the basic character of the nuclear histones.

L-Lysine is essential to human and animal nutrition. It, and threonine, are unique among the amino acids in that they apparently do not participate to a significant extent in reversible transamination; rather, the loss of the α-amino of lysine to give α-keto-E-aminocaproic acid is irreversible and the metabolism of lysine then proceeds, via a number of intermediates, to α-ketoglutaric acid, which is also formed in the reversible transamination of glutamic acid. The formation of hydroxylysine from lysine occurs by a different route which is also irreversible; hydroxylysine cannot substitute for lysine in the diet. In general, cereal proteins, e.g., wheat, rice, corn, etc., are deficient in their lysine content and therefore have a poor nutritive value. The protein value of the inexpensive, and widely available cereal grains can be improved by the addition of small quantities of L-lysine, or of animal proteins which do contain adequate amounts of lysine. This approach is being taken to improve the protein nutrition of people living in many areas of the world.

Methionine (α-amino-γ-methylthiolbutyric acid), $[CH_3SCH_2CH_2CH(NH_2)COOH]$. Mol. wt. 149.22, $[M]_D$ of L-antipode, −14.9° in H_2O +34.6° in $5N$ HCl, and +29.8° in HAc; c = 1 − 2, T = 25°. Methionine was discovered by Mueller in 1921. Its structure was established in 1928 by Barger and Coyne, using the Strecker reaction. The sulfur of methionine is alkali-stable, as contrasted to the labile sulfur of the other sulfur-containing amino acids found in proteins, cystine and cysteine.

L-Methionine is an essential amino acid in both human and animal nutrition and serves many important functions. Along with cystine, it is a source of sulfate, used in detoxication and in the formation of the sulfated polysaccharides. The body cannot synthesize methyl groups in adequate amounts; the S-methyl group of methionine, via transmethylation, thus serves as the donor of the methyl group found in a host of body constituents, e.g., the chlorine present in phospholipids, creatine, N-methylnicotinamide and adrenaline. Methionine is also converted to cystine, via homocysteine.

Phenylalanine (2-Amino-3-phenylpropanoic acid).

—$CH_2CH(NH_2)COOH$

Mol. wt. 165.20, $[M]_D$ of L-antipode, −57.0° in H_2O, −7.4° in $5N$ HCl, and −12.4° in HAc; c = 1–2, T = 25°. Schulze and Barbieri described, in a preliminary report in 1879, the isolation of a new amino acid from ethanolic extracts of lupine seedlings. They later obtained this amino acid from plant protein hydrolyzates, and established its identity with a product synthesized by Erlenmeyer and Lipp, which the latter investigators had named phenylalanine.

L-Phenylalanine is an essential amino acid in both human and animal nutrition. In normal individuals it is mainly metabolized to tyrosine, which is then transformed to a variety of substances. Some persons suffer from a hereditary defect in the enzymatic hydroxylation of phenylalanine to tyrosine. In this condition, termed phenylpyruvic oligophrenia (phenylketonuria), abnormally large amounts of nonhydroxylated phenylalanine metabolites are present in the blood and urine. The brain damage present in persons with phenylpyruvic oli-

gophrenia is probably due to high blood levels of phenylalanine or its metabolites.

Proline (2-Pyrrolidinecarboxylic acid). Mol. wt.

115.14, $[M]_D$ of L-antipode, −99.2° in H_2O, −69.5° in $5N$ HCl, and −92.1° in HAc; c = 1–2, T = 25°. Proline was synthesized in 1900 by Willstatter from α, δ-dibromopropylmalonic ester. A year later Fischer isolated it from acid hydroylzed casein and later, in much higher yield, from gelatin. Proline is readily racemized in hot aqueous solutions, particularly in the presence of alkali. Like hydroxyproline, proline is present in highest concentration in the proteins of connective tissue, such as collagen and gelatin. Unlike most amino acids, it is soluble in alcohol.

L-Proline is not a nutritionally essential amino acid. It is formed from ornithine, and after incorporation into protocollagen it serves as the precursor of hydroxyproline.

Serine (2-Amino-3-hydroxypropanoic acid)

$$[HOCH_2CH(NH_2)COOH].$$

Mol. wt. 105.10, $[M]_D$ of L-antipode, −7.9° in H_2O, and +15.9° in N HCl; c = 2, T = 25°. Cramer in 1865 isolated a propionic acid derivative containing both hydroxy and amino groups from acid hydrolyzates of the silk protein serecine; he named this new amino acid serine. The structure of serine was established synthetically in 1902 by Fischer and Leuchs. Serine also occurs in proteins as the phosphate ester, phosphoserine, and as such apparently forms a part of the active site of some enzymes.

L-Serine is not an essential amino acid in nutrition. It is formed reversibly from glycine and its metabolism was briefly outlined in the discussion of that amino acid.

Threonine (2-Amino-3-hydroxybutanoic acid) $[CH_3CH(OH)CH(NH_2)COOH]$. Mol. wt. 119.12, $[M]_D$ of L-antipode, −33.9° in H_2O, −17.9° in $5N$ HCl, and −35.7° in HAc; c = 1–2, T = 25°. During studies on the nutritional requirements for individual amino acids Rose and his coworkers noted that a mixture of all of the known amino acids would not support the growth of rats, and that this deficiency could be relieved by adding a specific fraction from protein hydrolyzates. Following this lead, Rose isolated and identified a new amino acid which he named Threonine. This is the only amino acid sought on the basis of nutritional studies, and the last one to be discovered having importance in animal nutrition. Carter synthesized threonine in 1935. Like serine, threonine also exists in proteins as the σ-phosphate ester.

L-Threonine is essential in the nutrition of animals and man. Like L-lysine, it apparently does not participate to a significant extent in reversible transamination. At least two pathways for the metabolism of threonine in animals have been identified, one resulting in the formation of α-aminobutyrate and the other in the formation of glycine and acetate, the latter compound presumably via acetaldehyde.

Tryptophan   (2-Amino-3-indolepropanoic acid).

$$-CH_2CH(NH_2)COOH$$

Mol. wt. 204.23, $[M]_D$ of L-antipode, $-68.8°$ in H_2O, $+5.7°$ in N HCl, and $-69.4°$ in HAc; c = 1–2, T = 25°. Hopkins and Cole in 1902 isolated an amino acid containing an indole nucleus from an enzymatic digest of casein and named it tryptophan, following an earlier suggestion of Neumeister. Its structure was established by synthesis in 1907 by Ellinger and Flammand.

The dietary requirement for L-tryptophan is lower than for any other essential amino acid. Despite this, it is most often the limiting amino acid since its concentration in proteins, especially those of vegetable origin, is very low. Tryptophan undergoes extensive metabolism in the body. Perhaps its most interesting, though by no means major, metabolites are the vitamin nicotinic acid and the neuro-hormones serotinin (5-hydroxytryptamine) and melatonin (N-acetyl-5-methoxytryptamine).

Tyrosine   [α-Amino-β-(p-hydroxyphenyl)propionic acid]. Mol. wt. 181.20, $[M]_D$ of L-antipode, $-18.1°$

$$-CH_2CH(NH_2)COOH$$
$$HO-$$

in $5N$ HCl; c = 2, T = 25°. Liebig first obtained tyrosine in 1846 as a product of the alkaline degradation of casein. Erienmeyer and Lipp synthesized tyrosine in 1883. The extreme insolubility of tyrosine in water aids its isolation from protein hydrolyzates. The o-sulfate ester of tyrosine has also been reported to occur in proteins.

L-Tyrosine is a nonessential amino acid but can spare a portion of the body's need for phenylalanine. Its metabolic formation from phenylalanine his been mentioned. Tyrosine is transformed metabolically to a number of important products:

skin and hair pigments
(melanin)
↑
L-dopa
↑
phenylalanine → Tyrosine → thyroid hormones
                           (diiodotyrosine → thyroxine
fumarate⎱   ⎧phenol   ↓
acetoacetate⎰   ⎨pyruvate   triiodothyronine)
                    ⎩NH_3

adrenal hormones
(nonepinephrine → epinephrine)

Valine   (α-Aminoisovaleric acid), $[(CH_3)_2CHCH(NH_2)COOH]$. Mol. wt. 117.15, $[M]_D$ of L-antipode, $+6.6°$ in H_2O, $+33.1°$ in $5N$ HCl, and $+72.6°$ in HAc; c = 1–2, T = 25°. Valine was probably first isolated by van Gorup-Besanez in 1856 from extracts of glandular organs. Emil Fischer later identified it in casein hydrolyzates and determined its structure.

Like the other branched chain amino acids leucine and isoleucine, L-valine is an essential amino acid in human and animal nutrition. Also, like

them, the first step in the metabolism of valine is transamination to the corresponding α-keto acid, α-ketoisovalerate, which is further metabolized to acetoacetate and acetate. The absence of valine in the diet of rats, in addition to the usual cessation of growth, results in an unusual syndrome which, among other symptoms, is characterized by a lack of muscular coordination. The mechanism of this effect is unknown.

DUANE G. GALLO

Cross-references: *Polypeptides, Proteins, Foods, Nutrition, Asymmetry, Optical Rotation.*

AMINOPLAST RESINS

Aminoplast resins are a family of resins resulting from the reaction of an aldehyde with an amino compound having a functionality greater than 1. In this context the concept of functionality may be defined as the number of labile hydrogens present in the compound. In the case of simple aliphatic amines containing one primary amino group, the functionality would be two which would correspond to only the two labile amino hydrogens. At the present writing, there are only two chemical types of aminoplast which are of chemical importance. These are condensation and free radical addition cured aminoplasts. The first class comprises the vast majority of the commercially available materials while the second is a relative newcomer.

Condensation Aminoplasts. The condensation class of aminoplasts is prepared by the reaction of the aldehyde and amine at near neutral pH's. The product is a liquid or readily fusible polymeric composition of hydroxyalkyl substituted amines and higher polymers:

$$RNH_2 + R'CHO \rightarrow RNHC\!-\!R' + R\!-\!N\overset{CHOHR'}{\underset{CHOHR'}{\mid}}$$
$$\overset{}{\underset{OH}{\mid}}$$
$$+ R\!-\!\overset{H}{\underset{H}{N}}\!-\!\overset{}{\underset{R'}{C}}\!-\!\overset{}{\underset{R}{N}}\!-\!CHOHR' \quad x$$

The most commercially important resins of this type are the ones prepared from melamine or urea and formaldehyde. These have found wide application as adhesives and binders for molding materials, enamels, and foundry resins. Because of the large volume of these two resins as compared to the volume of the other less widely known aminoplasts, they deserve special attention.

Urea Resins. In a typical urea adhesive formulation, the urea and formaldehyde are combined in a mole ratio of from 1.5–4.0 moles of formaldehyde per mole of urea at a pH of 7.5–8.0. This mixture is refluxed for several hours to achieve optimum uptake of formaldehyde by the urea. The pH is then lowered to 5.5–6.0 to allow the solution viscosity to rise until the desired value is reached. The pH is then quickly raised again to approximately 8.0 to quench the reaction and also to provide increased storage stability to the formulation.

Melamine Resins. Melamine resins are prepared at reflux by the reaction of 2.1 to 3.1 moles of formaldehyde per mole of melamine. This reaction is conducted at a pH generally between 8.5 and 10.0. The initial reaction generally occurs quickly to form dimethylol- or trimethylolmelamine. Trimethylolmelamine may be easily isolated in excellent yield in a very pure state by cooling the reaction mixture to room temperature, shortly after complete solution of the melamine is achieved. The sugar-like crystals may be collected by filtration after overnight crystallization. Hexamethylolmelamine may be prepared in a similar fashion employing the appropriate quantity of formaldehyde.

If resin is desired, the solution containing dimethylol or trimethylolmelamine may be converted to resin by prolonging the reflux period until a drop of the reaction mixture, when placed in cold water, forms a slightly milky cloud which indicates conversion of methylol derivatives to polymer. The solution may then be used directly as an impregnating resin or glue. Extensive use of these resins is made in the manufacture of kitchen counter tops where material with good abrasion and water resistance is required. If a solid fusible resin is desired, the water may be removed by vacuum distillation at temperatures below 212°F or by spray drying. Granular resins prepared in this manner are widely used in glues and in a variety of molding materials.

Although urea resins are sold in very high volume as inexpensive glues for use in interior grade plywood, their water resistance is generally far inferior to melamine-based resins. Both urea and melamine resins may be cured to the infusible thermoset state by exposing the resin or resin-containing matrix to temperatures in the range of 250–450°F. In both cases the addition of acid or acid-liberating material will increase the cure speed considerably. However, only in the case of urea resins, is the addition of acid effective enough to render the resin capable of room-temperature cure. Because of this capability, urea resins have become widely used in conjunction with phenolic resins in the foundry industries as a sand binder in the "Hot-Box" process.

The hydroxymethyl melamines and melamine resins in general are considerably stronger bases than the analogous urea derivatives. Because of this basic nature, melamine resins when combined with strong acids at room temperature, and at pH's below 3.5, form hydrophilic colloids having particle sizes in the neighborhood of 150 Å. These colloids exhibit the opalescent blue color attributable to the Tyndall light-scattering effect. At these low pH's the resin chain is believed to exist with an enormous net positive charge. Consequently, when these solutions are placed in a slurry of cellulose fibers, which possess a sizable net negative charge naturally, the resin particles are almost quantitatively deposited on the cellulose. Papers treated in such a fashion possess improved wet tensile strength and have become the basis of most wet strength paper towel formulations in use today.

Urea and melamine resins can be and are modified by various means in commercial quantities to extend their useful properties beyond their normal range. Among the most useful modifications of aminoplasts are the methylated and butylated melamines and ureas, and to a lesser degree the similarly alkylated benzoquanamines (1, 3-diamino-5-phenyl-s-triazine). In large measure the alkylation procedures for all aminoplasts are nearly the same. The bis or tris (hydroxymethyl) derivatives of the amine or amide are combined with approximately a tenfold excess of the alcohol corresponding to the alkyl group to be attached. An azeotropic agent such as benzene, toluene, or xylene and an acid condensing catalyst are then added, and the mixture is held at reflux, under a Dean-Stark trap, until the appropriate quantity of water has been collected which corresponds to the desired degree of alkylation. The solution is then cooled and used as is, in conjunction with alkyd resins in the manufacture of various enamels and glues.

Perhaps the largest use of urea and melamine resins is in the manufacture of molding material. Urea molding materials have found wide use in a large number of interior applications where the moldings do not require resistance to severe weathering conditions. Examples of such applications are buttons, decorative bottle caps, and elec-

TABLE 1.

	Urea	Melamine
Specific gravity	1.47–1.52	1.47–1.52
Flammability	Self-exting.	Non-burning
Heat distortion temperature, °F	260–300	350–370
Water absorption, %	0.4–0.9	0.1–0.6
Impact strength, Izod, ft-lb/in.	0.25–0.40	0.24–0.35
Flexural strength $\times$ 10^{-3}, psi	10–18	10–17
Hardness, Rockwell	M110–M120	M115–M125
Arc resistance, ASTM D495, sec	80–150	110–140
Dielectric constant, 60 cycles	7.9–9.5	7.9–9.5
Chemical resistance		
Strong acids	Decomposes	Decomposes
Strong bases	Decomposes	Attacked

trical housing devices. On the other hand, because of their good water resistance, melamine compounds are used in many applications requiring good resistance to rather severe conditions, such as tableware and electrical switch gear and numerous other household uses.

Urea and melamine resins are nearly or completely colorless. The refractive index of urea resins, and to a somewhat lesser degree melamines, approximates that of many of the common fillers used in molding material. This makes possible the production of translucent and in some cases almost clear moldings.

Melamine resins may be easily differentiated from urea resins by the observation of a strong band in their infrared spectrum at 800 cm⁻¹. This band is present only in melamine materials and is rarely, if ever, shifted. Urea resins exhibit absorption in the 650 cm⁻¹ region, which is very characteristic of the urea moiety.

Addition-cured Aminoplast. To date only one member of this family has reached commercial acceptance with any degree of success. This is the polymer prepared from diacetoneacrylamide (N-(1, 1-dimethyl-3-oxobutyl)-acrylamide) and formaldehyde. Because of the residual unsaturated acrylamide groups present in the resin after formaldehyde condensation, it may be thermoset by the action of many free radical initiating catalysts such as benzoyl peroxide or tertiary butylperbenzoate.

Diacetoneacrylamide resins are reportedly being used in surface coatings where good grain-crack resistance is required and as a textile treatment and adhesion polymer in adhesive formulations.

PHILLIP A. WAITKUS

AMMONIA

Derived from the name of the Temple of Jupiter Ammon in ancient Libya, ammonia is a colorless alkaline gas (NH_3, m.w. 17.03) at one atmosphere with a penetrating odor, lighter than air. The word is derived from sal ammoniac (ammonium chloride), said to have been first obtained from camel's dung near the temple of Jupiter Ammon in Egypt. It is formed by destructive distillation of hartshorn (hoofs, horn) (Spirits of hartshorn).

Physical Properties. B.P. −33.35°C; F.P. −77.7°C; critical temperature 133.0°C; critical pressure 1657 psi. Specific heat at constant pressure for pressure of one atmosphere: 0°C 0.5009 cal/g; at 100°C 0.5317; at 200°C 0.5029. Heat of formation: near 0°K, 10,329 cal/mole; 700–1000°C 12,000–12,800 cal/mole. Solubility in water at one atmosphere, NH_3 % by weight: 0°C 42.8, 20°C 33.1, 40°C 23.4, 60°C 14.1. On compressing and cooling NH_3 condenses to a liquid approximately 60% as heavy as water. Because of the high vapor pressure of the liquid at room temperature, it is shipped in pressure cylinders.

Chemical Properties. *Oxidation reactions.* Ammonia does not support ordinary combustion but it does burn with a yellow flame in air or oxygen (ignition temperature 780°C): $4NH_3 + 3O_2 \rightarrow 2N_2 + 6H_2O$. An ammonia and air mixture will explode under certain conditions; high temperature and pressure greater than atmospheric. Ammonia

is also oxidized by many oxides: $3CuO + 2NH_3 \rightarrow 3Cu + N_2 + 3H_2O$.

Under the proper conditions ammonia may be oxidized to nitric acid (platinum gauze catalyst):

$$4NH_3 + 5O_2 \rightarrow 4NO + 6H_2O$$

$$2NO + O_2 \rightarrow 2NO_2$$

$$3NO_2 + H_2O \rightarrow 2HNO_3 + NO$$

In general reducing agents do not affect NH_3. Ammonia is stable at ordinary temperatures but it decomposes into nitrogen and hydrogen at 450–500°C under atmospheric pressure. In the presence of catalysts this decomposition may take place at 300°C.

Liquid ammonia is an important ionizing solvent (dielectric constant equal to 16.9 at 25°C) which has been used in many electrolytic solution investigations. Sodium and potassium dissolve in liquid ammonia to form highly conducting colored solutions. Potassium dissolves slowly forming the amide: $2K + 2NH_3 \rightarrow 2KNH_2 + H_2$. Chlorine, bromine, and iodine all react:

$$3Cl_2 + 8NH_3 \rightarrow 6NH_4Cl + N_2$$

$$2NH_3 + 3I_2 \rightarrow NI_3 \cdot NH_3 + 3HI.$$

Phosphorus reacts with ammonia at red heat to form phosphine and nitrogen: $2NH_3 + 2P \rightarrow 2PH_3 + N_2$. Sulfur vapor reacts with ammonia to form ammonium sulfide and nitrogen.

Ammonia forms a great variety of addition compounds which are often called ammoniates in analogy with hydrates.

The water solution of ammonia is distinctly alkaline. The reaction between water and ammonia may be represented by:

$$NH_3 + H_2O \rightleftharpoons NH_3 \cdot H_2O \rightleftharpoons NH_4^+ + OH^-.$$

If an aqueous solution of NH_3 is neutralized with an acid, nitric (HNO_3), hydrochloric (HCl), or sulfuric (H_2SO_4), the corresponding salt is formed: NH_4NO_3, NH_4Cl, $(NH_4)_2SO_4$. The radical NH_4^+ is referred to as the ammonium radical.

NH_3 forms many coordination compounds, which are often called ammines, e.g., $[Cu(NH_3)_4]SO_4$.

The double decomposition in which NH_3 is a reactant is known as ammonolysis in analogy with hydrolysis. The substitution of an —NH_2 (amine) group for Cl, OH, SO_3H, is known as amination by ammonolysis.

Production. Ammonia is one of the by-products in the coking of coal. About 5.5 to 6.5 lbs. of ammonia per ton of coal coked is recoverable as ammonia. Dobereiner was the first to synthesize ammonia from nitrogen and hydrogen. Le Chatelier was the first to recognize that to produce ammonia commercially, high pressure was needed. Haber studied the thermodynamics of the nitrogen-hydrogen equilibrium. Lewis and Randall calculated the change of free energy, and the change of specific heat for the equilibrium as a function of temperature.

The process used to produce the largest quantities of ammonia today is the Haber-Bosch process, which consists essentially of the following: water gas (CO, H_2, CO_2) is made from coke, air, steam.

The mixed gas (which also contains N_2) passes through a scrubber cooler where dust particles and undecomposed material are removed with water. Most of the CO is converted to CO_2 and removed with a carbon dioxide purifier. The rest of the CO is removed with an ammoniacal cuprous solution. The pure gases ($3H_2$ and N_2) pass over a catalyst in a high pressure (up to 1000 atm) and high temperature (near 700°C) ammonia converter. The ammonia generated is absorbed and removed by water. Modifications of this process are used widely. The modifications consist of differences in sources of the N_2 and H_2, the methods of purifying the catalysts, the temperature and pressure and the methods of ammonia recovery.

Storage. Anhydrous ammonia is usually stored at pressures up to 40 psi. Cylinders containing up to 150 pounds and 26 ton tank cars are usually used for shipping it. Anhydrous ammonia is marketed under two specifications: (1) commercial grade—ammonia content not less than 99.5%, (2) refrigeration grade 99.95% NH_3. The nonbasic gas content is not more than 0.20 ml/gm NH_3, moisture content less than 0.01 ml per 100 ml NH_3, pyridine content none.

Toxicology. There are four hazards connected with ammonia: (1) pressure, (2) thermal, (3) physiological, (4) explosive. Solutions of ammonia are much less dangerous to handle than the pure substance. There is no pressure hazard; the physiological hazard is eliminated. Liquid ammonia severely irritates the skin. Gaseous ammonia intensely irritates moist tissue. High concentrations of ammonia cause cessation of respiration.

Uses. Ammonia is oxidized in great quantities to make nitric acid, which is used to make TNT, nitroglycerine, nitrocellulose, and ammonium nitrate. In the textile industry ammonia is used in the production of synthetic fibers, e.g., nylon, cuprammonium rayon. Ammonia is also used in the dyeing and scouring of cotton, wool, rayon and silk. The principal nitrogen carriers in fertilizers are anhydrous ammonia, ammonium nitrate, urea, and calcium cyanamide. Anhydrous ammonia is also used to ammoniate superphosphates in the preparation of mixed fertilizers.

Ammonia is used as a catalyst in the phenolformaldehyde (Bakelite) condensation and also in the ureaformaldehyde condensation to make synthetic resin. The melamine component of melamine-formaldehyde resins is produced by the polymerization of dicyanodiamide in the presence of ammonia. The sulfa drugs, sufanilimide and sulfapyridine, as well as many vitamins and antimalarials, use ammonia in their synthesis. In the petroleum industry, ammonia is used as a neutralizing agent to prevent corrosion due to acidic components in the petroleum products.

Ammonia is a commonly used refrigerant to produce ice, to air condition, for cold storage. Its characteristics of high latent heat of vaporization, low density, high stability and low corrosion make it valuable in this respect.

A recent development is the substitution of NH_3 for Ca in the bisulfate process for pulping wood. This improves the yield and quality of the pulp. NH_3 is also used as a solvent for casein in the coating of paper.

Chemicals. Ammonium salts are produced in the inorganic chemical field by the neutralization of ammonia with an acid. The following equation represents the production of sodium cyanide:

$$NH_3 + Na \rightarrow NaNH_2 + \tfrac{1}{2}H_2,$$

$$\text{sodium amide}$$

$$NaNH_2 + C \xrightarrow{800°C} NaCN + H_2$$

The nitric oxide required for the conversion of sulfur dioxide to sulfur trioxide in sulfuric acid manufacture is obtained from the oxidation of ammonia. The bulk of nitric acid used in the United States is obtained from ammonia oxidation. The Solvay ammonia-soda process for the production of soda ash from sodium chloride uses large quantities of ammonia.

In the organic chemical field amines, amides, nitriles are produced with ammonia, e.g., aniline:

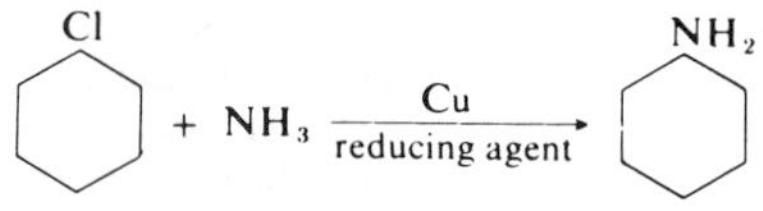

Metallurgy. Ammonia is a processing agent to recover copper, molybdenum, and nickel from ores. Anhydrous ammonia, after it has been dissociated into N_2 and H_2, is used in nitriding alloy sheets to impart a hard wearing surface. In powder metallurgy, ammonia is used as a protective atmosphere in sintering operations. Metal oxides are reduced with ammonia gas in atomic H_2 welding. Since ammonia can be decomposed to give hydrogen it is an economical means for transporting hydrogen.

Water Purification. In conjunction with chlorine, ammonia is used to purify municipal and industrial water supplies. The ammonia and chlorine are metered into the water in the desired proportions. The chloramines formed are water purifiers.

DANIEL BERG

Cross-references: *Bases, Amines, Amides, Noxious Gases.*

ANALYTICAL CHEMISTRY

The terms "instrumental analysis" or "instrumental methods of analysis" are difficult to define exactly, since many chemists maintain that all measurements are made with instruments of some kind. General usage, however, is to divide analytical methods into two rough classifications, "wet" or classical methods and "instrumental methods." Volumetric and gravimetric analyses fall into the first classification and other methods which employ some instrument other than the buret or balance for the final measurement are known as instrumental methods. Instrumental methods can be classified in the following manner:

Optical methods (methods dependent on radiant energy measurements)
  1. Absorption of radiant energy
     a. absorption of gamma rays
     b. absorption of x-rays
     c. ultraviolet, visible and infrared spectrophotometry

d. atomic absorption spectrometry
e. colorimetry
f. absorption of microwaves
2. Diffraction of x-rays and electrons
3. Emission of radiation
   a. spectrography
   b. flame photometry
   c. fluorescence
   d. Raman effect measurements
4. Dispersion of radiant energy
   a. nephelometry
   b. turbidimetry
5. Polarimetry, i.e., rotation of plane of polarized light
6. Refractometry
7. Interferometry
8. Nuclear magnetic resonance
Electrical methods
1. Potentiometry, including pH measurements
2. Conductiometry
3. Coulometry
4. Chronopotentiometry
5. Polarography
6. Amperometric titration methods
7. Magnetic susceptibility measurements
8. Dielectric constant measurements
9. Mass spectrometry, i.e., ratio of charge to mass measurement
Mechanical properties
1. Specific gravity or density measurements
2. Viscosity
3. Velocity of sound in a gas
4. Rate of diffusion of a gas
5. Gas chromatography
Thermal methods
1. Heat of reaction
2. Thermal conductivity of a gas
Nuclear methods
1. Radioacitvity

Besides the methods mentioned above, which are in fairly general use, many others have been devised, primarily for specific analytical problems.

Advantages

1. Instrumental methods are generally much faster than wet methods. This is especially true when a large number of samples have to be analyzed and the calibration time for the method can be prorated among many samples. One notable example of this is the determination of as many as eleven constituents in steel samples in 45 sec. by use of a recording spectrograph. The calibration time for the instrument may be several hours to several days. However, a "wet" analysis would require the time of a single analyst for an hour or more, for each sample, regardless of the number of samples to be run.

2. Instrumental methods are often more precise than ordinary volumetric or gravimetric methods. When one considers that the analytical balance is an extremely precise instrument, being capable of detecting differences of 0.1 mg in loads of around 100 g, it is apparent that an instrumental method gains its precision, if any, from some part of the procedure other than the final measurement. Frequently, it is that the instrumental method requires

little or no preliminary chemical separation before measurement.

3. Instrumental methods may require no separation of the desired constituent from the substrate. As an example, infrared methods can determine the amount of 1,2-dibrompropane in 1,3-dibrompropane without preliminary separation.

4. Instrumental methods often leave the sample unchanged. This is usually the case when a physical property is being measured. When the sample is expensive, rare or small in amount, this advantage is of great importance.

5. Instrumental methods may be extremely sensitive. The most notable examples of this are the methods employing radioactivity. Amounts smaller than 10^{-10} g can often be detected.

6. Many instrumental methods can be adapted to give a continuous recording of the composition of the sample. This makes these methods extremely useful in industrial plants and for following the kinetics of reactions.

7. Many instrumental methods can be adapted to control a process. The future will undoubtedly see many manufacturing processes automatically controlled by instrumental means. It is possible to feed the results of many instrumental measurements into high-speed computers which will then compute the changes necessary to give optimum yield and, in turn, feed this information to control mechanisms to effect the required changes.

Disadvantages

1. Instruments are generally quite expensive and, unless a large number of samples is to be run or a great saving in time and labor is achieved, the cost may be prohibitive.

2. Instruments require highly skilled technicians for repair and maintenance. Although the operators of the instruments may often be less skilled than ordinary analytical technicians (using classical methods), every instrument will eventually require the attention of a highly skilled chemist, physicist or electrical engineer.

3. Most instruments require calibration which may be time-consuming and may limit the application of the method to cases where several samples are to be run.

4. Since most instruments must be calibrated and since the calibrating samples are often analyzed by classical methods, the accuracy of most instrumental methods is limited to that of the classical methods used for determining the composition of the standard samples.

5. Instruments may not recognize changes in the properties of the system under investigation which would be immediately apparent to an experienced analyst. For example, a photoelectric colorimeter will compare a greenish-purple permanganate solution containing iron with a purple permanganate standard solution and, unless very narrow band filters are used, an incorrect determination of manganese will result. The difference in color of the two samples would be immediately apparent to the eye.

LYNNE L. MERRITT, JR.

Cross-references: *Analytical Chemistry, Optical Spectroscopy, Raman Spectroscopy, Polarography.*

Analytical Chemistry of Radioactive Elements

The field of analytical chemistry was revitalized upon the advent of atomic energy. Three factors were responsible: (1) the need for determining traces of many elements in high-purity uranium; (2) the need for separation and characterization of the many nuclides resulting from nuclear fission; and (3) the availability of radionuclides as tracers for studying analytical processes.

Uranium was considered a rare and relatively useless metal until the Manhattan Project was begun. Since that time, it has been intensively studied and there are more data available on the analytical properties of uranium than on any other element. In addition to analysis for uranium itself, the high purity requirements for reactor uranium meant that suitable methods had to be found for determination of traces of almost all the known elements in a uranium matrix. Sensitivity requirements were such that most of the determinations had to be carried out instrumentally after suitable chemical separations. Therefore, basic research was carried out on emission and absorption spectra, electrode potentials, diffusion potentials and the like as well as on chemical separation procedures such as extraction, paper chromatography and ion exchange.

The advent of atomic energy brought several new elements into being, especially the transuranic elements. The need for analysis of small quantities of these elements required improvements in many microchemical techniques. Seaborg in 1946 described chemical analyses involving only 0.1 microgram in volumes as small as 10^{-5} ml, and physical measurements were made on quantities of the same order. Although larger amounts of these elements are now available, the high toxicity of most of them tends to limit the amounts handled to the microgram range.

The mixture of reactor fission products, largely the more common elements plus the rare earths, offers a large source of radionuclides for chemical research. This, in turn, requires the development of methods for separating the individual elements. The techniques are largely those of analytical chemistry, since clean-cut, small-scale separations are required. While the standard precipitations and solvent extractions have proved quite valuable, the synthetic resin ion-exchange column is a major tool. The preferential absorption and elution of various elements in different media provides remarkably sharp separations of even individual rare earths.

Other radionuclides have become available by neutron bombardment in reactors or charged particle bombardment with accelerators. The short radioactive half-lives of some of these has required development of very rapid separation techniques, but allows tracer studies of many elements that do not have fission product isotopes.

Radioactive tracers have been a particular boon to the analytical chemist for the development of new methods and the testing of old ones. In many of the older procedures that have now been tested with radioactive tracers, it was found that correct results had been obtained by compensating errors. Tracer methods have shown the actual contribution of the factors concerned.

Radioactivity has two other major applications; radiometric analysis, and the isotope dilution technique. In radiometric analysis, the element to be determined is precipitated with a radioactive reagent. The classic example is the precipitation of traces of chloride with Ag^{110}. The minute quantity of radioactive silver chloride may be collected by coprecipitation with ferric hydroxide and counted. In isotope dilution, a known amount of a radioactive isotope of the element determined is added to the original sample. When the final precipitate is collected and weighed, the measure of its radioactivity compared to the added activity gives the yield of the chemical process and therefore allows determination of the total amount of the element initially present.

A corollary of the isotope dilution procedure is applied to analyses for traces of radionuclides. In this case, the minute quantities of the element cannot be handled physically and may even give deviations from expected chemical behavior. By dilution with a known amount (usually milligrams) of the inactive form of the element sought, the final quantity of the element may be measured and the per cent recovery of the process may be estimated. This recovery factor may be applied to the measured radioactivity of the product to calculate the total radionuclide initially present. Originally recovery was measured by gravimetric techniques, but more specific analyses by flame photometry, atomic absorption or fluorescence are preferred.

Two additional techniques allow the application of radiometric measurements to inactive elements. In activation analysis, a sample is subjected to neutron activation in a nuclear reactor or with a neutron source. Traces of certain elements may then be determined from the radiometric or radiochemical properties of the nuclides produced. This system now relies heavily on high resolution gamma spectroscopy. X-ray fluorescence spectroscopy uses a radionuclide source to provide low-energy gamma radiation or x-rays to irradiate the sample. Characteristic x-rays from elements in the sample are then measured with a high resolution gamma detector.

Atomic energy has presented the analytical chemist with new problems and also with new tools. Both of these have promoted the rapid advance of basic research in analytical methods, and have opened many new fields of development, as well as making possible accurate evaluation of older methods.

JOHN H. HARLEY

References

NAS—National Research Council Publications. NAS also publishes a series of monographs on individual elements and techniques in their Nuclear Science Series (Reports NAS-NS 3001 ff).
Cross-references: *Radioactivity; Spectroscopy, Gamma-ray.*

Kinetics in Analytical Chemistry

The role of kinetics in analytical chemistry is a multiple one and includes, at the least, the following major aspects:

(1) Study of the rates and mechanisms of ana-

lytical reactions as a means of elucidating deviations from stoichiometry, detecting interferences, and providing optimum reaction conditions.

(2) Direct use of reaction rate measurements for purposes of analysis.

(3) Utilization of catalytic and inhibition phenomena for selective and trace analysis.

(4) Evaluation of new reagents of possible use in analytical chemistry.

A less tangible, but no less important, benefit is that the methods and terminology of kinetics permit the analytical chemist to organize and communicate analytical information in a non-empirical manner and to eliminate much of the trial and error approach which often characterizes classical topics in analytical chemistry. It is no exaggeration to say that the impact of kinetics on analytical chemistry may be comparable to the influence of the equilibrium approach which has been so beneficial to the treatment of analytical problems. The role of kinetics in analytical chemistry is, therefore, not only one of application, but also of viewpoint.

An example of this and of (1) above may be found if one considers how a "new" reaction becomes part of accepted analytical procedure. In the past, the reaction of analytical interest would probably have been carried out under a variety of experimental and environmental conditions until the appropriate conditions for reproducible stoichiometry, convenient rate, and minimum interference have been found; a procedure, incorporating these experimental details, is then published. Other analytical chemists may then make use of this reaction—preferably with a minimum of alteration of the recommended procedure. The kinetic approach to the same situation is quite different, however; the reaction would be investigated systematically and its characteristics expressed in terms of kinetic generalizations (such as rate laws and reaction mechanisms) so that the potential user could have sufficient information to decide for himself what exact conditions and modifications would be appropriate to the specific circumstances of his individual application. The main difference between the two approaches is that the first encourages experimentation by rote while the second permits the logical exploitation of scientific information for practical ends.

The direct use of reaction rates in quantitative analysis (2) is enjoying considerable popularity. Two different approaches may be distinguished: the determination of the concentration of a single reacting material via the well-known proportionality between the concentration of a reactant not in excess and the observed reaction rate and, secondly, the measurement of differential reaction rates to distinguish between two or more components which are chemically similar but react at different rates. The primary advantage of the direct reaction rate method is that it permits rapid estimation of the concentration of many materials (even unstable substances) without the need for detailed information regarding the reaction employed. The method suffers, however, from lack of specificity and shows only moderate accuracy. The differential rate method also does not yield highly accurate results but finds important application in systems where

isomeric and other highly similar components must be resolved. The utility of the method improves as the difference in reactivity of the components in question increases, but, under otherwise favorable circumstances, permits the resolution of reactants whose reaction rates differ by less than 1%. Future application of these reaction rate methods will probably be in the areas of organic and biochemical analysis.

Catalytic methods (3) are among the most sensitive analytical techniques yet devised and have great potential for selective as well as sensitive analysis. They have their basis in the fact that the rates of chemical reactions are often related, in a simple manner, to the concentration of catalytic material present. Again, it is possible to use catalytic methods for analysis without a detailed knowledge of the mechanisms of either the uncatalyzed or the catalyzed reaction because the relationship between the reaction rate and the concentration of the catalyst may be empirically established. In favorable cases, such as the thiosulfate-catalyzed iodine—azide reaction, as little as 10^{-13} gram of the catalyst may be measured with reasonable accuracy. The major experimental problem to be overcome is elimination of unwanted catalytic impurities and other chemical interferences. Similar considerations apply to enzyme catalysis.

Many of the new reagents (4) proposed for analytical purposes either react slowly with the sample component or are unstable with respect to time in their active state. The investigation of the kinetics of such processes often yields information valuable in the selection of appropriate reaction conditions or suggests means by which the reagent might be stabilized, e.g. the spontaneous decomposition of silver (II) is inhibited by the presence of excess silver (I) because of the displacement of a disproportionation equilibrium. Furthermore, such kinetic information may be employed to set the effective limits for the use of specific reagents and, in addition, may demonstrate whether or not a given reagent will react stoichiometrically with the desired constituent in the presence of other components.

The future role of kinetics in analytical chemistry is dependent not only upon the further development of the areas of investigation outlined above but also upon the willingness of analytical chemists to make use of kinetic information. Conversely, modern analytical chemistry may benefit the field of chemical kinetics in many ways, principally by the development of new techniques to follow the progress of chemical reactions and by establishing ever higher standards of chemical purity for both inorganic and organic reagents.

G. A. RECHNITZ

Reference

Mark, H. B., Jr., and Rechnitz, G. A., "Kinetics in Analytical Chemistry," John Wiley and Sons, Inc., New York, 1968.

ANESTHETICS

Anesthetics are agents which, when suitably applied, cause a general or localized loss of feeling or sensation. General anesthetics exert their action on the higher nerve centers and produce involuntary loss of consciousness. When a suitable agent is applied to peripheral nerve endings (topical anesthesia), by infiltration into the nerve fiber (intra-neural anesthesia), around the nerve sheath (para-neural anesthesia), or into the spinal canal (intra-spinal anesthesia) at various levels, a loss of sensation is produced within a restricted or readily predictable large segment of the body. This type of anesthesia is called local, block or regional anesthesia. It does not produce loss of consciousness but effectively prevents the transmission of certain kinds of pain sensations from any region which lies farther from the brain than the site of application of the anesthetic agent and which is served by the anesthetized nerve.

General anesthetics may be classified according to their physical properties as volatile (gases or low boiling liquids) and nonvolatile (high boiling liquids and solids). Several chemical classes are either potentially or practically valuable as general anesthetics. These include the following useful individuals.

Hydrocarbons. Gaseous and volatile members of the aliphatic series, while potentially anesthetic, have such a low degree of activity that effective concentrations produce dangerous anoxia. None has been found useful.

Cyclopropane or trimethylene may be made from propane by chlorination, fractional separation of 1,3-dichloropropane, b.p. 119°C, from other isomers and cyclization with zinc in the presence of iodide to yield the hydrocarbon. It was first used as an anesthetic in 1929. The anesthetic grade is not less than 99% by volume of C_3H_6. It is soluble in 2.7 parts of water at 15°C and is freely soluble in fatty oils. It is marketed in cylinders of compressed gas which must comply with the standards described in the United States Pharmacopeia. It is highly flammable and forms explosive mixtures with air or oxygen. One distinct physiological advantage is that it exerts its anesthetic action in the presence of substantial proportions of oxygen. Anesthesia is quickly achieved and consciousness rapidly recovered when the administration is stopped.

Ethylene. The chemical was described as early as 1795 but its anesthetic properties were not reported until 1923. The classical synthesis is achieved by the dehydration of ethanol with concentrated sulfuric acid at about 180°C. Phosphoric acid may be used in a similar reaction as well as the dehydration of ethanol over aluminum oxide at 360°C. Large quantities of ethylene result from catalytic cracking of petroleum. Medicinal grade ethylene is marketed as a compressed, liquefied gas in steel cylinders. It must comply with the standards described in the United States Pharmacopeia. At atmospheric pressure it is soluble (1:4 v/v) in water at 0°C and (1:9 v/v) at 25°C; it is very soluble in alcohol and in ether. It forms explosive mixtures with air or oxygen but can be administered under suitable safeguards with substantial volumes of oxygen. Anesthesia is quickly induced and recovery is rapid.

Ethers. *Ethyl ether* was discovered as early as 1540 but its first recorded use as a surgical anesthetic was in 1842. Anesthetic grade ether contains 96–98% of $CH_3 \cdot CH_2 \cdot O \cdot CH_2 \cdot CH_3$, the remainder being a mixture of alcohol and water. It is marketed as a colorless liquid, boiling point 35°C, in tight containers. It should be used promptly after the container is opened before any substantial amount of ether peroxide may form. This toxic and highly explosive oxidation product of ether results when ether is exposed to air and light for any considerable time. The medicinal grade of ether must meet the tests described under that title in the United States Pharmacopeia.

Vinyl ether resulted from a calculated effort to prepare a general anesthetic having many of the physiological properties of ethylene and ether. It was synthesized and pharmacologically tested in 1930. The product is marketed as a liquid, b.p. 28–31°C which comprises about 96% vinyl ether and about 4% anhydrous alcohol. A trace of an antioxidant preservative is permitted.

Other ethers which are analogous to or isomeric with ether have been used experimentally, among these may be mentioned methyl ethyl ether, methyl propyl ether, methyl cyclopropyl ether, ethyl cyclopropyl ether. None of these has achieved wide use.

Halogenated hydrocarbons containing one to three carbon atoms and from one to five atoms of halogen, chlorine, bromine, or fluorine, or mixtures of these in the same molecule, have been shown to have potential and useful anesthetic activity. All of this group are characterized as toxic when administered over a long period of time or for repeated short periods. Some of them are used as solvents in dry-cleaning fluids and in technical adhesive products, which spreads the health hazards beyond their use as anesthetics, e.g., "glue sniffing." The first of this group to be used as an anesthetic was chloroform.

Chloroform was first prepared in 1832, but its use as a general anesthetic dates from 1847. The marketed product is 99 to 99.5% pure with alcohol as the permitted impurity. Chloroform boils at about 61°C and is nonflammable and nonexplosive. The periods of induction and recovery are short.

Ethyl chloride was prepared in 1845 and achieved its first use as an anesthetic in 1847–1848. Ethyl chloride is a gas at ordinary room temperature and pressure (b.p. 12–13°C). It liquefies under relatively low pressures and is frequently marketed in specially designed glass ampuls with a valve for withdrawal of the liquefied substance. In this way a stream of liquid may be directed on tissue to anesthetize it by lowering the temperature. Ethyl chloride, and its bromine analog, are sometimes used as inhalation type general anesthetics. Induction and recovery are rapid. Ethyl chloride is flammable.

Trichloroethylene was prepared in 1864 but was first used as an analgesic and general anesthetic about 1933 to 1934. It is a colorless liquid (in commerce sometimes colored blue with an approved medicinal dye) boiling at 88°C. It is not flammable. Induction and recovery periods are short.

Examples of compounds of this type which have been experimentally used as general inhalation anesthetics but which have not attained major importance are: *methylene chloride*, methyl chloride, carbon tetrachloride, ethyl bromide, ethylene chloride, ethylidene chloride, Halothane (1,1,1-trifluoro-2-bromo-2-chloroethane), Teflurane (1-bromo-1,2,2,2-tetrafluoro-ethane), Haloprane (1-bromo-2,2,3,3-tetrafluoropropane).

Halogenated Ether. One example of this type which has been tested but has not attained major importance is Methoxyflurane (2,2-dichloro-1,1-difluoroethyl methyl ether).

Halogenated alcohol. Alcohols containing one hydroxyl group and short hydrocarbon chains are potentially anesthetic, but their potency is largely increased by replacement of several of the hydrogen atoms of the hydrocarbon chain with chlorine or bromine. The latter is particularly effective and is exemplified in tribromoethanol.

Tribromoethanol is a low-melting (79–82°C) crystalline compound. The substance is somewhat soluble in water but is usually marketed as a solution in tertiary amyl alcohol (amylene hydrate) of approximately 1 g in 1 ml of solution. The solution is suspended in warm water and administered as a retention enema. It is recommended for use as a basal anesthetic and when complete anesthesia is desired, it is usually supplemented with an inhalation type of general anesthetic.

Nitrous oxide was first prepared by J. B. Priestley in 1772. Its anesthetic properties were observed by Humphry Davy about 1799–1800 but it did not achieve use in surgery until after the dramatic failure of a public demonstration by H. Wells in 1845 had called the attention of dentists and surgeons to its possibilities. It is usually prepared by the thermal decomposition of fused ammonium nitrate at about 240°C.

The gas is freed from possible contamination with nitric oxide (NO) by passing it over iron powder and is then compressed and cooled in three stages to a liquid. The liquefied gas is marketed in steel cylinders. It is required to be not less than 95% pure N_2O with the allowable impurity being nitrogen. It must be administered with some oxygen or air to avoid anoxia. It produces rapid anesthesia and the recovery period is equally short. It is not suitable for prolonged anesthesia nor for surgery which requires profound relaxation.

Barbituric acid derivatives. When administered in the usual manner by mouth and in therapeutic doses, the derivatives of barbituric acid are sedative and hypnotic but not anesthetic. However, when the soluble salts of these compounds are administered intravenously, they produce typical anesthesia. Many of these barbituric acid salts have such a prolonged action that they are not usable for anesthetics, since one of the desirable properties is that the anesthetist should have rapidly adjustable control over the depth and duration of anesthesia. Those barbiturates which act promptly and are destroyed in the body rapidly make useful general anesthetics. Among those which are frequently used for this purpose are some which contain a sulfur atom in the place of C_2-oxygen and others which have a methyl group in place of the N_1-hydrogen.

Examples which have been used are: Thiopental Sodium (sodium 5-ethyl-5-(1-methyl butyl)-2-thiobarbiturate); Hexobarbital Sodium (sodium N-methyl-5-methyl-5-Δ^1-cyclohexenyl)-barbiturate);

Methohexital Sodium (sodium-1-methyl-5-allyl-5-(1-methyl-2-pentynyl)-barbiturate); Thiamylal Sodium (sodium-5-allyl-5-(1-methyl butyl)-2-thiobarbiturate); Methitural Sodium (sodium-5-(1-methyl butyl)-5-(2-methylthioethyl)-2-thiobarbiturate); Thialbarbital Sodium (sodium-5-allyl-5-(cyclo-Δ^2-hexenyl)-2-thiobarbiturate); Thioethamyl Sodium (sodium-5-ethyl-5-isoamyl-2-thiobarbiturate).

Other compounds having analgesic and general anesthetic properties when administered intravenously have been proposed and tested from time to time. Among them may be mentioned: Dolitron (5-ethyl-6-phenyl-metathiazane-2,4-dione); a steroid derivative named Hydroxidione (21-hydroxypregnane-3,20-dione sodium succinate); a phenoxyacetamide, Propanidid (4-(diethylcarbamoylmethoxy)-3-methoxyphenyl acetic acid propyl ester).

Local Anesthetics

It has long been known that lowering the temperature of a portion of the body will make the perception of pain originating in the area of lowered temperature less apparent. Likewise, it is known that pressure applied to a nerve fiber or trunk will interfere with the transmission of pain impulses originating peripherally to the site of the pressure. The first principle is still occasionally used to produce local anesthesia. Advantage is taken of the low temperatures produced by the vaporization of such compounds as ethyl chloride, ether and solid carbon dioxide to produce local anesthesia in restricted segments of the body.

The beginning of the modern period of local anesthesia by physiologically active compounds can be traced to the isolation and characterization of the alkaloid cocaine. This alkaloid was isolated from the leaves of *Erythroxylon coca* by Niemann in 1860. Its local anesthetic effect on the tongue was noted by Wöhler. Its use in surgery dates from the work of Koller who, in 1884, described the use of cocaine in surgery of the eye, nose and throat. Koller also used cocaine by infiltration techniques. (See **Alkaloids**).

A systematic investigation of the chemical nature of cocaine led to its synthesis by Willstätter in 1902 and to the development of a large number of synthetic compounds which have valuable local anesthetic action, for example, the procaine group (Novocaine). As a result of much work relating to the chemical constitution of cocaine analogs and local anesthetic activity, a number of general relationships were traced. It was found that alkyl esters of aromatic acids such as benzoic and naphthoic acids had potential local anesthetic activity. This is enhanced by the presence of substituent groups in the aromatic ring such as alkyl, amino, hydroxy, alkoxy and alkylthio. Isosteric compounds such as substituted amides and amidines, as well as thiocarboxylic esters and urethanes retain the anesthesiophore.

Many of the simple alkyl esters of aromatic acids were only useful as topical anesthetics, since their water solubility was too low to permit their use in parenteral solutions. Of these, Benzocaine (ethyl-4-aminobenzoate) is a prototype with the *n*-propyl and *n*-butyl esters known respectively as Raythesin, and Butesin. The isobutyl ester of benzoic acid is known as Isocaine and the methyl ester of 3-amino-4-hydroxy-benzoic acid as Orthoform.

It was found that the inclusion of a tertiary or secondary aliphatic or cycloalkylamine in their structure made it possible to prepare water-soluble salts with acids which could be administered by parenteral pathways. This was also possible in the case of amides, thiocarboxylic esters and urethanes. Amidines which were of moderately high molecular weight are sufficiently basic to form salts which can be dissolved and used. A large number of individual compounds have been synthesized and a considerable number have been marketed.

Some of these compounds, like cocaine, are effective when applied topically or parenterally. A sampling of such compounds would include: Beta-Eucaine (2,6,6-trimethyl-4-piperidylbenzoate); Naepaine, (n-amylaminoethyl-4-aminobenzoate); Tetracaine, (4-n-butylamino-diethylaminoethyl benzoate); Piperocaine, (3-(2-methyl-piperidino)-propyl benzoate); Dibucaine, (2-n-butoxy-4-(β-diethylamino-ethylcarbamyl)-quinoline); Phenacaine, (N,N'-diethoxyphenyl acetamidine); Diothane, (1-piperidino-propane-2,3-di-N-phenyl carbamate).

A number are poorly active topically but adequately effective when injected. Among these may be metioned: Procaine, (diethylaminoethyl-4-amino benzoate); Butacaine, (γ-dibutylaminopropyl-4-amino benzoate); Lidocaine, (2,6-dimethyl-α-diethylamino-acetanilide); and Thiocaine, (diethylaminoethyl thiobenzoate).

Many experimental compounds are rejected because of their locally irritant effects or systemic toxicity or both. Other compounds which, structurally, are potentially local anesthetics have found their greatest clinical use for other effects such as antispasmodics.

The local anesthetic effect of quinine salts and Eucupin (isoamylhydrocupreine) should be mentioned. Both quinine hydrochloride, rendered more soluble with urea, and Eucupin have been used where local anesthesia of long duration was desired.

A number of simpler compounds such as Chlorbutanol (1,1,1-trichloro-2-methylpropanol-2), benzyl alcohol and salicyl alcohol have mild local anesthetic effects when topically applied to mucous membrane.

G. L. WEBSTER

Cross-references: *Ethers, Oxides, Alkaloids.*

ANHYDRIDES

The term "anhydride" means "without water." Thus an acid anhydride is a compound that lacks the water necessary to make it become an acid.

Inorganic Anhydrides. Oxides, usually those of nonmetals, that react with water and form acids are called acid anhydrides or acid oxides. Sulfur dioxide (SO_2), carbon dioxide (CO_2), phosphorus pentoxide (P_2O_5), and sulfur trioxide (SO_3) are examples of such oxides. $SO_3 + H_2O \rightarrow H_2SO_4$ and $P_2O_5 + H_2O \rightarrow 2HPO_3$ are typical reactions of acid anhydrides with water, forming acids.

Oxides, usually those of metals, that react with water and form bases are called basic anhydrides or basic oxides. Calcium oxide (CaO), sodium monoxide (Na_2O), and chromous oxide (CrO) are examples of such oxides. The reaction CaO +

$H_2O \rightarrow Ca(OH)_2$ (the slaking of lime) is one of the most definite examples of the reaction of a basic anhydride with water forming a base. A simple definition of the terms acid and base is assumed in this description.

For some metals that have several states of oxidation, chromium and manganese being the most common examples, the oxides in which the metal has the lower oxidation values are basic anhydrides; the oxides in which the metal has intermediate value are amphoteric; and the oxides in which the metal has higher oxidation values are acid anhydrides. The explosive compound, manganese heptoxide (Mn_2O_7), and manganese trioxide (MnO_3), which has doubtful existence, are acid anhydrides that form permanganate ions (MnO_4^-) and manganate ions (MnO_4^{--}), respectively, when they react with water. Manganese dioxide (MnO_2) is somewhat amphoteric. Manganese sesquioxide (Mn_2O_3) and manganous oxide (MnO) are basic anhydrides. In these five oxides, the oxidation number of the manganese is respectively 7, 6, 4, 3, and 2. A similar pattern exists for the oxides of chromium. Chromous oxide (CrO) is a basic anhydride, and chromium trioxide (CrO_3) is an acid anhydride.

Organic Anhydrides. Organic anhydrides seem to have been made by combining the residue of two carboxyl (·COOH) radicals by means of an oxygen atom when water is eliminated between the two carboxyl radicals. Thus two molecules of acetic acid form acetic anhydride. A molecule of water is eliminated from two molecules of acetic acid. The use of anhydrous calcium sulfate ($CaSO_4$) aids the dehydration. Anhydrides generally hydrolyze slowly in cold water, more rapidly in warm water, and readily in alkaline solution.

Acetic anhydride, boiling point 139.6°C, is the most important of the organic acid anhydrides. It is an aliphatic anhydride that has a sharp penetrating odor. Production each year in the United States is about one-half million tons. Synthetic methods are used; one involves ketene.

Acetic anhydride is used to introduce the acetate radical into organic compounds, and is called an acetylating agent. About three-fourths of the acetic anhydride made is used to make cellulose acetate. Cellulose acetate in turn is used to make fibers, plastics, photographic films, and transparent wrapping sheets. Acetic anhydride heated with salicyclic acid (C_6H_4—OH—COOH) forms acetylsalicylic acid (aspirin). In general, alcohols react with acetic anhydride forming an ester and acetic acid. For example:

$$C_2H_5OH + (CH_3CO)_2O \rightarrow$$
$$CH_3—COOC_2H_5 + CH_3COOH.$$

Phthalic anhydride, a colorless solid, melting point 130.8°C, is the leading aromatic anhydride. Its annual production in the United States exceeds 100,000 tons. It is the anhydride of o-benzenedicarboxylic acid. Commercial manufacture is carried out by careful partial oxidation of naphthalene ($C_{10}H_8$) in the presence of vanadium pentoxide (V_2O_5) as a catalyst. An anhydride of a polybasic acid, it polymerizes with polyhydroxy alcohols and forms resins of the alkyd type.

Glycerol $[C_3H_5(OH)_3]$ and phthalic anhydride form the well-known thermosetting "Glyptal" resins noted for their resistance to electricity and their use in paints and other surface coatings. Phthalic anhydride is also used to make phthalate plasticizers, anthraquinone, phthalamide, anthranilic acid, phenolphthalein, xanthene dyes, and benzoic acid.

In a manner similar to the oxidation of naphthalene yielding phthalic anhydride, the controlled catalytic partial oxidation of benezene yields maleic anhydride, another cyclic anhydride.

Maleic acid $(COOH\!-\!CH\!=\!CH\!-\!COOH)$ is changed to maleic anhydride by heating. In this reaction the ring is closed by means of an oxygen atom. Such ring formation takes place when the two carboxyl $(-\!COOH)$ groups are separated by two or three carbon atoms. Malonic acid $(COOH\!-\!CH_3COOH)$ forms no anhydride of this type, but merely heating succinic acid $(COOH\!-\!CH_2\!-\!CH_3\!-\!COOH)$ changes it into succinic anhydride, thus forming a ring compound. Maleic anhydride is used to make alkyd resins.

ELBERT C. WEAVER

Cross-references: *Acids, Bases, Oxides.*

ANTIBIOTICS

Definition. Antibiotics are chemical substances produced by microorganisms; in dilute solution, they have the capacity to inhibit the growth of and even to destroy other microorganisms. Numerous attempts have been made, unjustifiably, to broaden the definition of antibiotics so as to include antimicrobial substances of plant and animal origin, on the one hand, and synthetic material, on the other.

Antibiotics are produced by various groups of microorganisms, particularly bacteria, fungi, and actinomycetes. The ability to produce antibiotics is characteristic not of the genus, nor even of the species, but of strains of particular organisms. Among the actinomycetes, as many as 50 per cent of all cultures isolated from the soil or other substrates are able to produce antibiotics when grown on suitable media and under favorable conditions. Identical antibiotics are sometimes produced by different species. Some cultures can produce more than one antibiotic.

Specific media and special conditions of cultivation are required for the production of antibiotics. Most of the media favorable for antibiotic formation contain complex organic substances.

Antibiotics vary greatly in chemical nature, physical properties, selective activity upon bacteria and other microorganisms, toxicity to animals, and *in vivo* activity. The selective action of antibiotics is spoken of as the *antibiotic spectrum.* Some microbes are sensitive to a given antibiotic and others are resistant. The degree of sensitivity is qualitative and quantitative in nature. Some are active largely upon bacteria and some upon fungi, others are active on both bacteria and fungi. Still others are also active upon the so-called large viruses. Some are active upon protozoa. Some antibiotics have found extensive application as chemotherapeutic agents in the treatment of numerous infectious diseases in man, animals, and plants.

Historical background. The ability of certain microorganisms, especially bacteria and fungi, to inhibit the growth of other microbes has been known since the latter part of the last century. The activity of green molds, belonging to the genus Penicillium, upon various bacteria was first demonstrated in 1874. This is also true of the effect of various non-spore-forming and spore-forming bacteria upon other bacteria, including disease-producing organisms. Attempts were made to utilize the products of such microorganisms in the treatment of various infectious diseases, including tuberculosis. A degree of success was attained with certain bacterial preparations, such as pyocyanase, and with certain mold preparations. The results were not sufficiently clear, however, to justify broad generalizations.

In 1939, R. Dubos isolated from a spore-forming soil bacillus certain polypeptides, described as gramicidin and tyrocidine (tyrothricin), which were found to possess remarkable antimicrobial properties and were active both *in vitro* and *in vivo.* These could be used in the treatment of certain infectious diseases. This discovery was soon followed (1940) by Florey and Chain's study of the formation of penicillin (originally observed and named by Fleming in 1928) by a culture of *Penicillium notatum,* and Waksman and Woodruff's isolation, the same year, of actinomycin, the first true antibiotic produced by a culture of an actinomyces.

The three major groups of antibiotic-producing organisms were thus recognized. These have found extensive application in the treatment of numerous infectious diseases. Their discovery has led to tremendous developments in the field of chemotherapy and other antibiotics soon isolated caused a revolution in medical practice. They have also found extensive application in veterinary medicine, in the treatment of certain plant diseases, in animal feeding, and in the preservation of various biological materials.

These antimicrobial substances were first designated as lysins, toxins, antibacterial agents, bacteriostatic and bactericidal substances, lethal or staling principles. The word *antibiotic* was first used in this sense by Waksman in 1942.

Production. The most important groups of antibiotic-producing organisms are the bacteria, lower fungi or molds, and actinomycetes. Members of the genus Bacillus, among the spore-forming bacteria, and of the genus Pseudomonas, among the non-spore formers, are extensive producers of antibiotics. Among the fungi, the genera Penicilium, Fusarium, and Cephalosporium are particularly important; the ability to produce penicillin belongs to a few strains of certain species of these genera. Other strains and even other groups of molds may be able to produce penicillin-like substances, but the yields are so low that they cannot be considered on a par with the important commercial producers. Formation of streptomycin and of other important antibiotics is characteristic of certain species of Streptomyces.

Some microbes are capable of forming more than one antibiotic. *Pseudomonas aeruginosa* produces pyocyanase, pyocyanin, pyolipic acid, and certain pyo-compounds. *Streptomyces griseus* produces not only the antibacterial substances streptomycin, mannosidostreptomycin, and grisein, but also the antifungal substances actidione and candicidin, the antiprotozoan agent streptocin, and certain other antibiotics. Another actinomyces, *Streptomyces fradiae,* produces several forms of neomycin, and fradicin. *Streptomyces rimosus* produces oxytetracycline and rimocidin.

Some antibiotics have been modified chemically to yield substances with somewhat different, perhaps more desirable, properties, as in the formation of dihydrostreptomycin from streptomycin, the semi-synthetic penicillins, and cephalosporines. Only a few antibiotics, notably penicillin and chloramphenicol, have been synthesized, chemically.

Most of the antibiotics formed by bacteria are polypeptides. The most important of these are tyrothricin, bacitracin, subtilin, nisin and polymyxin. The fungi have yielded the penicillins, cephalosporines and griseofulvin. The other antibiotics of fungi include clavacin (claviformin, patulin), citrinin, viridin, fumagillin, and gladiolic acid. Actinomycetes are now known to produce nearly 500 compounds or preparations that possess remarkable antimicrobial properties of which 75 have found practical applications. These include the actinomycins, the streptomycins, chloramphenicol, the neomycins, the tetracyclines, erythromycin, carbomycin, novobiocin, lincomycin, cleomycin, viomycin, trichomycin, nystatin, candicidin, and numerous others.

Isolation. Once an organism has been selected for the production of a particular antibiotic, the next step comprises the development of suitable media and of conditions favorable for growth and the production of the antibiotic. The antibiotic-producing culture is transferred through a series of stages from the test tube to small fermenters and finally to large tanks. The final stage of growth usually lasts 2 to 4 days, when the concentration of the antibiotic reaches a maximum. The assay of the concentration of the antibiotic is carried out by suitable microbiological or chemical methods against a known standard. When the antibiotic has finally been isolated and purified, a registered standard is established by the Antibiotics Control Division of the Food and Drug Administration in Washington.

Chemical Nature. Antibiotics vary greatly in chemical composition. On the basis of their elementary chemical composition they have been grouped as follows:

I. Compounds containing C, H, and O, such as clavacin ($C_7H_6O_4$).

II. Compounds containing C, H, O, and N. These include such compounds as pyocyanin ($C_{13}H_{10}O_2N_2$), streptomycin ($C_{21}H_{39}O_{12}N_7$), and actinomycin ($C_{41}H_{56}O_{11}N_8$).

III. Compounds containing C, H, O, N, and S. Here belong the penicillins (R—$C_9H_{11}N_2O_4S$).

IV. Compounds containing chlorine, including chloramphenicol ($C_{11}H_{12}N_2O_5Cl_2$) and chlortetracycline ($C_{22}H_{23}N_2O_8Cl$). They vary greatly in chemical structure. New types of chemical compounds never known before have been found among the antibiotics.

Antimicrobial Activities. Antibiotics vary greatly in their ability to act upon different groups of microorganisms. Some are active only upon bacteria and actinomycetes but not on fungi. Others are active only upon fungi but not upon bacteria and actinomycetes. Some have a very wide spectrum, being active against bacteria as well as upon fungi, or upon bacteria, rickettsiae, and other intracellular parasites. Others, like viomycin, have a very narrow spectrum, being active only upon mycobacteria.

The antimicrobial spectrum is of great importance in characterizing an antibiotic and in deciding upon its importance as a chemotherapeutic agent. Penicillin is active upon cocci and various gram-positive bacteria and spirochaetes, but it has little activity against the gram-negative rod-shaped bacteria and the acid-fast bacteria. Streptomycin is active against gram-negative and the gram-positive bacteria, including those causing tuberculosis, but not against most fungi and viruses. The tetracyclines are highly active against many bacteria and rickettsiae. Some antibiotics, like candicidin, nystatin, trichomycin, and ascosin, are active primarily against fungi. Certain microorganisms are also capable of producing substances that inhibit the growth of or destroy viruses, including phages and tumor cells. Among the antibiotics active upon neoplasms, one may include the actinomycins, azaserine, sarkomycin, cleomycin and carzinophilia.

Mode of Action and Development of Resistance. Antibiotics are largely bacteriostatic (growth-inhibiting) agents, although some are markedly bactericidal (cell-destroying). Some antibiotics are highly toxic to animals; others, notably penicillin, are relatively nontoxic. Some like tyrothricin, are hemolytic, others are not. Some are readily absorbed from the digestive tract, and others are not.

Upon continued contact with a given antibiotic, originally sensitive bacteria become resistant to it. As a result of this, freshly isolated cultures of micrococci or staphylococci which originally were highly sensitive to penicillin may have gradually become resistant, largely because of the constant use of this antibiotic. *Mycobacterium tuberculosis,* the causative agent of various forms of tuberculosis, originally sensitive to streptomycin, may develop resistance to it when in contact with it or when isolated from patients treated with it. A culture of an organism that has become resistant to one antibiotic may remain sensitive, however, to other antibiotics, or certain synthetic compounds. The problem of transfer of resistance from some bacteria to others has recently received considerable attention.

There is a certain degree of cross resistance among certain groups of antibiotics, such as the tetracyclines. As a result, an organism made resistant to chlortetracycline automatically becomes resistant to oxytetracycline. To overcome the development of resistance, two antibiotics may be combined. A combination of streptomycin and penicillin is used in the treatment of certain mixed infections. Sometimes an antibiotic is combined with a chemical compound, as the sulfa drugs or isoniazid.

Antibiotics as Chemotherapeutic Agents. During the last 30 years several hundred antibiotics have been isolated. Only about 75 of these have found a place as chemotherapeutic agents. These include penicillin, streptomycin, and their semisynthetic forms, chloramphenicol, chlortetracycline, neomycin, oxytetracycline, viomycin, tetracycline, erythromycin, carbomycin, bacitracin, tyrothricin, polymyxin, nystatin, trichomycin, candicidin, and a number of others previously listed.

Some of these antibiotics, like penicillin, are used against a large number of infections caused by various cocci, gram-positive bacteria, and spirochetes. Others, like streptomycin and the tetracyclines, are used against diseases caused by gram-negative bacteria, such as infections of the urinary tract, or those caused by gram-positive bacteria that have become resistant to penicillin. Some antibiotics are specific for certain diseases, such as streptomycin for tuberculosis, chloramphenicol for typhoid fever, the tetracyclines for typhus fever and other rickettsial diseases. Some antibiotics are given by injection, others by mouth. Some are used for skin, eye, and ear infections, others for generalized infections.

As a result of the introduction of antibiotics, most of the diseases of childhood have virtually disappeared, many of the infections of adults have lost their dangerous implications, and the average life expectancy of man has been increased by 10 to 20 years. Tuberculosis, which only 25 or 30 years ago was the number one killer of the human race is gradually coming under control. The effectiveness of antibiotics in the treatment of true virus diseases and tumors is still uncertain.

Many antibiotics cause allergic and other reactions. In some patients streptomycin may cause dizziness, loss of balance, or loss of hearing. This can usually be avoided by reducing the dosage or by combining streptomycin with certain synthetic compounds, such as PAS and INH. The tetracyclines, taken by mouth, may cause nausea, development of fungus infections, and certain other undesirable reactions. These reactions can be largely overcome by switching to another antibiotic.

Antibiotics have found many other applications, principally in the treatment of diseases of animals, such as mastitis in cattle and various infections in poultry; certain bacterial diseases of plants, such as apple and pear blight; feeding of nonherbivorous animals, and the preservation of biological materials, such as virus vaccines and semen of animals.

Manufacture. The manufacture of antibiotics has increased from a few dollars a year in 1940 to a tremendous industry amounting to nearly a billion dollars a year in the U. S. alone. Penicillin is the leading antibiotic, streptomycin takes second place, and the tetracyclines and chloramphenical third place. Neomycin, bacitracin, polymyxin, erythromycin, fumigallin, and tyrothricin are produced on a much smaller scale.

More recently, the semisynthetic penicillins, the cephalosporins, griseofulvin, lincomycin, and several others have come to occupy important places as chemotherapeutic agents. The production abroad is probably as great as, if not greater than, in this country. It has been said that nearly 50 per cent of all prescriptions sold over the counter in drug stores throughout the country are antibiotics.

SELMAN A. WAKSMAN

Cross-references: *Chemotherapy; Antibiotics by Fermentation.*

ANTIBODIES AND ANTIGENS

The name *antibody* implies that a body is acting against an introduced substance. The introduced substance which generates the antibodies is called the *antigen*. Animals have the ability to resist certain infections when they have been previously exposed to them. The existence of antibodies in the blood was originally postulated to account for this immunity to repeated bacterial infection.

Antibodies are serum proteins of the globulin fraction and generally with the physicochemical properties of the γ-globulin of the species producing it. In the rabbit and man they have molecular weights of about 160,000. Antigens are always molecules of at least 10,000 molecular weight and almost always proteins. However, some lipides and polysaccharides can induce antibody formation under certain circumstances. Antigens lose their antigenicity when they are administered orally and, consequently, they are injected parenterally to produce antibodies. In general, the antibodies produced in response to one antigen are specific in that they are capable of reacting only with the homologous antigen or those with great similarities in molecular structure.

By means of this action of the antibody the animal has the capacity to destroy the virus or the bacteria. The principle of antibody response it is hoped, but without success thus far, will be of benefit in the fight against cancer, just as it can be used to destroy the viruses causing polio. The work of Karl Landsteiner is responsible for our understanding of human blood groups which arise from antigenic considerations. This work demonstrated that the specificity of the antibody-antigen reaction is not due to the antigen molecule as a whole, but only to certain chemical groupings. He discovered that an animal can manufacture antibodies with the power to combine with well-defined chemical groups. He used the technique of coupling diazotized amines to proteins. One which has been used extensively to *p*-aminophenylarsonic acid, whose diazonium salt has been coupled to foreign serum proteins. Landsteiner found that antibodies manufactured against the D-tartaric acid grouping do not react with the L form. He also found that proteins coupled to *o-, m-,* and *p*-aminophenylsulfonic acid produce different antibodies.

Not all types of groups are capable of inducing antibodies against them. Only polar groups, such as —COOH, —SO$_3$H, —PO$_3$H$_2$, and quaternary ammonium groups appear to have this capacity. It is probable that antigen specificity arises from a particular configuration of a single chemical group like the above, or a given arrangement of several of them. There may be many such groups incorporated into the natural protein, making up the so-called antigen valence. It is believed that the antibody molecule has one or more regions with structures complementary to these, and consequently, that the forces responsible for combination of antibody with antigen are of the electrostatic, van der Waals, and hydrogen bond type. The antibody may be formed by a unique folding of the peptide chains which makes available certain chemical groupings on the resultant molecule.

Although the site of formation of the normal serum globulins is not known, it is believed that they arise in the reticuloendothelial cells of the liver, bone marrow, and spleen. There have been many theories proposed for the mechanism of formation of the antibody, and it is generally accepted that the antibody is a new globulin, synthesized under the influence of a nearby antigen. Pauling has described a theory in which the antibody is formed from the precursor of the globulin molecule—that is to say, the amino acid sequences are all the same, but the antibody differs in the way in which the molecule is folded. As the folding occurs, two parts of the chain are in direct contact with the antigen molecule and fold in stable configurations which would be different if the antigen were not present. Consequently, Pauling postulates bivalency for antibody molecules. According to this concept it may be possible to manufacture antibody molecules *in vitro*. This has not yet been accomplished.

Immunochemists have studied the formation of the antibody and its reactions with the homologous antigen to gain understanding of the immunity mechanism. As is not uncommon, their efforts have probably contributed as much to the knowledge of the physical and chemical nature of proteins as to their original thesis. Chemists are presently studying the reactions of antibodies with antigens, using for this purpose the earliest tool, the precipitin reaction, as well as light scattering, ultracentrifuge, and electrophoresis techniques.

The precipitin reaction can be studied by observing the results of adding a constant amount of antibody or antiserum to each of a series of test tubes containing increasing amounts of antigen (usually increasing by a factor of 2). The reverse can be accomplished by adding constant amounts of antigen to varying amounts of antibody. If the proper relative amounts are chosen, the former test will show maximum precipitation in the center tube and decreasing to either side. Real inhibition to precipitation has occurred on the antigen excess side of maximum precipitation, while the decrease on the other side may be the result of inhibition due to antibody excess or merely lack of sufficient antigen to cause precipitation. To find out which is the case, one performs the latter test, and if real inhibition exists, precipitation will rise through a maximum and then decrease as the amount of antibody increases. Equine antibodies show a pronounced antibody excess inhibition, whereas rabbit antibodies do not. All systems show antigen excess inhibition. The precipitates can be evaluated to determine the amount of antibody and antigen present. The precipitation and inhibition phenomena have been the subject of a great deal of work in recent years in the attempt to reveal the mechanism of the precipitin reaction. It is hoped that this information will eventually lead to an understanding of antibody manufacture and its duplicating mechanism which may not be unlike those which produce replicas of complex biological molecules, for instance viruses and genes.

Ultracentrifugation and electrophoresis have been used to study the antibody-antigen system in the antigen excess inhibition region. The effects of ionic strength and pH have given indications of great usefulness in determining the nature of the forces involved in the antibody-antigen bond.

RICHARD J. GOLDBERG

Cross-references: *Lipides, Proteins.*

ANTIDOTES

An antidote is an agent which prevents or counteracts the effects of a poison. Since emergency treatment is of paramount importance in many, if not all, cases of acute poisoning, it constitutes an important phase of antidoting. Such emergency treatment, rather than specific antidotes, often prevents or decreases the damage which might be produced by the toxic material and may be a life-saving event.

One of the perplexing problems which confronts the physician is the proper antidoting procedure for many commercial products which may be toxic. It has been estimated by the American Medical Association that there are some 250,000 brand name products which may be potential hazards; many of these carry inadequate information on the container, making proper treatment extremely difficult and often delaying it.

Information about noxious substance(s) in a commercial product may be obtained from a number of sources. To the writer's knowledge, the most recent and most complete information on the noxious substances contained in commercial products and industrial chemicals can be found in "Clinical Toxicology of Commercial Products" by Gleason *et al.* (Williams and Wilkins Co., 1969), and "Dangerous Properties of Industrial Materials," edited by Sax (Reinhold, 1968). A useful pamphlet is "Clinical Handbook on Economic Poisons, Emergency Information for Treating Poisons" by W. J. Hayes, Jr., USHEW, PHS, 1963. However, since new brand names and new chemicals are continually being marketed, no single printed source is completely adequate.

The pharmacist plays an important role as a source for such information. Other important sources of information for toxic substances and their antidotes are the recently established poison centers located throughout the various states. In certain localities, hospitals, consulting laboratories, and pharmacists privately operate a poison information center at no cost. A knowledge of the nearest of these centers might prove extremely useful, since information from such centers is usually available on a 24-hour basis. A unit known as the National Clearinghouse for Poison Control Centers under the Department of Health, Education, and Welfare, in Washington, D.C. may be called upon in certain situations.

Since space does not provide for the listing of diagnostic features and antidotes for even the most important poisons, certain emergency measures designed to save lives and diminish serious damage from noxious materials are outlined below. These measures are essentially those recommended by the American Medical Association Committee on Toxicology and also by leading textbooks of toxicology; they are designed for the layman who has little knowledge of the more complicated and specific antidoting procedures.

Time is of paramount importance in the institution of first aid and follow-up treatment of accidental poisoning; therefore, emergency telephone numbers for physician, pharmacist, hospital, resuscitator and/or rescue squad, etc. should be readily on hand. Although the first aid treatment of the victim should receive immediate attention, some thought should be given to possible detection of source or type of poisoning unless this is obvious. Attending to such matters as saving the poison container or the vomitus may decrease the delay before specific antidotes can be administered.

The measures that can be taken before the arrival of the physician are aimed towards the delay of absorption and/or removal of the poison whether ingested or otherwise applied. It is to be emphasized here that many medications and household products, for example, baby aspirin, waxes and polishes, although not labeled "Poison," are potential hazards, particularly for the young.

(1) Ingested Poisons. In the case of orally ingested poisons, materials may be administered to protect the lining of the gastrointestinal tract and/or to retard the absorption of the poison. Common household ingredients such as, starch, flour, or mashed potatoes, suspended in water may be used for these purposes. Recent research indicates that activated charcoal is the most valuable single agent available for emergency treatment in cases of ingestion of a wide variety of potential poisons. A dose of activated charcoal, no less than five times the dose of the poison, should be suspended in a glass of water and administered as soon as possible after the ingestion of the poison. Activated charcoal *should not* be administered in combination with other substances that may interfere with its adsorptive power (e.g., universal antidote, second edition). Also, it should be pointed out that burnt toast, which has been recommended as a source of charcoal, is not a satisfactory substitute for activated charcoal since the former has no adsorptive capacity. Activated charcoal is of little value in cases of poisoning with caustic alkalies or acids and suggested emergency treatment for these materials is mentioned below.

Vomiting should not be induced if the patient is unconscious or in a comatose state or showing signs of convulsions. In the latter situation, induction of vomiting may precipitate a serious convulsive seizure. If petroleum products or solutions made with petroleum products have been ingested, e.g., kerosene, vomiting should not be induced, since these compounds do not readily stimulate the cough reflex and, hence, may be aspirated into the lungs and produce disastrous results. *In all cases of ingestion of corrosive poisons* which is always accompanied by severe pain and a burning sensation in the mouth and throat region, the induction of vomiting is contraindicated. Violent contractions of the stomach may perforate the stomach. A physician should be called immediately. *If the corrosive is acid* in nature, eg. toilet bowl cleaners and rust removers, and if the patient is not unconscious or comatose, administer milk, milk of magnesia (3–10 ounces) or any non gas-producing mild alkali (do not give sodium bicarbonate) or demulcent. *If the corrosive is alkaline* in nature, e.g., household bleaches, ammonia water, drain cleaners, etc., demulcents such as milk or any mild neutralizing substance such as household vinegar

(3–6 ounces diluted 4 to 1 with water), lemon juice (3–6 ounces) or orange juice (3–10 ounces) should be administered. *In all cases* the patient should be kept warm. *Alcohol* should NEVER be administered.

Vomiting should be induced when noncorrosive poisons have been ingested. This can be done by administering milk, starch suspension, water, etc., and then introducing a blunt instrument or finger (caution) into the mouth and gently stroking the pharynx. Often a glass of warm salt water (2 tablespoonfuls of ordinary table salt in a glass of water) may automatically induce vomiting. When vomiting begins the patient should be placed abdominal side down and the head below the rest of his body. This may prevent the aspiration of the vomitus into the lungs.

(2) Skin Contamination. Certain substances are absorbed through the skin as rapidly as when they are ingested. This is particularly true of organic phosphate compounds (insecticides). All contaminated clothing should be removed immediately and the area flushed copiously with water.

(3) Eye Contamination. The mucous membrane in this region is extremely sensitive to irritation and damage (which may be permanent). The eyelids should be held open and the eye bathed in running water. Permanent damage may be greatly reduced by the rapidity with which this treatment is instituted. *Chemical antidotes* should be avoided. These may produce additional damage by liberating heat, or by producing irritation themselves.

(4) Inhaled Poisons. These poisons cause one form or another of oxygen lack. In order to minimize the danger from hypoxia, the patient, even if able to walk, should be carried out of the contaminated atmosphere. The patient should be kept quiet and warm and all clothing should be loosened. Artificial respiration should be instituted if necessary. A resuscitator should be called for immediately. In this connection, everyone should become familiar with the mouth-to-mouth resuscitative procedure which has been shown to be the most effective.

(5) Injected Poisons. (snake bites, etc.). Bites from poisonous snakes are characterized by two puncture wounds. Nonpoisonous snakes show more than two teeth marks—usually a semi-circular row. Keep the patient quiet. Apply a tourniquet between the bite and the heart. The pulse below the tourniquet should not disappear; a sensation of throbbing should not be felt. Loosen for one minute periods every fifteen minutes. *Do not* immerse the limb in cold water or ice since this treatment has recently been shown to be detrimental to recovery from snake bite. Carry, *do not walk* the patient to a place where medical attention can be obtained. Incision and suction should not be performed unless local reactions are severe and develop promptly. *The administration of alcohol is absolutely contraindicated.*

(6) Burns. If the burn is from a chemical source (i.e. not from a flame) wash immediately with copious amounts of water. An exception to this are burns produced by phosphorus. The burned area should be loosely covered with clean cloth. Do not apply *any medication* since quite often burns may be contaminated with debris which is more

difficultly removed after application of creams, ointments, and the like. Such first aid measures can actually be detrimental. Keep the patient warm and transport him carefully to the hospital or await the physician.

Medication for specific antidoting usually is not immediately available. Even if they are available, these antidotes may do more harm than good if improperly administered. Hence, emergency measures become extremely important. Specific antidoting depends on the knowledge of the poison and its effectiveness often depends on how quickly correct emergency measures are instituted.

J. H. MENNEAR

Cross-reference: *Toxicology.*

ANTIFERTILITY AGENTS

The increased awareness in the past decade of the geometrical increase of the world's population sharply focused attention on the need for a simple and reliable biochemical means for the control of conception. Studies in the endocrinology of reproduction had shown that the steroid hormones progesterone and estrogens such as estradiol were intimately involved in the regulation of the human menstrual cycle.

Experiments in animals suggested that administration of progesterone and estradiol could inhibit ovulation, and thus in effect prevent conception. These findings were, however, of limited utility since neither of these agents is sufficiently active orally in the human. The development of processes for obtaining steroids from abundant natural sources such as Mexican yam roots and soybean sterols led to the preparation of a staggering array of synthetically modified steroid hormones. Biological evaluation of these analogues revealed potent orally active progestins such as norethynodrel, norethindrone and medroxyprogesterone acetate and estrogens such as ethynylestradiol.

Work at the Worcester Foundation, spurred largely by Dr. G. Pincus and Dr. J. Rock of Harvard, showed that a combination of norethynodrel and the 3-methyl ether of ethynylestradiol (prepared in the laboratories of Searle) was effective in inhibiting ovulation. There then followed a series of extensive human trials of this combination in which it was found to be close to 100% effective as an oral contraceptive. In 1960 sale of this combination as a contraceptive was approved by the FDA. Subsequently, many other such agents were developed for use as oral contraceptives; there are today sold in the U. S. about a dozen drugs which are combinations of either ethynylestradiol or its methyl ether with a variety of synthetic steroidal progestins. As currently employed, these agents are taken by the woman daily from the 5th to the 25th day of the menstrual cycle. Ovulation which would normally occur in the vicinity of the 14th day is suppressed. Soon after discontinuance of treatment on the 25th day, menstrual bleeding takes place.

The growing suspicion that the estrogenic component of the "pill" was responsible for the antiovulatory action led to development of the "sequential" contraceptives. With these, the woman typically takes a tablet consisting of only estrogen for the first twenty days of the cycle and a combination of estrogen and progestin for the last five days. The effects are much the same as above.

Considerable further work is underway in this area, spurred particularly by unease over potential problems arising from chronic administration of hormonal agents. Especially noteworthy are recent reports of the efficacy of prostaglandins—a series of ubiquitous biologically active fatty acids isolated from mammalian tissues—in bringing about abortions in early pregnancy. Reports have also appeared of other nonsteroid agents active as antifertility agents in female as well as male laboratory animals.

DANIEL LEDNICER

References

"The Control of Fertility," G. Pincus, Academic Press, N. Y. 1965.
"Contraception: The Chemical Control of Fertility," D. Lednicer, Ed., Marcel Dekker, N. Y. 1969.

Cross-reference: *Prostaglandins*

ANTIFOULING AGENTS

Nature of Fouling. Fouling is defined here as an accumulation of either macro- or microorganisms on submerged surfaces. The macroorganisms are frequently referred to, in a general way, as barnacles. They include other species of molluscs such as mussels and oysters, and such other organisms as encrusting and filamentous bryozoa, annelids, tubularia, and tunicates. Coral deposits and calcareous and siliceous sponges also are encountered. Sea grasses and algae are most common in areas reached by sunlight near the surface. The microorganisms appear as slimes which frequently precede the larger fouling organisms.

Practical Importance. Fouling on ships' hulls greatly increases "drag" and must be controlled to avoid excessive consumption of fuel to maintain speed. It is necessary, also, to control fouling of intake tunnels for power plants. Fouling as slimes can interfere with heat transfer from either salt water or fresh water.

Accumulations of marine growths may be detrimental to the performance of underwater vessels by increasing their weight and by interfering with the functioning of instruments or other devices.

Fouling organisms such as barnacles can promote pitting by setting up crevices which lead to corrosive concentration cells and active-passive cells on alloys that depend on oxygen induced passivity for their resistance to corrosion.

Heavy accumulations of fouling can reduce general corrosion by acting as barriers to diffusion of oxygen to corroding surfaces. They also can eliminate aggravating effects of high-velocity flow by reducing the velocity to zero at the surface under the organisms.

The barrier effect of the accumulated macroorganisms is offset, to some degree, by the corrosive effect of microorganisms such as sulfate reducing bacteria that can thrive in the anaerobic conditions that prevail under the heavy growths.

Hydrogen sulfide resulting from the decay of fouling organisms can promote pitting and other corrosion of metals.

Fouling Conditions. Fouling can occur throughout the natural range of temperature encountered in rivers and oceans. A principal difference between frigid and tropical waters is in the duration of the seasons in which fouling can occur and in the number of species that may be involved. This leads to heavier growths of more organisms in the warmer waters.

Fouling occurs at great depths but is relatively sparse in terms of species and extent, in line with the low oxygen content and low temperature of very deep waters. Reduction of oxygen by organic pollution in surface waters restrains fouling.

Fouling organisms become attached to surfaces while in the embryonic stage. There is a finite time during which the embryonic organisms must find a home. Consequently, fouling occurs to the greatest extent near shore where the sources of embryo and surfaces to which they may become attached are greatest.

A relative velocity of movement in excess of 2 to 3 ft. per sec. will prevent attachment of fouling organisms to smooth surfaces. However, once attached during periods of rest, the organisms can adhere and flourish at very high velocities. Temperatures much in excess of the customary ambient temperatures will prevent the attachment and growth of fouling organisms. A temperature rise of as little as 20°F for as little as one hour once a week will restrain fouling.

Control of Fouling. Copper and its alloys which can release copper ions in corrosion products to the extent of about 5 mg of copper per sq. decimeter of surface area per 24 hours can be expected to remain free from fouling.

Unalloyed copper, brasses, bronzes, other than aluminum bronzes, and copper-nickel alloys containing up to 30% nickel without modification by iron over about 0.05%, or up to 10% nickel with the addition of iron, e.g., 1.5%, have useful antifouling properties. Galvanic effects or applied cathodic protection which suppress corrosion of copper and copper alloys will permit them to foul.

Silver also has antifouling characteristics and zinc often remains free from fouling for several weeks or months.

Other metals and alloys, wood, concrete, elastomers and plastics can be expected to become fouled.

It has been observed in experiments with plastics that the accumulation of fouling organisms is delayed considerably on the lighter colored surfaces, this effect being most noticeable on clear and white surfaces.

Preventive Measures. The most common antifouling coatings take advantage of the antifouling characteristics of copper used as such in the form of powder or more often as cuprous oxide in appropriate vehicles. To extend the protective effect there must be a proper balance between the rate of degradation of the vehicle and the rate of release of copper from the pigment so that the source of copper ions will not be exhausted too fast or that the surface layers will become completely leached before underlying sources of copper can become exposed by removal of the exhausted outer layer.

Mercury compounds are sometimes added to antifouling coatings to increase their effectiveness towards algae and grasses that occur near the water surface.

Organic tin compounds have been found to be effective in antifouling coatings. They are particularly desirable for use over aluminum to avoid the corrosive effect of copper leached from a cuprous antifouling coating.

When coatings containing copper or cuprous oxide are used on metals they must be applied over an effective protective coating system to avoid acceleration of corrosion by the copper from the antifouling coating. Coatings containing mercury must not be used over aluminum or copper alloys subject to stress corrosion cracking by mercury.

Fouling of intake tunnels and heat exchanger surfaces can be controlled by treatment of the water with chlorine. As little as 0.5 ppm of available chlorine as measured with ortho-tolidene will suffice. Tests have shown that continuous treatment to achieve such a level of chlorine is likely to be more effective than intermittent dosages at higher levels.

Agitation as with air bubbles and high-frequency vibrations will also restrain fouling. Periodic heating of surfaces of the water can also be used effectively.

Applications of direct or alternating currents through a wide range of voltages and frequencies have not been found useful beyond effects of chlorine generated by electrolysis at anodes.

For an extensive discussion of this subject see "Marine Fouling and its Prevention" prepared by the Woods Hole Oceanographic Institution and published by the United States Naval Institute, Annapolis, Maryland.

FRANK L. LaQue

ANTIFREEZE AGENTS

Water is an excellent heat-transfer agent since it has a high specific heat and is readily available at virtually no cost. However, it has disadvantages: namely, a high freezing point, about a 9% expansion in volume on freezing, and corrosive action toward common metals.

The 32°F freezing point of water is lowered by the addition of various compounds. The extent of the lowering is generally indicated by Raoult's law. Thus, it is possible to employ an aqueous solution as a coolant below the freezing point of water. These solutions serve as means of transferring heat from one part of a system to another without a change of state over the operating temperature of the system. The properties of the antifreeze agent-water system determine the operating temperature range for a particular aqueous solution. The critical properties of the system are freezing point, boiling point, chemical stability, specific heat and viscosity. The most important of these are the freezing point and boiling point composition curves, since these establish the operating temperature range of a particular solution.

The earliest antifreeze agents were aqueous brine solutions such as sodium chloride, magnesium chloride and calcium chloride. The eutectic tempera-

FREEZING PROTECTION PROVIDED BY HYDROXY ANTIFREEZE AGENTS

	Molecular Weight	Conc. for 0°F Protection	Max. Usable Concentration	Protection Provided
Methanol	32	27% (by vol.)	50% (by vol.)	−54°F
Ethanol	46	34	72	−50
Isopropanol	60	42	81	−50
Ethylene glycol	62	33	58	−56
Proylene glycol	76	36	52	−32
Glycerine	92	42	65	−42

ture for aqueous sodium chloride solutions (23%) is given as −6°F while the corresponding value for magnesium chloride solution(21%) is −22°F. The eutectic temperature for calcium chloride solutions(30%) is −50°F. However −40°F is about the lowest practical operating temperature, since the brine solutions tend to precipitate solid salt if the concentration approaches the eutectic value. The precipitated solid will clog pipes and pumps with a subsequent decrease in heat transfer.

A shortcoming of the low-cost salt solutions, which seriously limits their use, is the corrosive action on metals due to the electrolytic nature of the aqueous solutions. Both calcium and magnesium chlorides undergo hydrolysis resulting in an acid reaction, which increases the corrosiveness of these salts. Further, unless deionized water is used in the preparation of the calcium salt brines, a precipitate will be formed by the components of hard water. While sodium chloride does not hydrolyze, it is a highly corrosive material. The brines may be inhibited partially to reduce their corrosive nature in laboratory corrosion tests, but they still present very serious corrosion problems in actual use. Sodium dichromate is commonly used as a brine inhibitor. Also, inorganic phosphate salts such as tripolyphosphate have been used. However, chemical corrosion inhibitors have been ineffective in decreasing the corrosion of brine solutions on common metals.

The most useful antifreeze agents consist of organic compounds such as mono(lower boiling aliphatic alcohols), di(ethylene, propylene glycols) and polyhydroxy(glycerine, sugar) derivatives. The effectiveness of a specific antifreeze agent is inversely related to its molecular weight, since the lowering of the freezing point of water is determined by the number of moles of antifreeze agent present in a given solution. The following table gives the molecular weight, the amount of antifreeze agent required to lower the freezing point of water to 0°F and the maximum usable concentration for several common antifreeze agents. These data rate the antifreezes on their ability to decrease the freezing point of water. Methanol is the most effective, while glycerine is the least efficient. Of the higher-boiling di- and poly- derivatives, ethylene glycol is the most effective and is widely used to protect water in automotive cooling systems.

These aqueous solutions will not precipitate crystalline salt deposits at low temperatures. Certain glycol solutions may even be pumped through pipes at temperatures 10°F below their freezing points without mechanical damage to the system. Higher molecular weight polyhydroxy derivatives (cane sugar, honey and glucose) have been used, but their high molecular weights make them inefficient depressants. Further, they are not thermally stable and their aqueous solutions are too viscous for satisfactory pumping.

In general, the pure hydroxy antifreeze agents are flammable or combustible, unlike inorganic solid salts. However, aqueous solutions of the di- and polyhydroxy derivatives are nonflammable at normal operating temperatures. Aqueous solutions with boiling points lower than water (methanol, ethanol and isopropanol) tend to lose the antifreeze agent when exposed to overheating; aqueous solutions of polyhydroxy derivatives (glycols, glycerine) will only lose water, leaving the freezing point depressant in solution. The latter class is known as permanent or nonvolatile antifreeze, while the former is referred to as the volatile type. The table below compares the boiling points of aqueous solutions of both types at one and two atmospheres pressure.

With the higher horsepower engines now in use, more heat must be dissipated through the cooling system. This is accomplished by using the higher-boiling ethylene glycol and raising the boiling point by installing a 13–17 psi pressure cap on the radiator. The volatile type coolant is practically eliminated in a pressure system because of its low boiling point.

Antifreeze Additives

The hydroxy-containing antifreeze agents are relatively noncorrosive to the common metals, but when dissolved in water, the solutions are corrosive and inhibitors are needed. This is particularly true in the automotive cooling system because of multi-metal combinations, consisting of copper,

BOILING POINTS OF AQUEOUS SOLUTIONS OF ANTIFREEZE AGENTS

	Methanol		Ethylene Glycol		
Concentration (% by vol.)	33	50	33	50	58
1 Atmosphere	181°F	171°F	220°F	227°F	232°F
2 Atmospheres	226	216	265	272	277

solder, brass (radiator metals); steel, cast iron and aluminum alloy (engine block metals), and high temperature and high coolant flow rate. More than one corrosion inhibitor is required to provide protection to the cooling system. In general, about 1–4% of the formulated concentrate consists of corrosion inhibitors. These may be inorganic or organic compounds. For the nonvolatile product there are two general types of inhibitor systems, single or two phase. The volatile products are usually produced in the single phase form although some additives may form a second phase on dilution with water. The mono phase inhibitor system which is completely soluble in the glycol or alcohol, consist of borates, phosphates, nitrites, arsenites, molybdates, sodium mercaptobenzothiazole and tolutriazole. In general, these compounds protect the metal surfaces by forming a protective film. A two-phase inhibitor system usually contains an oil phase of about 1–2% (by vol.) and consists of sulfonated oil, mineral or vegetable oil derivatives. These hydrophobic materials protect the metallic surfaces by coating them with an oil film, although there is always the possibility that the film may not be uniformly distributed throughout the cooling system.

Aqueous antifreeze solutions have a tendency to foam in cooling systems which have not been properly maintained (air or exhaust gas leaks; rusty or dirty system). Therefore, antifreeze products usually contain a defoamer such as silicones, higher aliphatic alcohols, organic phosphates or polyglycols at low concentration (0.2%).

Antifreeze solutions creep or seep from under radiator hose, gaskets, small cracks, etc. more readily than water because of the difference in surface properties of the hydroxy derivatives. This is especially true in a pressurized cooling system. Therefore, the concentrates contain an anti-leak additive, consisting of a petroleum base derivative, which functions by preferentially wetting the metal surface. Recently solid plugging agents have been added to the nonvolatile concentrate. These consist of polystyrene polymer with particle size varying from less than 1 micron up to 500 microns, powdered ginger root or asbestos.

The various antifreeze manufacturers have developed additive combinations which produce a satisfactory product. Since the additive systems vary between products, it is not advisable to mix them because the additives can react and nullify each other. Also this may result in the formation of an undesirable precipitate.

Selection and Use of Antifreeze

The glycol and methanol type antifreezes are usually available, but for some time the trend has been toward the use of a glycol-type coolant. This product now comprises about 99% of an annual multimillion gallon antifreeze market. The preference for ethylene glycol is due to its high boiling point and superior corrosion characteristics. Today's automotive engines require ethylene glycol solutions to operate efficiently for twelve months. The higher boiling point is especially important for cars with air conditioning units to prevent boil over in the summer months. Car manufacturers are installing ethylene glycol antifreeze on the assembly lines at concentrations of 44–55% (by vol.) which is sufficient to prevent freezing in most sections of the country. These coolants have a service life of at least one year. Some are even recommended for two years. These coolants contain an additive system which enables the antifreeze to function for a specified service period. Car manufacturers' specifications for these long-life products require the coolant to pass various simulated service tests such as circulating radiator, cavitation-erosion, engine-dynamometer and vehicle tests.

With these extended-life coolants it is necessary to check the concentration of the antifreeze agent over the life of the coolant. This may be done by means of a specific gravity hydrometer. Charts giving the freezing points as a function of specific gravity are available for ethylene glycol and methanol aqueous solutions. An inexpensive refractometer, giving the freezing point of ethylene glycol solutions directly, is the best of the field testers.

Other Applications

Brine solutions are still frequently used as refrigerents in ice skating rinks, plant processing and snow-ice melting equipment. Hydroxy compounds (glycols) are also employed for the same purpose to avoid corrosion. A glycol solution is used to remove ice, formed by passage of moisture over the cooling coils of large industrial air conditioning units. In food processing plants, such as dairies, breweries and packaging plants, inhibited propylene glycol is preferred because of its low toxicity. Antifreeze agents are added to prevent freezing and possible destruction of emulsions, used in pharmaceutical and cosmetic creams.

Aqueous nonflammable industrial hydraulic fluids for hydraulic lifts, die casting machines, ingot manipulators, etc. contain glycols to depress the freezing point of the water. A somewhat similar composition operates the catapult mechanisms on aircraft carriers.

Several deicing compositions have been developed for removal of ice and frost from such exposed surfaces of aircraft as leading edges of the wings, fuselage, etc. They consist of inhibited alcohols or alcohol-glycol combinations. The problem has become more acute with long flights at high altitudes when the humidity may vary through a wide range. The deicing fluid should not only melt ice but also prevent its formation for several hours. These deicers consist of a blend of ethylene and propylene glycols, wetting agent and corrosion inhibitor. Automotive deicing fluids in aerosol containers are widely used to remove ice from windshields. These products consist of a blend of lower aliphatic alcohols (ethanol, propanol or isopropanol), glycols (ethylene or propylene), water, wetting agent and corrosion inhibitor.

Antifreeze compounds have been added to gasoline to prevent freezing of condensed moisture during the winter. These usually are methanol, isopropanol and ethanol. Some petroleum refiners add isopropanol to the gasoline at the refinery for the same reason.

Cities utilize solid inorganic salts, such as sodium chloride and calcium chloride, to melt ice and snow on city streets. They are very effective, but are highly corrosive to vehicles. Chemical inhibitors have not been successful.

CHESTER M. WHITE

References

Anderson, "Consumer Chemical Specialties," *C&E News*, p. 37, Dec. 12, 1969.

ASTM Standards, "Engine Antifreezes," Part 22 (1972) American Society for Testing and Materials, Philadelphia, Pa.

Bergman, "Corrosion Inhibitors," Chapters 4 and 3, The Macmillan Co., New York, N. Y., 1963.

Howard et al., "Automotive Antifreezes," U. S. Nat. Bur. of Stds. Circ. #576, U. S. Dept. Commerce, Washington, D. C. (1956).

Mellan, "Polyhydric Alcohols," Spartan Books, Washington, D.C., 1962.

ANTIHISTAMINES

The powerful physiologically active compound histamine is present in an inactive form in most tissues of the body. It is liberated by antigen-antibody reactions, causing the symptoms of allergic reaction, which may vary from nasal rhinitis to anaphylactic shock. Histamine may also be responsible for symptoms of other pathological conditions, such as the inflammation of burns and radiation sickness, and pregnancy toxemias. In 1937, Bovet and Staub, at the Pasteur Institute, reported that certain compounds, including the ethylaniline derivative F-1571, protected guinea pigs against lethal

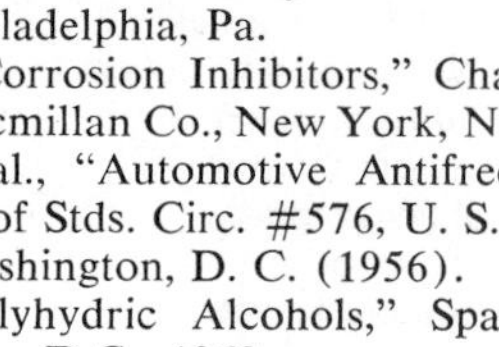

$$HC\!\!=\!\!=\!\!CCH_2CH_2NH_2$$

Histamine

doses of histamine, and also had a protective action in anaphylactic shock. These compounds were too toxic for human use, but the discovery stimulated a great deal of research, first in France and, following the war, in the United States, for new compounds having specific antihistaminic action. Bovet was awarded the Nobel prize in 1957 for his work in chemotherapy, particularly on the antihistaminic compounds.

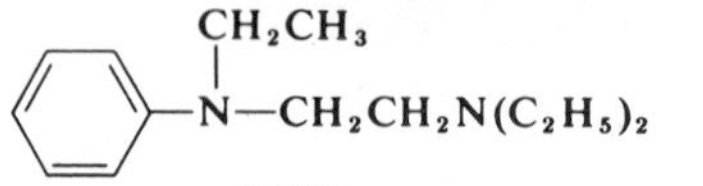

F-1571

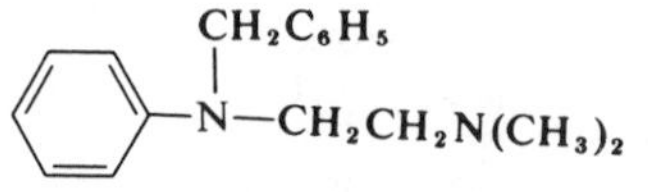

(RP-2339)Antergan

In 1942, Halpern, of the Rhone-Poulenc laboratories, reported on a series of compounds related to F-1571 which included the first compound used

in clinical work, Antergan. This compound was nearly 100 times as active as F-1571 against histamine-induced bronchiospasm, but caused gastric distress, including nausea, emesis and diarrhea. Shortly thereafter, the benzhydrol ether, Diphenhydramine (Benadryl), was introduced in the United States. This drug was about equal to Antergan in antihistamine action, but proved to be a potent sedative in some cases, making its administration to ambulatory patients, especially operators of motor vehicles, dangerous. The high incidence of side reactions showed a need for a better

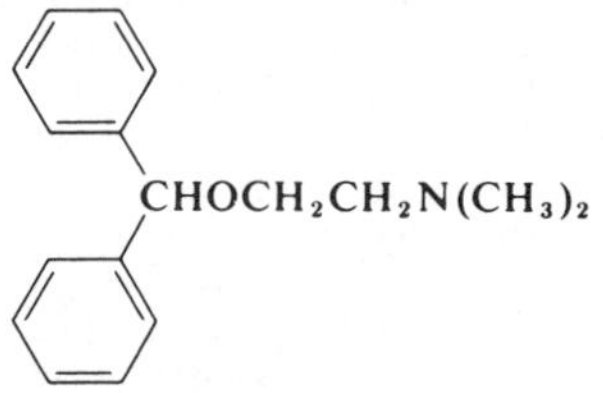

$$CHOCH_2CH_2N(CH_3)_2$$

Diphenhydramine

drug, and stimulated further research which produced hundreds of compounds, among which were several having activities ten or more times that of Antergan or Diphenhydramine, or more than one thousand times that of Bovet's original discoveries. Presently, the U. S. Pharmacopea and New and Non-Official Remedies list about twenty different antihistaminic drugs, each with its uses and specific side reactions. The wide variation of susceptibility to the side effects among individuals has made it desirable for the physician to have a choice of drugs in this field.

Most of the active compounds in this group are closely related to a general formula of ethylamine derivatives, having large bulky groups at one end. In this formula, Ar is usually phenyl or pyridyl, R is commonly an aromatic, aromatic methyl or

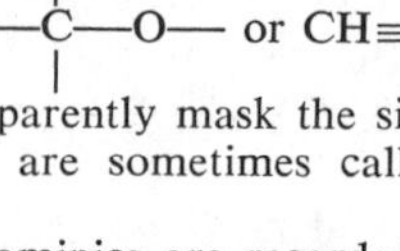

$$\begin{array}{c} R \quad\quad\quad R' \\ Z\!-\!C\!-\!C\!-\!N \\ Ar \quad\quad\quad R' \end{array}$$

heterocyclic methyl group, while R′ is usually methyl but may be more complex, as in some heterocyclic systems. The trivalent function Z may be nitrogen, $H\!-\!C\!-\!O\!-$ or $CH\!\equiv$. Compounds of this class apparently mask the site of action of histamine, and are sometimes called "umbrella" molecules.

True antihistaminics are regarded as substances which specifically antagonize the action of histamine, without other pharmacological response. Thus epinephrine and its analogs, although having opposite effects to histamine, are not antihistaminics. True antagonism has been demonstrated for several of the clinically useful antihistaminic drugs by the fact that the Law of Mass Action is obeyed when various concentrations of the drug are plotted against concentrations of histamine required to maintain a constant response in test systems, such as intestinal strips. The inhibition ratios are different in different systems, but the dose-response curves follow the form of the Langmuir adsorption isotherm in all cases. This indicates competition for a receptor-substance, which may be histaminase. At any rate, antihistamines do not inactivate histamine *in vitro,* nor do they prevent the antigen-antibody reaction *in vitro.* They do not prevent the release of histamine during anaphylactic shock. It should be noted that rather diverse structures exhibit the competitive inhibition of histamine action, in contrast to other metabolite-antimetabolite systems, such as *para*-aminobenzoic acid-sulfanilamide, where the structural requirements of the antimetabolite are very exact.

The principal use of antihistaminic drugs is for their therapeutic effect on nasal allergies. Relief is greatest from mild attacks of hay fever evidenced by sneezing symptoms. These agents are also useful in prevention and treatment of systemic allergic reactions, such as those caused by the use of penicillin, the sulfonamides, or other drugs and antibiotics. They have a palliative effect on nasal rhinitis and urticaria, and use is made of the strong sedative properties of some of these drugs in proprietary preparations designed to both relieve hay fever symptoms and induce sleep.

E. CAMPAIGNE

Cross-references: *Antibodies and Antigens; Chemotherapy*

ANTIHYPERTENSIVE AGENTS

The measurement of blood pressure with a stethoscope and occluding arm cuff is an established routine clinical procedure, and yet it was not until the early part of this century that the technique was devised. Subsequent use of the procedure over the passing years has revealed the widespread occurrence of elevated blood pressure and its serious consequences in those afflicted with the disorder. One of the more significant achievements stemming from the quest for more effective therapeutic agents has been the discovery of several substances capable of lowering the blood pressure in human essential hypertension. The following compounds and groups of antihypertensive compounds are discussed in the order of the discovery of their therapeutic usefulness.

Adrenergic Blocking Agents. Historically, substances which impede transmission at the junction between the terminal sympathetic nerve and the blood vessel were the first to be tried in the modern treatment of hypertension. The mechanism of action of these blocking compounds was based on a competition with the norepinephrine liberated by the nerve for a binding site on the receptor substance of the smooth muscle cells of the blood vessels. As a result the vascular muscle relaxed and peripheral resistance decreased. Among the older compounds of this type were ergotoxine and yohimbine, but the former exerted a vasoconstrictor action of its own on the blood vessels and the latter caused central nervous stimulation as an undesirable side effect. Phentolamine, 2[N-(m-hydroxyphenyl)-p-toluidinomethyl] imidazoline and piperoxan, 2-(1-piperidylmethyl)-1,4-benzodioxan, were typical of the more recent synthetic attempts in this direction. They were free of the side effects of

the earlier agents but, unfortunately, they did not block the cardioaccelerator action of the sympathetic nerves. The blood pressure could be lowered, but the concurrently developing reflex speeding of the heart tended to be excessive, caused subjective discomfort, and placed an increased work load on the heart. As a consequence this group of compounds, though of considerable theoretical interest, has not been strikingly successful from the therapeutic standpoint.

Ganglionic Blocking Agents. These derivatives prevented the passage of nervous impulses through autonomic sympathetic ganglia by a competitive antagonism with the acetylcholine liberated by the preganglionic nerves. As a result, the postganglionic nerves were not depolarized by the acetylcholine, the nervous impulses were blocked, and the caliber of the arteries on which they impinged was widened. This group comprised the quaternary ammonium derivatives (tetraethylammonium, hexamethonium, chlorisondamine, pentapyrrolidinium chlorides) and the amines (mecamylamine and pemipidine). These substances were not selective in their action, affecting the autonomic parasympathetic as well as the sympathetic ganglia. Blockade of the parasympathetic segment, unfortunately, led to blurred vision, dry mouth, constipation and difficult urination. The quaternary derivatives especially were irregularly and unpredictably absorbed from the intestine, providing a therapeutic hazard in maintaining a uniform antihypertensive effect. Thus, at times, the control of the hypertension was inadequate while on other occasions it was excessive, leading to postural hypotension characterized by weakness, faintness and dizziness, especially when assuming the erect position after sitting or lying.

The ganglionic blocking agents succeeded clinically where the adrenergic blockers failed, chiefly because they blocked the sympathetic cardioaccelerator fibers at the ganglia and thus prevented reflex tachycardia as the blood pressure fell. Despite their annoying side effects, the ganglionic blocking agents represented the initial laboratory success in the search for a chemical sympathectomy, thus serving to establish control of hypertension by drugs as a feasible procedure.

Veratrum Compounds. The veratrum alkaloids were the next group of substances to be tested as antihypertensive drugs. It had long been known that these compounds lowered the blood pressure in experimental animals, but they had never been employed in humans. Detailed study of their mechanism of action showed desirable characteristics which led to their clinical use in 1950 based on a dual action on peripheral nervous receptors and upon centers in the brain. By the stimulation of nervous receptors in the heart and lungs, impulses were sent to the brain which caused reflex dilation of peripheral vessels, together with a slowing of the heart caused by reduced sympathetic nervous activity. This action produced a significant lowering of the blood pressure. These compounds were free of the side effects of the ganglionic blocking agents but, unfortunately, they caused emesis at a dose very close to that affecting the blood pressure, a finding which has limited the application of these potentially useful hypotensive agents.

Hydralazine. This substance, which is 1-hydrazinophthalazine, caused a blood pressure fall in experimental animals which was more gradual in onset and more sustained than that brought about by ganglionic blocking agents. When administered to humans for the first time in 1952, it was found to lower the blood pressure in essential hypertension. In contrast to the ganglionic blocking compounds, this substance was well absorbed by the oral route and was virtually free of the tendency to cause postural hypotension. Some increase in heart rate was experienced in the human, but on continued administration the rate tended to gradually decrease toward the pretreatment level. The primary site of action of hydralazine was thought to be upon sympathetic centers in the brain, but more refined experiments have shown it to exert also a direct relaxant effect upon the smooth muscle of the arterioles. An interesting aspect of its action was its capacity to dilate the renal blood vessels, thus increasing the blood flow through the kidney. This property was of considerable import since it was known that an embarrassed blood supply to the kidney could lead to hypertension. It was of interest in this connection that hydralazine prevented the damaging effect of renin, a substance produced by the ischemic kidney, upon the vascular system. Hydralazine had adrenergic blocking properties of a moderate degree and also antagonized to some extent the constrictor effect of hypertension on the blood vessels. It is a fair statement that its beneficial action in essential hypertension is based on the fortuitous concurrence of several separate pharmacological actions, which yield an effective chemical sympathectomy.

Reserpine. This alkaloid, known as 2,4,5-trimethoxybenzoyl methyl reserpate, isolated from the root of Rauwolfia serpentina in 1952, differed from the previously studied antihypertensive drugs in possessing a quieting or sedating effect in addition to a subtle hypotensive effect. In both laboratory animals and in man, the central nervous and circulatory effects of reserpine appeared gradually and persisted for some days after discontinuing the use of the drug. The tendency toward a cumulative effect and a prolonged action provided the means for the maintenance of a uniform control of the blood pressure level. Early experiments suggested that reserpine suppressed vasoconstrictor activity by an inhibitory action on the sympathetic center in the midbrain, but later observations suggested that it also blocked the passage of impulses from the peripheral sympathetic nerves to the blood vessels. In fact, actions at both points are probably involved, since it was subsequently shown that the compound had the capacity to release serotonin and norepinephrine from their binding sites in the midbrain and to release norepinephrine from the peripheral sympathetic nerves. Depletion of the available norepinephrine from the sympathetic nerves reduced their effectiveness in maintaining a high level of constrictor tone to the arterioles, since norepinephrine was the neurohumor responsible for the constriction of the blood vessels. Thus, it was generally accepted that the relaxation of the blood vessels and resultant fall in blood pressure was due to depletion by reserpine of norepinephrine from the nervous system. However, there was not

complete agreement as to whether serotonin or norepinephrine release was the factor responsible for the sedative effect of reserpine. As a practical matter, it was quite likely that the sedative effect played some part in the antihypertensive action especially when anxiety was an undesirable complication in the patient with essential hypertension. Reserpine had the definite advantage of slowing the heart and was also practically free of the tendency to cause postural hypotension or to lower the blood pressure in the absence of hypertension. As a matter of fact, when reserpine was administered for about ten days before the start of hydralazine treatment, the speeding of the heart caused by hydralazine alone was prevented. It was thus possible to secure the added effects of the two agents on the blood pressure while their opposing actions on the heart rate were neutralized.

In view of the mechanism of action of reserpine whereby the sympathetic nervous system is literally deprived of its neuroeffector substance, a specific reversible and practically complete chemical sympathectomy can be achieved with reserpine. In actual practice a dosage is selected which produces an optimum rather than a complete sympathectomy.

Syrosingopine (methyl carbethoxysyringoyl reserpate). This synthetic derivative of reserpine, made available for clinical use in 1959, represents an attempt to eliminate the sedative and retain the hypotensive properties or reserpine. In the dog it was found that syrosingopine and reserpine had comparable hypotensive effects while syrosingopine had but one-tenth the sedative action of the latter; syrosingopine caused marked depletion of norepinephrine from both the central and peripheral sympathetic nervous systems but had a considerably reduced tendency to liberate serotonin centrally. Such biochemical evidence served to provide an explanation for the observed pharmacological actions of syrosingopine on the hypothesis that the hypotensive action was due to catecholamine depletion and the reduced sedative effect was due to a reduced effectiveness in liberating bound serotonin.

On this premise, syrosingopine was theoretically capable of effecting an even more specific chemical sympathectomy than reserpine. The results of human clinical studies appear to substantiate such a premise, since less sedation was produced by effective hypotensive doses. Its use would be indicated when sedation was neither needed nor desired.

Guanethidine: [2-(octahydro-1-azocinyl) ethyl] guanidine sulfate. This guanidyl derivative, first employed in the treatment of hypertension in 1959, possessed a novel chemical configuration and a unique inhibitory action on the sympathetic nervous control of the cardiovascular system. Guanethidine released the norepinephrine present in the peripheral sympathetic nerves but, in contrast to reserpine, it did not alter the catecholamine content of the brain or of the adrenal. Thus its antihypertensive effect was not accompanied by any sedation. The characteristic sympathetic nervous inhibition produced by guanethidine was gradual in onset but was prolonged. The fall in blood pressure and the reduction in heart rate appeared about six hours after an adequate oral dose and usually persisted for several days. Interestingly, from a practical therapeutic standpoint, guanethidine produced a considerably greater hypotensive effect in experimentally-produced neurogenic or renal hypertensive dogs than in normal animals. After the initial stage of amine depletion, guanethidine impeded the release of norepinephrine from sympathetic nerves and thus permitted a relaxation of the peripheral arterioles. In contrast to the adrenergic blocking agents such as phentolamine, this agent augmented rather than blocked the effect of injected norepinephrine upon the blood vessels. Since it was not a ganglionic blocking agent, it did not depress the parasympathetic system with attendant blurred vision and dryness of the mouth. It caused some postural hypotension which was usually controlled by proper adjustment of dosage.

From a practical standpoint, the high potency and prolonged action of guanethidine, coupled with a more intense effect in hypertension than in normotension, were the most intriguing aspects of its action.

Part of the antihypertensive action of guanethidine was related to a reduction in cardiac output, but fundamentally its primary action was based on the chemical sympathectomy which it brought about.

Alpha Methyl Dopa. This substance, more properly known as α-methyl-3,4-dihydroxyphenlalanine, was reported to possess sedative, antihypertensive and decarboxylase-inhibiting properties in 1960. Its activity was found to reside solely in the levo isomer. It was originally thought that its blood pressure lowering capacity was related to interference with the decarboxylation of dopa to dopamine, the immediate precursor of norepinephrine. Reduced synthesis of this neurohormone would in turn diminish the vasoconstrictor tone of the peripheral sympathetic nerves and lower the peripheral resistance. Subsequent studies showed, however, that alpha methyl dopa produced an active depletion of norepinephrine from the sympathetic nerves, heart and brain in a manner resembling that of reserpine. Unlike reserpine, alpha methyl dopa did not remove the catecholamines from the adrenal medulla, and, unlike guanethidine, it did remove the catecholamines from the brain stem. The latter effect may explain the sedation following alpha methyl dopa. As in the case of guanethidine and reserpine, the amine-depleting action of alpha methyl dopa was gradual in onset but persistent. Certainly all three compounds are capable of effecting a chemical sympathectomy, and though there are certain differences in their patterns of action, it would seem reasonable to ascribe the similar influence where they exert on the blood pressure to a final common action.

Monoamine Oxidase Inhibitors. Some substances of this type which have been used as psychic stimulants in humans have also been reported to lower the blood pressure in essential hypertension. The most interesting of these is pargyline, N-benzyl-N-methyl-2-propynlamine. The precise manner in which the disturbances of amine metabolism produced by these inhibitors results in hypotension is not clear. The matter is further complicated by the observation that certain potent monoamine oxidase inhibitors are not hypotensive agents. The

final answer must await further research. The hypotensive action of pargyline appears to be due primarily to a reduction in peripheral arterial resistance by an action which is not completely understood.

Diuretics. The organic mercurial diuretics were originally used for the removal of edema fluid by the urine by interfering with the reabsorption of NaCl and water by the kidney tubule. In 1958 it was learned that the mercurial diuretic, mercaptomerin, displayed a definite hypotensive action in patients with essential hypertension even in the absence of edema. When the sulfonamide thiazide diuretics were introduced into therapeutic use, it was found that they also had similar hypotensive properties coupled with very low toxicity. As a result, a number of these, typified by chlorothiazide, 6-chloro-7-sulfamyl-2H-1,2,4-benzothiadiazine-1,1-dioxide, and hydrochlorothiazide, 6-chloro-3,4-dihydro-7-sulfamyl-2H-1,2,4-benzothiadiazine-1,1-dioxide, have found a place in the therapeutic management of hypertension over the past several years.

Clinical Efficacy of Antihypertensive Agents

During the past five years no new type of compound has been made available for the clinical management of hypertension. Rather this has been a period of assessment during which long-term controlled studies have demonstrated that not only is it possible to maintain a reduced blood pressure for many years through the judicious use of drugs, but also that the procedure greatly reduces the medical complications resulting from hypertension, and in so doing improves general well being and increases life expectancy.

One of the more significant long-term studies of this nature was designed by Doctor E. Freis of the Veterans Administration Cooperative Study Group on Antihypertensive Agents. In a well-organized program 143 male hypertensive patients were divided randomly and equally into a placebo control and a drug test group. Freis chose a combination of reserpine, hydralazine and hydrochlorothiazide as the drug to be administered. Blood pressure fell promptly in the drug-treated patients and remained elevated in the control group. Over a four-year period 27 severe complications including strokes, kidney deterioration, coronary attacks, degeneration of the blood vessels of the eyes and progressive elevation of the blood pressure occurred in the control group, while a single case of stroke was the only untoward circulatory involvement noted in the drug-treated group. There were four deaths among the placebo-control group and none in the drug-treated group. From these results and those of other investigators it is apparent that the use of drugs in the management of hypertension—and there are approximately twenty-two million cases in the United States alone—should provide many added years of well being to such patients.

Although the ideal drug or combination has probably not yet been discovered, it is certain that the reward has been worth the effort and has provided a motivation to continue the search for even more effective substances.

ALBERT J. PLUMMER

ANTIKNOCK AGENTS

An antiknock agent is a chemical which, when added in small amounts to gasoline, raises its octane number. An increase in octane number is useful because it enables engines to operate without knock at higher compression ratios, with attendant advantages in thermal efficiency and power output. Although many compounds possess antiknock properties, they all are compounds of rather few elements. These are: nitrogen (as aromatic amines), iodine (elemental or combined), and several metallic or semimetallic elements.

Gasoline composition, methods of test and antiknock concentration affect both the absolute and relative effectiveness of different compounds, even of the same element. Moreover, some compounds can function as an antiknock under some conditions but promote knock under others. Therefore, the following comparison of active elements in effective compounds is representative but subject to considerable variation:

RELATIVE EFFECTIVENESS OF ANTIKNOCKS

Compound	Relative Effectiveness, per Atom of Metal or Nitrogen
Tetraethyllead	100
Methylcyclopentadienyl manganese tricarbonyl	65
Thallium naphthenate	64
Iron carbonyl	43
Copper methylaminomethyleneacetone	40
Nickel carbonyl	30
Cyclopentadienyl cobalt dicarbonyl	28
Diethyltellurium	23
Triethylbismuth	20
Tetrahydronaphthalene chromium tricarbonyl	10
Diethylselenium	6
Tetraethyltin	3
N-methyl aniline	1
Ethyl iodide	1

The effectiveness of an element can vary greatly dependent on the specific compound considered. Some aromatic amines are not antiknocks while others are twice as effective as N-methyl aniline, but nitrogen in other forms is not effective. Most of the metallic compounds that possess antiknock activity have metal-to-carbon bonds, but exceptions exist. Minor variations in organometallic structure may profoundly influence antiknock effectiveness.

Knock is caused by the sudden, spontaneous ignition of the unburned mixture ahead of the flame front advancing through the combustion chamber. Partial oxidation reactions of the hydrocarbons in this mixture produce reactive intermediates that lead to the spontaneous ignition. Reaction of these intermediates with the antiknock agent reduces their ability to propagate the oxidation that leads to knock. In the case of organometallics, it is believed that products released by the thermal decomposition of the organometallic at the appropriate stage of the precombustion process are the active antiknock species.

The most widely known antiknock, tetraethyllead (TEL), has been in commercial use since 1923, and is a component of nearly all automotive and aviation gasolines. Concentrations greater than 4.23 grams Pb per gallon are not currently used, largely for economic reasons. Incremental effectiveness decreases with increased concentration, a characteristic common with antiknock agents. The average TEL usage in present motor gasoline (2.3 g Pb per gallon in regular, and 2.7 g Pb in premium) usually adds from 6 to 10 octane numbers. Tetramethyllead (TML), mixtures of TML and TEL, and the intermediate lead alkyls produced by redistribution of TEL with TML also are commercially used. These higher-volatility alkyls, in comparison to TEL, offer advantages of better vaporization and effectiveness in some multicylinder engines. The choice of alkyl for use depends largely on economic considerations and the relative responsiveness of different fuels to the different alkyls.

To prevent accumulation of lead oxide in engines from the combustion of organolead compounds, commercial antiknock formulations contain organic halides such as ethylene dibromide and ethylene dichloride. On combustion of the fuel, these organic halides enable lead halides to form which, in turn, can vaporize from hot surfaces in the combustion chamber. In aviation gasoline two moles Br are used per mole of TEL (a ratio expressed as "one theory" based on the formation of $PbBr_2$). In most motor gasolines one mole Br and two moles Cl are combined per mole of TEL.

Concentrated lead alkyls are hazardous to handle. Therefore, these antiknock agents are blended with gasolines in refineries under safe conditions, and the resulting leaded gasoline can be handled safely as a motor fuel.

As to the question of restricting the use of lead antiknocks, those favoring restrictions contend that antiknocks will interfere with catalytic devices that may be developed to control automobile emissions, and that the use of antiknocks might be harmful to public health. Those opposing restrictions contend that automobile emissions can be controlled by means that are compatible with antiknocks, that studies made during many years of antiknock use have detected no potential harm to the public health, and that there are important advantages to the use of antiknocks. Such advantages include conservation of crude oil resources, lower cost gasolines, and the prevention of valve wear in engines. It is also asserted that antiknocks permit gasolines to be formulated with lower concentrations of aromatic hydrocarbons, and that this reduces the smog-forming potential of the exhaust products.

Since the development of pi-complex (or coordination) principles of metalorganic chemistry in the 1950's, many such compounds of the transition metals have been patented as antiknocks. Only one, however, has been commercialized—methylcyclopentadienyl manganese tricarbonyl. It may be used in admixture with TEL to provide about 5 wt. per cent of the antiknock metal. The manganese compound not only is an active antiknock alone, but also acts as a synergist with TEL. Other forms of antiknock synergy exist. Some organic acids and esters (such as t-butyl acetate) which pyrolyze easily to acids under precombustion conditions, improve the effectiveness of lead alkyls but, unlike the above manganese compound, have no antiknock action of their own.

The many qualities that a commercial antiknock must possess have prevented other organometallics from reaching commercialization. These requirements include, (1) high effectiveness per unit cost, (2) volatility comparable to that of gasoline, (3) good solubility in gasoline, with resistance to extraction by water or hydrolysis, (4) stability and compatibility with gasoline, and (5) minimum undesirable side effects in engines. The last of these is the principal reason why inexpensive iron carbonyl has not found use as a primary antiknock. On burning, it forms excessive amounts of iron oxides which cause engine wear and greatly reduce spark plug life.

D. A. HIRSCHLER

Cross-references: *Additives, Gasoline.*

ANTIMALARIALS

Malaria is certainly one of the most, if not the most, widespread of human diseases. Because of the outstanding progress made in the early years of the global malaria eradication program coordinated by the World Health Organization and launched in 1955, it was the expectation of malariologists that the disease would be eliminated in the foreseeable future. In the last few years, however, the program appears to be showing diminishing returns. One reason for this could be the increasing appearance of drug-resistant, especially chloroquine-resistant, malaria although obviously the effect of such drug resistance on malaria eradication programs is difficult to assess. For practical purposes the problem of drug resistance is, at this time, confined to *P. falciparum* malaria. Resistance to proguanil was noted as early as 1948 and to pyrimethamine in 1953. Such reports did not cause undue alarm because of the availability of chloroquine to fill the breach. The report in 1961 of chloroquin-resistant malaria in the Magdalena Valley of Colombia dispelled this complacency. This report of abnormal drug response was followed by others from Brazil, Cambodia, South Vietnam, Thailand, the Philippine archipelago and the western coast of Colombia just south of Panama. The problem of malaria became so acute for the U. S. combat forces in South Vietnam that it was affirmed as the single most important medical problem encountered. This crisis resulted in a large U. S. Army program to uncover and develop new and more effective chemotherapeutic and chemoprophylactic antimalarial drugs.

There are four types of human malaria which result, respectively, from infections produced by *P. vivax, P. falciparum, P. malariae* and *P. ovale.* The two first named are by far the most important and most prevalent. In order to understand the chemotherapy of malaria it is necessary to be aware of the various stages in the life cycle of the parasite. Part of the life cycle, an asexual development

(schizogony), occurs in man and another part which is sexual (sporogony) occurs in the mosquito. When an infected female *Anopheles* mosquito bites to obtain a blood meal, sporozoites from the saliva of the mosquito are released into the blood stream of the vertebrate host. In less than an hour the sporozoites are arrested by the liver and disappear entirely from the blood stream. Within the liver an asexual development occurs, probably exclusively in the parenchyma cells. In from one to two weeks the parasites are released into the surrounding tissues and the blood stream where they shortly invade erythrocytes. During this first stage of the life cycle in the liver, the preerythrocytic or primary tissue stage, no symptoms of the disease are apparent. Having invaded the red cell, the nucleus of the parasite undergoes multiple division (schizogony). When division is complete, the red cell ruptures and the parasites, now called merozoites, are released into the blood stream where they invade new red cells in a cyclical process. The mass release of the merozoites along with the cellular debris into the blood stream of the host brings on the fever and other well known symptoms of an attack. The rhythm of the attacks depends upon the particular parasite involved. During this cyclical multiplication sexual forms called gametocytes, both male and female, are also produced. At the same time that the mosquito takes its blood meal and injects sporozoites, it sucks up gametocytes which conjugate in the stomach of the mosquito and after some further development, penetrate the stomach wall and eventually end up as sporozoites in the salivary glands ready to begin the cycle again. With the exception of *P. falciparum,* there is an additional phase to the life cycle of the parasites in man. When the parasites are released from the hepatic cells after the preerythrocytic stage a small number invade other hepatic cells (secondary tissue forms) and begin a second developmental cycle. When this second generation of parasites is released from the hepatic cells the bulk of them invade erythrocytes to begin a second erythrocytic cycle, but again, a small number invade other hepatic cells to begin a third preerythrocytic phase. It is these subsequent cycles that bring on additional overt attacks or relapses; such relapses may occur for many years.

In discussing antimalarial drugs the multiple phases of the life cycle of the parasites must be kept in mind because the various drugs are not equally effective against all of the human malarias nor against all stages of any given species of parasite. For example, because of the absence of a secondary tissue phase in *P. falciparum* infection, a drug capable of destroying asexual erythrocytic forms will effect a cure, whereas with *P. vivax,* the drug will eradicate the immediate clinical disease but not effect a cure because the parasites resulting from the cyclical development of the secondary tissue forms in the liver will periodically reinvade the blood. A drug that is gametocytocidal will, by destroying the sexual forms of the parasite, prevent the spread of malaria by the mosquito vector. Such specific drug activities will be discussed in more detail shortly.

The characteristics of the test systems to which a compound is subjected for evaluation after it has been synthesized by the chemist as a candidate agent are extremely important and the results obtained determine whether the compound will ever progress to the point of testing in man.

Since just before World War I the primary screening of potential antimalarials has been carried out using a variety of avian malarias, for example, *P. gallinaceum* in the chick, *P. lophurae* in the duck, and *P. cathemerium* in the canary. Today avian screening is largely confined to the use of *P. gallinaceum* in chicks. Just over 20 years ago a new rodent malaria, *P. berghei,* was uncovered. The use of this malaria as a screening method has steadily developed. At first, *P. berghei* could be used only to estimate suppressive activity because of technical difficulties in obtaining sporozoite-induced infections. These difficulties have been overcome and today, despite some well-known idiosyncrasies, the superiority of this test system is well established. Regardless of the primary testing, advanced testing is based on the effectiveness of the compound against an infection in a simian host. A strain of the simian plasmodium, *P. cynomolgi,* whose characteristics in many ways parallel those of the human parasite, *P. vivax,* is usually used with the rhesus monkey as the host. The simian parasite *P. knowlesi* is also used to some extent. A recent and outstanding achievement in experimental malaria, which should have profound effects in the area of malaria chemotherapy and the testing of compounds, is the establishment of human malaria (*P. falciparum, P. vivax* and *P. malariae*) in the owl monkey (*Aotus trivergatus*).

The drug Ch'ang Shan, consisting of the powdered roots of the plant Dichroa Febrifuga, Lour. (Saxifragaceae), has been used as a remedy for malaria in China since antiquity. Later, cinchona bark, introduced into Europe about 1640, and subsequently the alkaloid quinine (isolated by Caventon and Pelletier in 1820), constituted essentially the only remedy for the disease until the beginning of WW-II. Indeed, the armamentarium of synthetic antimalarial drugs in use today is largely the result of research carried out shortly before, during and shortly after that war. Although a number of drugs with different structures have been evaluated, pyrimethamine, introduced in 1951, is the last antimalarial of lasting importance that represents a new structural class.

The currently available drugs for the prevention and treatment of malaria are those that have been used for many years. Some modifications in the methods of use of the drugs have evolved in an attempt to cope with emerging parasite resistance. In this respect, there is a trend toward the use of new combinations of established drugs; these have often proved to be very effective. The following is a brief summary of drugs that are presently in general use:

Quinine: Although quinine is not suitable as a routine suppressive it is effective against the asexual blood forms and is useful for emergency treatment in acute attacks. It has gametocytocidal action against *vivax* malaria but not against *falciparum* malaria; it has no action against primary (preerythrocytic) or secondary tissue forms of any of the human parasites.

The 4-aminoquinolines (e.g., chloroquine and amodiaquin) and the related 9-aminoacridine (mepacrin) as a class are highly effective and rapid blood schizontocides. They inhibit development of the gametocytes of *P. malarie* and *P. vivax* but not of *P. falciparum.* They are not effective against either primary or secondary tissue forms. In most countries mepacrine has been replaced by the 4-aminoquinolines which are, today, the most widely used antimalarials and the drugs of choice against susceptible strains.

The 8-aminoquinolines (e.g., primaquin, quinocide and pamaquin) demonstrate a relatively poor effect against asexual blood forms at doses that can be generally tolerated. They are, however, highly active against the primary and secondary tissue forms of *P. vivax* and the preerythrocytic forms of *P. falciparum.* Thus, this class of compounds shows true prophylactic and radical curative activity. However, they are normally not used alone for malaria prevention but rather in combination with a 4-aminoquinoline.

The biguanides (e.g., proguanil and chloroproguanil), *cycloguanil,* and *pyrimethamine* have slow blood schizontocidal activity. They are effective in inhibiting the development of the primary tissue forms of *P. falciparum* but much less effective against these forms of *P. vivax.* These compounds are not gametocytocidal but do interfere with the development of the parasite within the mosquito. It is now generally agreed that the activity of proguanil, N^1-(p-chlorophenyl)-N^5-isopropylbiguanide, can be largely attributed to the metabolite 2,4-diamino-5-(p-chlorophenyl)-6,6-dimethyl-5,6-dihydro-1,3,5-triazine (cycloguanil). The latter compound, as the embonate, has been tested widely in field trials as an injectable, long acting drug.

The sulfonamides (sulfadiazine, sulformetoxine, sulfadimethoxine and sulfalene) and *sulfones* (e.g., diaphenylsulfone) have moderate but slow and incomplete action as blood schizontocides. These drugs are generally not used alone but rather in combination with a potentiating drug such as pyrimethamine. Surprisingly, such a combination has often proved to be effective against pyrimethamine-resistant strains.

The problems associated with drug-resistant malaria have prompted investigations of combinations of drugs; these studies have met with considerable success. The following are some combinations that have been explored:

Chloroquin and primaquin
Chloroquin and pyrimethamine
Chloroquin, pyrimethamine and sulfisoxazole
Quinine and pyrimethamine
Quinine, pyrimethamine and diaphenylsulfone
Pyrimethamine and sulfonamides or sulfones

Although the armamentarium of antimalarial drugs that is now available is almost universally effective, providing adequate facilities are available for their effective application, the increasing loci of drug-resistant *P. falciparum* malaria throughout the world is disquieting and emphasizes the need for continuing research to develop new and more effective antimalarials.

T. R. SWEENEY

References

"Chemotherapy of Malaria," World Health Organization Technical Report Series, No. 375, 1967.

Peters, W., "Chemotherapy and Drug Resistance in Malaria," Academic Press, Inc., New York, 1970.

ANTIMONY AND COMPOUNDS

Antimony with atomic number 51 and atomic weight 121.75 appears in Group VA of the Periodic Table along with nitrogen, phosphorus, arsenic, and bismuth. It is represented by the symbol Sb (L. *Stibium*). Chemically antimony may exhibit a valence of +3, +5, or −3 and for this reason it may be classified as both a nonmetal and a metal.

The use of antimony in the form of the natural occurring sulfide and as the metal has been traced to early Biblical times. In order of terrestrial abundance antimony ranks 58th. More than 100 minerals containing antimony have been observed in nature. While native antimony is found in isolated instances, the metal and its compounds are derived chiefly from the ore stibnite or antimony glance, Sb_2S_3. Complex sulfide ores containing lead (Jamesonite), mercury (Livingstonite), or copper (Tetrahedrite) and oxysulfide type ores are other sources of interest. Small quantities of antimony mineralization disseminated with lead, and copper ore bodies are another important source of recovery.

Antimony is recovered from stibnite ores by first roasting and converting it to a volatile oxide which is then reduced to the metal with carbon. In the case of some rich stibnite ore, it is reduced directly to metal with scrap iron under a sodium sulfide cover which forms a fusible matte with the iron sulfide produced. The antimony in complex sulfide ore is extracted by leaching with sodium sulfide and recovered as the metal by an electrowinning process.

In the smelting of lead ores, the antimony is reduced along with the lead and during the softening operation it may be removed with caustic as a slag from which sodium antimonate is subsequently recovered and reduced to metallic antimony or it may be removed by air oxidation as a slag which is reduced to produce an antimonial lead. The antimony associated with copper ores is recovered as a by-product which is sent to a lead smelter for processing.

Another important source of antimony is from the treatment of antimony bearing scrap. This antimony is referred to as secondary antimony. Lead storage battery scrap is one of the prime sources of secondary antimony. The United States production of secondary antimony normally exceeds that of primary antimony and for the past several years has averaged around 21,000 tons compared to an average of 12,000 tons for primary antimony.

Normally in the elemental form, antimony is a silvery white, brittle, crystalline metal, however, under certain conditions it can exist in an unstable black or yellow allotropic modification.

The physical properties of antimony include: melting point, 630.5°C; boiling point, 1640°C; density 6.618 g/cc (20°C); specific heat, 0.0494 cal/g/°C (20°C); latent heat of fusion, 38.3 cal/g; latent heat of vaporization, 373 cal/g; vapor pressure, $\log_{10} p(mm) = -9871.5/T + 0.051$ (1070–1325°C) where $T = °K$; surface tension 383 dynes/cm (635°C) in H_2; thermal coefficient of expansion 8.5–10.8 micro-in./in./°C depending on crystalline orientation; electrical resistivity, 39.1 microhm-cm (0°C); magnetic susceptibility at 18°C, -99×10^{-6} cgs; thermal conductivity, 0.045 cal/sec/cm²/°C/cm (20°C); Brinell hardness, 42; modulus of elasticity, 11.3; thermal neutron absorption cross section, 5.7 barns/atom (121.75 at. wt.); isotopes 121,123 at. wts.; crystal system-rhombohedral; contraction in volume on solidification, 0.6%.

Metallic antimony is quite stable under ambient exposure. When heated in air metallic antimony will be converted to the oxides Sb_2O_3, Sb_2O_4, Sb_2O_5. When heated with chlorine it will ignite to form $SbCl_3$. Antimony is not readily attacked by hydrofluoric, dilute hydrochloric or sulfuric acids. It will dissolve in a mixture of concentrated hydrochloric and sulfuric acid as well as a mixture of nitric and hydrofluoric acid. Nitric acid will also attack antimony. It combines directly with sulfur, phosphorus, arsenic, tellurium, selenium, and with many metals such as indium, sodium and zinc to form definite compounds. Under ordinary conditions antimony does not react with carbon, silicon, boron and nitrogen.

Chemically antimony, like arsenic and bismuth, will react to form both trivalent and pentavalent compounds with the former being more common. All of the halogens with the exception of bromine form compounds which can exist in either valence state. Most of the halogen compounds are readily hydrolyzed, e.g., $SbCl_3$, will form SbOCl and HCl. Antimony hydride ((stibine), a toxic gas is formed by the hydrolysis or the action of a reducing acid such as hydrochloric on a compound such as Zn_3Sb_2 or AlSb. It is a colorless gas with a garlic-like odor. Stibine will react with solutions of certain metallic ions such as Ag+ to form antimonides. The trioxide of antimony is amphoteric. The trioxide is soluble in concentrated acids, however, only basic salts such as $(SbO)_2SO_4$, $(SbO)NO_3$ are ordinarily crystallized from these acids. Salts containing the basic radical SbO are referred to as antimonyl compounds. Potassium antimonyl tartrate (tartar emetic) is used medically. It is formed by boiling a solution of potassium hydrogen tartrate (cream of tartar) with antimony oxide. Salts of antimonous acid are formed when Sb_2O_3 is dissolved in alkalies. From a sodium hydroxide solution, sodium meta-antimonite may be crystallized. Salts of ortho- and pyroantimonic acids are also known. Many insoluble compounds of antimony are soluble in hydrofluoric acid.

The United States consumption of primary antimony in 1969 was 20,000 tons which was divided almost evenly between metal and oxide. Battery grid metal (3–6% Sb) accounted for 8,000 tons of the primary metal consumption while ammunition, bearing alloys, type metal and smaller applications accounted for 2,000 tons. Of the 24,000 tons of secondary antimony produced, a substantial amount was consumed as antimonial lead for automotive battery use.

The oxide when formulated with chlorine or one of the other halogens finds wide use as a flame retardant agent in plastic materials, and outdoor textiles such as tarpaulins, tents and military supplies. About 6000 tons of antimony oxide are used in flame retardant applications. It is also used as an opacifier in porcelain enamels. When added to glass it acts as a decolorizer and also prevents a change in the color of the glass when exposed to the sun, improves the light transparency near the infrared end of the spectrum, and acts as a fining agent in optical glass and ruby glass manufacture. It also finds application as a catalyst in the polymerization of polyesters.

In the form of the trichloride and pentachloride it is used as chlorinating agents in organic syntheses. The trifluoride is used in dyeing and as a fluorinating agent. The pentasulfide is used in rubber vulcanizing while the trisulfide is a constituent of matches, percussion caps and pyrotechnics, and a pigment for glass and paints. The trisulfide paints are used for camouflage purposes since the infrared reflecting properties are similar to vegetation.

In recent years high-purity antimony (99.999+%) has been used in the production of the semiconductor compound indium antimonide and as a dopant for producing n-type germanium. Antimony with a purity of 99.99+% is used in the formulation of bismuth telluride-type compounds used for thermoelectric applications.

The price of antimony for a number of years was relatively stable. In January of 1969 the price was around 45 cents per pound, however, in January 1970 the price soared to $4.00 per pound due to a longshoreman's strike and increased demand. By the spring of 1971 the price of antimony had fallen well below $1.00 per pound. The price of high-purity antimony depending on quality varies from $12 to $90 per pound.

S. C. Carapella, Jr.

References

Wang, C. Y., "Antimony," Charles Griffing Co., London, 1952.

Carapella, S. C. Jr., "Antimony," pp. 22–25, in "Encyclopedia of the Chemical Elements," C. A. Hampel, Editor. Van Nostrand Reinhold, New York, 1968.

Sneed, M. C., and Brasted, R. C., "Comprehensive Inorganic Chemistry," Vol. 5, D. Van Nostrand Co., Inc., Princeton, N. J., 1956.

ANTIOXIDANTS

Rubber

Rubbers of all kinds undergo gradual deterioration during aging because of attack by oxygen and/or ozone. The attack by oxygen may be increased by light, heat, presence of certain metal ions as Fe, Mn, Cu, Ti and by mechanical working or flexing. To protect rubber against these various agents requires different materials or a material capable of reacting in several ways. Thus, materials to protect against ozone are called antiozonants; against ultraviolet light, U.V. absorbers; against metals,

metal deactivators; against flex-cracking, antiflex-cracking agents, and against oxygen, antioxidants.

Rubber needs protection against oxidation, both in the raw and vulcanized states. Natural crude rubber contains a very effective but as yet unknown antioxidant. However, after compounding and vulcanization the naturally occurring antioxidant no longer gives sufficient protection. Synthetic antioxidants must be added to insure adequate performance.

Synthetic polymers must have stabilizers added sometime before drying, otherwise the rubbers may catch fire during drying or storage. The stability of these unprotected polymers may vary with their composition but particularly depends upon the amount of unsaturation. The greater the unsaturation the greater the reactivity of the polymer with oxygen. The amount of stabilizer commonly employed varies from 0.5 to 1.25 parts on 100 parts of dry polymer.

Both natural rubber and SBR (styrene-butadiene rubber) can unite with at least 15 to 20% oxygen when unvulcanized. In so doing they become orange-colored and hard, losing all rubber properties. Small quantities of oxygen cause large changes in the Mooney Viscosity, gel content and molecular weight of the raw polymer. In vulcanized stocks, oxidation results in changes in tensile, elongation and modulus as well as surface crazing with oxidation. Any rubber which has combined with more than 1% oxygen has lost most of its desirable elastic properties.

The evaluation of antioxidants is quite important if new and better materials are to be developed. Accelerated tests are in wide use in the rubber industry. For raw polymers these tests involve aging at elevated temperatures. Deterioration is followed by measuring Mooney Viscosity, gel content and DSV (dilute solution viscosity). The rate of oxygen absorption at elevated temperatures is quite useful here.

For vulcanizates, tests involve aging at elevated temperatures in both air and oxygen bombs, and also in circulating air ovens. Resistance to flex cracking before and after aging is determined by a number of devices. The correlation between these accelerated tests and actual service is many times quite poor. Hence these tests are followed by actual service tests before a new antioxidant is accepted for commercial use.

Antioxidants are of two types. Those designed for use in light-colored rubber articles are called nonstaining or nondiscoloring antioxidants. These compounds are usually composed of hindered phenols, phenolic sulfides and phosphite esters. Staining and discoloring types of antioxidants are usually used in black stocks. This type of antioxidant is usually more potent than the nondiscoloring type. Discoloring antioxidants generally contain a secondary amino group —NH—.

Desirable properties of an antioxidant are that it should be odorless, tasteless, nontoxic and nonirritating to the skin. It should emulsify easily if it is to be used as a polymer stabilizer and should be readily soluble in rubber if it is to be used as a compounding ingredient. Solubility is necessary to avoid blooming. It is desirable that the antioxidant have little effect upon cure. Cost is also an important consideration. Today the most widely used antioxidants cost from \$0.50 to \$1.60 per pound.

There are several recent trends in the field of antioxidants. Increasing service requirements have caused emphasis on different types of products. In the nondiscoloring field this has meant shifting to higher molecular weight products such as bis- and poly-hindered phenols or to complex high-molecular weight phosphites.

In the discoloring field there has been a move to replace the old "work horses" of the secondary amine or ketone-amine types with the more potent diaryl or alkyl, aryl-*p*-phenylenediamines. These latter compounds are very effective antioxidants as well as excellent antiflex-cracking agents. The *p*-phenylenediamines are being used both as polymer stabilizers and as compounding ingredients. Along with the use of the more potent materials has come the use of greater quantities of these materials. While rubber articles may contain as little as 0.25 part per 100 parts of rubber of an antioxidant, it is not unusual to find 5.0 or more parts per 100 of rubber for severe service conditions.

Use of synergistic mixtures is gaining acceptance. The most effective of these consists of a hindered phenol as a chain stopper, an ester of thiodipropionic acid and/or an organic phosphite as a peroxide decomposer. More sophisticated mixtures contain metal deactivators and UV absorbers in addition. Widespread use of synergistic mixtures is common in polypropylene and in ethylene-propylene terpolymers (EPDM).

Oxidation of rubber is autocatalytic. Antioxidants function to prolong the onset of the autocatalytic stage. The time required to reach the autocatalytic stage is called the induction period, which can serve as a measure of the stability of a rubber or the effectiveness of an antioxidant. Presence of metals as Fe, Cu, Mn, shortens the induction period. The induction period is also quite dependent upon temperature; a rise of 10°C is sufficient to cut the induction period to one-half.

Oxidation is thought to take place by a free radical mechanism. In simplified form this consists of three stages—initiation, propagation, and termination. If RH is the rubber hydrocarbon and AH is the antioxidant the oxidation may be pictured by the following equations:

$$RH + energy \rightarrow R° \; \textit{Initiation} \tag{1}$$

$$R° + O_2 \rightarrow RO_2° \; \textit{Propagation} \tag{2}$$

$$RO_2° + RH \rightarrow R° + RO_2H \; \text{Chain reaction} \tag{3}$$

$$RO_2H \rightarrow RO_2°, RO°, OH° \; \text{etc.} \tag{4}$$

$$RO_2° \rightarrow \text{stable products } \textit{Termination} \tag{5}$$

$$RO_2° + A° \rightarrow RO_2A \tag{6}$$

$$RO_2H + AH \rightarrow \text{stable products} \tag{7}$$
$$\text{(peroxide destruction)}$$

The key to the prevention of oxidation of rubber by an antioxidant depends upon its ability to stop the chain reaction of steps (2) and (3) or upon its ability to destroy hydroperoxides, reaction (7). Phenolic antioxidants are thought to be chain stoppers, while amine antioxidants function both as chain stoppers and peroxide decomposers. Thiodipropionate esters and organic phosphites function as peroxide decomposers. It is interesting to note that the antioxidant does not protect by combining directly with oxygen. As rubber ages, however, the antioxidant is consumed. Once it is gone autocatalytic oxidation sets in.

Estimated annual U.S. production of all antioxidants (rubber, gasoline, oils, etc.) is 240 million pounds with a value of \$150,000,000. The average annual expansion is estimated at 8%. By 1975 the total production of antioxidants in the U.S.

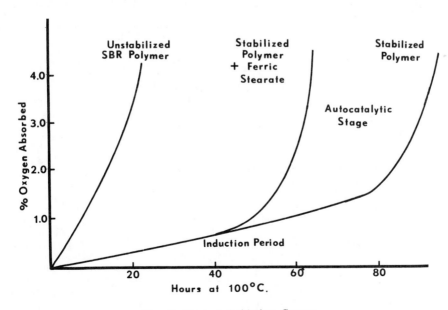

FIG. 1. Typical Oxidation Curves

could reach 350 million pounds with a value of $250,000,000.

New and more effective materials must be discovered to meet increased service requirements for rubber articles. Specifically for tires, formulations must be found to withstand higher speeds, higher temperatures, greater loads, higher ozone content and also give longer service. Age resisters must be found that possess greater stability, are less volatile and are more resistant to extraction by water and other solvents.

CHEMICAL CLASSIFICATIONS OF ANTIOXIDANTS

Classification	Examples
p-Phenylenediamines	N,N' Diphenyl-p-phenylenediamine
	N-Isopropyl-N'-phenyl-p-phenylenediamine
Secondary Aryl-amines	N-Phenyl-2-naphthylamine
	Dioctyl diphenylamine
Alkylarylamines	N,N' Diphenylethylene-diamine
Ketone-Amine	R.P. of acetone and diphenylamine
Dihydroquinolines	Polymerized 2,2,4 tri-methyl-1,2-dihydro-quinoline
	2,2,4-trimethyl-6-ethoxy-1,2-dihydroquinoline
Alkylated Phenols	2,6 di-t butyl-4-methyl-phenol
	2,2'-methylenebis (4-methyl-6 t butyl-phenol)
	Butylated octyl phenol
	4,4'-methylenebis (2,6 di t butyl phenol)
Phosphite ester	Tri (nonylphenyl) phosphite
Alkylated phenol sulfides	4,4'-thio bis (6 t butyl-3-methylphenol)
Thiodipropionates	Dilaurylthiodipropionate

R. B. SPACHT

References

Scott, G., "Atmospheric Oxidation and Antioxidants," Elsevier Publishing Company, New York, 1965.
Reich, L., and Stivala, S. S., "Autoxidation of Hydrocarbons and Polyolefins—Kinetics and Mechanisms," Marcel Dekker, Inc., New York, 1969.
Rubber World, "Materials and Compounding Ingredients for Rubber," 1970 Edition, "Protective Materials—Antioxidants, etc.," pages 100 to 132.

Cross-references: *Antiozonants*

Lubricants

Since almost all lubricants consist wholly or in large part of organic compounds, they oxidize in the presence of air. This oxidation results in deterioration of the lubricant with a marked change in viscosity (usually an increase), the precipitation of insoluble materials and/or the corrosion of metal parts in contact with the lubricant. Small amounts of certain additives, known as antioxidants or inhibitors, greatly decrease the rate of oxidative deterioration. The amount added usually lies between 0.01 and 3 per cent, depending on the antioxidant used. The total amount of lubricant antioxidants used in the United States in one month is about twelve million pounds.

Most lubricant antioxidants fall into the following classes: (11) aromatic amines, (2) phenols, (3) compounds containing sulfur or selenium, (4) compounds containing phosphorus. Phenyl-α-naphthylamine, alkylated diphenylamines and unsymmetrical diphenylhydrazine are typical amine-type antioxidants. Tetramethyldiaminodiphenyl-methane is an effective antioxidant at temperatures above 100°C where many other amines are ineffective. Among the phenolic compounds, hydroquinone, β-naphthol and alizarin show inhibitory action. However, the most commonly used antioxidants are substitution derivatives of phenol, such as 2,6-di-*tert*-butyl-p-cresol, o-cyclohexylphenol, and p-phenylphenol.

Lubricating oil containing a small amount of dissolved sulfur is effectively inhibited, but is quite corrosive toward copper and its alloys. However, if an organic compound of sulfur is added, inhibition may be achieved without corrosion. Suitable compounds can be produced by reacting sulfur with unsaturated esters such as sperm oil, or with other unsaturated organic compounds such as terpenes and polybutenes. Similar compounds may also be prepared by the reaction of chlorinated wax with sodium sulfide. The products of these reactions are complex, and the sulfur is present in a number of forms such as sulfides, disulfides, etc. Aromatic and aliphatic sulfides, such as dibenzyl sulfide and alkylated diphenyl sulfide, are sometimes used.

Some selenium derivatives are also excellent antioxidants. Dicetyl selenide gives much greater inhibition than the corresponding sulfur compound. Dilauryl selenide is particularly useful for temperatures above 100°C.

Like sulfur, elementary phosphorus is an effective antioxidant, but is too corrosive for actual application. The most common phosphorus-containing antioxidants are alkyl and aryl phosphites such as tributyl phosphite and tris (p-tert-amylphenyl) phosphite. Naturally occurring phosphorus compounds such as lecithin are also used.

Compounds containing both sulfur and phosphorus have been used extensively as antioxidants. In general, inhibitors containing both elements are definitely superior to those containing only one. Most of the phosphorus-sulfur inhibitors are produced by the reaction of high molecular weight alcohols or unsaturated organic compounds with phosphorus pentasulfide. The alcohols (such as lauryl alcohol, cyclohexanol or butyl phenol) yield dithiophosphoric acids, which are used in the form of their barium, calcium or zinc salts. Phosphorus pentasulfide reacts with terpenes, polybutenes, unsaturated fatty acids and esters, and so forth to give an almost unlimited number of complex addition products.

There are also some organic compounds containing sulfur and nitrogen which are excellent antioxidants. A number of polyvalent metal dithiocarbamates give good inhibition. Phenothiazine is at present the antioxidant most generally employed in diester synthetic oils.

Naturally, the effect of any antioxidant will vary greatly with the nature of the lubricant in which it is used.

Since autoxidation apparently proceeds by a free radical chain mechanism, antioxidants are considered to function by breaking the chain. Thus, the reaction of one molecule of antioxidant with a chain carrier serves to prevent the oxidation of hundreds or even thousands of lubricant molecules. When the inhibitor reacts, it may be oxidized to a compound which no longer will inhibit oxidation, it may be oxidized to a compound which is a less potent antioxidant, or it may be regenerated. The latter type is, of course, the most desirable. Phenothiazine apparently owes its great inhibitory power to a regenerative reaction which converts active peroxide oxygen to a less active form.

ALAN DOBRY

Food Products

Practically all food products can react with atmospheric oxygen under various processing, storage, and use conditions, with resultant degradative changes in flavor, odor, color, and possibly other characteristics of the food. Such oxidative changes can be related to carbohydrate, protein, or fat, which are the prime building blocks of all foodstuffs; but generally, the oxidative rancidity problem in foods results primarily from autoxidative degradation of fatty (glyceridic) components.

A number of substances are known to be effective, in varying degrees, as inhibitors of the harmful oxidation processes which can occur in food materials. These substances (referred to collectively as food antioxidants) may be naturally occurring (such as tocopherols found in plant and animal tissues), or synthesized chemical compounds (such as butylated hydroxyanisole and butylated hydroxytoluene which are manufactured in modern chemical plants). Whether natural or synthetic in origin, however, food antioxidants usually are phenolic compounds and their antioxidant activity is believed to arise from their ability to inactivate free radicals which are formed as an early step in the oxidation of fat molecules. These free radicals will, if not inactivated, catalyze oxidation of other fat molecules and thus propagate the oxidation chain reaction. It is important to realize that the oxidation reaction is only temporarily inhibited by the antioxidant and will proceed when the antioxidant has been used up or destroyed. Also, the antioxidant will exhibit maximum effectiveness only if added prior to initiation of the oxidation chain reaction.

Although various amines and some sulfur-containing compounds with structures similar to phenols are also known to be effective inhibitors of the type of oxidation which occurs in foods, they are generally not acceptable as food antioxidants because of questionable toxicity characteristics and other problems, such as discoloration tendencies or off-flavors and off-odors associated with their use. Also, other compounds such as organic acids (citric and ascorbic, for example) and phospholipids (lecithin, for example) exhibit apparent antioxidant effectiveness and are often referred to as food antioxidants or synergists. The principal activity of compounds such as these in improving oxidative stability of foods probably stems from their ability to inactivate (chelate) prooxidant metals or to magnify the effectiveness of prime antioxidants which are in the system.

Substances intended for use as food antioxidants must be shown to be safe for such use and must be cleared under pertinent government regulations. Those antioxidants finding some use in foods are:

> Butylated hydroxyanisole (BHA)
> Butylated hydroxytoluene (BHT)
> Dilauryl thiodipropionate
> Glycine
> Gum guaiac or resin guaiac
> 4-Hydroxymethyl-2,6-di-*tert* butylphenol
> Lecithin
> Nordihydroguaiaretic acid
> Propyl gallate
> Thiodipropionic acid
> 2,4,5-Trihydroxybutyrophenone
> Tocopherols

Of these, the most widely used are BHA, BHT, and propyl gallate.

Food antioxidants are commonly used in the concentration range of 0.005–0.10% in fats and oils or in fatty portions of food products. Since the various individual antioxidant compounds exhibit significantly different degrees of effectiveness in various substrates, it is common practice to use them in combinations to take advantage of their different beneficial effects or to minimize certain problems which might be encountered with the individual compounds. Also, the prime antioxidants are often combined with the various synergising compounds, which are known to enhance their effectiveness in foods.

A variety of techniques may be employed in adding antioxidants to foods. Most commonly, they are dissolved or dispersed directly in fats and oils. In treating complex, multi-component food products, the antioxidant may be added to one of the ingredients which serves to carry it into the finished food product. In some instances, treatment of nut meats, for instance, an antioxidant solution may be sprayed onto the food product. Other examples of techniques for applying antioxidants are the use of antioxidant-treated spices and seasonings and the addition of antioxidants to food packaging materials.

In the past two decades, food antioxidant technology has developed to a point where the majority of oxygen-sensitive foods can be effectively protected by the food antioxidants currently available. However, there is still need for better techniques for incorporating them in complex types of food products where the oxidizable fraction is dispersed throughout some other medium which presents a barrier to the antioxidant and its protective effect. Also, with the use of more unsaturated types of oils in food products, there is a pressing need to develop more potent food antioxidants which will adequately protect these more readily oxidized food products. Much laboratory work is being conducted to elucidate further the antioxidant mechanism and to develop improved antioxidant systems for foods.

E. R. SHERWIN

References

Blanck, F. C., "Handbook of Food and Agriculture," Chapter 8, Reinhold Publishing Corporation, New York, 1955.

Furia, T. E., "Handbook of Food Additives," Chapter 5, The Chemical Rubber Company, Cleveland, Ohio, 1968.

Lundberg, W. O., "Autoxidation and Antioxidants," Interscience Publishers, New York, Volume I, 1961, and Volume II, 1962.

Scott, Gerald, "Atmospheric Oxidation and Antioxidants," Elsevier Publishing Company, New York, 1965.

ANTIOZONANTS

Antiozonants impart to natural and synthetic rubber compounds various degrees of resistance to atmospheric ozone. The cracking of rubber on exposure to the atmosphere is caused by two separate mechanisms. Light-activated oxidation results in a resinification of the surface; the surface, being brittle, breaks in an irregular crisscross pattern as a result of surface shrinkage or movement caused by contraction and expansion or other disturbances. The second mechanism is the attack of ozone on strained rubber which takes place in the dark as well as in light and results in cracks running at right angles to the strain. A complex strain pattern will result in a complex crack pattern, but cracking resulting from ozone can usually be identified as being distinct from cracking due to light-activated oxidation. Ozone cracking occurs only in rubbers containing chemical unsaturation; saturated rubbers do not crack on exposure to ozone.

The atmospheric ozone effect is very often visible in a matter of days, and severe, damaging cracks may develop in weeks. Its action is progressive and once started is not reduced or halted until stress relaxation occurs or the ozone concentration is reduced.

The light-activated oxidation effect is a surface effect which tends to lend protection to the underlying material and is often not particularly disruptive of the rubber's useful life, providing that the article is not thin in cross-section and a high ratio of volume to surface exists.

Petroleum waxes used in rubber were in the past often referred to as "sunproofing agents" because the cracks in rubber caused by ozone were formerly considered by many to result from exposure of rubber to the sun, and the addition of waxes to rubber compounds tends to reduce or prevent this type of cracking. They function by being incorporated in the rubber in amounts exceeding their solubilities in the compound so that they migrate to the surface (bloom). This surface wax layer prevents ozone from contacting the rubber and thus ozone cracking does not result. A small amount of wax may be more harmful than

beneficial, as the surface is spottily coated with wax, and where the wax is thin or missing, ozone attack results in large, deep cracks. With no wax, the surface is often uniformly covered with small, shallow cracks which stress-relieve the surface as they form. Wax blooms are often protective for static exposure, but if such blooms are lacking in adhesion or flexibility, they will rupture or break away from the rubber surface during dynamic use. This again results in larger and deeper cracks than if no wax was present.

Waxes used for imparting ozone protection are usually mixtures of microcrystalline and amorphous waxes, some of which are designed to form flexible blooms at winter atmospheric temperatures and yet produce sufficient blooms at the higher temperatures of summer. As temperature rises, the solubility of waxes in polymers increases, and thus less wax is available to form the protective bloom. Higher quantities of wax generally help at elevated temperatures, but they may form heavy, crusty, undesirable blooms at normal room temperatures. Paraffin is often effective, but paraffin blooms are generally brittle.

Chemicals developed specifically as antiozonants are principally dialkyl and alkyl-aryl derivatives of p-phenylenediamine, such as, N,N'-dioctyl-p-phenylenediamine and N-isopropyl-N'-phenyl-p-phenylenediamine. Certain antioxidants of the type which provide resistance to flex cracking also impart added ozone resistance. Examples of these are (1) quinoline derivatives such as 6-ethoxy-1,2-dihydro-2,2,4-trimethylquinoline and polymerized 1,2 dihydro-2,2,4-trimethylquinoline, and (2) mixtures containing N,N'-diphenyl-p-phenylenediamine. Concentrations of 2 to 3 parts on 100 parts of rubber hydrocarbon are normal, although up to 5 parts may be used for severe applications. Often some wax is desirable in the compound to assist in carrying the antiozonant to the rubber surface and to act as an inert barrier on the surface.

Some of these materials impart dynamic as well as static protection which is of concern in such things as tire sidewalls. The word "antiozonant" usually refers specifically to these and related chemicals, but as already explained, waxes alone in many cases provide adequate protection against ozone cracking, especially in static usage, and may be considered as antiozonants. The commercial chemical antiozonants are discoloring and can be used only in black compounds. They also tend to migrate from rubber and stain light-colored materials which may be in contact with rubber containing them. They have accelerating and scorching properties and must be carefully evaluated for their effect on the processing properties of rubber compounds in which they are used. The exact mechanism of how chemical antiozonants protect rubbers from ozone cracking is not known, but it is known that they readily react with ozone, and that the reaction products, possibly combined with those from the rubber hydrocarbon, form a protective surface layer which shields the underlying rubber from attack by ozone. If this layer is mechanically removed, ozone cracking develops in the stretched rubber.

Reduction of light-activated oxidation effects can be obtained in two ways. Incorporation of opaque ingredients in the rubber will lessen penetration of ultraviolet energy into the rubber and thus reduce the thickness of the oxidized layer. Carbon blacks, certain colors and titanium dioxide act as light shields in this manner. Black compounds resist light-activated oxidation fairly well, but nonblack compounds will generally deteriorate on exposure. The incorporation of antioxidants in rubber compounds will reduce light-activated oxidation effects to some extent. The use of nickel dibutyl dithiocarbamate appears helpful in nonblack neoprene compounds for reducing and delaying the formation of an oxidized skin.

C. V. LUNDBERG

Cross-references: *Antioxidants, Waxes.*

References

Ambelang, J. C., et al., "Antioxidants and Antiozonants for General Purpose Elastomers," Rubber Chem. and Technology, **36**, No. 5, 1497–1541 (Dec., 1963).

Biggs, B. S., "The Protection of Rubber Against Atmospheric Ozone Cracking," Rubber Chem. and Technology, **31**, No. 5, 1015–1034 (Dec., 1958).

Lake, G. J., "Ozone Cracking and Protection of Rubber," Rubber Chem. and Technology, **43**, No. 5, 1230–1254 (Sept., 1970).

Loan, L. D., et al., "Ozone Attack and Antiozonant Action in Elastomers," Journal of the IRI, **2**, No. 2, 73–76 (April, 1968).

ANTISEPTICS

Antiseptics are agents which kill or prevent the growth of microorganisms on living tissue. In general a less efficient antibacterial action is expected of antiseptics than that required for germicides, sterilants or disinfectants. The latter terms imply the destruction of all or almost all microorganisms present. Compounds, designated as disinfectants are usually employed on inert objects, while antiseptics are designed to produce their action on living tissue. Antiseptics are usually employed in the first aid treatment of cuts or abrasions or to inhibit bacterial growth on mucous membranes. For the latter purpose they are incorporated into mouthwashes, toothpastes or gargles. Agents employed as antiseptics should not be confused with antibiotics or internally administered antibacterials which inhibit the growth of microorganisms by interfering with their metabolism, processes of cell division or protein synthesis.

Besides their medicinal uses antiseptics are employed to reduce or eliminate odors resulting from bacterial decomposition on the body or in the mouth. For such use they are commonly incorporated into deodorants, foot powders and breath purifiers. They are often added to various pharmaceutical and cosmetic preparations as preservatives to prevent bacterial growth.

Many of the less irritating or toxic chemicals used as disinfectants are also employed as antiseptics. For this use their concentration is often reduced. In the following discussion compounds with antiseptic properties will be discussed by the chemical classes to which they belong.

Halogens. Iodine and compounds which liberate chlorine (hypochlorites) were among the first widely used antiseptics. Although they are efficient in destroying bacteria they are highly irritating and cause destruction of tissue. Iodine is primarily employed as Tincture of Iodine, a hydroalcoholic solution containing 2% iodine and 2.4% sodium iodide. For many years this was considered to be the antiseptic of choice for the first aid treatment of cuts and abrasions despite its irritating properties and the unattractive coloration it produces. It has been replaced to a considerable extent by the less irritating and less toxic organomercury antiseptics. While the iodide ion is practically devoid of antiseptic activity a number of preparations containing it have been marketed as antiseptics. Newer organic compounds which can combine loosely with iodine and slowly liberate it with minimum irritation offer promise as antiseptics. One such product combines iodine with N-ethylpyrinium acetamide alkaonate:

$$+ \; Cl^- \cdot I_2$$

Mercury and Silver Compounds. Such inorganic mercury and silver salts as mercuric chloride and silver nitrate destroy most bacteria, but are both corrosive and toxic. While a 1% solution of silver nitrate has been routinely used in the eyes of newly born infants as a prophylaxsis against eye infections, it is gradually being replaced by less toxic and irritating compounds, such as the antibiotics.

Attempts to retain the antiseptic properties of the silver and mercury ions but to decrease their toxicity and irritating properties by incorporating them into organic molecules have been successful. While combinations of silver with protein (Mild Silver Protein) were used for many years for nasal and throat infections, the greatest success has been achieved with such organomercury compounds as phenylmercuric nitrate, phenylmercuric chloride, merbromine (disodium 2'7'-dibromo-4'(hydroxymercuri)fluorescein), thimerosal (sodium ethylmercurithiosalicylate) and nitromersal (4-nitro-3-hydrosyamercuri-o-cresol anhydride) which have largely replaced the inorganic mercury compounds. These organometalics are less toxic, less irritating and are not as easily inactivated by organic material. As a result they are often included in first-aid kits for the treatment of minor wounds.

Phenolic Compounds. Most of the common phenols and cresols possess the ability to destroy or inhibit the growth of microorganisms. While their usefulness as disinfectants has been limited by their solubility in water and therefore the ability to satisfactorily formulate them, a 1% solution of phenol is a useful antiseptic. Resorcinol (1,3-dihydroxybenzene), being water soluble, is used in dermatological preparations for its antiseptic properties. The toxicity of phenols and cresols has been decreased and their antiseptic potency increased by introducing alkyl and chloro groups into the molecule. While these have been used chiefly as disinfectants, Hexylresorcinol (4-hexyl-1,3,dihydroxybenzene) has been incorporated into mouthwashes, gargles and throat lozenges.

In the search for a more antiseptic soap, attempts have been made to incorporate many antiseptics into such formulations. Some antibacterial compounds such as the quaternary ammonium salts are inactivated by soap. Others are physically incompatible with the necessary formulations. While phenols have been successfully added, their pronounced odor has been objectionable to many users. Development of the bisphenols such as hexachlorophene [bis-(3,5,6-trichloro-2-hydroxyphenyl)methane] led to antiseptics which were compatible with soap formulations. The resulting products are supposed to reduce or eliminate objectional odors resulting from bacterial decomposition of organic materials on the skin. Hexachlorophene has the disadvantage of being inactivated by sebum and of having little effect on gram-negative organisms. Bithionol [bis-(3,5-dichloro-2-hydroxyphenyl)sulfide] is somewhat analogous to hexachlorophene in structure and can be incorporated into soap to add antiseptic properties. Both compounds may also be applied by other means. m-Cresolacetate, an ester of a phenol, is used as an antiseptic on the mucous membranes of the nose, throat and ear.

Quaternary Ammonium Compounds. Although used chiefly as germicides for the disinfection of skin and surgical instruments such compounds as alkyl dimethylbenzylammonium chloride (Benzalkonium Chloride), dodecyldimethyl-(2-phenoxyethyl)ammonium bromide (Domiphen Bromide) and cetylpyridinium chloride are the active ingredients of many antiseptic mouthwashes, gargles and ophthalmic preparations. The manner in which they exert their antiseptic properties is not entirely understood, but it is at least partly related to their ability to reduce surface tension. While many primary, secondary and tertiary amines show antiseptic properties, the stability and solubility of the quarternary ammonium compounds in acidic, basic and neutral solutions have favored the use of the quarternary compounds. It is reported that bacteria can develop resistance to the quaternary salts in much the same way they develop resistance to the antibiotics. Cetylpyridinium chloride is essentially free from irritating properties and can be used on mucous membranes. Because of this property it has been incorporated into throat lozenges. These compounds are commonly employed as preservatives in numerous pharmaceutical products.

Acids, Acid Derivatives and Alcohols. Benzoic acid has long been employed as a preservative for syrups and other preparations which can serve as a media for mold and bacterial growth. When combined with salicyclic acid in an ointment base (Whitfield's Oinment) it has been employed for the treatment of such fungal infections as athletes foot. Because of the antiseptic and kerolytic properties of salicylic acid this was considered to be a most efficient treatment. More recently the use of this combination of acids has been surplanted by sodium propionate and zinc undecylenate which have excellent antifungal properties.

Boric acid, as a saturated or more dilute solution, is often employed as an antiseptic on the skin and in the eye. Because of its toxicity it should not be used on broken skin and should be kept away from children. Methylparaben (methyl p-hydroxybenzoate) and Propylparaben (propyl p-hydroxybenzoate) possess antiseptic properties and are employed in ophthalmic solutions to inhibit the growth of molds and bacteria.

While ethyl and isopropyl alcohols are usually employed to disinfect skin, they are often employed as first aid treatment for wounds. They are, however, most irritating.

Peroxides. The germicidal value of oxygen is well known and compounds capable of serving as oxidizing agents or liberating oxygen (chlorates, permanganates, perborates and peroxides) have been employed to kill or inhibit bacteria. Solutions of hydrogen peroxide (3% or less) are used for cleaning wounds or for infections of the mouth. Sodium perborate has been incorporated into tooth powders or dissolved in water as a mouthwash for infections of the oral cavity. Glycerin solutions of urea peroxide are applied to wounds. Moisture from the skin decomposes the peroxide with the liberation of oxygen. These solutions have also been employed for treatment of ear infections. Zinc peroxide is also employed as an antiseptic.

JAMES E. GEARIEN

ANTISTATIC RESINS

Despite many technical advances, there is limited knowledge and agreement about the mechanisms of conduction in organic matter. The same is true with plastics. For example, several authorities believe electrostatic buildup is a surface phenomenon; others contend that it occurs as a volume distribution throughout the plastic.

The problems resulting from electrostatic buildup on plastics, unfortunately, are readily observed. Since thermoplastic resins are usually fine insulators, and therefore, poor conductors, they retain charges easily. The magnitude of these charges is partly a function of polymer composition and geometric shape of the plastic object. Equally important generative factors are climatic conditions to which the object is exposed, and the degree of its physical contact with other objects and with air. Although static propensities and the resulting problems differ from resin to resin, most thermoplastic matter is affected by static, either while at rest or while in motion.

Electrostatic charges on thermoplastic objects can be generated spontaneously by air currents even if such objects are otherwise left undisturbed. These charges then attract and tenaciously hold dust and dirt particles to plastic surfaces. The resulting unsightly appearance is detrimental to the esthetic appeal of plastic merchandise itself, e.g., appliance housings, housewares, toys, furniture, and phonograph records, as well as to nonplastic goods packaged in plastic.

In packaging, where a dirty package appearance can detract from a potential sale, static is a particularly troublesome problem. Polyethylene bottles and squeeze tubes are cases in point. Generally, packaging for durable merchandise, i.e., hardware and soft goods, is more adversely affected by poor appearance than packaging for nondurable merchandise, i.e., fresh fruit and vegetables.

Electrostatic problems with plastic objects in motion are different from those at rest; they are primarily manufacturing problems and usually involve electrostatic magnitudes of higher orders; they occur throughout production, converting, printting, and packaging operations.

As unsupported film, sheeting, and extrusion coatings move along at high speeds during processing, frictional forces generate powerful electrostatic charges. The magnitudes of these charges are commonly measured in many thousands of volts.

The charges cause many manufacturing and handling problems. For example, they cause improper film-to-machine alignments which can jam equipment. Some relief may be achieved by installing destaticizing equipment or by the uneconomic alternative of slowing down high-speed production lines. Less frequent, but more serious, are strong electrostatic discharges which have resulted in severe shocks, and even fires and explosions.

To solve these problems, external antistatic agents have been used on plastic objects in a limited manner for years, but since they wash off, wear off, and rub off, only short-term protection is provided. Three approaches have been used to attack static problems:

(1) *Conductive Polymers.* One is the development of new resins which would have inherently better conductivity and therefore would not easily develop electrostatic charges. These resins will necessarily be of a different polymer configuration than those which now comprise mass markets. A step in this direction is the electrically conducting polymer, TCNQ, tetracyanoquinodimethane, which has had resistivities recorded as low as 0.01 ohm-cm. By comparison, present conventional thermoplastic resins have resistivities which range from 10^{10} ohm-cm upward.

(2) *Internal Antistatic Additives.* In the second approach, thermoplastics can be made antistatic by physically admixing nonreactive antistatic additives throughout the plastic matrix. Such an additive must have the capacity to exude uniformly to the plastic object's surface and to change the plastic's electrical properties. It has been hypothesized that antistatic additives function by one or more of three possible mechanisms.

(a) A strongly polar additive after exuding will form a continuous, conductive thin film on the plastic object's surface. As a result, surface resistivity is lowered markedly and any static buildup that does occur is instantaneously drained off. Examples of such polar or ionic additives which have been successfully used commercially are fatty quaternary ammonium compounds, amines, diamines and imidazolines, ethoxylated fatty amines, and phosphate esters.

(b) In a similar way, hygroscopic additives have been used successfully in thermoplastic resins; they also migrate to the surfaces of the plastic object to form a continuous film. In the presence of high relative humidity, a thin continuous film of water forms on the surface. Water, itself a polar compound, will act as an ionic compound to dissipate rapidly any induced charge. Examples of antistatic additives which are thought to function by

their hygroscopic nature are polyethylene gycols, and polyethylene gycol fatty acid esters.

(c) Finally, it is believed that certain hydrophobic slip additives function by first exuding to the plastic surfaces, but instead of lowering surface resistivity appreciably they function by dramatically increasing the slip properties of the plastic surface, so that static charges cannot be so easily induced. Examples of hydrophobic antistatic additives are oleylamide, erucylamide, ethylenebis-stearamide and methylenebis-stearamide. It should be pointed out that certain additives, such as ethoxylated fatty amines and quaternaries, can employ all these hypothesized mechanisms concurrently.

A patent and literature search for the period 1965 through 1970 indicates more than 300 new references to antistatic experimentation, primarily from research laboratories in the United States, Germany, and Japan. The great preponderance of this research investigated such conductive organic additives as alkyl sulfonates, amines, quaternary ammonium compounds, glycols, alkanolamines, fatty amines, polyphosphites, and polyphosphates.

(3) *Copolymer Antistatic Agents.* In the third approach, the antistatic compounds contain a reactive site that can be copolymerized with the static-prone polymer. Such copolymers would also have to possess functional groups which impart antistatic properties by the hypothesized mechanisms. Little has been accomplished to date with copolymer antistatic agents, probably because few such compounds have been made commercially available at reasonable cost. One potential advantage from the copolymer approach is that truly "permanent" protection will result; that is, since the copolymer antistatic agent is actually part of the polymer, it cannot be leached out by solvents during the service life of the plastic.

H. S. HOLAPPA

ARCHAEOLOGICAL CHEMISTRY

The most obvious and frequent application of chemistry to archaeology is the exact identification of materials found in excavations. Errors of identification, some of them rather glaring from the chemical viewpoint, are rather frequent in the older archaeological literature and are by no means absent from some of the more recent literature. Such confusions have arisen either because no correct identification was made in the first place or because the archaeologist did not appreciate the chemical distinction between certain materials similar in appearance and put to similar uses.

The problem of making a correct identification of ancient materials from excavations may be very different from that of identifying the same materials in a fresh condition. It is usually less important to determine what it is now than to establish what it was originally before it underwent extensive chemical change. Though the original nature of inorganic materials may usually be determined with little difficulty, it is otherwise with organic materials, for these are likely to undergo complicated chemical changes during long burial in the ground.

Simple qualitative chemical tests are often sufficient to establish the identity of a material, but sometimes rather extensive tests are necessary, including complete quantitative analyses and measurements of physical properties. Microanalytical methods are especially useful because it often happens that only small amounts of materials are available for investigation. For the examination of rare or valuable objects, which cannot be altered in any way even to the extent of taking samples for microanalysis, nondestructive physical methods must be used.

The exact identification of a large number of specimens of materials, and especially a chronological series of specimens, from a particular site or group of sites, may yield valuable information about the cultural status or development of a people. The absence of metal among the remains of a given people obviously indicates a low state of technical development and a low state of general culture. The greater the variety of metals and alloys utilized by a people, the greater their technical development, and, as a rule, the higher their general cultural development. The same appears to be true of certain kinds of nonmetallic materials such as glazes and glasses.

Some materials which serve as an index of cultural development may, of course, be recognized without chemical aid, but the application of chemical methods may make their recognition more certain. On the other hand, some kinds of materials which are indicative of cultural level are not easily recognized for what they are without chemical aid. For example, the introduction of lead into glazes, or the introduction of less common elements, such as cobalt, into glasses, represents technical advances which may be, and in fact sometimes have been, overlooked in the study of ancient remains.

Chemical methods are apparently not generally applicable to the important problem of determining the chronological sequence or absolute age of materials and objects, though they are applicable to a few kinds of materials and objects, especially when supplemented by other methods of relative or absolute dating. Ancient remains such as wood, textiles, and bones that contain carbon derived from plants or animals may now be dated absolutely by a measurement of the radioactivity of the carbon they contain, a method developed by W. F. Libby and co-workers at the University of Chicago. Its basic principle is that the carbon in the carbon dioxide of the atmosphere taken in by plants, both now and in ancient times, contains a fixed proportion of C^{14} with a half-life of about 5580 years.

During long burial in the ground a great variety of reactions occur between the material of most objects and the surrounding soil and ground water. (An example of this is the petrefaction of wood, wherein the organic components of the wood are gradually replaced by soil silicates.) The appearance and general condition of objects of the same material and of the same age taken from various excavation sites may be very different. A bronze from one site may be coated with a hard coherent layer of corrosion products of the type commonly called a patina, whereas a similar bronze of the same age from another site may be coated with a loose porous mass of corrosion products. Archaeologists observed that bronze found at Corinth was almost always in a severely corroded state, whereas similar bronze of equal or greater age found at certain other sites in Greece was not so corroded. It was found that the ground water at Corinth contains a high enough concentration of chloride to account for the severe corrosion of the buried bronze.

The deterioration of buried objects is predominantly the result of chemical change, and it is therefore reasonable to expect that the restoration of such objects should usually be best effected by chemical treatment. The deterioration of bronze, for example, is primarily the result of oxidation, and the reverse process of reduction should tend to restore corroded bronze to its original condition. This principle has been applied with much success to the restoration of metal objects by various electrolytic reduction procedures.

The possibilities in the application of chemistry to archaeology have by no means been fully explored, and much of the work up to the present has been of a fragmentary and unsystematic nature. Enough has been done, however, to indicate clearly that this middle ground between chemistry and archaeology is capable of being developed into a distinct and systematic branch of applied chemistry with its own special data, techniques, and general rules.

EARLE R. CALEY

Cross-references: *Chemical Dating, Microchemistry.*

References

Caley, E. R., "The early history of chemistry in the service of archaeology," *J. Chem. Education,* **44,** 120–123 (1967).

Levey, M., "Archeological Chemistry," Philadelphia, 1967.

Plenderleith, H. J., "The conservation of antiquities and works of art," London, 1956.

ARGON

Argon, chemical symbol Ar., atomic number 18, atomic weight 39.948, is one of six elements, helium, neon, argon, krypton, xenon, and radon, which comprise Group 0 of the Periodic Table. At ordinary temperatures it exists as a colorless, tasteless, odorless inert (or "noble") gas. Dry atmospheric air contains about 0.94% by volume of argon in the form of three stable isotopes of mass number 40, 38 and 36 which occur in the ratios of 99.26%, 0.05% and 0.33%. In addition five unstable isotopic species of mass number 35, 37, 39, 41 and 42 have been produced artificially. Ar^{35} which emits positive electrons (positrons) has a half-life of 1.8 sec; Ar^{37} decays with a half-life of 35 days by capture of an electron from the K or L shell of the atom; Ar^{39}, Ar^{41}, and Ar^{42} emit negative electrons with 260 year, 1.82 hour and 3.5 year half-lives.

The presence of an inert gas constituent of air was suspected by Henry Cavendish as early as 1785, when he found that a small fraction (less than 1/20th) of the air remained after removing the oxygen and nitrogen. The nitrogen was oxidized in an arc process and the oxides of nitrogen formed were adsorbed in caustic soda. Over a century later in 1894 Lord Rayleigh and Sir William Ramsey, using a somewhat similar method, isolated an inert constituent from air which they identified as a monatomic gas of atomic weight about 40. The name Argon given to the element, comes from the Greek word meaning "inactive." By 1898 the other four noble gas constituents of the atmosphere, helium, neon, xenon and krypton, which together constitute only 22 parts per million of dry air were isolated and identified in Ramsey's laboratory. Radon, a decay product of radium, was subsequently identified to complete this group of the Periodic Table.

As in the case of the other noble gases, argon is chemically inert, with zero valence. The 18 electrons of the atom form stable closed shells of 2, 8, and 8 electrons around the nucleus. However, certain complexes of argon and boron trifluoride of composition $Ar.nBF_3$ where $n = 1, 2, 3, 6, 8$ and 16 do exist as unstable crystalline solids under limited conditions. Argon can also be trapped in the holes or cages formed in the clathrate compounds. An example of such a "compound" is the quinol clathrate $3C_6H_4(OH)_2 . 0.8 Ar$.

The normal boiling point at 1 atmosphere of argon is $-185.7°C$ (above that of nitrogen and slightly below that of oxygen). The freezing point at 1 atmosphere is $-189.2°C$; the critical temperature $-122.5°C$; critical pressure 48.0 atm.; critical density 0.531 g/c.c.; first electron ionization potential 15.7 volts.

Argon is extracted from the atmosphere in commercial quantities by careful fractionation of liquid air. Since 1920 it has been widely used as a fill-gas in incandescent lamps. It is now extensively used in arc welding to prevent undesirable oxidation by providing an inert gas shield or mantle around the work and weld metal. Millions of cubic feet are used for this purpose annually. It is used in fluorescent lamps and in some gas-filled electronic tubes. It is one the gases used in Geiger-Mueller tubes and in ionization chambers used in the detection of ionizing radiation. In 1971 production approached 4 billion cubic feet in the U.S.A.

H. A. FAIRBANK

Cross-references: Clathrate Compounds, Noble Gas Compounds.

AROMATIC COMPOUNDS

Compounds that contain a ring structure composed of one sort of atoms only are called homocyclic. These are divided into two groups: (1) a relatively small number retain the essential characteristics of aliphatic compounds and are called alicyclic compounds; (2) most of the other ring compounds are related to benzene (C_6H_6) or its derivatives, and are called aromatic compounds. The name is associated with the penetrating odor of the first few compounds of this group that happened to be investigated; this odor was described as aromatic. The classification, however, is not limited to an odor or type of odor. Today all derivatives of benzene generally are classed as aromatic compounds, whether they have an odor or not. The benzene ring

is usually abbreviated by a simple hexagon. Thus

Benzene has been studied by electron diffraction and x-ray diffraction techniques. Evidence points to a ring of six carbon atoms that lie in one plane. One hydrogen atom is attached to each carbon atom. The spacing between adjacent carbon atoms in the benzene ring (1.39 Å) is less than that for a single bond space (1.54 Å), and close to that for a double bond space (1.34 Å). Benzene does not react, however, as if it had typical double bonds, nor does it show the typical reactions of an alternate double and single bond conjugated system. The chemical reactivity of benzene, however, is far greater than that of n-hexane (C_6H_{12}). Theorists have concluded that the correct structure of benzene probably resonates among all possible structures. The two forms that probably contribute most to the structure of benzene are

A small circle placed within a benzene ring is a symbol sometimes used to represent all contributing forms of benzene.

If one carbon atom in benzene is labeled 1, then the others are called 2 through 6 successively clockwise around the ring. If the ring has one hydrogen only substituted by halogen, for example, then only one such compound is known. One and only one chlorobenzene is known. This is evidence that all the hydrogen atoms in benzene are equivalent. Three dichlorobenzenes $(C_6H_4Cl_2)$ are known. These correspond to positions of the chlorine atoms: 1,2-(ortho); 1,3-(meta); and 1,4-(para).

1,2-Dichloro-
benzene; ortho-
Dichloro-
benzene

1,3-Dichloro-
benzene; meta-
Dichloro-
benzene

1,4-Dichloro-
benzene; para-
Dichloro-
benzene

In the case of three substituents, the naming is as follows:

1,2,3-Tri-
chlorobenzene;
vicinal-Tri-
chlorobenzene

1,2,4-Tri-
chlorobenzene;
asymmetrical-
Trichlorobenzene

1,3,5-Tri-
chlorobenzene;
symmetrical-
Trichlorobenzene

Benzene (C_6H_6) is a colorless liquid that boils at 80°C. It has an aromatic odor, and it is immiscible with water. Like all hydrocarbons, it burns readily in air. Benzene, like acetylene, burns with a sooty flame, a characteristic attributed to the high percentage of carbon in the compound.

Benzene is obtained from the destructive distillation of coal, being fractionated from coal tar. It can also be made synthetically from petroleum or acetylene.

Benzene reacts with halogens and forms products of varying halogen content:

$$C_6H_6 + Br_2 \rightarrow C_6H_5Br + HBr.$$

In a similar manner, substitution in the benzene ring can be accomplished by fuming nitric acid (in the presence of a strong dehydrating agent):

$$C_6H_6 + HONO_2 \rightarrow C_6H_5NO_2 + H_2O,$$

and by concentrated sulfuric acid:

$$C_6H_6 + H_2SO_4 \rightarrow C_6H_5SO_3H + H_2O.$$

Halogens can also add to benzene as well as substitute in the ring: $C_6H_6 + 3Cl_2 \rightarrow C_6H_6Cl_6$. 1,3-Cyclohexadiene, a compound that shows the reactions typical of a conjugated system, forms when benzene adds hydrogen in the presence of nickel as a catalyst:

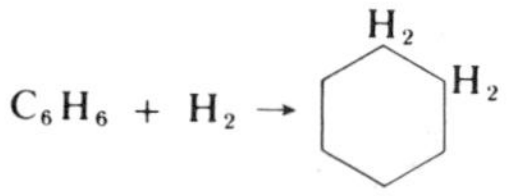

Benzene is a useful solvent and it is an additive in motor fuel because of its high antiknock values. It is the starting material from which a number of useful substances is made, including styrene, clorobenzene, phenol, nitrobenzene, biphenyl, phthalic anhydride, and others. The vapors of benzene are toxic.

Many compounds are known in which the benzene ring is attached to other benzene rings. This attachment can occur end-to-end as in biphenyl.

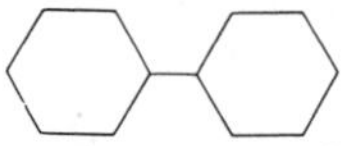

$$C_6H_5\!-\!C_6H_5$$

Biphenyl is formed when benzene vapor is partially decomposed by heating it above 600°C in the presence of iron. This colorless crystalline compound melts at 70°C and boils at 255.9°C. It is stable enough to be used successfully as a heat-transfer agent.

Side-to-side attachment (fusion) of two benzene rings results in *naphthalene*

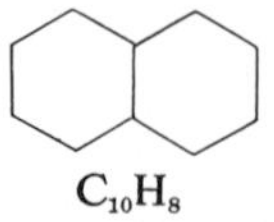

$$C_{10}H_8$$

a compound obtained by the fractionation of coal tar; three fused benzene rings give *anthracene*

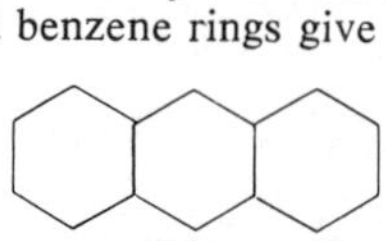

$$C_{14}H_{10}$$

also from coal tar fractionation, and *phenanthrene,* the basis of the steroid family.

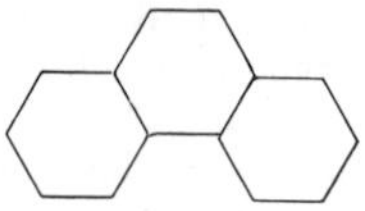

$$C_{14}H_{10}$$

Some of the compounds that contain multiple benzene rings are known to be carcinogenic (cancer-producing). Methylcholanthrene

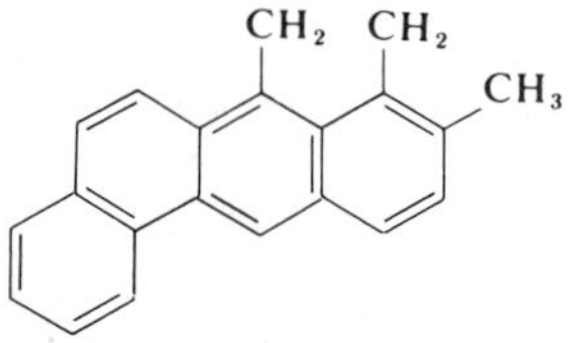

is one prominent example.

The substituent $CH_3\!-$ in the benzene ring forms *toluene* ($C_6H_5CH_3$) a well-known hydrocarbon. Although some toluene is made from coal tar, most of it is manufactured from petroleum. Toluene is a colorless aromatic liquid that boils at 110.6°C. When strongly nitrated, it forms the well-known TNT, (symmetrical trinitrotoluene) that is a powerful explosive that can be handled with relative safety.

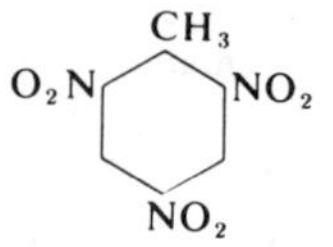

The next hydrocarbon in the series, is *ethylbenzene* ($C_6H_5C_2H_5$).

Three dimethylbenzenes or xylenes are known [o-, m-, p- $C_6H_4(CH_3)_2$]. They differ slightly in boiling points, and are used as solvents, often mixed in commercial preparations.

Styrene ($C_6H_5CH\!=\!CH_2$) is a liquid that boils at 146°C. It is an example of an unsaturated aromatic hydrocarbon. Styrene is the base material for the manufacture of many thermoplastic polymers (polystyrene).

All the classes of compounds known in aliphatic chemistry have their counterpart in aromatic chemistry. A few illustrations of each sort will be given.

Aromatic hydroxides are called *phenols.* Ordinary phenol (C_6H_5OH), which has been described as a pink slush, melts at 43°C and boils at 182°C. The compound has a characteristic penetrating odor. It is widely used in the manufacture of resins. Some phenol is obtained from coal tar, but most of it is manufactured synthetically.

1,2-Dihydroxybenzene (ortho) is called catechol, the meta compound (1,3-dihydroxybenzene) resorcinol, and the para compound (1,4-dihydroxybenzene) hydroquinone. Catechol is used in the synthesis of adrenaline, a powerful stimulant. Resorcinol is used to make hexylresorcinol, a urinary antiseptic. Hydroquinone is used as a developing agent in photography. Trihydroxybenzenes are also known.

Phenols are not similar to alcohols in properties. An example of an aromatic alcohol is benzyl alcohol ($C_6H_5CH_2OH$). This compound, alpha hydroxytoluene, and its esters are used in perfumes.

Diphenyl ether (phenyl ether ($C_6H_5)_2O$) is a colorless solid that melts at 28°C. It has a geranium-like odor, and it is used as a heat-transfer agent because of its stability.

Aromatic aldehydes are represented by *benzaldehyde* (C_6H_5CHO) which is found in nature and is called oil of bitter almonds. Another important aldehyde is *vanillin* [4-hydroxy-3-methoxybenzaldehyde, $1,3,4-C_6H_3(CHO)(OCH_3)OH$], which is found in nature and which is manufactured synthetically.

Like aliphatic ketones, aromatic ketones may be simple or mixed. The least complicated of the simple ketones is diphenyl ketone ($C_6H_5COC_6H_5$), or *benzophenone*. Its melting point depends on the form it takes. It boils at 305.9°C.

The simplest aromatic amine is *aniline* ($C_6H_5NH_2$). Many derivatives of aniline are known. Aniline can be made by the reduction of nitrobenzene, or by the reaction of chlorobenzene with ammonia. Aniline is the parent compound of many dyes. Aniline tends to destroy red blood cells in the body, so it should be handled with care.

Aromatic acids of the carboxylic type are represented by *benzoic acid* (C_6H_5COOH) as the simplest member. Benzoic acid or its sodium salt may be used as a food preservative. Three phthalic acids (dicarboxylic) are known: phthalic anhydride [$C_6H_4(CO)_2O$] is made from the *ortho-* form; it is important in the manufacture of resins, together with glycerol. Other important aromatic carboxylic acids include *cinnamic acid* ($C_6H_5CH=CHCOOH$), an unsaturated acid which occurs in nature, and *salicylic acid* ($C_6H_4 \cdot OHCOOH$) which is both a phenol and an acid. Salicylic acid (2-hydroxybenzoic acid) derivatives are widely used. Methyl salicylate is oil of wintergreen, and acetylsalicylic acid is aspirin.

The great variety of aromatic compounds is more fully appreciated when one realizes that substitution in the benzene ring by halogen, NO_2, SO_3H, NH_2, and other groups is possible. In addition, a great variety of side chains can be used, and also combinations of them. Many of the aromatic compounds are extremely valuable as medicines, dyes, antiseptics, anesthetics, and flavorings.

ELBERT C. WEAVER

Cross-references: *Aliphatic Compounds, Alicyclic Compounds; Resonance.*

AROMATICITY

Chemical experiments have led to the rather surprising fact that monocyclic conjugated polyenes containing a certain number of π-electrons are particularly stable. This favored π-electron configuration is given by the formula $4n + 2$, where n is zero or an integer. Benzene with six π-electrons was the first of these "aromatic" molecules encountered by chemists. More recent observations have revealed that $4n + 1$ and $4n + 3$ electron molecules

(radicals) tend to form a "more stable" system through the gain or loss of an electron to form the $4n + 2$ configuration. Physically, the aromatic system is one that can be written in at least two equivalent classical (Kekulé) structures; it is coplanar and its C—C bonds are of equal length, intermediate between single and double bonds. Chemically, aromaticity is distinguished by a preference toward substitution rather than addition reactions. The modern fundamental definition of aromaticity is that of complete electron delocalization and bond equivalence as measured by ability to sustain a magnetically induced ring current. This latter property is determined from the nuclear magnetic resonance spectrum. Although the realization of the unusual stability of a cyclic system of π-electrons was of experimental origin, it can be shown following the principles of molecular orbital theory to be a theoretically predictable fundamental property of such systems. Our aim in this discussion is to gain an understanding of the Hückel $4n + 2$ rule as it was developed within the framework of LCAO-MO theory [linear combinations of atomic orbitals —molecular orbital].

When carbon is hybridized to its trigonal sp^2 state, it can form three equivalent sigma bonds, leaving one *p*-orbital, containing a single electron, perpendicular to the plane of the three sigma bonds. The overlap of this *p*-orbital with an identical orbital on an adjacent carbon atom forms a π-bond. A conjugated system consists of a series of sp^2 hybridized carbon atoms linked together by sigma bonds and oriented in such a manner that the *p*-orbitals are all perpendicular to the plane of the molecule. The *p*-electrons are free to move over the entire carbon core system. This electron delocalization results in a lower total electronic energy, compared to the system in which each electron is assigned to a particular carbon atom or to a particular pair of carbon atoms.

Consider a planar, cyclic system of sp^2 carbon atoms bound together by sigma bonds. The single electron in the *p*-orbital on each carbon atom is free to move throughout the potential field of the entire carbon core ring. Following the Hückel LCAO-MO method the energy of the π-molecular orbitals for cyclic molecules composed of 3 to 10 sigma-bonded carbon cores can be calculated and considered in terms of the empirical resonance integral, β. The number of molecular orbitals obtained is equal to the number of carbon atoms in the molecule. The energy scale increases in the upward direction. A positive energy is bonding, since β is negative and energy is released on forming a bond. The negative energy levels are called anti-bonding, as an electron in a negative energy orbital reduces the energy gained through the delocalization of the electrons. When an electron is in a zero orbital, it neither adds to nor decreases the π-electron energy. This orbital is termed non-bonding. Each orbital can, following the Pauli exclusion principle, contain a maximum of two electrons. In Table 1 these are represented by arrows with the sense of the arrow indicating electron spin. The total π-electron energy for each molecule is obtained by putting the available *p*-electrons (one from each carbon atom) into the lowest energy molecular orbital etc. until no electrons are left. The total π-energy is the sum of the energies of the occupied orbitals times the number of electrons in each orbital.

Looking first at the 6-carbon-atom molecule, benzene, we see that all six π-electrons can be accommodated in the bonding orbitals. Also, each electron is paired, i.e., each orbital is doubly filled; therefore, benzene has a closed-shell singlet ground state. Looking at the five- and seven-membered rings, which of course represent cyclopentadienyl and cycloheptatrienyl free radicals, we see that each has an unpaired electron. The striking feature is that in the five-membered ring, the unpaired electron is in a *bonding* orbital. The lowest empty orbital available for another electron is this bonding orbital; thus, the addition of one more electron to this system increases the total π-bonding energy—at the same time forming a $4n + 2$ electron system. This accounts for the unusual stability observed for the cyclopentadienyl anion. In the seven-membered ring, the unpaired electron is an *anti-bonding* orbital. The loss of this electron increases the total π-bonding energy—again forming a $4n + 2$ electron system. Thus, the remarkable stability observed for the cycloheptatrienyl positive ion is readily understood. Notice that in the $4n + 2$ electron system all the bonding orbitals are filled and each electron is paired with another of opposite spin.

In the three-membered ring and the nine-membered ring, the pattern repeats itself, the gain of an electron in an unoccupied bonding orbital or the loss of an electron from an antibonding orbital resulting in the more stable anion or cation having the "preferred" $4n + 2$ number of electrons. The ten-membered ring having just the number of electrons necessary to completely fill all its bonding π-molecular orbitals would have a closed shell singlet ground state. However, it has not yet been prepared and is expected to be highly destabilized by steric conflict between the internal hydrogen atoms. Naphthalene and azulene, in which the bridgehead bond is only a minor contributor to the total resonance stabilization, resemble such a ten carbon atom aromatic molecule.

The point to remember is that the π-electron molecular orbitals of conjugated monocyclic molecules varies with the number of carbon atoms such that $4n + 2$ number of electrons completely fill the *bonding* molecular orbitals and produce the largest electron delocalization (resonance) energy.

[Delocalization energy is the stability gained through electron delocalization in excess of that of a comparable system of alternating single and double bonds. In terms of β it is calculated by summing the occupied energy levels of the molecule times the number of electrons in each energy level and subtracting 2β for each ethylene unit in one of its Kekulé structures. An odd electron localized on a particular carbon atom, as it must be in a

TABLE I

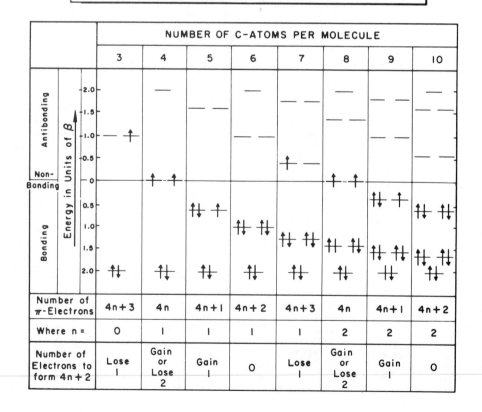

π-ELECTRON MOLECULAR ORBITALS of MONOCYCLIC CONJUGATED POLYENES

	NUMBER OF C-ATOMS PER MOLECULE							
	3	4	5	6	7	8	9	10
Number of π-Electrons	4n+3	4n	4n+1	4n+2	4n+3	4n	4n+1	4n+2
Where n =	0	1	1	1	1	2	2	2
Number of Electrons to form 4n+2	Lose 1	Gain or Lose 2	Gain 1	0	Lose 1	Gain or Lose 2	Gain 1	0

Kekulé structure, has an energy of zero β as its contribution to the resonance energy of the molecule.]

The four-membered ring and the eight-membered ring each contain $4n$ π-electrons. As before, we put the available p-electrons into the π-molecular orbitals. The result is that the lower energy orbitals are all doubly occupied. The highest occupied π-orbital which in this case is the doubly degenerate non-bonding orbital, can accommodate four electrons, but only two electrons remain to occupy these *two* energy levels. The most favored distribution is one electron in each orbital with their spins parallel, i.e., a triplet state. Very close in energy are three singlet states, one in which the electrons are opposed in spin but in different orbitals and two where they are opposed in spin and in the same orbital. The result of these four close-lying electronic states is that these $4n$ molecules do not possess a closed shell ground state in the planar delocalized configuration. Molecules not having closed shell ground states are known to be very reactive or unstable.

These $4n$ π-electron molecules may be stabilized by losing two electrons, thus producing the di-cation or gaining two electrons to form the dianion. In this manner they form a closed shell and acquire the $4n + 2$ electron configuration.

There is another way this unstable situation can be relieved. The calculation of the π-electron molecular orbitals assumed two things; a planar molecule, and that the electrons were delocalized (i.e., free to move over the entire molecule). If the electrons were to become localized, to form a system of alternating single and double bonds, this energy level scheme would no longer apply and the molecule might become more stable. This has proved to be the case for the eight-carbon ring cyclooctatetraene, which exists in a puckered form consisting of an alternating single and localized double bonds, thus having a closed shell ground state. Also because of the localization of the electrons, cyclooctatetraene has only a small fraction of the resonance energy (3–4 kcal) that would be expected from the π-electronic energy levels calculated above assuming planarity and delocalization. Furthermore, its chemical reactivity is that of a typical unsaturated system and not that of an aromatic system, i.e., cyclooctatetraene readily undergoes addition reactions.

The four-membered ring, cyclobutadiene, has as yet resisted all efforts toward its preparation. The simple MO treatment, as depicted here, predicts a triplet ground state for cyclobutadiene, but more comprehensive theoretical treatment predicts that if it is prepared, its most stable form will be rectangular having alternating single and double bonds (i.e., localization of the π-electrons as in cyclooctatetraene). Its first electronically excited state would be the square shape triplet. Because of the presence of nearly degenerate low lying excited states of different symmetry $4n$ molecules are expected to distort in an unsymmetrical manner and in the extreme be vibrationally unstable. Cyclobutadienes, therefore, may be capable of only transient existence in an excited state. The theoretical predictions of stable cyclobutadiene-transition metal complexes has been verified by isolation of several such complexes.

The fundamental theory of $4n$ π-electron molecules has been discussed by Craig. He concludes that in such "pseudoaromatic" molecules the notion of resonance is of dubious significance, and they are expected to show marked unsaturation and unequal bond distances.

The enhanced stability of $4n + 2$ π-electron molecules is not expected to continue *ad infinitum* as the ring size increases. As the rings become very large, theory indicates that the most stable configuration is that of alternating bond distances, regardless of the number of π-electrons. Monocyclic polyenes as large as $C_{18}H_{18}$ [which has $4n + 2$ π-electrons] have been prepared and shown to be aromatic in the sense of having equivalent bonds and sustaining a ring current, but not to have enhanced stability. Annalenes which do not have $4n + 2$ π-electrons, i.e., $C_{24}H_{24}$, $C_{16}H_{16}$, do not show "aromatic" electron delocalization.

In summary, the calculation of the π-electron molecular orbitals for planar monocyclic conjugated polyolefins gives an explanation for the observation that the $4n + 2$ number of π-electrons forms a highly stable aromatic molecule. The aromatic stability results from the molecule having its *bonding* molecular orbitals completely filled, thus forming a closed shell ground state. Odd numbered conjugated molecules (which must be radicals) not having this number of electrons tend to stabilize by gaining or losing electrons to attain the $4n + 2$ number, which produces a closed shell and the greatest π-resonance energy. In the $4n$ π-electron molecule, electron delocalization results in a degenerate ground state and so is not expected to enhance its stability. The most favored molecular structure for $4n$ molecules is one of alternating non-conjugated double bonds.

RICHARD WAACK

Cross-references: *Aromatic Compounds, Orbitals, Bonding.*

ARSENIC AND COMPOUNDS

Arsenic with atomic number 33 and atomic weight 74.9216 appears in Group VA of the Periodic Table. It is represented by the symbol As (L. *Arsenicum*). It is classified as a nonmetal and may exhibit a valence of +3, +5, or −3.

Arsenic is a rather ubiquitous element being widely disseminated throughout the earth's crust and in sea water. It ranks 47th in the order of the occurrence of the elements in the earth's crust. Arsenic has occasionally been found in the natural state, however, most arsenic is found in nature in a combined form. The native sulfides realgar, As_2S_2, and orpiment, As_2S_3, have been known since ancient times. Most commonly, it occurs as arsenides and sulfarsenides of heavy metals such as silver, lead, copper, gold, cobalt and iron. Over 150 minerals of arsenic have been reported. Metallic arsenic may be recovered directly by sublimation of the mineral arsenopyrite, FeAsS, however, most of it is produced by the direct reduction of As_2O_3 with carbon.

Most of the arsenic of commerce is sold in the form of As_2O_3 which is derived as a by-product from the treatment of copper and lead ores. The world production of arsenic trioxide in 1968 was in excess of 65,000 tons. The American Smelting and Refining Company in early 1971 was the only domestic producer of arsenic trioxide. At that time, the availability of arsenic trioxide exceeded the demand. Arsenic trioxide is sold in a crude form 95% As_2O_3 and as "white arsenic" 99.0% As_2O_3. The price of the crude oxide in 1971 was $94 per ton, while that of the white oxide was $120 per ton. Metallic arsenic is imported chiefly from Sweden. The price of the metal in 1971 was about 56 cents per pound.

Metallic arsenic is a steel grey brittle crystalline material with poor heat and electrical conductivity. It can exist in other allotropic forms such as amorphous or vitreous arsenic and yellow arsenic. The metal does not exhibit a melting point under normal conditions but sublimes on heating.

The physical properties of arsenic are as follows: density, crystalline, 5.72 g/cc (20°C); black amorphous, 3.7 g/cc; yellow cubic, 2.0 g/cc; melting point, 817°C (28 atmospheres); boiling point, (v.p. 1 atm.) 613°C; latent heat of fusion, 88.5 cal/g; latent heat of sublimation, 102 cal/g; specific heat, 0.082 cal/g/°C (20°C); linear coefficient of expansion, 4.7×10^{-6} in./°C (20°C); electrical resistivity, 33.3 microhm-cm (20°C); magnetic susceptibility, -5.5×10^{-6} cgs (18°C); thermal neutron absorption cross section, 4.3 barns; Brinell hardness, 147; crystal system, rhombohedral.

Metallic arsenic when exposed to the atmosphere will tarnish and the surface will develop a black coating. When heated in air or oxygen, it will oxidize to form arsenic trioxide, As_4O_6. Arsenic due to its electronegative nature unites directly with many metals to form arsenides of definite composition, e.g., Zn_3As_2, Mg_3As_2, $CoAs_2$, etc.

It is oxidized to orthoarsenic acid, H_3AsO_4, by concentrated nitric or sulfuric acid. Arsenic will be attacked by hydrochloric acid only in the presence of an oxidant. The halogens will react with it directly to form trihalides; the only pentahalide known is AsF_5. When fused with alkalis like NaOH, arsenic forms arsenites, $AsO_3\equiv$. It reacts with sulfur to form di- tri- and pentasulfides: As_2S_2 (realgar), As_2S_3 (orpiment) and As_2S_5. Trivalent and pentavalent arsenic oxides, oxygen acids and salts, and thio salts are known. Arsenic trioxide is the most important compound of arsenic. It is oxidized by HNO_3 or Cl_2 to orthoarsenic acid from which arsenate salts may be produced by neutralization. Arsine, AsH_3, an extremely toxic gas, can be produced by the hydrolysis of compounds such as Al_3As_2. When heated in the absence of air, AsH_3 dissociates into arsenic and hydrogen. The detection of arsenic by the thermal decomposition of arsine is the basis of the Marsh test.

Because of its brittle nature, arsenic has no use as a pure metal, however, it is often used as an alloying ingredient. When added in amounts of up to 3% to lead bearing alloys, it increases the hardness. The sphericity of lead shot is improved with quantities of ½ to 2%. A small percentage of arsenic is also added to lead base battery grid metal and cable sheathing to increase hardness. Minor additions of arsenic to copper increases the corrosion resistance and toughness.

High-purity arsenic metal has an important use as a constituent in the production of III-V type semiconductor compounds such as gallium and indium arsenides, gallium arsenic phosphide, and indium gallium arsenide. It is also used as a doping agent to form n-type germanium and silicon semiconductors.

Arsenic trioxide often referred to in commerce as arsenic not only finds use as such but is also the starting material for the preparation of many arsenical compounds. Arsenic trioxide is used as a fining agent in the manufacture of glass. In formulations with copper and chromium, it is used as a wood preservative. Since many arsenical compounds are biologically active, major applications are in the area of pesticides and herbicides. Some of the familiar pesticides are Paris green or copper acetoarsenite, lead and calcium arsenate, and cacodylic acid. Sodium arsenite is used in cattle dips to control infection by ticks. Sodium arsenite in addition to disodium methylarsonate and cacodylic acid are used as herbicides. Arsenic acid is used as a cotton dessicant and defoliant prior to harvesting. Arsenilic acid is used as a feed additive to chicken, turkey and hog rations. Other compounds of arsenic, such as arsenic sulfide (realgar), is used for tanning hides and in pyrotechnics; arsenic trisulfide is used in infrared lens applications.

Metallic arsenic and arsenic trisulfide may be handled but, in general, skin contact with arsenical compounds should be avoided. An effective antidote for arsenic poisoning due to accidental ingestion is freshly prepared ferric hydroxide (prepared from ferric chloride and milk of magnesia) which reacts with the arsenical forming a nontoxic compound.

S. C. CARAPELLA, JR.

References

Mellor, J. W., "Comprehensive Treatise on Inorganic and Theoretical Chemistry," Vol. 9, Longman, Green and Co., New York, 1930.

Sneed, M. C., and Brasted, R. C., "Comprehensive Inorganic Chemistry," Vol. 5, D. Van Nostrand Co., Princeton, N.J., 1956.

Carapella, S. C., Jr., "Arsenic," pp. 29–33, in "Encyclopedia of the Chemical Elements," C. A. Hampel, Editor, Van Nostrand Reinhold Co., New York, 1968.

ASPHALT

Asphalts are black to dark brown highly viscous mixtures of paraffinic, naphthenic and aromatic hydrocarbons together with heterocyclic compounds containing sulfur, nitrogen and oxygen as well as small amounts of metals. These components are now known to cover the molecular weight range of about 300 to 5000, although small amounts of lower molecular weight material are frequently present. Asphalts are primarily employed in applications which make use of their adhesive quality, waterproofing ability, relative chemical inertness, and high viscosity at ambient temperatures.

Over 95% of the asphalt used in the U.S. is produced from petroleum as the residue from vacuum distillation or steam distillation or from vacuum flashing operations. About 4% of the asphalt used occurs in natural deposits, of which Bermudez Lake

(Venezuela) and Trinidad Lake (British West Indies) are the most important examples. A number of very hard naturally occurring asphalts which have special uses in paints and printing inks include gilsonite and glance pitch.

Production and Major Uses. About 34 million tons (about 164 million barrels) of asphalt of all kinds were consumed in the United States in 1970. About 72% of the asphalt from petroleum is of the viscous Newtonian fluid type and is used in the construction and maintenance of roads and airports. About 19% is processed by air-blowing to render it non-Newtonian with a definite yield stress and is used to produce roofing and floor-covering materials. The remaining 9—10% is specially processed and formulated for use in the construction of hydraulic works such as canals, dikes, dams, and erosion-control membranes and in the manufacture of paints, electrical insulation, waterproof paper, joint fillers for concrete pavements, battery seals, acidproofing for the interior of storage tanks, wood preservatives, gear lubricants and other items. About half the asphalt used for road construction (the "penetration grades") is sufficiently viscous to require heating in order to mix with mineral aggregates. The other half is blended with petroleum fractions such as naphtha, kerosene or furnace oils to yield low viscosity "cutback grades" or is emulsified in water. These products are sufficiently fluid to apply without heating.

Chemical Composition. Chemical studies show that asphalts obtained as residues from vacuum distillation, steam distillation, vacuum flashing or solvent deasphalting operations on petroleum normally consist of carbon (70–85%), hydrogen (7–12%), nitrogen (0–1%), sulfur (1–7%), oxygen (0–5%) and generally small amounts of metals, some of which are present in the form of finely dispersed oxides and salts and some in metal-containing porphyrin compounds. The content of elements other than carbon and hydrogen varies greatly with the source of the crude oil. Asphalts are composed primarily of high-boiling compounds, and their fractionation for purposes of composition studies without thermal cracking has been accomplished by the use of molecular distillation under high vacuum with short contact time at high temperatures to yield distillate cuts of successively higher molecular weight up to a molecular weight level of about 1500. The color of these distillation fractions progresses from light yellow to green to brown to very dark greenish black and the logarithm of their viscosities at a given temperature is a linear function of their molecular weight. The hard residue from such a distillation has an average molecular weight in the range of about 1500–3000. The molecules of which asphalts are composed are relatively small compared with those of polymeric materials such as rubber and plastics.

By chromatography over silica gel, distillate fractions of an asphalt may be separated into three general classes of compounds: (1) saturated hydrocarbons, (2) aromatic hydrocarbons, and (3) a fraction containing heterocyclic sulfur, nitrogen and oxygen compounds of various types which for convenience are called "resins." In general, the saturated hydrocarbon fractions are white to light yellow and may amount to as much as 35% of the lower molecular weight distillate fractions; their proportion decreases progressively with increasing molecular weight. By chilling a solution of the saturated hydrocarbons in acetone, a filterable wax is obtained which consists primarily of paraffins (predominantly normal) and secondarily some alkyl-substituted naphthenes. The normal paraffins in the wax may be removed by complexing with urea, leaving isoparaffins and naphthenes. The remaining oil is composed largely of naphthenes with normal or isoparaffinic side chains.

Aromatic hydrocarbons may constitute about 50% of the lower molecular weight fractions; their proportion increases with increasing molecular weight of the distillate fraction. At high molecular weight, the aromatics may subsequently decrease due to the rising amounts of heterocyclic and other compounds containing sulfur, nitrogen and oxygen. The aromatics may be separated further into monocyclic and dicyclic fractions by chromatography over alumina. Ultraviolet absorption spectra show the presence of only small amounts of aromatics with three or more aromatic rings condensed directly together. These aromatic compounds consist primarily of benzene and naphthalene structures combined and substituted in a variety of ways. Both normal and isoalkyl and cycloalkyl substituents are present on the aromatic rings.

The resin fraction increases in amount with increasing molecular weight and in many asphalts at very high molecular weights the amounts of sulfur, nitrogen and oxygen are sufficient to provide each molecule with several hetero atoms. Sulfur is present primarily in thiophene and benzothiophene structures, which may be isolated by oxidizing them under mild conditions to sulfoxides that are adsorbed on silica gel during chromatography of the resultant mixture. Because of the complexity of the system, little is known definitively concerning the relative amounts of the various oxygen compounds in asphalt. Oxygen is present in hydroxyl, carbonyl, carboxyl and ester groups. Nitrogen occurs in basic and nonbasic forms, primarily as pyridine and pyrrole types of structures, respectively. The pyrrolic structures are important in connection with the occurrence of the metals magnesium, iron, nickel, vanadium and others in petroleum residues. Four pyrrole rings linked together by methylene bridges encircle various metal and metal oxide ions to form the porphyrin group of compounds apparently derived from the hemin of animal blood and from the chlorophyll of plants. The surface-active properties of asphalt, and hence its adhesive characteristics, are due to its heterocyclic and polar constituents.

Less is known of the details of chemical structure of components with molecular weights above about 1500 which are insoluble in pentane and hexane (generally referred to as "asphaltenes"), but on the basis of their chemical analyses and mass spectra they are known to be a continuation into higher molecular weights of the types of compounds which constitute the aromatic and heterocyclic fractions. Ultracentrifuge studies confirm that the molecular weights of the molecules in asphaltene fractions are at the most a few thousand.

Chemical Reactivity. From what is known of the composition of asphalt, it is evident that the chemical reactivity of such a mixture will be extremely complex. Present knowledge of asphalt chemistry is mostly limited to the effect of a chemical reagent on the rheology and solubility of the asphalt and to speculation on the nature of the reactions which may have taken place. Asphalts generally are resistant to many chemical reagents, and it is for this reason that they are widely used as protective coatings.

Oxidation. The reactivity of asphalt toward oxygen under a variety of conditions is one of its most important chemical characteristics. About 25–30% of all asphaltic products are produced by blowing air through distillation residues from petroleum at temperatures of about 300–500°F. Air-blowing increases the hardness of the asphalt and converts it from a Newtonian material with the properties of a viscous liquid into a non-Newtonian material with the properties of a plastic. During air-blowing, oxygenation and dehydrogenation reactions take place followed by others including dehydration and decarboxylation, which result ultimately in the removal of most of the reacted oxygen in the form of water, carbon dioxide and some low boiling carbonyl compounds. Only a small fraction of the reacted oxygen remains in the asphalt, primarily in the form of carboxyl, hydroxyl and ester groups. At lower air-blowing temperatures the amount of reacted oxygen remaining in the product increases. High molecular weight components are produced from those of lower molecular weight during air-blowing as a result of carbon-carbon coupling reactions made possible by removal of hydrogen atoms and by coupling through oxygen atoms following the same initial step. The number, size, configuration and position of ring substituents will govern the rate of oxidation of aromatics and to a lesser extent that of naphthenes, as well as the solubility of the oxidation products. The presence of small amounts of unsubstituted aromatics, phenols or amines may greatly retard oxidation of naphthenes and paraffins. Hence, it is not surprising that asphalts from different sources respond differently to air-blowing.

Oxidation of asphalts at atmospheric temperatures is of importance in two general classes of applications which differ primarily in whether the material is exposed during oxidation to the light or not. When asphalt is exposed to the air at room temperature in the dark, oxidation takes place at the surface to produce a surface film of very high viscosity which presents an effective barrier to the diffusion of more oxygen into the material. The rate of oxidation is then controlled by the very low rate of oxygen diffusion, and this accounts for the excellent resistance to oxidation of asphalts in the dark.

Atmospheric oxidation of asphalt at ordinary temperatures in sunlight proceeds much faster than in the dark; peroxides, aldehydes and ketonic acids are among the oxidation products. Also, much more oxygen remains combined with the asphalt than in air-blowing at high temperatures. The oxidized surface becomes hydrophilic and more soluble in oxygenated solvents; eventually, water-soluble compounds are produced. During this rapid oxidation in the presence of light, mechanical strains develop which very frequently lead to cracking of the hardened surface with resultant exposure of unreacted asphalt. This progressive degradation is prevented in practice by shielding the asphalt layer from sunlight by a thin layer of rock granules or by incorporation of additives which absorb at specific wavelengths in the ultraviolet region. The reactivity of the various chemical type fractions of asphalt to oxygen has been shown to be: saturates < aromatics <resins < asphaltenes. The hardening of paving asphalts during application and use has been studied extensively as a function of mixing temperature, air content of the compacted mixtures, thickness of the pavement and chemical nature of the asphalt. A simple and rapid laboratory procedure employing a sliding plate microviscometer has been shown to predict accurately, in advance, the hardening from oxidation to be expected over a period of years for paving asphalts in field installations.

Action of Acids at Ordinary Temperatures. Asphalt resists attack by dilute sulfuric acid and by hydrochloric acid in all concentrations; it is attacked by concentrated sulfuric and by nitric acid in all concentrations. Concentrated sulfuric acid converts the aromatics and heteroaromatics into water-soluble sulfonic acids. The combined oxidizing and nitrating power of nitric acid leads to products containing nitrogen (probably nitrated aromatics). Paraffins and naphthenes generally are not attacked by sulfuric acid. The passage of hydrogen chloride gas through the material results in appreciable hardening, the extent depending on the amount of nitrogen bases converted to hydrochlorides or polymers.

Action of Alkalies. Asphalt is not attacked by concentrated alkaline solutions at ordinary temperatures, but dilute alkaline solutions react with acidic constituents to form salts such as sodium naphthenates which serve as excellent emulsifying agents for the asphalt. The interfacial tension of asphalt-water systems is generally very low at high pH.

Action of Sulfur. Reactions between asphalt and sulfur take place at temperatures above about 350°F to give products of high molecular weight and low solubility. This is not exactly analogous to the vulcanization of rubber, since olefins usually are not found in asphalts. Rather, the reactions of sulfur with asphalt are similar to those of oxygen, wherein most of the sulfur is released as hydrogen sulfide and only a small part remains. Dehydrogenation-coupling reactions probably are important as in the case of oxygen.

Action of Halogens. The reactivity of halogens and halogenated compounds with asphalt decreases in the order, chlorine, bromine, iodine. Chlorination at 200°F produces high molecular weight material of low solubility, which on further heating gives off hydrochloric acid. At higher temperatures, 400–500°F, little of this chlorine reacted remains in the asphalt but appears primarily as hydrochloric acid. High temperature chlorination produces a material quite similar to that which results from air-blowing at the same temperatures. Treatment of asphalts with boiling halogenated solvents generally results in some loss in halogen from the solvent and in hardening of the asphalt.

Trends in Research. Although much technological activity continues to be directed to the solution of specific problems in asphalt manufacture and applications, it is clear from a careful study of the literature that the chemical constitution of asphalt and its relationship to broad rheological types, the complex stress/strain characteristics of the material as a function of loading time and temperature, and the major factors controlling durability are now reasonably well understood. New developments in this subject are emerging from the modification of the composition, rheology and chemical characteristics of asphalt as a result of mechanical and chemical combination of this material with elastomers, plastics and resins.

W. C. SIMPSON

ASSAY

The term "assay" as used today, implies analysis for only a certain constituent or constituents of a mixture; the others usually are neglected. The term is generally applied to ores, alloys and pharmaceuticals. The analysis of the principal component as given in the specifications of some industrial chemicals is often referred to as assay.

Chemical assay methods are divided into two classes, dry and wet. The assay of an ore for gold is the best known of the dry or fire methods. The ore is finely ground and mixed with charcoal and an appropriate flux, the composition of which depends on the acidic or basic nature of the ore. The mixture is slowly brought to red heat. The gold, lead, and silver separate as a free metal button in the bottom of the crucible and is isolated from the slag by breaking the mass with a hammer. The lead and silver are separated by various methods and the free gold is weighed. Among the oldest of the wet methods is the assay of certain plants for their alkaloid content. The dried plant material is extracted with a solvent and the active constituent separated from the extract by chemical means.

Pharmaceutical preparations generally contain one or more active ingredients mixed with inert fillers, binders, etc., or dissolved or suspended in a liquid. The finished product must be assayed to determine whether or not it contains the stated amount of the active constituent(s) per unit (capsule, ampule, tablet or unit of volume).

The newer field of assay, *biological assay,* generally referred to as *bioassay,* is the test of living organisms to a substance. The organisms used include single and multicellular plants and animals. Assay using single cell organisms is called *microbiological assay.* The most familiar technique is the "plate assay" used for testing antibiotics against various strains of bacteria.

Not only is bioassay used to control the quality of many pharmaceutical preparations and animal feed supplements, but it is an important research tool. The search for new antibiotics, vitamins and hormones depends on this method of analysis for detection and initial evaluation. The feed supplements for cattle, hogs and chickens that promote more rapid growth and immunity to certain costly diseases are developed by feeding tests with the animal concerned. The screening of chemicals as insecticides is done with insects. A household insecticide for flies and mosquitoes, for instance, is evaluated for its effectiveness by the percentage of "kill" of these insects under a given set of conditions.

JOHN A. RIDDICK

ASSOCIATION

Association was first thought of as a reversible reaction between like molecules that distinguished it from polymerization, which is not reversible. Association is characterized by reversibility or ease of disassociation, low energy of formation (usually about 5 and never more than 10 kcal per mole), and by the coordinate covalent bond which Lewis called the acid-base bond. Association takes place between like and unlike species. The most common type of this phenomenon is hydrogen bonding. Association of like species is demonstrable by one or more of the several molecular weight methods. Association between unlike species is demonstrable by deviation of the system from Raoult's law.

The strength of the coordinate covalent bond is a function of polarity of the associating molecules. Hence, associated molecules vary in stability from very unstable to very stable. The argon-boron trifluoride complex is quite unstable, whereas calcium sulfate dihydrate (gypsum) is very stable. The bond strength associated with stability has been measured for a number of combinations. The strengths of some hydrogen bond types decrease in the order FHF, OHO, OHN, NHN, CHO but they are dependent upon the geometry of the combination and upon the acid-base characteristics of the group. Steric effects can have a marked effect on the strength of the coordinate covalent bond. This was demonstrated by a study of the strength of the series NH_3, $C_2H_5NH_2$, $(C_2H_5)_2NH$, $(C_2H_5)_3N$ as bases toward an acid in solution and in the gaseous state and the comparison of the base strength of triethylamine and quinuclidine. The latter is, in effect, triethylamine in which the 2-carbon atoms of each ethyl group are tied together by another carbon. The geometry of the ethyls around the nitrogen is drastically changed, and the cyclic is a stronger base than the triethyl compound. The factors affecting the strength of the hydrogen bond also influence the degree of association.

Association within the same species accounts for the high boiling points of, for example, water, ammonia, hydrogen fluoride, alcohols, amines and amides. Ethyl ether and butanol contain the same number of atoms of each element but butanol has a boiling point 83° above that of ethyl ether as a result of more extensive hydrogen bonding. Some substances associate completely to two or more formula weights per molecule. Carboxylic acids, by a hydrogen-to-oxygen association, form dimers with a six-membered ring. N-Unsubstituted amides dimerize in the same manner, whereas N-substituted amides dimerize in a chain form in a *trans* configuration (see **Solvents**).

Hydrogen bonding is so common that coordinate bonds between other elements are sometimes overlooked. Antimony (III) halides form very few complexes with other halides, whereas aluminum halides readily form complexes. The octet of electrons is complete in all atoms of the antimony halides but is incomplete in the aluminum atom of aluminum halides:

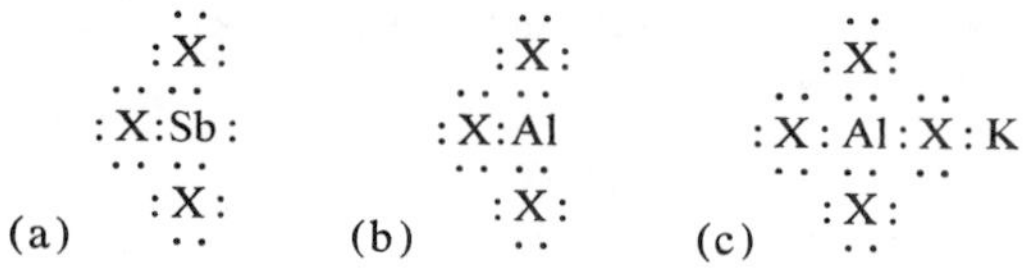

(a) : X : Sb : with :X: above and :X: below

(b) : X : Al with :X: above and :X: below

(c) : X : Al : X : K with :X: above and :X: below

Aluminum can accept two electrons to complete its octet. The pair of electrons is available from the halogen. An alkali halide can supply the electrons and form a complex (c, above); or the electron pair may come from the halogen of another aluminum chloride. Association with other aluminum halides accounts for the higher melting point of aluminum halides over antimony (III) halides which have a formula weight of 95 or more. The association of aluminum sulfate, alkali metal sulfate and water to form the stable alums is one of the more complex examples.

The formation of solvates is association between unlike species. Solvation is more frequent between substances of high polarity than those of low polarity. This is illustrated by the decrease in the tendency to form solvates with decrease in dipole moment and dielectric constant (in parentheses) for N-methylacetamide (3.59, 172), to water (1.84, 78.4), to ethanol (1.70, 24.6), to ammonia (1.48, 17.8), to methylcyclohexane (0, 2.02) for which few associations are known.

JOHN A. RIDDICK

Cross-references: *Solvents, Complex Inorganic Compounds, Chelation, Sequestering Agents.*

ASTATINE*

Astatine, element 85, is the heaviest member of the halogen family of elements. Unlike other members of the family, however, it has no stable or long-lived isotope. Although minute amounts of the very short-lived At^{215}, At^{218} and At^{219} are present in nature in equilibrium with long-lived radioactive elements, the only practical way of obtaining astatine is by synthesizing it through nuclear reactions.

Element 85 was first characterized in 1940 by Corson, MacKenzie, and Segré, who synthesized the isotope At^{211} by bombarding bismuth with alpha particles. They observed chemical behavior somewhat similar to that of other halogens and named the new element "astatine," from the Greek "astatos," meaning "unstable." Since that time a number of astatine isotopes have been identified, with masses ranging from 200 to 219 and half-lives varying from 2 microseconds to 8 hrs.

The least unstable astatine isotope, At^{210}, has only an 8.3 hr half-life. This corresponds to a specific activity of two curies per microgram! Because of its short half-life and formidable radioactivity, studies of ponderable amounts of astatine have not been possible, and nothing is known of the bulk physical properties of the element. The largest preparations of astatine reported to date have been about 0.05 microgram.

The atomic absorption spectrum of astatine has been measured. Two lines at 2244.01 Å and 2162.25 Å have been assigned respectively to the transitions $^2P_{3/2}{}^0 - {}^4P_{3/2}$ and $^2P_{3/2}{}^0 - {}^4P_{5/2}$ between configurations $6p^5$ and $6p^47s$.

Although astatine behaves much like iodine, there is to date no evidence for the existence of the molecule At_2. Mass spectral studies of astatine and its compounds have failed to reveal any sign of such a molecule. The elemental astatine species may be atomic At resulting from dissociation of At_2 at the very low concentrations involved in all the research carried out so far. It is also likely that in many cases the species thought of as elemental astatine actually consists of a mixture of organic astatine compounds formed by reaction of the astatine with impurities in the experimental systems.

Most investigations of astatine chemistry have been carried out by tracer techniques, in which the element is detected only by its radioactivity. However, the compounds HAt, CH_3At, AtI, AtBr, and AtCl have been identified by direct measurement of their masses in a mass spectrometer.

Tracer studies have shown the existence of at least five oxidation states of astatine in aqueous solution:

The "zero" or "elemental" state is characterized by extractability into nonpolar organic solvents, by volatility, and by a tendency to adsorb on various surfaces. The species At and At_2 have usually been assumed to be present. However, various organic astatine compounds may behave similarly, while in the presence of iodine the species AtI is formed.

The −1 state is characterized by coprecipitation with insoluble iodides. It is formed by reducing higher states with strong reducing agents, such as stannous chloride. The species present is presumed to be At^-, the astatide ion.

The +5 state is characterized by coprecipitation with insoluble iodates. It is formed by oxidizing lower states with such strong oxidants as Ce^{+4} or H_5IO_6. Presumably the astatine is present as the astatate ion AtO_3^-.

One or more intermediate positive oxidation states of astatine result from oxidation of lower states with mild oxidants, such as bromine or dichromate. After such oxidation the astatine does not extract into nonpolar organic solvents, nor does it usually coprecipitate with either insoluble iodides or iodates. Species present may be intermediate oxides or oxyacids of astatine, polar interhalogens such as AtBr, or even polar organic astatine compounds. Cationic astatine species also appear to be present in some cases. Under certain conditions polyhalide complex ions are formed, such as AtI_2^-, $AtIBr^-$, $AtICl^-$, $AtBr_2^-$, and $AtCl_2^-$.

Evidence has recently been found for the existence of perastatate—astatine in the +7 oxidation state. It is prepared by oxidation of lower states with xenon difluoride in alkaline solution, and is characterized by coprecipitation with KIO_4 and $CsIO_4$.

The aqueous chemistry of astatine may be summarized by the following approximate electrode potential diagram for solutions in 0.1 M acid:

$$At^- \text{----} 0.3v \text{----} At(O) \text{---} 1.0v \text{---} At(+X)$$

$$\text{---} 1.5v \text{---} AtO_3^-$$

Tracer techniques have been used to characterize a number of organic astatine compounds, among

---

* Written under the auspices of the Atomic Energy Commission.

them: C_6H_5At, HOC_6H_4At, $AtCH_2COOH$, p-AtC_6H_4COOH, p-$AtC_6H_4SO_3H$, $(C_6H_5)_2$ $AtCl$, $C_6H_5AtO_2$, $At(C_5H_5N)_2ClO_4$, $At(C_5H_5N)_2NO_3$, and several alkyl astatides. Astatine has also been successfully incorporated into protein molecules.

When injected as At^- into experimental animals, astatine concentrates in the thyroid, though not quite as selectively as does iodine.

EVAN H. APPELMAN

References

Aten, A. H. W., Jr., "The Chemistry of Astatine," *Advances in Inorganic and Radiochemistry,* **6,** 207 (1964).

Appelman, E. H., "Astatine," MTP International Review of Science, Inorganic Series One, Vol. 3, V. Gutmann, Ed., Butterworth and Co. Ltd. London, 1972, pp. 181–198.

ASYMMETRY

Asymmetry involves the presence of four different atoms or substituent groups bonded to an atom. Its existence was discovered in 1815 by the French physicist J. B. Biot (1774–1867). He found that oil of turpentine, solutions of sugar, camphor, and tartaric acid all rotate the plane of plane-polarized light when placed between two Nicol prisms. This phenomenon is called optical rotation and is indicated in symbols in the following manner: $[\alpha]_D^{20°} = +53.4$ aq., meaning that the substance gives a rotation of 53.4° to the right (clockwise, or plus) in water solution at 20°C using the sodium D line as the light source. Substances in solution that rotate light to the right are designated d and called dextrorotatory; substances rotating light to the left are designated l and called levorotatory.

In 1848, Louis Pasteur (1822–1895), working with the sodium ammonium salt of optically inactive racemic acid, an acid isolated from winery sludges, made a remarkable discovery. On recrystallizing the salt in a cool room, he found that it separated into two crystalline forms, differing only in that they had hemihedral faces inclined in opposite directions. He found that a solution of those crystals having right-handed faces was dextrorotatory, while the left-handed crystals gave a levorotatory solution. A mixture of equal parts of both crystals was optically inactive, as the original acid had been. The d crystals were shown to be identical with the then-known d-tartaric acid, while the l-acid was a new compound. This was the first separation of such a dl mixture, and such mixtures are now called *racemic* for this reason. J. Wislicenus (1835–1902), in 1873, working with the naturally occurring d and l lactic acids, showed that the two optical isomers have identical physical and chemical properties, except for the direction of rotation of polarized light. If the l form has a rotation of $-3.3°$ then the d form will have a rotation of $+3.3°$.

These results were puzzling to the organic chemists of the time, because they did not have a three-dimensional concept of molecular structure. It was the brilliant postulate of the tetrahedral carbon atom, arrived at independently in 1874 by Jacobus H. van't Hoff (1852–1911) and J. A. Le Bel (1847–

1930) that solved the problem and made the existence of optical isomers seem reasonable. The postulate that the four valences of carbon point to the corners of a regular tetrahedron implies that optical isomers can exist, and their well authenticated existence can be taken as evidence for the postulate.

Consider the molecule formed of carbon and four different atoms or groups D, E, F, G.

$$\begin{matrix} & D & \\ & | & \\ G & \!\!-C-\!\! & E \\ & | & \\ & F & \end{matrix} \qquad (I)$$

Now, keeping the line DCF fixed, rearrange the other two groups to get

$$\begin{matrix} & D & \\ & | & \\ E & \!\!-C-\!\! & G \\ & | & \\ & F & \end{matrix} \qquad (II)$$

It can be readily seen that the two molecules are related to each other as right and left hands; they are mirror images, but not superimposable if left in the plane of the paper.

$$\begin{matrix} & D & \\ & | & \\ G\!-\!C\!-\!E & & (I) \\ & | & \\ & F & \end{matrix} \qquad \begin{matrix} & D & \\ & | & \\ E\!-\!C\!-\!G & & (II) \\ & | & \\ & F & \end{matrix}$$

It must be added that they can be superimposed simply by rotating structure II 180° about the axis DCF. This is a fault of the planar representation of a three-dimensional atom. If a model of the two atoms is constructed in three dimensions, it can be seen that they can in no way be superimposed. In considering whether two optical isomers are identical from their planar projections, it must be remembered that rotations of any kind in the plane of the paper are allowed, while lifting the projection off the paper and inverting it is not allowed. Compounds related to each other as the two model compounds (image-mirror image) are called *enantiamorphs* or optical antipodes.

Carbon atoms with four different groups attached to them are called asymmetric carbon atoms, and in chemical symbolism are quite frequently marked with an asterisk (*). It is possible for a compound to have more than one of them. Tartaric acid has two, while naturally occurring compounds such as proteins, with one asymmetric carbon per peptide unit, may have millions. Tartaric acid, with two asymmetric carbon atoms, has a possibility of having four optical isomers, of which three are known.

In order to consider asymmetric compounds without writing out a complete formula each time, a conventional way of writing the planar projection has been adopted. The molecule is represented with the carbon atoms in a line, with the groups falling to the right or left as if one were sighting down the length of the carbon chain. As a further guide, the configuration of the first carbon atom in the chain is referred to one of the two glyceraldehydes.

```
        CHO                         CHO
         |                           |
   H—C*—OH   (D)           HO—C*—H   (L)
         |                           |
       CH₂OH                       CH₂OH
```

Compounds that have the same configuration as D-glyceraldehyde are designated D, and those with other configuration L. It must be added that configuration as represented in a structural formula and direction of rotation are not necessarily related. D-Glyceric acid is dextrorotatory, while D-lactic acid is levorotatory.

With three, four, and more asymmetric atoms in a molecule, the number of possible isomers becomes tremendous. The maximum number of isomers for n asymmetric atoms is 2^n, but the actual number may be smaller because of mesomeric pairs. Compounds having the same configuration on all but one carbon atom are called *epimers*. Epimers are quite frequent in the hexoses—6-carbon sugars that have four asymmetric carbon atoms. All the 16 possible optical isomers are known in this series.

Asymmetry of this kind is not, however, limited to carbon compounds. Any atom with tetrahedral or octahedral valences may be asymmetric. If three different groups are present in the octahedral coordination sphere of cobalt, it can be seen that mirror images and optical isomers will exist.

```
   NH₃  Cl  Cl        |        Cl  Cl  NH₃
     \  |  /          |          \  |  /
       Co             |            Co
     /  |  \          |          /  |  \
   NH₃ NH₃ Br         |        Br  NH₃ NH₃
```

Asymmetric atoms of sulfur, tin, nitrogen, selenium, and other elements have been observed in compounds of those elements.

Asymmetry has been a very useful tool in the detailed study of organic reactions. The question of which bond is broken in the hydrolysis of an ester was one of the first questions to be solved by using pure optical isomers. Other methods, such as radioactive tracer studies, have been used to decide questions of mechanisms, and they are useful supplements to the classical methods of stereochemistry.

ELBERT C. WEAVER

Cross-references: *Optical Rotation, Polarimetry, Stereochemistry.*

ATOMS

The word "atom" is used universally in chemistry and physics to denote the smallest particle of an element which can exist either alone or in combination with other atoms of the same or of other elements and still have the properties of the given element. Breaking up an atom would yield particles that would not have the same properties as the original element. The radius of an atom is about 2 or 3×10^{-8} cm. Atoms are the building blocks from which molecules are constructed and they are the particles which occupy regularly spaced positions in the lattices of crystals. The mass of the lightest atom, the hydrogen atom, is 1.67×10^{-27} kilogram. This means that its mass is about 1840 times the mass of an electron.

Historical Development. The name "atom" was derived from the Greek word "atomos" which means uncut or indivisible; the Greek philosopher, Democritus, was the first to propose the existence of such elementary particles. The idea was revived late in the nineteenth century in an effort to explain experimental results. Dalton's Law of Partial Pressure showed that in any mixture of gases or of vapors or of both, each constituent exerted its pressure as though the other constituents were not present. Other studies, such as the development of the kinetic theory of gases and measurements on the combining weights of elements, brought out a more complete atomic theory. In many experiments atoms behaved as though they were tiny solid spheres, in accordance with the ideas of Democritus. However, later developments indicated that the atom was anything but solid and indivisible and further that it had associated with it some waves. These results showed that the atom was far more complex than a simple solid particle.

Many important discoveries in the last half of the nineteenth century had a direct bearing on beliefs about the structure of matter. Studies of electrical discharges through gases and in particular of cathode rays led to the discovery of electrons, and J. J. Thomson and others showed that electrons are constituents of atoms. Later, Lord Rutherford carried out elaborate experiments on the scattering of radioactive emanations as these particles passed through metal foils. The results convinced him that the positive charges in matter were concentrated in very tiny regions whose size was about 10^{-12} cm or a millionth of a millionth of a centimeter. This positively charged core or nucleus was surrounded by an equal amount of negative charge.

The Orbital Model. Albert Einstein invoked the idea of light quanta. This theory proposed that the emission of light energy occurred in certain discrete amounts called quanta or in even multiples of these amounts, but not in other amounts. The Danish scientist, Niels Bohr, tied this concept to processes in individual atoms by developing his orbital model of the atom. He postulated that the negative charge in the atom was carried by electrons which rotated in stable orbits around the positive core or nucleus of the atom. His hypothesis for the hydrogen atom was that there was one stable orbit for the single electron which rotated about the nucleus of this atom. The angular momentum of the electron in this orbit was assumed to have a value equal to $h/2\pi$ where h was the universal constant known as Planck's constant. Bohr also assumed the existence of other orbits in each of which the angular momentum was an integral multiple of $h/2\pi$. If the electron should get out into one of these orbits the atom was in an excited state which was not a stable configuration. All other orbits were considered to be impossible arrangements.

This meant that the atom could exist (at least temporarily) in states in which it had certain definite amounts of energy, but states corresponding to electrons in other orbits or other amounts of energy were ruled out. Hydrogen atoms which were not combined into molecules were found in electrical discharges, and in the discharges the excited states corresponding to electrons in other than stable orbits also appeared. When an electron dropped from one orbit to another in which it was closer to the nucleus, Bohr's theory stated that energy was emitted and the atom lost this same amount of energy. Since only certain orbits were considered possible, only certain definite amounts of energy were emitted and these amounts corresponded to Einstein's "packets of energy." Also, energy could be absorbed by the atom in amounts equal to the amounts needed to move the electron out to specific orbits.

Atoms of elements other than hydrogen have more than one electron and these are found in specific orbits when the atoms are in their normal states. Each element is characterized by the number of electrons which makes up its normal complement. This number is also equal to the amount of positive charge which is contained in the nucleus when that amount is expressed in multiples of the charge on one proton. These numbers, which are called atomic numbers, range up to 105. For each kind of atom there are certain stable orbits and others which correspond to excited states of the atom. If an atom has either more or less than its normal complement of electrons in orbits around it, the atom is said to be ionized.

Extensions of Bohr's model were proposed which used elliptical orbits or shells of negative electricity to account for the results of some precise experiments. None of these have been entirely successful and Bohr's theory is not considered to be a complete picture of the atom. None the less, the model has retained its usefulness because it provided a picture of the atom which could be visualized and the model is frequently used to describe the conclusions reached by theoretical calculations based on much more complicated models of the atom.

Atomic Weights. The atomic weight of an element is the average mass of the atoms making up a sample of the element when these masses are expressed on a proportional scale on which the average mass of the atoms of a sample of ordinary oxygen is taken as exactly 16. On such a scale, hydrogen has an atomic weight of 1.008. When using this scale, the whole number or integer nearest to the mass of a particular atom is called the mass number of that atom. This scale is sometimes called the "chemical" system. Another scale called the "physical" system is also used, especially where single atoms are to be compared. Some types of oxygen atoms are heavier than others though all have the properties of oxygen. The physical scale selects the mass of the most common form (isotope), calls it 16, and uses that as a standard. Under this system the atomic weights come out about 1.0002 times those based on the chemical scale.

A "unified" scale to replace the other two was adopted by the International Union of Pure and Applied Physics and the International Union of Pure and Applied Chemistry. For this scale a standard value of 12 is used for the most abundant type of isotope of carbon.

Classification of Elements. Since all the atoms of any one element have the same amount of positive charge on each of their nuclei and this amount is different from the amount on the nuclei of any other element, the elements can be arranged in order of increasing positive charge. When this is done, it is found that these amounts are all exact multiples of the charge of the first element on the list—hydrogen. Each element is assigned an atomic number which denotes its position on this list. This number also indicates the number of electrons in the orbits about the nucleus. The arrangement of the electrons, particularly the number of electrons in the outermost orbit which is occupied, determines the way in which the atom will combine with other atoms to form molecules. In an attempt to classify elements, Mendeléeff found that they could be placed in groups such that those in any one group had similar chemical and physical properties and that various properties showed a continuing trend throughout the group. Names of the elements can be arranged in rows with atomic number increasing from left to right, starting a new row whenever an element is reached which is similar to hydrogen in its chemical properties. This results in a periodic chart of the elements similar to the one proposed by Mendeléeff. Elements which are in any one column are those which are similar in their chemical properties. These characteristics are determined mainly by the number of electrons in the outermost orbit which is occupied.

Electron Distribution. The electrons in any atom fall into groups or classes in accordance with the amount of energy needed to remove them from the atom and these groups are called the K shell, the L shell, the M shell, etc. Experiments with x-ray photons are used to determine the energies necessary to remove electrons. For each kind of atom the electrons which are hardest to remove are the K electrons, those in the next group are L electrons, etc. The maximum number of electrons in each shell is 2 for the K shell, 8 for the L shell, 18 for the M shell, and 32 for the N shell.

If a start is made with the lightest atom, the hydrogen atom, and each element is considered in turn throughout the Periodic Table, we find that more and more orbits are occupied. The second element, helium, has two electrons for each atom and these fill the K shell. The next atoms have some electrons in the L shell. It is not always found that any given shell is filled before the next shell receives any electrons. Sometimes a shell is partially filled to a convenient semi-complete stopping place with 8 or 18 electrons in the outermost occupied orbit. Then in the following elements some electrons are placed in the next orbit before any more are placed in the semi-complete orbit. Later, this orbit is filled in. When the outermost orbit which is occupied is full, the atom is nearly inert. There are also nearly inert atoms when the M shell reaches its semi-complete point with 8 electrons and when the N shell has 18 electrons.

The atoms of elements just beyond the nearly inert ones have one and only one electron in the outermost shell which is occupied. This last electron revolves in a field which is similar to that of the one electron of the hydrogen atom. This results in these elements having spectra which are similar to that of hydrogen, and these atoms all have similar chemical properties. Throughout the Periodic Table there are correlations between the chemical properties and the number of atoms in the outer orbit.

One might expect that the atoms which have several occupied shells would be considerably larger than those which have fewer occupied shells. However, the larger attractive forces, which the more strongly charged nuclei of these heavier atoms exert on the electrons, result in the stable orbits being drawn inward so that there is only a comparatively slight variation in size among all the known atoms.

The Nucleus. The nucleus or core of the atom contains all the positive charge associated with that particle. For a long time it was believed that nuclei consisted of protons and electrons. This gave way to the belief that nuclei consisted of protons and neutrons. Elements differ from each other in the number of protons in their nuclei. The lightest element, hydrogen, has only one proton in each nucleus. The next element in the Periodic Table, helium, has two, and so on throughout the table. In addition to the protons present all atoms, except some of those of the very lightest elements, contain one or more neutral particles or neutrons. A neutron has almost exactly the same mass as the proton but differs in that it is uncharged.

It is possible for two atoms to contain the same number of protons in their nuclei and thus be atoms of the same element, but to have different numbers of neutrons and thus have different atomic masses. Such atoms are called *isotopes*. Some elements are known to exist in as many as seven or eight isotopic forms. The mass number of an atom of a particular isotope can be obtained by adding the number of neutrons in its nucleus to the number of protons in that same nucleus. Different isotopes of an element may differ widely in the stability of their nuclei.

The way in which the components of a nucleus (the nucleons) are arranged is at the present time the object of much study. One hypothesis is that these particles are in much the same form as they are when existing separately but are closely packed together in the atom. Another theory states that they exist more in the form of shells one inside the other. The nucleus is held together by forces which are extremely great when the separations are very small, but which fall off very rapidly as the distances between particles increase. The Japanese scientist, Yukawa, developed a mathematical meson theory which deals with these forces.

The nuclei of all the very heavy elements are unstable in varying degrees and decay spontaneously with the emission of radiations. This process, called *radioactivity*, was first discovered by Becquerel; it transforms an atom of one element into an atom of another element, the disintegrations following statistical laws. Some isotopes decay so slowly that it takes thousands of years for one-half of the atoms of a given sample to decay and another equal period of time for one-half of the remainder to disintegrate. Others decay so rapidly that one-half of the atoms in a given sample will decay in a tiny fraction of a second. Some isotopes of the lighter elements are also radioactive. The half life for radioactive decay is the length of time it takes for one-half of the atoms in a given sample to decay. At least one radioactive isotope of every element has been found to occur naturally or has been prepared artificially.

Theoretical Interpretation of the Atom. Powerful mathematical methods give quantitative treatments of atomic and subatomic processes. These theories are called quantum mechanics, matrix mechanics, or wave mechanics and are primarily the results of investigations by Heisenberg, Dirac, and Schrödinger. The background for these theories comes from Louis de Broglie's realization that the circumference of the circular orbit of Bohr's atom model for hydrogen in the normal state is equal to the wave length of the waves which can be associated with the moving electrons. This leads to the idea that in the atom there is a standing wave associated with the electron as it moves in its orbit. Erwin Schrödinger proposed that this wave length be substituted in a classical wave equation and from this beginning he derived a wave equation for the hydrogen atom. This pictures the negative charge of the electrons as a standing wave about the nucleus. The square of the amplitude of the wave represents the probability that the electron can be found at that point.

For other atoms wave mechanics indicates a method for finding the energy values and the electron distributions for stationary states of the atoms. The mathematical difficulties, where many particles are involved, are stupendous, but successes are being achieved. This development is generally known as the orbital theory of atomic structure.

Atomic Energy. Much is being written at the present time about atomic energy, using the expression to refer specifically to the energy obtained from certain changes in nuclear structure. The foundation of this idea of obtaining energy in usable form from atoms can be traced to Albert Einstein who showed that matter seemed to be a form of energy and that it could be changed to other forms of energy. The amount of energy obtained is given by the equation $E = mc^2$. The energy E is given in ergs if m is the mass in grams and c is the velocity of light in centimers per second. One gram would correspond to 9×10^{20} ergs. At the present time only a small fraction of the mass of any given sample can be converted, but this fraction still yields a vast quantity of energy.

Considering two atoms of deuterium, the heavy hydrogen isotope, each nucleus contains one proton and one neutron. At high temperatures and high pressures these will combine into one nucleus. This new nucleus has two protons and two neutrons. Thus, it is the nucleus of an atom of helium. However, if we add up the masses of the starting atoms very carefully and compare the sum with that of the helium nucleus, we find that a small amount of mass is missing. It has been converted into other forms of energy. The loss of mass is often referred to as "mass defect."

Considering any other combinations of very light nuclei which result in heavier nuclei, we find that some mass is lost in the process of *fusion*. This mass appears as some other type of energy. On the other hand, if we consider the heaviest atoms in the Periodic Table, we find that they can be broken into two nearly equal parts plus some very light particles. The sum of all the masses of the resulting particles is less than that of the original materials. This process is called *fission,* and again we have the conversion of mass into other forms of energy.

Chain reactions occur when some of the particles produced in the fission process are ones which are capable of breaking other atoms, and the geometry of the arrangement is such that there is a great enough chance of their breaking other atoms before they are absorbed by competing processes, or are lost outside of the mass of fissionable material, or are slowed down to the point where they can no longer trigger the fission process.

Atoms should be thought of as entities which are very small but none the less so complex that they stagger the imagination and offer rich fields for further research and speculation.

ROBERT M. BESANCON

Cross-references: *Molecules, Protons, Electrons, Elements, Radioactivity, Periodic Law, Nucleonics, Molecular Weight, Carbon-12, Fission, Fusion.*

References

Stearnes, Robert L., *Atomic Physics,* 1970, Barnes and Noble, New York, N. Y.

Oldenberg, Otto and Holliday, Wendell G., *Introduction to Atomic and Nuclear Physics,* 4th Edition, 1967, McGraw-Hill, New York, N. Y.

AUTOCATALYSIS

Autocatalysis is a term used to describe the experimentally observable phenomenon of a homogeneous chemical reaction, showing a marked increase in rate with time, reaching its peak at about 50% conversion and then dropping off. The temperature has to remain constant and all ingredients mixed at the start for proper observation.

This definition excludes those exothermic reactions which show an increase in rate with time (like explosions) caused by the rapidly rising temperature.

Autocatalysis can be readily observed, for example, in some thermal decompositions in which a gas is released. A simple case of autocatalysis involves three essential, consecutive kinetic steps:

(1) $A \xrightarrow{k_1} B + C$ Start or background reaction

(2) $A + B \xrightarrow{k_2} AB$ Complex formation

(3) $AB \xrightarrow{k_3} 2B + C$ Autocatalytic step

Compound A decomposes with rate k_1 into products B and C, the background reaction. The autocatalytic agent B forms a complex AB with rate k_2. Next, the complex AB decomposes with rate k_3, releasing $1B$ in addition to forming $B + C$.

Thus, reactions (2) and (3) together form the path by which most of A decomposes. Reaction (1) is the starter, but continues competitively with (2) and (3) as long as there is any A around.

A mathematical analysis of steps 1, 2, and 3 readily leads to curves representing the rate of formation of C versus time. These curves, calculated for various values for k_1, k_2, and k_3, give a peak reaction rate at about 50% conversion, in good agreement with experimentally determined curves. The mathematical analysis also demonstrates that k_3 should be much larger than k_1 and that k_2 should be of the same order as k_3 in order to show autocatalysis. If we define α as $\alpha = k_3/k_1$, we can call α the *degree* of autocatalysis.

One can then classify various autocatalytic reactions as:

$\alpha = 1$ to 10   Barely noticeable autocatalysis

$\alpha = 10$ to 100   Mildly autocatalytic

$\alpha = 100$ to 1000   Strongly autocatalytic

During most kinetic investigations, cases with $\alpha = 1$ to 10 would tend to escape being detected as autocatalysis, and in cases with $\alpha = 100$ to 1000, the background reaction would tend to go by unnoticed. This condition is responsible for some inaccurate statements in the literature regarding autocatalysis. Autocatalysis is very common, and not at all a rare phenomenon.

Precursor autocatalysis is a special form in which a precursor E is necessary to get a reaction to start, has no further role. For example, a trace of iodine, of a heavy metal, or of a base frequently acts as a precursor.

The kinetic steps of precursor autocatalysis involve 4 essential reactions:

(1) $A + E \xrightarrow{k_4} AE$ precursor-complex

(2) $AE \xrightarrow{k_5} E + B + C$ precursor complex decomposition

(3) $A + B \xrightarrow{k_2} AB$ complex formation

(4) $AB \xrightarrow{k_3} 2B + C$ autocatalytic step

Thus reactions k_4 and k_5 replace or overshadow k_1. Conditions for clearly noticeable precursor autocatalysis are that k_4 is about equal to (or faster than) k_5, k_2 is about equal to (or faster than) k_3 and $\alpha = k_3/k_5$ between 10 and 1000, as discussed before.

Inhibition. Autocatalytic reactions are especially susceptible to inhibition. The inhibition step in autocatalysis is simply

$$B + F \xrightarrow{k_6} BF.$$

Thus, a small amount of inhibitor can effectively block autocatalysis, but not the starting reaction k_1 which will go on unchanged.

Reaction k_6 has to be about 10 times as fast as k_2. Inhibition of precursor autocatalysis would involve either reaction k_6 or:

$$E + F \xrightarrow{k_7} EF$$

Here it is a condition that k_7 is about 10 times as fast as k_2. One can see that in the above reactions the inhibitor is slowly used up as a result of the slow background reaction k_1 (or k_4 and k_5 in precursor autocatalysis). After inhibitor F is used up, the reaction will resume its normal autocatalytic character.

More variations of autocatalysis and inhibition are possible. One of them, called superautocatalysis, is intimately tied to inhibition. It can be observed in the thermal decomposition of various oxazolidinones inhibited with cyanuric acid as an abnormally rapid autocatalytic decomposition which sets in after the inhibitor is exhausted (a "cold explosion"). A characteristic of superautocatalysis is, that there is a proportionality between the amount of inhibitor used and the rate increase of the superautocatalysis. This proportionality gives the key to the kinetic understanding of this interesting phenomenon.

During the period of superautocatalysis, there is an autocatalytic release of the autocatalytic agent. Instead of the formation of one extra B per reaction sequence, the example quoted has a release of 3 B per reaction sequence. This increases the concentration of B exponentially from: $1 - 3 - 9 - 27 - 81$, etc. This increase goes on until all B stored by the inhibitor is released.

The concept of superautocatalysis can serve as a model for certain biochemical reactions in which a sudden release of a chemical occurs.

W. E. WALLES

Cross-references: *Catalysis.*

AUTOXIDATION

The term "autoxidation" is applied to those spontaneous oxidations which take place with molecular oxygen or air at moderate temperatures (usually below 150°C) without visible combustion. Autoxidation may proceed through an ionic mechanism, though in most cases the reaction follows a free radical-induced chain mechanism. The reaction is usually autocatalytic and may be initiated thermally, photochemically or by addition of either free radical generators or metallic catalysts. Being a chain reaction, the rate of autoxidation may be greatly increased or decreased by traces of foreign material.

Most organic and a variety of inorganic compounds are susceptible to autoxidation. The mild conditions under which many compounds react with oxygen presents a serious problem of deterioration though, on the other hand, autoxidation offers wide possibilities in the field of synthetic chemistry. Several industrial processes such as the manufacture of *o*-phthalic anhydride from naphthalene and more recently of phenol and acetone from cumene are based on autoxidation. In the paint industry many of the finishes employing drying oils are dependent on autoxidation as well as polymerization for their durable qualities. The reaction has found limited use in development of analytical techniques.

Degradative processes involving autoxidation are very numerous. The rancidity developed by edible fats and oils on standing in the presence of air is due to autoxidation, with formation of acidic products. Gum formation in lubricating oils and fuels may be attributed largely to autoxidation, as may the deterioration of many polymers, especially on exposure to sunlight. In the field of metallurgy the corrosion of certain metals may be considered in part an autoxidative process.

The chemistry of autoxidation has been studied extensively in the hydrocarbon field and the mechanism elaborated for this class of compounds shown to be characteristic of free radical type autoxidation. At temperatures of 100°C or less or on exposure to ultraviolet light at lower temperatures many hydrocarbons react with oxygen rapidly to produce a variety of products. It is now generally agreed that reaction is initiated by removal of hydrogen from a molecule of the substrate to form a free radical, $R\cdot$. The following steps represent the overall reaction in its simplest form.

Initiation $\quad RH \xrightarrow[\text{or Catalysis}]{\text{Activation}} R\cdot$
$$+ \text{atomic hydrogen}$$

Propagation $\quad R\cdot + O_2 \rightarrow ROO\cdot$

$\qquad\qquad ROO\cdot + RH \rightarrow ROOH + R\cdot$

$\qquad\qquad ROOH \rightarrow RO\cdot + OH\cdot$

$\qquad\qquad RO\cdot + RH \rightarrow ROH + R\cdot$

$\qquad\qquad OH\cdot + RH \rightarrow HOH + R\cdot$

Termination $\quad R\cdot + R\cdot \rightarrow$ disproportionation

$\qquad\qquad R\cdot + R\cdot \rightarrow RR$

$\qquad\qquad ROO\cdot + OH\cdot \rightarrow ROH + O_2$

Kinetic studies with tetralin have shown that formation of hydroperoxide is almost quantitative with respect to oxygen absorbed in the initial stages of reaction. After a critical concentration is reached, homolytic cleavage of peroxide commences and the resulting $RO\cdot$ and $OH\cdot$ radicals may then initiate new chains—this is the autocatalytic phase of the reaction. In the case of aldehydes it has been shown that peracids rather than hydroperoxides are the first products of reaction; though the peracids do attack the aldehyde, reaction is through a heterolytic mechanism and hence these autoxidations do not become autocatalytic.

Ease of removal of hydrogen to form free radicals is a function of molecular structure. Groups such as carbonyl, carboxyl, phenyl or the ethylenic double bond activate hydrogen on adjacent carbon atoms so that autoxidation occurs more readily when these groupings are present in the molecule. Thus olefins are more susceptible to autoxidation than paraffins. Autoxidation may be inhibited by addition of antioxidants (HA) which are attacked by the propagating radicals breaking the oxidative chain:

$$ROO\cdot + HA \rightarrow ROOH + A\cdot$$

An effective antioxidant must form a radical ($A\cdot$) which is incapable of chain propagation. In this way autoxidation is inhibited as long as unreacted antioxidant is available. In contrast, addition of foreign peroxides or other generators of active free radicals affords a source of additional radicals capable of initiating autoxidation or increasing the rate. Also metals having variable valence, such as copper, cobalt, manganese, etc. can promise autoxidation through catalytic decomposition of hydroperoxides.

W. LINCOLN HAWKINS

Cross-references: *Antioxidants, Oxidation.*

AZIDES

Azides comprise a group of chemicals of characteristic formula $R(N_3)_x$. R may be almost any metal atom, a hydrogen atom, a halogen atom, the ammonium radical, a complex ($[Co(NH_3)_6]$, $[Hg(CN)_2M]$ with M = Cu, Zn, Co, Ni), an organic radical (e.g., methyl, phenyl, nitrophenol, dinitrophenol, p-nitrobenzyl, ethyl nitrate, etc.), and a variety of other groups or radicals. In the inorganic series, and according to most authorities in the organic as well, the azide group has a chain structure (N═N═N) rather than a ring structure. All the heavy metal azides, hydrogen azide, and most if not all of the light metal azides (under appropriate conditions) are explosive. Many of the organic azides are also explosive, especially those containing in addition to the azide group a nitro group ($—NO_2$). The alkali and alkaline earth metal azides are the least explosive of the metal azides some of which, e.g., $Ba(N_3)_2$, NaN_3, are incapable of propagating a detonation wave in small quantities but would probably propagate in large charges.

Hydrogen azide (NH_3), the chemical properties of which are indicative of azide chemistry in general, is a very sensitive explosive with a heat of explosion between 1400 and 1550 kg-cal/kg. This heat of explosion is comparable with that of the powerful explosives RDX, PETN and nitroglycerin. Hydrogen azide is a colorless, highly volatile liquid of boiling point 37°C and melting point −80°C. It is a protoplasmic poison resembling in this respect hydrogen cyanide (HCN). Hydrogen azide is soluble in water, alcohols, and ether and is itself a solvent for many substances. In the pure state it exhibits slow decomposition at room temperature after four to five days. In aqueous solution (hydrazoic acid) it is stable, but catalyzed by active platinum, it is decomposed into ammonia and nitrogen.

Hydrazoic acid is a monobasic acid with properties resembling the halogen acids. It reacts with acids to liberate nitrogen gas among other compounds and decomposes upon electrolysis with the evolution of nitrogen and hydrogen, probably with $(N_3)_2$ as an intermediate. It reacts readily with oxidizing agents to give a variety of products one of which is generally nitrogen. It is reduced by the action of reducing agents probably through triazene (H_3N_3) and triazane [$(H_2N)_2NH$] to hydrazine (NH_2NH_2) and finally to ammonia. Tetrazene (N_4H_4) may also be produced by reduction of hydrazoic acid. Hydrazoic acid is a powerful nitridizing as well as oxidizing agent.

Sodium azide is important in the preparation of heavy metal azides, principally lead azide. It is prepared in a variety of ways, principal among which are the reactions of nitrous oxide and sodium amide ($N_2O + NaNH_2$) and the alkyl nitride-hydrazine synthesis, both of which have been adapted to laboratory or larger scale production of sodium azide. Solutions of sodium azide react metathetically with soluble copper, silver, lead, mercurous and thallous salts to precipitate the corresponding (insoluble) azides. The soluble metal azides are obtained from hydrazoic acid by metathesis of barium azide with soluble sulfates, or with potassium azide and soluble perchlorates. Ammonium, ethylammonium and diethylammonium azides may be obtained from reactions of solutions of hydrogen azides in ether with the corresponding anhydrobases in ether or alcohol.

Some metal azides, including those of silver, mercury (mercurous), lead and sodium, are sensitive to visible or ultraviolet light, resembling the silver halides in this respect; they are decomposed by light to form the metal and nitrogen. All metal azides are characterized by thermal instability at temperatures starting in the range 100 to 200°C, depending on the particular metal azide. At higher temperatures the heavy-metal azides may explode violently. Few if any metal azides are stable at their melting points. The thermal decomposition of the heavy-metal azides is autocatalytic, (q.v.) being promoted by metallic nuclei which form during decomposition, or by the action of light of appropriate wave length following a mechanism similar to that involved in the development of photographic emulsions. The thermal decomposition of metal azides has been used to prepare the corresponding metals in pure form. This method is inapplicable, however, to the alkali metals owing to the relative stability of the alkali nitrides.

In general, the explosive azides, particularly the heavy-metal azides, are very sensitive and may be detonated directly by shock, friction, heat, electrical discharge, and other energy sources. They must therefore be prepared and handled with extreme caution. Their shipment and storage are carefully regulated by law.

The most important commercial metal azide is *lead azide,* used extensively as the *primary explosives* in commercial and military detonators. Lead azide is one of the most effective primary explosives available for use in the "composition detonator" which contains several specialized elements: a bridge wire, an ignition agent, a primary explosive, and a base charge. Application of sufficient current heats or melts the bridge wire, igniting the igniter element, which delivers a hot intensified flame to the primary explosive. The initial combustion reaction is transformed suddenly into a detonation in the primary explosive. The detonation wave formed in the primary explosive is intensified in the base charge.

Mercury fulminate, for many years the most useful primary explosive in detonators, is more easily ignited and develops a greater detonation pressure or priming impulse than lead azide. However, the transition from an explosive deflagration to a detonation takes place more readily in lead azide than in mercury fulminate, especially at high densities where the latter may "dead press." Where the primary explosive itself is required to perform all the functions of a detonator (e.g., an "ordinary" fuse cap) mercury fulminate is apparently unsurpassed. A "composition cap," however, may make use of the most efficient elements for each separate stage of reaction in which few, if any, substances are more suited as the primary explosive than lead azide in all necessary requirements (surveillance, cost, sensitivity, ease of preparation, and effectiveness to carry out the primary purpose of the detonator—the creation of the detonation wave). Other azides have proved either too sensitive and expensive (e.g., silver azide, cadmium azide, copper azide) or too insensitive (e.g., the alkali and alkaline earth azides) to replace lead azide as the primary explosive in a "composition detonator." Lead azide is, however, not without its limitations, one of which is the tendency in the presence of moisture to attack copper and brass to form the extremely sensitive cuprous azide. Modern detonators based on lead azide, however, are designed effectively to overcome some of the undesirable properties of lead azide. Even the extreme hazards of the detonator have been minimized by ingenious design of the "composition cap," but a detonator remains a highly hazardous device and should be handled cautiously at all times.

M. A. COOK

Cross-reference: *Explosives.*

B

BAKING CHEMISTRY

Wheat flour is a heterogeneous mixture of proteins, carbohydrates, lipids, minerals, enzymes, and related compounds. It appears to be unique in that one of its proteins, called gluten, has the property of being stretched into a film capable of retaining gas. This attribute accounts in great part for the economic importance of wheat. The term "baking" means the cooking of food in dry heat; it is the conversion of doughs or batters containing wheat flour into light, palatable, and nutritious products such as bread, cakes, and cookies. Under the influence of heat, the gases generated by leavening expand and form a semi-dry foam that is the crumb, enclosed within a crust.

Two principal types of products are recognized, depending on the type of leavening used. The first consists of the yeast-leavened bread and rolls, usually manufactured from higher-protein "hard" winter and spring wheats. Flour constitutes the majority of the basic ingredients in bread doughs, along with sugar, shortening, milk, yeast, yeast food, oxidizer, and enzymes in proper proportions. The dough is mixed until full development of gluten is attained. Then it is permitted to proof, or rest, while enzymatic activity proceeds.

Amylases inherently present in the flour, and those added to the dough and elaborated by yeast, accelerate the hydrolysis of some of the starch to maltose. The sugars are then converted by yeast enzymes to carbon dioxide and small quantities of lactic and other acids, alcohols, aldehydes, and other compounds which give flavor to the finished product. The mechanism of glycolysis is essentially that of Embden-Meyerhof-Parnas. Both the straight dough (one-stage mixing of ingredients) and sponge (two-stage mixing) methods are used; the latter is probably more common in commercial circles. In either case, the dough is occasionally punched, or squeezed, to drive off large gas pockets, distribute the gas nuclei more uniformly, and renew the yeast environment. When the dough is heated, the gas cells expand but gas is prevented from escaping by films formed by proteins and starch. With evaporational loss and binding of water, the films "set."

It is commonly agreed that product quality is determined to a large extent by the quantity and quality of protein in the flour. In general, flours with greater protein contents produce larger loaves with silky crumb. Protein quality is usually associated with two factors—genetics and environment. Since wheat is propagated by self-fertilization, genetic characteristics bred into a variety are retained from generation to generation; hence quality is identified with varietal designation to a considerable extent. Some of the attributes thus inherited are: dough water absorption; mixing time; mixing tolerance; dough oxidizer requirements; and bread characteristics such as size potential, external characteristics, and internal crumb properties. When the wheat plant is exposed to excessively high temperatures during the critical grain-forming period, the protein in the grain appears to become damaged, resulting in poorer bread than would be expected on the basis of variety and flour protein content.

Recent research in the chemistry of bread flours has centered about both the chemical and physical aspects of the protein fraction. Electrophoretic and chromatographic studies indicate that there are several albumins and globulins in wheat flour. In addition, the gluten or insoluble protein of flour which is considered to be most important in baking is also believed to consist of a number of fractions characterized by differences in charge group, molecular weight, and migration velocity. However, it has not been possible to relate directly any specific protein to baking quality. Amino acid composition of gluten does not appear to differ significantly among flours of different varieties or even among wheat classes. Physical properties of flours which contribute to dough rheology have also been the subject of study. Equipment and apparatus have been developed which aid in the determination of dough elasticity, extensibility, and changes during relaxation, and of the influences on these properties induced by dough oxidation or reduction, electrolytes, enzymes and similar agents. Enzyme systems present in fermenting doughs have also been extensively studied.

The second type of product, manufactured by using chemical leavening rather than yeast, includes cookies, cakes, sugar wafers, cake doughnuts, and similar pastry baked goods. For these items, "soft" rather than "hard" wheat flour is used. These wheats are characterized by their generally low protein contents as well as by the fine granulation and low water absorption of their flours.

Relatively large quantities of sugar and shortening (the former at times exceeding the weight of flour), and of other ingredients such as milk, eggs, carbohydrates, and flavoring are incorporated in the dough or batter. Leavening is provided by chemicals such as baking powder (sodium bicarbonate plus potential acid source), and ammonium bicarbonate or similar salts, sources of carbon dioxide and ammonia gases which provide the lightness of the finished product.

After mixing, the dough or batter is baked without the waiting period necessary for yeast-raised doughs. Enzymes present in the flour are not believed to play a major role in product development. Leavening systems may be simple or complex, depending on whether single or multiple-stage gas generation is desired and on the temperatures of optimum activity. As is true of bread and rolls, the objective of leavening is to form a foam which dries or sets when sufficient water has been evaporated or otherwise removed from the sphere of action by binding. Because of the great diversity of products baked from soft wheat flours, it is not yet possible to ascribe soft wheat flour quality in general to any specific component, but it is believed that proteins are not as important in pastry products as they are in bread. The preference of commercial pastry manufacturers for flours low in protein would tend to support this belief.

The basis of baking chemistry of "soft" wheat flour products appears to be associated with the competition between flour components and sugar for the limited quantity of water available. Since sugar is a hydrophilic agent, it tends to sorb water, the extent of this sorption depending on the quantity of available water and on the presence of other hydrophilic materials. It appears probable that in some products such as cookies there is not enough water in the dough to bring about any extended starch gelatinization. If the flour were less hydrophilic, the syrup in the dough would tend to maintain a fluid medium, permitting expansion of the product through continued leavening action and the relatively low viscosity of the dough. In cake batters, on the other hand, there appears to be sufficient water for starch gelatinization to take place. The juxtaposition of temperatures of starch gelatinization, protein denaturation, leavening action, and loss of water to a critical level are factors believed to contribute to cake quality as expressed by volume and internal texture. Cake baking potential of a flour is improved by treating it with chlorine gas; this brings about hydrolytic and oxidative modifications of starch. In general, the mechanisms of pastry-baking appear to be associated with the physical aspects of flour quality such as hydrophilicity, gelatinization, denaturation, emulsification, and surface forces that influence the rheological properties of doughs and batters.

Recent Technological Developments. Two recent developments which have made some impact in the field are continuous bread-baking and the air-classification of flour. In the first, a pre-ferment is made by permitting yeast action on a nutrient medium or solution; thus the metabolic products which contribute to bread flavor are generated. The solution is then mixed with flour to form a dough which is baked without the lengthy fermentation time that was formerly required. The influence of such a change on the production schedule is obvious.

Air classification involves the separation of flour particles by size and density through the use of air centrifuges which can be adjusted to various cut-off points. Since flour composition is not uniform for all particles, separations result in fractions rich in protein or in starch. Judicious reblending of fractions can result in flours which are better suited to specific purposes than was the parent flour. One application of this process has been in the fractionation of low-protein hard wheat flours; thus a low-protein fraction is extracted and the protein content

of the residue is elevated to improve its bread-baking potential. The separated fraction has been found suitable for certain products formerly made exclusively from soft wheat flours.

W. T. YAMAZAKI

Cross-references: *Brewing, Foods, Yeast.*

BARBITURATES

The barbiturates are derivatives of barbituric acid (2,4,6-trioxohexahydropyrimidine) which produce central nervous system depression and result in a condition of sedation or hypnosis when administered to man and most other animals.

Literally thousands of barbituric acid derivatives have been synthesized and tested but a relatively few, perhaps no more than thirty, have been used extensively in medicine. This is due primarily to the similarity of action which is characteristic of the series. Wide variations in hypnotic potency are prevalent, but the ratio between potency and toxicity is relatively constant; thus there is no clear-cut advantage in the margin of safety of any particular barbiturate. Differences between barbiturates in rapidity of onset and length of action are pronounced. These factors, along with variation of dosage, make it possible to produce any degree of depression from mild sedation to complete surgical anesthesia. The broad spectrum of action in conjunction with proved safety in clinical application accounts to a large extent for their widespread use.

Barbiturates are habit-forming under certain conditions of continued usage, most frequently in the absence of proper medical supervision. Addicts are usually stimulated rather than depressed and symptoms are similar to those of chronic alcoholism. Withdrawal reactions vary considerably but they can be severe to the extent of requiring hospitalization for special treatment. Poisoning may occur accidentally or with suicidal intent, the symptoms simulating alcoholic inebriation in many respects, including mental confusion, ataxia, disturbed vision, shallow respiration, weak pulse, slow reflexes and coma. Death occurs from respiratory depression. In order to control improper use, the law requires that barbiturates be sold only on prescription and that prescriptions cannot be refilled without consent of the physician.

The basic requirement for depressant activity in barbituric acid derivatives is replacement of both hydrogens in the 5-position. The substituents may be almost any combination of alkyl, alkenyl, aryl, cycloalkyl or certain modifications of these groups. Further substitution in the side chain groups usually, although not universally, will reduce or destroy depressant activity. The most active com-

TABLE 1.

Name	Substituents	Duration of Action
Amobarbital	5-ethyl-5-isoamyl	Moderate
Butabarbital	5-ethyl-5-(1-methylpropyl)	Moderate
Butalbital	5-allyl-5-isobutyl	Moderate
Hexobarbital	5-(1-cyclohexenyl)-1,5-dimethyl	Ultrashort
Methohexital	5-allyl-5-(1-methyl-2-pentynyl)	Ultrashort
Pentobarbital	5-ethyl-5-(1-methylbutyl)	Moderate
Phenobarbital	5-ethyl-5-phenyl	Long (anticonvulsant)
Secobarbital	5-allyl-5-(1-methylbutyl)	Short
Thiopental	5-ethyl-5-(1-methylbutyl)-2-thio	Ultrashort

pounds, with some exceptions, contain two groups in the 5-position of different size, structure, or configuration, the sum of the carbon atoms in both groups totaling 7 or 8. Activity is modified by substitution in the 1- or 2-position. For example, replacing oxygen by sulfur in the 2-position shortens the duration and increases the intensity of action. A methyl group in the 1-position may shorten the duration of action or modify it to increase anticonvulsant activity. Additional replacement in other positions of the barbiturate ring will usually destroy activity or modify it to such an extent that central nervous system depression is no longer the predominant action. Some of the barbiturates and their characteristic actions are listed in Table 1. Barbiturates are used either as free acids or as salts, usually the sodium salt, which are water-soluble and more rapidly absorbed.

Synthesis. Barbiturates are prepared by the condensation of a disubstituted malonic ester with urea in the presence of a sodium alcoholate in an alcohol solvent. Numerous variations of the general procedure are possible. For example, a substituted cyanoacetic ester or malononitrile may be used in place of the malonic ester or a urea derivative such as thiourea, guanidine or dicyandiamide may replace the urea. Ethanol and methanol with the corresponding sodium alcoholate are the most commonly used solvents but other alcohols and other solvents have been used. Direct alkylation in the 5-position of the barbiturate ring is generally not successful with the exception of the allyl group which may be introduced by reaction of the sodium salt of a 5-monosubstituted barbituric acid with an allyl halide in the presence of a trace of a copper salt as a catalyst.

Properties and Reactions. Barbiturates are usually odorless, white, crystalline solids, sometimes occurring in more than one crystal form. They are poorly soluble in water and soluble in a variety of organic solvents. Water solubility is highest in compounds with the smaller substituents, while solubility in organic solvents increases as the size of the substituents increase. The barbiturates are weak acids due to their ability to react in enol form. They form salts with alkalies that are hydrolyzed in aqueous solution and decompose through rupture of the barbiturate ring. They are stable in water and in dilute acids.

Uses. The barbiturates have many clinical uses all of which are associated with their ability to produce central nervous system depression. Proper dosage as directed by a physician will induce a nor-mal, refreshing sleep with relatively little "hangover" if a compound with a moderate duration of action is used. Small repeated doses at intervals during daytime activity induces sedation that is useful in controlling various nervous disorders. When administered prior to surgery, they tend to allay fear and induce relaxation. The ultrashortacting derivatives are administered intravenously to produce general anesthesia for surgical procedures of short duration. Many barbiturates have anticonvulsant properties that are widely utilized, phenobarbital being outstanding in this respect in the control of epilepsy. The barbiturates also have a specific prophylactic action against toxic reactions from local anesthetics.

ROBB V. RICE

References

Blicke, F. F., and Cox, R. H., Editors, "Medicinal Chemistry," Vol. IV, "Barbituric Acid Hypnotics," John Wiley & Sons, New York, 1959.
"Hospital Formulary," Section 28.24, "Sedatives and Hypnotics," Amer. Soc. of Hospital Pharmacists, Washington, D.C., 1972.

BARIUM

Barium, symbol Ba, atomic number 56 and atomic weight 137.34, is located in Group IIA of the Periodic Table. It is the heaviest of the three alkaline earth elements: calcium, strontium and barium, and is the least volatile of the three.

In 1774 Scheele distinguished barium oxide from lime and recognized that a new element was present. The element was first prepared by Davy in 1808 as an amalgam by the electrolysis of a soluble barium salt on a mercury cathode. It was named barium from the Greek *barys* (heavy) which had also been applied to the oxide, baryta, and the sulfate, barite or barytes.

Occurrence. Barium constitutes 0.4–0.5% of the earth's crust. The most common barium minerals are barite or barytes, $BaSO_4$, and witherite, $BaCO_3$. The former is the chief source of barium compounds in the United States. Of the estimated world production of 3 million tons annually of barite, the United States production is about 800,000 tons, chiefly from Missouri and Georgia.

Barium minerals are heavy, with a specific gravity of 4.4–4.5 compared with 2.71 for calcite, $CaCO_3$.

The greatest tonnage of natural or crude barite is used as such, principally as a constituent of oil well drilling mud, without further processing except for washing, gravity separations to remove lighter material present in the as-mined ore, and grinding.

Derivation. Metallic or elemental barium cannot be made by the reduction with carbon of barium oxide at elevated temperature because the acetylide, BaC_2, is formed rather than the metal.

The most effective method of making barium is the reduction of barium oxide with aluminum or silicon in a high vacuum at elevated temperature. The reactions involved are: $4BaO + Si \rightarrow 2BaO \cdot SiO_2 + 2Ba$ (gas) and $4BaO + 2Al \rightarrow BaO \cdot Al_2O_3 + 3Ba$ (gas).

Useful alloys of barium and aluminum or magnesium are readily made by the reaction of aluminum and magnesium, respectively, with barium oxide. Fusion electrolysis with a heavy metal cathode, such as zinc, lead, antimony, tin and bismuth, also produces barium alloys with the respective cathode metals.

Barium is produced by Dominion Magnesium at Haley, Ontario. The market price is some $7–10/lb.

Physical Properties. Barium is a silver-white metal slightly harder than lead that is malleable, extrudable and machinable so that it can be made into rods, wire and plate. It is the densest of the alkaline earth metals. Barium and barium compounds give green colors in flames.

Values of various physical properties are given in Table 1.

Chemical Properties. Barium forms divalent compounds since its valence electrons are $6s^2$. It is an extremely reactive element and the free energy of formation of barium compounds is very high. Barium reacts directly with water, oxygen, nitrogen, ammonia, the halogens, phosphorus, sulfur and most acids. The chief use for it as a "getter" or degassing agent in vacuum tubes indicates its reactivity with gases.

When barium is heated in hydrogen at about 200°C a vigorous reaction occurs to form barium hydride, BaH_2, a solid compound readily decomposed by water and acids. Barium nitride, BaN_6, decomposes violently when heated.

Barium reduces tthe oxides, halides and sulfides of less reactive metals to produce the corresponding metal. However, it is not more effective than calcium in such reactions and is not used for this purpose because of its much higher cost per equivalent weight of reducing agent.

Compounds and Their Uses. Barium forms compounds analogous to those of calcium and their properties, including color or lack thereof, closely resemble those of the corresponding calcium compounds. One prominent exception is barium hydroxide which is quite soluble in water, forming an almost 50% solution at 100°C.

Commercially, most barium compounds are derived from chemical grade barite, $BaSO_4$. In the production of many barium compounds, the $BaSO_4$ is first reduced to the sulfide by carbon in a furnace, and the soluble BaS then is reacted to form other compounds.

Barium carbonate is probably the most important barium compound insofar as chemical applications are concerned. Precipitated $BaCO_3$ is made by the reaction of sodium carbonate or CO_2 with barium sulfide. One of its most important uses is the treatment of salt brines to remove sulfates from the solutions fed to chlorine-alkali cells; the reaction involved is $CaSO_4$ (or Na_2SO_4) + $BaCO_3 \rightarrow BaSO_4 + CaCO_3$ (or Na_2CO_3). Barium carbonate is also used as a raw material for other barium chemicals, as a flux in ceramics, as an ingredient in optical glass and fine glassware, and in the preparation of barium ferrite ceramic permanent magnets used in loud speakers, a major use.

Barium chloride and *barium nitrate* are made by the action of hydrochloric and nitric acid, respectively, on barium carbonate or barium sulfide. Barium nitrate, $Ba(NO_3)_2$, is used to produce a green color in flares, pyrotechnic devices and tracer bullets.

Barium oxide, BaO, is made by the ignition of barium nitrate and by the decomposition of barium carbonate at high temperature in the presence of

TABLE 1. PHYSICAL PROPERTIES

Atomic number	56
Atomic weight	137.34
Atomic volume, cc/g-atom	39
Melting point, °C	729
Boiling point, °C	1637
Density, g/cc, 20°C	3.6
Crystal structure	body-centered cubic
Lattice constant	a = 5.015 Å
Latent heat of fusion, kcal/g-atom	1.83
Latent heat of vaporization, kcal/g-atom	41.74
Specific heat, cal/g/°C, 20°C	0.068
Electrical resistivity, microhm-cm	50
Electron work function, eV	2.5
Surface tension, dyne/cm	195
Vapor pressure, 10 mm	1049°C
100 mm	1301°C
400 mm	1518°C
760 mm	1637°C
Thermal neutron absorption cross section, barns	1.2

carbon. The major use of BaO is in the manufacture of lubricating oil detergents.

Barium peroxide, BaO_2, a stable material at room temperature when dry, is readily formed by heating BaO in air at about 1000°F. Barium peroxide when added to dilute sulfuric or phosphoric acid yields hydrogen peroxide solution which can be separated from the precipitated barium sulfate or barium phosphate: $BaO_2 + H_2SO_4 \rightarrow BaSO_4 + H_2O_2$.

Barium sulfate, prepared chemically as a fine white powder by the metathetical reactions of various barium compounds with sulfates, is one of the most insoluble salts known and is widely used as a filler in paper, leather, rubber goods, etc. A small but vital use is in x-ray photography of the gastrointestinal tract where it provides a fine opaque contrasting medium when ingested prior to the x-ray examination.

The brilliant white paint pigment, lithopone, is a mixture of zinc sulfide and barium sulfate prepared by the joint precipitation of these compounds by the reaction: $BaS + ZnSO_4 \rightarrow BaSO_4 + ZnS$.

Toxicology. All of the soluble compounds of barium are poisonous when taken by mouth. This includes barium carbonate, which although insoluble (0.002 g/100 ml water at 20°C), is dissolved by the hydrochloric acid in the stomach if ingested. The fatal dose of barium chloride is 0.8 to 1.0 g; larger amounts of less soluble compounds, such as the sulfide, may be tolerated. However, very few cases of industrial systemic poisoning have been reported.

The barium ion is a muscle stimulant. It is very toxic to the heart and may cause ventricular fibrillation.

The antidote for ingested poisonous barium compounds is to drink a solution of sodium sulfate or Glauber's salts which converts the barium ion to the insoluble harmless barium sulfate.

As mentioned previously, nontoxic barium sulfate from which all soluble barium compounds have been removed is widely used as an opaque medium for radiography of the gastrointestinal tract.

CLIFFORD A. HAMPEL

References

Pidgeon, L. M., "Encyclopedia of Chemical Technology," Kirk, R. A., and Othmer, D. F., Editors, 2nd Edition, Vol. 3, pp. 80–82, New York, John Wiley and Sons, 1964.

Priesman, L., *ibid.,* pp. 82–98.

"Minerals Yearbook," Vol. 1, Washington, D.C., U.S. Bureau of Mines, issued annually.

Light Metals, **34,** 77 (1944).

Kroll, W. J., U.S. Bur. Mines Inform. Circ. 7327, 1945.

Mantell, C. L., "Rare Metals Handbook," Hampel, C. A., Editor, 2nd Edition, pp. 25–28, 1961.

Hampel, C. A., "Encyclopedia of the Chemical Elements," pp. 43–46, Van Nostrand Reinhold Corp., New York, 1968.

BASES

The subject of acids and bases has been one of the most controversial in chemistry, and has led to the development of an interesting series of theories. In the 17th century, during the infancy of experimental chemistry, acids and bases were defined or described on the basis of their behavior. Thus, bases were substances that neutralized acids, turned plant dyes blue, had a bitter taste, and had a smooth or slippery feeling to the skin.

In the 18th century, following the discovery of oxygen by Priestley, Lavoisier advanced the idea that oxygen was the acidifying principle of all acids. Thereafter, the experimental approach was largely abandoned and emphasis was placed on the composition of substances instead of the phenomenological properties. The development of the hydrogen theory of acidity and Faraday's studies of electrolytic conductance in the early 19th century led logically to the water-ion theory proposed by Arrhenius. By this concept a base may be defined as any hydroxy compound which gives hydroxyl ions in water solution. Neutralization then involves the combination of hydroxyl ions with hydrogen ions formed by the acid, producing water and incidentally a salt. The role of solvent as an ionizing medium for acid-base reactions was emphasized. Although the theory was very useful and adequate for many reactions in aqueous solution, its many limitations soon became apparent. The theory excludes basic substances that are not hydroxy compounds, does not provide for the amphoterism exhibited by many oxides and salts, and limits the field of acid-base reactions to aqueous solutions in spite of many known typical neutralization reactions in nonaqueous solutions.

These objections led to two more or less conflicting theories: the protonic theory advanced by Bronsted and Lowry in 1923 and the older solvent system theory advanced by Franklin in 1905 and extended by Germann, Cady and Elsey, Smith and others. In terms of the protonic concept, a base is any substance, molecule or ion, which accepts a proton. Thus, Base + H^+ ⇌ Acid. Obviously, the proton must be supplied by an acid which is stronger than the acid which is formed. A reaction between a base and an acid may be independent of the solvent, and the formation of a salt as a primary criteria for the reaction is eliminated by this theory. The formation of a salt is incidental to the fundamental proton transfer in the reaction. The reaction of a base with a proton will occur only if a substance (an acid) is present to supply the proton; every acid-base reaction may be considered as a competition of two bases for the proton, as illustrated in the following reactions:

$$Acid_1 + Base_2 \rightleftharpoons Acid_2 + Base_1$$

$$HCl + H_2O \rightleftharpoons H_3O^+ + Cl^-$$

$$H_2O + NH_3 \rightleftharpoons NH_4^+ + OH^-$$

$$HCl + NH_3 \rightleftharpoons NH_4^+ + Cl^-$$

The reaction will proceed in the direction of the formation of the weaker base and acid. The protonic theory eliminates the objection to the Arrhenius theory in limiting bases to hydroxy compounds, but places undue emphasis on the proton by ignoring nonprotonic acidic substances.

In turn the theory of solvent systems eliminates the objection of limiting acid behavior to protonic substances but suffers from the limitation of such behavior to ionic reactions in solution. Ionic dissociation into solvent cations and anions, as water forms hydronium and hydroxyl ions, is a phenomenon exhibited by many solvents. For example:

Solvent	Solvo-positive Ion	Solvo-negative Ion
H_2O	H_3O^+	OH^-
NH_3	NH_4^+	NH_2^-
$HC_2H_3O_2$	$H_2C_2H_3O_2^+$	$C_2H_3O_2^-$
SO_2	SO_2SO^{++}	$SO_3^=$
$COCl_2$	$COCl^+$	Cl^-

A base in this system is defined as a solute which produces an increase in the concentration of solvo-negative ions. Neutralization is the union of solvo-negative ions from the base with the solvo-positive ions from the acid to form solvent molecules. Thus, NH_4Cl behaves as an acid and KNH_2, as a base in liquid ammonia solution. A salt may be regarded as any substance that gives a solution of higher conductivity than the pure solvent and yielding at least one ion different from those characterizing the solvent.

A more general concept, the electronic theory, was proposed by G. N. Lewis in 1923 but gained little attention until 1938 when it was revived and extended. It is noteworthy that Lewis returned to experimental behavior as a basis for the theoretical interpretation. He rejected such criteria as the presence of a given element, a solvent, or ions as being essential for a theory of acids and bases. A substance, in order to be classified as a base, must display, according to Lewis, each of the following experimental criteria:

(1) *Neutralization:* The base will combine rapidly with an acid.
(2) *Titration with indicators:* The base may be titrated with an acid using a substance, usually colored, for an indicator.
(3) *Displacement:* The base will replace a weaker base from its compounds.
(4) *Catalysis:* The base will promote a chemical reaction requiring a base catalyst.

Structurally, a base is any substance containing an atom which is capable of donating a pair of electrons to an acid, the acceptor. Neutralization is the formation of a coordinate covalent bond between the base and acid. The typical acid-base reaction is $A + :B \rightarrow A:B$, where the species $A:B$ may be called a coordination compound, an adduct, or an acid-base complex. This fundamental reaction may or may not be followed by dissociation or rearrangement. For example:

Base		Acid	Neutralization Product
H_2O	$\rightarrow$	SO_3	H_2SO_4
NH_3	$\rightarrow$	BCl_3	$H_3N:BCl_3$
Cl^-	$\rightarrow$	$SnCl_4$	$SnCl_5^-$ or $SnCl_6^=$
NH_3	$\rightarrow$	Ag^+	$Ag(NH_3)_2^+$
Cl^-	$\rightarrow$	Mg^{++}	$MgCl_2$

The strength of the Lewis Electronic Theory probably rests on its experimental basis and the fact that all the other theories presented here may be interpreted as special cases of this inclusive Lewis

concept. It is applicable not only to all solvent systems, protonic and nonprotonic, but also to reactions occurring between solids and liquids not in solution. The Lewis Theory has been used extensively in the interpretation of the formation of coordination compounds in which the ligand is the base and the metal ion or ligand acceptor is the acid.

Pearson has proposed a principle of hard and soft acids and bases which has value in extending the Lewis concept to an understanding of a wide variety of chemical phenomena. A soft base is one in which the valence electrons are easily distorted, polarized, or removed. A hard base holds its valence electrons tightly, is not easily polarized, and is hard to oxidize. A hard acid is one of small size, high positive charge, and contains no valence electrons that are easily distorted or removed. A soft acid contains an acceptor atom of large size, small or zero positive charge, and has several valence electrons which are easily distorted or removed. The general statement which sums up the principle of hard and soft acids and bases is: Hard bases coordinate preferentially with hard acids and soft bases with soft acids.

Audrieth and Moeller extended the Bronsted-Lowry Theory to reactions of dry and fused "onium" salts at elevated temperatures. The Lewis Theory was used successfully to explain the high temperature nonprotonic reactions encountered in the fields of ceramics and metallurgy. For example:

Base	Acid	Neutralization Product
$O^=$	SiO_2	Silicates
from metal oxides, hydroxides, carbonates or sulfates	Al_2O_3	Aluminates
	B_2O_3	Borates
$O^=$	$S_2O_7^=$	$SO_4^=$
from metal oxides, etc.	HSO_4^-	$SO_4^=$
F^-	AlF_3	$AlF_6^=$
from alkali fluorides		

The Lux-Flood concept is a more limited view in which a base is any material which gives up oxide ions and an acid is any material that accepts oxide ions.

An even more comprehensive theory of acid and bases is the positive-negative theory proposed by Usanovich in 1939. A basic is defined as any substance which will give up anions or electrons, or combine with cations. In addition to the reactions included in all the other theories, all oxidation-reduction processes are included as acid-base reactions. Sodium is considered to be a base in reacting with chlorine, an acid, to form sodium chloride. The theory is too broad to be generally useful.

Although the Lewis Theory with its extensions and the Bronsted-Lowry Theory are currently most widely applied to interpret acid-base reactions, it frequently is desirable to use the simplest theory to explain adequately the phenomena observed under a given set of experimental conditions.

It is apparent from the foregoing discussion of several theories that the terms "base" and "acid" are not absolute but relative terms to be used with reference to a particular environment. An amphoteric substance may behave either as an acid or a base depending upon the solvent and reactant. Nevertheless, since many substances are considered to have an intrinsic basicity, an effort will be made to list several common bases and mention their properties and industrial uses.

Alkali Metal Hydroxides. Both NaOH and KOH are white solids which absorb CO_2 and H_2O rapidly from the air and are soluble in water, alcohol and ether. Principal uses are in the manufacture of chemicals, rayon and cellulose film, pulp and paper, lye and cleansers, and soaps by saponification of fats, in petroleum refining to neutralize acids, and in reclaiming of rubber to dissolve the fabric.

Alkali Metal Carbonates. Na_2CO_3 and K_2CO_3 and their hydrates are white solids, soluble in water, insoluble in alcohol. Large quantities are used in the manufacture of glass, and other chemicals, chiefly NaOH and $NaHCO_3$. Other uses include manufacture of soap, cleansers, pulp and paper, water softeners, and petroleum refining.

Borax and Sodium Phosphates. The chief uses for borax and the sodium phosphates such as their use in cleaning compounds depend primarily on their basicity in aqueous solution.

Quicklime and Hydrated Lime. CaO, a white solid, absorbs H_2O and CO_2 from the air to become air slaked lime. $Ca(OH)_2$ also absorbs CO_2 from the air. Principal uses are in the building trade for making mortar, agriculture for controlling the acidity of soils, metallurgy for a steel flux and ore concentration, water purification, tanneries for dehairing hides, and manufacture of calcium carbide, cyanamide, and paper.

Ammonia. NH_3 is a colorless gas with a pungent odor, very soluble in water and readily liquefied at moderate temperature and pressure. It is marketed in the anhydrous form or aqueous solution. Important uses are for fertilizers, either by direct addition to the soil or conversion to an ammonium salt by neutralization with H_2SO_4, H_3PO_4 or HNO_3, for extracting plant dyes, and for the manufacture of aniline and other amines. It is used in refrigeration because of its physical properties, as an intermediate in the preparation of nitric acid by oxidation, and as a solvent for many chemical reactions. [See further article on **Ammonia**].

Aliphatic Amines. Methyl-, ethyl-, butyl-, and amylamines, and ethylenediamine are used in manufacturing chemical intermediates, dyestuffs, insecticides, synthetic detergents, corrosion inhibitors, and as solvents. Monoethanolamine is used for absorption of CO_2 from combustion gases. Diethanolamine is used for the absorption of CO_2 and H_2S.

Aniline, 2-Naphthylamine, and Other Aromatic Amines. These are used for manufacture of dyes and intermediates, and synthetic rubber additives such as diphenylamine and cyclohexylamine.

Pyridine and Its Homologs are used in the manufacture of vitamins, sulfa drugs, and fungicides.

There has been a tremendous increase in interest in reactions in nonaqueous solvents in recent years for use in both synthesis and analysis. Ammonia and derivatives of ammonia, the simple aliphatic amines, ethylenediamine and pyridine are all basic solvents which have been used for reactions for which it was desirable to enhance the acidic properties of the solutes.

WILLIAM F. WAGNER

References

1. Audrieth, L. F., and Moeller, T., Acid-Base Relationships at Higher Temperatures, *J. Chem. Ed.,* **20,** 219 (1943).
2. Hall, N. F., Systems of Acids and Bases, *J. Chem. Ed.,* **17,** 124 (1940).
3. Luder, W. F., and Zuffanti, S., *The Electronic Theory of Acids and Bases.* (New York: John Wiley and Sons, Inc. 1946).
4. Pearson, R. G., Hard and Soft Acids and Bases, *J. Chem. Ed.,* **45,** 581, 643 (1968).

BATTERIES

History. While the science of electrochemistry began with the invention of batteries, the fields of chemistry, physics and electricity too owe much of their beginning to Allessandro Volta's announcement of his voltaic pile and crown of cups in 1800. Until that time, electricity was a little understood subject which could only be feebly demonstrated in the laboratory as a static charge.

Within a few weeks of the announcement of Volta's discovery, Nickolson and Carlisle used this new source of electrical energy to decompose water into hydrogen and oxygen. By 1807 Davy had used Volta's battery in his celebrated experiment to decompose fixed alkalies. This was soon followed by the classic work of Faraday and Daniel which did so much to establish the relationship between electricity and chemistry. During the first half of the 19th century many other workers attempted to improve this method of generating power, since it had inspired vast areas of experimental work which had never before been possible. As a result, many of the fundamental laws of electricity, physics and chemistry were established during this period.

While many types of primary batteries were developed during this period, they were so low in capacity that their use in experimental work was quite limited. Another very important step was made in 1860 when Planté invented the lead-acid storage battery, which not only could be recharged but was capable of supplying much stronger currents. This new electrochemical source of power made it possible to carry out almost unlimited experimental work and initiated the separation of the physical sciences into fields now recognized as electricity, physics, chemistry and electrochemistry. Soon after this, direct current generators powered by steam or water were developed, making it possible for light and power to be supplied to many cities. These power systems further increased the demand for the lead-acid storage batteries for supplying peak and emergency loads. Thus, during the last half of the 19th century, secondary or storage batteries came into extensive use. By the end of this period, both Edison and Junger had developed several types of alkaline storage batteries which have since been produced in many forms and in large quantities.

At the beginning of the 20th century, methods were developed for generating and transforming alternating current so that electric power could be transmitted at high voltages from city to city without prohibitive losses. This meant that batteries no longer played such an essential role in everyday life. In the meantime, however, the dry battery was developed from Leclanche's invention into a practical item that is widely used in both private and public life. The lead-acid storage battery likewise has received extensive technological development and in addition to its common use in automobiles, has found considerable application in electric trucks, the telephone service, switch gear control, submarines and in many emergency standby sources of power.

While batteries are poor sources of electrical energy for their weight and volume when compared to other sources, there are many important applications where nothing else can be used. For this reason, considerable effort has always been directed toward the improvement and development of increased life and capacity in most types of batteries. This has been particularly true during and since World War II because of the many military applications and more recently because of space applications.

Principles Involved. Batteries make use of spontaneous chemical reactions in which the oxidation and reduction reactions are arranged to take place at separate electrodes. If these reactions are not readily reversible the battery is referred to as a primary battery. If they are reversible, the battery is called a secondary or storage battery. If the two reactants are continuously supplied to the electrodes and their products can be simultaneously eliminated, the battery then becomes a continuously acting primary battery or fuel cell. Although a large variety of primary batteries have been developed, only a few are capable of being reversed, thus severely limiting the number of possible storage batteries. Likewise, many combinations of electrochemical reactions are possible in continuously acting primary batteries, but only a few have so far been successful.

Theoretically, almost any of the elements or their compounds should be capable of forming part of a battery when suitably coupled with another in an appropriate electrolyte, provided that (1) the anodic reactant has at least one higher state of oxidation and can lose electrons in the reaction; (2) the cathodic reactant has at least one lower state of oxidation and can gain electrons; and (3) the sum of the free energy changes for the two reactions is negative. From the free energy change of the net reaction it is possible to calculate the open-circuit voltage from the equation

$$\Delta F = -n\mathcal{F}\mathcal{E}$$

provided the net reaction is at equilibrium. ΔF is the change in the free energy, n is the number of equivalents evolved in the reaction, $\mathcal{E}$, the open-circuit voltage and $\mathcal{F}$, the Faraday equivalent in coulombs per gram equivalent. To be useful as a battery, except in very special cases, the potential should be at least one volt and preferably much higher.

A compromise must usually be made between the need to obtain as high a voltage as possible to get maximum power and at the same time have a stable system which is compatible with a suitable electrolyte. During the 170 odd years of existence of batteries, only five anodes have so far been commonly employed. These, in the order of their frequency of use are lead, zinc, iron, cadmium and magnesium. Recently, some use has been made of hydrogen, sodium and a few hydrocarbons or their partially oxidized products in fuel cells. During this period only a few cathodes have found any appreciable application and they are, listed in an approximate order of their use, the oxides or salts of lead, nickel, manganese, silver, mercury and copper, and oxygen, itself. More recently, some use has been made of partially oxidized organic compounds. There are a few types of special cells and, of course, fuel cells where oxygen, hydrogen peroxide, the halogens, and SO_3 have been used. It is of interest to note, that Grove first demonstrated the operation of a fuel cell with hydrogen and oxygen at platinum electrodes in an acid solution in 1839.

The relationship between the potentials of these few anodes and cathodes presently in use is shown in a simplified version of the Pourbaix type of diagram (Fig. 1). While complete diagrams of this kind show the thermodynamic relationship between the reversible potentials, $\mathcal{E}°$, of all the various oxidation states of a given element and pH, this simplified version gives only the relationship between the $\mathcal{E}°$ of the electrode reactions at pH's of 0, 7 and 14, covering the range of most battery electrolytes. The relationship between these and any other electrode reaction may be compared by referring to a table of standard potentials in acidic, basic or neutral solutions.

Other types of batteries use ions in solution which are spontaneously oxidized or reduced at an inert electrode. These batteries have had limited

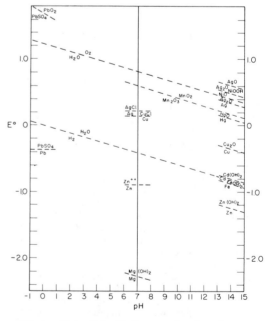

FIG. 1. Potential pH diagram for the electrode reactions involved in most of the practical batteries.

use and are mentioned only because of their historical interest and also to illustrate the many ways in which an electrode reaction can be utilized in a battery.

Two other types of batteries have recently been developed which use either a fused salt or a solid electrolyte as a medium of ion transfer. The former is referred to as a thermal battery and the latter as a solid-state battery. Both may employ the same couples used in conventional batteries and are different only in the fact that water does not enter into the electrode reactions.

There are few successfully developed batteries, in spite of the literally thousands of theoretical possibilities and the never-ending search for new types. This can best be explained by describing the role each part of a battery plays in determining its feasibility of manufacture or in its operational characteristics.

Electrolyte. The battery electrolyte must be an ionic conductor which carries the current between the electrodes by means of ions, and at the same time, it must be chemically stable in contact with the electrodes.

Except for a fortunate quirk of nature, the water in all batteries using an aqueous electrolyte at voltages higher than 1.23 v would theoretically decompose. Thus, electrodes in a successful battery using aqueous electrolyte must have hydrogen and oxygen overvoltages which are high enough to avoid this decomposition. This means that many electrode materials with high potentials cannot be used directly. For example, the alkali metals ordinarily cannot be used because their low hydrogen overvoltages result in a spontaneous reaction with water to evolve hydrogen. In some cases, however, it is possible to use amalgamation to raise the hydrogen overvoltage of these anode materials. Overvoltage plays an even more important role in the recharging of storage batteries since a higher than open circuit potential must be used which further increases the possibility of producing hydrogen and oxygen at the sacrifice of recharging the electrode material.

If high discharge rates are not required, it is sometimes possible to use a salt solution or even an electrolyte which has been immobilized as a gel. A number of batteries have been developed using sea water as the electrolyte and in other cases the products of the electrode reactions are used to impart conductivity to water. Considerable effort has recently been made to utilize the low atomic weight metals in nonaqueous electrolytes in order to avoid their decomposition in water and at the same time take advantage of their much higher energy density.

Electrode Reactions. While many substances are theoretically capable of being utilized as one or the other of the electrodes in a battery, several factors limit their use. The transfer of current from the electrode reaction to the external leads requires that either the electrode material itself must be a good conductor or be made so by the addition of conducting materials or by physically subdividing it in such a way that there are many low-resistance paths to the supporting grid or plate. At the same time this electrode material must have as large a surface area as possible to minimize the current

density and yet retain enough strength to maintain its shape during manufacture and use. In many cells this material is often referred to as the paste, and its method of preparation in most batteries is an art rather than science. Thus, most electrodes are fabricated from a very finely-divided material which can be formed into a suitable paste and then applied to a supporting and conducting grid in such a manner that its surface area is high, its resistance low, and the products of its discharge do not form high resistance films which would retard further discharge. In a storage battery the discharge products must be retained on the plate without being redistributed. Thus it is readily seen why an empirical approach has been necessary in the development of reasonably satisfactory battery electrodes and that the ultimate success of any battery depends primarily on the physical properties of electrode active material.

Separators. Most batteries require some method of physically insulating the two electrodes without appreciably reducing the conductivity of the electrolyte. Separators, in addition to their simultaneous electron-insulating and ionic-conducting properties, must be capable of withstanding the high oxidizing potentials, high temperatures and strong electrolytes which are encountered in many batteries. Materials that have been used include wood, microporous rubber, porous plastic polymers and very thin ion exchange resin membranes.

Side Reactions. Another important factor is that all components of construction, as well as any of the accidental or unavoidable impurities, are subject to the reducing potential of the anode and/or the oxidizing potential of the cathode. If either of these electrodes is sufficiently negative or positive to reduce or oxidize these extraneous materials, a side reaction will take place at one or both of the electrodes which could lead to self-discharge and eventual deterioration of the battery.

For example, the high positive plate potential in the lead-acid storage battery is responsible for a difficult corrosion problem associated with the supporting grid for the reactants of this plate. This battery has been in use for over a hundred years only because of a protective oxide film that develops on the grid and yet is conducting enough to allow current to pass.

Polarization. Most of the factors which influence the discharge rates of a battery are usually grouped together under a term commonly referred to as polarization. It is possible to isolate a number of these factors and determine those which are important in controlling the overall electrode reaction. These include the resistance of the electrolyte, the active material and the supporting grid; concentration polarization resulting from the reactants and the products of the electrode reaction; and the rate of the over-all electrochemical reaction, which depends on the activation energy of the reaction. It is often possible to reduce the resistance effects by a physical modification. The activation energy, however is basic to the initiation of a given reaction and depends on the amount of energy required to transfer an electron through the double layer that is established between the electrode and its ion in solution. If the activation energy is high, the kinetics of the reaction are quite limited and the reaction is probably not suitable for use in a practical battery.

Except for fuel cells and very low rate batteries, most of the successfully used anode materials have been reasonably good conductors and are used in a finely-divided state so that the current density is low enough to avoid excessive polarization. Of the metallic oxides which have been successfully used as cathodes, only the higher oxide of lead, PbO_2, has a reasonable conductivity. In order to utilize nickel oxide, either carbon or nickel flake must be added or else the oxide must be supported on a sintered nickel plaque. While the oxides of silver are poor conductors, the metallic silver produced during the discharge reaction reduces electrode resistance.

In charging a storage battery it is customary to refer to the anodic active material and its support as the negative plate and the cathodic active material and its support as the positive plate since the terminals for these plates are connected respectively to the negative and positive poles of a direct current source.

Cost and Availability. From a practical standpoint, cost and availability of materials are of paramount importance, regardless of how well a given type of battery may perform. This automatically eliminates many possible combinations, except where the need is essential or the application so unique that cost is immaterial. This is illustrated by the use of silver in the silver oxide-zinc battery for a number of military and satellite applications or, if the world production of a required item is limited, availability becomes the controlling factor.

J. C. WHITE

References

1. *J. Electrochem. Soc.,* **99,** No. 8, August, 1952.
2. *J. Electrochem. Soc.,* **99,** No. 9, September, 1952.
3. Vinal, George Wood, "Storage Batteries," 3rd Ed. rev., Wiley, New York, 1940.
4. Vinal, George Wood, "Primary Batteries," 1st Ed., Wiley, New York, 1950.
5. Latimer, Wendell M., "Oxidation Potentials," 2nd Ed. rev., Prentice-Hall, New Jersey, 1952.
6. Pourbaix, M. J. N., Translated by J. N. Agar, "Thermodynamics of Dilute Aqueous Solutions," Arnold, London, 1949.
7. Heise, G. W., and Cahoon, N. C., Editor, "The Primary Battery: Vol. I," Wiley Interscience, New York, 1971.
8. Falk, S. V., and Salkind, A. J., "Alkaline Storage Batteries," Wiley Interscience, New York, 1969.

Storage Batteries

Lead Batteries. The lead battery requires a large number of thin plates (from 12 to 17) to supply the heavy currents required for automobile cranking. Other lead batteries are made with fewer and thicker plates where heavy currents are not required but a longer life is desired. The manufacturing process usually begins with the casting of a framework or grid. A paste consisting principally of lead oxide is applied to this grid and dried. The plates are then charged, assembled and put into a container. The grid is a lattice work which supports the active material and conducts electric current to the battery terminals. These grids are usually cast from antimonial lead (lead containing 7% to 12% antimony, and smaller amounts of tin, copper and arsenic).

The paste has in the past been made from a wide variety of lead compounds. Most paste is made by mixing a litharge containing 20 to 30% metallic lead with dilute sulfuric acid to give a paste of the desired density and consistency. The paste is then applied to the grids in automatic machinery and passed through an oven which partly dries the plates. Finally, the plates are allowed to stand or cure for several days while the residual moisture dries, and most of the metallic lead is oxidized. This gives plates of good mechanical strength without introducing foreign "bonding agents" or requiring precisely controlled drying conditions.

The oxide used for the negative plates contains about 1% of materials called *expanders.* These comprise carbon black, barium sulfate and an organic material usually derived from the lignin fraction of wood. They prevent a deterioration of plate capacity and a shrinking of the sponge lead, and also improve the capacity during a high rate low temperature discharge.

After the plates have been cured, they are assembled and given their first charge. This charge is usually carried out in dilute sulfuric acid at a specific gravity of about 1.100, since a higher strength acid retards charge. After this first, or forming charge, the acid is replaced by a stronger acid. Maximum initial capacity for cold weather operation is attained with 1.300 acid. For high-temperature operation, a much better life is attained with 1.220 acid. Most batteries are made with 1.250 to 1.275 acid. This is measured at 80°F compared with water at 60°F. The charge process converts the negative plate to sponge lead and the positive plate to lead dioxide.

The ordinary automobile battery weighs about 40 pounds, contains six cells, will deliver about 12 volts, and has a capacity of about 50 ampere hours at the 20 hour rate. It will deliver 150 amperes for about 3.5 minutes at 0°F and at an initial voltage of about 8.6 volts. It is kept fully charged by a generator which may deliver as much as 40 amperes to a discharged battery. The amount of charge current is controlled by a voltage regulator. Whenever the battery voltage rises to about 15 volts, the charge current is automatically reduced to avoid excessive overcharging. The chemical reaction normally occurring in a battery is:

$$Pb + PbO_2 + 2H_2SO_4 \rightleftarrows 2PbSO_4 + 2H_2O + 2e$$

when e is 96,500 ampere-seconds if molecular weights are in grams. This reaction proceeds in either direction readily without the development of much heat, and without the consumption of water or production of gas. When a battery is given an overcharge, water is decomposed by the reaction.

$$2H_2O + 4e \rightarrow 2H_2 + O_2.$$

This reaction produces considerable heat, consumes water, and because of the high voltage and heat, tends to shorten battery life. With a good voltage regulator, water consumption of a battery is small and water is added to the battery not more than four times per year. Excessive water consumption, or zero water consumption, indicate something wrong with the electrical system of the automobile. The water added to the battery is usually specified as distilled water, but almost any water that is safe to drink is also harmless to the battery.

The life of a battery depends on the service it receives. It is customary to guarantee a battery for 18 months or 18,000 miles, whichever occurs first. The average battery in private automobile service will last 24 to 36 months. The most frequent cause of failure is corrosion of the positive grid. No method of extending this life without sacrifice of other important properties has been proved by laboratory and field test.

Industrial lead-acid batteries are used in fork-lift trucks. These batteries differ from automobile batteries in the use of thick plates and an envelope to keep softened active material in the positive plate. They have a normal life of five to ten years. Telephone and switchgear batteries serve as an emergency power supply; and when made using grids of a lead-antimony alloy, they are float charged at 2.15 to 2.17 volts per cell. New cells require a current of 20 to 40 milliamperes per 100 ampere hours of rated capacity. As they age, they require more current, and more frequent equalizing charges at 2.35 volts per call. Such batteries have a life of 10–15 years when pasted plates are used, and 20–30 years when Plante type plates are used. Standby batteries are also made using a calcium-lead alloy as the grid metal. These are usually floated at 2.17 to 2.25 volts per cell at a current of about 4 milliamperes per 100 AH of rated capacity. They use very little water, they give off almost no hydrogen, and require practically no equalizing charges. When properly made and operated, a life of more than 20 years has been obtained, and recent designs are expected to do considerably better than this.

Nickel-Iron Batteries. Between 1900 and 1910, Thomas A. Edison developed a nickel-iron battery which has found extensive use. It is mechanically more rugged than the lead battery. It uses an alkaline electrolyte which is less corrosive, and its active materials withstand electrical abuse in the form of overcharge, incomplete charging or prolonged idle periods. It also withstands high temperatures better and in general has a much better life. Its higher cost and different electrical characteristics have limited its use to electric truck propulsion and related heavy duty services. In general, it is not used where exceptionally heavy currents are required, such as for automobile cranking, or where low temperature operation is required.

The chemistry of the nickel-iron battery is not fully worked out. The reaction has been written:

$$Fe + NiO_2 \rightarrow FeO + NiO.$$

The voltages obtained indicate that probably two different higher oxides of nickel are present and are mutually soluble. Other studies indicate that the nickel oxides are hydrated. The electrolyte is potassium hydroxide containing lithium hydroxide, and the specific gravity is 1.21.

The negative plates are prepared from iron oxide containing a small amount of mercury oxide. These are surrounded by a nickel plated steel pocket pierced by many small holes to allow electrolyte and current to reach the active material. The positive plate consists of tubes of nickel plated steel with reinforcing bands. The active material is nickel oxide with alternate layers of metallic nickel flake to supply electrical conductivity.

The cell delivers an average of 1.25 volts during discharge at the five hour rate. Consequently, it requires approximately five cells to do the work of three lead-acid cells. The cells, however, are smaller, and lighter so that the overall battery weight is slightly less and the volume somewhat higher than a lead battery of the same power output. The battery is usually rated at the 5 hour discharge rate, that is 500 ampere hour battery will deliver 100 amperes for 5 hours. The normal charge is 100 amperes for 7 hours with this battery. This overcharge makes it necessary to ventilate the battery compartment sufficiently to carry off the heat developed, as well as the hydrogen and oxygen.

Nickel-Cadmium Batteries. If cadmium is substituted for the iron of the negative plate of the nickel-iron battery, it becomes possible to float charge the battery, and also to discharge it at low temperatures. The result is the "tubular" type nickel cadmium battery. Only the positive plates are of tubular construction. This difference in negative active material gives a battery which has achieved considerable success in Europe.

If the tubular positives of such a battery are replaced by pocket type positives, the battery has a better high rate performance, but a somewhat inferior cycle life. The pocket type positive contains nickel oxide but uses graphite for electrical conductivity instead of the nickel flake used in tubular positives. These pocket type nickel-cadmium batteries are manufactured in various plate thicknesses depending on the service desired. Thin plate batteries have a good high rate performance, but sacrifice low rate performance and are quite bulky.

A third type of nickel-cadmium battery uses sintered plates. For these batteries a special grade of nickel powder is heated in a mold until the particles just begin to adhere. The pores of these plates are then impregnated with a solution of the nickel nitrate to make positive plates and with cadmium nitrate to make negative plates. If these plates are made quite thin, a battery is obtained which has excellent high rate performance, even at low temperatures.

The sintered plate type of nickel cadmium battery is finding increasing usefulness because it can be hermetically sealed. This makes it adaptable to portable equipment where the absence of spilled electrolyte or periodic watering is important. It has found applications in electric toothbrushes, electronic flashguns, electric shavers and portable radio and television sets where a small amount of power is required.

Silver-Zinc Batteries. A still more recent development is the silver-zinc battery. This is an extremely expensive battery of relatively short life, but with a tremendous power output during a high rate discharge. The chemical reaction is: $AgO + Zn \rightarrow ZnO + Ag$.

The discharge voltage is 1.4 to 1.3 volts. The positive plate is a porous silver peroxide sponge, usually with a support of metallic silver. The negative plate is spongy zinc.

This couple has been known for many years, but it was not a successful storage battery because zinc oxide is soluble in the potassium hydroxide electrode. The discharge product would, therefore, settle to the bottom of the cell and during charge form spongy zinc not available for further useful work. Also a part of the silver would go into suspension and contaminate the zinc plate with metallic silver, where it promoted rapid self discharge. The development of a separator which would prevent the migration of silver and zinc was needed to make the battery useful. Such separation has been developed, and the silver-zinc battery is now a commercial article.

Silver-Cadmium Batteries. If cadmium is substituted for zinc in the silver-zinc battery, there is a very considerable improvement in life at the expense of some reduction of capacity. These batteries are still expensive, but can be used in some applications where neither the nickel cadmium, nor the silver zinc batteries are suitable.

EUGENE A. WILLIHNGANZ

Cross-references: *Electrochemistry, Ions.*

BERYLLIUM

In 1797 Vauquelin discovered beryllium to be a constituent of the minerals beryl and emerald. Soluble compounds of the new element tasted sweet, so it was first known as glucinium from the corresponding Greek term. Quarrels over the name of the element were perpetuated by the simultaneous and independent isolations of metallic beryllium in 1828 by Wohler and Bussy. Both reduced beryllium chloride with metallic potassium in a platinum crucible. The name beryllium has been adopted universally.

Hope for the emergence of beryllium beyond the laboratory curiosity status resulted from publication of the work of the French scientist Lebeau in 1899. His paper described the electrolysis of fused sodium fluoberyllate to produce small hexagonal crystals of beryllium. Lebeau also reported the direct reduction of a beryllium oxide-copper oxide mixture with carbon to yield a beryllium-copper alloy. In 1926, Lebeau's alloy was rediscovered and found to have remarkably good mechanical properties. There emerged an industry based on the beryllium-copper alloy market which remains important today.

Commercial development of beryllium in the United States was begun in 1916 by Hugh S. Cooper with the production of the first significant metallic beryllium ingot. This was followed by formation of the Brush Laboratories Company, which started its development work under the direction of Dr. C. B. Sawyer in 1921. In Germany, the Siemens-Halske Konzern began commercial development work in 1923.

Occurrence. Occurrences of beryllium in the earth's crust are widely distributed and estimates of the amount vary from 4 to 10 ppm. There are over thirty recognized minerals containing beryllium. Only two are commercially important— beryl, $3BeO \cdot Al_2O_3 \cdot 6SiO_2$, for its high beryllium content and bertrandite, $Be_4Si_2O_7(OH)_2$, for its large quantities located in the United States.

In 1959 beryllium was found in the rhyolitic tuffs of Spor Mountain, Utah, containing from 0.1 to 1.0% beryllium oxide. The practical processing limit is now around 0.6% beryllium oxide content of the ore. Deposits of this processable grade are adequate for the industrial requirements of the United States for several decades at present levels of consumption. Although the ore grade is much lower than that of beryl ore, the beryllium values in the rhyolitic tuffs are acid soluble and recoverable by established processing technology.

In pure form beryl mineral contains approximately 14% beryllium oxide, as found in its precious forms, emerald and aquamarine. Industrial grades of the mineral contain 10 to 12% beryllium oxide. Beryl occurs as a minor constituent of pegmatic dikes and is mined primarily as a by-product of feldspar, spodumene, and mica operations. Only the relatively large crystals are recovered by hand-picking.

Hand-sorted and cobbed beryl ore has been supplied to meet industrial demands of 3–5000 tons/year. In 1969 the principal producers were Brazil, the Union of South Africa, India, and Uganda. Nearly all of the beryl ore consumed by the beryllium industry in the United States has been imported. The opening of mines and mills to process bertrandite-bearing ores in Utah has decreased beryl requirements to 1–2000 tons/year with further reductions expected.

Extractive and Process Metallurgy. The production of metallic beryllium, its alloys, or its ceramic products centers around the recovery of an intermediate partially-purified concentrate from ore processing. The usual intermediate is beryllium hydroxide or basic carbonate.

Two processes are used to extract the low content of beryllium from its ores. One is based on sulfuric acid and the other on sodium fluosilicate. Each process yields beryllium hydroxide as the end product which is then converted to beryllium fluoride by reaction with ammonium bifluoride. Thermal reduction is commonly carried out with magnesium metal to form beryllium pebbles. Final purification is accomplished by vacuum melting the beryllium pebbles to remove fluorides and magnesium impurities and casting into graphite molds. Standard powder metallurgy processes are generally used to convert the cast billets to solid shapes. The prevalent final consolidation step is hot pressing.

Properties

Beryllium has several unique properties which have given it a position of commercial significance. Its low atomic mass, low absorption cross section, and high scattering cross section are neutronic properties which spurred the expansion of beryllium production beyond the pilot scale immediately after the formation of the United States Atomic Energy Commission and the initiation of nuclear reactor development programs. Structural applications using beryllium began about 1960 to utilize its modulus of elasticity of 42,000,000 psi, its low density, and its relatively high melting point. Beryllium has good thermal conductivity and excellent thermal capacity properties which gave rise to its use as a thermal barrier and heat sink for re-entry vehicles and other aerospace applications. The latter properties have been coupled with favorable ductility properties at elevated temperatures for the development of aircraft brakes.

Physical Properties. Beryllium, with atomic number 4, has an atomic weight of 9.0122. It has a valence of 2 corresponding to the arrangement of its four electrons as $1s^2$, $2s^2$. The crystal structure is close-packed hexagonal. The average values of lattice parameters at 25°C are a = 2.2856Å and c = 3.5832A. The density calculated from these average lattice spacings is 1.8477 g/cc. Beryllium products generally have a density around 1.85 g/cc or higher because of impurities, such as beryllium oxide and other metals. The alpha-form of beryllium transforms to a body-centered cubic structure at a temperature very close to the melting point.

Beryllium melts at 1285° ± 10°C. The vapor pressure at the melting point calculates to be 0.04 torr from the equations for the vapor pressure:

$$\text{Solid } \log P_{(atm)} = 6.186 + 1.454 \times 10^{-4}T - 16,734T^{-1}$$

$$\text{Liquid } \log P_{(atm)} = 6.494 - 11,710T^{-1} \, (°K)$$

Considerable variation exists in the reported specific heat data. The following appear to be representative:

$$C_p = 4.322 + 2.18 \times 10^{-3}T \text{ cal/deg/mole}$$
$$(T = °K; \quad 600–1560°)$$

$$C_p = 6.079 + 5.138 \times 10^{-4}T \text{ cal/deg/mole}$$
$$(T = °K; \quad 1560–2200°)$$

Similar scatter exists in the reported thermal conductivity data; an average value lies around 0.35 cal/cm²/cm-sec-°C.

The thermal neutron absorption cross-section is 0.0090 barns/atom.

The electrical conductivity of beryllium is dependent on both temperature and metal purity. It varies at room temperature between 38 and 42% International Annealed Copper Standard. Beryllium also has been reported to show superconductivity at temperatures below 11°K.

Chemical Properties. Many chemical properties of beryllium resemble aluminum, and to a lesser extent, magnesium. Notable exceptions include solubility of alkali metal fluoride-beryllium fluoride complexes and the thermal stability of solutions of alkali metal beryllates.

All the common mineral acids attack beryllium metal readily with the exception of nitric acid. It is also attacked by sodium hydroxide and potassium hydroxide, but not by ammonium hydroxide.

Beryllium interacts with most gases. Polished beryllium exposed to air quickly acquires a nonporous protective oxide film. The oxidation rate increases parabolically at temperatures around 800 to 825°C, above which a sharp increase occurs.

Compounds of Beryllium. Basic beryllium acetate, $Be_4O(C_2H_3O_2)_6$, is soluble in glacial acetic acid and can readily be crystallized therefrom in very pure form. It is also soluble in chloroform and other organic solvents. It is used as a source of pure beryllium salts.

Beryllium hydroxide, $Be(OH)_2$, is precipitated as an amorphous gelatinous material by addition of ammonia or alkali to a solution of a beryllium salt at slightly basic pH values. All forms of beryllium hydroxide begin to decompose to beryllium oxide at 190°C.

Beryllium sulfate, $BeSO_4 \cdot 4H_2O$, is an important salt of beryllium used as an intermediate of high purity for calcination to beryllium oxide for ceramic applications. A saturated aqueous solution of beryllium sulfate contains 30.5% $BeSO_4$ by weight at 30°C and 65.2% at 111°C.

Beryllium chloride, $BeCl_2$, with a melting point of 440°C, is used as a component of cell baths for electrowinning or electrorefining of the metal. The compound hydrolyzes readily with atmospheric moisture, evolving HCl so that protective atmospheres are required during processing.

Beryllium fluoride, BeF_2, is readily soluble in water, dissolving in its own water of hydration as $BeF_2 \cdot 2H_2O$. The compound cannot be crystallized from solution and is prepared by thermal decomposition of ammonium fluoberyllate $(NH_4)_2BeF_4$.

Beryllium nitrate, $Be(NO_3)_2 \cdot 3H_2O$, is produced commercially by dissolving beryllium hydroxide or basic carbonate in nitric acid and crystallizing out the nitrate salt on cooling.

Ammonium beryllium carbonate solutions are prepared by dissolving the hydroxide or the basic carbonate in warm (50°C) aqueous mixtures of NH_4HCO_3 and $(NH_4)_2CO_3$. Heating the solution above 88°C evolves NH_3 and CO_2 and precipitates a basic beryllium carbonate.

Basic beryllium carbonate, $BeCO_3 \cdot xBe(OH)_2$, is variable in composition. The value of x depends on the concentration of NH_3 and CO_2 in solution at the time of precipitation and on the temperature of hydrolysis of the ammonium beryllium carbonate solution. Basic carbonate is readily dissolved in all mineral acids, making it a valuable starting compound for laboratory synthesis of beryllium salts of high purity.

Applications

In the above sections on properties, the early applications dependent on the nuclear characteristics of beryllium were described. Structural uses of beryllium in the aerospace industry have been realized because no other known material exceeds its modulus-to-weight ratio while supplying significant ductility.

Examples of applications which fully exploit the nuclear, rigidity, and high-temperature properties include heat shields, guidance system parts, such as gimbals, gyroscopes, stable platforms and accelerometers, housings, mirrors, aircraft brakes, and formable grades of beryllium used in drawing and related methods of fabrication. Recent applications have been in structures which are loaded in compression. Wrought products are being developed with yield strengths approaching 100,000 psi and 10% elongation at room temperature.

Additions of beryllium to commercial copper-base alloys enable these materials to be precipitation-hardened to strengths approaching those of heat-treated steels. Yet, beryllium copper retains the corrosion resistance, electrical and thermal conductivities, and spark-resistant properties of copper-base alloys.

The oxide of beryllium, when fabricated into finished shapes, has a unique combination of properties unavailable in any other design material. The neutronic properties of beryllium are combined with high-temperature resistance (melting point above 2500°C) which permit use in high-temperature reactor systems where the metal would deform or melt. Beryllia ceramic materials also combine extremely high electrical resistivity and dielectric strength with extremely high thermal conductivity.

Biology and Toxicology

Under some circumstances beryllium dust, mist, fume, or vapor, when inhaled, may be hazardous to health. In addition, some beryllium salts and compounds may produce a dermatitis on contact with the skin. There is, however, no ingestion problem.

The hazards are generally classified as (a) the acute respiratory disease, (b) the chronic disease (berylliosis), and (c) dermatitis.

The acute disease is principally a concern of the basic extractor. With proper medical attention, it resolves itself in complete clinical and x-ray recovery.

The chronic disease is a more important industrial hazard, since it has a latency period marked by a remarkable variance from a few weeks after exposure to as long as twenty years. Investigations by medical, toxicological, and engineering personnel led to the promulgation of safe limits of exposure in 1949 by the AEC, adopted in 1957. It should be noted that no disabling cases of chronic berylliosis from exposure were documented after 1949. The disease is believed to be avoidable when air counts are held within limits of 2 micrograms per cubic meter average for an 8-hour exposure, with a maximum at any time of 25 micrograms per cubic meter of air.

Dermatitis is produced by skin contact with soluble salts of beryllium, especially the fluoride. It is controlled by a program of good personal hygiene, frequent washing of the exposed parts of the body, as well as by a clothing program where clothing is laundered on the plant site.

Local exhaust ventilation is the major engineering control used controlling concentration of airborne beryllium. Modern air cleaners allow control within recommended outplant levels of 0.01 micrograms beryllium per cubic meter, averaged over a period of one month.

Permissible emission and exposure limits to airborne beryllium are currently being reviewed by the Environmental Protection Agency in compliance with recent legislation.

KENNETH A. WALSH

References

"Beryllium in Aerospace Structures," Brush Wellaman Inc., Cleveland, 1963.

Busch, Lee S., "Beryllium," pp. 49–56, "Encyclopedia of the Chemical Elements," Hampel, C. A., Editor, Van Nostrand Reinhold Corp., New York, 1968.

Darwin, G. E., and Buddery, J. H., "Beryllium," Vol. 7, New York, Academic Press, Inc., 1960.

Gonser, B. W., "Modern Materials," "Advances in Development and Applications," Vol. 6, "Beryllium," by N. P. Pinto and J. Greenspan, New York, Academic Press, Inc., 1968.

Hampel, C. A., "Rare Metals Handbook," article on "Beryllium," New York, Reinhold Publishing Corp., 1961.

Hausner, H. H., "Beryllium, Its Metallurgy and Properties," Berkeley and Los Angeles, University of California Press, 1965.

Kirk-Othmer, "Encyclopedia of Chemical Technology," 2nd Ed., Vol. 3, "Beryllium and Beryllium Alloys," by C. W. Schwenzfeier, Jr., New York, Interscience Publishers, Inc., 1964.

Stokinger, H. E., "Beryllium, Its Industrial Hygiene Aspects," New York, Academic Press, Inc., 1966.

BET THEORY

The data of adsorption can be most conveniently represented by adsorption isotherms, i.e., plots of the amount of material adsorbed against its partial pressure, concentration or activity after equilibrium is established in the system at a constant temperature. In the literature of the adsorption of gases and vapors, five different types of isotherms have been found, as shown in Fig. 1. A complete theory of these isotherms would furnish a complete understanding of the phenomenon; however, such theory does not exist. The BET theory offers a qualitative explanation for all five isotherm types but only a partial solution of the quantitative aspects of the problem.

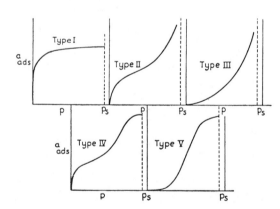

FIG. 1. Types of physical absorption isotherms.

The first theoretical treatment of the adsorption isotherm was advanced by Langmuir in 1915. He reasoned that at equilibrium the amount adsorbed on a given adsorbent was constant at a given temperature and pressure; consequently, the rate of adsorption must be equal to the rate of desorption. He assumed that (a) adsorption took place only on the bare surface of the adsorbent, i.e., only a single layer was adsorbed; and (b) the rate of desorption was directly proportional to the amount adsorbed. The second assumption implies a constant heat of adsorption throughout the entire range of adsorption.

On the basis of these two assumptions Langmuir

derived the equation

$$\theta = \frac{v}{v_m} = \frac{bp}{1 + bp} \tag{1}$$

where θ is the fraction of the surface covered with adsorbed molecules, v is the volume of gas adsorbed, v_m is the volume of gas necessary to cover the entire surface with a unimolecular layer of adsorbed gas, p is the equilibrium pressure, and b is a constant at constant temperature. Eq. (1) gives a good representation of the Type I isotherm, shown in Fig. 1.

The adsorption of vapors in most cases gives Type II, or S-shaped, isotherms. In 1938 Brunauer, Emmett and Teller published an equation that accounted for this type of isotherm. They discarded the idea that vapor adsorption was unimolecular, and used, instead the following assumptions: (a) adsorption was multimolecular, and each separate layer obeyed a Langmuir equation; (b) the average heat of adsorption in the second adsorbed layer was the same as in the third and higher layers, and it was equal to the heat of condensation of the vapor; and (c) the average heat of adsorption in the first adsorbed layer was different from that of the second and higher layers. The equation, in its linear form, is

$$\frac{p}{v(p_s - p)} = \frac{1}{v_m c} + \frac{(c-1)}{v_m c} \frac{p}{p_s} \tag{2}$$

where p, v and v_m have the same meaning as in Eq. (1), p_s is the saturation pressure of the vapor (or the vapor pressure), and c is a constant at constant temperature. Harkins named Eq. (2) the BET equation, from the initials of its authors.

Most vapor adsorption isotherms plotted according to Eq. (2) give straight lines in the range $p/p_s = 0.05$ to 0.35. The constants v_m and c can be calculated from the slope and intercept of the straight line. From v_m the surface area of the adsorbent can be evaluated. The number of molecules that cover the adsorbent with a complete monolayer is given by v_m, and the area covered by a molecule is obtained (a) by calculating a spherical volume for the molecule from the density of the liquid, and (b) by assuming hexagonal close-packing of spheres on the surface. An alternative method of obtaining the area of a molecule is to use a two-dimensional analog of the van der Waals equation. The two methods give approximately the same molecular surface areas.

The surface area values obtained for a variety of adsorbents by means of the BET equation agreed in all cases within a few per cent with the surface areas determined by direct visual means (e.g., by electron microscope), where such determinations were possible. For porous adsorbents a number of indirect surface area measurements confirmed the BET values. Thus, at present, the method is regarded as the most reliable one for the determination of the surface areas of finely divided substances, and largely because of this the BET equation is the most widely used isotherm equation in the field of adsorption.

The constant c is given by

$$c = c_1 e^{(E_1 - E_L)/RT} \tag{3}$$

where c_1 is another constant, E_1 is the average heat of adsorption in the first layer, E_L is the heat of liquefaction of the vapor, R is the gas constant and T is the absolute temperature. The difference $E_1 - E_L$ is called the net heat of adsorption. Cassie and Hill independently derived an equation, formally identical with the BET equation, by statistical mechanics and found that

$$c_1 = j_a/j_e \tag{4}$$

where j_a and j_e are the pressure-independent parts of the partition functions of the adsorbate and the liquid, respectively.

From the constant c one can obtain an approximate estimate of E_1, the average heat of adsorption in the first layer, and the values so obtained are in semiquantitative agreement with the experimental results. Eq. (2) also correctly describes the temperature dependence of the adsorption isotherms.

The equation gives too low adsorption values below $p/p_s = 0.05$ and too high values above $p/p_s = 0.35$. The former discrepancy is caused by the heterogeneity of the surface of the adsorbent, the latter by the pore structure of the adsorbent.

The BET equation was derived for adsorption on a free surface, and it was assumed that at p_s an infinite number of layers could build up on the surface of the adsorbent. In a porous adsorbent even at saturation pressure only a finite number of adsorbed layers can build up. For this case the isotherm equation is

$$\frac{v}{v_m} = \frac{cx}{1-x} \frac{1 - (n+1)x^n + nx^{n+1}}{1 + (c-1)x - cx^{n+1}} \tag{5}$$

where x is the relative pressure (p/p_s), n is the maximum number of layers that can be adsorbed, and the other terms have the same meaning as in Eq. (2). For $n = \infty$, Eq. (5) reduces to Eq. (2), and for $n = 1$ it reduces to the Langmuir equation, Eq. (1).

The three-parameter BET equation gives a better fit at higher relative pressures than the two-parameter equation, but it does not fit the data up to saturation pressure. Pickett, Anderson, Cook, Dole, Hüttig and others introduced semi-empirical modifications of the BET equation and obtained better fits at higher relative pressures for certain isotherms. There exists, however, no isotherm equation that can give good agreement with experiment throughout the entire pressure range, from $p = 0$ to $p = p_s$.

The theoretical interpretation of the Type III, IV and V isotherms of Fig. 1 was given by Brunauer, Deming, Deming and Teller. Eq. (2) represents a Type III isotherm if $c = 2$ or smaller. The data of Reyerson and Cameron for the adsorption of bromine and iodine on silica gel were well fitted by the equation. Also, the surface area of the adsorbent, the average heat of adsorption and the temperature dependence of the isotherms were accurately given by the equation.

In the Type IV and V isotherms the attainment of saturation adsorption below saturation pressure was attributed to capillary condensation. A four-parameter equation was developed, in which the fourth parameter was dependent on the heat of capillary condensation. The equations fitted certain experimental data fairly well; the surface area and the temperature dependence were accurately obtained, and reasonable values were found for the heat of adsorption and the heat of capillary condensation.

STEPHEN BRUNAUER

Cross-references: *Adsorption, Catalysis.*

BILE ACIDS AND ALCOHOLS

Vertebrates convert cholesterol in the liver to a variety of saturated C_{27} alcohols belonging to the 5α-cholestane or 5β-cholestane (coprostane) series. The lower vertebrates excrete such bile alcohols as sulfates in their bile. As an example, Fig. 1 shows the 27-desoxy-5β-cyprinol sulfate, found in toad bile. In a further step the alcohols are metabolized to C_{27} bile acids, which are conjugated with amino acids. The taurine conjugate of 3α, 7α, 12α-trihydroxycoprostanic acid, shown in Fig. 1, was isolated from frog bile. Hydroxylation of the steroid nucleus occurs preferentially in 7α- and 12α-position, but in most species 12-hydroxylation does not take place after the oxidation in the side chain. Permutation of hydroxyl substituents in the 3α-, 3β-, 7α-, 12α-, 26- and 27-position of C_{27} steroids with 5α- and 5β-configuration results in a large number of derivatives, many of which have been identified in the bile of different species and named, usually after the species in which they were first encountered.

Vertebrates above the evolutionary level of the amphibians have the ability to degrade the side chain of the C_{27} acids to produce C_{24} acids belonging to either the 5β-series (cholanic acids) or the 5α-series (allocholanic acids). These are conjugated in the liver with taurine or glycine, but only mammals can produce glycine conjugates, and in man this is the predominant form of conjugation.

Again, hydroxyl groups may be found on C-7 and C-12 and the 12-hydroxylation is inhibited when the side chain is oxidized. Fig. 1 shows the structures of the two bile acids that predominate in the human bile, chenodesoxycholic acid and cholic acid, the latter being shown in the form of the glycine conjugate, glycocholic acid. Microorganisms in the intestine remove the 7α-hydroxyl group of bile acids, converting chenodesoxycholic acid to lithocholic (3α-hydroxycholanic) acid and cholic acid to desoxycholic (3α, 12α-dihydroxycholanic) acid.

The conjugated bile acids are produced in the liver, collected in the biliary passages, stored and concentrated in the gallbladder, and intermittently discharged into the common bile duct. In the duodenum, where the pH is about 6.5 in man, the conjugated bile acids exist in the form of the anions and are referred to as bile salts. There the bile helps to neutralize the gastric chyme and to emulsify the dietary lipids. The bile salts have a powerful detergent effect and promote the contact between the pancreatic lipase and the fats. Their emulsifying action enables the intestinal mucosa to absorb lipids, including cholesterol and the fat-soluble vitamins.

In the ileum the bile acids are reabsorbed and are then transported to the liver via the portal system. Their synthesis in the liver is probably regulated by this enterohepatic circulation. Gallstones in man may contain up to 1 per cent bile salts. Human enteroliths are almost always made up of choleic acids. These are complexes of desoxycholic acid and fatty acids in a mole ratio of 8:1 (in the case of the higher fatty acids). In obstructive jaundice bile acids appear in the peripheral circulation and cause hemolysis.

ERICH HEFTMANN

Cross-references: *Steroids, Cholesterol, Lipids.*

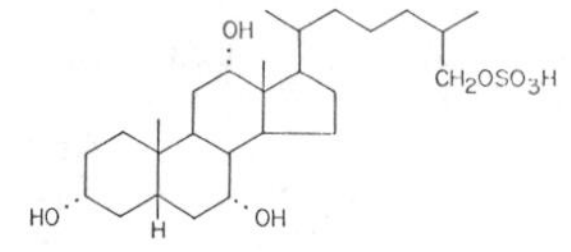

References

Heftmann, E., "Steroid Biochemistry," Academic Press, New York, 1970.

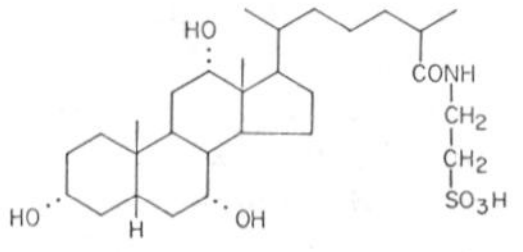

Schiff, L., Carey, J. B., Jr., and Dietschy, J. M., eds., "Bile Salt Metabolism," C. C. Thomas, Springfield, Ill., 1969.

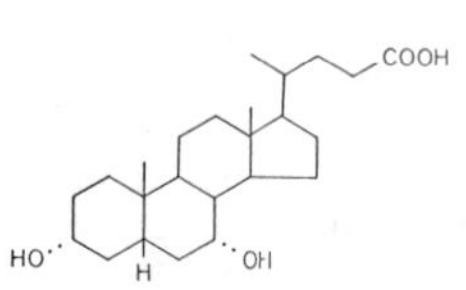

Haslewood, G. A. D., "Bile Salts," Methuen, London, 1967.

Bergström, S., "Metabolism of Bile Acids," *Federation Proceedings,* **21**, Suppl. 11,28 (1962).

27-Desoxy-5β-cyprinol
Sulfate

Taurine Conjugate of 3α,7α,12α-
Trihydroxycoprostanic Acid

Chenodesoxycholic Acid   Glycocholic Acid

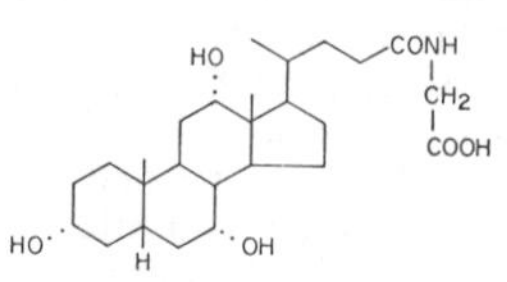

FIG. 1. Bile acids and alcohol.

BIOCHEMISTRY

The term "biochemistry," which deals with the chemistry of living things, means now about what "organic chemistry" meant before Wohler, whose synthesis of urea (1828) marked a turning point. Up to that time there was "mineral chemistry" which dealt with minerals, and "organic chemistry" which dealt with the chemistry of organisms. Organic chemistry has now come to have a much more restricted, and in a sense an expanded, meaning—the chemistry of carbon compounds—and biochemistry encompasses not only the organic chemistry of living things, but the inorganic chemistry as well—the latter a very important subdivision.

Biochemistry includes in its purview all the major branches of chemistry—as they are related to living things. Thus a biochemist may be interested in any one of a number of broad topics—the physical chemistry of living things, the colloidal chemistry of living things, the electrochemistry of living things, catalysis in living things, photochemistry as it applies to living things, e.g., photosynthesis, etc. He may use all sorts of tools: microanalysis, spectroscopic analysis, nuclear magnetic resonance, electron spin resonance, x-ray diffraction, electron microscopy, mass spectometry, isotopic tracer techniques, in short, almost any technique that is used by chemists in any area. Preparation for biochemistry therefore demands a sound training in all the fundamental branches of chemistry and a good foundation in mathematics and physics. Since the material studied is biological, a grasp of the essentials of biology is also needed.

Biochemistry has become so broad that it is no longer merely a separate discipline; it permeates all disciplines that are related to biology. Molecular biology is currently a branch of learning which would be impossible and meaningless without biochemistry. It often deals with the fundamental phenomena associated with genetic inheritance.

The immediate forerunner of biochemistry as it is known today is the "physiological chemistry" as it has been taught for several decades in medical schools. This, of course, has its center of interest in mammalian and human biochemistry with more or less emphasis on clinical and pathological findings. Plant Physiology is another field closely related to biochemistry, but its development from the chemical standpoint has been relatively slow, partly because there has been no one kind of plant organism on which everyone in the large field would willingly focus attention.

In the earlier development of biochemistry the emphasis was necessarily upon the topic: the composition of living matter. This, of course, was and is fundamental, and until a substantial storehouse of information and insight accumulated in this area, there was no opportunity to study some of the more absorbing problems related to how organisms live. The development of biochemistry in this area is therefore intimately tied to advances in the fields of protein, carbohydrate and lipide chemistry as well as the chemistry of numerous miscellaneous substances which also enter into the make-up of living things.

Earlier books and treatises on biochemistry (whether called by this name or not) were often content to deal with the "straight chemistry" of cellular constituents and excretion products and leave to physiologists or others speculation regarding how these constituents functioned and why they were present. In the course of the development, however, particularly in the past 25 years, there has been an increasing interest in the topics: Nutrition and Metabolism.

Having learned in earlier decades something about the composition of living matter (there is, of course, still much to be learned in this area), biochemists have become engrossed in the question: What do organisms need in their food? For different types of organisms the answer to this question may be very different. Some organisms such as single-celled algae, for example, are able to utilize the energy of sunlight for building up organic matter and do not require in their food any energy-yielding organic material whatever. Others such as molds, as well as certain bacteria, have relatively very simple organic requirements. Sometimes a form of carbohydrate is the only organic substance needed. In all such cases it is a mistake to think of the organisms as simple. Indeed, they must possess a most intricate mass of synthetic machinery because they contain in their make-up a wide assortment of organic materials—carbohydrates, lipids, proteins, including those in cell nuclei, vitamins, etc.—all of which must be built up from relatively simple starting materials.

Such organisms, however, have mineral requirements which are absolute and fairly numerous. Some elements are needed in relatively large amounts, others in minute traces. No organism so far as we know is able to bring about a transformation of one element to another; if molybdenum, for example, is an essential cell constituent for a particular organism, it must be supplied by food. Investigations of trace element requirements can be carried out with greater ease in organisms which have simple organic requirements than with those (e.g. mammals) which require complex organic food. In general, insofar as evidence is available, there is a high degree of unity in the biological kingdom, and a number of the trace elements are known to be needed alike, for a wide spectrum of organisms from single-celled plants to the most complex multicellular animals. At the other end of the biological scale, so far as *organic* food requirements are concerned, are such organisms as mammals which need to be supplied a multiplicity of organic materials of diverse nature without which they cannot live.

It is well to call attention also to the fact that various organisms which have a complex life history may have very different nutritional requirements at different stages in their development. For example, insects may, at the larva stage, have very exacting and complicated requirements for growth and development, but after they become adult flies, for example, they may be able to subsist on a very simple energy-yielding fare. Similar observations apply to typical seed plants. We may think of green plants, since they are capable of carrying on photosynthesis, as having very simple nutritional requirements. In the life history of a seed plant, however, there is a time—before its photosynthetic apparatus is developed—when its requirements may be highly complex. A typical seed, besides carrying a partially developed embryo, carries a store of organic material, carbohydrates, fats, proteins and vitamins, which supply the developing embryo with what it needs for early development. The needs at this stage may be comparable in complexity to those of mammals.

Nutrition has been studied most thoroughly as it applies to young mammals, notably weanling rats. Of the various types of organic compounds found in the tissues, it is notable that certain ones must be supplied in the foods, while others need not be. Thus while carbohydrates are from the quantitative standpont a most important food for mammals, there appears to be no obligatory requirement for carbohydrate. Fats and phospholipides, particularly the latter, are highly important tissue constituents, yet young mammals can thrive without anything of this nature being supplied in the diet, with the exception of small amounts of certain unsaturated acids: linoleic or linolenic acids. Proteins as such are not required by young mammals, but certain essential building stones are: histidine, lysine, tryptophan, phenyl alanine, leucine, isoleucine, threonine, methionine, valine. Nucleic acids, while playing an extraordinary role in cell reproduction and activity, are quite dispensable nutritionally for young animals. At this point it may be well to mention that requirements for growth in mammals and fowls are more exacting and deficiencies are easier to demonstrate in the early stages of growth than at or near maturity. It cannot be assumed that in mammals the nutritional needs are the same throughout the life span.

A vast new area in nutrition has developed, mostly in the last half century, in the discovery of vitamins. These nutritionally required substances, which often play catalytic roles, are discussed more fully in a section devoted to that subject.

A major part of the subject of biochemistry is that devoted to *metabolism*. This has to do with all the chemical events which transpire from the time food material is taken in by the organism until it is made use of and the waste products are eliminated. In mammals, for example, digestion and absorption are preliminary steps, followed in the body proper by an intricate series of degradations, syntheses, transformations, and oxidations, all of which may involve cyclical processes. These events are distinctive for each substance metabolized, and are often localized with respect to organs, tissues and types of cells involved. Since there are numerous substances entering into metabolism, the subject as a whole is vast.

Chemical reactions taking place within organisms are most often mediated by specific catalysts built up by the organism for the purpose. Even such a commonplace reaction as the decomposition of carbonic acid into carbon dioxide and water is mediated by a specific enzyme catalyst, carbonic anhydrase. Into the make-up of these enzymes go the essential protein building stones, and often vitamins and trace elements. Carbonic anhydrase, for example, contains zinc, and one basis for the need of this element is its presence in this indispensable enzyme. The number of enzymes produced by mammalian organisms is large. For example, instead of two or three enzymes being required for protein digestion, there are perhaps dozens. While we often think of enzymes as being present in minute amounts, this is not always the case, and it is probable that a substantial part of the total protein in a mammalian body has enzymatic functions. Many enzymes and apoenzymes (discussed elsewhere) have been isolated in the form of purified proteins.

One of the large areas of metabolism which is as yet relatively obscure has to do with the interrelations between *hormones* and enzymatic processes. Hormones (which are discussed more fully elsewhere) are produced in various localities and are transported by the blood to their site or sites of action. Though some starts have been made toward elucidating how they perform their functions, this section of biochemical history is largely yet to be written. It may be presumed in many cases that they play a catalytic role, but this is difficult to demonstrate. Enzymes can be shown to be catalysts by the most rigorous type of chemical experiment. Hormones, however, usually exhibit their effects within living organisms and when they are studied for direct chemical effects *in vitro* the results are generally negative.

A relatively new area of investigation which impinges directly upon problems of nutrition and metabolism is that of biochemical genetics. It is now apparent, on the basis of numerous studies, that the ability to produce the multiplicity of specific enzymes contained in an organism is transmitted through the genes, just as are the capabilities for producing specific morphological structures. Different species of animals, for example, are different with respect to their metabolisms, because of differences in their inherited genes. These in turn also make them different in their nutritional needs. Guinea pigs, monkeys and humans require ascorbic acid in their food because the guinea pig, monkey and human genes do not transmit the capability of producing the enzymatic machinery necessary for its synthesis. Rats and dogs, for example, have transmitted through their genes the ability to produce ascorbic acid as needed. In view of the complete dependence of enzyme production on genic transmission, it becomes increasingly certain that substantial differences in metabolism and in nutritional needs exist not only between different species of animals, but between different strains and individuals within the same species. This finding has far-reaching potentialities in relation to both the susceptibility to and the treatment of disease.

Biochemistry is rich in intriguing problems. How is stored chemical energy transformed into mechanical energy when muscles work? Just what are nerve impulses and what chemical reactions underlie their variable speed of propagation? What chemical mechanisms are involved in the process of photosynthesis and can these be duplicated artificially? What chemistry is involved in the process of vision? What are the details of the enzymatic process whereby bioluminescence (cold light) is produced? What chemically is back of the process of differentiation, whereby parent cells produce progeny which are entirely different from the parents? How and why do cancer cells arise, and what keeps them from arising much more often than they do? When specific drugs act upon the body, what phases of metabolism and what enzyme systems do they influence?

Answers to these questions and many others are being continually sought, and the successes and partial successes which have resulted make the field of biochemistry attractive to larger and larger numbers of scientifically inclined men and women.

ROGER J. WILLIAMS

BINDERS, PAINT

A coating consists of a binder that becomes the continuous phase of the final film and gives it physical integrity; a combination of color pigments and colorless extenders in amounts to give the desired color, opacity and rheological characteristics;

and additives in minor proportion to impart necessary properties at each stage of the life of the coating, from the moment production is started through its hopefully long subsequent existence, such as dispersants, thickeners, biocides, defoamers and the like. In addition, most coating compositions initially include a volatile liquid to reduce the consistency to a brushable, rollable or sprayable level. This may be a solvent for the several solid ingredients of the binder, or a nonsolvent, such as water in a latex base paint, which is a carrier only and on evaporation forces the binder particles to coalesce into a continuous film.

After a paint is made, there are four stages in its life cycle in which the nature of the binder exerts a major influence. First, the paint must be applied to the substrate by one of a variety of methods. Each of these requires a consistency in its own narrow range, to which the viscosity of the binder makes a major contribution. The consistency may be varied by the choice of solvent, in kind and in amount, or by modifying the internal structure of the paint, as with surface active agents that alter the degree of flocculation of the particulate ingredients. However, coatings need not be applied in a wet state only. Thus, a finely powdered pigmented resin may be induced to adhere to the article to be coated by electrostatic attraction from a cloud of the powder, or by dipping the preheated article into a vessel in which the powder is suspended in a fluid state in a stream of air bubbled up from the porous bottom, followed by a bake which fuses the powder into a continuous film.

Secondly, the binder must assist in keeping the liquid paint on the substrate. The paint must wet the surface and tend to spread over it. Yet it must not remain so fluid as to sag on a vertical surface. In paints containing a solvent, its rate of evaporation governs this factor.

Thirdly, the dry film may require cure. In one class of coatings, that is, lacquers, the film is practically cured after solvent evaporation, and here the fixation and cure stages merge into one. The binder selected must therefore be one that has the desired durability as soon as the solvent has left. However, most binders must be altered chemically during the cure to achieve the necessary resistance properties. One type of cure is the oxidation, catalyzed by drier additives, of the binder in a house paint based on a drying vegetable oil, like linseed oil. Another type of cure is chemical reaction between components of the binder that ignore each other in the package at room temperature but react in minutes at an elevated temperature to leave a tough, insoluble, highly resistant thermoset polymer of the original binder mix. Still another type is one in which the binder components are too reactive to permit a single package at room temperature, and two separate components must be intimately mixed and applied, generally the same day, after which cure is effected by chemical reaction. In the case of powder coatings, the fusion step cures the film, which may be thermoplastic because the chemical nature of the film is not changed by the bake, or thermosetting because the heat induces a chemical cure as well as fusion. Latex-base paints on drying by water evaporation leave a coating of nonvolatile ingredients dispersed between minute

globules of latex binder. These globules coalesce at ambient temperatures to give a continuous film with physical integrity and water resistance. The coalescence of latex-base paints and the fusion of powder coatings are not unrelated.

Fourthly, the durability of the cured film depends primarily on the nature of the binder. To the extent that the pigment shields the binder behind it or ultraviolet absorbers divert harmful actinic rays, the binder is assisted in its battle with the elements. A durable binder can be formulated into a poor paint by ignorance or design but a durable paint cannot originate with a poor binder.

The drying vegetable oils, among which linseed oil is the most important, constitute a class of liquid binders that has been longest in common use. They are, in the main, mixtures of the triglyceryl esters of unsaturated fatty acids. Under the catalytic influence of driers, that is, the cobalt, manganese, lead and other metallic salts of organic acids, and after an induction period, these esters absorb oxygen at the unsaturated sites in their molecules and combine to form much larger molecules, yielding the leathery coats we know as house paint films, for example. The major drying process occurs in hours and days and persists slowly for a long time thereafter, often to the detriment of the film. Vegetable oils are processed and heat treated to enhance properties that increase their utility as binders.

Phenolic resins are among the oldest binders in current use. They are the products of phenols combined with formaldehyde and further condensed. Some are soluble in vegetable oils; others are not. Some need baking to bring out their best properties; others are ingredients of air-dry finishes such as spar varnishes.

Another class of binders, basic to many lacquer formulations, is the cellulose esters, such as nitrocellulose and cellulose acetate, and the cellulose ethers, such as ethyl cellulose. These solids must be dissolved in suitable solvents to be used. To give them acceptable binder characteristics, they are modified with resins and plasticizers.

During the last several decades, many new classes of synthetic paint binders have appeared on the scene at an increasing pace. Historically early are the alkyd resins, which in the simplest sense are the glyceryl esters of a combination of phthalic anhydride and drying or nondrying fatty acids, or, more broadly, equivalents in which any of a number of organic polyols replaces the glycerol in whole or in part and any of a number of polybasic organic acids replaces the phthalic anhydride. If no fatty acid is included, the composition is a polyester rather than an alkyd. Alkyds are valuable as binder components in combination with a host of other film-forming resinous materials, to give hybrid properties no component by itself possesses.

The polymers of the alkyl acrylates and methacrylates, generally modified with other unsaturated monomers, constitute an important class. They excel in light-fastness and durability. The simplest give thermoplastic lacquer-type binders. However, proper modification of the molecular architecture yields thermosetting compositions.

The preparation of these acrylic esters, not in solution in organic solvents, but in an aqueous carrier in which the resin globules are insoluble, that is, in latex form, has had a tremendous impact on the coatings industry. The broader field includes latexes based on vinyl acetate polymers and copolymers and a number of other monomers and monomer mixtures that lend themselves to latex technology. The ease of clean-up with warm soapy water and the decrease in air pollution through the elimination of organic solvents are among the advantages.

A commercially significant class of binders is the aminoplast group. They started historically with the methylol derivatives of ureas and melamines. These may be condensed under the influence of heat to give highly durable but brittle products. When suitably plasticized, as with other compatible binders, a group of baking finishes results, of the type very useful in appliance coatings.

Catalyzed epoxy binders are a class of two-package compositions which react when mixed to give highly resistant finishes that suffer only from a slight tendency to chalk on exterior exposure and to yellow. The epoxy components are long-chain ethers with spaced hydroxyl groups that combine readily with polyamines, aminoplasts, organic acids, isocyanates and the like. The reaction rate at room temperature may be accelerated by heat.

In addition to the vinyl latexes already mentioned, vinyl polymers in organic solvent solution find wide application as binders in coatings, usually in conjunction with other binder components. To obtain particularly tough films, the polymer may be processed to a relatively high molecular weight. It is at this stage no longer soluble but may be mixed in fine powder form with an organic plasticizer. When the applied mixture is baked, polymer and plasticizer fuse into the intended coating. These plastisols may be thinned, if necessary, with a suitable solvent for the plasticizer for greater ease of application, in which case the mixture becomes an organosol.

The isocyanate group is the reactive unit in one of the youngest classes of binder, the urethanes. With two or more isocyanate groups in the molecule, it can condense with a wide selection of other compounds containing active hydrogen loci to form resins. Some of these are then no longer reactive. When the active hydrogen resides in glycerol which is partially fatty acid esterified, the result is in effect an alkyd in which the phthalic anhydride is replaced by the diiosocyanate. Several other types of urethane coatings are possible when residual active groups remain. In one, the finish dries rapidly on exposure to air by reacting with the water in the atmosphere. In another, the isocyanate is blocked with a transient bond to a third ingredient in a package also containing the component supplying the reactive hydrogen sites. Heat after application frees the isocyanate to combine with this chemical partner. The most durable type is a two-package system in which the polyisocyanate and a polyester as the second component are supplied separately, isolated because of their mutual reactivity, to be mixed on the day of application.

The siloxane group $\equiv Si - O - Si \equiv$ is a building block that is compatible with carbon chemistry in its chain-forming ability and thereby originates the silicone class of resins. These excel in heat resistance and durability, so much so that they can be copolymerized and even blended with more conventional components to retain a portion of their excellence at reduced cost. The simplest are lacquer types but heat curing types have also been synthesized. It should be mentioned that the extraordinary properties of the silicones adapt them also to other uses than as binders.

There are additional useful classes of paint binders, such as the hydrocarbon resins, the chlorinated rubbers, and others of minor volume but major significance when their peculiar combinations of characteristics are required. New classes are being continually unveiled as a direct result of binder research and as spin-off from allied industries like plastics. It is probable that the best are yet to come.

DAVID M. GANS

References

Myers, R. R., and Long, J. S., Editors, "Treatise on Coatings," Volume 1, Parts I (1967) and II (1968), "Film-Forming Compositions," Marcel Dekker, Inc., New York.

Martens, C. R., Editor, "Technology of Paints, Varnishes and Lacquers," Van Nostrand Reinhold, New York, 1968.

Taylor, C. J. A., and Marks, S., Editors, "Paint Technology Manuals," Parts 1 (1961), 2 (1961) and 3 (1962), Chapman and Hall, Ltd., London.

Payne, H. F., "Organic Coating Technology," Volume 1, John Wiley & Sons, Inc., New York, 1964.

BIOMATERIALS

The recognition of biomaterials as an identifiable discipline within the broader field of biomedical engineering has occurred relatively recently. Although materials scientists and engineers have made sustained contributions to the development of dental restorative and orthopedic materials since the turn of the century, the interest in new biomaterials has had increasing impetus with the development of soft tissue implants and artificial organs since World War II and with the achievements in biomedical engineering during the past two decades.

The definition of a biomaterial is still evolving, but is generally considered to encompass any material, usually synthetic or semisynthetic, that is implanted subdermally by a surgeon. Ideally, the implant is expected to closely duplicate the mechanical, physical and chemical properties of the replaced tissue while achieving unlimited compatability and nondegradability with the host tissues and body fluids. Currently, the application of biomaterials is broader than this simple definition implies, and the ideal expectation for biomaterials is still a distant goal. The above definition does not encompass those biomedical materials developed for surgical tools, instrumentation, synthetic ligatures, blood storage and transfusion containers and conduits, probes, drains, external braces and artificial limbs, pharmaceutical packaging, and general health care equipment.

However, biomaterials are considered to include dental restoratives, membranes, adhesives, supporting structures, percutaneous leads, and devices where cellular or tissue interfaces are sustained.

The scope of the biomaterials field has been described by S. N. Levine[1] to include the following topics:

"• performance of materials in prosthetic applications,
• new techniques for the *in vitro* and *in vivo* evaluation of prosthetic materials,
• chemistry and metallurgy underlying the preparation and behavior of biomaterials,
• stability and corrosion of materials in physiological environments,
• effects of materials on mechanisms of blood coagulation,
• influence of implanted materials on living tissues and related aspects of toxicology and histopathology,
• progress related to standards and specifications for prosthetic materials,
• research on reprocessed animal bone and tissue in prosthetic applications————, and
• fundamental investigations on the biophysical and biochemical aspects of materials."

These activities involve the close collaboration of metallurgists, ceramists, polymer chemists and engineers, biochemists, biomechanists, physiologists, surgeons, dentists, pathologists, and veterinarians, among others. Consequently, new curricula are being developed to train surgeons in physical science subjects and to train materials scientists in biological and medical subjects in order to satisfy demands for new competencies in this discipline. Such competencies include quantification of biomaterial requirements on the part of the surgeon and the specification, selection, and testing of biomaterials to satisfy biochemical and physiological criteria on the part of the materials scientist and engineer.

Biomaterials are selected from several broad groupings of materials to include metals, ceramics, polymers, elastomers, carbons and graphites, composites, bone, bone derivatives, and reprocessed elastins. With some important exceptions, the compositions, formations, and designations of implant materials have corresponded to those developed for other technical and industrial applications due to requirements for standardization, availability, and low cost. However, through the collective efforts of the medical profession, investigators, manufacturers, standardization committees (e.g., Committee F-4 of the American Society for Testing and Materials), the Department of Health, Education and Welfare, and the Pure Food and Drug Administration, materials processes and quality control standards for biomedical applications and biomedical grades of materials are being established. Specifications for dental materials have long been in existence and have been an important factor in the success enjoyed by restorative dentistry. The recent classification of implantable materials as drugs is also placing important restraints on the selection and clinical application of biomaterials.

Available experience from clinical evaluations and surgical procedures has established a number of technical, functional, and compatibility requirements for implant materials; they should (1) be reproducibly obtainable as an uncontaminated material, (2) be easy to fabricate into the derived form at relatively low cost, (3) not be degraded or adversely changed during storage, sterilization, and insertion, (4) have the needed chemical, physical, and mechanical properties for performing their function in the body, (5) not be degraded or adversely modified by the tissue, (6) not induce blood trauma or alter cellular or soluble materials in blood, (7) and be nontoxic, noninflammatory, nonimmunogenic, and noncarcinogenic to the host tissues. In addition to these general requirements R. Johnsson-Hegyeli[2] lists additional factors which need to be considered in selecting an implant material. Among these are: (1) etiology of the defect requiring implant, (2) anatomy and pathology of the site of implantation, (3) factors of importance to successful implantation at the site, (4) suitable form, size, and nature of the intended implant, (5) method of sterilization, (6) consideration of possible interactions between implant material and therapeutic drugs, (7) surgical techniques required, and (8) follow-up of patient and implant. In addition to the above, cosmetic factors such as resiliancy, softness, texture, and color are important considerations in many soft-tissue implants and dental restorations. Unfortunately, materials that ideally satisfy the above requirements do not exist, nor can the requirements be adequately defined in quantitative terms.

After systematic study of the above factors, the biomaterials investigator in collaboration with life scientists establish testing procedures and protocols for evaluating the properties of candidate materials under simulated body environments. Among the properties of interest are hardness (or softness), ductility, strength, and elastic or viscoelastic properties under static and dynamic loading; notch sensitivity; fatigue or flex life; surface charge, chemisorption, structure, and friction; corrosion and erosion resistance; electrical and thermal conductivity, thermal expansivity, heat capacity and dielectric constant; hydrophobicity; porosity and permeability; molecular fraction (polymers); and long-term microstructural and phase stability under typical service conditions of temperature and stress. At appropriate stages in the evaluation program, tests are conducted *in vitro* in common solvents, organic acids, and isotonic salt solutions to evaluate electrochemical responses to body and oral environments. Toxicology responses are determined by *in vitro* and by *in vivo* tissue culture and histopathological methods which measure acute and chronic antigenic, systemic, carcinogenic, and localized material-cellular reactions. Animal tests are conducted to statistically evaluate (1) the release of ions and other ingredients from the implanted material to the biological environment, (2) alteration of the implant properties by the biological environment, (3) sequestration, biodegradation, or fixation of the implant by the host tissue, (4) the response of the host tissue to the material (e.g., the development of pseudoendocardium or blood trauma in cardiovascular devices), and (5) the mechanical response of the structural implant to stresses, wear, and flexure. The final stage, after the biomaterial has satisfactorily passed the above tests, is clinical evaluation in humans. The relatively recent application of such techniques as electron microprobe analysis, scanning electron microscopy, chromatography, acoustic holography, ultrasonic detection, telemetry, neutron activation analysis, radiosotope tracer analysis, and *in vivo* corrosion analysis, to mention only a few, are adding new dimensions and precision to the *in vivo* evaluation of biomaterials.

Biomaterials can be divided into three general, but nonrigorous, categories: structural biomaterials, soft-tissue substitutes, and implantable electrodes and leads for muscle, nerve and bone-growth stimulation. Under structural biomaterials can be listed prosthetic and restorative materials for bone, restorative dental materials, and adhesives and hemostatic agents. Under soft-tissue substitutes can be listed artificial hearts and assist devices, membranes for oxygenator elements in heart-lung machines and dialysis elements in artificial kidneys, artificial skin, cannulae, corneas, bulk tissue substitutes, joints and facial prostheses. Since multicomponent devices, such as total hip joint replacements or artificial heart pumps can involve both structural and soft-tissue elements these classifications must be considered tentative.

Although a vast variety of materials have been used in the past by orthopedic surgeons for internal bone fixation and bone replacement, the metals and alloys presently used for bone fixation and prostheses are almost exclusively limited to types 316, 316L, and 317 stainless steel (described in ASTM Specifications F55-66 and F56-66), wrought (HS-25) and cast (HS-21) cobalt-chromium alloys (described in ASTM Specifications F75-67 and F90-68), and unalloyed titanium (described in ASTM Specification F67-66). Tantalum has also been used since the 1920's for cranial plates, sutures and clips, foil and wire for nerve repair, and as plate, sheet and woven gauze for abdominal muscle repair.

Among the shortcomings of the alloys leading to service failures are inadequate strength, limited fatigue life; fabrication inhomogeneities; mislabeled and mismatched materials resulting in galvanic corrosion; crevice corrosion at countersunk screw holes or where differential oxygen activities occur along varying surfaces; pitting corrosion and corrosion fatigue in type 316 stainless steel containing less then the minimum 2.00 wt. % molybdenum specified; and stress corrosion cracking in highly stressed components. As alternatives to autogenous, homogeneous, and heterogeneous bone replacements, porous metals, ceramics, and carbons are being evaluated. Although these implants have insufficient strength for normal body load applications, they have the advantage of closely matching the elastic modulus of cortical bone, thereby reducing interfacial stress concentrations under load. In addition they provide a scaffolding for tissue fixation, ingrowth, and calcification under normal bone healing processes. Consequently, porous materials are also being evaluated for tooth and mandibular replacements. Current developments of orthopedic implants are underway to improve wear resistance and to reduce friction at articulating joint surfaces, to increase alloy strength and fatigue resistance, to reduce metal ion release rates during initial surface passivation, and to apply composite material technology to model and simulate the composite nature of cortical bone.

A variety of metals, ceramics and plastics have been evaluated for dental restorative materials and applications; however, the use of these materials is encumbered by a number of problems inherent in material preparation, the nature of enamel and dentin, constantly changing chemical, thermal, and biological characteristics of the oral environment, compatibility requirements, and clinical performance. Standard Ag-Sn dental amalgam is inherently weak in tension and can undergo undesirable embrittling phase transitions during long-term service. Similar deficiencies occur in other amalgam systems containing varying amounts of silver, tin, copper, and zinc. However, there is some expectation that the strength and toughness of these amalgams can be improved further by reducing porosity, increasing the strength of individual phases and improving bonding between the matrix and unreacted particles. Fluoride-containing silicate cements and porous oxides and carbons are among the ceramic materials being investigated, and acrylic resins show promise among the polymers because of their acceptable mechanical properties and good resistance against discoloration and deterioration by heat and light. However, the lack of satisfactory initiators, differences in coefficients of thermal expansion, and volume and wetting changes upon solidification indicate that direct bonding of the resin to a vital tooth will be unsatisfactory. A current objective is finding a material which is adhesive to tooth structure and which will eliminate leakage and discoloration around the margins of the restoration. The use of elastomeric adhesive liners between the resin and tooth surface may overcome these problems, and efforts are being made to analyze and synthesize barnacle cement to provide such a nonwetting adhesive.

A significant advance in the development of cardiovascular devices and artificial blood vessels occurred when woven Dacron® velour was discovered to stimulate the development of an autogenous tissue layer, thereby providing a partial solution to the tissue-interface problem. Subsequently, Dacron® velour of varying weave designs has been used as backing for steel medullary pins, silicone rubber implants, artificial skin, percutaneous leads, and artificial tendons to promote tissue adhesion. Blood trauma, particularly hemolysis and clotting, has also been a major problem in heart pump and valve design. Efforts by materials scientists to develop a nonthrombogenic surface have led to the use of low-energy hydrophobic polymer surfaces and coatings containing collodial graphite-benzalkonium chloride, and heparin (benzalkonium chloride greatly increases the sorption of heparin on hydrocarbons). Recent *in vivo* tests have shown, however, that impermeable, isotropic carbons are significantly thromboresistant without the exogenous application of heparin, and such carbon valves are currently under clinical evaluation.[3]

Medical-grade silicone rubber has had remarkable success in subdermal soft-tissue replacements (millions of implants) because of its almost negligible tissue reaction and the wide spectrum of structural forms of properties that can be formulated. It is currently the most widely used material for soft-tissue prostheses and is being applied as shunts, ligatures, ureters, bladders, tracheas, larynxes, and scleral bucklers, and for ear and breast augmentation, among many others. However, undesirable properties of silicone rubber for some applications include a lack of tissue response, low flexural strength, rapid lipid absorption causing hardening, hydrophobicity in most physical forms, and thrombin promotion under some circumstances.

The classes of polymers that are being used or have potential use as biomaterials by virtue of their useful properties for specific applications, good resistance to adverse tissue reactions, and specific clot-inhibiting properties are numerous, and the reader is referred to other reviews[4,5] for details. However, D. J. Lyman[6] lists the following as promising: polyethylene, polypropylene, polytetrafluorethylene (Teflon®), polydimethyl siloxane, polyparaxylylene, polycarbonate (Lexan®), polyethylene terephthalate (Dacron®), polyhydroxyethyl methacrylates (Hydron®), cellulose, epoxides, polyurethane (spandex elastomers), polyelectrolytes, polypeptides, (including collagen), block copolymers, and grafted and composite materials. In applying these materials, however, accurate control of polymer molecular weight and molecular weight distribution must be exercised and rigid quality control is required to guard against undesirable additives which might include unreacted monomers, catalyst residues, solvents, contaminants, antioxidants, plasticizers, fillers, etc. Examples of polymer applications in biomedical applications involving a specificity in properties include the use of silicone polycarbonate block polymers and dimethyl silicone rubber for membranes (high O_2 and CO_2 permeability), alpha cyanoacrylate adhesives for hemostasis and for holding skin grafts in place (rapid polymerization, good wettability, and biodegradable) poly-α- amino acids for temporary burn wound covers (biodegradable and wettable), high-density polyethylene as self-lubricating sockets (high lubricity, resistance to wear) and polymethylmethacrylate as a self-curing cement and grouting compound (rapid self-curing and strength) in the total hip prosthesis and as a construction material for denture plates.

Among the greatest current challenges in the field of biomaterials are the solution of tissue interface problems; the measurement and simulation of biomechanical properties of hard and soft tissues; the development of incentives, controls, and standards for the adequate supply of medical-grade biomaterials, but at nonprohibitive costs; and the understanding of long-term surface behavior of implants under dynamic and aggressive body environments. Some of these problems go to the heart of the ultimate distinctions between living and nonliving matter and provide an exciting stimulus for both scientists and engineers.

ARDEN L. BEMENT, JR.

References

1. Levine, S. N., "Editorial," *J. Biomed. Matls. Res.,* **1,** No. 1, 1962.
2. Johnsson-Hegyeli, R., "The Physicological and Biomedical Basis for Soft Tissue Implants," in "Bioengineering Applied to Materials for Hard and Soft Tissue Replacement," Editor, A. L. Bement, Jr., U. of Washington Press, Seattle, 1971.
3. See also "Proceedings of Artificial Heart Program Conference," Editor, R. Johnsson-Hegyeli, Washington, D.C., June 9–13, 1969, National Heart Institute, U. S. Govt. Printing Office.
4. "Proceedings of Symposium on Medical Applications of Plastics," Editor, H. P. Gregor, Toronto, Canada, May 25–26, 1970, Interscience Publishers, 1971. (*J. of Biomed. Matls. Res.,* **5,** No. 2, 1971).
5. Sanders, H. J., "Artificial Organs," *Chem. Eng. News,* April 5, April 12, 1971.
6. Lyman, D. J., "Biomedical Polymers: Their Problems and Promise," in "Bioengineering Applied to Materials for Hard and Soft Tissue Replacement," Editor, A. L. Bement, Jr., U. of Washington Press, Seattle (1971).

BISMUTH AND COMPOUNDS

In the middle ages Europeans had recognized bismuth as a specific element and referred to it as *wismut* which was later Latinized to *bisemutum.* Bismuth occurs in the earth's crust in about the same abundance as silver. Although it is found as native bismuth, the major portion is combined in ores. The minerals are widely distributed in the world but in small quantities. An estimated 2,000,000 to 4,000,000 pounds of bismuth were produced annually in the world over the last decade. 40–50% of this is refined in this country.

The primary source of bismuth in the western hemisphere is as a by-product of copper and lead smelting and refining. The bismuth occurs in ores with these metals and remains with the metals after smelting. Bismuth is recovered from copper in the anode slimes during the electrolytical refining and the procedure for handling the slimes is such that the bismuth is collected in lead. Bismuth is removed from lead during the refining by the Betterton-Kroll process, which depends upon the formation of calcium-bismuth and magnesium-bismuth compounds whose high melting points and lower density permit them to be liquated from the lead bath. The enriched bismuth dross is freed of calcium, magnesium and lead by chlorination.

Bismuth will fracture as a brittle, crystalline metal having a high metallic luster with a pinkish tinge. As a member of Group VA of the Periodic Table, it is in the same subgroup as arsenic and antimony.

Atomic no.	83
Atomic weight	209
Density—20°C g/cm³	9.80
Density—271.3°C g/cm³	10.067
Density—20°C lb/in³	0.354
Atomic volume $\dfrac{\text{cm}^3}{\text{g-atom}}$	21.3
Melting point—°C	271.3
Boiling point—°C	1420
Specific heat—20°C cal/g/°C	0.0294
Heat of fusion—cal/g	12.5
Vapor pressure—540°C	10^{-3} mm
840°C	1 mm
1200°C	100 mm
Crystal form—rhombohedral	
A = 4.7361 A°	
Axial angle—57° 14.2′	

Bismuth is one of the two metals (gallium the other) which increases in volume on solidification, this expansion being 3.32%. The thermal conductivity of bismuth is lower than any metal with the exception of mercury. The most diamagnetic of all metals, bismuth has a mass susceptibility of -1.35×10^6.

The low absorption cross section of bismuth for thermal neutrons has focused attention upon it as a possible coolant for nuclear reactors.

Bismuth alloys with a number of other low-melting elements to form a group of alloys commonly referred to as "fusible alloys." Table 1 gives a number of the common eutectic compositions and melting temperatures. Fusible alloys are used in a number of ways for such safety devices as safety plugs in compressed gas cylinders, automatic fire sprinkler systems and fire door releases. Some of the bismuth alloys have other unusual properties such as low liquid-to-solid shrinkage and expansion in the solid state. These properties have led to their use for gripping tools, punches and parts to be machined. Bismuth may be melted and cast in the same manner as lead, but its low ductility does not permit working at ordinary temperatures. At higher temperatures (above 225°) it becomes more plastic, permitting a limited amount of shaping.

TABLE 1.   FUSIBLE ALLOYS

System	Composition	Eutectic Temperature
Cd—Bi	60 Bi	144°C
In—Bi	33.7 Bi	72°C
	67.0 Bi	109°C
Pb—Bi	56.5 Bi	125°C
Sn—Bi	58 Bi	139°C
Pb—Sn—Bi	52 Bi-16 Sn-32 Pb	96°C
Pb—Cd—Bi	52 Bi-8 Cd-40 Pb	92°C
Sn—Cd—Bi	54 Bi-20 Cd-26 Sn	102°C
In—Sn—Bi	58 Bi-17 Sn-25 In	79°C
Pb—Sn—Cd—Bi	50 Bi-10 Cd-13.3 Sn-26.7 Pb	70°C
In—Pb—Sn—Bi	49.4 Bi-11.6 Sn-18 Pb-21 In	57°C
In—Cd—Pb—Sn—Bi	44.7 Bi-5.3 Cd-8.3 Sn-22.6 Pb-19.1 In	47°C

In other metallurgical applications bismuth is used as a carbide stabilizer in the manufacture of malleable iron. When added to aluminum and low carbon steels, it improves the machinability of those materials.

In electronic applications bismuth is an important ingredient for counter-electrode alloys used on selenium rectifiers. It is also an important constituent in thermoelectric materials.

Two sets of compounds will form in which bismuth is trivalent and quinquevalent. The trivalent compounds are the more common. Bismuth does not readily form oxides at ordinary temperatures. The metallic luster is retained for a long period of time. Bi_2O_3 is the best defined of the oxides, the existence of others being questioned. The compounds of bismuth with the halogens are of the form BiX_3. They may be readily formed by dissolving the bismuth in nitric acid and adding a soluble halogen salt. The salts of trivalent bismuth hydrolyze in water to insoluble basic salts precipitating the oxysalt. The best solvent for bismuth is nitric acid. From the concentrated solution the bismuth nitrate pentahydrate is formed. Concentration of this solution by evaporation will form the nitrate. The subnitrate is prepared by the hydrolysis of bismuth nitrate between temperatures of 30 to 70°C. Bismuth subcarbonate is prepared from the bismuth subnitrate by adding sodium carbonate. Both bismuth subcarbonate and bismuth subnitrate are used for medicinal purposes.

Bismuth oxychloride is used as a pearlescent ingredient in cosmetics. Bismuth phosphomolybdate is used as a catalyst in the production of acrylonitrile fibers.

H. E. HOWE, REVISED BY S. C. CARAPELLA, JR.

BLOCK AND GRAFT POLYMERS

Definitions. A *block* polymer is a macromolecule comprised of long chains of a particular chemical composition, the chains being separated by (a) a polymer of different chemical composition or (b) a low molecular weight "coupling" group; a stereoblock is derived from a single monomer and contains alternating sequences of different steric conformations. A *graft* polymer is made up of a high molecular weight "backbone" to which are attached chains of a different polymeric species; the backbone may be homopolymeric or copolymeric with pendant groups of either type. Solution and bulk properties of blocks and grafts reflect the composition and structure of these species.

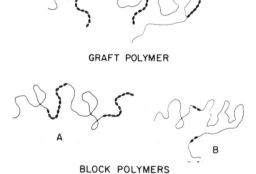

GRAFT POLYMER

BLOCK POLYMERS

Most synthetic routes to blocks and grafts depend on the presence of an active site on a polymer chain capable of initiating the chemical reactions leading to the desired product; the ingenious techniques employed permit precise control of the properties of the product, and impressively establish the potential of "molecular engineering" of polymer systems.

Chain transfer. In a system containing polymer P and growing chains of monomer M, if chain transfer to P (i.e., abstraction of hydrogen or halogen atom from P) occurs by the growing chain of M units, polymerization of free monomer can be initiated by these newly formed free radicals; the product is a graft copolymer. In practice both homogeneous and emulsion systems have been employed successfully. For

$$
\begin{array}{c}
\wr \\
\text{M.} \\
\sim \text{P—P—P—P—P} \sim \ \rightarrow \\
\quad\quad \sim \text{P—P—P—P—P} \sim + \cdot \sim \text{M} \\
\quad\quad\quad\quad\quad \cdot \\
\quad\quad\quad \downarrow \text{ monomer} \\
\quad\quad \sim \text{P—P—P—P—P} \sim \\
\quad\quad\quad\quad | \\
\quad\quad\quad\quad \text{M} \\
\quad\quad\quad\quad \text{M} \\
\quad\quad\quad\quad \text{M} \\
\quad\quad\quad\quad \wr
\end{array}
$$

efficient grafting, it is essential that the radical transfer step occur exclusively with the backbone, and that the favored reaction of the resulting polymer radical be chain initiation of the second monomer. Often, it is convenient to modify the structure of the backbone to favor chain transfer by incorporating susceptible carbon-halogen or sulfur-hydrogen bonds in the starting polymer. The common monomers such as styrene, vinyl acetate, methyl methacrylate and acrylonitrile have been grafted *via* chain transfer to polyacrylamide, polyacrylonitrile, polyethylene and polymethyl methacrylate.

The unique structure of a graft is reflected in the polyelectrolyte properties of polyampholytes prepared by polymerizing 2-vinyl pyridine in the presence of polyethyl acrylate, followed by hydrolysis of the ester group to the free acid. In aqueous base, the specific viscosity of the graft is less than that of a random copolymer of about the same composition, because of differences in the distribution of the functional groups in the graft and in the random copolymer. The latter is soluble because it forms a polymeric anion which is extended in solution as a result of the repulsive forces present in the chain; in the graft, the effect of the repulsive forces between the ionic groups in the backbone is attenuated by the pendant chains so that the normal coiled configuration is least disturbed.

Direct radical attack. If the starting polymer is unsaturated (polyisoprene, for example), in addition to hydrogen abstraction, direct addition of the initiating radical to the backbone double bond occurs; the macroradical so formed can initiate vinyl polymerization. Grafts of methyl methacrylate to natural rubber, of structure similar to a number of commercially useful high-impact polyblends, illustrate that the properties of the graft both in solution and in the solid state reflect the different properties of the component polymers: when a graft polymer containing, for example, 30% methacrylate is isolated by evaporation from benzene-petroleum ether, the product is rubbery, while when benzene-methanol is used, the product is a glossy, hard plastic surface exhibiting the properties of the rigid methacrylate.

Electromagnetic radiation. Radicals on a polymer chain can be produced by bombardment with electromagnetic radiation of energies greater than bond energies, i.e., greater than about 3 eV. Ultraviolet radiation results in cleavage only of a particular bond in an absorbing molecule, thus leading to grafts the structures of which can be deduced from the chemistry of the starting polymer. For example, polystyrene brominated in the α-position, when photolyzed with light of about 4000Å wavelength decomposes into the macroradical and bromine atoms; methyl methacrylate has been grafted to polystyrene in this manner. Carbon-chlorine bonds, hydroperoxy, amino, keto- and thioketo linkages are photosensitive groups which have been incorporated into polymers for subsequent block and graft syntheses.

Higher energy radiation (up to several million ev, obtained from radioactive isotopes or machine sources) interacts randomly with polymers to form radicals. Irradiation in an inert atmosphere of a polymer swollen with monomer permits graft formation to an extent dependent on the relative radiation sensitivity and diffusion characteristics of the chemical species present, and on the radiation dose and dose rate employed. Grafts of styrene to polyethylene exhibit elastic moduli between those of the component polymers; sulfonated grafts are cation exchange membranes which may be converted to anionic membranes by replacement of the acid group with an amine. The polymeric peroxides and hydroperoxides formed when the backbone is irradiated in air subsequently may be decomposed in the presence of a vinyl monomer, in the manner identical with conventional initiators, to form grafts. This "preirradiation" technique affords grafts of acrylonitrile to polyethylene: the product contains 220% acrylonitrile, with densities, softening points, and ultimate strength properties of the latter polymer, but with the film-like characteristics of polyethylene.

Radiation polymerization of unsaturated monomer-polymer mixtures (acrylics, silicones, e.g.) leads to complex reaction products which contain graft structures; and irradiation of such syrups on wood or plastic substrates produces grafts to that organic substrate.

Mastication. When a polymer such as natural rubber is masticated or milled, the applied shear forces may break primary bonds to form macro-radicals capable of reacting with vinyl or oxygen compounds, although it is difficult, if not impossible, to isolate and identify pure products from milling reactions. The best known milled systems are the commercially important rubber blends with the styrene and with acrylonitrile-styrene copolymers, in which incorporation of a small amount of rubber provides considerable improvement in toughness with small sacrifices in mechanical properties and thermoplasticity; increasing the rubber content imparts greater impact strength to the system, although the surface hardness, rigidity, and tensile strengths are lowered.

Living polymers. For some monomers, with sodium-napththalene as initiator, in the absence of a suitable terminating agent (e.g., water) the growing chain ends remain active when the monomer is consumed so that addition of a second monomer to this "living" chain produces a block polymer. Living polymers of styrene, isoprene and methyl methacrylate have been used to initiate block formation of the common vinyl monomers.

Unvulcanized blocks of polyisoprene on polybutadiene between polystyrene units exhibit rubber-like properties at ambient temperatures, similar to conventional vulcanized elastomers, but can be molded as a normal thermoplastic resin at elevated temperatures. At room temperature, a stable network is formed as a result of phase separation of the dissimilar blocks, and so a two-phase system exists where the glassy polystyrene agglomerates are embedded in a polydiene matrix, and function as multifunctional crosslinks.

Functional group reactions. A polymer containing chemically reactive groups at the ends of the backbone or along the chain, in suitable environments, can initiate block and graft formation. For example, linear and isotactic polymers and copolymers of the α-olefins ($RCH\text{-}CH_2$) contain tertiary carbon atoms which may be air oxidized at low temperatures; under these conditions, attack occurs only on the surface of the material or in the amorphous regions next to the surface. Exposure of the macroperoxide to a suitable monomer yields grafts. Grafts of vinyl chloride to polypropylene and polybutene exhibit unique properties: The plasticizing action of the hydrocarbon portion of the graft is different from that of conventional polyvinyl chloride plasticizers, since the latter contain polar groups which associate with the polar groups of the polymer, thus behaving as solvents. Poly-α-olefins, on the other hand, while not compatible with polyvinyl chloride, permit the latter chains to associate; consequently, lowering of the second-order transition temperature is not noted as it is with conventional plasticizers, and the resilience and ultimate elongation of the graft are higher than those of the plasticized material.

Nonionic detergents have been developed based on a starting, hydrophobic, polyoxypropylene chain containing hydroxy endgroups which initiate block formation of ethylene oxide:

$$HO\text{---}(CH_2CH_2O)_m\text{---}(C_3H_6O)_n\text{---}(CH_2CH_2O)_mH$$

The structure, and therefore the properties, of the copolymer can be varied conveniently over a wide range by altering molecular weights and ratios of the components. The starting polyoxypropylenes are essentially water insoluble at room temperature; as the fraction of polyoxyethylene in the copolymer is raised, the water solubility of the product increases; by varying the ethylene oxide fraction one can obtain blocks ranging in physical form from mobile liquids to waxy solids to hard materials which can be flaked. Low values for the surface tensions of aqueous solutions of the blocks are realized as the weight of the hydrophilic group in the copolymer decreases, or as the molecular weight of the hydrophobic fraction increases.

The synthesis of high molecular weight of blocks connected by a urethane bond can be effected by reaction of diisocyanates with high molecular weight compounds containing two labile endgroups, e.g., a polymeric glycol or diamine. If either or both molecules contain three or more functional groups, branched networks result. Although polyurethane can be synthesized from a variety of active hydrogen containing compounds, the ones employed commercially are the hydroxyl terminated liquid polyethers, polyesters, and polyester amides. Industrially useful products are obtained from starting materials which range in molecular weight from 1000 to 2500, since, under these conditions, the physical configuration of the system can be designed to ensure the internal mobility and crosslink density required for rubbery characteristics, and to impart outstanding tear and abrasion resistance.

Stereoblocks. In heterogeneous polymerizations (in the presence of a catalyst complex derived from a transition metal such as $TiCl_3$ or VCl_3 in combination with aluminum or beryllium alkyls) normally leading to tactic polymers, inversion of configuration occurs occasionally during chain growth so that the product contains chains of differing tacticities. The melting points of isotactic polypropylene blocks separated by atactic segments are lower than those of equivalent mixtures of atactic and isotactic polymers; blocks of low isotacticity (15–30% crystallinity) are highly elastic; in fact, reversible elongations up to 200% have been observed. At higher elongations, the stress increases sharply, probably because the isotactic segments are able to align and crystallize, in a manner similar to the stress-induced crystallization of some rubbers.

W. J. BURLANT

References

Burlant, W. J., and Hoffman, A., "Block and Graft Polymers," Van Nostrand Reinhold, New York, 1960.
Ceresa, R., "Block and Graft Copolymers," Butterworth, Inc., Washington, D.C., 1962.
Battaerd, H., and Tregear, G., "Graft Copolymers," Interscience Publishers, New York, 1967.

Polyorganosilicate Graft Polymers

Concept. Polyorganosilicate graft polymers are organic polymers chemically bonded to silicates, especially clays. The clays are usually montmorillonite and kaolinite (see **Clays**). Organoclays are hybrid unions of organic and inorganic materials, and thus may be expected to have properties characteristic of both classes. In fact it is the growing needs of technology for new materials having new combinations of engineering properties at increasingly higher temperatures that underlies much of the interest in this field. Hence, the aim of organoclay chemistry is to make mineral-organic materials with properties as variable and controllable as organic chemistry itself. Clay minerals are layer-lattice polysilicates of very high molecular weights and with dimensions approximating those of organic high polymers. Essentially they are polymer-size rocks with chemically active surfaces. The surfaces are negatively charged due to charge imbalance in the ionic lattice and, therefore, are able to attract and hold both inorganic and organic cations. Once the surfaces become clad with reactive organic radicals, all the type-reactions of organic chemistry become possible, at least theoretically.

There is an important chemical distinction, however, between polyorganosilicate graft polymers and filled polymers containing a mineral extender. The latter disintegrate in solvents which dissolve the polymer, but the former do not because the organic polymer is anchored by strong chemical bonds to the insoluble silicate sheets.

Preparation. Clay-polymer complexes can be made by at least three general reactions: (a) *organoclay + monomer,* e.g., vinyl-pyridinium montmorillonite + styrene; (b) *organoclay + active polymer,* e.g., vinyl-pyridinium montmorillonite + polymer containing active double bonds; (c) *clay or organoclay + polymer + ionizing radiation.* In the (a) graft, polystyrene chains grow from the silicate surface; in a (b) graft, isoprene-terminated polyisobutylene is an example of an active polymer; an excellent example of a (c) graft is the bonding of polyethylene to polyvinyl alcohol montmorillonite by gamma rays or 2 MeV electrons. The (a) and (b) grafts are made using standard polymerization techniques. Experimentally, the irradiation technique is much simpler: organoclay and polymer are dispersed homogeneously in the absence of any added chemicals and irradiated. In practice, if the irradiated mixtures prove to have superior properties, an effort to prepare similar materials by possibly more economical copolymerization reactions may be justified.

Properties and Applications. In addition to exhibiting improved thermal stability the polyorganosilicate graft polymers and simple organoclays are insoluble in organic solvents. This is because the organic part is chemically bonded to the insoluble silicate lattice. Thus, a fully bonded irradiated clay-polyethylene containing as much as 50-weight per cent silicate lattice retains the essential properties of polyethylene but is insoluble in boiling toluene, a solvent in which it is normally soluble. Other property improvements have been recognized for this relatively new class of materials. For example, improved ion exchange sorbents which do not swell in water have been made by irradiation grafting of styrene and acrylic acid to silica followed by sulfonation. There is possible use of clay-organic compositions as ablatants in nose-cones, reinforcing fillers in elastomers and as structural hydraulic fluids in shock absorbers.

The future for these materials seems bright in view of their potential for chemical novelty, the ready availability of the extremely low-cost, high-purity clay minerals which makes the grafting of expensive polymers economically feasible, and finally the newness of the field in which there is as yet little research competition.

PAUL G. NAHIN

Cross-references: *Block and Graft Polymers; Clays; Silicates*

BLOOD

Blood is best described as a heterogeneous system in which a variety of cells is suspended in an aqueous solution (plasma) composed of protein, glycoproteins and lipoproteins in colloidal suspension and organic and inorganic metabolites in solution.

The blood circulates to each organ in the body within a closed system of vessels. The blood volume, pressure and biochemical composition is maintained at a constant level within the limits of

biological variability despite the fact that there are positional changes, and functional and dietary stresses during the work, recreational, and/or resting periods of the day. The composition of the blood, morphologically, is unique to the biological species; biochemically, it is unique to the biological individual. As a consequence, the assessment of the biochemical composition of the blood and the morphological study of blood cells may be used to differentiate between states of health and disease. Composite data on the biochemical composition of the blood, statistically assessed, accounting for the biological parameters of age, sex, work and/or recreational activities and habits, previous illnesses, dietary history, genetic endowment and environmental exposure, can be used with discretion to assess the state of health, provided the physician does not forget that these biological parameters establish the subject as a biological individual within a select biological population.

Function. *(in relation to physiological economy of the body).* The blood is an aqueous dispersion containing emulsified lipids, soluble electrolytes and nonelectrolytes (organic metabolites) and a colloidal suspension of proteins. Its prime function is to provide a medium for metabolic exchanges between the cells, tissue fluids and blood, and to serve as a transport medium for the complex biochemical metabolites occurring in the body: (1) all nutrients ingested and absorbed across the intestinal wall; (2) O_2 and CO_2 transported from the lungs to the cell to the lungs, respectively; (3) waste metabolites synthesized in the cells and transported to the kidney and skin; and (4) hormones, enzymes, vitamins, antibodies, clotting factors and other metabolites synthesized inside specific cells and transported to sites of biochemical activity. An equally important function is the role played by the blood in maintaining the constancy of the internal environment, i.e., water, protein and electrolyte content (osmolarity); anion-cation composition (relative constant concentrations of the major electrolytes); acid-base equilibrium (pH); oxidation-reduction potential (ratio of reduced to oxidized forms —SH/S—S compounds, lactate (hydroxyacids)/pyruvate (keto-acids)); and temperature. This is accomplished by the selective movement of metabolites across cell membranes, sometimes against concentration gradients; this is known as active transport because of the energy required to achieve the transport. The unequal distribution of electrolytes is in part also governed by the Donnan Equilibrium.

The Biochemical Anatomy of The Blood. The formed components of blood are red cells (erythrocytes), white cells (leukocytes) and platelets (thrombocytes). These are present in relative amounts of 500:1:30. The cells may settle out of solution on standing or can be separated mechanically and differentially by techniques of differential centrifugation, which separate cells on the basis of molecular size and weight. Cells may be classified also by their morphology (size and shape) and by their unique biochemical nature as demonstrated by staining reactions.

Red Cells. The red cell as a mature cell does not contain a nucleus; is red by virtue of the fact that it contains a substantial amount of hemoglobin

(28%), (along with lipids, 7%, carbohydrates, proteins and salts, 3%, and water 71%) and has a unique structure in the shape of a biconcave disk 8.4μ in diameter, with a thickness varying from 1.0μ to 2.4μ. The primary function of the red cell is to transport hemoglobin which in turn transports oxygen to all the cells of the body. The hemoglobin contained within the red cell membrane has a half-life of 50–65 days, whereas hemoglobin circulating free in plasma has a half-life of 200 minutes. The normal limits for red cell content in blood is between 4.5–5.5 × 10⁶ cells/cu.mm. The approximate hemoglobin composition of each cell is 30μg/erythrocyte, which leads to a composition of 13.5–16.5g of hemoglobin/100 ml of blood. Variations from the norm may be indicative of illness. Among the methods used to assess the normalcy of the red cell and its hemoglobin content in blood are (1) to count the cells present per cubic millimeter; (2) to determine the packed cell volume after centrifugation under regulated conditions of time and speed (hematocrit); (3) to determine the hemoglobin content per 100 ml of blood by chemical and/or spectrophotometric techniques; (4) to determine blood oxygen content, since hemoglobin is the specific biochemical substance involved in oxygen transport; (5) to determine serum iron and iron-binding capacity, since only the porphyrin containing iron can transport oxygen; (6) to determine the electrophoretic mobility of hemoglobin of the cell, since the protein rather than the porphyrin component is the portion of the molecule which is improperly synthesized in genetic diseases of hemoglobin synthesis; (7) to determine aspects of red cell membrane structure in order to ascertain hemolytic disorders by means of morphologic studies, osmotic fragility studies, red cell survival times (the usual red cell survival time is about 120 days), the concentration of bilirubin, bilirubin-glucuronide and other intermediates in the hemoglobin catabolic sequence; and (8) to determine aspects of the red cell as a dynamic biochemical functional unit by means of studying the enzyme activities of the anaerobic (Embden-Meyerhof) and the aerobic (Pentose Phosphate) metabolic pathways.

White Cells. The mature white cell as distinguished from the red cell is nucleated, colorless, can reproduce itself, moves spontaneously with amoeboid-like character, and is capable of phagocytizing foreign solid particles. Since the specific gravity of white cells is less than that of red cells, the white cells collect as a thin buffy layer above the red cells after centrifugation. Among the cells characterized as leukocytes are granulocytes 40–75%, (containing neutrophils 35–70%, eosinophils 0–5% and basophils 0.3–1.0%), lymphocytes 20–45% and monocytes 2–10%. The major function of leukocytes is to provide a defense mechanism in the circulating blood against bacterial infection by phagocytosis and stimulation of the synthesis of antibodies. The white cell contains a considerable number of lysosomes, a subcellular organelle that contains proteolytic and hydrolytic enzymes important in the phagocytic process. It is well to note that the number of leukocytes per cu.mm in blood and the qualitative distribution of the kinds of leukocytes vary physiologically based on age. Biochemically, leukocytes contain

metabolites similar to those found in other cells which use as their major energy source the anaerobic and aerobic glycolytic metabolic pathways. The major technique used to assess alterations from the norm in respect to white cells is a total count of white cells (4–11 × 10⁹ cells/cu.mm) and a quantitative study of the differential morphology of white cells.

Because the methods of differential centrifugation and the isolation of intact white cells are difficult compared to the isolation of red cells, the application of biochemical techniques to study white cell metabolites which accumulate in quantity, such as ascorbic acid and histamine, or white cell enzymes present in significantly high concentration, such as acid and alkaline phosphatase, are not yet routine techniques for the laboratory assessment of alterations due to disease.

Platelets. Platelets are formed in the megakaryocyte, which is the largest of all the blood cells with a diameter of 30–90μ. The thrombocytes or platelets are 2–4μ in diameter and like the mature red cell are nonnucleated. They are usually present in blood in the amount of 150–500 × 10⁹ bodies per cu. mm. The primary role of platelets is hemostasis. They contain virtually all the blood content of 5-hydroxytryptamine, along with substantial amounts of epinephrine and norepinephrine. These amines cause a constriction of the blood vessel at the site of an injury. The platelets themselves agglutinate and adhere to the site of injury of the blood vessel. When the platelets undergo dissolution, thromboplastin is released, which in turn initiates the blood coagulation mechanism.

The biochemical assessment of platelet function has been developed around this fact, and is commonly known as the Thromboplastin Generation Test. This technique as well as all techniques used to assess blood coagulation factors requires the finite care and control of any enzymatic technique in which one seeks to determine that activity of one selected enzyme in a metabolic sequence of many steps, this specific case involving twelve currently known steps.

Plasma. Plasma is the solution which remains after the cells have been removed from the blood. It differs from serum in that it contains the protein fibrinogen. It is an amber opalescent solution which contains proteins in colloidal suspension and solutes (electrolytes and nonelectrolytes), either emulsified or in true solution. The proteins, which are of varying size, shape, electrophoretic mobility and immunochemical specificity, can be separated from each other and from the other solutes by the usual techniques used to separate macromolecules. The techniques are ultrafiltration and ultracentrifugation (size), electrophoresis (mobility in an electrochemical field at a specific pH), viscosity and streaming birefringence (shape), and immunochemical techniques which require the preparation of a pure protein to be able to synthesize in an experimental animal a specific antibody without interfering cross reactions.

Electrophoretic techniques permit qualitative detection of the presence or absence of normal serum protein fractions and the presence of abnormal fractions. This technique in connection with photometry or immunochemical procedures also permits quantitative estimation of each of these components. Immunochemical techniques alone are used to identify and quantitatively assess blood group specific polysaccharides.

Electrolytes in plasma may be defined as mineral present in concentrations expressed as milliequiv./L (Na^+, K^+, Ca^{++}, Mg^{++}, Cl^-, HCO_3^-, $HPO_4^=$, $SO_4^=$), mineral present in concentration of $\mu g/L$, (I^-, Fe^{++}), and trace minerals which include all other cations. The amino acids, lipids, carbohydrates and other nitrogenous compounds in solution are either nutrients which are ingested, metabolites which are intermediate in a metabolic scheme, or metabolites which are the end product of a metabolic scheme. These substances are all found in the blood because the blood carries the entering nutrient to the site of cellular metabolic activity and transports the final product to the kidney or skin. Intermediates are present because of the role the blood plays as a transport medium carrying the circulating pool of metabolites. Consequently, a genetic defect in which an enzyme is deficient results in production of an incomplete metabolic sequence, with the piling up of an intermediary metabolite which can be detected in the blood. By the same token, deficiency of a nutrient may also result in an impaired metabolic sequence, with accumulation of an intermediate which is not usually present in relatively high concentration. In addition to the determination and quantitative assessment of known metabolic intermediates, it is often of interest to detect the presence of drugs (foreign organic molecules) or environmental pollutants which are hazardous to life.

It should be kept in mind that the techniques used in biochemical analysis of blood for specific components attempt to detect and quantitate them with accuracy and speed. Consequently, techniques have been sought which are specific, yet do not require the ultimate in the isolation of the pure substance. The use of spectrophotometric, electrometric, chromatographic techniques is now commonplace. For the specific and detailed biochemical composition of the blood the reader is referred to the following sources.

CARL ALPER

Cross-references: *Iron and Compounds in Biochemistry and Physiology; Porphyrins.*

References

Altman, P. L., and Ditmer, D. D., "Blood and Other Body Fluids," Washington, Federation of American Societies for Experimental Biology, 1961.

Boyd, M. J., and Alper, C., "Blood (Chemical Composition, Vertebrates)," in Williams, R. J., and Lansford, E. M., Jr., "The Encyclopedia of Biochemistry," Reinhold Publishing Company, New York, 1967.

Faulkner, W. R., King, J. W., and Damm, H. C., "Handbook of Clinical Laboratory Data," 2nd Edition, The Chemical Rubber Company, Cleveland, 1968.

Ferguson, J. H., "Blood and Body Functions," F. A. Davis, Philadelphia, 1965.

Seitz, J. F., "The Biochemistry of the Cells of Blood and Bone Marrow," C. C. Thomas, Springfield, Illinois, 1969.

BOND ENERGIES AND BOND DISSOCIATION ENERGIES

The terms "bond energy" and "bond dissociation energy" are often confused and used interchangeably in the classroom as well as in the literature. The bond energy concept is usually introduced in the teaching of elementary physical chemistry during the discussion of thermochemistry as a means of estimating heats of reaction. This, however, is usually different both in character and numerical magnitude from the term used in chemical kinetics and free radical chemistry which is properly referred to as the bond dissociation energy.

Bond energy is defined as the strength of a bond as it exists in a molecule, or the contribution of the bond between a particular pair of atoms A-B in a molecule to the total binding energy present in that molecule. The proper symbol is $E(A$-$B)$. This definition allows precise values of bond energies to be assigned only for diatomic molecules and for polyatomic molecules of the type AB_n, where all of the A-B bonds are identical. In the case of diatomic molecules the bond energy is the heat of atomization of the molecule. For polyatomic molecules of the type AB_n the bond energy is $1/n$th of the heat of atomization of the molecule, i.e., $1/n$ times the heat necessary to convert the molecule AB_n to $A + nB$ with the molecule and the atoms formed all in the ground state. Thus, $E(A$-$B) = Q_a/n$.

Polyatomic molecules with structures other than AB_n usually contain a variety of bonds, and it is not possible to make precise assignments of bond energies. A precise definition would necessitate the assumption of a theory of local pair bonds with no interaction among the various bonds. Taking methane as an example, as the four hydrogen atoms are brought toward the carbon atom, most of the energy liberated is from the overlap of electron "clouds" in the carbon-hydrogen bonds that are formed. However, some of the energy is due to overlap of "clouds" other than in the carbon-hydrogen bond. In a like manner, the force constant of a bond is affected by the neighboring bonds, and the bond cannot be considered as completely isolated. For many molecules these non-localization corrections are small; therefore, the various descriptive qualities of a bond are still useful in most instances.

In the case of polyatomic molecules with structures other than AB_n, the heat of atomization or the atomic heat of formation must arbitrarily be divided into reasonable contributions from the several bonds. There are a number of ways to do this which are commonly used. For example, where a series of similar molecules can be examined, like bonds can be assigned constant values in such a way as to reproduce best the heats of formation of the whole series. In the case of the paraffin hydrocarbons, $E(C$-$C)$ and $E(C$-$H)$ are assigned constant values, and the energies of other bonds are deduced by an obvious extension of the method. Values of bond energies derived in this manner are "average" bond energies, and their numerical values will depend on the compounds used in the assignment process.

More refined thermochemical estimates may sometimes be obtained by using a "true" value for E

such as might be obtained from E(C-H) in methane. Many attempts have been made to resolve the ambiguities inherent in such procedures, including the correlation of the structural features of molecules (such as bond lengths) with bond energies. Also, there are a number of "mathematical" definitions of bond energy, but these go beyond the scope of this article.

Bond dissociation energy is defined as the difference in energy between the parent molecule and the two fragments after bond breaking. The molecule and its fragments are all assumed to be ideal gases in their equilibrium configurations, and the reference temperature is 0°K (although 298°K values are often reported). The proper designation is $D°_0$, where the superscript refers to the reactant and products being in their ground states and the subscript designates that the temperature reference is 0°K. For convenience, the change in enthalpy at some standard temperature is often used. Actually, $\Delta H°_{298}$ does not often differ greatly from $D°_0$, and much of the enthalpy data is not precise enough to justify a correction. In a dissociation reaction $\Delta C°_p$ is generally positive but not large, so that $\Delta H°_{298} > D°_0$; the difference is rarely more than 1 kcal/mole.

What, then, is the relationship between the bond dissociation energy and various other properties of the molecule? The heat of atomization or the atomic heat of formation (Q_a) is equal to the sum of all the bond dissociation energies involved as the molecule is stepwise degraded into its atoms. The dissociation energy of the bond between two atoms A and B generally depends not only on the bonded atoms (A and B) but also on the other atoms or radicals (if any) attached to A and B as well as the configuration of the molecule as a whole.

The dissociation energy is the difference between the heats of formation of the fragments and the heat of formation of the parent molecule:

$$D(AB\text{-}C) = \Delta H_f(AB) + \Delta H_f(C) - \Delta H_f(ABC).$$

(For convenience, the subscripts and superscripts are omitted with the tacit understanding that the parent molecule and fragments are all ideal gases in their ground state configurations at 0°K.) This can be turned around to give the heat of reaction when the dissociation energies are known or can be measured. The heat of reaction is the difference between the sum of the dissociation energies of the bonds formed and the sum of the dissociation energies of the bonds broken. Thus, for the reaction $AB + C \rightarrow A + BC$:

$$\Delta H = D(B\text{-}C) - D(A\text{-}B).$$

In general, a chemical reaction consists of breaking one bond and forming another bond simultaneously. The activation energy for such a reaction is a very complex function of the dissociation energies of the bonds broken and formed. In a unimolecular reaction where bond breaking produces two atoms or radicals, however, the activation energy will be almost if not exactly equal to the bond dissociation energy, because a radical combination reaction (which is the reverse of a bond breaking dissociation) requires little, if any, activation energy. For most exothermic reactions between radicals or atoms and molecules, the activation energies are also small. Thus it is clear that bond

energies should not be used in estimating activation energies; only the specific dissociation energies of the bonds in the specific molecules in question should be used, and even then with caution.

A direct measure of the heat of recombination can sometimes be obtained by calorimetry. By means of the van't Hoff equation, the energy of dissociation can be calculated from known dissociation equilibrium data covering a range of temperatures.

Using kinetic methods and assuming that the activation energy for radical recombination is zero or unimportant, the dissociation energy can be equated to the activation energy for dissociation.

Appearance potentials $A(R^+)$ can be obtained directly from electron-impact studies. The dissociation energy is the difference between the appearance potential and the ionization potential $I(R)$ of one of the fragments, assuming that there is no activation energy or excess energy involved:

$$A(R^+) = D(R\text{-}R') + I(R).$$

If the ionization potential is known precisely, then the dissociation energy can be calculated directly. If $I(R)$ is not known, then D can be obtained indirectly by measuring several more appearance potentials of appropriate species.

If the states of the dissociation products are known, spectral analysis is the most accurate method of measuring the dissociation energies of diatomic molecules. The spectra of polyatomic molecules are extremely complex, however, and analysis to determine dissociation energies is nearly impossible. Statistical mechanics can be used to relate molecular spectra to thermodynamic quantities but is of no value for making independent calculations of dissociation energies. Quantum mechanical calculations are capable of yielding exact values of dissociation energies; however, these have been of limited use because of their very great mathematical complexity.

Where the proper measurements can be made, these methods, either individually or in combination, can give reliable bond dissociation energies in most cases. However, different methods sometimes give different answers or at least lend the results to different interpretations by different observers. There are many subtleties in the computations such as choosing the proper basic thermochemical quantities. For example, there are two values for the heat of sublimation (or heat of atomization) of graphite which are prominent in the literature before about 1960. The value chosen will affect every calculation involving a carbon atom. Some investigators have reported values for this of about 140 kcal/mole, while others have presented values of about 170 kcal/mole. The currently accepted value for the heat of formation of C_1 is $\Delta H°_0 = 169.98 \pm 0.15$ kcal/mole. There are other pitfalls to beware of, but careful examination of the literature and careful experimentation can yield satisfactory results in most cases.

BRUCE E. KNOX

BONDING, CHEMICAL

All atoms and molecules consist of one or more positively charged nuclei surrounded by a cloud of negatively charged electrons. When two or more such particles (atoms or molecules) come near to one another, electrostatic interactions between particles will include (1) repulsions between electron clouds, (2) repulsions between the nuclei of the separate particles, and (3) attractions between the nucleus or nuclei of each particle and the electrons of the other particle. Since an electronic cloud is both nebulous and mobile, it must adjust to the influence of an external electrical field, such as that of another atom or molecule. There are two principal ways in which this adjustment can lead to a net attraction.

The first way is observed when neither particle possesses any low-energy electron vacancies in the outermost part of its electronic cloud. Even though under such circumstances electrons of one particle are prevented from becoming closely associated with the nucleus or nuclei of the other particle, two adjacent clouds exert an influence on one another that must inevitably lead to net attractive forces. By correlation of the electronic motion around the nuclei, significant reductions in the repulsions between clouds are produced. Coulombic forces vary inversely with the square of the distance between the charges. Therefore repulsive forces tend to increase the distance, thus becoming weaker, whereas attractive forces tend to decrease the distance and grow stronger. The net force between any two such particles therefore tends to hold them together. Such forces, which may be thought of as attractions between very rapidly oscillating induced dipoles, are called "van der Waals forces." They are relatively weak between small atoms and molecules, commonly much too weak to hold against the disruptive influence of kinetic energy at ordinary temperatures. Reduction of the kinetic energy by cooling, however, allows these van der Waals attractions to become dominant, and the fundamental particles of all substances tend to condense to liquid and solid state when the temperature is sufficiently low. Although van der Waals attractions are generally thought of as relatively very small, they increase significantly as the molecules become larger and so structured that close intermolecular contact is possible. For example, although the cohesive energy in liquid nitrogen and oxygen is less than one kilocalorie per mole (kpm), it is about 6 for liquids boiling around 25°C, and for liquids boiling about 300°C, the van der Waals forces produce a cohesive energy of about 20 kpm.

The second way by which net attractions between atoms or molecules can result is observed when at least one and more commonly both of the approaching particles have low-energy electron vacancies in the outermost part of the electron cloud. A low-energy electron vacancy represents a region around a nucleus wherein the positive nuclear charge is not well shielded from any electron that might occupy that region. In other words, it represents a region wherein an electron from another atom can be accommodated in stable association with the nucleus. Thus it permits the same electron(s) to be attracted to two or more different nuclei, the one with which it was originally associated, and one or more outside nuclei. Such attractions are normally many times stronger, per atom-pair, than van der Waals forces. They are called "covalent bonds." Helium, with a filled electronic shell of two electrons, and

neon, argon, krypton, xenon, and radon each with an outermost shell of 8 electrons, are without low-energy vacancies and therefore are incapable of forming covalent bonds. (See Noble Gas Compounds for apparent exceptions.) All the other chemical elements have fewer than 8 outermost shell electrons, and thus vacancies that can accommodate electrons jointly with other atoms to form bonds.

The most familiar type of covalent bond is that formed when each of two atoms possesses one half-filled orbital (one unpaired electron and one vacancy) in its outermost shell. The electron of each is shared with the other which accommodates it in its vacancy. Alternatively, in what is called "coordinate covalence" or simply "coordination," one atom provides a vacant orbital and the other, an electron pair. The two electrons, called "valence electrons," are thus mutually attracted to both nuclei, holding them together. This mutual electrostatic attraction of two nuclei for the same electron pair holds the atoms together in a more stable system than that of the two separate atoms. The distance between the two nuclei is called the "bond length" and the energy required to separate the joined atoms completely is called the "bond energy." If two atoms at wide separation are considered to possess zero potential energy, then allowing them to approach will result in a decrease in potential energy (negative value) until a minimum is reached at a certain distance, closer than which the potential energy rises rapidly as the atoms resist further approach. This energy minimum corresponds to the bond energy and occurs at the bond length.

From the observed requisites for a covalent bond, namely a half-filled outer shell orbital on each atom, the number of covalent bonds an atom can form is easily seen to be equal to the number of such half-filled orbitals. In general, the number of bonds is limited by the number of outermost electrons when the number of vacancies is greater, and by the number of vacancies when the number of outer shell electrons is greater. For example, across the periodic table, lithium with but one outermost electron and seven vacancies can form but one covalent bond. Beryllium, with two electrons and six vacancies, can form two bonds. Boron with three electrons and five vacancies can form three covalent bonds. Carbon with four of each can form four covalent bonds. Nitrogen with five electrons but only three vacancies can form only three covalent bonds. Oxygen with six electrons but only two vacancies can form only two covalent bonds. Fluorine with seven electrons has but one vacancy and can form only one covalent bond. The formulas of compounds are thus often predictable (and nearly always understandable) from the structure of the component atoms. For instance, the hydrogen atom having only one half-filled orbital can form but one covalent bond, but a carbon atom can form four, so it combines with four hydrogen atoms producing methane, CH_4.

The number of possible covalent bonds can sometimes be increased, two at a time, if one of an outer shell electron pair can be promoted to an otherwise vacant outer d orbital. This process, called "expanding the octet," creates two new half-filled orbitals at once, and allows, for example, the number of covalent bonds formed by sulfur, which normally has an outer shell of two half-filled orbitals and two filled orbitals or lone pairs of electrons, to be increased from 2 to 4 and then to 6. Such promotion only occurs when withdrawal of some of the electron cloud by other atoms allows the nuclear charge to become more effectively sensed in the outer d orbitals, increasing their stability. Fluorine, being more electronegative than sulfur, accomplishes this electron withdrawal, and thus allows the formation of SF_4 and SF_6. Under some conditions, not yet thoroughly understood, an increase in the normal number of bonds can also be brought about through the formation of half-bonds not requiring the promotion of lone pair electrons to outer d orbitals. For example, ClF_3 appears to consist of one normal covalent bond and two half-bonds which are longer and weaker.

Usually an atom of a major group element uses all possible electrons in bonding, but toward the bottom of certain groups is observable a tendency for two electrons to remain aloof. This is known as the "inert pair effect," as though the s orbital of the outermost principal quantum level tends to retain its electron pair as in the ground state instead of permitting one electron to move into a p orbital for maximum covalence. It leads to compounds in which thallium, for example, forms only one bond instead of its possible three, lead two instead of four, and bismuth three instead of five. In the transitional elements, which characteristically use underlying d orbitals as well as outermost shell orbitals in their bonding, the tendency to use fewer than the maximum possible half-filled orbitals is common, and variable valence is the rule.

The bonding orbitals of a combining atom have a strong tendency to equalize even though different in the isolated atom. This phenomenon is called "hybridization." It leads, for example, to two equal sp hybrid bonding orbitals in zinc instead of one s orbital and one p, to three equal sp^2 hybrid bonding orbitals in boron instead of one s and two p, and to four equal sp^3 hybrid bonding orbitals in carbon instead of one s and three p.

The geometry of atomic orbitals permits more than one electron pair, up to three pairs, to become involved in a given bond under certain conditions. A four electron bond is called a double covalent bond and a six electron bond is called a triple covalent bond. Such bonds are said to have a bond order of 1, 2, and 3. Owing to the difficulty of concentrating so many mutually repelling electrons between the same two nuclei, a double bond is not ordinarily twice as strong as a single bond nor is a triple bond three times as strong. However, in elements of Groups VA, VIA, and VIIA, the presence of lone pair electrons in the valence shell appears to exert a weakening effect on the covalent bond energy. This effect is reduced by formation of a double bond and eliminated by formation of a triple bond. The effect is especially large in nitrogen, oxygen, and fluorine, making their single covalent bond energies abnormally weak. As a consequence, the triple bond between two nitrogen atoms (226 kpm) is indeed more than three times as strong as the weakened single bond (39 kpm), and the double bond between two oxygen atoms (119

kpm) is also more than twice as strong as the weakened single bond (34 kpm).

Bonds of nonintegral order are common. They demonstrate two important points. One is that although extremely useful, the concept of the two-electron bond, or integral multiples of it, has limited applicability. The second is a very important principle of bonding. When an atom might equally well provide multiple bonding electrons to any one or some of several atoms to which it is bonded, but not all at once, it tends to provide these electrons partially but equally to all rather than just to one or some, excluding the others. Similarly, when integral covalent bonding would leave one or more outer level orbitals unoccupied, the tendency is for available bonding electrons to occupy all vacancies partially rather than one or two exclusively and the others not at all. These tendencies often result in nonintegral bond order, and this phenomenon of partial utilization of electrons among several bonds in preference to localization of electrons between a particular pair of atoms is called *resonance*. A familiar example of a compound involving resonance is benzene, C_6H_6, in whose six-membered ring each carbon atom could form a double bond to one or the other of its two neighbors but not to both. Instead it shares equally, giving a ring containing six equivalent bonds of order 1.5.

The resonance principle assumes especially great importance in bonding between atoms in each of which the number of outer vacancies exceeds the number of electrons. The tendency toward covalence would suggest that such atoms should unite to the limit of their covalent capacity and no further. But this would leave low energy orbitals unoccupied. Accordingly, the electrons spread out through all available orbitals, becoming delocalized instead of paired between specific atom pairs. This results in much closer packing of atoms, in fact, most commonly the closest possible packing of uniform spheres, in which each atom on the interior is in contact with 12 neighbors. This is called *metallic* bonding. It may be regarded crudely as the electrostatic attraction between the metal cations and the continuum of valence electrons in which they are imbedded. Such bonding is usually quite strong, and imparts unique characteristics to the metallic crystal, including high lustre, electrical and thermal conductivity, and malleability and ductility. In contrast, covalently bonded substances with their relatively localized bonding electrons tend to be electrical and thermal insulators, and brittle or hard.

Being delocalized, the bonding in metals is all-directional. Not so in the localized covalent bond, which makes a definite and predictable angle with each other bond formed by the same atom. Such bond angles determine the molecular structure that results from a given combination of atoms. Sophisticated explanations of the geometry of molecules involve calculations by wave mechanics of the directional nature of the pure and/or hybrid orbitals that may be involved in the bonding. A much simpler and at least equally useful explanation can be based on the recognition that electron pairs, or electrons in bonds, tend to locate as far apart as possible to minimize the electrostatic repulsions. For example, if all the electrons in the outer shell of an atom are used in only two bonds, these two almost invariably are directed opposite to one another, giving a bond angle of 180°. If the atom forms two bonds but has one lone pair of electrons left over, then there are three locations, and the farthest apart these can be is at corners of an equilateral triangle, making bond angles of 120°. Similarly, if all the electrons are involved in three bonds, the bond angles are 120°. If all the electrons in the outer shell of an atom are used in four bonds, or in three bonds with one lone pair of electrons not used, or in two bonds with two lone pairs left over, the bond angle is close to that of a regular tetrahedron, 109°28′. Five exactly equivalent positions on the surface of a sphere do not exist, but the best separation of five groups of electrons, whether used in bonding or not, is at the corners of a trigonal bipyramid. Six equivalent positions for electrons, whether involved in bonds or not, are at the corners of a regular octahedron, which makes the angle 90° between adjacent bonds. In general, a lone pair is expected to exert somewhat stronger repulsion than a shared pair, and bonds exert higher repulsion the higher the bond order. Such factors may modify the bond angles predicted above. Practically all molecular geometry can be explained, at least approximately, on the basis of this simple picture of electrostatic repulsion among groups of electrons surrounding the atomic nucleus.

In many chemical combinations, especially between metal and nonmetal, the simplest molecules that might be predicted from the electronic configurations of the component atoms appear to have at most a very transitory existence at ordinary temperatures. They rapidly condense to crystalline solids in which individual molecules no longer exist. In such nonmolecular solids, the exact relationship between the packing of atoms and their normal valence as exhibited in molecules is not yet clear, but the simple electron pair repulsion rules appear uncertain if not invalid. However, in "network" solids where the atoms are covalently bound in indefinitely extensive aggregates or giant molecules, the bond angles are those predictable, or at least can be rationalized, from the repulsion concept.

The formation of coordinate covalent bonds most commonly supplements normal covalence. For example, boron trimethyl, $B(CH_3)_3$, uses only three boron orbitals and thus has one vacant orbital left. Trimethylamine, $N(CH_3)_3$, has a lone pair of electrons on the nitrogen left uninvolved in the bonding. When these two molecules, each capable of stable independent existence, come together, they join by sharing the lone pair of the nitrogen with the vacant orbital of the boron. The product molecule is called a "molecular addition compound." Sometimes, however, a neutral atom can coordinate without first forming its normal covalent bonds. For example, an atom of iron can coordinate five molecules of carbon monoxide forming a liquid compound, iron pentacarbonyl, $Fe(CO)_5$. Such coordination frequently brings the number of electrons up to that of the next higher noble gas element. For example, if each CO molecule contributes two electrons to the iron atom's 26, the final number is 36 like that of krypton.

Most common among acceptors in coordination are positive ions. Thousands of different coordination complexes are known, in which the donor or "ligand" may be either a molecule or an anion, and the acceptor a cation.

Of central importance in theoretical chemistry is the exact mathematical description of chemical bonding, especially to permit the accurate prediction of bond length and bond energy. In *principle,* this can be accomplished by applying the Schrodinger wave equation and the methods of wave mechanics. The Schrodinger equation is

$$\frac{\partial^2\psi}{\partial x^2} + \frac{\partial^2\psi}{\partial y^2} + \frac{\partial^2\psi}{\partial z^2} + \frac{8\pi^2 m}{h^2}(E - V)\psi = 0$$

where x, y, and z are coordinates of the particle, m is its mass, h is Planck's constant, E is the total energy, and V the potential energy. Ψ is called the "wave function," and has physical significance in that its square is proportional to the probability of finding the electron in the position designated by the coordinates. Unfortunately, the mathematical complexities increase so rapidly as the number of particles increases that only in the simplest cases, such as the hydrogen molecule ion, H_2^+, have rigorous solutions been obtained. For applications to systems having even a few more interacting particles, grossly simplifying assumptions are essential, and the results are inevitably less reliable.

Two principal types of approach to the wave mechanical description of the covalent bond are used. One, the *valence bond* method, considers electron sharing to involve an overlap of orbitals, one from each atom, so that the atoms retain their individuality but occupy an intermediate region in common. For example, if the wave function for hydrogen atom A which holds electron (1) is $\Psi_A(1)$, and the wave function for hydrogen atom B which holds electron (2) is $\Psi_B(2)$, then the total wave function for the two atoms is $\Psi = \Psi_A(1)\Psi_B(2)$. But since electrons are indistinguishable, from one another, the total function could equally well be $\Psi_A(2)\Psi_B(1)$. The true function may then be some combination of the two:

$$\Psi = C_1\psi_A(1)\psi_B(2) + C_2\psi_A(2)\psi_B(1)$$

The parameters C_1 and C_2 are mixing coefficients giving the relative contributions of each function. The weight of each coefficient is proportional to its square. In the symmetrical example of the two identical hydrogen atoms, $C_1^2 = C_2^2$ and $C_1 = \pm C_2$. Thus two wave functions are possible:

$$\Psi_+ = \psi_A(1)\psi_B(2) + \psi_A(2)\psi_B(1)$$

$$\Psi_- = \psi_A(1)\psi_B(2) - \psi_A(2)\psi_B(1)$$

Application of the Schrodinger wave equation to calculation of the energy shows that Ψ_- corresponds to the state of parallel electron spins, leaving only to increased repulsion as the internuclear distance is decreased. Ψ_+ gives an energy minimum at 0.80 Å corresponding to sharing an electron pair with opposed spins, and a bond energy of 72 kpm. The observed bond length and energy in H_2 is 0.741 Å and 103.2 kpm (at 0°K), showing that Ψ_+ accounts for about two-thirds of the energy.

Another possibility not yet considered is that both electrons may happen near the same nucleus at the same time, creating in effect the ions H^+ and H^-, for which ionic wave functions $\Psi_A(1)\Psi_A(2)$ and $\Psi_B(1)\Psi_B(2)$ should be included. The total wave function thus becomes

$$\Psi = \psi_A(1)\psi_B(2) + \psi_A(2)\psi_B(1) +$$
$$\lambda'[\psi_A(1)\psi_A(2) + \psi_B(1)\psi_B(2)]$$

or $\Psi = \Psi$ covalent $+ \lambda'\Psi$ ionic

The parameter λ' gives the relative contribution of the ionic wave function, which for H_2 is somewhat less than 6 kpm. Consideration of additional factors improves still more the agreement between calculated and experimental values. James and Coolidge performed laborious calculations using a wave function containing 13 terms, and found the bond energy to be 102.8 kpm at the bond length 0.74 Å as observed. Recent extension to a 100 term wave function has led to computer results in exact agreement with experiment. In this sense the H_2 molecule may be regarded as completely "understood."

The second principal method of describing the covalent bond is in terms of molecular orbitals. According to this concept, polynuclear "molecular" orbitals may replace atomic orbitals, resembling them in possessing specific energy and spatial characteristics and in having a capacity of two electrons each. For example, in the hydrogen molecule, the two 1s orbitals are replaced by two molecular orbitals, different in energy. The one lower in energy concentrates the electrons between the two nuclei. It is called a *bonding* molecular orbital and designated as σ. The one higher in energy would keep electrons away from the region between the nuclei, and the molecular system with the electrons in this orbital would be less stable than the separate atoms. Such an orbital is called an *antibonding* orbital and designated as σ^*. Just as with atomic orbitals, electrons fed into molecular orbitals go into the most stable orbitals available to them, and occupy molecular orbitals of equal energy singly, with parallel spins, before pairing. Each electron in an antibonding orbital cancels the effect of one electron in a bonding orbital, so that the bond order is taken as half the difference between the number of electrons in bonding orbitals and the number in antibonding orbitals. Molecular orbitals having cylindrical symmetry about the bond axis are called sigma, σ. If they lead to nodal planes through the bond axis they are designated as pi, π.

A molecular orbital (MO) wave function is usually written as a *linear combination of atomic orbital* (LCAO) wave functions, on the basis that in the region of each nucleus, the molecular orbital must closely resemble the atomic orbital which it replaces.

$$\Psi_{M.O.} = \psi_A + \lambda\psi_B$$

λ denotes the degree to which one atomic orbital is favored over the other, and of course in homonuclear bonding as in H_2, must equal 1. Just as in the valence bond method, the potential energy change with distance can be calculated, leading to a minimum at the bond length.

Each of these two approaches to the mathematical description of bonding has certain advantages over the other, but neither is suitable for dealing simply with the complex interactions involved in most chemical bonding. The difficulties involved in the wave mechanical treatment of bond energy are especially troublesome because of the fact that the energy of interaction *between* atoms is very much smaller than the total energy *of* the individual atoms or of the molecular system. For example, if one could calculate by wave mechanics the total energy of the oxygen molecule, and subtract from this the total energy of two oxygen atoms, the difference would be the equivalent of what is called "bond energy." By summation of the individual successive ionization energies of an oxygen atom determined experimentally, the total energy is found to be about 47,000 kilocalories per g-atom. Twice this is 94,000 kilocalories per mole of O_2, to be compared with the dissociation energy of O_2 of only 119 kpm. Even very good approximations of the total energies would thus be inadequate to provide accurate bond energies, for the latter are relatively small differences between very large numbers. Although extensive calculations for a few small molecules have been carried out with a considerable degree of success, in general, wave mechanics in its present form does not provide a practical means of determining bond energies and thus heats of formation of compounds and heats of reaction. For a practical understanding, a more approximate and intuitive approach is an essential supplement to wave mechanical theory.

The most recent and successful such an approach is based on avoidance of the insoluble many-body problem by acceptance of certain atomic properties as representing the resultant of all the incalculable interactions of the atomic components. These properties are the nonpolar covalent radius, the electronegativity, and the homonuclear single covalent bond energy. All three are the result of the same basic but not easily measurable property, the effective nuclear charge that can be sensed by an electron occupying a vacancy at the covalent radius distance from the nucleus.

In bonds between like atoms, equal electronegativities (which see) result in equal sharing of the valence electrons, and the covalent bonds are nonpolar.

For heteronuclear bonds, electronegativity assumes especially great importance, for different kinds of atoms differ in their initial attraction for the valence electrons, and thus differ in electronegativity. Unequal attraction results in uneven sharing, which causes the bond to be polar. It is assumed that a heteronuclear covalent bond can be divided, for purposes of calculation, into two contributions, one covalent, the other ionic. In a simple diatomic molecule, the model is that of a bond which is a composite of a covalent form, where the two bonding electrons are exactly equally shared, and an ionic form, where the two bonding electrons are held exclusively by one atom, making it a negative ion and leaving the other atom a positive ion.

The energy, E_c, of the covalent form can be calculated by assuming that each atom contributes to a nonpolar covalent bond with an unlike atom exactly as it would to a like atom. The energy is taken as the geometric mean of the two homonuclear covalent bond energies, corrected for any difference between the sum of the two nonpolar covalent radii, R_c, and the actual bond length, R_o:

$$E_c = (E_{AA}E_{BB})^{1/2}(R_c/R_0)$$

The energy of the ionic form, E_i, can be calculated by Coulomb's law, assuming that the charges on the ions are centered at their nuclei:

$$E_i = 332e^2/R_0$$

The factor 332 converts the energy to kilocalories per mole, and e^2 is the charge product, where the charge is unity for one electron.

The total heteronuclear bond energy is then calculated as the weighted sum of the covalent and ionic contributions:

$$E = t_c E_c + t_i E_i$$

The weighting coefficients, t_c and t_i, total 1.00, since the bond is considered to be of one form or the other all the time. The critical part of this calculation is the evaluation of t_i, the ionic weighting coefficient. It is determined simply as the average of the partial charges on the two atoms:

$$t_i = \frac{\delta_A - \delta_B}{2}$$

The heteronuclear covalent bond is thus described quantitatively as a consequence of superimposing a part-time ionic character upon a part-time covalent character, such that the individual contributions can be calculated. Unlike the early concept of ionic energy as *additional* to covalent energy, this method treats ionic energy as *substituting for part* of the covalent energy. In other words, when two atoms initially different in electronegativity form a heteronuclear covalent bond, they equalize their electronegativity by uneven sharing of the valence electrons. This causes a reduction in the nonpolar covalent energy, but ionic energy is substituted for the lost covalent energy. Because the factor 332 is large compared with any homonuclear single bond energy, this substitution of ionic for covalent energy inevitably produces an increase in the total bond energy. Polarity always increases the strength of a bond. Electronegativity differences are thus a principal contributor to bond strength, and in fact exert a dominant influence on the direction of chemical reaction.

Two factors are recognized to lead to stronger bonds than would be calculated for a single heteronuclear bond as described above. One is bond multiplicity. It is found empirically that a double bond energy is obtained by multiplying the total single bond energy by 1.50. A triple bond energy is obtained by multiplying the total single bond energy by 1.77. The other factor is reduction in the normal homonuclear single bond weakening effect of lone pair electrons. Such reduction occurs whenever double and triple bonds are formed. It occurs also when the other atom possesses low energy outer shell orbitals which might accommodate the lone pair electrons in such a way that their weakening effect is reduced or eliminated. The latter effect, usually called "pi bonding," is not thoroughly understood but there is some evidence that it strengthens the bond not by increasing the number of shared electrons, but rather by reducing the lone pair weakening effect. For example, the normal weakened single bond energy of nitrogen is only 39 kpm but the unweakened single bond energy is about 95 kpm.

The concepts of polar covalence which lead to the successful interpretation and quantitative calculation of bond energies to molecular compounds can with only slight modification be applied equally well to nonmolecular compounds such as the binary compounds of metals with nonmetals. Before describing these modifications, however, it may be useful to review briefly the conventional older treatment of such compounds, according to the "ionic model."

Binary solids formed by the combination of metal with nonmetal are usually regarded as ionic. The nonmetal atom is considered to have acquired one or two electrons from the metal atom, thus becoming a negative ion and leaving the metal a positive ion. The ions are imagined to be "hard spheres" having their charges centered at their nuclei. Because of their charges, the ions tend to join together so that each positive ion is surrounded by negative ions and each negative ion by positive ions. In other words, greatest stability of the crystalline solid is achieved when opposite charges are as close together as possible, consistent with keeping like charges as far apart as possible. The bonding energy in the crystal is electrostatic, and is generally called "lattice energy" or "crystal energy," which is the energy which would be released by forming the crystal from the gaseous ions.

In the crystal, each positive ion has negative ions as its nearest neighbors, then other positive ions a little farther away, then other negative ions still farther away, and so on throughout the crystal. To evaluate the electrostatic energy over the whole crystal, it is necessary to add up these successive attractions and repulsions. Summation of all the interactions within the crystal leads to a crystal energy equal to the coulombic energy between two neighboring ions multiplied by a constant. This is called the Madelung constant. It is a characteristic of the crystal structure. The Madelung constant has been calculated for most of the common crystal structures. To an extent it evaluates the advantage of condensation to a crystalline solid over remaining as ion pairs in the simplest possible metal-nonmetal combination. For example, the Madelung constant for sodium chloride type structures is 1.75, which means that the electrostatic energy of the crystal is 1.75 times as great as the electrostatic energy in a pair consisting of one sodium ion and one chloride ion.

However, it is recognized that the point charges of the ionic model could not come into stable equilibrium when separated as in the crystal. They are prevented from coming closer together by the fact that the ions are not actually point charges but rather, surrounded by electron clouds. The repulsion of the electronic cloud of a positive ion for the electronic cloud of the adjacent negative ions prevents closer approach. On this basis, in the ionic model, the electrostatic energy is considered to be balanced against the repulsions among the electronic clouds. By equating the two, it is possible to arrive at an expression for the extent to which the attractions are weakened by the inter-cloud repulsions, which may be summarized as a "repulsion coefficient." Values can be determined by experimental studies of the compressibility of the crystal. They ordinarily are in the range of about 0.8 to 0.9. In other words, the total electrostatic energy of cohesion of the crystal is reduced by a factor of about 0.85. The equation for calculating the crystal energy, U, usually called the Born-Mayer equation, is then:

$$U = 322(z+)(z-)e^2 Mk/R_0$$

M is the Madelung constant, k the repulsion coefficient, $(z+)e$ the charge on the positive ion, $(z-)e$ the charge on the negative ion, and 332 again the factor for converting to kpm.

The Born-Mayer equation appears to work very successfully for the alkali metal halides, but less well for most other so-called "ionic" compounds. In a general way, the magnitude of error appears related to the probable degree of covalence in these compounds. In fact, of the many criticisms that might be leveled at the ionic model, perhaps the most cogent is that every positive ion is a potential electron pair acceptor, since it has both positive charge and vacant orbitals, and every negative ion is a potential electron acceptor, since it has both negative charge and lone pair electrons. The ionic model requires that each potential electron pair acceptor be surrounded by potential electron pair donors in close contact, and each potential electron pair donor in close by potential electron pair acceptors, without any coordination occurring. This in fact seems so unrealistic as to cast grave doubts on the validity of the ionic model.

In recognition of the probability of coordination under the conditions prevailing in a crystalline compound of metal with nonmetal, an alternative model of such nonmolecular solids has been proposed, based on the concepts of polar covalence described earlier. Here too the total bonding energy is considered divisible, for purposes of calculation only, into a covalent contribution and an ionic contribution. The covalent contribution is again the geometric mean of the homonuclear single covalent bond energies, corrected for any deviation between actual bond length and covalent radius sum. However, it must now be recognized that most or all of the outermost shell electrons are now involved in the bonding, including the electrons that would have remained as lone pairs on the nonmetal in the simplest molecule of metal-nonmetal compound. Therefore the covalent bonding contribution must be multiplied by n which is the number of electron pairs involved in the bonding, per formula unit. For the ionic energy contribution, the Born-Mayer equation for crystal energy is used. The ionic weighting coefficient, t_i, is simply the partial charge on an atom in the crystal, divided by the number of charges on its ion. The total energy of atomization is the weighted sum of the covalent and ionic contributions:

$$E = \frac{nt_c R_c (E_{AA} E_{BB})^{1/2}}{R_0} + \frac{332 t_i Mk(z+)(z-)e^2}{R_0}$$

This alternative model of nonmolecular solids is called the *coordinated polymeric* model. It works very accurately for the alkali metal halides, and in addition is successful for a complete range of

bond polarities, in contrast to the ionic model which has much more limited applicability. The principal complication in the use of the coordinated polymeric model arises when the nonmetal is divalent. Then, probably because of incorrect evaluation of the repulsion coefficient, k, the ionic contribution must be reduced by an empirical factor which is about 0.63 for oxides and 0.57 for sulfides.

The usual dichotomy of covalent and ionic bonding is thus largely replaced by a more comprehensive theory in which polar covalence in both molecular and nonmolecular compounds can be quantitatively analyzed according to the same general principles and concepts. There do exist, however, certain solid compounds wherein coordination between oppositely charged ions appears to be an impossibility. An example is $(CH_3)_4N^+B(CH_3)_4^-$, which therefore must be ionic in the usual sense that there is no electron sharing but the ions are held together by the attraction between their opposite charges. Other compounds exist wherein equalization of electronegativity cannot occur. For example, compounds such as sodium sulfate, Na_2SO_4 are probably best considered composed of Na^+ ions and $SO_4^=$ ions because the maximum electronegativity possible for sodium, that of sodium ion, is still lower than the electronegativity of the sulfate ion. Sodium ions are held in the crystal by their attractions for the electrons of the sulfate ions, but the latter electrons probably do not penetrate toward the sodium nucleus at all.

Clearly, much remains to be learned about the nature of the chemical bond, and perhaps especially in the solid state. At the present time, however, it is possible to apply reasonably simple concepts toward a quantitative understanding of bond energies based on the nature of atoms. It is thus becoming possible to calculate heats of formation and reaction in a manner which can lead to a better understanding of chemistry.

R. T. SANDERSON

Cross-references: *Electronegativity; Electronic Configurations*

BORIC ACID

Boric acid (H_3BO_3), is made by treating borax ($Na_2B_4O_7 \cdot 10H_2O$) with sulfuric acid. It is a solid, and may be obtained as a very fine, air-borne powder, as a granular powder, or in the form of crystals. It is made in a technical grade and a U.S.P. grade. The U.S.P. grade is a pharmaceutical; its saturated solution is a mild antiseptic, or at least, a bacteriostatic agent. The technical grade finds use in ceramics, and in other industries. Borax is obtained from the brine in Searles Lake, but very especially from rasorite, a mineral ($Na_2B_4O_7 \cdot 4H_2O$); the latter gives by solution, filtration, and recrystallization, the desired borax.

ELBERT C. WEAVER

Cross-references: *Acids (General); Nitration; Nitrates; Fertilizers.*

BORON AND COMPOUNDS

Boron is the fifth element of the Periodic Table, having an atomic weight of 10.82, with stable isotopes of mass numbers 10 and 11. It is essentially metalloidal in character (in this respect differing from all the other, strictly metallic elements of Group III). In the form of borax ($Na_2B_4O_7 \cdot 10H_2O$), it has been an article of commerce for hundreds of years. The element itself is difficult to isolate, and many of the characteristics assigned to it by early investigators were due to highly impure and oxygen-containing material, and even to metallic borides and borocarbides, as made by reduction of oxygenated compounds. The element can be isolated in highest purity, by pyrolysis of diborane, B_2H_6, or boron halides with hydrogen generally by deposition on a heated tungsten wire. Major commercial production has involved reduction of boric oxide by magnesium, followed by leaching out impurities. Electrolysis of potassium fluoborate, KBF_4, in fused potassium chloride, with or without boric oxide additions has also been used. The element appears in several crystalline forms and also microcrystalline (so-called amorphous). The crystalline modifications and several borides are all based on B_{12} icosahedra packed in near close packed arrangements and with or without interstitial boron or other atoms. Boron has a density of 2.24–2.34, and its melting point is in the 2000–2300°C range. Its electrical conductivity is very low at room temperature, but rises phenomenally with temperature; impurities have a great influence on the conductivity, carbon being extraordinarily effective even at 0.1% concentration.

In alloy form, boron has proved of considerable worth in degassing and deoxidizing metals, copper base in particular; similarly, boron aids in the grain refinement of aluminum. World War II brought boron steels rapidly to the fore, since quantities of the order of .0005 to .005% have proved adequate to increase hardenability very sharply. Here boron is usually added as ferroboron or as manganese-boron.

Boron is of value in atomic work because it occupies top rank of all the elements as a neutron absorber. Boron has the third highest heat of combustion per unit weight of all the elements (25,120 BTU/lb) and consequently it has been seriously considered as a fuel booster in jet engines. Boron is a good reducing agent for many refractory oxides, and when its cost is lowered it could replace silicon and aluminum for this purpose. An important use for boron is in aluminum wire of high electrical conductivity. In certain other aluminum base alloys, it is used in minute amounts to confer exceptional strength, especially to those alloys which are sand cast, and require no heat treatment.

Elemental boron unites readily with all the halogens, but it is unaffected by boiling HCl or HF; conversely, with HI it reacts explosively. Concentrated nitric acid or hot sulfuric acid reacts slowly on massive boron; caustic alkalies, either as aqueous solution or in fused form, have substantially no effect. Fused sodium peroxide or a fused mixture of sodium carbonate and potassium nitrate, on the other hand, react vigorously. In air, the massive element is unaffected at 750°C, and action is slow even at 1000°C; in oxygen, attack at 1000°C is quite rapid. Surprisingly, finely divided material, of 0.1–10.0 micron range size, oxidizes slowly even at room temperature; in general, the finely divided material is far more reactive than the massive.

A remarkable compound of boron is the nitride, BN, "white graphite" with a hexagonal graphite-like platelet structure. Of very low apparent density, significant oxidation resistance (up to about 650°C), very high melting point (about 3000°C), and extremely high electrical resistivity, the material has been used as a special lubricant and for crucible purposes. A cubic, diamond-like form "borazon" has been prepared under very high pressures. It is as hard as diamond.

The hydrides of boron or boranes have unusual structures. Their boron skeletons are generally segments of the icosahedron found in crystalline boron. They have received considerable attention as high energy fuels. They have negative heats of formation, with consequent unusually high energy liberation on oxidation. Sodium borohydride, $NaBH_4$, and the amine-boranes, $R_3N:BH_3$, are useful reducing agents.

Boron is also an important trace element in plants and serves as a significant nutrient factor.

Borates

The principal natural borates of commercial importance today and their sources are:

Name	Approx. Composition	Deposits
Tincal (Natural Borax)	$Na_2O \cdot 2B_2O_3 \cdot 10H_2O$	U.S., Tibet
Kernite (Rasorite)	$Na_2O \cdot 2B_2O_3 \cdot 4H_2O$	U.S.
Colemanite	$2CaO \cdot 3B_2O_3 \cdot 5H_2O$	U.S.
Ulexite (Boron-atrocalcite)	$Na_2O \cdot 2CaO \cdot 5B_2O_3 \cdot 16H_2O$	U.S., Chile, Argentina, Boliva, Peru
Priceite (Pandermite)	$5CaO \cdot 6B_2O_3 \cdot 9H_2O$	Turkey (Asia Minor)

The water-soluble borates have been most studied. They are limited to the salts of alkali metals. From the systems M_2O—B_2O_3—H_2O the following salts separate at 30°C:

Metaborates	Tetraborates	Pentaborates
$LiBO_2 \cdot 8H_2O$	$Li_2B_4O_7 \cdot xH_2O$	$LiB_5O_8 \cdot 5H_2O$
$NaBO_2 \cdot 4H_2O$	$Na_2B_4O_7 \cdot 10H_2O$	$NaB_5O_8 \cdot 5H_2O$
$KBO_2 \cdot 2 \cdot 5H_2O$	$K_2B_4O_7 \cdot 4H_2O$	$KB_5O_8 \cdot 4H_2O$

The borates react with strong aqueous acid to precipitate orthoboric acid.

$$2NaBO_2 + H_2SO_4 + 2H_2O \rightarrow Na_2SO_4 + 2H_3BO_3$$

The alkali metal borates react with the soluble salts of other metals to precipitate their borates. This reaction is partially responsible for the softening action of borates in hard water. The sodium borates react with aqueous ammonium chloride to precipitate ammonium pentaborate:

$$5Na_2B_4O_7 + 10NH_4Cl \rightarrow$$
$$10NaCl + 4NH_4B_5O_8\downarrow + 6NH_3 + 3H_2O$$

Boric oxide may be prepared by heating ammonium pentaborate and driving off ammonia and water.

$$2NH_4B_5O_8 \xrightarrow{\Delta} 2NH_3 + 5B_2O_3 + H_2O$$

Colemanite is commonly converted to boric acid by adding SO_2 or H_2SO_4 and floating the boric acid from the gangue.

$$Ca_2B_6O_{11} + 2SO_2 + 9H_2O \rightarrow 2CaSO_3 + 6H_3BO_3$$

Colemanite is converted to borax by boiling with soda ash, filtering from the calcium or magnesium carbonate, and allowing the borax to crystallize.

$$2Ca_2B_6O_{11} + 4Na_2CO_3 + H_2O \rightarrow$$
$$4CaCO_3\downarrow + 3Na_2B_4O_7 + 2NaOH$$

Upon fusion of transition metal salts with borax, metaborates, often distinctively colored, are formed. These have traditionally been used as qualitative tests.

Scandium, yttrium, and indium borates are the only known orthoborates and have structures similar to those of $CaCO_3$ and $NaNO_3$ with planar BO_3 groups. A unique common structure in most borates is the boroxin (B_3O_3) ring. In the alkali metal metaborates the anions are cyclic $(BO_2)_3^{-3}$ trimers. Both borax and colemanite have polymeric anions containing boroxin rings. The pentaborate anion has two boroxin rings sharing a tetrahedrally coordinated boron. However, in calcium metaborate the anions are linear $(BO_2)_n^{-n}$ polymers.

Perborates. Sodium "perborate," obtained by the action of hydrogen peroxide and sodium hydroxide, or sodium peroxide, on cooled borax solution, formerly considered to be a perborate $NaBO_3 \cdot 4M_2O$, is probably a borate-containing hydrogen peroxide of crystallization, $NaBO_2 \cdot 3H_2O \cdot H_2O_2$. It does not liberate iodine from concentrated potassium iodide solution. The compound ("perborax") is stable in the dry state and only sparingly soluble in water. The solution has bleaching and antiseptic properties. It is stable at room temperature but evolves oxygen when heated. The solid loses $3H_2O$ at 50–55°C, and, if it is then heated in a vacuum at 120°C, it loses another molecule of water, leaving a yellow solid formulated as $(NaBO_2)_2O_2$, which evolves oxygen in contact with water, but does not liberate iodine from concentrated potassium iodide solution. The crystalline perborate $NaBO_2 \cdot H_2O_2 \cdot 3H_2O$ is also obtained by the electrolysis of a solution of borax and sodium carbonate with a platinum gauze anode.

ROY M. ADAMS

References

Adams, R. M., Editor, "Boron, Metalloboron Compounds and Boranes," John Wiley & Sons, Inc., New York, 1964.

Lipscomb, W. N., "Boron Hydrides," W. A. Benjamin Inc., New York, 1963.

Muetterties, E. L., "The Chemistry of Boron and Its Compounds," John Wiley & Sons, Inc., New York, 1967.

Steinberg, H., "Organoboron Chemistry," John Wiley & Sons, Inc., New York, Vol. I, 1964, Vol. II, 1966.

Steinberg, H., and McClosky, A. L., Editors, "Progress in Boron Chemistry," Pergamon Press, London, Vol. I, 1964, Vol. II, 1970, Vol. III, 1970.

Boron 10

Boron consists of two isotopes, one of atomic mass 10 and the other of mass 11. The B^{10} isotope, which constitutes between 18.5 and 19% of natural boron, has a large capture cross section (affinity) for neutrons of low energy; i.e., "slow" or "thermal" neutrons which possess energies of the order of 0.01 eV. The capture of such a neutron results in the expulsion of an α-particle:

$$B^{10} + n^1 \rightarrow Li^7 + \alpha^4 + 2.5 \text{ MeV}$$

The high energy of this reaction (2.5 MeV) is shared by the lithium atom and the α-particle and consequently both cause considerable ionization along their recoil tracks. As with other nuclear processes the $B^{10}(n, \alpha)Li^7$ reaction is independent of the chemical state of the boron-10 atom.

An obvious and well-used outlet for this unusual property is in the manufacture of neutron absorbing materials, e.g., for control rods in nuclear reactors and for constructing neutron shields for personnel. Elemental boron itself cannot be used for these purposes because, being a typical nonmetal, it has no ductility, tensile strength or other useful fabricating properties.

A semimetallic neutron absorbing material called "boral" is made by suspending boron carbide, B_4C in molten aluminum. Ingots of this mixture, after being wrapped in an aluminum jacket, can be rolled out, at 700°C, into sheets only a quarter of an inch thick. In this way a thin, lightweight neutron shield is formed; the production of such shields will probably be a major consideration in the design of nuclear-powered aircraft of the future. The α-particles, which are doubly charged helium nuclei He^{++}, emitted after neutron capture by the boron-10 nuclei rapidly lose their energy by collisions with surrounding molecules and pick up stray electrons to become atoms of helium gas:

$$He^{++} + 2\epsilon \rightarrow He$$

After a long period of time the accumulation of helium may cause slight distortion of the shields but for most purposes this would be of little consequence. Pressure-molded bricks of finely powdered boron carbide and water, when baked in air at 600–1000°C also provide useful neutron absorbing material. During this baking period, some of the boron is oxidized to boric oxide, B_2O_3, which is liquid at these temperatures; on cooling the boric oxide solidifies and binds the brick together.

A similar technique involving the pressure molding at 1500°C of mixtures of 5–20% boron with one of several metals, such as copper or nickel, produces coherent slugs suitable for making the control rods of nuclear reactors. The role of these rods is to "soak up" spare neutrons in the reactor which would otherwise allow the nuclear fission processes to get out of hand. By either inserting or withdrawing these boron-containing absorbers it is possible to control the power of the pile. The efficiency of these absorbers is of course greatly increased (as is the cost) if the boron source is artificially enriched in B^{10} nuclei before fabrication.

Normally, neutrons, due to their lack of charge, pass unchecked through vast amounts of matter before being captured by a nucleus and problems therefore arise when scientists try to "count" them. However, efficient neutron counters may be manufactured from simple Geiger counters by filling the counter tube with a mixture of argon and boron trifluoride (or boron trimethyl) enriched with boron-10. As a slow neutron passes through the tube so it is captured by a boron-10 nucleus and an α-particle is emitted which, being charged and having high energy, causes ionization of the argon and triggers the counter in the usual way. Another, possibly less sensitive, instrument has been made in which the boron-10 source is mixed with activated zinc sulfide and spread uniformly on the surface of a photomultiplier tube. The α-particles emitted from the boron nuclei on neutron capture cause scintillations in the zinc sulfide which are detected and amplified by the photomultiplier.

These nuclear properties of boron-10 have also found some application in research into the treatment of brain tumors.

A. G. Massey

Borides

Several borides were prepared before 1900 and by 1935 some of the crystal structures had been determined. Recently, however, because of a growing interest in high-temperature materials, the transition metal borides have been the subject of intensive study by L. Andrieux, L. Brewer, R. Kieffer, R. Kiessling, J. T. Norton, P. Schwarzkopf, their co-workers and many others.

The borides do not conform to the simple valency rules which determine the composition of many chemical compounds. The borides, as well as certain other interstitial compounds such as the carbides and nitrides of the transitional metals, defy interpretations in terms of any rational valency rules. Further examination shows that for many of the borides, the stoichiometric laws are no longer rigorously valid. Hence, the borides can be called compounds only in the sense that they exist as a solid phase of characteristic structure, which generally conforms closely to a simple chemical formula, but which may exist over a limited range of chemical composition. In view of this, the borides will logically be discussed in terms of their crystal structure. Although it is not yet possible to state the principles governing the boride structures, one feature is outstanding. This is the tendency for the boron atoms to be linked together to form zigzag chains, 2-dimensional nets or 3-dimensional frameworks extending throughout the whole crystal. The boron-boron bond plays a priminent part in the boride structures and their formulas are generally quite different from those of the interstitial carbides and nitrides. This linking together of the boron atoms seems to preclude the applicability of the Hägg rule. Consequently, there are no discontinuities in the properties of the borides when the critical ratio of the radius of the boron atom exceeds 0.59. The Hägg rule would appear to apply only when the metalloid atoms are in isolated positions in the lattice as in certain carbides and nitrides.

Borides may be obtained by sintering mixtures of the powdered metal with boron at temperatures of 1,800 to 2,000°C. The material thus obtained may be compacted and purified by further sintering compressed bars in a vacuum or low pressure inert atmosphere at temperatures near the melting point. Most of the impurities, being more volatile than the refractory borides, will be volatilized away. Very pure boride samples have been prepared in this way.

Borides are also prepared by the thermite process in which a mixture of the metal oxide and boric oxide are reduced by Al, Mg, Si or C. The intermediate products react to form the boride. The boride crystals must then be separated from the by-products. This is difficult and usually results in impure products.

Several borides have been prepared by the electrolysis of fused salts. A molten mixture of the borate and fluoride of the metal is electrolyzed between graphite electrodes. The boride is deposited at the cathode and must be chemically separated from the mixture.

Borides have also been deposited from the vapor phase. A tungsten filament is heated in mixed vapors of boron tribromide, hydrogen, and a volatile metallic halide. The reaction, proceeding at the surface of the incandescent filament, deposits a coating of the boride. Large single boride crystals have been prepared in this way.

The borides are most remarkable for their thermal and chemical stability, hardness, and true metallic properties. The hardness of the borides lies in general between the diamond (Mohs 10) and the topaz (Mohs 8). The metallic character of the borides is evident from their low electrical resistivities and high thermal conductivities. The borides also have positive temperature coefficients of resistivity, showing that true metallic conduction is involved. Superconductivity has been reported for some of the borides. In several cases, however, these results have not been substantiated when very pure samples were examined. The borides are very refractory, with melting points in most cases between 2,000 and 3,000°C. The thermal stability of the diborides of the metals of Group IVB are exceptionally high. The borides are very stable chemically. They are not attacked by moisture or air at moderate temperatures and in nearly all cases do not react with HCl or HF. All the borides, however, are readily dissolved by molten alkali hydroxides. The transition metal diborides and all of the borides of Group VIB are stable at high temperatures in the presence of carbon or carbides.

The transitional metal borides are gray in color with a metallic luster. The alkaline earth borides are black and brownish-black. The rare earth borides are various shades of blue and purple. Lanthanum boride turns a deep red when moist. Thorium hexaboride is red and thorium tetraboride is yellow.

The Me_2B borides crystallize into structures which are isomorphous with $CuAl_2$-type structure shown in Fig. 1. The small black spheres represent the boron atoms and the large white spheres represent the metal atoms. In this tetragonal structure there are four boron atoms and eight metal atoms in one unit cell. The boron atoms are arranged in layers with the metal atoms interleaved between them. Although the boron atoms lie in layers, they

FIG. 1. The Me$_2$B boride crystal structure.

are apparently isolated with no boron-boron binding in these layers. However, each boron atom does have two close boron neighbors, directly above and below it in adjacent boron planes. The metal atoms form pairs, as shown, and the distance between the atoms of such a pair is shorter than in the metal crystal. These pairs, one above the other and rotated 90°, form tetrahedra of metal atoms. The isolated boron atoms then fit in the holes between the tetrahedra.

The MeB$_2$ borides have a hexagonal AlB$_2$-type structure as shown in Fig. 2. In this structure the arrangement again consists of alternate layers of boron atoms and metal atoms. The boron layer consists of hexagonal meshes like that of the graphite layer structure. Each boron atom is located at the center of a triangular prism of metal atoms. The metal atoms form a simple hexagonal structure. Each metal atom has twelve equidistant boron neighbors, six in the plane above it and six in the plane below. In the boron planes each atom is equidistant from three other boron atoms. Each boron atom has six equidistant metal atoms at the apexes of the triangular prism.

The alkaline earth metals, the rare earth metals and thorium all form hexaborides with the CaB$_6$-type structure. This crystal structure is shown in

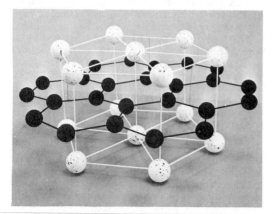

FIG. 2. The MeB$_2$ boride crystal structure.

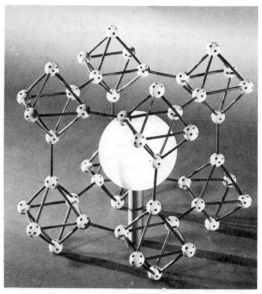

FIG. 3. The MeB$_6$ boride crystal structure.

Fig. 3. The small boron atoms form a three-dimensional framework structure which surrounds the large metal atoms. The boron framework is made up of octahedra, one at each corner of the cube. These are bonded together at their apexes. Each boron atom has four adjacent neighbors in its own octahedra and another neighbor in the direction of one of the cubical main axes.

The monoborides have at least two different crystal structures characterized by zigzag chains of boron atoms extending indefinitely throughout the crystal. Because of their low scattering powers for x-rays, the positions of the boron atoms are not always readily determined and frequently some doubt exists as to their exact locations in some of these structures. The compound ferric boride, FeB, has an orthorhombic crystal structure which is isomorphous with CoB. In this structure, each boron atom is surrounded by six iron atoms at the apexes of a trigonal prism. However, the nearest neighbors to each boron atom are two other boron atoms, thus forming an infinite boron zigzag chain. Several other monoborides form variations of this structure, some tetragonal, but all have boron chains.

The dodecaboride, ZrB$_{12}$, is isomorphous with UB$_{12}$. This structure consists of a three-dimensional framework of boron atoms with metal atoms located in the interstices of this skeleton. A face-centered cubic cell contains four ZrB$_{12}$ units.

The AlB$_{12}$ compound has three structures: a tetragonal diamond-like structure and graphitic structures of both teragonal and monoclinic systems.

The tetraborides of Ce, Th and U form tetragonal structures with the metal atoms in sheets and the boron atoms in positions resembling both the diboride and hexaboride structures. With the intensive study now being devoted to refractory materials it is most likely that many more boride compounds will be reported in the near future.

The high-melting borides of the transition metals in Groups IVB and VIB are considered most promising as high-temperature materials for such applications as turbine buckets and rocket nozzles.

Lanthanum hexaboride is used as a thermionic cathode material because of its metallic properties and high chemical and thermal stability.

Small quantities of boron are frequently added to steel in the form of borides for improving its ability to be hardened.

Excellent references to the literature on the transition metal borides may be found in Schwarzkopf and Kieffer's book "Refractory Hard Metals."

J. M. LAFFERTY

Cross-references: *Boron, Crystals and Crystallography.*

References

Schwarzkopf, P., and Kieffer, R., "Refractory Hard Metals," p. 271, Macmillan Company, New York, 1953.

Aronsson, B., Lundström, T., and Rundqvist, S., "Borides, Silicides and Phosphides," Methuen & Co. Ltd., London, 1965.

Samsonov, G. V., and Paderno, Yu B., "Borides of the Rare Earth Metals," U. S. Atomic Energy Commission, Div. of Tech. Info., Translation Series, AEC-tr-5264, 1961.

Lafferty, J. M., "Boride Cathodes," *J. Appl. Phys.,* **22,** 299, 1951.

Beck, P. A., "Electronic Structure and Alloy Chemistry of the Transition Elements," p. 179, Interscience Publishers, New York, 1963.

Brewer, L., "The Chemistry and Metallurgy of Miscellaneous Materials: Thermodynamics," McGraw-Hill Book Company, New York, 1950.

Carboranes

The first carborane was isolated and its structure was deduced in 1957. Since then, primarily under the sponsorship of the United States Office of Naval Research, the discovery rate (and structural characterization) of new carboranes has been explosive. Fortunately, a thread of regularity running throughout carborane chemistry allows one to assimilate it fairly rapidly, though in a somewhat cursory fashion.

Although composed of boron, carbon and hydrogen, the carboranes have distinctively different structural and chemical characteristics and are not to be confused with mere hydrocarbon derivatives of the boron hydrides. They may be divided into three groups: $closo\text{-}(C_{0-2}B_nH_{n+2})$, $nido\text{-}(C_{0-4}B_nH_{n+4})$ and $arachno\text{-}C_{0-6}B_nH_{n+6}$, the prefixes being derived from the Greek words for *cage, nest* and *web,* respectively. When the carbon subscript is zero the empirical formulae denote the familiar boranes (boron hydrides), and indeed the borane structural geometry is dominant even in compounds where most of the skeletal atoms are carbon.

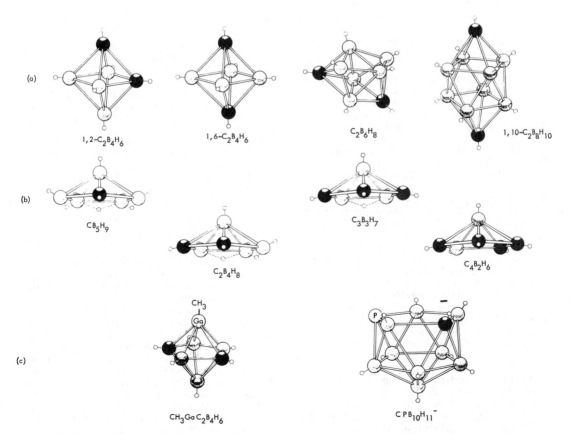

(a)

$1,2\text{-}C_2B_4H_6$ $1,6\text{-}C_2B_4H_6$ $C_2B_6H_8$ $1,10\text{-}C_2B_8H_{10}$

(b)

CB_5H_9 $C_2B_4H_8$ $C_3B_3H_7$ $C_4B_2H_6$

(c)

$CH_3GaC_2B_4H_6$ $CPB_{10}H_{11}^-$

FIG. 1. Structure of (a) *Closo*-carboranes, (b) Nido-carboranes, and (c) Heteroelement carboranes.

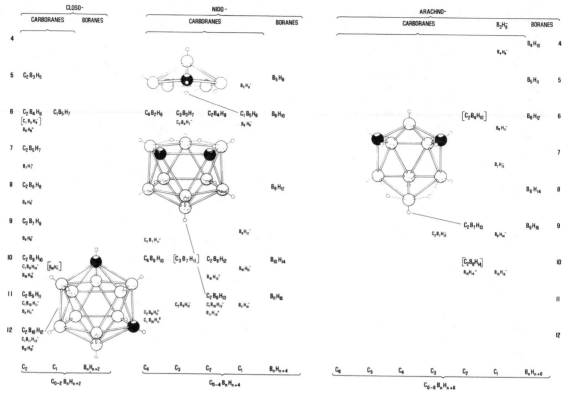

FIG. 2. Classes and families of carboranes.

All the carboranes (including boranes) are most readily visualized as being built about the vertices of a unique and complete series of triangular faced polyhedra containing from five to twelve vertices (e.g., *closo*-compounds, Fig. 1a) or fragments of that same unique series of polyhedra generated by the sequential removal of high coordination vertices (e.g., *nido-* and *arachno*-compounds, respectively, Figs. 1b and 2). The examples shown in Fig. 1 (a and b) represent sets of closely related carboranes, and in general the most stable species (e.g., $C_4B_2H_6$ and 1, $10\text{-}C_2B_8H_{10}$) do not have any bridge hydrogens. When a choice exists the carbons prefer to locate in low coordination sites, as in 1, $10\text{-}C_2B_8H_{10}$, the most stable carborane thus far discovered, while bridge hydrogens associate with low coordination edge positions. The many and varied carboranes thus far discovered may be correlated with related (isoelectronic) species, as catalogued in Fig. 2; the blank space reveals that many more carboranes remain to be discovered.

Syntheses. Several synthetic routes have been utilized for the preparation of the carboranes. In general the reaction of an acetylene and a borane at either elevated temperatures in the gas phase or in the presence of a Lewis base is involved. German workers additionally have prepared many alkylated derivatives by various other routes.

Derivatives. The derivative chemistry of the *closo*-carboranes has generally involved the icosahedral *closo*-carborane $C_2B_{10}H_{12}$ because it has been available in quantity for the longest period of time. The hydrogens on the carbons of the *closo*-carboranes $C_2B_{10}H_{12}$, $C_2B_8H_{10}$, $C_2B_5H_7$, etc., may be replaced with Li, which in turn leads to a great number of C-attached derivatives including S, N, Sn, P, As, C and Si, etc.; most substituents that can be attached to benzene may also be attached to the carborane cages, including alkyl groups, halogens, and so on. The smaller *closo*-carboranes $C_2B_5H_7$, $1,6\text{-}C_2B_4H_6$ and $1,5\text{-}C_2B_3H_5$ are now becoming available in developmental quantities, and the rich derivative chemistry heretofore restricted to $C_2B_{10}H_{12}$ will apply directly to these smaller *closo*-carboranes.

Commercially available polymers which are very useful at elevated temperatures have been prepared from $C_2B_{10}H_{12}$, $C_2B_8H_{10}$ and $C_2B_5H_7$ which conform to the general formula

$$\{-CB_nH_nC[Si(CH_3)_2O]_mSi(CH_3)_2-\}_x.$$

When $m = 2$ or 3 and when $C_2B_{10}H_{12}$ is utilized elastomeric silicone derivatives result; because of its smaller size, $C_2B_5H_7$ produces elastomeric materials even when $m = 1$, and about 20% of $C_2B_8H_{10}$ or $C_2B_{10}H_{12}$ is utilized randomly to break up any predilection to crystallinity.

Other heteroelements are also found substituting for boron and/or carbon in a number of carborane derivatives. Those elements in the upper right hand corner of the Periodic Table seek low coordination vertices, (e.g., N, P, S) thus substituting for carbon, while other elements to the left of boron and carbon in the Periodic Table seek out high coordination vertices (substituting for boron) when a choice exists. Most transition metal elements in addition to Be, Al, Ga, In, etc. have been introduced into the icosahedral framework in place of boron.

ROBERT E. WILLIAMS

BROMINE AND COMPOUNDS

Bromine is a nonmetallic element of the halogen family, with chemical properties resembling those of chlorine and iodine (Group VII of Periodic System). At room temperature bromine is a dark red-brown liquid which vaporizes readily. Both liquid and vapor are very corrosive and the vapor has an intensely irritating odor. The element has atomic number 35, and consists of two stable isotopes, Br^{79} and Br^{81}, present in nearly equal amounts so that the atomic weight is 79.904. The liquid and vapor are diatomic over a wide range of temperature.

It was discovered independently by A. J. Balard in France and C. Löwig in Germany, in 1825. Balard obtained it by chlorinating sea water bitterns and liberating the element by distillation, whereas Löwig treated salt spring brine with chlorine and extracted the bromine with ether. Balard selected the name from the Greek *Bromos,* meaning "stench."

Occurrence. The occurrence of bromine generally parallels that of chlorine. In the earth's crust, the latter is about 300 times as abundant as the former. Neither element occurs free in nature, but is always found as a halide. With the exception of some rather rare silver salts, no natural mineral contains bromine as an essential constituent. Alkali and alkaline earth halides, because of their solubility, are susceptible to leaching from rocks and soils by rain water and are carried to the ocean where they accumulate. Average ocean water contains 67 mg of bromine per liter. In various parts of the world there are salt deposits or brines where bromine has been enriched by evaporation of water from prehistoric seas or salt lakes. Bromine is extracted commercially from the ocean, from underground brines in Michigan and Arkansas (0.2 to 0.5% Br), from saline basins such as Searles Lake, California (0.085%) and the Dead Sea, Palestine (0.56%), and from solid salt beds at Stassfurt, Germany. In the latter, bromine is present in the form of chlorobromide mixed crystals with the more soluble components (carnallite, $KCl \cdot MgCl_2 \cdot 6H_2O$, and tachhydrite, $2MgCl_2 \cdot CaCl_2 \cdot 12H_2O$).

Preparation. Bromine is prepared in the laboratory by reaction of a bromate with a bromide and acid, in water:

$$NaBrO_3 + 5NaBr + 3H_2SO_4 \rightarrow$$
$$3Br_2 + 3Na_2SO_4 + 3H_2O$$

The bromine is distilled out and condensed together with a little water, from which it may be separated by gravity, and is dried by treatment with anhydrous calcium sulfate.

Commercially, bromide-containing brines are treated with chlorine and the bromine is swept out by steam:

$$2Br^- + Cl_2 \rightarrow Br_2 + 2Cl^-$$

The mixture of steam, bromine, and some chlorine is condensed and the halogen layer is fractionally distilled to obtain pure bromine. In the recovery of bromine from ocean water the use of steam is not economically practical, so the bromine is blown out with air. The halogens are removed from the air stream either by scrubbing with sodium carbonate solution or by introducing sulfur dioxide and scrubbing with water:

1. $3Br_2 + 3Na_2CO_3 \rightarrow 5NaBr + NaBrO_3 + 3CO_2$

2. $Br_2 + SO_2 + 2H_2O \rightarrow 2HBr + H_2SO_4$

In the former case, bromine is regenerated by acidification as in the laboratory preparation above, whereas in the latter case, the acid mixture is rechlorinated and distilled as in the steam process. The waste sulfuric acid is used for acidifying the incoming ocean water, which must be brought to a pH below 4 before treatment with chlorine.

Properties. Bromine freezes at $-7.2°C$ and boils at $58.8°C$. The density of the liquid at $25°C$ is 3.104 and the specific heat is 0.107 calorie per gram. Heats of fusion and vaporization are 16.1 and 44.8 calories per gram, respectively. The solubility of bromine in water at $25°C$ is 3.35 grams per 100 grams of solution. In the presence of alkali halides, particularly potassium bromide, the solubility is increased. Presumably a complex polyhalide (KBr_3) is formed. Bromine is completely miscible with many of the common organic solvents such as carbon tetrachloride and benzene, though in most cases bromination occurs and the solutions are not stable.

In reactivity bromine is similar to chlorine, though its normal oxidation potential is somewhat lower (-1.087 volts for the aqueous system $Br_2 + 2e \rightarrow 2Br^-$). Bromine attacks most metals. Aluminum reacts with it vigorously, with emission of light, and potassium reacts explosively. On the other hand, lead, nickel, and magnesium are not attacked and may be used as containers for the liquid. Even sodium does not react with dry bromine below 300°C. Iron and zinc are corroded rapidly if moisture is present.

Bromine hydrolyzes slightly in aqueous solution, producing hypobromous acid which is unstable:

$$Br_2 + H_2O \rightarrow HBrO + HBr$$

$$2HBrO \rightarrow 2HBr + O_2$$

The active oxygen is responsible for the bleaching action of bromine water. Hypobromite solutions formed by neutralizing bromine water with alkali are also strong oxidants, capable of oxidizing ammonia to nitrogen, sulfur compounds to sulfates, and various metals to their higher valences. The hypobromite is also unstable, disproportionating to bromate and bromide.

Unsaturated organic compounds form *addition* products with bromine, in some cases almost quantitatively:

$$\begin{array}{ccc} & H\ \ H & & H\ \ H \\ & |\ \ \ | & & |\ \ \ | \\ RC&=CR + Br_2 \rightarrow & RC&-CR \\ & & & |\ \ \ | \\ & & & Br\ \ Br \end{array}$$

The *substitution* reaction of phenol with bromine is also practically quantitative:

$$C_6H_5OH + 3Br_2 \rightarrow C_6H_2Br_3OH + 3HBr$$

Uses—Organic Compounds. Most of the bromine that is produced is converted to ethylene dibromide. A small proportion is sold as liquid bromine for use in organic syntheses, as an analytical reagent, and for miscellaneous oxidizing purposes. Bromine is an effective antiseptic but except in specialized cases cannot compete economically with chlorine for water sterilization. The usefulness of bromine in synthesis stems from the ease with which it may be introduced and replaced in organic compounds. Since bromine may be replaced more readily than chlorine, the reaction conditions are frequently less drastic. Selective reactions are also possible so that, for example, two different groups may be introduced into a molecule by successive operations upon a bromochloro intermediate. Bromine also modifies the shades and solubilities of indigos and other colored compounds, thereby finding uses in the dye industry. Ethylene dibromide (1,2-dibromoethane) is used principally as an ingredient of antiknock fluid for motor fuels, but has an additional important use for insect control in grains and soils. Methyl bromide and 1,2-dibromo-3-chloropropane are also useful agricultural fumigants. Bromochloromethane is an effective fire extinguisher fluid, while a number of bromine compounds, such as tetrabromobisphenol, pentabromochlorocyclohexane and tris (2,3-dibromopropyl) phosphate, serve to impart fire retardant or self-extinguishing properties to plastics in which they are incorporated.

Bromides

Inorganic bromides are salts of hydrobromic acid, HBR. Their properties are intermediate between those of the chlorides and the iodides. Most metallic bromides are quite water-soluble, those of silver, lead and monovalent mercury, thallium or gold being exceptions. In the absence of chloride and iodide, a bromide can be detected and determined either volumetrically or gravimetrically by precipitation as silver bromide. Interference from iodide is avoided by first boiling with nitrous acid (sodium nitrite plus dilute sulfuric acid), whereby iodine is expelled. Bromide in the presence of chloride may be detected by oxidation to bromine (amber color) with chlorine water. It may also be determined by oxidation to bromate with sodium or potassium hypochlorite. The excess hypochlorite is decomposed with sodium formate and the bromate is determined by treatment with potassium iodide in acid solution, followed by titration of the liberated iodine.

Alkali and alkaline earth bromides are prepared most simply by treating the corresponding hydroxides or carbonates with hydrobromic acid, followed by evaporation and crystallization. Sodium and potassium bromides may also be recovered from mother liquors obtained in the preparation of bromates from the carbonates and bromine. Bromides are used in the preparation of light-sensitive emulsions for photography, as bleach modifiers, and pharmaceutically as mild sedatives. Calcium and lithium bromides are used as dehumidifiers in air-conditioning.

Bromates

Bromates are salts of the very unstable acid $HBrO_3$ (bromic acid), which exists only in aqueous solutions. The salts, though quite stable at ordi-

nary temperature when dry or in neutral or alkaline aqueous solution, are strong oxidizing agents in acid solution. Like the chlorates, they react vigorously with organic matter when heated or subjected to shock.

Alkali bromates are prepared by the electrolytic oxidation of bromides or by the reaction of bromine with an alkali hydroxide or carbonate, in water:

$$KBr + 3H_2O - 6e \rightarrow KBrO_3 + 3H_2$$

$$3Br_2 + 3Na_2CO_3 + 3H_2O \rightarrow$$
$$NaBrO_3 + 5NaBr + 3H_2CO_3$$

The bromates are generally less soluble than the bromides and may be crystallized from the mixture.

Uses for the bromates are based upon their oxidizing properties. The baking characteristics of wheat flour are improved by addition of 5 to 10 parts of bromate per million; apparently the salt oxidizes sulfhydryl groups in the protein, which otherwise would affect the dough viscosity adversely. Bromates are also used as analytical reagents and in some hair wave preparations. "Mining salts" or "bromine salts" are mixtures of bromates with bromides, which yield bromine upon acidification and have some uses as brominating agents.

V. A. STENGER

BUFFERS

When acid is added to an aqueous solution, the pH falls; when alkali is added, it rises. If the original solution contains only typical salts without acidic or basic properties, this rise or fall may be very great. There are, however, many other solutions which can receive such additions with only a slight change in pH. The solutes responsible for this resistance to change in pH, or the solutions themselves, are known as *buffers*. A weak acid becomes a buffer when alkali is added, and a weak base becomes a buffer on the addition of acid. A simple buffer may be defined, in Brönsted's terminology, as a solution containing both a weak acid and its conjugate weak base. Buffer action is explained by the mobile equilibrium of a reversible reaction:

$$A + H_2O \rightleftharpoons B + H_3O^+$$

in which the base B is formed by the loss of a proton from the corresponding acid A. The acid may be a cation such as NH_4^+, a neutral molecule such as CH_3COOH, or an anion such as $H_2PO_4^-$. When alkali is added, hydrogen ions are removed to form water, but, as long as the added alkali is not in excess of the buffer acid, many of the hydrogen ions are replaced by further ionization of A to maintain the equilibrium. When acid is added, this reaction is reversed as hydrogen ions combine with B to form A.

The pH of a buffer solution may be calculated by the mass law equation

$$pH = pK' + \log \frac{C_B}{C_A}$$

in which pK' is the negative logarithm of the apparent ionization constant of the buffer acid and the concentrations are those of the buffer base and its conjugate acid.

A striking illustration of effective buffer action may be found in a comparison of an unbuffered solution such as $0.1M$ NaCl with a neutral phosphate buffer. In the former case, 0.01 mole of HCl will change the pH of 1 liter from 7.0 to 2.0, while 0.01 mole of NaOH will change it from 7.0 to 12,0. In the latter case, if 1 liter contains 0.06 mole of Na_2HPO_4 and 0.04 mole of NaH_2PO_4, the initial pH is given by the equation:

$$pH = 6.80 + \log \frac{0.06}{0.04} = 6.80 + 0.18 = 6.98$$

After the addition of 0.01 mole of HCl, the equation becomes:

$$pH = 6.80 + \log \frac{0.05}{0.05} = 6.80$$

while after the addition of 0.01 mole of NaOH it is

$$pH = 6.80 + \log \frac{0.07}{0.03} = 6.80 + 0.37 = 7.17.$$

The buffer has reduced the change in pH from ± 5.0 to less than ± 0.2.

Figure 1 shows how the pH of a buffer varies with the fraction of the buffer in its more basic form. The buffer value is greatest where the slope of the curve is least. This is true at the midpoint, where $C_A = C_B$ and $pH = pK'$. The slope is practically the same within a range of 0.5 pH unit above and below this point, but the buffer value is slight at pH values more than 1 unit greater or less than pK'. The curve of Fig. 1 has nearly the same shape as the titration curve of a buffer acid with NaOH or the titration curve of a buffer base with HCl. Sometimes buffers are prepared by such partial titrations, instead of by mixing a weak acid or base with one of its salts. Certain "universal" buffers, consisting of mixed acids partly neutralized by NaOH, have titration curves which are straight over a much wider pH interval. This is also true of the titration curves of some polybasic acids, such as citric acid, with several pK' values not more than 1 or 2 units apart. Other polybasic acids, such as phosphoric acid, with pK' values farther apart, yield curves having several sections, each somewhat similar to the graph in Fig. 1. At any pH, the buffer value is proportional to the concentration of the effective buffer substances or groups.

The following table gives approximate pK' values, obtained from data in the literature, for several buffer systems:

Constituents	pK'
H_3PO_4, KH_2PO_4	2.1
HCOOH, HCOONa	3.6
CH_3COOH, CH_3COONa	4.6
KH_2PO_4, Na_2HPO_4	6.8
HCl, $(CH_2OH)_3CNH_2$	8.1
$Na_2B_4O_7$, HCl or NaOH	9.2
NH_4Cl, NH_3	9.2
$NaHCO_3$, Na_2CO_3	10.0
Na_2HPO_4, NaOH	11.6

Buffer substances which occur in nature include phosphates, carbonates and ammonium salts in the earth, proteins of plant and animal tissues, and the carbonic acid-bicarbonate system in blood.

Buffer action is especially important in biochemistry and analytical chemistry, as well as in many large-scale processes of applied chemistry. Examples of the latter include the manufacture of leather and of photographic materials, electroplating, sewage disposal and scientific agriculture. See also pH.

DAVID I. HITCHCOCK

References

Bates, R. G., "Determination of pH," Ch. 5, New York, John Wiley & Sons, 1964.

Clark, W. M., "The Determination of Hydrogen Ions," Chapters I, II, IX, Baltimore, Williams and Wilkins Co., 1928.

Kolthoff, I. M., and Laitinen, H. A., "pH and Electro Titrations," Chapters I, III, New York, John Wiley & Sons, 1941.

MacInnes, D. A., "Principles of Electrochemistry," pp. 275–278, New York, Reinhold Publishing Corp., 1939.

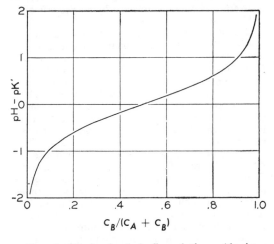

FIG. 1. pH of a simple buffer solution. Abscissas represent the fraction of the buffer in its more basic form. Ordinates are the difference between pH and pK'.

C

CADMIUM AND COMPOUNDS

The element cadmium was first discovered by Strohmeyer in 1817. It is a relatively rare element; its abundance in the lithosphere is estimated in the order of 0.5 gram per ton of the earth's crust. Cadmium minerals are rarely found alone; they are usually associated with zinc minerals.

Cadmium is in Group II, Period V of the Periodic Table. It has eight isotopes ranging from 106 to 116 in mass. Other physical constants are presented below:

Atomic number	48
Atomic weight	112.41
Color	Silver-white
Crystal structure	Hexagonal pyramids
Hardness (Mohs)	2.0
Ductility	Considerable
Density (g/cc)	
20°C (68°F) (s)	8.65
330°C (626°F) (l)	8.01
Melting point	321°C (609.8°F)
Boiling point	767°C (1412.6°F)
Specific heat (g-cal/g)	
25°C (77°F) (s)	0.055
Electrochemical equivalent Cd^{++} (mg/coulomb)	0.582
Electrode potential Cd^{++} ($H_2 = 0.0$ volt)*	−0.40 volt

*National Bureau of Standards nomenclature.

The major sources of cadmium are from zinc ores in various parts of the globe. The more important deposits are in Australia, Tasmania, Belgian Congo, Canada, Mexico, Peru, Southwest Africa, Western United States, and the Tri-State District (Missouri-Oklahoma-Kansas).

Mexico is probably the largest primary source of cadmium-bearing ores on the basis of quantities mined. It should be emphasized that practically all the cadmium obtained is as a by-product of zinc recovery and the supply of cadmium is dependent on the amount of zinc which is extracted; however, not all zinc ores contain cadmium.

The present world production of cadmium is in the order of 15 million pounds, of which a large proportion is recovered in the United States. At the present time, cadmium is quoted at $3.50 per pound. Most cadmium is recovered from primary sources. Only small amounts are recovered from secondary metals.

A large proportion of the cadmium is recovered from zinc sulfide ores. These ores are roasted to crude oxide, mixed with coal or coke and sodium or zinc chloride and passed over a sintering machine. Combustion of the coal provides the necessary temperature in the sintering bed to allow reaction of the chlorides with cadmium, lead, and some other impurities; these impurities are volatilized and collected, usually in an electrostatic precipitator. These chlorides are then reacted with sulfuric acid, solubilizing the cadmium. The cadmium is then removed from solution by careful selective precipitation with zinc dust; the cadmium sponge from this operation is then compressed (to prevent excessive oxidation) and distilled. This crude cadmium is then redistilled and sometimes chemically purified. It is then cast into slabs or special shapes.

Complex ores containing cadmium (almost invariably associated with zinc) require special treatment but use the same general principles. Another source of cadmium is in the purification of zinc sulfate liquor for electrolytic recovery of zinc. Prior to the electrolysis of the zinc sulfate solution, cadmium is removed by selective precipitation with zinc dust and the cadmium is then purified as previously described. In some cases the final step in recovering cadmium is effected by electrolysis of a cadmium sulfate solution.

Cadmium is almost always divalent. It slowly oxidizes in moist air at room temperature. At higher temperatures, the oxidation is more rapid. Cadmium reacts with the halogens to form the corresponding halides, chlorine being the most reactive. Cadmium is soluble in most acids but, unlike zinc, is not soluble in alkalies. It has good corrosion resistance except in acid environment. It is this resistance which makes cadmium coatings so effective in retarding corrosion.

The fumes of cadmium and its compounds, as well as solutions of its compounds, are poisonous. The toxicity of cadmium has not been fully appreciated and unwitting exposure to cadmium fumes is probably responsible for many cases of poisoning. Adequate precautions, such as the use of good respirators in the presence of cadmium fumes, will prevent the vast majority of poisonings. It cannot be too strongly emphasized that cadmium has highly lethal potentialities.

The major use of cadmium is for plating articles to give a protective coating (mainly for iron and steel); most of this coating is done by electrodeposition. Practically all commercial plating is done from cyanide baths which are essentially a solution of cadmium oxide and sodium cyanide in water. Cadmium coatings have good resistance to atmospheric and galvanic corrosion and to alkalies, but their resistance to acid attack is poor.

A wide variety of both small and large parts for

many uses are cadmium-plated for such protection. However, cadmium is almost never used in equipment or in containers for food or drinks, in view of its toxicity.

Cadmium alloys with a large number of other metals. This property is responsible for another use of cadmium although smaller quantities are used than for electroplating. Cadmium alloys are used for high-temperature, high-speed bearing metals. Other alloys are used for high-temperature solders, and still another series of alloys is used for low-temperature applications.

Cadmium is one of the more efficient absorbers or capturers of neutrons and, as such, finds utility in atomic work. It is used in the control of nuclear reactions and can also be used as a shield to prevent escape of the neutrons. The isotope CD[113] is especially effective for such purposes.

A small but important scientific use of cadmium is in the Weston standard cell, which is the working standard for the United States in maintaining the value of the volt. Another electrolytic couple is the nickel-cadmium storage battery in which the negative is cadmium and the positive is nickel oxide, the electrolyte being an aqueous solution of potassium and lithium hydroxides. Nickel-cadmium cells are finding uses as power sources in space satellites and in cordless appliances.

The most important cadmium compound is cadmium sulfide, usually prepared by precipitation of cadmium in solution with the sulfide radical. Cadmium sulfide is a brilliant yellow and is used as a pigment. Colors varying from yellow to a brilliant red can be secured by replacing an increasing proportion of the sulfide radical by selenium. Still another cadmium pigment is cadmium lithopone, prepared by co-precipitation of barium sulfide and cadmium sulfate; proper proportions of these two materials give pigments from light yellow to deep red. The cadmium pigments range in price from about 50 cents to $1.50 per pound. Ultrahigh-purity cadmium sulfide is being used in solar cells and photosensitive devices.

Cadmium sulfate and its hydrates are made by reaction of the metal or oxide with sulfuric acid. It is used in medicine, for fluorescent screens, and as an electrolyte in the Weston standard cell.

Cadmium oxide is commercially prepared by oxidizing cadmium vapor with air. Its major use is as an ingredient of electroplating baths and in the manufacture of pigments.

JOHN R. MUSGRAVE

CALCIUM AND COMPOUNDS

Calcium is a white, silvery alkaline-earth metal, in Group II of Periodic Table, with atomic number 20 and atomic weight 40, boiling point 1170°C, melting point 810°C (pure, 851°C), density 1.48 to 1.55 grams per centimeter; face-centered cubic crystals. It tarnishes in air with the formation of thin bluish gray films of oxide. These films are tight, adherent and protective. Calcium may come into contact with the skin without danger, and can be machined, cut, extruded or drawn. It is commercially available as a remelted or a sublimed material.

Calcium is made by thermal processes under high vacuum from lime reduced with aluminum. The Downs cell for the electrolysis of fused salt produces a sodium metal containing calcium. The calcium crystallizes out of solution and is filtered from the liquid sodium.

Calcium metal is an alloying agent for aluminum, bearing metals of the lead-calcium or lead-barium-calcium type; a reducing agent for beryllium; an alloying agent and a deoxidizer for copper; an alloying agent for the production of the age-hardening lead alloys for cable sheaths, battery plates, and related uses; a modifying agent for magnesium and aluminum; a debismuthizer for lead; a carburizer and desulfurizer, as well as a deoxidizer for numerous alloys such as chromium-nickel, copper, iron, iron-nickel, nickel, nickel-cobalt, nickel-chromium-iron, nickel bronzes, steel and tin bronzes; an evacuating agent; a reducing agent in the preparation of chromium metal powder, thorium, uranium, and zirconium; and a separator for argon from nitrogen.

Calcium oxide is commonly prepared by burning the carbonate in large vertical shaft kilns at temperatures below 1200°C. It is a white solid, often called quick-lime. It slakes or reacts with water to form the hydroxide [$Ca(OH)_2$], a common constituent of mortars and plasters, and an industrial alkali and neutralizer for acids.

Limestone is *calcium carbonate* in an indistinctly crystalline and massive form. It is found throughout the United States, but extensive deposits exist in Indiana. All varieties of calcium carbonate are almost insoluble in pure water but dissolve appreciably in the presence of carbon dioxide because of the formation of calcium hydrogen carbonate (bicarbonate):

$$CaCO_3 + H_2O + CO_2 \rightleftharpoons Ca^{++} + 2HCO_3^-$$

It is by this reaction that limestone is dissolved, often resulting in the formation of caves. Natural waters containing $CaCO_3$ are called hard waters or limestone waters. The action is reversible; and in many regions the underground waters, carrying large quantities of the carbonate, lose carbon dioxide on exposure in caves and deposit limestone. Travertine, used as a building stone, is a white concretionary calcium carbonate deposited by some springs when the pressure on the water is suddenly released, permitting the rapid escape of carbon dioxide. It occurs chiefly in Italy.

Calcium fluoride, as the mineral fluorite or fluorspar, is the commercial raw material for hydrofluoric acid and fluorinated organic compounds such as the "Freons."

Calcium chloride occurs in nature as tachhydrite [$CaCL_2 \cdot 2MgCl_2 \cdot 12H_2O$], and in some other minerals and also to the extent of about 0.15% in sea water. A considerable amount is removed from salt brines. It is a byproduct of Solvay sodium carbonate production and of other industrial processes. Upon evaporation of its aqueous solution the hexahydrate [$CaCl_2 \cdot 6H_2O$] is obtained, and by partial dehydration it is converted into a porous mass which is used for drying gases and liquids. Calcium chloride is very soluble in water; for this reason it gives with ice an excellent freezing mixture. With the hexahydrate and crushed ice a temperature as low as -50°C can be reached. A solution of the salt is used as a refrigerating brine in cold-storage plants and in the manufacture of ice.

Because of its deliquescent property it is sprinkled on roads to prevent ice formation, and (in solution) in mines to decrease the danger of explosion from dust.

Anhydrous *calcium sulfate* occurs in nature as the mineral anhydrite, crystallized in rhombic prisms. The dihydrate, or gypsum [$CaSO_4 \cdot 2H_2O$] is, however, more common and plentiful. Three general groups of gypsum products—uncalcined, calcined building, and calcined industrial—are sold. Uncalcined gypsum is used as a portland cement retarder to prevent too rapid hardening, and as a soil corrector in agriculture. Calcined gypsum is employed in making tile, wallboard, lath, and various kinds of plasters. When gypsum is heated to about 125°C it loses three fourths of its water of hydration and forms the hemihydrate ($2CaSO_4 \cdot H_2O$) or plaster of Paris. The resulting product is ground to a fine white powder. When this is mixed with water it forms a plastic mass which quickly sets to a coherent white solid consisting of small tangled crystals of more highly hydrated calcium sulfate.

Calcium phosphate is a tonnage mineral, primarily used for fertilizer after reaction with sulfuric acid to convert it into superphosphate.

The principal constituents of the bony skeleton of vertebrate animals are $Ca_3(PO_4)_2$, $Mg_3(PO_4)_2$, calcium carbonate, alkaline salts, and fatty and cartilaginous matter. The inorganic constituent of bone, is a hydroxyapatite in which a little magnesium is substituted for calcium and some carbonate for phosphate. The phosphate minerals or rock are the geologic residues of the decomposition of animal skeletons. Apatite is a calcium phosphate mineral crystallizing in the pyramidal group of the hexagonal system. Fluorapatite is $Ca_4(CaF)(PO_4)_3$, and chlorapatite is $Ca_4(CaCl)(PO_4)_3$; hydroxyapatite would be $Ca_4(CaOH)(PO_4)_3$.

When the apatites or their beneficiated products are reduced with carbon in an electric furnace, phosphorus is distilled and then or later burned to phosphoric anhydride (P_2O_5), which is converted to drug- and food-grade phosphoric acid and chemicals.

Calcium metaphosphate is widely used as a mild abrasive and neutralizer in tooth pastes.

CHARLES L. MANTELL

CALORIMETRY

Calorimetry is the science of measuring the quantity of heat released or absorbed by matter when it changes its physical state, participates in a chemical reaction, or interacts with its environment in any way. The increase in internal energy of a system, ΔE, is related to the heat absorbed, Q, and the work done by the system, W, by the equation

$$\Delta E = Q - W$$

Most calorimetric measurements are made under constant-pressure conditions, where the work done is equal to $P\Delta V$. In this case, the heat absorbed is equal to the increase in enthalpy.

$$Q = \Delta E + P\Delta V = \Delta H$$

Calorimetric measurements yield such useful thermodynamic functions as heat capacity at constant pressure, (C_P), heat capacity at constant volume (C_V), entropy (S), free energy (F), and the changes in these quantities which accompany chemical reactions.

In principle, nothing could be more fundamental than the relationship between temperature and internal energy for a given material, and calorimetry is obviously the most direct means of extracting such information. However, calorimetry is an unusually demanding science, characterized by a variety of experimental techniques, each one optimized for a particular application; almost all calorimeters of the adiabatic or isothermal-phase-change type operating today were designed and constructed by the people who use them.

Most calorimeters can be divided into two classes. In adiabatic calorimeters, the sample and its container are thermally isolated from the remainder of the apparatus, and the heat flow rate from the sample is inferred from the temperature rise of the sample container. The bomb calorimeter is semi-adiabatic, in that the thermal resistance between the sample container and the environment is not as high as in the adiabatic type; the measurement principle, however, is the same. In these instruments, the heat content of the sample is the independent variable, while its temperature is the dependent variable.

In isothermal calorimeters, the thermal resistance is as small as possible, to facilitate heat flow into or out of the sample, and this heat flow is measured directly; in these instruments, therefore, the temperature of the sample is the independent variable. The scanning calorimeter falls into this category, since the sample is coupled through a small thermal resistance to a controlled-temperature source.

Adiabatic Calorimetry. The experimental difficulties associated with all types of calorimetry are best appreciated with reference to the adiabatic calorimeter, which is used from cryogenic temperatures to about 600°K. The sample material is placed in a container provided with a heating element and a temperature-measuring device such as a platinum resistance thermometer. This assembly is suspended inside a temperature-controlled enclosure. The relationship between the heat flow rate into the sample $\frac{dH}{dt}$, the sample temperature T, and the enclosure temperature T_A is given by the equation

$$W = \frac{dH}{dt} + \frac{T - T_A}{R_T} + C \frac{dT}{dt}$$

where W is the power input to the heating element, R_T is the incremental thermal resistance between the sample container and the enclosure, and C is the thermal capacity of the container. Integrating once, we have the equation

$$\int W \, dt = \Delta H + \int \frac{T - T_A}{R_T} \, dt + \int C \, dT$$

where ΔH is the quantity to be determined. The left-hand side of the equation can be determined electrically with as much precision as necessary, and the value of C can be found by measurements on the empty container. The remaining term, which describes the heat exchange between the sample container and the enclosure, represents the principal source of error in adiabatic calorimetry. In fact, control of this heat exchange is the major problem in all types of calorimetry.

There are three mechanisms by which heat may be transferred to the enclosure if T and T_A are not equal. The most complicated mechanism is convection in the medium filling the space between the sample container and the enclosure. Heat transfer by convection is a nonlinear function of the temperature differential and the absolute temperature, and is geometry-dependent. It can be minimized by operation at reduced pressure.

Heat is also transferred by conduction through the supports and electrical connections for the sample container, and through the medium surrounding it. Thermal conductivities are temperature-dependent, and therefore these losses also are nonlinear with respect to temperature and differential temperature.

Finally, heat may be transferred by radiation. This effect may be minimized by using low-emissivity surfaces on the sample container and on the enclosure walls, and is not difficult to suppress at low temperatures. However, radiation heat transfer is the principal limitation in high-temperature calorimetry.

These heat-transfer processes together define R_T, the incremental thermal resistance, for a given calorimeter design. In a true adiabatic calorimeter, heat is supplied to the enclosure to keep its temperature as close as possible to the sample temperature. The temperature differential $(T-T_A)$ is recorded during an analysis so that the data may be corrected for heat exchange which takes place due to imperfect tracking.

Bomb Calorimetry. The bomb calorimeter is used for measuring heats of combustion of organic compounds. It is of great commercial importance, and the apparatus design and operating procedure have been so standardized that a precision of 0.01% can be realized.

The sample rests in a stout metal container, which is filled with oxygen to a pressure of 30 atmospheres. The container is immersed in 2.5 liters of water in the calorimeter vessel, which is supported inside a constant-temperature water jacket. The heat of combustion of the sample is determined from the rise in temperature of the calorimeter; the instrument is calibrated by burning a sample of benzoic acid, supplied by the National Bureau of Standards, under the same conditions. Unlike most calorimetric measurements, bomb calorimetry takes place at constant volume, and the data must be corrected accordingly.

Isothermal Calorimetry. In isothermal calorimetry, the sample material is surrounded by an enclosure whose temperature is independent of the thermal behavior of the sample. The heat content of the sample, therefore, becomes the dependent variable.

The greatest precision is achieved with the isothermal-phase-change calorimeter, invented by Bunsen in 1870. The sample rests in a well inside a closed vessel filled with a working fluid such as water. If the reaction to be measured is exothermic, the calorimeter is prepared by withdrawing heat from the well so that a mantle of ice forms around it. As the water freezes, the volume of the water-ice system increases, displacing mercury from a pool below the water to an external precision-bore capillary tube where the height of the mercury column can be measured.

When the reaction is initiated, the process is reversed and the drop in the mercury level is proportional to the heat released. The instrument is calibrated by dropping a known weight of a standard material, at a known temperature, into the calorimeter well. Errors due to heat exchange are minimized by placing the calorimeter inside a second vessel containing the same fluid, also at the phase-change temperature.

Frequently, the Bunsen calorimeter is used with diphenyl ether instead of water; the working temperature, 300.02°K, is conveniently close to room temperature and the displacement of mercury per calorie is three times as great.

If the sample temperature is to be changed, as in a heat content measurement, the sample is rapidly transferred into the calorimeter from a high-temperature furnace; the entire apparatus is referred to as a drop calorimeter. The sample is usually enclosed in a thin-walled platinum container, and the measurement is repeated with the empty container to correct for heat exchange during transit.

Drop calorimetry is the only method routinely used for heat content measurements above 1000°K.

Scanning Calorimetry. In this recently developed technique, the temperature of the sample is varied in a reproducible manner, while the heat flow rate into the sample is measured. In the usual case of a linear variation of temperature with time, the heat flow rate is proportional to the instantaneous specific heat of the sample.

Two identical sample holders, one containing the material under study, are mounted inside a constant-temperature enclosure. Each sample holder is equipped with a platinum resistance thermometer and a heating element. The sample holder temperatures T_S and T_R are continuously measured and compared with an indicated temperature T_P. The temperature differences are amplified and used to control the heating elements so that the sample holder temperatures are equal to each other and to T_P, while the difference between the power requirements of the two heating elements is recorded.

The sample holders are very small, with a capacity of approximately 50 μl; the temperature can therefore be varied at rates as high as 1 degree per second without introducing calorimetric errors due to dynamic temperature gradients.

To make a specific heat measurement, the sample material is placed in one of the sample holders and the temperature is programmed through the interval of interest. The sample is then removed and the procedure is repeated with empty sample holders; the thermal capacity of the sample is given by the change in differential power divided by the program rate. The instrument is calibrated with a specific heat standard such as aluminum oxide.

Scanning calorimetry is uniquely suitable for the examination of reversible transitions and strongly exothermic behavior; since the sample holders are hotter than the enclosure, and thus transferring heat to it, it is possible to program the sample temperature in either direction, and to remove heat from the sample as fast as it is generated.

Descriptions of many other calorimeters are to be found in the literature, but the four types described here are of most importance at the present time. Generally, the trend is toward instrumentation for high-speed small-sample calorimetry, such as the stopped-flow calorimeter, or the scanning calorimeter described above. Calorimetry is coming to be regarded as an essential analytical method for routine work, rather than the difficult and time-consuming laboratory procedure that it used to be.

M. J. O'NEILL

Cross-references: *Thermochemistry, Thermodynamics*

References

Porter, R. S., and Johnson, J. F., *"Analytical Calorimetry,"* Plenum Press, New York, 1968.

Gray, A. P., *American Laboratory,* **3,** 43 (1971).

Westrum, E. F., Jr., Hatcher, J. B., and Osborne, D. W., *J. Chem. Phys.,* **21,** 419 (1953).

CARBIDES

General Binary Compounds of Carbon. The term "carbide" has a content of popular meaning which does not correspond to any natural classification.

For the purpose of this article the term "carbide" is defined as a binary compound of carbon and any other element. The elements of the major portion of the Periodic Table form carbides according to this definition. Broad as this definition is, it is still too narrow in certain instances. Thus the compound cohenite, $(Fe,Ni)_3C$, is isomorphous and closely related to cementite, Fe_3C, but definitely is not included in the above definition because this compound is ternary. In other cases the definition may seem too broad, as it regards liquid carbon tetrachloride, CCl_4, or gaseous carbon monoxide, CO, as carbides.

Two general factors have central importance for the determination of the chemical and physical properties of the carbides. The first is the electronegativity difference between carbon and the element in question, as defined by Pauling. The second factor is the presence or absence of an uncompleted electronic "d" or "f" shell; i.e., whether or not the element in question belongs to one of the transition series. These two factors form the basis for the method of classification here employed. It emphasizes the characteristic differences among the carbides and groups together those carbides with common properties.

Salt-like carbides. These carbides are characterized by a large electronegativity difference between carbon and the element involved, ranging between

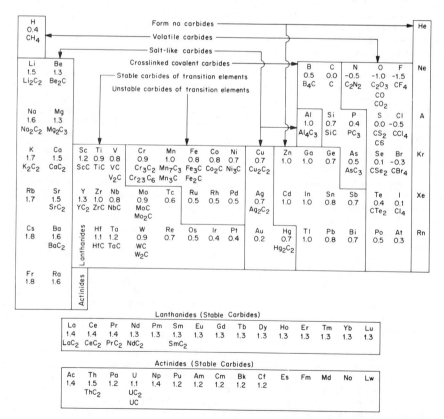

FIG. 1. Classes of the carbides, characteristic formulas, and the electronegativity difference of the carbon-element bond as related to the Periodic Table. (Courtesy of Bureau of Mines, U.S. Department of the Interior.)

+ 1.8 to + 1.0. They react with water to form the hydrocarbon corresponding to the chain length and to the valence of the carbon assembly within the crystal. Thus we have the methanides Be_2C and Al_4C_3 containing methanide ions which react to form pure methane according to the reactions:

$$Be_2C + 4H_2O \rightarrow 2Be(OH)_2 + CH_4\uparrow \quad (1)$$

$$Al_4C_3 + 12H_2O \rightarrow 4Al(OH)_3 + 3CH_4\uparrow \quad (2)$$

The acetylides containing acetylide ions in the crystal structure are widely represented in the Periodic Table among the alkali metals and alkaline earth metals. They have the formulae Me_2C_2 and MC_2. The reactions of sodium carbide and calcium carbide with water occur violently:

$$Na_2C_2 + 2H_2O \rightarrow 2NaOH + C_2H_2\uparrow \quad (3)$$

$$CaC_2 + 2H_2O \rightarrow Ca(OH)_2 + C_2H_2\uparrow \quad (4)$$

Finally, there is the very remarkable compound *magnesium carbide* which presumably contains the C_3^{+4} ion since on hydrolysis it yields methyl acetylene according to the reaction:

$$Mg_2C_3 + 4H_2O \rightarrow$$
$$2Mg(OH)_2 + CH_3{-}C{\equiv}CH\uparrow \quad (5)$$

When acetylene is passed through a solution of sodium in liquid ammonia a sodium acetylide is formed according to the reaction:

$$2Na + 2C_2H_2 \rightarrow 2NaHC_2 + H_2\uparrow \quad (6)$$

Similar reactions occur for the liquid ammonia solutions of potassium and lithium. On heating the sodium acid acetylide to 180°C, sodium acetylide is formed:

$$2NaHC_2 \rightarrow Na_2C_2 + C_2H_2\uparrow \quad (7)$$

The action of the electric arc on lithium carbonate yields lithium acetylide. The other alkali metals do not form acetylides by the direct reaction with carbon even at high temperatures.

The acetylides of the alkaline earth metals are in general prepared by the direct action of the element or one of its compounds on carbon at high temperatures in an electric arc or blast furnace. Specifically carbon and calcium oxide form calcium carbide at 1800°C. Calcium acid acetylide can be formed by the action of acetylene on the solution of calcium in liquid ammonia. It too will decompose into the dibasic acetylide by heating. Magnesium carbide, MgC_2, is obtained as the product of the reaction between magnesium and ethane at 450–500°C, but the carbide of the formula Mg_2C_3 is best obtained as the product of the reaction between magnesium and pentane at 650°C.

The unstable salt-like carbides are represented by the three compounds of Cu_2C_2, Ag_2C_2, and Hg_2C_2. They can be regarded as salts of acetylene and are synthesized by the reaction of acetylene with aqueous solutions of cuprous, argentous, and mercurous salts. By hydrolysis the acetylene can be again regenerated. These carbides differ from the other salt-like carbides in that they are explosive and cannot be prepared by the direct action of carbon on the metal. The difference may be ascribed to the low electronegativity difference between carbon

and copper, silver and mercury which is of the order of 0.7. The bonds thus have very little ionic character. Silver acetylide and cuprous acetylide are used commercially as detonators.

Cross-linked Covalent Carbides. Carbides falling within this category have electronegativity differences of 0.7 to 0.0. They are represented by boron carbide B_4C, silicon carbide, SiC, diamond, and possibly the carbides of arsenic and phosphorus PC_3 and AsC_3. The first three at least are composed of crystals which are covalently linked together so that every crystal represents a single giant molecule in the same sense that the word molecule is used in organic chemistry. This fact accounts for the tremendous hardness of these compounds. Thus B_4C has a hardness of 9.5 on the Mohs scale of hardness, and SiC, 9.0. On the same scale sapphire or corundum has a hardness of only 9.0.

Boron carbide reacts rather readily at 900–1000°C with chlorine gas to form boron trichloride. This has made it useful in the past as the starting point of total syntheses of boron compounds. Recently elemental boron has become commercially available for this purpose. Boron carbide can be synthesized by the interaction of boron oxide, B_2O_3, and carbon at 2500–2600°C. *Silicon carbide* is made commercially from silica and carbon at temperatures of 2200 to as high as 3000°C. Until recently *diamonds* were obtained only as a natural mineral. Small diamonds have been produced synthetically under controlled and reproducible conditions at pressures of 800,000 psi and advanced temperatures. Unusual hardness makes these compounds very important commercially as cutting, wear resistant, and abrasive materials. (See **Boron and Compounds**).

Volatile Covalent Carbides. The volatile covalent carbides are closely related to the crosslinked covalent carbides.

Cyanogen, C_2N_2, is a colorless poisonous gas. The chemistry is very similar to that of the halogens. At 400°C it polymerizes to form a solid paracyanogen, $(CN)_x$. This tendency to polymerization shows the relation of cyanogen to the crosslinked covalently bound carbides.

Carbon monoxide. (See **Carbon Monoxide**).

Carbon dioxide. (See **Carbon Dioxide**).

Carbon disulfide is a heavy, colorless, flammable, poisonous liquid. It reacts with chlorine to form carbon tetrachloride in commercial quantities. It is formed by the direct union of carbon and sulfur in the electric furnace.

Carbon diselenide has been prepared by the action of hydrogen selenide on carbon tetrachloride, forming carbon diselenide and hydrochloric acid.

The evidence for carbon ditelluride is somewhat unsatisfactory. The compound is thought to be formed when an arc is formed between tellurium and graphite electrodes under carbon disulfide.

The carbon tetrahalides are characterized by chemical inertness which can be related to their approximation of the electronic configuration of the rare gases. The stability of these halides decreases as one passes from CF_4 to CI_4 in the Periodic Table. *Carbon tetrafluoride* is a colorless inert gas. Commercially it is formed in the preparation of aluminum during the electrolysis of cryolite. *Carbon tetrachloride* is a colorless, heavy, inert liquid widely used as a solvent and fire extinguisher. *Carbon tetrabromide* and *carbon tetraiodide* are heavy inert liquids often used as standards for the refractive index measurement of high-refractive index numerals.

Stable Carbides of the Transition Elements. These carbides possess metal atom arrangements of either the face-centered cubic or hexagonal close-packed type. The carbon atoms in these compounds are located in the interstices between the metal atoms. The carbides of the formula MC generally possess the former structure and those with the formula M_2C the latter. In addition to these, there are carbides of the formula MC_2 which seem to form

free carbon or to the more stable but more complex carbides such as the Hägg iron carbide and cementite. The complexity of the latter carbides is shown by their large lattice parameters which are measures of the distance which must be traced through the crystal before the motif repeats itself. The reactions of the iron carbides are essential to the tempering reactions in carbon steel. The iron carbides tend to form as the solid solution of carbon in face-centered cubic iron (austenite) becomes unstable and changes to a solid solution of carbon in a distorted lattice of body-centered cubic iron (martensite) at 723°C. The sequence of tempering in carbon steel may then be regarded as

$$\text{austenite} \rightarrow \text{martensite} \tag{8}$$

$$\text{martensite} \rightarrow \text{epsilon iron carbide } (\epsilon Fe_2C) + \text{alpha iron } (\alpha Fe) \tag{9}$$

$$\text{epsilon iron carbide } (\epsilon Fe_2C) \rightarrow \text{chi iron carbide } (\chi Fe_2C) \tag{10}$$

$$\text{alpha iron } (\alpha Fe) + \text{chi iron carbide } (\chi Fe_2C) \rightarrow \text{theta iron carbide } (\theta Fe_3C) \tag{11}$$

only with elements of either the lanthanide series (rare earths) or actinide series of elements. The MC_2 carbides have crystal structures closely related to that of calcium carbide in that each occupied interstice contains two carbon atoms.

The stable carbides are characterized by exceedingly high melting points, unusual tensile strength and hardness, and considerable inertness to acids. The type of bonding responsible for these properties has been the subject of considerable speculation.

The actinide and lanthanide carbides, UC_2, ThC_2, LaC_2, NdC_2, and probably others, upon treatment with water or acids produce some acetylene but mainly hydrogen and more saturated hydrocarbons. The C_2^{++} group in these compounds cannot therefore be regarded as a true acetylide ion, or alternately the hydrogenation power of the rest of the compound must be such that acetylene cannot be liberated as such. The fact that divalent ions of the rare earth elements are not ordinarily stable in aqueous solutions may also be involved.

The uranium carbides, UC and UC_2, can be synthesized by the action of carbon on uranium metal at 2100 and 2400°C, respectively. UC can also be formed by the action of methane on very finely divided uranium (prepared by the decomposition of uranium hydride) at 450°C. Most of the other refractory carbides can be formed by the direct combination of the metal and carbon at about 2200°C.

Commercially, the hardness of some of these compounds is employed in carbide cutting tools where they are generally used in a matrix of cobalt metal. These compounds are also formed in tool steels where their presence insures the hardness and the hot-work characteristics of these steels.

Unstable Carbides of the Transition Elements. Most of these carbides have unusual and complex structures, although a few have the same structures as the stable carbides. Examples of the latter are the hexagonal close-packed carbides of iron and nickel, Fe_2C and Ni_3C, respectively. The instability is the result of the interstices between the metal atoms being too small to accommodate the carbon atoms. As might be expected, these are thermally highly unstable, and decompose either to the metal and

This sequence of reactions may also be observed in the tempering of finely-divided carburized iron formed in catalytic reaction and in the corrosion products produced by liquid sodium on various steels.

Nickel carbide and cobalt carbide appear to be formed as epitaxial inclusions in synthetic diamond when the elements are used as catalysts in the diamond synthesis. Iron and manganese carbides are found in the reaction products along with synthetic diamonds when these elements are used as catalysts or carbon carriers. Palladium and platinum have also been used successfully in the diamond synthesis. These elements are, therefore, also capable of forming carbides at high pressures as suggested by the classification in Fig. 1.

Graphitic Compounds. While not true carbides in the sense that no chemical bond of either the ionic or covalent type exists between the carbon and the other element, graphitic compounds should receive mention here. Graphite itself consists of sheets of catacondensed aromatic rings. Between these sheets potassium metal can intrude causing a swelling of the graphite in the direction perpendicular to the sheets. The bonding between the carbon and the potassium is of Van der Waals type. The graphite can be regenerated by distilling off the potassium. Graphite is known to form other similar interlayer compounds with other reagents.

L. J. E. Hofer

References

Sneed, M. Cannon, and Brasted, Robert C., "The Metallic Borides, Carbides, Silicides and Related Compounds. Chapter VIII., Volume VII., The Elements and Compounds of Group IV." "A Comprehensive Inorganic Chemistry." D. Van Nostrand Company, Incorporated, Princeton, New Jersey, 1958.

Anderko, K., and Hansen, M., "Constitution of Binary Alloys," 1958. First Supplement, R. P. Elliott, 1965. Second Supplement, Francis A. Shunk, 1969, McGraw-Hill, Incorporated., New York.

Kieffer, Richard, and Schwarzkopf, Paul, "Refractory Hard Metals; Borides, Carbides, Nitrides, Silicides," The Macmillan Company, New York, 1953.

Miller, S. A., "Acetylene, Its Properties, Manufacture and Uses," Academic Press, New York and London, 1965.

Hofer, L. J. E., "Nature of the Iron Carbides," Bureau of Mines Bulletin, **631,** 60 pp. (1966).

Refractory Carbides

These are characterized by great hardness, thermal stability, and chemical resistance. They include the carbides of boron and silicon with those of the Group IVB, VB, and VIB transition metals. The latter are interstitial compounds, the carbon atoms occupying interstices of the metal crystal lattice. Thus, chemical valence and stoichiometry considerations do not apply. The compositions and attendant properties, particularly of the Group IVB carbides, can vary over broad ranges. Other defined carbide phases of this class are formally V_2C, Nb_2C, Ta_2C, Cr_7C_3, $Cr_{23}C_6$, Mo_3C_2, W_2C, and W_5C_3.

Due to their high melting points, the refractory carbides are generally synthesized by solid state reactions, and are obtained as metallic appearing powders which conduct electricity. The purest specimens are prepared by heating precisely proportioned mixtures of carbon and metal powder at high temperatures for prolonged periods in vacuum or inert atmosphere. This process is used industrially for WC and Mo_2C. The other carbides are manufactured by reducing the metal oxide with excess carbon at electric furnace temperatures. Silicon carbide, which sublimes by dissociation above 2700°C, forms macro crystals from the vapor phase in this reaction. Protective carbide coatings can be applied by the thermal reduction of metal halide vapors with methane and hydrogen. Carburization rates are increased and reaction temperatures are lowered by employing a menstruum, solid or molten, in which both reactants are soluble. Thus, carburization of a ferroalloy, e.g., ferrotantalum, followed by acid leaching to separate the iron yields carbide powders in impure state. Aluminothermic and fused salt electrolytic processes have also been used.

Although relatively inert chemically, the carbides are all decomposed by fusion with alkali, and they are dissolved by mixtures of nitric and hydrofluoric acids. Chlorination at red heat yields the anhydrous metal chloride. Only SiC and Cr_3C_2 exhibit substantial oxidation resistance at high temperatures.

Cemented Carbides. Cutting edges of tools and wear-resistant parts of great strength and hardness are produced by consolidating the refractory metal carbide powders with bonding metals of the iron group, e.g., Fe, Co, and Ni. The exceptional industrial importance of WC is due to the unique properties which are attained by bonding with cobalt. The WC powder is milled with Co powder, and the mixture is pressed to shape in steel dies. The compacts are then heated in hydrogen or vacuum for an hour at temperatures of the order of 1400°C. Porosity is eliminated and the WC grains are recrystallized through the medium of a molten Co–WC eutectic. Various grades may contain from 3% to 25% of cobalt, the toughness increasing with cobalt content while hardness decreases. Compressive and transverse rupture strengths of 750,000 and 250,000 pounds per square inch respectively are readily obtained. Such cemented carbide tools permit cutting speeds in rock or metal 100 times as great as can be obtained with alloy steel tools. Enhanced resistance to cratering, the erosion of the tool by metal chips, is provided by small additions of TiC and NbC. Such multicarbide compositions are solid solutions of the component carbides.

Consolidation of carbide powders by hot pressing in electrically heated graphite dies produces hard but brittle parts. Gage blocks of B_4C are produced this way. Porous, sintered shapes can be strengthened by infiltration with molten metals. Thus, TiC compacts are bonded with an Fe-Ni-Cr alloy to provide high-temperature strength and oxidation resistance. TiC and Cr_3C_2 are important constituents of ceramic-metal structures (cermets).

M. L. FREEDMAN

Cross-references: *Carbon, Carbon Dioxide, Carbon Monoxide, Acetylene, Refractories.*

CARBOHYDRATES

Carbohydrates are the most abundant class of organic compounds. They constitute three-fourths of the dry weight of the plant world and are widely distributed, often as important physiological components, in animals and lower forms of life. In plants and animals they serve mainly as structural elements or as food reserves. Plant carbohydrates, in particular, represent a great storehouse of energy either as food for men and animals, or after transformation in the geological past as coal and peat. Large industries process such carbohydrates as sucrose, starch, cellulose, pectin and certain seaweed polysaccharides.

The term carbohydrate originated from the belief that this class of compounds consisted of hydrates of carbon because elemental analysis of common carbohydrates, such as lactose, sucrose, starch and cellulose, led to the empirical formula $C_x(H_2O)_y$. Although the formula represents the majority of carbohydrates, many have compositions which do not fit such a simplified generalization. While it is not possible to give a simple yet comprehensive definition of such a broad group of compounds, one fairly good definition describes them as compounds of carbon, hydrogen and oxygen which contain the saccharose group

$$\left[\begin{array}{c} H \\ | \\ -C-C- \\ | \ \ || \\ OH\ O \end{array} \right]$$

or its first reaction product, and which usually have the hydrogen and oxygen in the ratio found in water.

Most carbohydrates are hydroxyaldehydes or hydroxyketones or substances producing them by hydrolysis. Glycoaldehyde ($HOCH_2CHO$) is considered the simplest carbohydrate, but glyceraldehyde, (glycerose, $HOCH_2 \cdot CHOH \cdot CHO$), is more representative because it contains an asymmetric carbon atom and is optically active. All other aldehydic carbohydrates contain one or more asymmetric carbon atoms and influence a beam of polarized light according to the rules of stereochemistry. Because glycerose contains an aldehyde group it is classified as an *aldose*. Since it contains three carbon atoms it is an *aldotriose*. There are similar aldose molecules with three, four, five or more alcohol groups for each aldehyde group. Such molecules have four, five and six or more carbon atoms respectively and are therefore aldotetroses, aldopentoses, aldohexoses, etc. Common aldopentoses are xylose and arabinose while common aldohexoses are glucose, mannose and galactose. Another group of carbohydrates is ketones with dihydroxyacetone ($HOCH_2 \cdot CO \cdot CH_2OH$) as the simplest member. These substances are *ketoses*. Dihydroxyacetone is a ketotriose. There are also ketotetroses, ketopentoses, ketohexoses and higher members. All these, except the ketotriose, contain asymmetric carbon atoms and are optically active. Common ketohexoses are frustose and sorbose.

All the above carbohydrates contain one saccharose group and are classified as *monosaccharides.*. More complex carbohydrates that break up on hydrolysis to produce two monosaccharides are called *disaccharides*. Still larger carbohydrates are classified, depending on the number of monosaccharides each molecule produces, as *trisaccharides, tetrasaccharides,* and so on. *Oligosaccharides* is a collective term given to carbohydrates which contain two to ten monosaccharide units. All carbohydrates which contain more than ten monosaccharide units are classed as *polysaccharides;* typical examples are starch, pectin, glycogen and cellulose.

All the monosaccharides and many of the oligosaccharides are called *sugars*. Frequently the monosaccharides are called "simple" sugars. The sugars are readily soluble in water but vary greatly in their sweetness. The sweetest, fructose, is about 1.5 times as sweet as sucrose (table sugar) and about 3 times as sweet as glucose (corn sugar). Other sugars are less sweet, with some exhibiting a barely noticeable sweetness.

When the sugar structural formulae are written in the vertical position with the aldehyde (or ketone) function at the top, half will have the hydroxyl group next to the primary alcohol group lying on the right side of the molecule as with dextrorotatory glycerose, and half will have this penultimate hydroxyl group lying on the left as in levorotatary glycerose. The former are therefore classified to the D-series and the latter classified as L-series. It should be understood that these symbols denote configurational relations only and do not indicate the optical rotation of the molecule.

Because of the presence of asymmetric carbon atoms, numerous stereoisomeric monosaccharides exist. Although the structures of carbohydrates are often written as acyclic compounds, they usually occur as cyclic compounds formed by the carbonyl group condensing with the hydroxyl group on carbon C_4 or C_5. These forms are designated as furanose and pyranose, respectively. Because of the cyclization, carbon atom Cl is an asymmetric center which gives rise to two anomers. When the hydroxyl group of the anomeric carbon atom is in the same direction as the penultimate hydroxyl group the sugar is designated as the α-form. On the other hand, when the hydroxyl group is in the opposite direction the sugar is the β-form. All these forms, together with the acyclic form, coexist in an equilibrated, aqueous sugar solution.

Monosaccharides. The aldotrioses, D-glycerose, and dihydroxyacetone are very important biochemically since they occur as integral parts of the glycolytic cycle of carbohydrate metabolism in both plants and animals. None of the aldotetroses, D- or L-threose or D- or L-erythrose is common or of much importance.

D-Xylose is rarely found free in nature but is abundant as units in the polysaccharide xylan found in various plants. It is freed by subjecting corn cobs or seed brans to hydrolysis by hot dilute mineral acid. Concentration of the hydrolyzate and purification with carbon causes the sugar to crystallize. Subjecting corn cobs, brans or other plant tissues high in xylan content to hot 12% HCl causes conversion of the D-xylose and some other substances (glycuronic acids) to *furfural*. The process can be used for quantitatively measuring the amounts present of pentose containing polysaccharides (pentosans). It is also used commercially for manufacture of furfural.

L-Arabinose occurs in nature, sometimes in polysaccharides combined with D-xylose units but more often in polysaccharide gums and as a polysaccharide component of pectin. It may be prepared by mild acid hydrolysis of such plant gums as mesquite, arabic or cherry. D-Arabinose does not occur in nature but may be prepared by appropriate degradation of D-glucose. D-Ribose is an important constituent of nucleic acids which are found in all plant and animal cells. Ribonucleic acids are usually found in the cytoplasm whereas 2-D-deoxyribose nucleic acids are found in the nucleus. 2-D-Deoxyribose has the hydroxyl group on carbon atom C2 replaced with a hydrogen atom. Lyxose does not occur in nature.

D-Glucose is the most abundant and important of the aldohexoses. It occurs free in fruits, plant juices, honey lymph, cerebrospinal fluid, urine and blood. Normal human blood contains 70 to 100 mg. of D-glucose per 100 ml. D-Glucose is taken into the blood from the intestines as a result of digestion of sucrose, starch and other carbohydrates. It is delivered to the body cells and muscles to supply energy through enzymatic glycolysis and the citric acid cycle. Solutions of D-glucose are sometimes used for intravenous feeding of hospital patients. The sugar is produced commercially in large amounts by hydrolysis of corn starch. It is crystallized and sold as "dextrose" but the largest amount is sold in solution as corn sirup. For this purpose the hydrolysis of the starch is stopped short of completion so that the hydrolyzate contains besides D-glucose various oligosaccharide fragments, resulting from incomplete depolymerization of the starch molecules. Their presence in the sirup prevents crystallization of the D-glucose and hence promotes sirup stability. D-Glucose is also a constituent of sucrose (table sugar).

D-Galactose is a constituent of the disaccharide lactose, from which it may be prepared by hydrolysis. The sugar may also be obtained in good yield by hydrolysis of numerous polysaccharides which contain D-glactose in varying amounts. Some such as agar and carrageenan are composed almost entirely of D-glactose units. From such polysaccharides L-glactose may sometimes be obtained in small amounts.

D-Mannose is widely distributed in nature as a repeating unit in polysaccharides known as mannans.

Among the ketohexoses D-fructose (levulose) is most abundant and important. It occurs with D-glucose in honey and in "invert sugar" which is the product of sucrose hydrolysis. Sucrose is dextrorotatory and the hydrolyzed mixture is levorotatory; hence the name *invert sugar*. D-Fructose can be separated from invert sugar, but in the laboratory it is made by hydrolysis of one of its numerous polymers, polysaccharides termed fructans which are commonly found in plants as reserve foods. Inulin from dahlia tubers is a convenient source. L-Sorbose, an intermediate in the synthesis of ascorbic acid (vitamin C), is prepared by the bacterial oxidation of sorbitol.

Commercial D-fructose is now being produced in large quantity by isomerization from D-glucose using the enzyme isomerase. It is mainly sold in its equilibrium mixture with D-glucose as a sweet sirup for use in the food industry and is widely used because it is cheaper than invert sugar.

Of the four naturally occurring "methylpentoses" or 5-*C*-methylaldopentoses, L-rhamnose or 6-deoxy-L-mannose is the commonest. It is obtained by hydrolysis of quercitrin.

Disaccharides. Disaccharides are formed by condensation of two monosaccharides through one or both of the carbonyl groups. The new linkage is labile to acid and often to alkali. Acid hydrolyzes the disaccharide to its constituent monosaccharides. Disaccharides which have a free carbonyl group show reducing properties. The more important members of this group are maltose, cellobiose and

lactose. Maltose is readily prepared by the partial acid or enzymatic (diastase) hydrolysis of starch. Cellobiose is obtained by the partial acid hydrolysis or acetyolysis of cellulose, whereas lactose, sometimes called milk sugar, is usually obtained by evaporation of whey whereby crystalline lactose is deposited.

Those disaccharides in which both of the carbonyl groups are involved in the linkage have only slight reducing properties. The two most important nonreducing disaccharides are sucrose and trehalose. *Sucrose* is obtained commercially from sugar cane or sugar beet. The sucrose is extracted by crushing and extraction with water. Purification is by heating the extract with calcium hydroxide which causes the impurities to separate either as a scum or a deposit. Concentration of the clear solution yields crystalline "raw" sugar. Table sugar is obtained by recrystallization of the raw sugar. The mother liquors from this process deposit less pure crystals known as "brown" sugar. α,α-Trehalose is found in rye ergot and young mushrooms, but the best sources for its isolation are trehala manna, an exudant of certain insects found in Syria, and the *Selaginella Cepidophylla* plant of southwestern United States.

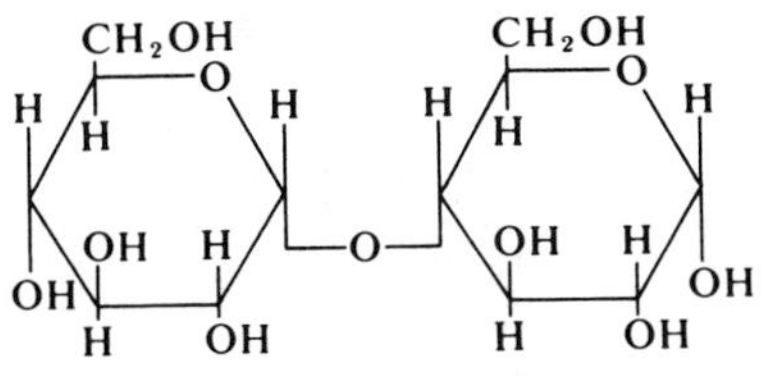

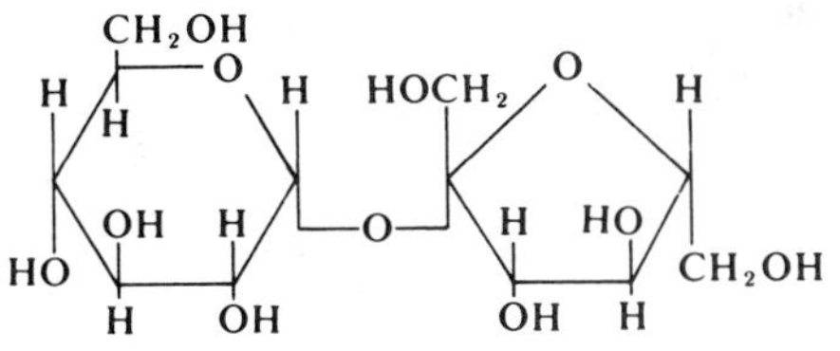

Sucrose

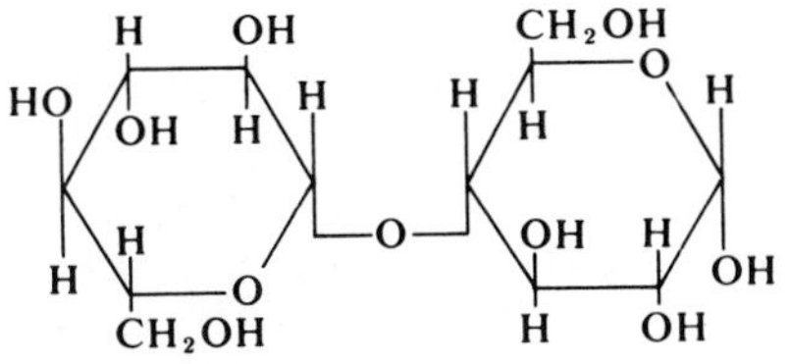

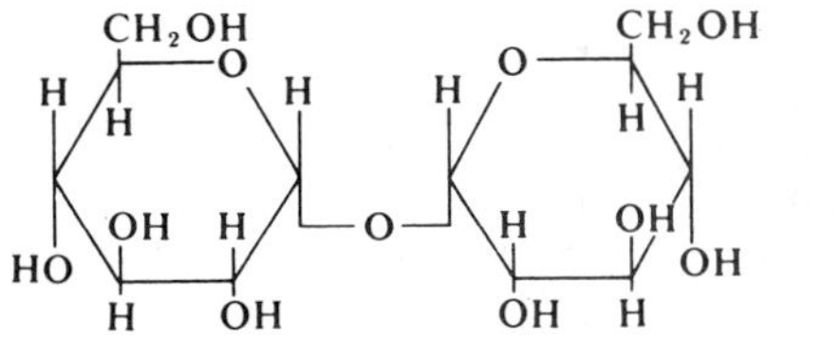

α,α-Trehalose

Oligosaccharides. The commonest oligosaccharides found in nature are derivatives of sucrose.

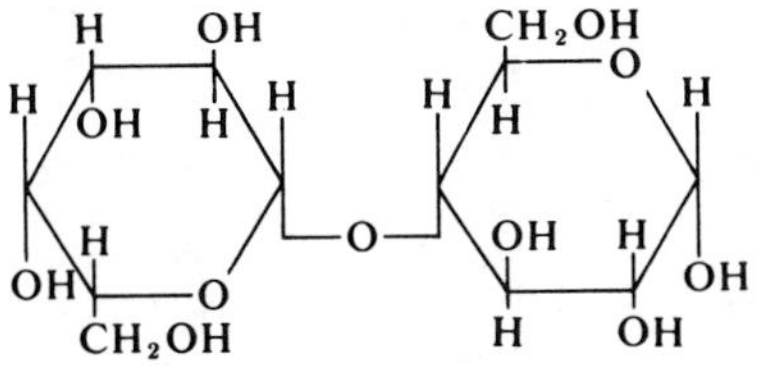

Polysaccharides. The polysaccharides are widely distributed in the plant and animal worlds, serving as food reserve substances and structural material. The two best known polysaccharides are starch and cellulose, which consist of D-glucopyranosyl units linked by α- and β-1 → 4 bonds, respectively. Similar to cellulose is xylan, a polysaccharide consisting of β-1 → 4 linked D-xylopyranosyl units. This polysaccharide occurs in practically all land plants and in some marine algae. Although xylan is difficult to extract from cellulose, there is no positive evidence for covalent linkages between the two polysaccharides.

Also similar in many ways to cellulose is *chitin,* the most abundant of the polysaccharides containing aminosugars. It is the principal structural component of the shells of insects, crabs, and lobsters, and is also found in lower plants such as the mycelia and spores of fungi.

Maltose

Cellobiose

Lactose

Polysaccharides may also be built of uronic acid units. For instance alginic acid, a constituent of the cell walls of most brown algae, is a linear polymer made up of D-mannuronic and L-guluronic acid units united by β-1 $\to$ 4 linkages. The presence of the uronic acid group stabilizes the glycosidic linkage to acid hydrolysis. Treatment with 19% hydrochloric acid at 145° for 2 hours causes decarboxylation and the theoretical amount of carbon dioxide is evolved. During the last decade alginic acid has become an important commercial product and is used extensively in the food industry. Since the calcium salt of alginic acid can form strong fibers which are soluble in mild alkali, it is used in the textile industry for weaving special fabrics such as imitation lace and furs. The alginic acid fibers are used as a frame-work during weaving and then are dissolved out by alkali.

Certain polysaccharides occur as esters of sulfuric acid. The most important, commercially, is the polysaccharide mixture carrageenan, extracted from red seaweed and consisting mainly of kappa and lambda carrageenans containing 1→3 and 1→4 linked D-galactopyranosyl and in the former polymer 3,6-anhydro-D-galactopyranosyl units with most units having a monoesterified sulfuric acid group. The kappa component forms a gel with potassium ions. Carrageenan is widely used as a stabilizer in the food industry.

Reactions of Carbohydrates. Carbohydrates which contain a free carbonyl group have reducing properties typical of those of the simple aldehydes. The carbonyl group will also undergo condensation with amines, hydrazines, mercaptals and hydrogen cyanide to form derivatives which are useful for identification. Aldehyde groups are quantitatively oxidized to a carboxyl group by bromine or alkaline iodine. This reaction is used extensively for the determination of aldoses in the presence of ketoses and for molecular weight determinations. Carbonyl groups may also be reduced to alcohols.

The polyhydric nature of the carbohydrates is illustrated by the formation of esters and ethers. Esters of carbohydrates are usually prepared by the action of inorganic acids, organic acids or anhydrides of the latter in the presence of an impelling agent such as sulfuric acid, phosphoric acid or pyridine. The presence of ester groups reduces the solubility of the carbohydrate in water but increases the solubility in organic solvents. Carbohydrate ethers are usually prepared by the action of alkyl halides or sulfates on a sodium hydroxide dispersion of the carbohydrates. Complete etherification is achieved only after repeated treatment. The ethers and esters of polysaccharides, especially of cellulose and starch, are of fundamental commercial importance.

Hydroxyl groups of carbohydrates are readily oxidized to acids, aldehydes and ketones. The mechanism of oxidation is quite complex and only with lead tetraacetate and periodic acid is the reaction specific. In these cases vicinal hydroxyl groups (α-glycol groups) are cleaved to give carbonyl groups. Glycol groups containing a primary alcohol group give rise to formaldehyde whereas three vicinial secondary hydroxyl groups give formic acid. α-Hydroxycarboxylic acids react slower and produce carbon dioxide. The quantitative

study of lead tetracetate and periodate oxidation of carbohydrates is most valuable in structural determinations. Hypochlorite and hydrogen peroxide are used commercially for modifying the properties of polysaccharides such as starch. The mechanism of oxidation and the type of products produced are still unknown. A general characteristic of oxidized carbohydrates is their sensitivity to alkali.

ROY L. WHISTLER

Cross-references: *Asymmetry, Sugars, Cellulose, Starches, Foods, Nutrition.*

CARBON

The element carbon, C, atomic number six, atomic weight 12.011, is the first of five elements located in Group IVA of the Periodic Table, the other members being silicon, germanium, tin and lead.

Carbon is unique among the elements because it forms a vast number of compounds, more than the total of all other elements combined, with the exception of hydrogen. It exists in three allotropic forms, namely, diamond, graphite and amorphous carbon. Diamond and graphite are naturally occurring crystalline solids possessing widely divergent properties, whereas amorphous carbon comprises a comparatively large variety of carbonaceous substances not classified as either diamond or graphite.

Prevalence. Carbon ranks nineteenth in the order of abundance of the elements comprising about 0.2% of the earth's outer crust. In the earth's atmosphere carbon is present in amounts up to 0.3% by volume as carbon dioxide. Though it is widely distributed in nature, mainly in the combined form, only minor amounts are found in the free or elemental state. It is present as a principal constituent of all animal and vegetable matter. Coal, petroleum and natural gas are also composed essentially of carbon. Various minerals, such as limestone, dolomite, and marble, as well as certain marine deposits such as oyster shells all contain carbon in the form of carbonates.

Carbon plays a vital role in what is known as the carbon or life cycle. Carbon dioxide from the air, together with water, is absorbed by plants and converted into carbohydrates in the process of photosynthesis. Animals consume the carbohydrates, returning the carbon dioxide to the atmosphere by the process of respiration, excretion, fermentation and decay under bacterial action.

Diamond. Though diamonds have been discovered on all the major continents, over 90% of the world's natural diamond production is from Africa. Other significant producers are the South American countries of Brazil, British Guiana and Venezuela, and most recently Siberia of the U.S.S.R.

Diamonds are most frequently found imbedded in "volcanic pipes" of a relatively soft, dark, basic peridotite rock called "blue ground" or "kimberlite," from which they are mined. While this is considered the primary source, diamonds are also found in alluvial deposits. To recover the diamonds from the blue ground or alluvial gravel, a series of gravity separations and flotation operations is first performed on the crushed ore to concentrate the heavier diamond. Final separation is accomplished on a grease table, the diamonds adhering to the grease. As found in nature, diamonds range widely in size, shape, color, purity, state of aggregation and crystal perfection. The largest known diamond weighs 3024¾ carats or a little over 1¼ pounds.

At least five basic varieties of diamond are recognized: (1) diamond proper; (2) macles; (3) boart; (4) ballas; and (5) carbonado. The term "diamond proper" includes single crystals and gem stones occurring in the octahedral (8-sided), dodecahedral (12-sided), and tetrahexahedral (24-sided) crystal habits. Gem diamonds are colorless or pale shades of pink, blue, yellow, green or brown, of good crystal soundness, with a minimum of flaws or inclusions. Macles, an industrially useful form, are triangular, pillow-shaped stones consisting of twin crystals. Boart, also called bort or bortz, refers to minutely crystalline, grey-to-black, translucent-to-opaque pieces of diamond used industrially. The term is also applied to diamond fragments useless as gem stones. Ballas is a name given to dense spherical masses of randomly oriented crystallites. It is extremely tough and hard, and therefore suited for industrial applications. Carbonado is a cryptocrystalline material composed of diamond crystallites, graphite and other impurities.

Diamonds have been produced synthetically since 1955 when the General Electric Company first announced the successful development of a reproducible process. (See **Diamond Synthesis**).

Graphite. Deposits of natural graphite are located in virtually all continents. Of the major producers, the Republic of Korea is currently the world's largest, followed by Austria, North Korea, the U.S.S.R., China, Mexico, Malagasy Republic (Madagascar), West Germany, Ceylon and Norway.

Natural graphites are classified into three physically distinct varieties based upon geological occurrence—lump (from vein deposits), amorphous (from metamorphosed coal beds) and crystalline flake (from layered metamorphosed rocks). Depending upon the nature of the deposit, both underground and surface methods are employed in mining the graphite. The purity of the deposits vary widely, and where necessary, the graphite bearing ores are generally beneficiated by flotation techniques.

Manufactured or synthetic graphite represents more than 70% of the total graphite consumption in the United States today. The various geometrical shapes which comprise the bulk of the manufactured graphite products are produced primarily from petroleum coke and a coal tar pitch binder. The raw materials are thoroughly mixed at about 150°C, cooled to about 100°C, formed by extrusion or molding into desired shapes, and baked to around 950°C in a nonoxidizing environment. Heating to temperatures of about 2800°C in special graphitizing furnaces converts the baked carbon body into a polycrystalline graphite article.

Pyrolytic graphite, another form of manufactured graphite, is deposited on surfaces when low molecular weight hydrocarbons are pyrolyzed at low pressures (4–6 mm Hg) in the temperature range of 1700–2500°C.

Fibrous forms of graphite—thread, yarn, felt and cloth—are produced by controlled carbonization of natural and synthetic organic fiber products followed by heating to temperatures of around 2500°C. Principal applications for pyrographite and the fiber forms of manufactured graphite are found as components for rockets, missile and other aerospace vehicles.

The manufactured forms of graphite find extensive use in a wide variety of applications, of a chemical, electrical, metallurgical, mechanical and physical nature. The largest single use is in the form of electrodes for electric open arc furnaces producing regular and alloy steels. Electrolytic anodes for electrolysis of brine to produce chlorine and caustic, and of molten salts to produce magnesium and sodium constitute another major use of manufactured graphite. Because of its excellent resistance to corrosion by many chemicals, particularly acids, alkalis, organics and inorganic compounds, graphite is used for process equipment in the chemical process, steel, food and petroleum industries. Corrosion resistance combined with a good thermal conductivity make it useful as a material of construction for heat exchangers and tower packings. Principal metallurgical uses for manufactured graphite are as molds, dies and crucibles. Important but lesser volume uses are those of moderator and reflector for nuclear reactors and rocket and missile components. (see also **Graphite**).

Amorphous Carbon. Natural organic matter, such as coal, petroleum, gas and timber, constitutes the primary source of amorphous carbon. All are in abundant supply in various countries throughout the world. Coal deposits are usually mined, although other special mining techniques, such as coal gasification by partial combustion, are sometimes employed to obtain the carbon values from the underground deposits. Petroleum and natural gas are recovered from subterranean deposits by drilling wells, while timber is derived from the forests.

The processes by which the various forms of amorphous carbon are obtained from these carboniferous materials generally involve some type of thermal decomposition or partial combustion. The so-called coking coals and mixtures thereof are converted to coke in beehive and slot-type ovens. Anthracite, a "hard" coal, is calcined to remove noncarbon constituents. Residual oils resulting from the refining of petroleum crudes are converted to petroleum coke by the delayed coking and fluid coking processes, both being forms of destructive distillation. The carbon blacks are produced by vapor phase decomposition of hydrocarbons in an open flame, in a partial combustion chamber or in a thermal decomposition chamber in the absence of air. They are classified, respectively, as lampblacks, furnace combustion blacks and thermal blacks. Charcoal, a generic term, is obtained from the destructive distillation of wood, sugar, blood and other carbonaceous materials. The so-called activated carbons are produced by gas (selective oxidation) or chemical treatment to create a very large surface area. Two distinct types are generally recognized: liquid phase or decolorizing carbons, and gas phase or vapor adsorbent carbons.

Physical Properties. The atomic weight of carbon is 12.011, its atomic number 6, and atomic radius (for single bonding) 0.77 Å. Six isotopes of carbon are known: C^{10}, C^{11}, C^{12}, C^{13}, C^{14}, and C^{15}. Isotopes 12 and 13 are stable; the others are radioactive, having half-lives as follows; C^{10}, 20 sec.; C^{11},

20.5 min.; C^{14}, about 5560 yrs.; and C^{15}, 2.4 sec. The isotope C^{12} is the current base to which the atomic weights of all the other elements of the Periodic Table are referred.

The thermal-neutron-absorption cross section (σ_a at 2200 m/sec) for carbon is 3.4 millibarns. The electronic configuration or orbital arrangement of electrons is $1s^2 2s^2 2p^2$, where the superscript indicates the number of electrons in that particular energy level.

Some physical properties of the allotropic crystalline forms of elemental carbon are given in Table 1. Since the amorphous forms of carbon are not capable of rigorous definition, a tabulation of corresponding physical property data is quite impossible. On the other hand, the specific classes or types of amorphous carbons, such as the carbon blacks, activated carbons and anthracite coals have been extensively characterized and the references should be consulted for specific information of this nature.

Specific mention must also be made with reference to the physical properties of the manufactured forms of polycrystalline graphite. In general, these properties are a function of the raw materials and the processing techniques employed; consequently, the range of physical properties attainable is very broad. Again, the references should be consulted for further information.

Of the two allotropic crystalline forms of elemental carbon, graphite is the more thermodynamically stable at atmosphere pressure. Diamond is transformed into graphite above 1500°C. Because of the formidable experimental difficulties associated with the extremely high temperatures and ultra high pressures required in the study of the carbon

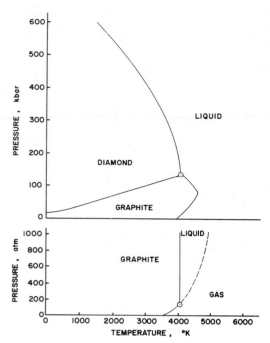

Fig. 1. Phase diagrams of carbon (1 kbar = 987 atm)

system, the phase diagram for carbon is incompletely known. Recent proposed diagrams of selected regions are shown in Fig. 1. At atmospheric

TABLE 1. PROPERTIES OF THE CRYSTALLINE FORMS OF ELEMENTAL CARBON†

	Diamond	Natural Graphite
Density, g/ml	3.15	2.26
Melting point, °C	3700	—
Boiling point, °C	4200	—
Sublimation point, °C	—	3620 ± 10
Mean coefficient of thermal expansion, per °C	$3\text{--}4.8 \times 10^{-7}$ @ 100°C	
a axis (<383°C)	—	$-.15 \times 10^{-7}$
c axis (15–800°C)	—	238×10^{-7}
Thermal conductivity, cal/sec/cm/°C	1.27	
a axis	—	0.6
c axis	—	0.2
Specific resistance, ohm-cm	5×10^{14}	
a axis	—	$\sim 1 \times 10^{-4}$
c axis	—	~ 1
Specific heat, C_p, cal/mole/°C	1.462	2.038
Magnetic susceptibility, x, emu/g	52×10^{-6}	$\sim -6.5 \times 10^{-6}$
a axis	—	-22×10^{-6}
c axis	—	-0.5×10^{-6}
Compressibility, cm²/kg	0.18×10^{-6}	3×10^{-6}
Young's Modulus, dyne/cm²	—	1.13×10^{15}
Shear Modulus, dyne/cm²	—	2.3×10^{10}
Hardness, Mohs	42	0.5–1.5
Refractive index	2.4173	1.93–2.07
Heat of formation, $\Delta H°_{298}$, kcal/mole	0.4532	0
Free energy of formation, $\Delta F°_{298}$, kcal/mole	0.6850	0
Entropy, $S°_{298}$, kcal/mole	0.5829	1.3609
Latent heat of vaporization @ b.p., kcal/mole	—	170

†Values listed are those measured at room temperature unless otherwise indicated.

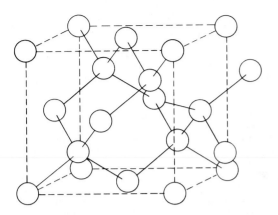

FIG. 2. Crystal structure of diamond.

pressure, graphite sublimes at 3640 ± 25°K. The triple point (graphite-liquid-gas) is 4020 ± 50°K at 125 ± 15 atm. A second triple point (diamond-graphite-liquid) occurs at about 130 kbars at 4100°K. Regions involving metastable phases of either diamond or graphite have not been included in the figure.

The marked differences in the physical properties of diamond and graphite are readily understood from their crystallography and the nature of the interatomic forces within the crystals. The diamond crystal lattice consists of two interpenetrating face-centered cubic lattices so arranged that each carbon atom of one is surrounded by four carbon atoms belonging to the other (Fig. 2). Each carbon atom is covalently bonded to each of its four neighboring carbon atoms in the form of a tetrahedron. This type of carbon-to-carbon bonding is described in terms of the hybridization of one $2s$ and three $2p$ atomic orbitals of the carbon atom. Such bonds, termed sp^3 hybrids, are effectively localized, thereby greatly restricting electron mobility within the crystal. The extreme hardness and the low electrical conductivity of diamond are consistent with this picture.

The idealized crystal structure of graphite is hexagonal (Fig. 3), with the stacking arrangement of

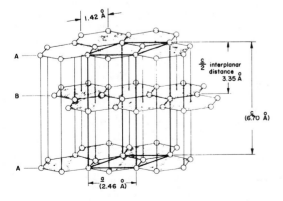

FIG. 3. Hexagonal form of graphite.

the parallel layers of carbon atoms in ABABAB . . . . form.

Bonding between carbon atoms within the graphitic planes is different from that found in diamond, and results from a second type of combination of the atomic orbitals of the carbon atom known as trigonal or sp^2 hybridization. Only three of the bonding orbitals are effectively localized in this instance, leaving a fourth whose nature is such that the electrons are quite mobile within the layer of carbon atoms. Because the bonding electrons are all used in forming carbon-to-carbon bonds within the carbon layers, the layers themselves are held together only by weak van der Waals forces. The bonding energy between planes is only about 2% of that within the planes. These structural characteristics explain the highly anistropic and widely divergent properties of graphite compared to diamond.

Chemical Properties and Reactions. At room temperature, elemental carbon in any of its allotropic forms is relatively unreactive. It is insoluble in water, dilute acids and bases and in organic solvents. At elevated temperatures, however, it becomes highly reactive, combining directly with other elements such as oxygen and sulfur. When carbon reacts with oxygen or air, carbon monoxide or carbon dioxide is formed, depending upon the relative amounts of carbon and oxygen present in the reaction environment. With carbon in excess, carbon monoxide is formed:

$$2C + O_2 \rightarrow 2CO$$

With oxygen in excess, the reaction continues and carbon dioxide is formed:

$$2CO + O_2 \rightarrow 2CO_2 \qquad \text{and}$$

$$C + O_2 \rightarrow CO_2$$

The temperature at which these reactions take place will vary greatly depending upon the form in which the carbon is present. Thus, while diamond does not oxidize in air until about 800°C, natural graphite oxidizes slowly above 450°C. Some forms of amorphous carbon may react with oxygen at temperatures even lower than 450°C because of their greater surface area and the presence of certain non-carbon elements and impurities.

Carbon reacts with water vapor to form carbon monoxide, carbon dioxide and hydrogen. The reaction rates and product compositions are dependent upon the form of the carbon employed and the reaction conditions. Diamond is virtually unreactive even at white heat, whereas graphite reacts at temperatures above 800°C. Certain forms of amorphous carbon will react at temperatures as low as 600°C.

Most metals react directly with elemental carbon to form the respective carbides. The temperature at which carbide formation takes place varies widely depending upon the metal involved. Cobalt forms the metastable carbide, Co_3C, at a temperature of 218°C, whereas hafnium carbide, HfC, is formed at 2000°C.

In reacting with oxides and oxygen-containing salts, carbon acts as a reducing agent. Depending on the temperature and other conditions of reaction, metal oxides are reduced to the free metal or converted to the carbide:

$$Me_xO_y + yC \rightarrow xMe + yCO$$

The reduction of calcium phosphate rock with carbon in the presence of silica is the basis for the commercial process for the production of phosphorus:

$$Ca_3(PO_4)_2 + 3SiO_2 + 5C \rightarrow 3CaSiO_3 + 2P + 5CO$$

Certain chemical reactions of carbon are indigenous to the specific crystalline form of the element. Graphite, for example, forms a series of compounds known as lamellar compounds with a large number of salts, a few oxides, fluorine, chlorine, bromine, potassium, rubidium and cesium. Heating with a mixture of potassium chlorate and concentrated nitric acid converts graphite to graphite oxide. With hot oxidizing agents such as nitric acid and potassium nitrate, mellitic acid, $C_6(COOH)_6$, is formed.

The unique electronic band structure of the carbon atom is responsible for its ability to combine not only with itself but also with a large number of other elements, principally hydrogen and oxygen, to form the vast system of chemical compounds known as organic.

In bonds to carbon a unique concept of bonding is involved, namely, hybridization. Three basic types of hybridization of the bonding orbitals of the carbon atom are possible: the tetrahedral or sp^3 type; the trigonal or sp^2 type, and the digonal or sp type. Each type is stereospecific, or in other words, has directional properties. The first or tetrahedral type describes the nature of the bonding in organic molecules of the paraffin class as well as in diamond. The second or trigonal type describes the nature of the carbon-to-carbon double bonds in unsaturated organic compounds of the olefinic type. The third or digonal type describes the nature of the carbon-to-carbon triple bonds in unsaturated organic compounds of the acetylenic type. Such hybrid types form what is termed localized or σ type bonds. Still another type of bond is required in describing carbon bonding in organic molecules; namely, the π bond. This type of bond is used in describing the nature of the chemical bonding between carbon atoms in the so-called aromatic hydrocarbon series as well as in graphite. This versatility of the carbon atom with respect to bond formation serves to explain the prevalence and endless multiplicity of organic compounds, both natural and synthetic.

Principal Compounds. The principal compounds of carbon are its oxides, carbon monoxide and carbon dioxide; carbon disulfide; carbon tetrachloride; and the carbides, notably silicon carbide, calcium carbide and the heavy-metal carbides.

Carbon monoxide is an extremely poisonous, colorless, odorless, tasteless gas, stable at room temperature. Industrial quantities of carbon monoxide are derived from processes involving the partial combustion of a carbonaceous fuel or the water gas reaction. Blast furnace gas, methane partial-combustion gas and oil partial-combustion gas are examples of sources of carbon monoxide from partial-combustion processes. Industrial preparation of carbon monoxide by the water-gas reaction involves the passage of steam over a bed of hot coke or coal at 600 to 1000°C:

$$C + H_2O(g) \leftrightharpoons CO + H_2$$

The reaction is endothermic and air is alternated with steam to maintain the temperature above 600°C. (See **Carbon Monoxide**).

Carbon dioxide is a colorless gas with a faintly pungent odor and acid taste. It is produced commercially by separation from flue gases resulting from the combustion of carbonaceous materials and as a by-product of synthetic ammonia production, fermentation and lime kiln operations. It is used in solid (dry ice) liquid and gaseous form in such widely diverse applications as carbonation of beverages, chemicals manufacture, fire extinguishing, food preservation and mining operations. (See **Carbon Dioxide**).

Carbon disulfide, a low-boiling, highly toxic and flammable organic solvent is made by the catalytic reaction of natural gas, CH_4, with sulfur vapor at 500–700°C. Carbon disulfide is used mainly in the manufacture of viscose rayon and cellophane, and as a raw material for the production of carbon tetrachloride.

Carbon forms compounds with the halogens according to the general formula CX_4, where X is fluorine, chlorine, bromine, iodine and mixtures thereof. By far the most important is carbon tetrachloride, a colorless, heavy, nonflammable, low-boiling liquid (b.p. 76.4°C) manufactured by passing chlorine gas into carbon disulfide containing catalytic amounts of metal chloride such as $SbCl_3$.

The metal carbides, an important class of carbon compounds, are formed by heating the metal or metal oxide with carbon or a carbonaceous material. Calcium carbide, produced by heating calcium oxide and coke in an electric arc furnace, has industrial importance as a source of acetylene. The carbides of titanium, hafnium, tantalum and tungsten are very refractory and find important use as cutting tools. Silicon carbide and boron carbide are extremely important compounds of carbon, inasmuch as they constitute the backbone of the entire abrasives industry. Silicon carbide and boron carbide are produced by reacting sand, SiO_2, and boric acid, respectively, with coke in a special type of electric furnacing operation.

L. H. JUEL

References

Kirk, R. E., and Othmer, D. F., "Encyclopedia of Chemical Technology," Vol. 4, Second Ed., Interscience Publishers, New York, 1965.

Juel, L. H., "Carbon," pp. 106–115, "Encyclopedia of the Chemical Elements," Hampel, C. A., Editor, Van Nostrand Reinhold Co., New York, 1968.

Walker, P. L., Jr., Editor, "Chemistry and Physics of Carbon," Marcel Dekker, Inc., New York, 1965.

"Minerals Yearbook," Vol. 1, Bureau of Mines, U.S. Dept of Interior, Washington, published annually.

CARBON, ACTIVATED

Activated carbon is a manufactured form of carbon having a large specific area and designed for adsorption from either the gas or liquid state. It is made by destructive distillation of carbonaceous material under controlled conditions. The specific area of an activated carbon may range from 600 to 2,000 square meters per gram. In both granular and powdered carbons, this area is almost entirely internal; it is the area of the pore structure created by the two steps of driving off volatile constituents from the carbonaceous raw materials, and oxidizing the residue. Such oxidation is accomplished by means of steam or carbon dioxide, at temperatures approximating 1000°C.

Carbons activated for adsorption of gases and vapors differ markedly from those designed for adsorption from liquids. The former class is characterized by a preponderance of small pores (under 20 Å in radius); they are in general harder and denser than the socalled "decolorizing" carbons. The fine pore structure required for gas-adsorbent carbon may be secured either by choice of suitable raw material or (within limits) by choice of the method of manufacture. Coconut and other nut shells are ideal raw materials for producing such carbons.

The "decolorizing" carbons for liquid phase use (which adsorb many noncolored substances as well) are divided into the animal and vegetable carbons. The former, and historically older, type is made from animal bones. The finished bone char ranges approximately 12 by 20 mesh, and consists of about 90+% of hydroxy apatite, the balance being carbon. The vegetable-activated carbons of commerce are made from lignite, paper mill waste liquor, coal, peat, and wood. They are sold in powdered and granular form. Activated carbons of both the gas-adsorbent and decolorizing types can also be made by chemical activation. Here the pore structure is initiated by destructive dehydration instead of destructive distillation. The agents used are zinc chloride, phosphoric acid, or sulfuric acid.

Activated carbons are usually leached to extract inorganic contaminants, if they are to be used for purifying food or chemical products. Such leaching is done with mineral acid, and is followed by water washing to eliminate excess acid and salts. A good grade of activated carbon as sold will contain well under 1% of soluble impurity.

The military use of the gas-adsorbent carbons in gas masks is well known. There are a number of sizable peace-time uses as well. Important among these is the recovery of volatile solvents in industrial processes. Air or gas streams containing such solvent vapors, even in very low concentrations, are effectively stripped by passage through a bed of granular gas-adsorbent carbon. After a suitable adsorption period, the solvent is desorbed, usually with steam, and recovered by conventional condensation.

The same principle is used, but at higher concentration levels, in the petroleum industry for separation of hydrocarbons by fractional adsorption. This is useful with lower members of the aliphatic series, where low boiling points cause complications in separation by distilling and subsequent condensation.

Other applications of gas-adsorbent carbon include the removal of oil vapor, sulfur compounds and other contaminating agents from such industrial

gases as carbon dioxide, carbon monoxide, hydrogen and acetylene. Objectionable odors or irritating vapors are removed from air to be circulated in theaters, restaurants, offices, railroad cars, airplanes, and other places for human comfort. Both gas-adsorbent and decolorizing carbons are used as catalysts or catalyst carriers for various commercial organic chemical reactions such as hydrogenation and synthesis of vinyl chloride.

The "decolorizing" carbons, both animal and vegetable, find a major use in the refining of raw cane sugar. Bone char dates back well into the nineteenth century. It is used in large char towers through which sugar liquor is percolated for removal of color as well as inorganic constituents. The spent bone char, after "sweetening-off" with water to recover occluded sugar liquor, is regenerated by a heat treatment in char kilns; this process is very similar to the original activation. Some of the original carbon is burned off, but is replaced by secondary carbon from the organic nonsugars adsorbed in the char tower.

Vegetable carbons have replaced bone char in corn syrup refining. In cane sugar refining, they are used both as the sole purification agent and also as an adjunct to bone char treatment. U.S. beet sugar refineries also use these carbons.

Granular vegetable carbons are used by the cane, corn, and beet sugar industries in fixed, moving, or pulse beds. The spent carbon is thermally regenerated; losses approximate 5%.

Another major field of use for the vegetable carbons is in the treatment of municipal water supplies. The objective here is the adsorption of impurities producing odor and taste. These may arise from industrial pollution, from decay of vegetation, or from algae. The carbon is applied in powdered form at the filtration plant, usually during the coagulation and sedimentation steps, in dosages averaging only two to five parts per million. These low levels are effective because the odors and tastes are caused by such minute concentrations of organic compounds.

A similar function is performed by granular activated carbon on industrial water supplies. In the soft drink and brewing industries taste-free and substantially sterile water is essential; sterilization is effected by heavy chlorine treatment. Carbon is used to remove residual chlorine, as well as odors and tastes, in a single operation. Other large fields of use for the vegetable carbons are the treatment of glyceride-type oils and fats, waxes, plasticizers and other esters, the purification of nylon monomer, polyhydric alcohols such as glycerine, the glycols, pentaerythritol and sorbitol; food products such as gelatin, pectin, vinegar, and monosodium glutamate; photographic chemicals and pharmaceuticals such as sulfa drugs, antibiotics, blood plasma extenders and synthetic vitamins; organic acids including acetic, benzoic, citric, fumaric, maleic, and many others; natural and synthetic caffeine, dye intermediates, and a great host of miscellaneous products.

In recent years concern with pollution of lakes and streams has led to extensive investigation of activated carbon for municipal and industrial waste water treatment. Granular carbon has proved effective, and powdered carbon will probably find application also.

There are two rather unusual uses for activated carbon. One is the immobilization of a troublesome liquid in an essentially solid-phase system. When reclaimed rubber is used in the production of white side-wall tires or other light-colored rubber products, some of the oils used in the reclaiming process tend to migrate from the back ply to the light outside ply and cause a yellow stain. When activated carbon is milled with the reclaimed rubber, these oils are immobilized on the carbon and staining is prevented.

The other is the use of carbon to adsorb impurities causing haze in a liquid product on long storage or on chilling. This may be the case with some types of sugar used in the soft drink industry, and in the past it has also caused much concern in the brewing industry. Such haze is caused by the presence of an impurity at or near the point of insolubility; substances like this are readily adsorbed. Activated carbon is a valuable aid to the brewing industry where it supplements the enzymes employed to offset chill haze caused by protein precipitation.

WALTER A. HELBIG

References

Deitz, V. R., "Bibliography of Solid Adsorbents," (2 Volumes) Government Printing Office Vol. I 1900–1942, Vol. II 1942–1953, Washington, D.C.

Hassler, J. W., "Active Carbon," Chemical Publishing Co., New York, 1963.

CARBON BLACK

Carbon black is the generic name for a group of intensely black, submicron size pigments composed of essentially pure carbon and made by vapor phase decomposition of hydrocarbons in an open diffusion flame, in a partial combustion chamber or in a thermal decomposition chamber in absence of air. Common soot is a true carbon black; commercial carbon blacks are the result of making soot under exactly controlled process conditions.

Carbon black is formed in a fraction of a second in a flame having a mass temperature ranging from about 2100 to 2700°F; at this temperature level all hydrocarbons decompose rapidly to molecular fragments which in turn condense or polymerize to polynuclear aromatics. These aromatic "compounds" grow rapidly and dehydrogenate almost completely by the time the final carbon particle is formed. Evidently nuclei, which have never been isolated and identified, are formed by *thermal degradation of the hydrocarbon raw material*. The nuclei grow almost instantaneously into the final carbon particles which, in turn, have a tendency to aggregate. The 3-dimensional branched aggregates are held together with varying degrees of bond strength, known as "structure," depending on fuel and production conditions.

X-ray analysis indicates that all carbon blacks are made up of crystallites containing parallel layer groups similar to graphite except that the layers are randomly oriented about the layer normal and the layer spacing is slightly greater than in graphite. Prolonged heating at 2700–3000°C graphitizes carbon black with considerable growth in crystallite size.

The range of mean particle size of commercial carbon blacks is from about 10 to 500 mμ with the distribution about the mean usually broadening as particle size increases. Channel blacks may contain up to about 11% oxygen and 0.8% hydrogen; oxygen contents above 4% are usually obtained only by secondary aftertreatment in an oxygen-containing atmosphere at 800–1000°F or by prolonged high-temperature exposure on the channels. Channel blacks contain negligible amounts of benzene soluble matter; ash is usually 0.10% or less.

Furnace and thermal blacks contain about 0.3–0.5% each of hydrogen and oxygen, a trace of benzene extractables up to a maximum of a few tenths of a percent and ash contents of 0.1 to 1.5% depending upon the solids content of the water used in cooling. Sulfur content of black depends on sulfur content of the fuel; in commercial furnace blacks sulfur varies from 0.1 to 1.5%. Neither sulfur nor ash have important primary effects in carbon black applications but oxygen content (or some variable related to oxygen content) has a strong influence on the dispersing properties of blacks in fluid systems.

The pH of aqueous suspensions of ash-free carbon black varies from about 3.0 for aftertreated channel blacks to 7.0 for furnace black as oxygen content varies from 11% to under 0.5%. In practice furnace blacks are alkaline (up to pH 10) due to contaminants from cooling water.

The most important present day commercial carbon black process is the so-called furnace process. This involves non-pre-mix partial combustion of atomized liquid or vaporized hydrocarbon fuels, with natural, refinery or coke oven gas as auxiliary fuels, and with air as an oxygen source in a refractory enclosure or furnace. Black is collected from the smoke in plants of the latest design by cyclones followed by bag filters. Use of electrostatic precipitators and wet scrubbers is obsolete in new plants.

The channel black process, in use since about 1880 and intensively developed in the period 1920–30, is now virtually nonexistent, accounting for only about 4% of U.S. consumption in 1970. Before 1940 channel black constituted over 90% of the world's supply of carbon black.

In the thermal black process natural gas is decomposed or "cracked" by contact with hot silica checker brick in a cyclic process employing 2 chambers alternately used on heating and cracking cycles. Collection of black from the gas stream is usually by bag filters with or without cyclones in series. Net yields are in the range of 25–40% when the by-product hydrogen is used as heating fuel. U.S. thermal black consumption in 1970 was more than 11% of total carbon black production.

Particle Size

Over 95% of all carbon black produced is used in natural and synthetic rubber as a reinforcing and filling agent; the largest single use is in motor vehicle and airplane tires. The most important variable affecting reinforcement is *particle size;* in general, basic reinforcing properties—tear resistance, tensile strength, and abrasion resistance—increase as particle size decreases. Particle size is determined from electron microscope pictures in which only the outline or the external surface area is considered. Total surface area, which includes the area of pores and surface irregularities, is determined by nitrogen adsorption. The nitrogen adsorption surface areas of commercial blacks vary from 1 to 5 or 6 times that of the external area determined from electron microscope counts. Rubber grades of black have a ratio of nitrogen area to electron microscope area not over about 1.5. Generally the higher the ratio of nitrogen surface area to the external area determined by the electron microscope the slower the vulcanization rate, evidently due to greater adsorption of accelerators or intermediates formed in curing. Stiffening effects —modulus, viscosity, hardness—increase in proportion to structure and inversely with particle size. The mechanism of reinforcement of rubbers by black is only partially understood.

Uses

The growth of the carbon black industry here and abroad parallels that of the rubber industry; for the U.S. an average growth rate of about 4% to 5% per year seems indicated as long as new competitive nonblack fillers do not develop. For the rest of the world a growth in demand of 7 to 8% per year can be anticipated for the next few years.

The current annual production of carbon black in the U.S. is about 3 billion pounds.

About 90% of the blacks sold in the U.S. are in the price range 6 to 10¢/lb. f.o.b. plant in bulk. Carbon black has a low specific gravity (1.86) compared to most filler pigments; its price is low in relation to its reinforcing power in rubbers. Synthetic white reinforcing fillers, principally silicates and hydrated silicas, are finding increasing use in rubber products where color is critical, e.g., in shoe soling. Clay and ground calcium carbonate, the cheapest nonblack fillers, are not gaining in use relative to carbon black. It is difficult to see how carbon black, which is made from the heavier refinery fractions falling in the industrial fuel category, can be surpassed in reinforcing power per dollar by products made from chemical raw materials.

The second largest use of blacks is in inks, news ink accounting for most. Carbon black for use in black inks is not likely to be supplanted by other black pigments or dyes. The trend is towards methods of masterbatching black with vehicles to increase its density, reduce transportation costs, cut down dust and speed dispersion at the ink factory. Growth in ink use of blacks can be expected to follow the growth of newsprint, book and magazine papers at 3 to 4% per year. The use of carbon blacks in plastics is constantly increasing and may exceed the use in inks in 5 to 10 years.

C. A. STOKES

CARBON, INDUSTRIAL

Industrial carbons can be considered broadly as all forms of carbon used for industrial purposes excluding carbon used simply as a fuel, but including coke used for reduction of ores. Under this broad definition carbon becomes one of the largest volume items in the entire field of chemical products. Very few chemical materials are produced in the U.S. in quantities exceeding one million tons. Industrial

carbon is far up on the list at roughly 100 million tons, worth over 2 billion dollars.

Of minor volume tonnage-wise, a significant dollar volume (about $50 million per year) of industrial carbons is consumed as diamond, ballas, bort, and carbonado, whose use in abrasives, cutting operations, and die-orifices is indispensable.

Naturally occurring graphite is used for a wide variety of applications. The principal ones are lubricating, refractories, marking instruments, electrical products, corrosion resistant paints, and carbonizing of steel. Crude graphite ore is usually processed by air and/or froth flotation to remove impurities. Final processing can include sizing and/or grinding depending on the applications for which it is to be used. The 1968 consumption was about 39,000 tons valued at $6 million. Consumption has been declining in recent years.

Manufactured industrial carbons are usually primary products or byproducts of fuel processing. They are used, either as is or after calcining, for a variety of purposes.

Synthetic graphite is an important manufactured carbon made by electric furnace heating of selected pure solid carbons compacted with a pitch binder to various shapes near that of the final graphite articles, usually furnace electrodes or anodes for electrolytic cells. In 1968 the U.S. produced over 210,000 tons of synthetic graphite.

By far the greatest use of industrial carbon, as defined, is in the iron and steel industry, where nearly 64 million tons of beehive and slot-oven coke were consumed in 1968. The major portion of this coke is used in the reduction of iron ore in blast furnaces at the average rate (1970) of about 0.6 ton per ton of iron. Coke produced by slot-oven coking of bituminous coal is the principal source of this form of industrial carbon. High strength, large lump size, low ash, low sulfur, low phosphorus, and good reactivity are the major requirements for coke used in the manufacturing of iron in the blast-furnace process. There is a trend toward preforming carbon for this use but it will be many years before such a product can compute on large scale with the coke oven based on raw coal feed.

Anthracite, calcined anthracite, coal coke, and petroleum coke from the destructive distillation of residual oils are all used, in combination with coal tar pitch, for refractory pot linings, furnace linings, carbon bricks, and the like in such industries as the electrowinning of aluminum, magnesium, and for lining of equipment to withstand unusual corrosion conditions. In this latter application, carbon or graphite pipe is often used for handling such chemicals as hot aqueous HCl, HF, etc.

Low-grade coal-cokes and chars of noncoking coals are used somewhat interchangeably with calcined and green petroleum coke in electrothermal processes for producing elemental phosphorus, silicon carbide, and calcium carbide, and in the reduction of zinc ore to the metal. In addition, the aluminum industry consumes large quantities of high-grade calcined petroleum coke in the form of electrodes in the electrolytic reduction of alumina to aluminum.

Specifications of coke for the production of silicon carbide are not stringent, but those for coke for producing calcium carbide are somewhat more exacting. Low moisture ($<1.5\%$), volatile ($<2.0\%$), and phosphorus ($0.<1\%$) are the most important. Ash content of coke for this use can run as high as 4%. Apparent density is limited to 0.84 maximum while lump size should be between ⅛ inch and 1 inch. It is also desirable that the coke or char be reasonably strong. Coke consumed by this industry was about 360,000 tons in 1971 or about 0.6 tons per ton of calcium carbide produced.

Typical analysis of coke used by elemental phosphorus producers is: 2% volatile; 3 to 5% ash; $-1''$ $+ ⅛''$ lump size. Again, strength and apparent density are important considerations although no specifications are set on these qualities. Coke consumed by this industry is over ½ million tons per year at the rate of about 1 ton of carbon per ton of phosphorus.

Zinc producers use foundry coke to reduce zinc oxide to metallic zinc in the retort process. About 1.36 tons of coke are used to produce one ton of retort zinc.

Specifications for petroleum coke used by the aluminum industry are very stringent. Moisture and volatile content of calcined coke should be less than 1.0%. Ash should be less than 0.8%, while silicon, iron, calcium, nickel, vanadium, and titanium content of the coke are all specified at levels less than 0.05%. Sulfur content should be less than 2%; for calcined coke, a minimum specific gravity of 2.0 is specified. Actually, carbon is not the primary reducing agent in this application. However, due to the proximity of the carbon electrodes to the high temperature reaction zone, the carbon in the electrodes is oxidized in this process to the extent that coke consumed by the aluminum industry was about 2 million tons in 1969 or 0.5–0.6 pound per pound of aluminum. World consumption of carbon for aluminum is 3 times U.S. consumption nearly all of which comes from petroleum coke. With diminishing supplies of low-sulfur, low-ash crude, a world-wide shortage of carbon for aluminum is in prospect.

Charcoal from the destructive distillation of hard and soft woods is characterized by high surface area. This property is the basis for many of its industrial uses. This can be either for difficult reactions involving carbon, where charcoal presents more surface to react than other forms of carbon, or in cases where high surface for adsorption of colors or catalysts is desired.

The carbon disulfide industry consumes about 50,000 tons of charcoal yearly, mainly of the hardwood type. Specifications for charcoal used in the carbon disulfide industry are: low ash (2–2.5%), volatile (16–18%), and moisture ($<4\%$) content, good lump strength and high bulk density. Newer processes are based on sulfur or hydrogen sulfide reacted with methane as the source of carbon.

Activated carbons consume about 70,000 tons of carbon per year. (See Carbon, Activated).

Approximately 3 million tons of various carbon blacks worth about $400 million are sold annually to the rubber, paint, ink, and plastics industries. Ninety percent of this volume of carbon goes into rubber (e.g., tires) where it acts as a reinforcing agent, increasing abrasion resistance, and in general improves the physical properties of the finished product. Major applications for these carbons are in tires, mechanical goods, shoe soles, and heels. (See Carbon Black).

C. A. STOKES

Cross-references: *Coal and Coke, Carbon (Activated), Calcination, Carbon Black.*

Diamond Synthesis

Carbon can crystallize in at least five distinct structural modifications. The most abundant is graphite, with atoms normally arranged according to a simple hexagonal unit cell. Dimensions are $a_0 = 2.456$ Å and $c_0 = 6.696$ Å. A second less common variety of graphite possesses a rhombohedral lattice with unit cell parameters of $a_0 = 3.635$ Å and $\alpha = 39°\ 30'$. If indexed according to a simple hexagonal cell, dimensions would be $a_0 = 2.456$ Å and $c_0 = 10.044$ Å. The third modification is cubic diamond, with $a_0 = 3.567$ Å. The fourth polymorph, a new high pressure-high temperature phase is named lonsdaleite. It often is called hexagonal diamond since it has a wurtzite type structure with $a_0 = 2.52$ Å and $c_0 = 4.12$ Å. The fifth and most recently discovered crystalline form of carbon is named chaoite. It has a primitive hexagonal lattice with $a_0 = 8.948$ Å and $c_0 = 14.078$ Å. It has a density of 3.43 g/cc in contrast to 2.25 for graphite and 3.51 for diamond and lonsdaleite.

In addition to the five established polymorphs, there is indirect evidence suggesting a possible sixth metallic phase of carbon. Reasoning is based on a probable analogous behavior with the other Group IV elements which do exhibit metallic forms. It appears as though pressures in excess of 600 kb would be required for its existence. As yet, however, no confirmation has been reported. The only structure of carbon that is thermodynamically stable at standard conditions is the simple hexagonal variety, graphite.

Attempts to synthesize diamond had been made for a long time. The role of pressure and temperature in its genesis was recognized as early as 1896. One of the earliest phase diagrams for the carbon system was that proposed by Roozeboom in 1901. The first quantitative determination of the relationship of pressure and temperature to the graphite-diamond equilibrium was made in 1938 by Rossini and Jessup. In 1939 Leipunskii proposed a more comprehensive phase diagram for carbon, and in 1955 Berman and Simon published an improved equilibrium curve for graphite and diamond. The first successful synthesis of diamond was reported in 1955 by the General Electric Company.

The major obstacle to the synthesis of diamond had been difficulties in developing apparatus capable of generating sufficiently intense pressures and temperatures. Once this was overcome, success was achieved. There are several types of apparatus now developed that are capable of diamond synthesis. Examples are the General Electric Belt, the Hall Tetrahedral Anvil, the Cubic Anvil, the Stromberg Girdle, and the Bridgman Anvil.

The chemistry which first yieded diamond, namely, Fe + C, was not far removed from some of the metallic diamond-bearing meteorites. Later reactions included Ni, as well as combinations of Fe and Ni with carbon. Each of the elements, Cr, Mn, Fe, Co, Ni, Pt, Ta and Nb, either alone, with each other, or with other metals, have been found to yield diamond when reacted with sufficient carbon at sufficiently intense pressures and temperatures. These metals are similar with respect to electron vacancies in the outer d shell. All react readily with carbon, especially at elevated pressures. The lower limits of pressure and temperature at which diamond growth has been observed coincide with the graphite-diamond equilibrium boundary above a pressure of about 45 kb.

The metal plus carbon reactions that produce diamond have been variously referred to as catalytic, catalytic solution, and simply as solution of carbon in the metal. Based on analyses of residual reactants and products, it appears that in some cases reactions proceed with increasing temperature by a sequential formation of carbon compounds, each having a higher carbon content until a limiting composition is reached. Beyond this temperature, spontaneous diamond crystallization occurs.

The spontaneity may be explained on the basis of free carbon released by the decomposition of unstable carbide or carbonyl compounds. Reactions usually are carried out in the presence of excess carbon. The sequence of events leading to diamond formation is readily determinable in those reactions that yield residually stable carbides in the reaction matrix. Two examples are the Fe + C and Mn + C systems.

Residually stable free carbides of cobalt have not been observed in diamond synthesis reactions. No positive evidence is available, therefore, to define events that might occur in the Co + C system. The only residual products are recrystallized elemental cobalt and carbon. These are arranged, however, in three distinct microstructural patterns that infer a possible existence of three carbon compounds of cobalt prior to diamond crystallization. Further indirect support for this hypothesis can be drawn from the known existence of the metastable carbides, Co_3C and Co_2C, and of the existence within diamond synthesized from the Co + C reaction of finely dispersed and systematically oriented cobalt carbide inclusions. The stoichiometry of this included carbide is uncertain. The assigned formula is Co_xC. Available evidence indicates that $x > 4$.

The nickel-carbon system is similar to that of cobalt in that no residually stable free carbon compounds are found. Here, however, the residual microstructure is a simple and uniform one of recrystallized nickel and carbon in complete segregation. The only evidence of possible compounds having a high carbon content is the ratio of recrystallized nickel and carbon. The latter yields averaged compositions as high as NiC_3. As in the case of cobalt, finely disseminated oriented inclusions of Ni_xC with $x > 4$, also are persistently observed within the diamonds. Whereas two metastable cobalt carbides can be produced at near normal conditions of pressure and temperature, only one carbide of nickel is known, and this is even less stable than those of cobalt. If carbides of nickel do exist during the course of the reaction leading to diamond formation, their stability must be such that complete decomposition occurs beyond narrow limits of high pressure and high temperature, except in the case of entrapment within diamond.

The significance of the included carbides, Co_xC and Ni_xC, to diamond crystallization is uncertain. Both have stable crystal structures within diamond and their stoichiometry appears to be relatively consistent. Their concentration within diamond shows an inverse relationship to the temperature at which the diamond is synthesized. Included carbides are not observed in other metal-carbon systems. It has been shown that both Co_xC and Ni_xC have crystal structures that very closely match that of diamond except for slightly smaller unit cell dimensions. Similar compatibilities were not observed with residual carbides of other systems. All possess unit cell dimensions larger than diamond. The presence of Co_xC and Ni_xC are known only as epitaxial inclusions within synthetic diamond.

The function of catalytic carbon compounds in diamond synthesis remains to be determined. In cases such as the Fe + C system, carbides appear to constitute the primary activation mechanism. In other cases, experimental evidence ranges from meager in the system Ni + C, to nonexistent for Pt + C where nothing but recrystallized platinum and carbon are found. The latter may represent carbon recrystallization directly from solution. Another possibility is the catalytic action of unstable compounds such as carbonyls as suggested by Bokii and Volkov.

Diamond growth on diamond substrates by low pressure thermal decomposition of carbon monoxide and of methane has been demonstrated by Eversole, Deriaguin, and others. The amount of growth appears to be limited, and is observed to occur as whiskers and other discrete shapes at surface junctures of dislocations. The mechanisms of growth by this process is not well understood.

A direct conversion of graphite to diamond could not be achieved in the 100 kilobar range of early static pressure vessels. In 1960, DeCarli and Jamieson succeeded in recrystallizing graphite to diamond by subjecting it to intense shock pressures calculated to be of the order of 300 to 400 kb with concurrently generated temperatures of 1000 to 1500°K. The degree of conversion was small and appeared to be related to the percentage of rhombohedral graphite in the starting material. A logical mechanism proposed for the conversion was a strong compression along the c axis of graphite. This would yield the necessary atomic rearrangement for coincidence with the structure of diamond.

In 1962, Bundy modified his static pressure apparatus to permit generation of pressures to about 200 kb concurrent with transient temperatures up to approximately 5000°K. A direct conversion of polycrystalline graphite to diamond was achieved with this vessel at a pressure and temperature of about 130 kb and 3500°K, respectively. The temperature for transformation was found to decrease with increasing pressure.

Later, in 1967, Bundy and Kasper discovered that compression of single crystal graphite to about 140 kb and 1300°K yielded a new diamond-like material with a wurtzite-type crystal structure. This "hexagonal diamond" was found to be similar to diamond in density and hardness, and was shown by Hanneman, Strong and Bundy to occur naturally in diamond-bearing metallic meteorites. This form of carbon has been named lonsdaleite.

At about the same time that Bundy and Kasper succeeded with static techniques, Cowan, Dunnington and Holtzman also synthesized "hexagonal diamond" by subjecting cast iron that contained nodules of well-crystallized graphite to shock pressures in the range of 1 megabar. Conversion products were mixtures of cubic and hexagonal diamond in single crystal and aggregate form in sizes ranging to tens of microns. The material has been found suitable for industrial polishing use.

In 1968, El Goresy and Donnay discovered a new white-colored form of carbon in the graphitic gneiss of the Ries meteorite crater in Bavaria, Germany. This polymorph has a simple hexagonal crystal structure, and a density of 3.43 g/cc. It occurs with graphite, and is believed to have been transformed by the impacting shock of the meteorite. In 1969, Whittaker and Kintner found that this same form of carbon could be synthesized by low pressure sublimation of graphite at temperatures above 2550°K. This phase of carbon has been named chaoite.

In 1967, Kalashnikov, Vereshchagin and colleagues succeeded in synthesizing diamond in ballas form; i.e., as fine-grained ball-like polycrystalline aggregates similar to natural ballas. In 1969, Vereshchagin and associates described the synthesis of very fine-grained diamond aggregates similar to natural carbonado in texture. In 1970, Hall, and Stromberg and Stephens, using pressures and temperatures above 60 kb and 2100°K, successfully sintered diamond powders in predesigned shapes to bulk densities approaching that of single crystal diamond. Also in 1970, the General Electric Company succeeded in growing gem quality diamond single crystals in the weight range of one carat.

A. A. Giardini

Graphite

Structure. The solid forms of elemental carbon, excluding diamond, consist of flat, parallel planes of carbon atoms. Within each layer the atoms form an ordered "chicken-wire" pattern of condensed planar C_6 rings; a two-dimensional ordering. When the individual crystals (crystallites) exceed 200 Å in diameter and the parallel layers are spaced 3.354 Å apart, the "chicken wire" layers then orient to form a third dimensional ordering. Carbon of this three-dimensional ordering is called graphite.

When the crystallite diameter is less than 200 Å, the pinning forces weaken and the parallel layers further separate and rotate at random so as to destroy the 3-three-dimensional ordering of the carbon atoms. Carbon of this two-dimensionally ordered structure is called "amorphous" carbon—turbostratic is the more descriptive, scientific term. The interlayer spacing (d spacing) measures 3.44 Å or greater.

The bonding force between planes in graphite is ca. 2% of that within layers. This bonding structure results in a marked anisotropy of most properties.

Carbons whose d spacing average between 3.354 Å and 3.44 Å are termed graphite or carbon depending on the preponderance of one structure over the other. Natural graphite of 3.3538 Å d spacing is 100% graphitic. Manufactured graphite of 3.358 Å d spacing has about 15% disordered structure and is called graphite. A carbon product of 3.43 Å d spacing exhibits 7% graphitic (ordered) structure and is called carbon.

Natural. Natural graphite (also known as black lead and plumbago) occurs widely distributed over the world as a silvery-gray mineral and in three forms: (1) beds that once were coal; (2) veins, possibly of carbonaceous gas origin; (3) flakes disseminated in metamorphitized rock. The first form, commercially (and misleadingly) known as "amorphous" graphite, customarily analyzes ca. 80–85% C—the associated minerals being predominantly clay-like hydrated aluminum silicates. The second form, vein (or lump) graphite, sometimes occurs almost pure and is completely graphitized; associated minerals usually are aluminum silicates, some silica and often small percentages of iron sulfides. The third form, flake graphite, favored for refractories, is beneficiated by flotation; commercial grades range from 85–99% graphitic carbon.

Graphite processors further beneficiate, grind, sieve, and blend graphites for the market.

Manufactured. H. Y. Castner and E. G. Acheson in 1895 independently discovered that heating some forms of amorphous carbon above 2200°C changed the carbon to graphite. Thus started the manufactured graphite industry (also known as artificial graphite and synthetic graphite). A further development consisted of bonding petroleum coke and other carbons with coal tars and pitches, baking the piece to ca. 900°C to form a baked carbon product (See **Carbon, industrial**) and then heating at 2200–3000°C to graphitize the carbon. The baked piece may be impregnated with a light coal-tar pitch before graphitizing. Nuclear reactor graphites require further purification, by gases, to reduce impurities below 1 ppm. Recent developments include hot pressing for increased densities and include further heat treatment to control grain structure.

Pyrolytic. Pyrolytic graphite is formed on a substrate by carbon deposition of pyrolzed carbonaceous gases such as natural gas, methane, or carbon suboxide, C_3O_2, at 25–150 Torr and 1700–2500°C. The deposited crystallites align with their carbon layers parallel to the substrate. These carbons exhibit increasing densities with increasing temperatures and range from 1.8 g/cc to 2.2 g/cc (maximum theoretical density 2.268 g/cc). Their unusual characteristics lie in their strongly anisotropic properties parallel to and perpendicular to their substrate. The ratio of thermal conductivity $\parallel / \perp$ is in order of 100; ratio of electrical conductivities in the order of 1000. The anisotropy results from alignment of crystallites much more than from ordering of the carbon planes within crystallites. The d spacing of lower temperature products approach 3.44 Å, indicating random arrangement of the parallel carbon planes. Graphitizing at higher temperatures produces a well ordered crystallite approaching 3.354 Å d spacing. Alloys may be formed by introducing low atomic weight elements (e.g., boron) into the process.

Colloidal. Colloidal graphite may be either natural graphite or artificial graphite ground to ca. 1 micron particle size, coated with a protective colloid, and dispersed in a liquid. One selects the liquid carrier—water, oils, or synthetics—for the use intended. The two outstanding properties of colloidal graphite dispersions are: (1) the particles remain in suspension; (2) the particles "wick"—are carried by the liquid to most places the liquid penetrates.

Fibers. Graphite not only is the strongest material above 2000°C but is the stiffest and is comparatively light—giving it exceptionally high specific strength and modulus. Whiskers, fibers, and filaments come the closest to the calculated theoretical values of ca. 8×10^6 psi tensile strength and 80×10^6 psi elastic modulus. Whiskers, scrolls of single crystals, exhibit ca. 4×10^6 psi T.S. and other graphite fibers—much less costly—in the order of 0.5×10^6 psi T.S.

Fibers in composites provide stronger, stiffer, and lighter materials of construction. The most effective fibers have the higher T.S./E ratio. The desirable ratio of per cent strain to failure is unity; ratios above 0.5 are acceptable.

Graphite cloths are made from filaments, the precursor of which may be rayon, acrylics, or other synthetic yarns; or may be strands of continuously extruded tars or resins.

Basically the process of producing filaments is one of oxidizing, carbonizing, and graphitizing under tensile constraint.

Properties. The properties of graphite give rise to a variety of commercial uses. It marks (pencils); it is soft and unctuous (solid lubricant); it conducts electricity (motor and generator brushes, electrodes); it is a refractory, strength increases with temperature (crucibles and graphite refractories, aerospace technology); it is inert to most chemicals (chemical equipment); it is opaque (paint pigment); it also slows fast neutrons and has small cross section for neutron absorption (nuclear moderator for atomic reactors).

The physical strength of graphite *increases* with temperature as frozen-in stresses are relieved; an unusual property. Above 2500°C plastic deformation begins.

Sp. gr. ca. 2.25; sublimation point 3350°C; Mohs hardness 0.5–1.5; strongly diamagnetic; hydrophobic; strongly anisotropic. Forms both covalent compounds and intercalation (adduct) compounds. Heat of sublimation 170 kcal/mole.

SHERWOOD B. SEELEY

Cross-references: *Allotropes; Carbon; Diamond Synthesis.*

CARBON-12 SCALE OF ATOMIC MASSES

The Concept of Atomic Weight. The chemical method of determining the atomic weight of any element is, in many cases, dependent on a previous knowledge of the molecular weight of compounds in which the element is a constituent part. The determination of molecular weights is based fundamentally on Avogadro's hypothesis. The application of this hypothesis can yield only relative molecular weights and not absolute values; therefore, a defined standard is required on which one can base the atomic weights of other elements. Since hydrogen is the lightest element, it appeared best to choose this element as standard and assign the value of unity to its atomic weight. This would have been an adequate selection if the atomic or molecular weights alone were to be considered. But since these values must be used in connection with combining weights, oxygen serves as a better standard for comparison because of its abundance and reactivity. Since oxygen is a better working standard for chemically determined ratios, a scale based on oxygen gradually came to be preferred. It was natural to use an integral value for the atomic weight of this element. The choice of O = 16 had the fortunate coincidence of making nuclidic weights nearly identical with their mass numbers.

Existence of Two Scales. The scale based on oxygen seemed to provide a satisfactory atomic weight scale. It remained a happy choice until it was discovered that some elements consisted of isotopic mixtures. The discovery of isotopes would not have altered the situation at all were it not for the fact that ordinary oxygen, the standard itself, was found to consist of a mixture of isotopes. The physicists, working with mass spectrometers in their study of isotopes, naturally chose for their standard system the oxygen nuclide of mass 16, while the chemists continued to use 16 for the atomic weight of the isotopic mixture.

The unit of the physical scale is called the "absolute mass unit" (amu) or "isotopic mass unit" and the unit of the chemical scale is called the "atomic weight unit" (awu). Atomic masses are given on the physical scale while atomic weights are given on the chemical scale. For isotopic elements it would be more rigorous to use the term "mean atomic mass or weight," but there is no evidence of confusion resulting from the shorter designation used by chemists. The unit of the physical scale is $1/16$ the mass of O^{16} atom. The unit of the chemical scale is $1/16$ the weight of the mean mass of an atom of natural oxygen and is therefore slightly greater than the physical unit.

The quantitative relationship between the two scales is established in the following manner:

Let $M_c(X)$ = atomic weight, on the chemical scale, of any natural element X

$M_c(O^*)$ = atomic weight, on the chemical scale, of natural oxygen

$M_p(X)$ = atomic mass, on the physical scale, of element X

$M_p(O^*)$ = atomic mass, on the physical scale, of natural oxygen

By definition of the chemical scale, $M_c(O^*) = 16$ exactly. The following relation exists among the above four quantities:

$$\frac{M_c(X)}{M_c(O^*)} = \frac{M_p(X)}{M_p(O^*)} \qquad (1)$$

$M_p(O^*)$ is the average atomic mass of natural oxygen referred to O^{16}, and may be expressed in terms of mass, $M_{p,i}$, of each isotope and its abundance (atom fraction) A_i,

$$M_p(O^*) = \,< M_p(O^*) > \, = \sum A_i \times M_{p,i} \qquad (2)$$

By definition, the conversion factor is

$$r = \frac{M_p(O^*)}{M_c(O^*)} = \frac{\text{physical mass (amu)}}{\text{chemical weight (awu)}} \quad (3)$$

Hence from (1)

$$M_c(X) = \frac{M_c(O^*)}{M_p(O^*)} M_p(X)$$

$$= \frac{1}{r} M_p(X) \quad (4)$$

The conversion factor, r, may also be defined in terms of the ratio of the units of atomic weight (chemical scale) to the units of atomic mass (physical scale).

$$r^* = \frac{\text{unit of the atomic weight}}{\text{unit of the atomic mass}} = \frac{\text{awu}}{\text{amu}} \quad (5)$$

The numerator and the denominator are magnitudes of the units when referred to the same reference base scale. We could say that the ratio is defined as a pure number. Thus, one may write for r and r^* the following identity

$$r = r^* \frac{\text{amu}}{\text{awu}}$$

The value of r can be calculated from a knowledge of the masses and the abundance ratios of oxygen nuclides O^{16}, O^{17}, and O^{18},

that its mass should be intercomparable with the masses of other nuclidic species. Requirement (1) is clear in the light of what has already been discussed. Requirement (2) is closely related to the mass-spectroscopic method used by physicists in mass determinations.

Of course, even if the indefiniteness of the chemical scale could have been removed by another, better definition, there would still be two lists of atomic weights and also two sets of values for those constants that are dependent on the mole. To appreciate the seriousness of the problem, let us assume that the chemical scale was to be abandoned in favor of the physical scale. On the basis of this scale, $O^{16} = 16$ exactly, the atomic weight of natural oxygen would be 16.00440. This would mean a change of 275 ppm in the atomic weights and related values, an amount which cannot be regarded as negligible. The need would then arise for the immediate revision of all mass and related quantities reported in the literature. It is well to note that abandoning the chemical scale would have involved the greater burden and difficulties involved in the revision process. It was on such grounds that alternatives to abandoning either of the scales were sought, both by chemists and physicists.

At this time the necessity for giving up both the physical and the chemical scales and the need for adopting a "universal" scale were receiving more urgent attention. Several proposals were made. Among these a scale based on $H^1 = 1$ exactly was

$$r = \frac{\text{mean atomic mass of natural oxygen}}{\text{atomic weight of natural oxygen}}$$

$$\frac{A(O^{16})M(O^{16}) + A(O^{17})M(O^{17}) + A(O^{18})M(O^{18})}{16} > 1$$

Using values for the isotopic abundances and isotopic masses from the literature,

$$\frac{(99.7587)(16) + (0.0372)(17.004533) + (0.2039)(18.00487)}{16} \qquad 16$$

$$= 1.000273$$

The disparity between the chemical and the physical scales was further heightened when it was observed that the isotopic abundance of oxygen in nature varies from source to source. This led to the value of r varying between 1.000268 and 1.000278. In 1940 the IUPAC Commission on Atomic Weights adopted a value of 1.000275 for r corresponding to a "typical" natural oxygen. We thus note that the physical scale differs from the chemical scale by 275 ppm.

The Avogadro number, defined as the number of oxygen atoms in 16 grams of oxygen (now defined as the number of atoms in 12 grams of carbon-12), also depends on what scale is chosen. So also do other properties which are functions of the mole, for example, the Faraday constant, or the gas constant, R. The value based on one scale differs from the value based on the other scale by the factor r.

It is immediately clear that the key to the solution of the problem of two scales would be to agree on a single reference substance and assign a value to its mass. However, there are at least two requirements that this reference substance should meet: (1) that its mass should be an invariant, and (2)

suggested. The great disadvantage of such a scale was that on this basis the masses (on the chemical scale) of all atoms would have had to be reduced by 7870 ppm. This, indeed, is too great a change to be ignored. Hence it was argued by some that a large change of this magnitude might provide the spur for the total revision of all mass and related quantities.

When the Commission on Atomic Weights of the IUPAC met in 1957 it was obvious that certain steps would need to be taken in order to clarify the situation. The ideas proposed at that time aimed to solve the chemical atomic scale problem so as to avoid the need for the great task of revision of the already existing chemical data. The suggestions were the following:

(1) to refer the scale of 16 (exactly) as the atomic weight of a defined mixture of oxygen isotopes;

(2) to adopt a defined ratio of the atomic weights on the chemical scale to those on the physical scale; and

(3) to define the chemical scale such that the mass of O^{16} will be $16/r$, where r is the conversion

ratio discussed above and currently taken as 1.000275.

It should be noted that actually none of the above suggestions provided any means for the unification of the two scales. On the other hand, these suggestions and discussions served to emphasize the need for perhaps an entirely new scale instead of modifications of the existing ones.

Adoption of Carbon-12 Unified Scale. The Commission on Atomic Weights of the International Union of Pure and Applied Chemistry (IUPAC) had, prior to the meeting of the Union in 1959, proposed the adoption of a reference mass scale on C^{12} (carbon-12 isotope = 12 exactly) to which the atomic weights of all elements would refer and which would serve as a common scale for use by both chemists and physicists. In August, 1959, the IUPAC approved the recommendation for a unified atomic weight scale providing that similar action would be taken by the physicists. The corresponding representative organization for physics, the IUPAP, at its meeting in 1960, approved the adoption of the carbon-12 atomic weight scale, permitting the chemists to take final action at the following meeting of the IUPAC in 1961. This meeting was held on August 2–5, 1961, in Montreal; the reference scale, based on $C^{12} = 12$ exactly, was formally adopted.

Having approved the adoption of the scale based on $C^{12} = 12$ exactly, the IUPAC recommended universal use of the new scale as of January 1, 1962. It changed all chemical atomic weights by about 40 ppm, which is well within the limits of accuracy and precision of present-day chemical atomic weight determinations. However, these small changes still need to be taken into consideration whenever one is reporting critically selected physicochemical data or constants of the highest precision.

It might be stated that while carbon-12 as a reference standard does not operationally lend itself to techniques of chemical atomic weight determinations as satisfactorily as oxygen does, yet it has inherent advantages. These advantages result from the fact that the nuclidic mass of C^{12} can be very accurately related to the atomic weights of elements which in turn can be used as reference standards in the chemical atomic weight determinations.

A. LABBAUF

Cross-references: *Atoms, Periodic Law, Molecular Weight, Isotopes, Gas Laws.*

CARBONATES

The carbonates are of considerable industrial importance. The largest tonnages are consumed in the ceramic and metallurgical industries. For example, limestone (essentially $CaCO_3$) as a source of CaO, is used extensively for fluxing and refining purposes in iron and steel making. CaO, derived from $CaCO_3$ in sea shells, is used for the recovery of Mg from sea water. Dolomite [$(CaMg)(CO_3)_2$] and magnesite ($MgCO_3$) as sources of MgO, are used for the manufacture of refractories. Limestone, as a source of CO_2, is used in the Solvay process for the manufacture of Na_2CO_3 and $NaHCO_3$. Thus, both products of decomposition of abundant limestone are important. Na_2CO_3, as a source of Na_2O, is used for the manufacture of glass.

Most metals form carbonates, and a great many of these are found native. The metals which do not form carbonates are those whose hydroxides are very weak bases, for example Al and Cr. The carbonates can be classified as ionic compounds. The carbonate ion is planar and since the three C—O distances are equal, it is believed that resonance of a double bond occurs among the three C—O positions.

Metal carbonates are very slightly soluble in water, those of the alkali metals and Tl_2CO_3 being the only important exceptions. Therefore, the metallic carbonates can usually be prepared by precipitation by the action of a solution of a carbonate on a solution of a soluble salt of the metal. $(NH_4)_2CO_3$ and NH_4HCO_3 are also soluble in water at room temperature but decompose in hot water. The action of excess CO_2 on carbonates in solution results in the formation of bicarbonates, which are more soluble. "Temporary hard water" contains bicarbonates in solution which decompose on heating to form the less soluble carbonates. The latter carbonates precipitate out of solution and yield the deposit often observed on the walls of teakettles and boiler pipes. Carbonates are decomposed by acids with evolution of CO_2. CO_2 is slightly soluble in H_2O, and this solution exhibits feeble acid properties. It contains carbonic acid, H_2CO_3, which is unstable and cannot be isolated.

The thermal decomposition rates of the carbonates increase with increasing temperature. For some, rapid decomposition starts at comparatively low temperatures (Ag_2CO_3, 200°C, for example) whereas for others, rapid decomposition starts at comparatively high temperatures ($CaCO_3$, 880°C, for example). In addition, the alkali metal carbonates melt before rapid decomposition occurs.

There exists at a given temperature a minimum pressure or partial pressure of CO_2 in the environment which will prevent a carbonate from decomposing. This pressure is known as the equilibrium decomposition pressure for the carbonate, and it increases with increasing temperature. At a given temperature, this pressure is different for the various carbonates existing in nature. The above statements can be placed on a thermodynamic foundation by considering the thermal decomposition of the carbonate MCO_3, where M is a bivalent metal cation. The decomposition reaction is:

$$MCO_3(\text{solid}) \rightarrow MO(\text{solid}) + CO_2(\text{gas})$$

The equilibrium constant, K, for this reaction is $(P_{CO_2})_e$, where $(P_{CO_2})_e$ is the equilibrium partial pressure of CO_2. The activities of MO and MCO_3 are unity, since they are in their standard states. Thus, the free-energy change for this reaction is

$$\Delta G = -RT \ln (P_{CO_2})_e + RT \ln P_{CO_2} =$$
$$\Delta G° + RT \ln P_{CO_2}$$

where R is the gas constant, $\Delta G°$ is the standard free-energy change, and P_{CO_2}, is the partial pressure of CO_2 in the environment at a given absolute temperature T. When $P_{CO_2} = (P_{CO_2})_e$, this reaction is at equilibrium or no net change occurs and $\Delta G = 0$. When

$$P_{CO_2} < (P_{CO_2})_e$$

ΔG is negative and the reaction proceeds continually to the right provided this condition is maintained. When

$$P_{CO_2} > (P_{CO_2})_e$$

ΔG is positive and no decomposition can occur. In fact, if free MO is present under this latter condition, it will convert to MCO_3. Therefore, the driving force or tendency for decomposition to occur is determined by the equilibrium partial pressure of CO_2, analogous to the use of vapor pressures of substances as an indication of their vaporizing or escaping tendency.

The equilibrium partial pressure of CO_2 for $CaCO_3$ at 880°C is one atmosphere, whereas for Li_2CO_3, Na_2CO_3, and K_2CO_3 at 1200°C, the equilibrium partial pressures in mm Hg are 300, 41, and 27, respectively. The melting points of the latter three carbonates are 618, 851, and 890°C, respectively. $CaCO_3$ melts below 1100°C if the CO_2 pressure in the environment is maintained above the equilibrium decomposition pressure. It is apparent that the atomic binding forces are different in the various carbonates.

An equilibrium decomposition temperature is ascribed to the various carbonates. It is the temperature at which the P_{CO_2} required for equilibrium is one atmosphere. In this case, the standard free-energy change, $\Delta G°$, is also zero. For the carbonates of Mg, Ca, Sr, and Ba, the equilibrium decomposition temperature increases with increasing atomic number of the cation or with increasing cation radius. The heat of decomposition can be obtained by measuring the equilibrium decomposition pressure as a function of temperature and utilizing the Gibbs-Helmholtz equation. The equilibrium partial pressure of $\Delta G°$ for a given temperature can also be calculated from thermochemical data provided the solid reactants and products are at unit activity or if activity data are also known.

The decomposition of a carbonate particle starts at the surface and proceeds inward. It is known that the zone of decomposition increases linearly at a given temperature for $CaCO_3$, $MgCO_3$, (CaMg) $(CO_3)_2$, and (Ag_2CO_3). This rate of zone increase is constant at a given temperature regardless of the diameter of the particle. It appears, at least for these carbonates, diffusion of CO_2 from the decomposing interface is not the limiting process. Any mechanism of decomposition must include the above observation and the fact that the resulting oxide usually has a different crystal structure than the parent carbonate. The overall process can be represented by two consecutive reactions: the first, decomposition; and the second, a phase change.

The thermal decomposition rate of the carbonates can be increased by the presence of other gases in the environment, notably H_2O and NH_3. These gases have large dipole moments in the liquid state and are dipoles in the gaseous state and can catalyze the decomposition.

The initial thermal decomposition rate of $CaCO_3$ is influenced by line imperfections (dislocations) and area imperfections (twin boundaries).[1,2]

J. H. WERNICK

Cross-references: *Thermodynamics; Crystals (Dislocations); Heterogeneous Reactions.*

References

1. Thomas, J. M., and Renshaw, G. D., *J. Chem. Soc.* (A), 2058 (1967).
2. Stern, K. H., and Weise, E. L., "High Temperature Properties and Decomposition of Inorganic Salts," Part 2, "Carbonates," *N5RDS-NBS-30*, U.S. Dept. Commerce, National Bur. Standards, Washington, D.C., Nov., 1969.

Organic Carbonates

Organic carbonates are esters of carbonic acid. Based on the formula, $\begin{matrix} HO \searrow \\ HO \nearrow \end{matrix} C{=}O$ for carbonic acid, the normal esters are formed by the replacement of both H-atoms by aliphatic (alkyl) groups or by aromatic (aryl) groups so that the typical formula for the esters is $\begin{matrix} RO \searrow \\ RO \nearrow \end{matrix} C{=}O$. Other carbonic acid esters contain chlorine or an alkali metal instead of one —OR group. For example, chloroethyl carbonic ester or ethyl chlorocarbonate is $Cl \cdot CO \cdot OC_2H_5$ or $\begin{matrix} C_2H_5O \searrow \\ Cl \nearrow \end{matrix} C{=}O$, and potassium ethyl carbonate is $KOCOOC_2H_5$ or $\begin{matrix} C_2H_5O \searrow \\ KO \nearrow \end{matrix} C{=}O$.

If both OH— groups of carbonic acid were replaced by Cl—, the product would be phosgene, $COCl_2$, and phosgene is actually a raw material for the preparation of many organic carbonates. For example, phosgene and methyl alcohol form dimethyl carbonate, by the reaction: $COCl_2 + 2CH_3OH \rightarrow CO(OCH_3)_2 + 2HCl$.

In some cases chlorocarbonic ester, $\begin{matrix} RO \searrow \\ Cl \nearrow \end{matrix} C{=}O$, is made as a first step by the action of phosgene on alcohol, such as ethyl alcohol which yields $ClCOOC_2H_5$. In a second step, the chlorocarbonic ester is reacted with more alcohol to form normal carbonic acid ester. Further, two of the methods of forming polycarbonate resins depend on the reaction of bisphenol A [2,2'-bis (4-hydroxyphenyl) propane] with phosgene (see **Polycarbonate Resins**).

The normal alkyl carbonic acid esters are colorless sweet-smelling liquids, but diphenyl ester is a solid melting at 78°C. The dimethyl ester boils at 90°C, the diethyl at 125.8°C, the dipropyl at 168°C, the diisobutyl at 190°C, the diisoamyl at 229°C, and the methyl ethyl at 109°C. All are insoluble in water, are readily hydrolyzed by caustic alkalis, and react with ammonia to form carbamic esters and urea.

The aliphatic chlorocarbonic esters are decomposed by water with the formation of alcohol, HCl and CO_2, and have lower boiling points than the corresponding normal carbonic acid esters. The alkali metal aliphatic carbonic acid esters, such as $KOCOOC_2H_5$, are also decomposed by water to yield alcohol and the alkali metal carbonate.

Another type of organic carbonate is the tetraalkyl orthocarbonate based on the hypothetical orthocarbonic acid, $C(OH)_4$. The tetraethyl ester, $C(OC_2H_5)_4$, also called tetrahydroxymethane, boils at 159°C, and the tetra-*n*-propyl ester also named tetrapropoxymethane, boils at 224°C. The former is prepared by the reaction of chloropicrin and sodium ethoxide: $CCl_3NO_2 + 4C_2H_5ONa \rightarrow C(OC_2H_5)_4 + NaNO_2 + 3NaCl$.

CLIFFORD A. HAMPEL

CARBON CYCLE

Biochemical. In bright sunlight the rate of the photosynthetic reaction is about 15 to 30 times the respiratory rate. A sunflower leaf may gain 9% of its dry weight per hour. Considering that respiration proceeds continuously while photosynthesis occurs only during the hours of light, the net rate of the latter is about five times greater. This means an excess of carbohydrate, which is partly stored and partly used for the growth of the plant. Non-green plants and animals, which constitute the heterotrophic organisms, depend upon this excess for their organic food.

Land plants, which utilize the carbon dioxide present in the air, grow faster if supplied with more carbon dioxide than they can now find in nature. This averages only 0.03% of the gas volume at the earth's surface. Aquatic plants thrive on the carbon dioxide dissolved as gas or as carbonates in the water. The best estimate for the rate of the total carbon dioxide reduction on earth is a turnover of 10^{10} tons of carbon per year, more than two-thirds of which is contributed by the flora of the oceans. An amount of carbon dioxide equal to the total readily available reservoir passes through the life cycle in about 350 years. The cycle of carbon requires also cycles of oxygen and hydrogen. Since both these elements are available in larger quantities, it takes about 2,000 years to renew all the oxygen in the air and 2,000,000 years to decompose all the water on earth.

There was probably more carbon dioxide in the atmosphere of the earth at earlier times. The enormous deposits of coal and oil, the so-called fossil fuels, are the remnants of once living plants and microorganisms. These products of ancient photosynthesis (which, since the beginning of the Industrial Revolution, man has been returning to the air in ever increasing amounts as carbon dioxide) have been out of circulation for many millions of years.

(Reprinted with permission from Encyclopaedia Britannica, Copyright 1970.)

HANS GAFFRON

Cross-reference: *Photosynthesis.*

Stellar. Thermonuclear reactions involving carbon in the role of a catalyst play an important role in the conversion to hydrogen to helium, hence in energy generation, in the interiors of stars. The basic reaction sequence, first proposed independently by Bethe[1,6] and by von Weizacker[2] in 1938, proceeds as follows:

$$C^{12} + H^1 \rightarrow N^{13} + \gamma$$
$$N^{13} \rightarrow C^{13} + e^+ + \nu$$
$$C^{13} + H^1 \rightarrow N^{14} + \gamma \qquad (1)$$
$$N^{14} + H^1 \rightarrow O^{15} + \gamma$$
$$O^{15} \rightarrow N^{15} + e^+ + \nu$$
$$N^{15} + H^1 \rightarrow C^{12} + He^4$$

Here four proton (H^1) capture reactions, interspersed by the decays of N^{13} and O^{15}, leave a residual C^{12} nucleus and an alpha-particle (helium nucleus). The net result of this sequence is therefore the transmutation of four hydrogen nuclei (protons) to a single helium nucleus and two positive electrons or positrons (e^+), with the release of two rather low energy neutrinos (ν) and approximately 25 million electron volts of energy in the form of photons or gamma rays (γ). This reaction sequence, involving primarily carbon and nitrogen isotopes, is variously referred to as the Carbon-Nitrogen cycle or simply the Carbon cycle. The carbon and nitrogen nuclei assist in the transformation of hydrogen to helium, but they are not themselves consumed in the process once equilibrium in their relative abundances is established under the prevailing temperature and density conditions: they are true nuclear catalysts.

The interaction of a proton with N^{15} does not lead in every case to the products He^4 and C^{12} as indicated above. Occasionally (about one time in 2500), under typical burning conditions, the proton is captured by N^{15} to form O^{16} with accompanying gamma-ray emission. The ensuing reaction sequence is given by

$$N^{15} + H^1 \rightarrow O^{16} + \gamma$$
$$O^{16} + H^1 \rightarrow F^{17} + \gamma \qquad (2)$$
$$F^{17} \rightarrow O^{17} + e^+ + \gamma$$
$$O^{17} + H^1 \rightarrow N^{14} + He^4$$

In this case two successive proton captures result in the formation of F^{17} which undergoes position decay to O^{17}; the interaction of a proton with O^{17} then gives a helium nucleus and N^{14}, which can feed back into the previous cycle. This entire sequence (Eqs. 1 and 2) is generally referred to as the Carbon-Nitrogen-Oxygen cycle, or simply the CNO-cycle.

The CNO-cycle defined above does not contribute significantly to hydrogen burning in our sun. The central temperature of the sun is slightly too low for these reactions to proceed sufficiently rapidly. In more massive stars, however, higher temperatures accompany the higher pressures required to support the heavier outer regions; under these conditions hydrogen burning by means of the CNO-cycle will generally dominate.

An important consequence of this CNO-cycle hydrogen burning is the synthesis of N^{14}. Detailed calculations of CNO-burning[3] reveal that, generally, N^{14} will be the most abundant nucleus under equilibrium burning conditions; this is due to the fact that the probability that a given N^{14} nucleus will capture a proton is relatively low. Most of the initial carbon and oxygen present in the hydrogen-burning region will therefore be converted to N^{14} by this mechanism. This is extremely important, as the initial formation of C^{12} and O^{16} by means of helium burning reactions[4] is not accompanied by the production of amounts of N^{14} consistent with the solar system abundances[5]. The formation of N^{14} is therefore a "secondary" process of nucleosynthesis, requiring C^{12} and O^{16} nuclei formed in a previous generation of stars to have been mixed into the interstellar gas from which the CNO-burning star has since formed.

There is rather convincing observational evidence for CNO-cycle burning in stars.[10] Wallerstein, Greene and Tomley[6] have analysed the star HD 30353, which shows a very high abundance of nitrogen and a deficiency of hydrogen. They interpret the nitrogen as N^{14} which has been formed in CNO-burning in the interior and carried to the surface of the star by convective motions. They have determined the ratio of the abundance of nitrogen to that of oxygen in this star to be approximately 50. This is in marked contrast to the ratio $N/O \simeq$ 0.1 characteristic of solar system material[5]; it is quite consistent, however, with calculations[9] of the ratios expected ($N/O \simeq 60$) for CNO-burning under typical hydrogen-burning conditions in stars. Similarly, the ratio of C^{12} to C^{13} observed in carbon stars[7,8] is about 3 or 4 (as compared with roughly 90 for solar system abundances), which is almost exactly the inverse ratio of the interaction probabilities for the $C^{12}(p,\gamma)N^{13}$ and $C^{13}(p,\gamma)N^{14}$ reactions. This material is again assumed to have participated in equilibrium CNO-burning in the interior of the star, and then to have been transported to the surface by convection.

J. W. TRURAN

References

1. Bethe, H. A., Physical Review **55**, 434 (1939).
2. von Weizacker, C. F., Physikalische Zeitschrift **39**, 633 (1938).
3. Caughlan, G. R., and Fowler, W. A., Astrophysical Journal **136**, 453 (1962).
4. Burbidge, E. M., Burbidge, G. R., Fowler, W. A., and Hoyle, F., Reviews of Modern Physics **29**, 547 (1957).
5. Cameron, A. G. W., in *The Origin and Distribution of the Elements*, L. H. Ahrens (Ed.), Pergamon Press, New York (1968).
6. Wallerstein, G., Greene, T. F., and Tomley, L. J., Astrophysical Journal, **150**, 245 (1967).
7. McKellar, A., in *Stellar Atmospheres*, J. L., Greenstein (Ed.), University of Chicago Press, Chicago (1960).
8. Wyller, A. A., Astrophysical Journal **143**, 829 (1965).
9. Bethe, H. A., Science **161**, 541 (1968).
10. Wallerstein, G., Science **162**, 625 (1968).

CARBON DIOXIDE

Carbon dioxide, CO_2, formula weight 44.01, is a colorless gas with a pungent odor and acid taste. The first recognition of carbon dioxide as a gas distinct from other gaseous substances is attributed to J. B. Van Helmont (1577–1644), who detected in the products of combustion of charcoal the same gas as that given off in the process of fermentation. Carbon dioxide is widely produced today as a by-product from synthetic ammonia production, fermentation, or lime kiln operations, and by separation from flue gases or certain natural gases by absorption processes.

It occurs in the combustion products of all carbonaceous fuels and may be recovered from them in various ways. Carbon dioxide is also a product of animal metabolism and is important in the life cycle of animal and vegetable matter on earth. It is thus present in the atmosphere in small quantities (about 0.03% by volume or 0.0474% by weight) but may not be recovered economically from this source, although the total available amount is about 2,750 billion tons.

Properties. Sublimation point, $-78.5°C$ at 1 atm; triple point, $-56.6°C$ at 5.11 atm (1 kg per cm²); critical temperature 31.0°C, critical pressure 72.80 atm or 1070.16 psia; latent heat of vaporization 149.6 Btu per lb or 83.12 gm cal per gm at the triple point, and 101.03 Btu/lb at 0°C; gas density, 1.976 g per liter at 0°C and 1 atm; liquid density, 0.914 kg per liter at 0°C and 34.4 atm; solid density, 1.512 kg per liter or 94.39 lb per cubic foot at $-56.6°C$; solubility in water, 1.713 vols. per vol. at 0°C and 0.759 vol. per vol. at 25°C and 760 mm partial pressure of carbon dioxide. The heat of formation of carbon dioxide is 94.05 kg cal per gram mole (or 169,450 Btu per lb mole), showing the oxidation of carbon to be a highly exothermic reaction. Carbon dioxide will not support combustion and is used as a fire-extinguishing agent.

Chemical Reactions. Carbon dioxide is not a chemically active compound as such and high temperatures are generally required to promote its reactions. In water solution, however, the opposite is the case, as the solution is acidic in nature and many reactions take place readily. A few of the more important of these are included here.

Carbonic acid, H_2CO_3, is formed by the reaction between water and carbon dioxide and is a highly dissociated acid. The pH of saturated carbon dioxide solutions in water varies from 3.7 at 1 atm and 25°C to 3.2 at 23.4 atm and 0°C. A hydrate, $CO_2 \cdot 8H_2O$, is formed on cooling at elevated pressures. CO_2 is generally stable and does not break down under normal conditions. At high temperatures the reaction

$$2CO_2 \rightarrow 2CO + O_2$$

will take place. This reaction is also assisted by ultraviolet light, high pressure, and electric discharge but only a small percentage dissociation occurs.

Carbon dioxide may be reduced by several means. The most common of these is the reaction with hydrogen:

$$CO_2 + H_2 \rightarrow CO + H_2O$$

This is the reverse of the "water gas shift" reaction which is used commercially in the production of hydrogen and ammonia. CO_2 may also be reduced catalytically with various hydrocarbons and with carbon itself at elevated temperatures. The latter reaction occurs in almost all cases of combustion of carbonaceous fuels and is generally employed as a method of producing carbon monoxide:

$$CO_2 + C \rightarrow 2CO$$

Carbon dioxide will react with ammonia, as in the first stage of urea manufacture, to form ammonium carbamate:

$$CO_2 + 2NH_3 \rightarrow NH_2COONH_4$$

Carbon dioxide is an important factor in plant growth through the process of photosynthesis. This reaction takes place in the green coloring material in the leaves of plants and trees where the energy obtained by absorption from the rays of the sun causes a reduction of the carbon dioxide in the chlorophyll-bearing cells to carbohydrates and oxygen. The oxygen is released to the atmosphere thus completing the carbon dioxide-oxygen cycle between animal and plant life.

Radioactive Carbon ^{14}C. A radioactive carbon isotope of mass 14, with an estimated half-life of 5568 years, is present in trace amounts in atmospheric carbon dioxide and in natural carbon-containing substances, such as wood. Procedures have been developed for dating such substances, by comparing the relative amounts of ^{14}C in them with that in newly formed natural organic substances. Artificially produced $^{14}CO_2$ is being used as a tracer to study various chemical reactions and biological processes.

Commercial Production. Of the various sources the following are most important in the commercial production of carbon dioxide:

(1) Recovery from flue gases resulting from the combustion of carbonaceous materials according to the following formula:

$$C + O_2 \rightarrow CO_2$$

(2) Byproduct from synthetic ammonia production in which hydrocarbons are converted to hydrogen and carbon dioxide:

$$C_nH_{2n+2} + 2nH_2O \rightarrow nCO_2 + (3n + 1)H_2$$

This has become the largest source of commercial carbon dioxide.

(3) By-product of the fermentation industry in which a sugar such as dextrose is converted to ethyl alcohol and carbon dioxide by an enzyme reaction as follows:

$$C_6H_{12}O_6 \rightarrow 2C_2H_5OH + 2CO_2$$

(4) By-product of lime kiln operation in which naturally occurring carbonates are thermally decomposed:

$$CaCO_3 \rightarrow CaO + CO_2$$

Although the carbon dioxide produced and recovered by the methods outlined above has a high percentage of purity, traces of hydrogen sulfide and sulfur dioxide are frequently present which give the gas a slight odor or taste. The fermentation gas recovery processes include a purification stage but carbon dioxide recovered by other methods must be further purified before it is acceptable for beverage or dry ice use. This applies particularly to dry ice which is to be used for refrigerating food. The most commonly used purification methods are treatments with (1) potassium permanganate, (2) potassium dichromate, and (3) activated carbon.

Before the liquefaction of carbon dioxide, the water with which it has become saturated during the various recovery and purification operations must be removed. This may be accomplished by any one of several commercial methods such as treatment with calcium chloride, silica gel, activated alumina or bauxite, or refrigeration.

Carbon dioxide may be liquefied at any temperature between those at its triple point ($-70°F$) and its critical point ($87.8°F$) by compressing it to the corresponding liquefaction pressure. Two conditions are employed in commercial practice, the first near the critical temperature, with water being employed for cooling, and the second at temperatures in the neighborhood of 0 to 20°F, with ammonia or other refrigerants being utilized for cooling. In the first method, liquefaction of the carbon dioxide is accomplished by compressing and cooling the gas until it is slightly below the critical temperature of 87.8°F at a pressure of 73 atm or more. Low-temperature liquefaction of carbon dioxide is usually employed in cases in which the liquid carbon dioxide is to be used for the production of dry ice.

Dry Ice. Solid carbon dioxide, popularly known as "dry ice," accounts for about one third of the carbon dioxide produced. In most of the larger installations in this country, solidification takes place directly in a hydraulic press where the blocks of dry ice 20×20×10 inches are formed. The blocks are subsequently cut up into 10 inch cubes weighing approximately 50 pounds each. In one method of dry ice press operation, carbon dioxide snowing takes place at the triple point and thus the evaporated gases are available for recycle at elevated pressures. After the snowing operation is completed, the press chamber is vented down to atmospheric pressure thus releasing more carbon dioxide gas for recycle at low pressure.

The major use for dry ice is for preserving food by refrigeration. The advantages of carbon dioxide over water ice may be summarized as follows: (1) higher heat of fusion per pound, (2) lower storage temperatures are possible, (3) no residual water or brine results from melting. Other uses for solid carbon dioxide are more or less specialties and do not account for a large percentage of its production. They are: shrink fitting of machine parts, chilling of aluminum rivets before use in aircraft manufacture, chilling golf ball centers before winding, laboratory uses for cooling baths and sample freezing; refrigeration of serum and blood banks in hospital uses; "seeding" of clouds to precipitate rainfall.

While the production rate of dry ice has declined slightly in recent years, the production of liquid carbon dioxide has been increasing at a rapid rate. Carbon dioxide manufacturers have developed efficient systems for shipping and storing liquid carbon dioxide in bulk, and are supplying refrigerated storage tanks to large users which they refill by tank truck or tank car shipments. New uses for liquid carbon dioxide include rapid chilling of trucks and rail cars, chilling of meat in hamburger grinding machines, chilling of test chambers for research purposes, and fracturing and acidizing of oil wells. It is used in considerable quantities in mining operations using the Cardox method of controlled vaporization blasting. Miscellaneous uses include: operation of bell buoys and railroad signals by the power produced from the expanding gas, inflating life rafts and many others.

The largest user of gaseous carbon dioxide is the carbonated beverage industry. Considerable quantities are also used in fire-fighting equipment. It is used by the chemical industry as an inert gas blanket over reaction and storage vessels to prevent oxidation. Other chemical uses are mentioned in the section on reactions above. Miscellaneous uses include: air displacement in drying electrical cables, hardening concrete specialty products, scale removal from water pipes, as a respiratory stimulant, testing gas masks, humane killing of animals, shielded arc welding, hardening of sand cores and molds in foundries, and as an atmospheric additive in greenhouses. Carbon dioxide has become important in gas lasers. It is being used as a laser power source with electrical, thermal and chemical pumping. Power outputs up to one kilowatt have been realized, with applications being developed in drilling and cutting metals, welding, and bonding operations.

Production Data. The annual production in the United States, as obtained from the U.S. Department of Commerce, is given below (in thousands of short tons):

	1963	1965	1967	1969
Liquid and Gas	649	665	725	790
Solid	434	421	368	395

The total production of all three forms in 1971 amounts to 1,258 thousand tons.

R. M. REED
N. C. UPDEGRAFF

Physiological Aspects

Carbon dioxide, a by-product of the metabolic activity of all cells, is one of the most important chemical regulators in the human body. It can be truly said that without carbon dioxide human life would be impossible. In less specialized forms of life carbon dioxide is merely a waste product. In the more highly evolved animals, such as man, it serves to regulate the activity of the heart, the blood vessels, and the respiratory system.

Normal air contains about 0.03% by volume of CO_2. A poorly ventilated room may contain as much as 1%. Concentrations of the gas from about 0.1 to 1% by volume induce languor and headaches; 5% CO_2 greatly stimulates respiration; 10% induces severe distress; over 30% produces unconsciousness, above 40% causes coma and death.

Up to 10–15% CO_2 in air progressively causes an increase of respiratory rate; maximum stimulatory effect is approached at this CO_2 range. In concentrations of CO_2 above 7% subjective effects in man are pronounced: dyspnea, restlessness, fainting, severe headache. Between 20–30% CO_2 general convulsions are observed.

Resting men who are transferred from breathing room air to a mixture of 4.1% by volume of CO_2 in room air show an increase in expiratory minute volume from an average of 8 to about 15 liters; the respiratory rate increases from 14 to about 18 per minute; and the tidal volume from 0.500 to almost 0.900 liters. If the amount of CO_2 is increased, much greater increases (percentagewise) in the above physiological measurements follow; expiratory minute volume, 420; tidal volume, 260; respiratory rate, 160. Acclimitization to elevated CO_2 concentrations occurs in man.

As a general rule, the respiration of individual cells decreases as the concentration of CO_2 in the medium increases. Fish show a lessened capacity to extract oxygen from their environment when increasing amounts of CO_2 are present. On the other hand, many invertebrates show marked increases in respiratory rate (or ventilation) with increased amounts of the gas in their surroundings.

Photosynthetic and autotrophic bacteria reduce carbon dioxide, which is assimilated into complex molecules for use in synthesizing various cellular constituents. The gas is apparently assimilated, at least to a small extent, by the heterotrophic bacteria. Certainly it is required for any growth in these forms. Many pathogenic bacteria require increased carbon dioxide tension for growth immediately after they are isolated from the body. The production of hemolysins and like substances is greatly enhanced by adding 10 to 20% of CO_2 in the air which comes in contact with the cultures.

In men the average amount of CO_2 in the alveolar air is about 5.5% by volume; during the breathing cycle this concentration varies only slightly. In women and children somewhat lower mean values obtain.

The oxygen dissociation curve for blood is shifted to the right when the partial pressure of carbon dioxide in air is increased. This is referred to as the "Bohr Effect." It means that for a given partial pressure of oxygen, hemoglobin holds less oxygen at high concentration of CO_2 than at a lower. It is evident, then, that the production of CO_2 by actively metabolizing tissues favors the release of oxygen from the blood to the cells where it is urgently needed. Moreover, at the alveolar surfaces in the lungs the blood is losing CO_2 rapidly, which loss favors the combination of oxygen with hemoglobin.

In every 100 ml of arterial blood there is a total of 48 ml of free and combined CO_2. In venous blood of resting man there is about 5 ml more than this. Only about 1/20 is uncombined, a fact which indicates that there is a specialized mechanism, aside from simple solution, for the transport of CO_2 in the blood.

About 20% of the CO_2 in the blood is carried in combination with hemoglobin as carbaminohemoglobin. The balance of the combined carbon dioxide is carried as bicarbonate. A CO_2 dissociation curve for blood can be prepared just as for oxygen, but the shape is not the same as for the latter. As the partial pressure of CO_2 in the air increases, the amount in the blood increases; the increase is practically linear in the higher ranges. Oxygen exerts a negative effect on the amount of CO_2 which can be taken up by the blood.

In working muscles large amounts of CO_2 are produced. It causes local vasodilation. The diffusion of some of it into the blood stream slightly raises the concentration there. It circulates through the body and the capillaries of the vasoconstrictor center, where it excites the cells of the center, resulting in an increase of constrictor discharges. If one recalls the stimulating effect of CO_2 on cardiac output, it is evident that a most effective mechanism exists for increasing circulation through active muscles. More blood is pumped by the heart per minute and the arterial pressure is increased by the general vasoconstriction; blood is forced from the inactive regions, under increased pressure, through the widely dilated vessels of the active muscles.

Narcosis due to CO_2 is characterized as follows: mental disturbances which may range from confusion, mania, or drowsiness to deep coma; headache; sweating; muscle twitching; increased intracranial pressure; bounding pulse; low blood pressure; hypothermia; and sometimes papilloedema. The basic mechanism by which carbon dioxide induces narcosis is probably through interference with the intracellular enzyme systems, which are all extremely sensitive to pH changes. Carbon dioxide at partial pressures far below those associated with CO_2 narcosis acts synergistically with certain inert gases to produce narcosis at hyperbaric pressures.

CHARLES G. WILBER

CARBONIUM IONS AND CARBANIONS

Carbonium ions are characterized by the structure R_3C^+ in which the carbon's valence shell contains but 6 electrons. They can be named either as alkyl cations or as alkyl carbonium ions. The ion $(CH_3)_3C^+$ would be named tertiary butyl cation (t-bu$^+$) or trimethyl carbonium ion. Unfortunately, this ion has often been called the t-butyl carbonium ion, a name applicable to $(CH_3)_3CCH_2^+$. For this and other reasons, the alkyl cation system of naming is recommended. It is further recommended that RCO^+ species be named acyl cations, for example, CH_3CO^+ is the acetyl cation.

Between 1900 and 1950, only the triphenylmethyl cation, $(C_6H_5)_3C^+$, was known as a stable species. This stemmed from the work of Hantzsch on the freezing point depression of triphenylmethanol in 100% H_2SO_4. These studies were extended in 1950–1955 to other triarylmethyl cations as well as a limited group of Ar_2CH^+ and $ArC^+(CH_3)_2$ ions.

With the introduction of nuclear magnetic resonance spectroscopy, the stable existence of a wide variety of alkyl, alkenyl, acyl, arylalkyl, and polyenyl cations was quickly proved. The most successful method of producing such stable cations was the addition of alkyl fluorides, alcohols, or alkenes plus HF to SbF_5. However, other acid systems such as H_2SO_4, SO_3—H_2SO_4, $ClSO_3H$, and CH_2Cl_2—$AlCl_3$ were also used. At present, even such simple examples as $(CH_3)_2CH^+$, $(CH_3)_3C^+$, CH_2=CH—CH_2^+ and CH_3CO^+ have been extensively studied and their various spectra recorded.

Despite the paucity of direct knowledge (1900–1960), carbonium ions were proposed as transient intermediate at an increasing rate. By 1950, they had become accepted along with free radicals and carbanions as the three dominant intermediates in organic reactions. The many ingenious experiments leading to this acceptance are documented in standard texts of physical-organic chemistry.

The outstanding chemical characteristic of carbonium ions is their ability to share electron pairs or even abstract electron pairs from relatively inert compounds. Thus they may react with olefins to produce dimeric and polymeric cations; they may abstract hydride (proton plus an electron pair) from a hydrocarbon; and of course they react with all types of electron pair donors.

Related to the above chemistry is the prevalence of internal rearrangements. These are most commonly the shift of CH_3^- to the carbonium carbon from an adjacent alpha carbon. Alkyl groups larger than methyl can presumably also undergo such 1,2-alkyl shifts, but there is little definitive evi-

dence. The possibility of 1,3- and more distant alkyl shifts has been much debated, but it is only recently that definitive evidence has appeared which establishes 1,3-methyl shifts. Hydride shifts of the 1,2- and 1,3-type have been found.

Carbonium ion chemistry is often of great complexity in terms of the great rapidity of the reactions, the extensive structural rearrangements taking place, and the large variety of products formed. Despite such complexity, many industrial processes involve carbonium ion reactions, e.g., catalytic cracking of petroleum, production of isooctane and other components of high octane gasoline, and most aromatic substitution reactions.

Carbanions

Carbanions are characterized by the structure $R_3C:^-$, which is negatively charged and in which the carbon has a completed octet of electrons. Along with carbonium ions and free radicals, they compose the three dominant types of reactive intermediates in organic chemistry. Many base-catalyzed condensations, such as the Claisen, Michael, Mannich, Knoevenagel, and aldol, are believed to proceed via carbanions. Organic sodium, lithium, magnesium, etc. compounds can be sources of carbanions (R^-Na^+) and many of their reactions are conveniently so formulated. However, there is usually some degree of metal-carbon bond in such organometallics, which usually persists into the transition state in chemical reactions.

Carbanions are generated by abstraction of a proton (H^+) from an organic compound by the action of a strong base such as $NaNH_2$, NaH, buLi, t-buOK, or NaOH. They also arise (in the form of organometallics) by the action of Na, Li, Mg, etc. on organic halides.

Some carbanions that are highly resonance-stabilized, such as $^-CH_2NO_2$ and $(CH_3CO)_2CH^-$, can exist in high concentration in water solution. Simple alkyl carbanions, such as CH_3^-, abstract H^+ instantly from water, alcohols, and other protonic solvents to form the alkane.

The chemistry of carbanions is dominated by H^+ abstractions, additions to the carbon of C=O bonds, and the addition to certain electron poor C=C bonds. They are notoriously reluctant to rearrange internally and in this regard they contrast strongly with carbonium ions.

NORMAN C. DENO

CARBON MONOXIDE

Chemical Aspects

Carbon monoxide is a colorless, odorless, toxic gas. Its toxicity results from preferential reaction with hemoglobin. Carbon monoxide is formed when combustion of carbonaceous material occurs in an insufficient supply of air or oxygen. Carbon monoxide is a component of some manufactured gases; it occurs in exhaust gas from automotive engines and some household heaters, in certain industrial by-product gases, in some volcanic gases, and in tobacco smoke. It is used as an intermediate in the synthesis of various organic compounds, as a reducing agent, and as a fuel along with other components of some manufactured gases.

Preparation. In 1766 Lassone obtained a flammable gas by heating charcoal with zinc oxide. The identity of this gas was in dispute until 1800 when Cruickshank showed it was an "oxide of carbone."

The oxidation state of carbon in carbon monoxide (CO) is intermediate between elemental carbon and carbon dioxide; hence, CO can be prepared by high-temperature reduction or partial oxidation. High-temperature reduction includes reduction of (1) metal oxides with charcoal or carbon (2) some carbonates with a metal or carbon, (3) carbon dioxide with metals, carbon, or methane, and (4) steam with coal, char, coke, or methane. Partial oxidation of carbon, coal, char, coke, or methane yields carbon monoxide among other products. Some reactions are used to manufacture CO; others function as an intermediate step in preparing CO and converting it into various products.

Carbon monoxide can be prepared also by degradation of certain compounds. In the laboratory, CO can be obtained by dropping formic acid into concentrated sulfuric acid at 120–150°C. Passage of vaporized methanol over catalysts of various metal oxides at about 300°C gives CO and H_2. Heating methyl formate with sodium methoxide (as catalyst) at 60–100°C yields CO as well as methanol, which can be removed by chilling the effluent vapors.

Carbon monoxide is a component of some by-product and manufactured gases. The by-product gases of special interest are coke-oven gas (6% CO, 53% H_2) and blast-furnace gas (26% CO, 3% H_2). These gases, after removing some other components, often are used as fuel in the respective plants. Of the various manufactured gases, the group called synthesis gas is most relevant to the chemical aspects of carbon monoxide. Synthesis gas consists of mostly CO and H_2.

Water gas, one of the earliest manufactured gases, results from passing steam through coke previously made incandescent by a strong blast of air. Among other components, water gas contains 37–43% CO and 49–51% H_2. This conventional water-gas process has undergone many improvements, chiefly by operating under pressure with coal, coke, or char and by using oxygen instead of air. One aim of these changes is to increase the yield of hydrocarbon by-products, which raise the product's fuel value. Making enriched water gas under pressure is an integral step in some processes for converting coal into a gas-substitute for natural gas.

The conventional process for making water gas

has long been the source of synthesis gas used for making various products. But after rapid growth of the petroleum industry, much of the synthesis gas has been made from petroleum fractions.

Generally two methods are used to make synthesis gas from petroleum fractions—steam reforming and partial oxidation. Steam reforming is done with a nickel catalyst at pressures up to 600 psi and at 700–950°C, depending on the fraction used. Partial oxidation involves reacting natural gas, naphtha, or heavy fuel oil with oxygen and steam at pressures up to 600 psi and 1200–1500°C for the final temperature.

In steam reforming of methane, the main reaction is $CH_4 + H_2O \rightarrow CO + 3H_2$. Steam, however, can transform CO into carbon dioxide: $CO + H_2O \rightarrow CO_2 + H_2$. When a product rich in CO is desired, carbon dioxide is used along with the methane and steam to retard conversion of CO to CO_2. Steam reforming of light naphtha yields synthesis gas and methane; the resulting product normally is passed through a secondary reformer to convert methane into CO.

A promising commercial source of CO is the huge quantity produced by the basic oxygen process for making steel. Oxidation of the carbon in the iron (made in the blast furnace) yields off gas rich in recoverable CO.

Purification. Gases rich in carbon monoxide undergo varying degrees of purification. Gases intended for consumption as fuel usually receive less purification than a synthesis gas destined for chemical processing.

The purification method depends somewhat on operating cost and the efficiency sought. For raw gases made from coal or coke, the simplest choice is first to scrub with water at atmospheric pressure or about 15 atm. This removes suspended solids, tar, ammonia, and much of the carbon dioxide and sulfur compounds. Then the resulting gas is passed through high-boiling oil to remove entrained low-boiling oil products. Similar treatments are sometimes done for petroleum-derived synthesis gas.

Water and oil scrubbing are often performed merely as preliminary steps. For greater efficiency in removing carbon dioxide, synthesis gas is further purified by passage under pressure through an aqueous solution of potassium carbonate or through an organic solvent—ethanolamine, propylene carbonate, certain glycols, and other types; some methods employ mixtures of solvents. Many solvents can also remove sulfur compounds, but not well enough for using synthesis gas in some catalytic processes. Practically sulfur-free gas is se-

TABLE 1. COMPOSITIONS OF SYNTHESIS GAS

Source Process	Methane Steam-reforming	Methane Partial-oxidation	Naphtha Partial-oxidation	Heavy Oil Partial-oxidation
Components	Volume percent			
Carbon monoxide	16–20	34–35	42–45	47–48
Carbon dioxide	4–7	2–3	3–5	4–6
Hydrogen	75–76	61–62	51–53	46–47
Nitrogen	—	0.4–1.4	0.1–1.4	0.2–1.4
Methane	0.1–2	0.2–0.5	0.3–0.7	0.1–0.5

cured by flowing prepurified gas through a hot bed of alkali-treated iron oxide.

When pure CO is sought, purification and isolation is accomplished by selective adsorption or by cryogenic (low-temperature) separation, which is favored for large-scale operations.

Cryogenic separation involves liquefaction and subsequent fractionation of the condensate. But first it is necessary to remove impurities—particularly carbon dioxide and moisture, which will solidify and plug the equipment. Typically, synthesis gas under pressure is supercooled to condense all gaseous components except most of the H_2. Dissolved H_2 is removed by flash evaporation, and the condensate is fractionally distilled to get CO of over 98% purity. At least 95% pure CO often can be obtained from partial-combustion product gas by merely vaporizing the liquefaction condensate.

Selective adsorption when used to get pure CO from manufactured gas mainly consists of two steps: The prepurified gas is passed while under pressure through an ammoniacal copper solution which absorbs mostly CO; then the pressure is removed to release pure CO. The CO obtained by this method may need scrubbing with water to remove ammonia, and scrubbing with aqueous sodium hydroxide to remove carbon dioxide that builds up from oxidation of some CO by the copper solution.

Detection. Ampules or tubes with color-forming reagents can be used for estimating the CO content of ambient air by intensity of the color produced. For best results, a sample must be analyzed by methods specially designed to eliminate interferences. The air sample, after preliminary removal of interfering substances, can be analyzed gasometrically by adsorption in acidified cuprous chloride solution (or in a commercial reagent known as "Co-sorbent") in an Orsat or Bureau of Mines apparatus. CO can also be determined by other methods.

In areas subject to CO accumulation, the ambient air can be monitored by an automatic, continuous recording instrument. Some instruments have a device for sounding an alarm when the CO concentration reaches a specific level. Analysis by one instrument is based on the heat generated by catalytic oxidation of CO to CO_2. Another instrument determines CO by adsorptivity of infrared radiation. Even a gas chromatographic assembly can be used for automatic semicontinuous determination of CO.

Physical Properties. At 18–19°C, water-saturated mixtures containing air and 12.5–74.2% CO (carbon monoxide) are flammable; for dry mixtures, the lower limit is about 15% CO.

For CO gas, the autoignition temperature is 652°C; density at 0°C and 1 atm., 1.2504 g/l; specific gravity at 15–20°C, 0.968 (air = 1); enthalpy at 25°C, 2072.6 cal/mole; entropy at 25°C, 47.301 cal/(°C)(mole); free energy of formation at 25°C, −32.8077 kcal/mole; heat capacity at 25°C, 6.965 cal/(°C)(mole); heat of combustion at 25°C, 4343.6 Btu/lb or 2414.7 cal/g; heat of formation from the elements at 25°C, −26.416 kcal/mole; heating value (gross) at 30 in. pressure and 60°F (15.6°C), 322.6 Btu/ft³. CO is virtually insoluble in water; the solubility ranges from 0.0044 g per 100 ml water at 0°C to 0.0000 g at 100°C; its sol-

ubility is only several times greater in common organic liquids. The thermal conductivity as 10^{-6} cal/(sec)(cm²)(°C/cm) at given °C is 47.94, −40°; 55.87, + 4.4°; 57.86, 15.6°; viscosity in micropoise at °C is 127, −78.5°; 166, 0°; 172, +15°; 210, 100°.

For liquid CO, the normal boiling point, b.p., is −191.5°C; d_4^{-195}, 0.814; latent heat of vaporization at b.p., 1443.6 cal/mole; vapor pressure in mm Hg at °C is 1, −220°; 10, −215°; 40, −210°; 400, −196.3°.

Solid CO exists in two forms with a transition temperature at −211.6°C; latent heat of transition is 151.3 cal/mole; m.p. −205°C; heat of fusion, 199.7 cal/mole at −210°C.

The triple point for CO is −205.01°C at 115.14 mm Hg; critical temperature, −140.23°C; critical pressure, 34.53 atm; critical density, 0.301 g/cm³.

Chemical Properties. The inertness of carbon monoxide at ordinary temperatures and in the absence of catalysts or light is not surprising, since its electronic structure is similar to that of the nitrogen molecule. At elevated temperatures CO is a potent reducing agent. Its availability at low cost, its reducing power and its versatility in catalytic addition reactions make CO a valuable chemical in many commercial operations.

Stability. Carbon monoxide is stable with respect to decomposition into carbon and oxygen. But it disproportionates into carbon dioxide and carbon at as low as 35°C over palladium deposited on silica gel. At 400–700°C, almost any surface is sufficiently active to give copious deposits of carbon. Above 800°C, the equilibrium favors formation of CO.

CO burns with a bright blue flame to produce carbon dioxide. The reaction with oxygen is slow below 650°C in the absence of catalysts. At higher temperatures or when initiated with a spark, mixtures of CO and oxygen containing trace quantities of water will explode. A catalyst containing oxides of copper and manganese, known as "Hopcalite," has been used in gas masks to oxidize carbon monoxide to the harmless dioxide.

CO reacts reversibly with steam to produce CO_2 and H_2. The equilibrium constant of this water gas reaction, $K_p = (P_{CO} \times P_{H_2O}/(P_{CO_2} \times P_{H_2})$, varies with temperature, °C, thus: 225°, 0.007; 425°, 0.109; 625°, 0.455; 825°, 1.08; 995°, 1.76.

Reduction of metal oxides. At 300–1500°C, CO reduces many metal oxides or metal carbides. Among the metals whose oxides are so reduced are cobalt, copper, iron, lead, manganese, molybdenum, nickel, silver and tin. An interesting synthesis of $AlCl_3$ comprises reaction of CO and Cl_2 with Al_2O_3 in the presence of alkali. CO is one of the active agents that reduces iron ore in the blast furnace. Above 900°C, iron carbide, Fe_3C, is formed by the reaction of CO with Fe_2O_3.

Miscellaneous inorganic reactions. Chlorine and bromine react with CO under the influence of light or of charcoal catalysts to produce phosgene, $COCl_2$, and carbonyl bromide. No reaction occurs with iodine, but carbonyl fluoride can be prepared by the action of AgF_2 on CO. The phosgene synthesis is highly exothermic, and is therefore favored by temperatures below 350°C.

CO reacts slowly with liquid sulfur and rapidly with the vapor to form carbonyl sulfide, COS.

Aqueous bases absorb CO at ordinary and elevated temperatures with formation of formate salts. Sodium, potassium and ammonium formates are produced in this manner. The reaction is rapid at 200°C and high pressures.

Metal Carbonyls. Most of the transition metals from carbonyls: $Cr(CO)_6$, $Mn_2(CO)_{10}$, $Fe(CO)_5$, $Co_2(CO)_8$, $Ni(CO)_4$, $Mo(CO)_6$, etc. Several of these, especially the cobalt and nickel carbonyls, are useful as homogeneous catalysts for organic reactions.

Reactions of Carbon Monoxide and Hydrogen (Synthesis Gas). A wide variety of paraffinic and oxygenated compounds result when synthesis gas is treated at 250–350°C with iron or cobalt catalysts (the Fischer-Tropsch reaction). Methanol is synthesized in large amounts by heating synthesis gas at 320–380°C in the presence of chromium and zinc oxides at around 300 atm.

$$2H_2 + CO \rightarrow CH_3OH$$

A low-pressure version of the methanol synthesis operates at a CO partial pressure below 20 atm.

Many processes for producing synthetic pipeline gas involve hydrogenation of CO to methane:

$$3H_2 + CO \xrightarrow[400°C]{nickel} CH_4 + H_2O$$

The oxo reaction, used industrially to synthesize aliphatic aldehydes and alcohols, involves reaction of olefins with synthesis gas (see Oxo Process) in the presence of soluble metal carbonyls or other transition metal complexes.

Reactions of organometallic compounds. Rather complex reactions occur with Grignard reagents and CO. Acyloins, α-diketones and olefins are isolated. But organoboranes, synthesized *in situ* in a suitable ethereal solvent, react smoothly with CO at atmospheric pressure to give carbonylation products convertible in good yields to tertiary alcohols, ketones, aldehydes and other derivatives, depending on reaction conditions.

Catalysis by Transition Metal Complexes. Many new syntheses involving transition metal complexes of organic compounds have been discovered in recent years. Soluble transition metal complexes are good catalysts for syntheses of various organic compounds from carbon monoxide and an organic substrate: acetylene, olefins, alcohols, allylic compounds, amines, etc. Compounds containing active hydrogen and CO can be added to acetylenic compounds to produce acrylic acid, acrylates, and derivatives:

$$HC \equiv CH + CO + H_2O \rightarrow CH_2{=}CHCOOH$$

Nickel compounds are normally used as catalysts in this reaction. Acetylenes and CO can also give quinones and hydroquinones. Olefins, CO and water give saturated acids; while with alcohols, saturated esters are obtained. Dimethylamine and CO yield dimethylformamide. Allylic compounds react with CO to yield unsaturated acids and their functional derivatives, ketones, and cyclic compounds incorporating the CO molecule.

Acetic acid is being made industrially by a new process in which methyl alcohol is reacted with CO over a rhodium catalyst; a large plant using this process started in 1971.

$$CH_3OH + CO \xrightarrow{Rh} CH_3COOH$$

Miscellaneous organic reactions. Many organic compounds can form carbonium ion intermediates in the presence of acids (H_2SO_4, BF_3, etc.). These carbonium ions react with carbon monoxide and active hydrogen compounds. Substrates include olefins, alcohols, ethers, formals and organic halides. Active hydrogen compounds include water, alcohols and amines. Products include acids, hydroxyacids, esters, amides and their derivatives.

One such reaction that has achieved industrial importance is the synthesis of branched-chain acids from olefins, CO and water. High olefin conversion can be obtained at mild conditions, if the reaction is carried out in two stages:

$$CH_3{-}\underset{\underset{CH_3}{|}}{CH}{=}CH_2 \xrightarrow[\text{2. } H_2O]{\text{1. } H_2SO_4,\ CO} CH_3{-}\underset{\underset{CH_3}{|}}{\overset{\overset{CH_3}{|}}{C}}{-}COOH$$

Nearly all olefins react between −20° and +80°C and CO pressures of 1–100 atm. to give carboxylic acids in 80–95% yields; with alcohols instead of water, esters are obtained.

Copolymers of ethylene and CO are prepared at high pressures and temperatures in the presence of free radical initiators. The ratio of ethylene to CO in the polymer can be varied widely.

Uses. The major chemical use of carbon monoxide is in the catalytic synthesis of methanol; lesser amounts of CO are employed for manufacturing oxo-chemicals (alcohols, esters, etc.), phosgene, acrylates, and other products. Synthesis gas is used to make various chemicals and fuels, and to reduce metal-oxide ores to the metal, e.g., in producing iron without using the blast furnace.

Carbon monoxide has long been used in the Mond process to recover nickel from low-grade ores. In this process, CO reacts with nickel to form a volatile carbonyl, which on thermal decomposition gives pure nickel. Several metal carbonyls, particularly iron carbonyl, are manufactured for conversion by thermal decomposition to pure metal powders, which then are molded into complex articles by powder metallurgy.

The ability of CO to engage in various reactions, and the likelihood of its expanding availability from industrial operations indicate that CO will become increasingly important as a progenitor of many chemical products.

IRVING WENDER
CHARLES ZAHN

References

Falbe, J., "Carbon Monoxide in Organic Synthesis," Charles R. Adams, Tr., Springer-Verlag, New York, 1970, 219 pp.

Hust, J. G., and Stewart, R. B., "Thermodynamic Property Values for Gaseous and Liquid Carbon Monoxide From 70 to 300°K With Pressures to 300 Atmospheres," National Bureau of Standards, Technical Note 202, Washington, D.C., 1963, 109 pp.

Wender, Irving, and Pino, Piero, Eds., "Organic Syntheses via Metal Carbonyls," Vol. 1, Interscience, New York, 1968, 517 pp.

Physiological Aspects

Carbon monoxide is toxic to warm-blooded animals. Because of its extremely faint odor and taste, its lethal capacity can be insidious. The ordinary charcoal-filled gas mask is useless for filtering out carbon monoxide from contaminated air. Persons who are required to enter areas contaminated with carbon monoxide (firemen, rescue workers, maintenance men) must be provided with closed circuit breathing apparatus which delivers oxygen through a mask to the wearer. This is essential in atmospheres which contain more than 2% by volume of carbon monoxide. In atmospheres which contain less, an ordinary gas mask can be used for short periods, if it is fitted with a special canister filled with Hopcalite, a mixture of metallic oxides which serve to catalyze the oxidation of carbon monoxide to carbon dioxide. The reaction is exothermic and such canisters become very hot in use.

Carbon monoxide is physiologically quite inert, except for its strong combination with hemoglobin in the blood. It has no unique toxic action on any of the bodily tissues. As Henderson and Haggard point out: "Were it not for this one reaction carbon monoxide would be classified with nitrogen and hydrogen as a simple asphyxiant." The affinity of carbon monoxide for hemoglobin is about 300 times that of oxygen.

The reaction between carbon monoxide and hemoglobin is reversible:

$$HbO_2 + CO \rightleftharpoons HbCO + O_2$$

"Carbon monoxide displaces oxygen from hemoglobin, and in turn oxygen may displace carbon monoxide from its combination. Red corpuscles, in which the hemoglobin has been joined to carbon monoxide and then freed from the combination by means of oxygen, are not injured; they are capable of transporting oxygen as if they never had been exposed to the other gas. But so long as the combination with carbon monoxide continues they are incapable of fulfilling their respiratory function." Consequently, they cannot transport adequate oxygen to the various bodily tissues. Progressively severe anoxia results. Unfortunately, the victim is all too often unaware of his danger. Mechanical efficiency, e.g. driving a car, may persist until poisoning has advanced almost to the possibility of unconsciousness.

Death from inhalation of carbon monoxide can be summarized as follows: (1) reduction of the oxygen-carrying capacity of the blood due to the formation of HbCO; (2) tissue anoxia, especially in the brain, which is very sensitive to lack of oxygen; (3) consequent depression of respiratory center in the brain and decrease in respiration; (4) failure of the heart due to inadequate oxygen supply.

Carbon monoxide is absorbed into the body only through the alveoli in the lungs. It does not enter thru the eyes, mucuous membranes, cuts, or upper respiratory tract.

Inhalation of cigarette smoke is the most important source of CO intake in the United States. The one-pack-a-day inhaler smoker is exposed to a CO concentration of nearly 6%, great enough to pose a serious threat to health.

Air pollution also increases CO concentrations and the results are aggravated at high altitudes.

Tolerance Limits. On the basis of numerous experimental studies, the tolerance limits for the average man have been established. The following series of equations gives a ready method for estimating the safety of any carbon monoxide-air mixture under conditions of rest. Time is given in hours and concentration of carbon monoxide in parts per 10,000 of air:

a. Time $\times$ concentration = 3 (no perceptible effect).

b. Time $\times$ concentration = 6 (a just-perceptible effect).

c. Time $\times$ concentration = 9 (headache and nausea).

d. Time $\times$ concentration = 15 (dangerous).

Muscular activity or increased respiratory minute volume reduces the value in equation *a* to 1, 2, or less; it influences the other equations in like manner.

Drinker gives data on which the following table of allowable concentrations for carbon monoxide in air is based:

Concentration of Carbon Monoxide in		Effect
Per cent	Parts per 10,000	
0.01	1	No symptoms for 2 hours
0.04	4	No symptoms for 1 hour
0.06–0.07	6–7	Headache and unpleasant symptoms in 1 hour
0.1–0.12	10–12	Dangerous for 1 hour
0.35	35	Fatal in less than 1 hour

As a safe rule, based on sound experiments and experience, concentrations of carbon monoxide above 0.01 per cent, or 1 part per 10,000 (100 parts per million) should not be permitted in houses, garages, laboratories, or industrial plants where prolonged exposure to the gas may be experienced.

Chronic Effects. There is no such physiological entity as "chronic carbon monoxide poisoning." The gas is not a cumulative poison; it is readily removed from the blood when the victim is exposed to pure air or oxygen.

Older persons take longer to eliminate CO than do younger persons under comparable conditions.

After a severely acute exposure, the victim usually dies in about 36 hours or he recovers completely after a few days. The alleged chronic damage to man from carbon monoxide poisoning stems from prolonged cerebral anoxia severe enough to cause permanent brain damage but not severe enough to kill; it is not caused by retention of carbon monoxide in the body.

Lesions found in the various tissues of persons killed by CO are nonspecific and do not permit one to diagnose CO poisoning.

Treatment. The treatment of carbon monoxide poisoning depends on removal of the victim from the contaminated atmosphere, administration of artificial respiration, and inhalation of pure oxygen by the patient. If a good mechanical respirator is available, it can be used to advantage. In the absence of such device, air can be pumped into the victim's lungs by "rescue breathing," a mouth-to-mouth resuscitation procedure. It is now the medically accepted method for administering artificial respiration in the Army, Navy, and Air Force.

A normal individual breathing air will wash out about half his blood CO in 240 minutes; the breathing of oxygen reduces this time to 40 minutes. Hence, the use of oxygen is essential for effective treatment. From the purely academic point of view, it might be argued that the addition of 5 to 7% carbon dioxide to the oxygen will result in more efficient resuscitation. On the practical level, however, there is little to justify the use of carbon dioxide during resuscitation. Hyperboric oxygen (2 atmospheres absolute) administered as soon as possible to persons poisoned with CO is strongly recommended.

Drugs are of little use and may even be dangerous. Under no circumstances should a patient who is recovering consciousness after carbon monoxide poisoning be permitted to arise and walk about. He must be kept in a prone or supine position; every effort must be made to keep his oxygen requirements at a minimum.

CHARLES G. WILBER

Cross-reference: *Noxious Gases.*

CARBONYL COMPOUNDS

The compounds containing the divalent C=O group, or *carbonyl group,* if considered together solely on this basis, undoubtedly comprise the largest and most important single class of organic compounds. Not only is a very large number of such compounds known, but these include many diverse types of substances, such as the proteins and amino acids, sugars, many perfumes and flavoring materials, medicinals and antibiotics, plastics like "Bakelite," "Lucite," and "Vinylite," paint resins such as the alkyds, fibers such as silk, wool, rayon, nylon, and "Dacron," and many solvents and chemicals of industrial importance.

The carbonyl group may be considered as a product of the oxidation of a hydrocarbon unit, $-CH_2-$ or $-CH_3$, although for economic reasons the introduction of a carbonyl group into an organic molecule generally is accomplished through the

oxidation of an alcohol group $-\overset{|}{\underset{|}{C}}OH$ rather than

by the direct oxidation of a hydrocarbon. This oxidation of alcohol groups to carbonyl groups may be accomplished directly with oxygen at elevated temperatures in the presence of catalysts, by chemical oxidizing agents such as the permanganates, or by bacterial processes as in the production of vinegars. A few processes involve the reduction of more highly oxidized carbon compounds to carbonyl compounds, as in the chlorophyl-catalyzed photochemical reaction of carbon dioxide and water in living plants to form sugars. It should be remembered, however, that many important carbonyl compounds contain other functional groups in ad-

dition to the carbonyl group, and that the syntheses of such compounds often will employ simpler carbonyl compounds as starting materials.

We may divide compounds containing the carbonyl group into a few very general classes, depending on the nature of the two groups attached to the carbon atom of the carbonyl group. In all except a few compounds, the carbonyl group is attached to at least one organic group (designated generally as R in the examples below). For convenience, we refer to a functional group comprising a carbonyl group attached to one organic group [R(C=O)—] as an *acyl* group, and to specific examples such as the *acetyl* group [CH_3(C=O)—], or the *benzoyl* group [$C_6H_5 \cdot$(C=O)—].

In the *aldehydes,* at least one of the attached groups is a hydrogen atom, and the group $-\overset{H}{\underset{}{C}}=O$ is known as the *aldehyde group.* In all aldehydes except formaldehyde, HCHO, the carbonyl group also is attached to an organic group, as in acetaldehyde, CH_3CHO, or benzaldehyde, C_6H_5CHO.

The *ketones* are compounds in which the carbonyl group is attached to two organic groups. These may be the same, as in acetone, CH_3COCH_3, or different, as in acetophenone, $C_6H_5COCH_3$. When present in a ketone, the carbonyl group is sometimes referred to as the *keto* group.

In more highly oxidized derivatives, the carbonyl group is attached not only to an organic group but also to oxygen, through which it may be linked to other groups. These include hydrogen as in the *organic acids* [R(C=O)—OH], acyl groups as in the *acid anhydrides* [R(C=O)—O—(C=O)R], or organic groups as in the *esters* [R(C=O)—OR']. In formic acid, HCOOH, and its derivatives, R is hydrogen rather than an organic group. Common examples of compounds of these types are acetic acid, $CH_3 \cdot COOH$, benzoic acid, C_6H_5COOH, and ethyl acetate, $CH_3COOC_2H_5$. The —COO— group is known as the *carboxyl group.*

Other derivatives of the acyl group include the *acid* (or *acyl*) *halides,* such as acetyl chloride, CH_3COCl, and the *acid amides,* such as acetamide, CH_3CONH_2. A very important compound related to the amides is urea, $(NH_2)_2CO$.

It should be appreciated that organic compounds can contain more than one carbonyl group, and that such groups may differ in functionality. For example, the common compound *aspirin* is *acetylsalicylic acid,* which is at once both an acid and an ester. In *oxalic acid,* $(COOH)_2$, both carbonyl groups have the same type of functionality.

The chemical behavior of the carbonyl group depends upon the nature of the groups attached to it. In some types of compounds the carbonyl group generally is an active participant in chemical reactions, as in the aldehydes and ketones, while in others such as the organic acids and their derivatives, its primary action is to modify the functionality of neighboring reactive groups.

In the aldehydes and ketones, the carbonyl group is characterized by its unsaturation, and it is able to add a variety of reagents. This sort of reactivity resembles that of the carbon-carbon double bond, or olefinic linkage, but is somewhat greater because of the polar nature of the oxygen atom. Owing to the presence of the hydrogen attached to the carbonyl carbon, the aldehydes are considerably more reactive in general than are the ketones.

The aldehydes are very readily oxidized to the corresponding acids, and are intermediates in the oxidation of alcohols to acids. A very sensitive test for the aldehyde group, and one which often is used to indicate the presence of sugars (some of which are aldehydes), is the oxidation reaction with cupric ion in Fehling's solution. Ketones are less reactive toward oxidation and require more vigorous reaction conditions; they do not react with Fehling's solution, for example. In the oxidation of a ketone, two or more different organic acids will be formed depending on the nature of the groups originally attached to the carbonyl group. The reduction of a ketone leads to a secondary alcohol (RCHOHR'), whereas the reduction of an aldehyde leads to a primary alcohol (RCH$_2$OH). The reduction of an organic acid leads first to an aldehyde, then to a primary alcohol.

Of great importance are the polymerization and condensation reactions exhibited by the aldehydes and, to a lesser extent, the ketones. In condensations of the *aldol* type, there is a transfer of hydrogen between molecules, from the carbon adjacent to the carbonyl group on one molecule (or from the carbonyl group in formaldehyde) to the carbonyl group on an adjoining molecule of aldehyde, thus forming an alcohol group and a carbon-carbon bond between the two molecules. The new molecule thus formed is both an aldehyde and an alcohol; hence the name *aldol*. This process can be repeated to build up more complex molecules. It is believed that the synthesis of sugars in living plants occurs by the condensation of simpler units, such as formaldehyde, through processes of this sort. The sugars then are converted into more highly condensed products such as the starches and cellulose. The simpler sugars include both keto and aldo types. In different types of reaction, the condensation of aldehydes, and formaldehyde in particular, with phenols, with urea or derivatives thereof, or with casein, occurs readily to form resinous materials that are widely used in the plastics industries. One of the oldest and most common of these is "Bakelite." In the presence of strong acids, aldehydes may condense to form cyclical polyethers; some of these, such as paraformaldehyde, paraldehyde, and metaldehyde, are well-known articles of commerce. A number of industrial products and important chemical intermediates are prepared by the rearrangement of ketones.

In many of the reactions of aldehydes and ketones, and to a lesser extent organic acids, the reactivity of hydrogen atoms on carbon atoms adjacent to the carbonyl group becomes important through a process known as *enolization,* in which a hydrogen transfers from the adjacent carbon atom to the oxygen of the carbonyl group, leaving a double bond between the two carbon atoms. In the *enol* form, the hydrogen is highly reactive and very readily replaced by a variety of reactants; a molecular rearrangement follows such substitution in many cases. Aldehydes and ketones which have no hydrogen atoms on carbon atoms adjacent to the carbonyl group must remain in the *keto* form; such compounds are considerably less reactive and do not exhibit many of the reactions ordinarily characteristic of aldehydes or ketones.

In the organic acids and their derivatives, the primary function of the carbonyl group is to modify the reactivity of other attached groups. Thus, for example, the functional hydrogen atom in acetic acid, CH$_3$COOH, is moderately acidic owing to the presence of the adjacent carbonyl group, and exhibits a number of reactions characteristic of acids, such as ionization in water and the formation of salts, whereas the functional hydrogen in ethyl alcohol, CH$_3$CH$_2$OH, is essentially neutral. Similarly, the acyl halides, RCOX, are very reactive compounds, being readily hydrolyzed by water to form organic acids and hydrogen halides, whereas most of the alkyl halides, RCH$_2$X, are practically inert to water under ordinary conditions.

One of the more important reactions of acids, and also of acid halides, is *esterification.* The reaction of an acid or acid halide with the hydrogen of an alcohol group results in the elimination of water or hydrogen halide and the formation of an ester, e.g. RCOOH + R'OH → H$_2$O + RCOOR'. This sort of reaction is used on a large scale industrially in the preparation of polyester resins and plastics, such as "Mylar," and synthetic fibers such as "Dacron" as well as for the preparation of a variety of chemical specialties and intermediates. Esterification may be considered as analogous to the neutralization reaction of an acid with a base.

In the amino acids, which are of the greatest importance since they are the basic units in the structures of animal and plant proteins, the amine group —NH$_2$ is present in place of one of the hydrogen atoms on the carbon atom adjacent to the carbonyl group, e.g. glycine, or aminoacetic acid, NH$_2$CH$_2$·COOH. Since the amino acids are amphoteric, i.e. both acids and bases, they can undergo an unusual variety of chemical reactions, the amine group and the carboxyl group acting either separately or together. One of the basic linkages in the structure of proteins is the peptide, or amino, linkage —NHCO—, which joins amino acid groups together into large polymeric molecules.

DALLAS T. HURD

Metal Carbonyls

Many of the transition metals form volatile molecular coordination compounds with carbon monoxide in which the oxidation state of the metal atom is zero. The carbonyls may be divided into two classes: mononuclear, containing only one metal atom per molecule and polynuclear, containing more than one metal atom per molecule. The properties of the metal carbonyls are typical of covalent substances. In general, they are easily distilled or sublimed, soluble in nonpolar solvents and insoluble in polar media. Most of the carbonyls and their derivatives are toxic and care should be used in handling them. In the human system the carbonyls decompose to give carbon monoxide along with other products. The carbon monoxide combines with the iron in the hemoglobin of the blood, thus making it impossible for the hemoglobin to combine with oxygen, which is a necessity for the life process.

Numerous methods are available for the preparation of metal carbonyls. Nickel tetracarbonyl and iron pentacarbonyl can be prepared by the direct reaction between carbon monoxide and the finely divided metal. Other carbonyls are prepared by treating the metal halide in a suspension of an organic solvent such as tetrahydrofuran with carbon monoxide at 200–300 atm pressure and temperatures up to 300° in the presence of suitable reducing agents such as sodium, aluminum, or magnesium. Oxides, sulfides, or other salts of the metal have also been used for the preparation of carbonyls by the reaction with carbon monoxide in the presence of suitable reducing agents at elevated temperatures and pressures. Polynuclear carbonyls are often obtained from the decomposition of mononuclear compounds, e.g.,

$$\text{Fe(CO)}_5 \xrightarrow{\text{light}} \text{Fe}_2\text{(CO)}_9 \xrightarrow{\text{heat}} \text{Fe}_3\text{(CO)}_{12}$$

Except for vanadium hexacarbonyl, metals with even atomic numbers form mononuclear carbonyls. The metal atom coordinates to that number of carbon monoxide groups necessary to attain the configuration of the next higher rare gas. In these cases it is assumed that each carbon monoxide molecule donates two electrons. The metals with odd atomic numbers have an unpaired electron after combining with carbon monoxide. The unpaired electron is used in bonding by combining with another metal atom to give a metal-metal bond. This combination give rise to the various polynuclear carbonyls and accounts for their diamagnetism. In the polynuclear carbonyls, the carbonyl group may be a terminal carbonyl, or it may be used to bridge two or more metal atoms. When the carbonyl group is used to bridge several atoms, the bonding is through the carbon atom only.

In many of the carbonyls, carbon monoxide may be replaced by other ligands. These may be strong electron donor molecules such as nitric oxide, amines, aromatic isonitriles, ethers, phosphorus trihalides, or unsaturated organic molecules such as benzene. Carbonyls may also undergo reactions with aqueous base resulting in the formation of carbonylate anions. When iron pentacarbonyl is dissolved in basic solution, the carbonylate anion, HFe(CO)_4^-, is formed which upon acidification gives the corresponding weak acid, $\text{H}_2\text{Fe(CO)}_4$. The hydrides are volatile, unstable liquids, notable for their strong reducing action and the acid character of the hydrogen atoms. The carbonyl halides, e.g., $\text{Mn(CO)}_5\text{Cl}$, are obtained either by direct interaction of the metal halides with carbon monoxide at high pressure or in certain cases by the cleavage of polynuclear carbonyls by halogens. Carbonyl halide anions e.g., $[\text{Cr(CO)}_5\text{Cl}]^-$ are also known.

THEODORE M. BROWN

CARBOXYLIC ACIDS

Carboxylic acids are those organic compounds which contain the carboxyl group, $-\overset{\displaystyle O}{\underset{\displaystyle \|}{C}}-\text{OH}$. These compounds are acidic by virtue of the ionization of the carboxyl hydrogen. Although carboxylic acids are relatively weak in acid strength, their acidity is enhanced by the presence in the molecule of other carboxyl groups or negative substituents close to the carboxyl group, particularly in the alpha position in aliphatics, and the ortho position in aromatics.

Many carboxylic acids are found in nature, as free acids or in the form of esters or salts. Most natural aliphatic monocarboxylic acids are straight-chain acids containing an even number of carbon atoms. Those of higher molecular weight usually occur in the form of esters, in fats, oils and waxes. Among the most abundant acids are the saturated acids, palmitic and stearic, and the unsaturated acid, oleic. Some dicarboxylic acids, including oxalic, succinic, and fumaric, also occur in nature, as do the hydroxy acids, glycolic, lactic, ricinoleic, malic, tartaric and citric.

A limited number of aromatic acids are found in natural substances, notably benzoic and the hydroxybenzoic acids, salicylic, protocatechuic and gallic; and cinnamic and the hydroxycinnamic acids, p-cumaric and caffeic.

Some natural alicyclic acids are abietic, chaulmoogric, hydronocarpic, and the bile acids.

There are several general methods for synthesizing carboxylic acids: (1) oxidation of a primary alcohol; (2) Grignard synthesis; (3) nitrile synthesis; (4) malonic ester synthesis; the monoalkyl malonic esters can also be used to prepare branched chain acids; (5) acetoacetic ester synthesis.

These methods are generally applicable to the preparation of aliphatic acids. Formic acid, however, is prepared by the addition of alkali to carbon monoxide at high temperature and pressure, followed by acidification of the resulting salt. Acetic acid can be made by the oxidation of acetaldehyde prepared from either acetylene or ethyl alcohol; and it can be isolated from the products of the destructive distillation of wood. Dilute acetic acid (vinegar) is made by fermentation of fruit juices to form alcohol, followed by bacterial oxidation. A number of high-molecular weight fatty acids are obtained commercially by hydrolysis of their naturally occurring glycerides. An acid hydrolysis is employed, in contrast to the caustic hydrolysis which is the basis of soapmaking.

The higher dicarboxylic acids are generally prepared commercially by oxidizing certain alicyclic and aliphatic compounds. The acids having an even number of carbon atoms are more readily prepared by this method. Oxalic acid can be made by heating sodium formate, and also by fermentation. Catalytic oxidation of benzene yields maleic anhydride, which can be converted to succinic acid.

Alpha-hydroxy acids are conveniently prepared from the corresponding halogen compounds. Esters of beta-hydroxy acids are produced by the Reformatsky reaction. The hydroxy acids which can be made by fermentation include lactic, citric and gluconic. Tartaric acid is obtained from potassium acid tartrate, a by-product of winemaking.

Some aromatic acids can be made by oxidizing aromatic hydrocarbons (e.g. benzoic from toluene, phthalic from naphthalene) or substituted acetophenones. Benzoic acid is prepared by the decarboxylation of phthalic acid and phthalic anhydride. Salicyclic acid is made by the Kolbe synthesis.

Straight-chain aliphatic monocarboxylic acids containing fewer carbon atoms than capric are liquids, water-miscible up to butyric and thereafter decreasingly water-soluble. The acids beginning with capric are solids almost insoluble in water. All are quite soluble in most organic solvents. The acids containing an odd number of carbon atoms have lower melting points than the 'even acids' immediately preceding them in the series; boiling points increase regularly with increasing molecular weight.

Aliphatic polycarboxylic acids are crystalline solids, those of low molecular weight being relatively water-soluble. Dicarboxylic acids containing an odd number of carbon atoms have lower melting points and higher water solubilities than the 'even acids' immediately preceding them in the series. Alicyclic and aromatic acids are generally crystalline solids of low water solubility.

The following reactions are typical of carboxylic acids: (1) reaction with bases to form salts (RCOOM); (2) reduction to aldehydes (RCHO) and alcohols (RCH$_2$OH); (3) formation of acid anhydrides (RCOOCOR) and acyl halides (RCOX); (4) reaction with alcohols to form esters (RCOOR'); (5) amide formation (RCONR'R''); amides can be formed by heating ammonium or amine salts of the acid, or by reaction of an acid anhydride, acyl chloride or ester with ammonia or a primary or secondary amine; (6) decarboxylation to the hydrocarbon (RH), often achieved by fusing a salt of the acid with an alkali; (7) formation of ketones (R$_2$CO) by dry distillation of metal salts.

Formic acid undergoes several unique reactions. It is easily oxidized to water and carbon dioxide; heated with sulfuric acid it decomposes to water and carbon monoxide.

Aliphatic acids can be chlorinated and brominated in the alpha position in the presence of sunlight or an added catalyst.

Halogenated aliphatic acids react in different ways with alkali: α-halo acids are converted to α-hydroxy acids; β-halo acids form α,β-unsaturated acids with elimination of hydrogen halide; γ-halo acids form lactones (intramolecular esters).

Aliphatic hydroxy acids are dehydrated on heating: α-hydroxy acids undergo bimolecular dehydration to give lactides; β-hydroxy acids form α,β-unsaturated acids; γ- and δ-hydroxy acids form lactones; where the hydroxyl group is removed by 5 or more carbon atoms from the carboxyl group, intermolecular esters or polyesters may be formed.

Dehydration of polycarboxylic acids leads to the formation of cyclic anhydrides where a 5- or 6-membered ring is possible. Higher molecular weight dicarboxylic acids yield polymeric anhydrides on dehydration. Dry distillation of some salts of dicarboxylic acids (adipic and higher) yields cyclic kenotes.

The reaction of dicarboxylic acids with glycols forms polyesters, and with diamines, polyamides.

Oxalic and malonic acids undergo certain unique reactions. Oxalic acid on heating decomposes to carbon monoxide, carbon dioxide, formic acid and water; oxidation of oxalic acid gives carbon dioxide and water. Malonic acid on heating decarboxylates to form acetic acid and carbon dioxide; on dehydration it yields carbon suboxide. Maleic anhydride and other α,β-unsaturated anhydrides and acids undergo the Diels-Alder reaction with 1,3-dienes.

A few uses of some important carboxylic acids are listed below.

Formic acid is used in textile treatment and as an acid reducing agent.

Acetic and *propionic* acids are used in the production of cellulose plastics and esters. Calcium propionate is used as a mold inhibitor in foods.

Stearic acid finds application in rubber compounding as a dispersing agent and activator of accelerators. Stearic and *palmitic* acids and derivatives are used in the manufacture of soaps, candles, cosmetics, pharmaceuticals and protective coatings.

Oleic acid is employed in the manufacture of soaps and detergents, and in textile applications. Derivatives of oleic, *linoleic* and *linolenic* acids are constituents of paints and drying oils.

Acrylic, methacrylic, maleic, fumaric and *itaconic* acid derivatives are used in the preparation of a wide variety of polymers.

Oxalic acid is useful in rust-removal, cleaning and bleaching. Other dicarboxylic acids, notably *phthalic, adipic* and *sebacic*, are employed in the preparation of plasticizers, alkyd resins, polyesters and polyamides. *Succinic* acid derivatives are used in a number of medicinals.

Citric and *tartaric* acids are used in foods, pharmaceuticals, and metal cleaners. Citric, tartaric and *gluconic* acids are sequestering agents. Citric esters find application as plasticizers in food wrappings.

Salicylic and *benzoic* acids and their sodium salts are antiseptics and preservatives. Acetylsalicylic acid (aspirin) and sodium salicylate are analgesics and antipyretics. Methyl salicylate is used in pharmaceuticals, flavors and toilet goods. Benzyl benzoate is an insect repellent and miticide.

Abietic and related acids are the chief constituents of rosin, used in the manufacture of paper, resins and varnishes.

Naphthenic acid salts, derived from crude oils, are used in paint driers, greases and soaps.

Some important carboxylic acids for which U.S. production data were listed by the U. S. Tariff Commission for 1968 are given in the following table.

Acid	Production, million lb.
acetic, syn., 100%	1,738
propionic	38
chloroacetic	79
phthalic (anhydride)	744
salicylic (tech.)	30
acetylsalicylic	31
2,4-dichlorophenoxyacetic	79
benzoic (tech.)	22
acrylic	82
gluconic (tech.)	4
adipic	1,163
maleic (anhydride)	182
fumaric	43

Production data for 1971 for several of these acids, in millions of pounds, are: acetic acid, 2,050; adipic acid, 1,250; phthalic anhydride, 766; and maleic anhydride, 227.

C. J. KNUTH

Cross-references: *Acids, Aliphatic Compounds, Aromatic Compounds.*

CARCINOGENIC SUBSTANCES

Strictly defined, a carcinogen is a compound which can produce a malignant tumor. However, the term "carcinogenic substance" is generally used in a broad sense to refer to a compound which can initiate either a benign or a malignant tumor in a variety of tissues in plants, animals, and in some instances in man. The tumors are composed of neoplastic cells characterized by unrestrained growth in size and number; transplantability; invasiveness; and metastasis. These cells have abnormal nuclei, bizarre mitosis, differing cytoplasm with varying amounts of enzymes, although they obtain their nutriments from the same blood and lymph which support normal cells. Carcinogenic substances *initiate* the production of these abnormal cells but do not actually become part of the subsequently growing cells.

Contact of the carcinogens may be:

1. by repeated topical application to the skin over a long period of time (months, years) leading to benign papillomas and then to malignant carcinomas.

2. by inhalation (vapor, dust, mist) leading to adenomas of the lung.

3. by implantation in connective tissues leading to fibromas, then to malignant sarcomas.

4. by injection of a carcinogen in special tissue, e.g., brain cells (neuromas and gliomas); myeloid tissue of bone leading to bone tumors or myeloid leukemia; lymphatic glands and/or spleen leading to lymphatic leukemia.

5. by metastases: migration of neoplastic cells from one tissue or organ to others.

Thus carcinogenic substances may exert their effects on many different tissues in a variety of ways.

Carcinogens in Man. A limited number of compounds have been found to cause cancer in man by (1) clinical studies of the tumors in men and women whose occupations exposed them to continued exposure to the compounds, and (2) by subsequent studies which demonstrated that administration of these same compounds by proper routes to suitable experimental animals caused the development of tumors. Some examples are:

1. *Polycyclic aromatic hydrocarbons.* In the 18th century it was noted that soot was a factor leading to high incidence in cancer of the scrotum in chimney sweeps. After prolonged exposure, workers in the coal tar industry developed skin cancers, and tests of coal tar on rabbit skin caused skin cancers. Much later, the polycyclic aromatic hydrocarbon 3,4-benzpyrene was isolated from coal tar and shown to cause cancer in animals. This observation triggered the synthesis of a large number of similar types of compounds such as, 9,10-dimethyl-1,2-benzanthracene, and 20-methyl cholanthrene, which initiated skin papillomas and carcinomas. When injected they caused tumors in the tissues of contact.

Closely related to the above studies is the fact that 3,4-benzpyrene and others have been isolated from tars of cigarette smoke and from soot of heavily polluted atmospheres. Since 1930, an increased incidence of lung cancer in men and women has been observed in many countries and associated epidemiologically with both increased cigarette smoking and atmospheric pollution.

Some animal experiments suggest that application of 3,4-benzpyrene or related compounds in amounts too small to cause cancer of the skin may result in a "latent effect," which may be activated by later application of a noncarcinogenic compound with subsequent formation of tumors. These matters of sensitization and tolerance to carcinogens are very complex and quite unsolved.

2. *Aromatic amines.* Workers in certain dye production plants had a high incidence of carcinoma of the bladder. This was traced to absorption of arylamines such as β-naphthylamine, biphenylamines, benzidine, substituted anilines and aminoazodyes. Subsequent studies in dogs and rats demonstrated the carcinogenicity of these compounds. Many synthetic arylamines have been made and studied for their carcinogenic activities in animals.

3. *Radiation.* Workers who were applying luminous paint to hands and dials of watches and clocks used small camel hair brushes dipped in a paint containing thorium, mesothorium I, mesothorium II and zinc sulfide. They pointed the brushes by means of their lips. After several years, it was noted that cancer of the mouth, gums, and bones developed. This was due to the radiation given off by the radioactive elements. Also certain chemists working with radium compounds and other radioactive elements developed skin cancers and other tumors. It was established that these were also due to prolonged exposure to β-rays, alpha rays and gamma rays. Moreover, during the early use of x-rays for diagnosis, radiologists developed carcinomas of the skin and leukemia. These resulted from continued daily exposure to x-rays.

Survivors of the atomic explosions at Hiroshima and Nagasaki who were far removed from the central thermal effects but were exposed to high levels of instantaneous radiation have shown higher incidence of leukemia and some other cancers.

The above three examples of carcinogen induced tumors in man are of historical interest only. People in potentially hazardous occupations have been taught to wash frequently with soap and water. Modern industrial plants have controlled ventilation with modern instrumental detection devices, and provide respiratory masks and daily laundered work clothing. At present, new types of x-ray tubes giving directed beams of short exposure times, plus proper shielding of the patient and physician have prevented any carcinogenic action. Likewise modern dental practice now poses no hazards to x-ray of teeth or gums.

Moreover, controlled use of beams of high-intensity x-rays, radiations from cobalt-60, or radium salts, constitutes one of the therapeutic methods for destroying tumors in man and animals. Surgical removal of tumors is sometimes followed by controlled radiation to prevent metastases.

Carcinogens in Animals. Over 500 compounds have been shown to produce various types of tumors in many different species of experimental animals. The polycyclic aromatic hydrocarbons, mentioned above, have been studied in mice, rats, guinea pigs, rabbits, hamsters, dogs, cats, monkeys, birds, and amphibians. Likewise, the carcinogenic activity of coal tar and arylamines has been demonstrated in rats and dogs.

The following constitutes a partial list of the numerous compounds which have been found to produce some type of cancer when given to animals:

Ethyl carbamate (urethan), 2-acetylaminofluorene, carbon tetrachloride, alkaloids of *Senecio,* arsenic compounds, chromates, nickel carbonyl, asbestos, beryllium and selenium compounds, alkylating agents such as nitrogen mustards, *bis*-epoxides, ethylene imines, cellophane, N-nitrosodimethylamine and many other N-nitrosamines, N-hydroxylamino compounds.

The mold, *aspergillus flavus,* produces aflatoxin B_1 and G_1 which are highly toxic to chicks, turkeys, ducks, pigs, calves, and rats. These mycotoxins are very potent carcinogens, only 10 micrograms being sufficient to produce liver cancer in a rat. The mold grows on improperly stored cottonseed meal, raw peanuts and related products. There are a few reports that groups of people who have eaten foods contaminated with the mold show higher incidence of liver cancers.

Viruses. The status of these substances as carcinogens is unsettled. J. J. Bittner, in studies using inbred strains of mice, found evidence for a milk factor transmitted to suckling mice which led to mammary cancer. R. E. Shope described an agent which induced papillomas in the skin of rabbits. Some of these growths became malignant. Extracts of mouse leukemic tissue injected into day-old mice generated a high incidence of leukemia. The Stewart-Eddy polyoma virus was shown to be capable of growth in tissue culture and of causing tumors in mice. These various agents are termed oncogenic viruses. They are very different from well-characterized viruses (combinations of nucleic acids with proteins) which have been shown to cause *infectious* diseases: such as common cold, influenza, mumps, measles, chicken pox, smallpox, rabies, and polio. However, none of the forms of cancer are transmissable from one person to another. Cancer is not an infectious disease and no authentic human carcinogenic viruses have been separated and characterized as of 1970. This phase of cancer research is under intensive study.

Steroidal Hormones. The structures of the sex hormones, estrone, estrol, estradiol, progesterone, and testosterone have been established. Also the adrenal cortical hormones, cortisone, corticosterone and others have been characterized. All these possess portions of the ring structures of cholesterol and the bile acids; cholic, deoxycholic and chenodeoxycholic acids. Investigators have considered the questions concerning the possibility of carcinogenicity of these hormones, *if* excessive amounts were produced, or, *if* insufficient amounts were produced for normal functions. Also, are there possible faulty metabolic processes which could convert these steroidal ring systems to compounds related to the carcinogenic polycyclic aromatic hydrocarbons? The protein and polypeptide hormones, insulin, oxytocin, vasopressin, ACTH; and the secretion of thyroxin and adrenalin are also involved in the normal regulatory mechanisms of the body. The problems concerning these hormones plus that involving the numerous cell enzymes are being studied intensively to clarify their relationship to carcinogenic initiation.

Finally, it should be pointed out that the purpose of the studies of carcinogenic compounds is not only to learn why and how cancerous cells develop, but the results may serve as a warning of danger so that exposure to such compounds can be avoided. The studies serve to perfect the best possible test systems in suitable control animals so that it will be possible to advance the search for preventive and curative drugs, radiation, and surgery.

RALPH L. SHRINER

References

Braun, A. C., "The Cancer Problem," Columbia University Press, New York, 1969.

Sutton, P. M., "The Nature of Cancer," English University Press, London, Great Britain, 1969.

Weisburger, J. H., and Weisburger, E. K., "Chemicals as Causes of Cancer," *Chem. Eng. News,* **43,** 124–142 (1966).

Hartwell, J. L., "Survey of Compounds Tested for Carcinogenic Activity," U.S. Public Health Service (1951); and Supplement (1957).

"The Merck Index," By Merck and Co., Rahway, N.J. (8th Ed. 1968); should be consulted for structural formulas and properties of compounds mentioned above.

CAROTENOIDS

The carotenoids represent the most widespread group of naturally occurring pigments. They are present without exception in photosynthetic tissues and occur with no definite pattern in nonphotosynthetic tissues, such as roots, flower petals, seeds, pollen, etc. They are again found sporadically in lower plant forms which include the pigmented yeasts, molds, mushrooms, bacteria and algae. Although the carotenoids are found in all animal phyla, the animal is thought not capable of a *de novo* synthesis of the basic carotenoid structure; when ingested by the animal, they may be absorbed unchanged or undergo some modification. Certain of the carotenoids can be split into Vitamin A by animals.

The carotenoids are chemically and biochemically related to the more general class of compounds known collectively as the terpenes and terpenoids. These are compounds composed of repeating isoprenoid (methyl-butadiene) like units. The isoprenoid carbon skeleton is also found in such diverse compounds as sterols, bile acids, squalene, sex hormones, ubiquinones, natural rubber, essential oils, the phytol side chain of chlorophyll and the side chains of Vitamins E and K.

Mevalonic acid, which can be derived from acetyl CoA, has been shown to be an obligate precursor to the biosynthesis of the terpenoids. Mevalonic acid can be converted to the C-5 compound isopentenyl pyrophosphate. The latter compound is isomerized to dimethylallyl pyrophosphate, which condenses with a second isopentenyl pyrophosphate to form the C-10 compound, geranyl pyrophosphate. In a similar fashion this compound is converted to the C-15 farnesyl pyrophosphate and then to the C-20 geranylgeranyl pyrophosphate. Condensation of two C-20 compounds, tail to tail, yields phytoene which, though colorless, is generally thought to be the main precursor of the C-40 carotenoids. The polyenes phytoene and phytofluene (3 and 5 double bonds in conjugation) are usually considered in discussions of carotenoids, although a strict adherence to the definition of carotenoids as pigments would exclude these compounds.

Phytoene has been shown to undergo a series of dehydrogenation steps to form more unsaturated polyenes. The introduction of four such double bonds yield lycopene—the major pigment in tomato fruit. In the process, the absorption characteristics of the compounds change from a colorless phytoene through yellow intermediates to the red compound —lycopene.

The polyene chain may be wholly aliphatic, ending with structures resembling the pattern for pseudoionone, of which lycopene would be an example. The carotenoids may also be alicyclic, with one end or both ends cyclized.

Beta-carotene (Fig. 1) is an example of a carotene containing two rings, both of which have the unsaturation in the β form. The methyl groups are 1:5 relative to each other, except at the center where the two C-20 pyrophosphate groups were joined and the separation of the methyl groups is 1:6. If, as in the case of β-carotene, both halves are identical, a center of symmetry exists at the 15-15′ position.

Vitamin A($C_{20}H_{29}OH$) is one-half of the β-carotene molecule with a primary alcohol at the C-15 position. Beta-carotene could theoretically yield two molecules of Vitamin A, and any other carotenoid with one-half identical to β-carotene could yield, in theory, one molecule of Vitamin A (e.g., γ-carotene, α-carotene, etc.). Recent studies have shown that dietary β-carotene is cleaved into two molecules of retinal (Vitamin A aldehyde) in the intestine.

The retinal is then mainly reduced to retinol (Vitamin A_1 or Vitamin A alcohol), which is esterified with fatty acids and transported from the intestine via the intestinal lymphatics. Retinol and retinal have a dual function: in vision, and systemically as in the growth of a young animal. Retinoic acid (Vitamin A acid) administered to a Vitamin A-deficient animal can alleviate the systemic symptoms but does not serve as a vision pigment. Retinal is pale yellow; it was first identified as a component of visual purple (rhodopsin), the coloring matter of rods, in combination with a protein opsin. Visual purple is lost on exposure to light, causing night blindness. Visual purple is regenerated after a period of time in a dark environment.

Carotenoid Types. The carotenoids have traditionally been classified as carotenes (hydrocarbons) or xanthophylls (containing oxygen in some form). The main carotene found in higher plants is β-carotene. The main oxygen containing carotenoids in plants is lutein (3,3′ dihydroxy-α-carotene). Other functional groups include esters, aldehydes, ketones, epoxides, furanoids, ethers and acids. In addition to the usual double bond system, some carotenoids with triple bonds have been isolated. All new structures were variations of the basic C-40 structure and, although synthetic carotenoids had been prepared with greater than 40 carbon chains, these were thought to be of academic interest only, due to the "C-40 limit imposed by nature."

With use of the high-resolution mass spectrometer, however, the analysis has rapidly changed. Natural carotenoids whose structure had been "proved" were reexamined and found not to be limited to 40 carbons. Thus the impurities that were hitherto discarded from microorganism extracts on the one hand to citrus extracts on the other were found to contain many new carotenoids of a variety of carbon skeletons.

In addition, carotenoids with aromatic rings have been isolated. Carotenoids have also been shown to occur in glycosidic linkages to such sugars as glucose, rhamnose, mannose and gentiobiose.

Properties. With the exception of the carotenoid glycosides or protein complexes, the carotenoids are extracted with organic solvents of varying polarity. Most can be crystallized, although some of the more saturated polyenes tend to oil out on crystallization. Melting points are not as important as a criterion of purity since the actual melting point or, in some cases, the decomposition range, may vary with the rate of heating.

Preliminary identification of a carotenoid might be based on its relative position on TLC or column

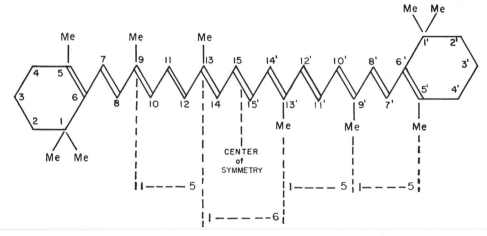

FIG. 1. β-Carotene.

chromatography, its absorption spectrum and its partition coefficient. Structural determinations of new pigments require the use of such instruments as NMR, ORD, IR and the mass spectrometer. A large number of the carotenoids have been synthesized, and thus direct comparisons with authentic samples can be made.

The most characteristic feature is the light absorption spectrum. This may vary from a sharp, 3-peaked spectrum to a single broad peak with examples in between the two extremes. The positions of the maxima in a given solvent, the shape of the curve and the extinction coefficient are a consequence of the basic structure and the type and position of functional groups. Each additional double bond in conjugation results in a displacement of the absorption 15–30 nm toward longer wavelengths. Introduction of a *cis* double bond to the all *trans* system or of a double bond out of plane, as in ring formation, generally results in a shift to shorter wavelengths with loss of extinction. Any functional group that can extend the chromophore, such as a carbonyl in conjugation, will shift the spectrum to longer wavelengths. The chromophore in a polyene chain arises from the possibility of charge separation. If there is a ketonic oxygen that can assume a negative charge, the presence of a proton acceptor, or electron donor such as a protein would stabilize the form. Such a situation exists in the lobster, and the color of the complex is shifted to longer wavelengths and is then blue. On boiling, the complex is broken and the lobster assumes a red color due to the liberation of the red keto-carotenoid astaxanthin.

Cis-trans Isomerization. Most of the natural carotenoids are in the all *trans* form, although the natural isomers of phytoene and phytofluene probably contain a central mono-*cis* configuration.

Although the most stable isomer is the all *trans* form, stereomutations can occur *in vivo* and *in vitro*. Carotenoids show *cis-trans* isomerism in solution, particularly at elevated temperatures. Light with or without a catalyst (iodine) is quite effective in causing *cis-trans* isomers. It has been calculated that lycopene with 11 conjugated double bonds could in theory exist in 1,056 different forms. Fortunately only 72 sterically unhindered isomers exist (overlapping of hydrogen atoms vs the overlapping of hydrogen-methyl groups). Some sterically hindered forms have been prepared and have been shown to occur in the metabolism of Vitamin A.

Prolycopene, proneurosporene and more recently a pro-ζ-carotene have been isolated and shown to be poly *cis*-carotenes.

Function. It is tempting to seek a single universal function for the carotenoids since they are found in such diverse biological forms. Failing this, one is tempted to ascribe some function for these pigments wherever they are found. Thus they have been implicated as an accessory pigment in photosynthesis, as being involved in phototropism in higher plants and in phototaxis in motile cells. They have been suggested as having a function in the reproduction of some fungi and in the protection of other cells, such as bacteria, from absorbed light.

The protective function of the carotenoids in photosynthetic tissues has been well-documented.

In a few cases the carotenoids have been shown to be able to pass on light energy to chlorophyll. Thus, in the diatom *Nitzchia,* light absorbed by the carotenoid fucoxanthin caused fluorescence in chlorophyll a.

While the function of the carotenoids has been proven in a few cases, their function, if any, remains to be determined in a large number of other organisms.

KENNETH L. SIMPSON

References

Goodwin, T. W., Editor, "Chemistry and Biochemistry of Plant Pigments," Academic Press, Inc., London, 1965; cf. chapters by B. C. L. Weedon, T. W. Goodwin, B. H. Davies, C. P. Whittingham, J. H. Burnett, and C. O. Chichester and T. O. M. Nakayama.

"Carotenoids Other Than Vitamin A," 2nd International Symposium, Las Cruces, N.M., IUPAC, Butterworth & Co., Ltd., London, 1969.

Mackinney, G., "Carotenoids and Vitamin A," in *Metabolic Pathways,* Vol. 2, Ed. D. M. Greenberg, Academic Press, N.Y., N.Y., 1968.

Porter, J. W., and Anderson, D. G., "Biosynthesis of Carotenoids," in *Annual Review of Plant Physiology,* **18,** 197–228 (1967).

Simpson, K. L., Phaff, H. J., Chichester, C. O., "Carotenoid Pigments in Yeast," in *The Yeasts,* Vol. 2, pp. 493–515, Eds. A. H. Rose and J. S. Harrison, Academic Press, London, 1971.

CATALYSIS

For many years it has been recognized that certain substances by their very presence are able to alter the rate of chemical reactions. It was not until 1835, however, that this phenomenon was given a name. Berzelius called these substances "catalysts" and named the phenomenon "catalysis." The name appears to have been well chosen. It was derived from two Greek words: "kata," meaning *entirely* and "lyo" meaning *loose.* The implication of the name is to the effect that a catalyst loosens the bonds of the reactant substances in such a way as to greatly alter the rate of reaction. At the time, it was decided that catalysis was due to some special "catalytic force." It is now generally recognized, however, that the forces involved are probably those of ordinary chemical reactions. The exact mechanism by which catalysts operate is still not certain. There certainly would be no disagreement among chemists, however, that catalysts do in some way "loosen up" the bonds of reactants and profoundly alter reaction rates.

Strictly speaking, catalysts can either increase or decrease the rate of a reaction. A very large fraction of the literature on catalysis is devoted to systems in which the catalyst *increases* the rate. The retardation of reactions, called negative catalysis, is known to exist, however. One common theory of the action of negative catalysts is that they combine with and remove from the system traces of positive catalysts, or that they combine with intermediates in a chain reaction in such a way as to break the reaction chain. In the remainder of this description the discussion will be limited to the action of catalysts in accelerating chemical processes.

It is frequently stated that a catalyst is a substance capable of altering the speed of a reaction without itself necessarily undergoing any chemical change. It must be recognized, however, that according to present points of view a typical catalyst as it operates in a reaction may well have been altered considerably from the form in which it was added to the reactants. The changes are especially notable in the composition of the surface layer of a catalyst, though in many instances changes throughout the body of the catalyst may also take place. For example, if an iron catalyst is placed in contact with a mixture of hydrogen and nitrogen at 450°C, the iron immediately becomes covered with a chemisorbed layer of nitrogen, hydrogen and perhaps even ammonia molecules. Furthermore, the catalyst becomes saturated with dissolved nitrogen and dissolved hydrogen. It may even be true that these added gases influence the electronic characteristics of the solid catalysts in such a way as to be controlling factors in the activity of the iron as a catalyst.

A more extreme instance of a change in the catalyst during use is afforded by the iron catalysts that are active in the synthesis of hydrocarbons from mixtures of carbon monoxide and hydrogen. Such a catalyst, which initially consists mostly of iron, changes very rapidly to a mixture of carbides and Fe_3O_4. A certain small amount of the original alpha-iron also usually persists. It is important to know, however, that during the various changes in the solid phase of the catalysts, the activity often continues at a fairly steady level. It might perhaps be more accurate to say that the solid which is initially added to a reaction and which is commonly called a catalyst undergoes rapid changes in the presence of reacting gases to some form in which the surface becomes a combination of catalyst and reactant atoms in a ratio that is often not known.

The persistence of activity for long periods of time indicates, however, that this surface layer presumably reaches a steady state and does not change extensively, even though the underlying bulk phase, may become altered as a function of time.

Another important characteristic of a catalyst is the fact that a formula weight of catalyst will usually be effective in participating in the transformation of many formula weights of reactants. For example, iron synthetic ammonia catalysts are known to have operated effectively for periods of several years, in the course of which a million or more formula weights of ammonia per formula weight of catalyst were produced. Eventually, catalysts lose their activity due to gradual sintering, accumulation of poisons, the occurrence of side reactions between the catalyst and one or more of the reactants, or the accumulation of products. In practice they may have to be regenerated every few minutes or not for years, depending upon the reactants, the catalyst, and the operating conditions.

Catalysts merely speed reactants toward their normal chemical equilibrium but do not actually alter the position of equilibrium. Thus, for example, if a mixture of three parts hydrogen and one part nitrogen is placed in contact with an iron catalyst at 450°C and a total pressure of one atmosphere, the final equilibrium amount of ammonia will correspond to 0.23% of a gaseous phase. This figure therefore represents the upper limit of the per cent ammonia that can be produced in a stream of 3:1 hydrogen to nitrogen gas over an iron catalyst at this particular temperature and at atmospheric pressure. It must be kept in mind that in complex reactions a catalyst frequently can yield many products. It does not always follow that the products formed will be those which would be obtained if equilibrium existed among all the various products. For example, cetane can be cracked over a silica-alumina catalyst to a variety of hydrocarbons. It is well known that the ratio of isobutane to normal butane obtained as a reaction product at a given temperature is always considerably in excess of the ratio that would exist if these gases were in contact with the catalyst long enough to permit sufficient isomerization of iso- to normal butane. It so happens in this instance that under operating conditions the cracking reaction occurs at a relatively faster rate than the isomerization reaction among the reaction products.

Catalysts may be solids, liquids or gases. If the reactants and catalysts comprise two separate phases, as in the case of gases reacting over solids, or liquids reacting in the presence of finely divided solids, the phenomenon is frequently referred to as *heterogeneous* catalysis. On the other hand if the catalyst and the reactants are all dispersed as a homogeneous phase, the process is usually designated as homogeneous catalysis. Iron catalysts for the synthesis of ammonia represent a good example of heterogeneous catalysts; on the other hand, acids in solution may act as catalysts for different components of a liquid phase by a process of homogeneous catalysis. Most of the present discussion is devoted to heterogeneous catalysis but a few brief remarks will also be made relative to some of the essentials of homogeneous catalysis.

Preparation and Nature of Solid Catalysts. Catalysts may consist of elements, compounds or amorphous mixtures of complexes or compounds. Among the elements, the metals are particularly useful as catalysts. Among compounds, metallic oxides and metallic sulfides are outstanding. Probably the principal example of an amorphous mixture of complexes or compounds is the silica-alumina catalyst used in cracking hydrocarbons.

One important characteristic of catalysts is that they are usually highly specific in their activities. For example, a catalyst may well be active for the hydrogenation of certain bonds of organic compounds and yet completely inactive for the hydrogenation of other bonds. Thus, for example, copper chromite will readily hydrogenate carbonyl groups on organic molecules but is relatively inactive for the hydrogenation of carbon-carbon double bonds and completely inactive for the hydrogenation of benzene. Metallic nickel on the other hand will hydrogenate all three types of bonds. Again, an alloy containing 80% copper and 20% nickel will rapidly hydrogenate ethylene but not styrene. This property of a catalyst—to be specific in regard to its action—is extremely important. In those systems in which a multiplicity of activities might be involved, this specificity of catalysts is for the most part an inherent characteristic of a given surface. Moreover, it is also dependent to a certain extent upon the pore size and pore distribution of the catalyst. For example, catalysts for partial oxidation should presumably have a pore size sufficiently large to lower the probability of oxidation of the desired intermediate product to carbon dioxide and water while it is passing out of the pore structure of the catalyst into the main gas stream.

Traditionally, catalysts are prepared in such a way as to produce large surface areas. They are therefore usually either finely divided or porous or both. For the maintenance of a large surface area of the catalytic component, use is frequently made of substances which in themselves are inert, but which are capable of supporting the active catalyst in a form that resists sintering. Common supports are alumina, kieselguhr, silica gel, and even silica-alumina catalysts.

Evidence very strongly suggests that the activity of a catalyst is often centered in only a small fraction of the catalyst surface. The nature of these active points or active regions is still very much a matter of dispute. Sometimes they are created by adding impurities known as *promotors*. For example, the addition of about 1% of potassium oxide and 1% of some inert oxide such as aluminum oxide to an iron oxide catalyst yields on reduction a porous iron solid partially covered with these added promoter materials. Such catalysts are much more active at high pressure than those produced by the reduction of pure iron oxide or those catalysts containing, for example, only aluminum oxide as promoter. It thus appears that one must attribute some intrinsic activity to the interface between the promoter components and the metal surface.

Another proposal is that only certain planes on the metallic catalysts are active for a particular reaction. For example it has been suggested that ammonia synthesis takes place on the 111 planes of iron rather than the 100 or 110 planes.

Regardless of the nature of the active regions, there is no question but that the surface of most solid catalysts is very nonuniform in activity. The actual preparation of catalysts frequently involves the reduction of metallic oxides or compounds to the metal form after the addition of necessary promoters. The oxides themselves are prepared in an amorphous or finely crystalline condition by precipitation under proper conditions and in the presence of suitable promoters.

Catalysts are usually subject to poisoning. This involves the deposition of certain impurities in such a way as to render the active centers or active portion of the catalyst surface inactive. For example, traces of hydrogen sulfide in a stream of hydrogen will usually cause a rapid decrease in the activity of a metallic hydrogenation catalyst. These poisons may be temporary and capable of being removed by some suitable treatment of the catalyst for a short period of time by one or both of the pure reactants; or they may be incapable of being removed by such procedure and will become permanent. For example, traces of oxygen or water vapor in a hydrogen-nitrogen mixture will serve as temporary poisons for an iron catalyst. However, treatment of the catalyst at normal operating temperature by a stream of pure hydrogen or pure hydrogen and nitrogen will rapidly remove this oxygen poison and thus regenerate the catalyst. In contrast to this, during the cracking of hydrocarbons over silica-alumina catalysts, carbonaceous deposits are built up on the catalysts. These deposits have to be periodically removed by combustion in order to restore the catalyst to its initial activity.

Exact details of the nature of the catalyst surface cannot at the present time be specified. A recent trend tends to classify catalysts as being metals, semiconductors or insulators. This classification throws emphasis on the electronic structure of the solid as a very important one in determining the nature and extent of catalytic action. It is also currently popular to interpret the "active points" or "active regions" of catalysts in terms of lattice defects that have been built up in a catalyst by the addition of impurities, by the removal of certain atoms from the original compounds to produce non-stoichiometric compounds frequently characterized by high conductivity, or by the preparation of the catalyst in such a way as to produce numerous lattice irregularities or dislocations. Much more experimental work will be required before the exact nature of the catalyst surface can be described with any certainty.

To illustrate the types of catalysts and types of reactions that are especially important, a limited number of examples is itemized in the accompanying table. Attention is also called to current commercial use that is being made of certain of the catalysts and reactions.

Mechanism of Catalytic Reactions. From what has so far been stated, it is evident that the mechanism of catalytic reactions must be as obscure as the nature of the catalyst surface itself. Progress is being made in elucidating both the mechanism of the reactions and the nature of the catalyst surface. It seems to be generally agreed that solid catalysts invariably combine chemically at the surface with one or more of the reactants. This combination is referred to as *chemical adsorption* or *chemisorption*. For example, both nitrogen and hydrogen are capable of being chemically adsorbed on iron catalyst at the temperature at which these gases are capable of combining to form ammonia. Silica-alumina cracking catalysts for cracking hydrocarbons in the temperature range of 400° to 500°C are sometimes cited as an exception to this rule. The most recent data, however, seem to indicate that chemical interaction between the reactants and this catalyst surface does actually occur, but takes place on only a very small fraction of the surface of the catalyst.

This has been found true for both the amorphous silica-alumina catalysts and for the silica-alumina molecular sieves that have recently been found to be much more active than the amorphous catalysts and that are being adopted as standard components in commercial cracking catalysts.

The action of the metallic catalyst has for many years been related to the geometric spacing and arrangement of the metal atoms with respect to the molecules of the reactants. Thus, for example, Beeck found that a plot of the spacing of the atoms of pure metal films against the activity of these films for the hydrogenation of ethylene yielded a curve with a maximum corresponding to activities larger by several orders of magnitude than the activity of some metals lower on the curve. Thus, a thin film of rhodium was, per unit area at a given temperature, about 1000 times as active as a thin film of nickel.

Type of Catalyst	Typical Catalysts	Reactions Caltalyzed
Acid	Amorphous silica-alumina and crystalline silica-alumina (porous zeolites)	Cracking of hydrocarbons
	HF	Alkylation
	H_2SO_4, H_3PO_4	Isomerization of hydrocarbons
	H_3PO_4	Polymerization of olefins
Hydrogenation-dehydrogenation	Ni	Oils to fats
	Fe	Ammonia synthesis
	Fe, Co	Hydrocarbon synthesis
	Pt, Pd, Rh	Hydrogenation of double bonds and other carbon linkages
	Fe_2O_3, MoO_3, Cr_2O_3	High-temp. dehydrogenation
	ZnO, Cr_2O_3	Methanol synthesis
Cyclization and aromatization	Pt, MoO_3, Cr_2O_3	Heptane to toluene Straight-chain hydrocarbons to cyclic and aromatic hydrocarbons
Oxidation	V_2O_5, MoO_3, WO_3	Partial oxidation of organic compounds
	Bi_2O_3-MoO_3	Partial oxidation of propylene to acrolein
	Ag_2O	Ethylene to ethylene oxide
	Fe_2O_3, Cu_2O-CuO, Pt, MnO_2, Bi_2O_3	Complete oxidation
Hydration-dehydration	Al_2O_3, ThO_2	Alcohols to olefins and water vapor, and the reverse
Halogenations	Metallic halides	Deacon process
Dual Type	Pt, MoO_3, or Cr_2O_3 on acid type supports	Hydroreforming of hydrocarbons
Polymerization	MoO_3 on Al_2O_3 CrO_3 on SiO_2-Al_2O_3 Aluminum alkyls plus titanium halides	Polymerization of olefins to solids

More recently, this difference in activity has been attributed to the electronic characteristics of the individual metals. Beeck showed that his data can be represented as a smooth curve if the logarithm of the activity per unit area of catalyst is plotted against the per cent d-character of the metal, as interpreted by the Pauling hybridized bond theory of metals. In a very spectacular demonstration of the way in which the electronic characteristics of a metal may influence activity, Dowden and Reynolds showed that adding copper to nickel gradually filled the "d-band vacancies" in the nickel and at the same time lowered the activity of the catalyst for the hydrogenation of styrene. As a matter of fact, the activity dropped to approximately zero when enough copper had been added to make the alloy consist of 40 atom % copper and 60 atom % nickel. This result was interpreted as indicating that one or both of the reactants on the catalyst surface tended to transfer electrons ino the lattice of the solid. It must be pointed out, however, that for the hydrogenation of other molecules, catalysts containing certain amounts of copper are definitely more active per unit area than nickel itself. For example, the rate of hydrogenation of ethylene to ethane has been reported to be manyfold greater for catalysts containing 20 to 80 atom % copper than for pure nickel.

For oxide catalysts, such as vanadium pentoxide, molybdenum trioxide, zinc oxide, and chromium oxide, evidence is accumulating to indicate that the steady state composition of the solid which cata-lyzes a reaction is definitely different from that of the original stoichiometric compound. It is also known that the electrical conductivity of these compounds changes by many orders of magnitude as lattice defects are built up. These defects, in some instances, are produced by adding excess oxygen (p-type semiconductors) and in some instances by removing atoms from the initial stoichiometric compound (n-type semiconductors). Although definitive data for correlating the conductivity of a semiconductor with its activity as a catalyst are for the most part still lacking, there seems to be little doubt that the creation of lattice defects is important in and perhaps essential to the catalytic action of solids. Some workers in the field believe that all catalytic reactions are controlled essentially by the electronic characteristics of the solids. Others admit that certain reactions are so controlled, but that some reactions may take place by mechanisms that are not concerned with the conductivity of the solids.

In conclusion, it may be well to mention a few of the newer research tools that are now available to help unravel the factors that are important in producing active catalysts. These include methods for measuring the surface area of a finely divided or porous catalyst (see **BET Theory**); for measuring the pore size and pore distribution of catalyst; and for obtaining values of the electrical properties of the catalyst particles.

In addition x-ray and electron diffraction, electron microscopy, field emission and ion emission microscopy, x-ray absorption edge and electron probe analyzers, electron spin resonance, nuclear magnetic resonance, and infrared, visible, ultraviolet and Mossbauer spectroscopy are combining to give valuable information about the catalyst itself and in some cases about the detailed nature of the chemisorbed reactants.

Finally the use of radioactive and nonradioactive tracers is helping to elucidate the way in which certain catalytic reactions take place. For example, tracer experiments employing radioactive alcohols, aldehydes and other oxygen compounds seem to establish that the synthesis of hydrocarbons over metals such as iron takes place through formation of an oxygen complex on the surface by the interaction of carbon monoxide and hydrogen. These various tools and approaches should in the years immediately ahead furnish a very much better picture of the nature of catalyst surfaces and catalytic action than is available at present.

Homogeneous Catalysis. The action of a homogeneous catalyst is, in a sense, less complex than that of heterogeneous catalysts. For the most part catalysis in homogeneous systems seems necessarily to involve the formation of intermediate chemical complexes. For gaseous reactions catalyzed by gases this is usually particularly clear-cut. For example, nitrogen pentoxide is known to catalyze the decomposition of ozone. This catalysis apparently takes place as a result of nitrogen pentoxide decomposing into lower oxides of nitrogen plus oxygen. The lower oxides of nitrogen then react rapidly with ozone to produce oxygen and to regenerate nitrogen pentoxide. This process repeats itself, until all the ozone is exhausted. The rate of ozone decomposition under these circumstances will be governed by the intrinsic rate at which nitrogen pentoxide decomposes.

In homogeneous catalytic reactions occurring in solution, the interpretation of results is frequently much more difficult than for homogeneous gas catalysis. This is due to a number of factors. To begin with, catalysts and reactants in solution are subjected to all the numerous variables that characterize reactions in liquid phase. These include phenomena of ionization in solution, complex formation, salting out effects, activity coefficients of reactants and catalytic components, and specific effects to be associated with the solvent medium and the presence of various added substances, which in some way affect the properties of the solutions.

Most examples of homogeneous catalysis may for convenience be grouped into three classes. These may be designated as acid-base catalysis, oxidation-reduction catalysis and transition metal complex catalysis. In the first category are the many reactions in solution that apparently are catalyzed by protons or hydronium ions on the one hand or by hydroxyl radicals on the other. A typical example is the acid catalyzed inversion of sucrose.

In the second category are reactions such as the decomposition of hydrogen peroxide capable of being catalyzed by the various metallic ions, such as those of copper, nickel, cobalt and iron. In this connection, it should be pointed out that fantastically small amounts of copper ion have been shown to have positive effects on certain reactions. Concentrations as low as 10^{-19} molar for copper ions in solution will produce a definite catalytic effect on the catalytic oxidation of sulfite ion to sulfate. It should perhaps also be pointed out that for just such systems the action of negative catalysts is most pronounced. Any substance capable of combining with some of these minute traces of positive metal ion catalysts would cause enormous changes in the rate of reaction, even though the negative catalyst might be present in quantities comparable in magnitude to the traces of positive ion that are effective for the reaction.

The third category, transition metal complexes, has come into prominence during the last ten years. In this short time, however, it has developed into one of the most important fields of catalysis. In fact, the claim has been made by some enthusiasts that every known heterogeneous catalytic reaction can also be catalyzed homogeneously, mostly by transition metal complex catalysts. Like enzyme catalysts (See **Enzymes**) the transition metal complexes are often very specific. A single example will perhaps suffice. It has recently been shown that tris (triphenylphosphine) chlororhodium, $RhCl(PPh_3)_3$, is an excellent hydrogenating catalyst for olefins and aceylenes at room temperature. If, however, one of the triphenyl phosphine groups is replaced by a CO to form $RhCl(CO)(PPh_3)_2$, a catalyst is produced that is excellent for hydroformylation but is no longer active for hydrogenating olefins and acetylenes at room temperature and atmospheric pressure.

                                                    PAUL H. EMMETT

References

Emmett, P. H., Editor, "Catalysis," Vols. I to VII. Reinhold Publishing Corp., New York, 1954–1961.

"Advances in Catalysis," Vols. 1 to 22, Academic Press, New York, 1949–1971.

Storch, H. H., Columbic, N., and Anderson, R. B., "The Fischer Tropsch and Related Syntheses," J. Wiley, New York, 1951.

Evans, D., Osborn, J. A., and Wilkinson, G., "Hydroformylation of Alkenes by Use of Rhodium Complex Catalysts," *J. Chem. Soc. (A),* **1968,** 3135–42.

"Homogeneous Catalysis," Advances in Chemistry Series, No. 70, American Chem. Soc., Washington, D.C., 1968.

Shape-selective Catalysis

A most important property of a catalyst is the *selectivity* with which it directs the transformation of certain molecular species into products of a specific structural identity. The need for high selectivity often exceeds that for high *activity*. In biochemical processes, molecules generated with a structure other than that of the required specific metabolic intermediate or product may alter or poison the entire chain of normal processes. In technology, undesired side products complicate process design and cause loss of materials.

Nature has evolved highly selective catalysts in the enzymes. But, at present the powers of selectivity of man-made catalysts are confined to an ability to discriminate between classes of reactive chemical groups. For example, catalysts containing transition metals can be made to act upon unsaturated bonds between carbon atoms, but not on saturated ones. These catalysts will cause hydrogenation reactions, but will not induce hydration reactions. Within a single class of chemical reactivity more detailed powers of discrimination are usually difficult to attain. For instance, the selective preference for hydrogenating olefinic rather than aromatic carbon-carbon bonds is generally achieved only by way of a difference in degree rather than in kind of catalyst reactivity. Going a step further, it is virtually impossible for conventional catalytic materials to selectively hydrogenate an olefinic group on molecules with a straight-chain carbon structure (hexane, for example) without also causing the analogous reaction of molecules with branched structure (methylpentene, for example).

Discrimination between molecules differing in shape rather than reactivity of chemical groups has been achieved in specially "designed" catalytic solids. In these solids the catalytically active centers, with their essentially chemical class-selective reactivity, are placed within the crystallite cages of certain aluminosilicates. In this way, the discriminating activity of the molecular sieve is combined with the catalytic activity of the active sites so that these sites are accessible only to molecules with a shape and size such that they can pass through the narrow port openings connecting the crystal cages.

The crystalline aluminosilicates (zeolites) known as A-zeolite, chabazite, gmelinite, offretite, and erionite (the latter two names have for some time been used for the same materials) are examples of zeolites with effective openings near 5 Å, that can serve as bases for highly shape-selective catalysts. When transition metals are introduced into the crystal cages, and proper care taken to eliminate such materials from the external crystal faces, linear molecules can be subjected to reactions (such as hydrogenation, hydrogenolysis, oxidation, etc.) in the presence of branched molecules that are left unreacted. Because of the unvarying dimensions of the crystalline port, discrimination can be so complete that even such a notoriously unselective reaction as combustion can proceed with discrimination; for example, *n*-butane alone can be oxidized from a mixture of *n*-butane and isobutane.

The above reactions illustrate what has been termed reactant-selective reactions in shape-selective catalysis. Still another process makes use of a different kind of selectivity. For, although the passageways to and between cavities are presumed to allow penetration by only a single molecule at a time, the crystal cavities are often large enough to accommodate a fairly large number of molecules. Thus we may visualize the intracrystalline space as a catalytic laboratory in which various products may be generated in a manner characteristic of the nature of the catalytic sites *per se*. Escape, however, and thereby a net rate of productivity, are attained only for those reaction products capable of passing through the ports. This process has been termed product-selectivity.

The cracking of *n*-paraffin hydrocarbons using small-pore crystalline aluminosilicates provides one example of product selectivity. The normal spec-

trum of products of carbon-carbon bond scission reactions on acidic cracking catalysts includes many branched hydrocarbon products. If the cracking reaction is carried out on an acidic gmelinite catalyst, however, these branched chain hydrocarbons are missing from the product.

The introduction of shape-selectivity into heterogeneous catalysis greatly increases the flexibility available in the fabrication of catalytic systems and processes. The first large-scale application of shape-selective catalysis was made recently. The antiknock quality of gasolines can be raised by selective removal of octane-number inhibiting straight-chain paraffins. Such shape-selective catalytic reforming processes have been placed in commercial operation in at least three petroleum refineries. The shape-selective zeolite catalysts continuously remove these components from the gasoline charge stocks by conversion of most of the material to propane.

P. B. WEISZ

Cross-reference: *Catalysis.*

CATIONIC SURFACTANTS

Surface-active agents or surfactants fall into four classes: anionic, cationic, nonionic and amphoteric. Cationic agents exhibiting surface activity are chemicals which ionize in solution with a positive charge residing with the lipophilic portion of the molecule, which is frequently a long-chain hydrocarbon. The hydrophilic or solubilizing portion containing the positive charge and attached to the hydrocarbon chain includes primary, secondary, tertiary and quaternary amino nitrogen, as well as nonnitrogenous compounds such as phosphonium, sulfonium, arsonium, etc., salts. Associated with the positive charge or cation is a negative charge, (an anion) most commonly a halide, acetate, sulfate, metho or ethosulfate or hydroxyl. The quaternary ammonium compounds are opposite in character to anionic agents such as sodium soaps, and the term "invert soap" has been applied to describe these cationic chemicals.

Surfactants, including cationic analogues, are characterized by solubility in at least one phase of a liquid system and are composed of groups of opposing solubility behavior. The surface activity of any molecule is dependent upon the relative strength and distribution of these dissimilar functions. Surfactants form oriented monolayers at phase interfaces and aggregates of molecules or ions, called micelles, at concentrations peculiar to the system. This latter phenomenon is called the critical micelle concentration or CMC. Cationic surfactants similar to other surfactants exhibit in solution one or more properties which include wetting, emulsification, solubilization, detergency and foaming. Surfactants can migrate to a surface such as the boundary between two liquids, a liquid and a solid, or a liquid and a gas. Cationic chemicals have the outstanding characteristic of being strongly adsorbed on solid surfaces and are held in place by strong chemical forces. The polar nitrogen group is adsorbed on the surface, and the hydrocarbon group is oriented away from the surface.

Cationic surfactants are incompatible with anionic surfactants, forming insoluble precipitates in water, but are compatible with the nonionic species. The cationic surfactants are readily exhausted from solution by surfaces which are negatively charged.

Amino structures of commercial interest include aliphatic, aromatic, heterocyclic compounds, and mixtures thereof. These include primary amines, symmetrical and unsymmetrical secondary amines, symmetrical and unsymmetrical tertiary amines, ethoxylated amines (tertiary), aliphatic quaternary ammonium salts, pyridinium salts, amido amines, and imidazolines [4,5-dihydroimidazoles].

Long-chain aliphatic amines are prepared by reacting fatty acids with ammonia to form nitriles, which are then reduced with hydrogen over a suitable catalyst to primary or secondary amine. Diamines result from cyanoethylation of primary amines followed by catalytic reduction. Tertiary amines are produced by reductive alkylation of primary or secondary amines with formaldehyde or by the reaction of a suitable alkyl halide or sulfate with dimethyl amine.

Amines range from volatile pungent liquids to high melting and high boiling solids. A peculiar property of long-chain aliphatic amines is their very low solubility in water, but the very high solubility of water in the amine. This solubility decreases with increasing temperature and is believed to be due to hydration. The salts of the amines behave as colloidal electrolytes in solution, tend to undergo micelle formation, and are generally polymorphic.

Secondary amines are stronger bases than primary amines. Ammonia has an ionization constant of 0.18×10^{-4} at a concentration of 8×10^{-4} mole per liter; octyl to octadecyl amine at the same concentration ranges from 4.5 to 4×10^{-4}, and dioctyl to dioctadecylamine from 10.2 to 9×10^{-4}. In general, tertiary amines are weaker bases than primary amines.

Amines for the most part are primary skin irritants and somewhat toxic. Acute oral toxicities of 200–7000 mg/kg have been reported. No toxic effects with octadecylamine were observed with dogs receiving 0.6–3.0 mg/kg for one year.

Amines are reactive compounds readily forming salts with organic and inorganic acids. Both primary and secondary amines can be reacted further with organic acids to form substituted amides. Primary amines form carbamates and substituted ureas with carbon dioxide, thiourea and dithiocarbamates with carbon disulfide. Ammino-metal complexes can be readily made. The growing urethane field is partially based on the isocyanates made from the reaction of primary amines with phosgene. Most amines can be readily reacted with ethylene or propylene oxide, and alkyl chlorides or sulfates to form quaternary ammonium compounds. Tertiary amines can also be reacted with hydrogen peroxide to form amine oxides.

Quaternary ammonium compounds are made by the reaction of a primary, secondary or tertiary amine in an appropriate solvent with the alkylating agent of choice. Sodium hydroxide and/or bicarbonate is used to neutralize by-product mineral acid and to maintain pH control.

Quaternary ammonium salts can be considered salts of strong bases and strong acids. These chemicals are soluble and ionize readily in water and are insoluble in nonpolar solvents. The exceptions to this are the long-chain dialkyl quaternaries which are hydrocarbon soluble.

Large variations in the anion and cation are possible so that wide differences in physical and chemical properties can be obtained. Accordingly, quaternaries can be obtained as crystalline solids to viscous liquids. They are hygroscopic, decompose on heating to tertiary amines and form micelles at specific concentrations. The hydroxides of these compounds are strong bases comparable in strength to the inorganic bases.

At high concentrations quaternaries are irritating and toxic although they are considered nonirritating at concentrations of 1000 ppm or less. Acute oral toxicities of 200–7000 mg/kg have been reported.

Primary and secondary amines are reacted with ethylene and propylene oxide usually catalyzed by a base such as sodium acetate or hydroxide. The moles of oxide added affects the cationic strength of these adducts. The cationic characteristic decreases with increasing moles of ethylene oxide. The long chain adducts can be considered to be either nonionic or, at best, weakly cationic. These products can be oxidized to amine oxides or alkylated to form quaternaries.

Ethylene oxide condensates of amines are also primary skin irritants and mildly toxic. Acute oral toxicities of 500 to 7700 mg/kg have been reported.

Amine oxides, prepared as previously stated, can be best described as nonionic compounds possessing a strong dipole nitrogen-oxygen bond. However, when protonated and at pH 3, the cationic form predominates. Amine oxides are generally hygroscopic solids, unionized in solution, decompose at elevated temperature into olefins and substituted hydroxylamines, and are biodegradable.

Amine oxides have been reduced to tertiary amines, alkylated, acylated or formed into adducts with SO_2, SO_3, BF_3, SF_4, and PCl_3.

Amine oxides in concentrated solutions are primary skin irritants but essentially nonirritating to the skin and eye at concentrations below 2%. Acute oral toxicities of 1850–6150 mg/kg have been reported. A medically interesting observation has been made that amine oxides possess all the therapeutic and physiological values of the parent amine but are much less toxic.

The family of amidoamines and imidazolines are closely related. The amidoamines are produced from the reaction of fatty acids or esters with polyamines, such as diethylene triamine. The amidoamines, upon further heating at higher temperatures, can be cyclyzed to the imidazoline. The commercial products do not have well-defined chemical and physical properties and will vary from liquid to solids, depending upon the type and the degree of saturation or unsaturation in the hydrocarbon chain. These products are primary skin irritants, and no oral toxicity is available.

Long-chain aliphatic amines and derivatives were commercially virtually unknown prior to 1940. According to U.S. Tariff Commission Reports 1968 and other more recent sources, these cationic surfactants are now sold in volumes of more than two hundred million pounds annually.

As previously mentioned, most surfaces are electronegative. The ability of cationic surfactant to deposit on these surfaces accounts for the growth of these chemicals for widely diversified uses in many fields, such as mining, plastics, textiles, protective coatings, metal working, automotive, dyeing, pigments, road construction and maintenance, etc.

One of the first major uses for nitrogen chemicals was in the field of mining or mineral benefaction by flotation. In flotation, the problem is to develop a hydrophobic coating on the mineral particle. The coating then permits an air bubble to become attached to the mineral particle which then rises to the surface. The crystal structure and particle size of the mineral and flotation reagent specific for the mineral are required for a successful operation.

Amine salts are used to float silica, to separate KCl (potash) from NaCl, and to float metal oxides such as iron, phosphates and feldspar. These agents are used at very low levels in the order of 0.25 to 0.5 pounds per ton of ore.

Amine salts have also been found to be of value in prevention of caking and dusting of bulk chemical agents such as KCl, NH_4NO_3, urea and mixtures of inorganic salts used as fertilizers. Caking is a problem with salts that tend to be hygroscopic or brought about by moisture traveling from the core of the salt to the surface. Chemical make-up, conditions of production, storage and physical shape all contribute to the problem. The amines are used at very low levels to form a hydrophobic coating to prevent caking but do not interfere with normal usage of the salt. Amine derived from tallow is used for sylvite and $NaNO_3$; hydrogenated tallow amine is used for NH_4Cl. Special amine formulations are recommended for NaOH, portland cement, sodium metasilicate, fertilizer salts and NH_4NO_3.

Road construction and maintenance practices are turning to the use of asphalt emulsions in an increasing volume because of problems associated with hot asphalt and because of cost and air pollution problems with solvent reduced cut-back asphalts. Asphalt emulsions made with cationic emulsifiers are particularly suited for these needs as the asphalt is chemically deposited upon the negatively charged aggregate. In road maintenance programs, such as seal coatings, this characteristic results in minimal closure of roads to traffic and permits this operation to be carried out under adverse conditions of wet aggregate, high humidity and threatening weather. Water is displaced from the aggregate surface by the hydrophobic film of the emulsifier, assuring a good bond between the asphalt and the aggregate. In contrast, anionic emulsions cure by the evaporation of water and are more susceptible to attack by water.

There are a multitude of uses for catonic asphalt emulsions for nonroad building use. Some of these include pond sealants for reduction of water loss through seepage, as an agricultural mulch, in protective coatings, cold applied roofing formulations, sound deadeners, tile adhesives, sealants, undercoatings and flashing cement. These applications with asphalt use principally diamines, ethoxylated diamines, selected quaternary derivatives, imidazolines and amido-amines.

The ability of these cationic surfactants to attach to a surface (substantivity) is used to great advantage in processing of pigments and related functions in the protective coating field. Monoamine and diamine salts are the principal cationic chemicals used for pigment dispersing, softening and flushing. The changes in the characteristics of pigment surfaces induced by the action of cationic chemicals improve their dispersion, reduce the problem of pigment flotation, flooding and striking, and also result in reduction in grinding time. Better dispersion gives better color, gloss, greater opacity and tinting strength. This property has also been applied advantageously to the production of textile softeners. Dimethyldi(hydrogenatedtallow)ammonium chloride and related quaternaries are enjoying widespread use for this purpose and are estimated to be consumed in excess of 25 million pounds annually. The molecule orients itself on the fiber, leaving two fatty tails exposed to give surface softness and lubricity. Besides imparting antistatic properties, clothes dry quicker with less wrinkles and iron easier. Many quaternaries and ethoxylated amines are used for imparting antistatic and softness characteristics to nylon, rayon, wool and polyester fibers, and, to a minor degree, in paper.

A major use for quaternary ammonium compounds centers about their ability to inhibit or kill organisms. They are extremely efficient with phenol coefficients in the 300–400 range. Killing efficiency increased with pH and optimum chain length of 14–16 carbon atoms, although other functional groups and organisms will affect this efficiency. The aliphatic tertiary amines alkylated with benzyl chloride are those found most useful for bactericidal activity.

At high pressures encountered in metal-working applications, mineral oils and synthetic lubricants fail to lubricate because they are forced from the surface leading to metal to metal contact and welding. An extreme pressure additive will overcome this deficiency, and primary amines and diamines are usually selected. The dioleate salts of diamines are used for both copper tube and iron wire drawing. Low viscosity polybutene with this agent has been found to be very efficient.

There are many other uses for these cationic agents. These include catalysts for urethane foams, latex rubber sensitizers, grease additives, in agricultural herbicidal formulations, fat liquoring of leather, as demulsifiers, flocculating agents, in secondary oil recovery, water treatment, and many more too numerous to mention. Their versatility and properties will continue to bring about new applications and increased output.

S. H. SHAPIRO

Cross-references: *Detergents, Synthetic.*

References

Ralston, A. W., "Fatty Acids and Their Derivatives," John Wiley & Sons, Inc., New York, 1948.

Pattison, E. Scott, Editor, "Fatty Acids and Their Industrial Applications," Marcel Dekker, Inc., New York, 1968.

Markley, Clare S., Editor, "Fatty Acids—Their Chemistry, Property, Production and Uses—Part 3," Chap. XVI Nitrogen Derivatives, p. 1551, Interscience Publishers, a Division of John Wiley & Sons, New York, 1964.

Astle, Melvin J., "Industrial Organic Nitrogen Compounds," Reinhold Publishing Corporation, New York, 1961.

Smith, Peter A. S., "Chemistry of Open Chain Nitrogen Compounds," Vol. 1 and Vol. 2, W. A. Benjamin, Inc., New York, 1966.

Millar, I. T., and Springall, H. D., "Sidgewick's Organic Chemistry of Nitrogen"—3rd Ed., Clarendon Press, Oxford, 1966.

CELLULAR PLASTICS

Products with low densities are produced by incorporating gas in a liquid polymer and retaining the foam-like structure during the solidification process. Such foams are comprised of many small, gas-filled cavities, each completely enclosed by a polymer film which is either flexible or rigid. They may be prepared from aqueous dispersions by adding thickening agents and "whipping" the mixture to incorporate air, as in whipped cream. The water is removed by heating. Cellular plastics may be formed by the addition of a gas such as "Freon" or by the incorporation of a low-boiling solvent during the molding or extrusion process. Organic nitrogen compounds which decompose to form nitrogen during heating or reactions which produce gases may also be used. For example, cellular urethane polymers are formed as a result of the carbon dioxide evolved when small amounts of water react with organic isocyanates.

ing hemicelluloses, and with ligneous and other noncellulosic components. Wood is composed of 40–50% cellulose, while the major nonprotein natural fibers of commerce, cotton, ramie, flax, linen, hemp, sisal, and jute are predominantly cellulose. Cotton contains as little as 3% of impurities other than moisture. Isolation of cellulose from bacteria and members of the animal kingdom has also been reported.

Cellulose has no nutritive value for humans and most animals and serves only as roughage. However, the alimentary tract of certain ruminants and insects contains microorganisms which break down cellulose to sugars utilizable by the host. These microorganisms excrete the enzyme cellulase, which breaks down cellulose to oligosaccharides including cellobiose (two glucose units) and another enzyme cellobiase which completes the degradation to simple, digestible sugars. Cellobiase cannot attack cellulose directly.

The most convenient source of pure cellulose is cotton fiber, but wood is an alternative. Purified cellulose is composed of 44.4% carbon, 6.2% hydrogen and 49.4% oxygen, corresponding to a hexose anhydride with the formula $C_6H_{10}O_5$. Acid hydrolysis provides D-glucose in greater than 95% yield. In cellulose, it has been shown that these glucose units are 6-membered (pyranose) rings in the chair form, linked together by β-1,4 glucosidic bonds. The conformational structure is shown by the following formula:

$$n = \frac{DP-2}{2}$$

Cellular plastics are widely used for buoyancy, insulation and resiliency. Over 300 million pounds of cellular plastics were produced in the U.S.A. in 1964, about 700 million in 1969, and will approach 1 billion pounds in 1972. Most of the commercial plastics can be converted to cellular plastics.

RAYMOND B. SEYMOUR

CELLULOSE

Cellulose, a naturally occurring high polymer, comprises more than one-third of all vegetable matter and is the most abundant of natural organic compounds. As the chief structural element of the cell walls of higher plants, it has been isolated and modified in a large variety of ways for use in commerce. Although it has been the most widely studied of all compounds, many detailed chemical and physical features of cellulose remain obscure.

In nature, cellulose is found imbedded in, or in part combined with, other polysaccharides, includ-

Cellulose differs from amylose, the straight-chain fraction of starch, and from glycogen only in the configuration at carbon-1, where the glucose units are joined together. The aldehyde group at carbon-1 is tied up in acetal formation in the glucosidic bond and cellulose is generally nonreducing. A single, potential aldehyde occurs at the chain end. This aldehyde group has been used to measure the average molecular weight, or chain length, of the polymer.

Each anhydroglucose unit within the chain has three hydroxyl groups which are available for chemical reactions common to alcohols, namely, the secondary hydroxyls at carbons-2 and -3 and the primary hydroxyl at carbon-6. This inherent reactivity makes readily available a large number of cellulose derivatives and this has been a major contributing factor to the wide commercial utilization of cellulose.

Cellulose is polydisperse and the average polymer chain length depends on the source and the method used in the purification. Typically, average chain length (degree of polymerization) varies from

200–600 glucose units for commercial regenerated cellulose, as it used to make rayon or cellophane, to 700–1300 in wood pulps, to more than 3500 in native cotton cellulose. The viscosity method is the simplest and most used method for molecular weight determination, but this must be standardized against osmotic pressure, light scattering, ultracentrifuge or other suitable methods.

Cellulose dissolves, with molecular weight reduction, in concentrated solutions of strong mineral acids (e.g., 72% H_2SO_4, 40% HCl, or 85% H_3PO_4) and in concentrated aqueous solutions of some salts, such as 72% $ZnCl_2$. The best solvents are solutions of cations such as copper or cadmium complexed with ammonia or ethylene diamine.

Physical Structure. In both its natural state and in the regenerated form, cellulose has a high degree of crystallinity, or lateral order. The joining of anhydroglucose units in the β-configuration and the trans, equitorial disposition of the pendant groups give a linear and stiff molecule capable of close packing. This provides for a high degree of association of polymer chains through hydrogen bonding. This peculiar feature explains the fact that cellulose is insoluble in water and in common organic solvents in spite of its large number of hydroxyl groups. It also contributes to the nonmelting character of cellulose; on rapid heating, it remains intact until thermal rearrangement occurs at ca. 320°C with the formation of the glucose anhydride, levoglucosan.

The unit cell of the crystalline portion of cellulose can exist in at least four known forms and these have been well described by x-ray crystallographic studies. Native cellulose (Cellulose I) has a monoclinic unit cell $10.3 \times 7.9 \times 8.35$ angstrom units which has the length of a cellobiose unit. Chemical treatment of Cellulose I with strong caustic soda (mercerization) or with liquid ammonia brings about a rapid change in orientation in the unit cell and gives Cellulose II and III, respectively. Heat treatment of swollen cellulose produces Cellulose IV, which is similar to Cellulose I. Regenerated cellulose has the Cellulose II crystal structure.

The actual *degree* of crystallinity continues to be a subject of debate. X-ray diffractograms for cellulose of different origins or histories give results which, although in relative agreement, differ widely in absolute values with those obtained by other methods of investigation, which include deuterium isotope exchange, moisture sorption, acid hydrolysis, density measurements, and chemical reactivity. Measurements of absolute degree of crystallinity must be interpreted in light of other structural features depicting the next higher level of order, namely, the size and orientation of the crystallites within the matrix. X-ray diffraction and DP and electron micrograph examination of micelles produced by controlled acid hydrolysis give a measure of crystallite size. X-ray diffraction and infrared dichroism measure the degree of axial orientation of crystalline domains, and optical birefringence registers the total orientation. Crystallites are of the order of 100 and 250 glucose units in length in regenerated and in native cellulose, respectively, and range from 50 to 100 angstrom units in thickness.

At the next higher level, molecular aggregates of cellulose are arranged in threadlike strands called *fibrils,* oriented parallel to each other and wound in successive spiral layers to form the secondary wall and major portions of a natural fiber. The outer or primary wall has a netlike pattern and contains the major portion of noncellulosic matter. Natural cotton fibers (seed hair from the *Gossypium*) are about 1–2 inches in length and about 5–20μ in diameter.

Chemical Reactions and Utilization. Numerous esterification and etherification reactions customary of alcohols have been utilized commercially in the production of cellulose derivatives which find application as fibers (textiles), films, coatings and lacquers, and in molded plastics. The reactions are topochemical in nature and carefully controlled reduction in chain length is generally necessary to facilitate handling. These derivatives are soluble in the more common organic solvents.

Cellulose is susceptible to oxidative degradation with all known oxidizing agents, being particularly sensitive at alkaline pH. Acidic degradation can occur involving hydrolytic cleavage of the glucosidic bond. In addition, cellulose is subject to actinic degradation during exposure to sunlight.

Controlled degradation with caustic soda followed by treatment with carbon disulfide gives the water-soluble xanthate derivative, cellulose-O-CSS^-Na^+, and this forms the basis of the viscose process used to manufacture cellophane and rayon. The pure cellulose is regenerated by extruding into an acid bath, where the xanthate decomposes.

In recent years, the development of durable press garments, largely comprised of polyester/cotton blend fabrics, has been based on chemical reactions of the cotton cellulose with aminoplast resins. These formaldehyde-based reactants introduce chemical crosslinks between the cellulose macromolecular chains, which maintain the fabric in the desired shape, flat or creased. Crosslinks severely reduce the fiber strength and polyester is mixed with cotton to provide satisfactory wear life.

The major processes used in chemical pulping in paper manufacture are the sulfite and the sulfate or kraft processes. The sulfite process involves digestion of wood chips in pressure vessels in aqueous solutions of an alkaline earth bisulfate, usually $Ca(HSO_3)_2$, plus excess SO_2. Lignins are converted to soluble lignosulfonates. In the sulfate process, lignins are solubilized by digesting chips with an aqueous solution of NaOH plus Na_2S. Sodium sulfate is added to make up for sulfur loss, hence the name sulfate process.

The hydroxyl group reactivity of cellulose lends this substrate readily adaptable to anionic and free-radical grafting techniques. Although extensive investigations have been carried out in this field, commercial acceptance at any significant level is yet to be realized.

The principal cellulose derivatives of commerce are the esters, cellulose nitrate and cellulose acetate or acetate-butyrate. Common ethers are the methyl, ethyl, benzyl, hydroxylethyl and carboxymethyl derivatives. New and varied uses are continually being found for these derivatives, such as food additives, dispersing agents, thickeners and coatings.

JOHN J. WILLARD

References

Heuser, E., "The Chemistry of Cellulose," John Wiley & Sons, Inc., New York, 1944.

Whistler, R. L., Ed., "Methods in Carbohydrate Chemistry," Volume III, "Cellulose," Academic Press, New York, 1963.

Ward, K., Jr., Ed., "Chemistry and Chemical Technology of Cotton," Interscience Pub., New York, 1955.

Bikales, N., and Segal, L., Eds., "Cellulose and Cellulose Derivatives—Part IV," John Wiley & Sons, Inc., New York. In Press.

CELLULOSE ESTER PLASTICS

The major organic cellulose esters used for the production of plastics are cellulose acetate, cellulose acetate butyrate, and cellulose propionate. The plastics produced from them are thermoplastic and are commonly referred to as acetate, butyrate, and propionate, respectively. Acetate was the first of these to be manufactured commercially. It appeared in the form of sheets, rods, and tubes in 1927, becoming available in granular form for molding in 1929. Butyrate was first offered for sale in 1938, followed by propionate in 1945. Propionate was withdrawn from the market after a relatively brief period because of production difficulties, but it reappeared and became established in 1955.

The primary manufacturer of organic cellulose ester plastics in the United States is Tennessee Eastman Company, a Division of Eastman Kodak Company. The plastics are marketed by Eastman Chemical Products, Inc., a subsidiary of Eastman Kodak Company, under the trade marks "Tenite" acetate, "Tenite" butyrate, and "Tenite" propionate.

All three organic cellulose ester plastics in pellet form, as they are most commonly produced, find their widest usage in injection-molding and extrusion operations; but butyrate is also available as a powder and is used for rotational molding and fluidized-bed coating. In addition, large quantities of all three cellulose esters that have not been processed into plastics are dissolved and used in solution to produce plastic coatings, cements, and films, and very large amounts are used to produce fibers. The solvents used are ordinarily mixtures containing low-boiling ketones or esters as active components, alcohols as latent solvents, and aromatic hydrocarbons as diluents.

Manufacture. Organic cellulose esters are produced by the reaction between chemical cellulose and the appropriate acids and anhydrides in the presence of a chemical catalyst, usually sulfuric acid. The cellulose used is usually pretreated to make it more active chemically and thus shorten the reaction time. The reaction is allowed to proceed virtually to completion—the triester stage (all three hydroxyl groups in each pyranose ring of the cellulose chain substituted). At this stage, the product dissolves in the reaction mixture. Since the cellulose is fibrous, the reaction is not homogeneous until it is almost complete, and only by allowing it to proceed until the ester is totally dissolved can a uniform, homogeneous product be produced. If the ester is to be compatible with common plasticizers, however, it must contain some highly polar groupings. These are supplied by adding water to the reaction mixture and partially hydrolyzing the triester, so that some free hydroxyl groups are re-formed. This reaction occurs in solution and is therefore homogeneous. Commercial plastic-grade cellulose esters normally contain about one or two free hydroxyl groups for each four pyranose rings in the cellulose chain. Since the hydroxyl groups are formed by a homogeneous reaction, their distribution along the polymer chain is random, which is highly desirable. If the hydroxyl distribution is excessively nonrandom, as it would be if the reaction were not carried to the triester stage, the plastic will be hazy.

Plastics are manufactured from cellulose esters by blending the esters with such additives as plasticizers, heat stabilizers, antioxidants, ultraviolet inhibitors, pigments, and dyes. Fillers are ordinarily not used in cellulose ester plastics because their normal effect is to weaken the plastics severely.

The first step in the production of plastic is mixing the additives with the ester, which is in powder or flake form. The mixture is then heated to its softening point and kneaded, or worked, until it becomes a homogeneous plastic mass. This is done on roll mills, in an extruder, or in a Banbury mixer. Before the plastic cools and hardens, it is generally shaped into small solid pellets suitable for processing in an injection-molding machine or an extruder. For the production of the powder, the hot-kneading step can be omitted or the finished plastic pellets can be ground. If the plastic is to be used in solution, the cellulose ester and additives are simply dissolved together.

Characteristics of Cellulose Ester Plastics. Depending primarily upon the amount of plasticizer incorporated, organic cellulose ester plastics can range from hard, stiff, strong materials to soft, limber, extremely tough compositions. The plasticizer concentration determines the "flow temperature" of the plastic, which in turn determines the "flow designation." Flow temperature and flow designation, like nearly all other properties of cellulose plastics, are determined in accordance with a procedure published by the ASTM. Flow designations range from various degrees of hardness (H4, H3, H2, H) through medium-hard (MH), medium (M), and medium-soft (MS) to various degrees of softness (S, S2, S3, S4, etc.). At any given flow designation, the physical properties and processing characteristics of these plastics will vary to some degree with the identity of the plasticizer used. Some plasticizers, for example, produce unusually hard formulations; some, unusually tough plastics; some, compositions that are exceptionally easy to process. Complete identification of a cellulose ester plastic requires the name of the plastic, a formula designation (which varies with the additives and the cellulose ester used), a color number or color description or both, and a flow designation.

Cellulose ester plastics have no one outstanding property in which they surpass all other plastic materials. The selection of a cellulosic material for an application is generally based upon some combination of properties which is unique in the particular plastic or can be obtained at the least expense with that plastic. Three characteristics that nearly always enter into consideration are processability, toughness, and colorability. No other commercial plastic combines these three characteristics to as high a degree as the organic cellulose ester plastics. The added quality of weatherability makes butyrate and propionate even more exceptional.

The organic cellulose ester plastics are easily molded or extruded, and the products of these operations are well suited for secondary operations such as vacuum forming, machining, printing, lacquering, and cementing. The toughness of the cellulosic plastics is such that properly molded items can ordinarily withstand a considerable amount of maltreatment without breaking. And the natural transparency and freedom from color of the organic cellulose ester plastics allow them to be produced in practically any transparent, translucent, or opaque color.

Cellulose ester plastics in suitable flow designations have strength, hardness, and stiffness adequate for most plastics applications, and the fact that these properties can be varied by changing the flow is often advantageous. If trials on a formed article should show that greater stiffness or strength is needed, and the properties of the particular cellulosic used are otherwise satisfactory, it is often possible to make small changes in the design of the article to provide the additional strength or stiffness.

Properly designed and properly manufactured articles made of cellulose ester plastics have good dimensional stability. It is necessary only to have free flow of the plastic into the shape of the article to be produced. Proper manufacturing involves heating the material sufficiently for it to flow freely, but not enough to degrade it, applying pressure to cause it to assume its finished shape, and then cooling it fairly slowly. If half-melted plastic is forced by high pressure into a mold or through a die, it will freeze, or harden, with the polymer chains in distorted positions, and the article will contain frozen-in stresses. Shock-cooling the plastic will have the same result—the stresses will form as a result of differential shrinkage. Such stresses can result in the development of a rough surface, gross warpage of the article, and brittleness. Butyrate and propionate are generally considered to be easier to mold and extrude than acetate and therefore are less likely to develop internal stresses during processing. Their plasticizers are also higher-boiling and less water-soluble than those used in acetate. Acetate, over a long period of time, may lose some of its plasticizer by evaporation or leaching, and in doing so, it will shrink slightly; but the plasticizer loss from butyrate and propionate will be negligible, and the dimensions of articles made of these materials will not change significantly because of this factor. For these two reasons, butyrate and propionate usually have better dimensional stability than acetate.

Acetate, because of the relative volatility, water solubility, and vulnerability to oxidation of its plasticizers, often has an outdoor service life of only a few months; but special outdoor formulations of butyrate and propionate remain serviceable for several years in full exposure to weathering. According to manufacturer's literature, butyrate, either clear or pigmented, may be expected to remain useful for 5 to 8 years outdoors; pigmented formulations of propionate may be expected to withstand 3 to 5 years of weathering without objectionable deterioration in either appearance or performance; and unpigmented formulations of propionate have approximately half the useful life of pigmented propionate formulations. It is important to note that these statements apply only to specially stabilized, outdoor-grade materials.

The cellulose ester plastics have moderately low heat capacities and very low thermal conductivities. These two characteristics, combined with the smooth surface that these plastics can be given, impart to articles made of them a pleasant feel, which is important in many applications. Such items as adding-machine keys and screwdriver handles, for example, must be considerably hotter or colder than skin temperature before they will feel uncomfortably hot or cold; and the plastic surface warms or cools rapidly upon contact with the skin.

All three of the organic cellulose ester plastics have high electrical resistivities, high dielectric constants, moderately high dielectric strengths, and moderately high dissipation factors. These characteristics have led to the use of these materials for manufacturing such items as meter covers, handles for electrical probes, and housings for switches and various electrical appliances; but the high dissipation factor has almost precluded their use in such applications as coil forms for TV and other VHF and UHF equipment. The high dissipation factor has one practical advantage, however, in that it permits dielectric sealing of films.

Uses. The largest single application of cellulose ester plastics is the production of sheet, which is manufactured in thicknesses ranging from less than one mil to more than one-fourth inch. (Sheet not more than ten mils thick is called film). Sheet up to twenty mils thick is manufactured by both extrusion and solvent casting, but all heavier gauge commercial sheet is extruded. Solvent casting gives better control of surface finish and thickness, so all film for such items as photographic film and magnetic tape, where smoothness and uniformity are essential, is produced by this process.

Some sheet is used flat, as in windows of envelopes and packages, but most is vacuum-formed or shaped by some other variant of the thermoforming process before it appears in the consumer market. Some of the most common products made from cellulosic sheet are blister packages, skin packages, decorative items, and indoor and outdoor signs. Some of the largest items produced from the organic cellulose ester plastics are vacuum formed from sheet. Outdoor signs that measure several feet in each dimension, billboard-high replicas of soft-drink bottles and cones of ice cream, and large Christmas adornments are examples of such items.

Extruded applications of the cellulose ester plastics, other than the production of sheet, include the manufacture of such items as pipe, table edging and other decorative trim, tubes for pneumatic message-carrying systems, and a wide variety of sizes and shapes of thin tubing for packaging toothbrushes, candy, bathing suits, and a host of other products.

Some extrusion blow molding is done by standard procedures, and a variation of the blow-molding process wherein cut sections of pipe taken from stock are re-heated and blown has been used to produce items as large as streetlight globes.

Cellulose acetate plastic initiated the injection-molding industry—it was the first plastic to be processed in this manner—and the organic cellulose ester plastics are widely used in this type of molding. Some of their major molded applications are electrical-appliance housings, radio and TV knobs, pens, pencils, dolls and other toys, knife handles, screwdriver handles, and automobile steering wheels.

ROY O. HILL, JR.

Cross-reference: *Cellulose.*

CEMENTATION

Cementation is a process for modifying the surface composition of a metal to provide special surface properties, such as hardness, wear resistance and/or resistance to oxidation and corrosion. The modification is accomplished by exposing the work to a powder, gas or liquid containing the desired elements and heating to allow these elements to diffuse into the basis metal.

The coating elements may be metallic or nonmetallic. The coating, or case, formed consists of the alloys or intermetallic compounds characterizing the basis metal and coating elements, and is often referred to as a "diffusion coating." All cases require relatively high treating temperatures in order to provide the necessary rate of diffusion. The thickness of the cemented coating increases with time and temperature of treatment. In contrast to coatings formed by most other methods, the dimensional change associated with cemented coatings is only a fraction of the total coating thickness.

Carburizing and Nitriding. By heating a ductile low-carbon steel to a temperature of 870 to 900°C while it is surrounded by a carbonaceous packing material, the metal at the surface is converted to a considerable depth, sometimes 0.15 cm or more, into a high-carbon steel by combining with carbon that diffuses into it. Alternatively, the carbonizing medium may be a gas mixture, usually of carbon monoxide and hydrocarbons. In either case the high-carbon surface can subsequently be hardened by a suitable heat treatment, while the low-carbon core remains relatively soft and tough.

Analogous to the carburization of steel is the nitrogen case-hardening process in which the surfaces of steel parts are hardened by heating within the range of 925 to 1000°C in a nitrogenous medium, usually cracked ammonia gas. Nascent nitrogen reacts readily with aluminum, chromium and other steel constituents, forming dispersed nitrides which surface harden the steel.

In carbonitriding, the steel parts are heated in a gaseous carburizing atmosphere that contains ammonia in controlled percentages. Both carbon and nitrogen are added to the steel by this process, usually to a depth of 0.05 cm or less. Treatment of steel in molten sodium cyanide at 850 to 870°C also case-hardens steel through the addition of carbon and a lesser amount of nitrogen. In liquid carburizing, the mixture is a bath of sodium cyanide, sodium chloride and sodium carbonate at a temperature of from 840 to 950°C. The case so produced is low in nitrogen, but high in carbon.

The above processes of case hardening are used when a combination of a tough core with an exceptionally hard surface is required. They meet the needs of an impact-resistant, wear-resistant material for such applications as gears, pinions, shafts, piston pins and cam shafts.

Sherardizing. Sherardizing is the cementation process for forming a coating of zinc alloy on steel. The articles to be coated are packed in zinc dust in a tightly closed drum and heated while the drum is slowly rotated to insure intimate contact between the work and the zinc. At a coating temperature of 350 to 375°C a coating of 15 mg per sq cm is formed in a 2 to 3 hour treatment. A recent development which reduces the treating time consists in mixing about 90 per cent alumina with 10 per cent zinc dust to permit coating at temperatures above the melting point of zinc without caking.

The protective properties of sherardized coatings on steel are much like those for equal thicknesses of hot-dipped zinc coatings. The chief difference lies in the brittle nature of the sherardized coatings and in the fact that, because of the iron content, the surface will show rust staining relatively early in its life on exposure to the atmosphere. This is often misleading since it in no way signifies failure of the coating.

Because there is only slight dimensional change, sherardizing is often favored for coating threaded parts and for other applications where it is desired to provide corrosion protection and at the same time to maintain the contours of machined surfaces.

Calorizing. Calorizing is the process for providing aluminum-rich surfaces on iron or copper alloys. The work is packed in a drum in a mixture of powdered aluminum, aluminum oxide and a small amount of ammonium chloride under an inert atmosphere, usually hydrogen. For steel, a calorizing temperature of 850 to 950°C is used and for copper and brass, 700 to 800°C. Coalescence of the aluminum at such temperatures, which are above its melting point, is prevented by the alumina. After a coating of from 0.002 to 0.015 cm containing up to 60 per cent aluminum has been formed, further treatment for 12 to 48 hours at 815 to 900°C permits interdiffusion with the formation of a solid solution of aluminum in the basis metal. The final diffusion depth is from 0.10 to 0.06 cm and the aluminum content at the surface is about 25 per cent, with some ductility and toughness.

Calorized coatings are particularly favored for the protection of steel from oxidation at elevated temperatures, up to 800°C or more, including exposures to sulfurous atmospheres. They are used widely for furnace parts, diesel engine construction, in oil refineries and in heat exchangers.

Chromizing. The surface of steel can be enriched in chromium by cementation to provide a heat and oxidation-resistant coating resembling chromium stainless steel. The parts to be coated may be packed in a mixture of 55 parts chromium powder and 41 parts alumina by weight and heated to 1000 to 1400°C under vacuum or in an inert atmosphere. A treatment at 1000°C for one hour will produce a coating of about 0.01 cm. thickness.

The more recent "gas chromizing" is carried out at lower temperatures. Here the parts are enclosed in a hydrogen atmosphere with a powdered mixture containing chromium or ferrochromium, a diluent such as alumina to prevent sintering, and an ammonium halide. At the coating temperature, gaseous chromium halide is formed which decomposes on the steel surface. The chromium diffuses into the steel while the halide, which serves as a carrier, recombines with more chromium powder to repeat the cycle.

Salt bath chromizing, in which the work is immersed in a fused salt bath containing chromium chloride, barium chloride,, sodium chloride and chromium flakes under an argon atmosphere, obviates the tedious loading and unloading and the furnace heating and cooling associated with the powder and gas methods. Moreover, articles requiring different periods of treatment can be treated simultaneously.

Chromized steels are effective for service at elevated temperatures up to 850°C. They are used for the protection of valves, nozzles, pumps, gauges, nuts, bolts and tools where a combination of wear and corrosion resistance is required.

Other Cementation Treatments. Many different techniques have been developed for using diffusion to enrich the surface of many basis metals with a variety of coating elements in addition to those described above.

Coatings are produced by the corronizing process by heating at 370°C a duplex coating of tin or zinc over nickel, these having been deposited on the steel substrate by electroplating. The final coating consists of alloy layers that are smooth and ductile and provides a good base for paints. Corronized coatings are used for the protection of, for example, wire screening, oil cans and parts for outboard motors.

Siliconizing of steel provides a hard, wear-resistant surface that resists oxidation at temperatures up to 870°C by exposing the work to hydrogen containing at least 40% silicon tetrachloride vapor, or by heating in contact with silicon carbide and chlorine.

Silicon-bearing coatings have attained particular importance as a means for protecting refractory metals for aerospace applications where service temperatures in excess of 1100°C may be encountered. For example, disilicide coatings may be applied to molybdenum parts by the pack cementation process using a powdered mixture of silicon, aluminum oxide and a halide salt activator, and heating at 980 to 1200°C in vacuum or an inert gas for from 2 to 16 hours. During heating, the silicon forms a halide salt that dissociates on the molybdenum surface, releasing silicon that then diffuses inward. For large parts the work may be dipped, brushed or sprayed with a liquid suspension containing the silicon, packed in aluminum oxide with an activator present, then heated.

The metalliding treatment deposits the coating element electrolytically from a molten fluoride salt bath at temperatures of 400 to 1350°C. The element immediately diffuses inward to form a coating of an alloy or intermetallic compound. This technique can accommodate a wide variety of coating elements and basis metals, including, for example, beryllium on copper and titanium; boron or silicon on molybdenum, steel, vanadium, chromium, nickel and copper; titanium on copper; and tantalum on nickel. Typically, coatings of 0.002 to 0.015 cm. thickness are formed in 2 to 3 hours.

Diffusion coatings of nickel-phosphorus on steel may be formed by painting the surface with a water slurry of nickelous oxide and dibasic ammonium phosphate, drying, then heating in a reducing gas at 870 to 1100°C. Such coatings are effective in protecting pipe and fittings used in heat-exchanger systems.

W. W. BRADLEY

Reference

"Protective Coatings for Metals," 3rd Ed., Burns, R. M., and Bradley, W. W., Reinhold Publishing Corp., New York, 1967.

CEMENT, PORTLAND

In 1824 Joseph Aspdin was granted a patent in England upon a product he named *portland cement* because the appearance of hardened mortars made with this cement was like that of a building stone quarried on the Isle of Portland. Aspdin prepared his cement by calcining a mixture of slaked lime and clay at a higher temperature than was used for producing natural cement.

Much of the early chemical work on the constitution of portland cement clinker lead to incorrect conclusions about the essential constituents of portland cement, because of the inadequacy of analytical procedures that were then available. It was more than 50 years after Aspdin was granted his patent that LeChatelier published his first paper describing work in which he applied petrographic methods to the study of clinker (1883), and that J. Willard Gibbs published the work in which he developed the phase rule (1878), which was the foundation for the classic work of Rankin and Wright on phase equilibria in silicate systems (1915). These works were fundamental to understanding the chemistry of clinker formation.

The publications of LeChatelier were the first step in establishing modern concepts of the constitution of portland cement. Törnebohm, working entirely independently, confirmed LeChatelier's observations, and named four component "minerals" that each had observed in thin sections, thus providing the classification used in much subsequent work.

Alite, the most abundant phase, occurred as colorless, biaxial crystals with rectangular or hexagonal outlines. LeChatelier's belief that this phase was tricalcium silicate, $3CaO \cdot SiO_2$, was subsequently confirmed by later workers. Belite, later identified as β-dicalcium silicate, β-Ca_2SiO_4, occurred in small rounded grains with higher birefringence and a darker color. Another phase, also shown later to be dicalcium silicate, was named "felite" by Töorebohm. This colorless, biaxial crystal with high birefringence occurred as striated, rounded or elongated grains; it was not always present in clinker. A dark orange-yellow phase with high birefringence was named celite; it formed part of the material filling the space between the crystalline grains. It has been identified as a crystalline solid solution of dicalcium ferrite, $2CaO \cdot Fe_2O_3$, in a calcium aluminoferrite $6CaO \cdot 2Al_2O_3 \cdot Fe_3O_2$. Usually the aluminoferrite phase in clinker is nearly an equimolar solution of these two components. A second interstitial phase appeared as an isotropic colorless glass with a high refractive index. Later work has shown this isotropic material to be predominantly tricalcium aluminate, $3CaO \cdot Al_2O_3$, a cubic crystal.

All these crystalline phases are now known to contain small proportions of other oxides in solid solution. In addition, low percentages of uncombined calcium oxide and of periclase, MgO, are sometimes found as separate phases. Under certain conditions the clinker may also contain small quantities of sodium and potassium sulfates.

Usually the two calcium silicate phases comprise 75 to 80% and the tricalcium aluminate and the calcium aluminoferrite phases together comprise 15 to 20% of portland cement clinker.

The manufacture of portland cement consists of producing the clinker which is then interground with about 5% gypsum. In either the dry or wet process of manufacture a calcareous material is blended, then interground, with an argillaceous material in controlled proportions. The principal calcareous materials used are limestone, by-product calcium carbonate or lime from other industrial processes, marine shells, marl, or chalk. Suitable argillaceous materials are clay, shale, or slate. Cement rock can be used, sometimes with only minor adjustment of its composition.

Care must be exercized in the selection of raw materials and the adjustment of the composition of the raw mix. Sometimes minor components in the raw materials interfere with the formation of desirable phases, and small variations in the ratios of the principal components of the ground rock mixture may cause large changes in the temperature required to burn the clinker and in the properties of the finished cement.

In passing through the kiln the raw mix is dried, organic matter is destroyed, and decarbonization of limestone occurs. In the hottest region of the kiln, where temperatures may be 1450–1600°C, from 20–30% of the charge is liquefied, and the chemical reactions producing the cement compounds occur in the liquid and in the solid particles it engulfs. As the clinker cools the cement compounds crystallize.

When cement is mixed with coarse and fine aggregates and water the cement reacts with water to form a strong hard solid, which becomes the matrix of the concrete so produced. The cement phase that reacts most rapidly with water, and also has the highest heat of hydration, is tricalcium aluminate. Unless this reaction is retarded, initially, the concrete may "flash set," harden almost immediately. Aside from the fact that this time of set is too short to mix and place the concrete properly, flash setting inhibits the normal strength development. The necessary moderation of the hydration of $3CaO \cdot Al_2O_3$ is provided by the interground gypsum. It is important that the amount of gypsum be controlled carefully to assure the development of optimum physical and chemical properties in concrete.

Tricalcium silicate is the second most reactive phase in cement in terms of both rate and heat of hydration. Dicalcium silicate has the lowest heat of hydration but hydrates more rapidly than does the calcium aluminoferrite phase.

The hydration products of the silicates are a colloidal, poorly crystallized calcium silicate hydrate with calcium hydroxide crystals, which are too large to be classed as colloidal, engulfed in the colloidal mass. The alumina-bearing phases form well crystallized calcium aluminate hydrates that are colloids. The iron in the cement substitutes isomorphously for aluminum in these hydrates. The hardened paste is a porous mass having the properties of a limited swelling gel. Its strength and other mechanical properties are functions of the volume fraction of gel in the hardened paste, which in turn depends solely upon the original water: cement ratio of the paste.

Applications. The apparent use of cement in the United States, as estimated from shipments, has grown from about 11 million barrels (one barrel equals 376 pounds) in 1900 to over 418 million barrels in 1969. Figures for the world production of hydraulic cements, and for the ten countries having the highest production are shown in Table 1.

TABLE 1. HYDRAULIC CEMENT PRODUCTION IN THE WORLD AND TEN LEADING COUNTRIES (UNIT = 1000 BARRELS)

USSR	526,228
United States	416,652
Japan	301,122
W. Germany	205,147
Italy	185,176
France	161,402
United Kingdom	102,081
Spain	93,830
India	77,704
Poland	69,324
Total (10 Countries)	2,138,666
World total	3,168,951

In 1969 the production of cement was sufficient to produce about 1.5 tons of concrete for each man, woman, and child in the world.

An estimate of the comparative sizes of the major construction markets for cement in the United States is given in Table 2.

TABLE 2. CEMENT USE BY MARKETS (APPROXIMATE)

Residential construction	25%
Streets and highways	18
Other public works and buildings	28
Industrial commercial	20
Miscellaneous—farm	9

L. E. COPELAND

References

Lea, F. M., "The Chemistry of Cement and Concrete," First American Edition, Chemical Publishing Co., New York, 1971.

Taylor, H. F. W., "The Chemistry of Cements," Academic Press, London and New York, 1964.

"Proceedings of the Fifth International Symposium on the Chemistry of Cement," The Cement Association of Japan, 1969.

Troxell, G. E., Davis, H. E., and Kelley, J. W., "Composition and Properties of Concrete," Second Edition, McGraw Hill Book Company, 1968.

1970 Annual Book of ASTM Standards, Part 9. American Society for Testing and Materials, Philadelphia.

Powers, T. C., "The Properties of Fresh Concrete," John Wiley and Sons, New York, 1968.

CERAMICS

The origin of the term "ceramics" can be traced to a Greek word meaning "to burn." The word was originally applied to a product in whose manufacture a high temperature was used; the product was usually manufactured from raw materials of an earthy nature.

Today ceramics, or the broader area of ceramic engineering, embraces one of the larger manufacturing fields. Recent figures show that the total value of ceramic products is in excess of six billion dollars annually and it is continuing to show good growth potential.

An all-embracing definition is rather difficult because it includes what appears by a cursory examination to cover widely diverse areas. A committee of the American Ceramic Society has defined ceramics to include those industries which manufacture products by the action of heat on raw materials, most of which are of an earthy nature, in which the chemical element silicon together with its oxide and complex compounds known as the silicates occupy a predominant position. It is much easier to understand the scope of the ceramic industry by a consideration of specific products. It may be divided into seven major areas:

(1) Structural clay products which include all kinds of burned clay products such as common and face brick, roofing tile, drain tile, sewer pipe, terra cotta, and glazed architectural bricks.

(2) Whitewares which include all kinds of vitreous and semivitreous dinnerware, sanitary ware, wall and floor tile, art pottery, and chemical and electrical porcelain. One rapidly growing area in this field is electronic ceramics or ferroelectric and capacitive elements, ferritic spinel or ceramic magnets and special resistive and insulating materials.

(3) All varieties of glass products including window and plate glass, container glass, glass cloth, fiber glass, and the wide variety of related products, vitreous silica and special glasses are covered in the glass field. (See also **Glass**).

(4) Porcelain enamel products where a special glass or ceramic material is fused to metal at high temperatures. It is used on steel and cast iron for various appliances, architectural units, chemical and industrial applications, on aluminum for architectural purposes, and on stainless and highly alloyed steel for high temperature applications of rocket and missiles and other purposes.

(5) Refractories are special materials capable of withstanding very high temperatures. Extensive quantities of these special products are used in furnaces for the iron and steel industry, in kilns in all phases of the ceramic industry as well as other special high temperature operations involving nuclear energy and rocket and jet applications. Other new developments include a wide variety of insulating materials and special castable refractories. (See also **Refractories**).

(6) Cementing materials such as Portland cement, lime, plaster and a variety of magnesia and gypsum products whose constituents are of an earthy nature and after controlled heating or calcination acquire the property of setting when mixed with water.

(7) Abrasive materials such as fused alumina, silicon carbide, silica and emery, together with the products manufactured from them by bonding with a ceramic or other material. (See also **Abrasives**).

Ceramics is frequently referred to as the field of high-temperature engineering. In practically every case the ceramic product has been subjected to elevated temperatures either in the preparation of the raw material or in the manufacture of the final product. The temperatures involved may be a few hundred degrees for some calcination operations up to 6000°F for some special purposes. In recent years the demands for ever higher temperatures have been increasing rapidly.

R. L. Cook

References

Kingery, W. D., "Introduction to Ceramics."

Norton, F. H., "Fine Ceramics."

Singer, Felix, "Industrial Ceramics."

CEREAL CHEMISTRY

Cereal chemistry embodies a great deal more than the chemistry of the cereal grains. It starts with a study of the growth of the plants which produce the seeds, and continues with the production and the composition of the seeds, the chemical changes on storage, the refining of the grains for special purposes, the use of these products and their byproducts, the nutritional value of the whole kernel and its parts, and includes in its realm all products and processes which come in contact with the cereals. The work of a large number of cereal chemists is involved with the milling of wheat and with the use of the flour in the baking industry. To them, cereal chemistry involves the chemistry of yeast, milk, soybeans, bleaching, minerals, enzymes, etc., as well as of wheat. In the brewing industry, cereal chemistry must relate the chemistry of the barley to its ability to germinate, and the function of the malt, hops, yeast, and other ingredients to the fermentation and processing of the final beverage.

Of the various seeds, the cereal grains are perhaps the most important. There are a number of kinds of cereals and many varieties of each. It has been estimated that over 30,000 varieties of wheat exist—the United States Department of Agriculture has over 22,000 varieties in its world collection.

The production of the individual cereals is unevenly distributed. In 1968, the world production of all cereals was 1.18 billion metric tons, some 17% of which was produced in the United States. The United States produces 45% of the corn, China 35% of the rice, Russia 28% of the wheat, and Europe 30% of the rye. The production of wheat is the most universal and it is the most important cereal in international trade. Between 1948 and 1968, the world production of cereals almost doubled even though the acreage planted to grain only increased by one-sixth. Much credit to the increased yield must be given to Norman E. Borlaug who won the 1970 Nobel Peace Prize for the development of high-yield wheat.

The term "corn" is frequently misunderstood, as the word itself is a synonym for grain. It is applied to the most common cereal in a given locality. Thus in the United States, maize is the most common cereal and is called "corn." In many parts of Europe, "corn" refers to rye, and in Scotland to oats.

Like all forms of living matter, the constituents of the cereal grain are very numerous and many are very complex. The composition of the kernels is dependent upon their genetic make up, the climate, and the composition of the soil. For comparative purposes, the gross analyses are given in terms of the water, protein, fat, fiber, soluble carbohydrate, and mineral. These components vary widely for each of the cereal grains, but by compiling a large number of analyses, characteristic compositions for the different types of cereals are noted. Representative analyses are given in Table 1.

TABLE 1. REPRESENTATIVE ANALYSIS
OF THE IMPORTANT CEREAL GRAINS, % OF TOTAL

Cereal	Water	Protein	Fat	Mineral	Fiber	Carbohydrate
Wheat	10.2	13.1	1.7	1.9	3.0	70.1
Rice	12.5	6.4	2.1	5.9	7.8	65.3
Corn	11.0	9.0	4.0	1.8	3.0	71.2
Oats	10.0	12.7	5.3	3.0	7.0	62.0
Barley	10.8	11.0	2.2	2.5	3.8	69.7
Rye	10.0	12.3	1.7	2.0	2.3	71.7
Sorghum	11.6	11.3	2.9	1.7	2.2	71.3

The kernels of grain may be divided into several parts. The hull is a protective layer and consists largely of fiber. The bran is the outer layers of the seed proper and is rich in protein and mineral. The endosperm is the largest portion of the seed and it is the storehouse for most of the starch and protein. The germ is the vital center of the seed and is relatively rich in fat, protein, vitamins, enzymes, and minerals. The composition of the various products obtained from the cereals is thus dependent upon the degree of separation of the component parts. In the milling of wheat, the flour obtained has more starch and less protein, fat, and mineral than the whole grains. The reverse is true of the bran and germ which go largely into animal feeds.

Water. The water of the cereals is present in two forms. The most abundant is held in multilayers by adsorption on the free amino groups and peptide bonds of the proteins. It is easily displaced by heating, and its concentration is dependent upon the atmospheric temperature and humidity. A small amount of moisture is closely bound to the grain and it cannot be evaporated without causing marked alterations. The proper moisture content during storage of the seeds is important for the preservation of their vitality and usefulness.

Carbohydrate. Among seeds, the cereal grains are characterized by a relatively high carbohydrate content, divided into crude fiber and nitrogen-free extract. The former is that portion of the grain which is insoluble in 1.25% concentrations of both boiling sodium hydroxide and boiling sulfuric acid. The fiber content is highest in the seeds with large hulls, oats and rice (Table 1), and is composed of pentosans and cellulose. The oat hulls have a commercial value for the production of furfural by the reaction of sulfuric acid on the pentosans. The soluble carbohydrate or "nitrogen-free extract" is chiefly starch.

Protein. Though the cereals grains usually are not considered a source of protein, they furnish about one-third of the human dietary protein in the United States, and in some countries the percentage is much greater. The cereals contain relatively less protein than the legume seeds and the protein has a poorer biological value. The concentration of the protein is determined from the nitrogen content. Thus the per cent nitrogen is multiplied by the factor 6.25. A factor 5.7 would be better for many of the cereal proteins, and, although it is sometimes used for wheat flour, the factor 6.25 has a much wider usage. The proteins are not evenly distributed throughout the seed. This is especially true in corn as portions of the endosperm are chiefly starch.

As already noted, great advances have been made in increasing the quantity of cereal produced. Since in some areas of the world individual cereals form the bulk of the dietary, equal or greater efforts are needed to increase the quality of the cereals. Some attempts toward this have been made in the selection and breeding of varieties of wheat and corn that are more adequate in the deficient amino acids as lysine. Here the cereal chemist can help in developing rapid tests for the estimation of lysine and other amino acids.

The cereal proteins are classified by their solubility properties. Each class may represent one or, more probably, many individual proteins. The crude water-insoluble protein of the cereals is known as gluten, and it includes up to 90% of the total protein. It is composed of two types, a prolamin which is soluble in 70% alcohol and a glutelin which is soluble in dilute acid or alkali. The solubility properties are used in their purification. Among the prolamins, gliadin is probably the same in wheat and rye, but hordein of barley and zein of corn have different amino acid components. Rice contains no prolamin, and its chief protein is a glutelin known as oryzenin. In wheat the glutelin is called glutenin. The ratio of the prolamin to the glutelin may vary for a given cereal and these proteins do not have the same physical properties when obtained from different grains. The gluten of corn does not form a sticky mass as does that of wheat, and so cornbreads are granular rather than porous. In wheat the gliadin is soft and sticky and is responsible for the binding, but the glutenin gives solidity to the gluten. Albumins and globulins are present in relatively small amounts, usually under 20% of the total protein. In general the proteins of oats have the highest biological value of any of the cereals, and it is possible to obtain a positive nitrogen balance with this cereal as the sole source of protein.

Fat. The fat of the cereals is found chiefly in the germ, with some in the bran, but very little in the endosperm. About 75% of the fatty acid present is of the unsaturated type, which results in an oily consistency. The germ oils of wheat and corn are available commercially and are used in the medicinal and table oil fields, respectively. Attempts are being made to increase the germ oil content of corn through genetics. The acid number of the fat increases during storage and may be used as an index of the degree of soundness of the grain.

Minerals. The distribution of the minerals in the seeds approximately follows that of the oil. It was thought for a time that one of the advantages of whole wheat bread as compared to white bread was the larger amount of minerals it contained, especially phosphorus, calcium, and iron. It has since been shown that 75–85% of the phosphorus is present as phytic acid, the hexaphosphoric ester of inositol. The evidence indicates that the calcium and iron form insoluble salts with phytic acid so that these minerals cannot be utilized unless the complex is first hydrolyzed by phytase. Potassium, magnesium, and sulfur also form relatively large portions of the ash and many minerals are found in traces. The hulls of oats and rice are very high in silica, which accounts for the high ash content of these cereals (Table 1).

Enzymes. The cereal grains result as maturation of one generation of plant life. They are not inert. Their vitality may be expressed best in terms of the enzymes present. When called upon by favorable conditions, they initiate the changes which result in a new plant. Amylases, proteases, esterases, and oxidizing enzymes all play a role in supplying the new plant with food. The enzymic activities increase with the germination of the seed. Of the enzymes, the amylase, or diastase, is of commercial importance. The brewing industry depends upon this group of enzymes in malt to produce maltose from starch so that the enzymes of yeast may produce alcohol. A certain amount of diastase should be present in flour for optimum baking qualities. Flour from intact grains will differ considerably in diastase content. The desired level is obtained by blending flours of different diastatic values or by the addition of small amounts of germinated wheat or barley. A small amount of sugar is normally present in the flour, and the diastase slowly liberates additional amounts for the yeast to ferment to produce the necessary carbon dioxide.

Vitamins. Vitamin A, which is found in plants as carotene, is not abundant in cereals. With the exception of yellow corn, the carotenoids of the seeds are not precursors of vitamin A so that the bleaching of flour can not be objected to on the grounds that this vitamin is destroyed. Vitamin D is not found in cereals and because of the adverse effect of phytic acid on the mineral metabolism, it is essential that adequate quantities of this vitamin be obtained from other sources. Vitamin E in the form of tocopherols is found chiefly in the germ, and is available from wheat germ. Vitamin C is not present in normal sound kernels, but it is formed in relatively large amounts when the seeds germinate. The various B vitamins are the most important in the cereals. Large portions are removed in the extraction of flour and in the preparation of polished rice. Flour is now fortified with some of the vitamins lost in milling. The vitamins of rice may be partly transferred to the endosperm by parboiling the cereal before processing it.

ALBERT A. DIETZ

Cross-references: *Proteins, Carbohydrates, Foods, Nutrition, Enzymes, Brewing, Vitamins, Fats, Baking Chemistry.*

References

Matz, S. A., "Cereal Science," Avi Publishing Co., Westport, Conn., 1969.

Hlynka, I., Editor, "Wheat—Chemistry and Technology," American Association of Cereal Chemists, St. Paul, Minn., 1964.

CERMETS

Cermets are composite engineering materials composed of ceramic grains held in a metal matrix. The particle content usually runs up to 70% of the total volume and can be an oxide, carbide, nitride or any of the more conventional ceramic materials. In principle, any metallic element or alloy can form the metal matrix phase. However, chromium, nickel, cobalt and molybdenum are most commonly used.

In cermets, bonding between the constituents results from a small amount of mutual or partial solubility. Some systems, however, such as the metal oxides, exhibit poor bonding between phases and require additions to serve as bonding agents.

Cermets are produced by powder metallurgy techniques and can achieve a wide range of properties, depending on the composition and relative volumes of the metal and ceramic constituents. All such techniques are applicable, including cold and hot pressing and slip casting. Cermets can be produced in most of the shapes normally feasible by powder metallurgy methods, including standard shapes such as flats, rounds, rods and tubes as well as a variety of special shapes. Some cermets are also produced by impregnating a porous ceramic structure with a metallic binder.

Cermets can also be used in powder form as coatings, applied by flame-spraying. The powdered mixture, sprayed through a flame, fuses to the base material.

Although a great variety of cermets have been produced on a small scale, only a few types have significant commercial use. These fall into two main groups—oxide-base and carbide-base cermets.

A number of different oxide-base cermets have been developed, the most common having an aluminum oxide base. Al_2O_3 content ranges from 30 to 70%. Chromium or chromium alloys commonly make up the metallic phase. These cermets have specific gravities between 4.5 and 9.0, tensile strengths ranging from 21,000 to 39,000. Their modulus of elasticity runs between 37 and 50 × 10⁶ psi and their hardness range is A70-90 Rockwell.

The oxide-base cermets are used as a tool material for high speed cutting of difficult-to-machine materials. Other uses include thermocouple protection tubes, molten metal processing equipment parts and mechanical seals.

There are three major groups of carbide-base cermets—tungsten, chromium and titanium. Each of these groups is made up of a variety of compositional types or grades. Tungsten carbide, perhaps the most widely used of all cermets, contains up to about 30% cobalt as a binder. They are the heaviest type cermets (11.1–15.2 specific gravity). They are principally used for cutting tools; structural uses include gages and valve parts. The outstanding properties of tungsten carbide include high rigidity, compressive strength, hardness and abrasion resistance. Their modulus of elasticity ranges between 65 and 95 × 10⁶ psi. Compressive strength runs between 500,000 and 800,000 psi and they have a Rockwell hardness of about A90.

Most titanium carbide cermets include nickel or nickel alloys as the metallic phase and usually columbium carbide or chromium carbide for oxidation resistance. These cermets have relatively low density combined with high stiffness and strength at high temperatures (above 2200°F). Typical properties are: specific gravity, 5.5–7.3; tensile strength, 75,000 to 155,000 psi; modulus of elasticity, 36 to 55 × 10⁶ psi; and Rockwell hardness, A70 to 90. Typical uses have been gas turbine nozzle vanes, torch tips, hot-mill roll guides, valves and valve seats.

Chromium carbide cermets contain from 80 to 90% Cr_2C_3 and the balance nickel or nickel alloys. Their tensile strength runs about 35,000 psi and Rockwell hardness is about A88. They have superior resistance to oxidation, excellent corrosion resistance and relatively low density (specific gravity of 7.0). In addition, their high rigidity and abrasion resistance make them suitable for gages, valve liners, spray nozzles, bearing seal rings, bearings and pump rotors.

In addition to the oxide- and carbide-base cermets, several other types have some practical uses. Cermets, such as tungsten-thoria and barium carbonate-nickel are used in higher power pulse magnetrons. Some proprietary compositions are used as friction materials. In brake applications they combine the thermal conductivity and toughness of metals with the hardness and refractory properties of ceramics.

Finally, uranium dioxide cermets have been developed for use in nuclear reactors. Other cermets designed for this use include chromium-alumina, nickel-magnesia and iron-zirconium carbide.

H. R. CLAUSER

CESIUM

Cesium, symbol Cs, is a soft, ductile, silvery-white alkali metal belonging to Group IA of the Periodic Table. Its atomic number is 55 and its atomic weight 132.9054. With its low melting point of 28.5°C, cesium is one of the three metals (with mercury and gallium) liquid at room temperature. This rare element was discovered by Bunsen and Kirchhoff in 1860 with the spectroscope. It ranks 40th in order of prevalence in the earth's crust among the elements.

Cesium is found in uncommon minerals, such as pollucite ($2Cs_2O \cdot 2Al_2O_3 \cdot 9SiO_2 \cdot H_2O$), a hydrated silicate of aluminum and cesium occurring in Manitoba, Maine, South Dakota, Southwest Africa, and Southern Rhodesia. Cesium is also present in some lepidolites, a lithium ore, and is found in potassium brines and carnallite deposits, but these represent potential rather than current sources of cesium.

Cesium compounds can be derived from pollucite by acid digestion or by sintering with sodium carbonate. A major problem in the recovery of cesium compounds is the separation of them from other alkali metal compounds. Treatment of ground ore with sulfuric acid forms cesium alum which is separated from other alkali metal impurities by fractional crystallization. Use of concentrated hydrochloric acid yields cesium chloride which is separated and purified by fractionation of cesium antimony chloride, $3CsCl \cdot 2SbCl_3$, or cesium chlorostannate. Removal of the antimony or tin results in a pure cesium chloride. The alkaline extraction is conducted by sintering ground ore with NaCl and Na_2CO_3, leaching CsCl from the sintered mass and purifying this solution by solvent extraction. The process, developed at Oak Ridge National Laboratory, is also used to recover radioactive cesium isotopes from wastes generated in nuclear reactors.

Among the several methods of producing cesium metal are direct reduction of the ore with sodium followed by fractionation of the resultant alkali metals in a multiplate column. The more commonly used methods of preparing pure cesium metal utilize the reduction of pure salts, such as the chloride, with lithium or calcium in a heated closed reactor under high vacuum to prevent pickup of oxygen and water vapor by the cesium. Details of these and other methods of treating ore and of producing pure cesium metal are given by Mosheim.

Domestic production of cesium and its salts is only a few hundred pounds per year. The price of the metal is $100–375 per pound (depending on quantity) and that of the salts is correspondingly high, i.e., $35 per pound for CsCl and CsBr and $32.50 per pound for Cs_2SO_4.

The physical properties of cesium include: melting point, 28.5°C; boiling point, 705°C; density, 1.90 g/cc; specific heat, 0.052 cal/g (28.5°C); thermal conductivity, 0.044 cal/sec/cm²/°C/cm (28.5°C); latent heat of fusion, 3.766 cal/g; latent heat of vaporization, 146 cal/gg; electrical resistance, 36.6 microhm-cm (30°C); viscosity, 0.63 × 10⁻² poises (43.4°C); vapor pressure, 1 mm Hg at 278°C, 10 at 387°C, and 400 at 635°C; and thermal neutron absorption cross section, 29 barns. Cesium is the most easily ionized of the elements (ionization potential of gaseous atoms = 3.87 volts), and is one of the most electropositive (V° = −2.923 volts).

Chemically, the properties of cesium closely resemble those of potassium, differing in no outstanding respect. The hydroxide, CsOH, m.p. 272.3°C, is the strongest base known and must be stored in silver or platinum out of contact with air because of its reactivity with glass and CO_2. Soluble cesium compounds include the sulfate, nitrate, carbonate, hydroxide, sulfide and the halides, while $CsClO_4$, Cs_2PtCl_6, $CsMnO_4$ and $Cs_6SiW_{12}O_{42}$ are relatively insoluble. The metal reacts vigorously with oxygen and water and must be protected from them.

Of limited industrial importance, cesium is used in scintillation counters, in photoelectric cells and infrared-detecting instruments, such as the "sniper-scope," and as a getter in low-voltage vacuum tubes. In this last application the metal is usually produced in place by inserting a capsule charge of a salt plus a reducing agent, e.g., CsCl and Ca, into the tube and the heating the capsule with a high frequency source. Some cesium salts are used in the manufacture of mineral waters. Its use as a coolant in atomic energy applications is not promising because of its high thermal neutron absorption cross section, its scarcity, and its high cost, as compared with sodium and sodium-potassium (NaK) alloys.

The major potential applications for cesium lie in two fields: ion propulsion engines for space travel and the conversion of heat to electricity. For the first, the ease of ionization and the high atomic weight of cesium are favorable properties. For thermoelectric generators, such as those utilizing the magnetohydrodynamic (MHD) principle and the thermionic mechanism, the low thermionic work function (1.81 volts), the low ionization potential, the low melting and boiling points, and the high vapor pressure of cesium are valuable properties.

CLIFFORD A. HAMPEL

References

Mosheim, C. Edward, "Cesium," pp. 127–133, "The Encyclopedia of the Chemical Elements," Hampel, C. A., Editor, Van Nostrand Reinhold Co., New York, 1968.

Perel'man F. M., "Rubidium and Cesium," Macmillan Co., New York, 1965.

CHELATION

Chelation is the formation of a heterocyclic ring containing a metal ion, the metal being attached by coordinate links to two or more nonmetal atoms in the same molecule. The parent word "chelate" which is properly used as an adjective (e.g. chelate ring) and colloquially, as a verb and as a noun, is derived from a Greek word meaning "crab's claw," which refers to the tenacity with which the coordinating group holds the metal ion. The word was coined by G. T. Morgan and H. D. K. Drew. The terms "chelate" and "chelation" have been extended to refer to rings formed by hydrogen bonding, as in salicylaldehyde and acetic acid,

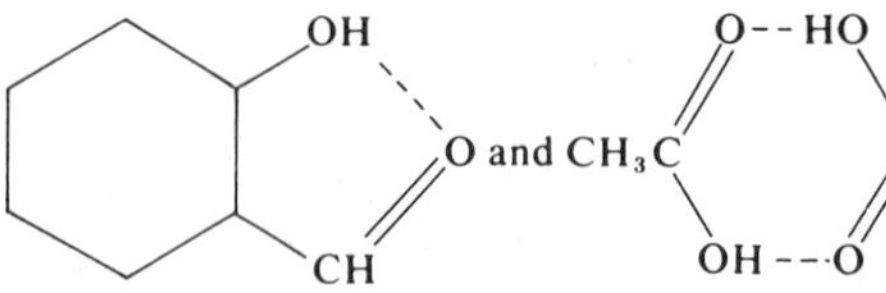

but this is not entirely justified, as the hydrogen atom does not form coordinate bonds with the two oxygen atoms simultaneously.

The importance of ring formation in metal coordination compounds was first established by Ley through a study of the copper (II) derivative of glycine. This compound is quite different from copper (II) acetate in that it is deep blue in color, and its solutions show little conductivity. These properties would seem to rule out the structure $Cu(OOCCH_2NH_2)_2$, and the structure $Cu(NHCH_2COOH)_2$ is ruled out by the fact that N, N-dimethylglycine gives a similar product. Ley therefore considered the structure to be correctly shown by

the formula

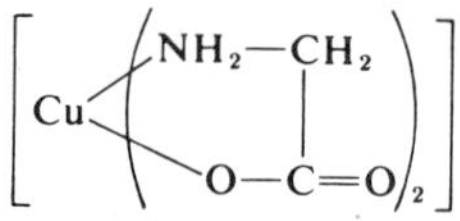

The most important property of chelation is found in the fact that it brings about a great increase in the stability of the bond between the metal atom and the coordinating group that forms the chelate ring. Monoamines are much poorer coordinators than ammonia, but complexes of the bidentate (literally, "two toothed" or "two clawed") ethylenediamine are many times more stable than those of ammonia, as shown below, where the logarithms of the stability constants of some ammonia complexes and ethylenediamine complexes are collected:

Metal Ion	Log. Stability Constant	
	Ammonia Complex	Ethylenediamine Complex
Co^{+2}	5.3	10.7
Ni^{+2}	7.8	14.1
Cu^{+2}	12.6	20.1
Zn^{+2}	9.1	11.1

Similarly, complexes in which both ammonia and acetate groups are coordinated to the same metal ion are very much less stable than those containing the aminoacetate (glycinate) group.

This increase in bond strength is largely an entropy effect, and is shown primarily by the coordinate bonds between the metal and the donor atoms in the chelate ring, other bonds in the complex being affected only indirectly. Thus, the cobalt-ammonia bonds in the partially chelated complex

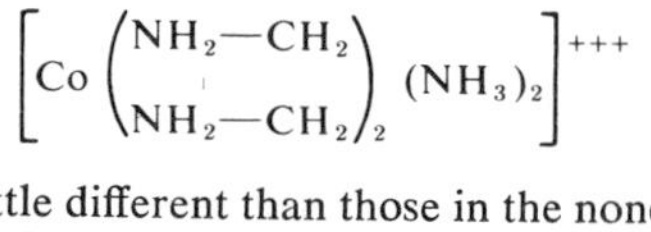

are very little different than those in the nonchelated $[CO(NH_3)_6]^{+++}$. However, if the chelate ring has unusual steric properties, it may distort the entire complex and change the thermal stability or the chemical reactivity of all of the bonds.

Stability of the chelate ring is attained only when the ring size and other steric factors are right. It has been shown in many ways that with chelating agents containing only single bonds, the greatest stability is achieved when the ring consists of five members. For example, the complex formed between platinum (IV) chloride and α,β,γ-triaminopropane might have either structure A or structure B:

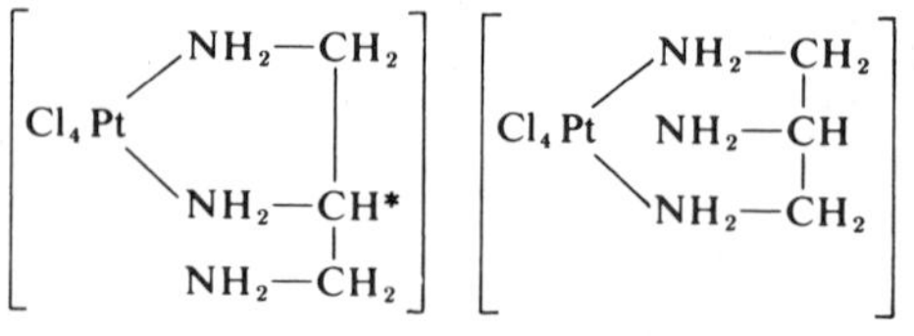

The fact that this complex is optically active indicates that the five-membered ring of structure A has been formed, the carbon atom marked with an asterisk thus becoming asymmetric.

The α-amino acids form much more stable complexes than do the β-amino acids; the corresponding γ, δ, and ϵ acids apparently do not form chelate rings at all. The α and β amino acids can be distinguished by the fact that the former react with cobalt (III) hydroxide to form deep-colored chelate complexes, while the latter do not. Ethylenediamine forms stable five-membered rings with many metals; trimethylenediamine $[NH_2(CH_2)_3NH_2]$ forms much less stable six-membered rings, and the higher homologs $[NH_2 \cdot (CH_2)_x NH_2]$ show almost no ability to coordinate. If two or three double bonds are introduced into the chelate ring, maximum stability seems to be shown when the ring consists of six members. Four-membered rings are also known, and are illustrated by the familiar carbonato tetraammine cobalt (III) complex $[CO(NH_3)_4CO_3]^+$. Rings of seven, eight, and more members have been described, but most reports of such complexes are not adequately supported by proofs of structure. As the distance between donor atoms in the coordinating group increases, the tendency to ring formation decreases, and a tendency to form linear polymers become evident.

Complexing agents containing three or four or more donor atoms, all of which can coordinate simultaneously, are referred to as tridentate, tetradentate, or polydentate coordinators. When such complexers form fused chelate rings, the stability of the metalligand bonds is increased far beyond that achieved by simple chelate formation. This so called "chelate effect" is illustrated by the increasing stabilities of the complexes of ethylenediamine (one ring), diethylenetriamine (two rings), and triethylenetetrammine (three rings), which is the more remarkable when it is recalled that secondary amines, generally form much less stable coordinate bonds with metals than do primary amines.

INFLUENCE OF NUMBER OF RINGS
ON STABILITY OF ZINC COMPLEXES

Coordinating Agent	Donor Atoms	Number of Fused Rings	Log. Stability Constants				% dissociation of 0.001 M solution
			k_1	k_2	k_3	k_4	
NH_3	1		2.4	2.4	2.5	2.2	7.7
$NH_2CH_2CH_2NH_2$	2	1	5.9	5.2			1.2
$NH(CH_2CH_2NH_2)_2$	3	2	9.0				
$N(CH_2CH_2NH_2)_3$	4	3	14.6				1.4×10^{-4}

The chelate effect is strikingly illustrated by the complexes of the sexidentate ethylenediamine tetraacetate ion,

$$^-OOCCH_2 \diagdown \qquad \diagup CH_2COO^-$$
$$NCH_2CH_2N$$
$$^-OOCCH_2 \diagup \qquad \diagdown CH_2COO^-$$

which can coordinate through both nitrogen atoms and all four carboxyl groups, forming five fused rings. In some cases it utilizes only two or three of the carboxyl groups, but even so, it is a powerful coordinator, and has found wide use as a sequestering agent.

The charge which a complex bears is the sum of the charges on the coordinating groups and the metal ion from which it was generated. It may be either positive or negative, or it may be zero. The zero charge is usually achieved by the combination of a bidentate ligand bearing a single negative charge with a metal ion for which the coordination number is twice the oxidation number. Fully chelated complexes of this sort comprise a special class of compounds known as "inner complexes." Such compounds show greatly enhanced stability. Some of them are volatile, and most of them are insoluble in water, but easily soluble in nonpolar solvents. They behave like organic substances. Nonchelated complexes of zero charge, such as trinitro-triammine cobalt (III) usually do not show these properties and are not considered to be inner complexes. The stability of the inner complex compounds is illustrated by tris-glycine cobalt (III), which can be recrystallized from hot, 50% sulfuric acid, and by bis-acetylacetone beryllium (II), which distills without decomposition at 270°C.

Many of the organic precipitants used in analytical chemistry are applicable because they form inner complexes; among these are 8-hydroxy quinoline, dimethylglyoxime, and nitroso-β-naphthol. In some cases, the metal in the resulting precipitates is determined by drying and weighing while in other cases the complex is extracted into an organic solvent and the metal is determined spectrophotometrically.

The fulfilment of the rules which lead to the stability of chelate rings is well illustrated by the amazingly stable copper phthalocyanine, which is not decomposed even at 500°C. In the phthalocyanine molecule we have an inner complex containing four resonating six-membered fused rings:

JOHN C. BAILAR, JR.

Cross-references: *Heterocyclic Compounds, Asymmetry, Sequestering Agents, Phthalocyanines.*

CHEMICAL DATING

The principal chemical methods of dating geological formations and artifacts are based on the measurement of the end products of radiochemical changes or on the degree of residual activity of a radioactive substance; they yield absolute dates and can be rather widely applied.

The oldest and most thoroughly investigated of the chemical methods is the estimation of the age of uranium minerals from a measurement of the amount of helium or lead they contain. Though a great deal of attention has been given to age determination from the measurement of helium content, it is now generally recognized that because of the tendency of this gas to escape, the helium method is always likely to yield results that are considerably in error. Hence the measurement of the amount of lead produced by radiochemical decomposition is the preferred method. If a pure uranium mineral has not been altered by leaching or weathering, the ratio of the percentage of lead it now contains to the percentage of uranium is an index of its age, which may be calculated from this ratio and the known disintegration rate of uranium. Likewise for a pure unaltered thorium mineral. If a mineral contains both uranium and thorium, the ratio of the percentage of lead to the percentages of uranium and thorium is still an index of its age, though now the calculation must take into account the different disintegration rates of uranium and thorium. A further complication and a possible source of serious error, arises if the mineral originally contained ordinary lead as an impurity. Fortunately, the presence of such lead may be detected by isotopic analysis with the mass spectrograph since ordinary lead contains the isotope Pb^{204} which is not a product of the radiochemical decomposition of either uranium or thorium, and a correction may be applied to the total percentage of lead in the mineral or to the percentages of the other isotopes of lead that are present. Instead of depending on the ratio of the percentage of total radiogenic lead to the percentage of uranium, thorium, or both, it is generally more accurate and more advantageous to depend on the ratio of the percentage of one or more of the individual lead isotopes to the percentage of the parent element or elements. However, because of the experimental difficulties of this method of analysis, most age determinations up to now have been based on the results of ordinary gravimetric analysis.

Although it is theoretically possible to measure the age of minerals containing a few naturally radioactive metals other than uranium or thorium, the only such measurement that has been applied practically as yet is based on the radioactivity of rubidium. Natural rubidium contains the isotope Rb^{87} which emits a beta particle to yield the strontium isotope Sr^{87}. The percentage of Rb^{87} in a mineral is determined by multiplying the percentage of rubidium found on analysis by the factor 0.27. The ratio of the percentage of strontium found on analysis to the percentage of Rb^{87} is the index of age, which may be calculated in the usual way from the known half-life of Rb^{87}.

For the determination of the time of occurrence of very recent geological events, for dating prehistoric human events, and for determining the age of certain kinds of artifacts, the radiocarbon (C^{14}) method of Libby is the most generally useful chemical method available.

A fundamental assumption of the method is that the amount of cosmic ray activity, the composition of the atmosphere, and other conditions under which C^{14} is formed have remained constant for the entire period of time over which this method is applicable; in other words that the proportion of C^{14} in the carbon of living tissues has remained the same over this whole period.

The careful selection and preparation of the sample is an essential step for reliable dating by the radiocarbon method. Organic matter of more recent origin than the sample, such as plant remains or soil containing humus, must be removed, and care must be taken to remove all mineral matter containing carbon. In general, the sample must be cleaned physically to remove foreign matter, and as an additional precaution soaked for several hours in normal hydrochloric acid. After treatment with the acid, it is thoroughly washed with water and dried completely in an oven. Sufficient purified sample must be prepared to yield at least ten grams of elementary carbon.

The prepared sample is completely burnt to carbon dioxide in a suitable train, or if it consists of shell, is treated with hydrochloric acid of sufficient concentration to yield carbon dioxide. For combustion, oxygen is passed over the heated sample and the combustion products are passed over hot copper oxide to complete the combustion. In either case the resulting gas is passed through a dry-ice trap and a Drierite tube to remove water and is collected by condensation in a nitrogen-cooled trap. Since the product thus collected is usually contaminated with other condensed gases such as oxides of nitrogen and radon in traces, it must be purified. This is done by warming the condensate, passing the gas mixture into ammonium hydroxide solution, and adding a solution of calcium chloride to yield a precipitate of calcium carbonate. After washing thoroughly with distilled water, this should contain all the carbon of the sample in the form of pure calcium carbonate. This pure product is then treated with hydrochloric acid, using the same train as in the combustion procedure, and the evolved pure carbon dioxide is dried as before and collected as a gas in storage bulbs.

The next step is the reduction of the carbon dioxide to carbon. This is done in an iron combustion tube containing magnesium turnings mixed with about one per cent of cadmium turnings as a catalyst. The gas is passed at a moderate rate into the tube containing the hot metals until all the carbon dioxide has reacted. After cooling, the mixture of carbon, magnesium oxide, and unreacted metal is removed from the tube and placed in a large beaker. This mixture is then treated first with water and then with concentrated hydrochloric acid to dissolve everything but the carbon. The resulting solution is separated from the carbon by the use of a glass filter stick, and the carbon is repeatedly washed with hot distilled water by decantation, and dried. The carbon is then again treated with hydrochloric acid, and filtered, washed, and dried as before.

The final step is grinding in an agate mortar to a uniform powder. The sample is then placed in a small bottle with a tight cap until the measurement of its activity can be made. In spite of the double acid treatment, the sample always contains some magnesium oxide. The percentage of this must be determined in a small part of the sample by combustion so that a correction may be made for this inactive material when the activity of the sample is measured and calculated.

In general, the error of age determination by the radiocarbon method is 5–10% in the range where reasonable accuracy may be expected. This degree of error is not serious in dating geological or prehistoric materials or events, but is usually too large to give useful results for periods within the limits of human history.

There are also a few methods of relative dating that depend on chemical reactions. The slow chronological increase in the fluorine content of bones in contact with ground water containing fluoride is the basis of one method. Another depends on the slow hydration of the surface of buried obsidian artifacts. Such methods are applicable only to limited areas or single sites.

Earle R. Caley

Cross-references: *Radioactivity, Archaeological Chemistry.*

References

Libby, W. F., "Radiocarbon Dating," 2nd Ed., Chicago, 1955.

McConnell, D., "Dating of fossil bones by the fluorine method," *Science,* **136,** 241–244 (1962).

Michels, J. W., "Archeology and dating by the hydration of obsidian," *Science,* **158,** 211–214 (1967).

Zeuner, F. E., "Dating the Past," 4th rev. Ed., New York, 1964.

CHEMICAL ENGINEERING

Chemical engineering is that branch of engineering which is concerned with the production of bulk materials from a few basic raw materials. The chemist works out the details of a given reaction in the laboratory, and the chemical engineer translates his work into large-scale plant operation. Typical examples of these bulk materials are industrial chemicals, gasoline, lubricating oils, rubber, soap, sugar, cement, metallurgical products, resins and plastics, synthetic fibers, and glass, to mention only a few. The producers of these materials are the chemical process industries, which now comprise the largest segment of all manufacturing enterprise in the United States.

Although chemical engineering is concerned primarily with the activities of the chemical process industries, in a broader sense it touches on all technical problems wherein chemical change or chemical character are important. For example, modern chemical engineering is increasingly concerned with biochemical and biomedical problems, where the skill of the chemical engineer in dealing with the coupling of physical and chemical processes is uniquely relevant.

Chemical engineering emerged as a separate branch of engineering when it was recognized that the chemical process industries employed a certain few basic operations in the manufacture of all their numerous products. These basic operations are called *unit operations,* and all chemical processing involves serial combinations of these unit operations together with some kind of chemical reactor. An understanding of these operations permits an intelligent attack on the technical problems relating to the production and refinement of any chemical product.

Unit Operations. The unit operations of chemical engineering are the basic physical operations common to all chemical processing. Among the more important are: fluid flow, heat transfer, filtration, evaporation, drying, distillation, mixing, gas absorption, solvent extraction, and adsorption. Fluid flow, heat transfer, and mixing are important in almost all chemical processing and are necessarily involved in the other unit operations, which are concerned primarily with separating constituents of mixtures. Fluid flow relates to the transport and metering of fluids; heat transfer relates to the transport of thermal energy between systems at different temperatures. Filtration is the operation of removing solids from a liquid by retention on a porous medium. In evaporation, solids are separated from liquids by boiling off the liquid. Drying is very similar, except that the proportion of liquid relative to solid is much smaller than in the case of evaporation.

Distillation is the vaporization and subsequent condensation of volatile liquids. When applied under proper conditions to mixtures of constituents which have different boiling temperatures, a separation of constituents results. Distillation is of particular importance in industries like petroleum refining, where the principal occupation is the separation of a complex mixture into a variety of products. Gas absorption is the removal of constituents from a mixture of gases by scrubbing the gases with a liquid which selectively dissolves out certain constituents. Solvent extraction is an analogous separation in which a liquid mixture is contacted with an immiscible liquid solvent. In adsorption a solid having an extended surface is added to a fluid mixture to adsorb selectively certain constituents in the mixture. Filtration is a mechanical separation, whereas all the other separations mentioned above involve the diffusion of a constituent from one phase to another. These *diffusional operations* may be treated by similar techniques, and for convenience they are often lumped together as *phase-change physical separations.* The latter name arises from the fact that the separations take place by a selective migration of constituents from one phase to another.

As an example of a typical process sequence, consider thermal cracking in a petroleum refinery. The preparation of the charge stock for the cracking involves distillation of the crude oil to separate it into various light cuts, intermediate cuts, and residuum. Selected high-boiling fractions make up the cracking feed stock, which is pumped to the cracking furnace through one or more heat exchangers wherein heat is exchanged with other oil streams. In the cracking furnace the charge stock flows at high velocities through many passes of pipes, which are heated by the flames in the furnace. After leaving the cracking still (furnace) the cracked oil is vaporized into a distillation column where it is separated into a variety of products including gas, gasoline, and recycle stock, i.e., stock having the same boiling range as the feed stock. In this particular process sequence the unit operations of fluid flow, heat transfer, and distillation are used both before and after the purely chemical step of cracking. It is typical of the chemical process industries that the unit operations comprise the bulk of the processing, while the purely chemical operations constitute only a minor, but critically important, part. (See **Cracking**).

Basic Principles. Although the introduction of the concept of the unit operations permitted the development of chemical engineering and the concomitant rapid technological advance in the process industries following World War I, recent trends in chemical engineering have made this concept unwieldy. Underlying the unit operations is a hard core of five basic principles which embrace every technical problem in chemical engineering. These principles may be identified simply as *states, conservation, equilibrium, kinetics,* and *control.*

States. The states concept must be invoked to identify the system under consideration. It affirms that the physical universe is not chaotic and that the specification of a very few properties of a system fixes all other properties of that system. Equations of state and the Gibbs phase rule are typical articulations of this concept.

Conservation. The conservation concept is an accounting concept which states that what goes into a system, if it is not transformed, must either accumulate in the system or come out of the system. Familiar examples are the law of conservation of matter and the law of conservation of energy. These laws as applied to problems of fluid flow are the basis of the equation of continuity and Bernoulli's theorem, respectively. The conservation of momentum is another powerful tool in flow problems.

In phase change separations material balances account for matter, and heat balances account for energy. In chemical reactions the material balances must be based on atoms or total masses.

The economic balance accounts for money and is of strategic importance in gaging actual performance or projected possibilities.

Equilibrium. Physical and chemical processes cannot go beyond the limits set by equilibrium conditions; hence a clear understanding of these conditions is essential to sound process engineering. In diffusional operations such as distillation, solvent extraction and the like, the degree of difficulty of separation and therefore some idea of the costs involved can be determined from the number of *equilibrium contacts* required for the separation.

The feasibility of chemical reactions as well as physical processes can be assessed from equilibrium relations. Furthermore, as will be seen, the driving forces for both physical and chemical processes are given by the appropriate displacements from equilibrium. Problems in kinetics can be handled meaningfully only when the equilibrium conditions are known.

Kinetics. By kinetics is meant time-dependent processes or changes. A synonymous term is rate process, and in chemical engineering only four rate processes are of any great importance at present, namely, the physical rate processes of *momentum transfer, thermal transfer,* and *mass transfer,* on the one hand, and chemical rate processes or chemical kinetics on the other. All these rate processes involve a transport or transformation resulting from a gradient in the appropriate potential. Momentum is transferred by velocity gradient; heat by a temperature gradient; mass by a concentration gradient; and chemical change by a gradient in chemical potential. In each case the total potential gradient is given by the total displacement from equilibrium, and in each case the rate equation is given by the product of a rate coefficient or reciprocal resistance and the driving force for the process.

Momentum transfer is the basic rate process in fluid flow; thermal transfer is the basic rate process in the various heat transfer operations; and mass transfer is the basic rate process in the phase change physical separations. However, in typical process contexts many, and often all, of the different rate processes occur simultaneously. A principal contribution of modern chemical engineering has been the development of practical strategies for dealing with coupled rate processes. This development has provided a surer basis for the design and analysis not only of industrial chemical reactors but also of devices like artificial kidneys, in which chemical and physical rate processes occur together.

Control. Intelligent application of the principles of states, conservation, equilibrium, and kinetics permits the design and performance analysis of processes and equipment insofar as sizes and capacities are concerned. However, these principles of themselves give no picture of the behavior of processes under dynamic operating conditions. Since the dynamic behavior of processes strongly affects the yield and quality of products, a fifth basic principle must be added to handle problems relating to quality of operation. This fifth principle is one of *control,* since it necessarily relates to the character of performance of regulated operations and equipment.

Control is the purposeful regulation of activity. In an ultimate sense it is regulation to achieve optimal performance. It is the concept of process units like fractionating towers and reactors, the management of chemical works and petroleum refineries and of corporations and national economies, and indeed the regulation of all living organisms and organizations of living organisms. All control systems, however disparate their natures, have certain common characteristics:

(1) They are integrated, information processing systems requiring an overall systematic (systems) approach for adequate appraisal.

(2) They are dynamic systems, and their temporal behavior is the measure of their performance.

(3) They necessarily incorporate feedback loops in their information circuitry, and as a consequence they have the potential for becoming unstable. Thus the limiting condition of stability provides an important first criterion of controlled performance.

As examples of how the five basic concepts of chemical engineering are utilized consider the following situations. In preliminary surveys the states concept permits identification of the system, and the equilibrium concept identifies feasible and unfeasible process schemes. In gross process evaluations the conservation concept gives the gross amounts of materials, energy, and money that may be involved. In more detailed evaluations the kinetic concept in conjunction with conservation and equilibrium fixes sizes and capacities of equipment. The final kind of operation achieved by the process units is determined by the control concept applied with all the other four basic concepts.

In all engineering problems economic practicality is the overriding consideration, ideally gauged not in narrow terms of profitability but in respect to overall social impact including environmental effects. While factors of human relations bear importantly on engineering decisions, the purely technical aspects of chemical engineering problems can be handled via combinations of the five basic concepts of states, conservation, equilibrium, kinetics, and control.

ERNEST F. JOHNSON

Cross-references: *Computer Simulations of Chemical Reactions.*

CHEMICAL LITERATURE

The literature of chemistry is enormous and continues to grow exponentially. The exact size can be estimated reasonably accurately from the number of abstracts of papers and patents published in *Chemical Abstracts.* In 1969, 249,777 abstracts were published. From 1907 (when about 8,000 abstracts were published) through 1969, the total number of abstracts is 4,435,451. Inasmuch as *Chemical Abstracts* completely covers the literature of chemistry and chemical engineering reporting new data and information (within Chemical Abstracts Service (CAS) policy for coverage), and since other abstract journals are later checked and abstracts copied through exchange agreements to ensure that no abstracts have been missed, counts of abstracts do provide the most accurate measure available for the chemical-research activity of the world. In recent years the increase in number of chemical papers has been about 10% per year, compounded. The annual output of abstracts is expected to double before 1980.

Of the very limited amount of information that can be remembered for effective use, it is highly desirable that part of it consist of a thorough knowledge of how to obtain the recorded information that is required from time to time. Skill in the library is turning out to be as valuable as skill in the laboratory. Information from the record can nearly always be obtained much more rapidly and inexpensively than can the same information from experiment. In chemistry more than in any other field, organized means for making the published literature available have been provided. Some of these means will be discussed briefly below. Pioneering in publication of abstracts, indexes, and the like has been extensive in the field of chemistry.

Before discussing the kinds of chemical publication and listing some specific books and periodicals, we should mention some developments stemming from the growing realization that the rapidly accumulating mass of documents must remain searchable.

Many universities now teach courses in chemical literature. Several textbooks and guides are available. Examples of these have been prepared by: Crane, Patterson, and Marr; Dyson; Mellon; Smith; Soule; and the American Chemical Society as the Advances in Chemistry Series, No. 30.

Organizations. Since 1949, the American Chemical Society has had a Division of Chemical Literature that meets at least twice yearly for presentation of papers and symposia. The Division has been commendably active in production of the *Journal of Chemical Documentation*—started in 1961. Organizations holding meetings devoted, at least in part, to problems of the chemical literature include: American Association for the Advancement of Science, American Association of Clinical Chemists, Inc., American Chemical Society, American Federation of Information Processing Societies, American Society for Information Science, American Institute of Chemical Engineers, American Leather Chemists Association, American Library Association, American Microchemical Society, Association for Computing Machinery, Inc., Association of Official Analytical Chemists, Association of Research Libraries, Battelle Memorial Institute, Chemists' Club (of New York), Drug Information Association, Electrochemical Society, International Council of Scientific Unions, International Federation for Documentation, International Federation for Information Processing, International Federation of Automatic Control, International Federation of Clinical Chemistry, International Federation of Library Associations, International Union of Pure and Applied Chemistry, Manufacturing Chemists' Association, National Academy of Sciences—National Research Council, National Federation of Science Abstracting and Indexing Services, National Science Foundation, Scientific Research Society of America, Society of Chemical Industry, Society of Cosmetic Chemists, Inc., Society of Industrial Chemistry, Society of Physical Chemistry, and Special Libraries Association.

Meetings. There have been many congresses and other meetings, some of them international, for discussion of problems of dealing with the chemical literature. The Association of Special Libraries and Information Bureaux, London (ASLIB) holds an Annual Conference. There are the Gordon Research Conferences in New Hampshire. Other conferences and symposia include: conferences of the International Association of Documentalists and Information Officers, International Computer Application Symposia, International Conference on Information Processing, National Colloquium on Information Retrieval, and meetings of the Society for International Development. Some of these have been sponsored by UNESCO, some by the International Union of Pure and Applied Chemistry (IUPAC) that has special committees on Machine Documentation and on Nomenclature and Symbols. There are still others by nations or national groups, such as the Royal Society of England. The International Council of Scientific Unions has organized an International Abstracting Board to aid abstracting services in science in areas where international cooperation is effective. Chemistry is included in this effort.

There are many national chemical organizations,

the publications of which constitute a very important source of original chemical papers. These organizations are listed in the UNESCO Directory of International Scientific Organizations and in other directories as well as in the guides to the chemical literature—mentioned above.

Periodicals. The journal literature constitutes nearly all the scientific source material and is by far the most important part. There were in 1970 at least 11,370 scientific, technical, and trade serials that publish original papers of chemical interest—the results of experimental investigation. The publication entitled *Chemical Abstracts Service Source Index* (*CASSI*—formerly *ACCESS*) is a union catalog of these serials that include such productive publications as the *Journal of the American Chemical Society* which carried (in 1969) 2,086 original chemical papers. Other publications in fields not primarily of chemical interest report valuable chemical information from time to time. Chemical articles in these are also covered by *Chemical Abstracts.*

In addition to original papers, the periodical literature of chemistry is a source of data. There are abstracts, indexes, book and other reviews, bibliographies, news, statistics, trade information, and descriptions of new products and services.

Abstracts and Indexes. About 140 years ago, it became evident that no scientist could read all the pertinent literature, even in a limited subject field such as chemistry. Indexed abstracts were developed to solve this problem. Thus, the abstract journals came to have special importance in dealing with scientific literature. Indexed abstracts are designed to keep scientists informed about current developments and to facilitate searches back through the literature, to help with the language problem, and to provide surrogates that overcome limitations of access to literature. The National Federation of Science Abstracting and Indexing Services, Report Number 102 lists, in 1963, 1855 titles of such services published in 40 countries. Malinowski states that there are over 3500 abstracting and indexing services throughout the world and publications covering all areas of thought.

Chemisches Zentralblatt, the first chemical abstract journal, started in 1830 under the name *Pharmaceutisches Centralblatt,* continues as *Chemischer Informationsdienst* from 1970. It is comprehensive and well indexed.

Abstracts in the English language appeared during the years 1871–1925 in the *Journal of the Chemical Society* (London), during 1882–1925 in the *Journal of the Society of Chemical Industry,* and from 1926–53 in a combination of these, first called *British Chemical Abstracts* and then, after being known as *British Chemical and Physiological Abstracts* during the period 1938–44 was called *British Abstracts* until it was discontinued at the end of 1953. These journals published excellent abstracts, but they were not quite complete in their coverage of the chemical field and their indexes were not so complete as others.

The principal abstracting journal for pure and applied chemistry now appearing in the English language is the American Chemical Society's *Chemical Abstracts.* It also publishes thorough keys to the information found in abstracts in the form of:

issue, volume, and cumulative author and patent-number indexes; issue keyword (subject) indexes; issue, volume, and cumulative patent concordances; volume and cumulative subject, molecular-formula, and organic ring-system indexes; and a volume Hetero-Atom-in-Context index. An index guide and *CASSI,* just mentioned, are also provided. In 1969, 22,100 pages of volume indexes appeared. The 24 volumes of the 7th cumulative indexes have 41,000 pages bearing 10 million entries that lead to 930,000 abstracts.

In 1956, Chemical Abstracts Service (CAS) became the designation of the activity of the *Chemical Abstracts* office and a research department on methods for handling chemical literature was established. Through this large and active research department, CAS has initiated a number of new services. In 1971, additional services include:

1. *Basic Journal Abstracts* from 35 journals and issued as computer tape plus corresponding printed issues 26 times a year.

2. *Chemical Abstracts—Section Groupings* in bio-, organic-, macromolecular, applied, physical, and analytical chemistry, and in chemical engineering. Groupings are distributed without volume indexes.

3. *CA Condensates,* a weekly, computer-readable service bearing titles, names of authors and assignees, citations, and entries from keyword indexes.

4. *Chemical-Biological Activities*—biweekly digests in printed and computer-tape form.

5. *Chemical Titles,* a prompt biweekly KWIC (Key-Word-In-Context) index of titles of chemical papers in about 650 journals.

6. *Plastics Industry Notes*—of abstracts from 30 journals.

7. *Polymer Science & Technology*—of digests.

8. *CA on Microfilm*—leased to save copying of heavy volumes.

9. *CASSI,* mentioned above, and its quarterly service.

10. The *Ring Index* and supplements.

11. Synthetic Organic Chemical Manufacturers Association (*SOCMA*) *Handbook*—of organic compounds in trade.

12. *Steroid Conjugates*—a bibliography of abstracts.

13. Compound searches—an experimental service based upon *CA.*

14. Reproductions of Soviet Chemical Papers—as photocopies.

15. Searches of Subscription Data Files—as computer listings from certain tape files produced at CAS.

16. Nomenclature reports.

17. Publications available from the U.S. Clearinghouse for Federal Scientific and Technical Information (CFSTI).

18. CAS Reports and Instruction Manuals.

19. CAS Information Centers for making searches.

20. License to reproduce material.

21. Back issues.

Some of the above services are provided in the form of leased or sold computer tape or searches thereof.

As early as 1858, *Bulletin de la société chimique de France* published abstracts. This service was discontinued during World War II and now in France the whole field of natural science is reporting briefly in sections of a journal formerly called *Bulletin analytique* and later named *Bulletin signalétique*. Sections 5, 7, 8, 11, 12, and 13 are most closely related to chemistry. Brief descriptive abstracts, similar to annotations, are published. There are no indexes.

Complete Abstracts of Japanese Chemical Literature (Nippon Kagaku Soran) are published in Japan. A series of abstract journals was started in the Soviet Union in 1953; one of these publications is *Referativnyĭ Zhurnal, Khimiya*.

There are many other sources of abstracts related to chemistry and chemical engineering. Some are for limited fields and for limited periods of time, and some for fields that are only partly of chemical interest. Some journals that publish original papers also carry abstracts for their fields as, for example, the *Journal of the American Leather Chemists' Association*. Both journals publishing original papers and those publishing abstracts are listed with descriptive matter in *CASSI,* mentioned above.

Patents. To many chemists, especially industrial, patent specifications and claims are an exceedingly valuable source of new information. (See **Patents**).

Other Sources. Other than periodicals and patents as sources of new information of interest to the chemist, there are numerous publications, such as: bibliographies; government bulletins, circulars, and reports; trade literature; annual reviews; reports by research organizations; dissertations or theses; privately published addresses or lectures; preprints, offprints, and reprints. Personal correspondence may be of importance.

Annual catalogs are often a very useful part of the trade literature. Good examples are "Chemical Engineering Catalog" and the "Chemical Materials Catalog," both published by the Reinhold Publishing Corporation.

Books. Chemical books, likewise important to the chemist, are discussed here last because, as a rule, they appear much later than do the journal and other literatures. While some books do report new information (the results of experimental investigation not previously published), ordinarily they are based upon information already in print, usually in journals. Authors of books customarily assemble published information on a given subject, organize and edit it, and present it in readable, more or less exhaustive form. Books are useful in teaching and learning, as a convenient means for reviewing a subject, and in some phases of literature searching. They serve well to introduce the user to a more or less general subject or field; often they explain new theories in the light of facts already known, and they help to coordinate and systematize knowledge. Popular books of science are helpful in widening interest and support, attracting students to the field, and improving public relations. Writers with long experience advantageously interpret investigational results scattered by chance throughout the journal and report literatures.

Many books offer convenient starting points for literature searches; reference books not infrequently provide ending points in certain kinds of searches, particularly when single bits of information (data) are needed. A good reference book cites its authorities and serves as a reliable guide to original articles.

Scientific books take various forms. They may be general books, such as textbooks, but they may also be encyclopedias, exemplified by this volume and others like it in the subfields of chemistry, treatises of a broad scope, indexes, dictionaries, pocket books, and guides to the literature of chemistry. Some, such as encyclopedias and broad treatises, may consist of many volumes.

General treatises or handbooks on chemical subjects are far too numerous to be listed here. The compilation and publication of books of the "Handbuch" type has been particularly common in Germany through the years. There is a growing tendency to publish annual review volumes in other countries, including the United States. Some examples of general treatises of greatest value to the chemist include: Gmelin's "Handbuch der anorganischen Chemie" (Verlag Chemie, G.m.b.H., Weinheim/Bergstrasse, Germany); and Beilstein's "Handbuch der organischen Chemie" (Springer, Berlin, Germany).

Gmelin, now in its eighth edition (started by R. J. Meyer and continued by Erich Pietsch), is a lineal descendant of Leopold Gmelin's "Handbuch der theoretischen Chemie" (1817–19). The eighth edition of the valuable publication, with supplements, exhaustively covers inorganic chemistry between 1750 and 1950. "Metallurgy of Iron" appeared in 1968.

Beilstein, now in its fourth edition, was originally compiled by Friederich Konrad Beilstein, a tremendous worker, who issued his first edition in 1881–3, his second in 1886–90, and his third in 1892–9. The main series of the fourth edition includes all organic compounds of known structure to 1910. The first supplement to the fourth edition covers the period 1910–19, the second, the years 1920–9; a third supplement, volumes of which are appearing regularly, covers a 20-year span, 1930–49. Since this is a most important series, the delay in appearance is unfortunate; the same is true for *Gmelin.*

Another series of volumes is *Mellor* ("A Comprehensive Treatise on Inorganic and Theoretical Chemistry") first published in 1922 with supplement on "Nitrogen (Part II)" coming out in 1967.

Chemical Abstracts started its annual formula indexes in 1920 to aid those experiencing difficulty with organic nomenclature; *Chemisches Zentralblatt* began its annual formula indexes of organic compounds in 1925 and produced a cumulative formula index for 1922–4. There is a cumulative formula index to *Chemical Abstracts* for the period 1920–46, a ten-year cumulative index for 1947–56, with five-year indexes of this kind thereafter.

"The Ring Index" by Austin M. Patterson and Leonard T. Capell is an example of an index in book form. The second edition of the Patterson-Capell-Walker "Ring Index" was published by the American Chemical Society in 1940. There have been three supplements covering through 1963. Supplement II features a cumulative index of Names and Hetero-Element Index.

Monographs (rather full treatises on single topics) usually give the detailed information needed by the research worker and are yet of convenient size.

These may be issued as parts of series. Of special interest to the American chemist are the Advances in Chemistry monographs on pure and applied chemistry published by the American Chemical Society. Number 96, "Engineering Plastics and Their Commercial Development," appeared in 1969. "Organic Syntheses," started in 1921 with volume 46 appearing in 1966, is another series. Yet others are: "Organic Reactions," "Heterocyclic Compounds," Advances in Protein Chemistry," and Sadtler's "Standard Spectra."

In the chemical field, there are several dictionaries and glossaries (word books) that consist of definitions of words or terms all in one language. Examples are: "Condensed Chemical Dictionary" (8th Ed., 1971, Van Nostrand Reinhold), "Hackh's Chemical Dictionary," (Fourth edition, 1969) and "A New Dictionary of Chemistry," by Miall and Sharp. Also, there are chemical dictionaries for two or more languages written for the purpose of helping in the reading and translating of scientific and technical publications. Patterson and De Vries are remembered in this connection for French-English and German-English dictionaries limited to chemistry and science, respectively.

Special Services. For limited subfields of chemistry, special services have been developed. For pharmaceutical research, for example, Derwent provides Ringdoc including Ringcode and Codeless Scanning in giving access to the literature through chemicals and their pharmacological effects. The De Haen card services extract pharmacological data and present it in convenient card form. Both of these services are costly, but very inexpensive when compared with duplicating results within individual industries.

Automation. For many years now, it has been apparent to chemists and chemical engineers that their professional reading was limited by time rather than by pertinency. They have realized that they are unable to keep up in reading their chosen subspecialties. Chemical engineers know that they waste time searching for valuable data in the wilderness of the primary and secondary literatures; they know that these data could have been organized and published in the form of convenient handbooks. There is a growing concern among all professional people about the adequacy of the services provided to them through their professional literatures. As a result of these problems and concerns, research has been stimulated to create vastly superior information services—knowledge-transfer systems. Problems of lack of time to read all believed necessary, lack of facility with about 70 languages, and lack of access to documents, have all received attention by the best minds in the field of information science. The results of this research effort over the last quarter century have been some very ingenious proposals, techniques, and devices, but nothing in the way of vastly improved services for all professional people. Use of edge-notched cards, optical-coincidence cards, character-reading machines, computers, microform equipment, and the like has been extensively studied and great ingenuity displayed. Study of machine translation, once extensive, has declined largely owing to semantic complexities. Automated abstracting, indexing, and question-answering have met similar fates. Computer-aided indexing, in which the computer aids the human indexer, is more promising, as is also associative retrieval, a technique for avoiding indexing altogether and for ranking references in response to questions. However, problems of too much to read in the limited time available, too many languages, and limited or delayed access to documents are still with us.

Since the literature of information science has been increasing at a rate even greater than that of chemistry, it is impossible to do justice to this subject in a few paragraphs. Several large information services have been established and are functioning. An example is the Medical Literature Analysis and Retrieval System (MEDLARS) of the National Library of Medicine, which has proven to be especially effective in generic or combination generic searches in which formulations comprising many terms correlated by the logical "or" are generated automatically. These systems function by human indexing, human search formulation, and machine searching by use of terms chosen by the formulator from a standard vocabulary or thesaurus. Terms are correlated by the human formulator by the logical "or" to make the search more comprehensive and to avoid invisible loss, or by the logical "and" to make the search more specific and exclude more irrelevancies, or by the logical "not" to exclude definitely unwanted material. Many smaller automated services for science and engineering are functional; about 300 are described in the Directory of Computerized Information in Science & Technology, edited by Leonard Cohan. Since natural English is unsuitable, at present, for direct input to these systems, standard vocabularies or thesauri are used to avoid scattering of like subjects, and enable indexing and searching. Examples of organized word lists are: The Thesaurus of Engineering and Scientific Terms (1967) of the Department of Defense and the Engineers Joint Council, MeSH (Medical Subject Headings) of the National Library of Medicine, and the NASA Thesaurus. Some computerized systems have been abandoned because of expense.

Systems for searching files of organic-compound structural designations are being developed; some are operational. These systems enable the searcher to locate references to specific organic compounds or to classes of compounds all having in common the same groups, combinations of groups, or other specific structural features. The inputs to these sysstems are: fragments of structures, notations embodying fragments, ciphers or codes, or atoms and their bonds. Output is in the form of references to specific structures.

The problem of keeping up in chemical fields ten to one hundred times broader than can now be read is becoming metamorphosed into a problem of providing, organizing, and distributing terse surrogates from each document. If we cannot read all of the words that we believe we should, then we must read fewer; and these fewer words should be chosen rationally rather than by chance. Short surrogates for documents provide rationally these fewer words. Surrogates even shorter than abstracts, e.g., terse conclusions, terse results, and handbooks of current data are found to be increasingly useful.

One of the major problems in obtaining authorization for implementation of superior information systems (automated or manual) is clear demonstration of a favorable cost-value relationship for bibliographic service. Cost of such services, including abstracting or extracting and indexing, ranges from 0.1 to 1% of the cost of obtaining and publishing the technical information in the first place. The value of technical information is unquestioned, but difficult to document in such a way that investment in vastly improved service is obviously justified.

CHARLES L. BERNIER

CHEMICAL NOMENCLATURE

The development of names for all that chemistry represents has been and still is one of its major concerns. It is obvious that every principle, fundamental concept, all elements, compounds and a multitude of other factors in the science must be labeled with a word or combination of words. To meet all these requisites there has been a constant attempt to make a coherent language of chemistry.

The word atom (Greek, uncut) is one of the oldest words of our present vocabulary and as far as ordinary chemical change is concerned (not counting nuclear fission) means about the same that it did when first used by Democritus c. 400 B.C. It is the least unit of matter involved in chemical reactions by means of which molecules or compounds are formed. The molecule is therefore a product made up of atoms of elements and represents a unit of matter able to exist alone and to exhibit a special set of properties of itself. The molecule is (except in the instance of certain gases) also a compound and as such has a chemical formula. This formula is so written to show the number of atoms of each element present. In other words, there are molecules of elements as well as molecules of compounds.

Each atom has what is known as a *symbol* made of one or two letters, the first of which is always the initial letter of the name of the element, either in English or some other language, *e.g.* Se, selenium, W, Wolfram (tungsten). The weight ascribed to the atom is called the atomic weight (or symbol weight) and represents a chemically determined relation of its weight to that of carbon 12. The sum of the number of the symbol weights in a molecule of a compound is called a *molecular* or *formula* weight. An example: KNO_3 (potassium nitrate) $= K = 39.09 + N = 14 + (3 \times 16) = 101.09$.

The term *substance* is applied to matter which has a uniform or constant composition with one set of chemical properties. Accordingly, only *elements* and *compounds* can be strictly called substances. *Mixtures* and *solutions* may differ in that their composition can be variable and their properties would depend upon it. One of the most used terms in chemical nomenclature is *valence*—a word developed to describe the relative combining capacity of the elements as they form compounds. On a chemical basis the valence numbers are reckoned in terms of a value of 2 for oxygen or 1 for hydrogen. Hence, since in water 2 H atoms combine with 1 oxygen atom the valence of each H is 1, because it takes two atoms of this element to be

chemically equal to 1 of oxygen. However, since the development of atomic structure the value of valence is usually given in terms of its valence electrons. The valence in respect to electrons is that number which represents the number of electrons gained or lost by an atom or a radical, or the number of shared pairs of electrons.

Inorganic Nomenclature

To demonstrate a naming system used in the chemistry of inorganic compounds it is necessary to introduce the words *ion* and *radical*. Ions in a general sense are electrically charged particles, and are usually capable of conducting electric current. Their main importance in chemistry relates to properties of solutions. A solution of an acid, base or salt has the property of electrical conductivity because of the presence in it of *ions*. They are either atoms of elements carrying an electric charge (*e.g.* H^+ hydrogen ion) or a number of atoms bound together in what is called a *radical* which is also charged, *e.g.* SO_4^{--} (sulfate radical). It is through these ions that acids, bases and salts have their existence as well as their properties.

A compound consisting of a H^+ (ion) in connection with a negatively charged radical, *e.g.* NO_3^-, is called an *acid*. In this case it would be nitric acid. The ions that carry positive charges are called cations (*e.g.* Cu^{++}) and those which are negative are called anions, (*e.g.* PO_4^{---}). A *base* is a compound made of a positively charged metal in conjunction with a negatively charged ion, called hydroxyl (OH^-). A typical case is K^+OH^-, potassium hydroxide. *Salts* are the product resulting from the combination of the positive ion of the base and the negative radical of the acid, *e.g.* $K^+NO_3^-$, named potassium nitrate. This general case, however, is not the only way salts can be formed.

The recommendations of the Committee of the International Union of Chemistry on nomenclature are perhaps the best source of systematic naming of inorganic compounds which has been provided. [See *J. Am. Chem. Soc.*, **63**, 889–897 (1941)]. There are still and probably will always be many names of chemical compounds that will be referred to with a colloquial term or expression that has little if any relation to any system. However, in the literature the only wise plan is to follow the recommendations of this committee for terms to use for inorganic and organic compounds. Even then many instances will arise that will present debatable problems.

Binary (two-element) compounds such as the common oxides, chlorides, sulfides, phosphides, etc. are usually named as follows:

MgO—magnesium oxide
KCl—potassium chloride
ZnS—zinc sulfide
Ca_3P_2—calcium phosphide

It is seen that the name of the most electropositive element is given first, and the name of the electronegative element is modified so that the last part of the name is changed to *ide*. Additional insight can be gained by a reference to:

HCl—hydrogen chloride
NaH—sodium hydride,

the hydrogen changing from the first element to the last because of its relative electropositive nature in the first to a relative electronegative in the second.

In the naming of binary acids a special syllable is used to indicate the oxygen content:

HBr—*hydro*bromic
H_2S—*hydro*sulfuric
HCN—*hydro*cyanic

The names here are instructive, in relation to the system given below for oxygen-bearing acids. *Hydro* means "without oxygen." The name of the nonmetal is changed to give it a characteristic syllable—*ic*. In the cyanide radical the system holds.

The oxygen-containing substances have names for special reference to the oxidation state of the central element—the main element in the compound besides hydrogen and oxygen. In those compounds the suffixes *ic, ous, hypo,* and the prefix *per* are used. Sulfuric acid will serve as a full example.

H_2S—in aqueous solution—*hydro*sulfur*ic* acid
H_2SO_2—in aqueous solution—sulfoxyl*ic* acid
H_2SO_3—in aqueous solution—sulfur*ous* acid
H_2SO_4—in aqeuos solution—sulfur*ic* acid
$H_2S_2O_5$—in aqueous solution—*per*sulfur*ic* acid

The highest oxidation state of the element is given a name with -*ic* as the characteristic syllable; one step lower is -*ous;* the lowest oxidation state is indicated by the prefix *hypo,* which means "less than." There is no assurance that all these can be prepared in the free form.

While the above syllables are still in general use, a strong recommendation is made that the Stock system should be used to indicate proportions by constitution. For example:

$FeCl_2$—iron(ll)chloride—not *ferrous* chloride
$FeCl_3$—iron(lll)chloride—not *ferric* chloride

The committee even goes so far as to say that the use of the older affixes should be discontinued in scientific literature. Some writers are already doing so. Compounds with an excess of oxygen are called *per* compounds, and there is usually a special way to account for the structure of compounds that can be so produced.

It should be observed at this point that the prefix *per* by no means tells the same thing about structure. For example, sodium persulfate is said to be a true *per* compound, but $KClO_4$, potassium perchlorate is not. This is a good example of how confusing words are still used in the nomenclature.

To continue the above system for acids as it is applied to salts, the following rules apply. The word acid is dropped, the name of the metal or radical is placed first, and the suffix syllables are changed: *ic* becomes -*ate* and -*ous* becomes -*ite.*

Na_2S—sodium sulf*ide*
Na_2SO_2—sodium (*hypo*sul*fite*) sulfoxylate
Na_2SO_3—sodium sulf*ite*
Na_2SO_4—sodium sulf*ate*
$Na_2S_2O_8$—sodium *per*sulf*ate*

Sometimes it seems desirable to use other roots to aid in meaning, such as *mono, di, tri, tetra,* and *penta* from the classical languages.

CO—carbon *mono*xide (meaning one)
CO_2—carbon *di*oxide
$AsCl_3$—arsenic *tri*chloride
CCl_4—carbon *tetra*chloride
P_2O_5—phosphorus *pent*oxide

Hydrates, as inorganic compounds, must be given special treatment, since they are common substances and at the same time more complex. Examples are copper sulfate, pentahydrate, $(CuSO_4 \cdot 5H_2O$, or sodium sulfate, decahydrate, $Na_2SO_4 \cdot 10H_2O$. Convention is not uniform either in formulation or naming. Some writers use a comma between the anhydrous salt and water, which has been suggested, while others prefer a center dot, *e.g.,* $CaCl_2 \cdot 6H_2O$. Some name a compound of this sort calcium chloride, hexahydrate while others suggest calcium chloride, 6 hydrate. The first seems more prevalent and acceptable.

Hydrates such as involved above introduce another common term, viz., *anhydrous.* A substance, usually a salt that can crystallize with water of hydration, is called anhydrous when it is free of water. The same compound when united with a definite number of molecules of water is said to be hydrated or is a hydrate.

To show how difficult a perfect consistency is: the case of the above-mentioned sulfur compounds compared to those of chlorine should be examined. The chlorine oxygen acids are:

HClO—*hypo*chlorous acid
$HClO_2$—chlorous acid
$HClO_3$—chloric acid
$HClO_4$—*per*chloric acid

There should be a "*hypo*sulfurous" acid, but it is recommended that the word sulfoxylic be retained.

Naming in respect to stoichiometric composition, the Greek numerical prefixes are recommended. These should precede, without hyphen, the constituent which they represent. This plan is applied chiefly to nonpolar compounds. For prefixes above 12, Arabic figures are used for simplicity. Examples are:

N_2O_3—dinitrogen trioxide
N_2O_4—dinitrogen tetroxide
$Fe(CO_4)$—iron tetracarbonyl

Intermetallic Compounds. No rigid system of naming is recommended for this class of compounds. The best usage is formulation, in which the exact distribution by atoms is shown. Example: $Cu_5 \cdot Sn \cdot Cu_3Al$.

Designation of atomic number, atomic weight, and state of ionization is as follows:

$$^{16}_{8}O_2^{-2}$$

These are read as follows:

Upper left—atomic weight or mass
Lower left—atomic number
Upper right—state of oxidation
Lower right—number of atoms per molecule

Naming within groups of elements: While the term halide for the halogens is widely used, it is strongly urged that *halogenides* replace it.

Names like alkali elements and alkaline earths

should be discontinued, but it is good form to retain "alkali chloride."

The prefixes *ortho, meta* and *pyro* are used in relation to the degree of hydration which an anhydride can undergo to form an acid or a salt:

$$P_2O_5 + H_2O \rightarrow 2HPO_3\text{—}meta\text{-phosphoric acid}$$
$$P_2O_5 + 3H_2O \rightarrow 2H_3PO_4\text{—}ortho\text{-phosphoric acid}$$
$$P_2O_5 + 2H_2O \rightarrow H_4P_2O_7\text{—}pyro\text{-phosphoric acid}$$

The accepted meanings are: *Ortho,* fullest possible hydration; *pyro* (Greek, fire) made from ortho by use of heat; and *meta,* lowest stage of hydration.

Salts containing hydrogen: The I.U.C. committee urges a change in the name for hydrogen-bearing salts to a system different from much of the common practice in current writing. Such compounds as $NaHCO_3$ are to be *sodium hydrogen carbonate,* not sodium bicarbonate. $KHSO_4$ is potassium hydrogen sulfate; Na_2HPO_4 is orthodisodium hydrogen phosphate. Salts containing a hydroxyl radical are to be called hydroxy compounds; $Ca(OH)Cl$, would be called calcium *hydroxy* chloride.

Complex Compounds. The system devised by A. Werner is to be followed as far as possible. Use is made in this plan for a letter to be used in place of a Roman numeral to indicate the valence of the central element. For example:

I = a
III = i K_2PtCl_6 = Potassium hexachloroplate*e*ate
IV = e

There seems to be a growing use of the Stock system of naming rather than that based on the Werner plan. This may be due in part to the lack of euphony in the words. Stock suggested the use of Roman numerals in parentheses after the name of the central element to refer to its oxidation state:

$FeCl_3$ would be iron(III) chloride
$FeCl_2$ would be iron(II) chloride

In the case of anions:

$KMnO_4$—potassium permanganate(VII)
K_2ReCl_6—potassium (hexa) chlororhenate(IV)

Coordination compounds can also be fitted into the Stock scheme easily and systematically. Examples:

$K_4[Fe(CN)_6]$—Potassium hexacyanoferrate(II)
$K_3[Fe(CN)_6]$—Potassium hexacyanoferrate(III)
$K[Au(OH)_4]$—Potassium tetrahydroxoaurate(III)

In regard to the recommended plan for naming isopoly, heteropoly, double salt compounds, hydrates, the use of hyphenation for hydrates and the ammoniates, see *J. Am. Chem. Soc.,* **63,** 895–897 (1941).

Due to extensive developments in recent years of the chemistry of coordination compounds, it seems in order to recommend other literature sources to deal with proper nomenclature in this area. It is too extensive to treat in a brief discussion. See references.

Organic Nomenclature

There are three essential facts with which to deal in the nomenclature of organic compounds: (1)

The common names of long and accepted usage: (2) the Geneva System, dating from the Congress held in 1892; (3) the report of the Commission on Reform of Nomenclature in Organic Chemistry. This study was in 1930 in Paris. The reference in English to the 68 recommendations is in *J. Am. Chem. Soc.,* **55,** 3905 (1933).

The I.U.P.A.C. system is now preferred and will in time be the plan adopted by most authors. Details of the system are included under organic chemistry.

The Geneva System is based on compounds derived from hydrocarbon as a starting point, and the names correspond to the longest straight carbon chain which is present. The position is usually indicated by numbers applied to the carbons of the straight chain. The numbering beginning with the end carbon nearest the substituent element or group.

The *normal* compound is one in which the carbons are linked one to another in a straight chain. These are named such as ethane, propane, or pentane. Normal propane would be:

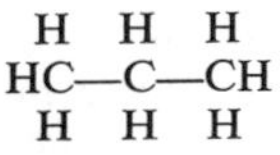

$$\begin{matrix} H & H & H \\ HC\text{—}C\text{—}CH \\ H & H & H \end{matrix}$$

It is necessary therefore to consider only the branched compounds, *e.g.:*

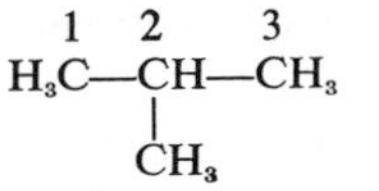

$$\begin{matrix} 1 & 2 & 3 \\ H_3C\text{—}CH\text{—}CH_3 & \text{(2-methyl propane)} \\ & | \\ & CH_3 \end{matrix}$$

To put the idea another way, it is a derivative of propane, not of methane. A certain pentane, for example, was once known as isopentane. Its structure is:

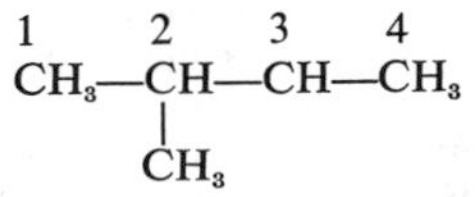

$$\begin{matrix} 1 & 2 & 3 & 4 \\ CH_3\text{—}CH\text{—}CH\text{—}CH_3 \\ & | \\ & CH_3 \end{matrix}$$

It is now named in accordance with the system 2-methyl butane. The 4 carbons in a straight line give rise to the name butane and the side group is attached to the *second* carbon atom in the chain. Another pentane has the structure.

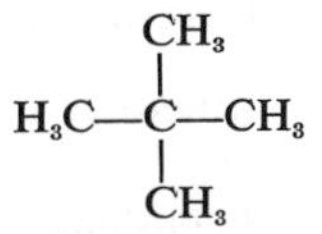

$$\begin{matrix} & CH_3 \\ & | \\ H_3C\text{—}C\text{—}CH_3 \\ & | \\ & CH_3 \end{matrix}$$

and was once named tetramethyl pentane. Under the Geneva System it is named in terms of 2 methyl groups being joined to carbon number 2 in a propane chain. It is thus 2,2-dimethyl propane.

An example from the hexanes ($C_{12}H_{14}$) can be studied as follows. It could be called (old name) tetramethyl ethane:

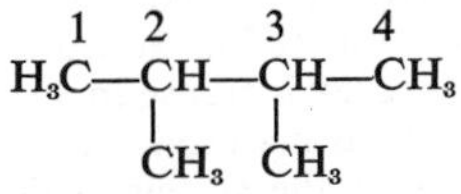

$$\begin{matrix} 1 & 2 & 3 & 4 \\ H_3C\text{—}CH\text{—}CH\text{—}CH_3 \\ & | & | \\ & CH_3 & CH_3 \end{matrix}$$

Its name is actually 2,3-dimethyl butane.

Some special recommendations which follow and extend the Geneva System define the use of certain syllables.

Open-chain hydrocarbons with one double bond should end in -*ene*.

Open-chain hydrocarbons with two double bonds should end in -*diene*.

Their names would now be alkenes, and alkadienes. Triple bond compounds end in -*yne* and -*dyne*. It is important to note that prop*yne* is a change from the Geneva System, which would call the same compound prop*ine*. It is suggested that acetyline will probably never become ethyne, but to illustrate the application and practice applied to higher hydrocarbons these examples will be instructive:

$$\overset{1}{H}C\overset{2}{\equiv}C\overset{3}{-}CH_3 \qquad \text{propyne}$$

$$\overset{1}{H}C\overset{2}{\equiv}C\overset{3}{-}CH_2\overset{4}{-}CH_3 \qquad \text{1-butyne}$$

$$\overset{1}{H_2}C\overset{2}{-}C\overset{3}{\equiv}C\overset{4}{-}C\overset{5}{\equiv}C\overset{6}{-}CH_3 \quad \text{2,4-hexadiyne}$$

Aromatic hydrocarbons should end in -*ene*, e.g., "phene" might be used in place of benzene.

Alcohols are to be named by use of the hydrocarbon name with -*ol* as the characterizing syllable: phenol, from phenene, and naphthol, from naphthalene. In order to extend this plan to polyhydric alcohols, a syllable such as *di, tri,* and *tetra* will be inserted between the name of the parent hydrocarbon and the suffix *ol*.

$$CH_2OHCH_2OH \qquad \text{1,2-ethane}diol$$

$$\overset{1}{C}H_2OH\overset{2}{C}HOH\overset{3}{C}H_2OH \qquad \text{1,2,3-propane}triol$$

The name *mercaptan* is to be discontinued and thiol used in its place. C_6H_5SH is benzenthiol.

Sulfides, disulfides, sulfoxides and sulfones will be named like ethers, the oxy- term being replaced with thio-, dithio-, sulfinyl and sulfonyl.

The acids are also named in accordance with the Geneva System. For a detailed discussion of the naming of acids the report on which this article is based should be consulted.

F. A. GRIFFITTS

Cross-references: *Chemical Nomenclature, Formulas, Atoms, Molecules, Acids, Bases.*

References

Cahn, R. S., "An Introduction to Chemical Nomenclature," Plenum Publishing Corp., New York, 1968.

Fernelius, W. C., "Nomenclature of Coordination Compounds: Present Status," *J. of Chem. Documentation,* **4,** 70 (1968).

CHEMICAL POTENTIAL

J. Willard Gibbs introduced the concept of the chemical potential in 1876. It arose from the need to describe the energy increase in a system upon the addition of chemical material. In particular if the equilibrium state of the system is described by the volume V, the entropy S, and the molar composition n_1 moles of species 1, n_2 moles of species 2 etc., then the increase in the energy of the equilibrated system is given by

$$dE = TdS - pdV + {}_i\Sigma\mu_i dn \qquad (1)$$

where the chemical potential of species i, μ_i, is determined by the relation

$$\mu_i = \left(\frac{\partial E}{\partial n_i}\right)_{S, V, n_j} \qquad (2)$$

and n_j denotes the moles of all species except i. In the expression for dE the μ_i's play a role analogous to that which temperature or pressure play in describing the equilibrium or nonequilibrium properties of the system.

We speak of a system in thermal equilibrium when the temperature is equal throughout, and in an entirely similar way we regard the system in chemical equilibrium when the chemical potential of each species is equal throughout. In this connection Lewis and Randall identified the chemical potential with the descriptive notion of the escaping tendency of a substance. Thus a solid in an evacuated container would tend to escape until an equilibrium concentration of the material was developed within the container.

From either equation (1) or (2) the chemical potential μ_i of a particular species i is equal to the increase in the energy per mole of that species, provided the entropy, volume, and number of moles of all other species are held constant, and the amount of the particular species added to the system has no measurable effect on the value of μ_i. The operational nature of this definition can be pictured by supposing that the dn_i moles of i added to the system carries a certain amount of entropy, volume, and chemical matter into the system. This added amount of entropy and volume must then be removed in order to place the system in a condition where the sole effect of the material addition upon the energy is present. This is conceptually feasible since it is possible to remove entropy by the process of heat exchange and volume by the process of compression.

Gibbs noted that the chemical potential could also be defined in terms of other thermodynamic functions, such as the enthalpy H, the Helmholtz free energy A, or the Gibbs free energy G. The appropriate expressions can be derived from equation (1) along with the definitions of H, A, or G. The results are

$$dH = TdS + VdP + \Sigma \mu_i dn_i$$
$$\text{so } \mu_i = \left(\frac{\partial H}{\partial n_i}\right)_{S, P, n_j} \qquad (3)$$

$$dA = -SdT - pdV + \Sigma \mu_i dn_i$$
$$\text{so } \mu_i = \left(\frac{\partial A}{\partial n_i}\right)_{T, V, n_j} \qquad (4)$$

$$dG = -SdT + Vdp + \Sigma \mu_i dn_i$$
$$\text{so } \mu_i = \left(\frac{\partial G}{\partial n_i}\right)_{T, p, n_j} \qquad (5)$$

The last two definitions of μ_i have the convenience of independent variables, either T, V, n_j or T, p, n_j that are readily accessible to experimental control. However a knowledge of A and G must now be at hand. The last definition is sometimes written as $\mu_i = \bar{G}_i$, where $\bar{G}_i$ is the partial molar free energy of species.

One of the fundamental relations between the chemical potentials of a system and the other thermodynamic properties is contained within the Gibbs-Duhem equation. It is obtained by recognition that the total energy of a system depends upon the extent of the system, i.e., the quantity of material within the system. From equation (1) the energy is thus construed to be

$$E = TS - pV + \Sigma \mu_i n_i \qquad (6)$$

However upon comparing the differential of this total energy with equation (1) a sum of terms as follows must be equal to zero:

$$SdT - Vdp + \Sigma n_i d\mu_i = 0 \qquad (7)$$

This is known as the Gibbs-Duhem equation. It points out that a change in the chemical potential must necessarily be related to changes in other parameters of the system. In a one component system at constant temperature $d\mu = (V/n)dp$.

The properties of the chemical potential can perhaps be more readily appreciated from the analogy of the behavior of other thermodynamic parameters. We have noted the analogy between thermal equilibrium as constant temperature with the concept of chemical equilibrium as constant chemical potential. It is also pertinent to recognize the role which other potentials play, for example the electrostatic potential ϕ, or the gravitational potential gh. These later potentials serve to define the amount of work of transporting either a unit quantity of charge or mass from one potential level to another. The difference in the value of the potential between the two states multiplied by the quantity transported is equivalent to the work of the transport process. The chemical potential describes the appropriate potential for the work of transporting chemical matter between two different chemical states. This interpretation of transport processes has been generalized by Brønsted to define all thermodynamic parameters. In this sense the work dW, as might be measured in a reversible transport of a quantity described by dK from potential P_i to P_f, is given by an equation of the form $dW = (P_i - P_f)dK$. A tabulation of typical potentials and their associated quantities is given as:

Work Process	Potential P	Quantity K
Weight transport (gravitational field)	gh (h, height)	m (mass)
Volume transport	$-p$ (pressure)	V (volume)
Surface area transport	γ (surface tension)	σ (area)
Charge transport	ϕ (electrostatic potential)	q (charge)
Thermal transport	T (temperature)	S (entropy)
Chemical transport	μ (chemical potential)	n (chemical matter) (moles)

One of the features to be noted in either the last tabulation or in equation (1) is that the terms which can be described as potentials are intensive properties (do not depend on extent of system) whereas those terms which have been denoted as quantities are extensive properties. The chemical potential is intensive with regard to the molar constitution of the system, or mathematically it is a homogeneous function of zero order.

The potentials of the various work processes can be used to interpret the direction of spontaneous processes. A mass falls spontaneously from one height to another, or a quantity of charge flows from a high electrostatic potential to a low electrostatic potential. Similarly spontaneous chemical processes involve the flow of chemical matter from a state of high chemical potential to a state of low chemical potential. In the case of a chemical reaction several components are necessarily involved in the process, and it becomes possible to define reaction chemical potentials.

The measurement of the chemical potential of a substance relies upon some form of the basic defining equations. There are two procedures; one might be called a direct method and the other an indirect method. In either case it is possible to measure only a difference in the chemical potential between two different states. In the direct method, the work of transport, or the difference in Gibbs free energy for a constant temperature pressure process, is measured. This type of measurement can be achieved in an electrochemical cell where the process is reversible. The work of the process taking place in the cell is determined by the potential difference of the electrons at the electrodes of the cell times the quantity of electrons transported. This work is equal to the opposing chemical work and provides a direct measure of the difference in the chemical potentials of the process taking place within the cell.

In the indirect procedure for determining the chemical potential, use is made of the Gibbs-Duhem equation, usually under the condition of constant temperature and with regard to the gaseous state. For example, the difference in the chemical potential of an ideal gas can be found by integrating the equation $d\mu = (V/n)dp$ between two different pressures at a given temperature:

$$\int d\mu = \int (V/n)dp = \int (RT/p)dp \qquad (8)$$

$$\mu - \mu^\circ = RT \ln p - RT \ln p^\circ \qquad (9)$$

$$\text{or} \qquad \mu = \mu^\circ + RT \ln p, \quad p^\circ = 1 \qquad (10)$$

It is customary to choose a reference pressure p° as a unit pressure, the μ° of course refers to this condition. In the case of nonideal gases knowledge of the equation of state provides the necessary information to place in the integral which determines the difference of the chemical potential. The chemical potential difference of a solid or a liquid can be determined by the appropriate equation of state, but it is of more general use to recognize that the chemical potentials of a given component in two phases must be equal. In the case of a liquid mixture which obeys Raoult's Law the vapor pressure of component 1 is given by $p_1 = p_1^\circ X_1$, where p_1° is the vapor pressure of the pure liquid 1, and X_1 is the mole fraction of component l in the liquid. Then at equilibrium the chemical potentials of liquid and gas phases for component l may be set equal and upon use of the ideal gas expression for the gaseous component

$$\mu_1^{liq} = \mu_1^{gas} = \mu_1^{\circ G} + RT \ln p = \mu_1^{\circ G} +$$
$$RT \ln p_1^\circ + RT \ln X_1 \qquad (11)$$

and the expression for the chemical potential of

pure liquid $\mu_1^{\circ L}$ is just $\mu_1^{\circ G} + RT \ln p_1$. In the case of solutions or gases which deviate appreciably from this simple ideal behavior, one still must make use of the phase equilibrium condition, usually using a phase which is simply characterized as either pure or gaseous.

The most important experimental methods for the determination of chemical potential differences in solutions involve the determination of vapor pressures, the osmotic pressure, the freezing point depression, and the boiling point elevation. In these situations it is convenient to express the chemical potential in terms of an activity a_1 of solvent, defined from the analogy of the equations for ideal gases or Raoult's law solutions:

$$\mu_1 = \mu_1^\circ + RT \ln a_1 \tag{12}$$

where μ_1° is the chemical potential of pure liquid l. In ideal solutions the activity a_1 might be identified as the mole fraction X_1. The osmotic experiment provides a determination of the osmotic pressure π that must be exerted to achieve equilibrium. The relation to the chemical potential of species 1, which is permeable to the osmotic membrane is given by

$$\mu_1 - \mu_1^\circ = RT \ln a_1 = -\int_p^{p+\pi} \overline{V}_1 dp \tag{13}$$

where $\overline{V}_1$ is the partial molar volume of the species 1 in solution. The freezing point depression experiment provides a similar result expressed by

$$RT \ln a_1 = \int_{T_0}^{T_0 + \delta T} (\overline{H}_1^L - H_1^{\circ S})/T^2 \, dT \tag{14}$$

where δT is the freezing point depression from pure liquid freezing point T_0 and $\overline{H}_1^L - H_1^{\circ S}$ is the heat of solution of solid species 1 to the concentration under investigation. The expression for the boiling point elevation is analogous except for the replacement of the heat of condensation from vapor to liquid.

The chemical potential plays a fundamental role in the discussion not only of phase equilibria, as we have seen, but also in the conditions of chemical equilibria. For a reaction such as

$$a A + b B = c C + d D \tag{15}$$

the condition of chemical equilibrium is

$$a \mu_A + b \mu_B = c \mu_C + d \mu_D \tag{16}$$

Substitution of the chemical potential expressions for the various terms establishes a relation with the equilibrium constant K of the reaction:

$$c \mu_C^\circ + d \mu_D^\circ - a \mu_A^\circ - b \mu_B^\circ = -RT \ln \frac{[a_C^c a_D^d]}{a_A^a a_B^b} = $$
$$-RT \ln K \tag{17}$$

Finally it is possible to obtain the chemical potential from a detailed expression of the partition function Q, for a canonical ensemble since $A = -kT \ln Q$. The appropriate result for an ideal monatomic gas of molecular mass m is

$$\mu = -RT \ln \left[\left(\frac{2\pi mkT}{h^2} \right)^{3/2} \frac{kT}{p} \right]$$

$$= -RT \ln \left[\left(\frac{2\pi mkT}{h^2} \right)^{3/2} kT \right] + RT \ln p \tag{18}$$

so that μ° is established by molecular parameters. This expression shows that the use of statistical mechanics provides the way by which the chemical potential can be established in an absolute manner, where the reference state is based upon the system being in its lowest possible energy level.

S. J. GILL

Cross-reference: *Thermodynamics.*

CHEMICAL RESEARCH

A continuing steady growth of chemical technology requires a proper balance between fundamental and applied research. The phenomenal growth of the industry during recent decades is certainly dramatic proof of the vitality of applied research during this period. A sustained growth in future decades will prove that our present fundamental research effort also is adequate.

There is a growing awareness on the part of industry that the twentieth century economy was born with a valuable inheritance of well-developed scientific understanding. In the field of chemistry proper, that is, molecular architecture, the basic laws of chemical reactions were quite well understood in terms of atoms and valence. Thermodynamics, the science which provides such powerful quantitative generalizations of myriads of chemical observations, had been completely developed. Classical physics had also reached a stage of considerable maturity, although its important contributions to chemistry by way of quantum theory and wave mechanics were to come during the first quarter of the twentieth century. Industrial research as we know it today did not exist, indeed, had not even been conceived. The word "research" needed no qualifying adjective such as "basic" or "fundamental," "developmental" or "applied" and "product," in order to explain what kind of research was meant. There was essentially only one kind of research, and this was what we now refer to as "pure" or "basic" or "fundamental."

The birth of modern industrial research, which is more inclined to applications, had necessarily to wait for pure research to reach a certain stage of maturity. It is being recognized, however, that the legacy of scientific understanding must constantly be extended if continued success in application research is to be expected.

Fundamental research is essentially a matter of inquiring into nature. The motivation for this activity is only imperfectly understood, but it is primarily an intellectual pursuit. Unlike applied research, the reward being sought is attained through the ability to understand and explain natural phenomena. The interest centers on elucidating the laws of nature, not on manipulating or exploiting them. Perhaps for this reason fundamental research has sometimes been referred to as "pure," presumably because of an emotional attitude on the part of those who use the term that is opposed to exploitive activity.

Liebig has been credited with the statement that, "To one man science is a sacred goddess to whose service he is happy to devote his life; to another she is a cow who provides him with butter." Value judgments of this kind are usually rejected as intellectual snobbery by present-day scientists, both fundamental and applied. Nevertheless, the fact remains that the interests and motivations of those engaged in research vary enormously, and that these temperament factors more than anything else determine whether an individual will center his interest in fundamental or applied research. In order to maintain an adequate supply of fundamental research scientists, it is necessary for society as a whole to appreciate and affirm the spirit of scientific investigation for its own sake.

Prior to World War II the chemical industry in the U.S. had already begun to recognize the necessity of making its own contributions to basic research rather than relying entirely upon the academic institutions, as had been done during the early part of the century. Du Pont, for example, pioneered in the chemistry of elastomers and polymers and quickly utilized the new learning to develop such products as neoprene and nylon. Characteristically, the relatively small amount of really new learning thus acquired opened up a possibility for new applications of such magnitude that it has supported many years of intensive exploitation in the form of applied research. The significance of the words of Goethe, "He who advances science may say to himself that he is laying the groundwork for unlimited results to come" becomes clear in the present age.

With the outbreak of the second world war and the interruption of German scientific contributions, the U.S. government very wisely recognized the need for large-scale sponsorship of scientific research. Naturally, much of the sponsored work was "applied," in the sense that it was directed either toward new weapons or products required by the wartime economy. Nevertheless, much fundamental work was also done, and for the first time in history the center of basic scientific investigation shifted from Europe to America. Many of the dramatic wartime developments were founded upon basic European discoveries, such as penicillin, DDT, polyethylene and the synthetic detergents. On the other hand, some of the synthetic rubbers and the whole of the complicated body of chemistry associated with the atomic energy development were almost entirely American.

It has often been said, on both sides of the Atlantic, that although the United States is the recognized leader in applied research, most of the fundamental research is contributed by European laboratories. This was undoubtedly true during the first half of the century, but there is some reason for believing that it is no longer so. While the question is a very difficult one to treat quantitatively, it may be reasonable to assume that the ratio of American to European Nobel Prize winners in physics and chemistry is a fair index of basic research activity. The percentage of Nobel Prizes in chemistry and physics awarded to Americans rose from 4.2% during the first decade to 9.1% during the third decade, and in the interval 1940–1950 mounted to 36.8%. This does not mean, however, that the ratio of basic to applied research has undergone any such change within the United States, but perhaps only that the total research activity has increased at a much greater rate than in Europe.

Another effect of the war was that a whole new methodology of research had to be developed because of the necessity for haste. It was no longer feasible to depend upon the success of individual investigators; instead, teams were encouraged to attack specific problems with definite objectives for each member of the team. This new, highly organized research method has been spectacularly successful in many cases, but probably not as economical of men and money as the more leisurely accomplishments of individuals. To some extent, the pattern developed by the government during the war is similar to the industrial research method which was already in the formative stage, but on a much larger scale. The justification for it in a peacetime economy, for anything other than problems very close to application, may be questionable.

At the conclusion of the second world war the contributions of scientific research had been so spectacular and had achieved such wide public recognition that the Federal government adopted a policy of continuing support on a broad basis. The Office of Naval Research was conspicuous among government agencies in the support of basic research, even though it might be entirely lacking in any forseeable application. The Atomic Energy Commission has also continued sponsorship of fundamental work, though in general the subject matter of the sponsored work has had a relation to matters of special interest to the A.E.C. Many of the other government agencies have also sponsored research, both chemical and physical, but in general it has been directed to specific problems. In the past few years, since the birth of the National Science Foundation, the ONR and other government agencies have been gradually withdrawing support of the more basic research and the National Science Foundation has in many cases been assuming these responsibilities. It is to be hoped that this far-sighted policy will be maintained and strengthened in the future. Sponsorship of only applied research by the government agencies tends to drain off talent from the basic science fields and may, in the long run, constitute a serious disservice to the nation.

Basic research in chemistry very often crosses over into fields variously known as physical chemistry, chemical physics or molecular physics. Thus, chemistry, the science, is in reality the physics of the extra-nuclear electrons of atoms and molecules. The elucidation of the Periodic Table of the elements with its many irregularities, the understanding of the various kinds of chemical bonds, and in fact all the nonempirical generalizations of chemistry owe their existence to the atomic theory of Rutherford and Bohr. On the other hand, the earliest and, in many respects, most convincing evidence for the existence of atoms was provided by chemistry rather than physics. The empirical laws of definite proportions and multiple proportions led directly to the concept of atoms. It is also true that the whole elaboration of structural chemistry occurred with relatively little aid from physics. Physics has not progressed sufficiently to be able to deal in any exact way with the complex molecular structures which are the everyday problems of organic chemists. Even in this field, however, the physical concepts derived from wave mechanics such as "resonance stabilization" and "hydrogen bond" are of great use to the chemist. Thus basic chemistry as a science merges with physics. Only in the application of the science is their a complete divergence in the subject matter.

Applied research has brought about astonishing changes in almost all fields of human activity during this century, and while it is impossible to separate out the importance of the contributions of the various sciences, chemistry has certainly accounted for a large part of the evolution. Agricultural productivity, for example, has been multiplied enormously, partly because of the development of new machines, but also because of the availability of chemical fertilizers, insecticides, soil conditioners and plant growth regulators of various kinds which were unavailable and even unknown fifty years ago. Medicine has been assisted immeasurably by the many pharmaceutical preparations, the antibiotics, vitamins and hormones. These developments are in large part responsible for increasing the life expectancy from 48 years in 1900 to 68 years, which is the present figure for Americans. At the same time, the cost of these precious substances has been constantly decreased by improved manufacturing processes.

The textile industry has experienced a virtual revolution because of the introduction of many new synthetic fibers, most of which have come within the last twenty-five years. At present the quantity of chemical fiber consumed in the United States, including viscose rayon, is twice as great as the amount of sheeps' wool. Perhaps equally important is the fact that entirely new properties are to be had in the synthetics which permit the development of new products and substantial improvements in old ones. Produced as continuous sheet rather than fiber, many of the polymers are making revolutionary changes in methods of packaging and distributing food products and other materials. These changes, together with the enormous improvements in transportation, have made it possible for the United States to surpass all other nations in the quality and selection of foods available everywhere in the country regardless of local climate or season. The polymers have, of course, made enormous contributions to the fields of protective finishes and as paints and varnishes, floor coverings, molded parts for appliances, automobiles and homes, and virtually every other industry.

Simultaneous advances in the chemistry of dyes and pigments have kept pace with the textile and molded plastic revolution so that our products are not only cheaper, but are also brighter and more colorful than ever before. Color in the graphic arts has reached a peak of perfection undreamed of in 1900, and even the amateur photographer may enjoy the remarkable achievements of the color chemists.

The profusion of new products is much too great to be enumerated, even by categories. Petroleum chemistry, as well as bringing us our modern fuels and lubricants, has made possible many of the excellent synthetic detergents, plastics and a host of other products. Silicone chemistry, still in its infancy, has already yielded improved waterproofing agents, electrical insulation materials, high temperature-stable elastomers, as well as better waxes and finishes. New advances in extractive and process metallurgy have made available as industrial materials metals which were only laboratory curiosities

a few years ago. The rapid proliferation of new materials is exemplified by the fact that over 50% of the present products of the du Pont Company have been developed within the last twenty-five years.

Even though great progress has been made in basic chemical research to date, there are still vast fields of chemistry for which little or no detailed understanding exists. Undoubtedly chemists of the future will not be content with the present state of knowledge in these areas. Virtually all organic chemical reactions proceed through the medium of free radicals and similar fragmentary molecules. Many of these substances have been detected spectroscopically and also chemically, and in some cases certain properties can be inferred, but in general very little is known of them. Catalysis, combustion, polymerization and all high-temperature reforming reactions must remain empirical until a better understanding is formed of free radicals and reaction mechanisms.

Ever since the synthesis of urea by Wöhler in 1828 the fundamental distinction between "vital force" and "inert" chemicals has been constantly receding. And yet the synthesis of life itself, even the lowly life of a virus, continues to elude the chemist. Careful and ingenious investigations have elucidated certain portions of metabolic chemistry of remarkable complexity, but the chemistry of self-duplication is still unknown. Closely related to this is the chemistry of immunology, enzymology and heredity.

The rapidly growing list of chemical elements which are becoming available as industrial materials will certainly attract the attention of applied research chemists. Organic compounds, including polymers, containing metals such as titanium, zirconium, molybdenum and tungsten, to mention only a few, show promise of providing new and interesting products. The combinations are almost inexhaustible. So long as a proper balance is maintained between basic and applied research it would appear that we need never fear a scarcity of new developments, the life-blood of an ever-expanding technology.

HOWARD O. MCMAHON

CHEMISORPTION

A substance is chemisorbed by a surface when it forms bonds with the surface atoms that are of comparable strength with ordinary chemical bonds. Chemisorption is distinguished from physical adsorption primarily by the strength of these bonds which in the case of physical adsorption are of the weaker van der Waals type. It is not always possible, however, to decide unambiguously that an observed adsorption is a weak chemisorption rather than a strong physical interaction.

Chemisorbed molecules are nearly always altered when they are chemisorbed. Hydrogen, for instance, is chemisorbed on metal surfaces as hydrogen atoms while chemisorption of hydrocarbons may result in the formation of adsorbed hydrogen atoms and hydrocarbon fragments. Chemisorption of hydrocarbons by oxide surfaces may result in the formation of adsorbed carbonium ions or carbanions plus water or surface hydroxyl complexes. Even when dissociation does not occur, the properties of the adsorbed molecule are changed in important ways by the surface.

The amount of a substance chemisorbed at a particular temperature varies with its concentration in the fluid phase. Attempts to describe the equilibrium achieved between a gas and its chemisorbed phase have yielded functional relationships between the fraction of the surface covered and the pressure of the gas. Three isotherms of general utility are described below.

The Langmuir isotherm is derived from the assumptions that adsorption occurs only if the gas molecules collide directly with the adsorption site, that the energy of each adsorbed species is the same everywhere on the surface independent of surface coverage, and that each surface site can hold but one adsorbed species. It is of the form: $\theta = aP/(1 + aP)$, where a is a function of temperature only, θ is the fraction of the surface covered (equal to the ratio of the volume of gas adsorbed to the volume of gas contained in one monolayer), and P is the pressure. A statistical derivation shows that $1/a = bkTe^{-q/RT}$, where b is a constant, k is Boltzmann's constant, T is the absolute temperature, R is the gas constant, and q is the molar heat of adsorption. Only a few examples exist which are strictly described by this isotherm, but many experimental results are in qualitative agreement with it.

The Freundlich isotherm results when the uniform surface and constant adsorption energy of the Langmuir surface are replaced by a heterogeneous surface on which the heat of adsorption varies logarithmically with coverage. It is of the form: $\theta = (a_0P)^{1/n}$ where n is a constant greater than unity, and a_0 is a constant. It has been shown that

$$\frac{1}{n} = \frac{RT}{q_m}$$

where $q_m \ln \theta = -q$, and q is the molar heat of adsorption at the coverage θ.

The Temkin isotherm, $\theta = (RT/q_0a) \ln (A_0P)$, is obtained if the heat of adsorption varies linearly with coverage. $A_0 = a_0e^{q_0 \cdot fRT}$, q_0 is the molar heat of adsorption at zero coverage, a_0 and a are constants. Because of assumptions made in its derivation, this isotherm is expected to be valid in the range, $0.2 < \theta < 0.8$ only.

Whereas the Freundlich isotherm refers unambiguously to adsorption on a heterogeneous surface, the Temkin isotherm may be shown to apply equally to a heterogeneous surface or to adsorption on a uniform surface with repulsion between adsorbed species. Each of the isotherms may be modified to account for dissociative adsorption, or for the presence of more than one kind of surface site.

The Freundlich and Temkin isotherms satisfy a larger amount of experimental data than the Langmuir isotherm. However, care must be taken before deciding that a particular isotherm describes the adsorption process. Not only must the experimental data fit the isotherm in the appropriate range, but also the heat of adsorption must vary in the appropriate way.

An additional factor complicating the application of equilibrium theories to the chemisorption process is the difficulty of obtaining true equilibrium data. Although the initial adsorption rate may be rapid at all temperatures, as is the case when hydrogen is chemisorbed by metals, the subsequent adsorption is likely to be a slow, activated process accounting for an appreciable fraction of the total adsorption. The adsorption at lower temperatures may be so slow that equilibrium is never reached, while at higher temperatures the accelerated rate may allow a larger amount of adsorption to occur before the process becomes very slow. Often only at high temperatures is true equilibrium established. Approaching equilibrium via a desorption process is predicted and found to be even slower than adsorption.

The slow chemisorption may often be described (neglecting back reaction) by the Elovich equation, an expression of the form: $-dP/dt = +d\theta/dt = ae^{-\alpha\theta/RT}$, or the integrated form $\theta = (RT/\alpha) \cdot \ln [(t + t_0)/t]$ where $t_0 = RT/a\alpha$, a and α are constants. This dependence of the rate on the fraction of the surface covered is expected in those cases where the activation energy for adsorption is a linear function of the amount adsorbed. This situation may occur if the chemisorption proceeds through a physically adsorbed state as shown by the shallow minimum in the potential energy curve in the attached figure. The heat of chemisorption varies with coverage and thus the activation energy will vary with the coverage. When the heat of chemisorption changes linearly with coverage, so also will the activation energy if the shape of the potential energy curves and the distance of their minima from the surface do not change with the amount chemisorbed.

In some cases, in addition to the strong chemisorption illustrated in the attached figure (Type "A" adsorption), an intermediate chemisorbed state (Type "C" adsorption) has been postulated, and the slow process described by the Elovich equation then would correspond to an activated transition

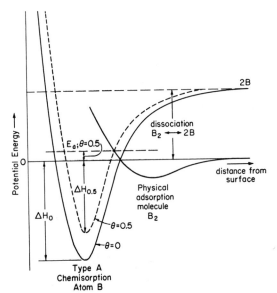

FIG. 1. Interaction between the surface and the chemisorbed species.

from the Type C state to the Type A state, the activation energy varying with the degree of coverage of Type A sites.

The desorption kinetics may be treated in an analogous manner, but the process is slower since the activation energy for desorption is the sum of the heat of adsorption and the activation energy for adsorption. The activation energy for desorption must always be larger than that for adsorption if the chemisorption is exothermic. In general this is expected because entropy usually is lost when the gaseous molecule is constrained to move on a surface.

The Elovich equation may describe other adsorption-desorption mechanisms than the one mentioned here, e.g., activated migration from surface sites of low adsorption heat and low activation energy to more energetic adsorption sites with a higher activation energy.

The study of chemisorption on clean, or rigorously reproducible surfaces has been made possible by relatively recently developed techniques of achieving ultra high vacuum routinely in the laboratory. Studies of chemisorption on carefully prepared surfaces have been carried out utilizing the techniques of flash filament desorption, contact potential difference measurements, low energy electron diffraction (LEED), field emission microscopy and field ion microscopy, as well as various combinations of these and other methods. These experiments have demonstrated that several distinct chemisorbed states may be formed by a single adsorbate on a particular surface. Thus when small amounts of hydrogen are adsorbed on a clean tungsten surface at low temperatures, desorption occurs from a single chemisorbed state with a maximum rate of desorption in the vicinity of 700°K. LEED studies show that under these conditions a characteristic, regular diffraction pattern is observed from the (100) crystal surface that is different from the clean surface pattern. When greater amounts of hydrogen are adsorbed additional lower temperature desorption peaks develop, and the LEED pattern also changes. If the tungsten surface is exposed to carbon monoxide before, or after, exposure to hydrogen changes in the flash desorption spectrum show that the CO competes strongly with hydrogen for clean-surface adsorption sites. Many other examples of this kind of phenomena are known.

Most of the common simple gases such as H_2, O_2, N_2, and CO are adsorbed rapidly by a clean metal surface.

The sticking coefficient, which may be defined as the ratio of the number of gas-surface collisions which lead to chemisorption to the total number of gas-surface collisions, often lies in the range 1.0 to 0.1 for these gases. Simple kinetic theory shows that if this ratio is 1.0, about one second at room temperature and a pressure of 10^{-6} torr will suffice to cover the surface with chemisorbed species. However, chemisorption also may occur on surfaces that are not atomically clean by displacement of a portion of the contaminating layer, or perhaps by formation of strong bonds with the layer itself.

Catalysis of chemical reactions by surfaces must proceed through the chemisorption of one of the reactants. Since the properties of chemisorbed mole-

cules are often considerably different from the properties of the normal molecule, it is not surprising that the rate of reaction of chemisorbed species may be much different from that of the normal molecules. Thus, information about the catalytic properties of a surface can yield information about the nature of the chemisorbed species. For example, the hydrogen-deuterium exchange, or ortho-para hydrogen conversion on surfaces demonstrates that only a fraction of the surface, and only a fraction of the hydrogen adsorbed participate actively in the reaction. This has been explained on the basis that more than one kind of chemisorption occurs, or that molecules adsorbed within a specific range of chemisorption energies only are active in the catalytic reaction. However, since the chemisorption energy is dependent on the degree of coverage, all adsorbed species may be available for reaction at a particular coverage. Lowering the coverage would then increase the chemisorption energy, and thus change the reactivity of all remaining adsorbate. Whether the surface is composed of equivalent sites whose energy depends on the degree of coverage, or whether the surface is heterogeneous with a few specific sites only active in the catalytic process, is a question that appears to depend on both the surface and the reaction studied. There is at present no a priori way of predicting the situation for a particular system.

Many strongly chemisorbed substances may poison the catalytic properties of a surface, presumably because strong chemisorption occurs on sites which are catalytically active, blocking access to them by the reacting species.

Other techniques have been used to study the nature of the adsorbed species. At least twelve different kinds of hydrocarbon fragments bound to metal surfaces have been identified by kinetic studies of various catalyzed reactions. These fragments range from simple monoadsorbed alkenes to pi-complexes of allylic species with surface atoms. If the adsorbed species absorb radiation, they may be studied by spectroscopic methods. Infrared, optical, EPR and NMR spectroscopy have all been applied to particular situations. In these cases the spectra obtained from the adsorbed molecules are sometimes similar to those of normal molecules, and thus allow identification by analogy.

L. E. CRATTY, JR.

Cross-references: *Activation, Adsorption, Catalysis.*

References

Trapnell, B. M. W., and Hayward, D. O., "Chemisorption," 2nd Edition, Butterworths, 1964.

Thomas, J. M., and Thomas, W. J., "Introduction to the Principles of Heterogeneous Catalysis," Academic Press, 1967.

Ehrlich, Gert, "Advances in Catalysis," (Academic Press, 1963) Vol. 14, pp. 280–427.

CHEMOTAXONOMY (BIOCHEMICAL SYSTEMATICS)

With the advent of analytical chemical techniques such as thin layer, paper, and gas chromatography, effective screening of natural products became possible. Paper chromatography, for example, revolutionized comparative studies of certain groups of compounds such as amino acids and plant pigments. Out of these advances it was not surprising that interest deepened in the possible biological applications of efficient chemical screening methods, and taxonomy (or, more broadly, systematics) represented a major application of these methods, especially the use of comparative distributions of secondary compounds. Although applications of chemistry to taxonomy accelerated in the 1940's, it is notable that periodically, either chemists or botanists had emphasized much earlier the possible utility of chemical data in taxonomic studies.

Early in the history of chemotaxonomy the major emphasis was upon broad and often rather crude surveys such as the presence or absence of alkaloids, steroidal sapogenins, cyanogenetic glycosides, anthraquinones and other classes. At this stage the taxonomic implications tended to be equally crude and often somewhat exaggerated. With the advent of paper, gas and now liquid—liquid chromatography, allowing the separation of complex mixtures, a qualitative element was added to such surveys. Initially, the protein amino acids were favorite subjects of comparative chemical studies, until it was obvious that these 20 compounds because of their nearly ubiquitous occurrence were of somewhat limited value. Later, when many additional, nonprotein, amino acids were detected from such surveys, these latter compounds (now numbering over 200) proved to be taxonomically more interesting. After amino acids, the much larger pool of flavonoid compounds (plant pigments) were studied extensively, and again many new compounds were disclosed often having quite restricted taxonomic distributions; e.g. isoflavones, found in only a few families; biflavonyls, found in gymnosperms plus *Casuarina;* rotenones found only in the Leguminosae. Simultaneously, gas chromatographic techniques combined with mass spectrometry allowed effective rapid screening of terpenoids and certain alkaloids, leading to further interesting taxonomic correlations. One of the most significant pieces of chemotaxonomic work was made possible by the disclosure of the structure of the betalain pigments by Dreiding, Mabry and coworkers at Zurich and Austin, Texas. These compounds erroneously widely regarded as nitrogenous anthocyanins, replace anthocyanins in a group of 10 families, each of which has been placed by one or more taxonomic workers in the Order Centrospermae. The betalains are completely unrelated to anthocyanins, and the chemical dissimilarity plus the mutual exclusion of the two groups of pigments has influenced some taxonomists to support the betacyanins as a phylogenetic link uniting the families of the Order Centrospermae.

Biologists interested in the field of molecular taxonomy have become cognizant of some special advantages in the study of chemistry which are not always inherent in other approaches to the study of evolution. For example, genetics and biochemistry have become interdisciplinary in the form of biochemical genetics. This field allows the possibility of approaching simple distributional data in considerably more intellectual depth. For example, one may now ask detailed questions concerning biosynthetic origins of the same or similar compounds. Recently, the basic amino acid, lysine, has been found to be synthesized by rather different routes in most fungi as opposed to bacteria and most green plants. Furthermore, certain mint species have apparently evolved a new series of mint oils following a change (presumably of genetic origin) in the position of an oxygen function in the monoterpene ring system. All these examples involve efforts to combine distributional studies with other biochemical considerations. Now, one is forced to face the question of biochemical or enzymatic homology versus parallel evolution when considering the possible systematic meaning of chemical distributions.

Effective screening techniques have also allowed the use of chemical data in the study of different types of systematic studies. Alston and Turner, for example, have utilized effectively chromatographic techniques in the study of 2-way hybrids in natural populations and even more complex natural populations of several parental species and their hybrids. They have emphasized flavonoid components of leaves and flowers in these studies and have noted that the chemical data are most effective when used in conjunction with other data, such as morphological data. One direct taxonomic application of this work was the discovery, utilizing chemical criteria, that several named species of *Baptisia* were actually interspecific hybrids.

While the lower molecular-weight secondary compounds have been traditionally the substances emphasized in chemotaxonomic studies, it is now evident that macromolecular compounds will be given increasing attention in such investigations, and indeed these compounds appear to have unique advantages in that they often reflect, more directly, individual and even sequential changes in the hereditary material. The most important advances in serology have come through the development of immunodiffusion and immunoelectrophoretic techniques. The later methods have resulted in more effective use of the criterion of protein relationships in the framework of chemotaxonomy. Despite the advantages of the serological approach, it is possible that simple electrophoretic separation of protein components will yield as many different components as does serology. Zymograms, as these patterns are called, have yielded extremely complex patterns from extracts from the seeds of both grasses and legumes. A combination of such techniques, plus standard methods of the *in vitro* characterization of enzymes such as phosphatases, other esterases, etc. will advance this form of chemotaxonomy rapidly.

The most complex and yet perhaps the most rewarding of modern chemotaxonomic techniques is the comparative study of the primary structure of similar and possibly homologous enzymes. Following the work of Sanger on insulin, and the pioneer contributions of Perutz and Kendrew on the structure of hemoglobin and myoglobin and more recently through the efforts of Ingram, a remarkable stock of data on the comparative chemistry of hemoglobins is now available. A hypothetical phylogenetic sequence for the alpha, beta, gamma and delta chains of hemoglobin is now supported by very strong circumstantial evidence, and impressive comparative structural studies of primate hemoglobin have been reported. Although primary structural data on proteins are acquired by slow and quite complex efforts, nevertheless, several enzymes have been analyzed successfully and compared among a variety of organisms. The investigations of Margoliash, Boulter and others on cytochrome c and those of Hirs, Smyth and others on ribonuclease have emphasized the comparative primary structure and additional data of this type are now being obtained for a number of organisms, both plant and animal.

Today more is known about the molecular evolution of cytochrome c than any other protein. The total structures for cytochrome c from some 39 organisms are now known and several phylogenetic trees have been erected utilizing these data. For example, the cytochrome c results suggest that fungi should be treated as a kingdom equivalent to those for plants and animals.

The latest technique, applicable to chemotaxonomy as broadly interpreted, involves the fundamental hereditary material deoxyribonucleic acid (DNA) or its transcribed copy, RNA which can now be compared directly at gross levels. Following development of techniques of denaturing (or unwinding) the helical DNA of viruses and bacteria and subsequent "hybridizations" of the derived single strands, similar techniques have been developed for use with DNA's of mammals and flowering plants. In principle, the method involves the extraction of denatured DNA strands and imbedding this single-stranded DNA in agar or trapping on nitrocellulose filters. RNA or sheared DNA from the same or another organism is used as a test against the long-strand DNA already present in the agar or on the filter. The RNA or sheared DNA has a tendency to pair with the original DNA, the affinity (or extent of pairing) presumably reflects some function of DNA similarity between the two interacting nucleic acids. At least, the reaction of DNA or RNA with its own type (homologous) always exceeds the reactions of different nucleic acids (heterologous). Although this method is still somewhat empirical and the precise significance of the binding is not as closely understood as one might expect, DNA and DNA–RNA hybridization studies represent another powerful tool for chemotaxonomy.

R. E. ALSTON*
T. J. MABRY
B. L. TURNER

Cross-references: *Biochemistry, Proteins, Amino Acids, Genetic Code, Nucleic Acids.*

* Deceased.

CHEMOTHERAPY

Chemotherapy is the treatment or prevention of a disease by administration of a chemical. As used originally by Ehrlich, the term was more restrictive, meaning the use of a chemical to bring about the cure of an infectious disease without injury to the host. The idea of preventive use of chemotherapeutic agents has now been firmly established.

The earliest chemotherapeutic agents were introduced in the sixteenth and seventeenth centuries. These were mercury for alleviation of symptoms of syphilis, cinchona (quinine) for the treatment of malaria, and ipecacuanha for the cure of amoebic dysentery.

Chemotherapy was launched as a branch of science and was given its name by Paul Ehrlich in the first decade of the twentieth century. In 1904 Ehrlich and Shiga showed that the dye "Trypan Red" had some controlling action on the course of trypanosomiasis in mice. However, this compound was inactive in man. In 1905 Breinl and Thomas found that "atoxyl" (sodium arsanilate) had value for the treatment of trypanosomiasis in humans. This led Ehrlich and his associated chemists, notably Bertheim, into an empirical search for a superior arsenical drug with a favorable therapeutic index (ratio of toxic level to effective level). Their work culminated in 1910 with the discovery of the effectiveness of arsphenamine ("Salvarsan;" "606;" 3,3'-diamino-4,4'-dihydroxyarsenobenzene dihydrochloride) as a chemotherapeutic agent for the cure of syphilis.

Ehrlich advanced the idea that chemotherapy results from the interaction of chemically reactive groups on drugs and of chemically reactive receptor groups on parasitic cells. He was the first to state that an effective drug must be of fairly low molecular weight. His co-workers, Franke and Roehl, discovered the phenomenon of drug resistance built up by repeated administration of sub-clinical doses of a drug. Cross-resistance was also observed with related compounds, and this was interpreted as chemical specificity on the part of the parasite.

The first antibacterial chemotherapeutic agents, acriflavine and proflavine, were discovered by Browning, a pupil of Ehrlich's, in 1912 and 1913. These substances were extensively used in World War I for sepsis of wounds, but were found to be without value in systemic control of bacterial infection.

The discovery of the antibacterial effectiveness *in vivo* of "Prontosil" (2',4'-diamino-4-sulfonamidoazobenzene) by Domagk in 1935 was the climax of a testing program involving more than a thousand azodyes. The goal was an antistreptococcal drug that would be active in the blood stream. Actually "Prontosil" in the test tube was without effect on bacteria. Trefouel and co-workers showed that this *in vitro* inactivity could be overcome by the addition of a reducing agent, and went on to demonstrate that in the body "Prontosil" was reduced to form sulfanilamide. Sulfanilamide, when administered alone, was equal in effectiveness to "Prontosil." It thus became apparent that sulfanilamide was the active antibacterial compound.

This discovery was followed by much activity in synthesizing "sulfa drugs." Among the approximately 7000 compounds made and tested during the period 1937–1946, several were found to be more potent than sulfanilamide, controlled more types of bacteria, or were more persistent in the blood stream. Some of the more important "sulfa drugs" discovered during this period were sulfapyridine, sulfaguanidine, sulfathiazole, sulfadiazine, sulfamerazine and sulfamethazine.

In 1940 Woods found a yeast extract which antagonized the antibacterial action of sulfanilamide. It was shown later that the active antagonist was *p*-aminobenzoic acid (PAB), a growth factor for most bacteria. Fildes developed this discovery into a useful theory having to do with the inhibition of enzymes. He stated that sulfanilamide, because of its steric and chemical similarity to PAB could be taken into the enzyme structure in place of PAB, but the enzyme, so modified, would be unable to function normally. The generalized concept of structural analogs inhibiting growth of organisms by competition with metabolites for the enzymes has become known as the Woods-Fildes Hypothesis, or the Anti-Metabolite Theory. (See **Structural Antagonism**).

The battery of useful chemotherapeutic agents has been greatly expanded by the discovery of the antibiotics, substances with antibacterial activity elaborated by certain fungi and bacteria. Penicillin was discovered by Fleming in 1928. Isolation on a practical scale and clinical utilization were significantly advanced by the work of Florey (1941). (See **Antibiotics**).

One of the most fascinating chapters in the development of chemotherapeutic agents has been the discovery of new antimalarial agents. Although quinine had been used to control malaria since the seventeenth century, it was ineffective against some forms and was available only in certain tropical areas. The first notable synthetic antimalarial was pamaquin ["Plasmoquine;" 8-(4'-diethylamino-1'-methylbutylamino)-6-methoxyquinoline]. Pamaquin was obtained in a synthesis program in 1926 by Schulemann, the object of which was to put various basic side chains on the quinoline portion of the quinine molecule; it was too toxic to be of more than limited usefulness. The German workers continued searching, and after testing over 12,000 compounds, came up in 1932 with mepacrine ["Atabrine," "Atebrin," 6-chloro-9-(4'-diethylamino-1'-methylbutylamino)-2-methoxyacridine], which was more successful than quinine in destroying the blood forms of malaria but which imparted a yellow stain to the skin of the user. During World War II a large-scale search by U.S. investigators for superior antimalarials resulted in the development of chloroquine [7-chloro-4-(4'-diethylamino-1'-methylbutylamino)-quinoline, first synthesized in Germany in 1934] and pentaquine [6-methoxy-8-(5'-isopropylaminopentyl)-quinoline]. Pentaquine was similar to pamaquin in action, but the therapeutic index was higher. Chloroquine was found to be more potent than mepacrine and did not stain the skin. Recently chloroquine has been found promising in the treatment of arthritis. (See **Antimalarials**).

Proguanil ("Paludrine;" chlorguanide; 1-p-chlorophenyl-5-isopropylbiguanide) was developed in 1945 by Curd, Davey and Rose after testing almost 5000 compounds in a search for a less toxic, less expensive, nondyestuff antimalarial molecule. It was decided to avoid quinoline and acridine nuclei, since these were foreign to the animal organism. Most of the early compounds examined were pyrimidine derivatives. Although proguanil itself does not contain a heterocyclic ring, it has been shown that the body converts it into a triazine derivative, 2,4-diamino-1-p-chlorophenyl-6,6-dimethyl-1, 6-dihydro-1,3,5-triazine, which is the active antimalarial agent.

Other useful chemotherapeutic agents are 4,4'-diaminodiphenyl sulfone and derivatives for treatment of leprosy and tuberculosis, isonicotinic hybrazide for tuberculosis, p-aminosalicylic acid for tuberculosis, and various antimony compounds for use in treating leishmaniasis, trypanosomiasis and amoebiasis.

There have been advances in the prophylactic control of systemic viral diseases using 1-adamantanamine hydrochloride against influenza and N-methylisatin β-thiosemicarbazone to control smallpox. In the important field of the chemotherapy of cancer, temporary remission of symptoms of leukemia has been obtained with nitrogen mustards, 6-mercaptopurine and 8-azaguanine.

JONATHAN W. WILLIAMS

CHLORATES AND PERCHLORATES

Chlorates

Chlorates are manufactured commercially by electrolysis of aqueous solutions of the chlorides. Sodium, potassium and barium chlorates constitute the most important chlorate compounds. The chemistry of electrolysis to form chlorates involves first the formation of chlorine, which then produces mixtures of HOCl and the metal hypochlorite. These then react to give the chlorate and chloride. Further oxidation yields the perchlorates. Of the two commercial processes the alkaline electrolytic process has a theoretical 100% efficiency and the acid one is characterized by high electrochemical efficiency. Other metal chlorates are prepared by double decomposition from barium chlorate and the metal sulfate.

The most characteristic property of chlorates is their oxidation potential caused by their relatively easily effected decomposition to the chloride and free oxygen. Mixtures of chlorates (in appropriate "oxygen balance") with organic materials, metals (Al, Mg, As, Cu, etc.), carbon phosphorous, and sulfur are generally powerful and more or less dangerous explosives. Chloric acid solutions are very reactive and are able to dissolve various metals (Cd, Cu, Fe, sodium amalgam, etc.) and decompose organic materials rapidly, often with flame. Esterification of AcOH in EtOH is greatly accelerated by $Ca(ClO_3)_2$.

The oxidation potential is responsible for the strong toxic effects of chlorate. Thus chlorate penetrates the erythrocrytes and attacks the hemoglobin, converting it to methemoglobin through an autocatalytic process in which reduction of the chlorate is a necessary step. Mere traces of chlorate in the soil are injurious to wheat and other plants.

While relatively high sensitivity toward explosion is achieved with chlorates by admixture with combustibles, the decomposition of the pure metal chlorate to the chloride and free oxygen is itself exothermic and may lead to explosion. Thus $KClO_3$ and $NaClO_3$ generate 16 kilocalories per mole in decomposition, and the spontaneous explosion of potassium chlorate is not uncommon. Decomposition of chlorates may be catalyzed or accelerated by contaminants of various types, for example, manganese dioxide and even fine sand or powdered glass.

The chlorates are more powerful oxidizing agents than the chemically similar bromates and iodates. Indeed, $KClO_3$ will oxidize iodine chloride to KIO_4 with evolution of chlorine and will oxidize bromine to the bromate stage. (Chlorine is not oxidized by iodates or bromates, but iodine is converted to iodate by chlorine). The power of chlorates as oxidizing agents is further illustrated by the fact that $KClO_3$ will oxidize ammonia to KNO_3 with the liberation of KCl and chlorine. It must therefore be a stronger oxidizing agent even than the corresponding nitrate.

Monovalent metal chlorates do not add water of crystallization, but the bivalent ones add from one to six molecules of water per molecule of chlorate. Tervalent chlorate salts, while almost unknown, probably have considerable water of crystallization in the solid state. On heating at sufficiently low temperatures, the hydrated salts lose their water of crystallization without further decomposition showing that they are salts of a strong acid. In all these respects the chlorates are very similar to the corresponding nitrates.

The primary uses of chlorates are in the manufacture of explosives, fireworks and matches. One of the oldest and still useful detonators (the fuse cap) is based on mixtures (90/10 and 80/20) of mercury fulminate and potassium chlorate. These detonators are more powerful, cheaper and safer than pure mercury fulminate. Chlorates are also used in insecticides and for medicinal purposes.

Like perchlorates, nitrates and other oxidizing agents, chlorates in combination with combustibles of all types constitute a serious fire and explosion hazard. Such mixtures are frequently quite sensitive to flame, shock and friction and must thus be handled cautiously.

Perchlorates

Chlorates and perchlorates are made commercially from the chlorides by electrolysis or from perchloric acid. Perchlorates are intermediates in the decomposition of chlorates and form also by oxidation of chlorates. Potassium and sodium perchlorates are found in small percentages in Chilean nitrates.

Perchloric acid forms normal salts [$M(ClO_4)_x$]. The perchlorates are generally more stable than the corresponding chlorates. Moreover, pure anhydrous perchloride acid exists whereas anhydrous chloric acid does not, although pure perchloric acid is a dangerous explosive, extremely reactive and injurious to the skin. Perchlorates decompose upon heating to the chlorides and oxygen. Carbonates accelerate the thermal decomposition of potassium and sodium perchlorates which decompose rapidly at 450 to 550°C. The constant-boiling perchloric acid (72.4%) is not an explosion hazard. Aqueous perchloric acid solutions are valuable acid reagents. Dilute perchloric acid is a powerful oxidizing agent. An important use of perchloric acid and perchlorates is in their reagent value where they constitute an extremely useful series of compounds.

As with the chlorates, "oxygen-balanced" mixtures of perchlorates with combustible materials are powerful though often dangerous explosives. This is a result simply of the high-temperature decomposition to form free oxygen, which reacts with the combustible. Perchloric acid is used extensively in destroying organic materials. Serious explosions have sometimes resulted in this application and many warnings and recommendations have been sounded regarding the use of perchloric acid with organic materials, bismuth and other combustibles.

Ammonium perchlorate (which is itself explosive) and other perchlorates have been proposed for use in commercial explosives, but they have not in general met with success owing to the adverse influence of perchlorates on stability and sensitivity. Ammonium perchlorate in percentages as small as 10% appreciably sensitizes ammonium nitrate-combustible mixtures and adds considerable explosive strength. But it may also promote instability, probably as a result of hydrolysis to produce perchloric acid. Metal perchlorates react with ammonia to produce metal ammine perchlorates [$M(ClO_4)_x \cdot yNH_3$] which are powerful but unstable explosives. The metal ammine perchlorates with the smallest cations have the largest heat of formation and are the most stable. Alkaline earth metal ammine perchlorates are more stable than the alkali metal ammine perchlorates. Ammonium perchlorate is the basic (oxidizer) ingredient of "composition" propellants, today among the most important solid rocket propellants.

Perchloric acid is an acetylation catalyst for cellulose and glucose and is used in the preparation of cellulose fibers. Also $Mg(ClO_4)_2$ is a catalyst in the preparation of cellulose ethers and ether-esters. The perchlorates of Be, Al, Mg, Zn, and Cu are used as catalysts in the preparation of artificial filaments, coated fabrics and plastic materials.

Perchlorates, like chlorates, are in general toxic. Sodium perchlorate appears in the urine of man within ten minutes and is largely eliminated in five hours. Seeds germinate in 0.2% but not in 0.5% solutions of sodium chlorate. A 2–4 g/kg dose by mouth is lethal to rabbits. Disease in corn and other plants, traced to the presence of small percentages of $KClO_4$ and $NaClO_4$, was encountered in the early days of use of Chilean saltpeter as a fertilizer.

The use of chlorates and perchlorates in commercial explosives, while prominent at one time, fell into disfavor owing to their great hazards caused by spontaneous ignition and often resulting in disastrous explosions. With the advent of slurry explosives, with their high water contents, it seemed reasonable to reinitiate the practical application particularly of the perchlorates as oxidizers for the slurry explosives. The phlegmatizing properties of water have largely removed the spontaneous ignition hazards of the perchlorate and chlorate when used in slurry explosives. The advantages of perchlorate slurry explosives in particular include increased density, higher bulk strength, and better performance. Sodium perchlorate contains the highest available oxygen concentration of any practical oxidizer for slurry explosives. Certain slurries made with sodium perchlorate in place of ammonium nitrate appear particularly attractive for secondary blasting and in specialty applications such as metal forming requiring propagation in thin films. The characteristic features of sodium perchlorate slurries are: (a) They are easily formulated on site, e.g., by simply mixing two stable, nonexplosive liquids, and (b) they are able to propagate in amazingly small sizes and thin films. A practical sodium perchlorate slurry is now in commercial use for secondary blasting in open pit mining.

M. A. Cook

Cross-references: *Explosives, Chlorine.*

CHLORIDES

The simple chlorides include that group of compounds which may be regarded as formal derivatives of hydrogen chloride, HCl, and which may be represented by the general formula MCl_n where M represents a metal, a nonmetal, an organic or inorganic radical, or a complex. Strictly speaking, M should have a lower electronegativity than chlorine. Thus, compounds having a great range of physical and chemical characteristics are included in this group. If M represents a metal of very low electronegativity, such as one of alkali metals, the resulting chloride is a typical ionic compound with a crystal lattice consisting of an indefinitely extended three-dimensional array of positive and negative ions but containing no finite molecular groups. Silver, thallous, and cuprous chlorides belong to this group also. A similar situation obtains when M represents a complex ion such as NH_4^+ or $Co(NH_3)_6^{+++}$, except that the positive ion positions in the ionic lattice are occupied by complex rather than simple ions.

The chloride ion is, however, sufficiently polarizable that when M becomes a metal of somewhat higher electronegativity than the alkali metals, covalent bonding may begin to play an important role in the crystal lattice. This effect manifests itself by the appearance of more or less covalent complexes indefinitely extended in one dimension (chain-type structures) or in two dimensions (layer-type structures). An example of the chain-type structure is palladous chloride ($PdCl_2$) which has indefinitely extended chain structures of the type

in its crystal lattice. Examples of chlorides having layer-type structures are the chlorides of Cd(II), Fe(II), Ni(II), Mg, Zn, Mn(II), Cr(III), and Fe(III). As the differences in electronegativity between M and Cl decrease, the trend toward covalent bonding increases. This factor, combined with steric and other considerations, causes many chlorides to have molecular type lattices composed of finite, discrete molecules. The configurations of these molecules depend upon the bonding orbitals available on the atom M. In some instances the molecular formulas do not correspond to the empirical formulas. Thus, aluminum chloride vapor is composed of Al_2Cl_6 molecules. Ferric chloride vapor is likewise composed of dimers, Fe_2Cl_6.

The molecular type chlorides are generally characterized by weaker crystal lattices, lower melting points and higher volatilities than the previously mentioned types of chlorides. Many molecular chlorides exist in the liquid phase at room temperature. The molecular chlorides include the organic chlorides such as CH_3Cl, C_6H_5Cl, and $CHCl_3$.

There are many chlorides in which more than one element is combined with chlorine in a given compound. Such *complex* chlorides may be represented by the general formula $A_xB_yCl_z$. This includes the situation where A and B refer to the same element but in different oxidation states; e.g. $TlCl_2$ is really $Tl^ITl^{III}Cl_4$. The variety of complex

halides is very great. There are four chief structural possibilities for the complex halides:

(a) A, B, and X forms an infinite 3-dimensional array of ions in which no finite or indefinitely extended chain-type or layer-type complex may be distinguished. There are few examples of this among the complex chlorides but it is quite common for the fluorides.

(b) B and X form layer complex anions. Again no structures of this type have been found among the complex chlorides.

(c) B and X form chain complex anions. Examples include NH_4CdCl_3, $K_2HgCl_4 \cdot H_2O$, $K_2SnCl_4 \cdot H_2O$, and $RbCdCl_3$.

(d) B and X form finite complex anions of the formula $[B_yX_z]^{(ny-z)}$ where n is the oxidation number of B. There are a great many examples of these among the chlorides.

The preparatory methods and reactions of the various chlorides cover an exceedingly wide range. It may be pointed out, however, that since chlorine in virtually all the chlorides is in its lowest oxidation state, the chlorides are in general susceptible to the action of strong oxidizing agents to yield molecular chlorine or in some instances even higher oxidation products.

Harry H. Sisler

CHLORINATED HYDROCARBONS

Chlorinated hydrocarbons are hydrocarbons in which one or more of the hydrogen atoms has been replaced by chlorine. Two general methods, chlorination and hydrochlorination, are used to prepare them.

Chlorination. Chlorine or chlorinating agents react with hydrocarbons either by substitution or by addition.

In most cases chlorination by *substitution* involves the reaction of elemental chlorine with hydrocarbons. The chlorination of methane is a good example of this process:

$$CH_4 + Cl_2 \rightarrow CH_3Cl + HCl$$

$$CH_4 + 2Cl_2 \rightarrow CH_2Cl_2 + 2HCl$$

$$CH_4 + 3Cl_2 \rightarrow CHCl_3 + 3HCl$$

$$CH_4 + 4Cl_2 \rightarrow CCl_4 + 4HCl$$

The composition of the reaction products varies with the conditions. The reactions are highly exothermic so good temperature control is essential. Other chlorinating agents besides chlorine can be used such as phosphorus pentachloride, sulfuryl chloride, thionyl chloride, sulfur dichloride, and antimony pentachloride. Some of these chlorinating agents also act as catalysts in chlorination with elementary chlorine.

In chlorination by *addition,* chlorine reacts with unsaturated hydrocarbons to form dichloro derivatives. The addition of chlorine to ethylene to form ethylene dichloride is a good example. Chlorine can also react with unsaturated chlorine compounds by addition; the formation of 1,1,2-trichloroethane from vinyl chloride occurs in this way.

A third method of chlorination has been developed, called oxyhydrochlorination. This is simply a variation of the chlorination by substitution process in which the by-product hydrogen chloride from this process is catalytically converted by air or oxygen in the vapor phase to additional chlorine, which is then returned to the process. In practice the hydrocarbon chlorination and the hydrogen chloride oxidation are carried out in a single simultaneous operation. This process results in some oxidation of the hydrocarbon, but the improved utilization of chlorine more than offsets the hydrocarbon losses.

Hydrochlorination. Hydrogen chloride can react with a number of organic compounds by *substitution*. In the majority of cases, however, the substitution is with alcohols. The reaction in the case of methyl chloride is as follows:

$$CH_3OH + HCl \longleftrightarrow CH_3Cl + H_2O$$

Since this reaction is reversible it is made to proceed to the right by removing the methyl chloride as fast as formed and by maintaining a high concentration of methanol and hydrogen chloride. Ethyl chloride and butyl chloride are made industrially by this process from their respective alcohols. Hydrogen chloride will also react by substitution with other organic compounds such as amines, ethers, esters, acids, aldehydes and ketones to give chlorinated hydrocarbons. Due to economic considerations, these reactions are seldom used industrially.

The *addition* of hydrogen chloride to unsaturated compounds to produce chlorinated hydrocarbons is of considerable importance industrially. For example, hydrogen chloride will add to ethylene to form ethyl chloride and to acetylene to form vinyl chloride.

Physical Properties. The stepwise substitution of chlorine for hydrogen in hydrocarbons in general increases the boiling point, melting point, density and refractive index. One outstanding exception to this generalization is the decrease in the melting point in going from benzene to monochlorobenzene. A reduction in molecular symmetry is believed responsible for this anomaly. The substitution of one hydrogen by chlorine in a hydrocarbon generally produces only a slight drop in flammability. Further substitution then produces a rapid drop. Fully chlorinated derivatives are nonflammable. The chlorinated hydrocarbons are generally good solvents for fats and oils and poor solvents for water and polar substances.

Chemical Properties. The chlorinated hydrocarbons are moderately reactive chemicals and show a wide variation in reactivity with structure.

Pyrolysis. The position of the chlorine in aliphatic chlorinated hydrocarbons has a marked effect in pyrolysis. The stability decreases in proceeding from primary to secondary to tertiary. The primary chlorides decompose only at temperatures in the range of 400–500°C. The completely chlorinated derivatives have about the same stability. The unsaturated chlorinated hydrocarbons show low stability at low chlorine content and high stability when completely chlorinated. The aromatic chlorides all show remarkable heat stability. For example a bright red platinum wire is required to decompose chlorobenzene. The chlorinated alicyclic hydrocarbons have somewhat the same stability as the primary saturated paraffin derivatives and

in both cases hydrogen chloride is nearly always produced as a decomposition product.

Hydrolysis. The rate of hydrolysis of saturated aliphatic chlorinated hydrocarbons increases from primary to tertiary derivatives. The primary chlorides are converted chiefly to alcohols. Secondary chlorides give some alcohols but mostly olefins. Tertiary chlorides give olefins almost completely. The chemical properties of the saturated aliphatic derivatives differ strikingly from the aromatic and unsaturated derivatives. This is illustrated in rates of hydrolysis. For example methyl chloride and benzyl chloride are readily hydrolyzed by boiling dilute sodium hydroxide solution, whereas chlorobenzene requires solid caustic at about 300°C. Any chlorine adjacent to a double bond shows increased chemical stability. In the methane series methylene chloride is considerably more stable than the other members. This increased chemical stability of methylene chloride, trichloroethylene and tetrachloroethylene has stimulated their use as industrial solvents. They show less hydrolysis, corrosion of metals and toxicity than such old chlorinated solvents as chloroform and carbon tetrachloride.

Reactions with amines. The reaction of chlorinated hydrocarbons with ammonia and amines is of considerable importance in synthetic organic chemistry. The usual reaction is one of addition, forming an amine chloride. The reaction is illustrated with methyl chloride and ammonia:

$$CH_3Cl + NH_3 \rightarrow CH_3NH_3Cl$$

If alkali is present the free amine is formed,

$$CH_3NH_3Cl + NaOH \rightarrow CH_3NH_2 + NaCl$$

which can react with more methyl chloride forming higher amines. The ultimate product, when excess alkali is present is tetramethyl ammonium hydroxide.

These reactions occur quite readily with most primary monochlorinated hydrocarbons. Secondary and tertiary derivatives react with ammonia and amines but generally by hydrogen chloride removal and olefin formation. Some polychlorinated hydrocarbons will react with ammonia to form amines if one chlorine is singly attached to carbon. The formation of ethylene diamine from ethylene dichloride and ammonia is an example.

Chlorinated aromatics will react with ammonia at high temperatures and pressures to form amines. For example chlorobenzene will react with ammonia at about 200°C, in the presence of a catalyst of copper oxide and chloride to form aniline.

Reactions with metals and metal salts. Primary monochlorinated hydrocarbons in general react with metals such as sodium, potassium, and lithium to form hydrocarbons, by a Wurtz reaction,

$$2RCl + 2Na \rightarrow R—R + 2NaCl$$

This reaction proceeds through the intermediate formation of a metal salt of the hydrocarbon,

$$RCl + 2Na \rightarrow RNa + NaCl$$

followed by reaction of this salt with more of the chloride

$$RNa + RCl \rightarrow R—R + NaCl$$

The primary monochlorinated hydrocarbons react quite readily with several types of metal organic compounds. Some examples are the reaction with sodium alcoholates to form ethers (Williamson synthesis of ethers):

$$RCl + NaOR' \rightarrow ROR' + NaCl$$

the reaction with the sodium salt of organic acids to form esters,

$$RCl + NaOOCR' \rightarrow ROOCR' + NaCl$$

and the reaction with sodium cyanide to form nitriles,

$$RCl + NaCN \rightarrow RCN + NaCl$$

Uses. The largest use for *ethylene dichloride* is as a coupling agent in antiknock fluid. *Ethyl chloride* has two important uses—the manufacture of tetraethyl lead and of ethyl cellulose. *Vinyl chloride* monomer, which includes some vinylidene chloride, finds its major use in vinyl plastics. The major captive uses for chlorobenzene are in the manufacture of phenol, aniline and nitrated chlorobenzenes.

Trichloroethylene and *tetrachloroethylene* are used almost entirely as solvents for dry cleaning, metal degreasing and extraction of fats and oils. The major use for *carbon tetrachloride* today is as a raw material for the manufacture of fluorochloromethanes, which in turn are used as refrigerants and aerosol propellants. It has been used in the past as a dry cleaning solvent and fire extinguishing agent, but better and less toxic materials are now available for both these applications. *Methylene chloride* is used mainly in paint and varnish removers. *Methyl chloride* has three important uses, one as a catalyst solvent in Butyl rubber production, another as a reagent in silicone production and the third as a methylating agent. *Chloroform* is used as a solvent and as a chemical reactant.

Benzene hexachloride is produced almost entirely for its gamma isomer content, which is an excellent insecticide and miticide. *Benzyl chloride* is used primarily as a chemical reactant. The chlorinated paraffins are important as plasticizers, mildewproofing and flameproofing agents.

Toxicology. The toxicity of chlorinated hydrocarbons varies widely. Dichlorodifluoromethane is considered nontoxic, whereas carbon tetrachloride is rated as very toxic. Some of the relatively nontoxic chlorinated insecticides, such as DDT, endrin and dieldrin have been found to build up in food chains to potentially toxic levels, due largely to their low solubility in aqueous systems and their high solubility in fatty tissue.

P. J. Ehman

Cross-references: *Solvents, Hydrocarbons.*

CHLORINE

History. Chlorine was discovered in 1774 by K. W. Scheele, as a reaction product of manganese dioxide and hydrogen chloride. Scheele noted the solubility of this new substance in water, its bleaching action on organic matter and reactivity with metals, including gold. However, to Sir Humphry F. Davy goes the credit of establishing in 1810 the elementary nature of chlorine to which he gave its present name, derived from the Greek word for greenish-yellow.

Industrial exploitation of this chemical was soon undertaken by Berthollet (1789) after finding that it can be fixed in stoichiometric amounts by an alkaline solution and that the latter thus has bleaching properties. In 1798, Charles Tennant replaced the alkali lye used by Berthollet with cheaper milk of lime; subsequently, on finding the possibility of absorbing chlorine in slaked lime, he originated the bleaching powder industry.

The industrial production of chlorine was initially achieved by adding sulfuric acid to a mixture of sodium chloride and manganese dioxide, to perform essentially the same reaction discovered by Scheele. A substantial economic improvement was reached about 1870 by Deacon's process, whereby chlorine is liberated from hydrogen chloride by treating the latter with atmospheric oxygen at about 450°C in the presence of a catalyst such as copper sulfate.

The invention of electromagnetic machines for producing electric power on an industrial scale made possible the development of electrochemical processes that, at the turn of the past century, gradually displaced the other methods of chlorine production.

Natural Occurrence. Chlorine abundance in the lithosphere, the ten-mile thick crust of the earth, is estimated at about 0.045% by weight. Due to its strong chemical affinity for other elements, it never occurs in the free state, except as a minor constituent of the gaseous output from volcanic eruptions, which, however, are often quite rich in hydrogen chloride. Under such conditions the presence of free chlorine can be easily explained as a consequence of thermal dissociation of chlorides at the high temperatures prevailing in the volcanic phenomenon.

Among the most common minerals consisting of chlorides are rock salt, or halite (NaCl), sylvite (KCl), carnallite ($MgCl_2 \cdot KCl \cdot 6H_2O$). Beside being dispersed in the lithosphere, chlorine salts are also dissolved in the hydrosphere. The average chlorine content in sea water, as Cl^- ion, is nearly 2%, although local departures from this value can be quite marked in the several oceans. Since the mass ratio of hydrosphere to lithosphere is about 1:13, the over-all chlorine abundance averages 0.19%.

In the hydrosphere as well as in the lithosphere sodium chloride NaCl is by far the most plentiful alkali metal salt. The average concentration in sea water is approximately 2.6%.

Physical Properties. Chlorine, symbol Cl, at normal temperature and pressure is a greenish-yellow gas. Its molecule is diatomic, so that the free state is usually designated by the symbol Cl_2. The atomic number 17 corresponds to the electronic configuration $1 s^2 2 s^2 2 p^6 3 s^2 3 p^5$. The wide departure of the atomic weight, 35.457, from an integer number is mostly due to the coexistence of the two isotopes Cl^{35} and Cl^{37}, which are present in naturally occurring chlorine compounds to the extent of 75.4 and 24.6%, respectively.

The density of chlorine gas is nearly 2.5 times that of air. This explains why any amount released in the atmosphere tends to form a cloud hovering close to the ground and hard to dissipate. Its physical properties have the following values[3]:

Boiling point   −34.5°C
Compressibility 0.0118% per unit vol. per
  of liquid   1 atm pressure at 20°C
Critical density 573 gpl
Critical
  pressure   78.525 atm
Critical
  temperature 144°C
Density   gas 3.214 gpl at 0°C,
              (1 atm); liquid 1468 gpl at 0°C
Freezing point −100.98°C
Heat of fusion 22.9 g-cal/g.
Latent heat of
  vaporization 68.8 cal/g.

Solubility in Water. The solubility at different temperatures under atmospheric pressure changes from 9.97 gpl at 10°C to 1.27 gpl at 90°C and becomes nil at boiling point of water. These values include, besides the chlorine dissolved in the molecular state, the amount required for the hydrolysis products in equilibrium with the former, according to the reaction

$$Cl_2 + H_2O \rightleftarrows HClO + H^+ + Cl^-$$

On account of hydrolysis, the dependence of chlorine solubility on pressure does not obey Henry's law.

The solubility in saturated sodium chloride is 1/7 to 1/8 of that in water at the same temperature and is further decreased by acidifying the solution. However, chlorine solubility is substantially higher in muriatic acid, and at 25°C approaches 10 gpl in 25% HCl; this fact is explained by assuming the existence of the compound HCl_3.

When chlorinated water or brine are cooled down to near 0°C, crystals of hexahydrate, $Cl_2 \cdot 6H_2O$, and octahydrate, $Cl_2 \cdot 8H_2O$, are formed. In an open vessel, i.e., under atmospheric pressure, chlorine hydrate dissociates into gaseous chlorine and water vapor above +9.6°C and into gaseous chlorine and ice below −0.24°C. Accordingly, chlorine solubility in pure water attains its maximum at +9.6°C, if chlorine hydrate is present as a solid phase.

The solubility of gaseous chlorine in organic liquids, notably chlorinated derivatives, is generally high. For instance, in carbon tetrachloride at 0°C and 1 atm, chlorine solubility is 250 gpl.

Liquid Chlorine. In the liquid state chlorine is a pale-yellow substance, whose color intensity decreases with a lowering in temperature.

In modern industrial plants the chlorine gas from the electrolytic cells is liquefied at the daily rate of hundreds of tons, either by compression at several atmospheres while keeping the temperature close to the ambient value, or by deep cooling with "Freon," under pressures slightly above 1 atm.

The mean specific heat at constant pressure and within the range from 0°C to 24°C is 0.2262 cal/g.

Liquid chlorine is a good solvent for a number of compounds, notably chlorides, such as CCl_4, $TiCl_4$, $SnCl_4$, $PbCl_4$, $POCl_3$, $AsCl_3$, SCl_2. Most of oxygenated organic compounds, notably alcohols, are soluble, with formation of addition compounds.

Oxidation States. Chlorine is a nonmetal that occupies the second place, after fluorine, in the group of halogens (salt producers). Its proximity to the right corner of the periodic table points out its outstandingly electronegative nature, that is, its marked electronic affinity, or oxidizing power, which is next only to that of fluorine and oxygen and imparts to its chemical bond with most other elements the total or partial ionic character required to form stable compounds.

As a consequence of the fact that chlorine and nitrogen have the same electronegativity value, the molecule of nitrogen trichloride, NCl_3, is kept in metastable equilibrium by a purely covalent bond; accordingly, this oily substance is so unstable that it can be made to decompose with explosive violence even by a moderate shock. On the other hand, although belonging to the same group, chlorine and fluorine are sufficiently apart on the electronegativity scale to form the stable compounds ClF and ClF_3, while the slight defference in electronegativity between chlorine and oxygen allows them to form several compounds, most of which are unstable, due to the weakness in ionic character presented by such bonds.

The oxidation states that chlorine may acquire, when combined with hydrogen and oxygen, range from −1 to +7, as follows:

− 1 Cl^-, HCl	+ 4 ClO_2
0 Cl_2	+ 5 ClO_3^-, $HClO_3$
+ 1 HClO, Cl_2O	+ 6 no compound
+ 2 no compound	+ 7 ClO_4^-, $HClO_4$, Cl_2O_7
+ 3 ClO_2^-, $HClO_2$	

Metallic Compounds. The greatest stability is shown by the oxidation state −1, in particular when the chemical bond structure is prevailingly ionic in character, as in alkali metal chlorides (LiCl, NaCl, KCl). In the molten state these salts are dissociated into alkali metal ions (Li^+, Na^+, K^+) and chloride ions, Cl^-, thus exhibiting electrolytic conductivity. The ionic dissociation of these chlorides, when dissolved in water, is almost complete, so that they are typical "strong electrolytes." Hydrogen chloride, HCl, is a colorless gas with a very pungent odor; it is very soluble in water, and its aqueous solution is called hydrochloric or muriatic acid. It is a strong acid, forming salts that are called *chlorides*. Metal chlorides are generally characterized by crystalline structure and high melting point (for example, NaCl 800°C; KCl 790°C; $FeCl_3$ 282°C).

Oxygen Compounds. Hypochlorous acid, HClO, is formed from free chlorine dissolved in water, as the result of a hydrolysis process, whereby Cl atoms are partly reduced from the original state +0 in the free molecule, and partly oxidized with the consequent production of hydrochloric acid and hypochlorous acid:

$$Cl_2 + H_2O \rightleftarrows 2H^+ + Cl^- + ClO^-$$

The bleaching and sterilizing action of chlorine in water, so widely used for sanitation purposes, is due to this buildup of hypochlorous acid, which can easily react with organic matter by giving up oxygen in atomic form:

$$ClO^- \rightarrow Cl^- + O^*$$

If the solution to which chlorine is added contains an alkali, such as sodium hydroxide, the reaction may be written as follows:

$$Cl_2 + 2OH^- \rightarrow Cl^- + ClO^- + H_2O$$

The reaction product is a bleaching and sterilizing solution of sodium hypochlorite, NaClO, with a substantially higher concentration of "available chlorine" (i.e., chlorine capable of releasing an equivalent amount of atomic oxygen) than is possible to achieve with simply chlorinated water.

Bleaching powder is produced by passing chlorine gas over slaked lime, i.e., $Ca(OH)_2$, so as to form a compound whose average hypochlorite content corresponds to the reaction:

$$Ca(OH)_2 + Cl_2 \rightarrow Ca(ClO)Cl$$

All these processes for obtaining of hypochlorite must be conducted under carefully controlled conditions of temperature and reactant concentration, to prevent disproportionation of hypochlorite to the more stable chlorate:

$$3ClO^- \rightarrow ClO_3^- + 2Cl^-$$

This reaction can be most simply achieved by electrolyzing a sodium chloride or a postassium chloride solution at relatively high temperature without taking any precaution to keep the hydroxide forming at the cathode separated from chlorine developing at the anode, as would be necessary if these two products were to be obtained separately.

Perchlorates, such as $KClO_4$, are obtained by electrolyzing a chlorate solution, and making use of an anodic material, such as platinum, presenting a sufficiently high overvoltage to oxygen discharge, so as to allow the anodic oxidation of chlorate ion to perchlorate ion.

Compounds with Nonmetals and Metalloids. In most other chlorine compounds the covalent bond character prevails and the intermolecular forces of attraction are comparatively weak, with correspondingly low boiling points and melting points.

A typical example is offered by carbon tetrachloride, CCl_4, melting at $-23\,°C$ and boiling at $77\,°C$.

Organic Chlorine Compounds. The high electronegativity value of chlorine enables it to form very stable compounds with a host of saturated and unsaturated hydrocarbons, as well as with more complex organic substances.

Chlorine Industry. Industrial chlorine production has been achieved for more than half a century by electrolysis of alkali metal chlorides. Well over 90% of this production is carried out by the electrolysis of sodium chloride solution, which yields caustic soda (sodium hydroxide or NaOH) as a co-product. As chemical raw materials, chlorine and caustic soda individually are exceeded only by salt, sulfuric acid and soda ash in the quantities consumed.

In a minor proportion, chlorine is also manufactured by fused salt electrolysis of sodium chloride and magnesium chloride; in such cases it is to be considered as a co-product in the electrowinning of sodium and magnesium (q.v.).

Sodium chloride or potassium chloride electrolysis in aqueous solutions can be achieved by making use of two different cell types, involving different processes[2,4]: they are the diaphragm cell and the mercury cell process. In both of these the anodic reaction is the same and proceeds under equal conditions:

$$2Cl^- \rightarrow Cl_2(gas) + 2e^-$$

The chloride ion Cl^- gives up its excess negative charge (electron) with the consequent formation of free radicals Cl; these combine by pairs to build up chlorine molecules that evolve in the gaseous state. The cathodic reaction, however, is accomplished in the diaphragm cell in quite different steps than in the mercury cell and establishes the main point of distinction between the two processes, which are briefly described in some of their most modern and outstanding examples.

Diaphragm Process. (Fig. 1) The anode assembly includes a number of parallel rows of vertical anode plates; a steel structure makes up the cathode and is formed by a set of double screens extending vertically between each pair of anode rows. The two elements forming a double screen are a fraction of an inch apart and are covered on their outer surfaces facing the anodes with an asbestos diaphragm. Each double screen thus builds up a narrow chamber or catholyte compartment, enveloped by the diaphragm, while the anodic compartment is confined between the diaphragm and the graphite anodes.

Sodium chloride brine at about 310 gpl is fed into the anolyte compartment and the effluent is continuously withdrawn from the catholyte compartment, into which it percolates through the diaphragm, while electrolysis proceeds. The catholyte, or cell liquor, contains the amount of sodium hydroxide produced by the cathodic reaction:

$$Na^+ + H_2O + e^- \rightarrow Na^+ + OH^- + \tfrac{1}{2} H_2(gas)$$

together with an unreacted residue of sodium chloride; a typical cell liquor concentration is 135 gpl NaOH plus 150 gpl NaCl. The average cell voltage, when using the newly developed activated titanium anodes instead of graphite, is about 3.5 V at the current density of 2000 A per square meter and the current efficiency is about 97%. Modern diaphragm cells are rated for current capacities up to 80,000 A.

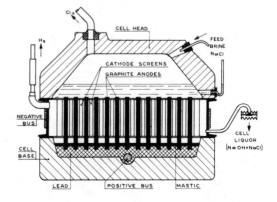

FIG. 1.

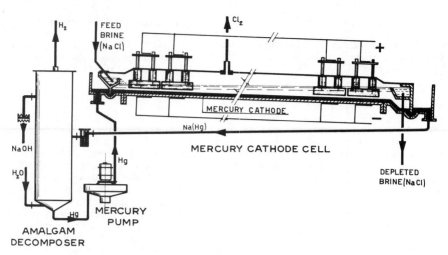

FIG. 2.

Mercury Cathode Process (Fig. 2). In this process there is no need for a diaphragm or for any "counterflow" device, since the alkali metal (sodium or potassium) is discharged on a cathode consisting of a flowing mercury layer and becomes dissolved in it, thus forming an amalgam that is continuously drained out of the electrolytic cell. Accordingly, the mercury cathode stream performs the function of a virtual diaphragm as well, in that the cathodic product is thereby impeded to diffuse back into the solution and bring about any side-reactions with chlorine discharging at the anodes. The amalgam is flowed into a reaction vessel, called *decomposer* or *denuder*, where it is reacted with water so as to liberate its alkali metal content, with production of caustic and hydrogen:

$$2Na(Hg) + 2H_2O \rightarrow 2NaOH(aq) + H_2(gas) + Hg$$

The reaction would not be viable without the presence of a suitable catalyst such as graphite, that performs as the positive metallic piece of a short-circuited battery, in which the mercury mass provides the negative pole and the caustic solution makes up the electrolyte. Hydrogen is thereby displaced from the water molecule HOH and replaced by sodium to form NaOH, as the consequence of an electromotive force that sets hydrogen free in the gaseous state, while its discharge is facilitated by the electrocatalytic material.

The caustic solution that can thus be obtained is characterized by a high-purity grade and a concentration that may reach 50% or more. Mercury cells can be operated at current densities four to five times higher than diaphragm cells, at a voltage of about 4.2 V. The current efficiency is nearly the same.

Modern mercury cells are rated for current capacities up to nearly 500,000 A.

Production Statistics and End-use Patterns. The production of chlorine increased in 1969 to a new record of 9,427,000 tons in the United States, with a gain of approximately 11.2% over 1968. This represents about a half of the overall world production. The 1969 production in Canada also reached a new high, 887,000 tons, corresponding

to 14.4% over 1968 figures. It is estimated that new projects completed in 1970 have increased the combined U.S. and Canadian capacity by about 6.5% or to approximately 12 million tons.

In 1970 the production of chlorine in the United States reached 9,755,000 tons, but in 1971 the output fell off the 9,340,000 tons. This decline coincided with a decline in the production of most of the basic chemicals in 1971.

The following percentages correspond to the estimated breakdown of end-uses for chlorine in the United States[2,4] organic chemicals 60%; inorganic chemicals 10%; pulp and paper 15%; sanitation (water treatment) 4%; others 11%.

Chlorine Institute. The Chlorine Institute[1] was founded in 1924, "to foster the industrial interests of those engaged in the chlorine industry; to promote a more comprehensive relationship between those engaged in the production and those engaged in the consumption of its products; and to engage in research work directed toward developing new uses and the more extended consumption of chlorine and its products."

Physiological Response and Toxicity. Chlorine gas has a characteristic, pungent odor with a detectability threshold of a few parts per million in air.

Liquid chlorine in contact with eyes, skin or clothing may cause severe burns; as soon as it is released in the atmosphere, it vaporizes with irritating effects and a suffocating action, which were exploited by using it as a war gas. The physiological response to the presence of any amount of chlorine gas in air may be evaluated from the following data published by the Bureau of Mines (Technical Paper No. 248). Concentration values are given in parts per million by volume.

Least detectable odor	ppm	3.5
Least amount required to cause irritation of throat	ppm	15.1
Least amount required to cause coughing	ppm	30.2
Least amount required to cause slight symptoms of poisoning after several hours exposure	ppm	1.0

Maximum amount that can be
breathed for one hour without serious
effects   ppm   4.0

Amount dangerous in 30 minutes to
one hour   ppm 40–60

Amount likely to be fatal after a few
deep breaths   ppm 1000

Consequently, chlorine must be considered as a hazardous chemical, to be handled in accordance with pertinent regulations and specifications.

Storage and Shipment. Before liquefaction, chlorine is thoroughly dried, so as not to be corrosive against the steel containers in which it is stored or shipped. Cylinders for chlorine transportation are of seamless construction and have a capacity ranging from 1 to 150 pounds. Ton containers are of welded construction and in their maximum size have a loaded weight of 3700 pounds. Tank cars may hold up to 55 tons and tank barges as much as 600 tons.

Patrizio Gallone

References

1. Chlorine Institute, Pamphlets, Drawings and Miscellaneous Publications, The Chlorine Institute, Inc., 342 Madison Ave. New York, N. Y. 10017.
2. Hampel, C. A., Ed., "Encyclopedia of Electrochemistry," Reinhold Publishing Corp., New York, 1964.
3. de Nora, O., and Gallone, P., article *Chlorine* in "Encyclopedia of Chemical Elements," Hampel, C. A., Ed., Reinhold Publishing Corp., New York, 1968.
4. Sconce, J. S., Ed., "Chlorine," Reinhold Publishing Corp., New York, 1962.

CHLOROHYDRINS

Aliphatic organic compounds which are both alkyl chlorides and alcohols are called chlorohydrins. Those most frequently encountered contain one chlorine atom and one hydroxyl group on adjacent carbon atoms.

Preparation. A general method of preparation is the addition of hypochlorous acid to alkenes. Addition to an unsymmetrical alkene yields predominately a product in which the hydroxyl group is attached to the carbon atom poorest in hydrogen. Thus, reaction of hypochlorous acid with 1-alkenes gives chlorohydrins of the structure $RCHOHCH_2Cl$. The initial step in this reaction is an attack on the electron-rich carbon atom of the double bond by an electrophilic reagent. The formed intermediate is subsequently decomposed by a water molecule or in alkaline medium by hydroxide ion. The active electrophilic reagent in hypochlorous acid, acidified with sulfuric or perchloric acids, is the chlorinium ion Cl^+ for reactive alkenes and the hypochlorous acidium ion $ClOH_2^+$ for very reactive alkenes such as 2-methylpropane. In weak acidic solutions the effective electrophilic reagent is hypochlorous acid and chlorine monoxide.

Chromyl chloride in carbon tetrachloride reacts with 1-alkenes to form chlorohydrins of the type $RCHCl \cdot CH_2OH$ and with cyclohexene to form trans-2-chlorocyclohexanol. Presumably the chromyl chloride donates a positive CrO_2Cl fragment to the alkene to form a three or five-membered cyclic intermediate which reacts in a subsequent step with chloride ion; hydrolysis of this product gives the chlorohydrin. Industrial processes for producing chlorohydrins from alkenes include the addition of hypochlorous acid indirectly by the use of monochlorourea or *t*-butyl hypochlorite, as well as the passing of chlorine and an alkene into water under controlled conditions of temperature and concentration.

Cyclic ethers may easily be converted to chlorohydrins by treatment with hydrochloric acid. The rate of reaction is proportional to the concentration of the cyclic ether, hydrogen ion and chloride ion. Inversion is observed to accompany the opening of the oxide ring. These kinetic and stereochemical results of opening the oxide ring by hydrochloric acid are best explained by a mechanism involving the addition of a proton to the epoxide oxygen to form the conjugate acid of the oxide, followed by a nucleophilic displacement by chloride ion. The opening of unsymmetrically alkyl substituted ethylene oxides by hydrochloric acid usually results in mixtures of the two possible chlorohydrins. In some cases particularly with aryl substituted ethylene oxides a single product is formed.

Chlorohydrins may also be formed by the treatment of various glycols with chlorinating agents such as hydrochloric acid, thionyl chloride or sulfur chloride. These methods are often complicated by the occurrence of disubstitution. Other methods of preparation of chlorohydrins include the reaction of Grignards with chloroaldehydes and chloroesters and the reduction of chloroketones with aluminum ethoxide. Chlorohydrins are also made successfully by the reduction of chloroacids, chloroesters, and chloroacid chlorides by lithium aluminum hydride.

Reactivity. Kinetic studies on the reaction of some chlorohydrins with water established that 4-chloro-1-butanol reacts 1000 times as fast as 2-chloroethanol and 200 times as fast as 3-chloro-1-propanol. Tetrahydrofuran was the main reaction product of 4-chloro-1-butanol with water, while ethylene glycol and trimethylene glycol are the only products in the hydrolysis of 2-chloroethanol and 3-chloro-1-propanol.

Treatment of 2-chloroethanol with sodium hydroxide yields ethylene oxide. The reaction is first order with respect to hydroxide ion and first order with respect to the chlorohydrin. A mechanism which accounts for the kinetics and the formation of the epoxide involves a two step process. The first step is a rapid equilibrium between the hydroxide ion and the hydroxyl group of the chlorohydrin followed by a rate-determining displacement of chloride ion from the chloroalcoholate ion. The rate of formation of ethylene oxide thus depends upon the acidity of the chlorohydrin and upon the reactivity of the chloroalcoholate ion formed. Reaction of hydroxide ion with 4-chloro-1-butanol also gives a cyclic ether, tetrahydrofuran, and follows second-order kinetics.

The reaction of 3-chloro-1-alkanols with hydroxide ion is quite slow compared to 2-chloroethanols and 4-chloro-1-butanols and the yields of trimethyl-ene oxides are usually low, the predominant reaction being elimination and some substitution. The elimination reaction is favored by the inductive effects of both the chlorine atom and the hydroxyl groups affecting the same beta carbon atom favoring the release of a proton. *Gem*-dialkyl substitution on the hydroxyl bearing carbon seems to aid trimethylene oxide formation. Yields of trimethylene oxides are also high in those 3-chloro-1-alkanols where elimination is impossible and where steric hindrance would prevent intermolecular substitution.

Substitution of the chlorine atom of chlorohydrins by nucleophilic reagents such as the amino or substituted amino groups, mercapto groups, cyanide ion, sulfide ion, phenoxide ion and iodide ion takes place easily.

Due to the displacement of electrons toward the chlorine, the alcohol group of α-chlorohydrins is less susceptible to attack by electrophilic reagents than unsubstituted alcohols. Thus, the chlorohydrins of the 2-butenes are reported to be inert to the action of fuming hydrochloric acid or to zinc chloride-hydrochloric acid—nor is there any reaction with fuming hydrobromic acid. The half-life of trans-2-chlorocyclohexanol with fuming HBr at 20°C is 85 days compared to a half-life of 20 minutes with cyclohexanol. On the other hand 3-chloro-1-propanol where the chlorine and hydroxyl groups are further removed from each other reacts readily with 48% HBr.

The hydroxyl group of chlorohydrins, on the whole, exhibit the characteristic reactions given by simple alcohols. Oxidation of chlorohydrins containing a primary alcohol group yield chloroaldehydes and chloroacids. Halides of phosphorus replace the hydroxyl group with ease and reaction with organic acids or acid chlorides forms esters.

Reaction of (+)-threo-3-chlor-2-butanol with thionyl chloride gives equal amounts of *dl*-2,3-dichlorobutanes but no meso dichloride. This sterochemical result can be accounted for by a mechanism involving the formation of the ester which subsequently decomposes to the chloronium ion due to the displacement of the negative ion $SOCl^-$ by the neighboring chlorine atom. A nucleophilic displacement by chloride ion yields the final products.

H. W. HEINE

Cross-references: *Chlorine, Alcohols.*

CHLOROPHYLLS

A group of closely related green pigments occurring in leaves, bacteria and organisms capable of photosynthesis. The major chlorophylls in land plants are designated *a* and *b*. Chlorophyll *c* occurs in certain marine organisms. Because of the overwhelming percentage of the total photosynthesis which is performed by marine organisms, it is possible that chlorophyll *c* is equivalent in importance to chlorophyll *b*. Chlorophyll *a* is several times as abundant as chlorophyll *b*. Bacteriochlorophyll contains two more hydrogens than the plant chlorophylls and has the vinyl group altered to an acetyl.

Formula I represents one of the canonical forms for chlorophyll *a*, II the similar form for chlorophyll *b*. These structures have been established by a long series of degradation studies mainly by R. Willstätter, Hans Fischer and their collaborators, and by synthetic studies in the laboratories of Fischer.

I R=CH₃
II R=CH

The laboratories of Woodward and of Strell announced the complete synthesis of chlorophyll *a* almost simultaneously.

The biological significance of the chlorophylls stems from their role in photosynthesis, the process by which plants fix the sun's energy in the form of organic matter. This process corresponds to the reversal of the combustion of hydrogen. The oxygen liberated is set free in the air. Under special conditions, some organisms are also capable of liberating the hydrogen, but usually this is used for chemical reductions in the plant. Atmospheric carbon dioxide is fixed enzymatically and is thus used as the source of the carbon in the synthetic process but is not reduced directly. The path of the carbon from carbon dioxide in photosynthesis has been elucidated largely by the studies of Calvin and his collaborators. While it is known that most of the energy fixed in photosynthesis is absorbed originally by the chlorophylls, the exact reactions which they undergo to initiate the process of reduction are not yet understood. It is known, however, that the photosynthetic sequence requires a high degree of organization within the plant cells where it occurs, and that destruction of the organization of the chloroplasts by processes like grinding are sufficient to bring photosynthesis to a stop, even when the chlorophyll and the soluble enzymes participating in the process are still presumably intact.

Weak acids remove the magnesium from the chlorophylls giving the pheophytins. Strong acids selectively hydrolyze the phytyl group, yielding the pheophorbides. Hydrolysis of the chlorophylls by the enzyme chlorophyllase in the absence of alcohols also removes the phytyl group, giving chlorophyllides. When the enzymatic process is conducted in the presence of alcohols, chlorophyllide esters, such as methyl or ethyl chlorophyllide, are produced by alcoholysis. The chlorophylls, because of the phytyl group, are microcrystalline waxes. The chlorophyllides, on the other hand, crystallize in visible crystals.

Hot, quick, alkaline saponification of the chlorophylls yields chlorophyllins, magnesium-containing pigments with three carboxylate ions. These result from the removal of the two alcohol groups and the cleavage of the five-membered isocyclic ring. The cleavage occurs readily because this ring contains a keto group β to a carboxylic ester group. Acidification of the product obtained in this manner from chlorophyll a removes the magnesium and gives chlorin e_6, one of the most readily obtainable and important degradation products of chlorophyll a. The corresponding degradation product of chlorophyll b is called rhodin g_7. The numerical subscripts in this and lower ranges refer to the number of oxygen atoms contained in the molecules. In the high ranges, for example pheopurpurin$_{18}$, the subscript refers to the so-called "acid number," the percentage strength of hydrochloric acid which will remove two-thirds of the substance from an equal amount of its ethereal solution.

When an ethereal solution of a chlorophyll, or of a derivative containing an intact isocyclic ring, is treated with cold alcoholic KOH, the green color is momentarily discharged to a yellow or brown, depending upon the derivative, and the green color then reappears. This is known as the "phase test" and was discovered by Molisch. When chlorophyll derivatives are exposed to air in the presence of alkali, oxidation accompanies the hydrolysis and ring cleavage and the product is said to be "allomerized." Such a product will no longer give a positive phase test, nor will it crystallize readily. The prevention of allomerization is of great importance in securing high quality chlorophyll derivatives. The complicated series of reactions taking place during allomerization was elucidated by the work of J. B. Conant and his collaborators. Equivalent oxidation can also be secured with quinone. Stronger oxidizing agents, such as ferricyanide or molybdicyanide, are capable of stripping off the extra hydrogens from the nucleus and converting members of the green chlorophyll series into the red porphyrins. Chromic acid degrades all these substances into maleic imide derivatives, together with other products.

Mild hydrogenation of chlorophyll derivatives saturates the vinyl group. Drastic hydrogenation is capable of discharging the color, with formation of leuco compounds, or even of cleaving the nucleus with the formation of pyrrole derivatives.

It is thus evident that the chlorophylls are sensitive to acid, alkali, reducing agents or oxidizing agents, whether weak or strong. This variety of sensitivities to chemical agents renders them hard to purify without change and makes chemical synthesis in the field difficult. In commercial practice, it complicates the problem of securing uniformity of product and gives rise to the possibility that each batch will contain different substances or different proportions of various products. The sensitivity of the b series to chemical reagents is generally greater than that of the a series.

Compared to the highly stable phthalocyanines, chlorophyll derivatives are pigments of relatively low stability to light, oxidizing agents and other common pigment destroyers. This property forms one of the serious limitations to their wider commercial exploitation. A moderate increase in stability is conferred by the substitution of copper for magnesium. This also produces a clearer shade

that is frequently more desirable. Numerous metals which have square, planar bonds are quite tightly bound by chlorophyll derivatives. This is especially noticeable in the case of copper, which is so tightly held in this configuration that it is deprived of its usual catalytic action. Even these complexes possess relatively low stability to light and oxidizing agents, however. An additional difficulty is introduced by the fact that the destruction of the organic portion of the molecule renders the metal available for new combinations.

Commercially, chlorophyll derivatives are usually assayed spectrophotometrically or colorimetrically. These assays serve as an index of tinctorial power but do not accurately measure the chlorophyll derivatives because of the interferences of other plant pigments and decomposition products. For reliable assays, preliminary fractionations must be performed.

The special techniques most frequently used in the fractionation and purification of chlorophyll derivatives are acid fractionation, column chromatography and paper chromatography. The acid fractionation procedure of Willstätter and Mieg relies upon the fact that porphyrins, chlorins and related substances are weak bases which can be extracted from ethereal solutions by varying concentrations of hydrochloric acid. The fractionations obtained are sharper than might be expected because small changes in acid concentration cause changes in the solubility of ether in acid and of aqueous acid in ether, thus producing greater effects than they would in a relatively invariant system. The method of chromatography was invented by Tswett expressly for the separation of chlorophylls and carotinoids but lay dormant in the literature for many years until rediscovered for the fractionation of carotinoids. Paper chromatography is a powerful tool for assay because of the small quantities of material needed for a satisfactory fractionation. Even these small quantities, however, are sufficient for quantitative purposes when eluted and assayed in a spectrophotometer.

Chlorophyll derivatives with the phytyl group intact are oil-soluble and form a series of green dyes which have found wide commercial application in the coloring of oils and waxes. The chlorophyll soaps, resulting from combined saponification and cleavage of the isocyclic ring, form valuable water-soluble dyes, useful in the coloring of soaps and similar products.

Both the medical and the cosmetic literature are replete with claims of therapeutic or physiological activity of "chlorophyll." The substances utilized in this work range from partially purified chloroplasts to mixtures of materials which have undergone deep-seated chemical alteration. Some of the types of activity claimed can be shown to be due to incidental impurities. The field for investigation of the action of pure chemical individuals produced by the action of various reagents upon chlorophyll or its derivatives is unexplored. It is known, however, that neither chlorophyll nor hemoglobin in the diet is utilized by the body in the formation of the physiologically active pyrrole pigments. These are derived, instead, from such simple building blocks as glycine and acetate ion. Only the iron in dietary blood pigment can be utilized by the body.

The work of Granick has shown that, in the physiological processes of plants, chlorophyll is formed from protoporphyrin, which can be obtained in the laboratory by the removal of iron from hemin. The pathways to heme and to chlorophyll diverge at protoporphyrin. To form heme, an organism introduces iron into protoporphyrin. To form chlorophyll from protoporphyrin, an oxidation, a reduction, a ring closure and esterifications are performed and the magnesium is introduced. The end product of the enzymatic synthetic chain is presumably protochlorophyll, the magnesium derivative of the porphyrin corresponding in structure to chlorophyll. The addition of the two hydrogens necessary to convert the red compound to the green is accomplished under the influence of light.

A. H. Corwin

Cross-references: *Photosynthesis, Phytochemistry, Flower Pigments.*

CHOLESTEROL

First isolated from gallstones in 1816 by Chevreul and named *cholestérine* (Greek *chole,* bile and *stereos,* solid), cholesterol has been the subject of innumerable investigations by many illustrious chemists. Berthelot, Windaus, Wieland, and Bernal have contributed to the elucidation of its structure. The brilliant research work of Bloch and Woodward, of Cornforth and Popják, of Lynen and of several other outstanding chemists has established the mechanisms whereby this substance is formed biologically. Total laboratory synthesis of cholesterol was accomplished by Woodward and by Robinson.

Cholesterol probably occurs in all living organisms and may be an essential cell constituent. Sterols such as cholesterol are required nutrients for certain bacteria, protozoa, fungi, arthropods, annelids, mollusks, sea urchins, sharks, and other creatures that are incapable of synthesizing them. Among mammalian tissues liver, skin, and intestines are most active in the biosynthesis of cholesterol. Generally, bacteria contain only traces of cholesterol, plants relatively low concentration, and certain animal organs rather high concentrations. Besides gallstones, which may consist almost entirely of cholesterol, the skin fat, brain, adrenals, eggs, and spermatozoa have a high cholesterol content. Eggs and butter are among the richest dietary sources.

The biosynthesis of cholesterol is outlined in Figs. 1 and 2. It involves the condensation of acetyl CoA with acetoacetyl CoA to produce 3-hydroxy-3-methylglutaryl CoA. This is followed by reduction to 3,5-dihydroxy-3-methylvaleric acid (mevalonic acid), which is phosphorylated. Mevalonic acid pyrophosphate is decarboxylated and dehydrated to isopentenyl pyrophosphate (C_5). The latter condenses with its isomer 3,3-dimethylallyl pyrophosphate to yield the monoterpene derivative geranyl pyrophosphate (C_{10}). Addition of another isopentenyl pyrophosphate molecule leads to the formation of the sesquiterpene derivative farnesyl pyrophosphate (C_{15}).

Condensation of two molecules of the sesquiterpene produces the acyclic triterpene, squalene (C_{30}). The latter then cyclizes via its 2,3-epoxide to the

$CH_3CO-S-CoA + CH_3COCH_2CO-S-CoA \rightarrow$

Acetyl CoA — Acetoacetyl CoA

$HOOCCH_2C(CH_3)(OH)CH_2CO-S-CoA \rightarrow$

3-Hydroxy-3-methylglutaryl CoA

$HOOCCH_2C(CH_3)(OH)CH_2CH_2OP_2O_6H_3 \rightarrow$

Mevalonic Acid Pyrophosphate

$CH_2=C(CH_3)CH_2CH_2OP_2O_6H_3 \rightarrow$

(Isopentenyl Pyrophosphate)

$CH_3C(CH_3)=CHCH_2OP_2O_6H_3 \rightarrow$

(3,3-Dimethylallyl Pyrophosphate)

$CH_3C(CH_3)=CHCH_2CH_2C(CH_3)=CHCH_2OP_2O_6H_3 \rightarrow$

Geranyl Pyrophosphate

$CH_3C(CH_3)=CHCH_2CH_2C(CH_3)=CHCH_2CH_2C(CH_3)=CHCH_2OP_2O_6H_3$

Farnesyl Pyrophosphate

FIG. 1. Biosynthesis of terpenes.

tetracyclic triterpene lanosterol. Subsequent demethylations yield the sterol zymosterol, which undergoes a series of transformations leading to desmosterol or 7-dehydrocholesterol and, ultimately, cholesterol (C_{27}).

Cholesterol has several important functions. In nerve tissues it may be involved in insulation and in blood it may act as a vehicle for the transport of unsaturated fatty acids. However, its main function is to act as a metabolic precursor of other steroids in plants and animals. In higher animals

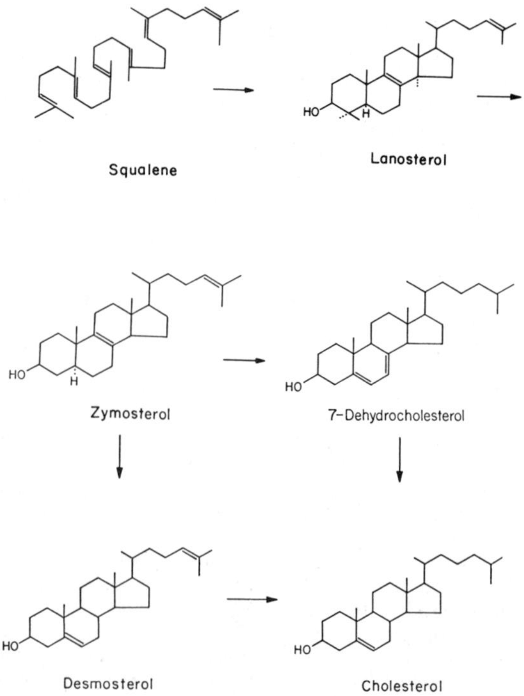

FIG. 2. Biosynthesis of cholesterol.

cholesterol is mainly metabolized to the bile alcohols and acids. Insects and certain plants convert it to sterols with molting hormone activity. Other plants alter the cholesterol side chain to produce C_{27} sapogenins and alkaloids. Degradation of the cholesterol side chain by various organisms leads to pregnane derivatives, including progesterone (C_{21}), which is a key intermediate in the biosynthesis of the other steroid hormones, of C_{21} alkaloids, and of C_{23} or C_{24} cardiac-active genins. Other steroids, such as the C_{28} and C_{29} sterols, the methyl sterols, cyclo sterols, and vitamins D are derived from various cholesterol precursors.

Dietary cholesterol is absorbed in the small bowel with the aid of bile salts and esterified in the gut wall. The thoracic duct lymph transports cholesterol, partly in esterified form, to the systemic circulation, where it migrates in combination with the α- and β-lipo-proteins. While there is very little esterified cholesterol in liver and practically none in mature neural tissue, 70–75% of the plasma cholesterol is in the ester form. Human plasma contains an average of 50 mg of free and 170 mg of esterified cholesterol per 100 ml. In the erythrocytes cholesterol occurs in the free form in the cell membrane and combines with hemolytic substances, such as saponins.

Statistically, the serum cholesterol of women is lower than that of men. It increases with age and is higher in obese persons than in lean ones and is higher in populations eating animal fats than in vegetarians. Hypercholesterolemia is often observed in nephrosis, diabetes, hypothyroidism, and atherosclerosis. In the latter condition cholesterol deposits may form in the walls of the aorta. Since hypercholesterolemia seems to favor the appearance of atherosclerotic plaques, factors which decrease the serum cholesterol level, such as estrogens, vegetarian or low-fat diets, or factors which interfere with cholesterol absorption, such as soybean sterols, have attracted some attention. In certain disorders of lipid metabolism (xanthomatoses) abnormal deposits of cholesterol are sometimes observed. Gallstones are formed when the bile loses its ability to keep cholesterol in solution.

ERICH HEFTMANN

Cross-references: *Steroids; Bile acids; Hormones, steroid; Vitamins; Glycosides, steroid.*

References

Heftmann, E., "Steroid Biochemistry," Academic Press, New York, 1970.

Briggs, M. H., and Brotherton, J., "Steroid Biochemistry and Pharmacology," Academic Press, New York, 1970.

Belle, H. van, "Cholesterol, Bile Acids, and Atherosclerosis. A Biochemical Review," North-Holland, Amsterdam, 1965.

Cook, R. P., ed., "Cholesterol," Academic Press, New York, 1958.

CHOLINESTERASE INHIBITORS

Cholinesterase inhibitors act by inhibiting the enzyme acetylcholinesterase, which catalyzes the hydrolysis of acetylcholine. Acetylcholine has the essential role of transmitting the effects of stimulation of certain nerves into functional changes in organs that are supplied by the so-called cholinergic nerves. Thus when the cholinergic nerves to the muscles and glands are stimulated they release acetylcholine which acts on those structures to produce increased muscular or glandular activity. The release of acetylcholine from the vagus nerves causes slowing of the heart rate. Acetylcholine also serves as the chemical transmitter of the nerve impulse in ganglia which function as relay stations to transmit impulses from one nerve to a second nerve in an autonomic chain. Acetylcholine also apparently plays a similar role in transmission of impulses from one nerve to another in the brain. The normal activity, which is referred to as normal tone, of various organs as the intestine, smooth muscles, skeletal muscles, and the heart and bladder is maintained by a constant release of acetylcholine from cholinergic nerves.

Acetylcholine was first synthesized in 1867. It consists of choline and acetic acid in an ester linkage. The component parts of the acetylcholine molecule are both normal constituents of the body. Acetylcholine has the following chemical structure:

$$CH_3\text{---}N\text{---}CH_2\text{---}CH_2\text{---}O\text{---}\overset{\overset{\displaystyle O}{\|}}{C}\text{---}CH_3$$

with three CH_3 groups on the N

Acetylcholine is stored in the tissues in a physiologically inactive form. It is bound to some constituent of the nerve cell, probably to fat or protein. Stimulation of nerves containing acetylcholine causes its release from the bound form to exert its action on cells, after which it is quickly rendered inactive by hydrolysis to choline and acetic acid. The inactivation or detoxification is catalyzed by enzymes called cholinesterases.

As early as 1941, it was suggested that an enzyme present in the blood brings about destruction of acetylcholine, and much attention was thereafter devoted to a study of the enzymatic destruction of acetylcholine. It is now well established that there is a specific enzyme located in nervous tissue that catalyzes the rapid hydrolysis of acetylcholine. It is called *specific* or *true* cholinesterase, and its concentration is highest at the nerve endings where acetylcholine is liberated. The rapid hydrolysis of acetylcholine catalyzed by cholinesterase is the reason that the acetylcholine liberated when cholinergic nerves are stimulated exerts only very short-lasting effects upon the cells. Furthermore, its rapid destruction prevents acetylcholine liberated at one site from being transported by the blood to other organs. This mechanism thus restricts the action of acetylcholine to a localized area of the body at the immediate site of liberation from the nerve endings.

In addition to the specific cholinesterase present in the nervous system, a variety of tissues contain enzymes capable of hydrolyzing acetylcholine. These enzymes are referred to as *pseudocholinesterases*. They are present in highest concentrations in the serum, the intestinal mucosa, and the liver. They hydrolyze acetylcholine and a variety of other esters. Normally they are not involved in the detoxification of acetylcholine. However, they do have an important role in protecting against acetylcholine poisoning under abnormal circumstances. For example, the pseudocholinesterase of the gastrointestinal tract rapidly hydrolyzes acetylcholine before it can be absorbed. This serves as a protective mechanism against poisoning from foods containing acetylcholine. The pseudocholinesterase of the serum also has a protective role in those instances where the specific cholinesterase is inhibited leading to elevated acetylcholine levels in the tissues. Under these circumstances the nonspecific enzymes in the serum and the liver hydrolyze acetylcholine and thus decrease its general effects.

The acetylcholine-cholinesterase system has served as the basis for the development of a number of drugs needed to alter the activity of the autonomic nervous system in certain disease states, and for the development of poisons for use as insecticides and chemical warfare agents. Inhibition of the cholinesterases results in preservation of acetylcholine with consequent elevation of the acetylcholine levels in the tissues, with the result that the tissues which respond to acetylcholine are excessively stimulated.

The first effective cholinesterase inhibitor to be discovered was the alkaloid physostigmine, commonly known as eserine. It is a constituent of the Calabar bean which grows in tropical West Africa. Physostigmine was isolated in crystalline form from this bean in 1864 but its chemical structure was not discovered until 1925. Shortly after that, its toxic action was shown to be due to inhibition of cholinesterase. Physostigmine is a complex derivative of carbamic acid (NH_2-COOH). A series of less complex phenyl carbamates was then synthesized and shown to inhibit cholinesterase, particularly when the amino group contained one or both hydrogens substituted by methyl groups. The carbamates are rapidly reversible inhibitors of cholinesterase. Some of them are used as drugs in diseases where elevation of acetylcholine levels is needed. For example, in a disease of skeletal muscles (myasthenia gravis) characterized by weakness of skeletal muscles, the carbamates elevate acetylcholine levels and restore muscular strength. Neostigmine (prostigmine) is the best known drug of the carbamate series. Some carbamates are used as insecticides, among which is carbaryl (1-napthyl N-methylcarbamate). They inhibit the cholinesterases of insects causing a marked increase in acetylcholine to the level that causes intense stimulation of various structures, convulsions, and death.

The other important cholinesterase inhibitors are organic esters of phosphoric acid having the following type structures:

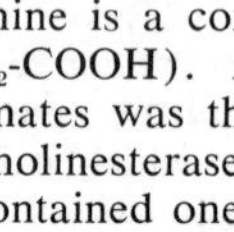

$$\begin{matrix} R^1 \\ R^2 \end{matrix} \!\!> \!\! P\!-\!X \atop \quad\|\atop\quad O$$

In one group of compounds of this class X is the fluoride or cyanide ion and R^1 and R^2 are alkyl or alkoxy groups. Members of this group are highly toxic and usually very volatile. These properties led to the development of several compounds of this class as chemical warfare agents. Two well-known "nerve gases" of this type are Tabun and Sarin:

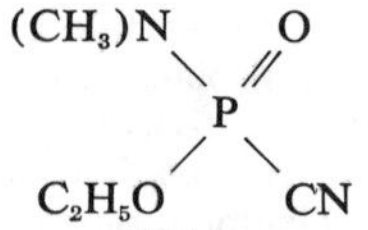

Tabun 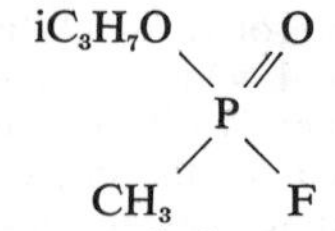   Sarin

The nerve gases are among the most potent synthetic toxic agents known. They are lethal to laboratory animals in minute doses. They readily enter the body through the skin, lungs, and intestinal tract.

Other organophosphorus anticholinesterase agents are widely used as insecticides. In this group the X attached to the phosphorus may be an alkyl, alkoxy or aryloxy group, a pyrophosphate linkage, or some other group. The R^1 and R^2 groups are usually alkyl, alkoxy, or dimethylamide linkages. Frequently sulfur replaces the oxygen of phosphoric acid because the sulfur analogues are more stable toward hydrolysis. The best known of these insecticides is parathion:

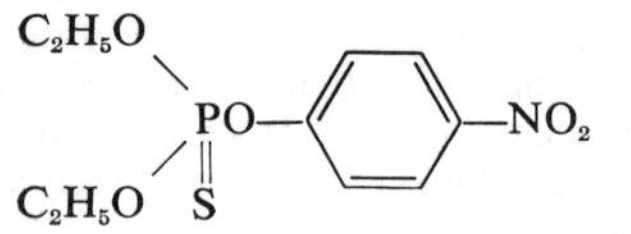

Parathion and related compounds containing sulfur in place of oxygen are not cholinesterase inhibitors themselves, but when they enter the body an enzyme system catalyzes replacement of the sulfur by oxygen. This reaction occurs in insect tissues and in the liver of man and animals. The resulting oxygen analogue is a strong cholinesterase inhibitor. The enzyme inhibition is due to phosphorylation of the active catalytic site of the enzyme. Specifically, the hydroxyl group of the amino acid serine in the enzyme molecule is phosphorylated.

The organic phosphates kill insects by raising the acetylcholine levels through inhibition of acetylcholine detoxification to the point where there is marked and excessive stimulation of various tissues. The same mechanism accounts for the lethal effects of these agents in animals and man. The estimated lethal dose for man is around 300 milligrams taken by mouth. Most of the widely used insecticides of this class have similarly high toxicity. Malathion, an organic phosphate derivative of succinic acid, is an exception in that it has low toxicity to mammals. The symptoms of poisoning by organic phosphates consist of bronchoconstriction, sweating, salivation and other glandular secretions, anorexia, nausea, vomiting, diarrhea, muscular twitching, convulsions, and paralysis of respiration. The inhibition of cholinesterase by organic phosphates is reversible after sublethal doses. The rate of reversibility and the consequent disappearance of symptoms varies for different compounds, but complete reversal usually occurs within a few days. Reversal of the inhibition can be accelerated by oximes and one of these compounds (pyridine aldoxime methiodide; 2-PAM) is widely used as an antidote for treatment of accidental poisoning by the organic phosphate insecticides. The alkaloid atropine is also used as an antidote; it does not reverse the inhibition of cholinesterase, but blocks the stimulating effect of acetylcholine on various organs.

KENNETH P. DUBOIS

References

Goodman, L., and Gilman, A., "The Pharmacological Basis of Therapeutics," Second edition, pp. 389–475, The Macmillan Co., New York, 1955.

DuBois, K. P., in "Cholinesterases and Anticholinesterase Agents," in "Handbuch der Experimentellen Pharmakologie," pp. 89–128, Springer-Verlag, Berlin, 1963.

CHROMATOGRAPHY

Theory. Chromatography is a method of separating substances that is widely used in analytical and preparatives chemistry. Although zone refining and zone electrophoresis may also be considered as chromatographic processes, chromatography is commonly understood to involve the flow of a liquid or gas over a solid or liquid stationary phase. Thus, we may distinguish, on the basis of the mobile phase, *liquid chromatography* and *gas chromatography*, or on the basis of both phases, *liquid-solid, liquid-liquid, gas-solid,* and *gas-liquid chromatography*. Liquid chromatography is applicable to soluble substances and gas chromatography to volatile substances.

As the mobile phase flows past the stationary phase (*chromatographic development*), a solute will undergo repeated sorption and desorption and move along at a rate depending, among other factors, on its ratio of distribution between the two phases. If their distribution ratios are sufficiently different, components of a mixture will migrate at different rates and produce a characteristic pattern (*chromatogram*). Differential migration is usually the result of a combination of sorption processes, but for convenience we may distinguish, on the basis of the stationary phase; *adsorption chromatography*, if it is an adsorbent; *partition chromatography*, if it is a liquid supported on a solid carrier; *gel filtration* and *ion-exchange chromatography*, if it is one of several special polymers.

Under a given set of conditions, the position of a substance in the chromatogram is constant relative to the mobile phase. The ratio of the distance traveled by the substance to the distance traversed by the mobile phase, the *R value*, is a constant related to the distribution ratio and hence to the chemical nature of the substance. It is therefore valuable for the identification and even for the structural analysis of substances, if used *judiciously*. Except in rare instances (*double zoning*), each pure substance gives rise to a single chromatographic zone, and substances located in different zones of the same chromatogram may be considered nonidentical. Analogous and homologous organic compounds often have *R* values that are related in a predictable manner.

A mixture applied to a sorbent in the form of a *narrow zone* will be resolved into a series of zones of similar shape, except for gradual spreading due to diffusion and distortions due to nonequilibrium conditions. Some factors which promote *spreading* and *distortion* of chromatographic zones are: time, large and irregular particles, rapid flow rates, and nonlinear sorption isotherms. The efficiency of a chromatographic system is expressed in terms of the number of *theoretical plates, N*. The average height equivalent to a theoretical plate, H, of a column or other chromatographic sorbent bed is obtained by dividing its length by N.

Nonlinear isotherms may be responsible for trailing (*tailing*) and consequently, overlapping of zones. Partition isotherms are usually linear over a much wider concentration range than adsorption isotherms. These factors must be considered if chromatography is to produce complete separation. However, for analytical work, the sample may also be applied in a *wide zone* and the chromatogram then interpreted by *break-through* or *frontal analysis*.

Techniques. *Liquid chromatography* may be performed either in a tube filled with a sorbent (*column chromatography*), on a sheet of paper (*paper chromatography*), or on a layer of sorbent spread over a solid support (*thin-layer chromatography*). *Gas chromatography* is column chromatography with a gaseous mobile phase. (See section on *Gas Chromatography*).

A *chromatographic column*, usually made of glass, is held vertically and is uniformly packed with a sorbent. The sample is usually applied to the top, and the chromatogram is developed by passing one or several solvents through the porous bed from top to bottom. The developing solvent (*developer*) may produce a series of zones in the sorbent. Earlier, the separated components were recovered by extruding the sorbent and cutting it into disks. In the currently used *elution* technique, the fluid leaving the bottom of the column (the *eluate*) is analyzed. This may be accomplished by continuously monitoring the effluent stream, usually by a suitable physical method. The eluate may also be collected in fractions—preferably with the aid of an automatic *fraction collector*—and either used for preparative work or analyzed by various physical, chemical, or biological methods.

During elution, the developing solvent (*eluent*) may be changed either stepwise or gradually in order to elute more strongly sorbed substances successively. *Gradient elution*, i.e., elution by a gradual increase in eluting power, is particularly advantageous for automatic operation and also because it reduces tailing. A recent technical improvement permits the use of finely divided sorbents in very long columns and affords high resolution. However, high pressure must be applied to accelerate the solvent flow. With the aid of instrumentation adopted from gas chromatography liquid column chromatography may ultimately offer the same speed and convenience as gas chromatography, but height load capacity and greater flexibility in the choice of chromatographic systems.

In *paper chromatography* the sample is concentrated in a small spot near the edge of a sheet of filter, modified cellulose, or glass-fiber paper. The paper is placed in a closed chamber and the edge is then dipped into a solvent. By a variety of techniques the paper is held in such a way that the solvent flows past the spot either upward, downward or horizontally. The solvent may also be made to enter a horizontal, circular sheet at the center and spread out radially past a spot near the center. This process can be speeded by rotating the sheet.

One advantage of paper chromatography is that a number of samples can be developed *simultane-*

ously under identical conditions on the same sheet. Another advantage is the possibility of *multiple development*. Repeated development of the same sample with the same or different solvents may increase the resolution. Another way of increasing resolution is *two-dimensional chromatography*. The sample is placed near one corner of a square sheet and developed first in one direction and then in a second direction, perpendicular to the first. Instead of paper chromatography, paper electrophoresis may be used for separation in one of the two dimensions.

Continuous development not only increases the resolution, but also permits the use of paper chromatography for the elution technique. Variants of paper chromatography that have been devised for preparative uses include a pad of rectangular sheets (*chromatopack*), a stack of disks (*chromatopile*), and a stick of tightly wound paper.

Colorless substances may be located on paper chromatograms by physical methods, e.g., in ultraviolet light or by *radioautography;* by biological methods, e.g., *bioautography;* or by chemical treatment. Zones of colorless material become visible after the chromatograms has been sprayed with, dipped into, or exposed to vapors of appropriate reagents. For quantitative work, the zones may be compared with standard zones, measured by planimetry, or scanned by a *densitometer.* Separated substances may also be recovered by excision and elution of individual zones, and the eluates may be used for analytical or preparative purposes.

Technically, *thin-layer chromatography* differs from paper chromatography mainly in requiring the preparation of a uniform layer of sorbent. A powder, either dry or—more frequently—in a slurry, is spread on a plate of glass or other solid supports, such as plastic or aluminum foil. Many types of mechanical spreading devices have been described, but by now ready-made thin-layer plates and sheets are commercially available. Usually, a *binder,* such as plaster or starch, is incorporated in the powder to give it mechanical stability.

Thin-layer chromatography has several advantages over paper chromatography: The sorbent particles are smaller and more regular than paper fibers and produce less zone distortion. They are also less permeable and concentrate the solutes on the surface. This increases the sensitivity of the method and gives smaller and more compact zones. Moreover, reagents that are too corrosive for paper may be used on thin-layer plates. Advantages of paper chromatography are that filter paper is cheap and readily available and that it is easier to recover material by elution from pieces of paper than by scraping and eluting the sorbent layers on glass plates. However, eluates from thin-layer chromatograms are not contaminated by the materials usually washed out of the filter paper. For preparative work sorbent layers of 1 mm or greater thickness may be required.

Chromatographic Systems. The selective retardation of migrants in chromatographic separations is the result of a variety of sorptive processes. Although a number of different mechanisms may be operative simultaneously, sorbents and conditions may be chosen which favor a particular type of sorption.

Chromatographic systems are composed of the sorbent, solvent, and solute. In *adsorption chromatography* the sorbent is an *adsorbent* and differences in adsorbability at the solvent-sorbent interface are exploited to separate the solutes. Adsorbents vary in strength, depending on their water content. They are activated by heating and deactivated by rehydration. In a general way, the commonly used adsorbents may be arranged in decreasing order of adsorptive power as follows: charcoal, alumina, silica, magnesia, magnesium silicate, calcium carbonate, calcium phosphate, talc, diatomaceous earth, cellulose, starch, sugar. In *partition chromatography* the stationary phase is a relatively hydrophilic *solvent,* such as: water, buffer solutions, dilute alcohols, acids, bases, glycols, amides, etc., that is partially miscible with a more hydrophobic mobile phase. In *reversed-phase* partition chromatography the more hydrophobic solvent, such as paraffin, silicone, etc., is held stationary on some hydrophobic support, while the more hydrophilic solvent is the mobile phase.

Solvents may be arranged in a series of increasing *polarity* (i.e., hydrophilic character, dielectric constant, etc.) to furnish a guide in selecting suitable chromatographic conditions. Such an *eluotropic series* is, e.g.: paraffin hydrocarbons, benzene, chloroform, ether, ethyl acetate, *n*-butanol, acetone, methanol, water. Intermediate polarity is obtained by mixing two or more solvents in proper proportions.

Gel filtration (gel permeation chromatography) requires natural or artificial polymers that sorb substances smaller than a given size. When a mixture of solutes is filtered through a column of such polymers, the latter will be retained or retarded, while larger molecules will pass through. Suitable sorbents for liquid-solid and gas-solid chromatography are the natural and artificial zeolites (*Molecular sieves*), and cross-linked dextrans (for hydrophilic solutes) or polystyrene resins (for hydrophobic solutes). The degree of cross-linking determines the upper limit of molecular size retained by a particular polymer and thus gel filtration may be used to classify solutes according to their molecular weight.

Ion-exchange chromatography is based on the ability of natural and synthetic polymers to sorb ionized solutes reversibly. The most commonly used ion exchangers are cross-linked resins with exchangeable hydrogen or hydroxyl ions. Strong *cation exchangers* contain sulfonic acid groups; weak cation exchangers, carboxylic acid groups; strong *anion exchangers,* quaternary ammonium groups; and weak anion exchangers amino groups. Although usually performed in columns, both gel filtration and ion-exchange chromatography can also be carried out on sheets. Sorbents have also been prepared that have specific affinities, such as steric specificity, electron-exchange properties, immunological specificity, etc.

History. Although the use of certain sorbents in analytical procedures akin to chromatography goes back to the 19th Century, the potential of chromatographic processes in the separation of solutes was not recognized until the beginning of the 20th Century. About 1903 the American petroleum chemist David T. Day and the Russian botanist Mikhail

Tswett independently described liquid-solid column chromatography. However, it was not until the Thirties that adsorption chromatography was applied systematically.

After the invention of paper chromatography by A. J. P. Martin and R. L. M. Synge in 1944, many chemists became interested in extending the applicability of the technique. In the Forties ion-exchange resins became generally available and gas-liquid chromatography was first described in 1952 by A. T. James and A. J. P. Martin. Two chromatographic techniques were introduced toward the end of the Fifties, thin-layer chromatography and gel-filtration.

Largely as a result of efforts by the various manufacturers of scientific instruments and supplies, the use of chromatography has become a routine operation in many laboratories. Gas chromatographs are particularly versatile analytical instruments. Further efforts will be required to make liquid chromatographs as convenient and dependable, to extend the use of chromatography to separations on an industrial scale, and to adapt it to continuous rather than batchwise operation.

ERICH HEFTMANN

Cross-references: *Gas chromatography, Ion exchange, Adsorption.*

References

Smith, I., ed., "Chromatographic and Electrophoretic Techniques," 3rd ed., Interscience, New York, 1968.
Heftmann, E., ed., "Chromatography," 2nd ed., Reinhold, New York, 1967.
Giddings, J. C., and Keller, R. A., eds., "Advances in Chromatography," Dekker, New York, 1965.
Giddings, J. C., and Keller, R. A., eds., "Chromatographic Science Series," Dekker, New York, 1965.
Lederer, M., ed., "Chromatographic Reviews," Elsevier, New York, 1959.

Gas Chromatography

Gas chromatography (sometimes inaccurately called vapor phase chromatography) encompasses both gas-solid (or adsorption) chromatography and gas-liquid (or partition) chromatography. It is principally a method for the separation and quantitative determination of gases and volatile liquids and solids. A. J. P. Martin and R. L. M. Synge in 1941 first suggested that a gas-liquid partition technique for the separation of volatile substances should be possible in a column in which the volatile substances are made to percolate through a solid impregnated with a nonvolatile liquid solvent. Despite the clarity of this suggestion, other investigators were slow to pursue it. Ultimately, in 1952, A. T. James and A. J. P. Martin investigated gas-liquid partition chromatography and established the method which has since enjoyed widespread application by chemists for the separation and analysis of complex mixtures of organic compounds.

Apparatus and Procedure. The separation and analysis of a mixture is carried out in the following manner. A thermostated oven containing the chromatographic column is adjusted to a temperature near that of the highest-boiling component of the mixture. A pre-heater, or injection-port flash vaporizer, is set for a temperature high enough so that the sample will vaporize rapidly when injected into the stream of heated carrier gas. The carrier gases most frequently used are helium or nitrogen. When thermal conductivity cells are used as sensing devices, helium has the advantage of higher sensitivity. Nitrogen is often recommended with the flame ionization detector.

A manometer in the inlet system is optional but is useful in determining the column inlet pressure and in detecting changes in column characteristics. An inlet pressure of 400–600 torr is typical.

The sample to be analyzed may be injected through a silicon rubber septum into the sample injection port by means of a calibrated hypodermic syringe. The sample size used is determined by the sensitivity of the detector employed. Typical sample size ranges from 0.2 to 2 ml. for a gas or from 0.001 to 0.010 ml. for a liquid or volatile solid when thermal conductivity detectors are used. An amount as small as one-thousandth of this is sufficient with the more sensitive ionization detectors.

The column is the "heart" of the instrument and separations depend to the greatest extent on the column packing material used. In gas-liquid partition chromatography the column packing consists of an inert solid support coated with a nonvolatile stationary liquid. The most commonly used solid supports are diatomaceous earths, the siliceous skeletons of diatoms. Commercial products are available through Johns-Manville Corp. under the trade names, Chromosorb-P, -W, and -G. "Chromosorb-P" has characteristics similar to ground particles of Johns-Manville C-22 "Sil-O-Cel" firebrick while "Chromosorb-W" is similar in properties to size-graded "Celite 545," a commonly used filter aid. "Chromosorb-G" is a newer product specially prepared for use in gas chromatography. A uniform particle diameter of 60-80 mesh (0.25–0.18mm.) gives high column efficiency without excessive back pressures.

The stationary liquid adsorbed on the inert solid support should be essentially nonvolatile at the column temperature. A few of the stationary liquid phases most commonly used at the present time include: (1) the relatively polar liquid phases for good retention of polar solutes, diethylene glycol succinate polyester (200°C); the polyethylene glycol, "Carbowax 20M" (220°C); dioctyl phthalate (160°C); and (2) the relatively nonpolar liquid phases for good retention of nonpolar solutes, silicone rubber gum, SE-30 (250°C); Dow-Corning high vacuum grease (250°C) and silicone oil 550 (190°C); Apiezon M (275°C), and "Squalane," 2,6,10,15,19,23-hexamethyltetracosane (80°C). The temperatures listed above are the generally accepted maximum useful column temperatures for the liquid phases when a sensitive ionization detector is in use. With the less sensitive thermal conductivity detector slightly higher column temperatures may be used without noticeable interference with the recorded chromatogram due to liquid phase bleeding. A complete discussion on choosing the correct liquid phase may be found in the comprehensive text by Dal Nogare and Juvet listed below.

The percentage of stationary liquid adsorbed on

the inert solid is usually in the range of 15–30 weight per cent. The column may be straight, U-shaped, or coiled and is normally 4–16 feet in length. The inside diameter of most analytical columns is in the range 4–6 mm.

The ideal-sensing device should meet the following requirements: high stability, high sensitivity, rapid and linear response, simplicity, and usefulness for all types of compounds. Several sensing devices have been used. These include thermal conductivity cells (both hot-wire and thermistors) and the flame ionization detector, as the most commonly used detectors, as well as the electron capture detector, flame photometric detector, gas density balance, coulometric detector and piezoelectric detector in less common use.

Injection of a mixture of components produces a series of gaussian-shaped peaks. Each peak corresponds to the elution of a pure substance from the column. At constant column temperature and flow rate, the area under any given peak (and to a good approximation the height of the peak) is proportional to the amount of substance present. The time required for each component to pass through the column is its *retention time, t_R*. The volume of carrier gas required for the elution of each component is the *retention volume*. The retention volume is characteristic of a compound for any given column at any given column temperature. Thus the instrument may be used for both qualitative and quantitative analysis. It is interesting to note that for a *homologous series,* a plot of the logarithm of retention volume (or log retention time) *vs.* the number of carbon atoms in each member of the series yields a straight line. This fact is often of value in qualitative analysis when a sample of a pure material is not available for comparison of retention volumes.

Principles. A 4-foot analytical column of 4 mm. internal diameter may have the equivalent of 700 to 4000 *theoretical plates*. The number of theoretical plates, *N,* in a column is a measure of the column's efficiency and is given by the expression,

$$N = 16(x/y)^2$$

where x is the distance to the peak maximum from the point of injection and y is the peak width measured between the intersection with the base line of tangents drawn through the inflection points of the peak. van Deemter and co-workers first suggested that column efficiency or band broadening is a function of a multiple path term, A, a molecular diffusion term, $B/\bar{u}$, and a mass transfer term, $C_i\bar{u}$, and devised the expression,

$$\bar{H} = L/N = A + B/\bar{u} + C_i\bar{u}$$
$$= 2\lambda d_p + 2\gamma D_g/\bar{u} + (8/\pi^2)[k/(1 + k)^2]$$
$$(d_i{}^2/D_i)\bar{u}$$

where H is the average height equivalent to a theoretical plate; L is the column length; λ and γ are functions of the uniformity in size and shape of the particles used to pack the column; d_p is the particle diameter; D_g and D_i are the diffusivity of the solute in the gas and liquid phases, respectively; $\bar{u}$ is the average linear gas velocity; k is the partition ratio, related to the thermodynamic partition coefficient; and d_i is the average thickness of

the liquid layer adsorbed on the solid support. Although several modifications and extensions of this original expression have been made, consideration of this expression alone shows that the best column efficiency (smallest value of H) is obtained by use of particles of small diameter, uniform in size; a carrier gas of relatively high molecular weight at an optimum velocity of $\bar{u}_{opt} = (B/C_i)^{1/2}$, often 3 to 10 cm/sec; a nonvolatile liquid phase of relatively low viscosity which will retain the solute well on the column; and a uniform distribution of the liquid phase in a thin layer over the surface of a solid support possessing a large specific surface area.

Gas-Solid Adsorption Chromatography. Turner, Claesson, Turkel'taub, Cremer and, more recently, Kiselev have adapted several gas-solid adsorption techniques to the separation of gaseous mixtures. Although poor separation of adjacent peaks is common in adsorption methods because of nonlinear adsorption isotherms, recent developments show promise of improved resolution. In adsorption chromatography the column packing material consists simply of a solid of very large surface area per gram (e.g., activated charcoal or alumina). Elution development, frontal analysis and displacement development adsorption techniques have been used. Of these, elution and displacement methods are more useful for analytical purposes. However, the long tailing edges often associated with elution adsorption chromatography may cause overlapping of adjacent peaks and add complications to analysis and separation.

Successful analysis of mixtures has been conducted using the displacement technique. After the sample is injected onto the column, the components of the sample are pushed through the column by the carrier gas presaturated with a substance (the *displacer*) which must be more strongly adsorbed by the column material than any of the components of the mixture. Each component of the sample acts as a displacer for the next most strongly adsorbed component so that pure zones of each substance move down the column one after another. A plot of recorder response *vs.* time results in a step-like curve. The recorder response is proportional to the thermal conductivity of each component, and the length of the step along the time axis is proportional to the moles of compound present. The disadvantage of this technique lies in the fact that the column must be replaced after each analysis since it is saturated with the strongly adsorbed displacer. At the present time, adsorption techniques have not found as much favor among chemists as the partition technique.

Applications. Recent work has emphasized speed in analysis. By optimizing experimental conditions as many as 37 compounds have been eluted in a period of less than 10 seconds. Other workers have replaced the packed columns with long lengths of capillary tubing coated on the inside with liquid phase and have achieved very high efficiencies. In one study a mile-long coil of coated capillary tubing had an efficiency of over one million theoretical plates, and very difficult separations could be performed. Gas chromatography has been used for the analysis of complex mixtures of volatile substances at column temperatures of less than −200°C to greater than 1000°C. An accuracy of 1% or better may be achieved. Sample volumes larger than 20 ml have been used in connection with large-diameter columns for the separation of relatively large amounts of very pure compounds. The preparation of high purity standards for mass spectrometry has become an important application. Gas chromatography is an excellent and rapid analytical tool for obtaining rates of reactions and thermodynamic data. The determination of trace amounts of pesticides on food products and air pollution control are other important uses of this versatile technique. Industrial process monitoring continuous stream analyzers employing gas chromatography are now commercially available.

RICHARD S. JUVET

Reference

Dal Nogare, S., and Juvet, R. S., "Gas-Liquid Chromatography. Theory and Practice," Interscience, New York, 1962.

Gel Filtration Chromatography

The gel filtration method is a relatively new fractionation procedure in liquid chromatography for the separation of molecules according to the differences in size and shape and can be considered as a kind of "molecular sieve."

This method has also been referred to in some publications as exclusion chromatography, restricted diffusion chromatography, gel chromatography and gel permeation chromatography, just to mention a few.

A number of various substances such as starch grains and polystyrene have been used as bed material for gel filtration. However, a major breakthrough occurred in this area with the introduction of the cross-linked dextran and polyacrylamide gels available from Pharmacia Fine Chemicals, Inc., Uppsala, Sweden and Bio-Rad Laboratories, Richmond, California under the trade names of Sephadex© and Bio-Gel©P.

While gel filtration has been used in aqueous and organic solvent systems, the largest percent of all chromatographic work has been on gels made from dextran and polyacrylamide used in the aqueous systems. Therefore, the following discussion will concentrate on these.

Dextran gels are produced by cross-linking the linear macromolecules of a modified dextran, creating a three-dimensional network of polysaccharide chains. Polyacrylamide gels are produced by the polymerization of acrylamide with the crosslinking agent, N,N′methylene-bisacrylamide. Under suitable conditions, the degree of cross-linking can be controlled to produce several types of gels, each insoluble, but capable of swelling in an aqueous solvent. Upon swelling, pores appear, the diameters of which are determined by the number of cross-linkages. The size of the pores decreases as the number of cross linkages increases.

Ideally, a gel should be insoluble, but should swell in an appropriate solution so that the pore sizes compare with the sizes of the molecules to be separated. To obtain a maximum molecular sieving effect, it must be possible to vary the pore sizes without losing the identity of the gel. In addition, the gel should not be adsorptive or have any property tending to interrupt movement of the molecules according to their relative size. The shape of the gel particle itself is important and should be such that good flow rates can be maintained without danger of compacting in any section of the gel bed. Spherical particles appear to have these properties and seem to be preferred to those produced by crushing, grinding or chopping.

At present, dextran and polyacrylamide gels are produced and supplied as a white powder in 8 and 11 different types. Choosing the appropriate type of gel depends on the molecular size of the substances to be separated. With the above types, fractionations can be accomplished over a wide molecular weight range; up to about 200,000 for polysaccharides and up to 800,000 for proteins.

Dextran and polyacrylamide gels are supplied in a dry powder; therefore they must be treated properly before they become functional as a gel filter. All types must be allowed to swell for a time sufficiently long to establish equilibrium and it must be remembered that complete swelling can occur only in aqueous solutions or other polar solvents such as formamide, dimethylsulfoxide and glycol. They will not swell in glacial acetic acid, methanol or ethanol; however, in some cases water mixtures of these solvents have been used effectively but complete swelling was not accomplished.

The agrose gels supplement the fractionation range of gel filtration for molecules in the molecular weight range of approximately 0.5×10^5 to 150×10^6 as determined with globular proteins and viruses. Agrose is the neutral portion of the agar and is a mixture of linear polysaccharide mainly composed of alternating D-galactose and 3,6-anhydro-L-galactose residues. The agrose gels differ from other gel types as the macromolecules in gel matrix of agar are believed to be held together by hydrogen bonds rather than by covalent bonds.

Agrose gels are produced in bead form from 2–10% agrose solutions now available from Litex, Glostrup, Denmark; Bio-Rad Laboratories and Pharmacia under the trade names of Gelarose, Bio-Gel©A and Sepharose©, respectively.

In a molecular separation resulting from restricted diffusion through a column of cross-linked gels, a liquid inside the swollen gel is available as solvent to a certain degree and will depend on the porosity of the gel particles as well as the different sizes of the molecules to be separated. The distribution of solutes inside and outside the gel beads is determined by the volume available as solvent inside and outside the gel particles and not be differences in solubility.

For every substance applied to a specific gel type, there is a distribution between the inner and outer liquid which can be defined by a distribution constant (K_d). When small molecules can penetrate the gel particles unimpeded this K_d equals 1 and when large molecules are completely excluded from the gel particles, K_d equals 0. Since the pore sizes inside the gel grains vary considerably, the K_d of intermediate solute molecular sizes becomes a measure of that part of the stationary phase available to the specific molecule. It follows then that K_d can be calculated from various known volumes.

For molecules completely excluded from the gel bed ($K_d = 0$), the volume of solvent necessary to elute these molecules is equal to the volume of the solvent outside the gel particle (V_o). For small molecules able to penetrate the gel bed ($K_d = 1$), the elution volume (V_e) includes the inner or absorbed volume (V_i). The K_d values can then be calculated according to Eqs. (a) and (b):

$$\text{(a)} \quad V_e = V_o + K_d \cdot V_i$$

$$\text{(b)} \quad K_d = \frac{V_e - V_o}{V_i}$$

V_o is obtained by passing through a column a high molecular weight substance that is known to be completely excluded from the gel particles at the same time collecting and measuring the volume of eluant required to move the high molecular weight substance from one end of the gel bed to the other. The absorbed or volume inside the gel grains (V_i) is calculated from the dry weight in grams (a) of the column matrix and the water regain (W_r) of the particular gel type. The W_r value is shown on the label of the containers in which the dry powder is supplied. From these values, the internal volume can be calculated according to the following equation:

$$V_i = a \cdot W_r$$

If the dry weight of the gel powder is not known, V_i can be calculated from the wet density (d) of the swollen gel particles.

$$V_i = \frac{W_r \cdot d}{W_r + 1} (V_t - V_o)$$

where V_t equals the volume of the gel bed. Another way to determine V_i is to measure the elution volume of a substance with a K_d very nearly 1. Two such substances are potassium chloride and tritiated water (THO). The equation can then be written:

$$V_i = V_e(\text{KCl}) - V_o$$

where V_e is that volume of buffer required to move KCl through the column. A buffer is recommended when determining V_i experimentally. As indicated earlier, V_o can be determined by measuring the elution volume of a substance known to be totally excluded from the gel particles. Blue Dextran 2000 serves very well for this purpose because its color renders it visible and having a weight average molecular weight of 2 million it is completely excluded from all the gels. Therefore, the equation can be shown as $V_i = V_e(\text{KCl}) - V_e$ (Blue Dextran).

To operate a column efficiently, it is sometimes important to know what maximum sample volume can be applied and still obtain complete separation of the two solutes in the sample. If on a certain gel type, two substances with different molecular weights have K_d values, K_d' and K_d'', the elution volumes will differ by the separation volume (V_s). The separation volume is a measure of the separating efficiency of a column and can be calculated by the following equations:

$$V_s = V_e' - V_e'' = (V_o + K_d' \cdot V_i) - (V_o + K_d'' \cdot V_i)$$

$$V_s = (K_d' - K_d'') \cdot V_i$$

For complete separation of two materials, the sample volume must not be larger than V_s. For example, if a substance, which has a K_d value of 0.8, is to be separated completely on a certain gel type from a substance which has a K_d value of 0, then V_s must be smaller than 0.8 V_i. If the V_i of the particular gel is say 40% of the bed volume, the maximum sample size is 32% of the bed volume. The above equations can be used also as a rough estimation of the column size needed for a particular separation.

The gel filtration method has found numerous uses in analytical, preparative and industrial scale applications. Its uses in these areas are too numerous to be listed in detail; therefore, only a few significant examples will be given, such as the determination of molecular weight distribution for characterization of natural and synthetic polymers, of equilibration constants and molecular weight determination. Gel filtration has also found wide application for the separation of salts from virus suspensions, of nucleic acids from nucleotides and phenols. In the clinical area, it has been widely used for quick determination of free iodide in the presence of sugars, for protein binding of drugs and steroids, for radioimmunosorbent assay for proteins and for estrogens in the urine of pregnant women.

In purifying and isolating substances from natural products, gel filtration becomes an extremely useful tool. Amino acids can be separated from peptides, as well as fractions from mixtures obtained by enzymatic or chemical hydrolysis. Plasma and serum proteins can be divided into useful fractions by gel filtration. Since the introduction of a more highly cross-linked gel, the heretofore arduous task of separating amino acids now becomes rapid and relatively simple.

In industrial processes, gel filtration has been used for the preparation of protein-enriched milk, for purification of diptheria and tetanus toxoids, for desalting of albumin, for the removal of allergy-causing penicillaylated protein impurity from penicillins and for antiflu vaccine.

Other broad applications for gel filtration include concentration of high molecular weight substances, partition chromatography, zone precipitation and zone electrophoresis.

WARREN G. WATREL

CHROMIUM

Chromium is element 24 in the Periodic Table occurring in the subgroup of Group VIB that contains molybdenum and tungsten. It is a hard, blue-white metal crystallizing in the cubic system. The mineral, chromite, is its only important source.

Vauquelin discovered chromium in 1797 in a new red lead mineral known as crocoite ($PbCrO_4$) from Siberia. The name chromium is derived from the Greek word "chromos" meaning color because chromium compounds are highly colored.

Chromium is never found in the free state in nature, and most ores consist of the mineral, chromite. The ideal formula for this ore is $FeO-Cr_2O_3$ containing 68% Cr_2O_3 and 32% FeO, but the actual composition of the higher grade ores varies between 42–56% Cr_2O_3 and 10–26% FeO with varying amounts of magnesia, alumina and silica also present.

The chief sources of chromite are the U.S.S.R., Union of South Africa, the Philippines, and Southern Rhodesia, in that order.

Chromium ores are divided into three groups: (1) metallurgical, (2) refractory, and (3) chemical. About 52% of the ore used in the United States is metallurgical, 32% refractory, and 16% chemical grade.

Derivation. There are two classes of chromium available to industry: (1) ferrochromium and (2) chromium metal. Ferrochromium is produced by the direct reduction of the ore and will be discussed later. Chromium metal is produced electrolytically or by the reduction of chromium compounds, generally Cr_2O_3 produced from the ore by processing to remove iron and other impurities.

Commercial chromium metal has been prepared in large amounts by the reduction of Cr_2O_3 by aluminum.

$$Cr_2O_3 + 2Al \rightarrow 2Cr + Al_2O_3$$

A silicon reduction of Cr_2O_3 also produces chromium metal:

$$2Cr_2O_3 + 3Si \rightarrow 4Cr + 3SiO_2$$

This reaction is not self-sustaining and is carried out in an electric arc furnace. The product is similar to that from the aluminothermic process, but the aluminum content is lower and the silicon may run as high as 0.8%.

Chromium metal may also be produced by the reduction of Cr_2O_3 with carbon at low pressures to cause the chromium to volatilize.

There are two commercial processes used for the electrowinning of chromium: (1) chrome-alum electrolytes and (2) chromic acid electrolytes. For details of the processes see Ref. 1.

Physical Properties. The physical properties of chromium are given in Table 1.

Chemical Properties. Chromium is characterized chemically by the several valences it exhibits, chiefly 2, 3 and 6, although it may be 4 and 5 in chromium phenyl compounds and probably zero in chromium carbonyl, $Cr(CO)_6$; the intense and varied colors of its compounds; and the two forms in which the metal exists: active and passive. In the active form the metal reacts readily with dilute acids (with evolution of hydrogen) to form blue solutions of chromous, Cr^{++}, salts which absorb oxygen rapidly from the air and change to green solutions of chromic, Cr^{+++}, salts. If the metal is treated with oxidizing agents, such as nitric, chromic, phosphoric, chloric and perchloric acids, a thin oxide layer forms on the surface which passivates the metal so that it is then unreactive toward dilute acids. Halogen acids and halide salts must be absent to maintain the passive condition in acidic solutions. This passive film is responsible largely for the fine corrosion resistance of chromium and its alloys.

The dominant factors in the chemistry of chromium are the chromic compounds, including chromic oxide, Cr_2O_3, in which chromium has a +3 valence, and the chromates and dichromates, including chromium trioxide, CrO_3, in which the

TABLE 1. PHYSICAL PROPERTIES

Atomic number	24
Atomic weight	51.996
Electronic structure	$(1s)^2, (2s)^2(2p)^6, (3s)^2(3p)^6(3d)^5, (4s)^1$
Isotopes	50 (4.31%), 52 (83.76%), 53 (9.55%), 54 (2.386%)
Crystal structure, 20°C	Body-centered cubic, $a_o = 2.8844–2.8848$ Å
Density, 20°C	7.19 g/cc
Melting point	1875°C
Boiling point	2199°C
Heat of fusion	3.2–3.5 kcal/mole
Heat content solid and liquid chromium	H_t-H_{298} 400°K = 595 cal/mole
	1000°K = 4640
	2000°K = 14,220
Vapor pressure	log $P_{atm} = -20,473/t + 7467$
Entropy	5.58 cal/g-atom
Latent heat of vaporization at b.p.	76.635 kcal/mole
Specific heat	5.55 cal/mole or 0.11 cal/g/°C
Linear coefficient of thermal expansion, 20°C	6.2×10^6
Thermal conductivity, 20°C	0.16 cgs units
Electrical resistivity, 20°C	12.9 microhm-cm
Magnetic susceptibility, 20°C	3.6×10^{-6} emu
Total emissivity at 100°C nonoxidizing atm	0.08
Reflectivity, λ Å	3000 5000 10,000 40,000
%R	67 70 63 88
Refractive index	$\mu = 1.64–3.28$; $\lambda = 2,570–6,082$
Standard electrode potential valence	0 to + 3; 0.71 volt
Valence	+ 2, + 3, and + 6
Superconductivity	0.08°K
Electrochemical equivalent (hexavalent)	0.08983 mg/coulomb
Electrochemical equivalent (trivalent)	0.17965 mg/coulomb
Thermal neutron absorption cross section	3.1 barns

valence is +6. Depending on conditions, the chromic compounds are green or blue or violet, the chromates are usually yellow, although the insoluble silver chromate is red, and the dichromates are orange-red or red. The chromic ion also forms many coordination compounds having a coordination number of 6. The +3 and +6 chromium compounds are oxidizing agents and are in general stable in both solid and solution form, unless a reducing agent is present. The interesting chemistry of chromium compounds is treated extensively in many inorganic chemistry books.

Elemental chromium reacts with anhydrous halogens, HCl and HF. Aqueous solutions of hydrohalogens will dissolve chromium, as will sulfuric acid. It is not affected by phosphoric acid, fuming nitric acid or aqua regia at room temperature. At 600–700°C chromium is attacked by alkali hydroxides; at 600–700°C sulfides are formed when it is exposed to sulfur vapor or hydrogen sulfide; it also reacts with SO_2 in this temperature range. Oxidation of the metal occurs at about 1000°C in carbon monoxide, and it is attacked by phosphorus at about 800°C. Ammonia reacts with chromium at 850°C to form a nitride, and hot nitric oxide forms both a nitride and an oxide with chromium.

Chromium has a high oxidation resistance at elevated temperatures due to the formation of a tight oxide film on its surface.

The first and second ionization potentials of chromium are 6.74 and 16.6 volts, respectively.

Chromium Compounds. The principal industrial compounds of chromium are sodium chromate and dichromate, potassium chromate and dichromate, chromium trioxide, basic chromic sulfate, and chromic oxide.

The first two compounds mentioned are produced by roasting chrome ore with soda ash or soda ash and lime in a kiln:

$$2Cr_2O_3 + 4Na_2CO_3 + 3O_2 \rightarrow 4Na_2CrO_4 + 4CO_2.$$

Sodium dichromate is made from the chromate by the following reaction:

$$2Na_2CrO_4 + H_2SO_4 \rightarrow Na_2Cr_2O_7 + Na_2SO_4 + H_2O$$

These salts are used in leather tanning, the textile industry, for passifying the surfaces of metals, as catalysts and in organic oxidations.

Potassium chromate is made in the same manner as the sodium salt but using K_2CO_3 instead of Na_2CO_3.

Chromium trioxide which is used in chromium plating is prepared from sodium dichromate by the following reaction:

$$Na_2Cr_2O_7 + 2H_2SO_4 \rightarrow 2CrO_3 + 2NaHSO_4 + H_2O$$

Basic chromic sulfate used in leather tanning is prepared by reducing sodium dichromate by a carbohydrate or SO_2.

Chromic oxide, Cr_2O_3, is used as a green pigment and as a starting material for aluminothermic chromium metal. It is prepared from sodium dichromate by reduction with sulfur if pigment grade is required and with carbon if starting material is desired. The reactions are:

$$Na_2Cr_2O_7 + S \rightarrow Na_2SO_4 + Cr_2O_3$$
$$Na_2Cr_2O_7 + C \rightarrow Na_2CO_3 + Cr_2O_3 + CO$$

Chromic oxide is used as a pigment when chemical and heat resistance are required. It is also used as a ceramic color, for coloring cement, for green granules for asphalt roofing, in camouflage paints and in the production of chromium metal and Al-Cr master alloys.

Chromium Coatings. Chromium surfaces are produced on other metals by decorative or hard electroplating and chromizing. Decorative plate varies in thickness between 0.00001 and 0.00002 inch and is usually deposited over a layer of electrodeposited nickel. "Hard" plating is used because of its wear resistance and low coefficient of friction.

For decorative and hard chromium plating, solutions of CrO_3 are exclusively used. A small amount of another ion (sulfate) is necessary to cause chromium deposition and H_2SO_4 in the ratio CrO_3/H_2SO_4 of 100/1 by weight is used but this ratio may vary.

Ferrochromium. The various grades of ferrochromium are the alloys most used for adding chromium to steels.

To produce high-carbon ferrochromium coke and chromite are fed into the top of an open-top submerged-arc furnace and the molten alloy is collected at the furnace bottom and cast into chills. It is used to produce steels in which both chromium and carbon may be present or where blowing the bath with oxygen to reduce the carbon content is feasible.

Low-carbon ferrochromium is made by the silicon reduction of chromite in a two-stage process. First, a high-silicon ferrochromium, practically carbon free, is made in a submerged-arc furnace. This product is then treated with a synthetic slag containing Cr_2O_3 in an open-arc furnace. This grade of ferrochromium is used in the production of steels in which the presence of carbon is harmful.

Chromium Alloys. Although pure chromium has been produced in the form of ingots, rod, sheet, and wire, its use as metal is limited due to its low ductility at ordinary temperatures. Most pure chromium is used for alloying purposes such as in the production of nickel-chromium or other nonferrous alloys. The chief use of ferrochromium is as an addition agent to steel alloys. Chromium is used in combination with molybdenum, nickel, manganese, and vanadium in steels for high-strength applications.

Nickel or nickel plus manganese are added to high-chromium steels to make them austenitic with improved corrosion and oxidation resistance.

Chromium is added to cobalt-base alloys in amounts up to 25% to give corrosion resistance and hardness. Cobalt-chromium-tungsten combinations are used for cutting tools and hard facing.

Nickel-base alloys with up to 20% chromium have high heat and electrical resistance.

Chromium Refractories. Chrome refractories are neutral or sometimes considered basic in character. This important group of refractories is made from mixtures of magnesite and chrome ore. Magnesite consists chiefly of periclase or crystalline magnesia, MgO, and the chrome ore consists of several spinels, such as $FeCr_2O_4$ or $FeO\cdot Cr_2O_3$ and $MgAl_2O_4$ or $MgO\cdot Al_2O_3$.

Turkish chrome ore was used for the first chrome refractories in the United States. Since domestic sources have never been adequate or suitable, ores from Greece, South Africa, and the Philippines have been used.

In one method for making refractory brick, a mixture of chrome ore and magnesite is fused in an arc furnace and cast into brick. The chemical analysis is 62.5% MgO, 18.5% Cr_2O_3, 6% Al_2O_3, 11% FeO, 1.0% SiO_2, 0.8% CaO. The bricks so formed have high transverse strengths at elevated temperatures. The bricks are high density with low voids and their slag resistance at high temperatures is also high.

Typical applications include roofs of open hearths with high oxygen-input hearths, lower sidewalls of vacuum degassing vessels, and the sidewalls of electric furnaces and copper converters.

F. E. BACON

Reference

Bacon, F. E., "Chromium," pp. 145–154, "Encyclopedia of the Chemical Elements," Hampel, C. A., Editor, Van Nostrand Reinhold Co., New York, 1968.

CLATHRATE COMPOUNDS

This term is derived from the Latin word *clathratus,* meaning "enclosed by cross bars of a grating." In its narrowest definition, a clathrate is a single-phase solid in which atoms or molecules of a guest component are completely surrounded and kept as in a cage by the crystalline lattice of a host component without any sensible interaction between them except van der Waals forces. Many authors have broadened the above definition to include compounds in which the guest component is held as in an open channel or even as in a sandwich by the crystalline lattice of the host component. [Sister Martinette Hagan, "Clathrate Inclusion Compounds," Reinhold Pub. Corp. New York, 1962; V. M. Bhatnagar, "Clathrate Compounds," S. Chand & Co., Ram Nagar New Delhi, India, 1968]. The crystalline lattice of the host component is generally much more expanded in its clathrate form than in its pure form. The stability of many clathrate compounds is due to the formation, in the expanded lattice of the host itself, of strong interactions such as hydrogen bonds which do not exist to such an extent in the normal crystalline lattice of the host alone; furthermore, van der Waals forces and even sometimes weak charge transfer interactions contribute to the stability of clathrates. Steric factors are always of prime importance because the clathrated molecule must adopt to the size and the shape of the rigid cage formed by the crystalline lattice of the host. A few clathrates that have historical, theoretical or practical importance will be used to illustrate the subject.

Hydroquinone Clathrates. When hydroquinone is crystallized under a high pressure of argon, there is obtained a compound which is perfectly stable at room temperature in dry air, although it contains up to 70 times its own volume of argon; because of its inertness, argon can only be held by a purely mechanical effect in the crystalline lattice of hydroquinone. Powell in England [*J. Chem. Soc.,* **1948,** 61], resolved the structure of this compound by

x-ray diffraction and coined "clathrate" as a generic name for such inclusion compounds. In the hydroquinone clathrates all the hydroxyl groups are tied together six by six through hydrogen bridges, forming hexagons with oxygen atoms at each angle ·and the benzenic rings pointing alternatively up and down; the interlocking of all hydroquinone molecules forms an infinite three-dimensional cagework, the voids of which have a diameter of 7.9 Å, big enough to entrap argon atoms. A total filling of the cages (1 argon for 3 hydroquinones) is never reached, but only approached when a high pressure of argon is applied during the crystallization of hydroquinone. Helium and neon are not clathrated because their atoms are too small to be retained by the cage; but krypton and xenon are easily clathrated. The radioactive krypton (Kr-85)-hydroquinone clathrate has been proposed as a safe and convenient source of beta radiation. Hydroquinone is able to clathrate many other compounds and its hydrogen sulfide clathrate was prepared as early as 1849 by Wohler [*Annalen,* **69,** 297 (1849)].

Dianin's Compounds ($C_{18}H_{20}O_2$). This compound is obtained by condensing mesityl oxide and phenol in the presence of dry HCl. It forms clathrates with most of the solvents from which it can be recrystallized. Here again, as for hydroquinone, 6 molecules of Dianin's compound are tied together by hydrogen bonding through their hydroxyl groups, forming an aggregate looking like two cups joined through their bottoms; the stacking of those aggregates forms closed cavities big enough to hold from one to three foreign molecules depending on their sizes, e.g., one molecule of ethylene dichloride, two of carbon tetrachloride and three of methanol. These clathrates have little practical value because they do not discriminate enough between molecules of different sizes or shapes.

Phenol also forms clathrates of this type with a wide variety of compounds such as hydrogen sulfide, hydrogen chloride, carbon dioxide, xenon, etc.

Gas Hydrates. Water forms true clathrates by crystallization in the presence of several gases; these are commonly called gas hydrates; their structure was elucidated by von Stackelberg [*J. Chem. Phys.* **19,** 1319 (1951)]. There are two types of gas hydrates differing by the size of the cubic unit cell: the first type has a cell of 12 Å made of 46 water molecules; it is formed in the presence of small molecules which can be accommodated in 6 cages of 5.9 Å and in 2 of 5.2 Å; the second type has a cell of 12.3 Å made of 136 water molecules; it is formed in the presence of larger molecules (e.g., $CHCl_3$) which are accommodated in 8 cages of

6.9 Å; and there are 16 smaller cages of 4.8 Å which can only be filled by small molecules or atoms (e.g., Kr). Compounds such as $8H_2S \cdot 46H_2O$; $6Br_2 \cdot 46H_2O$; $8C_3H_8 \cdot 136H_2O$ or $CHCl_3 \cdot 16Kr \cdot 136H_2O$ have been well characterized. An important feature is that these clathrates melt well above 0°C: this accounts for plugging of pipelines fed with wet natural gas even several degrees above freezing temperature. Several processes based on the formation of gas hydrates (mostly propane hydrate) have been studied for the desalinization of of sea water.

Nickel Werner Complexes. Three principal types of nickel Werner complexes form clathrates, but only with aromatic molecules. This fact suggests that van der Waals forces are not the only ones to contribute to the stability of such clathrates. In some cases, interactions of the charge transfer type between the host and the guest molecules have been clearly demonstrated; nevertheless, the clathrate character remains the most important because, for molecules having the same charge transfer potential, it is their shape (position isomerism) which is decisive for clathrate formation.

Cyano monoammine nickel (II). This complex form clathrates of formula $Ni(CN)_2NH_3 \cdot M$ wherein M can be benzene, thiophene, furan, pyrrole, aniline or phenol. In its clathrate form this complex crystallizes in planar arrays of nickel cyanide; the ammonia molecules, bonded to nickel atoms, project above and below each plane, delineating cavities between them; molecules larger than that mentioned above cannot be accommodated in the cages, offering a purification process for the latter.

Thiocyanato tetrakis (4-methylpyridine) nickel (II). Schaffer [*J. Amer. Chem. Soc.,* **79,** 5870 (1957)] has described many clathrates of this complex with benzene derivatives; they show a striking selectivity for para isomers; pure meta-xylene has been prepared on a semicommercial scale from a meta-para eutectic mixture by clathration of the para isomer. This complex forms a mole per mole clathrate by simple contact with para-xylene. This fact indicates that this clathrate could be of the channel type, although its structure is not yet known.

Thiocyanato tetrakis (alpha-phenylalkylamine) nickel (II). Hanotier [*Bull. Soc. Chim. Belges,* **74,** 381 (1965)] has prepared more than 70 complexes of the type $[Ni(NCS)_2(Y-C_6H_4-CH(Alk)NH_2)_4]$ in which alk is a primary alkyl group and Y can be a wide variety of substituants on any position of the benzenic ring; it is almost always possible to select one of those complexes for clathrating selectively any desired isomer in a mixture of aromatic molecules. Table 1 illustrates selectivities which are attained for the separation of diethylbenzene (D.E.B.) isomers.

Many of these complexes do not form clathrates by simple contact but must be prepared in the presence of the compound to be clathrated; clathrates having such behavior are probably of the cage type. It has also been shown that a factor contributing to the stability of these clathrates is a weak electron donor-acceptor interaction between the aromatic rings of the host and of the guest molecules. These clathrates have been used to bring about the resolution of optically active amines [*Nature,* **215,** N. 5100, 502 (1967)].

Urea Adducts. In the presence of straight-chain compounds such as paraffins, fatty alcohols, fatty acids, fatty amines, ketones or *n*-alkylhalides, urea crystallizes in an hexagonal system in which its molecules are hydrogen-bonded and form a spiral around the guest molecules; urea adducts are definitely of the open channel type and fit only in the broadened definition of clathrates. These compounds have been found in 1940 by Bengen in Germany and closely studied by Schlenk [*Liebigs Ann.,* **565,** 204 (1949)]. The channel has a free diameter of 5 Å and there is about 0.7 urea molecule per clathrated methylene group. Several factories, especially in U.S.S.R., use urea adduction for dewaxing petroleum products and for producing pure straight-chain paraffins. Thiourea forms clathrates of the same type with more bulky molecules like branched paraffins; the channels have a free diameter of 7 Å.

Other Clathrates. Several compounds have a macrocyclic structure having in its center a hole which can accommodate foreign molecules, e.g., tri-*o*-thymotide, cyclodextrins, desoxycholic acid, cycloveratryl and few others. Graphite gives sandwich-type inclusion compounds with alkali metals and with fluorine. Purely inorganic compounds, such as zeolites and montmorillonites, form with many organic molecules clathrates of the channel type for the former and of the sandwich type for the latter; both have found important industrial applications (see Molecular sieves).

P. DE RADZITZKY

Reference

Mandelcorn, Lyon, "Clathrates," *Chem. Rev.,* **59,** 827 (1959); Cramer, Friedrich, "Einschluss-Verbindungen," Springer Verlag, Berlin, 1954.

CLAYS

The term "clay" has decidedly different meanings to technologists in different fields, but has been well

TABLE 1. SEPARATION OF D.E.B. ISOMERS BY $[Ni(NCS)_2(Y-C_6H_4CH(ALK)NH_2)_4]$

Amine		Mole % D.E.B. isomers in						Weight % D.E.B. in clathrate
		feed			clathrated phase			
−Alk	−Y	o	m	p	o	m	p	
−CH₃	−H	32	31	37	93	4	3	8.6
−C₄H₉	p−F	5	63	32	1	99	0	11.4
−CH₃	m−Br	32	31	37	3	4	93	10.8

defined by Keller[1] as "a naturally occurring sediment (including that obtained by alteration in situ by supergene and hydrothermal processes) or sedimentary rock composed of one or more minerals and accessory compounds, the whole usually being rich in hydrated silicates of aluminum, iron or magnesium, hydrated alumina, or iron oxide, predominating in particles of colloidal or near-colloidal size, and commonly developing plasticity when sufficiently pulverized and wetted."

Geologists and soil scientists emphasize the fine particle size aspect and use the term clay to describe rock particles of any kind with diameters finer than 1/256 mm or materials which consist predominantly of particles with dimensions less than two microns. Ceramists, on the other hand, generally use the term to refer to aluminosilicates which exhibit plasticity in aqueous suspension.

The clay minerals have been classified by Grim[2] as shown below:

Classification of the Clay Minerals

I. Amorphous
   Allophane group
II. Crystalline
   A. Two-layer type (sheet structures composed of units of one layer of silica tetrahedrons and one layer of alumina octahedrons)
      1. Equidimensional
         Kaolinite group
         Kaolinite, nacrite, etc.
      2. Elongate
         Halloysite group
   B. Three-layer types (sheet structures composed of two layers of silica tetrahedrons and one central dioctahedral or trioctahedral layer)
      1. Expanding lattice
         a. Equidimensional
            Montmorillonite group
            Montmorillonite, sauconite, etc.
            Vermiculite
         b. Elongate
            Montmorillonite group
            Nontronite, saponite, hectorite
      2. Nonexpanding lattice
         Illite group
   C. Regular mixed-layer types (ordered stacking of alternate layers of different types)
      Chlorite group
   D. Chain-structure types (hornblende-like chains of silica tetrahedrons linked together by octahedral groups of oxygens and hydroxyls containing Al and Mg atoms)
      Attapulgite
      Sepiolite
      Palygorskite

It should be mentioned that a new classification scheme for the phyllosilicates is under consideration.[2]

Clays originate by weathering (and related reactions) of rock masses, or may be produced by hydrothermal reactions, which may produce clay deposits at considerable depth. The so-called primary clays are those found in their place of formation (e.g., the kaolin deposits found in the Cornwall district of England), while the secondary, or sedimentary deposits have been transported after weathering by water, wind, ice, etc. and have then been deposited in beds, lenses or pockets with other sedimentary rocks, where they may have been further altered by the action of ground water or other agencies. The kaolin deposits found in Georgia are of such origin.

Clays are used in very large volume in the manufacture of a variety of products. Factors influencing the properties and therefore the use of clays are the identity and abundance of the clay mineral component, the amount and type of nonclay material (including any organic material), the particle size distribution and particle shape of the clay mineral and the kind and amount of exchangeable ions and soluble salts.

Clays and shales suitable for use in the manufacture of brick, tile and other heavy clay products are found in every state of the Union and some 40 to 50 million tons of such material are used each year.

Clay suitable for use in paper and paint manufacture, in the production of ceramics and in drilling mud manufacture are much less widely distributed and suitable deposits for such application are limited. Kaolin clay of the quality and volume required for paper coating and filling use has been found in only two areas of the world—from an area in Central Georgia and from the Cornwall District in England. Ball clays, those exhibiting good bonding power and high plasticity (properties of importance to the ceramist) are found primarily in Tennessee and Kentucky and in parts of England. Most of the swelling type of bentonite, used in preparing oil well drilling fluids, comes from Wyoming deposits.

The largest volume use of clays is in the manufacture of ceramics—porcelain, pottery, chinaware, as well as glass, enamels, bricks, refractories. The important clay properties for such use are plasticity, firing properties, shrinkage on drying, and color.

Some eight or ten million tons per year of white kaolin clay are used in the production of paper and paperboard and in the production of paints. Clay for such use must be virtually free of discolorants, and must be of fine particle size. A variety of grades are produced, varying from ultrafine, very white clay for use in producing high-gloss coated paper and paperboard to relatively coarse grades which are used as paper fillers. The important properties of clays used in paper and paint manufacture are whiteness, particle size, particle shape, uniformity and predictable rheological properties of aqueous suspensions of the clay.

Kaolin clays are also used to fill rubber goods and plastics, in the preparation of catalysts, in medicines, pharmaceuticals and cosmetics, and are potential ores of aluminum.

About three million tons per year of bentonite (composed essentially of montmorillonite) are used in catalyst manufacture, in the preparation of drilling muds, in iron ore pelletizing, in foundries and steelworks (for preparing molding sands) and in a variety of other applications. The swelling properties of the clay, to permit formation of thixotropic systems, is one of its important attributes.

The term fullers earth (from the cleaning or fulling of wool whereby oil and dirt particles are removed by application of an aqueous slurry of earth) now refers to clays composed of attapulgite and some montmorillonites, although some other clays also exhibit superior decolorizing powers. Some one million tons per year of the material are used to decolorize mineral and vegetable oils, as insecticide and fungicide carriers, etc.

The combination of properties that make clays of unusual industrial interest are their extreme fineness of individual particles, their diversity of particle shape, varying from platy to tabular, the relatively open crystal lattice of many of the clays, yielding useful adsorptive and ion exchange properties and their relatively wide and plentiful distribution, with consequent low cost.

H. H. MORRIS

Cross-references: *Colloid Chemistry, Ceramics, Drilling Fluids.*

References

1. "Kirk-Othmer, Encyclopedia of Chemical Technology," Vol. 5, (2nd Ed.), John Wiley and Sons, Inc., New York, 1964.
2. Grim, R. E., "Clay Mineralogy," 2nd Ed., McGraw-Hill, New York, 1968.

CLEMMENSEN REACTION

Several efficient methods are available for reduction of carbonyl compounds, including the ketones produced by the Friedel-Crafts synthesis. One, the Clemmensen method of reduction (1913), consists in refluxing a ketone with amalgamated zinc and hydrochloric acid. Acetophenone, for example, is reduced to ethylbenzene. The method is applicable to the reduction of most aromatic-aliphatic ketones and to at least some aliphatic and alicyclic ketones. The reaction apparently does not proceed through initial reduction to a carbinol, for carbinols that might constitute intermediates are stable under the conditions used. When substances are sparingly soluble in aqueous hydrochloric acid, improved results sometimes are obtained by addition of a water-miscible organic solvent such as ethanol, acetic acid, or dioxan. Particularly favorable results are obtained by addition of the water-insoluble solvent toulene (Martin, 1936), stirring the mixture vigorously, and using for amalgamation zinc that has been freshly melted and poured into water (Sherman, 1948). The ketone is retained largely in the upper toulene layer and is distributed into the aqueous solution of hydrochloric acid in contact with the zinc at so high a dilution that the side reaction of bimolecular reduction is suppressed.

The Clemmensen method of reduction is applicable to the γ-keto acids obtainable by Friedel-Crafts condensations with succinic anhydride (succinoylation) and to the cyclic ketones formed by intramolecular condensation.

This and the Wolff-Kishner method of reduction supplement each other; the Clemmensen method is inapplicable to acid-sensitive compounds and the Wolff-Kishner method cannot be used with compounds sensitive to alkali or containing other functional groups that react with hydrazine.

L. F. FIESER
MARY FIESER

CLINICAL CHEMISTRY

Clinical chemistry is the branch of chemistry which deals with the composition and measurement of the secretions, excretions, concretions and fluids of the human body in health and disease, and the chemical composition and metabolism of cells and tissues. It is particularly concerned with the detection of substances (or their derivatives) for diagnostic or therapeutic reasons and the detection of poisons (or their derivatives). Clinical chemical analyses performed by, or under the direction of, a clinical chemist requires a written request from a physician, and the test results are reported only to him.

In the latter part of the nineteenth century, scientists in many parts of the world began to identify the various chemical constituents in blood, urine, and other body fluids, and to elucidate their function and role in various metabolic processes. After development of the venipuncture (withdrawal of blood from a vein using needle and syringe), quantitative chemical analysis of the various biochemical components of blood became practical. In the second and third decade of this century researchers began to develop a relationship between various pathologic processes and alterations in the concentration of blood chemical components. Laboratory methods became particularly useful in the diagnosis and treatment of diabetes mellitus, and liver and kidney diseases. As the common diseases were better understood, clinicians became increasingly interested in rarer diseases. The clinical chemist is often able to develop specific tests for these rare diseases, thus increasing the frequency of their detection.

Among the earliest methods were Bang's technique for blood sugar (glucose) in 1907 and the determination of creatinine in urine by Folin in 1904. Bang introduced several methods for analyzing milligram quantities of blood constituents, including a micro Kjeldahl analysis involving steam distillation of ammonia, and the technique of separating and washing quantitative precipitates by centrifugation. Bang was not primarily interested in methodology, but developed these methods to assist in the study of various metabolic and other biochemical problems.

If modern clinical chemistry could be traced to a single investigator, it would have to be Folin. His introduction of the use of a protein-free filtrate, providing a water-clear fixed dilution of blood, enabled new methods to be developed and led quite naturally into the use of colorimetry as an analytical technique.

Many other investigators played important roles in the development of clinical chemistry, but the work of a few was outstanding. These include Van Slyke, in the early 1920's developed volumetric and manometric methods of gas analysis, and made possible a whole system of analysis for constituents in blood, urine, and organic compounds. Significant also were Bloor's contributions in lipide chemistry, the work of Benedict and Nash on ammonia formation and numerous analytical methods, V. C. Meyers and his "Practical Chemical Analysis of Blood" which first appeared in the early part of this century, Tiselius and Svedberg and their development of the electrophoresis and ultracentrifuge, and Hawk's classic book "Practical Physiological Chemistry" first published in 1907.

Today, well over 100 blood and urine constituents are routinely analyzed in clinical chemistry laboratories. The more common tests include blood glucose, urea nitrogen, uric acid, creatinine, cholesterol, proteins (and their fractionation), bilirubin, electrolytes (sodium, potassium, calcium, inorganic phosphate, carbon dioxide, and chloride), protein-bound iodine, iron and lipids. Also common are tests for drugs such as salicylates and bromides in blood, and numerous enzyme assays, including amylase, lipase, alkaline and acid phosphatase, the transaminases, lactic dehydrogenase, and creatine phosphokinase. Increased interest in the levels of blood and urine enzymes followed introduction of the concept of biochemical biopsy. This concept states that any tissue, normal or abnormal, which can be distinguished morphologically should also be distinguishable by its biochemical constituents and milieu, particularly in terms of enzymes. Since the enzymes from various organs and tissues are secreted into body fluids, such as blood or cerebrospinal fluid, assays of these enzymes should permit location of the source and type of abnormality. These fluids are generally much more accessible than the tissue or organ itself. Modern research is confirming this theory, and as the enzymes and their iso-enzymes are being delineated their estimation is becoming a useful diagnostic tool. Iso-enzymes (or isozymes) are various molecular forms of enzymes differentiated from each other by use of chemical inhibitors, determination of Michaelis' constants, thermal inactivation, electrophoretic mobility, column chromatography, immunochemical properties, and in the case of enzymes requiring coenzymes, by kinetic studies with coenzyme analogues.

Assays of hormones and their metabolites and research investigations in endocrinology are also an important part of clinical chemistry. Included in the compounds of clinical interest are steroid, peptide, and protein hormones.

Chemical toxicology also plays an important role in clinical chemistry. Some clinical chemists, such as those connected with Medical Examiner's offices, are exclusively concerned with this type of analysis. The causative factor in accidental deaths, death by suicide, and cases of murder may be chemical compounds, and the chemist must be able to identify the compound, determine its level in blood or tissue, and be familiar with its mode of action in humans.

Ultramicro analysis of blood constituents is widely used in clinical laboratories. Most of the more commonly analyzed blood compounds can be determined using specimens of 10–25 microliters (0.010–0.025 ml). Clinical chemists usually define ultramicro in terms of specimen size, rather than the concentration of the compound being analyzed because they are often severely limited in the amount of sample available. Ultramicro specimen sizes are no greater than 50 microliters. Sample sizes for all techniques have decreased over the years because of the increasing multiplicity of tests being ordered on a single individual.

Techniques used in modern clinical chemistry include gas chromatography, radioimmunoassay, liquid scintillation spectrometry, and atomic absorption.

Clinical chemistry has contributed in no small measure to the lowered mortality and increased longevity of man, serving as a bridge between the academic and basic research investigators who study the nature of disease, and the clinicians who apply this knowledge to the diagnosis and treatment of human subjects.

R. P. MacDonald

Automatic Clinical Chemical Analysis

Although clinical chemistry is not readily defined as a discretely delimited area of analysis, as many as 160 to 200 different tests may be requested for analysis on body fluids. This, in general, makes it somewhat more complicated than an industrial process. The demands for more, faster, and more accurate information from blood have laid the foundation for acceptance of automatic analytic aids in clinical chemistry. This acceptance is facilitated by the knowledge that in the future, even greater amounts of information will be obtainable from body fluids than is presently available. The field is not static. In order that more information be obtained from a single specimen, it has been necessary to reduce the aliquot allotted for each analytic test result. There is not a comfortable amount of testing material available. The current goal then of the hospital affiliated clinical laboratory is to attain a rapid, repetitive, accurate system of microanalysis, the results of which are presentable within a matter of minutes.

Two distinct automated systems of analysis have been formulated . . . flow stream and robot. Chronologically, the flow system analysis unit was the first to be used by clinical chemists. The differences between the flow stream and robot units are mainly in the methods of sample handling . . . i.e. the functions normally described as pipetting, reagent addition, transfer and presentation for measurement.

In the flow stream technique, the multichannel peristaltic pump creates a flowing stream into which are pumped sample and reagents at appropriate points downstream. Dialysis has been employed to separate protein from nonprotein constituents by creation of a separate stream flowing on the opposite side of a dialysing membrane to act as recipient for diffusible components. Any of the streams may be further subdivided and each subsidiary branch used for determination of one or more of initial sample constituent. Heat may be applied to one or more of the flow streams with the length of time for heating and reaction being a function of stream velocity and tube length. Ultimately the streams enter some measuring device: absorptiometer, fluorimeter, flame photometer, etc. and results presented in chart or digitized form.

A recent hybrid between flow stream and robot has used centrifugal force for metering, reagent addition and transfer, moving from center to periphery. Multiple samples may be run simultaneously and during rotation, with oscilloscopic display, absorptiometric measurements on each sample at one second intervals are averaged using as many as 20 readings per sample. Computer linkup for data reduction is also a feature.

In the robot approach, manipulations historically performed by hand have been mechanized (and miniaturized) and arranged in a sequence controlled by a programmer. Electrical and pneumatic actuation has provided for the necessary operations of metering and transfer with reaction generally proceeding in a receptacle on a turntable. After sample presentation to the appropriate measuring unit, the flow stream and the robot become similar in terms of data reduction and presentation.

Inadequate information exists for supplying relative merits to the different types of approach since the robot and centrifugal hybrid have entered the field too recently. Suffice to say that the flow stream analyzer has already altered the entire complexion of clinical chemical analysis and continues to show vigorous progress.

Since the robot approach is the summation of a number of individual steps arranged in a sequence, it should be mentioned that commercially available units allow for automation of individual steps as automatic aids to analysis. This means that partial or complete automation can be employed for over 90 percent of all clinical chemical procedures.

Following the acquisition of automatic instrumentation, the field at present is examining the feasibility of computer technique for control, data reduction, storage and retrieval purposes. Although this aspect is still an infant in this field, a few pilot studies have indicated that many clinical chemical laboratories in the near future will have quality control of routine analytical processes under computer direction and that data handling will also be facilitated and less subject to human errors.

WALTON H. MARSH

COAL AND COKE

Coal is a solid, combustible carboniferous substance formed by the decomposition of vegetable matter without free access of air. The plant debris from which coal had its origin accumulated in peat swamps and, in the presence of stagnant water and buried under rapidly increasing deposits of vegetable matter, was decomposed in almost complete absence of air. Assisted by microorganisms, a chemical transformation took place resulting in the formation of peat. The conversion of peat into coal occurred after the disappearance of most of the water and under conditions of increased pressure and temperature. This conversion, extending over many millions of years, was progressive, leading first to lignite and then to higher rank coals, such as bituminous coal and anthracite.

In color, lignite may be brown but all coals of higher rank are black. The usual range of specific gravity is from 1.15 to 1.5, with 1.3 a typical figure for bituminous coal. When heated out of contact with air, decomposition begins at 300–400°C, and gas and usually tar are evolved. Chemically, coal is composed chiefly of condensed, aromatic ring structures of high molecular weight. Detailed, standard methods for the analysis and testing of coal and coke have been published by ASTM.

Coal is widely distributed throughout the world, although relatively small reserves have been found in Africa, South America, and Oceania. Recoverable reserves of coal in the United States have been estimated at almost one trillion tons, which is probably about one-third of the world's total. Other large reserves exist in certain parts of Asia and Europe. In general, anthracite and the high rank bituminous coals underlie the Eastern part of the United States, while the lower rank bituminous and sub-bituminous coals and lignite are found in the West.

The majority of coal in the United States is produced from underground mines which are almost completely mechanized, including the growing use of continuous-mining machines. A minor but increasing portion of coal is being produced by "strip mining" from the surface. As a result of these improved mining methods, productivity of coal mines in the United States has increased to almost 20 tons per man-day in 1968 as compared to about 5 tons per man-day in 1940. After the "run-of-mine" coal has been brought to the surface, it is usually screened into various sizes, and is then often cleaned by a process based on differences in specific gravity, in order to reduce the ash and sulfur content.

The United States, with 18% of the world's coal production, is exceeded only by Russia (21%). Statistics for 1969 are summarized below (U.S. Bureau of Mines):

	Bituminous	Anthracite
Production, million tons	561	10
Percent from strip mines	35	44
Value at mine, per ton	$4.99	$9.62

Due to increasing efficiency of fuel utilization, as well as growing competition of other fuels, production of coal in the United States has shown almost no increasing trend over the last few decades.

Approximately 73% of the bituminous coal and substantially all of the anthracite in the United States are burned to furnish power and heat and 10% is exported. The remaining 17% of the bituminous coal is processed in coke ovens to make coke and, usually, coal chemicals.

USE-PATTERN OF BITUMINOUS COAL (U.S.)

Electric utilities	55%
Coke ovens	17%
Other industries	15%
House heating	3%
Export	10%

The use of coal for generation of electricity has increased substantially, despite the fact that increased efficiency in utilization has decreased the average consumption of coal from 3.20 lb. per kwh in 1919 to 0.87 lb. in 1968. Some other uses, particularly by railroads, have been adversely affected by competitive fuels.

In the United States a considerable volume of research and development work is aimed at finding new uses for coal. Much of this has to do with the synthesis of liquid or gaseous fuel and/or chemicals by the application of hydrogenation and other techniques. Other studies include the use of coal in fuel cells, magnetohydrodynamic generators, gas turbines, etc.

Coal Carbonization. Carbonization is the process of heating coal to relatively high temperatures out of contact with air in coke ovens or other equipment. Thermal decomposition of the large molecules of the coal substance results in the evolution of volatile compounds in the forms of gas, tar and water vapors, and the formation of a carbonaceous residue of coke. Commercial coal carbonization today is confined almost entirely to the high-temperature process (900–1150°C) in batteries of chemical-recovery coke ovens. Most coke ovens today have a capacity of about 16 tons of coal per charge, but recently ovens of double this capacity have been installed. The coal is heated for about 16 hours, during which period the gas and vapors leave the oven chamber to be separated into the various coal chemicals and other products. At the end of the coking period the red-hot coke is "pushed" out of the oven and is quenched by a water spray. Most of the bituminous coals have the property of forming coherent coke on heating, and can usually be used in the process, but all other ranks of coal are unsuitable. In contrast to chemical-recovery ovens, "beehive" coke ovens produce only coke, since all the gases and vapors are burned. Comparatively few of these beehive ovens remain in operation today. Another process, which is operating on a limited scale in some other countries, is low-temperature carbonization (500–750°C), which yields products differing considerably in character and amount from those produced at higher temperatures.

In 1971, 82,000,000 tons of coal were carbonized in chemical-recovery coke ovens in the United States, leading to the following average quantity of products per ton of coal (U.S. Bureau of Mines):

Coke	0.691 ton
Coke breeze	0.054 ton
Tar	8.28 gal.
Ammonium sulfate	14.07 lb.
Gas	10,470 cu. ft.
Light oil	2.46 gal.

In addition, 1,200,000 tons of coal were carbonized in beehive ovens, with an average yield of 0.600 ton of coke per ton of coal. Total production of coke in the United States in 1971 amounted to 57,400,000 tons. The tar and light oil are prolific sources of benzene, naphthalene and other prime chemicals used in the synthesis of a broad range of aromatic organic chemicals.

The coke made in chemical-recovery coke ovens is in the form of dark gray blocks of irregular size and shape. Due to its cellular structure, coke is quite light, with apparent specific gravity usually in the range of 0.8 to 1.0. After quenching, coke is screened into various sizes depending on the intended use. Very small coke (under ½ inch) is known as "breeze." The ash content of coke is approximately ⅓ greater than that of the coal from which it is made, since all the coal ash remains in the coke. The volatile matter in high-temperature coke is usually under 1.5%.

The major use for coke in the United States is the reduction of iron ore to pig iron in blast furnaces. Much smaller amounts are utilized for the melting of iron or other metals in foundry cupolas, for the manufacture of water gas and producer gas, and for house heating. Since the steel industry uses 90% of the coke produced in the United States, production of coal chemicals is dependent to a very large degree on the rate of steel manufacture. The economics of chemical-recovery coke

ovens rest very largely on the financial return from the coke.

ALFRED R. POWELL

Cross-references: *Fuels, Aromatic Compounds, Coal Tar.*

COAL TAR

Coal tar is one of the products obtained in high-temperature carbonization of coal, carried out in chemical-recovery coke ovens. Coking coal is charged into the ovens and, after a coking period of about 16 hours,, the hot coke is removed. During this operation gas and vapors are evolved which, upon cooling, condense to liquid tar and water. Since the tar is heavier than water, it settles out and is then ready for processing. Coal tar is a black, viscous liquid having a specific gravity of about 1.17. The average yield of tar is 8 to 9 gallons per ton of coal carbonized, or a total production of 679,000,000 gallons in the United States in 1971.

Distillation of Tar. Most of the coal tar produced in the United States is subjected to a fractional distillation process to recover a variety of products. However, an average of about 15% of the tar is used as a fuel for open-hearth furnaces by the steel plant producing it. Various types of batch and continuous stills are used for distillation. Some of these are equipped with fractionating columns while others are not, depending on the kind of products desired. There is a general tendency recently to install continuous stills, especially in the larger plants.

Tar Products. Distillation procedures are usually determined by the quantity and quality of the products that are desired. The following yields of the various fractions are typical:

Light oil (up to 200°C)	2%
Chemical oil (200°–250°C)	15%
Creosote oil (250°–350°C)	21%
Pitch (residue)	62%

Light Oil. This fraction contains some benzene and its homologues, as well as the lower-boiling tar acids and bases. The volume of this light oil fraction is so small that it is usually added to the larger volume of light oil removed from the coke-oven gas.

Chemical Oil. Chemical oil contains tar acids, tar bases, and naphthalene. The tar acids, which include phenol and its homologues, are removed from the oil by treating with a solution of sodium hydroxide. The phenols are recovered from the resulting sodium phenolate solution by "springing" with carbon dioxide. The tar bases, such as pyridine and its homologues, may or may not be recovered depending on the demand for these products. They are removed from the oil by washing with dilute sulfuric acid, followed by springing with ammonia or sodium hydroxide. Naphthalene comprises 50% or more of the chemical oil. It is recovered by distillation or crystallization or a combination of the two, and may be further purified if necessary. Following this the residual oil is usually added to the creosote.

Creosote Oil. The principal use of creosote oil is for wood preservation. It is the most widely used agent for this purpose since it has high toxicity toward fungi and insects that destroy wood, is easily applied, and maintains its protective action for many years. Wood preservation equipment usually consists of large horizontal retorts. The utility poles, railroad ties, or other material to be treated are loaded onto cars and then hauled into a retort. After the end doors are tightly closed hot creosote oil is pumped into the retort and the pressure is raised to as high as 250 psi. In normal practice the wood is impregnated with 6 to 8 lbs. of creosote oil per cubic foot of wood, but may be higher if required. In some cases the retort is first put under vacuum to get more penetration of the oil into the wood.

The highest-boiling fraction of creosote oil is anthracene oil, which is usually collected along with the creosote oil. Sometimes it is processed separately for recovery of anthracene, phenanthrene, and carbazole.

Pitch. Pitch has many uses, for each of which a different consistency is required. The desired consistency may be obtained by stopping the distillation at the proper point or by "cutting back" hard pitch with some of the distillate. The various consistencies are listed below together with the corresponding softening points.

Very soft	Below 27°C
Moderately soft	27°–49°C
Medium hard	49°–71°C
Hard	71°–100°C
Very hard	Above 100°C

Very soft pitch is used in protective coatings and paints and in saturating felts. It also serves as a dust-laying agent for roads. Moderately soft pitch is often used for waterproofing purposes and in road surface binders. Medium hard pitch has a large variety of uses. It serves as a binder for bituminous, concrete, and macadam pavements, and as a joint filler in block pavements. It is also used in felt roofing and in laminated membranes used for waterproofing foundations. This type of pitch is a constituent of the better grades of bituminous paints. Hard pitch serves as a briquette binder. It may also be used as a fuel after pulverizing. Very hard pitch may be carbonized in coke ovens to produce pitch coke from which carbon electrodes are made. It is also used as a binder for sand cores for foundry castings and as a filler for rubber goods.

ALFRED R. POWELL

Cross-references: *Coal and Coke; Wood Preservatives.*

COBALT

The discovery of cobalt as an element is credited to Brandt, who isolated the metal in 1742. The word "cobalt" (German Kobelt) is from the Greek "cobalos," and its German form denotes gnomes and goblins. Cobalt is widely distributed in the earth's crust, making up only 0.0023% compared with 0.008% for nickel, and ranks thirty-fourth in order of abundance.

Metallurgy. The processing of cobalt ore comprises gravity or flotation concentration followed by:

(1) *Smelting* in the blast furnace, reverberatory furnace, or electric furnace, to obtain a product high in metal values for further processing by roasting, leaching, and chemical precipitation of the cobalt as cobaltic hydroxide, which is heated to convert it to oxide. The washed oxide is reduced to metal by means of charcoal.

(2) *Electrowinning* as practiced by Union Minière du Haut Katanga, at Jadotville in the ex-Belgian Congo. The cobalt is rendered soluble by the conditions for electrolysis of copper. It is recovered from the leach plant washing section and from the copper electrolyte enriched with cobalt. These solutions are treated with milk of lime for the removal of copper and iron. The cobalt is precipitated as hydroxide, which is dissolved in the sulfate solution at a pH of 7.0. This cobalt is melted and purified in a Héroult-type electric furnace. It is then granulated at 1600°C. The finished cobalt analyzes 99% Co, 0.5% Ni, 0.1% Si, 0.1% C, 0.15% Fe, 0.02% Cu, and 0.02% Mn.

(3) *Treatment of Copper-Cobalt* sulfide ores is illustrated by the processing of Rhokana Corporation's copper ores at N'kana. The concentrate containing 32% copper, 3.2% cobalt, 13% iron and 23% sulfur is given a sulfatizing roast and cobalt sulfate is water-leached from it.

The cobalt sulfate solution is treated by air oxidation and milk of lime at 5.5 pH for the removal of iron and copper to 0.2 and 0.02 gpl, respectively. The clarified solution is treated with annealed cobalt granules to remove copper to less than 0.001 gpl. The purified cobalt solution is treated with milk of lime to precipitate cobalt as $Co(OH)_2$ at pH 8.2. The cobalt hydroxide, 40 to 45% solids, is added to the spent electrolyte circuit. The purified electrolyte is heated to 70°C and passed through the cells at a rate to give a final acidity of 1.8 pH. The cobalt metal is stripped from the steel cathodes, melted in an electric furnace and granulated in water. Cobalt metal analyzes 99.89% cobalt.

Physical Properties. Cobalt metal is a hard, magnetic metal, silvery white on fracture, and resembling nickel and iron in appearance (see Table 1.)

Chemical Properties. Cobalt is very similar to iron and nickel in its chemical properties, being in Group VIII of the Periodic Table.

The simple compounds of cobalt are bivalent. The trivalent cobaltic compounds are practically confined to oxides, sulfides, sulfates, fluorides and acetates. This is in contrast to the cobalt complexes, where the trivalent state is the stable and predominating form. The simple cobaltic salts are produced from the cobaltous by the action of the strongest oxidizing agents, and are decomposed by water even at low temperatures with formation of the cobaltous salt. The simple cobaltic ion, Co^{+3}, is a powerful oxidizing agent.

Cobalt has less affinity for oxygen than has iron but more than nickel. Like iron, cobalt has three well-known oxides: the monoxide or cobaltous oxide, CoO; cobaltic oxide, Co_2O_3; and cobaltosic oxide, Co_3O_4. The first is formed when the carbonate is calcined in a nonoxidizing atmosphere and is the stable oxide film in contact with cobalt. The others are formed when cobalt compounds are heated in an excess of air.

The halides of cobalt are made by treating the

TABLE 1

Atomic number	27
Atomic weight	58.9332
Isotopes	57, 59, 60 (artificial)
Density, g/cc, at 20°C	8.85
Melting point	1495°C
Boiling point	2900°C (approx)
Latent heat of fusion, cal/g	58.4
Latent heat of vaporization, cal/g	1500
Transformation temperature, beta to alpha	417° ± 7°C (783° ± 13°F)
Heat of transformation, cal/mole	60
Curie point	1121°C (2050°F)
Specific heat, cal/g/°C	
15°–100°C	0.1056
Liquid	0.265
Coefficient of thermal expansion, 40°C	13.8×10^{-6}
Thermal conductivity, 20°C, cal/°C/cm/sec	0.165
Hardness	
Cast, Brinell	124
Electrodeposited, Brinell	300
Tensile strength, psi (99.9%)	
Cast	34,000
Annealed	37,000
Wire	100,000
Compressive yield strength, cast, psi	42,200
Young's modulus, psi	30,000,000

hydroxide or carbonate with the corresponding hydrohalide acid. Each is soluble in water. $CoCl_2 \cdot 6H_2O$ is pink and changes with dehydration to a blue color; it is useful as an indicator of humidity and moisture.

There are many cobaltous organic salts, e.g., the acetate, formate, citrate and oxalate; and the largest use of cobalt compounds is based on the linoresinates, naphthenates and octoates used as driers in paints and varnishes.

Cobalt forms a ferrite, $CoO \cdot Fe_2O_3$; possibly three phosphides, CoP_2, CoP and CoP_3, none of which has been isolated; two selenides, $CoSe$ and $CoSe_2$; silicides such as $CoSi$ and $CoSi_3$; four sulfides, CoS, CoS_2, Co_2S_4 and Co_2S_3; and two tellurides, $CoTe$ and $CoTe_2$.

Carbon dissolves in cobalt above 1300°C in a manner similar to the dissolution of carbon in iron to form the carbides Co_3C, Co_2C and CoC_2.

While nickel forms gaseous nickel carbonyl when CO is passed over it at 40–50°C, cobalt does not form a carbonyl at ordinary pressures. This fact is the basis of the Mond process of nickel refining to produce cobalt-free nickel. At 30–350 atm and 90–200°C a volatile cobalt carbonyl is formed with CO which decomposes at 52–60°C to the black crystals of cobalt tricarbonyl, $Co_4(CO)_{12}$.

Other soluble salts of cobalt are the fluosilicate, $CoSiF_6 \cdot 6H_2O$, used in the ceramic field; cobaltous nitrate, $Co(NO_3)_2 \cdot 6H_2O$; and the sulfate, $CoSO_4 \cdot 7H_2O$.

Among the insoluble cobalt compounds are the carbonate, $CoCO_3$; cobaltous hydroxide, $Co(OH)_2$, which results when an alkaline hydroxide is added to a solution of a cobaltous salt, and a hydrated cobaltic oxide, $Co_2O_3 \cdot H_2O$, formed by the oxidation of $Co(OH)_2$; cobaltous phosphates such as the commercial form, $Co_3(PO_4)_2 \cdot 8H_2O$, used in pottery production; and the silicates, $2CoO \cdot SiO_2$, and the intensely blue $CoO \cdot SiO_2$ used to obtain the mazarine blue or royal blue in pottery glazes.

Cobalt, particularly the trivalent ion, is one of the most prolific complex-formers known. The important donor atoms (in order of decreasing tendency to complex) are nitrogen, carbon in the cyanides, oxygen, sulfur, and the halogens. The most numerous are the complexes of ammonia and amines. Divalent cobalt exhibits a coordination number of either four or six, while that of the trivalent ion is invariably six.

Uses. Some 80% of the cobalt consumed is used in the manufacture of alloys, and the remainder in the form of compounds is used in pigments, salts, driers and ceramic frits.

About 20% of all cobalt consumed enters into several kinds of permanent and soft magnets. Cobalt is an essential constituent of tungsten tool steels, several ferrous and nonferrous alloys which exhibit high corrosion and oxidation resistance and excellent wear resistance, hard-facing alloys, and high-strength alloys for use at elevated temperatures. Cobalt powder is used to manufacture sintered alloys, chief of which are the tungsten carbide-cobalt alloys that permit the valuable cutting and abrasive properties of tungsten carbide to be used. The cobalt acts as a binder for the tungsten carbide particles in these cemented carbides.

In animal nutrition, cobalt in the form of chloride, acetate, sulfate or nitrate has been found effective in correcting mineral deficiency diseases such as bush sickness, Mairoa dopiness, nutritional anemia and Morton Mains disease. A healthy soil contains 0.13 to 0.30 ppm of cobalt. Cobalt has not yet been proved successful to plant growth, although present in many plants.

Cobalt-60 and other cobalt isotopes have been used in chemical, physical and biological research and as a radiotherapeutic agent. Cobalt-60 emits two gamma rays accompanied by beta radiation. The isotope's half-life is 5.3 years. Other applications are feed sterilization, food preservation, analysis, metal technology, and the initiation of chemical reactions of all sorts.

CARL R. WHITTEMORE

References

Battelle Memorial Institute, "Cobalt Monograph," Centre D'Information du Cobalt, 35, Rue des Colonies, Brussels, Belgium, 19.

Whittemore, C. R., "Cobalt," pp. 114–148 in "Rare Metals Handbook," Hampel, C. A., Editor, Reinhold Publishing Corp., New York, 1961; and "Cobalt," pp. 154–163 in "Encyclopedia of the Chemical Elements," Hampel,, C. A., Editor, Van Nostrand Reinhold Corp., New York, 1968.

Kirk-Othmer, "Encyclopedia of Chemical Technology," Vol. 5, pp. 716–736, Interscience Publishers, 1967.

Mellor, J. W., "A Comprehensive Treatise on Inorganic and Theoretical Chemistry," Vol. XIV, pp. 419–859, Longmans, Green and Co. Limited, London, 1935.

Young, R. S., "Cobalt—Its Chemistry, Metallurgy and Uses," Reinhold Publishing Corporation, New York, 1960.

American Society for Metals, "Metals Handbook," Vol. 1, 8th Ed., Metals Park, Ohio, 1961.

COLLAGEN AND GELATIN

Collagen

Collagen is a fibrous protein of animal origin. It is found in representative species of all animal phyla excepting the Protozoa. The primary functions of collagen are to provide mechanical support, and to form selectively permeable membranes. It also has a role in maintaining the water balance and in the binding of metals. Collagen is involved in the aging process and in many so-called collagen diseases. It is the basic raw material for the leather, gelatin, and glue industries, and for the manufacture of surgical sutures, "gut" strings, sausage casings, and many other products. The function of collagen in the living organism, and its usefulness as a raw material for a sizable industrial complex are based on the unique characters of collagen, some of which will be described and discussed in this article.

Collagen is produced by fibroblasts which have their origin in the mesodermal germ layer. A possible exception is the collagen in cuticles of invertebrate animals such as annelid worms. The formation of a collagenous tissue may be visualized as follows: the fibroblast forms monomeric collagen macromolecules, which are transported to the surface of the fibroblast, or are discharged into the intercellular spaces, where environmental conditions are such as to favor end-to-end polymerization to form macromolecule subunits and concurrent lateral association of the macromolecule subunits to form larger structural entities. The entities are termed fibrils, fibers, fiber bundles, and organized tissues, which are differentiated from one another by size. Collagen fibril bundles are also

termed fibers. The dimension of the width or diameter of the cross section is more easily defined than the length. The lengths of the entities are indefinite and very, very long compared to the widths. In general, the width-to-length ratios range from 1:40 to 1:500. The approximate widths are listed in Table 1.

TABLE 1. WIDTH OF STRUCTURAL ENTITIES OF COLLAGEN

Entity	Width (centimeters)
Macromolecular subunit and Tropocollagen macromolecule*	$1.5 \cdot 10^{-7}$
Fibril	$3.0 \cdot 10^{-7}$ to $2.0 \cdot 10^{-5}$
Fibril bundle**	$2.0 \cdot 10^{-5}$ to 10^{-1}

*The tropocollagen macromolucle is the only collagen entity that has a finite length which is 2800Å.
**Fibril bundles and fibers are synonymous terms.

Collagenous tissues are fabrics made up of collagen fiber bundles organized in definite patterns according to the requirements of the animal. Examples are diagrammatically sketched in Fig 1.

It is generally considered that collagen is insoluble in water, dilute acids, alkalies, and is not affected by proteolytic enzymes excepting the collagenases. However, treatment of collagen with any of the type reagents noted above (even storage in water) for prolonged periods, will induce degradation of the collagen. Pretreatment of the collagen by heating the collagen in water or with hydrogen bond splitting agents such as $6M$ urea hastens the degradative precess. Thus by appropriate treatments collagen may be converted into a series of degradative products: gelatin, glue, poly-

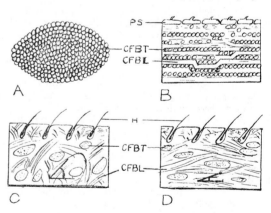

FIG. 1. Representative fabrics weave patterns of collagenous tissues.
A. Packed parallel array as in tendon.
B. Lamellar arrangement as in sharkskin.
C. High angle weave pattern as in the backbone portion of cattle hide.
D. Low angle weave pattern as in the belly portion of cattle hide.

H = hair; PS = placoid scale; CFBT = collagen fibril bundle (transverse section); CFBL = collagen fibril bundle (longitudinal section).

peptide chains of varying lengths, and amino acids. It is important to note that collagens derived from different sources, e.g., sharkskin and cowhide, or from a single animal species at different stages of development, e.g., embryonic and mature skin, will react to the same experimental treatment at different rates. Thus sharkskin collagen is more readily soluble in dilute acetic acid than is cowhide collagen. Similarly young rat tail tendon is more soluble than tail tendons from old rats. Briefly, some collagens are more stable than others, and this may be ascribed to variations in the intra- and intermolecular organization of different species of collagen.

When a piece of collagenous tissue, such as skin or tendon, is placed in water and the water is gradually heated, the tissue will shrink markedly and quite suddenly within a limited (2–3°C) and sharply defined temperature range termed the shrinkage temperature. The shrinkage temperature is an index of the hydrothermal and chemical stability of collagens. Leather, which is tanned collagen, also shrinks when heated in water, and in the case of leather the shrinkage temperature is an index of the degree of tannage, the higher the shrinkage temperature the greater the degree of tannage. Different species of collagens have characteristic shrinkage temperatures. Thus sharkskin will shrink at about 42–45°C, whereas cowhide will shrink at about 63–65°C. The fact that different collagens have characteristic shrinkage temperatures also may be ascribed to variations in the intra- and intermolecular architecture of different species of collagens.

The amino acid glycine constitutes about one third of the amino acid content of all collagens. This suggests that the amino acids sequence in the polypeptide chains may have repetitive units that are triads wherein glycine occupies one of three positions in the triplet of amino acid residues; further, proline and hydroxyproline making up from 10 to 25 per cent of the amino acid content of collagens, will occur in a significantly large number of the triads, and so on with the remaining amino acids. Analyses of peptides and polypeptides isolated from partially hydrolyzed collagen provide evidence to support the above observations.

Collagen is unique in that it contains substantial amounts of hydroxylysine and hydroxyproline. The hydroxyproline content of tissues and organs is a vary useful indicator for the qualitative and quantitative assay for collagen. Collagen contains relatively small quantities of tyrosine, methionine and histidine, and lacks tryptophan; therefore collagen may be considered to be an incomplete protein for nutritional requirements. The sum of the proline and hydroxyproline residues in a collagen is an index of the hydrothermal stability in that species with greater amounts of imino acid residues shrink and are denatured at higher temperatures than those with lesser amounts. This may be because the polypeptide chains with large amounts of pyrrolidine rings are in tighter helical configuration than those with small amounts of pyrrolidine rings; the tighter the helices the more stable the collagen.

It was noted previously that the length of the structural entities of collagen were indefinite. The 600–700 Å axial periodicities exhibited in the low angle x-ray diffraction patterns and electron micrographs of collagen fibrils, and studies of defects in collagen fibrils indicate that the length of the collagen macromolecule is of the order of 600–700 Å. However, many investigators believe that the primary monomeric collagen macromolecule is the tropocollagen unit which is 2800 Å long. The tropocollagen unit is an entity that is isolated or extracted from collagenous tissues by treating them with dilute acids, various buffer systems, or solutions of neutral salts. From such solutions one can reconstitute native type collagen fibrils, and a number of variant forms of fibrils with long (multiples of 600–700 Å) axial periodicities, and segments with similar periodicities.

In summary it may be stated that collagen is a generic term for a group of proteins, the species of which differ from one another in respect to variant properties that may be ascribed to variations in the intra- and intermolecular organization of the different collagens. Further the unique characters of a specific collagen are in accord with the functions and environment of the living tissue from which it is derived.

R. Borasky

Reference

Ramachandran, G. N., and Gould, B. S., eds. 1967/1968. *Treatise on Collagen,* 1, 2A, 2B. London, Academic Press, Inc.

Gelatin

Gelatin is a mixture of simple proteins derived from the collagen of animal connective tissues through a series of degradation or hydrolytic steps. Perhaps it is most frequently encountered in homemade soup stock into which it imparts the tendency to form a gel in the refrigerator. Collagen, one of the most abundant proteins found in the animal body, is the major component of the white fibers of connective tissue and is present in skin, sinews, hides and ossein, the connective tissue protein of bones. Through a number of hydrolytic steps effected by heat and/or acids and alkalies collagen is "cured" and "plumped" into a translucent mass of swollen fibers from which the gelatin can be extracted with hot water. When all of the raw materials, chemicals, equipment and procedures are maintained under strictly controlled conditions so as to insure compliance with regulatory food standards the product is called gelatin. Crude less refined products are commonly called technical gelatins or even glue.

In any case the raw material is almost always pork skins, calf skins (tannery by-product) or ossein from the demineralization of bones. Dependent upon the raw material, acids or alkalies are used to "cure" the collagen preparatory to extraction of the gelatin. Pork skins obtained from the packing industry are treated in dilute hydrochloric acid. Calf stock from tanneries has already been lime-cured or if not is so treated at the gelatin factory. Ossein, as the residue remaining from the acid leaching of the mineral calcium phosphate from bone, is acid-cured in the process of demineralization.

Irrespective of the acid or alkaline cure the collagen is thus prepared for hydrolytic cleavage with hot water after exhaustive washing to remove excess mineral acid or alkali. Since conversion of the hot water insoluble collagen into soluble gelatin is a time-temperature reaction a series of extractions are made by soaking in hot water until maximum solids content is reached in the liquor. Subsequent extractions or "runs of liquor" are made, each at a somewhat higher temperature than the one before. This is continued until all gelatin is extracted in separate runs each representing a stage in the hydrolysis of the parent collagen into a group of gelatins. For this reason the average molecular weight will decrease with each extraction and will be manifested by a gradual decrease in jelly strength and viscosity. Fresh from the extraction kettles these "light liquors" are quite dilute and lend themselves readily to subsequent steps of settling, centrifugation, filtration, and deionization.

Concentration is effected by vacuum evaporation first, in triple effect Swenson type and finally by turbulent film vacuum concentrators sometimes with a second intermediate filtration step if required. In final concentrated form the "heavy liquor" sets rapidly to a gel when cooled and is usually so converted on a chill roll from which sheets or strings of gel are conveyed to a low temperature drier. Low humidity cool air is passed over or through the product until skinning prevents agglomeration. The temperature is then raised to effect complete drying. Final steps include grinding and blending to the desired specification.

Photographic gelatin is prepared from lime-cured stock by the same overall process but numerous purification and other steps are mandatory in order to achieve the very special properties required in the manufacture of photographic film.

Throughout the processing of gelatin liquors various adjustments are made to control the pH since time and temperature effect a certain degree of hydrolysis which is exaggerated by extremes of acidity or alkalinity. For this reason one works at minimum temperature, minimum times, maximum concentration, and minimum extremes of pH in order to prevent as much hydrolysis as possible.

Commercial gelatin as produced in the United States is almost colorless and virtually free from odor. Its solutions are practically water-white until concentrations are raised to extremely high levels. As can be inferred from its method of manufacture gelatin is a complex mixture of collagen degradation products which range in molecular weight from about 30,000 to 80,000 or even higher depending upon the hydrolytic conditions to which it has been subjected. Little is known about the exact sequence of amino acids in the molecular chain, but almost half the gelatin molecule comprises glycine, proline and hydroxyproline. Irrespective of the degree of degradation found in a gelatin sample it contains about 45 milliequivalents of amino functions and about 70 milliequivalents of carboxyl functions per hundred grams. These values remain substantially unchanged unless gross hydrolysis is effected by acids, alkalies or proteolytic enzymes.

The three outstanding properties of gelatin are its ability to form reversible gels, viscosity and the extreme strength of its films. These properties explain the many uses for which this protein is suitable. As a dry granulated product gelatin appears lustrous and crystalline even though it is quite amorphous. It is not soluble in cold water but swells rapidly until it has imbibed about 6 or 8 times its weight. The swollen gelled particles then melt to a viscous solution when warmed above the melting point (40–45°C).

In almost all respects gelatin resembles many other dry proteins in that it does not melt and functions as an amphoteric electrolyte. Gelatin from acid-cured stock (Type A) has an isoelectric point of about 8.3–8.5 and alkali-cured gelatin 4.75–5.0 (Type B). Either type is most difficult to precipitate at its isoelectric point unless powerful protein precipitants are used. Various aldehydes and heavy metal salts exert a tanning action on gelatin so as to produce insoluble material. The tendency to gel can be sharply decreased by a number of peptizing agents in much the same way as salts depress the freezing point of water.

Commercial gelatins are graded according to the viscosity and gel rigidity of their aqueous solutions. Since these easily measured constants do describe the potential suitability of gelatin for a given use the commercial product is sold on the basis of its viscosity and Bloom test (jelly strength). These two physical constants are reproducible within 1% and indicate not only the suitability of the product for a given use but also the type of gelatin being tested. For example, gelatin from an acid-cured material will have a jelly strength-to-viscosity ratio of 4 or even 5 to 1 or 300 gms and 60 mp, approximately. From alkaline curing of the same stock a gelatin would result with a 2½ or 3 to 1 ratio, namely, 250 gms. and 100 mp, approximately.

Large quantities of gelatin are used in hard and soft capsules to contain drugs, vitamin preparations, etc. The photographic industry is another large consumer of gelatin; it requires a product having satisfactory characteristics for use in the preparation of the light-sensitive coating for plates and films, viscosity and jelly strength being of secondary importance.

By far the largest amount is used in the production of flavored powders for the preparation of desserts and salads. This is followed, on a tonnage basis, by use in ice-cream production where its characteristic protective colloid activity is put to work to prevent or reduce the growth of ice crystals.

Next in order of volume usage is that of the confectionary and baking trade, where it is an essential ingredient of marshmallow. Another important food use of gelatin is in a wide variety of meat-base products generally referred to as jellied meats. While the predominant characteristic made use of here is the jelly-forming property, color and clarity of the gelatin are also of special importance.

H. H. YOUNG

Cross-references: *Collagen, Foods, Proteins.*

COLLOID CHEMISTRY

Colloids are usually defined as disperse systems with at least one characteristic dimension in the range 10^{-7} cm to 10^{-4} cm. Examples include *sols* (dispersions of solid in liquid), emulsions (dispersions of liquids in liquids), aerosols (dispersions of liquids or solids in gases), foams (dispersions of gases in liquids or solids) and gels (systems such as common jelly in which one component provides a sufficient structural framework for rigidity and other components fill the space between structural units).

Thomas Graham's investigations of diffusion (1861) led him to characterize as crystalloids substances such as inorganic salts which in water solution would diffuse through a parchment membrane, and as colloids (after a Greek word for glue) substances such as starch and gelatin, which would not. Sols with a given weight percent dispersed material scatter light more strongly than a solution with the same weight percent dissolved inorganic salt. The Tyndall effect, in which the path of a beam of light through a turbid solution (or through dusty or smoke-filled air) is clearly defined through scattered light, is characteristic of sols. The slow diffusion and strong light scattering, together with the fact that the boiling point elevation, freezing point depression, and osmotic pressure caused by a given weight percent of dispersed material in sol form were much less than the corresponding magnitudes caused by the same weight percent of common inorganic salts, all indicated that the particles dispersed in the sol must be much larger than those resulting from dissolving inorganic salts in water.

The invention of the ultramicroscope by Siedentopf and Zsigmondy (1903) permitted particles substantially smaller than the wavelength of light to be observed by scattered light, and so counted. The invention of the ultracentrifuge by Svedberg (1924) made it possible to cause particles in sols to sediment at observable rates, to measure these rates with considerable precision, and to infer particle sizes accurately from these rates. The ultramicroscope and ultracentrifuge permitted validation of the early conclusions that colloidal particles were much larger than ions resulting from dissolving metal salts in water.

Svedberg further found that in some sols the particles were highly uniform in size. For example, he found that the gram particle weight of insulin, a protein, was 40,900 and that apparently all insulin particles had this gram particle weight. This made it extremely likely that the insulin particles were either single molecules, albeit giant ones, or aggregates of a very definite number of smaller (but still quite large by ordinary standards) molecules. The work of Staudinger beginning in 1920 and of Carothers beginning in 1929 opened up the field of macromolecular chemistry, leading to the recognition that giant molecules were not only abundant in nature, but could be made by well-established principles of chemistry.

It is convenient to classify sols in three types: *lyophilic* (solvent loving) colloids (examples: solutions of gelatin or starch in water), *association colloids* (example: a solution of soap in water at moderate concentration), and *lyophobic* (solvent hating) colloids (example: a sol of sulfur in water). The first two types can be prepared in thermodynamic equilibrium, so that when solvent is removed and then returned to the system, the original properties of the system are regained. The third

type is not (or at most rarely) an equilibrium system; when solvent is removed and then returned to the system the original dispersed material fails to redisperse, and it is usually convenient to regard this system as one in which the dispersed particles are continually aggregating. A lyophobic sol then appears to be stable if the aggregation rate is slow, and to be unstable if it is fast. The words lyophilic and lyophobic entered the literature before the characters of these systems were well understood, and are somewhat anachronistic but nonetheless well-established.

Lyophilic sols are true solutions of large molecules in a solvent. Solutions of starch, proteins, or polyvinyl alcohol in water, of rubber (not cross-linked) or polystyrene in benzene, and of polyvinyl acetate in acetone are examples. Properties of these solutions at equilibrium, for example density and viscosity, are regular functions of concentration and temperature independent of the method of preparation. The solvent-macromolecular compound system may consist of more than one phase, each phase in general containing both components. Thus if a solid polymer is added to a solvent in an amount exceeding the solubility limit, the system will consist of a liquid phase (solvent with dissolved polymer) and a solid phase (polymer swollen with solvent, i.e., polymer with dissolved solvent).

All these characteristics are also found with solutions of small molecules. Properties of solutions one of whose components is macromolecular differ from those having only small molecule components in quite understandable ways. Benzene, C_6H_6, is a small molecule hexagonal in shape; the hexagon may pucker or otherwise distort as its environment is changed but the distortions are minor. Quite generally the shapes of small molecules are little affected by environment unless the small molecules react chemically. Polyolefins, for example $\phi(CHRCH_2)_n\phi$ in which ϕ is a small end group, n may be in the range 1000 to 50,000, and R characterizes the polymer (polyethylene if R is H, polystyrene if R is C_6H_5, polyvinylchloride if R is Cl) can be thought of as long rope-like compounds with considerable flexibility because of the possibility of rotation about the carbon-carbon bonds in the polymer backbone. In contrast to small molecules, there is hence a considerable variation in polymer conformation with environment.

A polymer dissolved in a good solvent will tend to stretch out, and the resulting entanglement of polymer chains and interference with solvent movement will lead to a high viscosity. If the solvent is a poor one, the polymer molecule will tend to form a little ball, and the viscosity for a given weight percent polymer will be much less. The side group R may be ionizable, as in the case of polyacrylic acid (R is a carboxyl, i.e., COOH, group). Ionization of this group (for example, by partially neutralizing polyacrylic acid with base, changing the COOH group to a COO^- group) distributes a charge along the backbone, and charge repulsion causes the macromolecule to tend toward a rod shape. If there is a moderate salt concentration in the solution, the backbone charge will be partly shielded by ions from the salt of opposite charge, so the tendency toward rod formation will be less pronounced. The tendency of oppositely charged macromolecules to aggregate is much greater than the tendency of oppositely charged small ions to pair simply because the charges involved are greater in the former case.

The foregoing special properties of solutions of large molecules are relatively easy to understand qualitatively. A difference of more subtle (but still understood) origin occurs when the system forms two liquid phases. In a macromolecular solution both phases tend to be rich in the (small molecule) solvent, whereas in systems formed from two molecules of comparable size one phase is rich in one component, the other phase rich in the other component. The formation of two liquid phases from a solvent-macromolecule system is sometimes called coacervation, and the phase with the higher percentage of macromolecule is sometimes called the coacervate.

Association colloids are generally encountered in solutions of soaps and detergents in water. A typical soap (e.g., sodium stearate, $C_{17}H_{35}COONa$) or detergent (e.g., sodium dodecyl benzene sulfonate, $C_{12}H_{25} \cdot C_6H_4 \cdot SO_3Na$) consists of a long hydrocarbon tail and a polar (in the examples cited ionizable) head group. The solubility of the soap in water is largely conferred by the head group. As the soap concentration is increased, the soap molecules tend to cluster in aggregates called micelles, with hydrocarbon tails in the interior of the micelles and polar groups in contact with water. The formation of micelles is favored by the interaction between hydrocarbon tails (each of which replaces an "unfriendly" water environment with a "friendly" hydrocarbon one) and is opposed by charge repulsion of the polar groups which are placed close together at the micelle surface.

Micelle formation becomes pronounced at soap concentrations exceeding the critical micelle concentration. As hydrocarbon tail length is increased the interaction of tails is increased, and as salt concentration is increased the repulsion of head groups is reduced because their charges are partly shielded by ions of the salt. Both of these factors favor micelle formation, causing micelles to be larger and the critical micelle concentration to be smaller. Typically, a micelle might contain about 50 soap molecules. The micelle interior is hydrocarbon, and as such is receptive to other molecules soluble in hydrocarbons. Hence a soap solution can "dissolve" such molecules (taking them up in the micelle interiors) even if the molecules are quite insoluble in water. This phenomenon is called solubilization and is a factor in detergency.

Lyophobic sols and aerosols can be viewed most simply and in most cases with sufficient accuracy as two-phase systems in which the dispersed particles are steadily and irreversibly aggregating according to a second-order rate law. Thus where C is the number of particles per cc (an aggregate of many primary particles being counted as one particle) at time t, C_o the number of particles per cc at zero time, and K is a constant, C depends on t according to

$$\frac{C_o}{C} - 1 = KC_o t$$

and will be one-half its value at zero time when

$KC_o t = 1$. The time required for this is longer, the smaller K and the smaller C_o; if the time required is weeks, the sol will appear quite stable over a period of days. If there is no barrier to aggregation, so that the particles aggregate as fast as diffusion brings them in contact, the rate constant K can be calculated approximately from diffusion theory and is $8kT/3\eta$, where k is Boltzmann's constant, T the absolute temperature, and η the viscosity of the medium. Initial sol concentrations (particles per cc) giving one minute half-lives at room temperature are 1.4×10^9 in water and 2.7×10^7 in air in the absence of aggregation barriers.

The numbers appear large at first sight but actually correspond to quite small volume percentages of dispersed particles; a particle of radius 5×10^{-6} cm is at the upper limit of the colloidal range, and 1.4×10^9 such particles per cc occupy 0.07% of space. The behavior of smokes, fogs, and many dispersions of uncharged particles in water accords well with the rate equation and theoretical rate constant given. On the other hand, the dispersed particles in many sols are electrically charged, manifesting this charge through electrophoresis (motion of colloidal particles under the influence of an electric field).

Indeed Tiselius developed electrophoresis to a high degree, successfully fractionating and classifying proteins thereby. Evidently like charges on two colloidal particles will contribute to a repulsion between them, which will be greater, the greater the charge on each particle and the smaller the concentration of salts in solution (since ions from the salt will tend to mask the charges on the particle).

The theory of interaction between colloidal particles with a surface electrostatic potential (due to surface charges) surrounded by an electrical double layer (one layer of which is the layer of surface charges, the other a diffuse cloud of charges of opposite sign due to ions from salts in the solution) was developed by Derjaguin and Landau and independently by Verwey and Overbeek, and is generally known as the DLVO theory after the first letters in the names of these scientists. The DLVO theory shows how a barrier sufficient to reduce the rate constant K (and so to increase the half-life at a given initial concentration) by many powers of ten may arise from the interaction of charged particles in a solvent, and the magnitudes calculated agree rather well with experiment.

It is then plain why the properties of lyophobic sols depend so critically on the chemistry of the interface between dispersed particle and solvent, for this chemistry establishes the means by which the surface charge can be established or altered. Particles of silver halide dispersed in water will acquire a positive charge if silver ion is in slight excess in the water, because the silver ion can readily add to the silver halide lattice. A negative charge is similarly acquired if the halide ion is in slight excess. The silver ions and halide ions are called potential-determining ions for the silver halide sol; they can lose their waters of hydration and adsorb on the particle side of the electrical double layer, conferring a charge on the particle. Hydrogen ion and hydroxyl ion are similarly potential-determining ions for many oxide sols (e.g., silica, alumina) including particles such as carbon and many metals which are ostensibly not oxides but in fact usually have oxidized surfaces. Finally, charge can be conferred by the adsorption of charged macromolecules such as gelatin, which are called protective colloids. Salts added to the sol form ions which tend to mask the particle charges and so tend to promote flocculation. The ion whose charge is opposite to the particle charge (the counter ion) is of particular importance, and the greater its charge the lower the concentration at which its flocculating effect is evident.

Emulsions are dispersions of one liquid in another; most commonly one phase is water or an aqueous solution and the other phase an oil which is at most slightly miscible with water. The disperse phase can either be oil (an oil-in-water emulsion) or water (a water-in-oil emulsion). For apparent stability an emulsifying agent is almost invariably required. This agent is frequently a soap or detergent, a molecule with an oil soluble tail and a polar head. The emulsifying agent concentrates at (adsorbs at) the interface between oil and water, lowering the interfacial tension and frequently conferring a charge on the dispersed droplets. The film of emulsificant thus formed is usually only one molecule thick, but it is essential to emulsification. Mixed emulsificants, e.g., a mixture of sodium stearate and octadecyl alcohol, may be more effective in emulsification than either component alone, and there is a great deal of art in emulsion formulation. Lecithins and some proteins are effective natural emulsificants, and a mixture of lecithin and cholesterol is an effective natural mixed emulsificant. Emulsification of oily soils is a critical part of laundering, many liquid floor waxes are oil-in-water emulsions and milk is a fat-in-water emulsion stabilized by proteins and lecithins. Many cold creams and pharmaceutical ointments are water-in-oil emulsions. It should be noted that the difference between solubilization (discussed with association colloids) and emulsification is not sharp particularly insofar as large extensively swollen micelles and ultrafine emulsions are concerned.

Gels involve the formation of a three-dimensional structure. A simple illustration of such a structure arises if groups are introduced in a polyolefin chain $\phi(CHRCH_2)_n\phi$ permitting chains to be tied together. Polystyrene is the polyolefin with $R = C_6H_5$, and is obtained by polymerizing styrene, $C_6H_5CH=CH_2$. If a small percent of divinyl benzene, $CH_2=CH-C_6H_4-CH=CH_2$, is added with the styrene, a cross-linking of chains will occur, and this can lead to an infinite network and gelation. In the case just cited the gelation is irreversible, but if the chain cross-linking is accomplished by weaker bonds (hydrogen bonding, interaction of hydrated ions) gelation may be reversed by heating and adding solvent. A gel will tend to swell when exposed to solvent and the extent of swelling can be used to characterize the cross linking frequency. Gelatin desserts and jellies are familiar food gels, cross-linked sulfonated polystyrene gel furnishes a common ion exchange resin, and very large volumes of gasoline cracking catalysts arise from silica and alumina gels subsequently dried to give extremely large surface areas.

Foams are dispersions of gases in liquids, and an interface stabilizing agent such as a soap or detergent is again essential to a well-developed foam. The most familiar foams consist of polyhedron cells; each polyhedron wall is a sandwich-like structure with a soap (or equivalent stabilizing agent) monolayer on each side and solvent (e.g., water) in the wall interior. These polyhedral foams usually originate from an initial foam of spherical dispersed bubbles as the liquid drains from between them. As the polyhedron walls get thinner, the rate of drainage of the water in the film interior decreases markedly; in terms of the stresses placed on it by the draining liquid the surfactant monolayer is quite rigid, so the drainage law is substantially that for liquid between two rigid plates comparably spaced. The thinning of these films is beautifully illustrated by the thinning of free soap films, as in soap bubbles. Beautiful interference colors are observed as the film thickness approaches the wavelength of light, and the films become invisible as the thickness reduces to much below the wavelength of light. When the films thin sufficiently for charge repulsion between monolayers of soap on opposite faces of the film, a barrier to further thinning occurs, and the film collapse is a rate process somewhat analogous to the flocculation of lyophobic colloids with charge repulsion. These soap films have also provided a major model for biological membranes.

ROBERT S. HANSEN

Cross-references: *Aerosols, Aggregation, Emulsions, Foams, Interfaces, Micelles.*

COLORIMETRY

Colorimetry is a quantitative method of measuring the amount of a particular substance in solution by determining the intensity of its color. For example, the yellow carotene content of butter can be determined by saponifying a sample of butter in an alkaline solution, extracting the carotene with ether, and measuring the intensity of yellow color in the ether extract. The procedure is not actually as simple as that, but that is the principle. What is measured is the amount of colored light that passes through the solution.

Colored liquids appear colored because although they absorb a certain amount of light, they also transmit a certain amount. The kind of color such as we describe as yellow, red, or blue, depends on the wave-length of the light which passes through the liquid. The maximum amount of light which a yellow solution transmits has a wave length of 580–595 mμ, that of red 610–750 mμ, and that of blue 435–480 mμ. In each case, most of the rest of the color spectrum—the range of colors which we see when the sun strikes a prism—is absorbed by the solution.

Certain laws—named for the person who first expressed them—frame this in quantitative terms. According to *Bouguer's law*, each layer of a homogeneous colored solution absorbs the same fraction of light reaching that layer. The first mm. of solution will absorb a definite fraction of the light which strikes it, the second mm. would absorb the same fraction of the light passed through the first mm.; the third mm. would in turn absorb the same fraction of the light that passed through the second mm., and so on. Another form of expressing this is *Lambert's law,* which states that at constant concentration, the color intensity is proportional to the depth of the liquid. Thus, a short column of a dark solution may be matched by a longer column of a similar but lighter colored solution. *Beer's law*—the one most frequently referred to in colorimetry—is that the color intensity of a homogeneous solution is proportional to the concentration of the colored substance at the same depth. In other words, the darker the color, the greater the concentration; the lighter the color, the less the concentration of colored substance.

Not frequently, but sometimes, the natural color of a product can be used directly, as in the color of butter to measure its carotene content. More often, the compound to be determined is made to react with a reagent which by the combination gives a colored product. For example, benzoic acid is often added to food products as a preservative. The amount present should be limited to a certain level. The quantity can be determined by treating the food product in such a way as to separate the benzoic acid, and then measuring the amount of this after its reaction with ferric chloride to give a green compound. Similarly the vitamin nicotinic acid, or niacin, can be determined in beef by a suitable process to extract the niacin, then causing this to react with cyanogen bromide and *p*-aminoacetophenone to give a violet color. The intensity of the violet color shows how much niacin is present.

Several methods for measuring a particular color intensity are in use. One method much used in the past but not as much now, is comparison of the color of the sample with that of a series of standards. A series of tubes is prepared, containing known concentrations of the substance to be determined, over a definite range, with the color developed in each. Adjacent standards should differ by no more than 20% in concentration if accuracy of 5% or better is desired. The color of the sample or unknown is then developed in the same way in a similar tube and can be examined to see where it fits best in the scale of known concentrations. Several types of equipment are available for ease of comparison. This method can be used only over the visual part of the spectrum. The colors we can see are limited to wave lengths of about 400–700 mμ.

The Duboscq colorimeter, long used in this field, is an instrument for matching the color of the standard and unknown by placing each in a separate cup or tube, then varying the depth of one by sliding it up or down by means of a rack and pinion, and looking through an eyepiece to bring the colors side by side in each half of a circle. This also is limited to the visual part of the spectrum. Sometimes artificial standards are used, such as a colored glass, or stable liquids having the desired color range; such standards have a degree of permanence. Biochemical laboratories find the Duboscq type of colorimeter very useful. The color of oils was long measured by comparison with glass standards, but other methods have been developed. Glass standards permit very rapid reading which is important when the color developed is only transitory.

Most of the colorimetric methods currently presented in the chemical literature—and they are many—are photometric, where the work of the human eye is replaced by a photoelectric cell. An advantage is that readings can be made in the ultraviolet and infrared, at wavelengths invisible to the eye. The photoelectric cell operates over 200–1000 mμ. Under proper conditions of use, the sensitivity of the photoelectric cell over that of the eye is important.

In photometric methods, more complicated apparatus is required than the Duboscq or balancing type. This is because a source of approximately monochromatic light is necessary. In the simpler instruments, various filters are available for the purpose. In using such an instrument a calibration curve must be prepared for the particular filter under the conditions of use. In other words, each laboratory should prepare its own calibration curves.

A more expensive instrument contains a grating or a prism to transmit monochromatic light. This means light of a very limited wave-length range. In some instruments this may be as little as 1 mμ, in others it may be over 35 mμ, more usually the light band is 10–30 mμ. Several spectrophotometers are on the market, but the instruments are so complicated, and the possible difficulties so varied, that a skilled person should be in charge of its use. Employment of monochromatic light eliminates interference from substances which may be present which transmit light of quite different wave length from that of the substance being determined. The Beckman quartz spectrophotometer is an example of this type of instrument.

Many times the preparation of the sample, in order to get the substance to be determined into a form suitable for reading, is a long, time-consuming process, constituting the major part of the manipulations to be made. For example, in order to determine chromium in iron ore, the ore is first fused with sodium peroxide, taken up in water, and the solution filtered to eliminate insoluble material, mostly iron oxide. An aliquot of the solution is boiled to destroy the peroxide, then is made approximately neutral. The amount of alkali to be added is determined by titrating a separate aliquot of sample solution. If more vanadium than chromium is present, the sample solution is treated with 8-hydroxyquinoline in acetic acid, then extracted with chloroform. The chloroform contains the vanadium and is discarded.

The solution containing chromium is made acid with sulfuric acid, then treated with diphenylcarbazide and phthalic anhydride to develop a colored compound, which is read at 540 mμ. The color is sensitive to 0.008 ppm. From the reading, the amount of chromium originally present can be calculated.

This illustrates the fact that colorimetric methods can be used for determining very minute amounts of substances. Once the colored compound has been developed, the final reading takes very little time. For routine work colorimetric methods are extremely useful, as for determining small proportions of a particular substance in steel.

With most colorimetric methods, accuracy is to 2–5%, but with the spectrophotometer, it may be a fraction of a per cent if sufficient care is taken in the preparation of the sample. Colorimetric methods are applied in every field of analytical chemistry, for general work, and in many specialized laboratories. These cover hospital laboratories for blood and urine analysis, food laboratories for determinatin of vitamins, preservatives, coloring matter, etc. In metallurgical laboratories, traces of metals in raw materials and finished products are determined.

New colorimetric methods appear constantly in the chemical literature. They cover the whole range of inorganic, organic, and biological applications.

By reading in the ultraviolet, the corresponding determination can be carried out where there is no visible color to the human eye. Then it is essential to use quartz cuvettes to contain the solutions being tested. This field of colorimetry has had much attention in recent years. Often a reagent is not needed.

CORNELIA T. SNELL

Cross-reference: *Analytical Chemistry.*

COLOR VISION CHEMISTRY

Visual Pigments. In the eye as in photography, light acts by being absorbed by a photosensitive substance and causes chemical change. This substance known as a *visual pigment* has been extracted from the eyes not only of vertebrates but of animals as different as molluscs and arthropods. Its chemical constitution is always the combination of the aldehyde of Vitamin A (retinaldehyde) with a special eye protein (opsin). In fresh water fish the carotenoid is from vitamin A_2, in other animals vitamin A_1.

The most abundant source of visual pigment is from animals that see well in the dark, where the pigment (rhodopsin) is concentrated in the photoreceptors for twilight vision (the rods). But twilight vision is colorless, and we must look not to the rods but to the cones to find the pigments underlying color vision. Most mammals have predominantly rods in their retinas and no cone pigments have ever been demonstrated in their extracts. But in fowls (whose twilight vision is so poor that they go to roost at sundown) the retina contains mainly cones. Even here, however, it is so hard to extract the cone pigment, *iodopsin,* that only Wald, Brown & Smith (1955) have done it well. They found that (like rhodopsin) iodopsin was a compound of retinaldehyde$_1$ and an opsin, but this protein was different from that of rhodopsin since the maximum absorption lay at 560 nm (nanometer = 10^{-9}m) instead of at 500.

If birds have only the one cone pigment, iodopsin, it might be thought that all cones should exhibit the same spectral sensitivity and lack all color discrimination—a conclusion irreconcilable with their notorious sharp eye for ripe fruit. In fact fowls can discriminate color with only iodopsin rather as some photographic films can register color with only AgBr.

In these birds, each cone contains a minute colored oil droplet situated between the light and the visual pigment. In those cones that contain a (photostable) red droplet, for instance, mainly red rays will reach iodopsin hence this cone will be red-sensitive. Similarly other cones will be more sensitive to other colors, according to the oil drops they contain. But *our* eyes are organized differently. We do not contain stable filters; so we must seek three different visual pigments in the human cones.

Human Cones. Rushton (1955) first published photocell measurements upon the pigment in living human cones. Advantage was taken of the fact that the central 1 mm of the human retina (the precious *fovea centralis*) contains only cones and no rods. If measurements are restricted to this region, we shall avoid contamination with rhodopsin and study only cones in the human eye.

The technique for making measurements in this way on the living eye is to use the principle of the eye-shine seen when a cat is caught in the headlamps of a car. The light reflected from the eye has passed through the cat's retina and back, and must have suffered absorption by retinal pigments in this double passage. Naturally any change in the amount of cone pigment as the result of bleaching and regeneration will be shown by the change in reflectivity from the eye, and this reflectivity may be measured in lights of various wave lengths, using an intensity so weak that it does not produce appreciable bleaching.

The retinal densitometer (which has to be very sensitive and well compensated) has been described elsewhere (Rushton, 1958). Here we are concerned, not with the technique but with the cone pigments that it can measure in the red-green spectral range.

Color Blindness. About 1/20 of the male population has difficulty in distinguishing red from green. The condition is a sex-linked recessive character and hence only $(1/20)^2$ of the female population shows it. Half these defectives (those called 'protanope') cannot see the red end of the spectrum at all; the other half (deuteranopes) see it as bright as normal subjects can. Over a century ago (1855) Clerk Maxwell at the age of 23 proved by color matching that vision in the normal depended upon red, green and blue primaries, but protanopes lacked the red. Exactly 100 years later Rushton (1955) showed by retinal densitometry on the fovea that protanopes lacked the red-sensitive cone pigment *erythrolabe* that is present in normal eyes. Deuteranopes, on the other hand possess erythrolabe but lack the green-sensitive pigment *chlorolabe* present in the protanope.

It may be concluded that in the red-green range, the normal eye responds to two visual pigments erythrolabe and chlorolabe and in this range is dichromatic. Protanopes lack erythrolabe, deuteranopes lack chlorolabe and hence each is monochromatic. All subjects have in addition a blue-sensitive cone pigment, but *cyanolabe* cannot be measured with sufficient accuracy by retinal densitometry in living men.

Kinetics. The kinetics of bleaching and regeneration of cone pigments have been studied in color defectives and normals (Rushton, 1958). The rate of bleaching as would be expected is proportional to the rate at which quanta are absorbed. The rate of regeneration of the pigments is found to be always proportional to the amount of free opsin. These two rates, moreover, have been shown to be independently additive.

Thus

$$- \frac{t_0 dp}{dt} = \frac{lp}{I_0} - (1 - p)$$

where

p is the fraction of the pigment in the unbleached state

I is the incident quantum flux

t_0 is the time constant of regeneration = 2 min

$[I_0 t_0]$ is the incident instantaneous quantum energy that will change p from 1 to e^{-1} = [photosensitivity]$^{-1}$

After the eye has been exposed to a bright light it becomes much less sensitive, and takes some minutes to recover. The recovery is linked to the regeneration of the visual pigments that were bleached by the exposure. This is not simply due to the fact that if the pigment is half bleached away there will only be half left to absorb the incident light. That would only double the threshold; in fact the threshold is increased 30-fold. The threshold rise appears to be related to $(1 - p)$ the free opsin and is given by the expression $A \times 10^{3(1-p)}$ where A is the fully dark adapted *absolute threshold*.

Single Cones. Within the last few years the spectral sensitivity of single cones has been studied by two techniques requiring very great skill.

(a) *Microdensitometry*. Marks, Dobelle and MacNichol, Liebman and others have focused light onto single cones under a microscope and measured the pigment density at various wave lengths. With the excised retinas from man and monkey three types of cones were found, one containing erythrolabe and one chlorolabe coinciding well with reflexion densitometry in the living human eye. They have also found some cones containing the blue-sensitive pigment which had not previously been measured.

(b) *Microelectrode recording*. The spectral sensitivity of single cones has been measured by recording intracellularly with fine micropipettes. The reciprocal of the energy at each wave length necessary for a fixed response is found to be proportional to the absorption as measured by method (a).

W. A. H. Rushton

COMBINING NUMBER

The combining number of an element or a radical is a small whole number used in writing formulas. It is sometimes called the valence (number). Hydrogen is usually taken as the standard of combining numbers with value equal to 1.

(1) The number of hydrogen atoms that an element combines with directly is called its combining number. The compounds hydrogen chloride (HCl) and sodium hydride (NaH) are relevant examples. The formulas are derived from experimental data. In hydrogen chloride, one atom of hydrogen combines with one atom of chlorine. The combining number of chlorine is 1. Using similar reasoning, sodium also has combining number 1 because it combines with one atom of hydrogen. The compound between sodium and chlorine is written NaCl, each element having combining number 1. This formula (NaCl) for sodium chloride is verified by analysis. It contains 39.3% of sodium and 60.7% of chlorine—the proper amounts if this compound has a 1-to-1 relationship between the number of atoms. From the formulas of the compounds water (H_2O), ammonia (NH_3), and methane (CH_4), the combining number of oxygen is 2, nitrogen 3, and carbon 4.

Hydrogen is believed to have one electron in its K shell. It may fill the shell by borrowing or by sharing electrons as it does in NaH (borrowing); or it may lose its electron as it does in hydrogen chloride, where the electron pair that constitutes the bond probably is closer to the chlorine than to the hydrogen nucleus, electrically considered. In the first case the combining number of hydrogen is 1— and in the second case it is 1+. In general, negative combining numbers refer to elements that have gained electrons, and positive numbers to elements that have lost electrons. If a covalent pair is nearer one nucleus than another, the nearer nucleus is considered to be negatively charged.

(2) Certain combinations of elements remain together during many chemical changes, for example, OH, COOH, SO_4, NH_4, etc. Such combinations are called *groups* when they are uncharged and *radicals* when they acquire a charge in ionization.

The number of hydrogen atoms with which a radical combines directly is its combining number. From the formulas HNO_3, nitric acid; H_2SO_4, sulfuric acid; H_3PO_4 phosphoric acid, it can be seen that the combining number of the nitrate group (NO_3) is 1, that of sulfate group (SO_4) is 2 and that of phosphate group (PO_4) is 3.

(3) Combining numbers may be inferred by substitution. Not all elements readily combine directly with hydrogen. Silver is such an element. The formula of silver nitrate is $AgNO_3$. When this formula is compared with that of nitric acid (HNO_3), it is evident that one atom of silver replaces one atom of hydrogen. Since silver has replacing value equivalent to hydrogen atom for atom, the combining number of silver is therefore 1. Similarly, when $CuSO_4$ and H_2SO_4 are compared, it can be seen that a copper atom has taken the place of two hydrogen atoms, and therefore copper has combining number 2. In a similar manner the combining number of every element or radical may be determined.

According to theory, the combining number of elements in ionic compounds is the number of electrons transferred. It has the same value as the charge on the ion of that element. Calcium chloride ($CaCl_2$) is such an ionic compound in which the two chlorine atoms have gained one electron each from the calcium atom, which is thereby charged two units positive, or Ca^{2+}. Both of the chlorine atoms gain one electron each, becoming chloride ions (Cl^-). The combining number of calcium is 2, and that of chlorine is 1.

For covalent compounds, the combining number is the same as the number of shared electron pairs. In carbon disulfide (CS_2), carbon shares its four electrons with the two sulfur atoms, two pairs to each. The combining number of carbon is 4 (four shared electron pairs), and the combining number of sulfur is 2 (two shared electron pairs).

It is well known that many elements exhibit more than one combining number, nitrogen, chlorine, manganese, molybdenum, and chromium being among those that are especially noted for this property. The combining number depends on how many electrons in the outermost shell of an element are involved, or how many electrons it can draw up from or depress to lower levels.

ELBERT C. WEAVER

Cross-references: *Bonding, Radicals, Valence, Combining Weight, Formulas.*

COMBINING WEIGHT

The combining weight (or equivalent weight) of an element is found as one step toward determining its atomic weight by a chemical method. Most of the atomic weights today are found by use of the mass spectrograph, which also gives information about the isotopes present and their relative abundance. The chemical methods of finding atomic weights are of historical interest only.

Although any weight of an element may be selected as a standard for combining weights, the standard commonly accepted is exactly 7.9997 (approx. 8) grams of oxygen. Oxygen combines directly with many metals and nonmetals. The combining weight of an element may be defined as the weight of that element that combines with 8 g of oxygen, or its equivalent. Since 8 g of oxygen combines with 1.00797 (approx. 1.008) g of hydrogen, the latter is considered equivalent to 8 g of oxygen. When the value of 8 g is selected for the combining weight of oxygen, no element has a combining weight value less than 1.

The task of finding combining weights consists of measuring with precision the weight of the element that combines with 8 grams of oxygen. For elements that do not combine readily with oxygen, the weight of that element that combines with 1.008 g of hydrogen, or replaces that weight of hydrogen from an acid, may be found. In many cases, different weights of an element may combine with 8 g of oxygen. In such cases, the element has more than one combining weight.

The facts needed to find a combining weight may be measured in the laboratory. For example: 8 g of oxygen combines completely with 12.156 g of magnesium when the two elements combine. The value 12.156 is accepted as the combining weight of magnesium, and it is one-half of its atomic weight.

In another experiment more frequently performed, the combining weight of magnesium is found by measuring the volume of hydrogen that magnesium replaces from acid, finding the weight of the hydrogen, and calculating the weight of magnesium needed to produce 1.008 g of hydrogen. In this case, a piece of magnesium is used that weighs 0.0960 g. The metal is immersed in dilute acid and all the hydrogen released is collected in a gas-measuring buret. The volume of gas at standard temperature and pressure, after the necessary corrections have been applied, is 88.4 ml, or 0.0884 liter. Hydrogen weighs 0.0899 g/liter (its density at STP), so the weight of hydrogen is 0.0884 liter × 0.0899 g/liter, or 0.00795 g.

If 0.0960 g of magnesium released 0.00795 g of hydrogen, then the weight of magnesium that can release 1.008 g of hydrogen is found by the proportion

$$\frac{0.096 \text{ g}}{x} = \frac{0.00795 \text{ g}}{1.008 \text{ g}}$$

The value for x, the combining weight of magnesium, is 12.17 from the solution of this proportion. The accepted value is 12.156.

The combining weight is used to find the atomic weight of magnesium as follows: According to the law of Dulong and Petit, the atomic weight × specific heat = 6.4. The results from the use of this law are necessarily approximate because the specific heat varies with the temperature, state of matter, and other factors. For metals, application of the law gives an approximation that shows the order of magnitude of the correct atomic weight.

The specific heat of magnesium is 0.25 (cal/g/°C). Atomic weight of magnesium = 6.4/0.25, or 25.6, that is, the atomic weight of magnesium is approximately 25.6. Since 25.6/12.17 is approximately 2, the combining weight of magnesium should be multiplied by 2 to find the atomic weight. In this case 12.17 × 2 = 24.34, atomic weight of magnesium. The accepted value in the International Table, based on more exact work, is 24.305.

According to the experiments of Richards and Cuhman, an average of 8 trials, 24.28947 g of highly purified nickel bromide produced 6.52235 g of metallic nickel when the bromide was reduced with hydrogen. The weight of bromine present is found to be 17.76612 g. The ratio of bromine to nickel is 17.76612/6.52235. If we accept 79.909 as the combining weight of bromine, then the combining weight of nickel is found by the proportion 17.76612/6.52235 = 79.909/x; x = 29.336, the combining weight of nickel.

The specific heat of nickel is about 0.11. Applying the law of Dulong and Petit; atomic weight × 0.11 = 6.4. The atomic weight of nickel is about 58.2. Since 58.2/29.336 is approximately 2, 2 × 29.336 = 58.672, atomic weight of nickel. The accepted value in the International Table is 58.71.

ELBERT C. WEAVER

Cross-references: *Combining Number, Formulas.*

COMBUSTION

Combustion is a branch of science and technology that deals with the liberation and use of energy evolved during the reaction of chemical species. In its most common usage, combustion describes the process of rapid heat liberation, and is commonly associated with appearance of luminous flames.

The diversity of phenomena during combustion can be illustrated by the reaction of hydrogen and oxygen. A mixture of these two substances is stable at room temperature in the absence of catalysts. A slow reaction sets in at about 400°C. At somewhat higher temperatures, mixtures of hydrogen and oxygen can react very quickly (explode) to form water. The explosion is caused by free radical chain carriers (H, OH, HO_2) that are formed at a faster rate than they are destroyed, leading to a rapid increase in reaction rate and temperature.

An alternate way of accomplishing the rapid conversion to water is to initiate the reaction locally by means of a spark. Even at temperatures where the homogeneous reaction is slow or nonexistent, a flame will now propagate through the gas mixture, the formation of water occurring in the comparatively narrow reaction zone of the flame front.

If the initiating agent is sufficiently powerful (such as a small explosive charge) the reaction wave will propagate supersonically through the mixture at velocities about a thousand times faster than in the previous case. This process is known as *detonation* and is due to an intimate coupling of shock wave and chemical reaction. In contrast, the flow speeds in flame reactions at no time exceed the velocity of sound.

In addition to these examples, which may occur either in a static or a flow system, two additional cases of combustion have general application in flow systems: (1) *Diffusion flames,* wherein the reactants are not initially premixed. The reactions proceed as the reactants interdiffuse, the heat release being principally controlled by diffusion rates rather than by the speed of the chemical reactions. (2) The *"highly stirred reactor,"* wherein reactants are thoroughly stirred into the hot reaction products within a confined volume. The reaction occurs at high temperature, but below the equilibrium combustion temperature. The rate of mass addition of new reactants determines the temperature level at which the reaction proceeds, which, in turn, determines the heat generation rate.

It is apparent from the examples cited that combustion includes a large number of diverse phenomena even if one considers gaseous reactions only. Each of the cases described above has important technical applications. Explosions due to chain branching are believed to be responsible for the knock phenomena in internal combustion engines where the charge of fuel and air has been compressed and heated to a temperature at which the spontaneous reaction is rapid. Flame propagation through a premixed system occurs in engines (internal combustion, turbojet, and ramjet) where large rates of heat release are desirable. Detonations occur in high explosives. Combustion in diffusion flames is found in two distinct areas: (1) When it is desirable to produce strongly luminous flames for the sake of good heat transfer; hydrocarbons are used as fuels and the oxidation is made to proceed via strongly radiating carbon particles. (2) When the time for evaporation of the fuel and oxidizer is insufficient to produce a homogeneous gaseous mixture or when the vapor pressure of the fuel or oxidizer is too low for complete vaporization. Each drop or particle will be surrounded by a combustion zone based on interdiffusion of oxidizer and fuel. The "stirred reactor" has promise in application where precisely controlled residence times and compositions are of importance, as in the synthesis of chemical compounds at elevated temperature. The fuel and oxidizer supply the heat necessary for the initiation of the reaction as well as the atomic species required in the synthesis.

Full understanding of the details of the combustion process in terms of the pertinent chemical and fluid dynamic variables is hampered by the complexity of the interrelated phenomena. Nevertheless, improvements in the theoretical analysis of a variety of flame models and development of new techniques for the experimental study of flame front structure give promise for a more fundamental insight into the processes underlying combustion. One of the most pressing needs for the better understanding of flame reactions is an accurate description of the composition and temperature variations throughout the reaction zone, together with knowledge of the diffusion coefficients of the various species which are formed or destroyed in this zone.

Combustion reactions involving flame fronts and laminar flow conditions can be quantitatively specified as possessing a unique flame speed, defined as the velocity of flow of the unreacted mixture normal to the flame front. The magnitude of the flame speed depends on the particular reactants and the mixture composition. Flame speeds have been measured at room temperature ranging from a few centimeters per second to about 1000 cm/sec. Since they involve chemical reactions, flame speeds are also sensitive to pressure and inlet temperature.

The ignition limits of various mixtures are of considerable practical significance, for self-sustaining combustion reactions cannot be carried out beyond these limits and, therefore, do not present a safety hazard. One limit occurs when fuel is deficient, and another limit exists in the absence of sufficient oxidizer. These limits are usually referred to as the lean and rich (lower and upper) limits. Wide variations are found as a function of the chemicals involved. The precise location of the ignition limits depends on the means of initiating the reaction. The limits quoted are obtained in apparatus and with ignition sources of such magnitude that further increase of scale has no significant influence on the results. Several endothermic substances (acetylene) and monopropellants in which the fuel and oxidizer are part of the same molecule (such as nitromethane) may be unstable toward decomposition if the initiation source is sufficiently powerful.

Another quantity of considerable practical significance is the ignition or flame stabilization requirement. Flames in potentially combustible mixtures will not propagate unless an adequate initiation source is available. In view of the complexity of the initiation step the precise physical meaning of the terms is difficult to define. However, under controlled conditions minimum spark energies can be specified beyond which flame propagation will proceed. Similarly, the anchoring of flames in the high velocity flow field of a combustion chamber depends on the availability of a continuously initiating source whose function it is to supply the required ignition energy to the unreacted gas mixture. The flow in the wake behind bluff bodies is frequently used to serve this function.

Heat release rates have practical significance since they determine the size of the combustion volume in which the energy transformation takes place. In laminar premixed flames the heat release is confined to the combustion wave and depends on the point-by-point balance of heat release due to chemical reaction and heat gained or lost by diffusion. The highly stirred reactor permits a definition of a heat release rate in terms of the chemical reaction rates. In most cases of practical interest, the term is more loosely used and defines the heat output in unit volume of reaction space, even though the volume under investigation may not be of uniform composition. It permits the description of applications in which nonuniformities of composition or of phase (droplets or solids) exist. In such cases the fluid flow conditions are of very great significance in determining the magnitude of the heat release rates.

WALTER G. BERL

Cross-reference: *Flame Chemistry.*

Combustible and Flammable Materials

A combustible material is any substance that will burn, regardless of its autoignition point, or whether it is a solid, liquid, or gas. Although this definition logically includes all flammable materials as well, this fact is ignored by official classifications. In the latter, the term "combustible" is restricted to those materials that are comparatively difficult to ignite and that burn relatively slowly. For any solid material, the rate and ease of combustion depend as much on its state of subdivision as on its chemical nature; thus cellulose is combustible in the form of paper or a textile fabric, but flammable in the form of fibers (e.g., cotton linters). Solid magnesium will not ignite at less than 1200°F, whereas thin flakes or powder will burn easily and rapidly (as will other metal powders).

The definitions of combustible liquids offered by various organizations that are active in the field of hazardous materials are unfortunately far from uniform. For example, the National Fire Protection Association considers them as liquids having a flash point above 140°F (Tag Closed Cup). The Interstate Commerce Commission uses a flash point of 80°F as the dividing line between flammable and combustible. The International Air Transport Authority defines a combustible liquid as one having a flash point (Tag Open Cup) that is above 80°F but less than 101°F. The Manufacturing Chemists Association defines it as having a flash point between 80° and 150°F (Tag Open Cup).

A flammable material is any solid, liquid, vapor, or gas that will ignite easily and burn rapidly. Flammable solids are of several types: (1) dusts or fine powders (metals or organic substances such as cellulose, flour, etc.); (2) those that ignite spontaneously at low temperatures (white phosphorus); (3) those in which internal heat is built up by microbial or other degradation activity (fish meal, wet cellulosic materials); (4) films, fibers, and fabrics of low-ignition point materials.

The National Fire Protection Association divides organic liquids into three classes: (1) those having a flash point (Tag closed cup) below 100°F (flammable, dangerous fire hazard); (2) those having a flash point at or above 100°F and below 140°F (flammable, moderate fire hazard; (3) those having a flash point at or above 140°F (combustible, slight fire hazard). Thus the critical flash point temperature for a flammable liquid is 140°F (Tag Closed Cup). However, shipping regulatory authorities (ICC, CG, IATA), as well as the Manufacturing Chemists Association, use 80°F (Tag Open Cup) as the critical flash point temperature.

Flammable gases are ignited very easily; the flame and heat propagation rate is so great as to resemble an explosion, especially if the gas is confined. The most common flammable gases are hydrogen, carbon monoxide, acetylene and other hydrocarbon gases. Oxygen, though essential for the occurrence of combustion, is not itself either flammable or combustible; neither are the halogen gases, sulfur dioxide or nitrogen. Flammable gases are extremely dangerous fire hazards, and require precisely regulated storage conditions.

The terms "flammable," "nonflammable," and "combustible" are difficult to delimit. Since any material that will burn at any temperature is combustible by definition, it follows that this word covers all such materials, irrespective of their ease of ignition. Thus the term "flammable" actually applies to a special group of combustible materials that ignite easily and burn rapidly. Some materials (usually gases) classified in shipping and safety regulations as nonflammable are actually noncombustible. The distinction between these terms should not be overlooked. For example, sodium chloride, carbon tetrachloride, and carbon dioxide are noncombustible; sugar, cellulose, and ammonia are nonflammable.

G. G. HAWLEY

Cross-reference: *Extinguishing Agents.*

COMPLEXING AGENTS IN CHEMICAL ANALYSIS

Many types of complexes are relevant to the practice of chemical analysis. This article is restricted to complexes formed by metal ions.

A metal complex consists of a metal ion acting as an electron acceptor (i.e., a Lewis acid), coordinated with one or more electron donor atoms (or donor groups) of a second species, which is described as a complexing agent, ligand, or a Lewis base. An atom or group directly bonded to the metal is termed a ligand atom or ligand group. A complexing agent containing only a single ligand atom or group per molecule, is termed unidentate; if it contains two such atoms or groups, bidentate, and if many, multidentate. Elements serving as ligand atoms in complexing agents of analytical interest are few. The ligand atoms in metal complexes formed with common *inorganic* complexing agents are the halogens (as halide ions), carbon (in cyano complexes), nitrogen (in ammine, nitro, and some thiocyanato complexes), oxygen (in aquo, hydroxo, peroxo, nitrato, carbonato, sulfato complexes as well as complexes of various phosphate species), and sulfur (in sulfido complexes and some thiocyanato complexes). In the *organic* complexing agents now receiving attention as analytical reagents, ligand atoms are restricted to oxygen, nitrogen, and sulfur.

A complex formed by a bidentate or multidentate ligand may contain one or more rings in which a metal ion is bonded to two donor atoms. Such a ring was originally termed a chelate—a term which has now been extended to the entire complex, which can be described as a chelate complex, a metal chelate, or simply a chelate. Some chelating agents possess two types of ligand groups: one coordinating with neutralization of charge ("salt" formation) and the other without such neutralization. An electrically neutral chelate formed between a metal ion and a chelating agent, is sometimes called an inner complex compound or inner complex salt.

A metal complex may be positively charged, negatively charged, or electrically neutral. These three possibilities are illustrated by $[Cu(NH_3)_4]^{2+}$, $[Fe\cdot(CN)_6]^{3-}$, and $HgCl_2$, respectively. A complex may also become electrically neutral by association with a counter ion, e.g., $Pb_2[Fe(CN)_6]$ and $H[FeCl_4]$. The fact that some metal ions form anionic complexes and others remain as (solvated) cations, permits the separation of the metals on columns of ion exchange resins. Cations are retained by a cation exchanger and anionic complexes appear in the column effluent; anionic complexes are retained by anion exchangers and cations are passed. Complexing agents are frequently employed in separations of metal ions based on ion-migration (including paper, column, and thin-layer chromatography, electrophoresis, and ion exchange) since the migration rates and tendencies are strongly affected by complexation.

An electrically charged complex is more soluble in a polar solvent such as water. A neutral complex, or one made neutral by effective ion-association, is less likely to be soluble in a polar solvent. If the complex is relatively insoluble, it is, of course, a precipitate and the complexing agent is termed a precipitant. Such complexes are often extractable from an aqueous medium into a suitable immiscible solvent; in such a case the complexing agent may be termed an extractant.

If the reaction of a metal ion with a complexing agent is sufficiently complete, that is, the stability of the complex formed is great, the possibility of a titration of one of these species by the other is possible. If the complex is insoluble, the method is a precipitation titration. If the complex is soluble, the method is termed a compleximetric titration. For example, the titration of chloride ion with mercury(II) ion, is often used for the determination of the chloride content of biological fluids. The success of a compleximetric titration depends not only on the stability of the complex formed but also on the number of steps involved. This can be appreciated by the observation that the end point "break" is usually associated with the coordination of the last ligand. Consequently, compleximetric titrations have come of age only since 1946 with the availability of aminopolycarboxylate chelate agents that form highly stable, 1:1, water-soluble complexes with many metal ions. Of such agents EDTA, that is, (ethylenedinitrilo)tetraacetic acid, as the disodium salt, has received prominent attention. Titrations with EDTA and related compounds and polyamines, are variously termed complexometric titrations, chelatometric titrations, or chelometric titrations. By application of EDTA titrations to a visual end point, either directly or by indirect methods, over seventy of the chemical elements can now be determined.

Complexes of metals are often colored, that is, they show spectral absorption in the visible region. Two modes of color formation may be differentiated. A colorless cation may form a colored complex with a colorless complexing agent. Here the absorption of light is associated with an electronic transition between two orbitals for which the electron is located largely on different atoms (charge-transfer spectra). The red-orange complex formed by the iron(III) ion, Fe^{3+}, and thiocyanate ion, SCN^+, offers a familiar example. A second mode involves the electronic interaction between the metal ion and the resonating system of the ligand. When such ligands are already intensely colored in their noncomplexed (unmetallized) form, the interaction with the metal acts to extend the conjugation, which includes the chromophore groups. As a result, an absorption maximum of increased intensity is obtained, that is, displaced to a longer wavelength, and a pronounced change in color occurs. Color formation and changes (or, in general, changes in absorbance in the ultraviolet, visible, or near infrared regions) may be used for the purposes of detection and identification tests for the metal or ligand, of indication of the end point in precipitation and compleximetric titrations, and also of colorimetric and photometric determinations.

Where the cations of more than one metal are present in a sample, analytical selectivity becomes a primary concern. Some intrinsic selectivity is secured by the different tendencies of metal ions to react with a given family of complexing agents. For example, metals giving ions having an inert gas structure (Periodic Groups IA, IIA, B and Al in Group IIIA, Sc, Y, and La in Group IIIB, and Group IVB except thorium) show a preference for complexing agents containing oxygen as the ligand atom. In aqueous media, further selectivity can often be secured by adjustment of the pH. At an appropriate value, protonation of the ligand, when it is a weak acid, may proceed to an extent that one or more of the metal ions may not undergo complexation. For example, iron(III) can be selectivity titrated with EDTA at pH 1–2 without interference from most dipositive metal ions, including zinc, cadmium, and the alkaline earths.

Masking offers a further approach to gaining selectivity and may be defined as the process in which a species, without the physical separation of it or its reaction product, is so transformed that it ceases to enter into a reaction. The species is said to be masked toward the principal reaction or reagent. The masking of metal ions commonly involves complex formation. Naively, masking can be viewed as a tug-of-war for the species, between the principal reagent and a second agent. If two species are present, and the masking agent "wins" one and the principal reagent the other, then analytical selectivity or even specificity is secured. For example, the masking of copper(II) by the addition of cyanide ion, permits the selective precipitation of yellow cadmium sulfide. The introduction of organic complexing agents, including EDTA and related compounds, has greatly increased the possibilities of masking.

Complex formation often has a pronounced influence on a redox potential since, in brief, the stability of one of the two oxidation states involved may be affected more than the other. For example, the iron(III)-iron(II) couple in water has a standard redox potential of $+0.771$ volt; in the presence of EDTA, which forms a more stable complex with iron(III) than with iron(II), the potential is shifted to $+0.14$ volt. Consequently, in the presence of EDTA, iron(II) is a relatively strong reducing agent. Complexing agents therefore receive diverse use in various electroanalytical methods, including electrodeposition and polarography.

A. J. BARNARD, JR.

Cross-references: *Chelation, Sequestering Agents, Photometric Analysis, Ion Exchange, Titration, Gravimetric Analysis, Colorimetry, Electrochemistry, Polarography.*

COMPOSITE MATERIALS

Composites are composed of a mixture or combination of two or more macroconstituents that differ in form and/or material composition and that are essentially insoluble in one another. In principle, composites can be constructed of any combination of two or more materials, whether they be metallic, organic or inorganic. While the materials combinations possible in composites are virtually unlimited, the constituent forms are more restricted. The major constituent forms used in the structuring of composite materials are fibers, particles, laminas or layers, flakes, fillers and matrixes. The matrix is the "body" constituent. It serves to enclose the composite and to give it its bulk form. The fibers, particles, laminas, flakes and fillers are the "structural" constituents that determine the character of the internal structure of the composite. They are generally, but not always, the "additive" phase.

Perhaps the most typical composite is that composed of a structural constituent embedded in a matrix. However, many composites have no matrix, but rather are composed of one (or more) constituent form(s) consisting of two or more different materials. Sandwiches and laminates, for example, are composed entirely of layers. The layers taken together give the composite its form.

Because the different constituents are intermixed or combined, there is always a region contiguous to them. This region can be simply an interface, i.e., the surface forming the common boundary of the constituents. This interface is in some ways analogous to the grain boundaries in monolithic materials. In some cases, however, the contiguous region is a distinct added phase, which might be called an "interphase." Examples are the coating on the glass fibers in reinforced plastics and the adhesive that bonds together the layers of a laminate. When such an interphase is present, there are two interfaces—one between each surface of the interphase and its adjoining constituent.

Composites can be divided into five broad classes, based on the five structural constituents, as follows:

1. Fiber composites, composed of fibers with or without a matrix.
2. Particulate composites, composed of particles with or without a matrix.
3. Laminar composites, composed of layer or laminar constituents.
4. Flake composites, composed of flat flakes with or without a matrix.
5. Filled (or skeletal) composites, composed of a continuous skeletal matrix filled by a second material.

Fiber Composites. The most widely used composites are those composed of an inorganic fiber, in an organic matrix. There is considerable choice of materials for the two constituents making up such fiber composites, but to date no combination has proved as successful as glass-fiber-reinforced plastic composites.

Although many types of plastic resins, both thermosetting and thermoplastic, are being reinforced with glass fibers, polyester resins are the most widely used, especially for low-performance applications. Epoxy resins, although more expensive than polyesters, are being used in many high-performance applications. Thermoplastic resins, such as nylon, polystyrene, polycarbonate and polyvinyl chloride, show improved mechanical properties when reinforced with glass fibers. In addition, when reinforced with glass fibers these normally unstable plastics have excellent dimensional stability and improved heat resistance.

Metal-fiber-reinforced plastics, besides being mechanically strong and tough, can be designed to be good conductors of heat and electricity. By incorporating short, random metal fibers in epoxy resins, systems have been made with impact strengths, heat distortion points and thermal conductivities greatly exceeding those of glass-fiber-reinforced composites. Continuous metal fibers also have been investigated as a reinforcement for plastics. Steel wire of high tensile strength—on the order of 600,000 psi in 0.004-in. dia.—has been used to reinforce epoxy and polyethylene plastics.

Continuous metal fibers also are used to reinforce metals. Tungsten wire-copper matrix composites are considerably stronger than either material in bulk form. Molybdenum fiber-titanium matrix composites have shown substantially better high-temperature strength than titanium alloys. Steel-wire-reinforced aluminum has strengths over twice that of standard aircraft aluminum.

In recent years major attention has been focused on boron fiber and graphite-fiber-reinforced epoxies and metals such as aluminum. The great interest in boron and graphite composites stems from the combination of high strength, high modulus and low density, and the retention of strength at elevated temperatures. Whisker-reinforced metals are also in development. The whiskers are essentially single crystals ranging from about 0.5 to 10 microns in diameter and up to around 75,000 microns long. Alumina (sapphire) and silicon carbide whiskers seem to be receiving most attention. Whisker-reinforced matrices have outstanding strength properties. For example, the tensile strength of a silver matrix with alumina whiskers at 1400°F is nearly 45,000 psi, which is far greater than that of pure silver and more than twice that of dispersion hardened silver at 1000°F.

Particulate Composites. Particulate composites are composed of an aggregate of particles in a matrix. The two principal types of particulate composites are *cermets* and *dispersion-hardened alloys*. The major differences between the two types is that the size and volume of particles is larger in cermets than in dispersion-hardened alloys. Cermets are discussed under that title, and will not be repeated here.

Dispersion-Hardened Alloys. Dispersion-hardened alloys generally are composed of a hard particle constituent in a softer metal matrix. The dispersed particle constituent is only a small proportion of the total, ranging from 2 to 15% volume. The particles themselves often range down into the submicron sizes. Cold working is required to develop high strength levels. The dispersion-hardened alloys differ from precipitation-hardened alloys in that the particle is added to the matrix, usually by nonchemical means. Precipitation-hardened alloys derive their properties from compounds that are precipitated from the matrix, rather than added artificially.

There is a rather wide range of dispersion-hardened alloy systems. Those of aluminum, nickel and tungsten, in particular, have proved commercially significant. Tungsten-thoria, a lamp filament material, has been in use over thirty years. Perhaps dispersion-hardened aluminum alloys are finding the widest commercial use today. Known as SAP (aluminum-aluminum oxide) alloys, they have a unique combination of good oxidation and corrosion resistance plus hot strength and high-temperature stability considerably greater than that of conventional high-strength aluminum alloys.

Metal in Plastic. A number of useful particulate composites consist of metal particles in a plastic matrix. Such filled plastics may contain up to 90% by volume of metal particles. The range of metal-filled plastics is very broad. Aluminum, as a filler, has applications ranging from a decorative finish to the improvement of thermal conductivity.

Copper particles are used in plastics as a coloring agent. Generally, the alloys are used in conjunction with thermoplastic solutions, e.g., cellulose nitrate or vinyl lacquers. Epoxies predominate as the binder in thermosetting plastic systems because of the inhibiting action of copper on vinyl-type polymerization and the curing of thermosetting polyesters. Three properties of lead are used in lead particle-plastic composites: its ability to dampen sound and vibration, its effectiveness as a barrier against gamma radiation, and its high density.

Laminar Composites. Laminar, or layered, composites consist of two or more different layers bonded together. The layers can differ in material, form and/or orientation. For example, clad metals are made up of two different materials. In sandwich materials, such as honeycomb, the core layer differs in form from that of the facings, while the materials of which the layers are composed may or may not be different. Similarly, in plywoods, though the layers are often of the same type of wood, the orientation of the layers differs.

Laminar composites can be divided into two major classes: *sandwiches* and *laminates*. Laminates are defined as composite materials consisting of two or more superimposed layers bonded together. Sandwiches, a special case of laminates, consist of a thick, low-density core (such as honeycomb or foamed material) between thin faces of comparatively higher density.

Flake Composites. Flake composites could be considered as being in the particulate class, since flakes are a form of a particle. However, they are sufficiently different from the cermets and dispersion strengthened composites to warrant a separate class. A flake composite consists of flakes held together by an interface binder or incorporated into a matrix. Depending on the material's end use, the flakes can be present in a small amount or can comprise almost the entire composite.

The number of materials that are used in flake composites is limited. Most metal flakes are aluminum, with silver being used to a minor extent. The other important flake materials are mica and glass. In almost all cases the flakes can be used with a wide variety of organic or inorganic binders or matrixes, provided that the material has chemical, mechanical and processing compatibility with the flakes.

Filled Composites. In its simplest form a filled composite consists of a continuous, three-dimensional structural matrix that is infiltrated or impregnated with a second-phase filler material. The filler also has a three-dimensional shape that is determined by the voids in the matrix. The matrix itself may be an ordered honeycomb, a group of cells, or a random sponge-like network of open pores.

In effect, both the matrix and the filler exist as two separate constituents that do not alloy and, except for a bonding action, do not chemically combine to a significant extent. The matrix is always continuous, but the filler may be either continuous (as in an impregnated casting) or discontinuous (as in a filled honeycomb).

The types of skeletal structures that are most receptive to impregnation are an open honeycomb and a sponge-like network of open pores. In general, the open-pore structure presents more processing problems than the honeycomb because it is randomly oriented and it is not usually possible to introduce filler materials in their solid state.

Most filled composites in use today consist of a matrix formed from a random network of open passages or pores. Unlike the matrix in a filled honeycomb, which is specifically designed to a given shape, the open matrix in a sponge or pore composite is formed naturally during processing. Typical of materials that have this kind of structure and which lend themselves to filling are metal castings, powder metal parts, ceramics, carbides, graphite and foams.

The open network of a sponge-like structure can be filled with a wide range of materials including metals, plastics and lubricants, depending on the end properties that are desired. Metal impregnants, for example, can be used to improve the strength of a matrix or to provide better bearing properties. Plastics can be impregnated into metals to make them pressure-tight or—like lubricants—to provide special bearing properties. They can also be incorporated into porous ceramics and graphite as a structural binder.

H. R. CLAUSER

COMPUTER SIMULATION OF CHEMICAL REACTIONS

The process development problem—that of determining rapidly and economically, the best reactor design and the best reactor operation program to fit a new reaction just out of the laboratory—is one of the major tasks which has faced industry over the years. How can computers be best used to reduce the time and cost of converting the chemist's data to the chemical engineer's final plant design?

It is well known that the course of the simple chemical reaction

$$A \rightarrow B \qquad (1)$$

is readily expressed by the differential equation,

$$-\frac{d(A)}{dt} = \frac{d(B)}{dt} = k(A), \qquad (2)$$

and that

$$(B) = \int_0^1 k(A)\,dt \qquad (3)$$

This latter equation is of course readily solved by hand. However, let us use it as the first example of a computer solution. Its solution on the analog computer is as shown in Fig. 1. In the circuit of Figure 1.1, the triangles represent computing amplifiers and the circles adjustable potentiometers. The potentiometer marked A, or initial condition, sets the starting value of the amount of component A present. The second pontiometer, labeled k, represents the rate of reaction and sets the integration rate of the problem.

1. The Computer Circuit

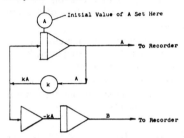

2. The Required Solution

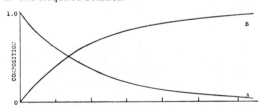

3. Manipulation of *k* to obtain the Required Solution

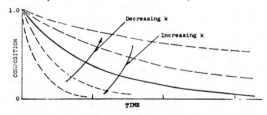

FIG. 1. A simple example to show the use of the analog computer for the mathematical analysis of a chemical reaction.

The circuit of Figure 1.1 is wired on the computer by connecting the components in the manner shown. The output leads A and B are connected to a plotter or recorder. The computer is then turned on and proceeds to solve Equation 3 by drawing the graphs of Figure 1.2 on a plotter. If the setting on the potentiometer representing *k* had been incorrect, one of the sets of curves shown in Figure 1.3 would be obtained. By readjusting the value of *k* and repeating the computer run as often as necessary, one could readily find by trial and error the right setting of *k* needed to duplicate a

desired curve such as one which had been obtained from laboratory data.

This then is the key to our thesis: The analog computer, by the very nature of its method of solution of the differential equations describing a reaction, gives us an easy procedure for the mathematical analysis of the reaction. Newer techniques for the use of digital computers give them the same flexibility in their use with the additional capability of automatic adjustment of the coefficient *k*.

More complex reaction schemes will of course require more elaborate wiring diagrams, but their analysis proceeds in exactly the same manner as above. For example the second order reaction,

$$A + B \rightarrow C + D \qquad (4)$$

requires the differential equation,

$$-\frac{d(A)}{dt} = -\frac{d(B)}{dt} = \frac{d(C)}{dt} = \frac{d(D)}{dt} = k(A)(B) \qquad (5)$$

which when put in integral form gives

$$C = D = \int_0^1 k(A)(B)\,dt \qquad (6)$$

Equation 6 requires the computer wiring diagram of Figure 2 for its solution. The resulting plots will closely resemble those of Figures 1.2 and 1.3 and one has the same facility for adjusting the reaction coefficient, *k*, as before.

If isothermal laboratory batch reaction data were available for the reaction of Equation 4 for a succession of temperatures, one could repeat the analysis of Equation 6 and Figure 2 and as a result obtain the value of *k* over the complete range of temperatures involved. When the resulting *k* values are plotted on a graph of reciprocal absolute temperatures versus the logarithm of *k*, the plot of

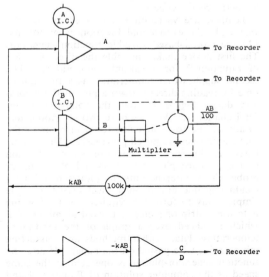

FIG. 2. Computer wiring diagram required for simulation of a second order chemical reaction.

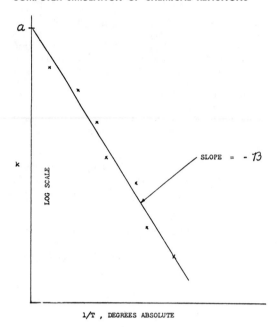

Fig. 3. Variation of reaction coefficient, k, with temperature.

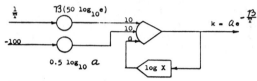

where

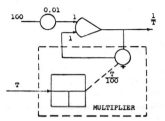 $\boxed{\log X}$ is an exponential function generator whose output is the logarithm of the input voltage.

G represents an input directly into the grid of the first state of the amplifier, bypassing the regular input elements.

are high gain amplifiers, that is, computing amplifiers used without their normal feedback elements.

Fig. 4. Computer circuit using the exponential function generator, a temperature variable reaction rate term.

Figure 3 results. From this figure the familiar Arrhenius equation

$$k = \alpha e^{-\beta/T} \tag{7}$$

$$= \alpha e^{-\epsilon/RT} \tag{8}$$

results. Here α represents a "frequency factor" related to the number of molecular collisions occurring per unit time per unit concentration and ϵ is a measure of the activation energy required to cause the reaction to occur. We can thus readily determine the value of k required to simulate the behavior of the reaction of Equation 4 at any temperature we choose.

Temperature variation is corrected by two devices: (1) the exponential function generator and (2) the plotter-based variable function generator. The first device makes possible the direct solution of Equation 7 for a variable temperature. The second supplies that temperature as a function of time by reading directly from a plot of the laboratory data. Figure 4 shows the computer circuit for Equation 7. We now have two potentiometers to adjust, one for α and one for β in place of the single adjustment required previously. The plotter-based variable function generator is a modification of the standard plotter to carry a high-frequency probe. This latter permits the plotter to follow a conducting line drawn on a sheet of paper. The simplest way to form the required conducting line is to use a strip of adhesive backed aluminum foil which is placed over a graph of the laboratory temperature data. Other methods are conducting ink or a fine wire pasted on the paper. If now the function generator-plotter is operated at the same speed as the computer solution of Equation 4 and if the circuit of Figure 4 along with another multiplier is substituted for the k potentiometer of

Figure 2, we can readily simulate a variable temperature curve by manipulating the values of the potentiometers representing α and β.

Needless to say, the same procedure can work in a plant reactor design problem. A series of possible reactor profiles may be placed on the plotter based variable function generator. Running each of these in turn will readily show the engineer which of the possible profiles will give the highest product yield or highest production from the reactor.

Complicating Factors. The effect of a catalyst on a chemical reaction is generally one of varying the rate of a reaction without in any way changing the final equilibrium products. Thus the presence of the catalyst changes the value of α and β of Equation 7 to a new α^* and β^*. We also have the problem of accounting for the effective concentration of the catalyst in the reaction. By convention on computers, a homogeneous catalyst is generally treated as a component in the reaction, because its concentration can vary radically during the reaction especially if a continuous stirred tank type reactor is used. Therefore

$$\frac{dC}{dt} = k^*(A)(B)(K) \tag{5a}$$

Its active concentration, K, is usually directly related to its molar concentration in the solution, and thus readily treated as in Equation 5a. Since K is treated as a component there must also be a differential equation written for it to define its

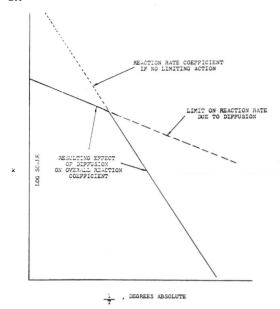

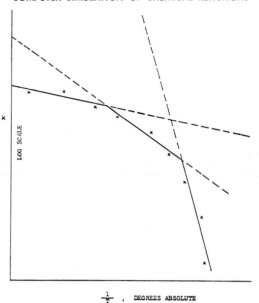

FIG. 5. Effect of diffusion on limiting reaction rates.

FIG. 6. Fitting of possible multi-step reaction rate data, three reaction steps assumed.

concentration just as for each of the other components in the reaction.

A heterogeneous catalyst, because of the problems of particle sizes, active sites, etc., is not so readily handled. However, its concentration is not usually subject to change during the reaction. Thus an empirical constant factor modifying k^* is commonly used

$$\frac{dC}{dt} = \overline{K}\, k^*(A)(B) \qquad (5b)$$

Equation 5a requires an additional multiplier to form the product $(A)\,(B)\,\overline{(K)}$ in the simulation. Equation 5b requires $\overline{K}$.

Other complicating factors commonly found are diffusion-limited reactions and limiting rate of solution of gases and of solids. As shown in Figure 5 this shows up as a change in the appearance of the k vs. $1/T$ graph and can often be approximated as shown here.

However, even when diffusion is not important, the procedure of Figure 3 may often lead to a set of data resembling Figure 6, or the use of a single α and β in Equation 7 may not give a satisfactory duplication of the composition curve over a temperature range. In many cases such action may be due to the formation, of short-lived chemical complexes or of free radicals and a resulting chain reaction as part of the reaction mechanism. This is especially so if a polymerization reaction is involved.

In such a case the data can be approximated as in Figure 6, where one or the other of the reactions becomes dominating as the temperature varies. One possible modification of Equation 4 to fit data such as those in Figure 6 might be

$$A \rightarrow \overline{A}, \qquad (9)$$

$$\overline{A} + B \rightarrow \overline{AB}, \qquad (10)$$

$$\overline{AB} \rightarrow C + D. \qquad (11)$$

where $\overline{A}$ refers to a possible free radical species and $\overline{AB}$ is a molecular complex. The mathematical model of this reaction is

$$\frac{d(A)}{dt} = -k_1(A)$$

$$\frac{d(\overline{A})}{dt} = k_1(A) = k_2(\overline{A})(B) \qquad (12)$$

$$\frac{d(B)}{dt} = -k_2(\overline{A})(B) \qquad (13)$$

$$\frac{d(\overline{AB})}{dt} = k_2(\overline{A})(B) - k_3(\overline{AB}) \qquad (14)$$

$$\frac{dC}{dt} = \frac{dD}{dt} = k_3(\overline{AB}) \qquad (15)$$

The computer plotted composition pattern of such a reaction as postulated here is shown in Figure 7. Intermediate concentrations, particularly of such components as $\overline{AB}$, may be very small at any one time but still be quite vital in the reaction scheme.

More complex reactors than the simple batch reactors just considered can also be readily handled by the methods illustrated here. Tubular reactors can be simulated provided only that lateral mixing can be considered very large compared to longitudinal mixing. The tubular reactor equations are then solved in the same manner as for the batch reactor except for the substitution of the length variable for the time variable of the batch reactor.

A continuous stirred tank reactor is more difficult to simulate than the others, but only in that the effect of a continuous flow of material into and out of the reactor must be taken into account. For the well-mixed tank this comprises the addition of flow terms into the reactor mathematical models.

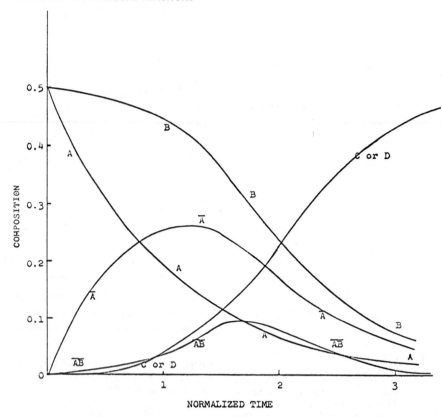

FIG. 7. Intermediate compositions for a possible multistep free radical or activated complex type mechanism of an over-all second-order reaction.

Equation 5 for example. Equation 5 would thus appear as:

$$\frac{dA}{dt} = -(A)(B) + \frac{f}{V}(A)_i - \frac{f}{V}(A)_o \quad (5c)$$

Such reactor operation has a decided influence on the appearance of the computer derived graphs but does not affect their interpretation or use as described here.

It can be readily seen that the use of computer simulations as an aid to the reactor design procedure is effectively the reverse of the method of attack required to devise and prove a chemical reaction mechanism and its associated reaction rate coefficients. Our task now is to take the mathematical model of the chemical reaction and modify it by the appropriate flow, heat balance and mixing terms as necessary. We then solve the resulting reactor model for a succession of assumed reactor temperatures or temperature profiles until that combination of sizes, flows and temperatures is obtained which gives the desired reactor output or the nearest possible approach to it.

Limitations. It will be noticed that each of the examples listed above requires the existence of a finite reaction rate for its solutions by computers. Thus this method of studying chemical reactions is best adapted to organic type reactions which usually have relatively slow apparent rates. Practically instantaneous reactions, such as ionic re-actions, which are completely diffusion limited, are difficult to represent by a computer simulation like that described here. This is due to the approximations which must be made for the complex phenomena of mixing and of diffusion.

Even though the computer provides a rapid and accurate solution of even very large families of ordinary differential equations, the procedure outlined here still requires much judgement by the chemist or engineer operating the computer in order to choose the best values of a large number of k factors or, even more difficult, to pick the best from several different possible mechanisms such as when free radicals or complexes may be present but whose identity is unproved.

Nomenclature:

A A component in a chemical reaction.
       As a subscript refers to a particular reaction rate coefficient.

α Frequency factor in Arrhenius reaction rate coefficient.

B A component in a chemical reaction.
       As a subscript refers to a particular reaction rate coefficient.

β Activated energy factor in modified Arrhenius reaction rate coefficient evaluation.

C A component in a chemical reaction.
       As a subscript refers to a particular reaction rate coefficient.

D A component in a chemical reaction.

ϵ Activation energy in Arrhenius reaction rate coefficient.

e Base of the natural logarithms.

f Flow rate of reactants in a continuous reactor.

i As subscript refers to input flow stream composition.

K A homogeneous catalyst.

$\overline{K}$ Constant expressing the effective concentration of a heterogeneous catalyst.

k Reaction rate coefficient.

L Length in appropriate units.

0 As subscript refers to reactor output composition.

R Thermodynamic constant.

T Absolute temperature, °R or °K.

t Time in appropriate units.

V Volume in appropriate units.

I.C.   Initial condition, concentration of reactant at beginning of reaction being simulated.

( )   Signifies concentration of component whose symbol is enclosed.

*   Signifies activated value due to catalysis.

¯   Indicates a free radical or activated complex type of chemical species (Overscript).

T. J. WILLIAMS

Cross-references: *Chemical Engineering*

CONCRETE, see CEMENT, PORTLAND

CONDUCTANCE

When a strong electrolyte is dissolved in water, the solute consists of electrically charged molecules, called ions. If the solution is sufficiently concentrated, some of the oppositely charged ions associate giving, in effect, a neutral molecule of the dissolved electrolyte. Weak electrolytes behave as though a major part of the solute is in the molecular state while only a small fraction is ionized. When an electrical potential difference exists between electrodes in the solution, the positively charged ions (cations) flow toward the negative electrode, the cathode. The anions flow toward the anode. Thus a flow of electricity occurs in the solution and Ohm's law, $R = E/I$, applies. E is the difference in potential, I is the current strength and R is the resistance of the solution. The resistance of 1 cc of a solution between electrodes 1 cm apart is the specific resistance, R, of the solution. The magnitude of the specific resistance is a function of several variables: the nature of the electrolyte, the concentration of the solution, temperature, etc.

Resistance of solution measurements are usually made by means of a Wheatstone bridge, one arm of which is a conductivity cell containing the solution. Several characteristics of the bridge should be taken into consideration, depending upon the accuracy which is desired. A pure sine-wave, low-voltage and moderate frequency alternating current is generally employed. Six volts and 1000 cycles are satisfactory. At high voltages (20,000 volts) the Wien effect and at high frequencies (30,000 kilocycles) the Falkenhauser effect are encountered, giving extraordinary results. Alternating current is used to minimize polarization effects at the electrodes. Polarization effects are further minimized by using platinum electrodes coated with platinum black. Such electrodes have large surfaces and low over-voltage. The conductivity cell is designed so as to minimize capacity effects due to the filling tubes and the lead-in wires. Residual capacity, effects are balanced out by including a variable capacitor across the decade resistance box which serves as one arm of the Wheatstone bridge. The resistance spools in this box should have a bifilar winding.

The *conductance, L,* of a solution is defined as the reciprocal of the resistance, (1), $L = 1/R$. The specific conductance of a solution, (2) $\overline{L} = 1/\overline{R}$, is the conductance of 1 cc of the solution between electrodes 1 cm apart. It is a function of the same variables as is the resistance. The equivalent conductance, Λ, is of more theoretical significance than the specific conductance. The equivalent conductance is related to the specific conductance by the equations, (3) $\Lambda = VL = 1000L/c$, where V is the volume of the solution containing one equivalent weight of electrolyte and c is the normality of the solution. It is defined as the conductance of a solution which contains one equivalent weight of electrolyte between electrodes one cm apart.

Kohlrausch, who pieneered in the field of electrical conductance measurements, established the relationships, (4) $L = l_+ + l_-$ and (5) $\Lambda = \lambda_+ + \lambda_-$; that is, the conductance of a solution of an electrolyte is the sum of the conductance of the ions in that solution. The l's are specific ion conductances and the λ's are equivalent ion conductances. Ionic conductances are functions of transference numbers and of ionic mobilities. They may be evaluated from equivalent conductance and transference number data by means of the relationship (6) $l_+ = t_+\Lambda$. Tables of ionic conductances are found in handbook literature.

For theoretical considerations the equivalent conductance at infinite dilution, Λ_0, is an important function. This is defined as the equivalent conductance of a solution which is dilute enough so that further addition of solvent does not affect the conductivity. Its value cannot be measured directly because the concentrations of these very dilute solutions are so low that the physical measurements have a high degree of uncertainty. Its value is obtained by extrapolating values of Λ versus $\sqrt{c}$ to zero concentration. These values of Λ's are obtained by measurements on solutions which are sufficiently concentrated so that accurate data can be obtained. The Debye-Hückel theory, which applies to the dilute region in which the conductivity measurements are uncertain, shows that a straight-line extrapolation is valid. Conversely, conductance measurements on fairly insoluble salts have been fruitful in confirming the validity of the Debye-Hückel theory (which see).

The extrapolation method does not apply to solutions of weak electrolytes. In this case, not only does the uncertainty in the measurements occur at higher concentrations, owing to the low values of Λ, but also the values of Λ are changing with dilution so rapidly that the destination of the curve at zero concentration cannot be estimated. A second method is available for these cases as well as for the case of the strong electrolyte. This is the application of Kohlrausch's law, (7) $\Lambda_0 = \lambda_{0+} + \lambda_{0-}$. The values of the ion conductances at infinite dilution have been determined by measurements of Λ's for strong electrolytes and the value of the transference number of some strong electrolyte by means of the equation (8) $\lambda_{0+} = \Lambda_0 t_{0+}$.

The Onsager equation—a modification of the Debye-Hückel equation—is (9)

$$\Lambda = \Lambda_0 - (A + B\Lambda_0)\sqrt{c},$$

where A and B are constants depedent on the solvent and on the temperature. A is a function of the dielectric constant and of the viscosity of the solvent and B is a function of the dielectric constant of the solvent. This equation is applicable to dilute solutions only. Some factors which cause deviations from the equation in more concentrated solutions are incomplete dissociation, yielding simple ions, intermediate ions, ion-pairs and neutral molecules, and the solvation of these charged and neutral molecules, thus changing the nature of both the solute and the solvent. These factors have not yet been satisfactorily evaluated.

Conductivity of solutions has been of practical value in several fields. One of the most important methods for determining the solubility of slightly soluble electrolytes is by measuring the conductance of their saturated solutions. In this case it is assumed that $\Lambda = \Lambda_0$ and equation (3) is applied to calculate the concentration. Conductance measurements are made to determine the end points in titrations. This is possible because of the difference in the mobilities of the ions formed or added as the titration proceeds. Conductance measurements are used in controlling the concentration of solution in chemical processes. In electrochemical processes the voltage drop across the electrolytic cell is composed of reversible and irreversible potential drops at the electrodes and of the potential drop required to overcome the ohmic resistance of the bath. Substitution of one electrolyte for another may substantially change the magnitude of this last component.

W. W. EWING*

Cross-references: *Debye-Hückel Theory, Electrolysis, Ionization, Solutions, Electrochemistry.*

* Deceased.

CONFORMATIONAL ANALYSIS

By "conformations" of a molecule are meant the (infinite) number of spatial arrangements which can normally be obtained by rotation about single bonds. Thus, in the case of butane, CH_3—CH_2—CH_2—CH_3, different families of conformations can be obtained by rotation about any of the three C—C bonds shown. When attention is focused on the central bond, six singular conformations may be discerned, three in which the substituents on carbons No. 2 and No. 3 are "eclipsed" and three in which they are "staggered." These conformations are best visualized with aid of molecular models. Fig. 1 shows the conformations of butane in an end-on or "Newman" projection: 1, 3 and 5 eclipsed, 2, 4 and 6 staggered. The diagram below the Newman projections represents (as ordinate) the potential energy of the various conformations. It will be seen that eclipsed conformations represent energy maxima whereas staggered conformations correspond to energy minima. Of the three staggered conformations, the one in which the

methyl groups are at a maximum distance (the so-called "*anti*" or "*anti*-periplanar" conformation) has maximum stability whereas the other two, mirror-image, conformations (so-called "*gauche*," "skew" or "*syn*-clinal" conformations) are less stable than the *anti* by about 0.6 kcal/mole. The abscissa in Fig. 1 represents the so-called "angle of torsion" or "dihedral angle," in this case the angle—positive if clockwise, negative if counterclockwise—between the planes defined by Me_1—C_2—C_3 and C_2—C_3—Me_2. Conformations are sometimes described according to the approximate angle of torsion between two salient groups (e.g. the two methyl groups in butane); the appropriate descriptions are indicated below the abscissa.

By "conformational anlysis" is meant an analysis of the physical and chemical properties of a compound in terms of the preferred conformation or conformations of the ground, transition and excited states, as well as the converse: an analysis of conformation from observed physical or chemical properties. For example, in 1,2-dichloroethane, $ClCH_2$—CH_2Cl, the *anti* form has zero dipole moment wheras the calculated dipole moment for the *gauche* form is *ca.* 3.2 D. The small observed dipole moment at 32°C, 1.12 D, indicates preference for the *anti* conformation. Similarly, the infrared spectrum of solid 1,2-dichloroethane shows relatively few lines as might be expected of the centrosymmetric (*anti*) form, indicating that the molecules crystallize in this conformation. As the substance is melted, additional infrared bands appear and are assigned to the *gauche* conformation; the enthalpy difference between *gauche* and *anti* can be calculated from the observed change of intensity of the pertinent infrared bands with temperature.

By far the most fruitful application of conformational analysis is in cyclohexane chemistry. This work advanced tremendously in 1950 when the salient differences between axial and equatorial substituents in cyclohexane were first pointed out by D. H. R. Barton (who along with O. Hassel, received the 1969 Nobel Prize in Chemistry for pioneering work in the field of conformational analysis). An equatorial substitutent is less crowded and therefore more stable than an axial one. Thus *trans*-1,2- and 1,4-dimethylcyclohexane (both methyl groups equatorial) are more stable than their *cis* epimers (one methyl group equatorial, one axial), but in 1,3-dimethylcyclohexane, the *cis* isomer (diequatorial) is more stable than the *trans* (equatorial-axial). Chemical equilibration will convert an axial isomer to an equatorial one, thus diethyl cyclohexane-*trans*-1,3-dicarboxylate (one $COOC_2H_5$ group axial) is epimerized by base to the *cis* isomer (both $COOC_2H_5$ groups equatorial).

As implied in Fig. 2, a monosubstituted cyclohexane may, by simple ring inversion, alternate between the conformation in which the substituent is equatorial and the one in which the substituent is axial. Since the barrier to ring inversion is only about 10 kcal/mole, inversion is rapid at room temperature and such a "mobile" system is not suited to demonstrate the chemical differences in stability, reactivity and signal position in NMR between equatorial and axial substituents. Systems in which the substituent occupies a well-defined

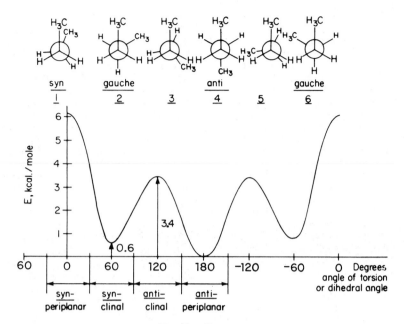

FIG. 1. Butane.

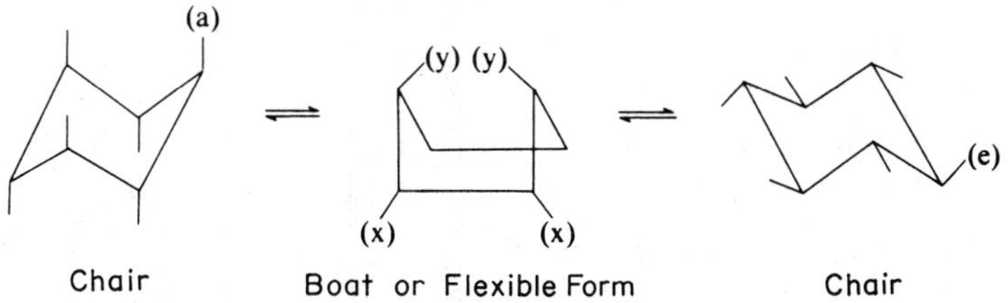

Chair   Boat or Flexible Form   Chair

FIG. 2. Cyclohexane.

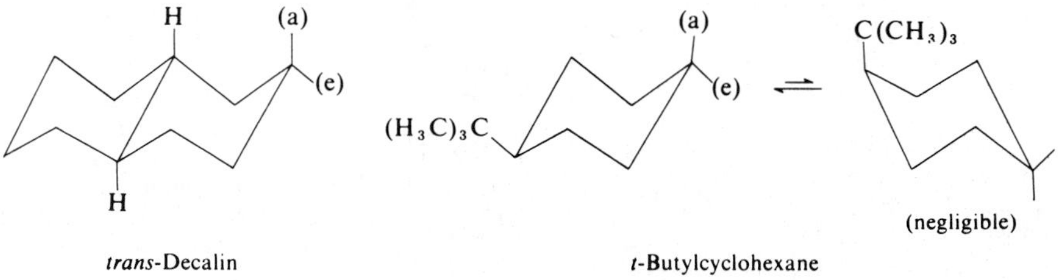

trans-Decalin   *t*-Butylcyclohexane

FIG. 3. Rigid and Biassed Systems.

equatorial or axial position are of two types: "rigid" and "biassed" (Fig. 3). *trans*-Decalin exemplifies a rigid system: inversion of the ring junction from the diequatorial to the diaxial conformation is sterically impossible and a substituent (e.g. at position 2) is therefore either rigidly equatorial (e) or rigidly axial (a). *t*-Butylcyclohexane represents a biassed (or "anancomeric") system: the *t*-butyl group is so large that (unlike most other atoms or groups) it will necessarily occupy the equatorial position; in the axial position it would be excessively crowded. Consequently, the second substituent in a *trans*-4-*t*-butylsubstituted cyclohexyl compound will be equatorial, whereas in the *cis*

isomer it will be axial (and hence less stable). By way of a typical example, epimerization of ethyl *cis*-4-*t*-butylcyclohexanecarboxylate (axial COOEt) by ethoxide leads to a mixture containing 85 per cent of the more stable *trans* isomer (equatorial COOEt). In a saponification reaction, the *trans* isomer reacts faster than the *cis* (even though the *cis*, being less stable, starts from a higher ground-state energy level) because the tetrahedral transition state for saponification is more space-consuming than the starting state and therefore relatively more crowded in the axial conformation. There are, however, also reactions in which the stability of the ground state determines reaction rate, thus *cis*-4-*t*-

butylcyclohexanol (axial hydroxyl) is oxidized faster than the *trans* (equatorial OH) because the hydroxyl group wants to escape the crowded axial position. The transition state in this case is sufficiently ketone-like to have lost much of the unfavorable axial interaction of the *cis* alcohol. Both saponification of esters and oxidation of alcohols with chromic acid illustrate steric factors: the former steric hindrance, the latter steric assistance.

Another conformational factor on reaction rate is the stereoelectronic one, i.e., the requirement of a certain preferred relative position of the electron clouds in the transition states. This factor is observed in bimolecular elimination reactions where the groups to be eliminated generally have to be *anti*-periplanar; thus *cis*-4-*t*-butylcyclohexyl *p*-toluenesulfonate (axial tosylate next to axial hydrogen) undergoes bimolecular elimination with ethoxide readily whereas its *trans* isomer (equatorial tosylate) does not. Similarly, $3\alpha,4\beta$-dibromocholestane eliminates its axial bromines with KI whereas the diequatorial $3\beta,4\alpha$-dibromide does not.

Physical properties, such as refractive index, density, infrared and NMR spectra, acidity, etc. are also strongly dependent on conformation. The isomer of lower enthalpy, usually the equatorial isomer, generally has the lower density and refractive index. It has the higher stretching frequency in the infrared for the bond to the substituent. Axial protons resonate at higher field in the NMR than equatorial ones, axial-axial coupling constants are larger than axial-equatorial or equatorial-equatorial ones, axial acids are weaker than their equatorial epimers (because of lesser solvation of the anions), etc.

The physical and chemical properties of monosubstituted cyclohexanes ("mobile systems," Fig. 2) must be considered in terms of the existing conformational equilibrium; in the case of reaction rates and NMR spectra, the properties of the mobile system are the weighted average of those of the two (equatorial and axial) prototypes. Conversely, knowing the properties of the mobile system and of the prototypes enables one to calculate the mole fractions of the mobile system existing in the two possible conformations and the corresponding free energy difference ("conformational energy"); typical values are listed by Hirsch in Vol. I of "Topics in Stereochemistry" (Wiley, 1967).

Cyclohexanone is a flattened chair and cyclohexene and cyclohexene oxides are half-chairs. Considerable work has been done on the conformation of substituted cyclohexanones, including important work involving their optical rotatory dispersion spectra.

Conformational factors are of great importance in fused cyclohexanoid systems (such as steroids, di- and higher terpenes, many alkaloids) and also in bridged systems (such as mono- and sesquiterpenes). Rings smaller or larger than six-membered ones have been studied; although four- and five-membered rings are not planar, the different conformations of the substituted rings (e.g. the "envelope" and "half-chair" conformations in substituted cyclopentanes) are not separated by sufficiently high energy barriers to show clearly distinct equatorial and axial substituents. Rings larger than six-membered ones have been discussed by

Sicher in Vol. 3 of "Progress in Stereochemistry" (Butterworths, 1962), and by Dunitz in Vol. 2 of "Perspectives in Structural Chemistry" (Wiley, 1968). The conformational analysis of saturated heterocyclic rings (cf. Havinga *et al.,* "Topics in Stereochemistry," Vol. 4, 1969) has thrown light on steric requirements of unshared electron pairs, dipole-dipole interactions and solvent effects thereon, conformational requirements for intramolecular hydrogen bonding and other interesting new factors.

Elucidation of the conformation of macromolecules, notably polypeptides and nucleic acids, by x-ray diffraction studies, is providing an important framework for the understanding of the biochemical function of these species.

E. L. ELIEL

References

Eliel, E. L., Allinger, N. L., Angyal, S. J., and Morrison, G. A., "Conformational Analysis," John Wiley & Sons, Inc., New York, 1965.

Hanack, M., "Conformation Theory," Academic Press, New York, 1965.

McKenna, J., "Conformational Analysis of Organic Compounds," The Royal Institute of Chemistry Lecture Series, London, 1966, No. 1.

CONJUGATION

The long known relationships and effects of the presence in organic molecules of single and multiple bonds between carbon atoms are now being subjected to intensive reexamination and interpretation in terms of the modern quantum theory of molecular orbitals. The chemical properties of dienes differ according to the positions of the double bonds. If the 2 double bonds are isolated, or separated by 2 or more single bonds, each double bond reacts independently and the reactions are no different from those when only a single double bond is present. If both double bonds are attached to the same carbon atom in the *allenes,* from the first member of the series $CH_2\!=\!C\!=\!CH_2$, they readily undergo rearrangement. If the 2 double bonds are separated by a single bond they are *conjugated,* and react still differently. The most important conjugated diene is 1,3-butadiene, $CH_2\!=\!CH\!-\!CH\!=\!CH_2$. A characteristic behavior is preferential 1,4 addition (with Br_2) to

$$CH_2\!-\!CH\!=\!CH\!-\!CH_2$$
$$||$$
$$BrBr$$

and of course polymerization by 1,4 addition to rubber-like products with free-radical catalysts.

The importance of conjugated molecules arises in connection with the present arguments as to whether bonds are formed, and so long pictured, by the sharing of *localized* pairs of electrons. This concept, of course, arose from the discovery of definite molecular geometries, implying rigid bonds between atoms. The pair of electrons forming a bond or a special component (σ or π) of a multiple bond occupies a 2-center molecular orbital formed from 2 atomic orbitals of the bonded atoms. The properties of molecules containing such bonds can be represented as additive functions of bond properties—bond energies, bond lengths, bond moments, etc., as in saturated compounds (single bonds between carbon atoms). However, as Dewar points out, additivity relationships break down in the case of conjugated molecules, where π-bonds are adjacent to another π-bond as in 1,3-butadiene mentioned above.

It is theorized the 2 π-molecular orbitals probably interact with each other, or with unused atomic orbitals to give many-center molecular orbitals; consequently the electrons occupying these areas are no longer localized. Similar though smaller effects are observed in systems where multiple and single bonds are in 1,3 relation as in $H_3C—C≡CH_2$. Again it is proposed that the effects are caused by similar interactions between a π-molecular orbital and a σ-orbital in 1,3 relation to it—an interaction to which the term *hyperconjugation* is assigned. (In the earlier literature hyperconjugation or no-bond resonance is used to explain how in phenylethylene the vinyl group $CH_2=CH—$ is activating and *ortho,para*-directing in the addition of a substituent group such as NO_2 because of a resonance effect. This is also true of the CH_3 group in toulene, to explain the marked increase in substitution of toulene over benzene—much greater than would be expected from the small dipole moment of toulene). On this basis Dewar and others maintain that in *all* cases where there are additivity relationships and conventionally accepted localized bonds from the sharing of electrons between atoms, such a concept is wrong, and that the electrons in any molecule containing 3 or more atoms are delocalized over all the atoms forming the molecule, as a necessary consequence of quantum theory and experimental measurements of spin-coupling in NMR, and wide variations for the paraffins of ionization potentials and excitation energies.

G. L. CLARK

Cross-references: *Bonding, Orbitals (Molecular), Nuclear Magnetic Resonance.*

COORDINATION COMPOUNDS

(COMPLEX INORGANIC COMPOUNDS)

The association of a metal ion with other ions or molecules called ligands produces a composite complex compound often called a coordination compound. Every metal ion of all metals in the Periodic Table is capable of being converted to coordination compounds but many of these are relatively weak thermodynamically. The complexes which are best known and which have been studied in the greatest detail are those of the transition metals. Because of the availability of d-orbitals for bonding, the transition metal complexes are of the highest stability and of the greatest number. Ligands may be positively or negatively charged or they may be neutral molecules. Examples are $NH_2NH_3^+$; Cl^-, NO_3^-, SO_4^{2-}; H_2O, NH_3, NH_2OH. The charge on the complex ion is the sum of the charges of the metal ion and the ligands surrounding it. Thus the addition of $6Cl^-$ to a Pt^{4+} ion gives the ion $[PtCl_6]^{2-}$ and $4NH_3 + 2Cl^- + Co^{3+}$ gives $[Co(NH_3)_4Cl_2]^+$ or $2NH_3 + 2Br^- + Pt^{2+}$ gives $[Pt(NH_3)_2Br_2]^0$. The brackets enclose the metal ion and the coordinated ligands.

Since the complex ion may have a positive charge a formula of a complex may contain a negative ion in two forms—one complexed and the other not, i.e., $[Co(NH_3)_5Cl]Cl_2$. Different chemical and physical properties are expected and found for the two types of Cl^-.

Complex compounds are often classified as being either *labile* or *nonlabile*. A labile complex is one which usually replaces its ligands rapidly by others while a nonlabile complex is slow to be converted to another complex. This property is related to the metal, its oxidation state and the ligand(s). Nickel(II) is generally labile while Co(III) is generally nonlabile and Fe(II) is variable. The addition of excess acid to $[Ni(NH_3)_6]^{2+}$ and $[Co(NH_3)_6]^{3+}$ gives $[Ni(OH_2)_6]^{2+}$ immediately but no reaction occurs for months with the Co(III) complex ion. Also the addition of excess acid to $[Fe(CN)_6]^{4-}$ and $[Fe(NH_3)_6]^{2+}$ rapidly produces $[Fe(OH_2)_6]^{2+}$ only with the ammonia complex ion. Thus, $[Ni(NH_3)_6]^{2+}$ and $[Fe(NH_3)_6]^{2+}$ are labile while $[Fe(CN)_6]^{4-}$ and $[Co(NH_3)_6]^{3+}$ are referred to as nonlabile. Most complexes of the first transition group are intermediate in behavior.

The coordination number (C.N.) is defined as the total number of metal-ligand bonds. Some ligands make more than one bond to the metal ion. This number is a characteristic of the metal ion in a particular oxidation state. In certain cases the coordination number may change with differing ligands or modified experimental conditions. For example, nickel(II) has a C.N. of 6 with H_2O or NH_3 as ligands, but only a C.N. of 4 CN^-. Common coordination numbers are 2, 4 and 6 as illustrated by the ions $[Ag(NH_3)_2]^+$, $[Ni(CN)_4]^{2-}$ and $[PdCl_6]^{2-}$. However, it should be noted that values of 3, 5, 7 and 9 have been shown in certain cases (usually in the solid state). Certain ligands coordinate through more than one atom. Thus, bidentate (2 bonding positions), tridentate (3 bonding positions) and tetradentate (4) ligands etc. are common. Oxalate ion :OOC-COO: and ethylenediamine (en): $NH_2-CH_2-CH_2NH_2$: are bidentate ligands and form with Ni(II) a complex with a coordination number of six and the formula $[Ni(en)_3]^{2+}$.

In complex compounds, the ligands are arranged in a definite geometrical fashion. These arrangements are present in solution as well as in the solid state and are responsible for many of the interesting properties of these molecule ions.

All ligands have electron pairs on the coordinating atom which are capable of being donated to or shared with the metal ions. The bonding in these complexes may be considered in two ways: 1) that the ligands share their electron pairs with the metal ions, 2) that the bonding is essentially electrostatic in nature. Neither picture agrees with all the known facts and it appears that the bonds are somewhat intermediate between these pictures. In the latter bonding scheme each ligand "shows" a negative charge to the positive metal ion and attraction occurs. Simultaneously another change occurs: the five metal ion d-orbitals become unequal in energy: two higher, three lower in octahedral geometry. The metal electrons occupy the lower energy orbitals. Then with $[Co(NH_3)_6]^{3+}$ for example, the six d-electrons populate the lower state and the complex is diamagnetic (no unpaired e's). In this treatment the ligand electrons provide the negative charge to split the metal ion d-orbital energies but do not enter the metal ion orbitals. Bonding "pictures" for both approaches to the bonding in $[Co(NH_3)_6]^{3+}$ are:

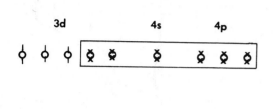

1.

The thermodynamic stability of the complex as measured by the equilibrium constant K_{eq} is always the difference between the stability of the aquo ion and that of the complex formed. The K_{eq} for the overall formation of $[Ni(NH_3)_6]^{2+}$ is

$$[Ni(OH_2)_6]^{2+} + 6NH_3 \overset{K_{eq}}{\rightleftarrows} [Ni(NH_3)_6]^{2+} + 6H_2O$$

The magnitude of K_{eq} depends primarily on the metal ion and on the ligand. The larger the charge and the smaller the size of the metal ion, the larger is K_{eq}. For the first transition elements in the +2 oxidation state the following order is generally observed.

$$Cu > Ni > Co > Zn > Fe > Mn$$

Ions from the 2nd and 3rd group elements form considerably stronger bonds than the related elements in the first transition series. In general, the stronger the base strength of the ligand the larger K_{eq} for a particular metal ion. For neutral organic ligands $NH_3 > CH_3NH_2 > (CH_3)_2NH > (CH_3)_3N$ and $H_2O > CH_3OH > (CH_3)_2O$. With negative ions usually $NO_2^- > F^- > NO_3^- > Cl^- > Br^- > I^- > ClO_4^- = BF_4^-$.

The complexes are generally formed in a stepwise manner with $K_1 > K_2 > K_3 \longrightarrow K_n$. For example, with $[Ni(H_2O)_6]^{2+}$ and en log $K_1 = 7.60$, log $K_2 = 6.48$ and log $K_3 = 5.03$. A few examples are known where the reverse is true, in the Fe(II)-dipyridyl system $K_1 > K_2 \ll K_3$. This is thought to be due to the change in magnetic properties.

Many geometrical and optical isomers have been prepared and studied. In the planar geometry two geometrical isomers of $[Pt(NH_3)_2Cl_2]^0$ have been prepared which have different chemical and physical properties. The *cis* structure has 2Cl⁻'s adjacent to each other while the *trans* compound has them across the plane. $[Co(en)_2Cl_2]^+$ exists in the *cis* and *trans* form but in addition the *cis*-form consists of two optical isomers, d and l. The d-l compounds have identical chemical and physical properties except for the direction they rotate a plane of polarized light.

Recent advances have included "sandwich" compounds such as *ferrocene* $[Fe(C_5H_5)_2]^0$ and dibenzene chromium $[(C_6H_6)_2Cr]^0$. They are called sandwich compounds because the metal is between two planes of the organic rings and equidistant from them. They are remarkably stable and react like aromatic hydrocarbons.

Another important region of investigation is the use of molecular N_2 and O_2 as ligands. Complexes like $[Ru(NH_3)_5N_2]^{2+}$ can be extremely important in producing a low-temperature "nitrogen fixation" process for producing ammonia type fertilizers for the farm.

Only recently has the importance of metal-to-metal bonds in complexes been realized. In Re_3Cl_9 there are three bridging chloride ions but also three Re-Re bonds of unusual strength. A large number of similar compounds have been found possessing similar strong metal-to-metal bonds.

The role of metal complexes in biochemistry is of great importance. The action of many enzymes is dependent on coordinated metal ions incorporated in their protein structure. Often simple metal ions will perform the same reactions as enzymes but at a reduced rate. A similar mechanism for the two processes is proposed. Model biochemical systems have been prepared which simulate an enzyme or vitamin but are simpler to study. Several "simple" systems similar to vitamin B-12 which contain a coordinated cobalt ion are presently under study in order to discern the underlying principles of vitamin B-12's unique reactivity.

R. KENT MURMANN

Cross-references: *Chelation, Sequestering Agents, Ligand Field Theory, Isomers, Complexing Agents, Ion Exchange, Electrochemistry, Stability Constants, Transition Metals.*

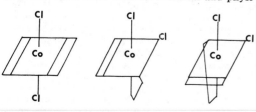

| trans | d-cis | l-cis |

COPOLYMER

A copolymer is a high molecular weight substance containing several types of repeating structures. A styrene-methyl methacrylate copolymer, for example, is obtained by polymerizing styrene and methyl methacrylate together. Copolymers can also be prepared by polycondensation techniques.

$$CH_2{=}CH \underset{\underset{C_6H_5}{|}}{} + CH_2{=}C{-}CH_3 \underset{\underset{COOCH_3}{|}}{} \xrightarrow{\text{polymerization}}$$

styrene   methyl methacrylate

$$....CH_2{-}CH{-}CH_2{-}C(CH_3){-}CH_2{-}CH.... \underset{\underset{C_6H_5 COOCH_3 C_6H_5}{|||}}{}$$

Portion of a styrene-methyl methacrylate copolymer

Copolymers are to be distinguished from mixtures of homopolymers (called polyblends). Some polyblends (rubber-polystyrene mixtures or mixtures of butadiene-styrene and butadiene-acrylonitrile copolymers) have valuable impact resistance properties and are important commercial plastics. Homopolymers are usually incompatible, however, and this makes it difficult to modify the properties of plastics, films, fibers or elastomers by blending polymers. Since monomer units are incorporated into common molecules in copolymers, such incompatibility difficulties are not encountered in a copolymer.

The properties of copolymers often vary progressively from one homopolymer to another, as copolymer composition is varied. For example, poly(vinyl chloride) is a high melting, hard, horny material which is poorly compatible with plasticizers and is difficult to process. Poly(vinyl acetate), in contrast, is a soft rubbery polymer. Copolymers of vinyl chloride with about 15 mole percent vinyl acetate are softer and more easily processed than poly(vinyl chloride); they are produced in large quantities for use in films, industrial fibers, plastics, phonograph records, floor tile, and many other applications.

Copolymers are usually prepared to accomplish any of the following objectives:

(1) *Preparation of low-melting materials of low crystallinity from monomers whose homopolymers are high-melting and highly crystalline.* The softening point of a polymer containing crystallizable monomer units is determined by the temperature at which the crystallizable units melt in the polymer. The softening point is the lowest temperature at which the polymer can be processed in bulk. When above their softening points, polymers and copolymers are generally transparent and they behave as elastomers; they become opaque (if crystalline), and hard and are sometimes brittle when cooled below their softening points.

When noncrystallizable units are present in a copolymer, they tend to lower the melting point of crystallizable units and they prevent the crystallizable units from crystallizing completely. This causes the copolymer to be softer, more transparent, more compatible and more easily processed than the corresponding crystallizable homopolymer.

Copolymers of ethylene with propylene, vinyl acetate or alkyl acrylates and also chlorinated polyethylene are elastomeric, although polyethylene is a crystalline plastic. Similarly, vinylidene fluoride-chlorotrifluoroethylene copolymers are elastomeric. For the same reason, silicone resins containing di-

phenylsiloxane and dimethylsiloxane repeating units have better low temperature properties than resins containing only dimethylsiloxane units. Also tetrafluoropropylene (FEP) copolymers are more easily processed than polytetrafluoroethylene ("Teflon").

(2) *Preparation of materials with low glass transition temperatures from monomers whose homopolymers have high glass transition temperatures.* Homopolymers and copolymers which do not contain crystallizable groups must be processed above their glass transition temperatures (Tg). Above Tg, polymers are soft and they behave as elastomers; below this temperature they are hard, brittle materials, useful as plastics. The glass transition temperature of a copolymer lies between the Tg's of the corresponding homopolymers. Copolymers of high Tg type monomers (e.g., vinyl chloride) with small amounts of low Tg type monomers (e.g., vinyl acetate) are often prepared to make materials which can be molded or extruded at relatively low temperatures.

(3) *Copolymerization of monomers with acrylonitrile or vinyl pyridine to obtain solvent-resistant materials.* Polymers based on hydrocarbon monomers such as styrene or butadiene either dissolve in or swell badly when exposed to hydrocarbon oils. Copolymers of these monomers with acrylonitrile are not swollen by hydrocarbons, however. Acrylonitrile-butadiene copolymers are oil-resistant, tough elastomers (Nitrile rubber); styrene-acrylonitrile copolymers are used in hair curlers and other plastic items where oil resistance is required.

(4) *Introduction of reactive groups into polymer chains.* Small amounts of monomers containing reactive groups are often incorporated into polymers to impart chemical reactivity. Vinyl pyridine and vinyl pyrrolidone are used in acrylonitrile polymers to give dye receptivity. Comonomer units are often incorporated into elastomeric polymers to provide sites for cross-linking (vulcanization). Thus, isoprene and cyclopentadiene units in isobutylene-isoprene copolymers (Butyl rubber) and in ethylene-propylene-cyclopentadiene terpolymers (EPT rubber) are required for the vulcanization of such polymers with sulfur. Similarly, vinylsiloxane units present in some silicone elastomers and β-chloroethyl acrylate or β-chloroethyl vinyl ether units in ethyl acrylate copolymers provide sites for the vulcanization of such elastomers. Small amounts of maleic anhydride or maleic acid units in polymers used for surface coatings react with metal surfaces and thereby cause the coatings to have good adhesion to metals.

(5) *To provide sites for cross-linking polymers by physical methods.* Strong intermolecular forces operate between certain types of monomer units. When such units are present in copolymers, the interactions tie the polymer chains together, acting as physical cross-links. These are temperature-sensitive. The copolymers can be processed using procedures applicable for thermoplastic materials even though they behave as cross-linked materials at moderate temperatures.

Ionomers are copolymers of unsaturated hydrocarbons and salts of unsaturated acids. The salt units aggregate at moderate temperatures and the aggregates behave as cross-links. Since the aggregates break down at elevated temperatures, the copolymers can be processed as thermoplastics. In addition to providing sites for reversible cross-linking, the salt units render the copolymers somewhat oil-resistant and they prevent the hydrocarbon units from crystallizing. Ionomers are consequently transparent materials.

(6) *Utilization of monomers with low reactivity or reduction of the cost of polymers.* Some readily available, inexpensive monomers such as maleic anhydride are slow to homopolymerize, but they copolymerize readily with other monomers. Maleic anhydride-styrene and maleic anhydride-methyl vinyl ether copolymers are prepared commercially and are converted by hydrolysis or alcoholysis into water soluble materials which find use as thickening agents, protective colloids and textile sizing agents. Styrene-methyl methacrylate copolymers have optical and weathering properties similar to poly(methyl methacrylate), but they are much less expensive. The copolymers are therefore replacing the homopolymer in many applications.

(7) *Improvement of heat stability of polymers.* The presence of a comonomer unit can sometimes improve the thermal stability of a polymer. Polyformaldehyde, for example, decomposes by successively breaking formaldehyde units from its ends. This process, called unzipping, can continue until the polymer is completely decomposed. In formaldehyde-ethylene oxide or formaldehyde-β-propiolactone copolymers, however, unzipping ceases when a comonomer unit is reached. This causes the copolymers to have better thermal stability than the parent homopolymer.

(8) *Special effects obtained with block and graft copolymers.* Block copolymers contain several long sequences of different monomer units. They have interesting properties when the monomer blocks are very different in character and are incompatible. Poly(ethylene oxide) is a water soluble polymer, whereas poly(propylene oxide) is hydrophobic. Ethylene oxide-propylene oxide block copolymers are surface active and are used extensively as wetting agents. Other block copolymers are used to stabilize polymer-oil dispersions.

Most applications of block copolymers depend on the blocks separating into different phases. When small amounts of polycarbonate-polysiloxane block copolymers are added to polycarbonate molding resins, the polysiloxane blocks tend to accumulate at the surfaces of the moldings. This causes the moldings to have the physical properties of polycarbonates but the surface characteristics of silicones.

When their blocks have vastly different softening points, some block copolymers behave as thermoplastic elastomers. The polymers can be molded at temperatures above the softening points of both phases and they can often be spun from solution to obtain fibers. At temperatures where one of the blocks (phases) is hard and the other soft, the materials behave like cross-linked elastomers, since the rubbery phases are connected to the hard phases. Typical thermoplastic elastomers are polystyrene-polybutadiene-polystyrene block copolymers and polyester-polyurea or polyether-polyurea block copolymers. The latter are prepared by polyurethane technology and are very useful elastomeric fibers.

Graft copolymers have properties similar to block copolymers. They contain one type of repeating unit in their main chain. Chains containing a different type of repeating unit are present as branches (grafts) to the main chain.

Copolymer Structure—Sequence Distribution

The physical and chemical properties of copolymers are dependent on the arrangement of monomer residues in their chains. The monomer units may be arranged randomly along the polymer chain, they may tend to alternate or they may tend to cluster in blocks of like units. These various possibilities are illustrated below for copolymer sections having equal numbers of A and B units.

$$\cdots\cdots \underline{A}\underline{B}\underline{A}\underline{B}\underline{A}\underline{B}\underline{A}\underline{B}\underline{A}\underline{B}\cdots\cdots$$
Alternating type

$$\cdots\cdots \underline{A}\underline{B}\underline{A}\underline{A}\underline{B}\underline{B}\underline{B}\underline{A}\underline{A}\underline{A}\underline{B}\underline{B}\cdots\cdots$$
Random type

$$\cdots\cdots \underline{A}\underline{A}\underline{A}\underline{A}\underline{A}\underline{B}\underline{B}\underline{B}\underline{B}\underline{B}\cdots\cdots$$
Block type

A given arrangement may be characterized by the run number, R, which is defined as the average number of monomer sequences (runs or blocks) which occur in a copolymer chain per 100 monomer units. The alternating, random, and block copolymer structures shown above (runs underlined) have run numbers of 100, 60 and 20, respectively.

Although run numbers can be predicted for copolymers if the mechanism of the copolymerization reaction and the monomer concentrations at the start of a copolymerization experiment are known, much attention is currently being given to the experimental characterization of sequence distribution in copolymers. Information about sequence distribution (run numbers) in copolymers can help our understanding of copolymerization reactions. Sequence distribution information is also needed to establish quantitative copolymer structure-physical property relationships.

Run numbers of copolymers can be determined by measuring some aspect of copolymer structure, such as the percentage of AA, AB or BB linkages or the percentage of A type units which have only A(P_{AAA}), only B(P_{BAB}) or both A and B(P_{BAA}) neighbors. The relationships of these quantities to run number and the percentages of A and B units in the copolymer, are very simple, as the following examples illustrate.

%AA linkages $= \%A - R/2$

%BB linkages $= \%B - R/2$

%AB linkages $= R$

$$P_{AAA} = 100(1 - R/2\%A)^2$$
$$P_{BAA} = 100(R/\%A)(1 - R/2\%A)$$
$$P_{BAB} = 100(R/2\%A)^2$$

Chemical methods can be used to estimate the percentages of AA, AB or BB linkages in copolymers, although such methods have yielded only approximate results at present. Examples of such methods are the selective degradation of specific types of linkages (e.g., AA), followed by examination of the degradation fragments or the cyclization of monomer pairs (e.g., AA or AB), followed by measurement of the maximum extent of cyclization obtainable. Physical techniques such as infrared, ultraviolet or nuclear magnetic resonance spectroscopy or dipole moment determinations can sometimes be used to estimate P_{AAA}, P_{BAB}, etc. Measurements of heats of copolymerization or heats of formation of copolymers can sometimes provide a measure of the percentages of AB linkages in copolymers. The melting behavior of copolymers also affords information about sequence distribution, even though the melting behaviour of polymers is very dependent on their thermal history.

Since stereoregular polymers can be considered to be copolymers of d and l-type monomer units, many of the methods used to characterize sequence distribution in copolymers can also be used to characterize stereoregularity in homopolymers.

H. JAMES HARWOOD

Cross-references: *Polymerization; ABS Resins; Rubber; Fibers, Synthetic; Elastomers; Cross-linked Polymers, Block and Graft Polymers.*

COPPER AND COMPOUNDS

Metallic Copper

Copper, the leading nonferrous metal, has been known since prehistoric times, and is used today in greater amounts than ever. The world's production of 6,854,000 short tons in 1970 surpassed that of any previous year. Its combination of high electrical and thermal conductivity, resistance to corrosion, ductility, and suitable strength, as well as its many valuable alloys, make it an extremely useful metal.

Occurrence. Copper is widely distributed in many parts of the world, and is found in a variety of ore minerals. The ores, however, are for the most part of low grade. Higher-grade ores once available have been exhausted, and those being mined at present in the United States average less than 1% copper. The ores are classified in three groups: sulfide, oxidized, and native copper. The sulfide ores are by far the most important. Chief among the sulfide minerals are chalcocite, Cu_2S, and chalcopyrite, $CuFeS_2$. Other principal minerals in this group are covellite, CuS; bornite, Cu_5FeS_4; enargite, $Cu_3(As, Sb)S_2$; tetrahedrite, $Cu_{12}Sb_4S_{13}$; and tennantite, $Cu_{12}As_4S_{13}$. The oxidized ores include the copper oxides and combinations of these with CO_2, SO_3, SiO_2, and H_2O. Native copper ores, containing uncombined copper, at one time occurred in numerous localities, especially in the State of Michigan.

Refining. Specifications for copper for electrical uses call for a minimum purity of 99.90%. Electrolytic refining is required to achieve this standard, as well as to recover gold and silver present in the ore and which remain in solution in the matte and blister copper. The preparation of anodes suitable for electrolytic refining necessitates a preliminary furnace refining, which is essentially a process of oxidizing those impurities more readily oxidized than copper. The anodes are suspended in an electrolyte of copper sulfate and sulfuric acid. Copper of 99.98% purity is deposited on the cathodes, which are usually remelted and cast into wire-bars, cakes, billets, and other shapes suitable for rolling, drawing, extrusion, and other fabricating methods.

Some of the impurities contained in the anodes dissolve in the electrolyte without plating out at the cathode, while others remain insoluble at the anode. The latter, constituting the "anode slime," include the gold and silver, which are recovered by further treatment of the slime.

Chemistry and Properties. Copper is the first element of subgroup IB of the Periodic Table. Its atomic number is 29 and the electron configuration is 2:8:18:1. The crystalline structure is face-centered cubic, with the cube side dimension a = 3.6080 kX at 20°C. The minimum interatomic distance is 2.551 kX.

The density of copper is 8.94 at 20°C, while that of liquid copper at the melting point is 7.93. The melting point is 1083°C and the normal boiling point about 2595°C.

The atomic weight of copper is 63.546. The natural element is a mixture of the two isotopes Cu^{63}, with 29 protons and 34 neutrons, and Cu^{65}, with 29 protons and 36 neutrons. Unstable isotopes of mass numbers 58, 59, 60, 61, 62, 64, 66, 67, and 68 have been found; of these Cu^{67} has the longest half-life, 61.88 ± 0.14 hours.

The electrical conductivity of copper is commonly stated in terms of a standard adopted by the International Electrotechnical Commission, which assigns a percentage value of 100 to copper having a mass resistivity of 0.15328 ohm (meter, gram) at 20°C. This value, called the International Annealed Copper Standard (IACS), corresponds to a volumetric resistivity of 0.0000017241 ohm for a cube measuring 1 cm on each side.

Specifications require that copper for electrical use meet the IACS conductivity of 100, but pure copper (polycrystalline) has been made having a conductivity of 102.3, and most commercial copper falls between 100.5 and 101.8. The conductivity of commercial annealed oxygen-free copper is 101. The volumetric conductivity is 94% of that of silver, while that of the next highest metal, gold, is only 66% of that of silver. The thermal conductivity at 20°C is 0.934 cal per sq cm per cm thickness per °C per sec.

The heat of fusion of copper is 50.6 cal per gram and the heat of vaporization about 1150 cal per gram. The specific heat of the solid is 0.092 cal per gram at 20°C and increases between 0 and t°C according to the expression $0.092 + 0.0000250t$ cal per gram. The specific heat of the liquid is 0.112 and of the vapor approximately 0.08. Copper is diamagnetic, having a susceptibility of -0.080×10^{-6} cgs units per gram at 18°C.

The ultimate tensile strength of hot-rolled copper is about 32,000 psi, yield strength 10,000 psi, elongation 45 to 55% in 2 inches. The tensile strength of cold-worked copper increases to about 57,000 psi after 70% reduction in area, with decrease of elongation to 4%. The elastic modulus in tension is 17,000,000 psi for annealed copper.

The mechanical properties stated above are those of electrolytic tough-pitch copper, which is purposely cast with an oxygen content of 0.03 to 0.04%. Some of the properties of oxygen-free copper will differ slightly from those of tough-pitch copper.

The standard potential of bivalent copper is +0.34 volt at 25°C. Its excellent corrosion resistance is due both to its relative nobility in the electropotential series of metals and to the formation of a protective film. In weathering, the coating is composed of hydrated copper carbonate, which is mixed with basic copper sulfate in industrial atmospheres. The coating is green and while somewhat powdery at the surface is dense and adherent at the junction with the underlying metal. Under other conditions, and especially when heated, copper receives a coating of black oxide, CuO, when freely oxidized, or of brilliant purplish red Cu_2O when oxidation is restrained, as by spraying with water while hot.

Uses. About 55% of copper is used for electrical purposes. About 15% is used in building construction, including pipes and plumbing, roofing, gutters, leaders, hardware, etc. The automotive industry takes about 12%, not including the electrical parts. Machinery and industrial equipment consumes 9%. These figures include the copper contained in brass and other alloys as well as pure copper.

Copper Alloys

A notable feature of copper is the number, variety, and usefulness of its alloys. The principal classes of alloys with respect to composition are as follows:

Copper-zinc (binary brasses)
Copper-tin (binary bronzes)
Copper-zinc-tin (special brasses and bronzes)
Copper-zinc-lead and copper-tin-zinc-lead (leaded brasses and bronzes)
Copper-zinc-nickel (nickel silvers)
Copper-zinc-manganese plus tin, iron, aluminum (manganese bronzes)
Copper-tin-phosphorus (phosphor bronze)
Copper-aluminum and copper-aluminum plus iron, nickel, or manganese (aluminum bronzes)
Copper-silicon plus manganese, tin, iron, or zinc (silicon bronzes)
Copper-nickel (cupronickel)
Copper-beryllium and copper-cobalt-beryllium (beryllium copper)

Copper alloys are classified as the wrought alloys and the casting alloys. The wrought alloys are for the most part compositionally simpler, including a number of binary alloys. Several of the important casting alloys contain both zinc and tin and in many cases lead also. Most of the alloys which do not contain lead are difficult to machine, and lead may be added to many of the alloys to improve their machinability.

Brass is designated according to zinc content as *low brass* (up to 20% zinc) and *high brass* (30% zinc and above). It is also designated as *red brass* (up to 20% zinc) and *yellow brass* (30% zinc or more). These terms are also applied to specific compositions.

Hardness and strength reach a maximum at about 40 per cent zinc, brass of this composition being known as *Muntz metal*. Brasses containing more than 40% zinc are rarely used owing to decreasing ductility and lowered corrosion resistance.

Copper Compounds

Copper forms both cuprous (Cu$^+$ and cupric (Cu^{++}) compounds; and Cu^{+++} occurs in a few unstable compounds. Among the more important cuprous compounds are *cuprous cyanide,* CuCN, which is used in the double-cyanide electroplating baths and as a catalyst for various organic reactions; *cuprous chloride,* CuCl, which has many catalytic applications; and *cuprous oxide,* Cu_2O, used in rectifiers and as a fungicide.

Important cupric compounds are more numerous. The sulfate, $CuSO_4 \cdot 5H_2O$, is the leading industrial copper compound. It has extensive agricultural applications as a fungicide and as a soil additive to prevent copper deficiencies in crops or animals or improve crop yields. It forms the copper plating bath of largest use and has many minor applications. Direct use as a fungicide has decreased greatly, but it is used to prepare other fungicides. Mixed with lime it forms *Bordeaux mixture,* the reaction on mixing forming $Cu(OH)_2$ and $CaSO_4$; this mixture is used to control plant diseases in many crops. Cupric acetoarsenite, *Paris green,* approximating the composition $(CH_3 \cdot COO)_2Cu \cdot 3Cu(AsO_2)_2$, and cupric arsenite, *Scheele's green* (variable composition), are used as wood preservatives and as larvicides for mosquito control. They were formerly used as pigments and insecticides.

Basic cupric carbonates and cupric oxide, CuO, are used as coloring agents, and the former also for their insecticidal and fungicidal properties. *Cupric chloride,* $CuCl_2$ is employed as a catalyst, as a deodorizing and desulfurizing agent in petroleum refining, and for many other purposes.

A number of organic copper compounds are in use. The most important is *copper naphthenate.* It is used for preservation of fish nets and fabrics exposed to weathering, as a wood and fabric preservative, in sterilizing wells, and as a fungicide.

ALLISON BUTTS

References

American Bureau of Metal Statistics, "Year Book," New York, annual.

Butts, A. (editor), "Copper, the Science and Technology of the Metal, Its Alloys and Compounds," American Chemical Society Monograph Series, No. 122, Reinhold Publishing Co., Inc., New York, 1954; reprinted 1970 by Hafner Publishing Co., Inc., New York, by arrangement with Van Nostrand Reinhold Co.

"Metals Handbook," American Society for Metals, Novelty, Ohio, 8th ed., 1961–69.

"Minerals Yearbook," U.S. Bureau of Mines, Washington, D.C., U.S. Govt. Printing Office, annual.

Newton, J., and Wilson, C. L., "Metallurgy of Copper," John Wiley and Sons, Inc., New York, 1942.

Biochemical Behavior

Plants. The activity of copper in plant metabolism manifests itself in two forms: synthesis of chlorophyll and activity of enzymes. In leaves, most of the copper occurs in close association with chlorophyll, but little is known of its role in chlorophyll synthesis, other than that the presence of copper is required.

Copper is a definite constituent of several enzymes catalyzing oxidation-reduction reactions (oxidases), in which the activity is believed to be due to the shuttling of copper between the $+1$ and $+2$ oxidation states. *Ascorbic acid oxidase* catalyzes the reaction between oxygen and ascorbic acid to give dehydroascorbic acid. This oxidase occurs widely in plants. *Tyrosinase*, also known as polyphenol oxidase or catechol oxidase, occurs in potatoes, spinach, mushrooms, and other plants. It catalyzes the air oxidation of monophenols to ortho diphenols, and the oxidation of catechol to dark-colored compounds known as *melanins*. Laccase also catalyzes the oxidation of phenols, and is fairly widely distributed. The cytochrome enzymes are also found in plants.

Traces of copper are required for the growth and reproduction of lower plant forms, such as algae and fungi, although larger amounts are toxic.

The effects of copper deficiency in plants are varied and include: die-back, inability to produce seed, chlorosis, and reduced photoysnthetic activity. On the other hand, excesses of copper in the soil are toxic, as is the application of soluble copper salts to foliage. It is for this reason that copper fungicides are formulated with a relatively insoluble copper compound. Their toxicity to fungi arises from the fact that the latter produce compounds, primarily hydroxy- and amino-acids, which can dissolve the copper compounds from the fungicide.

Animals. Copper is also a necessary trace element in animal metabolism. The human adult requirement is 2 mg per day, and the adult human body contains 100–150 mg of copper, the greatest concentrations existing in the liver and bones. Blood contains a number of copper proteins, and copper is known to be necessary for the synthesis of hemoglobin, although there is no copper in the hemoglobin molecule.

Copper in plasma is mainly present in the blue protein ceruloplasmin, which is thought to be responsible for the transport of copper in the body. Copper has been shown to be a constituent of some of the cytochrome enzymes, which are catalysts for the main respiratory reaction chains, involving transfer of electrons from various carbohydrates to oxygen. Copper is also required for the synthesis of a number of enzymes, and is involved in the glycolysis or breakdown of sugars.

The blue copper protein hemocyanin occurs in the blood of certain lower forms of animal life. This compound performs the oxygen-carrying function for these species. This protein is believed to be a polypeptide containing $+1$ copper. It is not, however, as efficient an oxygen carrier as hemoglobin. The enzyme tyrosinase is found in many animals, being mainly responsible for skin pigmentation and for hardening of fresh tissue in molting species.

Copper is also found in bacteria; in the diphtheria bacillus copper is necessary for the production of toxins.

Anemia can be induced in animals on a low copper diet, such as milk, and appears to be due to an impaired ability of the body to absorb iron. This anemia, however, is rare, because of the widespread occurrence of copper in foods. In some places, e.g. Australia and Holland, diseases of cattle and sheep, involving diarrhea, anemia and nervous disorders, can be traced either to a lack of copper in the diet, or to excessive amounts of molybdenum, which inhibits the storage of copper in the liver.

Ingestion of copper sulfate by humans causes vomiting, cramps, convulsions, and as little as 27 g. of the compound may cause death. An important part of the toxicity of copper to both plants and animals is probably due to its combination with thiol groups of certain enzymes, thereby inactivating them. The effects of chronic exposure to copper in animals are cirrhosis of the liver, failure of growth, and jaundice.

R. R. GRINSTEAD

Cross-references: *Metals, Corrosion, Chlorophyll, Toxicology.*

CORROSION

Corrosion is defined as the destructive alteration of a metal by reaction with its environment. Its economic importance is indicated by estimates that the annual cost of corrosion due to losses that result from it and the cost of combating it amounts to at least $6,000,000,000 per year for the world at large.

The basic cause of corrosion and the force that drives it is the difference in free energy between refined metals and the ores from which they have been derived or the compounds which they form during the processes of corrosion; the latter frequently are the same as the former. This is illustrated by iron where the oxides (ore) from which it is refined are almost identical in composition with the rusts formed by corrosion. The processes of corrosion, therefore, represent a retreat of the refined metals to their original compounds.

In corroding, metals become ionized and acquire positive electrical charges in accordance with the valence forces involved. Differences in potential and flow of current between discrete areas on a single metal surface or between dissimilar metals in a corrosive environment can be measured. There is ample evidence in support of the theory that corrosion is essentially an electrochemical process. This involves the presence of an anode where current leaves the metal, a cathode where current enters the metal, an electrolyte to conduct the ionic current between these anodes and cathodes, and a metallic or semimetallic path to conduct the electrons in that portion of the circuit which is outside the electrolyte.

The free energy relationships between most metals and their possible corrosion products are such that, in most cases, corrosion reactions should be spontaneous and proceed at high rates. Fortunately, however, these possible rates of reaction are reduced greatly by opposing influences, such as the formation of adherent corrosion products which are insoluble in the environment in which they are formed or in others that may be encountered subsequently. These serve as a barrier, and establish rates of further attack that are determined by the relatively low rates of diffusion of metal through the corrosion product to the corroding liquid or vice versa. This is exemplified by aluminum which should corrode at a high rate in moist air, but which is prevented from doing so by the formation of a protective film of aluminum oxide. Much thinner films involving oxygen are responsible for the phenomenon of passivity exhibited academically by iron after immersion in concentrated nitric acid and practically by chromium and the high chromium iron and iron-nickel alloys known as the stainless steels. Protective oxides and passivity are most important in determining the corrosion resistance of metals and alloys made from them that are listed above hydrogen in the standard electromotive series. Those that lie below hydrogen, e.g., copper, are less dependent on passivity for corrosion resistance.

The relationships between metals and hydrogen in the electromotive series are important because in the electrochemical processes of corrosion the discharge of hydrogen ions and the evolution of hydrogen as a gas is one of the principal cathodic reactions. The facility with which this can occur is determined by such factors as the hydrogen ion concentration (pH) of the electrolyte, the electrical potential of the corrosion cell, and the over-voltage characteristics of the cathodic surface.

Hydrogen evolution is not the only possible cathodic reaction; others include a possible reaction between atomic hydrogen and dissolved oxygen or, more likely, the direct reduction of oxygen in water to form hydrogen peroxide as a first step, or hydroxyl ions as a final product.

Thus, the electrochemical reactions in corrosion of a divalent metal may be written:

Anodic reaction

Anodic reaction

$$M^0 \rightarrow M^{++} + 2 \text{ electrons}$$

At the cathode

A. $2H^- + 2 \text{ electrons} \rightarrow H_2 \text{ gas}$

B. $\frac{1}{2}O_2 + 2H^- + 2 \text{ electrons} \rightarrow H_2O$

C. $O_2 + 2H_2O + 2 \text{ electrons} \rightarrow H_2O_2 + 2OH^-$

D. $\frac{1}{2}O_2 + H_2O + 2 \text{ electrons} \rightarrow 2(OH)^-$

It is evident that oxygen as well as hydrogen plays an important part in corrosion. It can accelerate corrosion by participating in cathodic reactions, or it can retard corrosion by forming protective oxides or passive films. This dual effect of oxygen is one of the things that complicate corrosion processes, the interpretation of observations and the steps to be taken to avoid corrosion difficulties.

Forms of Corrosion. (1) Pitting resulting from local action currents, as at discontinuities in protective or passive films or under or around deposits that set up concentration cells. (2) Stress corrosion cracking resulting from the combined effects of corrosion by a specific environment and either applied or internal static tensile stresses; depending on the metal and the environment the cracks may be either intercrystalline or transcrystalline. (3) Corrosion fatigue, resulting from the combined effects of corrosion and cyclic stresses; these cracks are characteristically transcrystalline. (4) Intergranular corrosion resulting from preferential attack on, or around, a phase or compound that occupies grain boundaries. (5) Erosion resulting from the combined effects of corrosion and either abrasion or attrition. The mechanism usually involves local or general removal of otherwise protective corrosion product films. Particular forms are impingement attack due to effects of high velocity or turbulence in flowing liquids, e.g., salt water

in steam condensers or other heat exchangers, in piping systems, valves, pumps, etc. A particularly aggressive form is associated with the severe mechanical forces that are characteristic of cavitation phenomena. (6) Uniform attack or general wastage, such as may be caused by the action of strong acids as used for pickling (scale removal) or etching. This is also characteristic of the slow corrosion of durable materials in appropriate environments, such as copper roofs in suburban atmospheres, cupronickel tubes in ships condensers, Monel—nickel copper alloy—racks for pickling steel in sulfuric acid, or stainless steel columns handling nitric acid.

Preventive Measures. (1) Use of the right metal in the right way in the right place. (2) Protective coatings—paints, enamels, other metals, oils, greases, etc. (3) Inhibitors, i.e., compounds added to the environment in small concentrations to form protective films which increase anodic or cathodic polarization, or both, or neutralize some corrosive constituents. (4) Neutralizing agents added to adjust acidity or alkalinity to a desired level. (5) Removal of dissolved oxygen or other corrosive gases by 'deaerators' or the addition of chemicals to react with oxygen (oxygen scavengers). (6) Drying of air or other gases to keep humidity below level where corrosion can occur. (7) Design of hydraulic systems to avoid excessive velocities or localized turbulence or to maintain a velocity high enough to prevent the accumulation of corrosion products or other deposits that would promote localized corrosion. (8) Various features of design and operation of structures or equipment to favor rapid drainage and drying, prevent accumulation or concentration of corrosive chemicals in crevices or low spots, hold operating stresses and temperatures within desired limits, eliminate fabricating stresses by appropriate heat treatment, avoid galvanically unfavorable combinations of different metals, provide protection against stray electrical currents by appropriate insulation and electrical bonding. (9) Heat-treating metals to leave them in optimum condition to resist corrosion. (10) Applying protective electrical currents (cathodic protection) from sacrificial metals (galvanic anodes) such as zinc, magnesium or aluminum or from some external source through a graphite, platinum or other appropriate anode receiving current from a rectifier, generator, or battery. The location of the anodes, the magnitude of the current and the applied voltage must be engineered so that without wasting current all surfaces that require protection will receive enough current to achieve this effect. Too much current may cause damage by the alkali generated by a cathodic reaction or by hydrogen evolved at the cathode which can destroy protective films or embrittle metals.

Corrosion can be suppressed, also, by the controlled application of current to the metal as an anode. This is called anodic protection. Passivity is induced and preserved by maintaining the potential of the alloy at, or above, a critical potential in what is called the range of passivity in a potentiostatic diagram. Such diagrams are based on the relationship between applied anodic current density and the corresponding potential in the environment of interest.

Continued research, especially of a fundamental kind, is necessary to learn more about the basic nature of corrosion as a guide to better alloys, better testing methods, better interpretation of data and improved means of preventing corrosion.

FRANK L. LAQUE

Cross-references: *Metals, Electrochemistry, Electrode Potentials.*

COSMETICS

Cosmetics are preparations applied to the surface of the body for the purpose of enhancing its appearance. They may be (1) make-up preparations, applied to bring about temporary effects, lasting only so long as the preparations remain on the body surface, or (2) treatment preparations, which effect no immediately noticeable change but which after repeated use are expected to have a beautifying effect.

The skin, on its outer surface, is an inert, rather tough material well suited to protect the delicate inner tissues from injury. The dead cells forming the outer layers are hard, rather dry, and expendable. In deeper layers of the skin, the cells are softer, alive, and vulnerable. An essential characteristic of the living cells forming the inner layer of the epidermis is their ability to divide and produce new cells which continually push the older cells upward. These older cells gradually toughen as soft proteins are changed to hard, horny keratin.

The living cells of the inner epidermal layer are believed to contain "tonofibrils" of composition similar to keratin, which may serve as crystallization centers for deposition of more keratin which is formed from the globular proteins of the cytoplasm as the cell ages and dies. This transformation of protein may be considered a denaturation process, with the long coiled chains of the globular proteins unfolding and assembling into the parallel bundles found in keratin.

Keratins are fibrous proteins, insoluble and relatively inert to chemical agents. The lengths of the molecules are over 100 times their thickness. Bundles of such molecules form fibers. In hair, the long axes of the keratin molecules lie roughly parallel to that of the hair; in nails, the fibers run transversely; in epidermal cells they run in all directions. Molecules are held together in the fibers (1) by linkage of the SH groups of their cysteine residues to form common disulfide bonds; (2) by salt linkages between free carboxyl groups of one molecule and free amino groups of another; and (3) by hydrogen bonds between NH groups of one molecule and CO groups of another. These linkages must be broken if the keratin is to be softened or altered.

Under the epidermis, the dermis contains blood vessels, lymph spaces, nerve endings, sebaceous and sweat glands, hairs in their follicles, and erector muscles for the hairs, all distributed in a matrix of connective tissue. This connective tissue contains a network of fibrous proteins (collagen, reticulin, elastin) filled in with a ground substance which appears to be semi-sol, semi-gel. Collagen, forming about 72% of the weight of dried, fat-free skin, has a high content of glycine, proline, and hydroxy-proline; very low tyrosine, methionine, and histidine; and no cystine or tryptophan.

The collagen molecule has the appearance of a rigid ribbon about 2800 Å long, 14 Å in diameter, with molecular weight about 300,000. The structure is formed of three different polypeptide chains interwoven into a triple helix, and held together about a common axis by H bonds. Along the chain are alternating polar and apolar regions. The apolar regions are made up of the repeating triplet glycine-proline-X, where X can vary; the less well known polar regions may be acidic or basic.

Carbohydrates are intimately associated with collagen, in the proportion of two molecules of glucose and two of galactose to one molecule of chrondroitin sulfate. Other mucopolysaccharides are present, probably bound to proteins. The structural unit of the collagen fiber is the microscopic fibril of diameter 0.3 to 0.5 micron. The fibers are bundles of these microfibrils. Reticulin fibers are branched, and are found mostly at the boundaries of the connective tissue. Collagen and reticulin fibers can be regenerated by living organisms, but elastin cannot. Elastin differs from collagen principally in having more alanine and valine, and less glutamic acid and hydroxyproline. Elastin is unique in containing the amino acids desmosine and isodesmosine.

The skin contains varying amounts of fat, depending upon nutritional conditions. Intercellular depositions of glycerides (fuel reservoirs) are very variable; intracellular lipides (mostly sterols and phospholipides) are more constant.

It is a well-known fact that the skin of elderly persons is generally wrinkled, dry, and in general less attractive than that of younger individuals. Some of the changes taking place as skin ages are: (1) components become less highly hydrated; (2) cholesterol content decreases; (3) calcium content increases; (4) fat content decreases; (5) globular proteins decrease; (6) phospholipide content decreases; (7) sebum excretion drops (in women only); (8) tension and elasticity decrease; (9) tensile strength increases.

Treatment cosmetics, if they are to be effective, should be designed to prevent, counteract, or compensate for one or more of the above changes.

If cosmetic preparations are to influence the living cells in the deeper layers of the skin, and possibly in the subcutaneous tissues, they must be able to reach those cells. The horny layer presents a barrier to such penetration, but not one which is insuperable. Knowledge of the composition and properties of the skin makes it possible to select cosmetic ingredients which will promote absorption. Oils and oil-soluble substances are absorbed principally by way of the sebaceous glands. Selection of solvent, or of emulsion type, can have an important effect upon extent of absorption.

Cosmetic make-up is generally designed to impart a desired color or texture to some part of the body surface. The skin of the face may be powdered to reduce gloss, or covered with a more adherent make-up which remains fixed all day. Face powders contain white pigments with high covering power, such as zinc oxide and titanium dioxide; colored pigments such as iron oxides; ingredients to give desired slip, such as talc; and materials to improve adhesion to skin, such as zinc stearate. In the more lasting make-up, the pigments are mixed with oils, waxes, etc., to improve adherence.

Rouge for the cheeks is similar to face powder, but contains bright red lakes of organic dyes in high enough concentration to redden the skin noticeably. Rouge for the lips is almost always prepared in the form of lipsticks, which have a base composed of wax and oil in such proportions as to remain stiff in hot weather but to be applied easily to the lips. Beeswax, carnauba wax, amorphous hydrocarbon waxes, castor oil, lanolin, butyl stearate, and polyethylene glycol and its esters are among the commonly used ingredients of lipsticks.

Nail "polishes" are generally colored lacquers with a nitrocellulose base. They are more lasting than most make-up preparations for the skin, and may stay on the nail until growth exposes a noticeable expanse of uncovered nail, necessitating removal and reapplication of the lacquer.

For all cosmetic preparations, ingredients must be free of any irritating or injurious effect under conditions of use. The U. S. Government has taken notice of the need for standardized coloring materials by providing that all coal-tar dyes for use in foods, drugs, and cosmetics must be certified by the U.S. Food and Drug Administration as harmless and suitable for such use. Manufacturers of cosmetic colors must submit to Washington a sample of each batch manufactured, and must obtain a certificate that the sample tested there was found to be composed of a dye on the approved list, and contained impurities in quantities below the specified limits. A certification number is assigned to each batch, and must be shown on the label of each container of color shipped to a cosmetics manufacturer.

For dyeing hair, phenylene diamine and similar compounds are much used. These amines are allowed to penetrate into the hair shaft; then an oxidizing agent is applied which converts the amine to a colored, insoluble compound which remains in the hair.

Bleaching of the skin with weak acids such as buttermilk and lemon juice has been practiced for centuries but is not very effective. Creams containing mercury salts, generally ammoniated mercury, are claimed to be effective bleaches, but their use involves some danger, and the mercury content is held to a low level by law. The mercury probably inhibits one of the enzymes involved in the production of melanin, the black pigment of the skin, from tyrosine. Hair is commonly bleached by hydrogen peroxide.

Since hair is a dead tissue, it may be subjected to rather drastic treatment. The salt links and disulfide links between adjacent keratin molecules may be broken by use of alkaline solutions of various sulfur compounds. Metallic sulfides are not convenient to use, but the ammonium salt of thioglycolic acid is well adapted to such application. The hair is wound on spindles, soaked with the thioglycollate solution until it is made soft and nonelastic, then rinsed or oxidized until all excess thioglycollate is removed. The disulfide bonds are supposedly reformed at new positions so that the restiffened hair thereafter retains its imposed spiral shape.

The same agents that soften hair may be used in greater strength to remove unwanted hair. Metallic sulfides were once used, but have been largely replaced by ammonium thioglycollate, which is less irritating and gives off a less objectionable odor.

In addition to improving visual appearance, cosmetic preparations may be designed to enhance olfactory attractiveness. Most cosmetics are perfumed attractively by addition of odorants, but deodorants and perfumes are specifically intended for odor improvement. Deodorants must counteract the tendency of sweat and sebum to form ill-smelling materials on the skin. They may do this either by inhibiting perspiration or by preventing bacterial decomposition of the excreted products upon the skin. Antiperspirants generally contain aluminum salts which exert astringent action. The antibacterial deodorants contain germicides, such as hexachlorophene, which cling to the skin and inhibit bacterial action for long periods.

The deodorant properties of chlorophyllins have been the subject of much controversy. It appears that they are able to neutralize or destroy some odorants by adsorption of both a physical and a chemical nature.

Reference was made above to the changes occurring in the skin as an individual grows older. Treatment cosmetics are designed to prevent, counteract, lessen, or compensate for, the effects upon the skin of age and other damaging agents. Knowledge of the biochemistry of the skin is fragmentary, and we do not yet know how to make completely effective treatment cosmetics.

Excessive exposure to sunlight may cause serious damage to skin, such as premature appearance of the wrinkles generally associated with advanced age, and even malignant growths. The development of effective sunscreens, applied to the skin in suitable cosmetic creams or lotions, has been an important step in combatting such damage.

Creams containing suitable oils have long been used to counteract the dryness (lack of oil) attendant upon age. Among the most effective agents for increasing hydration of skin cells are the sex hormones and related steroids; their skin-hydrating effect has recently been found to be independent of their sex activity. These steroids are reported to make senile skin plumper and firmer, and thus to make it look younger. The treatment problem is largely one of stimulating the cells of skin and underlying tissues, so that they will be more active and behave more like younger cells.

New chemical products with potential application in cosmetics are appearing rapidly. They include solvents, surfactants, germicides, hormones, and substances of many other classes. They increase tremendously the range of effectiveness to be reached by the cosmetic chemist. Their actual use in cosmetics must be preceded by careful tests for safety, first on animals, then by patch tests on humans, and finally under actual conditions of proposed use. By such a procedure preparations that are safe and harmless, as well as effective and attractive, will be produced.

PAUL G. LAUFFER

Cross-references: *Amino Acids, Proteins, Carbohydrates, Collagen, Keratins, Adsorption.*

CRACKING

Cracking refers to the decomposition of organic compounds, particularly of petroleum hydrocarbons, into compounds of lower molecular weight. Cracking is the major petroleum refining operation for the production of gasoline. The first method used was called thermal cracking, in which decomposition is achieved by heating the charge stocks to quite high temperatures for varying periods of time. Typical temperatures are 450 to 550°C for liquids, and 600°C or more for gaseous charge stocks. The thermal process still finds use in the large-scale production of gaseous olefins, ethylene and propylene, as monomers for the plastics industry.

Visbreaking and coking are variations in thermal cracking, but with heavy reduced crudes as charge stocks. In visbreaking, highly viscous residual fuel oils are cracked under mild conditions to reduce the viscosity of the residual feed stock without much formation of gas oil and lighter products. Coking extends the time of cracking so that coke yield is maximized, with accompanying upgrading of the residual feed into gasolines and gas oils. About half the by-product coke is used for aluminum cell electrodes and for metallurgical purposes, if sulfur and metals contents are low enough; the rest is burned as fuel. Today's petroleum coke production in the United States is 38,000 tons per day, equivalent to 940,000 barrels per day of residual fuel. The total daily capacity of thermal cracking in the United States including coking and visbreaking, is 450,000 barrels.

For gasoline production, the use of catalytic cracking has fully replaced thermal processing. The use of catalysts reduces the amount of raw material converted to methane, ethane, and other gases as compared to liquid, gasoline boiling range products, and produces a mixture of chemical compositions more desirable in quality, particularly in antiknock quality.

In the cracking processes, heavy hydrocarbons are not only broken down into lighter compounds, but their chemical structure is altered in the process of molecular weight reduction. In addition, condensation reactions occur which produce some high molecular weight hydrocarbons. The material remaining unconverted in terms of lowering of boiling range therefore tends to contain yet heavier components of lower hydrogen/carbon ratio. It is sometimes recycled for further cracking, processed separately such as by coking (see above), or marketed as heavy fuel oil. One end product of these condensation reactions is "coke," a highly condensed polycyclic hydrocarbon which accumulates on the catalyst. As the coke accumulates on the catalyst the rate of cracking slows down appreciably and the catalyst must be regenerated by burning off the coke with air. The catalytic cracking process is therefore a cyclic process, wherein catalyst is alternately exposed to the petroleum charge (cracking) and to air (regeneration).

The early catalytic crackers used several reactors with catalyst packed and remaining stationary therein. Cracking and regeneration would proceed simultaneously but in separate sets of reactors, the petroleum feed stock and regeneration being switched from one set to another. This fixed-bed processing system has been completely displaced by the more efficient and less complex continuous crackers, fluid- and moving-bed types.

In the moving-bed type (Thermofor Catalytic Cracking or TCC, and Houdriflow) the catalyst moves downward through a reactor concurrently with the feed. The deactivated catalyst passes into a steam purge zone and then into the regenerator where coke is burned off with air. From the regenerator the cleaned catalyst is carried back to the top of the reactor by air or steam lift.

In the fluid-type cracker, oil vapor in the reactor and air in the regenerator are fluidizing mediums for the fine powdered catalyst. The cracking and regeneration occur under fluidized turbulent catalyst conditions. The cracked vapors and regenerator flue gas pass through cyclone separators to remove any catalyst carry-over. The separation of cracked products in both the moving-bed and fluid crackers is carried out in conventional distillation units. Fluid-type crackers use catalyst in the form of a powder of some 20 to 100 micron particle size, while the moving-bed type uses bead or pelleted forms of catalysts of about 2 to 5 mm diameter. The fluid-type system is likely to be the major mode of use in the future.

In a typical catalytic cracking process, the temperature in the cracking reactor is 425 to 525°C, the pressure is slightly higher than atmospheric, the petroleum vapors contact the catalyst for a few seconds, catalyst remains in the cracking zone for several minutes, and a longer time in the regenerator.

The early cracking catalysts were mostly either synthetic silica-alumina composites or natural clays which had been activated. These catalysts are amorphous with an extensive random pore structure to create some 100 to 400 m^2/g of internal surface area on which most of the cracking action takes place. To be effective in cracking these surfaces must have acidic properties. One method used to demonstrate the acidic nature of the active surfaces has been to show the almost complete suppression of activity in the presence of high-boiling organic bases, such as quinoline.

Within the last decade a new family of cracking catalysts has profoundly influenced the practice and expansion of commercial cracking. These catalysts contain from a few per cent to about 15% of crystalline aluminosilicates, most frequently referred to as zeolite, in a porous matrix. These zeolites have an ordered structure of silicon oxide tetrahedra, with aluminum sites substituting some silicon positions, arranged in configurations that form intercrystalline channels and cavities sufficiently large to permit passage of the molecules of a chemical reactant, including petroleum fractions. The materials are related to the molecular sieves in geometric structure, but are modified, in chemical properties, to attain acidic catalytic properties.

Just as the discovery of certain amorphous solids to catalyze petroleum cracking rapidly displaced the practice of thermal cracking, the crystalline zeolite composite catalysts have quickly taken over dominance of the technology, because of the further dramatic improvement in conversion selectivity. Some 10 to 50% more gasoline is produced for the same amount of catalytic cracking charge stock. Conversely, the introduction of more selective cracking catalysts has significantly reduced the petroleum natural resources required to meet current gasoline demand.

The molecular processes of catalytic cracking are believed to follow certain rules of carbonium-ion reactions, while thermal cracking occurs through free radical reactions. The total of all reactions involved in catalytic cracking is very complex. However, catalytic cracking results in greater preference for bond rupture on such positions of the molecule that relatively few small (gaseous) molecular fragments are produced; isomerization reactions, i.e., skeletal rearrangements of molecules occur in catalytic cracking. These enhance antiknock quality of the gasoline, for example, by producing a relatively high concentration of branched molecular products.

Another class of reactions specific to acidic cracking catalysts involves hydrogen transfer between molecules, e.g., between naphthenes and olefins, transforming six-membered naphthenes into aromatics, and olefins into paraffins. This type of reaction is particularly pronounced in the zeolitic cracking catalysts; catalytic gasolines are therefore more saturated and more aromatic.

The size of catalytic cracking units has grown steadily, most modern installations having a daily capacity of near 100,000 barrels or more of charge stock. The total capacity in catalytic cracking has reached 5.8 million barrels per day in the United States, and 7.8 million barrels per day for the free world (1969).

P. B. Weisz

Cross-reference: *Gasoline.*

CROSSLINKING OF POLYMERS

A sample of a polymer such as nylon, polyethylene, or unvulcanized rubber normally consists of an aggregate of large molecules which, although they may interact with one another in the bulk, are nevertheless independent of each other in the sense that no molecule is tied to any of the others by primary valence bonds. A number of physical properties arise as a result of this independence. For example, these polymers can be made to flow like a viscous liquid upon melting. This implies that the molecules are able to glide past each other without the necessity of making or breaking any chemical bonds. Another consequence is that the polymer sample will dissolve completely if the proper solvent and thermal conditions are provided.

It is possible to picture a molecule such as one of those above becoming attached to another by a chemical bond that joins them together at some point other than the ends of the molecules. The two molecules are no longer independent of each other. The parts at some distance from the new bond may still engage in segmental motion as previously, but the molecules as a whole can no longer be separated without expenditure of the relatively large amount of energy necessary to break chemical bonds. The two distinct molecules are now considered to be crosslinked.

A crosslink may be looked upon as a more or less permanent connection between two polymer molecules binding them together through a system involving primary chemical bonds. In its simplest form it may be just a single valence bond such as occurs in the radiation crosslinking of polyethylene:

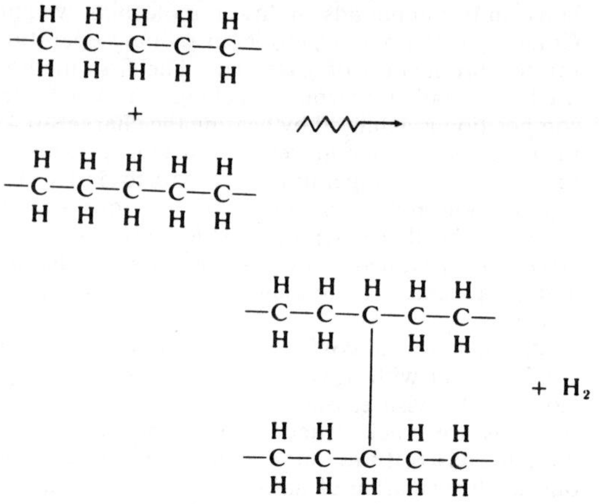

The sulfur vulcanization of rubber involves a somewhat more sophisticated form of crosslinking. Here the sulfur atoms become attached to the rubber molecules during a complex reaction and form the links between them. The crosslinking agent can also be a relatively complex molecular group such as in the *p*-dinitrosobenzene crosslinking of rubber:

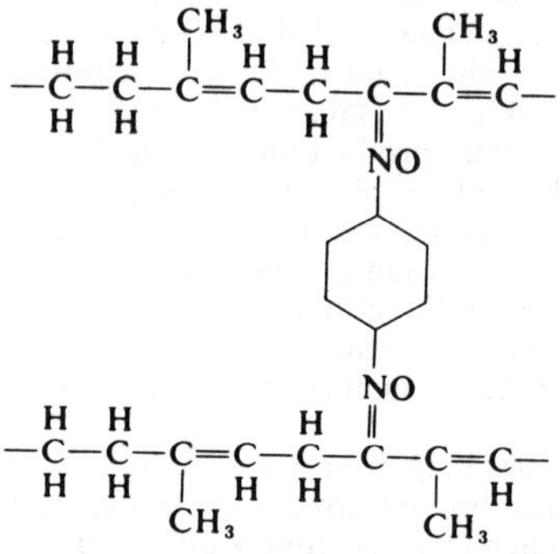

It is even possible to have a short segment of a polymer molecule connecting two other molecules so that the whole forms a sort of H shaped affair. Nor need the bonds involved be covalent. Ionic groups can also participate. For example, a polymer containing carboxylic acid groups may react with polyvalent cations to form "salt bridges":

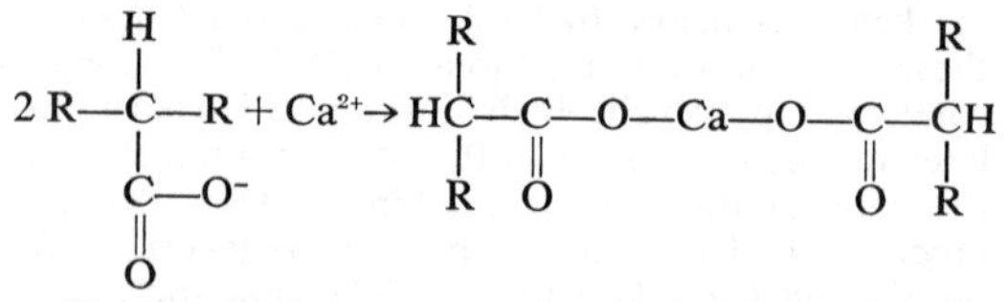

Here R designates the rest of the polymer molecule.

The above examples illustrate the crosslinking of already formed polymer molecules. In practice, this crosslinking is achieved by a variety of processes such as treatment with ionizing radiation or with sulfur and heat as in vulcanization. Since the crosslinking reaction frequently has a free radical mechanism, peroxides are often used. With some polymers simply heating the material is sufficient. However, many polymers are crosslinked during the polymerization itself.

A monomer molecule must be capable of reacting with two other monomers to be incorporated into polymer. The polymer molecule then contains a linear sequence or chain of monomer units. If some proportion of the monomer molecules is capable of reacting with more than two others, chain branching and crosslinking occur; thus the incorporation of some divinyl benzene in styrene polymerization produces a crosslinked polystyrene:

A common example of a polymer which is extensively crosslinked during polymerization is phenol-formaldehyde.

It should be noted that the crosslinks in the latter group of polymers do not differ in principle from those of the former. The distinction is solely whether or not they were formed during polymerization.

So far, we have considered the crosslink as binding two molecules together. These can also be linked to a third, the third to a forth, and so on. In the limiting case, the crosslinks can bind the entire polymer mass together into a single network-like structure which is, in principle at least, a single huge molecule.

The polymer mass now acquires new physical properties. The glass transition temperature of the polymer becomes higher. If it is heated above what was formerly the melting point, it will no longer flow like a liquid. Instead, it will exhibit rubber-like elasticity if the crosslink density is not too high. The higher the crosslink density, the higher the elastic modulus becomes. The elastic modulus can, in fact, be used as a measure of the crosslink density under these conditions. If the amount of crosslinking becomes very large, as in phenol-formaldehyde resins, then the polymer becomes hard, brittle and infusible. It simply cannot be melted without decomposition. It is no longer a "thermoplastic" material but has become "thermosetting." Similarly, a crosslinked polymer will not be completely soluble under the usual conditions, because the molecules are tied to each other and cannot be completely separated by solvent. Instead, the solvent will simply cause the mass to swell to some limiting value. This limiting or equilibrium swell value can also be used to measure the amount of crosslinking. The more extensive the crosslinking, the smaller the equilibrium swell.

Mention should also be made that a molecule can crosslink with itself if it twists and doubles back on itself. Here the crosslink joins two widely separate parts of the some molecule. A common example of this is the disulfide linkage in proteins.

It is the change in properties that makes crosslinking so useful. For example, the crosslinking of polyethylene prevents the molecules from flowing past each other. The material can now maintain its form at temperatures above its usual melting point. It is now useful at higher temperatures than formerly. The tensile strength, and elastic recovery of rubber-like materials depend upon crosslinking. The rubber-like property occurs when the segments of the molecules between crosslinks are long enough and flexible enough to assume a new conformation under stress but the molecules cannot slip past each other and tend to return to their former conformation when stress is released. Finally, plastic flow is inhibited by extensive crosslinking so that the shape and form of a polymer mass are more easily preserved, especially under stress.

Mention should be made of some related phenomena. A partially crystalline, partially amorphous, polymer may show limited swelling in a solvent adequate for the amorphous and not the crystalline polymer, thus showing a few of the characteristics of crosslinked polymer. A polymer having polar or hydrogen bonded groups may also show limited swell in some nonpolar solvents. Sometime polymers containing carbon black will exhibit a few of the properties of crosslinked materials. In fact, the term crosslinking has occasionally been used to describe these situations. This is an incorrect use of the term since crosslinks in the sense described above are not involved here.

GEORGE ADLER

References

Flory, P. J., "Principles of Polymer Chemistry," Cornell University Press, Ithaca, N. Y. 1953.
Meares, P., "Polymers: Structure and Bulk Properties," Van Nostrand, New York, 1965.

CRYSTAL-FIELD THEORY

Crystal-field theory, developed by Bethe in 1929, has become prominent in recent years in the interpretation of physical and chemical properties of compounds of transition elements. The theory describes the effects of perturbation of d orbitals of transition-metal ions in a crystal lattice. To a first approximation magnetic and exchange forces are disregarded.

A transition-metal cation in a crystal is subjected to the electric field of surrounding, negatively charged, anions or dipolar groups (ligands) which, in this theory, are represented as point negative charges. This "crystalline field" destroys the spherical symmetry possessed by an isolated transition-metal ion, and the nature and magnitude of the changes induced within the central cation depend on the type, symmetry, and distances of surrounding ligands.

In an isolated transition-metal ion, the five d orbitals are energetically equivalent (5-fold degenerate), and d electrons occupy singly as many

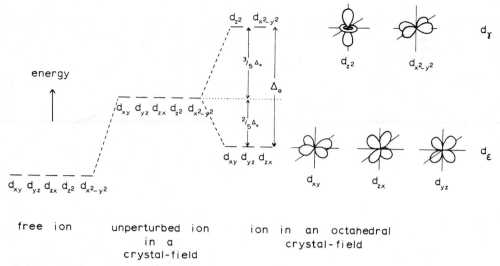

energy

d_{z^2} $d_{x^2-y^2}$

$\frac{3}{5}\Delta_o$

Δ_o

d_{xy} d_{yz} d_{zx} d_{z^2} $d_{x^2-y^2}$

$\frac{2}{5}\Delta_o$

d_{xy} d_{yz} d_{zx}

d_{xy} d_{yz} d_{zx} d_{z^2} $d_{x^2-y^2}$

d_{z^2} $d_{x^2-y^2}$ d_γ

d_{xy} d_{zx} d_{yz} d_ϵ

free ion   unperturbed ion   ion in an octahedral
                 in a   crystal-field
           crystal-field

FIG. 1. Energy level diagram illustrating the splitting of d orbitals by an octahedral crystal-field.

orbitals as possible with spins oriented in the same direction so as to minimize interelectron repulsion (Hund's Rule). However, the five d orbitals possess different spacial configurations. One group of orbitals, the d_γ orbitals (alternatively, e_g or Γ_2 orbitals) consisting of the d_{z^2} and $d_{x^2-y^2}$ orbitals, have lobes which are directed along three cartesian axes (Fig. 1). A second group of orbitals, the d_ϵ orbitals (alternatively, t_{2g} or Γ_5 orbitals) consisting of the d_{xy}, d_{yz}, and d_{zx} orbitals, possess lobes which project between the cartesian axes (Fig. 1). In a crystal-field, the d orbitals are no longer degenerate and some are lowered in energy relative to others. By preferentially filling low energy orbitals, d electrons stabilize certain transition-metal ions.

When a transition-metal ion is in octahedral co-ordination with six identical ligands, while electrons in all five d orbitals are repelled by the negatively charged ligands, electrons in the two d_γ orbitals are repelled to a greater extent than are those in the three d_ϵ orbitals (Fig. 1). The energy separation between the d_ϵ and d_γ orbitals is termed "crystal-field splitting" and is denoted by Δ_o (alternatively, $10\ Dq$). Each electron in a d_ϵ orbital stabilizes a transition-metal ion by $2/5\ \Delta_o$ whereas every electron in a d_γ orbital diminishes stability by $3/5\ \Delta_o$. The resultant net stabilization energy is called "crystal-field stabilization energy" (designated by CFSE).

The distribution of d electrons in a given transition-metal ion in a crystal field is controlled by two opposing tendencies. Coulomb and exchange interactions between electrons cause them to be distributed over as many orbitals as possible so that there is a maximum number of unpaired electrons with parallel spins ("Hund's stabilization energy"), but the crystal-field splitting favors the occupancy of the group of orbitals with the lowest energy (crystal-field stabilization energy). In an octahedral crystal-field, ions possessing one, two, or three d electrons (for example, Ti^{3+}, V^{3+}, and Cr^{3+}, respectively) can each have only one electronic configuration and d electrons occupy different d_ϵ orbitals with spins parallel. However, ions posses-

sing four, five, six, and seven d electrons (for example, Cr^{2+} and Mn^{3+}, Mn^{2+} and Fe^{3+}, Fe^{2+} and Co^{3+}, and Co^{2+} and Ni^{3+}, respectively) have a choice

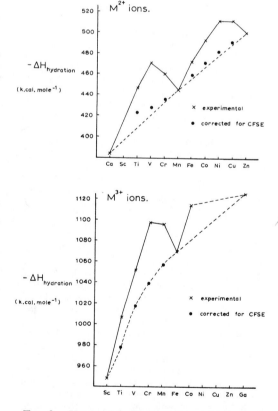

FIG. 2. Heats of hydration of transition-metal ions. Experimental values line on double humped curves. Corrected values, obtained by deducting crystal-field stabilization energies derived from absorption spectra, lie on smooth curves.

of electronic configuration. If the crystal-field splitting is small (the "weak-field" case), d electrons occupy both d_ϵ and d_γ orbitals singly to the maximum extent. The CFSE is reduced by electrons entering the d_γ orbitals, but energy is not expended in pairing of electrons in d_ϵ orbitals already half filled. Alternatively, if the crystal-field splitting is large (the "strong-field" case), it is energetically more favorable for d electrons to fill low-energy d_ϵ orbitals. In this situation, the gain in CFSE outweighs the electron pairing energy. Finally, ions possessing eight, nine, and ten d electrons (for example, Ni^{2+}, Cu^{2+}, and Zn^{2+}, respectively) can each possess only one electronic configuration and d orbitals are filled to completion. In a weak crystal-field, an ion generally possesses more unpaired electrons (the "high-spin" state) than it does in a strong crystal-field (the "low-spin" state). The distinction between the "low-spin" and "high-spin" configurations is fundamental in understanding magnetic and optical properties of transition-metal compounds. Note that ions possessing three, eight, and six ("low-spin" configuration) d electrons acquire high stabilizations in octahedral crystal-fields, whereas ions possessing zero, five, and ten d electrons have zero CFSE in weak octahedral fields.

When a transition-metal ion is in terahedral coordination, the d_γ orbitals become the more stable group, but the magnitude of the tetrahedral crystal-field splitting parameter, Δ_t, is smaller than that of the octahedral parameter, Δ_0. If the cation, ligands, and cation-ligand internuclear distances are identical in both octahedral and tetrahedral coordinations, $\Delta_t = 4/9 \, \Delta_0$

The magnitude of the crystal-field splitting parameter, Δ, may be estimated from measurements of absorption spectra. The energy to excite an electron from one d orbital to a vacant position in another of higher energy corresponds to the visible or near infrared region of the spectrum, and absorption of this radiation is the most general origin of color in transition-metal compounds. Certain generalizations may be made about the dependence of the numerical value of Δ on the valence and atomic number of the metal ion, the symmetry of the coordinated ligands, and the nature of the ligands.

(1) Δ values are higher for trivalent ions than for divalent ions. For the first transition series, the ranges are $7500 - 12500 \text{ cm}^{-1}$ for divalent ions and $14000 - 21000 \text{ cm}^{-1}$ for trivalent ions.

(2) The values are about 30% higher for ions in each succeeding transition series.

(3) Δ depends on the nature of ligands coordinated about the transition-metal. Ligands may be arranged in order of increasing Δ, the "spectrochemical series," an abridged version of which is

$$I^- < Br^- < Cl^- < F^- < OH^- \leq \text{carboxyanions} <$$

$$H_2O < NH_3 < SO_3^{2-} < NO_2^- \ll CN^-$$

Ligands at the beginning of the series generate weak crystal-fields, whereas those at the end of the series produce strong crystal-fields and "low-spin" electronic configurations in the central transition-metal ion. The cross-over point from "high-spin" to "low-spin" configuration varies from one cation to another and may be ascertained from magnetic

measurements and crystal structure analyses of internuclear distances. Note that, when an ion changes from a "high-spin" to "low-spin" configuration, internuclear distances are drastically reduced.

(4) Δ depends on the symmetry of the coordinated ligands, the crystal-field splitting for tetrahedral coordination being 40–50% of the values for octahedral coordination.

Crystal-field stabilization energies of ions produce observable effects on thermodynamic properties (for example, lattice energies and heats of hydration) of transition-metal compounds. If the ions were spherically symmetrical and no preferential filling of d orbitals occurred, a given thermodynamic function would display smooth periodic variation across a row of transition elements concomitant with contraction of the ions. However, observed values frequently fall on characteristic two-humped curves with maximum values for ions possessing three and eight d electrons (Fig. 2). Furthermore, the values lie above a smooth curve through those of ions possessing zero, five and ten d electrons, for which there is zero CFSE. When crystal-field stabilization energies, estimated from absorption spectra of hydrated ions, are deducted from observed heats of hydration, the corrected values conform closely to the smooth curve (Fig. 2). Such an analysis provides convincing evidence in support of the d orbital energy separation proposed by crystal-field theory.

A further consequence of great importance in the thermodynamics of heterogeneous systems, arising from the existence of crystal-field stabilization energies, is that solid solutions containing as one component a compound of a transition element cannot conform to ideal solution behavior. A necessary condition for a solution to be ideal is that the heat of mixing of components is zero. This criterion is not fulfilled, however, whenever differences exist between crystal-field stabilization energies of ions in different environments.

Although configurations in which identical ligands are in regular octahedral and tetrahedral coordination about a transition-metal ion exist in aqueous solutions and melts, such ideal configurations are rarely found in crystal structures. Frequently, cations are located in distorted environments in which metal-ligand internuclear distances are not constant, or surrounding ligands are not identical. Distortion of coordination polyhedra about a transition-metal ion is to be expected for theoretical reasons (the Jahn-Teller effect). For example, if one of the metal d orbitals is empty while another of equal energy (such as the d_γ orbital group) is half filled, the compound is predicted to distort spontaneously to a different geometry in which a more stable electronic configuration is achieved by making the occupied orbital lower in energy. Six-fold coordination sites in compounds of Cr^{2+} and Mn^{3+}, ions which possess the electronic configuration $d_\epsilon^3 d_\gamma^1$, and $Cu^{2+}(d_\epsilon^6 d_\gamma^3)$, are invariably distorted from regular octahedral symmetry as a result of the Jahn-Teller effect.

Crystal-field theory provides an insight into cation distribution in crystal structures and fractionation processes in heterogeneous systems. For example, cations in spinels, $X^{2+}Y^{3+}O_4$, occur in octahedral and tetrahedral coordination. In "normal" spinels, divalent ions (X^{2+}) occupy tetrahedral sites and trivalent ions (Y^{3+}) fill the octahedral sites. "Reverse" spinels, however, contain half the trivalent ions in the tetrahedral sites, and the remaining trivalent ions plus divalent ions in the octahedral sites. In this and similar cases it is possible to predict the distribution of cations between octahedral and tetrahedral sites, and to account for the type of spinel formed by each transition-metal ion, by crystal-field theory. The trends are related to the magnitude of the octahedral "site preference energy" parameter, which is the difference between octahedral and tetrahedral crystal-field stabilization energies of a cation. The tendencies for divalent and trivalent ions to fill octahedral sites and form "reverse" and "normal" spinels, respectively, are related to the magnitudes of the octahedral site preference energies of the ions. Similarly, crystallization behavior of transition-metal ions in melts (for example, silicate melts in nature) may be correlated with the relative values of site preference energies of cations in both liquid and solid phases.

WM. S. FYFE
ROGER G. BURNS

Cross-references: *Crystals and Crystallography; Ligand Field Theory.*

CRYSTALS AND CRYSTALLOGRAPHY

Crystal Optics

Chemists who are interested in the properties of solids have become increasingly concerned with crystal optics. From the practical viewpoint, this necessity arises in connection with many types of instruments, from fairly simple spectrometers to lasers. Furthermore, chemists are becoming increasingly interested in optical properties as a rapid, easy method for identification and comparison of inorganic substances through the use of the petrographic (polarizing) microscope.

The fundamental principles of the use of the microscope, including the universal stage, have been outlined, and several compilations on properties of inorganic crystals are available. In general, besides being useful for quickly observing qualitative crystallographic properties, the "immersion" method consists of measuring the principal refractive indexes—three (α, β and γ) for triclinic, monoclinic and orthorhombic crystals, two (ω and ϵ) for tetragonal and hexagonal, or one for cubic crystals—by matching the solid against liquids of known refractive indexes. In addition the relations of the principal optical directions to geometrical aspects of the crystal should be determined. Excellent graphic methods are available for recording even the most complex interrelations.

Optical properties are extremely sensitive to minute differences in crystal symmetry. For example, although the x-ray powder diffraction pattern of perovskite seems to be cubic, the twinning (as revealed by its optical anisotropism) indicates that the true symmetry probably is no higher than monoclinic.

The critical nature of optical properties serves another important function: that of revealing small differences in composition brought about through isomorphic substitution. Although changes in color

and/or pleochroism may be obvious, comparatively minor differences in composition are frequently indicated by quantitative consideration of the refractive indexes.

Knowledge of the influences of various elements (or ions) upon optical properties has resulted in the development of certain empirical relations which apply with astonishing reliability in some situations. The more frequent applications involve either the relation of Lorentz and Lorenz or that of Gladstone and Dale, although the latter was devised primarily for liquids.

Through the use of Avogadro's number, one of these empirical relations readily can utilize the unit-cell volume as determined by x-ray diffraction, and thus circumvent an accurate determination of the density. The Lorentz-Lorenz relation then can be written:

$$\frac{n^2 - 1}{n^2 + 2} = \frac{m_1 R_1 + m_2 R_2 + m_3 R_3 + \ldots}{V \cdot N}$$

where n is the refractive index, V is the volume (in Å), N is Avogadro's number (m_1, m_2, etc.) are numbers of atoms of each elemental kind per unit cell, and (R_1, R_2, etc.) are the characteristics of these atoms known as *ionic refractivities*.

The above relation does not, of course, take into account the asymmetric environments of the ions, and consequently is capable of yielding merely a mean index of refraction. However, Bragg has demonstrated that the structural orientation of such groups as CO_3 and NO_3 can be taken into account because of their pronounced vectorial properties.

Nevertheless, under favorable circumstances, such as exist for an isostructural series of cubic crystals (the garnets), the empirical relation of Lorentz and Lorenz leads to surprisingly good correlations through the use of values for ionic refractivities given in Table 1. Empirical values for alkali halides (due to Wasastjerna) are available. Jaffe has discussed the use of the Gladstone-Dale relation in connection with the refractive indexes of minerals.

TABLE 1. EMPIRICAL IONIC REFRACTIVITIES*

Ion	R	Ion	R
Al	0.65	Ca	2.08
Fe^{3+}	5.90	Mg	0.36
Cr^{3+}	5.20	Fe^{2+}	2.29
Si	0.18	Mn^{2+}	2.33

*Assuming a value of 3.5 for each oxygen of SiO_4^{4-} and sodium light.

In summary, optical characteristics may supply important clues to structural orientations of groups having pronounced polarization; mean refractive indexes can be calculated from other crystal-chemical data; refraction is frequently a sensitive criterion for recognizing compositional differences among isomorphic variants; the polarizing microscope is invaluable in the hands of a skilled operator; and crystal optics is assuming ever increasing importance in terms of its applications.

DUNCAN MCCONNELL

Cross-references: *Refractive Index, X-Rays.*

References

Bloss, F. D., "An Introduction to the Methods of Optical Crystallography," 294 pp., Holt, Rinehart & Winston, New York, 1961.

Hartshorne, N. H., and Stuart, A., "Crystals and the Polarizing Microscope," 614 pp., American Elsevier Inc., New York, 1970.

Wahlstrom, E. E., "Optical Crystallography," 3rd ed., 356 pp., John Wiley & Sons, New York, 1960.

Johannsen, A., "Manual of Petrographic Methods," 2nd ed., 649 pp., 1918. Hafner Publ. Co., New York, reprinted 1968.

McConnell, D., "Refringence of Garnets and Hydrogarnets," *Canadian Mineralogist,* **8,** 11–22 (1964).

Dislocations

Dislocations are one-dimensional crystal defects. By their motions, dislocations cause plastic deformation of crystals. Their great importance lies in the fact that they determine the mechanical properties of crystals and crystalline aggregates. Any crystals which are free of dislocations are brittle as glass. Since all metals are crystalline, this means that, in a very real sense, our culture would be impossible if it were not for dislocations. In addition, dislocations play an important role in some forms of crystallization as well as in other phenomena.

In order to understand the fairly complicated geometry of dislocations it is best first to consider the typical mode of deformation of crystals, which is called "translation," "slip" or "glide." It consists of a sliding motion along the directions of close-packed rows of atoms, i.e. along specified low-index crystallographic directions, named the "slip directions." Geometrically this type of deformation is the same as if the top part of a stack of corrugated iron sheets were pushed to move it relative to the lower part, in the direction of the corrugations. Commonly, low-index crystallographic planes act as preferred "slip planes" which in the model just used would correspond to the planes parallel to the sheets in the stack, but the shape of the surfaces of relative displacement can be arbitrary, provided only that they are everywhere parallel to the slip direction. During the deformation of a crystal, several combinations of slip directions and slip planes (named "slip systems") may act simultaneously; five independent slip systems being the lowest necessary number to allow any arbitrary deformation.

Dislocations arise because any two crystal parts which are displaced relative to each other during slip do not move rigidly. Instead, slip usually progresses in increments of one atomic diameter. Areas on active slip planes or surfaces over which a displacement of one atomic diameter along the slip direction has already taken place spread at the expense of the remainder of the slip surfaces. Within the described regions over which slip (or additional slip) by one atomic diameter along a close-packed direction has taken place, crystallographic order exists, just as crystallographic order exists at all parts of the slip surface, over which slip has not yet occurred. However, a narrow region of severe atomic misfit exists all along the boundary between the slipped and the not slipped area of the slip surface. This strongly disturbed region, with a cross sectional diameter only a few atoms wide, is a "glide dislocation."

By its nature, then, a glide dislocation is the one-dimensional center of a system of strong internal stresses and strains which, through its motion, causes slip. Since dislocation motion can be triggered by stresses—which follows directly from the fundamental property of glide dislocations that their motion causes glide—and since the dislocations themselves are centers of internal stresses, dislocations interact with each other strongly.

A glide dislocation as described above is but one of several types of dislocations. The common property of all dislocations is that they are one-dimensional crystal defects of a kind which *could* have been (but, of course, never actually are) produced by cutting, shifting and rejoining operations as illustrated in Fig. 1: (1) Slice a crystal partway through, generating a cut of arbitrary shape, planar, corrugated, or curved in three dimensions, which either may be completely confined within the crystal or which may end on free surfaces. (2) Along the edge of the cut, which eventually will form the dislocation line, drill a hole of roughly circular cross section but following the whole extent of the outline of the cut. This tunnel-like hole will be closed upon itself if the generating cut was confined within the crystal; it will end on free surfaces if the cut intersected free surfaces. (3) *Rigidly* displace the two sides of the cut relative to each other. No restriction applies to the magnitude or orientation of the relative displacement vector b, commonly called the Burgers vector, except that it must be small enough so that the deformation everywhere is elastic.

To perform this relative displacement, it will generally be necessary to remove or add material in places, so that gaps or overlaps are eliminated. Only in the particular case that the cut was made parallel to the vector everywhere, will it be unnecessary to cut away or add material. (4) Rejoin the two sides of the cut and fill in the tunnel-like hole.

With these operations a dislocation has been generated along the outline of the generating cut. The stress and strain distributions of stress systems generated in the manner described can be mathematically analyzed. These had been investigated in part from the mathematical viewpoint long before it was realized that dislocations play an important role in crystals.

It should be apparent that the glide dislocation described before, represents the particular case that the Burgars vector is equal to the distance between any two equivalent nearest neighbor atoms in the crystal; "equivalent" means atoms which are not only chemically alike but whose surroundings are identical in geometry and orientation. Slipping motion of such dislocations usually occurs at low to moderate stresses on their slip surface, i.e. the surface generated by all lines intersecting the dislocation axis and parallel to the Burgers vector. In other words, the slip surface is the surface on which the dislocation can glide without the transfer of material normal to the Burgers vector; it is that surface on which no transfer of material would be required if the generating cut according to step (1) above would be made coincident with it. It may be added here that a dislocation glides only in response

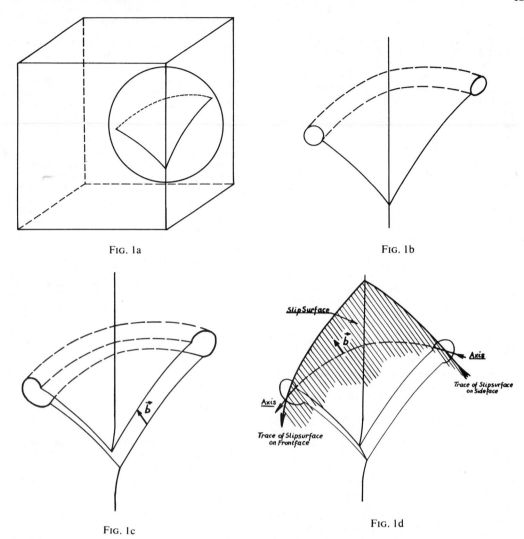

FIG. 1a   FIG. 1b

FIG. 1c   FIG. 1d

FIG. 1. The generation of a dislocation in four steps as explained in the text.

FIG. 1a. A crystal is sliced partway through. In the subsequent figures, only the area included in the circle is redrawn.

FIG. 1b. A hole is drilled along the outline of the cut.

FIG. 1c. The opposing sides of the cut are rigidly displaced relative to each other by the relative displacement vector b.

FIG. 1d. The hole bordering the cut, and the gap which arose when the rigid displacement was made according to Fig. 1c above, is filled in again. The line coinciding with the original outline of the cut is now the dislocation axis. The surface on which the dislocation can move in glide is generated by all lines parallel to the Burgers vector and intersecting the dislocation axis.

to one stress component, namely the shear stress acting on its slip surface in the direction of the Burgers vector.

In a crystal, the Burgers vector cannot be arbitrary in magnitude and direction. So-called "perfect" dislocations are bounded on all sides by ordered material. Their Burgers vectors must be lattice vectors: vectors linking equivalent lattice sites occupied by like atoms having identical surroundings, in geometry as well as orientation. Moreover, *stable* dislocations, such as are observed in actual crystals, virtually always have the smallest possible

Burgers vectors. The reason for this is that the energy of a dislocation is proportional to the square of its Burgers vector (amounting to a few electron volts per atomic plane normal to the dislocation line), so that dislocations with large Burgers vectors would dissociate by glide into two or more, mutually repelling dislocations with smaller Burgers vectors.

Burgers vectors which are not lattice vectors but lead from or to lattice sites which either are not equivalent sites or which are metastable positions, generate "imperfect" dislocations. Such dislocations border planar defects, named stacking faults. These arise because, in the imagined cutting and shifting operation, the atoms have been replaced in incorrect positions. Depending on whether the stacking fault(s) bordering the imperfect dislocations do or do not lie in the dislocations' slip surfaces, the dislocations can or cannot move by glide. Correspondingly, they are called "glissile" or "sessile." Gliding imperfect dislocations drag a stacking fault out behind them, or conversely remove it. Dislocations forming the juncture of two non-coplanar stacking faults are descriptively named "stair rod dislocations." These are necessarily sessile.

Particular classes of perfect dislocations have also been given descriptive names. As we saw already, every perfect dislocation is by its nature glissile. The name "glide" dislocation, introduced at the very beginning, should be reserved to dislocations whose Burgers vector is a shortest lattice vector, and whose slip surface is nearly planar, coinciding with a low-index crystallographic plane which commonly serves as a slip plane. A "prismatic dislocation" is a dislocation which has a general prism as its slip surface. By definition it is thus closed in itself and does not lie in one plane with its Burgers vector. Prismatic dislocations may be perfect or imperfect.

Since the Burgers vector over the whole length of any one dislocation is constant (commensurate with the prescription above that the relative displacement be *rigid*) while the dislocation line itself can be arbitrarily curved, either on a slip plane or on an arbitrary slip surface, the angle included between the Burgers vector and the dislocation line can assume any value. In the particular case that the said angle equals $\pi/2$, one speaks of an "edge dislocation." Such a dislocation can be visualized as generated by a planar cut into which a uniformly thick slice of material has been inserted before rejoining. The *edge* dislocation then coincides with the *edge* of this extra slice, be it straight or curved in the plane of the cut. Conversely, a "screw dislocation" is a dislocation whose Burgers vector is parallel to the dislocation line. By definition it is straight. It derives its name from the fact that a set of parallel imaginary or atomic planes, spaced one Burgers vector apart, is transformed into a continuous screw surface when intersected by a screw dislocation normal to the planes. The screw dislocation is of particular interest because any plane belonging to the zone defined by its Burgers vector can act as its slip plane. "Mixed dislocations" are those for which the angle between Burgers vector and dislocation line assumes arbitrary values. "Extended dislocations" arise when a perfect dislocation splits into mutually repelling "partial" dislocations, separated by intervening ribbons of stacking faults. A "partial" dislocation, thus, is an imperfect dislocation which, together with the other partial(s), makes up a glide dislocation. The extended dislocation assumes an equilibrium width because the force of repulsion between the partials decreases with the inverse of their distance of separation while the intervening stacking fault acts as a surface of constant surface tension.

Besides the glide motions of dislocations on their slip surfaces, which have been discussed so far, motions with a component normal to the slip surfaces are also possible. These are referred to as "climb." Unlike slip, dislocation "climb" requires the generation or destruction of point defects. Conversely, then, dislocations can act as sources and sinks of point defects and, when doing so, execute climb motions. For this reason dislocations sometimes play an important role in diffusion as well as processes involving diffusion.

The question as to the origin of dislocations has no single answer. They form by a variety of mechanisms during virtually every type of solidification and crystallization. In fact, their presence is of over-riding importance in solidification from vapors and diluted solutions, when they cause the crystallization rates to be orders of magnitude larger than they would be otherwise. Dislocations also form as a result of the condensation of thermal vacancies into planar arrays. This is a most important mechanism, leading to subboundaries during slow cooling from the melt and to prismatic dislocation rings after faster cooling and in quenched crystals.

Once present, dislocations can readily multiply through glide motions. In this case, glide dislocation segments which are joined to dislocations in other planes at either end, bow out in their slip plane under the influence of a stress, then double back in themselves, thereby restoring the original dislocation segment, which then may repeat the motion, and so on. For this reason a dislocation segment operating in the described manner is called a dislocation source. Dislocation sources of much the same type but emitting imperfect instead of perfect dislocations and modified so as to move up or down by one atomic plane for each revolution, are believed responsible for twin formation during straining, and may perhaps cause some transformations.

Over the course of the past twenty years, it has become increasingly possible to observe dislocations directly. Years before any direct evidence of dislocations had ever been obtained, most of the types and properties of dislocations described above had been derived from basic principle. Experimental evidence has confirmed these to a very gratifying extent. Current research on dislocations is largely concerned with unraveling the details of the interactions between dislocations and other types of crystal defects as well as with understanding the collective behavior of large numbers of dislocations. Only when this has been achieved to a considerable extent will the ultimate goal be reached, namely to predict accurately the mechanical properties, and other properties influenced by dislocations, of crystalline aggregates of all types, on the basis of dislocation theory.

DORIS KUHLMANN-WILSDORF

Cross-references: *Crystals and Crystallography.*

Color Centers in Crystals

The term "color centers" comes from the German "Farbzentrum," a word which was applied to a specific defect (localized imperfection) responsible for the characteristic coloration of alkali halide crystals which had been subjected to ionizing radiation or to alkali metal vapors. Before any such treatment, insulating solids of high purity show little absorption of radiation as one scans from the infrared into the ultraviolet regions until the energy of the incident quantum approaches that required to bridge the gap between the valence band and the conduction band. After treatments which generate defects in a crystal, one may detect bands of absorption, typically bell-shaped, which are responsible for an observable color if they occur in the visible region. A specific defect may give rise to one or more absorption bands. If defects are interconvertible, one may watch the growth of some bands at the expense of others. Although many impurity atoms (especially those of the transition metal or rare earth elements) may show characteristic absorption bands, we shall consider primarily the intrinsic defects—i.e., those involving components of the crystal—positive or negative ions, electrons, positive holes (sites at which an electron is missing) and vacancies (sites at which atoms or ions are missing). Two or more vacancies may aggregate; some of these vacancies may trap compensating charges. It is necessary that overall charge neutrality be maintained, but in local regions there may be a small excess positive or negative charges. In the absence of impurity ions with charge different from those of the host ions, one would expect equal concentrations of anion and of cation vacancies. The atomic fraction would be of the order of 10^{-6} at $300°K$. The fraction of one type of vacancy may be increased to as much as 10^{-3} by doping with ions of different charge, heating in metal or nonmetal vapors, irradiation with UV light, x-rays, fast electrons, protons or neutrons. The effectiveness of any of these treatments in generating vacancies depends greatly upon the host crystal.

Anion vacancies behave respectively as centers of charge $+1$ in the alkali halides M^+X^- or alkaline earth fluorides $M^{2+}F^{2-}$ and of charge $+2$ in the oxides $M^{2+}O^{2-}$. Except for cesium salts, the alkali halides and the alkaline earth oxides have the rock salt structure, while the MF_2 fluorides have the fluorite structure. Anion vacancies in any of these crystals (and in ZnO and NaN_3 as well) may trap either one or two electrons (F and F' centers respectively). The cation vacancies are sites of effective charge -1 in the alkali halides and of -2 in the others; they are potentially capable of trapping positive holes on an atom adjacent to the vacancy, and this trapping occurs in the oxides. However, in the halides, as soon as an electron is removed from an X^- ion, leaving an X^0 atom, it quickly unites with an adjoining X^- ion to form the more stable center X_2^-. The electrons or holes referred to here are generally produced by the same treatments which generated the vacancies.

In some hosts, one of the intrinsic atoms after displacement from its normal site may reside at an interstitial site. Such an atom may be located either in the center of the void or else in close association with another ion. For example, in alkali halide crystals irradiated at $20°K$ or lower, an X_2^- unit may symmetrically occupy *one* anion site (H center). The anion vacancy left behind by the interstitial X atom is populated by the electron removed from it; this is deduced from the one-to-one correspondence between the number of H- and of F centers generated in the low-temperature irradiation.

The details of composition of color centers, their orientation and their environment are primarily the result of electron spin resonance (ESR) and electron-nuclear double resonance (ENDOR) studies. While these techniques are applicable only to systems with unpaired electrons, the majority of known color centers fall into this category. Some centers which are diamagnetic in the ground state may be studied in a paramagnetic excited state after optical excitation. Most of the details derived from ESR or ENDOR studies are based on narrow-line spectra hyperfine splittings from nuclei within the center or those adjoining it. In contrast, optical lines from defect centers are usually so broad that they convey little detailed information.

F Centers. One electron trapped at an isolated anion vacancy is the description of the center responsible for the prominent F bands found by early investigators of alkali halide crystals; this structure was one of two alternative structures advanced. ESR and ENDOR studies of F centers in the alkali halides (as well as the psuedo-halides LiH and NaH) provide extreme detail of the distribution of the wave function of the unpaired electron over its surroundings. ENDOR spectra detect interaction out to the ninth shell of surrounding nuclei. Roughly speaking, the F center may be regarded as having a hydrogenic $1s$ ground state, with the first excited state $2p$-like. In the alkaline earth oxides, sulfides and selenides, the trapped electron is much more localized, owing to the higher charges of the host ions. The spectroscopic splitting factor (g factor) is generally close to the free-electron value 2.0023. One may gain more insight into the nature of the first excited state of the F center by measurement of the rotation of the plane of polarized light when the crystal is in a magnetic field and the exciting light is scanned through the F center absorption band.

Two electrons may occupy a single anion vacancy to give a diamagnetic center referred to as the F' center in the alkali halides. For the alkali or alkaline earth halides, two adjacent anion vacancies each containing one electron are referred to as an M center, which has a singlet ground state; however, it may be excited into the triplet state (spin $S = 1$) by irradiation in the M band. The R center consists of three anion vacancies lying in a [111]-type plane, with each containing one electron. Since the R center has a doublet ground state, its ESR absorption may be detected (at low temperatures); however, it may also be detected in its quartet state after excitation in the R optical band. For these systems with $S \geq 1$, the splitting of spin energy levels in zero magnetic field gives rise to multiple lines ("fine structure"), the anisotropy of which enables the principal axes of the defect to be established. In the alkali halides, as well as in MgO and CaO, an electron may be trapped at the anion vacancy of an anion-cation vacancy pair. Electrons trapped at next-nearest-neighbor anion vacancy sites are also observed in MgO and in CaO. There are a number of "F aggregate centers," mostly of unknown structure, identified in various hosts by their unusual very sharp peaks superimposed upon a broad band. One of these corresponds to the zero-phonon transition, that is, one which occurs without absorption or emission of vibrational quanta. Observation of the splitting of the zero-phonon lines under uniaxial stress, as well as the polarization of the components enables one to establish the axis of an aggregate center, which may be diamagnetic and therefore not amenable to ESR observation. By contrast, the absorption band of most F centers is very broad as a consequence of absorption of many vibrational quanta, being far more strongly coupled to the lattice than is a typical impurity ion.

Trapped-Hole Centers (V Centers). A positive hole generated in an undistorted region of an oxide crystal is free to move until it is trapped at a site of negative charge (e.g., a cation vacancy) or by an impurity of variable valence. The resultant O^- ion adjacent to a cation vacancy in the oxides MO (V^- center) undergoes a small relaxation away from the vacancy. This is sufficient for the ESR spectrum to display tetragonal symmetry; however, the lifetime of a hole upon a particular oxygen atom is too short to prevent migration among equivalent positions about the vacancy. This color center is responsible for the purplish appearance of irradiated MgO; the broad absorption peak is centered at 580 nm. In the alkali or alkaline earth halides, the rapid self-trapping of a hole to give X_2^- is a consequence of the binding energy of the order of 1.5 eV. The self-trapped holes or V_K CENTERS (after W. Känzig) can be generated in much increased numbers if there are sufficient Ag^+ or Tl^+ ions to accept the electrons liberated from X^- ions. Subsequent irradiation of the crystal with polarized light in the V_K band (~ 3.5 eV) can lead to preferential orientation; e.g., if the axis of propagation is [001] and the electric vector is along [0$\bar{1}$1], most of the V_K centers in an alkali halide crystal will have their axis along the [011] direction, as may be shown by either ESR or optical dichroism studies. By contrast, the axis of the V_K center in MgF_2 or CaF_2 is along a [001]-type direction; the fluorine ions at either end of the X_2^- unit are attracted inward, as revealed by an extra hyperfine splitting.

Hydrogen is one of the most fascinating of the many impurity color centers.

JOHN E. WERTZ

References

Fowler, W. B., "Physics of Color Centers," Academic Press, New York, 1968.
Seidel, H., and Wold, H. C., *phys. stat. solidi*, **11**, 3 (1965).
Henderson, B., and Wertz, J. E., *"Adv. Phys.,"* **17**, 749 (1968).

Crystal Defects

Outer Appearance. There are no two crystals of the same substance which are exactly alike: they may differ in their dimensions, in the smoothness and extent of their crystalline faces or in their habit, in the degree of deformation, if any, in the extent of twins formed, etc. Thereby crystals of organic substances show a larger degree of variation than those of a certain inorganic, which is related to the dimensions of the structural entities. If the latter are atoms, ions, or small molecules, as in inorganic structures, the crystals of the same substance differ less than those of certain organic compounds, usu-

ally consisting of much larger molecules. However, there is a relation which is common to all crystals of the same element or chemical compound: it is the rule of Nicolaus Steno (1669) about the constancy of angles between the faces of crystals of the same element or compound. Therefore, with the exception of the cubic system, many crystalline substances can be identified from the angles between the faces of single crystals. Nevertheless, precision goniometric measurements always show a departure from the angles supposed to be constant for a certain substance, frequently exceeding the limits of error of the measurements.

This deviation is a consequence of the *imperfect structure* of crystals caused by the presence of lattice imperfections, foreign atoms and stresses. A mosaic structure may be the result of such a condition in the crystal.

Point Defects. A structure (tridimensional) is perfect or sound, if all the positions in the respective lattice, provided for a certain kind of atoms or ions, are occupied, and only by them. For instance, if on all positions of a face-centered lattice there are Al atoms, the structure of the respective Al crystal (in absence of strains) is *perfect,* ideal, or sound. The number of atoms in such a crystal (face-centered) is exactly 4. However, there are probably no such crystals at all. Real crystals, even the purest, always contain admixtures, the atoms of which may be not dissolved (present in clusters), or they may be dissolved or be in equilibrium with the atoms of the parent crystal.

That there must be various kinds of deviations from the perfect structure in real crystals was followed first from the discrepancy between the experimental and the theoretically calculated strength of NaCl. Hence various proposals for the defects, occurring in real crystals, were made in order to explain the properties observed.

The simplest assumption explaining some deviations of a real structure from the ideal is that of the presence of *point defects* in all real crystals. There are two main kinds of such defects: interstitials and vacancies. To them may be added foreign atoms or ions including isotopes, causing disorder or changes in properties. Interstitials by definition are atoms which are not on the regular lattice sites but are squeezed into the lattice (Fig. 1). Depending upon

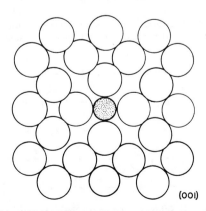

(001)

FIG. 1. Interstitial atoms in a cubic face-centered lattice.

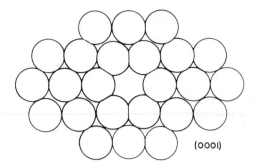

FIG. 2. Vacancies in a hexagonal close-packed lattice.

the size of the interstitial atoms, the resulting lattice will be more or less strained. When atoms are missing on the regular lattice sites, vacant sites remain (Fig. 2). The point defects may be evenly distributed throughout the structure or they may form clusters. Especially in case of vacancies they may assemble to di-, tri- and polyvacancies (holes).

The same kind of defects may also exist in all crystals of chemical compounds, e.g., with ionic constituents. Only here the electroneutrality of the resulting lattice has to be considered. Thus an ion can not be removed or squeezed into the lattice. It can only be displaced to an interstitial position (leaving a vacancy instead). Ions of various valencies but of opposite and equal charge (total sum = 0) can be removed, leaving positive (cationic) and negative (anionic) vacancies behind. The first kind of disorder is called the "Frenkel defect" (Fig. 3) and the second the "Schottky defect" (Fig. 4).

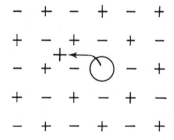

FIG. 3. Frenkel defects involving cationic and anionic (anti-Frenkel) interstitials and vacancies.

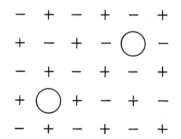

FIG. 4. Schottky defects: Simultaneous anionic and cationic vacancies; or simultaneous cationic and anionic interstitials (anti-Schottky defects).

A further type of disorder also affects the perfection of nonionic crystals: the substitutional or antistructural disorder, where the A and B atoms of a compound AB are misplaced; that is, part of the A atoms may occupy B sites and vice versa. There may also be various combinations of all 3 kinds of defects.

According to the theory developed by Schottky and Wagner the reasons for formation of imperfections in crystals are thermodynamic. From theoretical considerations it follows that crystalline materials with the above mentioned defects show a lower free energy than crystals with ideal structure. The defects are in equilibrium with the lattice in accordance with the law of active mass action and their concentration is relatively small. The concentration of the defects, e.g. the formation of vacant sites, increases with temperature and can be determined experimentally. The concentration of point defects, assuming that a certain kind of disorder is present, can be calculated theoretically.

M. E. STRAUMANIS

References

Mott, N. F., and Gurney, R. W., "Electronic Process in Ionic Crystals," Dover Publ., New York, 1964.

Hauffe, K., "Reactionen in Unst an festen Stoffen," Springer Verlag, New York, 1966.

van Bueren, H. G., "Imperfections in Crystals," Interscience, New York, 1960.

Weertman, J., and Weertman, J. R., "Elementary Dislocation Theory," Macmillan, New York, 1964.

CYTOCHEMISTRY

The microscopic cells of which all plants and animals are composed are essentially similar in that they consist of living protoplasm and, in most instances, a limiting membrane. The study of the variations in form, function and life-histories of many different types of cells has developed into the important branch of biological science known as *cytology,* which deals with individual cells and their intrinsic characteristics. The chemical identifications of the materials which make up the cells constitute that branch of cytological investigation known as *cytochemistry.* In its broadest sense cytochemistry includes the determination of the chemical constitution of all protoplasmic and cell membrane constituents at all stages of development of the cell. These determinations are carried out at suitable magnifications, with the least possible damage to the structural organization of the cell.

The substances identified by the earlier workers in the field of cytochemistry were often those which were either most conspicuous or most abundant. The carbohydrate *starch* was observed to exist in the form of comparatively large grains and to be located in special plant cell organs named *chloroplasts* and *leucoplasts;* found to be a temporary storage product produced from *sugar;* and later reconverted to sugar and translocated, from the cell in which it was formed, either to growing regions of the plant or to organs, such as tubers, where it was again converted to starch for permanent storage. *Proteins,* important constituents of all protoplasm, were found to be of many types and to occur in varying concentrations in the nucleus and other cell organs as well as in different regions of the cytoplasm. Such phenomena involve chemical processes which are not fully understood even today.

Starch grains have been found to contain only 80 to 85% "starch substance"; the additional components consist of such materials as fats, tannin, phosphates, water and even hemicellulose. The main component has been further subdivided into mucilaginous *amylopectin* and *amylose,* a denser, less reactive material. The chemical constituents and physical components of the starch grain lend themselves to cytochemical investigation and starches of many types are being studied intensively. Proteins are formed by the union of *amino acids.* Numerous qualitative cytochemical methods for the identification of specific amino acids or amino acid groupings have been developed, some of the more common of which are the Glycolic Reaction, Voisinet's Reaction, Xanthoproteic Reaction and Biuret Reaction.

The requirements of modern cytochemical research are rigorous. Standards of precision are far beyond those required in earlier cytochemical investigations. Optical principles involved in the formation of images in the microscope must be recognized in greater detail than was required for the microscopical examination of fixed and stained materials; artifacts must be avoided; diffusion of the substance under investigation from its normal location in the living cell must be detected and, if possible, prevented. Freeze-drying and special methods of fixation have contributed to a more normal condition of the cellular materials to be studied. The polarizing microscope, the ultraviolet microscope, the fluorescence microscope, the infrared microscope, the electron microscope and the microspectroscope now supplement the ordinary light microscope in obtaining cytochemical data.

In view of the highly specialized nature of many of these methods of analysis, a cytochemical study is now frequently conducted by a group of highly trained specialists rather than by a single individual. Within the last 25 years, important advances have been made in cellular chemistry by investigators such as Casperson, Linderstrøm-Lang, Lison, Feulgen, Commoner and their associates. Improved methods are now available for the cytochemical determinations of such a wide variety of substances as nucleic acids, pigments, aldehydes, enzymes and bacteriophages.

Recent progress in cytochemical research has been concerned with the development of more precise methods of analysis of the complex biological systems of plant and animal cells. The identifications and localizations of cellular constituents which have been made over many past decades with standard cytochemical methods serve as a background for detailed analyses made possible with improved instrumentation and new biophysical and biochemical techniques. Fluorescence microscopy is now supplemented by quantitative ultraviolet microspectroscopy; infrared microscopy with infrared microspectroscopy; interference microscopes have become generally available and the quality of electron micrographs has been enhanced by improved methods in sample preparation. The scanning electron microscope provides recordings of surface details of cells and tissues of superb quality. Autoradiography, a unique tool for the study of membrane permeability as well as the localization of adsorbed isotopes in the cell protoplasts, is now combined with a scanning device for the microscope which, in turn, can be replaced by a spectrophotometer to obtain microspectrophotometric measurements of the same specimen.

Many of these refinements of, and additions to, existing cytochemical techniques are adaptations from physical principles which are applicable to living cells and tissues without injury. The data obtained fulfill, therefore, a primary objective in cytochemical research—the identification and localization of cellular components without disturbance of the cell organization. Applications of these and other related techniques to the study of biological phenomena long under investigation serve to indicate their contributions to our understanding of the life processes of plants and animals.

Enzyme Activity and X-Ray Diffraction[1]. The mechanism of enzyme activity involving enzyme-substrate relationships continues to hold a permanent place in cytochemical research. The usefulness of x-ray diffraction in these experiments is usually influenced by the degree of three-dimensional disorder within even a crystalline state of the enzyme. The consequent reduction in intensity of the x-ray beam as it passes through the specimen and the low degree of resolution of the diffraction pattern produced are factors which are shared with many proteins. A review of the efforts to determine the nature of enzyme-substrate interaction is covered, in detail, by Blow and Steitz[1]. The low degree of resolution obtained with such enzymes as lysozyme and the value of the information obtained in spite of the limitation serves to indicate the potential value of x-ray diffraction analyses in current and future research with the mechanisms of enzyme activity. The unique features of x-ray diffraction are the detail, precision and certainty with which structure can be determined in favorable cases.

Photosynthesis (Cell-Free)[1]. Cell organs such as chloroplasts, isolated and cultured *in vitro* have led to the new field of investigation known as "cell-free photosynthesis." In the early effort by Hill (1939) to measure the physiological activity of isolated chloroplasts, rapid oxygen evolution was found but no measurable assimiliation of carbon dioxide. Improvements in technique which involved a rapid method of isolating the chloroplasts in a sugar-alcohol/sorbitol medium led to a rate of carbon dioxide assimilation which fell within the range of 80–120 μ moles/mg chlorophyll which is achieved by the average green plant under normal conditions. In isolating the chloroplasts for studies such as this, Bucke and coworkers (1966) found that when the outer membrane ("envelope") of the chloroplast is either broken or lost, leaving exposed the green protoplasmic masses (grana), the characteristic ability to assimilate carbon dioxide and evolve oxygen is lost. The ability of a chloroplast to carry on photosynthesis has been linked, through these same investigations, with the presence of "internal disc membranes." These findings with free chloroplasts have changed the conception of their membrane systems as structural elements and inert barriers to one of dynamic functioning in photosynthesis.

Plastids (self-duplication)[2,3,4]. The hypothesis that a vital function such as photosynthesis with its complex enzyme system may have been segregated during evolution into self-duplicating organelles, such as plastids, is often discussed[2]. The direct division of chloroplasts takes place at an early stage of their development in the living cell protoplasts and shows no relation to mitosis. The colorless amyloplasts which produce storage starch in tubers also increase in number by direct division at an early stage before the conversion of sugar to starch is in evidence.

The formation of cellulose in the chloroplasts of two marine algae, *(Halicystis ovalis and Valonia ventricosa)*, and also in the colorless plastids of the living cotton fiber was described in a series of papers between 1940 and 1950. The cellulose-forming plastids also increase in number at an early stage of development comparable to that of the starch-forming chloroplasts and amyloplasts. The successive stages of cellulose formation in the plastids, the discharge of the microscopically visible particles into the cell cytoplasm and their final deposition in the secondary lamellae of the algal and fiber cell walls, were followed by means of their cytochemical reactions to sulfuric acid and iodine, their double refraction in polarized light and their x-ray diffraction patterns.

WANDA K. FARR

Cross-reference: *Histochemistry*.

References

1. Annual Review of Biochemistry, Vol. 39, 1970.
2. Annual Review of Plant Physiology, Vol. 19, 1968.
3. Farr, W. K., Technical Documentary Report AMRL, TDR, 61-151, 1965, Wright Patterson Air Force Base, Ohio.
4. Paech, K., and Tracey, M. V., Modern Methods in Plant Analysis, Vol. II, 1955. Springer Verlag, Berlin, New York.

D

DEBYE-HÜCKEL THEORY

It was recognized shortly after Arrhenius' discovery of electrolytic dissociation that the observed freezing point lowering of dilute electrolytic solutions is smaller than that expected for an ideal solution of completely dissociated ions. Explanations of this deviation from ideality based on the assumption of the incomplete dissociation of strong electrolytes were unsatisfactory. In 1923 P. J. W. Debye and E. Hückel developed a theory for predicting quantitatively the deviations from ideality of dilute electrolytic solutions. Their basic assumptions are (a) strong electrolytes are completely dissociated into ions in dilute aqueous solutions; (b) all deviations of dilute electrolytic solutions from ideality result from the electrostatic interactions between the ions, *i.e.*, a solution of discharged ions is ideal; (c) the solvent is a structureless dielectric continuum; (d) each ion can be treated as a spherical cavity in the dielectric medium with radius a and there is no space charge inside this cavity. The constant a is often called the distance of closest approach of the two ionic species in solution.

Since like charges repel and unlike charges attract each other, there will be on the average more ions of opposite than of like sign in the vicinity of any ion. Thus in the Debye-Hückel theory every ion is considered to be surrounded by an ion-atmosphere of opposite charge. The electrical potential ψ at a distance r from a given spherical ion α is given by Poisson's equation as

$$\frac{1}{r^2} \frac{d}{dr}\left(r^2 \frac{d\psi}{dr}\right) = -\frac{4\pi}{D}\rho \qquad (1)$$

where D is the dielectric constant of the solvent and ρ is the charge density. But the charge density at any point in solution is given by the Boltzmann distribution as

$$\rho = \sum_i n_i z_i \epsilon \, exp\left(-\frac{z_i \epsilon \psi}{kT}\right), \qquad (2)$$

where n_i is the number of ions of species i per cc, z_i the charge in electronic units, ϵ, of ion i, k, the Boltzmann constant, and T the absolute temperature. Expanding the right-hand side of equation (2) into an exponential series, we find that all the terms following the second can be omitted without appreciable error because for dilute solutions $\dfrac{z_i \epsilon \psi}{kT}$ $\ll 1$. The first term in this expansion, $\sum_i n_i z_i \epsilon$, should also vanish because the solution is electrically neutral. Thus equation (2) reduces to

$$\rho = \frac{\epsilon^2}{kT} \sum_i n_i z_i^2 \psi. \qquad (3)$$

If the value of ρ given by (3) is substituted into equation (1), the latter can be solved for ψ. The appropriate boundary conditions are (a) $\psi = 0$ when $r = \infty$; (b) both ψ and $d\psi/dr$ must be continuous at $r = a$, where a is the distance of closest approach of the other ions in solution to the central ion α. The solution under these boundary conditions is

$$\psi = \frac{z_\alpha \epsilon}{Dr} \frac{exp[\kappa(a - r)]}{1 + \kappa a}, \qquad (4)$$

where $\kappa = \sqrt{\dfrac{4\pi\epsilon^2}{DkT} \displaystyle\sum_i n_i z_i^2}$ and z_α is the charge in electronic units of the central ion α.

At $r = a$, the potential given by (4) becomes

$$\psi = \frac{z_\alpha \epsilon}{Da} \frac{1}{1 + \kappa a} = \psi_0 + \psi_\alpha, \qquad (5)$$

where

$$\psi_0 = \frac{z_\alpha \epsilon}{Da}, \qquad (6)$$

$$\psi_\alpha = -\frac{z_i \epsilon}{D} \frac{\kappa}{1 + \kappa a}. \qquad (7)$$

Here ψ_0 is the self-potential of the ion α, and ψ_α is the potential due to its ion-atmosphere.

The activity coefficient, f_α, of the ion α can thus be found by calculating the additional electrical work required to charge the ion α due to the presence of its ion-atmosphere. Thus

$$kT \ln f_\alpha = \int_0^{z_\alpha \epsilon} \psi_\alpha d(z_\alpha \epsilon). \qquad (8)$$

Substitute (7) in (8) and carry out the integration, we obtain

$$-kT \ln f_\alpha = -\frac{z_\alpha^2 \epsilon^2}{2D} \frac{\kappa}{1 + \kappa a}. \qquad (9)$$

The mean activity coefficients of salts can be readily obtained by combining the corresponding single-ion activity coefficients given by (9).

If we introduce the ionic strength, μ, defined by

$$\mu = \frac{1}{2} \sum_i c_i z_i^2, \qquad (10)$$

equation (9) may be written in the familiar form

$$- \ln f_\alpha = \frac{z_\alpha^2 A \sqrt{\mu}}{1 + B a \sqrt{\mu}} \qquad (11)$$

where A and B are constants, and c_i is the concentration of ionic species i in moles per liter.

Results deduced from the Debye-Hückel theory have been extensively tested and verified by measurements in very dilute solutions of strong electrolytes. Although this theory is valid only for very dilute solutions, it is of great practical importance in providing us with a reliable method of extrapolating to infinite dilution the thermodynamic properties of electrolytic solutions. The Debye-Hückel theory also stimulated the rapid growth of theoretical studies along similar lines in many related fields. Generalized and modified treatments of ion-atmospheres by later workers have produced fruitful results in studying the rate of ionic reactions, polyelectrolytes, protein chemistry, the stability of lyophobic colloids, and various transport processes involving ions such as conductance, diffusion, viscosity, etc.

JUI H. WANG

DELIQUESCENCE

Deliquescence is a phenomenon exhibited by certain soluble solids. These solids become liquid (a solution) by absorbing moisture when exposed to air. Some solids show this effect readily, others not at all. Calcium chloride ($CaCl_2$), ferric chloride ($FeCl_3$), calcium nitrate [$Ca(NO_3)_2$], magnesium chloride ($MgCl_2$), sodium hydroxide (NaOH), and aluminum chloride hexahydrate ($AlCl_3 \cdot 6H_2O$) are among those that do.

Deliquescence is explained as follows: Moisture is absorbed from the air onto the surface of the compound. A saturated solution forms. This solution has a vapor pressure less than the partial pressure of the water vapor in the air. This concentrated solution continues to absorb moisture and to dissolve more solid. The absorption of moisture continues until the solution becomes so dilute that its vapor pressure is in equilibrium with the partial pressure of the water vapor in the air.

Different compounds deliquesce at different rates. Obviously the rate of deliquescence depends on the surface exposed and upon the relative humidity of the atmosphere, deliquescence being more rapid in a moist atmosphere than in a dry one. In the case of sodium hydroxide, carbon dioxide as well as water vapor is absorbed from the air, and a crust of sodium carbonate may be noticed.

Deliquescence is a phenomenon often observed, and because of it many compounds must be kept in airtight containers. Not only does slight deliquescence cause caking of some compounds themselves, but deliquescent impurities in compounds may cause caking. Common salt, chiefly sodium chloride (NaCl), contains impurities of deliquescent magnesium chloride ($MgCl_2$) and calcium chloride ($CaCl_2$) Because these deliquescent impurities are present, common salt does not pour freely, especially in damp weather.

Deliquescence has been used as one means to dry damp cellars. Racks of calcium chloride are exposed to the moist air in the cellar. The solution of calcium chloride that forms is led off to a drain. Moisture is removed from the air by the calcium chloride, which must be renewed from time to time.

ELBERT C. WEAVER

Cross-reference: *Absorbents.*

DESALINATION

Whenever the supply of fresh water has become scarce, man has turned to desalination of salty waters for supplemental supplies. This need was felt first on ships, where centuries ago man learned to supplement his supply with water desalted by boiling sea water and condensing the vapor. This is still the method most widely used at present to desalt water, practiced in increasingly sophisticated forms.

Water was desalted by natural processes long before man felt the need. Using membrane processes by mechanisms which are still under intensive study, fish can extract fresh water for their tissues from salty water ranging from brackish to sea water, and several species of marine birds concentrate salt in the avian salt gland. Important modern methods of water desalting are based on membrane processes, using membranes synthesized by man for this purpose.

On earth, fresh water is continuously generated primarily by the direct energy of the sun, which evaporates immense quantities of water from the oceans to be precipitated in the form of rain. Unfortunately, vast areas receive very little rain, resulting in arid areas and deserts, mainly in the U.S. Southwest, Northern and Southwest Africa, the Middle East, the Australian Southwest and Central Asia.

A growing need for water results from increasing population pressures, industrial needs and pollution of existing sources. Although much emphasis has been placed on water for agriculture, the economics of desalting up to now and for the immediate future favor primarily the municipal and industrial water markets, and special requirements such as military and remote outposts. The United States alone required approximately 500 billion gallons of fresh water in 1970. This is estimated to increase to 1 trillion gallons by the year 2000.

Fresh water contains less than 1000 parts per million (ppm) salts and preferably less than 500 ppm, which is the U.S. Public Health Service maximum limit. Sea water contains on the average 35,000 ppm, the salts consisting of 80% sodium chloride, 8.5% magnesium chloride, the balance made up of calcium, potassium, sulfate and bicarbonate, with many other ions in small quantities. Brackish waters by definition encompass waters less salty than sea water, and are easier to desalt. The salts in brackish waters are usually quite different in composition from sea water, consisting primarily of sodium, calcium, magnesium, chloride, sulfate and bicarbonate, with small amounts of iron, silica and other elements in solution. The proportion of calcium, sulfate, and bicarbonate to other ions in inland brackish waters is usually significantly higher than in sea water.

Sea water as a water supply is so tantalizing because there is so much of it. Approximately 97% of the 1.5 billion cubic kilometers of water in the world is in the ocean, 2% in the glaciers and ice caps, and the remaining 1% in ground water, lakes, rivers and the atmosphere. In desalting circles water is usually measured in gallons or cubic meters (there are a billion cubic meters or 264 billion gallons in a cubic kilometer). The projected U.S. consumption of nearly 4 cubic kilometers per year by the end of the century is literally a drop in the bucket compared to the 10 million cubic kilometers in the world's ground water and the 6,000 cubic kilometers annual precipitation over the country.

Sea water desalting will benefit only coastal areas, because of the prohibitive cost of transporting fresh water uphill from the coast any significant distance. Therefore the importance of brackish ground water and surface sources to inland areas must not be overlooked, since so many of the world's arid areas are inland.

All desalting processes accomplish the separation of water from the salts to yield product water and a brine waste, with recovery of 50 to 80% of the feed water as product. The brine waste or concentrate is usually short of saturation, although sometimes the brine will be supersaturated with respect to calcium sulfate or calcium carbonate. In large desalination plants disposal of the brine is a problem which is now under investigation by the Office of Saline Water (OSW). Ultimately the brine could be concentrated to give solid salts and achieve total recovery of the water, but this is not justified economically except to facilitate disposal of the salts. The market value of salts so produced is questionable, and is undergoing study by the OSW and others.

As in all separation processes, by the laws of thermodynamics there is a minimum amount of energy that must be expended to accomplish the desired separation. Barnett F. Dodge has written on this subject and his article is highly recommended. An understanding of the thermodynamics of desalting, which give the minimum energy required for any process to produce the same product and brine, is essential to avoid pitfalls such as selection of one process over another "because it requires less energy." In any process the energy requirement can be reduced to approach the minimum energy only at the expense of plant size and capital investment. A balance must always be struck between the investment cost and the energy cost to arrive at the minimum total cost. This optimum cost usually lies at energy requirements 10 to 20 times over the minimum thermodynamic energy. The larger the capacity of the plant, the lower the energy requirement per unit of water desalted at the optimum total cost.

The economics of size dictate very large plants, in the 10-100 million gallon per day (MGD) range, to realize the full promise of low-cost water by desalting. According to Dodge, the minimum work of separation to get 50% water yield from sea water is 4.15 KWH per 1000 gallons of product water; however, for distillation processes which use heat as the form of energy to do the work, the second law of thermodynamics indicates 3.94 times the minimum work if the heat is available as 250°F steam. To this approximately 0.4 KWH must be added for pumping. For brackish water the energy requirements are substantially lower, the minimum reversible work is 0.71 KWH per 1000 gal product for a 5000 ppm feed for 50% recovery compared with 4.15 KWH for sea water.

It is evident that brackish waters are less expensive to desalt than sea water, strictly on the basis of the reversible energy required for the separation. Of course, the minimum work increases as the percent water recovery increases, so that, for sea water, at 90% water recovery the minimum work is twice that at 50% water recovery. The implications to the ultimate separation of sea water into its salts and fresh water are self-evident. Fortunately, in the case of sea water, disposal of the brine in the ocean should not pose a serious problem, at least for medium-sized desalting plants. Very large plants may be faced with possible ecological effects caused by the brine discharge.

Existing desalting processes can be classified into distillation, membrane, freezing, ion exchange, and other miscellaneous processes in the research stage. The research and development effort in these processes has been concentrated in the following areas: (1) Achieving more efficient use of energy, as by multiple effects in distillation processes, and the search for more efficient membranes for membrane processes. (2) Overcoming mass transfer limitations, as in turbulence promotion for membrane processes. (3) Overcoming energy transfer limitations, as in vertical falling film evaporator tubes. (4) Controlling insoluble salt scale formation in distillation and membrane processes. (5) Finding and developing better materials of construction and increasing reliability to reduce maintenance costs.

Distillation. All these processes take advantage of the fact that the water vapor evaporated from salt water is essentially free of solids. It can derive its energy from any suitable source of heat, such as fossil fuels, industrial waste heat, nuclear fission or solar energy. Only the first two have been applied commercially, although nuclear fission looms large for future plants. Solar energy is very limited in application because of its diffuse nature, even in the desert.

In addition to designing for maximum economy, distillation processes must cope with scale formation and corrosion. Calcium carbonate, magnesium hydroxide, and calcium sulfate tend to form a hard scale on the heat-transfer surfaces, thus reducing the heat transfer rate. Methods of scale control include addition of acid or surface active agents, pretreatment of the feed water to remove or replace scaling salts with nonscaling salts, and improved methods of operation and improved equipment designed to minimize scale. Corrosion of the evaporator parts, especially the heat transfer surfaces, is reduced or prevented by selection of the proper materials of construction, removal of oxygen from the feed water, pH adjustment, corrosion inhibitors, and avoidance of dissimilar metal couples which encourage galvanic action. Copper-nickel alloys are excellent for this service, but the brine waste becomes contaminated with copper which could in turn contaminate shellfish in the out-fall area. Other alloys and metals such as aluminum, aluminum brass, Armco iron, mild steel and titanium are undergoing testing to determine their suitability as materials of construction for distillation equipment. For other parts of the system concrete polymers are also under study.

Multi-Stage Flash Distillation (MSF) is the most successful and widely used method in use to date for sea water desalting. Sea water is heated under pressure, then flowed into a chamber at reduced pressure to flash off some steam. Multiple stages are used for energy economy in a manner analogous to multiple-effect evaporators, except that more stages are generally required in the MSF process for the same steam economy. Commercial plants exist with a capacity up to 7.5 million gallons per day (MGD) and with as many as 40 stages. The advantages of MSF are better control of scale-deposition and cheaper construction than the boiling type evaporator, and less heat transfer surface area required for the same over-all coefficient of heat transfer. According to the U.S. Office of Saline Water, "a 50 MGD MSF plant can be built with confidence from the data obtained from the MSF module" at San Diego, California. For a plant of 100 MGD, in combination with power generation, projected costs are of the order of 40¢ per 1000 gallons of product water and investment costs are expected to be of the order of $0.60 to $1.00 per gallon per day. For 10 MGD plants the investment costs projected range from $1.00 to $1.50 per gallon per day. Water costs for existing or projected plants in the 2.5 to 7.5 MGD range from 65 to 85¢ per 1,000 gallons, and for 1 MGD are approximately $1 per 1,000 gallons.

Reductions in cost are expected from combination of the MSF process with the vertical tube evaporator distillation process, by reductions in capital investment, and lower operating and maintenance costs.

In *Vertical Tube Evaporation* (VTE) the salt water falls through long vertical metal tubes to exchange heat with steam in multiple effects for steam economy. The OSW has a water desalting test vehicle to provide engineering and operating data on the VTE process. The OSW projects capital investment reductions of the order of 30% and lower water costs by as much as 15–20% for large plants of 250 MGD capacity.

In the MSF, VTE and any distillation process the thermodynamic efficiency and thus steam economy are improved by increasing the operating temperature, at the expense of worsened corrosion and scaling aggressiveness. A high-temperature unit at the OSW Clair Engle Desalting Plant at the San Diego Test Facility is designed to operate up to 350°F to gather data at these higher temperatures.

Vapor Compression Distillation (VC) consists of compressing the vapor evaporated from the salty water to raise its condensation temperature and condensing indirectly against the feed so that the heat of condensation can be used to heat the salty water feed. The net energy supplied to the process is essentitally that required to drive the compressor. Small VC stills have been in use commercially on shipboard for many years. The capacity of VC stills is limited by the size of compressors, therefore attention has been directed by the industry to compressor size. A 1 MGD plant requires a 500,000 cu. ft. per min. compressor. Such a plant is in operation at Roswell, N.M. If feasible, for large

plants operating costs are comparable to MSF and VTE costs, but the investment cost is higher for a VC plant.

Other Distillation. Combinations and variations of the above methods aimed at increasing the heat transfer coefficient, have been investigated and piloted, for example, the Hickmann Still and the General Electric wiped-film evaporator. It is not expected that these will find application in large desalting plants.

Membrane Processes. These depend on the selectivity of synthetic membranes for either salts or water to achieve a separation between the two. In electrodialysis (ED) an electrical potential is used to drive the positive (cation) and negative (anion) salt ions separately through cation-selective and anion-selective membranes (usually made from reinforced ion exchange resin materials) away from the product water and into the brine waste. ED is the oldest membrane desalting process, having been commercialized in the mid-1950's.

In reverse osmosis (RO) the product water is driven under pressure away from the salt water through water selective membranes usually made of cellulose acetate, or similar materials. Its name is derived from the fact that the osmotic pressure must be overcome in the process before any desalted water rate is obtained. RO has been under the most intensive development of any process by the OSW and private U.S. corporations and has recently entered the commercial desalting market.

A third membrane process, still under development, is piezodialysis (PD). In PD a concentrated brine is driven under pressure from the salt water feed through a salt enriching membrane, leaving behind desalted water. The salt-enriching membrane is made in a mosaic structure containing both cation-selective and anion-selective synthetic resin materials. PD holds promise as a membrane process in the range of high salinity brackish waters up to sea water concentration,

Electrodialysis is more suitable to desalting brackish water than sea water, because it removes the salts from the water, and the desalting cost increases with salt concentration. However, research aimed at increasing the range of economic desalting up to sea water is underway. This will be accomplished by using thinner membranes and higher temperatures than are presently used in commercial equipment. Whereas distillation is heat transfer limited, ED is mass transfer or current density limited, and the equipment is designed to achieve high mass transfer rates through hydraulic turbulence promotion. When either brine saturation or the limiting current density is approached, calcium sulfate or calcium carbonate scales can form on the membrane surfaces. This is controlled by proper design, electrical polarity reversal, addition of acid for brine pH control, or pretreatment of the salt water feed. Another problem sometimes encountered is fouling of the membranes by iron, or organic matter. Pretreatment of the water to remove iron by manganese zeolite and carbon filtration to remove organics have been used to pretreat such waters prior to desalting.

The ED brackish water desalting market has been primarily in the small to medium sized plants primarily for the Middle East, and medium sized plants for municipalities in the United States: Coalinga, Cal.; Buckeye, Arizona; Dell City and Port Manfield, Texas; and most recently, a 1.2 MGD plant, expandable to 2 MGD, for Siesta Key, Florida, which cost $450,000 and produces 500 ppm water at a total cost of 30¢ per 1,000 gallons from a feed containing approximately 1500 ppm solids.

The largest ED plant in the world is in Benghazi, Libya, to produce 5.1 MGD of 850 ppm water from a feed containing 2200 ppm solids, at an anticipated cost of 42 cents per 1,000 gallons. The plant cost $2.1 million. Electrodialysis plants in the 2 to 10 MGD capacity range are expected to cost 20 to 40¢ per gallon per day and desalt water at a total cost of 30 to 40¢ per 1,000 gallons.

Reverse Osmosis is the newest of the commercial desalting processes and has been heavily backed by the Office of Saline Water in the last decade. This has resulted in the process reaching the commercial stage in less than 10 years since S. Loeb and S. Sourirajan prepared an asymmetric selective cellulose acetate membrane in 1961. Since, a virtual stampede has occurred involving not only the OSW, but several private concerns to develop the best RO membranes and equipment. Membranes have been made not only from cellulose acetate but other cellulose derivatives as well, and in the form of "hollow-fiber" membranes, from nylon and most recently from asymmetric aromatic polyamides.

The process is limited by the flux rate through the membranes and by mass transfer in the liquid-membrane interface. About 95% of the salts are removed by passage of the water through the RO membrane under a pressure differential of several hundred psi for brackish waters and about 1500 to 2000 psi for sea water. The process has been most effective for brackish water and much developmental work is in progress for sea water. These is such a proliferation of designs and membranes that these are available in flat sheets, tubes, spiral-wound modules, and bundles of hollow fibers which pack the most membrane surface area per unit volume of equipment.

It is usually necessary to pretreat the water and adjust the pH of the water to prevent scaling and deterioration of membrane properties (fouling). Pretreatment for iron removal and activated carbon required with present membranes may cost from 2 to 10¢ per 1,000 gallons.

The latest RO plants to enter the market are hollow fiber aromatic polyamide, brackish-water desalting units: one for Greenfield, Iowa, rated at 150,000 gallons per day, cost $85,000 and is expected to operate at a total cost of 45¢ per 1,000 gallons (not clear whether membrane replacement is included) on a 1700 ppm feed water; the other is a 300,000 GPD unit for the Bahamas, which cost $300,000, to desalt a feed water containing approximately 10,000 ppm solids.

The OSW is building a portable spiral-wound desalting unit rated at 250,000 gallons per day. Total water costs of 40–60¢ per 1,000 gallons have been predicted for 2 to 10 MGD RO plants at investment costs comparable to ED plants.

Freezing or Crystallization. Another desalting method worth mentioning is the freezing process which depends on the fact that the ice formed by freezing salt water is in theory free of salt. In fact brine is trapped between the ice crystals and much of the research and development effort in the freezing process is concerned with washing of the ice crystals. In the refrigeration cycle the ice is used to cool the compressed refrigerant for energy economy. Several variations of the freezing process exist: indirect freezing, direct freezing to eliminate the heat transfer area, vacuum freezing which combines evaporation and freezing, and the hydrate process which combines a hydrating agent such as propane with the water. Freezing as a desalting process has not progressed as fast as distillation in the United States. The OSW 200,000 GPD direct freezing process prototype, and the 10,000 GPD secondary refrigerant pilot plant at Wrightsville Beach, N.C. ran into problems and were dismantled in 1969. A commercial 100,000 GPD vacuum freeze vapor compression unit in St. Croix, V.I., has had problems of low compressor efficiency and carry-over of brine into the product.

It was recently announced that at Ipswich, England, the first large scale freezing plant is scheduled to begin desalting 1 million gallons per day in 1973. *n*-Butane will be used as a direct refrigerant to yield 100 ppm water containing 0.1 ppm residual butane starting with 2 million gallons of sea water. The plant is projected to cost over $2 million and desalt water for 48 cents per kilogallon.

Ion Exchange. Special ion-exchange resins have been developed recently, which reduce the chemical regeneration cost and open the possibility of economic brackish water desalting by increasing the economic range for ion exchange over the 1,000 ppm levels. The most promising new IE system is based on weakly ionized resins operating on a bicarbonate cycle.

Future of Desalting. There is no question that rapid progress will continue to be made in desalination, or that the world-wide market for such equipment will continue to grow. In 1968 the world desalting capacity was approximately 250 MGD, by 1975 it is expected to quadruple to 1,000 MGD. Both the OSW and private companies will continue to improve existing processes, and universities and other research institutions will continue to do research on the problem and on the fundamental properties of water and salt water. The OSW budget for fiscal year 1972 is nearly $27 million, divided as follows:

	$ Millions
Research and development	15.7
Test beds and facilities	7.4
Modules	1.4
Administration and coordination	2.5

Within this budget there is a significant increase from $500,000 to 1 million since FY1971 for environmental research, indicating increasing concern with ecological effects of brine effluents from desalting plants.

Much work will be done to improve existing processes and reduce costs further. In distillation, combinations of processes such as MSF/VTE, new materials of construction, higher operating temperatures and combination of large distillation plants with nuclear plants are foreseen.

Improvements in membrane processes will be accomplished by development of better membranes and operation at higher temperatures, up to 180°F. In RO, development efforts to make membranes that can desalt sea water in one pass will continue. Piezodialysis will emerge as the newest membrane process once the difficult task of making suitable membranes is solved.

When some of the problems in the freezing processes are solved we will also see a resurgence of these processes.

Esoteric desalting processes may enter the picture, such as polymer solvent extraction, an old idea; or the new and controversial "polymer" form of water, which is already being studied for its possible role in desalting. An ice "sandwich" which functions as a semipermeable membrane and an "air gap" membrane which does the same have been reported.

Interest has developed in geothermal water in the U.S., specifically in the Imperial Valley of California. There will probably be efforts to take advantage of the energy content of this water supply for desalting.

Finally, there are other applications for desalting processes which are worthy of mention because they apply to the total picture of water recycle and reclamation from industrial waste, sewage treatment, or acid drainage from coal mines. Many of these are under study and at least one has moved to commercialization. A 5 MGD ($14.2 million) acid mine water distillation desalting plant will be built at Wilkes-Barre, Pa., to reclaim water contaminated with iron sulfate by coal mining.

EDGARDO J. PARSI

References

"Saline Water Conversion Report, 1969–70," Office of Saline Water, U.S. Department of the Interior, U.S. Government Printing Office, Washington, D.C. 20402.

"Saline Water Conversion—II," "Advances in Chemistry Series," American Chemical Society, 1155 Sixteenth Street, N.W., Washington, D.C. 20036, 1963.

Popkin, Roy, "Desalination, Water for the World's Future," Frederick A. Praeger, New York, 1968.

"Manual of Water Quality and Treatment," American Waterworks Association, McGraw-Hill (publication imminent).

"Desalting Digest," Scope Publications, Inc., 1120 National Press Building, Washington, D.C.

"Water Desalination Report," Richard Arlin Smith, Publisher-Editor, 1732 Church Street, N.W., Washington, D.C. 20036.

"Water—1968," American Institute of Chemical Engineers, Lawrence K. Cecil, Editor, Vol. 64, 1968.

DETERGENTS, SYNTHETIC

Broadly defined, detergents are all substances that help remove dirt. Most important are those, like soap, which are water-soluble and greatly improve the cleaning ability of water. Synthetic detergents, sometimes called syndets or surfactants are generally considered as the man-made products, other than soaps, intended to combine with water to wash out dirt. After the introduction of syndets in the U.S. in the early 30's, their consumption grew so that in 1950 it was about a billion pounds and by 1970 had expanded to over 5 billion pounds, involving various product types and formulations for the many cleaning operations in the home and in industry. Synthetic detergents, thus, have a place among the most successful industrial developments of modern chemical technology, and command about 85% of the total soap and syndet market.

All household and some industrial syndets are made up of several components including one or more organic surface-active agents. Some other industrial detergents may be composed solely of one or more surface-active agents. The surface-active agent or surfactant is the essential and most important component. Surfactants have the ability to concentrate at interfaces and to reduce to varying degrees the surface tension of water. (See **Soaps**).

In the technical literature and elsewhere the words "detergent," "synthetic detergent," and "surfactant" are frequently used interchangeably. However, greater clarity results if "detergent" and "synthetic detergent" are applied only to complete cleaning products.

Household cleaning and washing products, either soaps or detergents, which contain significant amounts of surfactants are classed as (1) general-purpose or heavy-duty and (2) as light-duty. Products in which soap is the sole or predominant surface-active agent are considered soap products and not synthetic detergents. Synthetic detergents generally contain only surfactants other than soap, although soap is used in small quantities in some products. Both heavy- and light-duty types may be either liquid or solid. The solid or spray-dried granule type is by far of greatest importance in tonnage consumed. "Light-duty" detergents are intended primarily for hand dishwashing and fine fabrics, and like soaps are high sudsers. Heavy-duty or general-purpose detergents designed for the family wash, dishes, and other general cleaning chores, are of major importance. The "low-sudsers" both liquid and granule designed especially for certain automatic washing machines are members of the heavy-duty category. Other types of washing and cleaning products which utilize synthetic surface-active agents are toilet-bars, hard surface cleaners, shampoos, and products for various industrial purposes.

Typical all-purpose or heavy-duty synthetic detergent formulations of the granule type would contain not only surface-active agents but in addition, detergency builders to enhance cleaning ability, the most important of which are the moderately alkaline inorganic polyphosphate salts; fluorescent whitening agents (FWA) present in minute amounts, which by depositing on the fabric contribute to the appearance and whiteness; antiredeposition agents such as carboxymethyl cellulose to hold the dirt in suspension in the wash water after it is removed from the fabric; organic compounds to improve, stabilize or suppress sudsing; agents to inhibit tarnish or corrosion of metal surfaces, such as silicates; perfumes and colors. Also present in some products are chemical bleaching agents of oxygen or chlorine types to aid whiteness maintenance and stain removal; antibacterial agents for germ control; and bluing to further contribute to appearance.

In 1970, tonnage of synthetic detergents was over 5 billion pounds, and surfactant tonnage (active ingredient basis) over 4 billion pounds. A large portion of the surfactant tonnage includes food emulsifiers such as glycerol monostearates, lignin sulfonates, and other materials not used at all or used in insignificant amounts in synthetic detergents. Also, some surfactants are used for both detergent and nondetergent purposes. Commercial surfactants are usually mixtures of closely related isomers and homologs. Trade estimates usually assume that detergent surfactants represent about half of the total surfactant tonnage.

Early man-made surfactants developed in Europe during the 1914–18 war and used almost exclusively in the textile industry were anionic. Also anionic are the sodium alkyl sulfates which became available in the early 30's and which were prepared by the sulfation of higher alcohols derived from coconut oil. Their improved washing and cleaning ability stimulated the marketing in the U.S. of products such as light-duty household synthetic detergent granules, shampoos, liquid dentrifices and industrial synthetic detergents. The alkyl sulfates, including those from tallow fatty alcohols, continue to make up an important part of the surfactants going into synthetic detergents. The light-duty synthetic detergents did not wash as clean as the heavy-duty soap powders whose hard water performance still left room for improvement. The key to the invention of the heavy-duty spray-dried synthetic detergent granules was the basic development of new polyphosphate builders which gave unique cleaning ability when combined with detergent surfactants. Heavy-duty granules introduced in 1946 became the most important single type of synthetic detergent and remain so today.

The next advance in detergent surfactants were the alkyl benzene sulfonates. These anionic surfactants are members of a broad family of alkyl aryl sulfonates which combine one molecule of an alkyl side chain, an aryl ring structure and a sulfonate group. In the manufacture of most of the tonnage of alkyl benzene sulfonates, alkyl benzenes, known as "alkylates" are produced by the chemical and petrochemical industries and are sulfonated by the manufacturer of the finished syndet products. The first commercial alkyl benzene sulfonate, introduced in the middle 30's, was derived from keryl benzene alkylate. By the early 50's, the keryl chain of the alkylates had lost its market to branched chain hydrocarbons in the C-10 to C-15 range derived from tetrapropylene. Other closely related variations in the polypropylene hydrocarbon chain have been utilized commercially. The selected hydrocarbon chain is catalytically combined with benzene to produce the alkylate which is sulfonated to yield tetrapropylene (or polypropylene) benzene sulfonates called "TPBS," or commonly "ABS." Control of manufacturing conditions gives highly reproducible products. Because the water-repelling portions of the alkyl benzene sulfonates were petroleum derivatives independent of fats and oils (unlike the alkyl sulfates) they could be produced at lower costs. Capitalizing on this low cost and also on good cleaning ability, ABS became the most important detergent-surfactant in household washing and cleaning products, representing in the early 60's about 70% of the tonnage used.

ABS was replaced with alkyl benzenes made from linear alkylates to solve the biodegradability problem discussed below. In order to eliminate the tetrapropylene-based branched-chain alkylate, supplies of straight-chain alkylates of the proper molecular weight had to be developed. Linear alkylate is obtained from two primary sources: paraffin and ethylene. The paraffin is processed either by (1) thermal cracking to produce primarily alpha-olefin or by (2) dehydrogenation or chlorination followed by dehydrochlorination of paraffin feed stock. The desired molecular weight paraffin is made possible by the use of molecular seive technology. Linear alkylate from ethylene through the ethylene build-up process provides alpha-olefin. The resulting olefins, or in some instances the linear paraffin, are then used to alkylate benzene catalytically to produce a linear alkyl benzene (LAB). The LAB is sulfonated by the detergent manufacturer to yield linear alkyl sulfonate (LAS) which remains the most important surfactant today. With LAS first in importance and alkyl sulfates second, anionics have remained the most important class of surfactants used in household syndets.

Tonnage of the nonionic surfactants used in detergents is relatively small compared with anionics. Nondetersive uses of nonionic surfactants are greater than detergent uses and tonnage for industrial syndets is greater than for household syndets. On the other hand, nonionic tonnage has grown rapidly and its growth should increase. As their name implies, nonionic surfactants do not ionize in solution. Many nonionics are characterized by a long polyoxyethylene (ethoxylate) chain which accounts for their solubility in water. The most important commercial examples of nonionics used as detergent surfactants are the ethoxylates of alkyl phenols and the ethoxylates of primary and secondary fatty alcohols. By sulfating nonionic ethoxylates, the corresponding anionics are produced. The low sudsing character of most nonionics at concentrations normally used for washing and cleaning purposes made possible the "low sudsing" syndet category.

Biodegradability. Because of the increasing importance of water pollution control and abatement, the concept of biodegradability was added as a new requirement for surface-active agents during the decade or so ending in 1965. Biodegradable surfactants are those which are decomposed easily or readily by the biochemical processes typical of sewage treatment and surface streams. Microorganisms can destroy most, if not all, organic compounds, utilizing them as food to provide energy and growth requirements. However, the rate and completeness of destruction will vary. A yardstick considered reasonable by sanitary engineers is that detergent surfactants are "biodegradable" if they are as readily decomposable as the soluble organic matter of sewage.

The impact of the biodegradability development had its greatest effect on the polypropylene benzene sulfonates because this surfactant was used more extensively than any other and also was one of the substances in wastes which is relatively resistant to destruction by microorganisms. Furthermore, residues of TPBS found in surface waters (generally less than 0.5 ml/l) and in wastes were one of the factors involved in foams on streams and on some sewage treatment plants. On the other hand, detergent surfactant residues are not hazardous. They are permitted by the Public Health Service Drinking Water Standard with a limit of 0.5 mg/l of ABS set for aesthetic rather than health reasons. The branched character of the alkyl chain in polypropylene benzene sulfonate retards its destruction by microbial action. In studying biodegradability, the detergent industry found that alkyl benzene sulfonates made from straight chain (linear) alkylates (LAS) were easily destroyed and that they were excellent replacements for polypropylene benzene sulfonates from the viewpoint of performance and costs to the consumer.

As discussed above, the introduction of LAS provided the answer to the major surfactant biodegradability question. The detergent industry voluntarily replaced branched-chain alkylates in its production of syndets with straight-chain or linear alkylates. Ethoxylated primary and secondary alcohols are also biodegradable, and their use has expanded greatly. By the mid 60's it had become a voluntary prerequisite for the detergent industry that raw materials be biodegradable.

The detergent industry faces an even more difficult problem in the coming decade. The question of the role of phosphorus in the eutrophication, or overfertilization, of lakes has brought detergent phosphates into prominence. Despite continuing technical debate as to the role of phosphorus, it has become generally accepted that a reduction in detergent phosphates is desirable. The final answer to the problem will require removal from the environment of all man-made nutrients by proper sewage treatment and improvements in the management of run-off from agricultural lands. However, the possibility that a large source of phosphorus entering the nation's waters could be controlled by legislation generated nationwide pressure to have the phosphates in detergents reduced and ultimately replaced. In 1970 approximately two billion pounds of phosphate were required in detergent products. Alternates to phosphates were being sought by the entire industry and its major suppliers. Synthetic detergents require a builder to provide effective and economical fabric and surface cleaning.

In an effort to develop builder substitutes the detergent industry has evaluated several hundred materials. Possible substitutes must be safe for humans and safe for the environment. The first of these to gain significant acceptance was nitrilotriacetic acid (NTA) in 1970. The introduction of NTA into many detergent products was preceded by human and environmental testing that had no precedent among new materials of this type. In spite of this work additional technical questions have resulted in suspension of NTA usage in detergents in the United States (use continues in Canada and Scandinavia). Because of the massive quantities involved and ubiquitous nature of these products, this kind of thoroughness is essential. Many other materials, in addition to NTA, are being examined, so far without conclusive results.

Recent research in sewage treatment processes has yielded economical processes for removal of phosphorus from human and food wastes, as well as detergent sources in municipal sewage. These processes involve precipitation of phosphates by using lime, alum, iron compounds and other materials. Such processes will have to be used in many areas whether or not detergent phosphates are present. The growth in population alone will require improvements in waste treatment. The detergent industry includes compatibility with waste treatment systems as one of the technical criteria new materials must meet.

MARTIN CANNON

Cross-references: *Interfaces; Surface Chemistry; Soaps; Waste Treatment; Cationic Agents.*

DEVELOPERS

Development is a key to photography; it tremendously magnifies the effect of exposure, producing literally millions of atoms of silver for each quantum of light absorbed. It is a complex catalyzed heterogeneous reaction between a solution of a selective reducing agent and "grains" or crystals of silver halide (mostly the bromide) suspended in gelatin, or more rarely, other colloids. The catalyst in this reaction is the latent image which is believed to consist of a "concentration speck" of silver atoms formed by the action of light upon the grain. The visible image, following development, comprises black, finely divided silver which has been produced wherever the emulsion was exposed to light. It is made permanent or "fixed" by dissolving out the unreduced silver halide with sodium or ammonium thiosulfate.

In addition to (1) the selective reducing agent or "developing agent," the photographic developer usually contains (2) an accelerator or alkali to activate the agent, (3) a preservative to prevent atmospheric oxidation, and (4) a restrainer to increase its ability to differentiate between exposed and unexposed grains. It may also contain sequestering agents to keep the solution clear, wetting agents, bactericides, etc.

The Developing Agent. The most important class of developing agents consists of benzene nuclei substituted by two or three hydroxy or amino groups. Hydroquinone is a particularly important developer, being used to some extent alone but much more commonly in combination with N-methyl-*p*-aminophenol ("Metol"). The latter is sometimes used by itself in low contrast and fine grain developers (with considerable sulfite). This dual developing agent system (hydroquinone plus N-methyl-*p*-aminophenol) produces a large range of usable contrasts with good emulsion speed and clear differentiation from unexposed grains (fog). Catechol finds use as a tanning developer, since its oxidation product hardens the gelatin.

Diethyl-*p*-phenylenediamine is a very important developer in the color field, since its oxidation product readily couples with a large number of phenols and reactive methylene compounds to form indophenol and indoaniline dyes. This is the basis of most of the current color processes, which nearly all employ dialkyl *p*-phenylenediamines of this type. The ethyl group may be further substituted, as for example by hydroxy or sulfonamido to increase solubility and decrease toxicity.

The amino and hydroxy groups may be incorporated as part of a heterocyclic nucleus. The 4-amino-5-pyrazolones are of this type. 1-Phenyl-3-pyrazolidone, "Phenidone," has a rather different mechanism of oxidation. It is very useful when combined with hydroquinone and some other developing agents, forming superadditive mixtures. That is, more silver is developed by the combination of agents than would be developed by either alone. These developers now have considerable commercial importance.

The reaction of photographic development is typified by that of hydroquinone, which may be summarized as:

$$C_6H_4(OH)_2 + 2AgBr \rightarrow$$
$$C_6H_4O_2 + 2Ag + 2H^+ + 2Br^- \quad (1)$$

In the absence of sulfite, the quinone may react to tan the gelatin or to form colored humic acids, but in the presence of sulfite it forms hydroquinone-monosulfonate:

$$C_6H_4O_2 + Na_2SO_3 + H_2O \rightarrow$$
$$C_6H_3(OH)_2SO_3Na + NaOH \quad (2)$$

Another reaction of practical importance is the atmospheric oxidation of the developing agent:

$$C_6H_4(OH)_2 + \tfrac{1}{2}O_2 \rightarrow C_6H_4O_2 + H_2O \quad (3)$$

Development is controlled to considerable extent by the rate of diffusion of the solution into the gelatin layer, which will be increased by agitation. The reaction requires considerable time, practical development taking usually a few minutes, although it may be occasionally lengthened to an hour or reduced to a fraction of a second. Developing rate is increased by making the solution more concentrated and highly alkaline, and by developing at elevated temperature. As development progresses, the silver image becomes more and more dense and the contrast (difference in density between regions receiving much exposure and little exposure) of most emulsions, is also increased. However, if development is prolonged excessively, the unexposed grains are also reduced, producing fog and decreasing contrast.

In the diffusion transfer process, the developing agent, along with silver thiosulfate complexes formed by dissolution of the silver halide grains of the emulsion, diffuse into a receiving sheet. The developing agent then acts to reduce the silver thiosulfate complex, depositing metallic silver on nuclei which have been placed in the receiver. In a color diffusion transfer process the developing agent is itself a dye; for example a hydroquinone molecule linked to an azo dye. When the hydroquinone is oxidized by developing silver, the molecule becomes insoluble and only the unoxidized dye is transferred to the print, to form the colored image.

The Accelerator. The common organic developing agents reduce at reasonable rate only as the solution is made alkaline, and the rate increases with pH through the range of about 8 to 12. This is due principally to ionization of the phenolic agents to the active form, and possibly to the removal of the oxidation products of the amino agents. The magnitude of the rate increase varies with the agent but, for MQ developers functioning in the pH range of 9 to 12, the rate is increased about 70% for each 1.0 unit increase in pH.

Theoretically a wide choice of alkalies is possible, but practical considerations greatly limit the selection. The primary function of the alkali is to establish the degree of alkalinity or pH. However, the total alkalinity or buffering capacity is also very important, since the acid generated in reaction (1) will otherwise rapidly lower the pH.

Accelerating action is also obtained by the addition of quaternary ammonium salts, polyglycols, thallous salts, and thioureas under special conditions.

The Preservative. Alkaline solutions of the developing agents oxidize extremely rapidly on exposure to the air, darkening and losing activity. Sodium sulfite is added to reduce this oxidation, and is so effective that high-sulfite developers may be kept for periods of weeks under conditions of mild exposure to the air. This action is mutual, the sulfite preserving the developing agent and the developing agent preserving the sulfite.

The sulfite concentration employed varies markedly. Print developers, which are usually used for a single day, will contain only 10 to 20 grams of sodium sulfite per liter of working strength solution. Negative developers usually contain 30 to 50 grams of sodium sulfite, and fine grain developers as much as 100 grams per liter. In this case the sulfite is used for secondary effects, especially for its solvent action on silver bromide, which tends to reduce graininess.

The colored quinones formed by atmospheric oxidation of the developer react with the sulfite present to form colorless hydroquinone monosulfonate, according to equation (2). This greatly reduces the hue of the oxidized developer, which in any case is only a secondary indication of oxidation. Another effect of oxidation is the generation of alkali, which may cause a high sulfite-low pH developer to actually increase in activity on partial oxidation. Sodium sulfiite is a mild alkali and is the only alkali employed in some fine grain developers and amidol paper developers.

The Restrainer. Reduction of silver halide grains which have not been exposed to light is known as the formation of "fog." The developing agents employed are chosen to be as selective as possible, but their selectivity may be enhanced by adding certain antifogging agents or restrainers. Potassium bromide is the most important of these restrainers. Certain organic antifoggants, particularly the benzotriazoles, are finding extensive use for specialized problems. These may be used alone, but more commonly in addition to the bromide. The concentration employed is much lower, in the order of 0.01 to 0.1 gram per liter. These also affect image tone, tending to produce blue-black hues. Many other heterocyclic antifoggants have been employed, particularly in emulsions. Of these 5-phenyl-1-mercaptotetrazole is probably the most important.

Miscellaneous Additives. Calcium sequestering agents, including principally polyphosphates such as sodium hexametaphosphate, are added to developers to prevent precipitation of calcium sulfite and carbonate as sludge, scum or scale The calcium is introduced from the water supply and from the emulsion. Silver halide solvents are occasionally added to reduce graininess. Potassium thiocyanate is the most important of these. Sodium sulfate is added in large quantities to prevent the swelling of the gelatin emulsion when processing at high temperatures.

Bactericides are sometimes added to prevent the putrefaction of the gelatin in used developers and the formation of sulfides. Resorcinol, aminoacridine, and pentachlorophenol have been mentioned for this purpose.

Wetting agents are occasionally added to developers but it is difficult to demonstrate that these have any effect on either rate or uniformity.

Additives are often made to control image tone, particularly in the diffusion transfer process. These are closely related to the organic antifoggants, and are most commonly mercapto-substituted heterocyclic compounds.

Emulsion Incorporation. Rather than employing the developing agent in its own bath it is incorporated in the silver halide emulsion in some photographic papers. It is activated when the paper is immersed in an alkaline bath. Alternatively activation may be induced by heating.

R. W. HENN

Cross-references: *Emulsions, Silver, Photography.*

References

1. Mees, C. E. K., and James, T. H., "The Theory of the Photographic Process, 3rd Edition, MacMillan, 1966, Chapters 13–17.
2. Mason, L. F. A., "Photographic Processing Chemistry," Focal Press Limited, 1966.

DIAZOALKANES

The diazoalkanes comprise the class of compounds having the general structure R_2CN_2 where R may be hydrogen, alkyl, aryl, acyl, carboalkoxy, etc. Diazoalkanes are of interest from a theoretical standpoint because of their linear structures (i.e.,

$$\overset{-}{R_2C}-\overset{+}{N}\equiv N \leftrightarrow \overset{+}{R_2C}-\overset{-}{N}=N \leftrightarrow \overset{+}{R_2C}=\overset{-}{N}=N)$$

and from the practical standpoint because of their utility as synthesis reagents. The aliphatic diazo compounds (diazoalkanes) are quite different in chemical properties from the aromatic diazo compounds (known principally in the form of diazonium salts).

The lower molecular weight diazoalkanes are gases at room temperature; the higher molecular weight members are liquids or solids. All are colored, ranging from pale yellow to deep purple, and many are explosive under conditions of rapid heating. The parent member of the series is *diazomethane* which is a yellow gas boiling at $-23°C$ and melting at $-145°C$. In the vapor state it is highly explosive, although in solution it can be handled without undue difficulty. In spite of its drawbacks in being explosive and toxic it is a widely used reagent.

Methylation of Compounds Containing Active Hydrogens. Acidic hydrogens are replaced by methyl groups upon reaction with diazomethane, the facility of the replacement being directly related to the acidity of the hydrogen. Thus, carboxylic acids are rapidly and quantitatively converted to methyl esters, this providing a particularly useful method when the acid is sterically hindered. Although phenols are less reactive they, also, usually provide a quantitative yield of product, *i.e.*, the methyl ether, upon treatment with diazomethane. Only very weakly acidic phenols, such as those containing several alkoxyl groups, may fail to undergo the reaction smoothly. Other diazoalkanes react in a similar manner, although the yields may fail below those obtained with diazomethane.

Reaction with Unsaturated Compounds. Diazoalkanes are reactive toward most types of unsaturation. Diazomethane, for instance, interacts with carbon-carbon and carbon-oxygen unsaturation in the following ways:

(1) *Carbon-carbon unsaturation.* Diazomethane reacts with isolated double and triple bonds to form pyrazolines and pyrazoles, respectively. The reaction occurs much more readily, however, if the carbon-carbon unsaturation is in conjugation with another center of unsaturation as, for instance, in $CH_3CH=CHCO_2C_2H_5$. The pyrazolines are, themselves, useful synthesis intermediates, for they lose nitrogen upon heating and form cyclopropane derivatives, olefins, or mixtures thereof.

(2) *Carbon-oxygen unsaturation.* Sufficiently reactive carbonyl groups such as those present in most aldehydes, ketones, and acid chlorides will react with diazomethane. Aldehydes and ketones generally produce a mixture of the oxide and the homologous carbonyl compound, the relative proportion depending on the structure of the carbonyl compound and the reaction conditions. Cyclic ketones react in a comparable fashion, the carbonyl products in this instance being ring-expanded compounds. Catalysis of these reactions with alcohols and Lewis acids has been found to be effective. Acid chlorides react in another fashion, yielding diazoketones when an excess of diazomethane is used. The diazoketones are especially useful synthesis intermediates; they may undergo thermal decomposition to cyclopropane derivatives, react with halogens to form α,β-dihaloketones, react with halogen acids to form α-monohaloketones, react with organic acids to form esters of α-ketols, react with water to form the ketol itself, and undergo rearrangement in the presence of certain catalysts to yield the next higher homologous acid or its derivative. This last reaction provides one of the best methods for increasing the length of an organic acid by one methylene group and is usually referred to as the Arndt-Eistert synthesis.

Decomposition to carbenes. When heated, irradiated, or treated with catalysts such as copper and copper salts, diazoalkanes lose nitrogen and form divalent species called carbenes, i.e., R_2C:. Carbenes are exceedingly reactive entities and are incapable of being trapped and studied except under very special conditions. They add to double and triple bonds to give cyclopropanes and cyclopropenes, respectively, and they add to benzenes to give the ring expanded cycloheptatrienes. They are even reactive enough to insert into C—H bonds. For example, irradiation of a sample of diazomethane in cyclohexane yields a mixture of bicyclo[4.1.0] heptane (i.e., the cyclopropane compound from addition) along with the C—H insertion products 1-methylcyclohexene, 3-methylcyclohexene, and 4-methylcyclohexene. Many other types of bonds are also susceptible to insertion by divalent carbon generated from diazoalkanes.

Other types of reactions. Diazomethane reacts with Grignard reagents to form substituted hydrazines, with triphenylmethyl to form hexaphenylpropane, with halogens to form methylene dihalides, and with halogen acids to form methyl halides. Alkyl and aryl-substituted diazomethanes react, in most instances, like diazomethane itself. Acyl derivatives (i.e., diazoketones), however, may show such reduced reactivity in certain types of reactions as to appear qualitatively different. These several types of diazoalkanes have been arranged in the following order with respect to increasing thermal stability and decreasing reactivity toward acid: R_2CN_2, $ArCHN_2$, $(RCO)_2CN_2$, $(ArCO)_2CN_2$.

C. D. GUTSCHE

DIAZOTIZATION

Diazotization can be defined as the reaction of a primary aromatic amine with nitrous acid in the presence of excess acid to produce a diazonium compound. This is one of the oldest known reactions in organic chemistry having been discovered in 1858 by Peter Griess and for which he was granted German Patent No. 3224. Few chemical reactions can be relied upon with more certainty and few have more practical applicability, for the high reactivity of the resulting diazonium compound is the basis of most colors manufactured in the dyestuff industry.

The chemical reaction may be generalized as:

$$Ar—NH_2 + HX + HNO_2 \rightarrow$$
$$Ar—N{=}N—X + 2H_2O$$
diazonium salt

Ar is a mononuclear or polynuclear aromatic radical.

HX is a strong monobasic acid.

There are several methods of producing diazonium compounds. Factors which might influence the choice of one method over another would include: the solubility and basicity of the arylamine to be diazotized; whether or not the diazonium compound is to be isolated before further reaction; the source from which the nitrous acid is to be generated. Diazotization can be performed in aqueous solutions, mineral acid suspensions and solutions or in organic solvents.

By far the most frequently used method, termed the "Direct Method," of diazotization involves the *in situ* generation of nitrous acid by the action of a mineral acid, usually hydrochloric acid, on sodium nitrite. The nitrous acid immediately attacks the soluble arylamine salt.

$$Ar—NH_2 + 2HX + NaNO_2 \rightarrow$$
$$Ar—N{=}N—X + NaX + 2H_2O$$

An excess of acid is required, since one mole is required to form the soluble amine salt, one mole to produce the nitrous acid, with an excess needed to maintain the necessary overall acidity. The reaction should proceed as rapidly as possible, within the necessary temperature limits, usually around 0–10°C. Too slow a diazotization increases the chances of undesired side reactions, in particular the formation of diazo amino compounds:

$$Ar—N{=}N—X + ArNH_2 \rightarrow$$
$$Ar—N{=}N—NH—Ar + HX$$

Since most diazonium compounds are unstable at room temperature, and their isolation being not required, this "Direct Method" is usually used. The method is widely used with strongly basic amines and sulfonated amines that have solubility in aqueous mineral acid.

If the arylamine is not soluble in aqueous acids because of its weak basicity, only highly concentrated or anhydrous acids will effect diazotization. Strong sulfuric acid is usually used to dissolve the amine with the nitrous acid being introduced either as powdered sodium nitrite or a nearly saturated solution of nitrosyl sulfuric acid.

Another method for diazotizing weakly basic amines, as well as monoamino sulfonic acids and monoamino carboxylic acids that are insoluble in water is the "Indirect Method." Here the weakly basic amine is suspended in H_2O or the amino acid is dissolved in an alkaline solution. To either of these sodium nitrite solution is added. This mixture is then run into cold acid solution.

Once the diazonium compound has been generated, there are three general pathways further reactions may follow. The first involves reactions in which the diazonium group remains functionally intact. This would include the formation of diazonium salts; triazenes or diazoamino compounds; metallic diazotates; diazosulfonates; diazo oxides.

Many diazonium salts are dangerous to handle in their dry, unadulterated state because of their tendency to decompose, sometimes explosively. This is especially true of diazos containing nitro groups such as *o*-nitraniline and 2,4-dinitro-6-chloroaniline. However, it is possible to decrease their instability by converting them to $ZnCl_2$ complexes or diazonium aryl sulfonates which can then be precipitated out of solution.

The diazoamino compounds are very useful in the developed dye field. Certain diazoaminos, depending upon their structure, are easily split to the original diazo by acetic acid fumes. They can therefore be used in admixture with naphthols to generate a dyestuff directly in the fiber. These are the so-called Rapidogen® colors which are very popular for printing. The diazoaminos are also capable of rearrangement, thus offering a route to the important amidoazo compounds, i.e.,

as catalyst

The second pathway involves reactions in which nitrogen remains in the molecule, but not as a diazonium moiety. This would include:

Production of azo compounds by coupling the active diazonium salt to a compound possessing an active hydrogen atom bound to a carbon atom (phenols, naphthols, aromatic amines, active methylene compounds containing an enolizable carbonyl group such as acetoacetyl compounds and pyrazolones).

$$Ar—N{=}N—X + H—Ar' \rightarrow$$
$$Ar—N{=}N—Ar' + HX$$

Production of aryl hydrazines by reduction:

$$Ar—N{=}N—X + 2H_2 \rightarrow ArNHNH_2 + HX$$

Production of aryl nitroamines by oxidation:

$$Ar—N{=}N—X + [O] \rightarrow ArNHNO_2$$

Production of diazocyanides by reacting with cyanide. Production of diazoethers by reacting metallic diazotates with alkyl halides.

The third pathway includes reactions where nitrogen is eliminated. This would include various replacement reactions in organic synthetic routes starting with aromatic amines. For example the Sandmeyer reaction in which an aromatic amino group is replaced by a halogen.

$$Ar—N{=}N—X \rightarrow Ar—X + N_2$$

Replacement by nitrile:

$$Ar—N{=}N—X + KCN \xrightarrow{Cu} Ar—CN + N_2 + KCl$$

(The Gattermann-Sandmeyer Reaction)

Replacement by hydroxyl:

$$Ar—N{=}N—X + H_2O \rightarrow Ar—OH + HX + N_2$$

Replacement by hydrogen:

$$Ar—N{=}N—X + C_2H_5OH \rightarrow$$
$$Ar—H + CH_2CHO + N_2 + HX$$

Many of these reactions have commercial significance especially in the dyestuff industry, making the diazonium intermediate one of the most important in the chemical industry.

Certain types of diazos, especially those from N,N-dialkyl-*p*-phenylene diamines, are relatively stable to heat but are very sensitive to light. These have important commercial application in the field of diazotype copying, such as in the Ozalid® reproduction process. When a paper coated with such a light-sensitive diazo is exposed to light under a pattern, only those areas under the pattern will still retain the active diazonium compound, while in the exposed area there will be decomposed diazo. When the paper is developed by exposure to ammonia fumes, only the protected area under the pattern will couple to form a duplicate image.

L. N. Stanley and R. E. Farris

DIELECTRIC MATERIALS

Dielectric materials may be grouped into three categories—gaseous, liquid and solid. Of these, the solid type is looked upon as the "backbone" of electrical machine design. Such materials are utilized for their dielectric properties and as a mechanical means of conductor spacing and support. The gaseous and liquid dielectrics are used as impregnants of the solid insulation. The successful operation of commercial electrical equipment depends significantly on the dielectric and chemical stability of the gaseous or liquid impregnant used.

The use of gaseous dielectrics has been traditionally confined to low-voltage applications because of their characteristic dielectric strength, which is substantially lower than that of an insulating liquid. Air has been most generally used. More recently, however, gaseous dielectrics have been applied in high-voltage equipment. In these applications air has been largely replaced by nitrogen and in some instances by the synthetic gases ("Freon," sulfur hexafluoride, and the like). In order to increase the dielectric strength, such gases are frequently used at pressures higher than atmospheric. Pressures as high as 200 psi are applied in certain of the highest-voltage equipment. Gas pressure cables have been designed and operated at voltages of 200 kv and higher. The electronegative gases are characterized by a dielectric strength substantially higher than that of nitrogen gas. When used under pressure the dielectric strenght of the electronegative gases is increased to an extent that values approaching the dielectric strength of mineral transformer oil are possible.

Mineral oil still maintains a foremost position in its use as a dielectric and cooling insulant in high-voltage transformers, electric cables, circuit breakers and the like. Its application in electrical equipment at lower voltages has been replaced in part by the dielectric gases operating under pressures moderately above atmospheric. The application of hydrostatically applied pressure and the use of an oil impregnant in combination with a high gas pressure have advanced the utility of the high-voltage electric cable. Continued refining studies and the development of oxidation inhibitors have produced a mineral oil for transformer use which is characterized by a remarkable dielectric and chemical stability.

The most important contribution of synthetic chemistry in the field of dielectric liquids has led to the introduction of the so-called Askarel liquids. These liquids are nonflammable and nonoxidizing. Their use as a dielectric and cooling medium in transformers has resulted in a diminution of fire and explosion risk associated with the use of mineral oil. The Askarel liquids are chlorinated aromatic hydrocarbons which contain a chemical equivalency of chlorine and hydrogen in the molecule. The most prominent members of this group are the chlorinated diphenyls and trichlorobenzene. The use of pentachlorodiphenyl as an impregnant for paper-spaced, foil capacitors has produced a capacitor having about 60% of the physical volume per unit of capacitance of that normally associated with a similar capacitor impregnated with mineral oil. Other synthetic liquid dielectrics which have been used to a lesser extent include the silicone oils, which possess excellent stability at temperatures higher than is practical in the use of mineral oil, and ester liquids such as dibutyl sebacate, which is used as the sole dielectric in capacitors applied in circuits operating at frequencies up to about one megacycle. In these applications, the higher dielectric constant of the dibutyl sebacate presents a distinct advantage as compared to mineral oil. Its greater chemical stability and lower dielectric losses at rated frequency are the basis for its use in preference to castor oil. Both mineral oil and caster oil have been replaced for application in electronic heater circuits by the use of synthetic ester liquids such as dibutyl sebacate.

Cellulose in the form of paper tapes, sheets and boards is still the most largely used type of solid insulation. In high-voltage, oil-filled equipment such as the transformer and electric cable, manila and kraft papers are used almost exclusively. Kraft paper tissue is used in liquid-filled capacitors. The dielectric properties of the oil-treated cellulose insulation are very largely determined by the efficiency of the vacuum drying and oil impregnation treatment by which the insulation is freed from its normal content of moisture and air, the presence of which promotes increased dielectric loss and decreased dielectric strength. The presence of incompletely oil-saturated areas leading to air pockets or voids within the dielectric assembly results in the formation of ionization and corona at or below the operating voltage of the equipment with resulting dielectric and chemical deterioration and the ultimate failure of the apparatus. Other impregnants than mineral oil commonly used for treatment of cellulosic insulation include the mineral waxes, synthetic waxes, such as chlorinated naphthalene, shellac and the synthetic resins. The dielectric application of synthetic resins has been rapidly expanding. Synthetic resins of the phenolformaldehyde and "alkyd" types have largely replaced shellac in the treatment of coils and laminated resin-bonded dielectric structures. Such insulations have been limited to moderate temperatures, generally not exceeding 100°C. To meet the urgent demands for operation even at temperatures as high as 200°C, exceedingly stable resins of low loss and high dielectric strength have been developed. Among these, silicone resins and polytetrafluoroethylene are of particular utility.

Because of the higher losses associated with the use of impregnated cellulose insulation at high-frequency voltages, mica has been the traditional insulation for such applications. However, synthetic resins such as polyethylene and polystyrene have replaced mica in many of its high-frequency applications where low dielectric loss at relatively low operating temperatures is desired. Another type of synthetic resin which in sheet and tape form is replacing cellulose sheets and tapes in capacitors and motor windings is the terphthalate resin, e.g. "Mylar." The excellent dielectric properties of these products coupled with their chemical stability permit electrical application at temperatures substantially higher than are normally used with cellulose. *Polyvinyl chloride* is widely applied especially in the insulation of household wiring, electrical appliances and the like, where its low cost coupled with good electrical and chemical stability at moderate temperatures of use are advantageous.

In the resin molding and casting operations, the use of the *polyester resins* and *epoxy compounds* is increasing rapidly. The low-pressure molding techniques used with these resins are a distinct advantage in many applications. The low order of thermal expansion and the good dielectric stability associated with epoxy resins have promoted their use in the manufacture of resin-cast instrument transformers, insulators, cable terminations and the like. Their high resistance to moisture and moisture transfer, together with their excellent adhesion qualities have promoted the application of these resins as coil impregnants for motors and other types of rotating equipment.

In the form of an enameling composition, polyurethane resin has replaced a large part of the oleo resins heretofore used in the manufacture of enameled wire for communication and electronic equipment, because of its excellent stability and dielectric properties and its ease of manufacture and application. Its resistance to the effects of moisture, excellent adhesion properties and the wide variety of formulated products available for application as wire enamels, adhesives, waterproof coatings, insulating film and elastomeric compositions are the basis for the exploitation of this type of resin in its dielectric and structural application to a wide variety of electrical products.

Natural rubber has been almost entirely displaced by butyl rubber for insulating wires and cables. Properly compounded butyl rubber has good chemical and dielectric stability and a high order of resistance to the effects of corona. Silicone rubber and "Vulcolan," the polyisocyanate elastomeric composition, are also used. Silicone rubber has the widest range of possible temperature application and the greatest resistance to corona deterioration. "Vulcolan" has the greatest resistance to mechanical wear and abrasion and in this respect offers strong competition to other synthetic elastomers such as neoprene, for use as abrasion-resistant coatings for electric cables and wires.

The development of resin materials capable of continuous electrical operation at temperatures of 200°C and higher has served to emphasize the temperature limitations which must be maintained in the dielectric applications of cellulose products. Motor windings, transformer conductors, capacitors and the like which heretofore have depended almost entirely on the use of resin-treated paper tapes and resin-laminated cellulose boards and cylinders to meet high-temperature requirements demand the replacement of the cellulosic constituent by a material which in itself can match the stability of the resin impregnant at higher temperatures. Two types of products have been developed to meet this demand. These consist chiefly of glass and of mica.

Glass cloth, glass roving and thin glass films are being widely applied as a substitute for cellulose in resin-bonded coils, boards and the like designed for application at temperatures of 200°C or more. Glass-insulated, resin-treated wires are of equal utility.

Mica paper, mica mat, "Samica" and other commercially designated materials have been developed as a replacement for cellulose sheets and also as a superior replacement material for the resin-bonded mica tape which has been so widely used in the past. These recent mica materials are made from exfoliated mica prepared in the form of a sheet using the usual paper manufacturing technology. Possessing a somewhat low order of mechanical strength, these new mica sheets and tapes are usually treated with a bonding resin. Silicone bonding resins are used where the highest order of thermal stability is required. Tapes and sheets of this type of product are used in commutator construction and as coil insulation for motors and generators. Thin sheets have been used in certain types of capacitors. Coupled with the advances in the technology and application of glass cloth and other glass products, these materials offer an excellent solution to the problems presented by the instability of cellulosic insulations at high temperatures.

F. M. Clark

Cross-references: *Cellulose, Polyester Resins, Epoxy Resins, Polyurethane Resins.*

DIELS-ALDER REACTION

In 1928, Otto Diels and Kurt Alder reported their discovery of a reaction which now bears their name. They received the Nobel Prize in 1950 for their efforts on this reaction. It is one of the most carefully examined reactions in organic chemistry.

The reaction may be defined chemically as a 1,4-addition of an unsaturated molecule (called the dienophile) to a conjugated diene. The products always contain an unsaturated six-membered ring. The general reaction may be illustrated:

If the dienophile is an alkene, the product will be a cyclohexene derivative; if the dienophile is an alkyne, the product will be a 1,4-cyclohexadiene.

The utility of the reaction is illustrated by a few examples. It is possible to prepare cyclohexene itself in the simplest possible Diels-Alder reaction, although vigorous conditions must be employed. More esoteric molecules may be formed readily. For example, 1,3-butadiene reacts with *p*-benzoquinone to yield 5,8,9,10-tetrahydro-1,4-napthoquinone, and with maleic anhydride to yield cis-1,2,3,6-tetrahydrophthalic anhydride; cyclopentadiene reacts with acrolein to yield 6-formylbicyclo(2,2,1) heptene-2. Even anthracene may act as a diene with the powerful dienophile, tetracyanoethylene, forming an adduct across the central ring of the anthracene molecule. Conditions vary from room temperature within minutes with or without a solvent for very reactive reagents to temperatures of over 100° for several hours under pressure. The usefulness of the reaction to the synthetic organic chemist is emphasized by the role it has played in the synthesis of such interesting substances as cortisone, codeine, morphine, reserpine, yohimbine, and cantharidin.

The reaction is strongly influenced by the substituents on both the diene and the dienophile. Electron-donating groups (such as alkyl) on the diene and electron-withdrawing groups on the die-

nophile (such as cyano or carbonyl) increase their reactivity. Thus very reactive dienophiles include tetracyanoethylene, maleic acid, maleic anhydride, acrolein, and the acrylates; reactive dienes include butadiene itself, cyclopentadiene, cyclohexadiene, furan, or some of the methylbutadienes. The substituents on the diene must not be so numerous or of such size as to sterically hinder adduct formation.

In spite of intense scrutiny, the mechanism of the reaction is not yet clear. It is bimolecular in the rate-determining step. It has been speculated that the rate determining step involves the formation of an acid-base "pi complex" between the electron-rich diene and the "electron-deficient pi cloud" of the dienophile.

The Alder-Stein endo-cis principle sums up a great deal of the stereochemistry involved in the reaction: (1) the diene must be in the cis configuration in order to react; (2) the transition state must be one in which the substituents of the dienophile lie directly under the residual unsaturation of the diene. This indicates that, if a Diels-Alder reaction is accomplished between a substituted dienophile and a cyclic diene, the product will have the endo configuration. The following example illustrates this phenomenon:

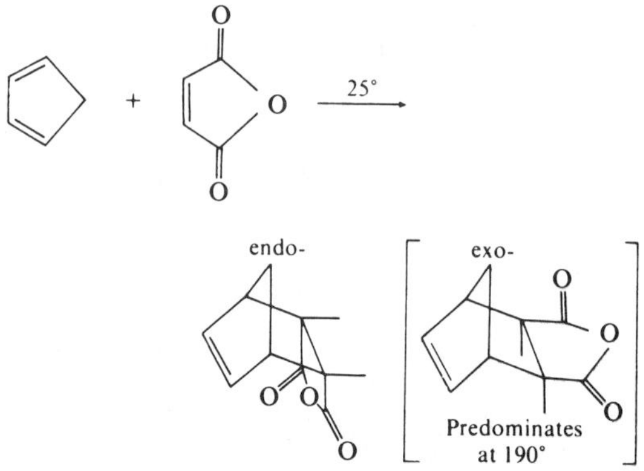

In actual practice, most Diels-Alder reactions do yield the endo-adduct. However, reactions requiring long reaction times or high temperatures often yield predominantly the exo isomer. It has been established that endo-exo isomerization may occur thermally and that the Diels-Alder reaction is thermally reversible. Therefore it is apparent that high temperatures or long reaction times may result in predominance of the thermodynamically more stable exo isomer as a result of isomerization of the initially formed endo isomer or the accumulation of more stable exo isomer produced directly from starting material.

J. C. Kellett, Jr.

Cross-reference: *Dienes.*

DIENES

While the term "diene" may be applied to all those compounds which contain two aliphatic C—C double bonds, e.g., butadiene, methyl linoleate, cyclopentadiene 1-cyanobutadiene, chloroprene, vinylcyclohexene, etc., it is generally limited to diolefins, hydrocarbons containing the above defined structure, and their simple substitution products. Dienes occur mainly in the products of high-temperature treatment of hydrocarbons, primarily natural gas and petroleum hydrocarbons, in thermal decompositions of other organic materials, and in various plant materials, particularly turpentine, rosin, etc. They vary from colorless gases through liquids, to solid materials. Generally they have characteristic strong odors, and may be highly colored. Most of the dienes are chemically very reactive and so have lent themselves to many interesting and important commercial developments.

Among the dienes of greatest interest are both open-chain and cyclic types. An important distinction is also drawn between conjugated and nonconjugated dienes. When the two double bonds are separated from each other by one single bond, i.e., contain the structure —C=CH—CH=CH—, they are spoken of as conjugated or 1,3-dienes. They enter into many special reactions of considerable importance not displayed by the nonconjugated type in which the double bonds are either adjacent (1,2-) or more widely separated. The simplest open-chain conjugated diolefin is 1,3-butadiene, CH_2=CH—CH=CH_2. It has been the subject of extensive research and has become a product of large industrial importance because of its great reactivity. Other important dienes of this type include isoprene, 2,3-dimethylbutadiene, piperylene, and chloroprene. Allene, CH_2=C=CH_2, and 1,2-butadiene are examples of open-chain dienes with adjacent double bonds, while 1,4-pentadiene, 1,5-hexadiene and other open-chain dienes with more widely separated double bonds are known. Among the cyclic diolefins, cyclopentadiene is perhaps the most reactive and is used extensively in industry. Cyclohexadiene and cyclooctadienes in which the rings contain six and eight carbon atoms, respectively, are also of interest. Other cyclic dienes of commercial interest include dipentene and terpinene.

The nonconjugated dienes react primarily as olefins, each double bond acting independently; either or both may be involved in a reaction, depending on conditions used and on molecular structure. Through simple addition reactions such as those of halogens, hydrogen cyanide, water, etc., and other typical reactions of the double bond including hydroxylation, epoxidation, and oxidation, they serve as convenient sources for many chemicals, particularly polyfunctional products. The 1,2-dienes readily isomerize to acetylenes; and, as in the case of allene, often behave in reactions as if they had the acetylenic structure. Normally, nonconjugated dienes do not readily polymerize.

Conjugated dienes display a particular type of activity, 1,4-addition, which results in several very important addition reactions having wide application. Addition to a 1,3-diene may occur normally, first to one double bond (1,2-addition), but more frequently the entering atoms are added 1,4, the remaining double bond shifting to the 2,3-position. An important example of 1,4-addition lies in the reaction of conjugated dienes with activated double bonds, the so-called diene condensation or Diels-Alder reaction. In this reaction, a diolefin adds 1,4 across a double bond which is activated through attached negative groups, giving a six-membered ring. The reaction occurs with many anhydrides, ketones, aldehydes, nitriles, and quinones and leads to a wide variety of valuable and interesting products and chemical intermediates.

The most important reactions of conjugated dienes are those of polymerization. This may occur through both 1,2 and 1,4 condensations. Polymerizations occur by both purely thermal, uncatalyzed mechanisms, and in both ionic and free-radical generating systems. Thermal polymerization results primarily in dimerization through the diene condensation; in the case of 1,3-butadiene to vinylcyclohexene, with some other isomers and higher polymers. Linear and branched chain polymers, ranging from light liquids to tough rubbers and resins are obtained in palymerization systems catalyzed by such materials as alkali metals or metal alkyls. Linear and cyclic polymers result from ionic catalysts such as hydrogen fluoride, sulfuric acid, phosphoric acid, boron fluoride complexes, etc.

Free-radical generating systems, including especially those containing peroxides, persulfates, "redox" and organometallic systems, produce a wide variety of polymers from dienes. Copolymerization of different dienes and of dienes with active vinyl groups in monoolefins and unsaturated nitriles, esters and the like, makes possible production of elastomers, drying oils and resins with almost any desired combination of properties. Stereospecific polymers and copolymers with distinctive and valuable characteristics are readily made and widely used. Diene polymerization is responsible also for formation of undesired sludges and tars in many cases, such as gummy residues in gasoline.

Among the individual dienes of commercial importance, *1,3-butadiene* is of first importance. It is used primarily for the production of polymers, of which synthetic rubbers are by far the largest tonnage. An approximately 75/25 copolymer with styrene is used to produce SBR rubber, used in tire construction and for general purposes. Oil-resistant rubbers result when acrylonitrile is used in place of styrene. "ABS" resins are terpolymers of butadiene, acrylonitrile and a high percentage of styrene. They are manufactured in large tonnages for use in shoe soles and as rigid engineering plastics. *cis*-Polybutadiene, a stereospecific homopolymer of butadiene has come into wide use in the tire industry because of its superior combination of properties. The *trans* form is a hard material resembling natural gutta percha; either may be produced by the proper selection of complex organometallic catalyst system. Liquid butadiene polymers, including those terminated by hydroxy or carboxy groups, may be produced in mass polymerization processes with finely divided alkali metal catalysts. They have found some use as binders. Through the diene condensation with maleic anhydride and acrolein, tetrahydrophthalic anhydride and tetrahydrobenzaldehyde have been made available. With furfural, two moles of butadiene condense to give a product marketed as a general insect repellent. Butadiene is one source of hexamethylenediamine for nylon manufacture. Chlorine is added to butadiene to give the 1,4-dichloride, which is reacted with hydrogen cyanide and then hydrogenated to adiponitrile and finally the diamine.

Isoprene, 2-methyl butadiene-1,3, is a product of the pyrolysis of natural rubber, and is regarded as the fundamental building block of the rubber molecule. It is produced in the pyrolysis of various essential oils, and of turpentine. In the manufacture of butyl rubber, the polymer chosen for inner tube fabrication, about 5% of isoprene is used as comonomer with isobutylene. *cis*-Polyisoprene, and isoprene-containing copolymers, are now finding extensive use in the tire industry due to their close approach to the desirable properties of natural rubber. Isoprene is recovered from products of cracking heavy petroleum oils, and is manufactured by several processes including dehydrogenation of isopentene, pyrolysis of methyl pentene from propylene dimerization or of the products of isobutylene-formaldehyde condensation and by dehydration of methyl butenol from acetone-acetylene condensation.

Chloroprene, 2-chlorobutadiene-1,3, is another diene of industrial importance, used as a monomer for neoprene-type synthetic rubber. Its polymerization proceeds extremely rapidly. Chloroprene is obtained by the addition of hydrogen chloride to vinyl acetylene produced from acetylene by the action of a cuprous chloride catalyst.

Cyclopentadiene has come into widespread use as a base for certain important chlorinated insecticides. It is isolated as the more stable dimer, from which it is readily regenerated on distillation, from the products of heavy oil cracking and from the light oil from coke oven tar distillation. It enters into a wide variety of diene condensations yielding products of unusual structure, some of which have found commercial application. It forms "sandwich" type chelate compounds with iron and other metals, which have also served as the basis for valuable homogeneous catalysts.

Cyclooctadiene is manufactured in a large plant by catalytic dimerization of butadiene. The trimer cyclododecatriene is also produced in this plant. It may be reduced to the diene; both it and cyclooctadiene may be reduced to the monoolefins, which can be opened up to yield α-ω-difunctional compounds of potential commercial value.

The terpene hydrocarbons include several dienes of some importance. They occur in the volatile oils of plants, particularly *Coniferae.* Dipentene, its optically active isomers the limonenes, and *alpha*-terpinene are among them. Other dienes of some interest are indene, a product of coal tar and oil cracking, and 4-vinyl-1-cyclohexene from dimerization of butadiene.

Installed capacity for 1,3-butadiene was 1,750,000 short tons per year in 1970. Consumption was approaching this figure and growing steadily. While considerable amounts are recovered as by-product in ethylene-propylene plants, particularly in Europe, the greater amount is based on dehydrogenation. Isoprene production was over 200 million pounds in 1968 and expanding rapidly. Cyclopentadiene is recovered in large tonnages, and multimillion pound capacity for the cyclic butadiene products is installed.

J. C. HILLYER

References

"Butadiene—Four Carbon Building Block" Texas-U.S. Chemical Co.

"Encyclopedia of Chemical Technology," 2nd Ed., Interscience, New York.
Vol. 3, p. 874, "Butadiene," 1964.
Vol. 6, p. 688, "Cyclopentadiene," 1965.
Vol. 12, p. 63, "Isoprene," 1967.

DIFFUSION

Diffusion is transport of a substance in a medium when a concentration difference is the cause of the transport. If a drop of an aqueous solution (of a specific gravity exceeding that of water) be placed on the surface of pure water in a glass, the mixing is due mainly to gravitation and is an example of convection rather than diffusion. If a similar drop be placed on the bottom of the vessel and cautiously covered with pure water, the solute (i.e., the dissolved substance) will (ideally) move into and in the water layer only because its concentration in the drop (or near the bottom) is higher than outside the drop (or far from the bottom); this is diffusion.

Consider a horizontal plane x cm above the bottom of the vessel. As long as diffusion continues, the amount dm of solute crossing this plane in the upward direction during the time interval dt is such that

$$\frac{dm}{dt} = -DA\frac{dC}{dx};$$

A is the area of the plane (i.e., the horizontal cross-section of the vessel), C is the concentration of the solute, and D is diffusion coefficient. The amount of solute in a horizontal layer between x and $x + dx$ (that is, also the local concentration C) increases in time so that

$$\frac{dC}{dt} = -D\frac{d^2C}{dx^2}.$$

These relations (known as the first and the second Fick equations) are the main mathematical expressions of linear diffusion; the equations are a little more complicated if diffusion takes place in several directions at once.

Diffusion may be said to be caused by a difference in partial pressures (in a gas), or by a difference in osmotic pressures (in a liquid), but a more general explanation is based on probability. If atoms, molecules or particles are free to move everywhere in the vessel, it is improbable that they will occupy only one part of the latter.

When a solute moves in a medium, its particles must push the molecules of the medium aside. Consequently, diffusion is slow when the viscosity of the medium is high. Thus the diffusion coefficient D, which is usually of the order of 10^{-1} cm²/sec in gases at atmospheric pressure, is of the order of 10^{-5} cm²/sec in water and often is below 10^{-8} cm²/sec in solids. The larger the diffusing particle, the smaller the diffusion coefficient. For instance, D for water and ethanol vapors in air at 0°C is, respectively, 0.2 and 0.1 cm²/sec. For urea and sucrose in water at room temperature it is 1.2×10^{-5} and 0.3×10^{-5} cm²/sec.

In porous solids, gas molecules can diffuse in the pores (gas diffusion), along the pore walls (surface diffusion), or in the actual solid material. The most common values of the surface diffusion coefficient are between 10^{-5} and 10^{-6} cm^2/sec. If the solute is an isotope of the substance in which diffusion takes place, then an instance of self-diffusion is present; the coefficients of self-diffusion are not greatly different from the common D. Molecular transport of matter from a warm to a cool space often is referred to as thermal diffusion.

In numberless instances, diffusion determines the measurable rate of evaporation, of the dissolution of a solid in a liquid, of chemical reactions at an electrode, and so on; thus it is of prime importance for chemical kinetics, chemical engineering, and several other disciplines.

J. J. Bikerman

References

Jost, W., "Diffusion in Solids, Liquid, Gases," Academic Press, New York, 1960.

Tyrrel, H. J. V., "Diffusion and Heat Flow in Liquids," Butterworth, London, 1961.

Shewmon, P. G., "Diffusion in Solids," McGraw-Hill, New York, 1963.

DIGESTION, PHYSIOLOGICAL

The Ingestion of Food. The urge to eat is felt when a calorie deficit occurs in the body. The amount of food ingested is regulated by the body's expenditure of energy as muscular work and loss of heat. Mental work is done at insignificant energy cost. It is common experience to feel hungry after exercise or exposure to cold. Hunger is a conscious, nonspecific manifestation of energy deficit and is different from appetite, which is a craving for specific foods. Appetite varies with the experience and sophistication of the individual. A young infant feels only hunger but the adult gourmet not only feels hunger, but has also developed many appetites such as those for raw oysters, Burgundy wine, or Turkish coffee.

In the central nervous system the hypothalamus integrates the various nerve impulses associated with hunger into the mechanism that controls initiation and cessation of eating. A man's weight tends to remain constant for years which indicates that the neural controls are extremely accurate. A precise energy balance is maintained despite wide fluctuations in energy expenditure as muscular effort or heat loss. Two areas in the hypothalamus are dominant in the regulation of food intake. One is the "satiety center" which when adequately stimulated, acts to inhibit the other one, i.e., the "feeding center." The satiety center, therefore, is the governor in this system, and it is fully active when energy balance is restored. Satiety signals become adequate to inhibit feeding behavior long before ingested food is completely digested and absorbed. These signals are probably generated by the combined action of mechanoreceptors and stretch receptors in the gastro-intestinal tract and chemoreceptors in the lining of the intestine as well as in the satiety center itself. In the rat the feeding and satiety centers have been destroyed separately or together to produce animals that either ate constantly to become obese monsters or refused completely to eat and thus actually starved to death in the presence of ample food. Details of these regulatory mechanisms are still obscure.

Salivation, Chewing and Swallowing. Saliva is produced in 3 pairs of glands controlled by reflexes that involve the brain. Simple reflexes are initiated by direct chemical stimulation of certain receptors in the mucous membrane of the mouth. Nerve impulses travel from these receptors to the salivary center in the brain, which in turn sends impulses over the secretory nerves to the salivary glands. The simple reflex may not enter consciousness. The conditioned reflex, however, involves the higher centers and no direct stimulation of the oral mucosa is necessary to bring about a flow of saliva. This phenomenon is definitely related to past experience and hence to appetite. Anyone, especially if hungry, can easily bring the conditioned reflex into play and make his mouth "water" simply by thinking about a particularly delectable dish. Hormonal control of salivary glands has never been demonstrated.

The amount of chewing is probably not decisive in the digestion of most foods. For hard foods the molar mill is quite effective and is capable, in man, of exerting a crushing force up to 200 lbs. Aside from its grinding effect, the act of chewing helps to saturate the food with saliva, which initiates starch digestion and lubricates the bolus of food so that it is easily swallowed.

Swallowing is a highly coordinated act that transports food from mouth to stomach. In essence a ring of contraction travels the whole length of the gullet, sweeps the contained bolus before it, and leaves the lumen clean. Since the upper part of this passage is also an airway, respiration is stopped reflexly at each swallow.

Secretion of Gastric Juice. Like the mouth, the stomach also "waters" at the sight, smell or thought of good food. This first response of the gastric glands is called the "cephalic phase" of gastric secretion because it involves the brain in a conditioned reflex. As a consequence of the cephalic phase, a certain amount of gastric juice is present when the first bolus of food arrives in the stomach. Digestion begins at once and the products of digestion stimulate chemically the mucosa of the lower end of the stomach to produce "gastrin," a hormone that enters the blood and stimulates the upper portion of the stomach to continue its secretion of gastric juice ("gastric phase"). This is an example of dual, neural and hormonal, control of a digestive secretion. The small intestine also is capable of producing a "gastrin" ("intestinal phase"). The two main digestive components of gastric juice are HCl and pepsin. The acid is produced only by the parietal cell as an aqueous solution of approximately 0.16 N HCl. Pepsin digests protein in acid medium and is itself a protein produced by the chief cells of the gastric glands. Neural and gastrin stimulation result in copious secretion of both HCl and pepsin. A man may secrete 700 ml of gastric juice in response to ingestion of a single meal.

Evacuation of the Stomach. When the stomach is filled with a meal, gastric digestion begins at the periphery. Liquefaction and small particle size are achieved by action of gastric juice and muscular movements. The fluid or semi-fluid contents are forced into the first portion of the small intestine (duodenum) which is thereby slightly distended to set off a reflex that inhibits gastric movements momentarily until the duodenum has distributed part of its load to lower sections of the small gut. This negative feedback mechanism is an important factor in controlling the evacuation of the stomach and in preventing an overloading of the duodenum. The duodenum is very sensitive to distention and if it is overloaded accidentally or in disease (obstruction), vomiting, another protective mechanism, comes into action to relieve the pressure.

It is important to emphasize that the pylorus, a narrow muscular passage between stomach and duodenum, does not act like a stopcock to prevent contents from passing into the gut from the stomach. The pylorus is open more than it is closed, and it is closed in that part of the gastric cycle in which the duodenum is beginning to contract and the stomach is quiescent. Its main function, therefore, seems to be prevention of regurgitation of duodenal contents into the stomach. This short interval of pyloric closure allows the duodenum to propel its bolus downward and prepare for the next load from the stomach. These movements of stomach and gut walls are nicely coordinated both by extrinsic nerves connecting with the central nervous system and by intrinsic nerves that lie in two nerve plexuses embedded in the muscular walls of these organs.

Secretions in the Small Intestine. Bile, pancreatic juice, and intestinal juice are poured into the small intestine to complete the digestion of food begun in the mouth and stomach. Bile is secreted continuously by the liver and stored and concentrated in the gall bladder. The ingestion of protein stimulates the liver to produce more bile, and fat causes release of cholecystokinin, a hormone that stimulates the gall bladder to contract and deliver a charge of concentrated gall bladder bile into the duodenum. Bile is especially useful in emulsification, digestion and absorption of fats. The pancreas produces potent enzymes that hydrolyze protein, fat and starch. The inactive forms of these enzymes are carried to the duodenum in an alkaline juice. The gut wall secrets enterokinase, an enzyme that activates the pancreatic enzyme precursors.

The pancreas is controlled by both neural and hormonal mechanisms. The water and salts of the pancreatic juice are secreted in response to stimulation by secretin. The hormone is released into the blood from the mucosa of the small gut as soon as gastric contents enter the duodenum. Normally the contents of the upper small intestine remain acid (pH 4.5–6.5), as demonstrated by repeated pH measurements of intestinal contents during digestion. The secretion of pancreatic enzymes is excited by the action of the vagus nerves as well as by the hormone pancreozymin; like secretin, this is produced in the gut wall by contact of gastric contents with the intestinal mucosa. Cholecystokinin and pancreozymin (CCK-PZ) are now known to be two classical names for the same hormone, a peptide containing 33 amino acid residues. A man may secrete from 1.5 to 2 liters of pancreatic juice a day.

The intestinal juice is produced by countless simple tubular glands in the mucosal lining of the gut. This tissue produces enterocrinin, a hormone that stimulates both volume flow and release of enzymes. The enzyme release may be another effect of CCK-PZ. It is not certain whether extrinsic nerves play a role here, but the intramural nerve plexuses are probably concerned because local mechanical stimulation of the mucosa is effective. The mere passage of a bolus of food along the lumen excites the secretion by rubbing the sensitive mucosal lining. The intestinal juice contains more different enzymes than any of the other secretions and generally serves to complete many of the hydrolytic processes begun by the other digestive secretions. It does not contain an enzyme capable of attacking native protein. In addition to furnishing a large variety of digestive enzymes, the intestinal juice provides a large volume of aqueous medium in which degradation products of food digestion can be dissolved or dispersed prior to absorption into the blood stream. The daily volume of intestinal juice is unknown but may be 10 to 20 liters.

Muscular Movements of the Small Intestine. The two main types of muscular movement are peristalsis and segmentation. Peristalsis is a ring of contraction preceded by a ring of relaxation that travels away from the mouth. The effect is to move contents in the same direction and to distribute them over most of the available surface. Peristaltic waves may travel only a few centimeters and die out, or they may travel the whole length of the small bowel in a "peristaltic rush." The contracted portion of the traveling wave represents a shortening of the inner muscle coat in which the fibers are disposed to form a cylinder around the long axis of the gut. The peristaltic wave in the small intestine is essentially the same type of movement that moves food onward in the gullet and the stomach.

Absorption. Absorption of digested food is accomplished with the aid of several special processes, none of which has been fully explained. There is little doubt that many physical processes, such as diffusion, osmosis and hydrostatic pressure, may play a role in the absorption of many substances. The absorption of water and some of the univalent salts in solution may be explained, in part at least, by known physical mechanisms. The absorption of glucose, however, is called "active transport," implying absorption against a concentration gradient and one which requires energy. The L-amino acids, fatty acids, and the glycerides are probably also moved across the intestinal mucosa, and eventually into the blood stream, by means of "active transport."

The digestive tract is capable of controlling to a large extent the qualitative composition of intestinal contents with respect to amino acids. The molar ratios of the free amino acids present in the intestinal contents tend to remain fixed under varied conditions. The amino acid mixture present in the jejunum after feeding is qualitatively the same whether the test meal includes, (1) a complete protein, containing all of the amino acids, (2) an incomplete protein, devoid of one or more amino acids, or (3) no protein whatever. The explanation seems to be that any meal, regardless of its composition, is mixed with so much endogenous protein (enzymes, mucoproteins, sloughed cells) that the differences of the ingested proteins are obscured by dilution.

Functions of the Colon. The colon primarily collects waste, recovers most of the water and salts and discards the remainder as feces. It is a common, but mistaken, notion that feces are composed chiefly of undigested food residues. In general the type of diet does not alter the chemical composition of the feces. This obviously is not true if the diet is changed to include irritating or non-nutritive substances. Of the fecal dry matter 5 to 10% is composed of bacteria. The remainder is mostly indigestible residues of digestive secretions which are not recovered in the small bowel; they probably include organic solids from saliva, gastric, pancreatic and intestinal juices, and bile. When feces are sufficiently dehydrated they pass from the colon into the rectum, which normally is empty, and from here passed to the outside. The stimulus for bowel movement is often originated in the stomach as a result of eating breakfast and is part of the "gastrocolic" reflex.

E. S. NASSET

DIPOLE MOMENT

A dipole moment is a vector quantity measuring the size of a dipole. An electric dipole consists of a pair of electric charges, equal in size but opposite in sign and very close together. The dipole moment, sometimes called the "electric moment," is the product of one of the two charges by the distance between them. A magnetic dipole is a magnet of very short length and infinitesimal width and its moment, commonly called the "magnetic moment," is the product of the strength of one pole by the distance between the two.

The electric dipole moment is normally a molecular property arising from asymmetry in the arrangement of the positive nuclear charges and the negative electronic charges in the molecule. A very small moment is induced in a molecule by an electric field, which displaces the positive and negative charges relatively to one another. A permanent moment arising from permanent electrical asymmetry is usually much larger. Its order of magnitude is that of the product of an electronic charge, 4.80×10^{-10} electrostatic units, times an atomic radius, 10^{-8} cm, that is, 4.80×10^{-18}. The moment values are, therefore, expressed in 10^{-18} e.s.u. cm, which is sometimes called a "debye."

In polyatomic molecules, the molecular dipole moment is the vector sum of the moments of dipoles associated with bonds or groups of atoms in the molecule. For example, the moment of a dichlorobenzene molecule may be treated as the resultant of moments acting in the directions of the two C—Cl bonds. The values of these C—Cl moments were taken to be approximately the same as the moment of the chlorobenzene molecule and used to calculate the moments of various models for the three isomeric dichlorobenzene molecules, thus establishing the correctness of the Kekulé structure. Dipole moments are widely used in this way to determine the geometric structures of molecules and also to investigate the distribution of electronic charge.

Dipole moments are usually calculated from the results of dielectric constant measurements on gases, dilute solutions, or pure liquids. Accurate values have been obtained for simple molecules from microwave spectroscopic measurements in gases. A few, usually approximate, values have been obtained for molecular moments by means of molecular beam measurements and for bond moments by means of infrared intensity measurements.

In a molecular dielectric material, the permanent dipole moment is responsible for the dielectric loss and for any part of the dielectric constant in excess of the small contribution of the induced charge shift. It is, therefore, a quantity of theoretical importance in determining electrical forces in matter and of practical importance in determining the properties of insulating materials. The induction of small dipole moments by the electric field of a light wave is responsible for the refraction of light by matter and the induction of small dipoles in molecules by the electric fields of neighboring molecules gives rise to intermolecular attraction.

Magnetic moment is due to the spin and orbital motions of unpaired electrons in atoms, molecules, and ions. The experimental values are obtained from measurements of magnetic susceptibilities by means of the Langevin equation, on the analogy of which the Debye equation was developed for the calculation of electric dipole moment from dielectric constant. When the unpaired electrons are not screened by other electrons, the contribution of the orbital motion becomes unimportant in comparison with that of the spin. For many cases, the moment is then $\sqrt{n(n + 2)}$ Bohr magnetons, where n is the number of unpaired electrons. The magnetic moment is often particularly useful in the investigation of complex ions and coordination compounds of transition metals, where determination of the number of unpaired electrons may determine the nature of the bonds and even the molecular shape.

CHARLES P. SMYTH

Cross-references: *Magnetic Phenomena, Magnetochemistry, Polar Molecules.*

References

Earnshaw, A., "Introduction to Magnetochemistry," Academic Press, New York, 1968.

Hill, N. E., Vaughan, W. E., Price, A. H., and Davies, M., "Dielectric Properties and Molecular Behavior," Van Nostrand Reinhold, New York, 1969.

Selwood, P. W., "Magnetochemistry," Interscience Publishing, New York, 1956.

Smith, J. W., "Electric Dipole Moments," Butterworth's Scientific Publications, London, 1955.

Smyth, C. P., "Dielectric Behavior and Structure," McGraw-Hill Book Co., Inc., New York, 1955.

DISINFECTANTS

Disinfectants are agents used to destroy disease germs or other microorganisms on inanimate objects. They destroy or inactivate viruses but do not necessarily kill the spore forms of bacteria. Disinfectants differ from antiseptics in that the latter are usually considered to be agents applied to the body to inhibit or destroy harmful microorganisms. An ideal disinfectant is highly germicidal, readily soluble in water, stable, nontoxic, noncorrosive to metals and nonirritating to the skin. It should be capable of penetrating without being inactivated by organic material, and must be of low cost.

The chemicals used as disinfectants may be divided into the following classes: (1) mercury compounds, (2) halogens and halogen compounds, (3) phenols, (4) synthetic detergents, (5) alcohols, (6) natural products and (7) gases. The efficiency of chemical disinfectants is often expressed by a ratio known as their "phenol coefficient"—the ratio of the greatest dilution of the compound that will kill all of a given organism in a test tube, to the greatest dilution of phenol that gives the same results. The phenol coefficient is not always valid when comparing nonphenol compounds. Disinfectants are better evaluated under conditions approximating the use for which they are intended.

Mercury Compounds. Mercuric chloride, $HgCl_2$, was one of the earliest compounds to be employed as a disinfectant and for many years was considered ideal for many uses. Like most mercury salts, it exerts its germicidal action by competing for the sulfhydryl group of a bacterial enzyme with a metabolite. Unfortunately, this reaction is reversible and bacteria appearing to be destroyed by mercury compounds may be revived by the addition of a compound containing a sulfhydryl group. For this reason its activity is reduced by organic materials. Its value as a disinfectant is further reduced by its low sporocidal activity, its reactivity with metals, its high toxicity and its irritant action on the skin.

In the hope of overcoming the corrosive and irritating action of the inorganic mercury salts, many organic mercury compounds have been prepared and examined for germicidal activity. A number of these, such as phenylmercuric chloride, sodium ethylmercurithiosalicylate and 2,7-disodium dibromo-4-hydroxymercurifluorescein, are used as antiseptics and germicides. These are said to be less easily deactivated by organic material, less corrosive and less toxic than the inorganic salts of mercury.

Halogens. All the halogens are bactericidal and virucidal with fluorine possessing the greatest germicidal activity followed by iodine, bromine and chlorine. Because of its availability and economy, chlorine is used extensively for the large-scale disinfection of water, swimming pools and food processing plants. Iodine in alcoholic or aqueous solutions is used for the disinfection of surgical instruments, catgut sutures and other equipment easily damaged by heat sterilization.

Organic and inorganic compounds which liberate chlorine when in contact with water are often employed as home and commercial disinfectants. Calcium hypochlorite, $Ca(OCl)_2$, available as "chlorinated lime," is used for the disinfection of sewage effluent, swimming pools and the water supply of food processing and handling plants. Sodium hypochlorite, $NaOCl$, is marketed as a disinfectant in aqueous solution containing from 1 to 15% of the compound. Organic compounds which liberate chlorine or hypochlorous acid (e.g., p-dichloro-sulfamylbenzoic acid or halazone) are employed for the emergency or small-scale disinfection of water.

Phenols. A 1% aqueous solution of phenol kills most bacteria within 20 minutes and a 5% solution kills bacterial spores. Homologs of phenols (cresols) have been employed as household disinfectants. These compounds are both more effective and economical than phenol, but are less water-soluble and must be "solubilized" in their commercial preparations.

A number of synthetically employed phenols such as o-phenylphenol and p-tertiary-amylphenols are used as disinfectants and are rapidly replacing phenol and the cresols. These do not have the characteristic disinfectant odor of the cresols and are less toxic and irritating to the skin. They are incorporated in aerosols for household use. The bis-phenols (hexachlorophene and bithionol) have the added advantage that their disinfectant properties are not reduced by soap and can therefore be incorporated into solid soaps or soap solutions.

Synthetic Detergents. While soap alone is popularly considered to be germicide and disinfectant, it acts only mechanically to remove bacteria and microorganisms and does not necessarily destroy them. The newer synthetic detergents, however, possess definite germicidal activity. These are effective germicides in concentrations as low as one part in several thousand parts of water. They are odorless, noncorrosive, nonirritating and possess low toxicities, and are therefore useful disinfectants for dairy and food equipment as well as in households and hospitals.

Anionic detergents, as the sodium alkylsulfonates, are effective against gram-positive organisms but not against gram-negative ones. The cationic detergents, or quaternary ammonium compounds, are as effective as the anionic types. Since most of the compounds originally employed as detergents and disinfectants are not biologically degradable, their widespread use has created problems in water pollution.

Alcohols. The lower molecular weight alcohols, with the exception of methanol, possess excellent bactericidal activity. While they increase in germicidal activity as they increase in molecular weight, the low water solubility of the higher molecular weight compounds limits their usefulness. Ethyl and isopropyl alcohol are employed for skin antisepsis and for the disinfection of sick-room equipment. Trimethylene glycol and propylene glycol are used for air disinfection by incorporation into aerosol formulas. It is generally believed that alcohols act by denaturing the proteins of mocroorganisms.

Natural Products. Pine oil disinfectants in dilutions of approximately 1 to 100 are effective household germicides. They are stable, nonirritating and possess a good penetrating power. Their pine-like odor has increased their popularity. These preparations are mixtures of pine oil and saponifying agents. The fraction of pine oil boiling between 170 and 350°C, which consists of a mixture of terpenes and terpine alcohols, is usually employed.

Gases. While terminal fumigation has been practically abandoned for the sick room, gaseous sterilization or disinfection is still employed in laboratories or industrial areas where the bacterial count must be kept low. Sulfur dioxide and chlorine, while effective disinfectants, are seldom employed because of their bleaching action. Formaldehyde, however, is considered to be an effective gaseous disinfectant. Its hydroalcoholic solutions are often used to sterilize instruments. Ethylene oxide is a most effective germicide. Because of its explosive hazard it is usually mixed with carbon dioxide. Other gases such as ozone, methyl bromide, chloropicrin and ethylenediamine have also been used as disinfectants.

Physical Disinfectants. Physical agents such as heat or electromagnetic waves are employed for the destruction of microorganisms. Although there is some germicidal activity action in the visual spectrum, the ultraviolet spectrum is more effective in the destruction of microorganisms. Ultraviolet lamps are employed for the disinfection of rest rooms, sterile laboratories and barber equipment. X-rays, and alpha- and beta-radiations have been employed for sterilization.

Seed Disinfectants. A number of chemical compounds have been employed as seed disinfectants. Among these are mercurous and mercuric chloride, copper sulfate, copper carbonate and mixtures of zinc oxide and zinc hydroxide. A number of organic mercury and sulfur compounds, quinones, and halocarbons have also been employed. Recent findings indicate that mercury containing disinfectants and fungicides should be used with utmost care. While many of the compounds used are relatively nontoxic, they are altered chemically by animals into more toxic organometalic compounds (methyl mercury). If the flesh of animals feeding on material treated with these is eaten, sufficient quantities of the toxic compounds are present to penetrate the central nervous system and cause death or permanent damage.

JAMES E. GEARIEN

DISPERSING AGENTS

A dispersing agent is a material which, when added to a suspending medium, will promote the separation of the individual extremely fine particles of solids, usually of colloidal size. In general, the term "dispersing agent" is limited to promoting the deflocculation and separation of solid particles, although some confusion exists regarding the exact distinction between protective colloids, suspending agents, stabilizers and true dispersants.

The basic function of a dispersing agent is the reduction of cohesive forces between individual particles, so as to aid in breaking up flocs or agglomerates and thus to permit each particle to act as a separate entity. The mechanism of the action of dispersing agents is not yet completely understood. In many cases, dispersing agents function by imparting a similar electric charge on the surface of solid particles, causing them to reverse their tendency to form aggregates, and in some cases, actually to become mutually repellent.

True dispersing agents generally are polymeric electrolytes, such as the condensed sodium silicates, the polyphosphates, and various derivatives of lignin. In nonaqueous media, nonionic dispersing agents can be employed, such as sterols, lecithin, and fatty acids.

The industrially important applications of dispersing agents lie in the fields of handling suspensions of solids in liquid media. Dispersing agents prevent agglomeration and settling and reduce the apparent viscosity of solid suspensions. Frequently dispersions of semisolid paste consistency can be reduced to creamy liquid consistency by the addition of the proper dispersing agent (e.g., stearic acid). Dispersing agents frequently are used in the wet grinding of pigments, inks, and water-insoluble dyes, the dispersion of insecticides, the grinding of colloidal sulfur, the preparation of oil-well drilling muds, and the processing of clays, slips, and glazes in the ceramic industry.

The amount of dispersing agent used generally is based on the weight of solids to be dispersed. Usually amounts of from 0.1 to 2% by weight are effective in promoting dispersion. The effect of dispersing agents is the result of a specific interaction between the agent chosen and the material being dispersed; hence, few generalizations can be made. Typical dispersing agents include the following:

Agent	Effective for
sodium silicate	clays, slips, glazes
sodium tripolyphosphate	oil well muds, clays
sodium hexametaphosphate	dyes, pigments
sodium lignin sulfonate	iron pigments, water based paints
calcium lignin sulfonate	sulfur dispersions
desulfonated sodium lignosulfonate	carbon black; titanium dioxide, clays
stearic acid	carbon black, zinc oxide and similar pigments

HOWARD M. GADBERRY

Cross-references: *Colloid Chemistry, Detergents, Carbon Black.*

DISSOCIATION

Dissociation can be broadly defined as the separation from union or as the process of disuniting. In chemistry, dissociation is the process by which a chemical combination breaks up into simpler constituents due, for example, to added energy as in the case of the dissociation of gaseous molecules by heat, or to the effect of a solvent upon a dissolved substance, as in the action of water upon dissolved hydrogen chloride. Dissociation may occur in the gaseous, liquid or solid state or in solution.

Elementary substances, if polyatomic in the molecule, will dissociate under conditions of sufficient energy. Chlorine and iodine, which are diatomic, are half dissociated at 1700°C and 1200°C, respectively. Just above the boiling point the molecule of sulfur is S_8. Its molecular weight decreases from 250 at 450°C to 50 at 2070°C. Thus there are some monatomic sulfur molecules at 2070°C. The dissociation probably takes place in reversible steps and can be represented by the equation:

$$S_8 \rightleftharpoons 4S_2 \rightleftharpoons 8S. \qquad (1)$$

Many chemical compounds dissociate readily upon heating or otherwise supplying them with energy. Acetic acid vapor consists of double molecules just above the normal boiling point, but dissociates completely into single molecules at 250°C. Nitrogen tetroxide (N_2O_4) is a pale reddish brown gas at temperatures near its normal boiling point of 21.3°C. On heating the density of the gas becomes less and the color becomes darker until it is almost black. At 140°C the molecular weight is 46 which is that of NO_2 molecules. The dissociation can be written:

$$N_2O_4 \rightleftharpoons 2NO_2. \tag{2}$$

If one mole of gas yields ν moles of gaseous products, and α is the fraction of the one mole which dissociates, then the total number of moles present is:

$$1 - \alpha + \nu\alpha = 1 + \alpha(\nu - 1). \tag{3}$$

Now the density of a given weight of gas at constant pressure is inversely proportional to the number of moles, and if d_1 is taken as the density of the undissociated gas and d_2 that of the partially dissociated gas, then:

$$\frac{d_1}{d_2} = \frac{1 + \alpha(\nu - 1)}{1} \tag{3a}$$

or

$$\alpha = \frac{d_1 - d_2}{d_2(\nu - 1)} \tag{4}$$

Therefore the *degree of dissociation* of a substance can be found by measuring the densities of the undissociated and partially (or completely) dissociated substance in the gaseous state. Molecular weights may be substituted for densities giving

$$\alpha = \frac{M_1 - M_2}{M_2(\nu - 1)} \tag{5}$$

The degree of dissociation can be used to calculate the *equilibrium constant* for dissociation. The equilibrium constant may be expressed in terms of concentrations, for example, moles per liter (K_c), or in terms of partial pressures (K_p). The degree of dissociation and equilibrium constants are important theoretically and practically, e.g., the latter can be used to ascertain the extent of a chemical process.

The temperature dependence of dissociation is expressed in terms of the equilibrium constant and is

$$\frac{d \ln K_p}{dT} = \frac{\Delta H}{RT^2} \quad \text{or} \quad \frac{d \ln K_c}{dT} = \frac{\Delta H}{RT^2} \tag{6}$$

where ΔH is the heat of dissociation. Integrating between the limits T_1 and T_2 one obtains

$$\ln \frac{K_{p_2}}{K_{p_1}} = \frac{\Delta H}{R}\left(\frac{T_2 - T_1}{T_1 T_2}\right)$$
$$\ln \frac{K_{c_2}}{K_{c_1}} = \frac{\Delta H}{R}\left(\frac{T_2 - T_1}{T_1 T_2}\right) \tag{7}$$

Electrolytes, depending upon their strength, dissociate to a greater or less extent in polar solvents. The extent to which a weak electrolyte dissociates may be determined by electrical conductance, electromotive force, and freezing point depression methods. The electrical conductance method is the most used because of its accuracy and simplicity. Arrhenius proposed that the degree of dissociation, α, of a weak electrolyte at any concentration in solution could be found from the ratio of the equivalent conductance, Λ, of the electrolyte at the concentration in question to the equivalent conductance at infinite dilution Λ_0 of the electrolyte. Thus

$$\alpha = \frac{\Lambda}{\Lambda_0} \tag{8}$$

This equation involves the assumption that mobilities of the ions coming from the electrolyte are constant from infinite dilution to the concentration in question. From the degree of dissociation and the concentration, the ionization constant or protolysis constant of a weak electrolyte can be obtained.

Water is a weak electrolyte, ionizing according to the equation:

$$H_2O + H_2O \rightleftharpoons H_3O^+ + OH^- \tag{9}$$

The specific conductance L, of water at 25° is 5.5×10^{-8} mho cm^{-1}, and the equivalent conductance of water at infinite dilution is found from the equivalent conductance of its constituent ions (H_3O^+ and OH^-) to be 547.8 mhos. The equivalent conductance Λ of water at 25°C is LV, where V is the volume of water (18 ml) containing 1 gram equivalent of water. Hence $\Lambda = LV = 5.5 \times 10^{-8} \times 18 = 9.9 \times 10^{-7}$. Therefore $\alpha = \Lambda/\Lambda_0 = 9.9 \times 10^{-7}/547.8 = 1.81 \times 10^{-9}$. Now $C_{H_3O^+} = C_{OH^-} = 55.5 \times 1.81 \times 10^{-9} = 1.00 \times 10^{-7}$ and

$$K = \frac{C_{H_2O^+} \times C_{OH^-}}{C^2_{H_2O}} \tag{10}$$

but C_{H_2O} is a constant, namely 55.5 moles/1 and therefore

$$K_w = (55.5)^2 K = C_{H_3O^+} \times C_{OH^-}$$
$$= 1.00 \times 10^{-7} \times 1.00 \times 10^{-7}$$
$$= 1.00 \times 10^{-14}$$

The ionization constant of pure water varies with temperature as shown below.

Temperature °C	0	10	25	40	50
$K_w \times 10^{14}$	0.113	0.292	1.008	2.917	5.474

Inserting corresponding values of K_w and absolute temperature into Eq. (7) and solving for ΔH one finds the heat of ionization per mole of water to be 13.8 kilocalories.

Ionization or dissociation in general can be repressed by adding an excess of a product of the dissociation process.

The acid formed when a base accepts a proton is called the conjugate acid of the base and the base formed when an acid donates a proton is the conjugate base of the acid. Thus in the reaction

$$HA + H_2O \rightleftharpoons H_3O^+ + A^- \tag{12}$$

HA and A^- are conjugate acid and base and H_2O and H_3O^+ are conjugate base and acid, respectively.

The common ion effect then can be found as the following example shows. When using ammonium hydroxide to which the common ammonium ion in the form of ammonium chloride has been added, the ionization can be represented by the equation

$$NH_3 + HOH \rightleftharpoons NH_4^+ + OH^- \tag{13}$$

Ammonium ion NH_4^+ is the conjugate acid of the

ammonia molecule. The ionization constant can be written

$$K = \frac{C_{NH_4^+} \times C_{OH^-}}{C_{NH_3}} \quad (14)$$

and

$$C_{OH^-} = K \frac{C_{NH_3}}{C_{NH_4^+}} \quad (15)$$

Now the base NH_3 is such a weak base that in the presence of NH_4Cl the concentration of unionized base, C_{NH_3}, is equal to the total concentration of base represented by C_{base}, and $C_{NH_4^+}$ coming from the weak base is so small as to be negligible. Hence the NH_4^+ ions can be considered as coming exclusively from the NH_4Cl. Therefore Eq. (15) can be written

$$C_{OH^-} = K \frac{C_{base}}{C_{salt}} \quad (16)$$

or

$$pOH = pK + \log \frac{C_{salt}}{C_{base}} \quad (17)$$

Thus C_{OH^-} and hence the degree of ionization of the base NH_3 is decreased with increasing concentration of salt. The salt effect of adding electrolytes with no common ion to a solution of incompletely ionizable substance can be seen from the following considerations and using the equilibriums represented by Eq. (14) which in terms of activities becomes:

$$K = \frac{a_{NH_4^+} \times a_{OH^-}}{a_{NH_3}}$$
$$= \frac{C_{NH_4^+} \times C_{OH^-}}{C_{NH_3}} \cdot \frac{f_{H_3O^+} \times f_{OH^-}}{f_{NH_3}} \quad (18)$$

This ionization constant in terms of activities is called the true or thermodynamic ionization constant. It does not differ too much from the K in Eq. (14) for sufficiently low ionic strengths. The two differ more markedly for appreciable ionic strengths. Now suppose a salt with no common ion is added to the solution. The ionic strength of the solution will be increased. This increase in ionic strength causes a decrease in the activity coefficients of the ions except in very concentrated solutions. Thus for K of Eq. (18) to stay constant the concentrations of the ions must increase to offset the decrease in their activity coefficients. The ammonia must therefore increase in ionization and K as defined by Eq. (14) must increase. This is known as a salt effect.

Ampholytes in solution give equal concentrations of a weak acid and a non-conjugate weak base. The amino acids are ampholytes which contain within their molecules equal amounts of a weak acid, the COOH group and a weak non-conjugate base, the NH_2 group.

According to Arrhenius those substances which yield the hydrogen ion in solution are acids, whereas bases produce the hydroxyl ion. As long as water was considered the only "ionizing" solvent these definitions were relatively simple. In the case of nonaqueous solvent chemistry at least three other concepts have been advanced. These are: (1) Franklin's *solvent system concept,* first limited to water and ammonia but since extended to nonprotonic media and defining an acid as a substance yielding a positive ion identical with that coming from auto-ionization of the solovent and a base as a substance yielding a negative ion identical with that coming from auto-ionization of the solvent; (2) the *protonic concept* of acids as proton donors and bases as proton acceptors advanced by Brønsted and by Lowry; and (3) Lewis' electronic theory according to which an acid is a molecule, radical or ion which can accept a pair of electrons from some other atom or group to complete its stable quota of electrons, usually an octet, and forming a covalent bond, and a base is a substance which donates a pair of electrons for the formation of such a bond.

In liquid ammonia as in water auto-ionization takes place. Ammonium and amide ions are formed by the dissociation or protolysis according to the following equation

$$2NH_3 \rightleftharpoons NH_4^+ + NH_2^- \quad (19)$$

The acid and base analogs of ammonia as a solvent is specified by this equilibrium as NH_4^+ and NH_2^- ions. All substances which undergo ammonolysis and hence bring about an increase in the ammonium ion concentration yield acid solutions. Thus P_2S_5 dissolves in liquid ammonia to give an acid solution as follows.

$$P_2S_5 + 12NH_3 \rightarrow 2PS(NH_2)_3 + 3(NH_4)_2S. \quad (20)$$

The solution is acid since an ammonium salt is formed and also because a solvo acid is obtained.

Many substances dissolve in liquid sulfur dioxide to yield ionic, conducting solutions. It has been found that such conductance data extrapolated to very high dilution yield the limiting conductance of sulfur dioxide. Both the Ostwald dilution law and the law of independent mobility of ions hold for "strong" electrolytes in highly dilute solutions.

The order of increasing dissociation and conductivity of salts in liquid sulfur dioxide apparently parallel the order of increasing cationic size. Probably because of solvation effects a similar relationship does not hold with respect to anion size. The mobilities of various ions in liquid sulfur dioxide have been studied. The van't Hoff i factors or mole numbers have been obtained by the ebullioscopic method for a wide variety of solutes in liquid sulfur dioxide. For non-electrolytes the mole number is one within experimental error. In liquid sulfur dioxide, univalent electrolytes give mole numbers which indicate large effects of ion-association of some kind. As would be expected the mole numbers of these electrolytes approach two in very dilute solutions. See Jander and Mesech, Z. physik. Chem., *A183,* 277 (1939). The mole number in general can be found from the ratio of the value of a colligative property of the solute is solution to the value of the same colligative property for a normal solute such as sugar, both solutes being at the same molal concentration.

The protonic concept of acid and bases is applicable to many of these high temperature solvent systems such as the fused ammonium salts which possess the "onium" ion or solvated proton, and the fused anionic acids which are salts possessing a metallic ion and a hydrogen containing anion. One of the most useful of the anionic acids is KHF_2

which is used to dissolve ore minerals containing silica, titania and other refractory oxides.

In many high temperature reactions there is an absence of hydrogen-containing ions. The Lewis electron pair concept of acid and bases can be used to advantage in such systems. In such systems strong anion bases such as the $O^=$ ion coming from basic compounds such as metallic oxides, hydroxides, carbonates or sulfates react with acidic oxides such as silica through the intermediate formation of polyanionic silicate complexes. The average ionic size of these complexes depend no doubt upon the temperature and the amount of added base.

Anion bases include the sulfide and fluoride ions coming from the corresponding alkali metal compounds. Likewise, metaphosphate and metaborate melts are acid in nature. Also proton-like character can be ascribed to any positive ion. The smaller the positive particle and the higher its charge, the greater is its polarizing tendency in bringing about deformation of negative ions, and the more reasonably can such an ion be looked upon as an acid analog.

When the potential energy of a diatomic molecule is plotted versus the distance separating the nuclei in the molecule, the potential-energy curve shows a minimum of zero in the energy at the distance separating the nuclei where the molecule is most stable, that is, where the nuclei are at the equilibrium internuclear separation. Energy is required to force them closer together or to pull them farther apart. The energy required to separate the nuclei to an infinite distance is D', the dissociation energy measured from the minimum of the potential energy curve. The spectroscopic dissociation energy D is smaller than D' by the zero point energy $\frac{1}{2} h\nu_0$. This results in the relationship,

$$D' = D + \frac{1}{2} h\nu_0 \qquad (21)$$

The spectroscopic dissociation energy D is the dissociation energy of an ideal gas molecule at absolute zero, where all the gas molecules are in the zero potential energy level, h is Planck's constant (6.62×10^{-27} erg second), and ν_0 is the frequency of vibration of the nuclei at the lowest vibrational level, which is above the point of zero potential energy at the equilibrium internuclear separation. Thus, for the hydrogen molecule, $D = 4.476$ electron volts, $\nu_0 = 1.3185 \times 10^{14}$ sec^{-1}, and since 1 electron volt = 23.06 kilocalories per mole we calculate D' using Eq. (21) as follows:

$$D' = (4.47 \text{eV})(23.06 \text{ kcal mole}^{-1}\text{eV}^{-1}) +$$

$$\frac{6.023 \times 10^{23} \text{ mole}^{-1} \times 6.62 \times 10^{-27} \text{ erg sec} \times 1.3185 \times 10^{14} \text{ sec}^{-1}}{(2)(4.184 \times 10^{10} \text{ ergs kcal}^{-1})} = 109.5 \text{ kcal mole}^{-1}.$$

E. S. Amis

Cross-references: *Ions, Solutions, Activity, Electrochemistry, Conductance.*

DISTILLATION

The term "distillation" refers to a process in which a liquid is converted to vapor and the vapor then recondensed to a liquid. The latter is referred to as the distillate. The original liquid (before vaporiza-

tion begins) is called the charge, and any unvaporized portion is known as the residue. Distillation is thus a combination of evaporation, or vaporization, and condensation. The natural cycle of evaporation and condensation that produces rain, and the vaporization and condensation of steam from a tea kettle on a cold surface might be considered as examples of distillation.

The usual purpose of distillation is purification, or separation of the components of a mixture. This is possible because the composition of the vapor is usually different from that of the liquid mixture from which it is obtained. Alcohol has been distilled for generations to separate it from water, fusel oil, aldehydes, and other undesirable impurities produced in the fermentation process. Gasoline, kerosine, fuel oil and lubricating oil are produced from petroleum by distillation. Distillation is extensively used in chemical analysis, in laboratory research, and for manufacture of a very large number of chemical products.

There are a number of different ways of performing distillation. Simple distillation is carried out by vaporizing the liquid mixture, as by heating with an electric heater, or steam, a flame, or bath of hot sand or liquid, and conducting the vapor to a condenser without subjecting the vapor to any treatment or circumstances that will cause the vapor to change composition after it is once formed. In the laboratory this is usually done by using a simple glass flask with a side arm that is connected to a glass water-cooled condenser and receiver (for example, the well-known Liebig condenser).

Historically, a retort was often used to accomplish the same purpose.

Simple distillation is widely used in industrial practice, the equipment consisting of a heated kettle of metal or ceramic ware with a vapor opening connected to a large condenser which drains into a receiving vessel. Steam is commonly supplied to a coil or jacket to provide the heat. The heated kettle is often referred to as a boiler, or still pot.

The most volatile component of a mixture tends to distil off first, but unless the other components are nonvolatile or are very much less volatile, even the first portion of the distillate is contaminated with the less volatile components. The degree of contamination, and also the progress of the distillation, is often estimated by reading a thermometer whose bulb is placed just below the point where the vapor enters the side arm of the still pot. Theoreti-

cally, such a thermometer should indicate the boiling point of the vapor passing over, but practical operating factors often cause considerable error in such measurements.

Simple distillation is useful when a volatile material is to be separated from one that is nonvolatile, or when there are great differences in volatility. The term evaporation is sometimes used for such processes, especially when the desired product is the residue, and the distillate is of only nominal or negligible value. More complex forms of distillation are used when it is desired to separate components of moderate or small difference in volatility. In mixtures of substances whose boiling points are widely different, the component of lowest boiling point is almost always the most volatile, the component of highest boiling point the least volatile. When boiling points are close together this simple normal situation no longer holds. The volatilities in a mixture of close boiling substances are often caused to be abnormally high or low by the presence of the other substances. Volatility is best expressed in terms of the actual partial pressures of each substance. This is calculated by the formula $p = \gamma P x$, where p is the desired partial pressure or volatility, x is the mole fraction of the desired substance in the mixture, P is its vapor pressure when pure at the temperature under discussion, and γ is a correction factor called the activity coefficient. The value of the activity coefficient is determined by the forces between the molecules of the different kinds of substances in the mixture being distilled. An extensive portion of physical chemistry and chemical engineering deals with methods for experimentally measuring vapor pressures and for predicting activity coefficients.

Every liquid or liquid mixture will evaporate into a closed space until the pressure of the resulting vapor reaches a characteristic or equilibrium value. This is termed the vapor pressure of the substance or mixture. The vapor pressure rises as the temperature of the liquid is increased. A liquid boils when its temperature is great enough so that the vapor pressure just exceeds the pressure of the surrounding atmosphere. Thus the boiling point is the temperature at which the vapor pressure of a liquid equals the external pressure on the liquid. An equivalent definition of boiling point is: the temperature at which a liquid and its vapor will be in equilibrium if maintained in a closed system from which the vapor cannot escape. Equilibrium as used here means that vaporization and condensation are going on with equal rates, so that the relative quantities of vapor and liquid do not change as time goes on, as long as the temperature and pressure remain constant.

When a mixture whose components differ very widely in volatility and boiling point is subjected to simple distillation, most of each component distils over in comparatively pure form at its appropriate boiling point. When nearly all of a given component has distilled over, the boiling point begins to rise and the succeeding portions of distillate contain less of the more volatile, lower boiling component and more of the next less volatile, higher boiling component until the latter comes over in comparatively pure form. A graph of boiling point versus quantity distilled (the boiling point curve) consists of alternate plateaus and steep rises. The plateaus represents nearly pure components and the rises are intermediate fractions.

When a mixture of close boiling components is subjected to simple distillation, all or most of the distillate is usually an intermediate fraction containing two or more components, and the boiling point rises gradually throughout the distillation. Sharp separation is not achieved.

A pure liquid distils entirely at a single fixed temperature, depending on the external pressure. However, such distillation with unvarying boiling point at a fixed pressure is not in itself a proof that the substance distilled is pure. It may be an azeotrope (or azeotropic mixture) of two or more components. Azeotropes may have boiling points higher than their pure components, or lower. The latter are more common.

The preceding discussion has assumed that the components of the mixture being distilled are completely soluble or miscible with one another. When a mixture of two completely immiscible liquids is distilled, both components distil over in constant proportions determined by their individual vapor pressures, until one component is exhausted. The most common example is steam distillation, described later. Partially miscible liquids behave somewhat similarly as long as two immiscible layers are present. When one disappears, the remaining material behaves like a completely miscible mixture.

Since simple distillation is ineffective for separating close-boiling mixtures, modified procedures known as fractional distillation and fractionation are in wide use. These procedures involve distillation in an apparatus with a still head or column as well as collection of successive portions of the distillate as a series of separate fractions or cuts. This is an example of *fractionation*.

The simplest way to obtain fractionation is to use a still head; this consists merely of an elongated neck on a distillation flask, usually with some baffle-type arrangement to prevent direct straight-through flow of the vapor. In consequence of air cooling some of the vapor condenses in the still head and the resulting liquid (called reflux) flows back down toward the boiler or still pot. This reduces the proportion of higher boiling or less volatile components that continue with the vapor stream, and thus produces sharper separation, but lowers the production of distillate per unit time. The use of still heads has been largely superseded by an even more effective device, the fractionating column or tower. This is placed between the boiler and condenser so that vapors ascend through the column, and some of the liquid condensate from the condenser (reflux) flows back down through the column to the still pot. The column is filled with a baffle-type device that forces the rising vapor and the descending liquid reflux to follow a devious irregular path and come into intimate contact with one another, to give the improved separation already described in connection with still heads. Condensation of vapor in the column itself is minimized by insulation or other means. The separation that results from contact of the countercurrent streams of vapor and liquid is referred to as rectification or fractionation. Product is withdrawn either as vapor from the top of the condenser, or obtained by withdrawing part of the liquid from the bottom of the condenser, before it reaches the point where it is introduced as reflux to the column.

Industrial distillation columns may be from 6″ to 20′ in diameter. These columns use a plate-type construction to achieve contact between countercurrent liquid and vapor streams. The most common type of plate column has horizontal plates at about two-foot intervals. Liquid stands on these plates to a height determined by an overflow weir delivering to the next plate below. Vapor bubbles through the liquid on the plate by means of one or more short chimneys on the plate, each covered by a cap. This arrangement is referred to as a bubble cap plate. Other plate constructions use a screen, or grid, or perforations over which liquid flows while vapor bubbles through the liquid. Smaller diameter columns use a great variety of packings such as glass beads, metal chain, short tube sections known as Raschig rings, and various specially shaped devices known as saddles, rings, or helices, as well as shapes made of screen or protruded metal.

The effectiveness of a fractionating column varies with the packing or contacting means used, and is measured in terms of theoretical plates. A theoretical plate is any contacting device that produces the same degree of separation as a single simple distillation. Thus a column that produces the same separation as ten successive simple distillations is said to have ten theoretical plates. The HETP, or height equivalent to a theoretical plate, is the total column height divided by the number of theoretical plates in the column. In plate columns, effectiveness is also measured by plate efficiency, which is the ratio (in percent) of theoretical to actual plates.

The use of a column or tower, with reflux, makes it possible to achieve sharp separations of all but azeotropes and materials of very nearly identical volatility. Columns with 50, 100, or more theoretical plates are used in laboratories and industrial operations. The chief factors determining separation are the ratio of volatilities of the components being separated, the number of theoretical plates, and the reflux ratio. The latter is the ratio of liquid returned as reflux to the product removed, or alternatively the ratio of descending reflux to rising vapor in the column.

The preceding discussion has dealt entirely with distillation processes in which a quantity of the material to be distilled (the charge) was introduced into the boiler or still pot at one time, before the start of the distillation, and product then removed from the condenser until the boiler contents were so nearly exhausted that operation was discontinued. This is known as batch distillation.

When large quantities of the same material are regularly distilled (as alcohol, petroleum, etc.) continuous distillation is used. The charge (material to be distilled, also sometimes called the feed) is introduced as a stream into some point in the column, so that part of it descends with the reflux and the remainder is vaporized and passes upwards. The most volatile component is concentrated at the top of the column, and removed from the condenser as already described. The least volatile component or a mixture may be removed from the bottom of the column or from the boiler, which is frequently referred to as a reboiler. Intermediate fractions are sometimes withdrawn from various points in the column, but these are never pure and must usually be redistilled.

Distillation under reduced pressure (vacuum distillation) is used to distil mixtures that normally boil at very high temperature, or that undergo decomposition when distilled at atmospheric pressure. Operation and equipment are similar to that already described, except that a vacuum pump must be used to reduce the pressure. Since a given weight of any substance produces a much larger volume of vapor at low pressures than at normal pressure, a given diameter column has a much reduced capacity when used for vacuum distillation. When very low pressures are used in distillation operations, the process is designated as molecular distillation because somewhat different phenomena determine the separation achieved and the proper operating conditions.

Steam distillation involves passage of a current of steam through the liquid in the boiler during a distillation. This causes a reduction in the boiling point of the liquid. When gases or vapors other than steam are used to achieve this effect, the process is called carrier distillation.

In some cases an extraneous material is purposely added in order to aid in separating by distillation. Azeotropic distillation is one such process, in which the added substance is usually chosen because it forms a low boiling azeotrope with one of the components to be separated. This azeotrope distils off first and leaves the other component or components. Benzene is commonly added to alcohol-water mixtures to remove the last 5% of water, since the benzene forms a low boiling azeotrope with water.

Extractive distillation is a somewhat similar process in which a properly chosen, relatively non-volatile liquid is added to the reflux at the top of a column because the volatility differences of the substance being separated are thereby increased enough to permit separation. Toluene and benzene are recovered from petroleum by extractive distillation with phenol. It is also an important step in obtaining the hydrocarbons used in the production of synthetic rubber.

Normally gaseous mixtures may be liquefied and then distilled and fractionated at subnormal temperature. Large quantities of air are thus distilled to produce commercial oxygen, nitrogen, and argon. Some gaseous mixtures are liquefied by pressure, and then distilled at the elevated pressure.

Distillation in one or more of its various forms is used in processing or purifying coal-tar products, turpentine, glycerol, fatty acids, essential oils, perfumes, formaldehyde, acetone, camphor, phenol, acetic acid, and a great host of other organic and pharmaceutical products.

One of the best known applications of simple distillation is the production of distilled water, for medical and chemical purposes. The large scale production of potable water from sea water or brackish waters is also practical. Specialized procedures such as multiple-effect vacuum distillation are necessary to reduce energy consumption and lower costs. Many plants producing hundreds of thousands and in some cases millions of gallons per day of sweet water from salt water or brine are in regular operation in arid areas and on islands.

Destructive Distillation

This is a process in which coal, or less frequently wood or oil shale or other substances are heated in the absence of air or oxygen so that thermal decomposition occurs and gases and vapors of the decomposition products are passed to a collecting and condensing system. A solid nonvolatile residue remains when the process comes to an end. The term carbonization is often used for such a process.

Bituminous coal is subjected to destructive distillation in large quantities to produce coke, coal tar, and coal gas.

Large quantities of wood were formerly subjected to destructive distillation to produce charcoal, methanol, acetone and other chemicals. Such destructive distillation processes could not compete economically with synthetic methods, but some charcoal is still produced by destructive distillation.

The recovery of oils and chemicals from oil shale or tar sands requires that these be subjected to heating in the absence of air, and this process is therefore at least partially an example of destructive distillation.

The terms cracking, pyrolysis, and thermal decomposition have the same general meaning and are roughly synonymous with destructive distillation. The latter however applies particularly to coal, while cracking refers most often to petroleum. Pyrolysis and thermal decomposition are general in meaning.

ARTHUR ROSE

Cross-references: *Coal and coke, Cracking, Pyrolysis, Desalination.*

DOSIMETRY (RADIATION)

The broad definition of radiation dosimetry is the measurement of radiation energy absorbed in a unit mass of the material of interest. Thus dose has the fundamental units of ergs/gram. Dose must be contrasted with the concepts of radiation source strength (ergs/sec) and radiation energy flux (ergs/cm² sec). The term radiation dosimetry is most frequently applied in radiobiology and radiation chemistry involving ionizing radiation such as the electromagnetic x and gamma rays, alpha, beta and proton charged particles and neutrons. It does include, however, the measurement of absorbed energy in other regions of the electromagnetic spectrum, particularly in photochemistry of the visible and ultraviolet.

The measurement of dose is relatively simple when the material is air or another gas, but becomes increasingly complex for liquids and solids, particularly when animal tissue is involved. As its name implies, ionizing radiation gives up its energy by causing ionization in the absorbing material. This ionization energy is then degraded to heat as the final form, just as in the case with less energetic electromagetic radiation.

The absolute measurement of dose in air is normally made with an ionization chamber where the ionization produced in the gas is collected and measured. Absolute measurements in water and certain solids can also be made calorimetrically but this technique is not widespread. A number of practical systems have been devised to approximate the dosimetry of ionizing radiation in air or tissue. These require energy absorption and a readout method that can be calibrated in terms of dose. Electrical techniques usually depend on the ion chamber but proportional and Geiger counter

measurements can provide dose data. Photographic emulsions are blackened by ionizing radiation and may be used for the more penetrating radiation. Photoluminescence, thermoluminescence and exoelectron emission properties of certain materials have practical applications in personnel protection and in the measurement of high radiation fields. High doses, however, are usually measured by chemical means, most of which rely on reactions taking place in water or on the oxidation of multivalent ions. All of these measuring devices except the absolute systems are generally more sensitive to flux than is desirable in a true dosimeter. Calibrations must be done carefully if the results are to be meaningful for estimating dose in the material of interest.

Fundamental units of ergs/gram are seldom used in practice. Workers in the photochemical field tend to use watt secs/gram for low energy radiation or to use the number of chemical transformations per 100 electron volts (eV) for ionizing radiation. This latter yield factor, abbreviated G, is a measure of the efficiency of energy transfer and also the nature of the transformation. For example, a very high G value may indicate that a polymerization is being triggered by the radiation.

Scientists in the field of ionizing radiation have developed entirely different units. They defined the roentgen, R, as a quantity of X or gamma radiation such that the associated corpuscular emission (i.e. the ions and electrons) per cm³ of air at STP produces ions carrying one esu of electricity of either sign. This unit was satisfactory when only electromagnetic radiation was studied and the measurements were being made with ionization chambers. It is not satisfactory, however, for particulate radiation where the rate of energy loss along the radiation path is highly variable. After several false starts, the rad which is a unit of absorbed dose of 100 ergs/gram was adopted. This unit is entirely independent of the type of radiation and is preferred for expressing absorbed dose in any specified material.

The order of magnitude of the energy changes may be shown quite simply. An x-ray photon may have an energy of a few thousand eV while an alpha particle energy will be a few million eV. Ionization requires about 32.5 eV in air, while the oxidation of Fe II to Fe III requires 6.5 eV. From this, it may be calculated that one rad absorbed in a gram of material will oxidize 1.6×10^{-11} moles of iron. The doses required in radiation chemistry are thus in the millions of rads, usually in small volumes of material. In contrast, 600 rad whole body doses to man will kill half of those exposed. Generally, radiosensitivity decreases as the biological system becomes simpler and millions of rads are required for bacterial sterilization.

The application of dosimetry to animals and man introduces a further complication. The heavier particles, such as protons and neutrons tend to produce a greater biological effect than would be found with the same number of rads of electromagnetic or beta radiation. An empirical factor has been introduced to allow for this difference. The original treatment was to use a multiplier called the RBE or Relative Biological Effectiveness. This has been succeeded by the term QF or Quality

Factor. Commonly used values for QF are shown in the following table.

Radiation	Quality Factor
X and γ rays	1
β particles	1
α particles	10
protons	2
thermal neutrons	3
fast neutrons	10
heavy nuclei	20

The product of the rad dose measured in air and the QF is known as the dose equivalent and is expressed in rems. Equal rem doses should produce about the same biological effects regardless of the type of radiation. The quality factors given are for purposes of radiation protection and are not exact for radiobiological studies.

It must be pointed out that the dose rate is also important for biological work. An instantaneous dose of radiation produces much greater damage than the same dose spread out over a long time. Thus the instrumentation for dosimetry includes direct-reading dose rate meters as well as integrating dosimeters designed to accumulate readings over weeks or months.

The ultimate biological application of dosimetry is handicapped by the need for measuring the absorbed dose in very small units of matter such as the cell or even chromosomes within the cell. The experimental aspects of this problem are far from satisfactory for radiobiological studies even though dosimetry for personnel protection appears to be adequate.

JOHN H. HARLEY

Cross-references: *Radioactivity, Photochemistry.*

References

International Commission on Radiation Units and Measurements, Radiation Quantities and Units—ICRU Report 11, Washington (1968).

Recommendations of the International Commission on Radiological Protection, ICRP Publication 9, Pergamon Press, Oxford (1966).

DRIERS

Driers comprise a group of metallic soaps (e.g., cobalt linoleate, cobalt naphthenate) incorporated in paints and varnishes to accelerate the oxidation of drying oils. The addition of driers tends to introduce complications. Drying oils, being natural products, are variable in composition, the resins often introduced are of extremely variable nature, the driers used are made of various raw materials, using different manufacturing methods, and the films, deposited in different ways, are dried in varying environments. Under these circumstances, it is almost impossible to do more than establish approximations of the directing effect of driers.

Oxidation. The degree to which the oxidation of drying oils is modified by the presence of driers is substantial, and the modifications fall into the following phases:

Shortening of the Induction Period. The I. P. is the time before the drying oil combines with a measurable quantity of oxygen. The *pure* glycerides of unsaturated fatty acids are presumed to have zero induction periods as compared to the impure glycerides in natural drying oils in which the I. P. is attributed to natural antioxidants not removed during refining.

Drying oils have shorter induction periods when they contain drier than when they do not. To account for this it has been proposed that driers may be considered as materials which chemically precipitate the natural antioxidants from true solution and/or as positive oxidation catalysts which counteract the natural antioxidants present.

Acceleration of Oxygen Combination. When driers are absent, the film, at the end of the I. P. combines with the oxygen at a rate characteristic of the oil. When driers are present, the rate of combination is accelerated. This has been explained in the following ways:

(1) Driers act as true oxidation catalysts, promoting the reaction but not entering into the reaction.

(2) Driers enter into the oxidation reaction, acting as oxygen carriers because of their susceptibility to oxidation-reduction reactions. The effective drier metals are postulated to be those of multiple valence. Further it is proposed that the lower valence state must be more stable than the higher valence state. High metal concentrations cause oil films to reach a maximum weight increase in a shorter time and begin losing weight in a shorter time than lower concentrations. No evidence has been found to support the contention that driers concentrate in the surface of the drying film and thus serve as special oxygen carriers. Activation of films deposited on other dried films containing driers appears to occur because of the high peroxide concentration in dried films during certain periods of their life.

(3) Driers may combine with the double bonds of the drying oils to form new compounds more susceptible to aerial oxidation. Experimental evidence indicates that this mechanism is unnecessary and probably does not occur.

Polymerization. Since polymerized films are more desirable for many purposes than oxidized ones, it is desirable to promote polymerization as much as possible. Some of the observed occurrences where polymerization has been modified by driers follow:

Solidification of Film at Earlier Stage in Oxidation Process. When driers are present, oil films solidify when less oxygen has combined than when driers are absent. Such a phenomenon may be caused because polymerization as well as oxidation is promoted by driers, so that the film reaches a given degree of thickening at a lesser degree of oxidation than when driers are absent.

Decreases in the Maximum Oxygen Combined. Less oxygen finally combines with the oil when driers are present than when absent, and this has been thought to occur because polymerization of unsaturated systems is accelerated in the presence of driers so that fewer double bonds are available for oxidation.

Promotion of a Higher Degree of Polymerization. It appears obvious from physical and chemical data that polymerization involving carbon-to-carbon bridging at the double bonds is promoted by driers. To explain this, polymerization has been postulated to be catalyzed by the formation of an addition compound with the drier at the double bond. As evidence to support such a proposal, it is known that lead soaps, which are known to form varying complexes with drying oils, are very active polymerization catalysts in spite of the fact that lead does not change valence readily. Other work, in many instances, has shown that the function of positive polymerization catalysts involves the formation of a complex with the catalyst.

Association. Metallic soaps (driers) are strongly polar compounds which show pronounced tendencies toward formation of complexes as well as exhibiting a dispersing action toward many colloidal aggregates. The chemical composition of the metallic soap used largely determines its dispersing activity. So, depending on the soaps used, various effects may be noted:

Solidification of Film at Earlier Stage in Oxidation Process. Because of their high polarity, driers may be considered capable of actively orienting oil molecules into micelles so that gelation or solidification may occur more readily. Such orientation of oil molecules, normal to the surface during the final stages of drying has been shown to occur in the absence of driers, so promotion of such action is feasible. The magnitude of this action, as compared to polymerization as a factor leading to solidification of the film at an earlier stage in the oxidation process, is unknown.

From these comments on the mechanism of the action of driers, it is apparent that the effect of driers in promoting more rapid drying of oils is marked. It is also apparent that a great deal more is known about the changes occurring in the films than about the reasons for the changes. This general situation is true in many fields, but is especially obvious in the case of driers.

S. B. ELLIOTT

Cross-references: *Metallic Soaps, Polymerization, Drying Oils.*

DRILLING FLUIDS

Drilling fluids are used in drilling wells by the rotary method. In that process, a fluid, or "mud," is circulated downward through a hollow drill pipe, whereupon it issues through ports in the bit attached to the lower end of the drill pipe, and rises to the top of the well in the space between the drill pipe and the walls or casing of the hole. The "mud" is freed of any cuttings or drilling debris which it has brought to the surface, and recirculated into the drill pipe. The bore hole is maintained full of drilling fluid during drilling. It functions to raise cuttings, to keep formation fluids from issuing into the hole by exerting a higher pressure hydrostatically, to cool and lubricate the bit and drill string, to buoy the drill pipe, and to bring formation samples to the surface for inspection. Accordingly, the fluid must have a suitable density, 2.0 or higher often being necessary; it must have proper rheological characteristics, in particular an apparent viscosity at high rates of shear of ten or twenty times that of water and a low but appreciable gel strength; however, it must be subject to thixotropic increase upon standing, and be able to resist filtration into a porous medium.

The first muds consisted simply of clay in water. Only rarely, however, do clays alone permit fluid densities greater than about 1.25, so that finely ground, inert materials of high intrinsic density were later introduced. Barite, hematite, celestite, witherite, and other minerals have been used; now ground barite is used almost exclusively, in excess of 1,200,000 tons per year annually in the United States, by far the major use of this mineral.

Clays are commonly obtained locally, but only in certain areas, notably California, do clays occur which provide excellent drilling muds without further admixture. Bentonite was introduced as a drilling mud admixture in the 1920's, and about 600,000 tons per year are now so employed in the United States. It is used for its properties of imparting viscosity, thixotropy, and infiltrability. The thixotropic nature of bentonite suspensions is particularly useful where weighting material such as ground barite must be suspended. Attapulgite is widely used where consistency must be imparted to aqueous muds high in dissolved salts, a condition routinely encountered in drilling through salt-bearing formations.

Water-base muds are subject to increase in consistency from a number of causes. Large quantities of polyphosphates were formerly used in combatting thickening in drilling muds, particularly from flocculation. Polyphosphates are not well suited, however, to drilling muds used at extreme depths because bottom hole temperatures rapidly bring about reversion to orthophosphates. Of growing importance, therefore, is the use of organic mud thinners, the most widely used of which are lignite and lignosulfonate derivatives, especially ferrochrome lignosulfonate. Tanstuffs, such as quebracho extract, are of declining importance. Certain surfactants, as well as certain calcium salts are used to inhibit swelling and disaggregation of drilled-up shale while still maintaining dispersion of the mud particles. The clay-slip deflocculants used in the ceramic industry have almost no place in drilling mud technology as such, although large amounts of caustic soda are used in conjunction with tanstuffs and lignites to neutralize the tannic and humic acids contained therein; and sodium carbonate is occasionally used for the removal of calcium ions.

Oil-base muds have a fluid medium of oil instead of water. Semirefined petroleum oil such as diesel oil and certain types of crude oil are generally used. Thickening and antifiltration agents used include air-blown asphalt, alkali metal and alkaline earth metal soaps, particularly of tall-oil fatty acids, and acids derived from rosin; and long-chain onium clay complexes. Ground limestone is generally used to increase density, to assist in resisting filtration, and to impart some consistency to the mud. Barite is the most generally used weighting material for heavy oil-base muds, although more hydrophobic material is desirable. Oil-base muds are particularly useful where contamination of formations penetrated, particularly oil-bearing formations, by water from the drilling fluid must be avoided.

Emulsion drilling fluids are widely used, both of the oil-in-water and water-in-oil types. The former is generally a dispersion of 10% to 20% by volume of diesel or crude oil in an existing clay-water mud, to which has been added an emulsifying agent such as tanstuff, a lignite or a water-dispersible soap. The oil reduces the filtration of the mud, reduces the density somewhat, diminishes bit wear, and reduces the incidence of stuck drill pipe. Water-in-oil emulsion muds are increasingly common; they have many of the characteristics of oil-base fluids, with the advantage of lower cost and somewhat higher density. The aqueous phase may contain salts, especially calcium chloride, to reduce osmotic effects, particularly in shale drilling.

Infiltrability in drilling muds is a desirable attribute because of the universal tendency of muds to undergo self-filtration whenever in contact with porous, permeable formations, since the hydrostatic pressure of the mud column is normally maintained higher than that of formation fluids, to prevent entry of the latter into the bore hole during the drilling operation. Infiltrability is measured by actual filtration tests in a pressure filter, and the choice of clays, colloidal additives generally, chemical treatments of the mud, and of constituents of oil-base and emulsion muds is largely regulated by the resulting filtration qualities imparted to the mud. Such filtration tests preferably conform to bottom-hole conditions of temperature and pressure, which may be quite high.

Even with careful selection of clays and bentonites, the filtration properties of water-base muds may not be low enough, and large quantities of organic additives are currently used as a replacement or supplement for clays or bentonites, including gelatinized starch, water-dispersible gums, bacterial polysaccharides, such as gum dextran and Xanthomonas colloid, sodium carboxymethyl cellulose, sodium polyacrylate and other similar water-dispersible colloids. These organic materials are particularly suitable in briny muds where maintenance of clays in a state of deflocculation is generally difficult.

D. H. LARSEN

Cross-references: *Clays, Colloid Chemistry, Emulsions.*

References

Anon., "Principles of Drilling Fluid Control," Edition 12, Petroleum Extension Service, University of Texas, Austin, 1969.

Rogers, Walter F., "Composition and Properties of Oil Well Drilling Fluids," Edition 3, Houston, 1963.

Larsen, D. H., "Use of Clay in Drilling Fluids," California State Division of Mines Bulletin **169**, 269–281 (1955).

DRYING

Drying is one of the unit operations of chemical engineering and can be defined as the removal of a liquid, usually water, from a solid by the application of heat. The goal is to obtain a product which is essentially moisture-free. This definition distinguishes thermal drying from mechanical dewatering that occurs in filters and centrifuges. Drying may also be accomplished by adsorption using such materials as molecular sieves, silica gel, activated carbon or alumina; these are commonly used to dry gases.

In the majority of the process industries, the drying operations are carried out on solids and for the following reasons: (1) to facilitate handling of the product, (2) to reduce the cost of shipping, (3) to meet the requirements for use of the product, and (4) to prevent deterioration of the product in shipment and storage.

Theory. The drying of a solid material involves the simultaneous transfer of heat and mass. The heat required for evaporation of the liquid in the material can be supplied by conduction, convection, and/or radiation. In a few cases the energy can be generated by dielectric heating. The liquid to be removed is transferred by diffusion or capillary flow within the material to the evaporating surface and then into a carrier gas or vacuum space.

Experimentally, most materials exhibit two or more distinct drying periods characterized by the terms constant-rate and falling-rate periods. These periods can be identified by the controlling mechanism in each case and refer to a constant or decreasing rate of water removal per unit time.

Drying in the constant-rate period proceeds by diffusion from the saturated surface of the material across a stagnant gas layer into the environment. The movement of moisture within the solid is usually adequate to maintain a saturated surface condition. In this period the rate of drying is controlled by the rate of heat transfer to the surface which balances the mass transfer. The temperature of the solid tends to remain constant during this period, although the temperature level is dependent on the mode of heat transfer. With convective heat transfer, the solid temperature approximates the adiabatic-saturation temperature.

As the drying proceeds, the rate the liquid is supplied to the evaporating zone is inadequate to maintain a saturated surface. The falling-rate period is initiated and the average moisture content of the solid at this point is termed "the critical moisture content."

During the falling-rate periods the instantaneous drying rate continually decreases while the temperature continually increases. The initial portion of this period is termed the first falling-rate period and is essentially a transition period between the constant-rate period and the second falling-rate period. The decrease in the drying rate is due to a decrease in the amount of saturated surface. In the second falling-rate period the rate of drying is governed completely by the rate of internal moisture movement.

The internal moisture movement takes place by liquid diffusion, capillary flow or vapor diffusion. Liquid diffusion takes place under the influence of the concentration gradient. Capillary flow results from the suction created in the pores of the material by surface-tension effects. Vapor diffusion occurs when the material is heated at one surface and drying occurs at another surface. Vapor movement results from a vapor pressure gradient through the solid.

In general, it is possible to predict the drying

rates for the constant-rate period from correlations of heat and mass transfer. Prediction of the critical moisture content and therefore the duration of the constant-rate period is usually impossible. In addition, drying rates during the falling-rate periods can only be approximated. As a consequence of these uncertainties, information on drying a particular material is usually obtained by tests conducted in prototype equipment. Handling characteristics and drying conditions are determined simultaneously and these are used for selecting the appropriate equipment.

Drying Equipment. Drying equipment must provide for the transfer of heat required for vaporization, the removal of the vaporized liquid and the physical handling and conveying of the material during drying. In addition, design controls must be used to insure that the thermal treatment does not degrade the material, that it transfers heat efficiently, and that it satisfies the handling requirements of the many different materials involved. The paper and food industries require large-scale drying equipment, including steam-heated rollers, and vacuum and tunnel dryers.

CHARLES K. HERSH

References

Williams-Gardner, A., "Industrial Drying," CRC Press, Cleveland, Ohio, 1971.

Chem. Eng. **74**, No. 13, "Drying," 1967.

DUST, INDUSTRIAL

Dust is a gaseous suspension of solids as a result of a mechanical process of dispersion; fume, on the other hand, is a term applied to suspended solid particles which have been formed by a thermal or chemical process, and is composed of much smaller particles. All dust is characterized by a broad distribution of particle sizes, a quality which has important implications in considerations of toxicity, measurement and separation from the gaseous phase. Ordinary dust includes an insignificant mass of discrete particles smaller than about 0.1 micron and even in the case of fumes a large population of separate, uncoagulated particles of smaller size can be expected only in cases where rapid dilution occurs at the site of formation.

In the open atmosphere natural sedimentation operates to limit the relative number of particles larger than 5 or 10 microns, but on a bulk basis particles as large as 50 microns or greater may be present in significant amounts, depending on the nature of the dust source. Though relatively few in number, such large particles may constitute the bulk of the weight in a typical dust mixture and thus profoundly influence measurements of weight concentration.

The toxicity of industrial dust resulting from its inhalation is dependent on the rate of accumulation within the lungs (ingestion is of comparatively minor importance) and on the biochemical nature of the dust. The accumulation is determined by the concentration of particles in the air and by the fraction which can penetrate the upper respiratory tract to the terminal sacs of the lung. In the case of mineral dust having a specific gravity near that of quartz, it has been established that the upper

respiratory tract does not remove any significant portion of particles smaller than 1 micron, but does remove nearly all those larger than 5 to 10 microns. Deposition in the lungs of smaller particles is influenced by two opposing factors. As particle size decreases, depth of penetration increases, whereas efficiency of removal by sedimentation decreases. The net effect is that the optimum size for deposition in the deep spaces of the lung is in the range 0.5 to 2 microns; and 40 to 50% of particles of this size inhaled are deposited in the alveoli. These facts have an important influence on considerations of dust toxicity.

Silicosis is a disease resulting from deposition in the lungs of dust containing significant amounts of uncombined silica, such as quartz, cristobalite, or tridymite. None of the silicate dusts can produce silicosis. However, if mineral silicates are subjected to high temperatures for sufficient time, free silica in the form of cristobalite and tridymite may be formed and dust from such materials can produce silicosis. There is some question concerning the ability of some of the amorphous free silica dusts, such as silica gel, to produce silicosis.

A superficially similar dust disease, asbestosis, results from human exposure to excessive quantities of asbestos dust. This is the only silicate dust whose toxicity has been well established although questions have been raised about some others.

In mineral dust exposure studies for appraisal of silicosis hazards, measurement of dust concentration is based on microscopic count of particles that have been condensed by a suitable air sampling instrument. Conventional dust count procedures in routine dust measurements (U.S.) employ low magnification which excludes from the count those particles much smaller than 1 micron in diameter. Conversely, an insignificantly small portion of total particles found in atmospheric suspension are greater than 5 microns.

More recent developments in the U.S. (ca. 1960–1970) permit the determination of dust concentrations on a weight basis. Air sampling methods for the weight concentration basis require attention to the larger particles that are present in atmospheric suspension but not respirable into the lung. Accordingly, air sampling assemblies may include a "size selector," and elutriation devices of specified aerodynamic characteristics attached to the front end of the sampling apparatus to remove oversize particles. Sampling may also employ apparatus with no size selector. Applicable TLV values depend on which sampling method is used.

The threshold limit values on a dust count basis for various mineral dusts, promulgated by the Occupational Safety and Health Administration, U.S. Department of Labor (Federal Register May 29, 1971), as later amended (Federal Register Dec. 7, 1971) are summarized, in part, below. These tabulated values are based on air sampling by impinger, and evaluation by light-field dust count. (Values given are in millions particles per cubic foot).

Portland cement and various
  other inert dusts .... 50 mppcf
Diatomaceous earth, mica,
  soapstone, talc (i.e., less
  than 1% crystalline silica) ..   20   "

Dusts containing free silica, as quartz, have TLV's expressed as a function of per cent free silica .... $\dfrac{250}{\% \ SiO_2 + 5}$ "

Dusts containing free silica in the form of tridymite or cristobalite .... One-half the value of quartz-containing dusts, above.

Values of TLV expressed in terms of weight concentrations are dependent not only on per cent free silica, but also on whether the sampling apparatus includes a size selector.

For straight quartz dust, the TLV in weight units is 0.3 milligrams per cubic meter for gross samples, i.e., if sampled with no size selector for removal of larger particles; and 0.1 milligrams per cubic meter for respirable dust, i.e., if a size selector is used. The indicated ratio of 3 to 1 represents from experience the approximate proportions of gross airborne dust to respirable dust.

For coal dust (less than 5% SiO_2) the TLV is 2.4 milligrams per cubic meter, employing a particle size selector in the air pollution assembly; there is no TLV value applicable to gross sampling.

The TLV for asbestos dust is based on the number concentration of asbestos fibers longer than 5 microns as determined by the membrane filter method, at 400–450X magnification, phase contrast illumination. The TLV for an 8-hour time-weighted average is 5 fibers per milliliter of air.

Due to comfort considerations dust respirators for protection of humans in dusty atmospheres are useful only where the exposure is of short or infrequent duration. The most certain and satisfactory protection lies in engineering measures either (a) for prevention of dissemination, as by controlled addition of moisture to the parent material, complete enclosure or isolation or (b) by the most generally applicable technique of local exhaust ventilation in which the dusty air is withdrawn into a duct system at its point of origin for discharge either to the outdoors or to a suitable dust collector.

A dust explosion may involve any dust composed of a substance which reacts with oxygen. This includes a number of metals. Particle size distribution and atmospheric concentration are important factors determining the explosion potential. Minimum required concentrations are reported to range upward from 25 grams per cubic meter.

W. C. L. HEMEON

Cross-references: *Air Pollution, Toxicology.*

DYES

Dyes are intensely colored chemical compounds, which when applied to a substrate impart color to this substrate. Retention of color as well as stability are required functional properties and are accomplished by chemical and physical forces such as chemical bonding, hydrogen bonding, Van der Waals forces, adsorption, solution, electrostatic interaction and others. The color of dyes is due to the interaction of visible light with the electron system of the dye molecule. Several hundred thousand known compounds qualify as dyes based on their light absorption in the visible region of the electromagnetic spectrum. However, of these only about 1500 have proved to be of practical value and are being manufactured. Commercial uses of dyes include the coloration of textiles, paper, leather, wood, inks, fuels, food items, and metals. Dyes are used also in photographic paper, as indicators in analysis, and as biological stains.

The U.S. market for dyes has grown from approximately $170 million in 1952 to nearly $400 million in 1970. Four application classes of dyes—the vat dyes, the directs, the disperse and the acid dyes—account for more than 60% of this sum. Following World War II U.S. imports of dyes increased stetadily and in 1970 were nearly $40 million.

The functional properties required of commercially useful dyes vary markedly and depend to a great extent on the end use. For certain uses such as printing inks fastness properties are of secondary importance and the dyes are largely selected on the basis of economy. For other uses, e.g., carpets, resistance to light fading is more important than washfastness; the opposite is true for apparel, which is washed more frequently than it is exposed to an intensive source of light. A number of the synthetic fibers are dyed and finished at elevated temperatures (200°C) and all dyes present have to exhibit sublimation fastness. Since many of the fastness properties of the dyes are dependent upon the environment in which the dye molecule finds itself, dyes have to be carefully screened under many different dyeing conditions before they can be commercialized. With textile technology advancing towards more complicated blends of different fiber types, greater demands are made on new dyes. Often it is advantageous to dye each type of fiber in the fabric with a different kind of dye. It is of great importance in such a system to avoid cross staining (a fiber dyed with a dye not intended for it) and to work with dyes essentially free of impurities.

Dyes may be classified in various ways, according to color (yellow, red and blue, etc.), origin (natural or synthetic), chemical structure, substrates to which they are applied, and methods of application.

Classification of dyes based on their chemical structure is the most precise. However, for the user of coloring matters this is not always the best arrangement, since it makes no provision for dyes whose constitutions are unknown or have not been disclosed. The Colour Index [3rd ed., 1971, publ. jointly by AATCC and SDC (Engl.)], the latest and most complete compilation of dyes, therefore classifies them according to both structure and usage.

Nitroso Dyes (Quinone Oximes). The chromophore —N=O characterizes nitroso dyes. They are classified as mordant dyes and, except for the green iron lakes, have found very limited application in dyeing and printing. Usually they are prepared by reacting phenols or naphthols with nitrous acid. Other synthetic methods have also been reported, such as the simultaneous introduction of the nitroso and hydroxyl groups by the reaction of nitrosyl radical and an oxidizing agent.

Nitro Dyes. *o*- and *p*-Nitrophenols and *o*- and *p*-nitroanilines and their derivatives make up this class. Picric acid, one of the first synthetic dyes, is no longer employed as a textile colorant, but Naphthol Yellow S in form of its sodium salt is still used as a cheap dye for wool and silk giving pure yellow shades.

A general method of preparation of these dyes is by nitration of phenols, naphthols and diphenylamines. Many dyes have another chromophore in addition to the nitro group, but they are generally classed with the other dyes containing the second chromophore.

Wool and silk have been dyed with water-soluble acid nitro dyes in yellow and orange shades. Nitro dyes in the form of insoluble disperse dyes have served for the coloration of acetate, polyamide, polyester and similar fibers.

Azo Dyes. Azo dyes are characterized by the common chromophore the azo group —N=N— and generally have such auxochromic groups as hydroxyl, amine, and substituted amino groups. Azo dyes to a large extent are manufactured by reacting primary arylamines with nitrous acid in mineral acids (diazotization), yielding diazonium salts which are reacted, usually without isolation, with aromatic amines, phenols or enolizable ketones (coupling). The coupling with amines generally proceeds under acidic conditions, while phenols require alkaline media.

The process of diazotization and coupling can be repeated giving rise to disazo and polyazo dyes. The number of possible synthetic alternatives is so great that azo dyes easily form the largest class of synthetic dyes. Azo dyes have been developed for coloring every fiber, natural and synthetic, and for the coloration of solvents and a wide range of non-textile substrates.

The shade range of monoazo dyes covers all colors from greenish yellow to blue. However, there are only a few blue structures with the exception of metallized dyes, which generally show a bathochromic shift compared to the corresponding unmetallized dyes. Additional azo groups in the molecule produce a change to predominantly dark colors such as browns, navies and blacks.

Azoic Dyes. There is no fundamental chromophoric difference between azo and azoic dyes. The differentiation is made to characterize a group of azo pigments which are precipitated within the cellulosic fiber by carrying out the dye coupling on the fiber. Good functional properties and a wide range of brilliant shades can be attained with these dyes, but with the advent of the equally brilliant but more easily applied fiber-reactive dyes the azoics have lost some of their importance. A few of the naphthol azoics still find use in discharge and resist printing.

Stilbene Dyes. This group of dyes has a limited range of shades and is essentially confined to yellows, oranges and browns; they result from the condensation of 5-nitro-o-toluenesulfonic acid in alkaline medium with other aromatic compounds, generally arylamines. Most of the structures have not been elucidated and most commercial stilbene dyes are mixtures rather than single compounds.

The stilbene dyes are essentially all direct dyes with a few exceptions where these colors find use as acid and solvent dyes.

Diphenylmethane Dyes. Diphenylmethane dyes are basic colors which exhibit relatively poor fastness properties. The common feature is the chromophore $\overset{\backslash}{C}$=NH, which shows sensitivity to hydrolysis.

The best known members of this group are the Auramines, used on wool, cotton, paper, leather, silk and jute.

Triarylmethane Dyes. The triarylmethane dyes are among the oldest synthetic dyes. The chromophore of this class is the quinonoid grouping, which may appear as

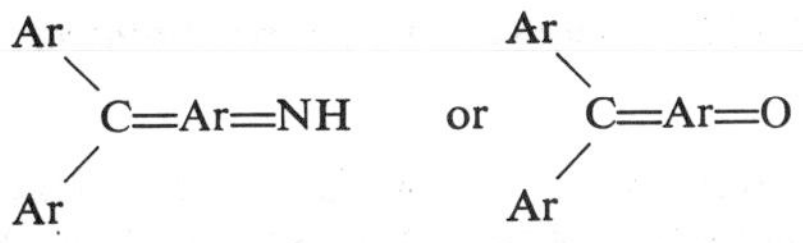

$$\overset{Ar\backslash}{\underset{Ar\diagup}{C}}=Ar=NH \quad or \quad \overset{Ar\backslash}{\underset{Ar\diagup}{C}}=Ar=O$$

where Ar symbolizes a substituted aryl group. The color and properties depend on the kind and numbers of the substituents. The presence of the sulfonic acid group confers water solubility and such dyes are applied as acid dyes. In the absence of acidic groups the dyes are called cationic or basic dyes and, where hydroxyl groups are present as auxochromes, adjacent carboxy groups confer mordant dyeing properties. Thus the class includes basic, direct, acid, mordant and solvent dyes. Nearly all are of brilliant hue, the range running from reds and violets to blues and greens. Fastness to light is generally poor to fair, except on polyacrylonitrile, where it is outstanding.

Xanthene Dyes. Xanthene dyes have the heterocyclic ring system of xanthene in common. The properties of the individual members depend largely on the substituents on the ring system and the class is subdivided into amino, aminohydroxy, and hydroxy derivatives. Some of the xanthene dyes resemble triarylmethane dyes in that they exhibit unusual brilliance and basic dye properties such as the following CI Basic Red.

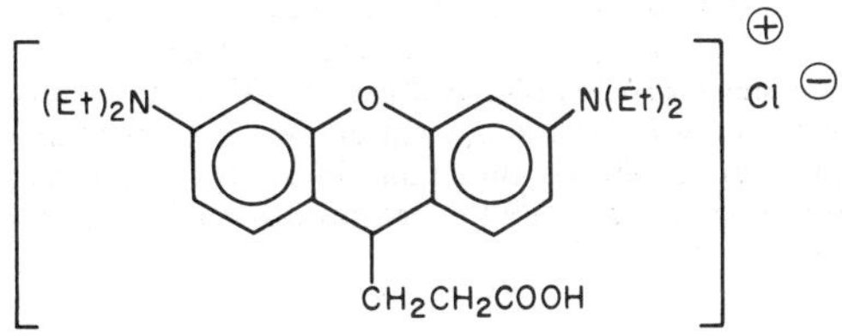

Several xanthene dyes are of commercial importance as cationics for polyacrylonitrile. Their use in the dyeing of wool and silk from weak acid baths and cotton on a tannin mordant is very limited because fastness properties on these substrates are generally poor, especially fastness to light.

Acridine Dyes. Acridine Orange R is a typical member of this class of dyes which incorporates mostly basic dyes of yellow, orange, red and brown shade. The dyes have found extensive use on leather. Their use on silk, wool or cotton, however, is restricted due to their lack of adequate fastness.

Quinoline Dyes. The condensation of methylquinoline and its derivatives with phthalic anhydride affords quinophthalones of the following structure:

2 - (2 - Quinolyl) - 1 , 3 - Indanedione

These dyes, mainly yellow or red compounds, may be applied as paper, food, solvent, basic and disperse dyes; when sulfonated they give excellent acid dyes for wool.

Methine Dyes. Methine dyes, also known as cyanines, are characterized by the presence of the methine group —CH= or a conjugate chain of such groups. The majority of the dyes contain heterocyclic systems such as quinoline, benzothiazole or trimethylindoline, and are formed by linking these heterocycles together by means of a chain of methine groups.

The main use of these dyes is as photographic sensitizers. In general they have poor fastness to light and therefore have limited use as basic dyes on textile fibers. One notable exception is CI Disperse Yellow 31, a dye of exceptional fastness to light and gas fume fading.

Thiazole Dyes. The parent substance of this group of dyes is dehydrothio-p-toluidine, which upon sulfurization at higher temperatures gives rise to the so-called Primuline Bases, which can be sulfonated to afford direct cotton dyes. They also can be diazotized and coupled with amines, phenols, and naphthols to produce azo dyes of yellow, orange and red shades. Primuline is used to color cotton where washfastness but not fastness to light is the primary consideration. Hypochlorite oxidation of the dehydrothio-p-toluidine, on the other hand, produces dyes of relatively high light fastness.

Indamine and Indophenol Dyes. Oxidation of mixtures of an aromatic p-diamine with an aromatic monoamine affords the indamines, while indophenols result from oxidation of mixtures of p-aminophenol and phenol.

Azine Dyes. Azine dyes are derivatives of the heterocyclic phenazine system. They are generally basic dyes, have been used on wool and silk, and dye cotton on a tannin mordant. Sulfonated derivatives are acid dyes and nonsulfonated compounds are used to color fats, lacquers, and oils. The most famous but now obsolete member of this dye class is Mauveine.

Oxazine and Thiazine Dyes. Oxazine dyes are derived from the heterocyclic system of the same name and are manufactured by condensing p-nitroso-N,N-dimethylaniline with suitable phenols. Most oxazine dyes are classified as basic colors for polyacrylonitrile, wool and cotton (on a tannin mordant). These dyes have also been used on leather and, after sulfonation, as direct dyes. Among the latter group are several brilliant blue dyes with excellent fastness to light.

The thiazine dyes are closely related, their structure including the thiazine ring. Several members of this group are used as stains for tissues in biology and medicine. As coloring matters for textile fibers they are of less importance.

Sulfur Dyes. Sulfur dyes are complex mixtures of uncertain composition and, while certain structural features such as the thioketone, the disulfide, the thiazole group have been identified, it has not been possible to assign precise chromophores. Sulfur dyes are used extensively on cellulosic fibers because of their low price and relatively good fastness. They are applied by a method similar to the vatting process. Since they are water-insoluble compounds, they are rendered water-soluble by reduction with sodium sulfide. In this form they are applied to the cotton and then are reoxidized (in air) to the original dye. Sulfur dyes are produced by heating relatively common aromatic compounds with sulfur or sulfur derivatives. The usual classification of sulfur dyes is based on the type of starting material such as amines, phenols, nitro compounds, and carbazoles.

The shade range provided by sulfur colors spans the spectrum from yellow to black. However, most of them are fairly dull and their great asset is their cheapness.

In spite of recent important technical advances in the production of sulfur dyes in the form of dispersions, stable reduced solutions, and nonsubstantive thiosulfonic acid derivatives, the use of these colors has decreased; in 1966 they represented

Indamine
(Phenylene Blue)

Indophenol
(CI Solvent Blue 22)

The functional properties of these dyes make application to fibers impractical. However, they have found use in photography as well as in the preparation of sulfur dyes.

nearly 9% of total United States production while in 1969 this percentage went down to about 6%.

Lactone Dyes. The group of lactone dyes is a very small one. They are prepared by oxidation or hydrolysis of polyhydroxy aromatic compounds, such as gallic acid to produce Alizarine Yellow (MLB).

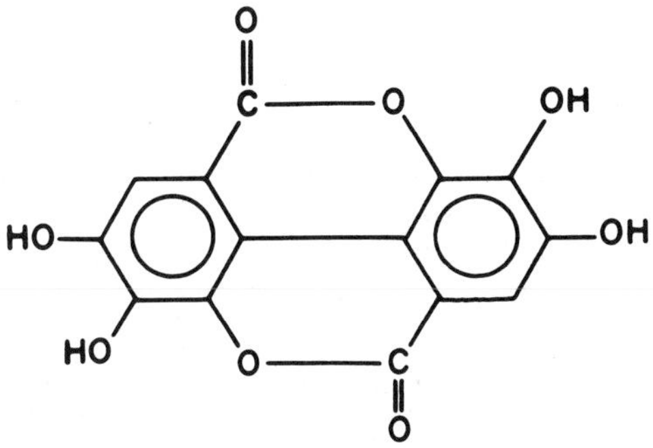

CI 55005 Alizarine Yellow(MLB)

The shade range of these colors is limited to yellows and olives and their application is restricted to use on chrome-mordanted wool.

Aminoketone and Hydroxyketone Dyes. These two groups of dyes derive their color from the carbonyl chromophore and amino and hydroxy groups as auxochromes. The amino-ketone dyes are generally arylaminoquinones and are applied to wool by a vatting procedure. The hydroxyketone dyes are mostly hydroxyquinones and hydroxy derivatives of aromatic ketones and are applied by mordanting techniques.

Anthraquinone Dyes. Anthraquinone dyes derive their color from one or more carbonyl groups in association with a conjugated system. Auxochromes include amine, hydroxyl, alkylamino, arylamine, and acylamino groups, as well as complex heterocyclic system. Anthraquinone dyes are found in many different usage groups, the more important ones being the acid, direct, disperse, mordant, solvent, vat, pigment and reactive dyes. The hydroxy derivatives of anthraquinone, such as alizarin (the coloring matter of Madder), exist naturally.

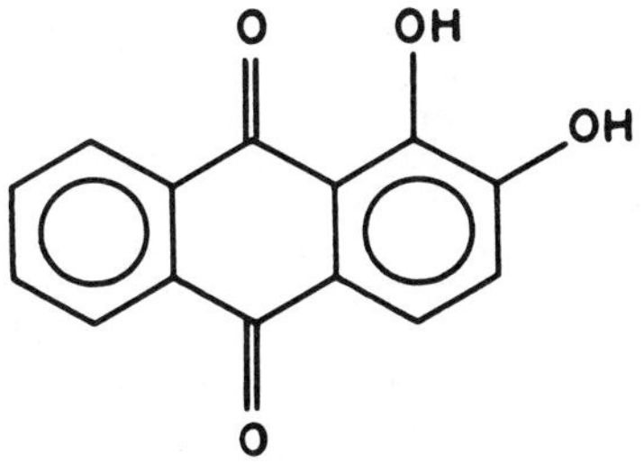

Alizarin

After the discovery of its constitution many other hydroxyanthraquinones and their derivatives were introduced as commercial mordant dyes. Anthraquinone acid dyes are usually sulfonated arylamino-anthraquinones. As a group they exhibit good fastness to light and wet treatment and provide bright shades on wool and synthetic polyamides. An important widely used example is CI Acid Blue 47.

Of ever-increasing importance are the anthraquinone disperse dyes. Structurally they are fairly simple, solubilizing groups such as the sulfonic acid group being absent. Initially these dyes were limited to use on acetate fiber; with the spectacular growth of synthetic fibers after World War II, however, production of disperse dyes increased rapidly. A well-known member of this group is CI Disperse Red 11

The first of the anthraquinone vat dyes was indanthrone, discovered in 1901. A wide range of vat dyes followed over the years, highlighted by the introduction of a bright green (Caledon Jade Green) in 1920 and the commercialization of stable leuco dye solutions in 1921. As a class, vat dyes exhibit excellent fastness to light, washing and bleaching. They are applied in the reduced form which is soluble under strong alkaline conditions and then are reoxidized to their water-insoluble colored pigment form by use of air or other mild oxidizing agents. The alkaline dyeing conditions have confined the use of vat dyes essentially to cellulosic fibers. However, the development of stable, neutral-dyeing leuco esters has broadened the applicability of the vat dyes to include wool, silk and nylon.

Indigoid Dyes. This class of dye is characterized by the chromophore

$$-\overset{\displaystyle O}{\overset{\|}{C}}-C=C-\overset{\displaystyle O}{\overset{\|}{C}}-.$$

The auxochromes of these dyes are —NH— groups or —S— atoms. The dyes may be symmetrical or unsymmetrical. The symmetrical dyes are generally produced by oxidative coupling, while the unsymmetrical ones result from condensation reactions of suitable molecules. Most indigo dyes are readily reduced in mildly alkaline medium and this makes them applicable not only to cellulose but also to wool and silk. The introduction of halogen, alkyl and alkoxy substituents gives rise to numerous derivatives of indigo and its analogs, many of which have gained commercial importance.

Phthalocyanine Dyes. The tetrabenzoporphyrazine nucleus is one of the more recently discovered chromophores and represents the common feature of the phthalocyanines. This group of dyes has produced the most brilliant known blues and greens. Many phthalocyanines are metallized such as in CI Pigment Blue 15 or copper phthalocyanine

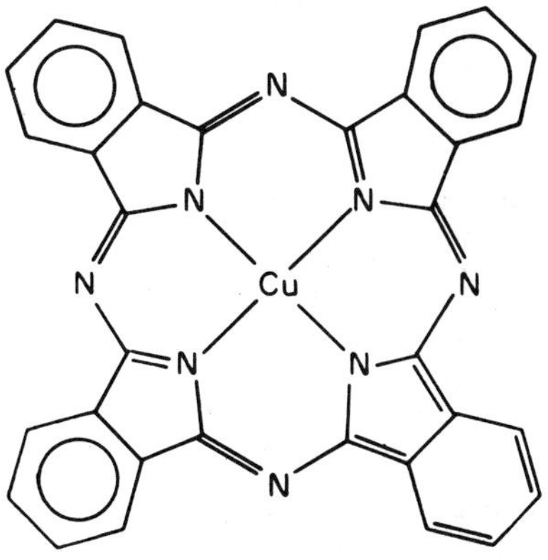

Copper Phthalocyanine

and have set new standards for brilliance and fastness properties. While most of the phthalocyanines are pigments in respect to their application properties, dye chemists have been successful in synthesizing a few vat, direct, solvent and fiber-reactive dyes based on the same chromophore.

The parallel development and interdependence of synthetic dyes and organic chemistry are well known. Many of the reactions pioneered in dye chemistry have proved fruitful in such areas as heterocyclic and steroid chemistry. Today interest in dyes and their synthesis is almost entirely confined to the research departments of large dye manufacturers.

Advances in organic synthesis and reaction mechanisms, as much as development of new or modified fibers, will determine the direction the research on synthetic dyes will take in the future.

HERMANN W. POHLAND

DYEING

Dyeing is the art of applying coloring matter to a substrate in such a way that the color appears to be a property of the substrate not readily removed by rubbing, washing or exposure to light. The term dyeing is usually understood to be confined to the application of colors to textiles and such related materials as furs, leather, plastics and paper, and is used to differentiate between other coloration processes such as printing, painting or staining. Printing and painting involve the application of colored pigments in a resinous vehicle to a textile or paper surface, while in staining small amounts of color are held on the surface of the substrate. True dyeing implies an interaction of the dye with the substrate on a molecular basis and involves chemical and physical principles which are responsible for bringing about a permanent union between the material to be dyed and the coloring matter applied.

General Principles. There is no uniform theory of dyeing, and different chemical and physical principles are involved in the dyeing of different substrates. In cotton dyeing with direct dyes, the nature of the dye-fiber interaction is predominantly physical, while the application of fiber-reactive dyes to the same substrate involves primarily chemical and to only a small degree physical phenomena. Another example is the case in which the same dye (magenta) may be used for the dyeing of two very different fibers such as cotton and wool. The cotton, however, has to be prepared prior to the application of the dye (i.e., with a mordant such as tannic acid), while wool acts itself as the mordant.

The simplest dyeing procedure is the one in which the fiber or fabric is placed in a solution or dispersion of the dye which is then absorbed upon increasing the temperature of the system. The ease with which the coloring matter is taken up by the substrate varies widely, and has made it necessary to devise dyeing processes which accelerate the penetration of the substrate by the color by controlling temperature, pH, salt concentration and other variables of the system. The trend in modern industry is to develop more rapid and economical dyeing processes, characterized by a greater emphasis on the utilization of scientific methods. Improved tools such as instruments for color measuring and matching, are now widely used.

Dyeing Processes. Different classifications have been suggested for dyeing processes referring either to the type of substrate or to the method of application of the dye. The classification of dyeing processes chosen in this article is based on the method of application.

Direct Dyes on Cellulosic Fibers. Dyeing of cotton, rayon and other cellulosic fibers with direct dyes is carried out in a neutral or mildly alkaline bath with additions of common salt or sodium sulfate at temperatures near the boil. The dyeing is facilitated by the use of surface-active compounds, which are used extensively in the trade for this purpose. Direct dyes are generally linear azo and polyazo dyes, but dioxazine, phthalocyanine and other structures are also encountered. The common features of these dyes are good water solubility, relatively high molecular weight and fiber affinity, i.e., they transfer directly (hence the name) from the aqueous solution to the cellulosic medium. They differ from acid dyes in that they are substantive and are fixed on the fiber by the addition of only NaCl. Best affinity is shown by dye molecules of linear and planar configuration. A full spectrum of shades is available from direct dyes, varying from dull to relatively bright. Wash and light fastness of direct dyes are not very good. However, the wash fastness can be improved by after-treatments. These treatments involve cationic agents which form water-insoluble complexes, or crosslinking agents such as formaldehyde which often adversely affect the light fastness. Other after-treatments such as diazotizing and coupling on the cloth aim at increasing the molecular weight of the dye without adding further solubilizing groups, thus decreasing the diffusion rates of the dye from the fiber during wet treatments.

The light fastness of o,o'-dihydroxyazo derivatives can be improved markedly by metallization with copper or nickel. This treatment often causes a broadening of the peaks in the absorption spectrum with resulting dullness, and is therefore restricted in use.

Fiber-Reactive Dyes on Cellulosic and Polyamide Fibers. Dyeing with reactive dyes is accomplished from aqueous solution by controlling pH and temperature. These dyes represent one of the most important developments in recent years. Among the properties which have been responsible for their outstanding commercial success are ease of application, brilliance of shade, and fastness to wet treatment. The common feature of all fiber-reactive dyes is a reactive group capable of forming covalent bonds with the substrate under the dyeing conditions. Many such reactive groups have been suggested and patented throughout the world over the past 17 years. Among them are the mono- and dichlorotriazinyl groups, the di- and trichloropyrimidyl groups, the 2,3-dichloro-6-quinoxalinecarbonyl and the 2-sulfatoethylsulfonyl group. A wide range of shades is offered in this class and different lines have been designed for cellulosic fibers, for wool and for polyamides. One of the still unsolved problems with reactive dyes is the accomplishment of quantitative fixation. Usually only 80–85% of the dye is chemically linked to the substrate, while the remainder is lost due to premature hydrolysis or to side reactions in the dyeing step. An interesting compromise between the fiber-reactive concept and mordanting is the fixation of dyes to the substrate in the presence of crosslinking agents. In this case no premature hydrolysis can occur since the reactive group is not covalently bound to the dye molecule. Fiber-reactive dyes readily lend themselves to continuous application.

Acid Dyes on Wool, Silk and Polyamide Fibers. The dyeing of natural and synthetic polyamides is achieved by salt formation between the basic fiber (free $-NH_2$ groups) and the acid groups of the dye. It is accomplished from aqueous medium by control of pH and temperature. Chemically, acid dyes are mainly sodium salts of sulfonic acids. Except for the common feature of the sulfonic acid, acid dyes vary widely in their chemical structure and derive from azo, anthraquinone, quinophthalone, nitroarylamine and triarylmethane chemistry. They vary widely in fastness to light, brilliance and tinctorial properties. They are usually applied to the fiber from an aqueous solution containing some acetic, formic or sulfuric acid; a few neutral-dyeing acid dyes are also known. These differences are reflected by the common grouping into:

(a) Level-dyeing types, which require sulfuric acid to exhaust.
(b) Neutral-dyeing or milling colors, which exhaust with acetic or formic acid.
(c) Metal-complex colors derived from *o,o'*-dihydroxyazo compounds and chromium or cobalt, which require relatively high amounts of sulfuric acid.

The higher concentration of acid for leveling the metallized dyes on wool is necessary to avoid demetallization of the dye through fiber-metal interaction. On nylon metallized acid dyes tend to accentuate barré and their exhaust rate, therefore, must be controlled carefully.

Nonionic surfactants facilitate level dyeing of many acid dyes by forming a complex (aggregate) with them, which releases the dye slowly. Anionic surfactants also assist in leveling acid dyes, but they function differently; they rapidly form salts with the basic sites in the fiber and are replaced only gradually by the dye molecules.

Disperse Dyes on Acetate Rayon, Polyester, Polyamide and Acrylic Fibers. Disperse dyes are generally applied in the form of dispersions from an aqueous medium and are "developed" by thermal means. The mechanism of dyeing is considered to involve a solid solution of the dye in the hydrophobic fiber.

Disperse dyes are compounds of low water solubility and are nonionic in nature. As a group they are largely anthraquinone, mono- and disazo derivatives; they provide a complete shade range for most of the hydrophobic fibers. At first restricted to the dyeing of cellulose acetate, their use has grown steadily since 1950. They now find extensive use on nylon, particularly since the advent of tufted carpets. Nearly all polyester fabrics are dyed with disperse dyes by wet and "Thermosol" application methods. The use of relatively high temperatures in the dyeing and durable-press application steps required the development of sublimation-fast disperse dyes.

The large number of end uses, the introduction of convenience finishes, and extreme processing conditions have imposed stringent requirements on these dyes. In automotive fabrics they have to show resistance to heat and light, and in the dyeing of polyester they must not be affected by the "carriers" (used in exhaust dyeing) or by heat applied in pleating, heat setting, or dyeing. They have to resist high pH conditions in carpet dyeing and must be fast to gas-fume fading and ozone bleaching. The latter requirement is especially important in hot and humid climates.

Disperse dyes are useful in light and medium shades on acrylic and modacrylic fibers. Even polypropylene is sometimes dyed with disperse dyes, for instance when used as primary carpet backing in place of jute. Elastomeric fibers are also dyed with disperse dyes.

With the continuing introduction of new antisoil, antistatic and fire-retardant finishes, the number of requirements will increase and many new dyes will have to be synthesized by the chemist.

Basic Dyes on Acrylic, Modified Polyamide and Polyester Fibers. Most manufacturers of acrylic fibers offer types that will dye with basic or cationic dyes. The mechanism by which these dyes color the substrate is believed to entail the adsorption of the dye on the fiber surface, neutralizing the negative potential of the fiber. Upon raising the temperature the dye diffuses into the fiber and combines with acid groups in the fiber. Basic dyes exhaust quickly within narrow limits of temperature and it is normally necessary to use cationic retarders to achieve level dyeings, particularly for light and medium shades. A wide range of shades can be dyed with basic dyes and in general the brilliance and fastness of these colors are outstanding. Chemically the basic dyes belong to two main types, (1) those in which cationic site and chromophore are insulated from each other, and (2) those in which the cationic site is a part of the chromophore.

Important members of this class are the triarylmethanes and amino salts of azo dyes, but xanthenes, acridines, azines, oxazines and thiazines also have found commercial uses. Cyanine derivatives are employed extensively as photographic sensitizers.

Basic dyes may also be applied to acid-modified polyester and polyamide fibers. These fiber types have found great acceptance in the carpet industry, where they permit the production of multicolored effects on tufted carpets.

Reduced Dyes: *Vat Dyes.* Vat dyes are applied by the so-called vatting procedure, in which they are first reduced in alkaline solution with sodium hydrosulfite to form a water-soluble "leuco" derivative, then applied to the fiber (the "leuco" form shows substantivity) and oxidized with air or other mild oxidizing agents to precipitate the colored insoluble species in the fiber. Vat dyes of the anthraquinone series generally exhibit outstanding fastness to washing, light and bleaching agents; their only drawbacks are a relatively dull shade and high cost. Vat dyes of the indigoid and thioindigoid type do not afford the outstanding fastness properties of the anthraquinone derivatives, but they do find extensive use in textile application where price is as important as dye performance. Vat dyes are readily applied by continuous dyeing methods and are used extensively in the dyeing of blend fabrics such as polyester/cotton. Vat dyes are primarily used for the coloration of cotton and viscose rayon, but since the indigoid dyes require only a weakly alkaline vat, they can also be used for the dyeing of wool and silk. A versatile subgroup are the soluble vat dyes, which are stabilized leuco salts (sodium salts of sulfuric acid esters of leuco vat dyes). In addition to their water solubility these dyes exhibit practically no color and possess a low affinity for cellulosic fiber but high affinity for wool and silk. They are generally applied by padding and developed by acid oxidation, using sulfuric acid in combination with mild oxidizing agents such as sodium nitrite or potassium dichromate.

Sulfur Dyes. Sulfur dyes are water-insoluble pigments which are applied by a vatting technique. By alkaline reduction with sodium sulfide the dye is dissolved and applied to the fiber at elevated temperatures. The affinity of the reduced dye is not great and salt has to be added to obtain acceptable exhaustion. The leuco dyes have good leveling properties but care has to be taken that they do not oxidize prematurely. After the dyeing is complete, the insoluble original form of the sulfur dye is then regenerated in the fiber by oxidation. Sulfur colors are made by treating a wide variety of organic compounds with sulfur or sulfur derivatives. The commercially important products are made from aminophenols, diphenylamines, indophenols, azines, carbazoles, and aminoaphthalenesulfonic acids. Structures of these dyes remain largely unidentified, but structural features such as mercapto, sulfide, disulfide and polysulfide groups have been identified and heterocyclic systems such as thiazole, thiadiazole and thiazine are often formed. The sulfur dyes tend to afford dull shades; because of their low cost and good fastness properties (except to chlorine) they have penetrated the market for dyeing cellulosic substrates and are produced in large amounts. Important technical advances have been reported in the production of sulfur dyes in the form of: (a) dispersions, (b) stable, clarified, reduced solutions, and (c) nonsubstantive thiosulfonic acid derivatives. These new types of sulfur dyes simplify the dyeing process and make it amenable to continuous application.

Wool and synthetic polyamides are not dyed with sulfur dyes, because the strong alkaline dyeing conditions would produce extensive chemical damage in the fiber by hydrolyzing amide linkages.

Ingrain Dyes. Ingrain dyes are all of those dyes which are formed *in situ* in the substrate by the development or coupling of one or more intermediate compounds which in themselves are not dyes. The principal subclasses are the azoic dyes and the oxidation dyes to be discussed below and the phthalocyanines, which are developed on the fiber by special treatments.

Azoic Dyes. The azoic dyes are water-insoluble mono- or disazo dyes produced *in situ* on the fiber. Because of their pigment-like structure they exhibit good washfastness and often very satisfactory light fastness. The azoic dyes are used primarily on cellulosic fiber but have found limited use also on wool, fur, acetate rayon, nylon and polyesters. A complete range of shades is available but the colors of greatest utility are the bright red, scarlet and orange shades. Commercial dyeing with azoic dyes is a two-step process. First a hydroxy coupling component is brought onto the fiber from an aqueous bath. The fiber subsequently is dried and then treated with a diazonium salt solution. In this latter step the azo pigment is formed *in situ* in the fiber. The coupling components primarily are naphthol derivatives. Because of the ease with which diazonium salts degrade they have to be applied at low temperature (ice colors is another name for azoics) or have to be employed in a "stabilized" form. These stabilized diazonium salts are easy to handle and have been used extensively in dyeing and printing.

The azoic dyes are less expensive than the quality vat dyes but their use is often difficult and this has limited their acceptance. Reproducibility of shade depth is a well-known problem with them and it is not practical to use mixtures of azoic combinations, since each diazonium salt in a mixture can react with every coupling component.

Oxidation Dyes. The principal dye of this class is aniline black, which is produced on the fiber by a two-step acid oxidation of aniline. In the first step emeraldine is generated by ageing or steaming, in the second step an ungreenable black is produced by further oxidation. The color has been used primarily on cotton and rayon and rarely on wool and silk.

Mordants. *Basic Dyes on Cellulosic Fibers.* Basic dyes are applied to cotton by a two-step process. The cloth is first mordanted with a solution of tannin, then dried and fixed (precipitation of antimony tannate). The basic color is now dyed on the fiber, the fiber's negative potential is neutralized on the surface allowing diffusion into the fiber and combination with the mordant. The common structural feature of the dyes is one or several positive charges, which are not necessarily localized on any specific site in the molecule (distributed charge).

The application of basic dyes to mordanted cotton today is essentially obsolete, since dyeings with equal brilliance and superior fastness can be obtained with fiber-reactive dyes.

Acid Dyes on Cotton. Acid dyes have little or no affinity for substrates such as cotton, yet may still be fixed if a mordant has been applied first. The mordanting (complex formation) is achieved with salts of metals such as aluminum and chromium which form a complex with the dye on the substrate. Often the self shade of the dye is changed slightly by the treatment and properties such as fastness to light and to washing are greatly improved.

The cheaper dyes of this type find use in paper dyeing, where they are applied to the surface of the finished sheet or by precipitation in the pulp with alum.

Mordant Acid Dyes on Wool. Mordant acid dyes (chrome dyes) provide excellent fastness on wool. They are applied by several methods. By one technique the wool is mordanted with sodium or potassium dichromate by boiling in the presence of the salt. After washing the wool, it is then dyed with the appropriate dye under acid conditions (bottom-chrome method). Another technique permits the simultaneous application of dye and mordant (autochrome method).

Future advances in fiber technology no doubt

will necessitate modifications of today's methods of application of dyes in textile dyeing. The trend will be towards faster, more economical and reproducible processes with still-higher requirements for improved fastness and tinctorial properties of the resulting dyeings.

HERMANN W. POHLAND

E

ECONOMICS OF THE CHEMICAL INDUSTRY

According to Webster, chemistry is "The Science that treats of the composition of substances and of the transformation which they undergo." The key word is transformation, since the change of basic matter from one form to another is the essence of chemical manufacturing.

The chemical industry broadly defined embraces a complex of subindustries. The borders of the industry and of its subindustries generally are indistinct. Large expenditures for research and development create new products and new technological processes which continually alter the product mix of the industry and create opportunities for companies already in the industry as well as for those in other industries.

A distinction may be made between heavy or bulk chemicals and fine chemicals. Bulk chemicals are produced in quantities of at least half a million tons annually and usually sell at a low price per unit. These products, which include sulfuric acid, alkalies and chlorine, ethylene, benzene, phenol, and vinyl chloride, are basic for many other industries and are key raw materials or intermediates for other chemical products; there is considerable captive use by the initial producer.

Fine chemicals ". . . include nearly all elements or compounds refined to high purity, or reagent grade; or frequently complex compounds made in relatively small volume to specification for specialized use." In terms of numbers, the overwhelming proportion of chemicals fall into this category. They do not benefit from the full economies of mass production and they sell at much higher prices per unit than the bulk chemicals.

Total shipments for the entire chemicals and allied products industry in 1970 were $49.3 billion of which $15.9 billion was accounted for in Standard Industrial Classification [SIC] 281 and $8.8 billion in SIC 282.

Economic Characteristics. The economic characteristics of the chemical industry include: (1) changing products and processes, (2) large numbers of by-products and joint products, (3) standardized products, (4) demand is derived, (5) multiple product companies, (6) intensity and diversity of competition, (7) high rate of innovation, (8) relative ease of entry, and (9) intensive use of capital. It is the combination of these economic characteristics that helps to explain the performance of the chemical industry.

The multiplicity of products manufactured by the chemical industry makes it difficult to draw a simple profile of their characteristics. Exceptions will be found to almost any generalization about the industry's products.

Contributions of Chemical Industry. The chemical industry has made impressive and vital contributions to all phases of the U.S. economy in peace and in war including: more efficient use of raw materials and manpower, advancement of space and missile technology, creation of new products and processes, stimulus to new industries, reduced dependence on foreign sources of supply, improved levels of living, lower costs to industry, improved health and longer life, increased food supply, national defense, and national economic growth.

We depend upon industrial chemicals for many everyday items including toothpaste, lipstick, color in our neckties, gasoline, antibiotics, paints, bleach in fabrics, color film, aerosol bombs, water repellents, milk containers, and a host of other products.

The major contributions made by chemicals and allied products are reflected in their high rank in the hierarchy of American industry and in their relatively large share of all manufacturing industries in terms of jobs, shipments, total assets, expenditures for research and development, and capital expenditures. More than one half of the 100 largest corporations in the United States and almost one-third of the five hundred largest industrial companies in the United States produce some chemicals although most of these companies are more heavily engaged in other types of manufacturing activity.

The changes in the nature of the chemical industry have been mirrored in the operations of many companies. The shift of Du Pont from a company specializing in explosives to a major chemical company after World War I provides a well-known and dramatic illustration. Moreover, the entrance into chemical production by companies with a primary interest in other industries also has been important in the post-World War II years. Illustrations include the large scale production of chemicals by petroleum companies, rubber companies and paper companies and by such large industrial companies as Pittsburgh Plate Glass, National Distillers, Eastman Kodak, Borden, and others.

At the same time, chemical companies have become increasingly interested in integrating forward into the industries producing consumer products which are built on chemical raw materials.

Growth of the Chemical Industry. The major factors contributing to the above average growth record of chemicals have included the large investment in research and development, the industry's technical know-how, the intense competition among chemical companies, the large-scale military and space demand for chemicals, the broadening of markets through reductions in price, the fact that many customers also are above average growth industries, and until 1966 the incentives created by a favorable rate of profits.

Test tube competition has been the vital stimulating and driving force in the chemical industry. Such competition is so intensive and so widespread that it has become indispensable to competitive success. The industry is constantly obsoleting its own products and processes. It is marked by intensive struggles for product leadership. Extensive research is undertaken not only by chemical companies but by the companies in the various chemical process industries thus broadening the base of research. The pervasiveness of research results in a fuller utilization of waste materials and increases the possibilities of discovery of new products and of the development of greater usefulness for older ones. Technical know-how and an insatiable drive for further knowledge have been among the industry's greatest assets.

The dramatic growth of the chemical industry provides an outstanding demonstration that extensive investment in research and development pays off. The recent history is one of inquisitiveness which has led to discovery, to changes, and to progress. An examination of the changing composition of output over time quickly reveals the dynamic contribution of R & D to expanding volume. Rayon and other cellulosic fibers in the 1920s and 1930s, synthetic rubber during World War II and the postwar years, plastics [e.g., polyethylene, polypropylene, styrene, urea, etc.] and noncellulosic fibers [e.g., polyesters and acrylics] in the 1950s and 1960s all represented major breakthroughs which resulted in the creation of major new markets for the chemical industry.

The industry also has been favored by a large rise in demand because of military and space needs, by the increase in effective demands as markets have broadened when prices for many new products have declined, by the substitution of chemicals for other products, and by many customers who are in industries with above average rates of growth. Finally, the favorable profit record throughout most of the chemical industry's history provided the incentive to make substantial investments in research and to manufacture and market the flow of products developed in its research laboratories: these profits also have helped finance the investment required. The economy has reaped high social dividends from these developments.

The growth of chemicals in dollar terms has developed overwhelmingly since 1939 when total shipments were a little less than $1.3 billion. By 1970, the total had increased to $24.7 billion. While increases were reported for all major product groups, the most outstanding growth rates developed for synthetic rubber, plastics materials, noncellulosic fibers, acrylics. These were mainly newly developed products. Since 1939, after eliminating the effects of higher prices, total shipments of chemicals have increased at an annual rate of 8.9% as compared with 4.3% for gross national product. The physical output of chemicals has increased more than twice as rapidly as that for the entire economy.

The volume of older products has increased as the national economy has grown in size. But it has been the sales derived from the new organics, particularly synthetic materials, which has made the difference between an about average growth rate and one which is more than twice as rapid as that recorded by the national economy. There has been no slowing down in the historic high rate of growth. However, a continuation of past trends will depend upon the continued high rate of innovation which has contributed so significantly to the brilliant record achieved by chemicals.

Competition in Chemical Markets. Chemical markets have been characterized by intensive competitive pressures. Of particular importance in this connection has been the ease of entry into many chemical markets—particularly the organics.

The number of companies producing specific chemicals has increased significantly as a result of the organization of new companies [often joint ventures], the entry of established companies from other industries, and/or the extension of product lines by chemical companies. Competitive pressures affecting many chemical products have been dramatically intensified by the large-scale entry of companies whose primary activity has been in other industries. The result has been persistent pressures upon existing producers, both large and small, to improve their products and services and to expand their own capacity in order to meet these competitive threats.

Most chemical markets are characterized by several large companies and a number of smaller ones. This is similar to the structure in other mass production industries. However, the combined volume of the biggest companies in the chemical industry accounts for a smaller share of the industry's total sales than is typically found in large manufacturing industries. On the other hand, the proportions of volume of specific products accounted for by the four largest companies are among the highest for any manufacturing industry. Concentration ratios in the chemical industry tend to be lower for the higher volume products than for those with smaller volume.

The high concentration ratios found for chemicals reflect the underlying economic characteristics of the industry, particularly the heavy investment required and the accompanying economies inherent in large scale production.

Fewness of sellers has not inhibited intensive competition for many chemical products. Nor has it prevented the development of excess capacity for many organic chemicals with the accompanying pressures on prices.

Substitute products add a significant dimension to competition in the chemical industry. Chemicals face competition from other chemicals [interproduct competition], from other processes [interprocess competition], and from products of other industries [interindustry competition]. The availability of these alternatives influences the environment within which decisions must be made concerning pricing, sales strategy, research and development, and various forms of nonprice competition.

Nonprice competition in the chemical industry emphasizes technical services, quality, and product improvement through research and development. The availability of technical services is the most important form of nonprice competition: brand names have been significant mainly for synthetic fibers.

Patents have been of considerable importance in the chemical industry. In some instances they have facilitated a limitation over the number of producers. However, judicial actions have modified some of these restrictions. The presence of a patent also tends to intensify research efforts to find substitute products or alternate processes, sometimes leading into entirely new areas of technology. Thus, because of the many substitute products available and the intensive research activities of most chemical companies, competition often develops despite the protection of patents. That patents do not inhibit price competition has been amply demonstrated by organic chemicals in recent years.

Chemical Prices. Chemical prices are determined by a number of factors including: competition, use value to the buyer, availability of excess capacity, and costs. However, costs generally do not play a major role in pricing. Rather, they are important primarily in determining the profitability of the product at a given level of prices.

The discretion in pricing available to a chemical company ranges over a wide spectrum, from substantial latitude for highly specialized products at one extreme to little or no discretion for bulk or commodity chemicals at the other. As a product moves from specialty to commodity status, the degree of latitude in pricing available to a company is steadily reduced. For new products, the extent of freedom in pricing depends on whether it is an entirely new product to the chemical industry or whether it is merely an addition to a company's product line of an already available product. In the latter case, the price must be slotted into the existing market and hence a company has little latitude in its pricing unless it can differentiate its product in some way. However, for an entirely new product, there may be considerable discretion and a company must decide at what price the product should be introduced. Where a natural product already is available, its equivalent cost is an important factor in the decision. Usually, the initial price is reduced if such an action will broaden demand.

Historically, the trend of the average level of chemical prices has been downward relative to the general price level. These relationships were evident during the 19th century and through the price inflations experienced during and after World Wars I and II.

In recent years, price cutting, often quite sharp, replaced price stability as a major characteristic of many chemical products, particularly the organics. Price reductions were associated with excess capacity, the entry of new producers, the expansion of foreign production, and the desire to broaden markets.

The fact that chemical prices tend to rise less and to decline more than all wholesale prices, affects the significance of comparisons of sales totals. Because of the more favorable price trends, a given percentage increase in dollar volume usually means a larger increase in physical output for chemicals than for the entire economy or for all manufacturing.

There has been a fairly steady increase in the real value received by purchasers of chemical products as compared with other products. These favorable price relationships undoubtedly have contributed to the enormous expansion in volume as chemical products were substituted for natural products and as the steady flow of new products has proved to be economically as well as technically acceptable. The reduction in prices has been accompanied by lower unit costs which have reflected the improvements in the manufacturing efficiency accompanying technological advances, expanding volume, and large scale capital investments.

Productivity and Profits. The chemical industry has experienced dramatic increases in productivity throughout its history. The annual rate of increase of 7% in the post-World War II period has exceeded by a wide margin the gains recorded in the entire economy, in all manufacturing industries, and in the allied industries.

The profits picture changed dramatically in the chemical industry after 1966. From a position with above average returns on net worth, the industry's return fell below the average in manufacturing industries. This loss of position developed despite a continued increase in production by about 10% a year, a rise in output per man-hour averaging about 7% a year, continued large expenditures for R & D, heavy capital investment, and further declines in unit labor costs.

Foreign Trade In Chemicals. Exporters of chemicals and allied products have been meeting increasing competition abroad from foreign companies which are growing rapidly in size. Thus, while total American exports of these products have continued to expand significantly, the world markets have grown so much more rapidly that the American share of those markets has declined from more than one quarter in 1956 to less than one-fifth in 1967.

The excess of exports over imports [export surplus] has been growing and has made significant contributions to the international balance of payments. In the thirteen years 1958 to 1971, chemicals and allied products earned a total export surplus of more than $22 billion. In 1971 the export surplus was $2225 million for chemicals and allied products, as compared with a deficit of $2689 million in the balance of trade. If it were not for the export surplus for chemicals and allied products, the trade deficit would have been almost $5 billion. Clearly, this group of industries has made a major contribution to the international balance of payments of the United States.

During the post-World War II period, the internationalization of chemical companies has been proceeding at a rapid pace. Activity by American companies overseas has been expanding steadily and is reflected in the significant annual increases in the volume of new investments, in licensing agreements, and in joint ventures. In 1970 direct investments overseas were in excess of $6 billion, exclusive of those made by petroleum companies. That overseas investments may be risky has been illustrated by the large losses reported by several American companies when they withdrew from some foreign markets.

JULES BACKMAN

ELASTOMERS

The term "elastomers" is a generic one intended to cover all polymeric materials exhibiting "long-range" elasticity, in other words, natural and synthetic rubbers. Because of the increasing variety of synthetic polymers now available which exhibit varying degrees of elasticity, the definition of an elastomer is not as clear-cut as desirable. However, the criterion of "long-range" elasticity is still quite adequate, signifying the ability of an elastomer to undergo stretching to at least twice its original length and, more important, to retract very rapidly to virtually its original length when released.

Originally only natural rubber had this remarkable property of high elasticity. Today, after something like a century of study and effort, a variety of synthetic elastomers are commercially available, including "synthetic rubber" itself, i.e., an exact molecular duplicate of natural rubber. All these owe their existence to the process of "polymerization," a chemical reaction whereby ordinary simple molecules are linked together into long chains, containing many thousands of these small molecules as units.

Natural rubber occurs in a milky fluid exudation, called latex, from such trees as *Hevea brasiliensis* (see **Latex**). It comprises about 35% of the latex, the remainder being mainly water. The rubber itself is a hydrocarbon polymer, consisting of long molecular chains of isoprene units. Furthermore, these isoprene units

$$(-CH_2-C=CH-CH_2-)$$
$$\underset{CH_3}{|}$$

all ($>95\%$) have the same configuration about the central double bond, viz. the *cis* configuration. Hence natural rubber is defined chemically as *cis*-1,4-polyisoprene. It is the stereospecific character of its chain structure that has made natural rubber so difficult to synthesize. In fact, the other isomer, *trans*-1,4-polyisoprene occurs naturally as gutta percha, a resinlike substance.

The molecular chains of natural rubber vary in length but their average molecular weight lies between 1 and 3 million. The presence of a double bond in each isoprene unit makes these chains capable of being interlinked by addition reactions, leading to the formation of networks. Unlike the linear molecular chains, such networks are insoluble in all solvents, and give rise to "gel," i.e. swollen, but undissolved, rubber. Hence it is not surprising that a portion of the natural rubber in latex is found in the form of gel, the remaining, soluble fraction being known as "sol." Latex may thus contain varying amounts of gel, and its gel content may increase on aging.

The elastic behavior of rubber is generally ascribed to the fact that its molecular chains, which are in a randomly coiled, entangled condition at rest, are capable of uncoiling when stretched, due to the possibility of rotation of any two carbon atoms comprising a *single bond*. When the force is released, this rotational energy results in the chain returning to its random coiled condition and the rubber quickly retracts. However, unless the chains are interlinked by actual chemical bonds, they will undergo a certain amount of slippage during stretching, and the rubber will not regain its original unstretched length. Hence the formation of a stable network, by vulcanization, is essential for true reversible elasticity.

Another property of natural rubber is the ability of its molecular chains to undergo partial crystallization, either upon cooling or upon high extension. Thus, when vulcanized rubber is stretched to more than 5 or 6 times its length, the chain molecules are closely aligned and form "crystallites" between well-oriented portions of their length. This partial crystallization resembles fiber formation and results in greatly increased tensile strength of the rubber. However, unlike true fibers, these crystallites revert to their coiled condition, when the stress is released, by virtue of the high kinetic energy of the chain carbon atoms. Thus it is not surprising to find that those elastomers which exhibit the phenomenon of crystallization on stretching also show far greater tensile strength than purely amorphous elastomers.

Ever since 1860, when the chemical composition of natural rubber first began to be established, continuous attempts were made to synthesize rubber from isoprene. It is not surprising that these early attempts failed, for a number of reasons. In the first place, the polymeric nature of the rubber molecule was not understood. In the second place, the stereospecific structure of the *cis*-1,4-polyisoprene was not appreciated. And finally, even later, when the molecular structure of natural rubber was fully elucidated (ca. 1930), there was no obvious method for such a stereospecific polymerization to be achieved.

In view of the above, attention was directed to the polymerization of other, more promising, monomers and combinations of monomers (copolymerization). Thus, at the start of World War II, world technology, mainly in Germany, Russia and, to a lesser extent, Japan and the United States, had developed practical and useful forms of synthetic rubber. During the war, the U.S. government built a giant synthetic rubber industry, which was finally turned over to private ownership in 1955. Thus, in today's world, there are available at least ten different classes of synthetic elastomers, with many more modifications of these. The main classes are described below.

Styrene-butadiene copolymer (SBR) is still the most important general-purpose elastomer for use in tires. It is a copolymer containing a mixture of butadiene units

$$(-CH_2-CH=CH-CH_2- \text{ or}$$
$$-CH_2-CH- \quad)$$
$$\underset{CH=CH_2}{|}$$

and styrene units ($-CH_2-CH-$) and is prepared
$$\underset{C_6H_5}{|}$$

by polymerizing a 3:1 mixture of butadiene and styrene in an emulsion polymerization system. The result is an amorphous, noncrystallizing elastomer, exhibiting poor strength by itself when vulcanized with sulfur, but showing a remarkable strength increase in the presence of large amounts of carbon black (30–40 parts per hundred parts of rubber). This reinforcement by the carbon black raises the tensile strength of SBR from 300–400 to 3000–4000 psi, equal to that of natural rubber. Other properties, such as abrasion resistance (tire tread wear) actually become superior to those of natural rubber. Hence SBR has been successful in displacing natural rubber in tire manufacture, in most cases, because of its lower cost. Its main shortcoming lies in its poorer dynamic properties, i.e., ability to recover from elastic deformation. Because of the higher "internal friction" of the SBR molecular chains, part of the deformation energy must be used to overcome these viscous forces, resulting in evolution of heat. This viscoelastic behavior (or higher "hysteresis") of the SBR therefore leads to a dangerous "heat build-up" in the bodies of large tires, where this heat is difficult to dissipate. Hence, although this elastomer can be safely used for passenger tires, it is still unsuitable for 100% use in the carcass of large truck tires, where a proportion of natural rubber is necessary.

Acrylonitrile-butadiene copolymer (*NBR*), also known as nitrile rubber, is a copolymer of butadiene containing from 10–40% of acrylonitrile units ($-CH_2-CH-$). It is prepared in emulsion poly-
$$\quad\quad\quad\quad\quad | $$
$$\quad\quad\quad\quad CN$$
merization systems similar to those used for SBR and yields a latex product. Its main use is as a solvent and chemical resistant elastomer, in which respect it is better than neoprene but not quite as good as Thiokol. It is vulcanized with sulfur, just like SBR, and also requires carbon black for strength.

Neoprene is the general name for polychloroprene rubber, prepared by polymerizing chloroprene (2-chlorobutadiene) in emulsion system. The chain structure is predominantly trans-1,4-polychloroprene ($-CH_2-C=CH-CH_2-$) and it does not
$$\quad\quad\quad\quad\quad\quad\quad\quad | $$
$$\quad\quad\quad\quad\quad\quad\quad Cl$$
easily lend itself to sulfur vulcanization. Instead vulcanization is accomplished by the reaction between metal oxides, such as ZnO and MgO, with the small proportion of reactive chlorine present in the polymer from occasional 1,2-polymerization. Neoprene is noted as a solvent resistant elastomer and was one of the first to be produced in this country. Like natural rubber, it shows the ability to crystallize on stretching, and hence has excellent tensile strength, even without carbon black.

Butyl rubber is a copolymer containing mostly
$$\quad\quad\quad\quad\quad\quad\quad\quad\quad CH_3$$
$$\quad\quad\quad\quad\quad\quad\quad\quad\quad | $$
isobutylene units ($-CH_2-C-$), with only 2–3
$$\quad\quad\quad\quad\quad\quad\quad\quad\quad | $$
$$\quad\quad\quad\quad\quad\quad\quad\quad\quad CH_3$$
per cent of isoprene units ($-CH_2-C=CH-CH_2$)
$$\quad\quad\quad\quad\quad\quad\quad\quad\quad\quad\quad | $$
$$\quad\quad\quad\quad\quad\quad\quad\quad\quad\quad CH_3$$
present to provide unsaturated sites for sulfur vulcanization. Unlike the preceding elastomers it is prepared by a cationic polymerization process, involving acid catalysts, which is extremely rapid. It is noted for its low gas permeability (about 1/10 that of natural rubber) which makes it ideal for inner tubes. In addition, its low unsaturation makes it exceptionally resistant to oxygen and ozone, properties of great importance in such uses as wire insulation.

"Thiokol," trade name of a polysulfide elastomer,

is synthesized by a condensation reaction between an alkyl dihalide, such as ethylene dichloride, $(CH_2Cl)_2$ and sodium tetrasulfide, Na_2S_4, resulting in a polysulfide chain unit ($-CH_2-CH_2-S-S-$).
$$\quad\quad\quad\quad\quad\quad\quad\quad\quad\quad\quad\quad\quad\quad\quad \| \quad \| $$
$$\quad\quad\quad\quad\quad\quad\quad\quad\quad\quad\quad\quad\quad\quad\quad S \quad S$$
The reaction is carried out in emulsion. This elastomer is noted for its extreme solvent resistance, although its mechanical properties are poor. Like neoprene, it was one of the first synthetic elastomers to appear, about 1930.

Silicone rubber is an unusual elastomer based on a polymer chain which does not contain chain
$$\quad\quad\quad\quad\quad\quad\quad\quad\quad\quad\quad\quad\quad\quad R$$
$$\quad\quad\quad\quad\quad\quad\quad\quad\quad\quad\quad\quad\quad\quad | $$
carbon atoms but a siloxane structure ($-Si-O-$)
$$\quad\quad\quad\quad\quad\quad\quad\quad\quad\quad\quad\quad\quad\quad | $$
$$\quad\quad\quad\quad\quad\quad\quad\quad\quad\quad\quad\quad\quad\quad R$$
where R generally represents a methyl group. The most common elastomer of this type is therefore a polydimethylsiloxane. The most unique characteristic of this elastomer is its ability to retain its flexibility and elasticity at extreme low temperature ($< -100°C$) where other elastomers become stiff and brittle. It is also oxidation resistant because of the absence of unsaturation, but cannot, therefore, be vulcanized with sulfur, requiring peroxides instead. A special type of sulfur-vulcanizable silicone rubber is available, having a few per cent of the methyl groups replaced by vinyl groups. In the gum state, it is essentially a viscous liquid and requires a filler, such as finely divided silica, for strength reinforcement.

The *urethane rubbers* comprise a whole new class of castable elastomers. These are based on a chain extension reaction, whereby low molecular weight fluid polymers, such as polyethers and polyesters, are interlinked to linear high molecular weight polymers. This is accomplished by the reaction of di-isocyanates with the hydroxyl terminal groups of the low polymers, hence the name "urethane." Because of their regular chain structure, these elastomers crystallize on stretching and hence require no reinforcing filler for high strength. An excess of the di-isocyanate can also be used to crosslink the chains into a network. These vulcanizates exhibit excellent abrasion resistance. If a little water is present during the chain extension reaction, it reacts with some of the di-isocyanate to liberate carbon dioxide, and a rubberlike foam is formed (urethane foam). This process also enables the formation of elastic or rigid foams *in situ* for many intricate applications.

"Stereo" Rubbers. The advent of the Ziegler-Natta catalysts (mixtures of aluminum alkyls and titanium halides), about 1955, enabled the synthesis of *stereospecific polymers.* including *cis*-1,4-polybutadiene and the long sought for duplication of natural rubber, *cis*-1,4-polyisoprene. These polymerizations are carried out in hydrocarbon solvents, in the absence of air or moisture. Lithium catalysts may also be used to synthesize *cis*-polyisoprene. The latter, as expected, has identical properties to natural rubber, but the *cis*-polybutadiene is an entirely new polymer, and exhibits entirely new properties, such as exceptionally high resilience (low hysteresis), excellent abrasion resistance and outstanding low temperature properties. These quali-

ties make it very attractive for use in tires, generally as a blend with SBR.

Ethylene-propylene elastomers (EPR). The Ziegler-Natta catalytic systems can also be used to prepare copolymers of ethylene and propylene (3:1), which are noncrystallizing elastomers exhibiting good mechanical properties for possible tire use. The absence of unsaturation makes these elastomers exceptionally suitable for age-resistant uses (wire insulation). However, in order to enable sulfur vulcanization, these copolymers are generally prepared with a few per cent of unsaturated units (nonconjugated dienes) and these are known as ethylene-propylene terpolymers (EPDM, for "ethylene-propylene diene monomer"). The low cost of the monomers makes these elastomers especially attractive commercially.

The 1969 statistics on production of natural and synthetic rubber are listed below.

	Long Tons	
	U.S.	Total World
Natural	——	2,895,000
Synthetic (all types)	2,232,260	4,867,500**

**Does not include USSR or China. Last available estimate for USSR production was 350,000 metric tons in 1960.

MAURICE MORTON

References

Alliger, G., and Sjothun, I. J., "Vulcanization of Elastomers," Van Nostrand Reinhold, New York, 1964.
Stephens, H. L., "The Compounding and Vulcanization of Rubber," Chapter II in "Introduction to Rubber Technology," (2nd Edition), M. Morton, Editor, Van Nostrand Reinhold Co., New York. (in production—1973)

Ozone-Resistant Elastomers

Although the concentration of atmospheric ozone is only about 7×10^{-5} mg/liter, its destructive action on elastomers is well known; this occurs primarily by the reaction of ozone at the double bonds. Damage may appear as surface cracking, hardening, or tack. In severe cases, complete failure or loss of elastic properties may occur. Surface cracking takes place only under stress. Hardening usually occurs in highly unsaturated rubbers, while tack formation is more prevalent in the butyl elastomer. Some of the factors that influence ozone activity are concentration and type of unsaturation, presence of chemical antiozonants, loading and plastization, temperature, rate of diffusion, stress, state of cure, and ozone concentration. Quantitative studies on the influence of these factors have led to a better understanding of the mechanisms involved.

When ozone reacts with rubber at rest, the reaction rate is slow and diffusion controlled. The surface is attacked but significant damage only occurs after long exposure. However, when ozone reacts with rubber under stress, as it usually is, cracking takes place rapidly. It is the stress required for crack propagation and the rate of crack growth that are of practical importance. More specifically, the elastically stored energy must equal or exceed the energy requirements for the creation of new surface resulting from crack formation. The energy required is about 60 ergs/cm² of newly formed surface. This value is extremely low, being comparable with the surface free energies of simple liquids. The critical stress is likewise quite small, of the order of 0.1 kg/cm². Mechanically, cracking occurs when ozone reacts with surface double bonds, resulting in chain scission. Having sufficient stored energy, the polymer chains move apart under stress to initiate the crack. The crack grows as ozone reacts with the newly formed surfaces at the tip of the crack where localized stress can be relatively high. It follows that protection at the surface to prevent the formation and growth of cracks is essential.

Chemical antiozonants, used for years to protect rubber vulcanizates, improve ozone resistance but have not eliminated the problem. The mechanism of this protection is not completely understood, but appears to differ for different materials. Most antiozonants retard the rate of crack growth, however, some of the best antiozonants, dialkyl-*p*-phenylene diamines, also increase the critical stress (and hence the critical energy) required for crack growth. This suggests reaction of the antiozonant with ozonized rubber to yield crosslinking at the surface.

The type and location of unsaturation can influence ozone resistance. Except for polychloroprene, all the unsaturated rubbers are derived from butadiene or isoprene. Most of the unsaturation results from 1,4-polymerization and is located in the polymer backbone. This structure, unlike the pendent double bond resulting from 1,2-polymerization, is subject to chain cleavage by ozone. Similarly, internal vinyl chloride unsaturation is indicated for polychloroprene. However, the low level of reactivity associated with this structure is reflected in its relatively superior ozone resistance. By the same token, vulcanization, at least in some formulations, is believed to occur through the more reactive allylic chloride groups resulting from small amounts of 1,2-polymerization.

A more direct approach to ozone resistance has been the synthesis of elastomers with little or no unsaturation. For example, butyl rubber, which is a copolymer of isobutene and 1 to 3% isoprene, contains only the unsaturation necessary for vulcanization. Its ozone resistance is greater than the highly unsaturated rubbers (natural rubber, SBR, ABR, etc.) by a factor of 10 to 100 times. Polyisobutene homopolymer, being completely saturated, is substantially unaffected by ozone. However, it must be used in applications where vulcanization is not necessary since suitable vulcanization systems have not been developed.

Elastomers prepared by copolymerization of ethylene and propylene or ethylene and butene-1 with Ziegler-type catalysts are also saturated. As such, they are resistant to the action of ozone. They are vulcanizable with peroxides and certain other special curing systems, but not with sulfur.

Recently, minor amounts of cyclic or unconjugated diene comonomers have been incorporated in experimental butyls and ethylenepropylene elastomers during synthesis. The products are sulfur-vulcanizable and ozone-resistant. It should be emphasized that these double bonds are no less subject to ozone attack than others, but their location is such that bond cleavage does not lead to chain scission.

Improved distribution of unsaturation and control can often be achieved during butyl polymerization by the use of a third comonomer. A diene, such as isoprene, known to give uniform distribution and with a relatively low poison coefficient, should be chosen. Terpolymers of cyclopentadiene, isoprene and isobutylene prepared in this manner are also highly ozone resistant.

An understanding of butyl terpolymer behavior involves two basic considerations: the cyclopentadiene component takes part in crosslinking; and ozone attack does not break the chain at this point. Therefore, even though the isoprene components are broken, the remaining cyclic components and associated crosslinks tend to maintain the surface network intact. This is equivalent to an increase in critical stress and results in a drastic reduction in the rates of crack initiation and growth. The extent to which the critical stress requirement is increased is primarily dependent on the relative amounts of the two diene residues present. The fact that relatively low concentrations of cyclopentene structures give large improvements in ozone resistance is in itself evidence for such a mechanism. Similar explanations apply generally in systems where other cyclic and unconjugated dienes are utilized.

A modification of the above approach combines the highly ozone-resistant butyl rubbers in blends with conventional isoprene-butyl rubber. The ozone resistance exhibited by the blend is roughly equivalent to that of a terpolymer prepared with the same total amount of protective diene component. For example, a 1:1 blend of a terpolymer containing 1% cyclopentadiene and an isoprene butyl would be equivalent, in ozone resistance, to a corresponding terpolymer containing 0.5% cyclopentadiene.

The synthetic approach to the problem of ozone degradation of rubber vulcanizates has been most fruitful. It is possible to design elastomers that are essentially free from the action of ozone or that possess intermediate degrees of resistance. Ethylene-propylene copolymer and terpolymers with excellent ozone resistance have been commercially available for some time.

L. S. MINCKLER, JR.

Cross-reference: *Antiozonants.*

ELECTROCHEMISTRY

Broadly interpreted, electrochemistry deals with all systems which involve chemical processes and electrical current or energy. This may include electrical production of high temperatures for chemical purposes (electrothermics, electrometallurgy); thermoelectric devices for heating or cooling; materials useful for electric insulation or for their dielectric properties; electrolytic decomposition or synthesis of organic and inorganic substances; electroplating or even "electroless" plating; all electric cells and batteries, whether rechargable or not; and many electronic materials such as semiconductors and phosphors, which require exacting chemical and electrical studies. In addition to research on industrial materials and processes, the science includes laboratory studies of importance in the theories of molecular structure, chemical reactivity and kinetics, energy relations; of electric cells and oxidation-reduction systems; of electrolytes and ionization in aqueous and nonaqueous solutions and melts; and of metallic corrosion. Electroanalysis is an important branch of electrochemistry (constant current and constant potential electrodeposition, amperometric and coulometric methods, polarography, electrochromatography). Electrobiology is essentially the electrochemical study of biological systems.

Modern electrochemistry may be said to have begun with the assembly of electric cells which can produce current in an outside circuit, due to chemical reaction within the cell. An example is the Daniell cell, which was widely used in the early days of telegraphy. It consisted of a copper electrode in saturated copper sulfate ($CuSO_4$) solution and a zinc electrode in dilute zinc sulfate ($ZnSO_4$), the solutions being separated by a porous membrane or by floating the lighter zinc solution on the denser copper one. On shorting the electrodes through an external circuit (copper wires and the telegraph assembly), the following reactions take place:

$$Zn \rightarrow Zn^{++} + 2\epsilon \quad \text{(anode, oxidation)}$$

$$Cu^{++} + 2\epsilon \rightarrow Cu \quad \text{(cathode, reduction)}$$

The symbol ϵ stands for the electron, which is regarded as the current carrier. In the early studies of such cells the electron was not known and it was postulated that positive charge flowed from copper to zinc.

Alessandro Volta is generally credited with the discovery of such cells, around the year 1790. While sometimes called voltaic cells, they are more commonly called galvanic cells after Luigi Galvani, whose experiments with metallic couples were initiated a few years earlier. Familiar examples of galvanic cells in modern use are the zinc-carbon "dry cell," the Mallory-Ruben cell which uses zinc and mercury-mercuric oxide electrodes in an alkaline electrolyte, and zinc-silver alkaline cells. For larger-scale power requirements rechargable "storage batteries" are used, including the alkaline nickel-cadmium battery (1.2 volts per cell), and more notably the lead-sulfuric acid batteries which are indispensable in automobiles and have many other applications.

Because of the extremely widespread use and importance of electric cells and batteries, a vast amount of developmental research has been, and is being done, on all types of cells. The familiar dry cell and its analogs have been tremendously improved in recent years with respect to electrical energy output per unit weight and volume, stability, shelf life, etc. Storage batteries last longer and are more dependable. The telephone systems have special needs for stand-by power, and their laboratories have devoted great efforts to design batteries for their own applications.

In the early 1800's Sir Humphry Davy was able to produce sodium and other active metals by electrolysis of their molten salts or hydroxides. His source of current was a battery of galvanic cells connected in series. In modern terminology we write the equations:

$$Na^+ + \varepsilon \rightarrow Na \quad \text{(cathode, reduction)}$$

$$2Cl^- - 2\varepsilon \rightarrow Cl_2 \quad \text{(anode, oxidation)}$$

Michael Faraday formulated his laws relating electrical and chemical quantities from experiments on electrolysis of solutions, again using simple galvanic cells as current source. If the symbol Na in the above equation stands for one gram-atomic weight of sodium (23 grams), the symbol ε stands for Avogadro's number of electrons, carrying 96,490 coulombs of charge.

Industrial Electrolysis. Many metals and nonmetals of commerce, as well as compounds of unusual nature, are obtained by electrolysis of solutions or melts. The alkali and alkaline earth metals are examples, as well as beryllium, indium, magnesium, aluminum; some zinc, copper and lead, and the "exotic" metals such as titanium, tantalum, hafnium and zirconium. Chlorine gas is obtained at the anode from molten or aqueous chlorides; fluorine is made in special cells from nonaqueous fluoride solutions. Caustic soda and other hydroxides are obtained from aqueous solution with hydrogen as a by-product. Chlorine bleaches, chlorates, perchlorates, peroxysulfates, perborates and other oxidizing agents are anodic products. Hydrogen peroxide, H_2O_2, is an indirect electrolytic product. Many other substances which could be made electrolytically are made more economically by direct chemical methods or even bacterial action.

Electrodeposition and Electrorefining. Electroplating is well-known as a means of depositing a thin layer of one metal on another either for decoration or for increased corrosion resistance. By special methods even plastics can be given a metal coating. Electroplating with zinc and tin has saved huge amounts of these metals over the older hot-dip methods.

Impure copper, nickel, tin and other metals are refined by anodic dissolution and simultaneous cathodic redeposition in suitable solutions of their salts; more noble metals do not dissolve as easily at the anode while baser metals do not deposit as readily at the cathode. In this process valuable metals are recovered, as silver and gold from copper.

Some low-grade ores can be leached with acids or other reagents and the metals recovered by electrodeposition (electrowinning). Sometimes heaps of tailings discarded in former years can profitably be reworked in this way.

Certain colloids whose particles are electrically charged can be electrodeposited; thin rubber articles such as gloves can be made in this way, and it is the basis of a method of applying a paint priming coat which is exceptionally free of pinholes.

The reverse of electrodeposition, i.e., anodic dissolution, has been developed into a valuable method of shaping metal parts (electroforming or electromachining).

Electrothermics. Since the first days of large-scale production of electric power, electrically heated furnaces have been used for chemical reactions which require high temperatures. Both resistance and arc furnaces are used, and in some cases, induction furnaces. "Carborundum" (SiC) and other abrasives, calcium carbide (CaC_2), calcium cyanamide ($CaCN_2$), and elemental phosphorus are electric-furnace products. Preparation of the so-called "hard metals" (carbides, borides, nitrides and silicides of many metals) often involves electric heating and sometimes electrolysis at temperatures of 2000°C and higher. Some metallurgical processes employ electric heating because of the excellent temperature control. Electric blast furnaces for the reduction of iron ores are not unknown.

Dielectric Materials. Electric insulating materials are of great importance in many industrial applications, from the coatings on conductive wires to the fluids or gases used in electric transformers. Much chemical research is involved in developing substances to meet requirements such as high or low dielectric constant, long-term stability against dielectric breakdown perhaps at high temperatures, or maintenance of electric charge separation. Silicones, fluorinated hydrocarbons, synthetic rubbers and other plastics are examples. In electrical condensers or capacitors both the electrode materials and the insulating medium must be well-chosen. In electrolytic capacitors true electrochemical processes are involved at the electrodes in forming and maintaining the oxide films (e.g., on aluminum, tantalum).

Semiconductors. Certain elements as germanium and silicon, which do not conduct electricity well and whose atoms have four electrons to share in covalent bonds with neighboring atoms, are called semiconductors. Normally a few electrons are not fixed in the crystal lattice and are free to conduct, and more are freed at higher temperatures (the resistance decreases in contrast to the resistance of metals, which increases with rising temperature). By refining these elements to unusual purity and adding controlled traces of elements with three or five valence electrons, the useful p- and n-type semiconductors are formed, which serve as the basis for transistors and other electronic units. The resulting *solid state* devices have revolutionized the entire electronics industry, from radio and television transmitting and receiving to computer design and operation, and in a host of other applications.

Phosphors. Phosphors are substances which become luminescent, i.e., emit visible light, when subjected to radiation which raises electrons to unstable energy levels, or when bombarded with electrons as on a television screen. The production of suitable phosphors for color television, phosphors which emit light on electric stimulation (electroluminescence), and materials and methods which convert heterogeneous white light into intense beams of "coherent" monochromatic light (lasers) has been the result of much intensive electrochemical research.

Theoretical Electrochemistry. Electrical studies of chemical materials have been powerful tools in elucidating the structure and properties of matter, and have led to many practical applications.

Studies of the electromotive force (emf) of reversible cells are of major importance in accurate determination of the energy produced in chemical reactions. Such cells are analogous to the cells used in flashlights, etc., but hundreds or thousands of reactions can be studied which are unsuitable for commercial cells. The emf is measured with an instrument (potentiometer) which requires negligible current from the cell (the electrodes do not become polarized). The free energy of the cell reaction is obtained from the equation $\Delta G = -nFE$, where E is the emf in volts, F is the Faraday constant (96,490 coulombs per chemical equivalent), and n is the number of equivalents appearing in the chemical equation. From the free energy the equilibrium constant of the reaction is deduced, and the possible yield of products. And just as chemical equations may be combined, so the energies may be combined to make predictions about reactions which have not been studied directly.

Electrolysis of solutions with direct current has led to measurement of relative and actual speeds of ions and the fraction of the current carried by individual ions (transference numbers). Direct current is used in studies of electrode kinetics and the molecular mechanism of electrode processes. Passing appreciable current invariably requires a higher voltage than the reversible emf; the electrodes become polarized because some step in the formation of a product requires more energy than the calculated minimum (the extra energy eventually appears as heat). It is of great theoretical (and practical) importance to understand what takes place at the electrode and what the slow steps are.

In polarography, electrolysis is carried out with the "falling mercury drop" as one electrode. A new mercury drop is presented every few seconds, with an uncontaminated surface for metal deposition or other process. Each reaction takes place at a characteristic potential and the components of a solution can be identified.

In electrophoresis, chemical compounds of high molecular weight can be separated by electrolysis, since their ions move at different speeds. The proteins of blood plasma have been separated and identified in this way, and the method has served as a guide in the large-scale separation of blood components.

Studies of the conductance of ionic solutions are made with alternating current of low amplitude, with special electrodes which do not polarize readily. Comparison of conductance with other physical properties of solutions led Svante Arrhenius in the 1880's to his ionization theory of acids, bases and salts. Conductance is used in titration analysis, in measuring reaction rates, to measure the purity of water and the contamination of industrial effluents, and for many other purposes.

The dielectric constant of a gas or liquid is measured in a capacitance cell using high-frequency current. From this quantity the dipole moment of the molecules is calculated, which gives a measure of their symmetry or dissymmetry of electron distribution. It is found, for example, that the water molecule (H_2O) is triangular with the hydrogen electrons drawn towards the oxygen; carbon dioxide (CO_2) has a linear molecule, etc.

Corrosion of Metals. Corrosion is in most cases an electrochemical process in that the metal is oxidized, losing electrons to something in the surrounding medium. For example, if a drop of water or salt solution is allowed to evaporate on a bare steel surface, a rust stain results. The initial anodic reaction is:

$$Fe - 2\varepsilon \rightarrow Fe^{++}$$

The cathodic reaction is the reduction of dissolved oxygen:

$$O_2 + 2\,H_2O + 4\varepsilon \rightarrow 4\,OH^-$$

The ferrous hydroxide formed, $Fe(OH)_2$, is further oxidized to hydrated ferric oxide (rust).

Such mechanisms suggest ways to prevent or delay corrosion: keep oxygen or other oxidants out, add inhibitors which are adsorbed on the metal and block the cathodic (or anodic) sites, make the metal cathodic in an electric circuit so that external electrons reduce the oxygen. Cathodic protection is widely used to mitigate the corrosion of pipe lines, water mains and other underground equipment, as well as ships' hulls (where the paint coating is never perfect) and many metal structures immersed in sea water. Current is supplied by generators or by attaching more active metals as magnesium or zinc to steel as sacrificial anodes.

Sometimes a metal may be protected by making it anodic in a circuit, thus forming a passivating oxide film; or by addition to the medium of an oxidizing agent (as chromate) to form the passivating film.

Electrochemistry in Life Processes. Ever since Galvani found that frog muscle contracted on application of an electrical potential, it has been apparent that electricity and current flow are intimately involved in life processes. Differential diffusion of ions through cell membranes results in electric potential differences, and due to the orderly arrangement of cells appreciable potentials are found between different parts of the body. In the extreme case of the electric eel, sufficient voltage and current are generated to stun or kill the eel's prey.

The heart beats as the result of periodic electric stimuli, the rate being controlled by the chemical balance and needs of the body. In case the stimuli are too weak they can be augmented by a battery-operated electronic "pacemaker" implanted in the body; more than 100,000 people now owe their lives to these devices. Unfortunately the pacemakers must be replaced in about two years, and much research is needed to make them last longer. In heart transplants, mild electric shock starts the new heart beating. A most desirable invention would be a mechanical pump to replace a worn-out heart, powered by current generated by electrochemical processes within the body.

Nerve impulses, while not carried by electronic conduction as in a wire, proceed along the nerve fibers by a series of chemical processes which transmit a potential difference. Storage of information in the brain and its retrieval (memory) are almost certainly electrochemical in nature. Electric stimulation at certain points in the brain can make the subject recall "forgotten" events, while stimulation at other points can result in pleasant, or unpleasant, feelings. There are suggestions that personalities can someday be drastically changed by implanting tiny electronic devices in the skull.

The study of electrochemical processes in living systems is evidently of utmost interest and importance, and holds great promise for the future.

CECIL V. KING

References

Bockris, J. O. M., and Reddy, A. K. N., "Modern Electrochemistry," Plenum Publishing Corp., New York.

Falk, S. Uno, and Salkind, A. J., "Alkaline Storage Batteries," John Wiley & Sons, New York, 1969.

The Bell System Technical Journal, Vol. 49, pp. 1249–1470, September, 1970.

Breiter, M. W., "Electrochemical Processes in Fuel Cells," Springer-Verlag, New York, 1969.

Hampel, C. A., Editor, "The Encyclopedia of Electrochemistry," Reinhold Publishing Corp., New York, 1964.

Electrochemistry in the Solid State

Electrochemical processes are those in which electrons are taken up from, or donated to, an electronically conducting phase by means of a chemical reaction occurring at a phase boundary.

Usually at least one of the phases comprising the circuit is solid (e.g., a metal serving as the electronic conductor). Purely electronic transport within a single phase is not of interest here. However, if two or more metals are used in a galvanic circuit, the electromotive force (emf) in the circuit, and hence the free energy change for the process accompanying charge transfer, may arise in large part from differences in electronic energies in the two solid metals. Frequently, also, an electrochemical reaction occurs in a solid phase or phases, or transport of matter takes place through a solid.

The Contact (or Volta) Potential. In a galvanic (emf) cell, the seat of the emf lies in the potential difference across phase boundaries. It can be shown that the minimum number of phase boundaries is three; commonly two of these are the boundaries between the electrodes and the electrolyte, which boundaries are the sites of reaction, while the third is the boundary between two metals comprising the electrodes.

Whether the emf of galvanic cells arises at the metal-metal interface or at the sites of electrochemical reaction was a question which gave rise to a lively controversy lasting well over a century. Volta discovered the potential difference which bears his name in 1797; on separating two pieces of metal which had been brought into contact, he found that they bore opposite charges. He believed that the contact potential between metals was responsible for the emf of his voltaic pile. Three years later Ritter found that the action of the pile was accompanied by chemical changes, which he concluded were the cause of the emf. Kelvin supported the contact theory of Volta and devised a way to measure contact potentials directly. Unfortunately, all early measurements were subject to serious errors from surface contamination; not until experiments were carried out with metals having really clean surfaces in high vacuum could it be proved that the contact potential was indeed a significant part of, but not solely responsible for, the emf of electrochemical cells.

The contact potential is due to a difference in Fermi levels. The Fermi level represents, approximately, the highest occupied electronic energy level in a metal; when two metals are brought into contact, electrons are transferred from the metal having the higher Fermi level to the other metal until the electronic energies are equalized, thereby rendering the former positive with respect to the latter. In order to establish the relative heights of the Fermi level, use is made of the work function, which is the difference in energy between that of an electron at rest in vacuum outside of a metal and that of an electron at the Fermi level in the metal. Consequently, the contact potential between two metals is given by the difference in their work functions.

Electrode Reactions. Faraday's Law of electrochemical equivalence is obeyed in the case of electrode reactions involving solids provided all of the current carried through the electrolyte is carried by ions. Apparent deviations from Faraday's Law are often noted, as in the case of the electrolysis of solid CuI, but these are satisfactorily accounted for on the basis that part of the charge transfer through the electrolyte is electronic.

The following examples illustrate the applications of solid-state theory to some characteristic electrode reactions.

Electron injection. When an NaCl single crystal is held between flat inert electrodes at elevated temperature, Na and Cl_2 are the electrode products. If, however, the cathode is a sharp platinum wire, heated and driven into the crystal, electrons are injected from the electrode into the crystal. These electrons cannot be accommodated in the valence band, but are trapped in negative-ion vacancies to form F centers. F centers impart a visible color, which is observed to drift toward the anode. Upon field reversal, the colored cloud is withdrawn at the platinum point. Since the F center is neutral, the drift of color implies that a portion of the centers are thermally ionized.

Reactions involving holes. When n-type germanium is made the anode in certain electrolytes a limiting current is soon attained as the anodic potential increases. The limiting current is not observed with p-type germanium in the same electrolytes. This is indication that holes, which are the minority carriers in n-type material, take part directly in the electrode reaction, the current being limited by the rate at which holes are generated at the surface or reach the surface by diffusion after generation in the interior of the electrode. Illumination, which generates holes at the surface, markedly increases the limiting current.

The reaction of holes with the electrolyte is equivalent to the injection of electrons from the electrolyte into the valence band of the electrode.

Oxide Layers. The formation of oxide layers on the metals aluminum, tantalum, titanium, tungsten, and zirconium is important in the technology and theory of electrolytic condensers, rectification, metal finishing, and corrosion prevention. When aluminum (for example) is made the anode in ammonium borate solution, an adherent resistive layer of aluminum oxide grows to a thickness limited by the applied voltage. Above about 600 volts dielectric breakdown occurs; at this voltage the film has reached a thickness of a few thousand angstroms.

Al_2O_3 is essentially an insulator; however, electronic conduction occurs and at sufficiently high fields Al^{3+} ions contribute to conduction. In fact, the layer could not grow, once it had been started, were it not for ionic transport through the layer. Thus the layer grows, not at the metal, but at the oxide-solution interface, where Al^{3+} ions and hydrated oxide ions neutralize each other. Electronic conductivity in the oxide is unimportant for this process.

Oxide "depolarizers." The cathodes (positive electrodes) of most useful voltaic cells are oxides of metals in one of their higher states of oxidation; the most common are MnO_2, PbO_2, and Ni_2O_3. These oxides are still referred to as *depolarizers,* because of the idea, prevalent at one time, that hydrogen gas was the product at the cathode and was depolarized by the action of the oxide.

All practical oxide depolarizers are semiconductors, whose conductivities may lie in the range 10^{-4} to 10^{-2} ohm^{-1} cm^{-1} at room temperature. The reduction of the oxide during discharge therefore takes place at the interface between the oxide and the electrolyte, not between the metal or carbon used as a connector and the oxide, since the field in the semiconductor is too low to cause appreciable ionic migration.

The behavior of MnO_2, during cathodic discharge is now reasonably well understood. The products depend upon current density and pH; at low current density and pH around 7 the principal reaction is

$$MnO_2 + H^+ + e^- = MnOOH.$$

During discharge, MnOOH accumulates as a solid solution in and near the MnO_2 surface, with a decrease in cell emf resulting from the lowered MnO_2 surface activity. When the discharge current is broken, the emf recovers to nearly its original value. This is believed to be due to the diffusion of hydrogen away from the surface into the bulk of the MnO_2, thus regenerating the MnO_2 surface.

Recent evidence indicates that the discharge and recovery of cathodes made from single crystals of V_2O_5 proceed by an analogous mechanism.

Ionic Conduction in Crystals. Frenkel, in 1926, and Schottky, in 1935, proposed that transport in crystals occurs as the result of defects in the otherwise regular structure. A Frenkel defect is a vacant lattice site plus an interstitial ion or atom. Either the interstitial, the vacancy, or both may contribute to conduction, which occurs as the result of ion-jumps from normal sites into vacancies or from one interstitial site to another. A Schottky defect is just a vacant site (the occupant having been taken to a new site on the crystal surface). In ionic crystals having Schottky defects, the concentrations of positive- and negative-ion vacancies are such that electroneutrality is preserved. Conduction may be by ion-jumps into vacancies on either the cation or the anion sublattice. Some solids, such as α-AgI and γ-Al_2O_3, have equivalent sites which are only partly occupied even in the pure perfect crystal. Such structures are called "defect structures."

Defects are produced as the result of thermal equilibrium in pure crystals, there being an entropy increase accompanying defect formation. Conductivity due to such defects is termed "intrinsic." The incorporation of foreign ions, of charge different from that of its host, may markedly increase the concentration of defects. For example $CaCl_2$, dissolved in NaCl, is accompanied by one positive-ion vacancy for each calcium ion, to compensate for the extra charge on calcium which occupies a sodium-ion site.

Ionic conductivity depends upon temperature according to

$$\sigma = \sum_i A_i \, e^{-E_i/RT},$$

where the summation extends over the types of charge carriers indexed by the subscript i; A is approximately constant, and E is an observed activation energy for conduction.

In many cases, E can be simply related to theoretical parameters. For example, for a binary salt at temperatures high enough that intrinsic conductivity predominates, $E = E_1 + \frac{1}{2}E_2$, E_1 being the activation energy for migration of the carrier and E_2 being the formation energy of a defect pair.

At lower temperatures, where the concentration of defects is established by the impurities present, $E = E_1$. At still lower temperatures, a vacancy may be bound by Coulombic forces to the foreign ion whose charge it compensates. Such complexes, if neutral, do not contribute to conductivity. When the degree of association into complexes approaches unity, $E = E_1 + \frac{1}{2}E_3$, where E_3 is the binding energy of the vacancy to the foreign ion.

In the alkali halides, in which the defects are of the Schottky type, E for the cation vacancy is sufficiently lower than that for the anion vacancy that essentially all of the conductivity is due to cation vacancies over a large range of temperatures. In this case the summation above reduces to one term; as seen above, E assumes different values in different regions of temperature. Near the melting point, there is evidence that anion vacancies may make a small contribution to the conductivity.

Recently, the computer has been used to fit the conductivity, written as a function of the parameters of the theoretical model, to accurate data taken over a wide range of temperature. For KCl, satisfactory values for the entropies and enthalpies of migration of both cation and anion vacancies, the entropy and enthalpy of formation of the defect pair, and for binding of the cation vacancy to foreign ions such as Sr^{2+}, have been obtained in this way.

Transport numbers are measured by means of a modification of the Hittorf method. Two or more single crystals, in series, are subjected to electrolysis. From mass changes in the crystals, total charge passed, and knowledge of the electrode reactions, the transport numbers are readily found. The cationic number is symbolized n_c, the anionic number, n_a. In most crystals ionic transport is either predominantly cationic or predominantly anionic. In some, for example γ-CuI and β-Ag_2S, transport is partly electronic and partly ionic over a considerable range of temperature.

Cells with Solid Electrolytes. Several schemes for high-temperature fuel cells have been devised in which the electrodes are oxides of iron or nickel and the electrolyte is a solid carbonate or vitreous "solid" through which transport is usually by sodium ions. The internal resistance is high, however, even at high temperature, and since the primary purpose of fuel cells is power generation, this presents a serious obstacle in the way of eventual application. A cell in which the electrolyte is ZrO_2 containing 15 mole per cent CaO has recently been described. At the porous platinum cathode, oxygen is converted to oxide ions. The electrolyte, because of the substitutional solution of divalent calcium ions on the tetravalent zirconium-ion sublattice, contains a high proportion of oxide-ion vacancies, so that oxide ions migrate easily to the porous anode, where the fuel, hydrogen, is oxidized. At 1000°, with a fuel containing 97 mole % hydrogen, cell voltage remains as high as 0.7 v at a current density of 50 ma cm^{-2}.

                                                    ALLEN B. SCOTT

ELECTRODEPOSITION

Electrodeposition, as distinguished from electroplating, is sometimes called electrophoretic deposition. It covers those electrolytic processes in which electrically charged particles other than ions are deposited at either the anode or the cathode. Although such processes are not so extensively employed as electroplating techniques, they have interesting industrial applications; they are also used in gravimetric analysis. Oxide coatings may be applied to the cathodes used in electron tubes by electrodeposition of carbonates of the alkaline earths precipitated from their nitrates by sodium carbonate and maintained in a slurry containing about ten grams of solid per liter. Electrolysis of the slurry will deposit the carbonates on the cathode in a uniformly thick film which can be baked to obtain the oxides, which comprise the active cathode of the electron tube.

The advantage of electrodeposition over spraying or other mechanical methods of application is the uniformity of thickness of the film obtained. An example from the field of organic chemistry is the deposition of rubber from an ammoniacal latex suspension. The electrically charged rubber particles are deposited at the anode, which is usually the shape of the product being manufactured (surgeon's gloves, for example). It is entirely practical to include in the latex suspension sulfur-bearing compounds, zinc compounds, accelerators, etc., which are required to produce desired properties in the cured rubber. These are prominent examples from among the many materials which may be peptized to a sufficient degree to permit electrolysis. Although it is usually desirable to work with negatively charged particles, it is possible in some cases to impart a positive charge to the material to be deposited and to obtain it at the cathode.

The conditions of electrolysis vary widely depending upon the material being deposited and the nature of the electrolyte used. The voltage required may range from 1.5 to more than 100 volts, although industrially a potential of over 15 volts is seldom encountered. In general, the lower voltages prevail in inorganic systems and the higher voltages in organic electrolytes. Current densities also are quite variable but commonly are about 10 to 20 amperes per square foot.

The current density which may be employed is determined mainly by side reactions and by-products at one or both electrodes. Excessive anodic current densities will result in the evolution of oxygen in aqueous electrolytes, usually with harmful results. The product formed at the anode may either be distorted by the evolution of the oxygen or it may be oxidized chemically.

Discharge of hydrogen at the cathode is commonly encountered in industrial processes, and usually causes foaming of the electrolyte. This is prevented ordinarily by interposing a semipermeable diaphragm between the electrodes. Although the voltage drop across the cell is increased, it is the easiest solution of the foaming problem.

In many cases the electrode at which the product is obtained must be selected with great care. Often a slight solubility of the anode will greatly affect the product; for example, a copper anode would never be used for rubber deposition because of the unfavorable effect of small amounts of copper on many rubber compositions. Stainless steel and lead are often used in practice.

In 1964 important developments in the application of paint by electrodeposition were announced. A dilute aqueous emulsion of resins and pigments is electrolyzed, with the workpiece to be coated as anode and the tank cathode (although auxiliary cathodes may be employed to improve current distribution). Deposition is effected by a combination of electrophoresis, the major process, and the discharge of some organic ions. Water is removed from the deposited film by electroosmosis. The process is used for the application of primer coatings, and the advantages are the uniformity of the coating thickness, particularly on sharp corners and edges, and the smoothness of the surface. Voltages of 50 to 100 are required, the current density is 2-4 asf, and the time is 1-2 min. Only primers are so applied.

                                                    HAROLD J. READ

Cross-references: *Electrophoresis, Electroplating.*

ELECTRODE POTENTIALS

Metals placed in contact with a solution containing ions of the metal tend to develop an electrical potential characteristic of the equilibrium between the metal and its ions. This potential can be measured by comparison with that of a reference electrode. Nonmetals can also develop such potentials when equilibrated with one of their ions in the presence of a catalytically active but chemically inert metallic conductor, e.g., platinum. For aqueous electrolytes, the common reference electrode is the *standard hydrogen electrode* (SHE) of platinized platinum in contact with an aqueous solution of hydrogen ion H^+ at unit activity (e.g., approx 1 M HCl) saturated with H_2 gas under 1 atm. fugacity (approx. 1 atm. partial pressure). The schematic diagram of the SHE is $(Pt)H_2/H^+$. Its potential V°(SHE) and isothermal temperature coefficient $(dV°/dT)_{iso}$ (SHE) are both assigned the value *zero* at all temperatures. Other reference electrodes with their potentials V in volts *vs.* SHE at t°C include: the saturated calomel electrode (SCE), Hg/Hg_2Cl_2, KCl(aq.sat.), V = +0.245 − 0.0007(t − 25); the normal calomel electrode (NCE), Hg/Hg_2Cl_2, KCl (1 N), V = +0.280 −

0.0003(t − 25); the decinormal calomel electrode (DCE), $Hg/Hg_2Cl_2,KCl(0.1$ N), V = +0.333 − 0.0001 (t − 25); the silver chloride electrode, $Ag/AgCl,KCl$(unit activity, V° = +0.2222 −0.00066(t − 25); the saturated copper sulfate electrode, $Cu/CuSO_4$(aq. sat.), V = +0.30 − 0.00009(t − 25). For the general electrode M/M^{z+} or M^{z+}/M, the electrode potential V is equal to the open circuit voltage V″ − V′ as measured, e.g., between two copper leads, on the isothermal cell: (V′)Cu/SHE//M^{z+}/M/Cu(V″), where // means that the liquid junction potential has been eliminated or minimized, e.g., by the use of salt bridges. The isothermal temperature coefficient is given by d(V″ − V′)/dT for the isothermal cell measured at a series of different temperatures. The thermal temperature coefficient (thermoelectric power) of any electrode is equal to $(dV/dT)_{180}$ + 871 microvolts per kelvin, where the latter constant is the thermal temperature coefficient of the SHE at 25°C across sat. KCl.

The potentials and isothermal temperature coefficients listed in Table 1 are given for the elements and their ions in their standard states of unit activity in water. For non-standard state electrodes, the Nernst equation applies: V = V° + (RT/zF) ln (Ox)/(Red). Here (Ox) and (Red) denote the activities, or activity products, of the electromotively active oxidized and reduced forms of the elements that appear in the balanced expression for the electrode reaction, z is the number of electrons transferred, R is the gas constant (8.3143 J/K.mole or 86.17 μVF/K.mole), F is the faraday (96487 A.s), and T is the absolute temperature in kelvins (K = °C + 273.15). For example, given the reduction of permanganate ion in acid medium: $MnO_4^- + 8 H^+ + 5 e^- = Mn^{++} + 4 H_2O$, the Nernst equation at 25°C is

V = +1.51 V + (0.05916 V/5) $\log_{10}$

$[(MnO_4^-)(H^+)^8/(Mn^{++})(H_2O)^4]$.

Two sign conventions have been used in the tabulation of electrode data. In Table 1, the potentials are given according to the international I.U.P.A.C.-Gibbs-Stockholm *electrode potential* sign convention, in which the electrode has the same algebraic sign as its observed D.C. polarity relative to SHE, e.g., V°(Zn/Zn⁺⁺) = −0.76 V means that Zn is the (−) terminal of the SHE//Zn cell. In the "oxidation potential" convention of Professor Latimer, extensively used in inorganic chemistry, Zn/Zn⁺⁺ is listed as +0.76V, corresponding to the fact that the tendency to undergo oxidation is stronger for zinc than for hydrogen. Latimer's "reduction potential" has the same sign as the *electrode potential*.

In every half-cell, an element appears in an oxidized form and in a reduced form, e.g., Zn⁺⁺ and Zn; H⁺ and H₂; O₂ and H₂O; F₂ and F⁻. The more negative (positive) the I.U.P.A.C.-Gibbs-Stockholm *electrode potential*, the stronger the reducing (oxidizing) power of the reduced (oxidized) form of the element and the stronger its tendency to undergo oxidation (reduction). In general, any oxidized form can react with a reduced form having a more negative electrode potential, but not vice-versa, e.g., H⁺ can attack Na, Zn, Fe, Pb, but not Cu, Ag or Au; Cu⁺⁺ can react with Zn,

but Zn⁺⁺ does not react with Cu. Electrode potentials provide thermodynamic information as follows: for the cathodic (reduction) half-cell reaction (coupled with the counter anodic reaction of the SHE): Ox + (z/2)H_2 = Red + zH⁺, ΔG° = −zFV°, ΔS° = +zF(dV°/dT)$_{180}$, ΔH° = −zFV° + zFT (dV°/dt)$_{180}$, ΔCp° = +zFT(d²V°/dT²)$_{180}$. Note that 1 volt-faraday(VF) = 96487 J = 23061 cal.

Electrode potentials provide a convenient way of representing the reducing (anodic or oxidation), and the oxidizing (cathodic or reduction) tendencies of the elements. With hydrogen selected as zero, potentials extend to almost equal voltages in both directions, from the extreme cathodicity of fluorine, with its high oxidizing power at +2.87 V, to the extreme anodicity of lithium with its high reducing power at −3.05 V. Table 1 illustrates the activity sequence of the metals from lithium to gold, and the inverse activity sequences of the non-metals F, Cl, Br, I and O, S, Se, Te, and includes the anode: $(-)Pb/PbSO_4$, and cathode: $PbSO_4/PbO_2(+)$, of the lead-acid storage battery. The potential shift of the hydrogen electrode from 0.000 V at pH 0 to −0.828 V at pH 14 illustrates the effect of pH which affects all electrodes involving H⁺ or OH⁻ in the half-cell reaction. Increasing *electrode potentials* (alt. decreasing *oxidation potentials*) can be used to construct oxidation chains such as those of iron in acid and alkaline media:

$$Fe \frac{(\text{Electrode Potential E.P.}) -0.44}{(\text{Oxidation Potential O.P.}) +0.44}$$

$$\xrightarrow{} Fe^{++} \frac{+0.77}{-0.77} Fe^{+++} \frac{+2.20 \text{ V}}{-2.20 \text{ V}} FeO_4^=$$

$$Fe \frac{(\text{E.P.}) -0.88}{(\text{O.P.}) +0.88} Fe(OH)_2 \frac{-0.56}{+0.56}$$

$$\xrightarrow{} Fe(OH)_3 \frac{+0.72 \text{ V}}{-0.72 \text{ V}} FeO_4^=$$

which define the domains of stability of the various oxidation states. When an inversion occurs in the potential sequence, e.g.,

$$Cu \frac{(\text{E.P.}) +0.521}{(\text{O.P.}) -0.521} \text{---}Cu^+\text{---} \frac{+0.153 \text{ V}}{-0.153 \text{ V}} Cu^{++}$$
$$\frac{(\text{E.P.}) +0.337 \text{ V}}{(\text{O.P.}) -0.337 \text{ V}}$$

the intermediate species Cu⁺ is unstable and disproportionates to Cu and Cu⁺⁺, so that the observed potential is that of the Cu/Cu⁺⁺ combination. In alkaline medium, the normal sequence reappears

$$Cu \frac{(\text{E.P.}) -0.358 \text{ V}}{(\text{O.P.}) +0.358} Cu_2O \frac{-0.080 \text{ V}}{+0.080 \text{ V}} Cu(OH)_2$$

and the copper-I oxide has a narrow band of stability as exhibited in the Pourbaix potential-pH diagram for this element.

Electrodes can also manifest non-equilibrium or irreversible potentials, marked by the phenomena of electrode polarization and overvoltage. These arise from current flows at the electrode surface, either local action currents between local anodic and cathodic spots (as in corrosion, these give rise to mixed potentials), or else external currents which shift the *electrode potential* in the positive (negative) direction as the electrode passes current in the

TABLE 1. SELECTED STANDARD AQUEOUS ELECTRODE POTENTIALS AND
ISOTHERMAL TEMPERATURE COEFFICIENTS AT 25°C

Reference Electrode: Standard Hydrogen Electrode (SHE)
I.U.P.A.C.—Gibbs-Stockholm Sign Convention
Potentials in volts, temperature coefficients in millivolts per kelvin.
The second isothermal temperature coefficient, $(d^2V°/dT^2)_{iso}$, where known, is given
in parentheses in microvolts per kelvin^2 and is designated by S.C.

Electrode	V°	$(dV°/dT)_{iso}$	Electrode	V°	$(dV°/dT)_{iso}$
	(V)	(mV/K)		(V)	(mV/K)
Li/Li$^+$	−3.045	−0.534	(Pt)Fe^{++},Fe^{+++}	+0.771	+1.188
K/K$^+$	−2.925	−1.080	Ag/Ag$^+$	+0.799	−1.000
Ca/Ca^{++}	−2.866	−0.175		(S.C. −0.924 μV/K^2)	
Na/Na$^+$	−2.714	−0.772	(Pt)Br$_2$(ℓ),Br$^-$	+1.065	−0.629
La/La^{+++}	−2.522	+0.085		(S.C. −6.210 μV/K^2)	
Mg/Mg^{++}	−2.363	+0.103	(Pt)O$_2$,H$_2$O,H$^+$	+1.299	−0.846
Al/Al^{+++}	−1.662	+0.504		(S.C. +0.552 μV/K^2)	
Mn/Mn^{++}	−1.18	−0.08	(Pt)Cl$_2$,Cl$^-$	+1.360	−1.260
Zn/Zn^{++}	−0.763	+0.091		(S.C. −5.454 μV/K^2)	
	(S.C. +3.84 μV/K^2)		Au/Au^{+++}	+1.498	—
Te/H$_2$Te,H$^+$	−0.718	+0.280	(Pt)MnO$_4^-$,Mn^{++},H$^+$	+1.51	−0.66
Fe/Fe^{++}	−0.440	+0.052	(Pb)PbO$_2$/PbSO$_4$,SO$_4^=$,H$^+$	+1.682	+0.326
Cd/Cd^{++}	−0.403	−0.093		(S.C. +2.516 μV/K^2)	
	(S.C. +2.2 μV/K^2)		(Pt)FeO$_4^=$,Fe^{+++},H$^+$	+2.20	−0.85
Se/H$_2$Se,H$^+$	−0.399	−0.028	(Pt)F$_2$,F$^-$	+2.87	−1.830
Pb/PbSO$_4$,SO$_4^=$	−0.359	−1.015		(S.C. −5.339 μV/K^2)	
	(S.C. −1.555 μV/K^2)				
Pb/Pb^{++}	−0.126	−0.451	**BASIC SOLUTIONS**		
(Pt)H$_2$,H$^+$ (SHE)	0.0000*	0.0000*			
	(S.C. 0.000*)		Fe/Fe(OH)$_2$,OH$^-$	−0.877	−1.06
(Pt)S,H$_2$S,H$^+$	+0.142	−0.209	(Pt)H$_2$,OH$^-$	−0.828	−0.834
(Pt)Cu$^+$,Cu^{++}	+0.153	+0.073		(S.C. −7.272 μV/K^2)	
Ag/AgCl,Cl$^-$	+0.2222	−0.658	(Pt)Fe(OH)$_2$,Fe(OH)$_3$,	−0.56	−0.96
	(S.C. −5.744 μV/K^2)		OH$^-$		
Hg/Hg$_2$Cl$_2$,Cl$^-$	+0.2676	−0.317	Cu/Cu$_2$O,OH$^-$	−0.358	−1.326
(calomel electr.)	(S.C. −5.664 μV/K^2)			(S.C. −6.828 μV/K^2)	
Cu/Cu^{++}	+0.337	+0.008	(Pt)Cu$_2$O,Cu(OH)$_2$,OH$^-$	−0.080	−0.725
Cu/Cu$^+$	+0.521	−0.058	(Pt)O$_2$,OH$^-$	+0.401	−1.680
(Pt)I$_2$,I$^-$	+0.536	−0.148		(S.C. −6.719 μV/K^2)	
	(S.C. −5.965 μV/K^2)		(Pt)Fe(OH)$_3$,FeO$_4^=$,OH$^-$	+0.72	−1.62

*Reference electrode, by definition.

anodic (cathodic) direction. An anode (cathode) passes positive current from (to), and electrons to (from), the external circuit, and is NEG (POS) in a galvanic cell and POS(NEG) in an electrolytic cell. The potential shifts caused by current flow can be subdivided experimentally into activation (reaction), concentration (transport), and ohmic (film) overpotentials. Electrode potentials are also referred to as *tensions* by analogy to the French: *tension d' électrode,* and the German; *Elktroden-spannung.*

ANDRE J. de BETHUNE

References

Christiansen, J.A., and Pourbaix, M., Compt. rend. 17e. Conf. I.U.P.A.C., Stockholm, 1953, pp. 82-84; J. Am. Chem. Soc., **82**, 5517 (1960).

de Bethune, A.J., and Swendeman Loud, N.A., "Standard Aqueous Electrode Potentials and Temperature Coefficients," C.A. Hampel, Pub., 8501 Harding Ave., Skokie, Illinois, 1964.

Gibbs, J. Willard, "Letter to Wilder D. Bancroft," May, 1899, reprinted in "Collected Works," Vol.

I, p. 429, New Haven, Yale University Press, 1949.

Hampel, C.A., Editor, "Encyclopedia of Electrochemistry," Articles on Electrode Potentials topics by S. Barnartt, A.J. de Bethune, S.N. Flengas, G.J. Janz, F.L. LaQue, J.Y. Lettvin et al., M.H. Lietzke and R.W. Stoughton, R.A. Marcus, J.M. Miller, R. Parsons, J.V. Petrocelli, F.A. Posey, M. Pourbaix, P. Ruetschi, P. Van Rysselberghe, New York, Reinhold Publishing Company, 1964.

Latimer, W.M., "Oxidation Potentials," 2nd Edition, New York, Prentice Hall, 1952.

Nernst, W., Z. physik. Chem. **4**, 129 (1889); Ber., **30**, 1547 (1897).

Pourbaix M.,"Atlas of Electrochemical Equilibria," Oxford, Pergamon; Brussels, Cebelcor, 1966.

ELECTRODES

Organic Electrochemical

Organic compounds which can either be oxidized or reduced offer varying degrees of resistance to this electron change. The amount of resistance offered depends to a great extent on the type of group to be attacked. Thus, for example, an aldehyde group generally requires less energy to be oxidized to an acid or carboxy group than would be required to oxidize a methyl group to a carboxyl group. The same holds true for reductions. If one desires to reduce an aldehyde or a ketone to the respective alcohol, less energy would be required than to cause the aldehyde or ketone to undergo a bimolecular reduction to form the pinacol. Therefore, an electrode material is used which will have the proper potential energy or overpotential required to accomplish the desired transformation. Generally this is not a great problem if there is present only one attackable group in the molecule and it is desired to cause a maximum transformation. In this instance any electrode of high overpotential can be used. However, if a partial reduction is necessary, then the choice of electrode material becomes limited. Thus, for example, N-methylphthalimide can be partially reduced to its respective hydroxy phthalimidine at a copper electrode. However, if complete reduction to the isoindoline is desired, a higher overpotential electrode, such as lead or mercury, is necessary.

The following table gives a group of metals in their approximate order of increased activity as oxidizing or reducing electrodes. This sequence may vary somewhat depending on the condition of the electrode surface, current density and the medium. As an example, the overpotential of a smooth surface lead electrode in an aqueous sulfuric acid medium is lower than mercury at current densities below 10^{-3} amp/cm². However, at higher current densities mercury is the lower overpotential electrode. It is evident that mercury, zinc, lead, or tin are the electrodes to use for difficult reductions and nickel or copper where milder conditions are required.

Oxidizing Electrodes	Reducing Electrodes
Nickel	Iron
Iron	Smooth Platinum
Lead	Silver Nickel
Silver	Copper
Cadmium	Cadmium
Palladium	Tin
Smooth Platinum	Lead
Gold	Zinc
	Mercury

The electrodes used for electrolytic oxidations are somewhat more limited because it is difficult to obtain a stable anode overpotential with many electrodes in the presence of an oxidizable compound. The potential generally rises rapidly from the value at which the anode dissolves, to the high value for passivity and oxygen evolution. However, since platinum and other platinum group metals are nearly always passive, it is possible to obtain graded potentials. Thus it is not possible to give definite rules concerning the efficiency of an electrode for anodic oxidation. If the oxidation is due to the presence of oxygen in an active form or is brought about by an oxide of the electrode material, as in the oxidation of iodate to periodate, the best results will be obtained with a lead dioxide or platinized platinum anode. In other cases, such as the oxidation of toluene to benzaldehyde or benzoic acid, best results are obtained with a smooth platinum electrode.

In general the overpotential of the electrode is a suitable indication of its reducing or oxidizing ability. There are, however, occasions in which the nature of the electrode material exerts a catalytic effect on the reaction to the point where it completely overshadows the potential effect. In these instances a low overvoltage electrode may be as effective or even more effective than a higher overpotential electrode. As an example one might consider the cathodic reduction of nitric acid to ammonia. If a relatively high overvoltage cathode such as amalgamated lead is used, a small percentage of ammonia is obtained, together with a large amount of hydroxylamine. However, when a low overpotential electrode such as spongy copper is used, high yields of ammonia are obtained. In the oxidation of iodate in an alkaline medium, quantities of periodate can be obtained using a smooth platinum electrode. A lower overpotential electrode, such as lead dioxide, results in far better yield than obtained with the platinum electrode. Methyl alcohol, when subjected to anodic oxidation in a dilute sulfuric acid medium at a smooth platinum electrode, yields formaldehyde with an 80% efficiency. When platinized platinum or lead dioxide anodes are used, the yield of formaldehyde decreases considerably, accompanied by an increase in the more highly oxidized products, formic acid and carbon dioxide even though the electrodes used were of lower overpotential.

It has also been found that alloy electrodes may be more efficient in certain instances. The reduction efficiency of nitrobenzene in an alcoholic alkaline solution is increased from 58 to 72% by the addition of 12% iron to the nickel cathode. The yield of pinacol from the electrolytic reduction of acetone is improved tremendously by the use of a lead-tin or lead-copper electrode. Thus it can be seen that although generally the overpotential of the electrode will give an indication as to the degree of attack which might be anticipated, the catalytic effect, which may very well be a surface phenomenon, must not be overlooked.

In view of the increasing interest in the measurement of electrochemical parameters of biological systems it is appropriate to include a brief discussion of the working electrodes used for such purposes. The most important criterion for such electrode materials is that they be chemically inert in the system being investigated. Electrodes with such characteristics are:

(1) Platinum, either as a shiny foil or wire has been the electrode of choice by most investigators. Although extremely inert under most conditions, it is attacked to an extent if used in the presence of halogen. Under these circumstances, the incorporation of about 1% iridium into the platinum will alleviate this difficulty.

(2) Gold electrodes have been used by some investigators. In the author's laboratory electrodes of this type were found unsatisfactory as it was ex-

tremely difficult to obtain the reproducible results observed with platinum.

(3) Iridium electrodes are suitable for use with biological systems but this metal offers no advantage over platinum.

Other electrodes made of graphite, tungsten palladium and mercury have all been used, but none have approached the suitability of platinum.

M. J. ALLEN

References

Allen, M. J., "Organic Electrode Process," Reinhold, New York, 1958

Allen, M. J., "Electrochemical Aspects of Metabolism," *Bact. Rev.*, **30,** 80 (1966)

Hewitt, L. F., "Oxidation Reduction Potentials in Bacteriology and Biochemistry," Williams & Wilkens, Balitmore, 1950.

Glass Electrodes

A thin glass membrane, when immersed in a suitable liquid medium, develops a measurable electrical potential which can be readily related to the activity of ionic species present in the solution. By appropriate manipulation of the glass composition, careful pretreatment of the glass surface, and reproducible experimental conditions, electrodes can be devised which not only yield meaningful information about the concentration of ions in solution but also display the ability to discriminate, in terms of a selective response, between a number of different ions of similar chemical characteristics. Because of their ability to give both qualitative and quantitative information about ions in solution, glass electrodes are widely used for purposes of chemical measurement.

Although a variety of special designs for unusual applications are available, glass electrodes are commonly formed as spherical or somewhat elongated bulbs. The interior of the electrode bulb, permanently sealed against contamination, is filled with a solution of invariable composition, often a buffer solution, in contact with a conducting reference element to provide a constant potential at the interior of the electrode. This arrangement is necessary because, experimentally, the potential difference between the interior and exterior interfaces of the glass membrane is measured. For practical purposes, the net potential of the glass electrode is measured against an external source of constant e.m.f. (reference electrode) with a device capable of measuring the resulting net e.m.f. with very little current drain (electrometer).

The composition of the glass is critical in determining the success of a glass electrode both as a means of accurately monitoring the relative activity of a single ion and distinguishing between different ions in solution. Most hydrolyzable glasses, including common window glass, show some electrical response to changes of ionic concentrations in solutions, but only glasses of carefully blended proportions of such materials as sodium, aluminum, and silicon oxides display the theoretical (Nernstian) response. The total number of possible combinations of ingredients which could be used to prepare glass electrodes is very large indeed, but because glass electrodes must also be sturdy, durable, and resistant to chemical attack the actual number of commonly used compositions probably does not exceed two dozen or so. Most of the current investigations of glass composition as a variable in manufacturing are concerned with the further improvement of pH type glass electrodes, on the one hand, and the development of so-called "specific ion" electrodes, on the other.

The latter category of glass electrodes represents an exciting and challenging advance in glass electrode development and is largely responsible for the currently revived interest in glass electrodes. In brief, these specific ion- or cation-sensitive glass electrodes show a more or less selective response to univalent cations (such as the alkali metal ions) in mixtures of such ions and in the presence of other (multivalent) cations (anions generally have little effect upon the glass electrode response). The selectivity of these electrodes is brought about by the substitution of certain ions (such as aluminum) into the glass lattice, so as to provide sites of excess negative charge which are "attractive" to cations in solution having a favorable charge-to-size ratio. Much of the recent work in this area has been semiempirical in nature and not all the observed effects have been rigorously explained.

The mechanism by which glass electrodes respond to ions in solution is itself of considerable interest. It is now known, with reasonable certainty, that glass electrodes immersed in aqueous solutions rapidly develop a swollen surface layer of hydrated glass in which ions can move with velocities intermediate between those in solution and in dry glass. The formation of this layer, accompanied by the slow penetration of ions from solution into glass, takes place during the preconditioning or "soaking" of the glass electrode, which is necessary for successful operation. When the preconditioned glass electrode is then transferred into the sample solution whose ionic composition is to be monitored, an electric potential is established at the swollen layer-solution interface as the result of a rapid exchange of ions between the solution and the hydrated glass layer. Tracer studies have shown that it is the properties of this ion-exchange process which determine the manner in which the glass electrode will respond in a given environment.

Applications of glass electrodes to chemical and biochemical problems have been wide indeed, ranging from the measurement of the salinity of blood streams to highly sophisticated investigations of ion-ligand interactions as correlated with nuclear magnetic resonance studies. Glass electrodes are routinely employed for a variety of clinical and process measurements and show every indication of further popularity as the principles and methodology of their functioning become more widely known.

G. A. RECHNITZ

Cross-reference: *Electrochemistry.*

References

G. Eisenman, editor, "Glass Electrodes for Hydrogen and Other Cations," Marcel Dekker, Inc., New York, 1967.

G. A. Rechnitz, "New Directions for Ion-Selective Electrodes," *Anal. Chem.,* **41,** (12), 109A (1969).

G. A. Rechnitz, "Chemical Studies at Ion-Selective Membrane Electrodes," *Acct. Chem. Res.*, **3**, 69 (1970).

ELECTROLYSIS

Electrolysis is a process by which a chemical reaction is carried out by means of the passage of an electric current. The process is generally used as a method of depositing metals from solution. The relation between the quantity of material undergoing reaction and the quantity of electricity used in the reaction was discovered by Faraday. The relation may be expressed by two laws: (1) The amount of chemical reaction produced by a current, e.g., the amount of any substance deposited or dissolved is proportional to the quantity of electricity passed; and (2) The amounts of different substances liberated at the electrodes, when the same quantity of electricity is passed, are proportional to their chemical equivalents. The proportionality constant, the Faraday, is equal to about 96,500 coulombs per gram equivalent weight.

Electrolysis is used for commercial purification of most metals and for winning some metals, especially sodium, magnesium and aluminum from ores or compounds. Sodium is produced commercially by electrolysis of molten sodium chloride, the other product being chlorine. Electrolysis of aqueous sodium chloride yields caustic soda. Electrolytic cells, such as Hooker, Vorce, and Allen-Moore, are used for this purpose commercially. Also, other halogens and metals may be produced by electrolysis of the appropriate metal halide. Aluminum is produced by electrolysis of alumina, Al_2O_3, dissolved in molten cryolite, $AlF_3 \cdot 3NaF$.

Copper is purified by using the impure copper as the anode and a bar of pure copper as the cathode of an electrolytic cell; the impurities silver, gold, et al., falling to the bottom of the cell as "slime." The impurities are later recovered from the slime. Hydrogen and oxygen are produced by electrolysis of water.

The waste sodium sulfate liquor from production of rayon is regenerated by electrolysis into sulfuric acid and caustic soda. Other materials produced electrolytically in commercial procedures are manganese and perchlorates.

Electrolysis is used to "anodize" metals such as aluminum or tantalum to provide decorative and protective coatings or coatings of high dielectric strength.

Because any electrolytic reaction requires some definite minimum potential, it is possible to selectively reduce one metal ion from a mixture of metal ions by keeping the potential high enough to carry out the desired reduction but low enough so that no other reaction occurs. By this means, i.e., controlled potential electrolysis, separations can be carried out or an unwanted reaction avoided.

Electrolysis is frequently used in chemical analysis. Coulometry, amperometric titrations, electrographic analysis and polarography, as well as simple electrodeposition depend on changes in chemical state due to the passage of an electric current. Electrolysis is also used to remove materials which interfere with determinations of other materials. This technique is especially useful in polarography.

Very little commercial production of organic materials is done by electrolysis. Two processes in use are the production of succinic from fumaric acid and the production of mannitol and sorbitol from glucose. There are, however, a large number of reactions which can be carried out by electrolysis, some of them with excellent yields.

PAUL D. GARN

Cross-references: *Electrochemistry, Dissociation, Electroplating, Chlorine.*

ELECTRONIC CONFIGURATION

The arrangement of electrons around the nucleus of an atom is called the "electronic configuration" of the atom. Since the electronic cloud is highly dynamic, the arrangement can have no static description with each electron occupying a definite physical position with respect to the nucleus and the other electrons. Instead, the situation of each electron is identified in terms of four quantum numbers which designate its quantized energy with respect to the other components of the atom. These quantum numbers may be crudely described as follows:

(1) *principal*: designated by n and of positive integral value: 1, 2, 3, 4, ... This number indicates very roughly the average position of the electron with respect to the nucleus. The smaller the principal quantum number, the closer the average distance between the electron and the nucleus, and correspondingly, the stronger the attraction tends to be. Very crudely, the electronic cloud of an atom may be imagined as composed of concentric spheres of electrons around the nucleus, increasing in energy with increasing radius as the value of n increases.

(2) *orbital*: designated by l and having some integral value 0, 1, 2, 3, ... up to $(n-1)$. Each electron is "confined" to a certain region within the principal quantum level, in the sense that at any given instant the electron is most probably somewhere within that region. This probability is represented by the square of the wave function of the electron. Such a region is called an "orbital." Orbitals may differ in their shape, which is specified by the orbital quantum number. Approximate shapes of the different kinds of orbitals possible are shown in Fig. 1. The values of l equal to 0, 1, 2, and 3 correspond to what are called s, p, d, and f orbitals, respectively.

(3) *orbital magnetic*: designated by m_l and having an integral value from 0 to $+l$. An electron moving around a nucleus is a charge in motion. Like all charges in motion, it creates an electromagnetic field capable of interacting with an external field. If there are more than one orbital of a certain kind (having the same value of l) within a given principal quantum level, these must be oriented differently. When the atom is placed in an external magnetic field, the energies of the different possible orientations are quantized, as represented by the vector of the magnetic moment of each orbital in the direction of the external field. Thus m_l can have only certain values, corresponding to the restricted number of similar type orbitals. The number of orbitals is given by $(2l + 1)$, which is

the number of m_l values possible if m_l can equal 0 up to $\pm l$.

(4) *spin magnetic:* designated by m_s, and of value either $+\frac{1}{2}$ or $-\frac{1}{2}$. Each electron, independent of its translational motion, acts as a magnet. The m_s value of $+\frac{1}{2}$ represents an electron magnet of poles opposed to those of another electron of value $-\frac{1}{2}$. The possession of magnetic properties is described as "spin." Parallel and opposed spins correspond to parallel and opposed magnets.

A given principal quantum level, or "energy shell" of an atom may then include several types of orbitals. The number of types is the same as the value of n, but commonly up to 4 only. Where $n = 1$, l can only have the value 0, corresponding to an s orbital. The number of orbitals of s type within a given principal quantum level is given by $(2l + 1)$ which can only equal 1. Thus only one s orbital can occur in any given principal quantum level.

When $n = 2$, l can be either 0 or 1. The value 0 corresponds to an s orbital. The value 1 corresponds to a p orbital. The number of similar p orbitals is given by $(2l + 1) = 3$. The second principal quantum level thus consists of four orbitals, one s and three of p type, the latter corresponding to m values of 0, 1, and -1.

When $n = 3$, l can have values of 0, 1, or 2. The values of 0 and 1 lead to one s orbital and the three p orbitals as before. When $l = 2$, the orbitals are called d orbitals. There are $(2l + 1) = 5$ of them, corresponding to m_l values of 0, 1, 2, -1, and -2. The third principal quantum level thus consists of one s, three p, and five d orbitals, for a total of nine.

When $n = 4$, l can have the values of 0, 1, 2, and 3. The first three values correspond to the s, p and d orbitals, nine altogether, as before. When $l = 3$, the orbitals are called f. Their number is $2l + 1) = 7$, corresponding to m values of 0, 1, 2, 3, -1, -2, and -3. The total for the fourth principal quantum level is thus sixteen orbitals.

In theory, higher principal quantum levels can contain additional orbitals, but in practice, stable atoms contain no more than sixteen orbitals in any principal quantum shell.

An important empirical principle of wave mechanics is that stated by Pauli and called the "exclusion principle": No two electrons of the same atom can have the identical four quantum numbers. Since each orbital is specifically designated by its values for n, l, and m_l, electrons occupying the same orbital must differ in m_s value. This value can only be $+\frac{1}{2}$ or $-\frac{1}{2}$, which means that no orbital can accommodate more than 2 electrons, and then only if their spins (magnets) are opposed. This establishes the capacity of each principal quantum level at twice the number of orbitals. The number of orbitals (through $n = 4$) is given by n^2: 1, 4, 9, 16. Therefore the shell capacity is $2n^2$: 2, 8, 18, 32.

In the building up of atoms (aufbau) by feeding electrons one by one into the region surrounding the nucleus, while simultaneously adding protons to the nucleus to balance them, each electron occupies the most stable position left available to it. When given a choice of orbitals within a given principal quantum level, the electron enters that type of available orbital that penetrates closest to the nucleus and therefore is most stable. The order of decreasing penetration and decreasing stability within a given principal quantum level is always $s>p>d>f$. When several orbitals of equal energy are available, as for example a set of p orbitals or d orbitals, the electron tends to occupy a vacant orbital if possible rather than overcome the repulsion necessary to become paired with another electron. A set of similar orbitals accommodates one electron in each, all of parallel spins, before pairing begins. The rules of separate orbital occupancy ("maximum multiplicity") and parallel spins are called "Hund's rules."

On the basis of these rules, the Pauli exclusion principle, spectroscopic data, and a knowledge of the general chemistry of the elements, all of the chemical elements have been assigned electronic configurations which appear to describe atomic structure with a minimum of ambiguity. The only significant source of controversy arises in the heavier elements, especially the transuranium elements, where the energies of $6d$ and $5f$ orbitals are so nearly alike that any specific assignment of electrons must remain somewhat uncertain.

The spectroscopic designation of electronic configuration involves giving the principal quantum number of each shell, followed by the orbital type (s, p, d, or f), with a superscript number (if greater than 1) to tell how many electrons are present in orbitals of that type. For hydrogen, $1s$ means that there is one electron in the s orbital of principal quantum level $n = 1$. For carbon, $1s^2$, $2s^2$, $2p^2$ means that the $1s$ orbital has its full complement of 2 electrons, as does also the $2s$ orbital. Two additional electrons occupy $2p$ orbitals, which are p orbitals of principal quantum level 2. The thirty electrons of zinc are distributed as follows: $1s^2$, $2s^2$, $2p^6$, $3s^2$, $3p^6$, $3d^{10}$, $4s^2$.

Within any given principal quantum level, the s orbital always fills before p orbitals are occupied, and these in turn become filled before d orbitals are occupied. The f orbitals accept electrons only after the d orbitals are full. Consequently if the total number of electrons in a principal quantum level is given, detailed spectroscopic designation is unnecessary. For example, the electronic configuration of zinc could be simplified to 2-8-18-2. One knows, for any principal level, that the first 2 electrons *must* be in an s orbital, the next 6 in p orbitals, the next 10 in d orbitals, and any beyond 18 therefore in f orbitals.

With only a few exceptions, each element has an electronic configuration exactly like that of the element one lower in atomic number, with then one additional ("differentiating") electron. The exceptions occur principally where a slight rearrangement can produce an especially symmetrical array that is energetically favorable. A set of three p orbitals, or five d orbitals, or seven f orbitals, all are spherically symmetrical with respect to the nucleus. This symmetry is lacking when the set is only partly filled with electrons, unless it happens to be exactly half-filled, with one electron in each orbital of the set. For example, an atom of chromium would normally have four electrons in its five $3d$ orbitals, and two electrons in the $4s$ orbital. However, it gains stability by transfer of one of the $4s$ electrons into the fifth $3d$ orbital, completing the half-filled set.

TABLE 1. ELECTRONIC CONFIGURATION OF THE ELEMENTS

n=	1	2	3	4	5	6	7
H	1						
He	2						
Li	2	1					
Be	2	2					
B	2	3					
C	2	4					
N	2	5					
O	2	6					
F	2	7					
Ne	2	8					
Na	2	8	1				
Mg	2	8	2				
Al	2	8	3				
Si	2	8	4				
P	2	8	5				
S	2	8	6				
Cl	2	8	7				
Ar	2	8	8				
K	2	8	8	1			
Ca	2	8	8	2			
Sc	2	8	9	2			
Ti	2	8	10	2			
V	2	8	11	2			
Cr	2	8	13	1			
Mn	2	8	13	2			
Fe	2	8	14	2			
Co	2	8	15	2			
Ni	2	8	16	2			
Cu	2	8	18	1			
Zn	2	8	18	2			
Ga	2	8	18	3			
Ge	2	8	18	4			
As	2	8	18	5			
Se	2	8	18	6			
Br	2	8	18	7			
Kr	2	8	18	8			
Rb	2	8	18	8	1		
Sr	2	8	18	8	2		
Y	2	8	18	9	2		
Zr	2	8	18	10	2		
Nb	2	8	18	11	2		
Mo	2	8	18	13	1		
Tc	2	8	18	13	2		
Ru	2	8	18	15	1		
Rh	2	8	18	16	1		
Pd	2	8	18	18			
Ag	2	8	18	18	1		
Cd	2	8	18	18	2		
In	2	8	18	18	3		
Sn	2	8	18	18	4		
Sb	2	8	18	18	5		
Te	2	8	18	18	6		
I	2	8	18	18	7		
Xe	2	8	18	18	8		
Cs	2	8	18	18	8	1	
Ba	2	8	18	18	8	2	
La	2	8	18	18	9	2	
Ce	2	8	18	19	9	2	
Pr	2	8	18	21	8	2	
Nd	2	8	18	22	8	2	
Pm	2	8	18	23	8	2	
Sm	2	8	18	24	8	2	
Eu	2	8	18	25	8	2	

TABLE 1. (Cont.)

n=	1	2	3	4	5	6	7
Gd	2	8	18	25	9	2	
Tb	2	8	18	27	8	2	
Dy	2	8	18	28	8	2	
Ho	2	8	18	29	8	2	
Er	2	8	18	30	8	2	
Tm	2	8	18	31	8	2	
Yb	2	8	18	32	8	2	
Lu	2	8	18	32	9	2	
Hf	2	8	18	32	10	2	
Ta	2	8	18	32	11	2	
W	2	8	18	32	12	2	
Re	2	8	18	32	13	2	
Os	2	8	18	32	14	2	
Ir	2	8	18	32	15	2	
Pt	2	8	18	32	17	1	
Au	2	8	18	32	18	1	
Hg	2	8	18	32	18	2	
Tl	2	8	18	32	18	3	
Pb	2	8	18	32	18	4	
Bi	2	8	18	32	18	5	
Po	2	8	18	32	18	6	
At	2	8	18	32	18	7	
Rn	2	8	18	32	18	8	
Fr	2	8	18	32	18	8	1
Ra	2	8	18	32	18	8	2
Ac	2	8	18	32	18	9	2
Th	2	8	18	32	18	10	2
Pa	2	8	18	32	20	9	2
U	2	8	18	32	21	9	2
Np	2	8	18	32	22	9	2
Pu	2	8	18	32	24	8	2
Am	2	8	18	32	25	8	2
Cm	2	8	18	32	25	9	2
Bk	2	8	18	32	26	9	2
Cf	2	8	18	32	27	9	2
Es	2	8	18	32	28	9	2
Fm	2	8	18	32	29	9	2
Md	2	8	18	32	30	9	2
No	2	8	18	32	32	8	2
Lw	2	8	18	32	32	9	2
(104)	2	8	18	32	32	10	2
(105)	2	8	18	32	32	11	2

Similarly, an atom of copper would be expected to have 9 electrons in its $3d$ orbitals and two in its $4s$ orbital. Transfer of one of the $4s$ electrons to the remaining $3d$ vacancy would provide spherical symmetry to the five $3d$ orbitals, which occurs because it also is energetically favorable.

The filling of successive principal quantum levels gives a periodicity to atomic structure as atomic number is increased. This periodicity of electronic configuration underlies the periodic law (which see). It is complicated, however, by an apparent overlapping of adjacent principal quantum levels, such that the most stable orbital of the next higher principal level seems more stable than the least stable orbitals of the level being filled. Specifically, the $4s$ orbital appears more stable than the $3d$, the $5s$ more stable than the $4d$, and the $6s$ more stable than the $5d$ or the $4f$. In other words, the elements build up in this order of orbital filling. (No such overlap is evident within any given element, where, for example, $4s$ electrons are less stably held than $3d$.) As a consequence of this overlap, no outermost shell ever contains more than 8 electrons (2 in the first shell). The ninth electron always starts a new outermost shell.

The electronic configurations of the chemical elements are tabulated in condensed form in Table 1. For the last ten elements listed, these are quite speculative owing to a lack of information. All the physical and chemical properties of all the chemical elements are the direct or indirect consequences of their atomic structure as revealed in these electronic configurations.

R. T. Sanderson

Cross-references: *Electrons, Periodic Law and Periodic Table.*

ELECTRONEGATIVITY

Electronegativity is that property of each atom that determines the direction and extent of polarity of the covalent bond that holds two atoms together. A difference in electronegativity between the two atoms causes the valence electrons to spend more than half time more closely associated with that atom which was originally of higher electronegativity and therefore attracted them more strongly. This increases the total average electron population around this atom, which therefore imparts a partial negative charge. The other atom is left with an average deficiency of electrons, corresponding to a partial positive charge since each atom was initially neutral. The magnitude of these partial charges depends on the initial electronegativity difference between the two atoms and also on the nature of other atoms in the compound.

Unfortunately, a precise, quantitative definition of absolute electronegativity is thus far lacking. However, the property certainly must originate with the nuclear charge and the negative charge on a valence electron. It may be regarded as proportional to the coulombic force between a valence electron and that fraction of the total nuclear charge which, despite the screening effect of the underlying electronic cloud, remains effective at that point. The problem of evaluating the effective nuclear charge as it would be felt by a valence electron is far too difficult for exact solution at present. However, approximate evaluation has been achieved by recognizing that although underlying electrons are quite effective in cancelling positive charge of the nucleus, outermost shell electrons are only about one-third efficient in this respect. In other words, one outermost shell electron cannot very effectively interfere with the coulombic attraction between the nucleus and another outermost shell electron. Consequently, in the building up of elements across a period of the Periodic Table, each successive proton added to the nucleus is only about one-third neutralized or cancelled by the corresponding successive electron added to the outermost shell. Thus with each unit increment in atomic number, the effective nuclear charge increases by about two-thirds of a unit protonic charge.

This progressive increase in the effective nuclear charge across a period has two highly significant results. One is a reduction of the atomic radius as

the electronic cloud is more effectively pulled together by the increased nuclear charge. The other is an increase in the electronegativity because of the combined effects of increasing the effective nuclear charge and decreasing the average distance between it and the valence electrons. This corresponds to a steady increase in electronegativity from one element to the next across the Periodic Table from left to right. Within each period, therefore, the lowest electronegativity is that of the alkali metal and the highest, that of the halogen. (The concept of electronegativity has only a very special and different application to the helium family elements that end each period.)

Relative values of electronegativity have been determined by a wide variety of methods. For example, electronegativities have been derived from ionization energies and electron affinities, from heats of reaction and bond energies, from calculations of the relative compactness of electronic clouds, from calculations of the effective nuclear charge at the surface of an atom, from work functions of metals, from force constants determined by infrared spectroscopy, and by miscellaneous other methods. When adjusted to the same arbitrary scale, that established by Pauling and ranging approximately from 1 to 4, the different methods show surprisingly good agreement, with only a few minor discrepancies that are still controversial. The general order of increasing electronegativities is Cs, Rb, K, Na, Li, Ba, Sr, Ca, Mg, Al, Be, Cd, Si and In, B and Hg, Zn, Tl, Pb, Sn, Bi, Sb, P, H, Ge and Te, C, I, As, S, Se, N, Br, Cl, O, and F. Reliable values for the transition elements are not yet available but they would probably be placed near to magnesium.

If it is assumed that the homonuclear single covalent bond energy of an element is proportional to the coulombic energy of the interaction between the valence electrons and the effective nuclear charge, then it follows that homonuclear single covalent bond energy E is quantitatively related to the electronegativity by the simple equation, $E = CrS$, where r is the nonpolar covalent radius, S the electronegativity, and C an empirical constant which can be derived from the electronic structure type. Future development of these concepts gives promise of the possibility of determining absolute electronegativity values from experimentally determined homonuclear single covalent bond energies.

The most useful practical application of electronegativities is to a quantitative evaluation of heteronuclear bond energies and heats of formation, as described under "**Bonding, Chemical.**" In such calculations, electronegatives afford the basis for apportioning the total bond energy between covalent and ionic contributions. Also extremely useful is the quantitative estimation of the relative condition of combined atoms made possible through the use of electronegativities. Much of chemistry becomes intelligible if one recognizes that the contribution made by an atom to the properties of its compound depends at least as much on the condition of that atom in the compound as it does on which element it is. The best available index of the condition of a combined atom is its partial charge, which results from the initial atomic electronegativities in the following way.

The electrons involved in a covalent bond must, in effect, be equally attracted to both nuclei. When these two atoms are initially different in electronegativity, their bonding orbitals must of necessity be different in energy. Therefore the process of forming the bond must provide some mechanism by which these energies can be equalized. Such a mechanism can be based on the fact that the electronegativity of an atom must decrease as the atom begins to acquire an electron and increase as it begins to lose an electron. An atom of fluorine has a very high electronegativity but a fluoride ion has none. An atom of calcium has relatively little attraction for electrons but a calcium ion attracts them strongly. Consequently, the energies of the bonding orbitals can be equalized through an equalization of electronegativity, which in turn can result from uneven sharing of the bonding electrons. When the bonding electrons spend more than half time more closely associated with the nucleus of the atom that initially attracted them more, they reduce its electronegativity, at the same time imparting a partial negative charge. By spending less than half time more closely associated with the initially less electronegative atom, they leave it with net partial positive charge and correspondingly a higher electronegativity. The final state is one of even attraction through uneven sharing.

Such adjustment is postulated by the *Principle of Electronegativity Equalization*, which may be stated: When two or more atoms initially different in electronegativity unite, their electronegativities become equalized in the compound. The intermediate electronegativity in the compound is taken as the geometric mean of the electronegativities of all the atoms before combination. An estimate of the relative condition of a combined atom with respect to partial charge can be based on the assumption that the charge changes linearly with electronegativity. Assignment of a particular ionicity to a particular bond (75% to Na-F) then permits calculation of the change in electronegativity per unit charge for any element of known electronegativity. The relative partial charge on any combined atom is then defined and estimated as the ratio of the electronegativity change undergone in forming the compound to the electronegativity change that would correspond to the gain or loss of one electron. The sign of the charge is negative if the electronegativity of the element has decreased on combination, and positive if it has increased.

The partial charge distribution in any compound composed of elements of known electronegativity can thus easily be estimated. Such information provides many valuable insights that contribute to a fundamental understanding of chemistry.

Theoretical chemists have recently found it advantageous to consider electronegativity as the property of a particular orbital rather than of the atom as a whole. For the purposes of practical interpretations of chemistry through knowledge of partial charge distribution and calculation of bond energy, however, the additional complication of this refinement has thus far seemed unnecessary.

R. T. SANDERSON

Cross-references: *Bonding, Chemical; Hydrogen Chemistry; Halogen Chemistry; Oxide Chemistry.*

ELECTRONS

The electron is an elementary particle with a mass of $(9.109558 \pm 0.000054) \times 10^{-28}$ gram and a negative charge of $(1.6021917 \pm 0.0000070 \times 10^{-19}$ coulomb $[(4.803251 \pm 0.000036) \times 10^{-10}$ electrostatic unit]. The name electron (from the Greek, elektron, amber) was proposed in 1891 by G. Johnstone Stoney for the unit of electricity which ruptures each chemical bond within an electrolyte solution during electrolysis. Hence one mole of electrons would be one faraday of electrical charge. However, the electron as a fundamental particle was not discovered until 1897 by J. J. Thomson in his studies of the effect of the passage of x-rays through gases. Thomson also measured the ratio of charge to mass (e/m) for the electron by observing the deflection of the path of a beam of electrons in an electric field. He found the value to be of the order of 10^{11} coulombs per kilogram; the presently accepted value of e/m for the electron is 1.7589×10^{11} coulomb/kilogram (5.273×10^{17} esu/gram). By observing the rate of fall of a cloud of negatively charged water droplets in air Thomson and Townsend obtained a preliminary value for e, the charge on the electron, of 6.5×10^{-10} esu. A better value of e was obtained by Robert A. Millikan, who in 1909 observed the rate of fall of charged oil droplets under the combined influence of gravity and an electric field. His result of 4.774×10^{-10} esu was slightly low, owing to his use of an erroneous value for the viscosity of air.

In 1921 Arthur H. Compton expressed the idea that an electron might possess an intrinsic angular momentum or spin and thus act as a magnet. In 1925 Wolfgang Pauli investigated the problem of why lines in the spectra of the alkali metals are not single, as predicted by the Bohr theory of the atom, but actually made up of two closely spaced components. He showed that the doublet in the fine structure could be explained if the electron could exist in two distinct states. G. E. Uhlenbeck and S. Goudsmit identified these states as two states of different angular momentum. They showed that the spectral multiplets could be explained by introducing a new quantum number s which could have either of two values, $+\frac{1}{2}$ or $-\frac{1}{2}$, so that the intrinsic angular momentum of the electron could be $\frac{1}{2}$ $(h/2\pi)$ or $-\frac{1}{2}$ $(h/2\pi)$. The spinning electron thus behaves as a tiny magnet with a magnetic moment of $eh/4\pi m_e c$, where h is Planck's constant, m_e is the mass of the electron, and c is the velocity of light. Electronic spin was confirmed experimentally by the famous Stern-Gerlach experiment. Later, when P. Dirac worked out a relativistic form of wave mechanics for the electron it was found that the property of spin fell out naturally from the theory.

The allowed stationary states of the electrons in even the most complex atoms can be referred to the same four quantum numbers n, l, m, and s, i.e., the principal (or shell), the azimuthal (or angular momentum), the magnetic and the spin quantum numbers, respectively. However, an important rule governs the quantum numbers allowed for an electron in an atom. This rule is known as the Pauli Exclusion Principle, which states that no two electrons in an atom can exist in the same quantum state, i.e., no two electrons in a given atom can have all four quantum numbers the same. A general statement of the Pauli Exclusion Principle is: a wave function for a system of electrons must be *antisymmetric* for the exchange of the spatial and the spin coordinates of any pair of electrons.

The Pauli Exclusion Principle provides the key to the explanation of the structure of the Periodic Table of the Elements. The most stable state, or *ground state,* of an atom is that in which all the electrons are in the lowest possible energy levels that are consistent with the exclusion principle. To explain the structure of the table arrange the different orbitals, each specified by its unique set of four quantum numbers, in the order of decreasing energy and then place the electrons, one by one, into the lowest open orbitals until all the electrons, equal in number to the nuclear charge Z of the atom, are safely accommodated. This process was called by Pauli the Aufbauprinzip. When this building process is carried out it is found that the number of electrons that can be placed in successive shells is *two* in the first, *eight* in the second, *eighteen* in the third, and *thirty-two* in the fourth. In general the number of electrons that can be placed in the *n*th shell is $2n^2$.

The tendency for atoms to form complete electron shells accounts for their valence states. For example, sodium, which has atomic number eleven, in the ground state has two electrons in the first shell, eight in the second, and one in the third. When electrons are removed from an atom, the atom is said to be ionized. Removal of the electron from the outermost shell of sodium, so that it assumes a configuration with the other two shells complete, accounts for its observed valence of $+1$. Similarly, fluorine has its nine electrons arranged with two in the first shell and seven in the second. The tendency is for the fluorine atom to accept one more electron to complete the second shell, forming a fluoride *ion* with a charge of -1.

Although the Bohr theory of the atom, which treated electrons as particles, had many successes, it also had troublesome failures. For example, the theory could not explain the spectra of helium and other more complex atoms. In 1923 it was suggested by Louis de Broglie that electrons, and in fact all material particles, must possess wavelike properties. For example, an electron that has been accelerated through a potential of ten kilovolts, according to de Broglie, should have a wavelength of 0.12 Å or about the same as that for a rather hard x-ray. The experimental demonstration that a beam of electrons may be diffracted and hence that electrons do indeed have wave properties was accomplished by C. Davisson and L. H. Germer in the U.S. and by G. P. Thomson and A. Reid in Scotland. The fact that electron beams may be diffracted is the basis for the electron microscope.

Free electrons for study can be produced from metal or other surfaces in various ways: heating the surface (thermionic emission), allowing light of the appropriate wavelength to strike the surface (photoelectric effect), placing the surface in a strong electric field (field emission) or bombarding the surface with charged particles (secondary electron emission).

                                M. H. LIETZKE AND R. W. STOUGHTON

References

Moore, W. J., "Physical Chemistry," 3rd Ed., Prentice-Hall, Inc., Englewood Cliffs, N. J., 1962.
Glasstone, Samuel, "Source Book on Atomic Energy," D. Van Nostrand Co., New York, 1950.
Stranathan, J. D., "The 'Particles' of Modern Physics," Blakiston, Philadelphia, 1946.
Taylor, B. N., *et al.,* "The Fundamental Physical Constants," *Sci. Am.,* **223,** 62 (1970).

ELECTROORGANIC CHEMISTRY

Electroorganic chemistry is the borderline field between electrochemistry and organic chemistry in which reactions of organic compounds are carried out by means of electrolysis.

Organic compounds themselves may be electrolyzed to form coupled products at the anode as in the case of some salts of aliphatic acids, $2RCOOM + 2F \rightarrow R{-}R + 2CO_2$ (Kolbe synthesis) or to form metal salts if the anode is soluble. Similarly, certain compounds may undergo electrolysis to give coupled products at the cathode, methylmercuric acetate, $2CH_3COOHgCH_3 + 2F$ to form dimethylmercury with free mercury, $2CH_3Hg \rightarrow (CH_3)_2Hg + Hg$, for example. Then, an organic compound may not carry current itself, but may react at either electrode with a substance resulting from the electrolysis of another conductor, at the anode, oxygen, halogens, thiocyanogen, etc., and at the cathode, hydrogen.

Still another type is the reaction of an organic compound with an inorganic compound as an intermediate. Thus a hypobromite from the electrolysis of a bromide may react with acetone to give bromoform, or may oxidize the aldehyde group of a sugar to form an aldonic acid; or a reversible oxidizing or reducing agent formed at the anode or cathode may react with the organic compound, the reduction of nitrobenzene to aniline by stannous chloride, for example, the resulting higher valent or lower valent salt returning to the anode or cathode to complete the cycle. Reactions may also take place through the concentration of substances at either electrode. Thus, the diazotization of amines may be carried out at the anode by nitrous acid resulting from the electrolysis of a nitrite. If a coupling agent is present, an azo dye may be prepared. An example of such a reaction at the cathode is the saponification of an ester by the alkali which is formed by the loss of hydrogen from the aqueous electrolyte containing sodium chloride.

Electrolytic reactions take place through several mechanisms. The compound may give up an electron to the anode or receive an electron from the cathode, or it may react with substancs formed at either electrode in atomic or molecular form. Thus, at the anode, oxidations may take place through OH radical, atomic oxygen or molecular oxygen formed by the discharge of OH^-. Substitutions may take place through the formation of an atomic or molecular species after the discharge of an ion, thus: $Cl^- \rightarrow Cl + Cl_2$. It will be noted that while oxidations and substitutions may be selective, they are not necessarily so. For example, organic compounds may be oxidized quantitatively to carbon dioxide, water and residual material. At the cathode, reducible groups may accept electrons, after which the resulting intermediate may react with H+, or they may react wih atomic hydrogen or molecular hydrogen.

In electrolytic reactions polarization or the back electromotive force developed when voltage is applied to the electrodes in solution plays an important part. To carry out an electrolysis it is necessary to apply an electromotive force exceeding the back electromotive force in order to cause current to flow through the solution. Theoretically, the minimum back electromotive force is developed by a cell with electrodes of platinum black. Other systems will develop higher voltages. The difference between the value of the back electromotive force of these systems and the theoretical value is called the polarization voltage. The polarization voltage may be subdivided into the overvoltages at the electrodes. Thus it may be seen that gases will be more easily evolved at certain electrodes than at others. In addition to relationship with the electrode material, the overvoltage increases with the current density and decreases with increasing temperature.

Reactions at the electrodes may take place considerably below the point of gaseous evolution or near it. In the first case the compound entering the reaction prevents the polarization voltage from being reached and is, therefore, said to behave as a depolarizer. As an illustration, a compound at a cathode may accept electrons at a potential lower than that for their acceptance by H+. As long as sufficient depolarizer is present the potential will remain below that for hydrogen evolution. Similarly, compounds may lose electrons to the anode considerably below the potential for oxygen evolution. When reactions take place near the potential of the discharge of the electrolytically formed reactant, the mechanism may or may not be electron transfer.

The potential is also important in controlling reactions which take place in steps. Thus by maintaining a low potential at the cathode it is possible to obtain phenyl hydroxylamine, C_6H_5NHOH, in the sequence $C_6H_5NO_2 \rightarrow C_6H_5NO \rightarrow C_6H_5NHOH \rightarrow C_6H_5NH_2$.

Control of the potential is particularly desirable when a slight variation will affect the type of product, e.g., certain reactions at the anode.

Apparatus for carrying out electroorganic reactions is very simple, a cell containing two electrodes immersed in the electrolyte, usually with means of agitation and of preventing contact of the products formed at one electrode with the opposite electrode. The electrode at which the reaction is to take place should either be surrounded by the other electrode and equidistant from it at all points, as a cylindrical cathode around a rod alone, or be placed between two electrodes of opposite sign, as a working electrode in the form of a sheet or plate between two nonworking plates. Agitation is most simply effected by a stirrer within the cell; the electrode itself may be used as stirrer. Such an arrangement is particularly advantageous when stirring at high speeds is desired. In most reductions it is necessary to prevent the products from making contact with the anode at which they may be reoxidized. Commonly this is accomplished by separating the anode and cathode with a porous diaphragm. On the other hand, the reduction of an oxidation product

may be prevented simply by using an inactive cathode, or by setting the current density at a level such that the compound is not reduced. In reactions involving the use of inorganic intermediates, such as hypohalites, the reduction may be prevented by the use of calcium or chromium salts which form hydroxides at the cathode. These precipitates behave as diaphragms.

Asbestos, "Alundum" and other ceramic materials commonly have been used as diaphragms. Now available with a controlled range of porosity are paper-resin compositions which can be formed to desired shape and then thermoset, also flexible all-resin sheet materials.

The use of a stationary porous electrode through which fluids can be passed has practical advantages for introducing depolarizers or removing products from a cell. Operation of such a device may be equivalent to stirring and sometimes is more effective. It also may make unnecessary the use of a diaphragm by removing products before they can diffuse back to the opposite electrode. Carbon and graphite long have been available with varying degrees of porosity. Numerous sintered metals now can be obtained.

Largely from the exploration of fuel cell systems during the last few years have come porous electrodes with bimodal structures. In these, part of the structure is relatively coarse and provides for easy passage of fluids. The rest of the pore volume conststs of relatively fine pores which provide extensive area, available through the coarse voids, on which electrode reaction can take place. Nickel and carbon have been made with bimodal structures which may prove advantageous in synthetic electrochemistry.

Recently there has been interest in the use of fluidized beds as cathodes. Their application seems to be limited to easily reducible substances such as nitro compounds.

The cost of current for an electrolytic operation will be at a minimum when the electrolyte is an excellent conductor, if a diaphragm is unnecessary and if the compound entering the reaction is a depolarizer. To obtain the maximum yield in an electrolytic reaction certain conditions must be obeyed. In the first place the electrode material must be chosen carefully. In potential-controlled stepwise reactions, the low potentials necessary for obtaining intermediates are obtained more easily at cathodes of low minimum overvoltage. In general at cathodes of high overvoltage the reaction will be more rapid and complete. There are, however, notable exceptions, making necessary a study of the relationship between the structure of the compound entering the reaction and the electrode material. Furthermore, the type of reaction will determine the nature of the latter. Thus, for example, cathodes of low hydrogen overvoltage apparently behave as catalysts for the hydrogenation of olefins, which are not reducible at cathodes of high hydrogen overvoltage at which the mechanism of reduction is by electron transfer or atomic hydrogen.

In the case of reductions which take place only at cathodes of high hydrogen overvoltage, the character of the cathode material may be of more importance than the overvoltage. Thus a typical aliphatic amide has been shown to be reducible only at a cathode of lead and not at other cathodes of high hydrogen overvoltage. Not only is the electrode material itself of importance, but also its physical structure. Thus, benzoic acid C_6H_5COOH in aqueous sulfuric acid is not reduced at a smooth cadmium surface, but is reduced to benzyl alcohol $C_6H_5CH_2OH$ at a surface which has been macroetched. It is likely that the crystal orientation plays an important part. Cathodic reactions are much more easily controlled than anodic reactions, particularly oxidations on account of the wide choice of electrode material. There are too few anodes which are not corroded and some of these are expensive, i.e. the platinum metals. Furthermore, it is sometimes difficult to prevent the combustion of an organic compound.

Until recently the choice of anodes in acid solution has been restricted to the platinum metals, lead peroxide and carbon, but it is possible that some of the materials coming from research in electroanalytical chemistry, such as boron carbide, may be suitable for certain reactions.

A few others, such as nickel and iron, may be used in alkaline solution. On the other hand, many metals may be used as cathodes, also carbon and possibly certain other nonmetals.

Another deciding element is the electrolyte, not only its pH but its kind. The course of a reaction is often decided by the pH of the electrolyte. Thus, aromatic nitro compounds are usually reduced to amines in acid solution and to azoxy, azo and hydrazo compounds in alkaline solution. Sulfuric acid is preferable to hydrochloric acid in many reductions, because it is less corrosive with certain cathodes. The concentration of conductor in the electrolyte is also important. Higher yields will often be obtained at higher concentrations. In aqueous solution the usual conductors are the commoner acids, bases and salts. The field is not limited to aqueous electrolytes, however. Reactions have been carried out in the lower alcohols, ethylene glycol, glacial acetic acid, acetonitrile and dimethylformamide for example. Nonaqueous media may be advantageous in solving the problem of the insolubility of many organic compounds.

The course of a reaction may also be affected by the particular medium used. When no intermediate reagent, such as a hypohalite, enters the reaction, an organic compound must make contact with an electrode. In an aqueous electrolyte this may be accomplished by adding a blending agent such as ethyl alcohol to increase the solubility of the organic compound or by emulsification either by stirring alone or in the presence of an emulsifier.

Another very important factor is the current density, except in reactions which are not potential controlled. The optimum current density will depend on the rate of reaction at the electrode and on the rate of diffusion to the electrode of the compound undergoing reaction. It is obvious, then, that normal stirring will often increase the optimum current density.

The temperature will also influence the current density by its effect on the diffusion. The higher the temperature the greater is the diffusion. As in chemical reactions in general, the rate may also increase with temperature. In certain reactions an increase in the temperature may have no influence

and in others even have an adverse effect, probably by disturbing the layer immediately adjacent to the electrode, as in the Kolbe synthesis mentioned previously.

For high yields it is often necessary to allow more than the theoretical amount of current to pass through the solution on account of the low efficiencies of certain reactions.

Finally, the current concentration or number of amperes per volume of solution may be of importance. Too high a current concentration will cause an increase in the temperature.

Electrolytic processes in organic synthesis are of decided advantage for reactions difficult to carry out by catalytic or ordinary chemical methods, and for obtaining precise control of reaction conditions or high purity of product, especially freedom from metallic contamination.

It should be pointed out, however, that there is still too little known about the field, particularly from an industrial point of view. In the laboratory many reactions may be carried out easily, but careful studies of optimum conditions are necessary for the development of an industrial process. An example is the reduction of acrylonitrile to adiponitrile.*

SHERLOCK SWANN, JR., AND N. M. WINSLOW

Cross-references: *Electrochemistry, Electrolysis, Organic Chemistry.*

* Baizer, M. M., *J. Electrochem. Soc.,* **111,** 215 (1964).

ELECTROPHORESIS

Electrophoresis involves the migration of charged particles, either colloidally dispersed substances or ions, through conducting liquid solutions, under the influence of an applied electric field.

Prior to about 1950, two methods were generally employed for studying the electrophoretic behavior of charged particles in a liquid. In the microscopic method, the migration of individual particles is observed in a solution contained in a glass tube placed horizontally on the stage of a microscope. The method is suitable for the study of relatively large particles such as bacteria, blood cells or droplets of oil. Its usefulness was extended somewhat by the finding that various finely divided inert materials, such as tiny spheres of glass, quartz or plastic can, in some instances, be so completely covered with adsorbed protein that they act as if they were large protein particles and respond to an electrical field in terms of the charge on the protein. The method is mainly of historical interest.

In the moving boundary technique of electrophoresis, the movement of a mass of particles is measured, thus obviating the necessity of observing individual particles. The displacement of the particles in an electric field is recorded photographically as the movement of a boundary between a solution of a colloidal electrolyte, such as a protein, and the buffer against which it was dialyzed. The material to be studied is poured into the bottom of a U-tube and on top of it, in each arm of the U-tube, a buffer solution is carefully layered in order to produce sharp boundaries between the two solutions. Electrodes inserted in the top of each arm of the tube are attached to a DC electric source. If the material under study is a protein bearing an excess of negative charges on its molecular surface, the boundary will move toward the positive electrode. Since the net electric charge on the protein molecule varies with the acidity of the buffer solution, the charge on the molecule, and hence its velocity, may be varied by changing the acidity of the buffer. As the hydrogen-ion concentration is increased or the pH lowered, the velocity of the protein is reduced until a point on the pH scale is reached at which it fails to move (isoelectric point or pI). If the pH of the buffer is further reduced, the protein will acquire a net positive charge and will move toward the negative electrode. The migration velocity of any particular migrant is, of course, directly proportional to the applied voltage gradient. Other important factors which may affect the observed velocity include the molecular shape and structure of the specific substance under study as well as the concentration of the buffer solution or, more specifically, its ionic strength, the temperature and electroosmosis. Electroosmosis refers to the constant flow of liquid relative to a stationary charged surface, for example, a strip of filter paper, under the influence of an applied electric field. The net movement in the case of aqueous solutions is generally toward the negatively-charged electrode. Owing to its experimental complexity and cost, this technique has been largely replaced as an analytical tool by the simpler method of electrophoresis described below.

The third technique for carrying out electrophoretic separations has been named variously as ionography, zone electrophoresis, electrochromatography, etc. Although the rootlets of the technique are discernible in the publications of Lodge dating back to 1886, the modern era began in 1950, when several papers on electromigration in paper-stabilized electrolytes were published from a number of countries. From that time on, the number of reports on various applications, modifications and limitations of the technique has grown phenomenally, and this procedure has become one of the important tools in biochemical research and in routine clinical chemical analyses. It lends itself equally well to the study of the electromigration of ionic substances of low molecular weight, such as amino acids, peptides, nucleotides and inorganic ions, and of colloidal materials such as proteins and lipoproteins. Under favorable conditions, substances are not only separated totally from mixtures but may be recovered almost completely. This aspect of the procedure is particularly valuable in work with radioactive materials. Rigid restrictions on the temperature, current and composition of the solutions, in order to minimize convectional disturbances, are largely removed when electrophoresis is carried out in a solution stabilized with a relatively inert material, originally paper but later on principally cellulose acetate, starch gel, polyacrylamide gel or agar, which permit sharper separations. The high resolving power of gel media is to a large extent a consequence of molecular sieving acting as an additional separative factor.

As an example, blood serum can be separated into about 25 components in polyacrylamide gel but only into five components on filter paper or by moving-boundary electrophoresis. Only minute amounts of material are required, the equipment is relatively simple and inexpensive, and the method can be utilized over a wide range of temperature. Some of the many uses to which electrophoresis in stabilized media has been applied are identification of the individual components of a mixture, establishing homogeneity, concentration and purification. In conjunction with other microanalytical techniques such as immunology and polarography, it may often be the means by which identification is ultimately established and quantitative assessment made. Adaptations of the technique, e.g., curtain and planar electrophoresis, provide for continuous operation thus permitting the separation of relatively large quantities of substances.

Normally, a micro amount of the mixture to be separated is streaked across the midpoint of a horizontal column or strip of stabilizing agent, e.g., a narrow, paper-thin cellulose acetate strip saturated with a buffer solution of known pH and ionic strength; a controlled DC source of electric potential is then applied to the ends of the column or strip. The substances under study, the migrants, begin to move and each rapidly reaches a constant velocity of electromigration through the stabilizing structure. The velocity depends, among other factors, upon the potential gradient, the charge on the substance, the ionic strength of the solution, the temperature, the barrier effect interposed by the stabilizing agent, the wetness, electroosmosis and hydrodynamic movement of the solvent through the stabilizing structure. In general, as the electromigration proceeds, the original zone separates into several discrete zones having different specific electromigration velocities. The distance of each separated spot, zone or band on the ionogram from the point of application provides a measure for determining the mobility of the particles making up the zone and the density and area of the spot provide an index to the quantity of material in the mixture. If the substances separated are colorless, the bands may be developed by the use of suitable dyeing reagents, e.g., bromophenol blue for blood serum proteins.

Several methods have been utilized for the quantitative determination of the dyed zones, e.g., protein fractions, on the ionogram. The strip may be cut into sections, the colored material eluted with suitable solvents and its concentration determined in a spectrophotometer fitted with small cuvettes. The most common method in use today involves direct determination by a transmission densitometer.

The charge on a particle may arise from charged atoms or groups of atoms that are part of its structure, from ions which are adsorbed from the liquid medium, and from other causes. It is evident that the behavior of colloidal particles, as compared to that of ions, in an electric field differs only in degree rather than in kind. Although a colloidal particle is much larger than an ion, it may also bear a much greater electrical charge, with the result that the velocity in an electric field may be about the same, varying roughly from $0-20 \times 10^{-4}$ cm/sec per second in a potential gradient of 1 volt per cm.

To understand the phenomenon of electrophoresis, let us suppose, for simplicity, that a nonconducting particle, spherical in shape, of radius r, and bearing a net charge of Q coulombs is immersed in

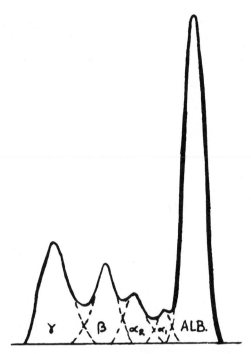

FIG. 1. A densitometer reading of an ionogram (optical density versus distance of migration along the ionogram) for normal human plasma. The major peak represents the albumin fraction; the lesser peaks represent the α_1, α_2, β, and γ globulin fractions.

a conducting fluid of dielectric constant D and a viscosity of η poises. Suppose, further, that the particle moves with a velocity of v cm/sec under the influence of an electric field having a potential gradient of x volts per cm. The force causing the particle to move, namely, $Qx \times 10^7$ dynes, is opposed by the frictional resistance offered to its movement by the liquid medium. From Stokes' law, the latter is given by $6\pi\eta rv$. Under steady-state conditions, and introducing the electrophoretic mobility, $u = v/x$; rearrangement yields the expression $u = Q \times 10^7/6\pi\eta r$. It is evident that if the electrophoretic mobility of a particle can be computed, it should be possible to determine Q, the net charge on the particle.

Electrophoresis in stabilized media is used mainly as an analytical technique and, to a lesser extent, for small-scale preparative separations. The most important applications are to the analysis of naturally occurring mixtures of colloids, often of plant or animal origin, such as various proteins, lipoproteins, polysaccharides, nucleic acids, carbohydrates, enzymes, hormones and vitamins. Electrophoresis often offers the only available method for the quantitative analysis and recovery of physiologically active substances in a relatively pure state. It provides a most convenient and dependable means of analyzing the protein content of body fluids and tissues and provides an important tool in most hospital laboratories. Like chromatography, electrophoresis in stabilized media is mainly a practical technique

and the most important advances have involved improvements in experimental procedures and the introduction and development of a wide range of suitable stabilizing media. The marked differences between normal and pathological serum samples are useful in the diagnosis and understanding of disease. Such changes in the electrophoretic pattern of blood serum are evident in diseases characterized by marked protein abnormalities such as multiple myeloma, nephrosis, liver cirrhosis and various parasitic disorders. Because of the small amount of fluid required, the method is applicable to the study of spinal fluids.

The electrophoretic pattern is not to be interpreted as specific for a given disease but rather as an index to the physiological and nutritional condition of the patient, to be combined with other information for a more complex diagnosis. If the electrophoretic pattern of a patient's blood plasma shows an excess of gamma globulin, the inference is that the body may be suffering from an infection, since most of the antibodies evoked by the presence of infectious microbes are gamma-globulin-like proteins. An increase in the alpha-globulin, a result of the breakdown of tissue proteins, may herald a fever-producing disease, such as pneumonia or tuberculosis. When the blood shows a decrease in albumin, the clinician looks to the liver as a possible seat of the disease because it is the main site of albumin production. When the liver fails, other tissues will often react to the lower albumin level by producing an excess of globulins.

Because fractionation of mixtures by ionography and quantitative estimation of the components has been of such great practical value, the potentiality of the technique for theoretical electrochemistry and colloid chemistry is often overlooked. Since substances can be characterized by their electrophoretic mobilities, phenomena such as the binding of various ions to proteins, which produce changes in the net charge of a molecule, can, in principle, be evaluated from the changes in the mobility. It has been possible to arrive at some conclusions as to relative binding strength and the charge on the chemical site involved. The isoelectric points of ampholytes, such as proteins and lipoproteins, can be determined and progress has been made toward determining such factors as ionization constants and associated thermodynamic quantities. Since these determinations can be made with less than a milligram of substance, the technique is often of considerable value for establishing the nature of functional groups.

HUGH J. McDONALD

References

1. Abramson, H. A., Moyer, L. S., and Gorin, M. H., "Electrophoresis of Proteins and the Chemistry of Cell Surfaces," Reinhold Pub. Corp., New York, 1942; reprinted by Hafner Pub. Co., New York, 1964.
2. McDonald, H. J., "Ionography: Electrophoresis in Stabilized Media," Year Book Medical Publishers, Chicago, 1955.
3. Shaw, D. J., "Electrophoresis," Academic Press, New York, 1969.

ELECTROPLATING

Electroplating is the deposition of a layer, usually thin, of some material on an object by passing an electric current through a solution containing the material in which the object has been immersed, the object being one of the electrodes. The purpose of electroplating is usually to protect and/or to decorate. Also by suitable procedures, shapes such as that of printer's type are reproduced.

Usually the object being plated is a metal; the thin coating deposited is a different metal, and the solution is an aqueous solution of a salt containing the element being deposited. The object being plated is the cathode, that is, it is the electrode to which an outside electric source delivers electrons. The anode often is composed of the metal being deposited, and is the electrode from which the outside electrical source accepts electrons. Ideally it dissolves as the process proceeds.

The current supplied is almost always steady D.C. but sometimes is made a pulsating one by superimposing an A.C. upon the D.C. The A.C. may be of the sine wave type, square wave type, etc.

The thin layer being deposited sometimes is composed of two or more metallic elements, in which case it is an alloy. The solution, or plating bath as it is called, contains dissolved salts of all the metals which are being deposited. It often also contains an appropriate acid, base and/or salt added for the purpose of holding the pH at a desired level.

Other substances, called addition agents, are often added to the plating bath for the purpose of giving the plate a desired texture, such as one which is strong, adherent and mirror smooth rather than rough, granular, loose and mechanically weak. Boric acid, glue, gelatin, urea, glycine are examples of such substances. Often, if not always, these are substances which when added to the solvent of the bath alone will form a solution having a higher dielectric constant than that of the pure solvent. Small amounts of these addition agents are likely to find their way into the plate itself which may or may not be desirable.

While the object undergoing plating usually is a good electrical conductor, it may be a nonconductor which has previously received a thin coating of some conductor. Graphite is often used for this purpose, but other materials, such as the gold paint normally used on chinaware will serve, or a thin silver film produced in a manner similar to that on the back of an ordinary mirror.

The anode must be an electrical conductor but may or may not be of the same chemical composition as the plate being deposited and may or may not dissolve during the electroplating process.

In addition to the plating of metals, colloids are sometimes electroplated. A positively charged colloid in a plating bath will plate or tend to plate on the cathode and a negatively charged colloid on an anode. Rubber is electroplated from latex by this means. This process is called *electrodeposition;* it permits rubber coating of complicated shapes and thin films such as surgeon's gloves.

There is nearly always a temperature range within which it is desirable to hold the plating bath and a range of current density within which it is desirable to hold the current. By current density is meant the current per unit area of the object being plated. This may be expressed in amperes per square centimeter, per square decimeter, per square foot, etc.

Ideally to plate monovalent ions such as Ag^+ from the plating bath would require one electron per ion, divalent ions such as Cu^{++} two electrons per ion, etc. Thus ideally one faraday of electricity would deposit one equivalent weight of any metal, i.e., 107.880 g of silver, 63.54/2 = 31.77 g of divalent copper, etc. Actually this ideal is almost never realized. The term "current efficiency" (of an electroplating process) is defined as the mass of a metal actually deposited divided by the mass of the same metal which would ideally have been deposited by the passage of the same quantity of electricity.

Objects to be plated are often irregular in shape and thus have recessed and remote areas. With some plating baths such areas may receive very little deposition while with other baths considerable deposition. The term "throwing power" is used to describe the ability of a plating bath to reach these remote or recessed parts of the object; the thicker they are plated, compared with the more exposed parts, the better the throwing power of the particular plating bath.

The voltage required for electrodeposition under ideal conditions would be given by the following equation:

$$E = E^0 + \frac{RT}{VF} \ln A$$

where E is the required voltage relative to the solution as measured by a hydrogen electrode in a unit molal activity hydrogen ion solution. E^0 is the electrolytic potential, in volts, of the metal being plated when immersed in a solution containing its ions at unit molal activity (approximately unit molal concentration) see **Electrode Potentials**; $R =$ 8.31 joules per degree mole; T is the Kelvin temperature; $F = 96,500$ joules per gram-equivalent; V is the valence of the ions which are depositing out; $\ln A$ is the natural logarithm of the activity of these ions (approximately the natural logarithm of their molality).

In actual practice the concentration and therefore the activity of the ions soon after the electroplating process starts is different in the solution just next to the cathode, called the cathode "film," than in the main body of the bath. The foregoing equation must in practice be modified to read:

$$E = E^0 + \frac{RT}{VF} \ln A - P,$$

where A is the molal activity in the cathode film of the ions being electrodeposited and P is the extra potential required to keep the plating going. A and P depend on temperature, current, density, concentration, valence, pH, and ion mobility.

Some of the metallic ions that are commonly electroplated and their electrolytic potential, $E^°$, are as follows:

Ion	E^0	Ion	E^0
Zn^{+2}	-0.762	Cu^{+2}	$+0.345$
Cr^{+3}	-0.71	Cu^{+1}	$+0.522$
Cd^{+2}	-0.402	Ag^{+1}	$+0.800$
Ni^{+2}	-0.250	Au^{+3}	$+1.42$
Sn^{+2}	-0.136	Au^{+1}	$+1.68$
$(H^{+1}$	$0.000)$		

As will be seen, in order to plate out the metals above hydrogen from an aqueous solution, the concentration of the hydrogen in the cathode film must be low. Hence a basic solution such as provided by a cyanide bath may be resorted to. A cyanide bath may also be used to give a smooth adherent plate.

To plate out an alloy, that is, to codeposit at least two kinds of metals, the plating bath must be so contrived that the electrodeposition potentials of the two metals in the cathode film are equal or nearly so.

Four typical electroplating baths are as follows:

Copper: CuCN 26 g/l, NaCN 35 g/l, Na_2CO_3 30 g/l $KNaC_4H_6O_6 \cdot 4H_2O$ 45 g/l, NaOH to give pH of 12.6

Copper: $CuSO_4 \cdot 5H_2O$ 188 g/l, H_2SO_4 74 g/l

Tin: Tin (as tin fluoborate concentrate) 60 g/l, free fluoboric acid 100 g/l, free boric acid 15 g/l.

Zinc: $Zn(CN)_2$ 60 g/l, NaCN 23 g/l, NaOH 53 g/l

The bright, hard, ornamental chrome so popular on automobiles and household and office equipage is produced by electroplating. In this electroplating process the chromium is not present in the bath as a positive metal ion but rather as part of the anion of chromic acid, H_2CrO_4. The object being plated is made the cathode. Usually it has already been electroplated first with copper and then with nickel. The pleasing to the eye, long lasting surface is the result of electroplating a coating of chromium only 0.00001 to 0.00005 inch thick on the nickel.

The plating bath is an aqueous solution of chromic trioxide, CrO_3, and sulfuric acid, H_2SO_4, with a ratio of approximately 100 to 1 by weight. The sulfuric acid acts only as a catalyst. The total concentrations may vary widely in different baths. Fluosilicate catalysts are also sometimes added to the bath.

As an example a chromium electroplating bath might contain 300 grams of CrO_3 and 3.0 grams of H_2SO_4 per liter, operate at 120°F with a current density of 2.6 amperes per square inch requiring a potential of 10 volts or more, with a current efficiency of around 20 per cent. It would have a nondissolving anode, lead, which would also serve to oxidize trivalent chromium, produced at the cathode, back to chromic acid.

Thicker coatings of chromium are electroplated onto wearing surfaces such as cutting tools and the walls of cylinders of internal combustion engines, with or without a previous plating with copper and nickel.

PENROSE S. ALBRIGHT

References

Raub, E., and Müller, K., "Fundamentals of Metal Deposition," Elsevier Publishing Company, New York, 1967.

Mohler, J. B., "Electroplating and Related Processes," Chemical Publishing Company Inc., New York, 1969.

Hampel, Clifford A., Editor, "Encyclopedia of Electrochemistry," Van Nostrand Reinhold Co., New York, 1964.

Arrnet, Rex Conde, "The Modern Electroplating Laboratory Manual," Robert Draper LTD, Teddington, 1965.

"Metal Finishing Guidebook," published annually by Metals and Plastic Publications, Inc. Westwood, New Jersey.

ELEMENTS

The chemical elements are the simplest entities into which all other substances and mixtures may be *decomposed* by chemical means. The recognition of the elements as the fundamental chemical constituents of all other substances came in the late eighteenth century through the work of Lavoisier, among others. In ancient and medieval times the elements were considered to be earth, air, fire, and water, the four simple substances of which all material bodies were supposed to be compounded.

Many attempts were made during the nineteenth century at a unified classification of the elements. These attempts culminated in 1869, when Dmitri Mendeleev enunciated the periodic law by which "the elements arranged according to the magnitude of atomic weights show a periodic change of properties." In the early twentieth century the x-ray spectrum work of H. G. J. Moseley removed certain anomalies in Mendeleev's table of classification and led to the true ordering of the elements in terms of their atomic numbers. The atomic number of an element is the ordinal number which is equal to the number of protons in the nucleus of the atom of the element; hence it is equal to the number of orbital electrons.

All atoms of a given element have the same number of protons in the nucleus, but the number of neutrons may vary. Atoms of an element containing different numbers of neutrons in the nucleus are called *isotopes* of that element.

There are ninety elements found in nature ($_1H$ through $_{92}U$, except for $_{43}Tc$ and $_{61}Pm$), although the shorter-lived elements $_{84}Po$ through $_{89}Ac$ and $_{91}Pa$ would not exist today if they were not continually being formed through radioactive decay from their long-lived parents: $_{90}^{232}Th$, $_{92}^{235}U$ and $_{92}^{238}U$. Isotopes of elements $_{93}Np$ through 105 have been made artificially by neutron or charged particle nuclear reactions.

The first twenty or so elements have approximately equal numbers of protons and neutrons in the nucleus. From that point on the ratio of neutrons to protons gradually increases so that isotopes of element 100 (fermium) have approximately 150 neutrons in the nucleus. All elements above bismuth (atomic number 83) are unstable with respect to radioactive decay.

Of the naturally occurring elements about 30 are

found chemically free on the earth; that is, not in chemical combination with any other element. These elements are obviously not very active chemically and include, for example, nitrogen, gold, platinum, copper, and the inert gases. However, oxygen, a very active element, occurs uncombined in great quantities in the atmosphere.

Chemically free elements in the vapor and liquid phases may exist as discrete atoms, as in the case of the noble gases He, Ne, Ar, Kr, Xe, and Rn; or as diatomic molecules, as in the case of hydrogen (H_2), oxygen (O_2), nitrogen (N_2), and chlorine (Cl_2). The molecules of phosphorus in the vapor phase contain four atoms (P_4) at fairly low temperatures and two atoms (P_2) at higher temperatures.

In the solid state there are a few elements whose crystals are composed of identical molecules in which the atoms are bound more strongly to each other than to those in adjacent aggregates, for example, sulfur (S_8) and iodine (I_2). Carbon has four valence, i.e., outer shell, electrons and each atom can form covalent bonds with adjacent carbon atoms. The diamond, one of the crystalline forms of carbon, may be regarded as one giant molecule. Solid metals are crystalline but contain no discrete groups of atoms.

Some elements exist in two or more forms or modifications, a phenomenon called polymorphism or allotropy. Allotropy may be due to the formation of different kinds of molecules, as in the case of oxygen: O_2 (ordinary oxygen) and O_3 (ozone); or to the formation of two or more crystalline forms, as in the case of carbon (diamond and graphite) or sulfur (rhombic and monoclinic).

The eight most abundant elements in the earth's crust are oxygen (46.6%), silicon (27.72%), aluminum (8.13%), iron (5%), calcium (3.63%), sodium (2.83%), potassium (2.59%), and magnesium (2.09%). Certain generalizations can be made concerning the abundance of the elements and the relative abundance of isotopes of an element: (1) elements of even atomic number are more abundant and have more isotopes than those with odd atomic numbers; (2) isotopes of an element which have an even number of neutrons are more abundant than those with an odd number of neutrons; (3) there are only four nuclides known to be stable which have an odd number of neutrons and an odd number of protons: ^2_1H, ^6_3Li, $^{10}_5B$ and $^{14}_7N$.

On the basis of their properties and structure the elements may be classified into three general groups:

(1) *The noble gases.* These elements have their outer valence shells completely filled and are for the most part chemically inert. A few compounds, such as XeF_6, have been characterized.

(2) *The metals.* The metals are the most numerous of the elements. In the free state they are lustrous and good conductors of electricity. Most of them are malleable, ductile, and with a few exceptions have high melting and boiling points. The metals have few valence electrons, which are more or less easily lost, yielding ions carrying a positive charge. The oxides and hydroxides of the metals are basic. With negative ions or radicals the metals form numerous salts.

(3) *The nonmetals.* The nonmetals have few physical characteristics in common. At ordinary temperatures some are gases, one (bromine) is a liquid, and the rest are solid. Although there are only about fifteen nonmetals they constitute the greater portion of the earth's crust and are essential constituents of living and growing things. The nonmetals, in general, have relatively larger numbers of outer shell electrons than do the metals. Those with nearly complete valence shells pick up extra electrons to form negative ions. Others, as for example carbon, with fewer valence electrons form strong covalent bonds with other elements. The oxides of the nonmetals are acidic and with water form the oxyacids.

A much more systematic grouping of the elements into families is made through the Periodic Table of the elements.

M. H. LIETZKE AND R. W. STOUGHTON

References

Hampel, C. A., "The Encyclopedia of the Chemical Elements," Van Nostrand Reinhold Co., New York, 1968.

Abundance of Elements

The relative abundances of the chemical elements in nature are of considerable importance to a wide variety of scientific fields of research. In addition to interests based on more fundamental studies, there are of course economic reasons for interest in abundances of elements such as precious metals. Most of the early data on elemental abundances were derived from geochemical investigations of the crust, oceans, and atmosphere of the earth. It was soon realized, however, that the earth is a highly differentiated cosmic body. Therefore, crustal elemental abundances can certainly not be taken as representative of the entire earth, much less as representative of the solar system or the cosmos.

Since it is not yet possible to sample the mantle and the core of the earth directly, compilations of average solar system abundances have been commonly derived from analysis of the meteorites and from spectrographic investigations of the light from the sun and the stars.

It was suggested by Urey and Craig (1953) that a certain class of stony meteorites called the chondrites (derived from the fact that they contain abundant spherical mineral inclusions known as chondrules) might represent the best average sample available for the relative solar system abundance of the nonvolatile chemical elements. Chondrites make up approximately 85% of all observed metorite falls and are believed to be fragments of parent meteorite bodies once present in the asteroid belt between Jupiter and Mars. One class of chondrites, the Type I carbonaceous chondrites, show little evidence of heating or alternation and are probably more representative of the primitive solar system materials than any other sample available to us for direct analysis. This postulate is supported by the rather good agreement in the comparison of their abundances of the nonvolatile elements with abundances derived from spectrographic analysis of the light from the sun and analysis of the

TABLE 1.   ABUNDANCES OF SOME SELECTED ELEMENTS IN THE EARTH, METEORITES, AND THE MOON*

Atomic Number	Element	Earth's Continental Crust	Type I Carbonaceous Chondrites	Apollo 11 Lunar Igneous Rock
8	Oxygen	464,000	460,000	393,000
14	Silicon	281,500	147,000	194,000
13	Aluminum	82,300	12,000	42,000
26	Iron	56,300	180,500	157,000
20	Calcium	41,500	11,000	74,000
11	Sodium	23,600	5,100	3,550
12	Magnesium	23,300	95,000	41,000
19	Potassium	20,900	430	1,800
22	Titanium	5,700	430	71,000
1	Hydrogen	1,400	21,000	—
15	Phosphorus	1,050	1,100	700
25	Manganese	950	1,900	1,770
16	Sulfur	260	55,000	2,300
6	Carbon	200	31,000	70
40	Zirconium	165	10	590
92	Uranium	2.7	0.012	0.8
79	Gold	0.004	0.14	0.0002

*All abundances in parts per million by weight.

components of the primary cosmic radiation. It is also interesting to note that complex organic compounds, such as amino acids, have recently been discovered in the carbonaceous chondrites.

Additional information on the composition of members of the solar system has recently been obtained by analysis of the returned lunar samples (Proceedings of the Apollo 11 Lunar Science Conference, 1970). While exhibiting structural features similar to some terrestrial rocks, the lunar rocks have a distribution of elemental abundances different from any known terrestrial rock, or meteorite.

Abundances in parts per million by weight in the Earth's continental crust, the Type I carbonaceous chondrites, and an Apollo 11 lunar igneous rock for some important elements are listed in Table 1.

The abundances in the Type I carbonaceous chondrites are generally regarded as the closest approach to true solar system abundances for nonvolatile elements. The first twelve elements listed are also the twelve most abundant elements in the Earth's crust.

Using compilations of abundance data, largely from analyses of chondritic meteorites, Suess and Urey (1956) plotted the logarithm of the relative atomic abundances versus atomic weight for the stable nuclei. From this plot it was obvious that there is an approximately exponential decrease in abundance with an increase in atomic weight until approximately mass 100. After this, there is a much slower decrease on which are superimposed distinct fluctuations. Other features noted include a depletion of Li, B, Be, and deuterium relative to neighboring elements, a large abundance peak in the region of Fe, enrichment of nuclei that are simple multiples in atomic number and mass number of the alpha particle, and a tendency for the heavy nuclei to be neutron-rich.

Observations of this type coupled with calculations using the data of nuclear physics have led to the development of theories of nucleosynthesis. Burbridge, et al. (1957) assumed that the elements are all synthesized from hydrogen and that this process is still taking place today in the sun and the stars. To account for the observed features of the experimental abundance curves, they postulated eight different processes in nucleosynthesis:

(1) Hydrogen burning to produce helium.

(2) Helium burning to produce carbon, oxygen, and neon.

(3) Alpha capture processes to produce the "alpha particle nuclei," such as Mg^{24}.

(4) An equilibrium process, involving a variety of simple nuclear reactions. Free protons and neutrons interact with the lighter elements produced above leading to the formation of elements principally in the region of iron. Steller temperatures at this point would exceed 10^9 °K.

(5) Neutron capture reactions on a slow time scale (s-process) in which neutrons are captured by seed nuclei in the iron region to form radioactive nuclei which decay by negatron emission. Many nuclei in the region between iron and bismuth are produced by this process. The rate of neutron capture is slow enough that the radioactive product nuclei have time to decay on the average before another neutron capture takes place.

(6) Neutron capture reactions on a fast time scale (r-process) in which a number of neutrons are captured in succession by seed nuclei before a negatron decay can take place. This process probably takes place in supernovae and leads to the formation of many heavy nuclei, including thorium and uranium.

(7) Processes probably involving spallation reactions to produce the light elements lithium, beryllium, and boron from heavier nuclei.

(8) Addition of protons to various heavy nuclei to produce certain proton-rich nuclei.

WILLIAM D. EHMANN

References

Cherdyntsev, V. V., "Abundance of Chemical Elements," The University of Chicago Press, Chicago, 1961.

Aller, L. H., "The Abundance of the Elements," Interscience Publishers, New York, 1961.

Aherns, L. H., "Distribution of the Elements in Our Planet," McGraw Hill Book Co., New York, 1965.

"Proceedings of the Apollo 11 Lunar Science Conference," Geochim. Cosmochim. *Acta,* Supplement I, Pergamon Press, New York, 1970.

EMULSIONS

An emulsion is an intimate mixture of liquids, one of which (the disperse phase) is distributed in large or small globules throughout the other (the continuous phase). A third component, the emulsifying agent, without which the emulsion would immediately break down into layers, is always present at the interfaces between the two phases. The properties of an emulsion depend on a complex interaction of the following factors: (a) emulsion type, (b) concentration, (c) degree of dispersion and (d) the nature of the emulsifying agent.

Emulsion Type. Since water-and-oil emulsions are the most studied, the terms *water-phase* and *oil-phase* are used to designate, in an emulsion of whatever composition, the phases whose relative behavior resembles that of oil and of water respectively. An emulsion in which water is the continuous phase is conveniently designated an *O/W emulsion*; and emulsion in which oil is the continuous phase is called a *W/O emulsion*. The gross properties of an emulsion are in general determined by the nature of its continuous phase: milk, an O/W emulsion, is watery; whereas butter, a W/O emulsion, is greasy. Since the water-phase of any emulsion is necessarily more *hydrophilic,* or water-soluble, than the *lipophilic* oil-phase, ease of dilution with water provides a handy test to show whether an emulsion is of type O/W or W/O.

Concentration of the Disperse Phase. A space completely filled, in closest hexagonal packing, with uniform spheres is 74% occupied, provided that the spheres are very small in relation to the total volume. The globules in an emulsion are seldom of uniform size; and, on paper, arrangements of nonuniform spheres can be worked out in which more than 99% of the space is occupied. In practice, however, emulsions of concentration above about 70% either contain deformed polyhedral particles in close contact, or tend to *invert* into emulsions of the opposite type, containing less than 30% of the disperse phase. If inversion occurs, physical properties that vary with concentrations, such as viscosity or electrical conductivity, will show a discontinuous function through the concentration range. If inversion does not occur, the particles come in contact and are deformed in shape from spheres to polyhedra. Resistance to flow is caused by mutual interference, and there is a marked increase in viscosity, accompanied by a departure from Newtonian behavior, made evident by a pronounced plasticity of the emulsion. Ostwald has called such concentrated and semisolid emulsions "liquid-liquid foams."

Degree of Dispersion. The stability of an emulsion depends partly on the size of the dispersed globules, partly on the nature of their surface. The globules may range in size from a few hundred Ångstrom units to a few microns. Large globules tend to rise or fall because of density difference, and may produce a concentrated "cream," or even a separated layer. The globules will coalesce if their surface is too weak to resist the forces of surface tension, tending to reduce the surface area. An *emulsifying agent* stabilizes the surface of each globule sufficiently to prevent coalescence.

Nature of the Emulsifying Agent. An emulsifying agent or *stabilizer* may act: (a) by supplying negatively charged ions for preferential adsorption on the globules; (b) by surrounding each globule with an adsorbed film, as of a protein, a soap, or a synthetic detergent; or (c) by coating each globule with a mechanically strong layer of fine solid particles ("Armor-plated emulsion"). *Bancroft's Rule,* which has been found valid for both types of emulsions and for all types of emulsifying agents, states that *the emulsifying agent is always more soluble in, or better wetted by, the continuous than by the dispersed phase.* Thus, for example, emulsions stabilized by electric charges on the surface of the globules occur almost exclusively with an aqueous continuous phase. Carbon black, which is lipophilic, emulsifies water in oil, whereas silica, which is hydrophilic, emulsifies oil in water. Benzene can be emulsified in water by water-soluble sodium oleate; water can be emulsified in benzene by oil-soluble magnesium oleate.

A surface-active emulsifying agent must be neither completely hydrophilic nor completely lipophilic. Sodium oleate, for example, while soluble in water by virtue of its polar group, has in its molecule a large organic and lipophilic group with a high surface energy in an aqueous medium. Molecules arriving at an oil-water surface arrange themselves in a low-energy orientation, with the hydrophilic group in the water and the lipophilic group in the oil. Since molecules do not migrate spontaneously from low to high energy positions, this surface orientation is stable. Sodium oleate and other soaps are particularly valuable in the stabilization of O/W emulsions, since the whole soap molecule acts as a bridge across the oil-water interface, reducing the interfacial energy, and the negative ions thus firmly attached to the surface cause mutual repulsion of the globules of oil. Even in the absence of an electric charge, a judicious balance of the *hydrophil-lipophil balance* (HLB) of molecular structure has made commercially available a large number of effective nonionic emulsifying agents; which do not have the harmful physiological effects of either the anionic or the cation-active soaps, and are suitable for use in foods and drugs. Since, according to Bancroft's Rule, an agent in which the hydrophilic character predominates will stabilize O/W emulsions, and one in which the lipophilic character predominates will stabilize W/O emulsions, an appropriate emulsifying agent can be selected in advance, on the basis of its HLB, for either type of emulsion.

Industrial Applications. Many emulsions are found in nature, such as milk, egg-yolk, and crude petroleum. Others are produced artificially, chiefly in the food, pharmaceutical, and cosmetic industries, by means of high-speed stirrers, colloid mills or homogenizers. It is found in practice that the stability and other properties of an artificial emulsion are affected by many, apparently trivial, details of operating technique—the order in which ingredients are added, the speed and duration of the process, intermittent or continuous operation of the stirrer or other machine. Emulsions are sensitive to temperature and to small changes of concentration of the emulsifying agent. The production of a stable emulsion therefore requires much patient experimenting, and an intuitive "feel" acquired only by experience and practice. Instability can be determined microscopically by periodic measurement of the size-frequency distribution of the globules. An increase in average globule size, or decrease of interfacial area, shows coalescence of the disperse phase.

In many industrial operations it is necessary to break stubborn emulsions—e.g. the W/O emulsions of crude petroleum, the wool-fat emulsions encountered in wool scouring, engine condensates containing oil, the emulsions obtained in steam distillation of organic liquids. Chemical means may be used, the point of attack being the stabilizing agent at the interfaces. If the agent's action is mainly electrical, polyvalent ions are added to neutralize it: if it is a soap or a synthetic detergent, it can be destroyed with acid. Addition of an agent of the opposite type may induce an emulsion to invert, thus passing through a period of minimum stability, at which it may break. Physical means, such as freezing, heating, ageing, centrifuging, or the application of high-potential alternating current, are also used.

SIDNEY ROSS

Cross-references: *Colloid Chemistry, Detergents.*

ENAMINES

Enamines are amino olefins; more specifically, the name enamine usually refers to α,β-unsaturated tertiary amines, of the general formula $R_2N-\overset{|}{C}=\overset{|}{C}-$, where R is any alkyl group. Although compounds of this type find little practical use as end products themselves, they are extremely valuable as intermediates for syntheses of a multitude of organic compounds.

Preparation. Enamines are usually prepared by the reaction of a carbonyl compound and a secondary amine with the loss of water. The water may be removed either by the action of some dehydrating agent such as anhydrous potassium carbonate or by azeotropic distillation with benzene. Thus, cyclohexanone may react with diethyl amine to give N,N-diethyl-l-cyclohexenyl amine.

$$\bighexagon\!\!=\!\!O \; + \; H-N(C_2H_5)_2 \; \xrightarrow{(-H_2O)}$$

$$\bighexagon\!\!-\!\!N(C_2H_5)_2$$

The water formed in the reaction must be removed as formed, since enamines are quite unstable and are readily hydrolyzed to give back the parent carbonyl compound and amine. This necessitates meticulous exclusion of water during the preparation of enamines, but provides the very important advantage that an enamine, once formed, may be employed in other reactions (providing water is not necessary in the reaction) and the product may then be hydrolyzed to regenerate a new carbonyl compound.

Enamines as Nucleophiles. By conjugation of the unshared pair of electrons of nitrogen with the double bond in enamines, a partial negative charge may be placed on the β-carbon atom; thus enamines may act as nucleophiles to form new bonds to the β-carbon or to nitrogen. Nucleophilic attack by the β-carbon of an enamine on active alkyl and aryl halides, electrophilic olefins or acid halides yields substituted ketones as the final products. Treatment of enamines with mineral acids gives imonium salts which may be converted back to the enamines upon treatment with nonaqueous base.

Alkylation with Alkyl Halides. Enamines displace halide ions from active alkyl halides. Apparently N-alkylation is reversible, and C-alkylation of an enamine followed by loss of hydrogen halide from the resulting imonium salt yields a new alkylated enamine which, by hydrolysis, may give the monoalkylated ketone. In this way, the pyrrolidine enamine of cyclohexanone may be alkylated with methyl iodide to form after hydrolysis, 2-methyl cyclohexanone. The new alkylated enamine is not alkylated a second time because of steric hindrance to approach of another alkyl halide to the enamine. Therefore, good yields of monoalkylated ketones may be obtained in this way.

Optimum yields are obtained with very electrophilic primary halides, but satisfactory results are also possible with unactivated primary alkyl bromides and iodides.

The mildness of the enamine alkylation reaction makes it possible to alkylate ketones with compounds containing groups that would be destroyed by using the conventional synthetic methods. For example, ethyl bromoacetate may be used to alkylate enamines to obtain γ-ketoesters.

Alkylation with Electrophilic Olefins. Enamines react with electrophilic olefins such as α,β-unsaturated nitriles, esters, ketones, etc. to give products which can be easily converted to the monoalkylated carbonyl compounds. The imonium salt formed in this reaction generates the alkylated enamine by transfer of a proton from the β-carbon to the carbon adjacent to the electron withdrawing group.

The reaction requires no catalyst such as the aqueous base needed for similar alkylations by other methods and therefore avoids base-catalyzed polymerization of the unsaturated compound and aldol condensation of the carbonyl compound.

The imonium salt resulting from addition of the enamine of an α,α-disubstituted aldehyde to an electrophilic olefin cannot become neutral by proton transfer but can become neutral by cyclization to form a cyclobutane product. Many cyclobutane derivatives and analogues have been prepared in this way.

Acylation. Enamines react with acyl halides to form 1,3-diketones and related compounds after hydrolysis. The reaction is useful for extending carboxylic and dicarboxylic acid chains by five or six carbons, because β-diketones such as 2-acetyl-cyclohexanone are readily cleaved by a strong base to give the ketoacid, in this case $CH_3CO(CH_2)_5COOH$, which can be converted by the usual ketone reduction procedures to the carboxylic acid, $CH_3(CH_2)_6\cdot$ COOH. In a similar way, dicarboxylic acids may be extended by five, six, ten or twelve carbons by using their mono- or dihalide forms to alkylate enamines of cyclopentanone or cyclohexanone. For example, the 20-carbon dicarboxylic acid may be made in 44% yield from one mole of the 8-carbon dicarboxylic acid dichloride and two moles of the morpholine enamine of cyclohexanone.

Reductions of Enamines. Enamines are readily reduced in a number of ways to yield saturated tertiary amines. Catalytic hydrogenation of enamines over active metal catalysts like platinum oxide probably involves the same kind of addition of hydrogen across the double bond as occurs in other olefins. For example, acetophenone may be converted to a N,N-disubstituted-1-phenylethyl amine by converting it to the corresponding enamine and hydrogenating:

$$C_6H_5\overset{\overset{\displaystyle O}{\|}}{C}CH_3 \rightarrow C_6H_5\overset{\overset{\displaystyle -N-}{|}}{C}=CH_2 \xrightarrow{\ H_2,\,PtO_2\ } C_6H_5\overset{\overset{\displaystyle -N-}{|}}{C}HCH_3$$

Enamines may also be reduced by formic acid and by sodium borohydride. Enamines themselves are not reduced by lithium aluminum hydride but salts of enamines are. Because enamines do not react with lithium aluminum hydride, the enamine function may be used as a blocking group to protect carbonyl functions during the lithium aluminum hydride reduction of other groups in a compound; subsequent hydrolysis, of course, regenerates the carbonyl function.

JOSEPH WEST

ENVIRONMENTAL CHEMISTRY

Environmental chemistry is a new and rapidly expanding science. Though numerous chemists who were active during the 1800 and 1900's could be described as environmentalists, the modern surge of interest in this field was due primarily to Rachel Carson and her book, "Silent Spring" (1962). Dr. Carson focused national attention on the widespread distribution of pesticides and their ramification with respect to nontarget organisms, including man. This motivated research by both those who agreed or differed with the conclusions in "Silent Spring."

Several spin-offs resulted from the above. The realization that minute quantities of toxicants could adversely affect an ecosystem motivated the development and refinement of numerous analytical tools including the liquid-gas chromatograph, atomic absorption spectrophotometer and infrared remote sensing. Many of these instruments had been employed initially in medical research. However, they were "borrowed" and modified in attempts to understand and remedy environmental problems. Also, interest in the detection and possible effects of other foreign compounds in the air, water and land led to research on heavy metals, such as mercury and lead, and numerous organic compounds such as polychlorinated biphenyl residues (a compound in some paints and sealents) and nitrosotriacetate (a substitute for phosphates in detergents).

In the early 1960's, the majority of those who called themselves environmental chemists probably were trained as toxicologists and/or analytical chemists. Currently, this area is attracting individuals from nearly every branch of chemistry—physical chemistry, biochemistry, chemical engineering, etc. Since a larger number of industries are being required to abate their effluents, as well as to report to governmental and other outside agencies on the impact of their discharges and products on the environment, opportunities for environmental chemists continue to expand.

The realization that traditional treament techniques for wastes discharged to the water or air were inadequate has resulted in research and development on these problems. For example, chemists are now contributing to the improvement of methods to remove nutrients from municipal sewage. Up until a few years ago, the design of such facilities and processes was done almost exclusively by sanitary engineers.

Reduction of the detrimental impacts of effluents has led to the development of a specialized field within environmental chemistry. This phase is concerned with the mixing of the discharges from one or more processes in order to neutralize these pollutants and/or to produce one or more usable materials. The United States, with approximately 5% of the world's population, uses nearly 40% of the annual global production of raw materials. With the reserves of several fuels and ores possibly becoming low and the reluctance of developing nations to continue to export these raw materials to the industrialized nations, it is anticipated that the chemistry of reuse or recycling—another component of environmental chemistry—also will be an area of expansion in terms of personnel and interest in the near future.

ROBERT A. SWEENEY

References

Anon., "Waste—Management and Control," National Academy of Science—National Research Council, Washington, D.C., 257 p., 1966.

Anon., "Cleaning Our Environment: the Chemical Basis for Action," American Chemical Society, Washington, D.C., 249 p., 1969.

Carson, Rachel, "Silent Spring," Houghton Mifflin, Boston, Massachusetts, 368 p., 1962.

Sayers, William T., "Water quality surveillance: the federal-state network," *Environmental Science and Technology,* **5**(2):114–119 (1971).

ENZYMES

Enzymes are proteins which function as catalysts in living organisms. They differ from other chemical catalysts in four important respects. First, enzymes selectively catalyze reactions. Both hydroxide ions and the enzyme α-chymotrypsin catalyze the hydrolysis of peptide bonds. Hydroxide ions will catalyze the cleavage of any peptide bond. α-Chymotrypsin preferentially cleaves bonds formed by the aromatic amino acids phenylalanine, tyrosine and tryptophan. Even more remarkable, chymotrypsin discriminates between L- and D- enantiomers. Only the former are substrates for the enzyme. Second, enzymes are more efficient than other catalysts. α-Chymotrypsin breaks peptide bonds ten million times faster than do hydroxide ions. A protein which requires one hour for digestion in the presence of chymotrypsin would require more than one thousand years in its absence. Third, enzymes may regulate their activity in order to maintain the proper balance of metabolites in a living cell. In one of the first examples studied, the production of the amino acid isoleucine from threonine is controlled by the enzyme threonine deaminase. If isoleucine is present in excess, the efficiency of threonine deaminase is depressed. A deficiency of isoleucine causes an increase in the activity of the enzyme until the physiological concentration of isoleucine is restored. The enzymatic mechanism resembles the feedback circuits used to control the output of electronic amplifiers. Finally, many different enzymes may collaborate in performing a metabolic function. A second digestive enzyme, trypsin, is required to cleave peptide bonds formed by the basic amino acids lysine and arginine. Threonine deaminase catalyzes only the first in a sequence of five reactions involved in the biosynthesis of isoleucine from threonine. Each of the subsequent reactions is catalyzed by a different enzyme. Virtually every reaction in the labyrinth of chemical transformations required for metabolism is specifically catalyzed and controlled by a different enzyme. Enzymes are responsible for the harmony as well as for the pace of life. A complete physical-chemical explanation of how enzymes perform their functions is still lacking, but astonishing progress has been made toward that goal. In the last decade, techniques have been developed which permit direct measurement of the fastest steps in enzymatic reactions. One enzyme, ribonuclease, has been chemically synthesized from its constituent amino acids. X-ray diffraction "pictures" showing the positions of the atoms in ten enzymes have been obtained. Pictures are available of substrate analogues bound to three enzymes—lysozyme, carboxypeptidase A and α-chymotrypsin.

Specificity. The ability of enzymes to selectively catalyze reactions is called specificity. Nearly one thousand different enzymes have been isolated from living organisms. All catalyze a single type of chemical reaction. They discriminate further among possible substrates for the reaction. Many distinguish between enantiomers of a molecule. A few enzymes interact with a single substrate. The specificity of enzymes provides the basis for their classification. X-ray crystallographic pictures of enzymes with substrate analogues bound at their active sites show how the substrate is recognized.

In 1964 a commission appointed by the International Union of Biochemistry recommended classification of enzymes according to the type of reaction catalyzed. All known enzymes can be placed in one of the six classes

1. Oxidoreductases
2. Transferases
3. Hydrolases
4. Lyases
5. Isomerases
6. Ligases or Synthetases

named by adding the suffix -ase to a root indicating the reaction catalyzed. For example, hydrolases catalyze reactions of the type

$$A—B + HOH \rightarrow A—OH + H—B \qquad (1)$$

in which a bond is cleaved by addition of a water molecule. This class is subdivided according to the type of bond cleaved. Carbon-nitrogen bonds other than peptide bonds are grouped in the fifth subclass (3.5). Each subclass is further subdivided according to the type of compound containing the bond. The first sub-sub class of hydrolases (3.5.1) consists of linear amides. Each enzyme is given an enzyme commission (EC) number, a systematic and a trivial name. The systematic name specifies the substrate and the type of reaction catalyzed. For example, the first enzyme listed in the sub-sub class of linear amide hydrolases (EC 3.5.1.1) is L-asparagine amidohydrolase. It catalyzes the hydrolysis of the side chain amide bond in L-asparagine. Its trivial name is asparaginase.

The relationship between the three-dimensional shape, or conformation, of an enzyme and its ability to discriminate among possible substrates was recognized as early as 1894 by Emil Fischer. He proposed a rigid "lock and key" complementarity between the shapes of enzymes and their substrates to explain enzyme specificity. A minimum of three points of contact between an enzyme and its substrate are required to explain the ability to distinguish between enantiomers. More recently the importance of subtle changes in an enzyme's conformation has been appreciated. In 1958 Daniel Koshland proposed that recognition involves alteration of the enzyme shape to fit the substrate ("induced fit").

The first picture of an enzyme with its substrate bond was published in 1966 by David Phillips. The enzyme, lysozyme, has a large cleft in its surface into which the substrate fits. The substrate, a polysaccharide or sugar, is held in the cleft by six weak bonds, called hydrogen bonds, between the substrate and groups on the enzyme. In addition there are a large number of interactions, called hydrophobic bonds, between apolar groups on the enzyme and the substrate. A comparison of the enzyme structure in the presence and in the absence of the substrate shows that the cleft closes slightly when the substrate binds, supporting the idea of "induced fit." The indole ring of the tryptophan residue moves 0.75 angstrom (less than three billionths of an inch) toward the opposite side of the molecule.

α-Chymotrypsin does not have a cleft in its surface. It does have a pocket, or slit, which is just large enough to accommodate the side chain of phenylalanine, tyrosine or tryptophan. The narrowness of the slit constrains the aromatic ring of the substrate to a single plane. A second interaction occurs between the imino hydrogen of the substrate and the carbonyl oxygen of a serine residue in the enzyme. These two interactions fix the carbonyl group of an L-amino acid near the hydroxyl group of a second serine residue of the enzyme. This serine hydroxyl is the attacking group in bond cleavage. Since interchanging the positions of the hydrogen atom and the carbonyl group to form the D-enantiomer of the substrate places the carbonyl group too far away for attack by the serine hydroxyl, only polypeptides of aromatic amino acids having the L-configuration are substrates for chymotrypsin.

A comparison of the structures of α-chymotrypsin and trypsin explains their different specificities. The amino acid sequences of the two enzymes are similar. So are their three-dimensional shapes, including the pocket at the active site. A striking dissimilarity is the substitution of a negatively charged aspartic acid residue deep in the pocket of trypsin for an uncharged serine residue in the same location of chymotrypsin's pocket. It is the formation of an ionic bond between this negatively charged group and the positively charged side chains of the basic amino acids lysine and arginine which accounts for the different specificity of trypsin.

Catalysis. The rates of enzymatic reactions usually depend directly on the concentration of catalyst, i.e., doubling the enzyme concentration (E) doubles the rate. The variation of the substrate concentration (S) with rate is more complex. The rate is directly proportional to low concentrations of the substrate, but it is independent of the substrate at high concentrations. To explain the appearance of a maximum rate, Leonard Michaelis and Maud Menten proposed in 1913 that an intermediate complex is formed between the enzyme and the substrate and that the rate of product (P) formation is proportional to its concentration (ES)

$$E + S \rightleftarrows ES \rightarrow E + P \qquad (2)$$

The resulting analytical expression for the rate

$$\text{Rate} = \frac{V(S)}{(S) + K_m} \qquad (3)$$

contains two experimentally determinable constants which characterize the reaction. V is the maximum rate. K_m is called the Michaelis constant. It is numerically equal to the substrate concentration at which the rate is half its maximum value. The ratio of the maximum rate to the enzyme concentration is called the turnover number. Turnover numbers for enzymes typically fall in the range 10^2 to 10^4 reciprocal seconds, which means that a single enzyme molecule can convert between one hundred and ten thousand substrate molecules to product each second. The Michaelis constant has no unambiguous physical interpretation, although it frequently indicates how tightly the substrate is bound by the enzyme.

It is clear from attempts to simulate enzymatic rates in model systems, that several reaction intermediates must be invoked to explain a ten million-fold rate enhancement. Nevertheless, most enzyme catalyzed reactions follow Michaelis-Menten kinetics, because the insertion of any number of additional intermediates in Eq. 2 results in the same analytical expression for the rate. The maximum rate and the Michaelis constant become complex combinations of rate constants for the individual steps in the reaction mechanism. To learn the details of enzymatic catalysis, it is necessary to investigate each step separately. The difficulty is that the intermediates are short-lived. A promising approach is chemical relaxation. By perturbing an equilibrium, or a stationary-state, faster than it can respond, chemical reactions with lifetimes as short as a millionth of a second can be studied. Using this approach five intermediates have been identified in the catalytic mechanism of the enzyme glutamic-aspartic transaminase. Other approaches to learning how enzymes accelerate reactions include chemical modification of the enzyme, isotope labeling, kinetic studies of the effects of pH and temperature, model studies and x-ray crystallography. Three important elements of catalysis have emerged from these studies—immobilization of the substrate on the enzyme surface, strain of the bonds to be broken and concerted attack by several functional groups on the enzyme.

Attachment of the substrate to the active site of the enzyme converts an intermolecular to an intramolecular reaction. It fixes the labile bond in correct juxtaposition to the attacking groups on the enzyme. The complex between lysozyme and its substrate places the glycosidic bond to be broken near a carboxylic acid group and a carboxylate ion which catalyze its cleavage. Estimates of the rate enhancement which results from immobilizing a substrate on an enzyme's surface range from five hundred to fifty thousandfold.

Enzymologists have long speculated that enzymes facilitate reactions by exerting a "rack" effect on their substrates. The complex between lysozyme and its substrate provides evidence for this picturesque theory. In order to fit the polysaccharide into the cleft of lysozyme, it is necessary to distort the sugar residue containing the labile glycosidic linkage from its more stable "chair" configuration into a partial "boat" configuration. Since this is the preferred configuration for the carbonium ion which results from bond cleavage, distortion of the sugar ring by the enzyme facilitates the reaction.

One of the intermediates formed in catalysis by α-chymotrypsin involves covalent attachment of the acyl part of the substrate to the enzyme. By labeling the substrate with an isotope of carbon, quenching the reaction and locating the label, it was established that the acyl moiety becomes attached to a serine hydroxyl group of the enzyme. Involvement of a histidine residue in catalysis had been inferred from the pH dependence of the reaction. At pH's below six the enzyme is catalytically inactive. It is fully active above pH 8. The imidazole side chain of histidine is the only group in chymotrypsin which titrates at pH 7. How a serine hydroxyl group, an imidazole group of histidine and a third group, the carboxylate ion of an aspartic acid residue, cooperate in breaking peptide bonds was revealed by the x-ray structure of the enzyme. The aliphatic hydroxyl group of serine is not effective in attacking peptide bonds, but the negatively charged anion is. In chymotrypsin a negatively charged carboxylate ion is connected through an imidazole ring to a serine hydroxyl by a string of hydrogen bonds. Negative charge is transferred from the carboxylate ion through these bonds to the oxygen atom of the serine hydroxyl, making it a good attacking group.

Control. One of the most remarkable properties of enzymes was discovered in the 1950's. An explanation, called the allosteric model, was proposed in 1963 by Jacque Monod, Jean-Pierre Changeux and Francois Jacob. In the biosynthetic pathway of every essential metabolite there is at least one enzyme which can regulate its activity in order to maintain the proper balance of metabolites. Regulation is accomplished by a "feedback" control mechanism. Five reactions, each catalyzed by a different enzyme, are involved in the biosynthesis of isoleucine. In the first step threonine deaminase converts threonine to α-ketobutyric acid. Its catalytic effectiveness is inversely related to the concentration of the final product of the biosynthetic pathway, isoleucine. The enzyme threonine deaminase is turned "on or off" by a deficiency or by an excess of isoleucine. Isoleucine regulates the activity of threonine deaminase by binding to the enzyme and inducing a shift between two conformations of the enzyme with very different catalytic properties.

Relaxation studies of the enzyme D-glyceraldehyde-3-phosphate dehydrogenase have provided evidence for the allosteric model of enzymatic control. This enzyme catalyzes the phosphorylation of D-glyceraldehyde-3-phosphate, one of eleven steps in the anaerobic metabolism of sugars. The enzyme is composed of four protein molecules, called subunits, each containing a catalytic site. Essential for its activity is a small, nonprotein molecule, nicotinamide adenine dinucleotide, called a coenzyme. The coenzyme binds cooperatively to the enzyme. It binds more readily at high than at low concentrations and can therefore turn the enzyme on or off. Kinetic studies of coenzyme binding revealed three processes. Two depended on the concentrations of both the enzyme and the coenzyme and were attributed to binding to two conformations of the enzyme with different affinities for the coenzyme. The third process did not depend on the enzyme concentration, but it did vary with coenzyme concentration and was associated with the change in conformation of the enzyme. At low concentrations the less affine form of the enzyme predominates. As coenzyme binding sites on the subunits become occupied, the conformer equilibrium shifts to the form which binds coenzyme molecules tightly and all the remaining sites are rapidly filled. The binding process resembles a zipper. Once started, it proceeds rapidly.

Organization. The poorest understood aspect of enzyme function is the ability of enzymes to cooperate with nonprotein molecules and with other enzymes in carrying out their metabolic role. At the simplest level of organization, nonprotein cofactors may be required for the activity of an enzyme. Zinc ions are required for catalysis by the enzyme carbonic anhydrase. Nicotinamide adenine dinucleotide is reduced when D-glyceraldehyde-3-phosphate dehydrogenase phosphorylates D-glyceraldehyde-3-phosphate. The conformational change involved in regulation of the phosphorylation rate results from a change in the weak interactions among the four subunits of the enzyme. At a higher level of organization a number of physically unassociated enzymes may catalyze the individual steps in the biosynthesis of a metabolite. In this case the intermediates must diffuse from one enzyme to the next. The biosynthesis of isoleucine from threonine provides an example. Alternatively several different enzymes may be physically associated in a multienzyme complex. At the highest level of organization the enzymes required for a metabolic transformation may be organized on the surface of an organelle. For example, the enzymes required for oxidative phosphorylation are attached to the inner membrane of the mitochondrian.

The pyruvate dehydrogenase system illustrates the complexity of enzyme organization. The oxidation of pyruvate to acetyl coenzyme A requires five different coenzymes and three different enzymes organized in a multienzyme complex large enough to be visible in the electron microscope. The molecular weight of the complex isolated from the bacterium Escherichia coli is over four million. The core of the complex is a molecule of dihydrolipoyl transacetylase which is a noncovalent aggregate of twenty-four identical polypeptide chains each of about 40,000 molecular weight. One molecule of lipoic acid is covalently bound to each of the polypeptide chains. Around the transacetylase molecule are arranged twenty-four pyruvate dehydrogenase molecules of molecular weight 90,000 and twenty-four dihydrolipoyl dehydrogenase molecules of molecular weight 55,000. A molecule of thiamine pyrophosphate is associated with each molecule of the former enzyme and a molecule of flavine adenine dinucleotide with each molecule of the latter. In addition, coenzyme A and nicotinamide adenine dinucleotide are involved in the catalytic mechanism. Astonishingly, the complex is self-assembling. If the completely dissociated components are mixed at neutral pH, the catalytically active complex, morphologically identical to that isolated from the bacterium, forms spontaneously.

LARRY FALLER

EPOXY RESINS

Liquid epoxy resins were first synthesized in Europe in the mid-1930's by reaction of a large excess of epichlorohydrin with bisphenol A. In 1939, higher-molecular-weight solid epoxy resins from the same intermediates were synthesized in the U.S. by using more nearly stoichiometric quantities of epichlorohydrin. The generalized reaction is as follows:

$$(1) \quad ClCH_2\overset{O}{\overset{\diagup\diagdown}{CH}}CH_2 + (H)\underset{x}{-}R \rightarrow$$

$$(\overset{Cl}{\overset{|}{CH_2}}\overset{OH}{\overset{|}{CH}}CH_2)\underset{x}{-}R$$

$$(2) \quad (\overset{Cl}{\overset{|}{CH_2}}\overset{OH}{\overset{|}{CH}}CH_2)\underset{x}{-}R + NaOH \rightarrow$$

$$(CH_2\overset{O}{\overset{\diagup\diagdown}{CH}}CH_2)\underset{x}{-}R + NaCl + H_2O$$

In the case of the diglycidyl ether of bisphenol A, R = $O\varnothing-\overset{CH_3}{\overset{|}{\underset{|}{C}}}-\varnothing O$ and $x = 2$. The higher molecular weight species are produced when the newly formed epoxy groups ($-\overset{O}{\overset{\diagup\diagdown}{CH}}-CH-$) compete with the ones supplied by epichlorohydrin.

From a base of zero in 1947, U.S. production of these "epi-bis" resins, solid and liquid, increased to over 150 million pounds by 1970, with the annual growth rate stabilizing at about 10%. The resins are also produced in most other industrialized countries.

The epi-bis resins account for the majority of the epoxy resins of commerce, although other types are being sold, including ones where R in Eq. (2) is the residue of a different phenol, of an aromatic amine, or of a short-chain alcohol. Too, a second large class of resins was introduced in the mid 1950's, these produced by the oxidation of di- or polyolefins with a peracid, usually peracetic acid. Resins from the peracid route may have vinyl precursors, or the unsaturation may be differently situated in the molecule, most frequently on otherwise saturated five- or six-membered rings.

The epoxy resins are characterized by having an average of two or more reactive epoxy groups per molecule. The reactivity of the epoxy group will depend on its location within the molecule and the nature of adjacent substituents. An "idealized" epoxy group will undergo reaction with well over 40 distinct reagents. In commercial practice, epoxy resins are either homopolymerized with tertiary amines or Lewis acids or are copolymerized with compounds containing hydrogens replaceable by sodium: most frequently primary and secondary amines, carboxylic acids, mercaptans, phenols, and alcohols. Alicyclic anhydrides are particularly valuable coreactants (curing agents) for the liquid species. They open on available hydroxyls to yield the reactive carboxylic acids. Reaction of the acids with epoxy groups is by direct addition, as in Eq. (1), no volatiles being given off.

With resins derived from epichlorohydrin, primary and secondary aliphatic amines are quite reactive, the reaction highly exothermic, and cure occurring in from 20-30 minutes at normal room temperatures in masses of 25-50 grams or more. With aromatic amines, carboxylic acids, and phenols, elevated temperature cures are normally required, with tertiary amines being used as accelerators for the latter two classes. The acids, in particular, cure with negligible exotherm. Mercaptans require accelerators or coreactants, and alcohols are at best sluggishly reactive except under special conditions. With resins containing ring-situated epoxy groups, amine reactivity is considerably reduced and acid reactivity somewhat enhanced.

In any event, since one reactive site on the coreactant molecule reacts with one epoxy group, the two materials must be combined to provide a reactive site on one species for each reactive site on the other, the exact ratio being somewhat critical in terms of ultimate cured properties. Often this will require use of substantial volumes of coreactant, such that the final product will partake of the properties of both major ingredients. In the selection of coreactants, it is necessary only to maintain system functionality in excess of two in order to achieve a crosslinked, thermoset network. Within this restriction, thousands and thousands of chemical entities are suitable, with some 25 or so being routinely encountered in commercial formulations.

For these reasons—the variety of basic epoxy resins available and the large number of suitable coreactants—the epoxy resins are extremely versatile. Depending on the specific selection of ingredients, cure may be made to occur at any convenient temperature from 20°C upward; and the finished product may range from soft and flexible to hard and rigid. Handling and cured properties may be further modified, in particular through the use of organic and inorganic fillers, and less frequently through the incorporation of chemical additives.

As a general rule, the more hard and rigid, the better the heat and chemical resistance and the better the electrical insulating properties. Thermal shock and adhesive strengths may be improved by building into the molecular network optimum degrees of flexibility. Systems may be produced which withstand temperatures up to about 250°C and which are virtually inert to all forms of chemical attack. Other systems will take repeated thermal shock from +180 to −80°C when cast around stress-concentrating metal inserts. Lap shear adhesive strengths to metals may exceed 4000 psi. In some cases, careful design and formulation may be required to exploit the inherent advantages of the epoxy resins and minimize their weaknesses—but under optimum conditions they are capable of functionally replacing, often to advantage, other rigid and semirigid thermosets in nearly all applications. Only the flexible species are inherently inferior to other available plastics and rubbers.

Because of this extreme versatility, it is not surprising that the epoxy resins have gained acceptance in a wide variety of often dissimilar applications.

The liquid epoxy resins are used most widely as potting compounds and casting resins, where tough, dimensionally stable, chemically resistant products with excellent electrical insulation properties can be produced by simple pour-and-cure techniques using low-cost molds. Particularly advantageous in these applications is the low shrinkage and the absence of volatiles. Major use areas include embedment of electrical and electronic equipment, preparation of low-cost molds and final shapes such as bowling balls, and casting of specialized flooring containing sand or marble chips. The liquid resins are also used in the formulation of adhesive systems, the epoxies being the first of the "miracle plastic adhesives" capable of supporting tons of weight in tension across very small bonding areas. But for their inherent expense, they would doubtless be more widely used as matrix materials for cloth, mat, or filament laminates. As it is, they are used with these reinforcements to make specialized tooling fixtures and for various hydrospace and aerospace constructions. Epoxy resin foams are used in a few areas but are of little commercial importance, foams either lower in cost or superior in properties being available.

Solid epoxy resins may be used in the formulation of molding powders, but more frequently these are made by grinding a B-staged formulation based on the liquid species. The molding compounds are most useful in low-pressure injection equipment. Some solid resins are also used in adhesive formulations.

The vast majority of the solid resins, however, are used in coatings—an application area that absorbs, year after year, fully 50% of the total production of the epi-bis resins.

The coatings may be solvent free, based on the liquid species, in which case they are applied in thicknesses ranging upward from 50-100 mil. These are used in dip coating applications and unreinforced over small areas for chemical resistance. More commonly, however, the coatings are based on the solid resins and contain from 20 to 80% solvent, based on the resinous vehicle. A range of progressively higher-mol.-wt. epi-bis resins are supplied for these coating applications. At the lower end, the resins melt at about 65-75°C and contain sufficient epoxy groups to react with the coreactant curing agents employed with the liquid species; at the upper end, the resins melt at up to 100°C higher and contain few unreacted epoxy groups. These high-melting species are valuable because of the numerous hydroxyls present along the thermoplastic chain. The hydroxyls are reacted with unsaturated fatty acids to produce air-dry ester coatings or with melamine, urea, or phenol formaldehyde resins to obtain long-pot-life baking coatings.

The epoxy-resin solution coatings are valuable as electrical varnishes, corrosion resistant paints, high-adhesive-strength primers, and durable appliance finishes.

The backbone of the epi-bis resin contains aromatic groups alternating with short chain aliphatic segments bearing polar hydroxyls and capable of rotation at repeated ether linkages. These properties are carried over in the cured network which also contains the hydroxyls, the aromatic/aliphatic segments, often an even increased number of ether links, and whatever specialized groups are contributed by the coreactant. This particular combination of chemical groups in the resinous backbone accounts for the toughness, chemical resistance, and adhesive strengths of the final polymer. No other type of epoxy resin possesses all the advantages of the epi-bis species, and in consequence the others are used only in specialized low-volume applications.

The valuable properties of the epi-bis backbone have been extended to produce phenoxy species. These are very high-mol.-wt. epi-bis resins which are used in their own right as thermoplastic coating vehicles, as well as being reacted with coreactants through the hydroxyls. The liquid epi-bis resin may also be reacted with methacrylic acid through the acid moiety to produce a molecule terminated with methacrylate groups, capable of polymerizing by the free radical mechanism and yet possessing, when fully cured, the combination of properties more normally associated with epoxy resins than with acrylics.

Thus, the epoxy resins may be of various types, each with their own unique set of inherent properties; they may be reacted with virtually any compound containing sufficient active hydrogens to provide crosslinking; they may be reacted to produce species with acrylic, hydroxyl, or conceivably other types of functionality; they may be modified with virtually any filler imaginable; and numerous diluents, flexibilizers, extenders, colorants, solvents, etc., may be added to the formulation to provide selective upgrading of properties. Because of this extreme versatility, they are supplied in commercial practice not only by the companies who synthesize the basic resins but by a variety of formulators who specialize in manufacturing adhesive, laminating, potting, molding, and coating systems to their proprietary specifications.

HENRY LEE
KRIS NEVILLE

References

Lee, H. L., and K. Neville, "Epoxy Resins, Their Application and Technology," McGraw-Hill Book Co., 1957.
Lee, H. L., and K. Neville, "Handbook of Epoxy Resins," McGraw-Hill Book Co., 1967.
Lee, H. L., (ed.) Epoxy Resins, "Advances in Chemistry" Series 92, American Chem. Soc., Washington, D.C., 1970.
Paquin, A., "Epoxydverbindungen und Epoxyharz," Springer-Verlag OGH, Berlin, 1958 (in German).
Schrade, J., "The Epoxy Resins," Dunod, Paris, 1957 (in French).
Skeist, I., "Epoxy Resins," Van Nostrand Reinhold Co., 1958.

EQUILIBRIUM

In many instances a chemical reaction which proceeds spontaneously in an isolated system does not go to completion. The products of the reaction in such a case themselves react to reform the initial materials as indicated by the following general equation:

$$aA + bB \longleftrightarrow cC + dD \qquad (1)$$

where a, b, c and d denote the number of moles of the species A, B, C and D that are involved in the reaction.

Under given conditions of temperature, pressure and concentration a state is reached wherein the overall reaction in effect comes to a halt. At this point the system is said to have reached equilibrium. From a microscopic point of view, however, the reactions involved have not stopped but are proceeding in their respective directions at equal rates.

In accordance with the Law of Mass Action, the velocities of the forward (V_1) and reverse (V_2) reactions are proportional to the active concentrations of their reacting substances, each raised to the power numerically equal to the number of molecules appearing in the balanced equation, as a, b, c and d in (1) above:

$$V_1 = k_1 A^a \, B^b \text{ and } V_2 = k_2 C^c \, D^d$$

At equilibrium $V_1 = V_2$ and

$$k_1 A^a \, B^b = k_2 C^c \, D^d$$

and hence

$$\frac{k_2}{k_1} = \frac{C^c \, D^d}{A^a \, B^b} = K_{eq} \qquad (2)$$

where K_{eq} is the equilibrium constant.

Approximate equilibrium constants may be derived employing molalities, mole fractions, partial pressures or the equivalent quantities that are usually measured. However, the activities or fugacities must be employed to derive exact constants.

The science of thermodynamics is particularly well fitted to the interpretation of equilibrium measurements. In the thermodynamic sense equilibrium is attained whenever the energy available to drive the reaction in one direction is equal to the energy available to drive it in the other. Under these conditions there would be no tendency for the over-all reaction to proceed spontaneously in either direction. This driving energy is called the free energy decrease accompanying the reaction. The free energy change which occurs when the reaction goes completely from left to right in equation (1) under specified conditions is readily expressed as:

$$\Delta F° = RT \ln K_{eq} \qquad (3)$$

where $\Delta F°$ is the difference in the standard free energies of the products and reactants at the temperature of measurement, R is the gas constant and T is the absolute temperature. In the standard state reactants and products have unit activity.

Equilibrium constants can be determined either from the measured equilibrium concentrations of reactants and products or from derived values of $\Delta F°$. The latter are usually calculated from measurements of electromotive force in galvanic cells or from calorimetric data.

It should be borne in mind that $\Delta F°$ as well as K_{eq} change with temperature. The exact relationship between temperature and equilibrium is defined by the formula commonly associated with the name of van't Hoff.

$$\frac{d \ln K_{eq}}{dT} = \frac{\Delta H°}{RT^2} \qquad (4)$$

where $\Delta H°$ is the change in heat content accompanying the process for the products and reactants in their standard states.

There are various kinds of chemical reactions to which these concepts can be applied. They can be broadly divided into those in which phase changes occur and those in which they do not. Processes involving a change in phase may occur for single component or multicomponent systems. Examples of the former are the melting of ice, the vaporization of water, the sublimation of CO_2 and the conversion among the allotropic forms of phosphorus and sulfur. The reaction

$$CaCO_3(\text{solid}) \longleftrightarrow CaO(\text{solid}) + CO_2(\text{gas}) \qquad (5)$$

is an example of the latter.

Processes which do not involve a change in phase include the chemically reversible gas reactions such as the dissociation of nitrogen tetroxide:

$$N_2O_4 \longleftrightarrow 2NO_2 \qquad (6)$$

and the formation of ammonia from nitrogen and hydrogen:

$$N_2 + 3H_2 \longleftrightarrow 2NH_3 \qquad (7)$$

The control of a chemical reaction by manipulation of its equilibrium conditions is very important in many technical processes. As an example, consider the synthesis of ammonia (Eq. 7) by the Haber process. Such a gas reaction which produces a different number of molecules from those which combine as reactants is subject to one of the most useful principles of equilibrium—that of LeChatelier. This principle states that every system in equilibrium, when subjected to a change in equilibrium

conditions, tends to react in such a manner as to restore the original equilibrium conditions.

If increased pressure is applied to the system (Eq. 7), the reaction will tend to occur so as to relieve the pressure—that is, so as to form more NH_3. Similarly, to counteract a rise in temperature, the system will react to use up heat. Since heat is evolved in the above process, low temperatures will favor the formation of NH_3.

Reaction (7) is very slow at room temperature but can be accelerated by employing catalysts and by increasing the reacting temperature. The above state of affairs indicates that the combination of a reaction temperature no higher than that necessary to provide a reasonable reaction rate while employing a suitable catalyst and a very high reaction pressure provide the most favorable conditions for the desired synthesis.

L. G. van Uitert

Cross-reference: *Thermodynamics.*

ESSENTIAL OILS

It is difficult to define essential oils precisely and concisely; for practical purposes they may be described as natural odoriferous compounds occurring in, or isolated from, plant materials. Normally liquid (in some instances semisolid; rarely solid), insoluble in water and volatile with steam, they evaporate at different rates under ordinary atmospheric pressure at room temperature. Hence the alternative term "volatile" or "ethereal" oils. (The common term "essential" originates from the Latin "essentia"—the "quinta essentia" which the medieval alchemists considered the characteristic and most important component of every natural substance.) It is in their relatively rapid evaporation and pronounced odor—aside from their chemical composition—that the essential oils differ fundamentally from the fixed, fatty oils. Of the many thousands of plant species known, a relatively small number yield essential oils. These may develop throughout the entire plant, or in certain parts only. Some oils occur solely in the root, or the wood, the bark, the leaves, flowers, or fruit. In some cases different organs of a single plant may contain essential oils of different chemical composition.

Several theories have been advanced to explain the biochemistry of the essential oils; however, none can be accepted as completely satisfactory. Perhaps the oils are merely elimination products in the life processes of the plant. If so, they resemble certain gums, balsams, and resins; some essential oils appear, indeed, to be precursors of such exudation products.

The essential oils vary widely in their physicochemical properties, and their chemical composition is usually complex. A few are composed almost exclusively of one component—for example, oils of wintergreen and sweet birch (methyl salicylate), and cassia oil (cinnamaldehyde). Most essential oils, however, contain a larger number of constituents, 50 and more not being unusual. These individual components belong to many classes of organic compounds, particularly the terpenes and sesquiterpenes, and their alcohols, esters, aldehydes, ketones, lactones, oxides, etc., some of them open-chained, many cyclic and bicyclic. Members of the aromatic series, too, are present (for example, phenylethyl alcohol and benzyl acetate), and from recent investigations it appears that the azulenes play an important role in essential oils.

Assay is accomplished by conventional physico-chemical tests, such as determination of specific gravity, optical rotation, solubility in alcohol, boiling range, etc.; determination of free acids, alcohols, esters, aldehydes, ketones, phenols and phenol ethers. Enormous progress in the examination of essential oils—particularly in the isolation and identification of individual constituents—has lately been made by the introduction of modern spectroscopic and chromatographic techniques, such as ultraviolet and infrared absorption, gas and thin-layer chromatography, nuclear magnetic resonance and mass spectrometry. Since odor will always remain an important criterion no assay of an essential oil will be complete without careful organoleptic tests which, however, require considerable experience.

The bulk of essential oils are isolated from the plant material (flowers, leaves, bark, wood, and roots) by hydrodistillation, partly in primitive, movable stills—in the interior of undeveloped countries —and partly in modern stationary distilleries. Only the citrus oils, which occur in the peel of citrus fruit, are obtained by the mechanical expression of the peel. Certain types of flowers are so delicate that their essential oils do not withstand hydrodistillation and are not amenable to expression; these must be isolated by extraction with volatile solvents (usually highly refined petroleum ether) yielding the so-called natural flower oils in **concrete**, solid form, which can be transformed into **absolute**, liquid form—jasmine, tuberose, acacia, mimosa, etc. Some flowers, rose and bitter orange blossoms among them, can be processed either by hydrodistillation or by solvent extraction.

The yield of essential oil differs with individual plant species—ranging in most cases from about 0.2 to 2.0%. As examples of extremes 0.025% for rose oil (otto of rose) and 17.0% for clover oil may be noted.

Essential oils are produced in many parts of the world, particularly in warm and temperate areas.

Less than one hundred essential oils have attained real commercial importance. These are employed widely for imparting odor and flavor to an almost unlimited variety of consumers' goods, such as, pharmaceutical and dental preparations, food products, beverages—alcoholic and nonalcoholic—, confectionery, chewing gums, soaps, detergents, room sprays and insecticides, cosmetics and perfumes; and for masking of odors in synthetic products, such as, plastics, artificial leathers and rubber goods.

ERNEST GUENTHER

Cross-references: *Drying Oils, Perfumes, Flower Pigments.*

ESTERS

An ester can be regarded as a compound formed by the replacement of the acidic hydrogen of an inorganic or organic acid by an aliphatic, aromatic or heterocyclic radical. The word "ester" generally has the connotation of a substance prepared from a carboxylic acid and an alcoholic or phenolic hydroxy compound. Organic esters of this type are represented by the general formula,

$$\underset{\text{R—C—O—R}}{\overset{\displaystyle O}{\overset{\displaystyle \|}{}}}$$

The common esters are named in terms of the acids and alcohols from which they are formed. Thus, the organic ester formed from acetic acid and ethyl alcohol would be named ethyl acetate. Similarly the inorganic ester formed from nitric acid and ethyl alcohol would be designated as ethyl nitrate. The term ethyl refers to the radical contributed by the alcohol. The second portion of the name, acetate or nitrate is formed by dropping the -ic ending of the acid and then adding -ate.

Esters are water insoluble but generally soluble in organic solvents. The simple aliphatic esters are generally less dense than water. Esters exhibit a characteristic absorption in the infrared at 1750-1735 cm^{-1} which is attributed to the carbonyl (C=O) group. The low molecular weight, simple esters are colorless, low-boiling liquids having pleasant odors. Almost all fruits contain numerous esters which contribute to their flavor. Of the 150 components identified in strawberry oil, 42 of them are simple esters. Some of the naturally occurring esters which are flavoring agents are menthyl acetate in peppermint oil, menthyl benzoate in clove oil, and methyl salicylate in oil of wintergreen. With a few notable exceptions, esters of short chain acids and alcohols have no recognized biological function. Such esters are toxic to higher organisms and are removed from the system by hydrolysis.

The high molecular weight aliphatic esters are solid substances having only faint odors, if any. The higher molecular weight esters of monocarboxylic acids and monohydric alcohols or alcohols of the sterol series are known as waxes. The aliphatic alcohols found in naturally occurring waxes are found to have only an even number of carbons, ranging from 16 to 36 (See Waxes). Fats differ from waxes in that glycerol replaces the higher molecular weight alcohols and sterols; an ester having glycerol as the alcoholic substituent is known as a triglyceride. Fats and fatty oils are composed of mixtures of triglycerides (See Fats). A variety of high molecular weight esters such as the triglycerides and phospholipids are important biologically active compounds.

Esters may be prepared by various methods. The most common is the reaction of a carboxylic acid and an alcohol in the presence of an acid catalyst (hydrogen chloride, sulfuric acid, boron trifluoride, or an aromatic sulfonic acid) with the elimination of water. The oxygen appearing in the water comes from the –OH group of the acid. The most direct proof of this was obtained by esterfying an organic acid with methyl alcohol containing isotopic oxygen, O^{18}. The water formed was free of O^{18}. This would be true only if the oxygen came from the carboxylic acid:

$$\underset{\text{acid}}{\overset{\displaystyle O}{\overset{\displaystyle \|}{\text{R—C—O}^{16}\text{—H}}}} + \underset{\text{alcohol}}{\text{H—O}^{18}\text{—CH}_2} \rightleftharpoons$$

$$\underset{\text{water}}{\text{HO}^{16}\text{—H}} + \underset{\text{ester}}{\overset{\displaystyle O}{\overset{\displaystyle \|}{\text{R—C—O}^{18}\text{—CH}_3}}}$$

This is an equilibrium (reversible) reaction and in order to increase the yield of ester either an excess of the starting alcohol must be used or the water or the ester formed during the reaction must be removed. The rate of esterification is decreased by sterically hindered (bulky) acids or alcohols.

Cation ion-exchange resins can be used to catalyze esterification. The advantage of ion-exchange catalysis is that the catalyst can be removed from the reaction by filtration. Acid sensitive materials which cannot stand strong mineral acids will undergo esterification in the presence of resin catalysts.

Esters can be prepared by treating an alcohol or phenol with acylating agents such as carboxylic acid anhydrides and acyl halides. Such reactions are not reversible, therefore, excellent yields can be obtained.

Alcoholysis (transesterification) has been used as a synthetic method for preparing esters, especially in the field of fats and polyesters. The process is either acid or base (alkoxide ion) catalyzed and is reversible. Good yields are obtained by forcing the equilibrium toward the desired product by employing a large excess of the alcohol (R^3—O—H).

$$R^1—\overset{\overset{O}{\|}}{C}—O—R^2 + R^3—O—H \rightleftharpoons$$

$$R^1—\overset{\overset{O}{\|}}{C}—O—R^3 + R^2—O—H$$

Methyl esters of complex acids can be prepared in excellent yield by treatment of the acid with diazomethane, CH_2N_2. Treatment of a salt (Na^+, K^+, Ag^+) of a carboxylic acid with an alkyl halide in N,N-dimethylformamide provides good yields of the ester. Olefins treated with carboxylic acids in the presence of palladium salts afford esters.

Hydrolysis of esters can take place in the presence of either an acidic or basic catalyst:

$$\underset{\text{ester}}{R—\overset{\overset{O}{\|}}{C}—O—R'} + \underset{\text{water}}{HOH} \rightleftharpoons \underset{\text{acid}}{R—\overset{\overset{O}{\|}}{C}—OH} + \underset{\text{alcohol}}{R'OH}$$

Alkaline hydrolysis is frequently referred to as saponification, because it is the type of reaction used in the preparation of soaps from triglycerides (fats). The base-catalyzed reaction goes to completion because the acid formed during hydrolysis reacts irreversibly with the alkaline catalyst. At least one equivalent of alkali is required for each equivalent of ester which is hydrolyzed. This hydrolysis can be carried out quantitatively to determine the equivalent weight of an ester. This is known as the saponification equivalent.

Amides are obtained from the reaction of esters with ammonia and primary amines:

$$\underset{\text{ester}}{R—\overset{\overset{O}{\|}}{C}—O—R} + \underset{\text{ammonia}}{NH_3} \rightleftharpoons \underset{\text{amide}}{R—\overset{\overset{O}{\|}}{C}—NH_2} + \underset{\text{alcohol}}{R—OH}$$

The formation of hydroxamic acids from hydroxylamine and esters provides the basis for the colorimetric determination of esters. The hydroxamic acids in the presence of ferric chloride form reddish blue ferric hydroxamate complexes which can be analyzed spectrophotometrically.

Reduction of an ester function can be performed catalytically in the presence of molecular hydrogen or chemically to yield primary alcohols. A copper chromium oxide catalyst at temperatures from 250–350°C in the presence of hydrogen and generally in the absence of a solvent are the usual conditions for the hydrogenolysis of esters. This process finds use in the preparation of long-chain alcohols from glycerides for use in the preparation of certain detergents. Chemical reduction can be carried out using lithium aluminum hydride, lithium borohydride, sodium bis(2-methoxyethoxy)aluminum hydride or sodium and an alcohol.

Esters which possess weakly acidic α-hydrogens are involved in many important synthetic reactions in which C-C bonds are formed. These condensation reactions are generally base catalyzed. An example of this type of reaction can be demonstrated by the self-condensation of ethyl acetate in the presence of sodium ethoxide to yield ethyl acetoacetate.

$$CH_2\overset{\overset{O}{\|}}{C}{-}OC_2H_5 + H{-}CH_2\overset{\overset{O}{\|}}{C}{-}OC_2H_5 \underset{\longleftarrow}{\overset{NaOC_2H_5}{\longrightarrow}}$$

$$\underset{\text{ethyl acetoacetate}}{CH_3\overset{\overset{O}{\|}}{C}CH_2\overset{\overset{O}{\|}}{C}OC_2H_5 + C_2H_5OH}$$

Uses. An important general use for low molecular weight esters such as ethyl acetate and butyl acetate is as solvents for lacquers, paints, and varnish. Pharmaceutically important esters are generally found in the aromatic series. On a tonnage basis the esters which are derivatives of salicylic acid are the most important. Aspirin, acetylsalicylic acid, is the most generally used analgesic in the world today. Methyl salicylate, oil of wintergreen, is used as a flavoring agent in medicinals and in the preparation of muscle liniments. The esters of p-aminobenzoic acid find extensive use as local and topical anesthetics. Ethyl p-aminobenzoate is known as benzocaine and N,N-diethylaminoethyl p-aminobenzoate is called procaine.

Esters find an economically important use in the preparation of cosmetics, perfumes, soaps, and detergents. High molecular weight aminoesters are used as low foaming, nonionic detergents; sucrose esters of fatty acids are used as emulsifiers and biodegradable detergents.

Synthetic low molecular weight simple esters are used commercially as the base for artificial food flavors. Ethyl formate is used as an artificial essence constituent in peach, raspberry, and rum flavors; ethyl acetate in apple, pear, strawberry, and peach; amyl acetate in banana and apple; isoamyl acetate in pineapple, pear, and raspberry.

Each year millions of pounds of high molecular weight esters such as di(2-ethylhexyl) phthalate are used as plasticizers in polymers such as polystyrene and polyvinyl chloride. Other important uses for the high molecular weight esters are as hydraulic fluids, greases, and lubricants. Synthetic diesters prepared from C_6–C_{10} diacids and C_8 to C_9 branched-chain alcohols are mostly commonly used.

A widely used diester in lubricants is di(2-ethylhexyl) sebacate.

Polymerization of methyl methacrylate affords poly(methyl methacrylate) which is a strong thermoplastic solid. It is known as "Lucite" or "Plexiglas" and is used in place of glass and for molding transparent objects. The automobile industry has found that thermosetting enamels consisting of poly(methyl methacrylate) and melamine, after baking to produce cross linking, afford a finish that does not require polishing. The acrylic ester polymers are also useful in modern dental technology in the preparation of dental plates and in filling minor cavities.

Polyesters have found important uses as fibers, films, laminates and chromatographic stationary phases as well as in urethan-polymer manufacture. These polyesters include poly(ethylene terephthalate) "Dacron," poly(1,4-cyclohexanedimethylene terephthalate) "Kodel" and poly(bisphenol A carbonate) "Lexan".

MARVIN R. BOOTS and SHARON G. BOOTS

References

Noller, C. R., "Chemistry of Organic Compounds," 3rd ed., W. B. Saunders Company, Philadelphia, 1965.

Patai, S., Ed., "The Chemistry of Carboxyl Acids and Esters," Interscience Publishers, New York, New York, 1969.

ETHERS

Ethers (oxides) are compounds of the type

$$\overset{|}{-}\overset{..}{\underset{..}{\text{C}}}\overset{|}{-}\text{O}\overset{|}{-}\overset{|}{\underset{|}{\text{C}}}-$$

in which the carbons may be included in all varieties of organic radicals. Types differing from the *simple* (symmetrical or *mixed* alkyl ethers, particularly in chemical properties, are: *acetals*, $>C(OR)_2$; *ortho esters*, $-C(OR)_3$; *orthocarbonic esters*, $C(OR)_4$; and *oxygen heterocycles*, such as, furan, $O-CH=CH-CH=CH$,

1,4-dioxane, $O-CH_2CH_2-O-CH_2-CH_2$, and

ethylene oxide, $O-CH_2-CH_2$. The wide occurrence of the ether linkage is seen by mentioning a few of the natural products which contain it, e.g., vanillin, saponins, anthocyanin and flavone pigments, morphine, terpenes, sucrose, starches, cellulose, and tannins.

Ethers are frequently regarded as derivatives of alcohols, but in properties they more closely resemble the corresponding hydrocarbon analog. Thus, ethers are much less reactive than alcohols but are more reactive than the equivalent hydrocarbon.

Ethers are similar to hydrocarbons in being relatively inert (except at high temperature) to alkali metals, alkalies and other strong bases, as well as resisting hydrogenolysis. The extra reactivity of ethers lies in the unbonded electrons and electronegativity of the oxygen atom. They may be split by the exceedingly strong bases, the alkali alkyls, with the formation of an alcohol and a hydrocarbon, usually an olefin. Basic displacements at the α-carbon only proceed readily with active ethers such as 2,4-dinitrophenetole, ethylene oxide or ethyl orthoformate. Halogens attack the α-carbon slightly more easily than in hydrocarbons. Ethers autoxidize slowly in air, leading to explosive peroxides. These should be reduced, e.g., by shaking with ferrous sulfate solution, before any ether is distilled. Also strong oxidizing agents split ethers readily into at least two fragments.

Most of the reactions of ethers occur after a primary step in which the oxygen acts as a Lewis base. In fact, the salts, *oxonium* compounds, e.g., $[R_2OH]^+X^-$, of some ethers with strong acids, especially hydrogen halides and boron trifluoride, are stable enough to be isolated. Such coordination compounds are necessary in the formation of Grignard reagents, and are generally soluble complexes, e.g., $(R_2O)_2MgXR'$.

The complexes with strong acids, particularly hydriodic acid, react further, especially if warmed, to split the ether, giving alkyl halides and alcohol or water:

$$R_2O + HX \rightarrow RX + ROH$$

$$R_2O + 2HX \rightarrow 2RX + H_2O$$

The ease of cleavage of methyl ethers to give methyl iodide is the basis of the *Zeisel* determination of methoxyl groups. Cleavage also occurs with other strongly acid reagents, such as phosphorus pentachloride, which freqently leads to polychlorinated fragments. Reagents such as benzoyl chloride or acetic anhydride which do not react readily alone, are catalyzed by acidic salts, e.g., zinc chloride or aluminum bromide. The reaction of cyclic ethers with hydrogen halides leads to open chain halohydrins or dihalides. Activated ethers such as ortho esters or ketals are split so readily with mild acid that they are often used to protect sensitive groups against basic or neutral reagents, after which they allow easy recovery of the original group.

The first method of ether formation and still an important one is *dehydration of alcohols* by means of sulfuric acid, probably through intermediate ester formation. This is not useful for most higher-boiling alcohols, especially when secondary or tertiary, because of the strong tendency to dehydrate to olefins. There are many modifications of this process using catalysts such as phosphoric acid and boric acid, or high-temperature vapor-phase dehydrating agents such as alumina or thoria.

Catalytic hydration of olefins is now the major source of most widely used ethers. As the process is usually carried out, the corresponding alcohol is the product while the ether is a by-product. The conditions can, however, be altered to make ethers the major product.

$$C_nH_{2n} + H_2SO_4 \rightarrow C_nH_{2n+1}HSO_4 + (C_nH_{2n+1})_2SO_4$$

$$(C_nH_{2n+1})_2SO_4 \text{ or } C_nH_{2n+1}HSO_4 + H_2O \rightarrow$$
$$C_nH_{2n+1}OH + H_2SO_4$$

$$(C_nH_{2n+1})_2SO_4 \text{ or } C_nH_{2n+1}HSO_4 + C_nH_{2n+1}OH \rightarrow$$
$$(C_nH_{2n+1})_2O + H_2SO_4$$

Although *simple* ethers are prepared in this way, the reaction occurs even more readily in the addition of alcohols or phenols to substituted olefins, leading to *mixed* tertiary ethers. In this manner many cheap industrial solvents of varying properties can be made.

The most versatile and definitive method for the preparation of ethers is the Williamson synthesis. Organic halides, sulfates or sulfonates are treated with sodium alcoholates or phenolates, leading directly to the ethers:

$$RX + NaOR' \rightarrow ROR' + NaX$$

This is the most common preparation of phenolic ethers, for example, the weed-killer, 2,4-D:

$$2,4\text{-}Cl_2C_6H_3ONa + NaOH + ClCH_2COOH \rightarrow$$
$$2,4\text{-}Cl_2C_6H_3OCH_2COONa + H_2O + NaCl$$

Activated aryl halides, such as 2,4-dinitrochlorobenzenes, may be used as the halide and even the relatively inactive chlorobenzene, under forcing conditions, can be used to give such products as diphenylether, b.p. 259°C, a valuable heat-transfer medium ("Dowtherm").

By far the greatest use of ethers as a class is as solvents, either as reaction media, extractants, or as plasticizers or vehicles for other products. For these purposes, cost rather than purity is generally the prime requisite, and the lower alkyl ethers far outbulk the higher. A broad spectrum of variations in solvent properties, solubility in water, etc., is obtained by changing alkyl groups, by use of chlorinated ethers, with hydroxyethers such as polyethylene glycols and their ethers ("Cellosolves" and "Carbitols"). Polyethers of polyalcohols give compounds of different properties, such as gums and plastics (polyoxymethylenes, polyvinylethers, cellulose ethers) and suspending and wetting agents. Ethers are also frequently used as protecting devices while carrying out reactions in other parts of a molecule. Ethers are also of importance in medicine and pharmacology, the major use being for anesthesia, but also because the activity of a drug may be modified in a useful way; codeine, for example, is the methyl ether of morphine. Some of the industrially important ethers are given below:

Ethyl Ether (ether) ($C_2H_5OC_2H_5$) is a highly volatile and flammable liquid, b.p. 34.6°C, flash point (closed) −40°C, prepared by the sulfuric acid process from ethanol, or by catalytic hydration of ethylene, large amounts being available as a by-product of ethanol production. It is the most important ether. United States production being about 100,000,000 kg/yr. Although it is not completely immiscible with water (it may contain 1.2% and dissolves to 6.8% in water), its greatest use is as a solvent for oils, fats, gums, resins, perfumes and essential oils, nitrocellulose, pharmaceuticals and botanicals, etc. It is important for extracting organic acids, especially acetic acid, from the aqueous solutions usually produced in their manufacture or use. As a reaction medium for fine organic chemicals, it is particularly useful for Grignard and other organometallic reagents where an ether is necessary for a reaction to proceed. Ethyl ether is sometimes used as a source of ethylene by dehydration, in areas where petroleum gases are not obtainable. The best known use is as a

general anesthetic. Some related ethers, e.g., vinyl ether and ethyl vinyl ether are also used for this purpose. Precautions should be taken against the presence of peroxides.

Methyl Ether (CH_3OCH_3), b.p. −24°C., shipped as a compressed gas, mainly used as a spray propellant and as a refrigerant which can freeze foods directly; *Isopropyl Ether* [(CH_3)$_2$CHOCH(CH_3)$_2$], b.p. 68°C., by-product from isopropyl alcohol production, mainly a cheap solvent and extractant, as is *n-Butyl Ether* [$CH_3(CH_2)_3O(CH_2)_3CH_3$], b.p. 142°C; *Bis(2-Chloroethyl) Ether* [Cl(CH_2)$_2$O(CH_2)$_2$Cl], b.p. 177° C., not a vesicant although mustard gas is its sulfur analog, used as a chemical intermediate and as a solvent combining the properties of ethers and chlorinated hydrocarbons, e.g. for degreasing.

Ethylene Oxide, b.p. 13–14°C., (ca. 2 billion kg/yr); and *Propylene Oxide* (CH_3CHCH_2—O)

b.p. 34°C. (500 million kg/yr) are colorless, highly volatile and flammable and have by far the greatest bulk production of any ether. Their heavy use is due to their being unlike other ethers (unreactive solvents) in having a strained 3-membered oxide ring which reacts readily with amines and alcohols to form hydroxy alkyl and polyoxyalkyl derivatives. Their use as fumigants is also based on their reactivity to kill living organisms.

Complex ethers made from the oxides include many commercially important products, generally of high water miscibility due to the large oxygen content. These are used as solvents, fabric lubricants and conditioners, binders, plasticizers, humectants, antifreeze and as intermediates (e.g. by esterification with fatty acids) to make nonionic detergents. Some are: *Diethylene Glycol* [(HOC$_2$H$_4$)$_2$O] b.p. 245°C., f.p.−8°, sol. water, insol. hydrocarbons (100 million kg/yr); *Dipropylene Glycol* [(CH$_3$CH(OH)CH$_2$)$_2$O] b.p. 232°C., sol. water and toluene (20 million kg/yr); and the corresponding polymers R(CH$_2$CHR'—O—)$_x$H, *Polyethylene Glycol* H(OCH$_2$CH$_2$)$_n$OH (20 million kg/yr) and *Polypropylene Glycol* HO(C$_3$H$_6$O)$_n$H 90 million kg/yr), both being nonvolatile, with M.W. ranging from 200 to 6000, changing from syrups to waxy solids. A wide variety of related products is available as R can be many different mono- and polyalcohols, amines, phenols and acids. A major example, with properties similar to the polyglycols, is the triol polymer, *glycerol tri(polyoxypropylene) ether* (R≡OH, R′= —OCH$_2$CH(O—)CH$_2$O—) (100 million kg/yr). An important nonionic detergent is *nonylphenol ethyl ether* (C$_9$H$_{19}$C$_6$H$_4$OC$_2$H$_5$) (60 million kg/yr). Ethers are also important in plastics, adhesives and coatings, epoxy resins being used to the extent of 80 million kg/yr.

DAVID H. GOULD

Cross-references: *Oxides, Solvent, Glycols, Cellulose.*

References

Patai, Saul, Ed., "The Chemistry of the Ether Linkage," Interscience Publishers, New York, 1967.

Gaylord, Norman G., Ed., "Polyethers," (3 vol.), Interscience Publishers, New York, 1962.

EXPLOSIVES

Explosives are substances capable of exerting, by their characteristic high-velocity reactions, sudden high pressures usually generating loud noise and more or less destructive action on surroundings. Explosive classification is based on the types of reactions that produce explosions, namely mechanical, chemical, and nuclear. This article is concerned only with the chemical explosives. They comprise two main types, the "low" or "deflagrating" (sometimes also called "propellant") explosives, and the "high" or "detonating" explosives. The latter are further classified as "primary" and "secondary" detonating explosives.

Pre-Explosion Reactions. Chemical explosives are based on exothermic (heat-liberating) reactions of the type that increase in rate (exponentially) with temperature. At ambient temperatures chemical explosives undergo negligible or no reaction. Indeed the reaction rate must not be appreciable if the explosive is to be considered safe for preparation and storage at temperatures up to 80–100°C, although many show evidence of slow decomposition at the upper limit of this range. At higher temperatures the characteristic reaction of the explosive takes place at rapidly increasing rates, and will transform suddenly into an explosion reaction at the "explosion temperature." Usually self-heating occurs at temperatures considerably below the "explosion temperature." Its extent depends on the heat balance ratio (fraction of the total heat of reaction retained in the explosive).

Reaction rates increase with temperature much more rapidly than heat dissipation to the surroundings (radiative, conductive and convective). While the "explosion temperature" may seem to be a sharply and distinctly defined value in any particular test designed to measure it, this is seldom, if ever the case; it is really a rather ill-defined property which depends critically on the many factors determining the heat balance ratio (confinement, size of charge, surrounding medium, rate of reaction, heat of reaction, and numerous other factors).

Sources of initiation include direct heating, flame, electrical discharge, impact or shock, contamination with incompatible chemicals which promote spontaneous decomposition, instability of the explosive itself, etc. Experimental evidence shows that all these sources act by producing sufficiently high temperature to initiate self-heating. Even those explosives which exhibit maximum shock or impact sensitivity, for example, are probably never exploded directly by shock or impact; shock of sufficient intensity to cause explosion merely generates hot spots or high-temperature regions which initiate self-heating, and the explosion results eventually from the latter.

While the pre-explosion process is characterized by a thermal chain branching or self-accelerating reaction, the explosion itself does not continue to accelerate indefinitely but quickly reaches a maximum rate, which may maintain until the explosive is entirely consumed, as in the detonating explosives, or the reaction may even decrease after a maximum rate is attained, as in some propellants. The conditions determining the limiting rate of reaction in explosion are the physical state of the explosive, the magnitude of exothermicity (or heat) of the reaction, and (in the low explosives) the heat balance ratio. The explosion proper is actually a very high-velocity reaction requiring total reaction times of the order of only milliseconds in the "low" explosives and microseconds in the high explosives. While reported "explosion temperatures" for useful explosives are in the range between about 150 and 350°C, no explosion ever occurs at these low temperatures, but requires a minimum temperature of perhaps more than 1000°C. However, this temperature range is bridged so suddenly that the temperature transient is seldom even observed by conventional experimental methods.

Explosive Deflagration. "Low" explosives are characterized by a reaction rate which increases nearly in direct proportion to the pressure (as a result of the influence of pressure on surface temperature), but always remains one or two orders of magnitude lower than in the detonating type. The limiting rate of reaction and pressure in propellant or granular "low" explosives is determined by the effective burning surface and the upper limit of surface temperature. The pressure-time (or p-t) curve of a propellant tends to exhibit a maximum usually below about 50,000 psi. By careful control of the geometry and size of the propellant grains one may control to a large extent the nature of the p-t curve associated with the explosion of a propellant. Indeed, this is one of the most important problems in propellant technology.

The relatively low rates of pressure development and peak pressures of the "low" explosives, in addition to rendering them useful in guns and rockets as propellants, give them a desirable "blasting action" for coal mining and other commercial blasting operations where "heaving action" (or sustained pressure) and controlled fragmentation are desired. Black powder (an intimate mixture of sodium or potassium nitrate, charcoal and sulfur) for centuries the sole source of explosive powder in both the military and commercial fields, has remained in use as a blasting explosive only because of its superior action for this type of blasting. Unfortunately, it is a very dangerous explosive owing to its extreme sensitivity to primary ignition sources, particularly spark and flame. Moreover, because of its relatively long reaction time, it is a very hazardous explosive for use in an explosive dust or gas environment, e.g., in dusty or gassy coal mines. Its chief present use in the miltary—as the igniter or fuse element—takes advantage of its ease of ignition and hot flame of relatively long duration.

The smokeless powder propellants, which represent the only other "low" explosives in extensive use today, are indispensable in the military field, but owing to their cost they have little commercial value. However, government-surplus smokeless powder is now being used extensively as a sensitizer for "slurry" explosives used in commercial blasting. Smokeless powders are the basis for most modern artillery, small arms, and rocket ammunition. Three general types are available, the "single base" powders, in which nitrocotton is the basic ingredient, the "double base" powders containing primarily nitrocotton and nitroglycerine, and the "triple base" powders made principally with nitroguanidine, nitroglycerine and nitrocellulose.

Detonation. Detonation is a process in which the (effective part of the) explosive reaction takes place within a high-velocity shock wave known as the "detonation wave" or "reaction shock." This wave generally propagates at a constant velocity from below 800 to above 8500 meters/sec, depending on the explosive, its density and physical state. Detonation velocity may be measured accurately by direct flame photography using "streak" cameras of various types and by electronic chronographs. Since the pressures generated by detonation may be controlled over a broad range as low as the maximum pressures of "low" explosives to about 100 times greater, the detonating explosives are more readily adaptable to most military and commercial demolition and blasting purposes than the low explosives. Even in operations where the low explosives exhibit better blasting action, they are less desirable than the detonating explosives because of their tendency to initiate explosions in explosive gas or dust-air mixtures. The commercial industry is today therefore based on the detonating type.

Owing to their great tendency to transform very suddenly into detonation from a very short duration pre-explosion reaction, the primary explosives are the basis of blasting caps and fuzes (military detonators) which are used to create the detonation wave in the less sensitive secondary explosives. Some of the most important primary explosives are mercury fulminate, lead azide, diazodinitrophenol, nitromannite, and lead styphnate.

In the applications of the secondary explosive one always requires the use of a detonator and in some cases also a booster in order that detonation may be produced easily and reliably. A booster is usually a small, pressed, cast, or low-density (loose-packed) charge of one of the more sensitive secondary explosives capable of being detonated readily by a detonator. It is used to reinforce the detonation wave from the detonator and thus deliver to the secondary explosive a more powerful detonation wave than is possible with a detonator alone. This makes possible the use of many relatively insensitive explosives that otherwise would have no direct application.

Even the least sensitive secondary explosive may explode in essentially the same type process involved in forming the detonation wave in primary explosives, except that the transition time and the amount of explosive required to go from the pre-explosion reaction to detonation may be very much larger in the secondary than the primary type. In some cases tons or even thousands of tons of the explosive are required for explosion to occur spontaneously following initiation. This is strikingly illustrated by the Texas City catastrophe and that of Brest, France, which involved one of the least sensitive of all secondary explosives, namely (fertilizer) ammonium nitrate, which in both instances ignited spontaneously, burned for some time, and then exploded.

Secondary explosives develop detonation pressures from a minimum of about 2,500 to a maximum of about 350,000 atmospheres. Even in a single explosive the detonation pressure may be varied over a wide range simply by varying the apparent density. The lowest detonation pressures are obtained by combining the effects of low density and dilution either by inert additives or incomplete chemical reaction accomplished by particle size control. Both means are used in the design of coal mining explosives ("permissibles") to produce lump coal where the lowest detonation pressures are required. The detonation temperatures of secondary explosives may range from as low as about 1500°C to 5,500°C or higher depending on the nature of the explosive.

Some of the most important military and commercial pure explosive compounds are RDX (cyclotrimethylenetrinitramine; HMX (cyclotetramethylenetetranitramine); PETN (pentaerythritol tetranitrate); NG (nitroglycerine); tetryl (trinitrophenylmethylnitramine); TNT (trinitrotoluene); AN (ammonium nitrate), picric acid and ammonium picrate. While pure AN can be detonated with heavy boostering only in large diameters (e.g., 5″ or greater) and/or under heavy confinement, it is an important source of explosive power in the commercial field because of its low cost. It is used, however, only in mixtures with combustibles and/or other explosives in which its strength is greatly enhanced. A suitable mixture of AN and combustible such as wax or wood pulp has approximately the same explosive potential as cast or pressed TNT. While RDX, HMX, and PETN are among the most powerful chemical explosive compounds known, they are never used in the pure state owing to their excessively high sensitivity, but instead are used in conjunction with other less sensitive explosives or nonexplosive ingredients in desensitized or phlegmatized condition.

In the past liquid NG, comparable in strength to RDX and PETN, was used extensively for oil well shooting, despite its exceedingly high sensitivity, but its use incurred many devastating accidental explosions. Today NG has been largely replaced in oil well shooting by far less sensitive types, e.g., non-NG high-AN explosives or desensitized liquid mixtures of NG, dinitrotoluene, and/or mononitrotoluene. However, oil well shooting itself has been largely replaced by other methods of sand fracturing.

Slurry explosives are replacing the older types of oil well explosives. The coarse TNT type slurry explosives are especially effective under very high hydrostatic pressures, but still better, more powerful slurry explosives capable of functioning at much higher temperatures are coming into use for deep, hot oil well blasting.

NG is also the basic ingredient of dynamites, and in rocket propellants. Dynamites are of two general classes, the "straight" dynamites containing "balanced dopes" consisting of mixtures of sodium nitrate and combustibles (wood pulps, meals, starch, sulfur, etc.), and the "ammonia" dynamites consisting of mixtures of ammonium nitrate and combustibles. These mixtures are carefully "oxygen balanced" (i.e., adjusted in composition to utilize all the available oxygen of the nitrate) for fume control and optimum strength. Calcium carbonate and other "antacids" are used in dynamites to prevent acids from accumulating during storage; dynamites are rendered unstable and dangerous in the presence of acid.

Numerous explosives based on AN (used either as the only explosive ingredient or sensitized by the use of a small amount of another explosive, e.g., TNT) have been used for commercial blasting, notable among which are the duPont "Nitramon" blasting agents. However, today these have been largely replaced by slurry explosives.

A decade ago two new types of commercial "blasting agents" came into extensive commercial use: "ANFO" and "slurry." ANFO, or "prills and oil," is a 94/6 mixture of porous prilled ammonium nitrate and fuel oil of average density about 0.85 g/cc. Owing to low cost and high weight strength, ANFO has moved ahead rapidly since its discovery in 1955 and today comprises about half of the commercial explosives industry. "Slurry Blasting Agents" or SBA are based on thickened or gelatinized aqueous ammonium nitrate slurries sensitized with TNT or other usually coarse solid high explosives, smokeless powder or aluminum. SBA, discovered by Cook and Farnam in 1956, is much higher in density and velocity than ANFO and is water resistant. SBA is intermediate in cost between dynamites and ANFO, but because of its superior properties is today gaining rapidly in commercial blasting even replacing ANFO in large open pit mines where the latter had been used very successfully. In hard rock blasting, the economic advantages of SBA are unsurpassed. Moreover, the aluminized slurry blasting agents are the safest, most powerful and economic modern explosives available, especially when handled by new field-mixing and loading system known as the "pump truck system."

The second most significant development in commercial explosives since the invention of slurry explosives by Cook and Farnam over a decade ago was the on-site mixing and loading system called the "pump truck," also a development of Cook and associates. First introduced in 1963, the "pump truck" has revolutionized open pit commercial blasting. Today commercial explosives are being bulk handled with an efficiency not even visualized a decade ago.

While slurry explosives first came into use in large diameter applications, they have since been perfected to the point that they may replace dynamites not only in large but also in small diameter, underground applications. Some slurry types are cap sensitive, i.e., able to be detonated with a blasting cap alone, without the use of the high pressure "booster" generally required for explosives of minimal sensitivity, the basic characteristic of slurry explosives.

Composition B (60/40/1 RDX-TNT-wax), cast from the slurry produced by melting the TNT, develops a detonation pressure of about 240,000 atm. Other secondary explosives of still higher detonation pressures are now in use. For example, the military explosive "PBX 9404" has a measured detonation pressure of about 350,000 atm. Another very useful military explosive capable of developing detonation pressures in the same range is cast 50/50 pentolite (PETN-TNT). These explosives are most useful where high "brisance" is needed, e.g., in military demolition operations and shaped charges. "Brisance" is identified with detonation pressure; before the development of the thermo-hydrodynamic theory it was an empirical somewhat ill-defined quantity interpreted as "shat-

tering action." The aluminized military explosives, while not highly "brisant," are characterized by high available explosive energy; they develop only moderately high but sustained pressures. Explosives of this type, of which Torpex (RDX, TNT and aluminum) is a notable example, are among the most powerful chemical explosives available.

A powerful slurry explosive known as DBA-22M was approved for military purposes and used extensively for clearing helicopter landing zones in the jungles in the spring of 1970 in American operations in Cambodia. At 15,000 pounds each, these bombs were the largest chemical bombs yet employed in field applications. However, several 40,000 pound bombs of DBA-22M were very effectively used in early 1969 in the Air Force's "cloud-maker" program for studying upper atmosphere air movements.

M. A. Cook

EXTINGUISHING AGENTS

Water. Because of its high specific heat, water is the most effective extinguishing agent on fires involving ordinary combustible materials. Water exerts its maximum cooling effect in fire extinguishment where it is applied cold and evaporated into steam. One gallon (8.336 lbs) of water applied at 50°F and evaporated into steam at 212°F has a cooling effect of 1350.4 Btu + 8088.4 Btu = 9438.8 Btu. If it is applied at 60°F and runs off after being heated to 80°F it has a cooling effect of 167 Btu. The superiority of water spray over solid streams in certain types of fires may in part be explained by the fact that the spray is largely evaporated to steam, thus exerting the maximum cooling effect, whereas water from the solid steam is merely heated, and only a small part evaporated.

Water spray extinguishes fires in kerosene, fuel oil, linseed oil, lubricating oil and other heavy viscous liquids. (Water applied to liquids heated to over 250°F—for example, quenching tanks in heat treating—may cause the liquid to foam over and spread fire.) Fires in gasoline, and similar light liquids are not readily extinguished by water spray, although the spray absorbs heat and tends to prevent fire from spreading to other combustibles. Fires in acetone, alcohol and similar water-miscible liquids may be extinguished if water is used to dilute the surface layer sufficiently.

Carbon Dioxide. Carbon dioxide fire extinguishing systems are effective primarily because they reduce the oxygen content of the air to a point where it will no longer support combustion. Under suitable conditions of control and application, a cooling effect is also realized.

A reduction of the oxygen content of the air from the normal 21 to 15% will extinguish most fires in spaces which do not include materials that produce glowing embers or smoldering fire. The amount of oxygen necessary to support combustion varies with different materials. Reduction to considerably less than 15% is required in some cases. To extinguish fires completely in areas or spaces containing materials which will produce glowing embers or smoldering fires, or fires involving highly heated metal containers, it is necessary to reduce the oxygen content to 6% or less and to maintain that dilution for more than the normal brief period. Otherwise, a reflash may occur. In the case of deep-seated fires, such as in cotton or paper, wherein the material acts as a thermal insulator, it may be necessary to maintain the dilution for hours to achieve cooling below the ignition point.

Dry chemical extinguishing agents are mixtures of dry chemicals in powder form expelled from extinguishers or extinguishing system nozzles by carbon dioxide, nitrogen or dry air. There are two different dry chemical extinguishing agents: one, composed primarily of sodium bicarbonate (or potassium bicarbonate), is suitable for fires in flammable liquids or electrical equipment; the other consists primarily of monoammonium phosphate and is a satisfactory extinguishing agent for fires in ordinary combustibles, flammable liquids, and electrical equipment. In both cases, materials are added to the extinguishing agent to produce water-repellency and free-flowing characteristics. The extinguishing action of the bicarbonate-base dry chemical is believed to be due primarily to a chemical reaction of the extinguishing agent with one of the intermediate products of combustion. The extinguishing effectiveness of the monoammonium phosphate-base dry chemical on flammable liquid fires is also thought to be due to a chain-interruption reaction in the combustion zone. One important factor in its effectiveness on fires involving ordinary combustibles is its tendency to form a coating on surfaces and thereby retard combustion.

Foam. Foam is employed primarily for extinguishment of flammable-liquid fires. It extinguishes by blanketing the liquid surface, obstructing access of the flammable vapors to the air, insulating the flammable liquid from the heat of the fire, and cooling the surface.

Fire-fighting foam consists of a mass of fine bubbles which must be relatively heat-resistant and long-lasting to be effective. It must be delivered to the fire faster than it breaks down under fire exposure. Poor quality foam may break down so rapidly that it may not extinguish a fire. The ability of foam to prevent reignition of flammables blanketed by it is an outstanding advantage of foam protection.

Chemical foam is produced by a chemical reaction which generates carbon dioxide bubbles in water solution containing a foaming ingredient; mechanical or "air" foam is produced by mixing and agitation of air with water containing foaming ingredients. Both are effective; the content of the bubbles, whether inert gas or air, is not a significant factor in extinguishment. Ordinary foam, whether chemical or mechanical (air) foam, is effective on hydrocarbons which are liquid at ordinary temperatures and pressures, but special foams are usually required for the more common alcohols, ethers, aldehydes, ketones and the more volatile esters, to resist their solvent action, which tends to break down ordinary foam.

C. S. Babcock

The use of halogenated hydrocarbons as extinguishing agents, such as carbon tetrachloride, methyl bromide, and chlorobromethane, is inadvisable because of their high toxicity. [Editor]

F

FATS

Natural fats comprise a group of lipid materials composed of glyceride esters of fatty acids and associated phosphatides, sterols, simple alcohols, hydrocarbons, ketones and related compounds which are obtained from plants and animals by pressing, cooking with steam or water, extraction with organic solvents or combinations of these processes.

The term "fat" is now generally applied technically to include fatty *oils*, since the two words refer merely to different physical states of the same substance. Mere change in temperature of the appropriate degree and in the appropriate direction will change a liquid *oil* into a solid *fat*, or a solid *fat* to a liquid *oil*. The two terms are frequently used interchangeably.

When freed of all associated lipids (a process exceedingly difficult if not impossible on a technical scale), fats consist of mixtures of glyceride esters of various fatty acids. The great variety of natural fats results from the variation in the number and proportion of different glycerides which may be mixed to form any given fat. Further variation results from the number and nature of the fatty acid residues (radicals) which may be attached to the glycerol skeleton to form the individual glycerides. The simplest fat which can be conceived would consist of a single glyceride ester derived from but one fatty acid and would be represented by the formula

$$
\begin{array}{l}
\text{H} \\
\text{HC—OR} \\
| \\
\text{HC—OR} \\
| \\
\text{HC—OR} \\
\text{H}
\end{array}
$$

where R would be an acid radical such as palmitoyl, $CH_3(CH_2)_{14}CO—$. No such fat is known in nature and only a very few in which the mixture of glyceride esters is composed predominantly of such a *homo*-glyceride. Most fats consist of mixtures of *hetero*-glycerides, i.e., of glycerides containing two or three different fatty acid residues.

The fatty acid radicals may be saturated or unsaturated; if unsaturated they may contain one to six double bonds. Two or more bonds may be in either isolated or conjugated positions. The carbon chain may vary in length from that present in butyric acid to 24 or more carbon atoms, and it may contain side groups such as hydroxy, keto, methyl, etc. Any given acid group may be attached to any one of the three carbon atoms in the glycerol molecule. It is thus apparent that fats having very different compositions and properties can, and actually do, exist in nature.

Sources. Since fats are found in almost all living organisms they may be obtained in some quantity from any source, but practically these are limited to a few plants and even fewer animals. The world production of fats in 1969 was estimated to be 41,420,000 short tons divided into 5 categories as follows: (all figures below represent 1,000 short tons), edible vegetable oils 19,992, palm oils 4,387, industrial oils 1,591, animal fats 14,070, and marine oils 1,380. The edible oils in decreasing order of magnitude were soybean 5,940, sunflowerseed 3,980, peanut 2,955, cottonseed 2,730, rapeseed 1,840, olive 1,487, sesame 600, corn 275, safflowerseed 185. The principal palm oils comprised coconut 2,242, palm pericarp 1,650, palm kernel 430, and babassu 60. Industrial oils were linseed 1,035, castor 405, tung 144, and oiticica 7. Animal fats were butter (fat content) 5,100, tallow and grease 4,645, and lard 4,325. Marine oils were fish (including liver oils) 1,125, sperm whale 165, and other whale oils 90. Source: U.S. Department, Foreign Agricultural Service, Statistical Report January 1970, p. 10. Minor amounts of vegetable fats are obtained from germs of cereals other than corn, cacao beans (cocoabutter), a wide variety of palm nuts, and edible nuts such as walnuts, pecans, almonds, Brazil nuts, etc., the seeds of the poppy, hemp, grape, tea, grapefruit, kapok, parsley, tomato, rice bran, etc.

Lard and white grease are obtained by rendering various parts of the hog; edible and inedible tallows are obtained principally from the cow, but also from the sheep; butter is produced by churning cow's milk and in some countries the milk of the water buffalo, goat, sheep and other animals; fish oils are derived principally from the sardine, pilchard, menhaden, and herring; fish-liver oils from the cod, halibut, tuna, dogfish and several species of sharks.

Classification. Fats may be classified in a number of ways, none of which is entirely satisfactory or rigid. One classification is based on the end uses of these products, which are broadly edible, inedible or industrial, cosmetic, and pharmaceutical. The inedible class may be further subdivided in soap and detergents, drying oil, and miscellaneous industrial. Fats are also classified on the basis of composition or some readily determinable property, such as iodine value, which reflects their composition. This classification is somewhat more satisfactory than by uses.

Physical and Chemical Characteristics. Many physical and chemical properties of fats have been investigated and a number of these have obtained importance for identifying individual products and in some cases have been incorporated in trading rules and contracts to insure delivery of acceptable materials. Others are useful in controlling quality or in the development of new products or processes.

Chemical Reactions. The oldest and best known reaction of fats is that of hydrolysis to form free fatty acids and glycerol and the reverse reaction of esterification. Analogous reactions include alcoholysis, in which the glycerol of the fat is exchanged for another alcohol; also acidolysis, in which one fatty acid radical is exchanged for another; and ester-ester interchange wherein the glycerides react to exchange their radicals to form new esters. Most of these reactions are carried out at elevated temperatures and in the presence of a catalyst. Monoesters, formed by some of the above reactions or by esterification of the free fatty acids, undergo condensation to form β-keto-esters, or can be reduced with sodium to α-hydroxyketones or alcohols. Many of these reactions are applied on a large scale to form soaps, fatty acids, monoesters, alcohols, etc.

Typical of a wide variety of reactions is the ability of fats to add various atoms and radicals to their olefinic bonds. These additions include hydrogen, halogens, thiocyanogen, hydrogen halides, sulfuric acid, formic acid, maleic anhydride, etc. The addition of halogens, hydrogen halides and thiocyanogen is used to measure the degree of unsaturation of fats. The reaction with maleic anhydride is applied for the same purpose to tung and other fats containing conjugated double bonds. Strong sulfuric acid reacts with the olefinic linkages to form sulfated oils, a reaction of particular importance in the case of castor oil.

The reaction of oxygen is of special economic importance. Reaction with atmospheric oxygen is detrimental in the case of edible fats, producing rancidity and other deleterious effects. With drying oils the same reaction is beneficial, because it is involved in polymerization and film formation. Other oxidizing agents such as potassium permanganate, hydrogen peroxide, ozone, peracetic acid, etc. are used in investigating the structure of fats and have some limited industrial applications.

Fats undergo isomerization, polymerization, and copolymerization. Polymerization is usually carried out at elevated temperatures and is oxygen-induced. It is the principal reaction occurring in preparing heat-bodied and blown oils.

The most characteristic and perhaps most useful reaction of fats is the addition of hydrogen to their olefinic linkages. It is carried out at elevated temperatures in the presence of catalyst. Millions of pounds of liquid oils are hydrogenated or hardened annually for the production of shortening, margarine oils, confectioners fats, soap fats, and for other purposes.

Physiology. Of the three primary food constituents fats represent the most concentrated form of energy, namely, about 9 large calories per gram compared to about 4 each for carbohydrates and proteins. Certain unsaturated acids which are constituents of many fats appear to be essential to the maintenance of good health, especially in infants and young children. Since they apparently cannot be synthesized by the body from carbohydrates or other fatty acids they must be ingested. Two of these acids appear to be linoleic and arachidonic, and possibly γ-linolenic. Claims have been made that other fatty acids possess growth-promoting or other physiologic activity but subsequent investigation has usually failed to confirm these claims.

Several vitamins (A, D, E, and K) are soluble in fats rather than water. Some of these vitamins are found in natural fats; others are conveyed in a fat medium when used therapeutically or when used to fortify food products. Some fats are nearly devoid of one or all of these vitamins, or they may be partially or wholly removed during processing. Such fats as well as other food products can be fortified to any desired degree. Much of the margarine is now fortified with vitamin A and D at levels in excess of those occurring naturally in the best summer butter.

Methods of Recovery. Methods of recovering fats vary with the nature of the raw materials and the state of the development of technology in a given country. Animal fats are recovered by relatively simple processes, which consist of cooking dry or with steam or water various fatty tissues trimmed from other portions of cattle, sheep and hogs and the whale. Fish oils are obtained by wet-rendering the whole animals, often continuously using centrifuges to separate the aqueous and oil phases. Recovery of additional oil from the rendered tissues is usually accomplished by pressing them in open or cage hydraulic presses or continuous screw presses; sometimes they are extracted with solvents.

Fruit pulps (olive, avocado and palm) are each recovered by processes especially adapted to the nature of the fruit. Olives are first crushed in edge-runner or other types of mills and then pressed in hydraulic or screw presses. The residual solids (marc) are often solvent-extracted to recover additional oil. Bunches of palm fruits are first sterilized to inactivate the fat-splitting enzymes, the fruits are stripped mechanically from the bunch, and digested with steam in open kettles. The digested fruits are then either pressed or centrifuged to separate the oil and water phases.

The fat from oilseeds and nuts are recovered by mechanical pressing or solvent extraction or a combination of these processes, but many of them must first be subjected to a certain amount of pretreatment. Nuts generally have to be cracked and the kernels separated from the shells. Cottonseed must be delinted and hulled. Some seeds like sunflower, and sometimes castor and soybeans are hulled because of the abrasive action of the hulls on presses. Other seeds like sesame, linseed, and sometimes soybeans are simply cleaned, cracked, or ground, cooked and pressed or extracted with solvent. In general, all kernels and seeds are ground and cooked to some degree prior to pressing.

Processing. Most fats used in the United States are subject to some form of processing to remove nonfat materials such as free fatty acids, pigments, odorous constituents, and extraneous solids. This is especially true of fats derived from plant sources. The degree of processing depends on the nature and quantities of nonfat constituents present in the crude fat, and the end use of the product.

Practices in different parts of the world are not uniform. In some countries fats and oils may be consumed even as food with little or no processing. In parts of Africa and Asia many people, through long conditioning and acquired taste prefer the natural flavor of oils like soybean, peanut, sesame and palm to highly refined and deodorized products.

Liquid edible fats may be processed further to yield hardened fats with a consistency similar to lard or margarine by the controlled addition of hydrogen to the double bonds of the unsaturated portion of the glycerides. They may be hardened also to produce soap fats or special products for the confectionery trade. The process termed hydrogenation is carried out in the presence of a catalyst and under slight pressure and at elevated temperature.

Castor oil may be dehydrated to produce a drying oil with properties intermediate between tung and linseed; hydrogenated for use in making metallic soaps for lubricants; sulfonated for use as a wetting agent; or cracked to produce dibasic acids, especially sebacic acid.

Many types of fats are saponified with alkalies or with steam or water under pressure with an organic catalyst to produce soaps or fatty acids. The resulting fatty acids may be distilled to remove impurities or fractionated to separate individual acids. Glycerol is usually recovered as a byproduct of the saponification.

KLARE S. MARKLEY

Cross-references: *Fatty Acids, Esters, Foods, Soaps, Hydrogenation, Fat Splitting.*

FATTY ACIDS

The term "fatty acids" was formerly restricted to a small number of saturated and unsaturated acids produced by the hydrolysis of glyceride oils and fats and natural waxes. Present usage has broadened the term to include the entire homologous series of normal aliphatic acids, homologs and isomers of the unsaturated acids, and various substituted acids which have been isolated from natural sources and/ or have been synthesized in the laboratory.

Fatty acids may, therefore, be considered to include all of the odd- and even-numbered aliphatic acids from acetic to at least *n*-octatriacontanoic, $C_{37}H_{75}COOH$. The general formula for the saturated fatty acids is $C_nH_{2n}O_2$ where *n* can be any even or odd integer. Since all members of this series except formic consist of an alkyl chain and a terminal carboxyl group, they may be represented by the formula RCOOH, and all members above acetic by $CH_3(CH_2)_nCOOH$. There are several systems of naming the saturated fatty acids of which the modified Geneva Nomenclature used by Chemical Abstracts is perhaps the best known to American chemists. In this system, these acids are considered as being derived from hydrocarbons having the same number of carbon atoms (CH_3 being replaced by COOH). The final letter of the hydrocarbon is replaced with "oic" to designate the corresponding acid; thus caproic acid is *n*-hexanoic. Many of the saturated acids also have well established common names which are used as frequently or more so than the scientific name.

No sharp dividing line can be drawn with regard to the occurrence of these acids in nature, but in general the lowest members are found in the fruit esters and essential oils, the intermediate members in the glyceride oils, and those of the highest molecular weight as constituents of natural waxes. However, butyric acid occurs in animal milk fats, and coconut oil contains all of the even-numbered acids from C_6 to C_{18}. Almost all the even-numbered acids from acetic to C_{18} or higher are produced by fermentation of sugars through the action of bacteria, yeasts and fungi.

The branched-chain fatty acids are isomeric with normal fatty acids, and contain one or more branching alkyl groups. Three of the better known of these acids are isomeric with valeric acid ($C_5H_{10}O_2$). They are α-methylbutyric; isovaleric or 3-methylbutanoic; and pivalic or trimethylacetic. The first two occur in the oil of valerian root. During the 1940's 2-ethylhexanoic [$CH_3(CH_2)_3CH(C_2H_5)COOH$] became industrially important because its metallic salts, called octoates, found considerable application as driers, stabilizers for polyvinyl resins, thickening agents in paints and lacquers, mineral oil sludge inhibitors, etc. Higher molecular weight branched-chain acids were unknown until 1929, when they were first isolated from the lipides of acid-fast bacteria. Since then many such acids have been isolated from similar sources, and from wool fat, and petroleum.

A considerable number of mono- and polyethenoic unsaturated fatty acids varying in chain length from C_4 to C_{30} and having one to six double bonds have been found as constituents of natural fats, as have a number of triply bonded or ethynoic acids. The general formula for the monoethenoic acids is C_nH_{2n-2}, for the diethenoic acids $C_nH_{2n-4}O_2$, and for the triethenoic $C_nH_{2n-6}O_2$. The scientific names are derived from the corresponding hydrocarbons with a numerical prefix to designate the position of the double bond(s). Since the double bonds can be in any position along the carbon chain it is evident that many positional isomers are possible, especially in the long chain acids. Furthermore, the presence of one or more double bonds leads to *cis trans* or stereoisomerism, thereby still further increasing the number of possible isomers. In recent years an increasing number of positional and stereoisomeric acids have been found in natural fats. A number of triply bonded or "ynoic" acids have been found as constituents of certain fats and others have been prepared synthetically.

A number of substituted acids, including hydroxy, keto and cyclic side chain fatty acids have been found as important constituents of natural fats. These include ricinoleic acid in castor oil, licanic acid in oiticica oil, and chaulmoogric and related acids from chaulmoogra oil.

The fatty acids of commerce are derived from a limited number of the total known vegetable, marine-and land-animal fats and oils. The United States is particularly fortunate with respect to the variety of fats and oils that it produces and as consequence with the variety of fatty acids, either mixed or relatively pure that are commercially available to the enduser. Fatty acids of different degrees of purity are produced by several different methods, namely (a) fractional crystallization and pressing, (b) crystallization from solvents and (c) distillation. Each of these processes can be varied to produce different products or grades of products from the same fatty acid stock.

Whatever the process of separation, it must first be preceded by hydrolysis of the glyceride fat to liberate the acids, followed by their separation from the glycerine. The hydrolysis is carried out with water or steam at or above atmospheric pressure, and in the presence or absence of a catalyst. The early catalyst was sulfuric acid which was used almost exclusively until 1890 when Ernest Twitchell introduced the sulfonic acid catalyst, originally a product of a fatty acid and benzene or naphthalene. The original Twitchell reagents have since been replaced by similar but more active compounds prepared by sulfonating certain petroleum fractions.

Sometimes the fat stocks are acid-refined with strong sulfuric acid to destroy nonfatty impurities before hydrolysis. The reaction occurs stepwise, with one fatty acid being replaced at a time and the intermediate formation of di- and mono-glycerides. The splitting is carried out in two or more stages, between each of which the liquor or "sweet waters" containing the liberated glycerine is drawn off. The glycerine is subsequently recovered from these collected "sweet waters." After the splitting is complete or nearly so, the fatty acids are then purified by one or another of the methods mentioned above.

Through the application of combinations of splitting, purification, and separation there are produced a variety of products, such as pressed stearic acid and red oil (oleic acid), mixed fatty acids derived from soybean, cottonseed, linseed, castor, coconut and other oils, as well as individual acids of various degrees of homogeneity.

Another source of fatty acids is tall oil, a byproduct of the manufacture of paper by the sulfate or kraft process. Tall oil contains 18-60 per cent of fatty acids together with rosin acids. Various processes have been applied for the recovery and purification of fatty acids from tall oil. Crude tall oil contains about 7 per cent of neutral materials and the balance is about equal parts of rosin and fatty acids. The fatty acid ratio may vary from around one-third to two-thirds of the total rosin-fatty acid content, depending upon where the pine trees are grown. A fatty acid fraction with not more than 2 per cent of rosin acids, low in saturated acids, and with equal amounts of oleic and linoleic and no linolenic acid can be produced from tall oil. The production of tall oil fatty acids has increased from 149.6 million pounds in 1959 to 337.2 million pounds in 1966. The saturated acids (particularly stearic and palmitic) are used mainly in soaps, detergents, cosmetics, candles, waxes, and many chemical intermediates. The unsaturated acids (including oleic acid or red oil) and tall oil fatty acids go primarily into protective coatings, inks, metallic driers, soaps, and detergents. Tall oil fatty acids are in strong demand and the price of tall oil has remained relatively stable, with a constant and dependable source of supply.

In addition to the processes mentioned above for obtaining fatty acids they are obtained as byproducts from degumming, refining, deodorizing, bleaching and other fat recovery and purification processes. The total annual production of fatty acids in the United States is not known. The largest amount of these products is consumed by the rubber industry (about ¼ of the total), followed by the synthetic resin industry. The metal working, protective coating, textile, synthetic detergent and candle industries also consume appreciable amounts of fatty acids. Other uses include insulation, leather, pharmaceuticals, toilet articles, paint and ink vehicles, disinfectants, glue and adhesives. Production and consumption figures do not include the amounts of fatty acids produced and consumed by soap manufacturers in their own operations.

Several solvent-crystallization processes for the separation of fatty acids have been described in the patent literature, but to date only two have become industrially important, namely, the Emersol Process and the Armour Process. Both processes are based on the separation of certain fractions by crystallization of mixed acids at a temperature of $-10°C$ or below depending on the solvent used and the type of fractionation desired. In the Emersol Process mixed fatty acids, usually derived from tallow or grease, are continuously mixed with 90% methanol and pumped to a multitubular crystallizer. The concentration of the acids in the methanol is normally maintained at 25-30%. A small amount of crystal promoter may be added during mixing of the acids and methanol. Each stainless steel tube in the crystallizer is fitted with stainless steel rotating scraper blades to remove the crystalline acids from the walls and promote uniform regulated transfer of heat. The crystallizer tubes are cooled by circulating refrigerated methanol through the jackets. The fatty acid-methanol enters the crystallizer at 27°C and flows countercurrently to that of the refrigerated methanol in the jackets. The solution is cooled to about $-10°C$ at which temperature most of the solid (higher melting and least soluble) acids crystallize from solution. The resultant slurry of crystalline acids flows to an enclosed rotary vacuum filter where the solids are separated and washed with cold 90% methanol. The filtrate containing the liquid acids is passed through a heat exchanger in which 50–65% of the total refrigeration is recovered. The filter cake containing 35% solid acids and 65% methanol, is melted and pumped to a solvent recovery still. The filtrate containing the liquid acids is preheated and pumped to a solvent-recovery still of unique design. The bulk of the solvent is removed from the dissolved acids at low temperature, and highly concentrated methanol (92–94%) is recovered in the operation. Various types of mixed fatty acids may be separated by this process. Mixed fatty acids of animal origin yield commercial stearic acids, iodine value 5.0–6.0 and oleic acid or liquid acids, titre 2–5°C. Vegetable oil acids (cottonseed or soybean) yield solid acids, usually high in palmitic acid, iodine value 5–10, and liquid acids high in polyunsaturated acids, iodine value 130–160. Commercial installations employing this process are operated in the United States, Great Britain, The Netherlands and Australia.

The Armour process of low-temperature separation of fatty acids dissolved in an organic solvent, sometimes referred to the Armour-Texaco process, is an adaptation to fatty acids of a process originally developed by The Texas Company for solvent-dewaxing petroleum refinery stocks. The process involves a controlled, low-temperature crystallization of the saturated acid fraction from an acetone solution of mixed fatty acids of natural origin. Almost any type of mixed fatty acids can be processed, but operations have been concentrated principally on the mixed acids obtained by splitting tallow and animal grease. With these materials as feed stock, the products of the solvent-crystallization process are red oil (oleic acid) and a mixture of stearic and palmitic acids. Approximately 4000 lb./hr. of distilled fatty acids are sent to the crystallizer. Acetone is introduced into the pipeline, and a mixture of acids and acetone is passed through a series of water-cooled tubular heat exchangers to precool the solution. The precooled mixture is then pumped through two chilling units connected in series. Each chiller contains eight 40-foot lengths of 4-in. pipe inside of 6-in. pipe, which corresponds to a total path (both chillers) of 640 ft. Spring-mounted scrapers operate continuously inside the 4-in. pipe in which the crystallization occurs. Additional acetone (called "secondary" acetone) is added part of the way through the first chiller which serves to improve the efficiency of the separation. The annular space between the concentric pipes in the chiller contains ammonia refrigerant for temperatures down to −18°C. A booster compressor is available when it is desired to reduce the temperature to −45°C. The lower temperature is used to separate two unsaturated acids, such as oleic and linoleic. The crystalline saturated acids are separated from the still liquid red oil from the chillers in an enclosed, insulated, rotary filter. Some of the filtrate is recycled to decrease the thickness of the filter feed material. The cake is washed in the filter by an acetone spray. The filter is maintained under slight pressure (2–4 in. water) of nitrogen, which is also the case with the acetone storage tanks. The peripheral surface of the filter is divided into sections its full length. Under each section is a shallow pan which receives the filtrate and wash liquid from that section. Outlets from these pans are piped (inside the filter drum) to a rotary valve arrangement which controls the filtration, washing, and blowing cycle. The filtrate and washing liquids are combined and the acetone is removed in a flash tower, followed by stripping the residue with steam. The cake (solid acids) is melted by steam coils, and the acetone stripped off in the same manner, after which the products are pumped to their respective storage tanks. When operating with distilled tallow acids, the yield of products are 53% red oil (commercial oleic acid) and 47% of solid acid cake (stearic acid). For the separation of oleic and linoleic acid, a crystallization temperature of −40°C is required and the products are oleic acid (cake fraction) and a liquid fraction which is a eutectic mixture of 84% linoleic and approximately 16% oleic acid.

One of the more recent processes for the separation of saturated (stearic, palmitic) and unsaturated (oleic) acids is known as hydrophilization. The separation is carried out in an aqueous medium containing a surface-active agent. The process was developed at Henkel & Cie. GmbH, Dusseldorf, Germany, prior to about 1960. Since that time approximately 8000 pounds per hour of fatty acids of vegetable or animal origin have been separated by the aforementioned process. The process is described in a series of patents issued to W. Stein and H. Hartmann (Henkel & Cie.) U.S. 2,800,493 (1957); H. Hartmann and W. Stein (Henkel & Cie. U.S. 2,972,636 (1961); H. Waldmann and W. Stein (Henkel & Cie.) U.S. 3,052,700 (1962). In addi-

tion to the original plant in Dusseldorf, Germany, another is in operation by Canadian Packers, Toronto, Canada, and a third plant is reported to be under construction in Chicago.

The aqueous separation of saturated and unsaturated fatty acids in the presence of a sulfo fatty acid was first described in U.S. Patent 918,612 issued to Ernest Twitchell April 20, 1909. In 1942 Prof. A. H. Schutte was granted a U.S. Patent 2,296,456 describing the aqueous separation of mixtures such as natural waxes, linoleic and oleic acids, and naphthalene from anthracene as well as stearic and oleic acids. Apparently neither the Twitchell nor the Schuette process was ever reduced to practice.

In 1965 Adolf Koebner and Thomas Thornton obtained a British Patent 1,002,609 assigned to Marchon Products Ltd., London covering the filtration separation of a mixture of solid and liquid fatty acids in an aqueous solution containing sodium sulfonates. In 1966 Adolph Koebner et al. obtained a Canadian Patent 738,448 assigned to Marchon Products Ltd., London covering an aqueous process for separating saturated from unsaturated fatty acids, citing particularly rapeseed and linseed fatty acids and also tallow fatty acids.

In 1970, George R. Payne et al. obtained a Canadian Patent 837,647 assigned to National Dairy Products Corp., New York for the separation of fatty materials by centrifugation of mixtures of fatty acids wherein cooling of a fluid mixture of fatty acids may be effected rapidly and still provide solid phase crystals that are separated from the liquid phase. The process is claimed to be applicable to a variety of vegetable oils, land animal fats and marine animal fats. Examples cited in the patent are aqueous solutions of tallow fatty acids in the presence of magnesium sulfate plus sodium lauryl sulfate; also cottonseed oil to produce a winterized product.

In addition to the patents cited above Werner Stein described the process as operated by Henkel & Cie. Ltd. in an illustrated article in the *JAOCS,* **45,** 471–474 (1968) and the process was described less briefly in *Chem. & Eng. News,* May 15, 1967.

KLARE S. MARKLEY

FERMENTATION

Antibiotics by Fermentation

The production of antibiotics by fermentation involves the cultivation of selected pure strains of microorganisms in sterilized liquid medium contained in large tanks provided with systems for aeration and agitation. Typical media are composed of an aqueous solution or suspension of 1–5% carbohydrate source (glucose, other sugars or starch) and 0.1–1.0% nitrogen source (meat extracts, peptones, nitrates or ammonia) with 1% or less of one or more inorganic salts such as potassium phosphate as sources of important elements and/or to control pH. Natural products such as soybean meal and waste products such as distillers solubles and corn steep liquor may serve as sources of carbohydrate and/or nitrogen. Control of the metabolism of the producing organism is maintained by manipulation of conditions such as temperature, pH, the rates of aeration and agitation, and the feeding of nutrients.

The fermentation is usually initiated by inoculation with several per cent of a vegetative growth of the microorganism previously developed in a smaller fermentor. Several stages of increasing size may be employed sequentially from a small shaken flask culture up through a final unit containing many thousands of gallons. Fermentations are allowed to proceed for a period of from 1 to 6 days depending upon the rate of growth of the producing microorganisms. After completion of the fermentation, antibiotic is recovered by chemical extraction and purification from the culture broth.

Penicillin. The first antibiotic to be produced on a large scale was penicillin. The organism, *Penicillium notatum* Westling, was grown on the surface of the nutrient medium, and relatively small yields of penicillin were obtained. Yields of penicillin have been increased to over ten grams per liter as a result of the development of submerged methods of fermentation, the use of more satisfactory media for growth, the discovery of more productive strains such as the substitution of *P. chrysogenum* for *P. notatum,* and the improvement of strains by x-ray, ultraviolet light, nitrogen-mustard and other mutation treatments.

Penicillin may be regarded as a metabolic product, but a minor one under natural conditions. A number of other organisms also produce the antibiotic. Penicillin as produced by mold cultures does not represent a single compound but, rather, a group of closely related compounds which have a common β-lactam-thiazolidine structure with different side chains. Penicillin G was the first penicillin to be isolated, and it possesses highly desirable characteristics for recovery, purification, and therapeutic use.

Factors contributing to high penicillin yields are adequate, effective aeration; the presence of a precursor at nontoxic levels, preferably phenylacetic acid or some derivative of it; and the restriction of growth during the later stages by use of a poorly fermentable carbohydrate such as lactose, or by continuous feeding of glucose in a limiting amount.

The carbon dioxide of the effluent air from the fermentor may be utilized as a guide to the amount of growth and metabolic activity of the fungus, and as a function of the amount of aeration supplied. The ammonium-nitrogen decreases during penicillin formation and rises slightly toward the end, lactic acid is consumed, the pH rises, the lactose disappears, and autolytic changes in the mycelium may take place. Synthesis of penicillin appears to be associated with an unbalance of the metabolism.

Through an enzymatic process for removal of the side chain and condensation of the resulting 6-amino-penicillanic acid with other structures, new penicillins with desirable properties have been synthesized. Some of these derivatives have broadened spectrum; others inhibit staphylococci which are resistant to penicillin G.

Streptomycin. Another genus of microorganisms, the *Streptomyces,* contains several important species that produce streptomycin, tetracyclines, chloramphenicol, erythromycin and many other useful compounds. In his early work on streptomycin, Waksman used glucose, peptone, beef extract, and sodium chloride to grow *Streptomyces griseus.* For large-scale production in fermentors, soybean meal has replaced beef extract and peptone as a source of nitrogen. Dextrose, distillers' soubles or corn steep liquor, and calcium carbonate are also employed, the latter to control pH; lard oil and other fats may replace glucose and starch. Actinophage-resistant strains were developed to prevent low yields resulting from actinophage infection. Aeration of submerged cultures is practiced in growing the organism.

Chloramphenicol. The constituents of the nutrient medium for growing *Streptomyces venezuelae* n. sp, the organism that produces chloramphenicol, in the laboratory include glycerol, meat extract, tryptones, soybean protein, distillers' soubles and molasses, with inorganic salts. It has been reported that a medium containing starch, peptone, and inorganic salts gives good results. Phenylalanine, tyrosine, and methionine are stated to enhance yields of chloramphenicol. Aeration and submerged growth are practiced. Production of mycelium and chloramphenicol run parallel. The chemical composition and structure of chloramphenicol have been established. It is produced commercially chiefly by chemical synthesis.

Chlortetracycline. Chlortetracycline is produced by *Streptomyces aureofaciens.* Commercially, it is stated, corn steep liquor, sucrose, $(NH_4)_2HPO_4$, KH_2PO_4, $CaCO_3$, $MgSO_4 \cdot 7H_2O$, and traces of Mn, Cu, and Zn salts may be employed. Media in which growth is limited due to low levels of phosphate favor chlortetracycline production. The pH of the medium initially varies from 6.0–6.4, and may fall to 4.5–4.8. Media containing cottonseed meal, peanut oil meal or peanut meal with a low fat content have given yields in excess of five grams per liter. Aeration and submerged growth are employed. The chemical structure of chlortetracycline was reported in 1952.

Oxytetracycline. Oxytetracycline is a product of *Streptomyces rimosus.* Information indicates that soybean meal, corn starch, N-Z-Amine B, $NaNO_3$, and $CaCO_3$ may be used as a nutrient medium. Aeration and submerged growth are practiced commercially. The pH of the nutrient medium is adjusted to 7.0 before sterilization and may rise to 8.0. The chemical structure of oxytetracycline was published in 1952.

Erythromycin. Erythromycin is an antibiotic of intermediate spectrum produced by *Streptomyces erythreus.* It is produced commercially by submerged fermentation in a soybean-corn steep-glucose medium. The range of pH is 6.4 to 7.2. The structure has been established as a macrocyclic lactone containing two unique sugars and having a molecular weight of 733.

Bacitracin. Bacitracin is produced by a sporeforming bacterium, *Bacillus subtilis.* The antibiotic is a polypeptide which accumulates in the liquor in which the cells grow. Aeration is employed when using submerged growth. Surface growth can be used when aeration cannot be practiced. Soybean or peanut meal, dextrose, sucrose, and starch may be employed as growth media.

Other antibiotics produced on a commercial scale. The genus *Streptomyces* has proved to be a fertile source of antibiotics. Many products, including lincomycin, neomycin, novobiocin, kanamycin, and rifamycin, are manufactured on a commercial scale in large, aerated fermentors, employing culture media rich in natural products. Gentamicin, an unusually broad spectrum antibiotic, is derived from a closely related genus of microorganisms, the *Micromonospora*. Other antibiotics produced commercially from molds include griseofulvin, for systemic treatment of chronic fungus infections, and the cephalosporins, a family of antibiotics somewhat related to penicillin, which are stable to penicillinase. From bacteria, the polymyxins have proved useful in treatment of resistant gram-negative bacteria, including pathogenic strains of *Pseudomonas;* tyrothricin is employed in treatment of topical infections.

The conditions for producing each of these antibiotics are similar to those discussed in more detail above. These include selection of a rich inexpensive nutrient medium, development of high-yielding strains by mutation, and growth of the producing organism in pure cultures under aerated conditions in large fermentors. The optimum conditions for antibiotic production must be selected for each producing culture. Frequently, the best yield of antibiotic is obtained with a balance of nutrients which is not optimum for maximum cell growth. A well developed commercial fermentation will yield several grams of antibiotic per liter of fermentation broth which represents from a hundred to a thousandfold increase over the initial culture and conditions.

Modified Fermentation Products. In recent years many antibiotics have been produced in chemical forms different from that originally discovered as natural products. Modified forms of the natural products may be achieved by the use of precursors, medium modification to eliminate or reduce precursors of unwanted forms, mutation to develop new cultures which produce a new form of the antibiotic, or production of an intermediate with final completion of the antibiotic by a process of chemical synthesis. The best example of the use of a precursor is that of phenylacetic acid for the formation of benzyl-penicillin cited earlier. Medium modification has been useful in the production of tetracycline. Variations in the medium to reduce the level of chlorine available result in production of more of the chlorine-free tetracycline molecule instead of the natural major product, chlortetracycline. The same end result has been obtained by the isolation of mutant cultures which produce tetracycline as the major fermentation product. Modification by chemical manipulation has resulted in a number of antibiotics with improved characteristics. In analogy with the studies done on penicillin, a series of cephalosporins have been prepared by chemical removal of the side chain of the cephalosporin molecule to result in 7-amino-cephalosporanic acid which can then be combined with a variety of other structures resulting in new cephalosporins with improved potency, broader spectrum or other desirable properties. Chemical modification to result in semisynthetic fermentation products has been very important for the antibiotic lincomycin, a product of *Streptomyces,* which has been converted to clindomycin which is reported to have higher antibiotic activity and be better absorbed than the natural product. Rifamycin, another *Streptomyces* product, has been converted

to rifampicin which has superior antibacterial properties especially for the treatment of tuberculosis.

EDWARD O. STAPLEY

References

Blakebrough, N., "Biochemical and Biological Engineering Science," Volume 1, Academic Press, London and New York, 1967.
Perlman, D., "Fermentation Advances," Academic Press, New York and London, 1969.
Hockenhull, D. J. D., "Progress in Industrial Microbiology," Volume 1, Interscience Publishers Inc., New York, 1959.

Yeast

Yeasts, molds and mushrooms are common names for fungi, the largest botanical group (about 100,000 species) sharing the earth's population of microorganisms with viruses, bacteria, actinomyces, algae, protozoa and lichens. The word "yeast" is a general descriptive term for a small group of microfungi currently classified in 39 genera representing 349 species. But only a few species of yeast are of economic and academic importance. Their worldwide cultivation and usage is due mainly to two outstanding biochemical properties: (a) Fermentation: the transformation of simple sugars to ethyl alcohol and carbon dioxide by yeast cells living anaerobically, and (b) respiratory (oxidative) metabolism: the great capacity of some yeasts for protein synthesis during growth in richly aerated media containing a wide variety of carbonaceous and nitrogenous nutrients. Thus, yeasts serve in many ways: (a) as living cells they are biocatalysts in the production of bread, wine, beer, distilled beverages and ethyl alcohol; (b) as dried nonfermentative whole cells or hydrolyzed cell matter yeasts contribute nutriment and flavor to human diets and animal rations; (c) as producers of vitamins and other biochemicals yeasts are a rich source of enzymes, coenzymes, nucleic acids, nucleotides, sterols and metabolic intermediates; and (d) as a versatile biochemical tool yeasts aid research studies in nutrition, enzymology and molecular biology.

The art of making wine, leavened bread and beer was practiced more than four thousand years ago. Phenomena producing these foods were attributed to yeast. And in many languages the word for yeast describes the visible effects of fermentation, as observed in the expansion of bread dough and the accumulation of froth or barm on the surface of fermenting juices and mashes. Historically, yeast cells were first seen in a droplet of beer mounted on a crude microscope by van Leeuwenhoek in 1680. He found globular bodies but was not aware that these were living forms. For nearly two centuries the theory of spontaneous generation dominated thought and research on the causes of fermentation and disease. In 1818 Erxleben described beer yeast as living vegetable matter responsible for fermentation. In the following 20 years, yeasts were shown to reproduce by budding, and in 1837 Meyen named yeast *Saccharomyces* or "sugar fungus." By 1839 Schwann observed "endospores" in yeast cells, later named ascopores by Reess. As early as 1857 Pasteur proved the biological nature of fermentation, and later (1876) he demonstrated that yeast can shift its metabolism from a fermentative to an oxidative pathway when subjected to aeration. This shift, named the Pasteur effect, is especially characteristic of bakers yeast, *Saccharomyces cerevisiae,* and is applied in the large-scale production of yeasts.

Botanically yeasts form a heterogeneous group of saprophytic forms of life occuring naturally on the surface of fruits, in honey, exudates of trees, and soil. They are disseminated by air-borne dust, insects and animals. Typical industrial yeasts are generally oval, microscopic, unicellular organisms lacking chlorophyll and means of locomotion. They reproduce vegetatively by budding and sexually by spore formation (ascospores) within the mother cell or ascus. These properties place them in the family *Endomycetaceae,* class *Ascomycetes,* phylum *Eumycophyta,* division *Thallophyta.*

The major industrial yeasts are classified with the family *Endomycetales: Saccharomyces cerevisiae,* the yeast for alcoholic beverages and bread; *S. cerevisiae* var. *ellipsoideus, S. bayanus* and *S. beticus* used in wineries; *S. uvarum* (formerly *S. carlsbergensis*) in brewing; *Kluyveromyces fragilis* (formerly *S. fragilis*) in whey disposal.

Food and feed yeast production employs several species in the family *Cryptococcaceae: Candida utilis, C. tropicalis* and *C. japonica* are cultivated on plant wastes (wood sugars, molasses, stillage); *C. lipolytica* converts hydrocarbons to yeast protein.

Yeast cell shape and size is influenced by the growth environment—temperature, nutrition, acidity, aerobiosis and age of culture. For the type species, *S. cerevisiae,* single oval cells occur most frequently. In bakers yeast they measure 3 to 8 microns (μ) in width and 5 to 12 μ in length. Single cells are generally smaller in the nonspore-forming yeasts, such as *C. utilis.*

The cell structure of *S. cerevisiae,* as observed in an optical microscope, reveals a rigid cell wall, a colorless, granular cytoplasm and one or more vacuoles. Electron microscopy of ultrathin sections of a yeast cell show the microstructures: birth and bud scars on the cell wall, plasmalemma (cytoplasmic membrane), nucleus, mitochondria, vacuoles, fat globules, cytoplasmic matrix and volution or polyphosphate bodies. The cell walls of bakers yeast contain 30 to 35% glucan (yeast cellulose), 30% mannan (yeast gum) which is bound to protein (about 7%), 1 to 2% chitin, 8 to 13% lipid material and inorganic components, largely phosphates.

Gross chemical composition of compressed bakers yeast is approximately 70% moisture, and in the dry matter, 55% protein (N × 6.25), 6% ash, 1.5% fat, and the remainder mostly polysaccharides, including about 15% glycogen and 8% trehalose. Food yeast, molasses-grown, is dried to about 5% moisture, has the same chemical composition as bakers yeast, and the following vitamin concentration, in $\mu g/g$ (micrograms/g) yeast: 165 thiamine (B_1), 100 riboflavin (B_2), 590 niacin, 20 pyridoxine (B_6), 13 folacin, 100 pantothenic acid, 0.6 biotin, 160 *p*-aminobenzoic acid, 2710 choline and 3000 inositol. Yeast crude protein contains 80% amino acids, 12% nucleic acids and

8% ammonia; the latter components lower the true protein content to 40% of the dry cell weight.

Yeast protein is easily digested (87%) and provides amino acids essential in human nutrition. Most commercial yeasts show the following pattern of amino acids, as per cent of protein: 8.2 lysine, 5.5 valine, 7.9 leucine, 2.5 methionine, 4.5 phenylalanine, 1.2 tryptophan, 1.6 cystine, 4.0 histidine, 5.0 tyrosine and 5.0 arginine. The usual therapeutic dose of dried yeast prescribed by the National Formulary (N.F. XIII) is 40 grams per day (one heaped teaspoonful weighs about 5 grams). This amount of food yeast supplies approximately 40% of the recommended daily allowance (RDA—National Research Council) of thiamine, 100% riboflavin, 80% niacin, 60% pyridoxine and 28% protein.

Ash content of food yeasts ranges from 6 to 8%, dry basis, consisting principally of calcium, phosphorus and potassium. Molasses-grown food yeasts contain, in per cent of dry matter, about 2 potassium, 1.1 phosphorus, 0.4 calcium, 0.2 magnesium, 0.4 sulfur, 0.2 sodium; and (in $\mu g/g$) 42 zinc, 92 iron, 21 copper, 2.5 lead, 4 manganese and 1.6 iodine.

Triglycerides, lecithin and ergosterol are the major constituents of yeast lipid. Oleic and palmitic acids predominate in yeast fat which resembles the composition of common vegetable fats. Ergosterol, the precursor of calciferol (vitamin D_2), varies from 1 to 3% of yeast dry matter.

Metabolic activity of yeast is generally associated with the well-known alcoholic fermentation in which, theoretically, 100 parts of glucose are converted to 51.1 parts of ethyl alcohol, 48.9 parts of carbon dioxide and a small amount of heat. In addition, however, the anaerobic reaction also yields minor byproducts in small amounts, mainly glycerol, succinic acid, higher alcohols (fusel oil), 2, 3-butanediol, and traces of acetaldehyde, acetic acid and lactic acid. Fusel oil is a mixture of alcohols: *n*-propyl, *n*-butyl, isobutyl, amyl and isoamyl.

Respiratory activity or oxidative dissimilation is characteristic of many species of yeasts. During their aerobic growth, sugar is oxidized to carbon dioxide and water with the release of large amounts of energy (about 680 kcal when complete oxidation occurs). Aerobiosis produces a variety of byproducts, some in unusually high concentration—acetic acid, succinic acid, zymonic acid, polyhydric alcohols (glycerol, erythritol, D-arabitol, D-mannitol), extracellular lipids, carotenoid pigments in shades of red and yellow, black pigment (melanin), and capsular polysaccharides (phosphomannan).

Saccharomyces cerevisiae is a major raw material for recovery of enzymes, especially invertase, coenzyme A, catalase, alcohol dehydrogenase, hexokinase, L-lactate dehydrogenase, diphosphopyridine nucleotides (NAD, NADH) and adenosine triphosphate (ATP). *Saccharomyces fragilis* is a potential source of lactase. *Candida utilis* is an excellent raw material for extraction of nucleic acids.

About 80,000 tons of yeast dry matter are produced annually in the United States; 75% of this is in the form of bakers yeast, the remaining 25% represents approximately equal amounts of food yeast and feed yeast. This production issues from four types of manufacture: (1) bakers yeast grown batchwise in aerated molasses solutions; (2) continuous propagation of *Candida utilis* in pulp mill spent liquid; (3) *S. fragilis* grown batchwise in cottage cheese whey; and (4) dried yeast recovered as spent beer yeast. World production of all types of food and feed yeast is estimated at more than 400,000 dry tons annually.

The process for growing bakers yeast is a model system for the propagation of microorganisms. It begins with a laboratory culture of a pure strain of *S. cerevisiae*. Seed yeast is developed in successively larger volumes of nutrient solutions, beginning with a Pasteur flask and ending in a trade fermentor containing as much as 40,000 gallons of sterilized and diluted molasses maintained at 30°C. During the highly aerated growth period, minerals are added, pH (acidity) is adjusted to 4.5, and diluted molasses is continuously fed in proportion to the increase in cell mass. Under ideal conditions yeast cells may double in number every 150 minutes, converting more than half (56.7%) of the sugar supplied to cell components. Biosynthesis of cell matter requires an equal weight of oxygen. To produce 100 lb yeast dry matter with 50% protein content requires about 400 lb molasses, 25 lb aqua ammonia, 15 lb ammonium sulfate, 7 lb monobasic ammonium phosphate and 75,000 cubic feet of air.

Only a few yeasts and yeast-like fungi are troublesome, either as contaminants and opportunists in foods or the cause of relatively mild forms of disease. Any yeast other than *S. cerevisiae* is undesirable in bakers, brewers and wine yeast production.

Yeasts may cause spoilage of fruits, juices, liquid sugar products, brined pickles and olives, and honey. Mild yeast infections may occur in cotton bolls, coffee berries, lima beans, tomatoes, pecans, and citrus fruits. As intestinal parasites, certain yeasts are harbored regularly in swine, cattle, horses, rabbits, mice and rats. Insects, beetles, birds and man are yeast carriers—but appear to suffer no ill effects. *Cryptococcus neoformans* causes a chronic and often fatal meningitis in man, dogs, horses and mice, and has been identified in mastitis in cattle. Otherwise yeasts play a minor role in medical mycology. A few species, notably *Candida albicans,* are associated with infections of the skin, mucuous membranes and urinary tract. In the fatal fungal diseases histoplasmosis and blastomycosis, a yeast cell phase occurs in the body of the host.

HENRY J. PEPPLER

References

Rose, A. H., and Harrison, J. S., Eds., "The Yeasts," Vol. 1, 2, 3, Academic Press, London, 1970.

Lodder, J., Ed., "The Yeasts—a Taxonomic Study," Vol. 1 and 2, North Holland Publishing Co., Amsterdam, 1970.

Phaff, H. J., Miller, M. W., and Mark, E. M., "The Life of Yeasts," Harvard University Press, Cambridge, Mass., 1966.

FERRITES

Ferrites are oxide compounds of iron that exhibit the property of magnetism due to the alignment of individual magnetic spin dipoles of atoms within the structure. In general, ferrites are limited to compounds of those elements that have unpaired electrons located in d and f suborbitals of the atom. These electrons are shielded from the forces of atomic bonding and participate in the energy exchange which gives rise to the observed macroscopic magnetism. As in the ferromagnetic metals, magnetism arises from an exchange of energy between elements which acts to align all the electron spins. In metals all spins are aligned parallel. In ferrites, however, the magnetic energy is divided into two or more magnetic "sublattices" within which all spins are parallel. The antiparallel array of spins between sublattices tends to cancel the magnetic contribution of each, and one necessary condition for magnetism in ferrites is that the magnetic moment in these sublattices must be unequal, so that a net magnetic moment will occur. This is most easily visualized by the example of a typical ferrite, the compound magnetite ($FeO \cdot Fe_2O_3$). In magnetite there are two magnetic sublattices which are indicated by parentheses( ) and square brackets [] in writing the chemical formula.

$$(\uparrow Fe^{+3})[Fe^{+3} \downarrow Fe^{+2} \downarrow]O_4$$
$$-5 \qquad\qquad +5 \qquad +4 \quad = 4\,\mu_B \text{ per formula unit}$$

The divalent iron in magnetite has four unpaired d electrons and trivalent iron has five. Each contributes a magnetic moment per atom proportional to this number of electrons. The magnitude of the magnetic contribution of each cation is indicated by the number appearing below, in this case five units (called Bohr magnetons) for Fe^{+3} and four for Fe^{+2}. Since the sublattices are aligned antiparallel (as shown by the direction of the arrows), the opposed moments of the trivalent iron exactly cancel, leaving a net moment of four Bohr magnetons per formula unit due to the divalent iron in the second sublattice.

A number of transition metal elements can substitute for iron in the magnetic structure. The location of these substituting cations and the number of unpaired electrons contributed by each alters the balance of magnetic intensity between sublattices and therefore the net magnetic moment. The magnetic moment of several typical ferrites at 0°K is shown in Table 1. The lack of exact agreement between simple theory and the experimental values is due in part to orbital contributions to the magnetic moment. It is also possible to substitute nonmagnetic elements such as Zn, Al, or Mg which again alters the balance of magnetic interaction between sublattices. There are, however, limitations to this procedure. Magnetism in ferrites depends on the maintenance of sufficient antiparallel electron spin interaction between sublattices to keep all spins within each sublattice aligned parallel. Without this antiparallel interaction of magnetic ions between sublattices the spins within each sublattice become random and magnetism disappears. This is shown by the compound $ZnFe_2O_4$ in Table 1 where nonmagnetic Zn(zero unpaired electrons) does not maintain parallel spins in the second sublattice and the net moment becomes zero due to a randomizing of the Fe^{+3} moments. As temperature increases, the alignment of spins within and between magnetic sublattices is progressively weakened by

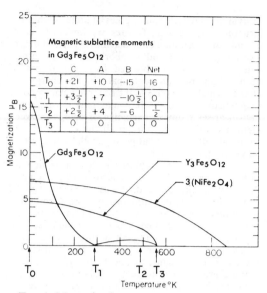

Fig. 1. Magnetization versus temperature curves for three representative ferrites.

thermal motion, and the net magnetization finally becomes zero at the Curie temperature. A plot of magnetization vs temperature for nickel ferrite is shown in Fig. 1.

In ferrites the free energy associated with the spin alignment is minimized by an antiparallel array of magnetic sublattices. In a similar way the demagnetizing energy in all magnetic material is minimized by a tendency to subdivide the magnetic solid into small volumes called domains, each having a net moment but randomly oriented such that the overall magnetic moment is zero. The domain pattern in a virgin sample can be altered by the application of an extetrnal magnetic field. In a constant dc field those domains that are oriented parallel or nearly parallel to the field grow at the expense of unfavorably oriented domains. The nonlinear growth and reorientation of domains with magnetic field intensity gives rise to a hysteresis curve in the plot of magnetization vs applied field. In ferrites, as in other magnetic materials, the relative ease of magnetizing the material and the degree of energy required to demagnetize the sample, i.e., to return it to the minimum energy random domain state, are important parameters.

One of the principal advantages of ferrites is the very high electrical resistivity of the oxides as compared with metals. Electrical conduction in ferrites causes electrical losses due to eddy currents, especially when used at high frequencies. Verwey had

discovered in 1947 that conduction in ferrites is due almost entirely to an electron transfer between Fe^{+2} and Fe^{+3}. The electron from Fe^{+2} transfers to an adjacent Fe^{+3} making the latter an Fe^{+2} ion. The electron transferred in this way can migrate to other Fe^{+3} ions in the structure giving rise to electrical conductivity. If, however, the Fe^{+2} is removed by substituting another divalent cation such as Ni^{+2} in $NiFe_2O_4$, this conduction process does not occur. The electron transference leading to conductivity is not the same as the electron spin exchange causing ferrimagnetism.

A second useful characteristic of the ferrites is the wide range of magnetic properties, which can be achieved by substitution of many different magnetic and nonmagnetic ions in the ferrite structure (see below). Thus, ferrites can be made with several substitutions in the basic composition, each designed to improve, in principle at least, a specific characteristic such as magnetic moment, dielectric loss, temperature stability, etc. Sintered oxide ferrites have inherent advantages in low cost and amenability to automated production and testing. Finally, these ferrites have high temperature stability and imperviousness to oxidation—properties that are essential under severe environmental conditions.

The growth of ferrite technology has occurred very rapidly in the period since World War II. The first systematic study of ferrites was reported by Verwey and co-workers in 1947 in papers dealing with a class of ferrites called spinels because of their structural similarity to the mineral spinel ($MgAl_2O_4$). This discovery was followed in 1948 by the development by L. Néel of a theory of ferrimagnetism explaining the magnetization process in these new oxide compounds and relating the phenomenon to the more familiar ferromagnetism in metals. Néel was awarded the Nobel prize in 1970 for this work. Snoek, in particular, was instrumental in the early development of the spinels for radio and other high-frequency applications. A second class of ferrite materials, similar in atomic structure to the mineral magnetoplumbite ($PbFe_{12}O_{19}$) and referred to as hexagonal ferrites, was announced by Went, Braun, and Wijn in 1952. This was followed in 1956 by the independent discoveries of Bertaut in France and Geller in the U. S. of ferrimagnetism is a family of rare earth oxide-iron oxides which are called garnets or garnet-ferrites because of the structural similarity to the mineral silicate garnets. Ferrimagnetism has been observed in other crystal structures, notably the perovskites, ilmenite-types, and certain sulfides, but the effect is weakened by tendencies toward covalent bonding and by lower crystal symmetry which decreases the

TABLE 1. MAGNETIZATION OF FERRITES OF THE TYPE (MFe_2O_4)

Composition	Sublattice (A)	Sublattice [B]	Theoretical Net Moment μ_B	Experimental Moment at 0°K
Fe_3O_4	$Fe^{+3}(-5)$	$Fe^{+3}(+5)Fe^{+2}(+4)$	4	4.1
$CoFe_2O_4$	$Fe^{+3}(-5)$	$Fe^{+3}(+5)Co^{+2}(+3)$	3	3.7
$NiFe_2O_4$	$Fe^{3+}(-5)$	$Fe^{+3}(+5)Ni^{+2}(+2)$	2	2.3
$CuFe_2O_4$	$Fe^{+3}(-5)$	$Fe^{+3}(+5)Cu^{+2}(+1)$	1	1.3
$ZnFe_2O_4$	$Zn^{+2}(0)$	$Fe^{+3}(+5)Fe^{+3}(-5)$	0	0

effectiveness of the magnetizing process. The term ferrite has, on occasion, been applied exclusively to the type first discovered, i.e., the spinel class, but the basic similarities between spinel, garnet, and hexagonal structures in chemical and physical properties indicate that the more general use of the term "ferrite" to include all three types is preferable.

Ferrites are generally prepared by ceramic processing techniques in which the raw materials in the form of oxides, carbonates, etc., are intimately mixed and ground together, prefired or calcined to form the ferrite powder, and then pressed or extruded into the desired shape and sintered by a high temperature firing to a dense polycrystalline piece. In each step of the processing, care must be taken to maintain thorough mixing and the proper particle size to obtain a homogeneous material with minimum internal porosity. Contamination is also a problem in the grinding and mixing of the powders since certain properties are sensitive to deviation from the ideal or stoichiometric composition. The microstructure of the finished ceramic piece, i.e., the size and distribution of grains and pores is influenced primarily by the condition of the fine powder before pressing.

New processing methods are constantly being sought which will improve chemical homogeneity, purity, and the reproducibility of physical characteristics, all of which profoundly effect sinterability of the ferrite powder. Spray drying and freeze drying of aqueous salt solutions are both possible contenders to replace conventional wet-milling processes for certain critical ferrite components. In both processes, solutions are broken into small droplets ($\wedge 20\mu$m) by atomization and either evaporated or frozen and subsequently dehydrated to produce a fine powder of the given salt. The salt must then be decomposed to form the desired ferrite powder, which is then pressed and fired as before. The advantages of these processes are lack of mill (impurity) pickup, and chemical homogeneity on an atomic scale because droplet size and drying time do not permit segregation of components as would occur in a normal batch precipitation. The disadvantages are the additional costs in raw material and special equipment which must be balanced against better performance.

In certain cases ferrites have been grown as single crystals for research purposes or for uses in devices where the strictest requirements of uniformity and crystal perfection are needed. These crystals are generally quite small and are usually produced by slow cooling a saturated melt of an oxide which will dissolve the ferrite but does not enter the ferrite structure during crystallization.

Thin films of single crystalline spinels and garnets are also grown by epitaxial deposition from vapor or liquid phase for use as magnetic bubble memories. The oxide thin films, typically 10μm in thickness, have the capability of storing digital information of the order of 10^6 bits/in^2 which far exceeds the storage density of present core and disk magnetic memories. The bubble memories, moreover, are less susceptible to radiation effects or accidental erasure than are plated wire (magnetic) and semiconductor (electrical) memory devices.

H.J. VAN HOOK

Cross-reference: *Magnetochemistry.*

FERTILIZERS

Fertilizers are materials that contain one or more nutrient elements essential for plant growth. They are used chiefly to compensate natural deficiencies in the soil and/or to replace nutrients removed in cropping regimes. In addition to carbon, hydrogen, and oxygen, which plants acquire from water and the air, the following groups of elements are considered essential: (1) nitrogen, phosphorus, and potassium, usually called primary nutrients because they are used in large amounts and are most likely to be limiting; (2) calcium, magnesium, and sulfur, called secondary nutrients, which also are used by plants in relatively large amounts but on most soils are not as likely to be limiting; and (3) iron, copper, manganese, boron, zinc, chlorine, and molybdenum, which are needed in relatively small amounts and therefore are called minor or micronutrients. Other elements are being scrutinized continuously for their influence on plants, and it is possible that the list of essential nutrients may be extended.

Materials used as fertilizers include manufactured chemicals, natural and refined minerals, byproducts of other industries, and organic materials derived from plants and animals. However, the growing demand for fertilizers has far outstripped the supply of byproducts and organics, and manufactured materials are the major type by far.

The nutrient elements cannot be supplied to plants as such since they are not readily assimilated (available); nitrogen is an inert gas and phosphorus and potassium are toxic. Therefore, they must be combined with other elements in the form of suitable compounds. Nitrogen is supplied in the ammoniacal (NH_4^+) or the nitrate (NO_3^-) form or as urea $[(NH_2)_2CO]$. Except in unusual situations, one form is as good as the other because in the soil the ammoniacal and urea forms are rapidly converted to nitrate. Phosphorus is supplied as phosphate compounds such as calcium and ammonium phosphates. Potassium is supplied chiefly as potassium chloride and, for a few crops, as potassium sulfate and potassium nitrate.

Most forms of nitrogen and potassium used in modern fertilizer manufacture are water-soluble and are considered fully available for plant growth. The availability of phosphorus is considered, in the United States and some other countries, to be the percentage of the total that is soluble in water plus that soluble in neutral ammonium citrate solution. Details of official procedures for determining nitrogen and potassium contents and total and available phosphorus are contained in "Methods of Analysis" of the Association of Official Analytical Chemists. In some European countries, the availability of the phosphorus is considered to be the amount that is soluble in water plus that soluble in an alkaline citrate solution. In other countries, the phosphorus value is considered to be only the proportion that is soluble in water.

It is customary to describe fertilizers by analysis, or grade, with respect to nutrient content. In some countries, the grade is reported as per cent of N (nitrogen), P (phosphorus), and K (potassium). In the United States and many other countries, these analyses are expressed in terms of N, equivalent P_2O_5, and K_2O. Thus, a 46-0-0 grade contains 46% of available nitrogen and a 14-14-14 product contains 14% nitrogen as N, 14% available P_2O_5, and 14% K_2O (potassium oxide).

Consumption. Since World War II world consumption of fertilizer has increased in yearly increments ranging from 6 to 10%. It may be expected that as tonnage increases during the 1970's the annual percentage gain may fall towards 5% per year. In 1970, the consumption of primary nutrients was about 62.8 million metric tons: 28.5 N, 18.5 P_2O_5, and 15.8 K_2O. In the United States, consumption in 1970 increased only about 3% over 1969 and totaled 39 million tons of materials, which represented 16 million short tons of plant primary nutrients (7.4 N, 4.6 P_2O_5, and 4.0 K_2O). Of this, 25% was in the form of fluids [aqua and anhydrous ammonia, nitrogen solutions, and mixed (multicomponent) clear liquids and suspensions], 35% as bagged solids, and 38% as bulk solids. Much of the solid fertilizer was produced in granular form, especially as compound products, which contain more than one nutrient.

Nitrogen. Over 95% of the nitrogen for fertilizer use is produced from synthetic ammonia, made by reacting nitrogen and hydrogen under high pressure and temperature. The nitrogen is derived from the air. The hydrogen may be derived from a number of sources such as natural gas, naphtha or other petroleum fractions, or coal. Generally, natural gas or naphtha is preferred because of their relatively high purity. As a source of nitrogen for fertilizers, anhydrous ammonia itself is applied directly to the soil in large tonnages. It also is used as a raw material for the production of nitrogen solutions, nitric acid and nitrates, urea, and ammonium salts of phosphoric acid.

In addition to ammonia, the major nitrogen fertilizer materials are ammonium nitrate, ammonium sulfate, and urea. Solutions containing nitrogen also comprise a very important class of fertilizer materials; they normally contain one or more of the following materials: ammonia, ammonium nitrate, and urea. Calcium nitrate, sodium nitrate, calcium cyanamide, and natural organics are minor sources of long standing which continue at more or less the same supply levels but which, as indicated earlier, contribute a steadily diminishing proportion of the total nitrogen consumed.

Phosphate. Phosphate is derived from phosphate rock, the principal mineral constituent of which is a complex fluorapatite generally indicated as $Ca_{10}(PO_4)_6F_2$. It is found as ore deposits in conjunction with clay, quartz, and other impurities. Twenty-eight countries reported some product in 1969. Major producing countries were the United States, USSR, Morocco, and Tunisia. Little of the phosphate in most of the phosphate ores is readily assimilated by growing plants until it is dissociated from the fluorine, which is accomplished by treatment with acids or heat. Before such treatment, the clay and quartz gangue are removed from the phosphate rock by beneficiation methods such as washing and flotation to conserve acid or heat requirements and to obtain more concentrated products.

Some consider the phosphate fertilizer industry to have originated nearly 130 years ago with the production of "normal" superphosphate (18-20% available P_2O_5), often also called single or ordinary superphosphate. The process involves mixing pulverized phosphate rock with sulfuric acid so that the calcium will react to form $CaSO_4 \cdot 2H_2O$ (gypsum). The available phosphate is in the form of monocalcium phosphate, $Ca(H_2PO_4)_2$, in admixture with gypsum and other impurities, some of which are valuable micronutrients. Although normal superphosphate still is the important fertilizer in some developing countries and in some areas of developed regions where markets are close to the phosphate fields, the trend throughout the world, because of economics in handling and transportation, now is toward the production of more concentrated materials such as triple superphosphate, sometimes called concentrated superphosphate (46–48% P_2O_5), diammonium phosphate (18-46-0), and nitric phosphate (20-20-0); potash and additional nitrogen may be added during or after manufacture to produce a variety of grades. Triple superphosphate is made from phosphate rock and phosphoric acid and contains about 2.3 times the P_2O_5 in ordinary superphosphate. Diammonium phosphate is made from phosphoric acid and ammonia. Production of nitric phosphates involves acidulating the rock with nitric acid. Subsequent processing may include partial removal of calcium before ammoniation.

Phosphoric acid can be made by several methods including (1) the "wet-process" method, in which phosphate rock is treated with sulfuric acid sufficient to convert all the calcium to gypsum, which is removed by filtration, and (2) the thermal method, in which the ore is smelted under reducing conditions to form elemental phosphorus, which is converted by oxidation and hydration to ortho or polyphosphoric acid. Method (1), through which impure orthophosphoric acid is obtained, is dominant. However, with recent increased interest in fluid fertilizers, some of the acid is concentrated to such a degree as to form polyphosphates, which are sequestrants for impurities and micronutrients and permit the production of high-analysis fluid fertilizers. Phosphoric acids produced by thermal reduction of rock generally are very pure; however, production of acid for fertilizer use by this method is limited by economic considerations.

Potassium. Potassium is found in several parts of the world. The largest known reserves are in the USSR, Canada, and Germany (Federal Republic and Eastern). Very significant deposits also are present in Israel and Jordan (Dead Sea), Spain, France, the United States, and the United Kingdom. Deposits are as water-soluble salts such as sylvite (KCl), langbeinite ($K_2SO_4 \cdot MgSO_4$), carnallite ($KCl \cdot MgCl_2 \cdot 6H_2O$), and kainite ($KCl \cdot MgSO_4 \cdot 3H_2O$). The major source of potassium for fertilizer use (over 90%) is potassium chloride (from sylvite); some potassium sulfate is used mainly for crops, such as tobacco and potatoes, whose quality may be affected by high levels of chloride; langbeinite is used where both magnesium and potash are needed.

Sylvite ore (about 20-30% KCl, usually mixed with NaCl) is mined and beneficiated by solid-flotation and solution-crystallization processes. The resultant products usually contain 60 to 63% K_2O, and are marketed on size bases as requested for use in granulation or dry blending with other fertilizer material or in fluid fertilizers. Early sources of potash, now used in very minor quantities, include cotton hull ashes, wood ashes, and potassium carbonate.

Trends. New technology has changed the fertilizer industry from a rock-acid mixing business for supplying phosphate and the use of sodium (Chilean) nitrate for nitrogen to a chemical industry in which fertilizer materials are produced in large complex plants. With increased demand for fertilizer, with development of marketing, and with increasing attention to fertilizer requirements as dictated by soil analyses, the industry, especially in the United States, can look forward to increased demand for more customer service and "prescription" type fertilizers. This demand probably can be met best through the use of dry blends and fluid fertilizers which, in 1970, represented about 40% of all multinutrient fertilizers in the United States.

There probably will be a continued increase in the proportion of nitrogen supplied as urea because of its high plant nutrient concentration (46% N) and the fact that the tremendously large plants (1500+ tons/day) being built to produce it probably will make it the lowest-cost nitrogen source other than ammonia. There is a very significant move toward use of polyphosphates which permit higher concentration of fluid fertilizers, sequester some of the troublesome impurities in wet-process acid, and have a high solubility for most of the micronutrient elements. New high-analysis NP and NPK products being studied and expected to move rapidly into industry are combinations of urea and ammonium phosphate and ammonium polyphosphate. Typical grades may be (N—P_2O_5—K_2O) 36-18-0, 29-29-0, and 19-19-19. Also, increases in the use of fluid fertilizers, both clear liquid and suspensions, are expected.

Controlled release fertilizers, made by coating fertilizer granules with relatively water-insoluble materials such as sulfur, probably will receive considerable attention in the near future, since such coatings appear to be the only low-cost practical means for controlling nutrient release to better match the requirements of growing plants.

JULIUS SILVERBERG

Cross-reference: *Soil Chemistry.*

References

"Fertilizer Technology and Usage," Second Edition. R. A. Olson, T. J. Army, J. J. Hanway, and V. J. Kilmer, Eds., Soil Science Society of America, Madison, Wisc., 1971.

"Micronutrients in Agriculture," J. J. Mortvedt, P. M. Giordano, and W. L. Lindsay, Eds., Soil Science Society of America, Madison, Wisc., 1971.

"Fertilizer Manual," United Nations Industrial Development Organization, New York, N. Y., 1967.

"Chemistry and Technology of Fertilizers." Vincent Sauchelli, Ed., Reinhold Publishing Corporation, New York, N. Y., 1960.

"Manual on Fertilizer Manufacture," Vincent Sauchelli, Third Edition, Industry Publications, Inc., Caldwell, N. J., 1963.

"Fertilizer Developments and Trends," A. V. Slack, Noyes Development Corporation. Park Ridge, N. J., 1968.

"New Fertilizer Materials," Y. Araten, Ed., Noyes Development Corporation. Park Ridge, N.J., 1968.

FIBERS, NATURAL

Natural fibers may be divided into three elementary classifications, animal, vegetable and mineral, a large number making up each category. Many are of commercial importance; others of only academic interest.

Classification

Animal Fibers. Wool, hair and fur are largely the protective hair or fur of animals. Of these, sheep's wool is by far the outstanding animal fiber of commerce. For normal use as textile materials, the fibers are spun into yarns for subsequent weaving or knitting into cloth. Other animal fibers used in significant amounts are alpaca, angora, badger, beaver, camel's hair, cashmere, llama, mohair, rabbit and vicuna. Fiber cost depends upon availability and general quality. The scarcer and finer luxury fibers, particularly cashmere and vicuna, command higher prices. The fibers are normally clipped from the live or "pulled" from the dead animal and thus exist as "staple," i.e., they have short discrete lengths ranging from one-half up to 15 inches or more. Sheep's wool fibers are 10 to 70 microns in width and 1½ to 15 inches in length. They are almost uniformly cylindrical except for the tapered natural ends. Cross sections are oval to round. In common with other animal hairs the surface of the wool fiber is made up of flat scales which overlap like shingles. The free ends of the scale cells project outward and point toward the tip of the hair, thereby giving the outline of the fiber a barbed appearance.

Other animal fibers of a coarser nature are cattle, goat and horse hair, and hog bristles. These are mostly used as resilient wadding or stuffing in mattresses and upholstered furniture.

The silkworm or *bombyx mori* extrudes long continuous filaments, in which it then wraps itself to form its cocoon. During yarn manufacture the cocoons are floated on warm water and the fibers are unwound and then rewound onto a reel. Silk is the only continuous filament (i.e., of essentially infinite staple length) natural fiber of commercial value. It is mostly used to make smooth, uniform yarns for apparel and decorative fabrics where surface luster and luxurious "hand" are desirable. Silk filaments can also be cut to any desired staple length and spun into yarns.

Vegetable Fibers. Vegetable fibers are divided into three sub-classifications:

(1) *Seed and Fruit Hair Fibers.* Cotton is the most well-known and widely-used fiber in this class. Cotton fibers are elongated single-celled outgrowths of the cotton seed. A mature fiber is a flattened tube with characteristic irregular twists or convolutions, and with a central canal or lumen extending throughout its length. Cross sections are often bean shaped, although configurations ranging from

U-shaped to circular are encountered depending upon the degree of maturity. Fibers from different varieties of plants vary considerably in length and fineness. Short staple fibers are about ½ inch long, but the long staple types may exceed 2 inches. The range of fiber widths is about 12 to 25 microns. Cotton types include Upland, American, Sea Island, Peruvian and Egyptian. Cotton fabrics of a large variety of constructions and weights are used for apparel, decorative, and industrial purposes.

Other fibers in the seed and fruit hair category, all having limited use as stuffing and wadding, are akund, kapok and milkweed.

(2) *Bast Fibers.* Bast fibers are taken from the stalks of dicotyledonous plants. They are found just under the outer bark and exist as bundles held together by cellular tissue, gums, and waxes. Over 100 botanical varieties are listed, but the most commercially important are flax, from which linen is derived, hemp (*Cannabis Sativa*), jute, ramie, kenaf and sunn. Of these, jute is probably used in greatest tonnage for bagging, cordage, and the reinforcing backs of carpets and linoleums. Linen has limited use for apparel textiles, and is especially suited for tablecloths, napkins and toweling. Ramie, in spite of the wide publicity which has been given it in recent years, remains as a fiber of rather restricted application for special or expensive apparel.

(3) *Leaf Fibers.* These are also called hard fibers because most of them are harder, stiffer, and coarser than the bast fibers. The two most important are abaca (*Musa Textilis*) and sisal (*Agava Sisalana*). Abaca, often inappropriately called manila hemp, is the most widely used and is probably the best of the hard cordage fibers. Staple lengths run to as much as 12 feet; it is processed into rope on heavy spinning machinery. Other useful cordage leaf fibers are henequen, cantala, maguey, phormium, istle, sisal, pineapple fiber, sansevieria and yucca. Palm leaf fibers are used for brushes and brooms.

Mineral Fibers. Chrysotile and chrocidolite asbestos are included in this category. The former, found in Canada and Russia, has a magnesium silicate base. The latter, found in Australia and South Africa, has an iron silicate base. Asbestos is primarily used in products where fire and heat resistance are needed. Because of its brittleness, 100% asbestos fiber yarns are not normally spun, but when mixed with more flexible fibers, usually up to about 20% cotton, excellent industrial heat resisting products are produced.

Chemical Properties

Animal Fibers. Commonly referred to as keratin, wool (and all other fibers in this class) is composed of a complexity of amino acids, one of the most important being cystine:

$$[COOH \; CH(NH_2)CH_2—S—S—CH_2$$
$$(NH_2)CH \; COOH].$$

Some 18 others have been identified, including glycine, alanine, tyrosine, histidine, aspartic and glutamic acids, arginine and lysine. The main molecular chains are composed of amino acid units (polypeptides) connected or crosslinked with relatively stable disulfide (—S—S—) bonds, and with less stable salt linkages derived from acid side chains of aspartic and glutamic acid, and basic side chains of arginine and lysine. The disulfide bonds are said to be responsible for the excellent ability of the wool fiber repeatedly to extend and contract, thus contributing to its good wrinkle and abrasion resistance, and dimensional stability. The acid and basic side chains are responsible for its amphoteric ability to combine with both acids and alkalies, more especially with "acid" and "basic" dyestuffs. Wool contains 3-4% sulfur, about 13% cystine, 16% nitrogen, and 0.2% ash. It is neither well oriented molecularly nor does it have a high degree of crystallinity. Rather its molecular structure is considered relatively amorphous or random as compared with silk, ramie, linen and cotton.

Silk is a similar natural protein composed of fibroin (the true fiber portion), and sericin, a natural gum which coats the two fibroin filaments extruded by the silkworm. The sericin is easily soluble and dispersible in slightly alkaline warm water. Silk is chiefly composed of the simpler amino acids; glycine, alanine and some tyrosine. It has no cystine-like crosslinks or other side chains. Thus the polypeptide chains can assume a well-oriented and close-packed structure. Chains cannot fold and unfold as in the case of wool, and this causes a lesser extensibility in silk. However, at low tensile stress levels the elastic deformation-recovery properties are good even though the magnitudes are less.

Cellulosic Fibers. These fibers in their purified state, free from natural waxes, pectins, and lignins are composed primarily of cellulose, the simple empirical formula being $(C_6H_{10}O_5)_x$. The end product of cellulose hydrolysis is glucose, and it has been established that cellulose is composed of long chains of two β-glucose units called cellobiose, bound together by a 1,4-β-glucosidic linkage. Polymeric cellulose chains are formed when cellobiose units condense with the splitting out of water between the hydroxyl groups attached to carbon atom 1 of one cellobiose unit and carbon atom 4 of an adjacent unit. Thus the long chains of anhydro-β-glucose units are built up via 1,4 oxygen bridges.

Other cellulosic fibers, such as flax and ramie, require considerable preparation to separate the useful fiber from the plant. When this is done mechanically, as in the case of ramie, it is called *decortication*, and involves the separation of the fine, long fibers from the ramie plant stalk. When separation is done chemically, as in the case of flax, it is called *retting*, and consists of immersing the stalks in stagnant river water. The microorganisms in the water attack the cementing media which fasten the fibers to the flax straw, thus permitting the fibers to be separated out.

Leaf fibers are normally used in their natural unpurified state for industrial purposes where the aesthetic or comfort properties of apparel and decorative textiles need not be taken into consideration. Thus, manila abaca fiber is used for cordage without separating the cellulosic constituents from the accompanying lignin.

Physical Properties

Wool is a relatively weak but highly extensible fiber. Its "tenacity" (breaking strength per unit weight per unit length) is 1–2 grams per denier (gpd)* as compared with 3–6 gpd for cotton and 4–9 gpd for nylon. Its rupture elongation ranges from 20 to 50% depending upon type. Not only does wool have a great capacity to absorb energy, it can do this repeatedly (at below rupture loads) because of its ability to deform under load and recover upon load removal. This combination of properties causes wool textiles to exhibit good crease retention, wrinkle resistance, abrasion resistance, and dimensional stability. Wool is hygroscopic, having a moisture regain at standard conditions (65% R.H. and 70°F) of 15%. When immersed in water its cross-section area swells 35%. Because of its ability to absorb and transfer moisture, it is accepted as being a particularly comfortable fiber for apparel fabrics.

Wool exhibits excellent resistance to weak and strong acids, but is severely damaged by alkalies. It is unaffected by most of the common dry-cleaning solvents. Upon prolonged exposure to heat or to sunlight it tends to yellow and to lose strength. It is attacked by moths and while it may be subject to mildew attack, the relative humidity and temperature must be high before mildew sets in. Wool is nonthermoplastic; it does not melt at elevated temperatures, but rather chars and decomposes. While it will burn when exposed to a direct flame, it will not support combustion and is considered relatively safe insofar as flammability danger is concerned.

One of wool's outstanding properties is its ability to felt. This is due to the fiber's scale structure, which imparts different friction coefficients in the root-to-tip and tip-to-root directions. The resulting "Differential Friction Effect" (D.F.E.), coupled with the ability of the wet fiber to extend and contract, causes the fiber masses upon mechanical agitation in the presence of heat and moisture, continuously to tangle and compact, thus producing a felted or "fulled" fabric without the need for spinning yarns and weaving fabric. Felting may also be a disadvantage in that wool textiles shrink and mat when wet-laundered. Much effort has been expended in developing the following antishrink treatments: (1) oxidation process, e.g., wet or dry chlorination or bromination, permanganate treatments or peroxides; (2) hydrolysis processes, e.g., alkalies, alcoholic alkalies, proteolytic enzymes; (3) resin applications, e.g., melamine formaldehyde, vinyls, acrylics and polyamides. Only the first and third methods, and combinations thereof, are now used commercially.

Silk. Silk has a tenacity of between 3 and 6 gpd, while nylon and polyesters can be manufactured with strengths up to 9 gpd. It is a nonthermoplastic fiber which exhibits excellent recovery from below-rupture strains. It has a moisture regain of about 9½% at 65% R.H., approximately three times the regain for nylon. For this reason it normally is considered to be a more comfortable apparel fiber than nylon and other synthetic hydrophobic fibers. Silk does not exhibit particularly good resistance to sunlight on prolonged exposure. It is more resistant to moths than wool, but it may be attacked.

It has generally high resistance to microbial attack, but it may be discolored by molds. Silk does not exhibit resistance to strong acids as does wool, and is very sensitive to alkalies.

Cotton. The cotton fiber has a tenacity range of from 3 to 6 gpd, the wet strength being from 110 to 130% of the dry strength. Its rupture elongation is in the order of 3 to 7%. Cotton has a higher tensile modulus and is a relatively brittle fiber as compared to wool and silk. Its large energy absorption capacity is attributable to its high strength rather than its high extensibility. Cotton does not exhibit particularly good strain recovery characteristics, and untreated cotton fabrics have poor crease retention and wrinkle resistance. Recently great commercial success has been achieved in developing "durable press" or wrinkle resisting-no ironing cotton apparel fabrics by application of resinous chemicals. These resins react via crosslinking with the cellulose. Typical reactants are dimethylolethylene urea (DMEU), dihydroxy-dimethylolethylene urea (DHDMEU) and various triazones. While these resins are effective with respect to improving wet and dry wrinkle resistance, they result in partial degradation and weakening of the cotton, reducing fabric abrasion resistance to an unacceptable level. To overcome this deficiency cotton fibers are now blended with synthetic fibers, particularly polyesters such as "Dacron," "Fortrel," or "Kodel." The resulting blended fiber fabrics are then treated with the durable press resins. The resin-treated cotton fibers and the inert polyester fibers both contribute high wrinkle resistance, while the polyester fibers contribute strength, toughness and abrasion resistance. Cotton-polyester blend fabrics have now supplanted all-cotton fabrics in a large number of apparel products.

One of cotton's outstanding attributes is that it is actually stronger when wet than dry. This condition, which holds for fiber, yarns, and fabrics, is particularly important for products which are wet laundered. Because of their higher wet strengths, cotton textiles become much more resistant to the tensions and strains which develop in the course of laundering, drying and ironing. Because of its ability to absorb large amounts of water, as well as its launderability and its nonthermoplasticity, cotton is particularly useful for toweling and diapers. It can be sterilized and so is used for all types of surgical and medical purposes.

Cotton will burn readily. It does not melt at high temperatures but chars and decomposes. Fire-retardant treatments that are fast to washing and dry cleaning are currently being developed. The most widely used chemicals are tetrakis hydroxymethyl phosphonium chloride (THPC) and tris 1-aziridynal phosphine oxide (APO).

Linen and Ramie. Most of the comments made above for cotton apply to linen and ramie. These fibers are more highly oriented than cotton and have somewhat higher strengths and lower elongations. Because of their greater cost they are utilized far less for both apparel and industrial purposes.

Hard Fibers. Abaca, sansevieria, sisal and henequen have respective tenacities of 6.5, 5, 4.5, and 3 grams per denier. The first three have 3% rupture elongations, while henequen is a lower modulus fiber having an elongation of about 5.5%. Abaca,

---

* If 9000 meters of filament or yarn weighs one gram, it is called "one denier."

having the highest fiber strength-to-weight ratio, produces the highest strength cordage. The hard fibers are considerably larger in diameter and denier than such "normal" textile fibers as wool, cotton and rayon. For example, the deniers of individual abaca, henequen and sisal fibers range from 300 to 500 while sansevieria has a denier of 75–100. This is in contrast to wool and cotton where fiber deniers are in the 3–10 denier class. Because of their larger size they are considerably stiffer, stiffness increasing as the fourth power of fiber diameter. Thus hard fibers are not used for apparel purposes but are confined primarily to cordage type end uses. In recent years synthetic fibers and monofilaments such as nylon and polypropylene have replaced the natural hard fibers for many cordage applications.

Economic Significance

One of the significant scientific achievements of the twentieth century is the development of man-made fibers. The production of rayon (regenerated cellulose) starting in the U.S. in 1910, and of nylon starting in 1939, has resulted in large changes in fiber consumption in the U.S. and throughout the world. Prior to 1940 more than 80% of all U.S. fiber consumption was cotton. While cotton consumption has remained reasonably constant at over 4 billion lbs per year from 1950 to 1970, the total consumption of man-made fibers has shown large yearly increases (498 million lbs in 1940 vs. 5.5 billion lbs in 1969). Thus cotton's percentage has progressively dropped so that in 1967, for the first time, U.S. consumption was less than 50%. Wool also has shown a progressive decline both in poundage and percentage. It remains, however, an important fiber for use in winter clothing, carpets, and certain industrial applications such as paper maker felts. With the introduction of nylon and other man-made fibers, the amount of silk consumed has significantly diminished (80 million lbs in 1930, 10 million lbs in 1950, and 2.5 million lbs in 1969).

Total pounds per capita of fibers consumed progressively rises, reflecting a continuous increase in our living standard. Cotton and wool per capita consumptions have remained fairly constant; the per capita increases have resulted almost entirely from the large yearly increases in the production of man-made fibers.

World fiber production figures tend to parallel U.S. figures, but with one important difference. The natural fibers, particularly cotton, as a per cent of total fibers consumed, still exhibit a somewhat more dominant position in the rest of the world than they do in the U.S.

Thus in 1967, the first year that the per cent of cotton consumed in the U.S. dropped below 50% (to 49%), world production of cotton was 57% (Table 3). In 1969, as man-made fiber consumption increased, U.S. cotton consumption dropped to 40% while world cotton production dropped to 54%.

Certain general statements may be made about the natural fibers. As a group, but excluding asbestos, they are hygroscopic, nonthermoplastic and are available in a variety of sizes with a variety of mechanical properties. Because of their hygroscopicity they usually afford excellent body comfort

in clothing and are useful wherever water must be absorbed. For the same reason they are easily dyed, producing a wide range of attractive colors that are "fast" to sunlight and laundering. In spite of the continued increase in synthetic fiber production, and the accompanying severe competition by the synthetic fiber producers, there is a continuing place for the natural fibers in our present civilization. This is because they have unique, desirable, and economical properties which man has not yet been able completely to incorporate into synthetics. In the opinion of the writer, cotton, wool, linen, jute, and abaca will continue to maintain significant fiber market positions for the foreseeable future.

ERNEST R. KASWELL

References

Hock, C., in "Harris' Handbook of Textile Fibers," Textile Book Publishers, Inc., New York, 1954.

"Textile Organon," Textile Economics Bureau, New York (a monthly journal).

FIBERS, SYNTHETIC

Most synthetic fibers are produced in both continuous filament and staple form. Some are supplied in the form of tow, which is later broken or cut into staple by the yarn manufacturer. The continuous filament and staple forms of a given synthetic fiber composition may have quite different physical properties, and the chemical properties, such as rate and extent of dye absorption, may also differ because of the higher state of orientation of the filament form. The cross-sectional shape of a synthetic fiber depends greatly upon the method of spinning employed—melt, wet, or dry spinning—as well as on the shape of the hole in the spinneret. Some synthetics are supplied with several different cross-sectional shapes for specific applications. Only those synthetic fibers which are in commercial production or show special long-range promise are discussed here.

Inorganic Fibers

Glass. Glass fiber, supplied in both continuous filament and staple forms, ranks high in production among the synthetic fibers; it is well-suited for electrical and thermal insulation, fireproof curtains, chemical filter fabrics, and plastic laminates. Glass single filament strength is very high, about 500,000 psi equivalent to a tenacity* of 15 grams per denier; glass filament yarns exhibit ultimate elongations of 3–4%. Recent development of yarn impregnants that provide interfiber isolation and glass-rubber adhesion have greatly improved the fatigue properties and glass filament yarns are finding increasing use in industrial fabrics and tire cord. Low-denier glass fiber yarns are also being used effectively in bedspreads and drapery fabrics. Glass fibers are dyed by the exterior application of pigments, either to the fabric or to the yarns. Because of its outstanding strength, wet and dry, high melting point, nonflammability, dimensional stability, outstanding electrical resistance, and excellent resistance to chemicals, mildew, heat, moisture, and sunlight, the glass fiber has a very bright future.

* Tenacity is defined as rupture load/linear density. Denier is the weight in grams of 9000 meters of fiber or yarn.

Fibers for Structural Composites. Rapidly advancing space age technology has generated the need for materials of construction having low specific gravity and exceptionally high strength and rigidity at very high temperatures. Such materials have been made through the use of fiber-reinforced composites. Several glasses with improved fiber properties, "S-glass" for example, have been developed for this purpose and quartz fibers are now commercially available. However, in order to attain still higher strength and modulus, a wide range of exotic fibers have been studied. Among those available in 1970 were carbon and graphite fibers obtained by pyrolysis of rayon or acrylic yarns, fibers formed by condensation of boron or silicon carbide from the vapor phase upon hot tungsten wire, and sapphire single-crystal fibers grown continuously by pulling a seed crystal from an alumina melt through a controlled temperature gradient. Other fiber-forming materials investigated include beryllium, boron nitride, boron carbide, spinel, titanium diboride, ceramics, metal alloys, and whiskers (single crystals formed by vapor phase growth). Fibers with fantastic physical properties have thus been obtained and, naturally, their prices are equally fantastic. However, these prices are justified in view of the very high operating cost of each pound of structural material built into a jet plane or other aero-space vehicle.

Metal Fibers. A number of metal fibers formed of a variety of alloys have been developed for both terrestrial and space applications. Such fibers are usually made by successive drawing through tungsten or diamond dies to a final diameter as small as 0.002 cm. Steel cord is commonly used in European-made radial tires and has recently been employed in both radial and bias-belted tires in the United States. A very fine stainless steel wire has been developed specifically for dissipation of static charges on carpets. It is reported to be effective when blended in miniscule proportions with other fibers in carpet yarns. These functional metal fibers should not be confused with the so-called "metallic" fibers which are formed by combining thin metallic foil or coatings with polymeric films such as cellophane, polyester, or cellulose acetate. They are used in lace, knitted, tufted, and woven fabrics to give decorative effects.

Regenerated Organic Fibers

Viscose Rayon. The viscose process for regenerated cellulose fibers is accomplished by the following major steps: (1) digestion of wood pulp or cotton linters cellulose in NaOH to form alkali cellulose; (2) aging of alkali cellulose to reduce cellulose D.P. (that is, the degree of polymerization) to optimum value for spinning; (3) formation of cellulose xanthate by reaction with CS_2; (4) dissolving xanthate in dilute NaOH; (5) "ripening," filtering, and deaerating xanthate spinning solution; (6) extrusion of xanthate through multiorifice spinneret into a coagulating and regenerating bath (acid solution of metal salts), stretching the filament yarn or tow to produce the desired orientation; and (7) washing, desulfurizing, and drying the yarn or tow. The result is a regenerated cellulose fiber or crenulated cross-section which may be supplied as a continuous filament yarn, or as a tow which may be cut or broken into staple fibers of any desired length. Viscose rayon cellulose differs from native cellulose essentially in that the D.P. is considerably lower, the crystallinity is lower, and the unit cell is slightly altered. The morphology of the native fiber is, of course, completely lost in the viscose process.

Viscose rayon is hydrophilic (13% moisture absorption at 65% R.H., 70°F). Consequently, it may be dyed without difficulty by several methods and presents no problems as regards generation of static electricity and body comfort. The traditional "regular tenacity" viscose does not have very impressive physical properties—tenacity 1.5–2.6 grams per denier "dry"* and 0.7 to 1.8 grams per denier wet. The corresponding ultimate elongations are 15–30% dry and 20–40% wet. It swells greatly in water and has been noted for its unfortunate progressive shrinkage during consecutive launderings. In spite of these deficiencies, it has long maintained the top position among man-made fibers.

The metamorphosis of viscose rayon began with its introduction into tire cord in continuous filament form. The severe demands of this application led to the development of "high-tenacity' filament rayon (3–6 grams per denier). This was achieved by a combination of increased fiber orientation and higher skin to core ratio. It was not long after this development that high-tenacity viscose rayon almost completely replaced cotton in tire cord. Rayon is still important in this field, once the greatest single market for cotton, and competes quite effectively against the aggressive encroachment of nylons, glass, and polyesters.

A more recent development is high wet-modulus rayon, in which the microcrystalline structure has been modified to give a marked increase in stiffness at low strains, both wet and dry. These fibers exhibit increased tenacity and somewhat lower swelling in water. They also are of higher D. P. and have a circular cross section. But the secret in achieving greatly reduced progressive laundering shrinkage, approaching cotton in this respect, is their high wet modulus. The acceptance of these new rayons has been spectacular and the future of regenerated cellulose as an apparel fiber, not long ago a subject of despair, has brightened considerably.

Cuprammonium Rayon. This regenerated cellulose fiber, sometimes referred to as "Cupra" or "Bemberg rayon," is produced by dissolving cellulose in cupric ammonium hydroxide and spinning into a water bath, followed by "hardening" in acid. The fibers thus produced resemble viscose rayon very closely in chemical properties, but they are circular in cross-section and very fine filaments may be produced. Cuprammonium rayon occupies a relatively minor position with respect to rayon made by the viscose process, but it is still produced in quantity and finds application in knitted and woven wearing apparel, upholstery, and decorative fabrics.

Derivative Fibers

Cellulose Acetate. Acetate ranks high in American consumption of synthetic fibers. The cellulose

---

* The term "dry" refers to the standard testing condition —65% R.H., 70°F.

acetate is prepared in the following major steps: (1) acetylation of wood pulp or cotton linters cellulose by soaking in glacial acetic acid; (2) formation of cellulose triacetate by reaction with acetic anhydride in presence of acid catalyst; (3) hydrolysis of triacetate in acetic acid; (4) precipitation of resulting "secondary acetate" in water; and (5) washing and drying of the flakes. The "secondary acetate" has a degree of substitution of about 2.3, as compared to 3.0 for theoretically complete substitution. It is dissolved in acetone to form the spinning solution which is extruded from a spinneret, downward into a warm air column where rapid evaporation of the solvent takes place and the fibers are oriented under gravity. Both continuous filament and staple fibers are produced.

Acetate may be considered the ancestor of the newer hydrophobic fibers, for it is thermoplastic and has a "livelier" handle than its cousin, viscose. Acetate is on the borderline between the hydrophilic and the hydrophobic fibers, absorbing only about 6% moisture under standard conditions. It is not particularly strong. The "regular" variety has a dry tenacity of 1.3 to 1.5 grams per denier and a wet tenacity ranging from 0.8 to 1.2 grams per denier. Its ultimate extension may be as high as 50% dry and 40% wet. Acetate is somewhat "wool-like" in mechanical behavior and yields better wrinkle recovery than untreated viscose in equivalent fabrics. It cannot be dyed by ordinary methods; in fact, an entirely new type of dyestuff, the so-called dispersed dye, had to be developed for acetate. It is subject to atmospheric fading and shade changes, and this has led to the "dope-dyeing" technique where the colorant is incorporated into the spinning bath. Acetate suffers from several minor weaknesses: it is soluble in acetone, creates static in processing, and has a low softening point. It has many compensating good properties, however, and because it can be produced at low cost, will continue to be an important apparel fiber. Vast quantities of staple acetate are used in cigarette filters.

Cellulose Triacetate. A cellulose triacetate fiber is considerably more hydrophobic than conventional cellulose acetate and has a much higher softening temperature. Thus, it has the wet dimensional stability of the other hydrophobic fibers, such as the polyamides, acrylics, and polyesters, but is much cheaper to produce. It is colored with acetate dyes. Triacetate is superior to ordinary acetate in that fabrics made from it are fast to machine washing, can be permanently pleated and can be ironed at higher temperatures. Its level of wash and wear performance is quite satisfactory, although somewhat below that of the polyesters. Both the filament and staple varieties of triacetate have enjoyed considerable success, but so, also, has ordinary acetate in recent years.

Synthesized Fibers

Polyamides. The first synthesized fiber to meet with commercial success was nylon 66, formed by condensation polymerization of the salt resulting from the reaction of hexamethylene diamine with adipic acid. It is melt-spun into continuous filament yarn or tow and oriented by cold or hot drawing. Highly oriented nylon 66 has a tenacity of 6–9 grams per denier, a high extensibility, and an unusually low modulus at low strains. Its resilience characteristics are good, although the time for recovery from large strains is rather high. It has a reasonably high melting point of about 264°C and a good resistance to alkalies. It is degraded by acids. It absorbs about 4% moisture at 65% R.H., 70°F.

The great strength and ability to absorb energy possessed by nylon 66 have made it useful in many industrial and military applications where it is employed in continuous filament form. The continuous filament found early application in women's hosiery yarns, where it has replaced silk. It is also used in tire cord, carpets, satins, taffetas, marquisettes, and women's lingerie fabrics. The staple form of nylon 66 is made with lower orientation and thus has a somewhat lower tenacity. It can be crimped by several methods and is blended with other fibers on the woolen, worsted, and cotton systems of spinning. It does not have the bulking power of some of the other fibers, in spite of its low specific gravity (1.14), for it has the circular cross-section characteristic of a melt-spun fiber. The shape of the cross-section can be varied, however, to give excellent carpet yarns.

Another very important polyamide fiber, nylon 6, had its early development in Europe. It is made by the polymerization of caprolactam involving the conversion of the caprolactam to amino-caproic acid which polymerizes to nylon 6. Nylon 6 is said to be somewhat cheaper to produce than is nylon 66. It has a lower melting point—about 223°C—but in other physical and chemical properties, is very similar to nylon 66. Nylon 6 is produced in quantity by a number of American and European firms. It has been successful in many applications ranging from apparel to tire cord.

The polyamides offer almost limitless possibilities for new fibers. Examples of higher nylons in commercial use are: nylon 610 for bristles, nylons 7 and 9 (Russia), nylon 11 (France, Italy and Brazil), and nylon 91 (Japan). Nylon 4, based upon 2-pyrrolidone, was produced in pilot plant quantities in 1970. It is unique among the polyamides in its high moisture regain (9%) and dyeability with cellulose dyes. "Nomex," a wholly aromatic polyamide fiber thought to be poly-m-phenylene isothalamide is now produced in both staple and continuous filament form. It does not melt and burns with difficulty, forming a char at temperatures above 370°C. Its great thermal stability, strength, and resistance to flexing, abrasion, and chemicals bring it into many demanding applications such as filters and flame-protective clothing. A recently developed dyeable form is being used in aircraft interiors. Still newer on the polyamide fiber scene is "Qiana," announced in 1968. Referred to by its manufacturer as an alicyclic polyamide, this fiber is claimed to have all the aesthetics of silk with the performance properties of nylon 66.

Acrylics. Several synthesized fibers based principally upon acrylonitrile are now in quantity production. American acrylic fibers are supplied only as staple. They are formed either by "dry spinning" (extrusion of the polymer in solution into a heated chamber in which the solvent evaporates) or by "wet spinning" (extrusion of the polymer solution into a coagulating bath). Various solvents have been used, prominent among them being dimethyl formamide. The earliest form of "Orlon" was essentially polyacrylonitrile, but because of dyeing and other difficulties, it now consists of a copolymer in which acrylonitrile is the major constituent (85% or higher). This is also true for the other acrylics and great secrecy surrounds the nature of the minor constituents. It is known, however, that vinyl pyridine and methyl methacrylate have been employed. "Zefran" differs from the other acrylics in that it is referred to as a "nitrile alloy" polymer. It has a very much higher tenacity—3.3 to 4.2 grams per denier compared to 2.0 to 3.0 grams per denier for the other acrylics and has different dyeing characteristics.

The acrylic fibers are quite hydrophobic with water absorption ranging from 1.5 to 2.5% under standard conditions. They are of low specific gravity (1.16–1.18) and possess unusual bulking power, either alone or blended with other fibers. They are relatively free of pilling and thus are suitable for knitted or loosely woven fabrics. A special bicomponent acrylic fiber with permanent crimp behaves very much like wool and is excellent for bulky sweaters; indeed its bicomponent structure is patterned after the natural bilateral asymmetry of the wool fiber cortex. The acrylics have also found large markets in blankets and carpets.

Modacrylics. The generic term modacrylic has been adopted by the U.S. Federal Trade Commission to describe a fiber composed of a substance in which less than 85%, but at least 35%, by weight of the material is acrylonitrile. Only two modacrylic fiber types are produced in the United States, "Dynel," which is a 40/60 copolymer of acrylonitrile/vinyl chloride and "Verel," which is composed of acrylonitrile and vinylidene chloride. "Dynel," the older of these fibers, is characterized by moderate tenacity (3.5 grams/denier), low water absorption (0.4% under standard conditions), and low softening temperature. "Verel" is less hydrophobic, absorbing 3.0% water under standard conditions, and has a somewhat higher softening temperature. Both fibers are unusual in that they do not support combustion and have excellent chemical resistance. They are dyeable with disperse dyes without carriers. Cationic dyes are also effective. The modacrylics have found volume only in the limited areas of work clothing, carpeting, synthetic furs, fire-resistant drapery fabrics, and in crease-retaining fabrics for men's and boys' slacks. "Dynel" is dominant as a fiber in wigs.

Polyvinyls. A great number of polyvinyl halide fibers have been developed in the United States, Europe, and Japan. Many of these have been based wholly upon polyvinyl chloride. Only a few of these vinyl halide fibers have survived the pilot plant stage for they are inherently low in strength and dimensional stability and have very low softening temperatures. They are useful for plastic molding purposes and have limited industrial application because of their chemical resistance and nonflammability. Vinyl halide fibers now commercially available include the French "Rhovyl" and the German "PeCe" polyvinyl chloride fibers. "Vinyon HH," an 86/14 vinyl chloride/vinyl acetate copolymer, is the only commercial survivor of the vinyl halide fibers in the United States.

A very different picture is presented by the polyvinyl alcohol fibers, which were developed in Japan. Insolubility of the polyvinyl alcohol fiber is achieved by reaction with formaldehyde. Fibers with a tenacity of 9.0 grams per denier and quite good abrasion resistance can be made. The ultimate elongation is rather low (10 to 20%). The dry melting point lies between 210° and 232°C. The fiber can be dyed to excellent fastness with vat dyes. The staple resembles cotton in handle while the filament is similar to silk. The fiber is commercially produced in Japan.

Polyvinylidenes. Several fibers based primarily on vinylidene chloride, but generally containing a minor component such as vinyl chloride to permit chemical and physical processing, are produced in sizable quantities. For example, saran and "Rovana," made in the United States, are extremely hydrophobic, chemically resistant, and moderately strong (2.5 grams per denier in highly oriented form). Their major application has been in rather coarse monofilament form for weaving seat coverings for public vehicles, nonstaining window screens, etc. They have shown considerable promise as carpet fibers. They offer difficult problems in dyeing and are generally supplied in dope-dyed form.

Polyesters. The first of this important class of fibers was developed in England in 1941. The U.S. patent rights were acquired in 1946 and the polyester fiber "Dacron" was introduced to the American market. It is composed of polyethylene terephthalate which can be formed by direct esterification of ethylene glycol or by catalyzed ester exchange between ethylene glycol and dimethyl terephthalate. The fiber is melt-spun and is drawn at an elevated temperature. The melting point is 264°C and the moisture absorption under standard conditions is about 0.4%. It is furnished in filament, tow, and staple forms. The tenacity of the staple fiber, wet or dry, ranges from 2.2 to 4.0 grams per denier. Tenacities up to 9.5 grams per denier are attained in the continuous filament variety. The outstanding characteristic of the polyesters is high resilience under wet conditions. Blended with wool, polyesters confer excellent wrinkle-recovery and crease-retention behavior upon worsted fabrics. Blended with cotton or rayon, polyesters yield fabrics with the so-called "durable press" or "minimum-care" behavior that have achieved wide consumer acceptance. Unlike the all-cotton "durable press" fabrics, polyester-cotton blends do not exhibit lowered strength and abrasion resistance; instead the polyester component improves fabric behavior in these respects. The early problems of pilling, static electrification, and dyeing, characteristic of hydrophobic fibers, have been largely solved and it may be said that introduction of the polyesters was the beginning of a renaissance in apparel. The continuous filament polyester yarns have also been very successful, particularly in industrial applications. Recent experience with polyesters in tire cords indicates that these fibers will largely replace nylon in this important market.

Polyolefins. The first of the olefin polymers, low-density polyethylene, was not successful in fibers.

Fibers were weak, impossible to dye, very low melting, and poor in sunlight resistance. With the advent of low-pressure polymerization and the resulting linear, high-density polyethylene, the fiber potential of polyethylene increased. Nevertheless, its use as a fiber has been extremely limited. It is available only in continuous monofilament form. It can be obtained in numerous pigmented colors and with either low or high thermal shrinkage values. Its low softening temperature and "waxy hand" make it unsuitable for apparel. Its unusual chemical resistance and low cost, however, have brought it into limited and very specialized fiber usage.

Polypropylene was hardly more promising as a fiber than polyethylene until the discovery of stereospecific catalysts by Natta and Ziegler. Professor Natta was the first to demonstrate the possibility of making isotactic polypropylene fibers with much higher melting temperature and strength. Primarily because of the very low cost of propylene, the chemical industry became very interested in its potential as a base for films and fibers. Enormous investments in facilities for producing isotactic polypropylene fiber and/or film have since been made in the United States. Several American firms are supplying polypropylene fiber in monofilament, multifilament, staple, and tow forms. Polypropylene monofilaments and multifilaments are being used in rope, cordage, webbing, and auto seat covers. Bulked continuous filament is being used extensively in carpets. Its low specific gravity (0.90) offers advantages in several marine applications and, combined with low monomer cost per pound, makes its use attractive in fabrics where the number of yards per dollar is the significant cost factor. Tenacities of the commercial fiber are 4.5 to 8.0 grams per denier for the multifilament; 3.0 to 6.0 grams per denier for the staple and tow varieties. The "recovery" properties are quite good at room temperature. It must be noted, however, that properties decay with increasing temperatures and that isotactic polypropylene melts at 165°C and shrinks about 4% at 100°C. This shrinkage can be greatly reduced by heat setting, however. The future of polypropylene as a textile fiber seems to hinge upon two factors: (1) whether the good mechanical properties inherent in its isotactic structure can be preserved in the course of modifying that isotactic structure to achieve dyeability, and (2) how much of a limitation will be its low softening and melting temperatures.

Polyurethanes. A number of commercially unimportant polyurethane fibers were developed as early as 1937. One such product was produced by the reaction of 1,4 butanediol and hexamethylene diisocyanate. The fibers, prepared by melt spinning, melt at 178°C as compared to 264°C for its nylon 66 cousin. Recently, the segmented polyurethane fibers with elastomeric properties have come into prominence and bid fair to revolutionize older concepts of garment manufacture and performance. Indeed, the long reign of the rubber elastomers in foundation garments and in the stretch portions of other textile items appears to be challenged by these new, dyeable polyurethane polymers, classified by the Federal Trade Commission as spandex fibers.

Important among the spandex fibers is a segmented polyurethane. It has an ultimate extension of about 600% and a reasonably good tenacity of 0.6 to 0.8 grams per denier. Obviously, the evaluation of elastomeric fibers depends importantly upon elastic reversibility and the time constants of elastic recovery. It is furnished in coalesced multifilament form. Other spandex fibers are supplied as multifilament yarns. The spandex fibers are superior to rubber in resistance to sunlight, abrasion, oxidation, oils, and chemicals; they have a far superior flex life; can be dyed in a full range of colors, and are offered in much finer fiber diameters.

Anidex Fibers. The new generic classification, anidex, was created by the U.S. Federal Trade Commission in 1969 to cover fiber-forming polymers composed of at least 50% by weight of one or more esters of a monohydric alcohol and acrylic acid, $CH_2 = CH-COOH$. The first elastomeric fiber of this class is a chemically cross-linked polyacrylate elastomer reinforced with a crystalline organic filler. It is supplied as a monofilament of circular cross-section and is spun from emulsion by a process claimed to be entirely new. It is similar to the spandex fibers in its high extensibility but is said to be superior in chemical stability and resistance to light, atmospheric gases, and dimensional changes in laundering. Its proponents appear to have good reason to expect that this new elastomeric fiber will generate renewed interest in "comfort stretch" fabrics for apparel.

Fluorocarbons. Only one fiber has thus far been developed which fits this Federal Trade Commission category. This is "Teflon," based on polytetrafluorothylene. It deserves special mention because of its unusual resistance to chemicals and heat. The monofilament melts at about 288°C. Its most outstanding characteristic is its low coefficient of friction—it is the slipperiest material known. It is furnished as monofilament, staple, tow, and flock. Its applications thus far have been for pump packings, bearings, gaskets, artificial arteries, and electrical insulation.

General

The foregoing summary includes the principal classes of synthetic fibers now in use. Each class is supplied in many special types which could hardly be enumerated in this brief article. It must also be recognized that there are many variations within each class; for example, the fibers employing the principle of bi-componency and the various bulking and stretch treatments. Several types of synthetic fibers that have been omitted from this discussion; for example, the polycarbonate fibers and the nonflammable organic fibers "PBI" and "Kynol." "PBI" (poly-1,3,4 benzimidazole), developed for protection of air crews, has 10% moisture regain, high abrasion resistance, and excellent strength retention in air at 430°C. Flight suits of "PBI" fiber have remained for 3 seconds in a jet fuel fire without burning. "Kynol" is an amorphous organic polymer consisting of cross-linked phenolic units. Its fibers exhibit a tenacity of 1.8 grams per denier and a rupture elongation of 30%. In a flame, they convert to carbon fibers with 60% yield and little shrinkage, giving off CO_2 and H_2O. It should also be noted that a viscose rayon fiber with a "built-in" flame retardant is in commercial production. Several synthetic fibers with internal antistatic agents were announced in 1970. The possibilities of fibers based on the so-called "ladder polymers" have also been omitted; such fibers may invade realms of high temperature never before considered appropriate for organic materials. One thing is certain, the field of synthetic fibers is dynamic. There appears to be no limit to the potential of present man-made fiber types and others yet to come.

J. H. DILLON

Cross-references: *Fibers, Natural; Textile Processing, Polyester Resins, Polyamide Resins, Cellulose Esters, Wood, Glass, Polypropylene.*

FILLERS (EXTENDERS)

A filler, or extender, as used in paints, plastics, and rubber, is a solid material which is added to the basic composition to modify the physical properties of the composition and to reduce costs. Pigments, on the other hand, are similar materials used primarily for their colorant and hiding powers. Fillers may be inorganic, such as clay, talc, and silica, or may be organic, such as woodflour, various fibers, and nut shells. The type of filler used is determined by the processing details and requirements of the finished product.

The general requirements for fillers to be used in plastic and rubber applications are: low moisture absorption; good wettability by resins; no effects on equipment; good electrical resistance; retention of color at high temperatures; and inertness to acids, alkalies, and solvents. Other requirements which are becoming of increasing importance are high heat resistance, nonflammability or low burning rate, and absence of odor. In paint compositions, fillers and extenders are used to add bulk and light weight, control gloss, increase hardness and toughness, remove stickiness, control body, and improve dispersion, to name but a few applications.

Organic fillers have more application in plastics and rubber than in paints. Some of the types used are various forms of cellulose derived directly from wood or vegetable fibers or obtained as by-products from the textile industry. These include woodflour, cotton floc, chopped fabric, twisted cord and walnut shell flour.

Woodflour fillers are perhaps the most commonly used materials in phenolic resin molding compositions. They produce good molding properties and fair physical properties in end products, but absorb moisture and tend to cause lack of uniform electrical properties in molded pieces. Woodflour is prepared by attrition grinding, which retains the maximum strength inherent in the wood fiber and yet gives a finely ground product. Woodflour produced today varies in particle size, but generally is ground so that maximum distribution is around 100 mesh.

Fibers and fabrics are used where strength and high impact resistance are wanted. Fibers of particular interest are asbestos and "Fiberglas." These inorganic materials have been found effective in conjunction with plastic resins, especially where high temperature resistance and impact strength are required. Inorganic fillers are used in paint formulations to: (1) control penetration; (2) maintain gloss at relatively high pigmentation; (3) improve washability; (4) adjust flow and leveling;

(5) eliminate flashing or ghosting; (6) induce self-colored chalking; (7) aid in the dispersion of other pigments; and (8) reduce costs.

One of the oldest and still important fillers for paint and rubber compositions is *whiting,* an inert white crystalline pigment composed principally of calcium carbonate. The physical properties vary, depending on its source, method of manufacture, and the care used in its production. Today there are several specially prepared precipitated calcium carbonates, some of which have been treated to render them more readily wettable by organic liquids. A calcium carbonate of very fine particle size —in the range of 0.03–0.04 micron—is also available. Its value is based on its ability to induce in wet and dry films behavior not normally considered as a function of a pigment. These materials are used to increase the consistency of paint products without adversely affecting other properties. Such calcium carbonates are termed "reagents" rather than pigments.

Diatomaceous earth is another important filler used in paints, plastics, and rubber. This form of silica occurs in nature in the skeletons of microscopic marine animals called diatoms, and is found in deposits laid down in past geologic ages on sea bottoms which have since become land masses (See **Diatomaceous Earth**). The uncalcined grades have a gray color characteristic of the crude rock, while the calcined grades are light pink, due to traces of oxidized iron. This filler improves the appearance, heat resistance, and electrical properties of molded plastic products without increasing their specific gravity. Its principal drawback is its abrasive character and tendency to wear molds and cause stickiness. In paint applications, diatomaceous silica adds bulk and is used as an inert extender for flattening, suspension, and body in inside flat wall paints; for durability, easy brushing, and whiteness in outside house paints; and for superior filling, sanding, and adhesion properties in primers and undercoats. It is especially effective in imparting durability, night visibility, quicker drying, and better suspension to traffic and zone paints.

Silica derivatives are widely used as extenders and fillers for paints, plastics, and rubber. The most common naturally occurring types are calcium and magnesium silicates. These materials have an added advantage as fillers, since they can be used as suspending agents in emulsions. Synthetic silicas made by burning $SiCl_4$ are finding wide use because of their larger surface area and uniform properties.

Mica is an important filler in molded plastics. The principal and most important attributes of this filler are its high dielectric strength and low power factor, dielectric constant, and loss factor. Because of the low water absorption of mica, molded parts maintain these electrical properties in use. Mica is also used in paint applications to improve brushing characteristics. Its disadvantages are that it is not wetted readily by resins, and, in molded articles, mechanical strength is lowered while the specific gravity is increased.

Natural barium sulfate, or barytes, is used where a high specific gravity is necessary. Since it is unaffected by acids and alkalies, it is used extensively as an inert filler in paints and as a base for concentrated colors.

Other fillers or extenders include alumina hydrate, bentonite, calcium sulfate, china clay, infusorial earth, pyrophyllite, and talc.

D. W. LOVERING

Cross-references: *Paints, Rubber, Clays, Diatomaceous Earth.*

FILMS, SURFACE

The unique properties of surface films are derived from the extension of the film-forming material into a layer of large area compared to its thickness. Films are formed spontaneously by the action of intermolecular and chemical forces at an interface existing between two immiscible, bulk phases. The property of spontaneous formation distinguishes a surface film, in the chemical view, from the popular classification of films that includes industrially important materials, such as paints, lubricating greases, thin plastic sheets, and foams or bubbles.

A true surface film cannot exist in the absence of a bulk substrate that supports it and influences its chemical and physical properties. Surface films, because of their property of spontaneous formation, exist at virtually all solid or fluid surfaces, unless extreme precautions are taken to prevent their formation. They may alter drastically the surface characteristics and the industrial usefulness of the substrate, by changing the wettability of a solid by water or organic liquids, by promoting or interfering with chemical reactions at a solid surface, by stabilizing dispersion of two immiscible liquids or an insoluble solid in a liquid. Comparison of the surface properties of the substrate with and without the film present is the basis for the experimental studies of surface films, which have given information on molecular dimensions, intermolecular forces, film structure, and orientation of film-forming molecules.

Studies of films of organic materials on water have been extensive and informative. In films of organic acids and alcohols on water, the carboxyl and hydroxyl groups have an affinity for the water phase, whereas the hydrocarbon chains to which these groups are attached are repelled by the water and forced into the gas phase. Thus, these materials tend to spread into a monomolecular oriented film on the water surface. When the lateral pressure exerted by the film is measured, using a Langmuir surface-pressure balance, it is found that the pressure varies with the area occupied by a known quantity of the film-forming substance. As the area is reduced by force, applied to the barrier confining the film, a sharp rise in pressure indicates the point at which the film is closely packed. The area occupied by each molecule and the thickness of the layer, equivalent to chain length, can then be calculated. The results are in essential agreement with determinations of molecular dimensions by other methods, and confirm the deduction that the film is oriented with the hydrocarbon chains above the surface. With longer chains the film thickness increases while the cross-sectional area per molecule is essentially constant. When more than one hydrophilic group is present, so that each is attracted into the water surface, the surface area occupied by each molecule increases proportionately.

Other experimental methods confirm the oriented structure of such films. It has been found that the rate of evaporation of water from the surface is affected by the length of the projecting hydrocarbon chains. Longer chains oriented side by side in the film are more difficult for the evaporating water molecules to penetrate, and the rate of evaporation is reduced proportionally. Minor traces of impurities in the film, such as nonoriented benzene, cause imperfections or discontinuities that permit the evaporation rate to rise remarkably. Pressure applied to the film tends to reduce the solubility of the impurity and to force it out of the oriented-film structure.

The physical behavior of surface films is often analogous in two dimensions to three-dimensional systems, that is, films can exist in the gaseous, liquid, and solid states. Gaseous films are compressible, as are bulk gases, and the product of area times pressure is approximately constant at constant temperature. Liquid films exist in a number of different types, one of which is virtually noncompressible, as are bulk liquids. Liquid films will flow through a gap in the compressing barrier, as liquids flow through an orifice, and a two-dimensional viscosity can be determined. Solid films are also resistant to compression, and will collapse when the applied pressure exceeds their mechanical strength. Phase changes may occur with rising temperature, analogous to the melting of a solid and the boiling of a bulk liquid.

Certain types of films on water, such as oleic acid, barium stearate, and proteins, can be transferred from the liquid surface to a preconditioned glass or metal plate dipped through the surface. Successive layers can be built up to a thickness that is measurable by optical methods. Variations in experimental techniques permit certain of the films to be transferred with either the water- or the air-contact side exposed on the solid, and the resulting surfaces are wettable or nonwettable, respectively, as would be predicted.

The study of surface films on solids is a much more difficult problem on which substantial progress was being made in the late 60's and early 70's by novel applications of experimental methods and by devising new interpretations based on extensive data and theoretical correlations (Reference). These methods included the experimental capability for studying solid surfaces at very low pressures in the range of picoatmospheres, low-energy electron diffraction (LEED), Auger electron spectroscopy, and to a lesser extent laser diffraction, field-electron or field-ion microscopy, and ellipsometry. Extremely low pressures enable the experimenter to clean and maintain a clean solid surface for reasonably long periods while he makes base-line studies of the surface without a film, before the firm-former is admitted to the system. Laser diffraction is a sensitive method for observing surface movement of the molecules of clean substrates as a diffraction grating ruled on the surface is distorted or blurred by heat-induced migration.

Field-electron microscopy is a means to observe the pattern of the surface atoms of a specially prepared substrate (e.g., a polished tungsten point) which can detect and locate the position of a single foreign atom, so that the interchange between the solid surface and atoms of the surrounding gas can be monitored. Both LEED and Auger spectroscopy reveal characteristic properties of atoms on the surface of the solid which can be correlated with chemical and physical status of surface films.

The surface of a solid is normally not uniform in chemical activity or in physical structure, even when clean, so that films that do occur are often not uniform. Films on solids may be sometimes mobile, as if they were gases or liquids, but are more often fixed on the solid surface. The immobility of the film is probably caused more often by attachment of the individual film molecules to the solid surface by either chemical bonding or electrostatic forces than by the film's behavior as a two-dimensional solid, in which the source of dimensional stability is mutual attraction of the molecules in the film.

The noncorroding character of stainless steels and metallic aluminum is attained by the spontaneous formation of oxide films that are dense and impermeable to atmospheric oxygen, so that further oxidation or corrosion is prevented. Freshly formed glass surfaces exhibit frictional properties that induce chipping or flaking on glass-to-glass contact, but this can be reduced by adding certain gases such as sulfur dioxide to the atmosphere during cooling. The implication is that a lubricating film is produced by adsorption of gas, or by reaction of the clean glass surface with the atmosphere. Many metal and glass surfaces normally nonwettable can be made highly wettable by heating to a relatively high temperature. High wettability may be retained for several minutes or a few hours after the surface cools; but almost invariably the surface reverts to a nonwettable condition during exposure to the atmosphere. It has been proposed that traces of greasy air-borne contaminants are always present, which spread spontaneously on contact with the clean solid surface. Other fragmentary evidence indicates that adsorbed gas films are responsible for alterations in wettability.

The electrical properties of surface films that form between a solid and a dilute salt solution have been studied extensively. Measurements of surface conductivity of glass-solutions interfaces, and also of organic acid films on pure water, show that the bulk resistance of the liquid phase is paralleled by a conductive layer of charged ions adsorbed on the surface, balanced by a diffuse cloud of ions of opposite charge gathered in the liquid phase near the surface. The measurements of surface conductivity and of related phenomena, such as the migration of suspended solids (electrophoresis) and the flow of solution through a porous solid (electroosmosis) under the influence of an imposed electrical potential, provide a means for the study of surface films on solids that has not yet been fully exploited.

J. F. FOSTER

Cross-references: *Emulsions, Foams, Adsorption, Monomolecular Films.*

References

E. J. Drauglis, "Molecular Processes on Solid Surfaces," R. D. Gretz, and R. I. Jaffee, Editors, McGraw-Hill, New York, 1969.

FISCHER-TROPSCH PROCESS

The Fischer-Tropsch process—reaction of hydrogen with carbon monoxide to form hydrocarbons and oxygenated organic compounds—is carried out in the presence of a catalyst and under suitable conditions of pressure and temperature. This reaction, also called the gas-synthesis, synthine, kogasin, or Fischer-Pichler process (depending on the variant of the method), may be used for the preparation of waxes, oils, liquid fuels, and organic chemicals. It is one of a number of catalytic processes in which these two gases react with each other or with a third component: Iso-, methanol-, and higher alcohol- or synol-synthesis, usually in the presence of difficultly reducible oxides; and methanation, Fischer-Tropsch, synthol, and oxo (or hydroformylation) reactions, usually in the presence of metals or compounds having certain metallic properties. Of these processes, only the Fischer-Tropsch reaction has been used for large-scale preparation of liquid fuels.

Raw Materials. The mixture of hydrogen and carbon monoxide may be prepared by reforming natural or refinery gas (methane) or by gasification of carbonaceous solids, particularly coal and coke. The ratio of hydrogen to carbon monoxide should be adjusted to correspond closely to their usage ratio in synthesis; for cobalt catalysts this ratio is about 2, for iron about 1. Complete conversion in a single pass through the reactor is usually impractical or undesirable, and unconverted gas is usually recycled.

Catalyst. Hundreds of catalysts have been prepared under a wide variety of conditions. Their main constituents have been iron, cobalt, nickel, or ruthenium. These metals may be supported on carriers like diatomaceous earth to obtain high ratios of surface to volume, and they may be aided by "promoters" (see **Catalysis**). Apart from meeting certain physical requirements such as mechanical strength, shape, and size, which differ with mode of operation of the synthesis, the catalyst must have satisfactory activity (extent of conversion of synthesis gas), selectivity (types and relative amounts of products), and stability (reasonably constant activity and selectivity for a given period). Correct composition and mode of preparation of catalyst are necessary but not sufficient to meet these requirements.

During synthesis, the catalyst may undergo changes. Metallic iron, for example, will be at least partly converted to carbide rather rapidly, and both carbide and metal will be oxidized at a slower rate. Iron nitride will be converted to carbonitride, carbide and oxide. (Four types of carbide and three of nitride have been identified in iron catalysts). In addition, free carbon is deposited, especially at higher temperatures, and may cause swelling, spalling, deactivation, and even mechanical stoppages of the reaction under severe conditions of operation. Deactivation may be caused by blanketing of surface and pores with wax; in this case, mild hydrogen treatment or solvent extraction may restore the activity.

Modes of Synthesis. Optimum temperatures are 160°–225°C for ruthenium, cobalt, and nickel, and 220°–325°C for iron; optimum pressures are 220–1,000 atmospheres for ruthenium and 1–20 atmospheres for iron, cobalt, and nickel. The heat liberated by the reaction is about 600–700 kcal per cubic meter of converted gas. Because of the large heat of reaction and the necessity of maintaining a constant temperature, the main engineering problem in carrying out the synthesis on a larger scale is adequate removal of heat. This may be accomplished by indirect or direct heat exchange. A fluidized-bed technique with indirect heat exchange has been developed largely in the United States ("Hydrocol" process). Oil- and gas-cooled fixed and moving beds and oil-catalyst slurries have been studied intensively here and abroad.

Products. The primary products of the synthesis are aliphatic. They are predominantly straight-chain compounds, though the amount of monomethyl isomers increases considerably with increasing molecular weight. More highly branched and cyclic compounds occur only in small amounts. Unsaturation varies with catalyst and increases generally in the order nickel, cobalt, iron, though specific catalyst composition and conditions of synthesis may cause wide variations. The average molecular weight of the products may be varied widely, so that more than 60% of the total primary C_{3+} product may be in the gasoline range or in the range of heavy fuel oil and wax. Theoretical considerations as well as earlier experiments showed that not more than 30% of the primary product is in the boiling range of diesel oil. Oxygenated materials amount to not more than about 6% of the product from cobalt preparations, up to about 17% from non-nitrided iron, and up to about 60% from nitrided iron catalysts.

Mechanism of Synthesis. The types and relative amounts of products obtained from the Fischer-Tropsch synthesis are not in thermodynamic equilibrium. For example, one would expect carbon dioxide as an end product, which is produced over iron, rather than water, which is produced over cobalt and nickel; free carbon and methane rather than higher hydrocarbons; and much larger fractions of branched paraffins and of olefins with internal double bonds than are actually found. The course of the over-all reaction is therefore determined kinetically.

Fischer's original theory, that the products are formed by way of alternate carburization and reduction of the catalyst, has now been abandoned. Cobalt carbide has been shown to decrease the activity of the catalyst greatly, and experiments with radioactive carbon indicate that not more than about 15% of the products can possibly be produced by way of iron or cobalt carbide. The distribution of hydrocarbon products, both isomeric and by carbon number, can be accounted for by a set of rather simple rules of addition of carbon to the chain. Such purely mathematical considerations imply the eventual termination of chain growth so that selective cracking of waxes need not be postulated. The over-all activation energy for the synthesis ranges from 15 to 30 kcal/mole depending on the conditions used.

ERNST M. COHN

[This process is not now of significant economic value in the U. S., but might become so in the event of a shortage of hydrocarbons. Editor]

Cross-references: *Carbon Monoxide, Fuels, Catalysis, Coal.*

FISH: CHEMISTRY AND PRESERVATION

Although there are tens of thousands of species of fish, relatively few are of commercial importance. In the United States one species, menhaden, amounts to more than ⅓ of the fish caught, and the five leading species (menhaden, tuna, shrimp, herring, salmon, and crabs) represent 60% of the total harvest. Species fall into several large categories, the most important of which are the herring-like fish (anchovies, herring, menhaden, and sardines) caught largely at the surface in seines, and the cod-like fish (haddock, hake, pollock, and cod) taken from the bottom by trawl gear.

Composition. The flesh of fish consists primarily of protein, oil, water, and mineral matter. The protein content generally ranges between 15 and 20%, with a few species, primarily shellfish, falling somewhat below this range and certain other species, notably tuna and halibut, lying somewhat above the range. The amino acid makeup of fish protein resembles closely that of mammalian flesh, so that consumption of fish provides a good balance of essential amino acids.

Fish oils occur in fish flesh from a minimum of about 0.6% up to a maximum usually of 30%. Occurring primarily as triglycerides, the fish oils contain a much greater variety of fatty acids than do animal oils. Usually about ⅓ of the fatty acids of fish oils contain 4 to 6 double bonds. A 22-carbon atom fatty acid with 6 double bonds and a 20-carbon atom one with 5 double bonds often amount to 20% or more of the total. On the other hand, completely saturated fatty acids also occur in considerable quantity usually totaling 25 to 40% with myristic (C_{16}) predominating. Odd chain-length fatty acids (C_{15}, C_{17}, C_{19}) occur to a significant extent, and with a few species, notably mullet, these unusual fatty acids may exceed 10% of the total. The iodine number of fish oil fatty acids is much higher than that of most other fats, usually ranging between 130 and 180. The polyunsaturated fatty acids of fish oils possess very high cholesterol depressant activity and fish oils, even at very low dietary levels, exhibit this effect. The oil content of livers of fish is generally inversely proportional to that of the flesh with species having low flesh-oil content possessing high liver-oil content and vice versa.

The moisture content of fish varies inversely with the oil content, the sum of the two approximating 80%. Moisture in fish starts to freeze at about −1.5°C with most of it being frozen at −8°C. Very small proportions, however, are unfrozen even at −20°C.

The mineral matter of fish flesh amounts to about 1.5%. Principal inorganic constituents expressed as mg % average: potassium 300, chlorine 200, phosphorus 200, sulfur 200, sodium 63, magnesium 25, calcium 15, iron 1.5, manganese 1, zinc 1, fluorine 0.5, arsenic 0.4, copper 0.1, and iodine 0.1. From a nutritional standpoint the iodine content is unusually high, exceeding the quantity found in most other foodstuffs.

Minor components of fish include vitamins of which vitamins A and D are high in fish containing considerable oil. Most B vitamins occur in quantities approximating those found in meats.

Up to 1/10 the nitrogen in true fishes and up to 1/3 that in shark and other elasmobranch fishes occur in forms other than protein such as trimethylamine oxide, urea, and free amino acids. Glycogen is the principal carbohydrate occurring in most fish to about 0.6%, and in shellfish and oysters to 2% or more. In fish which have struggled excessively at the time of capture, a large part of the glycogen may have been converted to lactic acid with consequent lowering of pH from initial values close to 7.0 to values of 6.0 or less.

The many species of fish show wide variation in chemical composition both from species to species and among individuals of the same species. The oil content shows widest variation, with season of catching and geographical area being important factors. Even the fatty acid makeup of oils of the same species shows large variations.

Deterioration. The high susceptibility of fish to decomposition is due primarily to the very great degree of polyunsaturation of fish oils. This causes high instability with a tendency toward oxidation and development of rancidity and other "fishy" flavors. Most bacteria which decompose fish protein are of psychrophilic type, which cause spoilage much more rapidly at ordinary temperatures than is the case with most other flesh foods; fish, therefore, require lower storage temperatures for successful storage.

Bacterial invasion of fish for the most part starts from the outside, and thus highest contamination occurs on surfaces. This facilitates to some extent application of preservatives in the form of chemical dips or treatment with radiation. Fish oils, while present in the flesh of fish, oxidize exceedingly slowly so long as bacteria are active but when the bacterial action is arrested as by freezing, or by application of ionizing radiation, the rate of oxidation increases. Thus rancidity is a much greater problem with frozen than with iced fish despite the much lower storage temperature in the former case. Most antioxidants which may be effective with fats of other foods are much less so with oils of fish. Keeping air or oxygen away from fish (such as by vacuum packaging) is the most effective way of minimizing oxidative rancidity.

Preservation and Processing. Although fish caught close to shore or in cold climates may be held aboard the fishing vessel without refrigeration of any kind, most often they are packed in the hold with crushed ice. Use of refrigerated sea water aboard the vessel allows storage temperatures a little lower than those of melting ice and prolongs keeping quality. Some species, notably tuna, are frozen in brine and held on the vessel until landed.

Fish to be marketed fresh are shipped iced as whole or partially dressed fish or may be cut into fillets (boneless sides of the fish). The latter are often packaged in moisture, vapor-proof materials, frozen, and marketed in this form. Whole fish, when held frozen, are, after freezing, dipped in cold water to form an ice glaze that protects them in frozen storage against dehydration and oxidation. Some species handled in this way, particularly halibut and salmon, may, after frozen storage, be cut into steaks (transverse slices with bone left in).

A considerable quantity of fish frozen in the United States is cut into portion pieces or sticks, then breaded, and sometimes cooked in hot oil before packaging and freezing. Such breaded fish sticks or portions, as well as breaded shrimp, make up over 85% of the frozen packaged fishery products produced in the United States.

Recent research has shown that nonoily varieties of fish can be irradiation pasteurized by exposure to low levels of gamma radiation, which destroys most, although not all, the bacteria and provides double or treble storage life at iced temperatures.

Salmon and herring are canned by placing the dressed portions in a can and heat processing for a sufficiently long time under pressure to cook the fish, soften the bone, and render the fish sterile. Some species, notably tuna, are first cooked, then separated from skin and bone, and finally heat processed in the can. Sardines may be lightly smoked before canning.

Curing which includes salting, smoking, pickling, and drying is carried out on fish to a larger extent in underdeveloped countries than is today common in areas possessing extensive refrigeration or mechanized canning facilities. In all these processes moisture is withdrawn from the fish in one way or another so that in most cases primary preservation is brought about by the reduced moisture content whereby bacterial action is retarded. With salt fish some preservative action is provided by the salt itself. In smoked fish, phenols, formaldehyde, etc., derived from the smoke, add to the preservative action. In pickled fish the lowering of the pH is an important factor in preserving the product.

Recent developments in freeze drying yield fishery products which will rehydrate to a form closely resembling the fresh condition. Such processing is successful only with fishery products very low in oil content such as shrimp and crab since the oil is rapidly oxidized; even so, packaging under vacuum is essential for adequate storage life.

Industrial Products. About 40% of the fish caught in the United States is used for manufacture of industrial products, principally fish meal and oil, rather than for human food. Usually the whole fish is cooked with steam and passed through a press to separate cook-water (fish solubles containing protein, amino acids, vitamins, and minerals), oils and cooked solids. The oil is separated and purified by centrifugation. It may be then sold as crude oil or be suitably refined. The fish solubles are condensed to 50% moisture in multiple-effect evaporators. The solids (press-cake) from the press are dried most commonly in a rotating flame dryer. The resulting meal usually has 60 to 70% protein, 5–10% moisture, 8–12% oil, and 15–20% ash. Sometimes the fish solubles, instead of being evaporated, are added to the meal before drying to produce a "whole meal" higher in vitamins and soluble amino acids than ordinary meal. Fish meal and condensed fish solubles are added to mixed poultry feeds. The fish meal protein is better balanced with respect to amino acids than most vegetable protein and can, in small quantities, upgrade the protein quality of mixed poultry feed. The fish meal also provides additional vitamins, minerals, and perhaps some unidentified growth factors. Recent research has shown that solvent extraction and other processes can be used to make a fish protein concentrate free of oil and that is of high nutritive value for use as a human food.

The largest market for industrial fish oils is for use after hydrogenation in margarine and shortening. Many fish oils are also used without hydrogenation for various purposes in which the properties of their long-chain, highly polyunsaturated fatty acids are important. These uses include incorporation in paints and varnishes, for making special chemicals and chemical intermediates, in printing inks, as core oils, and for treatment of leather. Fish-liver oils are still used in some countries as a source of vitamins A and D, although in the United States they cannot compete economically with synthetic vitamins.

Other fishery byproducts include ground oyster and clam shells used as poultry feed; pearl essence essentially a solution of guanine from herring scales and used as a spray or dip to impart an iridescent appearance to many items, especially artificial pearls; leather from sharkskins; glue from fish skins; and liquid fish fertilizers.

MAURICE E. STANSBY

Cross-references: *Foods, Vitamins.*

FISSION, NUCLEAR

The term "fission" was first used by Meitner and Frisch (1939) to describe the process of the disintegration of a heavy nucleus into two lighter nuclei of roughly equal size. The conclusion that this unusual nuclear reaction occurs was the culmination of a truly dramatic episode, and set in motion an extremely intense and productive period of investigation. After the discovery of the neutron by Chadwick in 1932, Fermi undertook an extensive investigation of the nuclear reactions produced by the bombardment of various elements with this uncharged projectile. He observed (1934) that at least four different radioactive species resulted from the bombardment of uranium with slow neutrons. These new radioactivities emitted beta particles and were thought to be isotopes of unstable "transuranium elements" of atomic numbers 93, 94 and perhaps higher. There was, of course, intense interest in examining the properties of these new elements and many radiochemists participated in the studies. The results of these investigations, however, were extremely perplexing and the confusion persisted until Hahn and Strassmann (1939), following a clue supplied by Curie and Savitch (1938), proved definitely that the so-called "transuranic elements" were, in fact, radioisotopes of barium, lanthanum and other elements in the middle of the Periodic Table.

Armed with the unequivocal results of Hahn and Strassman, Meitner and Frisch invoked the new liquid drop model of the nucleus (Bohr, 1936, Bohr and Kalckar, 1937) to give a qualitative theoretical interpretation of the fission process, and called attention to the large energy release which should accompany it. There was almost immediate confirmation of this reaction in dozens of laboratories throughout the world. These experiments confirmed the formation of extremely energetic heavy particles and extended the chemical identification of the products.

Most of the energy of fission is released in the form of kinetic energy of the fission fragments. The initial velocities of the separating particles are of the order of 10^9 cm/sec, and since this is greater than the orbital velocities of the outermost electrons, the latter are stripped from the fragments by interaction with the medium, leaving a positive charge of about 20 units. In passing through matter, the fission fragments cause intense initial ionization. As the fragment is slowed down, outer electrons are captured, decreasing the positive charge, and the ionization intensity decreases. The energy loss at low velocities is due mainly to elastic collisions with the nuclei of the medium. Ranges of from 2 to 3 cm of air are observed for the fission fragments. The ionization produced by fission fragments is many times greater than that due to the most energetic alpha particles, and hence they are readily observed in ionization chambers, cloud chambers, photographic plates and semiconductor detectors. The recoil of the fragments is sufficient to eject them from the surface of the fissioning material and they can be collected on "catcher foils" or on a water surface in close proximity and identified by their radioactivity and chemical properties. These methods have all been used in confirming and studying the fission process.

The chemical evidence which was so vital in leading Hahn and Strassmann to the discovery of nuclear fission was obtained by the application of the "carrier" and "tracer" techniques. Since invisible amounts of the radioactive species were formed, their chemical identity had to be deduced from the manner in which they followed known "carrier" elements, present in macroscopic quantity, through various chemical operations. Known radioactive species were also added as "tracers" and their behavior compared with that of the unknown species to aid in the identification of the latter. The wide range of radioactivities produced in fission makes this reaction a rich source of tracers for chemical, biological and industrial use.

Although the early experiments involved the fission of normal uranium with slow neutrons, it was rapidly established that the rare isotope, U^{235}, was responsible for this phenomenon. The more abundant isotope, U^{238}, could be made to undergo fission only by fast neutrons with energy greater than 1 Mev. The nuclei of other heavy elements, such as thorium and protactinium were also shown to be fissionable with fast neutrons, and other particles, such as fast protons, deuterons and alphas, as well as γ-rays proved to be effective in inducing the reaction. Bismuth, lead, thallium, mercury, gold, platinum, tantalum, and even medium weight elements such as copper, bromine, silver, tin and barium have been made to undergo fission by excitation with high energy projectiles (the order of 100 Mev or more.) Some other nuclides which do not occur in nature but have been produced by transmutation reactions also undergo fission. Among these, U^{233} and Pu^{239} are fissionable with slow neutrons, while Np^{237} requires fast neutrons.

The very interesting and rare occurence of spontaneous fission in uranium was first observed in 1940. In this reaction, the nucleus undergoes fission in its ground state, without excitation from external sources. Although the partial half-life for this process in uranium is about 10^{16} years, isotopes of the new "transplutonium elements" have been discovered in which spontaneous fission represents the principal mode of decay, with half-lives of the order of a few hours or less. This process probably sets a limit for nuclear stability.

The outstanding feature of nuclear fission is the tremendous energy release which accompanies it. A chemical recation such as the explosion of TNT releases about 1.5×10^{11} ergs per gram; nuclear fission releases approximately 8×10^{17} ergs per gram. Fundamentally, the source of this energy lies in the fact that the total mass of the final products of fission is appreciably less than that of the reactants. This loss in mass appears as energy in an amount given by the famous Einstein relation, $E = mc^2$, one atomic mass unit being equivalent to 931 Mev. (1 Mev $= 1.6 \times 10^{-6}$ erg.)

The nature of the fission process may perhaps be best understood through a consideration of the structure and stability of nuclear matter. Nuclei are composed of neutrons and protons, the total number of them being equal to the mass number. The actual weight of the nucleus is always less than the sum of the weights of the free nucleons, the difference being the mass equivalent of the energy of formation of the nucleus from its constituents. This difference is known as the "mass defect" and is a measure of the total binding energy of the nucleus.

The neutrons and protons of which nuclei are composed are bound by a short-range attractive force which acts only between nearest neighbors. As long as the total number of protons is small the long-range electrostatic repulsion between them will be insufficient to overcome the cohesive forces between all nucleons. As the number of nucleons increases, the fraction of them near the surface (and hence with fewer neighbors) decreases, and the average binding energy per nucleon increases. At about mass number 55, this trend reaches a maximum, and a further increase in the number of nucleons decreases the average binding energy per nucleon due to the repulsive coulomb force between protons. In fact, in order to maintain stability, the number of protons must be diluted with an excess of neutrons as the mass number increases. Since a decrease in binding energy per nucleon means a decrease in the mass defect or an increase in the average mass per bound nucleon, uranium (with a greater mass per nucleon than that for nuclei of elements of medium atomic weight) will be energetically unstable with respect to fission. Qualitatively, at least, the fission process is thus seen to be a consequence of the coulomb repulsion between protons.

Coulomb repulsion between protons causes heavy nuclei to have rather high neutron to proton ratios (of the order of 1.5) and when such a nucleus undergoes fission, the primary fragments formed will possess a similar ratio. For nuclei in the mass region of these products, however, such a ratio is higher than is consistent with stability. A ratio corresponding to stability is attained by the evapo-

ration of neutrons from the highly excited primary fragments (within about 10^{-14} sec of the fission event), and by conversion of neutrons to protons through the beta decay process.

The average number of neutrons emitted per fission in U^{235} fission is 2.5, and hence a chain reaction becomes possible, the excess neutrons causing fission in other U^{235} nuclei which, in turn, contribute more neutrons. A vast energy source is thus made available for utilization in a controlled form (nuclear reactor or "pile") or in an explosion (atomic bomb).

A typical fission event in U^{235}, for example, may be described by the following equation:

$$_{92}U^{235} + _{0}n^1 \rightarrow (_{92}U^{236}) \rightarrow$$
$$_{38}Sr^{95} + _{54}Xe^{139} + 2_{0}n^1 + \gamma + Q.$$

A slow neutron is absorbed by a U^{235} nucleus forming the excited compound nucleus U^{236}, which then splits into two fission fragments, Sr^{95} and Xe^{139}, and two neutrons. The subscript at the left of the chemical symbol indicates the atomic number (nuclear charge) and the superscript to the right indicates the mass number. The fission fragments possess about 20 Mev of excitation energy which will be emitted in the decay process they undergo in reaching the stable members of their respective "decay chains," $_{42}Mo^{95}$ and $_{57}La^{139}$. In addition, about 5 Mev of gamma radiation, represented by γ, is released at the instant of fission. (Some energetic alpha particles have also been observed in the fission act, but they are rather rare.) Q represents the kinetic energies of the fission fragments and neutrons and is approximately 170 Mev. Thus, the total energy of a fission event is close to 200 Mev. Both nuclear charge and mass number must be conserved in the fission process. Other combinations of primary fragments and number of neutrons are possible, but they are not all formed with equal probability. For example, the formation of the complementary fragments of masses 95 and 139 is about 600 times as probable as the formation of two mass 117 fragments and almost 10^6 times as probable as the formation of masses 72 and 162.

The probability distribution of masses in fission is an important feature of the process. The percentage of fissions that leads to a given mass is referred to as the "fission yield" of that mass. A fission yield curve is obtained by plotting the yields against mass number. Such curves have been obtained by radiochemical investigations for fission of many nuclides at low and high energy of excitation. Ionization chamber and velocity measurements indicate a spread in the kinetic energy of the fragments for a given mass split. This is associated mainly with a variation in the number of neutrons per fission. In general, fission induced by low energy particles (generally referred to for convenience as "low energy fission") is characterized by an asymmetric splitting into two main groups of mass numbers (the "light" and "heavy" groups) the most probable mass ratio of light to heavy product being about ⅔. As the mass number of the fissioning nucleus increases, the light group distribution shifts towards heavier masses while the heavy group remains relatively stationary.

The fission of lighter, less-fissionable elements

such as Pb or Bi results in a symmetric mass distribution of the fission products. In the fission of Ra^{226} a triple-peaked mass distribution is observed, the three peaks occurring with about equal yield. This has been taken as evidence for two distinct modes of fission—one leading to a symmetric mass split in fission, the other to an asymmetric one. The threshold for symmetric fission lies lower than that for asymmetric fission in the Pb-Bi region, about equal in the Ra-Ac region, and higher in the region of Th and above.

As the energy of excitation increases, the probability of symmetric fission increases, and at very high energies it becomes the most probable mode. The fission yield curves are not entirely smooth functions of mass, and regions of fine-structure have been observed. These are attributed to an enhanced probability for the formation of particularly stable nuclei having "closed shells" of 50 and 82 neutrons.

In addition to a distribution in mass, a distribution in nuclear charge for a particular mass split also occurs. The present data on charge distribution in slow-neutron induced fission indicate that the most probable primary charge of complementary fragments is that for which both fragments are equally displaced from the most stable charge for their respective mass numbers. Each primary fragment undergoes, on the average, about three beta disintegrations before achieving stability, each beta disintegration increasing the nuclear charge by one, but leaving the mass number unchanged. Since the primary fragments are highly excited, the decay energy of the first few members of a decay chain may be quite large. For some nuclides near closed-shells this decay energy may exceed the binding energy of a neutron, and the latter may be emitted in preference to a beta particle. Since these neutrons follow beta-emitting precursors in the decay chain, they are referred to as "delayed neutrons," thus distinguishing them from the "prompt neutrons" which are coincident with the fission event. Although delayed neutrons account for less than one percent of all fission neutrons, they are an extremely important factor in the control of nuclear reactors.

In higher energy fission the most probable charge for the primary fragments appears to be that which maintains the same neutron-to-proton ratio as the parent fissioning species. Fission competes with other modes of deexcitation of an excited nucleus, such as nucleon or gamma-ray emission, and if the excitation is high enough the nuclei left after nucleon emission may also fission. Thus at high excitation energy a variety of fissioning species may be present and this complicates the analysis of mass and charge distribution determined by radiochemical techniques from the accumulated fission products.

Many different techniques have been employed in an attempt to understand the very interesting and complicated phenomena associated with nuclear fission. Advances in instrumentation continue to open new avenues of approach, and the application of semiconductor particle detectors, solid state track detectors, multi-parameter pulse-height analyzers, computer data processing, etc. have given new impetus to the studies and yielded extensive new data. These studies concern such aspects of the reaction as the role of angular momentum (as evidenced, e.g., by the angular distribution of the fragments, the ratio of formation of isomeric states, and fissionability), the spectra and numbers of γ-rays, x-rays, and neutrons emitted in the process, the occurrence of ternary splitting, etc.

The new techniques permit an event-by-event analysis of correlated information on the fission act. Thus, the neutrons, x-rays, γ-rays, α-particles, etc. accompanying each fission can be correlated with the two major fragments in that event and the mass and nuclear charge of the latter can be identified by kinetic energy, time-of-flight, and characteristic x-ray analyses. Angular distributions and ternary break-up can also be examined from the particle tracks recorded in solid state track detectors such as mica or Lexan. The dependence of such data on the energy and type of particle inducing fission and the mass and charge of the fissioning species is of particular interest in establishing the parameters of fundamental importance to a theoretical description of the process.

Although an extensive phenomenology of nuclear fission has been accumulated, no comprehensive theoretical treatment has yet emerged. The liquid-drop model has been considerably extended and improved, and other approaches such as a statistical treatment, and the "unified model" (which combines the liquid-drop "collective" properties with the single-particle or "shell" model) have been invoked to interpret fission phenomena. Each has had significant, though limited, success in accounting for particular aspects of the reaction.

The recent observation of spontaneous fission from isomeric states of a number of isotopes of uranium and transuranium elements has led to further developments in the theory of nuclear fission. The inclusion of a deformation (or shape)-dependent correction to calculations of closed-shell effects on the nuclear binding energy indicates that a minimum in the potential energy surface may be present in the vicinity of the fission barrier, yielding a double-humped barrier. The spontaneous fission isomeric states lie in the potential well formed by the double-humped barrier. This theoretical development along with other extensive calculations of the potential energy surface based on a deformed liquid drop model with appropriate correction terms for the effects of nuclear closed-shells has stimulated a renewed interest in fission theory and given hope for a fundamental interpretation of this complex reaction. In addition to their applicability to an understanding of nuclear fission, these new developments in nuclear structure theory predict a region of relative nuclear-stability in the neighborhood of atomic number 110 (which should have chemical properties of ekaplatinum). The search for such "super-heavy" elements is an exciting new phase of nuclear research and a new generation of heavy-ion accelerators is being constructed in an attempt to produce them in the laboratory.] A complete description of nuclear fission phenomena may necessitate the adoption of several complementary theories, in analogy with the wave-particle dualism of electromagnetic radiation.

ELLIS P. STEINBERG

Cross-references: *Atoms, Isotopes, Transuranium Elements, Fusion, Nuclear.*

FLAMES AND FLAME STRUCTURE

A flame is a gas phase* combustion reaction which is able to propagate through space. Fire is the common term for combustion, especially when out of control. Combustion refers to exothermic reactions in any phase. It usually implies propagation and oxidation. Oxidation normally involves atmospheric oxygen, but many other oxidizers produce flames and some flames do not involve oxidation reactions. In most combustion processes the exothermic stages occur in the gas phase, regardless of the initial phases of the reactants. Therefore, flames are associated with most combustion processes. The appearance of combustion reactions can vary considerably. This is controlled by the initial phase and the dominant mixing process. The character of the associated flames also depends upon mixing. There are three common types, premixed, diffusion and turbulent. (Table 1) Diffusion flames are common in industry for safety reasons. Turbulent flow is often added to maximize the volumetric combustion rate.

TABLE 1. EXAMPLES OF FLAME TYPES

	Laminar Flow	Turbulent Flow
Premixed	Bunsen Burner	Laboratory Blast Lamp
Nonpremixed	Candle	Household Gas Furnace

Flame propagation results from strong coupling between reaction and transport processes of molecular diffusion and thermal conduction. In premixed flames propogation is controlled by the balance between reaction and transport. In Diffusion flames propagation is limited by diffusion. In Turbulent flames propagation is dominated by turbulent eddy transport. The fundamental processes and reactions are common to all flame types, but the importance of the individual processes varies from type to type. Therefore, discussion will center on the easily-visualized, one-dimensional premixed flames (e.g., the Bunsen burner). (Fig. 1) The dominant processes are: flow, chemical reaction, thermal conduction, and diffusion. Under some circumstances turbulent eddy diffusion, thermal diffusion and radiation become important.

A premixed flame can be visualized as a thin reaction surface propagating normal to itself. This propagation can either be through free space or balanced out by an opposing flow as in the laboratory Bunsen burner. (Fig. 1) The microstructure is identical for the two cases. The observed differences result from viewing the flame front in moving and fixed coordinate systems, respectively. Although the reaction zone is narrow, the study of flame microstructure has allowed the identification of the elementary physical and chemical processes. Flames provide a convenient environment for studying high-temperature reactions.

* Some flame systems occur in liquid and solid phase. These are usually detonations.

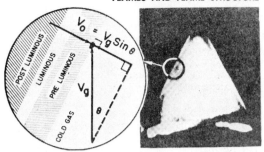

FIG. 1. Burning velocity—the rate at which a plane flame front advances into the unburned gas is called burning velocity. It depends principally on initial composition, temperature and pressure. At any point along a flame front the propagation velocity balances the cold gas approach velocity. For conical Bunsen flames there is a simple relation between cone angle, gas velocity and burning velocity. (See above).

The elementary flame processes are well understood individually and a set of "flame equations" can be written describing the flow; the constraints of conservation of energy and of atomic species (generalized mass conservation); and the differential equations of thermal conduction, molecular diffusion and chemical reaction. Momentum is neglected, since for subsonic flow the kinetic energy is negligible and the flame is effectively at constant pressure. In contrast, detonations are supersonic, show pressure discontinuities and require consideration of momentum and viscosity.

A one-dimensional form of the flame equations allows the quantitative description of many laboratory systems (e.g., the Bunsen, flat, and spherical flames). This is a boundary value problem. The eigen solution can be identified with the "burning velocity."

Simple flames (e.g., the H_2-O_2 and H_2-Br_2) can be simulated with computers, but the chemistry of most flames is too complex for this approach at present. A second limitation of this method is the lack of rate and transport coefficients for the calculation.

Size and stability limitations restrict flames to rapid reactions, most of which involve radicals or atoms. Thus, flame chemistry involves principally the high-temperature reactions of a few radical species. In oxygen supported flames they are: H, O, OH and CH_3. In some cases halogen atoms, HO_2 and nitrogen radicals can also be important.

Many flame systems are both qualitatively and quantitatively understood. Flame studies have made important contributions to kinetics.

There are two major divisions of flame chemistry: thermodynamics and reaction kinetics. The former is better understood.

The primary thermodynamic problem is: given a set of initial reactants, what will the final products and temperature be? Enthalpies are available for most species covering a wide temperature range (0–6000°K). In practice, the problem is complicated by the existence of radical species which make the normal stoichiometric relations poor approximations. Concentrations cannot be determined until the temperature is known, and vice versa. As a result, the calculation is a tedious iteration. Fortunately, it is well adapted to machine computation.

Some systems, e.g., rich hydrocarbon and boron hydride flames—stop short of final equilibrium. Therefore, caution should be used in interpreting such calculations.

Flame Systems

Some areas of flame chemistry are relatively simple because: (1) only rapid steps can be important, and (2) parallel reactions with identical rates are rare. A single scheme often explains most of the observed reaction.

Halogen Flame Systems. Hydrogen produces flames with bromine, chlorine, and fluorine. The absence of iodine flames probably results from the slow speed of the endothermic reaction $I + H_2 \to HI + H$. Burning velocity increases with halogen reactivity. The chemistry can be written by inspection. (Table 2)

TABLE 2. HYDROGEN-HALOGEN FLAME REACTIONS

$$M^* + X_2 \rightleftarrows 2X + M$$
$$M^* + H_2 \rightleftarrows 2H + M$$
$$X + H_2 \rightleftarrows HX + H$$
$$H + X_2 \rightleftarrows HX + X$$
$$M + H + X \rightleftarrows HX + M^*$$

Oxygen Flame Systems. Most common flames are oxygen supported.* Oxygen is a divalent radical and reactions of the radicals OH and HO_2 require consideration as well as H and O. HO_2 reactions become important in low-temperature hydrogen flames and ignition phenomena.

Oxygen-supported flames can be logically grouped into a hierarchy beginning with hydrogen, carbon monoxide, and passing through the homologous series of hydrocarbons. Each flame involves the reactions of the simpler systems in addition to its own. Oxygen disappears principally by reaction with hydrogen atoms. This branching reaction is responsible for the rapid buildup of radical concentrations. By contrast the branching step in hydrogen-halogen flames is dissociation.

The hydrogen-oxygen flame begins the hierarchy. It is assumed that the reactions of HO_2, O_3, and H_2O_2 may be neglected in high-temperature systems. This is attributed to their lack of thermal stability (H_2O_2, O_3) and relatively low reactivity ($HO_2 \ll OH$). The scheme can be written by inspection. (Table 3)

The carbon monoxide-oxygen system requires hydrogen (or deuterium) for flame formation. A trace of a hydrogen containing compound is sufficient. This suggests that the dominant reaction of CO is with OH rather than with O_2 or O. Therefore, the CO flame must involve the reactions of the hydrogen-oxygen system (Table 3), and characteristic reactions. (Table 4)

The scheme for the simpler hydrocarbon-oxygen flames can be partially outlined (see e.g., Table 5). The chemistry can be conveniently divided into fuel-

---

* Flame systems such as: the decomposition of hydrazine, nitric oxide, ozone, and hydrogen peroxide, as well as systems involving the nitrogen oxides as oxidizers also occur. These chemistries are less well understood.

lean and fuel-rich flames. Saturated and unsaturated hydrocarbons involve different reactions. Lean-flame gases are strongly oxidizing while rich-flame gases are strongly reducing. The transition between these limiting chemistries occurs on the rich side of stoichiometric. For saturated hydrocarbons the carbon oxidation sequence appears to be: hydrocarbon $\to$ radical $\to$ methyl + olefin $\to$ formaldehyde $\to$ CO $\to$ CO_2. These flames have been aptly characterized as hydrocarbon reactions feeding a $CO-H_2$ flame.

The two basic questions are (1) what is responsible for the initial hydrocarbon attack, and (2) what is the fate of the fragments from this reaction? In fuel-lean saturated-hydrocarbon flames, the initial attack is by OH. The corresponding hydrocarbon radical is formed. (Table 5)

Methyl is the most stable radical at flame temperatures so that complex radicals usually expel methyl leaving the next lower olefin. Even if fis-

TABLE 3. HYDROGEN-OXYGEN FLAME REACTIONS

$$H + O_2 \rightleftarrows OH + O$$
$$OH + H_2 \rightleftarrows H_2O + H$$
$$O + H_2 \rightleftarrows OH + H$$
$$2OH \rightleftarrows H_2O + O$$
$$H + H + M \rightleftarrows H_2 + M^*$$
$$O + O + M \rightleftarrows O_2 + M^*$$
$$H + OH + M \rightleftarrows H_2O + M^*$$
$$H + O + M \rightleftarrows OH + M^*$$

TABLE 4. CARBON MONOXIDE-OXYGEN FLAME REACTIONS

$$OH + CO \rightleftarrows H + CO_2$$
$$H + CO + M \rightleftarrows HCO + M^*$$
$$(H, O, OH) + HCO \to CO + (H_2, HO, H_2O)$$

TABLE 5. SOME HYDROCARBON FLAME REACTIONS

I. Saturated

  A. Fuel-Lean

$$OH + CH_4 \to H_2O + CH_3$$
$$OH + C_nH_{2n+2} \to H_2O + C_{n-1}H_{2n-2}(\text{Olefin}) + CH_3$$
$$O + C_nH_{2n} \to OCH_2 + \text{products}$$
$$O + C_nH_{2n-2} \to \text{products}$$
$$CH_3 + O_2 \to OH + OCH_2$$
$$OCH_2 + O \to (\ ) \to H_2O + CO$$

  B. Fuel-Rich

$$H + CH_4 \to H_2 + CH_3$$
$$H + C_nH_{2n+2} \to H_2 + C_{n-1}H_{2n-4} + CH_3$$
$$O + CH_3 \to H + OCH_2$$

II. Unsaturated

  A. Olefinic

$$O + R - CH = CH_2 \to (\ ) \to (\ ) \to R + CO$$

  B. Acetylenic

$$O + R - C = CH \to (\ ) \to (\ ) \to R + CO$$

sioning occurs, the ultimate product is methyl. The mode of oxidation of methyl radical is a subject of debate. Methyl could be attacked by either molecular or atomic oxygen. Formaldehyde is the product. This is attacked rapidly in flames, probably by OH radicals (possibly by O and H) with the formation of CO and intermediate formation of HCO radical.

In fuel-rich flames the hydrocarbon is attacked by hydrogen atoms forming the corresponding radical. Radical concentrations are high. This favors the formation of higher hydrocarbons and oxygenated intermediates similar to those found in low-temperature oxidations. In very rich flames, free carbon is formed and heterogeneous reactions may occur.

Unsaturated hydrocarbons burn more rapidly than saturated ones. The hydrocarbon reaction may involve atomic oxygen attack of the unsaturated bond as well as, or in some cases to the exclusion of, stripping by hydroxyl radical or hydrogen atoms.

R. M. Fristrom

References

Fristrom, R. M., and Westenberg, A. A., "Flame Structure," McGraw-Hill, New York, 1965.

Fenimore, C., "The Chemistry of Premixed Flames," Pergamon Press, New York, 1964.

Gaydon, A., and Wolfhard, H., "Flames," Chapman and Hall, Ltd., London, 1962.

Lewis, B., and vonElbe, G., "Combustion Flames and Explosions of Gases," Academic Press, New York, 3rd Ed., 1959.

Williams, F., "Combustion Theory," Addison-Wesley Publishing Co., Inc., Reading, Mass., 1965.

FLAME RETARDANTS

A flame retardant is a material used as a coating on or incorporated in a combustible product to raise the ignition point or to reduce the rate of burning of the product. Flameproofing agents and fireproofing agents are less accurate terms for these materials, since products treated with them will burn if the temperature is sufficiently high.

The products to which flame retardants are applied include apparel, carpets and rugs, construction materials (thermal insulation, wall coverings, panels, etc.), electrical materials, paper and wood products, transportation vehicle interior components, apparel fabrics, and home furnishings. Governmental regulations now require use of flame-retardant goods in many of these categories and more regulations are to be applied in the near future. In fact, the drive for fire safety is the underlying stimulus for the development and application of fire-retardant agents. The market for these agents is now estimated to be 200 to 300 million pounds per year and is expected to approach about 600 million pounds by 1977.

The materials involved in the products to be protected include fibers, fabrics, sheets, panels, structural forms, foams, insulation (thermal and electrical), padding, etc. made of natural or synthetic organic polymers.

The flame-retardant agents used vary with the physical and chemical properties of the material to be protected. Among them are:

(1) Inorganic salts, such as zinc borate, zinc carbonate, ammonium sulfamate, ammonium phosphate, ammonium sulfate, and mixtures of boric acid and borax.

(2) Antimony oxide and other antimony compounds like the oxychloride and the trichloride.

(3) Chlorine compounds, such as chlorinated paraffin, perchloropentacyclodecane, and hexachloroendomethylene tetrahydrophthalic (HET) acid and anhydride.

(4) Bromine compounds, among them tris(2,3-dibromopropyl) phosphate, bromine- and phosphorus-containing polyols, and derivatives of 2,3-dibromopropanol.

(5) Phosphorus compounds, such as phosphate and phosphonate esters, tris(beta-chloroethyl) phosphate, tris(dichloropropyl) phosphate, and tetrakis-(hydroxymethyl) phosphonium chloride derivatives.

Some flame-retardant agents are constituents of organic polymers, that is, they are part of the polymer molecule. For example, one flame- and heat-resistant nylon is a copolymer of metaphenylenediamine and isophthaloyl chloride. Other examples are polyester resins made from chlorendic (HET) acid, epoxy resins produced from tetrabromobisphenol A, and polyurethanes prepared from bromophenyl diisocyanate.

The flame-retardant agents can be classified by their permanence as (1) nondurable, consisting of water-soluble salts which are easily removed by washing or other exposure to water; (2) semidurable, which will be removed by repeated laundering; and (3) durable, which are not affected by laundering or dry cleaning.

The mechanisms by which flame-retardant agents inhibit the support or propagation of flame in the products to which they are applied are several. The agents may be effective by simply diluting the combustible material; they may form noncombustible residual coatings when heated during a fire; or they may decompose to yield noncombustible gases which inhibit flame propagation. One theory regarding cellulosic materials is that the flame-retardant catalytically controls the decomposition of cellulose when exposed to flame temperatures so that the pyrolysis is toward a carbonaceous char rather to flammable vapors.

Many unsolved problems exist in the flame-retardant field, chief of which is the development of satisfactory agents for application on synthetic fibers. In general such materials are much more difficult to protect than are natural fibers, where many simple chemicals are effective. Other problems are the effect of the agents on the properties of the materials being treated, durability of the agents, the basic nature of flame and combustion, and the cost of the agents and of applying them to the combustible materials. An excellent review of the commercial aspects of the flame retardant field is given in *Chem. Eng. News*, **49**, No. 43, 16–19 (Oct. 18, 1971).

Clifford A. Hampel

FLASH PHOTOLYSIS

Flash photolysis is an experimental method for investigating fast photochemical processes invented in 1950 by G. Porter and R. G. W. Norrish of Cambridge University.[1] The irradiation is performed with an intense light flash that generates in the reaction vessel a temporarily high concentration of the reaction intermediates. The photolysis flash can be obtained by discharging a capacitor through gas-filled lamps, although pulsed arc and spark discharges have been used. Typically, these sources provide polychromatic light pulses of 0.1 to 100 microseconds duration at power inputs up to 500 megawatts for xenon flash lamps and 30,000 megawatts in the case of open sparks. Since 1968 flash photolysis has been extended to Q-switched lasers that provide highly monochromatic irradiation in the visible and near-ultraviolet regions at output powers attaining several hundred megawatts within several nanoseconds. The short-lived chemical reaction intermediates produced by flash irradiation are studied before they disappear by making fast measurements of their optical absorptions spectra. This can be done with "flash spectroscopy" in which an auxiliary "spectroflash" lamp synchronized with the photolysis flash takes the absorption spectrum of the reacting system after a short time delay on a photographic plate spectrograph. Alternatively, in "kinetic spectrophotometry" a constant monitoring lamp is employed. The monitoring light beam passes through the irradiation cell to a monochromator with a fast photodetector at the exit slit, so that the transmission changes caused by the formation of intermediates may be followed during and after the irradiation flash. The two-lamp technique is usually employed for spectral identification of the intermediates while kinetic spectrophotometry is more accurate for measurements of reaction kinetics. Special detection methods are required to obtain the nanosecond time resolution available with lasers because of the poor signal-to-noise factor inherent in fast pulse amplifying systems. Successful procedures include the use of high-intensity, pulsed monitoring lamps with photoelectric detection, utilizing the short-lived fluorescence of a dye excited by the laser flash with optical path delay as the spectroflash source, or the use of an image-converter tube with a low-dispersion prism as a fast scanning spectrophotometer. Alternatively, the irradiation may be performed with a succession of short duration light pulses so that the signal-to-noise factor can be improved by means of an integrating detection system.[2]

One of the first flash photolysis accomplishments was the identification of unstable, inorganic free radicals postulated for many years as chemical reaction intermediates. This gas phase work requires high dispersion and multiple-traverse optics because of the low absorption coefficients. Radicals identified in this way include OH, SH, SD, SO, NH, PH, CH, CS, CN, PO, C_2, CH_2, CH_3, CD_3, HCO, ClO, HS_2, PH_2, and PD_2. In some cases vibrationally-excited or "hot" molecules have been observed including O_2^*, OH^*, BrO^*, ClO^*.[3] Inorganic radicals in solution are characterized by broad absorption bands of high molar absorbance. The flash photolysis of halide ions in water or alcohols led to the identification of the unstable dihalide ions I_2^-, Br_2^-, and Cl_2^-. These radical-ion spectra are accompanied by the broad, red absorption band of the solvated electron, because photoionization is the primary photochemical act in the ultraviolet irradia-tion of halide ions. The solvated electron has been observed by flash photolysis of other aqueous inorganic anions including OH^-, CNS^-, CO_3^{2-}, PO_4^{3-}, and $Fe(CN)_6^{4-}$.

Many unstable organic free radicals have been identified and studied with flash photolysis. In early work, resonance-stabilized aromatic free radicals were observed in gas phase and in solution, including benzyl, anilino, phenoxyl and thiophenoxyl radicals. Free radical intermediates have been obtained from many other aromatic molecules including hydrocarbons, quinones and carboxylic acids as well as synthetic dyes and biological molecules including chlorophylls, flavin pigments and retinal pigments. The optical generation of solvated electrons from aromatic molecules was first demonstrated with flash photolysis. It has been found that optically excited molecules utilize the excess energy by emission of heat, fluorescence, bond splitting, population of the triplet state by intersystem crossing, and ejection of a solvated electron. Electron ejection is most probable for molecules with low gas phase photoionization potential such as benzene derivatives with electron-donating substituents and heterocycles such as indole derivatives.[4]

Flash photolysis has made a significant contribution to the identification of metastable triplet states of aromatic molecules and the clarification of their role as reaction intermediates. It has been found that triplet states of aromatic molecules and dyes may be efficient oxidizing and reducing agents. Furthermore, energy transfer to molecules with lower triplet state energies has been demonstrated with many aromatics and dyes. For example, the triplet state of eosin dye is reduced to the semiquinone in the anaerobic oxidation of phenol, p-cresol, aniline, tyrosine, tryptophan, p-phenylenediamine, β-naphthol, and hydrogen peroxide, while triplet eosin is oxidized by reacting with ferricyanide ion and benzoquinone.

Early applications of flash photolysis to photobiology include studies on the combination of heme proteins with gases, the bleaching of rhodopsin, and the reversible bleaching of chlorophyll *in vitro*. Recently, the technique has been used to study primary photochemical mechanisms in the inactivation of the enzymes such as lysozyme. Flash photolysis has been employed in connection with the mechanism of photodynamic systems, i.e., the dye-sensitized photoautoxidation reactions of biological molecules. The formation of excited singlet oxygen as an oxidizing intermediary in this process has been the subject of current interest.[5]

The basic flash photolysis technique has been extended to the study of fast, thermally-induced reactions (flash heating) and to fast reactions initiated by ionizing radiations (pulse radiolysis). In some cases the same intermediates are formed with these techniques as in flash photolysis (e.g., the solvated electron), so that the availability of the different pulse irradiation methods provides flexibility in the investigation of fast chemical reaction mechanisms.

LEONARD I. GROSSWEINER

References

1. Porter, G., "Flash Photolysis and Some of Its Applications," *Science,* **160,** 1299–1307 (1968).

2. Boag, J. W., "Techniques of Flash Photolysis," *Photochemistry and Photobiology*, **8**, 565–577 (1968).
3. Norrish, R. G. W., and Thrush, B. A., "Flash Photolysis and Kinetic Spectroscopy," *Quarterly Reviews*, **Vol. X**, 149–168 (1956).
4. Grossweiner, L. I., "The Study of Labile States of Biological Molecules with Flash Photolysis," in *Advances in Radiation Biology*, (Ed. L. G. Augenstein, R. Mason, and M. R. Zelle), **2**, pp. 83–133, Academic Press, New York (1966).
5. Grossweiner, L. I., "Flash Photolysis Research in Photobiology," *Photophysiology*, (Ed. A. C. Giese.) **5**, pp. 1–33, Academic Press, New York (1970).

FLAVONES

The flavones are a class of organic compounds derived from 2-phenylchromone, which is flavone itself. Of even more importance are a series of hydroxyflavones, which constitute one class of dyes responsible for the colors of yellow flowers, roots, and woods.

Flavone exists as colorless needles, m.p. 99–100°C. It is insoluble in water and shows a violet fluorescence in concentrated sulfuric acid solution. It can be synthesized from *o*-hydroxybenzalaceto-phenone, which is obtained from *o*-hydroxyacetophenone and benzaldehyde. Treatment with al-

coholic alkali gives flavanone, which gives flavone on bromination and dehydrobromination. Thus, the flavanones are 2,3-dihydroflavones.

Flavone is decomposed by boiling alkali; the first product is *o*-hydroxydibenzoylmethane, which is then further decomposed partly into salicylic acid and acetophenone and partly into *o*-hydroxyacetophenone and benzoic acid.

Treatment of flavanone with amyl nitrite and hydrochloric acid gives the 3-isonitroso derivative, which can be hydrolyzed with acid to 3-hydroxy-flavone, or flavanol, from which yellow plant dyes are derived also. Flavanol crystallizes in yellow needles, m.p. 169°C. Its solution in concentrated sulfuric acid shows a violet fluorescence. It is a mordant dye and colors cotton mordanted with aluminum hydroxide a bright yellow. Some of the more important hydroxyflavones are:

Chrysin, m.p. 275°C, which occurs in poplar buds. It colors wool mordanted with alumina a bright yellow.

Apigenin, m.p. 347°C, occurs partly in the free state and partly as the glucoside in various flowers. It is also formed by the hydrolysis of the parsley glucoside, apiin. Its aluminum lake is pale yellow.

Luteolin, m.p. 328–329°C, is found in foxglove and colors fabric orange when mordanted with alumina.

Scutellarein gives a brownish-yellow color with an aluminum mordant.

Naturally occurring hydroxy derivatives of 3-hydroxyflavone or flavanol include:

Galangin, m.p. 217–218°C, the flavanol corresponding to chrysin, is found in the galanga root. It colors mordanted wool yellow.

Campherol, m.p. 247°C, the flavanol corresponding to agigenin, occurs in senna leaves, Avignon berries, and in various flowers; it gives a yellow aluminum lake.

Fisetin, m.p. 330°C, the coloring matter of fustic, is found in fustic wood, combined with glucose and tannin. It gives a brownish-orange color with an aluminum mordant.

Quercitin, m.p. 313–314°C, the flavanol corresponding to luteolin, is the most important and abundant of the flavanol dyes. It occurs in the bark of the American oak and is still used as a dye for wool and silk. Its aluminum lake is brownish-yellow in color.

Morin, m.p. 290°C, is contained in yellow wood (Morus tinctoria) and is still used in wool-dyeing and cotton printing, which it dyes yellow. It is a sensitive reagent for aluminum.

Myricetin, m.p. 355–360°C, occurs as a glucoside in various kinds of plants. The aluminum lake is brownish-orange.

R. D. Morin

FLOWER PIGMENTS

Flower pigments are composed of two groups: the water-soluble vacuolar flavonoid pigments which produce red, blue, or yellow colors and the fat-soluble yellow, orange, or red carotenoid pigments of the plastids. The plastids may also contain the chlorophylls, porphyrin derivatives, which are present in flower buds, but ordinarily disappear when the flower matures.

The flavonoid pigments have a 15 carbon skeleton composed of two benzene rings which are joined by a three carbon chain to give a C_6-C_3-C_6 frame. These three central carbons with an oxygen atom form a heterocyclic ring whose state of oxidation determines the class of pigments, such as the red to bluish anthocyanins and the ivory to yellow anthoxanthins which include the flavones, flavonols, aurones, flavanones, etc. The left ring is synthesized from three acetate radicals, while the right ring and the three-carbon fragment are produced through the shikimic acid pathway. Attached to various of the 15 basic carbon atoms may be—OH, —OCH_3, or —CH_3 groups which cause variations in the characteristics and color of the pigment. In anthocyanins an increase in the number of —OH groups causes the pigment to be more blue, while the substitution of an —OCH_3 for an —OH group causes the pigment to be more red. The elaboration of the molecule in this manner has been demonstrated to be under genetic control. The color of a pigment is also affected by factors in its environment such as pH, metallic ions, and the presence of other pigments with which it may form complexes. Most of the naturally occurring flavonoid pigments contain one or more sugar residues which increases their solubility. Often associated with anthocyanins are colorless compounds known as proanthocyanins and leucoanthocyanins which do not produce the reddish color of the anthocyanins until they have been boiled with a mineral acid.

A less common group of pigments are the water-soluble betalains found in the vacuoles of members of ten families of the order Centrospermae. The betacyanins are the red pigments of the beet, cacti, pokeberry, etc., plants which lack anthocyanins. Degradation of betacyanins give indole and pyridine derivatives. The yellow betaxanthins are less common.

The carotenoids, carotenes and xanthophylls, may be found in all parts of the plants and are obscured in the leaf by the chlorophylls. The carotenoids give the typical color to the carrot root, the fruit of the tomato as well as the petals of the sunflower. These yellow pigments insoluble in water are soluble in organic solvents. They are composed of eight isoprene-like units to give a total of 40 carbon atoms. The central carbons of the molecule are the same in all carotenoids. Since many aliphatic double bonds are present in this central portion, many *cis-trans* isomers are possible and several exist in nature. The variation between the pigments occurs in the nine carbons at either end of the molecule. The carotenes are hydrocarbons and are soluble in petroleum ether and only slightly soluble in alcohol, while the xanthophylls are insoluble in petroleum ether, but soluble in alcohol. The xanthophylls are similar to the carotene, but contain one or more oxygen atoms. Zeaxanthin, the yellow xanthophyll of corn grains, differs from β-carotene, the most universally present carotene, by two hydroxyl groups. The presence or absence or elaboration of the basic pigment molecules as with the flavonoids is under genic control. Beta-carotene serves as a source for animals of vitamin A which is equivalent to one-half of the molecule.

The function of flower pigments is probably to aid in assuring the pollination of the flowers. The flower is made conspicuous to the pollinator by the reflection of particular wave lengths of visible or ultraviolet light. Red pigments are common in flowers pollinated by birds, while yellow and blue ones are prevalent in flowers pollinated by bees and butterflies. The determination of which wave lengths of light are absorbed and which are reflected is due mainly to the pigment complement of the petals. Thus by conspicuous color patterns the pollinator is able to recognize the flowers and carry out pollination. Often both the carotenoids and the flavonoids make substantial contributions to the pigmentation of a particular flower.

SARAH CLEVENGER

FLUIDIZATION

Fluidization may be broadly defined as the unit operation which utilizes the properties of a bed of finely divided solids, suspended in an up-flowing fluid stream under such conditions that the solids are in turbulent, random motion within the bed, while the bed maintains a definite upper surface resembling that of boiling liquid. The major part of the development on fluidization has been devoted to gas-solids systems, and the subsequent discussion is limited to these systems. However, it should be noted that the same principles may be applied to liquid-solid systems when appropriate allowances are made for the differences in physical properties such as density and viscosity of the fluidizing medium.

A clear picture of the mechanism of the fluidization process is observed when a stream of gas passes upward through a bed of finely divided solids in such a way that the gas flow is uniformly distributed over the entire cross-section of the bed. As the gas flow is increased, a critical velocity will be reached wherein the bed starts to expand as the solids are lifted by the gas. This partially expanded but highly dense bed is sometimes called a "jiggled" bed. As the gas velocity is further increased, the bed will continue to expand and the movement of the particles increase until the particles are in free, random motion. Above this transition zone of fixed to random motion is the condition usually considered true fluidization.

Consider a bed of finely divided solids in a reactor or vessel which has provisions at the bottom for the distribution of gas uniformly over the cross-section of the vessel, and a considerable freeboard above the bed. When gas is introduced at a low rate it passes through the interstices of the bed without disturbing it. As the gas flow rate increases, the pressure drop across the bed increases approximately according to the equation $\Delta p = kv^2$, where Δp is the gas pressure drop, v is the superficial gas velocity and k is a constant characteristic of the system.

When the localized gas velocity becomes great enough to begin to lift the particles, the free area between them is increased and the bed expands. The velocity required to expand the bed is related to the free-falling velocity of the smaller particles. It is possible to obtain steady state conditions with a partially expanded bed. The rate at which the pressure drop increases under these circumstances is less than that calculated by the equation given above. It can be related to the gas velocity by $\Delta p = k'v^n$, where n is less than 2. The range over which the gas rate can be varied in this jiggled state depends on a number of factors which include the size distribution, density and shape of the particles. The particles in a jiggled bed are in constant motion; there is, however, relatively little migration of individual particles when the bed begins to expand. A limited amount of classification does take place, however, since the larger or heavier particles tend to migrate toward the bottom, while the smaller or lighter particles migrate upward. Particles reaching the upper surface of the bed, which are fine or light enough so that their free-falling velocities are less than the upward velocity of gas above the bed, are entrained with the outgoing gas stream.

As gas flow is further increased, the movement of the particles increases, the bed continues to expand, and a gradual transition takes place, culminating in a state where the solids are in very rapid random motion and migrate freely throughout the entire breadth and depth of the bed. The rate of increase in pressure drop with increasing gas flow becomes less and less until no appreciable increase results from the increase in gas flow. These conditions, which are those of true fluidization, produce a bed which is entirely supported by the gas passing through it. The pressure drop is then $\Delta p = \rho_b D$, where ρ_b is the bed density (bulk density) and D is the depth of the bed. All the solid particles are in rapid motion and the bed surface takes on the appearance of a violently boiling liquid.

At a gas rate corresponding to the minimum fluidization velocity a relatively small percentage of the solids are carried into the free space above the bed. Only those particles which reach the surface with sufficient velocity to carry them out of the bed and which have free-falling velocities less than the superficial gas velocity above the bed are carried out. As the gas velocity is further increased the entrainment of solids out of the bed increases more and more until finally all the solids are entrained into the gas stream and the upper surface of the bed completely disappears. This is the condition utilized when transporting solids in gas streams by pneumatic conveying systems.

Heat transfer between the gas and solids within the bed and between the bed and the vessel walls is greatly enhanced by the movement of the solid particles, increasing as the gas velocity is increased. In the fluidized bed, gas-solids heat exchange is such that the entire bed maintains a constant temperature even when highly exothermic or endothermic reactions are taking place.

The analogy between a fluidized bed of finely divided solids and a boiling liquid is well illustrated by the carryover of solids from the bed into the outgoing gas stream, which compares with the vaporization of volatile components from the boiling liquid.

For industrial uses, the application of fluidization depends on the intimate vapor-solids contact and the rapid dispersion of the solids through the bed. These bed characteristics produce extremely good heat transfer between the solids and the gas as well as a very uniform bed temperature. This latter characteristic is of special importance where the control of reaction temperature within very close limits is essential to the efficiency of the operation. These same physical characteristics of the bed are ideally suited for chemical interaction between gas and solid where the rapid random motion of the particles permits a high rate of mass transfer.

The original commercial application of the fluidization technique is the fluid bed Catalytic Cracking Process developed by the Standard Oil Development Company—a process which illustrates the application of all the above factors. The basic purpose of a catalytic cracking process is to present a catalytic solid surface to oil vapors at a carefully controlled temperature in order to produce the desired chemical reactions. In the fluid bed process, hot catalyst from the regenerator is introduced into the reactor bed. Heat is transferred to the oil vapor, bringing it to reaction temperature practically instantaneously and maintaining a very uniform temperature within the bed. The temperature can be accurately controlled by the flow rate and by the temperature of the catalyst entering the bed. Cracking, an exothermic reaction, takes place at the surface of the solid particles. The sensible heat of the catalyst particles supplies the heat of reaction, so that the cracking temperature is maintained. The heavy carbonaceous material produced in the cracking reaction is adsorbed on the catalyst—an example of mass transfer from the vapor to the solid phase. Spent catalyst, coated with carbon, is continuously removed from the cracking zone and transferred to the regenerator.

In the regenerator the fluidizing gas is air, which reacts with the carbonaceous material on the catalyst to burn it off—an example of chemical interaction between a gas and a solid. The catalyst is heated and revivified by the combustion reaction and is suitable for return to the cracking zone.

In the fluid bed catalytic cracking operation used in making gasoline for motor fuel, the heat produced in the regenerator is approximately in balance with the heat requirements of the cracking zone. Any difference in heat balance can be easily corrected by controlling the temperature of the oil entering the cracking zone and the incoming air to the regenerator zone. Where intense cracking is done, such as in the production of base stock for high octane aviation gasoline or in the cracking of heavy oils, the heat of combustion of the carbonaceous residue on the catalyst is in considerable excess of that required for the cracking operation itself. In such cases the excess heat is removed from the regenerated catalyst by passing it in a dense-bed, fluidized condition to a tubular type of heat exchanger to generate steam or to transfer the excess heat to some other medium. Although it has not been done in fluid catalytic cracking units, it is quite feasible to introduce the heat transfer surface within the regenerator bed itself.

Since the original development of the fluidizing technique for catalytic cracking, a number of other commercial applications have been evolved which utilize the unique features of the fluidized bed, for example, the oxidation of naphthalene to produce phthalic anhydride. Another example is the roasting of sulfide ores to produce the metallic oxides and sulfur dioxide. In this type of operation, developed mainly by the Dorr-Oliver Company, sulfide ore in the proper particle size range is introduced into a fluidized bed of essentially desulfurized material using air as the fluidizing medium. The fresh charge is almost instantaneously desulfurized as it enters the bed. The product is continually withdrawn from the bed at a rate corresponding to the feed. Fluidized bed roasting permits a better utilization of the oxygen in the air, so that a somewhat higher concentration of sulfur dioxide can be obtained than is possible when these ores are roasted in the conventional manner.

The thermal decomposition of limestone to produce lime and carbon dioxide is another application of the fluidized solids technique. In this operation the heat required to maintain the bed at the decomposition temperature and supply the heat of reaction is obtained chiefly by preheat in the fluidizing gas. Flue gas produced by burning fuel, itself containing carbon dioxide, can be used for this purpose.

Summarizing, the unique features of fluidization which make it attractive for a variety of industrial uses are: (1) extremely intimate contact between gas and solid within the bed; (2) extremely high heat transfer rates possible between gas and solid, and between the bed and heat exchange surface placed within the bed; (3) very close temperature control; (4) elimination of mechanical features within the reactor and over-all simplicity of operation.

The features which may in some cases be disadvantageous are: (1) solid material must be finely divided; (2) solid product from a fluidized bed is always withdrawn at the conditions existing within the bed, i.e., no concentration gradient is possible.

JOHN A. PATTERSON

Cross-references: *Cracking, Catalysis.*

FLUORESCENCE

Fluorescence is a process in which an atom or molecule emits radiation in a spin-allowed transition to the ground state. This transition is from the lowest excited electronic state which has the same spin as the ground state. The time interval between the acts of excitation and emission is short, of the order of 10^{-9}–10^{-6} sec. Fluorescence may be distinguished from phosphorescence, where the radiative transition to the ground state is spin-forbidden because it is from the lowest excited electronic state of spin different from the ground state. The time interval between absorption and emission is from 10^{-7} sec to several seconds.

Fluorescence is exhibited both by free atoms and by molecules; it can occur in the gaseous, liquid, and solid states, although not necessarily in all three phases of the same substance. It is observed in its simplest form in the classic experiment of Lord Rayleigh, in which sodium vapor, confined in a quartz vessel at low pressure, is exposed to the ultraviolet radiation from a spark between zinc electrodes. This radiation causes the sodium vapor to glow with a yellow fluorescence corresponding in wavelength to the D-line transition of sodium atoms falling from the first electronically excited state to the ground state. The absorption of the ultraviolet radiation excites the sodium atoms to high electronic energy levels. Part of this excitation energy is dissipated by collision with other sodium atoms, or the walls of the vessel, reducing the atom to its lowest electronically excited state, from which it subsequently drops to the ground state with the emission of the D-line radiation. In such excitation processes the wavelength of the fluorescent radiation is always longer than that of the exciting radiation—a generalization known as Stokes' law. Fluorescence can occur also in molecules, but the process is more complex, since the electronic excitation and de-excitation processes may be accompanied by secondary changes in the vibrational and rotational energy of the molecule.

The relationship between the molecular structure and the fluorescence of organic compounds is not completely understood. It is known that the molecule should contain a chromophoric system to absorb the exciting radiation. Also, the electronic system which excites should be fairly rigid in order to prevent too rapid a dissipation of the excitation energy into vibrational motion before fluorescence re-radiation can occur. Among organic compounds, brilliant fluorescence is associated particularly with phthalein structures, and also with aromatic structures such as anthracene and naphthacene. Few inorganic compounds fluoresce strongly in the liquid state or in solution, but in the solid state the fluorescence of certain uranyl salts and platinocyanides is outstanding.

In solids, fluorescence is often greatly modified by the presence of trace impurities.

As examples we may note that the blue fluorescence of pure solid anthracene changes on the addition of 10^{-4} mole of naphthacene, and the resulting green fluorescence is characteristic of the naphthacene and not the anthracene molecule. Similar effects are noted in solid inorganic systems, particularly zinc sulfide, the blue fluorescence of which is greatly intensified by the addition of one part in ten thousand of cupric chloride. The application of such effects to the analysis of trace amounts of the activating substances will be apparent.

Fluorescence can be induced by excitation mechanisms other than ultraviolet irradiation. The excitation of fluorescence by electron bombardment constitutes the basic process in the illumination of cathode-ray-tube screens, and the excitation of fluorescence by α-, β-, and γ-rays is employed in scintillation counters for the monitoring of radioactivity.

R. NORMAN JONES
JOYCE M. DUNSTON

Reference

E. J. Bowen, "Luminescence in Chemistry," 254 pp. D. Van Nostrand Co. Ltd., London 1968.

Fluorescent Protein Tracing

Fluorescent dyes may be conjugated to proteins, including serum antibodies, without material effect on their biological or immunological properties. Proteins labelled in this way can be injected into animals and traced directly in histological sections by fluorescence microscopy. Alternatively, labelled serum antibody may be employed in immunological tracing as a specific histochemical stain to locate the corresponding antigen in microscopic preparations. Fluorescent dyes are used for the labelling because they can be seen at much lower concentration than ordinary dyes.

Fluorescein, used as the isothiocyanate, is the fluorochrome in most general use for protein labelling and yields conjugates with brilliant green fluorescence. Sometimes rhodamine, commonly as the sulfonyl chloride, is employed to make orange fluorescent conjugates. Combination of fluorochrome with serum protein probably takes place largely through the ϵ-amino group of the lysine moities. Fluorescein thus forms a thiocarbamide linkage and rhodamine a sulfonamide linkage (Fig. 1).

An excess of dye is used in the conjugation and some of this is adsorbed by the serum proteins. Free dye in solution and this adsorbed dye may be removed by gel filtration with Sephadex or by shaking with powdered activated charcoal. The purpose of these procedures is to obtain a fluorescent solution in which all the dye is firmly bound to the serum proteins. For most applications, antiserum conjugates need further purification by absorption with tissue powders or homogenates or by fractionation on DEAE-cellulose columns to remove unwanted proteins which may be a source of nonspecific staining reactions.

Direct Tracing. This term is applied to the study of the distribution of fluorescent protein conjugates directly after injection into animals. The conju-

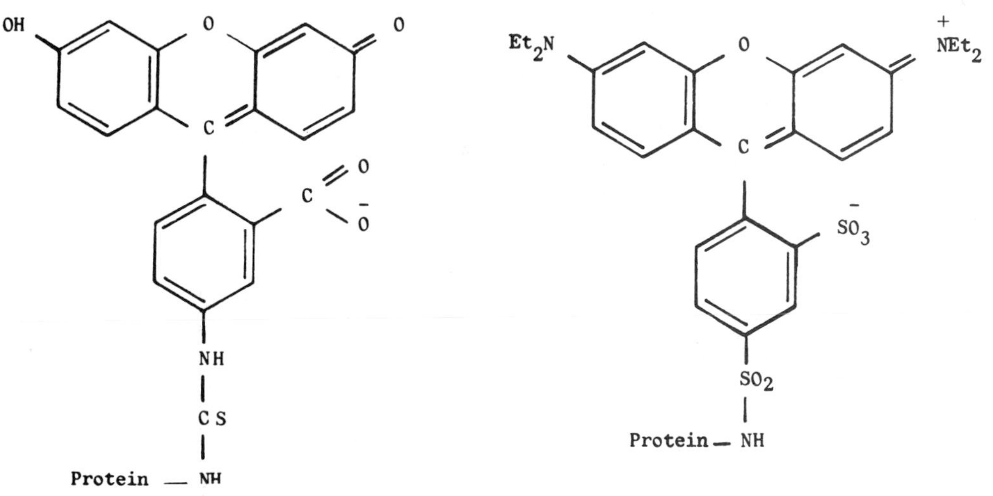

Fig. 1. Protein conjugates with (a) fluorescein isothiocyanate and (b) lissamine rhodamine B.

gates can be demonstrated in the animal tissues by fluorescence microscopy of histological preparations, or sometimes *in vivo* as in microcirculatory studies. The labeling procedure does not make the proteins toxic and the fate of injected or ingested native or foreign proteins can be investigated even in man. The technique is complementary to tracing proteins with radioactive labels.

Immunological Tracing. This is an immunohistochemical technique quite distinct from direct tracing and much more widely used. It depends on the fact that the serum antibody after the labeling still retains much of its immunological activity, usually more than 50 per cent. Such antibody can be used as a specific immunological stain for microorganisms, proteins, and other macromolecules, which can therefore be identified even in the presence of closely related organisms or substances. The principles of the method are illustrated by an experiment in which an antigen, such as a suspension of microorganisms, is injected into an animal to stimulate antibody production. The antibody, formed after two or three injections over a few weeks, is present in the γ-globulin fraction of the serum, which is conjugated with the fluorochrome. The conjugate and the corresponding antigen react with immunological specificity; the organisms, coated with fluorescent antibody, fluoresce brilliantly.

The method is applicable to any antigenic material provided it can be retained without denaturation in a microscopic preparation. The antigens that have been successfully studied include microorganisms and a wide variety of tissue components.

The unquestionable value of immunological fluorescent tracing in biology and medicine is underlined by the enormous increase in popularity of the method during the past ten years since reliable commercial reagents and equipment became readily available. Labeled antisera can now be bought for numerous research and routine diagnostic microbiological and immunological investigations, and methods of conjugating a worker's own antisera have been much simplified and standardized to bring the technique within the competence of any laboratory possessing a fluorescence microscope.

The chemical rationale of immunohistology is that the fluorescent label when conjugated to antibody globulin molecules does not affect the specific site of reaction with the corresponding antigen. Thus antibody-antigen interactions can still occur and the marker fluorochrome can be detected visually with the microscope. The exquisite specificity and sensitivity of immunology is thereby married to the precision of microscopy to provide the most elegant of all of the histochemical techniques.

Applications of Immunohistology. The simplest examples of immunofluorescent tracing are provided by the specific staining of microscopical preparations of microorganisms by conjugated specific antisera. Bacteria, viruses, rickettsiae, fungi, protozoan and metazoan parasites, indeed all varieties of microorganisms have been successfully studied in this way.

Less simple is the identification in tissue preparations of other kinds of antigenic material, especially native antigens. Pure foreign antigens such as ovalbumin injected into animals have been traced in histological sections by immunofluorescence in much the same way as microorganisms might be identified, the only special problem being the retention of the soluble foreign protein in the sections during processing. Native antigens may present a similar problem; another is that they can seldom be obtained pure, so that antisera prepared against them are likely to contain unwanted antibodies to the impurities. Staining by conjugates of such antisera is therefore not necessarily specific and the most acceptable test of specificity, namely inhibition of staining by neutralization of the serum with corresponding antigen only, is of diminished value if the latter is impure.

Future of the Method. Direct fluorescent protein tracing will no doubt continue to be increasingly used since it can provide histologically precise information which is complementary to the more quantitative information obtainable by radioactive tracing. Immunological fluorescent tracing has already expanded from microbiological research and applied medical and veterinary use, into fundamental studies of autoimmunity and of the immunological aspects of species- and organ-specificity, embryology, genetics, and carcinogenesis. Its widespread employment in botanical research may also be confidently expected. Biochemical applications are growing in importance particularly in currently topical fields such as the investigation of the kinetics of antigen-antibody reactions.

R. C. NAIRN

Cross-references: *Fluoroescence, Proteins, Immunology.*

References

Nairn, R.C., "Fluorescent Protein Tracing," 3rd Edition, Livingston, Edinburgh, 1969.
Nairn, R. C., Standardization in immunofluorescence, *Clin. exp. Immunol.*, 1968, 3, 465.

FLUORIDATION

Fluoridation is the addition of some fluoride salt to public drinking water for the purpose of improving the dental caries resistance of the members of the community. In 1939 an important study showed that drinking water containing 1.8 ppm of fluoride reduced dental caries in children, with virtually no enamel disfigurement.

There followed an amazing series of studies on the epidemiology of dental caries in relation to the fluoride naturally present in public drinking water. A good example is the comparison between the caries picture in the adults in Boulder (0.00 ppm fluoride) and in Colorado Springs (2.55 ppm fluoride). The average number of decayed, missing, and filled permanent teeth was 60% lower in the latter than in the former city. The results are even more striking in the many studies of the dental health of children. The literature is filled with evidence of the reduction of tooth decay in communities with fluoride-containing drinking water.

The logical result of the above studies was the introduction of fluoride into public drinking waters in many communities. Research teams continued to observe results in areas with newly installed water fluoridation equipment in their water source. In Newburgh, N.Y., in Brantford, Ontario, Canada, and in Evanston, Ill., the caries rate of the children was seen to descend when compared with the rate in control cities.

As a result of the many experiments, fluoridation of public drinking water has spread throughout the United States and other countries. As with all public health measures, care has been taken to estimate the possible injurious effects of fluoride at normal as well as at excessive levels in the water. Like many common chemicals, fluoride salts can be toxic in large amounts. However, one would have to drink in a day 400 gallons of water containing 1.0 ppm fluoride to receive a toxic dose. Daily ingestion of 0.5 to 1.0 mg of fluoride will result only in the decrease of dental caries with no systemic toxic effect.

The best proof of the safety of public water fluoridation is furnished by the reports of the good health of populations consuming fluoridated water. A detailed survey of the health of children from Newburgh and Kingston, N.Y., has been in progress since 1945. The water supply of Newburgh is treated to bring the fluoride content to 1.1 ppm while the water supply of the nearby city of Kingston is fluoride-free. The children of the two cities have been shown to be alike in height, weight, blood, and urine chemistry; condition of the nails, skin, and hair; eye and ear conditions; and joint conditions (i.e., wrists and knees) as evidenced by radiology. The only difference observed between the two groups was the better dental health found among the Newburgh children.

Five different chemicals are used to treat public waters with fluoride: NaF, Na_2SiF_6, H_2SiF_6, HF, and CaF_2. The latter was employed for the first time at Bel Air, Maryland, with aluminum sulfate used to liberate the fluoride ion from a water slurry of the CaF_2. Highly soluble sodium fluoride applied as a dry powder or in solution is one of the most common and earliest used fluoridation agents. Hydrofluoric acid is applied in solution as it comes from the manufacturer while the difficultly soluble Na_2SiF_6 is used in slurry form. Hydrofluoric acid has the advantage of a large fluoride availability tempered with obvious handling difficulties.

The cost of fluoridation naturally depends upon the choice of addition system and chemicals. An estimate of the annual cost per capita may be given as ranging from 4.6 cents for Na_2SIF_6 to 11.5 cents for H_2SiF_6. At this writing there are 32,000,000 people in 1500 communities throughout the United States drinking fluoride-treated water. In addition at least 3,000,000 persons are getting optimal quantities of fluoride for dental health in water where the ion is naturally present.

There have been other vehicles used for the ingestion of fluoride. Where water fluoridation is not feasible or available, fluoridation of table salt has been suggested. In Switzerland there is a commercially available table salt containing 245 mg of sodium fluoride per kilogram of salt. Another compound tried experimentally in table salt is Na_2PO_3F.

Toothpastes which contain fluoride compounds have been marketed for several years. Stannous fluoride and sodium fluoride are two of the substances added to different commercial dentifrices. Bottled water has been fluoride treated. Another method of fluoride treatment in current use is the application by the dentist of a solution of sodium fluoride to the tooth surfaces to give patients the benefit of fluoridation where other sources are not present.

Extensive research has been carried out on the effect of various chemical agents, fluoride salts, and others on experimental caries in animals. In general, reduction of caries given by fluoride metabolism depends upon the animal species and the availability of the salt to the animal. It has been demonstrated that certain fluoride salts are metabolically inert, that is, excreted without being used by the animals. Examples are KPF_6, $NaBF_4$, KBF_4 and $(C_2H_5)_4NPF_6$. On the other hand, NaF and Na_2SIF_4 commonly used for water treatment were found to be metabolically available to the animals used in the study.

The analysis of fluoride ion in public water presents certain problems due to interfering ions. Sulfate, phosphate, and chloride, commonly found in treated waters, all interfere with the color formation used in fluoride analysis. In order to avoid such interference, the fluoride ion is isolated by a distillation procedure involving the volatizing of H_2SiF_6 from a strongly acidified fluoride solution in contact with a source of silica. Fluoride is then determined colorimetrically or spectrophotometrically. A stable, colored lake is first formed between a metallic cation (Al, Zr, Fe, etc.) and an organic dye. The fluoride then complexes with the cation, shifting the color of the solution to that of the original dye. By comparing the color of the unknown solution with a series of standards, the analysis can be effected.

The earliest methods of analysis involved a zirconium-alizarin lake, while later methods employed lakes of aluminum-Eriochrome cyanine R, thorium-chromazurol-S, and zirconium-Eriochrome cyanine R.

When biological tissue is analyzed for fluoride content, the above methods are used after first ashing at high temperatures to convert the fluoride to an inorganic salt. Distillation and color formation procedures (as with water samples) are then utilized.

AARON S. POSNER

Cross-references: *Fluorine; Water.*

FLUORINE AND COMPOUNDS

Moisson first isolated elemental fluorine in 1886. His discovery was not utilized to any extent until the early 1940's when the requirements of the Manhattan project for chemically stable fluids and polymers fostered an investigation of the preparation of fluorine and its derivatives. Since then research in both university and industrial laboratories has been vigorous and productive, and today there are many fluorine-containing materials available commercially. These are finding widespread and varied applications ranging from refrigerants and aerosols to high-temperature and solvent-resistant elastomers.

Elemental Fluorine

Preparation. The element is obtained from the mineral ore by first forming hydrogen fluoride as:

$$CaF_2 + H_2SO_4 \rightarrow 2HF + CaSO_4$$

Hydrogen fluoride is then mixed with potassium fluoride (about 4:1) and electrolyzed as a melt (80–100°C) using a direct current to give fluorine as:

$$2HF \xrightarrow[\text{KF}]{\text{electrolysis}} F_2 + H_2$$

Properties. The element (at. no. 9) exists at room temperature as a greenish-yellow diatomic gas with the following properties:

Boiling point	85.02°K
Melting point	53.54°K
Density (at 85.0°K)	1.108
Critical temperature	144°K
Critical pressure	55 atm
Heat of vaporization	1564 cal/mole
Dissociation energy	37.7 kcal/mole
Electronegativity	4.0
Atomic weight	18.9984

Chemically, fluorine is considered to be the most re-

active known element and thus requires careful handling in specially designed equipment. It is available commercially in steel cylinders for laboratory use and small tank trucks (ca. 5000 lbs) for larger scale operations.

Inorganic Fluorides

A large number of inorganic derivatives are known and their chemistry has been thoroughly investigated. Recently it was discovered that fluorine would react with various of the noble gases to form a series of interesting new fluorides: XeF_2, XeF_4, XeF_6, and KrF_4. These may be prepared by reacting fluorine with either xenon or krypton under pressure at elevated temperature or in an electric discharge. Their properties are as follows:

KrF_4	Colorless crystal, m.p. 60°C (dec.)
XeF_2	Colorless crystal, m.p. 140°C
XeF_4	Colorless crystal, m.p. 114°C
XeF_6	Colorless crystal, m.p. 49.5C

They are metastable materials which can be hydrolyzed in the case of xenon to form oxygen-containing derivatives which act as powerful oxidizing agents. Through this discovery of rare gas fluorides, a whole new concept of chemistry is unfolding.

Organic Fluorides

The C–F bond is the one of most practical interest today. It is characterized by a high bond energy (110–120 kcal/mole), short bond length, low polarizability and small radius. These properties impart both chemical and thermal stability to a highly fluorinated hydrocarbon (loosely termed a fluorocarbon). Another characteristic arising from fluorine substitution is weak intermolecular forces in liquid fluorocarbons (sometimes referred to as low internal pressures). This property is reflected in low boiling points, low surface tension, decreased solubility, and low heats of vaporization.

Methods of Fluorination. The two general methods used in synthesizing a fluorocarbon are replacement of hydrogen in an organic molecule by fluorine or exchange of chlorine, bromine, or iodine by fluorine. The former technique requires the use of elemental fluorine while the latter employs an inorganic fluoride generally of antimony, sodium, or potassium.

Reaction with Elemental Fluorine. During fluorination, several chemical processes take place which include substitution of hydrogen, addition to an unsaturated C–C bond, fission of a C–C bond, and polymer formation. Since direct fluorination is highly exothermic, carbon bond rupture occurs readily and removal of reaction heat is of prime importance to achieve control of the reaction and good yields. An efficient reaction vessel has been designed which employs jets or other small openings into the reaction zone through which fluorine diluted with nitrogen is introduced. The reaction zone itself can be packed with a good heat conductor such as copper to aid in heat removal.

The fluorination reaction can best be explained on the basis of a radical chain process as the following:

$$F_2 \rightleftharpoons 2F\cdot$$

$$RH + F\cdot \rightarrow R\cdot + HF$$

$$R\cdot + F_2 \rightarrow RF + F\cdot$$

Initiation easily occurs by light or heat and the reaction can proceed rapidly at relatively low temperatures.

Addition of fluorine to an unsaturated C-C bond takes place in a similar manner as shown:

$$-\overset{|}{C}=\overset{|}{C}- + F\cdot \rightarrow -\overset{|}{C}F-\overset{|}{C}-$$

$$-\overset{|}{C}F-\overset{|}{C}- + F_2 \rightarrow -\overset{|}{C}F\overset{|}{C}F + F\cdot$$

Polymerization can also occur, leading to higher molecular weight fluorocarbons.

Reaction with Active Metal Fluorides. A second technique for substitution of hydrogen also results from the reaction of a higher fluoride of such metals as silver, cobalt, or lead with a C-H bond as:

$$2CoF_2 + F_2 \rightarrow 2CoF_3$$

$$RH + 2CoF_3 \rightarrow RF + 2CoF_2 + HF$$

The heat evolved for each C-F bond formed (46 kcal/mole) is much less than that obtained in direct fluorination (104 kcal/mole) and consequently the reaction yields are much higher.

Only a few metal fluorides are of practical use in this process. Their reactivity appears to be associated with the oxidation potential of the metal ion (Ag^{+1} or Co^{+2}) the greatest reactivity with the highest potential. In order of decreasing effectiveness as fluorination agents they are the following: silver difluoride, cobaltic fluoride, ceric fluoride, manganese fluoride, plumbic fluoride, bismuth pentafluoride, chromic fluoride, mercuric fluoride, ferric fluoride, cupric fluoride, and stannic fluoride.

The reaction is conducted by first passing fluorine through a bed of the metal salt at elevated temperature to form the active metal fluoride. The organic starting material is then passed through or over the fluoride at elevated temperatures to effect fluorination. Thus a semicontinuous process is realized. This technique is used principally for the fluorination of hydrocarbons yielding only saturated fluorocarbons. It may also be employed to fluorinate polychlorohydrocarbons, in which case replacement of chlorine as well as hydrogen by fluorine is accomplished.

Electrochemical Fluorination. A newer method of hydrogen substitution by fluorine comprises the electrolysis of an organic reagent dissolved in anhydrous liquid hydrogen fluoride. In practice, the solution is placed in an electrolytic cell constructed of steel or nickel with nickel plates serving as electrodes. A low voltage (5–6) direct current is applied and the cell contents cooled to minimize loss of hydrogen fluoride. Volatile reaction products are removed as gases, while the higher boiling products, which are usually insoluble in hydrogen fluoride, can be drained from the bottom of the cell.

The mechanism of this fluorination reaction remains obscure. Since the reaction potential is below that necessary for the formation of elemental fluorine, it is possible that the fluorination involves a higher fluoride of nickel formed on the surface of the anode. The conductivity of the solution directly affects yield and efficiency. The best results are obtained using organic acids, amines, or ethers as starting materials since they are usually quite soluble in hydrogen fluoride. Relatively insoluble compounds such as hydrocarbons require the use of a salt (usually lithium fluoride) to increase the solution conductivity.

Electrochemical fluorination works well with simple starting materials as acetic acid or tertiary alkyl amines. However, C-C bond rupture can occur readily resulting in lower yields of longer chain fluorocarbons.

Halogen Exchange. Replacement of halogen, primarily chlorine, by fluorine is the principal method for preparing aliphatic fluorocarbons. In particular the fluorides of antimony have found wide application and are the prime reagents used commercially. Antimony trifluoride can be used with compounds having the following structures:

$$RCX_2R \rightarrow RCF_2R$$

$$RCX_3 \rightarrow RCF_3$$

$$R\overset{|}{C}=\overset{|}{C}CX_3 \rightarrow R\overset{|}{C}=\overset{|}{C}CF_3$$

where R is hydrogen, alkyl, or aryl, and X represents chlorine, bromine, or iodine. Antimony trifluoride, however, is ineffective in replacing a single halogen atom on a carbon or a vinylic halogen.

A more effective and useful reagent is prepared from the pentavalent antimony chlorides. The fluorination of chloroform is illustrated as follows:

$$SbCl_5 + 3HF \rightarrow SbCl_2F_3 + 3HCl$$

$$SbCl_2F_3 + CHCl_3 \rightarrow SbCl_4 + CHClF_2$$

$$SbCl_4F + 2HF \rightarrow SbCl_2F_3 + 2HCl$$

In this manner continuous regeneration of the fluorination agent is possible.

Many other inorganic fluorides have been employed in halogen exchange, those of sodium and potassium being the most useful. In particular, potassium fluoride in the presence of polar media such as amides, sulfones, and nitriles can be used to replace vinylic halogen. In many reactions, the fluoride ion acts as a strong nucleophilic agent. With fluoroolefins, reversible addition may take place to produce a fluorocarbanion which may then add a proton from the solvent to yield a hydrogen fluoride addition product. The fluorides of lead and mercury, prepared *in situ* from the oxide and hydrogen fluoride, effect halogen exchange and also add fluorine to unsaturated linkages in halogenated olefins.

New Fluorination Agents. In the past few years other fluorides have been discovered which are useful fluorinating agents. While it is difficult to generalize on their chemistry, the following summarizes the present knowledge:

    (a) Iodine pentafluoride (IF_5). Halogen exchange agent.

    (b) Iodine heptafluoride (IF_7). Halogen exchange agent.

    (c) Iodine monofluoride (IF) and bromine monofluoride (BrF). Add readily to unsaturated bonds in olefins.

(d) Bromine trifluoride (BrF$_3$) and chlorine trifluoride (ClF$_3$). Replace hydrogen or halogen with fluorine and add to unsaturated linkages.

(e) Perchloryl fluoride (ClO$_3$F). Replaces active hydrogen with fluorine.

(f) Nitrosyl fluoride (NOF). When complexed with hydrogen fluoride it comprises a liquid halogen exchange agent.

(g) Sulfur tetrafluoride (SF$_4$). Replaces oxygen with fluorine.

(h) Xenon difluoride (XeF$_2$). Replaces aromatic hydrogen with fluorine.

Uses of Fluorocarbons

Organic fluorine compounds have found wide application in two general areas, as refrigerants and aerosol propellants and as polymers for elastomers or plastics. The former use, taking advantage of the chemical stability, low toxicity, and nonflammability of fluorocarbons, has developed into a major industry, producing several hundred million pounds per year of the various chlorofluoromethanes, ethanes, and ethylenes which constitute this general class of materials. Fluorocarbon polymers have found increasing applications based on their properties of chemical and thermal stability and low solubility in organic solvents.

OGDEN R. PIERCE

Cross-references: *Fluorocarbon Resins; Refrigerants; Fluorocarbons, Aromatic.*

References

"Advances in Fluorine Chemistry," edited by M. Stacey, J. C. Tatlow and A. G. Sharpe, Butterworths. (A series of reviews.)

"Fluorine Chemistry Reviews," edited by P. Tarrant, Marcel Dekker, Inc.

"Organic Fluorine Chemistry," W. A. Sheppard and C. M. Sharts, pub. by W. A. Benjamin, Inc. (1969).

Hexafluoride Molecules

In the hexafluoride molecules, the oxidation number and coordination number for the central element coincide and the hexafluorides are symmetrical compounds with very weak intermolecular binding and, therefore, high volatility.

The family of hexafluorides includes both very stable molecules and some of the most vigorous fluorinating reagents known. Since they all can be studied in great detail as monomeric vapors, systematic studies have been frequent and fruitful yielding considerable insight into subtle aspects of chemical bond formation and molecular structure such as the Jahn-Teller effect.

Eighteen elements form hexafluoride molecules. These may be divided into three categories, four group VI hexafluorides; Se$_6$, SeF$_6$, TeF$_6$ and PoF$_6$; thirteen metallic hexafluorides and one rare gas hexafluoride, XeF$_6$. The metallic hexafluorides may be arranged in four transition series: CrF$_6$($3d^0$): MoF$_6$($4d^0$), TcF$_6$($4d^1$), RuF$_6$($4d^2$), and RhF$_6$($4d^3$); WF$_6$($5d^0$), ReF$_6$($5d^1$), OsF$_6$($5d^2$), IrF$_6$($5d^3$) and PtF$_6$($5d^4$); UF$_6$($5f^0$), NpF$_6$($5f^1$) and PuF$_6$($5f^2$). The

parenthetic designation indicates the type and number of nonbonding electrons in the molecule, e.g., ($5d^2$) indicates two $5d$ electrons.

The discovery of fission for U^{235} and the attractive aspects of isotope separation with a stable volatile compound led to great interest in uranium hexafluoride and its industrial use in gaseous diffusion plants for concentration of U^{235}. The ensuing research led not only to the synthesis of the hexafluorides of the hitherto unavailable elements Tc, Np and Pu, but also to the development of enough expertise to permit the synthesis of the very reactive hexafluorides, PtF$_6$, RuF$_6$, RhF$_6$, and CrF$_6$ and the probable synthesis (still not adequately characterized) of the hexafluoride of intensely radioactive Po.

A number of the other hexafluorides merit special attention for their interesting history or somewhat surprising behavior.

The history of OsF$_6$, for example, is most unusual. This compound was first synthesized in 1914 but was incorrectly identified as OsF$_8$. As the only compound in nature known to have eight univalent atoms attached to a single central atom, "osmium octafluoride" held a celebrated position in chemical texts for over 40 years. However, this was an error; an octafluoride has not yet been prepared. The mistake was corrected when the systematic study of the hexafluorides showed the properties ascribed to the lower fluoride of osmium previously identified as OsF$_6$ (particularly its lower volatility) could not be those of a regular hexafluoride while those of the "octafluoride" fit the hexafluoride systematics.

Plutonium is ordinarily synthesized as a mixture of isotopes of different nuclear properties. A process for concentrating some of these has significant potential value and research on plutonium hexafluoride has been encouraged though the very poisonous α-active plutonium makes handling large quantities of a volatile compound rather hazardous. A few micrograms of plutonium hexafluoride may be a lethal dose. Plutonium hexafluoride is relatively unstable to dissociation to fluorine and the tetrafluoride, and both thermal and radiation decomposition have been studied. Solid plutonium hexafluoride decomposes under the influence of its own α-radiation, perhaps one to two per cent a day when stored in a representative container. In the vapor phase much of the energy may be dissipated in the walls and by storing the compound with fluorine, the net decomposition kept down to a negligible factor.

Chemical Properties. The Group VI hexafluorides are rather inert compounds. SF$_6$ is particularly unreactive, resisting hydrolysis at 500°C and finds use as an insulating gas for high-voltage electrical equipment. The metallic hexafluorides by contrast readily hydrolyze to form hydrogen fluoride, in some cases with explosive violence. Their order of reactivity as oxidizing and fluorinating reagents has been established. In general the reactivity increases with increasing number of nonbonding electrons, and hexafluorides formed from the $5d$ series elements are less reactive than the corresponding $4d$ series and $5f$ series. WF$_6$ will convert PF$_3$ to PF$_5$, but not fluorinate NO, MoF$_6$ yields NOMoF$_6$ but does not fluorinate AsF$_3$. UF$_6$ fluorinates the latter to AsF_5 but dissolves without reaction in BrF_3, while PuF_6 converts the latter to BrF_5 and PtF_6 and RhF_6 fluorinate PuF_4, molecular oxygen and xenon as well as all the stronger reducing agents. CrF_6 is synthesized with such difficulty that it is hard to be sure the experiments were really successful, but it would probably fluorinate even a lower platinum fluoride to PtF_6.

Molecular Structure. In the vapor, hexafluoride molecules have the structure of a regular octahedron with six fluorine atoms at the vertices equidistant from a central heavier atom. This structure was deduced from infrared and Raman spectra and confirmed by electron diffraction studies. The vibrational spectra of XeF_6 are dissimilar from those of the other hexafluoride molecules, and it may have a different structure.

Because of its peculiar properties, XeF_6 has been studied very extensively without completely elucidating the behavior. The abnormalities have been interpreted in terms of a seven coordination model with a "lone" pair of electrons in the seventh position rapidly oscillating so as to provide the average symmetry required by the zero dipole moment found experimentally. An alternative explanation interprets the complex spectra, incompatible with any simple symmetry class, in terms of an equilibrium mixture of a symmetrical octahedral ground state and electronic isomers whose rather unexpected stability is explained in part by the Jahn-Teller effect found for other hexafluoride molecules with one or two nonbonding d electrons.

The Jahn-Teller theorem states that symmetrical nonlinear polyatomic molecules with orbital electronic degeneracy will distort their structure to remove this degeneracy. Alternatively, a coupling of the electronic and vibrational motions may occur instead of a structural distortion. OsF_6, ReF_6, TcF_6 and RuF_6 have suitable electronic degeneracy and show the latter effect. Significantly, this effect is absent for eleven other hexafluorides, which should not exhibit it.

Electron diffraction measurements with hexafluorides were initially puzzling because they suggested unsymmetrical structures. Inclusion of an angle-dependent phase factor in the Born approximation was necessary to reconcile the diffraction data with the symmetrical structure. The metal-to-fluorine distances derived for WF_6, OsF_6 and ReF_6 are almost constant, being 1.833, 1.831 and 1.830 Å. The distances for UF_6, NpF_6 and PuF_6 are 1.996, 1.981 and 1.971 Å. This decrease in bond length with increasing atomic number is analogous to the "lanthanide" contraction observed for rare earth compounds.

UF_6 forms a close-packed orthohombic molecular solid in which the uranium hexafluoride molecule is slightly distorted. The other hexafluorides (except XeF_6) show x-ray diffraction patterns that are iso-structural with solid UF_6. Before melting, a transition occurs from this orthohombic phase to a less dense cubic form (except for UF_6, NpF_6 and PuF_6).

XeF_6 has at least three phases different from all other solid hexafluorides and interpreted in terms of specific polymeric arrangements. A fourth phase, described as a tetra-hexamer arranged as a cubic crystal appears to coexist with the other phases in a

very curious way and the published explanations have not gone unchallenged.

Some of the known physical properties are tabulated below. In spite of a lower molecular weight the $4d$ series hexafluorides are less volatile than the corresponding $5d$ series compounds, in good agreement with their greater reactivity and our ideas of chemical behavior which suggest this relationship. There is an interesting correlation between the polarizability of the molecule which governs the Van der Waals attraction between molecules and the ease with which the molecule can be ionized to initiate a reaction.

MELTING AND BOILING POINTS* FOR THE HEXAFLUORIDES

	M.W.	M.P. or T.P. °C	B.P. or S.P. °C
SF_6	146.1	−50.8	−63.7
SeF_6	193.0	−34.6	−46.6
TeF_6	241.6	−37.8	−38.9
PoF_6	324.9		
CrF_6	166.0		
MoF_6	209.9	17.4	34
TcF_6	212	37	~60
RuF_6	215.1	54	~70
RhF_6	216.9	~70	~75
WF_6	297.8	2.0	17.1
ReF_6	300.2	19	33.8
OsF_6	304.2	34	46
IrF_6	306.2	44	53.6
PtF_6	309.1	61.3	69.1
UF_6	352.0	64.0	56.5
NpF_6	351	54.8	55.2
PuF_6	356	51.6	62.2
XeF_6	245.3	49.5	75.6

*For many hexafluorides the vapor pressure of the solid is one atmosphere below the melting point and this sublimation temperature is given. Many melting points have been measured under the vapor pressure of the hexafluoride and are true triple points.

Electronic and Magnetic Properties. The hexafluorides containing one or more nonbonding electrons are colored compounds. Electronic spectra have been measured with many of their vapors and solids. Electron spin resonance and magnetic susceptibility measurements have also been made. The octahedral ligand field removes the ground state electronic degeneracy in $PtF_6(5d^4)$ and PuF_6 ($5f^2$). The spectra of the solids at low temperatures show further splittings arising from the small distortion of the molecular octahedron. The magnetic moment of NpF_6 first increases when diluted with UF_6 and then decreases upon further dilution; this is the only known example of this behavior.

B. WEINSTOCK
H. H. HYMAN

Dioxygenyl (O_2^+) Salts with Fluoroanions

Molecular orbital theory predicts a greater stability for the dioxygenyl ion than for molecular oxygen. Although this prediction is supported by spectroscopic data, no stable dioxygenyl salt had been reported up to 1962. The absence of stable dioxygenyl salts was attributed to the high ionization potential of molecular oxygen (283 kcal/mole). In 1962 Bartlett and Lohmann reported the synthesis of dioxygenyl hexafluoroplatinate (V), and shortly thereafter other investigators reported the existence of O_2BF_4, O_2PF_6, O_2AsF_6, and O_2SbF_6.

O_2PtF_6 may be obtained at room temperature as the result of the oxidation of molecular oxygen by platinum hexafluoride (Eq. 1). It is also obtained

$$O_2 + PtF_6 \rightarrow O_2PtF_6 \qquad (1)$$

when platinum or platinum salts are fluorinated in the presence of oxygen or some source of oxygen such as "Pyrex" or silica, at elevated temperatures (ca. 400°C).

In thin films, O_2PtF_6 is an orange-red solid but in mass it appears black. It may be sublimed at pressures below 10^{-2} mm at temperatures above 90°C. When heated in a sealed tube, O_2PtF_6 melts at 219°C with partial decomposition. Its density is 4.2 g/cc. Infrared absorptions observed for O_2PtF_6 sublimed onto potassium bromide windows were: 631(vs), 680(w), 425(vw), and 1308(w) cm^{-1}. Visible and ultraviolet absorption by a film of O_2PtF_6 on a quartz window rises steadily with decreasing wavelength and quite sharply at 4000 Å. A single maximum is exhibited at 3500 Å. O_2PtF_6 is paramagnetic, its effective magnetic moment (μ_{eff}) being 2.57 B. M. Its x-ray diffraction pattern can be indexed on a cubic unit cell with a = 10.032 Å, Z = 8, and it is isomorphous and nearly isostructural with $KSbF_6$. Dioxygenyl hexafluoroplatinate (V) behaves as a powerful oxidizing agent and as a derivative of platinum (V).

Other Dioxygenyl Salts. Dioxygenyl salts of complex fluoroanions having B, P, As, and Sb as the central element are obtained by the reaction of the appropriate fluoride with the thermodynamically unstable compound, dioxygen difluoride (Eq. 2).

$$O_2F_2 + MF_n \xrightarrow{-160°C}$$
$$O_2MF_{n+1} + \tfrac{1}{2}F_2, \; M = B, P^v, As^v, Sb^v \qquad (2)$$

Unlike the highly colored O_2PtF_6, these salts are white, although they tend to turn pink to violet on cooling to liquid nitrogen temperature. O_2BF_4 and O_2PF_6 are unstable at room temperature and decompose as shown in Eq. 3. The arsenic (V)

$$O_2MF_{n+1} \xrightarrow{25°C}$$
$$O_2 + MF_n + \tfrac{1}{2}F_2, \; M = B, P^v \qquad (3)$$

and antimony (V) derivatives are stable to above 150°C. The powder diffraction lines of O_2AsF_6 may be indexed on a face centered cubic unit cell with $a = 8.00$ Å. It is isomorphous with $NOAsF_6$.

All the salts are powerful oxidizing agents and react violently with organic solvents. They liberate oxygen and ozone upon reaction with water and are converted to the corresponding nitronium (NO_2^+) salts upon reaction with N_2O_4.

ARCHIE R. YOUNG II

Cross-references: *Fluorocarbon Resins; Refrigerants; Noble-gas Compounds.*

Fluorocarbons, Aromatic

Much of the earlier development of fluorocarbon chemistry involved aliphatic compounds. Aromatic fluoroderivatives containing one or two fluorine atoms have long been known, but highly fluorinated or fluorocarbon types have only become available since the mid-1950's. Since then much work has been done on their reactions and more recently on those of highly fluorinated heterocycles.

The standard route to partially-fluorinated aromatics has been from amines through diazo-compounds.

$$Ar.NH_2 \xrightarrow[\substack{\text{then} \\ \text{HBF}_4}]{\text{HNO}_2} Ar.N_2^+BF_4^- \xrightarrow{\text{heat}} Ar.F + N_2 + BF_3$$

Usually the fluorine in such compounds is stable. They undergo the usual general reactions with electrophilic (positive) reagents that are typical of aromatic chemistry. The synthetic routes to compounds such as these will not in general give full fluorination. Further, direct fluorination of an aromatic hydrocarbon with fluorine or any other drastic fluorinating agent does not give an aromatic fluorocarbon; fluorine is added to the unsaturated ring and alicyclic fluorides are formed, e.g., from benzene there would be obtained perfluorocyclohexane (C_6F_{12}) and polyfluorocyclohexanes ($C_6H_nF_{12-n;n=1-4}$). However, these compounds can be used as precursors of aromatic fluorocarbons. For example, compounds of formula $C_6H_3F_9$, if treated with strong bases, lose hydrogen fluoride to give hexafluorobenzene

the simplest member. Further, alicyclic fluoroderivatives with 6-membered rings will react with heated metals, such as iron or nickel, with loss of fluorine and re-creation of aromatic unsaturation. Since alicyclic fluoro- and fluorohydrocarbons are readily formed from aromatic hydrocarbons by fluorination with cobaltic fluoride or similar high-valency metal fluorides, these reactions provide a general synthesis of aromatic fluorocarbons.

Another route to aromatic fluorocarbons is by pyrolysis of simple aliphatic fluorides, for example:

$$CBr_3F \xrightarrow{\text{heat}} C_6F_6$$

A useful general route to both perfluoro-aromatic and -heterocyclic derivatives is provided by exchange of fluorine for chlorine in the analogous perchloro-compounds using potassium fluoride at elevated temperatures:

$$C_6Cl_6 \xrightarrow{\text{KF}} C_6Cl_2F_4 + C_6ClF_5 + C_6F_6$$

Octafluoronaphthalene, pentafluoropyridine and many related compounds have been made thus. This is probably the best route to perfluoro-heterocycles.

Saturated aliphatic fluorocarbons are characterised by an almost complete absence of useful chemical reactions. Aromatic fluorocarbons however are valuable synthetic intermediates from which can be made a whole vast range of new chemical compounds to parallel the enormous numbers of orthodox aromatic derivatives based on the hydrocarbon series. It will be recalled that aromatic hydrocarbons undergo substitution with electrophilic reagents (those deficient in electrons) e.g.

$$C_6H_6 \xrightarrow[\substack{\text{from} \\ HNO_3/H_2SO_4}]{NO_2^+} C_6H_5 \cdot NO_2 + H^+$$

From reactions of this sort followed by transformations on the functional groups introduced the whole field of orthodox aromatic chemistry has developed.

In an analogous way, aromatic fluorocarbons are substituted by nucleophilic reagents (those with excess negative charge, or regions of high electron density), fluorines are replaced and series of derivatives with different functional groups can be made.

The functional groups of these various derivatives have their characteristic chemical reactions, usually different in degree but not in kind from those of hydrocarbon analogues. A whole new area of synthetic chemistry is thus available for exploitation.

One of the major interests in hydrocarbon aromatic chemistry is the nature of the isomers formed, and the reasons for their formation, when a second attack occurs on a monosubstituted derivative of benzene. A similar problem arises in the hexafluorobenzene field, though, of course, the attacking species differ. Three isomers are possible, thus:

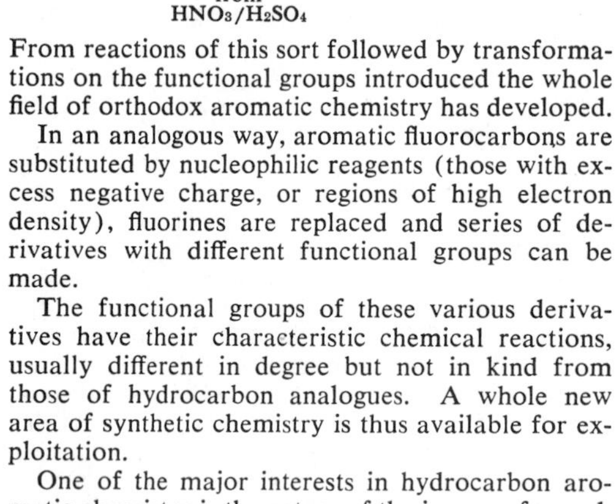

The product actually formed seems to depend on the character of the pentafluorophenyl derivative (C_6F_5X) attacked, and the reagent used (Y^-) is of secondary importance only. The same sort of isomer ratio is given by most nucleophilic substitutions on a given compound (C_6F_5X).

In the majority of cases, for example where X = H, CH_3, CF_3, C_6H_5, SCH_3, SO_2CH_3 or $N(CH_3)_2$, 90% or more of the para-isomer is formed, whatever the nature of Y. However, where X is a strong electron donor, substitution is very slow (as would be expected from attack by a negative species) and the meta-isomer is found to predominate. This occurs where X = NH_2 or OH; for example

$$C_6F_5 \cdot NH_2 \xrightarrow{NH_3} m-C_6F_4(NH_2)_2.$$

When X = OCH_3 or $NHCH_3$, mixtures of roughly equal proportions of meta- and para- isomers are formed; here again the groups X have some electron donor properties. An exceptional case arises when X = NO_2. Substitution is very easy (NO_2 is a strongly electron-withdrawing group) and usually gives rise to mixtures of ortho- and para-isomers.

Pentafluoropyridine and other nitrogeneous heterocycles, and their derivatives, also undergo nucleophilic substitution, some very readily.

The directional effects in these nucleophilic reactions are explained in terms of the most stable intermediate complex being formed, that is, the nucleophile attacks in that ring position which permits the negative charge in the adduct to occupy the least unfavourable situation.

By use of nucleophilic substitution reactions, and of classical reactions such as those involving Grignard reagents of the polyfluoroaryl series, to introduce functional groups, and by further transformations on the functional groups, whole families of polyfluoroaryl derivatives have been synthesized. All of the best-known types of aromatic compounds have now been paralleled in the polyfluoro series of which the general properties and reactions are now well defined. Simple physical properties are often surprisingly similar to those of hydrocarbon analogues and the extreme differences from general organic compounds shown by aliphatic fluorocompounds are not present.

Chemical reactivity is dominated by nucleophilic substitution of fluorine mentioned above. Functional groups usually show their fundamental properties and reactivities though these are often substantially modified by the electronegative fluorine substituents.

Interest in aromatic fluorocarbon derivatives lies not only in making great numbers of potentially interesting new compounds, and in studying the mechanisms of the many new reactions involved, but also in possible commercial exploitation of the series. Hydrocarbon aromatics are of course of immense industrial importance, and active work is in progress to try to develop new biologically-active compounds particularly, and also polymers, dyestuffs, etc., from highly fluorinated aromatics and heterocyclics. The introduction of fluorine into carbon compounds will always be expensive, and there will, therefore, be a delicate balance between special properties which some of these materials may be found to possess, and the higher costs associated with them.

J. C. TATLOW

Cross-references: *Fluorocarbon Resins; Fluorine and Compounds.*

References

Plevey, R. G., and Tatlow, J. C., *Science Progress,* 1970, **58**, 481.
Banks, R. E., "Fluorocarbons and their Derivatives," Macdonald, London, 2nd Edition, 1970.
Sheppard, W. A., and Sharts, C. M., "Organic Fluorine Chemistry," Benjamin, New York, 1969.

Fluorocarbon Resins

The fluorocarbon resins are among the most versatile of all thermoplastics. This versatility can be attributed to the following unique properties of this group: (1) inertness to almost all chemicals; (2) resistance to high and low temperatures; (3) excellent dielectric properties; (4) essentially zero moisture absorption; (5) nonflammability; (6) low coefficient of friction; (7) weather and oxidation resistance. There are several types of fluorocarbon resins in commercial use.

Polytetrafluoroethylene (PTFE or TFE Resins). Its basic unit consists of two atoms of carbon with fluorine. The strong chemical bond between the carbon and fluorine atoms is what gives PTFE its unique properties. These resins are trade-marked "Teflon" TFE by E. I. DuPont de Nemours and Company, Inc. and "Halon" TFE by Allied Chemical Corporation.

Properties of TFE. (1) Chemical resistance: within the limits of its thermal stability, TFE is only affected by molten alkali metals and elemental fluorine at high pressures. (2) Thermal stability: TFE is not affected by temperatures up to 500°F or down to −425°F. (3) Electrical properties: high dielectric strength, low dissipation factor even in high humidity and elevated temperature environments. (4) Low coefficient of friction: comparable to ice against ice. Few materials will stick to its slippery surface. (5) Zero water absorption. (6) Nonflammability. (7) Resistance to weathering and oxidation.

Since TFE is a relatively soft, waxy material, it is often combined with fillers to reduce creep under continuous load. The most common fillers are milled glass fibers, bronze powder, molybdenum disulfide or graphite. The choice of filled and concentration is naturally dependent on the intended application.

Processing: The techniques of processing TFE are unlike those for conventional thermoplastics due to its extremely high melt viscosity. The resin undergoes a transition from a powder to a gel state. It must, therefore, be processed with techniques like those employed in powdered metallurgy forming. Essentially, they consist of cold-forming a preform from the powder at pressures from 2,000 to 10,000 psi. The preform is sintered either in the mold under pressure (confined sintering) or out of the mold (free sintering). Sintering temperatures are generally 700° ± 50°F. The heating, sintering at temperature and rate of cooling must be carefully controlled, since they are important variables which affect the finished part's physical properties.

Extrusion of TFE is produced by compacting the cold powder with a reciprocating ram and then forcing the compacted resin through a heated die section for sintering.

For thin-wall extrusions or wire coating, fine-particle paste extrusion grades of TFE are compounded with a volatile lubricant such as naphtha or Deobase. The compound is then ram-extruded cold, and devolatizing and sintering is done in ovens outside the extruder.

TFE dispersions consisting of small particles of the coagulant of paste-type resins are available for the coating of metals and other surfaces. They are usually sprayed on, and the resin is fused by sintering the entire article in ovens at 700° ± 50°F.

Applications: The exceptional chemical inertness of TFE makes it useful for chemical process equipment, such as valve diaphragms, gaskets, seals, hoses, bellows, etc. Since TFE exhibits excellent electrical resistance over a wide range of temperatures and frequencies, it is often used for wire and cable insulation as well as stand-off and feed-through insulators, connectors, spacers, etc. Many mechanical goods such as piston rings, bearings, and housings are made of TFE to take advantage of its low coefficient of friction. For these applications, the filled resins discussed earlier are often utilized. The spray-on coatings are familiar on nonstick cookware.

Fluorinated Ethylenepropylene (FEP). The need for a conventional melt-processable fluorocarbon resin retaining most of the desirable properties of TFE fathered the development of FEP. It is essentially the copolymerization of TFE and hexafluoropropene. It is commercially available from E. I. DuPont de Nemours and Co. as "Teflon" FEP. Molding temperatures for FEP run from 625° to 750°F, highest of any of the true thermoplastics.

With the exception of continuous heat resistance limited to 400°F, FEP retains virtually all the properties of TFE. Typical applications are wire and cable insulation, corrosion-resistant pipe linings and fittings. The recent introduction of FEP films and sheets has opened the door for its use as a corrosion-resistant lining for chemical process vessels.

Chlorotrifluoroethylene (CTFE). This product should really be called a chlorofluorocarbon, since one of the fluorine atoms found in TFE is replaced with a chlorine atom. A great deal of the CTFE commercially used today, however, is in the form of copolymers. These are created by the addition of vinylidene fluoride and various other fluorine monomers to improve processability. Typical are "Plaskon" 2000 and 3000 series and "Kel-F82".

CTFE also has many of the properties of TFE, but with some limitations in dielectric properties, heat resistance, and chemical resistance.

The following, however, are properties in which CTFE surpasses TFE: (1) Physical properties: below 210°F, CTFE is stiffer and retains its impact strength. (2) Transparency: it can be made optically clear in thin sheets and films. (3) Moisture transmission: the lowest water vapor transmission of any known plastic.

Processing: CTFE resins can be molded using any conventional thermoplastic technique. Due to the high melt viscosity of these resins, close control over molding temperatures must be maintained to avoid thermal degradation. Generally, molding temperatures run approximately 500–550°F; the higher molecular weight resins will require temperatures 5 to 10% higher.

Applications: CTFE is used extensively in the manufacture of electronic components such as coil forms, tube sockets, and terminal insulators, and, of course, as wire and cable insulation. Its chemical resistance allows its use for lining valves, fittings, and pipe seals. Since at low temperatures it retains its flexibility, it is also used for seals and gaskets in cryogenic equipment where temperatures go down to −423°F.

Polyvinylidene Fluoride (VF₂). This homopolymer of vinylidene fluoride was introduced in 1961 by Pennsalt Chemical Corporation under the name of

"Kynar." VF_2 is characterized by: (1) Excellent chemical resistance within its temperature range. (2) Inertness to moisture, oxidation, and weathering. (3) Temperature range of $-80°$ to $300°F$, while retaining high tensile and compressive strength and low cold flow.

Processing: VF_2's crystalline melting point of $340°F$ allows fabrication on conventional thermoplastic equipment with relative ease.

Applications: Wire and cable insulation, chemical process equipment, and corrosion-resistant coatings for pipe and fittings.

E. F. EGGERT

Cross-references: *Fluorine and Compounds.*

FLUXES

The word "flux" is derived from the Latin word for "flow." A flux is any material which will lower the softening, fusion, or liquefying temperature of another material. Flux is a relative term and does not refer to any particular class of substances. Except in cases where two ingredients are soluble in each other in all proportions, small additions of any material to a pure element, oxide, or compound will reduce its melting temperature and, therefore, be a flux.

Some materials may act as a flux in one instance and, conversely, may be fluxed by that same material in another instance. For example, if a small amount of lime is added to sand, the lime will act as a flux on the sand and will reduce the melting point of the sand. On the other hand a small amount of sand may be added to pure lime, reducing the melting point of the lime; therefore, the sand becomes a flux. A similar example of this process in which mixtures rather than compounds are used is that of feldspars in porcelains and enamels. A feldspar is added to a porcelain body because it acts as a flux and reduces the refractoriness of the body. But when it is added to a sheet steel enamel, it is considered a refractory because it increases the melting point of the enamel.

A study of the many phase or equilibrium diagrams in common use illustrates graphically how many materials may act as fluxes or may be acted upon by fluxes. It is therefore necessary to determine the manner in which a material is used before it can be ascertained whether or not it is acting as a flux.

In the metallurgical field fluxes are introduced in the smelting of ores to promote fluidity and to remove objectionable impurities in the form of slags. Some of the common substances used are lime, calcium and barium sulfates, calcium phosphate, and silica. These substances unite chemically with impurities in melts to form slags, which can then be separated from the melts.

Sodium and potassium carbonates are valuable for fluxing off silica, while sodium carbonate and potassium nitrate, when mixed together, form an oxidizing fusion mixture. A valuable reducing flux, "black-flux", is composed of finely divided carbon and potassium carbonate, and is formed by deflagrating a mixture of argol with quarter to half its weight of niter.

Borax melts to a clear liquid and dissolves silica and many metallic oxides, making it very valuable in the manufacture of ceramic glazes and glasses. In connection with ceramic decorating, flux designates a prepared low melting glass, usually colorless, which may be mixed with pigments to produce vitrifiable coatings on clay wares.

Litharge and red lead are common fluxes in the ceramic industry and are also used in assaying. In the assaying of silver and gold ores they act as solvents for silica and any metallic oxides present.

M. A. HAGAN

FOAMS

A foam is a tightly-packed aggregation of gas bubbles, separated from each other by thin films of liquid. If foams were not so common, their existence would cause surprise. None of the obvious properties of a liquid would lead one to suppose that thin liquid films could sustain themselves for any appreciable time against the effect of gravity. The existence and stability of a foam depend, in fact, on a surface layer of solute molecules, which form a structure quite different from that of the underlying liquid inside the interbubble film.

At the surface of a liquid, molecules are in a state of dynamic equilibrium, in which the net attractive forces exerted by the bulk of the fluid cause molecules to move out of the surface; this motion is counterbalanced by ordinary diffusion back into the diluted surface layer. The equilibrium results in the surface layer being constantly less dense than the bulk fluid, which creates a state of tension at the surface. The tension can be somewhat relieved by adsorption of foreign molecules either out of the bulk solution, or out of the vapor phase. Soluble substances that have a strong tendency to concentrate in the surface layer are collectively known as surface-active agents: examples are soap, synthetic detergents, and proteins. The excess concentration of solute at the surface reduces the surface tension of water. The general relation, in the form of a differential equation, was first deduced thermodynamically by Gibbs; it is called Gibbs' adsorption theorem, and is

$$u = -\frac{c}{RT} \cdot \frac{d\gamma}{dc}$$

where u is the excess concentration at the surface, c is the bulk concentration, and $d\gamma/dc$ is the change of surface tension with concentration of solute.

An excess of solute at the surface, as measured by $+u$, can be termed *positive adsorption* to distinguish it from an excess of solvent at the surface $(-u)$, or *negative adsorption*. According to Gibbs equation, positive adsorption and the lowering of surface tension always appear simultaneously. When a fresh liquid surface is newly created, however, and before the excess solute molecules have had time to diffuse to the surface, the surface tension must remain high.

Lord Rayleigh showed, by means of his vibrating-jet experiment, that about 5 milliseconds is required for the surface tension of a fresh surface to reach equilibrium; during this time, the tension continuously declines from a high initial value of about 70 dynes/cm to a final equilibrium value of about 35 dynes/cm. The cause of the stability of a foam film resides in this effect. Should, for any reason, the equilibrium surface layer be disturbed, fresh surface is created and the tension immediately increases; a difference in surface tension cannot, however, be sustained for long because of the mobility of the liquid, which flows in response to the higher tension toward the area in which it has appeared. The first response of the liquid is no doubt just at the surface, but a considerable quantity of the bulk liquid is dragged along to the area of high tension. The following simple experiment beautifully illustrates the effect: pour a layer of water on to a thin metal plate, and touch the underside of the plate with a piece of ice: a high surface tension is created in the cold water just above the ice, and the motion of the surrounding liquid toward the colder area is to be seen immediately

On a foam film the stress that creates regions of higher surface tension is always present. The liquid film is flat at one place and curved convexly at another, where liquid accumulates in the interstices between the bubbles. The convex curvature creates a capillary force that sucks liquid out of the connected foam films (Laplace effects), so that internal liquid flows constantly from the flatter to the more curved parts of the films. As the liquid flows, the films are stretched, new surface of higher tension is created, and a counter-flow across the surface is generated to restore the thinned-out parts of the films (Marangoni effect). In this way the foam films are in a constant state of flow and counter-flow, one effect creating the conditions for its reversal by the other. Pure liquids cannot foam because of the absence of a Marangoni effect.

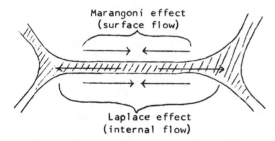

Fig. 1. Dynamic equilibrium in a stable foam film. The Marangoni effect reverses the destructive action of the Laplace effect.

The Marangoni effect maintains the stability of the foam films even against other disruptive actions, such as hydrodynamic drainage, that, like the Laplace effect, cause stretching of the films. To the Marangoni effect can be traced all the resilient ability of foam films for elastic recovery after external mechanical shock. Foam films sometimes have this property to a remarkable degree: lead shot, cork balls, mercury drops, and jets of water have all been dropped through foam films without causing rupture. In the fragility and brittleness of aged foams we see the effects of impaired resilience, probably due to the extreme depletion of solution from old films by prolonged drainage.

While the primary stabilizing factor in foam is the resilience of the film, provided by the Marangoni effect, in special cases additional surface-layer phenomena are significant: these include gelatinous surface layers and low gas permeability. Such effects can add enormously to the stability of the foam, resulting in such relatively stable structures as are exemplified in meringue, whipped cream, fire-fighting foams, or shaving foams.

Solutions of saponins or of proteins have surface layers that exhibit plastic viscosity, i.e., the surface does not flow until the shearing stress is greater than a characteristic "yield" value, which may be larger than the small gravitational stress inside a thin film. After the bulk of the liquid has drained away, the thin films remaining are gelatinous and resist rupture for long periods. Although carefully purified detergent solutions do not have surface plasticity, mere traces of "plasticizing" organic substances can cause surface plasticity to develop. The familiar foaming abilities of commercial soaps and synthetic detergents are often due to unintended traces of impurities, or to products of hydrolysis, such as aliphatic alcohols. These materials, however, are only indirectly responsible for the gelation of the surface layer. The evidence provided by x-ray diffraction studies of foam films is that a hydrous-gel structure extends from the surface to a depth of about 900 Å. Water makes up a principal portion of the surface film (at least 97% by weight); the water is oriented in an ice-like configuration, which acts as the linkage between solute molecules or micelles.

When a foam is first produced, e.g., by bubbling a gas through the liquid, each bubble is a little sphere, separated from its neighbors by thick liquid partitions. But soon after the foam is formed, a large amount of liquid has drained away by gravity, and the spheres of gas are now closer together. At this stage there is a passage of gas from one bubble to another, through the curved liquid convexities that separate them. Gas is confined in a bubble under a pressure greater than that outside, and described by the thermodynamic equation

$$p = 2\gamma/r$$

where p is the excess pressure, γ is the surface tension of the bubble wall and r is the radius of the bubble. In a foam, where γ is the same for every bubble, the pressure inside each bubble is inversely proportional to its size: i.e., the gas inside the smaller bubbles is at a higher pressure than the gas inside the larger bubbles. When, through drainage, the bubble wall becomes thin enough to be permeable, the gas in the smaller bubbles diffuses into adjacent larger bubbles to equalize the pressure. This spontaneous process increases the aevarage bubble size without any coalescence of bubbles taking place by film rupture. The final, stable equilibrium product is a fragile, honeycomb structure, in which the separating films have plane surfaces. At this stage of the life-history of a foam it is particularly vulnerable to external mechanical or thermal shocks, or air-borne contamination. Sir James Dewar kept plane soap films, in a horizontal position inside a closed bottle, for several months; in the open air they would have ruptured almost at once.

When comparing the foam stability of one solution with that of another, it is necessary to attend to the means by which foam is made, since foams of quite different characters can be obtained from the same solutions by different treatments. The area of interfacial surface, and the mechanical efficiency with which it is created, are the determining factors. The smaller the bubble the more persistent is the foam; and more foam can be produced if excessive agitation is avoided.

Different physical properties of foams are utilized in various industrial applications. In ore flotation, advantage is taken of the presence of an air-liquid interface and of the buoyancy of the bubble. Finely divided solid particles that are not wetted by aqueous solutions serve as the partitions between bubbles, and so rise with the foam. Thus hydrophobic sulfide particles can be separated from hydrophilic silica. In fire-fighting foams, use is made of the ability of a foam to retain a noncombustible gas (carbon dioxide), thus preventing air from reaching the fire. Foams in which the liquid phase is solidified are useful because of their insulating property and the low density conferred by the gaseous phase. Solid foams can be flexible, as in sponges, because the foam cells have all been ruptured in the course of production. Cellulose sponges, foam rubber, and polyurethane foam are examples of this type. Solid foams with closed cells are also made: they are rigid structures.

In many industrial processes excessive foaming of liquid causes waste and delays. Excessive foaming is particularly troublesome in the paper industry, the refining of beet sugar, the manufacture and the use of glue and many food industries. The foaming can often be inhibited by the addition of an insoluble liquid that is able to spread spontaneously, by virtue of surface-tension forces, over the surface of the foam films even as they are being formed. The spreading of the insoluble droplet is so violent, and the spreading liquid drags along with it so much of the underlying film, that a hole is "gouged" in the film, which is thus destroyed. Of such antifoaming agents the most versatile and the most effective are the silicone polymers.

SIDNEY ROSS

Cross-references: *Films, Colloid Chemistry, Surface Chemistry.*

FOOD

Food, the nutritive material taken into the body to maintain the vital processes can be classified in a number of ways. It may be divided into the two general classes, animal and vegetable, or into the commodity groups, or into classes based on the predominant organic substance in its composition, carbohydrate, lipid, or protein.

The carbohydrates found in foods are monosaccharides, disaccharides, polysaccharides, and pectic substances. Edible vegetation provides most of the carbohydrates in the human diet. In the occidental nations, cereals, vegetables and fruit supply most of the carbohydrates needed by the body as a source of energy. In the Orient, the chief energy-yielding food is rice. Throughout the tropics, cassava, grown for its fleshy edible roots, is a prime source of carbohydrate. Foods high in carbohydrates—wheat is 71% carbohydrate and corn (maize) 82% —are less expensive than high protein foods.

Fats or lipids are present in both plant and animal foods. Vegetable oils are derived from the seeds of plants for the most part, e.g., peanut oil, olive oil, corn oil, and cottonseed oil. Animal fats come from adipose tissue, e.g., lard. The properties of fats derive in considerable degree from the fatty acids. For the most part fats provide only energy but a few, notably, linoleic, arachidonic, and linolenic acids influence growth.

Although present in both plant and animal foods, proteins are usually thought of as being the characteristic organic substance in flesh foods. In recent years intensive study has been expended on protein, partly because of its key role in the life processes—growth, the repair of tissues, etc.—and partly because protein deficiency is a major problem in the developing nations and for that matter in poverty stricken urban and rural areas wherever located. As W.H. Sebrell of the Institute of Nutrition Sciences, Columbia University, has commented: "It is no coincidence that the underdeveloped peoples of the world today are those on poor protein diets."

The high cost of protein foods, a principal reason for protein deficiency in peoples of low economic status, has turned attention to the possibility of exploiting plant protein sources, not only conventional sources—beans, nuts, and cereals—but unconventional ones such as the leaves of trees, algae, and yeasts.

Since there is no possibility of maintaining the vital processes without them, it is not surprising that much research has been expended on the proteins, their chemical nature and their physical and nutritional properties. As to the general chemistry of the proteins it is not necessary to go further in this discussion than to say that the typical proteins are long chain molecules of amino acids and their elementary composition includes carbon, hydrogen, oxygen, nitrogen and in some cases sulfur. Of these elements, carbon is present in the highest percentage. The molecular weight of the typical protein is high, containing within it many different amino acid molecules. Egg albumin, a protein of high nutritional value, has a molecular weight of 34,000 or somewhat more.

Many years ago, W.C. Rose and his colleagues began studying the amino acids from the standpoint of those that are nutritionally essential, that is to say, those that must be in a food if young animals are to grow at a fully normal rate. Those found to be indispensable by the early workers were:

Arginine	Methionine
Histidine	Phenylalanine
Isoleucine	Threonine
Leucine	Tryptophane
Lysine	Valine

More recent work has shown that for children arginine is not essential and for adults histidine is not indispensable. The biologic value of proteins depends upon whether or not all the essential amino acids are present in the right quantities and whether or not, in the case of processed foods, their nutritive value remains unimpaired by the method of preservation used. In ranking proteins in terms of their biological value, it is now common knowledge that egg protein, casein, and the proteins of flesh foods are highest and the proteins of plant foods among the lowest.

It would be wrong, however, to jump to the conclusion that only foods of animal origin can supply the ideal diet. Where milk and foods of animal origin are in short supply, a mixture of plant foods can substitute. Thus, the Institute of Nutrition of Central America and Panama formulated a product called INCAP from corn, sorghum, cottonseed meal, yeast and vitamin A to combat the protein deficiency of that area. This it did, successfully and inexpensively.

This brief excursion into the composition of foods would not be complete without mention of the vitamins and minerals. The vitamins are essential to metabolism and to well-being, a fact that we have been made abundantly aware of in recent years by the media. Vitamins have found a secure place in human nutrition by virtue of the deficiency diseases which their absence causes and their presence prevents. The classic example is scurvy, caused by a deficiency of vitamin C. The administration of vitamin C restores the scorbutic patient to his normal health amazingly fast. Rickets, pelagra, and some skin diseases are symptomatic of vitamin deficiencies that can be cured by the administration of the vitamin absent, or present in low degree, in the victim. In addition to the vitamins, a number of inorganic elements—calcium, iron, copper, phosphorus, magnesium, sodium, potassium, sulfur, chlorine, iodine, cobalt, manganese, zinc, and fluorine—are needed by the body for the maintenance of normal life processes.

The food sciences are actually rather young. In fact only in recent years have they become a clearly identified set of disciplines. Perhaps enough has been suggested in the foregoing comments on the composition of foods and its relation to nutrition to indicate the complexity of the problems which the food sciences are seeking to solve. Food is taken into the body to maintain the vital processes. The vital processes are tremendously complex and in some degree mysterious. Equally complex— and in some areas, mysterious—is the role of food in supporting the vital processes. It is to be hoped that an observation by Seneca will one day be true: "The day will come when our descendants will be amazed that we remained ignorant of things that will seem to them so plain." Looking back, we can be amazed at some of the things that our ancestors were ignorant of—in many cases a result of food habits, a subject to which we shall now turn.

Food Habits. Ethnic factors, economic considerations, climate, the kind of agriculture pursued, and the state of food production, preservation, and distribution all play a part in the formation of the food habits that prevail in the various regions of the world. Religious restrictions or sanctions play a part and so too the life style of families. It is not surprising that the crops easiest to raise in the greatest abundance dictate the staple foods of nations—wheat and corn in North America, rice in the Orient, cassava in the tropics, fish in the coastal areas of many nations that border on the sea, the examples are endless.

The food technology of a nation may dictate the form a given food will take. It seems reasonable to infer that in Europe where the technology associated with the milling of grain developed rapidly, baked goods, from breads to cakes, early became the chief form, certainly the most popular form in which cereals were consumed. In the north of Europe the abundance of herring was instrumental in the perfection of the technology of preservation by salting, smoking, and drying. The religious observance of Lent in all Europe before the Reformation gave salt fish a prominent place in the diet of the ordinary family and this combined with the fact that in those days it was the only protein food available between Shrove Tuesday and Easter created and enforced a food habit that lasted until preservation methods, for example, canning, widened the choice.

Food habits play an important role in dictating the kind of food industry and the kind of food technology a nation or a group of nations will have. Social customs in the United States, for example, have decreed three meals a day, with breakfasts of a relatively standard type—a juice, cereals, toast, eggs, bacon, coffee, either one or two or all of these items depending upon the family or the individual; luncheons, a soup, sandwich, and beverage; and dinners—salad, a flesh food, vegetables, dessert and coffee. This pattern has influenced both the merchandizing and the manufacture of foods.

The behavioral sciences have been hard at work in recent years measuring, analyzing and evaluating the food habits of peoples, and their results have been widely applied. It is evident that in many of the western nations their findings have also influenced the merchandizing habits of the supermarkets and the manufacture of foods (for instance, the surge of convenience foods based on instantizing features) and, without question, food technology.

Food Technology. Although much indebted to that branch of engineering that has come to be classified as chemical engineering, food technology has become greatly diversified over the years. Nevertheless, and with full recognition of the fact that the food industry is not classified as one of the chemical process industries, the food industries are engaged in the production of finished products from raw materials of a chemical nature and in so doing rely upon unit operations that involve fluid flow, heat transfer, filtration, drying, distillation, mixing and conveying the raw product through a production line that ends with a finished product. Process sequences in the food industry are by no means all alike, however, and there is no one typical process. Processing operations, abstractly defined, proceed from preparing the raw material to the application of a preservation agent or method on to packaging or otherwise completing the manufacture of the food.

In addition to its concern with unit operations, food technology is involved in the business of applying the findings of the food sciences to product improvement over the whole spectrum of stability, palatability, nutritional adequacy and utility. The findings of the chemist and the bacteriologist are translated into increased stability. Advances in the sensory evaluation of foods are reflected in improvements in the flavor, texture and color of foods. Nutritional studies relevant to food carry with them suggestions for increasing nutritive values. With regard to the utilization of food, technology has long been concerned with making foods easier to prepare, better adapted to family or institutional use, and simpler to transport, store, and market.

It is manifest that food has many connotations. To the chemist or biochemist, a food is an assemblage of compounds that is distinct in varying degree from other foods in flavor, texture, color, and nutritive qualities. To the physiologist, foods are substances that furnish the body with materials that produce energy, enable the body to grow or to repair injured or depleted tissues, or to regulate its vital processes. To the psychologist, food carries the implication of an attitude toward individual foods, an emotional response of *like* or *dislike*. To the technologist, food is a product, the end result of a manufacturing process or, in the case of fresh foods, a product that requires techniques of harvesting, handling, storage and marketing that increase or at least maintain its quality.

As a result of the research and development effort of scientists and technologists at work on food, remarkable advances have been made over the years. As compared to the food resources of our not too remote ancestors, we have today a multiplicity of foods tailored to special purposes, e.g., baby foods, dietary foods, geriatric foods, convenience foods, seasonal foods available for all seasons, foods adapted to the needs of the military man, foods tailored to the needs of large groups of people, and foods designed for use of men engaged in voyages into space.

What undertakings are left for the future? It seems likely that the chemist and biochemist will be called upon, at least initially, to solve a problem that is intensifying as the years go by—namely, the problem of solving the world food crisis. It is clear that agriculture, unaided, cannot meet the demands of the future and that supplemental methods of food production, for instance, the chemical or biological synthesis of foods, will be required. These and a number of other unconventional methods of producing edible foods may soon become economic, especially so if waste products can be used in the process. It is now possible to convert cellulose to glucose by enzymatic means. It may be, in line with Seneca's thought, that "our decendants will be amazed that we remained ignorant of things that will seem to them so plain."

Flavor. Future generations are not likely to regard the phenomena of flavor as simple. Much work remains to be done, and it may take generations to do it, before an adequate understanding of the sensing mechanisms of taste and smell, the physical phenomena, and the complex chemical process that contribute to flavor are adequately understood. The enjoyment of food is the result of a dual stimulation of the senses of taste and smell, but enjoyment in the final analysis is a psychological response to food and is affected by past experiences, by food habits, by acquired attitudes and many subtle influences not easily identified.

The chemistry of flavor, by contrast, is yielding slowly to the studies of the experimentalists at work on its perplexities. The chemical substances which evoke the basic taste sensations—sour, salty, sweet and bitter—are somewhat more easily identified than those that contribute to flavor through the olfactory channel. Broadly stated, the sour taste is characterized by the taste of certain strong acids, although a weak acid (acetic) is readily identified when added to a food. Salty taste is represented by common table salt, sweet taste by sucrose, and bitter taste by a substance such as quinine sulfate.

Attempts to identify basic olfactory attributes have not been too successful. The German chemist, H. Henning, developed a conceptual frame of reference which included six types—flowery, foul, fruity, spicy, burnt, and resinous. A few years later, Crocker and Henderson, using four independent odors—fragrant, acrid, burnt, caprylic—established an intensity scale (0 to 8) for use in coding a flavor by means of digits, each digit representing the intensity of one of the four independent odors just listed. Compounds contributing to the characteristic flavor of food products, for instance, coffee, number in the hundreds. Unfortunately, when these compounds are synthesized in a neutral medium, the characteristic flavor is not reproduced; and this is true generally of other attempts to duplicate a flavor by synthesis of the known compounds. There are, however, a few compounds that have been shown to be major components of flavors, e.g., 2-methoxy-3-isobutylpyrazine, the major component of green bell pepper.

Flavor depends upon two classes of chemical substances—nonvolatile compounds sensed by the taste buds and volatile substances sensed by the olfactory receptors. The sense of smell is 10,000 times as sensitive as that of taste. It is therefore not surprising that the sense of smell is not only a monitor of what is fit to eat but also the arbiter of excellence in foods.

Martin S. Peterson

Food Preservation Methods

To preserve foods, it is necessary to stop or markedly retard deterioration. The four principal causes of deterioration are the following: (1) physical causes (such as loss of moisture which may cause wilting or other changes in texture, bruising, breakage or actions caused by rough handling); (2) chemical actions such as oxidation, which may cause change or loss of flavor (e.g., development of rancidity) and discoloration; (3) enzymatic actions, which accelerate physical and chemical changes (e.g., browning of apples and peaches); and (4) microbiological actions caused by the growth of bacteria, yeasts and molds.

Most physical changes are greatly retarded by proper packaging. For example, drying of chewing gum is largely prevented by packaging in aluminum foil, and/or plastic sheeting, thus preventing hadening and loss of flavor. Hygroscopic candies are preserved by packaging in moisture-proof sheetings to prevent them from becoming sticky and unattractive.

Deterioration because of chemical actions is also prevented or greatly retarded by proper packaging and storage at low temperatures. Thus, oxidation and hydrolysis of fats, which cause rancidity, are greatly retarded by packaging in an inert atmosphere in gas-tight containers substantially free from oxygen. If properly packaged fats and oils are held under refrigeration, their storage life is greatly increased. In fact, all chemical reactions are retarded by reducing the storage temperature. As a rule, reducing the temperature 18°F (10°C) reduces the rate of a chemical reaction by one-half.

The deteriorative action of enzymes may be prevented either by heating the food to inactivate the enzymes or by holding the food at such a low temperature that enzyme actions are greatly retarded. Fresh fruits and vegetables are alive and enabled through enzyme action to absorb oxygen and give off carbon dioxide. Reducing the temperature and/or reducing the oxygen content and increasing the carbon dioxide content of the storage greatly retards enzyme actions in stored fruits and vegetables.

Spoilage of foods by the action of microorganisms may be retarded by (1) drying or dehydration, thus reducing the moisture content of the food so low that bacteria, yeasts and molds cannot grow; (2) the addition of a chemical preservative such as sulfur dioxide which inhibits the growth of microorganisms; (3) salting and/or smoking, thus reducing the moisture content and adding substances which retard the growth of most microorganisms; (4) by storing the product under refrigeration thus greatly reducing the rate of growth of microorganisms; (5) by freezing the food and keeping it at a temperature sufficiently low to prevent enzymatic actions and the growth of bacteria, yeasts and molds; or (6) by placing the food in hermetically sealed containers and heating them sufficiently to sterilize the foods. Various combinations of these six methods are often used. Thus, halved apricots are treated with sulfur dioxide (a chemical preservative), dried in the sun, packaged and stored under refrigeration. Milk is reduced in moisture content by evaporation, filled into cans which are sealed, sterilized by heating and then stored under refrigeration. Herring are often first smoked and dried slightly, then canned or frozen.

Drying and/or Dehydration. Since prehistoric times, man has been preserving food by drying in the sun, and/or dehydrating it by heat from a fire. Millions of tons of grapes, apricots, peaches and other fruits are still preserved in this way. If the heat of the sun is used to reduce the moisture content of the food, the process is called drying. If "artificial" heat is employed, it is considered dehydration.

Removal of moisture has a two-fold effect: (1) at lower moisture levels the chemical reactions associated with food deterioration are retarded; (2) the growth of microorganisms is discouraged.

Dehydration may be brought about by the application of heat with or without a vacuum, by the use of low temperatures and vacuum, and by the use of a desiccant. Generally the removal of higher percentages of moisture is effected quite readily, but once the moisture level is lowered to about 3%, great care is necessary to prevent nutritional and physical damage taking place as it is lowered further. To minimize this effect, a judicious combination of various dehydration methods may be used.

Other methods of drying by heat involve the use of heated rollers over which a slurry of the material is fed; spray drying, in which the material is delivered by a nozzle into a chamber continuously fed by a blast of hot air; puff drying in a vacuum oven. Each of these methods results in a material characteristic of that method. Roller drying will produce a flaky material, spray drying a material in globular form, and vacuum drying an expanded or powdery type of material.

Canning. One of the most important methods of preserving foods is commonly known as canning. As ordinarily carried out, this involves the packing of cleaned and prepared food (usually hot) into cans or jars, then after covering the cans or jars with lids, conveying them through a steam chamber or exhaust to replace most of the air with steam (thus eliminating much of the oxygen in the headspace), crimping the lids on the cans and closing the jars, then heating the containers in retorts with steam under pressure (at 240 or 250°F) for a sufficient length of time to sterilize the food, then cooling, before placing in storage.

A relatively new canning procedure is the high-temperature, short-time method which involves pumping the food to be processed through a heat exchanger in which it is heated to 270°F or even higher in a matter of a relatively short period (10 or 12 seconds), and then passing it through a cooling section in which its temperature is dropped to about 212°F. It is then filled into hot, sterile cans which are immediately sealed. This process is applicable to products like cream-style corn which can be pumped through relatively small openings. Since products sterilized in this way are heated for less than a minute, the flavor and color are changed but little. The process works especially well for fruit juices and purées with a pH below 4.0, since the vegetative microorganisms are killed in a short time at 180–190°F and the spores remaining in the juice will not germinate during subsequent storage because of the acidity of the juice or purée.

Salting, Curing and Smoking. The salting of fish, beans and other foods is still common practice in various foreign lands. Further, this method of food preservation is often used in the United States in conjunction with other methods such as curing, smoking and drying. Cured meats constitute an important segment of the meat packing industry. The curing process involves the treatment of meat with salt plus other curing agents such as sugar, sodium nitrate, sodium nitrite and vinegar. Three general curing methods are recognized, namely, the dry cure, pickle cure, and injection cure. In the dry cure the curing agents are rubbed dry into the meat. When the pickle cure method is used, the meat is immersed in a solution of the curing agents. The injection cure process involves the use of a needle placed in the veins and arteries of the carcass so that the curing solution may be pumped into the arterial system. This type of cure minimizes the curing time. The use of nitrites and nitrates in pickles is of interest in that it has been shown that these salts stabilize the fresh meat color during subsequent processing.

The smoking of fish and meat often follows salting or curing. Smoke from such woods as hickory, juniper, maple and birch contains such components as formaldehyde, acetaldehyde, acetone, acetic acid, methyl and ethyl alcohol, creosote, phenols and pyroligneous acid. Of these formaldehyde is particularly important as a bactericidal agent. The effect of smoking is therefore twofold: the flavor of the food is enhanced and at the same time, food preservation is promoted.

Chemical Preservation. Many different chemicals are used to extend the storage life of various foods. In general, these are classed as food additives and the amount that can be used is subject to the regulations of the U.S. Food and Drug Administration and certain state agencies. The most commonly used bactericidal agents (substances which inhibit the growth of microorganisms) are sulfur dioxide, sodium benzoate and sorbic acid and their derivatives. Calcium propionate is also employed as a mold inhibitor.

Sulfur dioxide and sulfites are extensively used in the United States as an aid in preserving and to prevent darkening of dried or dehydrated apricots, peaches, pears, apples, certain grapes, and canned and frozen mushrooms. The British permit imports of heavily sulfured orange pulp and juice. During storage of sulfured foods sulfur dioxide is gradually oxidized, and, if sulfured dried fruit is stored for many months, additional treatment with sulfur dioxide may be required to keep the color bright.

Other chemical agents often used include the following:

Antioxidants (used to retard rancidity development)— propyl gallate; butylated hydroxyanisole (BHA); nordihydroguaiaretic acid (NDGA)

Emulsifying agents—lecithin, glyceryl monostearate and other monoglycerides

Bacteriostatic agents — quaternary ammonium compounds, sorbitol, and sorbic acid

Stabilizers—various gums, cellulose derivatives, algin and other seaweed products, starch, modified starches and dextrin.

Many other substances, too numerous to mention, are added to various products to extend shelf-life, etc.

Irradiation of Foods. Microorganisms can be destroyed by radiation with beta rays, cathode rays, x-rays and gamma rays. Radiation from both linear accelerators and atomic reactors has been used to preserve foods. Insects can be easily killed by low radiation dosages. Low dosages have also been used to treat potatoes to prevent sprouting during storage. Somewhat larger dosages are effective in "pasteurizing" fruits and vegetables. This "pasteurization treatment" kills nearly all of the vegetative microorganisms on a fruit (e.g., strawberries) so that the berries do not mold or rot quickly and consequently their storage life is considerably lengthened if the fruit is properly packaged and refrigerated. Heavier dosages of gamma rays will sterilize foods, and, if the products have been properly packaged, can be used as a means of keeping foods for many months free from spoilage because of the growth of microorganisms. Unfortunately, sufficient radiation required to kill the spores of certain putrefactive microorganisms (e.g., *Cl. botulinum*) often causes the development of undesirable off-flavors. For example, beef steak may take on the flavor and odor of scorched wool. Other meats, such as chicken and pork, may be sterilized without undesirable changes. In the United States, radiation has not been used as a commercial method of food preservation, because the U.S. Food and Drug Administration has not found that irradiated foods are completely safe for human consumption because of their possible toxicity.

Refrigeration. Cooling food retards deterioration since it reduces the rate of chemical actions such as oxidation and hydrolysis. It has a similar effect on reactions catalyzed by enzymes. Further, the rate of multiplication of all but a few microorganisms is greatly reduced by reducing the temperature of a food to the freezing point or below. Meats that will spoil in a few hours at 100°F can be kept for three weeks or even longer at 31–32°F. Apples and pears can be kept a few months at 32°F and if the proportion of oxygen, nitrogen and carbon dioxide is kept at the optimum level (under "gas storage") they may be kept in nearly perfect condition for a much longer period, provided they are at optimum maturity and they have not been bruised.

Refrigeration is an almost ideal means of greatly extending the storage life of almost all perishable and semiperishable foods. However, a few fruits (e.g., bananas and tomatoes) will not ripen properly if they have been subjected to temperatures near the freezing point. The skins of bananas will turn from yellow to black in a short time at 32°F.

Freezing Preservation. The frozen food industry has advanced very rapidly since 1925. Prior to that date, considerable quantities of fish, poultry, and meat were frozen, but the frozen products were considered to be inferior to strictly fresh foods. In the late 1920's, Clarence Birdseye showed that if strictly fresh foods were "quick frozen" (frozen rapidly), and subsequently held at 0°F or below, the products were comparable in quality to fresh foods that had never been frozen. Many theories have been advanced to explain why quick frozen foods are superior to those which have been frozen slowly. Probably, the great improvement in quality results from a number of factors:

(1) When quick frozen foods thaw, there is much less "drip," "weep," or "leakage."

(2) Quick freezing does not damage the texture of many foods as does "slow freezing," this may be because very small ice crystals are formed instead of larger crystals.

(3) When freezing occurs rapidly and the frozen product is held at 0°F or below, little or no deterioration or staling occurs before and during freezing and subsequent storage.

In order to obtain a frozen product of high quality (comparable to fresh), it is necessary not only to select strictly fresh foods of high quality but also to prepare and package the product carefully so as to avoid desiccation and enzymatic changes during freezing and storage. In general, foods are washed and prepared for the table just as they are preparatory for canning. Vegetables must be heated sufficiently to inactivate enzymes, then cooled, usually in cold running water, and either packaged in moisture-vapor-proof packages, each preferably holding not more than two pounds, and frozen at a low temperature either in an air blast or between metal plates, or frozen on a fluidized bed freezer and later packaged in moisture-vapor-proof packages. Freezing breaks down the cellular structure of fruits, therefore, nearly all fruits are frozen after the addition of either dry sugar or heavy syrup. Fish are commonly either filleted or cut into steaks and then packaged in moisture-proof packages. The cut surface of red meats may turn gray if packaged in plastic sheetings, consequently frozen steaks and roasts are not so appealing to the housewife as the unfrozen product; however, when cooked the thawed product cannot be told from the fresh. In recent years, many kinds of frozen precooked foods have become very popular. Frozen French-fried potatoes are sold in large quantities both to housewives and institutions. Many entrées, such as chicken pies and macaroni and cheese, are very popular. Several nationality foods, e.g., pizza, are frozen and sold in large quantities.

During the 1960's, ultrarapid or cryogenic freezing became of considerable importance. Cryogenic freezing may be carried out by the use of liquid nitrogen (boiling point-320.5°F) or Freon 12, (b.p.-21.6°F) or with powdered dry ice (which sublimes at −109°F). The food to be frozen may be sprayed with liquid nitrogen either before or after packaging, thus freezing the product in a minute or two. Cryogenic freezing yields a product of very high quality. Certain fruits and vegetables, which do not yield wholly satisfactory products when quick frozen, may be frozen cryogenically, thus obtaining superior foods. Italian-type tomatoes and certain varieties of strawberries and sliced mushrooms, which are of relatively poor quality when quick frozen, are usually cryogenically frozen commercially.

Packaging. Proper packaging is of great importance in food preservation and has made possible many of the great advances in the preservation, storage, and distribution of foods. Few persons have given much thought to the great importance of the sanitary "tin" can in our economy. Its use has permitted the preservation by heat sterilization of an astounding variety of foods. Special enamel-lined cans can be used as containers for products corrosive to ordinary tinplate. Aluminum foil is extensively used for packaging many kinds of foods of low moisture content and now an amazing variety of plastic sheetings have been perfected for use in packaging both frozen and fresh foods. Space will not permit a consideration of the great variety of packages used by the food processor, but it should be noted that without modern packages, few, if any, of the modern processes now used to preserve foods could be employed.

DONALD K. TRESSLER

FORMULAS

The representation in writing of a chemical element or compound is a formula for that substance. Simple examples are He for a helium atom which is also a helium molecule, H_2 for a hydrogen molecule composed of two atoms of the same sort, and HClO (or HOCl) to represent a compound composed of three atoms of different sorts, hypochlorous acid.

Subscripts in formulas represent the number of times each atom is represented. H_2O shows that one molecule of water contains two atoms of hydrogen and one atom of oxygen; CCl_4 represents one molecule of carbon tetrachloride containing one atom of carbon and four atoms of chlorine.

In the case of an ionic compound in which the bond between the atoms is electrovalent as it is in calcium chloride, the formula may be represented as $CaCl_2$. If attention is given to its ionic condition, then $Ca^{++}2Cl^-$ or $Ca^{2+}Cl^-$ represents the lattice. In water solution its ions have dissociated into Ca^{++} ions and Cl^- ions, each ion surrounded by molecules of water.

The complete meaning of a chemical formula can be seen from an example, KNO_3 (potassium nitrate).

(1) One mole of potassium nitrate. Two moles of potassium nitrate would be represented as $2KNO_3$.

(2) In the compound KNO_3, as shown by the subscripts written or understood, there are one atom of potassium, one of nitrogen, and three of oxygen. When parentheses appear in a formula, the subscript outside the parentheses applies to each symbol within the parentheses. For example, $Fe_2(SO_4)_3$ means that in the compound ferric sulfate there are two atoms of iron, three atoms of sulfur, and twelve atoms of oxygen.

(3) A definite weight is contributed by each element represented by the formula KNO_3. (Approximate atomic weights are used in this article.) One atom of potassium accounts for 39 parts by weight, one atom of nitrogen accounts for 14 parts by weight, and three atoms of oxygen account for three times sixteen or 48 parts by weight.

(4) The formula KNO_3 also means a definite weight of potassium nitrate. The sum of its atomic weights $(39 + 14 + 48) = 101$ atomic mass units. Often a gram-atomic weight or 101 grams (one mole) is indicated, but 101 pounds or any other weight unit may be used.

Derivation. A sample of a compound is analyzed, and by appropriate tests, chlorine, bromine, iodine, sulfur, nitrogen, and phosphorus are found to be absent. The compound burns, and it chars when heated with concentrated sulfuric acid. The compound mixes well with water in all proportions, and its solution is a very poor conductor of electricity. It is assumed to contain carbon and hydrogen, and possibly oxygen.

A two-gram sample is placed in a boat in a combustion tube and burned completely in a stream of dry oxygen. The resulting gaseous products of combustion are absorbed in a drying agent, and in a solution of potassium hydroxide successively. The drying agent absorbs moisture formed by burning hydrogen, and the potassium hydroxide solution absorbs carbon dioxide formed by burning carbon. Neither absorbing agent is affected by oxygen, so any excess oxygen bubbles through unchanged and does not affect the weight of the absorbing agents. The drying agent, weighed before and after the experiment, shows an increase in weight of 2.40 g, and the tube that holds the potassium hydroxide solution, similarly weighed, shows an increase of 4.40 g. Of the 2.40 g of water, 11.2% is hydrogen, or 0.267 g. The percent of hydrogen in the sample is $0.267/2 \times 100$, or 13.3%.

Of the 4.40 g of carbon dioxide, 27.2% is carbon, or 1.20 g. The percent of carbon in the sample is $1.20/2 \times 100$ or 60%. The balance, or 26.7%, is assumed to be oxygen supplied by the compound itself.

Because carbon, hydrogen, and oxygen atoms each have a different value for their weight, namely, that of their respective atomic weights, the relative number of atoms of each element in the compound is found by dividing each percentage by its respective atomic weight, as in the following example.

Element	%	Atomic Weight	Quotient	Ratio of Quotients
Carbon	60.0 ÷	12 =	5.0	3
Hydrogen	13.3 ÷	1 =	13.3	8
Oxygen	26.7 ÷	16 =	1.6	1

From the ratio of the quotients the empirical formula is C_3H_8O. The formula weight is therefore $(3 \times 12) + (8 \times 1) + (1 \times 16) = 60$.

In another experiment, three grams of the compound is added to 20 g of water and the freezing point of the mixture measured. It is found to be $-4.65°C$.

By general principle, one mole of an un-ionized (nonconducting) compound plus 1000 grams of water (a molal solution), freezes at $-1.86°C$, average value. This value is typical for the freezing point constant when water is a solvent. The depression of the freezing point by the presence of a dissolved substance depends on the number of particles of dissolved substance in a given weight of solvent. The number of particles in turn depends on the number of moles dissolved in a constant weight of solvent.

To resume with the example: 3 g in 20 g of water is a concentration equivalent to 150 g in 1000 g of water. Comparing the freezing point depression observed with the standard depression gives $-4.65°C/-1.86°C = 2.5$. Hence 150 g is 2.5 moles, or one mole is $150/ 2.5 = 60$ g. The correct molecular weight is therefore 60, and the simplest formula is also the correct formula.

From other experiments, such as when 60 g (one mole) of this compound reacts with nitric acid (HNO_3), 63 g of the pure acid, or one mole of that compound, is needed for a complete reaction (or amounts in the weight ratio of 60 to 63); thus it is assumed that one functional OH group is present. Hence a more descriptive formula is C_3H_7OH, and the compound is propyl alcohol.

Further experiments must be carried out to determine whether the carbon atoms are in a straight chain (normal propyl alcohol, 1-propanol) or in a branched chain (isopropyl alcohol, 2-propanol). One way to decide this matter is to measure the boiling point of the pure substance. Usually the boiling point of a straight-chain compound of this sort is higher than that of a branched-chain compound of the same number of carbon atoms. If the boiling point is 97.8°C, this compound is normal propyl alcohol; if its boiling point is 82.5°C, it is isopropyl alcohol.

Types of Formulas. There are several kinds of formulas used in chemistry; these are summarized below, as described in the "Condensed Chemical Dictionary":

(1) Empirical: Expresses in simplest form the *relative* number and the kind of atoms in a molecule of one or more compounds; it indicates composition only, not structure. Example: CH is the empirical formula for both acetylene and benzene.

(2) Molecular: shows the actual number and kind of atoms in a chemical entity (i.e., a molecule, group, or ion). Examples: H_2 (1 molecule of hydrogen), $2(H_2SO_4)$ (2 molecules of sulfuric acid); CH_3 (a methyl group); $Co(NH_3)_6^{++}$ (an ion)

(3) Structural: Indicates the location of the atoms, groups, or ions relative to one another in a molecule, as well as the number and location of chemical bonds. Examples:

$$CH_2{=}\overset{\overset{\textstyle CH_3}{|}}{C}{-}CH{=}CH_2 \quad \text{(isoprene)}$$

benzene

Since all molecules are 3-dimensional they cannot properly be shown in the plane of the paper. This is sometimes indicated by extra-heavy lines or 3-dimensional artwork (configurational formula).

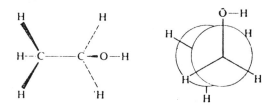

(4) Generic: Expresses a generalized type of organic compound, where the variables stand for the number of atoms or for the kind of radical in a homologous series (q.v.). Examples:

$$C_nH_{2n+2} \qquad C_nH_{2n}$$
a paraffin   an olefin

$$ROR \qquad ROH$$
an ether   an alcohol

(R = a hydrocarbon radical)

(5) Electronic: A structural formula in which the bonds are replaced by dots indicating electron pairs, a single bond being equivalent to one pair of electrons shared by two atoms. Example: the electronic formula for methane is:

$$\overset{\textstyle H}{\underset{\textstyle H}{H:C:H}}$$

ELBERT C. WEAVER

FRACTIONATION

A term with several related meanings, all referring to the separation of a mixture into fractions of different composition by systematic, repetitive use of distillation, crystallization, partial precipitation, or other separation processes. The most common use of the term, in this case synonymous with rectification, refers to distillation with a column or tower within which rising vapor and descending liquid are brought into intimate contact by use of baffle-type devices known as plates or packing, so as to maximize the separation and purification achieved. Such a column with plates or packing is equivalent to a sequence of simple distillation units. By extension of this meaning, the separation of the isotopic uranium fluorides by an ingenious arrangement of many diffusion barriers is also fractionation. In this case, each repetition of the basic diffusional separation is referred to as a stage, and the interconnected sequence of stages as a cascade. Fractionation is also carried out in equipment supplying a sequence of solvent extraction stages, and by extension of this concept fractionation can be achieved, at least in principle, by a connected sequence of stages of any separation process. In the case of distillation, the terms stage and plate are synonymous.

In a broader sense, fractionation can also be carried out by repeating the basic separation step with a single simple apparatus, or a few such units. Thus, a single simple distillation apparatus has been used to separate a sample into two or a few fractions, and then the same apparatus used to redistil and thus further separate each of these fractions. Some fractions will now be closely similar, and these are combined, and all fractions again redistilled, and the process continued. This type of method was used in the crystallization of rare earth salts to separate the individual rare earth elements from one another. Analogous systematic partial precipitation has been used on solutions containing complex mixtures of polymers to obtain homogeneous fractions or cuts.

ARTHUR ROSE

FRANCIUM

Francium, the element with atomic number 87, is the heaviest member of the alkali family of elements. In its chemical properties francium most closely resembles cesium. Owing to its great nuclear instability, all its isotopes have very short half-lives for radioactive decay, and consequently all knowledge of the chemical properties of this element comes from radiochemical experiments. No weighable quantity of the element can be prepared.

The element was discovered in 1939 by the French scientist Marguerite Perey, who established the presence of a 21-minute radioisotope with properties of a heavy alkali element among the decay products of actinium. The mass number of this isotope is 223 and, in the nomenclature of the natural radioactivities, it is called Actinium K. All isotopes of francium from mass number 203 through 224 are known as a result of their identification among the products of nuclear reactions. Some were identified in targets of thorium or uranium bombarded with high-energy protons, the others in targets of thallium, gold, lead, or bismuth bombarded with complex nuclear projectiles such as accelerated ions of carbon, nitrogen, or neon. All the artificially produced forms of francium have

half-lives for radioactive decay shorter than that of Fr^{223}.

Francium exists in aqueous solution as a large singly-charged ion with small tendency to form complex ions. Like other alkali ions, francium remains in solution when other elements are precipitated as hydroxides, carbonates, fluorides, sulfides, chromates, etc. Francium will coprecipitate with certain insoluble salts such as the alkali perchlorates, iodates, chloroplatinates and salts of heteropoly acids. One of the most selective and useful of the carriers for removal of francium ions from aqueous solution is silicotungstic acid precipitated from solutions saturated with hydrochloric acid.

Very few organic solvents immiscible with water are effective for the extraction of francium from aqueous solution, but there are a few exceptions. An example is a benzene solution of the sodium salt of tetraphenyl boron. Francium is readily adsorbed on synthetic cation exchange resins from neutral or slightly acidic solution and readily desorbed by solutions of higher acidity.

EARL K. HYDE

Reference

Hyde, Earl K., "Francium," pp. 222–225 in "Encyclopedia of the Chemical Elements," Hampel, E. A., Editor, Reinhold Book Corp., New York, 1968.

FREE ENERGY AND FREE ENERGY FUNCTIONS

The "free energy" of a system is a precise thermodynamic quantity which is used to predict the maximum work obtainable from a spontaneous transformation of the system. Furthermore it provides a criterion for spontaneity of a transformation (reaction) occurring and predicts the maximum extent to which the transformation can occur (maximum yield).

In order to understand the nature of the "free energy" it is useful to derive it from the internal energy and the entropy of a system. Transformation of a system can be brought about by thermal input (temperature change) or by mechanical work (volume-pressure change). In a cycle such that the system is alternately subject to a thermal change, dQ, and a mechanical change, dW, whereby after each cycle the initial state of the system is restored, the first law of thermodynamics postulates that:

$$\oint(dQ - dW) = 0 \qquad (1)$$

The quantity under the cyclic integral must be the differential of some property of state of the system. This property is called the *Energy, E,* (sometimes also internal energy) of the system. Hence

$$dE = dQ - dW \qquad (2)$$

which intergrates to

$$\Delta E = Q - W \qquad (3)$$

Only a difference in energy has been defined by Eq. 3 since one cannot assign an absolute value of energy to any state. It should be pointed out that dE is an exact differential, in contrast to dQ and dW, which are path dependent.

If one considers, for example, the expansion of a gas at constant pressure, one finds readily that the work gained can be expressed as:

$$\Delta W = p_2 V_2 - p_1 V_1 \qquad (4)$$

Since a temperature change during the expansion has not been excluded Eq. 3 can be written as

$$(E_2 + p_2 V_2) - (E_1 + p_1 V_1) = Q_p \qquad (5)$$

The function $E + pV$ is again a state variable and by definition called the enthalpy of the system:

$$H = E + pV \qquad (6)$$

Both E and H have important properties a discussion of which would exceed the scope of this paragraph. Suffice it to note here that:

$$\left(\frac{dE}{dT}\right)_{v=const} = C_v \quad \text{and} \quad \left(\frac{dH}{dT}\right)_p = C_p \qquad (7)$$

where C_v and C_p are the heat capacities for constant volume and constant pressure processes, respectively.

Again as in the case of the internal energy of a system it is not possible to assign an absolute value to the enthalpy. Hence it is customary to make a conventional assignment to the values of the enthalpy of the elements at a reference state (see later).

Just as the first law of thermodynamics leads to the definition of the energy of a system, the second law of thermodynamics leads to the definition of the entropy. It can be shown that for a Carnot cycle, or in fact for any cyclic engine, the cycle integral

$$\int \frac{dQ}{T} = 0 \qquad (8)$$

provided all transformations were carried out reversibly. Since the sum over the cycle of the quantity dQ/T is zero, this quantity is the differential of some property of state. This property is called the ENTROPY and is given the symbol S. Again, discussing the many properties of S would exceed the scope of this paragraph. Suffice it to say that in any real transformation in an isolated system S always increases, or:

$$dS > \frac{dQ_{irreversible}}{T} \qquad (9)$$

This prompted Clausius' famous aphorism that "the energy of the Universe is constant, the entropy strives to reach a maximum." Therefore the condition of equilibrium in an isolated system is that the entropy have a maximum value. On this basis then it is possible to formulate the various free energy relationships and to state the conditions for equilibrium or spontaneity of a reaction.

The basic condition for equilibrium or spontaneity is

$$TdS \geqq dQ \qquad (10)$$

where the equal sign stands for equilibrium condition. Using the above equations one can write:

$$-dE - P_{op}dV - dU + TdS \geqq 0 \qquad (11)$$

where $P_{op}dV + dU$ stands for the entire work (dW)

the system is capable of delivering. This fundamental relationship permits one to very rapidly formulate the three major energy relationships one unusually deals with in practical situations.

In an *isolated system* $dE = O$, $dW = O$, $dQ = O$ per definition, hence

$$dS \geqq 0 \qquad (12)$$

This means that a reaction in an isolated system can take place only if the entropy change of the system is positive. Equilibrium is attained when the entropy has reached a maximum.

If the *temperature* of the system remains *constant* during the transformation, then $dQ = O$ and:

$$-d(E - TS) \geqq dW \qquad (14)$$

The combination of variables E-TS is given the symbol A and is often called the "work function" of the system. However it is really the maximum amount of work (free energy) obtainable from the system under isothermal, reversible conditions.

Under constant pressure conditions Eq. 11 converts into:

$$-d(E + pV - TS) \geqq dU \qquad (15)$$

The combination $E + pV - TS$ is given the symbol G and is called the Gibbs free energy. (Often G is given other names and other symbols such as F. Therefore, when looking up free energy values in tables, one should always reassure oneself of the exact definition of this value.) It is good to remember, that in a certain sense S/T, A and G are all free energy terms although only G is specifically given this name. Finally by way of summarizing, the general relationship between G and A can be established:

$$G = E + pV - TS = H - TS = A - pV \qquad (16)$$

Free Energy Functions

All the above correlations would merely be of only theoretical interest if the various properties of state could not be evaluated numerically. Much of the research in the field of thermodynamics has been devoted to this problem. Originally properties of state such as the enthalpy were linked to easily measurable properties such as the heat capacity. Later however, an effort was made to arrive at more fundamental relationships.

The differential enthalpy can be written in the following manner:

$$dH = \left(\frac{\partial H}{\partial T}\right)_p dT + \left(\frac{\partial H}{\partial T}\right)_T dp \qquad (17)$$

and per definition

$$dH = C_p dT + \left(\frac{\partial H}{\partial T}\right)_T dp \qquad (18)$$

from where it follows that for any constant-pressure transformation the enthalpy differential is simply:

$$dH = C_p dT \qquad (19)$$

Integrated for a transformation from condition 1 to condition 2 we obtain:

$$\Delta H = C_p \Delta T \qquad (20)$$

Eq. 20 permits the calculation of enthalpy differences between two states by means of experimentally readily available data. In order to facilitate this job, an enthalpy standard was defined. The enthalpy at 298.15 °K, $H°$, for elements is zero. For compounds $H°$ then becomes the "heat of formation," and is tabulated in many reference books and tables. In order to calculate the enthalpy at a given temperature Eq. 20 can now be written as:

$$H_T = H° + C_p(T - 298.15) \qquad (21)$$

This equation is correct only if C_p is not a function of temperature, which is rarely the case. Early attempts to express C_p and C_v as temperature functions took the forms of power series:

$$C_p = A + bT - cT^2 + dT^3 \ldots \ldots \qquad (22)$$

with the coefficients A, b, c and d having been determined experimentally, and tabulated in standard reference works.

Similarly, an energy function for the entropy can be written and evaluated:

$$S_T = S° + \int_J^T \frac{C_p}{T}\, dT \qquad (23)$$

Being able to evaluate both the enthalpy and the entropy at any desired temperature enables one automatically also to evaluate the Gibbs free energy, G.

More modern approaches have been opened by statistical thermodynamics. If the energy levels of molecules composing a system can be obtained by solving the Schroedinger equation, the partition functions can be evaluated. This in turn permits one to arrive at any thermodynamic property.

Solutions of the Schroedinger equation are difficult and involve a number of approximations which may cast doubt on the validity of the result. The actual values of the energy levels may, however, be obtained experimentally from an analysis of the spectrum of the molecule. Again, inserting these experimental energy values into the partition function provides an approach to calculate any thermodynamic property of interest. This approach is particularly useful for high temperatures, since it may be much easier to determine a spectrum at 3000 or 4000°F than measure the heat capacity at these elevated temperatures. Actually a large amount of the thermodynamic data available now has been determined from spectral information.

R. H. HAUSLER

FREE ENERGY OXIDATION STATE DIAGRAMS. See FROST DIAGRAMS.

FREE RADICALS AND ATOMS

Free radicals may be defined as the fragments formed by breaking one or more bonds in a stable molecule. Thus by successively breaking C-H bonds in CH_4, free methyl (CH_3), methylene (CH_2), and methyne (CH) radicals would result. Free atoms such as H, O, or N are not radicals in the original meaning of the term, but behave chemically like free radicals and may be conveniently grouped with them.

By virtue of their free valence, free radicals are usually highly reactive and therefore shortlived under normal conditions. Lifetimes typically range from 1 to 100 milliseconds, but may be much longer or much shorter, since the lifetime of a free radical depends greatly on its environment, that is, on the rate at which it disappears by reaction with molecules, other free radicals, or surfaces which it encounters.

A monoradical such as methyl always has a single unpaired electron associated with its free valence. In a biradical such as methylene, there are two "free" electrons which may be unpaired (a triplet state) or paired (a singlet state). In the latter condition, it is still usually regarded as a free radical, and in fact is often more reactive than the triplet biradical in which the electrons are unpaired. Triradicals such as methyne in a similar way can exist in either doublet or quartet states. A few stable molecules, notably NO, NO_2, and O_2, have unpaired electrons and may properly be regarded as free radicals, although very unreactive ones. A number of organic free radicals, such as triphenyl methyl or diphenyl picryl hydrazyl, have been prepared which are also stable under normal conditions. The low reactivity of all these species arises from a reduced availability of the free electron due to such phenomena as delocalization, steric hindrance, or participation in odd-electron bonds.

The transitory existence of free atoms and radicals had long been established in spectroscopy. The idea that the reactions of free atoms may be important in chemical mechanisms first came from photochemistry. It became apparent that some sort of chain mechanism must be responsible for the very large quantum yields of reactions such as the photochemical synthesis of hydrogen chloride, and the so-called Nernst chain was suggested. It soon became evident that such atomic reactions are of major importance in photochemistry.

In 1925, H. S. Taylor first suggested that organic chain reactions involving free radicals might occur in a similar way. In 1929, Paneth and Hofeditz showed that free methyl radicals, formed by the thermal decomposition of lead tetramethyl in a flowing stream of inert carrier gas, could be detected by their reaction with a metallic "mirror" deposited some distance downstream from their point of formation, and therefore could exist for an appreciable time in such a system. Subsequently, F. O. Rice and his co-workers showed that free radicals could be detected in this way in the pyrolysis of a large number of organic compounds. On this evidence Rice and Herzfeld suggested that most organic compounds decompose by a free-radical chain mechanism. For ethane, for example, they suggested

$$C_2H_6 \rightarrow 2CH_3 \qquad (1)$$

$$CH_3 + C_2H_6 \rightarrow CH_4 + C_2H_5 \qquad (2)$$

$$C_2H_5 \rightarrow C_2H_4 + H \qquad (3)$$

$$H + C_2H_6 \rightarrow H_2 + C_2H_5 \qquad (4)$$

$$H + C_2H_5 \rightarrow C_2H_6 \qquad (5)$$

Reactions (3) and (4) propagate the chain, and for long chains, C_2H_4 and H_2 would be the only major products, as is observed.

Striking support for the occurrence of free-radical chain reactions was soon forthcoming. Various workers showed that it was possible to start chains in a system at temperatures below the normal decomposition range by adding radicals produced by the thermal or photochemical decomposition of suitable additives. Further evidence for the chain character of many decomposition reactions was the inhibiting effect of such compounds as nitric oxide and propylene which were thought to suppress chains by reacting with free radicals.

In the three decades or more following these early investigations, free-radical reactions have been widely studied. Rate constants and Arrhenius parameters have been measured for a large number of elementary processes such as reactions (1) to (5) by a variety of increasingly sophisticated techniques. Photochemical kinetic methods which permit radicals to be generated and studied under carefully controlled conditions have played a major role. Free radicals have been detected and measured quantitatively by physical methods such as absorption spectroscopy, mass spectrometry, and electron spin resonance; they have been studied in gases, liquids, and solids, in the latter often at low temperatures. Free radicals are recognized as important intermediates in a wide variety of systems, including photolysis, pyrolysis, radiation chemistry, polymerization, combustion, oxidation, chemistry of the atmosphere, and air pollution.

R. A. BACK

References

Steacie, E. W. R., "Atomic and Free Radical Reactions," 2nd Ed., Reinhold, New York, 1954.

Trotman-Dickenson, A. F., "Free Radicals," Methuen, London, 1959.

Dainton, F. S., "Chain Reactions," Methuen, London, 1956.

Bass, A. M., and Broida, H. P., (Editors), "Formation and Trapping of Free Radicals, Academic Press, New York, 1960.

Walling, C., "Free Radicals in Solution," Wiley, New York, 1957.

FRIEDEL-CRAFTS REACTIONS

In organic chemistry a class of reactions has been named after Charles Friedel, a Frenchman, and James Mason Crafts, an American, who reported mostly during 1877–8 the results of their joint research in France on reactions catalyzed by anhydrous aluminum chloride and other acidic metal halides. Although the original reaction was run on aliphatic halides alone to give a mixture of hydrocarbons, the Friedel-Crafts reaction is commonly considered to be a condensation of an alkyl or acyl halide with an aromatic compound in the presence of a strong Lewis acid*. More recently the definition of the Friedel-Crafts reaction has been extended to include any organic reaction brought about by the catalytic action of anhydrous aluminum chloride or related catalysts, including protonic acids.

$$ArH + RX \rightleftharpoons ArR + HX$$

* The term Lewis acid means any molecule capable of accepting a pair of electrons.

The RX reactant need not be a halide at all but can be any aliphatic compound capable of forming a carbonium ion with a Lewis acid. The more general equation may be written as follows:

$$ArH + R^+ \rightleftharpoons ArR + H^+$$

R^+ can be derived from compounds such as aldehydes, ketones, alcohols, esters, anhydrides, acids, olefins, etc.

The aromatic reactant can be taken from a large variety of substituted benzenes as well as aromatic heterocycles such as thiophene, pyridine and furan, and nonbenzenoid aromatics such as ferrocenes, fulvenes, azulenes, pseudoazulenes, pentalene, and heptalene derivatives.

The catalyst, though most often aluminum chloride, may include any electron-acceptor such as aluminum bromide, boron fluoride or zinc chloride as well as sulfuric acid, hydrogen fluoride, or solid acid catalysts such as supported phosphoric acid, the silica-alumina catalysts and cation-exchange resins. Frequently the latter group is not considered as typical Friedel-Crafts catalysts but the mechanism of reaction is essentially the same.

Sometimes cases where both reactants are aliphatic and the catalyst is aluminum chloride are called Friedel-Craft reactions. Such cases properly belong to the broader field of the Friedel-Crafts synthesis. Aluminum chloride catalyzed isomerizations, polymerizations, alkylations, as well as cracking reactions of aliphatic compounds do not belong in this category.

When the aromatic compound is substituted with typical meta-directing groups, the reaction is inhibited. Thus nitrobenzene can quite effectively be used as an inert solvent. Also, in the reaction with acyl halides only one group enters, deactivating the molecule and preventing further attack. The presence of ortho-para directing groups enhances reactivity at the ortho and para positions in a manner similar to other reactions involving electrophilic substitution in the aromatic nucleus.

In the case of alkylations the reaction does not stop at the monoalkylated stage, and mixtures are inevitable. Also typical carbonium ion rearrangements may be expected to occur. Thus, n-propyl halide will give isopropyl benzene.

In the case of acylations at least one mole of aluminum chloride must be used, although with other Lewis acids catalytic amounts are sufficient. Alkylations on the other hand need only small amounts of catalyst to complete the reaction.

The extended definition of the Friedel-Crafts reactions encompasses also the nitration, amination, sulfonylation, sulfonation, perchlorylation, and halogenation, as well as rearrangement, dealkylation, and fragmentation of aromatic compounds.

General Reaction Conditions. For alkylations the alkyl halide is added at room temperature with stirring to an excess of aromatic compound containing about five mole per cent of aluminum chloride under anhydrous conditions. When HCl is no longer evolved, the reaction is complete and the mixture is allowed to settle. The upper hydrocarbon layer is separated from the lower catalyst complex, and the product is obtained by distilling off the more volatile aromatic starting material. In the case of acyl halides the reaction product is complexed with a mole of aluminum chloride and must be decomposed with water before isolation of the product. When an olefin is used in place of the alkyl halide, a trace of HCl must be present to initiate the reaction and no HCl will be evolved.

Industrial Applications. Ethylbenzene, the key intermediate in the manufacture of styrene, is made on a large scale by reaction of ethylene and benzene in the presence of aluminum chloride and hydrogen chloride.

Isopropylbenzene, the intermediate in the manufacture of phenol and acetone, is commercially made from benzene and propylene.

Another similar alkylation which has achieved economic importance is the reaction of benzene with higher aliphatic halides or olefins to give alkyl benzenes. These are subsequently sulfonated to give synthetic detergents. In both of these cases predominant monosubstitution is achieved by using a large excess of benzene and recycling. These reactions are worked up by separating the catalyst complex from the hydrocarbon and refining the product by distillation.

The production of high-octane gasoline by alkylation of olefins as well as that of synthetic rubber and plastics by cationic polymerization of olefins fall within the scope of the extended definition of the Friedel-Crafts reaction.

In the field of pharmaceuticals a useful cathartic, phenolphthalein is made by condensing phenol and phthalic anhydride in the presence of zinc chloride and toluenesulfonic acid. In this instance phenol is a highly active aromatic compound and the relatively weak zinc chloride is sufficient for condensation.

The important insecticide DDT is made by a Friedel-Crafts type reaction.

$$CCl_3CHO + 2 \,\bigcirc\!\!-\!Cl \xrightarrow{H_2SO_4}$$

$$CCl_3CH \left(-\!\bigcirc\!\!-\!Cl\right)_2$$

In place of H_2SO_4 as a catalyst HF, $ClSO_3H$, and FSO_3H, have been successfully used. In all cases some ortho substitution occurs, but the crude mixture is usually sold without purification.

CHARLES ALLEN THOMAS

FROST DIAGRAMS

Pioneered by Professor A. A. Frost of Northwestern University, whose name provides an acronym for them. *FRee energy Oxidation STate diagrams* exhibit the oxidation-reduction states of the chemical elements with a plot of the oxidation number versus the free energy ΔG^0 of the ionic partial reaction which converts one gram-atom of an element to each of its possible redox states. The oxidation number is the number of electrons *lost* by one atom of the element X in going to a particular redox state (which is supposed, for present purposes, to contain only H, X and O), H_2O, H^+ (or OH^-), and e^- being supplied as needed to balance the half-reaction. If other ligands are involved, e.g., Cl^- as in AgCl or

Hg_2Cl_2, the ligand is supplied in the state of oxidation in which it appears in the complex. For nitrate ion, the ionic partial

$$\tfrac{1}{2}N_2 + 3H_2O = NO_3^- + 6H^+ + 5e^-,$$
$$\Delta G^\circ = +143.6 \text{ kcal} = +6.23 \textbf{ FV} \qquad (1)$$

shows that one atom of nitrogen loses five electrons in going to nitrate ion, with a standard free energy increase ΔG^0 of 6.23 faraday-volts per gram-atom. Nitrate ion is therefore placed at the point $+5$, $+6.23$ in the nitrogen Frost diagram (Fig. 1, acid curve). Note that 1 faraday-volt (FV) = 96.487 kJ = 23.061 kcal. In alkali, the corresponding reaction is

$$\tfrac{1}{2}N_2 + 6OH^- = NO_3^- + 3H_2O + 5e^-,$$
$$\Delta G^\circ = +29.07 \text{ kcal} = +1.26 \textbf{ FV} \qquad (2)$$

and the nitrate point in alkali is plotted at $+5$, $+1.26$ (Fig. 1, basic curve). All half-reaction free energies, both in acid and in alkali, are based on an assigned null value for the standard hydrogen electrode (SHE), i.e.,

$$\tfrac{1}{2}H_2 = H^+ + e^-,$$
$$\Delta G^\circ = 0.000 \textbf{ FV} \text{ (by definition)} \qquad (3)$$

For ammonium ion, the ionic partial is

$$\tfrac{1}{2}N_2 + 4H^+ + 3e^- = NH_4^+,$$
$$\Delta G^\circ = -19.00 \text{ kcal} = -0.825 \textbf{ FV} \qquad (4)$$

The oxidation number, for a gain of three electrons, is -3, and NH_4^+ has the coordinates -3, -0.825 (Fig. 1, acid curve).

Frost diagrams are illustrated for nitrogen (Fig. 1) and manganese (Fig. 2) in acid (pH 0) and alkali (pH 14) at 25°C. Also plotted (on a displaced scale) are the Frost lines for H_2/H_2O and O_2/H_2O in acid and alkali. Every tie-line connects a particular *redox pair* or *couple*, which consists of two oxidation states of the same element. The slope of the line is equal to the I.U.P.A.C.-Gibbs-Stockholm electrode potential V^0 at which the given redox pair is at equilibrium at unit activity; this is also equal to the reduction (or oxidizing) potential E^0_{red} for the given couple, and to the negative of the oxidation (or reducing) potential, $-E^0_{ox}$. The numbers shown on the tie-lines give in volts the electrode potentials V^0 for the particular couple, referred to $V^0(\text{SHE}) = 0.000$ V. The steepness of the line is a measure of the *tendency* of the couple to react in the direction of the state of lowest free energy (when paired with H_2/H^+).

The profile of the Frost diagram indicates the comparative thermodynamic stability of the various oxidation states of an element *at the potential of the SHE*. Thus Mn^{++} in acid and MnO_2 in alkali are the most stable forms of Mn at $V = 0$ (Fig. 2). If a sequence, such as $Mn^{++} - MnO_2 - MnO_4^-$ finds the middle state *below* the tie-line connecting the two outer states, the middle state can be formed thermodynamically by reaction of the two outer states. If the middle state is *above* the tie-line, e.g., $MnO_2 - MnO_4^= - MnO_4^-$, the middle states tends thermodynamically to disproportionate to the two outer states.

A shift to a new reference potential $V = V_1$ alters

the the profile of the Frost diagrams by raising the ordinate of every point by the amount $-nFV_l$. Alternatively, a family of lines of slope V_l can be drawn on the existing diagrams: these new lines become the isoergonic lines at the new reference potential. It can thus be shown that metallic Mn is the species of lowest free energy at potentials more negative (i.e., more reducing) than -1.18 V in acid (-1.55 V in alkali), and MnO_4^- is the species of lowest free energy at potentials more positive (i.e., more oxidizing) than $+1.70$ V in acid ($+0.59$ V in alkali).

The nitrogen diagram (Fig. 1) identifies hydroxylamine NH_2OH (or NH_3OH^+) and hydrazoic acid HN_3 as the only noteworthy reducers among the redox states of N. Hydroxylamine can act as an oxidizer and reducer both, and disproportionates, sometimes explosively, to a variety of N compounds. The oxides and oxyions of N are mild oxidizers less strong than oxygen. N_2 itself occurs at a dip in the free energy profile, corresponding to the remarkable stability of the elemental form, which can nevertheless be converted to NH_4^+ by nitrifying bacteria. The reduction of nitric acid almost never yields N_2, but rather a variety of reduction products going all the way down to NH_4^+, presumably passing through such intermediates as $H_2N_2O_2$ and NH_3OH^+.

To establish whether any particular oxidation-reduction reaction is thermodynamically possible, take the tie-lines for the oxidation and the reduction partial and bring their end-points, i.e., product points, together, keeping the slopes and lengths unchanged. If the product point falls *below* a new tie-line connecting the two reactant points, the reaction is thermodynamically possible, otherwise not. Thus H_2 can reduce MnO_4^- and MnO_2 in acid and base but cannot reduce Mn^{++} or $Mn(OH)_2$ to the metal. O_2 can oxidize the metal to the $+2$ state but not further, and MnO_4^- should thermodynamically be able to oxidize H_2O to O_2 since the MnO_4^-/MnO_2 line is steeper than the H_2O/O_2 line in both acid and alkali. Its failure to do so is attributable to kinetic factors, e.g., oxygen over-

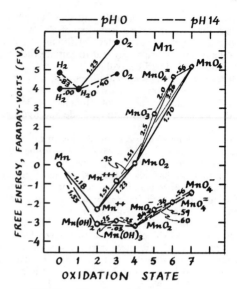

FIG. 2.   Manganese frost diagram.

voltage. To account for hydrogen and oxygen evolution overpotentials, the sets of H_2/H_2O and O_2/H_2O lines in Figs. 1 and 2 should be made steeper by about 0.5V—these would be "practical" Frost lines from which could be gauged whether any particular oxidizer or reducer is, in fact, able to bring about the decomposition of water. Wherever sufficient kinetic data are available, the free energy of activation for the reaction of any particular couple could be shown as a hump superimposed on the tie line.

ANDRE J. DE BETHUNE

References

Frost, Arthur A., *J. Am. Chem. Soc.,* **73,** 2680 (1951)

Phillips, C. S. G., and Williams, R. J. P., "Inorganic Chemistry," Oxford University Press, New York and Oxford, 1965, Vol. I; 1966, Vol. II.

Gray, H. B., and Haight, G. P., "Basic Principles of Chemistry," W. A. Benjamin, Inc., New York, 1967.

LEGENDS FOR THE FIGURES

Fig. 1: Nitrogen Frost Diagram

Fig. 2: Manganese Frost Diagram

Free Energy Oxidation State (Frost) Diagrams at 25°C.

Solid lines in acid (pH 0); dashed lines in base pH 14).

The free energies are given in Faraday-Volts (**FV**) for the ion-electron partial which converts one gram-atom of the element to the given redox state. The slopes, in volts, are equal to the I.U.P.A.C.-Gibbs-Stockholm *electrode potentials* for the redox couples formed by the pair of oxidation states at the two ends of any given tie-line.

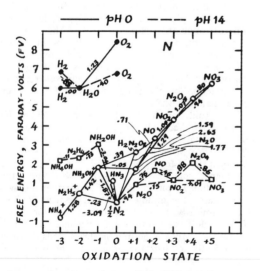

FIG. 1.   Nitrogen frost diagram.

FUEL

Fuels are materials that evolve usable energy when they combine with oxygen. Normally, the oxidation of fuels can be controlled so that the liberated energy may be used for heating or be converted to other forms of energy such as luminous, electrical, or mechanical. The most common fuels are solid, liquid, and gaseous carbonaceous materials. The extent of use of a particular fuel is determined by such factors as availability, cost, consumer preference, and status of technology. According to P. C. Putnam, more energy was obtained on earth from farm wastes than from all fluid fuels until 1940; dung is still India's main fuel. In the United States, wood was the chief fuel until about 1880, coal from then until about 1948, when fluid fuels assumed the lead.

The consumption of energy in the United States is increasing at a rapid rate. In 1970 the per capita annual use was over 300 million BTU and is expected to reach 390 million in 1980 when the population is estimated to be 235 million, a 17% increase over 1970. The fuel sources of energy at the present are about 40% from oil, 35% from natural gas, 21% from coal and 4% from water power, with a small but growing use of nuclear energy for generation of electricity. In addition to the influence of the depletion factor for oil and gas reserves, the fuel-energy source picture is increasingly affected by the steps to control environmental pollution from the use of fuels. Thus, the above ratios may change greatly in the future.

Currently energy usage is about 22% by households and commercially, 32% industrially, 24% for transportation, 20% for electricity and 2% for miscellaneous purposes.

While the per capita consumption of fuel in the United States has increased in the past 50 years, the efficiency of fuel utilization has jumped spectacularly: one kilowatt-hour of electricity required 3 lbs of coal in 1920, 0.9 lb or less in 1970; one ton of pig iron required 1.1 tons of coke in 1913, 0.6 ton in 1970; and one barrel of cement required 2 million BTU about 50 years ago, less than 1 million BTU (0.8 million at one large plant) in 1970. Future large increases like these are unlikely, unless methods of utilization are changed radically.

The economic value of a fuel is affected by the cost of mining, storing, transporting, preparing (cleaning, refining, etc.), cost and maintenance of combustion equipment, and disposal of waste, as well as by the quality of the fuel as judged by its performance in the burner. To predict performance, numerous standards for comparison have been developed. In the United States, the ASTM publishes standards annually; it contains standard definitions, specifications, and test methods for fuels. Typical of these are tests (on gaseous fuels) for calorific value, composition, and dew point; (on motor and aviation fuels) for octane number, distillability, gravity, aromaticity, unsaturation, color, gum, stability, and vapor pressure; (on heavier fuels, for example, diesel oil) for flash point, cloud and pour points, cetane number, neutralization value, sediment, and viscosity; (on coal and coke) for proximate and ultimate analysis, ash content, fusibility of ash, volatile matter, size, grindability, and swelling.

Carbonaceous fuels occur naturally in solid, liquid, and gaseous form. Although one kind of fuel may be plentiful in a geographic area, a prob-

Sources of Energy

Source	H*
Hydrogen	61.0
Hydrocarbons	
methane	23.9
ethane	22.4
propane	21.7
butane	21.3
benzene	18.2
acetylene	21.6
cetane	20.4
Alcohols	
methanol	10.3
ethanol	13.2
Carbon monoxide	4.4
Petroleum	
crude	18.8–19.5
gasoline	20.8
kerosine	19.8
gas oil	19.2
fuel oil	18.5–19.4
Coal, "as received"	
anthracite	9.0–14.1
bituminous and subbituminous	7.9–14.8
lignite	6.0–7.4
coke (average)	12.4
Wood	
typical nonresinous, seasoned	6.3
tanbark	2.6
pine	8.0
charcoal (willow)	13.5
Peat, air-dried	5.1–9.3
Dynamite, 75%	2.3
Gunpowder	1.3
Magnesium	10.9
Nuclear reactions	
U-235	3.5×10^7
D_2O	0.7×10^7
LiH	9.0×10^7

*Approximate amount of energy, 1,000 BTU/lb.

lem of rapidly increasing economic and technological importance is the provision of adequate supplies of the most useful types of fuels. The growth in energy demand during the last three decades has been absorbed largely by fluid rather than solid fuels. As coal and oil shale constitute more than 95% of the fuel reserves of the United States, their convertibility to liquids greatly increases the potential amount of domestically available fuels suitable for gasoline engines and diesel motors. The technical feasibility of obtaining oil from solid oil shale was demonstrated more than 100 years ago. Similarly, Bergius showed the feasibility of hydrogenating coal to produce oil. Coal can also be gasified with steam and oxygen to yield a gas which is reformed to methane (and steam) to be used as a gaseous fuel; or the impure gas, resulting from gasification of coal or controlled combustion of natural gas (mostly methane), can be purified and converted to liquid fuels. The thermal efficiency of liquefaction of bituminous coal is low, about 45% for the Fischer-Tropsch and 40% for the coal-hydrogenation process.

Ernst M. Cohn

Cross-references: *Coal, Petroleum, Fuel Oils, Fuel Gases Fischer-Tropsch Process, Natural Gas.*

Fuel Gases

All but one or two fuel gases are produced from materials which contain such large amounts of carbon (up to 90%) that they are used as fuels *per se*. But the cost and effort of deriving gaseous fuels from solid and liquid fuel materials is economically justified because fuel gases can be more quickly and conveniently delivered to a large number of users by mains, and can be more effectively utilized, than can solid or liquid fuels.

All compounds containing more than five carbon atoms are liquid or solid under normal conditions. Naturally occurring low carbon number hydrocarbons, e.g., natural gas, have become a major source of fuel gas. Fuel gas is also produced by a number of methods for converting large, complex, carbon-containing molecules to simple compounds containing only one or two carbon atoms in union with hydrogen (or oxygen). The basic actions used in various fuel gas production processes are:

(1) *Thermal decomposition:* removal by destructive distillation of volatile materials contained in coal, shale, wood, etc.
(2) *Thermal dissociation:* the "cracking" or pyrogenic fracture of large hydrocarbon molecules of oils.
(3) *Reaction with steam* at high temperatures: the "water gas" reaction for producing carbon monoxide and hydrogen from coke, coal, heavy oil, etc.
(4) *Semioxidation:* partial combustion producing carbon monoxide, usually in conjunction with and providing energy for one of the foregoing actions.
(5) *Physical partition:* fractionation by solution absorption or condensation, such as the removal of vapors and liquids from casing-head or refinery gases.
(6) *Biological decomposition:* as used in sewage and garbage treatments.

A few gases having fuel value (hydrogen, acetylene) can be produced by simple direct chemical reactions, but this method is used only where very pure products are needed for special purposes.

Composition. All fuel gases are mixtures in varying proportions of a limited number of constituent gases, which can be classified as follows:

(a) Combustibles (intentional): C_1 and C_2 hydrocarbons; CO; H_2
(b) Combustibles (incidental): small amounts of hydrocarbons above C_3
(c) Inerts: N_2; CO_2; O_2
(d) Impurities: H_2S; CN; NH_3

The most important combustible constituent is methane, which with ethane, ethylene, and hydrogen comprise the more desirable substances in fuel gases. In deriving these the simultaneous production of small amounts of propane, butane, propylene, butylene, benzene, toluene, etc., can seldom be avoided. The inerts content may be due to the use of semioxidation to promote molecular dissociations, may result from inadvertent inclusion of combustion products or air, or may be found in the natural gas.

The usual manner of indicating fuel gas compositions reflects the analytical procedure used in their analysis—in percentages by volume on a dry basis, in the order of their determination by chemical reactions:

CO_2 absorption by NaOH solution
"Unsats" reaction with Br, and solution in water
O_2 absorption by alkaline pyrogallate
CO absorption by ammoniacal cuprous chloride
H_2 oxidation by cuprous oxide
"Sats" by combustion and alkaline absorption of CO_2
N_2 by difference.

"Unsats" are primarily ethylene, but also include other olefinic gases, acetylene, or vapors of aromatic hydrocarbons which, being unsaturated, will react with bromine or fuming sulfuric acid. "Sats" include methane, ethane, etc. Where it is necessary to identify the constituents an ordinary chemical analysis must be supplemented by mass spectrometer, gas chromatograph or fractional distillation methods.

Classification. The calorific values of the combustible constituents of fuel gases range from 323 Btu/CF for carbon monoxide or hydrogen to 3260 Btu/CF for butane. The heating values of the gaseous mixtures most widely used as fuels range from 90 Btu/CF to about 3000 Btu/CF—or from 1150 Btu/lb to 21,000 Btu/lb as compared to 19,000 Btu/lb for fuel oils and 14,500 Btu/lb for coals.

Fuel gases may be divided into four general classes according to their heating values:

(1) *Producer or low Btu gases*, of less than 300 Btu/CF, usually by-products of metallurgical processes; used as fuels only near their origin, and only because of negligible cost and availability in large quantity. An example, steel-mill blast furnace gas.
(2) *Manufactured, or medium Btu gases*, of from 500 to 600 Btu/CF, produced from coal, coke or oil, and until recent years widely distributed by public utility companies. Examples are carbureted water gas and coke oven gas.
(3) *High Btu gases*, of from 900 to 1100 Btu/CF, such as natural gases and manufactured oil gases.
(4) *Heavy, vaporous gases*, of from 1800 to 3300 Btu/CF, mostly by-products of refining operations, such as "LP" or liquefied petroleum gases, cracking still gases, etc.

In recent years the use of high Btu fuel gases in America has increased rapidly, and intentional production of 500-600 Btu/CF fuel gases from solid materials has lost most of its importance. Only negigible amounts of carbureted water gas are now being manufactured, and although there has not been a significant decrease in the production of coke oven gas, it is no longer a public commodity. With a few isouated exceptions, all public utility companies now distribute gases containing 950–1100 Btu/CF, i.e., mostly "dry" natural gas supplemented by manufactured oil gas and dilutions of the LP gases. Thus, except where economic factors dictate that advantage be taken of low-cost by-products of smelting, retorting, or carbonizing operations, today the American gas industry is almost entirely dependent on high Btu gases (over 950 Btu/CF) derived from petroleum and natural gas as raw materials.

Until recently the transportation of fuel gas has been largely confined to locations that are connected by pipeline. The technology and equipment to liquefy and ship natural gas have been developed to the point that large quantities of LNG (liquefied natural gas) are being imported into Western Europe, Japan and the United States. LNG is the liquid resulting from the condensation of natural gas at low temperatures, i.e., −259°F and 1 atm. It is shipped under atmospheric pressure at temperatures ranging from −250 to −259°F. The heating value of LNG is 1000 to 1200 Btu/CF.

Another fuel gas shipped and marketed in liquid form is LPG, whose initials refer to "liquefied petroleum gas" whose heating value is 2000 to 3500 Btu/CF. It consists of readily condensible hydrocarbon gases of natural gas or refinery still gas (i.e., propane, butane, propylene, butylene or mixtures thereof). It is widely used for domestic heating in rural areas.

Royce E. Loshbaugh

Fuel Oil

Fuel oil may be defined as any liquid petroleum product that is burned in a furnace for the generation of heat, or used in an engine for the generation of power, excepting oils having a flash point below 100°F and oils burned in cotton or wool-wick burners. The three major classifications of fuel oil are related to the principal uses: burner fuel oil (frequently designated simply "fuel oil"), diesel fuel oil, and gas turbine oil. (In addition, certain aviation fuels can be considered as gas turbine fuel oil.) Each of these oils can be made from fractions distilled from petroleum, the residuum from refinery distillation, crude petroleum, or a mixture of any of the above, as long as it meets the appropriate specifications.

Burner Fuel Oil. Because fuel oils are used in burners of various types and capacities, different oils are required to meet specific needs. The ASTM has developed specifications for six grades of fuel oil as shown in Table 1. No. 1 oil is a distillate boiling approximately 350–550°F, equivalent to a heavy kerosine, intended for use in a vaporizing-type burner. No. 2 oil is a heavier distillate, boiling in the range of approximately 375–625°F, primarily used in atomizing-type burners for commercial and domestic heating. No. 4 oil is usually a light residual oil but may be a heavy distillate or a residual oil "cut" with a heavy distillate. The boiling range of a typical No. 4 oil is 425 to 675°F. The viscous residual oils, Nos. 5 and 6, referred to as bunker oils when used as ship fuel, usually must be preheated before burning. Under some climatic conditions, No. 5 (light) oil can be handled without preheating. No. 6 oil (Bunker C) is the most common heavy fuel, used to raise steam in power plants, ships, and industrial plants. It is also used to supply heat in the metallurgical, ceramic, and other industries.

The specifications given in Table 1 refer to values obtained by standard ASTM tests, as published an-

TABLE 1. BURNER FUEL OIL SPECIFICATIONS (ASTM D396)

Properties	No. 1	No. 2	No. 4	No. 5 light	No. 5 heavy	No. 6
Gravity, min, °API	35	30				
Distillation, °F						
10% pt, max	420					
90% pt	550 max	540–640				
Flash pt, min, °F[a]	100	100	130	130	130	150
Pour pt, max, °F	0	20	20			
Water & sediment, max, vol %	trace	0.05	0.5	1.0	1.0	2.0
Carbon in 10% residuum, max, %	0.15	0.35				
Ash, max, wt %			0.1	0.1	0.1	
Viscosity, Saybolt, sec[b]						
SU at 100°F		(32.6–37.93)	45–125	150–300	350–750	(900–9000)
SF at 122°F					(23)–(40)	45–300
Viscosity, Kinematic[b]						
cS, at 100°F	1.4–2.2	2.0–3.6	(5.8–26.4)	(32–65)	(75–162)	
cS, at 122°F					(42–81)	(92–638)
Copper strip corrosion	No. 3					
Sulfur, per cent[a]	0.5	0.5				

[a] Must meet legal standard.
[b] Figures in parentheses are for information only.

nually. For the lighter fuels, the important criteria are those related to safety in handling and storage (flash point), freedom from residue and corrosion, (carbon residue, copper test, and sulfur), and volatility (distillation range). For residual fuels, flow and atomizing properties (viscosity), water and sediment, and ash are important criteria. In some cases, the composition of the ash is important; alkali metal salts represent one problem, vanadium or nickel compounds another. The sulfur content is an important factor in residual fuel oils because of the air pollution caused by sulfur dioxide generated in burning. Many communities and states have put limits of 1.0 per cent, 0.5 per cent, or even lower on the sulfur content of fuel oils. This is not a problem with distillate oils since they typically have only 0.1 per cent sulfur.

Chemically speaking, fuel oils are a complex mixture of straight and branched paraffins, olefins, naphthenes (saturated ring compounds), and aromatic compounds. As in crude petroleum, small amounts of sulfur, nitrogen, and oxygen are found chemically bound to various hydrocarbon chains. The content of S, N, and O can be reduced by treating the oil with hydrogen under conditions of high heat and pressure over a catalyst. This is accompanied by some carbon-carbon bond breaking, which increases the amount of hydrogen used and creates molecules of lower volatility (lower flash point, viscosity). Costs of desulfurizing residual oils can be reduced by starting with lower-sulfur crude oil (North African or Nigerian instead of or blended with Venezuelan or Middle Eastern crude oil), blending desulfurized distillate with residual oil, or deasphalting before hydrodesulfurization. Number 2 oil typically has a percentage composition of C, 86.1; H, 13.2; S, 0.5; and N, 0.1. A typical residual oil (No. 6 from Venezuela) is C, 86.;3 H, 10.5; S, 2.1; and N and O, 1.1. Porphyrin complexes of vanadium and nickel are also found in residual oils. The vanadium content of a 900°F residuum may be as high as 2000 ppm, although 20-200 ppm is typical.

The heating value of fuel oils is of prime importance, but this property is not specified directly. The heating value per unit weight varies almost linearly with density in API degrees*, from 35° API (No. 1) at 19,600 BTU per pound to a typical "resid" of 10° API at 18,250 per pound. On a volume basis, which is the usual basis of sale, the heating value varies from 138,500 BTU per gallon (35° API) to 152,500 BTU per gallon (10° API). Thus the heavy oils have a heat content advantage on a volume basis.

Diesel Oil. Diesel oils are intended for use in diesel engines, hence the specifications reflect the special needs for this usage. As with furnace oils, a wide range of materials can be used, including light and heavy distillates, special residual oils, and even certain crude oils. Three ASTM grades have been designated, No. 1-D, No. 2-D, and No. 4-D. The physical characteristics correspond roughly to No. 1, No. 2, and No. 4 fuel oils, respectively. In addition, a measure of diesel performance, the

cetane number is specified (40 for 1-D and 2-D, 30 for 4-D). The cetane number is the per cent cetane (hexadecane) in a mixture of cetane and α-methylnaphthalene that gives the same performance in a test diesel engine as the fuel under test.

The requirements for volatility are set by engine design, load, nature of speed, and atmospheric conditions. For bus and truck use, the lighter fuels are best; for stationary and marine diesel engines, heavier fuels are used because of their lower cost and greater heat content.

Gas Turbine Oil. Distillate and residual oils intended for use in gas turbines are classified in a manner similar to burner and diesel fuels. No. 1-GT and No. 2-GT oils are similar to No. 1 and No. 2 fuel oils but have strict requirements for vanadium, sodium plus potassium, and calcium content. No. 3-GT corresponds to the heavier oils (Nos. 4-6), although the specific metal ion requirements virtually preclude anything except heavy distillate. No. 4-GT covers the same physical properties as No. 3-GT except that the allowed vanadium ion content can be much higher as long as the ratio of magnesium ion to vanadium ion is 3.0 to 3.5 (A magnesium-containing additive can be metered into the oil at the turbine if desired.) It is used principally in gas turbines in intermittent service where periodic cleanout is possible.

Aviation Fuels. A variety of light and middle petroleum distillate products are used in aircraft engines. Piston engines use aviation gasoline. Jet engines can use naphtha-type (light distillate), kerosine-type, or mixed types of fuel. ASTM and military standards for jet fuels have been set, an important factor being flame luminosity, which is related to the smoke point, and carbon-hydrogen ratio. Some of the kerosine-type fuels (e.g., JP-5) are essentially the same as No. 1 fuel oil. Some petroleum companies simplify their refining and storage problems by producing a single grade of kerosine that meets the requirements for jet fuel, No. 1 burner fuel, and No. 1-D diesel fuel.

Production and Marketing. The proportion of fuel oils produced by refineries from crude petroleum varies according to the type of crude, refinery processes used, and the economics of supply and demand for each product. In the United States, refiners have maximized the yield of gasoline and other high-value refined products and have minimized the yield of residual oil. In 1969, production of gasoline accounted for 47% of crude oil refined in the United States, distillate fuel excluding kerosine and jet fuel, 20%, and residual oil, 6.2%. As recently as 1963, the yield of residual oil was 14.0%, down from 23.4% in 1950. As the amount of residual oil produced in the United States declined, the oil import quota regulations were relaxed to allow importation of foreign residual oil. Countries with low production costs have maximized their yield of residual oil to get as much oil into the United States as possible. In 1969, about 60% of residual oil demand was satisfied by imports, chiefly from the Caribbean area.

The very great growth in the use of residual fuel oil by the electric utilities companies has come since 1967, as sulfur restrictions led to the conversion of many coal-fired power plants to residual oil.

W. J. Sheppard

*Degrees API $= \dfrac{141.5}{\text{Specific gravity at } 60°F} - 131.5$

References

Annual Book of ASTM Standards, Part 17
  Petroleum Products-Fuel, Solvents, Burner Fuel
  Oils; Lubricating Oils; Cutting Oils; Lubricating
  Greases; Hydrolyic Fluids, American Society for
  Testing and Materials, Philadelphia, Pennsyl-
  vania, ASTM D-396, D-975, D-1655, D-2880.
Guthrie, V. B., Ed., "Petroleum Products Hand-
  book," McGraw Hill, New York, 1960.
Sawyer, J. W., Ed., "Gas Turbine Engineering
  Handbook," "Gas Turbine Fuels" (H. R. Hazard)
  2nd Ed., Gas Turbine Publishers, Inc., Stamford,
  Connecticut, in press.
Gruse, W. A., and Stevens, D. R., "Chemical Tech-
  nology of Petroleum," 3rd Ed., McGraw-Hill,
  1960.
U.S. Bureau of Mines, Washington, D.C.
  Minerals Yearbook, Annual
  Mineral Industry Surveys
    Crude Petroleum, Petroleum Products & Natu-
    ral Gas Liquids, Monthly and Annual
  Mineral Industry Surveys
    Sales of Fuel Oil and Kerosine, Annual.

FUMIGANTS

Fumigants comprise the toxic agents which act in vapor form to destroy pests such as rodents, termites, grain weevils, nematodes, arachnids, bacteria, weeds and fungi. The chemical agents may be applied as solids, liquids or gases, but to be effective the toxicant must be readily volatile so the gas will permeate the treated area and reach the organism to be destroyed. Fumigants, most useful in closed spaces or inaccessible locations, are regularly used to control insects and rodents in grain storage, feed mills, flour mills, warehouses, greenhouses and holds of ships carrying agricultural products. In Southern California and Hawaii about 50,000 buildings are fumigated each year to control termites.

All fumigants exert their toxic effect by interfering with the metabolism of the pest. Therefore, fumigation is most effective when the pest is in an active metabolic condition, and is least effective when it is in a dormant state. Fumigants are generally most effective at temperatures above 70°F.

Several conditions influence the selection and usefulness of fumigants, particularly containment, flammability, toxic hazard and residues. The vapor must be confined sufficiently long to provide effective control. For soil fumigation in seed beds or greenhouses, plastic, canvas or paper covers are used. Buildings are sealed to insure thorough penetration of the agent. For those flammable agents for which the toxic concentration lies within the flammable limits, it is desirable to dilute the vapor with a nonflammable fumigant or carrier, such as carbon tetrachloride or carbon dioxide. Fumigants involve varying degrees of hazard to the user, as well as other persons or pets in the vicinity treated. For this reason, agents possessing good warning properties and relatively low toxic hazards, such as p-dichlorobenzene, formaldehyde and sulfur dioxide are used for household fumigation. Only professional operators equipped to employ adequate safety precautions should apply hydrogen cyanide, methyl bromide and equally poisonous agents. To minimize residues in the materials treated, agents which are rapidly absorbed and desorbed are preferred. Strict limits are placed upon the residues of chemical agents permitted in agricultural products.

The largest volume of fumigants is used for grain and flour treatment, and for soils. The table shows the properties of a number of common fumigants and includes the mean lethal concentration to kill the common flour weevil.

Agricultural crops, especially stored grain, are fumigated most commonly with carbon disulfide mixed with carbon tetrachloride. Other agents include acrylonitrile, dichloroethyl ether, ethylene oxide combined with carbon dioxide, hydrogen cyanide, calcium cyanide (Cyanogas), methyl bromide and phosphine (Phostoxin).

Household fumigants include para-dichlorobenzene against clothes moths and carpet beetles, and ortho-dichlorobenzene for termite control. Professional pest control operators frequently use sulfuryl fluoride, methyl bromide or hydrogen cyanide.

Soils are fumigated against nematodes, soil insects, microorganisms and weeds by means of chloropicrin; 1-2,dibromo-3, chloropropane (Nemagon); dichloropropene; and 3-5,dimethyltetrahydro-1,3,5,thiadiazene-2-thione (Mylone).

H. M. Gadberry

Cross-references: *Disinfectants, Sterilization.*

	BP °C	Sp. Gr. of Gas	Vapor Pressure mm Hg at 25°C	Flammability (Lower Limit %)	LD_{50} for Tribolium Mg/l at 25°C
Formaldehyde	−21	1.0	>760	—	—
Sulfur dioxide	−10	2.3	>760	N.F.	6
Methyl bromide	5	3.2	1824	13.5	11
Ethylene oxide	11	1.5	>760	3.0	18
Phosphine	−87	1.14	>760	1.8	1
Hydrogen cyanide	26	0.94	739	5.6	0.6
Sulfuryl fluoride	−55	3.44	>760	N.F.	—
Carbon disulfide	46	2.63	361	5.0	61
Carbon tetrachloride	76	5.3	114	N.F.	185
Acrylonitrile	77	—	—	N.F.	2
Chloropicrin	112	—	24	Flam.	58
Hexachloropropene	203	—	0.3	N.F.	1.1

FUNGICIDES

The fungicides are a heterogeneous group of chemicals that mitigate, inhibit or destroy fungi. Most chemicals, except the very inert ones, are fungitoxic if present in sufficient quantity, but they usually are not designated as fungicides unless they are effective at nominal dosages of 1000 ppm or less in aqueous suspensions. Those chemicals that inhibit spore germination or mycelial growth without destroying the fungous body are properly known as fungistats, although in common usage they are referred to as fungicides.

The commercial fungicides are indispensible to the welfare of mankind in preventing or curing the diseases of plants, man and animals and in suppressing the deterioration of stored agricultural produce, material and structures made of cellulosic, lignified or plastic materials. Such diseases as athlete's foot, skin mycosis and ringworm of the scalp in man, pulmonary infection of fowl, and moist eczema of dogs are amenable to control by fungicides. The major use of fungicides, however, is in preventing plant diseases such as the leaf blights, powdery mildews, downy mildews, rusts, anthracnoses, fruit rots, fruit scab and stem cankers by spray or dust application of protective or eradicant fungicides. They are also used as soil fumigants and ground sprays to destroy spores and mycelium in their natural habitat before they can attack plants, to disinfest seed known to bear spores or mycelium of smut and other types of fungi, and to protect seed from decay and damping-off organisms in the soil. Other uses include impregnation of fabrics to prevent mildewing and decomposition when in contact with moist substrates, impregnation of wallpaper, paints and leather goods to suppress mildew, treatment of structural timbers, piles, fenceposts, etc. to prevent dry rot and decay, and incorporation into bread to suppress mould growth.

About 100 fungicides are required for these various uses in the United States. The principal ones are sulfur; lime-sulfur (polysulfides of calcium); copper sulfate (or its equivalent in the oxides, basic sulfates, oxychloride and other relatively insoluble copper compounds); creosote products and zinc chloride, both used as wood preservatives; and a wide variety of organic compounds. Among the latter are several dithiocarbamates, such as ferbam and zineb, and other thio compounds, like N-(trichloromethylthio)-phthalimide (folpet); cis-N((trichloromethyl)thio)-4-cyclohexane-1,2-dicarboxyimide (captan); and 8-hydroxyquinoline.

Prior to 1939, the inorganic sulfur and copper compounds were used almost exclusively as spray and dust materials and the copper and organic mercury compounds as seed treatments. Sulfur had been used from before the time of the Romans in various plant prescriptions and had been used for powdery mildews since the beginning of the 19th century. Copper sprays were introduced in 1882 as Bordeaux mixture for control of downy mildew on grapes, and as a copper carbonate seed treatment in 1917. Lime-sulfur was developed as an apple spray in 1906. A new era in fungicides was initiated in the period 1934–1939 with the announcement of the dithiocarbamate and quinone fungicides which indicated the potentialities of organic compounds.

Mercurial fungicides were abandoned in 1971 because of environmental pollution and hazard of conversion into poisonous methyl mercury.

The intensive search for new organic fungicides has continued for over 30 years. The technique employed in most laboratories is to make an extensive survey of various structures by empirical methods to locate materials that suppress spore germination on glass slides or prevent mycelial growth on nutrient agar plates or rolled tubes. Those materials having an ED_{50} (effective dose for 50% inhibtion) in the order of 10 ppm are further tested in use applications.

The mechanism of action of the fungicides has been only partially solved. Sulfur deposits volatilize and within two to three minutes after coming into contact with a spore the sulfur begins to be released as hydrogen sulfide. It was formerly thought that hydrogen sulfide was the lethal agent, but it is now evident that it is the product rather than the cause of fungus destruction, probably because the hydrogen transport system of the sulfur-sensitive spore becomes overtaxed after about 10,000 ppm of sulfur has been reduced. In any event, sulfur appears unique in that its lethal effect does not depend upon its accumulation inside or on the spore

The organic fungicides have tremendous affinity for spores. For example, within 30 seconds after spores are placed in suspension containing 2 ppm of 2-heptadecyl-2-imidazoline they will remove up to 6,000 ppm of their body weight. There is reason to believe that this attribute may very well depend upon the compound's lipoid solubility induced by the 17-carbon chain in the 2-position. Homologs containing longer or shorter chains are less fungitoxic and fungitoxicity in this series is directly correlated with ability to fractionate into the lipoid phase of an aqueous-lipoid system.

Once the fungicides penetrate to the cell membrane or into the cytoplasm they may operate by devious means to disrupt vital functions. There is substantial evidence that the quinones immobilize the sulfhydryl and imino prosthetic group of enzymes. The 8-hydroxyquinoline and dithiocarbamate compounds are active against copper and other metallic members of an enzyme system, presumably by their ability to chelate metals. Heavy metals such as mercury affect certain enzymes such as amylases and may serve as general protein precipitants.

The best evidence available indicates that most, if not all, fungicides are not particularly specific in their action on vital cell systems and may react with nonvital molecules; thus a large percentage of them is detoxified before an essential biochemical process can be effected. The actual dosage required on a spore weight or mycelial basis for a lethal effect has been determined for very few compounds but most of those investigated by Miller and McCallan, who used radioactively labelled molecules, were not lethal until 5,000 to 20,000 ppm were accumulated by the spores.

There is appreciable specificity in the action of fungitoxicants on different types of fungi. Ferric dimethyldithiocarbamate, for example, is much more effective than sulfur against the cedar-apple rust fungus (*Gymnosporangium juniperi-virginian-*

nae) but has no particular advantage in control of apple scab (*Venturia inaqualis*). Ferbam is also superior to copper sprays against anthracnose of tomato (*Colletotrichum phomoides*) but is totally inadequate against *Alternaria* and *Phytophthora* leaf blights of this crop. Insofar as is known at present, specificity is only qualitative and not absolute so fungicides may be detected in a general group of candidate compounds with reasonable accuracy by measuring their effects on two or three indicator species of fungi.

George L. McNew

Cross-reference: *Pesticides*.

FURANS

The industrially important furfural, derived from annually renewable agricultural sources, is the best known member of the family of chemical compounds known as the furans. It is also the source of the other commercially available furan types, the family characteristic of which is the ring structure, generally represented as:

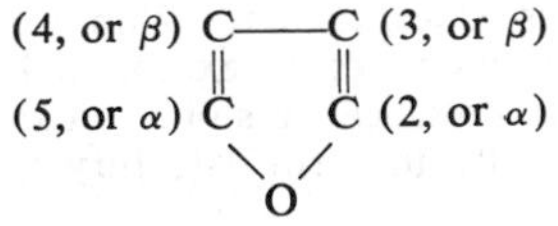

$$(4, \text{ or } \beta) \text{ C} \underline{\quad\quad} \text{C} (3, \text{ or } \beta)$$
$$(5, \text{ or } \alpha) \text{ C} \qquad \text{C} (2, \text{ or } \alpha)$$
$$\diagdown \text{ O} \diagup$$
$$(1)$$

From a chemical standpoint, the parent is furan (C_4H_4O), and the series develops by replacement of one or more hydrogens by other atoms or groups. Since the oxygen atom imposes a condition of asymmetry, two isomeric monosubstitution products can exist (e.g., 2- and 3-furancarboxylic acid); the number of isomeric polysubstitution products can be deduced readily from the structure shown.

In accordance with that structure, the furan nucleus is both a diene and a cyclic vinyl ether, but the activity of these functions is diminished as a consequence of resonance which, in furan itself, imparts a stabilization energy of 17–23 kcal/mole. Generally the furans are more reactive and less "aromatic" than their benzene analogs—they substitute more readily, exhibit a greater variety of addition reactions, and are more prone to cleavage leading to the formation of open-chain types. The common substitution reactions may be carried out successfully, orientation being almost exclusively alpha where such a position is available. Usually, however, it is necessary to select reagents and conditions carefully in order to minimize resinification and/or loss of identy of the furan ring.

That the furan ring lacks the integrity, as a unit, of the more aromatic ring systems is reflected in the fact that the extensive chemistry of the phenols and aromatic amines has no counterpart in the furan series. The simple furanols and furylamines appear to exist in the tautomeric dihydro configuration, and to exhibit a behavior more akin to the aliphatic keto-enol and ketimine-eneamine systems, respectively, than to their benzene analogs, phenol and aniline.

The literature is amply illustrated with 1,4-type addition reactions attesting to the butadiene structure in furans. Among others, Diels-Adler type syntheses involving the simple furans are com-

mon. However, this dienic activity of the ring is greatly diminished in those furans possessing such electron-withdrawing substituents as CO, COOR, NO_2, etc., and these furan types fail to react in the diene synthesis.

As vinyl ethers, furan and its homologs are susceptible to hydrolytic cleavage (generally accompanied by resinification) in acidic media. Furan, for example, may be split to succinaldehyde, although considerably more energetic conditions are required than in the related hydrolysis of ordinary vinyl ethers. Functional groups affect the ease with which hydrolysis occurs, and furfural, furoic acid, and particularly the nitrofurans are rather notably resistant to attack by acids. Conversely, however, the relatively electron-poor rings of these negatively substituted types are subject to attack by basic reagents.

Furfural. Furfural (2-furaldehyde, $2\text{-}C_4H_3O\cdot$ CHO) is a liquid aldehyde, colorless when pure, but subject to darkening in contact with air. It is manufactured on a large scale from agricultural materials, particularly corncobs, oat hulls, cottonseed hulls, and rice hulls. The production process, a single-stage acid digestion, involves hydrolysis of the pentosan component of these agricultural residues, and conversion of the resulting pentose sugars to furfural, which is removed continuously by steam distillation. The ease and economy of recovery of furfural with steam is a very important facet of its behavior, not only in connection with its manufacture, but also in many of its industrial applications. Furfural and water form a nonideal binary system which, at atmospheric pressure, gives rise to a minimum-boiling azeotrope (b.p. 97.9°C) containing 35% furfural by weight. At pressures up to 7–8 atm, the composition of the azeotrope is not materially altered.

Furfural is notably stable toward heat in the absence of catalytic materials and oxygen. Even at 230°C, several hours exposure is required to bring about detectable changes other than darkening. At 565°C decomposition gives rise to carbon monoxide, furan, and pyrolysis products of furan. With suitable catalysts, however, decarbonylation to furan occurs at considerably lower temperatures.

As an aldehyde, furfural takes part in the typical reactions of this group (Grignard reagent, bisulfite complex, acetal formation, condensation with suitably activated methylene groups, etc.). However, it may not always be substituted indiscriminately for another aldehyde, inasmuch as the reactive nucleus may introduce complexities which can affect the course of reaction. Thus, dimerization to furoin is complicated by simultaneous formation of a by-product of undetermined constitution. To a much lesser extent, the Cannizzaro reaction with furfural is also accompanied by competitive side-reactions although, in this instance, the difficulty can be eliminated by addition of a catalytic amount of silver. In alkaline media, oxidation of furfural can be restricted to the aldehyde group, giving rise to furoic acid in excellent yield. However, under acidic conditions, and on autoxidation in air, the ring is the major point of initial attack with degradation to formic acid and maleic acid or derivatives thereof.

The principal uses of furfural are (a) as a chemical intermediate in the manufacture of other furans, and tetrahydrofurans, and for phenolic resins; (b) as a selective solvent for the purification of petroleum oils, vegetable oils and wood rosin, and the extractive distillation of butadiene from other C_4 hydrocarbons; and (c) as a solvent and wetting agent in miscellaneous applications including the manufacture of resinoid-bonded abrasive wheels, and brake linings.

Furfuryl Alcohol. Until recently, furfuryl alcohol (2-furylcarbinol, 2-$C_4H_3O\cdot CH_2OH$) was manufactured by batch hydrogenation of furfural over copper chromite at 175–190°C, and 75–100 atmospheres. It is now produced by a continuous, low-pressure, vapor-phase hydrogenation process which permits better control and greater uniformity of product quality.

Furfuryl alcohol is a water-white liquid, soluble in water and in the common organic solvents except the paraffin hydrocarbons. It exhibits the typical behavior of a primary alcohol, and under carefully controlled conditions may be converted to the common functional derivatives (esters, ethers, etc.). In general, no difficulty is encountered in carrying out syntheses in neutral or basic media. Reactions in acidic medium, however, are accompanied, to a greater or lesser extent, by resinification, and it is difficult to prepare simple products in good yield under these conditions. Accordingly, esterification is best accomplished by the comparatively mild technique of ester-interchange, or by the Schotten-Baumann reaction.

In the presence of acid catalysts, furfuryl alcohol reacts with itself, exothermically, to form a dark-colored, thermosetting resin which cures to an insoluble, infusible state characterized by extreme resistance to attack by solvents, acids and alkalies. As such, the resin has found application in corrosion-resistant cements, asbestos-filled cast-molded items and other uses connected with the fabrication of a wide variety of chemical processing equipment. Furfuryl-alcohol-modified urea-formaldehyde resins are used extensively as foundry core binders.

Tetrahydrofurfuryl Alcohol. This saturated alcohol is currently manufactured by catalytic hydrogenation of furfuryl alcohol. The reaction is conducted in the vapor phase at low pressure over nickel; copper catalysts do not promote hydrogenation of the furan ring.

Tetrahydrofurfuryl alcohol is a colorless liquid of mild pleasant odor, soluble in water and in the common solvents. Unlike its unsaturated relative, it can be esterified directly by standard techniques. The chief commercial outlet is for the preparation of esters, as plasticizers for polyvinyl chloride films. The alcohol itself is a good solvent for cellulose esters and ethers, and vinyl polymers among others.

When heated to about 270°C, in contact with alumina or phosphated alumina, tetrahydrofurfuryl alcohol undergoes a rather remarkable dehydration with ring expansion to form 2,3-dihydropyran—an excellent starting material for a series of compounds (primarily 1,5-difunctional aliphatic types) of potential industrial interest.

Furan. Furan (C_4H_4O) is a colorless, highly flammable liquid of strong, ethereal odor. It is relatively insoluble in water (ca. 1% at 25°C), but miscible with the common solvents. It is produced commercially by passing furfural with steam over suitable catalysts (e.g. mixed chromite of zinc and manganese) at about 400°C. Though quite stable to heat, furan decomposes at about 700°C in a quartz tube; at 360°C, in contact with nickel. The pyrolysis products are carbon monoxide, gaseous hydrocarbons, and hydrogen. No bifuryl is formed, indicating that the C—H bonds are more resistant to rupture than the C—O linkages. With ammonia at 400–450°C, over alumina, furan is converted to pyrrole; thiophene is obtained when hydrogen sulfide is used in place of ammonia.

On exposure to air, furan forms an unstable peroxide, and therefore it should never be distilled without due precautions. Oxidation in the vapor phase with air over vanadium pentoxide gives rise to maleic anhydride. Carefully controlled alkoxylation, electrolytically or with chlorine or bromine, converts furan to acetals of malealdehyde.

Chlorination of furan may be conducted to give addition or substitution products. Acylation by the Friedel-Crafts reaction gives the appropriate 2-furyl ketones. With maleic anhydride, furan forms a normal Diels-Alder adduct. With weaker dienophiles no reaction occurs except under catalysis. Thus, in the presence of sulfur dioxide, acrolein adds abnormally to form 3-(2-furyl)propionaldehyde.

Tetrahydrofuran. The saturated, cyclic ether, tetrahydrofuran is a colorless liquid, miscible with most organic solvents, and with water. It is manufactured by catalytic hydrogenation of furan over nickel. Chlorination in the light at moderate temperatures gives primarily 2,3-dichlorotetrahydrofuran. With gaseous hydrogen chloride the ring is cleaved to form 4-chlorobutanol. Esters of the latter are obtained when the ring is cleaved with carboxylic acid chlorides. 1,4-Dichlorobutane is obtained by passing a mixture of tetrahydrofuran and hydrogen chloride through a reactor at 180°C and 15–20 atmospheres. With ammonia in the gas phase (400°C) over alumina, tetrahydrofuran is converted to pyrrolidine; with hydrogen sulfide, the product is tetrahydrothiophene.

The most active area of current interest is concerned with the cationic polymerization of tetrahydrofuran to polytetramethylene ether glycol,

$$HO-[(CH_2)_4-O]_n-H,$$

which is reacted with appropriate isocyanates for the production of polyurethane elastomers, fibers, and flexible foams.

Tetrahydrofuran forms an explosive peroxide in air, and for this reason the commercial product contains a stabilizer. Oxidation of tetrahydrofuran with air over a cobalt catalyst gives butyrolactone; with nitric acid, succinic acid may be obtained.

Other than its uses as a chemical intermediate, tetrahydrofuran finds application as a solvent for polyvinyl chloride and polyvinylidene chloride in the preparation of printing inks, adhesives, lacquers and other coatings. It is also an excellent medium for conducting Grignard reactions, and for carrying out reductions with metal hydrides.

A. P. DUNLOP AND F. N. PETERS

Cross-references: *Heterocyclic Compounds.*

Furane Resins

The furane resins occupy a unique position in the field of plastics in that they have been limited principally to industrial applications and derived from raw materials not used in the preparation of other plastics. The basic raw material is furfuraldehyde, extracted from vegetable product residues such as rice hulls, bagasse fibers, corn cobs, wood by-products, etc. Furfuraldehyde is hydrogenated to furfuryl alcohol, an unusual chemical intermediate which undergoes resinification under strong acid catalyst to yield a dark colored, highly cross-linked thermosetting polymer.

Resins from furfuryl alcohol are distinguished by their bonding qualities to mineral surfaces and their excellent chemical resistance. In consequence, one will find significant applications to diverse fields such as liners for chemical tanks exposed to both alkalies and acids; grouting compounds for quarry tile; utilization by foundries as binders for sand cores; gap filling adhesives for wood (in conjunction with urea formaldehyde); etc. Because the condensation reaction which ensues during the polymerization of furfuryl alcohol is highly exothermic, the reaction is not easy to control. The use of fillers is recommended to insure stability of the cured product by not only reducing shrinkage, but also providing viscosity control to an otherwise low viscosity product. It is also commercial practice to form more viscous prepolymers from furfuryl alcohol to aid application and processing of the furane resins.

Acidic catalysts for furfuryl alcohol include t-sulfonic acid, phosphoric acid, sulfuric acid, hydrochloric acid, maleic anhydride, and many others. Patent literature is replete with descriptions of many strong acids or acid-forming substances useful in the polymerization of furfuryl alcohol. These facts will influence the use of furfuryl alcohol resins because the presence of residual acid catalysts during cure may adversely affect some materials. One will find furane resins used in conjunction with mineral products, for example, rather than cellulosics.

Furane resins also include other condensation products, most notable of which are phenol-furfuraldehyde resins as well as furfuraldehyde-ketone resins. Furfuraldehyde imparts longer and easier flow to phenolic resins and may be used also as a partial substitute for formaldehyde in the manufacture of the resins. The furfural-ketones are an interesting group, though aside from their application in a few specialty coatings, they do not appear to have attained a prominent industrial position. Furfuryl alcohol, as a modifier of urea-formaldehyde reaction products has made valuable contributions to gap filling adhesives for bonding wood products. Small amounts of furfuryl alcohol have been reported as modifiers to epoxy resins, usually used in conjunction with curing agents for epoxies. Also, furfuryl alcohol is reported as a chain stopper in isocyanate terminated polymers used in urethanes.

Though not a resin-forming body in itself, tetrahydrofurfuryl alcohol should be noted for its significant use as a plasticizer, or esters thereof, as plasticizers for polyvinyl resins. Table 1 lists the basic physical properties of furfuraldehyde, furfuryl alcohol, tetrahydrofurfuryl alcohol, and tetrahydrofuran. The latter material is an outstanding solvent for many thermoplastics.

TABLE 1. PROPERTIES OF FURANE CHEMICALS

Name	Density g/cc	Boiling Point °C	Index of Refraction
Furfuraldhyde	1.16	161.7	1.5261
Furfuryl alcohol	1.13	171.0	1.4845
Tetrahydrofurfuryl alcohol	1.054	177.0	1.4517
Tetrahydrofuran	0.89	65.0	1.4076

Characteristics:

Color: Resins from furfuraldehyde and furfuryl alcohol are generally dark or black in color.

Odor: Pronounced, but not necessarily harmful.

Adhesion to: Minerals and many inorganic surfaces—good. Metal surfaces—poor, requires primer. Cellulosics—good, but embrittling tendency. Cured thermosets—usually good.

Physicals: Good when high-temperature (300°F) cure is used in addition to acid catalysts. Highly cross-linked.

Chemical Resistance: Excellent to many solvents, acids, and alkalies. Poor to oxidizing acids.

Electrical Properties: Poor, due primarily to ionic nature of catalysts.

Heat Resistance: Fair, up to 300°F service would be a conservative rating.

JOHN DELMONTE

FUSION, NUCLEAR

Nuclear fusion reaction modes are quite analogous to reaction modes in chemistry. Both chemical and nuclear combustion processes reveal cases of thermal (or thermonuclear) reactions, breeding reactions, propagation chain reaction cycles, chain branching reactions, and chain breaking reactions. The best example of the complex nuclear fire is the hydrogen or superbomb in which we follow Jetter as regards the basic processes for rapid, high-temperature burning of Li^6D. Jetter's paper sets the stage for an understanding of charged particle fusion chain reactions which may be of some interest in controlled fusion reactors. In general,* $n, p, d, t,$

* n = neutron, p = proton, d = deuteron, t = triton.

He^3 and He^4 are suprathermal chain centers or catalysts for the rapid combustion of otherwise difficult to burn nuclear fuels.

Three chemistry concepts are applicable to nuclear fusion systems: 1) the distinction between ignition and burning temperature, 2) the concept of a stable operating or burning temperature (negative temperature coefficient for the system), and 3) the importance of chain centers for self-multiplying properties and more efficient burning of nuclear fuel.

Thermal (Thermonuclear) Reactions. Bethe and Weizacker in the late 1930's proposed nuclear reactions which proceed at very low temperatures in massive stars like the sun. These reactions are known as the proton-proton cycle and the carbon cycle. In both cases four protons are fused to form a He^4 nucleus with the release of about 28 MeV (45 ×

10^{-18} joule) of energy per cycle. The reactions are thermal (i.e., thermonuclear), proceed at about 15,000,000°K and convert about 5 million tons of solar matter into energy every second, equivalent to about 10^{11} megatons of TNT exploding per second! Supernovae explosions involve more complex reactions at temperatures up to 10^{10}°K.

Thermal reactions were postulated for the superbomb by Hans Thirring in 1946 in terms of either $D \cdot D$ reactions or $Li^7 \cdot H$ reactions triggered by a plutonium or U^{235} fission detonator. The reactions suggested were

$$d + d \rightarrow n + He^3 + 3.3 \text{ MeV} \quad (1)$$

and

$$p + Li^7 \rightarrow 2\alpha + 17.3 \text{ MeV} \quad (2)$$

In reaction (1) the fast neutron (2.5 MeV) would produce suprathermal deuterons in close nuclear collisions and these would propagate the thermonuclear reaction. Reaction (2) would also propagate by thermallizing collisions between alpha particles and fuel nuclei. Both thermonuclear reactions can have a chain-like effect under proper circumstances. Thirring suggested initiation energies of 10^5 eV ($\sim 10^{9}$°K).

Fusion Chain Reactions Involving Neutrons and Tritons. In 1950 Ulrich Jetter proposed some dramatic ideas for the fuel and reaction mechanisms of the so-called hydrogen or superbomb. He selected Li^6D as nuclear fuel with Be^9 and possibly U as neutron multipliers. His reactions are the first to suggest a true nuclear fusion chain reaction which like the fission chain reaction has self-multiplying properties. The main difference is that fusion chain reactions are multistepped analogous to chemical chain reactions whereas the fission chain reaction gives multiplication of chain centers—the neutrons —in one step. Jetter took the neutron and triton to be the chain centers or catalysts and suggested several modes of reactions as discussed below.

Propagation Chain Reaction Cycle:

$$\begin{cases} n + Li^6 \rightarrow t + \alpha + 4.8 \text{ MeV} & (3) \\ d + t \rightarrow n + \alpha \; 17.6 \text{ MeV} & (4) \end{cases}$$

Here, the neutron and triton alternate as chain centers and so long as neither is lost from the system (or slows down too much in the case of the triton) the chain reaction will propagate through the fuel with a fusion energy release per cycle about 1/10 the fission energy release (~ 200 MeV). The net effect of the catalysts is to burn Li^6D to alpha particles indirectly but much faster than reaction (7) permits.

Thermal Reactions Giving Chain Branching:

$$d + d \rightarrow n + He^3 + 3.3 \text{ MeV} \quad (1)$$

$$d + d \rightarrow p + t + 4.0 \text{ MeV} \quad (5)$$

Deuterons are heated by thermalizing collisions with suprathermal fusion reaction products and then generate new n and t chain centers which can give an exponential growth to cycle I thus resulting in faster energy release. Three of these reactions (1), (4) and (5) have also been suggested for controlled fusion reactors.

Propagation Reaction with Energy Release:

$$t + t \rightarrow 2n + \alpha + 11.3 \text{ MeV} \quad (6)$$

① $t(d,n) \; \alpha + 17.6$ Mev
② $^3He(d,p) \; \alpha + 18.4$ Mev
③ $d(d,p) \; t + 4.0$ Mev
$+ d(d,n)^3He + 3.3$ Mev
④ $^6Li(p,^3He) \; \alpha + 4.0$ Mev
⑤ $^6Li(^3He, \alpha) \; p + \alpha + 16.9$ Mev
⑥ $^6Li(d,p) \; ^7Li + 5.0$ Mev
⑦ $^6Li(d,t) \; p + \alpha + 2.6$ Mev
⑧ $^6Li(d,\alpha) \; \alpha + 22.4$ Mev
⑨ $^6Li(d, d_1') \; d + \alpha - 1.5$ Mev
⑩ $^6Li(n,n') \; d + \alpha - 1.5$ Mev
⑪ $^6Li(p,p') \; d + \alpha - 1.5$ Mev
⑫ $^6Li(d,n) \; ^7Be + 3.4$ Mev
$^6Li(d,n)^3He + \alpha + 1.8$ Mev
⑬ $t(p,n)^3He - 0.8$ Mev
⑭ $^6Li(n,t) \; \alpha + 4.8$ Mev

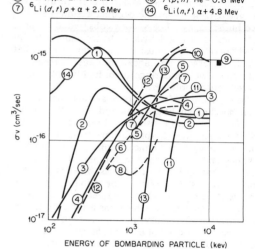

FIG. 1. Reaction rate parameter, σv vs. energy of bombarding particle for several light element reactions. Data only approximate in most cases.

Here, two tritons react to produce two alternate chain centers with a large energy release. The number of chain centers is essentially unchanged unless the α particle energy exceeds about 2.5 MeV. *Thermal (Thermonuclear) Reaction:*

$$d + Li^6 \rightarrow 2\alpha + 22.4 \text{ MeV} \quad (7)$$

Thermalizing collisions heat the Li^6D fuel to reaction energies and produce very energetic alphas which can heat more fuel material by nuclear elastic and inelastic collisions. This reaction has a small cross section but will contribute significantly to the energy release. *Multiplying or Chain Branching Reactions:*

$$n_{fast} + d \rightarrow 2n + p - 2.2 \text{ MeV} \quad (8)$$

$$n_{fast} + Be^9 \rightarrow 2n + 2\alpha - 1.7 \text{ MeV} \quad (9)$$

These endothermic reactions have the very important feature of doubling the neutrons with a consequent rapid increase in the reactivity. Jetter also pointed out that surplus neutrons from fusion would increase the efficiency of fissioning the trigger material of the superbomb thus releasing still more neutrons. He also suggested mixing fissionable nuclei such as U with the lithium deuteride to give additional neutron multiplication.

Charged Particle Fusion Chain Reactions. Various modes of charged particle nuclear reactions can also occur, some of which have self-multiplying properties. These electrically charged nuclei could hopefully be contained electromagnetically in a controlled thermonuclear reactor. Controlled fusion reactors might operate at temperatures up to 1

MeV (10^{10} °K) and fusion fuels include elements at least up through lithium ($Z = 3$) and possibly heavier elements especially if strong nuclear resonances could be found which would permit operating at lower temperatures.

The reaction usually considered for controlled fusion reactors is the $d + t$ reaction (4) because of its very large cross section at low energies (σv peaks at 60 keV due to a resonance in the compound nucleus He⁵*) and its very large energy release. It has several disadvantages: (1) tritium is highly radioactive, (2) tritium is initially very costly ($\sim 10^6$ \$/lb) so that tritium breeding must be eventually accomplished with the excess fusion neutrons in an external lithium blanket, (3) the reaction is purely thermal in nature, (4) only 20% of the reaction energy is available in the α particle for collisional heating of the confined plasma fuel, and (5) the 14 MeV neutron would produce serious radiation damage and pronounced radioactive afterheat problems in the containment structure.

The following sections discuss fusion fuels up to Li^6 only but the same types of reaction modes may be expected at least up through neon especially if one includes the neutron as a chain center. These latter reactions probably occur to some degree in supernovae explosions which have temperatures approaching 10^{10} °K. Only charged particle reactants are considered below since these can be electromagnetically confined in a potential fusion reactor.

Propagation Chain Reaction Cycles:

$$(a) \begin{cases} d + He^3 \rightarrow p + \alpha + 18.3 \text{ MeV} & (10) \\ p + Li^6 \rightarrow He^3 + \alpha + 4.0 \text{ MeV} & (11) \end{cases}$$

$$(b) \begin{cases} p + Li^6 \rightarrow He^3 + \alpha + 4.0 \text{ Mev} & (11) \\ He^3 + Li^6 \rightarrow p + 2\alpha + 16.8 \text{ MeV} & (12) \end{cases}$$

Propagation cycle (a) was first proposed by R. F. Post in 1961 and would use Li^6D fuel; cycle (b) would use Li^6 or Li^6H fuel and might lead to a "cleaner" reactor since fewer neutrons and radioactive products would be produced by D·D and other side reactions. Both cycles generate the costly ($\frac{1}{2} \times 10^6$\$/lb at present prices) He³ *in situ*, involve cheap, nonradioactive fuels, produce charged particle reaction products only, and are highly exothermic. In both (a) and (b) cycles the p and He³ alternate as chain centers or catalysts for the burning of Li^6 to alpha particles. The alphas produced in reaction (12) have energies up to 9 MeV and are active chain centers as in the next paragraph.

Multiplying Chain Reaction Cycle:

$$\begin{cases} \alpha_{fast} + Li^6 \rightarrow \alpha' + Li^{6*} - 2.2 \text{ MeV} \rightarrow \\ \qquad\qquad 2\alpha + d_{fast} - 1.5 \text{ MeV} & (13) \\ d_{fast} + Li^6 \rightarrow 2\alpha + 22.4 \text{ MeV} & (14) \end{cases}$$

This illustrates how an alpha induced disintegration reaction can produce a suprathermal deuteron by a nuclear inelastic collision. The fast deuteron can then give a doubling of the fast α particles (~ 11 MeV). Thus, even energetic alpha particles can be important as active chain centers as well as simply energy carriers in the reacting plasma.

Breeding Reactions (Thermal Reactions Giving Chain Branching):

$$d + d \rightarrow n + He^3 + 3.3 \text{ MeV} \qquad (1)$$
$$d + d \rightarrow p + t + 4.0 \text{ MeV} \qquad (5)$$
$$p + Li^6 \rightarrow He^3 + \alpha + 4.0 \text{ MeV} \qquad (11)$$
$$d + Li^6 \rightarrow p + t + \alpha + 2.6 \text{ MeV} \qquad (15)$$
$$d + Li^6 \rightarrow n + He^3 + \alpha + 1.8 \text{ MeV} \qquad (16)$$

Here, thermal reactions between cheap, nonradioactive fuel nuclei generate the costly t and He³ *in the plasma itself* and give a significant energy release with the production of only low energy neutrons. Reactions (5) and (15) also produce a suprathermal proton chain center. Successful breeding of the t and He³ would permit one to eliminate the lithium blanket in the reactor structure. Also, since there would be no need to conserve neutrons in the structure an optimized choice of structural materials could reduce the afterheat problems induced by the neutrons.

Thermal (Thermonuclear) Reactions:

$$d + t \rightarrow n + \alpha + 17.6 \text{ MeV} \qquad (4)$$
$$d + Li^6 \rightarrow n + Be^7 + 3.4 \text{ MeV} \qquad (17)$$
$$d + Li^6 \rightarrow p + Li^7 + 5.0 \text{ MeV} \qquad (18)$$

The first two reactions involve radioisotopes, t and Be⁷; however, both may be reacted efficiently in a high beta ($\beta = 8\pi nkT/B^2$), fusion reactor. Reaction (17) preferentially branches by $n + He^3 + \alpha$ production as in reaction (16). Reaction (18) produces an energetic (~ 4 MeV) proton chain center.

Chain Branching Reactions:

$$x_{fast} + Li^6 \rightarrow x' + d_{fast} + \alpha - 1.5 \text{ MeV} \quad (19)$$
$$\alpha_{fast} + Li^6 \rightarrow p + Be^9 - 2.1 \text{ MeV} \qquad (20)$$

Reaction (19) illustrates how nuclear inelastic collisions by fast particles can produce suprathermal deuteron chain centers analogous to Eq. (13); reaction (20) shows how fast alpha particles can be converted into more reactive proton chain centers with only slight energy loss.

Chain Breaking Reactions:

$$p + t \rightarrow He^3 + n - 0.8 \text{ MeV} \qquad (21)$$
$$x_{fast} + d \rightarrow x' + p + n - 2.2 \text{ MeV} \qquad (22)$$
$$p + Li^6 \rightarrow n + Be^6 - 5.2 \text{ MeV} \qquad (23)$$

In these endothermic reactions the neutron will escape from the plasma which reduces both the number of active chain centers and the heat content of the plasma. Thus, in reaction (21) two chain centers are reduced to one proton-rich chain center. In reaction (22) the proton is less active than the deuteron since there is only one exothermic reaction (11) possible for the proton compared to seven for the deuteron (1. 5, 14, 15, 16, 17, 18). In addition, the total reaction rate parameter, $\overline{\sigma v}$, for $d + Li^6$ and $d + d$ reactions is much greater than for $p + Li^6$ and the total energy release in charged particles is also greater. The endothermic nature of these reactions also leads to a cooling of the plasma which will hold down the operating temperature.

Summary. The nuclear burning of light elements is an extremely complex nuclear-chemical process.

The various chain reaction modes of burning suggest important fusion reactor criteria such as the all important negative temperature coefficient at the burning or operating temperature. Fuel and ash feed rates and losses, chain breaking reactions, endothermic reactions, $\overline{\sigma v}$ fall-off vs T, radiation cooling, and nuclear elastic and inelastic processes will all play important roles in determining the operating temperaure and the negative temperature properties of the system.

Whether fusion reactors will ever become feasible is still debatable. The first pilot plant experiment may involve a test of brute force techniques (θ-pinch or laser initiated microexplosions), empirical scaling laws (Tokamak), or exponential build-up (Mirrors) but hopefully will produce a plasma which exceeds the minimum Lawson number (density times confinement time $= 10^{14}$ sec/cm^3) with $T > 10$ keV ($10^{8\circ}$K). As in chemical and laser resonance processes the possibility of undiscovered low energy nuclear resonances lures one on to the prospect of even better fusion fuels.

J. RAND MCNALLY, JR.

References

See Bethe, H. A., *Physics Today,* **21,** 36 (1968).
Thirring, H., "Die Geschichte der Atombombe," Wien, 1946; *Bull. Atomic Scientists,* **6,** 69 (1950).
Jetter, U., *Physik. Blatt.,* **6,** 199 (1950); ORNL-tr-842.
McNally, J. Rand, Jr., *Nuclear Fusion,* **11,** 187, 189, 191, 554 (1971).

G

GALLIUM AND COMPOUNDS

Occurrence and Recovery. Gallium is a close chemical analog of aluminum and is widespread in the earth's crust, occurring in trace amounts as the hydrated oxide, $Ga_2O_3 \cdot H_2O$, in almost all aluminous minerals. There are also small amounts of gallium in some zinc blende ores, particularly in those from the tri-state area of Missouri, Oklahoma, and Kansas.

The metal is currently produced as a by-product of the aluminum and zinc industries. The trace of gallium (0.005 to 0.01%) that occurs in the aluminum ore, bauxite, passes through all the ore-refining and metal-extraction steps and appears in the final aluminum product, unless measures are taken to separate and recover it. Gallium values are recovered from the caustic liquor used to extract alumina from its ores in the Bayer refining process (See **Aluminum**). After the greater part of the aluminum is precipitated as $Al_2O_3 \cdot 3H_2O$, the gallium is concentrated and recovered from the aqueous caustic solution by electrolysis.

During the production of zinc, the small amount of gallium that occurs in the zinc sulfide ores follows the zinc through the air-roasting process that converts the sulfides to oxides, and into the zinc sulfate solution obtained by leaching the roasted ore with sulfuric acid. When this solution is carefully neutralized, the gallium accompanies the iron and aluminum impurities into the "iron mud" precipitate, from which it is separated, together with the aluminum, by leaching with caustic. This leach is treated with HCl, then the strongly acidic solution is extracted with an organic ether to separate gallium chloride. The gallium is recovered by electrolysis from a caustic electrolyte, and purified by recrystallization.

Physical Properties. Some of the physical properties of gallium are given below:

Symbol	Ga
Atomic weight	69.72 (60.2% Ga-69; 39.8% Ga-71)
Crystal structure	pseudotetragonal
Specific gravity,	
29.6° (solid)	5.904
29.8° (liquid)	6.095
Melting point	29.78°C
Boiling point	2403°C

As a solid, gallium strongly resembles zinc, showing a bluish luster and semibrittle fracture. The liquid metal, when free of oxide, is almost indistinguishable in appearance from mercury. However, the surface film that forms immediately on contact with air prevents it from coalescing. This rapid oxidation limits the number of applications that could be envisioned for a metal that is liquid near room temperature.

Gallium has a very open crystal structure and the metal expands on freezing. Bismuth is the only other element that exhibits this property, although it is familiar, of course, in the case of water. The unusual crystal structure of gallium also accounts for the marked anisotropy in some of its properties. For example, the difference in both electrical and thermal conductivity between two particular crystallographic directions in the solid is greater than for any other metal.

Gallium supercools greatly and can be maintained in the liquid state indefinitely at temperatures considerably below its normal freezing point. Although many substances can be supercooled to some extent, this property is generally associated with extreme purity. In the case of gallium, however, it appears to be due to the unusual crystal structure of the solid, which is not compatible with the various surfaces that ordinarily would be effective in nucleating solidification of other materials. Supercooled gallium begins to freeze immediately when brought into contact with a "seed" crystal of solid gallium, and solidification is generally initiated in this manner.

Chemical Properties. Gallium is in Group IIIA of the Periodic Table. It is normally trivalent, and forms the expected stable compounds with the halides and chalcogenides, and with phosphorus, arsenic, and antimony in Group VA. A nitride, GaN, can be prepared, but it is unstable at high temperatures. The gallium halides can be prepared by direct reaction of the elements but, because of the vigor of these reactions, preparation from the corresponding hydrogen halides is preferred. Gallium fluoride can be prepared by evaporation and dehydration of aqueous gallium fluoride solutions. The other halides require nonaqueous syntheses.

Gallium phosphide, arsenide, and antimonide are very interesting semiconductor compounds whose method of preparation is dictated more by requirements of purity and crystal perfection that are peculiar to semiconductor technology than by simplicity or economy. These materials will be discussed in the next section.

Gallium forms a stable oxide and all the other chalcogenides and halides, with the exception of the fluoride, hydrolyze to some degree to form a hydrated gallium oxide, $Ga_2O_3 \cdot xH_2O$. Gallium sesquioxide, Ga_2O_3, is the most common gallium compound. It precipitates in a hydrated form during neutralization of either acidic or basic solutions. Calcination of gallia hydrate at a high temperature yields beta-Ga_2O_3.

An interesting property of gallium is the ease

with which it forms the relatively volatile suboxide, Ga_2O, by reduction of the trivalent oxide by any of a variety of reducing agents. Gallium becomes clean in appearance and coalesces readily when heated above red heat in vacuum because of the reduction of the oxide skin to volatile suboxide by the underlying metal. Gallium also forms lower-valent halides. The monohalides, such as GaCl, can exist in the vapor state, but there is considerable doubt that they can exist as condensed phases. The dichloride, $GaCl_2$, is a stable liquid or solid (m.p. 170°C) in the absence of oxidizing agents.

Most of the other chemical properties of gallium are very similar to those of aluminum. Important exceptions are the absence of a gallium carbide and the stability of the complex chloride ion, $GaCl_4^-$. The latter affords a high solubility of gallium in organic ethers and permits almost quantitative separation from aluminum in acid solutions (after three successive extractions with equal volumes of isopropyl ether, only one part in a billion of the gallium remains in the aqueous layer).

Uses of Gallium. Gallium is relatively inert, nontoxic, and with a melting point only slightly above room temperature and a boiling point of over 2000 degrees it has one of the widest liquid ranges of any metal. It is not surprising, therefore, that the first uses that were envisioned for this unusual element exploited these properties. There were half-serious attempts to use it as the thermometric fluid in high-temperature thermometers of otherwise conventional design. It was considered as a liquid sealant in high-vacuum systems, as a heat-transfer medium in high-temperature engines, such as nuclear reactors, and as a component of dental alloys. Either because of cost considerations or unrealistic demands on the material, none of these applications became commercial.

It is somewhat paradoxical that the rather substantial uses of gallium that eventually evolved were based not on the unique physical properties of the metal, but on the specific chemistry of some of its compounds. The first significant use of gallium, and in fact the application that justified the first commercial production, was in the spectroscopic analysis of uranium oxide in conjunction with operations of the Atomic Energy Commission. Gallium oxide is added together with graphite to the powdered sample and, in the high temperature of the arc, the vigorous evolution of the volatile gallium suboxide formed carries impurities into the vapor and greatly enhances the sensitivity of their detection.

The application of gallium that has received the most attention, and possibly promises the largest market, is in the production of semiconductor compounds. For many years this technology was dominated by the elemental semiconductors, silicon and germanium. In 1952, German workers reported the achievement of semiconduction in compounds between elements in Group III and Group V of the Periodic Table, bordering the Group IV semiconductors. Of these, compounds of gallium with antimony, arsenic or phosphorus are the most important. The characteristic semiconductor behavior exhibited by gallium antimonide and gallium arsenide is being utilized in various electronic devices to perform functions that were previously in the domain of silicon, germanium, or conventional vacuum tubes. Among these are voltage rectification and amplification, and magnetic field and temperature sensing. New phenomena are observed in some of the gallium compound semiconductors that had no counterpart in the elemental semiconductors or conventional devices. Among these are semiconductor "lasing" and microwave generation in gallium arsenide and electroluminescent light emission in gallium arsenide (infrared) and gallium phosphide (visible).

Recent reports indicate that gallium compounds such as magnesium gallate, $MgGa_2O_4$, containing divalent impurities such as Mn^{+2}, will find a place among the growing list of commercial ultraviolet-activated powder phosphors.

For some years to come the uses of gallium will probably be restricted to these and similar applications where the importance of some specific phenomenon justifies the relatively high cost of the material.

L. M. FOSTER

GALVANIZING

The coating of ferrous materials by dipping them in molten zinc is known as *galvanizing* (after Galvani, an Italian physiologist). Products to be coated are commonly pickled in hydrochloric or sulfuric acid to remove scale and other surface contamination and are then introduced into the molten zinc through a layer of flux which floats on the surface of the zinc. The flux, usually a mixture of ammonium chloride and zinc chloride, removes the last traces of oxide and activates the metal surface. At the temperature of molten zinc, iron and zinc readily combine to give one or more of the many intermetallic compounds which these elements are capable of forming. When the work is removed from the bath, relatively pure zinc adheres to the outer surface and freezes quickly before it has a chance to react with the iron, from which it is separated by the compounds already formed. It is desirable to conduct the process so as to obtain a relatively thin layer of brittle intermetallic compound and a thick layer of the somewhat more ductile pure zinc.

The temperature of the molten zinc and the time of immersion are the variables which have the most effect on the nature and thickness of the coating. High temperatures increase the fluidity of the zinc and decrease the thickness of the pure zinc coating which will adhere to the articles as they are withdrawn from the bath. High temperatures and long immersion times increase the thickness of the layers of intermetallic compounds which form between the zinc and iron. It is desirable in general, therefore, to operate with a zinc bath whose temperature is only a little above the melting point and to use immersion times which are as short as possible. Commonly this would lead to a greater thickness of zinc than is necessary and it is common practice to remove excess zinc from the surface of the workpieces by wiping, squeegeeing or other similar process. In the case of small articles which are handled in a basket, centrifuging may be employed to remove excess zinc.

The attractive appearance of galvanized articles is obtained by controlling the size of the "spangle" formed by the crystallization pattern of the freezing zinc. A spangled pattern may be controlled by providing a special chemical atmosphere or by mechanical means. In some cases additions of metals such as tin or aluminum to the zinc are also employed.

Continuous coating of strip has almost entirely replaced hand dipping of single sheets, and in 1970 at least 90 per cent of the galvanizing capacity of the U.S.A. and Canada was represented by continuous lines. Not only are labor costs reduced, but a better product is obtained. It is possible to control the aluminum and lead contents of the pots as well as the operating variables so as to produce a very thin intermetallic compound layer; hence, the formability of the resulting sheet is superior to hand-dipped sheets. Nothing fundamentally new is involved in the continuous processes, and the differences from one plant to another are mostly engineering details. Perhaps the most significant difference centers about the choice of annealing the strip as part of the coating sequence or as a separate operation. The former appears to be the most popular procedure, probably because handling and coiling costs are reduced.

During the 1960's most galvanizers adopted some form of air or vapor knife for controlling the thickness of the coating. In this technique air or steam is blown from a narrow slot perpendicular to the moving sheet as it emerges from the galvanizing pot. Just the right amount of molten zinc is blown back into the pot to give the desired thickness. Not only is it possible to achieve more uniform thickness than by squeegeeing or wiping, but it is quite practical to produce a differential coating, that is, one with different thicknesses on the two sides of the sheet. Continuously reading thickness gages that move back and forth across the moving sheet are now in common use.

The corrosion-resistant properties of galvanized products are quite satisfactory in most outdoor environments, but the duration of the protection afforded is almost directly proportional to the weight of zinc per unit area. It is not important whether the zinc is free or combined with iron, although in the latter case the appearance of the product will be inferior. The rate of corrosion is much affected by the nature of the environment in which the products are used. A coating which will last for several decades in a dry rural climate may fail in less than five years in a wet industrial atmosphere. The thickness of the coating specified must be tailored, therefore, to the prospective application. The following table, taken from the work of Hubbell and Finkeldey, will summarize the relationship between thickness of coating:

Weight of Zinc on Steel Sheet in oz/sq ft	Expected Life in Years			
	Rural	Sea Coast	Semi Industrial	Heavy Industrial
1.25	30–35	20–25	15–20	8–10
1.00	20–25	15	10–15	6–8
0.60	8–10	5–8	5	3–5
0.2–0.3	3–4	2–3	1–2	1

HAROLD J. READ

Cross-references: *Protective Coatings, Zinc.*

GAS ANALYSIS

Though gas analysis is correctly defined as the detection and determination of constituents of gaseous mixtures, the scope customarily is broadened to include the determination of any desired property of a gas or gas mixture and also the detection and determination of liquid and solid constituents carried by a gas. In application, major interest is centered in three fields of technology: (1) combustion control and economy, (2) chemical control and investigation, (3) Biological and health control and investigation.

Gas analysis methods may be classified in several ways. A realistic and practical classification is according to purpose of the analysis:

A. Complete analysis of gaseous mixtures:
  (a) Standard chemical absorption and combustion methods
  (b) Physical measurements
B. Detection and estimation of individual components:
  (a) Chemical methods, volumetric, gravimetric and colorimetric
  (b) Physical measurements
  (c) Direct gravimetric
C. Measurement of properties of gas mixtures:
  (a) Density
  (b) Heating Value

No one method can be considered adequate for accurate, complete determination of all constituents in all gaseous mixtures. Despite the outstanding advances in modern physical methods, coal, gas, water gas and certain combustion gases are analyzed more satisfactorily by standard absorption and combustion methods. For natural gas and hydrocarbon analysis of refinery gases only certain physical methods, such as fractional adsorption, fractional distillation or spectrophotometry, are applicable.

Absorption and Combustion Methods. Essentially the equipment consists of a gas-measuring burette and a series of vessels or "pipettes" containing reagents specific to individual components. For gases requiring combustion technics, either pipettes incorporating an ignition mechanism or vessels containing a combustion catalyst are employed. The various types of equipment are (1) Hempel type; (2) Bunte type; and (3) Orsat type. In the latter, the burette is connected permanently through a manifold to a series of absorption pipettes containing appropriate reagents. The Barnhart arrangement employ a circular manifold with the pipettes grouped in a semicircle about the burette. All other Orsat forms employ a straight manifold, the number of outlets depending upon the number of pipettes employed, from three in the standard flue gas Orsat (CO_2, O_2, CO) to perhaps eight in the U. S. Steel modification. Although there are several classical modifications, the modern trend is to "build up" models in which number and arrangement of pipettes can be altered to suit the problem. Combustion equipment includes copper oxide for H_2, and CO and slow combustion, explosion or catalytic oxidation for general work.

Absorption Reagents. Properties desirable in a reagent are: (1) complete absorption, (2) high selectivity, (3) high capacity at high speed, (4) chemical and physical stability, (5) readily available or easily prepared. No reagent satisfies all these conditions. One shortcoming common to all is that solubility of gases, according to Henry's Law, can be minimized by preconditioning a reagent with the expected concentrations of the gas mixture. The usual constituents determined by absorption in the several standard apparatus types are CO_2, unsaturated hydrocarbons, O_2 and CO. CO_2 is universally determined by a potassium or NaOH solution of about 30%. The potassium solution is preferred. All other acidic gases are also soluble and must be separately determined. Unsaturated hydrocarbons may be absorbed in bromine or bromine water, or fuming or catalyzed H_2SO_4. Bromine cannot be used in contact with mercury. Catalyzed H_2SO_4 has less tendency to dissolve saturated hydrocarbons than fuming, and evolves no vapors. Oxygen is most frequently absorbed by alkaline potassium pyrogallate. Less well known but equally eeffctive are alkaline sodium hydrosulfite-beta anthraquinone sulfonate and acidulated chromous chloride. Yellow phosphorus also is employed. CO is usually absorbed by acid or ammoniacal cuprous chloride; a complex is formed which can evolve CO, necessitating fresh solution for complete removal. Cuprous sulfate-beta naphthol, although of lower capacity, forms a stable compound. Hydrogen is normally determined by combustion but occasionally absorption is desirable. Palladium chloride or organic oxidizing agents catalyzed by colloidal palladium are effective but rather easily poisoned.

Combustion Methods. Hydrogen and CO are simultaneously oxidized to water and CO_2 by slow passage through copper oxide at 300°C. Fast rates with high concentration may cause oxidation of hydrocarbons, if present. Hydrogen may be oxidized by pallidized asbestos at 100°C if accompanied by sufficient oxygen. CO and certain organic vapors are powerful poisons. Hydrocarbons, as well as H_2 and CO, may be oxidized over a platinum catalyst at 500°C or burned ("slow combustion") over a red-hot platinum spiral. Combustion by explosion, although still occasionally used, is the least satisfactory method.

Physical Measurement Methods. These methods depend upon property measurements and may be classified according to whether the property is specific or nonspecific for a particular constituent. The mass spectrophotometer is the only truly specific instrument; however, the infrared and ultraviolet spectrophotometer, fractional adsorption or desorption and fractional distillation methods are somewhat specific or may be quite specific, depending upon the constituents present. The nonspecific methods include thermal conductivity, velocity of sound, refractive index, density, and alpha particle ionization.

Specificity and precision may be increased by combining physical methods. Thus, gas chromatography, very popular at present, combines fractional desorption with detection by another physical method, usually thermal conductivity. A small gas sample is injected into a carrier gas, helium, argon, or hydrogen, from which it is adsorbed in a column of silica gel or diatomaceous earth, frequently treated with conditioning agents. It is fractionally desorbed quite quantitatively by the carrier gas, detection and estimation being made by thermal conductivity. An important limitation of all methods except possibly mass spectrometry is that qualitative identification of all constituents is essential for instrument calibration. (See **Mass Spectrometry.**)

Thermal conductivity is the most simple, inexpensive of the high precision methods. Although nominally restricted to binary mixtures of different conductivities, in combination with absorption or combustion its use may be extended. Usually, cells consisting of temperature sensitive resistances are incorporated in a Wheatstone bridge circuit and the difference between the conductivity of the sample and a reference gas is measured by the unbalance of the bridge. Methods involving velocity of sound, refractive index and density have rapid response but are also restricted to binary mixtures. Sometimes the second component may be a mixture of gases of similar properties. Thus density is frequently employed as a measure of CO_2 in flue gas.

Alpha particle ionization is a new interesting method similar to a Geiger counter except that the ionization is measured for a variable gas with fixed radiation source instead of a variable source with fixed gas.

Detection and Estimation of Individual Components. Usually this field is restricted to the determination of small concentrations of a substance because of its influence upon safety, health or the mechanism of a chemical process. With appropriate effort, specific methods unquestionably could be developed for any concentration of any constituent in any mixture.

Hydrogen sulfide is important because of its poisoning effect both human and catalytic, also its corrosive properties. The usual methods all involve chemical reactions and employ titration procedures (Tutwiler iodimetric chloride) or colorimetric (lead acetate or iodine).

Oxygen in small concentrations can be determined by several chemical methods and certain physical methods. The Shaw method treats the gas with fresh ferrous hydroxide, determining ferric ion colorimetrically. Oxidation of manganous hydroxide may be indicated with iodimetric titration or colorimetrically. The Texas Company method employs a copper-ammoniacal ammonium chloride reagent with subsequent iodimetric titration. An interesting physical method depends upon the relatively high magnetic permeability of oxygen, perhaps 200 times that of nitrogen or hydrogen.

Other gases commonly determined by chemical methods include carbon disulfide, nitric oxide, hydrogen cyanide, and water vapor. Methods for dusts, mists and condensible vapors have also been developed.

Measurement of Properties of Gas Mixtures. Aside from use in analysis, certain properties of a mixture are very important in themselves. Thus in calculations involving combustion, heat transfer and flow characteristics, the required values of such properties as heat of combustion, specific heat, density and viscosity are those of the gross mixture rather than individual constituents. However, only two of these, density and heat of combustion, are normally considered within the scope of gas analysis. Viscosities and specific heats of mixtures are usually calculated by the laws of mixtures from data for the individual components determined in basic physical investigations.

Density. Three principles are usually employed: gravimetric, effusion and kinetic energy. All methods except direct weighing determine specific gravity rather than density. Gravimetric methods may be at variable pressure or atmospheric pressure. The former employs the measurement of gas pressure necessary to balance a beam, the latter measures the displacement of the center of gravity of a beam.

The effusion type is typified by the Schilling apparatus. A fixed volume of gas is passed thru an orifice and the time measured. Specific gravity is the ratio of square of gas time to air time. Because of inaccuracies this method is limited to rough, routine tests.

An example of the kinetic energy type is the "Ranarex" differential torque apparatus. Impellers drive a gas stream and an air stream against vanes which are linked together. A displacement of the linkage results from a difference in kinetic energy of the streams. This apparatus is quite popular industrially.

Heating Value. Gas calorimeters may be of two types: constant volume or flow. The constant volume type are usually quite inaccurate because of the small amount of energy involved. Flow types may be intermittent (Junker, Sargent, etc.) or continuous (Thomas). In either case the heat of combustion of a known amount of gas is transferred to a known amount of heat absorbing medium and the temperature rise of this medium is measured. The Thomas meter is recording and is automatic but is quite expensive.

DONALD T. BONNEY

Cross-references: *Analytical Chemistry, Carbon Monoxide, Noxious Gases.*

GASES, HOMONUCLEAR

Only a limited number of elements are normally gaseous at room temperature. These include the "inert gases," which are atomic, and H_2, N_2, O_2, F_2 and Cl_2. The heavier halogens, bromine and iodine, are sufficiently volatile that their diatomic molecules are readily obtained at room temperature. However, among the other groups of elements in the periodic table only mercury is sufficiently volatile at room temperature (vapor pressure is $10^{-5.6}$ atm.) that its gas can be readily observed and even in this case the vapor pressure is quite small. Next come the heavier alkali metals and sulfur, but these vapors have equilibrium vapor pressures at room temperature in the range $10^{-8.2}$ to $10^{-8.9}$ atm and could be detected only with extremely sensitive instrumentation. The gases of most elements are, therefore, for all practical purposes nonexistent at room temperature.

At elevated temperatures these non-volatile elements do achieve significant vapor pressures and the question of whether gaseous molecules or simply atomic species are formed is a very interesting area of chemistry. Perhaps of equal importance is the question whether the gaseous homonuclear molecule, if it is formed at all, is simply diatomic or a more complex polyatomic molecule. The latter are considerably less common than the diatomic

molecules and have been found for elements comprising only three groups of the periodic table.

Main Series Elements. The alkali metals of Group IA all form stable homonuclear diatomic molecules at elevated temperatures. However the bond strengths, as exemplified by the bond dissociation energies at $0°K$, $D_0°$, are very weak, falling from 25.8 kcal/mole for Li_2 to only 10.4 for Cs_2. The latter is near the borderline for true chemical bonding. It is interesting that these bonds are weaker than any of those found for the permanent diatomic gases of the elements, but this is by no means true of all gases found only at high temperatures; there are a number of such gases that are indeed very strongly bonded, as will be described below.

The Group IIA alkaline earth elements have the helium-like valence-shell electronic configuration ns^2 in the ground state and thus form even weaker homonuclear diatomic molecules than do the IA elements. In fact the binding energies are so small that the bonds would have to be of the secondary or Van der Waals type. These elements form essentially atomic vapors although there has been one reported observation of the Mg_2 molecules in a nonequilibrium experiment.

The IIIA elements form homonuclear diatomic molecules with $D_0°$ values falling from ~66 kcal/mole for B_2 to ~14 for Tl_2. Because of their relatively high boiling points, as compared to those of the IA elements, and their low to moderate bond dissociation energies, these IIIA elements must form homonuclear diatomic molecules in detectible concentrations at quite high temperatures in saturated vapors.

With the IVA elements we come to more complex molecules. An entire spectrum of C_n molecules ranging from C_2 to C_7 has been found by optical and mass spectrometry, and even higher polymers have been predicted to be stable at very high temperatures. These C_n molecules are believed to be linear and bonded by double bonds. The odd-numbered molecules are more strongly bonded than the even-numbered ones. With the exception of lead, the other IVA elements—Si, Ge, Sn—have also been found to form a similar sequence of polymeric molecules.

The VA elements heavier than nitrogen also form polyatomic homonuclear molecules. However these do not constitute an almost continuous series of linear polymers, but rather only tetrahedral molecules of the type P_4. It is interesting that phosphorus forms molecules with 60° bond angles only in this instance, this being a rather strained configuration. For phosphorus, arsenic, and antimony, both tetrahedral and diatomic molecules are formed at elevated temperatures, the tetrahedral molecules predominating in the saturated vapors of the elements at lower temperatures and the diatomics at temperatures considerably higher. Bismuth apparently forms only Bi_2. It should be noted that the tetrahedral molecules are bonded only by single bonds while the diatomic molecules are bonded by triple bonds. The $D_0°$ values for the diatomic molecules of this series drop from 116 kcal/mole for P_2 to 47 for Bi_2.

The fact that at moderately high temperatures P_4, As_4, and Sb_4 are virtually the only species formed in the saturated vapors of these elements, over a wide temperature range, and that there is negligible dissociation to the atoms, differentiates VA molecules from the groups described previously, since in the latter the degree of dissociation generally tends to fall with increasing temperature in the saturated vapors of the elements. This difference stems from the dependence of dissociation on both the bond strengths of the homonuclear molecules and the vapor pressures of the respective elements. A high but steady mole fraction of molecules in the saturated vapor with increasing temperature is generally the result of strong bonding in the molecule coupled with a moderately low boiling point for the corresponding condensed element.

Stable polymeric molecules are also formed at elevated temperatures in the saturated vapors of several of the VIA elements. Sulfur and selenium form complex vapors of ring-shaped molecules with an entire spectrum of ring sizes, although the heavier VIA element tellurium forms only Te_2. In the sulfur vapor, gaseous species ranging from S_2 to S_8 have been found mass spectrometrically, with traces of S_9 and S_{10}. However the relative abundances of these species vary markedly with temperature, with S_8 predominating in the saturated vapor at moderately elevated temperatures. At very high temperatures the diatomic S_2 molecule is the stable molecular form of sulfur. Selenium differs from sulfur mainly in that Se_6 rather than Se_8 is the predominant form at moderately elevated temperatures. The $D_0°$ values fall from 101 kcal/mole for S_2 to 52.5 for Te_2.

Transition Metals. The IB elements Cu, Ag and Au form homonuclear diatomic molecules in the saturated vapors of these metals at high temperatures. These IB molecules are more strongly bonded than are any of the IA molecules. The difference is attributable to the d-hybridization energy possible with these transition elements. The homonuclear diatomic molecules of Zn, Cd and Hg have been observed by optical spectroscopy although not by mass spectrometry. Their $D_0°$ values are very low, ranging downward from 6 to 1.4 kcal/mole, as is predictable from the inert gas structure of the valence-shell electrons of the ground state atoms.

There are no known diatomic molecules for the transition metals of groups III through VII. Among the group VIII metals, only Ni_2 and Co_2 have thus far been observed mass spectrometrically.

Inert Gas Molecules. Although the inert gases are normally atomic and normal molecules of the type He_2 possess negligible binding energy, metastable molecules of this type have been observed spectroscopically. The latter are formed by excited state reactions via electrical excitation, but can be presumed to have very short life-times despite their appreciable $D_0°$ values.

BERNARD SIEGEL

Cross-references: *Bonding, Chemical, Heterogeneous Equilibria.*

GAS LAWS

The classical laws of gas behavior stand as a tribute to those figures prominent in the early history of chemistry. The kinetic theory of gases with its rational explanation of the relationships existing between gas volumes, pressures and temperatures did not appear until the middle of the nineteenth century. However, the laws governing these relationships were well charted by that period and, in some cases, had been in general use for well over a century. To a great degree, the development of the gas laws parallels the development of the science of chemistry itself.

As early as 1662, Robert Boyle described in detail the experiments which led to the formulation of the law which bears his name. Boyle's Law simply states that, *at constant temperature, the volume of gas is inversely proportional to its pressure.* Symbolically, it can be represented as: $PV =$ constant, or, in a more useful form,

$$V_1 = V_0 P_0 / P_1 \quad \text{(temperature constant)},$$

where

P_0 and P_1 = initial and final gas pressures, respectively.

V_0 and V_1 = corresponding initial and final gas volumes.

This relationship was independently discovered by Mariotte and, particularly in Europe, is sometimes referred to as Mariotte's Law.

The next major milestone in the history of the gas laws awaited the development of precise thermometry. In 1787 the French physicist, Jacque Charles reported the results of a series of experiments which led him to believe that, *at constant volume, the pressure exerted by a confined gas is proportional to its absolute temperature.* This is known as Charles' Law. In 1802, Gay-Lussac, using a somewhat different experimental approach, restated Charles' Law in the form more commonly used today, namely, that *at constant pressure, the volume of a confined gas is proportional to its absolute temperature.* Thus,

$$p_1 = p_0(1 + \alpha t) \quad \text{(volume constant)}$$

or

$$v_1 = v_0(1 + \beta t) \quad \text{(pressure constant)}$$

where

p_0 and v_0 = gas pressure and gas volume at 0°C.

p_1 and v_1 = gas pressure and gas volume at t°C.

α and β = temperature coefficients of pressure increase and of expansion respectively.

At low pressures, $\alpha = \beta = 1/273.1$ and Charles' Law can be expressed in terms of absolute temperature. Algebraic combination with Boyle's Law yields the following expression:

$$PV = kT$$

where,

P = pressure of confined gas

V = volume of confined gas

T = absolute temperature (K°)

k = a simple proportionality constant, the numerical value of which will depend on the units of pressure and volume used.

Although this expression adds no new facts, it is simpler and much easier to apply.

The enunciation of the atomic theory by Dalton in 1803 and the hypothesis put forth by Avogadro in 1811, stating that *equal volumes of gases, elementary or compound, when under the same conditions of temperature and pressure, contain the same number of molecules*, pointed the way toward universal relationships between gas masses and gas volumes. If Avogadro's hypothesis is correct, it follows that a gram-molecular-weight of any gas at a given temperature and pressure occupies the same volume as a gram-molecular weight of any other gas under identical conditions. The constant k in the equation $PV = kT$ can thus be evaluated experimentally for one mole of gas, leading to the Ideal Gas Law:

$$PV = g/M\ RT,$$

where

P, V and T = pressure, volume, and temperature (K°) of confined gas

g = weight of confined gas

M = molecular weight of confined gas

R = molar gas constant

It is usually convenient to express R either in liter-atmospheres/degree ($R = 0.08206$) or cc-atmospheres/degree ($R = 82.06$).

A gas which shows pressure-volume relationships in exact accordance with the Ideal Gas Law is called a "perfect" gas. Actually, careful experimentation, notably by Andrews and Amagat, soon revealed that no gas could be termed "perfect" and that observed deviations from this "Law" were in many cases quite significant. Beginning in 1869, Amagat reported a series of investigations on the compressibility behavior of nitrogen, hydrogen, methane and ethylene, and showed that each, particularly at elevated pressures, deviated markedly from its predicted behavior. Amagat deduced correctly that the deviations were due, at least in part, to the actual volume of the molecules themselves.

It remained, however, for van der Waal, then a student at the University of Leyden, to combine Amagat's findings with Maxwell's recently developed kinetic theory and to present in 1873 the following relationship:

$$(P + a/V^2)(V - b) = RT$$

In the van der Waal equation, a and b are constants having different values for different gases. The term a arises from the fact that the kinetic energy of molecules at the edges of a confined gas (e.g., at a manometer surface) is somewhat less than the average kinetic energy of molecules in the main body of the gas; the b term is a correction for the space occupied by the molecules themselves. It should be noted that these constants are usually reported for one *mole* of gas and that the van der Waal equation cannot be used directly for calculating the pressure-volume relationships of intermediate amounts.

The van der Waal equation is sufficiently accurate for exact experimentation at low pressures. However, the complexities of the intermolecular forces reflected in the a term usually make this equation inadequate for pressures above 10 atmospheres. Under these conditions, one must make use of empirical equations containing a number of constants which differ from gas to gas.

G. L. BARTHAUER

Cross-reference: *Molecules.*

GASOLINE

Gasoline is a mixture of hydrocarbons, boiling in the approximate range of 40–225°C (104–437°F), used as a fuel for internal combustion engines. Nearly all gasoline manufactured today is produced from petroleum and is used in automobile, aircraft, and marine engines, and small engines for miscellaneous applications. The composition and characteristics of gasoline vary widely with the source, manufacturing method and end use of the product. Meeting fuel specification requirements of different types of engines requires extensive processing, often involving steps of chemical synthesis. Total gasoline production in the United States in 1971 was about 92 billion gallons. This constituted about 52% of the crude oil charged to U. S. refineries during the same period.

Originally gasoline was produced by simple distillation of crude oil. The types of hydrocarbons found in unrefined petroleum and in "straight-run" gasoline (obtained by fractionation of crude oil) are paraffins, aromatics and naphthenes. The molecular size of the hydrocarbons in the gasoline boiling range is C_4 to C_{12}. More than 660 paraffins alone can exist in this range. Not all are found in each gasoline and not all hydrocarbons found in gasolines have been identified. Most of the paraffins in straight-run gasoline are of straight-chain (normal) or monobranched character.

Requirements. (1) *Antiknock characteristics (octane number).* Octane number of a gasoline is a measure of the ability to resist detonation caused by autoignition. When detonation occurs, it results in a rapid rise in combustion chamber pressure above that for normal combustion and causes high-frequency pressure fluctuations and an audible sound that is referred to as knock. When detonation (knock) occurs, heat is released which promotes further detonation, causes a loss in power and, if sustained, can result in engine damage. Antiknock requirements of spark ignition engines increase with increasing compression ratio and are affected by carburetion, spark timing, type of transmission and other engine/vehicle design characteristics. Antiknock requirement is also affected by changes in ambient conditions such as temperature, humidity and absolute pressure.

Octane number, an arbitrary scale of performance, is measured by standardized procedures and laboratory test engines by comparing a fuel's resistance to knock with reference fuels composed of mixtures of *n*-heptane (assigned a value of 0) and iso-octane (assigned a value of 100). The octane number of fuel is defined as the percentage of iso-octane in the reference blend giving equal resistance to knock as the test fuel. The Research Octane Number, RON (ASTM D-2699), is determined under mild test engine operation while the Motor Octane Number, MON (ASTM D-2700), is determined under more severe operating conditions (higher engine speed and temperature). Because of more severe test conditions, MON is generally lower than RON for commercial gasolines. In addition, a measure of the octane quality of the more volatile portion of gasoline, or front end octane quality, is often specified by Distribution Octane Number, DON (ASTM D-2886-70T). Because of the variations in engine and vehicle parameters and operating conditions on the road, no one of these individual octane ratings is synonymous with road octane quality; however, these laboratory octane ratings and their combinations are useful in predicting road performance of gasoline.

Normal paraffins have low octane numbers; naphthenes have intermediate ratings; isoparaffins (highly branched), aromatics and certain olefins have high Research Octane Numbers. Modern refining techniques are designed to produce predominantly the high-octane number components. Addition of lead alkyls (up to 4.0 ml/gal in motor gasoline) is also used to increase octane number of gasoline. Various lead alkyls are in use including tetraethyl lead, tetramethyl lead and physical and chemical mixes of the two. Research Octane Number of commercial motor gasolines in the U. S. has increased from a value of about 73-80 in 1937 to about 100 in the 1963-70 period through a combination of modern refining technology and use of lead alkyls.

(2) *Stability (low gum-forming tendency)*. Good stability, and protection against deterioration and the formation of gums in storage are important for all gasolines. Gums are sticky resins formed by polymerization and oxidation of certain olefinic (particularly diolefinic), and aromatic thiol compounds. Potential gum formers are removed during refining primarily by alkali treatment and catalytic hydrogenation. Antioxidants are usually added to gasolines to minimize oxidation and resultant gum formation. Metal deactivators are also often used to prevent oxidation, particularly that catalyzed by active metals, such as copper, present in the distribution system and in automobile fuel systems. (See **Antioxidants.**)

(3) *Low sulfur content*. Sulfur compounds are found in crude petroleum to varying degrees and occur in the gasoline fraction in concentrations from 0–2% sulfur, depending on the petroleum source. If left in the motor fuel, they cause corrosion in the engine and lower the octane number response to the addition of lead alkyl antiknocks. Current gasolines are treated by either chemical or catalytic methods to levels of less than 0.1% sulfur.

(4) *Volatility*. Varying amounts of C_4 and C_5 hydrocarbons are used to control volatility. Too high a volatility can result in excess formation of vapor in the fuel pump and/or carburetor and lead to vapor lock and hot-fuel handling problems, particularly in hot weather; too low a volatility results in poor starting and warmup in cold weather. Gasoline volatility is varied with the climate and seasons of the year to insure good vehicle performance.

(5) *Other considerations*. Additives are extensively used in gasoline to provide various benefits. Detergents have been generally used particularly to minimize carburetor deposits and also to effect cleanliness of other areas. Anti-icing additives are employed to reduce engine stalling in cool, humid weather due to ice formed in the carburetor from evaporative cooling. Corrosion inhibitors are generally used to prevent rusting and subsequent contamination of gasoline with rust fines. Many gasolines use phosphorus additives to minimize spark plug fouling.

Manufacture. Gasoline is produced in petroleum refineries by many different processes and methods. It is normally the end product of several different processes in each refinery. The various gasoline streams thus obtained may be pooled or they may undergo secondary treatment for special applications. Light liquid hydrocarbons separated from natural gas (natural gasoline) are also often blended into gasoline.

Fractionation of crude oil to produce straight-run gasoline is the oldest and simplest method. Because of the low octane of the heavy portion of straight-run gasoline, this is universally catalytically reformed to a higher octane. The light straight-run is usually blended directly to gasoline with just mild chemical or hydrogen treating.

Cracking and Reforming. *Cracking* converts higher molecular weight hydrocarbons, such as those in gas oils, to gasoline. Two general types of cracking are used: catalytic and thermal. *Reforming* is used to upgrade low-octane gasoline fractions into high-octane blending components, chiefly by use of a catalyst. These processes are discussed in articles under these headings and also under *Catalysis*.

Alkylation of C_3 and C_4 olefins with isobutane is an important source of high-octane number gasolines for aviation and motor fuels. Alkylation of C_2 and C_5 olefins is also feasible but not normally practiced. The synthetic gasolines produced by alkylation consist predominantly of branched isoparaffins and are more stable than the products of cracking or polymerization. Unleaded octane values are in the range of 91–97 for C_3–C_4 alkylates and can be as high as 102 for C_2 alkylates. Because of good lead alkyl susceptibility, leaded octane numbers for alkylate can exceed 100. Sulfuric acid and anhydrous hydrogen fluoride are used as catalysts in alkylating C_3, C_4 and C_5 olefins. The sulfuric acid process operates at about 0–15°C (32–60°F), the HF process at about 20–35°C (70–95°F). Aluminum chloride catalysts are used for alkylation of C_2 olefins with isobutane.

Isomerization. The light hydrocarbons presently blended directly into the gasoline pool have poor octane quality. This poor quality is primarily due to the presence of low-octane normal paraffins, principally normal pentane and normal hexane. These materials can be substantially upgraded by isomerization which converts the straight-chain normal compounds to branched-chain isomers. There are a number of commercial processes available for C_4, C_5, and C_6 isomerization. There are two principal types: (1) liquid phase isomerization with aluminum chloride catalyst, and (2) vapor phase hydroisomerization which employs a noble metal catalyst. Generally, the processes are designed for the lowest practical temperature since low temperature favors branching.

Both normal and isobutanes have a clear octane sufficiently high to be blended directly into gasolines. However, normal butane isomerization is sometimes practiced to furnish additional isobutane for alkylation. Both liquid and vapor phase processes are available for this purpose.

Pentane and hexane isomerization are generally accomplished in a vapor phase hydroisomerization process. Pentane isomerization, either single pass or with recycle, can be readily accomplished with a yield of about 99% isopentane. Hexane isomerization requires fairly precise feed preparation and becomes rather complicated if other than a single pass system is required. Hexane isomerization also gives high yields of the order of 95%.

Other Source Materials. While almost all gasoline now produced is derived from petroleum, other source materials have been developed and have at various times found, and in the future may find, large-scale application.

Oil shale is found in many parts of the world and constitutes a vast reserve supply of oil, which can be released by retorting and refined by methods developed in the petroleum industry. Oil content of the shale ranges downward from about 30 gallons per ton.

Tar sands exist in very large reserves in the northern portion of the Province of Alberta, Canada. The hydrocarbon portion has physical and chemical properties similar to some heavy crudes. The tar sands have to be mined and treated mechanically or chemically to separate approximately one part of hydrocarbon from seven parts of sand. The hydrocarbons can then be refined by conventional means to yield high-octane gasoline. Mechanical mining and washing were commercialized in 1967 in Alberta by Great Canadian Oil Sands Ltd. Discoveries of conventional petroleum reserves in Alberta and Alaska appear now to have delayed the need for further tar sand exploitation.

Coal has been used in the past as raw material in processes for making synthetic gasoline. Liquid hydrocarbon yields range from 1 to 3-barrels per ton of coal. During World War II Germany made 100,000 bbl/day of liquid hydrocarbons from coal using high-pressure hydrogenation, Fischer Tropsch synthesis and pyrolytic processes. The hydrogenation was conducted at high pressures (5,000 to 10,000 psig) in multiple stages using tin or iron catalyst. In the Fischer Tropsch process, synthesis gas (a mixture of carbon monoxide and hydrogen made by the high-temperature reaction of steam and coal) is converted to a mixture of hydrocarbons similar to crude oil. Cobalt or iron catalysts are used at pressures up to 300 psig and temperatures from about 170–180°C (340 to 360°F). Coal tars and other liquids produced by pyrolysis are upgraded by hydrogenation.

A large Fischer Tropsch plant has been operating for years in South Africa (the SASOL plant). Elsewhere high cost has prevented coal liquefaction from competing with crude oil.

Additional coal conversion processes are under development in the United States. In one of these processes, gas, gasoline, fuel oil and char are produced in fluidized-bed pyrolysis of coal. The liquid products are hydrogenated and further refined to adjust yields and improve quality. In another, coal is crushed, mixed with recycled gas oil and hydrogenated in an ebullating bed of catalyst. The prod-

uct from this reactor is fractionated and treated to produce finished products. Large-scale commercialization of these processes is unlikely in the near future.

C. B. Hood

Cross-references: *Cracking, Catalysis, Fischer-Tropsch Process, Reforming, Fuels, Petroleum, Antiknock Agents, Hydrocarbons.*

GATTERMANN SYNTHESIS

The Gattermann synthesis consists in use of liquid anhydrous hydrogen cyanide in place of carbon monoxide as a source of the formyl substituent; hydrogen chloride is required, usually in combination with aluminum chloride or zinc chloride, but cuprous chloride is omitted. The reaction formerly was regarded as proceeding through a transient addition product of hydrogen cyanide and hydrogen chloride (formimino chloride, $ClCH{=}NH$), but now appears to follow a more complicated course. In the absence of the hydrocarbon component the other reagents combine to form a molecular complex, $AlCl_3 \cdot 2HCN \cdot HCl$, possibly by addition of the intermediate formimino chloride to hydrogen cyanide. The complex condenses with the hydrocarbon component (2) with

1. $HC{\equiv}N + HCl \rightarrow$

$$[ClCH{=}NH + HC{\equiv}N \xrightarrow{\;AlCl_3\;}$$

$$ClCH{=}NCH{=}NH \cdot AlCl_3$$

2. $ArH + ClCH{=}NCH{=}NH \cdot AlCl_3 \xrightarrow{\;-HCl\;}$

$$ArCH{=}NCH{=}NH \cdot AlCl_3$$

3. $ArCH{=}NCH{=}NH \cdot AlCl_3 \xrightarrow{\;H_2O\;}$

$$ArCHO + 2NH_3 + HCOOH$$

elimination of hydrogen chloride and formation of an arylmethyleneformamidine complex, which subsequently is hydrolyzed to the aldehyde (3). The Gattermann synthesis has been improved by use of special solvents, chlorobenzene, *o*-dichlorobenzene, and tetrachloroethane, and by conducting the reaction at temperatures of 60–100° instead of at 40°; a distinct simplification is use of sodium cyanide in place of the hazardous hydrogen cyanide. Thus a simple procedure consists in passing hydrogen chloride into a suspension of sodium cyanide and aluminum chloride in an excess of the hydrocarbon component. Yields of aldehydic derivatives of hydrocarbons by the best procedures are generally low, however (benzaldehyde, 11–39%; *p*-ethylbenzaldehyde, 22–27%; 9-anthraldehyde, 60%; mesitylaldehyde, 83%).

Condensation with hydrogen cyanide or a metal nitrile, unlike that with carbon monoxide, is applicable to phenols and phenol ethers, often with considerable success. Thus anisaldehyde is reported to be formed in nearly quantitative yield by the Gattermann synthesis. Aluminum chloride is required as catalyst in the case of phenol ethers and some phenols; the less potent zinc chloride is generally adequate for phenols. The procedure has been modified to advantage (R. Adams, 1923) by substitution of zinc cyanide for hydrogen cyanide and a metal halide. When hydrogen chloride is passed into a mixture of the phenol or phenol ether and zinc cyanide in absolute ether or benzene, it liberates the hydrogen cyanide required for condensation and produces zinc chloride as catalyst. The method is illustrated for the reaction of thymol.

L. F. FIESER
MARY FIESER

GENETIC CODE

The "genetic code" is the name now given to the cellular information which is transferred to the daughter cells in the process of mitosis, or, by way of the fertilized ovum, to an entirely new organism. This information makes it possible for the new cell or the new organism to construct itself in a fashion identical, or nearly identical, with that of the parent cell or the parent organisms.

That such information must exist is obvious; the location of the information within the cell is less so. By 1904, Sutton specifically named the chromosomes as the site. The basic reasoning behind this lay in the fact that the chromosomes are carefully replicated in the process of cell division and as carefully shared among the daughter cells so that each has its full complement of different pairs. Furthermore, the sperm cell, carrying all the information contributed by the male parent, is a tiny bag containing virtually nothing but chromosome material.

The next problem lay in locating the information within the chromosome and here there seemed no doubt. The chromosomes contained complex protein molecules and protein molecules were the largest and most complex molecules in living tissue. Each protein molecule is made up of twenty different but related units, the amino acids, arranged in a chain, or in several connected chains. Each different order of amino acids produces a different protein molecule (with even trivial differences sometimes proving important) and the number of different orders possible is formidable. A medium-sized protein molecule, such as hemoglobin, can have its amino acids arranged in any of about 10^{600} ways.

The chromosomes also contained nucleic acids but through the first quarter of the twentieth century, these were viewed as relatively small molecules of relatively simple structure and were dismissed out of hand as possible carriers of genetic information. The fact that Kossel found, in 1896, that fish sperm contained only extraordinarily simple proteins, yet possessed a full complement of nucleic acid was puzzling but was, on the whole, ignored.

The turning point came in connection with two strains of pneumococci, the S strain ("smooth," because it formed a smooth pellicle) and the R strain ("rough," because it did not.) If dead S strain, or even an extract of the strain, were added to living R strain, the R strain became capable of forming the pellicle and changed over into S strain. Apparently, something in the R strain contained the information required to oversee the formation of the pellicle.

In 1944, Avery, MacLeod and McCarty produced the active extract in great purity and were able to show that it contained nucleic acid *only,* with no protein at all. For the first time, genetic information had been pinpointed and it was found in the nucleic acid component of chromosomes, not in the protein.

By then, nucleic acids were recognized as complex molecules after all, as large as, or larger than, protein molecules, and differing characteristically from species to species. In 1953, Watson and Crick had determined the structure of the variety of nucleic acid in chromosomes ("deoxyribonucleic acid" or DNA) and had worked out the manner by which it formed a replica of itself ("replication") in the course of cell division (see **Nucleic Acids.**)

But with the site of the information located, another problem was raised. How could the nucleic acids carry the necessary information?

The working of the cell depends on the nature and relative quantities of the thousands of different enzymes it contains. Each enzyme catalyzes a particular reaction and it is the network of reactions within the cell or cells of an organism that is responsible for all its characteristics. It is the nature of the enzymes, their relative quantities, the manner in which the working of one stimulates or inhibits the working of another, that is all the difference between a man and a mandrill, a cat and a catfish, a lion and a dandelion.

The essence of what now came to be called the genetic code, then, was the manner in which the structure of specific DNA molecules guided the production of specific enzyme molecules. The enzyme molecules were all protein in nature, made up of arrangements of twenty different amino acids. The DNA molecules were made up of arrangements of only four different nucleotides, and for all the great size of the molecules, that paucity in number of different units made DNA seem much simpler than proteins and too simple, perhaps, to carry the necessary information to guide protein-manufacture. (The four nucleotides, by the way, are adenylic acid, guanylic acid, cytidylic acid, and thymidylic acid, usually symbolized as A, G, C, and T, respectively. In another variety of nucleic acid, ribonucleic acid or RNA, thymidylic acid is replaced by the very similar uridylic acid, symbolized as U.)

In 1954, Gamow pointed out that the presence of only four different nucleotides did not matter. The code might involve groups of nucleotides. If a long chain, made up of four different nucleotides, is taken two at a time the number of different combinations is 4^2 or 16; if three at a time, the number is 4^3 or 64.

Investigation showed that it was indeed a chain of three adjacent nucleotides ("triplets" or "codons") in the nucleic acid molecule that represented a specific amino acid. The fact that there were 64 codons and only 20 amino acids, meant that several closely allied codons might all stand for the same amino acid, thus allowing redundancy, so that small errors in the replication of a nucleic acid often do not affect its working. One way of indicating that more than one triplet stands for a particular amino acid is to say that the genetic code is "degenerate." It is also "universal" for, as far as biochemists have been able to tell, the same triplet stands for the same amino acid in all organisms.

The next problem is to work out the code "dictionary," by determining which particular codon stands for which particular amino acid. The first breakthrough in this direction came in 1961 when Nirenberg and Matthaei made use of a synthetic RNA molecule made up of a long chain of a single nucleotide, uridylic acid. All the codons were therefore UUU. By adding to a solution of this RNA a supply of amino acids and all the cellular paraphernalia required for protein manufacture, a chain made up of a single amino acid, phenylalanine, was produced. Thus, the first item of the dictionary was determined: UUU = phenylalanine.

By 1967, the dictionary was completed. It was determined, for instance, that both UUU and UUC stood for phenylalanine. Similarly, GGU, GGC, GGA, and GGG all stood for the amino acid, glycine. The codons, AUG and GUG, were "capital letters" serving to indicate the beginning of a chain, while UAA was a "period" serving to end a chain.

There was also the problem of how the information originally contained in the DNA molecule, which existed in the chromosomes and never left the nucleus, reached the site of protein manufacture, which was in the cytoplasm. (In 1956 Palade had demonstrated the site of protein manufacture to be on tiny cytoplasmic particles rich in RNA, which he called "ribosomes." There are as many as 150,000 ribosomes to the cell.)

Clearly, the information had to pass from chromosomes to ribosomes and suspicion fell on RNA, which was similar enough in structure to DNA to carry the imprint of DNA information, and which existed both in nucleus and in cytoplasm. Thus, a section of DNA, using the process which ordinarily builds up another section of DNA of complementary structure (where A = T or U, and vice versa, and where G = C, and vice versa (see **Nucleic Acids**), builds up, instead, a section of RNA of that structure. The RNA molecule leaves the nucleus for the cytoplasm, carrying the message (so that it is called "messenger-RNA") to the ribosomes.

What is then needed is some sort of translating device to convert nucleotide codons into amino acids. The necessary device was located by Hoagland in the form of RNA molecules, small enough to be freely soluble in the cell plasma. One end of this small RNA molecule possessed a three-nucleotide combination that would attach itself only to a particular complementary codon on the messenger-RNA chain. At the other end of the small RNA molecule is a section that can combine only with a particular amino acid.

There are twenty different RNA molecules of this sort, one for each of the twenty different amino acids. Each one has a characteristic codon at the opposite end. Therefore, the small RNA molecules line up on the messenger-RNA according to the places into which their codons, at one end, fit; and on the other end their amino acids line up as well and are bound together to form an enzyme molecule. The structure of the enzyme molecule has thus been dictated by the structure of the messenger-RNA, the structure of which was in turn dictated by the structure of the DNA in the chromosomes.

Because the small RNA molecules transfer information from the messenger-RNA to the protein molecule, they are called "transfer-RNA."

In 1964, Holly and co-workers, determined the detailed structure of "alanine transfer-RNA," the one which combines with the amino acid, alanine. It was found to be made up of a chain of 77 nucleotides, bound together to form a three-leaf-clover structure. The structures of other transfer-RNAs were worked out in subsequent years and this three-leaf-clover effect seems common to all of them.

ISAAC ASIMOV

GEOCHEMISTRY

In a general sense geochemistry is the chemistry of the earth system, which as a result of the Apollo flights, now includes the moon. By combining aspects of both chemistry and geology this science attempts to determine and explain the distribution of elements in the various zones of the earth, in its waters, and in the atmosphere surrounding it. Geochemistry is concerned primarily with chemical properties whereas its sister science, geophysics, deals with the physical properties and forces of the solid earth, and its liquid and gaseous phases. Geochemistry is of relatively recent origin; most of the significant work has occurred in the last 50 years. The contribution of Frank W. Clarke, who carefully compiled thousands of rock and mineral analyses in his five editions of "The Data of Geochemistry," is a classic piece of work which is still a valuable reference source. The Russians, V. I. Vernadsky and A. E. Fersman, made important contributions to geochemistry and probably were the first to try to establish it as an independent science.

The discovery of x-ray diffraction and its application to the analysis of complex minerals represented the beginning of a new phase in geochemistry. Prior to this discovery geochemistry was a more or less incoherent collection of factual data. A system of fundamental laws and principles has been evolved which explains and predicts the distribution of many elements in the earth system. The initial applications of these fundamental principles of crystal chemistry are due largely to V. M. Goldschmidt and his co-workers. Goldschmidt recognized that in mineral formations ions of approximately the same size can substitute for each other in a crystal lattice (isomorphous substitution). Furthermore, this type of substitution can occur even though the ions differ in valence (e.g., Al^{+3} for Si^{+4}) provided that electrostatic neutrality of the system as a whole is maintained by the addition or omission of necessary counterbalancing charges in other portions of the crystal lattice. The additional charges can occur as further isomorphous substitution of other ions in the lattice (e.g., substitution of an Al^{+3}—Fe^{+3} pair for a Si^{+4}—Mg^{+2} pair is electrostatically equivalent), or additional charges may be fitted into empty spaces in or around the lattice as in the case of Na^+ and Ca^{++} in clay minerals. These principles brought order out of chaos in elucidating the complex analyses of minerals and rocks with their many "impurities" and trace constituents. Application of crystal chemistry gave new insight into the mechanisms by which complex silicates formed and accounted for many of the differentiations and fractionations of the elements which had occurred in rocks and minerals.

One of the important fields which concern geochemists is the elucidation of the transportation and migration of matter in the earth's crust. Rocks are decomposed by mechanical forces and by the action of water and its solutions to form sediments and sedimentary rocks. The transportation and redeposition of the elements released by this weathering action depends largely on the chemical properties of the systems involved, with special reference to aqueous solutions. Thus, although principles of crystal chemistry are of primary importance in understanding crystallization phenomena in molten rock systems, the weathering cycle is best described in terms of physical and inorganic chemistry.

Another problem with which geochemistry is concerned is the determination of the relative abundance of the elements in the earth's crust. Generally, this has been approached by careful chemical analysis of many rock specimens from all over the world. The composition of the earth's crust is represented to a first approximation by the composition of igneous rocks. There is no general agreement on the precise method of averaging analyses of various rocks, but the calculation of the composition of the earth's crust made over half a century ago by Clarke and Washington on the basis of more than 5000 analyses of igneous rocks is still of fundamental significance in geochemistry.

More recently studies by Goldschmidt and by Rankama and Sahama have confirmed in more detail this average composition.

The bulk of the earth's crust is composed of surprisingly few elements: oxygen, silicon, aluminum, iron, calcium, sodium, potassium, and magnesium make up over 98% of igneous rocks, and of these elements the first two comprise almost 75% by weight of the composition. The availability and usefulness of many elements is determined not by their true relative abundance, but by their properties, by the manner in which they have been concentrated or localized in certain ores, and by the ease with which they can be recovered from their native deposits. For example, gallium, lanthanum, and lead are about equal in true relative abundance in the earth's crust, but their commonness in technical use and products differs widely due to their properties and the fact that not all of them occur in concentrated deposits.

The earth consists of a core and several shells or layers which are predominantly of silicate composition. Therefore, the abundance of elements in the crust does not necessarily reflect the composition of the earth as a whole. Although there is some disagreement over the exact composition of the earth, there are considerable data which indicate that the earth has a very dense core probably composed of an iron-nickel alloy. Surrounding the core is a silicate mantle which may be largely a magnesium-iron silicate, and on top of the mantle is the relatively thin crust which is composed largely of igneous rocks. The crust is no more than 30 or 40 miles thick. The core and mantle make up almost 99 per cent of the earth's bulk. Thus the earth as a whole is composed of about one-third iron, one-fourth oxygen and one-fourth silicon and magnesium, the remaining one-sixth being made up of all the remaining elements.

The age of the earth has always been an intriguing problem to both geochemists and geophysicists.

Methods which utilize radioactive isotopes (Carbon 14) have been used most frequently in age determinations (see **Chemical Dating**). It is now generally agreed that the earth is about 5 billion years old. Study of the rocks obtained from the moon's surface may be expected to add considerably to geochemical knowledge.

FRED L. PUNDSACK

References

Fairbridge, Rhodes W., Editor, "Encyclopedia of Geochemistry and Environmental Sciences," Van Nostrand Reinhold Co., New York, 1972.
Cross-references: *Elements, Prevalence of; Chemical Dating; Ocean Water Chemistry.*

GERMANIUM AND COMPOUNDS

Germanium is a silvery-white metalloid in group IVA the Periodic System. Except for its appearance, it is unlike most metals for it is hard, brittle and a poor conductor of electricity. Present in the earth's crust only to the extent of about 10^{-11}% it occurs chiefly as the sulfide associated with other metal sulfides (e.g., Pb, Zn, Sb, Ag and Sn). At present the chief commercial source is from zinc sulfide ores of the Tri-State area of Missouri, Kansas and Oklahoma. In addition to small amounts of Cd, Ga and As these ores contain from 0.01 to 0.1% Ge. Other sources of germanium are flue dusts and coal ashes. Figures for the annual production of germanium are not readily available, but estimates are of the order of several tons.

Since the chief use of germanium is in the manufacture of electronic semiconductor devices, it must be of exceedingly high purity. Elements from groups IIIA and VA, present only to a few parts per billion, can produce detectable effects. Arsenic, the principal offender, is removed by refluxing the germanium tetrachloride condensate with copper turnings. Germanium dioxide is the chief article of commerce, selling for about $140 per pound. The dioxide is reduced in hydrogen at about 650°C to germanium powder, which is then fused to ingot form. Germanium currently sells for about $340 per pound in the United States. For use in semiconductor devices it must be further refined by fractional crystallization from its melt.

In addition to its semiconductor device applications, small amounts of germanium are used for magnesium germanate phosphors and in low-melting gold-germanium alloys which expand on freezing. Resistors made by depositing thin films of germanium have quite low temperature coefficients. Its optical transparency in the infrared is utilized in special filters. Glasses containing germanium dioxide have a high index of refraction and high dispersion.

Germanium is a typical covalent crystal having the diamond-type structure. The crystal is really a giant covalent molecule in which each atom is strongly bound to all its nearest neighbors. It is hard, has a high melting point and a large latent heat of fusion.

The metallurgy of germanium has been studied in many binary systems and in a few ternary cases. It enters into simple eutectic systems with Al, Ga, In, Tl, Bi, Pb, Sn, Sb, Co, Au, Ag, and Zn. With silicon it forms a continuous series of solid solutions. In many systems it forms compounds, examples of which are with As, Cu, Fe, Ni, Mg, and Na. The chemical properties of germanium are intermediate between those of silicon and tin, and indeed it may be regarded as a true "crossroads" element, exhibiting both metallic and nonmetallic characteristics.

Germanium oxidizes readily in air or oxygen at 600 to 700°C. It reacts only slightly with hydrogen peroxide, alkali hydroxide solutions, hydrochloric acid, and sulfuric acid. It is attacked by hydrofluoric acid, nitric acid, and aqua regia, and dissolves readily in molten alkali hydroxides.

In its compounds germanium predominantly forms covalent bonds and occurs in two oxidation states. The *germanic* compounds, in which the oxidation number of the element is +4, are in general more stable than the *germanous* compounds with an oxidation number of +2. Thus germanium is primarily quadricovalent as in the dioxide, disulfide, tetrahalides, hydrides and organic compounds. All the tetrahalides are known.

Germanium dioxide exists in two crystalline forms and as a glass. The *rutile* crystalline modification is relatively inert and insoluble in water. The *low-quartz* modification and the glass are soluble in water and in hot concentrated hydrochloric acid. Unaffected by nitric or sulfuric acids, germanium dioxide is readily soluble in alkali hydroxide solutions and in hydrofluoric acid.

Germanium forms a series of compounds with hydrogen similar to the methane series of carbon and further occurs in a variety of organogermanium compounds. Typical of these are compounds of the type GeR_4, where R refers to various organic radicals.

Germanium and germanium dioxide are not believed to be toxic. The tetrachloride irritates the nasal passages, but this is probably due to hydrochloric acid formed by hydrolysis.

HENRY E. BRIDGERS

Cross-reference: *Semiconductors.*

GETTERS

The term *getter* has been applied to solid substances used in vacuum devices to effect a variety of functions: removal of residual gas, maintenance of uniform filament diameter, and minimizing of darkening of bulbs in incandescent lamps; maintenance of substantially constant gas pressure in early gas-type x-ray tubes; ionic focusing of electron beams in early cathode-ray tubes; activation of photo, secondary, and thermionic emitters, and removal of excess activating material in electron tubes; purification of inert gases; production of very high vacuum; etc. However, today the term is applied primarily to substances placed in electron tubes to remove residual gas and gas evolved from the parts of the tubes during processing and throughout their life. This article is restricted to getters of this type.

There are three sources of gas in electron tubes: (1) *Residual gas,* usually air and H_2O vapor, not removed by the pumps during tube processing. (2) *Gases produced during processing* of the tube: occluded gas, mostly CO, H_2O, H_2, N_2, O_2, and hydrocarbons, expelled from the metal and glass; CH_4, other hydrocarbons, and H_2O produced by decom-

position of the nitrocellulose binder of the cathode coating; and CO_2 and CO produced by decomposition of the Ba-Sr-Ca carbonate cathode coating to oxide. (3) *Gases produced during life* of the tube by evolution from the metal and glass, by decomposition of metal oxides, carbides, etc. under electron bombardment, and by chemical reaction in the metal parts. Some authorities restrict the term *gettering* to "clean-up" of residual gas and gas produced during processing, and use the term *keeping* for clean-up of gas produced during life.

During manufacture, the gas pressure in "receiving type" and television picture tubes is reduced to the order of 10^{-3} torr or less by pumping and outgassing. Power tubes and photo, x-ray, and other special purpose tubes are pumped and outgassed to lower pressures. In all tubes it is the function of the getter to reduce the pressure further to 10^{-5} to 10^{-7} torr and to maintain such reduced pressure throughout the useful life of the tubes. Such high vacuum is required primarily to prevent ion bombardment of parts and impairment ("poisoning") of the electron emissivity of the cathode. Vacuum is required also to provide a long mean free path for the electrons, but even at 10^{-3} torr the mean free path of electrons is greater than the cathode-anode distance in most tubes. Getters assume added importance in the tubes produced for high-reliability use in military and industrial applications.

Probably no rigid classification of getters is entirely satisfactory, but in general, they may be classified in two groups which may be termed *bulk* (or *contact*) getters and *flash* getters.

Bulk getters consist primarily of elements which may be handled and stored in the open air and placed in tubes in elemental form. They require only outgassing in vacuum at elevated temperature to prepare them for gettering. The gettering mechanism of most bulk getters is essentially physical, involving adsorption or absorption of the gas, or formation with the gas of relatively unstable compounds. However, certain elements used as bulk getters form stable compounds with many gases and are exceptions to this generalization. Bulk gettering action is usually reversible, the gas being released at elevated temperatures. Clearly, alkali and alkaline earth elements are wholly unsuited for use as bulk getters.

Most metals, if clean and outgassed, are capable of absorbing a substantial volume of gas, and in fact, the electrodes of tubes act to a certain extent as bulk getters. However, some metals can absorb several hundred times their own volume of gas and, therefore, make excellent bulk getters. Among the metals that are used as bulk getters are Al, Be, Ce, Cr, Nb, Pd, Ta, Th, Ti, and Zr. Sometimes certain alloys are used, e.g., misch metal. Carbon, coated on anodes to increase heat radiation, also acts as a getter. For most such elements there is a temperature range for optimum adsorption below which the efficiency of adsorption is low and above which adsorbed gas is released.

Bulk getters are introduced in tubes in several ways. Anodes or other parts operating at suitable temperatures may be made of the getter metal, or a piece of wire or sheet may be welded to such parts. Powdered getter metal suspended in a binder may be applied to an electrode and later sintered, or a pyrolytic compound of the getter metal (e.g., ZrH_4) may be applied and decomposed. In some cases bulk gettering depends upon alloy formation; e.g., if Al-clad Fe anodes are processed at 700–800°C, Al-Fe alloys are formed which minimize evolution of oxygen from the anodes during life. The getter, in whatever form, is outgassed by heating in vacuum at a temperature above the optimum gettering range.

In general, small "receiving type" electron tubes do not operate at temperatures sufficiently high to permit efficient operation of bulk getters. Moreover, the relatively thin Ni and Fe electrodes used in such tubes cannot withstand the temperatures necessary to outgas such getters. Therefore, bulk getters are used primarily in power tubes in which the electrodes are of C, Mo, W, or other refractory elements which can operate at red or yellow heat.

Flash getters consist almost exclusively of alkali or alkaline earth elements which cannot be handled or stored in the open air and must be placed in tubes as a stable alloy, encapsulated in inert metal, or by chemical action. After the tube has been outgassed and evacuated, the getter is "flashed," usually by radio frequency ("rf") induction heating, and the active metal is deposited in the form of a "mirror" on the inside of the tube envelope. The gettering mechanism is essentially chemical, involving formation of quite stable chemical compounds, but certainly other mechanisms contribute to flash gettering. For example, the fact that a reduction in pressure is observed when a Ba getter is flashed in pure Ar or other noble gas is evidence that mechanical entrapment of gas occurs during flashing. Flash gettering action is substantially irreversible.

An ideal flash getter would have the following properties: (1) The vapor pressure of the (unflashed) getter must be low at temperatures suitable for outgassing the tube during manufacture. (2) The getter must volatilize easily at a temperature low enough to preclude melting, evaporation, or loosening of particles from the metal support from which the getter is flashed. (3) After flashing, the getter deposit must have a negligibly low vapor pressure ($< 10^{-7}$ torr) as the normal operating temperature of receiving tubes (~ 200°C). (4) The getter must be active during flashing and at all temperatures between room temperature and operating temperature for all gases encountered in tubes. (5) The products formed by reaction of the getter and gases must be porous, so as to permit penetration of gas to unused underlying portions of active metal. (6) The reaction products must be stable and of negligible vapor pressure. (7) The getter must not "poison" the alkaline earth oxide cathode.

At one time or another most of the alkali and alkaline earth elements, as well as Be, Al, Mg, and Zn, have been used as flash getters. Some of these elements (e.g., Mg) getter only while being flashed unless, after the metal is deposited, the gas is ionized by an electric discharge or other means. However, Ba has the properties listed above for a flash getter to a far greater degree than any other element, and is now used almost exclusively.

There are several ways to introduce Ba into tubes: (1) *Evaporation of pure Ba* from pellets crimped in Ni tabs. This method is no longer used because of the rapid deterioration of pure Ba. (2) *Thermal decomposition of a Ba compound,* usually Ba azide, $Ba(N_3)_2$. This method also is no longer used because of the explosion hazard. (3) *Evaporation from inert alloys,* usually with Mg, Al, or both. The powdered alloys, sometimes mixed with inert diluent (e.g., Fe powder), are held in metal channels with binder. Small lengths of alloy wire may also be held in such channels. (4) *Evaporation from self-supporting Fe- or Ni-clad Ba or Ba alloy wire.* This method is possibly the one most commonly used today. (5) *Production and evaporation of Ba by chemical action.* Powdered reactants, sometimes mixed with Ni powder, are held in a metal channel with binder. Upon "flashing," a chemical reaction is initiated which liberates free Ba and, if exothermic, provides heat for evaporation. The exothermic reactions provide a more rapid release and evaporation of Ba and are favored for modern high-speed, mass production processing equipment for electron tubes. (6) *Production of Ba in the alkaline earth oxide cathode* occurs during operation of the tube, and some of this Ba evaporates and condenses upon nearby structures. Consequently, the cathode surface and any surfaces which have received evaporated Ba act to a certain extent as getters.

The getter, in whatever form, is mounted in the tube, often with suitable shields to prevent the getter from depositing upon the tube electrodes, stem, or micas. After the tube is outgassed and evacuated, the getter is flashed by local heating, usually by rf, and the Ba film is produced.

At the operating temperatures of some large power tubes, the vapor pressure of alkaline earth metals, especially Ba, is high enough to cause detrimental effects, even evaporation of the metal film. Primarily, flash getters are used in tubes in which the active gettering surface area can be kept at temperatures below 200°C. Flash getters, therefore, are used in almost all receiving tubes, small power tubes, and cathode ray tubes.

EUGENE P. BERTIN
A. M. SEYBOLD

GIBBS-DUHEM EQUATION

The chemical potential, or free energy per mole, is an important attribute of the components in a homogeneous mixture. The chemical potential governs solubility limits, vapor pressures, phase separation, and other properties and processes of technical interest.

It is often useful to know how the chemical potentials change with composition, but this cannot be determined from thermodynamics alone. The Gibbs-Duhem Equation (GDE) is an exact thermodynamic relation that allows one to compute the changes of chemical potential for one component over a range of compositions provided the change of potential for each of the other components has been measured over the same range.

In a system of j components at constant temperature and pressure, let n_i designate the number of moles of the ith component and μ_i its chemical potential. If the amount of any one component is varied, the GDE states that

$$n_1 d\mu_1 + n_2 d\mu_2 + \cdots + n_i d\mu_i = 0 \qquad (1)$$

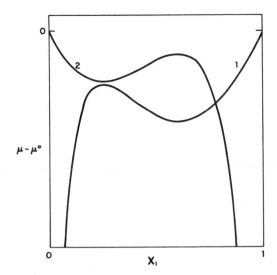

FIG. 1. Chemical potentials in a binary solution.

The GDE may be stated in terms of mole fractions, x_i; if this is done, the number of mole fractions that may be held constant is two less than the number of components. For systems of two components, a convenient form of the equation in terms of mole fractions is

$$\left(\frac{\partial \mu_1}{\partial x_1}\right)_{P,T} = -\frac{x_2}{x_1}\left(\frac{\partial \mu_2}{\partial x_1}\right)_{P,T} \qquad (2)$$

The figure exhibits the chemical potentials in a hypothetical binary solution. The ordinate, which applies to both components, is the difference between the potential at a given composition and the potential of the pure component; hence all the values are zero or negative. An important feature of equation (2) is that if one derivative vanishes, the other must vanish at the same composition. The result is that a minimum in the curve for one component must be accompanied by a maximum in the other. Since the potential of either component approaches minus infinity as the concentration of the component approaches zero, each curve must have both a maximum and a minimum, or neither. The type of solution illustrated exhibits phase separation at intermediate compositions.

The above remarks illustrate the qualitative conclusions that may be drawn with the aid of the GDE. If the curve for one component is known over a certain range of compositions, the curve for the other component may be derived graphically over the same range with the help of equation (2). The vertical placement of the second curve will not be known unless one point on it is determined by experiment. If an analytical expression is known for μ_1, one may differentiate, obtain an expression for $\partial \mu_2/\partial x_1$ by means of equation (2), and integrate to find an analytical expression for μ_2.

The composition variable need not be the mole fraction; the volume fraction is the most convenient independent variable in applications of the GDE to solutions of high polymers. The chemical potential is a partial molar quantity; by extension, the GDE may be applied to any partial molar quantity, such as partial molar volume, entropy, enthalpy, or heat capacity.

LAWRENCE F. BESTE

Cross-references: *Thermodynamics, Free Energy.*

References

Lewis, G. N., and Randall, Merle, "Thermodynamics," revised by K. S. Pitzer and Leo Brewer, 2nd Ed., pp. 210–217, McGraw-Hill Book Co., New York, 1961.

Glasstone, Samuel, "Thermodynamics for Chemists," pp. 213–216, 318–320, Van Nostrand Reinhold Co., New York, 1947.

Wall, Frederick T., "Chemical Thermodynamics," pp. 180–193, W. H. Freeman and Co., San Francisco and London, 1958.

GLASS

Glass is defined as an amorphous solid which is smoothly and reversibly converted to a liquid by the application of heat. If crystallization occurs during this conversion the glass is said to devitrify. Fundamentally the chemistry of glass is the chemistry of silica. In many respects vitreous silica is the glass par excellence and most commercial glasses contain 60% or more by weight of silica. The few unusual compositions made for special purposes which do not contain silica are based on boric oxide, phosphoric oxide or other glass formers and are similar to the silicate glasses in structure and behavior.

The physical chemistry of glass belongs in the field of liquids, where much of our knowledge is qualitative or at best semi-quantitative and empirical. Even in this complicated field, glasses constitute a problem of unusual difficulty. At ordinary temperatures, they are not in equilibrium but are potentially unstable with respect to crystallization. This tendency to crystallize or devitrify is of much practical importance in the manufacture of glass, since it becomes evident at elevated temperatures. Further complication is introduced by the fact that glass is not even in a state of metastable equilibrium at room temperature. Its properties are found to depend not only on temperature and pressure but also to an appreciable extent upon thermal history, that is, upon the way in which the glass was cooled through the temperature region in which it changed from a liquid to a rigid solid. It appears that at each temperature below the liquidus or crystallization point there is a state of metastable equilibrium characterized by definite physical properties which the glass approaches as long as crystallization does not occur. However, the rate of approach to this metastable equilibrium becomes progressively slower as the temperature drops and is vanishingly small before room temperature is reached. Thus at ordinary temperatures, glass is in a state of arrested equilibrium with properties depending upon the time allowed to approach the metastable liquid equilibrium at those temperatures where the rate of this change was appreciable during its previous cooling. As a consequence of this behavior, specimens of glass should be annealed by an adequate and carefully specified heat treatment before being used for any precise measurement of their physical properties.

Structure of Silica Glass. The key to the structure of silica glass lies in the fact that each oxygen is linked by strong bonds to two different silicon atoms. Thus each silicon, as appears from the formula SiO_2, must be bonded to four oxygens surrounding it in the tetrahedral arrangement, which is found by x-rays to be the characteristic structural element of silicate minerals and glasses. Study of the scattering of x-rays by vitreous silica established this structure, showing that each silicon is indeed surrounded by four oxygens on the average, and that the mean oxygen silicon distance is 1.62A, in good agreement with the distance observed in crystalline silicates. This short range order is not continued, however, and after a few steps from any particular atom the distances and orientations of the surrounding atoms become entirely random.

The bonds between oxygens and silicons in silica glass are regarded as partly ionic and partly covalent in character. Thus, a piece of silica glass is to be regarded as a network of SiO_4 tetrahedra linked together by strong chemical bonds having pronounced directional characteristics. The strength of the bonds results in a strong, hard, infusible material, while the random orientation of bonds with directional characteristics results in low density relative to the crystalline forms of silica and a low coefficient of thermal expansion. The strong, tightly bonded structure of vitreous silica is also reflected in its great resistance to chemical attack, its perfect elasticity and its excellent dielectric properties.

The random network which is characteristic of silica glass and of other glasses as well is usually obtained by melting crystalline raw materials and cooling the resulting liquid rapidly enough to avoid crystallization. This is possible in those compositions suitable for the manufacture of glass because the viscosity of the liquid increases rapidly with falling temperature and is sufficiently great at the liquidus, that is the point where crystals are thermodynamically stable, to prevent devitrification in the time required for cooling. At still lower temperatures, the viscosity increases enormously and the random glass lattice, although theoretically unstable, is for all practical purposes permanent and rigid.

The viscosity of glass, regarded as a function of temperature, is its most important property from the point of view of manufacture. Thus, pure silica glass has such a high viscosity at temperatures which can be conveniently reached that its manufacture is very difficult. A further complication arises from the fact that the vapor pressure of silica becomes appreciable at the temperatures required to break up its crystal lattice. The viscosity of liquid silica is so high at the melting point (1710°C) that the elimination of bubbles from the melt is almost impossible. Loss of material by rapid evaporation is avoided by using electric resistance or arc heating.

Alkali Silicates. For practical reasons it is necessary to weaken the structure of silica glass so that it can be manufactured economically. Experience shows that the oxides of the alkali metals are the most effective agents for doing this. Consequently,

they are the principal fluxes used in the manufacture of glass. Soda, the cheapest of these oxides, is used to the greatest extent, but potash has advantages in some compositions and lithia, the most potent of the fluxes, is also used in spite of its greater cost (see **Fluxes**).

The effect of the alkali oxides on the properties of glass may be understood in terms of the structure if we consider that the oxygen ions from the flux become part of the silica lattice. Since each silicon can be bonded to only four oxygens, these extra oxygens interrupt the silicon-oxygen-silicon links which give the silica glass its hardness and rigidity and produce a more mobile structure. The alkali metal atoms as singly charged positive ions lie in open spaces in the weakened silica network, loosely held by the negative charge contributed by the extra oxygen ions. By the addition of 25% of soda, the viscosity of silica glass is so far reduced that melting becomes quite easy. The thermal expansion is greatly increased, the electrical resistivity decreased and the density increased. However, sodium silicate glass is quite unsuitable for most of the usual uses of glass because it is rapidly attacked by water or even moist air. The resulting *water glass* finds applications as an adhesive, a detergent and a fire retardant.

Typical Glasses. By adding other oxides as well as the alkalies to silica the viscosity may be reduced sufficiently to make manufacture feasible without the disastrous loss of durability in aqueous solutions. For economic reasons, lime is the commonest of these stabilizing ingredients. Both calcium lime and dolomite are used in large quantities with sand and sodium carbonate in the manufacture of inexpensive soda-lime glass which is used for windows, bottles, electric light bulbs, etc.

Lead oxide is another useful stabilizing ingredient which enters the lead crystal or flint glasses as a major constituent. These glasses generally contain less total alkali than soda-lime compositions and usually include a considerable proportion of potash. They have a high refractive index and dispersion, which makes them useful in optical systems and also gives them a sparkle and brilliance desirable in art ware. The high electrical resistivity and dielectric constant of lead glasses makes them useful in vacuum tubes and electrical condensers. Their high density makes them useful in shields for x-ray tubes and radioactive materials. The long working range and low softening temperature of lead glass makes it a favorite for the fabrication of electric signs.

Boric oxide is a stabilizer which is very effective in lowering the viscosity of silica without increasing the thermal expansion. A small amount of alumina is desirable to prevent devitrification and to improve the chemical stability of these boron-containing glasses which are termed *borosilicates*. They are used in baking ware, laboratory apparatus and industrial piping because of their outstanding resistance to sudden temperature changes and excellent chemical durability. These are useful in power tubes, where their low dielectric loss is also desirable. Other compositions in this field transmit ultraviolet radiation unusually well and are useful in sun lamps, sterilamps and ozone generators. Glasses of very high chemical durability suitable for high pressure boiler gauges and ampoules are produced in the borosilicate field. Borosilicate compositions can also be made from which, after suitable heat treatment, the soluble fluxes may be leached with hot acid leaving a porous skeleton of almost pure silica. Upon drying and firing, this material shrinks to a dense clear glass containing at least 96% silica which is nearly as hard and as low in expansion as fused silica ("Vycor").

Alumina and lime in large proportions with about 10% of boric oxide and little or no alkali are used in glass for the manufacture of fibers. This is necessary in order to obtain the weather resistance required by the extremely large surface per unit weight of fibers. Similar compositions are used for combustion tubing and cooking ware where a very hard glass is required. They have a very rapid change of viscosity with temperature, making it possible to melt glasses which are sufficiently hard at service temperatures.

Optical glasses as a class have widely varied compositions. The common characteristic is that they must be of unusually high quality in respect to homogeneity and control of properties. They are characterized by their refractive index (N_D) and reciprocal dispersion (v) value. The original optical crowns were lime glasses and the optical flints lead glasses. However, in recent times, many other ingredients such as fluorides, boric oxide, zinc oxide, barium oxide, lanthanum oxide, tantalum oxide and thorium oxide have been used to obtain special values of N_D and v.

Colored glasses are produced by adding suitable coloring agents to the various types of glass which have been discussed. Blue may be produced by adding oxides of cobalt, cupric copper or ferrous iron; green by oxides of chromium, iron, copper, uranium or vanadium. Purple is produced by trivalent manganese, by mixtures of neodymium and praseodymium oxides and by nickel oxide in potash glasses. Cerium oxide produces a yellow color. These colors, which are believed to be due to the ions of the colorant subjected. Gold, cuprous oxide and cadmium sulfoselenide produce red colors in suitable base glasses.

Translucent white glass, known as opal glass, is usually produced by the addition of a sufficient quantity of fluoride to a suitable base glass. Phosphates, chlorides, sulfates, oxides of tin, zirconium, titanium and antimony are also used. The light diffusing properties of this kind of glass are due to minute inclusions, usually crystals, which separate during the cooling of the glass.

One of the most interesting of these colloidal glass systems is photosensitive glass. Here the concentration and state of oxidation of the colorant are adjusted so that exposure to ultraviolet light nucleates the precipitation of color or opal particles upon subsequent heat treatment while unexposed portions of the glass remain clear. Thus it is possible to develop a photographic image either in color or in white opal within a sheet of glass.

Glass ceramics are another important development in the field of glass technology. By adding suitable nucleating agents such as titanium dioxide it is possible to control the devitrification of glass to produce strong, well crystallized ceramic articles from shapes originally blown or pressed from molten glass. By choosing a suitable composition, glass ceramics may be obtained with a very low coefficient of thermal expansion and a high softening temperature which are excellent for cooking ware. Other glass compositions may be crystallized to yield ceramic materials with outstanding mechanical and electrical properties.

In certain lithium glasses silver particles, precipitated where the glass has been exposed to ultraviolet light, initiate devitrification of the glass. The white patterns, produced in this way, may be dissolved selectively by hydrofluoric acid leaving intricate perforated shapes termed "Fotoform" which find uses in the electronic and printing industries.

Strong Glass. Although traditionally a weak and brittle material, glass may be made strong and serviceable. The lattice of oxygen silicon bonds is itself very strong and it is only because of the stress concentrating effect of surface flaws that glass breaks so readily in tension.

Of the various means which have been found to throw the surface of glass into compression and thus improve its strength the simplest is a process known as thermal tempering. It is accomplished by heating the glass with suitable support until it is somewhat soft and then cooling the surface rapidly, usually with an air blast. The hard, rigid surface produced in this way is thrown into compression by the subsequent cooling and contraction of the interior so that the whole piece develops unusual mechanical strength.

In another process, known as chemical tempering, sodium ions in the surface layers of a piece of glass are replaced by potassium ions diffusing in from a molten nitrate bath. The larger potassium ions expand the surface and throw it into compression, again increasing the mechanical strength.

A third process for producing strong glass is known as "casing." Here the object is fabricated from two glasses. Its surface is made from a composition with a low coefficient of thermal expansion while the interior has a higher expansion. In such a compound article cooling from the solidification temperature produces strong compressive stress in the surface which must be overcome before a break can occur.

Chalcogenide Glasses. A new field of glass compositions of increasing technological importance is based on compounds of sulfur, selenium and tellurium, sometimes with additions of the heavier halogens. The arsenious sulfide may be cooled to a glass which has useful infrared transmitting properties. Other more complex mixtures have interesting semiconducting properties and may prove valuable in electronic devices.

CHARLES H. GREENE

Photochromic Glass

Color changes in many glasses have been observed under ultraviolet light or other high energy radiation. Under long exposure to intense sunlight, certain soda-lime-silica glasses, commonly used in bottles and windows, turn purple as a result of a change in the oxidation state of manganese, a minor constituent of the glass. Under higher energy electromagnetic radiation, such as X-rays, and γ-rays, silica-based glasses turn brown. These changes are typically very slow and are irreversible at room temperature. A wide range of special compositions of true photochromic glasses, which darken when exposed to light, and clear again when the light source is removed, has now been developed. These photochromic glasses do not fatigue—either with the passage of time or with the number of darkening/clearing cycles.

The active materials within these silicate glasses are colloidal silver halide crystals, formed by crystallization from the glassy matrix during initial cooling or subsequent heat treatment of glass into which silver and one or more of the halogens has been dissolved during the melting of the glass. In the absence of light, the silver halide glasses may be either transparent and essentially colorless, may have a residual and permanent color deliberately added, or may be translucent or opaque, depending on the composition of the glass and the particle size and concentration of the suspended colloid.

For those compositions which are transparent, the amount of silver is typically 0.5% or less, and the size of the crystals ranges from about 50 to 300 Å. A composition or treatment of the glass which permits growth of crystals to much larger sizes—sizes which are not small compared to a wavelength of light—results in an opal glass; the density of the opal may also range between wide limits. The average spacing between the randomly dispersed crystals in the transparent glasses is a few hundred Ångstroms.

Alkali borosilicates have been found to be, in general, the most suitable of several glass systems as a matrix for the photochromic sensitizers. The suspended crystalline silver halides absorb high-energy photons which dissociate the silver halide. Oxides of other metals, including arsenic and antimony, tin and lead, and copper, increase the sensitivity and the photochromic absorbance.

The reversible darkening of silver halide in these glasses results from the very small size of the crystals, and from the fact that the crystals are discrete and are separately embedded in rigid, impervious and chemically inert glass. This ensures that the photolytic color centers cannot diffuse away, or grow into stable silver particles, or react chemically to produce an irreversible decomposition of silver halide.

In the special case of silver chloride glasses that are darkened only by ultraviolet light, exposure to longer-wave visible and near infrared in the broad absorption band of the darkened color centers results in accelerated fading. In other words, such glasses fade more rapidly in visible light than in the dark. Similarly, glasses of this type become darker when exposed only to ultraviolet light than they do when exposed simultaneously to ultraviolet and visible light.

The optical transmittance of a transparent photochromic glass before darkening is similar to that of window glass. On exposure to light of a given intensity (temperature being held constant), the optical density measured at a given wavelength in the visible spectrum—and similarly, the total luminous absorption—increases exponentially, approaching a constant equilibrium value.

The rate of darkening depends on the composition and on the intensity of the activating light. In bright sunlight, maximum absorbance is approached in times of the order of a minute. Darkening to the same general level of transmittance occurs after exposure by a photographer's flash lamp, or after the much shorter duration, higher intensity illumination of a laser-activating xenon flash lamp.

The color of the glass in the darkened state is generally a neutral gray, or gray-brown with a broad absorption band extending through the near ultraviolet and the entire visible spectrum to the near infrared.

Potential uses for these glasses include control of light utilized by people, and control of light utilized by machines or equipment. In the former category are sunglasses, windows for buildings, land and air-borne vehicles. In the latter are variable density filters, compensating filters, elements of optical computers, temporary image storage in photography. The possibilities of modification of these chemically durable, reversible photochromic systems by their environment, i.e., by light and heat, suggests further applications.

G. P. SMITH

Cross-reference: *Liquid state.*

GLAZES

A glaze is a thin, impervious, glassy layer fused on ceramic bodies or whiteware materials. Glazes may be glossy or have a matte surface; may be colorless or exist in a wide variety of colors; and may be transparent, opaque, or translucent. A glaze is usually applied to ceramic bodies (a) to make the ware impermeable to liquids and gases; (b) to provide a surface that is hard and wear-resistant; (c) to obtain special colors or decorative effects; and (d) to provide an aesthetically attractive coating or surface. The glaze is usually applied as a slip or slurry by dipping, spraying or brushing on the surface of the ceramic body; it is then fired to bring about a fusion and interaction of the glaze materials to form the glassy or vitreous layer. The glaze slip may be applied to the unfired ware which is referred to as the one fire process, or it may be applied to previously fired bisque ware—the two-fire process. Decoration may be applied either on the bisque surface as an underglaze decoration, or on the surface of the fired glaze which is known as over-glaze decoration.

The term "glaze" is sometimes confused or associated with "vitreous enamel," "porcelain enamel" or "glass enamel." Technically speaking, the first two refer specifically to coatings fused on metals. The last refers to coatings usually colored, applied as thin films on container or other glass that mature at temperatures below the annealing range of the glass or around 1000°F. The term "glaze" applies specifically to glassy coatings fused on ceramic bodies.

The composition of glazes may be expressed as a direct percentage, as a batch weight ratio, or as is quite frequent in technical literature, as an empirical molecular formula. In order to understand the use of the empirical glaze formula it is helpful to consider that materials are classified on the simple oxide basis in the fused state, as illustrated in the following examples:

RO Bases or Fluxes	R$_2$O$_3$ Intermediates	RO$_2$ Acids
Li$_2$O	Al$_2$O$_3$	SiO$_2$
Na$_2$O	B$_2$O$_3$	TiO$_2$
K$_2$O	Cr$_2$O$_3$	SnO$_2$
MgO	Fe$_2$O$_3$	ZrO$_2$
CaO		
etc.		

The good glaze formula contains a proper balance among the RO, R$_2$O$_3$ and RO$_2$ groups. The empirical molecular formula not only expresses the relative proportion but also the specific function of a material in a glaze. As a uniform basis of comparison the RO group of all glazes is expressed as equal to unity.

A glaze may be very simple in composition, containing two or three materials, but is usually more complex, and may contain as many as twenty different materials. Glazes are conveniently classified into two main groups: raw glazes and fritted glazes. The raw glazes contain a wide variety of naturally occurring raw materials that are essentially insoluble. The fritted glazes contain as one of the materials a synthetically prepared special glass or frit. In order to incorporate certain of the soluble materials such as the alkalies and the borates it is necessary to make a special glass or frit so that these materials can be added in a relatively insoluble form.

Fritted Glazes. Fritted glazes differ from raw glazes in that they contain a synthetically prepared special glass or frit. Fritted glazes mature over a very wide temperature range from 600°C to over 1200°C depending on the specific composition. The principal reason for fritting is to make possible the use of water-soluble materials by melting them along with other materials to form a relatively insoluble glass. The molten glass is quenched in water to make the frit friable so that it can be readily ground. As an aid in the proper compounding of frits, certain empirical rules have been derived so that the frit will melt readily and be relatively insoluble.

The composition of fritted glazes varies quite widely. They may be characterized or classified as leadless or lead-containing glazes, as high borate or low borate glazes, or as high, intermediate or low alkali glazes or as combinations of these or other materials.

A wide variety of colors can be obtained with the fritted glazes by the use of color oxides or specially prepared color stains. The color stains are usually added to the final glaze batch along with the frit and other materials. By the addition of opacifiers such as zirconium oxide or tin oxide, opaque or white glazes are possible.

With all glazes, either raw or fritted, one of the important properties is the proper fit of the glaze on the body. In order for the glaze to exist as a continuous glassy layer on the body it is necessary that the coefficients of thermal expansion of the glaze and the body be closely related. If the glaze has a greater coefficient of thermal expansion than the body, crazing or a network of fine surface cracks may result. The most desired combination is for the glaze to have slightly less thermal expansion than the body so that the glaze will be in compression. It is to be noted that if the compressive stress becomes excessive it will produce shivering or rupture of the body. Proper selection of the glaze can increase the transverse strength of a glaze-body system by 15 per cent whereas improper use of glaze can decrease the strength by 75 per cent or more.

RALPH L. COOK

Cross-references: *Clays, Ceramics.*

References

Parmelee, C. W., "Ceramic Glazes."

GLYCERIDES

Glycerides are fatty acid esters of glycerol [HO—CH$_2$—CH(OH)—CH$_2$—OH]. Those in which all three of the hydroxyl groups are esterified are called *triglycerides.* These account for most of the substance of vegetable and animal fats. They occur in all forms of animal and vegetable life, and constitute one of the three major classes of food for humans and the higher animals. Triglycerides in which all the fatty acid radicals are alike are called *simple* triglycerides. Most natural fats contain only a minor percentage of simple triglycerides, and consist mainly of *mixed* triglycerides, which have two or three different kinds of fatty acid combined in a single molecule.

When only one or two of the hydroxyl groups of the glycerine molecule are esterified with fatty acid, the resulting glycerides are called *monoglycerides* and *diglycerides,* respectively. These occur in very small quantity in the fats as they are synthesized in nature, but are intermediate products of the hydrolysis of natural fats. Hence fat in the digestive tract contains a considerable proportion of mono- and diglyceride, and small quantities may be formed by incidental hydrolysis in the natural fats of commerce. Mono- and diglycerides are known mainly as laboratory products prepared for research or manufactured products for man's use. Basically, there are two types of monoglyceride, since the fatty acid radical must be attached either at the end or at the middle of the glycerol residue. There are four types of diglyceride structure, simple and mixed, with two possible positions in each case for the unesterified hydroxyl.

Man relies almost entirely on natural fats for his supply of triglycerides. They are transformed by blending, by frationation, by polymerization, by hydrogenation and by interesterification, but only a small amount of triglyceride fat is synthesized. When triglyceride is manufactured for some special reason, the reaction used is simple esterification, as follows:

$$C_3H_5(OH)_3 + 3R—CO—OH =$$
$$3H_2O + C_3H_5(O—CO—R)_3$$

If the stoichiometric quantities of glycerol and fatty acid are used, it is difficult to force this reaction to completion, and the resulting triglyceride is contaminated with free acid and with diglyceride. By use of an excess of fatty acid, diglyceride content can be held very low. By a combination of this precaution and repeated recrystallization, it is possible to prepare simple triglycerides such as tristearin of high purity, and to prepare mixtures of several mixed triglycerides practically free from other fatty material. For synthesis of any individual mixed triglyceride a more elaborate procedure is required.

If fatty acid is esterfied with glycerol in gradually increasing amounts over that shown in the above equations, the yield of triglyceride decreases at the expense of increasing amounts of di- and monoglyceride up to a limit fixed by the solubility of glycerine in the whole glyceride mixture. The esterification process is actually used in the manufacture of mono- and diglyceride mixtures for use in cosmetics and for other minor uses.

Mono- and diglycerides are manufactured mainly by reaction of glycerine with triglyceride fats, especially for their major use, i.e., in shortening and other edible products. For this reaction an alkaline catalyst is added, which can be an alcoholate, a soap or simply sodium hydroxide. The reaction product is an equilibrium mixture containing diglyceride, monoglyceride, and triglyceride. The production of mono- and diglyceride fat in the U.S. is about 100 million pounds a year.

Many of the physical and chemical properties of the glycerides closely parallel those of the fatty acids from which they are derived. Much of their behavior depends upon the hydrocarbon chains of their component fatty acids.

The most important reaction of triglycerides as such is hydrolysis. It proceeds stepwise, as follows:

$$C_3H_5(O{-}CO{-}R)_3 + HOH \rightarrow$$

$$C_3H_5(OH)(O{-}CO{-}R)_2 + R{-}CO{-}OH$$

$$C_3H_5(OH)(O{-}CO{-}R)_2 + HOH \rightarrow$$

$$C_3H_5(OH)_2(O{-}CO{-}R) + R{-}CO{-}OH$$

$$C_3H_5(OH)_2(O{-}CO{-}R) + HOH \rightarrow$$

$$C_3H_5(OH)_3 + R{-}CO{-}OH$$

Both of the final reaction products, fatty acid and glycerine, are commercially important and have many uses. One purpose of fat hydrolysis is the up-grading of the fatty acids by distillation, for use in soap. The hydrolysis is carried out at atmospheric pressure with use of a sulfonic acid catalyst or at higher pressure with use of a soap catalyst, e.g., zinc soap. Continuous countercurrent hydrolysis under high pressure is a particularly efficient process.

The outstanding property of monoglycerides is their emulsifying action. This was originally the sole basis of their commercial use. Today the chief value of fat containing monoglyceride is measured in terms of increased volume and tenderness of bakery products, but it is usually assumed, with or without proof, that monoglycerides act essentially as emulsifying agents.

A. S. RICHARDSON

Cross-references: *Fatty Acids, Vegetable Oils, Glycerol Fats.*

GLYCEROL

Glycerol (1,2,3-propanetriol, glycerin (U.S.P.), glycerine), the simplest trihydric alcohol, CH₂-OHCHOHCH₂OH, is one of the most important industrial chemicals; its production in the United States currently is at the rate of 360 million pounds a year, and it has thousands of industrial uses. The term "glycerol" is generally used for the pure chemical compound, whereas glycerine (or glycerin) commonly refers to commercial materials of whatever grade or purity. All three names, however, are often used without distinction.

Sources and Production. Glycerol occurs in nature esterified with fatty acids. These are the fats and oils of animal and vegetable origin and are called glycerides. Saponified with caustic to make soap or hydrolyzed with water to produce fatty acids, they yield glycerol as a by-product. The dilute glycerol is concentrated in vacuum evaporators to "crude glycerine," called soap-lye crude if from soap manufacture or saponification crude if from fatty acid manufacture. The crude glycerine is vacuum distilled (5 to 10 mm Hg) and steam is blown into the still to assist distillation. Vapors are condensed in a series of 2 or 3 hot condensers, and the distilled glycerine is further purified by bleaching with carbon. Glycerol is also obtained from fats as a by-product in the manufacture of higher alcohols. Synthesis of glycerol from propylene or allyl alcohol has become the major source in the past 20 years. This synthesis can follow any of several routes starting from propylene. There are three commercial routes now being used to prepare synthetic glycerine starting with propylene or propylene oxide. (See **Synthetic Glycerol**).

Another commercial route to synthetic glycerine is the hydrogenolysis of sugars.

Production of glycerine in the United States in 1971 was 326 million pounds of which approximately 160 million was from fats and the remainder synthetic.

Physical Properties. Colorless and practically odorless when pure. The taste is warm and sweet. Molecular weight: 92.09. Melting point: 18°C. Glycerol supercools very easily and is crystallized with difficulty. Flash point 177°C and fire point 204°C. Boiling points: 290°C at 760 mm; 222.4°C at 100 mm; 147.9°C at 4 mm Hg. Glycerol is about half as compressible as water.

Glycerol is hygroscopic, and is completely miscible with water, the lower alcohols, glycols, and phenols. It has limited miscibility with ether, acetone, ethyl acetate, and aniline. It is insoluble in hydrocarbons, chlorinated hydrocarbons and fats. Many organic and inorganic compounds are more or less soluble in it.

Physiological Action. Glycerol as it occurs in fat is a natural part of the diet. Taken as free glycerol, it is easily assimilated and forms glucose. In animal experiments ⅔ of the dietary starch was replaced by glycerol without ill effect. Tests with people have also shown it to be nutritious. Parenteral injection of glycerol may be harmful if it is not properly diluted to be isotonic. Glycerol is harmless to the skin and mucous membrane except that in high concentration it may be irritating because of its dehydrating action.

Chemical Properties. Glycerol has 3 alcoholic hydroxyl groups. Their positions may be indicated either by numbers or Greek letters. The β-hydroxyl is less reactive than the others. One or more of the hydroxyls may be replaced with other groups, such as ester or ether groups, or by halogens. A different type of group can be put in each position so that compounds of mixed character are obtained. Esters of the fatty acids are called mono-, di-, or triglycerides according to the number of acid groups and may also be named according to the acid groups present. Reaction with hydrochloric or hydrobromic acid yields mixtures of mono- and dichlorohydrins and mono- and dibromohydrins.

Glycerol reacts with alkalies, alkaline earths and certain metallic oxides or hydroxides to form glyceroxides which are analogous to the metallic alcoholates.

$$CH_2OH—CHOH—CH_2OH + NaOH \rightarrow$$
$$NaOCH_2—CHOH—CH_2OH + H_2O.$$

Reduction of glycerol with hydrogen yields propylene glycol.

Oxidation with different oxidizing agents yields a variety of aldehydic and acidic products and ultimately carbon dioxide and water. Glycerol is normally stable to atomospheric oxygen but is slowly oxidized in the presence of catalytic metals such as iron and copper.

When strongly heated, especially with a dehydrating agent (potassium acid sulfate is commonly used), acrolein is formed. This is a test for glycerol.

$$CH_2OH—CHOH—CH_2OH \xrightarrow{heat}$$
$$CH_2{=}CH—CHO + H_2O.$$

Chemical Derivatives. By far the most important derivatives of glycerol are its esters, and of these alkyd resins (principally esters of glycerol and phthalic anhydride) and ester gums (esters of glycerol and rosin) are of greatest commercial significance. (See **Alkyd Resins**).

Nitroglycerine (glycerol trinitrate), $C_3H_5(ONO_2)_3$, is a medically important ester. It is prepared by nitrating glycerol with "mixed acid," a combination of nitric and sulfuric acids, the latter functioning as a dehydrating agent (See **Explosives**).

The glycerol acetates or acetins are prepared by heating glycerol with acetic acid. Their primary uses are as solvents and plasticizers. Commercially available are monoacetin $[C_3H_5(OH)_2CH_3COO]$, a thick colorless hygroscopic liquid; diacetin $[C_3H_5(OH)(CH_3COO)_2]$, a colorless hygroscopic liquid; and triacetin $[C_3H_5(CH_3COO)_3]$, a colorless liquid with a slight fatty odor and bitter taste. The glycerol esters of fatty acids, the glycerides, occur naturally in all living matter, generally as triglycerides in fats and oils, but also as mono- and diglycerides in partially hydrolyzed fats with growing use in so called "convenience foods." (See **Glycerides**).

Glycerol can form mono-, di- or triethers with other alcohols or with itself (polyglycerols). It may also form inner ethers, such as glycidol.

$$CH_2—CH—CH_2OH$$
$$\diagdown \diagup$$
$$O$$

Epichlorhydrin, $CH_2—CH—CH_2Cl$,
with O bridge,

another inner ether, is prepared by reacting an alkaline substance with glycerol dichlorhydrin. (See **Ethers**.) The polyglycerols prepared by heating glycerol with an alkaline catalyst at 200–275°C range from viscous liquids to semisolids, and have uses similar to glycerol. Polyglycerols from diglycerol to triacontaglycerol have been prepared and are becoming commercially important. Ethers of aliphatic alcohols or phenols have not attained commercial importance, but are useful in synthesis and as solvents.

Glycerol amines have one or more of the hydroxyls of glycerol replaced by amine groups to yield products with properties of amines or an alcohol, depending upon the number of hydroxyls remaining. Their principal use is as intermediates.

Glycerol acetals and ketals, formed by condensing glycerol with aldehydes and ketones, are heterocyclic compounds showing structural, geometric and optical isomerism. The reaction, catalyzed by acid, is an equilibrium reaction sensitive to water. These acetals are useful solvents and plasticizers, and being fairly easily hydrolyzed, offer reactive groups for synthesis.

A number of substances that really form mixtures with glycerol are named as though they were derivatives. Mixtures of boric acid and glycerol, for example, are called glycerol borates or boroglycerides.

Uses. The wide applicability of glyceral is due to its unusual combination of properties. In addition to its utility as a chemical, it has many uses—as a plasticizer, humectant, solvent or emollient—based on its physical properties. Many applications of both classes, moreover, are possible because it is nontoxic and has no objectionable odor or taste. Chemical uses include the manufacture of alkyd resins and ester gums, nitroglycerine and monoglycerides. Principal physical uses are as a plasticizer for cellophane and certain types of paper; humectant in cigarettes; plasticizer for glue and gelatin in adhesives, capsules, and printers' rollers; vehicle, humectant, solvent, sweetening agent, and antifreeze in tooth pastes, pharmaceutical and cosmetic preparations; solvent and humectant in the beverage and food field, lubricant and humectant in the textile industry.

JOYCE C. KERN

Synthetic Glycerol

A wide variety of processes for synthesizing glycerol from propylene has been developed in the last 25 years. They are the subject of numerous patents and review articles.[1] Essentially two general routes to synthetic glycerol from propylene are known, a "chlorine" and a "nonchlorine" route (on a smaller scale, glycerol is also prepared by the hydrogenation-hydrogenolysis of sugar[2]).

I. In the "chlorine route," propylene is chlorinated at elevated temperatures to allyl chloride. The allyl chloride can be converted by either of two routes:
   a. It is hydrolyzed in the presence of a base to allyl alcohol, which is hydrochlorinated with a solution of chlorine in water to a mixture of glycerol monochlorohydrins (3). These, in turn, are hydrolyzed to glycerol.

b. It is hydrochlorinated with a mixture of chlorine and water to 1,2 and 1,3 glycerol dichlorohydrins. The dichlorohydrins are hydrolyzed in the presence of a base to glycerol.

II. Two major processes are known for the so-called "nonchlorine" route to glycerol:

a. Propylene is oxidized to acrolein over catalysts (3). Acrolein recovered by water extraction and distillation is either converted to glycidaldehyde, H_2C—CH—CHO, by ep-

$$\underset{\diagdown O \diagup}{}$$

oxidation with hydrogen peroxide or reduced to allyl alcohol. If allyl alcohol is the intermediate, it is converted to glycidol, H_2C—CH—CH_2OH, by epoxidation with hy-

$$\underset{\diagdown O \diagup}{}$$

drogen peroxide or peracetic acid; glycidol, in turn is easily hydrolyzed to aqueous glycerol. If glyceraldehyde is the intermediate, it is reduced to glycidol.

b. In a process brought on stream in 1969, (4), propylene oxide is manufactured by the reaction of a hydroperoxide (such as *tert* butyl hydroperoxide) with propylene. The propylene oxide is rearranged to allyl alcohol over trilithium phosphate catalyst. The allyl alcohol is converted to glycerol with peracetic acid.

A combination of some of the synthetic steps reported under either the "chlorine" or the "nonchlorine" route can also be used, since allyl alcohol can be the intermediate in either. It is interesting that the new processes have not displaced the older ones. Indeed the final choice of the desired process depends on the availability or on the price of the raw material (such as chlorine) and on the desirability of preparing the intermediate for its own sake. The "chlorine" route intermediate, epichlorohydrin is, for example, a fast growing product (3). Both acrolein (5) and propylene oxide (6) are valuable chemicals; allyl alcohol (7) and allyl chloride (8) are widely sold. Since glycerol, a high-boiling, heat-sensitive material, is subject to rigid specifications, the preparation of all precursors is carried out with the greatest of care to minimize impurities.

In the final step of either route I or II, an aqueous solution of glycerol is obtained by hydrolysis. As the first step in the purification, water is removed in evaporators. When salts are present in large amounts (such as in the "chlorine" route), they crystallize and are removed by filtration. Vacuum distillation then gives a crude colored product. Color bodies and some malodorous impurities can be now removed by extraction with a hot hydrocarbon solvent. The extracted glycerol still contains organic impurities such as polyols and ethers and some water, so the crude product is purified ("topped and tailed") by vacuum distillation. In certain cases where very high-purity glycerol is needed, the finished product can be treated with charcoal.

G. H. RIESSER

References

1. Kern, Joyce C., in Kirk-Othmer, "Encyclopedia of Chemical Technology," Second Edition, Volume 10, John Wiley and Sons, New York, 1967.

2. Lenth, C. W., and Du Puis, R. N., *Ind. Eng. Chem.*, **37,** 152 (1945).

3. Lichtenwalter, G. D., and Riesser, G. H., in Kirk-Othmer, "Encyclopedia of Chemical Technology," Second Edition, Volume 5, John Wiley and Sons, New York, 1964.

4. *Chem. Eng.* **75,** April 22, p. 78 (1968).

5. C. W. Smith, Ed., "Acrolein," John Wiley & Sons, Inc., New York, 1962.

6. Horsley, L. H., in Kirk-Othmer, "Encyclopedia of Chemical Technology," Second Edition, Vol. 16, John Wiley and Sons, New York, 1968.

7. Pilorz, B. H., in "Chlorine: Its Manufacture, Properties and Uses," ACS Monograph 154, Van Nostrand Reinhold Co., New York, 1962.

8. "Allyl Alcohol," Technical Publication SC; 46–32, Shell Chemical Corporation, San Francisco, 1946.

GLYCOLS

Glycols are generally accepted as compounds having two hydroxyl groups attached to separate carbon atoms in an aliphatic carbon chain. The term polyglycols is often applied to low and high-molecular-weight straight-chain, oxycarbon polymers having two hydroxyl groups. Glycols are also known as diols, although use of this Geneva designation is commonly limited to long-chain compounds of the aliphatic carbon series. Commercially available glycols find widespread application in the automotive, aviation, explosives, textile, surface coatings, food, cosmetic, pharmaceutical, tobacco, petroleum, and other industries.

The low molecular weight glycols are stable, odorless, water-white liquids that boil above 100°C and freeze or set to a glass below 0°C. They are hygroscopic and have a selective solvent action for dyes, synthetic resins, essential oils, and certain natural gums and resins. Their water solubility decreases with increasing molecular weight. Examples of these glycols are ethylene glycol, propylene glycol (1,2-propanediol), trimethylene glycol (1,3-propanediol), butylene glycols (1,2- and 1,3-butanediols), hexamethylene glycol (1,6-hexanediol), etc.

The polyglycols are stable, high-boiling or nonvolatile, odorless materials. Polyethylene glycols range from water-white liquids to wax-like solids. As molecular weight increases, specific gravity, flash point, and viscosity increase and the melting point increases to a maximum of about 64 to 66°C; vapor pressure and hygroscopicity decrease. Polyethylene glycols even of high molecular weights are completely soluble in water, although the degree of solubility in solvents generally decreases with molecular weight.

Commercial polypropylene glycols are colorless, viscous liquids with molecular weights up to about 5,000. They are more oil-soluble and less hygroscopic than the comparable polyethylene glycols. Higher molecular weight, crystalline polypropylene glycols, which are solids melting above 70°C, are known but are not commercially available. These special crystalline polymers have along the polymer chain successive asymmetric centers of identical configuration, molecular weights approaching a million, and are termed isotactic polymers.

The simple glycols have, in general, a relatively low order of toxicity and certain of them are considered safe for use in pharmaceuticals, cosmetics, and, in the case of propylene glycol, foodstuffs.

Glycols, like most simple alcohols, can be used to prepare esters, ethers, acetals, and other typical alcohol reaction products. Many of these derivatives particularly the esters and ethers are of major commercial importance.

Ethylene Glycol. Discovered in 1859 by Wurtz, this derivative first became commercially important about 1925 when Union Carbide went into large-scale manufacture at South Charleston, West Virginia, using the ethylene chlorohydrin process. It also has been made commercially from formaldehyde. The modern process is oxidation of ethylene to produce ethylene oxide which is subsequently hydrated to ethylene glycol. Annual U.S. production of ethylene oxide was over 3.6 billion pounds in 1971, part of which was used to produce over 3 billion pounds of ethylene glycol. The major producers of oxide and glycol are Union Carbide, Dow, and Jefferson. Permanent-type antifreeze is the largest end use of ethylene glycol, consuming 66% of the 2.4 billion pounds of ethylene glycol production in 1969. This glycol, containing rust and oxidation inhibitors, is marketed under several antifreeze trademarks. Of secondary but growing importance is its use as a polyester fiber intermediate. Significant amounts of ethylene glycol are consumed as hydraulic fluids, heat-transfer agents; and humectants for cellophane, fibers, paper, leather, and adhesives.

The monoethers of ethylene glycol are used as solvents and chemical intermediates, particularly in the protective coatings industry. Methyl, ethyl, and butyl derivatives constitute the bulk of these commercial products. Diethers are stable and useful as solvents and coupling agents. One diether, dioxane, is a cyclic compound which is an excellent solvent for many resins, waxes, dyes, oils, and organic and some inorganic compounds. It can be made by dehydrating ethylene glycol.

Acetals and ketals, which are essentially cyclic diethers, are formed by condensing aldehydes and ketones with ethylene glycol. Ethylene glycol is converted in the vapor phase to 2-hydroxymethyl-1,3-dioxolane, a substituted cyclic acetal. Ethylene glycol esters of organic acids can be prepared by reacting the glycol with acid, anhydride, or acid chloride, thus forming the mono- or diester depending on reactants mole ratio. The derivatives can also be obtained from the organic acid and ethylene oxide under acid or base catalysis. These esters find use as plasticizers, solvents, and heat-transfer media. Glycol diformate is useful in processes requiring a gradual release of formic acid; ethylene glycol diacetate is a solvent in printing ink, lacquers, fluorinated refrigerant gases, and for refining lubricating oils and removing free fatty acids from oils and fats. Ethylene glycol fatty acid esters are emulsifying, stabilizing, dispersing, wetting, foaming, and suspending agents. Polyesters result from reacting ethylene glycol with polybasic acids or their derivatives (see **Alkyd Resins, Polyester Resins**).

Diethylene Glycol. This glycol is the first member of a series of polyethylene glycols having the general formula $HOCH_2CH_2(OCH_2CH_2)_nOH$; for di-

ethylene glycol $n = 1$. Also known as diglycol, 2,2'-oxydiethanol, and bis-(2-hydroxyethyl) ether, it is obtained as a coproduct in the manufacture of ethylene glycol from ethylene oxide. In 1968, 226 million pounds were produced.

Diethylene glycol combines most of the properties of ethylene glycol and glycerol, but has added solvent powers imparted by the ether group. It is used in plasticizers, humectants for tobacco, in printing inks, glue, brake fluids, antifreezes, gas treating, explosives, resins, as a solvent, and in textile fibers. Diethylene glycol exhibits the chemical properties of a primary alcohol and an ether. The monomethyl-, monoethyl- and monobutyl ethers are excellent solvents and coupling agents in formulations for surface coatings, printing inks, hydraulic fluids, dry cleaning soaps, and other compositions.

Triethylene Glycol. The second member of the polyethylene glycol series is $n = 2$. Known also as triglycol, it is obtained as a coproduct in the manufacture of ethylene and diethylene glycols. The mole ratio of ethylene oxide and water determines in part the amount of these various coproducts obtained. About 73 million pounds were produced in 1968. Triethylene glycol does possess antibacterial properties that make it useful as an air disinfectant. This glycol is used in the manufacture of plasticizers for plastic interlayers in safety glass, in natural gas and oil processing, as a humectant, and solvent.

Polyethylene Glycols. Polyethylene glycols continue the series of ethylene oxide polymers with n running from 4 up to more than 100,000. The simple glycol derivatives up to tetraethylene glycol are usually isolated as distilled, pure compounds, but the polyethylene glycols sold, on the other hand, are composed of a molecular weight distribution of molecules, a consequence of the fact that they are derived by the base- or acid-catalyzed polymerization of ethylene oxide onto water or glycol. Union Carbide was the first to produce this family of commercial products in the U.S., and they are called "Carbowax" Compounds; the numerical designation (400, 600, 2000, 6000, etc.) refers to the average molecular weight of the derivatives. These compounds find wide use as lubricants, vehicles, solvents, binders and as chemical intermediates in the pharmaceutical, cosmetic, rubber, textile, petroleum, metal, electronic, agriculture, and other industries.

The highest members of this family are commercially available from Union Carbide under the trademark Polyox water-soluble resins. Polyox coagulant grade is reported to be over 5 million molecular weight; its melting point is 65°C. These unusual products have the extraordinary ability to gel water in as low as 1 to 2% aqueous solutions and to reduce very significantly the turbulent frictional drag of water in pipes and hoses when used in ppm concentration. The polymer can be extruded, cast, calendered, and molded, is an excellent binder, complexes with iodine and ureas, and has application in adhesives, paper coatings, toothpaste and water-soluble packaging films. Its friction-reducing properties make it useful in mining and in firefighting.

Propylene Glycol. Like ethylene glycol, propylene glycol (1,2-propanediol) was first prepared by

Wurtz in 1859, but it did not become commercially available until Union Carbide produced it in 1931. Current major producers are now Dow, Jefferson, Carbide, Olin-Mathieson, Wyandotte, and Celanese. The 1970 U.S. capacity was about 280 million pounds per year. It is generally produced by hydration of propylene oxide in much the same manner as ethylene glycol. The largest producer of propylene oxide is Dow, accounting for about half of the U.S. capacity of 1.7 billion pounds per year. Until recently propylene oxide came from propylene via the chlorohydrin route. A peroxidation route using propylene is now becoming commercially important and is practiced by Oxirane Chemical Company.

Propylene glycol is a colorless and odorless liquid with a slight sweet taste. It is miscible with water and many organic solvents and finds use in dissolving ingredients in medicinal preparations, essential oils for fine fragrances, and in combination with other solvents, some resins and paints. Propylene glycol finds extensive use in mold growth inhibition, food-color solvent systems, and in the food, cosmetic, and drug industries. The chemistry of propylene glycol nearly parallels that of ethylene glycol with the important exception that one of the hydroxyl groups in propylene glycol is secondary.

Dipropylene Glycol. This glycol is the first of a series of polypropylene glycols having the general formula $HO(C_3H_6O)_nC_3H_6OH$, n = 1. Dipropylene glycol and most of the other derivatives of propylene glycol are prepared from propylene oxide by methods quite similar to those used for ethylene oxide derivatives.

Dipropylene glycol exists in three structurally isomeric forms in which the two hydroxyl groups are primary-primary, secondary-primary, and secondary-secondary. Commercially, dipropylene glycol is a mixture of all three isomers with the composition of each isomer varying widely depending on the commercial process used to convert propylene oxide. One of the commercial processes produces dipropylene glycol that is about 70% primary-secondary isomer, (2-(2-hydroxypropoxy)-1-propanol) and 30% secondary-secondary isomer, (1,1-oxydi-2-propanol). Some commercial dipropylene glycol is about equal parts of primary-secondary and secondary-secondary isomers with traces of primary-primary isomer.

Dipropylene glycol is similar to other glycols in general characteristics. It is used as a water-miscible solvent in steam setting types of printing ink, in cosmetics, and as a chemical intermediate.

Polypropylene Glycols. Commercial polypropylene glycols are known in the polymer series mentioned above with n reaching a maximum of about 100. They are produced by base-catalyzed processes analogous to those used generally to produce polyethylene glycols. The commercial fluids are liquids reaching limited water solubility at about 700 to 900 molecular weight. The major producers of these polyglycols are Dow, Carbide, Jefferson, Olin-Mathieson, and Wyandotte. About 900 million pounds of propylene oxide were consumed in 1970 in the manufacture of polypropylene glycols.

The polyglycols are used as chemical intermediates for esters which find application as emulsifying agents, petroleum demulsifiers, wetting agents, mold lubricants for rubber, hydraulic brake fluids and as coupling agents in textile, cosmetic, and paper manufacture. A major use is in flexible and rigid polyether urethane foam manufacture.

Copolymers of propylene oxide and ethylene oxide, both block and random in configuration, find wide use as surface-active agents, industrial lubricants, hydraulic and heat-transfer agents, textile and rubber-processing and mold-release agents, engine lubricants, hair-dressings and leather processing. The copolymers are sold under tradenames such as Polyglycol 15 (Dow), Ucon brand (Union Carbide trademark), and Pluronic (Wyandotte trademark).

Higher Glycols. A few higher glycols are of commercial importance: 2-methyl-2,4-pentanediol; 2-ethyl-1,3-hexanediol; 1.3-butanediol; and 2,3-butanediol. 2-Methyl-2,4-pentanediol is produced by the liquid-phase hydrogenation of diacetone alcohol and is often called "hexylene glycol." It is used in the preparation of 2,4-dimethylsulfolane, a selective solvent of value in petroleum extraction processes. Hydrogenation of butyraldol leads to 2-ethyl-1,3-hexanediol, which finds wide use as an insect repellent. Other uses are as a solvent in inks, coupling agents, dry-cleaning, plasticizer, and in alkyd resins. 1,3-Butanediol is produced by reducing acetaldol and is used as a plasticizer, humectant, coupling agent, and in printing ink. 2,3-Butanediol exists as a crystalline solid or a viscous liquid depending on the isomeric form. It is obtained commercially from butylene oxide.

ROBERT K. BARNES

GLYCOSIDES, STEROID

In plants, steroids occur as glycosides, as acyl glycosides, as esters, and in the free form. Many of the steroid glycosides are important drugs or starting materials for the partial synthesis of drugs.

Sterolins. Although the sterols are largely in the free and esterified form, plants contain significant quantities of sterolins (steryl glycosides and acyl glycosides). As a rule, the 3-hydroxyl group of the sterol is linked to glucose or some other common sugar to form a heteroside, but other hydroxyl groups in the sterol molecule and higher saccharides may be involved. The two most common sterol aglycones in higher plants are sitosterol (previously called β-sitosterol) and stigmasterol, shown in Fig. 1. Cholesterol is usually present in very small amounts. (See **Cholesterol**). The biosynthesis of sitosterol differs from that of cholesterol in the alkylation of a precursor at C-24 and involves the successive introduction of two methyl groups. Stigmasterol is formed by the dehydrogenation of sitosterol.

Cholesterol and the C_{29} sterols are used by plants as the starting materials for the biosynthesis of other steroids with the same number of carbons, such as the insect-molting hormones and the steroidal sapogenins and alkaloids or they are converted to progesterone and other steroids with a lower number of carbons. Fig. 1 shows the structure of one representative of the insect-molting hormones, ponasterone A, which has been isolated from plants in the form of its 3-glucoside, ponasteroside A.

Steroidal Saponins and Glycoalkaloids. Certain

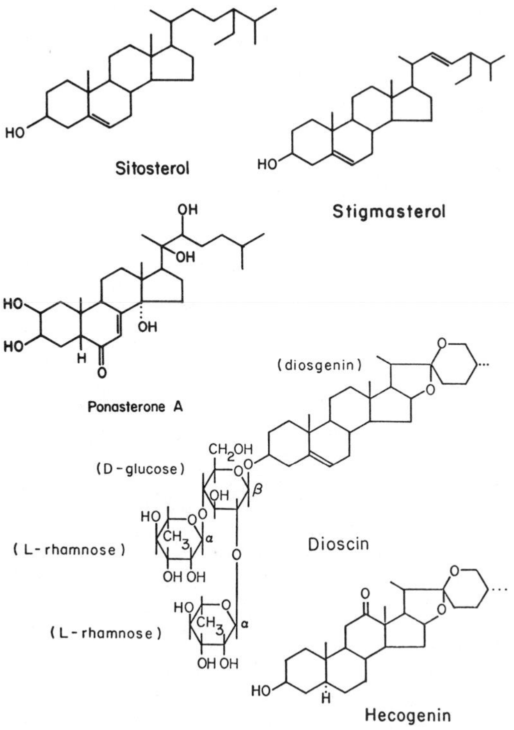

FIG. 1

in potatoes and other *Solanum* species in the form of various glycosides, the solanines and chaconines. For instance α-chaconine differs from dioscin (Fig. 1) only in the nature of the aglycone. The glycoalkaloids are likewise synthesized by plants from cholesterol.

Saponins and glycoalkaloids are characterized by their surface activity and hemolytic effect as well as their ability to form complexes with cholesterol and similar sterols. The best-known cholesterol-precipitating agent is digitonin, a saponin in *Digitalis*. While ingested saponins are nontoxic to warm-blooded animals, the glycoalkaloids are toxic. Tomatine and other glycoalkaloids have antifungal and cytostatic activity.

Pregnane Derivatives. The degradation of cholesterol in plants and animals produces pregnenolone (Δ⁵-pregnen-3β-ol-20-one), which is oxidized to progesterone (Δ⁴-pregnene-3,20-dione). (See **Steroid hormones**). Various plants contain neutral pregnane derivatives with C and D rings in cis-fusion in the form of glycosides. They have been called digitanol glycosides, because they were first isolated from *Digitalis* plants, and contain the rare hexoses otherwise only found in the cardiac glycosides. Fig. 2 shows an example of the digitanols, digipurpurogenin I, which occurs in *Digitalis purpurea* as digipurpurin, a glycoside in which three molecules of D-digitoxose are attached to the 3-hydroxyl group.

Plants belonging to the Apocynaceae and Buxaceae aminate the pregnane derivatives to produce alkaloids, which may be present as either glycosides or esters. Only one example of Apocynaceae alkaloid is shown in Fig. 2, funtuphyllamine A, which was isolated from *Funtumia africana*. The alkaloids of kurchi bark (*Holarrhena antidysenterica*)

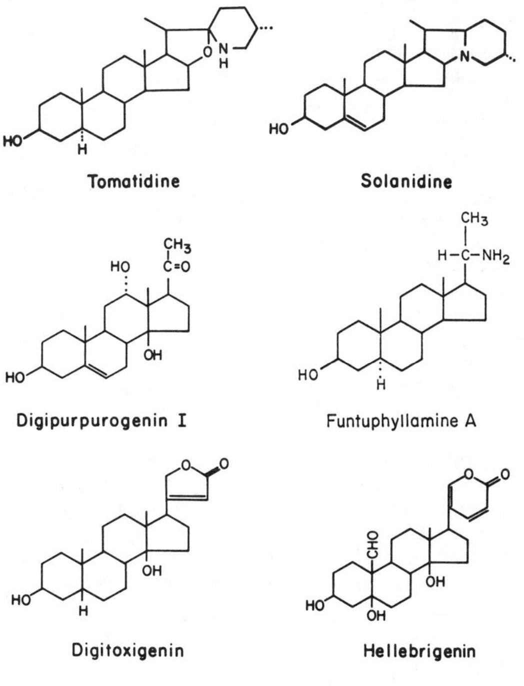

plants have the ability to hydroxylate cholesterol stereospecifically at C-26 or C-27. A glycoside of cholesterol with a glucose residue attached to this terminal hydroxyl group and chacotriose (2L-rhamnoses + D-glucose) attached to its 3-hydroxyl group is a biogenetic precursor of the saponin, dioscin, shown in Fig. 1. Steroidal saponins are glycosides of spiroketals, which form spontaneously, when the terminal sugar in an analogous 16-hydroxy-22-keto-steroid is enzymatically removed. While the configuration at C-22 is the same in all natural sapogenins, the orientation of the methyl group at C-25 depends on the position of the terminal hydroxyl group in the sterol precursor. In the D- or isosapogenin series the methyl group is α-oriented (equatorial), as in diosgenin (see Fig. 1), whereas in the L-, normal, or neosapogenin series it is β-oriented (axial).

Sapogenins are widely distributed in monocots belonging to the genera *Yucca, Trillium, Chlorogalum, Smilax, Nolina, Agapanthus, Agave, Manfreda*, and *Dioscorea*, and in dicots belonging to the genera *Digitalis, Solanum, Lycopersicon*, and *Cestrum*. Diosgenin (Fig. 1) from Mexican barbasco root (*Dioscorea* tubers) and hecogenin (Fig. 1) from wastes of African sisal fibers (*Agave sisalana*) are important starting materials for the commercial preparation of synthetic hormones.

The glycoalkaloids are nitrogen analogs of the saponins, occurring in the Solanaceae. Two of their aglycones are shown in Fig. 2. Tomatidine occurs in the form of a glycoside, tomatine, in tomato vines and may also be used for the partial synthesis of steroid hormones. Solanidine is found

FIG. 2

have many interesting pharmacological properties and one of them, conessine, is used as an amebicide and as a starting material for the partial synthesis of aldosterone.

Cardiac Glycosides. About eleven plant families are known to elaborate cardiac glycosides. Their genins have either 23 (cardenolides) or 24 (bufadienolides) carbon atoms and their sugars are not found elsewhere in nature. Cardenolides and bufadienolides have not been found together in the same genus. Both types of genins are synthesized in plants from a C_{21} steroid, usually progesterone, and contain a 14β-hydroxyl group.

Fig. 2 gives one example of a cardenolide, digitoxigenin, and one representative of the bufadienolides, hellebrigenin. *Digitalis* plants contain three cardenolides: digitoxigenin, gitoxigenin (16β-hydroxy digitoxigenin), and digoxigenin (12β-hydroxydigitoxigenin). These genins are combined with 2 molecules of digitoxose, 1 molecule of acetyldigitoxose, and 1 molecule of glucose to form the lanatosides (digilanides) A, B, and C, respectively. When the acetyl group is removed by mild alkaline hydrolysis, one obtains the corresponding purpurea glycosides (desacetyllanatosides or desacetyldigilanides). Enzymatic removal of the glucose unit, on the other hand, gives the acetyl derivatives of the digitoxose triosides. Combination of both hydrolytic procedures yields the digitoxose triosides digitoxin, gitoxin, and digoxin.

The bufadienolides occur in plants as well as animals, but only in plants are they in the form of glycosides. Their 3-hydroxyl group is attached to glucose, rhamnose, or thevetose. Hellebrigenin (Fig. 2), which is also known as bufotalidin, occurs in the rhizomes of the Christmas rose and other *Helleborus* species in the form of a rhamnoside. Bufadienolides have so far been found in plants in only two families, the buttercup and the lily family.

Crude leaf preparations of *Digitalis* have been in medical use since 1785. Pure cardiac glycosides are now available. These preparations in injectable tinctures or powdered leaf tablets are used extensively for the treatment of congestive heart failure. They increase the force of the heart muscle and the power of systolic contraction, apparently by inhibiting the active transport of K^+ and Na^+ ions through cell membranes.

ERICH HEFTMANN

Cross-references: *Steroids, Cholesterol, Steroid hormones.*

References

Manske, R. H. F., ed., "Alkaloids. Chemistry and Physiology," Academic Press, Vol. VII, 1960; Vol. IX, 1967; Vol. X, 1968.

Heftmann, E., "Biosynthesis of Plant Steroids," *Lloydia*, **31**, 293–317 (1968).

Heftmann, E., "Biochemistry of Steroidal Saponins and Glycoalkaloids," *Lloydia*, **30**, 209–30 (1967).

Reichstein, T., "Glycosides of Cardenolides and Pregnanes," *Naturwiss.*, **54**, 53–67 (1967).

GOLD AND COMPOUNDS

Gold (atomic weight 196.9665, atomic number 79, Group IB of Periodic Table) was probably the first metal known to man. It occurs in nature in the elementary state, has a characteristic yellow color, does not tarnish in air and goes through fire without oxidation or change in color. It is widely distributed but usually in too low a concentration to allow recovery.

Gold has a density of 19.32, a Brinell hardness of 58 in the hard and 25 in the annealed condition; its tensile strength, hard, is 32×10^3 psi and, annealed 19×10^3. Its coefficient of expansion is 0.000014 per °C; the electrical resistance is 2.06 microhm cm. It is extremely ductile, being easily drawable to fine wire and can be rolled or beaten to foil or leaf of a thickness of less than one micron. Its melting point is 1063°C and the boiling point is 2808°C.

Chemically, gold is one of the most stable ("noble") metals; it does not oxidize in air or oxygen at any temperature and resists almost all acids. It does dissolve readily in aqua regia and is also attacked by the halogens; hydrocyanic acid and the alkali cyanides dissolve gold in the presence of oxidizing agents. Gold readily forms alloys with most of the metals, the most important of which are the alloys with silver and with copper. Still more important are the ternary alloys of gold, silver and copper, which constitute the alloys used in jewelry; the alloy color depends on the proportions of the silver and copper. Where silver predominates the alloy is "green" gold; where silver and copper are present in approximately equal amounts the alloy is "yellow gold" and where copper predominates the alloy is "red gold" or "pink gold." "White gold" is an alloy of gold and nickel. In this, as well as in the vari-colored golds, zinc is usually added to improve the workability.

Gold forms two series of compounds, with valencies of 1 and 3. There is a strong tendency to form complex salts, derived from ions such as $[Au(CN_2)]_2^-$ and $[AuCl_4]$. Probably the earliest compound was the compound of gold and tin oxide known as "purple of Cassius" and much used by the ancients for dyeing of cloth. The most important compounds are, however, the trichloride and the cyanide; the former is the basis for many complex salts like potassium and sodium chloroaurates, widely used for photographic and medicinal purposes respectively; the latter is the salt universally used as the electrolyte in the electroplating industry.

For centuries gold has been used as the monetary standard in most nations of the world; its permanance and its rarity, compared with most base metals, make it suitable for this purpose. Extremely fine gold leaf is widely used for decoration or protection uses and in new surgical techniques; the gold is usually 22 carat and is obtained by the extremely old gold-beating process; since it permanently retains its color, gold leaf finds extensive use for sign writing, book-lettering, etc. "Liquid gold" is finely divided gold in suspension in certain vegetable oils, and is used for applying the pleasing gold designs on china, etc.; the liquid is simply painted on the porcelain or china which thereafter is fired, fixing the gold on the surface. Colloidal solutions of gold are composed of particles about 0.5 micron in diameter suspended in water; they are stabilized by electric charges (electrocratic solution). Industrially they are used to produce the characteristic color of ruby glass.

Gold is used increasingly in electronic and space

applications which consume over 1 million ounces annually. It is an important electrical contact material in computers, for example, where the thousands of sensitive contacts per unit must not suffer even slight oxidation. On other electronic components it provides excellent infrared reflectivity, corrosion resistance, solderability, and low noise contact reliability. The reflectivity of gold in the infrared region is about 97%. For this reason it is used in space applications, such as a coating on space ships, on the astronauts' face shields, and on the tether line (umbilical cord) connecting the astronaut to the space ship during space walks. It is used also for its heat reflectivity on windows in cars and buildings. For the above applications the gold film is deposited on surfaces by vacuum evaporation or by electrodeposition.

FRED E. CARTER
Revised by CLIFFORD A. HAMPEL

References

Robinson, H. W., "Gold," pp. 244–250, "Encyclopedia of the Chemical Elements," Hampel, C. A., Editor, Van Nostrand Reinhold Co., New York, 1968.
Jewelers' Circular-Keystone, CXXXVII, No. 9, Part II (June, 1967). (Editors' Note: This special issue contains 29 articles, 206 pages, covering the many facets of the technology, history and applications of gold.)

GRIGNARD REACTIONS

The Grignard reagents and their reactions are so called after Victor Grignard (1871–1935), who discovered them about 1900, and who for this achievement received a share (with Paul Sabatier) of the Nobel prize in chemistry for 1912.

The reagents are usually defined and represented as organomagnesium halides of the general formula $RMgX$, in which R signifies a carbon-linked organic radical, and X a halogen other than fluorine. Actually, Grignard systems, usually ethereal, are much more complex, comprising equilibrium mixtures of a variety of solvated components, some of which are indicated in the following simplified equations:

$$2RMgX \rightleftharpoons R_2Mg + MgX_2$$
$$2RMgX \rightleftharpoons [R_2MgX]^- + [MgX]^+$$
$$RMgX + MgX_2 \rightleftharpoons [RMgX_2]^- + [MgX]^+$$
$$RMgX + MgX_2 \rightleftharpoons [MgX_3]^- + [RMg]^+$$
$$RMgX + R_2Mg \rightleftharpoons [R_2MgX]^- + [RMg]^+$$
$$RMgX + R_2Mg \rightleftharpoons [R_3Mg]^- + [MgX]^+$$
$$R_2Mg + MgX_2 \rightleftharpoons [RMgX_2]^- + [RMg]^+$$
$$R_2Mg + MgX_2 \rightleftharpoons [R_2MgX]^- + [MgX]^+$$
$$2R_2Mg \rightleftharpoons [R_3Mg]^- + [RMg]^+$$
$$2MgX_2 \rightleftharpoons [MgX_3]^- + [MgX]^+$$

In solvents of such low dielectric constant as, *e.g.*, ethyl ether the degree of ionic association is presumably high.

The method of preparation employed by Grignard, and still predominant in laboratory practice, consists in the gradual addition of an ethereal solution of alkyl, aralkyl, cycloalkyl, or aryl halide to ether-covered magnesium in a suitable state of physical division. Stoichiometrically, the desired Grignardization may be represented by the equation:

$$RX + Mg \rightarrow RMgX.$$

Initiation of Grignardization of the more reactive halides may usually be effected by simple warming. A wide variety of "activators" has been recommended for use in small amounts for the facilitation of more difficult inductions, including iodine, iodinated magnesium, a more reactive halide, and a pre-formed Grignard reagent. Once started, Grignardization reactions in general tend toward self-acceleration. In some cases the exothermic process must be controlled by moderative cooling. Only for the least reactive halides is continued heating throughout the preparation necessary or desirable.

Although high yields (90–95%, or better) are attainable in favorable cases, the Grignardization process is usually accompanied by side reactions, of which the Wurtz and disproportionation reactions are those most often encountered. Without implication as to details of reaction mechanisms, over-all material diversions are indicated in equations (1) and (2a, b):

$$2RX + Mg \rightarrow [2R] + MgX_2 \qquad (1)$$
$$[2R] \rightarrow R_2(Wurtz) \qquad (2a)$$
$$[2R] \rightarrow R_{(+H)} + R_{(-H)}(disprop'n) \qquad (2b)$$

Whether the side reaction takes the Wurtz (2a) or the disproportionation (2b) course, exclusively or predominantly, depends primarily upon the nature of the radical R.

Under ordinary Grignardization conditions a few halides (notably the allylic and some benzylic) react readily with their own Grignard reagents to form Wurtz products:

$$RX + RMgX \rightarrow R_2 \rightarrow MgX_2$$

Special techniques have been developed for the mitigation of this and other atypical preparational adversities.

On the basis of (1) ease of initiation of reaction with magnesium and (2) rate of disappearance of magnesium after reaction begins, halides of like constitution decrease in Grignardization reactivity in the order $RI > RBr > RCl$. With the halogen remaining the same throughout, the decreasing order of halide reactivity conforms roughly to the series: trityl > benihydryl > alkyl >allyl and benzyl > cycloalkyl and s-alkyl > primary alkyl > aryl.

Of the methods of Grignard reagent preparation other than halide Grignardization those of greatest importance may be grouped under the general classification of displacement of carbon-linked hydrogen through the agency of an expendable Grignard reagent:

$$R'H + RMgX \rightarrow R'MgX + RH$$

Typical of the "active hydrogen" compounds amenable to this treatment are such hydrocarbons as acetylene and the monosubstituted acetylene-cyclopentadiene, indene, and fluorene, and such hetero-cycles as the pyrroles and indoles.

Of the various methods of Grignard reagent estimation that have been more or less thoroughly explored, only two are both convenient and reasonably accurate. The gas-analysis method is limited in applicability to reagents that yield gases on hydrolysis:

$$RMgX + H_2O \rightarrow RH_{(gas)} + MgXOH$$

The more generally applicable acid-titration method is based upon the basicity of the inorganic hydrolysis product of the Grignard reagent (MgXOH). In practice it is convenient to use a moderate excess of standard mineral acid (suitably sulfuric) and back-titrate with standard sodium hydroxide solution with phenolphthalein indicator. This method is reliable only if the Grignard reagent solution has been adequately protected from moisture and atmospheric contamination.

There are relatively few functional groups that do not react with Grignard reagents in some fashion. Among the classes of compounds that, in general, give sufficiently high yields of so-called "normal" products to constitute the basis for general preparative methods are: the aldehydes and ketones, the carboxylic esters, the carbonyl halides, the carboxylic anhydrides, the amides, the nitriles, the epoxides (notably ethylene oxide), the nocyanates, the aldimines, the boron and silicon halides and esters, and the phosphorus halides. The treatment of a Grignard reagent with carbon dioxide is a general method for the preparation of carboxylic acids, and alcohols may be obtained by the oxygenation of aliphatic Grignard reagents. Some organic halides (notably the allylic and some trylated methyl) couple with Grignard reagents to give good yields of hydrocarbons.

The over-all reaction of a Grignard reagent with a carbonyl compound is an essentially ionic addition at the carbonyl double bond.

$$R'R''C{=}O \xrightarrow{RMgX} R'R''RC{-}OMgX \xrightarrow{H_2O}$$
$$R'R''RCOH + MgXOH$$

In the special case of formaldehyde ($H_2C{=}O$) the "normal" end-product is a primary alcohol, with other aldehydes ($R'HC{=}O$) a secondary alcohol, and with ketones ($R'R''C{=}O$) a tertiary alcohol.

Kinetically, these reactions appear to be third-order, presumably involving one carbonyl-Grignard Werner complex aggregate and one molecule of Grignard reagent. It has been suggested that a quasi six-membered ring transition state plays a part in the reaction mechanism.

The chief reactions potentially competitive with the "normal" addition are reductions of the carbonyl compound and enolization (which may or may not lead to condensation).

Only Grignard reagents with a labile *beta* hydrogen atom are capable of effecting direct reduction of carbonyl compounds. The end products are an alcohol (corresponding to the carbonyl compound) and an olefinic compound (corresponding to the Grignard reagent). Such reductions appear to be bimolecular, and it has been suggested that they too may involve a quasi six-membered ring transition state.

In some cases at least, Grignard reagent reduction may be materially suppressed (with consequent increase in normal addition) by addition to the sanction system of a non-reducing complex-former, suitably magnesium halide. In terms of the hypothesis suggested, the trimolecular addition is thus favored over the bimolecular reduction.

The carbonyl compound may undergo a Meerwein reduction if it yields an alcoholate with a labile *alpha* hydrogen atom:

$$R'CHO + RMgX \rightarrow RR'CHOMgX$$
$$R'CHO + RR'CHOMgX \rightarrow$$
$$R'CH_2OMgX + RR'CO$$

Oxygen contamination of a "non-reducing" Grignard reagent (*e.g.*, $C_6H_5CH_2MgX$) to form the alcoholate (*e.g.*, $C_6H_5CH_2OMgX$) may also lead to Meerwein reduction. A bimolecular mechanism involving a quasi six-membered ring transition state (analogous to that suggested for Grignard reagent reductions) may be postulated for Meerwein reductions.

Ketones (and, very rarely, aldehydes) may undergo one-electron reduction to form pinacols when metallic subhalide free radicals are present, or readily available, in the reaction system. The usual cause of this phenomenon is residual metallic magnesium which, in the presence of magnesium iodide or bromide (but not chloride), behaves as though it were in equilibrium with magnesium halide:

$$Mg + MgX_2 \rightleftharpoons 2 \cdot MgX$$

Metallic subhalide free radicals combine with ketones to form ketyl free radicals which usually dimerize irreversibly to form pinacolates:

$$R'R''CO + \cdot MgX \rightarrow R'R''(XMgO)C \cdot$$
$$2R'R''(XMgO)C \cdot \rightarrow [R'R''(XMgO)C{-}]_2$$

Ketones with labile *alpha* hydrogen atoms are enolized by Grignard reagents if, for steric or other reasons, they undergo the "normal" addition with difficulty:

$$R'R''CHCOR'' + RMgX \rightarrow$$
$$[R'R''C{=}COR'']^-[MgX]^+ + RH$$

Many enolates behave like true Grignard reagents; for example, that of acetomesitylene

$$[CH_3COC_6H_2\text{-}2,4,6\text{-}(CH_3)_3]$$

acts in many respects as though it had the constitution $2,4,6\text{-}(CH_3)_3C_6H_2COCH_2MgX$. When an enolate is capable of reacting additively with its own ketone, ketolization (or condensation of higher degree) may occur.

OTTO REINMUTH

GUMS AND MUCILAGES (NATURAL)

Definition. Gums and mucilages are carbohydrate polymers of high molecular weight obtained from plants. They can be dispersed in cold water to give viscous or mucilaginous solutions which do not gel. They are composed of acidic and/or neutral monosaccharide building units joined by glycosidic bonds. The acid groups ($-CO_2H, -SO_3H$) are usually present as salts (Ca, Mg, Na, K, etc.) and in certain cases substituents such as acetyl (karaya gum) and methyl groups (mesquite gum) may also be present. Pyruvic acid residues, linked as ketals, are present in several cases (agar).

Plant Gums. Gums are the dried exudates from trees and shrubs produced with or without artificial stimulation by mechanical injury. Among those gums of commercial importance may be mentioned arabic, ghatti, karaya, and tragacanth. Gums have been subdivided for convenience into three groups: (1) those completely soluble in water, (2) those which swell but do not dissolve, and, (3) those composed of two parts, one of which dissolves completely while the other swells.

Plant Mucilages. Mucilages are not *extra*-cellular in origin but are obtained from seeds, roots or other parts of plants by extraction with either hot or cold water. While gums form tacky or sticky solutions, mucilages give slippery or mucilaginous solutions, e.g., those from guar bean, linseed, locust bean and other related leguminous plant seeds.

Generally plant gums and mucilages are completely insoluble in alcohol, but there are a number which are partly soluble in water and partly soluble in alcohol, e.g., gum myrrh and frankincense. In addition there are the so-called gum resins which are soluble in organic solvents but completely insoluble in water, e.g., gum benzoin, gum camphor. Bacterial extracellular polysaccharides with a wide variety of structures and physical properties are isolated from the culture media, which may have become quite viscous as a result of the secretion of the gum, e.g., *Xanthomonas campestris* NRRL B-1459. From various types of salt-water algae, the so-called seaweed mucilages or gums such as agar, algin and carrageenin, sometimes referred to as algal polysaccharides, may be obtained by extraction with hot water. Gum-like polysaccharides belonging to the hemicellulose group of carbohydrate polymers having a high pentosan content have been isolated from grains such as wheat, barley, rye, oats and maize.

Classification. Although the gums and mucilages are usually differentiated, the accumulation of factual evidence shows that the distinctions formerly drawn no longer apply. In the light of present knowledge, they may best be classified according to their chemical composition as follows:

Group I. Acid gums and mucilages. These contain acidic components (D-glucuronic, or its 4-*O*-methyl derivative, D-galacturonic, L-glucuronic, L-iduronic, or D-mannuronic acid) and neutral components (pentoses and/or hexoses). Certain seaweed mucilages contain the sulfonic acid grouping.

Group II. Neutral gums and mucilages. These are composed of the simple sugars (pentoses such as L-arabinose and D-xylose and/or hexoses such as D-glucose, D- or L-galactose and D-mannose). In certain gums methyl pentoses may also be present; thus L-rhamnose is present in gum arabic and L-fucose in gum tragacanth.

Origin and Function. Gums and mucilages may be found either in the intracellular parts of plants or as extracellular exudates. Those found within plant cells represent storage material in seeds and roots to tide the plant over a dormant period. They also serve as a water reservoir and as a protection for germinating seed.

The polysaccharides found as *extra*cellular exudates of higher plants appear to be produced as a result of injury caused by mechanical means or by insects. It is not known whether the exudates are formed at the site of the injury or whether they are generated elsewhere and then transported to the injured area.

Preparation. The true exudates, such as gum arabic and the East African and Indian gums, are picked by hand. Seldom are commercial samples pure. This is a serious disadvantage in product control and in constitutional studies. They are classified according to grade, which in turn depends on color and contamination with foreign bodies such as wood and bark. Tests are seldom available for judging quality. These exudates are processed merely by grinding, their only previous treatment being sorting and sometimes bleaching by the sun. In certain cases they can be purified by extraction with water and precipitated by alcohol. This procedure is usually adopted for research purposes but seldom if ever in commercial operations.

The gums and mucilages present in roots, tubers and seaweeds are usually extracted with hot water, dried and marketed as a powder. Those gums found on the inner side of the seed coat as a vitreous layer (e.g. locust bean, guar bean, etc.) are best obtained by a suitable milling process which first removes the seed coat and then makes use of the fact that the gum layer is very hard and tough as compared with the seed endosperm. For certain industrial purposes it is unnecessary to remove the seed coat prior to milling the outer vitreous carbohydrate portion of the seeds. These intracellular gums and mucilages can be purified by precipitation with alcohol from aqueous solution as in the case of the gum exudates or by a derivatization process such as acetylation. In a similar way, the bacterial polysaccharides can be precipitated from the cell-free culture fluid with alcohol or as the salt of a quaternary ammonium compound where acidic groups are present.

Constitution. The extracellular plant gums and mucilages, for example, gum arabic, karaya, tragacanth, etc., generally have a more complicated structure than the intracellular types. They are made up of a number of different sugar building units linked together by a variety of glycosidic bonds. They possess a central core or nucleus composed usually of D-galactose and D-glucuronic acid units joined by glycosidic bonds which are relatively stable to hydrolysis by acid. To this central nucleus there are attached as side chains those sugar units which are removed by mild acid hydrolysis. Thus in the case of gum arabic the acid resistant portion of the molecule is composed of D-glucuronic acid and D-galactose and to this nucleus are attached units of L-arabinose, L-rhamnose and D-galactopyranosyl $(1 \rightarrow 3)$ L-arabinose.

The neutral mucilages and gums such as mannans, galactomannans, and glucomannans extracted from seeds and roots have a relatively simple structure. The kinds of building units are fewer and the molecules are much less branched. The galactomannans are usually composed of a backbone of linear chains of D-mannose units joined by 1,6-glycosidic bonds, to which are attached at regular intervals side chains of D-galactose residues. The glucomannans are essentially linear polymers united by 1,4 linkages.

The algal polysaccharides resemble the relatively simplified structures of the neutral mucilages, as in the case of carrageenin. A wider spectrum of structures is found in the bacterial gums, which are generally of the highly-branched type exuded by higher plants.

Physical Properties. The industrial applications of gums and mucilages take advantage of their physical properties, especially the viscosity and colloidal nature. They are substances of high molecular weight, e.g., gum arabic has a molecular weight of 250,000–300,000. The gums and mucilages which possess a relatively linear molecule, e.g. gum tragacanth, form more viscous solutions, than the more spherically shaped gums, such as gum arabic, at the same concentration, and hence the former are more economical. Due also to the elongated molecular shape of the seed gums and mucilages the viscosity of their aqueous solutions varies widely with concentration; they exhibit structure viscosity. On the other hand, the gums and mucilages of more spherical shape, i.e., the exudates, give solutions whose viscosities do not depend so much on concentration.

Gums and mucilages influence each other, for the mixing of two gums of the same viscosity may result in a mixture with a different viscosity. The viscosity of solutions of gums and mucilages is dependent on the pH, especially for those containing acid groups, and in certain cases it decreases on standing due to enzymatic breakdown of the molecules. The molecules can undergo large changes in shape and size under the osmotic influence of opposing ions. Some of them, such as carrageenin from Irish Moss, can be fractionated by dilute salt solutions such as potassium chloride, and the poly-β-glucosan from barley grain may be precipitated with ammonium sulfate. Gum arabic shows the phenomenon of coaservation when mixed with gelatin.

One of the most useful physical properties of gums and mucilages is their protective colloidal effect. For this reason they are used in the refining of various kinds of ores where they coat impurities and enable them to be separated. Gums and mucilages can be distinguished by infrared studies. This might prove to be an excellent method for ascertaining the purity of gums and mucilages.

Industrial Applications. The diversity of applications of gums and mucilages is truly amazing. They have long been extensively used as adhesives and find use in increasing the strength of starch pastes. Gum tragacanth is used in toothpastes and for coating soap chips and powders to prevent dusting and lumping and also in hairwaving preparations. Emulsions with medicinals, insecticides, kerosene, paraffin oil, neoprene and natural latex can be stabilized by the addition of gums. Seaweed gums (e.g. carrageenin) and seed mucilages (guar gum) are said to be very good as stabilizers in dairy products such as ice-cream and cheese. They are used in confectionery, in making jams and jellies, in stabilizing citrus oil emulsions and salad dressings and are said to act as a fixative for 2,3-butanedione in the baking industry. The textile industry uses them in spinning and printing. Lithographic plates are often treated with gum arabic or synthetic gum-like compounds such as the carboxyethers of cellulose and starch. The electrolytic deposition of metals is often carried out in the presence of gums. Gums and mucilages, especially those from seeds, having essentially linear structures form excellent beater additives; by aligning themselves along the cellulose fibers, they act as a cementing medium and as a result paper strength increases. Gums can be used as soil flocculents in various ways, as adjuncts in the dressing of ores, and improving oilwell drilling muds. They act as good binders for example in thermite mixtures, in welding rods, and in explosives used under water.

REX MONTGOMERY

Cross-references: *Colloid Chemistry; Carbohydrates.*

H

HAFNIUM AND COMPOUNDS

Hafnium, symbol Hf, is a bright, ductile heavy metal of atomic number 72 and atomic weight 178.49 that was discovered in 1922 by Coster and von Hevesy. It is in Group IVB of the Periodic Table with titanium and zirconium. Little more than a laboratory curiosity until recent years, hafnium always occurs in zirconium minerals and current knowledge and interest about it have increased because of the demand for hafnium-free zirconium for atomic energy purposes. The very great tendency to absorb thermal neutrons makes the presence of hafnium in zirconium most objectionable.

Hafnium ranks 48th in order of prevalence among the elements in the earth's crust, but its constant association with zirconium, a much more prevalent element, which it resembles closely in chemical nature, offers a ready source of hafnium. However, this similar chemical nature makes the separation of hafnium and zirconium a most difficult practical operation.

The commercial production of hafnium metal is based on zircon ($ZrSiO_4$) concentrates which contain 0.5 to 2.0% hafnium. The zircon is heated with carbon in an electric arc furnace to yield an impure carbide or carbide-nitride, and this product is chlorinated to form a relatively pure zirconium tetrachloride which also contains the original hafnium as $HfCl_4$. The two chlorides are separated by liquid-liquid extraction of their thiocyanates with hexone (methyl isobutyl ketone) or of their nitrates with tributyl phosphate. In both cases hafnium free of zirconium and zirconium free of hafnium are obtained. The hafnium in the form of $Hf(OH)_4$ is calcined to HfO_2, and the latter is chlorinated in the presence of carbon to yield $HfCl_4$ by the reaction: $HfO_2 + 2Cl_2 + 2C \rightarrow HfCl_4 + 2CO$. The $HfCl_4$ is purified by sublimation (1-atm sublimation point = 317°C and melting point = 432°C) and then reduced with magnesium by the Kroll process in a furnace to form the metal in sponge form. Sodium can also be used for this reduction.

Ductile massive metal is made by arc melting and electron beam melting techniques, and purification is achieved by the latter and by the iodide, hot wire technique. Electrorefining techniques have also been developed. Special precautions must be taken throughout the processing to prevent air and carbon absorption, since oxygen, nitrogen and carbon impart permanent brittleness and cannot be removed without reprocessing.

The physical properties of hafnium include: density, 13.29 g/cc; melting point, 2230°C; boiling point, about 5400°C; specific heat, 0.0351 cal/g (25–100°C); electrical resistivity, 35.1 microhm-cm (0°C); coefficient of thermal expansion, 5.9×10^{-6} (0–1000°C); thermal conductivity, 0.0533 cal/sec/cm²/°C/cm (50°C); modulus of elasticity, 19.8×10^6 psi; hardness, 152 Vickers; and thermal neutron absorption cross section, 105 barns. Hafnium crystallizes in the close-packed hexagonal system and an allotropic transformation to body-centered cubic structure occurs at about 1760°C. The metal obtained after melting in an arc furnace under argon can be cold-rolled, swaged, hammered, and drawn into rod or wire, and has a brilliant luster.

The standard electrode potential for hafnium is −1.70 volts for the reaction: $Hf + 2H_2O \rightleftharpoons HfO_2 + 4H^+ + 4e$.

The remarkable chemical similarity of hafnium and zirconium, even though their atomic numbers differ greatly (72 and 40), results from their closely identical atomic radii (1.442 Å for Hf and 1.454 Å for Zr) caused by the lanthanide contraction among elements 57–71.

Hafnium in its compounds exhibits a principal valence of 4; unstable di- and trihalides have been produced by partial reduction of the tetrahalide. Very few hafnium compounds are soluble and stable in aqueous solutions. Among them are potassium and ammonium fluohafnates, K_2HfF_6 and $(NH_4)_2HfF_6$, hafnium oxyhalides such as $HfOCl_2$, and hafnium sulfate, $Hf(SO_4)_2$. Hafnium reacts with halogens to form tetrahalides which hydrolyze in water. At elevated temperatures hafnium forms HfO_2, m.p. 2775°C, with oxygen; HfN, m.p. 3300°C, with nitrogen; HfC, m.p. 3900°C, with carbon; and HfB_2, m.p. 3250°C, with boron. These compounds are among the most refractory ones known and their high melting points suggest a variety of high-temperature applications. At 700°C the metal absorbs hydrogen rapidly to form a hydride approximating HfH_2. However, the resistance of massive hafnium to attack by air, oxygen, nitrogen, high-pressure water, and steam is good to temperatures of several hundred degrees. In-pile tests show that hafnium's corrosion resistance to high-temperature water is unaffected by radiation, a valuable factor for its use as control rods in atomic reactors.

The massive metal is soluble in hydrofluoric acid and in concentrated sulfuric acid, is resistant to dilute HCl and to dilute H_2SO_4 and is unaffected by HNO_3 in all concentrations. Its resistance to alkalis, including boiling 50% NaOH, is good.

Hafnium is used chiefly as control rods in nuclear submarines and power reactors where its high thermal neutron absorption cross section, excellent mechanical properties and fine corrosion resistance are valuable. It has potential applications in alloys with other metals of high melting points, and some of its compounds have interesting properties at high temperatures. For example, hafnium carbide can be cast and is an extremely hard material, and hafnium nitride and hafnium boride are highly conductive at high temperatures. The high melting point and high electron emission of hafnium suggest uses in radio tubes, incandescent lights and rectifiers and as a cathode in x-ray tubes. Jewelry applications might take advantage of the density, tarnish resistance and beautiful luster of hafnium. Some hafnia, HfO_2, is used in special optical glasses, and hafnium halides can be decomposed at 800–1000°C to produce a corrosion-resistant hafnium film on base metals.

The metal is also used to a small extent as a component of alloys with other refractory metals, such as tungsten, tantalum and niobium. One such alloy, T-111, containing Ta-8W-2Hf, is used in space-power systems where good strength at high temperature is required.

CLIFFORD A. HAMPEL

HALL EFFECT

The Hall effect, one of several galvanomagnetic effects, was first described by E. H. Hall in 1879. If a bar of electrically conducting material is placed in a magnetic field and current is passed through the bar in a direction normal to the magnetic field, a potential is created in a direction perpendicular to both the current and magnetic field by deflection of the charged current carriers. The Hall effect is important because it provides a means for a direct estimate of the density of the charge carriers in the material and their sign, and the Hall coefficient can be combined with the result of a measurement of the material conductivity to yield a value for the mobility (velocity per unit applied electric field) of the charge carriers. In addition to its use for characterizing the electrical properties of a material, the Hall effect principle is applied in several practical devices.

Consider the bar shown in Fig. 1. The current density I (amps/cm^2) is proportional to the applied field E (volts/cm), $I = \sigma E$. The proportionality fac-

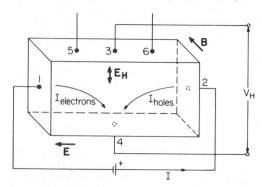

FIG. 1. Sample with electrical contacts arranged for measurement of Hall effect and conductivity. The Hall voltage V_H appears between contacts 3 and 4. The voltage drop between contacts 5 and 6 can be used to determine the conductivity of the material.

tor σ is called the conductivity; σ can be expressed in terms of the density of charge carriers n (cm^{-3}) in the material, the charge on each carrier e (coulombs), and the carrier mobility μ (cm^2/volt-sec), by the relationship $\sigma = ne\mu$. The carriers may be either negatively charged electrons or positively charged holes.

The magnetic field causes the charge carriers to be deflected by Lorentz force $\mathbf{F} = e\mathbf{v} \times \mathbf{B}$ toward one side of the bar. In this expression $\mathbf{B}$ is the magnetic field strength and $\mathbf{v}$ is the velocity of the charge carriers, $\mathbf{v} = \mu\mathbf{E}$. Note that carriers of either sign are deflected to the same side of the bar. The charge buildup creates a transverse field $\mathbf{E}_H$ which acts on each charge carrier in the steady state to balance the Lorentz force, $e\mathbf{E}_H = e\mathbf{v} \times \mathbf{B}$ or $\mathbf{E}_H = \mu\mathbf{E} \times \mathbf{B}$. The Hall coefficient R is defined by the relation $E_H = RIB$, so that $R = \mu E/I = \mu E/\sigma E = 1/ne$(cm^3/coulomb). Thus, the magnitude of the Hall coefficient is inversely proportional to the product of the density of charge carriers and the magnitude of charge on each. The polarity of the charge carriers is determined by the direction of the Hall field created in the sample. The sign of R is the same as the polarity of the charge carriers. If the conductivity σ of the material is known, the mobility of the charge carriers can be determined, since $R\sigma = \mu$.

In monovalent metals, the electron density calculated from the Hall coefficient agrees quite well with the density of atoms. In polyvalent metals, the results are more erratic. In semiconductors, the Hall coefficient for a single carrier must be modified to $R = r/ne$, where r is a correction factor dependent on the type of scattering the charge carriers encounter during conduction and the details of the energy band structure. The magnitude of r is rarely far from unity with the usual band structure; the value ranges from $3\pi/8$ for predominant lattice scattering to $315\pi/512$ for ionized impurity scattering. More complicated formulae must be used for the Hall coefficient when more than one type of charge carrier is participating in the conduction process.

At very low temperature, the conductivity of a semiconductor may be determined by "hopping" or tunneling of charge carriers from one atom to another. As the temperature is increased, impurities which may be present in the semiconductor become ionized and contribute charge carriers to the conduction process (extrinsic activation). The slope of the Hall coefficient measured as a function of temperature is in this case determined by the activation energy of the impurities. When the temperature is increased still further, conduction is enhanced by ionization of the semiconductor atoms themselves to produce pairs of holes and electrons (intrinsic activation). In the simple case of only one type of intrinsically activated carrier being mobile and the impurity contribution being negligible, the slope of the Hall coefficient as a function of temperature is determined by the energy gap between the valence and the conduction band, and this parameter of the substance may thus be computed simply. In general, since hopping, tunneling, extrinsic, and intrinsic activation effects may overlap in temperature, since any of several kinds of charge carrier scattering may be present, and since charge carriers of differing properties may be conducting current simultaneously, the interpretation of Hall effect data from semiconductors must be quite sophisticated.

In ferromagnetic and strongly paramagnetic materials, the magnetic field introduced into the formula is not the externally applied field H nor the internal field B; but an effective field must be used, $B_{eff} = H + 4\pi M\alpha$, where α is in general temperature and material dependent and can exceed unity, and M is the sample magnetization. In this case, $E_H = RIH + D\pi\alpha RMI$. R is the ordinary Hall coefficient and $4\pi\alpha R$ is called the extraordinary Hall coefficient. Other variations of the Hall effect include, for example, the photo-Hall effect, in which the density of carriers liberated by light may be measured in a photoconducting medium.

The Hall effect is used in some gaussmeters. A known current is passed through a small strip of a semiconductor which has a Hall constant large in magnitude and independent of temperature. The Hall voltage generated when the strip is inserted in a magnetic field is an indication of the magnetic field intensity.

Because the Hall voltage is proportional to the product of I and B, a Hall device can yield a voltage proportional to the product of the current in the device and the current producing a magnetic field surrounding the device. Such units are called Hall multipliers.

D. WARSCHAUER

Cross-reference: *Semiconductors.*

HALLUCINOGENIC DRUGS

There are many substances which, if taken in the appropriate quantities by normal subjects, produce distortions of perception, vivid images, or hallucinations. Most of these substances produce powerful peripheral as well as the central effects. Some few agents are characterized by *predominance* of their actions on mental and psychic functions. This group of drugs has been called hallucinogens, psychotomimetics, psycholytics, and psychedelics, among several ambiguous terms. None of the names which have been suggested to date is adequately descriptive of the compounds. The best known term is *psychedelic*, but *hallucinogen* and *psychotomimetic* are widely used.

The major hallucinogens of current interest[1] may be classed into six groups of chemically distinct compounds: (1) lysergic acid derivatives of which lysergic acid diethylamide (LSD-25) is the prototype; (2) phenylethylamines, the best known of which is mescaline; (3) indolealkylamines, which include psilocybin, psilocin, and bufotenine; (4) piperidyl benzilate esters, typified by Ditran (a 70:30 mixture of N-ethyl-2-pyrrolidylmethylphenyl-cyclopentylglycolate and N-ethyl-3-piperidylphenyl-cyclopentylglycolate); (5) phenylcyclohexylpiperidines (Sernyl); and (6) the tetrahydrocannabinols (THC) which are the active ingredients of marihuana. Chemical structures of these compounds are shown in Figs. 1 and 2.

Drugs representative of all but two of these classes have been isolated from natural sources. LSD-25 is a synthetic compound, but it is derived from a molecular component of ergot, a fungus

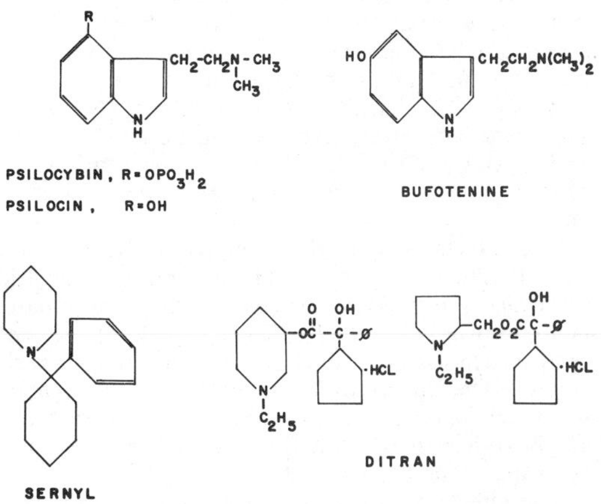

FIG. 1

which infects cereal grains. Mescaline, historically one of the older hallucinogens, was isolated from the Mexican cactus, peyote. Psilocybin and psilocin are constituents of the Mexican mushroom, *Psilocybe mexicana*. Bufotenine and related materials are found in psychoactive snuffs made by South American Indian tribes from indigenous plants. These indole derivatives are chemically similar to serotonin (5-hydroxytryptamine), a compound which plays an important, possibly a transmitter, role in the central nervous system. Tetra-hydrocannabinols are components of a resinous material secreted by the plant, *Cannabis sativa*. The piperidyl benzilate esters and phenylcyclcyclohexyl piperidines are synthetic compounds, and have not been shown to occur naturally

Clinical syndromes from LSD-25, mescaline, the indoleamines, and marihuana are very similar, differing mostly in frequency of occurrence and degree of severity. Somatic symptoms are nausea, dizziness, blurred vision, paresthesia, weakness, drowsiness, and trembling. These result frequently and are usually associated with sympathomimetic effects, such as increased pulse rate and slight temperature elevation. Perceptual and psychic changes are marked. Visual illusions and vivid hallucinations, decreased concentration, slow thinking, depersonalization, dreamy states, changes in mood, and often anxiety, are commonly found. Except for severe intoxication, consciousness and judgement remain intact. Experimental subjects are fully aware that their perception is distorted and the changes are drug-induced.

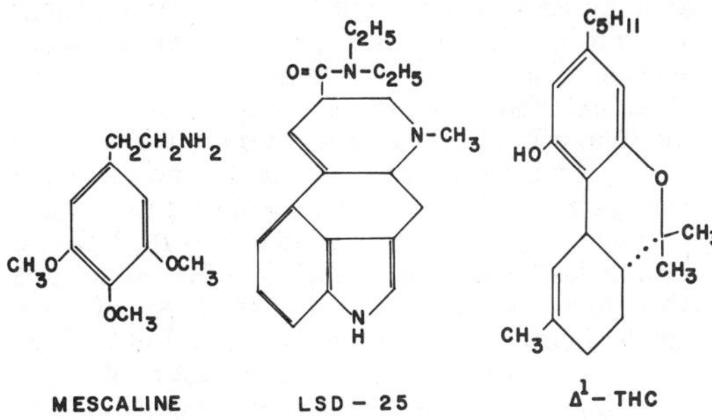

FIG. 2

Clinical syndromes from Ditran are different from those produced by the above drugs in some respects. Disorganization of thought, disorientation, confusion, mood changes, and visual and auditory hallucinations are observed. The piperidyl benzilate esters are central anticholinergics, and mental states produced by them are reminiscent of those from other anticholinergics, such as scopolamine.

The effects of phenylcyclohexyl derivatives are also distinctive. Comparatively minor somatic symptoms are evoked. Psychic effects predominate, being typically characterized by feelings of unreality, depression, anxiety, and delusional experiences. The effects of these drugs are said to be more analogous to natural psychoses than those of the other drugs; however the same claim has been made for Ditran.[2]

The effects of the hallucinogens on animals vary with the type of drug and the species tested.[1] Generally, the doses necessary to produce behavioral changes in animals are larger than those which cause symptoms in man. In humans, the effective dose for the various drugs are: (1) mescaline, 5000–10,000 μg/kg, of body weight; (2) psilocybin and psilocin, 100–200 μg/kg; (3) Sernyl and Ditran, 150–200 μg/kg; (4) LSD-25, 1-2 μg/kg, and (5) THC, 200–500 μg/kg, depending on whether it is smoked or taken orally. Except for THC and some of the indoles, the effective dose is the same regardless of the route of administration. Distribution studies indicate that in comparison to other tissues, brain does not accumulate these agents. There appears to be no clue to the mechanisms of hallucinogen action in the absorption, distribution, and elimination of the drugs.

LSD-25 is one of the most powerful drugs known to man. When it was discovered in 1943 by A. Hofmann, its extraordinary potency reawakened interest in the possibility of natural chemical activators in the schizophrenic process. The production of bizarre psychic phenomena, the lack of addictive and toxic properties, and the minimal side effects impressed those investigators who were concerned with mental illness, particularly schizophrenia. From this interest, a large research effort was directed toward known hallucinogens and toward the discovery of new ones. Among the concepts reintroduced was the possibility of using drugs to initiate so-called "model psychoses."[3] The use of drugs to produce disturbed mental states became a popular tool for investigators in several disciplines. A deluge of published work, much of it lacking scientific discipline and value, caused some authors to decry the lack of careful inquiry which invested "these agents with an aura of magic, offering creativity to the uninspired, "kick" to the jaded, emotional warmth to the cold and inhibited, and total personality reconstruction to the alcoholic or the chronic neurotic."[4] To date, the hallucinogens have proved to be of little therapeutic value.

Much attention was focused on the psychotomimetics by some clinicians and lay enthusiasts, who advocated unregulated dispension of these agents, in order that men may "transcend" or "expand their consciousness." As a result of garish publicity and notoriety, some elements of the population have been attracted to the use of these agents with detrimental consequences, particularly of a psychological nature. Legally, none of the hallucinating agents can be used, even for investigational purposes, without prior approval of the Food and Drug Administration.

F. CHRISTINE BROWN

Cross-references: *Psychotropic Drugs.*

References

1. Brown, F. Christine, "Hallucinogenic Drugs," Charles C. Thomas, Springfield, Illinois, 1971.
2. Gershon, S., and Olariu, J., "JB-329. A new psychotomimetic, its antagonism by tetrahydroaminacrin and its comparison with LSD, mescaline and Sernyl," *J. Neuropsychiat.,* **1,** 283, (1960).
3. Hollister, L. E., "Drug induced psychoses," *Ann. N. Y. Acad. Sci.,* **96,** 80, (1962).
4. Cole, J. O., and Katz, M. M., "The psychotomimetic drugs," *J. Am. Med. Assoc.,* **187,** 759, (1964).

HALOCARBON COMPOUNDS

Strictly speaking, the term halocarbon applies to compounds containing only halogen and carbon atoms. In the sense used herein some of the compounds will also contain hydrogen, the requisite being that they must contain carbon, a halogen including fluorine, and sometimes hydrogen.

The lower members of the various homologous series are used as refrigerants, aerosol dispersants, and fire extinguishing agents. They can be purchased in America under the trade marks, Freon, Genetron, and Isotron; in England, as Arcton and Isecon; and in Germany, as Frigen.

In 1893, the Belgian chemist Swartz started researches leading to the preparation and measurement of the properties of many of the simple compounds in this class. It remained for Midgeley and Henne, in the late 20's, to show that these compounds are inert, nontoxic, nonflammable, and otherwise well suited for use as refrigerants.

Some of the compounds can be made by starting with carbon tetrachloride, chloroform, or acetylene, and reacting them with anhydrous hydrofluoric acid. In some cases, antimony fluoride and cobalt fluoride are used as fluorinating agents.

An electrolytic process has been developed in which an organic substance is dissolved in anhydrous hydrofluoric acid, the electrolysis being carried out at about 0°C. The process is well suited for the production of such compounds as C_3F_8 and C_3F_7H. It may also be employed to make derivatives such as trifluoroacetic acid.

Because of the inconvenience involved in using the chemical names of these compounds, various systems have been devised in which numbers, along with a generic term, are assigned to the different substances. A few years ago the American Society of Refrigerating Engineers adopted numbering systems for all kinds of refrigerants, even including water to which the designation Refrigerant 718 was assigned.

As far as the saturated halocarbons are concerned, the number is to be considered as consisting of three digits. In the event there are actually only two digits as in "12," it is to be considered as "012." The number of carbon atoms is obtained by adding one to the first digit, the number of hydrogen atoms by subtracting one from the second digit, and the number of fluorine atoms by the "face value" of the third digit. All other valence positions are occupied by chlorine atoms, the total number of hydrogen, chlorine, and fluorine being $2n + 2$, where n is the number of carbon atoms.

An exception is that if any bromine is present, the number will be followed by the letter "B" with another digit following to show the number of bromine atoms. The Standard stipulates that the number may be used alone or following the word "Refrigerant." An abbreviation of the word is not allowed. Dichlorodifluoromethane CCl_2F_2 may be regarded as "Refrigerant 12" or simply "12." In the event of isomers, the number just described is given to the most symmetrical compound; the letter "a" follows the number for the next degree of symmetry; the letter "b" for the next, etc.

A numbering system used for Isecons in England is a little more straightforward; there is no necessity to add and subtract. In this system, the first digit is the number of carbon atoms, the second the number of chlorine atoms, and the third the number of fluorine atoms, the hydrogen being obtained by difference.

In refrigeration, the capacity depends upon several factors, but more especially upon the normal boiling point. The boiling points of some refrigerants are given in the following table along with the commercial status:

ASRE No.	B.P. (°F)	Status
11	74.7	C
12	−21.6	C
13	−114.6	C
14	−198.4	S
21	48.1	D
22	−41.4	C
23	−119.9	D
113	117.7	C
114	38.4	C
114a	38.5	C
152a	−11.5	C
142b	15.1	C

In the column headed "Status," C indicates that the compound is made commercially, S semicommercial, and D signifies that the refrigerant is in the development stage.

One can get a feel for the effect of boiling point upon refrigeration capacity by realizing that Refrigerant 22, with a boiling point of −41.4°F, has about 1.55 times the capacity of Refrigerant 12 whose boiling point is −21.6°F.

The very low boiling compounds can only be used advantageously at very low evaporator temperatures. Refrigerant 14, for example, cannot be condensed at a temperature higher than about −50°F. It can be used in a cascade system where the evaporator for the high temperature refrigerant, such as Refrigerant 22, takes the heat out of condensing Refrigerant 14, say, at −75°F. The evaporator temperature for the lower part of the cascade might well be −150°F.

An azeotropic refrigerant has been made commercial. Its ASRE designation is 500; and its trade name, Carrene 7. It consists of 26.2% (wt.) Refrigerant 152a and 73.8% Refrigerant 12. It boils at −28.0°F. This refrigerant gives an increase of 18% in capacity over that of Refrigerant 12 and serves to extend and fill in a line of reciprocating units.

If a concern makes the sizes 3, 5, 10, 25, 35, and 50 horespower units using Refrigerant 12, it automatically, through the use of Refrigerant 500, has the following additional sizes 3.5, 6, 12, 18, 30, 42 and 60 horsepower. Seven sizes are made on the assembly lines, yet, 14 are available for sale. Of course, where 500 is substituted for 12, the motor must be the proper size.

Refrigerant 500 is the only commercial azeotropic refrigerant. There are as many as a dozen azeotropes known in the field. From these, ASRE has selected two to put in their proposed Standard. Refrigerant 501 contains 75% (wt.) Refrigerant 22 and 25% Refrigerant 12; Refrigerant 502 contains 48.8% Refrigerant 22 and 51.2% Refrigerant 115. Refrigerant 501 has been challenged in the technical literature as to its very existence, evidence being given that the azeotrope within the system contains only about 2% (not 25) Refrigerant 12. Refrigerant 502 is thought to have a very low cycle efficiency, where compared with refrigerants like 12, 22, and 500.

Application has been found for halocarbons as pressurizing agents for dispersing insecticides from cans. The can contains a solution (aerosol) of the insecticide in a halocarbon or mixture of halocarbons having a pressure of about 35 psi at ambient temperatures so that an instantaneous discharge in the form of a fine spray can be obtained by opening a valve by merely pressing with the hand.

Four compounds 11, 12, 114, and 114a have already found wide use commercially as propellants; experimental work has been done with 21, 22, 113 and 152a.

There are a great many household and industrial applications for these propellants. Among the household, personal, and pharmaceutical may be listed adhesive balms, auto polishes, cold preventives, cleaners, deodorants, fungicides for garden plants, grease removers, hand cleaners, preparations for ladies' hair, insecticides, lubricants, paints, nail polish dryers, rust preventives, saddle soaps, sunburn creams, suntan sprays, inhalant sprays, weed killers, and window cleaners, just to mention a few. In industry, they are used for antifoam sprays, belt dressings, dye penetrants, fumigants, grease removers, livestock sprays, metal coating sprays, rust inhibitors, stencil inks, water repellants, and a host of other things.

A special class of halocarbons is the fluorocarbon which contains carbon and fluorine only. These compounds have a density of about 2 grams per milliliter. They are insoluble in water and in most organic compounds. They are only slightly soluble in hydrochlorocarbons or chlorocarbons but are readily miscible with chlorofluorocarbons and hydrofluorocarbons.

These compounds are resistant to concentrated nitric acid, oleum, strong alkalis, and virtually all other chemical reagents, an exception being molten sodium and the like. Because of their inertness, they have found limited applications as lubricants, dealing fluids, coolants, reaction media, and dielectric media, the use being limited by the high cost.

Two fluorine-containing polymers, polytetrafluoroethylene and polychlorotrifluoroethylene, are commercial products. The ASRE code numbers for the monomers are 1114 and 1113, respectively. Polytetrafluoroethylene can be bought as "Teflon"; polychlorotrifluoroethylene as "Genetron HL" plastic, "Kel-F," or "Fluoroethene."

13B1 and 12B2, which are quite stable and nontoxic, are being used instead of carbon tetrachloride and methylbromide as fire extinguishing agents. They may well be adopted for use with aircraft.

WILLIAM A. PENNINGTON

HALOGEN CHEMISTRY

In the successive building up of elements across the Periodic Table, only about one-third of the positive charge of each added proton is effectively neutralized at the surface by the outermost shell electron simultaneously added. The result is that the nuclear charge effective at the surface of the atom increases rapidly, producing both a contraction of the electronic cloud and an increase in the electronegativity of the atom. Consequently each halogen, being the last of the elements in its period before the outermost octet becomes completed, has smaller and more electronegative atoms than any of the elements preceding it in the period. The outermost principal quantum level, with seven electrons, has one half-filled orbital, thus providing the capacity to form one covalent bond. Since nearly all other elements are less electronegative, the halogen atom tends in most of its compounds to gain more than half share of the bonding electrons, giving it a partial negative charge and the characteristic oxidation number of -1. Such compounds are called "halides."

Electronegatives of the elements decrease in the order: F, O, Cl, Br, N, Se, S, As, I. . . . Although no reagent can attract electrons from fluorine, both fluorine and oxygen are seen to be potentially capable of attracting electrons from chlorine, bromine, and iodine, and other elements as well can attract electrons from bromine and iodine. Positive oxidation states are therefore possible for chlorine, bromine, and iodine but not for fluorine. Such states appear, however, to be formed only by the action of more electronegative elements. In some compounds, the effect of the more electronegative atoms may be to stabilize the outermost shell d orbitals by withdrawing some of the electronic cloud that otherwise shielded them from the nuclear charge. Promotion of a lone pair electron into a vacant d orbital may then create simultaneously two new half-filled orbitals, permitting the formation of two extra bonds. However, evidence from studies of bond energies suggests that the bonds formed beyond the one expected covalent bond may be only half-bonds, which may not require the use of the outer shell d orbitals. Further study of positive oxidation states of the halogens is clearly needed. Because of the apparent limitations imposed by the evident necessity of the presence of attached atoms of higher electronegativity than that of the halogen, the number of halogen compounds in which the halogen exhibits positive oxidation states is very much smaller than the number of halides.

Major group halides have the expected formulas: IA: EX; IIA and IIB: EX_2; IIIA: EX_3; IVA: EX_4; VA: EX_3; VIA: EX_2; VIIA: EX. In addition, some lower halides are known, especially where the "inert pair effect" is observed, as in the monohalides of thallium and the dihalides of lead. There are also some higher halides such as PCl_5, SF_6, and IF_7. The transitional metals also form a great number of halides, of lower as well as maximum oxidation states. Also, of special interest are the fluorides of krypton, xenon, and radon.

Direct synthesis from the elements is a common and widely applicable method of preparation of halides, especially in anhydrous condition. Reaction of the hydrohalic acids with metal oxides, hydroxides, carbonates, sulfites, or sulfides produces aqueous solutions of the halides. Anhydrous chlorides such as those of aluminum and silicon can be prepared by reduction of the oxide with carbon in the presence of chlorine. Anhydrous halides of many metals can also be prepared by heating the metal in an atmosphere of hydrogen halide gas or by a variety of special other methods.

In physical properties halides range from high melting, ionic solids to low melting, volatile liquids and gases. These differences correspond closely to the differences in states of aggregation as determined in part by the bond polarity. The higher the negative charge on the halogen, the better it can act as electron donor, and whenever the simplest molecule formed would contain vacant outer orbitals of low energy as well as the lone pairs of electrons on the halogen, condensation continues in the direction of utilizing these orbitals and electron pairs as thoroughly as possible. Lower negative charges on halogen correspond to weaker association, with lower melting temperatures and higher volatility, until volatile liquids and gases are formed in which only relatively weak van der Waals interactions among molecules are observed.

As indicated by heats of atomization, the thermal stability of fluorides exceeds that of chlorides, which in turn are more stable than bromides and iodides. Whenever a given element forms a series of halides representing different oxidation states of the element with the same halogen, the lowest halide is always the most highly associated, has the most polar bonds, and correspondingly is the most stable, with stability decreasing as the number of halogen atoms competing for electrons of the same central atom increases.

Being initially relatively high in electronegativity, halogen atoms tend to leave compounds in which they have not acquired much negative charge, in order to form more favorable bonds which permit them to become more negative. All halides in which halogen is not very negative are therefore potential halogenating agents, and may react with great vigor when in contact with better suppliers of electrons. On the other hand, halogen which has acquired fairly high partial negative charge has also lost its oxidizing or halogenating power to a large extent, and halide ions are not oxidizing agents at all.

No element that has acquired electrons avidly

will then give them up easily. Fluoride ion is therefore not a reducing agent at all. Chloride ion is a reducing agent only toward very strong oxidizing agents. Bromide ion is somewhat easier to oxidize, and iodide ion is a moderately effective reducing agent, in keeping with the lower initial electronegativity of the iodine atom.

Although the halide ions do not in general lose electrons easily, they are able to share their electrons by acting as donor in coordination complexes. In such action, fluoride ion seems sometimes superior by virtue of its higher concentration of each electron pair, resulting from its relatively small size, whereas iodide ion seems superior in other instances, presumably owing to its greater polarizability. Complex halides are extremely numerous, and simple halides that are relatively insoluble commonly dissolve in the presence of excess halide ion through the formation of soluble complexes. When halogen atoms are attached to certain Group IIIA elements, notably boron and aluminum, their withdrawal of electrons from the central element enhances its acceptor action using the fourth, otherwise vacant, orbital. Boron and aluminum halides, especially BF_3 and $AlCl_3$, are important Lewis acid type catalysts for industrial chemical processes.

Many halides are susceptible to hydrolysis, some extremely so. This property is closely related to the acid-base properties of the corresponding hydroxides. Hydrolysis of a halide, when complete, results in formation of the corresponding hydrogen halide, or hydrohalic acid if excess water is present, and the hydroxide of the other element. If the corresponding hydroxide is strongly basic, then both it and the halide are completely ionized and appreciable hydrolysis does not occur. If the hydroxide is capable of ionization as a weak base, then hydrolysis is reversible and the halide can be formed from the hydroxide using a high concentration of hydrohalic acid. But if the hydroxide is capable of ionization only as an acid, the hydrolysis is irreversible and must go to completion. An anhydrous halide can therefore be isolated from its aqueous solution by merely evaporating off the water only if the hydroxide of the element is a strong base. An equivalent statement is that the negative charge on halogen in the halide must be high. Otherwise, some hydrolysis occurs and hydrogen halide volatilizes with the water. Hydrolysis of halides is a useful method of preparing hydrogen halides, especially HBr and HI which are less easily prepared by other methods. As a class, fluorides are more resistant to hydrolysis than the other halides. Even when the hydroxide would be acidic, fluorine forms such strong bonds to other elements that it is not easily displaced, even as fluoride ion, by hydroxide.

Whereas the comparative chemistry of the halides is reasonably uniform and self-consistent, the halogens in their positive oxidation states resemble one another less. Fluorine, of course, has no positive oxidation state. With fluorine, chlorine forms ClF, ClF_3, and ClF_5 but the first is most stable. Bromine and iodine, on the other hand, form no stable monofluoride but preferentially BrF_3, BrF_5, IF_5, and IF_7. All three form the very unstable acids HOX, but of the anhydrides X_2O, only Cl_2O can be isolated at ordinary temperatures. (The fluoride, OF_2, is also known and more stable.) Other known halogen

oxides are ClO_2, ClO_3, Cl_2O_6, Cl_2O_7, but Br_3O_8, I_4O_9, and I_2O_5. Except for perchloric acid, $HClO_4$, the oxyacids of chlorine are too unstable to be isolated. They exist in solution, having the formulas $HClO$, $HClO_2$, and $HClO_3$. The acid strength increases markedly with increasing number of oxygen atoms attached to the halogen. Bromine forms similarly unstable oxyacids except for $HBrO_2$ which appears to be unknown. In contrast to $HClO_3$ and $HBrO_3$, which are too unstable to be isolated from solution, iodine forms HIO_3 as a stable polymeric solid. Similar irregularities are exhibited by the elements of the preceding groups in their positive oxidation states, but are not completely explainable at present.

R. T. SANDERSON

Cross-references: *Fluorine, Chlorine, Bromine, Iodine, Electronegativity.*

Halogen-Nitrogen Compounds

Compounds of the formula NX_3 where X = F, Cl, or I have been prepared. The bromide NBr_3 has probably not been prepared. In addition to the above compounds, ammoniates of nitrogen triiodide corresponding to the formula $NI_3 \cdot nNH_3$ where n = 1, 2, 3, and 12 have been reported. Likewise, $NBr_3 \cdot 6NH_3$ is reported to be formed when bromine reacts with ammonia under 1 to 2 mm pressure at $-95°C$. It explodes at $-70°C$.

Nitrogen trifluoride is obtained most conveniently by the electrolysis of fused anhydrous ammonium hydrogen fluoride, NH_4HF_2; small amounts of NHF_2 and NH_2F are obtained simultaneously. Nitrogen trifluoride can also be obtained by the direct union of nitrogen and fluorine or by the displacement reaction

$$2NCl_3 + 3F_2 \rightarrow 2NF_3 + 3Cl_2$$

Nitrogen trifluoride is a colorless gas which condenses to a colorless liquid at $-119°C$ under 1 atm pressure and freezes at $-216.6°C$. In contrast to the corresponding chloride and iodide, it is stable and nonexplosive; it does not undergo hydrolysis even in aqueous hydroxide solutions.

Nitrogen trichloride is formed almost quantitatively by the action of an excess of chlorine or hypochlorous acid on ammonium ion in acid solution (pH <3).

$$NH_4^+ + 3Cl_2 \rightarrow NCl_3 + 4H^+ + 3Cl^-$$

If pH = 3.0 to 5.0, dichloramine, $NHCl_2$, is obtained. If pH >8, only chloramine, NH_2Cl, is obtained. Nitrogen trichloride is a treacherously explosive, yellow, oily liquid which is irritating to the eyes. Contact with materials which can be chlorinated or heating to $95°C$ causes explosion to occur. Nitrogen trichloride boils below $71°C$ and is not frozen at $-40°C$. It undergoes hydrolysis to ammonia and hypochlorous acid.

The reaction of iodine with ammonia in alcoholic or potassium iodide solution yields a brown precipitate having the formula $NI_3 \cdot NH_3$. This substance, commonly known as "nitrogen triiodide" explodes at the slightest touch when dry. Nitrogen triiodide, NI_3, is formed by the action of dry ammonia on dibromoiodides such as $KIBr_2$.

$$3KIBr_2 + 4NH_3 \rightarrow 3KBr + 3NH_4Br + NI_3$$

The strong contrast in stability between nitrogen trifluoride on the one hand and the tribromide and iodide is worth noting. The instability of the latter two compounds results not from an exceptional weakness of N—Cl or N—Br bonds but rather from the exceptionally large energy of dissociation of the N_2 molecule. In the N—F bond the difference in electronegativity of the two atoms is sufficient to increase the N—F bond energy to a value large enough to counteract the high energy of formation of the N_2 molecule and thus to stabilize the NF_3 molecule.

Another series of binary halogen-nitrogen compounds of the general formula N_3X where X = F, Cl, Br, and I have been prepared. These compounds are halogen azides and contain the N_3 group characteristic of the covalent azides. Chlorine azide, ClN_3, is obtained by the reaction of hypochlorites with azides in acid solution

$$HOCl + H^+ + N_3^- \rightarrow ClN_3 + H_2O$$

Chlorine azide is a highly explosive, colorless gas. Bromine azide, BrN_3, is a volatile, orange-red liquid which freezes at $-95°C$ and hydrolyzes to hypobromous and hydrazoic acids. Iodine azide is a light yellow solid which is unstable above $0°C$. Fluorine azide, FN_3, a greenish yellow, explosive gas, is formed by the reaction of fluorine with hydrazoic acid

$$4HN_3 + 2F_2 \rightarrow 3FN_3 + NH_4F + N_2$$

Fluorine azide decomposes slowly at room temperature to yield the colorless, stable gas difluorodiazine, N_2F_2. Electron diffraction shows that this substance consists of equal amounts of the *cis* and *trans* forms of F—N=N—F.

More important than any of the binary halogen-nitrogen compounds, however, is the monohalogen derivative of ammonia known as chloramine. It was pointed out above that this compound is obtained by the action of chlorine or hypochlorous acid on ammonia in solutions of pH greater than 8. A better method consists of the gas phase reaction of chlorine gas with an excess of ammonia which under optimum conditions gives almost quantitative yields of NH_2Cl. Infrared spectroscopy of chloramine and deuterated chloramine has led to the conclusion that chloramine has a pyramidal structure. Similar results were obtained for dichloramine, $NHCl_2$.

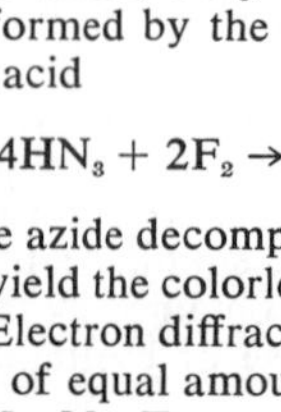

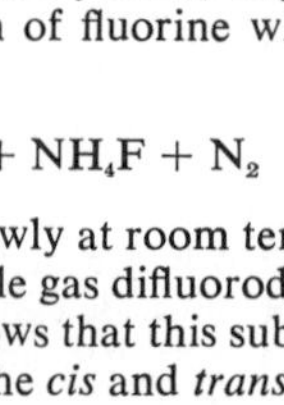

The preparation of pure, anhydrous chloramine by the vacuum distillation and dehydration of aqueous solutions of the substance obtained from the reaction of aqueous ammonia with hypochlorite has been reported. The pure substance is a colorless, explosive oil which decomposes at $-50°C$ and freezes at $-60°C$. Nonaqueous solutions of chloramine are obtained by solvent extraction of chloramine from dilute aqueous solutions.

Chloramine gives a wide variety of important chemical reactions. Among the more important of these are reactions of the SN_2 (bimolecular displacement) type; for example, we have the following reactions with ammonia and amines:

$$NH_3 + NH_2Cl \rightarrow$$
$$[NH_3NH_2]^+Cl^- \xrightarrow{NH_3} NH_2NH_2 + NH_4^+$$

$$RNH_2 + NH_2Cl \rightarrow$$
$$[RNH_2NH_2]^+Cl^- \xrightarrow{RNH_2} RNHNH_2 + RNH_3^+$$

$$R_2NH + NH_2Cl \rightarrow$$
$$[R_2NHNH_2]^+Cl^- \xrightarrow{R_2NH} R_2NNH_2 + R_2NH_2^+$$

$$R_3N + NH_2Cl \rightarrow [R_3NNH_2]^+Cl^-$$

These represent excellent methods for the synthesis of hydrazines and hydrazinium salts.

Other examples of this type of reaction include:

$$R_3P: + NH_2Cl \rightarrow [R_3PNH_2]^+Cl^-$$
$$2CN^- + 2NH_2Cl \rightarrow [2NH_2CN[+ CL^-$$
$$\downarrow$$
$$NH(CN)_2 + NH_3$$

$$NH_2Cl + RMgX \rightarrow RNH_2 + MgXCl$$

$$NH_2Cl + RO^- \xrightarrow{ROH} NH_2OR + Cl^-$$

It was pointed out above that mono- and difluoramine (NH_2F and NHF_2) are obtained as byproducts in the electrolysis of fused ammonium hydrogen fluoride. The mono- and dibromamines are formed by the action of bromine on ammonia in ether but have not been isolated. They behave somewhat like chloramines. Mono- and diiodoamines have not been prepared.

HARRY H. SISLER

Cross-references: *Nitrogen, Explosives.*

HEAT-TRANSFER AGENTS

The energy which is transferred under a temperature difference or gradient is designated as heat. The three modes of heat transfer are conduction, convection, and radiation, and applications may involve combinations of these modes. Primary sources of thermal energy are chemical and nuclear reactions, as illustrated by the burning of fossil fuels (coal, oil, gas) and fusion and fission processes. When one considers the continuous exposure to natural and man-made events which involve temperature changes, one begins to appreciate the vast number of applications of heat transfer. All heating and cooling operations involve heat transfer, covering an extreme range from near zero temperatures to millions of degrees. The applicable science at the low temperatures is termed *cryogenics*. At stellar temperatures, *fusion* reactions generate energy which, for example, the sun radiates to the earth. Thus, heat transfer manifested on both cosmic and terrestrial scales becomes essential for the preservation of life. Man utilizes heat transfer to meet individual and societal needs.

The transport of thermal energy through a material by the thermal motion or vibration of the molecules is called conduction. The basic equation for heat conduction is known as Fourier's law which relates the steady-state heat conduction per unit area to the temperature difference across a unit thickness, or temperature gradient. The constant of proportionality is the thermal conductivity. Fouriers' law is thus analogous to Ohm's law used in the conduction of electricity. Good heat transfer agents have large thermal conductivities, and good insulators have very low values. This transport property for a given material depends upon its state of aggregation and may be influenced by the temperature and other state properties. Materials with the lowest thermal conductivities are gases, and many common insulators involve porous solids which provide isolation of air cells. Use of evacuated regions, like in Dewar flasks, serves to eliminate thermal conductance as a mode of transfer, leaving radiation as the only mode. Metals, such as silver and copper, have the highest thermal conductivities. Materials with high thermal conductivity have high electrical conductivity and the Wiedemann-Franz theory assumes that heat transfer, like electricity, may involve conduction by free electrons. At normal conditions, the ratios of thermal conductivities of water to air and copper to water are approximately 25 and 660. At temperatures close to absolute zero, the phenomenon of superconductivity exists.

In heat exchangers in which the heating and cooling fluids are separated by metal surfaces, the heat transfer through the metal is by conduction. The efficiency of the transfer process may be increased through the use of metallic fins which provide extended surfaces. Insulators are used to minimize heat losses or heat gains, and examples include rock wool for insulating hot steam lines, bricks for kiln walls, an air layer between two panes of glass in thermopane windows, fiber-glass wall insulators for homes, and hot pad and cork holders for handling hot objects in the kitchen. The temperature of the earth's crust increases with depth and thus there is a continuous flow of heat to the surface by conduction, albeit relatively small compared to other modes of heat transfer.

Bodies emit radiant energy from its surfaces in the form of electromagnetic waves and this energy transport is called thermal radiation. The characteristics of the thermal radiation for a given material depend upon the surface conditions and the temperature. The surface is said to be "black" if it absorbs all incident radiation. For a black body, the Stefan-Boltzmann law states that the emission rate of thermal radiation per unit area varies with the fourth power of the absolute temperature. Thus the importance of heat transfer by radiation increases rapidly with temperature. Planck's law describes the distribution of the energy as a function of wave length. The radiant energy varies from wavelengths characteristic of infrared and longer, through the visible light spectrum, and to ultraviolet and shorter. A heated body glows red at about 1000°F and whitens at higher temperatures. The radiant energy of the sun corresponds to blackbody emissions at about 10,000°F. *Photosynthesis* involves utilization by plants of radiant energy in the visible and ultraviolet region of electromagnetic waves.

Whereas for a black surface, all incident radiation is absorbed, for most actual surfaces, a portion of the incident radiation may be reflected and a portion may be transmitted, with the remainder being absorbed. If the fraction of energy transmitted is zero, the body is said to be opaque. Further, the absorptivity may be dependent upon the wave length of the incident radiation as well as upon the surface temperature. For solids, the absorption is primarily at the surface. Some liquids and gases may be transparent, and for some materials, CO_2 and steam, absorption and emission are at selective wavelengths. Nonblack bodies, which implies all actual surfaces, emit and absorb less energy than that for a black body, and a factor called the emissivity is added to the Stefan-Boltzmann law. Polished metal surfaces may have low emissivities, for example 0.04 for polished aluminum, whereas oxidized metal surfaces have higher values, about 0.6 to 0.9. For an enclosed system of opaque surfaces at thermal equilibrium, Kirchhoff's law relates the emissive power to absorptivity, and the emissivity will be equal to the absorptivity. A surface is assumed to be gray if the absorptivity is not a function of the incident wavelength. The evaluation of the net interchange of radiant energy among surfaces requires the knowledge of the emissivities, absorptivities, and transmissivities, as well as geometrical factors relating how viewing of the surfaces are interrelated. Simplifying assumptions regarding gray bodies and view or exchange factors often may be justified in evaluating applications in radiant heat transfer.

High-temperature processes will generally involve significant contributions by thermal radiation, and examples include combustion of fuels for power generation and for metallurgical operations. A balance involving solar radiation and terrestrial radiation are major components in maintaining the earth's surface temperature. Whereas solar energy is more nearly transparent to our earth's atmosphere, terrestrial radiation, emitted at longer wavelengths, is partially absorbed and reradiated to the earth's surface by the CO_2 and water vapor in the atmosphere. This trapping of radiant energy is called the greenhouse effect. Radiation from the earth's surface on clear, still nights reduces the surface temperature below that of the ambient air permitting the nocturnal effects of dew and frost formation.

The third mode of heat transfer, termed convection, involves fluid motion. The fluid may be used as a heating or cooling agent, and for simplicity, the discussion is presented in terms of a coolant. The energy transfer from the source to the flowing or circulating medium is manifested most frequently as sensible heat, producing a temperature rise in the coolant. The magnitude of the coolant temperature rise, ΔT, is dependent upon the flowing heat capacity, wc (where w is the mass flow rate of coolant and c is the heat capacity of the coolant), and the rate of heat transfer, q, where in simple cases, $q = wc\Delta T$. In cases where the specific heat is a function of temperature and pressure, and/or where phase changes occur, reference should be made to the thermodynamic equations, charts and tables available for most common coolants for the evaluation of the coolant exit enthalpy and temperature rise. The Δt is called the transport temperature rise. Properties of the coolant other than the specific heat may be utilized, such as latent heat associated with melting, vaporization, sublimation, or other phase transitions. Flow systems with endothermic reactions can also be adapted as coolants. By far the most common coolants are water and air, and involve energy transfer simply as sensible heat.

The rate at which thermal energy is transferred to the coolant per unit area of heat transfer surface is directly proportional to the temperature driving force, ΔT, the transfer temperature rise, and usually is the difference between the surface temperature and the average or bulk temperature of the fluid. The relationship is referred to as Newton's law and the proportionality factor is called the heat transfer coefficient. The physical properties affecting the heat transfer coefficient are density, heat capacity, viscosity, thermal conductivity, and in the case of natural convection (movement of the fluid through density changes), the coefficient of volumetric expansion. Coolant flow rate is an important parameter as well as the geometrical configurations used for the heat transfer. Only under relatively simple conditions might a heat transfer coefficient be determined analytically, and for most applications, empirical correlations are available with the property, flow rates and geometry parameters presented in dimensionless groupings.

The power expended for pumping the coolant may be a factor involving coolant selection, and the two important physical properties in the pressure drop and flow work evaluations are viscosity and density. Aside from the usual requirements concerning availability and price, other coolant properties which need be considered are the phase pressure-temperature saturation curves, thermal stability, and component compatability covering scale, corrosion, and in some high-temperature cases, metal transfer. The choice of coolant for special purposes, such as involved in nuclear reactors, requires the evaluations of neutron absorption cross-sections, of possible induced radioactivity, and of radiation (exposure to fast neutron and gamma fluxes) damage. Some other special properties that have been emphasized include total weight of coolant system in airplane designs, good wetting and spreading action in water emulsion cutting coolants, along with lubricity to produce a fine finish, the use of hydrogen in cooling electric generators because of its high specific heat and low windage loss, and the use of natural gas as a cooling and reducing medium. A very common example of a gas coolant acting also as a carrier is in water cooling towers with forced or natural draft.

Low-temperature coolants include salt-water solutions, ethanol, methanol, glycols, glycerol; common industrial refrigerants such as ammonia, sulfur dioxide, the "Freons," methyl chloride, propane, butane, and propylene; and high-pressure refrigerants including methane, ethylene and carbon dioxide. Liquefied gases (air, nitrogen, hydrogen and helium) are used for very low temperature cooling.

Significant reductions in system operation pressures (and hence possible savings in equipment design) can be achieved by using coolants other than water for high-temperature heat exchange. For example, up to temperatures in the range of 700°F, coolants used include oil, "Dowtherm" (eutectic mixture of 73.5% diphenyl oxide and 26.5% diphenyl), and mercury. For somewhat lower temperature limits, chlorinated biphenyl and tetraryl silicates can be used. Above 700°F, special inorganic salt mixtures have been used, such as HTS (40% $NaNO_2$, 7% $NaNO_3$ and 53% (wt) KNO_3). The nuclear reactor program has accelerated the development of the technologies for handling molten metals, including sodium, and fused salts.

Solids may also be used to remove or transfer heat, and to stabilize temperature. For example, some designs of moving beds (pebbles of ceramic materials) represent significant improvements over the bulky regenerative type of heat exchange. In the catalytic cracking of petroleum, developments include the giant fluidized beds. Elemental solids with high melting points and relatively high volumetric heat capacities have been proposed as coolants for high temperatures.

Safety is a dominant factor which cannot be overlooked in handling potentially hazardous coolants. For example, for handling liquid metals and radioactive coolants, many new techniques are emerging for obtaining leak-tight systems. Special designs are used in liquid metal-to-water heat exchangers to detect and prevent leakage.

Heat pipes are special devices which transfer thermal energy under very little temperature difference over short distances by a self-contained, geometrical arrangement involving evaporation of the working fluid at the heat source and condensation at the heat sink, with return of the condensate by capillarity effects.

H. S. Isbin

HELIUM

Helium, chemical symbol He (atomic number 2), is the second element in the Periodic Chart of the elements and is the lightest member of the rare, noble or inert group of elements, all having zero valence. It is named from *Helios* the Greek word for the sun in whose atmosphere it was discovered by Sir Norman Lockyer and P. C. J. Janssen in 1868.

Helium exists in abundance in the sun and other stars, the sun's composition being approximately 80/20 hydrogen/helium with about 1% of heavier elements. The basic source of stellar helium is thought to be nuclear reactions such as occur in the sun and stars where hydrogen in a fusion process is converted into helium with a large release of energy. It is also a product of the natural radioactive decay of members of the uranium and thorium series. The alpha particle produced in this disintegration is a charged helium atom, and the helium found on earth is generally considered to be an accumulation of helium produced in this manner.

The earth's atmosphere contains about 5 ppm helium but the most prolific source of helium on earth is natural gas. Most natural gases contain 5,000 ppm (0.005%) or more, some 0.05 to 0.1% and a few, limited in number and low in volume, contain from 1 to 10% He.

Two stable isotopes of helium exist, He^3 and He^4,

but He^3 exists only in very minor amounts. Helium obtained from the earth's atmosphere contains about 1 ppm He^3 and helium obtained from natural gases, the principal source, contains only one tenth this amount, or 0.1 ppm. Thus the properties of helium of most concern are those of the heavier isotope He^4.

Both He^4 and He^3 are gases at ordinary temperatures and pressures and are colorless, odorless, tasteless, and nontoxic. Like all noble gases, their molecule is monatomic, so their atomic and molecular weights are equal, being 4.0026 for He^4 and 3.01603 for He^3 (Carbon-12 scale). He^4 has a critical temperature of 5.2°K above which the liquid cannot exist and a normal boiling point of 4.2°K. He^3 has a critical temperature of 3.35°K and normal boiling point of 3.20°K.

Gaseous He^4 has a density of 0.1785 gpl at 0°C and 1 atm., and 17.0 gpl at its normal boiling point, 4.2°K. Its index of refraction is close to unity, which makes it useful in optical equipment. The thermal conductivity of helium is greater than any other gas except hydrogen. It is less soluble in water than any other gas and its solubility in other liquids is low. Its rate of diffusion or permeation through tiny openings or pores is about three times that of air, a property that makes helium especially useful in leak detection.

In the usual sense of chemical reactions, helium is completely inert and forms no compounds, hence has a valence of zero. Only under extreme conditions—high temperatures, low pressures and ionizing conditions, such as are obtained in plasmas and glow and spark discharges (conditions for the production of spectra)—have combinations with other elements been observed. Being inert it is noncombustible, and is used in the interest of safety where hydrogen, another very light gas, otherwise might be used. As a lifting gas, helium has about 92% of the lifting power of hydrogen, though its density is about twice that of hydrogen. One thousand cut ft of helium lifts about 65.8 lbs at sea level. A striking feature of He^4 is the behavior of the liquid when cooled below its boiling pont of 4.2°K to a temperature of 2.19°K. Above this transition temperature the liquid behaves much like other low-boiling liquids and is called Helium I. At 2.19°K, a transition occurs which is completely unclassical. Below this temperature the liquid, which is called Helium II, exhibits amazing frictionless or "superfluid" flow properties. It will flow with ease through extremely small holes and capillaries through which no other liquid will pass. An extremely thin mobile surface-film of the liquid, only about 3×10^{-6} cm thick, forms on solid surfaces above the liquid and results in the transport of the liquid in an apparently frictionless manner. If a beaker is partially immersed in the liquid a surface-film of the liquid will form on the outer surface of the beaker, creep up and over the top of the beaker down on the inner surface and will fill the beaker to a level equal to the height of the liquid outside. Then if the beaker is raised above the level of the liquid in the helium bath, the flow is reversed and the liquid in the beaker will "drain" up and over the top, down the outer surface and fall off in droplets from the bottom of the beaker until it is empty.

Also, when liquid He^4 is cooled to 2.19°K, normally by pulling a vacuum on the system, it undergoes a striking change in appearance. All evidence of boiling ceases and it becomes quiescent. It becomes almost a perfect heat conductor, better than the best of metals, such as silver, and no temperature differences can be detected throughout the body of the liquid. It loses its viscosity and becomes a "superfluid," as just mentioned.

When He^3 is liquefied, it does not exhibit the strange phenomenon observed in liquid He^4. It does, however, have one characteristic in common with He^4, that is, it cannot be solidified or frozen by lowering its temperature alone. It remains a liquid down to absolute zero and can only be solidified by raising its pressure above its normal vapor pressure. Since the normal boiling point of He^3 is 3.20°K, lower than that of He^4, it is useful in obtaining the lowest temperatures. With it temperatures of the order of 0.00002°K have been reached.

In addition to the very limited supply of He^3 from natural resources, He^3 is produced in nuclear reactors and has been collected by the Atomic Energy Commission for sale to users. This is its principal source at present.

Because of the unusual characteristics of helium it has been found useful in many ways. Since it is very light, it is used as a lifting gas in blimps and in balloons used in aerology. Its low solubility in body fluids recommends its use in breathing mixtures for divers and manned under-sea chambers. Its inertness makes it useful in shielded-arc welding and in purging and shielding nuclear reactors. Because it can easily be detected by means of a mass spectrometer and because of its high leak rate through pin holes and crevices, it is universally used in leak detection. Its inertness and low liquefaction temperature makes it a choice in purging liquid oxygen and liquid hydrogen fuel lines and vessels in space program vehicles. Because of the low temperature and safety provided by liquid helium, one of its most important uses is in cryogenics. This undoubtedly will be one of its most useful fields in the future.

Although first discovered in 1868 as a constituent of the sun, helium was not discovered on earth until 1895. In that year, Sir William Ramsey discovered it in gases evolved from cleveite, a uranium ore. It was 10 years later that Cady and McFarland discovered it in natural gas at the University of Kansas in 1905.

Events in World War I created an interest in helium as a lifting gas for balloons for bomber protection of London. Between World Wars I and II helium was used, principally, as the inflating gas for dirigibles and blimps. Production of the gas was by the Department of the Interior, Bureau of Mines, from natural gases found in southwestern United States. During this period, however, uses for helium were found in preparing helium-oxygen breathing mixtures helpful in relieving respiratory ailments and in aiding deep-sea diving. Its use in shielded-arc welding also was developed.

Helium has been produced in commercial quantities since 1918. The quantities were relatively small until World War II. In 1948, a sharp upward trend began as new uses for helium were found in the U.S. nuclear and missile programs. In 1967, helium demands exceeded 900 million cu ft.

As all commercial helium is extracted from natural gas and this gas is being used for fuel purposes, the U.S. Government, in 1960, started a helium conservation program designed to save helium being lost when the fuel gas was burned. Contracts were made with commercial companies to purchase helium extracted from natural gas going to fuel markets. The helium purchased is transported by pipe line to a storage field where it is stored underground in a partially depleted gas field for future recovery and use. Under this program, about 3.5 billion cu ft of helium per year is being saved. At the end of 1970, 28 billion cu ft had been placed in underground storage and it is expected that by 1980 a total of more than 75 billion cu ft will have been recovered and stored.

WILLIAM M. DEATON

References

Keesom, W. H., "Helium," New York, Elsevier Publishing Co., 1942.

Lifshits, E. M., and Andronikashvili, E. L., "A Supplement to 'Helium'," New York, Consultants Bureau, 1959.

Cook, G. A., "Argon, Helium, and the Rare Gases," New York, Interscience Publishers, 1961.

Birmingham, B. W., Editor, "Liquid Helium Technology," Washington, D.C., U.S. Government Printing Office, April, 1968.

Deaton, W. M., "Helium: Yesterday, Today and Tomorrow," *Cryogenic Engineering News,* Cleveland, Ohio, May, 1968, pp. 22–27.

Brandt, L. W., "Helium," in "Encyclopedia of the Chemical Elements," Hampel, C. A., Editor, pp. 256–268, Van Nostrand Reinhold Co., New York, 1968.

HEMICELLULOSE

The term "hemicellulose" was first applied by E. Schulze in 1891 to those components of plant cell walls which are easily hydrolyzed by hot dilute mineral acids and are soluble in cold 5% aqueous sodium hydroxide. Today the term, as usually used, refers to the polysaccharide components of the cell wall other than cellulose. Most of these polysaccharides are of short chain length, as compared to cellulose, and behave according to Schulze's definition. However, some components of hemicellulose are quite resistant to mild acid hydrolysis or to caustic soda and are very difficult to separate from the cellulose.

The hemicelluloses are of indefinite composition and vary widely between different plant species. The term, therefore, does not refer to a specific chemical species, but to a group of polysaccharides, often of widely different structure and properties. The term, therefore, is mainly one of convenience.

Hydrolysis of the hemicelluloses of wood indicate that its building units are composed of D-xylose, D-mannose, D-glucose and, to a much smaller extent, L-arabinose, D-galactose and L-rhamnose. Besides these sugar residues, the hemicelluloses of wood when hydrolyzed, give acetic acid (originally present as acyl groups) and hexuronic acids. The latter frequently retain methyl ether groups, such as the 4-methyl ether of glucuronic acid. In the hydrolysis of hemicellulose, the uronic acid may be isolated as an aldobiuronic or aldotriuronic acid, which, in the case of aspen, pine and spruce wood, is 2-(4-methylglucuronosyl)-D-xylopyranose.

Hardwoods contain 25–35% of a partly acetylated 4-methylglucurono-xylan. Softwoods have about 10% of an arabino-4-methylglucurono-xylan and up to 20% of various galactoglucomannans. In heartwood of larch trees, the cells are filled with an arabinogalactan, usually 10–25% of the wood. Hemicelluloses present in only small amounts in hardwoods and/or softwoods are glucomannan, a 1,3-glucan, arabinan, and galactan. Bark contains more or less the same hemicelluloses as does wood.

Hemicellulose may be prepared by direct extraction from plant tissue by aqueous alkali. This method is often satisfactory for hardwoods and annual plants. In most cases, especially for softwoods, it is preferable to isolate the hemicellulose by alkaline extraction of holocellulose. Holocellulose is the total polysaccharide system of the cell wall and is obtained by removal of lignin from the plant tissue with acidic sodium chlorite or by alternate successive treatments with chlorine and alcoholic ethanolamine. The hemicelluloses are usually obtained from their filtered alkaline solutions by acidifying with acetic acid, followed by precipitation with an excess of ethanol.

There is substantial evidence that in xylan isolated from hemicelluloses the D-xylose units are joined through 1:4-glycosidic linkages. This is based on studies of isolated crystalline series of xylo-oligosaccharides, extending from xylobiose to xylohexaose. Some xylans also appear to contain some L-arabinose and, in other instances, a D-glucuronic acid.

There is a great gulf between the molecular magnitude of the hemicelluloses and that of cellulose. The latter may have an average degree of polymerization (D.P.) of 10,000 or more, whereas the average D.P. of isolated hemicelluloses is in the order of 200–300. Moreover, there is a wide distribution of molecular weights. The hemicelluloses probably form an interpenetrating system of polysaccharides, though evidence indicates that they are not evenly distributed throughout the cell wall. Presence of these short chain hemicelluloses decreases the crystallinity of the cell wall and increases the amount and areas of amorphous or poorly oriented material. The primary walls of wood fibers contain considerable amounts of hemicellulose. There is a controversial question as to whether or not part of the hemicelluloses are chemically combined with part of the lignin.

Hemicellulose has little or no direct industrial application. Indirectly, however, it is of considerable importance commercially. The hemicelluloses which are retained in paper pulps contribute greatly to the physical characteristics of paper. Only inferior paper can be made from pure cellulose or highly refined pulps. Such fibers, which contain little or no hemicellulose cannot make a strong paper without excessive mechanical beating in water or by the use of adhesive additives. The presence of hemicelluloses in pulps permits the fibers to "hydrate" readily upon beating and to form strong paper sheets. Hemicellulose contributes greatly to tensile strength, tear resistance, bursting strength and folding endurance of paper. The hemicelluloses appear to impart adhesive and bonding properties to paper fibers. Commercial coniferous pulps contain about 3–5% mannan, 5–8% xylan and 2% hexuronic anhydride. With increased amount of hemicellulose in pulp, particularly those containing uronic acids, the greater the improvement in the beating operation and the quality of the paper until a maximum is reached. Excessive amounts of hemicellulose may produce a brittle or glassine paper.

Hemicellulose may be detrimental in pulps made for purposes other than paper. This is especially true in cellulose acetate manufacture where small amounts of xylan or mannan cause hazy acetate solutions and the clogging of filters.

Another indirect use of hermicellulose is in the manufacture of furfural. This chemical is produced by the steam distillation of corn cobs and oat hulls. The pentosans and, to a lesser extent, the uronic acids, are the source of the furfural. The hemicelluloses of wood wastes are also a potential source of furfural.

EDWIN C. JAHN

HERBICIDES

The application of chemicals to kill weeds selectively has been practised for only about 75 years. In 1896, Bonnet, a French grape grower, observed that the leaves of a common weed growing with the grapevine turned black when Bordeaux mixture (a mixture of copper sulfate and hydrated lime still used principally as a foliage fungicide) was applied to grapevines as a protection against downy mildew. The weed killing properties of other inorganic salts such as ammonium sulfate, ferrous sulfate and other metal salts were soon discovered. The application of organic chemicals for weed control came in 1932 with the discovery of 2-methyl-4, 6-dinitrophenol (DNOC). Several years later it was found that chemicals structurally related to the plant hormones (auxins) could also be used as selective herbicides.

Today over 30% of the croplands in the U.S. are treated with herbicides. In 1965, U.S. farmers spent $494 million to treat over 120 million acres with 84 million pounds of herbicides. It is estimated that by 1975, 500 million pounds of herbicides will be used for farm and nonfarm use in the U.S. The widespread use of these chemicals is becoming of increasing concern both due to their potential toxicological hazard and their ecological effects which are still largely unknown.

Plants are considered weeds when they interfere with utilization of land and water resources or otherwise adversely intrude upon human welfare. Some plants are considered weeds because they are poisonous to livestock or because they otherwise affect the quantity and quality of animal products. Others, such as poison ivy and allergenic plants, are directly noxious to man. Water weeds clog irrigation and drainage canals, interfere with navigation and reduce the production and availability of fish and other wildlife. Weed control methods are classified as preventive, biological, managerial, physical and chemical. Chemical methods include the use of inorganic and organic chemicals as foliar sprays, soil and water treatments, fumigants, and stem application for selective or nonselective control of weeds.

The use of herbicides results in various types of injury to susceptible plants, often resulting in distinctive visible symptoms such as cessation of growth, desiccation, morphological aberration, chlorosis and death of the plant. Herbicides exert their effects by interfering with certain vital physiological processes in the plant.

Morphological and Anatomical Effects. Cell enlargement is an evident response to auxinlike herbicides such as 2,4-D. The symptoms seen are partly because of cell extension through changes in membrane permeability and cell wall plasticity. Certain herbicides such as the phenylcarbamates, maleic hydrazide, and DCPA cause severe stunting by arresting cell division, particularly in seedlings. Trifluralin inhibits root growth.

Effects on Uptake of Water and Nutrients. Certain herbicides, particularly the substituted phenylureas and s-triazines that inhibit photosynthesis, may lead to the closing of the stomata, inhibiting transpiration, thus causing water accumulation.

Effects on Photosynthesis. The triazines, substituted ureas and substituted uracil herbicides can be considered as exerting their action principally by inhibiting photosynthesis.

Effects on Respiration. The consumption of oxygen with concurrent evolution of carbon dioxide is a fundamental biochemical phenomenon. Nitrophenol-type herbicides such as dinoseb inhibit respiration to a moderate extent.

Interaction with Nucleic Acid and Protein Metabolism. Many of the actions of auxinlike herbicides are believed to be mediated by their effect on nucleic acid or protein metabolism.

Interaction with Plant Hormones and Growth Regulators. In many physiological processes, herbicides such as 2,4-D, 2,4,5-T, dicamba and picloram can replace the auxin, indoleacetic acid. The phytotoxicity of 2,4-D seems to some extent to result from its slow inactivation in susceptible plants, permitting a longer time for the chemical to exert its effect on cell growth and biochemical reactions in the plant.

Effects on Enzymes. A number of herbicides affect plant enzymes such as catalase (atrazine), peroxidase (2,4-D and atrazine), nitrate reductase and pectin methyl esterase. The arsenicals are relatively nonspecific inhibitors of plant enzymes containing sulfhydryl groups.

Following is a brief description of the inorganic and organic herbicides currently used for weed control. Although organic herbicides are discussed according to functional groups, there is no general theory relating chemical structure to herbicidal activity.

Inorganic Herbicides

Arsenic. The trivalent forms of arsenic, such as arsenic trioxide and sodium arsenite are effective herbicides. Arsenic trioxide has been used for soil sterilization but requires high application rates and remains for many years in the soil. Sodium arsenite is more effective than arsenic trioxide as a weed killer because it is translocated to some extent in plants.

AMS (ammonium sulfamate) is a water-soluble compound of low mammalian toxicity used for brush and poison ivy control.

Boron. A family of inorganic borates such as borax (sodium tetraborate), sodium metaborate and sodium borate enjoy wide usage for nonselective, long-term weed control. The borates have a low order of mammalian toxicity.

Sodium chlorate is used primarily as a soil sterilant and as a cotton and soybean defoliant. Effective application rates are high. This compound is a strong oxidizing agent and care should be exercised in its use.

Sulfuric acid has been used in the control of annual weeds in cereals and onions. Its usefulness is limited by its nonselectivity and by its corrosiveness.

Miscellaneous inorganic salts, such as ammonium thiocyanate, ammonium nitrate, ammonium sulfate, iron sulfate and copper sulfate have been used as foliar sprays at high dosages.

Organic Herbicides

Arsenicals. Two groups of organic arsenicals are widely used as herbicides; the arsonic and the arsinic acids. Cacodylic acid (dimethylarsinic acid) and its sodium salt and methane arsonic acid and its salts are widely used as contact herbicides. The pentavalant arsenicals used as herbicides such as DSMA (disodium methanearsonate), MSMA (monosodium acid methanearsonate), CMA (calcium acid methanearsonate), AMA (amine methanearsonate) and MAMA (monoammonium methanearsonate) are generally less toxic to animals than organic trivalent arsenicals such as arsenosomethane (CH_3AsO) and considerably less toxic than pentavalent or trivalent inorganic arsenicals. The methanearsonates are selective, whereas the cacodylates are not highly selective herbicides.

Phenoxyaliphatic acids. 2,4-D (2,4-dichlorophenoxyacetic acid) and its salts, esters and amides are used for broad spectrum post-emergence weed control in cereal grains, corns, pastures and lawns and aquatic weeds. The phenoxyaliphatic acid herbicides are relatively nontoxic to animals. Related compounds are: 2,4,5-T (2,4,5-trichlorophenoxyacetic acid); MCPA (2-methyl-4-chlorophenoxyacetic acid); MCPB (2-(2-methyl-4-chlorophenoxy)butyric acid); MCPP (2-(2-methyl-4-chlorophenoxy) propionic acid); 2-4-DB (4-(2,4-dichlorophenoxy)butyric acid) and its salts, amine salts and esters; Silvex (2-(2,4,5-trichlorophenoxy) propionic acid; and Dichlorprop (2-(2,4-dichlorophenoxy)-propionic acid). 4-CPA (4-chlorophenoxyacetic acid) is principally used as a plant growth regulator for bloom set on tomatoes and to thin peaches.

Aliphatic Acids. Two chlorinated aliphatic acids, dalapon (2,2-dichloropropionic acid) and TCA (trichloroacetic acid) are used as herbicides for the control of grasses and broadleaved weeds. Endothall (7-oxabicyclo-(2,2,1)-heptane-2,3-dicarboxylic acid) is used as an aquatic herbicide and as a selective herbicide in field crops.

Arylaliphatic acids. Representative arylaliphatic acids used as herbicides are 2,3,6 TBA (2,3,6-trichlorobenzoic acid) and its amine salt, fenac (2,3,6-trichlorophenylacetic acid), dicamba (3,6-dichloro-*o*-anisic acid), amiben (3-amino-2,5-dichlorobenzoic acid), DCPA (dimethyl-2,3,5,6-tetrachloroterephthalate) and TCBC (trichlorobenzylchloride) are used primarily for soil application against germinating seeds and seedlings.

Substituted Nitriles. Dichlobenil (2,6-dichlorobenzonitrile) is an effective herbicide against germinating seeds and young seedlings of both monocotyledonous and dicotyledonous species. Other compounds in this class are diphenatrile (diphenylacetonitrile), ioxynil (3,5-diiodo-4-hydroxybenzonitrile) and bromoxynil (3,5-dibromo-4-hydroxybenzonitrile).

Phenols. PCP (pentachlorophenol) and its salts and esters are nonselective herbicides used for weed control in certain crops such as soybeans and corn. Other phenols such as DNOC (2-methyl-4,6-dinitrophenol) and dinoseb (4,6-dinitro-*o-sec*-butylphenyl) are employed as foliar sprays and soil application herbicides.

Substituted Amides. Diphenamid (N,N-dimethyl-2,2-diphenyl-acetamide) is a selective pre-emergence herbicide for control of annual grasses and certain broad-leaved weeds. Related compounds are: CDAA (N,N-diallyl-2-chloroacetamide); propanil (3,4-dichloro-propionanilide); Solan (3-chloro-2-methyl-*p*-valerotoluidine); naptalam (N-1-naphthylphthalamic acid); and cypromid (3′,4′-dichlorocyclopropanecarboxanilide) is a corn herbicide which controls broad-leaved weeds and grasses.

Nitroanilines. Trifluralin (α,α,α-trifluoro-2,6-dinitro-N,N-dipropyl-*p*-toluidine) is a widely used pre-emergence herbicide for use on cotton, soybeans, sugar beets and other vegetables. It has a low order of toxicity to mammals and birds but is toxic to fish. Related compounds are nitralin (4-methylsulfonyl)-2,6-dinitro-N,N-dipropyaniline) and benefin (N-butyl-N-ethyltrifluoro-2,6-dinitro-*p*-toluidine).

Ureas. Diuron (3-(3,4-dichlorophenyl)-1,1-dimethylurea) is used for selective weed control on cotton, sugar cane and citrus and other crops and for general weed control on highways and industrial sites. Related compounds are: fenuron (3-phenyl-1,1-dimethylurea); monuron (3-(*p*-chlorophenyl)-1,1-dimethylurea); linuron (3-(3,4-dichlorophenyl)-1-methoxy-1-methylurea); chloroxuron (3-[*p*-(*p*-chlorophenoxy)phenyl]-1,1-dimethylurea); fluometuron (1,1-dimethyl-3(α,α,α-trifluoro-*m*-tolyl) urea); metobromuron (3-(*p*-bromophenyl)-1-methoxy-1-methylurea); norea (3-hexahydro-4,7-methanoinden-5-yl)-1,1-dimethylurea); and siduron (1-(2-methylcyclohexyl)-3-phenylurea).

Carbamates. The carbamates currently used as pre-emergence and post-emergence herbicides for the control of annual grasses and broad leaved weeds include propham (isopropyl N-phenylcarbamate), chlorpropham (isopropyl *m*-chlorocarbanilate), barban (4-chloro-2-butynyl-*m*-chlorocarbanilate) and terbutol (2,6-di-tert-butyl-*p*-tolymethylcarbamate).

Thiocarbamates presently used for selective pre-emergence control of weeds in croplands include CDEC (2-chloroallyl diethyldithiocarbamate), diallate (S-2,3-dichloroallyl diisopropylthiocarbamate), EPTC (S-ethyl dipropylthiocarbamate), molinate (S-ethylhexahydro-1H-azepin-1-carbothioate), pepulate (S-propyl butylethylthiocarbamate), triallate (S-2,3,3-trichloroallyl-diisopropylthiolcarbamate) and vernolate (S-propyl dipropylthiolcarbamate).

Nitrogen Heterocycles. The symmetrical triazines constitute a large group of widely used herbicides of low mammalian toxicity and a wide spectrum of herbicidal activity. Atrazine (2-chloro-4-ethylamino-6-isopropylamino-s-triazine) is widely used for season-long weed control in corn, sorghum and other crops and for noncrop and industrial sites. Other triazine herbicides in use are ametryne (2-(ethylamino)-4-(isopropylamino)-6-(methylthio)-s-triazine), prometone (2,4-bis(isopropylamino)-6-methoxy-s-triazine), prometryne (2,4-bis(isopropylamino)-6-methylthio-s-triazine), propazine (2-chloro-4,6-bis(isopropylamino)-s-triazine), simazine (2-chloro-4,6-bis(ethylamino)-s-triazine. Amitrole (3-amino-1,2,4-triazole) is used for control of grasses, broad-leaved weeds and certain aquatic weeds for nonfood products only. Pyrazon (5-amino-4-chloro-2-phenyl-3H(2H)-pyridazinone) is a selective herbicide used on beets. Picloram (4-amino-3,5,6-trichloropicolinic acid) is a highly active systemic growth regulating chemical effective against a broad spectrum of plants used to control brush or on utility rights-of-way, for noncrop use only. Terbacil (3-tert-butyl-5-chloro-6-methyluracil) and bromacil (5-bromo-3-sec-butyl-6-methyluracil) are two substituted uracils used as selective herbicides for the control of annual and perennial weeds.

Bipyridyliums. The two major herbicides in this category are diquat (6,7-dihydrodipyrido[1,2a:2',1'-c]pyrazinedium salt) and paraquat (1,1'-dimethyl-4,4'-bipyridinium salt). These materials are employed as contact herbicides and aquatic weed control.

Miscellaneous chemicals. Bensulide (S-O,O-diisopropyl phosphorodithioate) ester of N-(2-mercaptoethyl)benzenesulfonamide is a pre-emergence herbicide used for controlling crab grass and other weeds in turf and ornamentals.

Petroleum oils. Petroleum hydrocarbons are effective as contact herbicides for total vegetation control by probably preventing respiration of the plant.

JOHN L. NEUMEYER

References

1. Audus, L. J., "The Physiology and Biochemistry of Herbicides," Academic Press, London, 1964.
2. National Academy of Sciences, "Weed Control," Vol. 2, Washington, D.C., 1968.
3. Neumeyer, J. L., Gibbons, D., and Trask, H., "Pesticides," *Chemical Week*, April 12, 1969, pp. 38–68; April 26, 1969, pp. 38–68.

HETEROCYCLIC COMPOUNDS

If all the atoms in a ring compound are of the same sort, the compound is called *homocyclic*. If one or more of the atoms in the ring differs from the others, such a compound is called *heterocyclic*. Ordinarily rings are composed of carbon atoms plus attached hydrogen or other substitutents. In addition, one or more hetero atoms are included, usually nitrogen, sulfur, or oxygen, or a combination of them. Five and six-membered rings are

the ones most frequently met. One of the simplest heterocyclic compounds is furan which has the structure

The hetero atom is designated as number 1, and the other atoms in the ring are numbered clockwise 2 through 5. According to another system, the carbon atom to the right (as represented in the formula) is alpha and the next one is beta (lower right-hand corner). On the left the atoms are called alpha prime and beta prime (See **Furans**). A five-membered heterocyclic ring that contains nitrogen is *pyrrole*

This compound can be made by the electrolytic reduction of succinimide. It is both a weak acid and a weak base. The hydrogen attached to nitrogen can be replaced by potassium, forming potassium pyrrole. When potassium pyrrole is treated with an alkyl halide, an N-alkylpyrrole forms. The alkyl group, however, drifts to the 2 position, and some to the 3 position. Pyrrole cannot be nitrated or sulfonated because acids cause pyrrole to become polymerized. The pyrrole ring is found in natural materials including chlorophyll, hemoglobin, and some alkaloids.

Pyrrolidine is a secondary amine. It is a much stronger base than pyrrole, and it absorbs carbon dioxide from the air. It is a colorless mobile liquid, that has a penetrating amine-like odor. It freezes at −63°C and boils at 86.9°C. Its specific gravity 20/4 is 0.8618, its index of refraction N20/D 1.443, and its flash point 3°C. It mixes well with water and most organic solvents. It reacts with carbon disulfide, alkyl halides, and ethylene oxide, and it is useful for introducing the pyrrolidyl ring into synthetic products that have physiological activity.

An example of a five-membered ring that contains two hetero atoms is *antipyrine,* used somewhat in medicine. *Thiazole* is interesting because it contains two hetero atoms of different sorts.

Thiazole is a colorless liquid that boils at 116.8°C. It forms salts with acids. Its derivative, sulfathiazole, is well known in medicine.

Pyridine is a liquid that boils at 115.3°C. It mixes with water in all proportions, and its repelling nauseating odor is that smelled when bones are destructively distilled. Pyridine and its related compounds are found in coal tar. The nitrogen atom in pyridine is characteristic of a tertiary amine, but the ring structure is similar to that of benzene. The chemical reactions of pyridine are generally similar to those of benzene. Pyridine is a useful solvent. *Piperidine*, which is like pyridine except that hydrogen is attached to nitrogen, making it a secondary amine, is a stronger base than pyridine or ammonia. It is used as an accelerator in the vulcanization of rubber. Derivatives of a six-membered ring that contains one oxygen atom are known. They include some compounds such as tetrahydrocannabinol which have very high mari-huana activity.

A few other heterocyclic compounds will be mentioned. *Morpholine*, made by dehydration of diethanolamine, boils at 126–130°C. It is used in closed boiler systems to control corrosion, and it is a solvent for waxes, resins, and dyes. *Indole* consists of a benzene ring and a pyrrole ring. It has a fragrant odor, and can be used in perfumes when pure. The impure material, however, accounts for most of the odor of feces.

Among the heterocyclic compounds are purine, barbituric acid, uric acid, xanthine, theobromine found in cocoa beans, and caffeine found in tea and coffee. Interest in heterocyclic compounds is because some are related to the vitamins, e.g. riboflavin.

ELBERT C. WEAVER

HETEROGENEOUS EQUILIBRIA

A heterogeneous equilibrium is an equilibrium that involves substances in two or more phases of a heterogeneous system. The equilibria of interest to chemistry are those in which a condition of equilibrium is attained because the forward and reverse rates of a reversible process are equal. On a microscopic scale both processes continue to occur, but no net change takes pace in the amounts of the reactants and products at equilibrium. In heterogeneous systems the process that has attained a state of equilibrium can be a physical or a chemical change or a combination of physical and/or chemical changes. Examples of such physical changes are: changes of state, the physical adsorption-desorption of a gas, distribution of a nonreacting solute between two immiscible solvents, and all solution-separation processes of nonreacting solutes. Some examples of heterogeneous chemical changes are: the chemical adsorption-desorption of a gas, solution-separation processes that involve chemical changes, and all chemical reactions in which the reactants and products are present in more than one phase in a system. Two important types of the last example are gas-solid reactions and ion-solid reactions.

The existence of equilibria in heterogeneous systems can be predicted from kinetic or thermodynamic arguments. These arguments are more complicated than for homogeneous systems, but the reality of the equilibria are borne out by experiment. The simplest approach to the kinetic prediction can be illustrated by the solution-precipitation in water of the slightly soluble strong electrolyte, $BaSO_4$. The net reactions involved are

$$BaSO_4(solid) \rightleftharpoons Ba^{+2}(aqueous) + SO_4^{-2}(aqueous)$$

The solid $BaSO_4$ takes part in the reactions only at its surface. It is necessary to introduce some measure of the concentrations of the Ba^{+2} and SO_4^{-2} ions on the surface of the $BaSO_4$ since the usual concentration terms, molarity and molality, are not applicable. To achieve this, let A be the area of solid $BaSO_4$ exposed to the aqueous solution, and let X be the area covered by Ba^{+2} ions. Then the area covered by SO_4^{-2} ions is $A-X$. Neglecting interactions, the rate of solution of an ion is directly proportional only to the number of such ions on the surface which, in turn, is proportional to the surface area covered by that ion. The rate of precipitation of an ion is directly proportional to the number (molarity) of the ions in solution that can collide with the surface and to the number of sites on the surface that can accommodate the ion. The latter is directly proportional to the area covered by the ion of opposite charge. The essential feature of the kinetic derivation is to equate the rates of the forward and reverse reactions, which by definition, are equal when the process has attained equilibrium. In this case it is convenient to write the process as the sum of two steps.

$$1. \ Ba^{+2}(solid) \underset{r}{\overset{f}{\rightleftharpoons}} Ba^{+2}(aqueous)$$

$$2. \ SO_4^{-2}(solid) \underset{r}{\overset{f}{\rightleftharpoons}} SO_4^{-2}(aqueous)$$

Equating the rates for these two solution-precipitation reactions gives

$$k_{1f}(X) = k_{1r}(A-X)[Ba^{+2}(aqueous)]$$

$$k_{2f}(A-X) = k_{2r}(X)[SO_4^{-2}(aqueous)]$$

where the $k's$ are the specific reaction rate constants for the four reactions. The first equation can be reversed, divided by the second, and rearranged to give

$$[Ba^{+2}(aqueous)][SO_4^{-2}(aqueous)] = \frac{k_{1f}k_{2f}}{k_{1r}k_{2r}} = K_{eq}$$

Hence, when the process has come to equilibrium, the product of the concentrations of the two ions is equal to a constant, called the equilibrium constant. In general, for any slightly soluble strong electrolyte, A_xB_y, at equilibrium

$$K_{eq} = [A^{+y}]^x[B^{-x}]^y = x^xy^y[S]^{x+y}$$

where S is the molar solubility of the salt and K_{eq} is called the solubility product constant and designated K_{sp}. K_{sp} is a function of temperature and relates the maximum concentrations of the ions of a salt that can coexist together in aqueous solution at a given temperature.

For any heterogeneous reaction such as

$$aA(gas \ or \ solution) + bB(liquid) \rightleftharpoons$$
$$cC(solid) + dD(gas \ or \ solution)$$

it is possible to postulate a mechanism for the reaction that will allow the prediction that at equilibrium

$$K_{eq} = \frac{[D]^d}{[A]^a}$$

This equation is called the law of chemical equilibrium for the reaction and the right side is referred to as the equilibrium constant expression. This result is similar to the homogeneous case except that no concentration terms appear for those substances that are present in the form of pure solid or liquid phases. The laws of chemical equilibrium for heterogeneous reactions contain concentration terms only for gases and components of solutions and these concentrations are raised to a power given by the coefficient of the substance in the chemical equation.

The principle of LeChatelier can be applied to heterogeneous equilibria to predict the effect of a stress on the concentrations at equilibrium. This principle states that if a stress, such as a change in concentration, pressure, or temperature, is applied to a system in equilibrium, the system is shifted in a way that tends to undo the effect of the stress. In applying the principle to heterogeneous systems it is necessary to note that solid and liquid phases are essentially incompressible and a change in pressure will not affect the condensed phases. Therefore, an increase in pressure will shift the equilibrium in the direction of the smallest number of moles of gases. Changes in the amounts of solid and liquid phases will not affect the equilibrium concentrations, but a concentration stress can be applied by changing the partial pressure of a gas or the concentration of a species in a solution. The effect of a temperature change can be predicted by knowing whether the reaction as written is exothermic or endothermic. An increase in temperature favors the endothermic process. The quantitative effect of temperature on the state of an equilibrium is given by the relation

$$\frac{d \ln K_{eq}}{dT} = \frac{\Delta H}{RT^2}$$

where ΔH is the enthalpy change in the reaction. If this is known as a function of temperature, then the equation can be integrated to give a relationship between the value of the equilibrium constant and temperature.

The thermodynamic prediction or derivation of equilibrium constants follows quite simply from the definition of the term activity, a. The activity of a substance is a measure of the extent that its free energy per mole, G, deviates from its value in some reference or standard state, $G°$. The free energy and the activity are related by the equation that defines the activity.

$$G = G° + RT \ln a \qquad (1)$$

In the standard state where $G = G°$, $a = 1$. For any chemical reaction, regardless of the physical states of the reactants and products, the free energy change in the reaction is obtained by subtracting the total free energy of the reactants from the total free energy of the products

$$\Delta G = n_i G_i(\text{products}) - n_i G_i(\text{reactants}) \qquad (2)$$

where n_i is the number of moles of the i'th reactant or product of free energy G_i per mole. For any reaction indicated by

$$aA + bB \rightleftharpoons cC + dD$$

the free energies from Eq. (1) can be substituted into Eq. (2) to give for the free energy change

$$\Delta G = \Delta G° + RT \ln \frac{a_C^c a_D^d}{a_A^a a_B^b} \text{ where}$$

$$\Delta G° = cG_C° + dG_D° - aG_A° - bG_B°$$

The thermodynamic condition for equilibrium is that $\Delta G = 0$, hence

$$\Delta G° = -RT \ln \frac{a_C^c a_D^d}{a_A^a a_B^b}$$

and since $\Delta G°$ is a constant at constant temperature

$$\frac{a_C^c a_D^d}{a_A^a a_B^b} = K_{eq} \quad \text{and} \quad \Delta G° = -RT \ln K_{eq}$$

The thermodynamic derivation makes no assumptions about the nature of the process and gives a general law of chemical equilibrium that can be applied to any process if activities rather than concentrations are employed in the expression for the equilibrium constant. The experimental data show that if the chemical equation properly reflects the changes which occur for a system in equilibrium, then the equilibrium constant expression in terms of activities is always a constant independent of concentration. This is not always true if concentrations are employed in the equilibrium expression. Only under conditions where the rates of reaction show a simple direct dependence on concentration will the expression in terms of concentration be correct.

The law of chemical equilibrium in terms of activities can be simplified in its application to heterogeneous equilibria which involve phases consisting of pure solid and liquid substances. The standard states for these substances are chosen as the pure solid or pure liquid at one atmosphere pressure at each temperature. Under these conditions the activity is unity and no term for these substances will appear in the equilibrium constant expression. For the $BaSO_4$ example the equilibrium expression reduces to

$$K_{eq} = a_{Ba^{+2}} a_{SO_4^{-2}}$$

At sufficiently low concentrations of ions such as those resulting from the solution of slightly soluble electrolytes the activity becomes equal to the concentration and the expression reduces to that obtained from kinetic arguments.

The activity is constant at unity and does not appear in the law of chemical equilibrium only when a substance exists in the pure liquid or solid state. The activities for substances in solid and liquid phases have to be included in the equilibrium expression for heterogeneous processes when the solid and liquid phases exist as liquid or solid solutions. Although there is no such thing as complete immiscibility, if the solutions are sufficiently dilute, the activities of the solvents can be taken to be unity. Further, if the solute is also present as an essentially pure solid or liquid phase, at equilibrium the activity of the solute in the solution will be constant, equal to an equilibrium constant times the unit activity of the solute in the pure phase. The activity of the solute can be deleted from the equilibrium expression by combining this constant term with the equilibrium constant for the overall process.

One type of heterogeneous equilibrium constant

is of sufficient importance that it is given a special name, the distribution ratio. If a solute is shaken with two immiscible solvents it will distribute itself between the two solvents in a fixed ratio. If, for example, iodine is shaken with water and carbon tetrachloride, then at equilibrium $\Delta F = 0$ and the free energy of iodine per mole in the two solvents will be identical. Equating the free energy expressions for the iodine in the two solvents gives the result

$$K_{eq} = \frac{a(CCl_4)}{a(H_2O)}$$

where the K_{eq} is referred to as the distribution ratio. When a substance is present in more than one phase of a system the free energy per mole of the substance is the same in all the phases and the activities in the various phases are proportional to each other. At sufficiently low concentrations of solute where the solute molecules behave independently of each other, the activities can be replaced by concentrations to give

$$K_{eq} = \frac{[I_2(CCl_4)]}{[I_2(H_2O)]}$$

If the solute undergoes a reaction in one or both of the solvents then the simple distribution ratio is no longer valid. The usual reactions are association or dissociation of the solute. The proper equilibrium expression is obtained by writing a chemical equation for the net overall reaction and applying the law of chemical equilibrium to that reaction.

L. C. GROTZ

Cross-references: *Activity, Kinetic Theory, Thermodynamics, Electrochemistry.*

HIGH-TEMPERATURE CHEMISTRY

In the last 30 years a new area of chemical specialization has evolved—the study of high-temperature phenomena. Energetically, this realm is between "room-temperature chemistry" (up to $\sim 500°K$) and "low-energy physics" (down to $\sim 100,000°K$, i.e., a few volts). The evolution of industrial operations toward higher temperatures, the need for refractory and corrosion-resistant materials, the advances in space science and nuclear science and the almost limitless potential of chemical syntheses from electric arcs, flames, etc.—all have encouraged the expenditure of large amounts of money and scientific talent for the delevopment of a better understanding of high-temperature systems. The versatile laser can be used as a heating device and is providing many new opportunities for high-temperature research.

High-temperature systems can be complex. Currently, the complete characterization of a high-temperature vapor (for example, the equilibrium species over solid graphite at 2500°K; or in equilibrium with $B_2O_3(l)$ and $H_2O(g)$ at 1500°K) requires (1) classical vapor pressure measurements; (2) mass spectrometric examination of the vapors; (3) visible, ultraviolet, infrared and microwave spectroscopic studies; and (4) electron diffraction studies. In many cases, this detailed examination of high-temperature vapors has turned up new, complicated molecules and the old view that "at high temperature things will be simple" has had to be replaced by some more realistic laws of high-temperature chemistry, which may be stated as follows: (1) at high temperatures, everything reacts with everything; (2) the higher the temperature the faster the reaction; and (3) the products can be anything.

While these laws are somewhat too general, it is quite in keeping with them to note that high-temperature chemists ignore valence and octet rules; instead, they find the molecules AlF, Al_2O, Al_2O_2 and AlF_2 to be important Al-containing species in slightly reducing atmospheres or BO_2 and BS_2 to be important in slightly oxidizing situations. Furthermore, there is no simple relationship between a condensed-phase formula and the vapor species in equilibrium with it: $C_1, C_2, C_3, C_4, C_5, C_6$, etc. are all found over graphite; $WO_3(s)$ yields $(WO_3)_3(g)$, $(WO_3)_4(g)$ and $(WO_3)_5(g)$ on sublimation; $BeO(s)$ yields the elements, $BeO(g)$ and a series of polymers, $(BeO)_n$ where n = 1,2$\cdots$6; alkali halides exist in the vapor as monomers, dimers and trimers; etc.

One of the major problems of high-temperature chemistry has been the structural characterization of complex molecules, but infrared, Raman and ESR spectra of high-temperature species isolated in rare gas matrices can now provide reliable molecular parameters.

Similarly, the solid state is not adequately described by simple formulas like TiO_2 or TaC, but variable stoichiometry becomes the rule in high-temperature systems. Extensive research on lattice parameters, phase diagrams, optical, electrical and thermodynamic properties has been carried out on pure metals and alloys; on the various practical industrial refractory materials (Al_2O_3, ZrO_2, MgO, Cr_2O_3, TiO_2, silicates, etc.); on transition metal carbides, borides, silicides, nitrides, sulfides and phosphides; and on ternary systems with unusual properties.

Current research is mainly in the areas of (1) general and measurement of high temperatures; (2) high-temperature synthesis; and (3) property measurements at high temperatures. With special resistance heaters and induction furnaces, one may attain $\sim 4000°K$ in closed systems under reducing or neutral conditions. Chemical flames (H_2—O_2; B_2H_6—O_2; Al—O_2; H_2—F_2; C_2N_2—O_2; C_4N_2—O_2) and the atomic hydrogen or atomic nitrogen torches permit extension of this limit to nearly 6000°K; but shock tubes, electric arcs and lasers now offer practical opportunities for studies in the 6000°K range and higher. The extremely high translational energies, the presence of ions and the almost infinite number of thermodynamically acceptable ways to recombine ions and electrons from a plasma at 20,000°K make interest high in plasma syntheses. Workable schemes for preparation of NO, C_2N_2, C_2H_2, HCN, C_2F_4 and other halocarbons have been reported. Another synthetic technique of current importance is the "low-temperature co-condensation technique" developed by Skell, Timms and Margrave. Reactions of condensed gaseous atoms (C, Si, Fe, Pt, etc.) and of high-temperature molecules (SiF_2, SiO, BF, etc.) have been observed with a great variety of organic and inorganic species.

Property measurements on many high-temperature systems of industrial or defense significance are in progress, but the demand for data on literally all substances over the range 0-2000°K will not be met for several years, even with monumental support by governments all over the world. Heats of formation, entropies and free energy functions are still in doubt for many simple substances in spite of many thousand reports of thermodynamic data cited in the annual Bulletin of Thermodynamics and Thermochemistry. Surface tensions, viscosities, thermal and electrical conductivities and other transport properties are very incompletely studied at 1000°K and data are extremely rare at 2000°K and above.

In summary, the expansion of technology to allow convenient generation and measurement of temperatures in condensed systems up to $\sim$ 4500°K (and still higher by use of the rotating furnace technique proposed by Grosse) and in gaseous plasmas up to at least 100,000°K has created a new dimension to the Periodic Table. One has to consider the stability of various atom combinations as a function of temperature; as a function of various electronic configurations besides those of the ground states; and as a function of new rules of bonding, electron pairing, etc.—perhaps not formulated until actual high-temperature species are identified. Active research and data compilations are being carried out over a temperature spectrum, now routinely available, which is some 2-10 times broader than that of 30 years ago.

JOHN L. MARGRAVE

HISTOCHEMISTRY

Histochemistry is the study of the chemical composition of the minute structure of animal and vegetable tissues. Since the tissues of all plants and animals are composed of cells, another branch of chemistry known as *cytochemistry* (Gk. *kytos,* used biologically to designate cells) has much in common with histochemistry. Both involve the use of the microscope; both employ essentially the same chemical techniques; both strive to carry out chemical identifications with as little cell and tissue damage as possible. It is the function of histochemistry to identify chemically the constituents of fragments of tissues in which the cellular structure has not been destroyed.

Upon the foundations established by nineteenth century workers, histochemical methods have been developed for such a wide variety of substances as aluminum, calcium, iron, magnesium, manganese, potassium, sodium, ammonium, silicon, fats, waxes, proteins, glycosides, pigments, cellulose, pectic substances, chitin, latex, amyloid, hemicelluloses, lipoproteins, gums, enzymes, alkaloids, aldyhydes and ketones. Histochemical methods have involved the use of the microscope with ordinary light, polarized light, and ultraviolet light. X-ray diffraction has proved to be a valuable accessory technique. Electron microscopy and electron diffraction furnish direct information concerning fine structure and chemical composition. Sample preparations have included dissection of tissues, sectioning of tissues, freeze drying before sectioning, and electrophoresis, when quantitative as well as qualitative determinations were necessary.

Recent developments in histochemical research have not obscured the value of earlier methods. Among those in use before 1850 are the iodine reaction with starch; the xanthoproteic and Millon reactions with protein and the sulfuric acid-iodine reaction with cellulose. Before the end of the century, enzymes had been used to remove unwanted tissue from specimen of nerve fibers; differential solubilities of pectic substances had been determined, and the van Wisselingh method[4] of chitin identification in fungi and arthropods was in use. Refinements in earlier methods, new methods and combinations of methods have continued to expand the science of histochemistry. Developments in a wide range of biological disciplines (taxonomy, morphology, physiology, genetics) have been aided by established information concerning the composition, structure and arrangement of the constituents of cells and tissues. The prediction of Pearse[5] that histochemistry can transform the descriptive sciences of biology into dynamic functional sciences is beginning to be fulfilled.

Autoradiography[3], when used in the histochemical analysis of plants and animals, involves the incorporation of radioactive material in a tissue specimen and the photographic detection and localization of the ionizing radiations emitted by the radioactive material. The process is rapid and the photographic details clear in the absence of artifacts. Medical applications have included the use of radioactive iodine (I^{131}) in disorders of the thyroid gland. Detailed studies of bone development, using radioactive phosphorus and human bone marrow grown *in vitro* in the presence of radioactive strontium have been made. One of the earliest uses of the method was for the detection of radiation damage to bones produced by nuclear reactions.

Lipids[1]. Organisms which have been used for experimental purposes in the field of lipid chemistry and lipid metabolism are scattered throughout the entire plant and animal kingdoms. *Neutral lipids* include the hydrocarbons, polyphenols, fatty acids, waxes and glycerides. *Phospholipids* are phosphorus-containing lipids. *Glycolipids* are the phosphorus-free lipids which yield a fatty acid and a carbohydrate upon hydrolysis. Neutral lipids have been found in bacteria, yeasts, fungi, higher plants, invertebrates and vertebrates. They occur in liquid droplets in protoplasts of living cells, are abundant in the epicuticular waxes of the higher plants and in the common cricket. A number of lipids (squalene, cholesterol esters, cholesterol, glycerides, free fatty acids and phospholipids) have been found in fowl-pox virus, although their biological significance in this virus has not been determined.

Phospholipids are hygroscopic, swelling in water into a gluey, mucilaginous mass and then into a turbid emulsion. When a section of tissue containing phospholipids is mounted in water, the colorless colloidal masses collect under the cover-glass emanating mainly from regions where growth has been rapid. Below the swollen masses cellular detail is obscured. Histochemical techniques must be adapted, therefore, to the colloidal behavior of the phospholipid material. A combination of three methods serves to identify and localize the phospholipids in a given tissue section (a) staining with Sudan IV; (b) solubility in chloroform and insolubility in acetone; (c) the "phosphomolybdate reaction" which involves formation of ammonium phosphomolybdate crystals. The reduced state of ammonium phosphomolybdate is blue and the color is developed rapidly by the addition of a reducing agent.

Research in the chemistry of the glycolipids is active. Compounds obtained from *Mycoplasma* species are complex and those from the alga, *Chlorella*, even more so. Glycolipids obtained from bean leaves, oyster mantle, pig intestine, dog intestine, bovine brain and rat brain indicate their wide distribution. In medical research, Fabry's disease, a hereditary glycolipid lipidosis shows elevated glycolipid concentrations in kidney and other tissues; in plasma, but not in red cells and in muscular tissues, lymph nodes and arterial tissue.

Mechanisms of the Action of Herbicides[2]. The numbers of herbicides which have been formulated for use in weed control and in wide-spread defoliation serve to indicate the current importance of this branch of biochemical research. Abbreviations which have been approved by the Weed Society of America for the many herbicides now in use serve as a substitute for the elaborate chemical designations of most of them. The unwanted effects which the use of herbicides may have upon various types of conservation programs represent just another phase of this complex problem. Once a herbicide is formulated and prepared for use, the problem of application, penetration and physiological activity must be solved before the mechanism of action can be understood. Moreland[2] and others have discussed the environmental factors such as leaching, volatilization and biological degradation which render externally applied herbicides unavailable to the plant; anatomical, morphological, biochemical and physiological characteristics of the plant which control the entry of the herbicide and the complex conditions within the plant tissues which affect its distribution and location once the entry is accomplished. In order to achieve the desired degree of toxicity, it is assumed that vulnerable processes such as respiration, mitochondrial activity, photosynthesis and protein metabolism should be measurably affected. Histochemical techniques which have shown alterations in electron transport and energy-transfer inhibitors in mitochondria; alteration of the protein and lipoid components of the chloroplasts along with alterations in carbon dioxide absorption and oxygen evolution have been applied extensively to herbicide-treated plants and untreated controls. These have been supplemented by the effects of any given herbicides upon the alga *Chlorella* and various microorganisms. Influences upon protein synthesis and nucleic acid metabolism have been carefully explored. The complexity of the many biophysical and biochemical factors have so far prevented satisfactory explanation of the mechanisms of action of many herbicides. It is the opinion of many investigators in this field of research, however, that the combined efforts of organic chemists, biochemists, plant physiologists and histochemists will eventually produce the needed information.

WANDA K. FARR

Cross-reference: *Cytochemistry.*

References

1. Annual Review of Biochemisthy, Vol. 39, 1970. Annual Reviews, Inc., Palo Alto, Calif.
2. Annual Review of Plant Physiology, Vol. 18, 1969. Annual Reviews, Inc., Palo Alto, Calif.
3. Buller, J. A. V., and Randall, J. T., "Progress in Biophysics and Biophysical Chemistry," Academic Press, Inc., 1953.
4. Paech, K., and Tracey, M. V., "Modern Methods in Plant Analysis, Vol. II, 1955, Springer Verlag, Berlin, New York.
5. Pearse, A. G. E., "Histochemistry, Theoretical and Applied," Little, Brown and Company, 1953.

HOMOGENIZATION

The fat of normal cows' milk exists as a temporary emulsion of small globules suspended in the plasma or skimmilk fraction. The globules vary in size from about 0.1μ to about 15μ in diameter, averaging about 3.5μ with 80% in the 2.0 to 5.0μ range. On standing, especially at low temperature, the fat globules of milk first cluster through the influence of agglutinin, now identified as euglobulin, and then rise forming a cream layer. In this process the globules do not coalesce nor lose their identity and may be redispersed by simple stirring. The lack of coalescence, even when the milk is heated and the fat liquefied, is explained by the presence of an adsorbed layer or "membrane" of nonfat material adhering to the fat globule surface.

Recent studies on the formation of milk fat show the cells which synthesize fat droplets also produce Golgi vacules containing milk protein granules (casein). An osmiophilic membrane surrounds the Golgi vacuoles. Electron microscopy autoradiography experiments with labeled amino acids confirm synthesis of milk protein in the Golgi region and their conveyance to the cell membrane in Golgi vacuoles. It appears that a lactating cell secretes both milk fat and protein and that during the secretory process of the protein granules the membrane of the Golgi vacuole must become engaged with the plasma membrane. Thus, the Golgi vacuole membrane becomes continuous with the plasma membrane at the time the vacuole is secreting its content from the cell. It is now known that the milk fat droplet is secreted by being enveloped in the plasma (limiting) membrane of the cell. These cellular events indicate that the Golgi membrane, plasma membrane and milk fat globule membrane are closely related and derived sequentially in the milk secretion process.

The homogenizing process consists of forcing milk (or other fluids) through a small aperture at elevated pressures (500 to 4000 psi) resulting in high velocity, turbulence and great force of impingement as the product emerges. The effects of the process on the fat emulsion of milk have been ascribed to various causes, none of which has been fully substantiated: (a) fat globule shattering by impact, (b) expansion or explosion on release of pressure, (c) shearing between liquid strata and (d) cavitation.

When milk is homogenized the fat globules are broken up and reduced in size from an average diameter of about 3.5μ to an average of about 1μ.

This increases the number of globules some 100 times and expands the fat globule surface between six and seven times. In the violence of the homogenizing process much of the natural fat globule "membrane" material loses contact with the fat surface and becomes dispersed through the plasma. The new and smaller globules formed are almost immediately prevented from coalescing with other such globules because the great interfacial tension between the fat and the plasma is quickly reduced by the adsorption of surface-active materials which are abundant in the plasma. A new adsorbed "membrane" is thus formed, composed largely of the proteins and enzymes of the plasma proportioned according to their abundance, their mobility and their ability to reduce surface tension. The phospholipid-protein complex is largely absent from the new "membrane" because of its comparative scarcity. The fat globules have thus, in effect, been "resurfaced" with new and different substances. In the process, the euglobulin-agglutinin has been found to become denatured and nonfunctioning. When the fat content is high, as when cream is homogenized, the newly formed globules exhibit a tendency to clump or form bunches, readily visible under the microscope. This has been attributed to the slower stabilization of the extended fat surface and the proximity of the globules to each other, permitting an attempt at coalescence which, however, fails to materialize before sufficient surface adsorption takes place to prevent it. Almost all the changes in the characteristics of milk resulting from homogenization are caused by the alterations in fat dispersion and the associated shifts of substances from the fat surface to the plasma and vice versa. The homogenization of milk plasma or skimmilk causes no significant change of characteristics.

G. H. WATROUS, JR.

References

Hall, C. W., and Trout, G. M., "Milk Pasteurization," The AVI Publishing Co., Inc., Westport, Connecticut, 1968.

Patton, Stuart, 1969. "Milk," *Scientific American*, **221**(8), 58–68 (1969).

HORMONES, STEROID

Steroid hormones are a class of organic compounds containing C, H and O, possessing the cyclopentanoper-hydrophenanthrene nucleus. They are produced in endocrine glands such as the adrenal, gonads and placenta and they exert their action at a distant site. The steroid hormones may be subdivided into androgens, estrogens, progestational substances and corticoids on the basis of their physiological activity.

Androgens are produced in all the steroid-producing tissues, testis, ovary, adrenal and placenta, and have the specific function of maintaining the male sex characters of the mammal, such as the prostate and seminal vesicles. In man, other secondary sex characters such as facial hair and pitch of voice are controlled by androgens. In the fowl, androgens maintain the comb, wattles and spurs. In addition to sex-specific functions, androgens influence nitrogen metabolism and fat distribution and, to a limited extent, they exert control over

electrolyte balance. Testosterone is the most active naturally occurring androgen. Other less active androgens include androst-4-ene-3,17-dione. dehydroepiandrosterone and 11β-hydroxyandrost-4-ene-3,17-dione. Structurally, androgens possess 19 carbon atoms with oxygen substituents at carbons 3, 11 and 17. In certain tissues, such as the prostate, testosterone is converted to the 5α-dihydro derivative 17β-hydroxy-5α-androstan-3-one, which is the active hormone for this tissue.

The formation of androgens in the gonads is controlled indirectly by releasing factors from the hypothalamus and directly by the pituitary gonadotropic hormones, and the general sequence of biosynthesis in these tissues, the placenta and the adrenal, is (see **Steroids** for numbering system) cholesterol → 20α-hydroxycholesterol (or 22R-hydroxycholesterol) → 20α,22R-dihydroxycholesterol → pregnenolene → progesterone → 17α-hydroxyprogesterone → androst-4-ene-3,17-dione ⇌ testosterone. An alternate biosynthetic pathway for testosterone biosynthesis includes the conversion of cholesterol to pregnenolone and the pregnenolone so formed is converted to testosterone by the following sequence of reactions: pregnenolone → 17α-hydroxypregnenolone → dehydroepiandrosterone → androst-4-ene-3,17-dione → testosterone. A modification of this alternative pathway involves the transformation of dehydroepiandrosterone to androst-5-ene-3β,17β-diol by reduction of the 17-ketone and, finally, oxidation of the 3β-ol-Δ^5 grouping to the Δ^4-3-ketone, forming testosterone. Only the testis produces significant amounts of testosterone. Adrenal androgens are formed by the following two pathways: pregnenolone → 17α-hydroxypregnenolone → dehydroepiandrosterone → androst-4-ene-3,17-dione and cortisol → 11β-hydroxyandrost-4-ene-3,17-dione. Androgen biosynthesis also may proceed through pregnenolone-3-sulfate → dehydroepiandrosterone-3-sulfate → dehydroepiandrosterone.

Androsterone and etiocholanolone are the two principal catabolites of testosterone, androst-4-ene-3,17-dione and dehydroepiandrosterone. Four 11-oxygenated 17-ketosteroids, 11β-hydroxyandrosterone, 11-ketoandrosterone, 11β-hydroxyetiocholanolone and 11-ketoetiocholanolone, are the principal catabolites of the 11β-hydroxyandrost-4-ene-3,17-dione.

Estrogens are a class of C_{18} phenolic steroids where either one or two of the four rings in the structure is aromatic. Estrogens are produced in all steroid-producing tissues. The primary physiological action is maintenance of the female sex characters, including growth of the vagina, uterus, mammary glands and Fallopian tubes. Other effects include fat distribution, influence on electrolyte balance, calcium metabolism and blood clotting.

Estradiol-17β is the principal and most active estrogen and together with estrone, the corresponding 17-keto derivative, is widely distributed in body fluids. Equilin and equilenin are representative ring B unsaturated estrogens and are associated mainly with the pregnant mare.

Estrogen production in the gonads is controlled by pituitary gonadotropic hormones, while adrenal estrogen formation is regulated by the pituitary adrenocorticotropic hormone. The pathway testosterone → 19-hydroxytestosterone → estradiol-17β probably accounts for the major amount of estradiol-17β biosynthesized. Alternate pathways for estradiol-17β formation include: testosterone → 19-hydroxytestosterone → 19-oxotestosterone → estradiol-17β and testosterone → 19-hydrotestosterone → 19-oxotestosterone → 19-nortestosterone → estradiol-17β. Estrone may arise from the oxidation of the C-17 hydroxy group to the ketone or from the androgen androst-4-ene-3,17-dione by 19-hydroxylation and aromatization of ring A in reactions similar to those already detailed for the conversion of testosterone to estradiol-17β. By the biosynthetic route involving 19-hydroxylation and ring A aromatization, androst-5-ene-3β,16α,17β-triol is the substrate for estriol formation.

Catabolic reactions lead to the formation of a dozen or more products, of which estradiol-17β, estrone and estriol may be considered to be major metabolites. Other catabolites include 2, 6α, 6β, 15α, 15β, 16α, 16β and 16-keto derivatives of estrone and estradiol-17β.

Progestational hormones include progesterone and its 20-reduced derivatives, 20β-hydroxypregn-4-ene-3,20-dione and 20α-hydroxypregn-4-ene-3,20-dione. Of these three steroids, progesterone is the most active. These compounds are produced in all steroid-producing tissues and especially in the corpus luteum and placenta. Progesterone has a specific function on the vaginal and uterine epithelium and mammary glands and, in concert with estrogens, has a role in the maintenance of the female sexual cycle, whether estrus in the rodents or menstrual cycle in the primate. Progestational substances have a unique role in the maintenance of pregnancy. A second primary role of progesterone is to serve as an intermediate in the biosynthesis of all classes of steroid hormones.

The biosynthesis of progesterone in gonadal tissue is regulated by pituitary gonadotropins and by adrenocorticotropin in the adrenal. The pathway of formation is cholesterol → 20α-hydroxycholesterol (or 22R-hydroxycholesterol) → 20α, 22R-dihydroxycholesterol → pregnenolone → progesterone. Progesterone undergoes reductive catabolic reactions of the ketone groups and the nuclear double bond, forming the principal reaction products pregnanediol and pregnanolone.

Corticoids are essential to life and are classified on the basis of their metabolic function. Cortisol and corticosterone protect the organism against stress, form new carbohydrate from protein, influence fat metabolism, cause a dissolution of lymphatic tissue, and have minor influences on electrolyte balance. Aldosterone and deoxycorticosterone are particularly effective in causing sodium retention and potassium excretion.

The biosynthesis of cortisol proceeds by a series of hydroxylations from pregnenolone or progesterone at positions 11β, 17α and 21 and this process occurs in the fasciculata and reticularis layers of the adrenal. Corticosterone is produced in all three zones of the adrenal and is formed by 11β- and 21-hydroxylation. Aldosterone is formed exclusively in the glomerulosa and requires the hydroxylation of progesterone at positions 11β, 18 and 21. The 18-hydroxy group is oxidized to an aldehyde function.

The principal but not the only catabolites of cortisol include cortisone, tetrahydrocortisol (3α,5β), tetrahydrocortisone (3α,5β), tetrahydroallocortisol (3α,5α) and the 17-ketosteroids: 11β-hydroxyandrosterone, 11-ketoandrosterone, 11β-hydroxyetiocholanolone and 11-ketoetiocholanolone. The principal catabolite of aldosterone is the 3α,5β-tetrahydro derivative.

RALPH I. DORFMAN

Reference

Dorfman, R. I., and Ungar, F., "Metabolism of Steroid Hormones," Academic Press, New York, 1965.

Soffer, L. J., Dorfman, R. I., and Gabrilove, J. L., "The Human Adrenal Gland," Lea and Febiger, Philadelphia, 1961.

Young, W. C., Editor, "Sex and Internal Secretions," 2 vols., Williams and Wilkins Co., Baltimore, 1961.

Eichler, O., and Farah, A., Editors, "Handbuch der Experimentellen Pharmakologie Erganzungswerk," "The Adrenocortical Hormones, Their Origin, Chemistry, Physiology and Pharmacology," Vol. XIV (Helen Wendler Deane, Subeditor), Springer Verlag, Berlin, 1962.

HYDRATED ELECTRON

This newly discovered ion has profoundly altered research in radiation chemistry, and its further use will contribute to the development of other branches of chemistry and biology. The hydrated electron with its unit negative charge heads the series, e_{aq}^-, H$^-$, F$^-$, Cl$^-$, Br$^-$, and I$^-$. It is the most reactive of negative ions and in sufficiently pure water releases an atom of hydrogen by the reaction

$$e_{aq}^- + H_2O \rightarrow H + OH^-$$

It has a natural half-life of 800 μsec. But in spite of a short life, its chemical properties may be more easily studied than those of any other negative ion.

Formation. Electrons released in liquid water by ionizing radiations provide the most convenient means of preparation. Secondary, unhydrated electrons, e^-, produced via

$$H_2O + \gamma\text{-rays} \rightarrow H_2O^+ + e^-$$

rapidly thermalize. Then they polarize surrounding water molecules thereby becoming hydrated.

$$e^- + H_2O \rightarrow e_{aq}^-$$

Since the primary action of the radiation is on the water, foreign ions are unnecessary for e_{aq}^- formation. Photochemical ionization of neutral molecules or negative ions provides a second method of preparation. Simple inorganic ions such as Cl$^-$, Br$^-$, I$^-$, OH$^-$, and CNS$^-$, as well as many organic ions and molecules form e_{aq}^- on photolysis. The general reaction is

$$A^- + h\nu \rightarrow A + e_{aq}^-.$$

H-atoms, chemically generated in alkaline solution (i.e., OH + H$_2$ → H$_2$O + H) or bubbled in from the gas phase produce e_{aq}^- by

$$H + OH^- \rightarrow e_{aq}^-$$

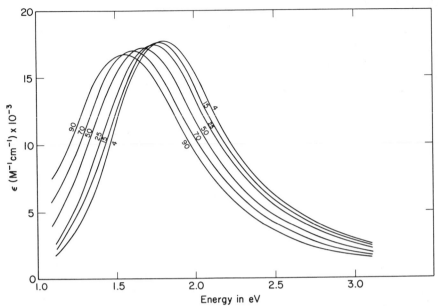

FIG. 1. Effect of temperature on the absorption spectrum of e_{aq}^- in the temperature range $-4°$ to $90°$.

Other possible methods of producing e_{aq}^- include photoionization of immersed metals, release from electrodes, and as an intermediate in electron transfer reactions.

Detection and Analysis. The hydrated electron was first suspected from discrepancies in the measurement of relative H-atom rate constants. However, the discovery of its optical absorption band not only settled the question of the existence and lifetime of e_{aq}^- but it also provided an unexcelled means of studying its properties. The broad intense band centered at 1.72 eV(7200 Å) at 25° is shown in Fig. 1, for temperatures in the range $-4°$ to 90°. The decadic molar extinction coefficient of this band at the maximum is 18500 M^{-1} cm^{-1}, placing it among the most intensely absorbing ions known. This spectrum was discovered first in a sodium carbonate solution and then in pure water by irradiating the solution with a high-electron current pulse accelerator. By synchronizing the electron pulse with light-flash absorption spectroscopy or spectrophotometry, the spectrum of e_{aq}^- was obtained.

Absorption spectrophotometry is the most convenient way to study the properties of e_{aq}^-. By pulse irradiating an aqueous solution and following the decay of the e_{aq}^- absorption at 5780 Å or 7200 Å, one may measure rates of reaction with ease and precision. From the optical density, OD, at 7200 Å, the concentration of e_{aq}^- is obtained from the equation

$$C = \frac{OD}{1.6 \times 10^4 l}$$

Since techniques for measuring optical densities of the order of 10^{-2} are available and multiple-reflection cell lengths, l, of 8–64 cm are in use, e_{aq}^- concentrations in the range of $10^{-5} - 10^{-8}$ M are generally employed.

Reactions. The redox potential, $E°$, of e_{aq}^- is + 2.7 volts and since $E°$ for the H-atom is +2.1 volts, e_{aq}^- is a stronger reducing agent than the H-atom by 0.6 volt. It is particularly reactive with simple free radicals and molecules or ions containing accessible vacant orbitals or positive zones.

Except for its reaction with water, rates with radicals are generally very fast, many of them being within the diffusion controlled range of 10^{10} M^{-1} sec^{-1}, where reaction occurs on every collision. Of especial interest in the radiolysis of pure water are the fast reactions of e_{aq}^- with H or with another e_{aq}^- to form H_2. Rapid also are reactions of e_{aq}^- with O_2 and CO_2 common gases normally present in water. Inorganic positive ions (such as Ag^+ and Cu^{++}) with stable lower valences are particularly reactive. However, because of its high redox potential some unusual e_{aq}^- reduction reactions occur. The doubly charged ions, Zn^{++} and Co^{++}, react quickly forming a transient ion of short life. These ions may exist as unstable complexes such as $Zn^{++}e_{aq}^-$.

The reaction rates of organic compounds vary from $<10^6 M^{-1} sec^{-1}$ for alcohols to $>10^{10} M^{-1} sec^{-1}$ for aromatic compounds with electron-withdrawing groups. General reaction types are:

$$e_{aq}^- + RH^+ \rightarrow R\cdot + H$$

$$e_{eq}^- + RX \rightarrow R\cdot + X^-$$

$$e_{aq}^- + R{=}Y \rightarrow \dot{R}{-}Y^-$$

$$\dot{R}{-}Y^- + H^+ \rightarrow \dot{R}YH$$

Structure. While the detailed structure of e_{aq}^- has not been definitely established, its properties, listed in Table 1, reveal it as a large highly mobile univalent negative ion. Furthermore, the close similarity of its optical absorption spectrum with that of electrons solvated in ammonia, alcohols,

TABLE 1. PROPERTIES OF e_{aq}^- AT 25°.[a]

Absorption maximum (nm)	715
Absorption maximum (eV)	1.73
ε (715 nm) (M^{-1}cm^{-1}10^{-4})	1.85
$dh\nu/dT$ (0 to 100°) (eV/deg 10^3)	−2.9
Half-width (eV)	0.93
Oscillator strength	0.71
ESR g-factor	2.0002
ESR line width (gauss)	<0.5
Charge	−1
Radius of charge distribution (Å)	2.5 to 3.0
Primary yield $g(e_{aq}^-)$ pH 7	2.65
$\tau_{1/2}(e_{aq}^- + H_2O)$ msec	0.78
$\tau_{1/2}$(neutral H$_2$O) msec	0.23
Diffusion constant (cm^{2}sec^{-1}10^5)	4.90
Equivalent conductivity (mho-cm^2)	190
Mobility (cm$^2V^{-1}$sec^{-1}10^3)	1.98
ΔF hyd (kcal/mole)	−37.4
ΔS hyd (cal/mole deg)	−1.9
ΔH hyd (kcal/mole)	−38.1
$E°$ $(e_{aq}^- + H^+ \rightarrow \frac{1}{2}H_2)V$	2.77

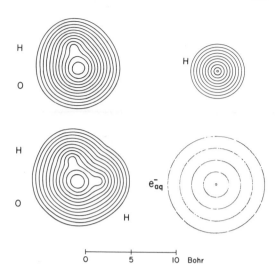

FIG. 2. Electron density contours for the polaron model of e_{aq}^-. Comparison with the hydrogen atom, hydroxyl radical and water molecule. In OH and H$_2$O the innermost contour corresponds to a probable electron density of $1 \cdot 0e^-$/bohr^3 ($0 \cdot 25$ in H), and each successive outer contour decreases by a factor of two down to $0 \cdot 00049$ e$^-$/bohr^3. In the e_{aq}^- diagram the innermost contour corresponds to an electron density of $0 \cdot 0077$ e$^-$/bohr^3 and the outermost to $0 \cdot 00049$ e$^-$/bohr^3.

amines, low-temperature glasses and F-centers supports the idea that e_{aq}^- is an electron trapped in water by its own polarization field. Its spectrum shifts toward lower energy (longer wavelength) with rising temperature, as shown in Fig. 1, and the similar areas and shapes of these spectra show that the number of electrons trapped in water is independent of temperature. Hydration even takes place above the critical temperature. Since the structure of water changes drastically over this temperature range it is clear that e_{aq}^- creates its own cavity and does not require preformed traps. Recent pulse radiolysis work has also shown that hydration occurs in less than 20 picoseconds, the resolving time of the pulse radiolysis unit. The hydrated electron is generally assumed to consist of an electron surrounded by a tetrahedron of H$_2$O molecules. And while the wave equations for this complex array have not yet been worked out, the simple dielectric model that accounts satisfactorily for the thermodynamic and optical properties of the ammoniated electron fit e_{aq}^- as well. Based on the wave equation for this simple approximate model, an insight into the structure of e_{aq}^- may be gained from a comparison of its electron density contour with those of the H-atom, OH radical, and H$_2$O molecule. These electron density contours appear in Fig. 2. The hydrated electron displays a much lower electron density than that of the other atoms and molecules associated with water. The similarities between these species are apparent even at this stage of development of the structure of e_{aq}^-. Further research must take into account the short range interactions between the electron and the atoms lining the cavity.

The hydrated electron, because of its simple means of formation, precise method of analysis, high reactivity and elementary structure, will find increasing applications in experimental and theoretical chemistry. It possesses considerable potential for the study of reaction mechanisms, structure of compounds and as an analytical tool. And because it can be generated photochemically from inorganic ions and organic molecules, it is likely

to be an important intermediate in photosynthesis and photobiology.

EDWIN J. HART

Reference

Hart, E. J., and Anbar, M., "The Hydrated Electron," Interscience, New York, 1970.

HYDRATION

Hydration is a reaction of molecules of water with a substance, in which the molecules of water act as a unit, that is, the H—OH bond is not split. The products of hydration are called *hydrates*. Hydration contrasts with hydrolysis, in which the molecules of water are split in their reaction with the substance. In hydrolysis, the reaction may be between ions and the solvent water in which the ions have dissolved. Thus the reaction:

$$CuSO_4 + 5H_2O \rightarrow CuSO_4 \cdot 5H_2O$$

is hydration, while the reaction of hydrated copper ions with water is hydrolysis:

$$Cu^{2+} + 2H_2O \rightleftharpoons CuOH^+ + H_3O^+$$

In water itself, both processes go on. Ionization takes place to the extent of one molecule in ten million. The resulting hydrogen ion is hydrated with at least one molecule of water.

$$2H_2O \rightarrow H_3O^+ + OH^-$$

Reactions similar to hydration can take place

with substances other than water. Alcohol, ammonia, and other compounds that add by means of a coordinate covalent bond react in a similar manner. In such cases the reaction is called alcoholism or ammonion respectively, and the product an alcoholate or an ammoniate. There is no essential difference between such reactions and hydration.

Reactions of water, especially with salts, produce hydrates of definite composition. Although the hydrates contain water, they are not ordinarily wet to the touch. Further, a given compound often forms more than one hydrate. For example, the hydration of sodium sulfate can form $Na_2SO_4 \cdot 10H_2O$ decahydrate, $Na_2SO_4 \cdot 7H_2O$, heptahydrate, and $Na_2SO_4 \cdot H_2O$, monohydrate. Hydrates often form when a solution of a salt is evaporated. Common salt, however, does not form a hydrate.

Strong evidence exists that cations in solution are hydrated. The Al^{3+} ion, effective ionic radius 0.50 Å, attracts six polar water molecules that cluster about it. The negative ends of the water molecules are toward the Al^{3+} ions, and the positive ends of the water molecules project like bristles. The hydrated unit $[Al(H_2O)_6^{3+}]$ is attracted toward the cathode in an electrolysis reaction. The attraction for water molecules seems to be less for monovalent ions than for ions that are divalent, trivalent, and beyond.

The number of molecules of water attached to an ion in a hydrate is the same as the coordination number of that ion. In the formation of hydrates, tetrahedral or octahedral structures are favored. Cations (positively charged) are usually smaller than anions (negatively charged) and they have stronger attractions for water molecules. Thus the magnesium ion, commonly written Mg^{2+}, is hexahydrated in water solution $[Mg(H_2O)_6^{2+}]$, and the chloride ion, commonly represented as Cl^-, is probably tetrahydrated $[Cl(H_2O)_4^-]$.

Alums are well-known hydrates, and common alum is $KAl(SO_4)_2 \cdot 12H_2O$. Other alums, such as the sodium aluminum alum used in baking powders, are dodecahydrates. Six of the water molecules are associated with the aluminum or trivalent ion, and the other six with the monovalent ion.

The act of hydration is associated with an energy change. In calories per mole, the fluoride ion (F^-) liberates 96,790 calories when hydration takes place, chloride ion (Cl^-) 64,540, bromide ion (Br^-) 57,610, and iodide ion (I^-) 48,000. The values for monovalent cations are somewhat larger, starting with 140,600 for lithium ion (Li^+) and reaching 80,690 for cesium (Cs^+). For divalent ions, the values are much greater, 894,400 for Be^{2+} to 731,400 for Ba^{2+}. For trivalent ions, the values are still greater.

Similarly, the act of dehydration requires that energy be supplied from some external source. A change in crystal structure also accompanies dehydration. In fact, some hydrates dehydrate spontaneously if their vapor pressure is greater than the partial pressure of water vapor in their environment. The dehydration of sodium carbonate decahydrate (washing soda) is spontaneous in most atmospheres. The act is called efflorescence. Efflorescence explains the crumbling that can be noticed when photographer's "hypo" ($Na_2S_2O_3 \cdot 5H_2O$) is left exposed to the air.

The obvious weight changes caused by hydration are of considerable industrial importance. Sometimes it is advantageous to ship a chemical compound in its anhydrous form, and to hydrate it at the point of its use if such hydration is needed.

In the manufacture of ethylene diamine tartrate (EDT), a crystal used for piezoelectric purposes, it was found necessary to keep the temperature above 40°C because of monohydrate formed on the surface of the crystal below that temperature. The monohydrate is cloudy, whereas the crystal above 40°C is clear and useful for manufacturing purposes.

The use of the term "hydration" is not as precise among organic chemists as the definition given at the beginning of this article. There are, however, numerous cases of hydration of organic compounds. Oxalic acid forms a dihydrate $(COOH)_2 \cdot 2H_2O$, as does glyoxal $(CHO)_2 \cdot 2H_2O$. Glyoxylic acid, made by treating glyoxal with nitric acid, crystallizes as a monohydrate $COOH—CHO \cdot H_2O$. In these cases the water seems to be combined with the $C{=}O$ group.

Some authorities call the act of adding water to chloral hydration. Others call it hydrolysis. According to the definition given here, it is hydrolysis. When water is added to chloral (CCl_3CHO), crystalline chloral hydrate forms. The hydrate is relatively stable. It melts without decomposition, and loses water at its boiling point, or by the use of a strong desiccant such as concentrated sulfuric acid. Chloral hydrate is known as a hypnotic. The anhydrous compound chloral is used to make DDT.

Two other cases are sometimes called hydration by organic chemists which by this definition are hydrolyses. They are the hydrolyses of unsaturated hydrocarbons, and of acid anhydrides. The well-known reaction of adding water in the presence of a catalyst to the ethylenic double bond of an alkene is illustrated in the case of ethylene: $C_2H_4 + H_2O \rightarrow C_2H_5OH$. Similarly, acetylene adds water and forms acetaldehyde: $C_2H_2 + H_2O \rightarrow CH_3CHO$. In adding water to acid anhydrides a splitting of the water molecules seems to be involved; this is properly hydrolysis, although sometimes it is loosely called hydration. The reaction with acetic anhydride is

$$(CH_3CO)_2O + H_2O \rightarrow 2CH_3COOH$$

ELBERT C. WEAVER

Cross-reference: *Hydrolysis.*

HYDRAULIC FLUIDS

Hydraulics involves transfer of pressure from one point in a system to another and is based on Pascal's Principle, which states that pressure on a confined liquid is transferred equally and undiminished in all directions. One pound acting on a 1 sq in. piston exerts 1 psi which, if transmitted by a fluid to a 10 sq in. piston, will exert 10 psi. For the larger piston to move 0.1 in. the smaller one moves 1 in. This simple system is the basis of the hydraulic brake. The brake cylinder pistons at the four wheels move a very short distance, compared to the smaller master cylinder piston, operated by the brake pedal. This is a static system with negligible fluid movement. Hydraulic fluids should be essentially incompressible, flow readily under operating conditions, provide lubrication and be sufficiently viscous to seal close piston tolerances. Each application has specific requirements which the hydraulic fluid must meet. While the same general properties are needed in some degree for all uses, a formulation developed for one use is not suitable for another because of differences in the system design and the operating conditions.

The principal **automotive** application of hydraulics is stopping the vehicle. Of some 15 requirements in the SAE brake fluid specification J1703, the reflux boiling point is the most critical from a safety standpoint. Heat from the braking operation raises the brake fluid temperature and at times causes the fluid in the system to boil. Brakes then are inoperative because vaporized fluid is compressible. This condition is known as "vapor lock." Recently, brake fluid temperatures have increased because of higher vehicle speeds, greater use of automatic transmissions (requiring more braking), less air cooling of the brakes because of body design changes, and the use of disc brakes, which conduct more heat to the fluid. As a result, boiling points of original equipment fluids were raised from about 400 to 450–550°F, depending on the manufacturer. The minimum boiling point of SAE J1703 was also increased from 302 to 374°F.

Brake fluids have three components: lubricant, solvent blend and additives. Aliphatic hydroxy derivatives are selected because they do not swell the rubber piston cups and seals used in the automotive brake system. The lubricant (20–30%) may be a castor oil derivative (early fluids used raw and blown castor oils), a polypropylene glycol with a molecular weight from 1000-2000, or a synthetic lubricant. A typical synthetic lubricant is a polymeric mixture of monobutyl ethers of 1,2 oxyethylene and 1,2 oxypropylene glycols. Suitable brake fluid grades have 100°F Saybolt viscosities of 170–660 SUS. Additives (1–3%) consist of an antioxidant (e.g., diphenylol propane), alkaline buffer (e.g., borax-ethylene glycol condensate) and a corrosion inhibitor system, developed for each formulation. Solvent blend (balance of the formula) is a combination of monoalkyl (methyl, ethyl or butyl) ethers of di-, tri- or higher alkylene glycols, mono- and/or poly- (ethylene, propylene or butylene) glycols, and limited amounts of higher-boiling aliphatic alcohols. The blend of solvents determines the boiling point because other components are nonvolatile. Fluids boiling from 450–500°F are blends of monoalkyl ethers of trialkylene glycols with smaller amounts of polyglycols (min. b.p. 450°F), combined with a lubricant and additives. Except for the additives, fluids, boiling over 500°F, are mixtures of monoalkyl (methyl, ethyl or propyl) ethers of tetra-, penta-, hexa- and higher alkylene (ethylene and propylene) glycols, depending on the boiling point. These are referred to as single-component brake fluids. Minimum boiling (ca. 385°F) replacement fluids employ the lower-boiling monoalkyl glycol ethers (b.p. under 400°F) with smaller amounts of the monoglycols and higher-boiling aliphatic alcohols (min. b.p. 260°F). All components (except castor oil derivatives) are hygroscopic and water-miscible.

Fluids readily pick up over 2% water from the atmosphere through the master cylinder and from the ground through the wheel cylinder seals after a relatively short service period.

Small amounts of water sharply reduce the boiling point of the fluid. For example, with 1% added water a 560°F fluid boils at 385°F, a 470°F at 380°F and a 385°F at 330°F. The minimum SAE boiling point is 374°F. The water has the greatest impact on the higher boiling fluids. Beside the boiling point, other SAE J1703 requirements are thermal and chemical stabilities, resistance to oxidation, protection of the brake system metals from corrosion, compatibility with rubber seals, fluidity at −40°F after 144 hours and at −58°F after 6 hours, and water miscibility at −40°F and 140°F after 24 hours. Test methods and limits for these properties are given in SAE J1703. The Department of Transportation now has the responsibility of establishing standards for this safety item. Initially DOT accepted SAE J1703 specifications but is now developing DOT requirements. For example, a minimum "wet" reflux boiling point will be specified. It is determined after the fluid has been exposed to a definite humidity and temperature. This requirement attempts to limit the hygroscopicity of the fluid and thus reduce the danger of "vapor lock."

A future automotive development is a central hydraulic system, which reduces the electrical demands by substituting hydraulic power. Possible automotive uses for this system include, in addition to brakes, power steering, window lifts, seat adjusters, windshield wipers, starting motor, etc. The hydraulic power circuit contains a hydraulic pump, driven by the engine, fluid reservoir, accumulator for storing fluid under pressure and relief valve. The system is capable of developing 2000 psi. Control valves direct fluid pressure to the various operating units and back to the reservoir. Unlike the static type brake system, the pump and not the brake pedal provides the pressure. This dynamic system is used for industrial and aeronautical hydraulic applications. SAE central-system fluid specifications, giving test methods and limits, have been published for synthetic (SAE 71R1) and petroleum (SAE 71R2) fluids. Car manufacturers have successfully tested the central system with both types. The synthetic type is basically a brake fluid with additional additives. SAE 71R1 requires that the fluid boil over 400°F and meet the requirements of SAE J1703, and that it also has shear stability, oxidation stability, and resistance to pump wear and foaming. Additives are needed to meet specific limits for these properties. The petroleum-type fluid (SAE 71R1) consists of highly refined base stock with a shear-stable viscosity index improver to meet the viscosity requirements. Additives are needed to satisfy all requirements of the SAE 71R1 specification. Petroleum fluid requires nitrile rubber seals and cups in contrast to the synthetic fluid which uses SBR or natural rubber. Since the central system will include the brake system which has always employed synthetic fluids, this type is favored for use in the automotive central system.

Industrial. Paraffin and naphthenic crudes serve as base oils for industrial hydraulic fluids. The petroleum component may consist of a blend of base stocks and possibly a viscosity-index improver, depending on the application. Antioxidant is added to slow the rate of oxidation and prolong the life of the fluid. A corrosion inhibitor is used to prevent rusting. For severe operating conditions an antiwear agent is added. A defoamer is included to cut foam buildup in the reservoir.

Fire-resistant fluids have been developed for hazardous operations. They resist combustion and, if ignited, will not propagate a flame. The fluid volume will increase because of the safety advantage. Applications consist of diecasting, heat treating, plastic fabrication and glass working equipment. Fire-resistant fluids are mainly water-base and phosphate ester types. Water is nonflammable and is used as a snuffing agent for flammable fluids. With sufficient water a flammable product can be made fire resistant. It may be a homogeneous solution or an emulsion, depending on the miscibility with water. The flammable part of the solution type is a blend of ethylene glycol, shear-stable polyglycol thickener and additives. About 35% water makes the solution fire resistant. By varying the molecular weight of the thickener different viscosity fluids are obtained. The additives control liquid- and vapor-phase corrosion, boundary lubrication and foaming. These aqueous solutions are used at below 0°F and have performed satisfactorily in pumps with no deleterious action on seals. The maximum operating temperature is limited to 150°F because of volatility. The water content must be checked regularly and maintained at a specified concentration. In the emulsion-type fluid the water is dispersed in petroleum fluid to yield a fire-resistant oil-in-water emulsion. An emulsifying agent is needed to form an emulsion, stable during the pumping operation. This type fluid adds fire resistance to the performance characteristics of the petroleum-base fluid.

Phosphate esters are not nonflammable, but their fire resistance is sufficient for use as hydraulic fluids in a hazardous environment. The physical properties of a tertiary phosphate ester (triaryl, trialkyl and trialkylaryl) are largely determined by the organic group attached to the phosphate radical. Their advantages are fire resistance with no adjustment problem, excellent lubricity, thermal stability at moderate temperatures, oxidation stability and a wide range of viscosities. Shortcomings are a tendency toward hydrolysis, swelling of standard seal materials (but an acceptable substitute is available) and corrosive attack on metals if hydrolysis or thermal decomposition occurs. Additives are employed to inhibit corrosion, to improve the viscosity-temperature characteristics and to reduce oxidation and foaming.

Aeronautical. As the aircraft became more complex, many operations, formerly handled manually, were performed hydraulically. As a result, the hydraulic system extended to all part of the plane. If a leak develops in the system, a flammable fluid in the form of a fine mist may be ignited if it strikes overheated brakes or a hot manifold. This could result in a serious fire. Early hydraulic fluids were flammable and consisted of a petroleum fluid or a combination of castor oil and solvent blend (higher aliphatic alcohols and monoalkyl

glycol ethers). During World War II the U.S. Navy developed a water-base fire-resistant fluid. The general composition was similar to the aqueous industrial fluids. Because of operational difficulties the aqueous fluid was eventually replaced by a phosphate ester formulation which meets requirements of the commercial (MS31500B) and the military (Mil-F-7100) fire-resistant fluid specifications. It is suitable for continuous service from −65 to 225°F.

The military has required a high-temperature but not fire-resistant fluid with an operating range from −65 to over 400°F. From a study of various structures hexa(2-ethylbutoxy)disiloxane showed outstanding high-temperature thermal stability, very low pour point and excellent viscosity-temperature characteristics. This siloxane-base stock with additives to improve oxidation stability and lubricity resulted in a fluid, usable from −65 to over 500°F. It is employed in military planes and missile installations where hydraulic power is used for handling and launching operations.

CHESTER M. WHITE

References

ASTM Standard "Hydraulic Fluids" Part 17, ASTM, Philadelphia, Pa., 1972.

Gunderson, R. C., and Hart, A. W., "Synthetic Lubricants," Reinhold Publishing Corp., New York, 1962.

Hattan, R. E., "Introduction to Hydraulic Fluids," Reinhold Publishing Corp., New York, 1962.

SAE Standards "Motor Vehicle Brake Fluid SAE J1703" and "Central Systems Fluids SAE J71," Soc. Auto. Eng. Inc., New York, 1972.

Spec. Tech. Publ. 267 "Symposium on Hydraulic Fluids," Am. Soc. for Testing and Materials, Philadelphia, 1960.

Spec. Tech. Publ. 406 "Fire Resistance of Hydraulic Fluids," Am. Soc. for Testing and Materials, Philadelphia, 1966.

HYDRIDES

Classification. There is some ambiguity about the meaning of the term "hydride." Since the addition of an electron to a hydrogen atom forms the anion commonly known as the hydride anion, perhaps the most consistent terminology would reserve the term "hydride" for compounds or molecules containing this anion. However, a large number of covalent compounds containing hydrogen are commonly called hydrides, despite the fact that they certainly do not contain hydride anions. To encompass these covalent hydrides one must enlarge the definition and refer to all binary compounds of hydrogen as hydrides (including those binary hydrogen compounds that are usually referred to by common names such as water, ammonia, hydrazine, etc.). Finally, we must consider the enormous number of compounds that are not binary hydrides but that do contain one or more hydrogen atoms. A number of these are ternary compounds containing two metals and hydrogen or, a metal, a nonmetal, and hydrogen.

One might make the observation that ternary or even more complex compounds containing hydrogen are usually referred to as hydrides if the element to which the hydrogen atom is bonded is less electronegative than hydrogen itself. On this basis compounds such as hydroxides, amides, phosphides and hydrosulfides are not hydrides, whereas the complex borohydrides and alumino-hydrides can properly be considered to be hydrides on the basis of the lower electronegatives of boron and aluminum, as compared to hydrogen. By this compromise only the binary hydrogen compounds of the electronegative elements are classified as hydrides, but not their derivatives, i.e., hydrogen sulfide but not NaSH. Because of the unusual ability of carbon to form long-chain molecules and its further ability to exhibit variable valences by the formation of single, double and triple bonds, the number of binary C—H compounds is immense, and it is impractical to consider the hydrocarbons as hydrides from the viewpoint of the systematic correlation of hydride properties. For convenience, therefore, only the inorganic hydrogen compounds are considered to be hydrides.

Binary Hydrides

Ionic Hydrides. The number of hydrides that form crystalline lattices containing discrete metal cations and hydride anions is quite small. Only the IA alkali metals and calcium, strontium and barium, among the IIA metals, are known to definitely form hydrides of this category. The IA hydrides are represented by the generic formula MX and the alkaline earth hydrides by MX_2. Each of these hydrides can be prepared by heating the metal in a hydrogen atmosphere, with the hydrogen pressure maintained at a level above the equilibrium dissociation pressure of the hydride at the temperature of the reaction. In each case the hydride formation reaction is exothermic and the resulting hydride is appreciably more dense than the metal from which it was formed. These hydrides are generally more thermally stable than the covalent hydrides discussed below. They have relatively high melting points, as one would expect from their salt-like structures. Usually the melting points are above the decomposition temperatures of the hydrides.

From the viewpoint of chemical reactivity, each of the ionic hydrides reacts readily with water to liberate H_2, as shown by the following equations. This property makes these compounds potentially good sources of hydrogen for specialized problems, and serves as a good analytical device for determin-

$$MH + H_2O \rightarrow MOH + H_2$$

$$MH_2 + 2H_2O \rightarrow M(OH)_2 + 2H_2$$

ing active hydrogen in ionic hydrides. The ionic hydrides are also very useful as selective reducing agents and as dehydrating agents. Their reactivity with water necessitates extra precautions in their use, since they do react with atmospheric moisture, and frequently they must be handled in an inert atmosphere, particularly the more reactive hydrides of this type.

Covalent Hydrides. The lightest IIA elements—Be and Mg—and the remaining main series elements of groups IIIB through VIIB form hydrides of predominantly covalent character. Such hydrides are also formed by the metals of the first two groups of transition metals, those of groups IB and IIB. Unlike the transition metal hydrides described in a subsequent section, these covalent hydrides are stoichiometric compounds of definite formulae, andare often quite volatile. They differ from the ionic hydrides in that the covalent hydrides are generally less thermally stable and usually cannot be prepared directly from the elements. Magnesium hydride, MgH_2, is one exception in that it can be prepared readily from the elements under high pressure, but this hydride is perhaps the least covalent of the covalent hydrides. It constitutes a bridge between the ionic and covalent hydrides. In contrast the lightest IIA hydride, BeH_2, is essentially covalent, as is evident from a comparison of the chemical and thermodynamic properties of the IIA hydrides.

The reactivity of the covalent hydrides toward water varies markedly. Many of these hydrides resemble the ionic hydrides in that they are readily hydrolyzed to liberate H_2. Examples of this high reactivity are the volatile boron hydrides, aluminum hydride and the silanes. Some covalent hydrides are not appreciably hydrolyzed and some, such as the hydrogen halides, are markedly dissociated in water. Generally the crystallized covalent hydrides are appreciably less dense than the element (other than hydrogen) from which they are derived.

The IIA hydrides, BeH_2 and MgH_2, are polymeric solids believed to be linked by three-centered hydrogen bridge bonds. This is particularly true for the more covalent beryllium hydride. These hydrogen bridge bonds are also found in the hydrides of the lightest IIIA element, boron, which forms a series of hydrides known as boranes. The simplest of these, BH_3, is not stable at ordinary pressures, dimerizing to diborane, B_2H_6, which is a gas at ordinary conditions. Condensation reactions convert diborane to higher boranes, in which the H/B ratio is less than the 3 in diborane, the extra valences being satisfied by three-centered boron to boron bonds. This series progresses through a number of well characterized crystalline compounds. The crystalline $B_{10}H_{14}$ was for a long time the highest of the known boranes but hydrides up to $B_{20}H_{26}$ have recently been discovered. The existence of such a series of boranes invites comparison with the hydrocarbons. However the similarity is very limited. Most of the boranes have positive heats of formation, are not very thermally stable and react readily with water to form H_2. Many of the lower boranes react violently with air. Further, the boranes are not chain compounds and do not have unsaturated linkages. Thus it is not possible to prepare an immense number of boranes comparable to the hydrocarbons. However, it should be noted that because of the high exothermicity of $B_2O_3(s)$ formation and the endothermicities of the boranes, these materials are potentially more important fuels than are the hydrocarbons.

Only aluminum among the heavier IIIA elements is definitely known to form a stable trivalent hydride. However unlike the boranes, volatile alanes are not ordinarily stable and the known aluminum hydride is a polymeric material, $(AlH_3)_x$. The monomeric AlH_3 and dimeric Al_2H_6 only exist as gaseous molecules at very low pressures.

Among the IVA elements, a series of tetravalent hydrocarbon-like molecules are formed by silicon and germanium. These silanes and germanes resemble the paraffins in structural formulae, although the maximum chain-lengths of the silanes and germanes are small by comparison with the known paraffins. Silicon and germanium do not form definite olefinic or acetylenic type compounds, although polymeric materials ranging from $(SiH)_x$ to $(SiH_2)_x$ are known, as well as $(GeH)_x$. In general, the silanes and germanes are less stable thermally than the analogous hydrocarbons. Tin forms the thermally unstable gases SnH_4 and Sn_2H_6, while only thermally unstable gaseous PbH_4 is known for lead.

The VA elements form MH_3 hydrides, all of which are quite volatile. The thermal stabilities of these hydrides decrease with increasing atomic weight for M, SbH_3 decomposing slowly at room temperature while BiH_3 is quite unstable at this temperature. Unlike the acidic trivalent IIIA hydrides these trivalent VA hydrides are basic, and unlike the covalent hydrides discussed above the more stable of the VA hydrides can be prepared directly from the elements. The more stable among the hydrides of this type can also be converted to the M_2H_4 hydrides which are less stable than the corresponding MH_3 hydrides because of the relative weakness of the M—M bonds.

The properties of the MH_2 compounds of the VIA elements and the MH hydrohalides are too well known to be described here. Less known are the MH hydrides of the IB metals and the MH_2 hydrides of the IIB metals. Crystalline CuH is known. It decomposes readily at moderately elevated temperatures and is one of the few metal hydrides that does not hydrolyze appreciably with water. Similar hydrides of silver and gold are unknown although diatomic gaseous hydrides of these metals were heated in hydrogen. The IIB hydrides ZnH_2, CdH_2, and HgH_2 are not very stable, the stabilities decreasing with increasing atomic weight of the metal to the point where HgH_2 decomposes readily at $-125°C$.

Nonstoichiometric Hydrides. The transition metals of groups IIIB through VIII form binary compounds with hydrogen that differ from both the ionic hydrides and the covalent hydrides. First, these transition metal hydrides are nonstoichiometric, in that the hydrides generally do not have definite formulas corresponding to integral ratios of hydrogen to metal. Upon heating of the metal in hydrogen the uptake of the latter at first resembles merely a solution of hydrogen in the metal. As the hydrogen contents increase distortions of the lattice require modifications of the original metal lattice and definite hydride phases are observed. However, the simultaneous presence of several of these, or a single phase with additional dissolved hydrogen, leads to nonstoichiometric formulas. These hydrides differ from the ionic hydrides in that the densities of the nonstoichiometric hydrides are lower than those of the corresponding metals. In this respect they resemble the covalent hydrides. However they are frequently far less reactive chemically than the covalent hydrides. The IIIB and IVB hydrides, although nonstoichiometric, are formed from the metal and H_2 exothermically.

This tendency is very much reduced for the VB hydrides, and the VIB through VIII hydrides are endothermic. It has been claimed that stoichiometric hydrides are formed for a number of the VIB to VIII metals by a Grignard-type procedure but this evidence has been disputed and it appears that organometallic hydrides are actually formed by these techniques. The evidence indicates that the binary hydriding tendency of the latter metals is considerably lower than those of the transition metals of lower groups of the Periodic Table. It might be noted that a number of the former do form definitive diatomic molecules in small concentrations when the metal is heated to very high temperatures in H_2. For example, CrH and MnH have been observed spectroscopically by this procedure.

Complex Hydrides of the Transition Metals

A number of transition metals form complex hydrides containing a metal to hydrogen bond that is stabilized by the presence of other ligands. One such class of complex hydrides is the metal carbonyl hydrides, in which the sigma M—H bond is stabilized by the electron withdrawing tendency of the CO ligand. Overlap of a filled d orbital of the metal with an empty pi orbital on the carbon atom reduces the electron charge on the metal and permits hydride formation. Examples of such carbonyl hydrides are the gases $H_2Fe(CO)_4$ and $HCo(CO)_4$ and the liquid $HMn(CO)_5$. Polynuclear carbonyl hydrides of iron are also known, corresponding to $H_2Fe_2(CO)_8$, $H_2Fe_3(CO)_{11}$ and H_2Fe_4-$(CO)_{13}$. These carbonyl hydrides are acidic and very poisonous. In a number of cases ligands such as OH, H_2O, OCH_3, and CH_3OH are substituted for some of the CO groups of the parent carbonyl hydrides.

Another type of complex hydride is one in which the cyclopentadienyl ion $C_5H_5^{-1}$ is the stabilizing ligand. It differs from the carbonyl hydrides in that the bonding between metal and ligand involves a pi molecular orbital of the entire ligand molecule. This is perhaps similar to the "sandwich" bonding in ferrocene, and the cyclopentadienyl hydrides are known as π complexes. Unlike the carbonyl hydrides the cyclopentadienyl hydrides are basic, for example forming $(\pi$—$C_5H_5)_2ReH_3^+$ from $(\pi$—$C_5H_5)_2ReH$. The cyclopendadienyl hydrides are generally less volatile than the corresponding carbonyl hydrides. Mixed carbonyl and cyclopentadienyl hydrides are known.

Mention might briefly be made of chelated hydrides in which chelates such as phosphines or arsines stabilize sigma bonds between transition metals and hydrogen. An example is $[C_6H_4(PEt_2)_2]_2$-FeH_2.

BERNARD SIEGEL

Cross-references: *Chelation, Bonding, Electronegativity, Carbonyls, Metal.*

HYDROCHLORIC ACID

Hydrochloric acid is the water solution of hydrogen chloride (HCl) and is generally referred to by that formula. Its most common form is the 20°Bé strength, with 31.45 per cent HCl. Some anhydrous HCl is available shipped in steel bottles.

Hydrochloric acid is made by one of three methods.

(1) By the action of sulfuric acid on salt (about 5%), on heating, the hydrogen chloride and such water as may be present pass over to the absorption vessel and towers, there to meet countercurrently a dilute solution of the acid or water. The reaction between nitre cake (NaHSO$_4$) and NaCl produces hydrochloric acid and sodium sulfate (Na$_2$SO$_4$), in the mechanical salt-cake furnace. In at least one large-scale operation, a stream of sulfuric acid itself is allowed to flow on a mass of salt cake to which salt is constantly added, in a Mannheim furnace, but the general practice is to use nitre cake wherever possible.

(2) By burning chlorine from an electrolytic cell in an excess of hydrogen, and dissolving the resulting hydrogen chloride in water. A mixture of hydrogen and chlorine in certain proportions is violently explosive, so that careful regulation is essential.

(3) By dissolving in weak acid or water the byproduct hydrogen chloride formed in the chlorination of hydrocarbons about 90%.

$$C_6H_6 + Cl_2 \rightarrow C_6H_5Cl + HCl$$

ELBERT C. WEAVER

HYDROFLUORIC ACID

Hydrogen fluoride is a gas, extremely soluble in water; the solution is called hydrofluoric acid. Hydrogen fluoride is liberated from its calcium salt by the action of concentrated sulfuric acid on the finely powdered mineral fluorspar (CaF$_2$). Hydrofluoric acid attacks silicates of any kind; with silica (SiO$_2$), it forms silicon tetrafluoride. The acid is shipped in 30% or 60% concentration in plastic bottles. It serves for making the soluble fluorides, for preparing glass-etching and glass-brightening baths, and for cleaning stone. The anhydrous acid is also prepared commercially, and is used for the alkylation of hydrocarbons in petroleum refineries.

ELBERT C. WEAVER

HYDROGEN AND COMPOUNDS

Hydrogen, H$_2$, atomic weight 1.00797, has a valence of 1. Under normal conditions it is a colorless, odorless, tasteless gas and is the lightest known substance. Free hydrogen occurs in nature in certain volcanic and natural gases in mixture with other gases. Its presence in the atmosphere, somewhat less than 1 ppm, has been attributed to these natural gases and to the decay of various organic bodies. The spectroscope shows that it is present in the sun, many stars, and nebulae. Our galaxy, which includes the sun and its planetary system containing the earth and the moon, plus the stars of the Milky Way, is presently considered to have been formed 12 to 15 billion years ago from a rotating mass of hydrogen gas which condensed into stars under gravitational forces. This condensation produced high temperatures giving rise to the fusion reaction converting hydrogen into helium, as presently occurring in the sun, with the evolution of tremendous amounts of radiant thermal energy, plus the formation of the heavier elements. Hydrogen gas has long since escaped from the earth's lower atmosphere, but is still present in the atmosphere of several of the other planets. In a combined state, hydrogen comprises 11.19% of water and is an essential constituent of all acids, hydrocarbons, and animal matter. It is present in most organic compounds.

Commercially, hydrogen is produced by (1) the steam hydrocarbon process (steam reforming), (2) noncatalytic partial oxidation of hydrocarbons with oxygen, (3) catalytic partial oxidation, (4) the steam-water gas process, (5) the steam-iron process, (6) the electrolysis of water, (7) the steam-methanol process, (8) ammonia dissociation, (9) the thermal dissociation of hydrocarbons, and (10) as a byproduct from certain petroleum refining operations. Steam reforming of natural gas (methane) has become the principal process used for the manufacture of hydrogen for its largest use, which is in the synthesis of ammonia. The second largest use, in petroleum refining operations, is supplied by (10) plus (1) and (2). Hydrogen of extremely high purity (99.99+%) is being produced commercially by diffusing hydrogen through films of palladium or palladium-silver alloys at elevated temperatures. These metallic films pass hydrogen and exclude all other gases.

At elevated temperatures (about 600°F) molecular hydrogen dissociates to atomic hydrogen on a palladium surface, dissolves in the palladium, passes to the opposite surface, recombines to molecular hydrogen, and escapes from the surface. This unique property of hydrogen is now being utilized commercially on a large scale for recovering pure hydrogen from gaseous mixtures. The ability of atomic hydrogen to dissolve in, or at least pass through, metals has given rise to hydrogen blistering of steel in which atomic (nascent) hydrogen produced on a steel surface by the corrosive action of acids such as HCl or H$_2$S, diffuses into the steel and collects at a discontinuity inside the steel. Within the discontinuity the atomic hydrogen recombines to form molecular hydrogen which cannot escape, but which builds up pressure enough to bulge or fracture the steel (embrittlement). This may be prevented by protecting the steel surface to prevent corrosion, or by using steel processed in such a way as to avoid the presence of discontinuities or laminations, so that any atomic hydrogen present may pass through the steel.

Hydrogen is used in conjunction with nitrogen in the manufacture of ammonia; in hydrogenation of vegetable and animal oils and fats, and hydrocarbons; with oxygen or air for welding and cutting metals; combined with chlorine to produce hydrochloric acid; with carbon monoxide or coal to form gasoline; and with carbon monoxide (or dioxide) to form methanol. Large quantities of liquid hydrogen are being made for use as rocket propellants, and the development of highly efficient insulations for tank trucks and rail cars has led to the commercial transportation and use of liquid hydrogen. Liquid hydrogen and oxygen were used in the second and third stage engines of the Saturn V launch vehicles in the Apollo moon flights. Liquid hydrogen is the sole propellant in the nuclear rocket system now under development. In this system liquid hydrogen pumped through a nuclear reactor is vaporized and heated to 3500°F to provide a high-velocity, low-molecular weight exhaust stream having a high specific impulse, with an estimated 50% performance gain over hydrogen-oxygen chemical rockets. Hydrogen slush, a pumpable mixture containing 50% or more of solid hydrogen crystals in liquid hydrogen at the triple point, 13.8°K, has been proposed as a replacement for liquid hydrogen in chemical or nuclear rockets, to give a 15% saving in storage volume, plus lower vaporization losses. Roasting oil shale in a retort at 900 to 1000°F with hydrogen present at 1000 to 2000 pounds per square inch pressure has proved to be a promising method for recovering hydrocarbons from oil shale, one of the world's largest fossil fuel reserves.

Constants. Molecular weight 2.0160; boiling point −252.7°C at atmospheric pressure; melting point −259.1°C at atmospheric pressure; critical data: temperature −240°C, pressure 12.8 atm, density 31.2 g/liter; gas density 0.0899 g per liter at 0°C and 1 atm; liquid density 70.8 g per liter at −253°C; solid density 80.7 g per liter at −262°C; specific gravity (air = 1.0) 0.0695; specific heat at constant pressure 3.44 (0 to 200°C), at constant volume 2.46 (0 to 200°C); ratio of specific heats 1.40 (0 to 200°C); gross heat of combustion 33.940 cal per g; latent heat of vaporization 107 cal per g at −253°C; latent heat of fusion 13.89 cal per g at −259°C; coefficient of thermal expansion 0.00356 per °C; thermal conductivity 0.00038 gram-calorie per sq cm per sec per °C per cm at 0°C; viscosity 0.0087 centipoises at 15°C and 1 atm; minimum ignition temperature 574°C. Flammability limits, 4 to 94% in oxygen, 4 to 74% in air.

Ortho- and Para-Hydrogen. The hydrogen atom consists of one proton, the nucleus, and one electron. Both have two possible directions of spin. However, only atoms with antiparallel electron spins can combine. When two such atoms are joined to make molecular hydrogen, the nuclear spins may then be either parallel or antiparallel. These have been designated as ortho- and para-hydrogen, respectively. Under normal temperatures a ratio of 3 ortho-: 1 para- is said to exist. At low temperatures equilibrium is on the side of the para-hydrogen. Para-hydrogen has been isolated by converting normal hydrogen at liquid air temperature over charcoal or a nickel catalyst. Ortho-hydrogen cannot be isolated in this manner as the equilibrium mixture at infinitely high temperatures is still a 3 to 1 ratio of ortho- to para-hydrogen. A physical method of separation is necessary.

The transition of liquid hydrogen from ortho- to para- is exothermic (168 cal/g), so that liquid hydrogen must be converted to the para- form before it can be stored as liquid. This is done by passing the liquid hydrogen over a suitable catalyst with refrigeration to remove the heat of transition.

Deuterium. Popularly known as "heavy hydrogen," deuterium is an isotope of hydrogen which is present in normal hydrogen to the extent of one part in 5000 to 7000 in rainwater and one part in 30,000 in electrolytic hydrogen. Deuterium differs from hydrogen in that its nucleus contains a neutron as well as a proton; it therefore has a mass of 2. The existence of deuterium was first indicated in the determination of atomic weight values by

mass-spectrographic methods. Aston obtained the value 1.00778 ± 0.00015 relative to oxygen 16.000 which agreed with chemical methods giving 1.00780. Compared with the standard chemical value of 1.00756 this gave a value of ordinary hydrogen one part higher in 5,000 than true hydrogen. Birge and Menzel in 1931 suggested that this might be due to a heavier isotope. This fact was established by Urey and co-workers in 1932.

Deuterium is principally found as deuterium oxide, or "heavy water," which is the analog of water. It may be separated from normal water by several methods including electrolysis of water, fractional desorption from charcoal, fractional diffusion, fractional distillation, and a catalytic exchange of deuterium between hydrogen gas and water. The last two methods mentioned were employed on a production basis for the atomic energy program during World War II, but are no longer in commercial operation.

Another process, the dual-temperature exchange process, employing hydrogen sulfide and water (the GS process) to separate heavy water from normal water, was used in two plants with a combined capacity of nine hundred tons per year of heavy water that were placed in operation in 1952 by the duPont Company for the U.S. Atomic Energy Commission. Other plants using this process have recently been built in Canada. A 28 ton per year plant was placed in operation in 1968 in France, using a catalytic exchange process between hydrogen gas and liquid ammonia, in which deuterium concentrates in liquid ammonia in the presence of potassium amide catalyst.

Tritium. Another isotope of hydrogen is tritium, which has an atomic mass of 3. It is radioactive, with a half life of 12.262 ± 0.004 yr, and decays into helium-3 with the emission of beta radiation. Traces of tritium are present in the atmosphere as a result of cosmic radiation. Tritium is produced by bombarding lithium with neutrons in a nuclear reactor to produce helium-4 and tritium. It is being used as a radioactive tracer in chemical research, and in other applications, such as luminous paints where low energy radioactivity is needed.

Reactions. Hydrogen will dissociate into atoms at high temperatures. The degree of dissociation at 1 atm has been given as 2.56×10^{-34} at 300°K; 1.22×10^{-3} at 2000°K; 0.9469 at 5000°K; and 0.996 at 10,000°K. The heat of dissociation is about 103 kg cal per g- mole. The reaction is endothermic.

Hydrogen will combine with oxygen to form water under the proper conditions according to the following equation:

$$2H_2 + O_2 \rightarrow 2H_2O$$

The reaction proceeds slowly at temperatures below 550°C but at high temperatures the reaction takes place with explosive violence. The reaction in burning hydrogen with oxygen is exothermic. At 25°C and 1 atmosphere the heat is given as 68.313 kg cal per g- mole. The temperature of such a flame in enclosed space may reach as high as 2800°C which is sufficient to melt platinum. Oxyhydrogen torches have been used in welding but have been largely replaced by oxyacetylene.

The reactions of hydrogen with carbon monoxide are of commercial interest and a variety of products can be made thereby. Unsaturated hydrocarbons will react with hydrogen to produce a more saturated product.

The reactions of hydrogen with metals vary considerably with the ones employed. The alkali metals, the alkaline earth metals (excluding Be and Mg) and many of the rare earth metals react directly with hydrogen to form hydrides. For the most part these are decomposed by water. With other metals the bond is more that of a hydrogen alloy rather than a chemical compound. Most of the metals take up hydrogen, particularly the rare earths. Copper hydride is a border line case between an alloy and a true hydride.

Hydrogen will react with all the halogens to form the corresponding halogen acids. The reaction is most violent and exothermic with fluorine, least active and endothermic with iodine. The reaction:

$$H_2 + I_2 = 2HI$$

which had been considered as a classical example of bimolecular reaction, has recently been found, as a result of photochemical and thermochemical data, to actually be the termolecular reaction:

$$H_2 + 2I = 2HI$$

R. M. Reed
N. C. Updegraff

Hydrogen Chemistry

An atom of hydrogen consists of one proton as a nucleus, surrounded by one electron in an s orbital which constitutes the first principal quantum level. The capacity of this orbital being two electrons, just one electron is insufficient to shield the nuclear charge from an outside electron. Consequently a hydrogen atom can form one covalent bond, using its orbital vacancy to accommodate an electron from another atom, and sharing its original electron similarly with the other atom. Having formed such a bond, hydrogen then is incapable of further bonding, having neither a lone electron pair nor a vacant orbital that might be used. In this respect, of all the other elements, only carbon resembles hydrogen. This is one reason for the uniqueness of carbon-hydrogen chemistry.

Hydrogen is usually placed in the Periodic Table both in Group IA with the alkali metals, in common with which it has one outermost electron and the capacity to exhibit an oxidation state of +1, and in Group VIIA with the halogens, in common with which it has one outermost vacancy and therefore the capacity to exhibit an oxidation state of −1. It belongs in neither group. It lacks both the extra vacant orbitals of the alkali metals and the extra electron pairs of the halogens. Hydrogen, having an outermost shell that is half full, resembles other elements with half-filled outermost shells by being intermediate in electronegativity. In the Periodic Table it would more logically be placed near Group IVA, therefore, than anywhere else, although it does not really belong in any of the established groups.

Hydrogen forms conventional, stoichiometric binary compounds only with major group elements. These compounds have the expected formulas: IA: EH; IIA and IIB: EH_2; IIIA: EH_3; IVA: EH_4; VA: EH_3; VIA, H_2E; VIIA: HE.

The stability of hydrogen compounds tends to diminish down each major group, consistent with the diminishing homonuclear bond energies in the group. Hydrogen compounds of mercury, thallium, lead, and bismuth are so unstable that knowledge of them is scant. Hydrogen telluride is least stable in its group (probably excepting hydrogen polonide), and hydrogen iodide is least stable in its group (probably excepting hydrogen astatide). The transition metals absorb hydrogen to a widely variable degree, forming solid interstitial compounds of stoichiometry depending on conditions, and having properties sometimes resembling, but generally unlike, those of the major group hydrogen compounds. Many complex molecules or ions also contain transition metal-to-hydrogen bonds, but the character of such combined hydrogen may differ unpredictably from expectations based on major group hydrogen chemistry. Indeed, the differences in their hydrogen chemistry are an important contributing argument for the separation of transition metals from the major group elements in systematic study. Compounds of hydrogen with major group elements are at present far more important.

The order of increasing electronegativity is: Cs, Rb, K, Na, Li, Ba, Sr, Ca, Mg, Al, Si-In, B-Hg, Zn, Tl, Pb, Sn, Bi, Ga, Sb, P; then H; then Ge, Te, C, I, As, S, Se, N, Br, Cl, O, and F. Because of its intermediate position, hydrogen can function both as an oxidizing and reducing agent, oxidizing the less electronegative metals and reducing the more electronegative nonmetals. The resulting partial negative or positive charges on hydrogen in its compounds have an unusually large influence on their properties because with only one electron in the neutral atom, relatively small electron gain or withdrawal makes a proportionately large change in the condition of the combined hydrogen. In the extreme condition (probably not actually attained stably), hydrogen may become hydride ion, H^-, or hydrogen ion, the bare proton H^+. Sometimes all binary hydrogen compounds are termed "hydrides" but properly the term should be reserved for those compounds in which hydrogen bears partial negative charge.

Binary hydrogen compounds of many of the elements, whether metallic or nonmetallic, can be made by direct synthesis from the elements. For example, sodium hydride, NaH, results from heating sodium metal in hydrogen gas, and hydrogen chloride, HCl, can be produced by burning hydrogen gas in an atmosphere of chlorine. One general method, suitable for the preparation of hydrogen compounds easily susceptible to oxidation and hydrolysis, is by metathetical reaction between an active hydride such as LiH, NaH, or $LiAlH_4$, and a halide or similar derivative of another element, such as BF_3 or $ZnCl_2$, relying on the higher bonding energy in the product halide of active metal to force the interchange, leaving the less active element to become attached to the hydrogen. When the hydrogen compound to be formed is not susceptible to hydrolysis, it can be made by hydrolysis of a binary compound, in which hydroxyl forms stronger bonds with the less electronegative component, leaving the more electronegative component to become attached to the hydrogen. For example, ammonia would be produced by hydrolysis of lithium nitride, Li_3N, the other product being lithium hydroxide.

Both physical and chemical properties of binary hydrogen compounds are closely related to the condition of the combined hydrogen, which can be evaluated usefully in terms of partial charge. Partial charge is defined as the ratio of the change in electronegativity undergone by the hydrogen in forming the compound to the change that would correspond to acquisition of unit charge. Partial charge values for hydrogen in a number of binary hydrogen compounds are listed in Table 1.

Hydrogen acting as an oxidizing agent acquires partial negative charge. In proportion to the amount of this charge, hydrogen compounds tend to become associated, with consequent reduction of volatility and increase in melting point. In combination with the highly reactive alkali and alkaline earth metals of very low electronegativity, hydrogen becomes sufficiently negative that it is convenient to treat these hydrides as essentially ionic. They condense in "ionic" lattices, becoming quite nonvolatile and melting only at high temperatures (unless they decompose first). When the charge on hydrogen is less negative, polymerization may occur through the formation of "hydridic bridges," in which each pair of metal atoms is joined by one or more hydrogen atoms. Each bridge seems to consist of one otherwise vacant orbital in the outer shell of each metal atom and the s orbital of the hydrogen, with one pair of electrons thus occupying a three-center orbital, holding the three atoms together. Only hydrogen appears capable of this kind of bridging, and only when it bears partial negative charge. Presumably as the result of such bridging, BeH_2, MgH_2, and AlH_3, for example, are polymeric solids, and BH_3 does not exist as such but dimerizes to diborane, B_2H_6, held together by a double hydridic bridge. Boron forms a unique group of hydrides in which hydridic bridges are a characteristic structural feature. Despite its negative hydrogen, SiH_4 exhibits no hydridic bridging, presumably because vacant orbitals of sufficiently low energy are absent in silicon. Consequently, SiH_4 is a very low melting substance, gaseous under ordinary conditions, and showing only very weak van der Waals intermolecular attractions.

When the hydrogen is neutral or nearly so in its chemical environment, the only intermolecular association is of a very weak, van der Waals type. The compounds so designated in Table 1 are all extremely volatile molecular substances of very low melting temperatures.

The electronegativity of hydrogen is somewhat above the median value, so that its bonds with more electronegative elements cannot be as polar as its bonds with less electronegative elements. For example, NaH is about twice as polar as HF. Intermolecular attractions are correspondingly weak and all binary compounds containing partially positive hydrogen, unless of unusually high molecular weight, are very volatile. However, attractions appreciably stronger than van der Waals forces or ordinary dipole-dipole interactions are possible. Hydrogen is unique among the elements in using all its electronic cloud for bonding. All other kinds of atoms retain electrons beneath the valence shell that serve to shield the nuclear charge from close approach of outside electrons, but when hydrogen is attached to an atom that withdraws part of the hydrogen cloud, the proton nucleus becomes exposed. In this condition it can exert a substantial attraction for a pair of electrons on another atom, provided that pair of electrons is available to it.

Two factors appear to govern the availability of a pair of electrons to such a positively charged hydrogen. One is the partial charge on the atom providing these electrons, for electrons become more readily available the better the supply, or in other words, the higher the negative charge on that atom. The other factor is the degree of concentration of the pair of electrons. A proton is essentially a point charge, and unless the electron pair is highly localized and concentrated toward it, the interaction must be weak. This requires that the electron-providing atom must be as small as possible, for on the surface of larger atoms the electron pairs can spread out too far. When the pair of electrons is part of a small atom having fairly high negative charge, then the attraction for the exposed proton of positive hydrogen becomes large enough to cause appreciable association among molecules. The hydrogen covalently bonded to one molecule serves as a bridge to another molecule. This kind of attraction is usually called a "hydrogen bond." The term "protonic bridge" seems better, for it reveals its dependence on partially positive hydrogen and distinguishes it from a hydridic bridge. The elements that most commonly fill the requirements for protonic bridging, and the only ones whose binary hydrogen compounds exhibit such association, are nitrogen, oxygen, and fluorine. Each of these elements is sufficiently electronegative to acquire substantial negative charge at the expense of attached hydrogen, and each has atoms small enough to require reasonably high localization of the electron pairs. Consequently NH_3, H_2O, and HF are much higher melting and less volatile than any of their congeners in the Periodic Table because their molecules, but those of none of their congeners, are highly associated through protonic bridging. Protonic bridging is an extremely important phenomenon in many areas of hydrogen chemistry, and has

TABLE 1. ESTIMATED PARTIAL CHARGE ON COMBINED HYDROGEN (in electrons)

Negative		Nearly neutral		Positive	
CsH	−0.59	SnH_4	−0.02	HI	0.04
RbH	−0.58	GeH_4	−0.01	H_2S	0.05
KH	−0.58	PH_3	−0.01	H_2Se	0.06
NaH	−0.50	SbH_3	−0.01	NH_3	0.06
LiH	−0.50	CH_4	0.01	HBr	0.12
BaH_2	−0.33	H_2Te	0.01	H_2O	0.12
SrH_2	−0.31	AsH_3	0.02	HCl	0.16
CaH_2	−0.29			HF	0.25
MgH_2	−0.18				
BeH_2	−0.13				
AlH_3	−0.12				
CdH_2	−0.09				
ZnH_2	−0.06				
B_2H_6	−0.05				
InH_3	−0.05				
SiH_4	−0.05				

special significance in the structure of complex biochemical species such as proteins.

The chemistry of *negative* hydrogen is more fully discussed under the subject Hydrides.

Binary compounds of hydrogen that is essentially *neutral* are relatively unreactive at ordinary temperatures, with respect to the hydrogen. They are neither acidic nor basic, nor do they exhibit oxidizing or reducing properties. At higher temperatures they can react in much the same way as molecular hydrogen, presumably by being sources of neutral hydrogen atoms.

Hydrogen compounds in which the hydrogen bears partial *positive* charge tend to participate readily in proton transfer reactions and thus are acidic. Since the other element is partially negative and possesses, commonly, lone electron pairs, the compounds can also be basic. In ammonia, for example, the basicity resulting from the lone pair on nitrogen tends to conceal the acidity of the hydrogen, but the existence of the latter is readily shown by the reaction with sodium to liberate hydrogen and form sodamide, $NaNH_2$. (See **Ammonia**.) When hydrogen compounds such as ammonia or water acquire protons, they form positively charged complex ions somewhat analogous to the negative complex ions formed by negative hydrogen and hydrogen compounds having vacant outer orbitals rather than lone electron pairs. Thus the reaction of ammonia with a proton to form the positive ammonium ion makes an interesting comparison with the reaction of aluminum hydride with a hydride ion to form the negative aluminohydride ion.

The strength of the acidity of positive hydrogen compounds is related more closely to the ability of the corresponding anion to donate electrons to a proton than it is to the actual partial positive charge on the hydrogen. For example, HF is a weak acid in water, and HI is a strong acid, despite the higher positive charge on hydrogen in the former. Fluoride ion is a much more effective proton attracter than is an iodide ion. The oxidizing power of positive hydrogen is not as conspicuous as the reducing power of negative hydrogen largely because the hydrogen can only become about half as positive (as with fluorine) as it can become negative (as with sodium). However, the activity of hydronium ion, H_3O^+, in removing electrons from metals more active than hydrogen is ample evidence of oxidizing power of positive hydrogen. Naturally, no reducing activity is shown by positive hydrogen.

In summary, binary hydrogen compounds across the Periodic Table exhibit a wide range in properties both physical and chemical that corresponds closely to the condition of the combined hydrogen with respect to partial charge.

R. T. SANDERSON

Cross-references: *Hydrides, Electronegativity, Periodic Law and Periodic Table.*

HYDROGENATION

Hydrogenation may be defined as the reaction of molecular hydrogen with an organic compound. Two major classes of hydrogenations may be distinguished: *addition* or *saturation,* in which hydrogen adds to a double or triple bond without cleavage of the bond, and *destructive hydrogenation* or *hydrogenolysis,* in which molecular cleavage occurs. Typical addition reactions are the hydrogenation of olefins to paraffins and of aromatics to naphthenes, and the reduction of aldehydes and ketones to alcohols. Hexamethylenediamine, which is copolymerized with adipic acid to make nylon 66, is formed by such an addition of hydrogen to the nitrile groups in adiponitrile. Hydrogenolysis includes such reactions as the hydrogenation of thiophene to form butane and hydrogen sulfide; this is one of the principal reactions occurring during the hydrodesulfurization of petroleum products. The hydrogenative splitting (or "hydrocracking") of large organic molecules, with the formation of smaller fragments which react with hydrogen, is another important example of hydrogenolysis. The reactions of molecular hydrogen with inorganic molecules for example, with nitrogen to form ammonia, and with carbon monoxide in the Fischer-Tropsch hydrocarbon synthesis, are not usually classified as hydrogenations and are not considered here.

Hydrogenations may be carried out at atmospheric pressure and room temperature or, more commonly, at elevated pressures and temperatures. The choice of reaction conditions depends on the nature of the compound to be hydrogenated, the catalyst, and catalyst poisons present in the system. For difficult hydrogenations, such as the conversion of coal to a distillable oil, pressures as high as 700 atm have been used, with temperatures up to 500°C. Serious equipment problems may be posed by such high-pressure, high-temperature reactions, since hydrogen embrittlement and hydrogen sulfide corrosion (in sulfur-containing systems) add special difficulties to normal equipment design. The feed may be in either the liquid or the vapor phase under reaction conditions. Hydrogen, in an amount exceeding that which is consumed in the reaction, is added to the feed prior to entrance into the reaction zone. For liquid-phase operation, intimate mixing or dispersion of hydrogen in the liquid is desirable in order to obtain optimum reaction rates.

The presence of a catalyst is vital in carrying out hydrogenation reactions. The metals of Group VIII, particularly nickel, platinum, and palladium, are very useful catalysts and are among the most active known. They suffer from the disadvantage, however, that they are very susceptible to poisoning, e.g., by sulfur compounds, especially when they are used at low temperatures. Less susceptible to poisoning are the oxides and sulfides of the Group VI metals, chromium, molybdenum, and tungsten, which constitute the second major class of hydrogenation catalysts. These catalysts are less active than the Group VIII metals and, in general, require the use of elevated hydrogen pressures. Their insensitivity to poisoning makes them particularly useful in such reactions as hydrodesulfurization. A number of other materials have been used as catalysts: tin compounds, for example, show an almost unique activity in coal hydrogenation. Hydrogenation catalysts are normally used either in granular or pellet form in fixed beds, or in finely dispersed form in a liquid feed. The fixed-bed arrangement is usually more efficient than the dispersion because it provides a higher catalyst-to-oil ratio in the reaction zone.

The addition of hydrogen to olefinic double

bonds is the basis of several reactions of commercial importance. One is the hardening of vegetable oils to make margarine and shortening, in which the principal reaction is the controlled addition of hydrogen to carbon-carbon double bonds in unsaturated glycerides. A dispersed nickel catalyst is used at pressure below 100 psig and temperatures of 120 to 200°C. After the hydrogenation has been carried to the desired state of completion, the nickel catalyst is removed from the hydrogenated product by filtration.

Nickel catalysts have also been used, in fixed beds, in the hydrogenation of isooctenes to produce aviation gasoline; this was an important process during World War II. Isooctenes result from the catalytic dimerization of isobutylene or from the reaction of butylene with isobutylene. The saturation of the olefins with hydrogen resulted in increased stability of the gasoline and in increased engine performance, after addition of tetraethyl lead. Because of the activity of the nickel catalyst, temperatures below 200°C and hydrogen pressures of only a few atm are again sufficient. However, it is necessary that both the isooctene feed and the hydrogen supply be carefully desulfurized to prevent catalyst poisoning. Carbon monoxide in the hydrogen supply also acts as a catalyst poison.

The production of liquid fuels by the hydrogenolysis of coal was practiced on a large scale in Germany during World War II. In this process the coal is dried, ground to a fine powder and made into a paste with heavy recycle oil. Small amounts (0.1 to 5%) of dispersed catalyst are added to the paste, and the mixture, along with the necessary hydrogen, is injected into the high-pressure system. Reaction pressures of 300 to 700 atm have been used, and temperatures of 460 to 490°C. The catalysts used in this step may be compounds of iron, tin, molybdenum, or nickel. The coal is converted, in this liquid-phase step, into (a) heavy oil, which, after centrifugal separation from catalyst, ash, and unreacted coal, is recycled, (b) distillable oil, of boiling range 200 to 325°C, and (c) some gasoline and hydrocarbon gas. The distillable oil is then converted to gasoline by a separate hydrogenolysis, this time in vapor-phase operation over a fixed-bed catalyst. In the over-all conversion of coal to gasoline, large quantities of hydrogen are consumed; gasoline produced from bituminous coal involves the consumption of over 20 wt % hydrogen, based on the weight of gasoline produced. To a large extent, this high utilization of hydrogen arises because coal is so hydrogen-poor relative to gasoline. The atom ratio of hydrogen to carbon in coal is in the neighborhood of 0.5 to 0.6; in a typical gasoline the ratio is 1.5 to 1.7. The hydrogen deficiency in coal must be supplied by the addition of molecular hydrogen. Light hydrocarbon gases, produced as unavoidable by-products during the hydrogenolysis, also account for a significant part of the hydrogen consumption.

Other by-products from the hydrogenolysis of coal are of considerable interest as chemicals. Phenol and its higher homologs are the most important of these, but a variety of nitrogen bases and of aromatic compounds, including such complex materials as carbazole, pyrene and coronene, are also produced in significant amounts.

The developing shortage of natural gas is also leading to increased interest in the direct hydrogenation of coal under pressure to form methane for pipeline gas. This direct hydrogasification process avoids the complicated series of steps of coal gasification in steam-oxygen, water-gas shift reaction, and final methanation in separate reactors.

Upgrading of gasoline, distillate oils, and residual fuel oils from petroleum by mild hydrogen treating is now practiced on a large scale. Relatively low temperatures and pressures are used. The most important result of the mild hydrogen treatment is desulfurization, the organic sulfur compounds being converted to hydrogen sulfide and hydrocarbon. There is also a significant removal of nitrogen and oxygen as ammonia and water, respectively, and olefins are at least partially saturated. Catalysts based on mixed cobalt and molybdenum oxides are most popular, though tungsten, nickel, and molybdenum sulfides are used in some processes.

Catalytic hydrocracking is extensively used to produce high-quality gasoline, jet fuel, and lubricants. Dual-functional catalysts are required. Sulfided nickel and tungsten on a silica-alumina support were initially most widely used, but of growing interest are the "shape-selective" catalysts employing a precious metal, typically palladium, on a synthetic zeolite support.

Also of increasing importance in petroleum processing is hydrodealkylation, in which alkyl-substituted aromatics are converted to unsubstituted aromatics (benzene, naphthalene) or to less-alkylated aromatics (e.g., xylenes to toluene). The reaction is conducted at elevated temperature and hydrogen pressure, either catalytically or thermally. A typical catalytic process to make benzene from toluene utilizes a transition metal oxide catalyst, hydrogen at about 35 atm, and temperatures around 600–650°C.

The availability of large quantities of hydrogen as a by-product of modern methods of gasoline reforming makes it certain that the utilization of hydrogenation processes will continue to increase.

SOL W. WELLER

HYDROLYSIS

Hydrolysis is the chemical decomposition or splitting of a compound by means of water, as in the following general equation:

$$R \cdot X + H \cdot OH = R \cdot H + X \cdot OH$$

Water, in the form of its hydrogen and hydroxyl ions, adds to the cleaved compound. The addition of water is generally catalyzed by ions and without added ions may be a very slow process. Consequently, the addition of acid or base increases the concentration of hydrogen or hydroxyl ions with a corresponding increase in the rate of hydrolysis. Certain enzymes also catalyze the hydrolysis of some organic compounds.

The most typical example of the hydrolysis of an organic compound is that of an ester, *viz*:

$$CH_3COOC_2H_5 + H \cdot OH \rightarrow CH_3COOH + C_2H_5OH$$

In inorganic chemistry the term is chiefly used in connection with the hydrolysis of salts of a weak acid and a strong base to give a basic solution or a strong acid and a weak base to give an acidic solution, e.g.,

$$Na_2CO_3 + H_2O \rightarrow 2Na^+ + HCO_3^- + OH^-$$

$$AlCl_3 + 2H_2O \rightarrow Al(OH)^{++} + 3Cl^- + H_3O^+$$

There are a great many important industrial processes involving the hydrolysis of organic compounds, one example being the base catalyzed hydrolysis of fats to produce glycerol and soap, according to the equation:

$$CH_2OCOC_{17}H_{35} + 3H_2O \xrightarrow{\text{(NaOH)}}$$
$$|$$
$$CHOCOC_{17}H_{35}$$
$$|$$
$$CH_2OCOC_{17}H_{35}$$

glycerol tristearate
(fat)

$$3C_{17}H_{35}COONa + CH_2OH$$
$$|$$
$$CHOH$$
$$|$$
$$CH_2OH$$

sodium stearate   glycerol
(soap)

The above hydrolysis reaction, as applied to fats, is termed *saponification*. (See **Soaps**)

The compound sugars, that is, the disaccharides, oligosaccharides and polysaccharides, may be hydrolyzed to their simple sugar components. Thus cane sugar (sucrose), when boiled in water containing a dilute mineral acid as a catalyst, yields glucose (dextrose) and fructose (levulose), *viz.*,

$$C_{12}H_{22}O_{11} + H \cdot OH \rightarrow C_6H_{12}O_6 + C_6H_{12}O_6$$

Similarly, maltose is hydrolyzed by acid catalysis or by the enzyme maltase into two molecules of glucose. Starch is hydrolyzed by diastase into maltose and dextrin.

In the field of cellulose and wood chemistry there are a number of examples of hydrolysis being practiced on a large scale industrially.

The development of cellulose acetate rayon would not have been possible without the discovery of the effect of partial hydrolysis on completely acetylated cellulose. This product was found to be insoluble in any available cheap and nontoxic solvents and so was impossible to spin from solution. However, the American chemist, Miles, found in 1903 that if the triester was allowed to partially hydrolyze, the product became soluble in acetone and could be spun economically into fibers. This procedure is still practiced on a very large scale for the manufacture of acetate rayon.

Another example is the acid hydrolysis of wood to produce sugars. The polysaccharide components of wood can be partially or completely hydrolyzed by acid catalyzed reactions by one of several different technical processes. The cellulose is converted to glucose and the hemicelluloses mainly to xyosle. The syrupy mixture (molasses) of wood sugar may be directly used as cattle food, or it may be fermented to alcohol, or the sugar glucose (dextrose) may be separated from the wood sugars and purified. Wood, however, is not yet competitive with other sources of sugars and alcohol in the United States.

A further example of an industrial process involving hydrolysis is in the production of regenerated cellulose which may be reformed into cellophane films or viscose rayon. In this process, cellulose is reacted with sodium hydroxide and carbon disulfide to form soluble cellulose xanthate (Cellulose − CS − S Na). This may be readily hydrolyzed back to cellulose and carbon disulfide by treatment with acids or acid salts. Industrially the cellulose xanthate (viscose) solution is extruded through fine pores or a thin slit into an acid bath where the cellulose is regenerated in the form of fibers or film. By using a tubular die, the cellulose may be regenerated in the form of sausage casing.

An interesting example of hydrolysis in the plastics industry is in the production of polyvinyl alcohol. Vinyl alcohol monomer is too unstable to exist and, therefore, the polymer is prepared industrially by the alkaline hydrolysis of polyvinyl acetate. In this way, mixed alcohol acetate copolymers can also be produced.

V. STANNETT AND EDWIN C. JAHN

Hydrolysis Polymerization

Hydrolysis polymerization is a novel technique for preparation of poly(ester-acetals) in which the ketal of a glyceryl monoester-acetal is converted to a polymer at an aqueous-organic interface. Specifically, isopropylideneglyceryl azelaaldehydate dimethyl acetal (I) hydrolyzes to give a poly(ester-acetal) of structure II.

This technique was discovered in an attempted preparation of α-glycerol azelaaldehydate by hydrolysis of I. Analogous isopropylideneglycerol esters of simple fatty acids readily hydrolyze under mild conditions to the corresponding α-monoglycerides. Attempted hydrolysis of I, however, resulted in recovery of starting material or, when concentrated acid was used, in formation of cloudy, viscous oils, many of which turned to gelatinous solids on standing. None of the products contained more than small amounts of α-monoglyceride. Even in reactions carried out at −40°C the free glycerol azelaaldehydate could not be isolated.

Since acetals in acidic media exist in labile equilibrium with their component aldehydes and alcohols, the conditions for acetal hydrolysis are essentially those for acetal formation. When the ester-acetal is hydrolyzed, the free glycerol ester-aldehyde apparently reacts rapidly with other liberated molecules of glycerol ester-aldehyde, and the product isolated is poly(ester-acetal) II.

Interfacial reactions are frequently encountered in biological chemistry. Heterogeneous ester hydrolysis at a hydrocarbon-water interface has been studied as a model reaction. The hydrolysis-polymerization technique might also be considered similar to interfacial polymerization because the initial hydrolysis and possibly the subsequent polymerization take place at an interface; however, the hydrolysis-polymerization is a homopolymerization with the monomer precursor entirely in the organic phase. The aqueous phase contains only the catalyst for hydrolysis and polymerization. The polymer remains in the organic phase and is recovered from it.

The products of hydrolysis-polymerization are low-molecular weight oligomers with maximum molecular weights of about 1500, representing about 7 repeating units. Hydrolysis of the glyceryl ester linkage terminates the polymerizing chain. The resultant free carboxyl group of the isolated oligomer will form salts and esters. The sodium salt of the oligomer is a water-soluble solid having surfactant properties. The methyl ester, formed by reaction of diazomethane with the oligomer, undergoes a typical polycondensation reaction.

Analogous polymers of molecular weight up to 14000 can be produced by polytranesterification of the glycerol acetal of methyl azelaaldehydate. This compound exists in four isomeric forms and the structure determines the physical properties of the polymers. Similarly, when hydrolysis polymers are prepared under different conditions, their properties vary, possibly because of structural changes.

W. R. MILLER

References

Miller, W. R., Awl, R. A., Pryde, E. H., and Cowan, J. C., *J. Polym. Sci.*, Part A-1, **8**, 415 (1970).

Lenz, Robert W., Miller, W. R., Awl, R. A., and Pryde, E. H., *ibid.*, **8**, 429 (1970).

Wittbecker, E. L., and Morgan, P. W., *ibid.*, **40**, 289 (1960).

Menger, F. M., *J. Am. Chem. Soc.*, **92**, 5965 (1970).

HYDROQUINONES

The hydroquinones are dihydroxy aromatic compounds with the two groups in positions corresponding to *ortho* or *para* substitution in the benzene ring. They are closely related to the quinones from which they can be obtained by reduction. Thus *o*-dihydroxybenzene (catechol) can be obtained from *o*-benzoquinone, and hydroquinone (*p*-dihydroxybenzene or quinol) from *p*-benzoquinone. Resorcinol (*m*-dihydroxybenzene) is not properly a hydroquinone since the corresponding *meta* quinone is not known to exist. Homologs of hydroquinone are usually named after the parent hydrocarbon. Thus toluhydroquinone is 2,5-dihydroxy-1- methylbenzene and naphthohydroquinone is 1,4-dihydroxynaphthylene. Unlike many of the quinones, the ring systems are fully aromatic and undergo substitution reactions common to phenols and other benzene derivatives. However they are easily oxidized by some reagents to the less stable quinones and degradation products frequently result. Thus treatment of hydroquinone with nitric acid yields oxalic acid while halogenation with sulfuryl chloride results in a mixture of chlorohydroquinones, quinone, quinone chlorides, and tetrachloro-*p*-benzoquinone. The formation of side and degradation products can be minimized if the molecule is protected against oxidation by acetylating or benzoylating at least one of the hydroxyl groups. For example 2-nitro-hydroquinone can be prepared in good yield by the nitration of monobenzoyl hydroquinone folowed by hydrolysis. Concentrated sulfuric acid gives hydroquinone-2.5-disulfonic acid directly, and tertiary amyl groups can be introduced into the ring in the 2 and 5 positions by treatment with amylene in the presence of sulfuric acid. The hydroxyl groups are weakly acidic and can readily be converted to ethers by treatment with alkyl halides or sulfates in the presence of alkali. A diacetate is formed on treatment with acetic anhydride. The most characteristic reaction of hydroquinones is their reversible oxidation to quinones.

Catechol, or 1,2-dihydroxybenzene, was first prepared by the dry distillation of catechin obtained from *Mimosa catechu*. It can also be formed by the hydrolysis of its methyl ether, guaiacol, which is a constituent of beechwood tar. It is prepared synthetically by fusing phenol-*o*-sulfonic acid with sodium hydroxide, or treating *o*-chlorophenol with aqueous alkali in the presence of copper at a high temperature and pressure. It crystallizes from benzene in colorless monoclinic plates which melt at 105°C. The lead salt can be oxidized to *o*-benzoquinone by a solution of iodine in chloroform. The ethers of catechol are of considerable importance and can be derived from a number of naturally occurring substances. The methylene ether of protocatechualdehyde is known as piperonal, and is closely related to various natural products including piperine, safrole, and isosafrole, from which it can be derived. These compounds have been used for the synthesis of pyrethrin synergists. *Vanillin*, the principal flavoring constituent of vanilla, is the 3-methyl ether of protocatechualdehyde.

Hydroquinone is found in nature combined in the glycoside arbutin, from which it can be released by hydrolysis with emulsin or dilute sulfuric acid. It is prepared commercially from *p*-benzoquinone by reduction with sulfur dioxide. It is a dimorphic solid with the stable form melting at 170.5°C. Hydroquinone is one of a number of compounds that possess the property of forming molecular compounds with gases such as hydrogen sulfide, sulfur dioxide, krypton, xenon, etc. These are known as clathrate compounds, and their existence is due to the entrapment of atoms or molecules of the gas in the crystal lattice of the hydroquinone. Three moles of hydroquinone can entrap one mole of gas, which is firmly held but which is liberated when the clathrate is dissolved in water. The most important commercial use of hydroquinone is for the development of photographic film. Its effectiveness is dependent on its ability to reduce the silver subhalide formed on exposure of the film to light to metallic silver. It gives films of high density and it is often necessary to reduce the harshness of contrast by using it in combination with other developers such as metol or paramidophenol. Hydroquinone and its derivatives are effective antioxidants for the preservation of fats, oils, and rubber. It has also been used as a short-stopping agent for controlling polymerization in the production of synthetic rubber of the butadiene-styrene type.

H. P. BURCHFIELD

Cross-references: *Quinones, Clathrate Compounds.*

I

IMPURITIES

In the broader chemical sense "impurities" are the extraneous or unwanted materials in a given preparation of a substance. A more precise definition can be given by first defining "purity." Impurities then represent the degree of departure from purity. A pure preparation of a chemical substance in the absolute sense is one in which all the molecules in the preparation are molecules of the substance. No other molecules differing in any way can be present. This ideal state of affairs never can be reached in actual practice and no experimental method for recognizing or proving that it has been reached is available.

From the practical standpoint, a preparation is "pure" when its characteristic physical and chemical properties coincide, within the experimental error of determination, with those recorded in the literature, and when no change in these properties can be found after application of the most selective fractionation techniques. We thus are forced to define purity in operational terms: a substance is pure when impurity can no longer be detected by any experimental procedure. The physical and chemical properties most commonly used in this connection are ultimate analysis, melting point, boiling point, refractive index, solubility, ultraviolet absorption spectra, colorimetry, x-ray analysis, mass spectrophotometry, radioactivity, behavior in an electric field, crystallography, optical rotation, partition ratios and specific gravity. Many less common but equally useful physical properties are important in specialized fields. For many substances quantitative bioassays are now available which can be very useful.

The fractionation techniques which have found widest usage for purity studies include fractional crystallization, fractional precipitation, fractional distillation, fractional adsorption (chromatography, paper chromatography, ion exchange chromatography), fractional extraction (counter-current distribution), sedimentation and electrophoresis (paper electrophoresis, electrofocusing). The whole concept and philosophy of chemical purity depends on the degree to which molecules can be sorted experimentally and preparations thereby achieved which approach the pure state as nearly as possible. It is only when they are used to indicate the absence of impurities by comparison with the purest preparation, that the physical and chemical properties have a real meaning.

A number of refinements of fractionation techniques specifically developed for the recognition of impurity deserve mention. The solubility method is based on the phase rule and has a very sound theoretical basis. However, it sometimes fails for practical reasons, such as the formation of solid solutions or the length of time required to reach equilibrium between a solid and liquid phase. Chromatography as refined by the procedure of Moore and Stein is a very selective and sensitive method of analysis for those impurities which can be detected in extremely dilute solution by some analytical method and for which the behavior on the column is known. When the impurities are of unknown nature the presence of impurities often may be missed entirely. In the search for impurities counter-current distribution has a number of very desirable characteristics. It is a very selective fractionation tool with wide applicability. The result can be evaluated by simple residue weight determination as well as by more specific methods. Impurities with unknown properties are less likely to be overlooked than with most procedures. When the search for impurities is concerned with volatile substances, the refinements of fractional distillation have been widely discussed. A recent and even more highly selective separating tool for this class of substance is that known as gas-liquid chromatography. A refinement of the determination of melting point is presented by cooling curves. This is a particularly good method for studying crystalline substances which are completely stable at their melting point.

Since impurities can never be entirely eliminated, a preparation must be accepted in spite of some degree of impurity. For this reason it is most important to know that the total impurity cannot be greater than a certain minimal percentage, and often even more important to know the nature of the impurity. Quantitative analysis for the solute of interest can often restrict the amount of impurity to a small percentage but fails when the total impurity becomes equal to or less than the experimental error of the analytical method used. One of the quantitative analytical methods mentioned above, such as spectroscopy or counter-current distribution, which can be made to magnify the impurity then becomes the tool of choice.

In present day technology certain impurities must be eliminated to an astonishing degree. Examples of this may be found in the production of transistors, certain intermediates for synthetic rubbers, pyrogenfree distilled water for use in parental preparations, conductivity water, many catalysts, certain enzyme preparations, antibiotics, graphite for reactor use, many drugs, and in countless other fields. Many metals have been found to have strikingly different properties after practically all the impurities have been removed.

LYMAN C. CRAIG

Cross-references: *Chromatography, Fractionation, Analytical Chemistry, Semiconductors.*

INDICATORS

Indicators are colored "signposts" conveying to an observer the idea that (1) a solution is acidic, basic, or neutral; (2) a precipitation is or is not complete, or (3) an oxidation-reduction reaction is or is not complete.

An acid-base indicator is generally a large organic molecule which is itself a weak acid or base changing color rapidly near the endpoint when the acidity (or basicity) of a solution containing the indicator increases (or decreases) only slightly. The choice of an indicator for a particular titration will depend on the character of the neutralization curve. The ranges of some common indicators are given below.

pH OF SOME INDICATORS

Indicator	pH Range*	Color in	
		Acid	Base
Methyl orange	3.1–4.4	red	yellow
Phenolphthalein	8.3–10.0	colorless	red
Phenol red	6.8–8.4	yellow	red
Litmus	4.4–8.3	red	blue
Congo red	3.0–5.2	blue	red
Bromthymol blue	6.0–7.6	yellow	blue

*By range of an indicator is meant the following: methyl orange is distinctly red at pH 3.1 and distinctly yellow at pH 4.4. Within the "range" pH 3.1–4.4, the color is changing.

A narrow range is a desirable characteristic for an indicator. Although litmus is a common indictator, it has a wide range. A sharp contrast in the two colors also helps. For example, the human eye can detect the transition from colorless to red easier than from yellow to red.

In practice the pH expected at the endpoint of a given titration may be calculated and an indicator chosen which changes near the calculated value. If more than one indicator is available, the best one may then be selected by experiment with standard solutions. A second method of choosing an indicator is to prepare the mixture from the pure substances which will be present at the equivalence point of the reaction. Then the pH of the solution may be measured and an indicator chosen as before or the behavior of the prepared solution may be tested with selected indicators experimentally by titrating back and forth through the endpoint.

One large class of indicators are azo dyes of which Methyl Orange and Congo Red are examples. These dyes change color because the color-forming group in base (azo, —N=N—) is changed to another chromophore (p-quinoid) when protons are added to the indicator.

A second class of indicators called phthaleins (derived from phthalic anhydride) also are colored because of a p-quinoid chromophoric structure. Phenolphthalein has no color-forming group in acidic solution (colorless) but takes on a p-quinoid structure is base (red). In very strong base, phenolphthalein undergoes a further structural change and again is colorless. It must therefore be used with discretion.

The coloring matter in many flowers exhibits the properties of an indicator. For example, the coloring matter in the corn flower (blue), the red rose, and the red dahlia is the same dye at different acidities in the three flowers.

The Volhard volumetric method for the determination of silver depends on the use of an indicator for the endpoint of the precipitation. If ferric nitrate is added to a solution of silver nitrate, the latter may be titrated with a standard thiocyanate solution. After the white silver thiocyanate precipitation is complete, a reddish brown color due to the formation of $Fe(SCN)^{++}$ appears. This is the endpoint.

The formation of a second precipitate which is colored may also be used as a precipitation indicator. In the Mohr method for chloride ion, chromate ion added to the solution allows the precipitation of red silver chromate after the precipitation of silver chloride is essentially complete. To be successful the solubility product of the compound used for the determination should be several orders of magnitude smaller than that of the precipitation indicator.

LEALLYN B. CLAPP

INDIUM

Indium was discovered in 1863 by F. Reich and T. Richter at the Freiburg School of Mines. In the course of examining local zinc ores spectroscopically for thallium, they observed a prominent indigo blue line which had never been reported before. Subsequently, they succeeded in isolating a new metallic element which they named indium from the characteristic indigo blue lines of its spectrum.

Indium does not occur in the native state. The element is widely distributed in nature. It is found in concentrations ranging up to 0.1% in ores of iron, lead, copper, and particularly zinc. During the recovery of some of these metals, indium becomes concentrated in various by-products. Commercially, it is most frequently recovered from lead-zinc residues, flue dusts, and slags.

Many processes have been developed for recovering indium from these by-products. The indium is usually extracted using acid, and separated either as crude indium sponge by cementation on zinc or aluminum, or as the insoluble hydroxide, carbonate, sulfite or phosphate. The sponge or precipitate is then redissolved in acid, the solution purified, and indium recovered electrolytically or by cementation. Although metallic indium can be obtained from the oxide by heating with hydrogen or carbon, these methods are not in commercial use.

Indium, atomic number 49, atomic weight 114.82, belongs to subgroup IIIA in the Periodic Table, together with boron, aluminum, gallium, and thallium. The common and most stable valence of indium is 3, although compounds of valence 1 and 2 are also formed.

Indium metal is unaffected by air at ordinary temperatures, but at red heat it burns with a blue flame forming the trioxide (In_2O_3). Indium forms compounds with other metals such as selenium, tellurium, arsenic and antimony, and with nonmetallic elements such as nitrogen, hydrogen, sulfur, and the halogens as well as oxygen. It amalgamates with mercury, and forms alloys with many other metals. Indium dissolves slowly in cold, dilute mineral acids and more readily in hot dilute or concentrated acids. The metal is not attacked by alkalies or by boiling water except in the finely divided state (sponge or powder), when the hydroxide is formed on contact with water.

Indium is a silvery-white metal with a brilliant luster. It is softer than lead, marks paper in the same way and can be scratched with the fingernail. It is malleable, ductile and crystalline. When a rod of the pure metal is bent, it gives a high-pitched "cry" similar to tin. Molten indium will wet smooth, clean glass.

Indium melts at 156.6°C and boils at 2000°C. It is less volatile than zinc (indium vapor pressure at 907°C, the boiling point of zinc, is 5×10^{-3} mm Hg). Some other properties are: density at 20°C 7.31 g/cc; specific heat at 20°C 0.058 cal/g°C; heat of fusion 6.8 cal/g; coefficient of linear thermal expansion (0–100°C) 24.8×10^{-6} cm/cm°C; electrical resistivity at 20°C 8.8×10^{-6} Ω cm; standard electrode potential −0.34V. Indium becomes superconducting at 3.38°K.

Structurally indium is a weak metal. In pure form its tensile strength is 380 psi and its compressive strength is 310 psi. Indium does not work-harden. Consequently its elongation is abnormally low (22% in 1 inch). It is a highly plastic metal and can be deformed almost indefinitely under compression. Freshly exposed indium surfaces cold-weld with great ease.

A large use of indium has been in high-performance sleeve bearings. After application by electroplating, the indium is diffused into the bearing surfaces to provide a low coefficient of friction, good resistance to corrosion and excellent anti-seizure properties. Indium is also used in low-melting alloys for surgical casts, foundry patterns, fusible safety plugs and links, in glass sealing and soldering alloys, as an additive to lubricating oils, and as a sensitive neutron counter.

At the present time the electronics industry is the predominant consumer of indium. The main use is in germanium device manufacturing; smaller quantities are used in making III-V compound semiconductors such as indium antimonide. For device manufacturing high-purity indium and indium alloys are supplied in form of tiny precision spheres, disks and washers. In many devices the indium serves both as the dopant which imparts the desired electronic properties, and as a low-temperature solder to make the electrical contacts. The properties of indium antimonide make it useful in detectors for infrared radiation. Development work is continuing to expand the use of indium in all areas of solid state electronics.

The experience of 20 years industrial handling of indium and its compounds, including long exposure periods for many employees, indicates that there is no significant toxicity or health hazard associated with the industrial use of this element.

H. E. HIRSCH AND A. G. WHITE

References

1. Liang, S. C., King, R. A., and White, C. E. T., "Indium" in "The Encyclopedia of Chemical Elements," Hampel, C. A., Editor, Van Nostrand Reinhold Co., New York, 1968.
2. Ludwick, M. T., "Indium," The Indium Corporation of America, New York, 1959.
3. Mills, J. R., King, R. A., White, C. E. T., "Indium" in "Rare Metals Handbook," Hampel, C.A., Editor, Reinhold Publishing Corporation, New York, 1961.

INDOLES

Indoles are weakly basic heterocyclic nitrogenous compounds in which a benzene ring and a pyrrole nucleus are fused in the 2,3 positions of the pyrrole ring. Indole is present in jasmine flowers, orange blossom oil, some citrus plants and coal tar. Its structure was elucidated by Adolf von Bayer while carrying out investigations on indigo:

Indole possesses aromatic properties and undergoes electrophilic substitution, chiefly at position 3. A colorless crystalline solid which melts at 52°C and boils at 254°C, indole is soluble in hot water, hot alcohol and certain organic solvents. It is stable under neutral and basic conditions, but polymerizes considerably in acids, forming polymers of unknown structure. Chlorine and bromine react violently with indole to form tars. Iodine reacts less vigorously, and it yields 3-iodoindole. Treated with sulfuryl chloride, indole yields 3-chloroindole and 2,3-dichloroindole. The pyrrole ring of indole is oxidized by permanganate with the formation of tars. The oxidation of indole with organic peroxide produces ring opening to yield o-formaminobenzaldehyde. Indoles are reduced to indolines by chemical, catalytic and electrolytic procedures. Acid anhydrides react with indole to form 1-acylindoles or 1,3-diacylindoles. Indole and its derivatives without substitution in the 1-position produce indolylmagnesium halides with Grignard reagents.

Indole is a chemical reagent in various syntheses, and is produced for use in the manufacture of tryptophan and indole-3-acetic acid. Perhaps the most broadly applicable synthesis of indole derivatives is the Fischer indole synthesis in which arylhydrazones of aldehydes, ketones, aldehydic acids, keto acids, or esters of such acids are converted into indole derivatives by heating with an acid catalyst. Indole itself cannot be prepared by this procedure, but can be made by the Reissert synthesis which involves condensing o-nitrotoluene with diethyloxalate in the presence of sodium ethoxide. The product is the ethyl ester of o-nitrophenylpyruvic acid which is first hydrolyzed to the corresponding free acid and then reduced to o-aminophenylpyruvic acid. Under the conditions of the reduction, cyclization occurs with the loss of water to form indole-2-carboxylic acid. Heating this acid yields indole with the loss of carbon dioxide. Substituted indoles are prepared by applying the Reissert synthesis to substituted o-nitrotoluenes. Another general synthesis of indoles is that of Madelung and Verley in which an intramolecular aldol condensation of an *N*-acyl derivative of *o*-toluidine is carried out under the influence of catalysts such as potassium alkoxides or sodium amide.

The indole nucleus is found in a large number of naturally occurring compounds including indigo, tryptophan, serotonin, lysergic acid and alkaloids from more than twenty genera of plants. Some of these natural products are biologically active. Tryptophan, an essential amino acid, is β-(3 indolyl) alanine and is present as the levo isomer in many proteins. Tryptophan can be synthesized nonphysiologically from indole and serine by an A-protein mutant of *E. coli*. In the large intestine, tryptophan and tryptophan-containing proteins are attacked by putrefying bacteria to form indole:

tryptophan → indole-3-propionic acid →
       indole-3-acetic acid → skatole → indole.

Indole is not formed by tryptophan which is absorbed into the blood. When indole is absorbed, it is detoxified, probably in the liver, by conversion to indoxyl, 3-hydroxyindole, which is excreted in urine as a sulfate.

Skatole, 3-methylindole, occurs in feces, in a secretion from the African civet cat, and in *Celtis reticulosa*, a Javanese tree. Tryptophol, 2-(3-indolyl) ethanol, is formed by the fermentation of tryptophan with yeast. L-2-Hydroxytryptophan is present in *Amanita phalloides*, a deadly mushroom. Indole-3-acetic acid, "heteroauxin," is a plant-growth hormone present in many higher plants and in the urine of humans. In extremely low concentrations, indole-3-acetic acid promotes plant stem elongation but not cell multiplication. Oral administration of indole-3-acetic acid has been shown to decrease the blood sugar of adult diabetic patients. Derivatives of this substance, however, are thought to produce mental disturbances. Gramine, 3-(dimethylaminomethyl) indole, is isolated from the germ of Swedish barley and from the Asiatic reed, *Arundo donax* L. Tryptamine, 2-(3-indolyl) ethylamine, results from the decarboxylation of tryptophan by putrefactive bacteria.

Serotonin, 5-hydroxytryptamine, is extremely interesting because it has a broad spectrum of biological activities. A mammalian hormone, serotonin is formed by the decarboxylation of 5-hydroxytryptophan, a reaction which occurs in the brain. Serotonin is located in other body sites, including the gastrointestinal tract, serum and platelets. It stimulates uterine contraction, facilitates intestinal peristalsis and serves as a precursor for the skin pigment, melatonin. Also a vasoconstrictor, serotonin participates in the regulation of blood pressure. It also plays unclarified roles in mentation, in psychoses and in sexuality. The brain level of serotonin is affected by light and darkness and fluctuates with circadian rhythm. LSD and inhibitors of monoamine oxidase increase serotonin levels by blocking its metabolism. Serotonin is also widespread among lower animals such as birds, crustaceans and fishes, and is even present in walnuts and bananas.

Many hallucinogenic (psychotomimetic) drugs contain the indole nucleus; examples are lysergic acid diethylamide (LSD), psilocybin (O-phosphoryl 4-hydroxy-N,N-dimethyltryptamine), bufotenine (3-dimethylaminoethyl-5-hydroxyindole) and N,N-dimethyltryptamine. LSD, synthesized by A. Hoffman in 1943, is one of the most potent drugs known to man. Lysergic acid itself occurs esterified with alkaloids in the rye mold ergot. Bufotenine occurs in secretions of the common toad. Similar indole bases found in other types of toad are bufotenidine and bufothionine. Psilocybin is found in the fruiting bodies of the Mexican hallucinogenic fungus, *Psilocybe mexicana*.

Reserpine, an indole-containing alkaloid of complex structure, is found in *Rauwolfia serpentina*, a member of the dogbane family. Reserpine reduces the activity of the central nervous system and is administered as a sedative to patients in psychotic states. This drug augments the urinary excretion of serotonin and of its main metabolite, 5-hydroxy-indole-3-acetic acid. Vinblastine and vincristine, alkaloids from the *Vinca* species, are used as chemotherapeutic agents for leukemia and Hodgkin's disease. Other alkaloids containing the indole ring system include harmine, harmaline, harman, yohimbine, evodiamine, rutecarpine, ergotoxine, ergotinine, ergometrinine, physostigmine, brucine and strychnine.

Indole chemistry owes much of its development to interest in indigo. A series of plants indigenous to Eastern Asia, the *Indigoferae*, contain the glucoside indican, which is hydrolyzed to glucose and indoxyl. The latter is immediately oxidized to indigo by atmospheric oxygen. Indigo was employed by the ancient Egyptians to dye mummy cloth. By 1866, chemists had converted indigo into indole by the following sequence of reactions: oxidation to isatin, reduction of isatin to oxindole, and reduction of oxindole to indole. In the early part of this century, synthetic indigo replaced the natural product as an article of commerce. Brom-indigo (6,6′-dibromoindigo, Tyrian Purple) is obtained from the glands of certain shellfish.

Indole solutions turn pink upon treatment with *p*-dimethylaminobenzaldehyde (Ehrlich's test). A pine splinter moistened with hydrochloric acid turns red in the presence of an indole. Alkaline solutions of indole become blue-green with sodium β-naphthoquinonesulfonate; this coloration is the basis of a quantitative method for indole analysis. Indole solutions become pigmented after the addition of concentrated sulfuric acid and an aliphatic aldehyde; the color produced depends upon the aldehyde employed.

Indolines are generally prepared by reduction of the corresponding indole derivatives. Oxindoles are synthesized by the reduction of derivatives of *o*-nitrophenylacetic acid, a reaction followed by ring closure. Some indoxyls are prepared by reducing the corresponding 3-ketoindoline. Many isatins are made by treating N-substituted aniline with oxalyl chloride and cyclizing the intermediate. Dioxindoles are formed by the reduction of isatin derivatives.

Carbazole (dibenzopyrrole, 9-azafluorene) is a colorless substance found in the anthracene fraction of coal tar. Carbazole is prepared by the Graebe-Ullman synthesis by treating N-phenyl-*o*-phenylenediamine with nitrous acid. In the Borsche synthesis cyclohexanone phenylhydrazone is cyclized to form tetrahydrocarbazole which is dehydrogenated to carbazole. Indolomorphinans have been synthesized as have indolopavine derivatives which are structurally related to the naturally occurring pavine-type alkaloids.

Indole derivatives of commercial value include: indole and skatole (perfume bases), indole-3-acetic acid and related acids (plant hormones), tryptophan (an essential amino acid), indigoid and thioindigoid compounds (vat dyes for fabrics; pigments for paints, printing inks and plastics), indigo Carmine (synthetic sodium indigotin disulfonate, a blue coloring for foods, drugs and cosmetics), phthalocyanines (complex synthetic blue-to-green pigments used in inks and paints), physostigmine (for the treatment of glaucoma and gastrointestinal atony), reserpine (a tranquilizing agent), strychnine (for destroying rodents) and ergotoxine (for oxytocic use). Other medicinals which contain the indole ring system are adrenoglomerolotropin (adrenal stimulating hormone, ASH) and indomethacin, 1-(p-chlorobenzoyl-5-methoxy-2-methyl-indole-3-acetic acid). ASH is found in extracts of pineal gland tissue and is employed for the stimulation of aldosterone secretion. Indomethacin is a well-known antiinflammatory agent. 2,3-Disubstituted indoles have been synthesized as potential nonsteroidal antifertility agents.

F. R. ZULESKI AND F. J. DI CARLO
Cross-references: *Hallucinogens; Psychotropic Drugs.*

INDUSTRIAL RESEARCH

Research is the systematic search for new knowledge. When conducted within an industrial establishment it is called industrial research. It consists of two different kinds: (1) research undertaken to meet the needs or to expand the opportunities of the enterprise itself and which is supported from its own resources and, (2) research performed as a service to a sponsor from outside the establishment, usually an agency of the government, and supported financially by the sponsor. In 1970, slightly over half of the industrial research done in the United States was of the first kind, and slightly under half was of the second kind. The industry classified as Chemicals and Allied Products accounts for about a tenth of the industrial research done in the United States and in that section of industry only about 10 or 15% of the research is federally sponsored.

Inasmuch as it is the larger of the two in total volume, let us first consider the research which industry does for its own account. Some industrial firms from their very beginning have done research to improve their operations, and it is now considered a necessary activity within any industrial enterprise. Even the proprietor of a one-man shop or business is likely to spend some part of his time experimenting, seeking new information for the purpose of increasing his efficiency in some way. Large industrial firms almost all maintain research departments staffed with technically trained professional people. They seek ways to reduce operating costs, improve the quality of the goods or services the firm produces, or invent new products or services for the firm to sell. The rapid pace of change in the industrial and commercial world,

caused partly by the progress of science and technology and partly by economic, social and political developments, imposes upon every company the need to maintain a high degree of adaptability in its methods, its products, and even in its conception of its mission, if it is to survive. This adaptability is based to an important degree on its ability quickly to gain whatever specific information it needs to identify its problems, recognize its opportunities, and to act effectively in dealing with them. Industrial research is an important part of the mechanism for survival of an industrial enterprise. Collectively, the research of all a nation's industrial enterprises provides the base for maintaining the vitality of an entire national industrial economy in competition with the rest of the world.

Research, in providing information, can only provide a basis for effective action. The action itself must be a product of managerial purpose, planning, and decisions which always involve risk-taking. The risk may be assumed by private investors, the government, or by both together. Industrial research was originally conceived largely as interdisciplinary study involving the natural sciences, such as chemistry, physics, biology, metallurgy, and engineering science. Such study has indeed led to useful ideas and discoveries. Action to take advantage of these may require the investment of capital at considerable risk because of uncertainty as to the acceptability of the proposed innovation within the whole system of people and operations involved, both within the company and among its suppliers and customers. For example, an improvement in a manufacturing process might be resisted by the plant labor force who might fear a loss of jobs, or it might require raw materials of a quality or uniformity which the supplier can not provide economically. Products offered to the building trade must satisfy a multitude of requirements set by building codes, fire underwriters, lending institutions, and labor unions, as well as to perform their intended function economically and please the ultimate user and owner of the building. Information concerning a very broad range of subjects is required to fit a research finding into an operating industrial or consumer system. Collecting this broad body of information is called development work and it commonly has been linked with research in the phrase "Research and Development," ("R & D").

The research part of R & D is normally considered to be the first phase of the innovating process. It consists of a very wide variety of activities. These may include theoretical studies and calculations; simulation studies on computers; basic or fundamental laboratory studies of natural phenomena; empirical laboratory studies; equipment development and assembly; testing of research products for physical, chemical, mechanical and biological properties; engineering tests to arrive at ideas for a manufacturing process or design concepts for process equipment; cost estimates to evaluate process alternatives, together with the literature and laboratory studies necessary for drafting patent applications.

The development phase of R & D usually results from and follows the research work on any given project. It is the additional work that must be done to transform the research discovery into an industrial innovation. It includes the scaling-up of a research idea to a workable demonstration, such as a prototype model of a machine or a pilot plant of a chemical process. It may involve trial production runs, test marketing, extensive biological testing, such as is required for clearance of a food or drug. It may involve test usage of a new product to establish and optimize its serviceability, working life, or value in use. It may include the preparation of engineering estimates for new plant or equipment and estimates of operating costs, distribution costs, selling expense and sales promotion expenses. It may include extensive experimental work to draft and prosecute patent applications to refine and elaborate the protection for the new venture. It deals with fitting the innovation into the entire industrial and commercial system for which it is intended.

Development work usually requires a much larger scale of operation than does research and is, therefore, much the more costly part of the R & D process. A research project must show considerable promise of success before development work on it is begun. Once the decision is made to undertake a very costly development program, it is likely that the need for basic understanding of the phenomena involved will be more clearly recognized than ever before. Small-scale basic laboratory studies are continued and often intensified as the project moves from the research to the development phase of its life. Whether the basic research, the search for fundamental scientific understanding, that is evoked by and accompanies the development effort should be called research or development work is a moot point.

All the diverse activities that blend together in any typical industrial R & D project are better classified, for managerial purposes, on the basis of their objective than on the basis of their apparent nature. Thus, while all the activities named as parts of the research and development phases of a project differ very greatly in nature, they all have in common the purpose of effecting some innovation. In the R & D work that industry does for its own account, the objectives are largely economic in character. The resources of the company limit the total amount of research it can finance, and thus the number of such objectives it can pursue at one time.

Few of these considerations apply to the federally financed research done by industry. In the first place, most of the objectives of this research are not economic in character. Defense of the nation, exploration of space, eradication of disease, and maintenance of a habitable environment are goals that account for the great bulk of this work and these are not undertaken for reasons which are primarily economic. Ways are being sought to apply industrial research techniques even more broadly to the amelioration of social conditions. As long as the credit of the government is not limited, there is no clear limit to how much total R & D activity may be sponsored.

As public zeal for pursuing noneconomic goals increases, money may even be appropriated for them in excess of the amount required to keep all the qualified research people available working effectively. This has the obvious effect of allowing less well-qualified people to make a living at R & D work and to raise the cost of R & D results of any kind. This reduces the amount of effective R & D effort that can be directed toward economic goals, both by increasing its costs and by reducing the number of qualified people available to do it. On the other hand, overzealous support of some branch of science or technology can lead to the training of more highly specialized workers than can find fruitful employment after a particular goal has been achieved or when public enthusiasm for it wanes. It is not easy for people highly trained and long experienced in one science or technology to achieve equal proficiency in another which may then be in greater demand by society, the government, or by industry.

Federally sponsored research is assigned to companies largely on the basis of their capability to do the desired work and of the attractiveness of the proposals they present. Proposals are judged for their technical merit as well as for the cost that is estimated for carrying them out. Companies undertake this research for many different reasons. One reason may be simple patriotism, together with the belief that the company is uniquely qualified to do a good job. Another may be that the company must participate to stay current in its field. Another reason may be the desire of a company to be represented in the forefront of advancing technology generally so that it can be alert to discoveries of potential importance to it in its commercial role. A manufacturing firm might expect that participation in the R & D work might give it some real and legitimate advantage in bidding for the government business if large-scale procurement should result from the development. Government work may enable a company to recruit and maintain a larger staff of highly trained professional employees than it could otherwise afford. This should increase the range of its capability to find and exploit new opportunities in its commercial business and give it a larger pool of candidates from which to select people for higher management positions.

Some industrial firms have been organized for the sole or principal purpose of conducting sponsored research. This has given rise to a research industry which produces only knowledge, the utilization of which is left up to others. All the many industrial firms that do some federally financed research may be said to be participating, to that degree, in the R & D industry.

The patent system was established as a national policy to increase the incentive for industrial research toward economic objectives because of the relationship that was recognized between industrial innovation, economic growth, and national strength. The acceleration of the accumulation of scientific and technical knowledge has created serious difficulties in the administration of the system and profound procedural changes are under consideration to cope with the problem. The very principle of the patent system is being reconsidered in another way in public debate relating to the administration of patents assigned to the government as a result of the R & D work it sponsors. One view is that most economic growth can be gained from government R & D only if the results and patents are open for all to use. Another view, more prevalent among business men, is that the government should license patents as another business would, granting enough exclusivity to encourage commercialization. They say that the effort, risk and financial investment required to develop an invention commercially are not likely to be forthcoming without some protection for the investor from imitative competition in the early years of the venture. The question is whether economic development should be sought by allowing everyone to use the patents which are public property or by offering these rights on a restricted or exclusive basis so as to motivate commercialization.

WM. A. FRANTA

Cross-references: *Chemical Research, Patents.*

INFRARED ABSORPTION SPECTROSCOPY

The absorption of electromagnetic radiation ranging in wave length from 2 to 40 microns is currently being applied in analysis by infrared spectrophotometry. Matter in all its states—gases, liquids, solids—absorbs energy in this region. One is generally limited in practice to substances in which the atoms are held together with covalent bonds since they give rise to numerous sharp bands. Thus organic molecules and inorganic ions such as phosphate, carbonate, sulfate, etc. are ideally suited for study by this means. Materials may be examined in solution providing the solvent does not absorb all the pertinent radiation. For this reason it is difficult to work in water solutions or in complex organic solvents which have many strong absorption bands. Common solvents are CS_2, CCl_4 and $CHCl_3$. Solids are commonly examined as mulls in Nujol and in fluorocarbon oil or pressed into disks with KBr or AgCl as a substrate.

The analytical uses of infrared spectroscopy may be described as identification, quantitative analysis and diagnosis. An infrared spectrum is a complex of absorption frequencies and intensities. As such it is a physical property more useful for the characterization of a molecule than boiling point, melting point or index of refraction. It has been used to prove the identity of two samples of a compound and to detect the presence of a specific substance in a mixture. It is a frequent criterion of purity. Minute quantities of some impurities may be detected while others are not noticed unless present to the extent of several per cent. One interesting application has been its employment for following the isolation and purification of substances—i.e., steroids and fluorohydrocarbons.

Quantitative analysis is a major industrial use of infrared spectroscopy. Mixtures of hydrocarbons, fluorocarbons, nitroparaffins, halogenated compounds, to mention only a few, have been analyzed. Hydrocarbon mixtures have been analyzed for as many as ten components. Methods for determining total aromatics in hydrocarbon mixtures and water in Freons or in bromine have been developed. Continuous monitoring of plant streams is carried out with either nondispersive or dispersive analyzers. Recently apparatus has been built whereby the spectrophotometer presents its data directly on punched cards. These may be run thru a digital computer with an appropriate calculation deck to

yield improved resolution or the composition of complex mixtures.

An infrared spectrum gives information about the atomic groups present in a molecule—an important analytical application.

Relation Between Absorption and Molecular Vibration. Some knowledge of the fundamental processes involved in infrared absorption is necessary for its successful application. When a quantum of infrared radiation is absorbed by a molecule, it excites vibrational motion. The frequency of the absorbed radiation (band center) concides with the frequency of molecular vibration. Since a molecule may vibrate in a number of different ways, each way with its own frequency, a spectrum consists of a number of absorption bands forming a unique set for each molecule.

In general all atoms move when a molecule vibrates. However, vibrations occur in which the major atomic displacements are found concentrated in a small group of atoms. The frequency of such a vibration is characteristic of the group involved plus the type of motion. Thus the "stretching" vibration of a methyl group appears near 3.3 μ while a "deformation" vibration is found near 7 μ. A change in the immediate environment of a structural unit generally results in a modest shift of the "group" frequencies. For the most part this reflects a change in the effective force resisting the vibrational motion.

Group frequencies make possible a qualitative analysis for the presence or absence of various structural units in a molecule. Noting the exact position of the band within the characteristic interval may yield information about the immediate molecular environment of the group.

Another type of molecular vibration takes place in which structural groups move as (approximately) rigid units relative to each other. These "skeletal" or "chain" frequencies yield bands which are characteristic of the molecule as a whole. Because of this they form the basis for a unique identification of the molecule and, in addition, make its quantitative analysis possible.

Identification and Quantitative Analysis. The multiplicity of bands and the specificity with which their locations and intensities are determined by details of molecular structure—especially for the longer wave length, skeletal vibrations—make the infrared spectrum a veritable "molecular fingerprint." Successful identification requires the *ready* accessibility of the infrared spectra of numerous pure compounds. One solution, frequently adopted, has been to prepare a catalog of spectra on McBee Keysort cards, which are punched with the positions of the major absorption bands, molecular formula, atomic groups present, physical constants and other pertinent information. The number of cards to be scanned is generally large and it is often advantageous to use IBM punched cards when sorting equipment is at hand.

The fundamental relationship for quantitative analysis is the Beer-Lambert law, relating the absorbance (d) to the concentration. The extinction coefficient (K) is dependent upon wave length and slit width. The components of a noninteracting mixture absorb independently and the absorbance is additive. The extinction coefficient of each substance is measured at each wave length and the equations are solved simultaneously for the concentrations. Difficulties are generally encountered unless a wave length exists for each compound where its extinction coefficient is large and that for all other substances present is small. When the Beer-Lambert law does not hold, modifications must be made in this procedure. If the deviations are small, empirical corrections may be applied. When the deviations are large, comparison is generally made with working curves prepared from synthetic mixtures having compositions in the range of that of the unknown.

Diagnosis. The use of infrared spectroscopy in structural diagnosis depends upon a theoretical and empirical correlation of absorption bands with structural units. A quick scanning of a spectrum for absorption bands in the regions listed in Table 1 provides a preliminary picture of the structural units present in the compound. Correction and refinement of this picture can be achieved by a detailed examination of the spectrum for band shifts and for confirmatory bands, especially in light of additional evidence as to the physical and chemical characteristics of the compound.

TABLE 1. PRELIMINARY SCANNING

Absorption Bands	Indicated Structural Unit
2.3–32. μ	OH, NH Groups; H_2
3.2–3.33	Olefins, aromatics
3.33–3.55	Aliphatics
4.5	Acetylenes, nitriles
5–6 (pattern)	Aromatics
5.4–5.8 (sharp)	Esters, anhydrides, acid halides
5.7–6.1 (sharp)	Aldehydes, ketones, acids, amides
7.5–10.0	Phenols, alcohols, ethers, esters
11.0–15.0	Aromatics, chlorides

Absorption bands due to O—H and N—H stretching occur in the 2.75–3.23 μ region; these bands are shifted to longer wave lengths and broadened by hydrogen bonding. Intermolecular hydrogen bonding is decreased in dilute solution, but intramolecular hydrogen bonding is not affected by dilution.

Aliphatic C—H stretching gives rise to multiple absorption in the region 3.37–3.52 μ; resolution in this region does not suffice for detailed diagnosis when instruments with rock salt prisms are used. C—H deformation bands of medium intensity have been observed at 6.80–6.99 μ for —CH_3 groups and at 6.73–6.92 μ for —CH_2—groups.

Olefins are characterized by C—H stretching bands at 3.29–3.32 μ and C=C stretching bands at 5.99–6.09 μ; the latter band is shifted to higher wave lengths by conjugation. The number and orientation of hydrogen atoms on the ethylene nucleus are revealed by the C—H deformation bands.

In addition to bands associated with the aromatic nucleus as such (3.30; 6.15–6.35; 6.56–6.88 μ), the infrared spectra of benzenoid compounds display absorption patterns in the 5–6 μ and 11–14.5 μ regions which have been correlated with the orientation of substituents. Patterns in the 5–6 μ regions are highly characteristic but are generally weak and can be obscured by strong carbonyl absorption; correlations here have been made only in the benzene series. The patterns at the longer wave lengths are very intense; C—Cl stretching can cause interference here. These patterns are related to the number of adjacent hydrogen atoms on a benzene ring; the correlations can also be used to determine orientation in naphthalenes and in the benzene rings of quinolines and isoquinolines. Similar, but displaced, patterns are produced by substituted pyridines.

The C=O group of aldehydes, ketones, acids, esters and acid halides gives rise to a single intense absorption band in the 5.5–6.1 μ region; anhydrides give two strong bands. The location of this absorption is highly characteristic of the type of functional group; small but regular shifts to higher wave lengths are produced by conjugation or chelate hydrogen bonding while shifts to lower wave lengths are produced by ring strain.

The use of infrared spectroscopy in structural studies is sometimes aided by the use of model compounds to duplicate electronic, geometric or strain effects which may cause band shifts. In other cases, the location of bands may indicate structural features but leave unsettled the question of whether the relative band intensities are consistent with the presence of such groups in a single compound. Here synthetic equimolar mixtures of simple compounds, each having one of those features, may permit a decision. The direct measurement of band intensities would appear to offer promise in the determination of the number of times a given structural unit appears in a molecule.

WALTER F. EDGELL AND JAMES H. BREWSTER
Cross-references: *Analytical Chemistry, Optical Spectroscopy.*

Infrared Emission Spectroscopy

Infrared emission, spectroscopy has played an important role in the evolution of modern physical science. However, because of its more limited areas of application, it has not achieved the popularity of infrared absorption spectroscopy. Recent developments have prompted scientists to return to the study of infrared emission spectra. Engineering heat transfer research, as in aerodynamically heated aircraft or missiles, requires information about the emission of infrared radiation. There is interest in the identity and quantity of chemical species in regions of high temperature, such as flames, combustion chambers, rocket exhausts and stellar atmospheres. Fast reactions and explosions are often followed with rapid-scanning infrared spectrometers. The superior sensitivity of interferometric spectrometers permits the study of chemical species in much cooler remote systems, including planetary atmospheres and atmosphere-polluting industrial stacks. Temperatures of hot gaseous systems can be determined by measuring the relative intensities of discrete emission lines originating in rotational or vibrational transitions. Luminescence radiaton from molecules excited by various nonthermal methods has been studied, as has recombination radiation from semiconductors, wherein the recombination of electrons and holes is accompanied by emission of infrared radiation. Under proper circumstances, luminescent systems can be made to achieve laser action, and numerous solid and gaseous infrared lasers have been devised.

Radiation emitted by an object as a consequence of the thermal agitation of its atoms or molecules is called thermal radiation. The radiation emitted by a hot glowing filament is thermal radiation. Radiation from atoms or molecules which have been excited by other than thermal means is called luminescence. The standard emitter of thermal radiation is the ideal black body, which by definition absorbs all radiant energy falling on it. The power radiated from a unit area of a black body at temperature T in the wavelength interval between λ and $\lambda + d\lambda$ is given by *Planck's radiation law*

$$J(\lambda, T)\,d\lambda = \frac{c_1}{\lambda^5(e^{c_2/\lambda T} - 1)}\,d\lambda$$

with $c_1 = 3.740 \times 10^{-12}$ watt-cm^2, $c_2 = 1.438$ cm degrees, and λ and T are in centimeters and degrees Kelvin. By a slight modification this and the following equations can be written for power radiated in a frequency interval. Planck's equation is closely approximated by the older *Wien equation* for sufficiently low values of λT

$$J(\lambda, T)\,d\lambda = \frac{c_1}{\lambda^5}\,e^{-c_2/\lambda T}\,d\lambda$$

and by the *Rayleigh-Jeans equation* for large values of λT

$$J(\lambda, T)\,d\lambda = \frac{c_1 T}{c_2 \lambda^4}\,d\lambda$$

The total power radiated by a blackbody is given by the *Stefan-Boltzmann law*

$$J(T) = 5.669 \times 10^{-12}\,T^4$$

which can be obtained by integrating the Planck law or from thermodynamics. The power radiated from a unit area of an object into a unit steradian of solid angle is the *radiance* of the object, and the radiance per unit wavelength increment at a specified wavelength is the *spectral radiance*. The power intercepted by a unit area of surface at some distance from a source is the irradiance at the surface, and the spectral irradiance is the irradiance per wavelength interval at a specified wavelength.

The spectral distribution of radiation from a black body has a maximum at a wavelength λ_m whose temperature dependence is given by the *Wien displacement law*

$$\lambda_m T = 0.2897$$

At room temperature, the maximum of the blackbody emission occurs in the infrared at about 10 microns. As the temperature of an object is raised, λ_m shifts to shorter wavelengths. At a sufficiently high temperature visible light is emitted and with further temperature increase the appearance of the object changes progressively from dull red to white. Just as an optical pyrometer takes advantage of this phenomenon to permit the measurement of the temperature of a hot object, an infrared pyrometer can be used to measure a cooler object.

The radiant power emitted per unit area by an ideal blackbody in a given wavelength interval cannot be exceeded by the *thermal* radiation emitted by a real body at the same temperature. The *emittance* of a body at a specified wavelength is defined as the ratio of the power emitted per unit area by the body to the power emitted per unit area by a blackbody at the same temperature. The emittance can vary with both temperature and wavelength. The emittance at specified wavelengths per unit wavelength increment is the *spectral emittance*. The characteristics of a blackbody can be closely approximated by an artificial blackbody which usually takes the form of an enclosure with a small opening through which radiation is sampled.

The absorptance of an object is the fraction of incident radiation which is absorbed, reflectance is the fraction reflected, and transmittance is the fraction transmitted. The sum of absorptance, reflectance, and transmittance is unity. *Kirchhoff's law* states that the emittance of an object is equal to its absorptance. Consequently, the emittance of a specimen which is not opaque depends on its transmittance, and therefore on its thickness. Emission spectra of thin layers of gas consist of narrow lines from discrete transitions separated by regions of relatively low emittance. But the transmittance of very thick layers of gas vanishes even at wavelengths of rather weak absorptivity. Since the reflectance of a gas is negligible, Kirchhoff's law tells us that thick layers of gas, such as the sun, emit as blackbodies. Lambert's law states that the power emitted per unit solid angle from a plane surface is proportional to the cosine of the angle between the direction of emission and the normal to the surface. This expression, derived geometrically, is true for most materials for angles not too large, but fails at large angles, partly because the reflectance increases at large angles of incidence. The surface condition of an object also plays an important role in determining its emittance, largely through its influence on reflectance.

There are a number of approaches to the measurement of infrared emission spectra. The most direct procedure is to place the specimen in a spectrometer in place of the usual source and record the signal obtained at various wavelengths. The record will not be a quantitative emission spectrum of the sample because of variations in the spectral bandwidth of the monochromator with wavelength, attenuation by the optical elements of the spectrometer and by atmospheric water vapor and carbon dioxide, and variations in detector sensitivity with wavelength. These difficulties can be avoided by recording the thermal emission from a blackbody at the same temperature and under the same instrumental conditions. The ratio of the spectra of the sample and the blackbody provides a true emittance spectrum, which is easily converted to spectral radiance by multiplication by $J(\lambda, T)$ from the Planck radiation law and division by 4π steradians.

The advantages of double-beam spectrophotometry can be exploited, in principle, by placing the specimen in the sample beam and a blackbody in the reference beam of a double beam instrument and recording the spectral emittance directly, with instrumental and atmospheric effects automatically canceled from the spectrum. Alternatively, if a means of programming the gain or slit width of a single-beam spectrometer is provided, the program can be adjusted to make the emission spectrum of the blackbody invariant with wavelength. The emission spectrum of another body, maintained at the same temperature as the blackbody, when recorded with the same program will be a true emittance spectrum. Or, if the program is adjusted to make the emission spectrum of the blackbody proportional to its spectrum radiance, the spectrum of another body will be its spectral radiance. In any work of this kind, great care must be taken to insure that the reference source and the body under investigation illuminate the spectrometer in exactly the same way.

It is often inconvenient to use a blackbody source in emittance measurement, and a secondary standard is substituted. The secondary standard is first measured relative to a blackbody, and rechecked occasionally as needed. Electrically heated silicon carbide rods (globar) have been used as reference sources in the 2 to 15 micron spectral region at temperatures up to 1300°K. The National Bureau of Standards certifies tungsten filament lamps as standards of radiance or irradiance. These lamps have quartz envelopes which limit their usefulness to about 2.5 microns in the near infrared. Emission studies are not limited to high temperatures. Engineers often require data for calculating radiant heat transfer between cold objects, such as artificial satellites and the sky. Laboratory measurements of low-temperature emittance are made with refrigerated samples and reference bodies. In astronomical and meteorological studies the sky might serve as the reference body.

The emittance of a specimen which is known to ʰe opaque can be determined indirectly by measuring its reflectance, from which its emittance can be deduced from Kirchhoff's law. This method does not require a reference blackbody. It is necessary to differentiate between reflected and emitted radiation from the sample. This can be done by recording spectra with and without the spectrometer source or by locating the radiation chopper between the source and the specimen, so the emitted radiation will not be modulated and therefore not detected. The reflectance method can be used for partially transmitting samples as well as opaque ones, provided the transmittance is determined and proper account is taken of multiple reflections within the sample.

Conducting or semiconducting solid materials can be heated electrically by conduction. Nonconductors can often be coated on conducting strips or held in mechanical contact with them. Materials which will not adhere to large heated surfaces can sometimes be applied to fine metallic screens either directly or indirectly. Fine metallic or ceramic screens can also be heated by a flame. It is often convenient to coat a combustible fabric with the specimen and then burn away the supporting material, leaving the specimen as a fine mesh. Welsbach mantles of thoria are prepared in this way. Induction heating in radiofrequency fields, heating by solar or other high-intensity radiation, and heating in a plasma jet or in an electron beam are also potentially useful methods for the production of emission spectra of solids. Gases can be heated in flames or other combustion or reaction systems, by shock waves, by induction heating, by heating in a confining oven, or by flowing a gaseous stream over a hot surface. Emission spectra of liquids can be observed by heating thin films between transparent windows.

JAMES E. STEWART

References

Rutgers, G. A. W., "Temperature radiation of solids," Handbuch der Physik, XXVI, Springer, Berlin, 1958, pp. 129–170.

Penner, S. S., "Quantitative Spectroscopy and Gas Emissivities," Addison-Wesley, Reading, Mass., 1959.

Blau, H. H., and Fischer, H., "Radiative Transfer from Solid Materials," Macmillan, New York, 1962.

Tourin, R. H., "Spectroscopic Gas Temperature Measurement," Elsevier, Amsterdam, 1966.

Stewart, J. E., "Infrared Spectroscopy: Experimental Methods and Techniques," Dekker, New York, 1970.

INHIBITORS

Inhibitors retard or eliminate changes which normally take place in matter. In a general sense inhibitors can be considered as negative catalysts. The use of inhibitors is a common practice in a number of fields, for an effective inhibitor in small amounts can minimize undesirable changes which may become severe if uncontrolled. The modes of inhibitor action are diverse and the successful application of inhibitors requires a specific additive for each change in question. The following selected examples attempt to illustrate inhibition practices and the chemistry involved.

Antioxidants. Organic materials in many cases are subject to attack by atmospheric oxygen, leading to development of undesirable products or properties. The ease of attack is dependent on a number of factors such as: the organic material (olefinic material is more vulnerable than saturated), the condition of storage, the presence of catalysts (e.g., metal salts) and the presence of effective antioxidants. The attack of oxygen leads to the initial formation of peroxides which can decompose to oxygen-containing products (alcohols, ketones, acids, etc.), The oxidation process itself involves chain reactions with the component parts of (1) initiation, (2) propagation and (3) termination. (For detailed discussion see **Antioxidants.**) (Foods, Lubricants, Rubber).

Corrosion Inhibitors. One of the means of controlling corrosion is the addition of inhibitors which under favorable conditions notably reduce the rate of corrosion. This effect is due to the changes brought about by the inhibitor at the interfacial areas where corrosion takes place. In corrosion, electrochemical processes occur, setting up local cells with anodic and cathodic areas. With iron the anode reaction is $Fe \rightarrow Fe^{2+} + 2e^-$ while at the cathode the reaction may be $H^+ + e^- \rightarrow H$. Inhibitors function by increasing the polarization at either anodic or cathodic areas. The polarization can be due to such factors as: changes in hydrogen or metal overvoltage values, absorption of a well oriented film on the metal, formation of a diffusion barrier, or satisfaction of metal surface affinities.

The addition of salts yielding nitrite, chromate or phosphate ions in amounts from 5 p.p.m. to 1% has been found effective. Certain salts of organic acids with polyamines or ethylene oxide addition products of amines are also effective at low concentrations (5 p.p.m.). (See **Corrosion**).

Foam Inhibitors. Foaming is another interfacial phenomenon controlled by the presence of certain agents at low concentrations. In foaming, the presence of a foreign material brings about surface tension conditions favorable to build up a stable film structure. In certain cases the foam inhibitor acts by displacement of the foam producing agent at the interfacial areas, leading to collapse of the film structure. The higher alcohols (e.g., 2-ethylhexyl alcohol) and polyalkylene glycols are useful foam inhibitors. Certain silicones in concentration of 10 p.p.m. and higher are foam control agents.

Ultraviolet Inhibitors. A stability problem is made more severe in the presence of light, particularly light in the ultraviolet range (2,000–3,500 Å). The absorption of the proper irradiation by a molecule results in an increase in energy content, and reactions are thereby initiated. Inhibitors can be added which strongly absorb the ultraviolet irradiation and do not pass on this energy to neighboring molecules but dissipate the energy by other means (as in the form of heat). Typical compounds which serve as ultraviolet inhibitors are phenyl salicylate, resorcinol dibenzoate and 2-hydroxy-4-methoxybenzophenone.

Vinyl Resin Inhibitors. High molecular weight vinyl resins are subject to deterioration, particularly at elevated temperatures (100–200°C). The incorporation of suitable inhibitors assists in the stabilization against polymer breakdown and color formation. Many of the additives used possess antioxidant properties and can be considered as oxidation inhibitors. Such compounds are p-t-butyl phenol sulfide, aldol-α-naphthyl amine condensation product, phenyl-α-naphthyl amine, dibutyl-p-cresol, and butylhydroxyanisole.

With vinyl chloride polymers, hydrogen chloride is liberated and acceptors must be present to avoid severe degradation. Basic materials (metal oxides, amines) can be incorporated in the resin, or ethylene oxide derivatives can be added to combine with the acid released. Special organic metallo compounds containing lead or tin, such as dibutyl tin dilaurate, are effective in stabilization.

Fuel Oil Additives. Those fractions of petroleum (boiling range about 400 to 800°F) used as fuel oils often deposit a sediment causing clogging in filters and burner areas. This process may involve oxidation by atmospheric oxygen, but the common antioxidants do not prevent sediment formation. The seriousness of this problem has been alleviated by the use of dispersing agents which suspend the insolubles and check deposition at critical points. Those materials which are effective in this dispersion are surface-active agents, such as the salts of high molecular weight amines, and a copolymer of lauryl methacrylate and diethylaminoethyl methacrylate.

R. H. ROSENWALD

Gastric Inhibitors

In the treatment of peptic ulcer it is of prime importance to reduce hypermotility and gastric secretion (hydrochloric acid) in order to promote healing of the lesions. Drugs which suppress one or both of these functions are commonly referred to as gastric inhibitors. They exert their effect by interrupting stimulus of the parasympathetic division of the autonomic nervous system. It is generally accepted that the principal site of action is the junction of the nerve and the muscle or secretory cell. Since such stimuli are mediated by acetylcholine, these drugs form a part of the general classification known as anticholergic agents. Among the many hundreds of compounds which have been studied as gastric inhibitors, the majority fall into one of three categories: (1) quaternary salts of carboxylic acid esters of amino alcohols; (2) amino alcohols of the type

$$\underset{\displaystyle >}{\overset{\displaystyle OH}{\underset{|}{C}}}-CH_2CH_2N(CH_2)_n \cdot MX,$$

where $n = 4$ or 5, $M = H$ or alkyl, and $X = $ halogen or sulfate; or (3) basic carboxylic acid amides.

In addition to these three types, there are a number of compounds of such widely varying structures as to preclude simple classification. It is noteworthy that, with few exceptions, the active compounds contain a common structural feature. A group consisting of $>N-(CH_2)_{2-3}$ is attached to a secondary or tertiary carbon atom which in turn holds either two aryl groups or a combination of one aryl and one alkyl or cycloalkyl groups. These two terminal moieties may be joined directly together or indirectly through a —CO— link or other

O

simple function. The apparent indispensability of the $>NCH_2CH_2$-linkage lends support to the concept that such compounds compete with acetylcholine at the receptor site of the cells effected.

The therapeutic potentialities of all known gastric inhibitors are limited in varying degrees by the incidence of side effects. Unfortunately, the antisecretory action affects not only the cells responsible for gastric secretion but also other secretory mechanisms controlled by the parasympathetic nervous system. Thus, to achieve the desired therapeutic effect, it may be necessary in some cases to employ a dose sufficiently high to cause dryness of the mouth. Also, suppression of gastric motility increases the gastric emptying time, which may disturb normal digestive and peristaltic processes. A new therapeutic approach to the management of peptic ulcer is now undergoing clinical trial. The substance in question is a sulfated amylopectin which is a potent inhibitor of pepsin secretion. It should be borne in mind that, from the evidence to date, none of these drugs influences the natural cause of the disease. They are of greatest value in promoting the healing of an existing ulcer, but apparently are incapable of preventing recurrences. However, when used in conjunction with proper diet and other adjustments by the patient, the gastric inhibitors are invaluable in the management of peptic ulcer.

R. R. BURTNER

Cross-references: *Antioxidants, Antizonants, Additives, Foams, Corrosion.*

INORGANIC CHEMISTRY

As the name implies, the field of inorganic chemistry was originally considered to consist of the study of materials not derived from living organisms. However, it is now viewed as including all substances other than the hydrocarbons and their derivatives. For purposes of discussion it will be convenient to subdivide inorganic chemistry into three parts: (a) synthetic, (b) theoretical, and (c) industrial.

One of the most outstanding developments in inorganic chemistry during the past half-century was the broadening of the field of inorganic synthesis attendant upon the study of reactions in liquid ammonia and the development of the concept of the nitrogen system of compounds. The work of the pioneers in this field (E. C. Franklin, H. P. Cady, and C. A. Kraus, and their successors) not only provided a new stimulus to the development of the chemistry of the amido, imido, and nitrido-derivatives of the various elements, but also developed the technique of using a basic solvent and the resulting very strong bases (stronger than OH^-) which are available in such a solvent (e.g., NH_2^-) to prepare salts of very weak acids.

The discovery of the fact that alkali metals, as well as calcium, strontium, and barium dissolve in liquid ammonia to yield solutions containing metal cations plus solvated electrons was another important factor in broadening the field of synthetic chemistry. These solutions in liquid ammonia are very strong reducing agents, and since many compounds are soluble in liquid ammonia, these solutions make it possible to carry out reductions with a strong reducing agent in homogeneous systems —a type of reaction impossible in aqueous systems.

In like manner the use of anhydrous acids such as acetic, sulfuric, hydrofluoric, and nitric as solvents made possible the use of very strong acids (stronger than H_3O^+) as reagents in inorganic synthesis.

Progress in the use of jet and rocket motors in aircraft propulsion have led to greatly increased interest in the synthesis of ceramic and metal-ceramic bodies capable of maintaining high strength and other desirable physical and chemical properties at high temperatures, such as required for the efficient operation of these new types of engines. This has in turn caused the inorganic chemist to turn his attention to the reactions of refractory materials at high temperatures and to the synthesis of new refractory substances.

There have also been a large number of interesting developments in the synthesis of complex compounds of metal ions with such electron donor groups as ammonia, carbon monoxide, cyanide ion, ethylene diamine, chloride and others. Studies of compounds of this type have been very fruitful in the field of structural and stereochemistry.

Developments in the field of nuclear processes have had a marked impact on inorganic chemistry. Of particular importance has been the fact that the fission of U^{235}, Pu^{239}, and other fissionable isotopes leads to isotopes of a number of elements in the middle of the Periodic System which had previously been little more than laboratory curiosities. Further, the search for elements of proper nuclear cross-section for use in nuclear reactors has led to the investigation of a number of elements previously little studied. Thus, the knowledge of the chemistry of the less familiar elements has been greatly increased. The field of inorganic chemistry has also been expanded by the synthesis by nuclear reactions of a number of new elements and the investigation of their chemical properties.

Theoretical Inorganic Chemistry. The underlying idea around which the theory of inorganic chemistry has developed has been the Periodic Law, first stated by Mendeleev in 1869 and modified by Moseley in 1913, which says "the properties of the elements are periodic functions of their atomic numbers (positive charges on their nuclei)." (See **Periodic Law**). The theory of inorganic chemistry has grown as a result of the need to explain the Periodic Law and of attempts to answer the following questions: (a) Why and how do atoms combine with each other to form molecules or ionic crystals? (b) What is the relationship between the properties of substances and their atomic and molecular structures?

The trend in inorganic chemistry has been away from the purely empirical approach to chemical data toward predictability based upon theoretical principles. This trend is by no means complete but successes enjoyed thus far have been impressive.

The Nuclear Atom. Far from being the solid spheres which the chemists of the 19th century visualized, atoms were shown by the work of physicists and chemists of the late 19th and early 20th centuries to possess complex internal structures and to be composed of still smaller fundamental particles. The results of the investigations of such eminent scientists as J. J. Thomson, Marie Curie, Ernest Rutherford, and H. G. J. Moseley have led to the formulation of the following picture of atomic structure:

(a) Almost all the mass of an atom is concentrated in a minute positively charged body at its center. This positively charged body is called the *nucleus.*

(b) Distributed about this nucleus at relatively large distances are negatively charged bodies of very small mass, called *electrons.* The charge on an electron is the smallest quantity of electricity ever observed, and is therefore considered as the unit of electrical charge. Since the atom as a whole is electrically neutral, the number of electrons in any atom is equal to the number of unit positive charges on the nucleus. The electrons may be thought of as moving about the nucleus at high velocities, the resulting centrifugal force balancing the electrostatic attraction of the positively charged nucleus for the negatively charged electrons.

(c) The radii of most atoms are of the order of 10^{-8} cm, but the nuclei of atoms are only about 10^{-13} to 10^{-12} cm in radius. From various considerations, the radius of an electron is estimated to be only 3×10^{-13} cm. Based on these values, calculations show that only about one-trillionth (10^{-12}) of the volume of an atom is actually occupied by material particles.

(d) The number of electrons in the neutral atom,

or the number of units of positive charge on the atomic nucleus, is the same for all atoms of a given element. This number is called the *atomic number* of the element. The simplest of all atoms is the hydrogen atom, which has an atomic number of 1, and which thus consists of a nucleus having one unit of positive charge about which one electron revolves.

(e) All atoms of a given element, though they have the same nuclear charge (atomic number), do not necessarily have the same atomic weight. Thus, whereas most hydrogen atoms have a weight of about one atomic weight unit, a few hydrogen atoms have a weight of about two such units, and a very few have a weight of about three units. Varieties of atoms of the same element (i.e., having the same atomic number) which have different atomic weights are known as *isotopes*. Isotopes of a large number of the elements are known.

Electronic Shells and Subshells. In answering the question, "Why do atoms combine with each other?" the characteristic of atomic structure with which we are most concerned is the arrangement of the electrons which move about the nucleus, for it is these electrons which are chiefly involved in the processes by which atoms react to form compounds.

Our knowledge of the electronic configurations of the atoms of the various elements has come chiefly from two sources: (1) the variation in chemical and physical properties with atomic number, and (2) the study of atomic spectra. The results of these studies indicate that the electrons in an atom are grouped into shells and subshells having certain definite energies and certain definite capacities for electrons. Each shell is usually denoted by a number $n (n = 1, 2, 3, 4, 5, \ldots)$, in order of increasing energy. The subshells within each shell are usually denoted by the letters s, p, d, and f (see table).

Shells	Subshells	Electron Capacity	
1	s	2	2
2	s	2	8
	p	6	
3	s	2	18
	p	6	
	d	10	
4	s	2	32
	p	6	
	d	10	
	f	14	
5	s	2	32
	p	6	
	d	10	
	f	14	
6	s	2	18
	p	6	
	d	10	
7	s	2	8
	p	6	

It will be noted that the first shell contains only an s subshell, the second an s and a p subshell; the third shell contains an s, a p, and a d subshell, whereas the fourth shell contains an s, a p, a d, and an f subshell. Theoretically the fifth shell should contain five subshells; the sixth shell, six and the

seventh shell, seven; but in known atoms in their normal, i.e. unexcited, states, no subshells other than those listed ever contain electrons. In any given shell the energies corresponding to the various subshells increase in the order $s < p < d < f$, so that in some cases overlapping occurs between the energies of subshells in adjacent shells.

Chemical Bonds and Properties. A study of the chemical properties of various elements soon shows that elements differ widely in the tendencies of their atoms to combine with other atoms. In general, chemical combination of atoms results in their achieving electronic configurations of increased stability.

(a) Electrovalent Bonds. In certain cases, stable electronic configurations are attained by *transfer of electrons* from atoms of one element to atoms of another. Thus, when metallic sodium and free chlorine are brought together, each sodium atom, which differs from neon in having one $3s$ electron, loses that electron to a chlorine atom, which differs from argon in having one less electron. The result of this transfer is that the sodium atom has attained the stable neon configuration; because it has lost a negative electron, however, it now bears a *positive* charge. This charged particle is called a sodium *ion*. The chlorine atom, on the other hand, now has attained the stable argon configuration, but, by virtue of the added electron, bears a *negative* charge; or, in other words, it has become a chloride *ion*. These two kinds of ions are held together by the electrostatic attraction of their opposite charges in the sodium chloride crystal lattice.

$$\text{Na} \quad + \quad \text{Cl} \quad \rightarrow \quad \text{Na}^+ \quad + \quad \text{Cl}^-$$
$$(2\ 2,6\ 1) \quad (2\ 2,6\ 2,5) \quad (2\ 2,6) \quad (2\ 2,6\ 2,6)$$

The reaction of calcium with fluorine is another example of electron transfer. Each calcium atom loses one electron to each of two fluorine atoms, producing a doubly positively charged calcium ion and two singly negatively charged fluoride ions; the calcium ion has the argon configuration and the fluoride ion the neon configuration:

$$\text{Ca} \quad + \quad 2\text{F} \quad \rightarrow \quad \text{Ca}^{++} \quad + \quad 2\text{F}^-$$
$$(2\ 2,6\ 2,6\ 2) \quad (2\ 2,5) \quad (2\ 2,6\ 2,6) \quad (2\ 2,6)$$

Electrovalent bonds are formed between elements one of which loses electrons readily and the other of which has a strong attraction for electrons. Electrovalent compounds include most of the simple binary compounds between elements of the two families immediately preceding the helium family in the Periodic System and the two families immediately following the helium family in the Periodic System.

Compounds which contain electrovalent or ionic bonds are composed of positive and negative ions, and such substances in the solid state consist of crystals made up of regular geometrical arrangements of these ions. The force which holds together the ions in such crystals is the electrostatic attraction which the oppositely charged ions exert on each other. Any given ion in a crystal exerts a similar force on all the oppositely charged ions in its immediate vicinity, and it is therefore impossible to select any group of atoms which might properly be called a molecule. Because the electrostatic forces are large, such crystals are hard and nonvolatile; ionic substances, therefore, have high melting points and their boiling points in most cases are above 900°C. Since the forces between the oppositely charged ions are greatly reduced in media of high dielectric constant, electrovalent substances tend to be more soluble in solvents having high dielectric constant than in solvents of low dielectric constant. The solutions thus formed are good conductors of electricity.

(b) Covalent Bonds. In a larger number of cases, however, bonded atoms do not differ sufficiently in their attractions for electrons to permit electron transfer. In these instances the bond is produced by the *sharing of a pair of electrons* between the bonded atoms. This means that the shared pair of electrons becomes in some rather complex way a part of the electronic configuration of both atoms.

Consider, for example, the chlorine molecule, Cl_2. Each chlorine atom has seven valence electrons (two $3s$ and five $3p$ electrons) and one more electron to attain the stable argon configuration. This is accomplished by the electron sharing process symbolized as follows:

$$\overset{..}{:}\underset{..}{Cl}: + \overset{..}{:}\underset{..}{Cl}: \rightarrow \overset{..}{:}\underset{..}{Cl}:\overset{..}{\underset{..}{Cl}}:$$

The type of bond formed by the sharing of a pair of electrons between two atoms is called a *covalent* or *electron-pair* bond. In some cases, two or even three pairs of electrons are shared between two atoms. When two atoms share two pairs of electrons between them, they are said to be connected by a *double covalent bond;* the sharing of three such pairs constitutes a *triple covalent bond.* Covalent bonds can also be formed in a manner somewhat different from that illustrated above: namely, one of the bonded atoms may furnish *both* of the shared electrons. For example, an ammonia molecule, which contains in the outer shell of its nitrogen atom a pair of electrons not yet shared with any other atom, may contribute this pair to form a bond with some other atom.

Crystal lattices of most covalent substances are made up of discrete molecules held together by forces which are relatively weak as compared to the electrostatic forces which operate between ions in crystals of electrovalent substances. This results in these molecular crystals having relatively low melting points and boiling points and being relatively weak and soft.

(c) In some cases crystals are made up of atoms held together by an indefinitely extended network of covalent bonds, making the whole crystal a giant molecule. Such crystals, e.g. diamond (carbon) and silicon carbide are exceedingly hard and have very high melting points,

(d) Metallic solids constitute a fourth type of crystal in which the bonding forces are different from those of the above three types. The hypothesis which has been generally accepted is that a metallic crystal is made up of a lattice of the positive kernels of the metal atoms, held together by the attraction of the valence electrons, which, being mobile, are free to pass from one atom to another. The valence electrons, therefore, belong not to any one particular atom but to the crystal as a whole.

Although the manner in which such a structure results in a binding force between the atoms is not obvious, it has been explained on a theoretical basis. The theory makes it possible, moreover, to account for many of the physical properties of the metals.

Since the valence electrons in the metallic crystal are free to move, the application of a difference of potential to a piece of metal produces a flow of electrons through the metal from the negative pole to the positive. This picture also explains why the electrical conductivity of most metals decreases with increasing temperature; as the temperature is increased the vibration of the atoms interferes more and more with the flow of electrons. This interference increases the electrical resistance. A few metals when cooled to within a few degrees of absolute zero lose practically all their resistance to electron flow. Displacement of the atoms in a metallic crystal relative to each other does not destroy the general attractive force which holds the crystal together; hence metals are malleable and ductile. In ionic crystals such displacements bring into such close contact ions of like charge so that their repulsion "breaks" the crystal. Ionic crystals, therefore, are not malleable or ductile.

Industrial Inorganic Chemistry. The production of inorganic chemicals in the United States is at the heart of chemical industry, for most of the "heavy" chemicals upon which the whole chemical industry depends come under this category.

The chief raw materials from which inorganic chemicals are manufactured include: (a) sulfur, from which sulfur dioxide, sulfuric acid, and various sulfates are produced; (b) natural gas, which is a source of hydrogen, carbon, carbon monoxide and dioxide; (c) salt, from which sodium carbonate, chlorine, hydrochloric acid, sodium hydroxide, sodium metal, and many sodium salts are manufactured; (d) phosphate rock, from which normal and triple superphosphates for fertilizer manufacture, elementary phosphorus, phosphoric acid, and many other phosphorus chemicals are obtained; (e) sand, which is the chief raw material for manufacture of glass, sodium silicates, silica gel, silicon, ferrosilicon, and silicon carbide; (f) air, from which oxygen, nitrogen, oxides of nitrogen, ammonia and nitric acid are manufactured; and (g) limestone, which is the source of lime, carbon dioxide, and is used in the manufacture of glass. Other important raw materials for inorganic chemistry include the aluminum minerals bauxite and cryolite, potash minerals (potassium chloride, potassium sulfate, potassium magnesium chloride, etc.), the titanium ores ilmenite and rutile, water, coke, and a variety of metal ores.

HARRY H. SISLER

Cross-references: *Bonding, Ligand Field Theory, Crystal-Field Theory, Chelation, Isotopes.*

INSECTICIDES

Insecticide chemicals are indispensable tools for food and fiber production and to protect man and domestic animals from annoyance and diseases resulting from insect attack. Although the use of chemicals for insect control spans several centuries, those currently in vogue are mostly synthetic organic compounds introduced within the past thirty years. Their use has revolutionized the concept of efficient agriculture, saving millions of people from starvation and helping to prevent rapid escalation in food prices. Millions of others have been spared debilitation or death resulting from insect-vectored diseases.

Large amounts of insecticide chemicals are used, i.e., about 300 million pounds annually in the United States, mostly applied to agricultural land. As their toxic effects are not always restricted to the insect target, careful consideration must be given to the short- and long-term effects of such use on man and his environment. Other control methods, such as biological control using predators, parasites or insect pathogens, are generally less effective in rapid disruption of insect populations, and their use to date is restricted to only a few pest problems. Where possible, both chemical and biological control procedures are used in an integrated control or pest management approach. It is thus necessary to balance the risks, benefits and alternatives in arriving at a judgment on when and how to use chemicals for insect control.

The chemicals used are often referred to generally as insecticide chemicals, but they may be designated more specifically on the basis of the organism to be controlled, i.e., insecticide for insects, acaricide or miticide for mites, scalecide for scales, aphicide for aphids, ovicide for eggs, etc. Thay may enter the insect through the spiracles in the vapor phase (fumigants), on ingestion (stomach poisons) or on contacting the insect integument (contact poisons). They may be absorbed and translocated by plants so that insects feeding far from the point of application are killed (plant systemic insecticides). Some are effective in control of internal and external parasitic insects when fed to or applied to the skin of domestic animals (animal systemic insecticides).

Insecticide chemicals are almost always marketed and used as formulated products. As the spray formulations must rapidly disperse in water, solids are treated with wetting agents to give wettable powders that disperse automatically as a suspension of fine particles. A few insecticidal chemicals are liquids soluble in water, but most of ʾhe liquids must be combined with a surface-active component (surfactant) and a solvent to give an emulsifiable concentrate which, on addition to water, spontaneously disperses as fine droplets to form an oil-in-water emulsion. The insecticide in a volatile solvent is sometimes dispersed from a pressurized can containing a propellant liquid to generate an aerosol. Not all insecticides are dispensed in water or other solvent or as sprays. Some are sufficiently volatile at room temperature to vaporize from a carrier into an enclosed space for insect control, while others are dispersed as "smokes" by applying heat to substrates containing the heat-stable toxicant. Dusts are prepared by intimate grinding of the insecticide chemical with an inert carrier or filler to obtain a free-flowing powder for application with a dusting machine or a dust that adheres to treated seeds. Systemic insecticides for protection of seedling crops are frequently formulated as small granules which gradually release the insecticide chemical.

The active component or ingredient of the for-mulated insecticide is the insecticide chemical. There are about 200 of these in use, representing many types of inorganic and organic compounds, natural products and synthetic chemicals. The use of inorganic insecticides (lead arsenate, paris green and sulfur) has decreased, because the synthetic organic compounds are usually more effective for the same purpose. Three botanical products are used, rotenone from Derris roots and a few other plants, nicotine from tobacco waste, and pyrethrum (mixed pyrethrolone and cinerolone esters with chrysanthemic and pyrethric acids) from the ground flower heads of *Chrysanthemum cinerariae-folium*. Although synthetic pyrethrum is not economically feasible, many synthetic analogs (pyrethroids—allethrin, phthalthrin, resmethrin and others) are used because of their rapid paralytic effect (knockdown) as contact insecticides. A noninsecticidal chemical (synergist—such as piperonyl butoxide, sulfoxide or Tropital) that enhances the potency of the insecticide is usually added to pyrethroids for greater economy. Each of these botanical chemicals and related compounds is unstable in light and in the animal body, rapidly undergoing oxidation or hydrolysis at several sites on the molecule; thus they are of short residual action and persisting residues are generally not a problem. Several organic thiocyanates (Lethanes and Thanite) are less expensive but much less effective knockdown agents than the pyrethroids.

Chlorinated hydrocarbons of several different types are extremely effective contact insecticides. DDT [*d*ichloro *d*iphenyl *t*richloroethane or 1,1,1-trichloro-2,2-bis(*p*-chlorophenyl) ethane] is generally the most effective (for susceptible strains) of a series of analogs (methoxychlor, TDE, etc.). The persistence of DDT and several other chlorinated hydrocarbons results from low volatility and water solubility and good stability; they provide protection of treated areas for several days or weeks after application. Hexachlorocyclohexane (normally referred to as benzene hexachloride) and particularly the purified gamma isomer (lindane) are highly effective insecticides that are more volatile and less stable than DDT. A number of related chlorinated methano-bridged cyclodienes (aldrin, chlordane, dieldrin, endosulfan, endrin and heptachlor) are extremely potent contact and stomach poisons, particularly against soil insects. Some are also very toxic to fish and other wildlife. They vary in persistence because of differences in volatility and ease of metabolic oxidation and photocatalyzed oxidation and isomerization but in general are of long persistence [as such or as toxic derivatives, such as the epoxides formed from aldrin (dieldrin) and heptachlor (heptachlor epoxide)]. A very complex mixture (toxaphene) resulting from chlorination of camphene is important in protecting many crops from insect attack.

Organophosphorus compounds are the largest group, by number, of insecticide chemicals. They are also used in the greatest variety of ways for insect control. Major contact insecticides include azinphosmethyl, carbophenothion, diazinon, dioxathion, Dursban, Dyfonate, EPN, fenitrothion, Imidan, malathion, methyl parathion, naled, parathion and trichlorfon, while the plant systemics include Azodrin, demeton, dimethoate, disulfoton, mevinphos, phorate and phosphamidon. Dichlorvos volatilizes from resin strips for household insect control and coumaphos, ronnel and Ruelene are animal systemics. The persistence varies from a few hours, for volatile compounds applied to crops, to several months, for some compounds used as soil insecticides. The metabolic breakdown, which normally occurs rapidly in mammals, involves hydrolytic and oxidative reactions at many sites on the molecule. Those containing thiono sulfur are oxidized to the oxygen analog which is the actual toxicant, i.e., they are biologically activated. Some but not all of the organophosphorus insecticides are highly toxic to man, mammals and birds and the more toxic compounds must be used with great care.

Methylcarbamates are the latest addition to the major groups of insecticide chemicals. The most important compound of this type is carbaryl, 1-naphthyl methylcarbamate, a contact and stomach poison. Other methylcarbamates have useful contact activity (Baygon, Meobal, mesurol, formetanate and Zectran), plant systemic activity (aldicarb and methomyl) or are effective in control of soil insects (BUX and carbofuran). These compounds are relatively unstable to light and are readily oxidized and hydrolyzed in mammals and other organisms; thus they are short residual materials except when incorporated in soil. A related compound, Padan, is a thiolcarbamate derived from a natural toxicant (nereistoxin), but this compound acts in a different way from the methylcarbamates.

The diverse group of acaricide chemicals includes not only organophosphorus compounds and methylcarbamates but also bridged bis(chlorophenyl) compounds (chlorobenside, chlorobenzilate, dicofol and tetradifon), 2,4-dinitro-6-alkylphenol derivatives (binapacryl, dinobuton, dinocap and DNOCHP), organotins (Plictran), benzimidazoles (Lovozal) and many others (such as Morestan and Galecron).

Fumigant chemicals include carbon tetrachloride, ethylene dichloride and dibromide, methyl bromide, p-dichlorobenzene, sulfuryl fluoride and carbon disulfide.

There are many ways to potentially achieve insect control with chemicals besides a direct lethal action. Aziridine derivatives and many other compounds are effective chemosterilants, i.e., they reduce or destroy the reproductive capacity of the insect, but their use is greatly restricted because of unfavorable toxicological considerations. (See Insect Chemosterilants). Insect growth and development are mediated by hormones including juvenile hormone (an epoxidized homolog of methyl farnesoate) for the development of larval characteristics and ecdysone or molting hormone (a ketosteroid) for the development of pupal and adult characteristics. These are not single hormones in insects but classes of related compounds, some of which have been characterized and synthesized. Similar materials of high potency occur in some plants. While it appears that the ecdysone-type compounds will not find major use in insect control, synthetic compounds with morphogenetic activity similar to that of juvenile hormone will probably find limited use because of their high potency and favorable toxicological picture in mammals.

Man and domestic animals are protected from insect annoyance with repellents (such as benzyl benzoate, deet, dimethyl phthalate and ethyl hexanediol). (See Repellents, Insect.) The pheromones or chemicals that mediate humoral correlations among individuals of a given species are of potential interest in control. The most potent are the sex attractants emitted by virgin females to attract the males for mating. Many of the natural sex attractants have been identified and synthesized; they are being tested for use in control in conjunction with insecticides. There are many other synthetic insect attractants, such as eugenol and geraniol, used in poison baits and to survey insect populations.

The mode of action of several types of insecticides is well understood at the biochemical level. For example, a number of compounds block cellular respiration in different ways. Rotenone inhibits the reduced nicotinamide-adenine dinucleotide (NADH)-dehydrogenase component of the electron transport chain. Dinitrophenol and benzimidazole derivatives uncouple oxidative phosphorylation, preventing synthesis of adenosine triphosphate. Organotins and organothiocyanates block other steps in respiratory metabolism. The organophosphorus compounds and methylcarbamates inhibit acetylcholinesterase in the nervous system, resulting in acetylcholine accumulation and conduction block. (See Cholinesterase Inhibitors). Atropine is the antidote used most commonly in cases of accidental poisoning resulting from the cholinesterase inhibitors and, in organophosphate insecticide poisoning, an aldoxime may also be useful to hasten reactivation of cholinesterase activity. The chlorinated hydrocarbons and pyrethroids act on the nervous system, disrupting ion transport, but the precise mechanism is not known. In most of these cases the mode of action appears to be the same in poisoning insects and mammals, and so selective toxicity is often not the result of differences in the nature of the target site. More often, selective toxicity is imparted by differences between insects and mammals in the ease of absorption or metabolism of the toxicant. The need for completely selective insect control agents, such as hormones or compounds acting on a system critical only to the livelihood of insects, continues to be an important goal for the future.

The synthetic organic insecticides used today resulted from a systematic search of hundreds of thousands of compounds for ones with a useful balance of properties. Obviously they must be highly toxic to insects. If used on crops, they must be nonphytotoxic. Their acute toxicity to mammals, when accidentally swallowed or spilled on the skin, and their chronic toxicity, when ingested in the diet, must be low. They must have no unfavorable side effects, such as carcinogenic, mutagenic or teratogenic activity. It is very necessary to use realistic doses in assessing possible hazards of insecticide chemicals on mammals because it is always possible to give doses sufficiently high that harm will result. This, in itself, does not indicate that levels normally contacted constitute any hazard.

Another characteristic needed is a suitable residual persistence. Until recently, long residual action was sought and valued by farmers and public health workers controlling pest species attacking man, because the compound did not have to be reapplied frequently to maintain insect control. Now, compounds that are easily degraded by metabolism or sunlight are favored. The hazard to nontarget organisms, i.e. pollinating, parasitic and predatory insects, and wildlife, particularly birds and fish, must also be weighed in the risk-benefit equation. The labeling and registration specifications for an insecticide define the conditions under which it is effective and can be safely used. There are also economic considerations. The compound must be inexpensive ahd readily available, limiting the use or complex molecules and natural products. Considering the unique requirements for a useful insecticide, it is not surprising that the cost of developing a new compound exceeds ten million dollars. Even then, the compound may be useful for only a few years in some areas of insect control because of the development of resistant strains.

Particular attention has been focused on insecticides as environmental pollutants because they are toxic to mammals, repeated applications are made, and there are very sensitive analytical methods for their detection (particularly gas-liquid chromatography with the electron-capture detector for chlorinated hydrocarbons). DDT and its derived products (DDE from dehydrochlorination and TDE from reductive dechlorination) are almost ubiquitous environmental contaminants because of their stability and the large amounts of DDT used since its discovery in 1939. DDT accumulates on passing through some food chains, leading to levels hazardous to certain species of fish and birds. It persists at a low level in the human body and passes into mother's milk. There are similar findings, although less dramatic, for a few other chlorinated hydrocarbons. Despite its persistence, there is no evidence of any fatalities or direct harm to man from the legitimate use of DDT. Some areas of insect control require persistent chemicals, but in others less persistent insecticides such as organophosphorus compounds and methylcarbamates are being used with increasing frequency. It is not reasonable to discontinue the use of persistent compounds that are well-proved tools in insect control without a suitable alternative that is as well understood in its long-term effects. A more discriminate use of the present insecticides while suitable replacements are developed will help preserve the very significant gains made in agriculture and public health in recent years.

JOHN E. CASIDA

INSECT CHEMOSTERILANTS

Chemosterilants are chemical compounds which reduce or eliminate the reproductive capacity of an organism to which they are administered. They may affect only one sex or both sexes, and their effects may be temporary or permanent. To eliminate an insect population by sterilizing it rather than controlling it with insecticides was first proposed by Knipling in 1955. For an insecticide to be effective it must come in direct contact with the insect it is designed to kill, and those insects that are not killed remain to propagate and reestablish themselves. Because the use of insecticides is simple and the action is rapid this method has wide appeal, but the problems associated with insecticides are well-known. The principle of the sterility control method is simple; if some members of a population of sexually reproducing organisms are rendered sterile, the reproductive potential of the population will decrease; if the ratio of sterile to fertile members remains sufficiently large, the population will die out. Detailed mathematical models have been constructed supporting this proposition by Knipling[1] and Borkovec, and the validity of the principle was demonstrated in practice (Knipling[3] 1960) with the eradication of the screwworm fly from the southeastern United States.

The first chemosterilants effective in male insects were biological alkylating agents, particularly the derivatives of aziridine, e.g., tepa (I) [tris(1-aziridinyl)phosphine oxide].

I

Alkanesulfonates, e.g., busulfan [1,4-butanediol dimethanesulfonate] and certain nitrogen mustards, e.g., mechlorethamine [2,2'-dichloro-N-methyl-diethylamine] were also effective in some insects. Unfortunately, because most alkylating chemosterilants are toxic and mutagenic in many organisms, including man, their practical application is severely limited.

In 1964, two classes of nonalkylating male chemosterilants were discovered by Chang, Terry and Borkovec: phosphoramides, e.g., hempa (II) [hexamethylphosphoric triamide] and melamines, e.g., hemel (III) [hexamethylmelamine]. Since the toxicity and scope of the sterilizing

II

III

activity of these compounds was usually much lower than that of the alkylating agents, new directions in the search for chemosterilants were indicated. Dithiazolium salts, e.g., (IV) [3,5-bis(dimethylamino)-1,2,4-dithiazolium chloride], dithiobiurets, e.g., (V) [1,1,5,5-tetramethyl-2,4-dithiobiuret], and anthramycin and some of its derivatives are examples of

IV

V

recently reported chemosterilants for male insects, but considerable advances have also been made in the search for compounds active exclusively in females. Antimetabolites, particularly the antagonists of folic acid, e.g., aminopterin and methotrexate, were the first specific female sterilants but other groups including 2,4-diamino-s-triazines, triphenyltin derivatives, and boron compounds were reported between 1960–1970. The highest degree of specificity coupled with high effectiveness was detected in certain analogs of insect juvenile hormones by Masner, Sláma, Zdarek and Landa. If similar chemosterilants active in males or in both sexes were available, the sterility control method could equal or surpass the use of insecticides in many control operations. For lists of active chemosterilants and the insects upon which they act the book by Borkovec can be consulted.

The physiological effects of chemosterilants are variable. In the male, the reproductive organs are usually affected only by high doses of the compound, but lower doses induce dominant lethal mutations in the sperm. Such sperm remains motile and capable of fertilizing the ovum but the embryo dies before reaching maturity. In the female, chemosterilants disturb the development and maturation of the eggs and even small doses cause atrophy and disfunction of the ovaries. Occasionally nonviable eggs are produced which may be a result of induced dominant lethals. At the molecular level, the mode of action of chemosterilants is unknown. Since alkylating agents, antimetabolites, and certain other compounds are known to interact with nucleic acids or to interfere with their synthesis, a direct effect of the sterilant on the chromosomes is a possible explanation. It is clear, however, that not all chemosterilants act by the same mechanism.

If insects are to be reared, sterilized and released, the sterile insects may be obtained by radiation or chemosterilants. Thus far, only radiation-sterilization has been used in large-scale control and eradication programs. In one instance chemosterilized Mexican fruit flies were used successfully for preventing infestations at Tijuana in northwestern Mexico. Which of the two methods of sterilization is to be used has to be decided individually on the basis of susceptibility of the insects, safety of the procedure, and cost. When the sterilization of natural populations is contemplated, however, only chemosterilants can be used. Because of the potential hazards involved, application of chemosterilants in the environment has so far been limited to strictly controlled field trial experiments, which Weidhaas[1] has discussed at length.

PAUL H. TERRY

References

1. LaBrecque, G. C., and Smith, C. N., Eds., "Principles of Insect Chemosterilization," Appleton-Century-Crofts, New York, 1968.
2. Borkovec, A. B., "Insect Chemosterilants," in "Advances in Pest Control Research," Vol. VII, Interscience Publishers, New York, 1966.
3. Knipling, E. F., "Eradication of the screwworm fly," Sci. Amer., 203, 54 (1960).
4. Chang, S. C., Terry, P. H., and Borkovec, A. B., "Insect chemosterilants with low toxicity for mammals," Science, 144, 57 (1964).
5. Masner, P., Sláma, K., Zdarék, J., and Landa, V., "Natural and synthetic materials with insect hormone activity, X. A method of sexually spread insect sterility," J. Econ. Entomol., 63, 706 (1970).

INTERFACES

An interface is the area of contact between two immiscible phases. Five types of interface are possible at equilibrium: solid-solid, liquid-liquid, solid-gas, solid-liquid and liquid-gas. As gases are completely miscible with one another there is no equilibrium gas-gas interface.

At a freshly prepared surface of either a liquid or solid the molecules experience a net inward pull, which is caused by the molecular attraction of their neighbors in the bulk phase. In a liquid, the surface molecules can readily respond to the inward pull by moving out of the surface. The lower concentration thus produced at the surface creates a counter-diffusion. When the two opposite rates of movement are equal there is a dynamic equilibrium, which sustains a layer of lower density at the surface of the liquid. The equilibrium surface layer of less dense liquid is a transition zone between the bulk liquid and the vapor phase; it is distended or extended liquid and is, consequently, in a state of tension. At a solid surface, however, the molecules do not have the same freedom to move, and the cohesive forces acting inward are balanced by elastic reactions set up within the solid, as well as by the adsorption of any polar or polarizable molecules from the adjacent liquid, vapor or gas phase. The characteristic phenomenon of a liquid surface is therefore surface tension, and that of a solid surface is adsorption; both phenomena arise from the same cause, namely, the inward cohesive forces acting on molecules at the surface.

At liquid-liquid interfaces, since liquids cannot sustain a shearing stress but readily change shape under the influence of surface forces, equilibrium is reached rapidly. The nature of the equilibrium between immiscible liquids is illustrated in Fig. 1. If a liquid (2) is dropped into a less dense liquid (1) the interfacial tension (γ_{12}) will cause the interface to assume the minimum possible area; hence, if gravity can be ignored, a spherical drop will be formed. If now the spherical drop reaches a second liquid-liquid surface (i.e., the surface of a third denser than either) it will assume a lenticular form,

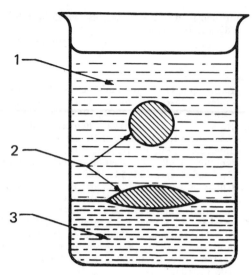

FIG. 1. Liquid 2 immersed in liquid 1 is a sphere; liquid 2 at the interface between liquids 2 and 3 is a lens, whose shape is determined by interfacial tensions and density differences. (*Courtesy American Mining and Metallurgical Engineers*).

whose shape is determined by the lowest total surface energy for all three interfaces. The angle of the lens is exactly the same as that established by three strings, tied to a common junction, each passing over a pulley and carrying a weight proportional to one of the three interfacial tensions, as shown in Fig. 2, together with the condition for mechanical equilibrium. This condition is often called Neumann's triangle. It has been applied not only to three immiscible liquids, as in the illustration, but also to the conjunction of liquid and solid phases (e.g., in studies of contact angle and degree of wetting); two crystals of the same solid phase (e.g., in studies of grain boundaries), and to interphase

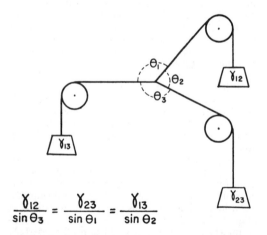

$$\frac{\gamma_{12}}{\sin \theta_3} = \frac{\gamma_{23}}{\sin \theta_1} = \frac{\gamma_{13}}{\sin \theta_2}$$

FIG. 2. A mechanical analogy of the three forces of interfacial tension that are in equilibrium around the periphery of the liquid lens shown in Fig. 1. (*Courtesy American Mining and Metallurgical Engineers*).

boundaries of solids (e.g., studies of microstructure of alloys).

Since the molecules at the surface of a solid do not move readily in response to the cohesive forces acting upon them, they remain in a nonequilibrium state with respect to the interior of the solid. Four factors can then operate to bring about changes in the surface structure that tend to equalize the chemical potentials of surface and bulk solid: (a) Polarization of surface ions of the solid; (b) Distortion of surface structure of the solid; (c) Nonstoichiometric excess of anions over cations at the surface; (d) Adsorption of polar or polarizable ions or molecules at the surface. Factor (b) is particularly important: as anions are generally larger and more polarizable than cations, any distortions of surface structure would tend to be of a sort in which the cations are screened by the anions, particularly if the cation is small, highly charged, and not readily polarized. In this way small particles tend to acquire a negative surface charge and to repel one another. The fluidization of catalysts such as MgO, Al_2O_3 and SiO_2, which depends on mutual repulsion of particles, is therefore possible only because of their surface distortion. Powdered quartz is an example of drastic distortion of surface structure; experimental evidence shows that the distortion of the crystal lattice at the surface of quartz particles extends to a depth of 0.3μ. The particles of many naturally occurring colloids (particularly, the clay minerals) are not large enough to include transition zones of this magnitude, and it must be concluded that small particles, thin plates, or fibers cannot possess such distorted surfaces as can more readily be produced on larger particles. The surface energy of such natural colloids is therefore relatively large. Some well-known applications of clays (i.e., as adsorbents, as plasticizing and thickening agents), depend on their high surface energy and small particle size.

Any reduction of the chemical potential of the surface of a solid, by one or more of the mechanisms mentioned above, results in a state of tension at the surface. It is doubtful, however, if the equilibrium surface tension of a solid is ever fully achieved. A real tension at a solid surface is only possible if displacement of matter can occur in the surface layer, whether by diffusion of atoms along interfaces or by plastic deformation occurring under the influence of unbalanced surface forces. The measurement of the angles established between interfaces provides a quantitative value for the relative surface forces involved; though when one or more of the phases are solids there is no assurance that equilibrium has been reached.

A freshly formed solid surface, such as might be produced by sudden fracture of a crystal, rapidly becomes contaminated on exposure to air, since chemisorption is one method by which the surface can get closer to equilibrium with the bulk solid. The influence of the chemisorbed films is often very marked. The following table reports typical values of the coefficient of friction of some clean solid surfaces in air and in high vacuum. Of these solids diamond and sapphire have high surface energies (several thousand ergs/cm²); graphite and carbon have intermediate surface energies (a few hundred ergs/cm²); and polytetrafluoroethylene has an extremely low surface energy (less than 100 ergs/cm²). The increase in the coefficient of friction of the first four solids *in vacuo* is caused by the loss of adsorbed oxygen or water vapor. Polytetrafluoroethylene has actually so low a surface energy that it is unable to chemisorb gases at ordinary temperatures, and consequently shows no increase in coefficient of friction when evacuated. A practical example of chemisorption is found in the boundary lubrication of moving metal parts in machinery. A film of oil forms a chemisorbed layer at the surface and prevents the enormous frictional forces that would otherwise oppose the relative motion of two clean metal surfaces in contact.

COEFFICIENT OF FRICTION OF SOME NONMETALS AT ROOM TEMPERATURE

Solid	Coefficient of friction, μ	
	In air	*In vacuo*
Diamond	0.05	0.5
Sapphire	0.05	0.6
Graphite	0.1	0.5
Carbon	0.12	0.6
Polytetrafluoroethylene	0.05	0.05

The solid-liquid interface is important for its relevance to the subject of adhesion, for adhesives are usually liquid when applied. The free energy of separation of two surfaces, which is the negative of the free energy of adhesion, ΔG^{adh}, is defined as

$$-\Delta G^{adh} = \gamma_{S/L} - \gamma_S - \gamma_L$$

where $\gamma_{S/L}$, γ_S, and γ_L are surface free energies per sq cm of the solid-liquid interface, the solid surface, and the liquid surface, respectively. If ΔG^{adh} is a large negative number, the work required to divide the solid-liquid interface into two separate surfaces is large.

The spreading or nonspreading of liquids on solid surfaces is another index of the adhesional forces. If the liquid spreads spontaneously, as for example, a drop of lubricating oil on the surface of a metal, the adhesional force is large. The spreading coefficient, S, is defined as

$$S = \gamma_{S/L} + \gamma_L - \gamma_S$$

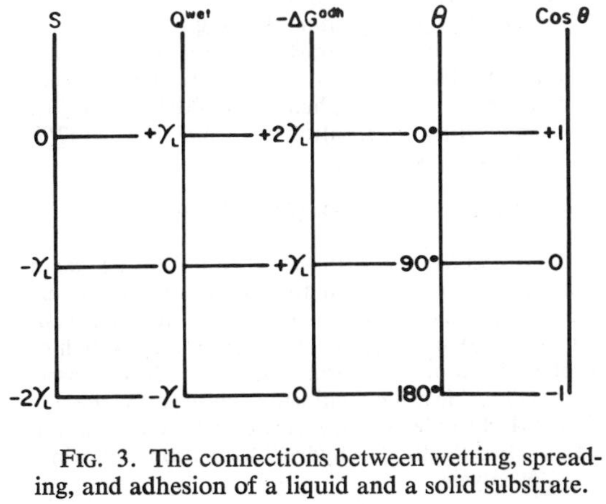

FIG. 3. The connections between wetting, spreading, and adhesion of a liquid and a solid substrate.

If S is positive, the liquid spreads as an ultrathin film over the surface of the solid; if S is negative, the liquid sits as a lens or sessile drop on the surface, making a definite angle of contact, θ, between the two phases. Contact angles have been established by the excellent work of W. A. Zisman as fundamental thermodynamic parameters; the measurement is conveniently made by the contact-angle goniometer. Contact angles are significant, however, only for substrates of relatively low energy, where both wetting and adhesion are difficult to achieve. When spreading is spontaneous, there is *no* finite contact angle, and the degree of wetting of the substrate must be obtained by another index.

This index is provided by the heat of wetting, which can be measured directly in a calorimeter. Wetting is effected in the calorimeter vessel by exposing the dried solid, usually a finely divided powder, to contact with a wetting liquid. The surface energy of the solid is replaced by the surface energy of the solid-liquid interface, and the difference appears as heat. The specific surface area of the powder can be obtained fairly accurately from a nitrogen-adsorption isotherm, and the heat of wetting per sq. cm., Q^{wet}, is found by dividing the experimentally measured heat by the area of solid substrate that was wetted in the calorimeter. In terms of fundamental parameters,

$$Q^{wet} = \gamma_S - \gamma_{S/L}$$

Only the difference between γ_S and $\gamma_{S/L}$ can be measured experimentally; the terms taken separately are not available by any known means of measurement.

In Fig. 3, all the indexes of adhesion, wetting, and contact are related. For nonspreading liquids, the measurement of γ_L and θ are enough to define any one of the indexes:

$$S = \gamma_L (\cos \theta - 1)$$

$$Q^{wet} = \gamma_L \cos \theta$$

$$-\Delta G^{adh} = \gamma_L (\cos \theta + 1)$$

For spreading liquids, the contact angle can not be measured, and the indexes are obtained by measuring Q^{wet} and γ_L:

$$S = Q^{wet} - \gamma_L$$

$$-\Delta G^{adh} = Q^{wet} + \gamma_L$$

SYDNEY ROSS

Cross-references: *Adhesives, Surface Chemistry, Colloid Chemistry, Soaps, Detergents.*

INTERMEDIATES

The term "intermediate" originated in connection with dyes nearly a century ago, when it became necessary to designate the products which were actually intermediate between basic raw materials and the finished dyes. For instance, in the sequence where benzene upon nitration furnishes nitrobenzene and upon reduction gives aniline, a constituent of many dyes, the nitrobenzene and the aniline are considered to be intermediates. Initially all the intermediates were based upon fundamental raw materials obtained from coal tar, such as benzene, toulene, xylene, naphthalene, anthracene, and the like. Most of these aromatic ring compounds are being supplied in two ways from the petroleum industry: (1) by direct distillation and (2) by chemical reaction, including ring closure and dehydrogenation. This greatly broadens the raw materials for cyclic intermediates.

These direct building units for the dyes were greatly extended in scope and importance after World War I; they became intermediate products not only for an increased number or a greater complexity of dyes but also for other chemical end products such as medicinals, plasticizers, resins, plastics, explosives, detergents, synthetic fibers, rubber chemicals, elastomers (synthetic rubber) and pesticides. The wider use and diversity of application of intermediates has had two main causes. (1) The chemist by his laboratory research ascertained more of the chemical properties, including reactivities, of both raw materials for intermediates and of the intermediates themselves. The chemist also determined new products that could be made therefrom by chemical change such as in the field of plastics, detergents and rubbers. (2) The chemical engineer applied some of the fundamentals the chemist had worked out, investigated certain of these himself, and decreased the cost of manufacture by increasing yields and improving equipment used for the manufacture of the intermediates.

Acyclic intermediates have outstepped the cyclic with a production of 66 billion lbs vs 25 billion lbs for cyclic in 1964. These intermediates are the foundation of the modern approach to organic technology. However, the cyclic intermediates like aniline, betanaphthol, o-benzoylbenzoic acid and even styrene are more generally employed in the original classification of a product *intermediate* between the cyclic raw hydrocarbon material and the finished dye. Acyclic intermediates, such as ethyl alcohol, methanol, and chloral, have widespread end uses in themselves as solvents or medicines, for example, as well as being intermediates for a chemical reaction to make end-use chemicals. On the other hand, butadiene is an acyclic intermediate for chemical change into synthetic rubber or synthetic fiber.

Chemicals can be classified by structure or by chemical conversion. In industry the latter is often called classification by chemical conversion or unit process. From an introductory educational viewpoint, classification by structure is most useful. However, as intermediates are essentially products of a *chemical* reaction, the classification by chemical conversion is much more pertinent. For example, in industry, different hydrogenations are often carried out in the same equipment or in analogous equipment in the same area of a factory. This is likewise true of nitrations, aminations, and many of the other chemical changes carried out by the chemical industry.

Intermediates by Chemical Conversion Classification. *Acylation* is employed to prepare *acetanilide* from aniline with glacial acetic acid. As such it finds use as a medicine, but much is nitrated and hydrolyzed to make *p*-nitroaniline as another intermediate step in the manufacture of dyes. Some *p*-nitroaniline is reduced to *p*-phenylenediamine.

Alkylation was one of the early chemical processes used by the Germans to furnish intermediates

for improved dyes, e.g., *dimethylaniline*. Other alkylation products are: cumene, dodecylbenzene, ethylbenzene, and nonylphenol.

Amination by ammonolysis has become a very important process for making many intermediates, since NH_3 is such a cheap chemical. The following can be manufactured by this process, all from the corresponding chloro- or occasionally sulfonic derivative: 1- and 2-aminoanthraquinone, 2-amino-1-naphthalenesulfonic acid (Tobias acid), aniline, hexamethylenediamine, and *p*-nitroaniline.

Aniline results in good yields and of excellent purity from chlorobenzene by this reaction:

$$C_6H_5Cl + NH_4OH \text{ (large excess)} \xrightarrow[850 \text{ psi}]{210°C}$$

Chloro-
benzene

$$C_6H_5NH_2 + NH_4Cl + H_2O$$

Aniline

Turbulent mixing is necessary to secure good yields and this is sometimes effected by a specially rotating jacketed steel autoclave.

Amination by reduction is the oldest method for making aniline, using cast-iron turnings. It is also of value for metanilic acid and *p*-phenylenediamine, both important dye intermediates:

$$4C_6H_5NO_2 + 9Fe + 4H_2O \xrightarrow[HCl]{100°C}$$

$$4C_6H_5NH_2 + 3Fe_3O_4 \quad \Delta H = -130 \text{ kcal}$$

The above reaction only expresses the starting and end materials.

Aniline is now manufactured on a large scale by continuous *vapor phase hydrogen* reduction of nitrobenzene:

$$C_6H_5NO_2(l) + 3H_2(g) \xrightarrow[270°C]{20 \text{ psi}}$$

$$C_6H_5NH_2(l) + 2H_2O(l) \quad -130 \text{ kcal}$$

Condensation as a reaction, leads to several important intermediates: o-benzoylbenzoic acid, phenylglycine, and hexamethylenetetramine.

The series of reactions from phthalic anhydride through o-benzoylbenzoic acid to anthraquinone are of great importance to the vat dye industry.

Hexamethylenetetramine has long been a urinary antiseptic but its large consumption is in the plastic field as an alkaline formaldehyde carrier in the second stage of making phenolformaldehyde resins. It is made with condensation of formaldehyde (37% solution) with NH_3.

$$6CH_2O + 4NH_3 \rightarrow (CH_2)_6N_4 + 6H_2O$$

Halogenation by virtue of cheap chlorine has a most important place in the preparation of intermediates, e.g., chlorobenzene, α-chlorotoluene (benzyl chloride), o- and p-dichlorobenzenes, 2,4-dichlorophenol, chloral, chloroacetic acid, chloroethane (ethyl chloride), vinyl chloride (chloroethylene).

Vinyl chloride is made cheapest by combination of the two methods: hydrochlorination; chlorination, pyrolysis:

(1) $CH\!:\!CH + HCl \xrightarrow{HgCl_2} CH_2\!:\!CHCl$

(2) $CH_2\!:\!CH_2 + Cl_2 \rightarrow$

$$CH_2Cl\cdot CH_2Cl \xrightarrow[\text{High temp.}]{\text{Pyrolysis}} CH_2 = CHCl + HCl$$

Use HCl in reaction (1)

Hydrogenation and dehydrogenation are employed for making certain largely consumed intermediates such as cyclohexane, methanol, styrene and butadiene. *Methanol,* by the hydrogenation of carbon monoxide, has generally replaced the product from wood distillation:

$$CO + 2H_2 \xrightarrow[\text{4500 lb.}]{300°C} CH_3OH; \Delta H = -24.6 \text{ kcal}$$

Butadiene, so essential for synthetic rubber and fibers, is an important intermediate petrochemical:

$$\underset{\textit{Butane}}{C_4H_{10}} \xrightarrow{-H_2} \underset{\textit{Butylene}}{C_4H_8} \xrightarrow{-H_2} \underset{\textit{Butadiene}}{CH_2\!:\!CH\cdot CH\!:\!CH_2}$$

Hydrolysis for many decades has been used to make phenol and β-naphthol from sodium benzenesulfonate and caustic soda, or from sodium 2-naphthalenesulfonate and caustic soda. It now serves also to make *phenol* from chlorobenzene. (See **Oxidation.**)

Nitration is one of the oldest of the chemical processes applied to intermediates such as *nitrobenzene,* nitrotoluene, and various nitronaphthalenesulfonic acids (H acid etc.):

$$C_6H_6 + HNO_3(H_2SO_4) \xrightarrow{30-60°C} C_6H_5NO_2$$

$$+ H_2O(H_2SO_4); \Delta H = -27.0 \text{ kcal}$$

The yield in this reaction is around 99%. As nitration is also employed for oxidation, exact conditions must be ascertained for the particular nitro product desired. Oxidation accompanies nitration especially for polynitro bodies and at higher temperatures.

Oxidation may be viewed as a controlled combustion, and this chemical change has been harnessed to supply phthalic anhydride, benzoic acid, terephthalic acid, adipic acid, acetic acid, acetaldehyde, ethylene oxide, and maleic acid. The chemical process for making *phthalic anhydride* intermediate is of greatest importance for dyes, plastics and surface coatings (80–85% yield).

Adipic acid for nylon is now made by nitric acid oxidation:

$$\underset{\substack{\textit{Cyclo-}\\\textit{hexanol}}}{C_6H_{11}OH} + \underset{\substack{\textit{Cyclo-}\\\textit{hexanone}}}{C_6H_{10}O} + \underset{(54\%)}{HNO_3} \xrightarrow{\text{heat}}$$

$$\underset{\textit{Adipic acid}}{HOOC(CH_2)_4COOH} + \text{Water} + \text{NO}$$

Phenol is also made by oxidation from cumene but the newest large-scale production is by *oxidation* of toluene.

$$\underset{\textit{Toluene}}{C_6H_5CH_3} \xrightarrow[\text{300°F, 30 psi}]{\text{Air, Co salts}}$$

$$\underset{\textit{Benzoic acid}}{C_6H_5COOH} \xrightarrow[\text{450°F, atm}]{\text{Air, Cu salts, H}_2\text{O}}$$

$$(C_6H_5COO)_2Cu \xrightarrow{H_2O}$$

$$\underset{\textit{Salicylic acid}}{C_6H_5COOH + C_6H_4OH\cdot COOH} \rightarrow \underset{\textit{Phenol}}{C_6H_5OH} + CO_2$$

Sulfonation was one of the earliest chemical processes employed in the dye industry, being a reaction leading to cyclic intermediates and dyes. It has been partially replaced by halogenation:

$$100 \text{ parts } \underset{\substack{\textit{Naph-}\\\textit{thalene}}}{C_{10}H_8} + H_2SO_4 \rightarrow \underset{(93\%)}{2\%} \underset{\textit{Sulfone}}{C_{10}H_7O\,C_{10}H_7} +$$

$$84\% \underset{\substack{\textit{2-Naphthalene-}\\\textit{sulfonic acid}}}{C_{10}H_7SO_3H} + 14\% \underset{\substack{\textit{1-Naphthalene-}\\\textit{sulfonic acid}}}{C_8H_7SO_3H}$$

The 1-naphthalenesulfonic acid is hydrolyzed to naphthalene by passing dry steam into the sulfonation mass at 160°C condensing the naphthalene and reusing it. The pure 2-naphthalenesulfonic acid upon alkali fusion ("hydrolysis") furnishes pure β-naphthol.

R. NORRIS SHREVE

Cross-references: *Dyes and Dyeing.*

References

Shreve, R. Norris, "Chemical Process Industries," 3rd Ed., McGraw-Hill Book Co., New York, 1967. See especially pages 682, 684, 689, 780, 781, 784, 787, 789, 799, 802, 804, 807, 809, 812, 815, 821, 823, 826, 827, 845.

IODINE AND COMPOUNDS

Iodine having the highest atomic weight 126.9045 of the halogens (excluding astatine) begins to exhibit some metallic properties. At room temperature, it exists as a bluish-black solid with a metallic luster and is classified as a semiconductor of electricity. Many of its chemical properties resemble those of chlorine and bromine (Group VII of Periodic System).

Iodine was discovered in 1811 by the French chemist, Bernard Courtois, who manufactured potassium nitrate for Napoleon's armies. In Courtois' process a crude ash obtained from kelp (seaweed ashes) was used to conserve potash. Courtois found while washing kelp with sulfuric acid that a black powdered precipitate was obtained which on heating gave a violet-colored vapor. Later, Gay-Lussac first recognized iodine as a new element and named it after the Greek word for violet.

Occurrence. Iodine is the 47th most abundant element in the earth's crust, counting the rare earths as a single element. It is widely distributed in nature, occurring in rocks, soils and underground brines in small quantities (20–50 ppm), while seawater contains about 0.05 ppm. Despite the low concentration in seawater, some seaweeds, notably those of the brown variety such as Laminaria family and the Fucus, are able to extract and accumulate up to 0.45% iodine on a dry basis. Less than a dozen minerals in which iodine is an essential constituent have been found; lautarite, anhydrous calcium iodate, is probably the most important of these since it is the form in which iodine is found in the large Chilean nitrate deposits.

Preparation. 1. A common method of preparing iodine in the laboratory is by reaction of an iodate with an iodide and acid:

$$IO_3^- + 5I^- + 6H^+ \rightarrow 3I_2 + 3H_2O$$

The element can also be obtained directly from the iodate or periodate using a reducing agent such as sodium bisulfite. In either case, the precipitated iodine can be separated and purified by sublimation.

2. The major world source of iodine is the Chilean nitrate deposits (50% of world production). The nitrate ore, or "caliche," contains between 0.05% to 0.3% iodine in the form of calcium iodate (lautarite). The element is recovered from nitrate leach solutions by reduction with sodium bisulfite.

3. Iodides occur in some oil-well and natural brines to the extent of 20–70 ppm. In the Dow process, hot Michigan brine is acidified with HCl and oxidized with chlorine to liberate the elemental iodine. The free iodine is concentrated by a blowing out process followed by addition of sulfur dioxide to reduce and absorb the iodine as an HI/H_2SO_4 solution. Iodine is precipitated by chlorination of the more concentrated solution.

4. Although sea water is too dilute for the direct extraction of iodine, certain seaweeds, like kelp, have the capacity of absorbing iodides selectively. The commercial production of iodine from ashes of seaweed continues to account for a portion of the iodine produced in Japan, France, Scotland, and eastern Russia.

Properties. Iodine melts at 113.6°C and its liquid boils at 185°C. The density of the solid at 20°C is 4.93 g/cc and the specific heat is 0.0518 cal per g. Heats of fusion and sublimation are 14.85 and 56.94 cal per g, respectively. Of the common halogens, iodine is the least soluble in water with 0.0162 part soluble in 100 parts water at 0°C and 0.45 part in 100 parts at 100°C. There are no known hydrates of iodine. The solubility of iodine in aqueous solutions is increased markedly by introducing iodide ion, owing to the formation of the polyiodide complex ions. Iodine dissolves readily in many other solvents such as alcohol, carbon tetrachloride, chloroform, carbon disulfide, and benzene. Solvents for iodine fall into two classes referred to as either the violet or brown solvents of iodine. The brown solvents are represented by the complexing of Lewis base compounds such as ethyl alcohol. The violet solvents such as carbon tetrachloride, on the other hand, show very little complexing and, subsequently, the absorption spectrum is shifted very little from iodine vapors.

Iodine, like the other members of the halogen family, is chemically very active in its elemental state, but is usually less violent than they in its action. Iodine has a much lower electronegativity (2.5) than other common halogens and considerably lower oxidation potential (−0.535 volt for the aqueous system $I_2 + 2e^- \longrightarrow 2I^-$). Its principal valences are −1, +1, +3, +5 and +7; examples of compounds having these oxidation states are KI, IBr, ICl_3, IF_5 and Na_5IO_6.

Iodine forms compounds with all the other elements except the noble gases, sulfur and selenium. However, it does not react directly with carbon, nitrogen or oxygen except at high temperatures requiring a platinum catalyst. With the exception of copper and silver, most metals which form volatile iodides react rapidly with iodine vapor, especially when finely divided and strongly heated. Iodine does not form compounds with CO, NO, or SO_2 corresponding to the carbonyl, nitrosyl or sulfuryl chlorides.

In general, iodine reacts with organic compounds in much the same manner as the other halogens; but, owing to the weakness of the carbon-iodine bond, the energy released is small and many reactions are readily reversible. However, iodination can, in some cases, be made to go to completion by preventing the formation of free hydrogen iodide by addition of a strong oxidizing agent such as nitric acid.

The chemistry of iodine in dilute aqueous solution has considerable practical significance in determining its effectiveness as a disinfectant. Unlike the other common halogens, iodine does not react with water, liberating oxygen and forming the hydrogen halide. The reverse reaction predominates in the case of iodine, that is, hydrogen iodide is readily oxidized to free iodine by oxygen:

$$I_2 + H_2O \leftarrow HI + HIO(HI + O_2)$$

The equilibrium constant for this reaction is only 3×10^{-13} at 25°C, showing that iodine is not hydrolyzed nearly as much as other halogens. There is a marked effect of pH on this reaction as shown by a total iodine concentration of 0.5 ppm having about 99% elemental I_2 and only 1% HIO at pH 5; whereas some 12% is present as elemental I_2 and 88% is HIO at pH 8. Another important reaction in aqueous solution is the formation of triiodide ion:

$$I_2 + I^- \rightarrow I_3^-$$

This reaction, which has an equilibrium constant of 1.4×10^{-3}, is important as a means of concentrating iodine in water.

Uses. One of the oldest and largest uses for iodine is in the disinfectant area. The once common alcoholic solutions or "tinctures" of iodine have been largely replaced by the "Iodophors" such as iodine complexed with polyvinyl pyrrolidinone. Complexes of iodine with nonionic surfactants have found considerable use in detergent-sanitizer formulations.

Iodine and some of its compounds have found use as catalyst systems for producing stereospecific polymers (TiI_4), dehydrogenation of alkanes and sulfuric acid oxidations. The importance of iodine in synthetic organic chemistry is exemplified by the well-known Hofman, Williamson, Wurtz and Grignard reactions. The use of "iodometry" in analytical chemistry and the application of radioactive isotope I^{131} are valuable tools for chemists. Other important uses of iodine and its compounds are food supplements, x-ray contrast media, photographic chemicals, cloud seeding, and electric light bulb additive.

Compounds of Iodine. Hydrogen iodide is prepared by: (1) direct union of hydrogen and iodine vapor in presence of a platinum catalyst; (2) treatment of an iodide salt with phosphoric acid; or (3) by the hydrolysis of phosphorus iodide. Because of the ease of oxidation, it cannot be made by the action of concentrated sulfuric acid on a metallic iodide. A solution of HI, called hydriodic acid, can be prepared by the action of hydrogen sulfide and iodine in water:

$$H_2S + I_2 \rightarrow S + 2HI$$

Hydrogen iodide is extremely soluble in water and forms a constant-boiling mixture containing 57% by weight HI and boiling at 127°C. Hydriodic acid is a stronger acid than either hydrochloric or hydrobromic acid. The concentrated acid is of sufficient strength that it will dissolve even metallic silver with the evolution of hydrogen. However, in the pure state hydrogen iodide is a nonpolar, covalent molecule that exhibits low electrical conductivity.

JACK F. MILLS

Physiological Aspects

Courtois' discovery of iodine in seaweed ash in 1811 or 1812 apparently constitutes the first reference to iodine as a plant substance. Much work has been done since that time on the content, distribution, and function of this halogen in the plant kingdom. In studies of the quantities of iodine or iodide in plants or plant products, because the amount present is exceedingly small, their detection requires highly sensitive chemical methods and skillful manipulation by the analyst. The analytical methods developed in the last quarter of a century may be considered much more accurate and reliable than those in use prior to that time.

The amount present in land plants varies from 10–$100\mu g.$ per 100 grams of dry matter. The seaweeds richest in iodine contain 10,000 times more iodine, in easily detectable form, than the land species highest in iodine, but no plant species so far has been found completely devoid of iodine. Neither has anyone succeeded in creating an environment for the growth of plants entirely deprived of every trace of iodine which at the same time provided all other nutrients in adequate amounts. Therefore, the question of the indispensability of iodine for any phase of plant growth, development, or metabolism remains unanswered.

The many instances in which traces of added iodine or iodide cause plant growth stimulation suggest that iodine is a trace element producing its effect in the most minute concentrations. The increased nitrogen assimilation frequently noted following application of iodine as iodides to plants may be linked to the promoting effect of iodine on nitrifying soil bacteria, rather than to any direct effect of the element.

It is thought that Chatin (1813–1901) was the first to point out a relation of iodine deficiency in man's environment to goiter. Further great impetus to studies of the iodine content of plant and animal foods arose from the discovery in 1895 that iodine is a normal constituent of the thyroid gland. Overwhelming evidence collected by numerous investigators all over the world during that last half century has proved that the major cause of goiter is iodine deficiency in soil, food, and water. In endemic goiter areas which have existed in many parts of the world the addition of one part of NaI or KI to 100,000 parts of NaCl used as table salt appears quite satisfactory in the prevention of the disease.

In spite of the addition of traces of iodide to table salt, endemic goiter is still prevalent in many parts of the world. Werner, Bora, Koutras and Wahlberg (*Science*, **170,** 1201, 1970) observed significant increased concentrations of immunogloblin M in many patients suffering with either endemic or sporadic nontoxic goiter. Their studies suggest that activation of this immunoglobulin response may occur in the thyroid during goitrogenesis.

From studies of the cause and prevention of goiter there has developed extensive knowledge concerning iodine and iodides and the prominent and essential role this element plays in the thyroid gland of animals and man. Three specialized functions performed by the thyroid are (1) the collection of iodide from the plasma (2) the transformation of iodide into organically bound iodine, and (3) the storage and release of the thyroid hormone. The unique property of the thyroid gland to concentrate iodide from the blood plasma constitutes the first step in the organic binding of iodine by the thyroid gland. This concentrating mechanism may increase the concentration of iodide in the gland several hundred times over that occurring in the plasma. Low dietary iodide intake increases the iodide concentration ratio between the thyroid and the blood serum. Conversely, increasing dietary iodide during low intake depresses the concentration ratio. The antithyroid drug thiocyanate also prevents iodide concentration by the thyroid and causes collected iodide to be discharged from the gland. Antithyroid drugs of the thiocarbamide type, however, produce their effect not by inhibiting the concentration of iodides by the gland but by preventing their organic fixation. Drugs of these two types have aided investigators to differentiate between various stages of iodide collection and fixation in the thyroid.

The oxidation of iodide to iodine constitutes the second step in the organic binding process, but since no free iodine has been unequivocally detected in the thyroid gland it is presumed that as rapidly as free iodine is formed substitution on the tyrosine molecule takes place. The oxidation of iodide *in vitro* in thyroid gland homogenates has been observed to occur in two ways: (1) Cell-free preparations of thyroid tissue either in the presence or absence of cupric ions (or some other oxidizing ions) form free iodine which is then combined with free tyrosine to form monoiodotyrosine in which process the enzyme, tyrosine iodinase, was identified; and (2) iodination of tyrosine within the protein molecule where the presence of the monoiodotyrosine formed could be demonstrated only after hydrolysis. The mechanism of tyrosine iodination occurring in the thyroid gland is only partially understood and requires further exploration.

Thyroid hormone released from the thyroid gland enters the circulation and is carried to the peripheral tissues where it controls tissue metabolism primarily through regulation of a large and various number of enzyme activities. It has long been accepted that thyroxine acts on fundamental chemical reactions normally taking place in the cell by increasing the rate at which the reactions are carried out, although recent work suggests that triiodothyronine rather than thyrosine may be the active biocatalyst and that a part, or the whole effect, ascribed to thyroxine may be due to its conversion to triiodothyronine. No clear choice exists, however, between (1) direct participation of thyroid hormone as a sort of coenzyme or (2) indirect release of metabolically active substances.

HAROLD P. MORRIS

Cross-references: *Hormones, Iodates, Iodides, Iodine Value.*

Iodine as Radioactive Tracer

There are 22 known radioactive isotopes of iodine; of these, I-123, I-124, I-125, I-126, I-128, I-129, I-130, I-131, I-132, and I-133 have been used as tracers. While I-128 (half-life of 25 min) is considered too short-lived to be useful today, it was used during the very early days (the mid 30's) of radiochemistry when nothing else was available. A number of short-lived nuclides, 4-day I-124, 12.5-hr I-130, 2.4-hr I-132, and 21-hr I-133, have been used for short-term diagnostic tracer applications, especially of the human thyroid function. Of the remainder, 8-day I-131 and 6-day I-125 have become far more commonly used than any others; 1.6×10^7-yr I-129 has utility for very long-term experiments or other special applications discussed below; 13-day I-126 has been useful as a third "everyday" tracer even though its production is not very economical; and a "newcomer," I-123 has a special attraction for nuclear medicine.

With its 8-day half life and ready availability, following the emergence of the nuclear reactor as a research and production tool, I-131 was the pioneer radioiodine tracer for the study and treatment of the thyroid function in the human being. Historically, it was actually in use before the invention of the nuclear reactor (the late 30's) when it was produced by cyclotron bombardments of tellurium. It has also been used for other organic and biological tracing experiments involving iodine and as an inorganic tracer, for example, in the form of the soluble salt, sodium iodide, for ground water tracing. I-131 is produced as a fission product (the mass 131 chain yield for thermal neutron fission is ~2.9%) or by the neutron irradiation of tellurium-130. The latter process produces tellurium-131 which subsequently decays by beta particle emission to yield carrier-free I-131. The chief advantages of I-131 as a tracer are: its economy, its readily measurable radiations, and, for certain purposes, its "short" half-life. Disadvantages include the inconvenience of its short life both for storage and for use in more lengthy experiments, and the detrimental effect of its energetic radiations upon the chemical stability of complex organic molecules.

The isotope, I-125, with its 60-day half-life, overcomes these disadvantages to a certain extent. The longer half-life permits the extension of experiments to several months and allows one the convenience of maintaining a stock of the tracer. On the other hand, its low-energy radiations, 27-35 Kev x- and gamma rays, are not easily measured and may require special precautions for sample preparation, while *in vivo* measurements may be

extremely difficult, if not impossible, to interpret. In spite of these difficulties, the use of I-125 as a tracer for complex biochemical systems has become more and more common as its availability (originally relatively poor) has improved considerably. It is now produced by the neutron irradiation of compressed xenon gas. The process produces xenon-125 (from the naturally occurring isotope xenon-124) which subsequently decays by electron capture yielding carrier-free I-125. A small amount (less than 3%) of 13-day I-126 is usually produced as an impurity but may be allowed to decay if a greater purity is desired.

The two isotopes discussed above have often been used concurrently in experiments where double tracing is desired (i.e., two iodine sources in the same system) and a third tracer is sometimes desirable for triple tracing experiments. Of the remaining iodine isotopes listed above, most have half-lives that are too short to be of practical use for the majority of applications or are not readily available in useful specific activities (a measure of radiation emission rate per unit mass of the element). The isotope, I-132 (2.4-hr half-life) for example, has been used for short term experiments (hours), being available as a daughter of 77-hour Te-132, from which it is "milked." The isotope, 13-day I-126, may be produced economically by high energy neutron irradiation of stable I-127 in a nuclear reactor, but the specific activities obtainable by this means are orders of magnitude below those generally required for biological tracers. Both I-126 and I-124 (4-day half-life) may be produced carrier-free by means of cyclotron bombardments of the appropriate tellurium isotope but such means of production are relatively costly (although justified under appropriate conditions). The identification of three iodine isotopes in a single experiment may also require sophisticated means of distinguishing between their irradiations and the availability of such means may well dictate the choice of tracers.

Recently, 13-hr I-123 has been produced in a very pure form by cyclotron bombardment of tellurium-122. This produces xenon-123 which subsequently decays to I-123. For applications in nuclear medicine in which the short half-life is not a drawback, this isotope has the advantage of giving only 1% as much radiation dose as I-131. This is due to its shorter half-life and its lack of particulate radiation.

A unique tracer for iodine is represented by 1.6×10^7-year I-129. While obviously not subject to the drawbacks of a short half-life, its specific activity is limited by its half-life to 1.6 microcuries per milligram as pure I-129 and so, for most tracer applications, can not be directly measured. It may, however, be determined, to a very high degree of sensivity, by neutron activation analysis. This results in the production of the easily measured isotope, 12.5-hour I-130. In effect, this allows one to achieve the same sensivity as with an ordinary radioactive tracer while exposing the subject of the experiment to a radiation dose approximately 10^4 times smaller. However, measurements cannot be made *in vivo*, and sample handling is always more involved than it is for ordinary direct methods of measurement. Indeed, the analysis may become quite involved when the utmost sensitivity of measurement is required (the equivalent in sensitivity to a few micromicrocuries of a directly measured tracer). The low dose rate experienced by the subject of the experiment allows the application of this method to experiments involving radiation-sensitive biochemicals over long periods of time as well as to experiments of long duration involving animal or human subjects. It has already been applied to large-scale (long-range) atmospheric tracing experiments with no danger to the population at large. Iodine-129 is readily available as a fission product in combination with the stable iodine also produced as a fission product. (The mass 129 chain yield for thermal neutron fission of U-235 is 0.9%). It is usually obtained as an 86%–14% mixture of I-129 and stable I-127.

BERNARD KEISCH

Iodates

Iodates have the general formula MIO_3, where M is a univalent metal or electropositive radical such as sodium or ammonium; they are derivatives of iodine pentoxide, I_2O_5, and of iodic acid, HIO_3. They are formally analogous to the chlorates and bromates and, like them, are powerful oxidizing agents. However, there are many points of difference. The iodates are not isomorphous with the others; they are generally more stable; they are less soluble in water; they form double and complex salts such as $NaIO_3 \cdot 4Na_2SO_4$ and $K_2Sn(IO_3)_6$; and acid and polyiodates, such as $KH(IO_3)_2$, $KH_2(IO_3)_3$, and KI_3O_8 ($=KIO_3 \cdot I_2O_5$) are known even though iodic acid is usually considered monobasic. Furthermore, iodic acid itself exists in two forms, the normal, HIO_3, and the pyro, HI_3O_8, both white crystalline solids; and it is only a moderately strong acid in aqueous solution.

Iodates of most of the metals and of a number of electropositive radicals are known. They are white except when the cation is colored. Those of the alkali metals yield iodides and oxygen at a low red heat; those of the other metals usually give oxide, iodine, and oxygen at lower temperatures. The salts of the alkali metals and of magnesium are soluble in water; the others are sparingly soluble or insoluble, but all except those of thorium and other tetravalent metals are soluble in dilute nitric acid.

Iodates can be prepared in a number of ways. When iodine is dissolved in a caustic alkali solution, ⅙ is converted to iodate, a large part of which can be recovered by simple filtration if sufficiently concentrated solutions are used. With sodium hydroxide care must be taken to avoid formation of a hydrate of the double salt $2NaIO_3 \cdot 3NaI$. Other methods include oxidation of iodine by concentrated nitric acid to iodic acid, followed by neutralization of the latter with an oxide or hydroxide; oxidation of an alkaline iodide solution with chlorine; and electrolytic oxidation of a neutral iodide solution containing chromate with a platinum anode and an iron cathode in a nondiaphragm cell. Sparingly soluble or insoluble iodates are precipitated when solutions of normal salts of the appropriate metals are mixed with those of the alkali iodates. The solid metallic iodates are stable and safe to handle when in the pure state, but they should be kept out of contact with organic substances and other combustible materials as such mixtures can be explosive. They are strong oxidizing agents in acid aqueous solutions. They liberate chlorine from hydrochloric acid, iodine from iodides, and oxidize formic acid to carbon dioxide and water. However, these reactions are not violent and some organic compounds are quite stable when dissolved in iodate solutions.

The alkali iodates are not toxic when ingested in moderate quantities, as they are quickly reduced to iodides in the stomach and their iodine is available to the thyroid gland. According to a recent report, mice tolerated single oral doses of 250 mg per kilogram of body weight and rabbits two doses of 10 mg per kg per week for six weeks without any ill effect. The iodates of toxic metals, such as lead, can be expected to be toxic, as iodic acid has no pronounced chelating action.

Since it is more stable than the alkali iodides under oxidizing conditions, sodium iodate is being recommended by the World Health Organization as an iodizing agent for salt for both human and animal consumption when oxidizing impurities such as iron are present or the salt is to be exposed to the weather. The alkali iodates are also used in small quantities as addition agents to certain types of flour to improve their bread making qualities.

The normal alkali iodates, potassium acid iodate, iodic acid, and iodine pentoxide are available commercially. The acid iodate is recommended as an alkalimetric standard, while the pentoxide is used for the analytical determination of carbon monoxide as it is one of the few known substances which reacts with that gas at ordinary temperatures.

A. C. LOONAM*

* Deceased.

Iodides

Iodides are generally considered to include those compounds of iodine in which the element iodine has acquired an electron, becoming an iodide ion (I^-). The size of the effective radius increases as a result, but less so than in the case of the other halogens, as shown below:

Halogen	Radius of Atom (Å)	Radius of Ion (Å)	Ratio Ion/Atom
Fluorine(F)	0.71	(F^-) 1.36	1.92
Chlorine(Cl)	1.00	(Cl^-) 1.81	1.81
Bromine(Br)	1.14	(Br^-) 1.95	1.71
Iodine(I)	1.33	(I^-) 2.16	1.62

Hydrogen iodide is a colorless penetrating gas. When liquefied, it boils at $-35.38°C$, and freezes at $-50.8°C$, values typical of a covalent compound. The gas dissolves in water 425/1 by volume at 10°C and 760 mm. A 57% solution of hydrogen iodide in water forms a constant-boiling mixture that boils at 127°C at 760 mm. Hydrates include $HI \cdot 2H_2O$ (m.p. $-43°C$), $HI \cdot 3H_2O$ (m.p. $-48°C$), and $HI \cdot 4H_2O$ (m.p. $-36.5°C$). All are colorless liquids.

Like hydrogen chloride, hydrogen iodide fumes in moist air. The reaction $HI + H_2O \rightarrow H_3O^+ + I^-$

accounts for the hydronium (hydrogen) ions present in a water solution of hydrogen iodide, commonly called *hydriodic acid*.

Hydriodic acid is a strong acid, and it reacts typically with metals, metallic oxides, hydroxides, carbonates, and sulfites. It is a powerful reducing agent, and is readily oxidized to water and free iodine by mild oxidizing agents, even air. With hydrogen peroxide, for example, the reaction is $H_2O_2 + 2HI \rightarrow 2H_2O + I_2$. Samples of the acid often become brown on standing.

In general, iodides are less readily fusible and less volatile than chlorides or bromides. Metallic salts of hydriodic acid are called iodides. Sodium iodide (NaI), colorless cubical crystals, sp, gr. 3.667, melts at 651°C and boils at 1300°C, values typical of an ionic compound. It dissolves well in water, and it is used to prepare silver iodide (AgI). Potassium iodide (KI), similar to sodium iodide in appearance, melts at 723°C and boils at 1420°C. It is also very soluble in water. Potassium iodide solution dissolves free iodine with the formation of the complex I_3^- ion, $KI + I_2 \rightarrow KI_3$. Ammonium iodide, sodium iodide, potassium iodide, calcium iodide, and strontium iodide are sometimes used in medicine. Potassium iodide in small percentage by weight is added to table salt to aid in formation of the hormone thyroxine.

Silver iodide is used in photography because it is photosensitive and as a cloud-seeding agent. It is very slightly soluble in water as are the following iodides: thallium (I), mercury (II), mercury (I), copper (I), and lead. The decomposition of titanium tetraiodide by heating it above 360°C has been explored as a method of preparing metallic titanium.

Polyiodides are formed by reaction with iodine or an iodo-halogen compound, such as ICl, ICl_3, or IBr. The number of iodine atoms in a polyiodide is almost always an odd number, although such compounds as $NaI_2 \cdot 3H_2O$ and CsI_4 have been reported. The tendency to form polyhalides is relatively slight with small cations such as H^+, Na^+, and Li^+, but greater with larger cations such as K^+, Rb^+, and Cs^+. The most stable polyhalides contain two different halogen atoms such as $RbICl_2$, deep orange, or $RbIBr_2$, bright red. Reactions such as $CsBr_3 + I_2 \rightarrow CsIBr_2 + IBr$ are known. $RbICl_4$, $RbI_7 \cdot 2C_6H_6$, and $RbI_9 \cdot 2C_6H_6$ are reported.

All polyhalides possess color, and as the number of halogen atoms increases, the color darkens, polyiodides being almost black. Polyhalides decompose when the temperature is raised and become simple halides, but the reaction of decomposition is not necessarily the reverse of that by which the polyhalide was formed.

Organic Iodides. Colorless iodine, sometimes used as an antiseptic, is diiodohydroxylpropane [$(CH_2I)_2$ CHOH]. Of the simple alkyl halides, the boiling point of the iodide is higher than that of the other halides. Also, as the length of the carbon chain attached to iodine increases, the boiling point increases. Primary alkyl iodides are characterized by the formula $R-CH_2I$, secondary by $RR'-CHI$, and tertiary by $RR'R''CI$, where R, R', and R'' are alkyl radicals, not necessarily different.

The alkyl iodides are insoluble in water, but soluble in organic liquids such as ether, alcohol, and benzene. They are used as alkylating agents.

The methods of preparing alkyl iodides, as well as their reactions, follow generally the standard pattern for alkyl halides. Iodides are as a rule more reactive than bromides which in turn are more reactive than chlorides. Reactivity of alkyl iodides increases with shortening of the carbon chain attached to iodine.

In the case of unsaturated organic iodides, reactivity is not much affected if iodine is attached to one of the carbon atoms involved in the double bond (ethylenic linkage), but reactivity is remarkably promoted if iodine is attached to the carbon atom one removed from the double bond. Hence

$$CH_3-C=CHI$$

is relatively less active than $CH_2I-CH=CH_2$.

Triiodomethane (iodoform, CHI_3) is formed by the reaction of free iodine with a compound of the type $RCOCH_3$ in the presence of alkali. Iodoform consists of yellow crystals that have a penetrating, lasting, antiseptic odor. The compound melts at 119°C, sublimes, and then explodes when the temperature is raised further. It is used slightly in external medicine. Diiodomethane (CH_2I_2) is noted because it is a liquid at 20°C that has sp. gr. 3.325. By mixing it with benzene (sp. gr. 0.879), mixtures that have different densities can be obtained through quite a range. Such mixtures are useful in the identification and separation of minerals. Carbon tetraiodide (CI_4) is also known.

Iodoethane (ethyl iodide, C_2H_5I) is a well-known alkylating agent. It melts at −108.5°C, and boils at 72.2°C.

Iodobenzene (phenyl iodide, C_6H_5I) is a colorless liquid that melts at −31.4°C and boils at 188.6°C. It is one example of an aryl iodide. 1,2-Diiodide benzene (ortho, $C_6H_4I_2$) melts at 286–287°C, the 1,3- (meta) compound at 284.8°C, and the 1,4- (para) compound sublimes at 285°C. Triiodo- ($C_6H_3I_3$), all varieties, tetraiodo- ($C_6H_2I_4$), all varieties, pentaiodo- (C_6HI_5), and hexaiodo-, C_6I_6, benzenes are all known. Iodine derivatives of phenol, aniline, and other cyclic organic compounds can be prepared.

"Polaroid" consists of two sheets of transparent plastic between which oriented crystals of an alkyloidal iodide are arranged.

ELBERT C. WEAVER

IODINE VALUE

Iodine value or number is the percent of weight of iodine absorbed by a sample of unsaturated organic materials. It is the principal method in common use for measuring the degree of unsaturation of fatty acids and their derivatives. Iodine values are actually determined by use of mixed halogen solutions, but the result is always calculated on the basis of the iodine equivalent of the total halogen absorbed.

Ideally iodine values should be determined by a procedure which results in addition of halogen to the point of total saturation, without side reactions with halogen such as substitution or oxidation. This ideal is closely approached in the case of fatty materials in which unsaturation is due solely to nonconjugated double bonds. Most of the double bonds in both vegetable oils and animal fats are in fact nonconjugated, which fact has enhanced the value of iodine values in research on these materials and in control of processing them for practical use. Fats and oils which owe their unsaturation to the commonly occurring oleic, linoleic, and linolenic acids show actual iodine values which correspond closely to the theoretical iodine values of these acids, as shown below:

Fatty Acid	No. of Double Bonds	Iodine Value
Oleic	1	89.9
Linoleic	2	181.0
Linolenic	3	273.5

Tung or Chinawood oil is outstanding among the few oils derived predominantly from conjugated unsaturated acids. Analysis of its mixed fatty acids indicates the presence of 85–90% of elaeosteric acid, a conjugated isomer of the common linolenic acid. Iodine values of tung oil tend to be erratic and are too low to serve as a true measure of its unsaturation.

Historically the most important procedures for determining iodine value were those of Hübl (1884), Wijs (1898), and Hanus (1901), all being of particular interest to chemists dealing with oils, fats and waxes. Hübl dissolved his sample in chloroform and reacted it with a mixed solution of iodine and mercuric chloride in 95% alcohol. The excess of halogen was then determined by titration with thiosulfate. The chief disadvantage of the Hübl method was instability of the alcoholic halogen solution.

In procedures generally similar to that of Hübl, the alcoholic solution of iodine and mercuric chloride was replaced by a glacial acetic acid solution of iodine monochloride in the Wijs method and of a mixture of iodine and bromine in the Hanus method. Both of these glacial acetic solutions of halogen had the advantage of greater stability and faster reaction with the sample, as compared with the alcoholic solution used by Hübl.

The procedures now used for determining iodine value are usually those prescribed by technical organizations and are based on collaborative research by many laboratories.

A. S. RICHARDSON

Cross-references: *Fatty Acids, Glycerides.*

ION EXCHANGERS

Ion exchangers are materials possessing the property that, when brought into contact with a solution, they do not dissolve in it but exchange can occur between ions initially present in the exchanger phase and ions initially present in the solution. This phenomenon (known as ion exchange) arises from the nature and the structure of the exchanger. Ion exchangers are essentially insoluble polyelectrolytes. Their insolubility arises from their chemical structure which consists of macromolecules often, though not always, crosslinked to one another to form a network or matrix. Their exchange properties arise from the fact that attached to the matrix by chemical (covalent) bonds are ionizable groupings. These may ionize into "fixed" ions, which remain attached to the network and "free" mobile counter-ions (so-called because their charge is of opposite sign to that of the fixed groupings). These counter-ions may exchange with any ions in a solution, provided that the charge of the ions in question is of the same sign as that of the counterions.

A large number of apparently unrelated types of substance behave as ion exchangers under appropriate conditions: synthetic ion-exchange resins; glasses; synthetic crystalline aluminosilicates; heteropolysalts; "hydrous" oxides and "mixed" hydrous oxides; natural alumino-silicates such as the zeolites and glauconites; natural organic materials such as coals, alginic acid, colodion and keratin; chemically modified natural substances such as sulphonated coals and sulphonated or phosphorylated paper and cotton. Recent work strongly suggests that ion-exchange processes may play an important part in the physiology of living things. This brings substances such as proteins and nucleic acids into consideration as ion-exchangers. Despite their apparent diversity, closer consideration shows that the chemical structures of all these types of substance possess the basic requirements of an ion-exchanger, namely, a macromolecular matrix with ionizable groupings.

Exchange groupings may be described as acidic or basic. Acidic groupings, when ionized, have a negatively charged entity attached chemically to the matrix and a free mobile cation. They therefore undergo cation exchange. Basic groupings, when ionized, have a positively charged entity attached chemically to the matrix and can therefore undergo anion exchange. Each of these can be conveniently subdivided into "strong" and weak." This subdivision, although a little arbitrary at times, is a useful qualitative way of describing the effect of pH on the degree of ionization of the grouping and is fairly analogous to the use of the same terms in describing simple acids and bases. There are therefore four types of grouping: strong acid, weak acid, strong base, weak base. Within each of these types, there may be more than one kind of grouping. Thus the following are different kinds of weak base anion-exchange grouping: $-NH_2$, $-N(CH_3)H$ and $-N(CH_3)_2$. An exchanger containing only one kind of grouping is said to be monofunctional. A polyfunctional exchanger contains more than one kind of grouping, though the different kinds of grouping may be of the same type. Many natural ion exchangers and some artificial ones contain both acidic and basic groupings, i.e., they are amphoteric.

The main consideration underlying the ion-exchange phenomenon is that electroneutrality must be maintained within the exchanger phase. For this reason, exchange is stoichiometric. Ion exchange is a reversible process and, in general, an equilibrium is set up. Where the competing counter-ions are of the same valency, the position of the equilibrium is determined by the relative *amounts* of the counter-ions in the system (exchanger and solution) and the relative affinity of these counter-ions for the exchanger. In this case, the total concentration of the counter-ions in the solution phase has little or no effect on the relative amounts of the counter-ions in the exchanger phase at equilibrium. Where, however, the competing counter-ions are of different valency, the situation is more complicated and the position of equilibrium is markedly affected by the total concentration of counter-ions in solution. A decrease of the total concentration (the relative amounts of the counter-ions remaining constant) always has the effect of shifting the equilibrium toward an increase in the amount of the counter-ion of higher valency in the exchanger phase.

Ion exchange selectivity or ion exchange relative affinity is the term used to describe the fact that, if an exchanger is in equilibrium with a solution containing two counter-ions of the same valency and in equal amounts (by equivalents), the amounts of these two ions in the exchanger phase will not, in general, be equal, i.e. the exchanger has a preference for one ion over the other. The problem of why exchangers exhibit ionic selectivity has received a lot of attention and ideas are still developing. However, opinion appears to be moving toward a consensus that, while a number of factors underlie selectivity phenomena, the two most basic are the interactions between fixed groupings and counter-ions (for the simplest counter-ions these interactions are mainly electrostatic in nature) and the interactions between counter-ions and solvent molecules (ionic solvation).

The kinetics of ion exchange processes have been studied and it is fairly generally agreed that, in all cases, the overall rate of exchange is determined by one or both of two kinds of diffusion process: (a) *particle diffusion,* meaning the diffusion of the exchanging counter-ions within the exchanger particle (b) *film diffusion,* meaning the diffusion of the exchanging counter-ions across a thin film of solution (the so-called "Nernst layer") surrounding each particle. Particle diffusion and film diffusion involve the simultaneous passage of the competing counter-ions in opposite directions. The "natural" rates of diffusion of these ions are, in general, different and could be characterized by different diffusion coefficients. Earlier treatments attempted to get round this difficulty by utilizing a "combined" (or "effective") diffusion coefficient, which was assumed to remain constant. Although this gave a fairly good representation of the experimental data, it was clearly an oversimplification. A more realistic theoretical treatment based on the Nernst-Planck flux equations has been more recently developed and its validity confirmed experimentally. A further complication arises when the exchanger is of the weak-acid or weak-base type, since then the degree of ionization of the exchange groupings varies with their position in the exchanger particle and also with time.

Although ion exchangers do not dissolve in aqueous solutions (or other solutions involving polar solvents), they swell in these media; i.e. the exchanger, if dry initially, imbibes solvent and increases in volume. This swelling is due primarily to the hydrophilic (polar) nature of the exchange groupings. The degree of swelling at equilibrium (i.e., the amount of solvent taken up per unit amount of exchanger) depends on a number of factors including the precise chemical structure and composition of the exchanger and the nature of the solvent.

The actual magnitudes of the three main properties so far discussed (selectivity, rates of exchange and degree of swelling) all depend on the precise chemical structure and composition of the exchanger. This structure is most easily discussed in terms of one particular type of exchanger, the ion-exchange resins. Ion-exchange resins are usually made by copolymerization of a suitable monomer with a suitable crosslinking agent, usually divinylbenzene (DVB). The exchange groupings are either introduced during the copolymerization (if they or their precursors are present in the monomer) or are introduced subsequently by chemical treatment of the copolymer. For most purposes, the structure and composition of the resin is defined by (a) the nature of the exchange grouping; (b) the degree of crosslinking, usually defined by the percentage of crosslinking agent used in the initial copolymerization; (c) the specific ion-exchange capacity. This last is defined as the number of milligram equivalents of fixed exchange groupings per gram of dry exchanger in a given ionic form. For cation exchangers, this is usually the hydrogen form; for anion exchangers, it is usually the chloride form. The number of milligram equivalents of fixed groupings is, of course, equal to the maximum number of milligram equivalents of counter-ions that can be exchanged without regeneration of the exchanger.

In general, the higher the degree of crosslinking, the lower the exchange rates and the higher the selectivity. Resins swell less in solutions of electrolytes than in the pure solvent, and this difference is the more marked with resins of low crosslinking. For most applications, resins are packed into columns and changes in volume (caused by changes in the concentration of the solution) disturb the uniform packing of columns. In choosing the appropriate degree of crosslinking for a particular purpose, one is therefore often forced to a compromise between the conflicting demands of a low swelling and a high exchange rate.

Exchange rates decrease with increase of particle size, since they are diffusion-controlled. However, if the exchanger is packed into a column of given dimensions, the smaller the particles the greater the hydraulic back-pressure associated with a given rate of flow of liquid. This is not usually a serious problem in laboratory applications but can become so with larger scale installations. Particle size, too, is therefore often a matter of compromise.

The effects of ion exchange specific capacity are complex and incompletely studied as yet. However, for most purposes it is usual to ignore refinements and aim at the highest practicable specific capacity. This is of the order of 5 mg equivs/gm for strong acid cation exchange resins, 10 mg equivs/gm for weak acid cation exchange resins and 4 mg equivs/gm for the usual strong base anion exchange resins.

The main reason (apart from convenience) why ion exchangers are usually used in the form of columns is the "equilibrium" or "mass-action" nature of the ion exchange phenomenon. Columns provide an almost ideal way of sweeping away the products of ion exchange and thus of driving the process to completion.

It is simplest (and perhaps most instructive) to

consider the applications of ion exchangers in terms of the phenomenon or effect being utilized rather than the field or scale of application. Many applications are based simply on the fact of ion exchange and its stoichiometry. Such applications may be called replacement applications. They include water-softening; water deionization; the removal of undesirable or interfering ions in various kinds of qualitative and quantitative analysis and in a number of medical applications; and the direct stoichiometric replacement of ions difficult to estimate by other more easily estimated (e.g., nitrate ions by hydroxide ions and alkali metal ions by hydronium ions).

Other applications are based on the phenomenon of selectivity. They consist mainly of various kinds of separation process, either for preparative or for analytical purposes. The separation is frequently enhanced by exploiting differences in the "solution chemistry" of the materials being seperated e.g., by the use of complexing agents. Such separations include those of rare earths and other groups of difficultly separable metals, nucleic acids, nucleotides, nucleosides, purine and pyrimidine bases, peptides and amino acids, sugar phosphates, sugars, aldehydes and ketones. In the last three cases quoted, the substances to be separated are not themselves ionic but can be converted to ionic species by complexing with borate or bisulfite ions. These separations are nearly always carried out using columns of ion exchange materials and with the usual techniques of chromatography. Ion exchange chromatography is, indeed, closely analogous to other kinds of chromatography but with a few special features deriving mainly from the stoichiometric nature of ion exchange.

Intermediate between these two types of application is a third, which may be called "concentration." In its simplest form it consists in the absorption of an ion by an ion exchanger from a dilute solution followed by its displacement from the exchanger with a concentrated solution of a suitable electrolyte. (In principle, an ion comes off a column at the same *equivalent* concentration as the total concentration of the solution displacing it.) An example of this kind of application is in the recovery of copper from waste rayon spinning solutions. It is often combined with some degree of separation e.g., in the recovery of uranium from leach liquors, of gold from cyanidation liquors, and of silver from photographic wastes.

Change of temperature usually causes little or no displacement of ion-exchange equilibria. However, it has recently been discovered that the uptake of hydrogen ions (competing with other cations) by weak acid cations exchange resins and of hydrogen ions (competing with other anions) by weak base anion exchange resins are both favoured by a rise of temperature. This raises the possibility of industrial processes involving the partial regeneration of weak acid and/or weak base exchangers using waste heat from other industrial processes. This kind of possibility is being explored, particularly in connection with water treatment.

Although ion exchangers are usually prepared and used in granular form (often in the form of spherical beads), in recent years increased attention has been given to ion exchangers in the form of sheets—usually known as ion-exchange membranes. Most of their applications (actual and potential) are based on their electrical conductivity, together with the tendency of all ion exchangers to exclude excess electrolyte. Work is proceeding towards developing suitable ion exchange membranes and cells for the electrolytic desalting of brackish water, for fuel cells, for various kinds of storage battery and for various kinds of electrolytic and electrophoretic separation.

D. REICHENBERG

Cross-reference: *Desalination.*

Ion Exchange in Inorganic Materials

Ion exchange was first demonstrated by Thompson and Way (1845) in an inorganic system; they showed that when ammonia solutions were percolated through soil ammonia was absorbed and displaced an equivalent amount of other bases such as lime and potash. This fact, which is relevant to the action of fertilizers on soils, is now known to be due to reversible exchange between ammonium ions in solution and calcium or potassium ions present in the clay minerals of the soil. Many aluminosilicates exhibit ion-exchange properties, and the first commercial ion exchangers, the zeolites used in water-softening, were aluminosilicates prepared by fusing together soda, potash, feldspar and kaolin. The absorption of cations and anions from solution by freshly formed oxide precipitates, which can be troublesome in analytical chemistry, is also due to ion exchange in many instances. Interest in inorganic exchangers has been re-awakened in recent years because many of them show greater stability at elevated temperatures ($\sim300°C$) than do the organic ion exchange resins, and it has been suggested that they might be used for cleaning up high-temperature water circuits in nuclear reactors. Their stability toward ionizing radiation offers promise of achieving economic separation processes under highly radioactive conditions, and some of them show greater selectivity toward certain ions than do their organic counterparts. Studies of the zeolites and of oxides suggest that these may even prove useful for separations in molten salt systems.

The inorganic ion exchangers may be either crystalline or amorphous, and chemically and structurally they fall into several fairly well-defined types. The clay minerals, which were the first to be studied, exist in many different classes, but all are characterized by layer lattice structures in which aluminate and silicate layers either alternate with each other in pairs or else form sandwich structures in which each silicate layer containing SiO_4 tetrahedra is situated between two aluminate layers containing AlO_6 octahedra. Pairs of layers in the first instance, or triple sandwich layers in the second, are stacked parallel to each other and perpendicular to the c-axis of the crystal. Strong covalent bonding within the composite layer units contrasts with weak van der Waals bonds between them, and the structure swells along the c-axis when immersed in salt solutions. The hydroxyl groups at the edges and corners of the sheets provide a source of roughly equivalent anion and cation exchange capacities via the reactions

$$R - OH + A^- \rightleftharpoons R - A + OH^-$$

$$R - OH + M^+ \rightleftharpoons R - OM + H^+$$

where R represents the clay backbone, A^- and M^+ being exchanging ions.

Such is the case in kaolinite (found in China clay), where both anion and cation exchange capacities are ~5 milliequivalents (meq) per gram. In other clay minerals, isomorphous substitution of Al^{III} for Si^{IV} in the tetrahedral layer, or of Mg^{2+} and Fe^{2+} for Al^{III} in the octahedral layer, produces a net negative charge on the sheets which is balanced by the entry of cations (Ca^{2+}, K^+ etc.) into the spaces between the layer units. The electrostatic forces between the layers and the balancing cations lead to stronger bonding between the units themselves than in kaolinite and so modify the swelling properties. When such clays are immersed in salt solutions the latter penetrate into the interlayer spaces and their cations exchange reversibly with the balancing cations present therein. The cation exchange capacity is thus much greater than the anion exchange capacity, the latter remaining relatively unchanged. The rate of exchange depends upon the rate of penetration of cations between the layer units; it is greatest for kaolinite, where only surface exchange is involved and the capacity is dependent upon particle size, and least for tightly bonded minerals such as illite and the micas, montmorillonite being intermediate between the two.

A wide range of cations are exchanged reversibly, the affinities of the minerals towards them generally following a pattern similar to that of the organic resins, e.g., $Li < Na < K < Rb < Cs$; thermodynamic studies have enabled heats and free energies of exchange to be determined. Clays cannot be converted to the hydrogen form by treatment with acid, which destroys them chemically, but hydrogen clays have been prepared by electrodialysis or by shaking with a suspension of mixed-bed resins in H^+ and OH^- forms. On exchanging long-chain alkylammonium ions, e.g., $C_{17}H_{35}\overset{+}{N}Me_3$, clays are converted to organophilic derivatives which readily absorb nonpolar compounds and are useful in chromatographic separations of hydrocarbons. On heating, clays lose water, reversibly at low temperatures but irreversibly above 300°C, the irreversible loss of water being accompanied by irreversible loss of capacity. At high temperature (900–1000°C) the lattice collapses to form a silica-like structure in which the trapped cations are very resistant to leaching, and it has been suggested that radioactive waste products could be exchanged on clay and then fixed in this way. Fixation also occurs at normal temperatures when small cations are exchanged on certain clay minerals in which collapse can occur by expulsion of interlayer water, the cations actually entering the layers instead of being situated between them. The phenomenon of "potassium fixation" by certain soils has been attributed to this cause.

In the *zeolites* aluminate and silicate tetrahedra are covalently bound together by sharing corners, edges or faces to form a series of three-dimensional structures possessing a net negative charge balanced by cations located in channels and cages in the lattice. These may also contain water molecules and simple anions such as sulfate, bicarbonate, etc. Many examples, both natural and synthetic, are known. When zeolites are immersed in salt solutions, ions from the latter which are small enough to enter the structure can exchange reversibly with the ions there present if the latter are in turn small enough to diffuse out into the solution. Since the rigid, three-dimensional structure cannot swell appreciably in solution as do the clays, exchange is strongly dependent upon the sizes of the two ions in question in relation to the structural parameters of the exchanger. In addition, therefore, to variations in affinity arising from thermodynamic factors, the zeolites, particularly those with cages and channels of small dimensions, can act as molecular sieves in excluding some ions completely. This property may be used to effect certain simple separations, e.g., analcite exchanges Rb^+ ions readily (radius 1.48Å) but not Cs^+ ions (1.63Å). In some of the synthetic zeolites, e.g., Linde Molecular Sieve 4Å, a complex ion sieve behavior results from the existence of more than one size of cavity in the lattice. The unit cell of this zeolite contains 13 sodium ions, 12 of which are located in cages linked by channels of diameter 4.2Å and the 13th in one with channels of diameter only 2.5Å. All 13 Na^+ ions may be exchanged for small cations such as Ag^+ (radius 1.26Å), but only 12 for larger cations such as Tl^+ (1.44Å). A volume sieve effect also exists for ions small enough to enter the cage but too large to enable the full number to occupy the available volume of the cage. The rates of ion exchange in zeolites are low compared to those for clay minerals and for organic resins, owing to the absence of swelling. Although the latter property generally results in considerable simplification of the thermodynamics of exchange compared with the clays and resins, this is not true when the end-members of the exchange system show limited solubility in each other. Hysteresis is then frequently seen in the exchange isotherms and the final position of "equilibrium" depends on the side from which it is approached.

The varying sizes of the channels and cages also enable the zeolites to act as molecular sieves towards gas molecules by excluding those whose dimensions are too great. Polar molecules are very readily absorbed if the pore size is adequate, and for this reason the zeolites are very effective in drying gases, even at high temperatures. The selectivity between two gases may often be varied by changing the temperature of absorption, and as the free channel diameter may be changed by altering the water content or by carefully selecting the particular cationic form of the zeolite (according to the cation radius) the zeolites find many applications in the chromatographic separation of gas mixtures.

The *ammonium salts of heteropolyacids* also exhibit ion-exchange properties, and NH_4^+ ions in ammonium 12-molybdophosphate, $(NH_4)_3PMo_{12}O_{40}$, may be replaced reversibly by K^+, Rb^+ or Cs^+ ions in order of increasing affinity. Many other heteropolyacid salts show this property, the cations in question forming sparingly soluble salts in which the cations are located in cavities within the unit cell lattice containing the polyacid anions. Ammonium molybdo- and tungstophosphates are particularly efficient at removing and separating the alkali metal cations even in acid solution, the separation factors being considerably higher than those for a resin such as Dowex-50, as the following figures show:

	$\alpha \dfrac{Rb}{K}$	$\alpha \dfrac{Cs}{Rb}$
"Dowex-50"	1.13	1.19
Ammonium molybdophosphate	68	260

Efficient column separations of the alkali metals have been carried out in acid solutions even in the presence of large concentrations of interfering ions; at higher pH the alkaline earth and trivalent cations may be separated into two groups, but separations among these latter elements themselves have not yet been achieved. Paper impregnated with ammonium molybdophosphate has also been used for separating alkali metals in tracer concentrations. Other ion exchangers of similar type, which have been less thoroughly studied, include the complex ferrocyanides such as $K_2[CoFe(CN)_6]$, which removes cesium very effectively from solution, and the curiously-termed "ferrocyanide molybdate," $[H_4Fe(CN)_6]_m \cdot [MoO_3(H_2O)_x]_n$ obtained by mixing solutions of sodium molybdate and ferrocyanic acid.

Most insoluble *oxides,* unless dried at temperatures where all their water is irreversibly lost, exhibit reversible ion exchange behavior when suspended in aqueous solutions; to this, which is due to hydroxyl groups present in the structure, one may attribute the absorption of cations or anions (depending upon the pH) when the oxide is first precipitated. While most oxides behave as anion exchangers at low pH and cation exchangers at high pH, due to the amphoteric properties of the OH group, the more feebly basic oxides such as hydrous silica show only cation-exchange properties. Cations which readily form metal-oxygen bonds (Fe^{+++}, Cr^{+++}, etc.) are readily absorbed by most oxides even at low pH. The difference in affinity among different cations is such that separations can be carried out by varying the pH or the eluant, e.g., alkaline earths and rare earths may be separated on ZrO_2 by elution with $1M$ NH_4Cl and $1M$ HCl respectively, while Cs^+ and Ba^{++} have been separated on UO_3 with $1M$ NaOH and $10M$ NH_4NO_3. Polyvalent anions are more strongly absorbed than monovalent ones, and those which form insoluble salts are irreversibly bound, e.g., phosphate on ZrO_2. A striking example of the selectivity of oxide exchangers is seen in the removal of uranium from sea water by hydrous titania or by basic zinc carbonate, where the exchanger not only removes uranium preferentially but also competes with the extremely strong bonding of the latter in the soluble tricarbonato complex $UO_2(CO_3)_3{}^{4-}$.

Zirconium phosphate, prepared by precipitation from a zirconium salt solution with phosphoric acid, and many other similar insoluble salts have been much studied as cation exchangers. By varying the conditions of preparation a wide range of crystalline or granular gel products may be obtained with P:Zr ratios up to 2:1 in the case of zirconium phosphate;

reversible exchange in these materials involves the hydrogen atoms of acid phosphate and similar groups in a manner analogous to the sulfonic acid groups of the strongly acid ion exchange resins. Although less acid than the latter, their functional groups are much more strongly ionized than those of the carboxylic resins. The capacities of these materials may be as high as 5 meq/g in alkaline solution, depending upon the mode of preparation and the temperature of drying, but in acid solution the capacity rarely exceeds 1.5 meq/g. They are very stable towards ionizing radiation and towards neutral or acid aqueous solutions at 300°C, but are hydrolyzed to form hydrous oxides in alkaline solutions. Rates of exchange are good, and for most exchangers of this type affinities towards the alkali metal cations increase in the order Li, Na, K, Rb and Cs. The hydrogen form may be used to effect group separations, elution proceeding in the reverse order to that on the oxides, viz. rare earths, alkaline earths, alkali metals; excellent separations among the alkali metals and alkaline earths are possible on zirconium phosphate, tungstate or molybate. The structures of a number of crystalline ion exchangers of this type have been determined by x-ray methods and correlated with their ion-exchange behavior; interesting ion-sieve properties are found, based on the nonswelling properties of the lattice. In some cases, e.g., cerium and thorium phosphates, the product may be formed in fibrous form and used to prepare paper or membrane for chromatography or dialysis.

Although up to the present time ion exchange has generally been limited to studies in aqueous and simple organic systems at modest temperatures, the growing interest in inorganic exchangers may lead to its extension to high-temperature aqueous systems and to fused salts, so extending its concepts into the realms of mineral chemistry and geochemistry.

C. B. AMPHLETT

Cross-references: *Soil Chemistry; Clays; Molecular Sieves; Ion Exchangers.*

References

Amphlett, C. B., "Inorganic Ion Exchangers," Elsevier, 1964.

Selected papers in "Molecular Sieves, Proceedings of Conference" organized by the Society of Chemical Industry, London, 1968.

Papers in Session 7, "Ion Exchange in the Process Industries," Society of Chemical Industry, London, 1970.

IONS

Atoms, or groups of atoms, which have either taken up or surrendered one or more electrons from their outer electronic shells are known as *ions.* These ions consequently bear positive or negative charges which are integral multiples of the elementary charge of the electron, 4.77×10^{-10} electrostatic unit. Positively charged ions are called *cations* while negatively charged ones are known as *anions.*

The charge carried by "simple" ions, i.e., those consisting of a single charged atom, is numerically equal to the valence of the elemental atom, or the number of electrons the neutral atom must gain or surrender in order to acquire a stable electronic configuration. Thus, the neutral sodium atom has one electron in its outer orbit and can acquire a stable configuration by losing this one to an atom lacking a single electron in its outer orbit, such as a halide atom. The resultant sodium ion has a positive unit charge and the halide ion a negative unit charge. Doubly charged ions such as the positive calcium ion may combine with a singly charged negative ion to form a singly charged complex ion—a class of ions which is very important in most natural geological and biological systems.

The charge carried by the ions is largely responsible for their unique properties in or out of solution, which differ markedly from the properties of the neutral atoms or groups of atoms. In solution the ions interact with the solvent to produce a kinetic entity, or solvated ion, whose properties are modified to a considerable degree by the solvent molecules attached in varying degrees to the ion. Out of solution, ions may exist in the solid crystalline state, as in the sodium chloride lattice, or in the gaseous state, where they are produced by the action of high energy radiation on neutral atoms or molecules.

A fundamental property of ions is the electrostatic attractive force between ions of opposite charge and the repulsive force between those of like charge. The former is largely responsible for the formation and stability of many crystalline and other type "ionic" compounds, while taken together these polar forces are responsible for the formation of the "ionic atmosphere" around each ion in solution, important in the Debye-Hückel theory of electrolytic solutions. In crystalline "ionic" compounds, the ions acquire their charge by a transfer of an outer orbital electron from one atom to the other, each atom thus acquiring a stable electronic configuration similar to that of the inert gases. The ions so formed are held together in a crystalline formation by the powerful electrostatic forces between them.

When ionic compounds are dissolved in a solvent medium such as water, the dielectric properties of the solvent reduce the electric field strength between oppositely charged ions, causing them to dissociate and behave as more or less independent kinetic entities within the solution. These dielectric properties of the solvent are the result of the action of ionic electric fields on the solvent molecules, giving rise to orientation polarization and distortion polarization. Thus the ions and the solvent interact to modify the properties and structure of both, with the strong electrostatic fields of the ions being the major cause of these interactions. Each dissociated ion tends to become solvated by attaching more or less firmly a number of solvent molecules, the number varying with external conditions of temperature and with the nature and relative concentration of all constituents present. The mobility of the ion is dependent on the extent of this solvation, which therefore greatly influences the electrochemical properties of the solution.

The extent to which a neutral compound dissociates into ions in solution depends on the nature of the compound and the nature of the solvent, as well as external condition of temperature, pressure, etc.

Under given external conditions, an equilibrium will exist between neutral molecules and ions, though some compounds designated as strong electrolytes are regarded as being completely dissociated even at high concentrations. The position of equilibrium constant if the activity coefficients of the ions are taken into consideration; these coefficients are a function of the concentration of the electrolyte in solution.

When an external electric field is applied across an electrolytic solution by means of suitable electrodes, the cations are electrostatically attracted to the negative electrode, or cathode, while the anions are attracted toward the positive electrode, or anode. If a chemical reaction occurs at both electrodes which transfers electrons from electrode to ion, or vice versa, the ions in the solution will move toward the electrode attracting them; and this movement constitutes the electric current in the solution. The electrical conductivity of solutions depends, then, on the concentration and mobility of the ions in the solutions. The mobility of an ion is defined as the velocity attained by that ion in cm per sec under a potential gradient of one volt per centimeter. Although the motion of a charged particle in a field is, in general, accelerated according to Newton's law that force equals mass times acceleration, ions are almost instantaneously accelerated to a limiting velocity determined by the viscous drag of the solvent, and thus move at a constant velocity under given conditions, which for ordinary fields is directly proportional to the applied field. Ohm's law is for this reason valid for electrolytes subjected to ordinary fields.

The mobility of an ion in a given solvent will depend on the size of the solvated ion, in other words, upon the ion-solvent interaction. Measurement of the mobility of an ion combined with Stokes' law of motion of spherical particles in a continuous medium provides a means for calculating a radius of the solvated ion, usually referred to as the "Stokes law" radius. Since electric current in solutions results from the motion in opposite directions of oppositely charged ions, each ion species can be considered to carry a certain proportion of the current. The fraction of the total current carried by a particular ion species is known as the transport (or transference) number of that ion. Evidently, it is proportional to the relative concentration of that ion in the solution, as well as to the magnitude of the charge on the ion and its absolute mobility under the given conditions.

In certain polymeric systems, as well as in various natural minerals and clays, one charge type of ions formed in the presence of a solvent is immobile, and fixed to the structure of the polymer or mineral. The mobile counter-ions (gegen ions) conduct all or nearly all the electric current which may be passed through the system, thus having a transport number near unity. Such systems form the basis for the process of ion exchange in which one species of ion replaces another in the polymeric or mineral system, as well as the basis for electrochemical operations involving permselective membranes.

The importance of ions in all natural processes can be understood by the realization that electrical conduction in all aqueous and other nonmetallic systems is almost entirely carried on by the movement of ions, and that most inorganic reactions involved in geological and biological systems are ionic in nature. The energy released in ionic reactions may take the form of electrical energy, as in storage and "dry" batteries, while the flow of ions in living systems due to concentration gradients gives rise to electrical potentials and impulses of great physiological significance, accounting in part for the great importance of ion concentration, particularly the hydrogen ion concentration, in living tissue. In soils and minerals, ion and ion exchange processes are of prime importance, determining in considerable measure the fertility of the soil and the durability and general utility of the mineral.

Ionization

The process by which a neutral or an uncharged atom or molecule acquires a charge, thus becoming an ion, is known as *ionization*. This process usually occurs by various mechanisms in liquid, gaseous, or solid media, acting on atoms or molecules constituting the medium, or dissolved in it. When a substance dissolved in a liquid, or a liquid itself, undergoes ionization, oppositely charged ions are formed, which may still interact with each other or the un-ionized solvent. When a substance in the gaseous state is ionized, usually by a collision process or by absorption of radiation, ions of positive charge only may be formed as a result of loss of an electron, and the electron is free as a result of the process. In all cases the ions formed are subject to motion or deflection in electric fields, and under suitable conditions, may recombine to form neutral atoms or molecules. Their motion in an electric field constitutes an electric current and, outside of metallic conductors, this is the mechanism of electrical conduction in most systems.

The atoms of a neutral molecule are held together by a number of molecular forces, among them the electrostatic attraction between charged portions of the molecule, which may be ions. If interionic attraction is the major force binding a molecule together, any significant reduction of this attraction will dissociate the molecule into two or more charged particles, each of which behaves as a more or less independent kinetic entity. This is the mechanism in the case of ionization of electrolytes in solution, where the solvent dielectric properties reduce the electrostatic forces between the ions and cause the molecule to dissociate. In a solvent such as water, some of the solvent molecules themselves are ionized, though to a very limited extent. In water the concentration of hydrogen or hydroxyl ions is 10^{-7} mole per liter. It is apparent that the ionization of a solute is dependent both on the properties of the solute and of the solvent, so that a substance may ionize in one solvent of high dielectric constant while remaining un-ionized in another solvent of low dielectric constant. Conditions of temperature and pressure will affect the degree of ionization of solute in a given solvent. This fact itself is dependent on volume or heat content changes which may occur during the ionization; the latter can be regarded as a reversible reaction involving an equilibrium between ionized and un-ionized solute. The rate of attainment of equilibrium is evidently very rapid for most solvents. Chromyl sulfate solutions are a noted exception in that weeks or even months are required for equilibrium attainment between the various ionized and un-ionized species.

Ionization of gases differs from ionization of solutes in a solvent both in the mechanism by which it occurs and the products formed. The atoms or molecules of gas lose an electron by collision with another electron, producing a positively charged ion and a free electron, rather than two oppositely charged ions. The ionization of gases occurs when an electric discharge is maintained in the gas at low pressure, as in a glass tube nearly evacuated of the gas in question. The electrons traveling from cathode to anode in the gas may, if their energy is sufficient, ionize the gas by detaching an electron from it. A particular voltage is required to do this, differing for each gas and known as the *ionization potential* of the gas. When the electrons emitted by the electrodes of the discharge tube strike an atom of the gas, the collision may result in excitation or ionization of the atom, depending on the magnitude of the voltage, or the energy of the electron. Excitation results when an electron of the atom is knocked to a higher energy level, without being ejected from the atom completely, while ionization results if the voltage is sufficient to cause the electron to be ejected completely. The frequency of the shortest wave length radiation the normal atom is capable of radiating can be used to calculate an energy of ionization, which should and actually does agree with experimentally observed values using the electric discharge method. The positive ions produced in the manner described above can be subjected to electric and magnetic fields to determine their mass and charge, which is the operating principle of the mass spectrograph.

Ionization of gases, liquids, or solids may also occur by the action of radiation on the atoms or molecules. The energy of the radiation, which may be particle or energy radiation, may be absorbed by the atoms encountered to knock an electron from them, forming an ion. α, β, and γ as well as cosmic rays may ionize a substance under suitable conditions, and the latter are considered responsible for the low level of ionization always present in the atmosphere. Conditions of extreme temperature may also ionize the gases existing at these temperatures.

ROBERT KUNIN AND ALVIN WINGER

Cross-references: *Ions, Electrochemistry.*

Ions, Gaseous

Chemists tend to think of ionization as exclusively a solution phenomenon, but this is not the case. In solution the solvent provides the energy necessary to bring about ionization, and at the same time exercises an influence upon the chemical behavior of ions existing in it. Energy must be provided from exterior sources if ions are to be formed in the gas, but the ions thus formed are isolated entities uninfluenced by any solvent; thus their chemistry is often quite different from that of ions in solution. The formation of an ion in the gas phase is usually brought about by impact with a rapidly moving particle, usually an electron, or by x-rays or ultraviolet light of short wave length. Studies of ions in gases usually require a mass spectrometer as an important part of the experimental equipment. In fact, a large part of the literature on the chemistry of gaseous ions has resulted from experiments in which a mass spectrometer was used to generate the ions, to provide a suitable reaction vessel and to yield a rapid analysis of reactants and products. A mass spectrometer is a device by which rapidly moving ions are separated according to the ratio of mass to charge, and the resulting separated ion beams are measured and recorded.

When a molecule is struck by an energetic electron or photon it may absorb sufficient energy to cause ionization. Further, the energy may be sufficient not only to ionize the molecule, but to bring about its dissociation into two or more fragments, at least one of which will be charged. This fragmentation, however, is slow compared to ionization, since the atoms or groups of atoms in the molecule necessarily move more slowly than electrons or light waves. Thus the process of fragmentation at low pressures, where most ionization phenomena are studied, results from the redistribution of energy through the various internal modes of the molecule, and occurs when enough energy is concentrated in a single mode to bring about bond rupture. This is, of course, a typical example of a first order rate process in an isolated system and the kinetics of such decompositions have been the subject of a great deal of theoretical and experimental research in the past 40 years. Eyring and his colleagues have modified the Rice-Ramsberger-Kassel theory of first order kinetics to apply to the decomposition of isolated ions, and the Eyring theory has been further refined to allow for the various quantum states of the ion. Although the resulting theory is not necessarily perfect, it clearly shows that mass spectra result from rate processes occurring in short but finite times.

The retention time of an ion in a mass spectrometer ion source is approximately a microsecond, and consequently any chemical behavior of ions that can be observed in the mass spectrometer must occur in times of this order of magnitude. Since mass spectra usually show that extensive decomposition has occurred often with quite profound rearrangement of the structure of the molecule, it is evident that many ions are very labile entities. Several examples of this will be given below; however, we should first explain certain measurements of ion energetics.

If a mass spectrometer is focused upon an ion of certain charge-to-mass ratio, and the energy of the ionizing beam is gradually reduced, the intensity of the ion will be similarly reduced and will eventually reach zero. This point is said to be the appearance potential of the ion and is in general a definite characteristic of the process by which the ion is formed. In many instances it is possible to treat this value as a heat of reaction and calculate from it the heat of formation of the ion. When this is done for the alkyl-carbonium ions, the following values are found: methyl, 260; ethyl, 220; s-propyl, 193; t-butyl, 170 kcal/mole. This order of increasing stability is just the one that would be expected from organic solution chemistry.

If one measures the appearance potential of an ion such as $C_3H_7^+$ (propyl) from n-butane, isobutane, propane, n-propyl chloride, and s-propyl chloride, and s-propyl radical, one obtains the same heat of formation from all; namely, 193 kcal/mole. The n-propyl radical, however, gives a heat for formation of about 220 kcal/mol. Thus, when propyl ion is formed by fragmentation it tends to assume the secondary structure. With n-butane and s-propyl chloride, rearrangement must occur during decomposition to yield the secondary propyl ion, and this rearrangement must occur without activation energy. Similarly, if one obtains the heat of formation of an ethylene ion from either ethylene by direct ionization or from ethane with the elimination of H_2 one obtains the same value. Thus one must conclude that the 4-center process of ethane ion dehydrogenation occurs without measurable energy of activation. This again illustrates the extreme lability of ions and should be contrasted with a similar process for dehydrogenation of neutral ethane which must involve at least 50 kcal/mole of activation energy above the endothermicity of the reaction.

It should not be inferred from the above that the appearance potential is always equal to the heat of reaction. In many instances it is not, but rather occurs with considerable excess energy so that the products of ionization are excited. When no fragmentation occurs this excitation will be either electronic or vibrational or both. When fragmentation occurs the fragments may also be formed with translational energy in excess of the normal thermal energy. When excess energy is involved it must be separately determined and deducted from the appearance potential before valid thermochemical calculations can be made with it. This is not a simple experiment and it cannot always be accomplished successfully. Specially equipped instruments are required to measure translational energy and the translational energy must be related to total excess energy. Encouraging progress has been made in developing this relationship but much more must be done.

In view of the great lability of isolated ions it would not be surprising to find that ions undergo extensive and rapid reaction upon collision. In recent years, extensive studies of reactions between ions and molecules have been made and all bear out this expectation. Many reactions occur, often with surprising results, and usually at rates that are very high compared to any known rates of reaction of neutral compounds. Thus it can be shown that the rate constant for the collision between an ion and a neutral molecule will obey the equation:

$$k = 2\pi e \left(\frac{\alpha}{\mu}\right)^{1/2}$$

where α is polarizability

μ is reduced mass

e is the unit electric charge.

In many cases reaction occurs at every collision so that the above equation also gives the reaction rate constant. The theory of ion-molecule reaction kinetics is being actively investigated by many chemists and physicists but as yet a completely satisfactory solution has not been obtained.

Among the various kinds of known ion-molecule reactions are those for the transfer of hydrogen atoms, of protons, and of hydride ions, as well as more profound bond-making and breaking processes. All proceed at comparatively high rates. Except for rather rare circumstances, the only ion-molecule reactions that have been observed are exothermic. It should also be noted that there are many instances in which several products result from the same pair of reactants. Thus, two major ions are formed from the reaction of $C_2H_2^+$ ion with ethylene, and in butadiene some five product ions are formed from the same reactants. Simple condensation reactions (sticky collisions) have been reported in a few instances, but they appear to be rare unless the pressure is sufficiently high to permit deactivating collisions to occur.

The ion CH_5^+ violates our concepts of valence, but is nevertheless a very real compound. Indeed, it is a very stable product, having been formed by the transfer of a hydrogen atom from methane, and thus has a C—H bond strength at least as great as that in methane. At pressures above 1 mm Hg, the CH_4^+ ion is almost completely converted to CH_5^+. The $C_2H_7^+$ ion, the next in the series, has also been observed, although in much less abundance than CH_5^+. The reason is that $C_2H_7^+$ apparently has a heat of formation only very slightly (3 or 4 kcal/mole) less than that of $C_2H_5^+$. Thus it can be observed only under conditions at which the ion is almost completely without excess energy, since even a small amount of excess energy would result in its decomposition to $C_2H_5^+$. It seems doubtful that the higher homologs are capable of existence.

If the pressure exceeds 50 to 100 microns, one observes many ions having pressure dependences of third order or greater. This is especially true in unsaturated systems. The normal paraffins show very few ions beyond 3rd order. In ethylene and acetylene, however, small amounts of ions showing pressure dependence of at least 6th order have been observed at pressures of a few hundred microns and, no doubt, far more extensive polymerizations would occur at either retention times or pressures greater than those available in ordinary mass spectrometry. It is of particular interest that with various polar compounds highly solvated protons have been observed in the gas phase. Thus with water, protons combined with as many as 8 molecules of water have been observed. With ammonia, solvation of the proton with up to 6 molecules of ammonia have been reported and similar results have been observed with several other polar compounds such as dimethyl ether, formic acid and acetonitrile.

The previous discussion has been directed toward the formation and reactions of positive ions. However, under the appropriate conditions, negative ions can also be formed. Photoionization can bring about a pair production process by which a positive and negative ion are formed simultaneously. Thus, ultraviolet light of the appropriate wave length will bring about the following reaction:

$$CH_3Cl + h\nu \rightarrow CH_3^+ + Cl^-$$

Electron impact can also cause pair production processes. In addition, electron impact can cause the formation of negative ions by either direct attachment to a sufficiently complex molecule or by dissociative resonance capture processes, thus;

$$SF_6 + e^- \rightarrow SF_6^-$$

and

$$CO + e^- \rightarrow O^- + C$$

$$Cl_2 + e^- \rightarrow Cl^- + Cl$$

Such processes occur over a relatively narrow range of electron energies and usually at quite low voltages. The attachment of an electron to SF_6 occurs at very close to 0 electron volts and the rate decreases rapidly above that value. Dissociative resonance capture processes involve the breaking of a bond but this is aided by the electron affinity of the species forming the negative ion. In the formation of O^- from carbon monoxide, this occurs at about 9.6 electron volts whereas the formation of Cl^- from chlorine occurs quite close to 0 electron volts. The latter, in fact, involves a considerable amount of excess energy which is observed as translational energy. Negative ions undergo collision reactions in the same fashion as do positive ions and at rates that are comparable to those of positive ions.

In addition to collisional reactions of ions there is a class of reactions known as chemi-ionization reactions by which ions result from the interaction of highly excited species. Thus, in helium the diatomic ion He_2^+ is formed at an electron energy about 1.4 electron volts below the ionization potential of helium. Obviously an electronically excited helium atom has reacted with an atom in the ground state with ejection of an electron to form the ion.

All the rare gases undergo reactions of this kind and in addition many other such chemi-ionization reactions have been reported. Several typical ones are given below:

$$Ne^* + Ne \rightarrow Ne_2^+ + e^-$$
$$He^* + Kr \rightarrow HeKr^+ + e^-$$
$$N_2^* + N_2 \rightarrow N_4^+ + e^-$$
$$Xe^* + CH_4 \rightarrow XeCH_4^+ + e^-$$
$$Xe^* + C_2H_2 \rightarrow XeC_2H_2^+ + e^-$$
$$Hg^* + CH_3OH \rightarrow HgCH_4O^+ + e^-$$
$$CO^* + CO \rightarrow C_2O_2^+ + e^-$$

Although it is difficult to make exact measurements of the rates of chem-ionization reactions a few attempts have been made to do this and, although the data are not very precise, they show that without question these reactions occur at rates comparable to those of ion-molecule reactions.

Although this discussion has referred only to results obtained with a mass spectrometer, it must be obvious that ionic reactions are not limited to the ion source of a mass spectrometer. Indeed, reactions of the same kind as those discussed above must occur in electric discharges, in radiation chemistry, and in flames. The latter have been the subject of extensive investigation because the presence of ions in rocket exhaust interferes with radio communication with the rocket. The origin of ions in most flames has not been established, but it is clear that the temperature is not sufficiently high to bring about ionization. It is now thought by many combustion scientists that some form of chemiionization reaction must produce the first ions in the flame and that these subsequently undergo a variety of ion-molecule reactions. One reaction that is widely thought to be the precursor of most ions in flames is $CH + O \rightarrow CHO^+ + e^-$ This, however, is largely speculative and has not been proved.

J. L. FRANKLIN

Cross-references: *Ions, Ionizations, Mass Spectrometry, Flame Chemistry.*

IRIDIUM AND ITS COMPOUNDS

Iridium is a member of the platinum-group elements. It was identified in 1803 by S. Tennant (England) who isolated it from the black residues remaining after treating native platinum from Spanish America with aqua regia. The name of the element is derived from the Greek *iris*—rainbow, in consequence of the variety of colors of its compounds in acid solutions.

This element is found as a minor component in native platinum, an alloy including most of the platinum metals (with platinum predominating) and some iron and copper. It also occurs in two native alloys with osmium—osmiridium ($<$ 32% osmium, cubic) and iridosmine ($>$ 32% osmium, hexagonal). In these natural occurrences the element is presumably uncombined, and indeed no mineral compound of iridium has been characterized. The abundance of iridium in the earth's crust has been estimated as 0.001 ppm.

The following are some of its important physical properties: atomic number 77; atomic weight 192.22 stable isotopes 191 (38.5%) and 193 (61.5%); density 22.65 g/cc (20°C); melting point 2447°C; hardness (annealed) 200–240 Vickers units; electrical resistivity 4.71 μohm-cm (0°C). The crystal form in the solid metal is cubic closest packing, with $\alpha = 3.8394$ Å, giving a metallic radius (for 12-fold coordination) of 1.354 Å. Iridium has the greatest density of all the elements, marginally exceeding that of osmium. It is also remarkable for its great mechanical strength which is sustained at high temperatures.

Iridium forms alloys with the other platinum metals as well as with many base metals. It is widely used to increase the hardness and corrosion resistance of platinum and palladium. In applications for hardening platinum, iridium is not preferred for high temperatures owing to appreciable losses in weight when this element is heated above 1000°C in an atmosphere containing oxygen; this phenomenon has been traced to formation of a volatile trioxide.

Of all the platinum metals iridium displays the greatest resistance to corrosion. At red heat it is attacked by oxygen to form the dioxide, though only superficially. At comparable temperatures it is attacked by the four halogen elements although to a less extent than the other platinum metals. Fluorine forms mainly IrF_6 with some IrF_5; chlorine produces $IrCl_3$. Iridium is not attacked by any acid, including aqua regia, and is scarcely attacked by molten alkali.

The common oxidation states of this element in its compounds are +3 and +4, though examples of all states from 0 to +6 have been described. The element shows no evidence of a simple aqueous cation, and in this it resembles the other members of the triad of heavier platinum metals.

Iridium dioxide is a black solid with the rutile structure. It may be formed by heating the finely divided metal in oxygen at 1000°C, or by hydrolysis of $IrCl_6^{2-}$ to produce an intensely blue hydrous oxide which may be dehydrated under nitrogen or carbon dioxide. The trichloride, when formed by direct union of the elements at 450–600°C, may be olive-green, brown, or black according to particle size, and is insoluble in water. A water-soluble hydrated form is made by the action of hydrochloric acid on the hydrous dioxide.

Like the other platinum metals, iridium tends to form a wide range of complexes. A common example is the dark-red chloroiridate ion, $IrCl_6^{2-}$. There are a number of salts of this ion which are sparingly soluble in water, e.g., those of the heavier alkali metals (K to Cs) and ammonium, and also of the protonated forms of many organic nitrogen bases; the latter are sometimes characterized by their chloroiridate derivatives. Chloroiridite, $IrCl_6^{3-}$, is formed by the fairly easy reduction of chloroiridate; it is a pale olive-green color in its salts and solutions. These complexes are comparatively inert to ligand substitution.

Iridium is used commercially mainly for alloying with platinum. It hardens and strengthens the latter, but also increases its resistance to corrosion. A drawback to its use for this purpose is that the alloys suffer losses in weight at high temperatures (above 1000°C) owing to formation of volatile oxides as mentioned above. Thus for high-temperature uses, rhodium is the preferable agent for hardening platinum.

Iridium-platinum alloys are used for electrodes, for vessels to handle corrosive chemicals, for electrical contacts exposed to corrosive atmospheres, for jewelry, for various medical devices, and so forth. Some pure iridium vessels have been fabricated to enable processes to be carried out at very high temperatures, e.g., the melting of materials for laser crystals. Also, for use at very high temperatures thermocouples have been developed in which the elements are pure iridium and an iridium-40% (or 50%) rhodium alloy.

W. A. E. McBRYDE

IRON

Iron, symbol Fe from the Latin *ferrum,* has an atomic number of 26 and an atomic weight of 55.847. The element is located in Group VIII of the Periodic Table as the first member of the triad: iron, cobalt and nickel.

Iron has four stable isotopes: 54 (5.90%), 56 (91.52%), 57 (2.245%), and 58 (0.33%). Its electronic configuration is $1s^2, 2s^2, 2p^6, 3s^2, 3p^6, 3d^6, 4s^2$. While iron can exist in the +4 and +6 oxidation states, they are rare and the common valences for it are +2 (ferrous) and +3 (ferric).

Iron is the most important metal known and used by man. The annual production of steel in the United States alone was 131,000,000 tons in 1970. This quantity exceeds the total production many-fold of all the other metals combined. Next to oxygen, iron is the element used in greatest amount in elemental form by man. This is due to its widespread prevalence, the ease with which its ores are reduced to the metal, and its many desirable structural properties. In addition, iron is found in all mammalian cells and is absolutely vital to the life processes of animals.

Occurrence

Iron is the fourth most prevalent element in the earth's crust, after oxygen, silicon and aluminum, and comprises about 5% of the earth's crust. It is a constituent of several hundred minerals, but fortunately, huge deposits of easily reduced iron oxide occur in many countries and these are the sources of most of the iron produced by man. The principal iron minerals are hematite, Fe_2O_3; magnetite, Fe_3O_4; limonite, $Fe_2O_3 \cdot xH_2O$, hydrated iron oxide; and the carbonate, siderite, $FeCO_3$. Minor use is made of iron pyrites, FeS_2, and of ilmenite, $FeTiO_3$, or $FeO \cdot TiO_2$.

By far the most important iron ore producing district in the world is the Mesabi range of northern Minnesota. More than 1.5 billion tons of high-grade (over 50% Fe) ore have been produced from the Mesabi and probably much less than one-half billion tons remain at this time. However, untold billions of tons of low-grade ore remain from which a high-grade product can be made. Much of this material is in the form of taconite, a very hard rock consisting of fine black crystals of magnetite intermingled with crystals of silica and containing 25–30% iron. It is now being upgraded to about 65% iron in huge plants located on the north shore of Lake Superior, and comprises about half of the total domestic iron ore shipments of about 90 million tons. Current iron ore consumption in the United States is about 140 million long tons annually, imports of ore chiefly from Venezuela meeting the demand.

Other iron ore deposits occur in most of the highly industrialized nations. Among them are those of the Lorraine area on the Franco-German border, Great Britain, Austria, Sweden, Germany and Russia. Future sources of iron ore, as yet largely untapped, are located in Brazil, Chile, Cuba, Venezuela and Canada.

Derivation

Only a minute portion of the iron produced is in the form of what might be called a pure metal, i.e., over 99.9% iron. The vast majority of the iron made and used is in the form of steel, an alloy of iron containing small amounts of carbon (usually less than 1%) that has properties much more attractive than those of pure iron. Actually, steel is a general term for hundreds of iron alloys containing, in addition to carbon, one or more additional elements, e.g., sulfur, silicon, manganese, chromium, nickel, vanadium, tungsten, molybdenum, columbium and titanium. The properties of a given steel are affected not only by the kinds and amounts of elements other than iron in it, but by the form of the iron and of the iron-carbon compounds in it as affected by heat treatment given the steel.

The first step in the extraction of iron from its ores is almost always the reduction of the iron oxide to pig iron in a blast furnace with carbon in the presence of limestone, whereby the noniron ore components, chiefly silica, are converted to a slag and the iron to a molten metal.

The average analysis of pig iron is about 1% Si, 0.03% S, 0.27% P, 2.4% Mn, 4.6% C (the solubility limit of carbon in iron), balance iron. It is a source of cast iron used in making a variety of products where the low tensile strength of cast iron is not objectionable.

Most of the pig iron is converted into steel, which requires the carefully controlled removal of sulfur, phosphorus and silicon; the elimination of most of the carbon but the maintenance of a definite carbon content; and the introduction of desired amounts of purifying and alloying elements. Steel is made from pig iron and scrap steel in open hearth furnaces, in Bessemer converters, in basic oxygen furnaces, and in electric furnaces. The open hearth process accounted for the greatest quantity of steel until just recently, but by 1970 half of the total steel production was by the basic oxygen furnace (BOF) process. Whereas the typical open hearth furnace produces 50 tons of steel in an 8-hour period, the BOF process requires less than an hour, chiefly because high-purity oxygen rather than air is used to oxidize the carbon in the charge.

High-Purity Iron. Pure iron, 99.9 + %, is a rare commodity for several reasons: the high chemical activity of iron which makes it difficult both to prepare and to maintain when made, the lack of attractive physical properties as compared with those of steel, and the high cost of pure iron. A small amount of quite pure, but not the 99.9 + % quality, is made for catalyst use and for incorporation in special magnets.

Among the methods used to prepare pure iron are: (1) thermal decomposition at about 250°C of iron carbonyl, $Fe(CO)_5$, a volatile compound made by the reaction at 180–200°C of carbon monoxide under pressure on iron powder; (2) hydrogen reduction of high-purity ferric oxide, or ferric oxalate, or ferric formate; and (3) electrolytic deposition from solutions of a ferrous salt.

Physical Properties

Most of the physical and mechanical properties of iron are altered by virtually all impurities, especially carbon, and many reported values of such properties are contradictory or in error because the iron tested was impure. An excellent compilation of physical properties is that of Moore and Shives of the National Bureau of Standards, published in the "Metals Handbook" of the American Society for Metals. Many of the data selected by them are listed in Table 1.

High-purity iron is attracted by a magnet, but unlike steel, rapidly loses its magnetism. When heated to the Curie point, 768°C, ferromagnetic iron becomes paramagnetic. No crystal structure change occurs at this point. However, iron has three allotropic forms: (1) alpha iron below 910°C is body-centered cubic; (2) gamma iron from 910 to 1390°C is face-centered cubic; and (3) above 1390°C delta iron is body-centered cubic.

Iron is the most tenacious of all the ductile metals at ordinary temperatures with the exception of cobalt and nickel, but it becomes brittle at liquid air temperatures. Iron softens at red heat, where it is easily forged, drawn, etc., and it can be welded easily at a white heat. However, most high-purity

TABLE 1. PHYSICAL PROPERTIES OF 99.9+% IRON

Atomic number	26
Atomic weight	55.847
Isotopes, natural, and abundance	54 (5.90%)
	56 (91.52%)
	57 (2.245%)
	58 (0.33%)
Electron configuration	$1s^2, 2s^2, 2p^6, 3s^2, 3p^6, 3d^6, 4s^2$
Density, 20°C, g/cc	7.8733
liquid, 1564°C, g/cc	7.00
Atomic volume, 25°C, cc/g-atom	7.094
Specific volume, 25°C, cc/g	0.12701
Melting point, °C	1536.5±1
Boiling point, °C	3000
Curie point, °C	768
Transformation points, °C	
alpha (bcc) to gamma (fcc)	910
gamma to delta (bcc)	1390
Specific heat, C_p, cal/g-atom/°C, at 25°C	5.98
at melting point	9.60
Latent heat of fusion, cal/g	65.5
Latent heat of vaporization, cal/g	1598
Heat of combustion (Fe to Fe_2O_3), cal/g	1582
Thermal conductivity, 0°C, cal/cm/sec/°C	0.2
Linear coefficient of thermal expansion,	
micro-in./in./°C	
alpha (20–100°C)	12.3
gamma (916–1388°C)	23.04
delta (1388–1502°C)	23.6
Electrical resistivity, 20°C, microhm-cm	9.71
liquid, at melting point	139
Temperature coefficient of electrical resistivity, % increase/°C	0.651
Electrode potential, Fe = Fe^{++} + $2e^-$, 25°C, volts	−0.4402
Hall effect coefficient, 13°C, volt-cm/amp/gauss	2.45×10^{-13}
Magnetic properties*, 25°C	
Permeability	88,400
Magnetic induction, gauss	
saturation	21,580
residual	11,830
Coercive force, H_c, oersteds	0.045
Hysteresis, egs/cc/cycle	150
*for iron of 99.99% purity, resistivity 9.71 microhm-cm	
Velocity of sound, 20°C, meters/sec	5130
Viscosity, 1743°C, centipoise	4.45
Surface tension, 1550°C, dynes/cm	1835 to 1965
Vapor pressure, 1600°C, atm	8.0×10^{-5}
Modulus of elasticity, psi	28.5×10^6
Shear or torsion modulus, psi	11.6×10^6
Poisson's ratio	0.29
Tensile strength, psi	30,000
Hardness, Brinell	60
Thermal neutron absorption cross-section, barns	2.62

irons are very ductile at room temperature and can easily be reduced or formed by any standard method.

Chemical Properties

Iron is a very reactive element and as indicated by its electrode potential of −0.4402 volts for Fe = Fe^{++} +$2e^-$, it is a strong reducing agent. Thus, it will displace hydrogen from water, slowly at room temperature and rapidly at temperatures above about 500°C. The reaction with water is associated with the atmospheric rusting of iron by oxygen, the most familiar chemical property of iron. Both oxygen and water or moisture are needed for rusting which is definitely an electrochemical reaction that can be prevented by application of a proper potential to the iron structure.

Solutions containing ions of many metals, such as gold, platinum, silver, mercury, bismuth, tin, nickel and copper, are reduced by solid iron to the corresponding metal and the ferrous ion, e.g., Cu^{++} + Fe → Fe^{++} + Cu.

Iron combines with most nonmetals directly and at moderate temperatures to form binary compounds; included are oxygen, carbon, sulfur, arsenic, phosphorus, all the halides, and silicon. Iron reacts only to a limited degree with nitrogen at elevated temperature. However, exposure to ammonia at 400–700°C results in the formation of Fe_2N; this reaction is used commercially to form *nitrided* surfaces on steel which are hard and abrasion resistant.

The most common valence states of iron are the +2 (ferrous) and the +3 (ferric), but it is possible to attain the +4 state in perferrite, $FeO_3^=$, and the +6 state in ferrate, $FeO_4^=$. Only a limited number of compounds of iron in the +4 or +6 state are known and the most important oxidation states by far are the ferrous and ferric, +2 and +3.

Iron dissolves in nonoxidizing acids, such as sulfuric and hydrochloric, with the liberation of hydrogen and the formation of ferrous ion. Since iron reduces ferric ion to ferrous ion, only ferrous ion is formed unless an oxidizing agent is present in excess.

It dissolves in cold dilute nitric acid with the formation of ferrous and ammonium nitrates and no liberation of gas. When hot acid is used or with stronger nitric acid, ferric nitrate is formed and nitrogen oxides are evolved.

The passivating effect of concentrated nitric acid on iron, whereby the iron will not subsequently react with acids or precipitate other metals from solution, was observed over a century ago by Faraday and others. The same effect is produced by anodic passivation or by other oxidizing ions, such as nitrate, chromate and chlorate, which result in the formation of an oxide film on the iron. This passivating film can be destroyed by scratching or hammering the surface, or by the action of reducing agents.

Principal Compounds

Iron forms two extensive series of compounds, the ferrous and the ferric, derived largely from the corresponding oxides, FeO and Fe_2O_3. A third oxide, magnetite, Fe_3O_4, is a mixed oxide, ferrous ferrite, $FeO \cdot Fe_2O_3$ or $Fe(II)Fe(III)_2O_4$.

Solutions of ferrous salts are difficult to maintain unless they are quite acidic, since the Fe^{++} ion is easily oxidized to the +3 state by oxygen in the air.

$FeSO_4$, one of the most important iron salts, is produced chiefly as a by-product of the pickling of steel with sulfuric acid to remove the oxide scale from the surface. It is used as a mordant in dyeing, as a disinfectant, as a reducing agent, in water purification and in the manufacture of ink and Prussian blue, ferric ferrocyanide, $Fe_4[Fe(CN)_6]_3$.

Ferrous chloride is made by the reaction of hydrochloric acid on iron or iron oxide. Since HCl is being used increasingly for pickling of steel, more and more $FeCl_2$ is being obtained as a by-product.

Hydrated ferric oxide is the basis, after dehydration, of a whole series of iron oxide pigments ranging in color from yellow (ochre) to red (Venetian red). Rouge is the red form and is used for polishing glass and plastics as well as for its cosmetic value.

Writing ink is prepared by adding ferrous sulfate to an extract of nutgalls which contains tannic and gallic acids. The ferrous salts that form are colorless, but they are rapidly oxidized by air to the intense black color of ferric gallate and tannate. To such an ink a colored dye is added to give the desired initial color before the ferrous to ferric change occurs.

Both ferrous and ferric iron form stable complexes with cyanides in which each iron atom is associated with six cyano groups as $Fe(CN)_6^{-4}$ and $Fe(CN)_6^{-3}$, the ferrocyanide and ferricyanide ions, respectively. These ions take part in some very complicated reactions to form useful compounds. Prussian blue, ferriferrocyanide, is the deep blue pigment of laundry bluing whose color covers the yellow stain on white goods to make them appear white.

Among the most interesting iron compounds are the carbonyls. Ferropentacarbonyl, $Fe(CO)_5$, is formed by the action of carbon monoxide on finely divided iron; it decomposes to Fe and CO when heated above about 200°C. $Fe_2(CO)_9$ and $Fe_3(CO)_{12}$ are also known.

Ferric thiocyanate, $Fe(SCN)_3$, is a compound whose bright red color makes it one of the most sensitive reagents for the detection of ferric ion which can be observed in concentrations as low as $10^{-5} M$.

Iron compounds are available as by-products of several industrial operations in quantities much greater than the amounts consumed by the rather limited number of applications for iron compounds. In many instances they represent waste disposal problems. Examples are the acidic pickle liquors of the steel industry which contain $FeSO_4$ or $FeCl_2$, and the "red mud" from the Bayer alumina process which contains hydrated ferric oxide from the bauxite raw material.

Applications of Iron

All but a minute fraction of the iron extracted from ores is used in the metallic form to make the multiplicity of devices and structures that characterize our modern civilization. In its applications iron is the major component of several types of alloys, the most prominent being those that contain carbon alone (up to 1.7%), the *carbon steels*. The properties of these steels can be varied not only by the amount of carbon present, but also by the heat treatment given the steel. Carbon has such potent effects that it is specified to hundredths of a per cent.

The next largest tonnage of iron alloys is found in the low-alloy steels containing, in addition to carbon, varying amounts and combinations of alloying elements (up to 5%). These, too, are heat treated to attain desired structure and properties. A four-digit numerical system is used to identify these steels.

Other important types of steel are tool steels, stainless steels, heat-resistant steels and maraging steels, a new group which is heat treatable to unusually high strength levels. All of these types of steels contain 10 to 20% or more of alloying elements, chiefly chromium and nickel.

Iron alloys not classified as steels include cast iron, wrought iron, silicon iron (e.g., "Duriron"), nickel iron, and malleable iron, each of which has a variety of important applications in modern life.

CLIFFORD A. HAMPEL

References

1. Moore, G. A., and Shives, T. R., "Metals Handbook," 8th Edition, pp. 1206–1212, Metals Park, Ohio, American Society for Metals, 1961.
2. Bain, E. C., and Paxton, H. W., "Alloying Elements in Steel," 2nd Edition, Metals Park, Ohio, American Society for Metals, 1966.
3. Sharp, J. D., "Elements of Steel-Making Practice," Long Island City, N. Y., Pergamon Press, 1966.

Biochemical and Biological Aspects*

Although considered to be a "trace" element, iron occupies a unique place in the metabolic processes of the animal body. It is a vital constituent of every mammalian cell. The role of iron in the body is closely associated with hemoglobin and the transport of oxygen from the lungs to the tissue cells. It is also concerned with cellular oxidation mechanisms which are catalyzed by iron containing enzymes.

It has been variously estimated that the total iron in the blood and tissues of a 70 kilo man is about 5 grams. All of it is bound to proteins in one form or another. These may be divided into two groups: (a) The iron porphyrin or heme proteins which include hemoglobin, myoglobin (muscle hemoglobin), and the heme enzymes—the cytochromes, catalases, and peroxidases. Blood hemoglobin, about 900 grams, contains 3 grams of iron and represents, therefore, from 60 to 70% of the total body iron; myoglobin from 3 to 5% and, the heme enzymes about 0.2% of the body iron. (b) Nonheme compounds, including siderophilin (transferrin), ferritin, and hemosiderin. In this group of proteins a significant quantity of iron, about 0.6 gram or about 15% of the total body iron, is contained in ferritin.

The iron porphyrin, or heme, proteins function together to bring about oxidation of cell metabolites with oxygen. This vital process is accomplished in a series of steps. The first involves the transport of molecular oxygen by hemoglobin contained in the red cells. Chemically speaking, hemoglobin is a conjugated protein, with a prosthetic group, heme, united to the protein globin. In the heme portion

HEME

of the molecule an atom of ferrous iron, with a coordination valence of six, occurs at the center of each 4 tetrapyrrole rings. Four of the valences bind the pyrrole nitrogens in the plane of the ring, while a fifth linkage above or below and perpendicular to the plane, is attached to globin through the nitrogen of the imidazole groups of histidine. This configuration permits the reversible combination with molecular oxygen, depending on the partial pressure, at the sixth coordination valence of the iron. In this manner the iron of hemoglobin facilitates the uptake of oxygen in the lungs where the partial pressure is high, and its release in the tissues where the partial pressure of oxygen is low.

Since the iron of functioning hemoglobin remains in the reduced form, the combination with oxygen is spoken of as "oxygenation," rather than oxidation. (Unfortunately the affinity of hemoglobin for carbon monoxide exceeds that of oxygen by a factor of about 210). In the tissues oxygen diffuses out of the red cells into the interstitial spaces and into the tissue cells. If the tissue is muscle, the oxygen may be combined temporarily with myoglobin, another iron-prophyrin-protein similar to hemoglobin with a greater affinity for oxygen. In a sense, myoglobin functions as a storage compound holding a reserve of oxygen in those muscles, e.g., heart muscle, which are required for emergency or for sustained work. The actual "burning" or oxidation of food fragments which takes place in the mitochondria of the cells is a complicated process catalyzed by the iron porphyrin enzymes: the cytochromes, cytochrome oxidase, catalase, and peroxidase. In these transformations a food fragment or metabolite, such as succinic acid from the citric acid cycle, is dehydrogenated (hence oxidized) with the loss of 2 hydrogen atoms and 2 electrons. The electrons are passed down through a series of cytochromes, b, c, a, in which the iron is alternately oxidized and reduced, and finally to cytochrome oxidase which activates oxygen to take up two electrons and hydrogen to form water. The net reaction is the removal of hydrogen from the metabolite and its combination with oxygen to form water. From the standpoint of energy involved in the reaction, the formation of 1 mole of water liberates 68,000 calories of heat. In this connection, it should be recalled that the opposite transformation takes place in plants. In the leaves, chlorophyll, a magnesium-porphyrin-protein, catalyzes the cleavage of water into hydrogen and oxygen.

Among the nonheme iron compounds of the body, ferritin, hemosiderin, and siderophilin (or transferrin) have been recognized as chemical entities. The first two are considered to be essentially iron storage proteins; the latter functions in the transport of iron in the blood.

In man, and in most other species investigated, the liver and spleen are the main storage organs for iron in the form of ferritin. Ferritin is also present in lesser amounts in the bone marrow, kidneys, and intestine. The protein devoid of iron is known as apoferritin, a colorless compound with a molecular weight of 460,000. Apoferritin is capable of combining with variable amounts of iron up to 23% of its dry weight, to form the brownish pigmented protein, ferritin. Both ferritin and apoferritin have been crystallized with cadmium salt. Since ferritin is not a definite chemical compound, it is assumed that the iron is present in the form of ferric hydroxide micelles or clusters, with an average composition, $[(FeOOH)_8 \cdot FeOPO_3H_2]$.

The transport of iron in the blood stream is accomplished by siderophilin, a serum globulin having a molecular weight of about 88.000 and capable of combining with a maximum of 2 ferric atoms per molecule. In normal individuals the siderophilin, or iron-binding protein, is about one-third saturated, which represents about 4 milligrams of iron. This permits varying amounts of dietary iron to be transported from the active absorbing areas in the intestine to the liver and spleen, and also from these storage depots to the specialized tissues for the synthesis of hemoglobin and the cellular cytochromes. The iron contained in the hemoglobin of the red cell is not transport iron in the sense that it can be made available for the synthesis of other iron compounds. This occurs only when the red cell ceases to function in the transport of oxygen and the hemoglobin is degraded.

The nutritional requirements for iron are exceedingly small and vary somewhat with age and sex. Fortunately, nature has provided the newborn infant with an adequate reserve of this vital element which is normally sufficient to tide him over a dietary period restricted to milk which is low in iron. The human adult male absorbs less than 5 mg of iron per day. This is a liberal estimate if one comparies it with the cumulative daily loss of 1 mg or less, in the urine, stool, and from the skin surfaces. Woman absorbs slightly more to replace loss through the menses or during gestation for foetal requirements. In growing children the demand for iron is geared to increased blood formation, but absorption perhaps does not exceed 10 to 15 mg per day. Iron absorption takes place primarily in the duodenal region just below the pyloric sphincter. It is generally agreed that dietary iron must be reduced to the ferrous state before it can be absorbed. Numerous observations have established the fact that the iron of different foods is not equally "available," and may be only a fraction of that determined by ash analysis. Hemoglobin-iron present in meat products is probably not available. Much of the iron in food is in the ferric form, as ferric hydroxide or as iron loosely bound in organic molecules. These compounds are acted on by gastric HCl so that at least a part of the iron becomes ionized. Ferric ions in contact with reducing agents in the food, such as ascorbic acid, cysteine, and other sulfhydryl compounds, are converted to the divalent state. Ferrous iron is combined with apoferritin, present in the mucosal cells, to form ferritin, which in turn passes the iron on the blood stream to be transported as sideraphilin. Thus, iron absorption is regulated by the ferritin mechanism of the mucosal cells to avoid both iron excess and iron deficiency.

Iron is frequently spoken of as a "one way" substance. Once it is incorporated into the body, whether by oral administration or by parental injection, there is no efficient mechanism for its excretion. Iron from outworn red cells is used over and over again as though it were a precious metal in the economy of the individual.

Among the foods rich in iron, meats—especially liver, fish, and egg yolk deserve prominent mention. Good vegetable sources are found in dried peas and beans, also in leafy green vegetables. Flour, fresh fruit, and milk are relatively poor in iron.

Since the availability of ingested iron is related to its solubility and its reduced state, ferrous salts (ferrous sulfate, ferrous gluconate) have proved to be effective in the treatment of anemia due to iron deficiency. It has been shown that ascorbic acid taken along with therapeutic iron increases the amount of iron absorbed. Iron replacement may also be accomplished by the transfusion of whole blood or by the use of a variety of organic-iron compounds which are available for parental administration. Perhaps the most widely used of the latter is saccharated oxide of iron. The therapeutic use of iron is a problem which should be left to the judgment of the physician.

*C. A. JOHNSON

ISOMERS AND ISOMERIZATION

Chemical compounds which have the same molecular weights and the same molecular formulas, but different properties, are called *isomers* (of each other) and the process of transforming one isomer to another is called *isomerization*. This phenomenon is accounted for by differences in arrangement (constitution) and spatial orientation (configuration) of atoms within the molecules. (But see *nuclear isomerism* below.)

Several examples of isomerism were observed in disbelief before 1824, but it was the work of Wöhler on silver cyanate (AgOCN) and of Liebig on silver fulminate (AgONC) which led Gay-Lussac to suggest that the atoms must be combined differently. Berzelius refused to accept this explanation but the discovery of new examples of isomerism caused him to change his views. In his own studies of tartaric and racemic acids (both $C_4H_6O_6$) he referred to isomeric bodies. In 1831 he introduced the term isomerism (composed of equal parts) for compounds with the same composition but different properties. At the same time he introduced the terms *metamerism* and *polymerism*. Metamers are isomers with the same molecular weight but with different distribution of atoms among radicals. Berzelius' example of tin sulfates was in error and the term has fallen out of use today (but see *functional isomerism* below for examples.) Polymers are compounds with the same percentage composition but different molecular weights; e.g., ethylene, C_2H_4 and butylene, C_4H_8. (See **Polymerization**) There are several major types of isomerism.

Structural isomerism, sometimes called *position isomerism,* results from differences in the arrangement of atoms or groups of atoms in the molecules. Variant forms of structural isomerism include skeletal and functional isomers. *Skeletal isomers,* including *chain* and *ring* isomers, differ in arrangement of atoms or groups around a skeleton of carbon atoms. Examples of chain isomers are *n*- and iso-butanes

$$CH_3—CH_2—CH_2—CH_3 \quad \text{and} \quad CH_3—CH—CH_3$$
$$\underset{\displaystyle CH_3}{|}$$

and *n*- and iso-propyl alcohols.

$$CH_3—CH_2—CH_2—OH \quad \text{and} \quad CH_3—CH—CH_3$$
$$\underset{\displaystyle OH}{|}$$

Ortho-, meta-, and para-xylenes are examples of ring isomers.

Still another type of ring isomerism is encountered in the case of multiple-ring compounds in which the rings are fused to one another at different positions as, for example, anthracene and phenanthrene.

It is readily observable that there would be three isomers of monosubstituted anthracene, but five isomers of monosubstituted phenanthrene. *Functional isomers* (*metamers*) have atomic arrangements resulting in different functional groups, for instance ethyl alcohol and dimethyl ether

$$\textbf{CH}_3\textbf{—CH}_2\textbf{—OH} \quad \text{and} \quad \textbf{CH}_3\textbf{—O—CH}_3$$

Stereoisomerism or *space isomerism,* resulting from differences in the arrangement of atoms in space (configuration), is represented by two classes: geometric isomerism and optical isomerism. *Geometric isomers* have similar atoms attached at similar positions but double bonds between carbon or nitrogen atoms prevent free rotation and two different arrangements in space are possible. When the double bond is between two carbon atoms the isomers are named *cis* and *trans;* for example, *cis* and *trans* crotonic acids are:

When the double bond is between a carbon and a nitrogen atom the terms *syn* and *anti* are used. For example, the benzaldoxime having the H and OH on opposite sides is *anti,* the other is *syn.*

$$C_6H_5—C—H \qquad C_6H_5—C—H$$
$$\underset{\displaystyle HO—N}{\|} \qquad \underset{\displaystyle N—OH}{\|}$$

Geometric isomerism is also encountered in saturated ring compounds with three or more members in the ring since the rigidity of the ring prevents free rotation of carbon atoms carrying different substituents.

In saturated ring compounds with six or more

ring atoms there is also the possibility for puckering of the ring in order to relieve bond strain. H. Sachse suggested in 1890 that cyclohexane might exist in two forms, commonly referred to as the chair and boat form.

Such isomers of cyclohexane have never been isolated since interconversion apparently takes place with ease. However, W. Hückel isolated *cis* and *trans* forms of dacalin (hexahydronaphthalene) in 1925, revealing that in fused ring systems bond rigidity was sufficient to prevent spontaneous interconversion. Since that time the existence of such isomers has been important in the understanding of saturated polysubstituted ring systems and fused-ring systems such as are found in sterols and triterpenoids.

Alfred Werner showed *cis-trans* isomerism in certain coordinated metal ions; e.g., dichlorobis (ethylenediamine) cobalt (III):

   cis   *trans*

Optical isomerism is observed in compounds containing one or more asymmetric carbon atoms. An asymmetric carbon atom is combined with four different atoms or groups of atoms. Van't Hoff, and independently Le Bel, showed in 1874 that four such atoms or groups can be arranged around the central carbon atom in two different ways, giving pairs of isomers which are mirror images. The members of such pairs are called *enantiomorphs.* An example is lactic acid:

$$CH_3—\underset{\displaystyle H}{\overset{\displaystyle OH}{\underset{|}{\overset{|}{C}}}}—COOH \qquad HOOC—\underset{\displaystyle H}{\overset{\displaystyle OH}{\underset{|}{\overset{|}{C}}}}—CH_3$$

One enantiomorph rotates the plane of polarized light to the right (+), the other to the left (−), a phenomenon known as *optical activity.* Atoms of various other elements are able to form asymmetric centers which are optically active. Optical activity has also been observed in compounds where there is no asymmetric atom but with a structure where hindered bond rotation results in the formation of two structures which are mirror images; e.g., allenes, spiranes, biphenyls with bulky ortho groups. When metal ions are coordinated with two or three bidentate groups optical isomerism is possible. For example, the *cis* dichlorobis (ethylenediamine) cobalt III ion shown above has an entantiomorph.

Dynamic isomerism is encountered in certain compounds where migration of a hydrogen atom can result in two or more structures. Frequently the two isomers are present in a state of equilibrium. The phenomenon is *tautomerism* and the isomers are *tautomers*. The earliest example was encountered with the discovery of acetoacetic ester in 1863 by Guether. This compound was found to show the properties of an unsaturated alcohol and of a ketone.

$$CH_3-\overset{\overset{\displaystyle OH}{|}}{C}=CH-COOC_2H_5 \rightleftharpoons$$

$$CH_3-\overset{\overset{\displaystyle O}{\|}}{C}-CH_2-COOC_2H_5$$

Such tautomers are named *enol* and *keto*. Pure samples of both isomers of acetoacetic ester were ultimately prepared. When the tautomers can be isolated they are sometimes referred to as *desmotropes* and the phenomenon is called *desmotropy*. (Tautomerism must not be confused with resonance where only a single structure exists but it is represented symbolically as an intermediate between two or more structures.)

The term *isomerization* is applied to any process whereby a compound is transformed into an isomeric form. The transformation may result from heat, pressure, catalysis, treatment with an appropriate reagent, or a combination of these. Isomerization seldom results in complete conversion but in the formation of an equilibrium mixture of two or more isomers. When an optically active compound is isomerized to its enantiomorph, equilibrium is generally reached when equal amounts of the (+) and (−) forms are present. This mixture is without optical activity and is called a *racemic mixture*.

In recent years isomerization has found an important practical use in the petroleum industry, namely, isomerization of normal paraffin hydrocarbons into their branched chain isomers. The importance of this reaction lies in the fact that the antiknock properties (as measured by octane number) of branched chain paraffins are much better than those at the normal chain isomers. In the case of butanes, of additional importance is the fact that the transformation of *n*-butane to isobutane leads to an increase of chemical activity due to formation of a tertiary carbon, which in turn makes isobutane available for the process of alkylation to highly branched liquid hydrocarbons, important components of aviation gasoline. Isomerization of *n*-butane to isobutane is carried out in the presence of various preparations of aluminum chloride as catalyst. In like manner, *n*-pentane can be isomerized to isopentane.

* * * *

Nuclear isomerism represents a variant use of the word isomerism since nuclear isomers are atomic nuclei with the same number of protons and neutrons but different energies. The higher energy forms decay to the ground state by one or more gamma ray emissions, a process called *isomeric transition*. For example, the zinc isotope, Zn^{69}, occurs in two isomeric forms, a 13.8-hour form emitting gamma radiation to reach the ground state.

AARON J. IHDE

Cross-references: *Asymmetry, Hydrocarbons, Gasoline, Polymerization, Alkylation, Reforming.*

ISOMORPHISM

Although the term "isomorphism" is attributable to Mitscherlich (1794–1863), its meaning has evolved to keep pace with modern crystallochemical concepts, chiefly through the efforts of mineralogists.

Two compositions are considered *isotypic*, such as AXO_4 and BYO_4, if they are isostructural; such a pair are quartz ($SiSiO_4$) and berlinite ($AlPO_4$). The axial dimensions of the unit cells are different, of course, depending upon the relative sizes of the atoms, Si, Al, and P. Were Al and P to substitute for Si atoms within a composition which is principally SiO_2 without alteration of the essential symmetry of the structure of the single phase, an *isomorphic variant* of quartz would result, and its composition could be represented as (Si,Al) (Si,P)O_4. Conversely, (Al,Si) (P,Si)O_4 would represent an isomorphic variant of berlinite. Were several intermediate compositions known to exist, the range of compositions would comprise an *isomorphic series*, having SiO_2 and $AlPO_4$ as *end members*. Actually, an isomorphic series with the quartz structure is unknown for such compositions, but substitution occurs to a limited extent for the *polymorphs*, cristobalite and tridymite.

When there is a significant difference between the cationic radii, ordering may take place. Dolomite [$CaMg(CO_3)_2$], for example, has a structure different from calcite [$CaCO_3$] and magnesite [$MgCO_3$], which are isotypic. Limited *isomorphic substitution* of Mg for Ca atoms produces magnesian calcite.

Other terms, such as "solid solution" and "mixed crystal," are encountered, particularly in connection with metallic structures where different types of "solid solution" are recognized: *substitutional* and *interstitial*. However, the bonding forces in metals are different from those encountered in oxygenated structures and probably the denotations for the several phenomena should be different also. "Mixed crystal" seems to be disappearing through disuse; it is essentially synonymous with "isomorphic variant."

Vegard's law is concerned with the unit-cell dimensional change as a function of composition, such as solid solutions of Au and Ag. The original data, according to Pabst, show better correlation with volume than with a linear dimension, and the relation becomes complex when nonisometric structures are considered.

In an attempt to relate the volumes of isotypic crystals and the radii of substituent atoms, McConnell (1966) devised the expression:

$$V_2 = V_1 + k(r_2 - r_1)NE \qquad (1)$$

where V_2 and V_1 are the predicted and known volumes, resp., in $Å^3$; r_2 and r_1 are radii (Ahrens') of cations in V_2 and V_1; N = the number of cations substituted per unit cell; E = the valence charge of the cations; and k = a constant, approximately 12.

In many situations the predicted volume is within a few per cent of the experimental value for complex structures, for which the relation seems particularly useful. From quartz, for example (taking E as 4 for both Al and P), one predicts the volume for berlinite to be 229 $Å^3$, as compared with the experimental value 232 $Å^3$.

For isometric crystals an additive expression interrelates the cube edge (a) and the radii of the constituent cations:

$$a = k + xr_1 + yr_2 + zr_3 + \cdots \qquad (2)$$

where k is a constant characteristic of the particular structure; x, y, z are dependent variables; and r_1, r_2, r_3 are radii of cations occupying equipoint structural sites, usually with different coordination numbers and valences.

For a fairly simple case (garnets), having the structural makeup $A_3B_2(SiO_4)_3$, with A = Ca, Mg, Fe^{++}, etc. and B = Al, Fe^{+++}, Cr, etc., the expression becomes:

$$a = 9.900 + 1.212\ r_A + 1.464\ r_B \qquad (3)$$

and the calculated a shows surprisingly little departure from the experimental value for any particular garnet species when the dependent variables and constant are obtained from six or eight synthetic *end-member* compositions, such as $Ca_3Al_2(SiO_4)_3$. The constant and dependent variables can be determined by multiple regression and will depend upon the ionic radii. Eq. (3) is based on radii of Ahrens with CN = 6, without correction for the actual, structural situations.

DUNCAN MCCONNELL

References

Strunz, H., "Mineralogische Tabellen," 5th ed., 621 pp., Geest & Portig K.-G., Leipzig, 1970.

Barrett, C. S., and T. B. Massalski, "Structure of Metals: Crystallographic Methods, Principles, and Data," 3rd ed., 654 pp., McGraw-Hill Book Co.. New York, 1966.

McConnell, D., "Calculation of the unit-cell volume of a complex mineral structure," *Z. Kristallogr.*, **123**, 58–66 (1966).

McConnell, D., and F. H. Verhoek, "Crystals, Minerals, and Chemistry," *J. Chem. Ed.*, **40**, 512–515 (1963).

ISOTOPES

An isotope is one of two or more nuclides of a given element which differ in atomic mass. Since they have the same atomic number and hence the same number of protons in the nucleus, the difference in atomic mass is due to varying numbers of neutrons in the nucleus. The term *isotope* (from the Greek meaning "same place") was coined by F. Soddy in 1913. Isotopes may be either radioactive, in which case they dissipate energy through the emission of radiation, or stable, i.e., nonradioactive.

At the time Soddy introduced the term isotope the discovery of the neutron was still nineteen years in the future. Hence his use of the term inferred nothing about the structure of the nucleus. Soddy proposed to place together in one position in the Periodic Table species which seemed inseparable by known chemical procedures even though their radioactive properties and even their atomic masses were different. This was a bold step since nearly forty species with different radioactive properties were known, but only twelve positions between thallium and uranium inclusive were available to accommodate them in the Periodic Table. Late in 1913 Soddy went even further and stated that he was of the opinion that isotopy might apply quite generally over the whole Periodic Table and therefore for any element the atomic mass was not a real constant but a mean value of much less fundamental interest than had hitherto been supposed.

In the same year that Soddy introduced the concept of isotopy he also enunciated his displacement laws for radioactive decay. These laws may be stated as follows: (1) the product resulting from an α-emission is shifted two places in the periodic chart in the direction of diminishing mass (i.e., a decrease of two in atomic number) from the place of the original substance; and (2) the product resulting from a β-emission is shifted one place in the direction of the heavier elements from that of the original substance. These considerations, used in conjunction with Moseley's correlation of the characteristic x-ray spectra of the elements with atomic number, led to the successful placing of all the natural radioactive species, some forty in number, between the atomic numbers 81 to 92, into what are known as the three natural radioactive series, designated as the uranium, thorium, and actinium series. With the evidence that these series all decayed to the single element lead it was pointed out by Soddy that lead from the uranium and thorium series should exist in forms having atomic masses 206 and 208, respectively. Experimental verification of this prediction was forthcoming within the next year, when it was shown that lead derived from uranium ores had an atomic mass close to 206, while the lead separated from thorium ores had an atomic mass close to 208. These results provided convincing support both for the theory of radioactive decay and for the existence of stable isotopes.

At about the same time J. J. Thomson, using positive ray analysis, obtained evidence that what had been called neon was not a single gas, but instead a mixture of two gases, one of which had an atomic mass of about 20 and the other an atomic mass of about 22. This result was given support by F. W. Aston, who, employing diffusion through pipe clay, was successful in separating neon into two fractions, one of which had an atomic mass significantly less than and the other an atomic mass significantly greater than the observed atomic mass of naturally occurring neon gas. Later, after he had developed the first mass spectrograph, Aston confirmed the existence of the two isotopes of neon with atomic masses of 20 and 22. The third isotope, with mass 21 and in abundance of 0.25% was not discovered until 1928.

By the end of 1920 Aston had examined nineteen elements in his mass spectrograph. He found that nine of them, including chlorine and mercury, consisted of two or more isotopes with masses which were close to integers. Furthermore, he concluded that elements like carbon, nitrogen, oxygen, fluorine, and phosphorus which have atomic masses close to whole numbers do not consist of isotopes. The isotopes which were later discovered in carbon, nitrogen, and oxygen occur in such small abundance that they do not sensibly shift the atomic mass of the elements from whole numbers.

On the basis of his observations Aston formulated his whole number rule which is essentially a modified form of Prout's hypothesis. According to his rule, all atomic masses are very close to integers, and the fractional atomic masses determined chemically result from the presence of two or more isotopes each of which has an approximately integral atomic mass.

In 1918, even before Aston had built his first mass spectrograph, A. J. Dempster designed and built the first mass spectrometer. In this instrument the ions were detected electrically rather than photographically, as in the mass spectrograph. With his instrument Dempster first measured the relative abundance of the potassium isotopes, and by 1920 he had discovered most of the isotopes of magnesium, calcium, and zinc as well as confirming, as Aston had reported, that lithium has two isotopes. By the end of 1922 thirty-four elements had been investigated and nineteen of them were found to have altogether over sixty isotopes.

The field of optical spectroscopy was also fruitful for the discovery of isotopes. In this method one examines the band spectra of isotopic molecules, i.e., molecules in which an atom of one kind is combined in turn with atoms of two or more isotopes of another element; for example, HCl^{35} and HCl^{37} are isotopic molecules. In band spectra the influence of nuclear mass is substantial, since it affects both the moment of inertia and the vibrational frequency of the molecule, which are predominant factors in the structure of spectral bands and band systems. Hence there is a considerable difference between the band spectra of two isotopic molecules. Isotopes discovered by the band spectrum method include ^{30}Si, ^{17}O, ^{18}O, ^{15}N, ^{13}C, ^{2}H, and ^{113}In. Studies involving the fine structure in optical line spectra led to the discovery of ^{203}Tl, ^{205}Tl, ^{204}Pb, ^{191}Ir, and ^{193}Ir.

The occurrence and degree of isotopy of the nuclides that occur in nature is very diverse. For our purpose "occurring in nature" means having existed over all geologic time. If the age of the universe is taken as 4.5×10^9 years, then, in order to have existed over geologic time and still be detectable by present counting techniques, a nuclide must have a half-life greater than about 10^8 years. If we impose this criterion and include the stable isotopes, then there are 283 isotopic species from hydrogen through bismuth (element 83). All the elements above bismuth are radioactive and only ^{232}Th, ^{235}U, and ^{238}U have half-lives sufficiently long to have survived over geologic time. Of the first 83 elements 20 have no isotopes; each consists of one kind of atom only. Of the 20 without isotopes all have odd atomic numbers except Be. The remaining elements possess isotopes ranging in number from two to ten (only Sn has ten) and occurring in widely different amounts in nature. For example, the relative amounts of ^3He and ^4He are of the order of 10^{-6} to 1.

Of the 283 isotopic species mentioned above, 165 may be classified as even-even (even number of protons and even number of neutrons), 56 as even-odd, 53 as odd-even, and 9 as odd-odd. These numbers display the pairing tendency of the nuclear constituents and show that there is no significant difference in stability between odd-proton and odd-neutron configurations. Eighteen of the naturally occurring isotopes of the first 83 elements are radioactive with half-lives greater than 1×10^8 years.

The years since World War II have seen an enormous advance in the instrumentation of mass spectrometers. Higher resolving power, increased sensitivity of ion detection, and improved stability of electronic power supplies are among the dominant characteristics of the new instruments. They have made possible the determination of isotopic abundances with very high precision.

One of the changes that has been made as a consequence of the improved knowledge of atomic masses and isotopic abundances is the adoption of a new and unified mass scale. Even after the discovery of the isotopes of oxygen with masses 17 and 18, chemists continued to use 1/16 weight of the natural mixture of the oxygen isotopes as the unit of atomic mass. In contrast, in the determination of atomic masses by means of the mass spectrograph Aston chose a mass scale based on the mass of the single most abundant oxygen atom taken as 16, and 1/16 of this as the unit of atomic mass. These two scales continued to be in use until the early 1960's. At that time, in order to obviate the confusion caused by the use of the two scales, both physicists and chemists agreed upon the adoption of an atomic mass scale based upon the ^{12}C isotope as having a mass of exactly 12.

Isotopes may be separated by a variety of techniques. For example, the pressing need for ^{235}U during World War II provided the impetus for the construction of large plants in Oak Ridge, Tennessee for the separation of ^{235}U from the much more abundant ^{238}U both by diffusion of uranium hexafluoride through a porous barrier and by an electromagnetic process which operated on the principle of the mass spectrograph. Many stable isotopes for research are still separated by the electromagnetic process. Other methods of separation include thermal diffusion, centrifugation, and chemical exchange.

Stable separated isotopes of many elements have been used as tracers in chemical and biological processes, with mass spectrometers being used for their detection. Outstanding examples currently being used in many studies include ^{17}O, ^{18}O and ^{15}N. The longest lived radioactive isotopes of these two elements have half-lives of only a few minutes.

The subject of radioactive isotopes is treated in the following section.

M. H. Lietzke and
R. W. Stoughton

Radioisotopes

A radioisotope is an isotope of an element which decays by particle emission or orbital electron capture. Each radioisotope may be characterized by its half-life, which is defined as the time required for one half of a sample to disintegrate. Half-lives for different radioisotopes may vary tremendously, from over 10^{17} years to a fraction of a microsecond. Radioisotopes may be either naturally occurring or artificially produced.

In 1913 there were about forty known radioactive species. One of the early triumphs of the radioactive displacement laws and the isotope concept, enunciated in that year by F. Soddy, was the proper placement of these species in the Periodic Table of the elements. It was soon shown that all these nuclides could be placed in three chains of successive decay. Each of these series is headed by a long-lived isotope and terminates in an isotope of lead. The chain known as the uranium series starts with ^{238}U as the parent substance and after fourteen major transformations (eight of them by alpha particle emission and six by beta particle emission) terminates in ^{206}Pb. This series is often designated as the $4n + 2$ series where n is an integer. The designation $4n + 2$ derives from the fact that division of the mass number of any member of this series by 4 leaves a remainder of 2. The actinium or $4n + 3$ series has ^{235}U as the parent and after eleven major transformations decays to ^{207}Pb. The thorium or $4n$ series starts with ^{232}Th as the parent and following ten major transformations decays to ^{208}Pb.

In 1934 I. Curie and F. Joliot announced that boron and aluminum could be made radioactive by bombardment with alpha particles. It was observed that both the boron and the aluminum continued to emit positrons after the removal of the alpha source and that the induced radioactivity in each case decayed with a characteristic half-life. This new phenomenon of induced radioactivity was quickly explained in terms of the production of new unstable nuclei.

At about the same time that artificial radioactivity was discovered machines were developed for the acceleration of the ions of hydrogen and helium to energies sufficient to produce nuclear transformations. With the discovery of the neutron in 1932 and the isolation of deuterium in 1933, there became available two additional particles which were especially useful for the production of induced activities. The growth in this field has been so rapid that by the end of 1970 about 1400 radioactive nuclides had been characterized. For every element there is now known at least one radioactive species.

As early as 1923 it was speculated that there should be a fourth radioactive series of the $4n + 1$ type. Although careful searches have been made, no members of such a series have been found in nature. However, during the course of research carried out in connection with the Manhattan Project, all of the members of the $4n + 1$ series below ^{237}Np and several above it were identified and the decay scheme was completely elucidated. Although the series includes several preneptunium members, the 2×10^6 y isotope ^{237}Np is usually considered the parent of the series.

Unlike each of the three natural radioactive series the neptunium series includes no isotope of radon. Also, unlike these series, there are representatives of the elements astatine (atomic number 85) and francium (atomic number 87). The series ends with the stable isotope ^{209}Bi instead of with an isotope of lead. Finally, there is present in the neptunium series the important isotope ^{233}U which undergoes fission with thermal neutrons and is hence a possible nuclear fuel. Its importance lies in the fact that since ^{233}U can be formed by the absorption of neutrons by thorium, nonfissionable thorium may be used as a fuel source.

In addition to the radioactive nuclides contained in the uranium, actinium, and thorium series there are other radioactive isotopes found in nature. Eighteen of these have half-lives greater than 1×10^8 years and have existed over all of geologic time. Included among these isotopes are ^{40}K, ^{50}V, ^{87}Rb, ^{115}In, and ^{123}Te. The heaviest representative of this group is ^{204}Pb.

Other shorter-lived radioisotopes are continuously produced in nature. Bombardment of stable isotopes by cosmic rays or by the particles (such as neutrons or protons) resulting from cosmic ray bombardment of air and earth is one such source. An example of such a radioisotope is ^{14}C with a half-life of 5600 years; it may be produced by an (n,p) reaction on atmospheric nitrogen. Absorption of the neutrons resulting from the spontaneous fission of ^{238}U may also result in the production of radioactive isotopes. For example, barely detectable amounts of ^{239}Pu have been found in some uranium-containing minerals. In this case the plutonium isotope is produced through neutron absorption by ^{238}U followed by beta decay (just as it is in reactors).

Some nuclides can exist in two or three energy states for times long enough for the half-lives of the upper states to be readily measured; these may be in the range of a fraction of a second to many years. The lowest state is called the *ground state* and any higher energy state is said to be *metastable*. Certain selection rules make immediate decay by gamma ray emission to the ground state *forbidden*. Usually the metastable state is shorter lived than the ground state (which may even be stable), but not always. The metastable state may decay to the ground state by nuclear photon or gamma ray emission (called *isomeric transition*), or both states may decay independently by electron capture or particle emission (which could include spontaneous fission for the heaviest nuclides). Often both isomeric transition and particle emission occur.

The relative amounts of two or three isomers formed depends on the mode of forming the nuclide. For example when ^{234}Pa is formed by beta decay of ^{234}Th the ratio of the 1.1 m isomer (of the former) to the 6.7 h isomer is a few hundred to one; when ^{234}Pa is formed by neutron capture of ^{233}Pa the ratio is roughly unity.

Radionuclides may be prepared by irradiating stable or other radioactive species with neutrons, charged particles or even high energy photons (x-rays or gamma rays). In general thermal neutron capture tends to give neutron-rich products; these, if radioactive, tend to emit (negative) beta particles, during which process a neutron in the nucleus is converted into a proton thereby raising the nuclear charge. Radioisotopes of elements from about 32(Ge) to about 64(Gd) can be made in large amounts by neutron fission in reactors. The products of fission are mostly beta emitters since the fissioning nuclei are more neutron-rich than the stable isotopes of the fission product elements. By the use of accelerators high energy charged particles may be produced and these used for irradiating various target materials. The compound nuclei formed have excess energy and immediately emit various numbers of neutrons, protons or alpha particles—depending on the energy of the incident particles. If the target element is heavier than about $_{82}$Pb and the kinetic energy of the particles is high enough fission can occur. X-rays are not very useful for producing radioisotopes, but where the upper energy limit is known they can be used for measuring the neutron binding energy which is the energy required to remove a neutron from a nucleus; they also can be used to measure fission threshold energies.

During the last few decades the large use of radioisotopes has greatly helped to revolutionize the analytical techniques used in medical, biological, chemical and technological processing studies. Activation analysis provides an extremely valuable tool in many types of research; an unknown sample is irradiated with neutrons after which the various gamma ray activities (accompanying beta decays) are determined with fine energy discrimination. These are interpretable in terms of known radionuclides which in turn are quantitatively interpretable in terms of stable elements in the original sample. The use of radioactive tracers simplifies and speeds up the detection of any reactant. For example, by the use of gamma-ray spectroscopy analyses can be made on solutions or solids in phase studies without further chemical separations. A technique of elucidating the various procedures and mechanisms in complicated chemical and biological systems involves the insertion of several different isotopes of an element each into different compounds. Then by following the path of each of these isotopes the chemical or biological history is revealed from which mechanisms may be deduced. Yields and losses in chemical technological procedures may be readily determined with the use of radioactive tracers.

In medicine radioisotopes are used as gamma-ray sources in the treatment of tumors (e.g., ^{60}Co) and in diagnosing certain abnormalities (e.g., ^{131}I is used in thyroid disease detection).

M. H. LIETZKE AND R. W. STOUGHTON

References

Hampel, C. A., "The Encyclopedia of the Chemical Elements," Reinhold, New York, 1968.

Weast, R. C., "Handbook of Chemistry and Physics," The Chemical Rubber Co., Cleveland, Ohio, 1970.

"Chart of the Nuclides," Tenth Edition—Revised to December 1968 (Revised about every 2 years: Eleventh Edition—Expected to be revised to April 1970; available to public about October 1972), Educational Relations, General Electric Co., Schenectady, N.Y. 12305.

"Neutron Cross Sections," BNL-325, Second Edition (1958) and Supplement No. 2 (1965-66), Sigma Center, Brookhaven National Laboratory, printed by Clearinghouse for Federal Scientific and Technical Information, National Bureau of Standards, U.S. Department of Commerce, Springfield, Virginia.

K

KERATINS

Keratins are natural fibrous protein materials which occur widely in nature, usually as protective coverings in vertebrates. They include such materials as feathers, hooves, claws, nails, hair and wool. The chief feature of their composition is their relatively high content of the amino acid cystine. They consist also of relatively high contents of serine and arginine. Generally, they have lower contents of proline plus hydroxyproline than the collagen family of fibrous proteins, but they have higher proline contents than most corpuscular proteins. Although the high cystine content is a significant feature, it is the most variable of the amino acids of keratins. The sulfur and cystine contents often differ considerably with species origin.

Two main types of keratins are distinguishable on the basis of physical characteristics, histology and chemical composition—the hard and the soft keratins. An example of soft keratin is the outer layer of the epidermis; examples of hard keratins are horn, nails, claws and hooves. The soft keratins are lower in cystine content, but higher in methionine. The hard keratins have low contents of the nutritionally essential amino acids, histidine, methionine and tryptophan.

Keratins are frequently cellular in structure. Wool and hair, for example, consist of an outer cuticular covering of overlapping cells. These cells have three distinguishable components: epicuticle, exocuticle and endocuticle. The epicuticle is a very thin outer layer which gives to wool and hair the characteristic property of shedding water even though these materials, taken as a whole, have rather high moisture absorption. The cuticle cells surround a cortex which comprises the bulk of the wool and hair fibers and consists of overlapping and highly elongated cells, commonly called spindle cells. These cells consist of fibrils and these, in turn, of microfibrils. The fibrils and microfibrils are held together by protein cementing material. Many fibers, particularly the coarser ones, have a central medullary core which consists of filamentous tissue and which is alkali resistant and low in sulfur. The medulla has a high proportion of air space.

Feathers are more complex in structure than hair. Consequently, they can show greater variation from species to species and even with respect to different areas of a single bird than does hair. A typical feather consists of a translucent calamus or quill, a shaft, and a vane. The calamus is typically one to three millimeters wide. An opaque, hollow, central shaft continues out of the calamus for 2 to 20 cm or more. Flattened or elongated branches, called barbs, branch from the shafts in regular rows on the opposite side of the shaft. The barbs are further subdivided into barbules, which may be flattened or elongated and have hooks, knobs, or other irregularities that are characteristic of the species or feather type.

Insofar as the constituent protein chains of keratins are concerned, x-ray and infrared evidence indicates that they occur in both the so-called alpha or folded configuration or the beta, or straight-chain form. Wool, taken as a whole, is generally regarded as a typical member of the alpha keratin group and feathers of the beta group, but there is increasing evidence that both groups can be present in the same keratin.

In wool the x-ray crystalline pattern is associated with the spindle cells and with the oriented, birefringent fibrils as first laid down. In feathers the outer portion of the quill exhibits crystalline structure.

The polar residues of the polypeptide chains of keratins include substantial amounts of both acidic and basic groups. Since over half the potential acid groups are amidized, the number of free acid groups from analytical evidence is only slightly less than equivalent to the basic groups present. The polar residues are responsible for the rather considerable water uptake by the keratins. Moreover they are sites for bonding of molecules other than water, for example, dyes and ionic detergents.

The physical properties of keratins differ, depending upon the degree of molecular consolidation and crosslinking. Keratins are generally described as insoluble in common solvents, and this property is ascribed as due mainly to the presence of a relatively high degree of crosslinking of the protein chains by the disulfide cross links of cystine. Some keratins have been solubilized, and this requires that the disulfide cross links be severed first. This may be accomplished either by reduction, hydrolysis or oxidation, following which the materials dissolve in such solvents as concentrated aqueous urea, dilute ammonia or aqueous solutions of anionic detergents.

The mechanical properties of keratins also depend upon the degree of molecular consolidation and cross-linking. In feathers the molecular chains are in more or less extended configuration and held together in rather extensive crystalline array. Consequently feathers are stiff and do not exhibit long-range extensibility which is typical of wool and hair. In wool and hair, not only is the degree of crystalline order low, but the chains appear mostly in the folded configuration so that when stress is applied the chains respond by unfolding and long-range extension may occur. The disulfide crosslinks assist in the recovery of the original length when applied stress is released.

HAROLD P. LUNDGREN

Cross-references: *Fibers, Natural; Proteins; Amino Acids.*

KETENES

Ketenes are members of a class of compounds which contain the functional group $=C=C=O$. Examples are ketene itself (CH_2CO), methylketene $CH_3CH=C=O$, dimethylketene (CH_3)$_2C=C=O$, diphenylketene (C_6H_5)$_2C=C=O$, and carbon suboxide $O=C=C=C=O$. Of these, ketene is by far the most important. It is a reactive, colorless gas of considerable industrial importance. Physiologically, it is extremely poisonous and care must be taken to avoid breathing it.

The availability of ketene by pyrolysis of acetone (or acetic acid) is the reason for the attention it has received, contrasted to other ketenes which are relatively unavailable. Ketene may be prepared also by pyrolysis of acetic anhydride or phenyl acetate or diketene. Other sources are quite unsatisfactory from a standpoint of yield. Small quantities may be made conveniently by heating acetone in a "ketene lamp." This is a glass apparatus containing a Nichrome filament, heated electrically to red heat. Larger amounts are made by passing acetone or acetic acid through a tube at 700°C. A very brief contact time is required, so that much of the acetone is undecomposed and has to be condensed and recycled. Also, it is imperative that the reaction tube be of inert material such as porcelain, glass, quartz, copper or stainless steel. A copper tube, if used, should be protected from oxidation by an iron sheath. Inert packing may be used (glass, vanadium pentoxide, porcelain), but just as good yields are obtained with empty tubes. No catalyst is known which accelerates this decomposition at significantly lower temperatures.

Methyl ethyl ketone is totally unsatisfactory as a source of methylketene by pyrolysis, but pyrolysis of propionic anhydride in a quartz tube at 400–600°C and low pressures does produce it in a stated yield of about 90%. Another synthetic approach is to prepare methylketene dimer by allowing a mixture of propionyl chloride and triethylamine to stand at 25° for 24 hours and then to pyrolyze the dimer.

The known disubstituted ketenes include dialkylketenes, diarylketenes, and the ester analogs. Dimethylketene may be made from α-bromoisobutyryl bromide by reaction with zinc in boiling ether. Diphenylketene may be made similarly, but the usual way to prepare it is to oxidize benzil hydrazone with yellow mercuric oxide to benzoylphenyldiazomethane which, on heating in benzene solution, decomposes into the ketene.

The best way to make carbon suboxide is a pyrolytic method, starting with tartaric acid. The latter is converted into diacetyltartaric anhydride and then pyrolyzed at 625–650°C (either in an empty tube or in a ketene lamp) into acetic acid and carbon suboxide, the latter in 35–50% yields.

Some recent synthesis of ketenes involve interesting chemistry. A butadienylketene is obtainable at −100 to −150°C by photolysis (mercury lamp) of the appropriate cyclohexadienone:

The reaction is reversed on warming, but the ketene may be captured by an amine to form an amide.

1-Ethoxy-1-alkyne, $R—C{\equiv}C—OC_2H_5$, pyrolyzes at 120° into ethylene and alkylketene, but the latter is consumed by the original alkynyl ether to form 2,4-dialkyl-3-ethoxy-2-cyclobutenone,

$$\begin{array}{c} R—CH—C=O \\ | \qquad | \\ C_2H_5O—C = C—R \end{array},$$

in high yield.

Ketene and diimide are formed during the alkaline decomposition of chloroacetic hydrazide. The diimide, however, spontaneously changes into hydrazine plus nitrogen, and the hydrazine consumes the ketene to form acetohydrazide, $CH_3CONHNH_2$. If an olefin is present in the reaction mixture it is required to a paraffin by the diimide, thus preventing hydrazine formation.

The formulas of acetic acid, acetic anhydride and ketene show that the anhydride differs from the acid by 0.5 mole of water, and that ketene differs by 1.0 mole. Ketenes, therefore, may be regarded as super acid anhydrides, and this viewpoint leads to a good appreciation of their reactions. Because of an original lack of understanding of this relationship, the monosubstituted ketenes were once classed as "aldoketenes," and the disubstituted ketenes as "ketoketenes." Such terms are quite misleading, however, and should be abandoned.

Ketene is absorbed in sodium hydroxide solution, yielding sodium acetate. Aniline adds to ketene to form acetanilide. Both of these reactions are quantitative and are used to assay the ketene in a gas stream.

Primary alcohols react readily with ketene to form acetic esters but tertiary alcohols require the catalytic help of sulfuric acid. Even with primary alcohols, as 1-butanol, it has been established that addition of ketene ceases at about the 75% conversion point unless a little sulfuric acid is present as catalyst. Phenol, which is inert toward ketene at ordinary temperature, may be converted into phenyl acetate by reaction at the boiling point of phenol or by reaction at room temperature if a trace of sulfuric acid is present.

An important industrial synthesis of acetic anhydride is via acetic acid and ketene. Mixed acetic anhydrides are made similarly by passing ketene into the acid in question: $RCOOH + CH_2CO \rightarrow RCO—O—COCH_3$. This is the basis of a good method of synthesizing symmetrical anhydrides in view of their formation from the mixed anhydrides on heating:

$$2RCOOCOCH_3 \rightarrow (RCO)_2O + (CH_3CO)_2O.$$

Comparable reactions of ketene are those with mercaptans to form thio esters (CH_3COSR), with amino acids (in water) to obtain N-acetyl derivatives ($CH_3CONHCHRCOOH$), with hydroxylamine to yield acetohydroxamic acid ($CH_3CONHOH$), dimethylchloroamine to form chloroacetic dimethylamide ($ClCH_2CON(CH_3)_2$), and Grignard reagents to form ketones ($RCOCH_3$).

Ketene adds to pyridine in a 4:1 ratio to form a yellow, crystalline compound,

Quinoline, isoquinoline, and phenanthridine resemble pyridine in reacting comparably, and diketene may be substituted for ketene.

Aromatic aldehydes take up ketene in the presence of potassium acetate in the manner of a Perkin reaction. The product is a cinnamic acid:

$$ArCHO + CH_2CO \xrightarrow{AcOK} ArCH{=}CHCOOH.$$

Friedel-Crafts catalysts are effective in converting formaldehyde and ketene into β-propiolactone. This is a process of industrial importance. Similarly, furfural and ketene give rise to 3- (2-furyl)-propionolactone.

When aluminum chloride is used as catalyst, ketene reacts with benzene to form acetophenone: $C_6H_6 + CH_2CO \rightarrow C_6H_5COCH_3$. Also, methyl chloromethyl ether under such conditions reacts to yield 3-methoxypropionyl chloride:

$$CH_3OCH_2Cl + CH_2CO \xrightarrow{AlCl_3} CH_3OCH_2CH_2COCl.$$

Ketones such as acetone, ethyl acetoacetate, ethyl levulinate or acetylacetone react smoothly with ketene if a trace of sulfuric acid is present. The products are enol acetates of the ketones, acetone giving rise to isopropenyl acetate, $CH_2{=}C(CH_3){-}OCOCH_3$, and ethyl acetoacetate changing into ethyl 3-acetoxycrotonate. Cyclobutanone derivatives are made quite easily by addition of ketenes to styrene or to vinyl ethers or to enamines:

$$\begin{array}{c} R_2C{=} \\ CH{-}NR_2 \end{array} + O{=}C{=}CH_2 \xrightarrow{0°} \begin{array}{c} R_2C{-\!-\!-}CH{-}NR_2 \\ | \qquad\qquad | \\ OC{-\!-\!-\!-}CH_2 \end{array}.$$

One of the most characteristic reactions of ketene or dialkylketenes is that of polymerization into dimers. Diarylketenes do not display this tendency. The dimer from ketene, or "diketene," is a liquid that boils without decomposition at 43° (28 mm), but the compound tends to decompose (into dehydroacetic acid and resinous substances) on distillation (b.p. 127°C) at atmospheric pressure. The structure of diketene was in doubt for many years, but recent critical chemical and physical evidence indicates the structure as 3-buteno-β-lactone. Diketene is an acetoacetylating agent. Thus, with aniline it yields acetoacetanilide, and with methanol (catalyzed by sulfuric acid) it produces methyl acetoacetate. These reactions are useful industrially.

CHARLES D. HURD

Cross-references: *Alcohols, Aldehydes, Aliphatic Compounds, Ketones, Organic Chemistry.*

References

Staudinger, H., "Die Ketene," F. Enke (Stuttgart), 1912.
"Organic Reactions," Vol. 3, pp. 108–140 (1946), Wiley (New York).

J. Am. Chem. Soc., **54,** 3427 (1932); **55,** 275, 757 (1933); **72,** 1461 (1950); **83,** 236 (1961). *Angew. Chem., Inter. Ed.,* **9,** 240 (1970).

KETONES

Ketones are a class of organic compounds, having as their generic formula, $R{-}\overset{\displaystyle O}{\overset{\|}{C}}{-}R'$, in which R and R' may be alkyl or aryl group. Examples of the possible types are as follows:

Type	Example
Dialkyl ketone	Acetone
Alkyl aryl ketone	Acetophenone
Diaryl ketone	Benzophenone
Cyclic ketone	Cyclohexanone

Acetone, methyl ethyl ketone, methyl isobutyl ketone and acetophenone are four ketones produced in relatively large quantities. Over 1.64 billion pounds of acetone were produced in 1971.

Preparation Methods. The more common ways of producing ketones are as follows:

(1) Hydration of an olefin to form a secondary alcohol, which is then oxidized to a ketone. A great proportion of all the acetone and methyl ketone prepared today is made this way.

(2) Dry distillation of the calcium salts of mono- or dicarboxylic acids. Symmetrical or mixed ketones can be prepared in this manner. Cyclic ketones result when dicarboxylic acids are used.

(3) Fermentation of certain carbohydrates by use of specific bacteria produces acetone.

(4) Alkyl aromatic ketones may be produced by reaction of an acid chloride on the benzene nucleus in the presence of aluminum chloride (Friedel-Crafts Reaction).

(5) Use of the acetoacetic ester synthesis, which involves reacting an alkyl halide with sodioacetoacetic ester, followed by alkaline hydrolysis.

(6) Isomerization of certain glycols with mineral acids. The best example of this is the pinacol-pinacolone rearrangement.

(7) Reaction of certain substituted ethylene oxides by heating in the presence of zinc chloride produces ketones.

(8) Hydrolysis of the sodium salts of secondary nitroparaffins with dilute mineral acids.

(9) Hydration of a monosubstituted acetylene in the presence of sulfuric acid and mercuric salts.

(10) Reaction of nitriles with Grignard reagents produces an intermediate, which on hydrolysis yields a ketone.

(11) Conversion of acids and anhydrides to ketones by passing the vapors over thorium oxide at 400°C.

Reactions. Ketones are similar to aldehydes in their reactions, both classes of compounds possessing the reactive carbonyl group and in most cases alpha hydrogens, which are also very reactive. Reactions of ketones may, in general, be divided into four classes: (1) those involving the keto group, (2) those involving the alpha hydrogens, (3) those involving both the keto group and the alpha hydrogens and (4) those not definitely in any of the first three classes.

(1) Reactions involving only the keto group.

(a) Reduction produces either a secondary alcohol or a substituted glycol, depending on conditions. By another type of reduction (Clemmensen), using zinc and hydrochloric acid, the oxygen of the keto group is replaced by two hydrogens.

(b) Reaction with some active hydrogen compounds produces a tertiary alcohol or an olefinic compound. An example of the former is the reaction of acetone with itself to form diacetone alcohol. An example of the latter is the reaction between acetone and malonic ester to form isopropylidene malonic ester.

(c) Reaction with primary nitroalkanes produces both nitroalcohols and dinitroalkanes. An example is the reaction of acetone and nitromethane to form 2,2-dimethyl-1,3-dinitropropane and a nitroalcohol.

(d) Reaction with substituted ammonia-type compounds, such as hydroxylamine, hydrazine and semicarbazide, to form oximes, hydrazones and semicarbazones, respectively.

(e) Reaction with Grignard reagents to form an intermediate, which on hydrolysis yields a tertiary alcohol.

(f) Reaction with certain acidic reagents to form addition compounds. An example is the reaction with hydrogen cyanide to form a cyanohydrin.

(g) Ammoniacal reduction produces an amine.

(h) Reaction with hydrogen sulfide yields a thioketone. Mercaptans react with ketones to give hemimercaptols and mercaptols.

(i) Reaction with sodium bisulfite results in crystalline bisulfite addition products.

(j) Reaction with phosphorus pentachloride results in the replacement of the keto oxygen with two chlorines.

(2) Reactions involving only the alpha hydrogens.

(a) Reaction with the carbonyl group of another compound to produce an alcohol or olefinic compound. See 1(b) above.

(b) Halogenation results in replacement of the alpha hydrogens.

(c) Reaction with a dialkyl carbonate produces a keto ester.

$$\overset{O}{\underset{\|}{}}$$

(d) Ketones with the structure, $—\overset{O}{\overset{\|}{C}}—CH_2—$, will react with nitrous acid to form alpha diketones.

(3) Reactions involving both the keto group and the alpha hydrogens.

(a) Acetone will react with itself to form mesitylene in the presence of sulfuric acid.

(b) Treatment of acetone with phosphorus pentachloride produces 2-chloro-1-propene, in addition to 2,2-dichloropropane.

(4) Reactions not in any of the above three classes.

(a) Pyrolysis of acetone at 700°C produces ketene and methane.

(b) Iodine and an alkali hydroxide on a methyl ketone gives iodoform. This is a test for methyl ketones.

Physical Characteristics. Ketones are liquids or low melting solids with pleasant odors. Certain physical characteristics of four common ketones are listed below.

Ketone	B.P. (760 mm)	M.P.
Acetone	56.2°C	−94.9°C
Methyl ethyl ketone	79.6°C	−86.4°C
Methyl isopropyl ketone	93.0°C	−92°C
Acetophenone	202.0°C	−20.5°C

Ketones are soluble in practically all common organic solvents. Acetone is completely miscible with water. As the chain length is increased the solubility decreases rapidly. Methyl ethyl ketone is only slightly soluble in water and aromatic ketones such as acetophenone and benzophenone are practically insoluble in water. The lower ketones have great solvent powers for many plastic and natural resinous materials such as cellulose nitrate, cellulose acetate, ester gum and the vinyl resins. This solvent characteristic makes the ketones unusually valuable in the explosive, lacquer and textile industries.

In addition to being an excellent solvent for organic materials, acetone will dissolve certain inorganic compounds such as potassium iodide and potassium permanganate. Waxes and pitches are soluble in certain of the lower ketones. Methyl ethyl ketone, like acetone, is a good solvent for cellulose nitrate, as well as Glyptal resins; however, in general, it is not a good solvent for cellulose acetate, carnauba, beeswax or paraffin wax. Methyl isobutyl ketone is also a good solvent for a number of organic materials, but it is considerably less volatile than either acetone or methyl ethyl ketone, thus decreasing explosion hazards.

Some ketones of increasing industrial importance are acetonyl acetone, cyclohexanone, diethyl ketone, diisobutyl ketone, dipropyl ketone, methyl isobutyl ketone, mesityl oxide, ethyl butyl ketone, isophorone and trimethylnonanone. All of these are commercially available.

Uses. The largest use of ketones takes advantage of their great solvent powers. In this category the largest amounts are used in the synthetic plastics, paint and allied fields. This, of course includes the textile industry, which, in addition to the use of ketones as solvents for synthetic material for fibers, uses them as leveling agents in dyeing and for delustering cellulose acetate fibers. The paint industry requires enormous amount of various ketones for use as lacquer thinners. Printing inks and polishes may have ketones in their composition. The manufacture of some photographic film uses acetone and some higher ketones. Other solvent uses are metal cleaning compounds, spot and stain removers, degreaser for metal parts and as an absorber for acetylene.

Relatively smaller uses of ketones are found in the tanning industry. Preservative for animal tissues is another use. Certain higher ketones find an application in the hydraulic fluid field. Generally mixtures of ketones with other materials, such as castor oil, are used to produce hydraulic fluids of the correct physical properties.

Although ketones find their greatest use as solvents of various sorts, their use as intermediates or starting compounds in the preparation of synthetic organic chemicals probably ranks second in importance. Alcohols can be produced by hydrogenation and amines by reductive amination. Many complex organic compounds are synthesized using simple ketones as building blocks. Examples of this are rubber accelerators, dyes, inhibitors of many types, insecticides and pharmaceuticals.

D. E. HUDGIN

Cross-references: *Solvents, Aldehydes, Hydraulic Fluids.*

KINETICS, CHEMICAL

The study of chemical phenomena can be made from two fundamental approaches. The first of these, known as *thermodynamics,* is a rigorous and exact method concerned with equilibrium conditions of initial and final states of chemical changes. The other method, known as *kinetics,* is less rigorous and deals with a more complex aspect of chemical phenomena, namely, the rate of change from initial to final states under nonequilibrium conditions. The two methods are related. Thermodynamics, which yields the driving potential—a measure of the tendency of a system to change from one state to another—serves as the foundation upon which kinetics is built. The rate at which a change will take place in general depends upon two factors: directly with a driving force or potential and inversely with a resistance. In chemical kinetics a measure of the tendency of the system to resist chemical change is the so-called activation energy, which is independent of the driving force or so-called free energy of the reaction.

In Fig. 1, a mechanical analogy is used to illustrate the difference between activation energy and driving potential. The chemical system is represented by a sphere resting in a valley between the hills. The initial equilibrium state A is at a higher elevation than the final state B. This difference in elevation between A and B is a measure of the free energy change of the reaction, that is, the tendency or driving force which will take the system from A to B. This quantity ΔG is determined by the classical methods of thermodynamics. Now both A and B are equilibrium states represented by the valleys between the hills. For the system to go from A to B it must first overcome the hill separating the valleys. The elevation of this hill from the valley of the initial state is a measure of the tendency to resist a change in the system in going from A to B. The quantity ΔG^*, known as free energy of activation, is determined by the methods of kinetics.

The system of molecules which is undergoing re-

action consists of these molecules in different energy states. If the temperature of a gas is raised, there is an increase in the energy of these states and hence an increase in the collisions of molecules which have the necessary activation energy, so that reaction will occur at a greater rate. Thus, in general the rate of reaction is increased with temperature. Also, if by means of a catalyst, the activation energy is decreased, more colliding molecules will react and again the rate of reaction will be increased.

Experimental Methods. Experiments to obtain data may be arranged as either *batch or flow* processes according to one of the following three types.

(1) Measurement of composition as a function of time in a *batch* reactor at a *constant volume* and temperature.

(2) Measurement of exit composition as a function of feed rate to a *constant volume* reactor at *constant pressure* and temperature under steady state *flow* conditions.

(3) Measurement of composition as a function of time in a *batch* reactor at *constant pressure* and temperature.

The experimental methods of analysis can also be divided into three types:

(1) Negligibly small samples are withdrawn successively, chilled and quickly analyzed.

(2) Multiple samples are started simultaneously and at given time intervals one of the samples is chilled and analyzed.

(3) Physicochemical methods are used to analyze the reaction mixture without withdrawing samples.

Classification of Reactions. Chemical reactions are broadly divided into *homogeneous* types, (single phase reactions) and *heterogeneous* (more than one phase reactions). In some catalytic reactions where permanent change may be only of a single phase, the overall chemical reaction is heterogeneous due to the interaction at the interface of the system and catalyst.

Reactions also may be classified as either a *flow* process, steady state condition where no change occurs with respect to time in constant volume reactors, or *batch* process where changes in the system occur with respect to time.

Conditions under which a reaction takes place may also serve as a means of classification as follows:

Isothermal (Constant temperature)

Adiabatic (No heat interchange with surroundings)

Constant Volume

Constant Pressure

An important distinction between chemical reactions is the question of reversibility. From a thermodynamic standpoint all reactions taking place in an isolated system are irreversible. When used in this sense, a reversible reaction could occur only if the isolated system were in equilibrium at all times. From the kinetic viewpoint, a reversible reaction is one that can take place in either direction whereas an irreversible reaction is unidirectional. This is an entirely different use of the term "reversible" than in thermodynamics. By isolating all the reactants in a closed system, every reaction can be theoretically made kinetically reversible. If

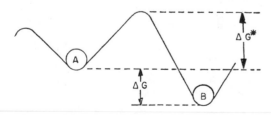

FIG. 1

one of the products can be removed from the reacting system as soon as it is formed, the reaction can be made unidirectional from beginning to end.

Rate Equation for Homogeneous Reaction. Many careful studies on chemical reactions coupled with thermodynamic considerations have shown that the rate of reaction is proportional to the product of the active concentration (that is activity). Taking the general reaction:

$$aA + bB + \cdots \rightleftharpoons rR + sS + \cdots$$

which states that a molecules of species A react with b molecules of species B, etc. to yield r molecules of species R, s molecules of species S, etc., then the rate of the forward reaction may be written as:

$$r = ka_A{}^a \, a_B{}^b \cdots = -\frac{dn_A}{V dt}$$

r = rate at which the molecules of component A are converted per unit volume of reacting system.

n_A = number of molecules of component A in the system at time t from the start of the reaction.

V = volume of reacting system.

k = proportionality constant, called the reaction velocity constant.

a_A = active concentration, called the *activity* of component A in the reaction mixture.

    = n_A/n_t for an ideal solution, where n_t is the total number of molecules of all kinds present.

In the above equation the active concentrations of each component are raised to a power equal to the number of molecules of that component taking part in the reaction, because the rate is proportional to the product of the concentrations of each molecule taking part. For the backward or reverse reaction, the rate is analogously given by

$$r' = k'a_R{}^r a_S{}^s \cdots$$

At equilibrium the forward and backward rates become equal so the net reaction reduces to zero. Equating rates gives:

$$\frac{a_R{}^r \, a_S{}^s \cdots}{a_A{}^a \, a_B{}^b \cdots} = \frac{k}{k'} = K$$

where K = the equilibrium constant defined by thermodynamics.

Reaction rates may be expressed in terms of concentration units. The reaction velocity constants so expressed are in general no longer true constants at a given temperature but vary with concentration. Thus, c will now replace a in the rate equation, where $c_A = n_A/V$. etc. and k now becomes k_c.

From thermodynamics the relationship between these reaction rate constants becomes:

$$k_c = k\gamma_A{}^a \, \gamma_B{}^b \cdots v^{(a+b+\cdots)}$$

where $\lambda_A = \dfrac{a_A n_t}{n_A}$ = activity coefficient of component

A in the mixture; $v = V/n_t$ = molal volume of the

reacting system. For gases this equation may be written as:

$$k_c = kv_A{}^a \, v_B{}^b \cdots (CRT)^{(a+b\cdots)}$$

where $v_A = \dfrac{\gamma_A}{P}$ (fugacity coefficient of component

A in gas mixture); C = compressibility factor of gas mixture; R = gas constant; T = absolute temperature; and P = total pressure.

For ideal solutions or ideal gases the coefficients in the above equations reduce to unity so that in these cases the equations simplify to the following forms:

$$k_c = kv^{(a+b+\cdots)}$$

for solutions and

$$k_c = k(RT)^{(a+b+\cdots)}$$

for gases.

Order of Reaction. The order of a chemical reaction is defined as the minimum number of molecules which appear to react simultaneously according to the kinetic equation which describes its mechanism. For a simple reaction the order is an integer which has never been found to exceed 3. For a complex reaction, which is really the result of several successive reactions, more than one kinetic chemical equation with integral coefficients is needed to describe the mechanism; although it may be described by a single kinetic equation with nonintegral coefficients. Because the probability of more than three molecules meeting simultaneously with the required activation energy to react is extremely small, no reaction has ever been found to exceed the third order.

The kinetic chemical equation as written may not agree with the thermodynamic or stoichiometric chemical equation, since the latter depends only upon the initial and final states which give the overall effect.

Change of Velocity Constants with Temperature. The relationship of reaction rate with temperature and activation energy was first discovered empirically by Arrhenius and is stated as follows:

$$k = Ae^{-E/RT}$$

where A = proportionality constant, which for bimolecular reactions is the collision number, and E = energy of activation.

From statistical mechanics, a new theory of absolute reaction rates has been developed by Eyring. This theory interprets the constants in the Arrhenius equation in terms of fundamental constants: thus,

$$A = \frac{1}{hv^*CN} \text{ and } E = \Delta G^* = \Delta H^* - T\Delta S^*$$

where: h = Planck's constant; v^* = fugacity coefficient of activated complex; C = mean compressibility factor of reaction mixture; N = Avogadro's number; and ΔG^* = free energy change in formation of the activated complex from the reactants; ΔH^* and ΔS^* = corresponding changes in enthalpy and entropy, respectively.

The new theory assumes that the reaction takes place by the formation of an activated complex from the reactants. The complex then decomposes to form the products. It is possible to predict reaction rate constants by means of this theory from a knowledge of chemical structure and energy distribution, but, unfortunately these predictions are still only of a qualitative nature giving orders of magnitude rather than true quantitative results.

Integrated Equations. The rate equations previously discussed are in differential form and give the instantaneous rate of reaction as a function of active concentrations at constant temperature. For batch processes at constant volume the differential or infinitesimal conversion of component A is given for the corresponding differential time interval. For flow processes at constant pressure, the differential conversion of component A is given for an infinitesimal increment in volume. In order to be of practical use, these equations must be integrated to yield conversions for finite changes in time or volume depending on whether the reaction is non-flow or flow. The integration of the general rate equation at constant volume yields the desired relationship between conversion and time for batch reactions.

The integration of the general rate equation at constant pressure for a flow process is carried out with respect to the volume of the reaction mixture, yielding the conversion of reactant A for steady-state conditions. The flow reaction is assumed to take place in a reactor in which no longitudinal mixing occurs. At any cross section in the reactor, the degree of conversion is constant and the same. The time variable is eliminated by incorporating it in the feed rate to the reactor. Thus, for any differential cross-sectional volume of reactor, a given mass passes through in a unit time at the feed rate. In this unit time a certain amount of reaction takes place in the differential volume at a constant rate and produces a differential amount of conversion. The number of molecules so converted disappears from the mass, leaving the differential volume. Equating this loss to the amount of reaction eliminates time as a variable and relates conversion to the volume of the reactor.

Simultaneous Reactions. When some or all of the reactants can undergo more than one reaction at the same time, the kinetic equation becomes more complex so that, in general, a direct mathematical solution is not possible. When the reactants for each of the simultaneous reactions are all the same and all the reactions are of the same order, simple integrated equations can be obtained.

Consecutive Reactions. In consecutive reactions the products of the reaction undergo further reaction to produce new additional products. Often an intermediate product may act as a catalyst in that it does not appear in the final product but serves to catalyze production of the final product. For example, the following simple set of reactions may be considered:

$$A \rightarrow B \rightarrow C$$

Normally A will not produce C, but by forming the intermediate B, which decomposes to form C, the final product is obtained. During this reaction B builds up to a maximum and then disappears with A until, upon completion, only C remains. This is a very important type of reaction which is some-times referred to as autocatalytic. In other words, the formation of B, which acts as the intermediary, comes from the reactant itself and therefore makes the reaction self-catalytic.

Chain Reactions. A series of consecutive reactions which repeats itself is known as a chain reaction. Sometimes the chain of reactions may be many thousand molecules long. For this reason, chain reactions are quite sensitive to negative catalysts or inhibitors, which break up the chain. Since only a single molecule may break up a chain of thousands of molecules, these inhibitors are effective in very small concentrations.

Empirically many chain reactions may appear to be reactions of simple order. This may be due to the fact that one of the reactions of the chain is dominating, that is, its velocity constant is considerably smaller than those of the remaining reactions. However, the fact that the reaction is susceptible to extremely small concentrations of inhibitor clearly indicates its chain nature.

Heterogeneous Reactions. In general, heterogeneous reactions are mostly found in the field of catalysis. However, reactions that are heterogeneous are not necessarily catalytic and a number of industrial reactions of the latter type are important. Most of these involve the interaction between gases and liquids. In the case of multiphase or heterogeneous reactions, the kinetic mechanism depends upon physical factors as well as chemical, and not just chemical as in the case of homogeneous reactions. Thus, the rate of transfer of material between the phases becomes important and complicates the mathematical treatment. In other words, diffusion becomes a most important factor in heterogeneous reactions.

The method of analysis of heterogeneous reactions involves the simultaneous effect of homogeneous reactions in each of the phases plus the added effect of interaction between the phases. Only in the very simple cases has it been possible to describe these reactions with mathematical rigor, and practically all cases of industrial importance require empirical methods of analysis.

ROGER GILMONT

Cross-references: *Thermodynamics, Activity, Catalysis, Chemical Engineering.*

References

Frost, A. A., and Pearson, R. G., "Kinetics and Mechanism: A Study of Homogenous Chemical Reactions," John Wiley & Sons, New York, 1953.

Glasstone, A. C., Laidler, F. J., and Eyring, H. E., "The Theory of Rate Processes," McGraw-Hill Book Co., New York, 1941.

Hinshelwood, C. N., "Kinetics of Chemical Change," Oxford Univ. Press, Oxford, 1940.

Hougen, O. A., Watson, K. M., and Ragatz, R. A., "Kinetics and Catalysis Part III of Chemical Process Principles," 2nd Ed., John Wiley & Sons, New York, 1947.

Gilmont, R., "Thermodynamic Principles for Chemical Engineers," Chapts. 12 & 17, Prentice-Hall, Englewood Cliffs, N. J., 1959.

Emmett, P. H., Ed., Vols. 1–7, "Catalysis," Vol. 1, Chapts. 3–5, Reinhold Publishing Corp, New York, 1954–60.

KNOEVENAGEL REACTION

Aldehydes, both aliphatic and aromatic, condense readily with malonic acid and other compounds having a methylene group activated by carbonyl, nitrile, or nitro groups under basic catalysis (Knoevenagel reaction, 1898). In the condensation with malonic acid the initially formed unsaturated malonic acid undergoes

$$ArCHO + CH_2(CO_2H) \xrightarrow{\text{Base}}$$

$$[ArCH{=}CH(CO_2H)_2] \xrightarrow{-CO_2} ArCH{=}CHCO_2H$$

decarboxylation during the condensation. Knoevenagel used ammonia or a primary or secondary amine to effect condensation, but pyridine, or pyridine with a trace of piperidine, is more satisfactory. With the latter combination piperonal, $C_7H_5O_2CHO$, is converted in 85–90% yield to β-piperonylacrylic acid, $C_7H_5O_2CH{=}CHCO_2H$; crotonaldehyde similarly affords sorbic acid (m.p. 134°), $CH_3CH{=}CHCH{=}CHCO_2H$ (30% yield). If such condensation is carried out in acetic acid solution rather than in the presence of a base the unsaturated malonic acid sometimes can be isolated.

L. F. FIESER
MARY FIESER

KRYPTON*

Krypton is the element with atomic number 36 and an atomic weight of 83.80 (Carbon-12 scale). It is a member of the noble gas family of elements and has the electronic ground state configuration $1s^2$, $2s^2p^6$, $3s^2p^6d^{10}$, $4s^2p^6$. At room temperature it is a colorless, odorless and nontoxic gas. The chief occurrence of krypton is in the atmosphere, where it is present in the rather low concentration of 0.000108% by volume of dry air. In addition, small concentrations of krypton have been found in gases from certain hot springs and volcanoes. The element was discovered in the residual liquid left after nearly complete evaporation of liquid air by Sir William Ramsay and M. W. Travers in 1898. The discoverers aptly chose the name krypton (from Greek $\kappa\rho\upsilon\pi\tau\sigma\sigma$-hidden) for the new element.

The present commercial production of the noble gas starts with the liquid oxygen fraction obtained by the large-scale liquefaction of air. The oxygen containing about 1% noble gases is evaporated and passed through an oxidizing chamber to remove organic impurities. Following this, the rare gases are recovered by low-temperature adsorption on silica gel. Finally, the separation of krypton from xenon is achieved by selective adsorption and desorption on activated charcoal. Krypton can be bought in Pyrex glass bulbs at 1 atmosphere pressure or in steel tanks at higher pressures. The current (1971) price is $20.00 per liter STP of gas.

Natural krypton is a mixture of six stable isotopes with the mass numbers 78 (0.354), 80 (2.27), 82 (11.56), 83 (11.55), 84 (56.90), and 86 (17.37). The numbers given in parentheses are the relative abundances in per cent. In addition 15 radioactive isotopes of the element are known. Certain of these occur among the fission products of heavy nuclei and have been injected in small quantities

*Written under the auspices of the U.S. Atomic Energy Commission.

into the atmosphere since the advent of nuclear weaponry.

Krypton can be condensed to form a white crystalline solid which melts at 116.0°K. It has a heat of fusion of 1636 J/mole. The crystal structure is face-centered cubic and the edge of the unit cell has a length of a = 5.69×10^{-8} cm at 88°K. The vapor pressure rises from 0.738 atmospheres at the melting point to 54.182 atmospheres (critical pressure) at the critical temperature 209.39°K. The critical density is 0.909 g/cc. At the temperature of the normal boiling point, 119.75°K, the enthalpy of vaporization is 9029 J/mole.

Gaseous krypton is monatomic. At 298.15°K the standard entropy of the gas, $S°_{298.15}$, is 163.97 J/deg/mole, and the Gibbs energy, $G°{-}H°_0$, has the value -42691 J/mole. The transport properties of gaseous krypton at 273.15°K are described by the following values for the respective coefficients: viscosity, 0.0002327 poise; thermal conductivity, 0.0000874 J/cm/deg/sec; self-diffusion, 0.0795 cm²/sec.

Of particular importance are the spectral properties of krypton. The first and second spectra of the gas, i.e., the spectra of the un-ionized atom and the one-fold positively charged ion, respectively, are readily obtained in discharge tubes. The lines are sharp and highly reproducible. They therefore serve as a class A secondary standard for the calibration of spectral apparatus. Since 1960 the orange-red line originating from the transition $6_{d\ 1/2,1} \rightarrow 5_{p\ 1/2,1}$ in the spectrum of the pure isotope Kr^{86} represents the fundamental standard of length and has replaced the meter etalon. By definition, one meter equals 1,650,763.73 wave lengths (in vacuo) of the orange-red line of krypton-86.

Being a noble gas, krypton is almost wholly inert chemically. However, two compounds of krypton have been made. They are the difluoride, KrF_2, an unstable white solid, and a complex of the difluoride with antimony pentafluoride. Krypton difluoride is best obtained by exciting an electrical discharge in a low-pressure mixture of krypton and fluorine at low temperatures (90°K). While it has not been possible to make other krypton compounds, it has been known for a long time that a hydrate can be obtained; and the formation of clathrates with hydroquinone and phenol has been described.

Because of its rarity krypton has found only limited uses. It is added to the filling gas of fluorescent lamps and has been employed in other lighting devices to obtain bright flashes of short duration. It has also been added to ordinary incandescent bulbs in order to lengthen the lifetime of the filament. Other uses include the embedding of the radioactive isotope Kr^{85} in solids for applications in analytical chemistry (so-called kryptonates), and the use of Kr^{86} to obtain the fundamental standard of length.

FELIX SCHREINER

References

Cook, Gerhard A., Ed., "Argon, Helium and the Rare Gases," New York, Interscience Publishers, Inc., 1961, 2 volumes.

Schreiner, Felix, "Krypton" in "Encyclopedia of the Chemical Elements," Hampel, C. A., Editor, Van Nostrand Reinhold Co., New York, 1968.

L

LACQUERS

The term "lacquer" has been used to designate a number of different types of coatings which vary considerably in composition and use. Modern lacquer can be defined as a protective or decorative coating whose primary film properties are formed by evaporation of solvent. The evaporation of the solvent is extremely fast resulting in rapid drying of the coating, allowing the coated ware to be handled within a very short period of time.

Lacquer is believed to have originated in China some time before the Christian era. It is not known exactly when lacquer was introduced to the Japanese by the Chinese, but it is believed to have become known to them just after the third century. Chinese and Japanese lacquers were natural products derived from the sap of a tree. The exact composition of the sap is not known but it is believed to have contained phenolic compounds, gum protein, volatile acids and water. This "raw" lacquer when exposed in a very thin film in moist air at ordinary room temperatures, hardens slowly; this is thought to be due to the catalytic action of an enzyme in the protein which brings about oxidation.

Another form of Oriental lacquer as used in India and Burma is derived from a resinous material (lac) secreted by tiny insects. In this case the sap of the tree has been transformed in the body of the insect into a resinous crust which the insects deposit on the branches and in which they become embedded. The resinous crust is gathered and ground between stones, then washed to remove the coloring matter. It is then melted to remove other impurities, cooled and formed into sheets which are broken into small flat pieces. The final product is dissolved in alcohol to form what is known as "shellac."

In much of the better grade of Oriental lacquer work, various colors were painted over a suitable background and then protected by several coats of clear lacquer. Many of the older pieces of this work are exquisitely decorated and are valued at fabulous prices, and the smoothness and uniformity of the various color coats show evidence of long hours of labor in the dispersing of the pigments and the rubbing of the finish coats of lacquer.

Modern commercial lacquers are entirely different chemically from the Oriental type, and resemble them only to a limited extent in drying characteristics and in final film toughness. They are highly specialized production finishes which originated during the latter part of the last century as clear lacquers with the advent of high-viscosity nitrocellulose. These lacquers were little known to the general public, but in the early twenties when low-viscosity nitrocellulose first became available,

lacquer was universally recognized as a revolutionary type of fast-drying coating in the paint industry. These original nitrocellulose lacquers were relatively simple formulas, since they were based on nitrocellulose modified with a resin such as ester gum, damar gum or rosin maleate, plus a chemical plasticizer such as dibutyl phthalate or tricresyl phosphate. Raw and blown castor oil were also sometimes used as plasticizers in these lacquers, particularly for flexible coatings.

The next distinct improvement in lacquer formulation came in the early thirties when the technique of blending nitrocellulose with alkyd resins was developed. This gave such an improvement in film properties and exterior durability over the old hard resin formulas that these lacquers are still being used successfully where a tough durable coating is required.

At about the same time, a large number of new synthetic lacquer polymers were introduced to the lacquer industry. In addition to cellulosic materials, as cellulose acetate, cellulose acetate butyrate, ethyl cellulose, etc., many entirely different polymers became available. It would be impossible to make a complete list of lacquer type film formers, but the list would include such materials as vinyl resins, acrylic resins, nylon resins, corn protein, chlorinated rubber, styrene-butadiene resins, polystyrene resins, allyl starch and silicone resins.

Nitrocellulose is still the most widely used lacquer component. It is made by the nitration of cellulose derived from cotton linters or wood pulp; most commercial grades have a nitrogen content of 10.7 to 12.2%. It has a broad solvency range in many different esters, alcohols and ketones and is compatible with a large variety of resins and plasticizers. Nitrocellulose is available in a wide range of viscosity grades which give the formulator considerable latitude in producing lacquers of any desired viscosity.

Modern nitrocellulose lacquers are used today to finish almost an unlimited number of material compositions, varying from metal and wood to plastic, paper and fabric. Both clear and pigmented lacquers are used in many large industries. The paper industry uses a very large volume of such lacquer to finish such items as labels, cellophane, book covers, foil, color cards and packages. Considerable quantities of nitrocellulose lacquer are used as fabric coatings and coatings for plastics.

Most modern furniture lacquers are still based on nitrocellulose and a chemical plasticizer blended with a nondrying alkyd resin to give the best color retention on furniture. The nondrying alkyds retain their resistance to cold checking on wood much better than the rosin modified hard resins used in furniture lacquers.

Automobile lacquers based on a vehicle of nitro-cellulose, coconut oil-modified alkyd and a chemical plasticizer are still widely used in the automotive refinish shops. Many new pigments have become available in recent years, which have made it possible to formulate automotive lacquers having excellent resistance to sunlight in almost any color. Some of the pigments involved are rutile titanium dioxide, indanthrene maroon, indanthrene blue, phthalocyanine blue and green, quinacridone red and maroon, indo yellow and orange and ferric hydrate. By the use of nonleafing aluminum and special dispersions of these bright-colored pigments in an extremely fine particle size, the modern "glamor" lacquers are produced.

Lacquers containing nitrocellulose continue to be popular because of their very fast solvent release. Nitrocellulose has the widest range of compatibility with resin modifiers of all the cellulosics and, therefore, gives the most latitude in the formulation of coatings for specific uses.

Another cellulosic that has found a limited but useful spot in the lacquer industry is ethyl cellulose. It is an ether made by treating cellulose with concentrated caustic soda and ethyl chloride and the commercial product has an ethoxy content varying from 45 to 50%. Ethyl cellulose is soluble in low-cost solvents such as alcohols and hydrocarbons and possesses heat resistance and flexibility superior to nitrocellulose. It is used in cable lacquers, paper lacquers, textile lacquers and plastic finishing lacquers.

Cellulose acetate is a specialized lacquer material which is utilized because of its behavior toward solvents as well as its heat and light resistance. It has limited application in the coating field as cable and wire coatings and the main use is confined to the production of synthetic yarn called "rayon."

Cellulose acetate butyrate is similar to the acetate in some respects but because of its butryl and acetyl content it has increased solvency and greater compatability with resins and plasticizers particularly in the low-molecular half-second grade. This grade is soluble in a wide range of ketones, esters, and alcohol-hydrocarbon blends. Some typical uses of cellulose acetate butyrate lacquers are metal coatings, plastic coatings, stripping lacquers, gel lacquers, fabric coatings and wood finishes.

Of the noncellulosic lacquers that have had general use in recent years those based on vinyl resins have probably had the widest usage. These resins contain the vinyl acetate, vinyl chloride and the vinylidine chloride groups and are characterized by the fact that they are water-white, thermoplastic, odorless, tasteless, nontoxic and chemically resistant. The copolymer vinyl acetate chloride resin has been used quite extensively as a chemical-resistant lacquer on metal. It has been modified with maleic anhydride to give a resin that has improved adhesion to metal surfaces even on low-bake or air-drying lacquers. Another modification of this copolymer is made by hydrolyzing part of the acetate portion to give a product that is compatible with a large number of alkyd and amine type resins. This property has extended the use of vinyl resins in maintenance and transportation finishes. Vinyl butyral resins, made by partially hydrolyzing polyvinyl acetate followed by reaction with butyraldehyde, are most useful in metal primers and specialized sealers. These resins have excellent heat and light stability and excellent adhesion to a variety of metals.

Thermoplastic acrylic resins have been known for many years, but were only a minor factor in the coating field until 1956 when they were first used for the finishing of automobiles. The acrylic lacquers used in the automotive field are based on resins that are polymerized ester derivatives of acrylic and methacrylic acid. All the resins in this group can be used as clear or pigmented coatings with very little modification, except the addition of chemical plasticizer. However, for many applications the acrylic resins are blended with cellulosics such as nitrocellulose and cellulose acetate butyrate to give the coating the desired properties. They are also sometimes blended with vinyl resins to get improved chemical resistance. The acrylic lacquers are widely used in the chemical coatings and transportation fields because of their outstanding color and gloss retention and excellent resistance to chalking on exterior exposure. Another area of interest for the acrylic polymers as well as their modifications is in reflow coatings. The coating is subjected to elevated temperature which causes the coating to melt or flow creating a glass-like, mirror-type coating. Most original automotive finishes are produced in this manner.

With all the new raw materials and new processing equipment available lacquer technology is a highly developed science. It has produced a variety of super finishes (catalyzed one- and two-package lacquers) whose properties far surpass those of the original nitrocellulose coatings.

FRANK J. STESLOW

Cross-references: *Paints, Enamels.*

LACTONES

The inner esters of carboxylic acids are known as "lactones." Appropriately hydroxylated or halogenated carboxylic acids will react intramolecularly with the elimination of water or hydrogen halide to form heterocyclic esters, the size of the ring depending upon the relation between the carboxyl group and the hydroxyl or halide groups in the original acid. Thus, the γ-hydroxyacids undergo this reaction very readily, and many of the corresponding gamma lactones are formed merely upon evaporation of the aqueous solutions of the hydroxyacids at only slightly elevated temperatures. The delta lactones are less readily formed and are more readily hydrated to the corresponding hydroxyacids, most of which can be isolated without any particular difficulties. The general reactions for the formation of lactones are as follows:

$$\gamma\text{-hydroxybutyric acid} \xrightarrow{-H_2O} \gamma\text{-butyrolactone}$$

$$\delta\text{-hydroxyvaleric acid} \xrightarrow{-H_2O} \delta\text{-valerolactone}$$

Not all hydroxyacids, however, form lactones on dehydration. For example, α-hydroxypropionic (lactic) acid loses water in two steps to form lactic anhydride and lactide, respectively; and substituted β-hydroxyacids lose water under appropriate conditions to form α,β-unsaturated acids. However, recently β-propiolactone has become available commercially. It is a stable liquid, boiling at 155°C, and its properties have been studied and reported in a series of articles by Gresham and co-workers.

The ε-hydroxyacids, and the ω-hydroxyacids from C_8 to C_{21} have been prepared, but upon hydration they do not yield lactones with large ring structures, but linear polyesters, instead. Many of the lactones with large ring structures have been prepared by other methods, however, and are well known. Upon the oxidation of sugars to sugar acids the corresponding γ-lactones are generally isolated. Compounds of this type are of importance in detoxifying a large number of phenolic substances in the organism. Thus, d-glucuronic acid can be isolated as the lactone from the urine of dogs that have been fed borneol, and this used to be a laboratory method for the preparation of glucurone. Better synthetic methods have been developed and glucurone is now commercially available.

A large number of lactones occurs in nature. The musklike odoriferous constituent of angelica root has been shown to be the lactone of ω-hydroxypentadecyclic acid, in which the ring consists of fifteen carbon atoms and one oxygen atom. Muskseed yields another similar lactone which, however, exhibits unsaturation in the ring. The odoriferous constituent of celery is the lactone of sedanonic acid, a derivative of cyclohexene. Coumarin, until recently used in flavorings, but discontinued on account of its toxicity, is a benzene derivative, while ascorbic acid (vitamin C) is a sugar acid lactone, as are the cardiac glucosides (aglucons).

Preparation. A large number of synthetic methods have been used for the preparation of lactones. Only a few of the more interesting methods will be considered. Thus, it has already been indicated that γ-lactones are readily formed, in many cases merely upon evaporation of the aqueous solutions of the corresponding hydroxyacids. The corresponding γ-halogenated acids will also yield lactones upon suitable treatment with alkalies, or by warming their silver salts.

The δ-lactones are somewhat less easily prepared by similar methods, while the ε-lactones are formed only by routes other than via the ε-hydroxyacids. The β,γ-unsaturated acids will form γ-lactones upon the acid-catalyzed addition of the elements of water. The tendency to form γ-lactones is so great that the double bond will move down a chain to the β,γ-position.

The substituted β-lactones can be prepared by treatment of the β-halogenated acids with bases, highest yields having been obtained with the chloro compounds and sodium hydroxide. Lactones are also formed by the reaction of enolyzable ketones with ketene in the presence of suitable catalysts (boric acid, clay, mercuric or zinc chlorides, etc.). Certain ketoacids, such as levulinic acid, will lose the elements of water on heating with dehydrating agents.

As has already been indicated, indirect methods have been employed to prepare the macro-ring lactones. Thus, lactones are obtained upon oxidation of macrocyclic ketones with Caro's acid. Lactones may also be prepared by the reduction of ketoacids, or of acid anhydrides. This may be accomplished, for example, by electrolysis in the presence of boric acid, or by sodium or aluminum analgams under mildly acidic conditions.

Chemical Properties. Lactones are not dissolved by sodium carbonate solutions, but do dissolve in sodium hydroxide with opening of the ring and formation of the sodium salt of the corresponding acid. As esters, they show reactions typical of the ester linkage. With ammonia they form amides of the corresponding hydroxy acids, pyrolysis under proper conditions may lead to loss of water to form unsaturated cyclic ketones, or they may lose carbon dioxide and be converted to ethylene derivatives. Catalytic reduction of lactones with hydrogen in the presence of nickel leads to the formation of saturated acids. β-Lactones, upon hydrolysis and subsequent dehydration, yield α,β-unsaturated acids. Glycols are formed when the butyrolactones react with Grignard reagents, while β-propiolactone and methyl magnesium iodide yield a mixture of β-iodopropionic acid, methyl vinyl ketone, and polymer.

Lactones can be reacted with aromatic hydrocarbons in the presence of anhydrous aluminum chloride to give aromatic derivatives of acids:

$$\begin{array}{c} CH_2 \cdot CO \\ | \quad\quad | \\ CH_2 \cdot O \end{array} + C_6H_6 \xrightarrow{AlCl_3} C_6H_5CH_2CH_2COOH$$

Biological Properties. Lactones show a wide range of biological activity. Unsaturated lactones have been shown to retard or inhibit the growth of plants and the germination of seeds. Lactones have been found to have cardiac, anthelmintic, hemorrhagic, and insecticidal activity. Some lactones are estrogenic, and recently some have been shown to have carcinogenic activity.

Uses. The lactones are useful as intermediates in chemical synthesis. They are used as solvents, drugs and detoxicants, textile assistants in dyeing, lubricating oil additives, and for plugging underground oil formations. They are also useful in the manufacture of perfumes and flavorings, and for epoxy, alkyd, and polyester resins.

CARL M. MARBERG

References

Moncrieff, R. W., "The Nature of Lactones," *American Perfumer and Essential Oil Review,* **48,** 47–8 (Nov. 1946; **48,** 44–6 (Dec. 1946); **49,** 42–4 (Jan. 1947); **49,** 148–51 (Feb. 1947). A review of lactones from the perfume and flavoring point of view.

Dickens, Frank, "Carcinogenic Lactones and Related Substances," *British Medical Bulletin,* **20,** 96–101 (1964).

Zaugg, H. E., "β-Lactones," "Organic Reactions," Vol. VIII, Chapter 7, pp. 305–363, John Wiley & Sons, Inc., New York, 1954. A review with 214 references to text and tables.

Etienne, Y., and Fischer, N., "β-Lactones," "The Chemistry of Heterocyclic Compounds," Vol. 19,

Part 2, Chapter VI, pp. 729–884, Interscience Publishers Div. John Wiley & Sons, Inc., New York, 1964. A review with 747 references.

LADDER POLYMERS

Ladder polymers are ordered networks comprised of double-stranded chains connected by hydrogen or chemical bonds regularly located along the chains. Naturally occurring ladder polymers include proteins, in which polypeptide chains are interlinked through hydrogen as well as through covalent disulfide bonds; the double chain polyester desoxyribonucleic acid in which two phosphate containing chains are hydrogen bonded between the purine and pyrimidine bases attached to the desoxyribose units; and several inorganic substances one of which is silicon disulfide

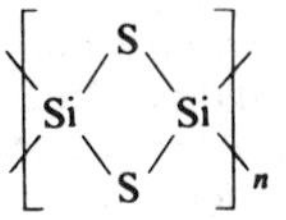

Partially ordered network formation can be induced in synthetic polymers either by (a) simultaneous ring closure and polymerization of suitably substituted molecules, exemplified by pyridine ring formation from the monoxime of 1,5-diketone to give

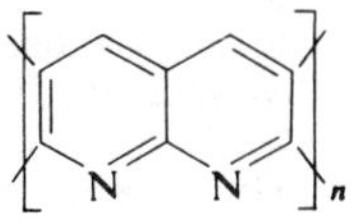

or (b) a two stage process, the first step of which requires formation of a single chain capable of ladder formation: cyclization by sulfuric acid of stereoregular polyisoprene or polychloroprene to give

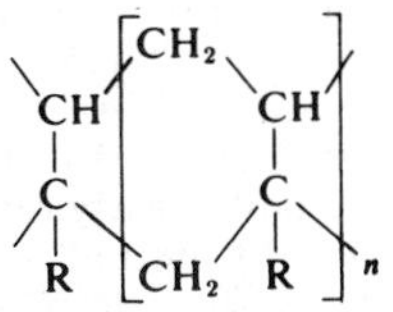

is such a synthesis.

The high degree of crosslinking in ladder polymers often leads to materials chemically and thermally more stable than their single chain analogs: for example, polyphenyl-silsesquioxane (from base catalyzed equilibrium of phenyl siloxane) is stable in air at 500°C.

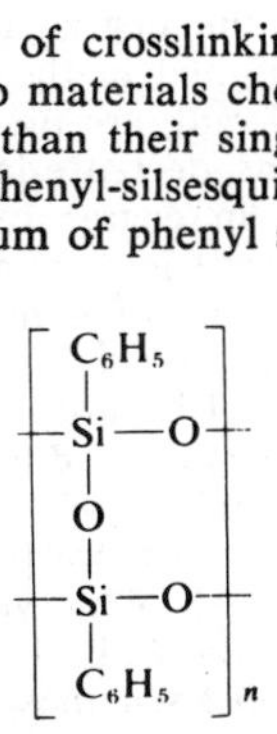

while the fused naphthyridine-ring type (pyrolysis product of polyacrylonitrile)

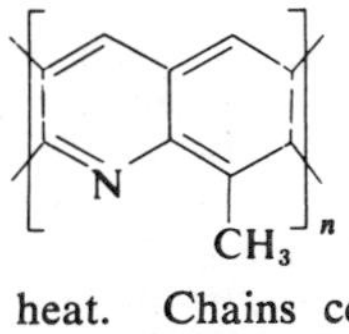

is stable at red heat. Chains containing highly conjugated groups, or fused aromatic rings with unpaired electrons, such as the one pictured above, exhibit strong electronic absorptions in the visible range, are metallic in appearance and semiconducting (electrical conductivity at 20° in the range of 10^{-10} to 10^{-13} ohm^{-1} cm^{-1}).

W. BURLANT

Reference

Overberger, C. G., and Moore, J. A., *Fortschritte der Hochpolymeren Forschung*, **7**, 113 (1970).

LAKES

The term "lake" as applied to an organic pigment color has, by virtue of usage, both a broad commercial and a narrow scientific definition. The broad, or commercially recognized general definition, is exemplified by that of the Dry Color Manufacturers' Association: "Lake: an organic pigment prepared either by precipitating a soluble dye on a reactive or absorptive substratum or by extending, blending, or diluting a full strength toner with a colored or colorless substratum." A *toner* is an organic coloring matter which does not require the assistance of an absorptive base to possess all the requisites of a pigment. The narrower scientific definition would be: "Lake: an organic pigment produced by the interaction of a soluble organic dyestuff, a precipitant, and an absorptive substratum." The discussion to follow is limited to those pigments which are in the scope of the latter category.

Essentially, there are as many fields of application for lakes as there are for toners. These include printing inks, metal decorative coatings, coated fabrics, rubber, plastics, interior oil and water paints, and many others. There are no lakes, and only a very few toners, such as the phthalocyanines, which have sufficient lightfastness for prolonged exterior application, especially in pastel or light tints.

The earliest known organic pigments were lakes formed from the naturally occuring dyestuffs obtained from either vegetable or animal sources. Some of the more important were the red Brazilwood lakes, the crimson Kermes lakes, the red and purple Cochineal lakes, the bluish-red Lac lake, the yellow Quercitron lakes, the brown and black Logwood lakes, the yellow Weld lakes, and the red Madder lake.

Madder lake, or Alizarin Red, because of its excellent soap and alkali fastness and good lightfastness, is still a very important organic pigment. It has been made from the synthetic dyestuff since Graebe and Liebermann discovered in 1868 that the main constituent of the madder dyestuff (alizarin) was 1, 2-dihydroxyanthraquinone. The formation of a commercially acceptable Madder lake involves the interaction of an aluminum hydrate base, alizarin, Turkey Red Oil, and calcium salt, followed by a period of boiling. Despite the age of Madder lake, there is still no general agreement among organic pigment chemists on the chemical constitution of the pigment, some believing that the Turkey Red Oil is an integral part of the pigment while others feel that it is present in intimate admixture as the insoluble metal salt, functioning principally as an oil wetting and dispersing agent.

A typical lake formed from an acid dyestuff can be illustrated by this practical procedure for manufacturing Persian Orange: 2000 pounds of aluminum sulfate are dissolved in 2400 gallons of water at 60°C in a large (7200 gallon) striking tank, which is a cylindrical wooden vessel equipped with a paddle agitator and either open or closed steam coils. 970 pounds of soda ash are dissolved in 1150 gallons of water which had been heated to 60°C. The soda ash solution is run into the agitating aluminum sulfate solution, thus precipitating the aluminum hydrate base. This base is filtered and washed in iron filter presses until essentially free of sulfate ions. The washed base is slurried with water till well dispersed and then brought to 1800 gallons at 30°C with additional water in the striking tank. 285 pounds of Orange II are dissolved in 1300 gallons of water at 35°C and added to the hydrate base. Finally a solution of 300 lbs of barium chloride in 325 gallons of water at 30°C is added to the dyestuff base slurry. This completes the formation of the lake and a water-insoluble pigment results.

Although the above are specific conditions governing the manufacture of a satisfactory Orange II lake, they demonstrate the general requirements for a lake of a soluble dyestuff, an adsorptive base or substratum, and usually a precipitant.

JOHN V. HALLETT

Cross-references: *Pigments, Dyes.*

LANTHANIDE ELEMENTS

The lanthanide elements (or rare earths) are those elements of atomic numbers 57 through 71 that lie between barium and hafnium. The fact that yttrium (atomic number 39, also in Periodic Group III) is a lathanide element in all of its characteristics except those that are completely dependent upon electronic configuration permits an operationally useful classification and consideration of this element as an additional member of the lanthanide series. The classic designation as rare earths, as based upon original recovery as oxides (earths) from relatively rare minerals, is unfortunate because of its connotation of scarcity. The more abundant lanthanides are as common in the crust of the earth as nitrogen, cobalt, or arsenic; the least abundant as common as silver, cadmium, or iodine. All of the lanthanides (plus yttrium) except promethium were painstakingly identified as components of the minerals gadolinite and cerite over the period 1794–1907. Promethium (atomic number 61) was first identified in the 1940's as a product of the neutron-induced fission of uranium-235. Promethium nuclides, formed by natural fission processes, do occur in minute traces in nature.

Atomic Structure. The marked overall similarities in properties that accounted for early difficulties of separation have their origins in the electronic con-

figurations of the atoms and derived cations. In their ground states, the Periodic Group III atoms have noble-gas cores plus the arrangement $(n-1)$ $d^n ns^2$ ($n = 4$ for Sc, $= 5$ for Y, $= 6$ for La). After lanthanum, the $4f$ orbitals lie at lower energy than the $5d$ orbitals, and the next fourteen electrons occupy these inner shells. Atoms of the resulting lanthanide elements have the electronic configurations [Xe core] $4f^1 5d^1 6s^2$ (Ce), [Xe core] $4f^{3-7} 6s^2$ (Pr-Eu), [Xe core] $4f^7 5d^1 6s^2$ (Gd), [Xe core] $4f^{9-14} 6s^2$ (Tb-Yb), and [Xe core] $4f^{14} 5d^1 6s^2$ (Lu). Whether the ground-state configuration is $4f^x 5d^1 6s^2$ or $4f^{x+1} 6s^2$ is of much less chemical than physical importance. Except insofar as electrons may be lost from $4f$ orbitals in the formation of cations of varying oxidation number, the $4f$ electrons are sufficiently buried within the atoms as to be largely unavailable for bond formation. The lanthanides are thus distinguished from the d-transition metals, in the atoms and ions of which the distinguishing d electrons are in the valency shells and participate in bonding.

Indirect consequences of electronic configuration are the *oxidation states* of the elements in their compounds and trends in the *crystal radii* of the ions. In the majority of their compounds, all of the lanthanide elements are in the +3 oxidation state as Ln^{3+} ions (Ln = any lanthanide plus yttrium). Only the terpositive ions are thermodynamically stable in aqueous solution. The non-terpositive ions Ce^{4+}, Eu^{2+}, and Yb^{2+} can exist in aqueous solution, but the Ce^{4+} ion is readily reduced and the other two readily oxidized, each to the +3 state. Quadripositive ions are found in solid compounds such as CeO_2, CeF_4, $(NH_4)_2Ce(NO_3)_6$, PrO_2, Na_2PrF_6, Cs_3NdF_7, TbO_2, TbF_4, and Cs_3DyF_7. Only the cerium(IV) compounds resist reduction in contact with aqueous systems. Bipositive ions are found in solid compounds such as $CeCl_2$, NdI_2, SmI_2, $EuCl_2 \cdot 2H_2O$, TmI_2, and $YbSO_4$. All are rapidly oxidized in contact with water except the europium(II) compounds. The ubiquitous +3 state apparently owes its stability in aqueous systems to a fortuitous combination of ionization and hydration energies. In the Ln^{3+} ions, the ground-state electronic configuration is [Ar core] for Y^{3+} and [Xe core] $4f^x$ for La^{3+} ($x = O$) to Lu^{3+} ($x = 14$).

Among the Ln^{3+} ions a nearly regular decrease in crystal radius is noted from La^{3+} (1.061 Å) to Lu^{3+} (0.848 Å). This ca. 20% decrease is termed the *lanthanide contraction*. It is primarily responsible for the trends in the degree to which particular properties are exhibited in the series. Paralleling decrease in crystal radii are increase in ease of hydrolysis of Ln^{3+} ions, decrease in basicity of the hydroxides, increase in degree of covalency with a given ligand, decrease in cordination number, and increase in the stability of complexes with a given ligand. The size difference is not sufficient to rule out many instances of isomorphism and similarities in solubilities. In crystal radius, the Y^{3+} ion (0.88 Å) is the same size as the Er^{3+} ion (0.881 Å) and not materially different from the ions Ho^{3+} (0.894 Å) and Tm^{3+} (0.869 Å). That yttrium is both found in nature in association with the heavier lanthanides and separates in most schemes with these ions are consequences of size similarities. Indeed, yttrium is so much more abundant than the heavier lanthanides that the minerals from which the latter are recovered are yttrium minerals, and these species are called the yttrium earths. The marked resemblances between the elements that immediately follow the lanthanides in atomic number and their lighter congeners (e.g., Hf and Zr, Ta and Nb, W and Mo, Re and Tc) are the result of decreases in crystal radii caused by the lanthanide contraction.

Direct consequences of electronic configuration are the *magnetic* properties and the *color* and *light absorption* properties of the ions. The ions Y^{3+}, La^{3+}, Ce^{4+}, Yb^{2+}, and Lu^{3+}, each of which contains no unpaired electrons, are diamagnetic. All other lanthanide ions are paramagnetic. The observed *magnetic moments* of the Ln^{3+} ions are determined by both the spin and orbital motions of the unpaired electrons and are, therefore, not directly relatable to the number of unpaired electrons, as are those of the $3d$ transition-metal ions, where only electron spin is significant. Two maxima in magnetic moment (ca. 3.5–3.6 BM for Pr^{3+} and Nd^{3+}, and 10.3–10.6 BM for Dy^{3+} and Ho^{3+}) are noted. The non-terpositive ions have roughly the same moments as the isoelectronic terpositive species (e.g., Eu^{2+} and Gd^{3+}), indicating comparable electronic configurations. A number of the Ln^{3+} ions are among the most strongly paramagnetic species known. The magnetic moments of the Ln^{3+} ions are essentially the same in the free state and in complexes, indicating minimal involvement of $4f$ electrons in bonding.

The striking colors that are characteristic of crystalline salts derived from a number of Ln^{3+} ions persist in aqueous and nonaqueous solutions and are largely unaffected by the anion present or the addition of complexing ligands. That the ions La^{3+}, Lu^{3+}, and Y^{3+} are colorless follows from the absence of unpaired electrons. However, the ions Ce^{3+}, Eu^{3+}, Gd^{3+}, and Yb^{3+}, each of which has one or more unpaired electrons, are colorless also. However, the first three have prominent absorption bands in the ultraviolet region, and the last absorbs in the infrared region. All of the other Ln^{3+} ions absorb both in the visible and the ultraviolet regions. The absorption bands of the Ce^{3+} and Yb^{3+} ions are broad and apparently result from transitions from the $4f$ to the $5d$ orbitals. Bands for the other ions are line-like in character and arise from transitions within the $4f$ level that are allowed by the changes in symmetry produced by surrounding ions or ligands. The wave lengths of these bands are so sharply and exactly defined (e.g., 4445, 4690, 4822, and 5885 Å for Pr^{3+}) that the lanthanide ions are used to prepare optical filters for standardization of instruments. For each absorbing ion, there are absorption bands that can be used for the spectrophotometric determination of that ion, either alone or in admixture with the other lanthanides. Fluorescence spectra are also characteristic of a number of ions. Europium(III)-activated phosphors for video receiver tubes are practical examples of this property. Certain europium(III) β-diketonate complexes emit absorbed energy coherently and thus show laser properties.

General Chemical Properties of the Oxidation States. The elemental lanthanides are comparable with magnesium as reducing agents under acidic conditions. Reactions at elevated temperatures may be summarized as: X_2 ($= F_2 — I_2$) $\rightarrow LnX_3$; $O_2 \rightarrow$

Ln_2O_3; $S_8 \rightarrow Ln_2S_3$; $N_2 \rightarrow LnN$; $C \rightarrow LnC_2$ or Ln_2C_3; $B \rightarrow LnB_4$ or LnB_6; $H_2 \rightarrow LnH_2$ or LnH_3; many metal oxides $\rightarrow Ln_2O_3$ + metal. Elemental europium and ytterbium, like calcium, dissolve readily in liquid ammonia to give blue, strongly reducing solutions which contain solvated electrons.

Reducing strength of the Ln^{2+} ions in aqueous systems increases as $Eu^{2+} << Yb^{2+} <<< Sm^{2+}$. Both the Yb^{2+} and Sm^{2+} ions are rapidly oxidized by hydrogen ion; Eu^{2+} ion is only slowly oxidized by hydrogen ion but rapidly oxidized if elemental oxygen is present. These bipositive ions have nearly the same crystal radii as the strontium and calcium ions (Yb^{2+}, 0.93 Å; Sm^{2+}, 1.11 Å; Ca^{2+}, 0.99 Å; Sr^{2+}, 1.13 Å), and as a consequence they form with given anions compounds of comparable crystal structures and solubilities (e.g., insoluble sulfates, carbonates, fluorides). These bipositive ions are colored (Sm^{2+}, reddish; Eu^{2+}, straw yellow; Yb^{2+}, green). Each absorption spectrum contains broad bands of considerable intensity. These ions are obtained by reduction of the anhydrous halides or chalcogenides with metals or hydrogen at elevated temperatures, by electrolytic reduction in aqueous solution (except Sm^{2+}), by chemical reduction in solution (Eu^{2+}, using Zn), or by thermal decomposition of the anhydrous iodides.

The terpositive ions form crystalline salts with most of the known anionic species. Where the anions can be decomposed thermally (e.g., OH^-, CO_3^{2-}, or $C_2O_4^{2-}$), these salts yield ultimately oxides when heated. Hydrated salts undergo thermal hydrolysis at elevated temperatures. Anhydrous compounds containing thermally stable anions (e.g., O^{2-}, F^-, Cl^-, Br^-) melt at high temperatures without decomposition. Crystal-structure determinations indicate the solid salts to be ionic. Aqueous solutions of soluble salts are highly ionic and hydrolyzed extensively only if strongly basic anions are present (e.g., CN^-, N_3^-). Highly stable complexes result only with strongly chelating ligands. In their complexes the Ln^{3+} species commonly have coordination numbers of 7, 8, or 9, and occasionally 10. Six coordination is uncommon. Because complexation has little effect upon the magnetic and light absorption characteristics, it cannot significantly involve $4f$ electrons. Bonding in complexes is predominantly electrostatic in nature.

Of the quadripositive ions, only Ce^{4+} appears in solution and in a series of compounds. Cerium(IV) is a strong and useful oxidizing agent. It is converted to cerium(III) by nearly all reducing agents, but the reverse process is effected in acidic solution only by such strong oxidants as the $S_2O_8^{2-}$ ion or ozone. Under alkaline conditions, hydrogen peroxide, elemental oxygen, or the OCl^- ion effects this oxidation. The other quadripositive species are stable only in crystals.

THERALD MOELLER

References

Moeller, T., "The Chemistry of the Lanthanides," Reinhold Publishing Corporation, New York, 1963.
Moeller, T., "Lanthanide Elements," pp. 338–349, "Encyclopedia of the Chemical Elements," Hampel, C. A., Editor, Van Nostrand Reinhold Co.,

New York, 1968. This book contains an article about each lanthanide element.

Spedding, F. H., and Daane, A. H., (Eds.), "The Rare Earths," John Wiley and Sons, New York, 1961.

Topp, N. E., "The Chemistry of the Rare-Earth Elements," Elsevier Publishing Corporation, New York, 1965.

LASERS IN CHEMISTRY

Laser, an acronym, is *l*ight *a*mplification by *s*timulated *e*mission of *r*adiation. This intense, coherent, monochromatic light has made great strides in the field of chemistry since its development about 15 years ago. The Laser Laboratory of the Medical Center of the University of Cincinnati, established by the John A. Hartford Foundation, has had a division of laser chemistry under the direction of Prof. R. Marshall Wilson for some time. Professor Wilson has established with Prof. Hans Jaffe, Chairman of the Department of Chemistry at the University of Cincinnati, a separate and comprehensive division of laser chemistry in the Department of Chemistry. This is the first such laser chemistry division in the United States.

In brief, according to Wilson, applications of the laser in chemistry are of two basic types. First, it replaces conventional light sources or ionization sources and enhances the sensitivity and the resolving power of conventional spectroscopic procedures; and second, the laser allows the observation of strange, nonlinear effects.

For the chemist, it is of interest to note that there are chemical lasers in which the stimulation for population inversion is the chemical reaction. These are becoming more popular because of their relatively high efficiency. In general, those chemicals which fluoresce may be capable of lasing. These include selenium oxychloride and deuterium among the recent developments. Europium and terbium chelates have used in pulsed liquid lasers at room temperatures. The input energy for these is 200–2000 joules. Europium and terbium chelate-doped polymer rods have also been developed. Whitaker has used recently a neodymium-containing chelate using polymethyl methacrylate (PMA) in a solvent.

One of the significant developments of modern laser technology is the tunable dye laser in which various dyes have been able in a single laser to develop many different frequencies. Dyes are also used in Q-switching techniques for nanopulse outputs. These include cells of nitrozene, phthalocyanine and kryptocynanine. Chemicals are used also to develop second and third harmonics. One crystal used often is potassium dihydrogen phosphate.

In photochemistry significant developments have also been made in studies of singlet oxygen, etc. A laser attached to a microscope has been used as a microprobe for microanalysis techniques. This analytical tool has now been perfected so that samples of a few microns in size can be examined for cations. Especially in biological materials, there is no complete destruction of the available sample. An example of the sensitivity limits indicates that laser microprobe detection can be made of Zn, Cd . . . 10^{-16}g, Al, Ca, Cu, Fe, Mn, No, Pb . . . 10^{-17}g, Ag, Cr, Mg, Sn, Sr . . . 10^{-18}g. The use of the laser microprobe has been extended to the field of ceramics and glass.

Lasers may be used for Raman and Brillouin spectroscopy as sources of illumination. Raman spectra can be shown by only a few milligrams of powdered materials. Work is being done currently to achieve Raman spectra from crystals. With lasers, even low-frequency Raman shifts can be measured. With a pulsed laser and a multi-channel photon counting detector, hyper-Raman effects have been detected in various systems. Laser Raman spectroscopy may be combined also with infrared spectroscopy.

In the field of the production of controlled nuclear fusion very high output lasers may be used to produce plasma confinement.

Lasers, especially CO_2 and Q-switched, have been used in air pollution studies for analyses techniques as well as for tracking air pollutants. SO_2, CO_2, NO, CO, H_2S, CH_4, etc. have been detected and measured as pollutants in air.

Chemists using lasers must be aware of the laser hazards. Laser safety programs in the chemical laboratory include the use of closed laser systems if possible, avoidance of specular reflectance from glassware and other highly reflectant surfaces. Protective eye glasses must be worn; each type depends upon the laser systems used; there is no single protective glass. Skin protection through gloves and protective creams is required not only to prevent acute burns but also chronic exposure to minimal laser irradiation. Air pollution may result from contamination from breakage of liquid chemical lasers or from lasers on various materials, such as toxic metals, plastics, malignant and infectious materials. With planned safety programs, laser chemistry may be expected to advance rapidly.

LEON GOLDMAN

Ultramicroanalysis with the Laser Microprobe

The focused energy from a medium to low powered solid state laser will vaporize a portion of any known material. The plasma generated by this process can then be analyzed by emission spectrography, mass spectrography, atomic absorption or gas chromatography.

In essence the laser head used for laser microprobe analysis consists of a cavity formed by a rod, such as ruby, neodymium-glass or yttrium-aluminum-garnet-glass (YAG), surrounded by a flash lamp and bounded by a totally reflecting back reflector and a partially reflecting front reflector which are parallel with each other and the front surface of the rod. The laser pulse emitted by this type of cavity is in the long-pulse mode, that is, a train of rather irreproducible spikes of energy spaced over a period of several microseconds. Although keeping the cavity at constant temperature improves the energy reproducibility, for quantitative ultramicroanalysis it is in addition preferable to incorporate a Q-switch into the cavity, which, when properly tuned, confines the laser pulse to a single giant spike of energy lasting between 15 and 25 nsecs. Four types of Q-switch are commonly used with the laser microprobe: a rotating front or back reflector, a dye cell sensitive to energy change at the wavelength of emission of the laser rod, a Kerr cell, or a Pockel cell. The latter two types generates r-f noise and should be heavily shielded, but have the advantage that they can be used as timers in the grating mechanism of automated direct readout systems. To increase reproducibility in microanalysis and act as coarse adjustment of volume of material evaporated from the sample, the laser head should also include an aperture within the cavity to restrict the laser beam to a single transverse mode. An infinitely variable filter, to act as a fine adjustment of volume evaporated and to keep the volume constant as composition of the sample is varied should also be available, as well as a beam splitter, to deflect off a fraction of the energy of the beam into a calibrated measuring device, thus monitoring the reproducibility of laser energy and allowing small corrections to be made to the sample volume evaporated.

The beam from the laser head is then, for reasons of mechanical stability, deflected through 90° with the aid of a prism or mirror into a microscope, which focuses it onto the area of the sample to be vaporized. In order not to degrade the quality of the beam it is preferable that the focusing be done by the microscope objective alone, and that all other parts required for viewing the sample under oblique, transverse or polarized lighting conditions be removable during the actual firing of the laser. Using long working distance 20 X and 40 X objectives, with a single transverse mode of the laser rod and suitable filters, craters less than 1 μ may be obtained in many types of sample. However, from a practical point of view, present analytical systems will require plasmas from craters between 3–10 μ in diameter to obtain quantitative data for minor and trace elements. The plasmas generated by the laser exceed 15,000 °K in temperature, often last for under half a microsecond if due to a 20-nsec Q-switched pulse, and are less than a millimeter in diameter. The plumes remain close to the surface of the sample if the resulting sample crater is less than 10 μ in diameter.

The most commonly used method for the analysis of the laser generated plume is emission spectroscopy. Light from the plume is relayed either by a parabolic mirror placed behind the plume, or a single large lens into a high-speed spectrograph (f/8 or preferably f/6.3), with mirrors and gratings of high efficiency. A photographic readout system using very high-speed Polaroid film is impossible when the plumes are generated from craters less than 25 μ in diameter. If the plumes are generated from craters greater than 8 μ in diameter, the plasma is large enough and far enough away from the sample surface to be further excited by two precharged spark cross electrodes. A Polaroid photographic readout system is possible for cross-excited sample sizes of 15 μ in diameter or greater. However, for sample sizes less than 25 μ without and between 8–15 μ with cross-excitation a direct readout system must be used. Because the plume is often less than half a microsecond in duration, the number of photons per trace element line generated from small craters being low, and the focal plane of high speed spectrographs often being 10 inches or less in length, it is imperative to use a direct out system with small-sized, high-gain, low-noise photomultipliers with rise times of less than 2 nsec. These must be followed by a fast rise time, fairly wide-band amplifier system, an integrator, an analog-to-digital converter, a multiplexing system and a printout system or computer input. It is also important to gate the system to exclude the high continuum occurring at the beginning of the laser-sample interaction and to minimize the electronic noise inherent in the readout system.

The laser plume contains electrons, positive ions and neutral particles, so isotopic ratios, as well as major, minor and trace elements, may be determined by mass spectrometry. Three types of instruments have been used so far: time-of-flight ion analyzers, magnetic mass analyzers and quadrupole mass filters. Although the first-named instrument has been used the most, a double-focusing magnetic mass spectrograph, with the latest ultra-sensitive electron multipliers coupled to a high-speed readout system, would seem to hold out the most promise for both isotope and trace element analysis.

For the future, analysis of the plume by atomic absorption would seem to hold much promise for the determination of traces of such elements as cadmium and zinc. In addition, laser microprobe pyrolysis may sometimes have more to offer than ordinary thermal pyrolysis prior to employing gas chromatographic procedures.

Laser microprobe spectroscopy has much to offer, with absolute sensitivities of about 10^{-16} g and probable 10^{-18} g or better in certain instances. It has no lower limit on atomic weight and for many methods, a vaccuum is not used: this is particularly useful where biological samples are concerned. Moreover, the sample need not be conductive. The most serious drawback to laser microprobe analysis remains adequate standardization, but with the advent of more ultramicro methods, this will doubtless be overcome.

The laser microprobe can be used to analyze a great variety of samples in many fields such as art, archeology, criminology, including ballistics, toxicology and forensic medicine; biology, including single cell analysis and studies of trace element localization and transport mechanisms; geology including zoning within grains; semiconductors, ceramics and the many facets of metalogy.

I. Harding-Barlow
K. G. Snetsinger

LATEX, RUBBER

The term "latex" was derived from an old Spanish word for milk, and was originally applied to the milky exudation from trees found in the New World, such as the *Hevea brasiliensis*, from which natural rubber was first obtained. To-day it may be applied to a variety of emulsions of polymeric materials, some of which are rubberlike and most of which are of synthetic origin.

Natural rubber latex occurs in the *cortex*, or inner bark, of such trees as the *Hevea brasiliensis* (over 400 plants are known to contain such rubberlike latex). As the name implies, this tree was indigenous to Brazil but was transplanted to the Far East (Malaya, East Indies, etc.) during the latter part of the 19th century, so that plantation rubber soon dominated the world scene. Latex is not the *sap*, but plays some other role, probably providing a self-healing mechanism for injuries. It can be tapped in small quantities periodically for many years without harming the tree. A tree thus yields from 5 to 10 lbs. of rubber per year, corresponding to 750–1500 lbs. per acre.

The latex itself is an aqueous emulsion containing about 35% (by wt.) of the rubber hydrocarbon, cis-1,4-polyisoprene, having a particle size of 0.5 to 3.0 microns and stabilized by a surface layer of proteins and fatty acid soaps. It also contains many complex organic substances. When freshly tapped, it is neutral (pH = 6.8) but, on standing for about 12 hours, putrefaction sets in and the latex shows an acid reaction, with resultant coagulation. Hence, for prolonged storage, it is stabilized with 0.5 to 1% ammonia, in which state it can be stored indefinitely. . To reduce shipping costs, this latex is generally concentrated to about 65% solids content.

The bulk of natural latex produced is intended for use as dry rubber. For this purpose, it is coagulated, shortly after tapping, by addition of weak acids, such as acetic or formic, washed and dried. To prevent putrefaction of residual traces of organic matter in the dry rubber, the drying procedure may involve smoking ("smokesheet"), or a powerful bleach (sulfur dioxide) may be introduced during the coagulation ("pale crepe").

Synthetic rubber latex now enjoys a wider use than the natural product, at least in the United States, in view of its lower cost and because it has proven satisfactory for most uses. The vast majority of synthetic latices, whether rubberlike or not, are actually prepared directly by the process of emulsion polymerization. In this process, the respective monomers are agitated in water, in the presence of soap and various catalysts (e.g. peroxides, etc.), the result being a latex of the synthetic polymer, similar in appearance to the natural product. Synthetic rubber latex can thus be handled and used in a similar fashion to natural latex.

It is now known, however, that synthetic latices generally have a much smaller particle size than natural latex, i.e., 500–1500 Å (0.05–0.15 micron). This is due to the mechanism of the polymerization, as elucidated by Harkins[1], and others, which involves the interaction, in the aqueous phase, between the catalyst and the dissolved monomer to form insoluble polymer particles. The latter, then, form the nuclei for further growth, by imbibing more monomer, which can then be further polymerized by interaction with the catalyst in the surrounding aqueous medium. In this way, the polymer particles are formed in very large numbers ($\sim 10^{15}$ per cc) by a nucleation process, leading to the observed small particle size. The relatively large monomer droplets, comprising the original dispersion, are not themselves polymerized, but simply act as a reservoir of monomer for the polymerization, which goes on in the many polymer-monomer particles.

It is this nature of emulsion polymerization which leads to its remarkable characteristics. Thus, since this process is known to occur by a free radical chain reaction, it is possible for growing chains to exist in *close proximity*, but in *different particles,* and not be subject to a mutual chain termination reaction, which would readily occur in a homogeneous medium. Hence, such "long-lived" radicals lead to the simultaneous attainment of rapid rates and high molecular weights[2], a feat impossible of attainment in homogeneous polymerization.

The main synthetic rubbers produced to-day by emulsion polymerization are as follows:

Styrene-butadiene copolymer (SBR)—a general purpose rubber

Acrylonitrile-butadiene copolymer (NBR)—a chemical resistant rubber

Polychloroprene (Neoprene)—a solvent-resistant rubber.

In addition, a wide variety of nonrubber-like polymers are also produced in this way to form synthetic latices for various uses. Furthermore, other synthetic rubbers, which are not prepared by emulsion polymerization, can be converted into latex form by mechanical dispersion, where latex technology is desirable. In those cases, however, the particle size is, of course, much larger, similar to that of *Hevea* latex.

Polymeric materials are used in latex form wherever technology demands the use of these materials in convenient liquid form, e.g. for coating and impregnating, and, in the case of rubber latex, for foam rubber. For these purposes, however, the latex must first be mixed (compounded) with all the necessary ingredients before its final application. In the case of rubber latex, both natural and synthetic, this requires incorporation of all the materials necessary for vulcanization, e.g. sulfur, accelerators, zinc oxide, etc., as well as the other ingredients such as antioxidants, fillers, pigments, etc. These are usually added in the form of aqueous dispersions. The compounded latex is then used for various purposes, such as coating or dipping to form various articles. The latex film is first dried and then heated to ensure vulcanization.

For the production of foam rubber, the latex must have a high rubber content, i.e. better than 60% by weight. This is easily attained in natural rubber, as mentioned previously, but, in the case of synthetic latex, the small particle size resulting from emulsion polymerization leads to very high viscosity latex at high rubber contents (>40%). Hence it is customary to prepare the synthetic latex at lower solids content, and then to subject it to a process of agglomeration (not to be confused with coagulation), so that the particle size is considerably increased, before concentration to a high solids content is carried out. The high solids latex is then compounded, mechanically foamed, generally by whipping, and coagulated in the foamed condition by the slow release of acid from chemicals incorporated in the compound. The foam is then dried and vulcanized.

A more recent use of synthetic latices has been as a vehicle for paints. A suitable latex is used for a particular type of paint, which is prepared by mixing and grinding in the required pigments and fillers. In this way a whole series of water-based paints have arisen, having many advantages, e.g. miscibility with water, lack of odor, rapid drying (since drying oils are absent) and excellent aging. Furthermore a specific latex can be used for a given performance requirement. A variety of latices are presently in use for this purpose, including styrene-butadiene copolymer, polyacrylates, polyvinylacetate, etc.

The latest available statistics on natural and synthetic rubber latex show that world consumption of natural latex in 1971 was 265,250 metric tons (dry rubber basis) while U.S. production of all types of synthetic latex in that year was 736,670 metric tons (dry rubber basis). No figures are available for the minor quantities of synthetic latex produced outside of the United States.

MAURICE MORTON

Cross-references: *Rubber, Vulcanization.*

References

1. W. D. Harkins, *J. Am. Chem. Soc.*, **69**, 1428 (1947).
2. W. V. Smith and R. H. Ewart, *J. Chem. Phys.*, **16**, 592 (1948).

LAURATES, PALMITATES, AND STEARATES

The laurates, palmitates, and stearates comprise two groups of compounds: the metallic salts or soaps, and the esters of monohydric or polyhydric alcohols of the corresponding fatty acids. Lauric acid, $C_{11}H_{23}COOH$, is the lowest saturated fatty acid found abundantly in nature. It is a principal constituent of palm kernel oils and is obtained commercially from coconut oil (50%). Palmitic acid, $C_{15}H_{31}COOH$, is the most abundantly occurring saturated fatty acid and is found in practically all vegetable and land- and marine-animal fats. It is a principal constituent of tallows and lard, bovine butter, palm oil, cocoa and other vegetable butters, and most fruit coat fats. Stearic acid $C_{17}H_{35}COOH$, occurs as a major constituent of animal fats (20 to 30%) and cocoa butter (34%) and is found in lesser amounts in nearly all fats and oils.

Commercial lauric acid is usually a mixture of solid acids obtained by saponifying coconut oil or it is the residue remaining after removal of caproic, caprylic, and capric acids from the mixed acids, by distillation. Although distilled acid of 90% purity is available, the commercial grade is used principally in manufacturing metallic salts. Commercial grade palmitic acid contains approximately 67% of this acid and commercial grade stearic acid contains about 55% palmitic acid and 45% stearic acid. For the manufacture of metallic salts the foregoing commercial grade acids are usually used. It should not be inferred that the foregoing are the only grades of these acids that are available.

The metallic salts of the aforementioned acids can be formed (1) by reaction of a base on the acid; (2) by double decomposition of a soluble salt of the acid and a salt of a mineral acid (precipitation process); or (3) by saponification of glycerides or esters of the acid with a base or metallic oxide. In general, commercial grade metallic laurates, palmitates, and stearates are produced by the precipitation process, which in most cases yields highly uniform light fluffy products. The raw materials of this process are an alkali, inorganic salt or hydroxide, and the acid of the desired metallic salt. The inorganic salt or hydroxide must be a water-soluble

sulfate, chloride, acetate, carbonate, etc., such as aluminum or zinc sulfate, calcium chloride, lithium hydroxide, etc.

The alkali metal salts of the above-mentioned fatty acids are soluble in water; the salts of other metals are not. The sodium and potassium salts of these acids are generally referred to as soaps, and the term is frequently applied to salts of other metals, whether soluble or insoluble in water.

With di- and polyvalent metals, lauric acid forms basic salts, e.g., $Al(OH)(OOCC_{11}H_{32})_2$. Unlike the metallic salts of palmitic and stearic acids those derived from lauric acid have only limited uses. Aside from its sodium salt only the zinc salt of lauric acid is commercially important. Commercial zinc laurate, $[Zn(OOCC_{11}H_{32})_2]$, is a colorless compound melting at 100°C and has a specific gravity of 1.10. It contains 14.2% metal, 0.5% free fatty acid. It is produced from lauric acid containing about 10% myristic acid and about 1% oleic acid. Zinc laurate has been used in the rubber industry as a softener to improve processing characteristics and as an activator to reduce the time of curing. Because lauric acid occurs to the extent of about 50% in coconut and various palm kernel oils, its sodium salt is an important constituent of soaps produced from stocks in which these oils predominate.

A large number of salts of palmitic acid and also of stearic acid have been prepared and their properties (solubility in various solvents, crystal structure, electrical conductivity, surface tension and viscosity effects, hardness, water resistance, plasticity, etc.) have been studied. Two important palmitates prepared from commercial grade palmitic acid (67% palmitic, 29% stearic, and 4% oleic acid) are aluminum (mono) and zinc palmitates. In general, metallic salts of palmitic or other higher fatty acids do not appear to volatilize when heated; instead they tend to pyrolyze with ultimate decomposition.

Despite the considerable accumulated knowledge of the reactions and properties of the salts of palmitic acid, the pure compounds have found little practical use. Because the palmitates are prepared principally from commercial palmitic and stearic acids, which are mixtures in different proportions of these two acids, their uses may be considered to correspond to those of the stearates. In general, commercial alkali metal palmitates admixed with other soaps, are employed as detergents, wetting and emulsifying agents; as intermediates in the preparation of water-insoluble metallic salts; and as saponification and ester interchange catalysts. Sodium palmitate and myristate, or a mixture of the two, are useful catalysts for polymerization of butadiene or copolymerization of butadiene and acrylonitrile, styrene, acrylic esters and the like to produce synthetic rubbers.

The sodium and, to a lesser extent, potassium salts of palmitic acid are important constituents of laundry and toilet soaps, the percentage depending on the particular mixture of fats and oils of the soapstock subjected to saponification. The lathering property of sodium palmitate in cold water is intermediate between the corresponding laurate (fast) and stearate (slow). Its action on the skin is also milder than that of the corresponding laurate.

In general, the industrial uses of metallic palmitates are the same as for metallic stearates, namely, in the manufacture of lubricating greases, pharmaceutical and cosmetic preparations, waterproofing agents, paints, printing inks, fungicides, rubber and plastics, etc.

In general, the primary uses of the stearates are the same as those of the palmitates. Although sodium stearate possesses poor lathering properties in cold water it imparts hardness to soaps. The commercially available stearates include Al, Ba, Ca, Li, Mg, Pb, and Zn. They are generally marketed in 25-lb. bags and 50-lb. cartons. The price varies with the price of tallow and the content and price of the metallic constituent. Annual production of metallic stearates has been estimated to exceed 20 million pounds of which the Al and Zn salts account for about 70% and together with the Ca salt for approximately 85% of the total production. Lithium stearate is used in the manufacture of lubricating greases as are a variety of other metallic stearates. The oldest metallic soap grease is the lime soap or calcium stearate grease. Aluminum stearate is also widely used in the manufacture of lubricating grease. These and other metallic stearates are also used in the manufacture of paints, printing inks, and other products enumerated under metallic palmitates above.

Esters. Simple esters, such as those of lauric, palmitic, and stearic acids, may be formed by the reactions of (1) an acid chloride and an alcohol or alcoholate; (2) an acid anhydride and alcohol; (3) salt of the acid and an alkyl halide or sulfate; (4) direct esterification of the acid and an alcohol; (5) interesterification of an ester with an alcohol or another ester. Esterification and interesterification are the most important methods for the commercial preparation of the esters of the aforementioned acids. For the preparation of some esters of the higher alcohols one of the indirect methods, such as (1) and (3), is generally used.

Most of the laurates, palmitates, and stearates of the monohydric alcohols from methyl to hexadecyl have been prepared and various properties such as melting point, boiling point, density, and refractive index of the respective esters have been reported. The lower members of the laurate series are liquids at room temperature but the corresponding palmitates and stearates are low-melting crystalline solids. Each series exhibits alternation in the melting points of the even- and odd-number carbon chains and at least the lower members exhibit polymorphism. When pure all of the aliphatic esters are colorless liquids above their highest melting points. In general, the vapor pressures in the liquid state are low; hence they are distillable without decomposition only at reduced pressures. The esters are insoluble in water but soluble in most of the common organic solvents. The lowest molecular weight esters of the above-mentioned acids have little or no commercial use. *n*-Butyl stearate and to a lesser extent methyl and ethyl stearate have had some use in compounding lubricating oils, in special lacquers, and in the manufacture of cosmetics. Production of butyl stearate has been reported to be about 2 million pounds per year.

There is an appreciable commercial production of laurates, palmitates, and stearates of polyhydric alcohols, especially of the various glycols. These esters are generally prepared by direct reaction of the acid with the polyhydric alcohol. Diethylene glycol monolaurate and 1,2-propylene glycol monolaurate are the most important glycol derivatives of lauric acid and are used principally as plasticizers. Commercially available stearates include the mono- and diester of ethylene glycol, monoester of propylene glycol, and a number of esters of polyethylene glycols. These esters vary in M.P. from 25 to 80°C. They comprise a group of nonionic surface-active agents used in the textile, petroleum, agriculture, food, pharmaceutical, and other industries.

Glycerol monostearate containing various proportions of di- and tristearate, depending on the process of manufacture, is used in the manufacture of a wide variety of products (cosmetics, pigment coatings, rubber and plastics, polishes, etc.), but its principal use is as an emulsifying agent in shortening and margarine. Tristearin (glyceryl tristearate), produced by the hydrogenation of vegetable oils, notably soybean oil, has a variety of uses and especially in the manufacture of shortening and margarine where it is used to control the hardness of the fat mixture. Commercial grade stearates produced from commercial stearic acid (mixture of palmitic-stearic-oleic acids) or by hydrogenation of natural fats and oils differ considerably in properties from the corresponding pure products.

KLARE S. MARKLEY

Cross-references: *Fats, Fatty Acids, Soaps.*

LEAD AND COMPOUNDS

Lead is one of the oldest metals known; references were made to it in the earliest world literature, and it is said to have been in use as early as 4000 B.C. Although lead occurs only to the extent of about 15 grams per ton of the earth's crust, its ores are concentrated in reasonably large deposits. Consequently, economic exploitation of these ores is much easier than for some of the other elements.

Occurrence and Production. Most of the lead deposits were formed as sulfides, and galena (lead sulfide) is by far the largest source of the metal. Most of the other lead minerals of economic importance were formed by alteration of the primary sulfides. Very frequently lead and zinc occur together; sometimes copper, gold, and silver are also associated with the lead minerals. Important deposits are located in the United States, Canada and Mexico in North America and on every continent. World mine production of lead in recent years has shown an average annual growth rate of about 3% and in 1970 amounted to about three and three-quarters million tons. The United States proportion of the total world mine production has increased from about 10% in 1967 to somewhat more than 15% in 1970 or almost 580,000 tons.

The United States produced about 700,000 tons of virgin lead in 1970 (the difference between mined and produced lead is made up by imports of lead ore and concentrates) and almost 600,000 tons of secondary lead (from scrap). The price of lead at New York averaged about 15.6¢/lb. On the basis of 1970 preliminary figures, the principal uses in order of decreasing importance remain as in recent preceding years: (1) lead storage battery components, antimonial lead and battery oxides, (2) metallic lead products such as ammunition, solder, cable covering, constructional purposes, etc., and (3) lead compounds such as tetraethyl lead, litharge, red lead, etc.

Concentration. The methods used in concentration of lead ores are dependent on their character. Massive lead sulfide, not mixed with other minerals, can be separated from gangue rock by suitable milling and gravity concentration. In some cases, mixed ores can be separated from gangue by differential density methods (sink-float); the minerals may then be recovered by jigging, tabling or flotation. Complex ores must usually be separated by selective flotation into appropriate fractions for smelting. Current milling and concentrating practices recover about 85 to 94% of the sulfide lead and up to about 88% of the oxidized lead.

Smelting. Lead concentrates are usually smelted in blast furnaces. The concentrates are first put through a roasting-sintering process to remove sulfur and produce a sinter physically suitable for charging the blast furnace. Coke and fluxes are also added to the furnace charge along with some lead-bearing by-products. A lead bullion results which collects also the silver and the gold and is tapped from the bottom or crucible of the furnace. Any iron, copper, nickel or cobalt in the ore combines with the sulfur left in the charge to form a matte which is tapped from the front of the furnace and given appropriate treatment. The zinc present in the charge enters the slag and together with any lead therein is recovered in a slag-fuming furnace. Lead recoveries in the blast furnace operation usually run about 97% or even higher and accordingly there is little opportunity for improvement.

Recently a blast furnace smelting process has been developed by the Imperial Smelting Corporation, Ltd. of England, according to which a bulk concentrate containing both lead and zinc is sintered and then reduced, permitting simultaneous recovery of both metals. The process is particularly valuble for treating a lead-zinc concentrate in which the metallic sulfides are so intimately associated that separation is practically impossible by the available milling and concentration methods. The products of this process are three in number, lead bullion containing the precious metals, zinc metal and matte. Furnaces employing this process are increasing in number and more are being constructed.

Refining. Since the lead bullion contains the silver and gold in the original ore (and other metals as well), it must be refined, the exact procedure employed depending on the characteristics of the bullion. "Softening" is the removal of copper, tin, antimony and arsenic in a drossing or refining kettle. The copper is first removed from the molten bullion by skimming off the copper as dross. Agitation with the addition of a small amount of sulfur serves to remove the remaining copper, as a copper sulfide dross which is also skimmed off.

The temperature of the bullion is then increased with or without the use of a stream of air, bringing about the oxidation of the arsenic, antimony, and tin which rise to the surface along with some lead oxide and are skimmed off.

Removal of the gold and silver is accomplished by the addition of a small amount of zinc to the molten bullion. The zinc alloys with the gold and silver in that order and the resulting alloys on cooling, float to the surface where they are skimmed off as gold and silver zinc crusts. Recently vacuum distillation has been employed with considerable success. If necessary, bismuth may be removed by several procedures, one being reacting the molten metal with calcium or magnesium, these metals forming high melting alloys with bismuth which can be removed as dross.

Occasionally pyrometallurgical methods are dispensed with and after "softening," electrolytic refining is employed. The anodes are of crude lead and the cathodes are made from electrolytic lead. The electrolyte is composed of fluosilicic acid and lead fluosilicate. As the lead is dissolved from the anode and deposited on the cathode, the impurities in the bullion drop to the bottom of the cell and are removed for subsequent treatment.

It should be emphasized that the sequence of steps in the metallurgy of lead is extremely varied and depends upon such factors as impurities present in the ore, location of plant, investment required, complexity of the various steps, etc.

Secondary Lead. As already pointed out, the production of secondary lead in the United States amounts to almost 600,000 tons annually, actually exceeding domestic mine production. It has its source in the salvage of discarded end-product items such as storage battery scrap, pipe, sheets, cable covering, etc., which are collected, remelted and refined in secondary plants to produce desired grades of refined lead or various lead base alloys.

Properties. Lead is in Group IVA, Period 6 of the Periodic Table. It has 29 isotopes with mass from 194 to 214, the important naturally occurring ones listed in decreasing order of abundance being 208, 206, 207 and 204. Other physical constants are presented below:

Atomic number	82
Atomic weight	207.2
Color	Bluish gray
Crystal structure	Face-centered cubic
Hardness, Mohs Scale	1.5
Density, g/cc	
20°C (s)	11.34
327°C (1)	10.69
Melting point	327.3°C
Boiling point	1744°C
Specific heat, g-cal/g	
20°C (s)	0.038
327°C (1)	0.039
Vapor pressure	
987°C, mm Hg	1.0
1167°C, mm Hg	10.0
Latent heat of fusion, cal/g	5.86
Latent heat of vaporization, cal/g	203.0
Electrochemical equivalent Pb^{++} (mg/coulomb)	1.073+
Electrode potential	
Pb^{++} ($H_2 = 0.0$ volt)*	−0.126
Pb^{++++}	+0.80

*National Bureau of Standards Nomenclature.

Lead is soft, malleable, of limited ductibility and

a poor conductor of electricity. It is the most corrosion-resistant of the common metals. It has two valences, two and four. It is relatively insoluble in sulfuric and hydrochloric acid and dissolves slowly in nitric acid. It is slowly oxidized and carbonated in moist air. Lead is amphoteric. It forms compounds with many inorganic ions which have definite utility in industry. Some of the organic compounds of lead are also quite important.

Lead and its compounds are highly toxic; however, if proper precautions are employed, little trouble need be experienced. Inhalation of lead fumes or ingestion of lead solids or solutions can lead to disabling intoxication. Workmen exposed to lead fumes must be protected by special respirators. Protection from organic compounds of lead is best accomplished by adequate control in the amount of vapor allowed in the air.

Initially, lead poisoning usually takes the form of colic; exposure over long periods of time causes chronic and acute intoxication which is characterized by weakness of muscles and a general feeling of poor health. Unless proper medication is administered death can result.

It should be emphasized that despite the poisonous characteristics of lead and its compounds, proper plant and process design together with good housekeeping and enforcement of necessary precautions have been extremely effective in combating this danger.

Lead is of interest in atomic work. The end product of each of the three series of naturally occurring radioactive elements is one of the three isotopes of lead (uranium series, Pb^{206}; thorium series, Pb^{208}; actinium series, Pb^{207}.) Lead has an extremely high density and is useful in protection from gamma rays. It is not feasible to present actual figures on the thickness of lead sufficient for shielding from radioactive rays since many factors are involved. For low-energy radiation, an inch of lead is equivalent to about 5 inches of iron, 15 of aluminum, or 36 of water. At higher energy levels, the differences are less. The linear absorption coefficient of lead per centimeter of thickness is approximately 0.5 for 2 Mev. (See also **Radiation.**)

Grades and Uses. Lead is furnished in seven standard specifications as set forth by ASTM, the maximum purity being 99.94% and the minimum 99.85% lead.

Because of its good corrosion-resistance lead is used in chemical construction, particularly in contact with sulfuric acid. Lead is extensively alloyed with other metals, particularly antimony and tin. Antimonial or hard lead alloys are used almost exclusively for lead acid storage battery grids and terminals. As a matter of fact, about one-quarter of the current lead demand is for this type of usage. Lead alloys have a wide range of uses.

Lead Compounds. Over one-third of the lead consumed in the United States goes into the manufacture of chemical compounds such as tetraethyl lead, litharge and red lead.

Tetraethyl lead, $Pb(C_2H_5)_4$, currently (1970) provides the second largest single use of lead (after that in storage batteries); almost 280,000 tons of lead were required for its manufacture. It is a colorless heavy liquid containing 64% lead, which when added to gasoline in amounts ranging from a few drops to 3 cubic centimeters per gallon, increases the octane rating, reduces engine knock and improves the general performance of high-compression internal combustion engines. The nonrecoverable nature of this use for lead should be emphasized (See **Antiknock Agents.**)

Other Organic Compounds. Lead salts of certain organic acids are used as driers for paints. Lead soaps find utility in lubricating greases. Lead azide is widely used as a detonator for high explosives.

Lead monoxide (litharge), PbO, is a most important compound of lead. It is manufactured by oxidizing molten lead in air at about 650°C. Different methods of manufacture and milling produce lead monoxides of varying characteristics. A large part of the lead monoxide manufactured is used as the active material in lead acid storage battery plates. Special blends of lead monoxide with small amounts of appropriate materials are available for both positive and negative battery plates. Lead monoxide is also the starting raw material for making many lead compounds. It is also used as an activator in rubber, in the manufacture of various ceramic products such as glass, enamels, etc., in paints and in petroleum refining.

Red lead, Pb_3O_4, an orange-red powder is prepared by the carefully controlled oxidation of lead monoxide at about 485°C. It is widely used as a rust-inhibitive pigment in paints, and for years has been the standard corrosion-resistant pigment. It finds use as an ingredient (with lead monoxide) in positive storage battery blends. It is also used in ceramics.

Lead dioxide, PbO_2, is a powerful oxidizing agent and finds utility in the dyeing and chemical industries. It is also used in the manufacture of lightning arrestors and as an accelerator for some synthetic polymers.

Lead silicates made by fusion of lead monoxide with the proper proportions of silica are used in ceramics, the ratio of PbO to SiO_2 depending on the particular application. A hydrated form of lead silicate is used as a stabilizer or inhibitor to retard the degradation of vinyl resins.

Basic carbonate white lead was formerly used in large quantities as an exterior paint pigment and although there still is some production, its use for this specific purpose has decreased markedly in recent years.

Basic sulfate white lead finds some use as an exterior paint pigment, particularly in mixed paints and also as the lead portion of leaded zinc oxide.

Lead chromates are used as both rust-inhibitive pigments and colored pigments. *Lead arsenates* are widely used as insecticides despite the advent of many organic materials for such purposes.

Lead and its compounds are so widely used that it is impossible to cover all applications in the space of this short review.

A. PAUL THOMPSON

References

Sections on Lead, Lead Alloys and Lead Compounds, "Encyclopedia of Chemical Technology," 2nd Edition, Interscience Encyclopedia, New York.

"Minerals Year Books," U.S. Bureau of Mines, Washington, D.C.

Chapter on "Lead," "Mineral Facts and Problems," U.S. Bureau of Mines, Bulletin 650, 1970.

"Year Book of the American Bureau of Metal Statistics," 50th Annual Issue, 1970, New York.

LEATHER

Leather is skin, permanently combined with a tanning agent so that its principal fibrous protein (collagen) is rendered resistant to hydrolysis by putrefactive bacteria, enzymes, and hot water, while the fibrous structure and desirable physical properties of the skin are retained. Leather is made mainly from skins and hides of domestic animals, though a few wild animals, such as kangaroos and reptiles, furnish significant amounts. Leathers are classified according to the species of hide, end use and method of tanning. The U.S. output in 1969 was valued at approximately 810 million dollars.

Structure and Composition of Skin. Skin conststs of two main parts—the epidermal area and the corium. A third layer of fat and flesh is present on the skins. The epidermis is composed of four layers of cells, three living layers and the outmost layer composed of dead cells. Mammalian epidermis produces hair follicles and hair which extend into the underlying corium. The epidermal area consists of the hairshafts and hair-roots, sweat glands, fat glands, small muscles, fibroblasts, small blood vessels and fine fiber bundles. The corium consists primarily of fiber bundles of collagen, made up of fibers and fibrils whose molecular structure consist of chains of aminoacids. The fiber bundles are interwoven as a three-dimensional fabric, which gives the skin its strength and elasticity. These fiber bundles become thinner as they approach the epidermis and form the "grain" immediately below the bottom layer of epidermal cells. The corium also contains large blood vessels and fibroblasts and in both epidermal area and corium area interfibrillary proteins (albumins, globulins, and mucoids) are present. The flesh layer is composed of fat cells and meaty muscle tissue which is removed prior to or during leather manufacture. Structure varies with species and age. In calfskin the fiber bundles are thin whereas in steerhide the fiber bundles are thick and closely packed; hence calfskin makes soft, flexible leather for shoe uppers, whereas steerhide makes strong, solid leather for soles and belting and when split for shoe upper leather makes sturdy, good foot supporting leather.

Curing. For preservation prior to tanning skins are "cured" mainly by immersion in saturated brine raceway systems until they are saturated with salt or in saturated brine in a concrete mixer. Disinfectant is added to the brine to limit bacterial action. The water content of the skin is reduced from 75 to about 40–48% and the residual water in the skin is saturated with salt. Fine salt is added to the flesh side of the skin before bundling prior to shipment. Properly cured skins can be stored in cool, dry storage areas for months without appreciable deterioration.

Soaking and Unhairing. Cured skins or hides are washed in a drum or paddle vat in cool water to remove salt, blood, and manure. They are soaked for 16-24 hours in water to rehydrate them and remove interfibrillary proteins.

Next the skins enter a depilatory bath containing a suspension of lime, to which a little sodium sulfide or sulfhydrate is added as a "sharpening agent." The cells of the epidermis are dissolved by mild alkali in a few days leaving the hairs loose in their follicles so that they can be scraped off. The hair is not damaged in such a process. When larger amounts of sulfide are used the hairs are pulped and destroyed in a few hours. A paste of lime and sulfide may also be used on the flesh side of skins to loosen the hairs. The sulfide penetrates to the hair roots and destroys them without damage to the hair or wool, which are valuable by-products.

Deliming and Bating. The alkaline unhairing liquors cause the collagen fibers to imbibe much water, so that the skin becomes swollen and turgid. Some opening-up of the fiber bundles occurs, mucoids are removed and collagen is partially deamidized. The unhaired skins are partially delimed by treatment with dilute solutions of ammonium salts, weak acids such as boric, or both, so that the pH value of the liquor in the skin is lowered to about 8-9. Simultaneously, most skins are "bated" with pancreatin or other tryptsin-like enzyme, which gives a softer and smoother-grained leather. The degree of bating varies from none at all for sole leather to a drastic overnight treatment for goat ("kid") leather that is soft and flexible.

Pickling. This process, applied to some types of skins after bating, may be done either for temporary preservation, or as a preliminary to chrome tanning. The skins are immersed in a solution containing about 6 to 12% NaCl, and enough H_2SO_4 to reduce the pH value to about 2.5 or lower. The salt prevents acid-swelling of the skin, which would otherwise occur.

Tanning. A tanning agent is a material that combines with skin more or less irreversibly, making it immune to bacterial attack, and generally raising the hydrothermal shrink temperature. When raw skin is heated in water starting at room temperature and heated at a rise of 3–5°C per minute it shrinks at about 60°C, and on further heating it dissolves as gelatin. Most tanning agents raise the shrinkage temperature. Some chromium-tanned leathers withstand boiling. The tanning agent also prevents the collagen fibers from glueing together on drying, so that the skin remains porous, soft and flexible. There are many tanning agents which range from simple molecules such as formaldehyde to large complex aggregates (vegetable tannins). Explanations of tanning action have ranged from formation of definite compounds to simple coating of the fibers by precipitated tanning agent. No mechanism suggested explains all types of tannage. The principal commercial tanning agents include: (1) vegetable tannins (aqueous extracts of barks, woods, leaves, fruit), (2) mineral tanning agents (complex salts of chromium, zirconium, and aluminum), (3) aldehydes, e.g., formaldehyde and glutaraldehyde, (4) oxidizable oils, and (5) "syntans" (various types).

Vegetable Tannage. This process is used for heavy leathers, particularly for soles. The delimed hides are suspended in weak infusions of the tannins, in a vat system. They are brought in contact, every day or two, with successively stronger liquors, until the tannin has penetrated to the center of the hide which takes several weeks. Hides for sole leather are then transferred to strong liquors in "layaway" vats, and left in the vats until they have taken up an amount of tannin nearly equal to the weight of the fibrous hide substance. The process is shortened from a few months to days by carrying out the later stages in revolving wooden drums, or by using warm liquors. Flexible vegetable-tanned leathers for belting, upholstery, luggage, etc. are much less heavily tanned than sole leather.

The mechanism of vegetable tanning is complex. The collagen molecule possesses numerous peptide linkages and side chains bearing NH_2, COOH, OH and other reactive groups. Tannins from different sources differ considerably in structure, but in general they are polymerization products of units containing polyphenolic rings and sometimes carboxyl groups. Probably most of the tannin is fixed by hydrogen-bonding between tannin phenolic groups and protein peptide groups, plus combination between tannin carboxyls and protein amino groups. In both reactions the tannin aggregates form cross links between neighboring polypeptide chains.

Mineral (Chromium) Tannage. Chromium (or "chrome") tannage generally is done with basic chromium sulfate made by reducing $Na_2Cr_2O_7$ with corn sugar or SO_2 plus enough sulfuric acid to give a chromic sulfate of the desired degree of basicity, from "⅓-basic" $Cr(OH)SO_4$ to "½-basic." The resulting solution is always acid (pH about 3) because of hydrolysis. Pickled skins, as described above are treated with the basic chromium sulfate solution, in a revolving drum equipped with pegs that lift and drop the skins or in hide processors equipped with spiral shelves for rotating action. The chromium compound penetrates very rapidly, and tannage is completed in 4 to 24 hours, depending on the procedure used. After tannage the pH is raised to about 3.5, either by careful addition of mild alkali or by adding the chromium liquor in increments of increasing basicity. The tanned leather contains somewhat more than 3% Cr_2O_3 and the Cr is present as a highly basic Cr sulfate. The main mechanism of chromium tanning is the coordination of carboxyl groups of collagen to the Cr atom.

Aluminum salts, which are weaker tanning agents than chromium are sometimes used when a white leather is wanted, and stability of tannage is unimportant. A better, but relatively expensive tanning agent for white leather is basic zirconium sulfate.

Aldehyde Tannage. Formaldehyde combines with collagen, best in slightly alkaline solution by reacting with the amino groups. Formaldehyde alone does not give a commercial leather but its use in conjunction with aluminum salts or with melamine resins results in fuller leather. Glutaraldehyde, as a tannage for preparation of washable sheepskin shearlings, and as a retanning agent for imparting perspiration resistance yields commercial leathers.

Oil Tannage. Chamois leather is made by milling wet sheepskin splits with cod or other oxidizable oil. On heating the oil is partly oxidized and aldehydes are liberated. It is thought that the oxidation products are responsible for the tannage which produces a leather that can absorb 3 to 5 times its weight of water which is important in filtering out

water from gasolines and for lining leathers in military uses.

Syntan Tannage. Because vegetable tannins are high molecular weight compounds containing many phenolic hydroxyls, attempts were made over a half century ago to make materials with tanning properties by condensing sulfonated phenol or naphthol with formaldehyde, the sulfonation being necessary for solubility. These early syntans were auxiliary tanning agents. During World War II and after these agents have been improved by modifying the manufacturing process so as to reduce the degree of sulfonation. Condensation products other than phenols, possessing strong hydrogen-bonding power, have also been developed as tanning agents. A number of commercial syntans produce salable leather but usually constitute only half of the total tannin mixture because of their high cost.

Combination Tannage. Chromium tanning and vegetable tanning frequently are combined to give a full tannage that has some advantages of each. The hides are first given a light chromium tannage to accelerate the absorption of the vegetable tannins and then treated with strong vegetable tan liquors. The degree of vegetable retannage varies from a mere surface treatment to full penetration. Chromium tanning and aldehyde tanning also are combined to give a full tannage that has some of the advantages of each.

Post-Tanning Operations. Only a general outline can be given on these operations because the use to which the leather is to be put dictates how it is to be treated for end-use products. For instance sole leather goes through several operations to introduce more tannins and other loading materials, to lighten the color by bleaching, and to lubricate the leather with oil. After these materials are incorporated into the leather it is compressed by rolling. Other leathers go through more extensive operations before the final product is ready for the market. Such stock is reduced to the desired thickness, and leveled by splitting and shaving. The grains and splits are made into lots of about 500 lbs or more which are washed, neutralized, mordanted, dyed, and "fatliquored." Dyeing of chrome leather is done with acid and direct dyes, sometimes topped with basic dyes; vegetable leather with basic dyes. The "fatliquor" is an emulsion of oil, very commonly a mixture of sulfated and raw neatsfoot or cod oil, though many other emulsifying agents and oils have been and are used. After dyeing, the skins are tumbled in the emulsion or may be fatliquored and dyed at the same time just so the emulsion does not break too rapidly. Finished upper leather after fatliquoring usually contains about 3–6% oil in the dry weight. When more oil is required, as in heavy vegetable-retanned leather, the skins or hides are "stuffed" by tumbling the damp leather in a molten mixture of greases or by applying silicone treatment or other surface treatments using surface impregnation to make it water-repellent. Such leather may contain as high as 20% of oil necessary for some end-uses of the product, or much less if given silicone impregnation.

Drying. Vegetable-tanned leather must be dried slowly, beginning at a temperature little above normal room temperature. Chrome leather is less sensitive, and usually is dried by increasing the temperature and decreasing the relative humidity. Much chrome or vegetable-retanned leather is dried by pasting the grain side onto glass plates which pass through a drying tunnel. Some leathers are dried using vacuum drying equipment to give special properties to the leather. The older drying process, which is still practiced, is to dry the skins in a loose condition, dampen them, and redry in a stretched condition, tacked out on boards, or toggled to perforated plates. Each type of drying gives special physical characteristics to the finished leather.

Finishing. Most leathers are given several coats of finish, to impart luster, bring out the natural grain pattern or correct the grain to hide minor defects, and render the surface somewhat water-repellent. Finishes commonly are water dispersions containing a protein binder (casein, albumins), resins (shellac or emulsified synthetic resins), waxes, dyes, and pigments. Protein binders are rendered insoluble by treatment with formaldehyde. Pigments are usually held to a minimum amount to avoid the painted look, except in those modern leathers which may require the painted finish look with the desired leather as the underlying comfort feature. Pigments are applied by the old-fashion methods of hand swabbing and brushing but at present mainly by spraying. Luster is developed by "glazing" in which a small glass cylinder is dragged across the skin under considerable pressure, plating (pressure between heated plates) or with luster imparted in the finishing coats. To prevent soiling of the finished leather and to impart water repellency a top coat of an emulsified lacquer or resin may be applied. The finish coats must have good adhesion, be elastic and flexible.

For those interested in more extended sources of information given in the article it is suggested that the references given below will give the details necessary.

H. B. Merrill
W. T. Roddy

References

O'Flaherty, F., Roddy, W. T., and Lollar, R. M., Vol. 3, A.C.S. Monograph No. 134, "Chemistry and Technology of Leather," Reinhold Publishing Corp., 1962.
Thorstensen, T., "Practical Leather Technology," Van Nostrand Reinhold Co., 1969.

LEGAL CHEMISTRY

Legal chemistry is a new subject but an old field of activity. It relates to chemical information, experiments, or services in connection with issues that may come before the courts, semijudicial tribunals and boards, or even lawyers in unofficial capacities for negotiation or determination.

Originally associated in the public mind with such fields as toxicology, specification writing and the meeting of specifications by materials sold or used, legal chemistry became extensively involved with pure food laws, quality control of milk products, water supplies and sanitation, and chemical patents. Now it is concerned also with marketing chemicals in such manner as to avoid running afoul of anti-

trust regulations, approvals required before offering new drugs to the public, federal and state provisions relating to sales of pesticides, transportation and sales of hazardous materials, patents, and crime detection.

Crime Detection. The information in this section has been abstracted from information furnished by Mr. John Edgar Hoover, Director of the Federal Bureau of Investigation, the parts used being mostly verbatim.

Examination for poisons should follow an autopsy by a competent medical doctor, preferably by a pathologist of experience in this type of work.

While there is no substitute for a complete and thorough post mortem, only chemical analysis offers conclusive proof of the presence of poison. Parts to be analyzed include the stomach and its contents, the liver, kidneys, heart, brain, all urine found in the bladder, an adequate sample of blood, and, in certain instances, sections of the intestines and the lungs.

Each organ or specimen should be placed separately in a clean, all-glass container, sealed and identified with the name of the victim, specimen enclosed, names of investigator and the autopsy physician. No chemicals or other preservative such as embalming fluids should be added to the specimens, since such materials often interfere with or prevent various reactions and tests used for identification purposes by the analyst. Steps must be taken, however, to preserve the tissue and halt the normal processes of putrefaction and decomposition which the body undergoes after death. This is best done by refrigeration with Dry Ice if the specimens must be shipped any distance. In this connection fluids such as urine and blood should be wrapped and protected from close contact with the Dry Ice to prevent them from freezing.

The chemical examination is accomplished by submitting the organs to certain procedures designed to separate the poisons from the body tissue. The classic procedures for toxicological analysis divide poisons into the following groups: (1) toxic gases; (2) poisons volatile in steam which are generally organic compounds; (3) nonvolatile organic compounds such as the alkaloids, barbiturates, and other drug products which are soluble to a greater or lesser extent in alcohol; (4) metallic poisons; and (5) other miscellaneous poisons.

In every instance the tissue is first minced or otherwise finely subdivided. A portion of the minced tissue is then made acid, placed in a suitable flask, and a current of steam passed through the flask. The vapors leaving the flask are led through a condenser, cooled, and the liquid is analyzed for poisons such as cyanide, the alcohols, ethers, chloroform, chloral hydrate, and many others which if present will be found in the liquid distilled from the tissues.

Another portion of tissue is placed in a separate flask and extracted several times with alcohol. The solutions of alcohol are filtered, pooled together and carefully purified in order to separate the poison from the normal tissue components which are extracted with the drug by the solvent action of the alcohol. When the poison is finally obtained in a satisfactorily pure condition, the poison is then identified by appropriate qualitative tests.

Some of the newer physical instruments are extremely useful in identifying unknown drugs. For example, substances obtained in crystal form may be identified by the characteristic patterns obtained when the specimens are analyzed on the x-ray spectrometer. This instrument is particularly useful in identifying such substances as the barbiturates which are closely related chemically and may be identified only by differences in their physical properties. Where colored derivatives can be prepared for identification tests, the electrophotometer and the recording spectrophotometer are also used to advantage.

Metallic poisons such as lead, arsenic, and others are identified by destruction of the organic components of the tissue and spectrographic analysis of the residues of inorganic matter which remain. The tissue may be destroyed by the action of acids and strong oxidizing chemicals or it may be ashed at a high temperature.

Miscellaneous poisons of group (5), such as fluorides, most commonly found in roach powders, cantharides (popularly believed to be an aphrodisiac), ergot (often used as an abortifacient), and many others are not isolated and detected by the methods of examination applicable to the four main groups of poisons discussed above. Since a routine examination cannot be made for all poisons, those of the miscellaneous group may be difficult to detect unless the toxicologist has reason, on the basis of a careful post mortem, to suspect the presence of an unusual poison. (See also **Toxicology**)

Bloodstains are first subjected to the benzidene test. This test, while extremely sensitive, is not a conclusive one for the presence of blood, but it is useful in eliminating those stains which are not blood.

Two confirmatory tests for the presence of blood are the microspectroscopic test and Teichmann's hemin test. Either of these tests shows the presence of hemoglobin and its derivatives, which are present only in blood, the hemin test being usually used when the quantity of blood being tested is small. Rust stains on iron, for instance, will be completely eliminated by the confirmatory test.

The precipitin test establishes the origin of a bloodstain. This test is based on the reaction of an antiserum with the protein in the stain. For example, a human bloodstain gives a positive reaction when tested with an antihuman serum and a negative reaction when tested with an animal antiserum.

The FBI Laboratory maintains antisera for most of the common domestic animals and some of the more common wild animals. By use of these sera it can be determined whether a bloodstain is of animal or human origin and, if animal, from which type of animal it came, i.e., dog, hog, cow, chicken, or rabbit. If a suspect claims that bloodstains on his clothing came from some animal for which the laboratory has a corresponding antiserum, then the truth or falsity of the claim can be readily determined.

When it has been established that a bloodstain is of human origin, grouping tests show to which of the four International Blood Groups it belongs, namely, O, A, B, or AB.

It is not possible by blood analyses alone to show that a bloodstain came from a particular person, but it is frequently possible to ascertain that the blood could not have come from that person.

For grouping purposes the stained portion of the specimen must be reasonably clean and, of course, there must be a sufficient quantity of blood in the stain. Usually a saturated stain which is about one-half by one-quarter inch is enough for a conclusive grouping test.

The blood on the clothing of the victim is often so dirty, contaminated, or putrefied as to preclude the possibility of making conclusive grouping tests. It is, therefore, recommended that, in crimes which involve blood examination, a liquid specimen of the victim's blood be taken so that it may be compared with bloodstains on the clothing of the suspect.

Drugs. Supplementing the discussion of the U.S. Food and Drug Administration, George P. Larrick, Commissioner of Foods and Drugs, summarizes clearly the legal aspects of offering drugs to the public as follows:

A manufacturer who introduces a drug to interstate commerce should conform with certain legal requirements in order to discharge his obligations to the public. The essential requirements are set forth in the Federal Food, Drug, and Cosmetic Act which is enforced by the United States Food and Drug Administration.

If the product is not a new drug it may be marketed without any formalities with the Food and Drug Administration. Its distribution, however, is subject to the provisions of the Act which require honest and informative labeling and certain standards of strength, quality and purity.

If the product is a new drug, definite requirements must be met before it may be marketed. A new drug is defined as one which is not generally recognized as safe, by experts qualified by scientific training and experience to evaluate the safety of drugs, when used as directed in its labeling. It continues to be a new drug, by definition, until it has been used to a material extent or for a material time. Under such conditions the distributor is required to submit an application for it to the Food and Drug Administration. The application should include full reports of investigations (animal and clinical) to show whether or not it is safe for use, a full list of the components used to make the drug, and a statement of its quantitative composition. In addition, the application must include a description of the manufacturing methods, facilities and controls for the product, such samples as may be required and specimens of the labeling. The drug may be legally marketed only if the application shows that the drug is safe when used as recommended in its labeling.

In order to obtain the data required in an application, the drug may be shipped for investigational use under certain conditions which are specified by regulation.

Some drugs, certain biological preparations such as serums, vaccines, antitoxins, etc. which are subject to licensing by the Public Health Service, and certain antibiotics for which certification by the Food and Drug Administration is required, are exempt from the new drug procedure described above. Certification of manufacturers' batches before distribution is also required for products containing insulin. New drug applications must be submitted for them if they contain new drugs.

Economic Poisons. The following private communication, slightly abbreviated, from Mr. Justus C. Ward, Head, Pesticide Regulation Section, Plant Pest Control Division, Agricultural Research Service, U.S. Department of Agriculture, states succinctly the phase of legal chemistry involving registrations and distribution of economic poisons.

The interstate shipment of insecticides, fungicides, rodenticides, and similar materials (known as "economic poisons" or "pesticides") within the United States and their importation and exportation is subject to regulation under the Federal Insecticide, Fungicide, and Rodenticide Act of 1947. Applications for the registration of economic poisons under the Act are made on appropriate forms indicating the information to be furnished by the registrant. These forms may be obtained from the Pesticide Regulation Section.

To be acceptable, the label for any economic poison must state the chemical composition of the formulation, including the percentages of active and inert ingredients; the net contents; adequate directions for use; and any caution or warning which may be required by the formulation and which will be adequate, if complied with, to protect the public.

In general, the claims made in advertising must agree with the claims made in connection with the registration of the particular product concerned. Literature which accompanies the product in the normal channels of trade and is distributed with the product, when it is sold, is regarded as labeling and must be registered. Newspaper, periodical, radio and TV advertising, while subject to the same requirements that it should not exceed the claims made in connection with the registration of the economic poison, is not subject to the registration requirements of the law. Such advertising, when it bears claims not acceptable for registration under the Federal Insecticide, Fungicide, and Rodenticide Act, is referred to the Federal Trade Commission for action.

The Department of Agriculture does not maintain or publish any listing of the products which have been registered under the act and which are being offered for sale in the United States. The Department does not approve specific proprietary products. Their registration means only that, on the basis of available information, they appear to meet minimum legal requirements. The manufacturer or registrant is responsible for the accuracy of all claims made in connection with the registration of any product, but no reference to the registration may be included in the labeling for the product.

The law is enforced through the collection of official samples of products which have been shipped in interstate commerce, their examination in the Section's laboratories and, when necessary, action in the courts. Work in these laboratories is limited to the examination and testing of samples which have been collected by the investigators in accordance with the provisions of the law. These samples are first analyzed in one of the chemical laboratories, and then tested, if necessary, in order to determine that the directions for use are adequate and that the product under examination will be safe and effective when used in accordance with the manufacturer's directions.

Any products not meeting the requirements of the law are brought to the attention of the manufacturer through correspondence, seizure or citation and prosecution whenever necessary.

When fertilizers are mixed with pesticides, in order to control certain types of soil insects, for example, the resultant pesticide-fertilizer mixture is then regarded as an economic poison subject to the registration requirements of the Federal Insecticide, Fungicide, and Rodenticide Act. (See also **Pesticides**).

Marketing Laws and Regulations. Alleged *discriminatory pricing* figured in the 1958 decision by the Supreme Court in the gasoline price war case involving the Standard Oil Co. (Ind.). The Court, in a divided opinion, held permissible the price differential on car lots of gasoline to four large Detroit dealers over tank wagon prices to others, even though the discount on the car lots was greater than the cost saving effected. The discount was held necessary to meet competition under the specific conditions shown in this case.

A circuit court decision in 1964, involving the Borden Co., turned on the issue of permissibility of charging a higher price for a product bearing the Borden trademark, with its implication of dependability of quality, than for material not so labeled. The court held that such variation of price was permissible.

Mergers including *purchases* of marketing outlets introduce new risks in building distribution of products or achieving vertical integration downstream. The decision in the DuPont case, in which the Supreme Court ordered divestment of DuPont's holding of General Motors stock, turned largely on this issue, namely, that DuPont might at some future date use the power of its stock holding to influence purchases of DuPont products even though such influence had not been exercised over the past 40 years or so of the stock ownership.

The following discussion of marketing laws and regulations is abstracted almost entirely and most of it stated verbatim from a recent comprehensive study by Henry H. Fowler (C. & E. N. **35**, 44, 1957).

Refusing to deal, by a seller of chemical products, with a particular class or type of distributor or with a distributor who follows certain practices is within the law. Refusal to sell, on the other hand, is illegal if the refusal to sell is made in concert or conspiracy with others such as rival sellers or even with distributor groups. Likewise, a refusal to deal, even if individually conceived and executed, may be illegal if it becomes part of a larger plan to restrain trade.

Tie-in agreements, such as requiring purchase of a complimentary product, run a serious risk of violating the Clayton Act.

Specifying territory and extracting a promise from the distributor not to sell outside the territory involves a substantially greater risk broadly than the promise by a chemical manufacturer that he will sell to no other distributor within any assigned territory, particularly as the chemical manufacturer's position in the market grows stronger.

Marketing affiliates and dealings between a chemical manufacturer and such affiliate or subsidiaries may lead to so-called "intra-enterprise conspiracy." To be avoided are coercion or unreasonable restraint of the trade of strangers or discriminatory treatment in prices or terms of trade.

Competitive pricing in the chemical industry may at times fall under the shadow of the doctrine of *"conscious parallelism."* The quoted market price of most homogeneous basic products inevitably seeks a common level since the product of one producer cannot be sold for more than the product of his competitors. Moreover, discounts and other terms of sale are often the result of custom and practice in the industry. The Supreme Court, in the Theatre Enterprise case (346 U.S. Sup. Ct. Reports 537), has said, however, that "This court has never held that proof of parallel business behavior conclusively establishes agreement or, phrased differently, that such behavior itself constitutes a Sherman Act offense." This reprieve does not mean that uniform pricing is without antitrust significance. Quite the contrary is true. Uniform pricing of chemical products by competing producers is relevant evidence of agreement to restrain trade. Its effect in establishing conspiracy will vary case by case.

Fair trade agreements setting up manufacturer-determined resale prices constitute a rather fundamental exemption from the antitrust laws but are under severe attack from time to time. The Miller-Tydings law in 1937 and the McGuire Amendment in 1952 exempted from the antitrust laws "contracts or agreements prescribing minimum prices for the resale of a commodity which bears, or the label or container of which bears, the trademark, brand, or name of the producer or distributor of such commodity and which is in free and open competition with commodities of the same general class produced or distributed by others, when contracts or agreements of that description are lawful in the state of resale."

Discriminatory pricing and sales practices in general are regulated through the Robinson-Patman Act.

[In 1965 the Supreme Court, in *U.S.* v. *Huck,* affirmed by a 4 to 4 vote a lower court decision that the licensor of a patent could fix the sale price of the invention by a *manufacturer-licensee*. This right would not extend to a resale or to multiple licensors.]

State laws govern the labeling of explosives, poisons, or other hazardous products and their transportation.

Patents. In the field of patents it is not improbable that a fifth to a third of the patent attorneys of the United States utilize chemistry or chemical engineering knowledge in their regular work. It has been estimated that in the Patent Office in Washington 277 Examiners or about 25 per cent of the total are trained chemists or chemical engineers. One of our largest chemical concerns estimates that it has in its patent division 45 persons who have received degrees in chemistry or chemical engineering; 40 of them have also received or are presently working for degrees in law. In addition there are the inventors who cooperate with the patent department in patent law matters and these in 1956 numbered approximately 575 as shown by the Patent Office Index of patents issued to this company in that year. The number of patents issued in the chemical classes in 1963 was 12,270.

There are reasons why chemical knowledge is required in this field of patents. Two questions recur in case after case: (1) Is the subject matter to be patented *new?* (2) Does it meet the statutory definition of invention? Only a chemically trained person is qualified to answer such questions on a broad range of chemical subjects in advance of a search on each specific question. Either before or after the search, no one is better qualified to know whether the chemical result obtained would have been obvious or *unobvious* in advance to one skilled in the art; the definition of invention depends upon this point.

Chemical patents require chemical knowledge also for recital or such features as the general properties required for each class of materials used, alternatives for the preferred materials, whether stoichiometric proportions are necessary or other proportions are permissible, the temperature, pressure, time period of the reaction, mixing conditions, and the end point or result which determines when the reaction, mixing, or other processing is completed and is to be discontinued. Without such information, the attorney preparing the chemical patent application must speculate on the basis of his own limited contact with the subject.

If the patent is issued and then sued upon, as is the case for about 1.2% of the currently issued patents, the chemist working with the lawyer frequently tests experimentally the procedures and compositions of the art relied upon to show whether the invention in suit is old and whether it gives a result so unlike what was known or forecastable before as to differ therefrom in kind rather than in degree only.

The chemically trained worker frequently becomes also the expert in the trial. As an expert, he interprets the other patents or publications that are relied upon to show the state of the prior art to which the invention relates. He presents his own data bearing on the invention and generally endeavors to impress the tribunal with his information and the reliability of his opinions. (See also **Patents**)

Water and Sanitation. (The following is a private communication from Professor William F. O'Connor of Fordham University, a consultant in the field of water and sanitation.)

Water is perhaps our greatest and cheapest natural resource. In lower New York State, for instance, a dollar will still buy about 80,000 lbs. of excellent drinking water. Possibly because of being constantly replaced, our water resources have been misused and abused. As a result, critical shortages are appearing in many sections of the country, thus creating a problem complicated by the fact that water supply, sewage disposal, and stream pollution are so intimately related that proper legislation for a single incident or location is extremely difficult to frame.

From the early solutions sought by individual communities, arrangements spread to several municipalities cooperating to form a sanitary district, as in Chicago or in Westchester County, New York. Later, still larger areas undertook joint action, as in the Central Valley Project in California and the

Miami Conservancy District in Ohio. Then the next larger scale appeared in the Tennessee Valley Authority, which has developed along many lines in addition to water and sanitation. The Ohio Valley Water Sanitation Commission covers over 150,000 square miles in Illinois, Indiana, Kentucky, New York, Ohio, Pennsylvania, Virginia and West Virginia. The Interstate Sanitation Commission, operating since 1941, represents New York, New Jersey and Connecticut. The much debated proposal to set up a Missouri River Basin Authority, if undertaken, would undoubtedly create the largest single development along these lines. It would involve over a dozen states.

Many state and local laws refer back to the common law principle of "riparian rights," under which a riparian owner has the right to full and uninterrupted use of the water adjacent to his property in quality and quantity to the same extent as all others on the stream. From this principle there have been developed by the courts many basic interpretations such as:

(1) Reasonable use of the waters by the riparian owner is subject to similar reasonable use by others.

(2) Drinking and other domestic purposes have the highest priority in uses of water.

(3) The principle of "superior use" favors municipal water supply over industrial use.

(4) "Natural and customary uses of water" enjoy prior rights over "unusual or novel uses."

(5) A preemptive right is said to exist when an action, even though it results in damage to the stream, has been permitted to exist continuously for some 20 or 30 years without being opposed.

(6) Public policy and the right of eminent domain constitute superior use.

The trend in complicated cases today leans toward sharing the benefits on an equitable basis by taking maximum advantage of the utilization potential of each stream. As this concept becomes more widely accepted, a more rational and individual approach will supplant the inflexible and arbitrary standards that have been applied indiscriminately in recent years to large areas, without regard to basic scientific or engineering features or sound economic considerations.

Every state has its own laws and regulations concerning drinking water and the disposal of wastes within the waters of the state, it being now generally recognized that the river basin is the most rational unit for correlating and integrating the various demands for water.

The authority of the federal government in this area stems from the Constitution. This grants to the government the power to provide for the common defense, promote the general welfare, and regulate commerce between the states.

Five Federal laws containing major provisions relating to control of water pollution have been enacted by Congress. The Rivers and Harbors Act of 1899 (Section 13) and the Oil Pollution Act of 1924 seek primarily to prevent damage to shipping. Their administration is by the Department of Defense. The 1899 Act prohibits depositing waste materials, other than that flowing from streets and sewers in a liquid state, into or upon the banks of navigable waters and their tributaries. The 1924 Act prohibits the discharge of oil into the coastal navigable waters of the United States. The Public Health Service Act of 1912 gives specific authority to the Public Health Service to investigate the pollution of streams and lakes by sewage and other causes. Investigations so made have provided the basis for much information and many consultative services to other agencies. The comprehensive Water Pollution Control Act of 1948 (P.L. 845, 80th Congress) expands activities by the Public Health Service and introduces the principles of State-Federal cooperative program of development, limited Federal enforcement authority, and financial aid. Finally, the new Federal Water Pollution Control Act (P.L. 660, 84th Congress), which was approved by the President on July 9, 1956, improves and extends the Federal effort to control water pollution. (See also "Encyclopedia of Patent Law and Invention Management")

ROBERT CALVERT

Cross-references: *Wastes, Industrial; Water Conditioning.*

LIGAND FIELD THEORY

The concepts of crystal field and ligand field theory have been used successfully during the past 15 years in discussing the magnetic and spectral properties of transition metal complexes. Crystal field theory treats the complex as if the only interaction between the central metal atom and its set of nearest neighbor molecules or ions—both of which are called *ligands*—is a purely electrostatic one. Thus, to a first approximation the orbitals of the central ion are considered as separated from the ligand orbitals.

The purely electrostatic concept, in which there is no orbital mixing or sharing of electrons, is never strictly true. In fact, recent electron spin resonance, ESR, and nuclear magnetic resonance, NMR, studies have shown appreciable mixing between the central metal's d-electrons and the ligand orbitals. Crystal field theory modified in such a way as to take account of the existence of moderate amounts of delocalization between the metal and ligand orbitals is called *ligand field theory*. When the amount of orbital overlap is excessive, e.g., in metal complexes of carbon monoxide or the isocyanides, the *molecular orbital theory* gives a more complete explanation of the metal-ligand bonding.

In ligand field theory one is concerned with the origin and the consequences of splitting the inner orbitals of the central metal by the surrounding ligands. The most satisfactory correlations have been demonstrated with the first transition series, in which the 3d-orbitals are split into different energy levels. To appreciate the effect of a ligand field, imagine that a symmetrical group of ligands is brought up to a charged ion from a distance. First, the electrostatic repulsions between the ligand electrons and those in the d-orbitals of the metal will raise the energy of all five d-orbitals equally. Then as the ligands approach to within bonding distances, the repulsion interactions will take on a directional character that will vary with the particular d-orbitals under consideration. This arises because of the different shapes and orientations of the five d-orbitals in space along a Cartesian coordinate system (Fig. 1).

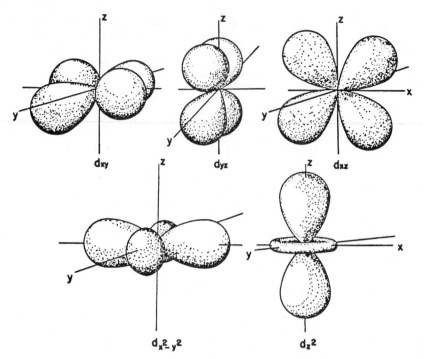

FIG. 1. Sketches showing the distribution of electron density in the five d-orbitals.

Consider the specific case of a metal ion, M^{n+}, at the center of an octahedral set of anions, X^- (Fig. 2). The orbitals which point directly along the Cartesian coordinate axes toward the ligands ($d_{x^2-y^2}$ and d_{z^2}) experience a greater repulsion interaction than those orbitals which point between the ligands (d_{xy}, d_{xz}, d_{yz}). In terms of orbital energy levels, the original set of five degenerate orbitals is split into two sets; the higher energy orbitals ($d_{x^2-y^2}$, d_{z^2}) are conventionally labeled e_g (or sometimes d_γ) and the lower energy orbitals (d_{xy}, d_{xz}, d_{yz}) are labeled t_{2g} (or sometimes d_ϵ). *Within each set*, the orbitals are all of equal energy; *i.e.*, the t_{2g} level is triply degenerate and the e_g level is doubly degenerate. These results are shown in Fig. 3.

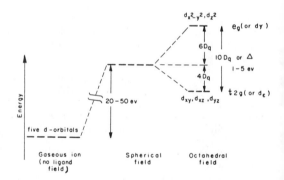

FIG. 3. Energy level diagram illustrating the splitting of the five d-orbitals in an octahedral ligand field.

The ligand field splitting is the energy difference between the t_{2g} and e_g orbital levels, and it is frequently measured in terms of the parameter Dq (or sometimes Δ; $10Dq = \Delta$). The magnitude of the splitting depends on the nature of the ligands and has been designated $10Dq$. The value of Dq in a given complex can be determined experimentally from its electronic absorption spectrum. Light absorption by the complex corresponds to the excitation of electrons from the lower energy levels to the higher levels and the frequency of the light absorbed is related to the energy-level difference by the equation $h\nu = E_2 - E_1$.

When an electron undergoes a transition from the ground state, e.g., the t_{2g} level, to an excited state, e.g., the e_g level, the complex must absorb energy. In the case of transitions involving the d-orbitals, e.g., $t_{2g} \rightarrow e_g$ electron transitions, the absorption

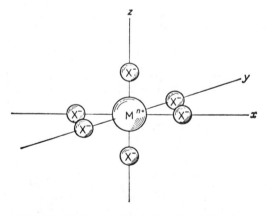

FIG. 2. Six X^- ions arranged octahedrally around a central M^{n+} ion.

Near Infrared
←
Red
Orange
Yellow
Green
Blue
Violet
Ultra-Violet
→
13,300 15,400 17,000 19,000 20,400 23,800 25,600 cm.

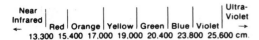

FIG. 4. The visible region of the spectrum.

bands occur in the near infrared, visible, or near ultraviolet region. One might infer from the scale in Fig. 4 that as the colors of a series of complexes of a given metal shift from red to violet, the $t_{2g} \rightarrow e_g$ splitting increases. However, we must recall that the color observed by the eye is the complement of the color that is absorbed. Thus, a complex that absorbs strongly at ~ 14,500 cm^{-1} appears blue and one that absorbs at ~ 20,000 cm^{-1} appears red to the observer. The shape of the bands and colors for some specific complexes are given in Fig. 5.

When the degeneracy of the five 3d-orbitals is split, electrons no longer will occupy all five members of the original set with equal probability. Instead, they tend to occupy the more stable ones, preferentially, subject to restrictions arising from

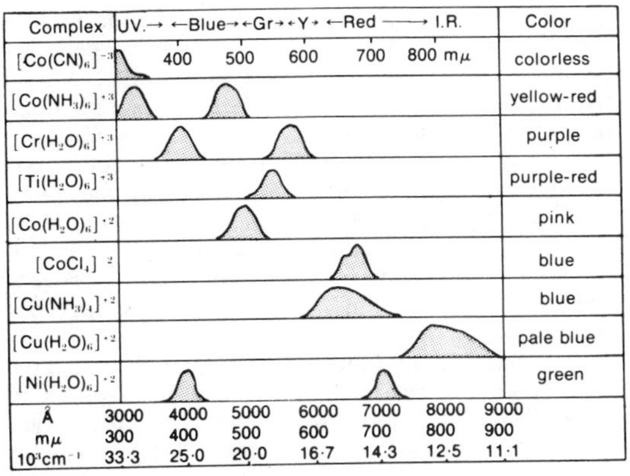
Complex
UV.→ ←Blue→←Gr→←Y→ ←Red → I.R.
Color
400 500 600 700 800 mμ
[Co(CN)6]-3
colorless
[Co(NH3)6]+3
yellow-red
[Cr(H2O)6]+3
purple
[Ti(H2O)6]+3
purple-red
[Co(H2O)6]+2
pink
[CoCl4]-2
blue
[Cu(NH3)4]+2
blue
[Cu(H2O)6]+2
pale blue
[Ni(H2O)6]+2
green
Å 3000 4000 5000 6000 7000 8000 9000
mμ 300 400 500 600 700 800 900
10³cm⁻¹ 33·3 25·0 20·0 16·7 14·3 12·5 11·1

FIG. 5. An illustration of the relationship between the colors and the positions of the electronic absorption spectra of some inorganic complexes. (Courtesy of Oxford University Press.)

Pauli's exclusion principle and from interelectronic repulsions. For example, a d^1 ion (Ti^{3+}) will preferentially fill one of the t_{2g} orbitals, and the ion is said to have a t_{2g}^1 configuration and to possess $4Dq$ of ligand field stabilization energy. Similarly, d^2 (V^{3+}) and d^3(Cr^{3+}) ions will have t_{2g}^2 and t_{2g}^3 configurations, with each electron in a different t_{2g} orbital and all electrons parallel (Hund's rule). These two configurations result in 8 and 12 Dq total CFSE, respectively. For a d^4 ion, there are two possibilities. All electrons may occupy the t_{2g} orbitals (t_{2g}^4), but then one orbital must be doubly occupied and the electron-electron repulsion energy must be overcome for the electrons to pair. Alternatively, we may have the configuration $t_{2g}^3\ e_g^1$, which does not require the pairing energy, but does require the expenditure of 10 Dq of energy to place one of the electrons in the unstable e_g orbital. The electronic configuration actually adopted in a given case is determined by which of the energies is smaller. If 10 Dq is less than the pairing energy the configuration is $t_{2g}^3\ e_g^1$, whereas if 10 Dq is greater than the pairing energy the configuration is t_{2g}^4. The configurations with the maximum possible number of unpaired electrons are called *high-spin* configurations (e.g., $t_{2g}^3\ e_g^1$), and those with the minimum number of unpaired electrons are called the *low-spin* or *spin-paired* configurations (e.g., t_{2g}^4). For convenience we now must distinguish the two limiting situations known as the *weak-field* case, (i.e., ligands which cause 10 Dq to be less than the pairing energy), and the *strong field* case (i.e., where 10 Dq is greater than the pairing energy). Electronic configurations for both the weak field and the strong field cases and the corresponding ligand field stabilization energies for d^0 to d^{10} ions are given in Table 1.

The splitting of the orbitals for a given central metal ion is dependent on the set of ligands. For example, the spectrum of a d^1 ion in a regular octahedral field should exhibit only one absorption band corresponding to excitation of the electron from the t_{2g} to the e_g orbitals; therefore, the energy change corresponding to this absorption represents 10 Dq. The spectrum of $[Ti(H_2O)_6]^{3+}$ (d^1) has a maximum at 20,300 cm^{-1}, so Dq = 2030 cm^{-1}.

TABLE 1. OCTAHEDRAL CRYSTAL FIELD STABILIZATION ENERGIES AND DIFFERENCES IN ENERGIES FOR WEAK AND STRONG FIELDS

Examples	Weak Field			Strong Field			Gain in Orbital Energy in Strong Field
	t_{2g}	e_g	Dq	t_{2g}	e_g	Dq	Dq
d^0 Ca^{2+}, Sc^{3+}, etc.	0	0	0	0	0	0	0
d^1 Ti^{3+}, V^{4+}	1	0	4	1	0	4	0
d^2 Ti^{2+}, V^{3+}	2	0	8	2	0	8	0
d^3 V^{2+}, Cr^{3+}	3	0	12	3	0	12	0
d^4 Cr^{2+}, Mn^{3+}	3	1	6	4	0	16	10
d^5 Mn^{2+}, Fe^{3+}	3	2	0	5	0	20	20
d^6 Fe^{2+}, Co^{3+}	4	2	4	6	0	24	20
d^7 Co^{2+}	5	2	8	6	1	18	10
d^8 Ni^{2+}	6	2	12	6	2	12	0
d^9 Cu^{2+}	6	3	6	6	3	6	0
d^{10} Cu^{+}, Zn^{2+}, etc.	6	4	0	6	4	0	0

Correlations of the spectra of a large number of complexes have demonstrated that generally the various ligands can be arranged in a series according to their capacity to increase the d-orbital splitting of a given metal ion. The spectrochemical series, for the more common ligands, is:

$$CN^- \gg NO_2^- > \text{dipyridyl} > \text{ethylenediamine}$$
$$> NH_3 > \text{pyridine} > NCS^- > H_2O$$
$$> OH^- > F^- > Cl^- > Br^- > I^-$$

The magnitude of the d-orbital splittings and, hence, the relative frequencies of visible absorption bands for two complexes containing the same metal ion but different ligands will generally vary according to the above series.

Metal ions in tetrahedral ligand fields may be treated by the same procedure outlined above for the octahedral cases. Note, by reference to Fig. 6, that in a regular tetrahedral complex, the e_g orbitals are lower in energy than the t_{2g} orbitals. That is, the d_{xz}, d_{yz}, d_{xy} orbitals experience the greater repulsive forces in a tetrahedral field. Also the magnitude of the orbital splitting is calculated to be only 4/9 as large as for an octahedral complex of the same metal ion and ligand with the same metal-ligand bond distance.

Distortions of the regular coordination polyhedron can cause important changes in the orbital energy levels and, consequently, on the magnetic and spectral properties of the complex. For example, for the d^8 configuration the strictly octahedral environment does not permit the existence of a low-spin state. However, tetragonal distortions of an octahedral complex will cause further splitting of the degenerate orbitals in each of the t_{2g} and e_g levels, and the splitting may become great enough to overcome the pairing energy and cause formation of a low-spin complex.

For the square-planar complexes (which may be considered to be the extreme case of a tetragonal distortion of an octahedron) the order of the d-orbitals going from lowest energy is generally represented as $d_{xz} = d_{yz} < d_{z^2} \ll d_{xy} \ll d_{x^2-y^2}$. In some complexes the d_{z^2} orbital may actually be at lower energy than the doubly degenerate d_{xz} and d_{yz} orbitals.

The inter-relationships of the orbital splittings for some different stereochemistries are shown in Fig. 6.

Applications of ligand field theory to many transition metal complexes have already played an important role in the interpretation of visible absorption spectra, magnetism, luminescence, and paramagnetic resonance spectra. In the future ligand field theory may be expected to be the basis for understanding redox reactions, photochemical reactions, and catalytic properties of metal ions.

D. W. MEEK

LIGNIN

Lignin is a major polymeric component of woody tissue in the higher plants, composed of repeating phenyl propane units and usually amounting to 20 to 30% of the dry weight of wood. About one-quarter of the lignin is concentrated in the middle lamella and serves to bind the individual cells together. The remainder serves as a diffuse matrix within the plant cell walls in which are imbedded the polysaccharide microfibrils consisting of cellulose and lesser amounts of the hemicelluloses. Physically lignin is an amorphous gel, the x-ray diagram of which gives no evidence of crystallinity. It exhibits little swelling or solubility in neutral solvents, but becomes soluble in polar hydrogen-bonding organic compounds after chemical degradation.

The classical methods of degradation that have been used to establish the structures of many natural products have been less successful when applied to lignin. None is known that gives yields of monomers much higher than 50%. Higher molecular weight fragments are present in such variety that in only a few cases has it been possible to isolate pure dimeric products. There are other complications in proving lignin structure. Lignin has a network structure and can only be dissolved and separated from carbohydrates with chemical degradation. All degradative methods involve side reactions and usually rearrangements.

Structure studies have shown, however that about one-third of the phenyl propane units are linked together through a phenolic oxygen to the beta position of an adjoining propyl side chain and the remainder by heterocyclic, biphenyl and other linkages. In softwoods essentially all the aromatic rings have one methoxyl group meta to the side chain and on an average there are at least three and one-half oxygens per monomer unit. In hardwoods there is in addition a second monomer with two meta methoxyl groups, and in grasses a third methoxyl-free monomer is found. The homogeneity or heterogeneity of lignin in wood, the kind of oxygenated functional groups on the side chain, the mode of linkage between monomers, a possible linkage to polysaccharides, and the macromolecular state of the native polymer are all under investigation at the present time.

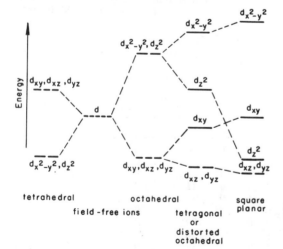

FIG. 6. Relative d-orbital splittings for ligand fields of different symmetries. (Diagrammatic only.)

From tracer studies it has been learned that lignin monomers are synthesized through the shikimic acid pathway. Coniferyl alcohol is quite probably the monomer precursor of softwood lignins. Further hydroxylation and methylation to yield sinapyl lignin units probably occurs at the monomeric state. Coniferin, the beta-glucoside of coniferyl alcohol, probably has no direct role in this process. Enzymes found in a number of plants convert phenylalanine and in grasses tyrosine to cinnamic acids. These acids are reduced to the alcohols and the latter polymerized through the action of peroxidases. Lignin is more resistant to biological attack than carbohydrates and is probably the main source of coal and the humic acids of the soil.

Chemical pulping processes degrade and solubilize the lignin in wood, freeing insoluble cellulosic fibers for paper making. The kraft process, using sodium sulfide and sodium hydroxide, liberates phenolic groups in the degraded lignin and these make the lignin alkali-soluble. Acid pulping methods degrade the lignin by acid-catalyzed hydrolysis and sulfonate it by bisulfite ion. The lignosulfonates are water-soluble. Semichemical and mechanical pulping methods permit fiber separation with less lignin loss but produce poorer quality pulps.

The alkaline liquors from the kraft and soda processes, containing the more soluble polysaccharides and lignin, are concentrated and burned to recover alkali and to obtain heat. The production of heat by its combustion is probably the most valuable use of lignin. The sulfite industry is in part replacing calcium bisulfite liquors with magnesium, sodium and ammonium bisulfite to alleviate scale formation in the concentration and combustion of their ligneous wastes, which are at present a serious disposal problem.

Lignin sulfonic acid is used, to some extent, as a low-cost wetting agent for industrial uses, and is the major raw material used for the production of vanillin. Dimethylsulfoxide, an industrial solvent and an experimental medicinal, is becoming the most important pure chemical made from lignin. Commercial lignins are also used for road and briquette binders, tanning agents, drilling mud additives, protein coagulants, as fillers for synthetic rubbers and phenolic plastics, as a component of linoleum, as soil conditioners and hardeners, and various other minor uses. Together these uses constitute a commercial outlet for only a minor portion of the lignin produced by the pulp industry.

CONRAD SCHUERCH

References

Sarkanen, K. V., and Ludwig, C. H., "Lignins: Occurrence, Formation, Structure and Reactions," Wiley-Interscience, New York, 1971; Browning, B. L., "Chemistry of Wood," Interscience, New York, 1963; Brauns, F. E., and Brauns, D. A., "The Chemistry of Lignin," Academic Press, New York, 1960.

Biosynthetic Lignins

Lignin is the three-dimensional, amorphous polymer that fills the spaces between cellulose and hemicellulose fibers in plant cell walls to make them

I R = R' = R'' = H, *trans-p*-Coumaryl alcohol
II R = R' = H, R'' = OCH_3, Coniferyl alcohol
III R = H, R' = R'' = OCH_3, Sinapyl alcohol
In glucosides, R =

woody. Lignin is important in agriculture because its presence in forage plants inhibits their digestibility by ruminants. Lignin is also an environmental concern since in modified form it is a troublesome by-product of the vast wood-pulping industry, contributing to air and water pollution. Much of our current knowledge of lignin structure and the mechanism of lignification, i.e., the processes involved in the conversion of monomeric precursors into the lignin polymer in plants, has come from simulation of the natural process in the laboratory by dehydrogenation of chemically synthesized *p*-coumaryl alcohols with phenol-oxidizing enzymes (*p*-diphenol:O_2 oxidoreductase, EC 1.10.3.2 = laccase, or donor:H_2O_2 oxidoreductase EC 1.11.1.7 = peroxidase), giving rise to buff-colored, insoluble amorphous powders with the appearance and properties of natural lignin. In recent work, peroxidase has been preferred because it allows better control of the extent of oxidation by exact dosage of the dilute aqueous hydrogen peroxide used as hydrogen acceptor. There is still controversy as to which oxidase is involved in natural lignification in plants, whether peroxidase, laccase, cytochrome oxidase, or other oxidizing system.

The lignin precursors are *p*-coumaryl, coniferyl, and sinapyl alcohols (I-III) whose $4O$-β-D-glucosides (glucocoumaryl alcohol, coniferin and syringin) occur in the cambial sap of growing trees. Suitable mixtures of the alcohols can be used to duplicate any lignin; e.g., 14% of I, 80% of II, and 6% of III for a typical softwood (spruce) lignin, or 5% of I, 52% of II, and 43% of III for a typical hardwood (beech) lignin. Forage plant lignins contain mainly units derived from I and have lower methoxyl contents (2-5%) than softwood (14-16%) or hardwood (20-22%) lignins. The mixture is dissolved in a large volume of water and added dropwise to an aerated solution of fungal (*Polyporus versicolor*) laccase or together with slow separate dropwise addition of very dilute H_2O_2 to horseradish peroxidase at pH about 6.5, the polymer gradually separating out as the reaction proceeds. The polymer can be freed from oligomeric materials by extraction with butanol or ethyl acetate, and purified from occluded enzyme by dissolution in dioxan/water (9:1 v/v), filtration, and precipitation in water or ether. Lyophilization from the filtered dioxan/water solution is also a convenient isolation method: the freeze-dried product is readily handled and easily redissolved for further investigation. The peroxide must be added very slowly, since excess reagent irreversibly inactivates peroxidase; about four equivalents of H_2O_2 are needed since some decomposes nonenzymically on the large active surface of the freshly precipitated polymer. No such difficulties are encountered with laccase, but this enzyme is harder to prepare and is not commercially available.

Chromatography of the low-molecular weight portion of this reaction or of the total reaction mixture at lower levels of oxidation has shown the presence of over 40 intermediates ('lignols'), containing 1–6 phenyl-propanoid units. Identification of their structures revealed how they had been formed and hence shed light on the complex mechanism involved in lignification. Abstraction of an electron from the phenoxide ion of the alcohols produces a free radical which is metastable on account of its ability to spread its unpaired electron by mesomerism over the conjugated system of the molecule. The radical becomes stabilized either by removing an H atom from another molecule to create other, less conjugated radicals which then disproportionate, or by radical coupling to form unstable methylenequinones (quinone methides). The first process causes oxidations of side chains in monomers and higher lignols; e.g., forming small amounts of coniferaldehyde, ferulic acid, vanillin, and vanillic acid from II. As phenols, these compounds are also condensed enzymatically into the lignin. The methylenequinones are stabilized by intramolecular rearrangements to phenoxides which either add a proton to form oligomeric lignols (dilignols, trilignols, etc.) or undergo further phenol oxidation or addition of nucleophiles (e.g., hydroxide or phenoxide ions) followed by coupling, etc., as before. Dilignols identified include dehydrodiconiferyl alcohol and its aldehyde, DL-pinoresinol and -epipinoresinol, guaiacylglycerol-β-coniferyl ether (IV) and its aldehyde, and the biphenyl derivative dehydrobisconiferyl alcohol. Trilignols include dehydrotriconiferyl alcohol, guaiacylglycerol-β-pinoresinol, -epipinoresinol, and -dehydroconiferyl ethers, and guaiacylglycerol-β,γ-diconiferyl ether. The last compound is formed by nondehydrogenative addition of coniferyl alcohol to the methylenequinone precursor of IV. Tetralignols, e.g., guaiacylglycerol-β-coniferyl-γ-dehydrodiconiferyl ether or bisdehydropinoresinol, and higher lignols are also formed by the mechanisms of phenol coupling and nucleophile additions to methylenequinones. In some dilignols, e.g., 1,2-diguaiacylpropane-1,3-diol and 4O-guaiacylconiferyl alcohol, one C_3 side chain has been eliminated; this is now known to be transferred to other units as glyceraldehyde ethers blocking terminal phenolic hydroxyls.

The structures of the lignols reveal that the units in lignin are bound together in a random fashion by a variety of strong carbon-carbon and ether bonds, sometimes with more than one bond between units. This explains why lignin, unlike other natural polymers such as proteins, polysaccharides or nucleic acids, is not depolymerized by hydrolysis with acids or enzymes. Strong acids transform lignin into a completely insoluble bakelite-like resin; this reaction forms the basis of lignin determinations performed by treatment of plant materials with strong acid. Mild treatments with acid cause cleavage of some weaker ether bonds while avoiding condensations; isolation of some lignols or related derivatives from such mild hydrolysis of lignin and wood has confirmed that the laboratory process for producing biosynthetic lignin does indeed closely simulate the natural process in plants.

Preparation of the artificial lignins in the laboratory in the presence of sugars produced fractions with sugar molecules covalently attached to lignols; the bonds can arise by free radical combinations or additions of the sugars onto methylenequinones and represent models of the bonding believed to exist between lignin and polysaccharides in wood.

Other extraneous acidic or phenolic substances can readily be condensed into artificial lignins in the same way. Some plants may behave similarly, condensing phenolic tannins or organic acids into their lignin, e.g., p-hydroxybenzoic acid in aspen, or ferulic acid in grasses.

Continued action of the phenol oxidases on the biosynthetic lignin polymer leads to its oxidative gradual degradation, converting it to humus-like material. This is in part what happens when plant residues are converted into humus in soil and may explain the presence of the phenol-oxidizing enzymes in soil microorganisms.

JOHN M. HARKIN

References

Harkin, J. M., "Lignin and Its Uses," USDA-FS Res. Note FPL 0206, 1969, 9 pp. Free copies available from Forest Products Laboratory, Madison, Wis. 53705.

Harkin, J. M., Taylor, W. I., and Battersby, A. R., "Oxidative Coupling of Phenols," pp. 243–321, Dekker, New York, 1967.

Pearl, I. A., "The Chemistry of Lignin," Dekker, New York, 1967.

Freudenberg, K., and Neish, A. C., "Constitution and Biosynthesis of Lignin," Springer-Verlag, New York, 1968.

Sarkanen, K. V., and Ludwig, C. H., "Lignins: Occurrence, Formation, Structure and Reactions," Wiley-Interscience, New York, 1971.

LIPIDES

Lipide (or lipid, lipin, fat, etc.) is a comprehensive term referring to simple or complex substances which are found in all living cells and which have either only a nonpolar hydrocarbon moiety or a hydrocarbon moiety with polar functional groups. Many lipides contain fatty acids or are capable of forming covalent bonds with fatty acids.

General Properties. (1) Most lipides are insoluble in water and soluble in fat solvents such as ether and chloroform. Some lipides, such as short-chain fatty acids, glycerides of short-chain fatty acids, and lysophospholipides, are appreciably soluble in water. The relative proportion of the hydrocarbon moiety (hydrophobic) and the polar moiety (hydrophilic) determines the physical properties of the lipides. Phospholipides and fatty acid soaps act as detergents and form micellar aggregates in water. Acetone dissolves the neutral lipides but precipitates the phospholipides. Ethyl alcohol dissolves one phospholipide (lecithin) and precipitates another phosphatidylethanolamine, (also known as cephalin).

(2) The fatty acid is the characteristic constituent of the lipide, but some lipides such as the sterols and the fat-soluble vitamins are not always found combined with the fatty acids.

(3) Many lipides have both hydrophobic and hydrophilic groups which cause them to orient at oil-water interfaces, the hydrophobic group in the oil phase and the hydrophilic in the water phase.

(4) Some lipides, especially the phospholipides, have a strong tendency to form complexes with each other and with various substances. Complex formation is due to the electrostatic attraction of polar groups and to the mutual solubility of the long hydrocarbon chains. Thus the lipoproteins and proteolipides are complexes of proteins and a variety of lipides such as cholesterol, phospholipides, glycerides, and glycolipides. The lipides are linked to the proteins by several types of forces. Electrostatic forces, van der Waals forces, hydrogen bonding, and hydrophobic bonding hold these complexes together. Because of their attraction for water, the polar groups of the protein and phospholipide arrange themselves on the outside of the complex, while the hydrocarbon groups of the lipides are folded into the center. Thus there is presented to the aqueous phase those groups which have an affinity for water. This arrangement accounts for the solubility of the complexes in water. The phospholipides, owing to their polar groups, act as water-solubilizers for the nonpolar lipides. The arrangement may be different in the proteolipides of the brain and nerves, since they are not soluble in water. In these complexes the lipides may completely envelope the protein.

(5) Some lipides, especially if isolated under conditions to minimize peroxidation, are colorless, odorless and tasteless. The color, odor, and taste which they may have are due to either the natural occurrence of certain functional groups (aldehydes, esters, ketones, conjugated double bonds) or to peroxidized products. Lipides, however, can solubilize certain colored pigments and other organic molecules and thus appear to be colored themselves.

Main Divisions of the Lipides. (I). *Simple Lipides.* (1) Fats or Neutral Fats. These are glycerol esters of the fatty acids which are chiefly palmitic, stearic, oleic, and linoleic, although many other fatty acids are found in nature. Most of the fatty acid combination is as triesters of glycerol. (2) Waxes. These are esters of fatty acids with long chain alcohols. Examples are cholesterol esters as in blood plasma, vitamin A esters, various plant waxes, beeswax, and animal skin waxes such as lanolin. In general the alcohols involved are large molecules and the waxes are very insoluble in water and serve as protective coatings.

(II) *Conjugated Lipides.* The complex or conjugated lipides contain other groups in addition to alcohols and fatty acids. (1) Phospholipides (phosphatides). These contain phosphoric acid and one or more other groups. Phospholipides are subclassified as: (a) Lecithins, which are glycerides containing two fatty acids with the third hydroxyl combined with phosphoric acid which in turn is combined with the base choline. Lecithin is now commonly called phosphatidyl choline. (b) Cephalins (sometimes called kephalins), in which the basic group is either ethanolamine (as in phosphatidyl ethanolamine), or serine (phosphatidyl serine). The term cephalin is seldom used today, since the meaning is vague due to its representing different types of phospholipides. (c) Sphingomyelin, which

contains a basic monoamino, dihydroxy alcohol called sphingosine, combined with a single fatty acid (amide linkage), and a phosphoryl choline group (but no glycerol). Sphingomyelin is elevated in Niemann-Pick's disease. (d) Phosphatidic acids, which are diacyl derivatives of glycerolphosphoric acid. (e) Aldehyde phospholipides, (plasmalogens) in which one of the fatty acid groups is replaced by a fatty aldehyde in the form of a vinyl ether linkage. The lipids containing vinyl ether linkages are also called alkenyl lipids. (f) Polyglycerolphosphatides. This group includes cardiolipin, phosphatidyl glycerol, bis-phosphatidic acid, amino acid esters of phosphatidyl glycerol, the lyso-compounds of these phospholipides, and their corresponding vinyl ether analogs. Cardiolipin is localized in the mitochondria of the cell. (g) Liponucleotides. Recent work has revealed that certain intermediates of cytidine nucleotides play a role in phospholipid biosynthesis. Two important intermediates are cytidinediphosphatediglyceride and long-chain fatty acyl adenylates. The fatty acyl adenylates are converted to fatty acyl coenzyme A thiol esters. (h) Lysophospholipides, obtained by the action of phospholipase A on lecithin and other phospholipides. This enzyme removes the beta-linked fatty acid, leaving usually a saturated acid combined with the glycerol. The enzyme is contained in snake venom, pancreatic juice and other sources.

(2) Cerebrosides contain the base sphingosine with one fatty acid linked by an amide group and a carbohydrate group which may be either galactose (galactolipide) or glucose (glucolipide). Cerebrosides are elevated in Gaucher's disease. (3) Sulfolipides are cerebrosides which contain a sulfate radical on the sugar moiety. (4) Gangliosides are complex lipides containing ceramide (lignoceryl sphingosine) hexoses, hexosamines, and neuraminic acid. They are found in ganglion cells of brain and are increased in Tay-Sach's disease. (5) Lipoproteins and proteolipides are combinations of protein, phospholipide, triglycerides, and cholesterol or its esters. Proteolipides, such as are found in brain and nerve, have the solubilities of lipides, i.e., soluble in fat solvents but not in water.

(III) Derived lipides—lipide products which may originate by breakdown of the more complex compounds mentioned above.

(1) *Fatty Acids.* Mostly long chain acids (16–18 carbon) with some representatives of shorter and longer chains. Fatty acids may be straight-chain, branched, or ring compounds but the straight chain acids with an even number of carbon atoms are the most abundant in animal tissues. (See **Fatty Acids**)

(2) *Alcohols* Those found in nature in combination with the fatty acids: (a) straight chain alcohols; (b) ring compounds such as inositol and cholesterol; (c) vitamin alcohols.

(3) *Hydrocarbons.* In this group are included hydrocarbons of the fatty series, the carotenoids and squalene. They form no esters with the fatty acids and are included here because they often occur in solution with some natural lipides and are soluble in lipide solvents.

(4) *Vitamins A, D, E, K,* some of which may occur as esters of the fatty acids.

Physiological Functions. (1) *Neutral glycerides.* These compounds are the most concentrated of the foodstuffs, yielding about 9 calories per g as compared to 4–5 cal/g for carbohydrates or proteins. They therefore yield the greatest quantity of energy per unit weight when burned in the body. They are economically valuable both for energy and storage because they are stored practically dry in the animal body while carbohydrate or protein is stored wet. They can serve as raw material for the formation of the phospholipides and other lipides. They are stored in adipose tissue.

(2) *Waxes.* Because of their chemical structure waxes are used largely as protective coatings to keep out bacteria and other invaders and to prevent excessive water evaporation. Common examples are the skin waxes and the waxy coating of fruits and leaves of plants.

(3) *Complex Lipides.* Phospholipides, glycolipides and cholesterol, in combination with proteins are integral structural components of cell membranes and are believed to play a role in cell permeability. Phospholipides are also believed to have a function in the mitochondrial electron transfer system and in nerve conduction. Other physiological processes in which lipides are believed to play a role include blood coagulation, relaxation of muscle contraction, and demyelination of nerve fibers. The grana of chloroplasts and the visual rods of the eye contain phospholipides and other lipides in a bimolecular leaflet array similar to that postulated in cell membranes.

Metabolism. Knowledge of lipide metabolism has greatly increased during the last decade. The detailed biochemical reactions whereby the fatty acids are synthesized and oxidized, how phospholipides, glycolipides and cholesterol are synthesized, and how lipides are absorbed and transported have been elucidated. Fatty acids are synthesized from acetyl coenzyme A and malonyl coenzyme A thiol esters. The vitamin biotin plays a vital part in the fixing of carbon dioxide to form malonyl coenzyme A, an important intermediate in fatty acid synthesis. The hormone insulin also favors fatty acid synthesis. The oxidation of fatty acids occurs as their coenzyme A esters in the Krebs cycle of the mitochondria. Cholesterol is biosynthesized from acetyl coenzyme A. Cholesterol in man is converted to bile acids, fecal sterols, and to steroid hormones. The synthesis of lecithin is mediated via phosphatidic acid and diglyceride precursors. Cytidine nucleotides play a role in the transfer of choline (as phosphorylcholine) to a diglyceride to form lecithin. Uridine nucleotides act to transfer sugar residues in the synthesis of glycolipides.

The transport of lipides in the blood plasma is effected by complex formation with proteins to yield lipoproteins. The liver is the major organ for the synthesis of the lipoproteins. Analysis of serum lipoprotein patterns is important in the understanding of vascular disease (atherosclerosis). The clearing of lipemic blood such as may occur after a heavy fat meal, is brought about by an enzyme known as lipoprotein lipase. This enzyme yields free fatty acids which combine immediately with the plasma albumin to form complexes known as NEFA (non-esterified fatty acids). NEFA act as important transport vehicles for transport of triglycerides and the levels of blood NEFA are very sensitive to hormonal control and neural control. Certain hormones such as epinephrine stimulate the membrane-bound adenyl cyclase which converts ATP to cyclic AMP. The latter stimulates adipose tissue lipase and mobilizes depot fat.

The excess utilization of lipides and excess oxidation of fatty acids causes an increase in acetoacetic acid in the body. This condition is known as ketosis and can lead to acidosis. This situation is common in severe diabetes and can occur whenever carbohydrate utilization is severely decreased.

Some of the major problems to be elucidated in lipide metabolism are as follows: (a) how are lipoproteins synthesized (b) how are lipides arranged and compared with proteins to form cell membranes (c) what specific role, if any, do lipides play in transport across cell membranes (d) how do hormones act to regulate lipide metabolism (e) what is the biochemical basis of such abnormal lipide metabolic states as Gaucher's disease, Niemann-Pick's disease, etc. (f) how do lipides themselves permeate cell membranes; (g) how many phenotypic lipoproteins occur in serum.

G. V. MARINETTI

Cross-references: *Fats, Fatty Acids, Phosphatides, Glycerides, Vitamins, Waxes.*

LIQUID CRYSTALS

Reinitzer observed in 1888 that cholesteryl benzoate melted to give a turbid but mobile fluid at 145°C, and only on heating to 179°C was the amorphous isotropic liquid formed. Other pure compounds were soon shown to behave similarly, and the turbid fluids were eventually recognized as representative of a new state of matter, neither crystalline nor liquid, but with properties intermediate between these commoner states of matter. This new state of matter was called the liquid crystalline state, and perhaps one in every two hundred organic compounds is capable of giving liquid crystals. With regard to terminology, some authors prefer to use the terms "mesophase" and "mesomorphic" instead of "liquid crystal" and "liquid crystalline," respectively. This alternative terminology emphasizes the intermediate nature of this state of matter without relating it too closely to either the crystal or the isotropic liquid.

The three-dimensional order of the molecules in a crystal is not therefore necessarily destroyed completely at a given temperature, but may be broken down in stages giving intermediate liquid crystalline states. These possess some of the properties of liquids—fluidity to a greater or lesser extent—and some of the properties of crystals—optical anisotropy.

Of great importance to progress in early studies of liquid crystals was the work of Vorländer and Lehmann in synthesizing a wider range of pure compounds of the liquid crystalline kind. Examination of the types of compound involved showed that they consisted of rod-like molecules possessing some degree of rigidity. Thus, compounds which give optically anisotropic liquid crystals are built up from geometrically anisotropic molecules. Friedel's detailed studies (1922) of the optical properties of liquid crystals were also notable in leading to the recognition of three types of liquid crystal and in stimulating studies of their other physical properties. Eventually it became possible to associate different arrangements of the elongated molecules with the different types described below.

(1) *Smectic Liquid Crystals.* In these viscous, turbid liquid crystals, the long molecules lie parallel, their ends in line, forming layers which are usually curved and distorted, but are capable of movement over one another. Only in thin unsupported films are the layers flat, as in soap films (Greek: *smectos*—a soap); in such plane films, layer flow may readily be observed. A high degree of molecular order therefore persists in the smectic state. Several polymorphic smectic forms have recently been identified, these varying in the degree of molecular organization within the layers and the angle which the long molecular axes make with the layer planes. An example of a smectogen is ethyl p-azoxybenzoate.

(2) *Nematic Liquid Crystals.* In these less turbid, more mobile liquid crystals, the long molecules are still essentially parallel, but there is no regular alignment of their ends. In thin films on surfaces, the degree of molecular orientation is higher than in bulk samples where we envisage a continuum produced through distortions of the parallel molecular alignment; under the polarizing microscope, thin nematic films often exhibit optical discontinuities called threads (Greek: *nematos*—a thread). A typical nematogen is p-azoxyanisole.

(3) *Cholesteric Liquid Crystals.* These are formed by optically active compounds which have the requirements for a molecular organization of the nematic type; common examples are the cholesteryl n-alkanoates. Thin films of cholesteric material in the Grandjean plane texture are best regarded as stacks of sheets, in each of which the molecular organization is nematic, but on passing from sheet to sheet the long molecular axes rotate through a small angle building up a supramolecular helix. The unique properties of cholesteric liquid crystals—optical negativity, very high optical rotatory power and the ability to selectively reflect and absorb colored light when the helix pitch is in the wavelength range of incident white light—are thus explained. Racemic mixtures of optical isomers are nematic, and a cholesteric liquid crystal can best be described as a twisted nematic state.

On heating a liquid crystalline compound, one of these three liquid crystal types may be formed, and only at a higher temperature is the amorphous isotropic liquid produced. If the compound gives two different liquid crystals, the lower temperature phase is always smectic; further heating then forms the amorphous liquid *via* a nematic phase—or a cholesteric phase if the compound is optically active. All these transitions are reversible on cooling. It should also be noted that an amorphous liquid may supercool below the melting point giving a monotropic liquid crystal.

Since liquid crystalline properties depend upon the stability of particular molecular arrangements, factors affecting molecular packing and molecular interactions are fundamentally important. Liquid crystalline behavior is therefore highly sensitive to change in molecular structure which may, for example, alter the thermal stability of the liquid crystal, alter the type of liquid crystal formed, or result in formation of two liquid crystal types rather than one.

Technological Applications. Such liquid crystals today assume considerable technological importance. For example, thin, clear films of nematic liquid crystals mounted between conducting glass surfaces may be rendered opaque by an applied voltage. The opacity is readily visible in natural light, and the film clears on removing the voltage. The construction of light shutters or curtainless windows is therefore possible. By addressing the film in segments, sequences of numbers or letters may be shown, or by etching the electrodes, a design may be produced. Such applications excite considerable interest in the field of electronic displays, *e.g.,* for alphanumeric indicators, and even have significance in the area of television. Adaptations of this effect may be used to produce information storage devices (using a nematic material containing 10% of a cholesteric material) or colored displays (using dichroic dyes dissolved in the nematic liquid or pure nematic liquid crystals in which the orientation of the long molecular axes with respect to the conducting surfaces is carefully controlled).

Secondly, thin films of cholesteric liquid crystals can give color maps of the temperatures of surfaces, and this application (surface thermography) is significant in medicine (in skin thermography and its applications in tumour detection, studies of cardiovascular diseases, *etc.*), in electronics (for temperature mapping of circuits) and in nondestructive testing of laminates, *etc.*

In all these applications, liquid crystals stable in the room temperature range are needed. Few pure compounds yet meet this requirement, and low melting mixtures giving nematic or cholesteric states are generally used.

Liquid crystals also have important applications in gas-liquid chromatography, nuclear magnetic resonance spectroscopy and electron spin resonance spectroscopy.

Lyotropic Systems. With suitable compounds, the ordered structure of the crystal may be broken down in stages by the action of a solvent giving smectic, nematic or cholesteric liquid crystals; these lyotropic liquid crystals give amorphous solutions with an excess of solvent.

Biological Systems. Although the full significance of liquid crystals in biological systems is imperfectly understood, it can be visualized that the ordered but deformable nature of liquid crystals may be very suitable for the construction and function of living matter. Thus, muscles, nerves and tendons can exhibit optical anisotropy, certain living sperms exhibit a liquid crystalline phase, and lyotropic liquid crystals are given by RNA and certain viruses.

G. W. GRAY

Cross-references: *Crystals and Crystallography.*

LIQUID METALS

The class of materials called liquid metals might be expected to include only those metallic elements which are liquid at room temperature. However, the term has come to include those which melt as high as aluminum (660.3°C). There are five other metallic elements that melt below the boiling point of water; there are a number of alloys which are molten at room temperature and even as low as −110°C. To a great extent the liquid metals have been utilized in heat transfer technology because they can remove heat rapidly from small volumes. However, there are some chemical uses which have developed.

The pure metals which comprise the liquid metal category are listed in Table 1. Aluminum has been historically chosen as the metal with the maximum melting point, although zinc could easily be chosen as the maximum.

TABLE 1. LIQUID METALS

Metal	M.P. (°C)	Group
Mercury	−38.87	IIB
Cesium	+28.5	IA
Gallium	29.62	IIIA
Rubidium	39.0	IA
Potassium	63.7	IA
Sodium	97.8	IA
Indium	156.4	IIIA
Lithium	179	IA
Tin	231.9	IVA
Bismuth	271.0	VA
Thallium	303	IIIA
Cadmium	321	IIB
Lead	327.4	IVA
Zinc	419.0	IIB
Antimony	630.5	VA
Magnesium	651	IIA
Aluminum	660.3	IIIA

The IA elements, the alkali metals, are quite uniform in their gradation of physical and chemical properties. They are known as the alkali metals because they form strong alkalis with water, are very reactive chemically and increase in reactivity with increasing atomic weight. All the elements are strong reducing agents, react with oxygen in the air and react violently with water; all will form relatively stable hydrides.

Group IIB elements show a decreasing chemical reactivity with increasing atomic weight (mercury behaves much like a noble metal). These elements are quite familiar because their purification-recovery is comparatively simple. Magnesium is generally classed with the alkaline earths but might be considered a light-weight element of the Group IIB series. It burns vigorously in air, whereas, although zinc will burn, it does so less readily than does magnesium.

Aluminum is not very representative of the IIIA group and is much like magnesium in many ways. In the molten state a protective oxide film forms which impedes further oxidation if the surface is unagitated.

In the IIIA group it is found that melting points increase with increasing atomic weight. The boiling points exceed those of most other liquid metals, giving them extremely long liquid ranges. Gallium is one of the few metals which expands on freezing. They are quite corrosive in that they are excellent solubilizing materials.

Group IVA again shows an increase in melting point with increasing atomic weight. Boiling points

are high and tin has the longest liquid range of the liquid metals. If the entire Group IVA series is considered (carbon, silicon, germanium, tin and lead) the progression from nonmetal to metal is quite apparent. There is a change in melting point from 3500°C for diamonds to 232°C for tin, evidencing the change from covalent to metallic bonding. Oxides of the metals are relatively insoluble in the pure metal, thus permitting the handling of molten tin and lead with the formation of an oxide protective layer. If this is stirred or removed the metals will burn rapidly.

In the VA elements antimony and bismuth are the only metallic elements having higher electrical resistance in the solid than in the liquid state. Both bismuth and antimony expand on freezing. There are several forms of antimony of which the normal is the gray or metallic form which is stable to the boiling point. There are two nonmetallic forms: an unstable yellow form and a black amorphous material.

Alloys of the metals are numerous, with some unusually low melting points being evidenced. The eutectic of sodium and potassium has a melting point of −12.5°C at approximately 78% potassium and 22% sodium by weight. Since the materials have been used in chemical and heat transfer applications, they are widely known and are commonly called NaK (pronounced nack). A sodium-potassium-cesium alloy has a melting pont of −79°C.

Nonchemical Uses. The primary application of liquid metals (as materials which are liquid) is in heat transfer or heat management. In this application heat is moved from one place to another by the liquid metal. Mercury boilers were the first notable use of a liquid metal and applications have increased considerably since that time. Characteristics which make them attractive as heat transfer agents are high thermal conductivity, good specific heat, low viscosity and the stability one would expect of an elemental material. Since the boiling points can range up to 2270°C for tin, heat can be transferred from extremely high-temperature heat sources. Until the liquid metals were applied to this use, materials such as water, Dowtherm, oils and fused salts had been used (and still are in many cases). However, the extremely high thermal conductivity of liquid metals makes it possible to transfer heat to the liquid metals at excellent rates. In fact, much experimental and developmental work was performed to determine applicable heat transfer equations for use in designing heat exchangers and heat transfer systems.

Applications which have been encountered include: heat removal from nuclear reactors, liquid metal-vapor turbine power plants, reaction vessel heating, consumable electrode crucible heat control, working fluid in high-temperature hydraulic systems, electrical switches, and thermometer elements.

The greatest interest in the use of liquid metals is as a coolant for nuclear reactors. Here the coolant must transport energy at high temperature levels at low pressures and not be damaged by radiation. Sodium and NaK have both been used in nuclear reactors. Although interest in very compact nuclear reactor power plants for submarines was the original impetus for consideration of liquid metals, only one was constructed, the Sea Wolf, which used sodium. The primary problem was corrosion of boilers on the water side of the system, not on the sodium side. The first electric power from a nuclear reactor was produced in the Experimental Breeder Reactor at Arco, Idaho, on 20 December 1951; this reactor used NaK as the heat transfer fluid. There is much interest in fast breeder reactors, which will use sodium as the coolant.

Since bismuth will dissolve uranium, a liquid metal-fuel reactor had been studied. Since the fuel is dissolved in the coolant it is possible to process the fuel continuously as the reactor operates.

An increasing interest in the application of sodium and NaK to heat management in reaction vessels has resulted in a number of heat control units in chemical processing.

Equipment to use the liquid metals has been developed. Pumps with no moving parts operate on the principle of an electric motor. Special centrifugal pumps have been developed. Flowmeters, liquid level gages and pressure gages are available and take advantage of the metallic properties of the coolant. Heat exchangers have been designed on the basis of the special properties of the alkali metals, and heat exchanger performance can now be predicted satisfactorily. Special valves have been developed for high-temperature work. While the liquid metals can be quite corrosive when improperly handled, they are among some of the purest materials and the Group I elements can be contained (except for lithium) in mild steels at temperatures up to 450°C and in stainless steels at temperatures up to ~ 1000°C. There are available handling techniques and purification techniques which will preclude corrosion problems. The major consideration is exclusion of oxygen from the air.

Chemical Uses. There are no large-scale chemical applications that are dependent upon the liquid property of the metals. However, in some chemical reactions the fluid state of the metals promotes intimate mixing, and reaction rates are enhanced.

Sodium-potassium alloy has been used in the directed interesterification of oils and fats to lower the melting point range of the treated material. This is a catalytic application and large quantities of liquid metal are not involved, although low-temperature fluidity is important. A large-scale use of sodium-lead alloy is in the production of lead tetraethyl.

Group I metals are of interest in some unusual reaction possibilities. Cesium combines the property of being strongly electropositive with a low melting point. Rubidium and cesium can promote polymerization of ethylene to branch products having up to 12 carbon atoms.

Some of the alloys promote reaction rates. For example, the alkylation of toluene with propylene goes about 500 times as fast with NaK as with sodium or lithium. Naphthalene can be hydrogenated with NaK as a catalyst. The polymerization of isoprene requires only a few hours with NaK, days with potassium, and weeks with sodium.

Mercury has been used as a reaction medium for inorganic reactions. Cadmium has been used as a menstruum to make nitrides of thorium, strontium and calcium. Titanium carbide is made using aluminum as a menstruum.

Pyrometallurgical processing of reactor fuel material has been performed to dissolve the irradiated fuels and separate the uranium and plutonium from fission products. Generally zinc, cadmium and magnesium are used in a series of crystallizations and progressive dissolutions by changing the solvent alloy composition, along with immiscible solvent extraction and chemical slagging.

J. W. MAUSTELLER

References

Jackson, C. B., and Mausteller, J. W., "Liquid Metals—Their Properties, Handling, and Applications," in "Modern Materials," Vol. 3, p. 401, Academic Press, New York, 1962.

Lyon, R. N., Ed., "Liquid Metals Handbook," 2nd Ed., NAVEXOS, p. 733 (rev.), AEC and Dept. of the Navy, Washington, D.C., 1952.

Mausteller, J. W., Tepper, F., and Rodgers, S. J., "Alkali Metal Handling and Systems Operating Techniques," AEC Monograph, Gordon and Breach, New York, 1967.

LIQUID STATE

Liquid is the term used for a state of matter characterized by that of a pure substance above the temperature of melting and below the vaporization temperature, at any pressure between the triple point pressure and the critical pressure (see Fig. 1). The liquid state resembles the crystalline in the relatively low dependence of density on P and T, and resembles the gas state in the inability to support shear stresses (see reference to glasses below). Structurally the molecules are relatively closely packed, but lack long-range crystalline order. The mutual solubility of different liquids is also intermediate between the complete mutual solubility of all gases, and the relatively rare appreciable mutual solubility of pure crystalline compounds. Two liquids of similar molecules are usually soluble in all proportions, but very low solubility is sufficiently common to permit the demonstration of as many as seven separate liquid phases in equilibrium at one temperature and pressure (mercury, gallium, phosphorus, perfluoro-kerosene, water, aniline, and heptane at 50°C, 1 atm).

Stability Limits. With the exception of helium, and certain apparent exceptions discussed below, Fig. 1

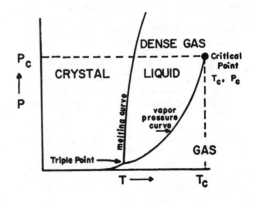

FIG. 1

gives a universal phase diagram for all pure compounds. The triple point of one P and one T is the single point at which all three phases, crystal, liquid, and gas, are in equilibrium. The triple point pressure is normally below atmospheric. Those substances, i.e., CO_2, $P_t = 3885$ mm, $T_t = 56.6°C$ for which it lies above, sublime without melting at atmospheric pressure.

From the triple point the melting curve defines the equilibrium between crystal and liquid, usually rising with small but positive dT/dP, and presumably always with positive dT/dP at sufficiently high P-values. The line is believed to extend infinitely without a critical point (it has been followed to $T \cong 16T_o$ for He, and calculations indicate that hard spheres would show a gas-crystal phase change). The gas liquid equilibrium line, the vapor pressure curve, has dT/dP always positive and greater than the melting curve. The vapor pressure curve always ends at a critical point, $P = P_c$, $T = T_c$, above which the liquid and gas phase are no longer distinguishable. Since the liquid can be continuously converted into the gas phase without discontinuous change of properties by any path in the P-T diagram passing above the critical point there is no definite boundary between liquid and gas.

The term *liquid* is commonly reserved for $T < T_c$, and "Dense Gas" used for $T > T_c$. However, certain properties, such as the ability to dissolve solids, change rather abruptly at the critical density. In many respects the dense gas resembles the low-temperature liquid of the same density more closely than it does the dilute gas.

The slope, dT/dP, of all phase equilibrium lines obeys the thermodynamic Clapeyron equation:

$$dT/dP = \Delta V/\Delta S = T\Delta V/\Delta H,$$

with ΔV, ΔS, and ΔH the differences, for the two phases, of volume, entropy, and heat content or enthalpy, respectively. The quantity ΔH is the heat absorbed in the phase change at constant P. Since always $S_{cr.} < S_{liq.} < S_{gas}$ and usually $V_{cr.} < V_{liq.} < V_{gas}$:, one usually has $dT/dP > 0$, the relatively rare cases, including water, for which $V_{liq.} < V_{cr.}$ at low pressures lead to $dT/dP < 0$ for the melting curve near the triple point.

Fig. 1 gives the P-T boundaries of the stable liquid phase. Clean liquids can readily be superheated or supercooled, and in vessels having walls to which the liquid adheres they can be made to support negative pressures of several tens of atmospheres. Thus the properties of the metastable liquid can be investigated outside the limits shown in the diagram.

Two apparent exceptions to the universality of the phase diagram of Fig. 1 deserve mention. First, many of the more complicated molecules decompose at temperatures below melting or boiling, and the diagram is unobservable. Secondly, some liquids, notably glycerine and SiO_2 and many multicomponent solutions, supercool so readily that crystallization is difficult to observe. In these cases there is a continuous transition on cooling to a glass, which has the elastic properties of an isotropic solid. The structure of the glass is qualitatively that of the high-temperature liquid, lacking long-range order. Since glass and liquid are not sharply differentiated, the term *liquid* is sometimes used to include glasses, although common parlance reserves liquid for the state in which flow is relatively rapid (see also **Glass**).

J. E. MAYER

Liquid Structures

The structures of liquids remain among the elusive secrets of nature. The liquid state represents an intermediate form of matter and it is thus difficult to delineate, since it has properties that suggest both extremes. At one extreme are solids which exhibit a very high degree of order, high viscosities and mean free paths that are small compared with those of the particles in the solid. At the other extreme are gases, which represent a chaotic state, have very low viscosities and large mean free paths. Liquids have intermediate properties such as short-range order but no long-range order, moderate viscosities and mean free paths; thus it is difficult to visualize their structure.

Clearly, in speaking of the "structure" of a liquid, a different meaning is implied from that normally used in reference to a solid. In a solid, or more precisely a crystal, the molecules (atoms) spend all their time vibrating about some point in space, called a lattice site. As a result, the structure of a solid can be precisely described in terms of the relative position of each molecule by means of some suitable experimental technique such as x-ray crystallography. In a liquid the molecules are continually milling about (as is evident from the relatively high diffusion coefficient) making a description in terms of the precise spatial position of each molecule impossible.

X-ray studies of a solid are able to determine the exact number of nearest neighbors and the distance between them. Studies of the liquid state yield more nebulous information—an average number of nearest neighbors and their average distance apart. In other words, experimental studies of the liquid state yield a picture that has been averaged over a time which is long compared with the time it takes for the particles to move about. For example, x-ray studies of a normal liquid such as argon indicate that near the melting point the number of nearest neighbors is between 10 and 11 with a distance of 3.8Å separating each particle. As the temperature is increased, the number of nearest neighbors decreases until, a few degrees below the critical point, it is about 4, while the distance has remained almost constant. At the critical point there are approximately 6 nearest neighbors, and the distance separating them is 4.5Å.

Since experimental techniques have not yet produced the desired understanding of the liquid state, great use has been made of inference: A model of the liquid state is visualized, this picture is translated into a mathematical equation (i.e., a partition function), and then thermodynamic and mechanical properties are calculated. Comparison with experimentally obtained properties allows one to infer whether or not the model is an adequate representation of the structure of the liquid. Naturally there is also a rigorous approach to the understanding of liquids in terms of radial distribution functions and intermolecular forces. This approach is formally correct, but solutions of the equations are intractable and thus require the introduction of well-defined mathematical approximations which correspond to unrealistic systems. Since the model approach offers a more visual representation of the liquid state, this article will concentrate on this method.

While there are many different types of liquids, they can be divided into two broad categories called "normal" and "associated." Normal liquids approximately obey certain empirical relations involving such quantities as vapor pressure, heat of vaporization, boiling point, and viscosity. Typical examples of normal liquids are argon, carbon tetrachloride, and most organic liquids. Associated liquids exhibit deviations from these empirical relations, and examination reveals that they are composed of substances which can form links such as hydrogen bonds with neighboring particles. Water is a good example of a strongly associated liquid, as are organic compounds containing hydroxyl or amino groups. Properties depend upon the number of effective links the particles can form with each other. Substances such as hydrogen fluoride apparently link into ring structures so that the liquid behaves as a normal liquid of higher molecular weight. Because additional forces are involved, the behavior of an associated liquid is more complicated than that of a normal liquid, and the situation is further complicated when one considers mixtures of various pure liquids. Since normal liquids are more uniform in behavior and thus easier to generalize about, the remarks below will, in general, be confined to normal liquids.

The most widely used description of the liquid state has been the cell model which visualizes the liquid as being very similar to a solid. Each particle in the liquid is assumed to be confined to a cell or cage whose walls are formed by the neighboring particles. The particle is free to move about in this cell and its motion is assumed to be independent of the motion of the surrounding particles, i.e., the motion of the particle in the liquid is uncorrelated. Numerous refinements have been made to the simplest cell models such as quantization of the cell energy levels, the possibility of multiple occupancy of the cell, and a variety of force fields governing the motion of the particle in its cell.

A slightly different approach to the liquid state is to visualize a liquid as a random mixture of particles and holes. A liquid at a temperature near the melting point has a volume 10 to 15% greater than that of the solid. It is assumed that this "excess" volume can be divided up into vacancies of about the same size as that of the particles in the liquid. It is apparent that if the vacancies are of less than unit size or are of unit size but several particles can share a vacancy, the model will possess no long range order. In the cell model, expansion of the liquid with temperature is accounted for by expanding the size of each cell, while the hole theory can account for the expansion by introducing additional vacancies into the liquid. An extension of this concept has been to "fluidize" the vacancies and to endow them with the properties of particles in the vapor phase which is in equilibrium with the liquid. The vacancies are fluidized by allowing the particles to jump into the vacancy, the vacancy then moving to the position formerly occupied by the particle. The vacancies move through the liquid in the same manner that posi-

trons move through an electron sea. This fluidization of the vacancies permits calculation of transport properties. By allowing the vacancies to be mirror images of the vapor particles (i.e., the holes possess the same degrees of freedom as the vapor phase particles and the concentration of the vacancies in the liquid is the same as the concentration of the particles in the vapor) an explanation may be given of such phenomena as the law of rectilinear diameters.

An entirely different way of understanding the liquid state is the geometric method. In this rather picturesque approach a three-dimensional array of spheres is shaken in a random fashion and then the position of each sphere is noted. The process is repeated a number of times and from the assembled data a radial distribution function of the spheres is constructed. The distribution function then allows calculation of the thermodynamic properties of the system. This type of procedure seems to indicate that the particles in a liquid are in the most random arrangement possible.

The geometric method may be treated mathematically by the use of high-speed computers and is then called the Monte Carlo Method. Here a number of particles are fed into a computer in some initial configuration. The properties of this configuration are determined, the configuration of the particles is changed according to a set of rules and then the properties recomputed. In this manner the properties of the system averaged over, say 50,000 configurations can be obtained. By the use of proper starting configurations, properties of both the solid and fluid states may be determined.

A variation of the Monte Carlo technique is the Method of Molecular Dynamics. In this method it is assumed that the liquid can be divided into cells containing, say, 100 particles (the number is limited only by the speed of the computer). The motion of the particles is simulated mathematically in a computer and the trajectory of each particle is followed through a great number of collisions. This information allows the calculation of various properties and, furthermore, it can be displayed on a picture tube making it possible to see the motions of particles in the liquid state. These studies indicate that a particle vibrates about some point and then wanders off and vibrates about some other point.

These models of the structure of a liquid have met with varying degrees of success, but none of them is entirely satisfactory; they are either too unrealistic or too empirical. One problem of trying to determine the structure of a liquid by inference is that the equilibrium properties of the liquid state are not greatly different from those of a superheated solid. Most of the models mentioned above are basically lattice models and so give results which do not violently disagree with the properties of a liquid. It is clear that for a model to become acceptable as a true representation of the liquid structure, it not only must yield correct values for such properties as volume, pressure, enthalpy, entropy, and coefficients of expansion and compressibility, but it must also be able to offer a quantitative explanation for the difference between a liquid and a solid. At this time only a broad description can be given of a normal unassociated liquid: A dense conglomeration of particles continuously changing from one lattice configuration to another and thus destroying any semblance of long range order.

R. P. MARCHI

Cross-references: *Water, Thermodynamics, Solid State.*

LITHIUM AND COMPOUNDS

Lithium with an atomic number of 3 is the lightest metallic element. It is a silvery white metal in Group IA of the Periodic Table with an atomic weight of 6.941, having two natural isotopes, 6 and 7. It has a specific gravity of 0.534, a melting point of 179°C and a boiling point of 1317°C±. It is derived primarily from the mineral spodumene occurring in considerable abundance in the United States and Canada. Lithium minerals are also found in Africa, South America, Europe, Australia and possibly in Asia.

The demand for lithium products calculated as lithium carbonate in the United States rose from less than 400,000 pounds per year prior to World War I to a peak after World War II of over 25,000,000 pounds per year; it is currently about 10,000,000 pounds per year.

Lithium has a hardness of 0.16 on the Mohs scale, and can be extruded, drawn and rolled with considerable ease. While it alloys readily with metals such as silver, aluminum, magnesium, cadmium and boron, the solubility is very low in copper, nickel and iron. In metallurgy lithium is used as a degassifier or scavenger in nonferrous metals and is used to remove nitrogen from noble gases.

Like the other alkali metals, lithium is a very reactive element, but is the least reactive of the alkali metals. In many of its chemical properties it is more like the alkaline earth elements than like an alkali metal.

Hydrogen reacts with lithium to form a stable hydride, LiH, and nitrogen forms a stable nitride, Li_3N. Lithium monoxide, Li_2O, is formed chiefly when lithium is burned in air, but the metal does not react with oxygen below 100°C.

The reaction with cold water slowly produces hydrogen and LiOH, the low solubility of the latter reducing the reaction rate. The halogens all combine vigorously and directly with lithium, and inorganic acids attack it rather violently to produce hydrogen. Molten lithium at high temperatures reacts with all known molecular gases and with phosphorus, arsenic, antimony, silicon, sulfur and selenium. The use of lithium in organic chemistry is described in the following article.

It has many industrial uses: lithium carbonate in ceramic and glass formulations; lithium fluoride in fluxes; lithium hydroxide in lubricating greases; lithium bromide and chloride in air-conditioning systems; lithium salts in fungicidal preparations and citrate beverages; lithium hydroxide in alkaline storage batteries; anhydrous lithium hydroxide for carbon dioxide absorption; butyl lithium in synthetic rubber and in polymerization reactions; lithium metal in silver solders, as a heat transfer medium, in underwater buoyancy devices, and as rocket fuel, as well as in synthesis of Vitamin A.

Lithium fluoride additives are used in fused salt electrolysis in the production of aluminum and magnesium.

Lithium hydride and lithium borohydride are important as a hydrogen source. Lithium aluminum hydride is used as a low-temperature reducing compound in organic reactions. In general, organolithium compounds are useful in organic synthesis.

Lithium metal was probably the first element fissioned to produce alpha particles (Rutherford *et al*) under the impact of highly accelerated protons. A significant nuclear reaction of fusion to produce tritium, H^3, through slow neutron irradiation, proceeds according to the following reaction:

$$3 Li^6 + 0n^1 \rightarrow 2 He^4 + 1 H^3$$

The isotope Li^7 may be used for control rods in fast reactions, due to the fact that any helium formed will not contaminate the pile. Because of its low melting point, lithium would have to be enclosed in a suitable thin tube, e.g., beryllium metal.

Lithium (Li^6) has important possibilities in applying controlled fusion reaction to power generation by nuclear means.

Lithium isotopes can be separated by electrochemical means whereby lithium-6 becomes enriched at the cathode of a mercury cathode cell electrolyzing LiCl solution or in an electromigration cell containing molten LiCl.

It has been discovered recently that lithium carbonate is remarkably effective in controlling the manic-depressive type of mental disease, a psychosis which afflicts millions of people. The lithium ion is thought to act by its effect on monoamine metabolism in the brain.

PERCY E. LANDOLT (deceased)
Revised by CLIFFORD A. HAMPEL

References

Landolt, P. E., and Hampel, C. A., "Lithium," pp. 367–374, "Encyclopedia of the Chemical Elements," Hampel, C. A., Editor, Van Nostrand Reinhold Co., New York, 1968.

Landolt, P. E., and Sittig, M., "Lithium," pp. 239–270, 2nd Ed., "Rare Metals Handbook," Hampel, C. A., Editor, Reinhold Publishing Corp., New York, 1961.

Lithium in Organic Chemistry

Among the alkali metals, lithium has earned the high regard of the synthetic organic chemist only in recent years. Certain characteristics of the free metal and its compounds indicate that lithium is superior to sodium metal both in reaction versatility and selectivity. The high solubility and the relative stability of associated, covalent organolithium compounds (RLi) in hydrocarbon or ether solvents contrast with the insolubility of the other ionic metal alkyls of Group IA in hydrocarbons and their facile decomposition in ethers. Moreover, although organolithium reagents generally undergo the addition and cleavage reactions displayed by Grignard reagents, the use of lithium alkyls often secures higher yields with fewer side reactions. Furthermore, the high solubility of lithium metal in ammonia or amines and of complexed lithium hydrides (*e.g.*, $LiAlH_4$ and $LiBH_4$) in ethers or pyridine permits the controlled reduction of many organic and inorganic systems. Finally, because of the relatively small size and high polarizing power of the lithium cation, many reactions both of lithium metal and of lithium alkyls proceed more rapidly than those of magnesium, on the one hand, and with greater stereoselectivity than those of sodium, on the other hand.

Examples bearing out this statement are the following: (a) lithium alkyls add to the azomethine linkage of pyridine and to the exocyclic double bond of 1,1-diphenylethylene under conditions where the corresponding Grignard reagents are unresponsive; (b) lithium metal forms adducts with conjugated systems in solvents where magnesium metal is inert; and (c) lithium metal or lithium alkyls in hydrocarbons promote the 1,4-*cis*-polymerization ($> 90\%$) of isoprene, whereas sodium leads to 1,4-*trans*- and 3, 4-polymer formation.

The particular uses of lithium and its compounds in organic chemical research stem from the foregoing considerations. Lithium metal suspensions or lithium-aromatic hydrocarbon adducts in amines or ethers (*e.g.*, tetramethylethylenediamine, tetrahydrofuran and the novel crown ethers) are able to cleave many covalent bonds, A—B (where A = C,O,Si,N,X and B = C,O,N,H,Si,X), thus providing a general method for the synthesis of organolithium compounds and the controlled degradation of organic systems:

$$C_6H_5\text{—}O\text{—}C_6H_5 \xrightarrow[\text{THF}]{2Li} C_6H_5Li + C_6H_5OLi$$

$$\downarrow H_3O^+$$

$$C_6H_6 + C_6H_5OH$$

Functioning as an electron source, lithium can cause the bimolecular reduction of certain alkenes:

$$2(C_6H_5)_3SiCH\text{==}CH_2 \xrightarrow[\text{2. } H_3O^+]{\text{1. Li/THF}}$$

$$[(C_6H_5)_3SiCH_2CH_2\text{—}]_2$$

as well as the stereospecific polymerization of conjugated dienes. Alternatively, the metal can effect the reduction of such dienes or α,β-unsaturated carbonyl systems, if a proton donor (R_3CH, NH_4Cl) is present at the time of metal dissolution.

Lithium hydride complexes ($LiBH_4$, $LiAlH_4$, $LiAlR_3H$ and $LiAl(OR)_3H$) and organolithium reagents are quite similar in their action providing a potential nucleophile (H^- or R^-) to various electron-deficient centers ($M\text{—}X_n$, $C\text{==}O$, $C\text{==}N$, $C\text{≡}N$ and, less rarely, $C\text{==}C$ or $C\text{≡}C$). Recently, complexes of lithium alkyls with copper alkyls have proved of value in effecting selective alkylations of organic halides, epoxides, and α,β-unsaturated carbonyl derivatives. Thus, these reagents permit the reduction and alkylation of many unsaturated organic compounds and the conversion of metal salts to metal hydrides:

$$R\text{—}COOH \xrightarrow[\text{2. } H_3O^+]{\text{1. } LiAlH_4} R\text{—}CH_2OH$$

$$(C_6H_5)_3SnCl \xrightarrow[\text{2. } H_3O^+]{\text{1. } LiAlH_4} (C_6H_5)_3SnH$$

In addition, lithium alkyls (*e.g.*, *n*—C_4H_9Li) undergo three ready exchange reactions which set them apart from Grignard reagents, namely (1) halogen-lithium exchange; (2) hydrogen-lithium exchange; and (3) metal-lithium exchange:

$$C_6H_5Li \xleftarrow{C_6H_5Br} C_4H_9Li \xrightarrow{(C_6H_5)_4Sn} C_6H_5Li$$
$$\text{(1)} \qquad\qquad \text{(3)}$$
$$\text{(2)} \downarrow (C_6H_5)_3CH$$
$$(C_6H_5)_3CLi$$

These interconversions permit the indirect synthesis of many valuable organometallic compounds for use in synthesis, mechanistic studies or structure elucidation. A variety of transient, synthetically useful intermediates, such as radical-anions, carbenoids, benzynes and polycarbanions, has been generated by means of lithium metal or organolithium reagents.

JOHN J. EISCH
HENRY GILMAN

Reference

Coates, G. E., and Wade, K., "Organometallic Componds, Volume I: The Main Group Elements," 3rd Ed., pp. 4–42, Methuen & Co., Ltd., London, 1967.

LUBRICATING GREASES

A lubricating grease is a solid or semisolid lubricant consisting of a thickening agent in a liquid lubricant. Other ingredients imparting special properties may be included. In such products the fluid is almost always the major component. This *liquid* is normally lubricating oil, which may vary in color, type and viscosity. For special applications, synthetic lubricating fluids are used in spite of the higher cost. Diesters, such as di-2-ethylhexyl sebacate, for a range of about -70 to $300°F$, certain silicone polymers for applications up to $600°F$ and unsubstituted polyphenyl ethers, for resistance to radiation oxidation and high temperatures, are illustrations.

Thickeners include soap and nonsoap types, the latter including inorganic or organic materials. Soap thickeners consist of compounds having cations of Al, Ba, Ca, Li, Na, Pb or Sr, most of which are formed *in situ* in the lubricating fluid by reacting a metal base and fatty acids or glycerides. Also, other acid-containing compounds, such as rosin, waxes, naphthenic acids or synthetic fatty acids, may be used. Complex salt-soap thickeners comprising the coneutralized products of low molecular weight acids, such as acetic, and intermediate or high molecular weight fatty acids have also been found of value as thickeners. A variation of this latter type of thickeners are aluminum complex soaps formed by the reaction of a mixture of straight chain carboxylic acids and aromatic acids with anhydrous aluminum compounds, like aluminum alcoholates. Preformed soaps, such as aluminum stearate or lithium 12-hydroxystearate, find some use as grease thickeners.

Nonsoap inorganic thickeners, having the widest application, include colloidal silicas or certain modified clays. Examples of the latter are bentonites which are rendered organophilic and water repellant by reacting the clay with amines containing radicals of at least 10 carbon atoms. Certain organic nonsoap thickeners, like arylureas, have excellent high-temperature stability and hence will find increasing use. Since "target" specifications visualize certain uses of lubricating greases up to at least 600°F, thickeners of such nature may be of value.

Minor components of lubricating greases may consist of additives or compounds resulting from reaction of ingredients. *Additives,* introduced in small concentrations as preformed materials, have the purpose of structure modifiers; to enhance the desirable properties; or to minimize the deleterious properties of the lubricant.

The main purpose of *structure modifiers* is to change the solubility characteristics of thickeners (soaps in most cases) or to control their crystallization during processing. The most widely used structure modifiers include water, high alcohols, fatty acids, glycols, and salts of low molecular weight acids. Other classes of compounds suggested and used for this purpose include amines, esters, phenols, and soaps of a second metal. While a single additive may serve more than one function, most of those to follow are not structure modifiers.

Most lubricating greases contain *oxidation inhibitors* which may consist of various amines; phenols; compounds of the oxygen and sulfur groups, organic phosphites, or dialkyl selenides or tellurides.

Other additives, present alone or in conjunction with others, may include: *metal deactivators,* which tend to precipitate undesirable dissolved metal ions out of solution or form inactive soluble complexes with the metal; *metal passivators,* which act by depositing a film over a metal surface; *corrosion* or *rust inhibitors,* which include some amines, amine salts, metallic sulfonates and naphthenates, etc.; *film strength additives* (more often referred to as extreme pressure (E.P.) compounds) which contain as active ingredients either chlorine, phosphorus, or sulfur compounds in such form that they will react with steel to produce a coating which will prevent welding of two metals; *wear prevention agents,* such as tricresyl phosphate; stringiness additives, such as isobutylene polymers; and *noise-reducing agents,* such as acryloid polymers. While the above is not a complete listing of additives, it does indicate some of the reasons for including such ingredients in lubricating greases.

Finely divided solids, functioning as either additives, like molybdenum disulfide or Teflon; or as fillers, like asbestos, metal flakes, metal carbonates, metal oxides, etc., are sometimes included in these lubricants. However, some such ingredients, for example certain grades or asbestos or of carbon black, actually provide a grease structure.

A satisfactory lubricating grease will flow into bearings by the application of pressure, remain in contact with moving surfaces, and not leak out under gravity or centrifugal action.

The fluid in lubricating greases is immobilized either by a mass of soap crystallites, which may be in the form of fibers from 0.2 to 1000 microns in length and from 0.01 to 1 micron in width, or by nonsoap particles of colloidal size but of varying shapes. Both composition and processing variables influence the character and geometry of soap fibers. The efficiency of thickeners depends upon the particle size and shape and the forces of attraction involved. Since the oil is held in the matrix either by absorption or by capillary attraction, the ability of soap-oil systems to resist oil separation depends largely upon the *surface area* of the soap fibers. From this standpoint, small fibers are preferable.

The *temperature stability* of soap-thickened lubricating greases depends primarily upon the phase behavior of the thickeners. Thus, soaps derived from different bases and fatty materials, as well as from complex soaps, exhibit phase changes at different temperatures. Additives may influence the temperature at which phase changes occur and loss of such additives may even be destructive to the body of the lubricant.

The value of a lubricating grease over a wide temperature range is dependent upon both the apparent viscosity at low temperature and the resistance to deterioration at high temperatures. The apparent viscosity in turn largely depends upon the pour point and the viscosity index of the fluid employed. By using very low-viscosity synthetic fluids, lubricating greases are manufactured which are useable at −100°F. Lack of stability at high temperatures can result from either collapse of the structure or deterioration of one or more of the ingredients. For the present, the upper temperature range for application of lubricating greases having general distribution is limited by the availability of fluids which will not deteriorate when subjected to temperatures above 400°F for prolonged periods. Target specifications visualize extension of use of such lubricants to 600°F.

During the last twenty years the properties of lubricating greases have been improved to the point that a large proportion of such lubricants now marketed can be classified as multipurpose greases. The availability and use of such products has decreased the inventories which consumers must stock and yet given better overall lubrication.

C. J. BONER

LUBRICATING OILS

Lubricating (or lube) oils are fluids whose function is the reduction of friction and wear between solid surfaces (generally metals), in relative motion. This function is accomplished in either of two ways: (1) by formation of adsorbed films on the two opposed surfaces, which can be more easily sheared than the solid substrate, or (2) by interposition of a fluid film between the two opposed surfaces. In the former case the shear strength of the film, and in the latter, the viscosity of the fluid determine the magnitude of the work which must be done to maintain the opposed surfaces in relative motion. In most cases, bearings are designed to operate under fluid film conditions and thus the viscosity of an oil is its most important property in classifying it for lubricating purposes. The range of viscosities required is approximately 1 centistoke at 100°F for high-speed, lightly loaded spindle bearings to approximately 4000 centistokes at 100°F for some gear units. The great majority of oils used fall within a much narrower viscosity range of about 20–200 centistokes.

Aside from the primary function of friction and wear control, lubricating oils are often called on to serve other purposes, such as corrosion prevention, electrical insulation, power transmission and cooling. This last is particularly important in metal cutting and grinding.

As petroleum consists of a highly complex mixture of hydrocarbons and other organic compounds, the raw material must be carefully processed to separate the fractions which are desired for lubricating purposes. This is generally accomplished by a combination of fractional distillation, solvent extraction and adsorption treatments to obtain various cuts which are blended if necessary, to produce the desired viscosity. A typical sequence of operations would be:

(1) Preliminary distillation of the crude petroleum to strip off volatile matter, i.e., dissolved gases and light hydrocarbons through the fuel oil range.

(2) Secondary vacuum distillation to separate the various lubricating oil fractions, which in some cases, may include the residuum. Since hydrocarbons tend to undergo thermal decomposition above 600°F, the distillation is generally carried out under vacuum to enable the higher-boiling fractions to be distilled.

(3) Solvent extraction to remove waxes, unsaturates, aromatics, asphalt, and nonhydrocarbon material.

(4) Contact with an adsorbent, e.g., activated clay, for final removal of polar impurities.

(5) Incorporation of special-purpose additives and blending to obtain the desired viscosity.

The lubricating oil fractions thus obtained consist of some of the largest and most complex hydrocarbon molecules to be found in crude petroleum. Their molecular weight ranges from about 250–1000 or more, based on structures containing 20–70 carbon atoms. They may be classified broadly as (1) straight-chain paraffins, (2) branched chain paraffins, (3) naphthenes (one or more saturated 5 or 6 membered rings with paraffin side chains), (4) aromatics (benzene ring structures with paraffinic side chains), and (5) mixed aromatic-naphthene-paraffin.

Of the above classes, the most desirable are the branched chain paraffins and the naphthenes. The refining processes are directed to increasing the concentration of these and removal of the others, as well as of organic compounds containing sulfur, oxygen, and nitrogen which are found in varying proportions in all crude petroleum.

Most hydrocarbons in the lubricating oil range are subject to low-temperature oxidation (180°F and above) by atmospheric oxygen, particularly under the influence of catalysts such as metals (copper is particularly active), moisture, and occasionally nonhydrocarbon impurities. The paraffinic types show the greatest resistance to oxidation, followed by the naphthenes and the aromatics. An empirical rule states that the rate of oxidation approximately doubles for each 20°F rise in temperature. Thus, in practice where long life of the oil is desired, bearing temperatures are held to around 150°F as a maximum. In an internal combustion engine where temperatures at the sliding surfaces of the order of 400–500°F are encountered, shorter life will result, e.g., 100 hrs. in an engine,

compared with several years in a water-wheel generator.

Besides viscosity, the viscosity-temperature variation of lubricating oils is a most important property. It has been found that this behavior can be accurately expressed by the relation: $\log_{10} \log_{10} (v + 0.8) = n \log_{10}(t°F + 460) + c$ where v is the viscosity in centistokes, t the temperature in °F, and b and c are constants. In most cases an oil with the smallest viscosity-temperature variation is considered the more desirable. As a generalization, it may be said that the oils with the highest paraffin-to-naphthene ratios have the smallest change, while low paraffin-to-naphthene ratios correspond to a higher viscosity-temperature change. However, it is possible to dissolve materials in oil (e.g., olefinic polymers) which will effect a decrease in this variation.

While most bearings depend on the viscosity of the oil for the formation of a lubricating film (.0001″ to .002″ thick), there are many cases where this is not feasible and successful lubrication depends on the formation of very thin films (10 to 10^3 Å thick). This was generally thought to occur through preferential adsorption by the solid surface of polar compounds (e.g., napththenic acids, sulfur compounds) present in small amounts in the oil.

Another factor possibly contributing to the lubricating ability of very thin films is the variation of viscosity with pressure. The viscosity-pressure relationship is approximately $\log_{10} v = kP + c$ where P is the pressure and k and c are constants. Thus, at very high unit pressures of the order of 10^5 psi, viscosity increases of the order of 1000–10,000 times are encountered. This can account for a significant increase in load-carrying capacity. More recent work on rolling contact bearings has shown that these extremely high viscosities may be partially responsible for the observed formation of true hydrodynamic lubricating films in the 10^{-6} to 10^{-5} inch range. (See **Lubrication**)

Though the major portion of industrial lubricants is derived from petroleum, a small but significant portion is being obtained from other sources.

(1) Natural liquid fatty esters, such as lard oil, palm oil, sperm oil, etc. These are good lubricants but have poor chemical stability.

(2) Synthetic hydrocarbons, prepared by polymerization of olefinic hydrocarbons; these have good stability when saturated and good viscosity-temperature coefficients.

(3) Polyalkylene glycol oils, made by reaction of alcohols with polymerized ethylene and propylene glycols. They are either water-soluble or -insoluble. Fair stability (improved with additives), good viscosity-temperature coefficient, and good lubricating qualities.

(4) Synthetic esters, (a) primarily esters of dibasic acids such as adipic and sebacic, though in Europe some monobasic acid esters have been prepared and used; (b) organic esters of phosphoric and silicic acid, which have some advantage of being more fire-resistant than the other organic compounds but which are subject to hydrolysis on exposure to water.

(5) Silicone oils, which are linear and cyclic siloxane polymers of the formula $(-SiR_2O)_n$. They generally possess good thermal and oxidative stabil-

ity, and good viscosity-temperature coefficients. When —R is a hydrocarbon substituent their lubricating ability is poor. This can be somewhat improved by the introduction of aromatic halogen in —R.

(6) Halogenated hydrocarbons (chlorinated or fluorinated). These have good lubricating properties but very poor viscosity-temperature relations. The fluorinated materials are extremely stable.

(7) Perfluorinated polyalkylene glycols which have better viscosity-temperature behavior and temperature behavior and temperature resistance.

(8) Polyphenyl ethers are very stable organic fluids which can be used in the 500–700°F temperature range although, like the silicones, they do not have good boundary lubrication properties. These reportedly, however, can be improved with suitable additives.

Generally speaking, these synthetic materials are at present used primarily as specialty lubricants where the need for a particular property outweighs high cost, e.g., diesters as military aircraft engine lubricants. As synthetic processes are improved more extensive application may be expected.

HENRY E. MAHNCKE

Synthetic Lubricants

Chemistry has been the guiding science in creating the various classes of chemical lubricants. Each class is named in terms of the chemical structure involved and each has at least one property that is unobtainable with naturally occurring materials. In varying degrees and combinations, synthetic lubricants provide excellent lubricity over extremely wide temperature ranges, high thermal and oxidative stability, fire-resistance, and outstanding resistance to nuclear radiation. Other noteworthy advantages of synthetic lubricants are the ease with which physical or performance properties can be altered by chemical modification or the incorporation of additives. The facility with which modifications can be achieved makes it possible for the chemists to tailor individual members of the family to meet exactly the requirements of a particular application. Compromises in properties are likewise made to obtain a satisfactory lubricant at lowest cost.

The primary disadvantage of synthetic lubricants is cost. Although production is measured in the millions of pounds, no single type is produced in sufficient quantity to rival the cost of petroleum or mineral oil lubricants. Costs of most commercial synthetic lubricants range from 2 to 40 dollars per gallon. Consequently, such lubricants are used only where mineral oils will not perform satisfactorily.

Polyglycols: $RO(CH_2CHR'O)_xR''$, where R's can be hydrogen and/or organic groups. Terminal groups determine the type of polyglycol such as diol, monoether, diether, ether-ester, etc. Various alkylene oxides including mixtures are the basic raw materials. The major attributes of the polyglycols are viscosity-temperature characteristics, volatility of products of decomposition, relatively low cost, and the wide variety of properties (water-soluble to organic-soluble) obtainable by structural variations. Major uses are as lubricants in auto-

motive hydraulic brake fluids and water-based hydraulic fluids, as compressor lubricants, textile lubricants and rubber lubricants, and in greases where the polyglycol is the carrier for solid lubricants.

Phosphate Esters: $R'OP(O)(OR'')(OR''')$ where at least one R represents an organic group while the remaining represent organic groups or hydrogen. These products are prepared from phosphorus oxychloride or phosphoryl chlorides plus phenols, alcohols or their sodium salts. The tertiary phosphate esters are often classified as triaryl, trialkyl, and alkyl aryl phosphates. The primary and secondary phosphates are used extensively as lubricant additives in various chemical forms. Phosphate esters are best known for their fire-resistant characteristics, and as a result have found extensive application as industrial and aircraft lubricants and hydraulic fluids.

Dibasic Acid Esters: These can include both simple and complex materials. The simple dibasic acid esters, $ROOCR'COOR''$ are made by reacting a dibasic acid, such as sebacic acid, with a primary branched alcohol, such as ethyl hexanol. Complex esters are prepared by reacting a dibasic acid with a polyglycol, such as polyethylene glycol, and capping the chain with a branched primary alcohol or a monobasic acid. The outstanding characteristics of the dibasic acid esters are favorable viscosity-temperature characteristics, excellent lubricating ability, and high stability. Because of this combination of properties, these products are now used as lubricants in almost all aircraft turbine engines.

Chlorofluorocarbons: The polymerization or telomerization of chlorotrifluoroethylene is the route to these relatively low-molecular-weight synthetic lubricants. Manufacture involves elaborate polymerization techniques and stabilization methods to eliminate all of the hydrogen and terminal chlorine introduced into the polymer chain by peroxide fragments or the chain transfer agent. The major characteristics of the chlorofluorocarbons are chemical inertness and thermal stability. Industrial and aerospace applications involving exposure to corrosive or oxidizing atmospheres are the largest uses for these lubricants.

Silicones:

$$\begin{matrix} R & & R & & R \\ RSiO & - & SiO & - & SiR \\ R & & R & & R \end{matrix}$$

where the R's may be the same or different organic groups. The properties are varied by the use of different types of organic substituents; the most popular are methyl, phenyl, and chlorophenyl groups. Recent advances involve fluorine-containing substituents. Manufacture entails the preparation of organochlorosilane intermediates, hydrolysis and condensation of these intermediates, and polymer finishing. In addition to good stability and low volatility, silicones have the best viscosity-temperature characteristics of any lubricant. Although they perform well under many conditions of lubrication, silicone lubricants are generally unsatisfactory for situations involving sliding contact of steel-on-steel. The many and varied uses include lubricating electric motors, precision equipment, plastic and rubber surfaces, and as greases for antifriction bearings. (See **Silicone Resins**)

Silicate Esters: ROSi(OR′)(OR″)(OR‴), where the R's may be similar or dissimilar groups. The best-known types are the tetraalkyl, tetraaryl, and mixed alkylaryl orthosilicates. The classic means of preparation is through the reaction of phenol or alcohol with silicon tetrachloride. A closely related group of products, the hexaalkoxy- and hexa-aryloxydisiloxanes, is also generally included in the silicate esters classification. These products, the so-called "dimer silicates," are conveniently made by the reaction of an alcohol or phenol with hexa-chlorodisiloxane. Notable characteristics of the silicate esters are low volatility, low-temperature fluidity, and thermal stability. The hydrolytic stability varies from poor to good, depending upon chemical structure. The products are used as high-temperature heat-transfer fluids, wide-temperature range hydraulic fluids, electronic coolants, and automatic weapon lubricants.

Neopentyl Polyol Esters: These polyesters are prepared by the esterification of 5-carbon polyfunctional alcohols with monofunctional acids. Because the beta carbon of the starting alcohol does not contain hydrogen, these esters are superior in thermal stability to the diesters. Most of the other characteristics are similar to those of the diesters. As a result of their superior stability, the neopentyl esters are finding increasing use as the lubricant for aircraft turbine engines.

Polyphenyl Ethers: Both alkyl-substituted and unsubstituted polyphenyl ethers are included in this class of synthetic lubricants. General preparation involves the Ullman ether synthesis. The unsubstituted polyphenyl ethers have outstanding thermal, oxidative and radiation resistance, however, poor low-temperature characteristics are a major drawback. Alkyl substitution improves low-temperature viscosity, but detracts from stability. Most lubricant uses are developmental in nature and involve aircraft and aerospace applications.

REIGH C. GUNDERSON

LUBRICATION

Lubrication is the process of separating two rubbing solid surfaces by means of a layer which is effectively "softer" than either surface. Depending upon circumstances, the "soft" layer (the lubricant) may be a gas, a liquid, a solid, or a combination of various phases. The many ways of accomplishing lubrication can be conveniently grouped into two categories: hydrodynamic and solid lubrication.

Hydrodynamic Lubrication. The separation of moving surfaces by a "fluid" (gas, liquid, or gel) is accomplished according to the laws of hydrodynamics. One of the surfaces "swims" on the lubricant, i.e., it is lifted through the simultaneous possession of velocity relative to the lubricating fluid and of an acute angle of attack. This so-called glider bearing is the basis of design of nearly all fluid-lubricated bearings, including the journal bearing, which is just a circular glider bearing.

The basic relation which expresses the load-carrying ability of a glider bearing can be given in the form:

$$W/\eta U = a(r/h) \qquad (1)$$

where W is the load on the glider, U the velocity of the glider relative to the stator, h is the minimum distance between the rubbing surfaces, r is a dimension characterizing the angle of attack, a is a numerical coefficient (somewhere between 1 and 10) and η is the viscosity of the lubricating fluid under the temperature and pressure conditions of the application. The energy expenditure required for the service performed by the lubricant film is usually expressed in terms of the friction coefficient, f, defined as the ratio of the frictional force required to move the glider (or the journal) to the load carried by it. For the case under discussion:

$$f = K \sqrt{\eta U/W} \qquad (2)$$

where K is a numerical coefficient which depends upon the geometry of the system and varies between about 2 and 6.

While Equations (1) and (2) provide only a qualitative guide—the detailed calculations for specific bearings are quite complicated—they clearly indicate that the only variable at the disposal of the chemist, the viscosity, will be chosen such as to give the maximum load-carrying ability of the bearing consistent with a reasonable amount of frictional energy lost in the lubricant.

The low friction losses (f is usually of the order 10^{-3}) and the naturally wear-free operation generally make the maintenance of hydrodynamic lubrication the primary aim of bearing design. This goal is attained perfectly only with journal and with glider (pad) bearings and to some extent in roller bearings. It can be achieved by a special mechanism (thermal expansion of the flowing fluid) in parallel thrust bearings. Its attainment is uncertain (and not readily subject to calculation) in ball bearings and in gear lubrication.

Since the viscosity of readily available fluid varies over a 10^{10}-fold range between gases and the thickest liquids, and the temperature and pressure coefficients of viscosity vary similarly over an about 10^4-fold range, the choice of a suitable lubricant is usually determined by the ancillary conditions of temperature, volatility, chemical stability, etc. The use of sulfuric acid as lubricant for oxygen compressors is a typical illustration of a choice dictated by chemical conditions. The evaporation and deterioration of liquid lubricants in high energy particle fluxes and/or in hard vacua ($<10^{-9}$ torr) has led to increased use and rapid development of gas lubricated bearings. These call generally for extremely high and therefore very costly standards of workmanship. Hence the use of gas lubrication is at present largely restricted to precision instruments and to military and space applications.

The most important source of lubricants is petroleum, from which about 10 million tons of lubricating oils and greases are manufactured per year. All other lubricants together aggregate probably less than ½ million tons per year. There is hardly a chemical species (esters, ethers, sulfides, metal-organic compounds, etc.) which has not contributed to the array of synthetic lubricants now available. Some chemicals, such as the perfluorocarbon compounds and the siloxane polymers were originally synthesized for just this service.

Elastohydrodynamic Lubrication. Well designed and equally well built machine elements with very smooth bearing surfaces permit the imposition of very high bearing loads. The resulting elastic deformation of the bearing may then change the geometry of the load-bearing surfaces substantially. Hence the elastic properties of the load-bearing materials enter the bearing calculations. The change of lubricant properties with pressure, temperature, and with the duration of exposure to the pressure and shear regime also enter the bearing calculations. The required viscoelastic properties of lubricants at high-frequency deformations are only beginning to be determined.

Hydrostatic Lubrication. Slow-moving or even static "gliders" can be separated from the bearing surface by a fluid film if the hydrostatic pressure required to carry the load is provided by an external source, as for instance by a pump. The very large journals of turbo generators, or heavy thrust bearing pads are generally lifted by these means before the onset of rotation in order to avoid scoring damage to the valuable bearings. Exceedingly low friction coefficients are obtained in these hydrostatic bearings, the most spectacular being that of the Mt. Palomar 200-inch telescope mirror pad bearings, where $f = 0.000004$, such that the 500-ton structure is easily moved by a $\frac{1}{2}$ HP clock motor.

Solid Lubrication. While fluid film separation of rubbing surfaces is the most desirable objective of lubrication, it is often unattainable, especially when bearings are too small or unsuitable for liquid lubricants. Even bearings built for full fluid lubrication during most of their operating periods experience solid-to-solid contact when starting and stopping.

Solid surfaces in rubbing contact are characterized by friction coefficients varying between 0.04 (Teflon on steel) and >100 (pure metals *in vacuo*). Solid lubrication, in contrast to fluid lubrication, is generally accompanied by a certain amount of wear of the rubbing parts. Optical inspection of the surfaces after rubbing reveals macroscopic (i.e., bulk) damage of the metal both when unlubricated and when lubricated. Quantitative differences between the two cases are easily measured radiographically when a radioactive glider has been used. In this manner it is found that effective solid lubrication can reduce the amount of wear—compared to the dry case—by a factor as high as 10^5.

Typical solid lubricants are the soft metals lead, indium, and tin, the layer lattice crystals graphite and molybdenum disulfide, many soft organic solids, such as metallic soaps, and waxes as well as the crystalline polymers Teflon (polytetrafluorethylene), polythene (polyethylene), and nylon. The integral bonding of these solids to the surface of the hard solid to be lubricated is essential for good performance. The bonding is accomplished either by alloying (copper-lead bearing metals), by flash coating, by introduction of the lubricating solid into the interstices of the sintered metal bearing (Teflon emulsion into sintered bronze), by chemical coating (phosphate coatings), or by anchorage to a phosphate or bonded plastic coating.

Special cases of solid lubrication are boundary and EP (extreme pressure) lubrication. In both cases the solid lubricant is formed by chemical reaction of special compounds, usually applied as oil solutions, with the metallic rubbing surfaces. Typical boundary lubricants are the fatty acids which react with the metal surface to form metallic soaps which then carry the load. Strongly adsorbed but nonreacting substances of linear structure, such as long chain fatty alcohols, can also act as boundary lubricants but only under very mild conditions.

Under the very severe conditions, sometimes encountered in automobile transmissions and especially in hypoid gear differentials as well as in machining operations, only those substances act as lubricants which contain chemically active chlorine, sulfur, or phosphorus to form the corresponding iron chloride, sulfide or phosphide by instantaneous attack on the surface hot spots resulting from the collisions of surface asperities. The chemical stability of these so-called E.P. agents is designed to permit activity at the temperature near the rubbing surface, say 200°C and above, but not be corrosive under normal conditions.

Mixed Film Lubrication. Mixed film lubrication is almost invariably the true state of affairs when boundary and EP lubrication are encountered, i.e., an appreciable fraction of the load is carried by the fluid film in the "valleys" of the surface while the asperities in contact are permitted to carry the balance of the load without seizure through the beneficent intervention of the boundary or EP lubricant. The very important break-in process of rubbing surfaces consists in the controlled reduction of the number and the size of the surface asperities so that fluid lubrication will prevail for most of the time.

"Real" Lubrication. In "real" lubrication of machinery, such as automotive engines, turbines, etc., the lubricating oil has to perform many functions besides lubrication. The most important of these is the cooling of the bearings. It also has to keep internal combustion engines clean by dispersing the partial combustion products of the fuel and its own degradation products, it has to carry chemicals to counteract wear, and—in common with many other lubrication applications—it must prevent corrosion of the equipment and be inhibited against its own deterioration in service. A relatively large amount of synthetic organic chemicals is therefore carried by many oils to perform the additional functions which one must expect from a modern lubricant.

Ideally one should always have full fluid separation of rubbing surfaces. But in inaccessible locations, or reactive environments, or under conditions of very slow motion or of intermittent operation, recourse must be had to solid lubrication.

A. A. BONDI

Cross-references: *Lubricating Greases, Fatty Acids, Lubricating Oils.*

LUMINESCENCE

Luminescence denotes an emission of light which is greater than could occur from a temperature radiation alone. Under various circumstances it is known as photo-, thermo-, tribo-, cathodo-, electro-, and chemi-luminescence. In all cases, it depends upon the conversion of a compound into an unstable, excited state; the light emission arises on the return of this phase to the normal one.

Its most noteworthy forms are photoluminescence that occurs under excitation by light of shorter wave lengths, and cathodo- and electro-luminescence that result from excitation by electrons.

The light that is confined to the period of excitation is called *fluorescence;* that which persists after excitation has ceased is the *phosphorescence*, or afterglow. *Phosphor* is the name given to those compounds capable of developing fluorescence. Its emission from inorganic compounds is associated with the presence, at a very low concentration, of an impurity feature called an activator, which may be a foreign element; silver or copper at about 0.05% in zinc sulfide, and manganese at 1–2% in the silicates and phosphates of zinc, cadmium, magnesium and calcium, in the borates of zinc and cadmium, and in zinc fluoride. It may also be a foreign condition, such as monovalent zinc in zinc sulfide, or perturbed groups of atoms in the self-activated tungstates. The effectiveness of monovalent activators in zinc sulfide requires the introduction of coactivators to secure charge compensation. Equivalent amounts of chlorine or aluminum serve this purpose.

The color of fluorescence ranges over the entire spectrum, and depends upon the activator and the composition of the compound. In the zinc sulfides, addition of cadmium sulfide shifts the color from blue to red for silver activation, and from green to infrared for copper. The fluorescence due to manganese ranges from green to red in different compounds. Lead, thallium, antimony, bismuth and titanium activators generally give rise to fluorescence in the ultraviolet or blue.

For photoexcitation to occur, there must be an absorption band at the spectral location of the exciting radiation, and this radiation must be of such wave length that its energy equivalent is sufficient to raise the activator ions to their excited state. In the absence of such an absorption band, the phosphor may still be photoexcited if another activator is added, the sensitizer, which introduces the needed absorption. The phosphor then emits two bands, one due to the sensitizer and the other to the activator proper. The latter receives excitation energy from the sensitizer by a resonance step.

Phosphorescence is generally of the same color as fluorescence. It occurs in two stages. The first has an exponential decay, lasting from a few microseconds to a few milliseconds and even to a second for zinc fluoride activated with manganese. It is independent of temperature below 100°C or so. The second stage of decay is bimolecular and its persistence ranges from seconds to hours, being longest for the bismuth-activated sulfides of calcium and strontium. It is strongly temperature dependent. At low temperature, it is frozen in and is released on warming—the phenomenon of thermoluminescence. At higher temperatures it is quenched and dissipated as heat to the lattice.

Phosphors are used for illumination and for television and radar screens. Fluorescent lamps utilize a low-pressure mercury discharge whose predominant emission lies in the 2537 Å line. Phosphors responding to this excitation include calcium and magnesium tungstates for blue, zinc silicate for green, calcium halophosphates for a combined blue and orange emission, calcium silicate for red, and calcium phosphate for a still deeper red. Combinations of these are used to give the color of light desired.

Bioluminescence is also a common and spectacular form of chemiluminescence. It is exhibited by living organisms, such as bacteria, glow worms, fireflies, and luminous fish. It is due to the admixture of two types of substances present in the organism; one of them, *luciferin*, is capable of oxidation in the presence of the second, an enzyme termed luciferase. The reaction produces an excited form of luciferase, and its return to the normal form is accompanied by the emission of light.

G. R. FONDA

Electroluminescence

Electroluminescence is light generated in crystals by conversion of energy supplied by electric contacts, in the absence of incandescence, cathodo- or photoluminescence.

It occurs in several forms. The first observation of the presently most important form, "radiative recombination in p-n junctions," was made in 1907 by Round, more thoroughly by Lossev from 1923 on, when point electrodes were placed on certain silicon carbide crystals and current passed through them. Explanation and improvement of this effect became possible only after the development of modern solid state science since 1947.

If minority carriers are injected into a semiconductor, i.e., electrons are injected into a p-type material, or "holes" into n-type material, they recombine with the majority carriers, either directly via the bandgap, or through exciton states, or via impurity levels within the bandgap, thereby emitting the recombination energy as photons. Part of the recombinations occur nonradiatively, producing only heat.

Exploitation of the effect was strongly dependent on progress in compound semiconductor crystal preparation and solid state electronics, since crystal perfection (absence of defects) is of prime importance. At present, single-crystalline p-n diodes made of gallium arsenide, GaAs, a "III-V compound" (from groups III and V of the Periodic System), yield the highest efficiencies (40% of the electrical power input converted into optical power output) in the near-infrared, and diodes made of gallium phosphide, GaP, yield red light with about 8% efficiency. Very important are alloys such as $In_xGa_{1-x}P$, $Al_xGa_{1-x}As$ and $GaAs_xP_{1-x}$ where the color of luminescence can be changed by changing the composition. Wave lengths of light emitted by III-V crystals range from 6300 Å to 30μ.

An important phenomenon, "injection laser action," was discovered in GaAs diodes in 1962. The crystal faces at the ends of a p-n junction are made optically parallel so as to form a Fabry-Perot optical cavity. Beyond a certain injection current density, (the "threshold current") the individual recombination processes no longer occur randomly and independent of each other, but in phase, so that a near-parallel beam of coherent light ($\sim$9000 Å) of enormous intensity (10^7 w/cm² in pulsed operation) is emitted. The efficiency has been improved by using graded bandgap $Al_xGa_{1-x}As$ heterojunctions and special doping profiles so that the lasing region near the p-n junction acts as a "light pipe," preventing light straying out sideways. The current threshold is now reduced to 2000 amp/cm² at room temperature (continuous operation).

These coherent or incoherent electroluminescent

p-n diodes are small point sources used for pilot lights, opto-electronic data processing, ranging systems, direct-sight communication, and as IR-lamps for night vision devices.

Another kind of electroluminescence, discovered by Destriau in 1936, uses inexpensive powders consisting of small particles of essentially copper-doped zinc sulfide, ZnS, a II-VI compound, embedded into an insulating resin and formed into a large flat plate capacitor with one plate transparent (e.g., SnO_2-coated glass). If an ac-voltage is applied, light is emitted (blue, green, red, depending on the exact material composition) twice per cycle, with brightnesses up to thousands of foot-lamberts. Brightness increases linearly with drive frequency, and exponentially with voltage, until saturation occurs. The efficiency is about 1% but it decreases with increasing brightness.

Microscopic examination of the interior of an individual particle reveals that the light is emitted inhomogeneously, in the form of two sets of comet-like striations which light up alternatingly, each set once per cycle. These comets coincide with long, thin conducting copper sulfide precipitates which form along crystal imperfections. The applied field relaxes in these needles and concentrates at the tips, so that electrons and holes are alternatingly field-emitted into the surrounding insulating luminescent ZnS. The holes are trapped there until they recombine with the more mobile electrons, emitting the typical luminescent spectra.

Among applications of this large-area, thin light source are safety lights for home use, luminous instrument faces, alphanumeric and other information display panels. The deterioration of these light sources is presently still a problem, but the time to half-brightness has been improved to useable intervals.

Compared to p-n junction recombination electroluminescence the main advantage of this type is low cost.

Large-area dc-electroluminescence of polycrystalline, faintly n-type films of ZnS on glass, doped with copper and manganese, has also been achieved. The mechanism involves high-field-aided hole injection.

Still another type of electroluminescence is acceleration-collision electroluminescence. In back-biased p-n junctions, Schottky barriers, and near conducting inclusions, the local field can become high enough (10^5v/cm or more) so that electrons acquire sufficient kinetic energy to impact-ionize luminescent centers or the host lattice, creating secondaries. Electroluminescence occurs upon recombination. The efficiency is very poor, because only a small fraction of the electrons can attain sufficient energy, the others create only heat.

This type of electroluminescence occurs in all materials displaying the earlier described types, and in materials such as ZnO, $BaTiO_3$, and Zn_2SiO_4. It has been optimized in large-area ZnS polycrystalline films, with a Cu_2S cathodic film electrode. Efficiencies are 10^{-2}% or less.

A. G. FISCHER

Luminescent Inorganic Crystals

General Luminescent Characteristics of Crystals. Luminescence is the phenomenon of light emission in excess of thermal radiation. Excitation of the luminescent substance is prerequisite to luminescent emission. Photoluminescence depends upon excitation by photons; cathodoluminescence, by cathode rays; electroluminescence, by an applied voltage; triboluminescence, by mechanical means such as grinding; chemiluminescence, by utilization of the energy of chemical reactions. Luminescent emission occurs with gases, liquids and solids—both organic and inorganic. The emission involves optical transitions between electronic states which are characteristic of the radiating substance. The luminescence of inorganic crystals involves in many cases the electronic states of impurities or imperfections; in some cases, the electronic bands of perfect crystals. Inorganic crystals which luminesce are called phosphors. Impurities responsible for luminescent emission are called activators.

Luminescent excitation and emission are separate processes, for example, photoluminescence is not a scattering phenomenon but consists of absorption of radiation followed by emission. The time delay between excitation and emission is long compared to the period of the radiation λ/c, where λ is the wavelength and c is the velocity of light. This time delay distinguishes luminescence from the Raman effect, from Compton and Rayleigh scattering, and from Cherenkov emission. For luminescent inorganic crystals the radiative lifetimes of the emitting states vary from 10^{-10} to 10^{-1} second, depending on the identity of the crystal and, if impurity luminescence, also on the identity of the activator. At ordinary intensities of excitation, the spontaneous transition probability predominates so that the luminescent radiation is incoherent; with high intensities of excitation, the induced transition probability may predominate for suitable phosphors, the emitted radiation is coherent and laser action is attained.

Excitation and Emission Spectra. For most phosphors the photoluminescent excitation and emission spectra are different. Normally, there is a Stokes' shift so that the emission occurs at a longer wavelength than the excitation. In other words, in contrast to Kirchhoff's law for incandescence, luminescent emission for most crystals occurs in spectral regions where the absorption is low. The spectral distribution of individual luminescent emission bands is normally the same with all types of excitation; for example, with photo-, cathodo-, or electroluminescent excitation. These effects are qualitatively understandable on the basis of the configuration coordinate model shown in Fig. 1. In this model the Born-Oppenheimer adiabatic approximation is used as a basis for plotting the energy of each electronic state of the activator as a function of the coordinates of the nuclei comprising the activator system. Photoluminescent excitation occurs vertically in accordance with the Franck-Condon principle, maintaining the nuclear coordinates of the ground state; emission occurs following a lattice relaxation to the equilibrium coordinates which are characteristic of the emitting state. Both the excitation and emission bands are broadened because of thermal vibrations at ordinary temperatures and the zero point vibrations at low temperatures.

Fluorescence and Phosphorescence. As noted earlier, the initial persistence of luminescent emission

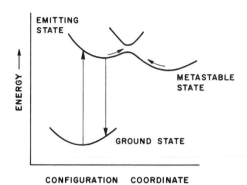

FIG. 1. Configuration coordinate model for activator systems.

following removal of excitation is a matter of the lifetime of the emitting state. This emission decays exponentially and is often called fluorescence. With many luminescent inorganic crystals there is an additional component to the afterglow which decays more slowly and with more complex kinetics. This component is called phosphorescence. The emission spectra for fluorescence and phosphorescence are the same for most phosphors; the difference in afterglow arises from trapping states from which thermal activation is prerequisite to emission. In some cases the trapping state is a metastable state of the activator; in other cases, is a state of another imperfection. A metastable state of the activator is included in Fig. 1.

Luminescence of Perfect Crystals and of Activator Systems. The luminescence of most phosphors originates from impurities or imperfections, however, there are also phosphors whose luminescence is characteristic of the perfect crystals. The latter include the alkaline earth tungstates whose luminescence is characteristic of the $WO_4^=$ group perturbed by the crystal field, rare earth salts which emit in narrow bands or lines characteristic of transitions within the 4f shell, and intrinsic semiconductors in which radiative recombination occurs between a conduction electron and a valence band hole coupled to form an exciton.

The impurities and imperfections which form activator systems in inorganic crystals are of diverse atomic and molecular types whose characteristics depend on the structure of the defect, on the electronic states of the defect and on the electronic structure of the pure crystal. Point defects which have been identified as capable of luminescence include lattice vacancies and substitutional impurities. Associated defects capable of luminescence include pairs of oppositely-charged point defects, that is, donors and acceptors. A donor-acceptor pair radiates following capture of a conduction electron and a valence band hole. In some cases the electronic states which participate in the luminescence of inorganic crystals can be described in terms of the energy levels of the impurity ion perturbed by the crystal field; in other cases, in terms of the crystal band structure perturbed by the impurity.

Ionic Crystals. The alkali halides are simple ionic crystals which become luminescent when doped

with suitable impurities. Tl^+ substituted at cation sites in alkali halides exhibits characteristic absorption and emission bands. The absorption bands correspond to the $^1S \rightarrow {}^3P, {}^{0,1}P^{\circ}$ transitions of the free ion perturbed by crystal interactions; the principal emission band, $^3P^{\circ} \rightarrow {}^1S$, similarly perturbed. The spectra can be understood qualitatively with the aid of Fig. 1 and with the configuration coordinate interpreted as the displacement of the nearest-neighbor halide ions of the Tl^+. It is the nearest-neighbor interaction which is strongly dependent on the electronic state of the Tl^+ and therefore is primarily responsible for the band widths and the Stokes' shift. In^+, Ga^+, Pb^{+2} and other impurities with electronic structures similar to Tl^+ have been shown to be activators in alkali halide crystals. In addition, alkali halides can be made luminescent by the introduction of imperfections. For example, the F-center which consists of an electron trapped at a halide ion vacancy has a characteristic infrared luminescence at low temperatures. Also, the self-trapped positive hole or V_k center has been shown to luminesce following electron capture.

Many inorganic crystals become luminescent when certain transiton metal ions are dissolved in them. The luminescence involves intercombination transitions within the 3d shell, therefore crystal field theory can be used to interpret the absorption and emission spectra. Divalent manganese is a common activator ion. Zn_2SiO_4, ZnS and $3Ca_3(PO_4)_2 \cdot Ca(F,Cl)_2$ are important phosphors activated with Mn^{+2}. The last, activated also with Sb^{+3}, is the principal fluorescent lamp phosphor. The excitation at 254 nm from the Hg discharge occurs at the Sb^{+3}, whose energy level structure is similar to that of Tl^+ modified by local charge compensation by an O^{-2} at a nearest-neighbor halide site; part of the energy is radiated in a blue band due to Sb^{+3}, and part is transferred to the Mn^{+2} which is responsible for an orange emission band. The emission spectrum of this important phosphor is shown in Fig. 2. The ruby laser involves the luminescence of Cr^{+3} in Al_2O_3. Excitation occurs in a broad absorption band, the system relaxes to another excited state from which emission occurs in a narrow band.

Rare earth ions in solid solution in inorganic crystals frequently exhibit the emission characteristic of transitions within the 4f shell. For examples, samarium, europium, and terbium give visible emission; neodymium, infrared; and gadolinium,

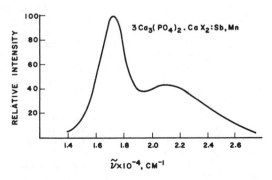

FIG. 2. Emission spectrum of manganese- and antimony-activated calcium halo-phosphate.

ultraviolet. Because of their narrow emission bands, the rare earth activated phosphors are of interest as lasers and photon counters. Crystal field theory can be used to explain the optical absorption and luminescent emission of the 4f transitions of rare earth ions in crystals. The rare earth ion is trivalent in most rare earth laser phosphors. Examples are Y_2O_3:Eu^{+3}, $Y_3Al_5O_{12}$:Yb^{+3} and $CaWO_4$: Er^{+3}. In the last material the trivalent rare earth is charge compensated by an equivalent concentration of monovalent cations such as Na^+. For some materials photo excitation or optical pumping involves transitions within the rare earth, i.e., transitions from the $4f^n$ to the $4f^{n-1}5d$ configurations, and for others, optical absorption by the matrix followed by energy transfer to the rare earth. Sensitized emission may also occur with suitable combinations of rare earths. An example is $CaMoO_4$:Er^{+3},Tm^{+3} where the Tm^{+3} emission is excited by energy transfer from excited Er^{+3}. In addition to crystals, glasses such as barium crown glass containing Nd^{+3} are capable of coherent luminescent emission and are used as high-power lasers.

Anti-Stokes' emission of visible radiation with infrared excitation occurs with suitable doubly rare earth doped crystals, for example, GdF_3:Yb^{+3},Tm^{+3}. Triple, consecutive energy transfer from Yb^{+3} to Tm^{+3} takes place so that the Tm^{+3} is triply excited and emits a single energetic photon in returning to its ground state.

Covalent Phosphors. The zinc sulfide phosphors, which are widely used as cathodoluminescent phosphors and are well-known for their electroluminescence, are large band gap, compound semiconductors. Two impurities or imperfections are essential to the luminescence of many of these phosphors: an activator which determines the emission spectrum, and a coactivator which is essential for the emission but in most cases has no effect on the spectrum. Activator ions such as Cu, Ag and Au substitute at Zn sites and perturb a series of electronic states upward from the valence band edge. In a neutral crystal containing only these activator impurities, the highest state is empty, that is, contains a positive hole, can accept an electron from the valence band, and therefore, in semiconductor notation the activator is an acceptor. In a similar way coactivators such as Ga or In at Zn sites or Cl at S sites are donors in ZnS. The simultaneous introduction of both types of impurities results in electron transfer from donor to acceptor lowering the energy of the crystal and leaving both impurities charged. The coulomb attraction of the donor and acceptor leads to a departure from a random distribution over lattice sites and to pairing. The electronic states, and the band-to-band intrinsic transition and some of the transitions of acceptors, donors and donor-acceptor pairs are shown in Fig. 3. The emission spectrum of ZnS:Ag,In is shown in Fig. 4. The longer wavelength emission band involves the transition from the lowest donor state to highest acceptor state (transition 5) in approximately fifth nearest-neighbor pairs; the shorter wave length emission corresponds more nearly to transition 3 of Fig. 3. In addition to luminescence due to donors, acceptors and the pairs, emission bands due to transition metals are well-known for zinc sulfide as noted earlier. In zinc sulfide crystals

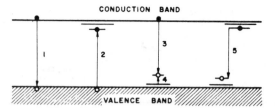

FIG. 3. Energy levels and transistor in semiconducting phosphors.

the donors which are unassociated with acceptors serve as electron traps and are responsible for phosphorescence.

The luminescence of Group III-Group V semiconductors has been extensively investigated during the past decade. In contrast to the broad gaussian emission of II-VI phosphors such as ZnS, the emission of III-V semiconductors such as GaP may consist of scores of lines. These lines have been identified as donor-acceptor transitions (transition 5 of Fig. 3) in pairs with different donor-acceptor distances. Two types of pair spectra are observed: those arising from donor and acceptor occupying equivalent sites, i.e., Si and S at P sites in GaP, and those arising from donor and acceptor occupying nonequivalent sites, i.e., Zn at Ga sites and S at P sites in GaP. Luminescent emission from the 11th to the 68th nearest-neighbor pairs are observed with theoretically predicted differences in their transition energies. In addition, luminescent emission at isoelectronic dopants (impurities with the same charge as those substituted for) is observed for III-V phosphors, for example GaP:N. The III-V semiconductors can be made either n- or p- type, that is, be doped so that electrical conduction is by electrons in the conduction band or by positive holes (empty electron states) in the valence band and therefore p-n junctions can be formed. With radiative re-

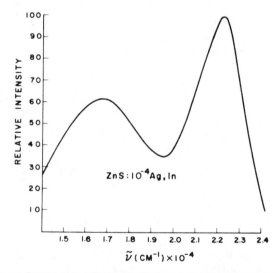

FIG. 4. Emission spectrum of zinc sulfide containing donors and acceptors.

combination centers such as donor-acceptor pairs in the junction electroluminescence occurs with forward bias. This is a light emitting diode. Both GaP and Ga(P,As) doped with pairs, and possibly isoelectronic dopants as well, are typical light emitting diodes.

Coherent luminescent emission is obtained from appropriate semiconductor phosphors with photo-, cathodo- or electroluminescent excitation. Lasers with electrical injection of charge carriers include In(P,As) and Ga(P,As), whereas cathodoluminescent lasers include CdS, ZnO and ZnS. The latter involve exciton emission (transition 1 of Fig. 3 modified for coupled electron and positive hole). Semiconductor lasers can also be pumped optically.

FERD WILLIAMS

References

Curie, D., (translated by G. F. J. Garlick) "Luminescence in Crystals," John Wiley and Sons, Inc., New York, 1963.

Goldberg, P., Ed., "Luminescence of Inorganic Solids," Academic Press, Inc., New York, 1966.

Williams, F., Ed., "Proceedings of the International Conference on Luminescence," North Holland Publishing Company, Amsterdam, 1970.

M

MAGNESIUM AND COMPOUNDS

Magnesium is in Group IIA of the Periodic Table of the elements, and has an atomic number of 12 and an atomic weight of 24.305. It has three isotopic forms with mass numbers of 24, 25 and 26.

With a specific gravity of only 1.74 it has long been recognized as the world's lightest structural metal. Aluminum, its nearest rival for lightweight honors, weighs 1½ times more while steel weighs approximately 4½ times more. Magnesium also has the distinction of being one of the most abundant elements in the earth's crust. It is the eighth most abundant chemical element, the sixth most abundant metal and the third most abundant structural metal, being exceeded only by aluminum and iron. Magnesium is readily available because sea water contains a substantial amount of magnesium chloride, and is one of the few materials that can be considered as being obtainable in virtually unlimited quantities. A practical process for the extraction of magnesium from sea water was developed by The Dow Chemical Company and a production plant was built at Freeport, Texas at the beginning of World War II. In addition to sea water and brines (Great Salt Lake) as a source of supply of magnesium and its compounds, there are several magnesium-rich minerals such as dolomite, magnesite and brucite which are very abundant in the U.S.

Production Processes. The electrolytic reduction of magnesium chloride has been and still is the principal method of producing magnesium in the U.S. The electrolytic method was the only means of producing magnesium from 1915, when the magnesium industry began in this country, until 1941 when a direct reduction process developed by F. J. Hansgirg of Austria was used in a plant built in Permanente, California. Shortly thereafter, another and more successful thermal reduction process using ferrosilicon to reduce MgO was developed and went into production. The inventor was L. M. Pidgeon of the Canadian National Research Laboratories.

The sea water process developed by Dow utilizes electrolytic reduction of magnesium chloride as the final step. In this method, ordinary sea water which contains approximately 0.13% magnesium is pumped into huge settling tanks where it is treated with lime. The calcium from the lime replaces the magnesium in the sea water forming magnesium hydroxide which settles to the bottom of the tanks. This magnesium hydroxide is then drawn off as a slurry which is filtered to remove the sea water. The magnesium hydroxide filter cake is then reconverted to a slurry with fresh water and transferred to a neutralizer where it is reacted with hydrochloric acid to produce a solution of magnesium chloride. The magnesium chloride solution is then transferred to evaporators which remove the water leaving dry magnesium chloride which goes to the electrolytic cells to be converted to metallic magnesium and chlorine gas. The chlorine goes back into the process in the form of hydrochloric acid and is used to neutralize magnesium hydroxide. Another firm, the American Magnesium Company, is using an electrolytic process to obtain magnesium from brines at Snyder, Texas. National Lead Company has a plant in Utah which employs an electrolytic process to obtain magnesium from the salt beds of Great Salt Lake. In 1972 the installed U.S. magnesium capacity was about 195,000 tons/year, and forecasts put the 1975 consumption at 145,000 tons.

Physical Properties. Magnesium, Mg, has a density of 1.74 g/cc; melting point, 650°C; boiling point, 1110°C; specific heat, 0.245 cal/g/°C (20°C); latent heat of fusion, 88 cal/g; latent heat of vaporization, 1260 cal/g; heat of combustion, 5980 cal/g; coefficient of thermal expansion, 26.1×10^{-6}; electrical resistivity, 4.46 microhm-cm (20°C); vapor pressure, 10 mm at 702°C, 100 at 909, and 400 at 1034; thermal conductivity, 0.37 cal/cm/cm²/sec/°C (20°C); surface tension, 563 dyne/cm (681°C); and crystal structure, close-packed hexagonal.

Chemical Properties. Magnesium forms only divalent compounds which are very stable with high heats of formation. Although it is very reactive at elevated temperatures and a powerful reducing agent when heated, magnesium at ordinary temperatures is not attacked by air, water and most alkaline substances. It reacts with most acids but displays excellent stability to chromic and hydrofluoric acids. It can be used with many organic compounds; however, it reacts chemically with methyl alcohol.

The reaction between magnesium and salt solutions is varied. There is little or no attack by alkali metal and alkaline earth metal chromates, fluorides and nitrates, but solutions of chlorides, bromides, iodides and sulfates of these metals usually corrode magnesium. Because of magnesium's electrode potential of -2.363 volts, it displaces heavy metals from solution with consequent corrosion of the magnesium.

Magnesium unites directly with many other elements, such as the halogens, sulfur, nitrogen, oxygen, phosphorus, boron and silicon, the rate of reaction depending on the temperature. It burns rapidly in air when heated, especially when in powder or ribbon form.

Magnesium Alloys. Magnesium, like most pure metals, requires the addition of alloying ingredients to strengthen and harden it for structural purposes.

Various alloy compositions have been developed for use in cast and wrought forms and for special applications. The principal alloying ingredients for most commonly used alloys are aluminum, manganese and zinc. For special properties, such as high creep strength or for extra strength qualities at room temperature, additional alloying ingredients are used. The rare earth metals and thorium provide added strength at elevated temperatures; the use of combinations of zinc and zirconium provide wrought alloys with improved strength at room temperature.

Mechanical Properties. The properties and characteristics of magnesium and its alloys determine the selection of the metal for a variety of commercial applications. Its light weight and relatively high strength make it a good construction material.

Magnesium possesses the advantage that it is worked by practically all methods known to the metalworking art. Magnesium alloys can be cast by sand, permanent mold and die-casting methods. They are readily extruded into a practically unlimited variety of shapes including round rods, bars, tubes, structural and special shapes. The extrusion process permits the efficient and economical production of shapes, many of which cannot be produced by other methods, thereby simplifying design problems. The excellent workability of magnesium allows it to be rolled on high production mills into plate and sheet ranging from heavy gauge down to 0.016″. Magnesium alloys can be readily forged to provide high-strength, lightweight parts with good pressure tightness. Both press and hammer forging methods are used, but press forging is by far the most commonly employed method. All forging operations are performed at elevated temperatures.

Uses. The uses for magnesium are conveniently divided into two types—structural and nonstructural. The structural uses include applications in the form of magnesium alloys in aircraft for fuselage, wing and empennage parts, landing wheels, engine parts and accessories. The portable tool industry utilizes magnesium castings for chain saws, drills, impact hammers, grease guns and many other manually handled articles of this type. Hand trucks, dockboards, grain shovels, gravity conveyors, foundry equipment and many other products in the materials handling field make good use of magnesium's lightness. Textile and printing machines and similar types of industrial machinery employ magnesium lightness to advantage. Household goods, office equipment, instruments of many kinds, sporting goods and transportation vehicles are other examples of well-known applications where magnesium is playing an increasingly important part.

Nonstructural uses of magnesium include its use as an alloying ingredient in aluminum, lead, zinc and certain other nonferrous alloys. It is also used as a deoxidizer and desulfurizer in the refining of nickel and copper alloys. Grey iron foundries make use of magnesium as an addition agent to produce the ductile cast iron. Other important nonstructural uses for magnesium are in the Grignard reaction of organic chemistry and in electrochemical applications, including batteries and the cathodic protection of underground pipelines, domestic water heater tanks, ship bottoms and similar installations where other metals must be protected

against corrosion. Magnesium also finds considerable application in certain metallurgical processes such as the Betterton-Kroll process for debismuthizing lead. Magnesium is used as a reducing agent in the production of metallic beryllium, titanium, uranium, and zirconium. It is also used in finely divided form in fireworks and specialized military devices.

Magnesium Compounds. Magnesium forms a series of compounds equal in variety to those of sodium and calcium. Almost all are soluble in water, the important exceptions being the oxide, carbonate, hydroxide, phosphate, pyrophosphate, arsenate, fluoride, stearate and silicate.

Compounds of magnesium are used commercially in a variety of industries. These include such manufacturing and processing operations as: the manufacture of iron and steel, cement, fertilizers, industrial insulation, rubber, paper, refractories, industrial chemicals; pharmaceuticals, paints, glass, ceramics; the processing of textiles, sugar, leather, and oil; and the production of magnesium and uranium.

The compounds of magnesium which find greatest commercial usage include the oxide (commercially referred to in the trade as magnesia), carbonate, hydroxide, chloride and sulfate (epsom salt). These represent the outstanding commercially used compounds and those which are produced in relatively large quantity. Numerous other magnesium compounds—both inorganic and organic are used by industry but in relatively small quantity compared to the compounds mentioned. Listed here are some typical specific uses of the common magnesium compounds:

Compound	Uses
magnesium carbonate	glass manufacture, inks, ceramics, fertilizers, insulation, industrial chemicals, rubber products
magnesium chloride	flocculating agent, catalyst, foliage treatment for fire resistance and prevention of drying, ceramics, paper and production of metallic magnesium
magnesium hydroxide	sulfite pulp, residual fuel oil additive, uranium processing, medicine, sugar refining
magnesium oxide (magnesia)	refractories, cement, insulation, rubber, fertilizers, rayon processing, paper manufacture
magnesium sulfate (epsom salt)	leather tanning, mordant assist in textile dying, paper sizing, cement, ceramics, explosives, match manufacture, medicines and fertilizers

WILLIAM H. GROSS

References

Beck, A., "The Technology of Magnesium and Its Alloys," translation of "Magnesium und seine

Legierungen," 2nd ed., 512 pp., F. A. Hughest & Co. Limited (London) 1941.

Gross, W. H., "The Story of Magnesium," 1st Ed., 258 pp., American Society for Metals, Metals Park, Ohio, 1959.

Roberts, C. Sheldon, "Magnesium and Its Alloys," 230 pp., John Wiley & Sons, Inc., New York, 1960.

Emley, E. F., "Principles of Magnesium Technology," 1013 pp., Pergamon Press Ltd., London, 1966.

Comstock, H., "Magnesium and Magnesium Compounds," a Materials Survey, 128 pp., U.S. Department of the Interior, Bureau of Mines, Washington, D.C., 1963.

Gross, W. H., "Magnesium," pp. 379–389, "Encyclopedia on the Chemical Elements," Hampel, C. A., Editor, Van Nostrand Reinhold Co., New York, 1968.

MAGNETOCHEMISTRY

Magnetism applied to chemical compounds and particularly to crystals is known as magnetochemistry. This discipline helps to characterize chemical compounds from both a structural and a compositional point of view. The analytical potential of magnetochemistry is, however, restricted almost entirely to compounds containing paramagnetic ions or atoms present in magnetic spin ordered and disordered systems. Although some diamagnetic organic crystals have strong diamagnetic anisotropic properties, nevertheless, it would be difficult to identify such radicals by using magnetic measuring methods.

Most of the elements in the Periodic Table are diamagnetic because they have closed electron shells. The magnetic susceptibility, χ, is defined as the ratio between the moment M and the intensity of the magnetic field, H, and is positive for paramagnetic and negative for diamagnetic moments. The latter are much smaller and virtually independent of temperature. The small magnitude of the diamagnetic moments makes them less attractive for magnetochemical studies.

The paramagnetic moments, however, are much larger, and in spin-ordered systems such as ferromagnets they can reach very high values. We shall review first the paramagnetic ions in condensed substances present as spin-disordered systems.

The Curie law states that χ is inversely proportional to the temperature. But χ is also related to the total angular momentum J and in some situations is given as: $\chi = \mathbf{N}g^2\mu^2_B [J(J + 1)]3kT$ where $\mathbf{N}$ is the Avogadro number; g is the spectroscopic splitting factor given as $g = 1 + [J(J + 1) + S(S + 1) - L(L + 1)]/2J(J + 1)$; μ_B is the Bohr magneton, a conversion constant equal to 9.22×10^{-31} erg/gauss, where gauss is the absolute unit for magnetic field or magnetic moment; k is Boltzmann's constant; and T is the absolute temperature. This value of χ is valid for the rare earth free ions in the gaseous phase, with the exception of Eu^{3+} and Sm^{3+}.

If the paramagnetic ion is measured in a crystal, the value of χ will change according to the following variation which distinguishes the transition elements $3d$, $4d$ and $5d$ from rare earth $4f$ and actinide $5f$ elements: (1) the magnitude of the spin-orbit coupling, λ, and (2) the electrostatic interaction between the ions and the crystal known as the crystalline field intensity, CF. For $3d$, $4d$ and $5d$ ions $CF > \lambda$, therefore the orbital magnetic moment L will have a very small contribution to J and the paramagnetic moment of these ions will have a value close to that of the "spin-only" S instead of the $J = L \pm S$ value. This statement is known as the "quenching" of the orbital moment. The value of $\lambda (5d) > \lambda (4d) > \lambda (3d)$ and is of the order of $10^2 - 10^8$cm^{-1}. Therefore, the orbital contribution of the $5d$ ions is much stronger than that for the $3d$ ion and the spin only approximation would fail to estimate the magnetic moment.

The magnitude of low-symmetry CF mainly for $4d$ and $5d$ ions is such that in certain cases even the spin moment are "quenched." In that case we speak about low-spin moments, where both L and S are quenched. The quenching of S is more complete for even than for odd numbers of electrons. For example, the even ion Rh^{3+} has a $4d^6$ configuration. According to Hund's multiplicity rule, the 6 electrons should have arranged themselves so as to yield a maximum value, i.e., $+5 -1 = 4\,\mu_B$. Instead, due to the high CF they arrange in the $+3 - 3 = 0\,\mu_B$ configuration and Rh^{3+} acts as a diamagnetic ion.

The $4f$ ions have $\lambda > CF$ and therefore χ is proportional to $J(J + 1)$. The CF, nevertheless, splits the J multiplet into $2J + 1$ sublevels lumped together as singlets, doublets, triplets or quartets. At very low T only part of the split levels of J will be occupied and contribute to χ which in certain cases will be much lower than its corresponding free ion value. The splitting of J is not very large and at room temperature or higher T, the χ of the rare earths will be close to that of the free ion.

The value of χ for Eu^{3+} does not follow the Curie law at low T. Eu^{3+} has a configuration of $4f^6$ and its ground state $J = L - S = 3 - 6(1/2) = 0$ should yield $\chi = 0$ at low T. Van Vleck was the first to realize that besides the first-order paramagnetism resulting from the J there is also a second-order paramagnetism κ which is independent of temperature and inversely proportional to the splitting Δ which separates the nonmagnetic ground state $J = 0$ from the first excited magnetic level $J = 1$. For Eu^{3+} this splitting is of only about 500°K, and it therefore yields a strong term κ independent of T up to about 100°K. The Sm^{3+} has also a strong κ term, since its ground state of $J = 5/2$ is separated by about 1500°K from $J = 7/2$.

Even numbers of unpaired $4f$ electron ions may have part of the ground-state multiplet split by CF so that the ground state is a nonmagnetic singlet. In such a case χ at very low T of about 10°K will also be independent of T and its magnitude will be proportional to Δ^{-1}. A typical value of Δ is of the order of about 20°K. At $T = 300$°K the value of χ even for these ions will be close to that of the free ion because the total splitting of the ground state J into its $2J + 1$ components is 100–500°K.

The light $5f$ actinide ions have strong $6d$ character, i.e., it seems that the $6d$ shell is partly occupied and the $5f$ shell is empty. This character changes for Np, Am and the heavier actinides which behave like $4f$ rather than $5d$ ions.

While paramagnetism is found in most oxidation

states of most transition ions, it also occurs in some molecular compounds. Examples of molecular paramagnetism are found in nitric oxide, nitrogen dioxide, triarylmethyls, and a surprisingly varied group of inorganic and organic substances. To this group we may add the superoxide ion O^-_2. All these substances owe their paramagnetism to the presence of the molecule of an odd number of electrons. The magnetic moment of such substances is, therefore, generally about 1.8 Bohr magnetons, although sometimes this is dependent on temperature.

Molecular paramagnetism has proved useful in the detection and estimation of free radicals and has many times helped to establish molecular configuration both in simple and in extremely complex systems. New developments in the technique of detecting paramagnetism have made the method one of quite extraordinary power.

A few molecular substances have an even number of electrons but, for one reason or another, two of the electrons remain unpaired and this gives rise to a fairly strong paramagnetism. This occurs in molecular oxygen and in a series of aromatic substances related to the chichibabin hydrocarbon. Such substances are known as biradicals. The existence of strong paramagnetism in molecular oxygen makes possible the use of magnetic susceptibility in the quantitative estimation of oxygen as in a variety of commercial oxygen meters.

If the concentration of the paramagnetic ions is high, then at low T the exchange forces can keep the spins of the paramagnetic ions ordered in the crystal. The simplest magnetic ordering is ferromagnetism, where the spins are parallel such as in the metals of Gd, Fe, Co, Ni and many of their alloys, in oxides or chalcogenides such as EuO, or EuS, $La_{1-x}Sr_xMnO_3$, Cd Cr_2Se_4 or CrO_2. Another possible ordering is collinear antiferromagnetism where the spins are aligned in an antiparallel arrangement cancelling their magnetic moments at $T = 0$ completely. Typical examples are the rocksalt oxides of MnO, CoO or FeO where the magnetic unit cell in the antiferromagnetic state is double that of the chemical one and the spins of one sublattice are oriented antiparallel to that of opposite one. Ferrimagnetism is the name given for uncompensated antiferromagnetism yielding a resultant magnetic moment. For example, Gd_3Fe_2 Fe_3O_{12} has three magnetic sublattices and the magnetic moment at $T = 0$ for Gd^3 and Fe^{3+} are 7 and 5 μ_B respectively. In a ferromagnetic arrangement they would yield a moment of $(3 \times 7) + (2 \times 5) + (3 \times 5) = 46 \mu_B$. Since, however, the Gd^{3+} sublattice is antiparallel oriented to that of the $3Fe^{3+}$ and parallel to the $2Fe^{3+}$ the moment is $(3 \times 7) + (2 \times 5) - (3 + 5) = 16 \mu_B$. The different temperature dependence of the three sublattices may lead to a temperature where the antiparallel magnetizations are equal and above that temperature, called *compensation point,* the resultant magnetic moment changes sign.

The three collinear spin ordered structures of ferroantiferro- and ferrimagnetism do also exist in nonlinear, screw-type arrangements or slightly canted antiferromagnetic systems. For example, hematite, α-Fe_2O_3 at room temperature has a weak ferromagnetic moment of about 1% that of similar ferrimagnetic compounds due to a small angle of canting of about 1° from perfect collinear antiferromagnetism.

All spin-ordered systems disorder at a critical temperature known as the Curie temperature for ferromagnetic and as the Neel temperature for antiferromagnetic compounds. The spin disordered state is the paramagnetic state discussed above.

The limitation of the total magnitude of the magnetic moment of the paramagnetic ion due to CF or λ effects, such as "quenching" of L or S is valid also in the ordered state. Therefore, a working knowledge of magnetochemistry is useful to characterize the valence state of paramagnetic ions or radicals in any compound which contains these ions.

For many useful references on the subject treated in this review, the reader is referred to the book: "Experimental Magnetochemistry" by M. Schieber, North Holland Publ. Co., Amsterdam, 1967.

M. SCHIEBER

Magnetic Phenomena

The measurement of magnetic susceptibility and the magnetic moment derived therefrom are widely used by the chemist in structural interpretations and in determining the concentrations of certain types of species. These magnetic quantities can be defined only in terms of certain basic concepts, which are given below.

Unit Pole. Unlike an electric charge of either sign ($+$ or $-$), a single magnetic pole (north or south) cannot be isolated. However, the purely fictitious concept of a unit pole helps to develop other useful quantitative aspects of magnetism. A unit pole may be defined as one which repels an equal and similar pole, placed 1 cm away *in vacuo,* with a force of 1 dyne. The force between two poles is governed by Coulomb's law.

Intensity or strength of a magnetic field. If a unit pole is placed at a fixed point *in vacuo* in a magnetic field, it is acted upon by a force which is taken as a measure of the intensity or strength of the magnetic field. It follows from the preceding definitions that unit magnetic intensity exists at a point where the force on a unit pole is 1 dyne. This unit magnetic intensity was formerly called the "gauss," and is so called even today by many manufacturers and users of magnets. According to the recommendations of the International Conference on Physics at London (1934), the term "Oersted" is used instead. Some writers use the abbreviation "oe." A smaller unit, the "gamma" (γ) is equivalent to 10^{-5} Oersted.

Magnetic field intensity or the magnetizing force is measured by the space rate of variation of magnetic potential. This unit is designated as the "gilbert per cm" and is the same as the oersted.

Magnetic moment. This is a term that is probably most widely known to chemists but one whose physical significance is least understood. Like the "moment of a force," the magnetic moment refers to the turning effect produced under certain conditions. When a magnetic dipole is placed in a magnetic field, it experiences a turning effect, which is proportional to a specific character termed the magnetic moment. If a field of strength H acts on a dipole N-S of length l and strength m, the N and S poles of the dipole will experience forces equal to +mH and −mH, respectively. These two equal and opposite forces constitute a couple whose turning moment M is given by

$$M = \text{force} \times \text{distance}$$
$$= mH \times l \sin \theta$$
$$= \mu H \sin \theta$$

where θ is the angle between the magnetic dipole and the direction of the applied field. Thus, the quantity $\mu(=ml)$ defines the magnetic moment and serves as a measure of the turning effect. It is expressed in dyne-cm/oersted or ergs/oersted. Although no practical unit for magnetic moment has been formulated, experiments with the basic electrical and magnetic properties of fundamental particles have revealed the existence of a fundamental unit of magnetic moment, the Bohr magneton. This is just as real a quantity as the charge of an electron, and may be placed among the "universal constants." Often abbreviated "BM" or given the symbol μ_B, this is equal to eh/4πmc, where e is the charge and m the mass of the electron, h is Planck's constant, and c is the velocity of light. Introducing the values of these quantities gives $\mu_B = 9.27 \times 10^{21}$ ergs/oersted. In early work the "Weiss magneton" was used; 1 Bohr magneton is equivalent to 4.97 Weiss magnetons.

Intensity of magnetization. The amount of pole strength induced over unit area represents intensity of magnetization. Thus,

$$I = m/A$$

where m is the induced pole strength over a total area of A cm². An alternative definition is obtained by multiplying both numerator and denominator by the distance l; this gives,

$$I = ml/Al = \mu/\text{volume}$$

or magnetic moment per unit volume.

Gauss' law and magnetic induction. According to the definition of field intensity, in a field of unity strength one line of force must pass through every square centimeter. If one considers a sphere of 1 cm radius (4π cm² surface area) enclosing a unit pole at its center, it follows that 4π unit lines of force emanate from a unit pole. Gauss' law states the total magnetic induction over a closed surface is 4π times the amount of pole enclosed. Hence, for a pole m, $4\pi m$ maxwells emanate from its surface.

A bar of unmagnetized material will become magnetized when placed in a uniform magnetic field. Consider a unit surface A within the material at right angles to the direction of the applied field H. If I is the intensity of magnetization induced, there will be $4\pi I$ unit lines of force across the unit surface. In addition, there will be H lines of force of the applied magnetic field superimposed on the induced magnetization. Therefore, the magnetic induction B, representing the total number of lines of force across the unit surface, is given by

$$B = 4\pi I + H$$

With a vacuum in place of the magnetic material, one would have simply $B = H$, because the mag-

netic permeability of vacuum is taken as 1. It also follows that, for a magnetic material of permeability μ, the magnetic induction B will be given by

$$B = \mu H$$

The cgsm (cm − gram- sec magnetic) unit of magnetic induction of flux density is the gauss; because of the dimensionless character of permeability, B, and H have the same dimensions in the electromagnetic system. As stated before, it is customary to express H in oersteds.

Electronic and nuclear magnetic susceptibility. The magnetic susceptibility observed in bulk matter represents the contributions from both the electrons and nuclei within the system (atoms, ions, molecules and free radicals). However, the contributions of the nuclei to the susceptibility ($\sim 10^{-10}$ cgs units) are negligible in comparison with those of the electrons, and therefore, the term "magnetic susceptibility" is usually taken to represent only the electronic property, since this results from the spin and orbital moments of electrons. However, some authors prefer the term "electronic susceptibility" to distinguish it from "nuclear susceptibility." The latter is usually encountered in discussions of nuclear magnetic resonance absorption spectroscopy, where the emphasis is naturally on the magnetic properties of nuclei.

Magnetic susceptibility. The intensity of magnetization, I, induced at any point in a body is proportional to the strength of the applied field H:

$$I = \kappa H \text{ (or } \kappa = I/H)$$

where κ is a constant of proportionality depending on the material of the body. It is called the magnetic susceptibility per unit volume, and may be defined qualitatively as the extent to which a material is susceptible to induced magnetization. For an isotropic body the susceptibility is the same in all directions. However, for anisotropic crystals the susceptibility along the three principal magnetic axes are different, and measurements on their powder samples give the average of the three values.

Magnetic susceptibility is obviously related to magnetic permeability, and the following relationships may be derived.

As was shown before,

$$B = \mu H = 4\pi I + H$$

Therefore

$$\mu = 4\pi I/H + 1 = \pi \kappa + 1$$

or

$$\kappa = (\mu - l)/4\pi$$

It should be noted that, since κ is a ratio of the intensity of magnetization, I, to the intensity of the applied field, H, the susceptibility κ should strictly be a dimensionless quantity if I and H are expressed in the same units. However, magnetic susceptibility is still expressed in terms of "cgs units" (or "cgs emu" units), more as a matter of convention than of scientific thought. This is so because there are uncertainties in both the measurement and the units of magnetic permeability which have led to confusion over whether or not B and H are quantities of the same kind. In order to simplify matters, the convention of expressing susceptibility in "cgs units" will be followed here.

Mass (or specific), Atomic, and Molar Susceptibilities.
If γ is the density of a material, then the susceptibility for 1 cm³ (or γ g) is equal to the volume susceptibility κ. Hence, the susceptibility per gram of the material, called the mass or specific susceptibility χ, is given by

$$\chi = \kappa/\rho$$

The atomic susceptibility χ_A and the molar susceptibility χ_M are simply defined as the susceptibility per gram-atom and per gram-mole, respectively. Hence

$$\chi_A = \chi \times \text{atomic weight}$$

$$\chi_M = \chi \times \text{molecular weight}$$

The "ionic susceptibility" is similarly defined as the susceptibility per gram-ion.

The magnetic susceptibilities are occasionally expressed in units of the rationalized Georgi system based on the mks (meter-kilogram-second) system. For volume susceptibility K, the ratio of units is

$$\text{Georgi/cgs} = 4\pi$$

For mass susceptibility χ, the ratio of units is Georgi/cgs = $4\pi \times 10^{-3}$ since the Georgi system employs density in units of kg/m³.

L. N. MULAY

Cross-reference: *Magnetochemistry.*

MANGANESE

Manganese, a hard brittle metal melting at 1245°C, was first recognized as an element by the Swedish chemist Scheele in 1774 while working with pyrolusite, the MnO_2 ore, and was isolated by his associate, Gahn, in the same year. Frequently found in conjunction with iron ores, the metal was named for the magnetic properties exhibited by pyrolusite from the Latin *magnes*, or magnet; the German equivalent is *Mangan* and the French, *mangané*.

Manganese is essential in steel manufacture for the control of sulfur content, and today this application accounts for the major portion of the manganese consumed in all forms in this country. Somewhat less than 14 pounds of manganese, chiefly in the form of ferromanganese, is used for each ton of steel produced.

The element, symbol Mn, atomic number 25, atomic weight 54.938, is located in Group VIIB of the Periodic Table between chromium and iron horizontally. Until rhenium, atomic number 75, and technetium, atomic number 43, were discovered in 1924 and 1937, respectively, manganese was the only known element in Group VIIB.

While manganese had been known and commonly used in alloy and compound form for a long time, the pure metal did not become an industrial metal until the late 1930's with the development of the electrolytic process for its recovery.

TABLE 1. PHYSICAL PROPERTIES OF MANGANESE

Atomic number		25		
Atomic weight		54.938		
Stable isotope		55		
Density, g/cc at 20°C	Alpha	Beta	Gamma	
	7.44	7.299	7.188	
Atomic volume, cc/g-atom		7.4		
Melting point		1244 ± 3°C		
Boiling point, 760 mm		2097°C		
Specific heat, cal/g/°C at 25°	Alpha	Beta	Gamma	Delta
	0.114	0.154	0.148	0.191
Latent heat of fusion, cal/g		63.7		
Latent heat of vaporization, cal/g (at BP)		997.6		
Linear coefficient of thermal expansion, Per °C (0–100°C)	Alpha	Gamma		
	22×10^{-6}	14×10^{-6}		
Electrical resistivity at 20°C				
Alpha, microhm-cm		185		
Beta, microhm-cm		44		
Gamma, microhm-cm		60		
Magnetic susceptibility, 18°C, cgs units		9.9		
Hardness, Mohs scale		5.0		
Vapor pressure, mm Hg °C				
1727		89		
1927		315		
2027		541		
2127		880		
Heat of transformation, cal/g-atom				
Alpha to beta, 727°C		535		
Beta to gamma, 1100°C		545		
Gamma to delta, 1138°C		430		
Standard electrode potential, $Mn = Mn^{++} + 2e^-$		+1.1 volts		
(referred to hydrogen electrode)				
Thermal neutron absorption cross section, barns		13.2		

Occurrence

Manganese is widely distributed in the combined state, ranking twelfth in abundance among the elements in the earth's crust. It is commonly found in association with iron ores in concentrations too low in most cases, however, to make its commercial recovery attractive from that source.

The principal ores are pyrolusite, MnO_2; rhodocrosite, $MnCO_3$; manganite, a hydrated oxide; and psilomelane, barium manganese oxide. Nodules found on the ocean floor and in Lake Michigan are important potential sources of manganese and other metals, and their exploitation is being undertaken.

The United States is a "have not" nation insofar as deposits of high-grade manganese are concerned. The known manganese deposits in this country are estimated to total 3,500 million long tons (2240 lb) of ore, most of it low-grade ore containing 75 million tons of manganese.

Over 90% of the country's consumption of manganese ore is imported. In recent years U.S. consumption of manganese ore has amounted to about 2.2 million short tons per year. Brazil provided the major portion of imports, followed in order by Gabon, Union of South Africa, India and Ghana. Ghana and Gabon supply most of the battery- and chemical-grade ore containing more than 47% Mn. At the present time imports offer a much cheaper source of manganese than do domestic deposits that would have to be upgraded to match foreign ores in quality.

Derivation

Manganese metal can be made by several reactions or processes:

Reduction of oxides with carbon; reduction of oxides with other metals; reduction of anhydrous halogen salts with metals; reduction of solutions of manganese salts with metals; and electrolysis of sulfate solutions of manganese. The last is the basis of the only commercial source of pure manganese and is used in this and other countries.

The American development is based on a process using the electrolysis of manganous sulfate and was extensively investigated in laboratory and pilot operations in the 1930s by Shelton and co-workers at the U.S. Bureau of Mines. After further development and engineering studies the process was brought to full-scale operation in 1939 by the Electro Manganese Corporation plant in Knoxville, now owned by Foote Mineral Co. While differing in many details, both operations used a diaphragm cell whose anolyte contained manganous sulfate, ammonium sulfate, and sulfuric acid to give a pH of about 1, and whose catholyte was maintained at a pH of 7.2 to 7.6 in the presence of added sulfur dioxide. Similar systems are currently used in electrolytic manganese plants.

A manganese plant cell room is shown in Fig. 1.

Process details, including flowsheets, are given in the references listed below.

The rise in production of electrolytic manganese has been great since the industry began. In 1941 some 600 tons were made, in 1952 over 3,500 tons, and in 1970 the market amounted to 33,000 tons. The price of pure manganese metal is 33¢/lb.

Electrolytic manganese of at least 99.9% Mn is produced by Foote Mineral Company, Union Carbide Corporation and American Potash and Chemical Corporation.

Ferroalloys. More than 90% of the manganese consumed is used in the form of ferroalloys by the metal industries, chiefly for steel manufacture. These are made by smelting operations in high-temperature furnaces, starting with suitable ores.

FIG. 1.   Cell room in Marietta, Ohio, electrolytic manganese plant. (*Courtesy Union Carbide Metals Corp.*)

The predominant type, ferromanganese (78–82% Mn, 12–16% Fe, 6–8% C, 1% Si), is made in blast furnaces and to a lesser extent in electric furnaces by the reduction with carbon of high-grade ores containing 48% or more Mn. Low carbon and other grades of ferromanganese are available.

Physical Properties. Manganese exists in four allotropic modifications, the alpha being the one stable at room temperatures. Alpha and beta manganese are hard brittle metals that will scratch glass. The pure metal cannot be fabricated.

The physical properties of manganese are summarized in Table 1.

Natural manganese is comprised of 100% of Mn^{55}. Artificial isotopes of Mn^{51}, Mn^{52}, Mn^{54}, and Mn^{56}, of which Mn^{54} has the longest half-life 310 days, have been prepared.

Chemical Properties. Although somewhat similar to iron in general chemical reactivity, manganese can exist in its compounds in the valence states of 2, 3, 4, 6 and 7, the most stable salts being those of the divalent form and the most stable oxide the dioxide, MnO_2. The lower oxides, MnO and Mn_2O_3, are basic; the higher oxides, acidic. The most stable compounds other than MnO_2 are those of valence 2, 6 and 7, exemplified, respectively, by the manganous salts, such as $MnCl_2$, $MnSO_4$; and $Mn(NO_3)_2$ the manganates, such as K_2MnO_4; and the permanganates, such as $KMnO_4$. Divalent manganese is a reducing agent, tetravalent manganese is a good oxidizing agent, and heptavalent manganese is one of the most powerful oxidizing agents.

These factors make manganese compounds useful for a variety of industrial applications and for analytical procedures.

Manganese metal oxidizes superficially in air and rusts in moist air. It burns in air or oxygen at elevated temperatures like iron; decomposes water slowly in the cold and rapidly on heating, and dissolves readily in dilute mineral acids with hydrogen evolution and the formation of the corresponding divalent salts.

Fluorine, chlorine, and bromine react with manganese when heated. When heated with nitrogen or ammonia various nitrides are formed. Manganese reacts with sulfur to form sulfides.

Fused manganese dissolves carbon, as does iron, ultimately forming a carbide. It reacts with carbon monoxide at temperatures above 330°C (626°F) and with carbon dioxide when strongly heated.

Hydrides of manganese have not been detected, but solid and liquid manganese dissolves appreciable quantities of hydrogen; electrolytic manganese normally contains 150 parts of hydrogen per million.

Principal Compounds and Uses

As might be expected for an element of five different valence states, a wide variety of manganese compounds is known. However, less than a dozen are made and used in more than minor quantities: the two oxides, manganous oxide, MnO and manganese dioxide, MnO_2; manganous sulfate, $MnSO_4$, and manganous chloride, $MnCl_2$; potassium permanganate, $KMnO_4$; and a few organic compounds, such as manganese naphthenate, oleate, linoleate, etc., used chiefly as dryers in paint and varnish.

Manganous oxide is more easily dissolved by acids than are the higher oxides present in ores and most ores destined for preparation of chemicals are first furnaced in a reducing atmosphere to reduce the oxides present to MnO. While most of the MnO is used as an intermediate to produce other manganese compounds, it is also currently being produced and used as a source of manganese in fertilizers. When MnO is treated with mineral acids the corresponding manganous salts are formed, e.g., $MnSO_4$ from which electrolytic manganese is made.

Manganese dioxide, MnO_2, is the manganese compound made and used in the greatest tonnage. A special grade of natural MnO_2 is used in the manufacture of the LeClanché and other primary batteries at the rate of about 40,000 tons/year. Most of this MnO_2 comes from mines in Africa and Mexico. It is characterized by a less well-defined gamma crystal structure than the pyrolusite form which is not suitable for batteries.

Synthetic MnO_2 is produced by chemical processes, but most of it is derived by an electrolytic process which is used to produce about 7,000 tons/year of the material in this country and over 30,000 tons/year in Japan. Manganese sulfate is purified and fed to a cell where MnO_2 is plated on the anode. Process details are given by Clapper. The electrolytic MnO_2 is of the gamma form and is eminently suitable for dry cell manufacture.

Manganese dioxide is used as the oxidizing agent in a number of chemical processes, for example, to prepare hydroquinone from aniline.

Manganous sulfate, $MnSO_4$, is a soluble, pale pink salt which forms several hydrates. Most of it is obtained as a by-product of processes using MnO_2 or $KMnO_4$ as oxidizing agents. Manganous sulfate is a good reducing agent and the MnO_2 which generally is formed by its use can easily be reconverted to $MnSO_4$ by the reaction: $SO_2 + MnO_2 \rightarrow MnSO_4$. It is used in the manufacture of paint and varnish driers; in the formulation of fertilizers to provide the small amount of manganese needed as a trace element; in the production of electrolytic manganese and manganese dioxide; in textile dyeing; and in ceramics.

Manganous chloride, $MnCl_2$, is a soluble, rose-colored salt usually obtained as the hydrate, $MnCl_2 \cdot 4H_2O$. It is formed by the action of HCl on manganous oxide or on MnO_2. Manganous chloride is added to molten magnesium to introduce the manganese present in most magnesium alloys.

Potassium permanganate, $KMnO_4$, is a dark purple compound that is a powerful oxidizing agent. It is produced by a multistep process wherein manganese dioxide or pyrolusite is fused with KOH in the presence of air or an oxidizing agent to form potassium manganate, K_2MnO_4, which is then converted to $KMnO_4$ by treatment with chlorine or carbon dioxide, or by anodic oxidation in an electrolytic cell.

When the purple $KMnO_4$ solution is concentrated by evaporation and cooled, slender opaque crystals of $KMnO_4$ are formed. Potassium permanganate is a disinfectant, deodorant and oxidizing agent. These properties make it useful for water purification, bleaching, air purification, descaling of steel, and preparing organic chemicals such as saccharine.

Applications for Manganese

Manganese is vital for sulfur control in steelmaking and is the most commonly and widely used deoxidizer of molten steel. The greater part of it ends up in the slag of the steelmaking process. Ferromanganese, the form in which about 90% of all manganese is consumed, is the standard additive agent for these purposes in the steel industry. However, pure manganese is used in many instances, especially where preparation of special steels is involved.

Manganese is present in several ferrous alloys of the chromium-nickel type. In some, most of the nickel is replaced by manganese to form alloys whose usefulness approaches that of the 17–7 stainless steels but whose corrosion resistance is somewhat less. Examples are the 16 to 17% chromium-14% manganese-1% nickel steels.

Practically all commercial alloys of aluminum and magnesium contain manganese. Corrosion resistance and mechanical properties, such as hardness, are improved by its presence. The amounts used are seldom above 1.2% for magnesium and 1.5% for aluminum. For use in aluminum alloys, electrolytic manganese competes with pure manganese oxides or carbonate. For use in magnesium it competes with pure manganous chloride that is added to the melting pots.

The instrument alloy, manganin, is comprised of 11 to 12% manganese-3 to 4% nickel-balance copper.

Biological, Biochemical, and Toxicological Aspects

Manganese is generally considered one of the five essential trace elements, along with boron, zinc, copper and molybdenum, for the vast majority of higher plants. At least one reason for their necessity is that they form essential constituents of certain enzymes. A number of enzymes concerned with the oxidation of carbohydrates in respiration are activated by manganese and for one of them, oxalosuccinic decarboxylase, it may be essential.

In plants a shortage of manganese first becomes evident in the form of an intervenal chlorosis (lack of chlorophyll) which results in the appearance of yellow or gray streaks between the veins of leaves or in a mottling.

The soil of at least 25 states, in particular, Florida, Michigan, Ohio and Indiana, is deficient in manganese. For this reason manganese is frequently added to fertilizers used in these areas, usually in the form of $MnSO_4$, but more recently as a soil-acid soluble MnO. About 10,000 tons/year of manganese compounds containing 54% MnO equivalent are used in mixed fertilizers and another 5,000 tons/year are directly applied to the soil or sprayed on leaves.

Manganese is widely distributed throughout the animal kingdom and may possibly be generally essential for the utilization of vitamin B_1. Deficiency of it in chickens results in deformity of the leg bones ("slipped tendon" or "broken leg"). Small concentrations of manganese compounds are added to many animal and chicken feed formulations to overcome deficiency of this element.

Many of the iron-depositing bacteria form deposits of manganese oxides as well as iron oxides.

This bacterial activity may account for the occurrence of manganese in the iron ores of the Lake Superior and other regions, and for many of the high-grade manganese ore deposits. The manganese nodules found in recent years on ocean floors are possibly of biochemical origin.

Manganous compounds are not in general regarded as poisonous due to the manganous ion alone, as compared to the toxicity of other metal ions such as mercury, cadmium, thallium, lead, etc. The strong oxidizing properties of manganates and permanganates can cause skin irritation. Chronic manganese poisoning is a disease which results from the inhalation of fumes or dusts of manganese and usually develops after one to three years of exposure to heavy concentrations of dusts or fumes. The central nervous system is the chief site of damage. However, companies which have produced manganese compounds for long periods of time have experienced few cases of toxicity as a result of exposure to manganese compounds. In the battery industry adequate precautions are taken in the handling of ores and oxides of manganese, particularly in milling operations, to provide ventilation and masks for workers, and few cases of damage to health have occurred over a period of several decades.

CLIFFORD A. HAMPEL

References

Bacon, F. E., "Manganese Electrowinning," in "Encyclopedia of Electrochemistry," C. A. Hampel, Editor, pp. 792–796, Reinhold Publishing Corp., New York, 1964.

Carosella, M. C., and Fowler, R. M., "A New Commercial Process for Electrowinning Manganese," *J. Electrochem. Soc.,* **104,** 352–356 (1957); for flowsheet, see *Chem. Eng.* **64,** No. 10, 136–139 (May 19, 1958).

Clapper, T. W., "Manganese Dioxide, Electrolytic," in "Encyclopedia of Electrochemistry," C. A. Hampel, Editor, pp. 789–792, Reinhold Publishing Corp., New York, 1964.

Hampel, C. A., "Encyclopedia of the Chemical Elements," pp. 389–399, Van Nostrand Reinhold Co., New York, 1968.

Jacobs, J. H., Hunter, J. W., Yarrol, W. H., Churchward, P. E., Knickerbocker, R. G., Lewis, R. W., Heller, H. A., and Linck, J. H., *U.S. Bur. Mines Bull.,* **463** (1946).

Mantell, C. L., "Manganese," in "Rare Metals Handbook," C. A. Hampel, Editor, 2nd Edition, pp. 271–282, Reinhold Publishing Corp., New York, 1961.

Sully, A. H., "Manganese," Academic Press, Inc., New York, 1955.

MANNANS

The mannan type polysaccharides are reserve polymeric carbohydrates which are found quite generally in the plant kingdom. One of the first to be studied in detail was salep mannan (or mucilage) derived from the *Orchidaceae*. Hydrolysis studies showed that D-mannose is the sole constituent of this polysaccharide. Structural studies by the methylation technique indicated that this is a linear polymer of 1,4-linked mannose residues; the sugar probably is present as the pyranose form and may be linked glycosidically through beta-links. The salep mannan closely resembles the ivory nut mannan in structure. This is found in the thickened cell walls of hard seeds and forms the chief component of the American ivory nut. The structure of ivory nut mannan may be depicted schematically as shown below.

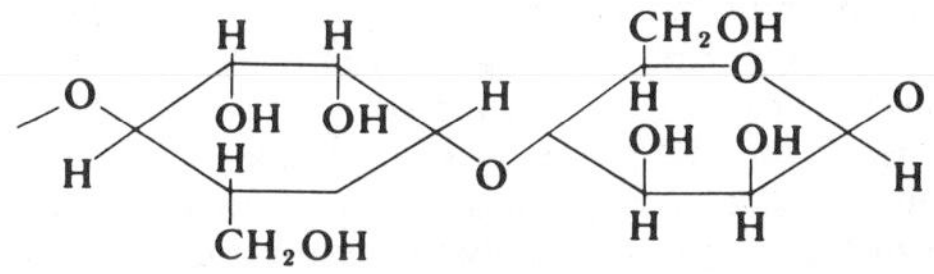

The mannans are found rather generally as components of coniferous woods, whereas deciduous woods contain either very small quantities or none. The mannan in coniferous woods is closely associated with cellulose, which also contains the 1,4-linkage system with glucose as the repeating sugar unit.

The galactomannans (mannogalactans) are found as the reserve carbohydrates in the endosperms of leguminous seeds such as guar, locust bean, honey locust, flame tree, foenugreek, Kentucky coffee bean, Paloverde, tara, huizache, lucerne seed, etc. These polysaccharides are called *galactomannans* since they are composed of galactose and mannose units, which, in turn, are the products of hydrolysis. Locust bean gum is the galactomannan polysaccharide obtainable from the locust (carob) bean. It is also known as Swine's bread, gum Hevo, etc. The bean is grown in the Mediterranean region and is utilized as a food called St. John's bread. In Europe and the United States the gum from the locust bean is used in the sizing of textiles, tanning of leather, sizing of paper, and as a mucilage in pharmaceutical products. Guar gum is the galactomannan polysaccharide obtainable from the guar seed. It was introduced into the United States in 1902 from India, where the beans are utilized for both human and animal consumption. It is noteworthy that in India guar is such an important dietary and feed factor that it is under government control. In the United States guar gum has industrial outlets principally in the paper, mining, textile, leather, and food industries. Careful and extensive structural studies have shown that 1,4- and 1,6-linkage systems are common to the galactomannans. Generally the mannose units are linked in a 1,4 fashion and constitute the main linear chain, whereas the galactose units occur as one unit branches with 1,6-linkage prevalent.

The *glucomannans* are reserve polymeric carbohydrates found in corms of plants of the *Aracae* family. Accordingly, *Amorphophallus oncophyllus* (found principally in Java) yields a product known as iles mannan flour, and konnyaku flour is obtained from *Amorphophallus rivieri*. These polysaccharides give glucose and mannose as the hydrolysis products. The glucomannan from *Amorphophallus rivieri* is an effective agent for the creaming of latex. The tubers of *Conophallus konnyaku,* a member of the *Araceae* family, contain a reserve glucomannan. This mannan forms the chief constituent of konnyaku, flour, which is utilized in Japan as a food. Ex-

perimental structural studies again indicate the presence of 1,4- and 1,6-linkage systems.

It is interesting to observe that nature has provided mannan type polysaccharides with 1,4-linkage systems (American ivory nut), and combination 1,4- and 1,6-linkage systems (galactomannans), which display amazing similarity to the common polysaccharides cellulose and starch, respectively.

                                                              OWEN A. MOE

Cross-references: *Carbohydrates, Sugars.*

MASS SPECTROMETRY

Mass spectrometry deals with the separation of gaseous ions of differing mass and charge by the action of electric and magnetic fields. By proper selection of experimental conditions, it is possible to measure precisely the mass of ions, prove the existence of mass isotopes, and measure relative abundance of ions in a mixture. Organic molecules produce, under proper conditions, a spectrum of ions corresponding to the parent molecule and most of the fragments that can be produced from it by bond rupture. Since mass spectra of even closely related organic molecules are quite different, mass spectrometry has become an important tool for the analysis of organic mixtures.

The mass spectrometer and the mass spectrograph, though based on the same principle, are quite different. The spectrograph is employed for precise measurement of mass, whereas the spectrometer is designed to measure precisely relative abundance of ions. Both are based on the observation of "positive rays" by Sir J. J. Thomson in 1912. The first spectrometer was built by Dempster in 1918; a spectrograph was built by Aston a year later. In the following 25 years, the instruments were developed slowly and their application was confined primarily to university laboratories. Early in World War II, need for precise, rapid analyses of complex mixtures of hydrocarbon compounds prompted development of spectrometers for this application. These commercially available instruments have been developed into versatile tools for the analysis of a wide range of organic and inorganic compounds.

In the last decade, more sophisticated instruments have been developed for special applications. The time-of-flight principle was used to decrease the time required to obtain a mass spectrum. Time-of-flight spectrometers and increased scanning speeds in conventional magnetic scanning instruments permitted coupling of mass spectrometers and gas chromatographs, greatly extending the complexity of organic mixtures which can be analyzed. High resolution (1 in 10,000 or greater) spectrometers have been developed which permit the definition of the exact elemental composition of ions. The technique has been applied to natural products with great success.

In a mass spectrometer, the sample to be studied is introduced into a vacuum bench, and, by suitable manipulations, reduced to low pressure. The vapor is passed through a beam of electrons, where ionization occurs. All possible ionization reactions occur, although the probability of some is much greater than others. The molecule may be ionized, or it may fragment in any possible manner, and the fragments be ionized. Molecules or fragment can add or lose one or more electrons, the loss of a single electron being most common. Ions thus produced are accelerated by an electric field and passed into the magnetic field. A charged particle traveling at high speeds follows a curved path whose radius depends on the speed of the particle and its mass-to-charge ratio (m/e). By varying the speed of the particles (by changing the accelerating voltage) or changing the magnetic field strength, ions of various m/e ratios can be focused on a collector plate and grounded. The resulting current is amplified and recorded by a suitable electrometer circuit. A plot of mass versus ion intensity is a mass spectrum.

No two molecules behave exactly the same on electron bombardment. Even closely related molecules or isomers show some differences in the manner in which various bonds react. Thus, mass spectra can be used for qualitative analysis. Most molecules give rise to some ions with the mass of the parent molecule, and the absolute molecular weight can be established directly. In mixtures, it is not always possible to identify "parent mass peaks," and other methods must be used. From the study of many spectra, an experienced spectrographer can form generalities that aid him in predicting the probable components in the sample. The final identification, as in most types of qualitative analysis, requires direct comparison of the spectra of known and unknown.

In quantitative analysis, the problem is somewhat different. First, the qualitative analysis of the sample must be known or determined. The mass spectrum of a mixture is the sum of the spectra of the components in the ratio of their partial pressures in the sample. The method of establishing the ratio of components depends upon the sample components. In the simplest case, each component gives rise to a characteristic ion not formed by any other component, and a direct breakdown of the mixture is possible. In the other extreme, each component gives rise to some of each of the observed mass peaks, and linear simultaneous equations are needed to make the analysis. The analysis of most mixtures is done by combining these two approaches. Some compounds in the mixture give rise to characteristic peaks, and their contributions can be subtracted directly. Linear equations may then be set up to handle the remaining components. By suitable combinations of technics, as many as 20 or more components can be determined in a single mixture.

For routine analyses of multiple samples, the mathematical details can be handled by a computer using signals fed directly from the spectrometer. In this case, the answer is available almost as soon as the spectral scan has been completed. The precision and accuracy of the instrument depend upon the type of sample being examined. In general, ±1 mole percent can be achieved.

R. E. KITSON

Cross-references: *Ionization, Isotopes.*

References

Ryland, A. L., "Mass Spectrometry," in Kirk-Othmer "Encyclopedia of Chemical Technology," 2nd Ed., Vol. 13, John Wiley and Sons, Inc., 1967.

Beynon, J. H., Saunders, R. A., and Williams, G. E., "The Mass Spectra of Organic Molecules," Elsevier Publishing Co., New York, 1968.

MECHANOCHEMISTRY

Energy must be supplied to stable substances before they will undergo chemical change. In a large majority of chemical reactions the necessary energy is supplied as heat. Heat is, however, only one of a number of forms of energy which should be regarded as equivalent to one another. To lead to chemical reaction, the molecules concerned must acquire at least a critical amount of energy to surmount the energy barrier between their original and changed state.

The difficulty with mechanical energy is that it is a blunt instrument compared with radiation, in which the individual bullets of energy strike and energize individual molecules. A large proportion of the mechanical energy is bound to be used up in merely moving the molecules about. If we take the stirring of a liquid as an illustration of applications of mechanical energy to a system, the molecules do not offer enough resistance to the shear forces to be able to absorb sufficient energy for chemical reaction. The energy is dissipated in the work of moving the molecules to new positions.

However, much greater mechanical forces are applied so as to energize molecules to the degree necessary for chemical reaction. Such a case is the rapid deforming of the big molecules of a rubbery or plastic high polymer. The molecules here are no longer the roughly spherical ones which move readily over one another, but are long chains made up of hundreds or thousands of small molecules strung together. These chains are highly entangled, and their disentanglement requires high energy. Some of the molecules may acquire enough energy to be ruptured.

Rupture of the chain requires breaking of one of the chemical bonds in its backbone. The broken ends are highly reactive and will combine with many substances to be stabilized again. Thus a chemical reaction can be induced mechanically.

It is only 35 years ago that such reactions were first contemplated, and only within the last 15 years have they conclusively been shown to occur. Their study is now going ahead actively, especially in the USSR, where the field of investigation has been aptly named "mechanochemistry," in line with "electrochemistry" and "photochemistry."

Thomas Hancock discovered empirically as long ago as 1819 that mechanical working of natural rubber caused it to soften permanently. His process of mastication was truly "the origin and commencement of India rubber manufacture." Hancock was soon operating industrial machines culminating in his Monster Masticator. Essentially the same process of working rubber, in mills and in mixers, remains one of the main operations in rubber technology. Mastication of rubber has now been recognized to be a mechanochemical reaction, and is probably the most important one at present in industrial use.

Natural rubber is made up of thousands of repeating units of chemical formula:

$$----\overset{\overset{\displaystyle CH_3}{|}}{C}=CH-CH_2-CH_2----$$

strung together into long chains. The units which are ruptured are near the center of the long chains and leave large segments of chain on either side anchoring into the entangled mass. The mechanochemical process in this case can be represented by as precise an equation as for a thermal reaction:

$$----\overset{\overset{\displaystyle CH_3}{|}}{C}=CH-CH_2-CH_2----$$

SHEARING $\downarrow$ FORCE

$$----\overset{\overset{\displaystyle CH_3}{|}}{C}=CH-CH_2\cdot + \cdot CH_2----$$

The broken ends readily react to form new chemical bonds. They normally combine with oxygen from the surrounding atmosphere. Other substances, so-called "peptizers," can be added to ensure this simple form of stabilizing the reactive ends of the broken chains. If the pure rubber is masticated in absence of oxygen, the broken ends unite in pairs, and superficially it appears as if no breakdown has occurred.

In the presence of oxygen and peptizers, the rubber molecules are stabilized at their reduced chain-length. The decrease in molecular weight is the cause of the softening first noted by Hancock. Most supplies of raw natural and synthetic rubbers have a wide distribution of lengths of their polymer chains. Mechanical working preferentially ruptures the chains near their middle. This is because the lengths of the molecule on either side act as anchors to avoid slippage and the relieving of the stress on the chemical bonds. For the same reason, rubber molecules below a certain molecular weight do not undergo rupture, depending on the particular stress. The result is that the average molecular weight is lowered on mastication and, in addition, the molecular weight distribution is made more narrow.

Other possible reactions explain otherwise mysterious products of the mastication of rubber. For example, the addition of only 1 or 2% of maleic anhydride to rubber being milled has long been known to change it abruptly to a biscuit-like product which sputters alarmingly from the mill. Here the additive is tying on to the broken ends of the rubber molecule to produce further reactive groups, but in this case with a great tendency to branch themselves on other rubber molecules; so making a highly cross-linked material, rather like a thermosetting resin, which has lost its characteristic rubbery properties. Again, the addition of carbon black may give a crumblike product unless conditions are arranged so that the broken rubber molecules do not tie themselves to the surfaces of the carbon black particles to give a cross-linked material.

The rubber industry is only one utilizing machines imparting considerable amounts of energy to high-polymer substances. In plastics technology, for example, the extrusion of PVC proceeds with continuous thermal formation of cross-links between the molecules and simultaneous mechanical

rupture of the network so formed; such a process is called "chemical flow." Grinding, rolling, extruding, mixing, homogenizing and stirring are practised in many branches of technology dealing with natural and synthetic high polymers, including cellulose, starch, proteins, synthetic fibers and resins. The available experimental results indicate that mechanochemical changes take place in these various mechanical processes, although up to the present they have not been very systematically and critically investigated from this point of view.

An interesting application of mechanochemistry is in the freezing and thawing of liquids. If polymer is dissolved in liquid or the liquid is absorbed by the polymer, mechanical forces come into play during freezing and thawing which can rupture the polymer molecules. Thus, the freezing of aqueous solutions of starch at temperatures from $-10°$ to $-70°C$ causes a sharp increase in the soluble fraction containing sugars and dextrins.

The milling of rubber, as we have seen, yields reactive ends of broken molecules. These free radicals, as they are called, have the same sort of reactivity as those produced by heating the catalysts used to produce polymerization of polystyrene, polymethylmethacrylate and PVC. Mastication would then seem a way to initiate polymerization. So it has turned out on trial. If a polymerizable substance is imbided (absorbed) by natural or synthetic rubber and the mixture is masticated in absence of oxygen, the original liquid polymerizes to a hard solid.

These products have one great difference from normal polymers. What starts the polymerization reaction is no longer a small piece of catalyst stuck on the chain end. It may be a polymer segment (of rubber, in this case) as long as or longer than that of the new polymer formed. It is specially interesting when the two parts of this so-called "block polymer" have quite different properties. Natural rubber, for illustration, can be suspended as a milky solution in acetone by grafting to it polymethylmethacrylate, and similarly the block polymer can be suspended by the rubber in petroleum ether in which the polymethylmethacrylate is quite insoluble. If this same block polymer is dried down from petroleum ether so that the rubber segments form a continuous mass and the other polymer is in the form of nodules, the dried material is quite rubbery. If the solvent is acetone, the dried material is tough and horn-like. So we have here a hard and a soft form for the self-same block polymer.

Mechanochemically initiated polymerization has been found to occur with a large variety of polymerizable substances using substrates other than rubber. The substrates may be natural products, including starch, casein, glue and shellac; modified natural polymers, including cellulose acetate and ethyl cellulose; amorphous synthetic plastics, including polythene; thermosetting resins; crystallizable polymers, including nylon. The substrate need not even be a polymer itself. Perhaps the ultimate has been achieved by the Soviet scientists who have, by the grinding of metallic powders and common salt, achieved thereby a firm union between crystal and polymer molecules.

Polymerization is only one example of many types of reactions involving free radicals which could be promoted by mechanical means. Mechanical forces may, with some materials such as silicates, lead to ionic reactions.

Biochemists appear to be studying related reactions. We can expect that the two independent approaches can be brought together and a more unified field of mechanochemistry to develop.

W. F. WATSON

Cross-references: *Rubber, Mixing, Polymerization.*

MELTING POINT

The melting point or freezing point of a pure substance is the temperature at which its crystals are in equilibrium with the liquid phase at atmospheric pressure. If this equilibrium is reached by cooling the liquid the temperature is called the freezing point and if it is reached by heating the solid it is designated as the melting pont. Though the terms are used interchangeably, freezing point ordinarily refers to temperatures below 0°C, the freezing point of water.

For ideally pure crystalline substances melting is accompanied by an abrupt increase in entropy, in heat content, and in volume. In a few exceptional cases; e.g., water, bismuth, and gallium, the volume decreases. The number of calories necessary to convert one mole of the pure crystals to the liquid state is called the *molar heat of fusion.*

The effect of pressure, *p,* on the freezing temperature of a pure substance in degrees absolute, *T,* can be expressed quantitatively by the Clausius-Clapeyron equation:

$$\frac{dT}{dp} = \frac{T(V_L - V_s)}{\Delta H}$$

where V_L and V_s are the molar volumes of the liquid and solid, respectively, and ΔH is the molar heat of fusion. Since with few exceptions the density of the solid at the freezing point is greater than that of the liquid, $V_L - V_s$ is positive, so that an increase in pressure almost always raises the freezing point.

The freezing point of water at 760 mm is 0°C. Its *triple point,* i.e., its freezing point when sealed in an evacuated vessel under its own vapor pressure (4 mm), is +0.0099°C. Thus, the freezing point of water decreases with a rise in pressure, as would be expected from the fact that $V_L - V_s$ is negative. At higher pressures, however, this term becomes positive because of the appearance of a much denser crystalline form of ice and the freezing point rises with further increase in pressure. At a pressure of 33,880 atmospheres the freezing point of water is 166.6°C. The freezing points of ethyl alcohol and carbon disulfide at this pressure are 109°C and 209°C, as compared to their normal freezing points, −117.3°C and − 111.6°C, respectively.

As a rule a soluble impurity always lowers the freezing point of a substance. Since the crystals which form on chilling such a liquid are those of the pure substance their separation causes an increase in the concentration of the impurity in the remaining liquid phase so that the freezing, or solid-liquid equilibrium, temperature is progressively lowered. In contrast, when the crystals are forming during the freezing of a pure substance there is no change in the freezing temperature of the resid-

ual liquid phase. Thus, a pure substance freezes completely at constant temperature and an impure substance freezes over a temperature range, the maximum being the *primary freezing point.* A curve showing the solubility of a solute in a given solvent at various temperatures can be constructed by plotting the primary freezing points obtained for a wide range of concentrations.

Though the freezing point of a pure substance is lowered by adding more and more of a second substance as an impurity a limit is finally reached (called the *eutectic point* or, when water is one of the components, the *cryohydric point*) beyond which the freezing point rises again and the crystals of the impurity rather than those of the original substance are the first to separate on chilling the liquid. The cryohydric temperature and cryohydric composition for the sodium chloride-water system are −21.1°C and 23.3 weight percent sodium chloride, respectively.

Freezing point depression is a colligative property; i.e., it is proportional to the number of dissolved molecules. It can therefore be used for determining molecular weights. The following approximate expression for the molecular weight, *M,* of a solute can be derived from the second law of thermodynamics:

$$M = F \frac{w}{\Delta T}$$

where *F* is a constant characteristic of the solvent, *w* is the number of grams of the unknown solute dissolved in 1000 grams of solvent, and ΔT is the experimentally observed freezing point depression; i.e., the freezing point of the solvent minus the primary freezing point of the solution. *F* is the *molal freezing point depression* of the solvent; i.e., the freezing point lowering per mole of solute dissolved in 1000 grams of solvent. For water *F* equals 1.86°C and for camphor it equals 37.7°C. This expression is valid for dilute solutions only and for solutes which do not dissociate, associate, or form solvates with the solvent.

EVALD L. SKAU

Cross-references: *Eutectic, Solubility, Solutions, Phase Rule.*

References

Skau, E. L., and Arthur, J. C., Jr., "Melting and Freezing Temperatures," in "Physical Methods of Chemistry," A. Weissberger and B. W. Rossiter, Eds., (Techniques of Chemistry, Vol. I), Wiley-Interscience, New York, Part V, 1970.

Utermark W., and Schicke, W., "Melting Point Tables of Organic Compounds," 2nd ed., Interscience, New York, 1963.

Stephen, H., and Stephen, T., "Solubilities of Inorganic and Organic Compounds," 2 vols., Macmillan, New York, 1963–1964.

Findlay, A., Campbell, A. N., and Smith, N. O., "The Phase Rule and Its Applications," 9th ed., Dover, New York, 1951.

MEMBRANE EQUILIBRIUM

This term is applied to a special type of osmotic equilibrium, first described in 1911 by Donnan and

Harris. They were studying the osmotic pressure of saline solutions of a dye, Congo red, which is the sodium salt of a high molecular weight sulfonic acid. The membranes of their osmometer were permeable to water and to ordinary salts, but impermeable to the dye. After osmotic equilibrium had been attained in their experiments, sodium chloride was present on both sides of the membrane, but its concentration was always higher in the external solution, which contained none of the large ions of the dye. To explain this unequal distribution, Donnan worked out a theory which he expressed in simple equations, based on thermodynamics and the laws of dilute solutions. He showed that diffusible ions tend to be unequally distributed in such a system whenever there is some constraint which prevents at least one kind of ion or charged particle from diffusing freely. Other investigators referred to the unequal distribution of ions as the Donnan effect, and called this type of equilibrium the Donnan equilibrium. After it was pointed out that Donnan might have based his theory on more general equations deduced by Gibbs in 1875–1878, the term Gibbs-Donnan equilibrium came into use.

A simple type of membrane equilibrium may be illustrated by the use of a diagram.

I	II
H_2O	H_2O
$z/n \ R^{n-}$	
$y + z \ Na^+$	$Na^+ \ x$
$y \ Cl^-$	$Cl^- \ x$

Here the vertical line represents a membrane impermeable to the ion, R^{n-} but freely permeable to water and to sodium chloride. The molar concentrations of ions, after equilibrium has been reached, are indicated by the small letters. The notation is consistent with the electroneutrality of each solution; z is the equivalent concentration of the anion of valence n and molar concentration z/n. According to Donnan's theory, equilibrium requires an equality of the products of the concentrations of the ions of sodium chloride in the two solutions. This may be expressed by the equation

$$x^2 = y(y + z)$$

which shows at once that x is greater than y, or that the concentration of diffusible salt is greater in the external solution, II. This unequal distribution may be very marked; for example, if z is equal to $100y$, the ratio x/y is 10.05. On the other hand, if y is equal to $100z$, the ratio x/y is only 1.005. It is characteristic of the Donnan equilibrium that a high concentration of any diffusible electrolyte tends to suppress the unequal distribution.

Equilibrium in such a case requires a difference in pressure between the two solutions, and this difference is the difference between their OSMOTIC PRESSURES. The observed difference, although it is often called the colloid osmotic pressure, may be largely due to the unequal distribution of diffusible ions. Donnan pointed out that it would approach that due to the ions of the whole colloidal electrolyte only if x and y were much less than z, while in the opposite extreme, it would approach that due to the colloidal ions alone.

Donnan also deduced the existence of an electric potential difference between the two solutions at equilibrium. Since this is a single potential difference, it cannot be measured directly; the best that can be done is to connect identical electrodes with the solutions on opposite sides of the membrane by way of salt bridges. Many measurements of this sort were made by Loeb in his work on the colloidal behavior of PROTEINS. It was later found that the electromotive force of such cells, of the order of 30 mv, was changed very little by the puncture or removal of the membrane after equilibrium had been reached.

The Donnan equilibrium in nonideal solutions has been treated mathematically by Overbeek.

The theory of membrane equilibrium has been especially useful in the study of proteins. Biological scientists have found it necessary to consider the Donnan equilibrium in trying to explain differences in ionic concentration, osmotic pressure, and electric potential across cell membranes.

DAVID I. HITCHCOCK

References

1. BOLAM, T. R., "The Donnan Equilibria," London, G. Bell & Sons, 1932.
2. DONNAN, F. G., "The Theory of Membrane Equilibria," *Chem. Rev.,* **1,** 73–90 (1924).
3. HITCHCOCK, D. I., "Proteins and the Donnan Equilibrium," *Physiol. Rev.,* **4,** 505–531 (1924).
4. HITCHCOCK, D. I., "Membrane Potentials in the Donnan Equilibrium. II," *J. Gen. Physiol.,* **37,** 717–727 (1954).
5. LOEB, J., "Proteins and the Theory of Colloidal Behavior," New York, McGraw-Hill Book Co., 1922, 1924.
6. OVERBEEK, J. TH. G., "The Donnan Equilibrium," *Progr. Biophys. Biophys. Chem.,* **6,** 57–84 (1956).

MERCURY AND COMPOUNDS

Mercury is a metallic element, among the first to be mined and utilized by man. Known also as quicksilver and hydrargyrum, from which Berzelius derived its symbol Hg, very little exists in the native state. Commercially, practically all is obtained from the pyrolysis of red cinnabar (mercuric sulfide), an ore found in at least fourteen countries at depths of less than fifteen hundred feet. Spain and Italy produce the most, annual output ranging from 50,000 to 80,000 flasks (76 lb net weight) each. Newly mined ore has a mercury content of from 0.3 to 20%; the poorer types are processed in the U.S.A. with high efficiency to compete at the present domestic market price of about $250–300/flask.

As the only metal which exists in the liquid state to temperatures below 0°C and the one with highest vapor pressures, it differs sharply in physical properties from all others. There are sixteen isotopes known with mass numbers from 189 to 205; the shortest-lived, Mass No. 194 and half-life 0.40 sec, is in contrast to the beta ray emitter, Mass No. 203, which has a half-life of 43.5 days and finds utility in tracer studies. A number of physical properties are tabulated[1]:

Atomic number	80
Atomic volume	14.81 cc/g-atom
Atomic weight	200.59
Atomic distance	3.005 Å
Density	13.546 g/cc at 20°C
Surface tension	475 dynes/cm at 20°C
Freezing point	−38.87°C
Boiling point	356.9°C/760 mm
Boiling point rise with pressure	0.0746°C/mm
Vaporization, heat of	14.69 kcal/g-atom at 25°C
Expansion coefficient of liquid	182×10^{-6}cc at 20°C
Heat capacity	0.0334 cal/g at 20°C
Heat of fusion	2.7 cal/g
Hydrogen overvoltage	1.06 v
Resistivity	95.8×10^{-6}ohm/cm at 20°C
Specific heat	0.033 cal/g/°C
Conductivity	0.022 cal/sec/cc/°C
Electrochemical equivalent	1.0394 mg/coulomb (univalent)
	2.0788 mg/coulomb (divalent)
Electrode potentials, normal	
$Hg^{++} + 2e^- = Hg$	0.85 v
$Hg_2^{++} + 2e^- = 2Hg$	0.79 v
$2Hg^{++} + 2e^- = Hg_2^{++}$	0.92 v

All physical states of the element are diamagnetic. The vapor is colorless and entirely monatomic, the only element other than the rare gases which forms monatomic vapor to such a degree at ambient temperatures.

The solubility in water is approx. 20–30 μg/liter, in benzene 20 mg/liter and dioxane 7.0 mg/liter. Interestingly, the element readily forms liquid metallic solutions or amalgams with other metals; at 18–20°C it dissolves 56.5 g of indium, 2.15 g of zinc and 0.042 g of silver/100 g of mixture. A variety of compounds results by reaction with other metals. Alkali elements interact exothermically and in the presence of air produce flame; examples include $NaHg_8$, Hg_2K, Hg_3Sb_2, Hg_5Ca and $Hg_{11}Rb$.

Energy absorption and radiation occur at wavelengths which are utilized in many devices. The mercury line at 2537 Å with an energy level of 4.86 electron-volts is the basis for fluorescent and germicidal lamps; that at 1849 Å and 6.7 electron-volts for ozone-producing equipment.

Oxidation of the element forms either mercurous, Hg_2^{2+}, or mercuric, Hg^{2+}, valence states. Pure metal is only slowly attacked by oxygen but is highly reactive with ozone, halogens, nitrogen dioxide, hydrogen peroxide, hot nitric or sulfuric acids. At elevated temperatures in air or oxygen red mercuric oxide forms, but at 440°C it dissociates completely to the elements. Ammonia gas containing traces of water causes one of the most hazardous reactions with liquid mercury, often leading to violent explosions.

Unique chemical properties set it apart from even the other two members of Group IIB of the Periodic Table, zinc and cadmium, and are attributable to having only two relatively inert valence electrons of $6s$ configuration.

Mercurous compounds contain primarily the diatomic mercurous ion in which one electron is removed from each atom and the remaining two valency electrons are shared by two atoms. In general, mercurous derivatives are insoluble, form no complexes, and differ from mercuric types in not being able to form stable covalent linkages with carbon or nitrogen. They convert fairly readily to mercury and mercuric compounds.

Mercuric salts compare with no element but hydrogen in dependence on anions for their degree of ionization. The least ionized are the halides, cyanide and thiocyanate. Industrially, inorganic salts are produced by direct reaction of the metal with chlorine to yield mercurous and mercuric chlorides; with hot aqueous nitric acid to form the nitrates; and with fuming sulfuric acid, the sulfates. Excess reagent completes the oxidation to mercuric salts.

High-purity metal is obtained by a combination of physical and chemical steps. The U.S. National Bureau of Standards utilizes a practical and efficient technique: dirt is removed by filtration through fritted glass, and base metals with a 12–16 hr air-agitated nitric acid wash (1 part conc. HNO_2 + 9 parts H_2O) followed with distilled water washing and drying. Repetition with the same procedure further lowers impurity levels. The metal is then distilled under reduced pressure in the presence of air, filtered through perforated filter paper, distilled under high vacuum in all-glass equipment and finally stored in acid-cleaned soft-glass bottles.

End uses are in a constantly changing flux as modern technology and increasing concern for ecological effects exert their influence. The consumption pattern in the U.S.A. for 1969* represents the most recent period prior to initiation of marked legislative restrictions and good recovery practices.

The high specific gravity, fluidity and electrical conductivity find continuing application in switches, rotating and stationary cathodes of electrochemical cells, relay systems and heat transfer equipment.

Compounds derived from the metal have served to inhibit the growth of deleterious bacteria, fungi and algae which otherwise unchecked cause economic losses estimated at billions of dollars annually. The major groups are alkyl and aryl mercury salts. For example, methylmercury acetate,

dicyandiamide, hydroxide, 2,3-dihydroxypropyl-
mercaptide, 8-hydroxyquinolinate, iodide, nitrile
and ethylmercurys are in worldwide use for the
control of seed and soil-borne fungal diseases of
cotton and cereal grains.

Outlet	Consumption (tons)	
	1969	1971
Agriculture	102	58
Amalgamation	7	withheld
Catalysts	112	43
Dentistry	116	91
Electrical equipment	710	644
Chlor-alkali plants	788	466
Laboratory uses	78	69
Industrial controls	265	185
Paints	370	327
Paper and pulp	21	withheld
Pharmaceuticals	27	26
Other uses	410	87
TOTAL	3006	1996

*U.S.Bureau of Mines.

Phenyl mercury salts, including the acetate, borate, dodecenylsuccinate, nitrate, oleate and propionate are in tonnage usage as bactericides and fungicides for preserving aqueous systems against
microbial spoilage; to prolong the useful service
life of paint films; and in the protection of textiles,
plastics, wood and other substrates against fungal
deterioration.

Inorganic compounds of importance include red
mercuric oxide, the major ingredient of the well-
known mercury battery and a key component in
the present trend of miniaturization. Mercury is
used as a catalyst in elemental form and as the sulfate and bichloride. Yellow mercuric oxide, mercurous chloride (calomel), ammoniated mercuric
chloride and mercuric acetate are employed in the
manufacture or formulation of pharmaceuticals.

Tremendous interest has developed in the widespread presence of mercury in the environment including the biological food chain. The sources appear to be from industrial and agricultural usage
and, possibly of greater importance, that from natural deposits. The element is distributed over the
earth's crust with a content estimated at $1\text{-}30 \times 10^{-6}$
weight %; among plant life one-tenth to one-hun-
dredth of this level appears. The concentration in
sea water is 0.00003 mg/liter; in a cubic mile 0.1
ton and in all the oceans, 46×10^{6} tons. Recent
estimates indicate there may be as much as 80,000
tons in the earth's atmosphere. Fuel oil and coal
release 0.1 lb Hg for each 500 tons burned. Vapor
equilibria with land and water metal contribute too.

Mercury absorption is via the respiratory tract,
skin and the gastrointestinal route. For an eight-
hour daily exposure to air containing 0.1 mg Hg/m^3
(the maximum allowable under present health
standards) and a mercury absorption efficiency of
25%, 2 to 4 micrograms Hg/kg body weight will be
taken in. Normally, this quantity is excretable in
24 hours, mainly in the urine and secondarily in
feces, tears, saliva, sweat and bile. Air analyses of
several industrial areas in New York State show a
mercury presence of less than 0.0001 mg/m³.

The most recent appraisal of mercury origins in
the tissues of humans is directed to dietary sources.
Although the estimated average background level
in foodstuffs is less than 0.2 part per million by
weight, it now seems that certain species of fish can
concentrate more in their tissues. It has been postulated with some experimental verification that
mercury in any form, i.e., elemental, inorganic or
alkyl and aryl, is converted to the highly toxic methylmercury salts by anaerobic bacteria in deep and
shallow waters. The ingestion of these microorganisms by higher organisms pursued up the food
chain to fish results in levels above 1 part per million in some tuna, swordfish, walleye pike and other
species. At present a mercury equivalent content
of 0.5 part per million (wet weight edible portion)
in fish means official regulatory rejection in Canada
and the U.S.A. In Sweden a permissible maximum
level of 1 part per million in fish has been set.
Otherwise it is feared that all coastal and lake fishing would have to be abandoned as food sources.

Elemental mercury has a relatively high lipid
solubility which assists in its widespread distribution to the tissues, including migration through the
blood-brain barrier similarly to the monoalkyl and
dialkylmercurials. The metal is oxidized after
transport *in vivo*. The kidneys and liver appear to
have highest concentrations with levels in the blood
and urine serving as a questionable diagnostic indicator. For periods as long as six months following
exposure mercury has been found in urine at concentrations of 0.03 to 0.07 mg/liter. Human finger-
nails and head hair contain approximately 1×10^{-4}
and 2×10^{-4} weight %, respectively.

The phenylmercurials are considered much less
toxic than the alkyl derivatives and may be compared to the inorganic mercurials. L. Goldwater
and associates found that workmen who were exposed for significant time periods to phenylmercury
compounds in air at concentrations equivalent to
0.1 mg/m³ of Hg showed no harmful effects. The
question still remains as to how much mercury can
be absorbed safely. It has been reported that the
ingestion of 500 g Hg (elemental) caused no poisoning; 2 g of metallic mercury taken intravenously
in a suicide attempt produced subacute poisoning
which cleared after three days.

The toxic signs of mercurialism are readily recognized. Vapor inhalation chronic toxicity is
marked by sore throat, listlessness, tremors, abdominal discomfort and other signs along with
mercury excretion in urine ranging from 300 to
1200 micrograms/24 hrs. Loss of kidney function
leads to death. High air concentrations damage
lung tissue. Acute poisoning is accompanied by
severe gastrointestinal symptoms.

Excellent analytical techniques have been developed for assaying concentrations as low as parts
per trillion. Atomic absorption, x-ray fluorescence
and ultraviolet absorption spectroscopy, neutron
activation and radiochemical isotope exchange procedures are available. A very useful method utilizes digestion of samples with oxidants to destroy
all organic material, reduction chemically to elemental mercury in a closed system, followed by
measurement of concentration in the vapor phase
using absorption spectroscopy at 2537 Å.

A major effort is under way to minimize mercury

pollution. Factory effluents may be treated by settling, physical precipitation with ferrous chloride and chemically as the sulfide, adsorption with carbon or resin binding and zinc amalgamation. Greater recycling of process liquors in closed systems reduces disposal problems. A search for nontoxic replacements in agricultural and industrial applications has proved partially successful.

The handling of mercury and its compounds requires segregation of materials with good ventilation. Protective equipment such as impermeable gloves, eye protection, and respirators with special absorbents are essential, especially when dusts and vapor inhalation are hazards.

Clinically employed detoxicants are based upon metal chelation for inactivation and excretion. N-Acetylpenicillamine, 2,3-dimercaptopropanol (BAL) and ethylenediaminetetraacetic acid are complexants of choice.

NATHANIEL GRIER

References

1. *Annals of The New York Academy of Sciences,* O. v. St. Whitelock, Editor-in-Chief, "Mercury and its Compounds," **65** (5), 357 (1957).
2. Bidstrup, P. L., "Toxicity of Mercury and its Compounds," Elsevier Publishing Co., New York, 1964.
3. United States Department of the Interior, Bureau of Mines, "Mercury in 1971," Mineral Industry Surveys, Washington, D.C.
4. "Hazards of Mercury," Report of a Study Group, Norton Nelson, Chairman, Pesticide Advisory Committee to Sec. HEW, Environmental Research, vol. 4, No. 1 and 2, 1971.
5. Grier, N., "Mercury," pp. 401–411, "Encyclopedia of the Chemical Elements," Hampel, C. A., Editor, Van Nostrand Reinhold Co., New York, 1968.

METALLIC SOAPS

Metallic soaps are compounds in which a metal has replaced the cation, the acid hydrogen or its equivalent, in a complex, monobasic acid. The metallic soaps divide rather sharply into two classes, the first being quite water-soluble and consisting of compounds based on the alkali metals (with the exception of some lithium derivatives). The second group consists of substantially water-insoluble and hydrocarbon-soluble compounds, and includes salts of both the alkaline earths and heavy metals. It is this latter group which is indicated, generally, when metallic soaps are mentioned.

As indicated, the anion cannot be too small, seven carbon atoms usually being taken as a minimum. The acid may be a fatty acid, either saturated or unsaturated, a rosin acid (e.g., abietic), a naphthenic acid (alkylated five-carbon ring with the carboxyl on a side chain) or similar compounds.

Various methods of preparation may be used, fusion being common. In this case an oxide or hydroxide may be fused into the acid, usually at a somewhat elevated temperature to complete reaction and remove water. Another popular technique is to metathetically react a water solution of the sodium salt of the acid with a solution of an appropriate metal salt, such as the chloride. In this instance, the impurities are washed away and the pulp dehydrated at an elevated temperature.

Though a very large number of metallic soaps are available commercially, the cobalt, lead, manganese, iron, zinc and calcium salts of napththenic, linoleic, 2-ethyl hexoic, rosin and talloil are the ones most used as driers by the paint and varnish, and printing ink industries. They are offered as solids, liquids and powders and the metal concentrations, the factor most affecting their catalytic activity, vary widely to answer technological requirements. The most widely used products consist of cobalt naphthenate 6%, a low viscosity, blue-purple liquid containing 6% of cobalt; manganese naphthenate 6%, a light brown liquid; and lead naphthenate 24%; a pale amber liquid.

Though the soaps really are oxidation and polymerization catalysts, the term "drier" was used early in the paint field and has clung. It is derived, of course, from the "drying" of unsaturated oils in which the catalyst has been included, the liquid gradually changing to the solid state commonly desired in protective and decorative coatings.

To be useful, the driers must be incorporated into vehicles (paints, varnishes, printing inks, etc.) which chemically have a polyfunctional structure. Polyenes, of course, as exemplified by the drying oils such as linseed, represent a common example of compounds in which two or more intermolecular linkages may be established. When these linkages are present, three dimensional growth becomes possible and gelable, "drying" type compounds may be produced. (See **Driers**).

Though the solidification of such films is extremely complex, oxidation, polymerization and association are usually involved. Oxidation usually occurs first; peroxide intermediates then form and initiate chain polymerization, as well as carbon-oxygen-carbon linkages and hydroxylated derivatives. Polymerization also progresses, a copious supply of free radicals being available from the oxidation process. Finally, because of the large, and frequently polar, molecules, which form as a result of the two steps mentioned, association of molecules through van der Waal forces, hydrogen bonding, etc., occurs.

The introduction of driers into such a system affects all three mechanisms, since the soaps used are not only oxidation and polymerization catalysts but nonaqueous wetting and dispersing agents as well. The most obvious effect of the drier addition is to shorten the induction period substantially, i.e., the time before oxygen combines with the film in measurable amount. This appears to depend on the precipitation or deactivation of the natural inhibitors present.

Once the induction period has been passed, oxygen combines with the material at a much greater rate when an oxidation catalyst, such as manganese or cobalt soap, is present. This may occur through true catalysis, in which the catalyst promotes the reaction but does not enter into it, or it may be that the soap functions as an oxygen carrier. Pointing in the latter direction is the fact that the effective drier metals are those which have multiple valence states.

An effective polymerization catalyst, such as a

lead soap, causes solidification of the film at an earlier state in the oxidation process, presumably because polymerization is progressing concurrently. Secondly, there is a decrease in the total oxygen combined with the oil when it has solidified, where driers are used. Further, polymerization of the film usually is more advanced at the gel point in the latter case.

No definite data are available as to whether the polar nature of metallic soaps assists in orienting the oil molecules during drying. If this occurs, it, of course, can substantially accelerate the reactions discussed above. At any rate, it is apparent that the whole film conversion is profoundly affected by the one or two per cent of soap customarily added in a judiciously chosen ratio of oxidation catalyst to polymerization catalyst.

In other fields metallic soaps have equally important functions. Thus, to prevent embrittlement and discoloration of halogenated plastics, such as polyvinyl chloride, small quantities of cadmium, barium, zinc, calcium and lead soaps are incorporated. By functioning as hydrogen chloride acceptors they markedly reduce the degradation rate of the polymeric system under both heat and light, making useful materials possible.

Huge quantities of copper soaps, especially the naphthenate, have been used as fungicides and have proved most effective for applications as demanding as timber preservation. Further, the lubricating grease industry uses large quantities of metallic soaps, especially those of the alkaline earths and of lithium, in the manufacture of all kinds of industrial lubricants. In short, the potential applications are manifold and increasing because of the unique combination of properties found in metallic soaps.

S. B. ELLIOT

METALLOCENES

The metallocenes constitute a class of neutral transition metal compounds containing two cyclopentadienyl (C_5H_5) ligands π-bonded to a central metal atom in a "sandwich" structure, as exemplified by ferrocene (see below).

The original metallocene, ferrocene, was first reported simultaneously and independently by two groups of workers in 1951. Miller, Tebboth, and Tremaine at British Oxygen Ltd. passed cyclopentadiene over iron powder and an ammonia catalyst; Kealy and Pauson at Duquesne University treated ferric chloride with cyclopentadienyl magnesium bromide. Both groups obtained the same orange, air-stable, hydrocarbon-soluble crystals (m.p. 173°) and proposed a sigma bonded structure for this first organometallic derivative of iron. Within a year groups at Harvard University including Woodward, Wilkinson, Rosenblum and Whiting revealed that $C_{10}H_{10}Fe$ exhibits unusual thermal stability, has no dipole moment, and undergoes typical aromatic substitution reactions; whereupon, they proposed the novel sandwich structure (Fig. 1) and advanced the name ferrocene to reflect both the iron content of the material and its aromaticity. X-ray crystallographic studies subsequently supported the sandwich structure, showing the preferred staggered, antiprismatic conformation in the solid state. In solution or the vapor phase the barrier to rotation of the rings has been shown to be very small.

One proposed bonding scheme, using a molecular orbital model, calls for utilizing the π orbitals of the $C_5H_5^-$ group and the d_{xz}, d_{yz}, s, p_x, p_y, and p_z iron orbitals to obtain a total of six bonding orbitals with the twelve π-electrons from the rings to fill them.

After the discovery of ferrocene many additional metallocenes and their derivatives were prepared, including titanocene (green), vanadocene (purple), chromocene (scarlet), cobaltocene (purple-black) and nickelocene (green). Dicyclopentadienyl manganese (amber) is ionic, and dicyclopentadienyl mercury, tin and lead are sigma bonded; their derivatives are not properly called metallocenes. In the second and third transition series, only ruthenocene and osmocene are known as simple, neutral metallocenes; the other metallocenes are known as cations (e.g. Cp_2Ti^+) or with additional ligands such as halide or hydride attached to the metal atom, as in Cp_2TiCl_2, Cp_2ReH, Cp_2TaH_3, etc. Though several preparative approaches are known, the most general reaction leading to metallocenes is the treatment of a metal halide with sodium cyclopentadienide in tetrahydrofuran. Ideally a +2 salt of the metal is used, although excess of the cyclopentadienide will often reduce a higher valent metal halide to a lower oxidation state during the reaction.

Chemically, ferrocene undergoes many typical aromatic substitution reactions such as Friedel-Crafts acylation or alkylation, sulfonation, mercuriation, lithiation, the Villsmaier reaction and the Mannich reaction (with dimethylamine, formaldehyde and acetic acid). Its reactivity is very great ("superaromatic") and is comparable in rate to that of phenol. Mono- and disubstitution on one or both rings can be realized, though some measure of control to predominately mono- or disubstitution can be exercised by adjusting conditions.

However, the central metal atom in ferrocene imposes some limitations on the chemistry of the system or provides in some cases additional reaction pathways. For example, the central iron atom is readily and reversibly oxidized from the Fe(II) state to Fe(III) in the form of the water-soluble red-blue dichroic ferrocinium ion. This oxidation occurs with halogens or with nitric acid so that direct aromatic halogenation and nitration cannot be realized; however, halo- and nitroferrocene have been prepared by indirect methods. Aminoferrocene cannot be diazotized, presumably due to oxidative destruction of the system. Coupling with diazonium salts is anomalous; ferrocene reduces diazonium salts to phenyl radicals in aqueous or nonaqueous solution to yield Gomberg (phenylation) products. Condensation with aldehydes in acid solution gives rearranged products due to a role that the iron atom can play. The central iron atom is readily protonated in strong acid media.

In the solid state ruthenocene and osmocene prefer the eclipsed, pentagonal prismatic structure (Fig. 1), and in solution exhibit a chemistry similar to that of ferrocene. The other metallocenes, on the other hand, are quite different chemically, none of them showing the typical aromatic substitution reactions of ferrocene.

Cobaltocene is rapidly oxidized in air or in solution (it liberates H_2 from water, slowly) to yield the very stable yellow +1 cation, which has been re-

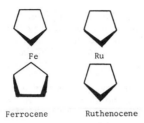

Ferrocene   Ruthenocene

Fig. 1

ported to be stable to aqua regia. Neutral co-baltocene reacts with alkyl and acyl halides to give adducts $(C_5H_5)(C_5H_5R)Co^+X^-$ in which the substituted ring is π-bonded to the cobalt as a diene.

Nickelocene, on the other hand, is more slowly oxidized and undergoes addition of suitable activated olefins to one ring converting it to a bicyclic ligand. Some ligand replacement reactions are also known for nickelocene.

Titanocene derivatives undergo substitution reactions at the metal atom:

$$Cp_2TiCl_2 + C_6H_5Li \rightarrow Cp_2Ti(C_6H_5)_2 + LiCl_2$$

$$Cp_2TiCl_2 + 2OCH_3^- \rightarrow Cp_2Ti(OCH_3)_2 + 2Cl^-$$

The utility of these materials is limited and disappointing. Ferrocene has been used to promote the burning of fuel oils and as a catalyst in rocket fuels. Its extreme thermal stability (to over 470°) spurred an interest in its derivatives as high-temperature fluids. Some polymers have been made, but they have found no substantial application. Titanocene dichloride has been shown to react with molecular nitrogen in systems containing added Grignard reagents or similar other materials.

It is generally agreed that the real importance of metallocenes lies in the effect of their discovery twenty years ago along with Ziegler's catalyst on the course and direction of organometallic chemistry. There has been a veritable explosion of interest and effort in this field during the past two decades that is unabated today. Many new materials and new insights into bonding of transition metals have evolved. Great strides in understanding of homogeneous catalysis and the design of new catalysts for hydrogenation, coupling, carbonylation, polymerization, etc., have been made. Even a closer understanding of the action of vitamin B_{12} has resulted from the new brand of organometallic chemistry, and we are moving closer to a practical method for fixation of nitrogen in solution.

WILLIAM F. LITTLE

References

Rosenblum, M., "Chemistry of the Iron Group Metallocenes," Part I, Interscience Publishers, 1965, New York.

Rausch, M. D., "Metallocene Chemistry–A Decade of Progress," *Canadian Journal of Chemistry,* **41,** 1289 (1963).

Little, W. F., "Metallocenes," *Survey of Progress in Chemistry,* **1,** 133 (1963), Arthur Scott, ed., Academic Press, New York.

METAL-METAL SALT SOLUTIONS

Solutions of metals in nonmetallic liquids are unusual. Most familiar are the solutions of the very electropositive metals in amines and ethers. Another important example is the solution of many metals in their molten salts. The few examples of metals dissolving in foreign salts are usually the result of oxidation-reduction reactions, and will not be considered proper solutions. By the latter part of the 19th century, many metal-metal salt interactions had been reported. A few investigators suggested that these solutions were the result of the formation of lower oxidation states and reported the isolation of such species as Na_2Cl, and CdF. Around 1900, Lorenz theorized that these solutions were, instead, colloidal suspensions of the metals in the molten salts ("pyrosols"). However, subsequent examination by ultra-microscopy, and cryoscopic measurements, indicated that these were true solutions.

A large number of metal-metal salt solutions are known today. Most involve halides, although a few chalcogenide and hydride systems have been reported.

The extent of solubilities of metals in their molten salts varies considerably. All of the alkali metal halide systems and most of the alkaline earth halide systems show complete miscibility. The consolute temperatture (the temperature at which complete miscibility first occurs) is observed to decrease with increasing weight of a halide for a given element, and is observed to decrease for increasing weight of an element within one family. Most of the transition metals show only slight solubility, while the post-transition elements vary from slight solubility of the Au-AuCl, Tl-TlCl, and Sn-SnCl$_2$ systems to complete miscibility of the Bi-BiCl$_3$ and Bi-BiI$_3$ systems. The lanthanide elements are somewhat intermediate in behavior. Some such as Nd show very extensive interactions with their halides while others such as Ce are quite limited in their interactions.

It has been convenient to divide these metal solutions into two main categories: The first consists of those solutions which show largely metallic character. Here, it is assumed that in solution of high salt concentration, the metal is solvated or ionized with the electrons occupying anion sites in the liquid salt structure. In the solutions of high metal concentration, the anions now behave as electrons in a metallic lattice.

The second category includes those systems where there is a strong interaction between the metal and its molten salt. Here, there appears to be loss of a metal electron(s) to form a lower valent state which may or may not be stable in the solid. For some systems such as Cd-CdCl$_2$, this lower valent state exists only in the liquid state, disproportionating to metal and dichloride on solidification, while in the system Cd-Cd(AlCl$_4$)$_2$, it is possible to isolate a solid compound containing the cadmium (I) ion.

Classification of the metal-metal salt solutions into these two categories is difficult since it is hard to differentiate between the species that results from a weak interaction which produces a solvated metal atom, M-M^{2+}, or from a strong interaction that may result in electron transfer to form M^+-M^+.

More recently, a more exact differentiation has been obtained by determining the standard free energies of solution for mtals in their molten salts. Those systems showing weak interactions, with free energies of solution less negative than -25 kcal, include most of the mono- and divalent metal ions; those showing strong interactions of more negative than -35 kcal are characteristic of the transition metals and the trivalent metals. The former is characteristic of a solvation process, while the latter is an indication of formation of a lower valence.

L. F. DRUDING

Cross-references: *Liquid State, Liquid Metals.*

References

Sundheim, B. R., "Fused Salts," McGraw-Hill, New York, 1964.

Blander, M., "Molten Salt Chemistry," John Wiley & Son, New York, 1964.

METALS

Occurrence and Properties

The word "metal" has definite connotations, but an accurate all-inclusive definition is impossible. However, by almost any classification roughly three--quarters of the elements are metals, including the Periodic Table Groups IA, IIA, all of the "B" groups, IIIA with the possible exception of boron, IVA with the exception of carbon and possibly silicon, VA with the exception of nitrogen and phosphorus, and VIA with the exception of oxygen and sulfur.

In general, they are solids with a metallic luster, conductors of electricity by electron flow, malleable, and of high physical strength. In compound form the metals have positive valences. Probably their most important characteristic is that when used *as metals* they are predominately in elemental form or alloyed with other metals. Another seldom-defined characteristic is the profound effect upon their properties caused by the presence in them of relatively small amounts of other elements. The great and extremely important differences between iron and steel caused by the small amount of carbon associated with iron in steel exemplify this effect (See **Impuriities**).

Great variations are found in the natural prevalence of metals in the earth's crust, as shown in the table on the following page.

Some metals which are low in absolute concentration in the earth's crust, such as copper, tin and antimony, are in common use, while many others in high absolute concentration, such as titanium, rubidium and zirconium, have been exploited to a minor degree. The metals long used by mankind—iron, copper, zinc, tin, lead, mercury, silver and gold—are those which exist as easily recognized minerals in large deposits and which are easily reduced from compound to elemental form. Fortunately, the most important metal, iron, is both prevalent and easily reduced to metallic form. In more recent times some of the most naturally prevalent metals that are difficult to reduce have become common and readily available due to the development of electrochemical processes for their production. These include aluminum, magnesium and sodium.

To be commercially attractive an ore must contain a high enough concentration of the desired metal to make the extraction of it technically and economically feasible. This varies greatly from metal to metal: for copper it may be 1% or less, but for iron it should be 30% or more; for magnesium it can be 0.13%, the amount in sea water, but for aluminum it is 30% or more.

The metals occur most commonly as oxides or sulfides in ores that contain variable amounts of gangue materials like clay, silica, granite, etc., from which the metallic compounds must be separated. Less commonly, metals exist in the natural state as chlorides, carbonates, sulfates, silicates, arsenides, and complex compounds of a wide variety. By far the greatest tonnage of metals is derived by the reduction of natural metallic oxides or ones formed by calcining ores, such as the sulfides, in air to convert their metallic content to oxides.

The majority of ores are given some type of beneficiation treatment prior to the reduction stage to increase the concentration of the desired component or to eliminate specific impurities. Such treatments involve the physical ones of hand picking, grinding and screening, flotation, magnetic separations, etc., or the chemical ones like the Bayer process to separate substantially pure Al_2O_3 from the iron and silica in bauxite. Many of the less common metals present in minute quantities in ores processed primarily for other metals are concentrated by these and other operations to values which permit their extraction and separation. If it were not for this fact, most of the rare metals would still be laboratory curiosities. With few exceptions, notably iron ore, most ores contain, and are mined and processed for the production of more than one metal.

Since only a few of the metals, such as copper, gold, silver, platinum and bismuth, exist naturally in elemental form, the chief problem is that of reducing them from compound to elemental form. Rarely can identical processes be applied to the production of more than one metal, and the great differences in chemical and physical properties among the metals demand almost as many processes for their isolation as there are metals. Because of these differences generalities are difficult, but the choice of process is dictated by the chemical and thermodynamic properties of the systems involved. Some of the more common reduction methods include:

A. Pyrometallurgical

(1) Heating sulfides in air

(2) Heating mixtures of oxides and sulfides

(3) Carbon reduction

By far the greatest tonnage of metals, chiefly iron, is produced by the high-temperature reduction with carbon or carbon compounds. This age-old technique is used to make iron, copper, zinc, cadmium, antimony, tin, nickel, cobalt, molybdenum, etc. from natural oxides or oxides formed by calcining ores like sulfides. It is limited to the production of metals which are weaker reducing agents than carbon at attainable furnace temperatures, and in many cases results in carbon or carbides being associated with the resultant metal. In the production of steel this is a desirable feature, but the presence of these materials is a drawback in many metals.

AVERAGE AMOUNTS OF THE ELEMENTS IN THE EARTH'S CRUST

From Brian Mason, "Principles of Geochemistry," John Wiley & Sons, Inc., New York, 1952.

Element	p.p.m.	Element	p.p.m.
O	466,000	Ge	7
Si	277,200	Be	6
Al	81,300	Sm	6.5
Fe	50,000	Gd	6.4
Ca	36,300	Pr	5.5
Na	28,300	Sc	5
K	25,900	As	5
Mg	20,900	Hf	4.5
Ti	4,400	Dy	4.5
H	1,400	U	4
P	1,180	B	3
Mn	1,000	Yb	2.7
S	520	Er	2.5
C	320	Ta	2.1
Cl	314	Br	1.6
Rb	310	Ho	1.2
F	300	Eu	1.1
Sr	300	Sb	1?
Ba	250	Tb	0.9
Zr	220	Lu	0.8
Cr	200	Tl	0.6
V	150	Hg	0.5
Zn	132	I	0.3
Ni	80	Bi	0.2
Cu	70	Tm	0.2
W	69	Cd	0.15
Li	65	Ag	0.1
N	46	In	0.1
Ce	46	Se	0.09
Sn	40	A	0.04
Y	28	Pd	0.01
Nd	24	Pt	0.005
Cb(Nb)	24	Au	0.005
Co	23	He	0.003
La	18	Te	0.002?
Pb	16	Rh	0.001
Ga	15	Re	0.001
Mo	15	Ir	0.001
Th	12	Os	0.001?
Cs	7	Ru	0.001?

(4) Hydrogen reduction
B. Electrolytic

Electrolysis offers the best means, and in some cases the only practical means, of producing many pure metals. It is also used to refine crude metals, such as copper, made by other reduction processes. Deposition from aqueous solution is used to prepare cadmium, copper, cobalt, gallium, indium, manganese, thallium and zinc. It is valuable in the application of many metals as electrodeposited coatings, among them cadmium, copper, cobalt, chromium, palladium, platinum, rhodium, indium, tin, tungsten, nickel and zinc. Deposition from molten salt baths has made possible the commercial production of aluminum, magnesium and sodium, and is used to prepare the other alkali metals, the alkaline earth metals, the rare earths, etc.

Electric furnaces are used in the production of the whole range of ferro alloys so vital to the preparation of alloy steels. Among the elements so combined with iron are manganese, chromium, vanadium, tungsten, molybdenum, nickel, silicon and titanium. In this case electrolysis is not involved but the electric furnace is the source of higher temperatures than can be reached conveniently by other means.

C. Metal replacement

The powerful reducing action of such metals as aluminum, calcium, sodium and magnesium is used in the preparation of other metals. The well-known Goldschmidt or thermite reduction of metal oxides with aluminum, and the Kroll process using magnesium to reduce metal chlorides are examples of this technique.

D. Halide decomposition

Certain halides of metals are decomposed at high temperature to yield pure metal and the halide or a lower halide of the metal. This is the basis of the DeBoer and Van Arkel process, whereby such metals as titanium, hafnium and zirconium are prepared in very pure form by the decomposition of the respective tetraiodides of these metals by a wire heated in an atmosphere of the iodides. It can be used to refine impure metals; the iodine liberated by the decomposition reacts with impure metal in the same container to form the iodide and allow the cycle to be repeated. Many metal halides disproportionate at high temperature to form the metal and a higher valence halide.

Many additional chemical operations are involved in the production of metals to effect the removal of undesired contaminants, to separate one or more valuable elements present in the mineral raw material, and to provide the correct environment for each of the several steps required for the preparation of metals. These reactions cover a great part of the field of inorganic chemistry and constitute one of the major commercial applications of chemistry.

Alloys. With the exception of the carbon steels, which may or may not be regarded as alloys, most of the metals used are in the form of alloys of two or more constituents. The whole field of alloy systems has grown in importance to both the supplier and the user of metals due to the variations in various properties of metals imparted by the presence of one or more alloying agents. The alloy steels, the bronzes and the aluminum alloys are well-known examples of such trends. Many applications, chiefly those where the metal is an electrical conductor, require the highest degree of purity in the metal, and in many cases the pure metal is desired, even if it is to be alloyed, to insure that only the desired constituents be present in the final alloy. Of special interest is the corrosion resistance of alloys, for example, the stainless steels and the copper-nickel alloys. Much of the chemical industry depends upon such special materials and the metals industry has permitted major chemical developments to be made by providing the required corrosion-resistant alloys. The advances in aircraft have been due in large measure to the development of strong, light-weight structural alloys as well as high-temperature alloys for propulsion systems.

For further information, see specific metals in this volume.

CLIFFORD A. HAMPEL

Classification, Theory and Structure

The word "metal" has two somewhat different meanings: (1) *Chemically* a metal is a chemical element which tends to form positive ions in solution and whose oxides form hydroxides rather than acids with water. (2) *Physically* a metal is a phase containing free electrons which give it certain characteristic properties such as high electrical and thermal conductivity, metallic luster, and, often, plastic formability. These phases need not be pure substances; when they are not, they are called *alloys*. The somewhat awkward term "metal and alloys" is often used when it is desired to clearly indicate the second meaning is intended.

There is no quantitative definition by either of these criteria which is without objection. The best single measure of chemical characteristics is probably the electrochemical series; of physical metallic character the best measure is probably the electrical conductivity.

Qualitatively, the two criteria agree rather well; elements which form positive ions are also conductors of electricity. Quantitatively, the correlation is extremely poor for several elements. Gold, for example, at the very bottom of the electrochemical series, is nearly the best conductor of electricity. In the present discussion attention will be focused on the metallic state; the metallic elements will be considered mainly as they serve as raw materials for making metals and alloys.

In metallic phases the atoms are held together by chemical bonds characterized by mobile or "free" electrons. These are responsible for their electrical conductivity, which is of the order of 10^{+5}, while nonmetals generally have a conductivity of less than 10^{-10}. A very few elements fall between; they are semiconductors (see **Semiconductors**). The free electrons are also efficient conductors of heat so metals are typically nearly a thousand times as good conductors of heat as nonmetals. For nearly all metals and alloys the conductivity of heat is nearly proportional to the conductivity of electricity. The ratio at room temperature (the Wiedemann-Franz ratio) lies between 100×10^{-8} and 200×10^{-8} for most metals and, probably, alloys. This ratio changes with temperature because electrical conductivity is almost inversely proportional to absolute temperature, while thermal conductivity is relatively constant with changing temperature.

Metallic phases may be recognized by a peculiar appearance known as metallic luster. This is due to the effect of free electrons in reflecting light rays. However, many nonmetals have striking metallic lusters. Iron pyrite (FeS_2) is known as "fool's gold" because of its resemblance to gold.

Modern industry is impossible without access to ample supplies of metals. Most great industrial centers are close to metal-producing districts. Economists recognize the per capita consumption of metals as an excellent index to a country's standard of living. The United States, for example, with one-fifteenth the world's population, consumes perhaps one-third of the world's metal.

The greatest source of iron ore in the world has been the Lake Superior region which produced 80% of the United States' supply. With the approaching exhaustion of the high-grade Lake Superior ores (more than 50% iron content) interesting changes have occurred.

New ore deposits have been found and are being exploited in Canada and Venezuela, so that the United States has become a net importing rather than an exporting country in iron ore. Concentration processes have been developed and applied to the enormous lower grade deposits of Lake Superior. These have been so successful that they are also applied to higher-grade ores formerly used without treatment. Over 75% of all blast furnace feed is now concentrate. The iron content now averages 56% instead of 51% iron formerly used from untreated ores. With the richer feed, blast furnace capacity is greatly increased.

Iron oxide minerals, placed in an ordinary bonfire, will be reduced to spongy iron which may be withdrawn and pounded into a solid mass. Left in prolonged contact, the solid iron will dissolve carbon, becoming steel. Until the fourteenth century all iron and steel was made this way as, for example, Damascus steel. To increase production, the fire was placed in a shaft with a blast of air admitted at the bottom. Larger shafts and greater blasts produced higher temperatures, eventually melting the iron. Large amounts of carbon, silicon, manganese, and other impurities are dissolved in this product, called *pig iron,* which is essentially the same as ordinary modern cast iron.

Under oxidizing conditions most of the impurities can be burned out of molten pig iron. Finally, deoxidizers and recarburizers are added which remove excess oxygen and bring carbon, manganese, and other components up to the percentages desired in the steel. The first mass production process for steel was the *bessemer process* invented in the 1850's. It is still very important, though at present most American steel is made in the open hearth and basic oxygen (BOF) furnaces, with electric steel an important factor.

The ancients knew seven metals, gold, silver, copper, iron, tin, mercury, and lead. With the passage of centuries a few more were added. In modern times progress has accelerated to a truly dizzy pace. The number of known metals has increased to 80 and none of them are so rare or difficult to extract that they cannot be considered for some of the increasing demands of modern technology. An insatiable demand exists for metals which can meet ever more rigorous conditions. Metals are badly needed which are stronger, lighter, or will operate at higher temperatures under more extreme conditions. The key problem in many fields, such as nuclear power, jet propulsion, or supersonic speed airplanes, is materials, for which metals supply the usual answer. The main research efforts of the great electric companies are devoted to materials, not to electricity.

During the twentieth century beryllium, magnesium, tantalum, niobium (columbium), titanium, zirconium, tungsten, and uranium have found important places in industry, though they were only scientific curiosities in the nineteenth. Of these, titanium and zirconium have become important only since the close of World War II. Zirconium's use, as yet, is specialized to atomic power; it has unusually low neutron capture characteristics.

Structure. Metallic elements, with few exceptions, crystallize in simple crystal structures: face-centered-cubic, body-centered-cubic, and hexagonal-close-packed. Metallic phases may be regarded as lattices of positive ions held together by free electrons which wander between the ions. Drude showed in 1900 this qualitatively explained the most important properties of ions, but it remained for wave mechanics to put this on a firm foundation quantitatively.

Recently, Pauling has developed a theory of metallic bonding in which the atoms are held together essentially by covalent bonds with, however, some good bonding orbitals available but unoccupied. By resonance the electrons move into and out of these, giving the "free" electron character to the metallic state. It probably can be developed to explain the same facts already well explained by the free electron theory while giving a physical picture easier to grasp. Pauling's theory meanwhile gives a better picture of strength of bonding and interatomic distances in metals.

The special nature of metallic bonding leads to principles of alloy formation which are inadequately appreciated by chemists. In metallic crystals there is a widespread indifference to the nature of the nearest neighbors to an atom and a tolerance for differences of atomic size. This makes extensive solid solutions possible as well as extensive ranges of composition over which intermediate phases are stable. Atomic distributions over lattice sites are in general disordered, rather than the ordered distributions characteristic of nonmetals. The factors which cause the formulas of nonmetallic compounds to be expressible in the ratios of small whole numbers of atoms are much less effective in metals, so that the effort to write compositions of intermetallic compounds as A_xB_y, with x and y as small whole numbers is often misleading.

The above factors are somewhat concealed by the tendency of intermetallic compounds to contain a large or small nonmetallic component in their bonding. Thus some of them are nonmetallic in properties and exist at a single composition only, at a simple atomic ratio.

The principles of alloy formation are not very well known and much experimental and theoretical work remains to be done. Several important crystal chemical principles have been established, however, especially by Hume-Rothery.

(1) *Atomic size factor.* Wide range of solid solubility can exist only if the sizes of the solvent and solute atoms do not differ too greatly. The limiting difference is about 15% in atomic diameter.

(2) *Electron concentration rules.* The energy levels of the valence electrons are clustered in bands called Brillouin zones, the capacities of which are determined by the crystal structure. When a zone is filled to capacity, new electrons can be introduced only by placing them in the next highest energy band, thereby raising the internal energy and making the phase less stable. New electrons are introduced when a solvent atom is replaced by a solute atom of higher valence, hence the solubility is limited by this factor. These rules have been applied especially to binary alloys of gold, silver, or copper (valence 1) with metals of higher valence (the valence taken to equal the number of the column in the Periodic Table). The copper-zinc system (brass) is a typical example.

Strength. Probably the most valuable single attribute of metals as structural materials is their ability to be plastically deformed, that is, to change shape under load without increase of stress. Obviously, this permits metals to be shaped freely and cheaply but it is much more important that by this means stress concentrations may be relieved.

Uneven loading conditions are difficult to avoid in service and cause stresses to be concentrated in the parts which bear the load, sometimes very unevenly. Brittle materials will fracture at the point of greatest stress, sometimes with rather small total loads. Ductile metals, on the other hand, will yield at the points of high stress, distributing the load more evenly to the rest of the cross section, resulting in a high total load before failure occurs.

Many types of laboratory tests are performed in the attempt to predict metal behavior under service conditions which are often very complex. Two of the most important are the *tensile test* and the *impact test.* Among other properties the tensile test determines the tensile strength and the elongation. The latter measurement gives the amount of plastic stretch before failure and is a good measure of the *ductility* of the metal. In the impact test the energy of fracture of a part which has been notched to produce stress concentration is measured. This value depends both on strength and ductility and is often called *toughness.*

Most pure metals used in structures are highly ductile but have only moderate strength. They may be strengthened by alloying in two different ways. The added element may enter into solid solution which strengthens the metal. Usually, but not always, this strengthening is at the expense of ductility. A much greater effect is obtained when the second metal causes the formation of a hard, brittle intermetallic compound which is dispersed in the first phase. Steel is an example of this. If the dispersion is fine enough, and if the brittle phase is not continuous, the dispersion will have satisfactory ductility and toughness with greatly increased hardness and strength. In some alloys, such as duralumin (aluminum with copper added), the greatest strengthening effect occurs when the second phase is on the point of precipitating, but has not actually separated. This phenomenon is called *precipitation hardening* or *age hardening.*

The science of the structure of metals and alloys is called *metallography.* It deals with the effects of structure on properties and with the manner in which structures arise in metallurgical reactions. Typical properties of some common metals and alloys are given in the table below. Commercially pure metals are there compared with a solid solution (alpha brass) of zinc in copper, with a dispersion of iron carbide in iron (steel), and with an age hardened alloy (duralumin).

The alloys given are examples only; a wide variety of properties may be developed at different compositions and heat treatments.

Cold work (plastically forming at room temperature) also increases strength at the expense of ductility, sometimes in a very desirable way. If a cold-worked metal is heated above its *recrystallization temperature,* a new crop of crystals will grow and the properties will revert to the state before cold working. The recrystallization temperature depends on the degree of cold work and the time allowed during the heating. It is low for low-melting metals; for lead and tin it is below room temperature.

Metal	Tensile Strength Pounds/Sq. Inch	Elongation Per Cent
Iron	40,000	45
Steel	140,000	17
Copper	32,000	50
Alpha brass	46,000	58
Aluminum	13,000	40
Duralumin	68,000	20

Metals are never very strong above their recrystallization temperature, and for a considerable temperature range below it will fail by *creep.* This is a failure under a stress too low to cause immediate fracture and occurs when the stress is prolonged; sometimes failure occurs only after weeks or months. The creep test is the only useful one under these conditions. Creep strength is the limiting factor in the high-temperature use of metals when they do not fail by oxidation or chemical attack. Solid solutions have much higher recrystallization temperatures than pure metals.

The science of the mechanical properties of metals had no scientific foundation before the development of the *dislocation theory,* which has become prominent only since World War II. Under stress the metal deforms principally by *slip* of one block on another along a slip plane. From known interatomic forces it is possible to calculate the stress required to move one block over another. The predicted strength is too large by a factor of 100 to 1000. Dislocation theory postulates the existence of defects in the crystal lattice, called dislocations. With new techniques of electron microscopy, these dislocations can be observed. Under stress it is possible, because of these dislocations, for one row of atoms at a time to slip instead of the whole block, though as the slips propagate, the net effect is to move the whole block. This accounts for the low strength of metals. So many phenomena have now been accounted for that it has become accepted that the dislocation theory is essentially correct.

R. R. HULTGREN

Cross-references: *Crystals, Steel, Iron.*

METHYLENES

Methylenes, also called *carbenes,* are organic compounds containing divalent carbon. While certain divalent carbon derivatives in which the carbon is joined by a multiple bond to oxygen (carbon monoxide) or nitrogen (isocyanides) are known as stable compounds, and will not be considered here, most are highly reactive species that are known only as reaction intermediates. Thus, methylenes,

with carbonium ions, carbanions and free radicals, comprise the four most important classes of organic reaction intermediates containing carbon in an unstable valence state.

Some of the first strong evidence for the formation of methylenes came from studies of the gas-phase photolysis and pyrolysis of diazomethane, CH_2N_2, and ketene, $CH_2{=}CO$, which lead to the parent member of the series, methylene itself. Later work showed that the decomposition of many ketenes and diazo compounds gives derivatives of methylene.

Methylenes may also be generated by the α elimination of two atoms or groups attached to the same carbon atom. In the dehydrohalogenation of haloforms this may take place by the stepwise mechanism shown below for bromodichloromethane

$$CHCl_2Br + OH^- \rightleftharpoons CCl_2Br^- + H_2O$$

$$CCl_2Br^- \rightarrow CCl_2 + Br^-$$

$$CCl_2 \xrightarrow[\text{several steps}]{\text{OH}^-, \text{H}_2\text{O, fast}} CO \text{ and } HCO_2^-$$

The nature of the first step was demonstrated and its rate measured by deuterium exchange experiments. The intermediacy of dichloromethylene, CCl_2, was demonstrated by capturing it with various nucleophilic reagents including bromide, iodide and thiophenoxide ions. Bromide ions have the effect of slowing the reaction, since the CCl_2Br^- ions produced are in equilibrium with $CHCl_2Br$, the starting material.

Reactivity studies indicate that as substituents on the methylene the various halogens facilitate dihalomethylene formation in the order $F \gg Cl > Br > I$.

Other α-elimination reactions take place when a compound with a metal and a halogen atom attached to the same carbon atom decomposes, e.g.,

$$C_6H_5HgCBr_3 \xrightarrow{80°} C_6H_5HgBr + CBr_2$$

Many compounds decompose photolytically to give methylenes. The strain in the three-membered ring probably facilitates the following reaction,

$$C_6H_5CH{-}CHC_6H_5 \xrightarrow{h\nu} C_6H_5CHO + H{-}C{-}C_6H_5$$

but even methane decomposes, to give methylene and hydrogen, when irradiated with light of wave lengths around 140 nm.

Flash photolysis of diazomethane produces methylene in high enough concentrations, for a few microseconds, to make spectral measurements. The spectra show that the methylene formed initially is a singlet (a species with no unpaired electrons) with an H-C-H bond angle of about 103°, but that this singlet is rapidly transformed to the presumably more stable triplet (having two unpaired electrons), in which the H-C-H bond angle is considerably larger. Electron-spin-resonance studies of the naphthylmethylenes, $C_{10}H_7$-CH, produced and trapped by the photolysis of the corresponding diazo compounds in a glassy matrix at very low temperatures, show that in the most stable state these species are triplets with nonlinear structures around the divalent carbon atom.

Perhaps the most useful reaction of methylenes is addition to multiple bonds to give products with three-membered rings. With singlet methylenes these additions are ordinarily stereospecific one-step reactions; e.g.,

$$\underset{H}{\overset{CH_3}{\diagdown}}C=C\underset{H}{\overset{CH_3}{\diagup}} \xrightarrow{CF_2} \underset{H}{\overset{CH_3}{\diagdown}}\underset{CF_2}{C-C}\underset{H}{\overset{CH_3}{\diagup}}$$

With triplet methylenes the additions involve intermediate diradicals that may undergo rotations around single bonds that make the addition non-stereospecific.

$$triplet\ CH_2 + \underset{H}{\overset{CH_3}{\diagdown}}C=C\underset{H}{\overset{CH_3}{\diagup}} \longrightarrow \underset{H}{\overset{CH_3}{\diagdown}}C-C\overset{CH_3}{\underset{\cdot CH_2}{\diagup}}{\overset{}{\diagdown}}H$$

$$\downarrow$$

$$\underset{H}{\overset{CH_3}{\diagdown}}C-C\overset{CH_3}{\underset{CH_2}{\diagup}}H \qquad \underset{H}{\overset{CH_3}{\diagdown}}C-C\overset{H}{\underset{\cdot CH_2}{\diagup}}CH_3$$

$$\downarrow$$

$$\underset{H}{\overset{CH_3}{\diagdown}}C-C\overset{H}{\underset{CH_2}{\diagup}}CH_3$$

A relatively unique reaction of singlet methylenes (the more reactive ones, at least) is one-step insertion into a bond, often a bond to hydrogen.

$$singlet\ CH_2 + CH_3SiH_3 \rightarrow 90\%\ (CH_3)_2SiH_2$$
$$+ 10\%\ C_2H_5SiH_3$$

Some methylenes rearrange to olefins by migration of a group adjacent to the divalent carbon atom,

$$CH_3CHN_2 \xrightarrow{h\nu} CH_3-C-H \longrightarrow CH_2=CH_2$$

and others give more complicated rearrangements, e.g.,

$$\underset{C}{\overset{CH_2-CH_2}{\diagdown\diagup}} \longrightarrow CH_2=C=CH_2$$

Such rearrangements may be accompanied by internal insertion or addition reactions, often to yield highly novel products that would be very difficult to synthesize in any other way, e.g.,

$$\begin{array}{c}CH_2\\CH_2\diagup\quad\diagdown CH\\|\qquad\qquad|\quad\diagup C\diagdown^{Li}_{Br}\\CH_2\diagdown\quad\diagup CH\\CH_2\end{array} \rightarrow \begin{array}{c}CH\\CH_2\diagup\ |\ \diagdown CH\\|\qquad CH\diagdown\ |\\CH_2\diagdown\quad\diagup CH\\CH_2\end{array}$$

$$+ \begin{array}{c}CH_2\\CH\diagup\quad\diagdown CH\\|\quad\diagdown CH\diagdown\ |\\CH_2\diagdown\quad\diagup CH\\CH_2\end{array}$$

Since a number of the reactions of methylenes are highly exothermic, their products are formed with large amounts of vibrational excitation. In the liquid phase the excess energy is ordinarily transferred to surrounding molecules, but in the gas phase, especially at low pressures, it may bring about the decomposition or rearrangement of the molecule before it can be dissipated.

Some of the reactions originally attributed to methylenes (carbenes) are now thought to involve species called *carbenoids,* which act as donors of methylenes without ever giving free methylenes. The reaction of α-lithiobenzyl bromide with isobutene, for example, appears to give the products shown in one step without the intermediacy of free phenylmethylene.

$$C_6H_5-\underset{Li}{\overset{}{\underset{|}{C}HBr}} + (CH_3)_2C=CH_2 \rightarrow$$

$$(CH_3)_2C\overset{\qquad}{\underset{\underset{C_6H_5\quad H}{C}}{-}}CH_2 + LiBr$$

As another example, the copper-catalyzed decomposition of diazo compounds gives intermediates that yield some of the products expected from methylates but that are often less reactive and more selective than the methylenes obtained by photolysis of the same diazo compounds; it seems likely that in many such cases the former reaction gives a copper carbenoid as the reaction intermediate.

Jack Hine

References

Hine, J., "Divalent Carbon," The Ronald Press Company, New York, N.Y., 1964.

Kirmse, W., "Carbene Chemistry," 2nd Ed., Academic Press, New York, N. Y., 1971.

Moss, R. A., *Chem. Eng. News,* **47** (25), 60 (1969); **47** (27), 50 (1969).

MICELLES

The term *colloidal micelle* (Latin *micula,* a tiny grain) was suggested by Duclaux in 1905 to designate the individual particle in an aqueous colloidal suspension, or sol, of a hydrophobic substance. The stability of such sols depends on a net surface-charge on each particle, which, by mutual electric repulsion, prevents flocculation. Examples of typical colloidal micelles, according to Duclaux's definition, are the particles of a gold sol, clay particles of colloidal size, and other truly colloidal hydrophobic sols.

The term *micelle colloid* was originally coined by H. Staudinger, referring to high-molecular-weight organic compounds, e.g. cellulose, natural and synthetic rubbers, proteins, and the like. Micelle colloids have the property of swelling in appropriate solvents and eventually exist in the sol as macromolecules of colloidal dimensions ($10m\mu$ to $1\ \mu$), so that the micelle in these solutions is the ultimate polymeric particle.

If a distinction between the particles of a lyophobic sol and those of a lyophilic sol had to be made, it was unwise to select terms that were so similar in form; inevitably, the distinction became less and less sharp, so that today all that remains is the use of the term *micelles* to designate colloid particles of *any* type of sol.

The most familiar context, however, in which the term is used is to designate the colloidal aggregates formed spontaneously by molecular association of surface-active solutes, such as soaps, synthetic detergents, and certain dyes. These agents are called *association colloids*. In extreme dilution the degree of association is negligible, but as the concentration is raised the solute ions and molecules combine with each other to form micelles of colloidal dimensions. The average size of these association micelles is approximately fifty molecules; their shapes vary from spherical to ellipsoidal. The concentration range in which the number of micelles in the solution becomes significant is often so narrow that it can be characterized by a single value, known as the *critical micelle concentration* (c.m.c.). The c.m.c. has in general a low value for the most highly surface-active solutes. The micelles of association colloids have the property of "solubilizing" materials that are normally insoluble in the solvent: thus, soap micelles in water will draw hydrocarbons into the aqueous phase; conversely, micelles formed by oil-soluble detergents will draw water into the nonaqueous phase. In each case the micelles either dissolve or adsorb the "solubilizate." The effect is utilized in emulsion polymerization (aqueous solvent) and in dry-cleaning (nonaqueous solvent).

SYDNEY ROSS

Cross-references: *Colloid Chemistry, Films, Surface.*

MICROCHEMISTRY

Early qualitative chemical work with small quantities of substances was mainly based on the use of the blowpipe and the microscope. Microchemistry as a field of scientific research has been recognized only since the turn of the 20th century. The methods developed by its pioneers, Friedrich Emich and Fritz Pregl, proved so useful that they replaced in a short time many of the standard procedures in all fields of chemistry. Work on problems, which had been impossible before that time on account of experimental difficulties, or scarcity of material, could now be undertaken. Quantitative microchemistry owes its greatest advances to the commercial availability of microchemical balances; these are small carefully constructed analytical balances with a maximum capacity of 5 to 20 grams on each pan, and a precision of ±1 microgram. The following table gives a widely used classification of chemical methods according to the size of the sample used.

Sample Size	Name of Method
100 mg and over	Macrochemistry
10 to 100 mg	Semimicrochemistry or Centigram
0.1 to 10 mg	Microchemistry or Milligram
less than 0.1 mg	Ultramicrochemisty or Microgram

This distinction is mainly used for the classification of analytical procedures. For preparative work with small quantities of material it has become customary to speak of procedures using up to 50 mg of substance as being of micro size, while those involving over 50 mg to about 2 g are called semimicro.

The recent development of chromatography, its extension to colorless substances, and its versatility in equipment and application has increased the working with "micro" quantities of material greatly. To "column methods" have been added the modifications of technique of paper, and of thin layer chromatography. Outstanding successes in the identification and quantitative determination of volatile or volatizable organic substances are achieved with gas—liquid chromatography, especially in combination with a suitable fraction collector, which allows the identification of the separated compounds by spectrographic methods in the various regions of the spectrum. The latter procedure is extremely valuable in the analysis for very small quantities of pesticides, insecticides, fungicides, herbicides, rodenticides, and other highly toxic substances used in agricultural crop protection.

Reference to small quantities of mass has been made more convenient by adding to the list of fractional units several new prefixes. The new arrangement is

d	deci	10^{-1}	n	nano	10^{-9}
c	centi	10^{-2}	p	pico	10^{-12}
m	milli	10^{-3}	f	femto	10^{-15}
μ	micro	10^{-6}	α	atto	10^{-18}

For the measurement of wave lengths the following units are in use: 1 m (meter) $= 10^{6}\mu$ (micron) $= 10^{9}$ mm (manometer or millimicron) $= 10^{10}$ Å (Angström) $\approx 10^{13}$ X-units (for x-ray).

Several micro balances of the torsion type for maximum loads in the milligram range are also commercially available.

Since the quantities of material used in this field today do not approach molecular dimensions, the chemical behavior is the same as when large amounts of substances are employed. Microchemistry is, therefore, essentially concerned with finding suitable reactions, developing techniques for the handling of small quantities of solids, liquids, and gases, and bringing the resulting physical phenomena to exact observation. The term "microtechnique of chemical operations" would describe correctly that phase of chemical work, for which the designation "microchemistry" is now in use.

Qualitative Microchemistry. Homogeneity, crystallographical and optical properties of a solid material are usually determined under a so-called "chemical," or better under a petrographic microscope, and this examination precedes wet methods. Micromanipulators allow in many instances to isolate under the microscope individual particles of the components of a mixture of solids for further identification.

On a small scale color reactions are in general more sensitive than precipitation reactions. Organic substances are often very valuable microchemical reagents due to color, high molecular weight, and molecular volume of the identifying compound produced by them. Specific reactions are under the experimental conditions used indicative of one substance or ion only. Selective reactions are characteristic for a comparatively small number of substances or ions. Selectivity often permits the simultaneous identification of several substances or ions, especially when the reaction is carried out on paper. Due to capillary and adsorption effects, and different rates of reaction and diffusion, clearly separated zones of reaction products may be observed. Schemes have been worked out for the systematic qualitative inorganic analysis of very small amounts of material, or of samples taken from large quantities, when the time element is important. Such analysis schemes are also available for semimicro procedures.

The "limit of identification" is the minimum quantity of material which can be clearly revealed by any reaction or method, independent of the volume used. The limiting concentration is the dilution prevailing at a given limit of identification (Feigl).

The sensitivity of a microchemical reaction is affected by the nature of the reaction product, its visibility, solubility product, and often by the quantity and color of the reagent used. Reaction time and method of observation (especially in precipitation reactions), presence of electrolytes, hydrogen ion concentration, colloidal properties, procedure and many other details must be carefully controlled in drop reactions. Equally important are the personal equation, and often the enthusiasm of the observer.

Physicochemical Methods. Many physical properties and constants of a substance can be determined qualitatively, and often quantitatively, by well-developed microchemical techniques and equipment. The melting point determination under the microscope on a hot stage is often combined with the simultaneous determination of the melting point of a eutectic mixture of the unknown with a known compound intentionally added. This method, and the determination of the "mixed melting point" usually result in positive identification. A very minute fragment of a solid is often sufficient for the determination of the refractive index. On a micro scale one may determine the visible spectrum (with a microspectroscopic ocular on the microscope), fluorescence spectrum (under quartz optics), and the spectra in the ultraviolet and infrared regions of the electromagnetic spectrum of a substance or solution.

Quantitative Analysis. Methods and apparatus were originally taken over from the macro technique and adapted to smaller quantities. Soon special equipment was designed: very small crucibles for gravimetry, burets to deliver 1 to 10 ml of 0.05 to 0.001 N solutions in fractions of a drop, gas analysis apparatus for the analysis of volumes of not more than 1 ml for possibly half a dozen component gases with the precision of the macro methods. Micro respirometers were originally designed for the study of biochemical processes (respiration of tissue fragments, fermentation, photoysnthesis), but may be conveniently used for the study of chemical reactions (e.g., in micro hydrogenations), and for micro gas analysis in general. Progress in instrumentation involves refinements in polarog-

raphy, amperometric and potentiometric titrations, and differential thermal analysis to allow the use of micro samples. Automation has been extended not only to individual pieces of equipment, like micro burets, but to complicated set-ups: thus the automatic simultaneous quantitative determination of elements like carbon, hydrogen and nitrogen in organic compounds requires not more than 8 to 10 minutes.

Due to its many attractive features, especially its economy of sample size, reagents needed, and time involved, microchemistry is today used in every branch of the physical sciences. Its techniques are of utmost importance in fields, where dangerous (radioactive or explosive) materials are examined, where the results must be available promptly, e.g., in clinical laboratories, or in the examination of rare or unique items (paintings, documents, antiques), when only a minimum of material may be used for testing. Microchemistry is invaluable in the examination of foods, drugs, and in criminology. Special micro instruments, devices as well as methods have been developed for securing samples by means of outer space vehicles, to collect information of chemical and "exobiological" nature about possible extraterrestrial life. In such cases special precautions must be taken to avoid contamination of the target planet by the sampling equipment. The automated biological laboratory "Multivator" measures about 25 cm in length, has about 8 cm diameter, and weighs only 400 g. It gathers particles, e.g. dust, of selected sizes. Vidicon apparatus will transmit a microscope image over a telemetry system via a relay satellite to earth. In the "Gulliver Experiment" the radioactivity of gases, which might be related to metabolic processes of organisms, is measured. Equipment for the determination of photosynthesis on planets has been developed; "sticky strings" will collect small samples of dust and organisms from a planet's surface, and will be reeled back into the device for later analysis.

Success in microchemical work requires the constant application of great manual skill and dexterity, of what has been termed "chemical asepsis"; it demands the special ability of critical evaluation of available methods and procedures for each task, and it challenges the chemist to solve a problem by adapting known operations and reactions in a novel way to small amounts of material.

PAUL ROTHEMUND*

*Deceased.

MILK AND MILK PRODUCTS

Approximately 120 billion pounds of milk are produced annually in the United States. It is utilized (1) as fresh milk and cream; (2) in various processed forms such as cheese, ice cream, butter, concentrated and dried milk; and (3) in pharmaceutical, industrial, and livestock feed products.

Intricate chemical and microbiological problems arise in the production, processing, and distribution of milk products. Milk is a complex mixture of fat (4%), protein (3.5%), carbohydrate (4.8%), and mineral components (0.7%) and is an excellent bacterial-growth medium; hence care is required in using it for food production. The excellent nutritional quality of milk, coupled with the high costs of producing it, have brought a gradual decrease in the quantities of milk diverted to industrial and feed use.

Milk Fat. Milk fat or butterfat is a mixture of triglycerides of various fatty acids. Milk fat also contains a small percentage of cholesterol (0.37%), a substance characteristic of fats and oils of animal origin in contrast to those of plant origin which contain plant sterols. The phospholipides, lecithin, caphalin, and sphingomyclin are present in milk within the range of 0.03–0.04%. These are fat-like substances containing phosphorus and nitrogen. They have emulsifying properties and are associated wth the fat-globule surfaces: hence their tendency to concentrate in butter and buttermilk.

Milk fat is distinguished from all other fats in that it is the only one containing butyric acid (C_4) as a component of glycerides. This acid occurs in the fat in a concentration of approximately 10 mole % and the total of C_6, C_8, and C_{10} fatty acids accounts for another 10 mole % of the component fatty acids. These acids (C_4–C_{10}) are often called the volatile fatty acids of milk fat. In the free form, they have a pungent characteristic flavor which is important in many types of cheese.

The deterioration of milk fat is an important cause of off-flavor development in dairy products and its control requires technical understanding of the processes involved. Three major types of deterioration associated with milk fat are recognized:

(1) *Rancidity,* due to free volatile fatty acids liberated from the glycerides by enzymic (lipase) hydrolysis. Lipases are normal components of raw milk, and are inactivated by the heat of pasteurization.

(2) *Tallowiness or oxidation,* due to autoxidation of unsaturated fatty acids with the production of flavorful unsaturated aldehydes. These reactions are accelerated by oxygen, high storage temperatures, and copper catalysts. Oxidation is usually the primary cause for spoilage of dried whole milk, cream, butter, and butteroil.

(3) *Heat-generated flavors,* due to the formation of lactones and methyl ketones from hydroxy and keto acid precursors, which occur in trace quantities in milk fat. These flavors are considered to be desirable in fried and baked goods and are partly responsible for the unique condiment properties of butter in food preparation. However, they are undesirable in dried whole milk and evaporated milk where the objective is to make a bland product as much like fresh milk as possible.

Proteins. The proteins of milk fall into two groups: casein, precipitable by both acid and proteolytic enzymes such as rennin; and whey proteins, which are acid soluble but heat-denaturable. There is about 3% casein in milk, the removal of which leaves a whey of approximately 1.0% nitrogenous matter. Of this 0.6% is heat-coagulable protein and 0.4% is nonheat-coagulable. The latter fraction is composed of protein-like fragments of a proteose or peptone nature plus other nitrogenous substances. Among these are small percentages of urea, creatin, creatinine, uric acid, and various forms of amino nitrogen.

Casein exists in milk as a calcium caseinate-calcium phosphate complex; the ratio of these compo-

nents is approximately 95.2 to 4.8. The dispersed casein particles appear to be spherical in shape and of various sizes. The size distribution of the casein micelles is not constant, but varies with aging, heating, concentration, and other processing treatments. Processing alters the water-binding of casein and this in turn affects the apparent viscosity of products that contain casein. Changes in hydration have not been measured quantitatively although the casein particles of raw milk appear to consist of one volume of water-free protein and three volumes of solvate liquid.

The whey or serum proteins have been partially resolved into three relatively homogeneous, crystallizable proteins: (1) β-lactoglobulin (50% of total serum protein), (2) an albumin resembling the albumin of bovine blood (5% of the serum protein), and (3) α-lactalbumin (12% of the serum protein).

On heat denaturation, the serum proteins show decreased solubility at pH 4.7 and in concentrated salt solutions. There is some variability in response toward heat treatment, but complete denaturation will occur during heating within the range of 60 to 80°C for periods of time up to two hours. There is practically no denaturation during normal pasteurization. Heat increases the activity of the sulfhydryl groups and the sulfhydryl titer can be employed as a measure of denaturation. The —SH groups are readily oxidized in liquid systems and consequently they appear to act as antioxidants to protect milk fat in dairy products. The fat of fluid milk, heated to produce a high —SH titer, shows increased resistance to oxidation and this carries through to the dried product which exhibits superior storage stability, if it is made from high heat milk. The high sterilization temperature to which evaporated milk is subjected and the low oxygen content in the can protect this product from development of oxidized and tallowy flavors during storage (See **Proteins**).

Milk is widely used as an ingredient for bread and other baked goods to which it adds substantial nutritional value. Milk is heated when used in bread to avoid softening of the dough and reduction of loaf volume. Why heat improves the baking properties of milk is not clear, but good baking properties have long been associated with low whey-protein-nitrogen values.

Lactose. The sugar of milk, lactose ($C_{12}H_{22}O_{11}$), occurs in the milk of all mammals. It is mildly sweet with a final solubility in water of 10.6% at 0°C, 17.8% at 25°, 29.8% at 49°, 58.2% at 89°. Lactose, on hydrolysis by acid or the enzyme lactase, yields a mixture of approximately equal parts of glucose and galactose, together with a small but variable quantity of oligosaccharides. The products of lactose hydrolysis are much more soluble than the original disaccharide. Lactose is a reducing sugar which is converted to lactobionic acid on mild oxidation. Two forms which differ in solubility and optical rotation are known. Alpha-lactose hydrate crystallizes at ordinary temperatures with one molecule of water, but this is lost with the formation of the anhydrous form during heating to a temperature between 149 and 200.3°F. Anhydrous β lactose, more soluble than alpha, crystallizes from supersaturated lactose solutions above 200.3°F. Solid beta-hydrate has never been prepared. The crystalline alpha-hydrate is stable in dry air at room temperatures, but both anhydrous forms readily absorb moisture and change to alpha-hydrate at ordinary temperatures. Alpha-lactose crystallizes out in some dairy products and because of the hardness of its crystals and their slow and limited solubility, "sandy" products may result.

The crystallization of lactose in frozen concentrated milk has been associated with a denaturation of casein which ultimately appears as a gel structure in the thawed product. Gelation in frozen milk can be retarded by enzymatic hydrolysis of part of the lactose before freezing or by addition of a polyphosphate salt.

Lactose, when fermented by lactic bacteria, is the source of the lactic acid formed in sour milk and whey. Lactose is helpful in establishing a slightly acid reaction in the intestine, which assists in calcium assimilation.

Mineral Components. When milk is heated to a temperature high enough to volatilize the water and oxidize the organic constituents, the residue of inorganic oxides that remains is called the milk ash; its major components are: K_2O, CaO, Na_2O, MgO, Fe_2O_3, P_2O_5, Cl, and SO_3. The calcium and phosphorus of the ash are of special interest because of their nutritional importance and because calcium phosphate is part of the casein micelle, influencing its physiochemical behavior toward coagulation with rennin, acid, and heat. Minor inorganic constituents are present in milk in trace amounts, i.e., iron, copper, zinc, aluminum, manganese, iodine, and cobalt.

Miscellaneous Components. The hydrogen-ion concentration of milk increases slightly with age, after milking, as natural carbon dioxide escapes. Most samples of cow's milk vary within the range of pH 6.5–6.7. Titratable acidity of fresh milk which may vary from 0.13 to 0.16%, expressed as lactic acid is an arbitrary measurement influenced by the protein and salt-buffer systems present in the particular sample. Citrates, phosphates, and carbonates are the principal buffers in milk.

Milk contains some important vitamins. The vitamin D content may vary from 30 I.U. per quart in summer to 6 in winter, depending upon the feed and the sunlight which reach the cow. Both pasteurized and evaporated milk are often fortified by the addition, on a fluid basis, of 400 I.U. of vitamin D per quart. Vitamins A, D, and E (alpha-tocopherol) are fat-soluble and stable at the heat treatments used in processing milk and milk products. The remaining vitamins are water-soluble and of varying stability. Vitamins B_1 (thiamine) and C (ascorbic acid) are partially destroyed by heat, while B_6 and B_{12}, are relatively heat-stable. Vitamin B_2 (riboflavin) is heat-stable but it is quickly destroyed by light. In spite of the varying sensitivity of the water-soluble vitamins toward heat, pasteurized milk is a good source of all the milk vitamins except C.

Two types of enzymes in milk are important: those useful an as index of heat treatment and those responsible for bad flavors. Phosphatase is destroyed by the heat treatments used to pasteurize milk; hence its inactivation is an indication of adequate pasteurization. Lipase catalyzes the hydrolysis of milk fat which produces rancid flavors. It must be inactivated by pasteurization or more se-

vere heat treatment to safeguard the product against off-flavor development. Other enzymes reported to have been found in milk include catalase, peroxidase, protease, diastase, amylase, oleinase, reductase, aldehydrase, and lactase.

BYRON H. WEBB

Cross-references: *Foods, Homogenization, Fats, Proteins, Sugars, Enzymes.*

References

"Fundamentals of Dairy Chemistry," 2nd ed., B. H. Webb and A. H. Johnson, Editors, Avi Publishing Co., Westport, Conn., 1973.

"Byproducts from Milk," B. H. Webb, and E. O. Whittier, Editors, Avi Publishing Co., Westport, Conn., 1970.

"Drying Milk and Milk Products," Carl W. Hall, and T. I. Hedrick, Avi Publishing Co., Westport, Conn., 1966.

"Ice Cream," W. S. Arbuckle, Avi Publishing Co., Westport, Conn., 1972.

"The Fluid Milk Industry," J. L. Henderson. 3rd Ed., Avi Publishing Co., Westport, Conn., 1971.

MINERALOGY, CHEMICAL

Chemical mineralogy pertains to the analysis and synthesis of minerals, and to the application of chemical theory to problems in mineralogy and in a more restricted sense to petrology. The alliance of chemistry and mineralogy is an ancient one and derives from a common source in alchemy. The earliest chemists used minerals as raw material for the preparation of their chemicals and as the basic substances from which to extract the then undiscovered chemical elements. Jöns Jakob Berzelius (1778–1848), the Swedish chemist, and Friedrich Wöhler (1800–1882), the German chemist, laid the basis for the analysis of minerals. Although Axel Frederich von Cronstedt, the Swedish Master of Mines, is generally regarded as the founder of chemical mineralogy for his *system of identification of minerals,* it was actually Berzelius who put blowpipe analysis on a sound basis. Berzelius and Wöhler together placed the qualitative and quantitative analysis of minerals on a firm foundation. The science, as the early workers developed it, was mineral chemistry.

Chemical analysis of minerals is a cornerstone of mineralogy for its results give the formulas of minerals. New methods of analysis for minerals are developed as needed in research and they are constantly revised as a result of the development of new tchniques and new reagents. Hillebrand's treatise as revised by Lundell, Hoffman, and Bright is a basic reference work. Microchemical tests, spot tests, spectrochemical analysis, staining and contact printing techniques, and thermal analysis are commonly employed in chemical mineralogy.

With the shift of emphasis in chemistry to synthesis, mineral synthesis based on empirical procedures (trial and error) was developed at the hands of many chemists, particularly by the French scientists Ferdinand Fouque (1828–1904) and Michel-Levy (1844–1911). Empirical synthesis reached a sudden maximum and declined rapidly before the turn of the nineteenth century, capitulating to the methods of physical chemistry. Empirical mineral synthesis is still done when making reconnaissance studies and in the search for unstable (and sometimes stable) polymorphic modifications of mineral phases.

The theoretical studies of J. Willard Gibbs (1839–1903), Jacobus H. Van't Hoff (1852–1911), Walther Nernst (1887–1941), and H. W. Bakhius Roozeboom (1854–1907) sounded the death knell of empirical mineral synthesis. Its place was taken by the rigorous physical and chemical approach developed at the Geophysical Laboratory of the Carnegie Institution of Washington under the directorship of the physicist Arthur Louis Day. The synthesis and stability of minerals as a problem in phase rule study and in other physical chemical studies, such as the determination of E_h and pH diagrams, is the major problem in chemical mineralogy.

The application of chemistry to mineralogical problems embraces a wide field of research and includes such topics as the weathering of feldspar to clay minerals; the polymorphism of $CaCO_3$ as aragonite and calcite; the influence of impurities on the morphology of crystals; the chemistry of hornblende (a complex aluminosilicate of calcium, sodium, iron, magnesium, and hydroxyl) crystals from igneous and metamorphic rocks.

GEORGE T. FAUST

MIXING

Mixing of fluids plays a part in almost all chemical research and processing. In exploratory work in the laboratory the effects of mixing may be very great, and it is essential that the desired type and amount of mixing can be reproduced or can be varied by known amounts. The primary purpose of mixing is to distribute components as uniformly as possible; temperature distribution is frequently a major purpose. These may be followed by a chemical reaction or a transfer of matter between phases, and by a transfer of heat for temperature control. The mixer produces mechanical effects only. Molecules of themselves will diffuse, but mixing impellers produce flow which results in forced convection and mixing. Hence, reactants can be brought to an interface as rapidly as desired by controlling the fluid motion. Most fluid mixing is done by rotating impellers.

Both large scale (mass flow) motion and small scale (turbulent) motion are ordinarily required to bring about rapid mixing. The discharge stream from an impeller initiates the large scale flow pattern. Turbulence is generated mostly by the velocity discontinuities adjacent to the stream of fluid flowing from the impeller, and also by boundary and form separation effects. Turbulence spreads throughout the mass flow and is carried to all parts of the container. Some mixing operations require relatively large mass flows for best results, whereas others require relatively large amounts of turbulence. There is usually an optimum ratio of flow to turbulence for a desired mixing operation, whether it is a simple blending of immiscible liquids or a mass transfer followed by chemical reaction.

In the research laboratory it is important to recognize the effect of mixing on reaction rate or on

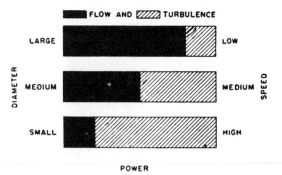

FIG. 1. Constant power, effect of impeller size, and speed on flow and turbulence.

other performance criteria. Energy must be supplied to produce fluid motion, thus, to compare mixing with different equipment or with different sizes of the same type impeller, it is essential that the comparisons be made on the basis of equal power input.

For the same power, the ratio of flow to turbulence from mixing impellers can be varied by changing the size and speed of the impeller. Figure 1 illustrates the differences in mass flow and turbulence which can be achieved for the same power input for dimensionally similar impellers. A large-diameter low-speed impeller produces a large ratio of flow to turbulence, whereas a small-diameter high-speed impeller will give a small ratio. Curve A, Figure 2, illustrates a reaction best accomplished by large flow and small turbulence. This curve, which is typical of blending operations, shows that the rate of blending increases to a maximum with a large impeller as impeller diameter is increased (and impeller speed is decreased) with power input constant.

Curve B of Figure 2 is typical of gas-liquid contacting operations. Here the rate of mass transfer between phases increases to a maximum at small impeller diameter and then decreases as impeller diameter is increased. The significance is that more turbulence is available with the small impeller and that turbulence is more important than flow in this operation.

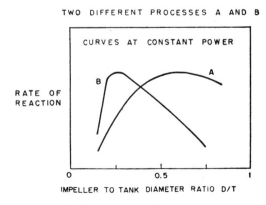

FIG. 2. Effect of impeller size on reaction rate at equal power output.

In all bench-scale and pilot plant work where mixing is important, the effect of the impeller diameter-turbine diameter ratio should be determined so that the type of flow motion best suited to the operation can be found. If an optimum ratio is found, it becomes the basis for larger scale design.

Mixing Vessels, Flow Patterns, and Impellers. Flow motion is dependent upon the shape and fitting of the container, the shape and position of the rotating impeller, and the physical properties of the fluid. The best mixing is usually one which produces lateral and vertical flow currents, and these currents must penetrate to all portions of the fluid; swirling motion should be avoided. Cylindrical vessels provide the best environment for mixing.

The most useful impellers are the simple flat paddle, the marine-type propeller, and the turbine. If any of these are on a vertical shaft rotating on the center line of a cylindrical vessel, the fluid motion will be one of rotation. A vortex forms around which the liquid swirls. A minimum of turbulence and of vertical and lateral flow motion will result. Very little power can be applied.

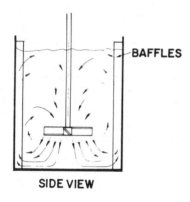

SIDE VIEW

BOTTOM VIEW

FIG. 3. Typical flow patterns from axial flow impeller in baffled tank.

Rotary motion (and surface vortex) can always be stopped by inserting projections in the body of the fluid; when these are at the side of the tank they are called baffles, and this is the method most commonly used to obtain good mixing in large industrial equipment. The propeller with baffles will produce an axial flow pattern, Fig. 3, and the paddle and turbine will produce radial flow, Fig. 4.

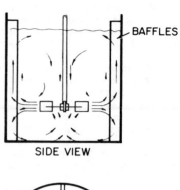

SIDE VIEW

BOTTOM VIEW

FIG. 4. Radial flow pattern for flat blade turbine positioned on center in baffled tank.

Laboratory. For this work round bottom flasks are the poorest type for mixing. Such flasks can, however, be vastly improved by having creases blown into the sides; such are now standard laboratory equipment. Cylindrical glass or metal containers should be provided with vertical baffles, or if a propeller is used in the "off-center" position, Fig. 5, the baffles may be omitted. In either case, the flow motion can be reproduced in large indus-

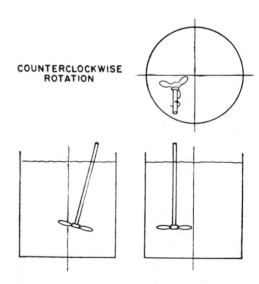

FIG 5. Position and resulting flow pattern for top-entering axial flow turbine or propeller, off-center, without baffles.

trial size. Swirling motion cannot be reproduced on scale-up with the same liquid.

Pilot plant work should be done with scale models of the type of impeller available for the large industrial size. The effect of impeller size and speed is a major objective so that scale-up can be accurate. If flow patterns are changed between pilot plant and production equipment, one can be certain that process results will not be the same for equivalent cost in power expended.

Power and Flow. Power imposed by an impeller is proportional to the density of the fluid, the cube of the speed and the fifth power of its diameter, in liquids of low viscosity. Thus, when impellers rotate with baffles or their equivalent, constant power can be achieved for different diameters according to $N_r = (1/D_r)^{5/3}$ where N_r is the ratio of speed and D_r is the ratio of the corresponding impeller diameters. For example, if one turbine is twice the size of a smaller geometrically similar turbine, the speed ratio to provide equal power input will be $N_r = (1/2)^{5/3}$, or 0.33. Proportionality constants are available in the references.

Flow from impellers is proportional to the speed and to the cube of the diameter. It is clear that for the same power there will be greater flow from the large-diameter low-speed turbine than from the small-diameter high-speed turbine.

Viscous Liquids. Power to move an impeller in high viscosity liquids is proportional to the viscosity, to the square of the speed and to the cube of the diameter of the impeller. Turbines and helical ribbon type impellers are in common use.

Very high viscosity liquids and pastes usually require a different technique for mixing than low-viscosity fluids. Special apparatus is necessary to provide for wiping, stretching, and squeezing, because turbulence cannot be generated in such fluids to provide for the small scale mass transfer necessary to cause interpenetration of particles. There are few quantitative data yet available relating performance to these types of equipment.

Mixing of Solids. Solids of different density and size are usually mixed in tumblers or rotating cylinders. Some data are available showing the effects of physical properties on the performance of rotating cylinder mixers. It should be noted that time of mixing is an important variable because classification and separation often follow the desired distribution if the operation is carried on too long. A systematic approach to the mixing of solids and pastes has not yet been developed.

Standard methods for calculating the extent of mixing for dry powders are available in the A.I.ChE. Standards.

J. H. RUSHTON

References

"Mixing of Liquids," J. H. Rushton and J. Y. Oldshue, Chem. Eng. Progress Symp. Ser., 55, No. 25 (1959). A.I.Ch.E. Standard Testing Procedure, Solids Mixing Equipment, 1961. Amer. Inst. of Chem. Engr., New York, N. Y.

"Liquid Mixing and Processing in Stirred Tanks," F. A. Holland and F. S. Chapman, Reinhold Publishing Corp., 1966.

"Mixing," Vol. I. and II, V. W. Uhl and J. B. Gray, Academic Press, 1966.

"Mixing for the Chemical Industries," J. H. Rushton, 1970 Institute Lecture, Amer. Inst. of Chem. Engrs., New York, N. Y.

MOISTURE CONTENT AND MEASUREMENT

The moisture content of a material is often considered to be the percentage of total water, including adsorbed and free water, as determined by analysis. A common method for determination of water is based upon measuring loss in weight of the specimen after heating at a specified temperature, e.g., in an oven at about 100°C, preferably in dry air, until the weight of the dried specimen is constant. The method will be in error if volatile substances other than water are present. Other methods depend upon (a) utilizing a quantitative chemical reaction in which a chosen reagent reacts selectively with the water, (b) measuring the weight or volume of water after removal from the specimen, or (c) determining physical properties that are related in a definable way to water content.

In a more limited sense, the term "moisture content" is restricted by hygroscopically bound water, the amount of which is responsive to changes in relative humidity of the surrounding atmosphere. The relative humidity is the ratio of the quantity of water present to the total quantity required for saturation at a given temperature. Under conditions of equilibrium, the moisture content of the material corresponds to a definite relative humidity in the surrounding atmosphere.

When the moisture content of a hygroscopic material is plotted against the equilibrium relative humidity, a sorption isotherm is obtained which usually has a characteristic sigmoid shape. Generally, the moisture content rises sharply with increase in relative humidity up to about 10% R.H. At intermediate relative humidities the moisture content increases more slowly. At very high relative humidities, as saturation is approached, the moisture content rises rapidly. At intermediate values of relative humidity, the moisture content of most substances depends upon whether the condition was approached from one of less or greater water content. Because of this hysteresis, the desorption curve normally lies above the adsorption curve.

The moisture content is important in the chemistry and technology of many products, including textiles, leather, wood, pulp and cellulose, paper, fuels, food, soils, seeds, pigment, and plastics; it influences many physical properties, such as strength, flexural rigidity, extensibility, dimensional stability, abrasion resistance, etc. The processing and application of many materials such as paper, textiles, and leather require control of moisture content to maintain dimensions and physical properties within required limits, as in fabrication, printing, etc. Many products are bought and sold on the basis of dry weight, and the moisture content of the moist or air-dried material is important.

The hygroscopic moisture at a given relative humidity depends upon the composition and previous history of the material. Typical figures (generally adsorption values) for a number of materials of general interest are listed in the following table.

Material	Temp. °C	Relative Humidity, %	
		30	70
Cotton	20	3.7	7.7
Mercerized cotton	20	5.9	11.9
Cellophane	25	7.2	14.4
Cellulose acetate	25	2.4	6.7
Wood, white spruce	25	5.4	12.0
Wood pulp	25	4.1	9.1
Fiberboard	23	5.6	10.9
Silk	36	5.0	10.4
Wool	36	7.6	15.0
Nylon	25	2.0	4.6
Starch (potato)	22	10.3	19.8
Leather (vegetable-calf)	25	12.6	19.0
Tobacco (Burley)	27	8.2	13.5
Activated Charcoal	25	7.9	34.5
Clay (Kaolinite)	—	0.5	0.9
Bentonite	—	12.9	25.5

B. L. BROWNING

MOLALITY

The concentration of a dissolved substance is usually expressed in one of three ways—molality, molarity, or mole fraction. These three kinds of units become proportional—hence, equivalent—in very dilute solutions. Molality is the number of moles of solute per 1000 grams of solvent. Since this unit defines the quantities of both solute and solvent, it provides a satisfactory basis for correlation of composition and thermodynamic properties of a solution.

Possibly the correlation of greatest theoretical interest at present is that between molality and activity coefficient of an electrolyte. The Debye-Hückel theory provides an explanation for the empirical discovery that the activity coefficient is proportional to the square root of the molality in very dilute aqueous solution. Much effort has been directed toward extending the theory to account for observed variations of activity coefficients with change in molality of electrolytes in less dilute aqueous solutions and in nonaqueous solutions.

The other correlations of major interest are the direct proportionality between molality and both the elevation of boiling point and depression of freezing point. There are important restrictions—the correlations are valid only for dilute solutions, the predicted boiling point rise is valid only for nonvolatile solutes, and the predicted freezing point depression is valid only for solutes which do not form solid solutions with crystallized solvent. These correlations were discovered and applied during the period when the concept of molecular weights was being established, and the unit of molality was defined in a way which avoided questions about the molecular weight of the solvent. The measurement of freezing point depression as a function of molality remains as one of the most useful techniques for determining molecular weights of new chemical compounds.

J. D. BUSH

Cross-references: *Activity, Debye-Hückel Theory, Mole Concept, Solutions.*

MOLE CONCEPT

The mole (derived from the Latin *moles* = heap or pile) is the chemist's measure of amount of pure substance. It is relevant to recognize that the familiar *molecule* is a diminutive (little mole). Formerly, the connotation of *mole* was a "gram molecular weight." Current usage tends more to use the term *mole* to mean an amount containing Avogadro's number of whatever units are being considered. Thus, we can have a mole of atoms, ions, radicals, electrons or quanta. This usage makes unnecessary such terms as "gram-atom," "gram-formula weight," etc.

A definition of the term is: *The mole is the amount of (pure) substance containing the same number of chemical units as there are atoms in exactly twelve grams of C^{12}.* This definition involves the acceptance of two dictates—the scale of atomic masses and the magnitude of the gram. Both have been established by international agreement. Usage sometimes indicates a different mass unit, e.g., "pound mole" or even a "ton mole"; substitution of "pound" or "ton" for "gram" in the above definition is implied.

All stoichiometry essentially is based on the evaluation of the number of moles of substance. The most common involves the measurement of mass. Thus 25.000 grams of H_2O will contain 25,000/18.015 moles of H_2O; 25.000 grams of sodium will contain 25.000/22.990 moles of Na (atomic and formula weights used to five significant figures). The convenient measurements on gases are pressure, volume and temperature. Use of the ideal gas law R allows direct calculation of the number of moles $n = (P \times V)/(R \times T)$. T is the absolute temperature; R must be chosen in units appropriate for P, V and T (e.g., $R = 0.0820$ liter atm mole^{-1} deg K^{-1}). It may be noted that acceptance of Avogadro's principle (equal volumes of gases under identical conditions contain equal numbers of molecules) is inherent in this calculation. So too are the approximations of the ideal gas law. Refined calculations can be made by using more correct equations of state.

Many chemical reactions are most conveniently carried out or measured in solution (e.g., by titration). The usual concentration convention is the *molar* solution. (Some chemists prefer to use the equivalent term *formal*). A 1.0 molar solution is one which contains one mole of solute per liter of solution. Thus the number of moles of solute in a sample will be

n = Volume (liters) × Molarity (moles/liter)

Another concentration expression used by chemists is the *molality* of a solution. This is defined as the number of moles of solute mixed with 1000 grams of solvent. Thus a one molal aqueous solution of sucrose, $C_{12}H_{22}O_{11}$, will contain 342.21 grams of sucrose in 1000 grams of water. The magnitude of the colligative properties of solutions (lowering of solvent vapor pressure, freezing point depression, boiling point rise and osmotic pressure) usually is considered to be proportional to the molality of the solution. Thus the depression of the freezing point of a solution can be calculated as $\Delta T_f = K_f \times m$. For the solvent H_2O, K_f has the value $-1.86°C$ when m is the molality of the solute.

The *mole fraction* is the most convenient expression of concentration to use whenever the properties of a solution reflect additivity of the properties of the components. This is especially true of solutions in which there is little interaction between components such as in mixtures of gases. Thus in a mixture of 3.0 moles of H_2 with 2.0 moles of N_2, the mole fraction of H_2 will be $3.0/(3.0 + 2.0) = 0.60$. The mole fraction identifies the relative numbers of chemical units (usually molecules) of each type present in the mixture. In the example given, the mole fraction 0.60 for H_2 means that 60% of the molecules present are H_2 molecules. If the mixture is confined at a pressure of 0.75 atm, the partial pressure of N_2 will be $0.40 \times 0.75 = 0.30$ atm.

The amount of chemical reaction occurring at an electrode during an electrolysis can be expressed in moles simply as $n = q$ (coulombs)$/z\,\mathfrak{F}$ where z is the oxidation number (charge) of the ion and $\mathfrak{F}$ is the faraday constant, 96,487.0 coulombs/mole. Thus the *faraday* can be considered to be the charge on a mole of electrons. This affords one of the most accurate methods of evaluating the Avogadro number (6.02252×10^{23}), since the value of the elementary charge is known with high precision.

Modern chemistry increasingly uses data at the atomic level for calculation at the molar level. Since the former often are expressed as quanta, appropriate conversion factors must involve the Avogadro number. Thus the *einstein* of energy is that associated with a mole of quanta, or $E = Nh\nu$. Thus light of 2537 Å wave length will represent energy of

talline zeolites to selectively separate molecules on the basis of critical diameter. A zeolite is a group of molecules characterized by the presence of structural SiO_4 and Al_3O_4 groups, cation(s) to balance the negative charge of the aluminosilicate structure, and water of hydration. They may be crystalline or amorphous. The former types are of particular interest as the molecular sieves. The cation(s) commonly found are sodium and calcium with barium, potassium, magnesium, strontium, and iron also possible. The natural zeolites are about 40 in number, the more familiar being chabazite, gmelinite, levynite, faujasite, analcime, erionite and mordenite; they have received considerable research by R. M. Barrer and his co-workers in England. These zeolites are of extreme interest from a physical chemical point-of-view, but have not been used industrially in any great quantity.

The Russians have had a large effort in the area of zeolites as exemplified by the work of Mirsky, Zhdanov, and Dubinin.

The first commercial molecular sieves were made available by Union Carbide Corp.'s Linde Division during late 1954 under the trademark "Linde Molecular Sieves." They were first used for drying refrigerants. In 1959, the Davison Chemical Co. Division of W. R. Grace introduced crystalline zeolites under the trademark "Microtraps"; Davison now uses the term Molecular Sieves. Early in 1962, the Norton Company introduced a zeolite under the trademark "Zeolon."

The physical and chemical properties of these materials are similar. In general terms, the cost varies with manufacturer and quantity and is in the range of $1 to 2 pound. The zeolites are available as powder, beads and pellets.

The zeolites have many uses, and the utilization

$$E = \frac{6.023 \times 10^{23}(\text{quanta/mole}) \times 6.62 \times 10^{-27}(\text{erg-sec/quantum}) \times 3.00 \times 10^{10}(\text{cm/sec})}{2.537 \times 10^{-5}(\text{cm}) \times 4.184 \times 10^{7}(\text{erg/cal}) \times 10^{3}(\text{cal/kcal})}$$

$$E = 113 \text{ kcal/mole}$$

Another convenient conversion factor is 1 eV/particle = 23.05 kcal/mole.

The chemist's use of formulas and equations always implies reactions of moles of material; thus HCl(g) stands for one mole of hydrogen chloride in the gaseous state. Thermodynamic quantities are symbolized by capital letters standing for molar quantities, e.g. C_v (heat capacity at constant volume in cal mole^{-1} deg^{-1}), G (Gibbs function in cal/mole), etc. At times it is more convenient to convert an extensive property into an intensive expression. This is especially true in dealing with multi-components systems. These are referred to as "partial molal quantities" and are given a symbol employing a bar over the letter. Thus the partial molal volume, $\overline{V}_1 = (\partial V/\partial n_1)$ is the rate of change of the total volume of a solution with the amount (number of moles) of component 1.

WILLIAM F. KIEFFER

MOLECULAR SIEVES

The term "molecular sieve" was originally used by McBain to describe the ability of dehydrated crys-

of these materials is widespread. There is still much to be learned in the application of these materials. They are currently used to dry gases and liquids, for selective separations based on size and polar characteristics of the molecules, as ion exchange materials, as catalysts, as chemical carriers, and as gas chromatography materials.

Physical and Chemical Properties. The molecular sieves designated as Type 3A, 4A, and 5A have what is termed the A crystal structure which is cubic ($a_o = 24.6$ Å, space group Fm3), characterized by a three-dimensional network which has cavities 11.4 Å in diameter separated by circular openings 4.2 Å in diameter. This latter figure is the so-called pore diameter. The removal of the water of crystallization leaves an "active" crystalline zeolite that has a void volume of 45 vol %. Adsorption occurs in these intracrystalline voids. The Type 4A unit cell has a chemical formula $8\ Na_{12}[(AlO_2)_{12}(SiO_2)_{12}] \cdot 27\ H_2O$. Types 3A and 5A are prepared from 4A by an ion-exchange process using salt solutions which replaces 75% of the Na ions present with potassium and calcium ions, respectively.

The Type 10X and 13X have the X-crystal structure which is also cubic, ($a_o = 24.93$ Å, space group Fd3m), characterized by a three-dimensional net work which has intracrystalline voids separated by pores which will admit molecules having critical dimensions of 10 and 13 Å, respectively. The void volume is 51 vol %. The Type 13X has a unit cell given by $Na_{86}[(AlO_2)_{86}(SiO_2)_{106}] \cdot 264\ H_2O$. The 10X is made by exchanging 75% of the sodium ions in 13X with calcium. The pH is 10 and they can be used environments of pH 5 to 12.

The internal surface area for molecular sieves is 650 to 800 sq.m/g and the external area is 1 to 3 sq.m/g. The average volume of voids is 0.27 cc/g for the Type A and 0.38 for the Type X. The material has a specific heat of 0.23 to 0.25, a bulk density of 33 to 45 lb/cu ft which is dependent on the physical shape of the bulk materials.

Linde has a zeolite designated as AW-500. This material has a pore diameter of 4 to 5 Å, is acid stable to pH 2.5. It has a bulk density of 42 lb/cu ft, a specific heat of about 0.2. It is particularly recommended for use in an acid environment.

The Linde zeolite designated Y has a cubic structure ($a_o = 24.67$ Å) and has a unit cell composition of $Na_{56}[(AlO_2)_{56}(SiO_2)_{136}] \cdot 250\ H_2O$. Zeolite Y has an effective pore diameter of 9–10 Å.

The Norton "Zeolons" are related structurally to the natural zeolite, mordenite. There are two forms, a sodium and a hydrogen. The effective pore diameter is 9 to 10 Å. The chemical analysis of the sodium form is 71.45% SiO_2, 12.05% Al_2O_3, 0.52% Fe_2O_3, 0.26% CaO, 0.31% MgO, 7.14% Na_2O and 9.15% H_2O. The hydrogen form is made by exchanging the sodium with hydrogen. These zeolites have a pore volume of 0.11 cc/g, a surface area of 500 sq.m/g and a porosity of 54%. They are acid resistant and can be used over the entire pH range and can be heated to temperatures as high as 800°C.

There are about 80 synthetic zeolites that have been synthesized by hydrothermal techniques.

Regeneration. The crystalline zeolites can be regenerated by simple heating supplemented by gas purging or pumping. Water is removable at temperatures between 150 to 350°C. In general zeolites should not be exposed to temperatures of 800°C. The Norton materials can be exposed to temperatures of 800°C. Other materials that have been adsorbed can be removed by a water or steam flush since in general these materials have a preference for water. In general a reverse flow of purge gas is recommended during the regeneration process. Care should always be taken to avoid hazardous or explosive situations in the selection of the purge gas. In liquid processes the liquid should be drained prior to regeneration.

Adsorbent Characteristics. These zeolites as a class are characterized by the ability to adsorb molecules that have critical dimensions less than the effective pore size of the zeolite. Type 4A will adsorb water, carbon dioxide, carbon monoxide, hydrogen sulfide, sulfur dioxide, ammonia, nitrogen, oxygen, methane, methanol, ethane, ethanol, ethylene, acetylene, propylene, n-propanol and ethylene oxide. The Type 5A will adsorb the molecules just listed as well as propane and the n-paraffins to C_{14}, n-butene and n-olefins, n-butanol and higher n-

alcohols, cyclopropane and "Freon-12." The Type 10X will adsorb molecules that have a kinetic diameter of less than 10 Å. They will adsorb with a slightly higher capacity than 4A or 5A. The Type 13X will adsorb all of the materials listed above at a slightly higher capacity than 3A, 4A, 5A and 10X. In addition, this type will adsorb the isoparaffins, isopropanol and other iso, secondary and tertiary alcohols, aromatics, cyclohexane, other cyclic compounds with at least 4 membered rings, carbon tetrachloride, hexachlorobutadiene, Freon 114 and 11, sulfur hexafluoride, and boron trifluoride. The AW-500 material will adsorb HCl, SO_2, and nitrogen oxides, molecules that have critical dimensions of about 4 to 5 Å.

The zeolite Y will adsorb molecules with kinetic diameters of 9–10 Å.

The Norton "Zeolon" adsorbs molecules, such as H_2O, CO_2, and H_2S. In addition, the hydrogen form adsorbs normal and cycloparaffins, and aromatics.

Drying. Because of their strong affinity for water these molecular sieve materials are extremely effective in the drying of gases and liquids in both static and dynamic applications. As a class they have good capacity; they do not deliquesce; they remain strong; they are not toxic; they are noncorrosive; and they are not explosive. In the drying of gases the synthetic zeolite can be used in drying low-humidity gases, they can be used at high temperatures, they can be used to dry gases without otherwise altering the composition, they can be used to remove selectively other impurities as well as water, they can be used to dry gases adiabatically, they are not damaged by liquid water, and they can dry gases more completely than other commercial adsorbents.

The drying of liquids can also be accomplished using the crystalline zeolites because they have a high capacity for water in liquids having only trace amounts of water; they give very dry liquids and they can dry polar liquids.

Chemically Loaded Zeolites. Linde scientists have explored the possibility of using the molecular sieves as carriers for a wide variety of materials. The adsorbed chemical is released by heating or by displacement using another material such as water. The volatility, toxicity, and odor of some compounds restricts their usefulness, and the molecular sieve can be used to depress the reactivity or undesirable properties. Materials such as anhydrides, amines, ethers, organic acids, alcohols, organometallic compounds, aldehydes, halogens, ketones, acid gases, esters, perfumes, peroxides, and hydrocarbons have been successfully loaded on molecular sieves. In all some 200 different chemicals have been "loaded" on Linde sieves.

Ion Exchange. The crystalline zeolites can act as ion exchange materials. The ability of the commercial zeolite to undergo ion exchange is evidenced by the fact that Type 3A and 5A are produced from 4A by introducing potassium and calcium for about 75% of the sodium ions present. The ion exchange capacity of the Linde materials has been studied with both the A and X structures. The following cations have been introduced into molecular sieves.

Li^+	Ag^+	Ba^{++}	Ni^{++}
K^+	NH_4^+	Hg^{++}	Cu^{++}
Rb^+	Mg^{++}	Cd^{++}	Al^{+++}
Cs^+	Ca^{++}	Zn^{++}	H^+
Tl^+	Sr^{++}	Co^{++}	Au^{+++}

This ion-exchange property while in itself very useful changes the pore diameter and hence the adsorption characteristics of these materials. The sodium in molecular sieve Type 13X has been replaced by Li^+, K^+, Mg^{++}, Ca^{++}, Zn^{++}, Cd^{++}, and Ba^{++}.

Catalysis. The literature during the past 10 years has heavily emphasized the use of molecular sieves in catalysis. The recent selection of Paul Venuto as Ipatieff Award Winner for 1971 points out the importance of this application. Two important uses are in catalytic cracking and hydrocracking. Some 60% of installed catalytic cracking capacity in the United States employs zeolite-based catalysts. They have several advantages: more active, cause less coking and increase production capacity.

In hydrocracking the zeolite catalyst is used to produce gasoline or fuel oil from heavy petroleum fractions. The catalyst is insensitive to sulfur and nitrogen containing impurities.

The catalytic application of zeolites is very wide —it can range from a mild catalytic effect (shifting of the carbon-carbon double bond) to very severe (fragmentation of a benzene ring). The applications cannot be listed here in detail but include dehydration, dehydrohalogeneration, additions to double bonds, alkylation, dealkylation, Cannizzaro reactions, aldol condensation, Beckmann rearrangement, water gas shift reactions, and inorganic reactions involving NH_3, SO_2, and NO.

CHARLES K. HERSH

Cross-references: *Adsorption, Drying, Catalysis, Ion Exchange, Zeolite, Crystal Structure.*

References

Hersh, C. K., "Molecular Sieves," Reinhold Publishing Corp., New York, 1961.
"Molecular Sieves," Soc. of Chem. Ind., London, 1968.
Venuto, P. B., Ipatieff Award, address preprint given at 1971 L.A. meeting of ACS.

MOLECULAR WEIGHT

The molecular weight of a chemical compound is the sum of the atomic weights of its constituent atoms. The molecule is the smallest weight of a substance which still retains all of its chemical properties. By convention, each atomic weight, and therefore molecular weights, are expressed relative to the arbitrary value, 16, for the oxygen atom. For example, the molecule of acetic acid, CH_3COOH, contains two atoms of carbon, four of hydrogen, and two of oxygen, so that its molecular weight is the sum of $2(12.01) + 4(1.01) + 2(16.00)$, which totals 60.06. This molecular weight is clearly in arbitrary units, but a related quantity, the gram-molecular weight or mole, is the molecular weight expressed in grams. One mole of any compound has been found to contain 6.023×10^{23} molecules, and this number is called the Avogadro constant.

The basis now accepted by all scientists for atomic and molecular weights is the carbon isotope of mass 12, whose atomic weight is assigned the exact value 12. This replaces earlier systems used before 1961 which were based on oxygen for which the small content (0.2%) of two isotopes heavier than 16 led to differences in scales used by chemists and physicists.

The weights of molecules range from a value of about two for the hydrogen molecule to several millions for some virus molecules and certain polymeric compounds. Molecular dimensions accordingly range from a diameter of about 4 Angstroms for the hydrogen molecule to several thousand Angstroms—which has permitted single large molecules to be resolved in the electron microscope. Molecular sizes are generally much smaller and are not measured directly, but are deduced from x-ray diffraction studies of ordered groups of molecules in the crystalline state or from the physical properties such as hydrodynamic behavior of molecules in the gaseous or liquid state.

Many methods for determining molecular weights depend fundamentally on counting the number of molecules in a given weight of sample. Since at least 10 trillion of the largest known molecules are present in the smallest weight measurable on a sensitive balance, an indirect count is made by measuring physical properties which are proportional to the large number of molecules present.

The term "molecular weight" is properly applied to compounds in which chemical bonding of all atoms holds the molecule together under normal conditions. Thus, covalent compounds, as represented by many organic substances, usually have the same molecular weight in the solid, liquid, and gaseous states. However, substances in which some bonds are highly polar may exist as un-ionized or even associated molecules in the gaseous state and in nonpolar solvents, but may be ionized when dissolved in polar solvents.

Truly ionic compounds, such as most salts, exist only as ions in the solid and dissolved states, so that the term "molecule" is not applicable. Instead, the term "formula weight" is used; this denotes the sum of the atomic weights in the simplest formula representation of the compound. If a broad definition of a molecule as an aggregate of atoms held together by primary valence bonds is adopted, then salts in the crystalline state would appear to have a molecular weight which is essentially infinite and limited only by the size of the crystal, since each ion is surrounded by several ions of opposite polarity to which it is attached by ionic bonds of equal magnitude.

Molecular Weight Distributions. There are many systems, particularly among the polymers and proteins, which consist of molecules of various chain lengths, and thus of various molecular weights— so-called polydisperse systems. In such cases, molecular weight values have an ambiguous meaning, and no single such value will completely represent a sample. Various techniques for measuring molecular weights, when applied to one of these materials, will produce values which often disagree by a factor of two or more. This disagreement arises from the different bases of the methods—for example, some methods yield so-called number-average molecular weights by determining the number-concentration of molecules in a sample, while other methods produce weight-average molecular weights which are related to the weight-concentrations of each species. Another common value is the viscosity-average molecular weight, which is related to the viscosity contribution of each species. Other bases are of importance for certain methods of study, and some of these are complex functions involving several averages. For some purposes, the determination of a single average molecular weight is sufficient for establishing relations between molecular weight and the behavior of polymers, but the type of molecular weight average must be so chosen as to have a close relation to the behavior property of interest. A more detailed knowledge of the constitution of a sample is sometimes required, particularly if several properties are to be considered, or if unusual forms of molecular weight distribution curve are present.

Uses. Molecular weight measurements, in conjunction with the law of combining proportions, have enabled the atomic weights of elements in compounds to be determined. When the atomic weights are known, molecular weight measurements permit the assignment of molecular formulas. Other applications to compounds of low molecular weight allow determination of the extent of ionization of weak electrolytes, and the extent of association of some uncharged compounds which aggregate. The study of molecular weights is becoming increasingly valuable in assessing the effects which various molecular species of a polymer sample have on the physical properties of the product. Through such knowledge, the synthetic process may be modified to improve the properties of polymers.

Methods of Measurement. Many physical and certain chemical properties vary substantially with the molecular weight of compounds, and these properties are the bases of all molecular weight methods. The summary given here includes principally the methods which are most frequently used or have general applicability. The choice of the most suitable method for a given sample depends on its state (gas, liquid, or solid), the magnitude of the molecular weight and the accuracy required in its determination, as well as on the stability of the compound to physical or chemical treatment.

Gases and Liquids. Avogadro's hypothesis (1811) that equal volumes of different gases contain the same number of molecules under the same conditions made it possible to determine how many times heavier a single molecule of one gas is than that of another. Thus, relative molecular weights of all gases could be established by comparing the weights of equal volumes of gases. The significance of the idea and utilization of this method were first clearly demonstrated by Cannizzaro in 1858, and this represents the first available method for determining molecular weights. With the additional information from chemical experiments on the number of atoms of each kind present in each molecule, the relative weights of each atom were obtained. The assumption of the integral value, 16, for the atomic weight of oxygen (to give a value close to unity for the lightest element, hydrogen) then enabled molecular weights of all gaseous compounds to be determined. The method obviously can be applied to other molecules which normally occur in the liquid

state but can be volatilized by heating. The Dumas and Victor Meyer methods are most used for molecular weight determinations of liquids by measuring the density of vapor produced by heating above the boiling point. These methods have been refined so that gas densities can now be determined with an accuracy of 0.02 per cent, and extremely small weights of material (about one microgram) can be similarly studied with somewhat less accuracy. High temperatures up to 2000°C have been used to study substances which are volatilized only with difficulty, provided decomposition can be avoided.

Solids Measured by Colligative Methods. It has been shown that nonvolatile molecules dissolved in a solvent affect several physical properties of the solvent in proportion to the number of solute molecules present per unit volume. Among these properties are a decrease of the vapor pressure of the solvent, a rise in its boiling point, a decrease in its freezing point, and the development of osmotic pressure when the solution is separated from the solvent by semipermeable membrane. Properties such as these, which are related to the number of molecules in a sample rather than to the type of molecule, are called colligative properties. They are the basis for some of the most useful techniques for molecular weight determination. The magnitude of the effects and the ease of measurement differ greatly, so that certain of the colligative properties are preferred for this purpose. For example, an aqueous solution containing 0.2 gram of sucrose (mol. wt. 342) in 100 ml has a vapor pressure 0.01 per cent less than that of the solvent, a boiling point 0.003°C greater, and a freezing point 0.011°C lower than the solvent, but will develop an osmotic pressure of 150 cm of water. Since the effects are related to the number-concentration of solute molecules, each method leads to a number-average molecular weight if the sample consists of a mixture of molecules of different sizes. Accurate results with any of the techniques are obtained only when measurements at a series of concentrations are extrapolated to infinite dilution where the system approaches ideality, i.e., is not affected by interactions between molecules.

Direct vapor pressure measurements with a differential manometer are generally limited to the larger depressions produced by low molecular weight solutes, while refined techniques such as isothermal distillation require the most exact control of conditions. Isopiestic methods allow the comparison of the vapor pressure of solutions of an unknown with those containing a known substance, and several modifications have been used more than other vapor pressure methods. Ebulliometric techniques which depend on the elevation of the boiling point of a solvent are often used for solutes of low molecular weight and find some use for large molecules. Since boiling points are highly sensitive to the atmospheric pressure, it is either necessary to control pressure very precisely, or more commonly to measure the boiling points of both the solvent and solution simultaneously. Often a differential thermometer is employed to determine only the difference of the two temperatures, and these devices have been made so sensitive that molecular weights as large as 30,000 have sometimes been studied. Techniques involving the lowering of the freezing point of a solvent (cryoscopic methods) are much used for rapid approximate determinations of molecular weights in the identification of organic compounds. For this purpose a substance such as camphor, which is a good solvent for many organic compounds and has a large molar depression constant, is often chosen to magnify the difference in freezing point of the solvent and the solution of the unknown. Since freezing-point depressions are not sensitive to atmospheric pressure, they are easier to measure accurately than the methods described above, and much use has been made of them for precise studies of solutes having low molecular weights. The possibility of association or ionization of the solute must be considered with any of these methods, since these effects will greatly influence the result.

Osmotic pressures are so much larger than any other colligative property that they are most widely used for molecular weight measurements, particularly for long-chain polymers were the high sensitivity of the method is required. For accurate measurements a membrane is required which permits the flow of solvent through its pores but completely holds back solute molecules. This condition is best satisfied where there are large differences in size of the solute and solvent molecules or of their affinity for the membrane. Membranes made from cellulose compounds are often successfully used for polymers which contain little material with molecular weights below about 10,000. Below this molecular weight the pore size of the satisfactory membranes is so small that solvent flow is very slow, and thus a very long time is required to reach constant osmotic pressure. In spite of this handicap, some of the most precise osmotic pressure measurements have been obtained with aqueous solutions of sucrose and similar small solutes by the use of membranes prepared by precipitating such materials as copper ferrocyanide in the pores of a solid support. The upper limit of molecular weights satisfactorily measured by osmometry is usually about 500,000, which is fixed by the lowest pressures that can be measured precisely and by the maximum concentrations of material which still give satisfactory extrapolations to infinite dilution. In comparing various colligative properties for the characterization of polymers, osmometry has the advantage that it is unaffected by the presence of impurities of very low molecular weight which will diffuse through membranes able to retain the polymer, whereas the other properties are greatly affected by the same impurities.

Modern instrumentation has provided commercial instruments utilizing several of these colligative properties for routine, accurate measurements in very short time and with small samples. This is true for boiling point, vapor pressure, and freezing point measurements of molecular weights up to several thousand, and for membrane osmotic pressure measurements of high molecular weight samples.

X-Ray Diffraction

X-ray diffraction analysis is a powerful method for determining exact molecular weight and structural characteristics of compounds in their crystalline state. However, the method is complicated and slower than many techniques which provide molecular weights of accuracy sufficient for many purposes, and so is usually employed only when the additional structural information is needed. The sample to be examined must have a high degree of crystalline order, and is preferably a single crystal at least 0.1 mm in size; such samples are prepared fairly readily from many inorganic and nonpolymeric organic compounds. Alternatively, crystalline powders of certain crystal types may provide suitable results. Diffraction patterns are then obtained by one of several methods, and the angular positions of the reflections are used to calculate the lattice spacings, and thus the size of the unit cell. This unit cell is the smallest volume unit which retains all geometrical features of the crystalline class, and it contains a small integral number of molecules. A rough estimate of the molecular weight of the compound is needed from a determination by an independent method in order to obtain this integral number. Finally, the resultant molecular volume is multiplied by the exact bulk density of the crystal and by the Avogadro number to yield the molecular weight.

Light-Scattering. Measurements of the intensity of light scattered by dissolved molecules allow the determination of molecular weights. Most commonly the method is used for polymers above 10,000 units, though under optimum conditions molecular weights as low as 1,000 have been determined. Since the intensity scattered by a given weight of dissolved material is directly proportional to the mass of each molecule, a weight-average of the molecular weight is obtained for a polydisperse system. An average dimension of the molecule can also be obtained by a study of the angular variation of scattered light intensity, provided some dimension of the molecules exceeds a few hundred Angstroms. The interaction between dissolved molecules substantially affects the intensity of scattered light so that extrapolation to infinite dilution of data collected at several polymer concentrations is required. The method has been so well developed in the last decade that it is now probably the most used method for determining absolute molecular weights of polymers. It is among the more rapid methods and provides information on sizes which is furnished by few other methods. The greatest problem encountered is in the removal of suspended large particles which otherwise would distort the angular scattering pattern of the solutions. This is rather easily accomplished by filtration in some cases, but may be a formidable difficulty for particles which are highly solvated or are peptized by the molecules to be studied. Auxiliary information is required on the refractive index increment of the sample, i.e., the change in refractive index of the solvent produced by unit concentration of the sample. This information is supplied by a differential refractometer using the same wave length of light as that employed in measurements of the intensity of scatter.

The Ultracentrifuge. The *sedimentation* of large molecules in a strong centrifugal field enables the determination of both average molecular weights and the distribution of molecular weights in certain systems. When a solution containing polymer or other large molecules is centrifuged at forces up to 250,000 times gravity, the molecules begin to settle, leaving pure solvent above a boundary which progressively moves toward the bottom of the cell. This boundary is a rather sharp gradient of concentrations for molecules of uniform size, such as globular proteins; but for polydisperse systems the boundary is diffuse, the lowest molecular weights lagging behind the larger molecules. An optical system is provided for viewing this boundary, and a study as a function of the time of centrifuging yields the rate of sedimentation for the single component, or for each of many components of a polydisperse system. These sedimentation rates may then be related to the corresponding molecular weights of the species present after the diffusion coefficients for each species are determined by independent experiments. Both the sedimentation and diffusion rates are affected by interactions between molecules, so that each must be studied as a function of concentration and extrapolated to infinite dilution as is done for the colligative properties. The result of this detailed work is the distribution of molecular weights in the sample which is available by few other methods. Extrapolation of diffusion coefficients to infinite dilution is difficult for high molecular weight linear polymers, and so alternate means are used to relate sedimentation constants to molecular weights in these important applications.

A modification of the sedimentation method which avoids the study of diffusion constants is the sedimentation equilibrium method in which molecules are allowed to sediment in a much weaker field. Under these conditions, the sedimenting force is balanced by the force of diffusion, so that after times from a day to two weeks molecules of each size reach different equilibrium positions, and the optical measurement of the concentration of polymer at each point gives directly the molecular weight distribution. However, again extrapolation to infinite dilution must be used to overcome interaction effects. The chief difficulty here is the long time of centrifuging required, and the necessary stability of the apparatus during this period.

The *Archibald method* utilizes measurements of the concentration changes near the cell ends during a sedimentation velocity experiment to determine the molecular weight of the sample. The result is obtained without dependence on separate measurements of diffusion coefficients, since it has been shown that these boundary conditions approach those at equilibrium. Both low and high molecular weights have been studied with some advantages shown over other methods.

Chemical Analysis. When reactive groups in a compound may be determined exactly and easily, this analysis may be used to determine the gram equivalent weight of the substance. This is the weight in grams which combines with or is equivalent to one gram-atomic weight of hydrogen. This equivalent weight may then be converted to the molecular weight by multiplying by the number of groups per molecule which reacted (provided they each are also equivalent to one hydrogen). If the number of reactive groups in the molecule is not known, then one of the physical methods for determining molecular weight must be used instead.

The chemical method is convenient and often used for the identification of organic substances containing free carboxyl or amino groups which can readily be titrated, and for esters which can be saponified and determinations made of the amount of alkali consumed in this process. The equivalent weights of ionic substances containing, for example, halide or sulfate groups may also be determined by titration or by gravimetric analysis of insoluble compounds formed with reagents which act in a stoichiometric fashion. In the titration of acids, the "neutral equivalent" is the weight of material which combines with one equivalent of alkali, and a similar definition applies to the "saponification equivalent" of esters. If only one carboxyl or ester group is present in the molecule, these values equal the molecular weight of the compound.

In a similar way, if the terminal groups on polymer chains can be determined by a chemical reaction without affecting other groups in the molecule, the equivalent weight or molecular weight of the polymer may be obtained in certain cases. For polydisperse systems, a number-average value of the molecular weight is obtained because the process essentially counts the total number of groups per unit weight of sample. Since the method depends on the effect of a single group in a long chain, its sensitivity decreases as the molecular weight rises, and so is seldom applicable above molecular weights of 20,000. Particularly at higher molecular weights, the method is very sensitive to small amounts of impurities which can react with the testing reagent, so that careful purification of samples is desired.

It is also important to know that impurities or competing mechanisms of polymerization do not lead to branching or other processes which may provide greater or fewer reactive groups per molecule. The analysis for end-groups must be carried out under mild conditions which do not degrade the polymer, since this would also lead to lower molecular weight values than expected. Labelling of end-groups either with radioactive isotopes or with heavy isotopes which can be analyzed with the mass spectrometer provides a rapid and convenient analysis for end-groups. This labelling can be accomplished with a labelled initiator if this remains at the chain ends, or after polymerization is complete, by exchange of weakly-bonded groups with similar groups in a labelled compound. Molecular weight determinations by end-group analysis are often used for condensation polymers of lower molecular weights, and are especially valuable in studying degradation processes in polymers.

GEORGE L. BEYER

Cross-references: *Molecules, Atoms, Osmosis.*

MOLECULES

Molecules are chemical units composed of one or more atoms. The simplest molecules contain one atom each; for example, helium atoms (one atom per molecule) are identical with helium molecules. Oxygen molecules (O_2) are composed of two atoms, and ozone (O_3) of three. Molecules may contain several different sorts of atoms. Water (H_2O) contains two different kinds, and dimethyl amine

$((CH_3)_2NH)$ has three kinds. Molecules of many common gases (hydrogen (H_2), oxygen (O_2), nitrogen (N_2), and chlorine (Cl_2)) consist of two atoms each.

Not all substances are molecular in structure. Some are atomic, many are ionic, and others have a partly ionic character.

Molecular substances are characterized by low boiling points and poor conductivity of electricity when dissolved or melted. Gases are generally molecular, and so are many liquids and some solids. All compounds of hydrogen and nonmetallic oxides are molecular. These compounds are considered to be bonded by one or more shared electron pairs, a force called covalent bonding.

Gases. According to Avogadro's principle, equal volumes of gases, regardless of composition, contain the same number of molecules at the same temperature and pressure. As a consequence of this principle, the gram-molecular mass (g-mole) of any gaseous substance occupies 22.4 liters at standard temperature (0°C) and pressure (760 torr). The number of molecules per g-mole has been calculated by different methods and is 6.02296×10^{23} (atoms per g-atom, or molecules per g-mole). For example, one mole of ammonia gas (NH_3) weighs 17.031 g, occupies a volume of 22.4 liters at standard conditions, and contains 6.02296×10^{23} molecules.

The size of molecules, especially of the smaller ones, is so tiny that it is difficult to make a meaningful comparison. Assume that the water molecules in a cup of water are dyed so that they can be identified and this cup thrown into the ocean 2000 years ago. These molecules would now be distributed evenly in all bodies of water on the earth. A cup of water, taken at random, would yield at least one hundred of the original dyed molecules! The exceedingly large number of molecules of water in a cup of water is, of course, directly related to the fact that each molecule is exceedingly small.

At the same temperature, all gases have the same kinetic energy. This means that molecules of denser gases have smaller average velocity and that hydrogen, the lightest has the greatest average velocity.

Liquids. At the same temperature, molecules of a liquid have the same kinetic energy but differ in potential energy from those in a gas. In a liquid, however, the extent of motion must be more restricted. Liquids flow as a stream and tend to form drops to a greater or smaller extent, thus giving evidence of the importance of the force of cohesion between the molecules in a liquid. When liquids are heated, as a rule they expand, an effect explained by the tendency of the molecules to occupy more space when they move at a faster rate. Also, increase in pressure has only a slight effect on the volume of a liquid. Liquids in general are but slightly compressible. From this evidence it is argued that molecules in a liquid are adjacent, close enough to flow in a continuous stream.

The molecules in a liquid, like those in a gas, are not all moving at the same velocity, but at the same average velocity at a given temperature. The molecules at the surface of a liquid, unlike those below the surface layer, have no force of attraction from molecules above. Some of the more rapidly moving molecules overcome the cohesive force of their neighbors and leave the surface. The tendency to leave the surface or to evaporate varies from liquid to liquid, and it increases when the temperature is raised. The pressure caused by the evaporation of molecules from a liquid, measured at equilibrium with the returning molecules at a given temperature, is called the vapor pressure. In general, vapor pressure increases when the temperature rises. With an increase in temperature the vapor pressure rises still more until the vapor pressure reaches the same pressure as the atmosphere above the liquid. Then evaporation goes on throughout the liquid, the liquid is boiling, and the temperature stays unchanged. Boiling can be accomplished either by raising the temperature of the liquid or by reducing the pressure of the atmosphere above the liquid.

Solids. Since solids have their own shape rather than that of the container (as for liquids and gases) and generally do not flow, the extent of molecular motion in a solid is even more limited than that in a liquid. True solids are crystalline, bounded by plane surfaces that meet in a definite dihedral angle, and have a characteristic melting point. The molecules in a solid have the same energy as those of a liquid or a gas at the same temperature, if they are molecules of the same substance. The motion of molecules in a solid must be confined, and probably it is a vibration or oscillation if more than one nucleus is present.

Crystals composed of molecules may evaporate in a manner similar to that of liquids. The phenomenon is called sublimation, and it may be noticed in solid carbon dioxide, paradichlorobenzene, camphor, and many odorous solids. Nonmolecular solids show little tendency to sublime. The van der Waals force between the particles in molecular solids is apparently less in general than the coulomb forces between ions in nonmolecular solids. As with liquids, solids vary greatly in their tendency to sublime, and the rate of sublimation varies with the temperature and inversely with the pressure.

In some solids the crystal is composed of molecules in a pattern that repeats. Graphite, for example, consists of atoms that are bonded in a hexagonal pattern in one plane, and the hexagons tie in with one another as if they were floor tiles. The properties of graphite are associated directly with this planar structure of the carbon atoms with only weak forces to hold the planes together.

The structure of the carbon atoms in a diamond is far different from that in graphite. Each atom is supported in a diamond crystal by a pyramid of three other carbon atoms, the four atoms spaced as at the vertices of a regular tetrahedron. The pattern continues as it does for graphite, but in three dimensions. One layer of graphite or a single diamond crystal can be considered a giant molecule.

Cellulose from wood pulp contains as many as 1800 glucose units, and cellulose from cotton linters may have 3500 units. These units are linked in a long chain, giving the fibers characteristic of cellulose, a giant molecule. Rubber latex contains about 6000 isoprene units, and its molecular mass varies from 130,000 to 400,000. Protein molecules have weights extending into the millions.

ELBERT C. WEAVER

Cross-references: *Atoms, Elements, Polar Molecules.*

MOLTEN SALT CHEMISTRY

A molten salt may be defined as an ionically conducting liquid which does not contain a large fraction of nonconducting constituents. In this definition a very large number of materials are included from such typical salts as NaCl to molten oxides. As early as 1807, Davy utilized molten salts for producing alkali metals electrolytically. The importance of molten salts in technology has long been illustrated in the electrolytic production of aluminum, steel-making and glass production.[1,2,3] In the past fifteen years many new technological uses have been developed including a molten salt thermal breeder nuclear reactor[4], high-temperature fuel cells, processes for the electrowinning of special metals, and for isotope separation.[5]

One of the most useful features of molten salts is the relatively large quantities of salts of interest which may be in solution. Thus, the volume of apparatus and equipment can be kept relatively small. This feature is utilized in many ways. Aluminum is produced in molten cryolite (Na_3AlF_6) in which aluminum oxide is soluble. The molten salt nuclear reactor is dependent on the relatively high solubilities of uranium, thorium and plutonium fluorides in molten fluoride solvents. $NaF-ZrF_4$ and $LiF-BeF_2$ are the two most important solvents. In electrodeposition and in fuel cells the high concentrations of ionic electroactive species lead to good current efficiencies. This new technology has spurred research activity on fundamental molten salt chemistry.

Pure one-component molten salts are characterized by high melting points, surface tensions which are generally higher than those for water but lower than those for metals, and viscosities in the range of 0.5 to 5.0 centipoises. The molten alkali nitrates or alkali halides are waterlike in appearance. The self-diffusion coefficients of individual ions are of the order of 10^{-4} to 10^{-6} cm²/sec. The most characteristic feature of molten salts is their high conductivity. The specific conductivity, κ, ranges from 1–10 Ω^{-1}cm.$^{-1}$ for molten alkali halides and is generally lower for salts containing polyvalent ions and for the more covalent salts.

The structures of pure molten salts are governed by the fact that they are dense collections of ions. The attractive forces between ions of unlike charge and the repulsive forces between ions of like charge lead to a charge ordering such that the first shell of nearest neighbor ions about any given ion is of the opposite charge and succeeding layers of ions about this central ion alternate in the sign of the charge. The structure of alkali halides upon melting changes so that both the coordination number and the average distance between nearest neighbor cation-anion pairs decrease whereas the average distance between ions of the same charge increases. Although there is a distinct disordering of the structure upon melting which is similar to other liquids, the long-range charge ordering of anions and cations has been demonstrated from x-ray and neutron diffraction measurements. The volume increase on fusion for the alkali halides ranges from about 10 to 30% and the entropy of fusion ranges from about 4.7 to 6.3 cal/deg-mole. Thus, aside from the conductivity and the charge ordering, the physical properties of most molten salts do not differ greatly from other liquids. The charge alternation has been important in developing theories of fused salts and the most significant result of these theories is that the influence of the electrostatic interactions on the equilibrium properties of pure fused salts is related largely to the sum of the radii of the cation and anion.

Binary mixtures containing one anion and two cations have been studied in many ways. The thermodynamic properties are in part related to the electrostatic interactions of the ions which are dependent on differences of the sizes and charges of the cations of the mixtures. Electronic absorption spectra of solutions in alkali halide solvents reveal some tendency towards low coordination numbers and a strong dependence of the spectra upon the size of the alkali cation. Solid-liquid phase equilibria exhibit such typical features as continuous series of solid solutions and simple eutectics. Many systems occur in which one or more solid binary compounds are formed.[5] Equivalent conductances of binary mixtures usually exhibit negative derivations from additivity.

Electronic absorption spectra of cations have been observed for many transition metal and rare earth ions dissolved in alkali salts. Although the ligands in some cases exhibit a high degree of symmetry, in many cases the spectra are consistent with an asymmetry caused by thermal motions of the ligands. Spectra of anions as NO_3^-, and $CrO_4^=$ have been observed. Vibrational spectra of ions such as NO_3^-, ClO_3^-, ClO_4^- and many salts which form stable ionic groupings reveal that for stable ionic groupings (e.g. NO_3^-) the spectra are usually not radically changed by changes in the solvent cations.

Molten reciprocal salt systems are perhaps the most interesting systems chemically. These are systems containing at least two cations and two anions. These systems provide a wide measure of control of the reactivity of individual components. Several useful theories for these systems have been developed. Solubility products for a slightly soluble salt may be estimated by the use of a simple cycle. For the solubility product of AgBr in molten KNO_3, for example, the sum of the standard free energy changes, ΔG, for the three processes

(1) AgBr (solid) + KNO_3(liquid)
$$\rightleftharpoons AgNO_3(liquid) + KBr(liquid)$$
(2) $AgNO_3$(liquid)
$$\rightarrow AgNO_3(\text{infinite dilution in } KNO_3)$$
(3) KBr(liquid)
$$\rightarrow KBr(\text{infinite dilution in } KNO_3)$$

is related to the thermodynamic solubility product (K_{sp}) of AgBr by

$$\Delta G_1^\circ + \Delta G_2^\circ + \Delta G_3^\circ = -RT \ln K_{sp}$$

Similar relations have been technologically useful for estimating solubilities.

In dilute reciprocal salt solutions many ions form associated groupings (complex ions). For example, in molten alkali nitrates:

$$Ag^+ + Br^- \rightleftharpoons AgBr$$
$$AgBr + Br^- \rightleftharpoons AgBr_2^-$$
$$AgBr + Ag^+ \rightleftharpoons Ag_2Br^+$$

The ion pair groupings, AgBr, in solution is characterized by an association constant K_1 given by

$$K_1 = Z(\exp(-\Delta E_1/RT) - 1)$$

where Z is a coordination number and ΔE_1 is an energy of association. Many other examples of complex ions in reciprocal salt systems have been observed.[5]

In concentrated reciprocal salt solutions the thermodynamic properties often exhibit large deviations from ideal behavior which in many systems (e.g., LiF-CsBr, AgBr-KNO_3) are manifested by the appearance of two immiscible liquid layers. Reciprocal systems are of special interest for chemical processing because of these immiscibility gaps and large deviations from ideal solution behavior.

One of the most interesting types of mixtures are the metal-metal salt mixtures. Even in the early experiments of Davy the deep color of these solutions was observed. Modern investigations have demonstrated that these are true solutions and not colloidal suspensions as had been believed. Very many metals are soluble in their own salts. In cases where there is complete miscibility of the molten salt with the molten metal one can obtain a continuous variation from ionic conduction of the pure salt to electronic conduction of the metal. A knowledge of these systems is important for the electrolysis of molten salts since the electronic contributions to conductivity may decrease current efficiency considerably.

The miscibility of alkali metals with their halides is smallest for lithium and greatest for cesium. Detailed investigations of many systems have been made to study their structure.

Of all of the properties which have been extensively investigated the transport properties (diffusion, conductivity, etc.) are least understood. The ionic mobilities appear to be related to ionic size and charge but no reasonable quantitative theory has been developed. Consequently, only a phenomenological description of results is possible at present.

Liquid-vapor equilibria of salts are quite complex because of the formation of many vapor phase compounds. All monovalent halides form dimeric species in the vapor as for example

$$2NaCl \rightleftharpoons Na_2Cl_2$$

Some highly associated vapor species such as Li_4F_4, Cu_3Cl_3 and Ag_3Cl_3 have also been observed. Many vapor compounds of alkali halides with polyvalent salts have been reported, e.g., $LiAlCl_4$, $KFeCl_3$, $LiBeF_3$, etc. The alkali halide compounds can be interpreted in terms of electrostatic forces whereas the cuprous, silver and other compounds are strongly influenced by other more specific interactions.

MILTON BLANDER

References

(1) P. T. Stroup, *Trans. AIME,* **230,** 356 (1964).
(2) G. R. St. Pierre (editor), "Physical Chemistry of Process Metallurgy," Interscience Publishers, New York (1961).
(3) W. Eitel, "Silicate Science," Vol. II Academic Press, New York, (1965); "The Physical Chemistry of Silicates," University of Chicago Press, Chicago (1954).
(4) W. R. Grimes *et al.* in "Fluid Fuel Reactors," ed. by J. A. Love *et al.* Addison-Wesley, Reading, Mass. (1958). "Chemical Aspects of Molten-Fluoride Reactor Fuels," p. 569.
(5) M. Blander (editor), "Molten Salt Chemistry," Interscience Publishers, New York (1964).
(6) B. R. Sundheim (editor) "Fused Salt Chemistry," McGraw-Hill, Inc., New York (1964).
(7) Iu. K. Delimarskii and B. F. Markov, "Electrochemistry of Fused Salts," The Sigma Press (1961) (Trans. from Russian by A. Peiperl).

MOLYBDENUM AND ITS COMPOUNDS

Molybdenum is a metallic element of Group VIB of the Periodic Table. It has an atomic number of 42 and an atomic weight of 95.94. Molybdenum was discovered by Scheele in 1778, but metallic molybdenum was not isolated until 1782.

The largest molybdenum deposits are found in the Western Hemisphere, the United States providing almost three-fourths of the free-world supply. Important deposits are also located in Canada and Chile. Molybdenum is obtained both from primary mining operations and as a byproduct in some copper mining operations.

The most abundant mineral in both primary and byproduct mining is molybdenite, MoS_2. Less important minerals are wulfenite, $PbMoO_4$, and powellite, $Ca(MoW)O_4$.

Molybdenite is concentrated by crushing and grinding the ore, then sending the finely-ground pulp through flotation cells. The cells contain a dispersion of oil and water, and the mineral's affinity for small oil globules allows it to be floated and spilled over into collecting troughs.

Most molybdenite concentrate is roasted to molybdic oxide, MoO_3, a product directly usable in some iron and steel-making processes. Several additional molybdenum-containing products are made available to end users. Roasted molybdic oxide may be converted to ferromolybdenum by an aluminothermic process. Roasted oxide may be purified by a sublimation process to provide a high-purity product used in the manufacture of chemicals and special alloys. By still another purification process, dissolving in ammonium hydroxide, filtering and crystallizing, molybdic oxide can be converted to ammonium molybdate, $(NH_4)_2Mo_2O_7$, a high-purity material used in the manufacture of chemicals.

Pure molybdic oxide or ammonium molybdate can be reduced in hydrogen to a high-purity metal powder suitable for use in preparing powder metallurgy or vacuum-arc-cast molybdenum metal. Molybdenum powder is also used in the manufacture of catalysts and as feedstock for molybdenum metal spraying.

Some molybdenite concentrate is purified to a grade of MoS_2 used in the manufacture of a dry lubricant.

Physical Properties. Molybdenum has a melting point of 2610°C and a boiling point of 5560°C. The heats of fusion and vaporization are 6.7 and 117.4 kcal/mole, respectively. The specific heats at room temperature and at 2000°C are, respectively, 0.064 and 0.105 cal/g/°C. The electrical resistivity is 5.78 microhm-cm at 27°C. Molybdenum is paramagnetic. The density is 10.22 g/cc.

Chemical Properties. Molybdenum has an atomic number of 42 and an atomic weight of 95.94. Natral isotopes have been observed at 92, 94, 95, 96, 97, 98, and 100. Artificial isotopes have been observed at 90, 91, 93, 99, 101, 102, and 105. Molybdenum forms compounds in which valence states of 0, +2, +3, +4, +5, and +6 are displayed. The ion radius of molybdenum is 0.92 Å in the trivalent state and 0.62 Å in the sexivalent state. The ionization potential is 7.2 eV.

Compounds. The chemistry of molybdenum is complex. It forms compounds in which valence states of 0, +2, +3, +4, +5, and +6 are displayed.

The most common oxides are MoO_2 and MoO_3 At least seven intermediate oxides have been identified, but they are largely mestastable. The reduction of the trioxide does not proceed stepwise through the intermediate oxides. The dioxide is readily converted to the trioxide by heating in an oxidizing atmosphere.

The trioxide is an acid anhydride forming molybdic acid, H_2MoO_4, on the addition of water. The salts of molybdic acid form an important family of compounds called molybdates. Acidification of normal molybdates results in polymerization. Heteropoly molybdates generally have very high molecular weights, ranging up to 4,000.

Halides representing a wide range of valency states and stabilities have been identified. Few molybdenum halides are considered monomeric. Many of the halides are volatile at atmospheric pressure, and many react with water and oxygen. Molybdenum pentachloride, $MoCl_5$, has been used in the vapor phase deposition of molybdenum coatings.

Molybdenum hexacarbonyl, $Mo(CO)_6$, represents a compound in which molybdenum exhibits a valence of zero. The compound can be prepared by reacting metal powder and carbon monoxide at high pressure. The compound decomposes at 150°C, but is stable at lower temperatures. The hexacarbonyl reacts with a variety of organic compounds to yield organomolybdenum compounds. $Mo(CO)_6$ has also been used as a source of molybdenum for vapor-deposited coatings.

Molybdenum forms a series of homologous compounds with sulfur, selenium and tellurium. The sesqui-compounds, Mo_2X_3, have been prepared by direct reaction of the elements at elevated temperatures. The di-compounds, MoX_2, are isomorphous. The disulfide is stable over a wide range of conditions, but it can be converted to MoO_3 under strongly oxidizing conditions.

In addition to the wide variety of complex ions derived from molybdic acid, a number of others are known which contain molybdenum in lower valence states. Complex cyanides and thiocyanates are included in this group.

Molybdenum forms many organometallic compounds because of its many valence states and great complexing power. Molybdenum forms chelates with nitrogen, sulfur and oxygen and a series of esters with alcohols, phenols and hydroxy acids. Molybdenum forms alkyl and aryl derivatives as well as complexes with cyanides, olefins and acetylenes.

Examples of organic compound applications include molybdenum acetylacetonate as a catalyst for the polymerization of ethylene and for the formation of polyurethane foam, molybdenum oxalate in certain photochemical systems, and molybdenum dithiocarbamate as a lubricant additive.

Applications. Approximately 85% of molybdenum production now goes into the manufacture of iron-base alloys. Included in this group of materials are alloy steels, tool and high-speed steels, stainless steel, and alloy cast irons.

Most alloy steels contain less than 1% molybdenum, but some of the recently-developed superstrength grades contain higher amounts. One of the important contributions of molybdenum additions to alloy steel is improved hardenability, that property which determines the hardness pattern throughout a part heat treated in a given manner. Other benefits are improved cold formability, toughness, machinability and weldability.

Molybdenum is added to tool steels in concentrations of 1 to 10% to obtain better hot strength, more resistance to softening and improved resistance to the effects of thermal cycling.

Molybdenum is added to several types of stainless steels to improve corrosion resistance, elevated temperature strength and weldability. The usual molybdenum addition to stainless steel is in the range from 1 to 4%. Certain special grades designed to handle specific corrosives may contain as much as 20% molybdenum.

Molybdenum is added to cast iron for applications such as automobile engine blocks primarily for strengthening purposes. Some cast irons (and some cast steels) used to combat very abrasive environments often contain molybdenum in concentrations up to 3%.

Superalloys, the materials used in the turbine wheels and blades of jet engines, usually contain molybdenum because of its contribution to hot strength. These materials may be based on nickel, iron or cobalt. The molybdenum content of superalloys varies over a wide range and can go as high as 30%.

Molybdenum metal and molybdenum-base alloys received considerable attention after World War II. The high melting point of the metal, about 2610°C, led to its consideration for a variety of missile and space applications involving exposure to very high temperatures. However at temperatures over about 550°C unprotected molybdenum oxidizes so rapidly in air that its use under these conditions is impractical. Further, the inability to find adequate oxidation-protection systems for molybdenum discouraged its use for many of these applications. Despite this setback, molybdenum and molybdenum-base alloys have found numerous applications that take advantage of properties such as strength, stiffness, electrical or thermal conductivity, and corrosion resistance.

Molybdenum-containing catalysts are now used in many petroleum and chemical processes. Molybdenum compounds may serve as principal catalyst or as activators and promotors of other catalysts. The ease of interchangeability of sulfur and oxygen in molybdenum compounds is a useful property in many catalyst systems.   Traces of sulfur often poison other catalysts but not those based on molybdenum.   Hydrocracking, alkylation and reforming are important processes utilizing molybdenum catalysts.

Purified molybdenum disulfide is an important dry lubricant that is used alone or in combination with grease or oil.   The compound functions in this application because its layered lattice structure favors easy cleavage between adjacent laminae.

Pigments containing zinc molybdate have been developed that have shown exceptional stability and anticorrosion properties and have a very low toxicity level.   A pigment of long-standing use is molybdenum orange.   This pigment is formed by the coprecipitation of lead chromate and lead molybdate.

Molybdenum is important in the life processes of both plants and animals, although its functions are not completely understood.   Large increases in crop yields, particular in legumes, have resulted from molybdenum applications in fertilizers, as foliar sprays or seed coatings.

Although the minimum requirements for molybdenum in humans have not been established in the United States, it has been added in trace amounts in vitamin supplements and specialized medications. Minimum molybdenum requirements have recently been proposed in the USSR.

Recent research has shown that molybdenum is an important element in promoting healthy teeth. For example, statistical studies have shown a direct correlation between dental health and molybdenum content of the soil in the subjects' geographical locale.   A molybdenum-containing dentifrice is now marketed in France.

JANET Z. BRIGGS
ROBERT Q. BARR

Reference

Herzig, A. J., and Briggs, J. Z., "Molybdenum," pp. 412–422, "Encyclopedia of the Chemical Elements," Hampel, C. A., Editor, Van Nostrand Reinhold Corp., New York, 1968.

MONOMERS

There exist a number of molecules (particularly organic) which have the capacity to react with each other in such a manner that very large molecules are formed.   Reactions of this type are called *polymerization* reactions.   The initial molecules which enter a polymerization reaction are called *monomers, monomeric units* or *base units,* all words which express the fact that they become a *part* of a large molecule, which in turn is called a *polymer.* Monomers have, in general, molecular weights between 50 and 250, whereas the polymers which are built up by the repetition of a monomer have molecular weights of many hundred thousands and, in certain cases, even several millions.

Because of their capacity to polymerize readily, monomers are, in general, very reactive materials, which must be handled with caution and, in many cases can be stored only in the presence of stabilizers under exclusion of oxygen or at low temperatures.

Monomers are produced from coal, oil, natural gas, farm and forest products and are used to synthesize a wide variety of valuable polymers covering the fields of fibers, plastics, rubbers, coatings and adhesives.   It can be estimated that the integrated value of all monomers used in the U. S. in 1975 will be close to 10,000 million dollars.   A special class of monomers are those which polymerize by *addition* polymerization; they contain one or more aliphatic double bonds and lead to a wide variety of synthetic rubbers and plastics.   Examples for this group are listed in Table 1.   Another class of monomers polymerizes by a *condensation* process with the aid of *functional groups* in such a manner that *two* monomers cooperate in the formation of the polymer molecules.   Examples for this type are given in Table 2.

TABLE 1.   MONOMERS CAPABLE OF UNDERGOING
ADDITION POLYMERIZATION

Ethylene	$CH_2{=}CH_2$
	Polyethylene
	"Alathon"
Vinylchloride	$CH_2{=}CHCl$
	"Geon," "Lucoflex"
Butadiene	$CH_2{=}CH{-}CH{=}CH_2$
	Buna
Styrene	$CH_2{=}CH(C_6H_5)$
	"Styron"
Acrylonitrile	$CH_2{=}CHCN$
	"Orlon"
Vinylidenechloride	$CH_2{=}CCl_2$
	Saran
Isobutylene	$CH_2{=}C(CH_3)_2$
	"Vistanex"

TABLE 2.   MONOMER PAIRS WHICH CAN
UNDERGO POLYCONDENSATION

Adipic acid	$HOOC{-}(CH_2)_4{-}COOH$
Hexamethylenediamine	$H_2N{-}(CH_2)_6{-}NH_2$
	Nylon 66
Phenolformaldehyde	C_6H_5OH
	CH_2O
	"Bakelite"
Glycol	$HOCH_2{-}CH_2OH$
Terephthalic acid	$HOOC{-}C_6H_4{-}COOH$
	"Dacron"
Glycol	$HOCH_2{-}CH_2OH$
Adipic acid	$HOOC{-}(CH_2)_4{-}COOH$
	"Paracon"

All monomers listed here and many others of somewhat minor importance are definitely organic molecules with known melting and boiling points and with a formula which is reliably established. They can be produced in a high grade of purity and must be used in this form in the course of polymerization reactions if one wants to arrive at reproducible results.

HERMAN F. MARK

Cross-references: *Polymers; Polymerization.*

MONOMOLECULAR FILMS

A monomolecular film or monolayer is a film one molecule thick; it is the ultimate thin film. Thin films that form at surfaces or interfaces warrant special attention. Such films reduce friction, wear, and rust, and stabilize solid dispersions, emulsions, and foams. Thin films on water surfaces reduce evaporation losses in arid regions throughout the world. The broad field of catalysis, which is basic to petroleum refining and many chemical industries, involves chemical reactions that are accelerated in the thin films of reactants at interfaces. Moreover, thin films containing proteins, cholesterol, and related compounds comprise the biological membranes or interfaces that control the complex processes of living matter. Recently, the removal of thin films of contaminants from the surface of water has become an important effort in pollution control.

In all these areas the monolayer is the most important layer. It is held to the adsorbing surface by forces stronger than those that hold any succeeding layer. On solid surfaces it is the only layer that can be chemisorbed. It is the last line of defense or protection.

Monolayers at a liquid-gas interface can be controlled, manipulated, and examined under far better conditions than at other interfaces. Most attention has therefore been focused on monolayers at the water-air interface. The pioneering studies of Adam and Rideal in England and Harkins and Langmuir in the United States have provided much basic information on such monolayers. Nevertheless, relatively little is known about their detailed structure.

A combination of film-balance and electron-microscope techniques has recently revealed some of the details of monolayer structure and behavior. The film balance probably provides the most reliable and versatile technique for studying the surface-pressure and surface-area relations for monolayers of polar organic molecules. Such film-forming molecules have a polar or water-soluble group and a hydrocarbon structure that is sufficiently large to make the molecule insoluble in water. Pressure-area data yield information on molecular geometry and orientation, location and strength of polar groups, and forces of cohesion and adhesion. Such measurements are extremely sensitive to small differences in molecular structure. The electron microscope provides details of size, shape, and thickness of film aggregates that are far beyond the range of earlier techniques. It reveals changes in film structure on compression and the nature of film collapse; it relates properties of films on liquids to those on solids.

A schematic drawing of the film balance is shown in Figure 1. The apparatus consists essentially of a long shallow trough filled with high-purity water on which the monolayer is spread, and a float system for measuring the surface pressure. The float is a strip of mica attached to the sides of the trough by thin flexible platinum foils, and to an aluminum stirrup by an unspun silk thread. The stirrup is fixed to a calibrated torsion wire that indicates the surface pressure. Small brass bars or barriers are used for sweeping the water surface free of contamination. A small amount (approximately 0.01 mg) of the film forming compound in a volatile solvent is spread between the float and the large or main barrier. The barrier is moved gradually toward the float to compress the film. Compression is continued until the pressure remains constant or falls. This procedure at constant temperature provides the data for plotting the pressure-area isotherms which characterize the films.

Before each experiment the entire apparatus— trough, barriers, float, and platinum foils—are thoroughly cleaned and coated with high-melting paraffin wax or with "Teflon." The trough is filled with water to a height well above the rim. This

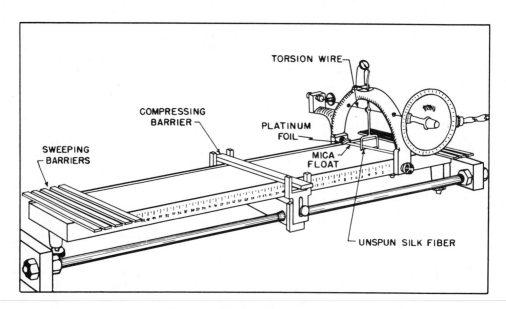

FIG. 1. Film-Balance Apparatus.

height is necessary if the sweeping procedure is to be effective and if the monolayer is to be contained and controlled.

Schematic drawings of three representative polar organic molecules oriented at the water-air interface are shown in Figure 2. Stearic acid, at the left, is the classical compound in monolayer studies; it is the simplest structure representative of thousands of important film-forming compounds. The stearic acid molecule consists of a long straight hydrocarbon chain and a polar group at one extremity. To the right of stearic acid is a very similar molecule, isostearic acid, and a very dissimilar molecule, tri-p-cresyl phosphate. The difference between isostearic and stearic acid is very slight—the displacement of a small group, the methyl group (CH$_3$), at the end of the molecule opposite the polar group. Tri-p-cresyl phosphate is greatly different; it has a bulky three-ring hydrocarbon portion attached to a strongly polar phosphate group.

Figure 3 shows the pressure-area isotherms for the three compounds. Area per molecule (in square Angstroms) is plotted against film pressure in dynes per cm. Extrapolation or extension of the steepest part of an isotherm to zero pressure gives the molecular area. The point at which the pressure falls or remains constant is called the collapse pressure. Compressibility of the monolayer may be calculated from the slope of the isotherm. Thickness of the monolayer, or the length of the vertically oriented molecules, may be estimated by assuming a density for the monolayer; the volume and area then yield the thickness.

Comparison of the isotherms for stearic and isostearic acids in Figure 3 demonstrates that the single small side chain of isostearic acid has increased the cross section from 20 Å^2 for the stearic acid molecule to 32 Å^2 for isostearic acid, an in-

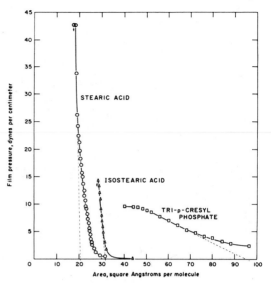

FIG. 3. Pressure-area isotherms.

crease of over 50%, Collapse pressure falls from 42 dynes to 14 dynes per cm. These are indeed striking differences between molecules that are extremely difficult to distinguish by most chemical methods.

The curve for the tri-p-cresyl phosphate reflects a very different molecular structure. The extrapolated area, 95 Å^2 per molecule, shows the bulkiness of the three-ring group held close to the surface. The low collapse pressure, nine dynes per centimeter, reflects the weakness of such a film. The gradual slope of the curve, or the high compressibility of the film, indicates poor packing of the molecules.

An example of the useful application of monolayer properties is in the rust-preventive area. Stearic acid is a fairly good rust preventive. However, stronger polar groups on straight-chain structures give stronger monolayers and better rust protection. Octadecyl-phosphonic acid collapses at 53 compared to 42 dynes per centimeter for stearic acid, and 2-hydroxystearic acid collapses at 57 dynes per cm. Both of these compounds give excellent rust protection at very low concentrations— far better than stearic acid. Tri-p-cresyl phosphate, which forms a weak film, gives essentially no rust protection even at high concentrations.

The monolayer of a very long chain acid, n-hexatriacontanoic (36 carbon atoms), is of interest because the film is twice as thick as that of stearic acid and it therefore facilitates electron-microscope study. Because its molecular cross-sectional area, 20Å^2, is the same as that for stearic acid, vertical orientation is clearly indicated. Its high collapse pressure of 58 dynes per centimeter, which is 16 dynes greater than that for stearic acid, demonstrates increased cohesion with increased chain length. As indicated before, one may also increase collapse pressure or film strength by increasing adhesion through the use of a stronger polar group.

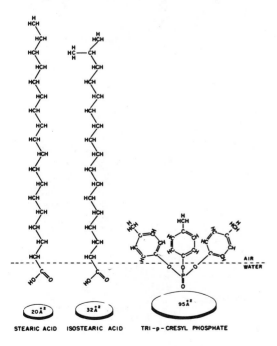

FIG. 2. Molecular orientation.

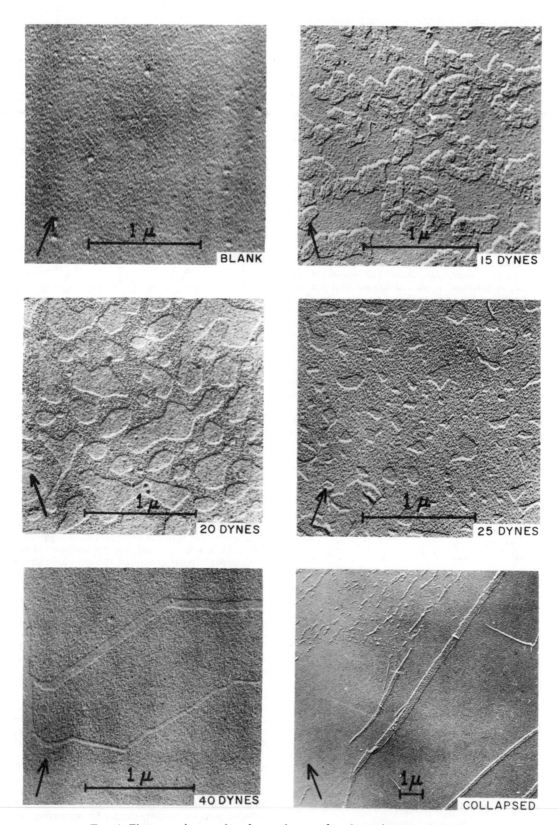

FIG. 4. Electron micrographs of monolayers of *n*- hexatriacontanoic acid.

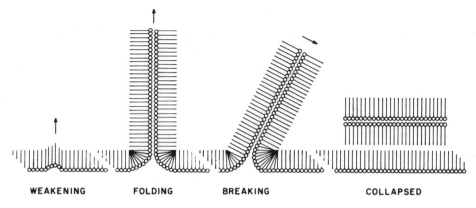

WEAKENING   FOLDING   BREAKING   COLLAPSED

Fig. 5. Mechanism of monolayer collapse.

A series of electron micrographs for *n*-hexatri-acontanoic acid is shown in Figure 4. Arrows indicate the direction of shadowcasting. The first of the series is simply collodion raised through a clean water surface on which no film had been spread. The smooth surface of collodion as well as some typical holes are apparent.

At low pressures (15 dynes per cm and below), islands of irregular size and shape appear; most are well under one micron, or 10,000 Å, in greatest dimension. Movement or rearrangement of such islands may account for unstable or nonreproducible surface pressures in the low-pressure region. The widths of the shadows correspond to a film thickness close to 50 Å—that expected for a monolayer of vertically oriented *n*-hexatriacontanoic acid. *Direct evidence that the film is one molecule thick is thus provided.* At 20 dynes, the film has become the continuous phase; bare portions are now discontinuous. At 25 dynes, the continuous monolayer occupies more of the available area, as might be expected. Large homogeneous areas of continuous monolayer are observed after transfer at 40 dynes. Uncovered areas at 40 dynes must result from strains or disturbances that crack the film; the matching contours of the film edges strongly suggest mechanical separation or cracking.

A sample of film transferred after the monolayer had collapsed is also shown in Figure 4. Long, narrow, flat structures appear. They are about 100 Å or two molecules thick, and they rest on a continuous monolayer substrate. Shadows at small breaks in the substrate establish that it remained 50 Å thick.

In the micrographs of Figure 4, the surface of the monolayer appears coarser in texture than that of the collodion support. Texture may reflect the structure of the monolayer. Perhaps the dark spots indicate micro-islands that form during the spreading process, persist at low pressures, and then consolidate. Areas that appear bare may contain smaller islands or even individual molecules that are scattered and beyond the range of the electron microscope used. High-resolution instruments have established the presence of micro-islands in the so-called uncovered areas.

Three widely different transfer techniques reproduce the main features of the micrographs. How-ever, one-to-one correspondence between the state of the film on the water and that observed in the electron microscope is not claimed. Nevertheless, the sequence of micrographs undoubtedly parallels a similar sequence of changes that takes place during compression of the film on water.

Electron micrographs of collapsed films suggest a mechanism for the collapse process. Figure 5 shows four probable stages in the collapse of a typical fatty-acid monolayer. In the weakening stage, some vertically oriented molecules are forced up from the water surface. The cohesion of the long chains and the mutual attraction of the polar groups are great enough that the molecules subsequently rise in closely packed folds, or ridges, two molecules wide. The double-layer ridge rises, bends, and apparently breaks. In the final stage, the collapsed fragment, two molecules thick, rests on the monolayer substrate. Presumably, this is the structure of the long, flat particles of Figure 4.

Combined film-balance and electron-microscope techniques have been extended to the study of mixtures because single-component systems are seldom encountered in practice. In film-balance studies, remarkable interaction effects of long polymer molecules and fatty-acid type molecules appear. Electron micrographs of such mixed films show structures entirely different from those of the individual components. Such studies may reflect the geometry of the complicated polymer structures so important in biology—but equally important in the broad fields of lubricants, fibers, and plastics. In addition, the use of radiotracer techniques is also shedding light on the formation, as well as the structure, of monolayers and mixed films under conditions in which the molecules are adsorbed from solution onto solid surfaces. Such new tools and techniques should permit more detailed study of thin films at other interfaces

Widespread interest in the removal of contaminant films from the surface of water has prompted considerable research on thin-film transfer by rotating cylinders. Remarkable effects are observed when a small cylinder is rotated in a monolayer held at constant surface pressure: the monolayer is transferred quantitatively from the water surface to the cylinder; fatty-acid and long-chain alcohol films differ markedly in behavior after the first layer is deposited; and well-defined flow patterns are produced when the cylinder is used as a surface pump for alcohol monolayers. In large scale experiments related to water-pollution control, thicker films varying widely in composition can be picked up by rotating cylinders or drums and continuously removed by scraper systems. Many related applications of monolayer techniques warrant further study.

HERMAN E. RIES, JR.

Cross-references: *Films, Surface; Emulsions; Colloid Chemistry; Adsorption; Surface Chemistry; Interfaces.*

References

Adamson, A. W., "Physical Chemistry of Surfaces," 2nd edition, Interscience Publishers, New York, 1967.

Gaines, G. L., Jr., "Insoluble Monolayers at Liquid-Gas Interfaces," Interscience Publishers, New York, 1966.

Ries, H. E., Jr., *Scientific American,* **204,** 152 (1961).

Ries, H. E., Jr., and Walker, D. C., *J. Colloid Sci.,* **16,** 361 (1961).

Ries, H. E., Jr., and Gabor, J., *Nature,* **212,** 917 (1966).

N

NATURAL GAS

Natural gas is a mixture of light aliphatic hydrocarbon gases, chiefly methane, occurring in porous formations in the earth's crust and frequently associated with petroleum deposits. It may also contain varying amounts of carbon dioxide, nitrogen and helium, and is the most important source of the latter valuable element. While used chiefly as a fuel, natural gas is a source of raw materials for the chemical industry.

Sources. Proved natural gas deposits are found in 34 states of the United States, but over 90% are located in Texas, Louisiana, Oklahoma, New Mexico and Kansas, which are also the leading producing states. Proved reserves in 1970 amounted to 291 trillion cubic feet (60°F, 1 atm.). This compares with a consumption of 22 trillion cubic feet in 1970, over 3.5 times the 1945 consumption. Additional potential reserves in this country totaled 1,178 trillion cubic feet in 1970 with more than 400 trillion located in Alaska.

Handling. Natural gas is withdrawn from deposits 1000 to 18,000 feet deep by wells drilled to the beds, and is processed to remove valuable by-products, such as natural gasoline, ethane, propane, butane, sulfur and, in some cases, helium. (See **Helium.**) It is also treated to remove unwanted materials, such as nitrogen, carbon dioxide, water and sulfur compounds that cannot be converted readily to sulfur. It is transported to the consuming centers through over 225,000 miles of large-diameter pipelines. To meet fluctuating demands natural gas is stored in more than 300 underground reservoirs, most of which are former producing gas or oil fields, whose capacity in the United States is over 5,000 billion cubic feet, about ¼ the annual consumption. They are located in 26 states, with four states having more than half of the total capacity: Pennsylvania with 696 billion cubic feet, Michigan 694, Illinois 526, and Ohio 502, all located nearer the northern cities than the producing wells of the southwest.

In recent years increasing numbers of storage plants for liquefied natural gas, LNG, have been built nearer cities than most of the underground storage reservoirs. Some 2 dozen are now in operation and have a storage capacity of about 27 billion cubic feet of natural gas. Tanks contain the LNG at atmospheric pressure and a temperature of -260°F; 1 cubic foot of LNG equals 630 cubic feet of gas. LNG is also shipped in ocean tankers; this permits the shipment of natural gas from such locations of huge reserves as Alaska and Algeria to consuming centers.

Uses. At first discarded as an unwanted adjunct of petroleum production, natural gas is now a major source of energy and a major raw material for the production of petrochemicals. Because it is low in sulfur content, demands are increasing drastically for natural gas as a fuel to replace coal which contains sulfur, whose combustion yields undesirable SO_2 air pollution.

In the United States in 1969 about 40% of the total energy came from petroleum, 35% from natural gas, 21% from coal and 4% from water power and nuclear sources. Since then the amounts from petroleum, natural gas and nuclear sources have increased with decreasing amounts from coal. Of all energy sources, the greatest increases in demand are for natural gas.

Although only about 5% of all natural gas consumed is used in production of chemicals, over 25% of all chemicals produced start with natural gas components as a raw material. Among them are methanol, ammonia, sulfur, acetaldehyde, acrylonitrile, carbon black, ethanol, chlorinated hydrocarbons, glycol, many plastics, acetylene and acetylene derivatives.

CLIFFORD A. HAMPEL

References

Springborn, Harold W., "The Story of Natural Gas Energy," American Gas Association, Inc., 605 Third Ave., New York, 1970.

"Gas Facts," American Gas Association, Inc., 605 Third Ave., New York, published annually.

NATURAL PRODUCTS

Although the term "natural products" literally includes all substances that occur naturally, in actual use it is restricted to organic compounds produced by living organisms; this is the original scope of organic chemistry, which began as a study of naturally occurring compounds. Until recently the main sources of natural products investigated were plants and microorganisms. Current sources also include plants and animals from the sea, insects, and food aromas.

The organic compounds that occur naturally have several distinctive characteristics. (1) They encompass a wide variety of chemical structures. (2) About half of all naturally occurring compounds contain heterocycles. (3) Many natural products have biological activity; in fact, one of the chief reasons for the selection of a natural product for study is an interesting biological action. (4) Many natural products are optically active. Usually only one member of a pair of optical isomers occurs naturally, and related compounds have the same configuration.

Classes and Functional Groups. Some of the more important classes of natural products are: alkaloids, amino acids, antibiotics, carbohydrates, enzymes, fats, fatty acids, flavonoids, hormones, proteins, quinones, steroids, and terpenes. Although most of these classes are based on chemical structure, a few (*e.g.,* antibiotics, vitamins, hormones) are based on biological activity and include very different structures.

A wide variety of functional groups has been observed in naturally occurring compounds. Even quite reactive groups, such as acid anhydride, acetylene, cyclopropene, and diazo, have been found; and it is likely that even less stable structures occur naturally but are not capable of isolation by techniques now available. The functional groups known to occur naturally include: olefin, acetylene, allene, cumulene; aromatic ring; alcohol, enol, phenol; ether, acetal, epoxide, peroxide; aldehyde, ketone; carboxylic acid, anhydride, ester, lactone, amide, lactam, imide; mercaptan, sulfide, disulfide, polysulfide, sulfoxide, sulfone, sulfate ester, sulfonic acid, sulfinic acid, isothiocyanate; selenide; $p, s,$ and t amines; amine oxides; nitro, nitroso, nitrile, hydroxylamino, diazo, azoxy; ammonium, oxonium, and sulfonium salts; phosphate ester, phosphonic acid; fluoro, chloro, bromo, and iodo. Both carbocyclic and heterocyclic rings occur ranging in size from three to approximately twenty-five members. The complexity of natural products ranges from simple hydrocarbons to compounds containing many functional groups and many rings.

Problems of Study. The study of natural products involves biology and biochemistry as well as chemistry and consists of the following fields of investigation.

(1) Isolation and purification of compounds. The method of isolation depends in part on the source (*e.g.,* bark, flowers, leaves, seeds) and on the type of compound; distillation, steam distillation, and extraction are the most common methods. The compounds are purified by standard methods of organic chemistry, especially the various types of chromatography, since a natural product often occurs as part of a complex mixture of which it may be a minor component.

(2) Elucidation of the structures of the compounds. This problem is more complex than that encountered with compounds of synthetic origin since no information about the natural compound is available and since it is frequently necessary to work with very small quantities. The physical data required for determination of the structure and stereochemistry of a natural product include elemental analysis and mass, ultraviolet, visible, infrared, and n.m.r. spectra. Chemical data are obtained by conversion to known compounds and/or degradation to smaller compounds, which are easier to identify. Additional data, such as optical rotatory dispersion, may be needed to determine the absolute configuration of an optically active compound.

(3) Synthesis. The synthesis of a natural product by an unambiguous method constitutes the final step in the structure proof, although synthesis is not always possible. The study of natural products often stimulates synthetic organic chemistry; some very unusual structures (*e.g.,* tropolone) have been found as natural products before they were synthesized.

(4) Function of the compounds. The biologist and biochemist are especially interested in the purpose which the compound serves in the organisms which produce it. The functions of enzymes and hormones in both plants and animals are well known; but the functions of the majority of compounds produced by plants are not known. Thus, the reasons why plants produce alkaloids, which are very useful to man, remain unknown despite more than a hundred years of study.

(5) Chemotaxonomy. A given natural product usually occurs in many sources. (α-Pinene is known to occur in more than four hundred species.) From a knowledge of the distribution of certain structures, it is possible to make correlations between chemical structures and groups of plants and animals. Such a study is known as chemical taxonomy, a new approach to biological classification. Information on the type of compounds produced by two species that appear to be similar (*i.e.,* similar external morphology) may greatly aid in solving the problem of whether the species are derived from a common ancestor (phylogenetic relationship) or have independently adapted in a similar manner (convergent evolution). Thus, the morphological evidence indicating a phylogenetic relationship among several families of flowering plants is supported by the occurrence of isoquinoline alkaloids in all of these plants.

(6) Biogenesis. The pathways by which compounds are synthesized within living organisms are known as biogenesis or biosynthesis. Several well-established theories account for the biosynthesis of large classes of natural products, such as alkaloids, flavonoids, terpenes, and steroids, starting from acetic acid. The oldest biogenetic rule, proposed by Wallach in 1887, is the isoprene rule that terpenes can be thought of as consisting of C_5 (isoprene)

$$CH_3$$
$$|$$

($CH_2{=}C{-}CH{=}CH_2$) units joined together. It has since been established that terpenes are in fact formed by the condensation of C_5 units, although the actual unit is not isoprene. Biogenetic rules often aid in the determination of structures of natural products. Thus, a compound containing 5, 10, 15, etc., carbon atoms may be a terpene and contain isoprene units. Similarly, fatty acids are expected to contain an even number of carbon atoms since they are formed from condensation of acetic acid, and hydrocarbons an odd number of carbon atoms since they are formed by decarboxylation of the fatty acids. Numerous exceptions to such rules are known.

Uses. It would be impossible to list all the uses of natural products, for one has only to think of the many articles of daily life that are not man-made. Some of the uses are as food, flavoring agents, perfumes, dyes, drugs, vitamins, hormones, oils, waxes, varnishes, glues, rubber, wood, paper, insecticides, and plant growth regulators. It is apparent that many of these are based on the biological properties of the natural products. As indicated before, the study of natural products has stimulated synthetic chemistry, and chemists have synthesized many new drugs, dyes, etc., patterned after the structures of the naturally occurring substances.

WILLIAM R. RODERICK

References

Alston, R. E., Mabry, T. J., and Turner, B. L., "Perspectives in Chemotaxonomy," *Science, 142,* 545–552 (1963).

Miller, M. W., editor, "The Pfizer Handbook of Microbial Metabolites," McGraw-Hill, New York, 1961.

Richards, J. H., and Hendrickson, J. B., "The Biosynthesis of Steroids, Terpenes, and Acetogenins," W. A. Benjamin, New York, 1964.

Robinson, T., "The Organic Constituents of Higher Plants: Their Chemistry and Interrelationships," Burgess Publishing Co., Minneapolis, Second Edition, 1967.

Roderick, W. R., "Structural Variety of Natural Products," *J. Chem. Educ., 39,* 2–11 (1962).

NEON

Neon with atomic number 10 and atomic weight 20.183 is the second lightest of the monatomic inert gases, helium, neon, argon, krypton, xenon and radon which comprise Group 0 of the Periodic Table. The ten electrons of the atom completely fill the K and L shells. Three stable isotopes of mass number 20, 21 and 22 exist and five radioactive species of mass number 17, 18, 19, 23 and 24 can be artificially produced. Ne^{17} has a half-life of 0.10 sec, emitting a positive electron (positron) followed by a proton. Ne^{18} and Ne^{19} emit positive electrons and have half-lives of 1.68 and 17.4 seconds, respectively. Ne^{23} and Ne^{24} emit negative electrons and have half-lives of 37.6 sec and 3.4 months, respectively.

Neon was discovered by Sir William Ramsay and M. W. Travers in 1898 as a volatile component of liquefied air. Dry air contains 18.18 parts per million by volume of neon and traces of the gas are widely found within the earth. The isotopic proportions of atmospheric neon are 90.92%, 0.257% and 8.82% for Ne^{20}, Ne^{21} and Ne^{22}, respectively.

Commercial quantities of neon are produced as a byproduct in the fractional distillation of air for the production of oxygen and nitrogen. A mixture of helium, neon, hydrogen and nitrogen, the most volatile components of air, collect at the top of the distillation column and can be withdrawn. The nitrogen can then be removed by further distillation and the last traces adsorbed on charcoal at liquid air temperature or on hot calcium or magnesium. The hydrogen is easily removed by chemical oxidation. The resulting mixture containing about 25% helium and 75% neon is sold as technical grade neon. Pure helium and pure neon can be obtained by condensing the neon from the mixture using liquid hydrogen or liquid neon as the refrigerant. The differential adsorption of helium and neon on cold charcoal can also be used to separate these gases.

Because of its stable electronic configuration the neon atom attracts other atoms with comparatively weak interatomic forces and forms no stable chemical compounds. However, it is known from optical and mass spectrometric studies that some short-lived ions $[(He\ Ne)^+, Ne_2{}^+, (NeA)^+, NeH^+]$ can be formed in electric discharge tubes. (See **Noble Gas Compounds.**) Neon is identified most easily by its spectrum: several intense red and yellow lines (e.g., wave lengths 6402. 2 Å, 5852. 4 Å, 5400. 5 Å) are prominent and account for the characteristic dazzling color of the neon gas electrical discharge. Neon has a normal boiling point at 1 atmosphere of 27.09°K; triple point at 24.54°K and 0.4273 atm; critical point at 44.40°K, 26.19 atm and 0.483 g/cc; latent heat of vaporization at normal boiling point of 429 cal/g-mole; heat of fusion at triple point of 80.1 cal/g-mole; first ionization potential, 21.563 eV; minimum excitation potential, 16.618 eV.

Neon with a liquid range of 24.54 to 44.50°K would be a convenient cryogenic fluid, except for its high cost. Because liquid neon is noninflammable and has a higher heat of vaporization and heat capacity per unit volume than hydrogen, it is used as a replacement for liquid hydrogen in some applications.

The principal commercial use of neon is in lighting devices, such as the familiar neon tubes used in advertising signs, fluorescent lamps, gaseous conduction lamps, etc. An electrical discharge in a tube filled with neon at a few millimeters of mercury pressure gives off a brilliant red-orange light. Commercial neon signs may contain other gases such as helium, argon or mercury and the color of the illumination is determined by the gaseous mixture used and the color of the glass envelope. In some fluorescent light tubes the purpose of the neon is to start and maintain an electric discharge. The ultraviolet light from the discharge excites a phosphor coating on the inside walls of the tube which provides the visible illumination. These neon lamps or tubes are highly efficient; the fraction of electrical energy converted to light is nearly 25%, or approximately four times that of the best metallic filament. Neon by itself or as a mixture with other gases is used in pilot lamps, regulating tubes, switching devices, starter switches for fluorescent tubes, Geiger-Mueller tubes for detecting and counting elementary particles, ionization chambers, proportional counters, spark chambers and other devices for detecting the passage of ionizing particles. It is also a component in the helium-neon gas laser.

H. A. FAIRBANK

NERVE GASES

Nerve gases comprise several severely poisonous compounds belonging to the group of organic esters of phosphoric acid derivatives. The general structure of members of this group is:

Diisopropyl phosphonofluoridate is a typical example. It is an odorless, colorless liquid with boiling point of 183°C (760 mm) and a freezing point of −82°C. It dissolves in water up to 1.54%, in which solution it slowly hydrolyzes; a 1% solution at 15°C is completely hydrolyzed in 72 hours to give diisopropyl hydrogen phosphate and hydrogen fluoride.

German synthetic chemists, under the leadership of Gerhard Schrader, are credited with opening up this field of biologically interesting compounds. Originally developed as possible insecticides, the incidental discovery of their amazing toxicity to man and animals resulted in their being seriously considered by the Germans as new and potent war gases. The following compounds were actually produced in industrial lots by the Germans during World War II: Tabun (ethyl phosphorodimethyl amidocyanidate) and Sarin (isopropyl methyl phosphonofluoridate), known as GA and GB, respectively, in American military nomenclature. Methyl pinacolyl phosphonofluoridate (GD) and methyl cyclohexyl phosphonofluoridate (GF) have been synthesized and studied.

Tabun may be prepared according to the following formulation.

1. $POCl_3 + 2(CH_3)_2NH \rightarrow$

$$(CH_3)_2NPOCl_2 + (CH_3)_2NH_2Cl$$

2. $POCl_3 + (CH_3)_2NH_2Cl \rightarrow$

$$(CH_3)_2NPOCl_2 + 2HCl$$

3. $(CH_3)_2NPOCl_2 + C_2H_5OH + 2NaCN \rightarrow$

$$\begin{array}{c} (CH_3)_2N \quad\quad O \\ \backslash \quad\quad // \\ P \quad\quad + 2NaCl + HCN \\ / \quad\quad \backslash \\ C_2H_5O \quad\quad CN \end{array}$$

Tabun is a colorless liquid with a faint fruity odor suggesting bitter almonds. It has the following characteristics: slightly soluble in water; readily soluble in organic solvents; MW, 162.13; D, 20°C, 1.077; n_D^{20}, 1.4250; mol. refr., 38.33.

Tabun hydrolyses in water solution, the rate increasing with the pH of the solution. In distilled water one half of the —CN in Tabun is split off in 9 hours at 20°C. The P—CN bond is not the only linkage split during hydrolysis; it is also known that very early in hydrolysis the P—N bond is attacked. Complete details of the hydrolysis of this compound are not available.

Sarin has the structure:

$$\begin{array}{c} C_3H_7O \quad\quad O \\ \backslash \quad\quad // \\ P \\ / \quad\quad \backslash \\ CH_3 \quad\quad F \end{array}$$

Corrosion is a critical problem in the preparation of Sarin. Heroic measures, such as lining reaction vessels with silver etc., are required to prevent speedy disintegration of pipe lines, vats, and other parts of the plant. A variety of corrosion-resistant plastic materials have been used in the U.S.A. manufacturing process. At the end of World War II, the Germans had building two plants aimed at a total production of 600 metric tons of Sarin per month.

British chemists have reported a general procedure for synthesizing this type of compound with an overall yield of 60 to 70%:

1. $PCl_3 + 3ROH \rightarrow (RO)_2POH + RCl + 2HCl$

2. $(RO)_2POH + Cl_2 \rightarrow (RO)_2POCl + HCl$

3. $(RO)_2POCl + NaF \rightarrow (RO)_2POF + NaCl$

The United States Army did produce nerve gas and load it into munitions at the Rocky Mountain Arsenal near Denver, Colorado. The Arsenal no longer produces the gas and has disposed of most of it by sea-dumping. No adverse effects have been reported. In sea water nerve gases, including the persistent, skin-penetrating VX, break down readily into relatively nontoxic constituents. Remaining supplies are scheduled for destruction by a form of pyrolysis.

Mechanism of Action. The nerve gases are anticholinesterase compounds, i.e., they inhibit the action of the enzyme cholinesterase. Acetylcholine is a compound produced in association with transmission of impulses along the nerves in animals and man. After the impulse has passed, the acetylcholine, having served its function, is hydrolyzed by cholinesterase and the cycle begins anew. Nerve gas inhibits the cholinesterase so that acetylcholine is not destroyed, but accumulates, resulting in continuous and uncoordinated bursts of impulses over the nerves. This situation brings about the appearance of characteristic symptoms which may be conveniently classified into several groups.

(1) Muscarine-like: Constriction of the pupils of the eye, increase of nasal secretions, tightness of the chest with occasionally wheezing during expiration, suggestive of bronchoconstriction or increased secretions in the air passages, nausea, vomiting, diarrhea, increased salivation, increased urination.

(2) Nicotine-like: Muscular twitching, difficulty in breathing resulting from weakness of chest muscles and diaphragm, pallor, sometimes increased blood pressure.

(3) Central nervous effects: Anxiety, headache, general convulsions, depression of respiration.

Death usually results from depression of activity of the respiratory center of the brain.

Toxicity to Animals. Nerve gases are the most toxic known substances to animals. They can enter the body by inhalation of vapor, by swallowing the liquid, through the cornea, or by percutaneous absorption of droplets on the skin. By all routes they are toxic but to different degrees (inhalation is the most toxic route). In cats and dogs an intravenous dose of 20 micrograms of Sarin per kg of body weight brings about respiratory distress in 15 to 30 seconds and death in 5 minutes or less. Rabbits given 40 micrograms per kg of Sarin intravenously stop breathing in about 10 seconds. The LD_{50} of diisopropyl phosphonofluoridate given to rats by intravenous injection is 0.5 to 0.75 mg per kg; for mice by subcataneous injection, 4 mg per kg. The LD_{50} for Tabun given intraperitoneally to mice is about 600 micrograms per kg. The toxic dose for rabbits given Tabun by slow intravenous infusion is 177 micrograms per kg; percutaneous toxic dose is 1.5 to 3 mg per kg.

Toxicity to Man. The toxicity of nerve gas to man is somewhat difficult to assess. The lethal dose is probably about 0.7 to 7.0 mg. For a 70-kg man this value would amount to 10 to 100 micrograms per kg to kill. The effects of nerve gas poisoning are prolonged and cumulative.

Persons who work with appreciable quantities of these compounds should be supplied with gas masks and protective clothing. Liquid spilled on the skin must be flushed away with copious amounts of cold water, and then the area should be sponged with sodium carbonate solution. Benches, floors, and glassware can be decontaminated by exposure to a solution of sodium carbonate.

Doses of VX (nonlethal level) given by vein cause a decrement in number facility suggesting mild mental impairment; vomiting occurs about 1 hour after dosing. Residual effects of nerve gases, apart from direct casualties produced by them, on man, domestic animals, wild animals, wild plants, and cultivated plants, result in relatively little suffering and damage. The medical and biological after-effects of the use of nerve gases are less important than other factors in arriving at a decision concerning the future use of these agents.

Chemical Structure and Biological Activity. There is a dearth of information on this topic. It is fairly clear that the activity of this group of compounds seems to be associated with the grouping R_1R_2—PO—X or R_1R_2—PS—X. R_1 and R_2 are alkoxy or alkylamino groups. X is an acid residue: —F, —OCOCH$_3$, or another R_1R_2—PO—group. The P—X bond has, therefore, the nature of an anhydride linkage.

Antidotes. Atropin in large quantities is the only known antidote for nerve gas poisoning. It serves to protect poisoned individuals from depression of the respiratory center in the brain; but it does not prevent the neuromuscular block which results in paralysis of the muscles of breathing. To overcome the latter difficulty, artificial ventilation of the lungs using a mechanical respirator or mouth-to-mouth resuscitation with a protective filter for the operator must be initiated. Persons who have been poisoned with nerve gas show a remarkable tolerance to atropine. Doses of atropine sulfate or tartrate many times the usual clinical amount must be repeated at relatively short intervals (often for several days after the poisoning). The proper dosage level is indicated by the initial signs of atropinization, e.g., dry mouth and throat, flushed skin, and some hesitancy of urination.

A variety of hydroxamic acid derivatives are useful as adjuncts to atropine in nerve gas therapy. Among these are 2-pyridine aloxime methiodide (a-PAM) and prolidoxime chloride (a-PAMCl).

Toxogonin given orally has shown therapeutic promise.

CHARLES G. WILBER

Cross-references: *Toxicology; Cholinesterase Inhibitors.*

NEUTRONS

The neutron, an important unit in the structure of atoms, is an uncharged particle with a mass of 1.008662 ± 0.000005 atomic mass units which is equivalent to $939.550 \pm .005$ MeV. The possible existence of the neutron was first considered in 1920 by W. D. Harkins in the U.S., by Orme Masson in Australia, and by E. Rutherford in England, all of whom believed that a neutral particle of mass unity on the atomic weight scale might be formed by the combination of an electron with a proton. Rutherford pointed out that since the external field of the particle would be practically zero it should be able to move freely through matter and hence difficult to detect. The word neutron (from the Latin, *neuter,* neither) was apparently first used to refer to such an uncharged particle by W. D. Harkins in 1921. Numerous unsuccessful attempts were made in Rutherford's laboratory to detect the formation of neutrons by the passage of an electric discharge through hydrogen.

In 1930 Bothe and Becker in Germany reported that when certain light elements, for example, beryllium, were exposed to the alpha rays from polonium, a very highly penetrating radiation was obtained which they thought might be a form of gamma ray of high energy. While repeating these experiments in 1932, I. Joliot-Curie and F. Joliot found that when a sheet of hydrogen-containing material, such as paraffin, was exposed to this new radiation, protons were ejected with a considerable velocity. The Joliots thought they had discovered a new mode of interaction of radiation with matter, whereby electromagnetic waves were able to impart large amounts of kinetic energy to light atoms. However, the results were not in accord with the accepted laws of mechanics. The explanation came in the same year from James Chadwick in England, who pointed out that the experimental results could be resolved if it were assumed that the radiation consisted of neutral particles with a mass nearly equal to that of the proton. In fact, his first experimental value was 1.16 times the proton mass. For his identification of the neutron Chadwick received the Nobel Prize in 1935.

Although the overall-charge is zero, the neutron shows evidence of a definite structure, having a positive core containing about one half of the total charge. This is surrounded by a more diffuse region, which is negative, and an outer shell containing a small net positive charge. The neutron has an anomalous magnetic moment in the sense that it is not zero as predicted by Dirac theory. In fact the neutron has a magnetic moment of -1.91354 ± 0.0006 nuclear magnetons. The minus sign indicates that the neutron behaves as a spinning negative charge. The spin quantum number of the neutron, like that of the proton or electron, may take on the values $\pm \frac{1}{2}$.

Free neutrons are unstable particles which decay into an electron, a proton, and an antineutrino with a half-life of 646 ± 10 seconds. In the process 0.782 MeV of energy is liberated.

Neutrons are often characterized as fast, epithermal, or thermal, depending upon their kinetic energy. Fast neutrons may be considered to have energies above about 0.1 MeV; epithermal neutrons to have energies in the range from some few tenths of an ev to about 0.1 MeV; while thermal neutrons have energy ranges of the same magnitude as those of ordinary gas molecules (modal energy of kT). At 300°K, $kT = 0.026$ eV.

Since the neutron is an uncharged particle, it is not repelled as it approaches a nucleus, even if its energy is very low. The interaction of a neutron and a nucleus (except for elastic scattering) can be represented by the intermediate formation of a compound nucleus which may then react in several different ways. If the neutron is released again, with the reformation of the original nucleus, the process is called *inelastic scattering.* If the neutron is retained for some time (which may be a minute fraction of a second), even though the compound nucleus may undergo subsequent decomposition into new products, the process is called *absorption.* Examples of subsequent nuclear processes include the emission of gamma rays, beta particles, alpha particles, two neutrons, or *fission,* which will be discussed later. If only gamma rays are emitted the process is called *capture.*

The yield of a nuclear reaction is expressed by its *effective cross section* which has the dimensions of an area. Hence the effective cross section may be considered as the stopping area which the nucleus opposes to incident neutrons. Since the diameters of heavier nuclei are of the order of 10^{-12} cm, effective cross sections are of the order of 10^{-24} cm² (at least for fast neutrons). This unit of area is referred to as the *barn.*

In examining the variation of nuclear cross sections with the energy, or velocity, of incident neutrons, it appears that three regions may be distinguished. There is first a low energy region, which includes the thermal range, where cross sections usually decrease steadily as the reciprocal of the velocity v. The total cross section is then the sum of two terms: one, due to neutron scattering, is small and almost constant, but the other, representing absorption by the nucleus, varies as $1/v$. The energy range in which this occurs is frequently spoken of as the $1/v$ region.

Following the somewhat indefinite $1/v$ region many elements, especially those of larger atomic weights, exhibit peaks, called *resonance peaks,* where the neutron cross sections rise more or less sharply to high values, for certain energies, and then fall again. Some elements, such as cadmium and rhodium, have only one peak, while others may have two, three, etc., or even tens of peaks. These regions of exceptionally high absorption may occur anywhere from about 0.1 eV to a few MeV. Absorption cross sections always vary as $1/v$ for neutron energies far from resonance energies in both directions and below the fast energy region. Except for some of the lighter and some of the heavier ($Z \geq 92$) nuclides, neutron absorption reactions occurring at low neutron energies are usually capture reactions. Other processes usually require more energetic neutrons.

With neutrons of high energy, in the MeV range, the cross sections are low, being a few barns, compared with possibly tens to millions of barns for the resonance peaks. There is usually a gradual decrease of cross section with increasing neutron energy in this range.

Like electrons and x-rays, neutrons exhibit wave characteristics. As in the case of electrons the equivalent wavelength λ is given by the de Broglie relation $\lambda = h/mv,$ where h is Planck's constant, and m and v are respectively the mass and velocity of the neutrons. Because of their wave characteristics neutrons may be diffracted by crystals and hence used in structure studies. X-ray diffraction intensities depend upon scattering from electrons; neutron diffraction intensities depend upon scattering from nuclei. Neutron diffraction offers an advantage over x-ray diffraction in structure studies where the sample contains elements which are either very close together in atomic number or very far apart. If they are very close, the x-ray scattering will be nearly the same for each; if far apart, the strong x-ray scattering of the heavier element may mask that of the lighter. Hydrogen shows a very small cross section for x-ray scattering and a very large one for neutron scattering; hence the positions of hydrogen atoms are readily determined with neutron beams in condensed phases.

Radioactive isotopes of almost all elements can be produced by neutron irradiation of stable or other radioactive nuclides. Advantage is taken of this fact in neutron activation analysis, in which a sample is irradiated with neutrons to produce radioactive isotopes, which may then be identified by counting techniques involving gamma ray energy discrimination. This method permits the identification of extremely small amounts of elements in a sample.

The neutron and the proton are the important building blocks in nuclear structure. Since the nuclei of atoms are stable there must be strong short range attractive forces between neutrons, neutrons and protons, and between protons. This force is probably of tensor type and will depend not only on the separation of the particles but also upon how the angular momenta of the particles are oriented with respect to the line joining them. When Heisenberg in 1932 suggested that atomic nuclei were built up of protons and neutrons, he tried to account for these intranuclear attractions by using the idea of wave-mechanical exchange forces. However, this theory encountered difficulties. For example, in considering the short-range attractive force between a proton and a neutron the electric charge exchanged was presumed to be an electron, either positive or negative. But this supposition was not in harmony with the idea that nuclear forces are of very short range, operating over distances of the order of 10^{-13} cm. The minimum effective range of the force involving a given particle is equal to $h/2 \pi m c,$ referred to as its Compton wave length, where c is the velocity of light. For an electron this distance turns out to be of the order of 10^{-11} cm, much too large for it to be applicable to proton-neutron forces.

In the course of developing a new theory of short-range intranuclear forces H. Yukawa in 1935 postulated the existence of a new charged particle as the basis of the exchange between the proton and the neutron. Taking the Compton wave length of this particle to be 2.8×10^{-13} cm he calculated the mass of this hypothetical particle to be about 140 times the mass of the electron. Further, he postulated that this particle was unstable, expelling an electron, either positive or negative, and having a mean life of 10^{-6} seconds. Further investigations, following the discovery of the presence in cosmic rays of charged particles, both positive and negative, having a mass approximately 200 times that of the electron, confirmed the existence of this particle of intermediate mass, now called a *meson,* which was shown to be unstable, decaying with a mean life of 10^{-6} second, in agreement with Yukawa's calculations. The meson theory of nuclear forces postulates that every nucleon is surrounded by a meson field, through which it interacts with other nucleons. The mesons themselves are regarded as the carriers of energy quanta in the meson field. Hence the strong attractive force between two neutrons at nuclear distances may be explained by the meson field theory.

As mentioned previously, an important nuclear reaction involving neutrons is nuclear fission, first reported in 1939 by O. Hahn and S. Strassmann in Germany. In the fission process an atomic nucleus bombarded with neutrons splits into smaller fragments. Nuclides fissionable by thermal neutrons include ^{233}U, ^{235}U, and ^{239}Pu. ^{238}U may be split by fast neutrons with energies greater than 1 MeV. Since secondary neutrons are produced in the fission process a chain reaction is possible. These secondary neutrons have high energies. However, the fission cross sections for ^{233}U, ^{235}U, and ^{239}Pu are much higher for thermal neutrons than for fast neutrons, and hence the neutrons produced in the fission process must be slowed down or *moderated* in order to sustain a chain reaction in a *thermal* nuclear reactor or *pile* using these fuels. In the slowing-down process the kinetic energy of the fast neutrons is transferred by collision to the atoms of the moderator. An effective moderator must have a high scattering and a low capture cross section for neutrons; otherwise, the neutrons will be captured during the moderating process and be unavailable for continuing the chain reaction. Also a good moderator must consist of light atoms so that a large decrease in energy can occur per collision; for hydrogen, with nuclear mass very close to that of the neutrons, the latter can lose as much as all of its kinetic energy in one collision. Good moderators include hydrogen, deuterium, beryllium, carbon, and compounds containing them (e.g., D_2O). In contrast to thermal reactors fast reactors require much more fuel (^{233}U or ^{239}Pu) because of the smaller neutron cross sections at high energies. They also require coolants which are poor moderators.

Whereas nuclear energy is released in a controlled fashion in a reactor, it must be released with great rapidity in an atomic bomb. The chain reaction must be maintained by fast neutrons, so the use of any material capable of slowing down the neutrons must be avoided. Hence virtually all the neutrons produced in fission will cause further fission, the multiplication factor will exceed unity, and the rate of energy release will increase with great rapidity.

Portable neutron sources for research may include alpha emitters combined with light elements (e.g., Ra-Be or Po-Be) where the neutrons are produced by (α,n) reactions and heavy nuclides (e.g., isotopes of Cm and Cf) which fission spontaneously at a rapid rate.

M. H. Lietzke and R. W. Stoughton

References

Moore, W. J., "Physical Chemistry," 3rd Ed., Prentice-Hall, Inc., Englewood Cliffs, N.J., 1962.
Glasstone, Samuel, "Source Book on Atomic Energy," D. Van Nostrand Co., New York, 1950.
Hughes, D. J., "Pile Neutron Research," Addison-Wesley Publishing Co., Cambridge, Mass., 1953.
Hughes, D. J., "Neutron Cross Sections," Pergamon Press, New York, 1957.

NEUTRON ACTIVATION ANALYSIS

Neutron activation analysis is a method of elemental analysis based upon the quantitative detection of radioactive species produced in samples via nuclear reactions resulting from neutron bombardment of the samples.

Types of Neutron Reactions. The neutron-induced reactions are of two main types: (1) those induced by very slow (thermal) neutrons, having energies of about 0.025 eV, and (2) those induced by fast neutrons, those having energies in the range of MeV.

The nuclei of all stable isotopes are capable of capturing thermal neutrons, but with characteristic reaction cross sections which vary widely from isotope to isotope, even of the same element. Promptly following the capture of a thermal neutron by a stable nucleus, the compound nucleus de-excites itself by the emission of one or more "prompt" gamma-ray photons. If the resulting product nucleus is a radionuclide, its later decay can be detected and can be of use in the activation analysis detection of that element. Thermal-neutron capture reactions are therefore referred to as "(n,γ)" reactions. For example, in the determination of vanadium, with thermal neutrons, some of the V^{51} stable nuclei present in the sample to be analyzed undergo the V^{51} (n,γ) V^{52} reaction. Vanadium-52 is radioactive, decaying with a half-life of 3.75 minutes, emitting a β^- particle and a 1.434 MeV gamma-ray photon.

Fast neutrons predominantly interact with nuclei by means of (n,p), (n,α), $(n,2n)$, and (n,n') reactions. Whereas the thermal-neutron capture reaction forms a nuclide of the original element, but now one mass unit higher, the $(n,2n)$ fast-neutron reaction forms a nuclide of the same element one mass unit lower. An example of this type of reaction is the N^{14} $(n,2n)$ N^{13} reaction. Nitrogen-14 is the abundant stable isotope of nitrogen, whereas N^{13} is radioactive, decaying with a half-life of 9.96 minutes by positron emission. The (n,n') fast-neutron type of reaction termed a "neutron inelastic scattering" reaction forms an excited state (nuclear isomer) of the original nucleus, with unchanged mass number (A), but a measurable half-life. The Se^{77} (n,n') Se^{77m} reaction is a good example of this type of reaction. Selenium-77 is one of the stable isotopes of selenium, whereas Se^{77m} is a radioactive isomer of Se^{77} that decays with a half-life of 17.5 seconds emitting an isomeric-transition 0.16 MeV gamma-ray photon. In (n,p) reactions, the product nucleus has the same mass number as the original nucleus, but is a different element, namely, lower by one unit in atomic number (Z). A widely utilized reaction of this type is the O^{16} (n,p) N^{16} reaction. Oxygen-16 is the principal stable isotope of oxygen; N^{16} is a radionuclide of nitrogen, decaying with a half-life of 7.14 seconds, emitting exceptionally high-energy beta particles and gamma-ray photons. In (n,α) reactions, the product nucleus has a mass number 3 units lower than the original nucleus, and a Z that is 2 units lower than originally. For example, in the fast-neutron detection of phosphorus, the P^{31} (n,α) Al^{28} reaction is often utilized. Normal phosphorus consists entirely of P^{31}; Al^{28} is a radionuclide of aluminum, decaying with a half-life of 2.31 minutes, emitting a β^- particle and a 1.780 MeV gamma-ray photon in each disintegration.

Theory of the Method. When a sample containing N nuclei of a given type (a particular Z and A) is exposed to a flux ϕ of neutrons, resulting in a particular nuclear reaction having a cross section σ, the rate of formation of product nuclei by this reaction, in nuclei per second, is simply $N\phi\sigma$. The units of ϕ and σ are neutrons per square centimeter per second and square centimeters per nucleus respectively. If the product nucleus is a radioactive species, some of these nuclei will be decaying while the irradiation is going on. If the irradiation of the sample with neutrons is continued for a long time, compared with the half-life of the radionuclide formed, a steady state, or "saturation" condition will be reached, in which previously formed nuclei are decaying at the same rate that new ones are being formed, thus with no further increase on the disintegration rate of this particular product with continued irradiation at that flux. Therefore, the saturation activity of a given species, at zero decay time (i.e., just at the conclusion of the irradiation), is A_o (satn) $= N\phi\sigma$.

At intermediate irradiation times (t_i), the activity of a particular induced species (A_o) expressed in disintegrations per second (dps) is equal to $N\phi\sigma S$, where S is a "saturation" term that is equal to $1 - e^{-0.693 t_i / t_{0.5}}$, where $t_{0.5}$ is the half-life of the radioactive species. The saturation term is dimensionless and ranges only from O (at $t_i = O$) to 1 (at $t_i = \infty$). It rapidly approaches a value of one, asymptotically, acquiring values of $\frac{1}{2}$, $\frac{3}{4}$ $\frac{7}{8}$, $\frac{15}{16}$, ... at $t_i/t_{0.5}$ values of 1, 2, 3, 4, ... Because of this rapid approach of S to its maximum value, it is pointless to activate a sample for a period of time longer than a few half-lives of the radioactive species of interest. Longer irradiation merely generates more interfering activities of longer half lives.

In the basic activation equation, σ is the isotopic cross section for the particular type of nuclear reaction, for neutrons of a specified energy. It assumes that the neutron flux, and energy, are constant throughout the sample. The N term is itself equal to $wfaN_A/AW$, in which w is the weight of the sample (in grams), f is the weight fraction of the element in the sample, a is the fractional abundance of the target stable isotope among all the stable isotopes of the element, N_A is Avogadro's number, and AW is the ordinary chemical atomic weight of the element. In actual analyses, of course, either f or the product wf is the unknown quantity.

Neutron Sources. In neutron activation analysis work, the most widely used neutron sources are (1) research-type nuclear reactors, and (2) small accelerators. Modern research reactors are mostly of the pool type and operate at power levels of 10 to 1000 kW, providing thermal-neutron fluxes of 10^{11} to 10^{13} n $cm^{-2}sec^{-1}$ and fission spectrum fast-neutron fluxes of about the same magnitude. The small accelerators used in neutron activation analysis work are largely low-voltage (100 to 200 kV) Cockcroft-Walton deuteron accelerators, capable of producing up to about 10^{11} 14-MeV neutrons/sec from a tritium target (with a 1-mA deuteron beam current), via the H^3 (d,n) He^4 reaction. Samples of typical size (~ 1 cm^3) can thus be exposed to a 14-MeV neutron flux of about 10^9 n $cm^{-2}sec^{-1}$. In a moderator, these can produce a thermal-neutron flux of the order of 10^8 n $cm^{-2}sec^{-1}$. Unfortunately, at full-power operation, the lifetime of the tritium target is only of the order of an hour to a few hours. Some work is carried out also with lower-energy neutrons generated by the Be^9 (d,n) B^{10} reaction with a 2-MeV positive-ion Van de Graaff accelerator, or by the Be^9 (x,n) Be^8 reaction with bremsstrahlung produced by a 3-MeV electron Van de Graaff accelerator. These produce thermal-neutron fluxes in the range of 10^8 to 10^9 n $cm^{-2}sec^{-1}$. Isotopic sources, such as Po^{210}-Be, Pu^{239}-Be, and Am^{241}-Be, also generate neutrons, but the maximum thermal-neutron flux attainable with such sources is only about 10^5 n $cm^{-2}sec^{-1}$. They are useful for teaching purposes, but not for real analytical work. Newly available Cf^{252} spontaneous fission isotopic sources can provide thermal neutron fluxes of about 10^7 n/cm^2-sec per mg of Cf^{252}.

Sensitivities for Various Elements. As the available neutron flux increases, the level of induced activity per unit mass of an element also increases; hence, the sensitivity of detection is improved, i.e., the limit of detection is lower. With a nuclear reactor thermal-neutron flux of 10^{13} n $cm^{-2}sec^{-1}$, a maximum t_i of one hour, and reasonable counting efficiencies, it is found that the median limit of detection for some 75 elements is about 10^{-3} μg. A few of these elements can be detected down to as low as 10^{-7} μg; a few only to about 10 μg. The method, with such high-neutron fluxes is the most sensitive known method for over half the elements of the periodic system. With 1-gram samples, the μg absolute sensitivities correspond to parts-per-million (ppm) concentration sensitivities. Samples ranging from minute samples up to 10 grams or somewhat more can be irradiated and analyzed. With longer irradiation periods, the detection limits for about half of the 75 elements (those forming longer-lived induced activities) can be reduced further. The most sensitively detected elements (limits $\leq 10^{-3}$ μg) are the following: Ag, Ar, As, Au, Br, Co, Cs, Cu, Dy, Er, Eu, Ga, Ge, Ho, I, In, Ir, La, Lu, Mn, Na, Nb, Pd, Pr, Re, Rh, Sb, Sm, Sr, U, V, W, and Yb. Some elements are more sensitively determined by activation with fast neutrons than with thermal neutrons.

At the lower (10^8 to 10^9) thermal-neutron fluxes attainable with the small accelerators, the limits of detection are of course 10^4 to 10^5 times higher than for the reactor 10^{13} flux. At a 10^9 thermal-neutron flux, the sensitivities for the same 75 elements thus range from 10^{-3} μg to 0.1 gram, with a median of 10μg. With a 14-MeV neutron flux of 10^9 n $cm^{-2}sec^{-1}$, a number of additional elements can be detected fairly sensitively, down to levels of 1 to 20 μg, e.g., N, O, F, Si, P, Cr, and Fe.

Forms of the Method. In practical analytical work, one does not employ the basic activation analysis equation ($A_o = N\phi\sigma S$), per se, but instead uses a comparator technique. When samples are to be analyzed for one or more elements, standard samples of these elements are activated at the same time as the unknowns and then are counted in an identical manner (counting efficiency ϵ). When the counting data are corrected to the same decay time, then A' (unknown)$/A'$ (standard) = grams element in unknown/grams element in standard, since ϵ, a, AW, ϕ, σ, and S are the same for both unknown and standard. In the equation, A' refers to the counting rate (rather than disintegration rate) of the radionuclide formed by the element in question —in unknown and standard respectively. Not only is the comparison technique simpler, but it removes any dependence upon literature values of σ, and experimental values of ϵ and ϕ, which often are not accurately known. At levels well above the limits of detection, careful application of this comparison technique results in precisions and absolute accuracies in the range of ± 1 to 3% of the value.

The method is employed in two different forms: the purely instrumental form and the radiochemical separation form. The instrumental form is fast and nondestructive, and is based upon the quantitative detection of induced gamma-ray emitters by means of multichannel gamma-ray spectrometry. It is the preferred method where it applies. Induced activities are identified by the energies of their gamma-ray photopeaks observed in the NaI(Tl) scintillation counter or Ge(Li) semiconductor detector pulse-height spectrum of the activated sample. The amount of the element present in a sample is computed usually from the photopeak (total absorption peak) area of its gamma ray, or one of its principal gamma rays, compared with that of the standard.

Where interferences from other induced activities are very serious, and cannot be removed adequately by decay, spectrum subtraction, or computer solution, one must turn to the radiochemical separation form of the method. Here the activated sample is put into solution and equilibrated chemically with measured amounts (typically 10 mg) of added carrier of each of the elements of interest, before chemical separations are carried out. The element to be detected needs then to be recovered in chemically, and radiochemically, pure form, but it need not be quantitatively recovered, since the carrier recovery is measured and the counting data are then normalized to 100% recovery. This form of the method is slower, but it applies to pure beta emitters, as well as to gamma emitters, and it does eliminate interfering activities. It is free of the usual complications of microconcentration analysis: high blanks from reagent impurity, and losses by adsorption and coprecipitation.

Neutron activation analysis is now a well-established method of elemental analysis, carried out in many laboratories and utilized by many more through available commercial activation analysis services. It is now widely applied in almost every branch of science, engineering, and medicine, where either its great sensitivity (at high fluxes) or its speed (with the instrumental form of the method), or both, are used to advantage. Interesting and important applications are also being made in such fields as archaeology, oceanography, geochemistry, crime investigation, and the environmental sciences.

VINCENT P. GUINN

References

De Voe, J. R., Editor, "Modern Trends in Activation Analysis," National Bureau of Standards Special Publication 312 (2 volumes), Washington, D.C., 1969.

Guinn, V. P., "A Forensic Activation Analysis Bibliography," Gulf General Atomic Report GA-9912, San Diego, California, 1970.

Guinn, V. P., Lenihan, J. M. A., and Thomson, S. J., Editors, "Advances in Activation Analysis. Volume 2," Academic Press, London, 1971.

International Atomic Energy Agency, "Nuclear Techniques in Environmental Pollution," Vienna, 1971.

Kruger, P., "Principles of Activation Analysis," Wiley-Interscience, New York, 1971.

Lederer, C. M., Hollander, J. M., and Perlman, I., "Table of Isotopes. Sixth Edition," Wiley & Sons, New York, 1967.

Lutz, G. J., Boreni, R. J., Maddock, R. S., and Meinke, W. W., Editors, "Activation Analysis: a Bibliography," National Bureau of Standards Technical Note 467 (2 volumes), Washington, D.C., 1968.

NICKEL AND ITS COMPOUNDS

Nickel, Ni, atomic number 28, atomic weight 58.71, is a ductile face-centered-cubic metal, essentially "white" in color. It occurs in the Periodic Table in the first triad of Group VIII after iron and cobalt and just before copper (Group IB).

It is estimated that the earth's crust contains an average of about 0.01% nickel. Nickel ranks twenty-fourth in the order of abundance of the elements in the earth's crust. The total amount of nickel is greater than that of copper, zinc and lead combined. However, there are relatively few known nickel deposits that are important enough to be economically worked.

Most of the world's present supply of nickel is of Canadian origin. Other deposits of importance occur in the United States, New Caledonia, Russia, Cuba, Australia and Indonesia. Recovery generally is from sulfide ores, however, the lateritic (oxide) deposits are gaining increasing importance.

Derivation. Nickel is recovered and refined from the crushed ore in one of several methods, the three most important being electrowinning, chemical reduction, and the carbonyl processes.

The *sulfide* ores, concentrated by usual mechanical methods, are roasted to form a matte. The sulfide matte is further roasted to an oxide which is then reduced in a pyrometallurgical operation and cast into nickel anodes which are electrically refined to produce high-purity cathode nickel. Alternately, the sulfide matte is melted and cast directly into anodes which are electrolytically processed to provide nickel of similarly high purity. In another refining process, nickel from sulfide ores is extracted utilizing the carbonyl process discovered in 1899 by Langer and Mond. In this refining process, nickel and carbon monoxide are reacted under certain conditions to form a volatile nickel carbonyl which, when heated, decomposes to form nickel and carbon monoxide. The nickel is deposited upon nickel nuclei or seeds in a "fluidizer." The product is a very high-purity nickel in which cobalt is essentially absent. (See Ref. 1.)

The *oxide* ores (which are high in iron) are treated somewhat differently since the usual mechanical methods of separation employed in the sulfide ores are not readily applicable and the entire ore must be treated by pyro-, hydro- or vapometallurgical procedures.

Hydrometallurgical refining of nickel usually involves the chemical leaching of the various ores. Metallic nickel is then precipitated from the aqueous nickel salt solutions by hydrogen reduction at appropriate pressures and temperatures. (See Ref. 1.)

General Characteristics. Commercially pure wrought nickel is a white metal having a slight yellowish cast. It is malleable, resistant to corrosion in many media and strongly magnetic at room temperature. It has good thermal conductivity, moderate strength and hardness, high ductility and toughness, and good electrical properties. It has good toughness and ductility at low temperatures. Nickel can be fabricated easily by all procedures common to steel and will take and retain a high polish.

Physical and Mechanical Properties. The normal crystal form of nickel is face-centered-cubic, and this structure persists to the melting point of 1453°C. The space group reprsentation is O_h^5; a = b = c = 3.517 kx units with four atoms per unit cell. Nickel is ferromagnetic and has a Curie temperature of 350°C with a saturation magnetization of 6100 gauss at 20°C.

The density of nickel is 8.908 g/cc. It boils at 2730°C and has a heat of vaporization of 89.4 kcal/g-atom. The heat of fusion is 4.23 kcal/g-atom and the mean specific heat is 0.108 cal/g/°C from 0–100°C.

Some additional physical properties are: thermal conductivity (20°C), 0.21 cal/cm²/cm/°C/sec; electrical resistivity (20°C), 6.84 microhm-cm; standard potential, bivalent nickel: −0.250 v at 25°C; and coefficient of thermal expansion (0–100°C), 13.3×10^{-6}/°C.

The ultimate tensile strength of commercially pure annealed nickel (99.5%) is 65,000 lbs/in² with 50% elongation. These values hold up to about 315°C after which tensile strength decreases and elongation increases.

Nickel Chemistry. Nickel has five stable isotopes with atomic weights of 58, 60, 61, 62 and 64.

It has a $3d^8 4s^2$ electronic configuration and forms compounds in which the nickel atom has oxidation states of −1, 0, +1, +2, +3, +4. By far the majority of nickel compounds are of the nickel(II) species. Powerful oxidants are required to generate the Ni(III) and Ni(IV) moieties. The only well established examples of stable crystalline derivatives Ni(III) and Ni(IV) are the complex fluoride anions $(NiF_6)^{-3}$ and $(NiF_6)^{-4}$. Even these materials are not stable in water solution yielding oxygen and Ni(II). In general, nickel(I) complexes are binuclear and diamagnetic, examples of which are $K_4(Ni_2(CN)_6)$ and $K_6(Ni_2(C_2H)_8)$. An example of nickel in the formal oxidation state of (−1) is the hydrocarbonyl $H_2Ni_2(CO)_6$ in which material hydrogen has no acid function. A number of derivatives of Ni(R) are known, the most common of

which is nickel carbonyl, $Ni(CO)_4$. Examples of other nickel (R) derivatives include $Ni(PF_3)_4$, $(\phi_3P)_2Ni(NO)_2$, $(\phi NC)_4Ni$ and the anion, $(Ni(CN)_4)^{-4}$; R = radical, ϕ = phenyl.

Nickel Oxides. Nickel oxide, NiO, is prepared by heating the metal in oxygen above 400°C. It is also prepared by calcining a number of salts including nickel nitrate. Nickel oxide displays no amphoteric properties, being insoluble in caustic solution and aqueous or liquid ammonia.

Nickel Hydroxide. Soluble alkalis give precipitates of green $Ni(OH)_2$ with nickel(II) salts. The hydroxide is a relatively strong base (pKb = 15) and is soluble in aqueous ammonia to give a blue solution containing the nickel tetrammine ion.

Nickel Carbonate. The addition of alkali carbonates to a solution of a nickel(II) salt precipitates an impure basic nickel carbonate. Nickel carbonate can be reduced with hydrogen at temperatures above 300°C to yield finely divided nickel that has good catalytic activity.

Nickel Halides. Hydrogen fluoride reacts with nickel powder at 225°C to give the fluoride NiF_2. The more conventional approach is to use hydrofluoric acid to dissolve nickel carbonate and nickel hydroxide. The anhydrous fluoride is stable and only slowly dissolved by water. Nickel oxide and nickel metal react with hydrogen chloride to form $NiCl_2$. Its major use is in nickel electroplating baths. Nickel chloride is generally less soluble than cobalt chloride in organic solvents and frequently provides the basis for separating these closely related compounds. Nickel(II) bromide can be obtained by the bromination of nickel powder in ether. Nickel iodide can be obtained directly from the combination of the elements. The anhydrous salt is black. The hydrate, $NiI_2 \cdot 6H_2O$, results from the neutralization of nickel hydroxide by hydriodic acid.

Nickel Sulfate, the most important nickel salt, is used in nickel electroplating baths. Dissolution of nickel carbonate or oxide in sulfuric acid followed by concentrating produces green crystals of nickel sulfate ($NiSO_4 \cdot 7H_2O$). Heating the heptahydrate in concentrated solution to approximately 50°C causes loss of one mole of water to form the $NiSO_4 \cdot 6H_2O$, the usual commercial form. It is also made by direct reaction of sulfuric acid on nickel metal.

Nickel Sulfamate. Nickel sulfamate, $Ni(SO_3NH_2)_2 \cdot 4H_2O$, is readily prepared by dissolving nickel carbonate in sulfanic acid. This salt is of commercial importance as a source of nickel for special sulfamate electrochemical plating baths.

Nickel Carbonyl. See **Carbonyl Compounds.**

Organonickel Compounds. Nickel, like the other transition metals, has the ability to react with functional organic molecules possessing acidic or basic character. The most commonly encountered acid functions are COOH, phenolic OH, SH, SO_3H and POH. The more common basic groups are trivalent: nitrogen, phosphorus, and arsenic; and oxygen in ethers and sulfur in thioethers.

The simplest derivatives are the nickel salts of organic acids. Nickel salts of organic acids or acid mixtures are prepared by metathesis between nickel sulfate and the sodium salts of the respective acids.

Nickel-π-Complexes. Nickel bromide reacts

with sodium cyclopentadienide, NaC_5H_5, to yield bis- (cyclopentadienyl) nickel (*nickelocene*), $\pi\text{-}C_5H_5)_2Ni$. Nickelocene is a diamagnetic, emerald green crystalline, sandwich compound in which nickel has a formal oxidation number of +2.

Another system of -complexes of nickel are those derived from allylic compounds. When nickel carbonyl is reacted with allyl bromide, the π-allyl system results with the formation of $((\pi\text{-}C_2H_5)NiBr)_2$. Again, no sigma-bonding exists and the three carbons are bonded to nickel through two bonds.

Corrosion. Commercially, pure nickel and most of the high-nickel alloys are used to a considerable extent for the fabrication of equipment in the chemical and process industries and in others where their corrosion resistance is an important consideration. Nickel does not discharge hydrogen readily from the common nonoxidizing acids. The presence of some oxidation agent, such as dissolved air, is necessary for corrosion to proceed at a significant rate. Reducing conditions usually retard the corrosion of nickel, while oxidizing conditions usually accelerate it. However, nickel has the ability to protect itself against certain forms of attack by development of a corrosion resisting, or passive, oxide film, and, consequently oxidizing conditions do not invariably accelerate corrosion.

Nickel Alloys. Nickel is one of the important alloying elements for steels and cast iron. Used with other alloying elements, such as chromium and molybdenum, it gives the strength, toughness and wear resistance that are essential for automative, heavy machinery, transportation and constructional applications.

Nickel is an ingredient of the popular 18% chromium-8% nickel stainless steels, particularly noted for their corrosion resistance, good formability, weldability, and desirable mechanical properties at both cyrogenic and elevated temperatures. They are widely used in the chemical, transportation and petroleum industries; for food processing and handling, dairy and pharmaceutical equipment; in hospitals and for domestic houseware applications; in paper making and textile processing, as well as in fields such as architecture and power generation.

Wrought and cast alloys containing about 25% nickel and higher, along with chromium, iron and a number of other elements, form the basis for several families of corrosion-resistant, high-strength and heat-resistant materials. Two are widely used for electrical resistance heating (60 Ni-15 Cr-Fe and 80 Ni-20 Cr). Others containing significant amounts of elements such as molybdenum, tungsten, copper and silicon are highly resistant to attack by acids and acid salts over wide ranges of concentration and temperature.

Wrought and cast nickel-copper alloys form two large groups, one containing about 10 to 30% nickel (the copper-nickel alloys) and the other containing about 60 to 85% nickel. Both are noted for their high strength properties and excellent corrosion resistance in a wide range of industrial, marine and atmospheric environments.

Catalysis. One of the most important chemical uses of nickel is hydrogenation catalysis. Nickel catalysts are used in three forms, finely divided nickel powder, supported catalysts and soluble nickel compounds. Nickel catalysts are extremely active toward hydrogenation and are poisoned by the presence of sulfur-containing compounds. Perhaps the most widely used form of nickel catalyst is Raney Nickel, prepared by extracting a finely divided nickel-aluminum alloy with caustic soda to remove aluminum. (See **Catalysis**.)

Nickel Plating. More nickel is plated than is any other metal. It is widely used as a coating for functional and decorative applications. All chrome plated objects actually have an underlayer of nickel many times thicker than the chrome plate. Nickel plating is done in electroplating baths containing a consumable nickel anode and solutions of various salts, e.g., nickel sulfate, or nickel sulfate-nickel chloride, or nickel sulfamate; or by electroless chemical reduction of its salts from aqueous solutions onto the object being coated.

THE INTERNATIONAL NICKEL CO., INC.

References

1. Rosenzweig, M. D., *Chem Eng.,* **76,** pp. 108–110 (April 7, 1969) and pp. 106–108 (May 19, 1969).
2. Adamec, J. B., and Springer, D. B., "Nickel," pp. 437–445, "Encyclopedia of the Chemical Elements," Hampel, C. A., Editor, Van Nostrand Reinhold Co., New York, 1968.

NIOBIUM AND COMPOUNDS

Niobium (Nb) is a metal with a melting point of 2467°C (4475°F) and a specific gravity of 8.57. Its most significant properties, other than its relatively high melting point, are its low neutron-capture cross section, its affinity for oxygen and nitrogen, its corrosion resistance, and its good mechanical characteristics.

It was discovered in 1801 by Charles Hatchett, an English chemist, who named it columbium. It is still referred to as columbium in most metallurgical literature in the United States but was officially designated as niobium by the International Union of Pure and Applied Chemistry in 1949, and this designation is now almost universally accepted.

Mineralogically, there are two important sources of niobium. One is the richer type ore, columbite or tantalite, in which the oxides of the metals "columbium" and tantalum occur together. The other is pyrochlore, which, although not so rich, represents by far the major source of known world reserves of niobium. Because of high percentages of tantalum contained in the richer columbite and tantalite ores, these are processed to separate the two similar metals, with the result that higher purity mill products are obtained. These are more expensive, and are used only where the higher purity material is essential, such as for pure metal production, or for certain impurity-critical high-temperature alloys. In the 1960's, development and application of niobium-bearing steels showed a pronounced increase. The raw material for this application, ferroniobium, is most economically won from the leaner pyrochlore ores, and as a consequence, during this decade, pyrochlores advanced from minor to major status, enjoying 85 to 90% of the market by 1971.

Chemical treatment is used to win niobium pentoxide from its ores. The pentoxide may be refined to niobium metal by carbon reduction. Consolidation and further refining of the metal and certain of its alloys is frequently done in electron-beam melting furnaces.

Niobium has excellent resistance to corrosion by aqueous solutioins of acids and alkalies, although its corrosion resistance is generally not so good as that of tantalum. As is characteristic of the refractory metals (Nb, Ta, Mo, W), niobium exhibits good resistance to molten alkali metals. At temperatures above about 600°C, niobium becomes very reactive. It has an affinity to absorb large quantities of interstitially soluble elements (O, N, C), and it readily reacts with most elements at elevated temperatures. Contamination resulting from the absorption of interstitial elements, particularly oxygen, severely reduces the ductility, thus the usefulness, of niobium and its alloys. Of particular importance is its poor resistance to attack by oxygen which in addition to dissolving in niobium results in rapid conversion of the metal to Nb_2O_5 when niobium is heated in air or oxygen-containing environments. So great is niobium's affinity for oxygen that it will reduce stable oxides such as MgO, BeO, and ThO_2 upon contact at temperatures of 1000°C and above.

Pure niobium is soft, ductile, and very malleable at room temperature, and it maintains excellent ductility down to at least −253°C. It is readily fabricated by conventional methods, and can be welded using GTA or electron-beam welding proccesses which exclude contaminating air from the hot-metal zone. Despite its high melting temperature, the strength of niobium at elevated temperatures is quite low. As a result of alloy development activities from the late 1950's to date, a number of niobium-base alloys possessing useful strength at temperatures up to 1400–1600°C are now commercially available. These alloys were developed to meet the mechanical requirements for aerospace hardware applications (e.g., leading edges, "hot" structures for re-entry vehicles, turbine blades and vanes) and substantial improvement in the strength of niobium at elevated temperatures was achieved by alloying, with but modest sacrifice in ductility and fabricability at low temperatures.

Despite the successful development of mechanically useful alloys, it has not been possible to improve the oxidation resistance of niobium by alloying sufficiently to permit its widespread use as an elevated-temperature structural material in oxygen-containing environments. However, several coatings (primarily silicide or aluminide compounds) for protecting niobium alloys from oxidation have been developed and brought to commercial status. In general, the better of these display protection reliabilities on the order of 99%. Although coated hardware is beginning to find some application in selected aerospace uses, lack of complete reliability understandably results in cautious use of coated niobium in critical areas such as man-rated aerospace vehicles.

The electric resistivity of niobium at room temperature is about eight times that of copper. Certain alloys of niobium with zirconium, titanium or tin exhibit excellent superconducting properties, and their potential as magnet winding appears good.

Such alloys exhibit relatively high threshold temperatures (e.g., 17°K) and maintain superconductivity in high magnetic fields. The particularly attractive niobium-tin alloys are brittle and require special fabrication techniques.

Niobium-base alloys, because of their high corrosion resistance at modestly elevated temperatures and their low cross section for thermal neutrons, are of interest as claddings for fuel elements and as a matrix for fuel material in nuclear reactors. Niobium-uranium alloys are used as nuclear fuel materials.

The principal commercial use for niobium is in the form of ferroniobium used as an alloying addition to "columbium" steels, stabilized stainless steels and welding rod, and certain high-temperature superalloys.

In carbon and low alloy steels, niobium acts as a grain refiner and stabilizer of carbides and other interstitial compounds. This results in steels with excellent toughness, cryogenic ductility, and formability. The use of ferroniobium has received particular attention in Western Europe, Canada, the United Kingdom, Sweden, and Japan, as well as the United States among the free-world nations.

Despite promise as a high-temperature structural material, reactor-core material, and winding for high-strength magnets, the use of niobium-base alloys in these fields has not yet attained a commercial or production status.

E. S. BARTLETT

References

Miller, G. L., "Metallurgy of the Rarer Metals-6-Tantalum and Niobium," Academic Press Inc., New York, 1959.

Douglass, D. L., and Kung, F. W., Editors, "Columbium Metallurgy," Interscience, New York/London, 1961.

Hausner, H. H., Editor, "Coatings of High-Temperature Materials, Plenum Press, New York, 1966.

"Columbium" (Annual Survey and Outlook), *Engineering and Mining Journal,* (annually, No. 3 issues).

"Strong Tough Structural Steels," Iron and Steel Institute Publication 104, pp. 1–10, 61–73, The Iron and Steel Institute, London, 1967.

NITRATION

Nitration is the process by which a hydrogen atom is replaced by a nitro ($-NO_2$) group through the agency of nitric acid or a derivative of nitric acid.

Nitration is one of the most general of all aromatic substitution reactions and is a reaction of great commercial importance. Currently, over 100,000 tons of nitrobenzene are prepared annually by the nitration of benzene, and yields of over 98% are obtainable in this industrial process. Most of the nitrobenzene is reduced to aniline for the production of dyestuffs. Nitration is also used in the production of most chemical explosives. For example, TNT is obtained by stepwise nitration of toluene, picric acid from nitration and hydrolysis of chlorobenzene, RDX and HMX from nitration of hexamethylenetetramine, nitroglycerin from nitration of glycerin, nitrocellulose from nitration of cellulose, etc.

The nitration of nitrobenzene to form 1,3-dinitrobenzene is the best understood of all nitration reactions. This reaction is irreversible under the usual conditions of nitration. In the water-sulfuric acid system, the rate of nitration is first order in respect to nitrobenzene and first order in respect to nitric acid. The rate increases rapidly with increasing sulfuric acid concentration up to a maximum of 90% acid. Above 90% sulfuric acid the rate slowly decreases with increasing acid concentration.

The rapid increase in rate with increasing acid concentration is related to shifting the equilibrium: $HNO_3 + H^+ = NO_2^+ + H_2O$.

It is probable that a mechanism similar to that described above obtains for most aromatic nitrations as evidenced by the fact that in several other cases studied, the dependence of rate on sulfuric acid concentration closely paralleled the behavior with nitrobenzene. Further support is found in the fact that the rate of nitration varies widely with the aromatic substance used. *Orthopara* directing substituents greatly increase the rate, while *meta* directing substituents decrease it. However, nearly all aromatic compounds can be nitrated if sufficiently vigorous conditions are chosen.

Aromatic nitrations are generally effected with a mixture of nitric and sulfuric acids in the temperature range of 0–120°C. Nitric acid-acetic acid-acetic anhydride mixtures and aqueous nitric acids have also been employed. The most recent development is the preparation and use of crystalline salts of the NO_2^+ ion.

An apparent exception to the above picture of aromatic nitration is the nitration of *p*-chloroanisole, which is strongly catalyzed by nitrous acid. In this case the aromatic compound is first nitrosated and the resulting nitroso compound is subsequently oxidized to the nitro derivative.

In contrast to aromatic hydrocarbons, the nitration of aliphatic hydrocarbons is generally unsuccessful. Tertiary hydrogen atoms on paraffin hydrocarbons have been replaced by nitro groups by refluxing with 30% nitric acid. The yields are low and the reaction has not found much use.

The nitration of alcohols and polyalcohols is a general reaction and can usually be effected by nitric acid-sulfuric acid mixtures. The compounds formed are esters of nitric acid and the nitration is reversible. However, in other respects the reaction closely resembles aromatic nitration and it is possible that the mechanism is the same. The nitration of polyhydroxy compounds is an important source of explosives and at one time was a method of solubilizing cellulose.

The preparation of the nitramines is effected in an indirect manner since direct nitration of basic amines is rarely satisfactory. Amides or amine derivatives of the same order of base strength as amides can be nitrated with nitric acid with or without sulfuric acid. Nitric acid-acetic acid-acetic anhydride mixtures have proved to be even more effective for this type of nitration. This latter reagent is explosive under certain conditions and should be used with caution.

Certain tertiary amines such as hexamethylenetetramine and dinitro-N,N-dimethylaniline cleave with nitrating agents to form nitramines, directly such as in the manufacture of RDX and Tetryl (2,4,6,N-tetranitro-N-methylaniline).

NORMAN C. DENO

Cross-references: *Nitroparaffins, Explosives.*

NITRIC ACID

Nitric acid is made by the catalytic oxidation of ammonia at the rate of about 500,000 tons a month in the U.S.

$$4NH_3 + 5O_2 \rightarrow 4NO + 6H_2O + heat \qquad (1)$$

The catalyst is a multiple screen of fine platinum wire held red-hot by insulation. The nitric oxide becomes dioxide on leaving the converter, reacting with the excess air present.

$$2NO + O_2 \rightarrow 2NO_2 + heat \qquad (2)$$

The nitrogen dioxide passes to a series of absorption towers where it travels countercurrent to a flow of acid, which increases in concentration in each tower.

$$3NO_2 + H_2O \rightarrow 2HNO_3 + NO + heat \qquad (3)$$

The nitric oxide reformed is oxidized again in the towers, so that finally almost all of it is absorbed. The acid discharged is 55 to 60 per cent HNO_3.

The main use of nitric acid is in nitrations, such as that of benzene.

$$C_6H_6 + HNO_3 \rightarrow C_6H_5 \cdot NO_2 + H_2O$$

The nitration is usually performed with mixed acid, H_2SO_4 and HNO_3. The function of the sulfuric acid is that of a catalyst. It also absorbs the water formed by the reaction, which otherwise would slow down the rate of nitration and finally prevent it altogether. Ethylene glycol, glycerin, and cellulose are other substances which may be nitrated.

ELBERT C. WEAVER

NITRILES

Nitriles are organic compounds of the general formula $R-C\equiv N$, where R is a univalent radical; compounds containing more than one nitrile group, such as dinitriles and polynitriles, also exist. Names for specific nitriles are derived from the names for the carboxylic acids having the same number and arrangement of carbon atoms, by replacing the ending **-ic acid** or **-oic acid** with **-onitrile**. For example: CH_3CN, acetonitrile; $CH_2=CH-CN$, acrylonitrile; C_6H_5CN, benzonitrile; and NC-$(CH_2)_4$-CN, adiponitrile (a representative dinitrile).

The nitriles are colorless liquids of pleasant ethereal odor, or colorless, odorless crystalline solids. Although acetonitrile is completely soluble in water, the higher molecular weight nitriles become increasingly less soluble. The nitriles have relatively high dipole moments and dielectric constants; consequently, the liquid nitriles are excellent solvents for a variety of substances. Acetonitrile, for example, is used commercially as a solvent for the separation and purification of butadiene during its manufacture.

The methods for preparing nitriles can be divided into two broad categories: (1) those methods in which the number of carbon atoms in the starting material remains the same, and (2) those methods in which the carbon chain of the starting material is lengthened. Examples of the former category include the dehydration of amides or oximes, the ammonolysis of acids or esters, the dehydrogenation of primary amines and the catalytic ammoxidation of activated hydrocarbons. Chain lengthening, on the other hand, occurs during the addition of hydrogen cyanide to the multiple linkages in olefins, acetylenes, aldehydes or ketones, during the double displacement reactions of organic halides or sulfates with inorganic cyanides and also during certain specific reactions of acrylonitrile, such as cyanoethylation of active hydrogen compounds and reductive coupling with activated olefins. The organic nitrile produced in the largest volume today is acrylonitrile, manufactured by the catalytic ammoxidation of propylene. More than 1 billion pounds was produced in the U.S. during 1970.

The chemical reactivity of nitriles can be described in terms of the chemical properties of the nitrile group itself and the effect of the nitrile group in altering the chemical properties of the rest of the molecule. The predominant chemical reaction of the nitrile group involves addition to the carbon—nitrogen triple bond of one or more reactive species such as water, alcohols, acids, ammonia, amines, hydrogen sulfide, hydrogen peroxide, carbonium ions, carbanions and molecular hydrogen. By these means many different compounds can be prepared. For example, the hydrolysis of acrylonitrile with 85% sulfuric acid gives acrylamide, which is an important ingredient of materials used for many water purification systems.

The presence of a nitrile group in a molecule can also increase or alter the reactivity of neighboring groups. For example, adjacent carbon-hydrogen bonds become acidic and various alkylations and condensations are possible[3,4]. Activation of the double bond in acrylonitrile leads to the facile addition of active hydrogen compounds (HX) to give cyanoethylated products ($X-CH_2-CH_2-CN$), and the electrolytic hydrodimerization of acrylonitrile to form adiponitrile, an important intermediate in the manufacture of nylons. In a similar manner the free radical polymerization of acrylonitrile gives polymers useful as textile fibers, oil-resistant rubbers, engineering plastics, disposable soft-drink bottles, and many other items.

Safety precautions in handling nitriles are largely dependent upon the nature of the organic radical R. The low molecular weight nitriles such as acetonitrile, acrylonitrile and benzonitrile are flammable and toxic; work areas should be adequately ventilated. Before any nitrile is handled in bulk, the data on its specific properties and toxicity should be thoroughly examined.

JANICE L. GREENE

References

1. Mowry, D. T., "The Preparation of Nitriles," *Chem. Revs.,* **42**, 189–283 (1948).
2. "The Chemistry of Acrylonitrile," 2nd edition, 272 pp., American Cyanamid Co., New York, 1959.
3. Cope, A. C., Holmes, H. L., and House, H. O., "The Alkylation of Esters and Nitriles," *Organic Reactions,* **9**, 107–331 (1957).
4. Jones, G., "The Knoevenagel Condensation," *Organic Reactions,* **15**, 203–599 (1967).
5. Krimen, L. I., and Cota, D. J., "The Ritter Reaction," *Organic Reactions,* **17**, 213–325 (1969).

NITROGEN AND ITS COMPOUNDS

Nitrogen, a colorless, odorless, tasteless diatomic gas constituting about four-fifths of the atmosphere by volume, was discovered independently in 1772 by both the Swedish druggist, Carl Wilhelm Scheele, and the Scotch botanist, Daniel Rutherford. It has an atomic number of 7, an atomic mass of 14.0067 and a molecular mass of 28.0134 and is a member of Group VA of the periodic system. Its specific gravity in the gaseous state is 0.96737, in the liquid state 0.804, and in the solid state 1.0265. Except when heated to a high temperature, where it combines with metals, forming nitrides (nitrogen boiling point $-195.8°C$, freezing point $-209.8°C$) is extremely inert. It is somewhat soluble in water and slightly soluble in alcohol. Nitrogen is an essential component of all living plant and animal matter, but cannot be used to form amino acids by the animal organism. This vital function is performed only by plants by a mechanism that is gradually being worked out. (See **Phytochemistry**.)

Commercially nitrogen is produced by fractional distillation of liquid air, with nitrogen boiling off at $-195.8°C$; or by the "Borsig" process in which methane is burned in air, producing carbon dioxide (which is absorbed under pressure in water) and nearly pure dry nitrogen. The element may also be produced by reduction from nitrogen-bearing compounds, such as ammonium nitrite. Nitrogen is important in the production of nitric acid, ammonia, cyanamide, cyanides and nitrides of metals; as an inert atmosphere for industry and laboratories; and as a source of cold when liquefied.

A nitrogen atom in an organic compound, in general, makes the substance more basic than it would be otherwise. For this generalization to have meaning, a comparison with substances of known basic strength is desirable. The most popular compound for comparison would, of course, be ammonia. Ammonia is itself a weak base in comparison with the hydroxides of the alkali metals.

If a short akyl group (1–4 carbons) is substituted for hydrogen in ammonia, the resulting primary amine is a slightly stronger base than ammonia itself. Further substitution of alkyl groups to secondary and tertiary amines enhances the basic strength slightly. Increasing molecular mass, however, decreases the solubility of the amine in water so that comparison is impractical in water solution beyond eight to ten carbons.

Substitution of an aryl group on nitrogen in ammonia decreases the basic strength rather sharply. Two aryl groups render an amine nearly neutral; diphenylamine, for example, is not soluble in dilute HCl. An acyl group has the same effect; amides are neutral in character. By putting two acyl groups on nitrogen in ammonia, the basic strength is reduced so far that substance is a weak acid instead. Phthalimide, e.g., will react with sodium hydroxide but not with the weaker base, sodium bicarbonate.

A sulfenyl or a sulfonyl group is even more effective than a carbonyl in reducing the basic strength of a nitrogen atom. One sulfonyl group in ammonia is enough to convert the compound into a weak acid.

Hydrazine (NH_2-NH_2) and hydroxylamine (NH_2-OH) are basic compounds of strength comparable to ammonia. Substitution of alkyl or aryl groups on nitrogen in these compounds has effects similar to those just described for ammonia. (See **Hydrazine**)

While ammonia and amines are electron donors, a nitro group is an electron acceptor. Consequently a nitro group in an aliphatic compound labilizes the hydrogen on the α-carbon and primary and secondary nitro compounds (RCH_2NO_2 and R_2CHNO_2) are weakly acidic.

In heterocyclic saturated compounds the nitrogen in general maintains its basic character and has the strength of a corresponding straight chain compound. But conjugation with the unshared pair of electrons on nitrogen reduces the basic strength to the neutral point or beyond. Pyrrole with a conjugated system of double bonds is slightly acidic. It forms a salt with solid potassium hydroxide but is hydrolyzed immediately in water.

Introduction of a second nitrogen in the six-membered ring makes a slightly stronger base in the o- and p- positions but pyrimidine displays little basic character. Some ring compounds occur as parts of more complex molecules in plants. Pyridine and piperidine rings occur in a number of alkaloids. Nicotine contains both the pyridine and

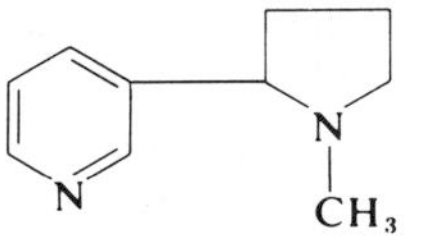

nicotine

the pyrrolidine rings. The piperidine nucleus is the principal structure in other alkaloids, for example, the hemlock (coniine) and the belladonna alkaloids (atropine, cocaine). Still other alkaloids contain the isoquinoline ring which is related to pyridine. The cactus (anhalamine), opium (morphine), and the curare alkaloids all embody the isoquinoline nucleus in complex structures. (See **Alkaloids**)

The pyrrole nucleus is found in the very important class of compounds called porphyrins, which occur widely in nature. The poryphyrins include chlorophyll, the green coloring matter in plants and hemin in blood. The well-known dye, indigo, is a derivative of indole (benzopyrrole).

Vitamins, found in plants in very small quantities, have in some cases turned out to be nitrogen containing compounds. As examples may be cited thiamine (pyrimidine derivative), riboflavin, and nicotinic acid and pyridoxine (pyridine derivatives). These structures are required to be preformed in the diet of man and other animals. (See **Vitamins**)

α-Amino acids are the important building blocks for proteins in both plants and animals. In plants

these nitrogen compounds are concentrated in the seeds. In animals, proteins occur in muscle, hair, tendons, skin, horns, feathers, nails, and nerve tissue. Putrefaction of histidine, an amino acid, leads to histamine which occurs widely in nature. Its release in body tissue is apparently responsible for some allergies and other body ailments. (See **Amino Acids**)

Enzymes (biocatalysts for specific body reactions) are nitrogen-containing compounds, some of which are simple proteins.

Nucleoproteins are also important nitrogen compounds found in the nuclei and cytoplasm of all plant and animal cells and more prominently in glandular tissues. The protein part of the nucleoprotein, of course, contains nitrogen, but so does the nucleic acid in the form of purine and pyrimidine bases. Purines are derivatives of the structure shown below, which itself contains the pyrimidine ring.

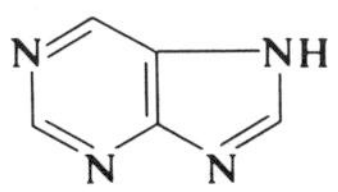

purine

Adenosine monophosphate (AMP), one of the ribonucleic acids has three components, adenine (a purine), ribose, and phosphoric acid combined in the structure shown.

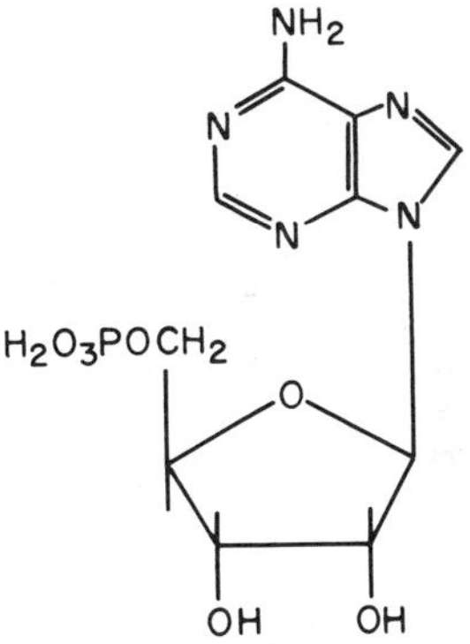

AMP

Hormones, the so-called chemical messengers in the body, are in some cases nitrogen compounds, e.g., thyroxine and epinephrine.

Synthetic Nitrogen Compounds. The wide variety of synthetic nitrogen compounds defies division into definite categories. A few representative groups of compounds will be described to indicate this variety.

Nitrogen compounds abound among therapeutic agents. Sulfa drugs are p-aminobenzenesulfonamide derivatives in which the R-group itself is also a nitrogen moiety, such as pyridine, pyrimidine, guanidine, etc.

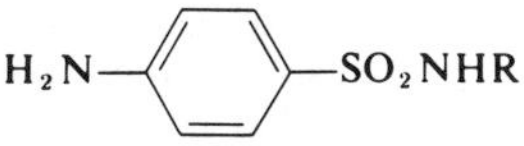

Sulfa drugs

Antibiotics generally contain nitrogen. For example, pyocyanine and aspergillic acid are piperazine derivatives and the penicillins have the ring structure shown for Penicillin G.

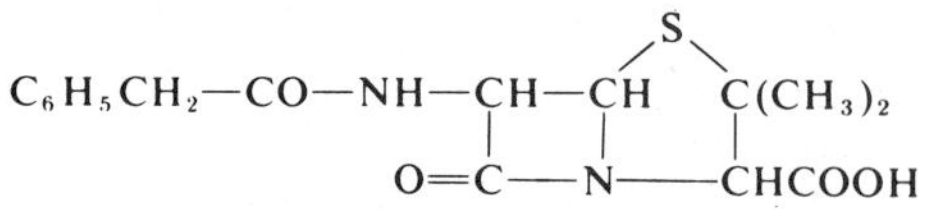

$$C_6H_5CH_2\!-\!CO\!-\!NH\!-\!CH\!-\!CH \quad \overset{S}{\diagdown} \quad C(CH_3)_2$$
$$O\!=\!C\!-\!\!-\!N\!-\!\!-\!\!-\!CHCOOH$$

Penicillin G

Many thousands of prospective antimalarials have been screened for medical use; some of the most promising are 8-aminoquinoline derivatives.

Organic fungicides have recently included nitroso pyrazoles, imidazolines, and triazines as well as the old stand-bys, dithiocarbamates.

In the polymer field, acrylonitrile ($CH_2\!=\!CH\!-\!C\!\equiv\!N$) is being used as a monomer and as a co-polymer in synthetic rubber, textile fibers, soil conditioners, and plastics. Nylon, a household word now, is a polyamide, man's superior creation of a protein-like molecule. The urea-formaldehyde and melamine-formaldehyde resins are also widely used in the field of polymers.

LEALLYN B. CLAPP

Nitrogen-Fluorine (NF) Chemistry

The term NF chemistry has been widely applied, since 1958, to the rapidly expanding field of chemistry involving both inorganic and organic compounds containing fluorine atoms bonded to nitrogen.

Prior to 1958 very little work had been done in this field, in fact only three fluorides of nitrogen were known prior to that time. Nitrogen trifluoride (NF_3) had been prepared by the electrolysis of molten ammonium bifluoride in 1928, while fluorine azide (FN_3) was prepared (1942) by the reaction of fluorine with hydrazoic acid. It was observed that fluorine azide decomposes (sometimes explosively) at room temperature. The nonexplosive decomposition of fluorine azide leads to the third nitrogen fluoride known prior to 1958, difluorodiazine (N_2F_2).

These compounds did not prove to be of significant synthetic use so the NF field languished until 1958 when tetrafluorohydrazine (N_2F_4) was first synthesized. With the synthesis of tetrafluorohydrazine and the subsequent discovery of its facile dissociation into the relatively stable difluoramino free radical, the field of NF chemistry underwent a very rapid expansion. Approximately 200 publications and numerous patents have appeared since 1958.

Tetrafluorohydrazine (N_2F_4). This compound was first prepared by the thermal reaction of nitrogen trifluoride with various fluorine acceptors, i.e., copper, bismuth, arsenic and antimony. Typical reaction conditions which led to yields of 60–70% are: temperature, 375°C; residence time, 13 minutes; reactor, copper-packed copper. Under these conditions it is really the difluoramino radical that is being formed. These radicals then dimerize to form tetrafluorohydrazine when the gas is cooled. There are now many synthetic methods for the preparation of tetrafluorohydrazine, most of which are based upon the original method, i.e., the abstraction of a fluorine atom from nitrogen trifluoride to

form NF_2 followed by dimerization of these radicals to form tetrafluorohydrazine.

The discovery, that permitted a logical straightforward attack on the synthetic aspects of the NF field was the observation in 1961 that tetrafluorohydrazine is in dissociative equilibrium with two di-difluoramino free radicals as shown in equation (1):

$$N_2F_4 \rightleftharpoons 2 \cdot NF_2 \qquad \Delta H = -19.8 \text{ kcal/mole} \qquad (1)$$

The heat of dissociation of this reaction is also the bond dissociation energy of the N—N bond in tetrafluorohydrazine. It has been determined by four independent methods and is now one of the best-known bond energies. This low bond energy (19.8 kcal) means that even at room temperature there is a significant concentration of difluoramino free radicals in tetrafluorohydrazine.

Difluoramino Free Radical $\cdot NF_2$. The difluoramino free radical is a typical stable radical in its reactions and this observation led to the synthesis of many new products containing the difluoramino functional group. A few of these reactions will be outlined below:

(a) *Hydrogen Abstraction.* The difluoramino radical is capable of abstracting hydrogen from molecules containing labile hydrogens to form difluoramine (HNF_2). If mercaptans or thiophenols are used the products are disulfides and difluoramine.

$$2\ RSH + 2\ NF_2 \rightarrow RSSR + 2\ HNF_2.$$

This method was used in one of the early preparations of difluoramine.

(b) *Radical Coupling.* The difluoramino radical can also couple with other radicals, either stable or reactive. NF_2 couples with nitric oxide to form the thermally unstable, highly colored nitrosodifluoramine (NF_2NO).

$$N_2F_4 + 2\ NO \rightleftharpoons 2\ NF_2NO \rightleftharpoons 2\ NF_2 + 2\ NO$$

Purple

The N—N bond strength in nitrosodifluoramine is only 10 kcals; consequently NF_2NO is only stable below −80°C.

The difluoramino free radical can also couple with very reactive alkyl radicals. These radicals may be generated either photochemically or thermally.

(c) *Combination of Hydrogen Abstraction and Radical Coupling.* In its reactions with aldehydes the difluoramino free radical first abstracts a hydrogen atom to form difluoramine and an acyl radical. The acyl radical then couples with a second difluoramino free radical to form a N,N-difluoramide.

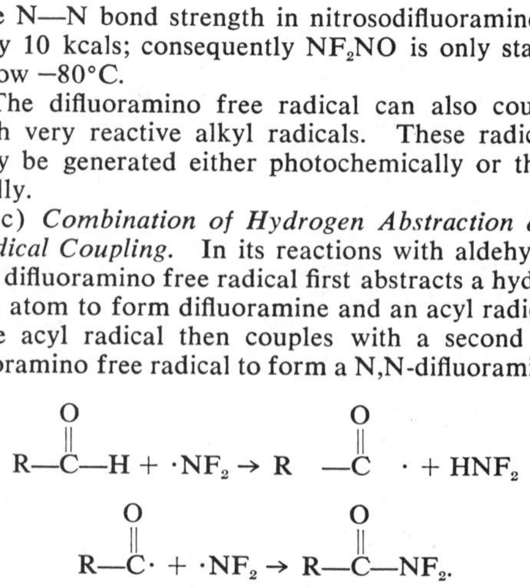

$$R\!-\!\overset{\overset{\displaystyle O}{\|}}{C}\!-\!H + \cdot NF_2 \rightarrow R\!-\!\overset{\overset{\displaystyle O}{\|}}{C}\cdot + HNF_2$$

$$R\!-\!\overset{\overset{\displaystyle O}{\|}}{C}\cdot + \cdot NF_2 \rightarrow R\!-\!\overset{\overset{\displaystyle O}{\|}}{C}\!-\!NF_2.$$

Additional radical coupling reactions take place between the difluoramino free radical and other reactive inorganic free radicals and atoms. For example there is an equilibrium between chlorine and tetrafluorohydrazine on the one hand and chlorodifluoramine on the other. This equilibrium is represented by:

$$Cl_2 + N_2F_4 \rightleftharpoons 2\ ClNF_2.$$

This reaction may be thought of as taking place by abstraction of a chlorine atom from Cl_2 by a difluoramino radical:

$$Cl_2 + \cdot NF_2 \rightarrow Cl\cdot + ClNF_2$$

followed by coupling of the chlorine atom with a second difluoramino radical:

$$Cl\cdot + \cdot NF_2 \rightarrow ClNF_2.$$

It has also been observed that the difluoramino free radical can couple with the fluorosulfonate free radical to form N,N-difluorohydroxylamine-o-fluoro sulfonate $[NF_2OSO_2F]$.

Disulfur decafluoride is believed to dissociate into sulfur pentafluoride free radicals. Certainly in its reactions with tetrafluorohydrazine to form difluoramino sulfur pentafluoride this presents an attractive mechanism:

$$S_2F_{10} \rightleftharpoons 2\ \cdot SF_5$$

$$N_2F_4 \rightleftharpoons 2\ \cdot NF_2$$

$$\cdot SF_5 + \cdot NF_2 \rightarrow F_5SNF_2.$$

In addition to chemical evidence for the existence of the relatively stable difluoramino free radical the following physical observations have been made on this radical. Its infrared, ultraviolet, epr and mass spectra have all been observed and are consistent with the difluoramino free radical species.

This free radical absorbs in the ultraviolet and the products of the interaction of the $N_2F_4 \rightleftharpoons 2\ NF_2$ system with ultraviolet irradiation are primarily difluorodiazine and nitrogen trifluoride. This reaction can best be rationalized as follows:

$$N_2F_4 \rightleftharpoons 2\ NF_2 \xrightarrow[uv]{} \cdot NF_2{}^*$$

$$NF_3 \xleftarrow{\cdot NF_2} F + NF$$

$$\downarrow NF$$

$$N_2F_2$$

Uses of NF Compounds. At present there are no commercial uses for NF compounds known to the author. Several potential uses have, however, been suggested. It has been suggested that NF compounds such as nitrogen trifluoride or tetrafluorohydrazine could replace oxygen as a filler for photographic light bulbs. The NF fillers would produce more light and a higher color temperature in the flash than does oxygen. Another investigator has suggested the use of nitrogen trifluoride with hydrogen as a torch. The torch can weld, braze and cut metals without additional flux since the nitrogen trifluoride acts as a gaseous flux. It has also been observed that difluorodiazine is a gaseous catalyst for the polymerization of various monomers. A recent study has appeared of the use of difluorodiazine as a vulcanizing agent for fluoroelastomers.

In addition, the difluoramino free radical itself has been reported to act as an initiator in free radical chain reactions and may be used as a polymerization catalyst.

Caution: Many NF compounds are extremely reactive; some are explosive while others form explosive mixtures with other materials. Extreme caution should be used in handling these compounds and reference should be made to the original literature before any experimental investigations are undertaken with these or similar materials.

CHAS. B. COLBURN

Hydrazine

Hydrazine (N_2H_4) is one of a homolgous series of chemical compounds known as the hydronitrogens. It was first identified by Fischer in 1875 in the form of organic derivatives and first isolated by Curtius in 1887. It is a hygroscopic, water-white liquid with an ammonia-like odor. It fumes in air reacting with both the oxygen and carbon dioxide present in the atmosphere. Hydrazine has a boiling point of 113.5°C, and a freezing point of 2°C. It forms a monohydrate $(N_2H_4 \cdot H_2O)$ whose boiling point is 120.1°C and whose freezing point is −51.7°C. Hydrazine decomposes thermally to yield ammonia, nitrogen and hydrogen; however, at sufficiently high temperatures nitrogen and hydrogen are the only products.

Hydrazine was first synthesized by Raschig in 1907 and practically all commercial hydrazine is produced by the Raschig process. This process consists of the following sequence of reactions:

a) sodium hydroxide + chlorine → sodium hypochlorite

b) sodium hypochlorite + ammonia → chloramine

c) chloramine + ammonia → hydrazine

The hydrazine produced in this manner is obtained as a dilute (ca. 2%) aqueous solution which is normally concentrated by distillation to the hydrazine monohydrate composition (64% N_2H_4). The hydrate may be converted to anhydrous hydrazine by azeotropic distillation with aniline. Mono- and unsymmetrical dialkylhydrazines $(RN_2H_3$ and $R_2N_2H_2)$ may be produced via the above process by substituting the appropriate amine for ammonia in step (c). However, the unsymmetrical dialkylhydrazines are best produced by catalytic hydrogenation of the dialkylnitrosoamine. Symmetrical dialkylhydrazines are prepared by the hydrogenation of azines. Aryl hydrazines are produced by the reduction of aryl diazonium salts or by the condensation of hydrazine with haloaromatic compounds.

Hydrazine is a base somewhat weaker than ammonia. It is capable of forming salts with both organic $(RCOOH \cdot N_2H_4)$ and inorganic $(N_2H_4 \cdot HX$ and $N_2H_4 \cdot 2HX)$ acids. Under anhydrous conditions, one of its hydrogens is somewhat acidic, and the preparation of metal hydrazides (MeN_2H_3) of Groups I, II and III of the Periodic Table has been described in the literature. These metal hydrazides can decompose explosively in contact with oxygen.

Hydrazine can react with certain metallic salts to form hydrazinates ($MeX \cdot N_2H_4$), double salts ($MeX \cdot N_2H_4 \cdot HX$) and complexes [$Me(N_2H_4)nX$]. Recent evidence has shown that in these metal complexes the hydrazine forms a bridge between two metal atoms. Previously it was suggested that both nitrogens were coordinated to one metal atom to form a three-membered ring. Caution should be taken in handling these metal complexes in which the anion is an oxidant (nitrate, perchlorate, etc.), as many of them are shock-sensitive.

Hydrazine undergoes autooxidation when exposed to oxygen with the reaction being catalyzed by traces of cupric ion, certain metal oxides and activated carbon. It has been shown that the autooxidation of hydrazine produces water and diimide and that diimide itself, rather than hydrazine, is the active agent in the hydrogenation of certain moieties such as carbon-carbon and nitrogen-nitrogen double bonds.

Hydrazine is a strong reducing agent and will reduce certain heavy metal oxides and metallic salts to the free metal. These reactions have been utilized in the silvering of mirrors and the plating of metal films on plastics. Hydrazine is also capable of reducing nitro groups to amino groups.

One of the most well-known organic reactions of hydrazine is its reaction with carbonyl compounds to yield either hydrazones ($R\overset{\overset{\textstyle H}{|}}{C} = NNH_2, R_2C = NNH_2$) or azines ($R\overset{\overset{\textstyle H}{|}}{C} = NN = \overset{\overset{\textstyle H}{|}}{C}r, R_2C = NN = CR_2$). Isohydrazones or diaziridines ($R_2C\overset{\diagup NH}{\underset{\diagdown NH}{|\;|}}$) are prepared by the interaction of a ketone with a mixture of ammonia and chlorine. These materials are hydrolyzed to hydrazine with acid, and this process has been patented as a method for the manufacture of hydrazine. The carbonyl group of aliphatic ketones can be reduced to a methylene group by forming the hydrazone followed by treatment with base. (See **Wolff-Kishner Reaction**).

Hydrazine reacts with organic acids, acid anhydrides, acid chlorides, esters, thioesters and amides to form the corresponding hydrazides. With acids, such as maleic and phthalic, cyclic hydrazides are produced. Nitrosation of a carboxylic acid hydrazide produces the corresponding azide which can be rearranged to the corresponding isocyanate via the well-known Curtius reaction. Aromatic hydrazides, when treated with benzenesulfonylchloride followed by base, undergo the McFadyen-Stevens reaction to form the corresponding aldehyde with the elimination of nitrogen.

Heterocyclic compounds including triazoles, thidiazoles, oxadiazoles, pyridazines, etc., are readily prepared from hydrazine and its derivatives. For example the heating of a carboxylic acid hydrazide at 150–200°C will form the corresponding 4-amino-1, 2, 4-triazole.

Numerous hydrazine-based polymers have been prepared and converted to films, fibers and resins. The most common of these include: polyhydrazides, polytriazoles, poly-4-aminotriazoles, polyazines, polyoxadiazoles and formaldehyde condensation products. In addition, the preparation of polyethylenehydrazine ($-CH_2CH_2 - \underset{\underset{\textstyle NH_2}{|}}{N} -$) is described in the literature.

Anhydrous hydrazine, unsymmetrical dimethylhydrazine, and monomethylhydrazine are being utilized as rocket fuels either in the main propulsion or in the attitude control systems. These fuels, either alone or admixed, have been selected for use in intercontinental ballistic missiles and for many outer space programs.

Because of its diverse activity, such as basicity, reducing power, biological activity, bi-functionality, etc., hydrazine and its derivatives are being commercially utilized in numerous applications. These include:

1. Pharmaceutical
   a) Antibacterials
   b) Coccidiostats
   c) Antihypertension drugs
   d) Antidepression drugs
   e) Antihistamines
   f) Treatment of tuberculosis
2. Agricultural
   a) Maleic hydrazide and amino-triazole are employed as growth regulators and herbicides
3. Polymer
   a) Blowing agents for vinyl and polyolefin plastics and rubber
   b) Curing agents for epoxy coatings
   c) Polymerization catalyst
   d) "Short stop" agent for synthetic rubber
   e) Extender for spandex fibers and synthetic leather
4. Petroleum
   a) Antioxidant
   b) Platinum catalyst recovery
   c) Mercaptan scavenger
   d) Carbon dioxide scavenger
   e) Corrosion inhibitor
5. Metals
   a) Metal plating of glass and plastics
   b) Processing of radioactive metals
   c) Copper soldering flux
   d) Antitarnish agent
6. Chemical Processing
   a) Chlorine scavenger in hydrochloric acid
   b) Polymerization inhibitor in hydrocarbon processing
7. Soap and Detergent
   a) Antioxidant
8. Energy
   a) Rocket fuel
   b) Fuel cells
   c) Primer in ammunition
   d) Boiler feed-water treatment in electric power generating plants.

PERRY R. KIPPUR

Cross-references: *Fuels, Propellants.*

Nitroparaffins

Nitroparaffins (or nitroalkanes) have the emperical formula $C_nH_{2n+1}NO_2$. The nitro group is attached through the nitrogen to a carbon atom, in contrast to the electronegative group ($-ONO$) of the alkyl nitrite esters which bond through the oxygen.

Commercial Solvents Corporation produced the first four members of the homologous series (nitromethane, nitroethane, 1-nitropropane and 2-nitropropane) by vapor-phase nitration of propane.

The nitroparaffins are colorless, mobile liquids with a characteristic pleasant odor. Some of their physical properties are listed in the following table.

This synthesis appears to be satisfactory for primary nitroparaffins only.

The Walden synthesis involves reacting a dialkyl sulfate with a concentrated alkali nitrite solution:

$$(C_2H_5O)_2SO_2 + NaNO_2 \rightarrow$$
$$C_2H_5OSO_2Na + C_2H_5NO_2$$

Some Physical Properties of the Nitroparaffins (1)

Property	Nitro-methane	Nitro-ethane	1-Nitro-propane	2-Nitro-propane
Boiling point, 760 torr, °C	101.20	114.07	131.18	120.25
Freezing point, °C	−28.55	−89.52	−103.99	−91.32
Density, 25°C, g/cc	1.13128	1.04464	0.99609	0.98290
Viscosity, 25°C, cp	0.610	0.638	0.790	0.721
Ionization constant, aci, 25°C	5.5×10^{-4}	3.93×10^{-5}	8.98 (2)	7.73×10^{-6}
Dielectric constant, 30°C	35.87	28.06	23.24	25.52
Solubility in water, 25°C, %w	11.1 (3)	4.68	1.50	1.71

(1) Selected from tables in Riddick and Bunger, *"Organic Solvents,"* Wiley-Interscience, New York, 1970. (2) pk_a nitro form. (3) Completely miscible at 105+°.

These liquids have an unusual combination of solvent properties. They have medium boiling points and a strong solvent power (solvent parameter for nitromethane is 10.25 and it is weakly hydrogen-bonded) for a wide variety of substances, such as numerous organic compounds, dyes, waxes, resins, gums, coating materials and many polymers. 2-Nitropropane has recently been found to have several unusual properties when used in the solvent formulation for flexographic and gravure printing inks. The 2-nitropropane inks give better wettability, sharper prints, and faster solvent release which permits faster running of the presses. The nitroparaffin reactivity is basic which limits their use to neutral or acidic systems. Nitromethane, and to a limited extent other nitroparaffins, has found limited use as explosives, propellants and fuel additives.

The laboratory methods for the preparation of nitroparaffins generally result in low yields. Primary nitroparaffins can best be prepared by the Victor Meyer reaction. An alkyl bromide or iodide reacts with silver nitrite to produce the corresponding nitroparaffin and alkyl nitrite.

$$RX + AgNO_2 \rightarrow RNO_2 + RONO + AgX$$

The yield of secondary nitroparaffins seldom exceeds 20% by this method. It cannot be used for tertiary nitroparaffins. Tertiary nitroparaffins may be prepared in high yields from the corresponding tertiary alkyl amine and potassium permanganate. The yields of the primary and secondary nitroparaffins can be increased to 55–62% by using dimethylformamide as the solvent instead of water. The yield can be increased further by using dimethyl sulfoxide as the solvent. The Kolbe reaction between a sodium salt of an α-monohalogen carboxylic acid and sodium nitrite provides an excellent laboratory method for nitromethane.

$$CH_2ClCOONa + NaNO_2 + H_2O \xrightarrow{\text{heat}}$$
$$CH_3NO_2 + NaCl + NaHCO_2$$

The higher members of the homologous series may be prepared by liquid-phase nitration. The reaction consists essentially of the replacement of a hydrogen by a nitro group. Considerable oxidation accompanies the slow and selective reaction. Poly nitrates are also formed (see **Nitration**).

The nitroparaffins offer the basis for the chemical synthesis of many interesting compounds. Some of the better known reactions are discussed below.

Action with bases. Primary and secondary nitroparaffins exist in equilibrium with the tautomeric *aci*-form. Nitromethane is uniquely sensitive to the action of alkali metals. Sodium or potassium converts nitromethane to sodium methylnitronate and then into the alkali metal salt of methazonic acid. This reaction is not as simple and straightforward as most of the literature implies. Further treatment with strong alkali converts the methazonate to alkali nitroacetate.

The dry salts of alkyl nitronic acids are sensitive to shock. Care must be exercised when handling.

β-Dioximes result from the condensation of three molecules of a nitroparaffin in the presence of a weak base such as an alkyl amine. The β-dioximes hydrolyze to form substituted isoxazoles.

Action with mineral acids. Strong mineral acids convert the primary nitroparaffins to the corresponding carboxylic acid and hydroxyl amine. The salts of the primary and secondary nitroparaffins warmed with dilute acids produce the corresponding carbonyl compound.

Reduction. Strong reduction with zinc and hydrochloric acid or hydrogen and Raney nickel results in the corresponding amine. Mild reduction with zinc dust and water results in N-alkylhydroxylamines.

Action with halogens. Halogenation studies of the nitroparaffins have shown some interesting phenomena. The three halogens (chlorine, bromine and iodine) react in the same manner in a basic solution. The halogen substitutes on the carbon holding the nitro group. The dichloro compound may be prepared if sufficient alkali is present to hold the monochloro compound in the *aci* form. The iodo compounds are unstable. Nitromethane reacts with chlorine in the presence of either an aqueous suspension of calcium carbonate or an alkali metal or alkaline earth hypochlorite to form chloropicrin.

Intense illumination catalyzes halogenation of nitroparaffins under acidic conditions. Bromine reacts in the same manner as in the presence of a base. Chlorine substitutes largely on a carbon other than that holding a nitro group. Nitroethane gives mainly 2-chloro-1-nitroethane; 1-nitropropane yields 2-chloro-1-nitropropane and 3-chloro-1-nitropropane.

Reaction with aldehydes. Nitroparaffins react with aldehydes in the presence of a basic catalyst to form nitroalcohols which are easily reduced to aminoalcohols.

Nitroalcohols react with amines to form nitroamino compounds which yield diamines on reduction.

Dehydration of certain aromatic nitroalcohols followed by reduction produces aromatic amines which do not have a hydroxyl group.

Reaction with ketones. The reaction of nitromethane with acetone produces 2,2-dimethyl-,3-dinitropentane (dinitroneopentane) which reduces to the corresponding diamine.

Toxicity and safety. The nitroparaffins act chiefly as moderate irritants when inhaled. The nitropropanes are somewhat more toxic than nitromethane and nitroethane. There is no evidence that absorption through the skin produces systemic injuries. The following threshold limit values are recommended: nitromethane, 100 ppm or 250 mg/m^3; nitroethane, 100 ppm or 310 mg/m^3, 1- and 2-nitropropane, 25 ppm or 90 mg/m^3.

Nitroparaffins form nitronates in the presence of alkalies. Evaporation of these mixtures should be avoided because some are explosive. Observe the special precautions for handling nitromethane recommended by the manufacturer.

JOHN A. RIDDICK

Technology of Nitrogen Compounds

In 1898 Sir William Crookes warned that the supply of fixed nitrogen for agriculture was rapidly reaching a point where it was insufficient to support an ever-increasing population. It is the basis of three important activities of man: (a) agriculture, (b) mining, (c) explosives. During the decade 1890–1900 three significant developments were taking place: (a) exploitation of the nitrate deposits in Chile, (b) the MacArthur-Forrest process for the cyanidation of gold and silver and (c) the work of Caro-Frank to produce calcium cyanide from calcium carbide and nitrogen.

When the Chilean natural nitrate deposits were opened up, a substantial but not an inexhaustible source of fixed nitrogen became available for agriculture and explosives. At about the same time MacArthur and Forrest developed the process for the cyanidation of gold and silver. This meant that gold and silver could be extracted from low-grade ores and since gold and silver were the basis of international trade, this trade expanded at an accelerated rate. Although nitrate from Chile had eased the situation insofar as agriculture and explosives was concerned, the problem of cheap cyanide still awaited solution.

In Germany Caro and Frank attempted to produce cyanide, but their reaction led to calcium cyanamide: $CaC_2 + N_2 \rightarrow CaCN_2 + C$. At about this time the Scheideanstalt in Germany developed the process for producing NaCN from metallic sodium, ammonia and charcoal.

However, it was found that when calcium cyanamide is treated with steam under pressure ammonia results: $CaCN_2 + 3H_2O \rightarrow CaCO_3 + 2NH_3$. The Germans also found calcium cyanamide to be an excellent fertilizer for direct application. Finally by 1916 Haber-Bosch developed the direct synthesis of ammonia from hydrogen and nitrogen and Ostwald at the same time developed the process for the oxidation of ammonia to nitric acid.

Calcium Cyanamide. The oldest fixed nitrogen process still practiced commercially is the Cyanamid Process. The reactions involved are:

$$Coal \rightarrow Coke$$

$$CaCO_3 \rightarrow CaO + CO_2$$

$$CaO + 3C \xrightarrow[\text{Furnace}]{\text{Electric}} CaC_2 + CO$$

$$CaC_2 + N_2 \rightarrow CaCN_2 + C$$

The Cyanamid Process requires cheap electric power and is having a difficult time in the United States competing with fertilizers based on ammonia. Pure calcium cyanamide contains 35% nitrogen, whereas the technical product contains 20–24%. This is because excess lime is used to lower the eutectic, and an 80–85% carbide is produced. To produce a higher grade carbide increases the power consumption to an uneconomic level.

Until about fifteen years ago calcium cyanamide was used only as a fertilizer and in the preparation of so-called black cyanide. However, in the last twenty years a considerable expansion in cyanamide derivatives has taken place.

Dicyandiamide. When calcium cyanamide is extracted with water it is hydrolyzed to calcium acid cyanamide and calcium hydroxide. The calcium is precipitated with CO_2 or sulfuric acid, filtered, the pH of the solution adjusted to 9.5 and heated. The cyanamide polymerizes to dicyandiamide.

Guanidine salts. When dicyandiamide is reacted with an ammonium salt a guanidine salt results:

$$(H_2NCN)_2 + 2NH_4X \rightarrow 2(H_2N-\underset{\underset{NH}{|}}{C}-NH_2 \cdot HX)$$

When ammonium nitrate is used guanidine nitrate is produced.

Nitroguanidine. When guanidine nitrate is dissolved in concentrated sulfuric acid nitroguanidine is formed:

$$(H_2N)_2C + NH \cdot HNO_3 \xrightarrow{H_2SO_4} H_2N \cdot \underset{\underset{NH}{|}}{C}NHNO_2$$

Considerable quantities were produced in Canada during World War II since it was used by the British to produce flashless and relatively cool explosions.

Melamine. When dicyandiamide is heated with anhydrous ammonia under pressure melamine results:

$$3(H_2NCN)_2 \rightarrow 2(H_2NCN)_3$$

Melamine when reacted with formaldehyde results in melamine resins. These resins are more water- and heat-resistant than the well-known urea resins. There are a great many other products which can be produced from calcium cyanamide, but the above are the ones which are produced in large quantity.

Urea(46.67N). After World War I Bosch (BASF) developed the process for the synthesis of urea from carbon dioxide and ammonia. Carbon dioxide and ammonia combine to form ammonium carbamate: $2NH_3 + CO_2 \rightarrow H_2NCOONH_4$. When ammonium carbamate is heated to 135–150°C in an autoclave at about 35 atmospheres pressure, one mole of water is eliminated and urea is formed:

$$H_2NCOONH_4 \rightleftharpoons H_2NCONH_2 + H_2O$$

The reaction takes place only in the liquid phase. Since water is formed, the reaction reaches an equilibrium which with excess of ammonia favors the formation of urea. With no excess of ammonia the yield per pass is about 42% of urea. The ammonia and carbon dioxide are recovered and recycled.

Cyanides. *Sodium cyanide.* In 1890 the world production of KCN amounted to 50–70 tons per year. Cyanide up to this time was produced from potassium ferrocyanide, which in turn was produced from the spent oxide of the coal gas industry. The preparation of KCN from prussiate was carried out according to the equation:

$$3K_4FE(CN)_6 + K_2CO_3 \rightarrow 14\ KCN +$$
$$Fe + 2FeO + CO + 4C + 4N$$

However, at the start of the twentieth century the process for producing high-grade sodium cyanide was developed in Germany. This process was unchallenged for the production of cyanide until World War I.

Black cyanide. Since Germany had a practical monopoly on the production of cyanide when this supply was cut off, it became imperative for the mining companies to obtain a source of supply of cyanide. In 1917 the American Cyanamid Company developed a process of producing a black cyanide from calcium cyanamide which contains approximately 50% sodium cyanide equivalent by fusing calcium cyanamide and salt at approximately 1000°C. When this melt is rapidly chilled to below 400°C, calcium cyanide is obtained, and is prevented from reverting to calcium cyanamide.

Hydrocyanic acid. Until the start of World War II, hydrocyanic acid was almost a rare chemical. However, the use of acrylonitrile (made from acetylene and HCN) for copolymerization with butadiene to produce Buna N required large quantities of HCN. It was soon found that polymerized acrylonitrile could be spun into fibers. Whereas most of the HCN used during the war was made from black cyanide and sulfuric acid, it now be-

came desirable to produce HCN directly without going through a cyanide salt. It is possible to crack methane in the presence of ammonia at temperatures of 1200–1300°C, but the heat required is extremely high, and since ceramic equipment is required, the heat-transfer problem is extremely difficult. The IG (Andrussow) developed a process by which the heat is furnished by reacting methane, ammonia and oxygen at a platinum gauze at about 1000°C according to the equation:

$$CH_4 + NH_3 + 3/2O_2 \rightarrow HCN + 3H_2O$$

The reaction is carried out with air and besides HCN and H_2O, H_2, CO, CO_2 and N are present in the exit gas. The use of air changes the reaction from an endothermic to an exothermic one. It is important to use a gas very high in methane and to have a slight excess of oxygen; otherwise carbon will combine with platinum and destroy the gauze. At present the production of HCN is about 400 tons per day and its use is growing.

Nitrogen Fertilizers. The three most important fertilizer elements are nitrogen, phosphorus and potassium, but nitrogen shows the largest and most consistent growth. Nitrogen is an important constituent of plant proteins and chlorophyll and lack of nitrogen can be detected by reduced growth and lack of healthy green color. Nitrogen should be supplemented by phosphorus to insure root development and early maturity.

The striking feature of the nitrogen fertilizer industry is that up to 1900 about 90% was supplied by packing house by-products, animal waste, tankage, etc., whereas at present hardly 1% of these materials furnish our present fertilizer nitrogen. (See **Fertilizers**)

L. J. CHRISTMANN

Cross-references: *Ammonia, Amines, Fertilizers, Heterocyclic Compounds, Amino Acids, Explosives, Dyeing, Nitriles, Nitroparaffins, Alkaloids, Proteins, Nitrogen Fixation.*

NITROGEN FIXATION

The term *fixation of nitrogen* denotes the chemical binding of nitrogen gas of the air to form a chemical compound. One of the simplest concepts in the fixation of nitrogen is the binding of the nitrogen gas of the air with the oxygen gas of the air. The final product is nitrogen dioxide, NO_2 or the dimer N_2O_4. Unfortunately, the intermediate reaction leading to NO_2 formation—namely,

$$N_2 + O_2 \rightarrow 2NO \qquad (1)$$

is endothermic by 42 kilocalories and does not proceed under normal conditions.

This article relates to the use of radiation to effect nitrogen fixation.

Radiation chemistry is the study of the chemical effects produced in a system by the absorption of ionizing radiation. Thus, when a system is irradiated with ionizing radiation the primary effects are ionization and excitation. High-energy radiations commonly used to produce these effects include alpha and fission recoil particles, beta rays, and x-rays or gamma rays, in addition to machine-made high-energy electrons and protons. Of the nuclear reactor radiations, the kinetic energy of the fission recoil particle is about 80 per cent of the total energy of fission. The remaining 20 per cent of the fission energy consists primarily of beta and gamma rays and high-energy neutrons, which contribute to a small extent to the primary effects of radiation, as compared with the kinetic energy of the fission recoil particles.

To interrelate the effects of the high-energy radiations of the chemical systems, a term *G value* is used, which denotes the number of molecules produced or decomposed per 100 electron volts. Since the band spectroscopy of nitrogen has been thoroughly investigated, it is possible to proportionate the 100 electron volts of energy absorbed by molecular nitrogen into excitation and ionization components. This is shown in Table 1. A similar apportionment and table could be made for oxygen. We can now examine Table 1.

In a mixture of nitrogen and oxygen subjected to ionizing radiation, the nitrogen and oxygen will be ionized and excited independently and proportionately to their relative stopping powers (which are about equal) and their concentrations. Immediately after the start of the irradiation and in less than 10^{-3} second, the following species will be formed and will be present in a stationary state: N_2^+, N, N^+, O_2^+, O_2^-, O, O^+, O^-, N_3^+, and N_4^+. These species may then interact among themselves, or in ion and atom reactions with O_2, or with products of the interactions. The predominance of one reaction over another will depend on many factors, such as temperature, pressure, mixture ratio, and radiation intensity. A knowledge of the various reactions and the effects of the enumerated variables upon them makes it possible to choose the variables in such a manner as to make the reaction yield the most favorable results.

The basic reaction, and the first step in the nitrogen fixation with oxygen via N atoms is

$$N + O_2 \rightarrow NO + O \qquad (2)$$

This reaction is exothermic and proceeds with a heat of activation of about 6 kilocalories. The NO thus formed will react in two different ways:

$$2NO + O_2 \rightarrow 2NO_2 \qquad (3)$$

and

$$N + NO \rightarrow N_2 + O \qquad (4)$$

Reaction 3 is the desired reaction. Reaction 4 is very fast (the reaction rate is $(2.6 \pm 0.6) \times 10^{-11}$ cm^3 per particle per second), so that in practice a very low steady-state concentration of NO is required to reduce G_N-fixed. The NO_2 formed in reaction 3 may react in three different ways:

$$NO_2 + N \rightarrow 2NO \qquad (5)$$

$$NO_2 + N \rightarrow N_2O + O \qquad (6)$$

$$NO_2 + N \rightarrow N_2 + 2O \qquad (7)$$

Of these three, only reaction 5 is favorable. Note that reactions 5, 6, and 7 consume the nitrogen atoms needed in the initial reaction 2. If the gas mixture of nitrogen and oxygen is irradiated longer than 10^{-3} second, not only are N_2, N_2^+, N^+, O_2, O_2^+, O_2^-, O, O^+, and O^- present but also NO, NO^+, NO_2, NO_2^+, NO_2^-, N_2O, O_3, NO_3, and N_2O_5, and possibly other species unknown at present. To a degree depending upon concentrations, the behavior of this mixture, when it is subjected to additional irradiation, will differ from that of the original mixture of nitrogen and oxygen. Therefore, yields or G values for the NO_2 and N_2O formation are comparable only if the overall mechanism is known.

With experimental information that nitrogen can be fixed through ionizing radiation, but with a yield limited to a G value of about 6, attention has been drawn to other techniques in an effort to increase the yields, or to determine why the yields converge at a value of 6. One method under study in our laboratories is a labeling technique. Since the basic reaction is reaction 2, the number of nitrogen atoms produced in the irradiation process must be known. By capsule irradiation of a mixture of $^{14}N_2$ and $^{15}N_2$ (mass units 28 and 30), the number of nitrogen atoms produced can be determined from the resulting concentration of $^{14}N^{15}N$ (mass unit 29). If ion recombination and excitation both produced nitrogen atoms, we see, from Table 1, that 8.3 nitrogen atoms should be formed per 100 electron volts in extremely pure nitrogen.

TABLE 1. Distribution of 100 eV absorbed in N_2 among its primary formed species, upon irradiation with ionizing radiation. Electron volts required to form one pair in N_2 is 36.3; or $G_{(ion)} = 2.86$. $G_{(N)}$ primarily formed = 2.6 ± 0.2; and $\Sigma G_{(N)}$ max = 6.0 ± 0.5 as fixed nitrogen in the presence of oxygen.

		E_{ev}	G	$G \times E$
1.	$N_2^+ + e^-$	15.5	2.70	42.1 ± 1.0
2.	$N^+ + N + e^-$	30.0	0.14	4.2 ± 0.2
3.	$N_2^{+*} + e^-$	18.5	0.03	0.6 ± 0.2
4.	N_2: $C^3\Pi\mu$, $B^3\Pi$, $A^3\Sigma^+\mu$ and other excited species	~9.0	~3	27 ± 4
5.	N + N	13	1.3	17.0 ± 1.5
6.	e^-	3.0	2.86	8.6 ± 3.0
	Total			99.5 ± 7.0

*Excited state of N_2^+; †Reactions 2 and 6 have the kinetic energy of the N-atom added to the dissociation energy (estimated from potential energy curves).

Experiments with a mixture of pure $^{14}N_2$ and $^{15}N_2$ (oxygen concentration less than 1/10,000) gave a result of 11.5, in excellent agreement with the prediction. The introduction of oxygen reduced this G_N value to about 6. The overall process is evidently complex. In all probability, an ion reaction of the type:

$$N_2 + (O_2^+) \rightarrow \text{fixed nitrogen} \qquad (8)$$

contributes to the overall process. The details of this important reaction are not known at present, but there is considerable evidence that the nitrogen fixation process involves O_2^+ ions. Note that NO and NO_2 consume N-atoms and also, because of their low ionization potentials, transfer charge with O_2^+ very rapidly. The latter inhibits the fixation of nitrogen via the O_2 mechanism.

In Table 1 a small fraction of the energy absorbed by the nitrogen molecule is shown as emitted light. This fraction was determined from experimental irradiation of pure nitrogen with polonium-210 alpha particles, in the course of which light was emitted. The spectroscopic analysis of this light showed which electronic levels of the nitrogen molecules were primarily excited and, from the intensity of the band spectra, the relative abundance of the species. These abundances were related to the overall excitation and ionization processes. However, these experiments show that, although energy is absorbed by a nitrogen molecule, not all of it is available for the dissociation of nitrogen and subsequent chemical synthesis; some of it is lost by light emission.

To reduce the negative effects of reactions 4, 6, and 7, the NO and NO_2 should be removed as quickly as possible. One method is to neutralize these acidic oxides of nitrogen with a base to form a salt that does not decompose readily. We conducted experiments on the irradiation of air in which the walls of the capsule containing the air were coated with solid KOH. In control experiments, noncoated capsules containing air were irradiated under identical conditions. The results were as anticipated. In the coated vessels practically all the oxygen was consumed and no N_2O was observed. In the noncoated vessel, a substantial quantity of oxygen remained, together with a similar amount of N_2O. These experiments confirmed the belief that in the coated vessels there was no steady-state concentration of NO and NO_2 for the nitrogen atoms to react with, by way of reactions 4 and 6. Also, in the coated vessels the NO and NO_2 appeared as KNO_2 and KNO_3.

The radiation chemistry of fixed nitrogen shows that the nitrogen may be fixed with atmospheric oxygen to yield nitrogen dioxide. Technical feasibility studies have been made by Brookhaven nuclear engineers using cores of fissionable material in the form of fibers, ribbons, or powder. Chemonuclear plants have been designed to produce nitrogen dioxide in quantities of 180 to 900 metric tons per day. Dual-purpose reactors have been designed for producing fixed nitrogen and electrical power as well. Economic feasibility studies based on various reactor designs show that NO_2 could be produced by chemonuclear reactors at a cost not very much higher than that of existing processes. With the population explosion and associated agri-cultural needs, the requirements for fixed nitrogen will continue to increase.

There are two other systems for fixing nitrogen: the nitrogen-hydrogen system for producing ammonia and the nitrogen-sulfur system for producing N_4S_4. The irradiation of a nitrogen-hydrogen mixture produces ammonia with a G_{NH_3} of about 0.7. Even if this value could be raised to 7, the high cost of hydrogen would still make the process unattractive economically compared with the Haber Process for NH_3 synthesis.

When the sulfur compound N_4S_4 was produced through irradiation of nitrogen and sulfur vapor mixture in a nuclear reactor at 300° to 400°C, the yield was poor. This poor yield results from the decomposition of N_4S_4 at temperatures above 200°C. However, temperatures above 300°C are required to vaporize the sulfur sufficiently to produce the proper partial pressure in the nitrogen-sulfur mixture.

In general, charge transfer reactions are much faster than ion-neutral reactions. In systems such as those involved in nitrogen fixation, where species with low ionization potentials are present, ion reactions involving O_2 will be inhibited, and the amount of nitrogen fixed will be limited to that primarily formed via N-atoms.

P. HARTECK AND S. DONDES

NOBLE-GAS COMPOUNDS

In 1962, Neil Bartlett at the University of British Columbia observed and published the first valid account of a true chemical reaction involving a noble gas—the reaction between xenon and platinum hexafluoride. This work was followed rapidly by the synthesis of a variety of xenon compounds and the demonstration that krypton and radon compounds could also be formed. Some rather simple generalizations appear to govern the reactions of these gases. Chemical bond formation is possible only for the heavier noble gases, most strikingly xenon, and only with strongly electronegative ligands, most effectively fluorine.

The most stable isotope of radon has a half-life less than four days and there has been little research on radon compounds. A krypton fluoride may be formed with much greater difficulty than its xenon homologue and no compound has yet been demonstrated for argon.

Xenon exhibits every even oxidation number from two to eight and stable compounds have been isolated for each of these. The compounds of xenon may be conveniently described in three categories: simple xenon compounds, mostly fluorides and oxide fluorides; complex fluorides containing xenon; and the xenon oxides, xenates and perxenates; although a few compounds do not fit easily into this classification.

Simple Xenon Compounds. The fluorides and oxide fluorides of xenon are the most studied and best understood group of noble-gas compounds. Some of their properties are listed in Table 1 together with those for the other simple compounds of xenon. The two lower fluorides form sparkling, large, colorless crystals which can be sublimed easily, even at room temperature.

Xenon difluoride is a linear symmetrical molecule in the gas phase while xenon tetrafluoride has the square planar structure. The solids are both molecular crystals in which the structure of the individual molecule is maintained. An equimolar mixture of these two compounds forms an interesting mixed crystal with the empirical formula XeF_3 but in which the crystal is made up of a regular array of equal numbers of xenon difluoride and tetrafluoride molecules.

TABLE 1. SIMPLE COMPOUNDS OF XENON

Oxidation number	Formula	Comment[a]	T.P.[b]°C	B.P.[b] (S.P.[b])°C
2	XeF_2	A, S	129	114
2	XeO	C, U		
2	$XeCl_2$	C, U		
4	XeF_4	A, S	117	116
4	$XeOF_2$	B, U		
4	XeO_2	D, U		
6	XeF_6	A, S	49	76
6	$XeOF_4$	A, S	−46	102
6	XeO_2F_2	A, U	−31	
6	XeO_3	A, U		
8	XeO_2F_4	A, U		
8	XeO_3F_2	A, U		
8	XeO_4	A, U		

[a]Comments cover the situation as of December, 1971. A—Prepared as a reasonably pure material in substantial quantities. (At least some milligrams.) B—Probably prepared, never adequately purified or characterized. C—Not yet isolated in ponderable amounts. Identified by optical and mass spectrometry. D—Only observed as a positive ion in a mass spectrometer with no evidence for long-lived independent existence. S—Thermodynamically stable. U—Thermodynamically unstable.
[b]T.P. = triple point, B.P. = boiling point, S.P. = sublimation point.

Xenon hexafluoride is a white solid which melts to a yellow liquid. High volatility and a short liquid range are generally attributed to the highly symmetrical structure associated with hexafluorides. In this respect, since xenon hexafluoride is the least volatile of the known hexafluorides and has a longer liquid range than any other, the physical properties tend to support the suspected lack of symmetry predicted by a simple valence bond description of the xenon fluorides. Alone among the stable xenon compounds and those hexafluorides for which data are available, xenon hexafluoride is extensively ionized in solution in anhydrous hydrogen fluoride.

Xenon hexafluoride shows a rich chemistry as a fluoride ion donor and acceptor and a rather bright yellow color in solution at elevated temperatures and in the irradiated vapor.

The symmetry of xenon hexafluoride has become one of the most controversial questions in simple inorganic chemistry. The complex vibronic spectra and the electron diffraction studies as well as x-ray and neutron diffraction studies of the solid rule out the octahedral model used to describe all other hexafluorides, the zero dipole moment seems to rule out many of the distorted structures suggested, and in fact the spectra and many of the other observations can best be interpreted in terms of a mixture of isomers.

Subtle changes have been observed in the spectra as a function of both time and temperature and not yet adequately interpreted. Electronic isomers may well be involved, and a symmetrical ground state has not been ruled out.

All the simple xenon fluorides may be made didirectly from the elements or by the exposure of xenon to a wide variety of fluorinating reagents and conditions. These include heating with fluorine or halogen fluorides, and with the use of the electric discharge even such mild fluorinating agents as carbon tetrafluoride or silicon tetrafluoride are effective. Mixtures of xenon and fluorine react even at room temperature at moderately elevated pressure while a photochemical reaction has been demonstrated for ordinary daylight in glass vessels. Ultraviolet light through silica or sapphire windows or even more energetic ionizing radiation are more effective, however, for a low-temperature synthesis.

At low pressures and particularly with an excess of xenon, the predominant product is xenon difluoride. As the pressure and fluorine excess are increased, the tetrafluoride and eventually the hexafluoride predominate in the direct combination.

All of the xenon fluorides may serve as fluorinating agents (for example with hydrogen). The difluoride is the least reactive while the hexafluoride is a most effective fluorinating agent. Krypton difluoride is an even more potent fluorinating agent than xenon hexafluoride.

Both krypton tetrafluoride and xenon octafluoride were reported in early synthetic work and these reports shown to be erroneous. Both are too unstable or reactive to be observed in any mixture examined to date, certainly not at room temperature.

In general, one oxygen may be expected to replace two fluorines in combining with xenon. The resulting xenon-oxygen bonds are probably somewhat weaker than a single xenon-fluorine bond but the difference in thermodynamic stability between the xenon fluorides and the corresponding oxygen compounds is largely due to the difference in stability between the fluorine-fluorine bonds in the fluorine molecule and the much stronger oxygen-oxygen bonds in molecular oxygen. As a result, except for xenon monoxide which appears to be formed in trace amounts in an electrically excited xenon-oxygen mixture, the compounds containing oxygen can be made only by reacting the fluorides with oxygen-containing materials, particularly water.

The only thermodynamically stable oxygen containing xenon compound is $XeOF_4$, formed by the controlled partial hydrolysis of the hexafluoride. This is a colorless compound with a long liquid range. The molecule has a square pyramidal structure with the oxygen at the apex and the xenon in the plane of the four fluorides much as in the xenon tetrafluoride.

A number of electronegative ligands may also be substituted for one or more fluorine atoms. The compounds are usually very unstable, often explosively so. Among those characterized most completely are $XeF(CF_3CO_2)$, $Xe(CF_3CO_2)_2$, $XeF(OSO_2F)$, $XeF_2(OSO_2F)_2$, $XeF_4(OSO_2F)_2$ and $XeF(OTeF_5)$ and $Xe(OTeF_5)_2$ which are more stable than the others.

Complex Fluorides of Xenon. The first compound of xenon actually prepared was a complex with platinum hexafluoride whose formula and properties are not yet well established. Depending on conditions, the mixture of products actually formed will include not only some of the simple fluorides, but such complexes as $XeF^+(PtF_6)^-$ and $Xe(PtF_6)_2$.

Similar reactions have been noted between xenon and rhodium, ruthenium and plutonium hexafluorides, but not with tungsten, molybdenum, uranium and neptunium hexafluorides.

Complex compounds of xenon have also been prepared by combining the simpler xenon fluorides with other metallic fluorides. With XeF_2 a variety of complexes have been prepared with 1:1, 2:1, and 1:2 stoichiometry and mostly with pentafluorides. They have been described formally as fluoride ion transfer ionic compounds, e.g., $(XeF)^+$ (MF_6^-), $(Xe_2F_3)^+(MF_6)^-$, $(XeF)^+(M_2F_{11})^-$, but such concepts as fluorine bridging and molecular association vs. ionic transfer may be important in describing many of the complexes actually observed.

Xenon tetrafluoride is a very poor fluoride ion donor or complex former, but as we have noted the hexafluoride has a rich chemistry. Xenon hexafluoride complexes have been studied both with fluoride ion donors such as alkali metal fluorides (e.g., $CsXeF_7$ and Cs_2XeF_8) and fluoride ion acceptors such as antimony pentafluoride (e.g., $XeF_6 \cdot SbF_5$). $XeOF_4$ forms rather weaker complexes than XeF_6 for both donor and acceptor molecules.

The strong, though readily reversible association with alkali metal fluorides has been used in the preparation and purification of xenon hexafluoride, much as in the purification of uranium hexafluoride.

Xenon Oxides, Xenates and Perxenates. The hydrolysis reactions of the xenon fluorides lead to stable aqueous solutions of xenon compounds with interesting and potentially useful properties. Xenon difluoride dissolves in water without ionization and reacts rather slowly at room temperature to yield xenon, hydrogen fluoride and oxygen. No intermediate oxygen compound has been isolated although, like the higher fluorides, the difluoride shows a transient bright yellow color when hydrolyzed in basic solution or when an aqueous solution is made basic.

Xenon tetrafluoride and hexafluoride react more rapidly with water to yield a variety of products which depend somewhat on the composition of the aqueous solution, but not on whether tetrafluoride or hexafluoride are involved. In neutral or mildly acid solution the predominant final product is XeO_3 which is appreciably soluble and appears to be neither ionized nor significantly complexed in solution. While the dry powder is a sensitive and powerful explosive, a solution of XeO_3 is reasonably stable, can be prepared free of other reagents, and is a good starting material for studying the aqueous chemistry of xenon. Rather unstable alkali metal xenates can be prepared at low temperatures, but in aqueous alkaline solution xenon (VI) typically disproportionates and perxenates are formed. Most are rather insoluble and the direct alkaline hydrolysis of the higher xenon fluorides yields perxenate precipitates heavily contaminated with fluoride.

When a XeO_3 solution is made alkaline with sodium hydroxide the XeO_3 disproportionates slowly and about half the xenon is lost while the remainder precipitates as the perxenate. In the presence of ozone, all of the xenon is converted to the Xe(VIII) salt, $Na_4XeO_6 \cdot xH_2O$ where, depending on the final conditions, x ranges from 8 to 2.2 (or less on intensive drying).

Potassium perxenate is more soluble and the pure white compound can only be prepared from a very concentrated alkali solution. In ordinary potassium hydroxide solutions, an intensely yellow, explosively unstable compound precipitates containing both Xe(VI) and Xe(VIII) constituents.

All aqueous xenon solutions are good oxidizing media; for example, in dilute acid sodium perxenate quantitatively oxidizes manganese to permanganate, thus providing what is perhaps the best analytical procedure for small amounts of manganese in metal alloys.

The oxidizing electrode potentials using standard conventions are 2.36 v for the VIII/VI couple, 2.12 v for the VI/O couple and 2.64 v for the XeF_2/Xe couple in acid solution and 0.94 v for the $HXeO_6^{-3}/HXeO_4^-$ couple in alkaline solutions.

Although in the hydrolysis of xenon fluorides an intense yellow color is often observed, the species best guess probably relates this color to a charge transfer complex involving both xenon (VI) and xenon (VIII) in the same molecular complex, as suggested for the potassium compound. An unstable xenon compound of lower oxidation number has been postulated during the disproportionation of the xenon(VI) fluoride solution, but a polarographic reduction of an aqueous solution of XeO_3 shows a 6-electron transition with no evidence for a xenon species of lower oxidation number.

Uses of Noble-Gas Compounds. The scarcity of xenon and even krypton tends to prevent their use in any industrial process consuming these materials. The fluorides may, however, play a role as highly specialized fluorinating agents where the noble gas is recovered and recirculated. A number of interesting syntheses of organic fluorochemicals have been demonstrated.

The compounds may be valuable in systems taking advantage of the nuclear properties of particular isotopes of the noble gases where a nonvolatile or concentrated source is needed. The water stable perxenates and related compounds have some specialized uses in analytical procedures where high price and relative scarcity are less important. It is very likely, however, that the most important use of noble gas compounds will be reflected in the stimulation of informed speculation on the nature of chemical bond formation, much as the discovery of the inert gases themselves stimulated similar speculation some years ago.

HERBERT H. HYMAN

NOMOGRAPHY

A nomograph (often called nomogram, alignment chart, or line coordinate chart), in its simplest and most common form, is a chart in which a straight line intersects three scales in values that satisfy an equation or a given set of conditions. This definition suggests the three principal advantages of the nomograph: its essential simplicity, the rapidity with which it can be used, and its accuracy. As for essential simplicity, the complexity of the calculation may be reflected to some extent in the design and to a greater extent in the construction of the chart, but to only a minimum degree in its use. Interpolations are usually made along closely graduated scales rather than vaguely in a confusing network of curves; and unskilled individuals, with no knowledge whatever of the underlying theory, can perform involved computations by nomograph as readily as can the trained chemist, chemical engineer, or the nomographer himself.

Usually a calculation can be made nomographically in merely the time needed to draw a straight line, with a straight edge, between two points on a sheet of paper; instances of calculations that require *minutes* by slide rule and *seconds* by nomograph are common. Nomographs are usually direct-indicating, a feature that contributes significantly to their accuracy of use, since it obviates outside operations and "pointing off." Almost any desired degree of precision can be attained when proper attention is given to the design of the chart, its size, and the type of graduations employed.

Although nomography is based upon simple principles of plane geometry that have been known for centuries, it seems to have awaited a demand from the technical world for its birth and development. Among the many texts on the subject are some that employ the determinant approach, so dear to the formal mathematician; some that appeal to the engineer because of their familiar geometric presentation; some that are dull, abstract treatises, impeccable mathematically, but of little real utility; and some "cook books" that concentrate on mere sets of directions, which cannot always be adequate. However, several texts achieve a neat balance between simple theory, practical working directions, and sound applications.

Construction of a nomograph requires (1) a knowledge of the equation that connects the variables or of linear relationships, of the same type, between two of the variables for each value of the third, (2) knowledge of the ranges covered by these variables, (3) identification of the particular case with one of the standard type forms, and (4) determination of suitable moduli or unit representations to be used in laying off the desired scales. Most of the equations of practical value to chemists and engineers are of the following forms:

Type	Equation
A	$f(x) = F(y) + \phi(z)$
B	$f(x) = F(y) \cdot \phi(z)$
C	$f(x) = \psi(x) \cdot F(y) + \phi(z)$
D	$f(x) = F(y) \cdot \phi(z)$
E	$\phi(z) = a + bF(y)$

where, in standard notation, $f(x)$, $\psi(x)$, $F(y)$, and $\phi(z)$ represent any functions of the variables x, y, and z, and where a and b depend upon x, but are defined by tables or plots rather than by a mathematical expression.

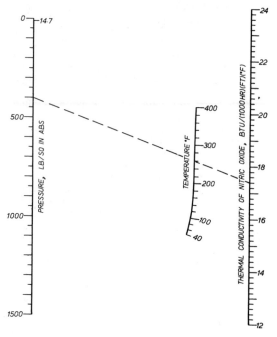

Fig. 1. Thermal conductivity of nitric oxide. Davis, D. S., *Brit. Chem. Eng.,* **15** (6) 805 (1970).

These forms are really closely related. Type **B** becomes the same as Type **A** when the equation is written in the logarithmic manner. When the equation for Type **D** is solved for $F(y)$, it is seen to be similar to that for Type **B**; when a and b are identified with $f(x)$ and $\psi(x)$, respectively, Types **C** and **E** are recognized as the same. The distinctions between the types are rather in the manner in which the charts are constructed. Type **A** usually calls for parallel, uniform scales; parallel, logarithmic scales characterize Type **B**; Types **C** and **E** frequently result in parallel y and z axes and in a curved x axis; parallel y and z axes connected with a diagonal x axis are characteristic of Type **D**. Numerous nomographs deal with four, five, or six variables instead of only three, so that combinations of these types are common.

In the accompanying typical nomograph the broken index line shows that the thermal conductivity of nitric oxide at 400 lb/sq in. absolute and 260°F is 17.4 Btu/(1000 hr) (ft) (°F).

For further essential theory and practical working directions to aid in designing nomographs, see "Nomography and Empirical Equations," Reinhold (1962). For collections of nomographs that pertain to chemistry and chemical engineering, see Peters, "Materials Data Nomographs" (Reinhold, 1965) and Davis, D. S., "Chemical Processing Nomographs," (Chem. Pub. Co., New York, 1969).

D. S. Davis

NONMETALS

Nonmetal is the term for a chemical element that does not exhibit metallic properties. An exact distinction between metals and nonmetals is not possible because a number of the chemical elements have properties that are somewhat intermediate between metallic and nonmetallic. Such elements are called "metalloid."* The most clearly metalloid elements are boron, siicon, germanium, arsenic, antimony, tellurium, polonium, and astatine. All the elements to the right of these in the Periodic Table, which include carbon, nitrogen, phosphorus, oxygen, sulfur, selenium, fluorine, chlorine, bromine, iodine, helium, neon, argon, krypton, xenon, and radon, are nonmetals, as also are hydrogen and the low-temperature form of tin. Of these, graphite, black phosphorus, and selenium have some electrical conductance which suggests metallic quality that might cause them to be called metalloid. But even if metalloid and nonmetallic elements are included together, as the nonmetals, there are still only 25 such elements. All the 80 other known elements are clearly metallic. Nevertheless the nonmetals have an importance in chemistry far out of proportion to their number.

When the atoms of an element possess more low-energy orbital vacancies than electrons in the outermost shell, the electrons are able to minimize repulsions by spreading out into orbitals that would otherwise have remained unoccupied. Thus there is brought about a delocalization of the valence electrons which produces those properties that are characteristic of the metallic state. In this state, the atoms are packed together much more closely than would be possible if they all had to be joined to their neighbors by two-electron bonds. Also there are no clearly definable bonds between neighbors. Rather, the delocalized valence electrons, through their attraction for the positive metal ions, serve as a cement to hold them together.

In contrast, the atom of a nonmetal has no low-energy orbital vacancies into which electrons can spread out. There are characteristically as many or more outer shell electrons than vacancies. Consequently, when nonmetal atoms combine with one another, any orbitals not directly involved in the bonding contain lone pair electrons rather than vacancies. The absence of alternative regions for electron occupancy restricts the bonding electrons to relatively localized regions between the nuclei which they hold together. The bonding in nonmetals being localized instead of delocalized is therefore covalent rather than metallic. Consequently the number of neighbors to which a given nonmetal atom can become bound chemically is limited by the number of vacancies in the outermost shell, for these determine the possible number of half-filled orbitals that are requisite for covalence. Under appropriate circumstances the lone pair electrons may then permit the already combined atoms to act as electron pair donors in the formation of coordination complexes.

In the pure nonmetallic element, there are two structural types of combination. First, bonding capacity may be fully realized in relatively small molecules, which then interact with one another only by relatively weak van der Waals forces. Sec-

ond, there may be bonding that is indefinitely extensive, so that either the element forms very large molecules, or all the atoms within a given chunk of the element are held together by a network of covalent bonds. Some examples of the first are H_2, N_2, O_2, P_4, S_8, F_2, Cl_2, Br_2, and I_2, (and of course the monatomic molecules of the helium group). Some examples of the second type are carbon in the form of diamond or graphite, silicon, germanium, red and black phosphorus, arsenic, and gray selenium.

In elements of the first type, the processes of mechanical breaking or thermal melting or vaporization require only the breaking of weak van der Waals forces among molecules. They are consequently relatively low melting, volatile, mechanically easily broken or crumbled substances. To melt, or deform, or break, or evaporate elements of the second type, however, relatively strong covalent bonds must be broken. This causes these nonmetals to tend to be hard, strong, high-melting, and relatively nonvolatile. In general, nonmetals, in striking contrast to metals, tend to be thermal and electrical insulators.

The efficiency of one outermost shell electron in blocking off another outermost shell electron from the nuclear charge is only about one-third. As a consequence, the successive filling of an outer shell with electrons, simultaneous with corresponding increase in the number of protons in the nucleus, results in a progressive increase in the effective nuclear charge felt at the periphery of an atom, by about two-thirds of a positive charge for each unit in atomic number. From left to right across the Periodic Table, therefore, the trend is toward smaller, more compact atoms of higher electronegativity. The nonmetals are all located toward the end of their respective periods and therefore have atoms that are relatively smaller and more electronegative than those of the metallic elements. In chemical combinations of nonmetals with metals, the bonds tend therefore to be highly polar and correspondingly strong. The nonmetal always acquires partial negative charge, leaving the metal with partial positive charge. In other words, the nonmetallic elements act as oxidizing agents toward metallic elements. When binary compounds of metal with nonmetal dissolve in water, the nonmetal tends to exist in solution as a negative ion isoelectronic with (having the same number of electrons as) the next "noble gas" element of higher atomic number. The metal exists in such a solution as a positive ion. Both type of ions are solvated by the water, the hydration energy being usually comparable to the energy needed to separate these ions from their crystal lattice. In general, the binary compounds of metal with nonmetal are called salts, except for those with oxygen which are called oxides. Such compounds are characteristically very stable, high melting, nonvolatile solids. However, as the positive oxidation state of the metal increases through increase in the number of nonmetal atoms per metal atom, the stability tends to decrease, the melting point to decrease, and the volatility to increase.

In chemical combination of different nonmetals with one another, the initially more electronegative nonmetal acquires partial negative charge, at the expense of the initially less electronegative nonmetal

---

*Editors' Note: The term "metalloid" is becoming obsolete in that the term nonmetal is more precise. The term semiconductor now is also applied to such nonmetal elements as silicon, germanium and selenium to more adequately define their properties as conductors.

which is left with partial positive charge. Under such conditions, the combining power of the initially more electronegative nonmetal remains just as predictable as ever, from the number of outermost shell electron vacancies. The combining power of the initially less electronegative nonmetal, however, frequently increases over the expected value. For example, sulfur can become SO_2 and SO_3 and SF_4 and SF_6; chlorine can become ClF_3 and ClF_5. It is only when the nonmetal atom is the more electronegative that its bonding capacity can be simply predicted from the number of available vacancies in the external octet. Since as a class, nonmetals are high in electronegativity, combinations of different nonmetals cannot have highly polar bonds because the competition between the two different nonmetals is too even. Therefore, such compounds tend to be less stable than the compounds of nonmetals with metals, lower melting, more volatile, and to retain the oxidizing power of the individual nonmetals so that the compounds themselves are effective oxidizing agents. In a very general sense, and especially in the form of their oxides, nonmetals tend to be acidic, metals basic. Most of the common acids are combinations of nonmetallic elements with one another. Most salts of complex anions, therefore, such as carbonates, nitrates, and sulfates, have anions that are combinations of nonmetals.

R. T. Sanderson

Cross-references: *Electronic Configurations; Metals.*

NONSTOICHIOMETRIC COMPOUNDS

The concept of stoichiometric compounds or compounds with integral combining ratios of the elements has played an important role in the development of the atomic theory of matter. The distinction between elements, compounds, and mixtures and the concept of pure substances was first clearly introduced by Boyle in the 17th century and further extended toward the end of the 18th century when Cavendish found simple integral combining volumes of hydrogen and oxygen gases in the reaction to form water and when Lavoisier established the law of conservation of mass and applied gravimetric analysis to determine the composition of water, carbon dioxide and other compounds. These observations were used by Dalton as the basis of his atomic hypothesis, but the experimental verification was not clear. Without the understanding that elemental gases could be diatomic, volumetric combining ratios of many gases did not work out properly. Although Avogadro recognized the existence of diatomic molecules in 1811, his hypothesis was not generally accepted until after 1860. Also the densities of some gases did not correspond to a simple formula and not until the work of Deville in 1864 was it recognized that the deviation was due to dissociation of the gases.

Thus the key step leading to the Dalton hypothesis was based primarily on the question of whether compounds contained fixed simple combining ratios of the elements, e.g., were stoichiometric. Berthellot vigorously attacked the assumption of constant combining ratios of reactants and was in fact the first to discover reversible reactions for which the yield depended upon the amount of excess of a reactant. Over the period 1799–1807 Berthellot's views were strongly refuted by Proust who did establish clearly, within the accuracy of measurements of those days, the Law of Definite Proportions or the existence of solid compounds that were stoichiometric. It was fortunate for the advance of science that stoichiometric compounds, now known as daltonides, were established. However, we now know that the concept of a daltonide is, in general, an approximation and that compounds must have a finite composition range and must therefore be berthollides, compounds with variable composition.

The basis for deviations from the laws of definite combining proportions and simple combining ratios can be illustrated by contrasting gaseous molecules and solid compounds. For the gaseous molecule NO, the smallest change that one can produce in this molecule is either the addition or removal of one atom. For example, one could add one nitrogen atom to produce N_2O or one oxygen atom to produce NO_2, both of which are compounds with properties drastically different from those of nitric oxide. As the atom is the smallest unit which can be added or removed from a molecule, it is clear that the smallest change that one can produce in a gaseous molecule will produce a large change in its properties. If, on the other hand, we consider solid sodium chloride, for example a single crystal weighing 10 g, its formula would be $Na_{10}{}^{23}Cl_{10}{}^{23}$. Here again, the smallest change that one can bring about is the addition or removal of one atom. If we remove one chlorine atom, the new formula becomes $Na_{10}{}^{23}Cl_{10}{}^{23}{}_{-1}$. When one is dealing with condensed phases with infinite lattices, the removal or addition of single atoms produces such minute changes in the formula that one can have essentially minute and gradual changes in the properties of the phase as one changes its composition. Thus it is not surprising that condensed phases have ranges of composition and are not restricted to definite compositions or simple combining proportions. In fact, one can prove thermodynamically that no condensed phase can be restricted to a unique composition except at a congruent melting or boiling point, a peritectic point, or at temperatures approaching absolute zero. Except for these singular points, condensed phases under all other equilibrium conditions must have appreciable ranges of composition or appreciable homogeneous ranges.

Consider the sodium chloride phase again. The phase in equilibrium with chlorine at 1 atm cannot be the same sodium chloride phase which is in equilibrium with sodium metal. It is impossible to have sodium metal and chlorine gas at 1 atm in equilibrium with one another. From available thermodynamic data one can calculate that the chlorine partial pressure at room temperature increases from less than 10^{-130} atm for the sodium chloride composition in equilibrium with sodium to a value of 1 atm for the sodium chloride composition in equilibrium with chlorine. Likewise the water to hydrogen ratio at equilibrium with alumina at 1000°K will lie between 10^{10} and 10^{-14} depending upon whether the composition corresponds to the oxygen-rich or the aluminum-rich end of the composition range.

The changes in properties of sodium chloride across its homogeneous range are quite substantial.

For example, its color changes from deep blue at the sodium-rich end to a white color at the chlorine-rich end (1). Likewise, its electrical and many other properties change quite markedly as one moves across the homogeneous range. For sodium chloride the range in composition is rather small, although it can be detected analytically without much difficulty. One finds that the homogeneous ranges will vary greatly from one phase to another. In some instances, a system of two components may correspond to a phase diagram with a complete homogeneous range from one component to the other. In other instances there may be a number of intermediate phases, some of which have moderately wide homogeneous ranges and others which have fairly narrow homogeneous ranges.

It is important to recognize that the compositions at the limits of these phases would normally have no particular significance of their own. These limiting compositions will be different at each temperature and are not determined alone by the properties of the phase in question but are equally well determined by the properties of the phases that exist in equilibrium with the phase in question. Thus, under some circumstances, when it is possible by proper seeding to have a choice of phases which might be saturating a given phase, then the limit of homogeneous range will be different under the two conditions. The reader may consult the literature (2) to become familiar with the plots of free energy versus moles of one component per fixed amount of the other component for a series of phases which illustrate that the limits of the homogeneous ranges occur when there is a common tangent to two free energy curves corresponding to equal partial molal free energies for the two phases. Such plots clearly show the influence of the saturating phases upon the homogeneous range of a given phase.

It is quite customary to designate a phase by a formula which corresponds to some simple combining ratio. For example, the sodium chloride homogeneous range is referred to as the NaCl phase region; the iron(II) oxide homogeneous range, or the wüstite phase, is referred to as the FeO phase region. The use of these names does not imply any commitment about the range of compositions nor that the simple composition chosen to name the phase has any special significance within this phase region. Thus in the instance of the iron(II) oxide phase region, the composition FeO does not exist in an equilibrium system (3). Nevertheless it is useful to refer to wüstite as the FeO phase region because of the fact that x-ray patterns for compositions in this phase region correspond to an ideal lattice of the sodium chloride type, which would have the composition FeO if it were an ideal lattice with no vacant lattice sites. Except at the absolute zero, it is not possible to have an ideal crystal for any substance at equilibrium and there will always be vacancies in the lattice. The vacancies in the cation sites need not be equal to those in the anion sites. In the instance of wüstite, any attempt to equalize the vacancies in anion and cation sites would result in a phase unstable with respect to disproportionation to metallic iron and an oxide phase richer in oxygen. Thus the terms FeO phase, TiO phase, NaCl phase, etc. do not imply that the phase has a fixed composition corresponding to the ideal crystal structure. No other commitment is made than the designation of the crystallographic arrangement. For those who are accustomed to dealing with intermetallic compounds, this is quite obvious. However, many people who have dealt with nonmetallic compounds under rather restricted conditions often have not been aware of the importance of recognizing the existence of homogeneous ranges for all compounds.

Just as the composition in the surface of a soap solution is not the same as in the bulk of the solution, the stoichiometric ratios will be different for the various crystal faces. The nonstoichiometry of surfaces has been demonstrated recently for alkali halides[4] and for the oxides Al_2O_3 and V_2O_5[5,6]. The nonstoichiometry of silica which has been treated in vacuum or in hydrogen is sufficient to be measured by titration of the surface with permanganate or dichromate solutions.[7]

There are many important respects in which not recognizing nonstoichiometry can greatly handicap the understanding of chemical processes. The designation of a vapor pressure for a phase causes difficulty unless one recognizes that the existence of a homogeneous range for a phase implies a range of vapor pressures as a function of composition. The designation of a single vapor pressure for a phase is normally meaningless unless one can characterize the composition of the phase and can use a method of vapor pressure determination that does not appreciably alter the composition; so that one can associate the vapor pressure with a specific composition within a homogeneous range. The phase will normally be saturated at either end of its range by some other phase. In a binary system at fixed temperature, the two condensed phases together with a gaseous phase constitute an invariant system with a unique characteristic vapor pressure. Thus if one does not measure the vapor pressure as a function of composition in the homogeneous range, one must normally insure that a saturating condensed phase is present to characterize the system.

In most systems this means that one does not wish to use a "pure" sample which would correspond to some composition in the single phase region, as the system would not be uniquely characterized and the vapor pressure measurement would have no significance. One of the two phases which could saturate the phase in question must be added in order to produce a completely characterized system before measuring the vapor pressure. In some instances a constant boiling composition exists within the homogeneous range and if one uses a method of vapor pressure determination which allows the system to approach this constant-boiling composition, there will be a third composition within the homogeneous range for which the vapor pressure measurement can be made without characterizing the composition of the phase. This is not general and one must verify that the constant-boiling condition is possible and insure that the method of determination allows one to reach this condition before attempting to make a measurement on a single phase system without carefully specifying the composition. In some systems the pressure may vary a millionfold across a homogeneous range, and it is quite meaningless to publish vapor pressures for a phase as is sometimes done unless one has made the system invariant by adding a saturating phase or by insuring constant boiling conditions. As the constant boiling composition as well as the limiting compositions will vary with temperature, one must be somewhat cautious in using the temperature coefficient of the vapor pressure to obtain heats of sublimation or vaporization as the heat may not correspond to any standard heat unless one has corrected for the changing composition with temperature.

The prediction of chemical behavior by means of thermodynamic data can be seriously in error if one does not take into account the homogeneous range of each phase. The thermodynamic data in the literature are normally given for one standard composition. The standard compositions may be hypothetical and need not correspond to any actual composition. In carrying out the calculations to predict actual behavior one must be prepared to calculate free energy changes for changes in composition from the standard state composition to the composition actually present at equilibrium. It is in this area that there is a most serious limitation to the use of thermodynamic data for prediction of chemical behavior, particularly in high temperature systems. This calculation of the free energy change in going from the standard composition to the actual equilibrium composition corresponds to knowing the activity coefficient as a function of composition within the homogeneous range. The few examples that have been studied indicate that the activity coefficients may vary in a complex manner in such homogeneous ranges. More data are needed to confidently predict activity coefficients in these homogeneous ranges and therefore reliably predict chemical equilibrium behavior.

Other properties can vary considerably across the homogeneous composition range of a compound. It has been noted earlier that color and vapor pressure change markedly. In addition, thermal shock properties, thermal conductivity, chemical reactivity, and, particularly, electrical conductivity can change very rapidly with changing concentrations of crystal defects due to changing composition. The two common types of crystal defects can be illustrated for a nonstoichiometric compound with excess metal. If the excess metal is at a cation site with an associated anion vacancy, the defect is termed a Schottky defect. Metal atoms in interstitial position in the crystal lattice are termed Frenkel defects. The names are those of the men who first proposed their existence.

The effect of nonstoichiometry upon electric conductivity is most marked for the poorly conducting semiconductors. If the crystal defects are due to atoms with excess electrons which can act as donors of electrons, the compound is designated as an n-type conductor. A deficiency of electrons or presence of acceptor sites results in a p-type conductor with conductivity due to migration of the positive defects. If the homogeneous range of a compound extends on both sides of stoichiometric, the conductivity will often change from n-type to p-type near the stoichiometric composition. Deviations from stoichiometry of as little as one part per million can make significant changes in conductivity of semiconductor materials.

Diffusion of reactants through a solid may depend strongly upon vacancies and the chemical reactivity of a solid can change drastically with composition. As both the equilibrium and kinetic behavior can vary strongly with composition, the degree of nonstoichiometry must be fixed to fix the chemical behavior of a compound. An excellent summary of recent work on the effect of nonstoichiometry upon the behavior of materials has recently been published[8].

LEO BREWER

References

(1) Rees, A. L. G., "Chemistry of the Defect Solid State," pp. 6–7, 57, 61, 77, John Wiley & Sons, Inc., 1954.
(2) Darken, L. S., and Gurry, R. W., "Physical Chemistry of Metals," ch. 13, McGraw-Hill Book Co., Inc., 1954.
(3) Darken, L. S., and Gurry, R. W., *J. Am. Chem. Soc.*, **67**, 1398–1412 (1945).
(4) Gallon, T. E., Higginbotham, M., Prutton, M., and Tokutaka, H., *Surface Science*, **21**, 224–232 (1970).
(5) French, T. M., and Somorjai, G. A., *J. Phys. Chem.*, **74**, 2489–95 (1970).
(6) Fiermans, L., and Vennik, J., *Surface Science*, **9**, 187–97 (1968).
(7) Ewles, J., and Youell, R. F., *Trans. Faraday Soc.*, **47**, 1060–64 (1951).
(8) Eyring, LeRoy, and O'Keeffe, Michael, Editors, "The Chemistry of Extended Defects in Non-Metallic Solids," North Holland Publ. Co., Amsterdam, 1970.

NORMALITY

Normality is one way of expressing the concentration of a solution that has definite concentration. In general, a solution that contains one equivalent weight of an ion, element, or compound dissolved with just enough water to make the volume become one liter is called a normal solution. Such a solution is designated by the letter N. A solution with concentration twice that of normal is $2N$; others are $12.0N$, $0.01N$, and so forth. Concentrated sulfuric acid, 95 per cent H_2SO_4, is described as 17.8 molar ($35.6N$).

The term equivalent refers to (1) the weight in grams of the element or ion capable of combining with or replacing 1.008 g of hydrogen (its atomic weight expressed in grams); (2) the weight of a metal replaced from a solution of a salt by an equivalent weight of a second metal; (3) the weight of a metal capable of combining with 8.00 g of oxygen (or 35.453 g of chlorine) with none of either element left over after the chemical change; (4) the weight of one compound that is changed into another compound that has known equivalence value.

Specifically, a normal solution of an acid contains 1.008 g of replaceable hydrogen ions (H^+) (19.024 g of hydronium ions, H_3O^+) per liter. For hydrochloric acid, formula weight 36.461 (35.453 + 1.008), 36.461 g of hydrogen chloride per liter makes a normal solution because 1.008 g of the 36.461 is hydrogen ions. Such a solution, therefore, contains 1.008 g of hydrogen ions per liter. For sulfuric acid (H_2SO_4, 2.016 + 32.064 + 63.998 = 98.078), formula weight 98.078, its formula weight divided by two, or 49.039 g, with enough water to make the volume of one liter contains 1.008 g of hydrogen ions per liter.

For soluble hydroxides, 39.998 g per liter makes a normal solution of sodium hydroxide (NaOH, 22.990 + 16.000 + 1.008 = 39.998); while a one-thousandth normal (0.001 N) solution of calcium hydroxide [$Ca(OH)_2$, 40.08 + 32.000 + 2.016 = 74.096] requires one-half of the formula weight of calcium hydroxide divided by one thousand, or 0.037048 g dissolved in one liter of solution.

The quantity 17.008 g of hydroxyl ions (OH^-) is equivalent to 1.008 g of hydrogen ions in reacting value

$$H^+ \quad + \quad OH^- \quad \rightarrow \quad H_2O$$
$$1.008 \text{ g} + 17.008 \text{ g} = 18.016 \text{ g}$$

It is not possible to make a normal solution of calcium hydroxide because such concentration exceeds the solubility limit of the compound. In other cases also, solubility limits concentration.

In order to make a normal solution of soluble salts, calculate the formula weight in grams of the salt, and divide this weight by the number of equivalents (x) involved. Then dissolve that weight of the salt in enough water to make one liter of solution. For sodium chloride (NaCl), $x = 1$; calcium chloride ($CaCl_2$), $x = 2$; aluminum chloride ($AlCl_3$), $x = 3$; sodium carbonate (Na_2CO_3) $x = 2$; potassium phosphate (K_3PO_4), $x = 3$; aluminum sulfate [$Al_2(SO_4)_3$], $x = 6$. In general

$$g = fw/x \times V \times N$$

where

g is the weight in grams of the substance under consideration,

fw is its formula weight in grams,

x is the number of equivalents per formula weight,

V is the volume of the solution in liters,

N is the normality of the solution.

Equal volumes of solutions of the same normality react completely, with none of either remaining unreacted if the compounds react. One liter of N hydrochloric acid (HCl) reacts with one liter of N sodium hydroxide (NaOH), calcium hydroxide [$Ca(OH)_2$], or silver nitrate ($AgNO_3$) solution. The same statement is true of a N solution of any other acid, provided such an acid reacts completely with the compounds listed.

In order to make a normal solution, the experimenter calculates the mass of the substance needed. The substance is weighed and quantitatively transferred to a volumetric flask. Distilled or deionized water is added, and the mixture stirred until the substance dissolves completely. The flask is then brought to the proper temperature, often 20°C, and water added to make the volume become the desired value by filling the flask to the mark etched on the neck of the flask. Sealed ampoules that contain a weighed amount of a given compound can be purchased from laboratory suppliers.

Such ampoules are used to make solutions of desired normality. The ampoule is placed in the flask, water added to the flask, the ampoule broken while it is under water, the contents of the ampoule dissolved and enough water added to make the desired volume.

Normality can express the concentration of solutions used in oxidation-reduction reactions. Potassium permanganate contains permanganate ions, a well-known oxidizer. It reacts as shown in the following partial equations:

$$MnO_4^- + 8H^+ + 5e^- \rightarrow Mn^{++} + 4H_2O$$

$$MnO_4^- + 2H_2O + 3e^- \rightarrow MnO_2 + 4OH^-$$

In the first, one-fifth mole of MnO_4^- ions involves one mole of electrons; in the second, one-third mole is involved.

A $1N$ solution of potassium permanganate ($KMnO_4$, formula weight 158) in the first reaction is 158 g $\div$ 5 or 31.6 g per liter; in the second 158 g $\div$ 3 or 52.7 g per liter for the same value, $1N$.

As shown, the concentration of normal solutions used in oxidation-reduction reactions depends on the reaction under consideration.

ELBERT C. WEAVER

NOXIOUS GASES

Noxious gases of current importance are those associated with occupational exposures and atmospheric pollution. The emphasis on the protection of the worker from the former has been greatly increased with the passage by Congress of the Occupational Safety and Health Act of 1970; that on the protection of the public from the latter has assumed immense proportions through increasing appreciation by Congress, industry and the public of the magnitude and complexity of the problem.

The toxic compounds discussed here are limited to those existing as gases at normal temperatures, and should not be confused with military "poisonous gases." These are mostly solids or liquids, not gases, the principal exceptions being chlorine and phosgene. Hydrogen cyanide is included because it is widely used as a gas. The considerable hazard attendant to the use of toxic liquids with high vapor pressure at normal temperatures and of toxic, low vapor pressure solids and liquids on being heated is not to be overlooked, even though such substances are not within the scope of this presentation. An extensive treatment of the noxious gases is available in Patty's "Industrial Hygiene and Toxicology."[1] The air pollution aspects of these gases are included in much detail in Stern's "Air Pollution."[2] The Hygienic Guides, Community Air Quality Guides, and Analytical Guides discussing industrial hygiene, air pollution, and analytical phases, respectively, of many of the noxious gases are published by the American Industrial Hygiene Association.[3] New Guides are continually being added.

"Threshold limit values" (TLV's) for occupational exposures, some 500 in number, are recommended by the American Conference of Governmental Industrial Hygienists.[4] A revised list is published annually by this unofficial organization. Excepting where the TLV is designated as a "ceiling" value, it refers to a time-weighted concentration for a 7 or 8-hour workday and a 40-hour workweek. The American National Standards Institute, Inc.[5] publishes "acceptable concentrations" of a number of gases, also for occupational exposure, and includes time-weighted averages, "ceiling" and "peak" values, together with information on physiology and analytical methods. Each publication is devoted to a single gas.

The TLV's of the ACGIH and the acceptable concentrations of ANSI have been given the force of law by their publication in the Federal Register, Vol. 36, pp. 10503–10506 (May 29, 1971) as Standards of the Occupational Safety and Health Administration of the Department of Labor.

Whereas the threshold limit values refer to occupational exposures of presumably healthy workers, community air quality standards are applied to the ambient (outside) air to which people, including those with cardiovascular, pulmonary, and other diseases, infants, and the aged, are continuously exposed. Several parameters are involved in air quality standards, namely, (1) concentration, (2) time over which results of sampling are averaged, and (3) frequency of occurrence of reaching the stated concentration. In the statement of regulations, the air quality standard is usually given as a basic standard of a concentration as an average over a stipulated period of time, often ranging from one hour to 24 hours, and also a permissible standard, defined as a concentration averaged over a briefer period of time and not to be exceeded more than a given percentage of time or days in a year. In some states, different air quality standards are promulgated for rural, urban, and industrial areas. In addition, emission standards may be set up to limit emissions from stacks and from motor vehicle exhaust pipes. The entire entity of air quality standards has been thoroughly discussed[2].

National Air Quality Standards for air pollution control to be met by July 1, 1975 have been stipulated by the Environmental Protection Agency for sulfur dioxide, particulate matter, carbon monoxide, photochemical oxidants, hydrocarbons, and nitrogen oxides.

In recent years colorimetric detector tubes have become commercially available for many of the noxious gases. A list of these and the suppliers is given in "Air Sampling Instrument Manual".[4]

Transportation of the noxious gases, usually in compressed or liquefied form, is regulated by the U.S. Department of Transportation under Public Law 86–710. The Hazardous Materials Regulations Board develops the format of the regulations, sets up the procedures for handling proposed additions or changes, and recommends adoption of those found to be satisfactory. The regulations for the transportation of poisonous gases are contained in a volume entitled "Code of Federal Regulations (CFR) Title 49-Transportation, Parts 1–199," available from the Superintendent of Documents, Government Printing Office, Washington, D.C. 20402. The regulations are also available in tariff form. Two of the most commonly used are: (1) T. C. George's Tariff No. 23 available from the Bureau of Explosives, Association of American Railroads, 1920 L Street, N.W., Washington, D.C. 20036: and (2) the ATA Motor Carriers Explosives and Dangerous Articles Tariff No. 14, available from the American Trucking Association, 1616 P Street, N.W., Washington, D.C. 20036.

WARREN A. COOK

References

1. "Industrial Hygiene and Toxicology," 2nd Ed., F. A. Patty, Editor, Vol. II. Toxicology, D. W. Fassett and D. D. Irish, Editors, Interscience Publishers, Div. of John Wiley & Sons, New York, 1962.
2. "Air Pollution," 2nd Ed., A. C. Stern, Editor, Academic Press, New York, 1968.
3. "Hygienic Guides," "Community Air Quality Guides," and "Analytical Guides," American Industrial Hygiene Association, Westmont, N.J., 08108.
4. "Threshold Limit Values of Airborne Contaminants" and "Air Sampling Instrument Manual," the former published annually and the latter in its 3rd edition, 1971, American Conference of Governmental Industrial Hygienists, Cincinnati, Ohio, 45201.
5. "Acceptable Concentrations of Selected Gases as American Standards" (Separate publication for each gas), American National Standards Institute, Inc., 1430 Broadway, New York, N.Y., 10018.
6. "Effects of Chronic Exposure to Low Levels of Carbon Monoxide on Human Health, Behavior, and Performance," a Report by the Committee on Effects of Atmospheric Contaminants on Human Health and Welfare, 1969. National Academy of Sciences, Washington, D.C.

NUCLEAR MAGNETIC RESONANCE

Nuclear magnetic resonance (NMR) is the term applied to a spectroscopy method and a physical phenomenon involving generally the absorption of applied radiofrequency (RF) energy by "magnetic" nuclei exposed to a strong uniform magnetic field. At the resonance condition, absorption occurs when the precessing magnetic nuclei "flip" their spin orientation, e.g., from $I = -\frac{1}{2}$ to $I = +\frac{1}{2}$. The magnetic properties of such nuclei are the spin number, I, and the magnetic moment, μ, both of which must be other than zero. The spinning and precessional motions are shown in Figure 1.

Historical Background. All atomic nuclei with finite spin numbers and magnetic moments, such as protons (hydrogen nuclei), fluorine, phosphorus, boron, nitrogen, carbon-13, and oxygen-17, have distinctive magnetic properties. Several experimental phenomena depend upon such properties. Hyperfine structure in optical atomic spectra arises from magnetic interactions. An NMR interaction was discovered during the course of some molecular-beam studies by Rabi, Stern, Gerlach, and Estermann in the 1930's. In 1945, Purcell, Torrey, Pound, Block, Hansen, and Packard detected NMR in bulk matter. Later it was shown that the molecular or chemical environment of the nucleus could produce characteristic shifts and fine structure in the NMR spectra. After this important discovery high resolution NMR rapidly evolved and was advanced by pioneers such as Gutowsky, Shoolery, Pople, Schneider, Bernstein, Tiers, Lauterbur, and McConnell.

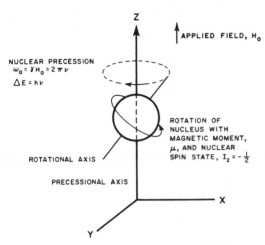

FIG. 1. Nuclear motion in magnetic field.

The shifts in the NMR spectra were named chemical shifts since they were found to depend primarily on the chemical or molecular environment. Because of such distinctive dependence on molecular architecture, NMR has become a fundamental structure tool for the research chemist. In fact, high resolution NMR analysis is the method of choice for a great many structure determinations. Used with vapor phase chromatography, infrared spectrophotometry, and mass spectroscopy, the NMR spectroscopy techniques have radically changed the practice of organic chemistry. Complex problems of organic isomer identification and structure elucidation are often solved in minutes. The usage of NMR in fields such as biochemistry, inorganic, physical, and analytical chemistry is increasing rapidly.

Both high resolution and wide-line NMR techniques are rapid, nondestructive, and apply to a wide variety of materials. High resolution techniques have been applied to certain solids, semisolids, liquids, solutions and gases. Wide-line NMR spectroscopy, using a magnetic field sweep method to present a derivative curve, is a powerful tool for many crystal structure studies and for routine quantitative analysis of liquids and solids.

Related Fields. A field somewhat related to NMR is electron paramagnetic resonance (EPR, or sometimes ESR) which is useful for detecting and studying free radicals, crystal F centers, transition elements, and other species containing unpaired electrons. EPR spectrometers are almost identical with wide-line NMR spectrometers except that microwave irradiation rather than RF is employed and the spin "flipping" of the electron instead of the nucleus is detected.

Another important related field, nuclear quadrupole resonance (NQR), is useful for many crystal structure and chemical bonding studies. The quadrupole properties exhibited by nuclei with spin equal to or greater than unity allow studies of electrical field gradients about such nuclei, covalent ionic character of bonds, crystallographic space groupings, and molecular structure. Other somewhat related magnetic resonance fields and pheno-

mena are spin echo NMR, ferromagnetic resonance, the Mössbauer effect, and dynamic polarization, e.g., electron-nuclear double resonance (ENDOR) and the Overhauser effect.

High Resolution Apparatus. In typical high resolution NMR analyses, liquid samples (*ca.* 0.5 ml) at room temperature in precision 5-mm glass NMR tubes are irradiated at a constant RF such as 100 Mc or 60 Mc for protons, 56.4 Mc for fluorine, and 24.3 Mc for phosphorus. Homogeneous magnetic fields at high field strengths such as 13 kilogauss at 60 Mc RF or 23 kilogauss at 100 Mc RF are used for proton work. A diagram of a simple single-coil NMR spectrometer is shown in Figure 2.

Chemical Shift. An NMR peak may be located with respect to a reference peak given by an internal or external standard, e.g., tetramethylsilane (TMS) for proton spectra and fluorotrichloromethane for fluorine spectra. The peak location or chemical shift, δ, for a given nucleus or group of equivalent nuclei can be defined as the difference in screening constants, σ, for the given nucleus and a reference nucleus: $\delta = \sigma - \sigma_r$.

Since the applied field H_0 is modified at the nucleus by swirling electrons which generate local magnetic fields, the local field H_{local}, is changed by an amount proportional to the screening ability of the local electronic environment: $H_{local} = H_0(1 - \sigma)$. Therefore, the chemical shift can be related to the magnetic field difference generated by the local fields:

$$\delta = \frac{H - H_r}{H_r}$$

where H is the actual magnetic field of the nucleus

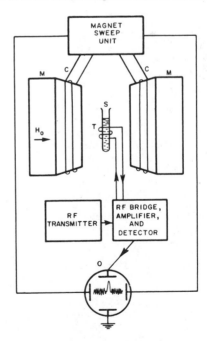

FIG. 2. Schematic diagram of single coil NMR spectrometer. M = magnet, C = sweep coils, S = NMR sample, T = transmitter-receiver coil, O = oscilloscope displaying NMR signal.

being observed and H_r is the field of a nucleus chosen for its reference signal.

Since frequency and field are directly related by $v = kH$ (where $k = \gamma/2\pi$ and γ is the gyromagnetic ratio), a simple substitution gives a more useful expression:

$$\delta = \frac{v - v_r}{v_r}$$

The actual value of the absorption frequency, v, analogous to an infrared absorption frequency, would be diagnostic for a given structural group or type of nucleus. In practice it is more accurate and convenient to measure chemical shifts and use these dimensionless values for chemical group identification. Many functional-group charts listing chemical shifts have been compiled. Since the fixed radiofrequency v_0 is virtually the same as v_r, one can obtain a very simple expression for chemical shift:

$$\delta = \frac{v - v_r}{v_0} = \frac{\text{frequency difference (cps)}}{\text{transmitter frequency (Mc)}}$$

Thus a frequency difference, $v = v_r$, of 60.0 cps for two compounds observed at a fixed transmitter frequency of 60.0 Mc would be designated as a chemical shift of 1.00 ppm (part per million).

Several chemical shift scales, conventions and standards are still being used but the major ones for proton spectra can be designated as δ (using tetramethylsilane, TMS, with downfield positive), δ TMS (with downfield negative(, and τ which can be defined as follows: $\tau = 10 - \delta$.

Coupling Constant. Another important quantity describing a feature of NMR spectra is the coupling constant, J, which represents the difference in allowed frequency transitions for a given nucleus interacting with another nucleus. The coupling constant can be measured or mathematically extracted from the fine structure lines within peaks. The fine structure or spin-spin splitting patterns give valuable information about the number, type, and proximity of neighboring magnetic nuclei.

Multiplets such as doublets, triplets (relative intensities 1:2:1) quartets (1:3:3:1), and quintets (1:4:6:4:1), are indicative of simple first-order spin-spin interactions of nuclei with discrete magnetic fields of neighboring nuclei. First-order coupling constants are independent of applied magnetic field. The separation or spacing of the two lines in a doublet is reported in cps, e.g., $J = 15.4$ cps. The center position (center of gravity) of the doublet is reported as a chemical shift, e..g, $\delta = 5.2$ ppm downfield from TMS.

A first-order approximation rule for nuclei with a spin of $\frac{1}{2}$ (H^1, F^{19}, P^{31}, Sn^{115}, Si^{29}, etc.) is that multiplets with $n + 1$ lines will be produced by n neighboring nuclei, if the latter are positionally equivalent or mode equivalent by rapid rotation. For example, *three* equivalent protons in a neighboring rotating CH_3 can produce *four* lines in a peak being observed as one from a CH_2 group. A compound such as CH_3OCHO would show a quartet and a doublet, an AX_3 type spectrum. The CHO proton "sees" three equivalent protons in CH_3 and is split into four lines, a 1:3:3:1 quartet. The CH_3 protons "see" one CHO proton and are split

into two lines, a doublet. Such first-order splittings arise when the chemical shift (in cps) between two groups (i.e., CHO and CH_3O) is large compared with the coupling constant: $|\delta_{H_A} - \delta_{H_X}| \gg |J_{AX}|$.

Spectral Analysis. A wealth of structural information can often be obtained from NMR spectra by completely analyzing the spectral features such as the chemical shifts, coupling constants, multiplicities, line and peak intensities, line widths, and changes in any of the above due to temperature effects. Computer programs for a variety of spin systems are widely used to facilitate spectral analysis, especially when second-order interactions are present. Also isotopic substitution and double resonance (homonuclear or heteronuclear spin decoupling) often facilitate analysis of spectra.

Fortunately, however, in the majority of cases, interpretations are readily made by comparing (fingerprinting) spectra with commercially available spectra compilations or by simply inspecting the first-order features of the spectra. Many new compounds have been discovered and many complex structures have been defined by a comparatively simple analysis of the first-order features of the NMR spectra.

Wide-Line Applications. When the magnetic field sweep is modulated a certain way, the first derivative curve is obtained from an absorption mode signal. Very broad signals arising from dipolar interactions in solids can be more readily detected and studied by wide-line techniques. Studies of crystal structure, particularly for hydrogen location, have proved useful. Hydrogen atoms which are impossible to study directly by x-ray diffraction methods can often be defined by wide-line studies of the absorption peak second moment (mean square width). The second moment is related to the sum of the inverse sixth powers of all appropriate internuclear distances in the crystal. The motion of molecules in the solid state, e.g., hexad rotation of benzene, can be studied by analyzing the second moment temperature variations. Studies of chain mobility, molecular diffusion, high pressure effects, transition temperatures, and changes in crystallinity have been carried out. Percentage crystallinity and first-order transition temperatures have been determined for many polymers. In many laboratories, quantitative analyses for percentage hydrogen and fluorine are routinely carried out. Studies of the spinlattice relaxation time, T_1, by standard procedures or spin echo methods, are proving useful for studies of molecular motion.

High Resolution Applications. It should be emphasized that the most important application of high resolution NMR is the elucidation of molecular structure, particularly of organic compounds. Molecular structure elucidation in the fields of biochemistry and inorganic chemistry is increasing. There is wide usage in chemical fields such as petroleum, pharmaceuticals, natural products, and fluoro-organics.

Since many NMR spectra are markedly time and temperature dependent, many chemical reactions and motion phenomena can be investigated. Chemical exchange, solvolysis, isomerization, tautomerization, hydrogen bonding, hindered rotation, ionization, and dissociation can often be studied by steady state high resolution techniques. Using intense RF pulses (spin echo) a closer study of many motion phenomena is possible.

The NMR absorption process for a given species may have a "shutter speed" of 0.01 second or even several orders faster. An average nuclear shielding or screening is obtained for systems undergoing faster chemical and physical processes. In many cases molecules move, rotate, exchange, etc., slower than once per 0.01 second and therefore give rise to separate peaks. However, peaks which are broadened and shifted by motional effects at room temperature can often be resolved at lower temperatures. Potential energy barriers and various thermodynamic quantities can be evaluated from temperature NMR studies. Studies of temperature, pressure, and solvent effects on NMR spectra have given valuable information about physical and chemical interactions. Fundamental studies of molecular diffusion, thermal relaxation, gas compressibility, basicity, and solvent bonding have been reported.

Since chemical shift, coupling constant, and line width data can be dependent in very specific ways to subtle features of the molecular environment, fundamental molecular effects can be studied. Atomic and group electronegativities have been estimated. Inductive, "resonance," steric, hyperconjugative, paramagnetic, and diamagnetic effects have been uniquely studied by high resolution techniques. Fundamental studies of electron densities and chemical bonding parameters have been reported. Diamagnetic susceptibilities have been determined by several NMR techniques.

In many laboratories, NMR elemental hydrogen analyses, which are rapid, accurate, and nondestructive, are routinely performed. The composition of mixtures, determination of percentage purity, and estimation of chemical yield can often be obtained. The determination of solvation numbers using a quantitative NMR technique has been reported. Future quantitative NMR studies combined with specific qualitative NMR studies may well solve analytical problems which are difficult or impossible by other methods.

BURCH B. STEWART

Cross-references: *Magnetochemistry; Stereochemistry; Resonance; Bonding; Crystals and Crystallization.*

Quantitative Analysis

While the use of nuclear magnetic resonance for qualitative analysis and structure determinations has been widespread in recent years, the application of this technique to quantitative analysis has been relatively slow in gaining acceptance. In part, this is due to the fact that early instruments did not possess the requisite magnetic field stability or the convenient electronic integration device now found on a number of commercially available instruments. Both broad band and high resolution nuclear magnetic resonance spectrometers have been applied in quantitative analysis. However, since the bulk of available instruments are of the latter type, this discussion shall consider only high resolution applications.

The basis of quantitative analysis by nuclear magnetic resonance stems from the fact that the intensity of the NMR absorption for a given nucleus is proportional to the number of such nuclei in the sample. This statement will only be true if certain precautions are taken, for the intensity of the signal (as given by the area under the peak) may be affected by field inhomogeneity, saturation effects at high r.f. power or slow field sweep rates, the operation of the instrument so that there is a contribution to the intensity from the dispersion mode, as well as certain other factors. While under certain conditions peak heights may be used to measure intensities, this method is often in error and measurements of peak areas are to be preferred since they are subject to less error. Most recent commercial instruments come with an electronic integration device which makes the problem of area determination a relatively rapid and accurate process.

The analysis of complex mixtures of organic compounds depends on being able to assign the observed absorption bands to the protons on the various substances in the mixture. Generally, it is helpful if one has the spectra of each pure component. However, since the positions of absorption for many type structures are now well known, this is not always necessary. In any event, successful analysis will depend on having one or more peaks clearly resolved from the others to serve as a reference.

Many other types of analyses by NMR spectroscopy have now been reported. Among these might be listed the determination of normal water in heavy water, the amount of deuterium substitution in a number of organic molecules, the composition of tautomeric equilibria of keto-enol forms, the moisture content of many products of interest in the agriculture, food, and paper industries, the chain sequence in copolymers, and the tacticity of homopolymers such as poly(methyl methacrylate), poly(vinyl acetate), poly (vinyl methyl ether), poly (vinyl trifluoroacetate), and poly(vinyl alcohol).

An example of the latter type of analysis is given by considering the NMR spectrum of a chloroform solution of poly(methyl methacrylate) which shows three bands at δ 0.91, 1.05, and 1.22 which have been attributed to the chain methyl groups on the polymer backbone in syndiotactic, heterotactic, and isotactic configurations, respectively. Determination of the areas of these three peaks allows a rapid determination of the contributions of each of these configurations to the tacticity of the polymer.

For the organic chemists, one of the most useful aspects of quantitative analysis by NMR is the determination of the per cent of hydrogen in an organic molecule. The method is nondestructive, and the sample can usually be recovered for other types of determinations with little trouble. Furthermore, once the operator has acquired skill in operation a complete analysis can be carried out in twenty minutes or less with a result which compares favorably with those obtained by the standard combustion technique. Sample sizes varying from 5 mg to 100 mg are usual for this type of analysis. The limiting factor is the width over which the spectrum is spread and the effect of the signal to noise ratio in obtaining a valid integration of the total hydrogen absorption of the sample.

Two techniques for obtaining the per cent hydrogen seem to have evolved. In the first of these, a known volume of a standard solution is integrated. A similar volume containing a known weight of the unknown is then integrated. It is necessary that both tubes be of the same diameter so that the same volume of material will occupy the sensing part of the probe. If this is not so, then a correction factor must be added in the proportionality constant of the following equation.

$$\text{Total integral area} = kHCV$$

Where k is a proportionality constant determined from the standard solution, H is the per cent hydrogen, C is the weight of sample per unit volume, and V is the effective volume of solution in the probe. The combined terms kV are determined from the standard immediately before or after (or both) from the standard. The unknown is then run separately, and the area used to calculate the per cent hydrogen. Obviously, this method requires magnetic field stability throughout the course of both determinations.

The second method obviates this difficulty by placing an internal standard within the same tube as the unknown. The two are then integrated on the same sweep. The internal standard must be chosen so that its absorption does not conflict with that of the unknown. The subsequent recovery of the unknown may be complicated by the choice of standard. This method also requires two weighings as opposed to one for the former method. In this case, the above equation may be arranged in the form

$$\%H = \%H_{ref}(A/A_{ref})(W_{ref}/W)$$

where A is the integral area and W is the weight per cent for standard and unknown as indicated.

W. B. SMITH

Cross-reference: *Analytical Chemistry*

References

Smith, W. B., *J. Chem. Ed.*, **41**, 97(1964).
Becker, E. D., "High Resolution NMR," Academic Press, New York, N.Y.

NUCLEAR REACTORS

A nuclear reactor is an assembly of special materials designed to sustain and control a neutron chain reaction with a fissionable fuel, such as plutonium-239, uranium-235, or uranium-233. The fission reaction releases neutrons and it is only through a delicate balance that the neutron population can be managed to operate a reactor. The energy released by fission is several million-fold greater per pound of fuel than that derived from the combustion of fossil fuels, and represents a potential reservoir of energy far exceeding that for fossil fuels. Since the fission reaction and the neutron flux produce radioactive fission products and activation products, redundant and diverse measures are provided to ensure the safety of the operations. With due regard for environmental effects, increasing use will be made of nuclear reactors for electrical power generation.

Significant components of the nuclear reactor unit include the fuel, moderator, reflector, control, coolant, shielding, and engineered safety features. Natural uranium contains only 0.7% U-235, and enrichments, achieved in the gaseous diffusion plants, of several per cent are necessary for the current nuclear power reactors, using water as the coolant. The form of the fuel is as uranium oxide and is contained in Zircaloy tubes. Generally, the fuel rods are about one-half inch in diameter with an active length of about 12 feet. The rods are closely spaced and water is forced to flow in the intervening spaces parallel to the rods. Tens of thousands of rods are used for the reactor core. Both U-238 and Th-232 are termed fertile fuels for upon absorption of a neutron, they undergo two successive beta decays and form the fissionable fuels Pu-239 and U-233, respectively. Since more than 2 neutrons are produced per fission and one is needed to continue the neutron chain reaction, it is possible to produce more fissionable fuel than is consumed. The term breeding had been used to designate the production of the same fissionable isotope as is consumed (for example, the use of Pu-239 as the fissionable fuel and producing Pu-239 from a U-238 blanket), but currently the concept is used in a general way to indicate successful breeding or conversion. In the United States, the major program to expand the fissionable fuel supply is through the use of liquid metal-cooled, fast breeder reactors, LMFBR. Although fast reactors and breeding have been already demonstrated, the current goal is to have an economically competitive LMFBR by the 1980's.

Neutrons released by the fission reaction have very large kinetic energies, generally expressed as 2 million electron volts (Mev) for the average. The kinetic energy of the neutron is dissipated as heat by successive scattering collisions of the neutron with the nuclides present, and this dissipation is called slowing down or neutron moderation. Materials which are effective in the slowing down process are called moderators. Good moderators provide high densities of light weight nuclides, but without much loss of absorption of the neutrons. Examples include heavy water, graphite, beryllium, and water. The low end of the kinetic energy of the neutrons is represented by an equilibrium in the energy exchange with the moderator and is called the slow or thermal range. Even in the thermal range, a distribution of neutron energy is found and can be approximated by a Maxwell-Boltzmann distribution. Under ideal conditions (no losses by leakage and absorption), the energy is 1/40 electron volt at room temperature, corresponding to the most probable neutron speed of 2200 m/sec. The probabilities for reactions of neutrons with nuclides often are tabulated at 2200 m/sec and corrections may be applied to relate to the temperature conditions of the reactor. Energy groups are also used to characterize the reactions of the neutrons at the higher energies. If no moderator is used, the energy spectrum of the neutrons is weighted at the high or fast energy range, and the reactor is called a fast reactor. At the low kinetic energy range, the probability that a neutron will produce fission with a fissionable fuel nuclide is greatly enhanced, and thus the concentration of the fissionable fuel can be reduced and still sustain the neutron chain reaction. Fast reactors thus require relatively high enrichments compared to that for thermal reactors.

Classification of reactors may be done in many ways, but usually distinguishes whether the neutron chain reaction is sustained primarily via fast or thermal neutrons, whether the fuel and moderator are physically separate such being heterogeneous or homogeneous, and may note the coolant. Most current nuclear power reactors are thermal, heterogeneous reactors using water as both the moderator and the coolant. In the BWR, boiling water reactor, the water coolant is allowed to boil to form steam at about 1000 psia; whereas in the PWR, pressurized water reactor, the pressure is raised to about 2000 psia and no steam is formed. The PWR coolant exchanges its heat in a steam generator which produces steam in a secondary loop. Just as with fossil power plants, the energy of the fuel is transferred to steam which in turn is passed through turbines driving the electrical generators. The steam existing from the turbines is condensed and recycled. Another reactor type used for producing power is the HTGR, a high-temperature gas-cooled reactor, utilizing a uranium-thorium fuel cycle, graphite as the moderator, and helium as the coolant to produce in a secondary loop superheated steam at a temperature of 1000°F. Power output of the largest operating BWR (in 1970) was 2527 Mwt (megawatts thermal) and 809 Mwe (megawatts electrical). Power ratings for both BWR and PWR under construction range up to 3400 Mwt and 1100 Mwe. Whereas nuclear power contributed less than 2% of the electrical generating capacity (in 1970), substantial growth will result from the more than 100 reactors, planned, ordered, and under construction.

In addition to the use of nuclear reactors for electrical power generation, many hundreds of reactors have been built for purposes which include research, development of prototypes, teaching, plutonium production, portable power sources, materials testing, and transuranic element productions. Numerous combinations of both homogeneous and heterogeneous reactor fuel-moderator configurations have been tested, and a variety of coolants has been used. Use of nuclear reactors for propulsion in naval submarines and ships is extensive; however, use of nuclear reactors in aircraft has not been adopted.

The loss of neutrons by leakage from the reactor core is minimized by the use of moderating materials which serve to scatter neutrons back into the core. Materials used to accomplish this function are called reflectors, and good moderators are also good reflectors. A reactor without a reflector is termed a bare reactor. With a reflector, the reactor core can be reduced in size, and thus while the over-all size of the core plus reflector is not appreciably affected, a substantial saving in fuel materials is achieved.

The sequence or cycle of events which on an average describes the behavior of a neutron from one fission to the next is called a generation. The corresponding neutron lifetime ranges in milliseconds for a natural uranium, graphite-moderated reactor, tens of microseconds for a water-cooled

reactor, and fractions of a microsecond for fast reactors. For thermal reactors, a convenient term characterizing the ratio of the number of thermal neutrons in one generation to that in the preceding generation is the effective neutron multiplication constant, k_{eff}. This constant is composed of factors which take into account the number of neutrons released per fission, $\nu = 2.4$ for U-235 and 2.9 for Pu-239; the number of thermal fissions per thermal neutron absorbed, f (thermal utilization factor); the correction for fast fission of U-238, ϵ (fast fission effect); the fraction of fast neutrons which escape capture during slowing down, p (resonance escape probability); the fraction of fast neutrons which do not leak from the reactor, $\bar{P}$; and the fraction of thermal neutrons which do not leak from the reactor $(1 + L^2B^2)^{-1}$, where L is the thermal diffusion length and B^2 is the buckling factor. The term k represents the thermal neutron multiplication constant in an infinite reactor (with no leakage terms).

$$k_{eff} = \nu \epsilon f p \, \frac{\bar{P}}{1 + L^2B^2} = k \, \frac{\bar{P}}{1 + L^2B^2}.$$

For $k_{eff} > 1$ the neutron density rises exponentially and the assembly is supercritical; for $k_{eff} < 1$, the neutron density decreases and the assembly is subcritical; and for $k_{eff} = 1$, the neutron chain reaction is just self-sustaining and the reactor is critical. Since the generation time is so short, only small excesses of $(k_{eff} = 1)$ are permitted, and generally this value is much less than β, the fraction of the neutrons produced per fission which are derived from the fission products. These neutrons are called delayed neutrons in that they are produced with a half-life of the radioactive fission product precursor. It is the presence of the delayed neutrons that augment the time available for control action that makes operating reactors possible.

The only critical homogeneous reactor combination possible with natural uranium is with heavy water as a moderator. The value of k, however, can be increased by lumping the fuel. Although the thermal utilization factor is decreased, given geometrical arrangements can be used to increase the resonance escape probability p to more than compensate the decrease in f. In this fashion, critical heterogeneous reactor combinations are achieved with a graphite moderator and natural uranium.

Nuclear reactors are generally controlled by neutron absorbers which may be moved in and out of the reactor or added to the coolant. Typical controls involve the use of boron. The neutron absorbers serve to regulate the value for k_{eff}. To shut a reactor down, it is necessary to make $k_{eff} < 1$, whereas to raise the operating level, k_{eff} is made just greater than 1. The safety of a reactor is assured by provisions for more rods called safety rods and other means that can be used to reduce k_{eff}.

For a given type and arrangement of fuel and moderator, the minimum size of the reactor core required to achieve $k_{eff} = 1$ is called the critical size. With a thermal, heterogeneous, natural uranium, graphite-moderated reactor, the critical size is of the order of magnitude of a 20-ft cube. Use of heavy water as a moderator, and of enriched fuel markedly reduce the critical size. Enriched, homogeneous, bare reactors may have critical sizes smaller than a 2-ft diameter sphere. One of the smallest cores reported was that for the fast reactor, EBR-1, which was the size of a football.

In addition to fuel, moderator, reflector, coolant and control designs, another important and essential feature of nuclear reactors is the shielding requirements. Most reactors are surrounded by massive concrete structures designed to reduce the gamma and neutron radiations to safe levels. Intense sources of gamma radiation are generated by the fissioning of the fuel and by the radioactive decay of the fission fragments. The fission products tend to build up to equilibrium concentrations so that even after a reactor has been shut down, considerable precautions need to be taken to protect not only the personnel from the radiation hazards, but also sufficient cooling must be provided for the fuel elements.

Nuclear reactor accidents cannot result in a nuclear explosion, but power excursions and release of radioactive fission products are possible. To prevent and reduce such occurrences and to minimize any resulting effects, a unique system of controls and regulations has been adopted by the federal government in the licensing of the construction and operation of the reactors. Intensive quality assurance programs for the design, construction, operation, and maintenance of all significant components and systems are being implemented. Research and development programs seek to confirm and establish improved safety features, including those involved with providing assured means to provide cooling of the fuel in the unlikely event that the main cooling system should be disrupted. The engineered safety features also include features which establish confinement of the radioactivity to the primary system, and further impose requirements of a secondary containment system with severe leak tightness provisions. Both preoperational and postoperational environmental monitoring programs are required to ensure that any release of radioactive wastes to the environment remain as insignificant health hazards. Heat releases (essentially the difference between ratings of the Mwt and Mwe) to the environment are also areas of concern and depending upon the location of the plant may require special treatment (cooling towers, cooling basins) as well as environmental monitoring.

H. S. ISBIN

NUCLEONICS

Nucleonics is the name proposed by Z. Jeffries of the Manhattan District in 1944 to describe the general field of nuclear science and technology, and is so used here.

Upon observing that in passing through matter alpha particles were scattered through larger angles than the then current concepts of the atom would predict, the English physicist Rutherford suggested that the atom actually consisted of a small, heavy, positively charged nucleus surrounded by negative charges of the same magnitude. Further observations and theoretical refinements have led to the now generally accepted concept that the

nucleus also contains elementary particles, i.e., neutrons, which are electrically neutral and have a mass of 1.00897, and protons, which are positively charged and have a mass of 1.00812. Stated in simple form, then, the atom consists of a positively charged nucleus, containing neutrons and protons, surrounded by a number of negatively charged electrons sufficient to provide electrical neutrality. Complete understanding of the forces which hold the neutrons and protons together has not yet been achieved.

In 1927 Aston found that experimentally measured isotopic weights differed slightly from whole numbers. From this he was led to the concept of the *packing fraction*, defined as the algebraic difference between the isotopic weight and the mass number divided by the mass number. Although the theoretical significance of the packing fraction is difficult to assess, it does lead to some interesting conclusions with respect to nuclear stability. A negative packing fraction derives from a situation where the isotopic weight is less than the mass number, implying that in the formation of the nucleus from its constituent particles some mass is converted into energy. Since an equivalent amount of energy would be necessary to break up the nucleus into its constituent particles again, a negative packing fraction suggests a high order of nuclear stability. By the same reasoning, a positive packing fraction indicates nuclear instability. As can be seen from Fig. 1, stable elements with mass numbers above about 175 and below about 25 have positive packing fractions. It is interesting to note that the packing fractions of both hydrogen and uranium are positive.

Actually, a comparison of the isotopic weight with the mass number as is done in determining the packing fraction is somewhat artificial. A rigorous determination of the mass-energy interconversion in the formation of an atom would seem to require a calculation of the difference between the sum of the masses of the constituent particles of the atom and the experimentally measured isotopic weight. The value of the mass difference thus obtained is the *mass defect*. The energy equivalent of this mass difference as derived from the Einstein equation yields a measure of the binding energy of the nucleus. Division of the binding energy of a nucleus by the number of nucleons (the

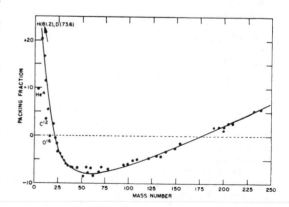

FIG. 1. Packing fraction curve.

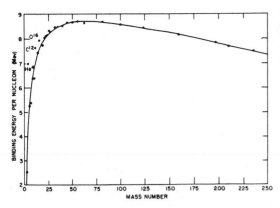

FIG. 2. Bending energy per nucleon.

total number of protons and neutrons) therein yields the binding energy per nucleon. As indicated in Fig. 2, in stable isotopes the binding energy per nucleon decreases with increasing mass number, a fact which is important in nuclear fission. Secondly, the binding energy per nucleon derived in the manner described above is an average value, whereas each additional nucleon added to the nucleus has a binding energy less than those which preceded it. Thus, the most recently added nucleons are bound less tightly than those already present.

Additional considerations regarding nuclear stability may be gleaned from a consideration of the odd or even nature of the numbers of protons and neutrons in the nucleus. According to the Pauli exclusion principle, no two extranuclear electrons having an identical set of quantum numbers can occupy the same electron energy state. The application of this principle to the nucleus leads to conclusions which at least are not at variance with observations of nuclear stability. Thus, one may infer that no two nucleons possessing an identical set of quantum numbers can occupy the same nuclear energy state. It would appear, then, that both protons and neutrons which differ only in their angular momenta or spins may exist in a nuclear state. The exclusion principle requires, therefore, that only protons having opposite spins can exist in the same state. The same consideration applies to neutrons. Accordingly, two protons and two neutrons might occupy the same nuclear energy state provided the nucleons in each pair have opposite spins. Such two proton-two neutron groupings are termed "closed shells," and by virtue of their proton-neutron interaction impart exceptional stability to nuclei which are made up of them. The nuclear forces in closed shells are said to be "saturated," by which it is meant that the nucleons therein interact strongly with each other, but weakly with those in other states. Since like particles tend to complete an energy state by pairing of opposite spins, two neutrons of opposite spin or a single neutron or proton also might exist in a particular energy state.

Any of the above conditions may be achieved when the nucleus contains an even number of both protons and neutrons, or an even number of one and an odd number of the other. Since there is an excess of neutrons over protons for all but the lowest atomic number elements, in the odd-odd situation there is a deficiency of protons necessary to complete the two proton-two neutron quartets. It might be expected that these could be provided by the production of protons via beta decay. As a matter of fact there exist only four stable nuclei of odd-odd composition, whereas there are 108 such nuclei in the even-odd form and 162 in the even-even series. It will be seen that the order of stability, and presumably the binding energy per nucleon, from greatest to smallest, seems to be even-even, even-odd, odd-odd.

Although the existence of binding energies holding the nucleus together has been demonstrated, the problem of defining the nature of these forces presents itself. Clearly, repulsive electrostatic forces must exist between protons. These are "long range" in effect. To achieve nuclear stability then, compensating attractive forces also must exist. It has been concluded that "short range" attractive forces exist between protons, neutrons, and protons and neutrons. The (p-n) attractive forces are considered to be of the greatest magnitude while the (n-n) and (p-p) forces are of lesser intensity, with the latter decreased by virtue of electrostatic repulsion. When the number of protons in a nucleus is greater than twenty, it is found that the ratio of neutrons to protons exceeds unity. The additional short range attractive forces provided by the excess neutrons, therefore, may be considered as compensating for the long range electrostatic repulsive forces between the protons. Nevertheless, when the number of protons exceeds about 50, the short range forces are insufficient to counteract the electrostatic forces completely, with the result that the binding energy per each additional nucleon decreases.

Unfortunately, the nature of the short range attractive forces between neuclons remains essentially unresolved. An interpretation of them has been presented by Heisenberg, however, in terms of wave-mechanical exchange forces. Thus, if the basic difference between the proton and neutron in a system composed of these two particles is considered to be that the former is electrically charged while the latter is not, then the transfer of the electric charge from the proton to the neutron results in an exchange of individual identity but not a change in the system. That is to say, the system still is composed of a proton and neutron, despite the fact that the particles have exchanged their identities. Since the system itself has the same composition, it must possess the same energy after the exchange as it did before. One of the principles of wave mechanics is that if a system may be represented by two states, each of which has the same energy, then the actual state of the system is a result of the combination, i.e., resonance, of the two separate states and is more stable than either. In the proton-neutron system under discussion, the energy difference between the "combined" state and the individual states may be considered as the "exchange energy" or "attractive force" between the particles. In an extension of Heisenberg's proposal Yukawa postulated that the exchange energy is carried by a particle which has been given the name *meson*. Particles having the properties attributed by Yukawa to mesons have been identified in cosmic rays.

Detailed discussions of fission and fusion phenomena will be found in the articles so entitled.

JOSEPH E. MACHUREK

Cross-References: *Atoms, Protons, Fission (Nuclear), Fusion, Plutonium, Uranium, Tritium.*

NUCLEIC ACIDS

Nucleic acids are macromolecules as large as, or larger than, the largest protein molecules, with molecular weights ranging into the hundreds of millions. They are as important as, or more important than, proteins to the working of living tissue.

Samples were first isolated in 1869 by Friedrich Miescher from nuclei in pus and in fish sperm and he therefore called the substance "nuclein." Some twenty years later, its markedly acid properties caused a change in name to "nucleic acid."

Complete hydrolysis of nucleic acid yields a mixture of breakdown products in the proportion of one mole of phosphate, one mole of a sugar, and one mole of a mixture of heterocyclic bases. Gentler hydrolysis shows that the larger units of nucleic acid structure are condensation products of a heterocyclic base and a sugar-phosphate, this condensation product being called a "nucleotide."

In 1911, Levene showed the sugar components of the nucleotides to be five-carbon monosaccharides, in some cases ribose and in others deoxyribose, the two differing in that one of the hydroxyl groups in ribose is replaced by a hydrogen atom in deoxyribose. When it became apparent that individual nucleic acid molecules never contained both sugars among their hydrolysis products, but always either one or the other, two classes of nucleic acids were recognized. One was ribosenucleic acid, usually abbreviated RNA, and the other was deoxyribosenucleic acid, usually abbreviated DNA.

The heterocyclic bases found in major quantities in nucleic acids are five in number. The purines, adenine and guanine, Fig. 1, are found in both DNA and RNA, as is the pyrimidine, cytosine. The pyrimidine, uracil, is found in RNA only while thymine (which differs from uracil by the presence of an additional methyl group) is found in DNA only. See Fig. 2. Minor quantities of other bases have been found in certain nucleic acids.

The nucleotides in RNA, then, are four in number: adenylic acid, guanylic acid, uridylic acid and cytidylic acid (often referred to as A, G, U, and C respectively). Removal of the phosphate group (Fig. 3) leaves "nucleosides" which, in RNA, may be referred to as adenosine, guanosine, uridine, and cytidine. The nucleotides and nucleosides in DNA are similarly named, with "deoxy" prefixed where it is necessary to distinguish it from the RNA component. Thus we can speak of thymidylic acid, but deoxyadenylic acid.

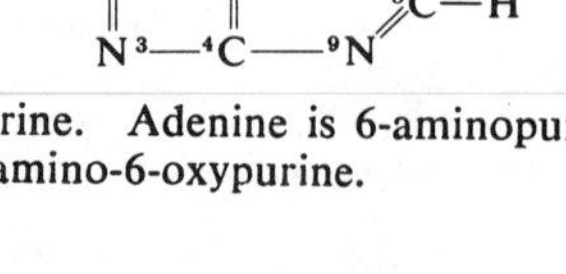

FIG. 1. Purine. Adenine is 6-aminopurine, and guanine is 2-amino-6-oxypurine.

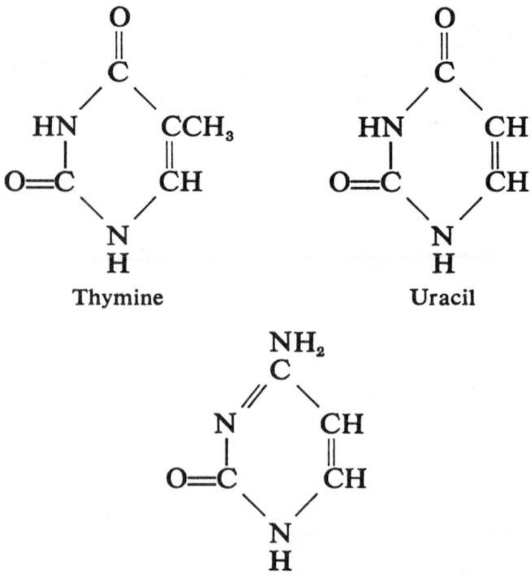

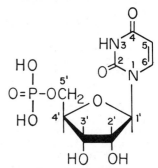

FIG. 2. Pyrimidines, heterocyclic bases found in nucleic acids. Ring members are numbered clockwise from the lower N.

Uridine 5'–phosphate (a nucleotide)

FIG. 3

The individual nucleotides are linked by way of the phosphate group, which is condensed at the carbon-5 position of one nucleotide and at the carbon-3 position of the neighboring one. Originally, hydrolysis methods were such as to reduce the nucleotide chain to small fragments and it was thought that the nucleic acid was a tetranucleotide containing one each of the four nucleotides. As new and ever-gentler methods for the isolation of the nucleic acids were employed, ideas as to the size of the molecule had to be escalated upward. By the 1940s, the macromolecular nature of nucleic acid was fully accepted.

Cell-staining techniques were used to localize the position of nucleic acids within cells. Robert Feulgen used reduced fuchsin to stain DNA but not RNA. · Pyronin will stain RNA but not DNA. Both classes of nucleic acid were found to be present, universally, within cells, but there were significant differences in their distribution. DNA was present primarily in the nucleus and, still more specifically, in the chromosomes, though some has been located in minor quantities in such cytoplasmic particulates as the chloroplasts of plants. On the other hand, RNA is present primarily in the cytoplasm, though it also occurs in the nucleolus, and even to a minor extent within the chromosomes.

Even in bacteria and in blue-green algae, which have no separate nucleus, there is DNA present, distributed in such a way as to mark the nuclear material that is distributed throughout the cell. As for those subcellular entities, the viruses, they, too, contain nucleic acids. Here a nucleic acid molecule is surrounded by a protein shell. In the virus, there may be only a single nucleic acid molecule, either DNA or RNA, and it is only in the viruses that one type or the other may be completely absent. There is no clear demonstration of any virus, however, in which *both* are missing.

The universal occurrence of nucleic acids and the close association of DNA, in particular, with chromosomes, could not help but raise the suspicion that these were somehow important in the economy of the cell. The fact that DNA seemed much simpler than protein in structure (being made up of four different nucleotides as opposed to the *twenty* different amino acids that contributed to protein structure) and that, at first, DNA seemed a relatively small molecule, obscured strong indications that nucleic acid importance was not only great, but overwhelming.

Thus, in certain fish sperm—which carry all the genetic information of the species, but which must pack that information into an unusually tiny volume—the proteins contained are extraordinarily simple in structure compared to those in ordinary cells of the same species, but the sperm DNA seems identical with ordinary cellular DNA. The sperm, it would seem, may economize on proteins, but never on nucleic acid. More startling still, Avery and coworkers found, in 1944, that certain cellular characteristics could be changed by a substance that was nucleic acid *entirely,* with no protein at all (see **Genetic Code.**)

Such information was conclusive, yet it could not be accepted as long as the notion of the complexity of protein structure and the simplicity of nucleic acid structure prevailed. Between 1950 and 1953, Chargaff studied the structure of DNA by paper chromatography and found that the simplicity of nucleic acids was a myth after all. The four bases present in the nucleic acid molecule were *not* present in equal quantities. The exact proportions differed, in some cases widely, from species to species. The base-proportions in the DNA from a particular species, however, were characteristic regardless of the tissue from which it was drawn and did not vary with age, development, nutritional state or any other physiological or environmental factor.

One could then visualize the DNA molecule as a long strand of nucleotide units built up by way of polyphosphate backbone, along which purine or pyrimidine deoxyriboside side-chains were spaced. These side-chains differed by the nature of the purine or pyrimidine they contained and Chargaff's work showed that their distribution along the chain was random in nature; that is, any distribution was possible.

If this is so, and if every different arrangement of bases produces what is, in essence, a different molecule, then the complexity of DNA is ample for any function demanded of it. Thus, imagine a DNA molecule of a thousand nucleotides (rather a small one) in which each nucleotide may be any of the four. The total number of different DNA molecules possible would then be 4^{1000}, or 10^{600}.

It would seem, then, that each species of organism might well have some specific DNA molecule characteristic of itself. Since the total number of species of plants and animals that have ever existed on earth can be numbered in the mere millions, we can see that the allotment of a different DNA molecule to each does not make the barest dent in the total number of DNA molecules possible. Indeed, it is perfectly easy to visualize a general molecule for the species—with a given length and a characteristic distribution of bases in the broad view—and with minor variations that could serve to allot a different molecule to every individual creature who has ever lived. The DNA potential would still be scarcely touched.

This view is made more reasonable by the care with which DNA molecules are multiplied and distributed in the process of cell division. Each daughter cell receives a replica of the one contained in the mother cell. Each infant receives (by way of sperm and ovum) replicas of those contained in either parent. It is almost as though accurate copies of a blueprint were being made, then passed from hand to hand.

But how is the enormously complex DNA molecule replicated? How is the copy made without unacceptable error? Chargaff had found certain regularities among the base distributions. In every sample, the total number of adenine molecules (with a double-ring structure) was about equal to the total number of thymine molecules (with a single-ring structure.) Similarly, the total number of guanine molecules (double-ring) was about equal to the total number of cytosine molecules (single-ring.)

Taking this into account, Watson and Crick studied the x-ray diffraction data which gave information concerning the general structure of the DNA molecule. These data could be interpreted in terms of a helical structure of the long molecule (such helical structures had earlier been successfully applied to protein molecules by Pauling.)

It occurred to Watson and Crick that the DNA structure was a *double*-helix; with two polyphosphate backbones twisting along the outside of the overall structure and the nucleotide side-chains jutting inward toward the cylindrical core. The purine or pyrimidine bases from one backbone approached those from the other and were held together by hydrogen bonds. To give meaning to Chargaff's data, it was assumed that an adenine would always have to be adjacent to a thymine group; a guanine to a cytosine. Models of such a double helix, taking into account all available information concerning bond-lengths and bond-angles, proved completely satisfactory.

By this "Watson-Crick structure," announced in 1953, the problem of DNA replication was solved. Each strand of the double-helix was a template for the other. At the crucial moment in cell division, the double helix would separate into individual strands. Each would then dictate the structure of the new strand formed next to it from the free-nucleotide stores of the cell. Every A along the chain would dictate the formation of a neighboring

T, and vice versa. Every G would dictate a neighboring C and vice versa, and where one double helix had existed originally, two (identical with each other and with the progenitor) would exist subsequently.

ISAAC ASIMOV

NUCLEOPROTEIN

Nucleoprotein is a generic term applied to various association products between nucleic acids or polynucleotides and proteins. These compounds were originally called *nucleins* by their discoverer Miescher (1844–1895). In later years, various more specific designations were introduced, such as *ribonucleoprotein* and *deoxyribonucleoprotein,* designating whether the nucleic acid is primarily of the ribose (RNA) or desoxyribose (DNA) types, and *nucleohistone* or *nucleoprotamine,* indicating the nature of the protein in the complex.

There is a growing tendency not to designate as nucleoproteins those proteins that are covalently linked to small nucleotides or nucleotide-containing structures, such as certain ones found in cell walls or enzymes which require a bound nucleoside phosphate for activity. In general, it is understood that nucleoproteins contain true RNA or DNA. From the conception of a nucleoprotein as a protein with a nucleic acid as a prosthetic group, the emphasis has been shifting toward considering it to be nucleic acid with associated protein. Nucleoproteins are, therefore, not now regarded as molecular substances in the usual sense but as highly specific molecular aggregates. Most early preparations of nucleoproteins were artifacts, in that they resulted from fragmentation, partial dissociation and unnatural reassociation of the components originally present *in vivo.* Nevertheless, most if not all of the nucleic acids of cells occur in association with proteins.

Covalent bonds appear to be lacking between the nucleic acid and the protein moieties in a true nucleoprotein complex. The two components are held together mainly by electrostatic bonds between the negative charges of the phosphates in the nucleic acid and the positive charges of basic proteins, as well as by hydrogen bonds between the nucleic acid bases and suitable structures in the protein. Nevertheless, the association between protein and nucleic acid is often very strong indeed, and rather drastic measures were often formerly employed to dissociate the complex. Subsequent procedures are based on the recognized necessity for breaking the electrostatic or the hydrogen bonds, or preferably both. The former are disrupted by strong salts and the latter by detergents, such as sodium duodecyl sulfate (SDS) or strong urea or guanidinium chloride solutions. In addition, it is desirable to separate the protein physically from the nucleic acid following dissociation, which is often accomplished by extraction with liquid phenol in the presence of SDS or certain salts. The protein is partitioned into the phenol while the more polar nucleic acid remains in the aqueous phase. Another method involves precipitation of the nucleic acid with strong salt (for example $2M$ LiCl) under conditions maintaining the protein in solution ($4M$ urea). Great care has to be taken in these separations if it is desired to recover both products, since the isolated nucleic acids are exceedingly fragile and the released proteins have a tendency to aggregate irreversibly.

Nucleoproteins are usually identified by their high absorption of light at 260 mμ, which is close to one hundred times greater than for proteins alone. As a result, the spectrum of a nucleoprotein resembles that of nucleic acid rather than that of protein. Deoxyribonucleoproteins are distinguished from ribo ones by color reactions characteristic of one or the other sugar. Deoxyribonucleoproteins are also usually much larger than the corresponding RNA structures and can normally be sedimented out of suspensions in low centrifugal fields, leaving ribonucleoprotein in the supernatant. The latter particles are in turn larger than most free protein and nucleic acid molecules and can be sedimented by high-speed centrifugation. Size separations can also be made effectively by filtration through columns of certain polymers capable of acting as "molecular sieves." A most useful general method for separating deoxy- and ribonucleoproteins from each other and from free RNA, DNA and proteins is based on the significant differences in density of these substances. During prolonged centrifugation, starting with a mixture of cell extract and a salt of high density such as cesium chloride, the salt forms a density gradient and each macromolecular component "bands" at a level where its density is matched. Free RNA has the highest density and free protein the lowest, with intermediate densities for DNA and the various molecular aggregates between the three.

Deoxyribonucleoproteins are usually found in the nucleus of a cell. The complex between DNA, various proteins and special kinds of RNA is called chromatin and it is the substance of which chromosomes, the carriers of heredity in higher organisms, are made. (See **Genetic Code**). The primary genetic information of chromosomes resides in the DNA, whereas the RNA and proteins, besides lending stability to the structure and protecting the DNA from enzymatic degradation, are thought to be regulators of the amount of information which can be transcribed into messenger RNA and eventually translated into proteins. The DNA is very large, double-stranded and helical. This helical structure has a minor and a major "groove" in which the proteins and the RNA are bound. A large part of the proteins in chromatin are very basic and are called *histones,* of which there are many different kinds. The function of histones in chromatin is not well understood, but, contrary to earlier beliefs, they do not appear to be concerned with the specific, gene-by-gene regulation of transcription. Among the nonhistone proteins there are the *repressors,* each of which specifically binds with a regulatory region of the DNA called the *operator,* thereby preventing the expression of a group of genes called a *cistron.*

The main ribonucleoprotein structures in the cell are the *ribosomes,* which are the site of protein synthesis. Active ribosomes (70S) are formed of two subunits, 30S and 50S, each of which is a complicated coil of one long RNA molecule bound to many proteins. The absence of covalent bonds between the RNA and the proteins is demonstrated by the fact that the ribosome can be taken apart to its constituents under mild conditions and then put together again in an active form. Both RNAs and most of the proteins are needed for protein synthesis. Furthermore, the exact folding of the nucleoprotein filaments is also essential, since certain changes in conformation lead to inactivation.

Most viruses (see Viruses), except for a few large ones, are formed exclusively of nucleic acid and protein, but only a certain class of small plant and bacterial viruses are nucleoproteins in the strict sense, i.e., possess a large nucleic acid molecule tightly and directly associated with proteins. In the larger bacterial and ainmal viruses the protein forms an outer protective coat or shell containing the coiled nucleic acid molecule inside. In every case, however, the nucleic acid, which can be double- or single-stranded DNA as well as double- and single-stranded RNA, is the chief functional unit.

Nucleoproteins are thus very important structures in the cell and information about one or the other kind is still pouring in at a tremendously rapid rate. The interested reader is referred to "Molecular Biology of the Gene" by James D. Watson, Second Edition, W. A. Benjamin, Inc., New York, 1970.

I. D. RAACKE

NUTRITION (HUMAN)

Nutrition is the physiological effect of foods on living things, including man, animals, and plants. This article concerns only the first of these. Specifically, nutritional studies can be described as determining the qualitative and quantitative needs of humans for the growth and maintenance of tissues, for the regulation of life processes, and for the production of energy. These needs are met by the basic constituents of food and are known as nutrients.

Between 50 and 60 nutrients are essential to mankind. These fall into six general classes: proteins, carbohydrates, fats, minerals, vitamins, and water. Most foods contain more than one kind of nutrient. Because the human body operates at peak efficiency only when all the required nutrients are ingested in optimum amounts, man must consume a wide variety of foods. If an essential nutrient is omitted from his diet over a period of time, man develops deficiency diseases, such as goiter, beriberi, or scurvy. The symptoms which develop depend on which nutrients are lacking and the severity of the deficiency.

Typical phases of nutrition deal with (1) the chemical determination of the nutrients in foods; (2) assays of growth and development of laboratory animals on varying levels of nutrient intake; (3) dietetics, the science of feeding people to attain and maintain good health through all stages of life (infancy through maturity and old age, as well as during pregnancy and lactation); (4) therapeutic dietetics, the assistance given medical science in the rehabilitation through special diets of bodies weakened by surgery, disease, or related conditions; (5) physiological studies of the effect of various nutrients on the body during conditions of stress, such as survival, aerospace travel, etc.; and (6) food habit surveys which contribute to our knowledge of psychological and sociological factors in the selection of foods. In its broader aspects nutrition encompasses and overlaps many other sciences, including biochemistry, genetics, microbiology, physiology, psychology, and sociology.

Proteins. Proteins are the basic material in every cell of the body, whether it be bone, blood, enzyme, or muscle. The daily diet must provide protein for growth, maintenance and repair of tissues, and many other body processes. During digestion, food proteins are broken down into simpler compounds called amino acids, 18 commonly occurring in foods. Combined into proteins, amino acids form compounds of high molecular weight: the protein gliadin, 27,500; zein, 50,000; and hemoglobin, 63,000. Food proteins contain 16% nitrogen on the average, the only significant source in the diet. They also contain carbon, hydrogen, and oxygen, and sometimes sulfur and iodine.

Certain amino acids which must be supplied in the diet—leucine, lysine, methionine, phenylalanine, threonine, tryptophan, and valine—are spoken of as essential. A protein food containing all of these is said to be complete, or to have high biological value. Such proteins are absorbed slowly by the body, with more uniform release and more efficient utilization of the essential amino acids.

Other amino acids are essential also, but can be synthesized by the body, if the necessary components are ingested. Many of the nonessential amino acids are readily interconverted in the body. However, all the essential amino acids must be available simultaneously. If the feeding of one or more is delayed, utilization of the others is markedly decreased.

Proteins of animal origin—meats, poultry, fish, eggs, milk, and cheese—contain 15 to 25% protein of high biological value. Plant proteins—cereals, legumes, nuts, and oil seeds—contain 10% protein of lower biological value. Plant proteins are frequently low in content of lysine, methionine, and/or tryptophan. They are spoken of as limiting, as they must be supplemented by dietary additions for utilization by the body (e.g., the addition of milk to cereals). Small quantities of protein are also found in fruits and vegetables; these do not contribute significantly to the protein requirements.

Protein intake varies considerably with environment and economic levels, sometimes being as high as 500 g per day. A level of 65 g per day has been established for 70 kg males, to meet dietary needs. Within rather wide limits man tends to adjust his rate of protein catabolism to his dietary intake. He is thus able to maintain a nitrogen balance (an equilibrium between intake and output of nitrogen) over a fairly wide range of protein intake. Proteins eaten in excess of body needs serve only as a source of energy (4 calories per g). They are deaminated, with their nitrogen eliminated chiefly as urea, and the organic residue used for energy in various chemical syntheses. The body uses protein most efficiently when there is a sufficient intake of carbohydrates and fats to meet all its energy requirements. This effect is known as the protein-sparing action of fats and carbohydrates, and has real economic importance when the relative cost of the three types of food is considered.

Carbohydrates. Carbohydrates, which supply energy for all man's activities and for maintenance of body warmth, make up about half of the usual caloric intake. Carbohydrates yield 4 calories per g. Thus a 70 kg man on a 2800 calorie diet would eat 350 g of digestible carbohydrate to supply 1400 calories. The chief sources of carbohydrates are cereals, 55%; refined sugars, 25%; vegetables, 10%; fruits, 5%; and dairy products, 5%. Most carbohydrates from natural sources also supply substantial amounts of minerals, vitamins, and small amounts of protein. Refined carbohydrates such as white sugar, polished rice, and cornstarch are almost pure carbohydrate, supplying only energy.

Carbohydrates are also important in the diet for other reasons. They greatly enhance the palatability of the diet, giving us a wide range of food textures and flavors ranging from bland to sweet. Some carbohydrates encourage growth of specific microorganisms in the intestine, thereby furthering digestive processes. Others pass through the body unchanged (celluloses) and aid in the elimination of waste materials. The body needs carbohydrates in order to use fat efficiently. When the supply is inadequate, fatty acids cannot be completely oxidized and ketosis—a condition of acidosis—results.

Fats. Fats and oils are the most concentrated source of food energy, providing 9 calories per g, and making up as high as 40 to 50% of the total intake in calories. However, it is now recommended by many research workers that the fat intake be restricted to a level lower than 20 to 25%. For a 70 kg man, intake should be not over 700 calories as fat, or about 65 g fat. Foods high in fat include cooking fats; salad oils; butter and margarine; the solid fat of bacon, salt pork, or other meats; nuts; chocolate; egg yolk; cream; and cheese. During the preparation of meats much of the fat cooks out and can be eliminated easily.

Fats are important in the diet for energy, but also as carriers of the fat-soluble vitamins. Certain fatty acids found in vegetable and marine oils are essential in the diet. One of these, arachidonic acid, or its precursor, linoleic acid, is essential to growth and for dermal integrity. Fats ordinarily contain lipides such as lecithin. Choline, present in the lecithin of most food fats, helps to prevent development of fatty livers. A diet low in fat tends to be dry and unappetizing, and has low satiety value. Fats improve the texture of foods, making them more palatable. The proportion of ingested carbohydrates and fats may vary widely, but if all the energy requirements are met by fat alone, ketosis may develop.

Minerals. All living things contain a wide variety of minerals, present largely in the form of complexes of carbohydrates, organic acids, and proteins. Minerals are a necessary part of the structure of bones and teeth. They are found in all tissues and body fluids, usually in aqueous solution, and are important in regulating vital body processes and functions.

About 4% of the dry weight of food is needed in the form of minerals. Required in macro amounts are calcium, chlorine, magnesium, potassium, phosphorus, sodium, and sulfur. Needed in trace amounts are cobalt, copper, iodine, iron, manganese, selenium, and zinc. In addition to these,

chromium, fluorine, and molybdenum appear to play a role in human metabolism. Milk furnishes about 40% of the dietary minerals, meat and eggs 25%, cereals 15%, vegetables 15%, and fruit and nuts 5%. Most of the sodium and chlorine is supplied as salt. Only three minerals—calcium, iodine, and iron—require planning to insure adequate intake. Other than these, a normal diet from both animal and vegetable sources contains adequate amounts of needed minerals.

Vitamins. Vitamins are organic compounds of widely varying composition which are essential in very small amounts for utilization of food and optimal functioning of the body. Each vitamin has a separate, specific function in the life process. All must be ingested, for they cannot be synthesized by the body. When omitted from the diet, each produces deficiency symptoms which become more severe with prolonged deprivation.

Vitamins are grouped according to their solubility: fat-soluble vitamins include A, D, E, and water-soluble vitamins include C (ascorbic acid), and the B vitamins—niacin, riboflavin, and thiamine. A well-balanced diet provides all normal body needs.

Water. Water is an essential part of all body tissues, aiding in the digestion and absorption of food, the transporting of nutrients to the tissues, the removal of body wastes, and the regulation of body temperature. Chiefly through the thirst mechanism, the body precisely maintains a salt concentration of 0.9%. Man can survive only a few days without water; normal replacement of the water in his tissues is about 6% a day. His body contains 60% water.

Variations in water requirements occur at high altitudes and high temperatures, during strenuous exercise, with fever, and with increased intake of salt. In addition to drinking water per se, water is obtained in beverages, in foods (many contain 80 to 90%), and from the water of metabolism, formed when foods are oxidized.

The Basic Four Foods. The recommended daily requirements for nutrients have been translated into a basic food pattern which will supply the body's needs:

Milk Group, including cheese and ice cream, and other milk-containing foods: Children—three or more glasses; teen-agers—four or more glasses; adults—two or more glasses.

Meat Group, including fish, poultry, eggs, and cheese, and occasionally legumes and nuts: two or more servings.

Vegetables and Fruits, including dark green or yellow vegetables, citrus fruits or tomatoes: four or more servings.

Breads and Cereals, especially enriched or whole grain: four or more servings.

Nutrition Problems Today. The most urgent health problem in the world today is probably malnutrition, due chiefly to insufficient supplies of complete proteins in underdeveloped countries. Children subsisting on such grossly inadequate diets are not only stunted in growth, but frequently suffer irreversible damage to their mental capacity. Because it would take generations to change the cultural patterns and the food habits of illiterate populations, the chemical addition of nutrients to their accepted foods has been recommended as the most efficient solution to the problem.

At the same time in highly developed countries malnutrition has taken the form of overindulgence and poor food habits, resulting in obesity and high incidence of degenerative diseases such as atherosclerosis, high blood pressure, and coronary heart disease. Intensified research throughout the scientific world is seeking to find contributory factors in the diet. Some areas being investigated are (1) the tendency among affluent people to eat more sucrose and less cereals; (2) consumption of caffeine in tea and coffee; (3) the higher level of fat consumption; and (4) the proportion of polyunsaturated to saturated fats in the diet. It seems probable that many of these factors may be involved.

LOUISE S. IRELAND

Cross-references: *Foods; Proteins; Amino Acids; Phytochemistry.*

References

"Present Knowledge of Nutrition," Third Edition, 1967. The Nutrition Foundation, Inc., 99 Park Avenue, N.Y., N.Y. 10016. Paperback, $2.00. (An excellent review and bibliography for each nutrient.)

"Recommended Dietary Allowances," Seventh Edition, 1968. A report of the Food and Nutrition Board, National Research Council. National Academy of Sciences, Washington, D.C. 20418. Publication 1694. Library of Congress Catalog Card No. 63-65472.

"Modern Nutrition in Health and Disease," Fourth Edition, 1968. Edited by Michael G. Wohl, M.D., and Robert S. Goodhart, M.D., Lea & Febiger, Philadelphia, Pa., $30.00.

"Normal and Therapeutic Nutrition," 13th Ed., 1967. The Macmillan Co., New York. Library of Congress Catalog Card No. 67-16055. Current revision by Corinne H. Robinson.

OCEAN WATER CHEMISTRY

In recent years sea water has been viewed as an aqueous solution in homogeneous equilibrium and in equilibrium with the solids within it and (or) transported into it. This article will explore some of the aspects of the thermodynamic approach to sea water chemistry.

Sea water is an electrolyte solution containing minor amounts of nonelectrolytes and composed predominantly of dissolved chemical species of fourteen elements O, H, Cl, Na, Mg, S, Ca, K, Br, C, Sr, B, Si, and F (Table 1). The minor elements, those that occur in concentrations of less than 1 ppm by weight, although unimportant quantitatively in determining the physical properties of sea water, are reactive and are important in organic and biochemical reactions in the oceans.

Dissolved Species. The form in which chemical analyses of sea water are given records the history of our thought concerning the nature of salt solutions. Early analytical data were reported in terms of individual salts NaCl, CaSO$_4$, and so forth. After development of the concept of complete dissociation of strong electrolytes, chemical analyses of sea water were given in terms of individual ions Na$^+$, Ca^{++}, Cl$^-$, and so forth, or in terms of *known* undissociated and partly dissociated species, e.g., HCO$_3^-$. In recent years there has been an attempt to determine the thermodynamically stable dissolved species in sea water and to evaluate the relative distribution of these species at specified conditions. Table 1 lists the principal dissolved species in sea water deduced from a model of sea water that assumes the dissolved constituents are in homogeneous equilibrium, and (or) in equilibrium, or nearly so, with solid phases.

Both associated and nonassociated electrolytes exist in sea water, the latter (typified by the alkali metal ions Li$^+$, Na$^+$, K$^+$, Rb$^+$, and Cs$^+$) predominantly as solvated free cations. The major anions, Cl$^-$ and Br$^-$, exist as free anions, whereas as much as 20% of the F in sea water may be associated as the ion-pair MgF$^+$, and IO$_3^-$ may be a more impor-

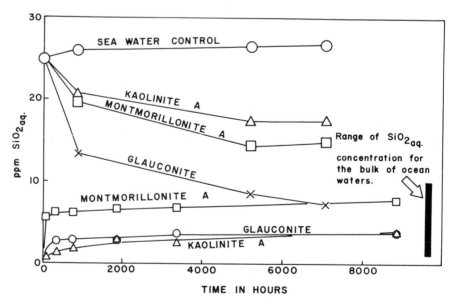

Fig. 1. Concentration of dissolved silica as a function of time for suspensions of silicate minerals in sea water. Curves are for 1–g (<62μ) mineral samples in silica-deficient (SiO$_2$ in water was initially 0.03 ppm) and silica-enriched (SiO$_2$ was initially 25 ppm) sea water at room temperature. Notice that the minerals react rapidly and that the dissolved silica concentration for individual minerals becomes nearly constant at values within or close to the range of silica concentration in the oceans (from Mackenzie, F. T., Garrels, R. M., Bricker, O. P., and Bickley, F., "Silica in sea water: Control by silica minerals" *Science,* **155**, 1404 (1967)).

TABLE 1. ABUNDANCES OF THE ELEMENTS AND PRINCIPAL DISSOLVED CHEMICAL SPECIES
OF SEAWATER, RESIDENCE TIMES OF THE ELEMENTS

Element	Abundance (mg/l)	Principal species	Residence time (years)
O	857,000	H_2O; $O_2(g)$; SO_4^{2-} and other anions	
H	108,000	H_2O	
Cl	19,000	Cl^-	
Na	10,500	Na^+	2.6×10^8
Mg	1,350	Mg^{2+}; $MgSO_4$	4.5×10^7
S	885	SO_4^{2-}	
Ca	400	Ca^{2+}; $CaSO_4$	8.0×10^6
K	380	K^+	1.1×10^7
Br	65	Br^-	
C	28	HCO_3^-; H_2CO_3; CO_3^{2-}; organic compounds	
Sr	8	Sr^{2+}; $SrSO_4$	1.9×10^7
B	4.6	$B(OH)_3$; $B(OH)_2O^-$	
Si	3	$Si(OH)_4$; $Si(OH)_3O^-$	8.0×10^3
F	1.3	F^-; MgF^+	
A	0.6	$A(g)$	
N	0.5	NO_3^-; NO_2^-; NH_4^+; $N_2(g)$; organic compounds	
Li	0.17	Li^+	2.0×10^7
Rb	0.12	Rb^+	2.7×10^5
P	0.07	HPO_4^{2-}; $H_2PO_4^-$; PO_4^{3-}; H_3PO_4	
I	0.06	IO_3^-; I^-	
Ba	0.03	Ba^{2+}; $BaSO_4$	8.4×10^4
In	<0.02		
Al	0.01	$Al(OH)_4^-$	1.0×10^2
Fe	0.01	$Fe(OH)_3(s)$	1.4×10^2
Zn	0.01	Zn^{2+}; $ZnSO_4$	1.8×10^5
Mo	0.01	MoO_4^{2-}	5.0×10^5
Se	0.004	SeO_4^{2-}	
Cu	0.003	Cu^{2+}; $CuSO_4$	5.0×10^4
Sn	0.003	(OH)?	5.0×10^5
U	0.003	$UO_2(CO_3)_3^{4-}$	5.0×10^5
As	0.003	$HAsO_4^{2-}$; $H_2AsO_4^-$; H_3AsO_4; H_3AsO_3	
Ni	0.002	Ni^{2+}; $NiSO_4$	1.8×10^4
Mn	0.002	Mn^{2+}; $MnSO_4$	1.4×10^3
V	0.002	$VO_2(OH)_3^{2-}$	1.0×10^4
Ti	0.001	$Ti(OH)_4$?	1.6×10^2
Sb	0.0005	$Sb(OH)_6^-$?	3.5×10^5
Co	0.0005	Co^{2+}; $CoSO_4$	1.8×10^4
Cs	0.0005	Cs^+	4.0×10^4
Ce	0.0004	Ce^{3+}	6.1×10^3
Kr	0.0003	$Kr(g)$	
Y	0.0003	(OH)?	7.5×10^3
Ag	0.0003	$AgCl_2^-$; $AgCl_3^{2-}$	2.1×10^6
La	0.0003	La^{3+}; $La(OH)^{2+}$?	1.1×10^4
Cd	0.00011	Cd^{2+}; $CdSO_4$	5.0×10^5
Ne	0.0001	$Ne(g)$	
Xe	0.0001	$Xe(g)$	
W	0.0001	WO_4^{2-}	1.0×10^3
Ge	0.00007	$Ge(OH)_4$; $Ge(OH)_3O^-$	7.0×10^3
Cr	0.00005	(OH)?	3.5×10^2
Th	0.00005	(OH)?	3.5×10^2
Sc	0.00004	(OH)?	5.6×10^3
Ga	0.00003	(OH)?	1.4×10^3
Hg	0.00003	$HgCl_3^-$; $HgCl_4^{2-}$	4.2×10^4
Pb	0.00003	Pb^{2+}; $PbSO_4$	2.0×10^3
Bi	0.00002		4.5×10^5
Nb	0.00001		3.0×10^2
Tl	<0.00001	Tl^+	
He	0.000005	$He(g)$	
Au	0.000004	$AuCl_2^-$	5.6×10^5
Be	0.0000006	(OH)?	1.5×10^2
Pa	2.0×10^{-9}		
Ra	1.0×10^{-10}	Ra^{2+}; $RaSO_4$	
Rn	0.6×10^{-15}	$Rn(g)$	

Adapted from Goldberg, E. D., "Minor elements in sea water," in "Chemical Oceanography," v. 1, pp. 164–165, J. P. Riley and G. Skirrow, Eds., Academic Press, New York, 1965.

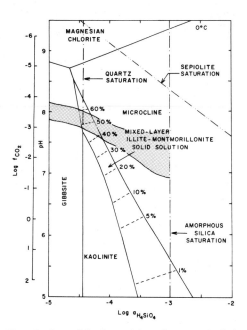

FIG. 2. Logarithmic activity diagram depicting equilibrium phase relations among aluminosilicates and sea water in an idealized nine-component model of the ocean system at the noted temperatures, one atmosphere total pressure, and unit activity of H_2O. The shaded area represents the compositional range of sea water at the specified temperature, and the dot-dash lines indicate the composition of sea water saturated with quartz, amorphous silica, and sepiolite, respectively. The scale to the left of the diagram refers to calcite saturation for different fugacities of CO_2. The dashed contours designate the composition (in % illite) of a mixed-layer illite-montmorillonite solid solution phase in equilibrium with sea water (from Helgeson, H.C. and Mackenzie, F. T., 1970, Silicate-sea water equilibria in the ocean system: Deep Sea Res.).

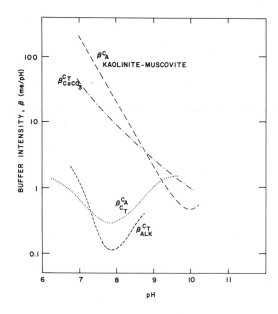

FIG. 3. Buffer intetnsity as a function of pH for some homogeneous and heterogeneous chemical systems. The buffer intensities are defined for C_A $\beta_{Kaolinite-muscovite}$, addition of a strong acid (or base) to sea water in equilibrium with kaolinite and muscovite; $\beta_{CaCO_3}^{C_T}$, addition of CO_2 in a sea water system of zero noncarbonate alkalinity in equilibrium with $CaCO_3$; $\beta_{C_T}^{C_A}$, addition of a strong acid (or base) to a sea water solution of constant total dissolved carbonate; and $\beta_{Alk}^{C_T}$, addition of total CO_2 to a sea water solution of constant alkalinity (data from Morgan, J. J., preprint, "Applications and limitations of chemical thermodynamics in natural water systems").

tant species of I than I⁻. Based on dissociation constants and individual ion activity coefficients the distribution of the major cations in sea water as sulfate, bicarbonate, or carbonate ion-pairs has been evaluated at specified conditions by Garrels and Thompson (1962).

About 10% each of Mg and Ca is tied up as the sulfate ion-pair. It is likely that the other alkaline earth metals, Sr, Ba, and Ra, also exist in sea water partly as undissociated sulfates; about 60% and 21%, respectively, of the total $SO_4^=$ and HCO_3^- are complexed with cations, and two-thirds of the $CO_3^=$ is present as the ion-pair $MgCO_3°$.

The activities of Mg^{++} and Ca^{++} obtained from the model of sea water proposed by Garrels and Thompson have recently been confirmed by use of specific Ca^{++} and Mg^{++} ion electrodes, and for Mg^{++} by solubility techniques and ultrasonic absorption studies of synthetic and natural sea water. The importance of ion activities to the chemistry of sea water is amply demonstrated by consideration of $CaCO_3$ (calcite) in sea water. The total molality

of Ca^{++} in surface sea water is about 10^{-2} and that of $CO_3^=$ is 3.7×10^{-4}; therefore the ion product is 3.7×10^{-6}. This value is nearly 600 times greater than the equilibrium ion activity product of $CaCO_3$ of 4.6×10^{-9} at 25°C and one atmosphere total pressure. However, the activities of the free ions Ca^{++} and $CO_3^=$ in surface sea water are about 2.3×10^{-3} and 7.4×10^{-6}, respectively; thus the ion activity product is 17×10^{-9} which is only 3.7 times greater than the equilibrium ion activity product of calcite. Thus, by considering activities of sea water constituents rather than concentrations, we are better able to evaluate chemical equilibria in sea water; an obvious restatement of simple chemical theory but an often neglected concept in sea water chemistry.

Constancy and Equilibrium. The concept of constancy of the chemical composition of sea water, i.e., that the ratios of the major dissolved constituents of sea water do not vary geographically or vertically in the oceans except in regions of runoff from the land or in semienclosed basins, was first proposed indirectly in 1819 by Marcet and expanded later by Forchammer and Dittmar. The concept was established on a purely empirical basis whereas in actual fact there is a theoretical basis for the concept.

Barth (1952) proposed the concept of residence (passage) time of an element in the oceanic environment and formalized this concept by the equation

$$\lambda = \frac{A}{dA/dt},$$

where λ is the residence time of the element, A is the total amount of the element in the oceans, and dA/dt is the amount of the element introduced or removed per unit time. Sea water is assumed to be a steady-state solution in which the number of moles of each element in any volume of sea water does not change; the net flow into the volume exactly balances the processes that remove the element from it. Complete mixing of the element in the ocean is assumed to take place in a time interval that is short compared to its residence time. Table 1 shows the residence times of the elements, and Table 2 compares the residence times of some elements on the basis of river input and removal by sedimentation. For the major elements the results are strikingly similar and suggest that at least as a first approximation sea water is a steady-state solution with a composition fixed by reaction rates involving the removal of elements from the ocean approximately equalling rates of element inflow into the ocean. Thus, as a first approximation the steady-state oceanic model implies a fixed and constant sea water composition and provides a theoretical basis for the concept of the constancy of the chemical composition of sea water. However, it is possible that at any time, t_o, for example, the present, the ratios of the major dissolved constituents in the open ocean may be nearly invariant simply

because the amounts of new materials introduced by streams and other agents to the ocean are small compared to the amounts in the ocean, and these new materials are mixed into the oceanic system relatively rapidly. But over time periods of 1000 to 2000 years or more the major ionic ratios can only remain constant if the ocean is a steady-state solution whose composition is controlled by mechanism(s) other than simple mixing.

Further insight into the constancy concept can be gained by exploring possible mechanisms governing the steady-state composition of sea water. The steady-state solution could be simply a result of the rates of major element inflow into the oceans being equal to rates of outflow by biologic removal, flux through the atmosphere, adsorption on sediment particles, and removal in the interstitial waters of marine sediments. For example, Ca^{++} carried to the oceans by streams is certainly removed, in part, in sea aerosol generated at the atmosphere-ocean interface and transported into the atmosphere, later to fall as rain or dry fallout on the continents. However, recent theoretical and experimental work suggests that sea water may be modeled as a steady-state solution in equilibrium with the solids that are in contact with it. Sillén has modeled the oceanic system as a near-equilibrium of many solid phases and sea water. Experimental work has shown that aluminosilicate minerals typical of those in the suspended load of streams and in marine sediments react rapidly with sea water containing an excess or deficiency of dissolved silica. Reactions involving these aluminosilicates may control on a long-term basis the activities of H_4SiO_4 and other constituents in sea water. Thus it has begun to emerge that the composition of the oceans represents an approximation of dynamic equilibrium between the water and the solids that are carried into it in suspension or are precipitated from it by the continuous evaporation and renewal by streams. Therefore, if sea water is a solution in equilibrium with solid phases, or even closely approaches such a system, then the *ion activity ratios* of the major dissolved species would be fixed and the chemical composition of the ocean would be "constant." Consequently, the activity of Ca^{++} in the ocean is not simply a result of removal processes involving sea aerosol, adsorption and so forth but is controlled by solid-solution equilibria. A model leading to nearly invariant ion activity ratios geographically and vertically at any time, t_o, in the oceans based on mixing rates alone may be sufficient to explain the constancy of sea water composition but is somewhat misleading and uninformative when considered in light of the recent advances in treating the oceans as an equilibrium system.

Some limitations of the equilibrium model of sea water do exist. Sillén has pointed out that based on equilibrium calculations all the nitrogen in the ocean-atmosphere system should be present as NO_3^- in sea water; however, most of the nitrogen is present as N_2 gas in the atmosphere. Also, the concentrations of the major alkaline earth elements, Mg, Ca, and Sr, in sea water may vary slightly with depth or geographic location.

Buffering and Buffer Intensity of Sea Water. The view has long been held that hydrogen-ion buffering in the oceans is due to the $CO_2 - HCO_3^- - CO_3^=$

TABLE 2. THE RESIDENCE TIMES OF ELEMENTS IN
SEAWATER CALCULATED BY RIVER INPUT AND
SEDIMENTATION

Element	Amount in ocean (in units of 10^{20} g)	Residence time in millions of years	
		River input	Sedimentation
Na	147.8	210	260
Mg	17.8	22	45
Ca	5.6	1	8
K	5.3	10	11
Sr	0.11	10	19
Si	0.052	0.035	0.01
Li	0.0023	12	19
Rb	0.00165	6.1	0.27
Ba	0.00041	0.05	0.084
Al	0.00014	0.0031	0.0001
Mo	0.00014	2.15	0.5
Cu	0.000041	0.043	0.05
Ni	0.000027	0.015	0.018
Ag	0.0000041	0.25	2.1
Pb	0.00000041	0.00056	0.002

After Goldberg, E. D., "Minor elements in sea water," in "Chemical Oceanography," v. 1, p. 173, J. P. Riley and G. Skirrow, Eds., Academic Press, New York, 1965..

equilibrium. Within recent years this view has been challenged, and the importance of aluminosilicate equilibria in maintaining the pH of sea water emphasized. The buffer intensity of a system is of thermodynamic nature and is defined as

$$\beta_{c_j}^{c_i} = \frac{dC_i}{dpH},$$

where $\beta_{c_j}^{c_i}$ is the pH buffer intensity for incremental addition of C_i to a closed system of constant C_j at equilibrium. Homogeneous buffer intensities are defined for systems without solid phases, e.g., the addition of a strong acid to a carbonate solution, whereas heterogeneous buffer intensities are defined for systems with solid phases, e.g., the addition of a strong acid to a solution in equilibrium with calcite, $CaCO_3$, or with kaolinite and muscovite. The homogeneous buffer intensities for the range of sea water and interstitial marine water pH values (7.0 to 8.3) are about 10- to 100-fold less than the heterogeneous intensities involving equilibrium between calcite and sea water or kaolinite, muscovite, and sea water. Both of these heterogeneous equilibria represent large capacities for resistance to sea water pH changes. Unfortunately, the kinetic aspects of these buffer systems have not been investigated quantitatively. However, it is apparent that aluminosilicate equilibria have buffer intensities equal to and perhaps greater than (the buffer intensities of most aluminosilicate equilibria in natural waters have only been qualitatively evaluated) the $CO_2 - CaCO_{3(s)}$ equilibria in sea water. Small additions of acid or base to the oceans could be buffered by the homogeneous equilibrium $CO_2\!-\!HCO^-_3\!-\!CO^=_3$. However, large incremental additions of acid or base or additions over a duration of time would involve the heterogeneous carbonate and aluminosilicate equilibria; the relative importance of each would depend on the buffer intensities of the various equilibria and the relative rates of aluminosilicate and carbonate reactions.

For geologically short-term processes on the order of a few thousands of years, it is likely that the carbon dioxide-carbonate system regulates oceanic pH. The long-term pH is controlled by an interplay of various near-equilibria involving carbonates and silicates.

FRED T. MACKENZIE

References

Goldberg, E. D., "Minor Elements in Seawater," in "Chemical Oceanography, 1," Academic Press, New York, 1965.

Garrels, R. M., and Mackenzie, F. T., "Evolution of Sedimentary Rocks," W. W. Norton and Co., Inc., New York, 377 pp., 1971.

Horne, M. K., "Marine Chemistry," Wiley-Interscience, New York, 568 pp., 1969.

Riley, J. P., and Skirrow, G., Eds., "Chemical Oceanography, 1," Academic Press, New York, 712 pp., 1965.

Riley, J. P., and Skirrow, G., Eds., "Chemical Oceanogrphy, 2," Academic Press, New York, 508 pp., 1965.

Sillén, L. G., "The physical chemistry of sea water," pp. 549–581, in "Oceanography," M. Sears, Ed., A.A.A.S. Publication No. 67, Washington, D.C., 1961.

ODOR

Although odor has long been classified as a chemical sensation, the precise relationship between the odorous properties of a substance and its molecular structure remains unknown. The remarkable advances made in conformational analysis in recent years will undoubtedly point the way to new understanding of structure-odor correlations.

In land vertebrates, odor is perceived when an odorant comes into contact with the olfactory epithelium, which covers an area of a few square centimeters (in the larger species) inside the nasal cavity. The nerve cells which are sensitive to odor are connected directly with the brain, thereby supplying the most direct, and the most primitive, of our senses.

Even in man, whose sense of smell is relatively dull, picogram quantities of many odorants suffice to produce an odor sensation. In terms of molecules, such quantities are of course not small. Skatole, a powerful odorant, must be at a concentration of 1.7×10^7 molecules per milliliter of inhaled air in order to be detected. On the other hand, calculation of the quantity of the sex pheromone bombycol adsorbed by the antennae of the male moth indicated that each olfactory sensor cell adsorbed only one molecule of the odorant.

The first step in the olfactory process appears to be adsorption of odorant molecules on the surface of the olfactory epithelium. The next step is unknown, but one theory (Penetration and Puncturing) assumes that olfactory nerve cell membranes differ in their lipid makeup, so that some are more easily penetrated by odorant molecules than others, and also more rapidly heal the puncture made by entry of such a molecule. Puncture of a membrane by an odorant molecule is proposed to allow penetration of the membrane by ions such as Na^+ and K^+, thereby changing conductance and resulting in propagation of an electrical nerve impulse. Free energies of desorption were found to differ according to odor type, being high for musky odorants, low for ethereal ones.

Another theory correlates odor type with vibrations of frequencies below 500 per second given off by the odorant molecule. Such vibrations are visualized as provoking resonance in a pigment molecule in the olfactory epithelium, thereby triggering the return to ground state of an excited electron and leading to a nerve impulse.

There is abundant evidence to show that type of odor is not referable to any one chemical grouping in the molecule, and that "molecular shape" appears to be an important factor. Molecular cross-sectional area and ratio of length to breadth have been seized upon as parameters relevant to odor. Methyl and other small alkyl groups appear to exert upon odor somewhat similar effects to those observed in certain other physiological processes. Tertiary carbons near to a hydroxyl or carbonyl group frequently produce a camphoraceous odor. The nitro musks depend on the presence of tertiary groups such as tert-butyl for their musk odor.

Such observations call to mind the strategic importance of alkyl groups in acetylcholine derivatives in determining intensity of reaction with cholinesterase, and the suggestion was made some years ago that cholinesterase may enter the olfactory process. Recent observations have confirmed the presence of cholinesterase in the nasal mucosa and in synapses of the olfactoy bulb. Application of acetylcholine to the nasal membranes of frogs caused the same nerve impulse as an adequate olfactory stimulus, and the reaction of olfactory receptors of fish to vanillin was blocked by preliminary treatment of the mucosa with specific acetylcholine-receptor blocking agents.

It therefore appears that further studies of molecular shape of odorants and olfactory cell membranes, particularly by the techniques of conformational analysis, should provide a key to the puzzling structure-odor problem. Odorants can then be tailored to specification, with their molecules designed to provide maximum intensity of a specific odor. Odor sensations are more quickly and simply measured than most other physiological effects, therefore information on odor-structure relationships may provide valuable insight into such correlations in other biochemical systems.

PAUL G. LAUFFER

OIL RESISTANCE

By "oil resistance" is meant the ability of an elastomeric composition to withstand the action of petroleum and coal-tar products, animal, vegetable and mineral oils, fatty acids, esters and various other organic compounds without decomposition, degradation of physical properties or serious change in volume. A number of rubbers and plastics have been used in service requiring contact with organic fluids and gases. The choice of a rubber or plastic must be based on the type of fluid with which it is in contact and the temperature of service. Obviously, if the elastomeric compound is brittle at low temperature or decomposes thermally at high temperatures, it is unsuited for use at either extreme; even though it may have excellent resistance to attack by a specific fluid at intermediate temperatures. In general, rubbers having polar groups are most resistant to attack by organic materials than nonpolar rubbers.

Natural rubber, butadiene-styrene copolymers (SBR) and chlorosulfonated polyethylene exhibit relatively poor resistance to most solvents and swell excessively in aromatic or ester type fluids. Butyl rubber (GR-I) is suitable for use in contact with animal fats and vegetable oils but is swelled excessively by many petroleum fractions. Polychloroprene elastomers (neoprene) find application in contact with animal and vegetable oils, fatty acids, gasoline, lubricating oils and aliphatic hydrocarbons but are deteriorated or swelled excessively by aromatic hydrocarbons, amines and esters. Neoprene WRT by proper compounding can be made serviceable up to 400°F while immersed in ester type hydraulic fluids. Neoprene is excellent in resistance to ozone, and this in combination with its oil resistance makes it a desirable elastomer for many applications.

The rubbers most widely used for oil-resistant service are the butadiene-acrylonitrile copolymers. By varying the acrylonitrile content of the rubber during its manufacture, the oil resistance is markedly altered. In addition to being satisfactory in animal and vegetable oils and petroleum products, this type of rubber is resistant to aromatic compounds, esters, alcohols, and lacquer solvents, as well as many other organic compounds. Amines and aromatic solvents cause moderate swelling, but ketones, chlorinated hydrocarbons and nitro-hydrocarbons swell the vulcanizates excessively. Acrylate ester-vinyl copolymers, commonly referred to as polyacrylate rubbers, are serviceable in most organic fluids and are especially recommended for service at high temperatures and in contact with oils containing different additives. Modified polyacrylates offer a higher resistance to certain oils.

Polysulfide ("Thiokol") rubbers offer maximum resistance to attack by aromatic compounds and are widely used in tank linings and other applications requiring maximum oil resistance. This type of rubber is harmed by contact with amines, acids and mercaptans and acquires high set under compression at elevated temperatures. Butadiene-vinyl pyridine copolymers are not highly oil resistant in themselves, but these copolymers can be quaternized with organic halides to yield oil resistance. By proper compounding, vulcanizates having excellent resistance to aromatic materials, organic or inorganic esters, ketones, glycols and alcohols can be produced and high temperature service is good. Polyurethane elastomers are reputed to possess moderately good oil resistance, but these materials can be hydrolyzed by alkali. Silicone rubbers are outstanding in high and low temperature service and are decidedly better than organic rubbers in resistance to deterioration by heat. Most silicone rubbers possess moderate resistance to aliphatic hydrocarbons but are highly swelled by aromatics. More recently developed fluorinated silicone rubbers possess greatly improved resistance to aromatic solvents. The inherent properties of silicone rubbers are not significantly altered by immersion in a number of liquids, and after being dried they recover their original properties. An example of extreme service conditions is the use of silicone in push rod seals operating at 230°C where, even though swelled somewhat by the oil, they outlast any other known rubber.

Oil resistance of organic rubbers is improved by adding fillers which themselves are not affected by oils and merely serve to dilute the rubber matrix. Glue is an additive which aids oil resistance without increasing hardness significantly. Mineral fillers and reinforcing agents also improve oil resistance, in addition to imparting greater strength to the vulcanizate. Carbon blacks reduce swell in a greater measure than can be attributed to the dilution effect alone; apparently the carbon black interacts with the rubber to give additional cross links which enhance oil resistance.

In many applications the rubber article is in contact with hot oils and must be able to withstand deterioration by heat as well as the solvent action of the organic fluid. Butadiene-acrylonitrile copolymers for high-temperature service are usually compounded with very little or no elemental sulfur and large amounts of thiuram disulfides. Polyacrylate rubbers, which are basically saturated, are exceptionally well suited for high-temperature applications and very resistant to oxidation. Basic compounding is essential to obtain heat-stable vulcanizates with these elastomers.

Low-temperature performance of organic rubbers is improved by the use of softeners. Ester type plasticizers are generally most effective; however, they are deficient in that they are usually extracted by the immersion fluid and the properties of the rubber article change after immersion.

Among the plastics, polyethylene exhibits only moderate oil resistance and can be used in contact with animal and vegetable oils but is attacked by petroleum products, aromatics, ketones and esters. Polyvinylchloride can be used in contact with some hydrocarbons and is stable toward amines and some alcohols. Vinyl chloride-vinylidine chloride copolymers excel the foregoing plastics in resistance to aliphatic hydrocarbons, gasoline and lubricating oils but are attacked by aromatic hydrocarbons, ketones and amines. Polyvinyl alcohol, while resistant to some hydrocarbons, has great water solubility.

The fluorocarbons are less susceptible to swelling and deterioration by oils than any of the other plastics. Polychlorotrifluoroethylene is excellent in contact with most solvents, but is degraded by esters. Polytetrafluoroethylene, the least thermoplastic of all vinyl plastics, is not attacked by any known solvent and is even stable in aqua regia and concentrated caustic acids. Polymers of 1,1-dihydrofluoro-butylacrylate reputedly possess excellent oil resistance and are easier to process than polytetrafluoroethylene.

The approximate practical operating temperatures within which the principal oil resistant rubbers and plastics find utility are given below:

	Temperature (°C)	
	Minimum	Maximum
Neoprene	*	160
Butadiene-acrylonitrile copolymers	−60	175
Polyacrylate rubbers	−60	185
"Thiokol"	−55	120
Epichlorohydrin rubbers	−40	175
Butadiene-vinylpyridine copolymers	−60	175
Silicone	−100	275
Polychlorotrifluoroethylene	−55	200
Polytetrafluoroethylene	−100	325

*Dependent on type—some types crystallize.

J. F. SVETLIK

Cross-references: *Elastomers, Rubber, Fluorocarbons.*

OIL ON THE SEA

The development of offshore oil resources and the rapidly expanding petroleum industry of the past two decades have resulted in steadily increasing quantities of oil on the sea, primarily a combination of deliberate disposal and accidental spills.

Crude oil, a common type in major oil spills, is a mixture of four major groups of substances: (1) aromatic, (2) olefinic, (3) saturated hydrocarbons and (4) various compounds of nitrogen, sulfur, oxygen, and trace metals, especially vanadium and nickel. Every crude oil source yields a product that is slightly different from all others so that it is possible to pinpoint the origin by means of chromatographic analyses. The well-field from which the sample came, or even the ship from which it was spilled or pumped, can be determined.

Crude oil on the sea releases water-soluble toxic substances for a time and then the remainder is relatively nontoxic, its deleterious effect on marine life a result of coating or smothering. Lighter fractions evaporate with time and a highly viscous residue will remain. Under some conditions, a water-oil emulsion wil form in which the material is as much as 80% water.

It has been observed that oil on the sea drifts at about 3.4% of the wind speed, so that it is possible to calculate the location of oil masses, following a spill, if meteorological conditions are known.

Natural Seepage. Oil on the sea is nothing new. It has been seeping into the sea from submarine oil reserves for thousands of years, probably millions, and is still doing so. Around North America it is oozing into the northern Gulf of Mexico and into waters along the coasts of California and Alaska. These natural sources of oil on the sea are, however, a drop in the proverbial bucket in comparison to the estimated 0.1% loss to the sea of the quantities consumed or transported upon the sea per unit of time. At Coal Oil Point near Santa Barbara, 50 to 70 barrels a day are seeping into the sea from depths of 40 to 100 feet.

Accidental Spills. Several oil spills of recent years from ships and offshore wells have received widespread publicity and have stimulated efforts and legislation to prevent spills and to contain and minimize the damage of those that do occur. Probably the most damaging of these was the wreck of the tanker *Torrey Canyon* off Lands End, southwest coast of England in March, 1967. Well over 100,000 long tons of Kuwait crude was spilled in three major batches during the following ten days. About 30,000 tons went ashore along the coast of Brittany in France and on the Isle of Guernsey; 20,000 tons along the northwest coast of Cornwall in southern England; 50,000 tons into the Bay of Biscay, west coast of France.

Subsequent major spills were those of the tanker *Ocean Eagle* near San Juan, Puerto Rico; the *General Colocotronis* near Eleuthera, Bahamas; the *Delian Apollon* in Tampa Bay, Florida; offshore well spills near Santa Barbara, California, and along the coasts of Louisiana, Delaware, and Connecticut.

The most obvious biological damage done by these oil spills has been destruction of sea birds, the smothering of some intertidal plants and animals (benthic algae, barnacles, oysters, limpets), damage to salt marshes and mangrove swamps. Direct economic losses were the temporary ruining of beaches, the coating of small boats and other marine installations, and the cost of clean-up.

Sea birds most commonly affected were mergansers, grebes, cormorants, loons, and scooters. Each of the major spills killed from a few hundred to an estimated 10,000. Fortunately, all these are abundant species and the numbers killed by oil has not significantly reduced the populations.

Although there has been much speculation, there apparently are no data to indicate damage to the sea fisheries either immediate or long-term. Oil is a naturally-produced substance and evolution has occurred in its presence.

Remedial Procedures. A variety of procedures have been tried in efforts to reduce or remove oil from the surface of the sea and to prevent its spread. Among these are absorption by polyurethane foam, gelation of oil, dispersion by detergents, burning, and confinement of an oil slick to permit removal by pumping or other means.

Polyurethane foam is easily generated by mixing two liquids with consequent hundred-fold expansion. It will absorb 90% of its volume of oil (about 100 times its weight). Water absorption is low and the absorbed oil is held until the foam is squeezed.

Gelation of oil in a tanker threatened with break-up is now possible, but further research is needed. A variety of detergents is available that will disperse an oil slick. They are toxic to various degrees and in many cases their use may be more damaging than the oil itself. In other cases, low-toxicity detergents may be a valuable remedy.

The burning of spilled oil is difficult and usually ineffective. However, the addition of fine particles of silica treated with silane to render it hydrophobic provides a wick-like structure and is said to promote combustion to the extent that about 98% of crude or bunker C oil on water can be burned.

Heavier types of oil can often be caused to sink by addition of mineral material such as sand or chalk, but it may be damaging on the bottom or rise to the surface later.

Many techniques have been tried in efforts to contain oil spills within a certain area and permit removal. One of these, the "pneumatic boom," is a submerged pipe with small holes through which compressed air escapes and forms a wall of fine bubbles which, under favorable circumstances, serve as a barrier to oil spread. This procedure is limited to calm waters of bays or harbors, as are most of the others. Physical booms are also useful. They usually involve some sort of floating wall around the oil. Barge-like vessels with absorbent rollers partially submerged have served to remove oil from harbors, but they are not effective where waves are as much as 12 inches in amplitude.

Deliberate Oil Dumping. Large accidental oil spills, while conspicuous and of much concern, are only one of the major sources of oil on the sea. The pumping of bilges and cleaning of tanks at sea is another major source. Recently an oceanographic vessel towing plankton nets at the surface in the Sargasso Sea near the center of the North Atlantic caught more oil and tar globules than *Sargassum,* the brown alga that is usually abundant in floating masses in this region. Other scientists working from oceanographic vessels have commented on the growing amount of oil on the sea surface in all parts of the world, a result of deliberate disposal. Even the U.S. Navy admitted deliberate dumping of quantities of oil off Jacksonville, Florida, late in 1970.

There are many species of marine and other bacteria that decompose petroleum hydrocarbons, ultimately to water and carbon dioxide. The natural process is relatively slow and oil is being added to the sea much faster than the bacteria can hydrolyze it. There is promise, however, of inoculating oil spills with carefully compounded assortments of bacteria plus stimulating substances with the result that decomposition of the oil can be accomplished at a greatly accelerated rate.

Concerted efforts by major world governments, perhaps through an up-dated International Convention for the Prevention of Pollution of the Seas by Oil (last held in 1954), may be the only answer to this progressively serious problem of global magnitude.

HAROLD J. HUMM

References

Hoult, D. P., Editor, "Oil on the Sea," Plenum Press, 227 W. 17th Street, New York, N.Y. 10011. 114 pages, 1969.

OLEATES

The oleates, like the corresponding saturated stearates, comprise two groups of compounds: the metallic salts or soaps, and the esters of oleic acid, $CH_3(CH_2)_7CH{=}CH(CH_2)_7COOH$. This acid is perhaps the most abundant in nature and occurs in the form of glycerides in substantial amounts in all land- and marine-animal fats and in all vegetable oils. It forms 40–50% of the fat in animal depot fats and in many vegetable oils it occurs to the extent of 30–85%. (See **Fatty Acids**).

The metallic oleates can be prepared by (1) reaction of a base (hydroxide or oxide) on the fatty acid (fusion process); (2) double decomposition of a soluble salt of the fatty acid and a salt of a mineral acid (precipitation process); or (3) saponification of glycerides or other esters of oleic acid with a base or metallic oxide:

$$2C_{17}H_{33}COOH + PbO \rightarrow Pb(OOC{\cdot}C_{17}H_{33})_2 + H_2O$$

$$2C_{17}H_{33}COOH + Co(Ac)_2 \rightarrow$$
$$Co(OCC{\cdot}C_{17}H_{33})_2 + 2HAc$$

$$2C_{17}H_{33}COONa + Pb(CH_3COO)_2 \rightarrow$$
$$Pb(OOC{\cdot}C_{17}H_{33})_2 + 2NaOOC{\cdot}CH_3$$

$$
\begin{array}{c}
\quad\quad\quad\quad H \\
\quad\quad\quad\quad | \\
C_{17}H_{33}COO\!-\!CH \\
\quad\quad\quad\quad | \\
C_{17}H_{33}COO\!-\!CH + 3NaOH \rightarrow \\
\quad\quad\quad\quad | \\
C_{17}H_{33}COO\!-\!CH \\
\quad\quad\quad\quad | \\
\quad\quad\quad\quad H
\end{array}
$$

Triolein

$$3C_{17}H_{33}COONa + C_3H_5(OH)_3$$

Sodium oleate   Glycerol

Several grades of oleic acid are commercially available for the preparation of metallic salts and esters. One of these is so-called "red oil," which was the best commercial grade available prior to 1949. It contains a maximum of 70% oleic acid; the rest is approximately equal parts of linoleic and saturated acids (palmitic and stearic). A purified technical oleic acid containing 90% or more oleic, 4% maximum linoleic and 6% maximum saturated acids is now available. Most of the physical data available with respect to the metallic oleates refers to products produced from red oil, either *per se* or after purification of the oil or the end product. Metallic oleates are produced principally by the precipitation process in the same manner as the corresponding saturated stearates. (See **Laurates, Palmitates, Stearates**).

The alkali metal salts of oleic acid are soluble in water, whereas the salts of the other metals are insoluble in this medium. The sodium and potassium oleates are referred to as soaps and those of the other metals as either soaps or salts. The acid salts (Na, K and NH_4) are equimolar mixtures of normal oleate and the free acid ($C_{17}H_{33}COOM \cdot C_{17}H_{33}COOH$). Basic salts of di- and polyvalent metals, e.g., ($C_{17}H_{33}COO)_2Al(OH)$, are also important. In fact, commercial aluminum trioleate probably corresponds to this formula as it usually reacts basic.

Many salts of oleic acid have been prepared and such properties as solubility in various solvents, surface tension, viscosity effects, electric and magnetic effects, spectral and crystallographic properties, etc. have been investigated. All the pure compounds are white or light cream-colored except those formed with cadmium (yellow), copper and nickel (blue), iron (green), manganese (light pink).

Uses. Since oleic acid is a constituent of all fats and oils used in the manufacture of soaps, sodium or potassium oleate is, therefore, a constituent of most of these products. The sodium soap is soluble in both water and alcohol, and it possesses good detergent, wetting, and emulsifying properties. Pure sodium oleate is a white powder which has been used in internal medicine in treating cholelithiasis. The potassium salt, which is softer, has been used as an emollient and as a cleaning and healing agent in treating abscesses. A number of the insoluble metallic oleates have a variety of uses.

Esters. Esters of oleic acid may be formed by the reaction of (1) the acid chloride and an alcohol or alcoholate; (2) the acid anhydride and alcohol; (3) salt of the organic acid and alkyl halide or sulfate; (4) direct esterification of the fatty acid and an alcohol; (5) interesterification of an oleate ester with an alcohol or another ester. Esterification and interestification are the most important methods for the commercial preparation of the oleates of the lower alcohols. For preparation of some of the oleates of the higher alcohols one of the indirect methods may be used.

Various oleates of the monohydric alcohols, from methyl to hexadecyl, have been prepared, but the series is very incomplete and there is a dearth of information concerning the properties of these compounds. At room temperature most of the oleates of the monohydric alcohols are colorless liquids.

Commercially available oleates include methyl, ethyl, butyl, and other oleates of monohydric alcohols, diethylene and polyethylene glycol monooleates, glycerol monooleate and trioleate, and oleates of various "Cellosolves." Total production

of oleates is in excess of 6 million pounds annually, of which about 10% each are butyl oleate and diethylene and 1,2-polyethylene glycol monooleates.

Methyl oleate, $C_{19}H_{36}O_2$, molecular weight 296.50, is prepared by refluxing oleic acid p-toluenesulfonic acid and methanol. Iodine value 85.6, density 0.8774 at 15°, melting point 19.9°, boiling point 166-2–167.4 at 2 mm, refractive index 1.44656 at 25°, insoluble in water, miscible with anhydrous ethanol, diethyl ether.

Ethyl oleate, $C_{20}H_{38}O_2$, molecular weight 310.52, boiling point 217–219 at 15 mm, density 0.8724 at 15°, 0.8671 at 25°, refractive index 1.4449 at 25°. Commercial grade ethyl oleate is a light yellow liquid, density 0.870^{20}_{20} and a distillation range 140–180° at 3–4 mm pressure; maximum f.f.a. 0.05% and saponification value about 180.

Butyl oleate, $C_{22}H_{42}O_2$, molecular weight 338.58, melting point $< -10°$, boiling point 227–8 at 100 mm, density 1.4480 at 25°, insoluble in water, soluble in alcohol, diethyl ether.

Ethylene glycol monooleate melts at 1°C, boils at 190–200°C at 0.05 mm and has a refractive index 1.4600 at 27°C. The dioleate boils at 183–5° at 3 mm and has a refractive index 1.4492 at 70°C. Polyethylene glycolmonooleate melts at 6°C; polyethylene glycol (600) monooleate at 16–20°C, and Carbowax dioleate at 25–30°C. The glycol oleates have a variety of applications in the textile, leather, petroleum, cosmetic, insecticide and other industries.

KLARE S. MARKLEY

Cross-references: *Fatty Acids, Metallic Soaps, Glycerides.*

OLEFIN COMPOUNDS

Definition and Nomenclature. The olefins are an important group of hydrocarbons which contain in their molecular structure at least one double bond between adjacent carbon atoms, usually represented by $-\overset{|}{C}=\overset{|}{C}-$.

Aliphatic olefins have straight or branching carbon chains, while alicyclic olefins have one or more rings of carbon atoms in the molecule. If the molecule contains only one double bond, it is called a monoolefin, while olefins with two, three and more double bonds are called diolefins, triolefins, and so on. In the Geneva system of nomenclature, the aliphatic olefins with one, two, three, and more double bonds are called alkenes, alkadienes, alkatrienes, etc., where the alka- prefix is that of the analogous paraffin (alkane in the Geneva system). The corresponding alicylcic compounds are called the cycloalkenes, cycloalkadienes, cycloalkatrienes, etc.

In order of increasing molecular weight, the simplest alkenes are ethylene (ethene), $CH_2{=}CH_2$, propylene (propene), $CH_3{-}CH{=}CH_2$, and the two butenes, $CH_3{-}CH{=}CH{-}CH_3$ and $CH_3{-}CH_2{-}CH{=}CH_2$.

The simplest alkadiene is butadiene,

$$CH_2{=}CH{-}CH{=}CH_2.$$

Note that two double bonds in an alkadiene may be arranged in any of three ways:

cumulated, $-\overset{|}{C}=\overset{|}{C}=\overset{|}{C}-$;

conjugated, $-\overset{|}{C}=\overset{|}{C}-\overset{|}{C}=\overset{|}{C}-$;

and isolated, $-\overset{|}{C}=\overset{|}{C}-\overset{|}{C}-\overset{|}{C}=\overset{|}{C}-$

(See **Dienes**).

Physical Properties. Individual alkenes are similar to the alkanes having the same carbon structure. Boiling points are only several degrees lower than the corresponding alkanes, while in the lower molecular weight members melting points are just a few degrees higher. At room temperature the lower molecular weight alkenes, up to 4 carbon atoms, are gases; the next higher members are liquids; and those above about 16 carbon atoms are waxy solids. Alkenes are usually colorless; they have densities of about 0.6 to 0.8 and are insoluble in water and soluble in chloroform, ethanol, hydrocarbons, and ether. The polarizability of an alkene is higher than that of the corresponding alkane.

Sources. Many alkenes are obtained from petroleum, either by direct separation or by treatment with heat or catalysts. Such materials are often very difficult to purify, however, and to obtain relatively pure alkenes some alcohols may be dehydrated over powdered aluminum oxide. Sometimes the elements of a halogen acid may be removed from an alkyl halide with potassium hydroxide dissolved in ethanol. Halogens on adjacent carbon atoms frequently may be abstracted with zinc powder. For some alkenes, an acetylene may be partially hydrogenated.

$$R-C\equiv C-H + H_2 \rightarrow R-\underset{\underset{H}{|}}{C}=\underset{\underset{H}{|}}{C}-H$$

Sometimes a solid quaternary ammonium hydroxide decomposes with heat to give an amine, water and the desired alkene. Grignard reagents can be used to lengthen the carbon chains of some bromoalkenes.

There are fewer general preparative methods for alkadienes, but often an alkene may be dehydrogenated or a glycol (dialcohol) may be dehydrated to give the alkadiene wanted.

Nature of the Double Bond. The double bond in alkenes consists of two pairs of electrons shared between the carbon atoms. Wave mechanics indicates that the sharing between the pairs is unequal; it seems as if two separate single bonds are present, one of higher energy (difficult to break) with its electrons tightly bound between the carbons (σ-bond), and another of lower energy with electrons less tightly bound (π-bond). The mobility of the π-bond electrons accounts for the high polarizability of the alkenes, and since these electrons are always available to convert to more stable σ-bonds in addition reactions, they are responsible for the very active chemical behavior of the alkenes.

Wave mechanics also predicts that the π-bond should allow the carbon-carbon double bond to rotate only with considerable difficulty.

Reactions. An alkene molecule can react at single bonds as the alkanes do, or at the double bond.

The alkane-type reactions are very feeble, however, and in nearly all cases are completely masked by the much more rapid reactions at the double bond. Many reactions in petroleum technology, such as pyrolysis, rearrangement of the molecules to give more desirable products, and catalytic removal of hydrogen proceed somewhat similarly with either alkenes or alkanes. Also with some alkenes, such as ethylene, one halogen atom can be substituted for a hydrogen at high temperatures much the same as in an alkane:

$$CH_2{=}CH_2 + Br_2 \rightarrow CH_2{=}CHBr + HBr$$

The most rapid and important reactions of alkenes, however, are the addition reactions, where the π-electrons share readily with a variety of reagents. Hydrogen adds either catalytically or with metal-acid combinations to give an alkane.

$$RCH{=}CHR + H_2 \rightarrow RCH_2{-}CH_2R$$

Halogens add rapidly to give dihalides at low temperatures.

$$RCH{=}CHR + Br_2 \rightarrow RCHBr{-}CHBrR$$

Ozone adds to give ozonides which then cleave at the previous location of the double bond into two carbonyl ($C{=}O$) compounds.

$$RCH{=}CHR' \xrightarrow{\;O_3\;} \text{Ozonide} \xrightarrow{\;H_2O\;}$$

$$R\overset{\overset{H}{|}}{C}{=}O + R'\overset{\overset{H}{|}}{C}{=}O$$

The carbonyl fragments are often used to identify the original alkene.

Unsymmetrical reagents (such as HBr) may add in two ways to a carbon-carbon double bond; thus

$$R_2-\underset{\underset{R_2}{|}}{\overset{\overset{R_2}{|}}{C}}{=}\underset{}{\overset{\overset{R_3}{|}}{C}}-R_4 + HBr \rightarrow R_1-\underset{\underset{H}{|}}{\overset{\overset{R_2}{|}}{C}}-\underset{\underset{Br}{|}}{\overset{\overset{R_3}{|}}{C}}-R_4$$

$$\xrightarrow{\text{or}} R_1-\underset{\underset{Br}{|}}{\overset{\overset{R_2}{|}}{C}}-\underset{\underset{H}{|}}{\overset{\overset{R_3}{|}}{C}}-R_4$$

The modern form of Markovnikov's rule predicts the products of this type of reaction. It states that when a reagent reacts with a double bond, the electron-poor (more positive) end will attach to the carbon atom with the greatest number of hydrogens. Thus HBr (the H is electron-poor) normally adds to an alkene in this fashion:

$$R-\underset{}{\overset{\overset{H}{|}}{C}}{=}\underset{}{\overset{\overset{H}{|}}{C}}-H + HBr \rightarrow R-\underset{\underset{Br}{|}}{\overset{\overset{H}{|}}{C}}-\underset{\underset{H}{|}}{\overset{\overset{H}{|}}{C}}-H$$

Organic peroxides, and often oxygen, which may convert an alkene to a peroxide, will cause the reaction to take place in a manner opposite to the predicted reaction; thus the abnormal addition of HBr to an alkene is:

$$R—\underset{\underset{H}{|}}{\overset{\overset{H}{|}}{C}}=C—H + HBr \xrightarrow{\text{Peroxide}} R—\underset{\underset{H}{|}}{\overset{\overset{H}{|}}{C}}—\underset{\underset{Br}{|}}{\overset{\overset{H}{|}}{C}}—H$$

Water adds to alkenes in the presence of sulfuric acid to give an alcohol (H is electron-poor):

$$R—\underset{\underset{H}{|}}{\overset{\overset{H}{|}}{C}}=C—H + H—OH \xrightarrow{\text{H}_2\text{SO}_4} R—\underset{\underset{HO}{|}}{\overset{\overset{H}{|}}{C}}—\underset{\underset{H}{|}}{\overset{\overset{H}{|}}{C}}—H$$

Sulfuric acid and sulfur dioxide will react with alkenes to give alkyl hydrogen sulfates and alkyl sulfonates, respectively, many of which are useful as detergents.

In the industrially important Oxo process the alkenes react catalytically with CO and H_2 to give aldehydes, according to

$$R—CH{=}CH_2 + CO + H_2 \longrightarrow$$

$$R—CH_2—CH_2\overset{\overset{H}{|}}{C}{=}O$$

Alkenes can be polymerized alone or copolymerized in mixture with other alkenes by heating with catalysts. Depending on the conditions, low molecular weight liquids, useful as high-octane gasolines, or high molecular weight solids, useful as rubber and plastics, are formed. Butyl rubber and polyethylene are examples of the latter. Also, alkanes will react with alkenes in the presence of catalysts to form good motor fuels in a process known as alkylation. (See **Alkylation**).

Alkadienes undergo many of the alkene reactions and, depending on the proximity of the double bonds to each other, will undergo some reactions peculiar to themselves.

The cumulated alkadienes, the allenes, with the grouping $—\overset{|}{C}{=}C{=}\overset{|}{C}—$, are relatively unstable, changing into acetylenes spontaneously, or polymerizing. They also add halogens, hypohalous acids and water.

Conjugated alkadienes ($—\overset{|}{C}{=}\overset{|}{C}—\overset{|}{C}{=}\overset{|}{C}—$) add halogens rapidly, but now in two stages. With excess alkadiene, both the 1,4 and the 1,2 additions occur in about equal amounts.

$$CH_2{=}CH—CH{=}CH_2 + Br_2$$
$$\xrightarrow{1,2} CH_2Br—CHBr—CH{=}CH_2$$
$$\xrightarrow{1,4} CH_2Br—CH{=}CH—CH_2Br$$

Note the shift of the double bond in the 1,4 addition. With excess halogen, only the tetrahalogenated product results.

Hydrogen usually adds to give completely saturated alkanes.

The Diels-Alder reaction is an extremely important reaction of conjugated alkadienes. In this reaction, a carbon-carbon double or triple bond (dienophile) will add to a conjugated alkadiene to form a six-membered ring compound. The dienophile multiple bond usually is conjugated with another multiple bond, such as $C{=}C$, $C{=}O$, $C{\equiv}N$,

although this is not always necessary. This reaction is very general and is used extensively in laboratory synthesis. Industrially, it is used for making insecticides and in some types of polymer production. (See **Diels-Alder Reaction**).

Hydrogen halides add 1,4 and hypohalous acids add 1,2 to conjugated alkadienes.

Alkadienes with isolated double bonds react much like alkenes.

Stereochemistry. The carbon-carbon double bond in alkenes will not rotate about its axis very easily. As a result, the four substituent groups are held rigidly in the same plane as the two carbon atoms to which they are attached. If each carbon atom has two different groupings attached to it, two different spatial arrangements of the molecule are possible, giving two different compounds. This is called *cis-trans* isomerism. Thus there are two different 1-chloropropenes:

$$\begin{array}{cc} CH_3—\overset{|}{\underset{||}{C}}—H & H—\overset{|}{\underset{||}{C}}—CH_3 \\ H—\overset{|}{C}—Cl & H—\overset{|}{C}—Cl \\ \textit{trans}\text{ isomer} & \textit{cis}\text{ isomer} \end{array}$$

Trans isomers usually have higher melting points, lower boiling points, are less soluble in a given solvent, have a lower heat of combustion, and are more stable than the corresponding *cis* isomers. Also, when a polymer has many double bonds in its structure, the *cis* or *trans* nature of the $C{=}C$ linkages may exert a strong effect on the physical properties of the material. In general, *cis-trans* isomers do not exhibit optical activity.

L. S. Nelson

Cross-references: *Aliphatic Compounds, Alicyclic and Cylic Compounds, Hydrocarbons, Paraffins, Acetylene Isomers.*

OPTICAL ROTATORY DISPERSION

All material media will interact with an electromagnetic light wave so as to change the speed of the wave from its value *in vacuo*, and diminish the intensity of the wave by absorbing energy from it. Both effects are frequency (wave length) dependent, and are commonly referred to, respectively, as *dispersion* and *absorption*. These effects are conveniently gauged in terms of an index of refraction n, the ratio of the speed of the wave *in vacuo* to its speed in the medium, and an index of absorption κ, defined by the equation $I = I_0 \exp(-4\pi\kappa l/\lambda)$. Here I_0 is the initial intensity of the wave, λ is its vacuum wave length, and I is its intensity after it has travelled l cm in the medium whose absorption index is κ. Substances for which the dispersive and absorptive effects are different for left and right circularly polarized light, even in the absence of external fields, are designated as naturally optically active. For such materials then, both $(n_L - n_R)$ and $(\kappa_L - \kappa_R)$ are in general different from zero. Here the subscripts L and R are used to designate quantities associated, respectively, with light of left or right circular polarization.

The condition that $(\kappa_L - \kappa_R)$ doesn't vanish has as its consequence that plane-polarized light, after traversing an optically active medium, emerges elliptically polarized, and the phenomenon is referred to as *circular dichroism*. In an elliptically polarized wave the electric field is constant neither in time nor in magnitude. Rather, if to indicate the magnitude and direction of the field one employs a vector whose length and direction are time-dependent, the head of the vector would trace out an ellipse in a plane perpendicular to the direction of propagation of the light wave. It is clear that a plane-polarized wave and a circularly polarized wave represent special forms of elliptically polarized waves: plane polarization corresponds to the case of the degenerate ellipse whose minor axis is of zero length, and circular polarization corresponds to the special circular ellipse whose major and minor axes are equal. If from the point of view of an observer looking toward the source (i.e., opposite to the direction of propagation) the electric field vector traces out the circle in a clockwise direction, then the polarization is termed right circular, and if in a counterclockwise direction, left circular.

Because of the differential dispersive condition that $(n_L - n_R)$ is not equal to zero, the major axis of the elliptically polarized wave emergent from the optically active sample is rotated through some angle with respect to the initial direction of the electric field vibration of the originally plane polarized impinging light wave. This rotatory phenomenon is termed circular birefringence. In actual practice, the ellipse is always extremely elongate (the major axis is much greater in length than the minor axis) and one speaks roughly of the rotation of the plane of polarization of the light.

A spectropolarimeter is a device which measures circular birefringence as a function of frequency, and a circular dichrometer (or dichrograph) measures circular dichroism, also as a function of frequency. In cgs units the rotation per unit length ϕ is given by

$$\phi = (\pi/\lambda)(n_L - n_R) \qquad (1)$$

The ellipticity is defined as the angle whose tangent is the ratio of the minor to the major axis of the ellipse. For the small ellipticities that are actually measured in practice, the tangent, in cgs units, is very nearly equal to the angle itself, and then the ellipticity per unit length Θ is given by

$$\Theta = (\pi/\lambda)(\kappa_L - \kappa_R)$$

Since n and κ are concentration-dependent, it is necessary to specify the concentration units. Hence one defines the specific rotation $[\alpha]$ as the rotation due to a unit weight concentration (one gm/cm^3) in a one decimeter tube; (no distinction is made here between 1 cm^3 and 1 ml):

$$[\alpha] = \alpha_{obs}/(C'l')$$

where α is in degrees, l' in decimeters, the subscripts obs mean "observed," and C' is the concentration in gm/cm^3. Actually, for making comparisons among different compounds, something proportional in a known way to the effect produced by a unit molar concentration is more useful. Hence one defines the molar rotation $[\phi]$(or frequently [M]).

$$[\phi] = [\alpha]M/100$$

where M is the molecular weight of the optically active material. Similarly, and for similar reasons, one defines the specific ellipticity Θ_s and the molar ellipticity $[\Theta]$:

$$\Theta_s = \Theta_{obs}/(C'l') ; [\Theta] = \Theta_s M/100$$

In homogeneous, isotropic media, both circular birefringence and circular dichroism are macroscopic manifestations of the interaction of the light wave with the individual optically active molecules. Associated with the energy level spacings ΔE of these molecules are the characteristic absorption

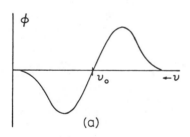

(a)

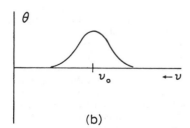

(b)

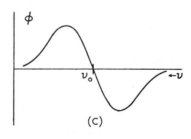

(c)

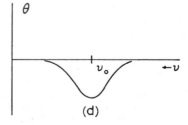

(d)

FIG. 1. Circular birefringence and circular dichroism in the vicinity of an absorption band at frequency ν_0 (idealized). Curves (a) and (b) correspond to a positive Cotton effect, and curves (c) and (d) to a negative Cotton effect.

frequencies ν, as given by the Planck condition $\Delta E = h\nu$, where h is Planck's constant. As the frequency of the impinging light wave approaches one of these characteristic frequencies of the molecule, the interaction between the wave and the molecule becomes more pronounced, and as a consequence, in the vicinity of these absorption frequencies, ϕ and Θ take on rather characteristic shapes. These shapes are sketched in somewhat idealized fashion in the accompanying figure. This characteristic behavior of ϕ and/or Θ in the vicinity of an absorption band is referred to as a *Cotton effect*. Because $(\kappa_L - \kappa_R)$ can be either positive or negative on the longwavelength side of the pertinent absorption band, one distinguishes between positive and negative Cotton effects. A contribution of the type shown in the figure will accrue from each possible transition in the molecule. Hence the experimental rotatory dispersion curve (ϕ vs. ν or λ) is the superposition of all such partial rotatory dispersion curves, and the observed circular dichroism curve (Θ vs. ν or λ) is a superposition of all such partial circular dichroism curves.

In order that a molecule be optically active, it must be nonsuperimposable upon its mirror image. Mathematically, this is equivalent to the statement that it must not contain any alternating axis of symmetry. However, for almost all practical purposes, this condition reduces to the absence of a center of inversion or of any mirror plane of symmetry for the molecule. An extremely large number of chemical functional groups do, of course, possess such elements of symmetry (e.g., the carbonyl group has two orthogonal reflection planes of symmetry). It follows that there are no Cotton effects associated with the absorption bands of such symmetrical groups when they are in a symmetrical molecular environment. For example, the electronic transitions of the carbonyl functionality are inactive in formaldehyde where the extrachromophoric hydrogen atoms are symmetrically placed with respect to the symmetry planes of the carbonyl chromophore. However, there are Cotton effects associated with the same set of electronic transitions, in, for example, the ketosteroids, where the molecular frame-work surrounding a carbonyl group is dissymmetrically disposed with respect to that chromophore's symmetry planes. The magnitude of the Cotton effect that is observed for a given chromophoric transition will then depend upon the nature of the extrachromophoric portion of the molecule and its disposition relative to the pertinent functional group. Hence, Cotton effect data for such chromophoric transitions can be interpreted in stereochemical terms. The type of chromophore just discussed is referred to as inherently symmetric, but dissymmetrically perturbed.

Of course, the intrinsic geometry of a chromophore may be of sufficiently low symmetry that no dissymmetric environment is necessary for the generation of optical activity. Such chromophores are designated as inherently dissymmetric. For example, inherently dissymmetric twisted diene chromophores exist in polycyclic systems, such as lumisterol or ergosterol. In these situations, the sign and magnitude of a chromophore's Cotton effect is determined to a large extent by the intrinsic geometry of the chromophore itself. Hence,

in such instances, analysis of the optical rotatory dispersion or circular dichroism curves can give information not only about the intrinsic geometry of the relevant chromophore, but also about any configurational or conformational consequences that derive from the pertinent dissymmetry of the functional group involved.

By substituting reasonable numbers into equation (1), one finds that for $\phi = 0.1745$ rad/cm $= 100$ degrees/decimeter, $\lambda = 5000$ Å, that $(n_L - n_R)$ is of the order of 10^{-6}. Since ordinary indices of refraction are of the order of unity, this means that the magnitude of $(n_L - n_R)$ is only an extremely small fraction of the value of the mean index of refraction $n = (n_L + n_R)/2$. Similarly, the ratio of $(\kappa_L - \kappa_R)$ to the mean absorption index rarely exceeds more than a few hundredths. These magnitudes provides a clue as to the comparatively subtle structural variations to which optical activity measurements are sensitive, and hence to the reason why comparatively detailed stetreochemical information can be obtained from measurements of optical activity as a function of wavelength. Both n and κ depend upon molecular geometry, and in spectropolarimetry and circular dichrometry one is, in effect, measuring small deviations from the mean values of the structurally sensitive parameters n and κ.

ALBERT MOSCOWITZ

Cross-references: *Asymmetry; Optical Rotation; Refractive Index.*

ORBITALS

Atomic

From spectroscopic studies we know that when an electron is bound to a positively charged nucleus only certain fixed energy levels are accessible to the electron. Before 1926, the old quantum theory considered that the motion of the electrons could be described by classical Newtonian mechanics in which the electrons move in well defined circular or elliptical orbits around the nucleus. However, the theory encountered numerous difficulties and in many instances there arose serious discrepancies between its predictions and experimental fact.

A new quantum theory called wave mechanics (as formulated by Schrödinger) or quantum mechanics (as formulated by Heisenberg, Born and Dirac) was developed in 1926. This was immediately successful in accounting for a wide variety of experimental observations, and there is little doubt that, in principle, the theory is capable of describing any physical system. A strange feature of the new mechanics, however, is that nowhere does the path or velocity of the electron enter the description. In fact it is often impossible to visualize any classical motion that could be consistent with the quantum mechanical picture of the atom.

In this theory the electron is viewed as a three-dimensional standing wave. The pattern of the wave is described by a wave function ϕ (analogous to the amplitude of a water wave). This one-electron wave function is called an atomic orbital. Since the wave function can be positive or negative (and real or complex) it does not describe an observable property of the electron. However, the square of the wave function (ϕ^2 or ϕ times its complex conjugate) is always positive and real, and can be identified with the probability of finding the electron at any point. This was first suggested by Born, but has now received ample experimental support. Hence, when the wave function is calculated for any electron we can determine the regions in space where the electron is most likely to be found, though we cannot say what type of motion results in that particular probability pattern.

Atomic orbitals are usually labeled by a set of designating numbers called quantum numbers. The one that determines the energy of the resulting state (for hydrogen) is given the symbol n and called the "principal quantum number." It assumes the values 1, 2, 3, 4, 5, . . . to infinity, with increasing electron energies. The second quantum number is given the symbol l. It can be identified with the angular momentum of the electron due to its orbital motion, and assumes values of 0, 1, 2, 3, . . . to $(n - l)$. For historical reasons the orbitals with these values are referred to as $s, p, d, f, . . .$ orbitals respectively. Hence a $3d$ orbital is one for which $n = 3$ and $l = 2$. The third quantum number is usually given the symbol m and is difficult to define in the absence of an external field. However, under all conditions it can assume $2l + 1$ values.

From the rules given in the previous paragraph it can easily be shown that for $n = 1$, 2 and 3 there are a total of 14 allowed orbitals. For an isolated hydrogen atom all orbitals must be spherically symmetrical and have the shapes shown in series (a) of Fig. 1. These are plots of the probability function (ϕ^2) for the orbitals and hence the darkest areas represent regions where the electron is most likely to be found. Though there are actually three $2p$ orbitals, three $3p$ orbitals and five $3d$ orbitals, in the absence of an external field the orbitals within a given set are identical in shape and energy, and

are said to be degenerate. If a direction is defined by the presence of a magnetic field or the approach of another atom, the degeneracy is removed. The shapes of the allowed orbitals under these conditions are shown in series (b) of Fig. 1. The quantum number m is well defined for these orbitals and its value is given beside each orbital. For the discussion of bonding in polyatomic molecules another set of orbitals is useful. These are shown in series (c) of Fig. 1. The quantum number m is not well-defined for these orbitals, and they are usually labeled according to the axis along which they lie.

In addition to the orbitals shown in Fig. 1 there are "hybrid" orbitals that are not stationary states for the electron in an isolated atom. They can be obtained by taking a linear combination of the standard orbitals in Fig. 1. Since the electron distribution is "off center" they are useful only for atoms that are perturbed by an electric field (Stark-effect) or by the approach of other atoms as occurs in chemical-bond formation.

In addition to the three quantum numbers discussed above, experimental evidence requires an additional quantum number m_s, which by analogy to classical mechanics is attributed to an intrinsic (i.e., position-independent) property of the electron called "spin." Unlike the other quantum numbers, however, it can assume only two values ($\pm\frac{1}{2}$). As we shall see, this fact determines the orbital population of the many-electron atom.

It is an unfortunate consequence of the mathematical complexity of the quantum mechanical equations that the hydrogenic atom (i.e., one-electron atom) is the only system for which an exact probability distribution (ϕ^2) can be obtained. Approximate methods must be used to calculate the wave functions for the many electron atoms. We begin with the natural assumption that the wave

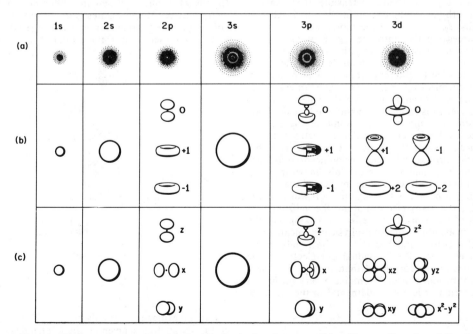

FIG. 1. Hydrogenic atomic orbitals (a) for isolated atoms; (b) with one direction defined; (c) with 3 directions defined.

function for such a more complex atom (ψ) can be obtained by taking a product of the appropriate one electron functions (ϕ). However, even if we ignore for the moment the coulombic repulsion between electrons, there are two fundamental postulates of quantum mechanics that complicate the picture. One is that electrons are indistinguishable. Hence, when their positions are interchanged, the probability function (ψ^2) must remain unchanged. This means that the wave function (ψ) must either remain the same or only change signs when electron positions are interchanged. A second postulate (called the Pauli principle) is that the total wave function must change signs when electron positions are interchanged. From these requirements it can be seen that when two electrons are placed in the same orbital (i.e., assigned the same orbital wave function ϕ) the total wave function will change signs (as required by the Pauli principle) only if the electrons have different spin quantum numbers ($+\frac{1}{2}$ and $-\frac{1}{2}$). It is hence a general rule that hydrogenic orbitals can only be occupied by a maximum of two electrons and these must have opposite spin.

If we begin with the most tightly bound orbital and add two electrons to each orbital the so called "ground state configuration" of the atom is obtained. For example the electronic configuration of the silicon atom is written $1s^2$, $2s^2$, $2p^2$; $3s^2$, $3p^2$. The orbital shapes and energies are, however, considerably altered by electron-electron repulsion, especially between electrons whose orbitals overlap appreciably. This has several marked effects. For a given value of n, all the orbitals no longer have the same energy. The binding energy now decreases with increasing values of the quantum number l. Secondly, because of the coulombic repulsion between electrons, they will tend to occupy separate orbitals whenever feasible (for example the $3p$ orbitals configuration in the silicon atom is actually $3p^1$, $3p^1$). Furthermore, when electrons are forced together into the region of one hydrogenic orbital it is quite likely that electron-electron repulsion (always greater than 25 kcal) leads, in effect, to slightly different orbitals for each electron.

For chemical purposes we are most interested in the shapes of orbitals in which the valence (outermost) electrons reside. By assuming that the inner electrons act only to screen some of the positive charge on the nucleus the valence electrons can be shown to assume the shape of the appropriate hydrogenic orbital, with the insignificant difference that the inner nodes in the orbitals shown in Fig. 1 are drawn in closer to the nucleus because the shielding is poorer in this region. It is worth noting that for the many-electron atom, the indistinguishability of electrons has the effect of making only the total probability function physically meaningful. For this reason and because of the repulsion between electrons, the individual one-electron wave functions are so correlated that, though the "independent orbital concept" remains a very useful approximation, it is not fundamental to the problem.

Molecular

By analogy with atomic orbitals (see above) the wave function (ψ) for one electron in a molecule is

called a molecular orbital, and the probability of finding the electron at any point is similarly given by the value of ψ^2 at that point. Just as in the case of atomic orbitals, an exact solution of the equations is possible only for the one-electron molecule (H_2^+) and it is only for this species that accurate molecular orbitals can be obtained.

The shapes of 10 of the most tightly bound orbitals of H_2^+ are shown in Fig. 2. Three quantum numbers can be used to label these orbitals. The one that is always well defined is given the symbol λ and can be identified with the component of the orbital angular momentum along the internuclear axis. It can take on values 0, 1, 2, 3 ... for which the orbitals are called σ, π, δ, ... respectively. The other two quantum numbers are defined differently depending on the nuclear separation. When the internuclear distance is short, as the case of H_2^+, it is convenient to consider the atomic orbital that would result from a given molecular orbital if the two nuclei were made to coalesce (in our imaginations). In the case of homonuclear diatomic molecules, for example, the molecular orbital designated $3d\,\sigma$ has $\lambda = 0$ and correlates to a $3d$ atomic orbital when the nuclei coalesce (i.e., the atomic quantum numbers n and l become well-defined at short internuclear distance). On the other hand, when the internuclear distance is large, a more useful and significant label is one which identifies the atomic orbitals with which a given molecular orbital correlates when the distance between the two nuclei approaches infinity. For this "separate atom" designation an additional symbol must be used to distinguish between the two molecular states that can arise from a given pair of atomic states. Chemists find it most useful to use the superscript * to indicate the higher energy (antibonding) orbital and the absence of * to indicate the lower energy (bonding) orbital. When the difference in symmetry between the two states is important the symbols g (gerade-symmetric) and u (ungerade-symmetric) are used. Thus, the orbital σ^* ($2p_z$) is an

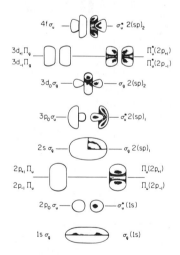

FIG. 2. Molecular orbitals of H_2^+. The "united-atom" designation is given on the left hand side and the "separate-atom" designation on the right hand side.

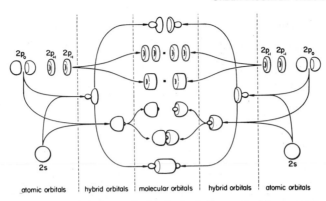

atomic orbitals | hybrid orbitals | molecular orbitals | hybrid orbitals | atomic orbitals

FIG. 3. L.C.A.O. method of obtaining molecular orbitals for N_2. Arrows indicate the combinations of atomic orbitals used to generate molecular orbitals. The stabilities of the orbitals increase from top to bottom.

antibonding orbital with $\lambda = 1$ that correlates with two $2p_\pi$ orbitals on the separated atoms. Similarly, a $\sigma_g 2(sp)$ orbital is a bonding orbital with $\lambda = 0$ that correlates with two atomic orbitals on the separated atoms. It is worth noting that the "antibonding" orbitals possess a nodal surface (i.e., a region of zero electron-probability density) between the two nuclei.

Exact solutions such as those given above have not yet been obtained for the usual many-electron molecules encountered by chemists. The approximate method which retains the idea of orbitals for individual electrons is called "molecular-orbital theory" (M. O. theory). Its approach to the problem is similar to that used to describe atomic orbitals in the many-electron atom. Electrons are assumed to occupy the lowest energy orbitals with a maximum population of two electrons per orbital (to satisfy the Pauli exclusion principle). Furthermore, just as in the case of atoms, electron-electron repulsion is considered to cause degenerate (of equal energy) orbitals to be singly occupied before pairing occurs.

It has not proved mathematically feasible to calculate the electron-electron repulsion that causes this change in orbital-energies for many-electron molecules. It is even difficult to rationalize the qualitative changes in sequence on the basis of the shapes of the H_2^+ orbitals. Greater success has been achieved by an approximate method which begins with orbitals characteristic of the isolated atoms present in the molecule, and assumes that molecular orbital wave functions can be obtained by taking linear combinations of atomic orbital wave functions (abbreviated L.C.A.O.). For homonuclear diatomic molecules, the atomic orbitals used are normally those with which a given molecular orbital correlates. An example of how molecular orbitals are manufactured in the L.C.A.O. approximation is shown in Fig. 3. According to this theory, the relative molecular-orbital energies will depend on (a) the spread of orbital energies on the individual atoms (since this determines the extent of hybridization) and (b) on the internuclear distance which is largely determined by the relative number of bonding and antibonding orbitals that are filled.

The L.C.A.O. approximation has also proved useful for the description of heteronuclear diatomic molecules, although valence-bond theory has been somewhat more successful in its quantitative calculations of bond energies. For these molecules the selection of appropriate atomic orbitals to be used in linear combination is governed by considerations of symmetry, energetics and overlap. These considerations also apply to the formation of molecular orbitals in polyatomic molecules by the L.C.A.O. approximation. In the case of polyatomic molecules it is always possible to develop either (a) molecular orbitals that extend between only two atoms localized M.O. s) or (b) molecular orbitals that extend over the entire molecule (delocalized M.O. s). In the case of saturated molecules, the two descriptions are practically equivalent. However, for molecules with conjugated double bonds and especially for aromatic compounds, delocalized molecular orbitals which extend over several atoms must be used. When the L.C.A.O. method is used to develop molecular orbitals in complexes between metal ions (possessing d orbitals) and ligands such as CO^-, OH^-, NH_3, CN^- etc., the procedure is called Ligand Field Theory.

A thorough test of the L.C.A.O. molecular orbitals has been possible only for H_2^+ where both the wave functions and orbital energies are known accurately. Such a comparison shows that though the shapes of the wave functions are reasonably represented by the approximation, their energies can be appreciably in error, especially for excited state. Nevertheless, the L.C.A.O. approximation has proved the most fruitful method of obtaining molecular orbitals that has been developed to date.

E. A. Ogryzlo

Cross-references: *Ligand Field Theory, Molecules, Atoms, Bonding.*

ORDER-DISORDER THEORY AND APPLICATIONS

Phase transitions in binary liquid solutions, gas condensations, order-disorder transitions in alloys, ferromagnetism, antiferromagnetism, ferroelectricity, antiferroelectricity, localized absorptions, helix-coil transitions in biological polymers and the one-dimensional growth of linear colloidal aggregates are all examples of transitions between an ordered and a disordered state.

The two quantities which apparently must be used to explain or describe these phenomena are the presence of a potential between the particles or spins and a small volume which must be assigned to each particle so that two particles or spins cannot occupy the same space. All the above phenomena may be semi-quantitatively treated by a single statistical model called the Ising model which consists of a lattice in space, each site of which possesses either a 0 or a 1. Thus two rows of a two dimensional lattice might look like

$$\ldots 011000011011110011 \ldots$$
$$\ldots 110110010100110111 \ldots$$

The 0's and 1's represent different particles or spins depending upon the system being studied and their definition will be given below for different systems. The disordered state corresponds to a random array of 0's and 1's. The ordered state is described by an ordered arrangement of the 0's and 1's. The total number of lattice sites equals N. The number of 0's and 1's on the lattice are respectively n_0 and n_1 and $n_1 + n_0 = N$.

The dimensionality of the lattice depends on the physical nature of the phase. For transitions in linear biopolymers and associative colloids, a one-dimensional lattice is used because these materials become ordered in a one-dimensional manner even though they actually exist in three dimensions. For absorption of gas onto a surface, a two-dimensional lattice is sufficient. For bulk phase changes, however, a three-dimensional lattice must be used.

The great advantage of this model is that it gives essentially the same explanation for a large number of seemingly completely diverse physical and chemical phenomena, often with quantitative success. This permits a deeper insight into the statistical thermodynamic behavior of different types of matter.

The calculation of the thermodynamic properties begins by selecting values for w_{00}, w_{11}, w_{10}, which are the potential energies between like and unlike particles or spins occupying nearest neighbor sites. Since two particles cannot occupy one lattice point, the energy of repulsion of two particles on the same site is considered infinite. The energy of a given configuration is then given by

$$E_{Conf} = n_{11}w_{11} + n_{10}w_{10} + n_{00}w_{00} \quad (1)$$

where n_{11}, n_{00} and n_{10} are the number of nearest neighbor pairs. The probability of a given configuration is given by Boltzmann's theorem as

$$P(E_{Conf}) = e^{-E_{Conf}/kT} \bigg/ \sum_{Conf} e^{-E_{Conf}/kT} \quad (2)$$

where the summation occurs over all possible configurations, keeping the number of 0's and 1's constant. The restriction of a constant number of 0's and 1's may be removed by letting the lattice interact with its surroundings. The probability that the lattice possesses a given number of 1's and a configurational energy, E_{Conf}, is given by

$$P(E_{Conf}, n_1 = \frac{e^{-Xn_1/kT} \, e^{-E_{Conf}/kT}}{\sum\limits_{n_1=0}^{N} e^{-Xn_1/kT} \sum\limits_{Conf} e^{-E_{Conf}/kT}} \quad (3)$$

where X is a suitable thermodynamic quantity such as the chemical potential or magnetic field, etc.

For a lattice gas, the 0 and 1 stand for an empty and filled site respectively. Consequently, w_{11} is the attractive potential energy between two gas molecules when they occupy adjacent sites on the lattice, and $w_{00} = w_{10} = 0$.

For binary solutions including linear associative colloids, the 0 and 1 represent solvent and solute respectively; w_{11} and w_{00} are the potential energies between like molecules and w_{10} is the potential energy between unlike molecules.

For a ferromagnet the 0 and 1 represent spins of $-\frac{1}{2}$ and $+\frac{1}{2}$ respectively. It is generally assumed that the potential energy between like spins is always the same, i.e., $w_{11} = w_{00}$. To show how the quantity, X of Eq. (3), is obtained, consider a ferromagnet interaction with an external magnetic field. The total energy is

$$E = (n_0 - n_1) \, H \cdot d + E_{Conf}$$

where H is the magnetic field strength, and d is the magnetic dipole moment of a spin. Thus for a ferromagnet $X = 2H \cdot d$ since $N - 2n_1 = n_0 - n_1$.

The above described order-disorder transitions are all three-dimensional phase transitions and occur with essentially infinite sharpness unless the condensed phase exists in a colloidally dispersed state. Recently, it has been shown that certain

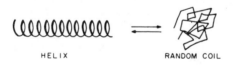

EXAMPLES

POLYLBENZY - L - GLUTAMATE
COLLAGEN
POLYGLUTAMIC ACID
DNA
AMYLOSE - IODINE HELIX

HELIX   RANDOM COIL

BENZOPURPURIN-4B
PSUEDOCYANINE CHLORIDE
SODIUM DEOXYCHOLATE
GUANOSINE MONOPHOSPHATE
FLAVONE - IODINE COMPLEX

HELICAL AGGREGATE   MONOMER

FIG. 1. Diagrammatic illustration of the helix = coil transitions in biopolymers (top). Helix = monomer transition in associative biocolloids (bottom).

polymers and associative colloids, particularly those of biological interest, have one-dimensional order-disorder transitions which may be explained in exactly the same terms that describe the three-dimensional phase changes discussed above. However, because the condensed state is colloidally dispersed and the ordering occurs only in one dimension, it may be shown that such transitions cannot be infinitely sharp.

Figure 1 illustrates the transition from an ordered helical to a disordered state which occurs in a large number of colloidal systems. For the helix-coil transition in polymers, the 1 and 0 represent, respectively, a hydrogen bonded turn of the helix and an unhydrogen bonded section of randomly fluctuating polymer. In an associative colloid, the 0 or 1 represent respectively a cell containing a solvent or colloid monomer. In either case, there exists an energy of attraction between 1's which leads to the formation of a one dimensional ordered helix. The fact that the transition tends to be rather sharp comes about because the 01 and 10 configurations which always occur at the beginning and end of a helical sequence are energetically unfavorable. This means that configurations with a lot of ends containing many short segments are suppressed since the configurational energy is large when n_{10} is large and this leads to a small Boltzmann factor for this configuration.

The fact that so many diverse phenomena can be correlated and explained by such a simple model makes it possible to develop tables of equivalent thermodynamic properties. This was first done by Yang and Lee who showed the equivalence between the properties of the lattice gas and Ising ferromagnet. Hill extended their analogies to the case of binary liquid mixtures. After the development of the helix-coil transition theory by Zimm and Bragg and others, Peticolas gave a corresponding table of equivalent thermodynamic properties between polymers and associative colloids. Thus the force-length curve for a helical polymer is the one dimensional analogue of the three dimensional pressure-volume curve for gas condensation and is equivalent to the chemical potential-mole-fraction curve for the associative linear colloid.

W. L. PETICOLAS

Cross-references: *Thermodynamics, Phase Rule, Colloid Chemistry.*

References

Hill, T. C., "Statistical Thermodynamics," Addison-Wesley, Reading, Mass., 1960.

Peticolas, W. L., *J. Chem. Phys.,* **37,** 2323 (1962); *ibid.,* **40,** 1463 (1964).

ORGANIC CHEMISTRY

Organic Chemistry is a study of the compounds of carbon. Usually excluded, however, are the metallic carbonates, cyanides, and other similar compounds, which are considered inorganic in nature. The name "organic" evolved from the original theory that any material derived from or produced by any living organism required a "vital force" identified only with life itself. All other compounds were considered to be of mineral origin and were termed "inorganic." The organic compounds were thought to be utterly complex and impossible of synthesis in the laboratory.

Lavoisier showed in the late 1700's by the analysis of many organic compounds that they were not hopelessly complex and that they all contained the element carbon, most commonly combined with hydrogen, oxygen, nitrogen, sulfur and phosphorus in decreasing order of occurrence. The first attack on the vital force theory was made when Scheele produced in his laboratory such organic compounds as lactic acid from sour milk, malic acid from apples and citric acid from lemons; but proponents of this theory claimed that the original source was still a living organism. In 1828 Wohler succeeded in the first laboratory synthesis of an organic compound when he fused the inorganic material ammonium cyanate and recovered urea, the organic waste product of nitrogeneous metabolism in the body, formed in this case by an intramolecular rearrangement. Soon many other of the simpler organic compounds were synthesized in the laboratory. Although the vital force theory was disproved, the name "organic" remains, today denoting the study of most of the compounds of carbon. The old scope of Organic Chemistry—the study of the chemistry of the living organism—is covered now in the field of Biochemistry, formerly called Physiological Chemistry. (See **Biochemistry**).

Over 500,000 different organic compounds have been isolated or synthesized and characterized, and the number of possible compounds is vastly larger. For example, the single organic molecule with the empirical formula $C_{40}H_{82}$ can form 6.25×10^{13} different and distinct compounds. This is possible because of the electrical neutrality of the carbon atom. Carbon lies midway in the periodic chart between the strongly electropositive elements of group I-A and the strongly electronegative elements of group VII-A. It thus shows little tendency to transfer its electrons from the last unfilled orbit to other atoms, but instead tends to share the four electrons in its last orbit with the formation of four electron pairs. It not only shares electrons with other species of atoms, but also with other carbon atoms to form either rings or chains. In addition to this, organic compounds can form isomers: that is, different compounds containing the same elements in the same proportions but with different structural or spatial arrangement of the atoms. Thus a study of Organic Chemistry would be hopelessly complex except for the orderly manner of arrangement and comparison of the various possible classes of compounds.

The field of Organic Chemistry is usually broken down into three divisions, namely: (1) aliphatic, (2) aromatic, and (3) heterocyclic.

Aliphatic Compounds. Aliphatic compounds are those in which the carbon atoms are linked together to form continuous or branched chains. Closely allied both physically and chemically are those in which the terminal carbon atoms of the chains are further linked so as to form a ring. These are often called the cyclo aliphatic or alicyclic compounds. The simplest aliphatic compounds are composed only of carbon and hydrogen, in which each of the adjacent carbon atoms in the chain shares only two electrons (one valence bond) while the other valence bonds are filled by sharing electrons with sufficient hydrogen atoms to give each carbon atom its normal valence number of four. These compounds are known for their relative chemical inertness, from which the common name "paraffin series" (Greek = little affinity) is derived. The most important reaction is that of oxidation to form carbon dioxide and water with the release of heat and possibly useful energy. Industrially long chains as found in petroleum are pyrolyzed or catalytically cracked to form more useful shorter chains and/or olefins. Saturated hydrocarbons also can undergo substitution reactions in which one or more hydrogen atoms is replaced by another atom or group.

When two adjacent carbon atoms share four or six electrons (two or three valence bonds) an unsaturated compound is formed. The double-bonded compounds are called "olefins" (Greek = oil producer) due to the property of forming oily liquids when combined with chlorine, while the triple-bonded compounds are called "acetylenes" after the first and most important member of that series. The physical properties of unsaturated compounds vary but little from the corresponding saturated derivative, but chemically they are much more active, the chief reaction being that of addition of other atoms to form less unsaturated or fully saturated compounds.

Other series of compounds are possible when one or more hydrogen atoms from a hydrocarbon is replaced by a functional group typical for each. These molecules are characterized by two parts, namely: (1) the hydrocarbon (R), and (2) the characteristic functional group. This results in the following important classes:

hydrocarbon	R—H
alcohol	R—OH
alkyl halide	R—X
amine	$R-NH_2$ and $R-\overset{\underset{\displaystyle \mid}{R}}{N}-H$ and $R-\overset{\underset{\displaystyle \mid}{N}}{\overset{\displaystyle \mid}{R}}-R''$
nitro compound	$R-NO_2$
nitrile	RCN
ether	R—O—R'
aldehyde	$R-\underset{\displaystyle O}{\overset{\displaystyle \parallel}{C}}-H$
ketone	$R-\underset{\displaystyle O}{\overset{\displaystyle \parallel}{C}}-R'$
carboxyl acid	$R-\overset{\displaystyle O}{\underset{\displaystyle OH}{C}}$
sulfonic acid	$R-SO_3H$
ester	$R-\overset{\displaystyle O}{\underset{\displaystyle O-R'}{C}}$
acyl halides	$R-\overset{\displaystyle O}{\underset{\displaystyle X}{C}}$

amides	$R-\overset{\displaystyle O}{\underset{\displaystyle NH_2}{C}}$
acid anhydride	$R-\overset{\displaystyle O}{\underset{\displaystyle O-C\overset{\displaystyle O}{\underset{\displaystyle R'}{}}}{C}}$
salts	$R-\overset{\displaystyle O}{\underset{\displaystyle O\text{-metal}}{C}}$ and R—SO_3-metal

These series are oxidation or replacement products of the hydrocarbon and can be prepared by a variety of relatively simple means.

Each series of compounds has a special functional group related to it with various methods of preparation and certain characteristic reactions. For example, ethers are characterized by the carbon-oxygen-carbon linkage, with a resulting relative inertness. Aldehydes and ketones are closely related, both having the carbonyl grouping $-\overset{\displaystyle O}{\overset{\displaystyle \parallel}{C}}-$ and the reactions are chiefly those involving this group. The carboxylic acids are characterized by the presence of the carboxyl group $-\overset{\displaystyle O}{\underset{\displaystyle OH}{C}}$ with the typical acidic reactions. All these series are more chemically reactive than the saturated hydrocarbon, undergoing reactions such as oxidation, reduction, hydrolysis, elimination or replacement with relative ease. Often two or more identical or different functional groups may occur in the same molecule, each group usually contributing its own chemical properties. This results in such important series as the amino-alcohols, hydroxy acids, halo-ketones, amino acids, dicarboxylic acids and many other similar compounds. (For further discussion, see **Aliphatic Compounds.**)

Aromatic Compounds. Aromatic compounds were so named because of the pleasant aroma, though many were not members of this series. Aromatic compounds are derivatives of benzene, a six-membered carbocyclic ring with bonding commonly represented by alternate double and single bonds connecting the carbon atoms and a single hydrogen atom attached to each carbon atom. The presence of multiple valence bonds between adjacent carbon atoms (unsaturation) usually imparts an increased chemical reactivity to the molecule; still the benzene molecule does not have the properties usually associated with unsaturation, but instead acts much more like the saturated molecule. This is due to the conjugated nature of the ring, that is the alternate double and single bonds, resulting in a resonance of the electrons between adjacent carbon atoms.

With the exception of addition; which can occur to the benzene ring under extreme conditions, there is only one type of reaction that occurs with relative ease—that of substitution of hydrogen atoms. This again shows the chemical resemblance to the saturated molecules. Direct substitution is limited, however, to four different groups, namely: (1) halogen atoms, (2) the nitro ($-NO_2$), (3) alkyl ($-R$)

$$\overset{\displaystyle O}{\underset{\displaystyle |}{}}$$

or acyl ($-\overset{O}{\underset{|}{C}}-R$) and (4) the sulfonic acid ($-SO_3H$) groups. Although the above four different types of groups may be substituted directly into the ring, still by indirect means any of the functional groups discussed under the previous heading can be attached to the aromatic ring. This results in a molecule having the chemical reactions typical of the aromatic ring, plus that of the functional group.

Just as carbon atoms can share electrons with other carbon atoms to form chains or rings, so carbon atoms in one ring can share electrons with atoms in another ring to form "fused rings;" that is, certain carbon atoms are common to two or more rings. The chemical properties are intermediate between those of aliphatic and aromatic compounds since in such systems the bond arrangement is not always uniformly conjugated. The simplest fused ring system is that composed of two rings (naphthalene), but much more complicated structures are found in nature or have been synthesized in the laboratory. Certain derivatives are known to be carcinogenic in nature when repeatedly applied to the skin of animals; others with functional groups attached are important in the normal functions of the body, while still others are the basis of many important dyes and drugs. (For further discussion, see **Aromatic Compounds.**)

Heterocyclic Compounds. Heterocyclic compounds are ring compounds in which at least one atom in the ring is not carbon. The most common hetero atoms are nitrogen, oxygen, and sulfur in decreasing order of occurrences. The most important series are those that contain either five or six atoms in the ring, as these are the most common and most stable. These are the only rings that show the common aromatic reactions, although they are chemically less aromatic than the benzene ring itself. Other rings with as few as three and as many as eight atoms are known, but they are relatively unstable and aliphatic in nature. Also heterocyclic rings are known that contain two and three hetero atoms; but except for a few compounds of physiological importance, the most common are those that contain a single atom other than carbon. Heterocyclic rings fused to aromatic rings are not uncommon but with the exception of quinoline, in general, they are of less importance than the simple heterocyclic ring.

The most common nitrogen heterocyclic rings are those of pyrrole, pyridine, and quinoline. These compounds are found in coal-tar distillate and are widely distributed as derivatives. Among the nitrogen heterocyclics found in nature are the alkaloids. (See **Alkaloids.**) These are basic compounds due to the presence of the nitrogen atom, thus "like alkali," from which the name is derived. They are usually found in combination with organic acids, as salts, and are characterized by powerful physiological activity. Alkaloids of importance include such compounds as quinine, codeine, morphine, nicotine and cocaine. Besides the alkaloids, nitrogen heterocyclics are found in such important compounds as chlorophyll, hemin of the blood, bile pigments and most of the B vitamins.

Oxygen heterocyclics include those of furan and pyran. The most common occurrence of these compounds is as the carbohydrates, one of the three major classes of foodstuffs derived mainly from the vegetable kingdom. The name "carbohydrate" is derived from an old concept that this class of compounds were "hydrates of carbon," since upon analysis they were all found to have the general formula $C_x(H_2O)_y$. Other organic compounds were found which had the same empirical formula but had none of the properties of the carbohydrates; thus the old term lost its meaning. It has survived, however, but now denotes a "polyhydroxy aldehyde or ketone or a compound that may form this on hydrolysis or rearrangement." It has been shown that carbohydrates through the addition of a molecule of water followed by elimination of a different water molecule, form either the furan or pyran ring. This class of compounds occurs most commonly as the simple molecule, a dimer (two molecules combined by the loss of a molecule of water) or as a polymer (many molecules combined). Examples of the simpler molecules are the sugars, while the best examples of the polymeric form are starch and cellulose. (For further discussion see **Heterocyclic Compounds** and **Carbohydrates.**)

ELBERT H. HADLEY

Reactions in Organic Chemistry

Organic reactions consist essentially of bond breaking and bond making processes. Bond breaking may involve: (1) homolytic fission, or homolysis, of an electron pair bond, leading to neutral free radicals, or (2) heterolytic fission, or heterolysis, of a bond leading to ion pairs. Bond making is the reverse of bond breaking and may precede, follow, or occur simultaneously with bond breaking.

Differences in ease of bond homolysis and bond heterolysis reactions may be predicted for organic substrates having an $X=Y-Z-A$ generalized structure. In such a system, bond $Y-Z$ is strong and bond $Z-A$ is weak. This generalization, originally called the Schmidt double bond rule, may be applied to a great variety of organic reactions because many of the complex and polar functions contained in organic compounds may be identified with the $X=Y-Z-A$ general structure.

In the $X=Y-Z-A$ general structure, the atom or group of atoms A is "active" and tends to dissociate as a radical, $A\cdot$, an anion, $A\!:^{\ominus}$, or a cation, $A^{\oplus}$. The manner by which atom A dissociates depends on the nature of atoms X, Y, Z, and A and on the conditions of the reaction. A driving force for the dissociation of atom A is the stabilization of the remaining $X=Y-Z$ radical, cation, or anion by solvation and/or charge or electron delocalization.

$$X=Y-Z-A \begin{cases} \xrightarrow{-A\cdot} X=Y-Z\cdot \longleftrightarrow \\ \xrightarrow{-A:^{\ominus}} X=Y-Z^{\oplus} \longleftrightarrow \\ \xrightarrow{-A^{\oplus}} X=Y-Z:^{\ominus} \longleftrightarrow \end{cases}$$

$$\cdot X-Y=Z \quad \text{or} \quad ---X\overset{...}{}Y\overset{...}{}Z---$$

$$\overset{\oplus}{X}-Y=Z \quad \text{or} \quad {}^{(+)}X\overset{...}{}Y\overset{...}{}Z^{(+)}$$

$$\overset{\ominus}{:}X-Y=Z \quad \text{or} \quad {}^{(-)}X\overset{...}{}Y\overset{...}{}Z^{(-)}$$

Some form of stabilization of the dissociated atom A may provide an additional driving force for the reaction.

Atom A may be monovaent or higher; it is usually found to be H, Cl, Br, I, O, N, or S.

The bond dissociation energies (BDE in kcal/mole) of C—H and C—C bonds in benzyl and allyl positions, which correspond to bond Z—A in the general system, are decidedly lower (C—H, 74–83; C—C, 11–65) than those for saturated alkyl (C—H, 89–103; C—C, 78–85) and vinyl (C—H, 102–121; C—C, 87–91) systems. The ease of the free radical abstraction or replacement of allyl or benzyl hydrogen atoms is thus expected and, indeed, is well known. In the light or peroxide induced halogenation of cyclohexene, the intermediate allyl radical is more highly stabilized, or is of lower energy, than any other possible radical and is therefore most favorably formed. *Alpha* hydrogen atoms in other olefins, in aralkyl hydrocarbons, ketones acids, and other carbonyl containing compounds are readily abstracted. Picrylhydrazine forms a stable free radical: N',N'-diphenyl-N-picrylhydrazine (DPPH) apparently exists only in the radical form. Peroxides, such as diacetyl peroxide (BDE$_{O—O}$ = 30), and azo compounds, such as azotriphenylmethane (Temp.$_{dec.}$ = 25°), capable of homolytic dissociation to form stabilized radicals of the X=Y—Z· type show typically low bond dissociation energies, low decomposition temperatures, low decomposition activation energies, and short half-lives.

The Schmidt double bond rule was used originally to predict the course of the pyrolytic dissociation of naturally occurring materials, such as limonene.

When of low electronegativity or when attached to an atom Z of high electronegativity, atom A in the general X=Y—Z—A system is likely to dissociate as a cation. The resulting anion, $^{(-)}$X$\cdots$Y$\cdots$Z$^{(-)}$, is capable of a maximum stabilization by charge delocalization if either X or Z is of high electronegativity. Prototropy is a special case of cationotropy in which atom A dissociates as a proton. Functional groups resembling the X=Y—Z—A general structure, and from which a proton would be expected to dissociate, include carboxylic acids, sulfonic acids, amides, imides, sulfonamides, sulfonimides, enols, phenols, oximes, C-nitroso and N-nitroso compounds, diazoic acids, alkyl olefins, alkyl benzenes, and inorganic acids such as carbonic, sulfuric, nitric, nitrous, sulfurous, phosphoric, and cyanic acids. A proton may thus dissociate from atoms (C, N, O) *alpha* to vinyl groups, phenyl groups, and unsaturated polar functional groups such as carbonyl, cyano, nitro, nitroso, imino, sulfinyl, sulfonyl, phosphoryl, and phosphonyl groups.

It is obvious from the measured relative acidities of carboxylic acids, phenols, amides, imides, ketones, and hydrocarbons that the acidity of a proton source is dependent in part on the electronegativity of the atom to which the dissociable proton is attached. Thus the relative ease of protolytic dissociation when the proton bearing atoms (C, N, O) are in similar structural environments is: O—H > N—H > C—H.

The acidity or the ease of heterolytic dissociation of a proton from an O—H bond in a carboxylic acid in aqueous medium is not related directly to the gas-phase homolytic bond dissociation energy of the carboxylic O—H bond. The O—H bond dissociation energy for most acids (102–112) are higher than most C—H bond dissociation energies (74–103) and as high as, or higher than, those of many alcohols (100–118). In aqueous solution, however, carboxylic acids lose a proton much more readily than do alcohols. The acidic properties of carboxylic acids are probably due primarily to the solvation energy of the carboxylate ion and to the high electron affinity of the carboxylate radical,

$$R—\overset{\overset{\textstyle O}{\|}}{C}—O\cdot,$$

which, in turn, depends on the extent of charge dispersal in the carboxylate anion,

$$R—\overset{\overset{\textstyle O}{\|}}{C}—O:^{\ominus}$$

Atom A in the general X=Y—Z—A system, when highly electronegative or when attached to an atom Z of low electronegativity, will tend to dissociate as an anion. The resulting cation, $^{(-)}$X$\cdots$Y$\cdots$X$^{(-)}$, in which atoms X and Z bear partial positive charges, is capable of the greatest degree of stabilization if both X and Z are of low electronegativity. In fact, the heterolysis of bond Z—A to form the anion A:$^{\ominus}$ is limited almost entirely to cases in which atoms X and Z are both carbon because of the inability of a more electronegative element, such as oxygen, to accommodate a positive charge. The lability of bond Z—A in anionotropic reactions is illustrated by the ease of formation of allyl and benzyl carbonium ions in the solvolysis reactions of the corresponding halides. Tropylium bromide exists as a completely dissociated water-soluble salt in which the tropylium cation is stabilized by solvation and by dispersal of the positive charge.

The relative ease of solvolytic anionotropic dissociation and the gas-phase homolytic bond dissociation energies of alkyl, allyl, and benzyl halides are parallel. The *gas-phase homolytic* bond dissociation energies, however, do not truly reflect the energy required to separate two particles of opposite charge. The *gas-phase heterolytic* bond dissociation energies of various alkyl hydrides (hydrocarbons), halides, and alcohols, and the ionization potentials of the corresponding free radicals more adequately illustrate that allyl and benzyl carbonium ions are more easily formed than vinyl and most alkyl carbonium ions.

The bimolecular displacement (S$_N$2) reaction is an anionotropic reaction in which atom A is displaced as A:$^{\ominus}$ by another nucleophilic atom or group of atoms, B:$^{\ominus}$. The relative rates of reaction of a series of halides under the conditions of the S$_N$2 reaction show the benzyl halides react about four times as rapidly as methyl halides. Moreover, *alpha*-halo acids, esters, amides, and nitriles react up to thousands of times more rapidly than the corresponding methyl halides.

The difficulty of breaking a vinyl bond, either homolytically (BDE$_{C—H}$ = 102–121; BDE$_{C—C}$ = 87–91) or heterolytically, may be due to a lack of appreciable stabilization of the resulting vinyl radical, anion, or cation. Thus, vinyl and aromatic halogen and alkyl groups, enolic, phenolic, and carboxylic hydroxy groups, and other atoms or groups of atoms occupying a vinyl position are not easily removed or replaced under any nonforcing conditions. The currently proposed transition state in the *forced* displacement of aromatic halogen has a structure resembling an X=Y—Z—A system in which the halogen being displaced occupies a position corresponding to atom A.

There are obvious extensions to the rule as formulated. Bond X—Y need not be double, but may be a triple bond, as in $C\equiv N$ or $C\equiv C$; or it may be any delocalized system of pi electrons as in the phenyl group. Atoms X and Y may be replaced by a single atom, such as sulfur, bearing at least one electron pair and capable of expanding its valence shell.

B. E. HOOGENBOOM

Cross-references: *Bonding, Electronegativity, Halogen Chemistry.*

ORGANIC SYNTHESIS

Organic synthesis may be defined as the formation of an organic compound by the chemical reaction of simpler compounds or by reaction of the elements. Sometimes the term "organic synthesis" is also applied to the formation of compounds via the breakdown of more complicated materials; in that case the process may also be called a degradation. The question why organic synthesis should be considered separate from inorganic synthesis merits some explanation. Compared to inorganic compounds, organic compounds usually have much lower melting points and boiling points. Organic compounds are decomposed by heat at much lower temperatures than inorganic. Profound differences in solubility behavior exist, inorganic compounds generally being water-soluble, while organic compounds are soluble in typical organic solvents. In contrast to inorganic materials, isomers are very common in the case of organic compounds. Typical inorganic reactions are instantaneous, while organic reactions are frequently slow. The chemical bonds of organic molecules are of a covalent nature, in contrast to the ionic bonds commonly encountered in inorganic materials. While inorganic compounds may be derived from any of the elements of the periodic chart, organic compounds generally contain only the elements carbon, hydrogen, oxygen, nitrogen, the halogens, sulfur and in exceptional cases a few others.

Carbon atoms have the ability of joining themselves to each other to form long chains or rings. These chains or rings are the backbone of the organic molecule. The ability of the carbon atom to form covalent bonds with other carbon atoms is responsible for the large number of organic compounds, which exceed the number of inorganic compounds by a large margin. It is also the cause of the frequent incidence of isomerism. Taking these differences into account, it is therefore not surprising that the organic chemist uses methods and tools quite different from those used for inorganic reactions.

Chemists once thought that the synthesis of organic compounds *in vitro* was impossible. The overthrow of this belief is generally credited to Wöhler. In 1824, he synthesized urea, long known to be a product of the animal metabolism, by the following reaction:

$$(NH_4)O\ CN \rightarrow NH_2 - \overset{\displaystyle O}{\overset{\displaystyle \|}{C}} - NH_2$$

At that time, this discovery appeared to be so radical that Wöhler repeated the synthesis several times, and did not make his findings public for several years. A few years later, Kolbe succeeded in forming acetic acid, a typical organic compound, from inorganic sources. From these humble beginnings, synthetic organic chemistry has grown so rapidly that our present civilization would be impossible without synthetic organic chemicals.

In the practice of organic synthesis, it is important to remember that with few exceptions, all organic compounds are flammable. Hence any exposure of the reaction mixture to flames, electric sparks and the like must be avoided. Nearly all organic compounds are toxic, at least to some extent, and the chemist must avoid breathing the vapors of the compounds, or the actual contact of the compound with his skin. For all but the simplest reactions, a preliminary literature survey is usually made. If the compound to be synthesized can't be found in the available literature sources, it is possible to try a procedure for the synthesis of a different but otherwise closely related compound.

Most laboratory syntheses are carried out in glass apparatus. For nearly all requirements "Pyrex" glass is satisfactory; however, at extremely high temperatures "Vycor" glass or even quartz may be used. While glass can be employed for work at atmospheric pressure or for vacuum work, steel vessels, such as bombs or autoclaves, must be employed when high pressures are to be applied. On an industrial scale, steel reactors are nearly always used. Sometimes corrosion becomes a serious problem; in that case one of the special alloy steels, steel lined with glass or ceramic, or more resistant metals, such as tantalum or titanium, may become the material of choice.

Many chemical reactions are carried out in the presence of a solvent. The polarity of the solvent often affects the course of the reactions. In addition, many reactions would be too violent without the modifying presence of a diluent. This diluting effect is sometimes also used to suppress undesirable side reactions. Typical syntheses are often carried out by refluxing the reactants in an excess of a particular solvent. Since under these conditions, the boiling point of the solvent becomes essentially the boiling point of the reaction mixture, the chemist is able to control the reactions temperature by a judicious choice of the solvent. Some frequently used solvents are methanol, ethanol, ethyl ether, acetone, petroleum ether, ethyl acetate, benzene, toluene, carbon tetrachloride and carbon disulfide. Sometimes an excess of one of the reactants may be conveniently used as the solvent.

The reaction generally is performed by adding the components and heating for a desired time. In place of adding the components at once, one of the reactants may be added in small portions as the reaction progresses. Most reactions proceed too slowly at room temperature; hence the reaction rate is usually increased by the application of heat. Many suitable heating devices are in use. Electrical heating mantles are very commonly used in the laboratory. Oil baths, sand baths, steam or water baths, baths of low melting alloys, or even open gas flames are also used. Superheated steam is often employed for industrial purposes. Many organic reactions are exothermic, and hence produce a considerable amount of heat on their own accord. It is therefore good practice to mount the heating source in such a fashion that it can be readily shut off and removed, should the reaction become too violent. Many organic syntheses require the use of homogeneous or heterogeneous catalysis. Generally speaking, catalysts are present in minor amounts only. They may be organic or inorganic. A number of processes are known where micro-organisms such as bacteria, yeasts or enzymes serve as catalysts. The speed of an organic reaction is increased greatly by the use of the proper catalyst. Stirring a reaction mixture is often beneficial, especially in heterogeneous systems. In that case stirring serves to effect a closer contact between the individual phases.

For many organic reactions, the equilibrium is unfavorable. To overcome this, the chemist frequently employs a molar excess of one of the reagents, usually the cheaper or more readily available one. Continuous removal of one of the products may eventually lead to good yields. In the case where one of the reactants is a gas, such as in hydrogenations, it may be advantageous to work under pressure. All these procedures have the effect of pushing the reaction equilibrium toward a more favorable direction.

On completion of a reaction, the crude product always has to be purified. Catalysts, unreacted starting materials, undesirable byproducts and solvents have to be removed. Inorganic drying agents may serve to remove water from the products. Commonly used drying agents are calcium chloride, sodium sulfate, magnesium sulfate, magnesium perchlorate and calcium oxide. In addition, water may sometimes be conveniently removed by azeotropic distillation.

The final purification step depends on the chemical and physical properties of the entities to be separated. Low-melting solids or liquids are frequently distilled. If one of the products tends to decompose at its atmospheric boiling points, vacuum distillation may be indicated. By this technique, the boiling point of the compound may be lowered considerably. By means of some of the newly developed laboratory columns, such as the spinning band columns, distillation efficiencies of over one hundred theoretical plates may be achieved. Such a column is capable of separating products boiling only a few degrees apart. When relatively high molecular weights are encountered, molecular distillation may be employed to advantage. As an intermediate purification step, steam distillation is a frequently employed procedure. In this instance, steam or superheated steam is passed through the reaction mixture. The compound to be removed is volatilized and subsequently condensed together with the steam. The organic material then forms a separated layer or it is insoluble in the condensate, and it is therefore easily removed. How-

ever, additional purification, at least a drying step, must follow a steam distillation. Occasionally an azeotropic distillation may be used to advantage in the purification of a compound.

Typical solids are most frequently purified by recrystallization. The choice of a proper solvent plays an important part in this process. The compound to be purified should be either more soluble or less soluble in the solvent than the impurities. Several recrystallizations may be required, especially if the impurities are very similar to the desired material. Occasionally several solvents may be combined in a crystallization, or, different solvents may be used in successive recrystallizations. Residual traces of colored materials can sometimes be removed by adsorption on special types of charcoal. For materials which are difficult to purify by other methods, chromatography is sometimes used. This procedure involves the selective adsorption and desorption of the various constituents of a mixture on a column of adsorbing material, such as alumina or silica. Another process which is capable of separating materials having almost identical properties, is the recently developed thermal diffusion method.

As criteria of the purity of an organic compound one compares the compound's physical and chemical properties with the properties recorded for that compound in the literature. In the case of a new compound, the material is usually purified until successive purification steps no longer alter the properties of the material. Properties which are often employed as indications of the purity of a compound are melting points, boiling points, refractive indices, elementary analyses, molecular weights and various spectrochemical characteristics, such as the ultraviolet, infrared, Raman or mass spectra. Theoretically both the melting and the boiling points should occur at a well defined definite temperature. If a temperature range is observed, the compound is impure. The chemist must exercise due care so that during the melting and boiling process a state of equilibrium exists; otherwise, his observations may be in error. For accurate work, the thermometer employed has to be calibrated, and the necessary stem corrections made. A powerful tool for determining the purity of organic compounds is the recently developed method of vapor phase chromatography. Only small samples are needed, and even minor amounts of impurities are easily detected.

Occasionally doubt exists regarding the identity of two solids having similar melting points. In that case the two substances are intimately mixed, and the melting point of the mixture is observed; if this shows an appreciable depression, the materials are different. The refractive index is often very useful as a criterion of purity, but it must be determined at a definite temperature and with reference to a particular wave length of light as source of illumination. Analytical data are often obtained by the combustion of the material. Good results agree to within .3% of the theoretical value for carbon, and to within .2% for hydrogen. Occasionally functional groups of the compound are used in analysis. Examples of this method are the neutralization equivalents of organic acids and bases, or the saponification equivalents of esters. When isomers are present in a reaction mixture, spectrochemical analyses often become particularly valuable, since all other properties of the compounds may be indistinguishable. Nuclear magnetic resonance spectroscopy is especially valuable for distinguishing between isomers. By this method, the different types of hydrogen in a molecule may be determined quantitatively. (See **Impurities**.)

The number of important organic syntheses, both on an industrial and on a laboratory scale, is very large. Hence any examples to be mentioned here obviously have to be chosen in a random manner. An important commercial reaction is the synthesis of methanol, which proceeds in accordance with the following equation:

$$CO + 2H_2 \xrightarrow[\text{Catalyst}]{\text{Pressure}} CH_3OH$$

Since there is a volume decrease in this instance, the use of pressure favors the establishment of a favorable equilibrium. This process has now almost completely replaced the older methanol synthesis, based on the destructive distillation of wood. Methanol is important as a solvent and as a chemical intermediate. Its dehydrogenation over copper or silver affords formaldehyde, another important chemical. Phenol is another compound of great importance. Several syntheses are known, the classical one being the alkali fusion of the sodium salt of benzenesulfonic acid. The first step of this synthesis is a *sulfonation*, a reaction of considerable commercial importance. Many important dyes and drugs are derived from aniline. This material is formed by (1) a nitration and (2) a reduction. Both *nitrations* and *reductions* are very important types of organic syntheses.

Styrene is an important raw material for both plastics and synthetic rubber. The first step in its synthesis is an *alkylation*, a type of reaction frequently encountered in the petroleum field. Styrene and butadiene form the starting materials for the synthesis of Buna S, an excellent synthetic rubber. Butadiene, the other component of Buna S is obtained by a two step dehydrogenation of butane.

The process whereby hydrogen is added to an unsaturated compound is called *hydrogenation*. It, too, is of considerable industrial importance. By means of a hydrogenation, liquid vegetable fats are turned into solid fats. This process may be formulated as follows:

$$-CH{=}CH- \xrightarrow[\text{Ni}]{H_2} -CH_2{-}CH_2-$$

A number of common household shortenings are prepared by partial hydrogenation of vegetable oils. (See **Hydrogenation**.)

The synthetic fibers have made great advances in recent years. The first materials of this class were derived from cellulose. At present, nylon is very popular as a synthetic fiber. This material is a polyamide; its synthesis may serve to illustrate the formation of amides:

$$x\text{HOOC}(CH_2)_4\text{COOH} + x\text{H}_2\text{N}(CH_2)_6\text{NH}_2 \rightarrow$$
$$-NH[CO(CH_2)_4CONH(CH_2)_6NH]_x\text{CO}-$$

A newer fiber, "Dacron," is based on *p*-xylene. This material is oxidized and then reacted with ethylene glycol. The resulting polymer is a polyester. Its formation is an example of *esterification*, an important type of organic reaction. (See **Esters**.) Ethylene glycol may be prepared from ethylene via ethylene oxide:

$$CH_2{=}CH_2 \xrightarrow[\text{Ag}]{O_2} CH_2{-}CH_2 \xrightarrow{H_2O}$$
$$\underset{O}{\diagdown\diagup}$$
$$HOCH_2{-}CH_2OH$$

Ethylene glycol has other extensive uses also. (See **Glycols**.)

Ethylene itself is the starting material for the synthesis of polyethylene, a very commonly encountered plastic:

$$x\text{CH}_2{=}CH_2 \xrightarrow{\text{Pressure}} -[CH_2{-}CH_2]_x-$$

This constitutes a typical example of a *polymerization*. Other polymers having large fields of application are made from acrylonitrile, methylacrylate, and methylmethacrylate.

Reactions whereby halogens are introduced into a molecule are called *halogenations*. Halogen may be introduced by addition or by replacement. The formation of ethylene dibromide is an example of the addition process:

$$CH_2{=}CH_2 + Br_2 \rightarrow BrCH_2{-}CH_2Br$$

Ethylene dibromide is used to improve the characteristics of automotive fuels. The chlorination of a pentane mixture to yield amylchlorides is an illustration of a halogenation by replacement or substitution:

$$C_5H_{12} + Cl_2 \rightarrow C_5H_{11}Cl + HCl$$

The amylchlorides serve as starting materials for the synthesis of the amyl alcohols, employed as solvents and plasticizers. Their formation is a typical *hydrolysis* reaction:

$$C_5H_{11}Cl \xrightarrow{\text{NaOH}} C_5H_{11}OH$$

Benzene, toluene, and the xylenes are familiar constituents of coal tar. However, a large proportion of these materials is now produced from petroleum by the *reforming* process. The synthesis of toluene may serve as an illustration:

$$C_7H_{16} \xrightarrow{\text{Catalyst}} \overset{CH_3}{\bigcirc} + 4H_2$$

Besides being valuable chemical intermediates, these compounds improve the octane number of gasolines. As a consequence, the reforming process is carried out on a huge scale in the petroleum industry.

Both naphthalene and *o*-xylene may be converted to phthalic anhydride. This compound is manufactured on a large scale for the production of plasticizers. The formation of phthalic anhydride is an example of an *oxidation* reaction.

Synthetic detergents have now replaced soap to a large extent. Their advantage lies in the fact that they do not require soft water. A typical example of the synthetic detergents are the alkyl aryl sulfonates. In order to minimize disposal problems, and to minimize ecological damage, newer types of detergents are now biodegradable. (See **Detergents**.)

Petroleum-deficient countries (Germany for instance) have attempted to overcome this drawback by synthesizing hydrocarbons on a large scale. In the Fischer-Tropsch process, carbon monoxide and hydrogen are reacted to yield the desired hydrocarbons:

$$nCO + (2n + 1)H_2 \xrightarrow{\text{Catalyst}} C_nH_{2n+2} + nH_2O$$

In another process, the Bergius process, finely divided coal suspended in heavy oil is hydrogenated. Neither of these processes is economical at present in the United States for the production of motor fuel. The Oxo process however, an outgrowth of the Fischer-Tropsch reaction, is now employed for the manufacture of alcohols and aldehydes:

$$RCH{=}CH_2 + CO + 2H_2 \xrightarrow[\text{Pressure}]{\text{Catalyst}}$$

$$RCH_2CH_2CH_2OH$$

In addition a coal hydrogenation plant is currently being operated for the production of aromatic chemicals. (See **Fischer-Tropsch Process**.)

A large number of organic chemicals may be made from acetylene. Two examples are vinyl chloride and acrylonitrile, both of which find extensive use in the polymer field:

$$N{\equiv}C{-}CH{=}CH_2 \xleftarrow{\text{HCN}} CH{\equiv}CH \xrightarrow{\text{HCl}}$$

$$CH_2{=}CH{-}Cl$$

While the examples cited above are mainly of industrial interest, there are a large number of well known reactions which are not well suited for large-scale applications, but are commonly employed in the laboratory. The various Grignard syntheses fall in this category. The various reductions with lithium aluminum hydride are also mainly laboratory syntheses:

$$RCOOH \xrightarrow{\text{LiAlH}_4} RCH_2OH$$

Prior to the discovery of lithium aluminum hydride the reduction of the carboxyl group or its derivatives was a difficult process.

Great contributions have been made toward the understanding of the steroids and the sex hormones. Many of these materials can now be prepared in the laboratory, but their syntheses are too complicated to be described here. Many other natural products such as amino acids, sugars, glycosides, porphyrins, alkaloids and the like can now be synthesized. Two famous examples in this field are the syntheses of morphine and quinine. Most of the vitamins are available on a commercial basis.

Many profound discoveries have been made in the drug field, so that only a few can be mentioned here. The drug made in the largest amount is probably aspirin. Salvarsan and Neosalvarsan proved to be the first effective drugs against spirochetal diseases. The sulfa drugs have saved count-less lives since their discovery. They are effective against a host of bacterial infections. Cortisone has brought relief in some cases of arthritis. Atebrin, an antimalarial, was hastily pressed into service when the quinine supply became inadequate during the last war. Since then, even better antimalarials have been discovered. Sufferers from allergies are often relieved by the various antihistamines. The synthetic analgesics have done much to relieve pain. Some frequently used analgesics are demerol, amidone and novocaine. In the field of hypnotics and sedatives, the barbiturates hold a prominent place. Veronal and barbital are two common drugs from this class. Many of the new antibiotics are of microbiological origin. The most famous drugs in this category are probably penicillin and streptomycin.

From the examples above it can be seen that the impact of synthetic organic chemistry upon our daily life is tremendous. If current predictions materialize, the importance of organic synthesis will become even greater in the future.

HENRY F. LEDERLE

ORGANOMETALLIC COMPOUNDS

An organometallic compound is an organic chemical that contains a metal bonded directly to carbon. Depending on whether the metal is of Groups IA-IIIA, of Groups IVA-VIA (Si, P and S families) or of the transition elements, these compounds are termed organometallics, organometalloids or transition organometallics, respectively. The first known organometallic compound was diethylzinc, obtained in 1849 by Frankland in his attempt to prepare a free organic radical from ethyl iodide and zinc. The first transition organometallic was isolated by Zeise as a complex from K_2PtCl_4 and ethylene, $K^+[C_2H_4PtCl_3]^-$.

Organometallic compounds (RM) of practically all the metals have been prepared. There are two main classes: "simple" and "mixed." A simple organometallic compound has only R groups attached to the metal, as R_4M^{IV}; whereas a mixed organometallic compound has both R and X groups attached to the metal, as $R_2M^{III}X$. The simple types may be further divided into symmetrical (as $C_2H_5HgC_2H_5$) and unsymmetrical (as $C_2H_5HgC_4H_9$) classes. Within these general groups there are many types of organometallic compounds. Those having but one metal may contain one or more R groups and one or more X groups, depending on the valence number of the metal and the stabilities of the organometallic compounds. Furthermore, two or more of the same or different metals may be present as in $CH_2(ZnI)_2$, $(C_6H_5)_3SiSn(C_6H_5)_3$, and $(C_6H_5)_3GeLi$. Transition organometallics can be classified by the periodic subgroup of the metal (e.g., Fe group: Fe, Ru and Os) or by the nature of the organic groups and their bonding (e.g., ethylene, acetylene, cyclopentadienyl, benzyl groups, etc, bonded in a sigma or pi fashion). Examples are $Fe(CO)_5$, $(\pi\text{-}C_5H_5)_2Ni$, dipositive $(\sigma\text{-}C_6H_5CH_2)Cr(H_2O)_5$ and $(\pi\text{-}C_5H_5)(CH_3)_2Co(Ph_3P)$.

Organometallic compounds may be classified as to reactivity in various ways, but one convenient system is by trends in the Periodic Table. In any group or subgroup the higher the ionization potential of the metal, the less reactive will be its organometallic compounds. Organometallic compounds may be grouped by relative reactivities on the basis of two reactions: (a) addition to an olefinic linkage; and (b) addition to a carbonyl group. The highly reactive compounds add to both; the moderately reactive add only to the carbonyl group; and the relatively unreactive add to neither functional group in reasonable time. A moderately reactive type is the Grignard reagent, RMgX, which is one of the important classes of organometallic compounds for synthetic purposes. Although more than thirty procedures, general and specific, have been used for the preparation of organometallic compounds, the following are of widest application for main group elements:

$$RX + M \rightarrow RMX \text{ or } 2RX + 2M \rightarrow R_2M + MX_2 \quad (1)$$

$$R_2M' + M'' \rightarrow R_2M'' + M' \quad (2)$$

$$R_2M' + M''X_2 \rightarrow R_2M'' + M'X_2 \quad (3)$$

$$n-\overset{|}{C}=CH_2 + MH_n \rightarrow [-\overset{|}{C}H-CH_2]_nM \quad (4)$$

Fundamentally, then, organometallic compounds derive from interaction of an RX compound with a metal, or its alloy, or amalgam. This applies particularly to the preparation of organozinc, –magnesium, and –lithium compounds which are among the most effective types for the transformations illustrated in Eq. (2) and (3).

There are currently four qualitative color tests for organometallic compounds. Of these, the one of greatest general value is Color Test I which depends on the formation of a dye when an RM compound reacts with Michler's ketone. It is of high diagnostic value and is particularly helpful in telling when a reactive or moderately reactive RM compound is formed, and when such RM compounds are used up in reactions.

At this point it might be convenient to make three rather sweeping generalizations: (1) organometallic compounds will be made from all metals and metalloids; (2) reactive and moderately reactive organometallic compounds will undergo reactions with all functional groups; (3) given sufficient time, organometallic compounds of a given structural type will probably show the same general chemical behavior. Because of characteristics due to structure and kind of metal, main group organometallics and transition organometallics form distinct classes both as to the kinds and the ease of their chemical reactions.

What are general reactions shown by all RM compounds? Two that stand out are oxidation, and cleavage by acids. (Incidentally these reactions underlie most procedures for the quantitative analysis of RM compounds.) The cleavage by acids may be selective, and also step-wise.

$$(C_2H_5)_2Sn(C_6H_5)_2 + 2HCl \rightarrow$$
$$(C_2H_5)_2SnCl_2 + 2C_6H_6$$

$$(C_6H_5)_4Sn \xrightarrow{HCl} (C_6H_5)_3SnCl \xrightarrow{HCl}$$
$$(C_6H_5)_2SnCl_2 \xrightarrow{HCl} C_6H_5SnCl_3 \xrightarrow{HCl} SnCl_4$$

Probably the most important reactions of main group organometallic compounds are those involving addition to an unsaturated linkage. In such reactions the R group attaches itself to the relatively less acidic element and the M to the relatively more acidic element. For example, in the reaction with a ketone the following takes place, leading to a tertiary alcohol:

$$R_2C{=}O + R'M \longrightarrow R_2\overset{|}{\underset{R'}{C}}{-}O{-}M \overset{(H+)}{\longrightarrow} R_2\overset{|}{\underset{R'}{C}}{-}OH$$

When the unsaturated linkage is made up of the same elements, then there are only two important types to consider. One of these is the olefinic or acetylenic linkage, to which the more reactive RM compounds add at an appreciable rate.

$$C_6H_5CH{=}CH_2 + R'Na \longrightarrow C_6H_5\overset{|}{\underset{Na}{C}}HCH_2R'$$

The other is the azo linkage, and here the course of reaction appears to be influenced by the relative reactivity of the RM compound. For example, a moderately reactive Grignard reagent like phenylmagnesium bromide reacts as follows:

$$C_6H_5N{=}NC_6H_5 + 2C_6H_5MgBr \longrightarrow$$
$$C_6H_5\overset{|}{\underset{BrMg}{N}}{-}\overset{|}{\underset{MgBr}{N}}C_6H_5 + C_6H_5{-}C_6H_5$$

However, a more reactive RM compound like phenyllithium adds as follows:

$$C_6H_5N{=}NC_6H_5 + C_6H_5Li \longrightarrow C_6H_5\overset{|}{\underset{C_6H_5}{N}}{-}\overset{|}{\underset{Li}{N}}C_6H_5$$

Some other reactions of RM compounds are the following:

$$RM + X_2 \longrightarrow RX + MX$$

(here the X_2 can be a halogen like bromine or a pseudo-halogen like cyanogen).

Cleavage by hydrogen occurs under varying experimental conditions:

$$RM + H_2 \longrightarrow RH + MH$$

With the more reactive RM compounds cleavage takes place readily. For example, phenylcesium undergoes prompt hydrogenolysis at room temperature, atmospheric pressure, and in the absence of a catalyst:

$$C_6H_5Cs + H_2 \longrightarrow C_6H_6 + CsH$$

It should be recalled in connection with the general rules for the relative reactivites of RM compounds that the organometallic compounds of the alkali metals form the most reactive group, and in this group the order of increasing reactivity is: Li, Na, K, Rb, Cs.

Somewhat related to the reaction with hydrogen is a reaction known as metallation. Inorganic salts may be prepared by interaction of an acid with a metal, a base, or a salt. The same general reactions can be used for the preparation of RM compounds from the very weakly acidic hydrocarbons.

Inasmuch as RM compounds have properties of salts, they react with an acid (RH) to form another salt and another acid:

$$RH + C_2H_5Li \longrightarrow RLi + C_2H_6.$$

Another reaction of broad interest is the cleavage of ether:

$$C_2H_5—O—C_2H_5 + RM \longrightarrow C_2H_5OM + RC_2H_5$$

This type of reaction takes place, as one might expect, more readily with the highly reactive RM compounds. For example, organometallic compounds of sodium, potassium, rubidium, and cesium cannot be prepared in a solvent like diethyl ether. This markedly limits the solvents or media for the more reactive RM compounds. Organolithium compounds can be prepared conveniently in ether, but they should be used reasonably promptly after preparation in such a solvent in order to avoid their loss by cleavage of the ether. On the other hand, ethers are the solvents of choice for the preparation of Grignard reagents. It has been shown that methylmagnesium iodide in ether solution, when not exposed to the atmosphere or direct light, is essentially unaffected after standing for more than twenty years. The cleavage of ethers, can in some cases, be a useful synthetic reaction:

$$\begin{matrix} CH_2 \\ | \\ CH_2 \end{matrix} \!\!\! \diagdown \!\! O + RM \longrightarrow \begin{matrix} CH_2OM \\ | \\ CH_2R \end{matrix} \xrightarrow{(H^+)} \begin{matrix} CH_2OH \\ | \\ CH_2R \end{matrix}$$

Obviously, solvents cannot be used for RM reactions where the solvent contains an "active" or acidic hydrogen, like RNH_2, ROH, etc. Compounds containing such hydrogens decompose moderately reactive and reactive RM compounds as follows:

$$R_2NH + R'M \longrightarrow R_2NM + R'H$$

Physical Properties. The physical properties of RM compounds vary over wide ranges. Some (like trimethylboron) are gases; many are liquids; but for the most part (particularly those having aryl groups) they are solids. The several rules or generalizations referred to earlier on the relative reactivities of RM compounds are not concerned with two highly obvious properties of some RM compounds: namely, thermal instability and spontaneous flammability. The highly unstable organosilver and organogold (RAu) compounds are of a relatively low order of so-called typical chemical reactivity; and ethylpotassium, which starts to decompose at room temperature, is extremely reactive. Trimethylboron and trimethylbismuth are spontaneously flammable but not particularly reactive otherwise; whereas the methylalkali compounds like methylsodium are spontaneously flammable and also highly reactive generally.

There appear, at this time, to be no correlations among thermal instability, spontaneous flammability, and other chemical transformations. The illustrations just given are of RM compounds which are either highly reactive or of a relatively low order of reactivity. Organomanganese compounds (having the C_6H_5—Mn type linkage) are of moderate reactivity, but they are not only thermally unstable but also spontaneously flammable. A typical criterion of relative reactivity is addition to the carbonyl linkage. This reaction is shown not only by the thermally unstable organocopper and organosilver compounds, but also by the spontaneously flammable organoberyllium and organoboron compounds, as well as by the relatively unreactive organomercury and organolead compounds.

As a general rule, the less reactive organometallic compounds are unaffected by water; but the moderately and highly reactive RM compounds are decomposed vigorously by water. Obviously water or other hydroxylated compounds cannot be used as solvents for the reactive RM compounds. Some of the less reactive RM compounds can be dissolved in water if the R group in the RM compound has functional polar groups like —COOH, —OH, and —NH_2 which impart water-solubilizing characteristics to the molecule. There are no good solvents for the very highly reactive RM compounds where M is Na, K, etc.

Although the *sigma*-bonded alkyl, aryl and alkenyl derivatives of the transition metals do adhere generally to the foregoing comments on stability, chemical reactivity and physical properties, *pi*-bonded transition organometallics do not. These latter compounds are generally more resistant to the thermal, hydrolytic and, often, oxidative disruption of their *pi*-carbon-metal bond than their *sigma*-counterparts. The *pi*-bonded organic groups, such as cyclopentadienyl, cyclobutadiene or trimethylenemethane, become much more stable than in the free state and often can undergo chemical reactions without detachment from the metal.

Physiological Properties. Organometallic compounds are relatively toxic. Little is known of the odor and taste of the moderately and highly reactive organometallic compounds, partly because of their slight volatility and more particularly because the aqueous medium presumably necessary for these sense perceptions is ideally suited for decomposing the reactive organometallic compounds. The odors vary over wide limits. Some have a pleasant fruity bouquet and others are highly obnoxious; some are without any appreciable odor and others have ill-defined "characteristic odors." There are highly purified RM compounds which develop an odor only after brief contact with the atmosphere. Thus, the sweet odor of air-sensitive $(C_2H_5)_3Ga$ has been shown to be due to its oxidation product, $(C_2H_5)_3GaOC_2H_5$.

Although it is true that organometallic compounds are generally quite toxic, it is equally true that there are marked variations in toxicity. The uncertain fate of the exhaust lead from the antiknock agent of gasoline, tetraethyllead, and the detection of methylmercury compounds in lakes and streams have raised serious alarm among ecologists. The variations are noted with different classes of organometallic compounds and also within the same class, depending on the alterations in structure of the R groups and the nature of X or the acid radical. This makes it understandable why some organometallic compounds have actually found application as therapeutic agents, and why others are being investigated for their possible curative effects.

If correlations may be drawn between metallic

hydrides and organometallic compounds, one may conclude that the physiological effects of organometallic compounds are not necessarily related to the physiological effects of the metal alone or to inorganic salts of the metals. For example, tin hydride is the most toxic of all hydrides so far investigated, whereas metallic tin and tin salts are apparently without any significant harmful effects on the organism. However, mercury and lead, as well as their many compounds, are generally toxic.

Indirect but significant contributions of organometallic compounds in biological problems have been their application in studies of reaction mechanisms and in procedures concerned with the structure and preparation of compounds like vitamins, hormones, carcinogens, and other biologically potent materials.

Applications. Although most valuable in fundamental synthetic and mechanistic studies, organometallic compounds recently have commanded wide and keen attention in industrial circles by their use as polymerization catalysts for the production of stereoregular polyalkylenes. Here transition organometallics, often with main group organometallics as co-catalysts, find their chief application. Among the more important technical applications is that of tetraethyllead as an antiknock compound. It is currently prepared by the following reaction:

$$C_2H_5Cl + (NaPb) \longrightarrow (C_2H_5)_4Pb$$

Figures on the production and use of organometallic compounds are difficult to obtain. There may be as much as 500,000,000 pounds of tetraethyllead used annually. Ecological concern for the potential hazards of this lead compound may force its discontinuance in gasoline, unless public opinion can be convinced of the safety of current practice. (See **Antiknock Agents.**)

Organomercury compounds find wide and varied uses, particularly as phenylmercuric acetate and ethylmercuric salts in microorganism control. The production of organomercury compounds may be as high as 4,000,000 pounds annually.

Organoaluminum compounds, principally, triethyl- and triisobutyl-aluminums, are produced on a large scale. The former is principally used in preparing by the Ziegler process long-chain alkenes and alkanols having an even number of carbon atoms:

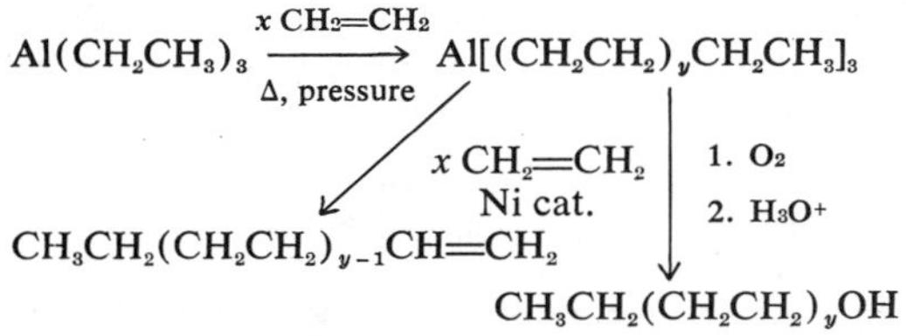

$$Al(CH_2CH_3)_3 \xrightarrow[\Delta, \text{ pressure}]{x \, CH_2=CH_2} Al[(CH_2CH_2)_yCH_2CH_3]_3$$

$$\underset{\text{Ni cat.}}{\overset{x \, CH_2=CH_2}{\swarrow}} \quad \underset{2. \, H_3O^+}{\overset{1. \, O_2}{\Big\downarrow}}$$

$$CH_3CH_2(CH_2CH_2)_{y-1}CH=CH_2$$

$$CH_3CH_2(CH_2CH_2)_yOH$$

The production of organotin compounds may be about 1,000,000 pounds annually. Among the more important ones currently prepared are dibutyltin salts and tetraphenyltin. Their chief use is as stabilizers and scavengers in chlorinated dielectric materials. For example, tetraphenyltin reacts by neutralizing small amounts of hydrogen chloride as liberated to give triphenyltin chloride and benzene:

$$(C_6H_5)_4Sn + HCl \longrightarrow (C_6H_5)_3SnCl + C_6H_6$$

The greatest industrial use of organosilicon compounds is in the field of silicones although the use of lithium, sodium, boron and aluminum alkyls is expanding rapidly. The number of siloxane units (Si—O—Si) and the nature of the R groups affect markedly the properties of the products, which in general are characterized by relatively high thermal stability and chemical resistance. All that can be said of production is that it is measured in millions of pounds.

HENRY GILMAN AND JOHN J. EISCH

Cross-references: *Antiknock Agents, Grignard Reactions, Lithium in Organic Chemistry.*

References

Coates, G. E., Green, M. L. H., and Wade, K., "Organometallic Compounds," 2 Volumes, 3rd Ed., Methuen, London, 1967 and 1968.

Eisch, J. J., "The Chemistry of Organometallic Compounds—The Main Group Elements," Macmillan, New York, 1967.

OSMIUM AND ITS COMPOUNDS

Osmium is one of the six platinum-group metals. It was isolated in 1804 by S. Tennant (England) during his examination of residues left when native platinum was treated with aqua regia. The name of the element is based on the Greek *osmé*—smell, in recognition of the strong odor of the volatile tetroxide which can almost always be detected in the vicinity of the metal.

The abundance of osmium in the earth's crust has been estimated as 0.001 ppm, and it is the platinum group metal of which the least amount is produced annually. In meteorites, however, it ranks second or third among the platinum metals in abundance. The chief source of osmium is native metal alloys especially with iridium; two of these are important, iridosmine ($>32\%$ osmium, hexagonal) and osmiridium ($<32\%$ osmium, cubic). Very little osmium is found in native platinum or palladium.

The following is a selection of physical properties: atomic number 76; atomic weight 190.2; stable isotopes 184 (0.018%), 186 (1.59%), 187 (1.64%), 188 (13.3%), 189 (16.1%), 190 (26.4%), 192 (41.0%); density 22.61 g/cc (20°C); melting point 3057°C; hardness (annealed) 300–670 Vickers units; electrical restitivity 8.12 μ ohm-cm (0°C). The metal crystallizes in a hexagonal close-packed lattice in which $a = 2.7341$ Å and $c/a = 1.5800$; this arrangement yields an average metallic radius of 1.35Å. The great hardness is remarkable, and militates against easy working with the metal.

Although osmium exhibits considerable miscibility with the other platinum metals, the resulting increase in hardness soon renders the alloys impractical for working. Thus osmium finds little application as an alloying element.

This element is unusual within the platinum group for the relative ease with which it reacts with oxygen. The compact metal, heated in air or oxygen to 200°C, or finely divided metal even at room temperature will form the volatile oxide OsO_4, which is toxic and, as mentioned, has a pungent odor.

Osmium is scarcely attacked by mineral acids or aqua regia even when these are hot. However, molten alkalis, especially with an oxidizing agent present, can convert the element to water-soluble compounds. This characteristic, which is shared with ruthenium, is employed for the separation of these two elements from iridium in the isolation and refining of the platinum metals. Fluorine and chlorine react with osmium at elevated temperatures, the products being OsF_4 and OsF_6, and $OsCl_3$ and $OsCl_4$, respectively. OsF_6 is a yellow green, volatile and quite reactive solid; for many years an octafluoride was described in textbooks, but this has proven to be nonexistent and the compound described is OsF_6. Of the chlorides $OsCl_3$ is best characterized; it is a brown solid, very soluble in water or alcohol.

Inorganic chemists have prepared osmium compounds with the element in any of nine oxidation states 0 to +8. The most commonly observed ones are +4, +6, and +8. The chemistry of the element does not include a simple aqueous cation. There are oxyanions in two oxidation states, $OsO_4(OH)_2^{2-}$ (yellow), and $OsO_2(OH)_4^{2-}$ (pink in aqueous solution); these species have no counterparts within the platinum group.

Osmium is expensive and difficult to work, and so has found few important uses. Some alloys containing about 60% of this element, with ruthenium and other platinum metals, are produced for tips of fountain pen nibs, long-life phonograph needles, or instrument pivots.

Osmium tetroxide finds a number of laboratory applications as an oxidizing agent and as a catalyst for oxidations. In particular it is very effective for hydroxylation of organic compounds at double bonds. It finds limited application as a histological stain.

W. A. E. McBRYDE

OSMOSIS

The term "osmosis" (from Greek ōsmos, impulse, push) refers to the flow of solvent through a porous membrane which separates two solutions of different concentrations from each other, and which is impermeable to the dissolved material (the solute). Such a membrane is called semipermeable or ideal. Under these conditions solvent tends to flow from the less concentrated solution—which may be pure solvent—into the more concentrated one. The driving force responsible for this flow of solvent is the difference of (thermodynamic) solvent activity between the two solutions: the solvent in the less concentrated solution has, generally, the greater activity, being less tied up by the solute molecule. The flow stops when equal concentrations are attained on both sides of the membrane, at which point solvent molecules pass through the membrane in both directions at equal rates. The flow of solvent can be counteracted by an excess pressure imposed on the more concentrated solution, and that pressure which just stops the flow is called the *osmotic pressure,* which in turn is a measure of the difference of solvent activities between the two solutions.

Osmosis and osmotic pressure are readily demonstrated by a simple experiment: a container, the lower end of which is closed off by a suitable membrane (see below) and whose upper portion is constricted to a thin vertical tube, is filled with a solution (eg., sucrose in water) and subsequently immersed in a larger container filled with pure water. The solution level in the vertical tube will rise until a certain hydrostatic pressure is reached. If the membrane is truly semipermeable, this hydrostatic pressure can be identified with the osmotic pressure for this particular system of pure water on one side and sucrose solution of that particular concentration on the other side of the membrane whereby the dilution of the original solution by inflowing solvent has to be considered.

As van't Hoff showed in 1887 for very dilute solutions, the osmotic pressure (π), at a given temperature (T), is determined exclusively by the number (n) of molecules of the dissolved substance contained in a given volume (V) of solution. This is expressed by

$$\pi = \frac{n}{V} \frac{RT}{N},$$

where R is the gas constant, and N Avagadro's number. This formula is completely analogous to the ideal gas law $PV = n \frac{R}{N} T$, and early investigators attempted to explain the osmotic pressure as arising from the solute molecules impinging on the semipermeable membrane, just as gas molecules striking the walls of a container are the cause for the gas pressure. This analogy, however, does not stand closer scrutiny; thermodynamic concepts rather than kinetic models of this kind are now favored for explaining the phenomenon of osmotic pressure.

It follows from van't Hoff's formula that the osmotic pressure is proportional to the absolute temperature. Raising the temperature for instance from 20°C to 30°C increases the osmotic pressure by 3.41%.

The formula is of great significance because it allows us to calculate the molecular weight of the dissolved substance from the osmotic pressure, since n = (weight of solute)/(weight of one molecule). Molecular weight determinations by means of the osmotic pressure are of special importance for substances with high molecular weights (up to one million) such as synthetic polymers (plastics) and biopolymers (proteins). The molecular weight is the weight—or more correctly the mass—of one molecule multiplied by Avagadro's number.

The crucial part of any osmometer (a device for measuring osmotic pressure) is the membrane. Porous membranes widely occur in nature where they play a most important role in the transport of nutrients and waste products across cell boundaries, but these membranes rarely have the proper size and shape for use in an osmometer. In early experiments pig's bladders have been used, but nowadays artificial membranes which are quite superior in their properties, are almost exclusively employed in the laboratory. The main problem is to find a membrane through which solvent flows at a reasonable rate while solute molecules are being fully retained. It is obvious that this condition becomes more difficult to fulfill the less the solvent and solute molecules differ from each other with respect to size and chemical properties. A membrane of very high retentiveness for use in aqueous solutions consists of copper ferrocyanide deposited as a thin layer in the walls of a porous pot, first used by Pfeffer in 1877. This membrane was found, for instance, to be impermeable to the comparatively small molecules of cane sugar (sucrose, molecular weight 342) dissolved in water. Today the interest in osmometry has shifted to larger molecules; for these, commercially available membranes made of cellophane, regenerated cellulose, cellulose derivatives, as well as porous synthetic plastics and porous glass are used. The pore diameters of these membranes are considered to be less than 100 Å in many cases, and such membranes are impermeable to substances with molecular weights of a few thousand and above. Membranes can only to some extent be compared with sieves, and it is not always a question of molecular size alone: electric charges and chemical properties play often a role in determining the selectiveness of a membrane. Thus, one has evidence that membranes sometimes act as dissolving media for the solvent while the second solution component, being insoluble in the membrane, is retained.

It should be mentioned that osmosis and experimental osmotic pressures can also be observed if the membrane is permeable to the dissolved substance as well and all solution components can pass through the pores, provided the membrane offers different resistances to the diffusing components. Here the measured osmotic pressure depends among other things very much on the design of the osmometer, and it is always found to be lower than the pressure measured on the same solution by means of a solute-retentive (semipermeable) membrane. Also, the observed pressures are not steady, but decline with time as the solute slowly diffuses across the membrane into the solvent container.

Osmotic pressures measured in the course of a molecular weight determination of a polymer are usually small but in other cases they can reach spectacular magnitudes. While a 1% solution of polystyrene of molecular weight 100,000 dissolved in toluene at room temperature gives a pressure corresponding to a hydrostatic head of 1.66 inches of water, osmotic pressures in some plants where the walls of the roots act as porous membranes reach pressures of 50 atmospheres (corresponding to 1,700 feet of water) and more. By comparison, a solution of 200 grams of sucrose in 1000 grams of water at room temperature generates an osmotic pressure of 15.4 atmospheres (524 feet of water). These experimentally determined values are greater than what one calculates from van't Hoff's formula (1.02 inches of water for the polystyrene solution and 439 feet of water for the sucrose solution) because the condition of a very dilute (ideal) solution is not fulfilled.

As already mentioned, the cell walls in living organisms are porous, but because of the smallness of the pores they are impermeable to the large colloidal protein molecules. Water, small molecules, and common ions can pass, but they encounter different degrees of resistance, depending on size and electric charge. The red blood cells are a classic example in this respect. Human red blood cells, when immersed in distilled water, burst within a very short time because the cell walls cannot stand up to the pressure developed by the inrushing water. This pressure would correspond to a hydrostatic head of approximately 260 feet of water which is about the same as the osmotic pressure generated by a 0.95% sodium chloride solution. As a matter of fact, blood cells suspended in such a salt solution remain unchanged. If the salt concentration is made higher, the blood cells shrink, owing to loss of water. It was recently found that the exchange of water takes place through pores which are approximately 8 Å wide, the cell membrane being about 100 Å thick. Water molecules which have a diameter of 3Å pass freely through these channels. Potassium salts inside the cell are mainly responsible for the osmotic pressure. The effective diameter of potassium ion in water is only about 4 Å; i.e., smaller than the diameter of the pores, but positive electric charges along the pore walls repel the likewise positive potassium ions. The membrane is not strictly impermeable to potassium salt, yet the resistance encountered by salts, or more correctly by positive ions, is so large by comparison with the uncharged water molecules, that the blood cell behaves almost like an osmometer with a solute-impermeable (semipermeable) membrane. It is clear that the serum surrounding the blood cells in the living organism must have a corresponding salt concentration to maintain the blood cells at their proper size.

Quite generally, many biological materials—such as plants, or only isolated cells, for instance—shrink when pickled in concentrated salt solutions, and swell when immersed in pure water as a result of osmosis: the walls are porous membranes, and water flows in the direction of the higher salt concentration.

HANS COLL

Cross-references: *Diffusion, Colloid Chemistry, Phytochemistry, Solutions.*

OXIDATION

The term "oxidation" originally meant a reaction in which oxygen was introduced into another substance, but its usage has long been broadened to include any reaction in which electrons are transferred. Oxidation and reduction always occur simultaneously, and the substance which gains electrons is termed the oxidizing agent. For example, cupric ion is the oxidizing agent in the reaction: $Fe(metal) + Cu^{++} \rightarrow Fe^{++} + Cu$ (metal).

Where the electrons are completely transferred from one molecule to another, without simultaneous transfer of atoms, identification of an oxidation process as such is usually relatively simple. However, electrons may be displaced within the molecule without being completely transferred away from it. Such partial loss of electrons likewise constitutes oxidation in its broader sense and leads to the application of the term to a large number of processes which at first sight might not be considered to be oxidations. Reaction of a hydrocarbon with a halogen, for example, $CH_4 + Cl_2 \rightarrow CH_3Cl + HCl$, involves partial oxidation of the methane from either of two points of view: (1) the chlorine atom, being more electronegative than hydrogen, has displaced a pair of electrons *away* from the carbon, or (2) the methyl chloride can be hydrolyzed to methyl alcohol which is an oxidation product of methane. By similar reasoning, halogen addition to a double bond may be regarded as an oxidation. Dehydrogenation, as in the reaction $CH_3CH_2OH \rightarrow CH_3CH=O + H_2$, is commonly regarded as an oxidation since two neutral hydrogen atoms each containing an electron have been removed.

Oxidation Number. Many elements can exist in more than one oxidation state. The oxidation number is the number of electrons which must be added to or subtracted from an atom in a combined state to convert it to the elemental form; for example, in barium chloride ($BaCl_2$) the oxidation number of barium is $+2$, that of chlorine is -1. This concept is a help in balancing equations and is most useful in dealing with ionic bonds as in much of inorganic chemistry; when an electron pair is only partly displaced toward one atom the bonding electrons are customarily assigned to the more electronegative atom. In some compounds of complicated structure the nature of the bonding is uncertain. If the compound contains oxygen, for the purpose of balancing equations, an oxidation number is sometimes assigned arbitrarily to an element by giving a value to oxygen of -2 (except in peroxy or super-oxide compounds, see **Peroxides**) and recognizing that the algebraic sum of the oxidation numbers of all the atoms present in the compound must equal zero. Thus in potassium permanganate ($KMnO_4$) manganese is assigned the oxidation number $+7$. No implication as to the actual character of the bonding should be read into the use of this formal procedure.

If reaction occurs very rapidly so that equilibrium is essentially established at all times, such as in some of the ionic reactions encountered in inorganic chemistry, the manner and extent to which an oxidation will proceed can be predicted by examination of the oxidation-reduction potentials of various alternate plausible reactions, provided that the reaction mechanism corresponds to that for which the potential is specified. However, the oxidation-reduction potential is not necessarily a measure of the *rate* at which a particular oxidant will react. The manner in which various oxidizing agents will attack the covalent bonds of organic substances is as yet primarily an empirical matter; the theories of structural chemistry are of help in organization, but guidance in prediction still comes largely from past experience.

Some oxidation-reduction potentials are determined by measurements with an inert electrode inserted in a solution containing both oxidized and reduced substances in equilibrium, in combination with a suitable reference electrode; in cases in which a solid is a participant in the reaction studied, the solid may constitute an electrode. Alternately the potential may be calculated from free energy data obtained in another fashion, e.g. from an equilibrium constant determined by direct measurements of equilibrium concentrations. (See **Electrode Potentials**.)

Autoxidation. Of particular significance because of their ubiquity and technical importance, are those oxidations involving molecular oxygen. The term "autoxidation" is applied to reactions of molecular oxygen at ambient temperatures. These reactions are generally slow, but occur with measurable rates with an enormous variety of organic and inorganic substances in addition to being an essential part of the biological processes on which life depends. Autoxidation is involved in the manufacture of a number of chemicals and is of technological importance in such diverse processes as the rusting and corrosion of metals, the polymerization of drying oils, the weathering of coal, the rancidification of fats and oils, the deterioration of rubber, and formation of undesirable gums and sludges in gasoline. (See also **Autoxidation**.)

Characteristically, the autoxidation of organic molecules is accelerated by light and by traces of various catalytic materials, particularly peroxy compounds and salts or oxides of the heavy metals such as cobalt and manganese. The autoxidation reaction is frequently preceded by an induction period, and the rate is strongly retarded or inhibited by traces of various easily oxidizable organic substances such as amines, phenols and alcohols. These characteristics are typical of free-radical reactions and it is now established that most organic autoxidations proceed in this fashion. The first step is the formation of a free radical by some (frequently unestablished) mechanism. (See also **Free Radicals**.) This is sometimes postulated to occur by the reaction.

$$RH + O_2 \rightarrow R + HO_2 \qquad (1)$$

Free radicals can also be formed by photochemical or radiochemical excitation. Most commonly the free radical R then reacts with oxygen:

$$R + O_2 \rightarrow ROO \qquad (2)$$

$$ROO + RH \rightarrow ROOH + R \qquad (3)$$

to form a peroxy radical and then, by hydrogen abstraction, a peroxy compound. It has been well established that reactions 2 and 3 occur in the autoxidation of a variety of substances and a peroxy compound is clearly the first molecular product formed. This in turn may decompose or react in various ways. The rate and mechanism of the decomposition are markedly affected by the medium, and it may be acid- or base-catalyzed in ionic media. Attack on the organic molecule by oxygen or free radicals will preferentially be at the weakest bond. In the autoxidation of a simple olefin or an ether at room temperature, attack is usually on an adjacent methylene group, and not on the double bond.

$$R-CH_2-CH=CHR' + O_2 \rightarrow$$

$$\underset{\underset{O-OH}{|}}{R-CH}-CH=CH-R'$$

This is probably the first step in the oxidative attack on natural rubber, which is an unsaturated hydrocarbon.

Substances which inhibit autoxidation contain reactive hydrogen atoms with which oxygen, or hydrocarbon or peroxy radicals readily react to form less active free radicals that are slow to continue the reaction chain. In some of these cases, such as the reaction of oxygen with aromatic polyphenols such as hydroquinones and the leuco forms of certain dyes, hydrogen peroxide is formed rather than a hydroperoxide. Usually the hydrogen peroxide reacts rapidly with the dehydrogenated substance; but in a few cases, as in the autoxidation of an anthrahydroquinone, the anthraquinone resulting from the oxidation is unreactive to hydrogen peroxide and this reaction is the basis for the principal industrial process for manufacture of hydrogen peroxide. Since oxygen is paramagnetic and the molecule contains two unshared electrons, it is sometimes termed a diradical. However, it is evident that its reactivity is far less than that of free radicals such as CH_3 or ROO and such characterization can be misleading.

The manner in which oxygen is first taken up to initiate an autoxidation is still rather speculative. The catalysis of autoxidations by peroxy compounds is clearly caused by the free radicals produced by their thermal decomposition. The role of metals and their salts is less clear; they appear to promote not only the formation but also the decomposition of hydroperoxides. The fact that all the metals that activate autoxidations can exist in more than one valence state suggests that some oxidation-reduction mechanism is involved. This promoting effect is desirable in some technical applications, such as in the oxidative hardening of drying oils, and undesirable in others, such as the deterioration of lubricating oils.

The induction period frequently observed in autoxidations may be caused by the time required to oxidize impurities which act as inhibitors or, alternately, to the time required to build up the hydroperoxide concentration to a level sufficient to maintain the chain reaction.

Oxygen may either markedly accelerate or strongly inhibit the rate of a polymerization process that occurs by a free radical mechanism. Polymerization can be accelerated if the oxygen reacts to form peroxides which in turn generate free radicals by thermal decomposition. Polymerization is inhibited if the oxygen preferentially reacts with the free radicals present to form other radicals or products that are less reactive. Oxygen or other oxidizing agents such as peroxy compounds can also act as depolymerizing agents; this will occur particuarly readily if the polymer contains α methylene groups or other bonds which are readily attacked. Likewise oxygen may facilitate branching or cross-linking in the polymerization process by promoting free radical formation.

High-Temperature Oxidation. As the temperature is increased, the character of the oxidation reactions will also change substantially, but only an elementary knowledge is yet available of how the reaction mechanisms are affected by temperature, as well as by pressure, structure of the oxidized material, etc. When the focus of attention is on combustion of a fuel to produce heat or power, high temperatures are required to obtain rapid rates of reaction and for high thermodynamic efficiency of the over-all process. At such temperature levels, the chemical reactions occur so rapidly that the limiting factor on the rate observed is usually no longer the rate of reaction of molecules but rather the rate at which the reactants can be mixed.

Electrolysis. Since removal of electrons constitutes oxidation, the passage of a current through an electrolytic cell causes oxidation to occur at the anode, where electrons are removed, and reduction to occur at the cathode, where the electrons are supplied. Under equilibrium conditions the oxidation process having the lowest oxidation potential will be the one to occur at the anode. However, if this process is relatively slow, a reaction requiring a higher potential may be made to occur preferentially if the cell potential is raised sufficiently. For example, consider the electrolysis of aqueous sulfuric acid to form peroxydisulfuric acid, used commercially in one of the older manufacturing processes for hydrogen peroxide:

$$2HSO_4^- \rightarrow H_2S_2O_8 + 2e^-$$

The undesired decomposition of water, which would instead occur exclusively under equilibrium conditions, is minimized by raising the potential by use of high current densities and the choice of an electrode material which has a high overvoltage.

The major industrial uses of electrolysis are in production of chlorine and sodium hydroxide from aqueous sodium chloride and in reduction of metals such as aluminum and sodium. Among the inorganic chemicals which may be industrially produced electrolytically, are hydrogen peroxide, sodium peroxyborate, potassium permanganate, chlorates and perchlorates. Very few organic oxidation processes are carried out electrochemically. Some of the difficulties stem from the fact that organic liquids are poor conductors, and they frequently are insoluble or insufficiently soluble in a suitable electrolyte. However, a large commercial plant in the U.S. manufactures adiponitrile, a nylon intermediate, by a rather unusual electrochemical process, the electrolytic hydrodimerization of acrylonitrile.

Biological Oxidation. Although it is obvious that oxidation reactions play the key role in the mechanisms whereby living organisms obtain energy, the series of reactions involved is extremely complex. In general three groups of compounds are oxidized, the carbohydrates, the amino acids and fatty acids. The oxidation of glucose to carbon dioxide and water has been studied most carefully as a model for carbohydrates and many of the steps are now established. In animal oxidations the oxygen does not usually react directly with the substrate but a variety of intermediate reactions occur including various processes other than oxidation.

Oxidizing Agents

Any substance which acquires electrons in a reaction acts in that circumstance as an oxidizing agent. In thermodynamically reversible systems, i.e., conditions under which electrons are very easily transferred, a quantitative measure of the "oxidizing power" of a substance is the oxidation-reduction potential. These systems are encountered principally in inorganic reactions occurring in ionizing media. Organic oxidation reactions are in general thermodynamically irreversible and here the oxidation-reduction potential offers but slight guide to expected behavior. Thus some substances of high oxidation-reduction potential, such as hydrogen peroxide, may sometimes be termed "weak" oxidizers because of the slowness of their reactions.

Many substances act either as an oxidizing or as a reducing agent, depending upon the other materials with which reaction is occurring. For example, hydrogen peroxide ordinarily acts as an oxidizing agent

$$H_2O_2 + Mn^{++} + 2OH^- \rightarrow MnO_2 + 2H_2O$$

but in contact with a strong oxidizing agent such as a permanganate, it acts as a reducing agent:

$$5H_2O_2 + 6H^+ + 2MnO_4^- \rightarrow 2Mn^{++} + 5O_2 + 8H_2O$$

The choice of an oxidizing agent rests primarily on its rate of reaction, selectivity, cost, and ease of handling. In organic chemistry oxidizing agents are usually used for determining structure or for chemical synthesis. The oxygen in the air, while the cheapest oxidizer, is relatively unselective and is seldom used for chemical preparation in the laboratory. However, a number of chemicals are made industrially by oxidation with molecular oxygen, frequently using a catalyst to control the reaction and moderately elevated temperatures to obtain suitably rapid rates. These processes include the liquid-phase oxidation of acetaldehyde or ethanol to acetic acid or the formation of a hydroperoxide from a hydrocarbon, the catalytic oxidation of methanol vapor to formaldehyde, and the gas phase oxidation of various hydrocarbons, with or without a catalyst, to form a variety of products.

The problem of what oxidant to choose to carry out a particular reaction, particularly in inorganic chemistry, must still be solved largely on an empirical basis, depending primarily upon the accumulated experience of the past. Following is a check list of the chemical oxidants commonly used. Their ranking for a given set of experimental conditions may be calculated from standard oxidation-reduction potentials, tabulations of which are generally available, for example, the one compiled by de Bethune and Loud.

(1) Ozone and oxygen. Ozone is one of the strongest oxidants known. (2) Hydrogen peroxide and various peroxy compounds. (3) Potassium permanganate. This is a powerful agent, its strength being controlled by the pH of the solution. (4) Chlorine and chlorine-containing compounds. These include elemental chlorine, hypochlorous acid and its salts, sodium chlorite, chlorine dioxide, chloric acid and its salts. (5) Other halogenated compounds, for example, periodic acid. (6) Metal oxides, including lead dioxide, manganese dioxide, mercuric oxide, selenium dioxide, silver oxide, and osmium tetroxide. The first are usually used in acid media. (7) Chromic acid, formed by dissolving sodium or potassium dichromate or chromic anhydride in an acid. (8) Mineral acids. The principal disadvantage of nitric acid, nitrous acid and nitrogen tetroxide is that they may introduce an undesired nitrate group. Likewise, fuming sulfuric acid may cause undesired side reactions. (9) Metal salts. Oxidized salts of metals which can exist in more than one oxidation state, for example, ferric chloride, ceric sulfate, mercuric acetate, may act as mild oxidizing agents.

Miscellaneous other oxidants which have had specific applications include lead tetraacetate, aluminum tert-butoxide, nitrobenzene, arsenic acid, potassium ferricyanide.

For the production of heat or power by combustion the usual oxidant is, of course, the oxygen in the air. In explosives, or where very rapid power production from a compact source is desired, or in locations where air is relatively unavailable, some other oxidant is commonly used. For example, in liquid-fueled rockets, liquid oxygen is a common oxidant. In most solid propellants an inorganic perchlorate is incorporated into a suitable combustible mixture.

Suitable storage and handling characteristics of oxidizing agents will vary greatly. They will react vigorously with organic materials at elevated temperatures and may create a fire hazard if stored in contact with or near organic substances. The specific precautions required for each individual oxidizer should be carefully investigated before use.

CHARLES N. SATTERFIELD

Cross-references: *Autoxidation, Oxygen, Peroxides, Electrode Potentials.*

References

Shtern, V. Ya., "The Gas-Phase Oxidation of Hydrocarbons," The Macmillan Company, New York, 1964. (English Translation).

Reich, L., and Stivala, S. S., "Autoxidation of Hydrocarbons and Polyolefins. Kinetics and Mechanisms," Dekker, New York, 1969.

Emanuel, N. M. (ed.), "The Oxidation of Hydrocarbons in the Liquid Phase," Pergamon, Oxford, England, 1965. (English Translation).

Scott, Gerald, "Atmospheric Oxidation and Antioxidants," Elsevier, New York, 1965.

de Bethune, A. J., and Swendeman Loud, N. A., "Standard Aqueous Electrode Potentials and Temperature Coefficients at 25°C," C. A. Hampel, Pub., 8501 Harding Ave., Skokie, Ill. 60076, 1964.

OXIDATION STATE DIAGRAMS. *See* **Frost Diagrams.**

OXIDE CHEMISTRY

An oxygen atom, of atomic number 8, has a kernel consisting of a nucleus and the $1s$ orbital containing two electrons. Outside this kernel is the valence shell, of principal quantum number 2, which contains 6 electrons, 2 short of its capacity of 8. These electrons are paired in the $2s$ and one of the $2p$ orbitals, leaving each of the other two p orbitals with one electron. Thus oxygen has the capacity of forming two covalent bonds, or one double bond. In either case, the oxygen atom still possesses two lone pairs of electrons which under favorable circumstances, principally involving a negative charge on the oxygen, can permit it to act as an electron donor in coordination bonding. If instead, the two outermost shell vacancies are combined in one orbital, then the oxygen atom can use this orbital to accept a pair of electrons from another atom, thus forming a coordinate covalent bond with oxygen acting as acceptor. The single bond so formed is commonly strengthened by the presence in the other, donor atom of otherwise unoccupied outer orbitals, usually d, capable of accommodating one or more of the oxygen lone pairs. Such strengthening is usually termed "pi bonding." These abilities suggest the most frequent behavior of oxygen in its compounds but there are also other possibilities that are less common.

Among all the chemical elements, only fluorine is more electronegative than oxygen. The high electronegativity of oxygen is consistent with the fact that most of its electrons are in the outermost shell where they are relatively ineffectual in shielding the nuclear charge from outside electrons. Consequently, in all its compounds except with fluorine, oxygen forms polar bonds in which it acquires partial negative charge. Its bonds to the least electronegative elements are generally described as "ionic" although the actual existence of oxide ions, $O^=$, seems very questionable. As the other element becomes one of higher electronegativity, the oxide bonds become less polar, until they are hardly polar at all. A consistent picture of oxide properties can be drawn largely from a consideration of the change in the nature of combined oxygen that is brought about by the acquisition of negative charge. In subsequent paragraphs the properties of binary oxides will be shown to exhibit a periodicity closely associated with the partial negative charge on the combined oxygen.

Oxygen combines directly with most of the chemical elements, especially when the reactants are heated. This constitutes a major preparative method for oxides. Simple binary oxides can also be prepared by thermal decomposition of complex oxides, when one of the product oxides is volatile. For example, both carbon dioxide and calcium oxide result by decomposition of calcium carbonate. Water is the volatile component when hydroxides are heated to decomposition. Finally, oxides are frequently among the most stable compounds that an element, especially a metal, can form, and oxygen will displace other elements from their combination with metals to form metal oxides. For example, metal hydrides, sulfides, nitrides, bromides, and iodides are commonly converted to oxides when these compounds are heated in air.

Oxides having the expected stoichiometry result from combination of oxygen with most of the major group elements and transitional elements as well. The generalized formulas for the major and subgroups are as follows: IA and IB: E_2O; IIA and IIB: EO; IIIA and IIIB: E_2O_3; IVA and IVB: EO_2; VA and VB: E_2O_5; VIA and VIB: EO_3 VIIA and VIIB: E_2O_7. Many lower oxides also exist. Transition metals characteristically form several different oxides each. For example, manganese forms MnO, Mn_3O_4, Mn_2O_3, MnO_2, and Mn_2O_7. The "inert pair effect" in the major groups is evidenced by such oxides as Tl_2O, PbO, and Bi_2O_3, and there are numerous other lower oxides of major group elements, especially of the nonmetals. Significant deviations from "ideal" stoichiometry are commonplace among solid oxide lattices. Many familiar solid oxides do not necessarily have the exact content indicated by their chemical formulas, but are stable over a range of compositions. Oxygen compounds also exist in which the oxygen molecule appears to have combined without splitting into its individual atoms. The best known examples of these are the alkali metal peroxides, M_2O_2, con-

sidered to contain $O_2^=$ ion, and the superoxides, MO_2, considered to contain O_2^- ion.

The physical properties of oxides are closely associated with their states of aggregation as determined, in part, by the partial charge on oxygen. When the oxygen is highly negative, as it is in combination with the active metals which are very low in electronegativity, the oxides are highly associated in what are termed ionic lattices. The coordination number of oxygen in those compounds varies with the partial charge on oxygen, increasing roughly by 2 for every -0.2 electron charge. Bonding energy is high, with the result that melting points are usually very high and volatility very low. With diminished bond polarity, and correspondingly lower negative charge on oxygen, the oxides are still highly condensed but tend to be polymeric rather than ionic, and to melt somewhat lower and be somewhat more volatile. As the negative charge on oxygen becomes less and less, the tendency is toward molecular oxides of relatively low melting point and high volatility, the least polar oxides being at ordinary temperatures either volatile liquids or gases.

The condensation of highly polar oxides corresponds to the fact that these are oxides of elements low in electronegativity, and such elements are always metals, having more outer shell vacancies than electrons. The simplest oxide molecule that might be formed thus contains oxygen which is highly negative and has extra electron pairs for donation, together with metal atoms that are positive and have extra orbitals capable of accepting electron pairs. As a general principle of bonding, condensation tends to continue whenever possible until all outer shell vacant orbitals are used in the bonding. In this sense, many solid oxides commonly regarded as ionic may usefully be considered as coordination polymers. At the other extreme, the molecular nature of relatively nonpolar oxides is not merely a reflection of the low bond polarity but also of the unavailability of vacant orbitals that could participate in further condensation.

The principal chemical properties of oxides include decomposition, oxidation-reduction, and acidity-basicity. Most solid oxides, having highly polar bonds, are very stable toward heat, but the molecular oxides, with bonds of low polarity, are commonly much less stable and easily decomposed. Partly because of this relative instability, but mostly because of the general tendency for a highly electronegative element to leave its combination with other elements that do not release electrons easily in favor of joining a new combination in which it can become more negative, binary oxides in which oxygen is only slightly negative are generally strong oxidizing agents. On the other hand, oxygen which is already quite negative has lost its oxidizing power. Oxides containing such oxygen cannot be expected to act as oxidizing agents under ordinary conditions. Oxygen, having sought electrons avidly, is not disposed to release them easily. Therefore even the oxides of most negative oxygen are not reducing agents. However, lower oxides of elements that have stable higher oxidation states can react as reducing agents. For example, CO readily becomes CO_2 and SnO is easily oxidized to SnO_2.

The most important general chemical property of oxides involves their relationship to the oxide of hydrogen, water, because of its ionizing ability. To a very limited but nevertheless very significant extent, water molecules interact to transfer a proton from one to the other, creating on the average, one hydronium ion, H_3O^+, and one hydroxide ion, OH^-, for every 556 million molecules at 25°. In the water molecule, hydrogen is partially positive and can be lost in the form of a proton to some other atom that can contribute an electron pair to hold it. Oxygen in the water molecule is partially negative, and its two lone pairs of electrons are therefore available to electron acceptors in the formation of coordination complexes (although most commonly just one electron pair is so used). The hydronium ion differs greatly from water by containing oxygen of much lower negative charge, such that its remaining lone pair is not available for coordination, and by having much more positive hydrogen, permitting the easier release of protons. Hydroxide ion, on the other hand, contains hydrogen which is no longer positive and cannot be released as a proton, together with oxygen which is much more negative than in water. This oxygen is therefore an excellent donor, much more effective than the oxygen in water. The two ions formed by water are thus quite opposite in nature.

An aqueous solution containing an excess of hydronium ions is called acidic. It readily releases protons to electron donating substances, oxidizes active metals, releasing hydrogen gas, and has a sharp and sour taste. An aqueous solution containing an excess of hydroxide ions is called basic. It readily accepts protons from substances that can release them, is in general an excellent electron donor, and has a flat and bitter taste and a soapy feeling. No aqueous solution can contain an excess of both hydronium and hydroxide ions because when these ions collide, a proton is immediately transferred from the hydronium ion to the hydroxide ion, and both become water molecules. Acids and bases thus neutralize one another by mutual destruction of the ions that characterize them, a process called "neutralization."

Almost all binary oxides react with water either to form definite hydroxides or as though they formed hydroxides, with two OH groups replacing each oxygen. Water itself, having the ability to provide both hydronium and hydroxide ions, is amphoteric, as are other oxides in which the partial charge on oxygen resembles that in water. As the partial charge on oxygen becomes more negative, the oxide loses acidity and increases in basicity, highly negative oxygen invariably corresponding to strong basicity. As the partial charge on oxygen becomes less negative than in water, the oxide loses the slight basicity of water itself and tends to become exclusively acidic, increasing in acidity with diminishng charge on oxygen. Since as a class, metals are low in electronegativity, their oxides contain oxygen of relatively high negative charge. Similarly, as a class, nonmetals are high in electronegativity and permit oxygen in their oxides to acquire only small negative charge. These are the underlying reasons for the validity of the common rule that metal oxides tend to be basic and nonmetal oxides, acidic.

When, as is common especially among the transition metals, several different oxidation states are known, the lowest oxide is always most basic and the highest oxide least basic and most acidic. Each successive oxygen atom increases the competition for electrons of the metal and diminishes the success with which any one oxygen acquires negative charge. In effect, the metal atom is made more electronegative and therefore less metallic and more nonmetallic, by each additional oxygen atom.

The acid-base properties of oxides do not depend on water as a medium, for the hydroxides are only a special case of complex oxides formed by combination of basic and acidic oxides. The aluminates, borates, silicates, sulfates, sulfites, carbonates, nitrates, nitrites, phosphates, and similar compounds are all examples of such combination, the most acidic oxide always acting as acceptor for ox-

TABLE 1. PROPERTIES OF SOME OXIDES

Oxide	Charge on O	Coord. No., O	Heat of atomization kcal/eq.	mp, °C	Acid-base*	Oxid. power of O*
Na_2O	−0.81	8	105.2	920	SB	none
Li_2O	−0.80	8	139.5	1727	SB	none
CaO	−0.57	6	126.7	2587	SB	none
MgO	−0.42	6	119.3	2802	WB	none
BeO	−0.35	4	139.7	2547	AB	none
Al_2O_3	−0.31	4	121.9	2027	AB	none
ZnO	−0.29	4	86.8	1975	AB	none
H_2O	−0.25	2	120.9	0	AB	none
B_2O_3	−0.24	2	127.7	450	VWA	none
SiO_2	−0.23	2	107.2	1700	VWA	none
SnO_2	−0.17	2	82.3	1927	AB	none
P_4O_{10}	−0.13	2,1	—	580	MA	none
CO_2	−0.11	1	96.0	−56.6	MA	W
SO_3	−0.06	1	55.0	17,28	SA	S
N_2O_5	−0.05	1	53.1	33	SA	VS
Cl_2O_7	−0.01	1	29.2	−81.5	VSA	VS

*V = very, S = strong, M = moderate, W = weak, A = acid, B = base.

ide ion and forming the anionic portion of the salt, leaving as cation the metal of the more basic oxide. In this sense an oxygen acid is merely a hydrogen "salt," although the relatively high electronegativity of hydrogen prevents this from being as ionic as are true salts of this acid.

A summary of some representative oxides and their properties is given in Table 1.

R. T. SANDERSON

Cross-references: *Bonding; Chelation; Electronegativity; Stoichiometry; Water; Acids; Bases.*

OXIDES

Oxides are chemical compounds obtained by the combination of oxygen with any other element of the periodic system. Also called oxides are substances like ethylene oxide (C_2H_4O), propylene oxide (C_3H_8O) and those formed by treating hydrogen peroxide with dialkyl sulfate in alkaline solution. Under these conditions dialkyl ethers [$(CH_3)_2O_2$, $(C_2H_5)_2O_2$] may be formed. These compounds on reduction give alcohols, so that their structure is Alk—O—O—Alk. Other organic compounds containing the O—O link can be formed from hydrogen peroxide (alkyl, acyl peroxides, percarboxylic acids and peranhydrides, etc.).

A nomenclature has been established to deal with cases of elements forming more than one oxide. Thus, the number of oxygen atoms which stoichiometrically combine with one atom of the element is recognized by introducing a numerical prefix: carbon monoxide (CO), carbon dioxide (CO_2), phosphorus pentoxide (P_2O_5), sulfur trioxide (SO_3), lead dioxide (PbO_2). Alternatively, the valence of the combining element is emphasized: cuprous oxide, copper (I) oxide (Cu_2O); cupric oxide, copper (II) oxide (CuO); ferrous oxide, iron (II) oxide (FeO); ferric oxide, iron (III) oxide (Fe_2O_3). For mixed oxides: ferrous-ferric oxide, iron (II, III) oxide (Fe_3O_4); plumbous-plumbic oxide, lead (II, IV) oxide (Pb_3O_4). Inorganic peroxides are compounds which, upon reaction with acids, produce hydrogen peroxide (BaO_2, but not MnO_2, PbO_2). Superoxides are the final oxidation compounds of heavy alkali metals (KO_2, RbO_2, CsO_2).

Due to the high electron affinity of this element and the relatively large amount of "free" electrons available in metals and elemental semiconductors, a great number of oxides, or their combinations, are found in nature. The most widely present, and essential for organized life, is water (hydrogen oxide). Oxides can be prepared by direct combination from the elements or by decomposition of suitable salts (nitrates, carbonates, hydroxides, formates, acetates, oxalates, etc.). Some oxides, however, cannot be formed by direct reaction. The oxides of bromine and chlorine have been obtained by reacting these elements with ozone. In the case of fluorine, an oxide has been formed by passing a mixture of F + O_2 through an electric discharge. Structurally, oxides form ionic and molecular lattices. Metal oxides contain anions O^{-2}, peroxides, O_2^{-2} (forming the so-called "oxygen bridge,"—O—O—), and superoxides, O_2^-. In some instances the bond is almost completely ionic (Li_2O, K_2O, Na_2O), while in other cases an appreciable contribution of

covalent bonding is present (ZnO, MgO, Al_2O_3, Fe_2O_3, Cr_2O_3, V_2O_3, Ga_2O_3).

Some oxides are good thermal and electrical conductors (Fe_3O_4), but most oxides are, at ordinary temperatures, either semiconductors or insulators. Their electrical conductivity increases with increasing temperature. Oxides can be classified as n-type (containing "free" electrons) and p-type semiconductors (containing "electron holes"). MgO, TiO_2, V_2O_5, MnO_2, PbO_2, In_2O_3, CeO_2, CdO, ZnO are n-type semiconductors, due to the presence in their lattice of cations in excess to the stoichiometric amount. This excess is accommodated into the lattice by the formation of anionic vacancies or cations in interstitial positions. NiO, CoO, FeO, MnO, Ag_2O, are p-type semiconductors, due to the presence in their solid phase of oxygen in excess to the stoichiometric amount. This excess is accommodated by the formation of cationic vacancies. Experimentally, this classification has been determined by a study of the sign of the thermoelectric effect and by following the relationship between electrical conductivity and oxygen partial pressure above the crystal. These studies have confirmed the fact that, if the oxide under consideration corresponds to the highest stable charge of the constituent metal ion, the tendency is to form an excess metal compound. On the other hand if higher oxidation states of the cations are possible, an oxygen excess compound is formed.

Metal oxides are essentially nonstoichiometric compounds, as compared to molecular oxides or Daltonide compounds. Stoichiometric deviations are possible in metal oxides, because at any temperature above 0°K the free energy of the system is found to be a minimum when there is a finite excess of one of the lattice constituents. Most metal oxide phases are therefore stable over a range of compositions. In some cases excess of both lattice constituents is possible.

The stability of oxides extends over a wide range of temperature. To a first approximation an indication of their stability can be obtained by noting the value of their heat of formation. When a classification of this kind is made, four different groups of oxides can be recognized. The first class consists of oxides with small heats of formation. These can be easily dissociated into metal and oxygen (oxides of gold, silver and mercury). The second class with heats of formation between 40–65 kcal/g-atom oxygen, is made up by oxides which are easily reduced with hydrogen or carbon monoxide (oxides of Bi, Pb, Ni, Co, Fe (ferric)). In the third group are those oxides which have heats of formation up to 100 kcal/g-atom oxygen (MnO, Cr_2O_3, WO_2, MnO_2, SnO_2). In the last group are the refractory oxides, with heats above 100 kcal/g-atom of oxygen (V_2O_3, TiO_2, BaO, Al_2O_3, BeO, MgO, CaO). It should be pointed out that these considerations apply to the bulk phase of oxides, since the surface values of thermodynamic quantities can be quite different from these of the bulk. The reduction of oxides of the third and fourth groups to the corresponding metal cannot be accomplished with CO or H_2. A more powerful reducing agent is necessary (carbon).

Metal oxides are soluble in acids with the formation of salts. For example, $ZnO + 2H^+ \rightarrow Zn^{++} +$

H_2O. Metal oxides therefore have basic properties, due to the relative ease with which metal atoms give up electrons. On the other hand, metalloid elements have a strong tendency to accept electrons with the formation of negative ions. Metalloid oxides, often called anhydrides, form with water compounds which have acid properties.

Calcium oxide is extensively used as a drying agent, because it can rapidly take up water to form calcium hydroxide. Commercial calcium oxide is prepared by roasting more or less pure calcium carbonate. It ranges in composition from almost pure calcium oxide to a material closely related to Portland cement. ZnO, which is industrially prepared by direct synthesis from metallic zinc, is used in the manufacture of paints, as a pigment for the preparation of white rubber, in pharmaceutical preparations (ointments, lotions), in the ceramic and glass manufacture and for the production of cosmetics. Carbon dioxide is a raw material for the manufacture of sodium and ammonium carbonates, sodium bicarbonates, for the preparation of mineral water, in sugar refining. It is the basic ingredient of many fire extinguishers, where it is generated by reaction between a carbonate and an acid. It is essential for plant life, where it is used for the formation of sugars. Lead oxides are used in the manufacture of flint glasses, glaze for earthenware, electric batteries (PbO), pigments and cement for stream joints (Pb_3O_4). MgO is used in the manufacture of refractory bricks, in basic steel processes and electrical furnaces. Manganese oxides are used in dry cells as depolarizers (MnO_2), for the manufacture of purple black enamels, for pottery and pigments.

G. PARRAVANO

Cross-references: *Semiconductors, Peroxides, Oxidation.*

OXIMES

The oximes are compounds containing the $>C=N-OH$ group and are commonly obtained in good yield by the condensation of an aqueous or alcoholic solution of an aldehyde or ketone with hydroxylamine. Thus, these derivatives of carbonyl compounds are conveniently designated as aldoximes, $RHC=N-OH$, or ketoximes, $RR'C=N-OH$. This condensation reaction proceeds rapidly with low molecular weight compounds, such as acetaldehyde or acetone, but with more complex compounds (acetophenone or benzophenone) more vigorous conditions, particularly prolonged refluxing of an alcoholic alkaline solution of the carbonyl compound and hydroxylamine, are required. A second general, although less frequently used, method of forming oximes involves the reaction of nitrous acid with compounds such as methyl ethyl ketone or benzoylacetone which have active methylene groups. Thus:

$$R-CO-CH_2-R' + HONO \rightarrow$$
$$R-CO-(C=N-OH)-R' + H_2O.$$

The oxime derivatives of aldehydes and unsymmetrical ketones may exist in two geometrically isomeric *syn* and *anti* forms analogous to geometric *cis* and *trans* isomers. With benzaldehyde and hydroxylamine in the presence of excess alkali, there is obtained the *syn*-benzaldoxime, m.p. 35°C,

(H and OH on the same side of the carbon-nitrogen double bond) which is stable to alkali but rapidly rearranged by acid to the *anti*-benzaldoxime, m.p. 130°C. Isolation of isomers of this type provides one of the strongest pieces of evidence for the indicated structure of oximes.

The oximes are frequently solids and for this reason have often been used as characterization derivatives for aldehydes and ketones. In this respect, however, the corresponding semicarbazones or phenylhydrazones are more generally useful, as they possess higher melting points. Nevertheless, oximes have often been used as an intermediate in the purification of carbonyl compounds; the aldehyde or ketone is recovered by acid hydrolysis.

Oximes exhibit properties of weak acids and weak bases to form salts of the following types: $[RR'C=N-O]^-Na^+$ and $[RR'C=NH-OH]^+Cl^-$. By the action of alkylating agents the O-alkyl ether or N-alkyl ether may be obtained having the alkyl group attached to the oxygen or nitrogen atom. Aliphatic and alicyclic ketoximes react readily with an alkaline solution of bromine or N-bromosuccinimide and sodium bicarbonate to form blue bromo-nitroso compounds ($RR'CBr-N=O$), which may be subsequently oxidized and then debrominated to produce secondary nitro compounds. An alternate and more general oxidation to the nitro compound may be accomplished using peroxytrifluoroacetic acid as oxidant.

A variety of reducing reagents convert oximes to primary amines and mild dehydrating agents (i.e., acetic anhydride) lead to formation of nitriles.

One of the most thoroughly studied rearrangement reactions has been the Beckmann rearrangement of oximes. When an oxime is treated with an acidic reagent (i.e., PCl_5 or H_2SO_4) in ether solution an N-substituted amide is obtained. It has been established that rearrangement in aldoximes and oximes derived from unsymmetrical ketones always involves exchange of the OH and the *trans* alkyl or aryl group. *Syn* and *anti* isomers of such oximes lead to structurally different substituted amides. In this way, the reaction has been invaluable in establishing the configuration.

Certain oximes form stable complexes with many metal ions. The use of dimethyl glyoxime as a qualitative and quantitative reagent for nickel is well known. Salicylaldoxime forms complexes with Mn^{++}, Cd^{++}, Ni^{++} and Cu^{++}. All these are precipitated in alcoholic solution. However, only the copper complex is insoluble in dilute acetic acid and this property enables this oxime to be used as a quantitative reagent for copper.

DON C. IFFLAND

OXOCATIONS

Oxocations are oxometal entities formed by many of the transition and a few early actinide elements in their higher oxidation states. Mononuclear species may be symbolically generalized as MO_x^{n+}, where M = a transition element in the Ti-Fe families or U-Am, x = 1, 2, or 3, and n = 0, 1, 2, 3, 4, or 5; and they are found as the central component of a larger unit complex molecule that may be neutral, anionic, or cationic. The oxocationic species are readily distinguishable from but are not that different in metal-oxygen bonding from oxyanions of the same metals, which may be generally formulated as MO_x^{n-}. Polynuclear species $M_xO_y^{n+}$ are also well known for a few transition elements and these may involve bridging (M—O—M) or terminal (O=M=O) oxygens or both (O=M—O—M=O). Bond strengths, orders, and IR stretching frequencies are all invariably lower and bond lengths longer for the bridging oxygens, although even for these there appears to be a substantial delocalization of oxygen pi electrons into available metal pi orbitals, resulting in partial multiple bond character.

Other ligands (besides oxygen) may be strongly or weakly attached to M, but the oxo oxygens are always very strongly bonded and generally persist during reactions involving the substitution of the other ligands or even redox. Oxygen appears to possess the capacity to form *multiple* (σ and π) covalent bonds with all nonmetals, except fluorine and the lighter noble gases, with all metalloids, and with almost all metals to which it can bond covalently. This strong tendency for multiple bonding on the part of oxygen is probably to be associated primarily with the high charge density of the relatively small O^{2-} ion (a Pearson hard base) and the resultant driving force for this species to transfer some of that charge via a pi bonding mechanism onto any species, with available acceptor (i.e., empty) pi orbitals of suitable symmetry and energy (a Pearson hard acid). Early transition metal ions and the heavier later ones seem to be particularly susceptible to this kind of bonding, especially when they are in high oxidation states so that they possess relatively few electrons in their d (or f) orbitals and high effective nuclear charges. Under these latter conditions the transition element ions possess empty pi bonding orbitals of appropriate symmetry and low energy.

There has been some recent theoretical work on bonding in oxocations and the electronic configurations of certain oxocations, but the experimental knowledge of the field is still far ahead of and increasing more rapidly than the theory at the present time. The experimental criteria most often used to establish a bond order greater than unity are these: (a) if a bond length is found to be shorter (by at least a few tenths of an angstrom unit) than the calculated or experimentally determined single bond distance (this is the best and most direct evidence); (b) if the bond angle *implies* a higher bond order, as for example in an M—O—M unit, if its value exceeds the tetrahedral angle by at least several degrees; or (c) if the bond stretching frequency (determined by IR or Raman studies) is considerably higher than is found (or predicted) for a single bond between the same atoms. This last method is the one most often relied upon because of its experimental simplicity, and it is often the only evidence available on the multiplicity of a metal-oxygen bond, although increasing numbers of single crystal x-ray structures are being reported for oxocation compounds. A magnetic criterion might also be employed in the case of bridging oxygens on paramagnetic metal ions since there is possible in this circumstance the canceling or lowering of paramagnetism in the metal ion by partial delocalization of electrons originally on the oxygen (superexchange).

The successive replacement of H_2O by OH^-, which in turn is replaced by O^{2-}, as the dominant species attached to metal ions, occurs in aqueous solutions with increasing pH or increasing ionic potential of the metal ion (charge-to-radius ratio), or increasing pi (acceptor) bonding capacity of the metal ion, or all of these. Indeed there are some metal ions which allow both H_2O and O^{2-} to co-exist within the same coordination sphere, as for example $UO_2(H_2O)_5^{2+}$ and $VO(H_2O)_5^{2+}$.

The most common oxocation species are UO_2^{2+}, VO^{2+}, MoO^{3+}, ReO^{3+}, OsO_2^{2+}, MoO_2^{2+}, CrO_2^{2+}, and ReO_2^+, with the number of examples and studies falling off roughly in this order. With the first two, many hundreds of complexes are known and many of these have been studied theoretically and by x-ray diffraction, visible-ultraviolet, infrared, ESR, NMR, equilibrium, magnetic, conductometric, and polarographic methods. A family by family survey of the oxocations leads to the conclusions which follow, some of which must still be considered tentative.

There are very few unambiguously established mononuclear MO^{2+} species (M = Ti, Zr, Hf, Th), as these metals, d^0 in their tetravalent state, have a much greater tendency to form either mononuclear hydroxo ions, $M(OH)_x^{+4-x}$, or chains or rings involving bridging by oxygen or hydroxyl ions. With vanadium the oxovanadium (IV), VO^{2+}, chemistry is voluminous and this ion is perhaps the most stable and persistent diatomic ion known. No analogous chemistry is known with Nb, Ta, and Pa, due in part no doubt to the much lower stability of the 4+ state. However in the pentavalent state oxo species MO^{3+} and MO_2^+ are found, and NbO^{3+} complexes are becoming more numerous. With V the VO^{3+} entity is found in many solid or liquid compounds whereas the dioxo ion VO_2^+ has been reported in relatively few compounds and is postulated to be the ion present in certain concentrated acid solutions. In Group VI (Cr, Mo, W) we find the +6 state represented by MO^{4+} and MO_2^{2+} species, and the latter are more commonly found, particularly with Cr. In the +5 state the MO^{3+} species are far more important than the MO_2^+ species, which in fact have never been observed for Cr. With Mo(V) there are dimers formed also, having both terminal and bridge oxygens:

$$(O=Mo—O—Mo=O)^{4+}, (O=Mo—(X)(X)—Mo=O)^{4+}, \text{ and } (O=Mo—(O)(O)—Mo=O)^{2+}.$$

and this dual role for oxygen within the same ion may be more widespread among the transition elements than is now realized. Mo(IV) forms several compounds containing the MoO_2^0 entity, shown recently to be a linear dioxo species, but there do not as yet appear to be any analogous Cr or W compounds. In Group VII, except for one or two Mn compounds and a few compounds of TcO_3^+, TcO^{4+}, and TcO^{3+}, all the oxocation chemistry is that of rhenium. Although oxo species abound here: ReO^{5+}, ReO_2^{3+}, ReO_3^+, ReO^{4+}, ReO_2^{2+}, ReO^{3+}, and ReO_2^0 it is only with ReO^{3+} and to a lesser extent ReO_2^+, that a large number of complexes have been prepared and studied. In Group VIII there do not appear to exist any oxocations with iron, cobalt, nickel, rhodium, palladium, iridium, or platinum. Ruthenium forms a few compounds believed to contain such species as RuO^{4+}; and RuO^{2+} is believed to exist in solution. Osmium forms a few compounds with such species as OsO_3^{2+}, OsO^{5+}, and OsO^{4+}, but most of its oxochemistry derives from OsO_2^{2+}. In the 5f series there are several very stable dioxo cations formed, MO_2^{2+} (M=U, Np, Pu, Am) and MO_2^+ (M=Np, Am); two unstable (toward redox), MO_2^+ (M=U, Pu); one recent monoxo species UO^{3+}; and two of very doubtful existence MO^{2+} (M=U, Pu).

The properties of the metal-oxygen multiple bond, such as bond length and IR stretching frequency, are not as sensitive to changes in the other metal-ligand bonds as might have been expected or hoped. However the following conclusions can be drawn from the accumulated data. The IR frequency range in which terminal metal-oxygen multiple bonds may be found is 1100–780 cm^{-1}, with the highest values being recorded for mononuclear *monoxo* species and the lowest values for *trans dioxo* species. Metal-oxygen bridge bonds generally give broader bands in a lower frequency range which extends approximately from 850–650 cm^{-1}, with most known values falling in the center of that range.

Enough good structural data on dioxo compounds in Groups V, VI, VII, and VIII are now available to permit some interesting generalizations to be made. The tendency for the metal-oxygen bonds to be linearly or angularly disposed seems to be determined by the formal valence state of the metal. Linear MO_2^{n+} groups are found only when an unshared pair of metal electrons is present. This then is the case for isoelectronic OsO_2^{2+}, ReO_2^+, MoO_2^0. UO_2^{2+}, which is 5f^{0}6d^0, is also linear as presumably are the other dioxoactinides, which are 5f^n electronic systems, and these therefore do not fit this generalization. The angular structure of the MO_2^{n+} group is predicted (and found in all cases studied) for the transition elements in their highest oxidation states, e.g., Os(VIII), Re(VII), Mo(VI) (the one studied the most and clearly possessing *cis* oxygens is MoO_2^{2+}) and V(V). Finally, judging from the available (but still meager) experimental data, the angular structure of the MO_2^{n+} group will be obtained also for intermediate oxidation states of the metals, e.g., in Re(VI) and Mo(V) compounds.

JOEL SELBIN

Cross-references: *Bonding, Oxide Chemistry.*

OXO PROCESS

The "oxo" or hydroformylation reaction consists of the synthesis of aldehydes and alcohols from olefins:

$$RCH{=}CH_2 + H_2 + CO +$$

$$Co \xrightarrow[125-175°]{200\ atm} \begin{array}{l} \rightarrow RCH_2CH_2CHO \\ \searrow RCHCHO \\ \qquad | \\ \qquad CH_3 \end{array}$$

As commonly practiced, a mixture of C$_7$-olefins obtained from propylene and isobutylene is treated with a mixture of synthesis gas (H$_2$:CO) at elevated temperature and pressure to produce a mixture of C$_8$-aldehydes. These aldehydes are then usually hydrogenated in a separate step to their corresponding alcohols which are then utilized to manufacture plasticizers. The total world production of oxo products in 1970 exceeded 5 billion pounds.

Cobalt salts are most frequently used as catalysts. Almost any form of cobalt may be used but frequently cobalt salts of organic acids are employed because they are soluble in the olefin feed. Under the condition of the reaction, the cobalt salts are transformed to dicobaltoctacarbonyl, $Co_2(CO)_8$, which reacts with H_2 to form the active catalyst:

$$H_2 + Co_2(CO)_8 \rightleftharpoons 2HCo(CO)_4$$

The hydrogen tetracarbonylcobaltate(-I), $HCo(CO)_4$, is an unstable gas at room conditions, quite soluble in organic solvents as well as soluble in water, in which it is essentially completely ionized to $H_3^+O + [Co(CO_4]^-$. $HCo(CO)_4$ can be extracted by aqueous alkali from the reaction mixture while it is still under pressure, and in some commercial processes, advantage is taken of this fact to recycle the cobalt. The sequence of reactions leading to aldehydes is most likely:

$$RCH{=}CH_2 + HCo(CO)_4{-}$$

$$\begin{array}{l} \rightarrow RCH_2CH_2COCo(CO)_3 \quad \xrightarrow{H_2} \\ \searrow RCHCH_3 \\ \qquad | \\ \qquad COCo(CO)_3 \end{array}$$

$$RCH_2CH_2CHO + RCHCH_3 + HCo(CO)_3$$
$$\qquad\qquad\qquad\qquad\quad |$$
$$\qquad\qquad\qquad\qquad\ CHO$$

If a pure terminal olefin is used, the product is principally a mixture of the two aldehydes shown above, with the straight chain product constituting about 70% of the mixture.

Recently the catalyst system has been modified by adding a trialkylphosphine, principally tri-*n*-butylphosphine, (n-C$_4$H$_9$)$_3$P. The modified catalyst permits the reaction to be carried out at higher temperatures and lower partial pressures of carbon monoxide because of the greater thermal stability of $HCo(CO)_3PR_3$ as compared to $HCo(CO)_4$. Under these conditions, the final mixture of alcohols can be produced in one step. The principal advantage of the phosphine-modified catalyst system, however, is the greater selectivity to straight chain product. With such catalysts it is possible to produce from a terminal olefin, a mixture consisting of about 90% straight chain and only 10% branched chain product. The enhanced selectivity may be due to steric effects in the conversion of alkyl to acyl cobalt compounds:

$$RCH{=}CH_2 + HCo(CO)_3PR_3 \rightleftharpoons$$

$$RCH_2CH_2Co(CO)_3PR_3 + RCHCH_3$$
$$\qquad\qquad\qquad\qquad\qquad\qquad |$$
$$\qquad\qquad\qquad\qquad\qquad\ Co(CO)_3PR_3$$

$$\qquad k_1{\downarrow}CO \qquad\qquad k_2{\downarrow}CO$$

$$RCH_2CH_2COCo(CO)_3PR_3 + RCHCH_3$$
$$\qquad\qquad\qquad\qquad\qquad\qquad\quad |$$
$$\qquad\qquad\qquad\qquad\qquad\ COCo(CO)_3PR_3$$

$$RCH_2CH_2CHO + RCHCH_3 \xleftarrow{H_2}$$
$$\qquad\qquad\qquad\qquad |$$
$$\qquad\qquad\qquad\ CHO$$

The possibility that $k_1 > k_2$ may account for the selectivity.

Although cobalt is used exclusively as the catalyst in commercial operations, rhodium catalysts are also effective and as a matter of fact have the advantage of faster rates and less side reactions. However, the prohibitive cost of rhodium precludes its commercial use even though it is marginally superior to cobalt.

MILTON ORCHIN

References

1. Falbe, J., "Carbon Monoxide in Organic Synthesis," Springer–Verlag, New York, 1970.
2. Chalk, A. J., and Harrod, J. F., "Advances in Organometallic Chemistry," Vol. 6, p. 119, Academic Press, New York, 1968.
3. Bird, C. W., *Chem Rev.,* **62,** 283 (1962).
4. Tucci, E. R., *Ind. Eng. Chem. Prod. Res. Develop.,* **9,** 516 (1970).
5. Wender, I., Sternberg, H., and Orchin, M., "Catalysis," Vol. V, 73, Reinhold Publishing Corp., New York, 1957.
6. Orchin, M., and Rupilius, W., *Catalysis Reviews,* **6,** 85 (1972).

OXYGEN

The total content of oxygen in the earth's atmosphere, crust, and oceans is about 50% by weight. Oxygen is by far the most abundant element, most of it due to the photosynthesis of both land and sea plants (algae). Oxygen removed from the air by respiration and combustion processes is restored by photosynthesis, in which carbon dioxide and water, energized by light, react in the presence of chlorophyll in green plants to form nutrients and liberate oxygen.

Oxygen, separated from air on a large scale by liquefaction and distillation processes, is of importance in the production, welding, and cutting of steel, as a raw material for chemical syntheses, in the propulsion of space rockets, and in several medical and other applications.

Occurrence. The element oxygen exists not only as O_2, but also to a slight extent as atomic oxygen, O, and as ozone, O_3 (see **Ozone.**)

Gaseous O_2 constitutes 20.95 volume % or 23.14 wt % of the earth's atmosphere. Oxygen is dissolved in the oceans and in most other natural waters. The respiration of fish and of most other marine animals depends upon this dissolved oxygen, which also serves the purpose of gradually oxidizing waste matter suspended or dissolved in the water.

Oxygen is a reactive element and is found in combined form in a great many chemical compounds, of which the most abundant are water (H_2O, 88.8% oxygen by weight), sand (SiO_2, 53.2% oxygen), and the silicates. Most of the solids in the earth's crust are essentially silicates.

Commercial Production and Distribution. For fifty or more years, after production of oxygen by the BaO_2 (Brin) process was discontinued, only one commercial process was employed for preparing oxygen: the fractional distillation of liquefied air. The year 1971 saw the introduction by the Linde Division of Union Carbide Corporation of an entirely new commercial process. This new process utilizes ambient temperature separation by means of a pressure cycle in which "molecular sieves" of the synthetic zeolite type are used to preferentially adsorb nitrogen from air. The product consists of about 95% oxygen and 5% argon, with substantially no nitrogen. The highest pressure required in this adsorption-desorption cycle is about 50 psi. The new process is expected to be economically attractive for oxygen uses with which the presence of 5% argon does not interfere, and when the oxygen is prepared in plants having capacities below about 40 tons per day of oxygen.

When high-purity (99.5%) oxygen is required, and in large plants, the rectification of liquefied air will continue to be used. In the older installations the incoming air, after removal of carbon dioxide and water vapor, was first compressed to about 3000 psi. The heat of compression was removed by heat exchange with water. The air was then cooled to its liquefaction temperature by expanding part of it through a Joule-Thomson valve and letting the remainder do work in a reciprocal expansion engine. This process was expensive, but supplied plenty of refrigeration, and was well adapted to producing liquid oxygen, nitrogen, and argon, which were then shipped as liquids in insulated railroad or truck trailer tanks.

In the last few years, however, the demand for oxygen has grown so large that a number of plants have been built on or near the property of large customers, oxygen being delivered to the customer as a gas in pipelines. In these "on-site" plants, only enough low-temperature refrigeration is produced to maintain liquid air in the distilling columns; the final products are gaseous and are passed through heat exchangers in order to recover as much refrigeration as possible.

Modern Cryogenic Plants. Practically all modern oxygen plants utilize the process described below:

Air is compressed, usually in a rotary compressor, to about 85 psig, cooled to room temperature, and passed through molecular sieve type adsorbent beds to remove moisture and CO_2. The clean air is then cooled in heat exchangers to near its dew point by means of the cold oxygen and the cold waste nitrogen produced in the main distillation column. A bed of silica gel or other adsorbent operating at this low temperature is then used to remove undesirable impurities such as acetylene and other hydrocarbons. About 85% of the air then enters the lower part of the double column (one column above another) which was invented many years ago for the efficient fractionation of liquid air. The upper (main) column, operating at about 10 psig, is connected to the lower by a heat exchanger which serves on one side as a condenser for the nitrogen vapor at the top of the lower column and on the other side as the reboiler for the upper column.

The lower column, which operates at a pressure of about 80 psig, serves to partially separate the air and to produce liquid nitrogen reflux for both columns.

Two liquid streams flow (by means of the available pressure difference) from the lower to the upper column: liquid nitrogen from the condenser and enriched (about 35% O_2) liquid air from the bottom of the lower column.

As mentioned above, about 85% of the intake air enters the lower column. The remaining 15% does work in an expansion turbine and then enters the upper distillation column. The refrigeration generated by the turbine is sufficient to cover all the unavoidable refrigeration losses of the entire plant.

The streams leaving the main (upper) column, both in the gaseous state, are (1) oxygen gas (about 99.5% oxygen and 0.5% argon) evaporated from the reboiler liquid and (2) waste nitrogen gas, about 99% N_2 and 1% argon, from the top. The liquid air phase is thus an intermediate state and never leaves the distillation column in liquid form.

Distribution. The gaseous oxygen is either (1) transported in pipelines to the point of use, (2) compressed into the familiar steel oxygen cylinders to about 3000 psi, or (3) liquefied in a separate liquefaction plant. Since low-temperature refrigeration is expensive, liquid oxygen costs a good deal more than the gas in a pipeline.

Consumption. It is estimated that during 1970, about 26 billion pounds of pure oxygen was consumed in the U.S., and this figure is expected to increase each year for some time to come.

Physical Properties. In its ordinary form, oxygen is a colorless, tasteless, and odorless gas which has the formula O_2.

Liquid oxygen is pale blue. When an open vessel of liquid oxygen is set beside a vessel of liquid nitrogen, it is easy to distinguish the liquids, because nitrogen is completely colorless.

For practical purposes, liquid oxygen may be considered to be a nonconductor of electricity. In both the gas and liquid phases, oxygen is paramagnetic. This property is fairly unique; nitric oxide is the only other diatomic molecule that is paramagnetic.

Other physical properties of oxygen are given in Table 1.

Chemical Properties. Oxygen can form compounds with all other elements except the lower-atomic-weight helium-group elements. It used to be thought that none of the helium-group elements would form oxides, but in 1961 Neil Bartlett prepared a compound, xenon trioxide, XeO_3. It forms colorless crystals which can be made to detonate easily.

Since oxygen and fluorine are both strongly electronegative, it might be thought that they would not combine with each other, but actually several oxygen fluorides are known. The most stable of these is OF_2.

Reactions at or Near Room Temperature. The oxygen molecule is not very reactive at room temperature, but it does react with strong inorganic reducing agents, such as ferrous sulfate in aqueous solution. Oxygen also reacts spontaneously with a number of organic compounds.

The most important reaction of oxygen that takes place in the neighborhood of room temperature is the metabolic process in which oxygen, taken from the air, oxidizes nutrients inside living cells with the help of enzymes. This process is the source of energy in all animals and in most plants.

Other examples of reactions at room temperature are the rusting of metals and the decay of wood. The presence of water is required to bring about these reactions. In the decay of wood, bacteria also play an important role.

TABLE 1. PHYSICAL PROPERTIES OF OXYGEN, O_2

Atomic number	8
Atomic weight	15.9994
Boiling point, 1 atm	$-182.97°C\ (=90.18°K)$
Melting point, 1 atm	$-218.79°C\ (=54.363°K)$
Triple point, pressure $= 0.0015$ atm	$-218.80°C\ (=54.353°K)$
Upper transition temperature for solid beta phase	$-229.4°C\ (=43.8°K)$
Lower transition temperature for solid beta phase	$-249.4°C\ (=23.8°K)$
Critical temperature	$-118.4°C\ (=154.8°K)$
Critical pressure	50.15 atm
Critical density	0.430 g/ml
Density of gas at 0°C and 760 mm pressure	1.429 gpl
Density of liquid at b.p.	1.142 g/ml
Density of solid average between 43.8 and 54.4°K	2.0 g/ml
Heat of vaporization of liquid at b.p.	50.94 cal/g
Heat of fusion at the triple point	3.3 cal/g
Solubility in water at 20°C, 1 atm	0.031 cc (STP) of gas/cc water
Heat capacity of gas, 25°C, 1 atm	7.02 cal/deg/mole
C_p/C_v for gas, 0°C, 1 atm (STP)	1.396
Viscosity of gas, 25°C, 1 atm	0.192 centipoise
Dielectric constant of gas, 0°C, 1 atm (STP)	1.0005233
Diffusion coefficient into air, 0°C, 1 atm (STP)	0.178 cm²/sec
Velocity of sound, 25°C, 1 atm	330 m/sec
Thermal neutron absorption cross section	<0.0002 barn

Reactions at Higher Temperatures. Most materials have to be heated before they will react with oxygen at an appreciable rate. Undiluted oxygen supports combustion at a much faster rate than does air. A wooden splint that is glowing in air relights in oxygen, iron wire burns vigorously, and in general, most types of combustion go very fast.

Most free elements react with oxygen at high temperatures to give the corresponding oxides. When these are water soluble, usually the oxides of metals form alkaline solutions, and the oxides of nonmetals form acids. Thus, sodium oxide dissolves in water to give sodium hydroxide, a strong base, and phosphorus pentoxide dissolves in water to form phosphoric acid.

Silicates and Silica. The greatest weight of combined oxygen exists in the form of silicates or silica. Silicates have complex polymeric structures subject to many variations. In nature they exist mostly in rocks and clay. Crystalline silicates have a basic structure of silicon-oxygen tetrahedra linked to metal-oxygen polyhedra by the sharing of oxygen anions. Common metal cations in silicates are those of sodium, potassium, calcium, aluminum, and iron.

Silica is an inorganic crystalline three-dimensional polymer, $(SiO_2)_n$. Silica occurs in nature as the mineral quartz. The most abundant form of quartz is sand.

Reaction of Oxygen with Organic Compounds. There are thousands of organic compounds which contain oxygen, but most of these are prepared indirectly rather than by the direct action of oxygen. However, there are a number of commercially important reactions of gaseous oxygen with organic compounds. An example is the direct catalytic oxidation of ethylene to ethylene oxide.

The production of water gas $(CO + H_2)$, from coal and steam, although strongly exothermic, can be made continuous by mixing oxygen with the steam. Enough of the coal then burns to supply heat for the reaction.

A mixture of carbon monoxide and hydrogen, called synthesis gas, is produced on a large scale by the partial oxidation of natural gas:

$$2CH_4 + O_2 \rightarrow 4H_2 + 2CO$$

While this reaction is being carried out, if the gases leaving the reactor are quenched by a spray of water, acetylene can be isolated from the off-gases and used on location as a raw material for organic synthesis.

Principal Uses. The three largest classes of uses for oxygen are (1) in the iron and steel industries, (2) in the large-scale manufacture of certain chemicals, such as synthesis gas and sodium peroxide, and (3) in rocket propulsion.

Iron and Steel. The oldest use for oxygen (as contrasted with air) was in the welding of steel. For this, a torch was used in which acetylene was burned with oxygen to give a very hot flame. This process is still in extensive use, although welding of aluminum is usually done by electric arc processes.

Steel in thicknesses up to two feet and more may be cut with a special cutting torch. The metal to be cut is first preheated by an oxygen-acetylene flame and then cut by a large stream of oxygen.

Heat is produced by the reaction of oxygen with the iron in a great shower of sparks.

A somewhat newer use for oxygen is the automatic scarfing of hot steel billets and slabs in steel mills. In this process, surface imperfections are removed by directing preheating flames and streams of oxygen gas against all four sides of the hot pieces of steel as they move through the rolling mill.

Rocket Propulsion. Large rockets are propelled upward ("boosted") from the launch pad by the combustion of a mixture of a kerosene-like fuel with oxygen. The liquid oxygen and the fuel are kept in separate tanks until the count-down is almost complete. At the moment of blast-off, streams of the liquids are mixed and ignited.

Mining. One method of drilling holes through exceptionally hard ore preparatory to blasting, is to spall the ore with heat from a special oxygen-kerosene burner. This "jet piercing" process is extensively used in the mining of taconite, an iron ore.

Physiological Uses. Oxygen is used in connection with space travel, in the decompression of divers who have been at great depths, and in medical therapy. Oxygen has been found helpful in the treatment of pneumonia, emphysema, some disorders of the heart, and a number of other diseases. Treatment may be carried out in a hospital room or even at home with the help of a tent or mask. Some hospitals have hyperbaric chambers in which patients may be treated with oxygen under pressure. Some diseases which have been successfully treated in these chambers are gas gangrene, carbon monoxide poisoning, and decompression sickness ("the bends").

Sewage Treatment. A new use for oxygen is to hasten the aerobic digestion of sewage solids. Use of oxygen instead of air greatly increases the rate of the bacterial action, so that equipment of a given size can handle far more sewage per day than when only air is supplied.

Acknowledgements

The help of Dr. L. C. Matsch of the Linde Division of Union Carbide Corporation in supplying information for this article and in reading it, is gratefully acknowledged.

GERHARD A. COOK

Reference

Cook, G. A., and Lauer, C. M., "Oxygen" in the "Encyclopedia of the Chemical Elements," edited by C. A. Hampel, Van Nostrand Reinhold Co., New York, 1968.

OZONE

Ozone is a blue unstable gas that has a characteristic pungent odor. Its chemical formula is O_3 and its molecular weight is 48. Ozone can be liquefied to form a deep blue liquid that has the following properties:

Normal boiling point	−111.9°C
Melting point	−192.7°C
Heat of vaporization	75.6 cal/g

Critical temperature	−12.1°C
Critical pressure	54.6 atm
Density at −183°C	1.571 g/cc
−195.4°C	1.614 g/cc
Viscosity at −183°C	1.55 cp
−195.4°C	4.20 cp
Surface tension at −183°C	38.4 dynes/cm

Liquid ozone is stable as long as it is kept below its critical temperature and free from trace quantities of reactive materials. Liquid ozone is soluble in liquid oxygen. At −183°C, a single phase exists to 29.8% by weight ozone, at this point a two-phase liquid region exists until the concentration reaches 72.4% ozone where a single phase exists. At −195.4°C the concentrations are 9 and 91% and at −180.3°C the liquid system forms a single phase across the entire concentration range. Ozone also forms mixtures with fluorine and halocarbons.

Ozone can be produced by several techniques. It is formed in low concentrations of a few ppm by absorption of ultraviolet radiation (2137Å) by oxygen. As a matter of fact ozone is formed in the upper atmosphere by this method. At 50–120,000 ft the concentration can be 5–10 ppm. This UV absorption by oxygen prevents such radiation from harming life on this planet. The amount of ozone at 30–40,000 ft varies with latitude, season of the year, atmospheric turbulence, and can be correlated to the height of the tropopause.

The common commercial method of ozone production utilizes an ozonizer or ozonator which gives a silent arc discharge. Numerous geometries have been used ranging from flat plates to cylindrical tubes. In general, all ozonizers consist of a pair of electrodes, a layer of insulation, an air space and a source of high-voltage alternating current. The discharge space must be cooled to produce ozone at a reasonable concentration and air or oxygen must be used to remove the ozone as it forms. A large part of the electrical energy supplied to the ozonizer is released as heat and must be removed usually by water or air, since ozone decomposition is accelerated by heat. Ozone can also be formed using microwave and nuclear energy.

Ozone may be quantitatively determined by several methods. The most common analytical method is the oxidation of potassium iodide solution. Ozone liberates iodine from this solution under neutral pH condition and the iodine is subsequently titrated with sodium thiosulfate. Ozone absorbs light at various wavelengths in the infrared, visible and ultraviolet regions; and this characteristic can be useful in photometric methods of analysis. Thermal decomposition and chemiluminescence can also be used to detect O_3.

Ozone is an excellent oxidizing agent. It oxidizes most inorganic compounds to their final oxidative state, e.g., ferrous, manganous and chromous ions are oxidized to their highest states of oxidation. Usually in this type of reaction only one atom of the ozone molecule enters the reaction and the by-product is oxygen. In organic reactions the ozone may behave in a similar fashion but usually the reaction proceeds to form an ozonide wherein all the ozone molecule is coupled to the organic compound. The ozonide is subsequently hy-drolyzed to give degradative products. Ozone is highly destructive to many microorganisms, such as bacteria, fungi and algae and has been used for water purification. In concentrations of less than 1 ppm it can be used to sterilize and deodorize water. In some instances it is more specific than chlorine. Ozone is also used commercially to ozonize phenols and olefins, in the production of metal oxides, in bleaching processes and recently has been used successfully for tertiary water treatment.

Ozone can also be reacted with alkali and alkaline earth hydroxides to prepare inorganic ozonides having the general formula MO_3. Tetramethylammonium hydroxide reacts in the same manner. These compounds decompose to give oxygen and the oxide.

The odor of ozone can be detected at 1–5 parts per 100 million. The maximum allowable concentration of 0.1 ppm/volume for an 8 hour day exposure was established in 1955 by the American Conference of Governmental Industrial Hygienists. In continued exposure (several parts per million) eye irritation and coughing may result. Headache, nausea and pulmonary edema can also occur if exposure is continued. An exposure for 3 hours to 12 ppm ozone is lethal to mice, 13 ppm to rats and 25 ppm to guinea pigs.

CHARLES K. HERSH

References

Hersh, C. K., "Production and Properties of Liquid Ozone and Liquid Ozone Oxygen Mixtures Progress in Astronautics and Rocketry," Vol. 2, p. 427, "Liquid Rockets and Properties," Academic Press, Inc., N.Y., 1960.

Jaffe, L. S., and Estes, H. D., "Ozone Toxicity Hazards in Cabins of High Altitudes Aircraft," *Aerospace Medicine*, **34,** 633 (1963).

"Ozone Chemistry and Technology," Adv. in Chem. Series, Vol. 21, American Chem. Soc., Washington, D.C., 1959.

Ozonolysis

The addition of ozone to olefins which leads to cleavage of the double bond is termed ozonolysis. Ozonization and ozonozation are general terms that include the reaction of ozone with other materials.

Schonbein (1855) was the first investigator to study the reaction of ozone with olefins. He reported on the reaction of ozone with ethylene to give formaldehyde, formic acid and carbonic acid. The initial ozonide to be isolated was that formed from the reaction of ozone with benzene by Houzeau in 1873. This material is a white amorphous sensitive explosive.

Ozonolysis is now recognized as a versatile technique for oxidative cleavage of the double bond. The interest in ozonolysis can be divided into three areas: (1) to locate the position of the double bond for structural identification, (2) to synthesize ketones, aldehydes, alcohols and acids, (3) to investigate the reaction mechanism.

In the ozonolysis reaction, the oxygen atoms of the ozone molecule are inserted into the carbon-

carbon double bond to form a cyclic compound, termed an ozonide. This ozonide contains one ether and one peroxy bridge between the carbon atoms. The mechanism of the reaction is complex and is not completely understood. However, ozone appears to add to the double bond to give an unstable adduct, sometimes termed a "molozonide." The molozonide then converts to the ozonide. The factors involved in this reaction are (1) the nature of the initial attack of ozone on the double bond, (2) the form of the initial adduct, (3) the mechanism by which the initial adduct is converted to the ozonide, (4) the influence of olefin geometry and substituents, solvent, temperature and concentration.

Some qualitative observations can be made regarding ozonolysis. Ozone reacts rapidly with isolated open-chain ethylenic linkages, but much less rapidly with those present in aromatic systems. The first double bond of a conjugated system reacts very rapidly; the second, more slowly. Double bonds with highly substituted carbons or in sterically hindered ring systems do not add ozone as readily as simple ethylenic double bonds. Polymeric ozonides can be formed under certain conditions. There is a definite solvent effect, e.g., polar solvents yield monomeric ozonides and nonpolar solvents form polymeric ozonides. Ozone concentration is also important. If the concentration is too high or the time of contact too long, the ozonide decomposes and the cleavage products are oxidized.

The most critical step in ozonolysis is the cleavage of the ozonides to the desired end-product. Where acids are the desired products, water alone or in the presence of oxidizing agents can be used. If aldehydes are to be recovered, it is usually necessary to perform the reaction in the presence of a reducing agent.

The first commercial use of ozonolysis was in the production of vanillin from isoeugenol prior to World War I. Similar processes were developed for the production of piperonal from isosafrol and anisic aldehyde from anethole. In more recent times the synthesis of cortisone and other hormones has involved use of ozonolysis techniques.

Ozone is being used commercially for the oxidation of oleic acids to azelaic and pelargonic acids. In this application it has a high specificity in formation of the ozonide and as a result higher yields of purer products are possible than with other oxidants.

CHARLES K. HERSH

References

ACS Division of Petroleum Chemistry Preprint, **16,** No. 2, 1971.

"Advances in Chemistry Series," **77,** "Oxidation of Organic Compounds-III," Washington, D.C., 1968.

P

PAINT

Paint is a mixture composed of solid coloring matter suspended in a liquid medium and applied as a coating to various types of surfaces.

The purpose of the coating may be decorative, protective or functional. Decorative effects may be produced by color, gloss or texture. A secondary decorative function of paint is lighting, as the color of the surface affects the reflectance. The proportion of light reflected by a surface is expressed as a percentage of complete reflectance. The following are approximate reflectance values for various colors.

White	90–80%
Very light tints	80–70%
Light tints	70–60%
Medium to dark tints	60–20%
Aluminum	45–35%
Deep colors	20– 3%
Black	2– 1%

An example of the functional properties of paint would be a traffic paint to mark the center or edge of a road. The protective properties of paints include defense against air, water, sunlight and chemicals (i.e., acids, alkalies, and atmospheric fumes) as well as mechanical properties such as hardness, abrasion resistance, etc. The protective coating may be the paint on a wooden boat for protection against rotting; the interior lining for metal cans or drums to prevent corrosion from foods or chemicals, coatings on electrical parts to exclude moisture; fire retardant paints to protect combustible surfaces, coatings on plaster and concrete for ease of cleaning, etc.

Between 600 and 700 million gallons of paint were produced in the United States in 1970 with a value of over 3 billion dollars.

Types. The paint industry serves two distinct types of markets; i.e., trade sales (shelf goods) and industrial (chemical coatings).

Trade sales is the large consumer-oriented portion of the business. Trade sales paints are house paints and other products marketed through wholesale and retail channels to the general public, professional painters and contractors for use on new construction or for the maintenance of old buildings. Also included in this category are the paints sold to garages and repair shops for automobile refinishing, marine finishes for the maintenance of pleasure boats, paints for graphic arts, such as signs, paints for refinishing machinery and equipment, and paints sold to government agencies, especially traffic paints used for road marking.

The other major market for paint products is industry for products finishing. These are the coatings sold directly to the manufacturer for factory application. Products include durable goods such as automobiles, appliances, house siding, etc. and nondurable goods such as coatings for cans for food and beverages, etc.

Two newer methods of applying coatings in industry are coil coating and electrodeposition. Coil coating or strip coating is a process in which long rods of metal strip, usually aluminum or steel, are continuously roller-coated with paint on one or both sides. The paint is then cured by passing through ovens. The coated coil stock is then fabricated into many types of finished products, such as venetian blinds, siding, shelving, awnings, roof decking, etc. Electrodeposition of organic coatings is similar to the electroplating of metallic films. The metallic object to be coated acts as the anode and the tank containing the coating material is the cathode. Anionic resins and pigments are dispersed or dissolved in water. When the current is applied, the negatively charged paint and pigment particles are deposited on the anode or metal surface and the paint is essentially dry as it leaves the tank. This process has the advantage of applying a uniform coating on irregular surfaces and inaccessible areas. This process is used on automobile bodies and parts.

Composition. All coatings contain a resinous or resin-forming component called the *binder*. This can be a liquid, such as a drying oil, or a resin syrup that can be converted to a solid gel by chemical reaction. In some instances where the binder is either solid or is too viscous to be applied as a fluid film, a volatile solvent or *thinner* is also added. This evaporates after a film is deposited and the evaporation causes solidification of the film. The binder plus solvent is known as the *vehicle*. Most paints also contain *pigments*. These are insoluble powders of very fine particle size; i.e., as small as 0.01 micron and usually no larger than 1 micron. Pigments impart color and opacity to the paint. Certain materials which do not contribute to the color and little to opacity, but impart other desirable properties to the paint, both in the wet and dry coatings, are known as extender pigments.

Binder. The protective properties of a coating are determined primarily by the binder. In the early days of paint technology, binders were limited to materials of natural origin, such as drying oils, congo resins, asphalts, etc. These still find some usage in the protective coatings industry. However, the chemicals and plastics industries during the past 60 years have supplied a large number of synthetic binders which make possible paints of better protective and decorative properties.

Coatings in liquid form can be divided into two types; i.e., solutions and dispersions. In solution systems the resin binder is dissolved in a solvent. In dispersion systems the resin is in the form of tiny spheres (usually 10 microns or less in size) suspended in a volatile liquid carrier. The liquid may be water or an organic material. When the liquid evaporates, a mixture of resin and pigment is left behind which fuses into a continuous film. If the liquid is water, the system is called an emulsion; if it is an organic material, it is called an organosol.

The table below shows a general classification of different resin types with their properties. Paints may be made up with mixtures of two or more different resin types to obtain optimum properties.

Oils: easy application, soluble in aliphatic solvents.

Alkyds: all-purpose, combined with other resins, most soluble in aliphatic solvents.

Cellulosics: (nitrate and acetate) used in lacquers, fast-drying.

Acrylics: good color and durability.

Vinyls: good durability, abrasion-resistant.

Phenolics: good chemical resistance, yellow color.

Epoxies: good chemical resistance.

Polyurethanes: good flexibility, abrasion resistance.

Silicones: good heat resistance.

Amino resins (ureas and melamines) blended with alkyd for baking finishes: tough, good color.

Latex form:

Styrene-butadiene: low cost, alkali-resistant.

Polyvinyl acetates: low cost, good color retention.

Acrylics: good color and durability.

Solvents. The type of binder determines the type of solvent or thinner used in the paint formulation. The preferred type of thinner is an odorless aliphatic hydrocarbon which can be used in all areas including the home. Unfortunately, aliphatic thinners do not dissolve many resins. In such cases, strong solvents such as aromatic hydrocarbons (toluene), esters, ketones, etc. are used. Strong solvents may be used in industrial applications where they are confined in spray booths, etc.

Solvents are generally classified as low- medium- and high-boilers, depending upon the evaporation rate. They may be classified by chemical composition, as follows:

Hydrocarbons: aliphatic, VM&P naphtha and mineral spirits, etc. aromatics, benzene, toluene and xylene, etc.

Alcohols: methyl alcohol, ethyl alcohol, butyl alcohol, etc.

Ethers: Dimethyl ether, ethylene glycol-monoethyl ether, etc.

Ketones: acetone, methyl ethyl ketone, methyl isobutyl ketone, etc.

Esters: ethyl acetate, butyl acetate, butyl lactate, etc.

Chlorinated solvents: tetrachlorethane, etc.

Nitrated solvents: nitromethane, nitroethane, 1-nitropropane, etc.

Latex or emulsion paints use water as the volatile which simplifies their use in the home. Latex paints are popular because cleanup after painting can be accomplished with water.

Pigments. Pigments contribute to the color, hiding power, mechanical strength, gloss and corrosion resistance of paints. They may be divided into the following classes:

Inorganic	Organic
White	Colors
Extender	Black
Colors	
Black	
Metallic	

White

Titanium dioxide: high hiding, nonreactive.

Lithopone: zinc sulfide crystals coprecipitated with barium or calcium sulfate.

Zinc oxide: fair hiding power, contributes to mildew resistance.

Antimony oxide: fire-retardant.

Extender Pigments

Extender pigments are also known by other terms such as fillers, inerts, and supplemental pigments. Extenders are lower in price than prime pigments and have little or no hiding power. Flat and semigloss paints are produced by the use of extender pigments.

Calcium carbonate (sometimes called whiting): poor acid resistance.

Clay: plate-like in shape, excellent for water paints.

Talc: fibrous and flake types, excellent for exterior paints.

Silicas: good durability, used in interior or exterior paints.

Inorganic Colors

Inorganic color pigments may be classified broadly as natural and synthetic colors.

Iron oxides: yellow, red, brown and black.

Red lead: corrosion-resistance.

Chrome yellows and orange: lead chromate.

Molybdate orange: mixtures of lead chromate, sulfate and molybdate.

Zinc yellow: basic zinc chromates.

Cadmium yellow, orange and red: cadmium sulfide with zinc sulfide and cadmium selenide.

Chrome oxide: greens.

Organic Pigments

Although there are a great many organic pigments available, they may be classified into about six groups based on characteristics of their chemical groups.

Insoluble azo pigments: toluidine, para, chlorinated nitronilines, naphthol reds, Hansa, benzidine, dinitraniline, dinitraniline orange.

Acid azo pigments: lithol, lithol rubine, BON colors, Red Lake C, Persian orange, tartrazine.

Anthraquinone: alizarine, madder lake, indanthrene, vat colors.

Indigoid: indigo blue and maroons.

Phthalocyanine: blues and greens.

Basic PMA, PTA—PTA toners and lakes, rhodamine malachite green, methyl violet, victoria blue.

Manufacture. The manufacture of paint involves mixing, grinding and thinning. The pigment is mixed with some of the vehicle to a heavy paste consistency. The paste is then put through a grinding mill. There are several types used, such as a three-roller mill, ball mill, and high-speed impeller type mills. The purpose of this operation is to disperse the pigment thoroughly in the vehicle. The term grinding is a misnomer as the particle size of the pigment is not reduced in this operation.

Equipment. Various types of equipment are used for grinding. These are:

Roller Mill. Three- and five-roll mills are used. These are water-cooled, hardened steel rollers turning at different speeds in opposite directions with very small clearance to the adjacent roll. A scraper blade on the last roll removes the well-mixed paste into which additional vehicle is incorporated.

Pebble Mill. The pigment and vehicle are placed in a porcelain-lined water-cooled mill which is about one-half full of pebbles the size of golf balls. The mill is rotated from 4 to 48 hours. When blacks or dark colored paints are ground, the pebbles may be replaced with steel balls.

Morehouse Mill. A high-speed continuous mill in which the pigment-vehicle mixture is dispersed by passing between a rotor or stator.

Sand Mill. Sand is rapidly agitated by steel disc to disperse the pigment.

Cowles Mill. Dispersion accomplished by high-speed blade agitator.

Application. Paint may be applied by many different methods. The application method determines the selection of many of the materials that go into the paint, particularly the solvents. Methods used to apply paints are:

Hand brushing
Hand roller coating
Spraying
    Conventional
    Hot
    Airless
    Electrostatic
    Steam
    Flame
Roller coating
    Direct
    Reverse
Knife coating
Curtain coating
Flow coating
Dipping
Electrodeposition

A paint should, upon application, show good leveling without sagging.

Performance. The average thickness of one coat film of a dry paint is 1 mil. Coverage of paint is expressed in square feet per gallon and refers to the number of square feet that can be painted to give satisfactory hiding and performance. For example, a coverage of 400 sq. ft. per gallon might be specified for application. To give satisfactory service, a paint must adhere to the substrate during the service life. For this reason surface preparation is

very important. The paint film should expand and contract with the substrate. For exterior use the paint film must be resistant to breakdown from moisture and sunlight.

C. R. MARTENS

References

Martens, C. R., "Emulsion and Water Soluble Paints and Coatings," Reinhold, New York, 1964.

Martens, C. R., "Technology of Paints, Varnishes and Lacquers," Reinhold, New York, 1968.

Roberts, A. G., "Organic Coatings—Properties, Selection and Use," U.S. Government Printing Office, Washington, D.C., 1968.

PALLADIUM AND COMPOUNDS

Palladium is a member of the platinum group of metals with which it is generally associated in nature. It was discovered in 1803 as a second element in native platinum by W. H. Wollaston (England), and named after the recently discovered asteroid *Pallas*. Its abundance in the earth's crust, estimated as 0.01 ppm, is probably the highest among the platinum group elements. However, the cosmic abundance, based on examination of meteorites, is relatively lower—fourth among the six platinum metals. Its common terrestrial occurrence is in native alloys, e.g., native palladium, native platinum, iridosmine, etc. However, probably because the element is the most reactive within the platinum group, it is known in more compound mineral forms than the others. These include potarite, PdHg; stibiopalladinite, Pd_3Sb; froodite and michenerite, $PdBi_2$.

Selected physical properties include: atomic number 46; atomic weight 106.4; stable isotopes 102 (0.8%), 104 (9.3%), 105 (22.6%), 106 (27.2%), 108 (26.8%), 110 (13.5%); density 12.02 g/cc (20°C); melting point 1554°C; hardness (annealed) 40–42 Vickers units; electrical restitivity 9.93 μ ohm-cm (0°C). The metal crystallizes with cubic closest-packed lattice, having $a = 3.8907$ Å, with metallic radius for 12-fold coordination = 1.375 Å. Palladium possesses a markedly higher magnetic susceptibility than the other platinum metals. It is the least dense and lowest melting of the six elements in the platinum group.

For many applications alloying elements are added to palladium. These tend to increase the resistivity, hardness, and tensile strength. Ruthenium is often used for this purpose, but other agents include copper, nickel, gold, silver, or iridium. Palladium exhibits complete miscibility with the other elements of Group VIII, save ruthenium and osmium, and with those of Group IB.

Palladium lies somewhere between silver and gold in resistance to tarnish and corrosion, and in general reactivity. The compact metal is attacked, and spongy metal dissolved, by hot concentrated nitric or sulfuric acid. In any form the metal is dissolved by aqua regia with formation of chloropalladic acid, H_2PdCl_6. Halogens attack the metal combining with it when hot to form compounds like PdF_3 or $PdCl_2$. The metal is superficially oxidized when heated in air above 350°C, but above 870°C the oxide decomposes. Many molten compounds such as oxides, hydroxides, cyanides, or nitrates of the alkali metals will attack palladium.

A remarkable characteristic of palladium is its uptake of hydrogen over a range of temperatures. As much as 800–900 times its own volume of this gas may be absorbed. At one time it was held that a chemical compound Pd_2H was formed, though now it is believed that two solution phases arise at temperatures below 300°C. Within each phase hydrogen atoms are held interstitially in such a way as to involve actual chemical bonding, as deduced from changes in electrical conductance and magnetic susceptibility. As a result of this permeability, hydrogen can diffuse through hot palladium sheets. A palladium-silver alloy utilizes this principle to produce very pure hydrogen gas from commercial hydrogen.

Palladous oxide, PdO, may be formed directly from palladium sponge heated in oxygen. Alternatively, it may be prepared by fusing palladous chloride with potassium nitrate and then leaching out the water-soluble residue. This black oxide is insoluble in water or boiling acids (including aqua regia). It can be readily reduced by hydrogen to form an active hydrogenation catalyst. Above 875°C it decomposes to the elements.

Palladous chloride is formed by direct combination of the metal and chlorine at 500°C. It is a red deliquescent solid. From its aqueous solution or from a solution of palladium in aqua regia the dihydrate, $PdCl_2 \cdot 2H_2O$, is obtained when solvent is evaporated. The corresponding iodide is black, insoluble in water but soluble in excess iodide as the species PdI_4^{2-}. The nitrate, $Pd(NO_3)_2$, may be formed by dissolving finely divided palladium in warm nitric acid. The salt may be crystallized from this solution, but may be contaminated with basic salts and is very hygroscopic.

As with all the platinum metals there is a strong tendency to the formation of complexes which are quite stable. In the majority of these palladium is bivalent; these are uniformly diamagnetic and square planar. The ligands are linked most commonly through nitrogen, halogen (other than fluorine), carbon, or heavy donor atoms like phosphorus or arsenic. Representative examples include $Pd(NH_3)_2Cl_2$, sparingly soluble and of importance in refining the metal; Pd (dimethylglyoxime)$_2$, important in analysis; $[Pd(CO)Cl_2]_2$, the dark color of which is the basis of a common method of detecting carbon monoxide with $PdCl_2$. Quadrivalent palladium complexes are mainly restricted to hexahalides, e.g., K_2PdCl_6, and ammine halides, e.g., $Pd(NH_3)_2Cl_4$.

Palladium finds numerous applications in preparative organic chemistry as a catalyst for hydrogenation or dehydrogenation reactions, especially the former. Modern practice favors adding palladium oxide as the catalytic material in the case of reductions because it is more conveniently prepared than the "black" metal. At the start of the hydrogenation the oxide is reduced to finely divided metal. Another practice is to prepare a catalyst of palladium supported on barium sulfate, calcium carbonate, or magnesium oxide. For instance, a palladous chloride solution may be treated with calcium carbonate which becomes coated with palladous hydroxide; the solid is then separated and heated in hydrogen to reduce the palladium. Activated charcoal is also used to support catalytic palladium. In some cases a used catalyst may be reactivated by air or oxygen, but in others this is not possible and the metal is recovered for use again. A representative application is the gas-phase hydrogenation of acetylene to ethylene; for this the catalyst is usually supported on silica gel. Another useful catalytic application is for the removal of traces of oxygen from electrolytic hydrogen by catalysis of the hydrogen-oxygen reaction on palladium suspended on alumina. Small cartridges for this purpose may be obtained for attachment to cylinders of hydrogen in the laboratory.

The greatest consumption of palladium in the U.S. is by the electrical industry. For example, the metal is used in an alloy for relays in telephone equipment; here its freedom from tarnish or corrosion as well as its lower cost compared to platinum contribute to its suitability. A silver alloy is used for resistance windings because it possesses an unusually low temperature coefficient of resistance.

Some palladium is used in jewelry, and for this purpose the metal must be hardened. Ruthenium is probably the commonest hardening element, but some rhodium is used for the same purpose. Also "white gold" is an alloy of gold decolorized by the addition of palladium.

The unusual dissolving of hydrogen in palladium has some practical applications. The solubility of hydrogen in the metal falls rapidly as the temperature is increased while the rate of diffusion increases. Advantage is taken of this to admit hydrogen into vacuum apparatus in the laboratory. Palladium thimbles are sealed into the apparatus to act as hydrogen valves. When these are heated in the presence of gases containing hydrogen, even by a burner flame, hydrogen, but no other gas, diffuses through the metal.

W. A. E. McBryde

PAPER

Paper is the term given to matted or felted, fibrous sheets formed on a fine wire screen from a liquid suspension. The fibers most commonly used in the United States are those derived from wood, while smaller amounts of cotton (rag or linters), straw, flax, other vegetable fibers, and certain mineral fibers are also used commercially. At present, papers made of glass and synthetic fibers have very specialized uses for chromatographic or filtration purposes.

Paper is made in a variety of sizes and shapes, from the very lightest tissue up to very heavy board. It is converted into many different kinds of products such as paperboard containers, bags, waxed papers and other laminated papers, building papers, writing papers, bonds and ledgers, and an infinite variety of other grades and classifications. During 1970, the paper industry of the United States produced approximately 53 million tons of paper and paperboard. The consumption of reused pulp fibers was 12 million tons brought about by a desire to minimize waste.

The isolation and preparation of vegetable fibers in an acceptable form for papermaking involves the processes of pulping, bleaching, and refining. The pulping processes depend for their success upon the significant differences in chemical and physical properties between the fibers themselves and the regions responsible for fiber bonding. In the case of raw vegetable material such as wood, this system consists of fibers (cells with thick cellulosic walls) chiefly laid parallel to the axis of the tree, possibly linked by tenuous carbohydrate membranes and filaments, and imbedded in a matrix of noncarbohydrate material termed lignin. In the case of waste paper and nonwoven fabrics adhesion is achieved by means of hydrogen bonding assisted by mechanical forces brought about by the presence of additives such as latexes, clays, fillers, and synthetic fibers. The process of pulping involves the separation of these fibrous elements, more or less completely, from one another by mechanical and/or chemical action.

The groundwood process is the commonest form of mechanical pulping. Here bolts of wood are converted to pulp by pressing them longitudinally against a rapidly rotating grindstone in the presence of water. Fibers and fiber bundles of irregular dimensions are torn from the log and further shredded by the grindstone, and the resulting pulp is used in newsprint and in many other grades of paper. Other mechanical processes have been devised using machines such as disk refiners to convert wood chips into pulp. The shredding action of the grinders may be modified by incorporating a mild chemical pretreatment of the logs to weaken the forces holding the fibers of the wood together. This treatment of chemical and shredding actions is particularly effective in weakening the bonds between fibers in reusable grades of waste paper. Unfortunately the fibers in paper impregnated with certain additives such as latex cannot be economically recovered by these processes.

The aim of chemical pulping is to remove sufficient lignin from wood to permit fiber separation, and depends for its success upon the differences in chemical reactivity of the lignin (which is concentrated chiefly in the regions between the fibers) and of the carbohydrate components (located chiefly in the fiber walls). The reagents used commercially in pulping include calcium bisulfite (or sodium, magnesium, or ammonium bisulfite) in the presence of a calculated excess of sulfur dioxide (the sulfite process); no excess sulfur dioxide (the bisulfite process); a mixture of sodium sulfide and sodium hydroxide (the sulfate or kraft process); or, more recently, a solution of sodium sulfite buffered with sodium carbonate or bicarbonate (the neutral sulfite semichemical process). Small amounts of pulp are also made with sodium carbonate, but the use of this process is on the decline. The desire to minimize noxious odors and eliminate recovery boiler explosions has sparked a growing interest in the previously obsolete soda (NaOH) pulping process and in novel alkaline sulfite processes. Pulps prepared by multistage processes, which involve mild alkaline or acidic pretreatment stages followed by terminal acid sulfite or kraft cooking processes, have been found to have unusual papermaking properties, and several mills employing these processes are in production in Europe and North America. High-yield pulps which may be prepared by most pulping processes have been found to have satisfactory characteristics for boxboard and corrugating medium.

The chemical reactions which occur in the digester during chemical pulping processes are only partly understood. Briefly, in the sulfite processes, lignin is converted to a soluble salt of lignosulfonic acid, and appreciable carbohydrate material is removed at the same time. The dissolved polysaccharides in the spent liquor are being utilized in some mills for the production of yeast. In the kraft process, the lignin is altered and converted to a soluble alkali lignin, while considerable polysaccharide material is converted to saccharinic acids. Useful chemicals such as dimethylsulfoxide may be isolated from these liquors.

The alkaline cooking processes and the sodium- and magnesium-base sulfite processes would be commercially impractical unless the alkali were recovered for reuse. The increasing concern for pollution abatement had led to more effective and "tighter" recovery systems. In the case of the kraft process, recovery is accomplished by collecting the waste liquor, evaporating it in multieffect evaporators, and burning out the organic matter in recovery furnaces. Because of the high organic solids content of the waste liquors, recovery furnaces, are frequently connected with boilers to utilize some of the heat produced during recovery. In the case of kraft recovery systems, the alkali remaining after combustion flows from the furnace as a red-hot, molten stream into a tank of water to give a solution of sodium carbonate and sodium sulfide. The liquor is regenerated for reuse by a treatment with slaked lime to convert soda ash to caustic soda. The recovery of cooking chemicals from the spent sulfite liquors is more complicated since the sodium-base liquors are converted to a mixture of salts composed chiefly of sodium sulfide. The recovery of the magnesium-base liquors, on the other hand, results in the isolation of magnesium oxide and sulfur dioxide, and the problem of utilizing sulfide is avoided.

The raw fibers produced during pulping may be used directly in the papermaking process after slight preparation, but frequently the pulp from the digester is first given a bleaching or brightening treatment. This is particularly the case when white, chemically stable pulps are required. Bleaching is less extensive in pulps used for containers and boards. The bleaching processes use oxidative or reductive chemicals to remove or convert to colorless forms the residual lignin, colored carbohydrate derivatives, and traces of other coloring matter present in unbleached pulps.

Generally pulps are bleached in several stages using a combination of bleaching agents, frequently with complementary effects. Chlorination is commonly employed as the first stage in such sequences. Chlorolignins are formed initially, but these are quickly converted to colored quinoid products which are soluble either in the chlorination liquors or, in the case of kraft pulps, in the alkaline washing which follows chlorination. The next step is to oxidize the washed, chlorinated pulp with an alkaline hypochlorite solution followed by a washing stage. The product may be satisfactory at this stage for many uses. Where a brighter pulp is required, there may be additional hypochlorite, chlorine dioxide, peracetic acid, or sodium peroxide bleaching stages, each followed by alkaline and water washes. No one common bleaching system is employed; on the contrary, modifications and combinations are developed by each manufacturer to produce the best pulp for his specific purpose. Groundwood pulp may be bleached by hypochlorite, peracetic acid, or peroxides, which decolorize by oxidation, or by hydrosulfite, which reduces the colored constituents. The exact chemical reactions which occur in bleaching are not too well understood, not only because of the complex nature of the bleaching agents but also because of the unknown nature of the chromophores.

The bleaching reactions are complicated by the fact that large volumes of water are employed which become contaminated with various halide and silicate ions. Since recovery is expensive and difficult, reuse of the water for other purposes in papermaking is discouraged. Therefore, considerable research is being devoted to high consistency bleaching to minimize water usage and to the use of less polluting chemicals such as oxygen as alternatives to existing bleaching stages.

The last step before papermaking involves "beating and refining," sometimes called stock preparation. Pulp fibers, as they emerge from the digester and bleach system, if run directly over the paper machine, form paper of low strength and high absorbency because of the stiffness and poor bonding of the fibers. This would be a satisfactory way to make blotting paper, but it is an unsatisfactory way to make paper strong enough for cement sacks or potato bags. Strength is developed in paper by the somewhat related processes of beating and refining, in which fibers in an aqueous slurry are swollen by mechanical treatment in various types of beaters and refiners. In earlier years, beating and refining were said to "hydrate" the pulp. Today they are considered to be more closely related to a partial breakdown of the fiber structure into fine fibrillae of varying dimensions which have the capacity of holding water by virtue of their increased surface area. The fibrillae and the fibrillated fibers likewise have an increased flexibility which increases the possibility of bonding the fibers in paper. All these factors favor a greater degree of bonding in the finished sheet, which in turn results in increased sheet strength to a value limited only by the strength of the fibers themselves.

The beater is also used as the place where other materials may be added to the pulp—such additives as rosin soap to impart water resistance, starch or galactomannan gum for strength, fillers of some type of clay for opacity, dyestuffs for color, and alum for the adjustment of pH and to assist in better retention of the additives in the sheet. After beating and refining, the pulp is now ready for conversion into paper.

The manufacture of paper from the stock suspension also requires several steps. The first of these is the formation of the wet web with its random orientation of the fibers, which is carried out by drainage on the "wet end" of the paper machine. In the next step, the wet sheet is pressed to remove still more water, a process which compacts the sheet and improves the physical characteristics of the paper. In the final step, much of the residual water is removed by heat.

The wet web is formed either by running a dilute water suspension of fibers over a moving, endless belt of wire (the fourdrinier machine) or by running the endless belt of wire through a fiber suspension (the cylinder machine). In the former case, part of the water drains from the fibers by gravity, part is taken off by suction, and still more by roll pressure. In the last-named case, a vacuum maintained below the stock level in the cylinder forms the sheet on the wire by suction. This process may occur at speeds of 2500 or even more feet per minute. Research indicates that rotary suction formers and two-wire formers will have greater flexibility of operation and control than the conventional machines.

From the wet end of the wire, the wet sheet containing about 80% water is taken off on a felt which carries it through several sets of press rolls. The sheet entering the press section with 80% or so of water leaves the section with a water content of 60 to 70%. Many factors control the amount the wet sheet is compressed in the press section.

The wet sheet of paper finally enters the dryer section of the paper machine with enough strength to bear its own weight. Here it first passes between a pair of smoothing rolls where surface irregularities are pressed out; then to a section of steam-heated dryer cylinders whose number varies from a dozen to more than one hundred and depends on the speed of the machine and on the amount of water to be removed. The dryer section reduces the moisture content of the sheet from 60–70 to 4–7%. The volume of water removed in the dryer is very large—roughly 200 tons of water are removed for every 100 tons of dry paper coming out of the end of the paper machine when the paper enters the dryer with 60% water and leaves with 7%.

The previous description of the pulping, stock preparation, and papermaking processes gives the reader very little idea of the immensity of the operation. In a sulfite pulp mill, for example, the digesters may be 16–18 feet in diameter and 40 or so feet high; 20 cords of wood may be cooked at a time in the form of chips. Paper machines (e.g., the fourdrinier) make paper up to 350 inches wide, although this top figure is rare. One machine can make between 500 and 600 tons of paperboard a day.

Norman S. Thompson

References

Rapson, W. H., Ed., "The Bleaching of Pulp," TAPPI Monograph Series No. 27, 1965.

Whitney, R. P., Ed., "Chemical Recovery in Alkaline Pulping Processes," TAPPI Monograph Series No. 32, 1968.

Rydholm, S. A., "Pulping Processes," Interscience Publishing Co., N.Y., 1965.

Macdonald, R. G., Ed., "The Pulping of Wood," Vol. I, 2nd Edition, McGraw Hill Book Co., New York, 1969.

PARACHOR

Attempts to correlate physical properties with the structure of organic compounds have included studies of molecular volume, critical volume, and "null-punkts-volume" (obtained by extrapolation of temperature-density curves to absolute zero). Generally, however, the results have not been satisfactory, principally because of the inability to compare molecules in the same or corresponding states. Perhaps the most successful attempt to attain comparable conditions for measuring molecular volume or a directly related property, has been by use of the *parachor*. The parachor is a function dependent on surface tension, density, and molecular weight. In 1923 MacLeod discovered empirically the relationship $\gamma = C(D - d)^4$ where γ is the surface tension of the liquid at a given temperature, D and d are the densities of the liquid and its vapor, respectively, at the same temperature, and C is a constant characteristic of the liquid. Sugden revised this equation in terms of molar proportions

and obtained $P = \gamma^{1/4} \dfrac{M}{(D - d)}$, where P is the par-

achor and M the molecular weight of the compound. If d is negligibly small compared with D and if $\gamma = 1$, the relation reduces to $P' = V_m$, where V_m is the molecular volume. The parachor may therefore be considered as the molecular volume of a liquid of surface tension equal to unity. Comparison of parachors may thus be regarded as equivalent to comparing molecular volumes when the surface tensions are the same. In order to adjust to such a condition, it would seem necessary to measure surface tension over a wide temperature range. This is usually not required, however, since for most unassociated liquids the effect of temperature on parachor is very slight.

The parachor was first considered to be primarily an additive function, though some constitutive features were recognized. Sugden calculated atomic and structural constants which gave values for the parachors of many compounds corresponding closely to those calculated from observed surface tensions and densities. Improved sets of fundamental constants have since been developed as a result of extensions and refinements of the original values. It is now possible to obtain agreement to within 0.1% between theoretical and experimental parachors. Vogel[2] has proposed a system of bond parachors instead of the atomic and structural parameters currently in common use. Though additive to a great extent, parachor is very sensitive to structural changes and its essential constitutive nature is now well recognized.

The parachor has been used as a tool in deciding among several possible structures for an organic compound. While there has been some success in this application, the use of the fundamental constants has, in some cases, led to incorrect conclusions concerning the structure of a given compound. Parachor should therefore not be offered as proof of a certain structure, but rather as supportive evidence in its favor. The parachor has also been used to indicate the extent of hydrogen-bond formation as well as supplying experimental evidence for the existence of the coordinate covalent bond and the triple-bonded isocyanide grouping. It has been employed to predict the surface tension of mixtures and to find suitable methods for predicting critical properties. A linear relationship between parachors and the logs of retention data of structurally similar compounds has been found useful in gas chromatographic separation and identification of components in mixtures of organosilicon compounds.[3] Relationships similar to parachor, but involving properties other than surface tension, have been developed, e.g., rheochor (intrinsic viscosity) and refrachor (refractive index).

Recently, Exner[4] has suggested that structural problems can be solved more simply on the basis of molecular volumes and with the same reliability as with parachors. He proposes that the concept of the parachor as a specially defined function of experimental quantities has no material significance and therefore, no reason for existence.

ROBERT FILLER

References

1. Sugden, S., "The Parachor and Valency," Routledge and Sons, Ltd., London, 1930.
2. Vogel, A. I., Cresswell, W. T., Jeffery, G. J., and Leicester, J., *Chemistry and Industry*, 358 (1950).
3. Wurst, M., *Mikrochim. Acta,* **1966** (1–2), 379 (German); *Chem. Abstr.,* **65,** 5346b (1966).
4. Exner, O., *Collect. Czech. Chem. Commun.,* **32** (1), 24 (1967) (English); *Chem. Abstr.,* **66,** 75582n (1967).

PARAFFINS

Paraffins are saturated aliphatic hydrocarbons having the empirical formula C_nH_{2n+2}. The term is derived from Latin (*parum affinis,* meaning too little akin) and signifies the inertness of the paraffin hydrocarbons to reaction with most reagents. However, it is now known that while methane, ethane and propane are relatively unreactive, the higher molecular weight homologs, particularly the branched-chain compounds which contain tertiary carbon atoms, are quite reactive.

Methane, ethane and propane have no isomers. Butane and the higher homologs exist in straight-chain and branched-chain isomers. Not counting optical isomers, there are 2 butanes, 3 pentanes, 5 hexanes, 9 heptanes, 18 octanes and 35 nonanes, all of which are known. While eicosan ($C_{20}H_{42}$) includes at least 300,000 isomers, a very minute fraction of these have been synthesized.

Nomenclature. The paraffins are more systematically known as alkanes. They are usually named in accordance with the rules of the International Union of Pure and Applied Chemistry (IUPAC) which adopted the ending "ane" for saturated hydrocarbons. A prefix indicates the number of carbon atoms in the hydrocarbon. The first four prefixes are meth-, eth-, prop- and but-. The succeeding prefixes are derived from the Greek (and, occasionally, Latin) word for the number; pent-, hex-, hept-, oct-, non-, dec-, hendec- (or undec-), dodec-, etc.

Branched-chain paraffins are named as derivatives of the longest straight chain in the formula, the location and names of the substituent alkyl radicals being indicated as prefixes. (For examples, refer to Table 1.)

TABLE 1. NOMENCLATURE OF PARAFFINS

Formula	IUPAC	Complete Skeleton or Simple Nucleus
CH_4	Methane	Methane
H_3CCH_3	Ethane	Ethane
$H_3CCH_2CH_3$	Propane	Propane
$H_3CCH_2CH_2CH_3$	Butane	n-Butane

H_3CCCH_3
   |
  CH_3
	Methylpropane	Isobutane

$CH_3CH_2CH_2CH_2CH_3$	Pentane	n-Pentane

$CH_3CHCH_2CH_2$
   |
  CH_3
	Methylbutane	Isopentane

$CH_3C(CH_3)_2CH_3$	Dimethylpropane	Neopentane
$CH_3CH_2CH_2CH_2CH_2CH_3$	Hexane	n-Hexane

$CH_3CHCH_2CH_2CH_3$
   |
  CH_3
	2-Methylpentane	Isohexane

$CH_3CH_2CHCH_2CH_3$
   |
  CH_3
	3-Methylpentane	Methyldiethylmethane*

$(CH_3)_3CCH_2CH(CH_3)_2$	2,2,4-Trimethylpentane	Isopropyl-t-butylmethane*

Formula	Type
C_7H_{16}	Heptane
C_8H_{18}	Octane
C_9H_{20}	Nonane
$C_{10}H_{22}$	Decane
$C_{11}H_{24}$	Undecane or Hendecane
$C_{12}H_{26}$	Dodecane
$C_{13}H_{28}$	Tridecane
$C_{14}H_{30}$	Tetradecane
$C_{20}H_{42}$	Eicosane
$C_{21}H_{44}$	Heneicosane
$C_{22}H_{46}$	Docosane
$C_{23}H_{48}$	Tricosane
$C_{30}H_{62}$	Triacontane
$C_{33}H_{68}$	Tritriacontane
$C_{40}H_{82}$	Tetracontane
$C_{50}H_{102}$	Pentacontane

*Rarely used simple nucleus name.

An alternate method for naming the paraffins is the complete skeletal system in which the type of branching may furnish a name for the compound. Unbranched (straight-chain) paraffins are called "normal" (abbreviated n-) paraffins as in n-butane. If the paraffin is straight-chain except for one methyl group on the second carbon atom, the compound is named with the prefix *iso-* (as in isobutane and isopentane). The prefix *neo-* is used to designate two methyl groups attached to the second carbon atom of a chain of 3 or 4 as in neopentane and neohexane, respectively; the term cannot be used for higher molecular weight hydrocarbons because they involve isomers. The petroleum industry has erroneously applied the name isooctane —correct for 2-methylheptane—to 2,2,4-trimethylpentane which is the reference standard (octane number, 100) for antiknock properties.

Occurrence. The major sources of paraffins are natural gas and petroleum. These contain both straight- and branched-chain isomers. Several normal paraffins (n-heptane, n-nonane, n-undecane, n-hexadecane, n-heptacosane, n-nonacosane and n-hentriacontane) have been isolated from vegetable products. n-Heptane occurs in the "petroleum nuts" of *Pittosporum resiniforum* of the Phillipines and in the oleoresin of two western pine trees, *Pinus jeffreyi* and *Pinus subiniana*. n-Undecane comprises up to about 5% of the turpentine from several types of pines.

Physical Properties. Paraffins containing up to 5 carbon atoms per molecule are gases at room temperature and atmospheric pressure. Those containing 5 to 16 carbon atoms are usually liquid; examples of exceptions include dimethylpropane which is a gas and tetramethylbutane which is a crystalline solid. The straight-chain paraffins of more than 16 carbon atoms are waxy solids; refined paraffin wax consists chiefly of n-alkanes containing from about 23 to 30 carbon atoms per molecule. The boiling points and melting points of the n-alkanes (which are higher boiling than any of their isomers) increase with increasing molecular weight.

The paraffins have lower refractive indices and densities than those of the other types of hydrocarbons (cycloparaffins, olefins and aromatics) having the same number of carbon atoms.

Chemical Properties. Unlike the olefins and the aromatics, the paraffins, particularly the normal isomers, do not react readily at mild temperatures with acids and oxidizing agents. Because they are saturated hydrocarbons, they react chiefly by substitution of hydrogen by other atoms. They also undergo reactions involving scission of carbon-carbon bonds (cracking and isomerization) or carbon-hydrogen bonds (dehydrogenation to olefins and aromatics).

The substitution reactions include halogenation to alkyl halides and polyhaloalkanes; for example, in diffuse light chlorine reacts with methane to form methyl chloride, methylene chloride, chloroform and carbon tetrachloride. Substitution of hydrogen by a nitro group occurs when paraffins are treated with nitric acid. Paraffins containing tertiary carbon atoms react with dilute nitric acid in the liquid phase at 105–110°C to yield nitroalkanes while straight-chain paraffins require higher temperatures; the vapor phase nitration of propane with nitric acid at 400–450°C produces nitromethane, nitroethane, and 1- and 2-nitropropane. While paraffins do not react with sulfuric acid at moderate temperatures, reaction with fuming sulfuric acid (oleum) yields alkanesulfonic acids.

Paraffins are used as fuels because they undergo a highly exothermic reaction with oxygen (air) to yield carbon dioxide and water. The oxidation can be controlled to form mixtures of alcohols, aldehydes, ketones and acids; oxidation of paraffin wax produces acids which may be reacted with alkali to form soaps or with glycerin to yield synthetic fats.

Condensation (alkylation) of paraffins with olefins to yield higher molecular weight branched-chain paraffins occurs under both thermal and catalytic conditions. Thus, the reaction of isobutane with ethylene at 500°C at 150–300 atm pressure yields chiefly 2,2-dimethylbutane. On the other hand, in the presence of aluminum chloride catalyst the principal product is 2,3-dimethylbutane.

LOUIS SCHMERLING

PARTICLES AND ANTIPARTICLES

The ultimate nature of matter is a problem which has always challenged the profoundest inquiry and the utmost theoretical and experimental ingenuity of scientists. Only 40 years ago, there seemed to be a triumphant solution of the structure of the atom in terms of motions of negatively charged electrons around positively charged nuclei. But as probing of the interior of the nucleus began, an unbelievably complicated jumble of new and unsuspected elementary particles was discovered. The fact that the nucleus is made up of protons and neutrons seemed simple enough, but nuclear properties could not be explained by these particles alone. For when man shatters nuclei with high-energy projectiles newly at his command, a bewildering array of very short-lived particles is created which simply do not exist in atoms of ordinary matter. It has taken quantum theory in its most powerful aspects to predict and account for these; but some fit nowhere and the fundamental concept of "strangeness," expressed numerically, has entered the concepts of nuclear physics. Gradually order is appearing out of chaos.

About 35 years ago the four elementary particles building blocks of atoms were the *electron,* the lightest particle, with a rest mass (the basic unit) and a negative charge (the base unit of electricity); the *proton* with a weight of 1836.1 electron masses and a charge of +1; and the *neutron* with a mass of 1838.6 and zero charge. The fourth particle was the *photon,* the quantum unit of radiation, or the building block of the electromagnetic field. Since the photon travels with the velocity of light, c, it can never be at rest or have a rest mass. Because of its motion, it possesses energy and therefore by virtue of its motion, it also has mass,

expressed by the equation $E = mc^2$. All four elementary particles have characteristic spins (the charged ones then become magnets) expressed by ½ for electrons, protons and neutrons, which are therefore called fermions (after Fermi), and 1 for photons, which are bosons (after Bose).

The understanding of necessary reactions, or coupling, of particles when they are close together has been a triumph of quantum electrodynamics in recent years. The idea of a virtual process was first exemplified by the coupling of the electron and photon; the continuous emission and absorption of photons by electrons is the means by which electromagnetic field and electron exert a force on each other. This "virtual" photon cannot be observed, and thus the law of conservation of energy is maintained; the virtual photon becomes real only by adding energy by accelerating the electron. Similarly, the proton emits and absorbs virtual photons. In any case, the four elementary particles, with these couplings, seemed sufficient for explaining atomic structures, and the masses and charges of nuclei. Dirac's theory of the electron had predicted some additional particles, including an electron with positive charge, the POSITRON. By this theory, a collision of a positive and negative electron would result in annihilation and their mass converted to photons with equivalent energy; or conversely, in a high-speed collision between two particles a positive and negative electron could be created. In this way Carl D. Anderson at the California Institute of Technology discovered the positron, the first anti-particle.

In 1956 came the discoveries at the University of California of the *antiproton,* with a negative charge and the *antineutron,* with no charge but with a magnetic moment opposite to that of the neutron. Neutron behavior led to the postulation and finally the experimental discovery of the *neutrino.* Within a nucleus the neutron is stable but outside, after an average of 18 minutes, it spontaneously ejects an electron (beta-ray) and turns into a proton. The proton and electron together are 1.5 electron masses lighter than the neutron, and this mass equivalent to 780,000 electron volts of energy is lost in the decay. To account for the discrepancy, Pauli suggested that another particle with zero rest mass is formed in the decay and carries the missing energy. Fermi named this the neutrino and constructed a theory that the neutron continuously loses and regains an electron and a neutrino by the virtual process which becomes real because the lost mass provides the energy.

There is some confusion in the use of terms, but most physicists prefer the *antineutrino* in this process accompanying electron emission, the antiparticle of the neutrinos which accompany positron emission. Next in order to hold protons and neutrons together in the nucleus, Yukawa postulated the same absorption-emission process by these nucleons for a particle called *meson,* because of intermediate rest mass. After 10 years, the pimeson or *pion,* positive, negative and neutral, with a weight of 270 electron masses was found, requiring 135 million electron volts of energy to turn a virtual into a real pion; the negative pion is the antiparticle of the positive pion (or vice versa) and the neutral pion, like the photon, is its own antiparticle. The positive pion decays into a neutrino and a positive muon; the negative pion into a negative muon and an antineutrino. The discovery of the *muon* weighing 207 electron masses defeated the so-called dozen-particle theory which was held until its discovery. The muon decays in a millionth of a second into electron (positron for the positive muon, electron for the negative) plus a neutrino plus an antineutrino.

Thus by analysis of particle interactions and the study of tracks made by charged particles in emulsions, Wilson cloud chambers and bubble chambers, the list of particles and antiparticles began to grow, first with *lambda* and *Kaon* particles and then with other "strange" particles which required theories hitherto undreamed of to account for them; for example, the law of the conservation of strangeness. In Table 1 appear the presently known "long-lived" elementary particles and antiparticles. Their lifetimes are long compared with a nuclear "year"—the time required for a nucleon to revolve about the center of the nucleus (10^{-22} sec). In Table 2 are listed the really "short-lived" particles which are called resonances, isobars and excited states. With the appearance of this host of particles blasted out of nuclei the challenge was to find interrelationships, and order in apparent chaos. Many of them showed enough similarities to be classified into families or "multiplets" whose members differed only in electric charge. They were designated by Greek letters. The further advances of extreme complexity have been due largely to

Gell-Mann and Ne'eman. Many of these particles have additional properties similar in some ways to electric charge (isotopic spin, hypercharge, etc.) that give rise to the strong nuclear interactions, just as electric charge itself gives rise to the electromagnetic force. Actually 8 properties, with 8 quantum number ratings, are necessary to account for the strong forces peculiarities. This "eightfold way," more recently called the SU (3)–SU (6) theory made it possible to arrange the particles in hexagonal or triangular patterns which together may form supermultiplets, analogous to the discovery of the Periodic Table of the elements. In the Table 3 triangle of a group of 10 particles in a decuplet the average mass of each submultiplet (horizontal rows) of SU(2) (isotopic spin) is 147 units heavier than the average mass in the multiplet below. At the apex, then, is a particle of predicted mass 1.676, entirely unknown, but from the beginning designated Omega minus. In November, 1963, scientists at the Brookhaven National Laboratory began to search with an alternating gradient synchrotron atom smasher half a mile in circumference, and an 80-inch hydrogen bubble chamber. To create Omega minus, Kappa minus particles of specific energy had to be selected from a mass of particles issuing from a tungsten target. On January 27, 1964, the photographed track of Omega minus appeared on the 97,025th expansion of the bubble chamber. After a 10-billionth of a second it decayed into a negatively charged pi particle, which turned abruptly downward and a neutral xi

TABLE 1. THE "LONG-LIVED" ELEMENTARY PARTICLES AND ANTIPARTICLES

			Particle name	Particle symbol and charge states	Mass (MeV)	Mean life (second)	Antiparticle symbol and charge states	Antiparticle name
FERMIONS (spin ½) Only one fermion can exist in a particular state at any given time	**LEPTONS** (light particles)		neutrino	ν_e	0	stable	$\bar{\nu}_e$	antineutrino
			neutretto	ν_μ	< 3.5	stable	$\bar{\nu}_\mu$	antineutretto
			electron	e^-	0.51	stable	e^+	positron
			muon (mu minus)	μ^-	105.66	2.2×10^{-6}	μ^+	muon (mu plus)
	BARYONS (heavy particles)	**nucleons**	proton	$p(=N^+)$	938.2	stable	$\bar{p}(-)$	antiproton
			neutron	$n(=N^0)$	939.5	1×10^{-3}	$\bar{n}(0)$	antineutron
		hyperons	lambda	$\Lambda(0)$	1115.4	2.5×10^{-10}	$\bar{\Lambda}(0)$	antilambda
			sigma — plus	Σ^+	1189.4	8×10^{-11}	$\bar{\Sigma}^-$	antisigma
			sigma — zero	Σ^0	1191.5	$\ll 1 \times 10^{-11}$	$\bar{\Sigma}^0$	
			sigma — minus	Σ^-	1196.0	1.6×10^{-10}	$\bar{\Sigma}^+$	
			xi — zero	Ξ^0	1315	3.9×10^{-10}	$\bar{\Xi}^0$	antixi
			xi — minus	Ξ^-	1321	1.7×10^{-10}	$\bar{\Xi}^+$	
BOSONS (spin 0 or 1) Two or more bosons can exist in the same state at the same time	**MESONS** (intermediate particles)		photon	γ	0	stable	γ	photon
			pion — pi zero	π^0	135.0	1.0×10^{-16}	π^0	pion
			pion — plus or minus	π^+ or π^-	139.6	2.6×10^{-8}	π^- or π^+	
			kaon — K plus	K^+	493.9	1.2×10^{-8}	K^-	antikaon
			kaon — K zero	K^0 K_1^0 K_2^0	497.8	1.0×10^{-10} 6.0×10^{-8}	$\bar{K}^0$	

TABLE 2. Short-lived particles—the so-called "resonances, isobars and excited states." The symbols in the last column indicate the decay products of these very brief phenomena. In some cases more than one kind of decay occurs. They are denoted by extra sets of products listed in order of importance.

	'Particle' name	Mass (MeV)	Charge states	Principal decay products
MESONS	η (eta)	548	0	neutrals $\pi^+\pi^-\pi^0$
	ρ (rho)	750	$-0+$	$\pi\pi$
	ω (ómega)	785	0	$\pi^+\pi^-\pi^0$ neutrals
	χ (chi)	1020	0	$K_1 K_1$
	ϕ (phi)	1019	0	$K_1 K_2 (K^0\overline{K}^0)$
	f	1250	0	$\pi\pi$
	K^*	888	$0+$	$K\pi$
BARYONS	N^*	1238	$-0+2+$	$N\pi$
	N^{**}	1512	$0+$	$N\pi$
	N^{***}	1688	$0+$	$N\pi$
	N^{****}	1920	$-0+2+$	$N\pi$
	Y_0^* (1405)	1405	0	$\Sigma\pi\,\Lambda\pi\pi$
	Y_0^* (1520)	1520	0	$\Sigma\pi\,\overline{K}N\,\Lambda\pi\pi$
	Y_0^* (1815)	1815	0	$\overline{K}N$
	Y_1^* (1385)	1385	$-0+$	$\Lambda\pi\,\Sigma\pi$
	Y_1^* (1660)	1660	$-0+$	$\overline{K}N\,\Lambda\pi\,\Sigma\pi$
	Ξ^*	1530	-0	$\Xi\pi$
	Ξ^* (predicted)	1600	-0	
	Ω (predicted)	1676	$-$	

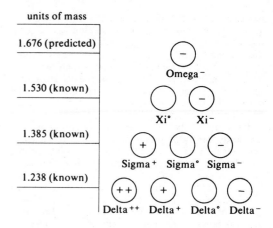

units of mass

1.676 (predicted)

1.530 (known)

1.385 (known)

1.238 (known)

Omega⁻

Xi° Xi⁻

Sigma⁺ Sigma° Sigma⁻

Delta⁺⁺ Delta⁺ Delta° Delta⁻

TABLE 3. DECUPLET OF PARTICLES, SPIN 3/2, BY WHICH OMEGA MINUS WAS PREDICTED AND FOUND.

ing beryllium with 30-GeV protons in the synchotron. The discovery is reassuring because it affirms that the strong forces which bind together neutron and proton in the deuteron are imitated in the mirror-image world of the antideuteron. The next step is the antitriton, which would require the sticking together of 1 antiproton and 2 antineutrons.

G. L. CLARK*

Cross-references: *Radiation, Radioactivity, Protons, Neutrons, Electrons, Nucleonics; Orbitals.*

* Deceased.

PARTICLE SIZE

Theory and Statistics of Analysis. The word *particle,* in its most general sense, refers to any object having definite physical boundaries, without any limit with respect to size. Particles varying from about 0.01 to 1000 microns are considered in this discussion. The term is often ambiguous unless carefully defined for a specific case. For example, the particle in a soil may be considered to be the loose aggregates, the "ultimate" particles of which the aggregates are composed, or even individual mineral grains in the ultimate particles.

The term *particle size* is also usually ambiguous unless defined for each type of application. Particle size is usually described in terms of rather artificially defined "diameters" which generally fall in one of two classes. Diameters of one class, called statistical diameters, are defined in terms of the geometry of the individual particles and are determined for a large number of particles. Definitions of some statistical diameters are:

(1) Martin's diameter; the distance between opposite sides of the particles, measured crosswise of the particle, and on a line bisecting the projected area.

(2) The diameter of the circle whose area is the same as the projected area of the particle.

(3) The shorter of the two dimensions exhibited.

(4) The average of the two dimensions exhibited.

(5) The average of the three dimensions of the particle.

particle whose track is invisible. Thus was verified a theory of order and symmetry in a seemingly hopeless confusion of about 150 particles, in one of the greatest theoretical and experimental adventures in the history of science. This is only a beginning and it will be a very long time before the particle physicist finds himself out of a job.

There remains the conjecture based on the antiparticles' existence concerning the possibility of the existence somewhere in the universe under stabilized conditions of "inverted" atoms, or antimatter, in which antiprotons and antineutrons and associated antiparticles constitute negatively charged nuclei surrounded by positrons in outer shells. The particles of normal matter combine to form nuclei of increasing size and complexity as they produce the heavier elements. But until June 1965 no one had discovered anything made out of antimatter more complex than a single particle. Then Professor Leon Lederman and his team from Columbia University and Brookhaven National Laboratory produced antideuterons, the nuclei of "inverted" deuterium, or heavy hydrogen, which consist of an antiproton coupled to an antineutron, by bombard-

(6) The distance between two tangents to the particle, measured crosswise of the field (for example, of a microscope), and perpendicular to the tangents.

Diameters determined by sieving are also statistical diameters.

Diameters of the other class are defined in terms of the physical properties of the particles. Examples are diameters determined by sedimentation or elutriation.

Numerous definitions of the mean particle size of a powder may be used, the choice depending largely on the use to which the data are to be applied. It is convenient to define such means in terms of data which have been classified into groups (classes) which are defined by means of particle size limits called class boundaries. The midpoint of each interval is called the class mark (d_i), the number of particles in each interval is called the frequency (f_i), and the total number of particles measured is denoted by n. The following table is a summary of the definitions of various means.

Name	Symbol	Definition
Arithmetic mean	$\bar{d}$	$\dfrac{1}{n}\,\Sigma d_i f_i$
Geometric mean	d_g	$(d_1^{f_1} d_2^{f_2} d_3^{f_3} \cdots d_n^{f_n})^{1/n}$
Harmonic mean	d_{na}	$\left[\dfrac{1}{n}\,\Sigma \left(\dfrac{f_i}{d_i} \right) \right]^{-1}$
Mean surface diameter	d_s	$\sqrt{\dfrac{\Sigma f_i d_i^{2}}{n}}$
Mean weight diameter	d_w	$\left(\dfrac{\Sigma f_i d_i^{3}}{n} \right)^{1/3}$
Linear mean diameter	d_1	$\dfrac{\Sigma f_i d_i^{2}}{\Sigma f_i d_i}$
Surface mean diameter	d_{vs}	$\dfrac{\Sigma f_i d_i^{3}}{\Sigma f_i d_i^{2}}$
Weight mean diameter	d_{wm}	$\dfrac{\Sigma f_i d_i^{4}}{\Sigma f_i d_i^{3}}$

The median and the mode are also often used as a measure of central tendency. The median is the middle value (or interpolated middle value) of a set of measurements arranged in order of magnitude. The mode is the measurement with the maximum frequency. The mode is sometimes reported with the arithmetic mean or the median to give an indication of the skewness of the distribution.

Some quantitative indication of particle shape is often desirable. Heywood has suggested the following method. The particle is assumed to be resting on a plane in the position of greatest stability. The breadth (B) is the distance between two parallel lines tangent to the projection of the particle on the plane and placed so that the distance between them is as small as possible. The length (L) is the distance between parallel lines tangent to the projection and perpendicular to the lines

defining the breadth. The thickness (T) is the distance between two planes parallel to the plane of greatest stability and tangent to the surface of the particle. Flakiness is defined as B/T and elongation as L/B.

Shape factors, which relate the surface (s) and volume (v) of a particle to its diameter, also provide a crude indication of shape:

$$s = \alpha_s d^2$$
$$v = \alpha_v d^3$$

For a sphere, $\alpha_s = \pi$ and $\alpha_v = \pi/6$. For particles other than spheres, the shape factors depend on the definition of diameter that is used as well as on the particle shape.

The sizes of the particles in a powder or other particulate system may be represented by some mean value, as indicated above. Usually some indication of size distribution or spread is desirable and can be provided by the standard deviation,

$$\sqrt{\frac{1}{n}\,\Sigma (d_i - \bar{d})^2 f_i},$$

or by means of quartiles if the median is used as the measure of central tendency. Quartiles are the values completing the first 25% and the first 75% of the values when they are arranged according to increasing size. However, it is often necessary to provide a more complete indication of the particle size distribution. Classified data can readily be represented by a bar type of graph called a histogram. The position of the bar locates the class interval and the length of the bar corresponds to the frequency.

Several methods have been proposed for determining the sizes of particles, but most of them are based on one or more of a small number of principles. Optical microscopy is often used for particle size determination. It has the advantage that the particles are observed directly and that the basic equipment is relatively inexpensive. However, it is a very tedious method if accurate results are to be obtained and is subject to large sampling errors unless adequate precautions are taken.

RICHARD D. CADLE

References

Allen, T., "Particle Size Measurement," Chapman and Hall, London, 1968.

Cadle, R. D., "Particle Size Determination," Interscience, New York, 1955.

Cadle, R. D., "Particle Size," Reinhold, New York, 1965.

Irani, R. R., and Callis, C. F., "Particle Size; Measurement, Interpretation, and Application," Wiley, New York, 1963.

Orr, C., and Dallavalle, J. M., "Fine Particle Measurement," Macmillan, New York, 1959.

PARTITION

Partition is a term used in the field of extraction to denote the distribution of a solute between two liquid phases. The ideal distribution law states that at equilibrium the ratio of the concentrations of the solute in the two phases is a constant for a given temperature. This may be expressed in equation form: $C_1/C_2 = K$, where C_1 = concentration of solute in the lighter phase, C_2 = concentration of solute in the heavier phase, K is usually referred to as the *partition coefficient* or *distribution constant*.

The numerical value of K depends upon the units used in expressing the concentrations. Any units may be used, but concentrations are given most frequently in moles per liter of solvent or solution.

This ideal law is exact only for ideal solutions. It is approximated most closely when the concentrations are dilute and the two phases are essentially immiscible. The mutual solubility of the solvents should not be affected by the presence of the solute. These conditions are not met by most commercial applications. The above equation is written for cases in which the molecules in each phase are in the same state of aggregation. If the solute is associated or dissociated, it must be applied to each individual state of aggregation. Solutes may be associated or dissociated in only one phase and not the other. Suitable modifications of the equation can be written if the type of molecular alteration is known.

Nearly all systems show a deviation from the ideal partition coefficient at some range of concentration or temperature. If the concentration of the solute in one phase is plotted against that in the other phase, the resulting figure will reveal deviations from ideal behavior. This type of plot is called a *partition isotherm*, and will yield a straight-line relationship for an ideal system. This two-coordinate representation is generally used for dilute solutions. Plots of more concentrated solutions generally are drawn on triangular coordinates.

Normally, extractions are concerned with the fraction of total solute in each phase. When the concentrations are expressed in terms of weight per unit volume instead of moles, the presence of ions, dimers, etc., does not affect the concentration values. This broader meaning has been referred to as *partition ratio* by Weissberger.

In commercial applications the attainment of equilibrium conditions is influenced by the problem of obtaining an appreciable rate of transfer across the boundary of the phases. If the rate of transfer is slow, it is not economical to reach equilibrium conditions. In continuous extractions, both phases are introduced into a contacting chamber continuously at one point and removed at the same rate at another point. Various factors affect the attainment of equilibrium in such an installation. These factors include the velocity of flow in the two phases, effective thickness of the boundary layers, diffusion rates, etc.

In laboratory applications discontinuous extraction is used more frequently. In this method a given amount of each phase is brought together for a sufficient time to establish equilibrium. Then the two layers are separated. The simplest form of equipment employed in single-stage extraction is the separatory funnel. Successive single-stage extraction with fresh solvent may be carried out with one separatory funnel, or multiple contacting can be accomplished by using several funnels. The use of successive contact with fresh solvent is per-

formed by a Soxhlet extractor in the case of solids.

Equipment has been developed where 100 tubes give simultaneous contacting of the two phases in each tube. The upper layers in all the tubes can then be transferred simultaneously to the next adjacent tube. A glass apparatus capable of giving 100 simultaneous transfers has been developed and is known as the Craig countercurrent distribution apparatus. This particular type of equipment is ideal for attaching an electric drive and clock mechanism for automatic operation.

If a solute is dissolved in the lower layer of one tube and fresh solvent added, the solute will be distributed between both layers according to the partition ratio, K. Transferring the upper layer to the second tube and adding fresh solvent to the first tube results in a distribution of solute in both tubes. This distribution is dependent upon the ratio of volumes of the layers and the partition ratio. If equal volumes for both layers are used and the partition ratio is equal to 1, both tubes will contain 50 per cent of the solute. A second transfer will move both upper layers one tube and result in a distribution in three tubes. This process can be continued until a countercurrent distribution curve is obtained which concentrates the solute in a few tubes. This greatly expands the use of extraction for analytical purposes.

Partition Chromatography is used in chemical analysis and in small scale preparative chemistry. The technique employs a small diameter column packed with an inert, porous substrate on which a nonvolatile, insoluble liquid is applied. The mixture to be separated is carried over the partition column by an inert gas and separation is effected by *partition extraction* or *partition absorption*.

FRANK C. FOWLER

PASSIVITY

In a general way, the term "passivity" refers to that behavior of a metal in which it exhibits a lesser reactivity than one would normally expect, a common example being that of iron dipped in concentrated nitric acid. The initial reaction of the iron with the acid quickly ceases and thereafter the metal remains unaffected. So many somewhat similar phenomena have been described as instances of passivity that it no longer becomes possible to formulate a simple definition and in the glossary of Uhlig's "Corrosion Handbook" two definitions are given, as follows: (1) A metal active in the *EMF Series* or an alloy composed of such metals is considered passive when its electrochemical behavior becomes that of an appreciably less active or noble metal. (2) A metal or alloy is passive if it substantially resists corrosion in an environment where thermodynamically there is a large free energy decrease associated with its passage from the metallic state to appropriate corrosion products.

The example of passivity involving iron and nitric acid can be considered to be wholly chemical, but there are many instances of electrochemical passivity. Copper anodes, for example, in a copper-cyanide plating bath will, under certain conditions, become insoluble, at least at applied voltages within the capacity of ordinary DC sources. Removal of the anodes and scraping of their surfaces destroys the passivity, indicating that it is caused by a surface film. Most authorities are agreed that a surface film of some sort, often an oxide or oxide mixture, is responsible for passive behavior, whether it be chemical or electrochemical. It is apparent, therefore, that passivity is basically a chemical phenomenon and the destruction or induction of passivity should be approached as a chemical problem.

Passivity is not an absolute property of a metal but depends to a great extent upon the chemical environment surrounding it. Oxidizing conditions tend to promote passivity, whereas reducing conditions lead to its destruction. Galvanic contact between two metals encourages passivation of the least noble of a given pair, but it must be realized that not every pair of metals will exhibit this behavior. If iron is coupled with platinum, it is more easily passivated, in a suitable environment, than iron alone. On the other hand, iron coupled to zinc cannot be passivated. In the latter case it is important to distinguish between passivation and cathodic protection.

The chemical environment for passivation or the preservation of a passive film is particularly important. Oxidizing conditions usually are required for passivation, and reducing conditions lead to its destruction. In addition, certain ions, particularly halogen ions, tend to destroy passivity even in oxidizing or neutral conditions. The passivity of stainless steel may, for example, be destroyed at small local areas by chloride ions, leading to the formation of small anodic areas surrounded by large areas of still-passive metal which will serve as cathodes for electrochemical corrosion. The result is disastrous pitting. Neutral or acid conditions are not necessarily required, for amphoteric metals such as zinc and aluminum may lose their passivity in alkaline environments, particularly in the presence of halogen ions.

In many cases it is desirable industrially to encourage passivity, usually to promote corrosion resistance. An outstanding example is that of stainless steel, for in the active condition stainless steel is only somewhat more corrosion-resistant than ordinary iron. A freshly exposed surface of stainless steel corrodes readily, but exposure to air will passivate the surface, although the process may be hastened by treatment with a suitable oxidizing agent, such as nitric acid. In other cases more drastic means may be required. In one sense, the anodizing of aluminum is a passivating process in which a relatively thick film of oxide is formed in a lengthy anodic treatment in an oxidizing environment.

Often the destruction of passivity is an important industrial process. The example of copper anodes given above is one problem which may be overcome by choosing a suitable chemical composition of the electroplating bath, e.g., increase of free cyanide concentration or lowering of pH. Another example is the nickel which is used as anodes in electroplating. Pure nickel passivates almost immediately in the usual plating solution, but the addition of controlled amounts of oxygen or carbon to the nickel during its metallurgical processing destroys the passive behavior and permits anodic solution at good efficiencies.

The study of passivity effects, particularly in corrosion processes, has been facilitated by the development of the potentiostat. If a gradually increasing potential is applied to a piece of iron in dilute sulfuric acid, the current increases in an approximately linear way until suddenly a slight increase in potential results in a remarkable drop in current (passivity). The exact characterization of the potential *vs* current behavior of various metals in many environments has not only led to a better understanding of passivity as a general phenomenon, but it has made possible the *anodic protection* of some large structures by polarizing them to a potential just large enough to cause the sudden transition from active to passive behavior but not sufficient to cause evolution of oxygen or other gas discharge. Although not so widely applicable as cathodic protection, the new technique is particularly useful for storage tanks and chemical processing equipment.

HAROLD J. READ

PASTEURIZATION

Pasteurization of food products is performed, primarily by the application of heat, to kill pathogenic microorganisms and to inactivate or destroy enzymes. Sterilization involves a more intense heat treatment and kills practically all (99.999%) of the organisms. Irradiation and chemical methods are additional means of pasteurization, but are not used extensively.

Pasteurization is used (1) to increase the storage life of the product; (2) to eliminate the transmission of communicable diseases; and (3) to provide a product with a uniform taste, which may differ from the new or original. Pasteurization processes are designed and controlled so as to minimize objectionable excessive heat treatment of products. Excessive heat treatment of some products would render them unfit for food purposes, particularly as beverages, which are often made to duplicate as nearly as possible the fresh product. More severe heat treatment may be used for liquids which are to be used for manufactured products.

Commercially, pasteurization must be carried out effectively and efficiently as the part of a processing system. Related processes—particularly in beverage operations—include clarification, vacuum treatment, homogenization, and separation. These can be designed as integral components of the pasteurization process or as separate operations.

Heat, the active agent in pasteurization, may be applied in a batch or holding operation, or to a continuous flow of the product. Regardless, a time-temperature relationship exists between the amount of the heating and the effect on the product. Effects on the product may be quantified as related to components such as enzymes and the associated organisms. As the temperature of treatment is increased the time for an equivalent effect is decreased, with a straight line relationship between the temperature and the log of the time at that temperature. The relationship is known as the thermal death time (T.D.T) curve for killing microorganisms. This relationship holds for enzymes and microorganisms in the product. In milk, for example, the enzymes which are present in small amounts are relatively unstable in heat, and the U.S.P.H.S. requirements for thermal pasteurization are sufficiently high to inactivate these enzymes. The inactivation of the enzyme phosphatase is a basis of determining whether pasteurization is complete. Phosphatase is more difficult to inactivate than the pathogenic organisms are to kill. Another enzyme in milk, lipase, must be completely inactivated before homogenization. Otherwise, homogenization activates lipase, and causes rancidity and poor keeping quality. Lipase, too, is inactivated at well below pasteurization thermal treatments.

For milk the standard heat treatment for batch pasteurization is 145°F for at least 30 min; for continuous pasteurization the process requires no less than 161°F for at least 15 seconds. The common continuous method, or short-time system, is called high-temperature short-time (H.T.S.T.) pasteurization. These minimum requirements are increased in either time and/or temperature for proper thermal treatment for products with an increase in fat, sugar, or other solids. With improved heat transfer methods and automatic controls the trend is to higher temperatures for shorter times. Those treatments above 200°F (for product exposure less than 1 second) are known as ultra high temperature (UHT) systems. These units must include equipment to cool the product rapidly after heating to avoid flavor impairment.

Plate heat exchangers are particularly adaptable for rapid thin film heating and cooling. Plate heat exchangers can be arranged for regenerative heating-cooling so that efficient heat utilization can be obtained.

Post-pasteurization contamination must be avoided. Thus, the filling-to-serving operations—container, sealing, storing, opening, and serving—must be done so as to maintain the quality of the pasteurized product.

Pasteurization of food is subject to control by the public health authorities. Standards of performance have been established.

Beverage foods—milk, fruit and vegetable juices, wine and beer—are products usually pasteurized. Pasteurized foods have a storage life exceeding the raw product. Normally, pasteurized foods are stored under refrigeration.

CARL W. HALL

References

Ball, C. O., and Olson, F. C. W., "Sterilization in Food Technology," McGraw-Hill Book Co., New York, 1957.

Hall, Carl W., and Trout, G. M., "Milk Pasteurization," AVI Publishing Co., Westport, Conn., 1968.

PATENTS

A patent, under United States law, is the grant to its owner of the right to exclude others from the use of an invention for 17 years from the date of issuance of the patent. In consideration of this grant, the patentee discloses the invention so clearly and completely that the public will know how to practice the invention after the patent expires.

Also he pays fees to the Government at the time of filing the patent application and of issuance of the patent, the minima being $65 and $100, respectively.

For *patentability* the invention (1) must not have been published in any country or in public use in the U.S., in either case for more than 1 year prior to the date of filing the application for the patent; (2) must not have been known in the U.S. before the date of invention by the applicant; (3) must not be obvious to an expert in the art; (4) must be useful for a purpose not immoral and not injurious to the public welfare; and (5) must fall within the five statutory classes on which only may patents be granted, namely, (a) composition of matter, (b) process (of manufacture or treatment), (c) machine, (d) design (ornamental appearance), or (e) plant produced asexually, tuber-propagated, plants being excluded.

Special regulations relate to atomic energy developments and subjects directly affecting national security. No U. S. patent on any subject is valid if the subject matter is disclosed in an application for patent filed abroad earlier than 6 months after filing the U. S. application, unless permission for export of the information is obtained from the Patent Office.

Systems of doing business and also *scientific principles,* as distinguished from the application of those principles, are not patentable.

The *test for invention* is unobviousness, at the time the invention was made, to a person having ordinary skill in the art to which the invention pertains and knowing all that has gone before, as shown by publication anywhere or public use in the U. S. When reliance for patentability is placed on a new mixture of components that have been used separately, it must be shown that there is some unexpected coaction between the ingredients and not just the additive effects of the several materials. It is not necessary, however, that the unobvious result be the principal result sought from the mixture; where the overall therapeutic effect, for instance, is evident in advance, a surprising result may arise in the preservation of one ingredient by another.

The *inventor* is the one who contributes the inventive concept in workable detail, not the one who demonstrates it or tests it, unless the experimenter must supply some inventive or unobvious step to make the concept operative. When error arises in the initial designation of inventor without deceptive intent, an inventor's name may be added or, in the case of joint inventors, subtracted, these changes being permissible after the application is filed or even after the patent is issued or suit filed thereon. Proper credit for inventions promotes morale in a research organization.

Patent rights of the employee are commonly stated in a contract. In the absence of such contract, the rights for various classes of employees under various circumstances will be understood to be as follows:

(1) The employee is not hired to make inventions. He makes an invention on his own time and with his own materials; the invention belongs to him. He makes the invention on his employer's time and with the employer's material; the inven-

tion belongs to the employee but the employer has a shop right to use the invention in his own factories without extra payment to the employee.

(2) The inventor is employed for research or under other circumstances showing that making invenions is a part of the work normally expected of him. The invention belongs to the employer if it is in the particular field of work assigned to the employee.

Data for the patent application, whether the data are presented orally or in writing, should include: brief discussion of the closest prior art; difference from it and the unobvious result of the change from what has been done before; the general properties necessary for each of the classes of components or reactants in a chemical case and at least three specific examples of each class; ranges of proportions for each class of materials that may be used without loss of operability of the invention; ranges of temperatures, times or other conditions, if any, that are critical; several examples illustrating in diverse manners methods of practising the invention successfully, including the best manner; and sketches in case a drawing is required and unless the inventor is to confer personally with the attorney or his draftsman. In all such presentations, the inventor should use the technical form and avoid unnecessary legalistic language.

In *interpreting patents,* claims are considered to cover all that they do not exclude. Adding to a process or composition a step or an ingredient not recited in a claim does not avoid infringement of the claim. Omitting an essential part of the claim, on the other hand, does ordinarily avoid the claim.

Interferences are declared by the Patent Office when two or more applicants seek patent on the same invention. The first inventor, to whom the patent will issue, will be the one who first reduced the invention to practice and tested it adequately for the intended use, if such test is required to show the utility. The inventor who first conceived of the invention (and can prove as much by having disclosed it to another who is not a joint inventor) may advance his reduction to practice to the date from which he began to exercise continuous diligence in work on the invention and continued diligent up to and including the time of actual reduction to practice.

Licenses may be exclusive, in granting to one party the sole right to an invention, or nonexclusive. An exclusive license excludes the patent owner himself from practice of the invention. Licenses may be for all or part of the U. S. License agreements should identify the parties to the agreement and the invention to be licensed, state the amount of royalty or other compensation and the periods of payment, and give the licensor the right to inspect the accounts to verify the reports of royalty due when calculated from the extent of operations under the invention and also the right on due notice to cancel the license (or convert it from exclusive to nonexclusive) if payments fall below a stated minimum sum.

Infringement is the unauthorized making, using, or selling of the invention within the U. S. Contributory infringement arises from supplying a part or component only of what is recited in the claims, provided it is a material part of the invention, the supplier knows that the part or component is especially made for the infringing purpose, and the part or component has no other commercial use for non-infringing purpose.

Validity of a patent is necessary for infringement, either primary or contributory. Validity will be assumed in the absence of proof to the contrary, the proof attempted being usually that the invention does not meet one at least of the requirements set forth above for patentability.

Liability for patent infringement when a patent is held valid and infringed shall be damages adequate to compensate the patent owner for the infringement. In no case will the damages be less than a reasonable royalty plus interest and costs as fixed by the court. In the absence of particularly unfavorable circumstances, such as continued infringement after a court holding of infringement and injunction thereagainst, the damages are ordinarily about the same as a reasonable royalty. A great disadvantage of infringing, as contrasted with taking a license in advance, however, is the possibility of being enjoined from continued use of the invention, regardless of the amount of royalty then offered, if the suit goes against the infringer. Also those who believe in patents should support the patent system by willingly compensating the patent owner for rights that have undoubtedly cost him money or research or both to acquire. Reputable companies generally seek to avoid knowingly infringing valid patents.

In general, patenting inventions is preferable by far to negotiating for license or standing the cost and hazard of suit after patents are taken by others. Patenting is one of the smallest items of research costs and should be carefully considered as the anchoring step for each successful phase of a research. Small companies sometimes overlook this step, although protection by patents is more vital to them than to large concerns which, in any event, derive some measure of protection from their size and reputation and from the amount of capital necessary for damaging competition with them.

ROBERT CALVERT*

* Deceased.

PECTINS

The pectins or pectic substances are a group of complex, high molecular weight compounds often regarded as polysaccharides or polysaccharide derivatives. They occur commonly in plants, particularly in succulent tissues, and are characterized by the fact that polygalacturonic acids make up their fundamental structure. The terminology which is now almost generally accepted for the pectic substances is that adopted by the American Chemical Society in 1944. Here the term "protopectin" is retained for the water-insoluble and as yet ill-defined parent pectic substances which occur in plants and which upon restricted hydrolysis, yield pectinic acids (pectins). The term "pectinic acid" designates polygalacturonic acids of colloidal nature in which some or most of the carboxyl groups are esterified with methyl groups. *Pectins* or *high-ester pectins* are pectinic acids that contain at least 7 or 8% methyl ester (expressed as methoxyl) while the *low-ester (low-methoxyl) pectins* contain less than 7% (usually 3–5%) methoxyl. The theoretical maximum methoxyl content of a pure pectinic acid is 16.35%. In *pectic acid* all carboxyl groups are free, or at least not present as the methyl ester. Under suitable conditions, pectins will form *jellies* with sugar and acid, whereas the low-ester pectins will form *gels* with traces of polyvalent ions. The term "pectins" is used in a generic manner to designate "pectic substances."

Chemical Composition and Structure. The basic structure of all pectins is made up of linear polygalacturonic acid chains. According to some authorities, arabinose and galactose units occur in the polygalacturonic acids at infrequent intervals. Others maintain that the small proportion of arabinose and galactose found even in purified pectins represented admixed polymers difficult to separate ("ballast"). Acetyl groups occur in some pectins as in beets and some fruits, for instance. Protopectin supposedly contains cellulose, but the ture nature of protopectin is so uncertain that it seems undesirable to discuss this matter here.

The anhydrogalacturonic acid unit building blocks in pectins are connected through 1,4-glycosidic bonds to form fibrillar molecules containing many units. The anhydrogalacturonic acids are present in the pyranose form. The pectinic acid and pectic acid macromolecules are thread-shaped and show typical and important colloidal properties. Apparently they have molecular weights between 100,000 and 200,000 but there is some variation in the values obtained by different methods. On account of the variability in composition as well as in the kind and distribution pattern of substitutions in the fibrillar molecules, it is most unlikely that any two macromolecules of a pectin preparation could be identical. Because of the wide variations in the kinds and distribution patterns of these macromolecules, it is equally unlikely that two "identical" samples of pectin would exist. Nevertheless, by the application of various criteria, pectin (pectinic or pectic acid) preparations can be fairly well defined.

Occurrence and Preparation. The location of various pectic substances in plant tissues is fairly well established. They make up most of the middle lamella in unripe fruit and are to be found in the cell walls and in small proportion in all plant tissues. The genesis and fate of pectins in plant tissues are not clear.

Citrus peel, apple pomace from juice manufacture, and beet pulp left over from the manufacture of sucrose are the common sources of commercial pectins. After some preliminary purification of the raw material, the extraction is usually performed with hot dilute acid (pH 1.0–3.5, 70–90°C). The pectin is then precipitated from the extract with ethanol or isopropanol or with metal salts (Cu or Al). The metal ions have to be subse-

quently removed by washing with water or acid ethanol. There are two formulas for denatured ethanol specifically authorized for pectin manufacture (Nos. 2B and 35A). The precipitates are purified, dried, and pulverized. Most pectins sold for commercial purposes are standardized on the basis of jelly-forming power by the addition of sugars and buffer salts. Purified but unstandardized pectins are available for pharmaceutical and research purposes. Since some pectin is extracted by manufacturers who subsequently use it in food manufacture, it is difficult to get an exact picture of the total production. The approximate amount of various pectins now produced in the U.S.A. is around 10 million pounds of dry pectin (standardized at 150 jelly grade) per year. The total world production may be about twice this quantity.

Properties and Characterization. Pectic substances in solution behave as typical colloids. Dry, purified pectins are light in color and soluble in hot water to the extent of 2–3%. The pH of pectin solutions is usually 3.0–3.5. Pectin is precipitated from solution by 50–60% ethanol or traces of heavy-metal salts. The free acid groups in pectins can be titrated directly in the usual manner, the titration curves resembling those of monobasic acids. Treatment with acid will progressively deesterify pectins and strong hot acid will cause decarboxylation. In alkaline solutions deesterification is rapid. As the methyl ester groups are removed by alkali, acid, or enzyme, the pectin becomes increasingly susceptible to precipitation with polyvalent ions. Pectic acid can be quantitatively precipitated with traces of calcium. The methods of colloid chemistry, particularly viscosity measurements, have been extensively applied in pectin research. Sodium chloride (0.8%) and polyphosphate (0.1%) (or Versene) are usually added in order to suppress electroviscous effects and the influence of traces of polyvalent ions. Low-ester pectins show a viscosity maximum at pH 6. With progressive deesterification the pH of maximum viscosity drifts towards the acid side.

Pectin preparations are usually characterized by the polyuronide content, by their solution viscosity, and by the proportions of free and esterified carboxyl groups present. The major method of evaluating the pectin of commerce is through its jelly-forming ability. The enzymes acting upon pectic substances have been extensively investigated. Some of the commercial pectic enzymes are used on a large scale in the fruit products industries and for other purposes.

Application in fruit products. The pectic substances in foods of plant origin are important since the firmness and texture of many fruits and vegetables depend on them. They also play a significant role in canning and freezing processes. In the presence of over 50% sugar and at pH values below 3.6, pectins will form firm jellies. The proportion of sugar which one unit weight of pectin will form into a firm jelly is the *jelly grade*. This can be determined by a variety of methods, of which the measurement of the breaking strength of the jelly, its elasticity modulus, or the extent of sag attained in two minutes of the turned-out jelly are most extensively used. All these tests must be performed under painstakingly exact and standardized conditions. The jellification (geling) in marmalades and jams is only partial, but is nevertheless important. In a jelly (or a jam or marmalade) the proportions of total solids of sugars, the pH, and the proportion and nature of the pectin used will determine the extent of jellification obtained. In commercial use, the kinds and proportions of these components as well as the fruit and juice which may be used are defined by government regulations. The use of added pectin in fruit jams, preserves, marmalades and jellies is permitted since the addition is thought to compensate for an incidental natural deficiency.

Low-ester pectins and low-ester gels. A new and significant development in pectin chemistry and technology is the manufacture and use of low-ester pectins. These will form gels with traces of calcium at sugar concentrations much below the customary 65% used in jams and jellies or, indeed, without sugar. The low-ester pectins are manufactured by a variety of methods, including the use of strong ammonia. In the latter case, apparently acid amide groups are introduced in addition to the deesterification. Low-ester pectins have rather exact calcium requirement for optimum gelation. Probably due to heterogeneity of the constituent pectinic acids, the gel formed often shows an undesirable extent of syneresis.

General. Over 75% of all manufactured pectin is used in fruit jams, jellies, marmalades and similar products, but pectin preparations are also used for a variety of other purposes, as thickening agents in malted milk beverages and other foods, in mayonnaise and salad dressing, in frozen dessert mixes, in frozen fruits and berries to prevent leakage, upon thawing, etc. The addition of a dilute pectin solution to milk will coagulate the caesin. In many food products the use of pectins as stabilizers is preferred, since they blend better into the flavor complex than do many gums, starches or carbohydrate derivatives. Pectin jellies do not melt at temperatures below 120°F, a distinct advantage over gelatin gels requiring refrigeration. Pectins have many nonfood uses, of which pharmaceutical and cosmetic preparations are the most important. The National Formulary lists several items made with pectin.

Z. I. KERTESZ*

Cross-references: *Gels, Colloid Chemistry, Molecules, Carbohydrates, Foods.*
* Deceased.

PENICILLINS AND CEPHALOSPORINS

Penicillin was first described in 1929 by Alexander Fleming, who gave the name to the culture filtrate of a mold, *Penicillium notatum,* which he had discovered at St. Mary's Hospital, London, as an accidental contaminant, on a plate of nutrient agar seeded with staphylococci. Fleming found that penicillin inhibited the growth in the test tube of a number of pathogenic bacteria and that it had a low toxicity to animals. He suggested that it might be used as a local antiseptic, but for ten years it aroused relatively little interest. Then, in 1939, H. W. Florey and E. Chain planned a systematic study, in the Sir William Dunn School of Pathology at Oxford University of antibacterial substances produced by microorganisms and chose penicillin

as one of the first substances for further investigation. They and their colleagues extracted and purified penicillin and demonstrated, first in mice and then in man, that it had remarkable chemotherapeutic properties when injected into the blood stream. These crucial experiments stimulated the pharmaceutical industry in England and the U.S. to attack what seemed to be the formidable problem of producing penicillin on a commercial scale. This problem was solved, largely through the efforts of groups in the U.S., by the use of deep aerated cultures, improved growth media containing corn steep liquor, and high-yielding mutant strains of *Penicillium chrysogenum.*

The chemistry of penicillin was studied by many groups of investigators in England and the U.S. during World War II and these groups maintained contact with each other through a system of confidential reports. At an early stage it became clear that the penicillins produced in the two countries differed in the nature of a side-chain attached to a characteristic nucleus. The two substances were called penicillin F and penicillin G, respectively. Chemical degradations and transformations led to two proposals for a general structure of the penicillin molecule. Structure I was shown to be correct by x-ray crystallographic analysis. In this structure, which contains a fused β-lactan-thiazolidine ring system, RCO is a variable side-chain

I

whose nature depends on the precursors present in the medium in which the mold is grown. In penicillin F R is Δ^2 pentenyl ($CH_3CH_2CH{=}CH{\cdot}CH_2$), in penicillin G it is benzyl ($C_6H_5CH_2$), and in penicillin V it is phenoxymethyl ($C_6H_5OCH_2$). These substances are called pentenylpenicillin, benzylpenicillin and phenoxymethylpenicillin respectively.

Penicillins are hydrolyzed by hot acid to D-penicillamine (D-β-thiolvaline), an organic acid (RCO_2H) from the side-chain and carbon dioxide from the C=O of the β-lactam ring. In cold dilute acid many penicillins isomerize easily to inactive penillic acids. However, this transformation does not occur readily with penicillin V. The latter can therefore survive the acidity in the stomach, a property which allows it to be used orally in smaller doses than benzylpenicillin. Hydrolysis of the penicillins with cold dilute alkali results in the opening of the β-lactam ring and the formation of inactive penicilloates, which can be split with mercuric chloride to give penicillamine, a penilloaldehyde ($R{\cdot}CONH{\cdot}CH_2CHO$) and CO_2. Penicilloates are also formed when penicillins are inactivated by penicillinase, an enzyme discovered by E. P. Abraham and E. Chain in 1940. Penicillinases, or β-lactamases, are now known to be produced by a number of gram positive and gram negative bacteria.

The penicillin nucleus, 6-aminopenicillanic acid, is found in cultures of *P. chrysogenum* to which no sidechain precursor is added. Evidence for the existence of this compound was obtained by Kato and by Sakaguchi and Murao in Japan, but it was first isolated in a pure state and characterized by F. R. Batchelor, F. P. Doyle, J. H. C. Nayler and G. N. Robinson in the Beecham Laboratories in 1959. 6-Aminopenicillanic acid (6-APA) is readily formed from benzylpenicillin (though not from penicillin N) by the action of penicillin amidases, enzymes which are produced by a number of bacteria. It can be acylated to yield a series of new penicillins which cannot be obtained by fermentation. Among the most important semisynthetic penicillins are methicillin, in which R is 2:6-dimethoxyphenyl, oxacillin, and cloxacillin, in which R is 5-methyl-3-phenylisoxazole and 5-methyl-3-o-chlorophenylisoxazole, respectively, ampicillin in which R is D-α-aminobenzyl, and carbenicillin in which R is α-hydroxybenzyl. The importance of methicillin, oxacillin and cloxacillin resides in their resistance to hydrolysis by penicillinase from *Staphylococcus aureus*. This is due in part to their bulky side-chains, which hinder their combination with the enzyme.

The introduction of members of the cephalosporin family of antibiotics into medicine derived from the discovery of cephalosporin C by G. G. F. Newton and E. P. Abraham at Oxford in 1953. This substance was found among the metabolic products of a *Cephalosporium* sp. isolated by G. Brotzu in Sardinia and was shown to contain a fused β-lactam-dihydrothiazine ring system (II) with an N-acyl side-chain (RCO) consisting of δ-(D-α-aminoadipyl), and an acetoxy group (X) attached to the dihydrothiazine ring. The nucleus of the molecule, 7-aminocephalosporanic acid (7-ACA) was obtained in very small yield on mild acid hydrolysis by Loder, Newton and Abraham. A chemical procedure which enabled 7-ACA to be produced in quantity from cephalosporin C was discovered by Morin, Jackson, Flynn and Roeske (1960) in the Lilly Research Laboratories and a chemical route from the penicillin to the cephalosporin family was later discovered in the same laboratories. These developments enabled a variety of cephalosporins to be produced which differed in the groups R and X.

Cephalosporins which are now used clinically include cephalothin (II, RCO = 2-thienylacetyl and X = 'acetoxy), cephaloridine (RCO = 2-thienylacetyl and X = pyridinium) and cephalexin (R = D-α-aminobenzyl and X = H).

A rational synthesis of penicillin V was achieved in 1957 by J. C. Sheehan and K. R. Henery-Logan. This involved the synthesis of the corresponding penicilloate and closure of the β-lactam ring by dicyclohexylcarbodiimide. A total synthesis of cephalosporin C was reported by R. B. Woodward *et al.* in 1966. However, processes for the production of penicillins and cephalosporins by total synthesis do not compete economically with the highly efficient methods involving fermentations with *P. chrysogenum* and enzymic removal of the side-chain.

L-Cysteine and L-valine are precursors of the penicillin and cephalosporin ring systems. In the latter case, valine must be oxidized further to yield a γ-hydroxy-αβ-dehydrovaline residue. A tripeptide, δ(α-aminoadipoyl)cysteinylvaline, appears to be produced by both *P. chrysogenum* and the *Cephalosporium* sp., and may be an intermediate in biosynthesis of these antibiotics.

The penicillins and cephalosporins are rapidly bactericidal to growing sensitive bacteria and act by interfering specifically with synthesis of the cell wall, so that the latter is weakened in the growing cell and the cell undergoes lysis. Benzylpenicillin, which has seen extensive use in medicine, shows a very high activity against many gram positive bacteria, but its effectiveness for the treatment of staphylococcal infections was seriously impaired by the selection and spread of resistant strains of *Staph. aureus* which produce penicillinase. The clinical problem presented by these staphylococci became less serious with the introduction into medicine of new penicillins and cephalosporins, such as methicillin, cloxacillin, cephalothin and cephaloridine, which show a high resistance to hydrolysis by the staphylococcal β-lactamase. The discovery of ampicillin and the cephalosporins has also extended the possibility of effective treatment of certain infections by gram negative bacteria. Furthermore, it appears that the cephalosporins cause no allergic reaction in many patients who are sensitive to penicillins.

E. P. ABRAHAM

Cross-reference: *Antibiotics.*

PERFUMES

Perfumes are pleasant smelling substances obtained from natural or synthetic sources. Because of their esthetic appeal, they are widely used either in the form of alcoholic solutions (perfumes, colognes or toilet waters), or are added to cosmetics, soaps, paper, etc. to which they impart an agreeable aroma.

A fine fragrance is the result of a great deal of work even by the most experienced perfumer, as well as a special type of creative ability. The perfumer must recall the odor of hundreds of compounds and be able to choose among them. For the originality required in a new perfume, it is necessary for him to discover a new note by an original combination of materials at his command and occasionally by newly developed compounds.

Before a perfume is commercially acceptable, it must also possess a high degree of stability. When applied to the skin or clothing, it must not only give an appreciable aura of fragrance to the wearer but must keep its character for several hours without changing. This lasting power of perfumes is very difficult to achieve, since perfumes consist largely of volatile materials and tend to evaporate. The compounding of a perfume must thus be carried out with an eye to the odor characteristics of each ingredient and to their physical and chemical properties.

Perfume consists of a judicious blend of odorous materials which can be roughly classified as follows:

(a) Flower oils, obtained from cultivated flowers either by steam distillation or solvent extraction.

(b) Essential oils, generally obtained by the steam distillation of a large number of plants, from roots, barks, leaves, fruits, etc.

(c) Isolates, consisting of chemical compounds of natural origin isolated from essential oils.

(d) Odorous materials manufactured synthetically. These may consist of compounds which occur in nature but are synthetically manufactured, or chemicals which do not occur in nature but are desirable in perfumes.

(e) Materials of animal origin, such as extractives from musk, ambergris, civet, castoreum, etc.

(f) Resinoids, balsams, etc.

Flower oils are usually very expensive and are used in varying proportions in fine perfumes. Their role is to give to the finished perfume a pleasant, natural tonality which is impossible to achieve by the use of various combinations of essential oils or synthetics alone. Rose and jasmine are by far the most important flowers yielding perfume oils. A rose (*rosa damascena*) cultivated in Bulgaria on steam distillation yields an oil known as Otto of Rose Bulgarian. In recent years, Turkey has also became an important producer of this oil. Roses (*rosa centifolia*) cultivated in France and North Africa are generally extracted with solvents to give waxy products known as *concretes,* which on de-waxing with alcohol give an oil known as rose absolute.

The jasmin flowers grown in the Mediterranean countries—notably in France, Italy, North Africa and Egypt—are extracted with solvents to produce concretes and absolutes. In spite of their high price, these floral oils find extensive usage in fine perfume compositions and many successful perfumes contain from several to fifteen or more per cent of jasmin or rose oils. Other flowers of lesser importance extracted for their essential oil content are violets, flowers of the bitter orange tree, tuberose, mimosa, etc.

Essential oils are obtained all over the world—lavender from France, rosemary from Spain, geranium from North Africa, sandalwood from India, vetivert from Java and Haiti, oakmoss from Yugoslavia, bergamot from Italy, rosewood from Brazil, patchouli from Indonesia, orris from North Italy, citrus oils from Italy and America, citronella from Guatemala and Formosa. The list may be lengthened to include at least one hundred essential oils which are used by the perfume industry.

Isolates are naturally occurring compounds obtained from relatively cheap essential oils. For example, geraniol and citronellol, which have sweet roselike odors, are extracted from the relatively inexpensive citronella oil. Or an isolate may be used for the manufacture of new types of compounds valuable to the perfume industry. Thus citral is isolated from lemongrass oils and used in the manufacture of various *ionones.* Isolates are limited in number but are indispensable to the industry.

Because the increasing demand for isolates in recent years can no longer be met from natural sources, synthetic methods have been developed for their manufacture. One approach to their synthesis starts with acetylene. Another process involves the use of readily available turpentine oil. Aside from their use in perfumes and flavors, very large quantities of isolates and their derivatives are used in the manufacture of Vitamins A, E and K.

Synthetics are odorous materials manufactured either from natural or purely synthetic raw materials. Since they are usually made from inexpensive raw materials and, under controlled conditions, are free from the vagaries of nature, their price is not only relatively low but also stable—an important factor in their application in soaps and for other industrial purposes. Moreover, constant improvements in the methods used in their manufacture has made them available in very pure form. The industry consumes large quantities of the many synthetics on the market.

Animal extractives, while expensive, are considered valuable in perfumes since they impart a desirable warmth and lasting quality. In addition, they exert a synergic effect on the odor of perfumes, rendering them stronger and livelier. In recent years, some of the constituents of these animal extracts have been synthesized and used with some success.

Resinoids such as styrax, benzoin, myrrh, olibanum, opoponax, galbanum and others, are moderately priced resins which find use in perfumes as time-honored fixatives.

Perfumes sometimes contain thirty or more ingredients. As some of the ingredients consist of natural essential oils which in themselves have a large number of constituents, the total number of chemical bodies in a perfume oil may well exceed a hundred. The complexity of a composition, has, of course, no bearing on the quality of the perfume.

In this country, *colognes* have attained much popularity, partly because they are more reasonably priced, and partly because being a diluted form of perfume, they are applied with greater ease. A modern cologne may contain 3 to 5% of perfume oil, the balance consisting of alcohol, and up to 15% of water to cut down the sharpness of the alcohol. *Toilet waters* as sold in the United States and Canada are about 2% solutions of perfume oil in alcohol and water. In all cases, the solutions are aged, chilled and filtered to obtain brilliant and stable liquids.

The soap industry is now the largest consumer of essential oils and aromatic chemicals, and it is largely responsible for the expanded cultivation of essential oil-bearing plants as well as the development of the synthetic aromatic industry. The cosmetic industry is of course another large user of perfumes.

In recent years, the perfumer has also been given the problem of odorizing or deodorizing various commercial products to make them more attractive to the consumer. Many rubber and plastic materials are deodorized or perfumed in order to make them more acceptable for wear or as household articles. Paints can now have a pleasant scent instead of objectionable odors. The air in industrial plants is scented for the sake of the workers and the neighbors. Industrial packages of household goods and even foods are sometimes scented to add to their consumer appeal.

PAUL Z. BEDOUKIAN

Cross-reference: *Odor, Essential Oils.*

References

Arctander S., "Perfume and Flavor Materials of Natural Origin," (1960). Published by Author.

Bedoukian, P. Z., "Perfumery and Flavoring Synthetics," Second Revised Edition (1967). El-sevier Publishing Co.

Jellinek, P., "The Practice of Modern Perfumery," (1954). Interscience Publishing Company.

Maurer, E. S., "Perfumes and Their Production," (1958). United Trade Press, Ltd.

PERIODIC LAW AND PERIODIC TABLE

Periodic Law

Comparison of the chemical elements in order of increasing atomic number shows many of their physical and chemical properties to vary periodically rather than randomly or with steady progression. The similar relationship between atomic weight and properties was recognized empirically a century ago by de Chancourtois in France, Newlands in England, Lothar Meyer in Germany, and Mendeleev in Russia. Since the work of Moseley, atomic number has been recognized to be more fundamental than atomic weight. The relationship between atomic number and properties is now understood to be the logical and inevitable consequence of a periodicity of atomic structure. The familiar statement of the Periodic Law is this: "The properties of the chemical elements vary periodically with their atomic number." A more informative statement of this same law is: *The atomic structures of the chemical elements vary periodically with atomic number; all physical and chemical properties that depend on atomic structure therefore tend also to vary periodically with atomic number.*

According to the modern concept, the atom consists of a positively charged nucleus imbedded in a cloud of electrons. The wave mechanical theory of atomic structure postulates that these electrons are distributed in "principal quantum shells" representing successively higher energy levels (1, 2, 3, 4. . . .). Within these shells, the electrons occupy regions of space called orbitals, each capable of holding two electrons, which are of different shapes designated as s, p, d, and f, having different energies increasing in that order. As the nuclear charge is increased one by one, with the simultaneous addition of balancing electrons, each successive electron goes into the most stable orbital available to it. The periodicity of atomic structure arises from the recurrent filling of new outermost principal quantum levels. This is complicated by the fact that although the principal quantum levels represent very roughly the general order of magnitude of the electron energy, the less stable orbitals of one level may be of higher energy than the most stable orbital of the next higher level. The result of this overlapping is that the outermost shell of an isolated atom can never contain more than 8 electrons. In the building up of successively higher atomic numbers, once the s and three p orbitals in a given principal quantum level are filled, electrons always find more stable positions in the s orbital of the *next higher* principal quantum level rather than the d orbitals of the *same* principal quantum level. When this s orbital is filled, electrons then go into the five underlying d orbitals until these are filled, before continuing to fill the outermost shell by entering p orbitals. The building-up of the atoms of successive atomic numbers may be represented by the following sequence: $1s$, $2s$, $2p$, $3s$, $3p$, $4s$, $3d$, $4p$, $5s$, $4d$, $5p$, $6s$, $5d$, $4f$, $6p$, $7s$, $6d$, $5f$. *The periodicity of atomic structure thus consists of the recurrent filling of successive outermost shells with from one to eight electrons that corresponds to the steady increase in nuclear charge.*

A *period* is considered to begin with the first electron in a new principal quantum shell and to end with the completion of the octet in this outermost shell, except, of course, for the very first period in which the outermost shell, having only one orbital, the s, is filled to capacity with only two electrons. From the order of filling given above, it should be apparent that periods so defined cannot be alike in length. The first period has only two elements, hydrogen and helium. The second period, beginning with lithium (3) and ending with neon (10), contains 8 elements, as does the third period that begins with sodium (11) and ends with argon (18). The fourth period begins with potassium (19), but following calcium (20), the filling of the outermost (fourth) shell octet is interrupted by the filling of the five d orbitals in the third shell. Thus 10 more elements enter this period before filling of the outermost shell is resumed, causing the total number of elements in this period, which ends with krypton (36), to be 18. In the fifth period, the first two outermost electrons are added in rubidium (37) and strontium (38), but then this outer shell filling is interrupted by the filling of penultimate shell d orbitals, which again adds 10 elements before filling of the outermost shell is resumed; this period, which ends with xenon (54), also contains 18 elements.

The sixth period begins as before with first one, then two electrons in the outermost shell (cesium (55) and barium (56)), but then interruption at lanthanum (57) occurs, where filling of the five $5d$ orbitals is begun. Here, however, occurs an additional interruption, in which 14 elements are formed through filling of the $4f$ orbitals, before the remaining $5d$ orbitals can be filled. Then the $5d$ orbitals must be completely filled before the outermost shell can receive any more electrons. Consequently here it takes 32 elements to bring the outermost shell to 8 electrons and thus end the period with radon (86).

The seventh period is believed to be similar but it is incomplete. In principle it would end with element 118, but artificial element 105 is the highest in atomic number known at the time of writing.

These elements which represent interruptions in the filling of the outermost s and p orbital octet are called "transition elements" (where d orbitals are being filled), and "inner transition elements" (where f orbitals are being filled.) This is to distinguish them from the other, "major group elements," in which underlying d or f orbitals are either completely empty or completely filled. A more fundamental distinction is based on the chemical behavior of these elements. Transition elements are capable of utilizing not only the outermost shell, but also the underlying d orbitals, in their bonding. Major group elements utilize only the outermost shell.

Physical properties of the elements that depend only on the electronic structure of the individual atom, such as the ionization potential and atomic radius, vary periodically with atomic number simply because of the recurrent filling of the successive outermost shells. Increasing the atomic number by increasing the nuclear charge while adding electrons to the outermost shell increases the attractive interaction between nucleus and outermost electrons more than it increases the repulsive forces among the electrons, with the result that the electronic cloud tends to be held closer (smaller radius) and more tightly (higher ionization energy and higher electronegativity), the greater the number of outermost shell electrons. For example, carbon (6) with half-filled octet has radius, ionization energy, and electronegativity intermediate between the larger lithium (3) atoms with their low ionization energy and low electronegativity, and the smaller fluorine (9) atoms with high ionization energy and electronegativity. Similar but smaller effects are observable for addition of d or f electrons to underlying shells. Transition elements differing in atomic number by one are therefore more alike than are corresponding neighboring major group elements. Changes in the number of underlying f electrons have an even smaller effect, resulting in still closer similarities among the inner transition elements. Throughout the periodic system, however, the differences among adjacent elements become smaller as the percentage change in atomic number becomes smaller.

The bonding properties of elements also depend on the electronic structure of the individual atom and therefore likewise vary periodically. For example, each period (except the first) begins with an alkali metal: lithium (3), sodium (11), potassium (19), rubidium (37), cesium (55), and francium (87). All of these show similar metallic bonding, crystallizing in a body-centered cubic lattice. Each has but one outermost electron per atom and can therefore form but one covalent bond. Each is very low in electronegativity and thus tends to become positive when bonded to another, more electronegative element. Crossing each period the elements become less metallic, more electronegative, and able to form a greater number of bonds until this number becomes limited by the number of outer shell vacancies rather than the number of electrons. The halogens, fluorine (9), chlorine (17), bromine (35), iodine (53), and astatine (85), each of which is next to the end of its period, are all nonmetals, higher in electronegativity than any preceding element in their respective periods and thus tending to become negative when bonded to other elements. Having seven outermost electrons, each has but one vacancy, permitting but one covalent bond. At the end of each period is an element having no low-energy orbital vacancy in its outermost shell and therefore usually showing no covalent bonding ability. These elements, helium (2), neon (10), argon (18), krypton (36), xenon (54), and radon (86), all occur as monatomic gases. Compounds of the first three are unknown. The electronic shells of the last three are more polarizable and under the influence of the most electronegative of all the elements, fluorine, the integrity of the octet can be broken to produce fluorides, and from these, oxy-derivatives. Despite this violation of the sanctity of the "inert elements," they remain extraordinarily unreactive and as deserving as ever as their position as cornerstone of the periodic system, terminating each period.

Physical properties of the elements that are of greatest practical interest are usually properties that depend indirectly on the atomic structure but directly on the nature of the aggregate of atoms which results from the atomic structure. Such properties are density, melting point, and volatility. They may tend to vary periodically, but the periodicity is not necessarily consistent or even evident, because of abrupt differences in the type of polyatomic aggregate. For example, the identical atoms of carbon may form soft, flaky, electrically conducting graphite or extremely hard, nonconducting, denser diamond, depending on the kind of bonding and the arrangement of atoms. Nitrogen follows carbon in atomic number, but because it forms N_2 molecules instead of giant 3-dimensional structures like diamond or graphite, it is a gas with physical properties strikingly different from those of either form of carbon.

Among the most useful applications of the periodic law is toward an understanding of the differences among compounds, whose properties may also vary in a periodic manner. For example, oxides of elements at the beginnings of periods tend to be very stable, high-melting, nonvolatile solids of strongly basic character and practically no oxidizing power. Oxide properties change progressively across each period until toward the end, the oxides tend to be unstable, low-melting, volatile compounds acidic in nature and of high oxidizing power. Such periodicity is recognizable throughout a very large part of chemistry.

Periodic Table

In order to erect a framework upon which the myriad facts that are in accord with the Periodic Law can be organized, the chemical elements can be arranged in an orderly array called a "Periodic Table." The Periodic Law is fundamental, but the Periodic Table is merely an arbitrary attempt to arrange the elements to represent their periodicity most usefully. Any such arrangement can be satisfactory if it organizes the elements in some order of increasing atomic number, showing the separate periods and at the same time grouping elements of greatest similarity together—in other words, placing the corresponding parts of the several periods together. Hundreds of variations, rectangular, triangular, circular, spiral, two-dimensional, and three-dimensional, have been proposed, with special merits claimed for each. None has universal appeal. The form currently in widest use is shown in Fig. 1. In it, elements are arranged in horizontal periods and in vertical groups bringing together corresponding parts of the periods. The groups are numbered by Roman numerals followed by A for the major group elements and B for the transition or subgroup elements. Elements of each group have an electronic similarity that is reflected in their properties, affording a considerable degree of predictability of the properties of one element if the properties of other elements in the group are known.

Unfortunately, the periodic table of Fig. 1 has serious defects, especially as a teaching aid, which

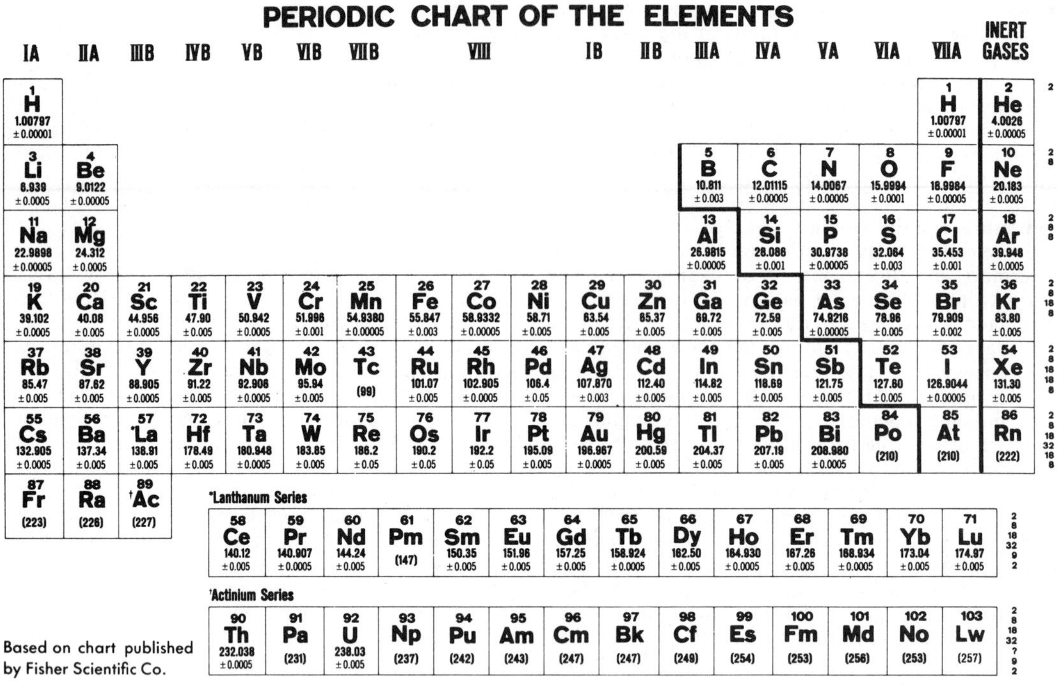

PERIODIC CHART OF THE ELEMENTS

IA	IIA	IIIB	IVB	VB	VIB	VIIB	VIII			IB	IIB	IIIA	IVA	VA	VIA	VIIA	INERT GASES
1 H 1.00797 ±0.00001																1 H 1.00797 ±0.00001	2 He 4.0026 ±0.00005
3 Li 6.939 ±0.0005	4 Be 9.0122 ±0.00005											5 B 10.811 ±0.003	6 C 12.01115 ±0.00005	7 N 14.0067 ±0.00005	8 O 15.9994 ±0.0001	9 F 18.9984 ±0.00005	10 Ne 20.183 ±0.005
11 Na 22.9898 ±0.00005	12 Mg 24.312 ±0.0005											13 Al 26.9815 ±0.00005	14 Si 28.086 ±0.001	15 P 30.9738 ±0.00005	16 S 32.064 ±0.003	17 Cl 35.453 ±0.001	18 Ar 39.948 ±0.0005
19 K 39.102 ±0.0005	20 Ca 40.08 ±0.005	21 Sc 44.956 ±0.0005	22 Ti 47.90 ±0.005	23 V 50.942 ±0.0005	24 Cr 51.996 ±0.001	25 Mn 54.9380 ±0.00005	26 Fe 55.847 ±0.003	27 Co 58.9332 ±0.00005	28 Ni 58.71 ±0.005	29 Cu 63.54 ±0.005	30 Zn 65.37 ±0.005	31 Ga 69.72 ±0.005	32 Ge 72.59 ±0.005	33 As 74.9216 ±0.00005	34 Se 78.96 ±0.005	35 Br 79.909 ±0.002	36 Kr 83.80 ±0.005
37 Rb 85.47 ±0.005	38 Sr 87.62 ±0.005	39 Y 88.905 ±0.0005	40 Zr 91.22 ±0.005	41 Nb 92.906 ±0.0005	42 Mo 95.94 ±0.005	43 Tc (99)	44 Ru 101.07 ±0.005	45 Rh 102.905 ±0.005	46 Pd 106.4 ±0.05	47 Ag 107.870 ±0.003	48 Cd 112.40 ±0.005	49 In 114.82 ±0.005	50 Sn 118.69 ±0.005	51 Sb 121.75 ±0.005	52 Te 127.60 ±0.005	53 I 126.9044 ±0.00005	54 Xe 131.30 ±0.005
55 Cs 132.905 ±0.0005	56 Ba 137.34 ±0.005	57 La 138.91 ±0.005	72 Hf 178.49 ±0.005	73 Ta 180.948 ±0.0005	74 W 183.85 ±0.005	75 Re 186.2 ±0.05	76 Os 190.2 ±0.05	77 Ir 192.2 ±0.05	78 Pt 195.09 ±0.005	79 Au 196.967 ±0.005	80 Hg 200.59 ±0.005	81 Tl 204.37 ±0.005	82 Pb 207.19 ±0.005	83 Bi 208.980 ±0.0005	84 Po (210)	85 At (210)	86 Rn (222)
87 Fr (223)	88 Ra (226)	89 Ac (227)															

Lanthanum Series

58 Ce 140.12 ±0.005	59 Pr 140.907 ±0.0005	60 Nd 144.24 ±0.005	61 Pm (147)	62 Sm 150.35 ±0.005	63 Eu 151.96 ±0.005	64 Gd 157.25 ±0.005	65 Tb 158.924 ±0.0005	66 Dy 162.50 ±0.005	67 Ho 164.930 ±0.0005	68 Er 167.26 ±0.005	69 Tm 168.934 ±0.0005	70 Yb 173.04 ±0.005	71 Lu 174.97 ±0.005

Actinium Series

90 Th 232.038 ±0.0005	91 Pa (231)	92 U 238.03 ±0.005	93 Np (237)	94 Pu (242)	95 Am (243)	96 Cm (247)	97 Bk (247)	98 Cf (249)	99 Es (254)	100 Fm (253)	101 Md (256)	102 No (253)	103 Lw (257)

Based on chart published by Fisher Scientific Co.

FIG. 1.

are nonetheless serious because most experienced chemists have become too accustomd to this chart to be sensitive to its faults. First, consistency in designating the groups as A or B is lacking. For example, in writing of the IVB elements some chemists would be referring to the carbon group and others to the titanium group. Second, the transition and inner transition elements differ sufficiently from the major group elements as a class that it is both convenient and customary to treat them separately. There is no advantage to disrupting the periodic arrangement of major group elements in the table to include the transition elements, thus introducing a meaningless and confusing space gap between beryllium and boron and between magnesium and aluminum. Third, the justification for separate treatment of the transition metals lies in their use of inner d orbitals in bonding. Since zinc, cadmium, and mercury use only their outermost shell orbitals in bonding, there is no basis for including them with the transition elements. Fourth, the assignment of only one group number, VIII, to three distinct transition element groups has no foundation except historical ignorance long since removed. Furthermore, it deprives the "noble gases" or "inert group" of this number. Fifth, the chart offers no distinction between 8-shell and 18-shell elements within the major groups, leading to erroneous expectations concerning the similarity of elements of the same group, or the trends down the group. Finally, the lack of extra vacant orbitals on the hydrogen atom disqualifies this element from membership with the alkali metals, and the lack of extra lone pairs disqualifies it from status as a halogen.

The periodic chart of Fig. 2 eliminates the defects of Fig. 1, as can be seen. The only radical change is in group numbers, the new chart designating the major groups by M and the transition groups by T, each followed by the arabic numeral corresponding to the electronic structure. For the major groups, this number is the number of outermost shell electrons. For the first two members of each transition group, the T group number is the number of electrons beyond the last M8 element. For the third member of each transition group, the number of electrons beyond the last M8 element is given by the group number plus 14. In this table, the iron, cobalt, and nickel groups become T8, T9, and T10 instead of being lumped together under VIII. The copper group becomes T11. Zinc, cadmium, and mercury are here properly recognized as major group elements, with the number M2′ to distinguish them from the alkaline earth elements of M2. Hydrogen has a place of its own just above and slightly to the left of carbon, since its chemistry despite its univalence more nearly resembles that of group M4. The "noble gases" are now appropriately M8. The change from 8-shell to 18-shell element down a major group is indicated by a gap, warning that a smooth trend in properties is not to be expected. The transition elements are grouped separately rather than squeezed into the major group table where they serve little useful purpose. Finally, the inner transition elements are shown as two separate series parallel but not adjacent, to emphasize the considerable differences observed, especially for the first half of each series.

The Periodic Table has long served two main functions: prediction and organization. Prediction can be made on two bases, comparison with horizontal neighbors and comparison with vertical

PERIODIC TABLE OF THE CHEMICAL ELEMENTS

MAJOR GROUPS

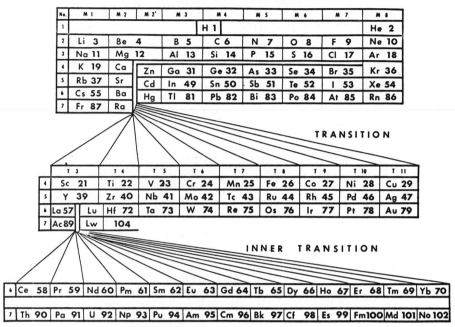

Fig. 2.

neighbors. Comparison with horizontal neighbors may not always be as precise, but in general it is more reliable. The trends across a period tend to be more consistent than they may be down a group. In general, the differences between elements one apart in atomic number are greater for small atomic numbers where the proportionate effect of unit change is greatest, and become progressively smaller as atomic number increases and the proportionate effect becomes less. For example, bismuth differs much less from lead than nitrogen from carbon. However, disproportionately large differences must be expected to accompany abrupt changes in electronic type. Comparison with neighbors in the group may be useful provided that electronic differences are properly taken into account. In the major groups, the first member differs from the second in having a penultimate shell of two electrons instead of eight. From Group M3 on, the third member differs from the second in having a penultimate shell of 18 electrons instead of 8. Although the elements of these groups use the same kind of bonding orbitals, it would be unreasonable to expect the third member to follow necessarily the trends set by the first two, because associated with the electronic differences, its atoms tend to be more compact, more resistant to the removal of electrons, and more electronegative. In the transition groups, $3d$ orbitals appear to be sufficiently different in behavior from $4d$ and $5d$ that the first member of each group is quite different from the other two.

As our understanding of the nature of individual atoms improves, we become less and less dependent on the Periodic Table as a basis for prediction, and one day soon we may dispense with it for this purpose. For no amount of organization or correlation will ever alter the fact that each chemical element is an individual and unique. Nor will the properties of any element be changed one iota by placing that element in any special position in a Periodic Table. Nevertheless, there is enough consistency to the structure and behavior of atoms to make any reasonable form of the Periodic Table an extremely and durable framework upon which to organize and correlate an enormous quantity of chemical information. As a basis for organization of chemistry, the Periodic Table will probably remain indispensable. The periodic law which it represents is truly one of the great generalizations of science.

R. T. SANDERSON

Cross-reference: *Electronic Configurations.*

PEROXIDES

A peroxide or peroxy compound is characterized by the presence of the —O—O— group with each oxygen atom having an oxidation number of −1. Such compounds may be regarded as derivatives of hydrogen peroxide, HOOH, in which one or both of the hydrogen atoms are replaced by other elements or groups. In comparison, a *superoxide,* such as potassium superoxide (KO_2) contains a pair of oxygen atoms each of oxidation number −½, whereas a normal oxide contains oxygen atoms of oxidation number −2. Substances of empirical composition, M_2O_3, (where M is an alkali element) have been reported but these may be mixtures of a peroxide and superoxide. However Rb_4O_6 and Ce_4O_6 are reported to exist from x-ray studies and inorganic ozonides having the formula MO_3 have been prepared. The perhydroxy radical, HO_2, is formed as a transitory intermediate in the gas phase oxidation of various organic substances.

Some normal metal *dioxides,* such as PbO_2, are still incorrectly referred to as peroxides. The use of the prefix *per-* of itself does not necessarily indicate a peroxy compound. Thus it may indicate an element in its highest state of oxidation, *e.g.,* potassium permanganate, or complete substitution, *e.g.,* perchloroethane. There also exist hydrogen peroxide addition compounds such as $2Na_2CO_3 \cdot 3H_2O_2$ or urea peroxide, $CO(NH_2)_2 \cdot H_2O_2$, analogous to a hydrate, which are termed hydroperoxidates or sometimes peroxyhydrates. Many peroxides also form true hydrates, *e.g.,* $Na_2O_2 \cdot 8H_2O$.

Several hundred different organic and inorganic peroxy compounds have been identified, and several dozen are articles of commerce.

Inorganic Peroxy Compounds. All the metals in groups I and II of the Periodic Table and many of those in the higher groups form peroxides. The most important industrially is sodium peroxide, Na_2O_2. Of the group II metal peroxides, those of barium and strontium are used in pyrotechnics, those of zinc, magnesium and calcium are used as bleaches and disinfectants and are incorporated in various pharmaceutical and cosmetic preparations. Calcium peroxide also aids in the mechanical handling of dough. In the nineteenth century barium peroxide was the sole starting material for the manufacture of hydrogen peroxide, but with the advent of the electrochemical and autoxidation processes, no hydrogen peroxide is made in this fashion commercially in the U.S. today.

Sodium peroxide, a pale-yellow solid, is manufactured by a two-step oxidation of metallic sodium with dry, carbon dioxide-free air, sodium monoxide being formed as the intermediate. The commercial material usually contains a minimum of 96% Na_2O_2. It is shipped in tightly sealed steel containers. Sodium peroxide is thermodynamically stable at room temperature if dry. If moistened, however, it hydrolyzes to form sodium hydroxide and hydrogen peroxide. Since the latter is relatively unstable in highly alkaline media, it will decompose with evolution of heat and oxygen. Sodium peroxide should therefore be kept from contact with oxidizable material such as wood or paper. Care should also be taken to keep from breathing the dust on handling, or allowing it to come in contact with the skin. Sodium peroxide is used as an oxidant and bleaching agent. In the laboratory it is used as a fusion reagent; in most industrial applications it is applied in the form of a dilute aqueous solution, usually partly neutralized, for many of the same purposes that hydrogen peroxide is used.

Strontium and barium peroxides may be manufactured by heating the oxide in air. As an alternate procedure, applicable to all of the alkaline-earth peroxides, the metal hydroxide may be reacted with hydrogen peroxide, or a soluble metal salt mixed with a base plus hydrogen peroxide. In some cases the metal peroxide is precipitated, in others it is recovered by evaporation. Barium, strontium, calcium, zinc, and magnesium peroxides are very stable at room temperature when dry. Since they are oxidizing agents, they should not be heated or intimately mixed with combustible material.

An aqueous solution of either *peroxydisulfuric acid* ($H_2S_2O_8$) or *ammonium peroxydisulfate* [$(NH_4)_2 S_2O_8$] is made as an intermediate in the various electrolytic processes for manufacture of hydrogen peroxide. Peroxydisulfuric acid has no commercial uses of itself. Ammonium peroxydisulfate and potassium peroxydisulfate, which is made from it, are used as polymerization catalysts, in soap and fat bleaching, and for other purposes. *Sodium peroxyborate,* which probably has the structure $NaBO_2 \cdot H_2O_2 \cdot 3H_2O$, is made by electrolysis of a solution of borax and sodium carbonate, or by treating a solution of borax with sodium peroxide or with hydrogen peroxide plus sodium hydroxide. The product is relatively stable at room temperature and, like other addition compounds such as urea peroxide, forms a convenient source of hydrogen peroxide in a "dry" form. It is used for oxidation of vat dyestuffs, and in bleaching preparations for the home laundry.

Analysis of inorganic peroxy compounds is usually by solution in water, using an acid if necessary, followed by iodometric or permanganate titration.

Organic Peroxides. A wide variety of organic peroxides have been isolated and identified. Most of them may be grouped by structure into the following categories: (1) The structure, ROOH, termed a hydroperoxide, where R may be an alkyl, aryl or other radical, e.g., C_2H_5OOH, ethyl hydroperoxide. If R is an acid radical (acyl) the compound is termed a peroxy acid, e.g., $CH_3C(=O)OOH$, peroxyacetic acid. (2) The structure ROOR, e.g., di-*tert*-butyl peroxide. A number of transannular peroxides are also known, such as ascaridole. (3) The structure $RC(=O)OOR'$, termed a peroxy ester. (4) The structure ($RC(=O)OO(O=)CR'$, such as dibenzoyl peroxide. (5) Various peroxy derivatives formed by condensation of aldehydes or ketones with hydrogen peroxide.

Hydroperoxides may be formed by alkylation of hydrogen peroxide in the presence of a strong acid, by oxidation of a hydrocarbon with molecular oxygen, or by addition of oxygen to Grignard reagents. The primary hydroperoxides, e.g., methyl or ethyl hydroperoxide, are very sensitive explosives. The stability increases with molecular weight and in the order primary, secondary, tertiary, and the same is true of the dialkyl peroxides. For example, di-*tert*-butyl peroxide is one of the most stable peroxides known and can be distilled at atmospheric pressure without decomposition. The alkyl peroxy acids, made by mixing an organic acid and concentrated hydrogen peroxide in the presence of a mineral acid, are among the least stable of the organic peroxides and, if concentrated, they explode violently on heating. Peroxyformic and peroxyacetic acid are powerful oxidizing agents and are used for causing epoxidation or hydroxylation of a carbon-carbon double bond, in various bleaching operations, and for other applications. A wide variety of organic peroxides are used to initiate polymerization reactions.

Organic peroxides are most commonly analyzed by iodometric methods, although the techniques may vary substantially for different types because of the considerable differences in their oxidizing powers.

Inhalation of vapors of most organic peroxides may cause headache and irritation of sensitive tissues. Many organic peroxides are highly sensitive to shock, friction, or heat, particularly the lower members of the diacyl and primary alkyl peroxide series. Various ethers and aldehydes may react with atmospheric oxygen to form sensitive and explosive peroxides.

Hydrogen Peroxide

Hydrogen peroxide, discovered by Louis-Jacques Thenard in 1818, is formed as an intermediate in the reaction of oxygen with a wide variety of substances including hydrogen, metals, and various organic compounds. It may also be prepared from various peroxy compounds. The principal manufacturing process involves the autoxidation of an alkyl anthrahydroquinone, such as the 2-ethyl derivative, in solution, in a cyclic continuous process in which the quinone formed in the oxidation step is reduced to the starting material by hydrogen in the presence of a supported palladium catalyst:

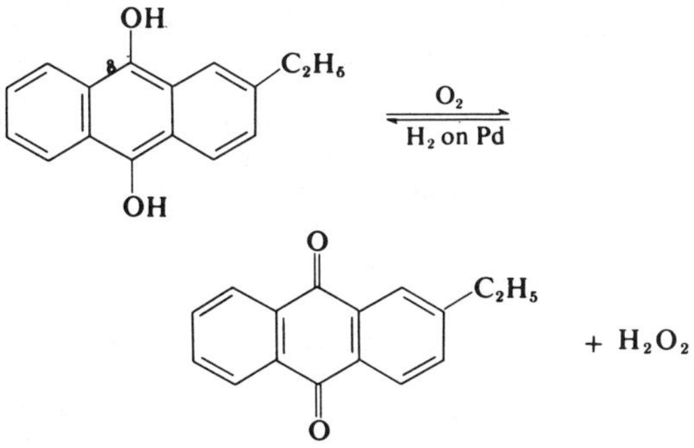

The hydrogen peroxide is removed by extraction with water and concentrated by distillation under reduced pressure. Some hydrogen peroxide is also manufactured by the autoxidation of isopropyl alcohol or, in countries other than the U.S., by the older electrolytic processes in which aqueous sulfuric acid or acidic ammonium bisulfate is converted electrolytically to the peroxydisulfate which is then hydrolyzed to form H_2O_2:

$$2H_2SO_4 \rightarrow H_2S_2O_8 + H_2$$
$$H_2S_2O_8 + 2H_2O \rightarrow 2H_2SO_4 + H_2O_2$$

U.S. annual production capacity in recent years has been about 100 to 120 million pounds, 100% basis. Hydrogen peroxide is manufactured and marketed as an aqueous solution; concentrations of from 3 to 90 wt. % or higher are commercially available. The price of the intermediate concentrations in commercial quantities is about $0.40 per pound of contained hydrogen peroxide.

Pure hydrogen peroxide solutions, completely free from contamination, are highly stable. However, a small amount of a stabilizing agent, such as sodium stannate, is usually added to help counteract the catalytic effect of traces of impurities, such as iron, copper, and other heavy metals. A relatively stable sample of hydrogen peroxide typically decomposes at the rate of about 0.5% per year at room temperature. Storage containers, which must always be vented, are usually made of high purity aluminum, although glass or polyethylene is sometimes suitable for small amounts of the lower concentrations. Organic material such as clothing or wood will frequently ignite on contact with *concentrated* hydrogen peroxide, and solutions or dispersions of organic substances with *concentrated* H_2O_2 may detonate on mechanical shock. A 90% aqueous H_2O_2 solution at room temperature cannot be detonated by mechanical impact. It is possible to obtain partial detonation, but only if highly confined and initiated by an explosive charge. It is important to avoid gross contamination of hydrogen peroxide solutions. In such a case the heat evolved on decomposition can lead to a self-accelerating reaction which can cause pressure rupture of the vessel even if vented.

Some of the important physical properties of pure, anhydrous hydrogen peroxide are as follows: Density of the solid, 1.71 g/cc, density of the liquid, 1.450 g/cc at 20°C, viscosity of the liquid, 1.245 centipoises at 20°C, surface tension, 80.4 dynes/cm at 20°C. Freezing point −0.43°C, (readily supercools), boiling point at atmospheric pressure, 150.2°C; heat of vaporization at 25°C, 12,334 cal/mole; heat capacity of the liquid at 25°C, 0.628 cal/(g)(°C). Changes of enthalpy and free energy on formation and decomposition of the vapor are, respectively:

$$H_2(g) + O_2(g) \rightarrow H_2O_2(g)$$
$$\Delta H°_{298} = -32.52 \text{ kcal/mole}$$
$$\Delta F°_{298} = -25.24 \text{ kcal/mole}$$
$$H_2O_2(g) \rightarrow H_2O(g) + \tfrac{1}{2}O_2(g)$$
$$\Delta H°_{298} = -25.26 \text{ kcal/mole}$$
$$\Delta F°_{298} = -29.39 \text{ kcal/mole}$$

Standard electrode potentials are:

$$H_2O_2 = O_2 + 2H^+ + 2e^- \qquad E° = +0.6824 \text{ volt}$$
$$2H_2O = H_2O_2 + 2H^+ + 2e^- \qquad E° = +1.776 \text{ volt}$$

Hydrogen peroxide has the structure H—O—O—H (bond angles not indicated here), and is unassociated in the gas and liquid phases other than the contributions from hydrogen bonding. Solutions of hydrogen peroxide in water are nonideal; there is an appreciable heat effect and volume change on mixing and the vapor pressures of the solutions do not follow Raoult's law. Hydrogen peroxide is catalytically decomposed by a wide variety of substances; particularly active are oxides and hydroxides of the metals of relatively high atomic weight such as manganese, iron, and silver.

The most commonly employed technique of analysis is titration of an acidified solution with standardized potassium permanganate solution; this is one of the most exact and reliable methods available. Probably the most specific and sensitive test for hydrogen peroxide is the yellow complex developed when it is allowed to react with an acidic aqueous solution of titanium sulfate.

Hydrogen peroxide is a strong oxidizing agent; about 70% of U.S. production is used for bleaching, accounting for practically all of the textile mill bleaching of wool and cotton, and a substantial fraction of that of synthetic fibers. It is also used in the bleaching of groundwood pulp and the final states of chemical pulp bleaching in addition to that of a wide variety of other substances. During World War II, particularly in Germany, large amounts of concentrated hydrogen peroxide were manufactured for use as a source of energy in various military devices such as torpedoes, rocket aircraft, submarines, and catapult launchers. Hydrogen peroxide is the starting material for preparation of many inorganic and organic peroxy compounds, and dilute solutions have long been used as a household antiseptic. As an oxidizing agent, it is used for epoxidation of unsaturated compounds, oxidation of vat and other dyed yarns, depolymerization of starch, adhesives, etc., and as an analytical reagent.

CHARLES N. SATTERFIELD

References

Vol'nov, I. I., *Peroxides, Superoxides, and Ozonides of Alkali and Alkaline Earth Metals,* Plenum Press, New York, 1966. (English Translation).

Tobolsky, A. V., and Mesrobian, R. B., *Organic Peroxides. Their Chemistry, Decomposition, and Role in Polymerization,* Interscience, New York, 1954.

Organic Peroxides. Their Formation and Reactions, by E. G. E. Hawkins, E. and F. F. Spon Ltd., London, 1961.

Davies, A. G., *Organic Peroxides,* Butterworths, London, 1961.

Machu, W., *Das Wasserstoffperoxyd und die Perverbindungen,* Springer Verlag, Vienna, 1951.

Schumb, W. C., Satterfield, C. N., and Wentworth, R. L., "Hydrogen Peroxide," Reinhold Publishing Corporation, New York, 1955.

Peroxides as Chain Initiators

The oxygen-oxygen bonds in inorganic and organic peroxides are much weaker than the normal single covalent bonds and hence undergo homolytic cleavage to produce free radicals that are able to initiate chain reactions. The general reaction is shown in the following equation:

$$—O:O— \longrightarrow —O\cdot + \cdot O—$$
peroxide   free radicals

The inorganic peroxides such as the potassium salt of peroxydisulfuric acid ($K_2S_2O_8$), sodium peroxyborates ($NaBO_3 \cdot xH_2O$), sodium peroxycarbonate (Na_2CO_4) and hydrogen peroxide (H_2O_2) are used as initiators for emulsion polymerizations.

As shown by the following equation, a water-soluble inorganic peroxy compound such as potassium persulfate may dissociate to produce two sulfate free radicals ($\cdot OSO_3^-$). Each of these free radicals may initiate a chain reaction by adding to a vinyl monomer ($H_2C=CHX$) to produce a new free radical capable of reacting with additional monomer molecules.

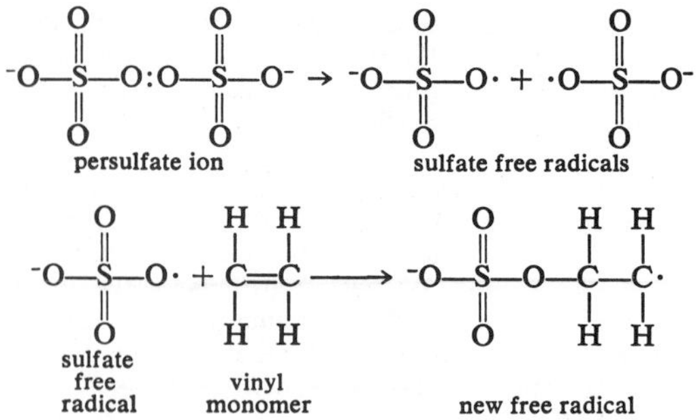

persulfate ion → sulfate free radicals

sulfate free radical + vinyl monomer → new free radical

Since the sulfate group is present in the final polymer, these peroxy compounds are not catalysts but this term is often used erroneously to describe both inorganic and organic initiators. The rate of formation of these free radicals may be accelerated in the presence of reducing agents, such as iron(II) salts in the so-called redox polymerization systems.

The most widely used organic peroxide initiator is benzoyl peroxide ($C_6H_5COO)_2$ which was first synthesized by Brodie in 1859. More than forty different organic peroxides are now available commercially. The most widely used organic peroxy compounds are benzoyl peroxide, lauroyl peroxide and cumene hydroperoxide. Since these compounds are readily decomposed by heat, mechanical shock, friction and by oxidizable contaminants, they must be used with caution. In spite of these hazards, the annual production of organic peroxides is about 20 million pounds.

As shown by the following equation, cumene hydroperoxide is produced by the autoxidation of cumene. Oxygen may be bubbled through liquid cumene, a solution of cumene or an emulsion of this arene.

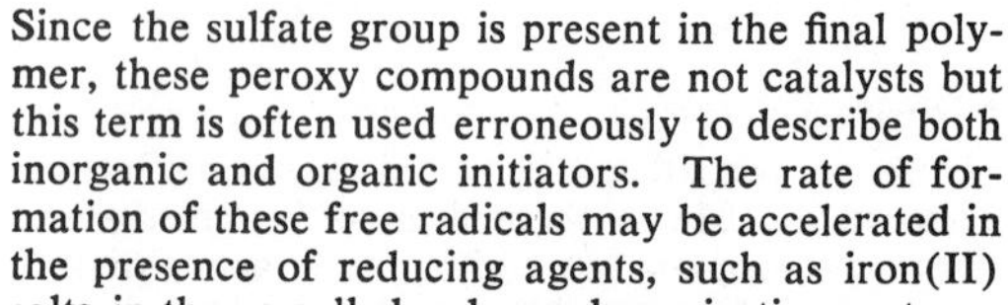

cumene   oxygen   cumene hydroperoxide

As shown by the following equation, acyl peroxides, such as benzoyl peroxide are produced by the reaction of sodium peroxide or hydrogen peroxide with the corresponding anhydride or acyl chloride. A base, such as pyridine, is used when acyl chloride reacts with hydrogen peroxide.

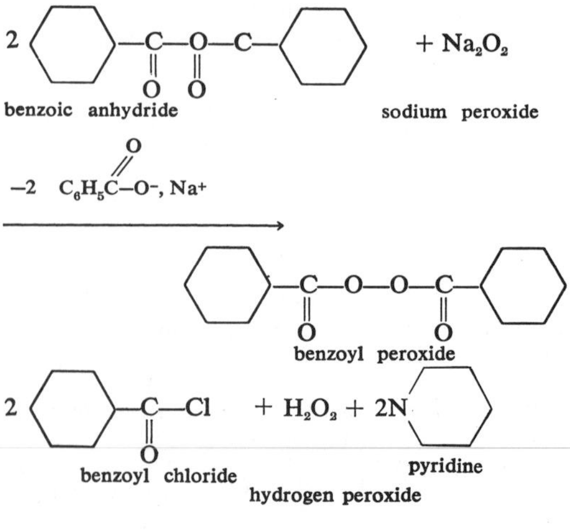

benzoic anhydride   sodium peroxide

benzoyl peroxide

benzoyl chloride   hydrogen peroxide   pyridine

−2 $C_6H_5NH^+$, Cl^-

⟶

benzoyl peroxide

Organic peroxides are often classified according to the temperatures at which they yield free radicals. These data are reported as half-lives or as the temperature for a half-life of ten hours. Thus cumene hydroperoxide has a half-life of 470 hrs at 115°C and a half-life of 10 hrs at 150°C. Benzoyl peroxide has a half-life of 2.1 hrs at 85°C and a half-life of 10 hrs at 72°C. Lauroyl peroxide has a half-life of 0.5 hrs at 85°C and a half-life of 10 hrs at 62°C. The rate of decomposition of these peroxides is decreased by the presence of electron withdrawing groups and increased by the presence of electron donating groups. The rate of decomposition is also affected by solvents and other substances present.

It is important to note that the rate of decomposition of an organic peroxide (R_d) is equal to twice the rate constant (k_d) times the concentration of the organic peroxide [I] times the fraction of free radicals (f) produced. That is, $R_d = 2f\,k_d[I]$. The rate of propagation of this chain reaction is proportional to the concentration of monomer and to the square root of the concentration of the organic peroxide.

RAYMOND B. SEYMOUR

PEST CONTROL BY ANTIMETABOLITES

Public opinion encourages substitution of conventional pesticides for chemicals which are less harmful to the environment. Antimetabolites (growth factor analogues) have proved their value in this endeavor.

Success with antimetabolites depends on knowing the nutritional requirements of insects to be controlled and then selecting inhibitors best suited to antagonize vitamins and amino acids most required by them. Candidate inhibitors are then incorporated into the insect's normal diet. Antimetabolites are more effective in controlling insects which thrive on a unidiet, for a single food supply eliminates the counterbalancing nutrients that would otherwise be derived from a varied diet. Fabric-feeding insects existing on wool alone are successfully controlled with antimetabolites.

Basic knowledge of differences between competitive and noncompetitive antagonism should be understood before experimenting with antimetabolites for pest control. The inhibition index is helpful in understanding the phenomenon of reversible antagonism as long as it remains constant. However, the index is constant only when antagonism is competitive, since the ratio between the contending pair of substances is constant. If change occurs as concentration of metabolite is raised, then antagonism is considered noncompetitive and the index is less helpful.

The classic method of demonstrating workable metabolic inhibition is by reversing the process involved—usually by putting back into the system

the same nutrient inhibited. Before attempting to reverse techniques, it is necessary to take into account the important factors of *time* and *rate of application*. If too much time elapses following intake of inhibitor, normalcy will never occur; nor will it occur if too much inhibitor is administered. Granted, early symptoms of dietary deficiency are corrected by the replacement of a deficient nutrient, no amount of corrective nutrient will restore a system already disrupted beyond repair. By this time secondary and tertiary symptoms will have set in, usually followed by a disease syndrome. Death is, therefore, attributed to complications beyond that of simple starvation.

Only small amounts of antimetabolite-treated wool fibers need be ingested by fabric-feeding insects before signs of dietary deficiency begin. Body weight losses and general decline lead to metabolic breakdown and eventual death, while fiber weight loss remains below that allowed by recognized standards.

Some antimetabolites are capable of inhibiting two or more essential metabolites. According to the favored hypothesis, a specified antimetabolite antagonizes a specific nutrient. However, in a living organism the selectivity observed may be the result of several factors. For example, a precursor which normally gives rise to two nutrients may be adversely affected. This results in greater disruption. A case in point is the antimetabolite imidazole, a recognized inhibitor of histamine. When histamine is immediately replaced in the system of carpet-beetle larvae already made deficient with imidazole, the larvae recover. On the other hand niacin also acts as a replacement and it too promotes recovery. Precise selectivity of action in this case is not clear, but it appears imidazole is capable of inhibiting both histamine and niacin as evidenced by the reversal effects of both substances.

Combinations of antimetabolites have proved valuable in controlling fabric-feeding insects. When DL para-fluorophenylalanine (inhibitor of the amino acid phenylalanine) is combined with imidazole (inhibitor of histamine and niacin), a triple blow is dealt the target insects. Decline is hastened and corrected only if the inhibited substance is replaced at first signs of deficiency.

Imidazole acts not only as a metabolic inhibitor, but also as a conventional insecticide when it is applied topically. When formulated in an oil spray, with a hydrophylic surfactant to improve penetration and internal distribution, imidazole demonstrates a good "contact" effect. Performance is not unlike that of some carbamate insecticides—yet the acute oral mammalian toxicity of imidazole is approximately ten times less than common aspirin.

Inhibition or alteration of essential nutrients is not restricted to antimetabolites alone. When some metabolites are ingested in excess, they "turn upon themselves" and create a condition of imbalance not always in keeping with the classical symptoms of hypervitaminosis. For example, when carpet-beetle larvae are fed more vitamin A than can be utilized, the surplus destroys certain gut bacteria which normally synthesize needed vitamin K. Importance of trace amounts of vitamin K to these insects has recently been shown. When vitamin K is inhibited directly, the insects decline and die.

When vitamin K is inhibited indirectly, due to upsetting the normal balance of symbiotic association, the insects also decline and die. Yet both direct and indirect deficiencies are corrected by vitamin K replacement if therapeutic amounts are given at first signs of decline.

Low concentrations of calcium phosphate integrated into the diet of stored grain insects have been shown to prolong the insect's total life span. Yet 2 and 3% of the amount of the same substance essential to human nutrition exhibits a marked toxic effect on important insect pests of stored grain. Symptoms of toxicity are characterized by retarded growth and inhibition of metamorphosis from larvae to pupae.

More in keeping with recognized symptoms of hypervitaminosis is the dual role of vitamin K. Trace amounts of this vitamin are beneficial. Slightly excessive amounts are lethal. 0.007% vitamin K_3 added to a carpet beetle diet of wool accelerates growth and development of larvae, yet as little as 0.125% is fatal. Pest control in these cases is realized by use of vitamins and minerals essential to man and higher animals.

Upsetting vital biological processes by any means other than metabolic inhibition is also a contribution to noninsecticidal pest control. Some compounds (chemosterilants) interfere with reproductive processes, others interfere with glandular secretion, hormonal discharge and in some cases cellular development. All are capable of inducing deleterious morphological changes. Work with a variety of alcohols applied topically and/or injected have exhibited interesting juvenile hormone activity.

Dodecyl alcohol, a harmless coconut oil derivative, and a few related higher alcohols and fatty acid derivatives, have recently been found to produce unusual morphological and physiological effects on six insect species of economic importance, e.g., termites, furniture carpet beetle, black carpet beetle, cockroach nymphs, and larvae of confused flour beetle and of the flesh fly. The most noteworthy of these effects is a progressive atrophy leading to complete loss of legs and other extremities. Other forms of abnormality include albinism, wing deformation, impeded blood circulation, and sterility. Of particular interest is dodecyl alcohol at 1 to 2% being immediately lethal to a number of insects while as little as 1/16% is lethal over a longer period of time. Fatal abnormalities are the result of prolonged exposure. In spite of dodecyl alcohol toxicity to insects, the substance is commonly used in manufacturing harmless soaps and detergents and as a preservative by confectioners. Currently, dodecyl alcohol is being combined with imidazole as an experimental substitute for DDT in drycleaning, and as a rug shampoo and "spray-on" for protecting woolens and furs against insect attack.

ROY J. PENCE

Cross-reference: *Insecticides.*

References

Wooley, D. W. "A Study of Antimetabolites," John Wiley and Sons Inc., London, 1952.

Bowers, W. S., and Thompson, M. J., "Juvenile hormone activity: Effects of isoprenoid and straight-chain alcohols on insects," *Science* **142**(3598):1469–70 (1968).

Majumber, S. K., and Bano, Athia, "Toxicity of calcium phosphate to some pests of stored grain," *Nature,* **202**(4939):1359–60 (1964).

Pence, Roy J., and Viray, Manuel S., "Inhibition vs excessive use of vitamin K and other nutrients for the control of carpet beetles," *Residue Reviews,* **12**:46-64 (1969).

Pence, Roy J., and Viray, Manuel S., "Atrophy and developmental abnormality in insects exposed to dodecyl alcohol," *J. Econ. Ent.,* **62**(3): 622–629 (1969).

PETROCHEMICAL FEEDSTOCKS

Products obtained from petroleum and natural gas were the starting materials for about 78 billion pounds or about 50% of the total United States organic chemical production in 1970. There are over 2,000 chemical end products derived from these feedstocks. Major categories of end products are plastics, elastomers, dyes, medicinal chemicals, surface active agents, pesticides, solvents, and a host of miscellaneous chemicals.

The large volume primary organic materials for chemical consumption include aliphatic hydrocarbons having one to five carbon atoms, benzene, toluene, xylene, naphthenic acids, cresylic acids, and inorganic materials include hydrogen, sulfur, and carbon black. Many of these primary materials are produced by more than one process, and alternative methods are also used for producing secondary products therefrom.

The quantities of chemical materials available from petroleum and natural gas vary with the chemical constitution of the oil or gas being used, and in the case of products from ordinary refinery processing of oil, with the process used.

Some material available for chemical conversion comes from gases from catalytic cracking and thermal reforming but an ever increasing share is provided from pyrolysis of hydrocarbons for olefin production, particularly from heavy feedstocks such as naphtha and gas oils. Aromatics are produced from UOP Platforming and other catalytic reforming processes, with increasing amounts from heavy hydrocarbon pyrolysis as noted above. Yields vary with charging stocks and process conditions.

Methane is the principal component of natural gas. Although methane is often extracted, natural gas is frequently used directly for chemical conversion. Chemicals are derived either by first producing synthesis gas (CO and H_2) or by direct oxidation or chlorination. The principal secondary products from methane or natural gas are methanol, methyl and methylene chlorides, ammonia, and acetylene.

Ethane is present in both natural gas and cracked gases from petroleum refining. In production of synthesis gas from either gas it is not separated initially. In pyrolysis, ethane is normally separated and fed selectively to the heaters for greatest efficiency and yields of olefin.

Propane is also present in both refinery and natural gas. It may be utilized without separation from natural gas, may be separated from either natural or refinery gases, or may be separated with the butane fraction. Much of the propane and butane fractions are marketed as liquefied petroleum gas (LPG). Propane is used largely for pyrolysis to ethylene and to some extent for dehydrogenation to propylene. Other secondary products include lower alcohols, aldehydes, ketones, and nitroparaffins.

Normal and isobutane are also components of both refinery and natural gases. Butane is used as LPG and to a considerable extent as a feedstock for pyrolysis heaters producing propylene and ethylene. Both butane and isobutane are also dehydrogenated to butylene and isobutylene with considerable amounts of butane also being dehydrogenated further to butadiene. Isobutane is also used in alkylation reactions to produce gasoline of high octane ratings. Alkylation reactions will gain increasing favor as the trend toward unleaded gasoline continues.

Pentanes are extracted from natural gasoline obtained from natural gas. They are converted to amyl alcohols, chlorides, and amines.

Higher paraffins found in kerosene and wax fractions of petroleum have some use as starting materials for higher alcohols, aldehydes, ketones, and chlorination products. A kerosene fraction was at one time the principal alkylating agent for preparation of alkybenzene for use in synthetic detergents. The kerosene product is now being largely replaced by dodecylbenzene from dodecene obtained by the polymerization of propylene.

Ethylene is the most important hydrocarbon raw material used for chemical production. While present in the gases obtained during ordinary cracking operations for the production of gasoline, by far the greatest quantity is obtained from pyrolysis of ethane, propane, and butane derived from natural gas. Naphthas comprise the major feedstock for ethylene producing facilities in Europe and are involved in most new, larger installations in this country. Produced at the rate of over 18 billion pounds in 1970 and an expected rate of 23 billion pounds in 1974, it is the starting material for many high-tonnage chemicals including ethyl alcohol, ethylene oxide, ethylbenzene, ethyl chloride, ethylene dichloride, and polyethylene plastics. These secondary products are basic for other important products such as acetaldehyde, ethylene glycol, ethanolamines, acrylonitrile, tetraethyllead, and vinyl chloride.

Propylene is present in the gases from cracking processes, and is also produced in large volume by cracking three- and four-carbon hydrocarbons from refinery gases and natural gas. It is used principally for production of isopropyl alcohol which is an intermediate for acetone. Current U.S. consumption of isopropanol is about 2 billion pounds/year. Commercial application of the substitutive chlorination reaction has led to the large scale use of propylene as a starting material for glycerine, allyl alcohol, epichlorohydrin, and many smaller volume chemicals. Polymerization of propylene is applied for the production of tetramer which is used for alkylaryl sulfonate detergents. Alkylation of benzene with propylene yields cumene, starting material for the production of phenol and acetone.

Both the polymerization and alkylation operations can be carried out over solid phosphoric acid catalyst in units used for production of polymer gasoline. Propylene is also used in increasing quantity to produce the plastic, polypropylene. Total 1970 production of propylene was about 8750 million pounds.

Butylenes (1- and 2-butene and methylpropene) are recovered from the gases produced during petroleum cracking operations. The principal uses of these products are in synthetic rubber. The normal butylenes are further dehydrogenated to butadiene, a major component of GR-S rubber. Butadiene is also used as starting material for adiponitrile, a component of nylon. Isobutylene is used mainly for butyl rubber in which it comprises 98.5% of the raw material. The remaining 1.5% is isoprene which is extracted from steam cracked naphthas. Other products of the butylenes are alcohols, ketones, and esters.

Heptenes are obtained by copolymerization of butylene and propene. These and other higher olefins are also extracted from fractions of cracked gasolines and oils. These compounds are starting materials for production of alcohols by the Oxo process. Isooctyl alcohol is the largest volume product.

Acetylene is currently produced from natural gas by partial oxidation of methane with oxygen and to a lesser extent by pyrolysis of natural gas in a regenerative furnace. Production from these processes is becoming increasingly important. A process for production of acetylene by converting petroleum coke to calcium carbide has also been offered. Acetylene is a versatile basic material for chemicals, including such important products as acrylonitrile, vinyl chloride, and acrylates, although it is gradually being supplanted by the general trend toward lower cost partial oxidation process using ethylene. (See **Acetylene and Compounds**).

Benzene of high purity is recovered from the products of UOP Platforming and other catalytic reforming processes. Starting material is the six-carbon fraction from natural or straight-run gasolines. Benzene for chemical conversion was first produced from petroleum in 1949. Production from this source had risen to the rate of 1280 million gallons per year or 90% of the total production by 1970. The highest volume uses for chemical grade benzene are in styrene, chlorinated intermediates, phenol, insecticides, nylon, aniline, synthetic detergents, and maleic anhydride.

Toluene was first produced commercially from petroleum by Hydroforming, an earlier catalytic reforming process, during World War II. UOP Platforming and other more recently developed processes are now used. The seven-carbon fraction from natural or straight-run gasolines serves as feedstock. Chemical-grade toluene is used mainly for trinitrotoluene (TNT), and saccharin and other chemical products are also derived from it.

Xylenes were first produced for chemical conversion after World War II. They are derived from the eight-carbon fraction of natural and straight-run gasolines processes. The isomers are separated by solvent extraction, distillation, and fractional crystallization. The ortho isomer is converted to phthalic anhydride. The para isomer is oxidized to terephthalic acid for use in Dacron fiber and Mylar film. Isophthalic acid is now derived from the meta isomer.

Naphthenic acids (higher fatty acids containing alkylcyclopentane groups) are obtained from the gas-oil fraction of crude oil by extraction with dilute aqueous caustic soda followed by acidification. They are converted to salts and used as paint driers, lubricants, greases, wood preservatives, and engine oil additives.

Cresylic acids (mixtures of alkylphenols, principally cresols and xylenols) are recovered from heavy cracked petroleum distillates by extraction with caustic soda. They are used principally in plastics and plasticizers.

The line between refining processes and chemical processes continues to blur as refiners are forced to create different yield patterns from the same barrel of crude oil. The trend to eliminate gasoline additives will cut severely into the petrochemical feedstock availability for a time.

Almost all petrochemicals have their origin in feedstocks from petroleum or natural gas. This will continue indefinitely until shortages force our technology to develop economical substitutes from coal and shale oil.

Douglas L. Allen

Cross-references: *Gasoline; Natural Gas.*

PETROLEUM

An essential natural resource, petroleum is an intriguing mixture of chemical compounds. Although it contains small amounts of sulfur, nitrogen, and oxygen compounds, petroleum is largely hydrocarbon material, with molecules ranging in size from 1 to 50 or more carbon atoms and comprising a diversity of molecular types, as paraffins, cycloparaffins, (or naphthenes), and aromatics (including benzene, toluene, naphthalene, and related compounds). Because petroleum contains many hydrocarbon compounds which are of themselves important chemicals of commerce, and because commercial chemicals derived from petroleum are increasing greatly in number and importance, petroleum has become known as nature's storehouse of chemicals.

Petroleum was found as an oily liquid exuding from the surface of the earth in tar springs, or on shores of lakes, or rising from springs beneath the beds of rivers. On the surface of water, petroleum appears as a thin film, with rainbow colors, or as dark globules that can be separated mechanically. In these forms, and in these limited quantities, petroleum was found in many places on earth—notably Rumania, Persia (now Iran), Italy, Trinidad, Cuba, India, and later the United States. In Western Pennsylvania, an enterprising businessman by the name of Kier put up petroleum in bottles, labeled Kier's Rock Oil, and sold it as a medicine.

In 1859, as the result of investigations of Professor Benjamin Silliman of Yale, the famous Drake well was completed, near Titusville, Pa. It was the first mechanically drilled petroleum well. This well, drilled to a depth of 69½ feet, produced about 25 barrels of crude oil per day—in those days an extremely large production. But the production

of petroleum in the continental United States since then has grown to astonishing proportions, as indicated by the following figures, in millions of barrels per year: 0.002 in 1860; 26 in 1880; 64 in 1900; 443 in 1920; 1300 in 1940; 2472 in 1960; and 3,327 in 1970. Since 1880, the United States production has been about 70 per cent of the world production.

Nonetheless, the proved reserves of crude oil in the United States have also increased to large proportions, being as follows, in billions of barrels; 3 in 1900; 7 in 1920; 19 in 1940; 32 in 1960; and 39 in 1970. For the continental United States, the ratio of proved reserves to the annual production has decreased somewhat over the years, being 21 in 1910, 16 in 1920, 15 in 1930, 14 in 1940, 13 in 1950, and 12 in 1970. It is seen from these figures that the ability of man to locate and recover petroleum in the United States is not keeping up with consumption.

Frederick D. Rossini

Cross-references: *Cracking, Catalysis.*

pH

The pH value of an aqueous solution is a number describing its acidity or alkalinity. The usual range of pH is from about 1 for 0.1 N HCl to about 13 for 0.1 N NaOH. The pH of a neutral solution is 7.2 at 15°, 7.0 at 25°, and 6.8 at 35°C.

The pH scale was first used in 1909 by Sørensen. He defined pH as the negative logarithm (base 10) of the concentration of hydrogen ions (equivalents per liter), and he also described an electrometric method for the measurement of pH. Modern chemists have abandoned Sørensen's definition but have retained, in all essentials, his method of measurement.

The approved definition of pH is an operational one.[1] The electromotive force, E_x, of the cell

H$_2$; solution X: KCl (saturated): reference electrode

and the electromotive force E_s of the cell

H$_2$; solution S: KCl (saturated): reference electrode

are measured with the same electrodes at the same temperature. The hydrogen electrode is now usually replaced by a glass electrode. The difference in pH between the unknown solution, X, and the standard solution, S, is defined by the equation

$$pH_x - pH_s = (E_x - E_s)/k$$

in which k is 2.3026 RT/F or 0.05916 for 25°C if E_x and E_s are expressed in volts.

To complete the definition of pH, values of pH$_s$ have been assigned to standard solutions. The National Bureau of Standards has recommended five solutions as primary standards.[2] One of these, 0.05M potassium hydrogen phthalate, has been adopted as the primary standard of pH in Great Britain and in Japan. For this solution pH$_s$ is 4.00 at 15°C, 4.01 at 25°C, and 4.03 at 40°C.[3]

An interpretation of pH is possible in special cases. The values assigned to the standards have been chosen in such a way as to make pH_s equal to $-\log C_H f_1$, where f_1 is an activity coefficient practically equal to that of sodium chloride. Approximate values of $-\log f_1$ are 0.04, 0.06, 0.09, and 0.11 for ionic strengths of 0.01, 0.02, 0.05, and 0.10, respectively. For solutions of ionic strength not over 0.1 and pH between about 2 and 12, these values may be employed in calculations involving ionic equilibria.

For example, the pH of a "standard acetate" buffer (0.1 M $HC_2H_3O_2$, 0.1 M $NaC_2H_3O_2$) is found to be 4.65. It follows that $-\log C_H$ is 4.65–0.11, or 4.54. For a 1:1 buffer the pH is equal to the pK' of the equation

$$pH = pK' + \log C_B/C_A$$

in which C_A is the concentration of the buffer acid and C_B is that of its conjugate base. The apparent constant, K', is not equal to the true constant, K, of the buffer acid because K' lacks a ratio of activity coefficients. In this case it may be assumed that $K = K'f_1$ or $pK = pK' - \log f_1$. On this bases pK (or $-\log K$) would be equal to 4.65 + 0.11, or 4.76, which is the accepted value.

A similar calculation may be made for a phosphate buffer in which the acid is $H_2PO_4^-$ and the base is HPO_4^{--}. In this case K_2 is equal to $K_2'f_2/f_1$ where f_2 is the activity coefficient of the bivalent anion. From the Debye-Hückel theory, it follows that $\log f_2 = 4 \log f_1$, and accordingly we have the relation

$$pK_2 = pK_2' - 3 \log f_1$$

For a buffer of 0.025 M KH_2PO_4 in 0.025 M Na_2HPO_4, the pH is 6.86 and the ionic strength 0.1. Accordingly, the value of pK_2 should be 6.86 + 0.33, or 7.19. The accepted value is 7.20. Again, the value of $-\log C_H$ is equal to $pH + \log f_1$, being 6.75 for this buffer.

Although the original definition of pH, in terms of the hydrogen ion concentration, has been modified, the following table of *approximate* relationship between pH and hydrogen or hydroxyl ion concentration is widely applicable in biological work:

	pH Value	Approximate Number of Times H^+ or OH^- Concentration Exceeds That of Pure Water
Acid side (excess of H^+ ions)	1	1,000,000
	2	100,000
	3	10,000
	4	1,000
	5	100
	6	10
Neutrality ⟶	7	
Alkaline side (excess of OH^- ions)	8	10
	9	100
	10	10,000
	12	100,000
	13	1,000,000

DAVID I. HITCHCOCK

References

1. Bates, R. G., "Electrometric pH Determination," New York, John Wiley & Sons, 1954.
2. Bates, R. G., "Revised Standard Values for pH Measurements from 0 to 95°C," *J. Res. Natl. Bur., Std.,* **66A,** 179–184 (1962).
3. Bates, R. G., and Guggenheim, E. A., "Report on the Standardization of pH and Related Terminology," *Pure Appl. Chem.,* **1,** 163–168 (1960).

PHARMACEUTICALS.

A pharmaceutical is a preparation of one or more medically useful compounds in a dosage form suitable for administration for medical purposes. Modern pharmaceuticals can be classified by the source of the pharmacologically active substance or by the dosage form in which it is administered.

(1) **Synthetic.** Thousands of novel synthetic organic compounds are prepared and tested annually for therapeutic value. Some rationale can be applied to the modification of a chemical structure to enhance a pharmacological activity already known to be present. However, the search for novel structures exhibiting a useful pharmacological activity remains largely based on the blind screening of large numbers of new compounds in small animal or *in vitro* test systems. In spite of the empirical nature of such an approach, many promising new compounds are detected by such primary screens and are then subjected to more extensive and sophisticated secondary testing. Such additional testing usually reveals unwanted side effects, toxicities, or an activity no better than a pharmaceutical already in use in medical practice. Only an extremely small number of these new compounds meet the rigorous criteria for a useful new drug substance. These few candidates are introduced into medical practice as new pharmaceuticals. Examples of synthetic compounds having important medical uses are analgesics (aspirin, meperidine, propoxyphene); local anesthetics (procaine, lidocaine); antihistamines (diphenhydramine, chlorpheniramine); ataractics (chlordiazepoxide, meprobamate, phenothiazines); diuretics (chlorothiazide); sedatives (barbiturates); and antidepressants (amphetamine, amitriptyline).

(2) **Natural.** Many useful pharmaceuticals are derived directly or indirectly from plant or animal sources. In this case the compound of interest is either so complex that it has resisted synthetic efforts or is economically impractical to synthesize in the face of large natural supplies. With natural products the problem is often one of isolation and purification of a small amount of a complex and unstable substance from plant or animal tissue. Sometimes an active derivative can be synthesized from a readily available but medically useless substance. Examples of naturally occurring compounds or derivatives in use as pharmaceuticals are peptide hormones (insulin, glucagon); alkaloids of widely varying pharmacological action (morphine and codeine as analgesics, reserpine as a hypotensive or ataractic); anticoagulants (heparin); and cardiotonic glycoside (digitoxin).

(3) **Microbial Fermentation Products.** The recent growth in importance of fermentation products in

the pharmaceutical industry and in human affairs is almost without parallel. Much research effort is expended in the screening for new antibiotic substances from unique microbial sources as well as in strain selection to improve production yields of antibiotics already in use or showing promise. Antibiotics are produced mostly by large scale fermentation procedures. Often an antibiotic having low potency or limited use can be greatly improved by the synthesis of derivatives having increased antibacterial potency, a broader antibacterial spectrum, or special properties useful in the treatment of specific diseases. Certain steps in the synthesis of some of the steroid hormones can best be carried out enzymatically in microbial fermentations; the metabolism of a steroidal raw material by the organism produces a desired intermediate. Examples of antibiotics produced by fermentation are natural and semisynthetic penicillins, tetracyclines, cephalosporins, erythromycin, griseofulvin, and neomycin. Examples of other important fermentation products are vitamin B_{12}, riboflavin, and certain steroid hormone intermediates.

(4) **Biologicals.** Biologicals are complex natural substances, often high molecular weight biopolymers, used to diagnose, treat, or impart immunity to a specific disease. The origins of such substances are varied since, for example, tetanus and diptheria antitoxin used in the treatment of the diseases are produced from the blood of immune horses or cattle while tetanus and diphtheria toxoid used to impart immunity are produced from chemically inactivated or modified toxin from growing virulent microorganisms. Measles, polio, and influenza vaccines are prepared from virus grown in tissue culture and may consist of killed virus for parenteral use or attenuated live virus for oral administration. Chemical manipulations on many of these preparations are limited because of the generally unstable chemical nature and high degree of biological specificity of the materials.

Pharmaceuticals may also be classified according to product form or dosage form. The dosage form chosen will depend both on the intended medical use of the compound and on the nature of the substance itself. Some drugs are poorly or unreliably absorbed from the gastrointestinal tract making parenteral administration necessary. Other compounds may be unstable in gastric acidity and require the protection of an enteric coating insoluble in the stomach but removed in the intestine. Still other drugs must be administered intramuscularly or intravenously because precise dosage or immediate onset of action is required. For obvious reasons of convenience and safety, oral administration is the preferred route of administration when possible. For most adults a capsule or compressed tablet is adequate. For children, however, liquids are usually a more acceptable dosage form and for this reason it is often necessary to synthesize a stable, soluble form of the drug, or to formulate a palatable, nontoxic and stable medium for suspending an insoluble drug. Often it is necessary to synthesize a tasteless or odorless derivative of a drug without sacrificing physiological activity. Some drugs are applied topically in ointments, creams, or lotions, while others are inhaled.

Governmental Regulations and Control. Several governmental agencies share responsibility for the enforcement of federal statutes dealing with the testing, approval, manufacture, and distribution of pharmaceuticals and biologicals for human and veterinary use. The Food, Drug, and Cosmetic Act and its many amendments is enforced by the Food and Drug Administration (F.D.A.), a branch of the United States Department of Health, Education, and Welfare. This agency not only has the responsibility for approving the clinical trial and marketing of a new drug, but also must approve manufacturing methods and sanitation, purity, safety, stability of the finished product, label and advertising claims, and, in many cases, therapeutic efficacy. In a few cases, particularly with antibiotics and insulin, this agency certifies each lot of a manufacturers drug as acceptable for sale based on uniform standards set by the F.D.A. and accepted throughout the pharmaceutical industry.

The Division of Biologics Standards, a separate division within the F.D.A., has authority to control the manufacture and sale of biological products (vaccines, antitoxins, blood products, and diagnostic agents). When standards of facilities, personnel and control are met, a manufacturer may be licensed to produce biologicals. Each lot must be accepted by the Division of Biologics Standards as meeting appropriate standards of safety and efficacy.

The Bureal of Narcotics and Dangerous Drugs (B.N.D.D.), a branch of the United States Department of Justice, controls the manufacture and distribution of narcotics and many other stimulant, depressant, and habit-forming drugs whose pharmacological action makes them objects of abuse. While all of these particular regulatory agencies apply to legal control only in the United States of America, almost all developed countries throughout the world have comparable bureaus for similar purposes.

Research and Development. Of the more than one-half billion dollars spent annually in the research and development of new pharmaceuticals, almost two-thirds of the expenditures are for anti-infective agents, antineoplastic agents, endocrine, central nervous system, and cardiovascular drugs.

A new pharmaceutical is developed through the joint efforts of many different scientific disciplines. A single compound from a promising series detected by primary screens and confirmed by secondary testing will be selected for development into a medicinal product. Years of effort by organic, analytical, and physical chemists, pharmacologists, toxicologists, and biochemists will then be required to generate the necessary information on toxicity, metabolism, side effects, and efficacy. At any time research may show the substance to be unsuitable for further development. The criteria for pharmaceuticals for human use are of necessity so exceedingly strict that most new compounds fail to survive these years. However, if satisfactory in all respects, pharmaceutical chemists will develop dosage forms suitable for clinical testing. If extensive clinical trials are successful, pilot plant and production engineers will scale up the laboratory processes to produce the finished pharmaceutical in large quantities. Marketing and distribution specialists will make the pharmaceutical and recommendations for its use available to the medical profession. Only then through the daily use of the preparation by thousands of physicians in millions of patients in an almost infinite variety of clinical situations will it be finally known whether the pharmaceutical that has shown so much promise really has an important place in the practice of medicine.

WALTER E. WRIGHT

Reference

"Remingtons Pharmaceutical Sciences," 14th Ed., Mack Publishing Co., Easton, Pa., 1970.

PHARMACOLOGY

Pharmacology has traditionally been considered, or at least so stated in text books, to consist of six subordinate disciplines: pharmacognosy, pharmacy, toxicology, posology, pharmacodynamics and pharmacotherapeutics. A contemporary view, however, is that pharmacology consists of the last four disciplines, pharmacognosy and pharmacy being identified as companion sciences, each contributing to the "science of drugs" which is a general definition of pharmacology (i.e., pharmakos = drug; logos = discourse or science).

Pharmacognosy, once called the "science of crude drugs," is more appropriately the "science of natural products" from animal or plant origins. Natural drug products originate from animal tissues (e.g., anticoagulants and the antihemophiliac factor) or from microorganisms (e.g., antibiotics and antitumor agents) as well as from plant sources. Pharmacognosists are no longer satisfied with crude plant extracts but, once biological activity has been confirmed by a pharmacologist, will attempt to isolate, purify and identify the active ingredients. The pharmacognosist is thus a trained chemist as well as an expert on the growth and development of medicinal plants. This interrelationship between medicinal chemistry and pharmacognosy explains the increasingly close cooperation now evidenced between these disciplines in the United States, although such cooperation has long been evident elsewhere in the world.

The trend in *pharmacy* from that of preparing, compounding, and dispensing medicines to that of largely dispensing is abundantly evident. A less obvious trend is to treat the patient as a complete medical entity: that is, total patient care means concern for the interactions of the patient's physical and psychological states with his drug regimen and with his environment. To perform this enlarged professional role, currently more visible in the hospital than in community practice, pharmacy curricula have been revised to include greater emphasis on the biological and social sciences, particularly pathology, clinical pharmacology and psychology. A new health team relationship is growing among the physician, nurse and pharmacist in which the pharmacist is emerging as the "pharmacologist," that is, the pharmacist knows best how to obtain maximum benefit from drugs administered in the presence of the diagnosed physical and psychological states.

Toxicology is still the science of poisons, although the supporting instrumentation now involves more elaborate and exacting techniques. A poison may be defined as a chemical substance which, when introduced into the body in a relatively small quantity, is capable of producing a harmful or toxic effect on the physical and/or psychological states of that species. A toxic response may be as minimal as a few minor side effects such as nausea and dry mouth, or as terminal as death. By the same token drug side effects need not always be toxic but in certain instances can be therapeutic objectives. For instance, accidental ingestion of belladonna plants usually induces dry mouth, dilated pupils and constipation, along with other symptoms of poisoning. However, for eye ground examinations the belladonna-induced dilated pupil is desired; dysentery and ulcers are relieved by reducing gastrointestinal motility; and in preparation for oral surgery a dry mouth is necessary. Therefore, what may be undesirable in one situation may be therapeutic in another.

Two of the more significant toxicologic advances which have been drawing attention to the significance to total patient care are drug interactions and drug susceptibility rhythms.

Drug Interactions are responsible for many unexpected clinical drug responses. The most frequently cited interactions are those of physicochemical interaction with the dosage form, or with the patient's environment or pathological state, or with concurrently administered drugs. Some common examples of each are: (1) the pH of a dosage form is such that, upon mixing with a solution for an intravenous administration, the drug precipitates out in the bottle; (2) if the solution were injected intravenously it might preferentially bind with plasma proteins and thus be inactivated before reaching its receptors; (3) the presence of pus in a wound inactivates some antibiotics and sulfa drugs; (4) persons with a goitre (i.e., hyperthyroid) may be more sensitive to digitalis preparations; (5) the induction of increased liver oxidation by barbiturates explains the more rapid destruction of concurrently administered drugs such as the antiepileptic diphenylhydantoin or the anticoagulant coumarin; and (6) the interaction of ethanol with sedative-hypnotics such as barbiturates and chloral hydrate, resulting in respiratory depression even to the point of death, has been known for years. Numerous clinical studies are now exhaustively cataloguing the various types of interactions; however, ready access to such information has not yet been achieved.

Drug Susceptibility Rhythms also account for variable clinical drug responses. Several drugs possess periods of peak sensitivity and resistance (i.e., morphine, lidocaine, pentobarbital, etc.). These periodicities or rhythms may be daily, weekly and/or annual depending on the drug, the species and the environment in which the species has been acclimated. Therefore the expected drug response is directly dependent on the time of administration as related to the patient's history. Drug responses may also depend on patho-physiological rhythms: daily rhythm to epileptic seizure thresholds, red cell regeneration cycle, asthmatic respiratory patterns, menstruation, etc. The rapid expansion of drug interaction, drug susceptibility rhythms and toxicologic data has provided considerable stimulation to develop computerized drug information centers.

The *Drug Information Center* is created so as to make pharmacologic and toxicologic data readily available in its most serviceable format, particularly at a moment's notice such as occurs in a hospital emergency room. Hand-sorting and searching of data files were adequate when the data bank was small and emergency demands few in number. However, as the services grew, especially as highly skilled drug information retrieval experts became available, the demands and the essential resource materials exceeded human physical capacities and computerized information retrieval systems were developed to service both the patients' and physicians' needs. This area of intercommunication amongst pharmacist, pharmacologist and physician can only expand, as more drug data banks are established.

The traditional role of pharmacology has changed from almost exclusively a basic medical science to an applied health team clinical and basic science. However, the basic pharmacologic principles remain the same: (1) a drug cannot initiate a new response but can only alter the rate of a pre-existing reaction or response; (2) for each drug there is an appropriate receptor to which the drug has affinity (binding ability) and, if an active drug, also efficacy (ability to produce a response); and (3) for each dose of a drug there is an appropriate response (the characteristic dose-response relationship so frequently referred to by pharmacologists). Therefore, a drug, to be clinically effective, must be able to (1) arrive at its "own" receptor in an active form (neither bound nor metabolically inactivated) (2) bind with that receptor (affinity); and (3) produce its characteristic response (efficacy), which is either an increase or a decrease in the rate of that tissue response (e.g., carcinogenesis or carcinostasis).

Any contemporary definition of pharmacology must differentiate between pharmacodynamics and pharmacotherapeutics since both relate to the mechanism of drug actions on living tissues; the former refers to actions on normal tissues whereas the latter is on diseased tissues. Numerous significant contributions to pharmacotherapeutics have been made by the clinical pharmacologist studying the mechanisms of drug actions in the clinical setting and by the clinical pharmacist determining optimal drug formulations for specific clinical situations. Enabling the drug to reach its receptor site within the patient in a maximal concentration in an active form is the role of the clinical pharmacist. Explaining the nature of the anticipated response and the mechanism by which that or any unexpected response is produced is the role of the clinical pharmacologist, who by training is also a physician.

Two other areas of interest for the chemist and pharmacist should be bioassay and drug screening. *Drug Screening* is the study of new compounds for biological activity of potential therapeutic merit. *Bioassay* is the quantitative determination of the doses of a drug required to produce therapeutic and/or toxic responses in a variety of animals. Bioassay also involves the biological standardization of chemicals for which no physical or chemical

analytical techniques exist. New drug development depends on knowledge of these areas and therefore, becomes essential to the chemist, who must synthesize or isolate new therapeutic agents, and to the pharmacist, who bases his professional judgment on knowledge of the quality and quantity of the drug in question.

An exciting neuropharmacologic accomplishment has been the rapid accumulation of biochemical data on the mechanisms by which drugs affect the central nervous system, particularly in the treatment of Parkinson's and several hereditary behavioral diseases. Notable biochemical advances have also been made with the treatment of leukemias, although more limited success has been achieved with the hard cell tumors.

Pharmacologists cannot discharge their responsibilities as scientists or instructors unless they are well versed in several areas of chemistry, including analytical, as well as physiology, biochemistry, anatomy, microbiology and pathology. In like manner most chemists and certainly medicinal chemists, should find considerable value in studying pharmacology, particularly chemical pharmacology which is the study of the relationships between chemical structure and biological activity. Analytical chemists should find the recent work of biochemical pharmacologists most helpful, particularly in the area of fluorescent and immunoassay techniques.

RALPH W. MORRIS

PHASE RULE

The phase rule is a general equation $F = n - r - 2$, stating the conditions of thermodynamic equilibrium in a system of chemical reactants. The number of degrees of freedom or variance (F) allowed in a given heterogeneous system may be examined by analysis or observation and plotted on a graph by proper choice of the components (n), the phases (r), and the independently variable factors of temperature and pressure.

Josiah Willard Gibbs propounded the rule about 1877 and H. W. B. Roozeboom about 1890 began pioneering in specific cases. This rule has been important in the development of metallurgy, the exploitation of salt deposits (e.g., Stassfurt, Germany and Searles Lake, Calif.), and the study of ceramic and mineralogic processes. Concepts recognized by experience with aqueous systems are readily transposed to nonaqueous systems at such temperatures that more than one phase exists, e.g., metal alloys at high temperatures, gases at low temperatures.

A phase is a homogeneous, physically distinct and mechanically separable portion of a system. H_2O has the three phases: water, vapor and ice. Each crystal form present is a phase. The relative amounts of each phase do not affect the equilibrium.

The one-component system, water-vapor-solid, is unary. The components of a system are the smallest number of independently variable constituents by means of which the composition of each phase taking part in the study of equilibrium can be expressed in the form of a chemical equation.

With the components fixed in a system, variance

—the degrees of freedom (F)—depends on the number of phases present. If water vapor alone is present, the system is bivariant since both temperature and pressure can vary within limits without affecting the number of phases; but if a second phase, liquid water, is present, the system is univariant and if either the temperature or pressure of the system in equilibrium is set, the other is automatically fixed as long as a second phase is present. A third phase, ice, makes the system invariant (the triple point), and any change in temperature or pressure, if maintained, results in the disappearance of one phase. Addition of another component forms a binary system, one degree of freedom is added, and the system is univariant until a fourth phase appears and the system becomes invariant (the quadruple point).

Schematic phase equilibrium diagrams outline experimental observation of physical and chemical changes in the system as the conditions of temperature, pressure and composition are varied. For unary systems, the diagram has two dimensions, for binary systems, three, and for ternary systems, four, etc. Binary systems are easily plotted with pressure or temperature constant, while ternary systems may be treated similarly as condensed systems with both constant if the vapor pressure is less than atmospheric. This added restriction reduces the variance by one and, at constant temperature, composition relationships may be plotted on a triangular diagram. Quaternary or quinary systems and ternary systems above atmospheric pressure may be treated by projections of surfaces of thermodynamic stability, but more complex systems require a mathematical approach.

The simple one-component system, water, plotted with rectangular coordinates, i.e., pressure and temperature, shows a variety of concepts which may be extended to more complex systems. Each area in the diagram is a bivariant, one-phase state. Each curve separating the areas is a univariant, two-phase state showing the conditions under which a transition of phase occurs. The fusion curve for the equilibria between the solid phase and the liquid phase, the sublimation curve for solid and vapor, and the vaporization curve for liquid and vapor meet at the triple point. The three distinct phases differ in all properties except chemical potential. The end of the vaporization curve is a singular point, the critical point where liquid and vapor become identical, a restriction which reduces the variance by one so that F becomes zero.

In a binary (or higher) system, a liquidus is a curve representing the composition of the equilibrium liquid phase and a solidus represents the composition of the solid phase. The conjugate vapor phase is represented by a vaporus with tie-lines or conodes to the liquidus or solidus points in equilibrium. A minimum point for the existence of a liquid is the eutectic, sometimes (in an aqueous system) called the cryohydric point. In a ternary system the eutectics of three binary systems initiate curves leading to a ternary eutectic. If two liquids form a miscibility gap in the system, multiple quadruple points are possible. At a peritectic a phase transition occurs at other than a minimum, i.e., one solid melts to another solid and a liquid. The solid may be a compound or a solid solution

(mixed crystals) in which the composition of the solid varies with the relative proportion of components and is shown by the solidus. A congruent melting point is a maximum in its curve, i.e., the solid melts to a liquid of the same composition. Where two phases become identical in composition, an indifferent or critical point exists. Where two conjugate phases become identical, the point may also be called a consolute point.

JOHN H. WILLS

Cross-reference: *Thermodynamics.*

PHENOLIC RESINS

Phenolic resins have been known since the 1870's when Baeyer first investigated the reactions of phenols and aldehydes. However, his findings were not commercially utilized until Dr. L. H. Baekeland disclosed his classic work in 1907. Through his use of high-pressure molding, he provided a solution to the problem of making quick-curing moldings which did not blister or crack. Contemporary with Dr. Baekeland was the work of Lebech and Aylsworth who provided the key to the application of large commercial quantities of phenolic Novolacs by suggesting the use of hexamethylenetetramine as a curing agent. Since that time, and because of their desirable price-to-property relationship, phenolic resins have enjoyed steady growth despite the encroachment into their areas of application by a few thermoplastics and other thermosetting materials.

Although in the pure state phenolic resins are quite weak and brittle, they are highly regarded among the plastic materials as being capable of producing very strong physical bonds with a large variety of materials at very low concentration. Consequently, phenolics have found use in many applications as binders. In addition to the strength they impart as bonding agents in matrixes, phenolic resins also possess resistance to chemical attack by all but the most polar organic solvents. The only inorganic reagents that have a deleterious effect on them are the strongest and most oxidizing of the acids and the strongest bases.

For many years prior to the development of high-temperature thermoplastics and thermosets, such as the polyimides, polysulfones, and epoxies, phenolic molding material dominated the high temperature-resistant market. This emphasizes their ability to resist temperature degradation in the 400–500°F range. Because phenolics were found to possess excellent ablative properties, it has been reported that both the American and Soviet space efforts used them in combination with certain other polymeric compounds in heat shield materials. Phenolics, like other aromatic hydrocarbon-based resins, possess excellent resistance to high-energy radiation degradation.

Unlike most thermoplastics, phenolic resins and moldings are characterized by high flexural modulus and good tensile strength while having relatively low impact resistance. In addition to their good physical strength properties, phenolic resins are used in the manufacture of many electrical devices where high dielectric breakdown strength and electrical resistance, in combination with excellent dimensional stability, are required. Table 1 com-

TABLE 1

	Phenolics	Polypropylene	Epoxy	Alkyd
Specific gravity	1.32–1.45	0.902–0.906	1.6–2.0	1.6–2.4
Tensile strength, psi $\times 10^{-3}$	6.5–10.0	4.3–5.5	10–30	3–9
Compressive strength, psi $\times 10^{-3}$	22–36	1.6–2.25	25–40	12–38
Impact strength, ft-lb/in	0.24–0.60	0.5–2.0	0.5–1.0	0.3–0.5
Hardness (Rockwell)	E64–E95	R85–R110	M100–M110	—
Flexural modulus, psi $\times 10^{-5}$	10–12	2.3–2.7	—	19–21
Dielectric strength, v/mil	200–400	500–660	300–400	—
Dielectric constant (60 cycles)	5.0–13.0	2.2–2.6	3.5–5.1	5.1–7.5

pares the properties of phenolic molding material with several of its more important contemporary plastics.

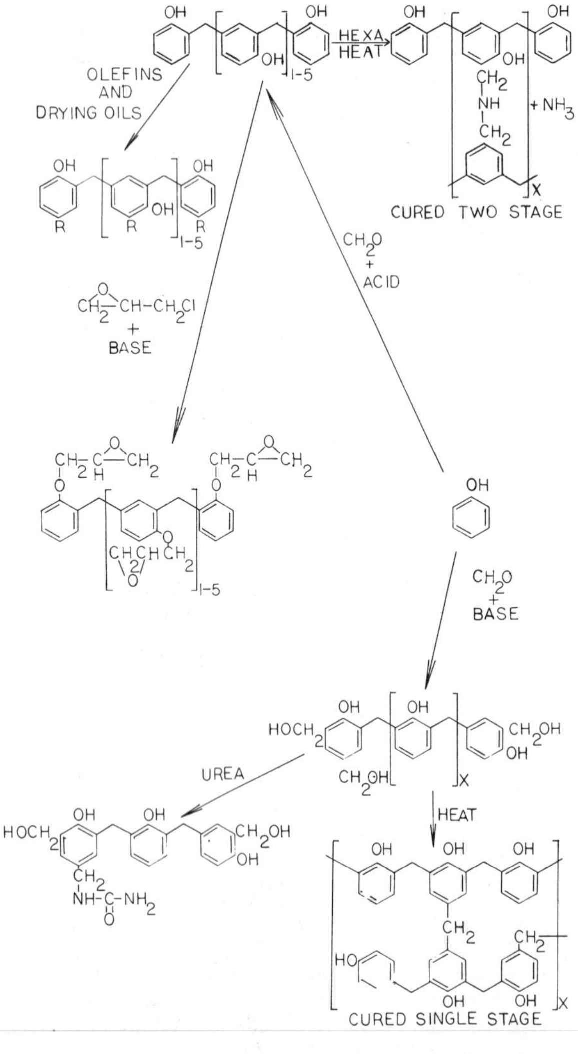

FIG. 1.

Chemistry of Phenolics. Phenolic compounds are capable of chemically combining with a large number of aldehydes and other compounds to yield an almost infinite spectrum of modified polymers. However, the reaction of a phenol with an aldehyde (most commonly encountered is that between phenol and formaldehyde) leads to the formation of only two classes of phenolic resins. These are Novolacs and resols. In general, these two classes of resins may be differentiated by the fact that Novolacs are prepared with an acid catalyst and substantially less than one mole of aldehyde per mole of phenol and require the addition of a curing catalyst to become thermosetting; while resols, or single-stage resins as they are commonly called, are prepared with from 1 to 3 moles of aldehyde per mole of phenol and employ a basic condensation catalyst, and are inherently thermosetting.

Novolacs. The aldehyde content of Novolac resins is insufficient to render the resin thermosetting, hence, they are true thermoplastics provided that a curing agent is not added. Novolacs may be stored indefinitely in the pulverized state at moderate temperatures even mixed with curing agent.

The final cure speed of simple Novolacs may be accurately controlled by the use of the proper condensation catalyst in the initial phase of the reaction. Structurally, a Novolac consists of a series of phenol nuclei joined by methylene ($-\overset{\text{H}}{\underset{\text{H}}{\text{C}}}-$) links at the o and p positions. Only two of the three possible o and p positions on each ring within the polymeric chain are substituted with a methylene group. Only one position on each of the two terminal phenol groups is substituted with a methylene group, hence, two positions on each terminal ring and one on each internal ring is available for future reactions, including the curing reactions. When the unsubstituted positions are predominantly the para positions, very fast-curing resins result. Slower-setting resins are obtained when the unoccupied positions are the ortho positions. When hexa is used as the hardener at approximately the 10% level, the reactive sites are joined by a $-\overset{\text{H}}{\underset{\text{H}}{\text{C}}}-\text{N}-\overset{\text{H}}{\underset{\text{H}}{\text{C}}}-$ linkage. One mole of ammonia is liberated for approximately every three of the above links formed.

Unaltered Novolacs and two-stage resins* find application in grinding wheel bonding, molding material, brake linings and clutch faces, foundry sand binding, premix and wood fiber bonding, thermal insulation. Modified phenolics are found in adhesives, coatings, and aerospace applications.

Resols. On the other hand, resols contain sufficient aldehyde to make them thermosetting without a curing agent. Consequently, they have only finite storage stability and care must be exercised to minimize both the length and the temperature of storage. The high aldehyde-to-phenol ratios used in the preparation of resols insure that a high percentage of the reactive *o* and *p* positions are utilized in either methylene links or are substituted by a hydroxymethyl group. It is these hydroxymethyl groups which function as crosslinking sites in the final curing reaction. In phenolformaldehyde resols, the ratio of methylene groups to hydroxymethyl groups is an important factor in determining the solubility of these resins. Low ratios insure water solubility in almost infinite proportions, while resins with high ratios can be dissolved in only low molecular weight alcohols, ketones, ethers, and esters. Solubility of resols in nonpolar solvents, such as hydrocarbons, is always very low.

In addition to acting as a crosslinking site, the hydroxymethyl group and unsubstituted *o* and *p* positions may be used as reactive sites to join numerous other compounds to the phenolic polymer. These modified resins often possess many properties normally not attributable to phenolics in the unaltered state.

Resols find application as impregnating resins, in laminating paper, cloth, glass and asbestos; as a pickup agent in grinding wheels, exterior and marine plywood, premix and granular molding material, adhesives, wood waste and particle board manufacture, and in coatings.

The final curing step of both classes of phenolics is accomplished by exposing the resin or the resin containing matrix to temperatures in the 190–450°F range for an appropriate length of time to render the resin infusible. In most applications high pressure is also applied concurrently with the heating cycle to eliminate blistering, which would normally occur if the trapped gases (ammonia in the case of Novolacs, and water in the case of resols) generated in the cure were allowed to escape unrestrained.

Comparative infrared analysis of the two classes of resins in their pure state show that only resols have strong absorptions in the 1,000 and 880 cm^{-1} range. As Novolacs require a curing agent, their presence may be inferred by a strong, sharp absorption at 510 cm^{-1} which is indicative of the most common curing agent used, hexamethylenetetramine. Unfortunately, hexa also has strong absorptions at or near 1,000 and 880 cm^{-1} which makes differentiation between resols and two-stage resins, by infrared spectroscopy, possible only for the experienced. In the cured state, it is very difficult to determine the class identity of an unknown sample.

PHILLIP A. WAITKUS

* By convention, two-stage resins are defined as a mixture of Novolac and curing agent, which is capable of thermosetting.

PHOSPHORESCENCE

Phosphorescence is the emission of electromagnetic radiation resulting from excitation of the phosphorescent material and occurring after such excitation. The excitation may be by electromagnetic radiation, by particle bombardment or by chemical or mechanical action. Phosphorescence is distinguished from fluorescence by the fact that there is an appreciable time lag between the excitation and the emission of phosphorescent radiation. There is no general agreement amongst authorities as to what is an "appreciable" time lag, some say 10^{-3} second, some 10^{-8} second. A more fundamental, but perhaps less practically useful distinction can be made on the basis that phosphorescence acts like radiation from a forbidden transition and has an appropriate time lag, whereas fluorescence has the characteristics of permitted radiation.

Materials which exhibit phosphorescence are termed *phosphors*. They may be conveniently considered in two groups: the mineral type and the molecular type. In the mineral type the emission of radiation is associated with energy levels, frequently called traps, which are produced in the assembly of molecules, rather than in the individual molecules. The molecular type comprises those in which the energy levels involved in the emission are characteristic of the individual molecules.

The mineral phosphors are normally crystalline. The existence of traps in which electrons displaced from normally occupied levels in the crystal may be held for a time and then released to return to the normal levels provides a mechanism by which delayed radiation, i.e., phosphorescence, may be produced. Such electron traps may be produced by the addition of small amounts of extraneous elements, called activators, to pure crystals. A very common example is zinc sulfide with copper as an activator. Mineral phosphors occur in nature, as the name implies, and also are readily synthesized in the laboratory. A wide variety of such phosphors is known. The most common ones are the alkalies and alkaline earths combined as halides, oxides, sulfides, silicates, tungstates, or borates, with activators of copper, silver, gold, or transition elements such as manganese, vanadium, etc. or rare earths.

To produce phosphorescence in this type of phosphor, the excitation process causes the electrons to be raised in energy from normally occupied energy states to unstable higher energy levels from which they then find their way into the electron traps from which they are slowly released. The return to the normal energy levels causes radiation. This phosphorescent radiation may have a half-life of milliseconds to hours. This release from the traps may be stimulated, i.e., accelerated, by increased temperature or by radiation with appropriate wavelengths or it may be quenched, i.e., retarded by radiation of other wave lengths.

The phosporescence emitted by mineral phosphors is generally spread over a wide wave length band with maxima at one or more wave lengths. Different phosphors will have different maxima, and these may be anywhere in the spectrum from ultraviolet through the visible range and into the infrared. The spectral distribution of radiation and the amount of radiation depend on the composition of the crystalline material, on the kind and amount of activators, on the presence of impurities, on the physical condition of the phosphor, such as the temperature, on the size of crystalline particles, on the vehicle, as in paints, in which the phosphor is held, and in some cases on the kind and amount of excitation.

Phosphors of this type are applied principally in "fluorescent" lamps (both fluorescence and phosphorescence are involved in the light produced) in illuminated clock and instrument dials and in cathode ray tubes including television screens.

Molecular phosphorescence occurs when an electron which has been raised from the normally occupied levels of a molecule, i.e., ground states, by excitation, finds its way into a metastable state and subsequently makes a forbidden, (i.e., forbidden by selection rules) radiation-producing transition back to the ground states. Many organic molecules exhibit this property. Usually the phenomenon has been studied by diluting the compound with appropriate solvents, then establishing a glassy supercooled structure by refrigerating to liquid air temperatures. Refrigeration tends to narrow the emission bands and to increase the amount of radiation by increasing the population of the metastable state. Both of these effects make the emitted radiation easier to detect and measure.

The radiation of molecular phosphorescence appears in wave length bands whose location in the spectrum and whose structure are determined by the energy levels of the individual molecule, usually only moderately modified by the environment of the molecule. These bands are generally narrower than those due to mineral type phosphorescence and they have a more pronounced structure of maxima and minima of radiation. This structure results from the super-position of vibrational energy on the electronic energy of the transition. Rotational structure is not usually observed.

In addition to the spectral distribution of radiation, molecular phosphorescence is characterized by a life time and by an efficiency of conversion of energy from excitation to phosphorescence. Each of these three features is specific to the particular molecule involved, and thus any or all characteristics may be used as the basis for qualitative and quantitative observations. The spectral distribution may be observed with low to medium resolving power apparatus. The life time (time required for the phosphorescent radiation to decay to $1/e^{th}$ of its first observed value) may be observed by electronic techniques or by a rotating shutter technique. The efficiency of conversion of energy may be observed by photometric procedures in which the energy absorbed by the phosphor is compared with the energy emitted as phosphorescence.

The first observation of *laser* action, that is, strong stimulated emission of radiation, at visible wave lengths was observed by making use of the long lifetime of the excited levels of the chromium ions in ruby material that give ruby its phosphorescent properties. Thus, although subsequent experience has taught us that long duration excitation levels are not essential to laser action at optical frequencies, the first successful steps in this new

exciting field were based on long-known facts about phosphorescence.

Fred W. Paul

Cross-reference: *Luminescence.*

References

"Fluorescence and Phosphorescence," Pringsheim, P., Interscience Publishers, Inc., 1949.

"Molecular Luminescence," Lim, E. C., W. A. Benjamin, Inc., 1969.

"Luminescence of Organic and Inorganic Materials," Kallmann, H. P., and Spruch, G. M., Editors, John Wiley & Sons, Inc., 1962.

"Fluorescence Analysis in Ultraviolet Light," Radley, J. A., and Grant, J., Chapman & Hall, Ltd., Fourth Edition, 1959.

PHOSPHORS

Phosphors are materials which emit ultraviolet (UV), visible or infrared (IR) radiation when exposed to, i.e., excited by, x-rays, cathode rays, UV, visible or IR radiation or when placed in an electric field or suitably connected to a battery. This emission is not due to incandescence as occurs in an electric light bulb and therefore is called "cold light" to distinguish the two processes. While many phosphors occur in nature, the most important ones are synthetic, not having a natural counterpart. Phosphors may be either organic or inorganic but the most interesting and commercially important ones are inorganic.

Phosphors are used in a wide variety of applications: radar; color and black-and-white television (cathode-ray excitation); the viewing screen in fluoroscopes (x-ray excitation); the coating of a fluorescent lamp (shortwave UV excitation); high-pressure color-corrected mercury lamps and fluorescent highway signs (longwave UV excitation); "safety" or "nite lites" which operate for about one cent a year (electric field excitation); and "solid state" lamps, i.e., luminescent diodes "excited" by the injection of suitable electric charge carriers from a battery or other power supply.

Phosphor Composition and Preparation. Conventional phosphors, that is, other than luminescent diodes, are either self-activated or impurity-activated. The former consist of two and sometimes three components: (a) the host which is the major component, (b) one or more activators (acceptors) and (c) a coactivator or charge compensator (donor) especially if the valence of the activator deviates from the valence of the host ion it replaces in the crystal lattice. The self-activated phosphors are not intentionally activated although they may contain a coactivator. The luminescent center (activator) in these phosphors appears to be a specific atomic group or individual atoms. Examples of self-activated phosphors and the presumed luminescent centers are:

Host	Luminescent Center
MWO_4 (M = Zn, Ca, Mg, Cd)	WO_4^{-2}
$MgMoO_4$, $CaMoO_4$	MoO_4^{-2}
$UO_2(C_2H_3O_2)_2$	UO_2^{+2}
ZnS	Zn vacancy
Rare earth salts	Rare earth ion

The impurity activated phosphors may or may not be photoconductive. The sulfides of the alkaline earths (Ca, Sr and Ba), Zn and Cd are photoconductive; oxide phosphors such as Y_2O_3, La_2O_3, Al_2O_3, $2.5Ca_3(PO_4)\cdot0.5CaF_2$, Zn_2SiO_4, Y_2O_2S and La_2O_2S are not photoconductive. The photoconductive phosphors are much more sensitive to impurities than the nonphotoconductive ones so that more care must be exercised in their manufacture to avoid accidental contamination. Also, the photoconductive phosphors are activated by lower concentrations of added impurities (activators) than the nonphotoconductive phosphors, 0.01 to 0.5 wt % vs. 0.1 to 5 wt % respectively. At concentrations much above these, the luminescent output decreases, i.e., is concentration quenched. While many transition elements can act as activators, the better ones are Cu, Ag, Mn and Eu^{+2} in the photoconductive phosphors and Tl, Pb, Sb, Mn and the trivalent rare earths in the nonphotoconductive phosphors. In the last several years, Eu^{+3} activated oxide and oxysulfide of Y and La have become very important as the red component in color television; the blue and green components are silver activated ZnS and (Zn,Cd)S, respectively. Several paramagnetic transition elements act as "killers," decreasing the fluorescence and/or phosphorescence at very low concentrations, ca. 1–10 ppm. In the ZnS phosphors, Fe, Co and Ni are particularly bad.

The mechanism of charge compensation can be illustrated by using silver activated ZnS phosphor. An Ag^+ ion enters the ZnS lattice by replacing one Zn^{+2} ion. This results in an excess of negative charge so that the phosphor no longer is electrically neutral. Neutrality can be restored by using a charge compensator according to the following schemes:

(a) $Ag^+ + M^{+3}$ replaces $2Zn^{+2}$ (Total charge +4)
(b) $2Ag^+ + M^{+4}$ replaces $3Zn^{+2}$ (Total charge +6)
(c) $Ag^+ + X^-$ replaces $Zn^{+2} + S^{-2}$

(M^{+3} may be Al, Ga, or In, M^{+4} Ti or Sn, X^- Cl, Br or I.)
In the case of the Eu^{+3} activated Y_2O_3, for example, no charge compensation is needed since Eu^{+3} replaces Y^{+3}.

The inorganic phosphors are prepared by intimately blending the proper amounts of the required components and heating at temperatures generally greater than 1000°C. The firing is necessary to promote diffusion of the activator into the host, creating the proper crystal structure as well as the required host compound.

Optical Properties. After excitation, a phosphor may exhibit fluorescence with or without phosphorescence. Fluorescence is the emission of radiation obtained while the phosphor is being excited; phosphorescence is the emission that occurs after the cessation of excitation. The arbitrary limit for the beginning of phosphorescence is any emission which lasts longer than 10^{-8} seconds after cessation of excitation. (10^{-8} seconds is the relaxation time of an excited sodium atom.) Phosphorescence may last from a fraction of a second to several days.

The emission from most UV or visible excited phosphors follows Stokes' law which states that the emitted radiation is at a longer wave length, lower energy, than the exciting radiation. Those IR excited phosphors emitting visible radiation are called anti-Stokes emitters since the emitted energy is apparently higher than the exciting energy.

The nature of the absorption and emission spectra as well as the phosphorescence decay gives an indication of the electronic processes that are occurring in the phosphor. In addition to being Stokes or anti-Stokes emitters, the spectra can be broad or narrow band(s) and/or lines. Their spectral location (wave length) and intensity largely determine the phosphor's application.

The emission and absorption spectra of most phosphors other than those activated by the rare earths consist of one or more bands. The trivalent rare earths give a series of lines of varying intensities whose wave length positions are relatively independent of the host crystal. This latter occurs because the electronic transitions responsible for the absorption and emission are between discrete 4f energy levels and are only slightly influenced by the lattice rare earth crystal fields. Eu^{+2} and the nonrare earth activators, however, have electronic transitions to levels that more strongly react with these fields and thus their wave length varies with a change in host lattice.

Some phosphors are very efficient emitters under one exciting wave length but not another desired wave length. By introducing a sensitizing activator which is excited efficiently by the desired wave length, it may be possible to transfer energy from the sensitizer to, i.e., excite, the emitting activator. This happens in some of the fluorescent lamp phosphors where Sb or Bi are the sensitizers in Mn activated alkaline earth halophosphate phosphors, i.e., $3Ca_3(PO_4)_2\cdot CaF_2$.

Luminescence Model. The Schon-Klasens model, which is based on the band model, is most often used to illustrate the luminescence mechanism. This is illustrated in Fig. 1.

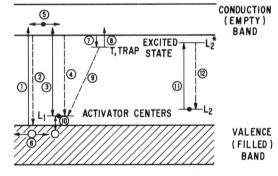

FIG. 1. Band model of a simple phosphor.

The activator as well as the impurity atoms introduce localized energy levels in the forbidden energy region between the lowest continuous empty energy level, the conduction band, and the highest filled continuous band, the valence band. L_1 is an activator (acceptor) level due to, say silver or copper; T, a trapping level due to the coactivator (donor) or some other impurity and L_2 and L_2^* are the ground the excited states of, for example, Mn^{+4} or a trivalent rare earth ion, respectively.

Excitation of the phosphor results in raising an electron from its ground state to a higher energy state, transition (1), (3) or (11). Transition (1) corresponds to the host absorption, transition (3) to absorption due to a host-activator interaction, and transition (11) to direct absorption by the activator.

Transitions (1) and (3) result in placing an electron in the conduction band, leaving behind an electron hole. The free electron in the conduction band and the electron hole in the valence band can move with or without the application of an electric field (5) and (6), respectively. Photoconductivity is the movement of the electron and/or electron hole caused by the electric field. On reaching the conduction band, the electron may immediately fall back to an empty luminescent center, transition (2) or (4), causing fluorescence or it may move in the conduction band and be trapped, transition (7). At some later time, the trapped electron will be returned to the conduction band, transition (8), after which it may return to an empty activator center, causing delayed fluorescence or phosphorescence, transition (2) or (4); the trapped electron may also go directly to the empty activator level resulting in "donor-acceptor emission." Usually, if excitation corresponding to transition (1) occurs, the electron hole left in the valence band may diffuse, transition (6), and can be trapped at an unexcited activator level, transition (10). The net result of this is that the electron returning from the conduction band can only make transition (4), thus producing emission characteristic of the activator. Excitation of the phosphor may also induce excitation of an electron from L_2 to L_2^*. The reverse transition can produce fluorescence or phosphorescence, depending on the lifetime of the excited L_2^* state. It is also possible for an electron to be transferred from the conduction band to L_2^*, and for a hole to be transferred from the valence band to L_2. This may then result in transition (12), producing the luminescence characteristic of the L_2 activator. Some type of energy transfer via the sensitizer is utilized in most fluorescent lamp phosphors.

The killers introduce other transitions which allow the radiationless return of an excited electron to its ground (unexcited) state.

PHILIP M. JAFFE

References

Leverenz, H. W., "An Introduction to Luminescence of Solids," Dover, New York, 1968.

Curie, D., "Luminescence in Crystals," Methuen, London, 1963.

Goldberg, P., Editor, "Luminescence of Inorganic Solids," Academic Press, New York, 1966.

Aven, M., and Prener, J. F., "Physics and Chemistry of II–VI Compounds," Wiley, New York, 1967.

PHOSPHORUS AND COMPOUNDS

Phosphorus atomic number 15, atomic weight 30.79376 is a common nonmetallic element, and is in Group VA of the Periodic System. H. Brand in 1669 heated white sand with evaporated urine and found that the resulting white solid glowed in the dark and ignited in air spontaneously. It was not until 1771 that K. W. Scheele prepared the element in large quantities by heating calcium phosphate (bone ash) with sand and carbon.

Occurrence. Phosphorus is widely distributed in the earth's crust, constitution about 0.13%; it is one of the few elements which is never found free in nature. The human skeleton contains 3 lbs. in the form of various compounds. Phosphorus is an essential constituent of vegetable matter and an important fertilizer. The chief sources of fertilizer are phosphorite, $[Ca_3(PO_4)_2]$ and apatite $[CaF_2 3Ca_3(PO_4)_2]$, which occur in many rocks and deposits located in the Gulf States, South Carolina, northern Africa, and Canada.

Preparation. Generally phosphorus is prepared in the electric furnace by the following reaction:

$$2Ca_3(PO_4)_2 + 6SiO_2 \rightarrow 6CaSiO_3 + P_4O_{10}$$

$$P_4O_{10} + 10C \rightarrow P_4 + 10CO$$

The vapors of the element together with the carbon monoxide leave the furnace and are condensed under water, while the calcium silicate melts and is drawn off at the bottom. Large quantities are produced in the Tennessee Valley where electricity is cheap.

Properties. There are two varieties of phosphorus. The white or yellow octahedral modification, formed when vapors of element are cooled very rapidly, is a wax-like substance of high volatility and low melting point (m.p. 44.1°C, b.p. 280.5°C, density 1.82). It ignites spontaneously in air, but does not burn in pure oxygen below 27°C. It is soluble in ether, carbon disulfide and turpentine, but insoluble in water. The white form is metastable; upon exposure to sunlight and heat in absence of air it turns yellow and then red, a transition which is catalyzed by iodine. For purpose of safety the white form is handled under water.

The other variety, red tabular, appears to be a mixture of the white and the stable form (violet phosphorus). This latter form is rather difficult to purify but can be obtained by crystallization from solution in molten lead. The properties of the red form are essentially those of the violet form. It is less active than the white, does not ignite in air below 240°C. It does not melt except at very high temperature and pressure. Below 1500°C phosphorus vapor has the formula P_4 and above 1700°C, the formula P_2. The yellow form ignites spontaneously in chlorine, while the red form has to be heated to start the reaction. A third, but much less known form of phosphorus is the black variety, formed when a pressure of 4000 atm is applied at 200°C.

Uses. Red phosphorus is one of the components of the box coating of safety matches. The element is used to a large extent in the manufacture of munitions, especially tracer bullets, smoke screens, and sky writing pyrotechnics. Certain rat poisons contain as their active ingredient white phosphorus incorporated into flour.

Phosphine (PH_3). Yellow phosphorus is soluble in hot alkalies with the liberation of the phosphine. Hydrolysis of calcium phosphide will also yield the hydrogen phosphide: $Ca_3P_2 + 6HOH \rightarrow 3Ca(OH)_2 + 2PH_3$. PH_3 resembles NH_3, as phosphorus occurs in the same position in the Periodic Table as nitrogen. It is classed as a noxious gas in concentrations of over .05 ppm. (See **Noxious Gases**)

Halides. Phosphorus exhibits both +3 and +5 valences in reacting with various halides, forming PF_3 and PF_5, which are gases: PCl_3 and PBr_3, both liquids, and PCl_5, PBr_5 and PI_3 which are solids. The pentahalides dissociate into the trihalide and halogen when heated. Hydrolysis produces phosphorous acid from PCl_3 and phosphoric acid from PCl_5, the oxychloride, $POCl_3$, being an intermediate in the latter reaction. *Phosphorus oxychloride* ($POCl_3$) is a colorless to slightly yellow liquid whose vapors are very irritating. It decomposes in water to yield phosphoric and hydrochloric acids. It is used as a chlorinating agent and production of intermediates of organic chemicals. Boiling range 106–108°C at 760 mm. *Phosphorus trichloride* (PCl_3) is a clear colorless liquid, whose vapors are very irritating. It decomposes in water to give phosphorous and hydrochloric acid. Uses are similar to those of $POCl_3$. Boiling range 75.5–77°C at 760 mm. *Thiophosphoryl chloride* ($PSCl_3$) is used as an intermediate in the production of some chemicals. It is a faint yellow to colorless liquid, very reactive toward water, alcohols, phenols and amines. Boiling range 120–125°C at 760 mm.

Oxides of Phosphorus. There are three oxides of phosphorus: the trioxide [$(P_2O_3)_2$], the tetroxide [P_2O_4] and the pentoxide [$(P_2O_5)_2$]. Phosphorus trioxide is the anhydride of phosphorous acid and the pentoxide the anhydride of phosphoric acid. Phosphorus trioxide, which is very poisonous has a melting point of 22.5°C and boils at 113°C, dissolves slowly in cold water forming phosphorous acid, while when added to hot water it reacts violently. At 440°C the P_2O_3 when heated in a sealed tube is converted to P_2O_4 and red phosphorus. The pentoxide (P_2O_5) appears to exist in a number of forms, one of which sublimes at 350°C. It is formed as dense white clouds by burning phosphorus in dry air or oxygen. Phosphorus pentoxide takes on water very readily and is used as a desiccating agent.

Hypophosphorous acid [$H(H_2PO_2)$], decomposes at 140°C to phosphine, hydrogen and phosphoric acid. The calcium salt [$Ca(H_2PO_2)_2$] is a white nonhygroscopic solid, soluble in water, insoluble in alcohol; pH of a 1% solution is 8.–8.1. Sodium hypophosphate, available as the monohydrate ($NaH_2PO_2·H_2O$), is a white crystalline solid; pH of a 1% solution is about 7–8. It is very soluble in water, only partially in alcohol. It decomposes at about 55–75°C and in hot alkaline solutions. All these compounds are strong reducing agents and used as ingredients of explosive mixtures. The acid is used in electroplating baths.

Phosphorous acid (H_3PO_3) may be formed by the following reactions: $PCl_3 + 3H_2O \rightarrow H_3PO_3 + 3HCl$ or $P_2O_3 + 3H_2O \rightarrow 2H_3PO_3$; it is a powerful reducing-agent.

Phosphoric acids. The three known phosphoric acids, meta, pyro and ortho, may be looked upon as one mole of P_2O_5 reacting with 1, 2 or 3 molecules of water to form HPO_3, $H_4P_2O_7$ and H_3PO_4. The meta acid (HPO_3) is obtained by dissolving P_2O_5 in cold water or by heating the ortho acid to 316°C. Commercially the meta acid is available in sticks (glacial phosphoric acid); it dissolves slowly in water to form the ortho form. Pyrophosphoric acid ($H_4P_2O_7$) is obtained by heating the ortho form between 213 and 316°C. It is a colorless solid

which melts at 61°C and dissolves in water to form H_3PO_4. The most important of these acids is orthophosphoric acid, (H_3PO_4), prepared by heating pulverized phosphate rock and sulfuric acid: $Ca_3(PO_4)_2 + H_2SO_4 \rightarrow 2H_3PO_4 + 3CaSO_4$. The ortho form is a white deliquescent solid and is soluble in water. Commercial "Syrupy Phosphoric acid" is an 85% aqueous solution containing chiefly $H_2PO_4^-$ ion and practically no $PO_4^\equiv$ ion. It is used in the manufacture of a great number of products: yeast, sugar, soft drinks, jellies, gelatin, pharmaceuticals, dental cements, glue, ceramics, explosives, fertilizers and numerous others. The meta acid coagulates egg white, while the ortho and pyro acids do not.

(For discussion of phosphates and their properties see **Phosphates.**)

Phosphorus in Animal Metabolism. The bulk of the phosphorus in the body, about 80%, is contained in bone and teeth. About half the remainder is combined with the organic constituents of the muscles, while the rest is distributed throughout the blood and other body tissues. Alkali phosphates play an important role as buffers in the blood. Phosphorus (as phosphate) is also a constituent of various organic compounds, such as lecithin, certain proteins, nucleic acids, and a series of other substances involved in the intermediary metabolism of fats and carbohydrates and muscle contraction. The energy used and released during these reactions is in part supplied by the substances containing a high energy phosphate bond. P^{32} (half-life 14 days) is one of the isotopes used in metabolic studies. (See **Phosphatides**)

The biochemical significance of finding naturally occurring C–P compounds has yet to be established. But since the isolation of two amino phosphonic acids

$$\left(\begin{array}{c} PO(OH)_2 \\ | \\ R-C-H \\ | \\ NH_2 \end{array} \right)$$

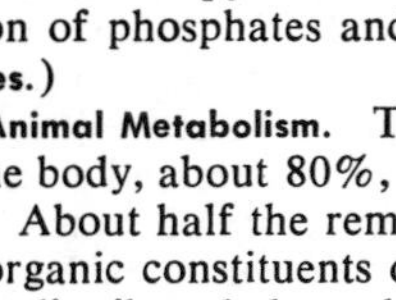

from certain sea animals, it would seem clear that new attitudes towards phosphorus metabolism—at least in lower animals—will result. To date about 26 amino phosphonic acids have been synthesized.

Toxicity. While the red form of phosphorus is essentially nonpoisonous, the white variety is extremely toxic, about 0.1 g being a fatal dose. Continued consumption of small amounts will lead to chronic poisoning. The symptoms of poisoning begin with anemia, loss of appetite and weight, albuminuria, and a tendency to bleeding in the mucous membranes. Characteristically there is a destruction of the jawbone which starts with a thrombosis of the bone vessels and is secondary to carious teeth. The bone becomes brittle, painful, and there is infection of the jawbone which can lead to a general sepsis. The infection of the jawbone must be treated operatively as early as possible.

Phosphorus poisoning occurs frequently in industry from accidental ingestion and from inhaling the fumes. A practice now prohibited by law, the tipping of matches with yellow phosphorus, resulted in poisoning of children rather frequently. At the present time, poisoning occurs as a result of accidental ingestion of phosphorus-containing fireworks, rat poisons, and during fireworks manufacture.

Post-mortem findings are fatty degeneration of the liver, heart muscles and kidney. Erosion of the gastro-intestinal tract occurs, and in a rapidly fatal case, the contents may be phosphorescent. The blood may not coagulate.

Insecticides and Nerve Gases. Certain organic phosphates have a history which almost parallels that of organic chemistry. Compounds of this type had been studied in Europe before the end of the 19th century; some of them have found use as lubricants, plasticizers and solvents. Recently interest in the compounds of the pyrophosphate type has turned to their clearly defined biological action, which makes them useful as insecticides. Tetraethyl pyrophosphate (TEPP) $[(C_2H_5)_4P_2O_7]$ is just one of the numerous tetraalkyl ester of this type used in the manufacture of miticides and aphicides. It is impossible to mention each of the other compounds because of the large number in use. From the original insecticide research several highly toxic organic phosphorus compounds resulted, generally referred as nerve gases, because they specifically inhibit cholinesterase, the enzyme responsible for monitoring transmission of nerve impulses through nerve endings either to muscle cells or to adjacent nerves. Nerve gases are toxic not only by inhalation but by absorption through the skin and eyes. (See **Nerve Gases**)

E. R. KIRCH

Black Phosphorus

Black phosphorus is a dense polycrystalline polymorph of phosphorus, black in color, not found in nature, and resembling graphite in appearance, feel, thermal, and electrical properties. Unlike white phosphorus, which is a waxy, white nonconducting molecular solid of high chemical reactivity, black phosphorus is stable at room temperature in a dry atmosphere, ignites only with difficulty, and is a semiconductor. White phosphorus can be transformed to black phosphorus only by high pressure at elevated temperature or by a slow catalytic method using elevated temperature and no pressure; but black phosphorus readily reverts to white or red phosphorus on heating.

Black phosphorus was first described by P. W. Bridgman in 1914. Bridgman found that white phosphorus subjected to a hydrostatic pressure of 12 kilobars at 200°C suddenly and irreversibly transformed to the black polymorph with a volume contraction of 30 per cent during transformation. He observed that the density of the black form is 2.69, that it is feebly diamagnetic, and that its compressibility at 25 kilobars is 2.6×10^{-12} dyne^{-1} cm^2. He found that red phosphorus can be transformed to black at 80 kilobars hydrostatically, but application of shear causes transformation at 45 kilobars.

Other workers have reported that the melting point at 18 kilobars is 1000°C, and there is no modification up to 20 kilobars and the melting point. Single crystals 4×6 mm of 0.1 to 0.2 mm thickness have been reported. Black phosphorus structure is side-centered orthorhombic with eight atoms per unit cell. An amorphous form, a black powder, can be obtained by pressure transformation. A form called vitreous phosphorus has also

been described. This is a nonconductive, dark grey, glassy phase of reactivity intermediate to black and white phosphorus. It does not appear to be related to amorphous black phosphorus.

Krebs, Weitz, and Worms showed that white phosphorus and mercury can be heated together for several days at 370°C to form an ingot of polycrystalline black phosphorus. Alloys or solid solutions of black phosphorus and arsenic have also been made this way. Brown and Rundqvist have obtained needle-shaped single crystals from solutions of phosphorus in liquid bismuth.

Black phosphorus is one of the few elemental semiconductors and, of these, one which is outside column four of the Periodic Table. The pressure-derived form has been found to be p-type with a room temperature resistivity of one ohm-cm. The apparent energy gap around room temperature is 0.35 eV and the carrier mobility is 9×10^5 $T^{-1.4}$ cm^2/volt-sec. An optical absorption edge is found in the 2 to 6 micron region of the infrared, consistent with the magnitude of the energy gap; photoconductivity has been observed. The material has a hydrostatic piezoresistive coefficient of 3.04×10^{-10} cm^2/dyne and a uniaxial coefficient $> 3 \times 10^{-9}$ cm^2/dyne, which is unusually high compared to many other strain-sensitive materials. Its thermoelectric power is about 370 microvolts per degree. Young's modulus is 6.89×10^{10} dyne/cm^2, and the ultimate tensile and compressive strengths are 3.4×10^7 and 3.5×10^8 dyne/cm^2, respectively.

Black phosphorus produced by catalytic conversion appears to be a matrix of phosphorus in which mercury is imbedded. Although the initial mercury content is about 25 to 30 per cent by weight, suitable processing can reduce it to 0.6%. The semiconducting properties of the material change from being near-metallic to being like those of the pressure-derived material as the mercury content is decreased.

D. WARSCHAUER

Phosphorus Sulfides

The phosphorus sulfides include a variety of crystalline, amorphous, polymeric, thermally stable and metastable compounds. They range in composition from P_4S to PS_6. Some are well defined; others need further study.

Reported compositions include P_4S_2, a crystalline material melting at 46°C; $(PS)_x$, apparently a mixed amorphous and crystalline polymeric composition; and P_2S_3 (or P_4S_6), variously reported as nonexistent, unstable at room temperature, and stable but melting with decomposition at 232°C. However, four crystalline phosphorus sulfides are well established and are named as follows (The correct technical name is listed first; the common name is designated by an asterisk):

P_4S_3 —tetraphosphorus trisulfide, or
       phosphorous tetritatrisulfide, or
       phosphorus sesquisulfide*

P_4S_5—tetraphosphorus pentasulfide*, or
       phosphorus tetritapentasulfide

P_4S_7 —tetraphosphorus heptasulfide, or
       phosphorus tetritaheptasulfide, or
       phosphorus heptasulfide*

P_4S_{10}—tetraphosphorus decasulfide, or phosphorus pentasulfide*

Preparation. P_4S_3, P_4S_7, and P_4S_{10} are industrial products. They are usually produced by the exothermic reaction of the molten elements at temperatures ranging from 180 to 450°C followed by cooling to the solid form. Hot operations are protected by an inert atmosphere such as carbon dioxide.

P_4S_3 can be purified by washing with water, which decomposes the other phosphorus sulfides, or by recrystallization from CS_2. P_4S_{10} is purified by distillation.

The reactivity of P_4S_{10} is enhanced by quick chilling of the melt, solidifying polymeric as well as crystalline forms.

P_4S_5 can be prepared by very slow cooling of the stoichiometric melt or by the reaction of P_4S_3 and S in CS_2.

Structure. Structures in Fig. 1 were determined by x-ray crystallography and confirmed by nuclear magnetic resonance and other techniques. Note symmetry differences and the variety of bond types in each: 3 adjacent (60°) P—P bonds in P_4S_3; one each in P_4S_5 and P_4S_7, none in P_4S_{10}; P—S—P bonds in all four compounds; and P=S bonds in all but P_4S_3. The P—P bond in P_4S_7 is uniquely long. These cage-like structures and bond varieties have a marked effect on physical and chemical properties.

Physical Properties. Characteristic physical properties are:

Compound	Color	m. p., °C	b. p., °C	Sol. (CS_2)
P_4S_3	yellow	174	407	100
P_4S_5	lt. yellow	162	—	~10
P_4S_7	near white	308	529	0.029
P_4S_{10}	yellow	288	514	0.222

Solubility is expressed in g/100 g CS_2 at 170°C. P_4S_5 melts with disproportionation to P_4S_3 and P_4S_7;

higher melting points have also been reported. Notice general trends in properties as sulfur content increases, with slight peaking at P_4S_7.

Reactions. Especially important are reactions of active hydrogen compounds with phosphorus sulfides by elimination of H_2S. In these reactions the nucleophilic group, originally bonded to hydrogen, bonds directly to phosphorus. Predominant products from P_4S_{10} are:

Reagents[1]	Products
H_2O^2	H_3PO_4
NH_3	$(NH_4)_2(PS_3NH_2)$, etc.[4]
ROH	$(RO)_2P(S)SH$
RSH	$(RS)_3PS$, $RSPS_2$
RNH_2	$(RNH)_2PSSH$
RNH_2^3	$(RNH)_3PS$
R_2NH	$R_2NP(S)(SH)_2$, etc.[5]

[1]R is aliphatic or aromatic.
[2]Hydrolysis in alkaline solution forms mixed thiophosphates, such as, Na_3PO_3S and $Na_3PO_2S_2$.
[3]High proportion of RNH_2 and higher temperature.
[4]Variety of products depending upon temperature.
[5]Excess amine forms ultimately $(RHN)_3PS$ and R_3N.

The lower phosphorus sulfides, P_4S_7, P_4S_5, and P_4S_3 also react with active hydrogen compounds but form other phosphorus derivatives often in addition to those derivatives formed by P_4S_{10}. For example, P_4S_7 and alcohols form $(RO)_2PSH$ as well as $(RO)_2P(S)SH$ and $(RO)_2P(S)SR$.

Primary amines react with the lower phosphorus sulfides to form derivatives containing PH as well as P—N and P—S bonds. Hydrolysis of P_4S_7 yields hypophosphorous and phosphorous acids as well as phosphoric acid. Hydrolysis of P_4S_3 yields PH_3 as well. Sequence of hydrolysis stability is $P_4S_3 > P_4S_{10} > P_4S_7 > P_4S_5$.

Perhaps the most remarkable reactions are those of cyclohexene, terpenes and aromatic hydrocarbons including benzene with P_4S_{10} when reaction temperature is carefully controlled. Thus, cyclohexene reacts with P_4S_{10} to form dimeric Δ^2-cyclohexenylthionophosphine sulfide, $(C_6H_8P(S)S)_2$, and anisole reacts to form p-anisylthionophosphine sulfide, $(CH_3OC_6H_4PS_2)_2$. Actual structures are uncertain.

The lower phosphorus sulfides, P_4S_3 and P_4S_7, do not react with boiling anisole.

The phosphorus sulfides also react with other classes of compounds. Grignard reagents react, followed by hydrolysis, with P_4S_{10} to form mono-, di-, or trialkylphosphine sulfides, $RP(S)(SH)(OH)$, R_2PSSH, or $R_3P(S)$, with P_4S_7 to form R_3P, R_3PS, and some R_2PSSH and with P_4S_3 to form a 3 to 1 mixture of R_2PH and R_3P.

P_4S_{10} reacts with K_2S to form K_3PS_4, with molten NaOH to form mixed sodium thiophosphates, such as $Na_3PO_2S_2$ and Na_3POS_3, and with KCl to form K_3PS_4 and $PSCl_3$. $PSCl_3$ can also form by the reaction of P_4S_{10} with PCl_5. In a similar reaction, a phosphorus oxysulfide, $P_4S_4O_6$, can be formed from P_4S_{10} and P_4O_{10}.

The lower phosphorus sulfides undergo related reactions. For example, P_4S_7 reacts with Na_2S to form Na_3PS_3 and undoubtedly other products.

Excess iodine reacts with P_4S_3 to yield PI_3 and

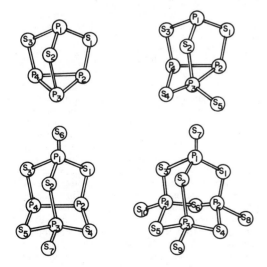

FIG. 1

P_4S_7. However, reaction of stoichiometric quantities of P_4S_3 and iodine in CS_2 at room temperature results in an interesting phosphorus thioiodide, $P_4S_3I_2$, the structure of which is related to that of P_4S_3 by breaking the bonds P_2—P_4 and interchanging the positions of atoms P_4 and S_3. Since these changes take place at room temperature, it has been inferred that the bonding in P_4S_3 is quite mobile.

Uses. The important industrial uses of P_4S_{10} involve the formation of dialykyl and diaryl esters of dithiophosphoric acid:

$$8ROH + P_4S_{10} \longrightarrow 4(RO)_2P(S)SH + 2H_2S$$

These esters in turn, in many variations, are neutralized with zinc or barium oxide to form oil additives, or with sodium carbonate to form ore flotation agents, or combined with various reagents to form powerful insecticides such as parathion and malathion.

P_4S_{10} is also a useful reagent in organic chemistry for converting OH, C=O, COOH, and $CONH_2$ groups into their corresponding sulfur analogs, and for preparing thiophene derivatives from 1,4 difunctional derivatives such as 1,4 diesters, 1,4 diketones and succinic anhydride.

P_4S_7, usually more active than P_4S_{10}, is available commercially on a small scale but is less popular due to greater difficulty of manufacture and greater tendency to form by-products.

P_4S_3, also capable of producing oil additives, is used primarily in the matchhead formulation of "strike anywhere" matches. In this use it capably replaces the toxic yellow phosphorus formerly used.

Economics. Production of P_4S_{10} in 1970 was approximately 124,000,000 lb sold at about 14¢/lb. Production of phosphorus sesquisulfide was (probably) less than 1,000,000 lb priced at about 38¢/lb.

T. H. DEXTER

References

1. Van Wazer, J. R., "Phosphorus and Its Compounds," Volume I, Interscience Publishers, New York, 1958.
2. Perot, G., *Chimie et Industrie*, **87**, (1), 83–86 (1962), a review (in French).

Organophosphorus Compounds

Organophosphorus compounds can be conveniently characterized by the number of groups attached to the P atom. Thus, there are found compounds with three, four and five groups bonded to P. The bonding is covalent and in the case of tetrasubstituted P compounds there are found tetracovalent materials, for example, phosphonium salts, and pentacovalent materials such as phosphates in which there is a semipolar double bond. The most common trisubstituted P compounds are the various phosphites, phosphines, halosubstituted materials and compounds bearing a combination of these groups.

The detailed structures of these compounds has been elucidated in only a few cases. In general they seem to have pyramidal arrangements of the atoms bonded to P. The bond angles between these groups vary from 90° to around 100°. Unlike amines, which undergo rapid inversion at room temperature, trisubstituted phosphines, and in at least some cases, phosphites are configurationally stable at room temperature. Optically active phosphines have been prepared and the stereochemistry of their reactions is a subject of considerable interest.

The synthesis of many of the trisubstituted P compounds proceeds from the trihalo compound, e.g., phosphorus trichloride. Phosphites are prepared by reaction with alcohols and phenols and phosphines by reaction with organometallics. Numerous variations on this general synthetic approach are employed.

Most of the trisubstituted P compounds behave as nucleophiles, entering into a wide variety of displacement reactions, including nucleophilic attack on carbon, oxygen, halogen, addition to carbonyl groups, etc. An important example is the Arbusov reaction in which a trisubstituted phosphite reacts with an alkyl halide to give a trialkoxyalkylphosphonium halide. These salts are generally unstable and decompose to an alkyl halide and a dialkoxy phosphonate.

In general the trisubstituted P compounds are only weakly basic. The trisubstituted phosphites and phosphines do act as bases toward many metallic ions and a wide variety of interesting complexes have been prepared by this reaction.

The conversion of tricovalent P compounds into tetracovalent or pentacovalent compounds is usually energetically favorable, and the reactions proceed readily and in high yield. Several of these reactions are of considerable economic importance.

Free radical reactions of trisubstituted P compounds have been described. Usually a radical adds to the P atom to give an unstable phosphoranyl radical with nine electrons in P orbitals. These intermediate radicals undergo further decomposition to give a tetracovalent or pentacovalent material.

By far the largest number of organophosphorus compounds are those in which four groups are bonded to P. These include such substances as phosphine oxides, phosphonium salts, phosphates, phosphonates, phosphinates and many others. Several classes of these compounds are valuable synthetic reagents and others are important biological substances.

Alkylidenephosphoranes, or Wittig reagents, are a class of organophosphorus compounds whose synthetic utility has only recently been recognized. These materials are prepared by allowing a phosphonium salt, which has at least one hydrogen on a carbon alpha to P, to react with an appropriate base. Loss of the proton yields the alkylidenephosphorane in which a semipolar bond now joins carbon to P. These materials act as nucleophiles and react with a variety of electrophilic reagents. By far the most important reaction of these alkylidenephosphoranes is the reaction with aldehydes and ketones. In this reaction the alkyl group is transferred specifically to the carbonyl carbon atom. A double bond is formed between this carbon atom and the entering alkyl group. The remainder of the alkylidenephosphorane is converted into a phosphine oxide. Important applications of this reaction include the synthesis of vitamins and many other natural products.

Quaternary phosphonium compounds can be obtained in optically active form and they serve as the source of the optically active phosphines. Extensive studies with the optically active phosphonium salts coupled with other mechanistic investigations have led to considerable understanding of the gross mechanistic features of many reactions of organophosphorus compounds. In general it is clear that pentasubstituted intermediates are often formed in many transformations of organophosphorus compounds. As yet little is known about the stability and structure of these intermediates; however it is important to note that they can lead to very complicated stereochemical and kinetic results. Knowledge of how these transformations occur is of great value in allowing the prediction of new reactions and an understanding of many old ones.

Of the many tetrasubstituted P compounds, the phosphate esters stand out in importance. They have many roles in nature. For example various monoesters of phosphoric acid such as sugar phosphates, glycerol phosphate and mononucleotides are found as are enol phosphates and phosphoramidates. Diphosphates and triphosphates occur in which two phosphate esters are bonded to each other. Examples of these are thiamine pyrophosphate and adenosine-5'-triphosphate. The nucleic acids DNA and RNA are polymers composed of a chain of phosphate-sugar-phosphate residues. The organic based are bonded to the sugars of the chains.

Other P- containing compounds which occur in nature are the phospholipides of which the lecithins and cephalins are examples. Several polymeric sugar-phosphate esters are also known. The profusion of compounds with phosphate ester groups and their tremendous importance to biosynthesis has created considerable interest in structural studies and laboratory syntheses of these materials. Real progress has been made but there is still much to be done.

The chemistry of pentasubstituted phosphorus compounds is under active study; perhaps the most interesting aspect of these molecules is the intramolecular transposition of groups bonded to phosphorus. This process is called pseudorotation and it has received considerable study.

D. B. Denney

References

Grayson, M., and Griffith, E. J., "Topics in Phosphorus Chemistry," Vol. 1–6, John Wiley and Sons, New York, New York, 1964–1969.

Kirby, A. J., and Warren, S. G., "The Organic Chemistry of Phosphorus," Elsevier Publishing Company, New York, New York, 1967.

Hudson, R. F., "Structure and Mechanism in Organophosphorus Chemistry," Academic Press, New York, New York, 1965.

Muller, E., "Methods of Organic Chemistry," G. Thieme, Stuttgart, Vol. 12, (1,2), 1963.

Phosphates

Originally obtained from calcined bone, phosphate is now derived almost exclusively from phosphate rock, fluorapatite, which in the United States is mined chiefly in Florida, North Carolina, Tennessee, Montana, Idaho and Utah. World sources include French Morocco, U.S.S.R., Tunisia, West Africa, Israel, Jordan, China, and Oceania. World production for 1969 totalled 77,400,000 metric tons of which 46.5% came from North America, 23.4% from Africa, 19.9% from East Europe, and 10.2% from Israel, Jordan, China, Oceania, etc.

The consumption pattern for phosphate rock differs from the production pattern as follows: North America consumed 40.9% in 1969, followed by 22.6% for West Europe, 21.1% for East Europe, and 15.4% for the rest of the world. The largest single use for phosphate rock is in fertilizers; only 15% of the world production finds its way into nonfertilizer fields. Current practice finds 28.8% of the phosphate rock converted to normal, or single, superphosphate by neutralization with sulfuric acid; the resulting product contains 16–18% P_2O_5 and was formerly the principal form of phosphate fertilizer. Double or triple superphosphate contains 40–48% P_2O_5 and is prepared by neutralizing phosphate rock with phosphoric acid; in 1969, 35.9% of the world's phosphate rock production was converted to phosphoric acid for producing fertilizers. The manufacture of elemental phosphorus accounts for another 4.8% of the world's phosphate, most of which is converted into alkali phosphates for use in detergents; in the U.S., about 600,000 short tons of elemental phosphorus are produced each year. Another major use for phosphate involves its addition to animal feeds as a nutritional supplement, particularly for poultry; the fluorine that is normally associated with phosphate in fluorapatite must be removed or greatly reduced for this application. In the U.S., this use represents about 5% of the total phosphate consumption.

The inorganic orthophosphates are prepared by neutralizing orthophosphoric acid. Although several others are commercially available, the sodium, calcium, ammonium, and potassium salts are the most important. Since all the hydrogens are replaceable, three salts of these, a mono-, a di-, and a tri-, are possible. The mono salts (MH_2PO_4) are water-soluble, yielding moderately acid solutions (1% solution pH = 4.4 to 4.5). The di salts (M_2HPO_4), excepting calcium, are also water-soluble, forming neutral to slightly alkaline solutions (1% solution pH = 8.0 to 8.8). Di- and tricalcium phosphates are relatively water-insoluble. The tri salts (M_3PO_4) of sodium and potassium are water-soluble, yielding strongly alkaline solutions (1% solution pH = 11.8 to 12.0). Triammonium phosphate is unstable under normal atmospheric conditions.

The commercially important polyphosphates (pyro, tripoly, and polymeta) are prepared by heating the proper orthophosphate or mixture. The metaphosphates, except tetrameta, are also prepared thermally. For example, dehydrating monosodium orthophosphate can yield any one of four condensed phosphates. Partial dehydration gives sodium acid pyrophosphate; controlled complete dehydration yields either an insoluble meta or the soluble trimetaphosphate, while rapid cooling of a melt forms the polymetaphosphate, frequently called "hexametaphosphate." Tetrasodium pyrophosphate is obtained by heating disodium phosphate. Controlled heating of an intimate mixture of one mole of mono with two moles of disodium yields the tripolyphosphate. Tetrapotassium pyro and potassium tripoly are prepared and are similar to the corresponding sodium salts. They are more soluble than the sodium salts but are not used as extensively because of their higher price. However, their high solubility often makes them the builder of choice in liquid detergents.

Dehydrating monopotassium orthophosphate does not yield an acid pyrophosphate quantitatively, since considerable conversion to meta takes place at the same temperature. Potassium metaphosphate is relatively water-insoluble but can be solubilized by adding sodium or ammonium salts. Sodium polymeta is very water-soluble, yielding approximately neutral solutions (1% solution pH = 6.2 to 6.8). Sodium and potassium tripolyphosphates ($M_5P_3O_{10}$) are water-soluble, forming moderately alkaline solutions (1% solution pH = 9.6 to 9.7). Tetrasodium and potassium pyrophosphates ($M_4P_2O_7$) are water-soluble, yielding alkaline solutions (1% solution pH = 10.0 to 10.2).

U.S. production of the more important industrial inorganic phosphates are shown below:

Alkali Phosphates	Approximate Production (Short tons, 1968)
Sodium tripoly	1,178,000
Tetrasodium pyro	107,500
Sodium meta	95,000
Trisodium ortho	63,500
Tetrapotassium pyro	53,250
Disodium ortho	22,900
Monosodium ortho	20,000

A large amount of phosphate is also consumed as ammonium or calcium salts. Approximately one-third of the phosphate fertilizer consumed in the U.S. is in the form of ammonium phosphate, particularly diammonium phosphate. Certain phosphate salts, such as monocalcium phosphate and sodium acid pyrophosphate, are used as leavening agents in baking; dicalcium phosphate has a major use as an abrasive in toothpaste or tooth powder.

Although many organic phosphates are known, only a few have achieved commercial importance as oil or gasoline additives, insecticides, plasticizers, etc. The largest volume in the U.S. goes into oil additives (125 million lbs/year) with the next largest use being insecticides (95 million lbs/year). The oldest large-scale use is as plasticizers (57 million lbs/year) and about 30,000,000 lbs/year are used as gasoline additives. The compound used in greatest volume in the U.S. is tricresyl phosphate (34.8 million lbs/year), followed by the insecticide methyl parathion (29.1 million lbs/year) and the plasticizer cresyl diphenyl phosphate (19.7 million lbs/year).

Modern Phosphate Chemistry. In contrast to the pioneering studies carried out many years ago on the reactions of inorganic phosphates and their interrelationships, described in the preceding section, more modern laboratory techniques have now been applied to the structural aspects of both in-

organic and organic phosphates. We are now able to describe phosphates as structures in which PO_4 groups—a phosphorus atom surrounded tetrahedrally by four oxygen atoms—are interconnected by the sharing of oxygen atoms. In solution it appears that only chains and rings of interconnected PO_4 groups can exist for a reasonable length of time. Two ring compounds are well known: the tetrametaphosphate, $Na_4(PO_3)_4$, and the trimetaphosphate, $Na_3(PO_3)_3$, and larger rings have also been found. Chains of all sizes up to 10^5 phosphorus atoms are known; the formula for the normal sodium salt of such chains is $Na_{n+2}P_nO_{3n+1}$.

Many of the techniques of high-polymer physics have been used to characterize and determine the size of long-chain phosphate anions, especially the anions obtained by dissolving the vitreous sodium phosphate known as Graham's salt and the crystalline potassium phosphate known as potassium Kurrol's salt. These techniques include (1) ultracentrifugation by both the sedimentation and equilibrium techniques; (2) relative and intrinsic viscosities both in the presence and absence of a swamping, low-molecular-weight electrolyte; (3) light scattering; (4) cryoscopy based on the transition point of sodium sulfate decahydrate; (5) end-group titration based on the weakly acidic hydrogen at each end of a chain phosphate; (6) flow birefringence both in the presence and absence of swamping electrolyte; (7) anisotropic electrical conductivity of the phosphates in flow; (8) solubility fractionation, and (9) nuclear magnetic resonance. Chromatography by both the paper and anion-exchange resin techniques has been used to identify chains as long as the nonaphosphate, as well as the two ring phosphates. Chromatography is now being employed to quantitatively carry out differential analyses of phosphate mixtures.

It has been known for many years that the polyphosphates are strong complexing agents, and this property has formed the basis of a number of industrial applications for these materials. Modern theory of polyelectrolytes, as supported by experimental studies on a number of such substances, indicates that all polyelectrolytes tend to bind their counter ions rather firmly. This action is working in the case of the polyphosphates, but in addition there also appears to be a specific complexing activity which can be attributed to one or two individual PO_4 groups. In general, it is found that the chain phosphates (especially the longer-chain materials) form complexes with a wide variety of cations, ranging from the alkali metals to the transition metals.

Phosphate Role in Biological Processes. As adenosine triphosphate (ATP) and other organic phosphates, the phosphate moiety plays an indispensible role in the biochemical processes that govern life itself. In photosynthesis, ATP is believed to be necessary for the fixation of carbon dioxide, and is probably produced by the endergonic reaction of adenosine diphosphate (ADP) with an orthophosphate group coupled with the photolysis of water to provide the necessary oxidation. In glycolysis and fermentation, phosphate transfer also plays a vital role; for example, glycolysis proceeds through the initial reaction of glycogen with an orthophosphate to form glucose-1-phosphate, which is converted to fructose-1,6-diphosphate through a series of enzyme-catalyzed reactions. Fermentation proceeds in a similar manner, except that ATP provides the initial phosphation of glucose.

Phosphate transfers identical to those that occur in glycolysis and fermentation also are part of the metabolism of the higher forms of life, although the metabolic pathways are more complex. Aerobic metabolism, osmotic processes, and the synthesis of nucleic acids are other biological processes that depend critically on phosphate. In fact, so many biochemical reactions depend on phosphate transfer that phosphates must be considered absolutely essential to life, just as are carbon, oxygen, sulfur, hydrogen, etc. Animals obtain their phosphatic requirements by ingesting plants, but the latter cannot obtain phosphate from the air as they do carbon dioxide and water. They must obtain it from the soil, which in turn receives it as the natural consequence of ecology or as the result of man's fertilization.

Finally, the biological process that has resulted in the large deposits of phosphate rock is the formation of mineralized tissues, teeth and bones, in animals by the deposition of hydroxyapatite in an organic matrix composed principally of the protein, collagen. At certain "calcification sites" in the body, the hydrolytic action of the enzyme phosphatase on a phosphate ester in the body fluid precipitates hydroxyapatite. This continuous deposition results in mineralized tissue that becomes teeth or bones. The accumulation of the bones of prehistoric animals in certain areas throughout the earth has formed the large phosphate rock deposits that now provide man with this all-important ingredient of life.

Ecological Developments. Concern with air and water pollution, and their effect on the ecology of planet Earth, is threatening to have a major effect on the use of industrial phosphates in certain countries (Canada, Sweden, and United States) where many bodies of fresh water are being choked with algae and other water plant life—a condition termed eutrophication. This is being blamed primarily on a high (>0.01 mg/l) phosphate content in these waters resulting from the extensive use of detergents containing large amounts of phosphate builders, particularly sodium tripolyphosphate; the phosphate finds its way into the sewer systems of cities and thence into nearby waters. Ecologists have concluded that this results in overfertilization of algae and other plant life in these waters, and have recommended a severe restriction on the use of phosphate builders in detergents as the most expedient answer to eutrophication. Many people disagree with the necessity for such action, pointing out that the eutrophication process is far too complex to be solved by banning phosphates from detergents and that other contributing sources must also be considered; these include other phosphate sources, such as agricultural runoff, and other nutrient sources such as suspended organic matter, nitrates and other nitrogenous matter, etc.

Meanwhile, researchers are busily engaged in searching for solutions to the eutrophication problem. Methods for removing phosphates from sewage and wastewater are being developed and a search for nonphosphatic builders for detergents has been initiated. At the present time, however, no completely satisfactory substitute for phosphate builders has been found. Salts of nitrilotriacetic acid (NTA) have been proposed and have been used to a limited extent; they are not without drawbacks, however, and do not appear to be an answer to the problem. Many detergent manufacturers have already reduced the phosphate content of their products and some have announced plans to market completely phosphate-free detergents based on such builders as sodium silicates, sulfates, and carbonates.

Whether or not the detergent industry is forced to alter the composition of its products, the net effect on the phosphate mining industry will not be crippling because the world-wide use of phosphate rock for sodium tripolyphosphate is only 8% of the total and part of this goes to countries where eutrophication is apparently not yet of major concern. There seems little likelihood that fertilizer phosphates will be restricted to an appreciable extent. However, the phosphate industry must continue to exercise control over fluorine emission in the manufacture of superphosphate fertilizers, phosphoric acid, and defluorinated feed phosphates as part of the battle against air pollution.

THOMAS P. WHALEY
JOHN R. VAN WAZER

References

Gmelin, "Handbuch der Anorganischen Chemie," Nummer 16, Achte Anflage, Teil A, B, and C, Weinheim/Bergstr., Verlag Chemie, Gmbh., 1965.

Van Wazer, John R., "Phosphorus and Its Compounds," Vol. 1 and 2, Interscience Publishers, Inc., New York, 1958.

Whaley, Thomas P., and Currier, James W., "Phosphorus," pp. 524–33, "The Encyclopedia of The Chemical Elements," Hampel, C. A., Ed., Van Nostrand Reinhold Co., New York, 1968.

Phosphatides and Phosphatidic Acids

All the naturally occurring phosphoglycerides appear to be derivatives of L-glycerol-3-phosphoric acid esterified with fatty acids at the 1- and 2-positions of the glycerol. The diacyl compound is called phosphatidic acid. Phosphatidic acid represents a relatively small part of the phospholipids of a tissue although many of the more complex phospholipids are phosphatidyl derivatives which contain alcohols esterified to the phosphate. For instance, choline forms phosphatidyl choline, commonly called lecithin.

Phosphatidic acid   *Phosphatidyl choline*

The unusually large amounts of phosphatidic acid described in early reports were found to be due to hydrolytic cleavage of the choline from phosphatidyl choline during the isolation process. Nevertheless, most tissues appear to use phosphatidic acid as an intermediate in the biosynthesis of the phosphatides so that it plays a vital role in lipid metabolism without normally accumulating to very high levels.

Other derivatives, such as phosphatidyl ethanolamine and phosphatidyl serine have been isolated from the aminophospholipid (cephalin) fraction of tissue lipids. Phosphatidyl inositol occurs in brain and nerve tissue along with its mono- and diphosphate esters and these three lipids have been called mono-, di-, and triphosphoinositides, respectively. Glycosides of phosphatidyl inositol containing mannose, with properties of both lipids and polysaccharides, have been isolated from mycobacteria. A closely related inositol glycoside has been found attached to ceramide phosphate in plant seeds.[1] Phosphatidylglycerol and diphosphatidyl glycerol are additional phosphatides found in many tissues. Appreciable amounts of the choline and ethanolamine phosphoglycerides are found in nearly all animal tissues and the general distribution of phosphatides in plants and animals was discussed in detail by Dittmer.[2] Orders of Insecta appear to differ among each other in the relative amounts of these two types of phosphatides (for example, ethanolamine phosphoglycerides predominate in Diptera).[3] In lipids of Gram-negative bacteria, choline is absent and ethanolamine is the principal amine.[4]

The phosphatides, as would be expected of any surface-active agent, are found to a large extent adsorbed in cellular membranes where they may play both structural and functional roles in maintaining the integrity of these membranes. The possibility that phosphatides may modify the surface membrane of a cell has been considered for erythrocytes and microorganisms, and this role may be important in the case of the amino acid esters of phosphatidylglycerol which are found in a variety of microorganisms. The content of the anionic detergent, diphosphatidyl glycerol (cardiolipin), in mitochondria has been shown to be related to the activity of the cytochrome electron-transport system. The action of cardiolipin as a hapten in the test for the antibodies developed in response to *Treponema pallidum* suggests further roles for this phosphatide. An interesting function of phosphatidic acid as a trans-membrane carrier for cation transport has been extensively investigated and may be significant in specialized salt excreting glands. Other phosphatides are now recognized as essential cofactors in restoring the activity of various enzymes that have been removed from their normal lipoprotein environment in subcellular membranes. Some enzymes can be treated to show a general requirement for phosphatides (glucose-6-phosphatase and adenosine triphosphatase from the endoplasmic reticulum), while other enzymes require a specific phosphatide for reactivation (β-hydroxybutyrate dehydrogenase from mitochondria).

Other closely related phospholipids that are not phosphatides in a strict sense, are the 1-alkyl and the 1-(1'-alk-1'-enyl) analogs in which the ester at the 1-position is replaced by an ether or alkenyl ether, respectively. The alkyl derivatives are particularly high in erythrocytes of some species, and the alkenyl ether derivatives are major components in the lipids of muscle and nerve tissue.

Essential fatty acids are esterified in phosphatides almost entirely at the 2-position and, in general, unsaturated acids (palmitoleate, oleate, linoleate, arachidonate, etc.) are esterified mostly at the 2-position with the saturated acids (palmitate and stearate) predominantly at the 1-position of the phosphatides. Phosphatidyl choline often contains more palmitate than stearate, whereas the reverse is often true for phosphatidyl ethanolamine from mammals. Each different phosphatide fraction isolated from a tissue generally contains a wide variety of fatty acids.

Because an individual phosphatide can contain only two of these fatty acids, there are different molecular species for each type of phosphatide, such as 1,2-dioleylglycerol-3-phosphorylcholine, 1-palmityl-2-oleyl-glycerol-3-phosphorylcholine, or 1-stearyl-2-arachidonyglycerol-3-phosphorycholine. In mammalian tissues, the palmityl-oleyl species is a predominant one recognized in the choline phosphoglycerides, and it appears to be formed by a pathway different from that for the stearyl-arachidonyl species.[5] These different species must be recognized when attempting to describe certain heterogeneous properties of some phosphatide fractions.

The interesting problem of separating and identifying the various species of phosphatides is developing in parallel with a similar situation for triglycerides. The existence of these species leads to speculation as to whether or not each species of glycerolipid could have a characteristic origin and function in a certain part of a cell. The manner by which the species of phosphatides originate is discussed in more detail in a recent review.[5]

The rapid development in the last 10 years of such sensitive and facile methods as thin-layer and gas-liquid chromatography for purifying and analyzing lipids[6] is quickly raising the level of sophistication at which phosphatide biochemistry can be dscussed, and the reader in this field should expect to check the useful annuals such as *Annual Review of Biochemistry* and *Advances in Lipid Research*.

WILLIAM E. M. LANDS

References

1. Carter, H. E., Johnson, P., and Weber, E., "The Glycolipids," *Annual Review of Biochemistry*, **34**, 109 (1965).
2. Dittmer, J. C., "Distribution of Phospholipids," in "Comparative Biochemistry," Vol. 3, pp. 231–264, Eds., M. Florkin and H. S. Mason, Academic Press, New York, 1962.
3. Fast, Paul G., "Insect Lipids: A Review," *Memoirs of the Entomological Society of Canada*, No. 37, 1964.
4. Kates, Morris, "Bacterial Lipids," *Advances in Lipid Research*, **2**, 17–90 (1964).
5. Hill, E. E., and Lands, W. E. M., "Phospholipid Metabolism," in "Lipid Metabolism," pp. 185–277, Ed. Salih J. Wokil, Academic Press, New York, 1970.

6. Marinetti, Guido V., "Lipid Chromatographic Analysis," Vol. I, Marcel Dekker, Inc., New York, 1967.

Phosphoric Acid

Phosphoric acid is a liquid (85 to 87%) and a solid (99%). It is prepared by either the wet or the dry method. In the former, dilute sulfuric acid is reacted with calcium phosphate in the form of ground phosphate rock or bone ash; 40% phosphoric acid is produced at first, which may be further concentrated. The concentrated grades are usually made by the dry method, which produces yellow elementary phosphorus at first. In this process a charge of calcium phosphate rock, carbon, and silica is fed into an electric furnace; phosphorus is evolved in vapor form, and is collected under water. The liquid solidifies easily and is kept under a blanket of water. It is shipped in specially equipped tank cars. The phosphorus is oxidized to phosphorus pentoxide, a white, extremely hygroscopic solid, from which acid of any concentration is made.

Phosphoric acid is used to make sodium phosphates, calcium phosphate, and other phosphates; to produce double and triple phosphates, which are double and triple strength because no inactive calcium sulfate remains in the product, as is the case when sulfuric acid is used; to substitute for tartaric and citric acid in preserves and soft drinks; to manufacture pharmaceutical chemicals and dental cements. It is used as a clarifying agent in the sugar house; in water softening; as a drying agent, and in numerous other industrial operations and preparations.

ELBERT C. WEAVER

PHOTOCHEMISTRY

Photochemistry is that branch of chemistry which deals with the permanent, chemical effects of the interaction of electromagnetic radiation and matter, as distinct from the usually temporary, physical effects. Upon absorption of a quantum of radiant energy a molecule is raised to an excited state, stable or unstable, in which, depending upon the particular wave length absorbed, the energy may be retained in rotational, vibrational and electronic degrees of freedom. Photoionization following the absorption of x- or γ-rays is usually referred to as Radiation Chemistry. (See **Radiation**) Excitation by longer wave lengths, the infrared, leads only to an increase in the population of the higher vibration-rotation levels of the ground electronic state; since no chemical effects are produced, such excitation is of little interest to the photochemist. Hence there exists, at ~ 7000 Å, a theoretical upper limit to the wave length region of interest. The lower limit of this region, which theoretically extends to the wave lengths of soft x-rays, is defined in practice by the availability of materials transparent to the far ultraviolet—~ 1800 Å for fused quartz and ~ 1300 Å for fluorite.

A molecule in an upper electronic state may undergo one of several processes. (1) Almost immediately, or after losing or gaining a number of vibrational quanta by collision, it may return to the ground electronic state with emission of light (fluorescence). (2) A major portion of the energy may be transferred to rotational, vibrational and translational degrees of freedom (rarely to electronic levels in molecular energy transfer) of the neighboring molecules or to the walls in collisions of the second kind (thermal degradation or collisional deactivation). (3) It may, within a period of one-half a vibration ($\sim 10^{-13}$ sec.) dissociate into mono- or polyatomic fragments, each of which may retain part of the excitation energy in any of its degrees of freedom (dissociation). (4) After a period of several vibrations in a vibrational level below the dissociation limit of the upper electronic state, but before a single rotational period ($\sim 10^{-11}$ to 10^{-12} sec.), it may, with no change in total energy, undergo a radiationless transition to a vibrational level in a second upper electronic state; if the second vibrational level is above the dissociation limit of the second state, dissociation will occur after one-half a vibration. The transition probability may be sometimes enhanced by collisions (predissociation and induced predissociation). (5) It may be transferred to an upper state of an isomeric molecule which is then usually stabilized by (2) (rearrangement). (6) It may react directly with other normal or excited molecules utilizing the excitation energy to overcome all or part of the activation energy barrier.

One or more of these processes must occur in any assembly of molecules absorbing light energy in the gaseous or liquid state, but not all need occur concurrently. The interpretation of the absorption spectrum of diatomic molecules and, to an increasingly greater extent, of polyatomic molecules may, in some cases, give definite information regarding (3) and (4). Advances in spectroscopy have been reflected in photochemistry; indeed, until the advent of quantum theory no satisfactory explanation of photochemical phenomena existed.

One of the major interests in photochemistry has been the production of free radicals by processes (3) and (4) and the study of their reactions. (See **Free Radicals.**) Studies of the photolyses of various aldehydes, ketones, alkyl halides, azo-compounds and organometallic compounds have been most fruitful in this respect. Intermittent illumination may be used to determine the lifetimes of the radicals in reactions in which the radical concentration is proportional to the absorbed intensity raised to a power n ($0.5 \leq n < 1.0$); this leads to the evaluation of rate constants for radical association reactions. A special case of (2) has also led to useful results. Metal vapors (Hg, Cd, or Zn) are raised to the excited $^3P_1$ and $^1P_1$ states upon absorption of resonance radiation from lamps containing these metals. Resonance fluorescence is efficiently quenched by the vapors of a large variety of compounds usually resulting in the production of atoms, radicals and metal atoms in the ground state. Such photosensitized methods are not confined solely to atomic vapours, but photosensitization by polyatomic molecules through collisions of the second kind has not been well developed.

The initial act of absorption and the various reactions of the excited state are usually referred to together as the primary process. The primary quantum yield, ϕ, i.e., the number of molecules changed per quantum absorbed, is given by the rate of processes (3) to (6) divided by the rate of absorption of light; in the absence of fluorescence and deactivation, $\phi = 1$. This is merely a statement of the Stark-Einstein Law of the Photochemical Equivalence, known as the first law of photochemistry. The concept of the quantum yield may be extended to include secondary processes (e.g., the reverse of (3)). Over-all quantum yields are measurable directly; ϕ must be inferred from kinetic studies of the secondary processes. Thus, any photochemical investigation is immediately separable into a chemical and an optical problem.

That fraction of the incident light which is absorbed by the reactant is readily measured by standard photometric procedures. The measurement of the absolute intensity is usually more difficult; all such determinations are based ultimately upon thermopile measurements of radiant energy, although a photocell may be calibrated as a more convenient secondary standard. To obviate the use of a thermopile or photocell, chemical actinometers have been used; these are photochemical reactions whose quantum yields have been established accurately as a function of wave length, etc. Hence the measurement of absolute intensity reduces to the determination of the amount of chemical change. Actinometric reactions in common use are the decomposition of oxalic acid sensitized by uranyl ions and the hydrolysis of chloracetic acid; recently the photolysis of malachite green leucocyanide and the reduction of potassium ferrioxalate have been advocated as suitable for actinometry. Gas-phase reactions which have been widely used are the photolyses of hydrogen bromide and of acetone. Since the response of all actinometric reactions and photocells is sensitive to wave length, monochromatic radiation is desirable for this reason as well as for the theoretical interest of the effect of various wave lengths on the photochemical reaction. This is obtained by removing the unwanted lines from sources of discontinuous spectra by means of monochromators or gaseous, liquid and glass filters; unfortunately the intensity of the wave length of interest is usually reduced by this procedure.

The last factor has a profound influence on the problem of determining the amount of chemical change. Under the most advantageous conditions with unfiltetred light, the rate of production of product is rarely as large as 10^{-4} mole/hr if the quantum yield is approximately unity; with filtered light and small fractional absorption, a few micromoles must be estimated. Inorganic compounds in solution can usually be handled in such quantities and techniques have been developed for the micromanipulation of gases in these amounts. Organic materials other than the gaseous hydrocarbons present much more difficulty, however. Since photochemical reactions are seldom carried to more than few per cent toward completion, the purity of the reactants, especially with respect to possible products and substances of high extinction, is of paramount importance. In certain instances (photopolymerizations, the photolysis of HCl, etc.) long reaction chains are set up with the quantum yield reaching 10^6 in the second example quoted. Measurements of physical properties (pressure of gases, density of solutions, etc.) then become practical and convenient methods of following the reaction.

K. O. KUTSCHKE

Cross-references: *Free Radicals, Radiation Chemistry, Flash Photolysis.*

PHOTOGRAPHY

The term "photography" comprises the process, the science, and the art of forming permanent images by means of the action of electromagnetic radiation cast, directly or indirectly, from the scene upon a photosensitive layer. In the most widely used photographic process the sensitive material is silver halide, as a suspension of microscopic crystals (grains) in gelatin, the so-called photographic emulsion, coated as a layer some 0.02 mm thick on a suitable support. In silver halide photography the effective wave lengths extend from the near infrared, through the visible and ultraviolet to the x-ray and γ-ray regions of the spectrum. Similar images can also be formed by high-energy particles, such as electrons, protons or α-particles.

To obtain a picture in black and white, a real image of the scene is projected by a lens arrangement onto the sensitive film placed in a camera. No visible change can be observed in the emulsion as a result of a normal photographic exposure, but some of the exposed grains are altered in such a way as to render them preferentially reducible to metallic silver by certain reducing agents called *developers*. The changed condition of exposed grains is referred to as the *latent image*.

It is now generally agreed that the latent image is constituted of colloidal particles of silver, containing three, four, or more silver atoms, usually mostly at the surface of the exposed grains.

The number of developable grains in any small area of the emulsion bears a regular relation to the intensity of light incident on the area; hence, on development, the photographic image appears as a silver deposit whose density and density differences correspond in some regular way to the brightness and brightness-differences in the original scene. The image is a *negative,* in the sense that the darkest parts of the image correspond to the brightest parts of the object.

The development process can be summarized as $AgBr + Developer \rightarrow Ag + Br^- + Oxidized Developer$. The components of typical developers are (1) the reducing agent, for example, a mixture of hydroquinone and *para*-methyl-aminophenol, (Metol, Elon) in aqueous alkaline solution, (2) a preservative sodium sulfite, added to inhibit aerial oxidation of the developer on standing and to decolorize oxidation products formed during development which might otherwise stain the emulsion, and (3), a restrainer such as potassium bromide. Without the latter, the developer may reduce some of the unexposed grains as well as the exposed ones, causing the superposition over the desired image of a uniform dark background called "fog." Antifoggants may also be added to the emulsion by the manufacturer.

Development leaves unreduced grains in parts of the film subjected to low light intensities during exposure, and to prevent darkening of the film by these residual grains, they are removed in the fixing process by a suitable solvent, usually sodium thiosulfate (hypo), which converts the silver halide to soluble complex ions. The film is then washed and dried. A positive image, that is one in which the tones of the image correspond directly to the brightness and darkness of the scene, instead of inversely, as in the negative, can now be formed by exposing a second sensitive layer, usually on paper, through the negative, followed by development, fixation and washing.

Emulsion making. The steps in the manufacture of a sensitive film are:

(1) Precipitation of silver halide by mixing an aqueous solution of silver nitrate with a slight excess of soluble halide, such as potassium bromide containing a little potassium iodide, in the presence of a portion of the gelatin finally required.

(2) A "first ripening," in which the growth of the silver halide crystallites is controlled by choice of temperature, time, and the concentration of silver halide solvent present (potassium bromide or, sometimes, ammonia), the larger grains growing at the expense of the smaller (Ostwald ripening).

(3) Washing of the emulsion, if it is to be coated on film base, to remove the bulk of the soluble salts and any ammonia. This step is omitted if the emulsion is to be coated on paper, since then soluble salts are occluded in the paper and do not crystallize on the surface of the coated emulsion.

(4) At this stage the light-sensitivity of the emulsion is very low, and it is now subjected to a second ripening or finishing process that increases sensitivity. The emulsion is digested, with additional gelatin, usually at about 50°C, until the desired sensitivity is attained. This sensitization process is known to be associated with the formation of small amounts of silver sulfide on the grain surface by chemical changes in labile sulfur compounds, such as sodium thiosulfate, which may be present in the gelatin or added at this stage to the emulsion.

(5) Before coating, small amounts of substances such as formaldehyde or chrome alum are added to the emulsion. This hardens the gelatin in the subsequent coating, thereby diminishing mechanical fragility of the coated emulsion and excessive softening in subsequent processing. Sensitizing dyes, if necessary, are added just before coating. In their absence the emulsion is sensitive only to light in the blue and shorter wave length regions of the spectrum; with sensitizing dyes sensitivity can be extended, according to the dye, over the green, yellow, red and near infrared regions of the spectrum. *Orthochromatic* emulsions are spectrally sensitized to green and yellow, while *panchromatic* emulsions are sensitive to all wave lengths from violet to red. The dye molecules add themselves to the grain surface as an incomplete layer one molecule thick, and the spectral sensitivity of the dyed emulsion corresponds to the absorption spectrum of the adsorbed dye, in addition to the sensitivity to blue and shorter wave lengths arising from direct absorption of light by the silver halide. Cyanine and merocyanine dyes form the most important groups of spectral sensitizers.

(6) The liquid emulsion is now coated on a support, usually a film of cellulose acetate, paper or glass, set by chilling and dried.

Manufacturers provide a great variety of emulsions designed for specific photographic needs. Emulsion types differ in the composition of the silver halide, the size of the grains and the addenda present, such as sensitizing dyes or antifoggants. The highly sensitive emulsions for the production of negatives contain silver bromide with a small proportion of iodide as relatively large, often plate-like crystallites of the order of 1 micron in diameter and some tenths of a micron in thickness, while the slower emulsions used in making positive prints contain smaller grains, usually of silver bromide, silver chloride or a mixture of both.

Besides materials for professional and amateur still- and motion picture making in black and white and in color, many types of special films are available for use, for example, in medical and industrial x-ray photography, in aerial photography, in graphic arts, in documentary copying and storage, in data recording, and in scientific applications such as micrography, spectroscopy, astronomy and the study of high-energy particles.

In addition to the conventional method of making a positive from a negative by contact printing or projection, a very useful new processing technique for the production of positives by *"diffusion transfer"* has been introduced within the last few decades. A negative film is exposed in the normal manner and developed more or less in the ordinary way in a developer which contains, in addition to the normal components, a solvent for silver halide, usually sodium thiosulfate. As development proceeds, the solvent dissolves the unexposed grains of the negative as a silver complex ion along with some of the grains in the lightly exposed regions, and very little in the highly exposed regions which are being rapidly reduced to metallic silver. The dissolved silver in the developer diffuses from the negative to a receiving sheet very close to the negative. The receiving sheet contains colloidal particles of materials such as silver sulfide, which catalyze the reduction of the dissolved silver complex by the developer. A deposit of silver metal is thus formed on those parts of the receiving sheet opposite the unexposed parts of the negative, corresponding to the dark parts of the original scene; little silver is deposited opposite the parts of the negative reduced to metallic silver, corresponding to the bright parts of the scene; and intermediate amounts of silver are deposited on the parts of the receiving sheet opposite the parts of the negative corresponding to the middle tones of the scene. A positive image of the scene is thus formed on the receiving sheet simultaneously with the development of the negative film. One application of the diffusion transfer process is to documentary copying, when relatively slow negative emulsions that can be handled in room light, containing a high percentage of silver chloride, are used.

The most spectacular application is the *Land process of rapid photography,* in which a fast negative roll film associated with a receiving sheet is exposed in a camera. When the film is wound after an exposure, a diffusion transfer developer from a pod attached to the film which is broken during the winding operation is spread as a viscous film in the narrow space between the negative and the receiving sheet. Development of the negative and the formation of a positive on the receiving sheet proceed, and after a short interval from the instant of exposure a positive of the scene is drawn from the camera, which has acted as exposure chamber and dark room.

Primary Photographic Process. During the exposure, a photon absorbed by the silver halide excites a bromide ion whereby a free electron and a bromine atom (positive hole) appear in the grain. In a perfect crystal, it is probable that the hole and electron would recombine after a period of free existence (demonstrated by the observation of photoconductivity in the illuminated crystal). Certain imperfections, however, are capable of trapping the electrons and positive holes separately and so retarding recombination. A trapped electron can combine with a mobile silver ion of the grain yielding a silver atom and a trapped positive hole may be inactivated chemically, or it may escape from the crystal as bromine. The process of chemical sensitization or finishing described above consists in the introduction of suitable traps for electrons and holes which will prevent their recombination.

A silver atom once formed at a trapping site is likely to be a more effective trap than the original trap, and a second electron is likely to be trapped at the same site, followed by combination with a mobile silver ion to form a two-atom silver center. A few repetitions of the electron capture-neutralization process builds up a latent image center as the result of the absorption of a few photons per grain. In the subsequent development, the latent image catalyzes the reduction to silver metal of all the silver ions in the exposed grain, some 10^9 in the grains of a high-speed negative emulsion, causing an enormous amplification of the initial photochemical change. It is this amplification of the exceedingly small initial photochemical formation of silver by development that is the basis of silver halide photography.

A developing silver halide negative can be used to form other images than silver images. For example, the oxidation products of certain developers, such as pyrogallol or hydroquinone, have the property of hardening gelatin (*"tanning developers"*), hence with such developers, the image-wise production of metallic silver is accompanied by an image-wise formation of hardened gelatin. On washing away the unhardened gelatin left in the unexposed parts of the negative, a relief image is left, which, after removal of developed silver by bleaching, can be dyed and used to obtain color prints.

Other Photographic Processes. Many photographic processes involving materials other than silver halides are used in special applications. They do not enjoy the very high amplification of the initial light reaction that distinguishes the silver halide process and are used in applications in which high light intensities and relatively long exposure times are feasible. Examples are processes depending on the photochemical formation or destruction of colored substances, such as the blue print and diazo processes, used in copying drawings, typewritten material, etc., or processes in which by various means light is used to cause image-wise insolubilization of a matrix, so that, by washing away the unexposed soluble portion, a relief image is produced, with applications in photomechanical printing processes, in decorating and marking ceramics, glass and other materials, in producing graticules, fine metal objects and printed circuits and in color printing processes. In *electrophotography*, widely used in copying documents and printed material, a latent image is formed by the light in the form of electrical charges on the surface of a semiconductor, such as a selenium plate or a zinc oxide coating, and a visible image is developed by the electrostatic attraction to the light-induced charges of minute particles of a suitable "toning" material.

W. WEST

Cross-references: *Developers, Color Photography, Sensitizing Dyes.*

Color Photography

In 1855, the English physicist, James Clerk Maxwell, first clearly stated the basic principles upon which all modern color photographic processes operate, and in 1861, he demonstrated the full-color photographic reproduction of a colored object before a gathering at the Royal Institution in London. His discoveries were based upon the earlier Young-Helmholtz theory of color vision, which stated that the brain perceives color because the eye is equipped with three different sets of receptors, sensitive to red, green, and blue light, respectively, and that the sensations corresponding to all discernible colors other than these so-called "primary" colors arise from the simultaneous stimulation of two or more of these selectively sensitive sets of receivers. Maxwell's demonstration showed that if a photographic system were arranged to produce separate records of the red, green, and blue rays, respectively, emanating from the subject, the three records could be recombined to yield a color photograph. For purposes of color photography, the entire range of wave lengths in the visible spectrum may be considered to be divided into three approximately equal regions: that from 400 to 500 nm, called blue; that from 500 to 600 nm, green; and that from 600 to 700, red. By properly mixing colored lights corresponding to sections of these three broad bands, it is possible to reproduce nearly all colors with a high degree of fidelity.

In the *additive* system of color photography demonstrated by Maxwell, photographs are taken through red, green, and blue filters to produce three separation negative images. From these, positive prints are prepared. These positives are placed in three separate projectors, each one equipped with its particular filter. The picture corresponding to the photograph taken with red light is projected through a red filter. Similarly, the green positive is projected with green light, and the blue positive with blue light. When the three projected images are superimposed on the screen, there is obtained a fairly faithful reproduction of the original colored object.

Many ingenious schemes have been suggested and tried for utilizing these principles of additive color mixture in color photography. However, the practical advantage of the so-called *subtractive* systems gradually became more apparent, with the result that today nearly all color processes are of this type.

In the various subtractive color processes, different routes and systems are used to reach the final dye images, but, in all cases, three such images are obtained, colored *cyan* (blue-green) *magenta* and *yellow,* respectively. Each of these three dyes absorbs approximately one-third and transmits most of the remaining two-thirds of the visual region of the spectrum. Thus, the cyan dye absorbs predominantly the red portion of the visual spectrum while transmitting freely the blue and green regions. The magenta dye absorbs predominantly the green region of the spectrum and transmits most of the blue and red regions. The yellow dye absorbs quite completely in the blue region of the spectrum and transmits the green and red portions. As a consequence of these absorption properties, when the three dye images are superimposed in the composite color photograph, each of them independently controls the transmission of one of the primary spectral regions: cyan dye for the red region; magenta for the green region; and yellow for the blue region. By using these "subtractive primary" dyes to make up the three separation images needed for color reproduction, it is possible to evolve color photographic systems which are much less cumbersome than the original additive systems demonstrated by Maxwell.

In practice, three light-sensitive photographic layers are coated inseparably in superposition, and each layer is selectively sensitized to one of the three spectral regions. A common arrangement has a red-sensitive silver halide emulsion layer next to the support, then a green-sensitive layer, and, finally, an emulsion layer which is sensitive to blue light only. After exposure of this multilayer material in a camera, a process is carried out to transform the photographic latent image in each layer into a dye image which is complementary in color to the light registered in that layer. Thus, the red portions of the object are reproduced in the bottom layer as a cyan dye image, the green objects are recorded in the magenta dye image in the middle layer, and the blue parts of the original object are recorded in the yellow image of the top layer. These three superimposed dye images constitute a photographic negative image, a negative both in the sense that light areas appear dark, and *vice versa,* and in the sense that the original colored object is reproduced in complementary colors.

This color negative is then printed onto a multilayer material containing the same arrangement of sensitized layers, using white light for printing. After exposure, the print material is again subjected to a process that produces in each layer a dye image complementary in color to the light registered in that layer. In this way there is obtained a positive color reproduction of the subject originally photographed.

In the so-called reversal type of color photographic process, separate negative and positive stages are not used. Instead, the multilayer photographic material is first subjected to normal black-and-white photographic development to produce a negative silver image. In a subsequent stage the silver halide not reduced in the negative developer is converted to the required dye images. Thus, the final color picture is composed of its three component dye images and is produced on the same piece of film that was exposed in the camera.

In order to make systems of this type work, it is necessary to have a means of imparting red and green sensitivity to the silver halide grains in the photographic emulsion. Silver bromide has an inherent sensitivity to blue light but none to red and green light. The required red and green sensitivity is achieved by means of *sensitizing dyes.* These are generally dyes of the cyanine series which have the special property of being adsorbed strongly to silver halide crystals and of transmitting the absorbed light energy to the silver halide to form a latent image. It is to be noted that these sensitizing dyes are present solely to confer the requisite sensitivities to silver halide in the separate layers but do not enter into the formation of the dye images which make up the final picture. Because silver halide sensitized with these sensitizing dyes retains its inherent sensitivity to blue light, it is necessary to insert a yellow filter layer above the red- and green-sensitive layers to prevent the blue light from forming a photographic image in these two layers and to restrict its action to the top emulsion layer. This yellow filter layer is decolorized during the processing of the film. (See also **Sensitizing Dyes.**)

In the evolution of color photography, one of the major problems from the start was to find methods for converting latent images in photographic silver halide emulsion layers into dye images. By far the most successful and, at present, the most widely used system is the process of *color development,* discovered in 1911 by Rudolph Fischer. In this process the exposed silver halide serves as the oxidizing agent to effect condensation of a *p*-phenylenediamine, the developing agent, with couplers which contain active methine or methylene groups, to produce *indoaniline* and *azomethine* dyes.

Thus, color development is actually the synthesis *in situ* of a dye within a photographic emulsion layer, the amount of dye deposited at any point in the image being a function of the degree of exposure of the silver halide. In a subsequent operation, the silver image is removed, leaving dye image alone as the photographic record. In most of the modern multilayer color photographic materials, the individual silver halide emulsion layers are only a few microns thick.

Accurate color reproduction in a subtractive system requires that each of the image dyes be formed exclusively within its proper layer in a multilayer system. Any departure from this selective deposition of the three dyes leads to serious losses in the quality of the color reproduction.

Three general methods are employed in commercial processes to achieve separation of the dye images. In the first method, employed for example with Kodachrome film, the multilayer coating contains only the light-sensitive silver halide emulsions. The couplers are introduced in the processing solutions. The three dye images are produced successively by utilizing the selective sensitivity to light of the separate emulsion layers.

In a second general type of color process based upon color development, the couplers are put into their proper layer during the manufacture of a multilayer sensitive material. Here the couplers must be essentially nondiffusing through gelatin so that none of them wanders out of its proper layer during either the coating operation or the subsequent processing. To meet these requirements, these couplers are synthesized with large "ballast" groups. Incorporation of the couplers during the manufacture of the multilayer material leads to great simplification in the processing procedure. The three dye images can be generated simultaneously in a single color-development step. The separation of the images is achieved solely by the ballasting of the couplers.

A third type of color development process in general use is exemplified by Polacolor. In this process the dyes, cyan, magenta and yellow, are present in layers of the film. The dyes are attached to alkali-soluble developing agents. The results of these unions, called dye-developers, are placed adjacent to their appropriately sensitized silver halide emulsions. After exposure, a pod is broken between the film and a sheet of paper that will ultimately contain the picture. The alkali released from the pod permeates the film and starts the dyes to move. When a cyan dye-developer encounters silver halide exposed to red light, it develops the silver halide and is, in turn, converted to an insoluble dye. Whenever there was no exposure to red, the cyan dye-developer diffuses to the paper receiver and is retained there by a mordant. Similarly, magenta and yellow dye developers form positive images on the receiver and a color picture is obtained.

The extensive growth of color photography during recent years has made it an important segment of the dye industry. The indoanilines, azomethines and azo dyes constitute the dyes in millions of color photographs produced each year, and although each photograph contains only a few milligrams of these dyes, a conservative estimate of the present annual production of the dyes would be many tons.

P. W. VITTUM

Cross-references: *Developers, Sensitizing Dyes.*

References

Mees, C. E. K., and James, T. H., Ed., "Theory of the Photographic Process," 3rd ed., Macmillan Co., New York, 1965.

Wall, E. J., "History of Three-Color Photography," American Photographic Publishing Co., Boston, 1925.

PHOTOMETRIC ANALYSIS

This term includes methods of chemical analysis based upon measuring the intensity (and/or spectral distribution of the intensity) of radiant energy in the extended optical spectrum. They may involve either the emission or absorption of radiation; examples include spectrochemical, spectrophotometric, atomic absorption, colorimetric and turbidimetric techniques.

The regions of the radiant energy spectrum of interest are the near ultraviolet (2,000–4,000 Å), visible (4,000–7,500 Å), near infrared (0.7–2.0 μ), fundamental infrared (2–24 μ), and far infrared (24–100 μ). Absorption or emission in the visible and ultraviolet regions are concerned primarily with transitions between energy levels associated with valence electrons in atoms and molecules, while in the infrared they are associated with changes in the vibrational and rotational energies of molecules (see **Photochemistry**).

Instrumentation used in photometric analysis generally includes a radiation source, refracting or reflecting optics, a dispersion system and a radiation detector.

For emission spectroscopy in the ultraviolet and visible regions electric arc, spark, laser or flame sources are used for excitation. For absorption spectroscopy in the visible region incandescent tungsten filaments are used as polychromatic sources and various special arcs and hollow cathode lamps for monochromatic sources (see **Spectroscopy, Absorption**). In the ultraviolet region hydrogen discharge lamps are most convenient. Nernst glowers or globars serve as sources in the infrared.

Dispersion systems may use prisms or gratings. Prisms and associated lens materials are selected for their dispersion and transparency in the spectral region of interest. In the visible glass or quartz optics may be used, in the ultraviolet quartz or fused silica, and in the infrared rock salt, potassium bromide, lithium fluoride or fluorite are generally used.

Photographic emulsions may be used in the visible and ultraviolet regions as radiation detectors. Infrared radiation is detected by its heating effect using thermocouples, thermistors, bolometers and Golay cells. In the visible the human eye, once the best detector, has been almost entirely replaced by more accurate and reproducible photoelectric cells. When very small amounts of radiation must be detected multiplier phototubes are available which provide detection and amplification in a single unit.

Spectrochemical Analysis. More than seventy elements can be qualitatively and quantitatively determined by spectrochemical analysis based upon the emission spectra of these elements when excited in an arc or high-voltage spark. The great value of this method is that many elements can be determined simultaneously. In qualitative analysis the speed and wide range is unmatched by other methods. Detection limits vary with technique from a few parts per billion to the parts per million level. The most useful range for quantitative analysis using this method is below 1%, and can be as low as a few parts per million.

Two types of instruments are used for spectrochemical analysis: prism and grating spectrographs. Prism instruments use quartz or glass optics with dispersing prisms of the same material. Grating spectrographs employ concave or plane reflection gratings to achieve dispersion. Resolution and dispersion vary with the number of lines on the grating surface. Concave gratings require no special optics while plane gratings use mirror optics. Photographic plates serve as radiation detectors and provide a permanent record of the spectra. Qualitative results are obtained from the position of spectral lines on the plate, relative to some "yardstick," usually the spectrum of iron. Quantitative data are gotten by comparing the measured density ratio of spectral line to background to that of known standards. Direct reading spectrographs utilize photocells to measure selected line intensity directly and the comparison is made electronically.

Such instruments are of great value where many repetitive analyses are required. While this method is quite rapid, it suffers from many interelement effects and care is required to compensate for them.

Absorption Spectrophotometry requires the measurement of radiation absorbed, as a function of wave length, upon passage through a given medium. Quantitative analysis is based on the Beer-Lambert law which relates the amount of radiation absorbed, at a particular wave length, to the length of the absorbing path, and the concentration of the material giving rise to the absorption. The method is applicable to a wide variety of samples and is of greatest use for low-level analysis of metals. Determinations can be made from a fraction of a part per million to several per cent.

Most instruments in use today use photoelectric detector systems. Single- and double-beam instruments, recording and nonrecording are available for use from 0.2 to 3 μ. These instruments employ a monochromator (prism, grating or filter) to isolate a narrow wave length band from a polychromatic source. They are equipped with a continuous source, cell holder and photoelectric detector. Double-beam instruments help to eliminate errors due to variations in the source, photocell response and interelement effects. Recording instruments give a permanent record of the entire spectrum over a given wavelength range.

Atomic Absorption is the absorption of radiant energy by atoms. As an analytical method it includes the conversion of elements combined in a sample into atoms and the absorption of radiant energy by those atoms. The conversion is accomplished by excitation in a high-temperature flame. Radiation from a hollow cathode discharge lamp, having the wave length of the element being determined, is passed through the flame. Atoms of that element absorb an amount of the incident radiation proportional to their concentration in the flame. This absorption is related to concentration by the use of known standards and the Beer-Lambert Law. The method, applicable to some 66 elements, is rapid, accurate ($\pm$ 1–2% of the amount present) and capable of determining amounts from a few parts per billion to several per cent. Perhaps its greatest advantage is specificity. Since iron radiation is absorbed only by iron atoms, interelement effects are greatly reduced. A hollow-cathode source for each element to be determined is generally required. In some cases certain compatible elements may be combined in a single multielement lamp.

Infrared Spectroscopy is mainly concerned with absorption spectra. In the range from 2–50 μ prism instruments are used, beyond 50 μ, or for high resolution, gratings are required. Single- and double-beam instruments of recording type are commonly available. Nernst glowers or globars are used as sources, thermocouples, thermistors or Golay cells serve as detectors. Windows and prisms are made of rock salt or special alkali halides.

Qualitative detection depends, in many cases, upon an adequate library of known spectra. Computer aided search and sort programs are available to speed spectra matching. Quantitative analysis requires the use of known standards and prior separation of complex mixtures. Water is opaque in this spectral region and must be removed except in certain special cases.

Raman Spectroscopy. The theory of Raman spectra is closely related to that of infrared spectra. The requirement for an infrared absorption band to appear is that there be a change in the dipole moment of the molecule as it is excited. The requirement for the appearance of the Raman effect is that there be a change in polarizability of the molecule. Since the two requirements are somewhat different, lines often appear in one spectrum and not in the other, although some lines may appear in both spectra. Thus, in part, infrared and Raman spectra complement one another and both are widely used in studying molecular structure.

To observe the Raman effect, the sample is illuminated by a high-intensity source of monochromatic radiation, frequently a mercury arc. (In recent years laser beams have been used extensively.) The scattered radiation spectra is recorded at 90° to the incident beam using a quartz spectrograph.

Chemical Colorimetry is principally concerned with measurements in the visible region and gets its name from the fact that most sample solutions are colored. However, absorption is actually measured rather than color. Colorimetric methods have been developed for most of the elements. They are widely used because of their high sensitivity, and relative simplicity.

The first instruments were simple visual comparators where the intensity of an unknown was compared directly to a standard identically illuminated. They have been almost entirely replaced by photoelectric photometers. These instruments have great advantage over visual comparators in both range and accuracy. Single-beam instruments consist of a cell, an incandescent tungsten source of variable intensity, and a photoelectric cell connected to a microammeter as a detector. Filters are used to isolate a narrow wave length band of radiation that matches the absorption maximum of the sample. Concentration is determined using known standards and the Beer-Lambert law. More sophisticated instruments use small replica gratings, or prisms, to achieve wave length selection and provide for a variable path length. Double-beam or split-beam instruments which measure sample and reference simultaneously are available.

Fluorimetry. Certain materials when exposed to radiation of a particular wave length absorb that wavelength and re-emit radiation at a longer wave length. This fluorescence, in many cases, is a function of concentration and can be used for analysis. The fluorescence is usually measured at right angles to the exciting beam. Instrumentation is similar to that used for colorimetry, but is less elaborate. The method is sensitive to interference from any source of scattered radiation and to many interelement effects.

Turbidimetry. Turbidity as used by the chemist is difficult to define. It refers to characteristic optical properties of dispersions and may be taken as the ratio of light scattered to that incident upon the dispersion. Other things being constant, the intensity of light scattered by a suspension is a function of concentration of the suspensoid.

Methods of measuring turbidity fall into three

groups: those which measure the ratio of scattered light intensity, Tyndall light, to that of the incident light; those which measure the ratio of transmitted light intensity to incident intensity; and those which measure extinction, that is, the depth at which a target vanishes beneath a layer of the turbid medium. Instruments which measure the Tyndall ratio are called tyndallmeters if the intensities are measured directly, and nephelometers when the intensities are compared with a known standard.

If turbidity is to be proportional to concentration the particle size must be constant. One of the most difficult aspects of these methods is the need to prepare reproducible suspensions. When the particle size is small compared to the wave length of the light, Rayleigh's law states that the Tyndall ratio is proportional to the cube of the particle size and inversely to the fourth power of wave length. If the particle size exceeds about one-tenth the wave length of the incident radiation, the Rayleigh law no longer applies, and the scattered radiation can exhibit a pronounced angular asymmetry. Light scattering, under these conditions, can be used to obtain information concerning the size, shape and molecular weight of polymers.

JACK P. WRIGHT

Cross-references: *Analytical Chemistry; Radiation; Spectroscopy; Spectroscopy, Infrared; Spectroscopy, Absorption.*

References

1. Mellon, M. G., "Analytical Absorption Spectroscopy," J. Wiley & Son, New York, 1950.
2. Slavin, Walter, "Atomic Absorption Spectroscopy," Interscience, New York, 1968.
3. Ahrens, L. H., and Taylor, S. R., "Spectrochemical Analysis," Addison-Wesley, 1961.

PHOTON

The photon is the quantum of the electromagnetic field. It is a particle with zero rest mass and spin one. For a photon moving in a specific direction, the energy E and momentum q of the particle are related to the frequency f and wave length λ of the field by the Planck equation $E = hf$ and the de Broglie equation $q = h/\lambda$. As for all massless particles, the energy and momentum are related by $E = cq$ and the photon can exist only when moving at the speed of light, c. Another property of all massless particles is this: given the momentum, the particle can exist in just two states of spin orientation. The spin can be parallel or antiparallel to the momentum, but no other directions are possible. The photon state with the spin and momentum parallel (antiparallel) is said to be right (left)-handed and is a right (left)-hand circularly polarized wave. By analogy with the neutrino, one can say that the state has positive (negative) helicity, and can call the right-handed particle the antiphoton, the left-handed particle the photon.

There is an operation, CP conjugation, that converts a photon state into an antiphoton state and vice versa. It is possible to superpose photon and antiphoton states in such a way that the superposition is unchanged by CP conjugation and so gives a type of photon that is its own antiparticle. The photons produced by transitions between states of definite parities in atoms or nuclei are their own antiparticles in this sense. As for all particles with integer spin, the photon follows Bose-Einstein statistics, which means that a large number of photons can be accumulated into a single state. Macroscopically observable electromagnetic waves, such as those resonating in a microwave cavity, for example, are understood to be large numbers of photons all in the same state. The photon, among all the particles, is unique in having its states macroscopically observable in this way.

The electric and magnetic fields $\mathbf{E}$ and $\mathbf{B}$ describe the state of the photon and make up the wave function of the particle. Maxwell's equations give the time development of the fields and state for the photon what Schrödinger's equation states for a nonrelativistic material particle. Many of the remarks above follow as direct consequences of Maxwell's equations. In Gaussian units, where both $\mathbf{E}$ and $\mathbf{B}$ are measured in gauss or dynes per electrostatic unit of charge, the equations for the free fields are

$$\epsilon_{jkl}\partial\mathbf{E}_l/\partial x_k + c^{-1}\partial\mathbf{B}_j/\partial t = 0, \tag{1}$$

$$\epsilon_{jkl}\partial\mathbf{B}_l/\partial x_k - c^{-1}\partial\mathbf{E}_j/\partial t = 0, \tag{2}$$

$$\partial\mathbf{E}_j/\partial x_j = \partial\mathbf{B}_j/\partial x_j = 0. \tag{3}$$

The particle aspect of the equations becomes evident when the equations are written in terms of the complex three-vector

$$\psi_j = \mathbf{E}_j - i\mathbf{B}_j \tag{4}$$

in which case they become

$$\epsilon_{jkl}\partial\psi_l/\partial x_k + ic^{-1}\partial\psi_j/\partial t = 0, \tag{5}$$

$$\partial\psi_j/\partial x_j = 0. \tag{6}$$

Equation (6) is to be considered as an initial condition rather than as an equation of motion since it follows from Eq. (5) that

$$\partial(\partial\psi_j/\partial x_j)/\partial t = ic\epsilon_{jkl}\partial^2\psi_l/\partial x_j\partial x_k = 0,$$

so if $\partial\psi_j/\partial x_j$ is zero at the start it is zero forever. Equation (5) can be cast into Hamiltonian form. One writes the three components ψ_j as a column matrix ψ and introduces three 3-by-3 matrices by

$$(s_k)_{jl} = i\epsilon_{jkl}. \tag{7}$$

With this notation Eq. (5) becomes

$$ic(s_k)_{jl}\partial\psi_l/\partial x_k = i\partial\psi_j/\partial t,$$

or

$$H\psi = i\bar{h}\partial\psi/\partial t, \tag{8}$$

where

$$H = -c s \cdot \mathbf{p} \tag{9}$$

and $\mathbf{p}$ is $-i\bar{h}\Delta$. The Hamiltonian for the photon is thus $-c s \cdot \mathbf{p}$. In detail the matrices that occur here are

$$\left(s_1 = \begin{matrix} 0 & 0 & 0 \\ 0 & 0 & -i \\ 0 & i & 0 \end{matrix} \right), s_2 = \left(\begin{matrix} 0 & 0 & i \\ 0 & 0 & 0 \\ -i & 0 & 0 \end{matrix} \right),$$

$$s_3 = \left(\begin{matrix} 0 & -i & 0 \\ i & 0 & 0 \\ 0 & 0 & 0 \end{matrix} \right). \tag{10}$$

They are Hermitian and, as is easily verified, they fulfil the commutation rules

$$[s_i, s_j] = i\epsilon_{ijk}s_k,$$

and so are a set of angular momentum matrices. Evidently each has eigenvalues 0, ± 1 so they are a representation of spin one.

Next consider the plane wave solutions. Let them be propagating in the 3-direction, so substitute

$$\psi = u \exp ih^{-1}(p_3 z - Wt)]$$

into Eq. (8). Here the same symbol p_3 is used for the eigenvalue as for the operator. The system reduces to the matrix eigenvalue problem

$$-c\begin{pmatrix} 0 & -ip_3 & 0 \\ ip_3 & 0 & 0 \\ 0 & 0 & 0 \end{pmatrix} u = Wu.$$

The eigenvalues are found to be $W = 0, \pm cp$, where p is $|p_3|$, and the corresponding eigenvectors are

$$u_0 = \begin{pmatrix} 0 \\ 0 \\ 1 \end{pmatrix}, u_\pm = \frac{1}{\sqrt{2}} \begin{pmatrix} \pm p_3/p \\ -i \\ 0 \end{pmatrix}.$$

The $W = 0$ possibility does not satisfy the initial condition, Eq. (6), and so must be discarded. The solutions $u_\pm$ are valid for either sign of p_3; choose $p_3 = \pm p$ so both waves are propagating in the positive z direction. The two solutions of the problem are then

$$\psi_\pm = \frac{1}{\sqrt{2}} \begin{pmatrix} 1 \\ -i \\ 0 \end{pmatrix} \exp [\pm iph^{-1}(z - ct)]. \quad (11)$$

The subscript $+1(-1)$ denotes a particle (antiparticle) solution with positive (negative) frequency of $W/h = +cp/h(-cp/h)$. Also $\psi_\pm$ are evidently eigenstates of the helicity operator $\mathbf{s} \cdot \mathbf{p}/p$ with eigenvalues ± 1. The electric and magnetic fields are the real and imaginary parts:

$$E_{\pm, x} = 2^{-\frac{1}{2}} \cos [ph^{-1}(z - ct)], \quad (12a)$$

$$E_{\pm, y} = \pm 2^{-\frac{1}{2}} \sin [ph^{-1}(z - ct)], \quad (12b)$$

$$B_{\pm, x} = \mp 2^{-\frac{1}{2}} \sin [ph^{-1}(z - ct)], \quad (12c)$$

$$B_{\pm, y} = 2^{-\frac{1}{2}} \cos [ph^{-1}(z - ct)], \quad (12d)$$

$$E_{\pm, z} = B_{\pm, z} = 0. \quad (12e)$$

Here it is seen that the $-1(+1)$ helicity solution is left (right)-hand circularly polarized with respect to the propagation direction.

The allowed states of the photon are eigenstates of the Hamiltonian H with eigenvalues $\pm cp$. Let $|H|$ be the operator which, applied to the same states, gives eigenvalue cp. The operators for the physical energy, momentum, and angular momentum of the photon are $|H|$, $(H/|H|)\mathbf{p}$, and $(H/|H|)(\mathbf{x} \times \mathbf{p} + \bar{h}\mathbf{s})$. One can understand these assignments for the energy and momentum by considering the plane wave states of Eq. (11). The states are eigenstates of the operator $|H|$ with eigenvalue cp and of the operator $(H/|H|)\mathbf{p}$ with eigenvalue p in the positive z direction. As further justification for these operator assignments, the expectation values of the operators are directly related to the

classical formulas for energy, momentum, and angular momentum in the electromagnetic field:

$$(\psi, |H|\psi) = (8\pi)^{-1} \int d^3x (E^2 + B^2),$$
$$\quad (13a)$$

$$\psi, \frac{H}{|H|}\mathbf{p}\psi = (4\pi c)^{-1} \int d^3x (\mathbf{E} \times \mathbf{B}),$$
$$\quad (13b)$$

$$\left(\psi, \frac{H}{|H|}(\mathbf{x} \times \mathbf{p} + \bar{h}\mathbf{s})\psi\right) = (4\pi c)^{-1} \int d^3x \, \mathbf{x} \times (\mathbf{E} \times \mathbf{B})$$
$$\quad (13c)$$

where the rule for taking the inner product is

$$(\psi_1, \psi_2) = \frac{1}{8\pi c} \int d^3x \, \psi_1^\dagger \frac{1}{p} \psi_2. \quad (14)$$

These equalities apply for any solution ψ of Eqs. (6) and (8). The dagger denotes the Hermitian conjugate. The operation $(1/p)\psi$ in Eq. (14) is to be carried out by expanding ψ in the plane wave components like $\psi_\pm$ and replacing the operator $(1/p)$ by the number $(1/p)$ in each component. Proofs of Eqs. (13) will not be given here; they can be made by expressing each side of the equations in terms of the plane wave expansion coefficients. Accepting these operator assignments, one sees that the helicity operator $\mathbf{s} \cdot \mathbf{p}/p$ is the component of the spin of the photon $(H/|H|)\mathbf{s}$ in the direction of its momentum $(H/|H|)\mathbf{p}$.

The CP conjugation operation is related to the space reflection covariance of Maxwell's equations. Consider a primed and an unprimed coordinate system such that the coordinates of any point in space referred to the two axes are related by $\mathbf{x}' = -\mathbf{x}$. Suppose the electric field is axial and the magnetic field is polar so that the functions describing the fields are related by $\mathbf{E}'(\mathbf{x}', t) = \mathbf{E}(\mathbf{x}, t)$ and $\mathbf{B}'(\mathbf{x}', t) = -\mathbf{B}(\mathbf{x}, t)$. It is evident that Maxwell's equations have the same form in both coordinate systems and that the transformation rule for ψ is $\psi'(\mathbf{x}', t) = \psi^*(\mathbf{x}, t)$ where the asterisk denotes the complex conjugate. The fact that the equations have the same form in both systems implies further that if $\psi(\mathbf{x}, t)$ is any solution then $\psi'(\mathbf{x}, t)$ or equivalently $\psi^*(-\mathbf{x}, t)$ is also a solution. The operation that carries $\psi(\mathbf{x}, t)$ into $\psi'(\mathbf{x}, t)$ is called CP conjugation, and one writes

$$\psi^{CP} = KP\psi, \quad (15)$$

where K is the operation "take complex conjugate" and the operator P changes $\mathbf{x}$ into $-\mathbf{x}$. If ψ is a solution of Maxwell's equations, so also is ψ^{CP}. However KP anticommutes with $\mathbf{s} \cdot \mathbf{p}$ so if the solution ψ has $\mp$ helicity then ψ^{CP} has $\pm$ helicity. The CP conjugation thus converts the particle into the antiparticle. The KP operator also anticommutes with the physical momentum operator $(H/|H|)\mathbf{p}$, so for a state $\psi_\pm(\mathbf{q})$ with definite helicity $\mp$ and physical momentum $\mathbf{q}$ one has

$$KP\psi_\pm(\mathbf{q}) = \psi_\mp(-\mathbf{q}). \quad (16)$$

Instead of the two states ψ_+ and ψ_- one may consider the superpositions

$$\psi_1(\mathbf{q}) = 2^{-\frac{1}{2}}[\psi_+(\mathbf{q}) + \psi_-(\mathbf{q})], \quad (17a)$$

$$\psi_2(\mathbf{q}) = 2^{-\frac{1}{2}}[\psi_+(\mathbf{q}) - \psi_-(-\mathbf{q})]. \quad (17b)$$

The reason for introducing them is the property

$$KP\psi_1(\mathbf{q}) = \psi_1(-\mathbf{q}), \qquad (18a)$$

$$KP\psi_2(\mathbf{q}) = -\psi_2(-\mathbf{q}). \qquad (18b)$$

Thus the KP operation applied to ψ_1 or ψ_2 reproduces the state, only traveling in the opposite direction and with a change of phase for ψ_2. The states ψ_1 and ψ_2 in this way are their own antiparticles. These self-antiparticle states are plane polarized in perpendicular directions. For the states with momenta in the positive z direction, as given by Eqs. (11) and (12), the fields are seen to be

$$E_{1x} = \cos [p\bar{h}^{-1}(z - ct)], \qquad (19a)$$

$$B_{1y} = \cos [p\bar{h}^{-1}(z - ct)]; \qquad (19b)$$

$$E_{2y} = \sin [p\bar{h}^{-1}(z - ct)], \qquad (19c)$$

$$B_{2x} = -\sin [p\bar{h}^{-1}(z - ct)], \qquad (19d)$$

with all other components zero.

The final point to be demonstrated here is that only a self-antiparticle type of photon is emitted or absorbed when a system makes a transition between states of definite parity. Consider for simplicity a spinless charged particle described by a Schrödinger wave function $\psi_m(\mathbf{x}, t)$. [The subscript m is used for the material particle, γ for the photon.] Suppose the particle is bound in some system and makes a transition from an initial state i to a final state f, both eigenstates of parity P, with emission or absorption of a photon. As is well known, the transition probability is determined by the interaction integral

$$I = -(e/Mc) \int d^3x [\psi_{mf}{}^*(\mathbf{x}, t)\mathbf{p}\psi_{mi}(\mathbf{x}, t)] \cdot \mathbf{A}(\mathbf{x}, t)$$

$$(20)$$

where e and M are the charge and mass of the particle and $\mathbf{A}$ is the vector potential of the photon in the Coulomb gauge,

$$\nabla \cdot \mathbf{A} = 0. \qquad (21)$$

Here and below the integrals extend over all space. The fields are found from the potential by the relations

$$\mathbf{E} = -c^{-1}\partial \mathbf{A}/\partial t, \qquad (22)$$

$$\mathbf{B} = \nabla \times \mathbf{A}. \qquad (23)$$

To make the argument, one first expresses the interaction explicitly in terms of the fields. The potential is found from the fields by integrating this way:

$$\mathbf{A}(\mathbf{x}, t) = \frac{1}{4\pi} \nabla \times \int d^3y \, \frac{\mathbf{B}(\mathbf{y}, t)}{|\mathbf{x} - \mathbf{y}|}. \qquad (24)$$

It is easily verified that this expression for $\mathbf{A}$ satisfies Eqs. (21), (22), and (23) by using Eqs. (1), (3), and the fact that $\nabla^2 |\mathbf{x} - \mathbf{y}|^{-1} = -4\pi\delta(\mathbf{x} - \mathbf{y})$. In the verification it is assumed that fields of interest will be zero outside a finite region of space so that in making partial integrations there are no contributions from infinity. Then by using Eq. (24) and replacing $\mathbf{B}$ by $\frac{1}{2} i(\psi_\gamma - \psi_\gamma{}^*)$ one can rewrite the interaction integral as

$$I = \frac{-ie}{8\pi Mc} \int d^3x [\psi_{mf}{}^*(\mathbf{x}, t)\mathbf{p}\psi_{mi}(\mathbf{x}, t)] \cdot \nabla \times$$

$$\int \frac{d^3y}{|\mathbf{x} - \mathbf{y}|} [\psi_\gamma(\mathbf{y}, t) - \psi_\gamma{}^*(\mathbf{y}, t)].$$

However, if i and f are eigenstates of parity, then, by changing integration variables from $\mathbf{x}$ and $\mathbf{y}$ to $-\mathbf{x}$ and $-\mathbf{y}$ in the $\psi_\gamma{}^*$ term, one sees that

$$I = \frac{-ie}{8\pi Mc} \int d^3x [\psi_{mf}{}^*(\mathbf{x}, t)\mathbf{p}\psi_{mi}(\mathbf{x}, t)] \cdot \nabla \times$$

$$\int \frac{d^3y}{|\mathbf{x} - \mathbf{y}|} (1 \mp KP)\psi_\gamma(\mathbf{y}, t),$$

where the factor is $(1 - KP)$ if i and f have the same parity, $(1 + KP)$ if i and f have opposite parity. Since

$$KP(1 \mp KP)\psi_\gamma = \mp(1 \mp KP)\psi_\gamma,$$

only a type of photon that is its own antiparticle can be involved in the transition in either case. As examples, the electric dipole radiation field has $KP = +1$ and the magnetic dipole field has $KP = -1$.

The following references provide further information:

1. Heitler, W., "The Quantum Theory of Radiation," Oxford University Press, London, 1954.
2. Good, R. H. Jr., *Amer. J. Phys.*, **28**, 659 (1960).
3. Good, R. H. Jr., and Nelson, T. J., "Classical Theory of Electric and Magnetic Fields," Academic Press, New York, 1971, Chap. XI.

Heitler discusses the properties of photons from different points of view than used here and especially shows various techniques for the quantization of the electromagnetic field. In Reference 2 a nonmathematical pictorial discussion of the different types of photons is given. A more complete treatment of the subject than given above is in Reference 3.

R. H. GOOD, JR.

PHOTOSENSITIZATION IN SOLIDS

Photosensitization is the enhanced response of a chemical, optical electrical or biological system to light by the addition of a sensitizing agent, usually a strongly colored substance that absorbs longer wave lengths than the base material or "substrate." An effective sensitizer is not destroyed during the photoprocess and may be incorporated within the substrate as a "dopant" or on the surface as singly adsorbed molecules or in layers. The best known sensitizers are colored inorganic ions and conjugated organic molecules including natural and synthetic dyes. The "action spectrum" of the photoprocess (i.e., the dependence of the optical efficiency on wave length) usually follows the absorption spectrum of the sensitizer. The mechanism by which the electronic excitation is utilized remains the subject of current research. In the Franck and Teller theory (1938) the excitation energy is transferred to a trapped electron in the substrate by a resonance process simultaneously

TABLE 1. IMPORTANT EXAMPLE OF SOLID STATE PHOTOSENSITIZATION

Substrate	Sensitized Photoeffect	Sensitizers	Application
amorphous selenium	photoconductivity	inorganic dopants	photocells, electro-photography (Xerox)
zinc oxide power in resin binder	photoconductivity	synthetic dyes	electrophotography (Electrofax)
silver halide powder in gelatin emulsion	reduction of Ag^+	synthetic dyes	photography
cadmium sulfide powder, sintered layer, in resin binder	photoconductivity	inorganic dopants	photocells
polycyclic aromatic compounds in resin binder	photoconductivity	synthetic dyes	electrophotography
aromatic polymers and copolymers	photoconductivity	synthetic dyes	electrophotography
plant chloroplast	reduction of CO_2 to carbohydrates	chlorophyll pigments	photosynthesis
rhodopsin in retina	photochemical isomerization	11-*cis*-retinal	animal vision

restoring the sensitizer and generating an additional conduction electron. The Gurney and Mott theory of 1938 postulates that the additional electron originates from the excited sensitizer molecule, so that restoration of the sensitizer requires transfer of an electron from the substrate to the oxidized dye. Although the net effect is the same in both theories, the mechanistic details and energetics are quite different. Factors considered in more recent work include changes in the energy levels induced by dye-substrate interactions, the effect of dye-dye aggregation, and the role of oxygen or other electron acceptor molecules present as impurities or added as dopants or activators.

Photosensitized systems of practical or natural importance are listed in Table 1. One of the first technical applications was the spectral sensitization of photographic emulsions discovered by H. W. Vogel in 1873. Dye sensitization was applied to zinc oxide electrophotographic layers by H. Grieg in 1959 and extended to other inorganic and organic layers by many workers since. Effective sensitization of silver halide and zinc oxide systems requires good adsorption between the dye molecule and the crystal grain and a proper relationship between the electronic energy levels. In general, the excitation energy of the adsorbed sensitizer must be higher than the conduction band of the substrate or sufficiently close that thermal motions can supply the required energy deficit. Stacking of dye molecules on the substrate grain leads to an intense adsorption band at longer wavelengths than the free dye adsorption (J band) which enhances the far red response. The combined effect of two dyes may exceed the sensitizing efficiency of either dye alone. This phenomenon is "supersensitization" and is particularly useful with "J aggregates." Sensitizing ability may be improved also by the presence of colorless "hypersensitizers." Hypersensitizers are electron accepting molecules that trap substrate conduction electrons lowering the dark electrical conductivity thereby enhancing the relative response to illumination. Organic layers may be n-type or p-type semiconductors and photoconductors. The p-type layers contain electron

donors (Lewis acid) and may be activated by adding electron acceptors (Lewis base) and *vice versa*. The mechanism of dye sensitization has not been established, but may involve electron transfer from the donor host to the excited dye molecule leading to enhanced p-type conductivity and the opposite for n-type conducting organic layers.

Some sensitization effects in solids are undesirable. In *phototendering* a dyed fabric disintegrates in the presence of light without noticeable fading of the dye. Anthraquinoid vat dyes are active in phototendering and the mechanism is believed to involve hydrogen abstraction from the fiber molecule by the dye triplet state. *Photodynamic action* refers to a large class of effects in which visible light irradiation of an organism in the presence of a colored drug or dye induces biological damage, mutagenesis or cell death. The phenomenon has been observed viruses, bacteria, single cells, and higher plants and animals. The presence of air is usually but not always required. The lowest triplet state of the sensitizer initiates photodynamic action, generated by intersystem crossing from the excited singlet state. When the triplet state oxidizes the substrate directly (dye-substrate mechanism), oxygen acts to restore the reduced dye. Alternatively, the triplet state may react first with oxygen (dye-oxygen mechanism) leading to several possible intermediates including excited singlet oxygen, an unstable dye-oxygen complex (moloxide) or the partially oxidized dye free radical plus the hydroperoxy radical, all of which might oxidize the substrate. The primary targets in photodynamic action have not been identified, but are believed to include DNA in the cell nucleus and possibly nucleoproteins and essential enzymes.

LEONARD I. GROSSWEINER

PHOTOSYNTHESIS

Photosynthesis is the process by which green plants harness the energy of sunlight as absorbed by chlorophyll to build organic compounds from carbon dioxide and water. The reaction is often referred to as assimilation or fixation of carbon.

Fundamental Role of Photosynthesis. For their subsistence, growth, and multiplication, all living beings, plants and animals alike, need organic food. The organic foodstuffs are utilized by living cells as building stones and as sources of energy. The energy stored in the foodstuffs is released mainly by respiration, the reaction in which organic matter combines with the oxygen of the air yielding carbon dioxide and water as final products. The rate at which living beings die and consume each other is so high that they would all disappear from the earth within the lifetime of a human generation if there were not a process providing for the re-formation of organic matter.

While there are many ways by which organic substances are decomposed in respiration and similar reactions, there is only one reaction, photosynthesis, that for millions of years has counterbalanced death and decomposition. In the course of photosynthesis the hydrogen of water is used to transform carbon dioxide into carbohydrate; simultaneously the oxygen of the water is liberated as free oxygen gas. Among the many pigments appearing in the plant kingdom only chlorophylls are known to convert sunlight into chemical energy. Chlorophyll is the green dye whose color is so characteristic of meadows and forests. It is located within green plant cells in special organelles called chloroplasts. Some yellow carotenoids and, in certain algae, blue or red phycobilins also are found in the same chloroplasts. These other pigments increase the efficiency of photosynthesis in some cases by transferring the energy they absorb to chlorophyll. Green plants further transform the carbohydrates produced by photosynthesis into fats, proteins, and many other substances. Thus, directly or indirectly, all the organic food of plants or animals depends upon the photosynthetic process. In the cycle of synthesis and death the basic materials, carbon dioxide and water, can be used over and over again, while the energy released upon the destruction of organic matter is lost forever; it is dissipated as heat in space and must constantly be replaced by the sun's radiation.

Origin and Evolution of Life. The chlorophyll systems are highly complex structures that cease to function whenever their protein components are damaged. Obviously, therefore, life cannot have started with the particular photosynthetic reaction that is its sole support at present. Life probably began with organic substances produced by ultraviolet radiation or by electrical discharges when the earth's atmosphere consisted mainly of hydrogen, methane, ammonia, and water vapour. Experiments have shown that such a gas mixture, if subjected to electrical discharges, yields many aliphatic and amino acids. Contemporary living cells still use these simple compounds as building stones, not only for proteins and fats but for the synthesis of even more intricate organic molecules. The most famous example is the formation of the porphyrin ring, the parent structure of chlorophyll and of heme, the red pigment in blood. It is formed by a succession of a few simple condensation steps. This synthesis starts with glycine and acetic acid, compounds now assumed to have been present in large quantities at the beginning of organic evolution on earth. It is likely, therefore,

that porphyrins and hence their iron and magnesium derivatives, *i.e.*, heme and chlorophyll, appeared early and took part in the emergence of the first living things.

Thus the complex mechanism of photosynthesis may have evolved as a part of the first organisms. Even at the initial stages of life the simplest photochemical reactions (reactions initiated by light) known to occur with porphyrins must have been instrumental in speeding up and selecting special types of synthetic reactions. Because the process of photosynthesis as found in present-day plants can be broken up into separate sets of partial reactions, it is assumed that, during evolution, these were added one by one, with the light-absorbing chlorophyll complex as the original centre. The last of these evolutionary steps was the capacity to release free oxygen. In this way photosynthesis has first furnished and then maintained the supply of oxygen in the atmosphere, which made the Darwinian evolution of respiring organisms possible. The preceding developmental step, the photochemical reduction of carbon dioxide at the expense of compounds other than water, seems never to have led beyond the level of unicellular microorganisms.

New understanding of the role of light in evolution has come from the discovery that the chloroplasts of higher plants contain their own nucleic acids, which have slightly different compositions from those in the rest of the cell. This agrees with the observations that chloroplasts grow and multiply independently of the cell, and that when the last chloroplast has by some circumstance been lost, the cell never makes new ones. The prime example is the case of the single-celled organism *Euglena,* which loses its chloroplasts easily while growing on organic media in the dark. Botanists therefore have long entertained the idea that chloroplasts are the descendants of green organisms that eons ago entered into symbiosis with nongreen organisms. The control of chloroplast behaviour by a double set of genetic factors lends the strongest support yet to this evolutionary hypothesis.

Efficiency of Photosynthesis. In land plants, between 1 and 3% of the light falling upon a green plant is transformed into chemical energy. Important as this answer may be for problems of agriculture, it is of no value in respect to the theoretical understanding of the process of photosynthesis. Most of the incident radiation is lost by reflection, transmission, and ineffective absorption by pigments other than chlorophyll. Only a fraction is absorbed by the active pigments. The efficiency with which this latter fraction is utilized in the course of the photochemical process is measured by the minimum number of light quanta necessary to bring one molecule of carbon dioxide to the energetic level of a carbohydrate. Ten quanta suffice: (1) to release one molecule of oxygen; (2) to furnish the energy stored as adenosine triphosphate (ATP) required for intermediate biochemical transformations; and (3) to reduce one molecule of carbon dioxide. Since ten mole quanta of red light amount to roughly 410 kilogram calories (kcal.) of energy, and while the organic food produced accounts for 112 kcal., the net overall efficiency of photosynthesis is about 30%. Much higher efficiencies are alleged to have been found in certain experiments where plants were kept under unusual conditions. In the unicellular green alga *Chlorella* the quantum yield remains essentially constant over the entire range of the chlorophyll absorption spectrum. This means that, expressed in terms of calories, the efficiency increases in the direction from blue to red in the spectrum, because the energy content of the light quanta decreases according to the laws of physics. Chlorophyll in the living cell obeys these laws exactly as do the chlorophyll-sensitized photochemical oxidations in vitro.

Artificial Food Production. Because the earth's human population is increasing so rapidly, with no corresponding increase in food production, scientists have wondered how much more food would be available if full use could be made of the great efficiency of photosynthesis which, as mentioned above, has been found to approach 30% of the intensity of the radiation absorbed by the active pigments. The average acre of land under cultivation yields hardly more than one ton of total dry organic matter per year because most of the light falling on the surface is not trapped by chlorophyll and thus is lost for photosynthesis. If agriculture were superseded by the continuous cultivation of tiny aquatic algae in thin layers of fertilized water, the same area could produce between 30 and 50 times more edible material. Before such a project can become practical, two other problems must be solved: how to maintain, easily and cheaply over large areas, conditions for optimal growth of algae and how to induce people to eat algae. A technological photochemistry for the synthesis of selected chemicals is certainly practicable. By contrast a completely artificial manufacture of food, imitating photosynthesis, does not appear to be a sensible aim to strive for. One cannot conceive of an artificial device as small and as efficient as a green plant cell for the purpose of producing all necessary nutrients.

The problem of how to nourish human beings and supply them with oxygen during long-lasting expeditions into outer space can be solved on the basis of natural photosynthesis. Oxygen for respiration would be provided by illuminated plants, which in turn would use the exhaled carbon dioxide and other waste products of man's metabolism as the raw material (carbon source, vitamins, and mineral nutrients) to grow and produce the necessary food. In short, the carbon cycle of the organic world on earth would have to be duplicated in miniature. The light energy to keep it running could be obtained either directly from the sun or indirectly from the power that propels the space rocket.

A third problem, the conversion of light energy into electricity for industrial purposes, has been solved in principle on the basis of purely physical reactions in "solar batteries," and a solution may be found for photochemical reactions also. Neither method of utilizing light, however, duplicates the photochemistry of chlorophyll in plants. (*See* also **Photochemistry.**)

Light Absorption and Energy Storage. Light energy may be used in several ways: (1) to speed up reactions that would proceed in the dark in the same way but at much slower rates; (2) for reactions that will not occur unless light energy is available; and (3) for a reaction in which the new products still hold as potential chemical energy a part of the light energy originally absorbed. In the course of organic evolution, photosynthesis reached the third stage and achieved an overall molecular efficiency of 30%. This efficiency is due to the incorporation of water as a reactant into the photochemical system and to the subsequent release of free oxygen arising from water. The photosynthetic reaction is summarized in the most general chemical terms by the equation:

$$n\text{CO}_2 \text{ (carbon dioxide)} + n\text{H}_2\text{O} \text{ (water)} + \text{light} =$$
$$n\text{O}_2 \text{ (oxygen)} + (\text{CH}_2\text{O})_n \text{ (carbohydrate)}$$

There are many other mechanisms in nature in which light plays a role (vision, for instance) and some in which carbon dioxide is transformed into organic matter without an overall gain in energy. A few species of autotrophic bacteria are known to grow in the absence of light with carbon dioxide as the sole source of carbon. They utilize the energy derived from the oxidation of inorganic compounds such as ferrous iron salts, sulfur, thiosulfate, hydrogen sulfide, and molecular hydrogen. This fact was generally ignored in earlier attempts to understand the mechanism of photosynthesis. The amount of organic matter formed by this means was considered insignificantly small as compared with photosynthesis, and the reduction of carbon dioxide in the dark appeared to be fundamentally different from photochemical reduction. In the decade between 1930 and 1940, however, it became evident that many organisms and tissues, from propionic acid bacteria to pigeon liver, include carbon dioxide among their metabolic substrates. In these cases, energy-yielding processes like respiration or fermentation are coupled to reactions in which carbon dioxide is bound by the cell and then reduced. The decisive difference between the organisms that assimilate carbon dioxide by means of their ordinary metabolism in the dark and the green plants lies, however, not in the way in which the carbon dioxide molecule is attacked but in the energy balance. No ordinary metabolic synthesis can end with an overall gain and storage of chemical energy, due to the fact that an excess of combustible material, previously synthesized, has to be sacrificed to make the reaction go.

In photosynthesis the energy of light furnishes, in the course of the reaction, the necessary combustible material or hydrogen donor by the "splitting" of water. The utilization of light energy, however, is not sufficient to guarantee a successful accumulation of chemical energy. This was shown by studies on the metabolism of purple bacteria, strongly coloured yellowish, red, or purple unicellular microorganisms. Because their growth depends on light, they were first believed to photosynthesize as do the green plants. Investigations proved, however, that purple bacteria are unable to produce molecular oxygen and unable to reduce carbon dioxide with water. In addition to radiant energy they need readily available hydrogen donors. According to the specificity of each species, the purple bacteria require for the assimilation of a certain amount of carbon dioxide an equivalent amount of either sulfur, hydrogen sulfide, thiosulfate, molecular hydrogen, aliphatic acids, or alcohols. In purple bacteria, therefore, the energy of light is not quite so successfully stored as in the green plants but is partially wasted. The light serves here only as a promoter, as it were, of reactions that might as well be carried on thermally in the dark. The overall gain consists quite directly in the appearance of complex living forms in place of unorganized molecules of "food."

During the 1960s it was found that there are at least two fundamentally different photochemical reactions in green cells not mediated by chlorophyll but indirectly influencing the outcome of photosynthetic processes. In far-red light around λ 7,500 Å, the pigment phytochrome not only determines photoperiodic responses of plant growth but also brings about changes in rates of respiration. In blue light between λ 3,700 Å (near ultraviolet) and λ 5,000 Å (blue-green), many organisms including green plants show a variety of growth responses. The responsible pigment, which is present in such small amounts that it is known only by its action spectrum, appears to be a flavin. The metabolic response to activation of this pigment, however, is anything but small. In algae it manifests itself by increasing the rate of respiration up to three times, a circumstance that must be taken into account when photosynthesis is studied at low light intensities.

These two different effects of light were discovered only recently because the intensities of either far-red or blue light needed to obtain maximal responses lie a hundred times below the average saturation intensity for photosynthesis. Both effects are examples of catalytic actions of light (like vision), since the light energy absorbed is far from commensurate with the metabolic energy consumed in the response.

(Reprinted with permission from Encyclopaedia Britannica, Copyright 1970.)

HANS GAFFRON

References

Clayton, Roderick K., "Light and Living Matter," Vol. I: "The Physical Part," McGraw-Hill Book Company, New York, 1970; Vol. 2, "The Biological Part," 1971.

"Energy Conversion by the Photosynthetic Apparatus," Brookhaven Symposia No. 19, 1966.

Fogg, G. E., "Photosynthesis," American Elsevier Publishing Co., Inc., New York, 1970.

Gaffron, Hans, "Encyclopaedia Britannica," Vol. 17, pp. 1002–1007, Chicago, 1971 Edition.

Rabinowitch, Eugene, and Govindjee, "Photosynthesis," John Wiley & Sons, Inc., New York, 1969.

Vernon, L. P., and Seeley, G. R., "The Chlorophylls," Academic Press, New York, 1966.

PHTHALOCYANINE COMPOUNDS

Phthalocyanines are tetrabenz derivatives of tetraazoporphyrins. The parent compound phthalocyanine, sometimes called metal-free phthalocyanine is:

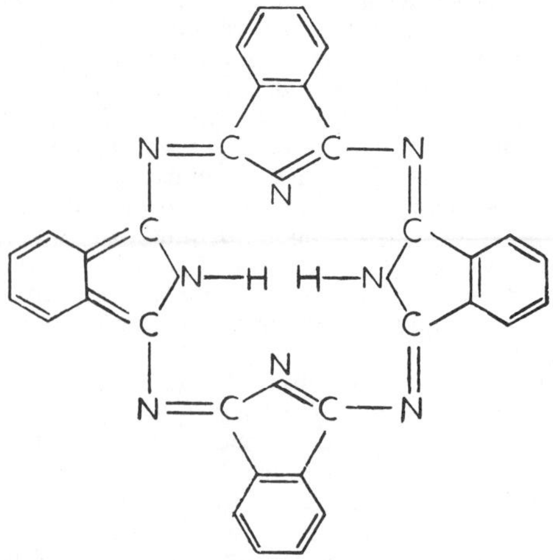

More than fifty metal derivatives of phthalocyanine have been made by substituting metals for the two central hydrogen atoms. The 16 peripheral hydrogen atoms may be replaced by numerous substituents and ring structures usual for an aromatic ring system.

The unusual stability of phthalocyanine and many of its derivatives has made it a subject of many investigations. The blue or green color of these compounds and the relative ease of their synthesis has provided new commercial colors for a range of shades where their combination of brilliance, color strength, and stability is unsurpassed.

The principal commercial use of a phthalocyanine compound is that of copper phthalocyanine and its derivatives as pigments and dyes. Copper phthalocyanine is used as a pigment in its two crystal forms. The beta-form is the more stable one and it is the so-called "peacock-shade" of commercial phthalocyanine blue. The pure alpha-form is readily converted to the beta-form by many solvents. Chemical and physical means are used to prevent the change to provide the redder alpha copper phthalocyanine blue in a tinctorially stable form. About 10,576,000 pounds of these two forms of copper phthalocyanine blue found use in the United States in 1969 in paints, printing inks, plastics, textile printing, and other pigment applications.

The highly halogenated derivatives of copper phthalocyanine containing 10 to 15 atoms of chlorine or bromine, or a combination of the two, provide the phthalocyanine greens of commerce. About 3,589,000 pounds of the greens were made and used in the United States in 1969.

The principal dye of phthalocyanine structure is the disulfonated copper derivative. It is used to dye cotton a turquoise blue shade.

Many complex derivatives of copper and other metal phthalocyanines have been patented for use as dyes. An oil-soluble form is used in making blue ball-point inks.

The unique properties of phthalocyanine compounds have contributed to their use in several noncolorant applications. Iron phthalocyanine has been used as a polymerization catalyst. Several metal phthalocyanines have been used as catalysts in the petroleum industry. One use is as a catalyst to oxidize the traces of mercaptan sulfur in petroleum distillates to a readily removable form in the "sweetening" of petroleum products. Metal-free phthalocyanine has been used as a metal scavenger in chlorinations where trace metals might initiate undesired side reactions.

Stable metal phthalocyanines have been used as thickening agents to make greases along with silicones, or other stable oils. These greases are used in very high- or low-temperature applications, where a stable product is needed.

The addition of 0.1 to 10% of phthalocyanine to polyphenyl coolants used in nuclear reactors improves their resistance to deterioration induced by exposure to neutron irradiation at high temperatures. Phthalocyanine (metal-free) can be prepared from o-phthalodinitrile by heating it in a high-boiling solvent such as quinoline, dimethylaniline or trichlorobenzene in the presence of ammonia or an amine catalyst. It can also be prepared by the removal of the labile metal with an acid from metal phthalocyanines.

Although there is no one method that may be used to prepare all metal phthalocyanines, some methods may be used to form a number of the metal complexes. Copper, cobalt, nickel, chromium, iron, vanadyl, chloroaluminum, lead and titanium phthalocyanines have been made by heating phthalonitrile in quinoline with the stoichiometric amount of metal chloride to $180-190°$ for two hours. A second well-known method uses phthalic anhydride, urea, a metal salt and a catalyst with or without a high-boiling solvent. Metal phthalocyanines may also be made by heating the metal-free phthalocyanine with a powdered metal in quinoline. Polyhalogenated metal phthalocyanines are generally made by halogenating the metal phthalocyanine.

Stable phthalocyanine compounds such as those of copper may be dissolved in concentrated sulfuric acid are reprecipitated by pouring the solution into water. The copper phthalocyanine is recovered quantitatively. Certain metal phthalocyanines are stable to acid, and to boiling caustic as well. Phthalocyanines may be sublimed in vacuum, or in an inert atmosphere at temperatures of about 550°C. Sublimation is sometimes used to purify the stable metal phthalocyanines. Halogen derivatives are stable to about 600°C but they do not sublime. They decompose above that temperature.

The stability of the metal phthalocyanines enabled Robertson to accomplish the first direct x-ray analysis of an organic molecule. The x-ray contour diagram not only confirmed the structure of the molecule but has also indicated the additional regularity of a complete resonance system. He also showed that the metal atom and the four nitrogen atoms lie in one plane. Metal-free copper, nickel and platinum phthalocyanine were shown to have planar symmetry. Cobalt, iron, manganese and beryllium phthalocyanines were the first examples of planar symmetry for those metals.

Single molecules of copper phthalocyanine on a tungsten surface have been photographed by Müller using a field electron microscope at a magnification of 2.8×10^6. The molecules appear as white discs on a black background. It was observed that bright quadruplet patterns, similar to the shape of

the molecules as usually sketched by the organic chemist, appeared on the screen. This is due to the planar nature of the molecule.

Absorption spectra for many metal phthalocyanines have been determined in both the visible and ultraviolet range and in the infrared. For the visible and ultraviolet range maxima appear in the $3300-3500$ Å, $5700-6000$ Å and $6250-6780$ Å ranges. The alpha and beta crystal forms can be distinguished by both x-ray diffraction and infrared spectrography. There is evidence that the so-called gamma type of copper phthalocyanine crystal is a finely divided alpha type. A delta-type crystal has been described.

Two types of bonding of the metal atoms are illustrated in metal phthalocyanine compounds. The metals sodium, potassium, lithium, magnesium, calcium, barium, cadmium, and antimony are held by electrovalent bonds. These metal phthalocyanines are insoluble in chloro-naphthalene and quinoline and they do not sublime. Bivalent nickel, copper, platinum, beryllium, manganese, iron, cobalt, tin, lead, zinc and vanadium and trivalent aluminum may be considered to be covalent salts. The electrovalent metals are removed upon treatment of their phthalocyanine molecules with acids. The covalent metal phthalocyanine complexes are stable to acids.

Phthalocyanine compounds are resistant to atmospheric oxidation at temperatures up to 100°C or higher depending on the particular metal complex. In aqueous acid solution strong oxidizing agents oxidize phthalocyanines to phthalic residues whereas in nonaqueous solution an oxidation product that can be reduced readily to the original compound is usually formed.

Copper phthalocyanine was observed to catalyze the formation of water from hydrogen and oxygen. Phthalocyanines have been observed to catalyze the decomposition of hydrogen peroxide, the oxidation of linseed oil and the oxidation of numerous other organic molecules.

Phthalocyanine and its metal derivatives were among the first organic compounds found to be intrinsic semiconductors when the unusual temperature dependence of their resistivities was discovered.

Electrical conductivity in the phthalocyanines can be induced by impingement of light as well as by application of an electrical field on a phthalocyanine substance. It has been shown that metal-free phthalocyanine has a greater photoeffect than magnesium, zinc or copper phthalocyanines.

Phthalocyanine compounds also participate in photochemical reactions, act as photosensitizers and may luminesce or fluoresce.

Knowledge of the photochemical properties of chlorophyll has stimulated interest in the photochemical properties of the phthalocyanines and of magnesium phthalocyanine in particular. The possibility of obtaining identical reaction products of chlorophyll, magnesium phthalocyanine, and their analogs by dark and photochemical methods has been demonstrated. There is a rapid drop in conductivity of a 10^{-5} M solution of chlorophyll and of magnesium phthalocyanine after cessation of illumination.

Photochemical reductions can be sensitized by

chlorophyll or by magnesium phthalocyanine such as the reduction of riboflavin by absorbic acid.

Chemiluminescence may be observed when magnesium or zinc phthalocyanines are added to boiling tetralin containing small amounts of benzoyl peroxide.

Phthalocyanine Blue (C.I. Pigment Blue 15, C.I. No. 74160.) copper phthalocyanine, is not approved as a food additive in the United States, but it may be used in rubber, urethane, thermoplastics, etc. used in contact with food, where it does not become a food additive as delineated in Federal Drug Administration regulations.

FRANK H. MOSER

Cross-references: *Dyes, Pigments, Organometallic Compounds, Organic Semiconductors, Catalysis.*

PHYCOCOLLOIDS

Agar. Agar is the cell wall polysaccharide (galactan) of certain marine red algae, including the genera *Gelidium, Pterocladia, Gracilaria, Acanthopeltis, Ceramium, Digenia,* and probably *Ahnfeltia.* It is regarded at present as a mixture of two polysaccharides, agarose and agaropectin. Agarose consists of 1,3-linked D-galactose (51 to 53%) and 1,4-linked 3,6-anhydro-L-galactose (about 46%). Smaller quantities of 6-O-methyl-D-galactose (1 to 20%, depending upon species), L-galactose (1 to 4%) and D-xylose (0.2 to 1.8%) are also present. Structurally agarose appears to be linear chain of alternating 1,3-linked β-D-galactopyranose and 1,4-linked 3,6-anhydro-α-L-galactopyranose with 3,6-anhydro-L-galactose at the nonreducing end, galactose at the reducing end of the molecule, and a half-ester sulfate on about every tenth galactose unit.

Agaropectin, the other constituent, is not as well known, probably because it is a mixture of polysaccharides that are difficult to separate. It consists mainly of D-galactose, 3,6-anhydro-L-galactose, 3.5 to 9.7% ester sulfate and D-glucuronic acid. It has some of the same structural characteristics as agarose but does not form as strong a gel. Agaropectin from some species of *Gelidium* is known to contain about 1% pyruvic acid. An agaropectin-like polysaccharide from *Ahnfeltia plicata* is about 20% L-arabinose, through its place in the structure is apparently unknown.

Agar forms a strong, thermally reversible aqueous gel. Normally 1.5% is sufficient to gel bacteriological media and permit inoculation with a wire loop without breaking the surface. It gels at about 38 to 42°C (although it will gel at several degrees higher temperature if kept long enough at that temperature) and melts at about 88°C. Agars from various sea plant sources exhibit considerable variation in these and other physical properties (syneresis, gel elasticity), a reflection of the proportion of agarose and agaropectin present, degree of hydrolysis during extraction (a function of time, temperature, pH), and of other unknown variables. Sea plant polysaccharide chemistry is considerably influenced by environmental conditions such as salinity, temperature, pH, nutrient salts, and polluting substances that may be present.

Though insoluble in cold water, agar absorbs as much as 20 times its weight, with much swelling—a characteristic related to its previous treatment (hysteresis). It dissolves readily in boiling water and has a relatively low viscosity as a sol. The viscosity changes little with fall in temperature until the temperature of gelation is approached. Acid hydrolysis is rapid when an agar solution is heated at a pH below 5.0.

Sea plant raw material is usually dried and then extracted in boiling water at a dry raw material/water ratio of 50:1000 (weight). This ratio varies with the species used. A pH between 5.0 and 6.0 speeds extraction but may involve risk of partial hydrolysis.

Freezing and thawing an agar gel is an effective means of dehydration and purification. About 85% of the water held by the gel will drip free as agar ice melts, and along with this water will go about 85% of the cold-water-soluble impurities such as salts, sugars, pigments, amino acids. This procedure is used in agar manufacture following removal of particulate matter by forcing the hot agar solution through a filter press with diatomaceous earth as filter aid.

Agar is a hydrophilic colloid of importance in foods, pharmaceuticals, cosmetics, and many other commodities. It is used in a dental impression compound and in other prosthetic work. It is an ideal laxative because it is indigestible (by humans), absorbs water, and provides harmless bulk. It is a stabilizer, gelling agent, emulsifier, controller of texture and moisture.

Agar-bearing sea plants are generally 20 to 60% agar on a dry weight basis. The wet weight/dry weight ratio ranges from about 5:1 to 20:1, depending upon species.

Carrageenan. Carrageenan is the term applied to cell wall polysaccharides of a number of marine red algae including the genera *Chondrus, Gigartina, Eucheuma, Agardhiella, Hypnea, Furcellaria, Iridaea,* and *Polyides.* Originally, the term was limited to extractives of *Chondrus crispus* and *Gigartina stellata* ("Irish moss"), but as knowledge of the chemistry of carrageenan developed, it became obvious that the term should be applied to the phycocolloids of a variety of marine red algae.

Carrageenan, like agar, is a mixture of polysacharides. Up to the present, six carrageenans of definite chemical structure have been recognized and designated by the Greek letters iota, kappa, lambda, mu, nu, and xi. Kappa and lambda, because of their abundance and economic value, are the best known. These two and iota may be the only ones of the six that are naturally-occurring end products in the plants.

Kappa carrageenan has 1,3-linked D-galactose 4-sulfate alternating with 1,4-linked 3,6-anhydro-D-galactose. ' Mu, regarded as a precursor of kappa, differs in having D-galactose 6-sulfate in place of the anhydride. Conversion of mu to kappa, catalyzed by an enzyme, involves removal of sulfate from C6 of the 1,4-linked groups with resulting closure of a ring to form anhydride. Iota carrageenan differs from kappa in that it has a sulfate at C2 on the 1,4-linked groups. Nu is thought to be a precursor to iota. In lambda about 70% of the 1,3-linked groups are sulfated at C2 and not at C4. Xi is poorly known, but there appears to be complete sulfation at C2 in the 1,3-linked groups.

In physical properties the carrageenans differ from agar in that agar forms thermally reversible gels, the properties of which are not significantly affected by solutes. The carrageenans will not gel at all in the absence of a solute (other than traces), and some carrageenans will not gel even in the presence of a solute. The properties of carrageenan gels (gel strength, temperatures of gelation and melting) are determined by the nature and amount of the solute or solutes present. Gel strength is also strongly influenced by concentration of the polysaccharide.

Considerable confusion exists because of the overemphasis in the literature on the influence of potassium ions as a gelling agent for iota and kappa carrageenan, a result of the use of KCl by the industry. All solutes cause gelation, not only inorganic salts but sugars, alcohols, and other organic compounds. Apparently the role of solutes in gelation has not been explained. Efforts to do so in terms of the potassium ion are not likely to yield the truth. Lambda carrageenan does not gel.

Some species of algae produce only one type, others two or more. The proportions in the plants vary considerably in response to environmental conditions in many species of algae.

Since a wide variety of polysaccharides, with reference to physical properties, can be prepared commercially from carrageenans these colloids find specific uses in a host of foods, cosmetics, pharmaceuticals, and other products. This is a sea plant industry that originated in the United States.

Algin. Algin, laminaran, and fucoidan are phycocolloids characteristic of the brown algae. Laminaran is a term applied to a variety of water-soluble β-1,3-linked glucans and is a food reserve along with mannitol, hence it is intracellular. Fucoidan apparently refers to several similar polysaccharides from the cell walls and consists mainly of fucose and ester sulfate, but xylose and glucuronic acid have also been obtained from it. These polysaccharides have not been commercially utilized, though laminaran may have medical applications.

Algin, a polyuronide, occurs in the cell walls of all brown algae and makes up 14 to 40% of the total dry weight. It is analagous to agar and carrageenan of the red algae. It apparently occurs in the plants in the form of salts of alginic acid, principally calcium alginate. It is apparently a polymer of anhydro-mannuronic acid and -guluronic acid residues with these two units varying in proportion in different species, in different parts of large plants, and within the same plant with season.

Alginic acid is insoluble in water. It is obtained by precipitation following the addition of strong acids to alginate solutions. Alginates of alkali metals and ammonia are water-soluble, as is magnesium alginate (in contrast to other divalent metals). As the concentration of alginate is increased, there is a tendency to change from a highly viscous sol to pastes and to gel-like solids. The viscosity of a 1% solution is usually between 500 and 3000 cps. Alginates of most divalent metals are insoluble in water and in organic solvents.

A variety of esters of alginic acid can be made, but the only commercially important one is the propylene glycol ester. It is not precipitated by acids and exhibits viscous sols at low pH. It is regarded as a safe food additive (in contrast to esters of ethylene glycol).

The algin industry originated in Great Britain. An estimated ten million pounds is produced annually with the United States and Britain the major sources. In California, the giant kelp, *Macrocystis pyrifera,* is harvested by barge-like vessels having an underwater cutter and a conveyor-ramp for loading. They can harvest 300 tons, wet weight, per day. In Great Britain the intertidal rockweed, *Ascophyllum nodosum,* is a major raw material along with species of the kelps in the genus *Laminaria* that grow well below low tide.

In the extraction process, the raw material is dried, pulverized, and washed. It is then extracted by steeping in a solution of dilute sodium carbonate with the result that ion exchange occurs in the raw material forming soluble sodium alginate which then diffuses out. Purification is accomplished by precipitation with mineral acid to form insoluble alginic acid. A second precipitation may be carried out.

Alginates are used in a wide variety of food, pharmaceuticals, cosmetics, and other products for their emulsifying, thickening, and suspending properties. In foods, they are important stabilizers of ice cream, sherbets, cheese spreads, fountain syrups, and toppings.

While uses of agar, carrageenan, and alginate are similar, usually one is superior to the other two in any specific application.

HAROLD J. HUMM

References

Lewin, R. A., Editor, "Physiology and Biochemistry of Algae," Academic Press, 111 Fifth Avenue, New York 10003, 929 pages, 1962.

Levring, T., Hoppe, H. A., and Schmid, O. J., "Marine Algae, A Survey of Research and Utilization," Cram, De Gruyter & Co., Hamburg 3, 421 pages, 1969.

Margalef, R., Editor, "Proceedings of the Sixth International Seaweed Symposium," Subsecretaria de la Marina Mercante, Direccion General de Pesca Maritima, Ruiz de Alarcon, Madrid, 782 pages, 1970.

Percival, Elizabeth, and McDowell, R. H., "Chemistry and Enzymology of Marine Algal Polysaccharides," Academic Press, 219 pages, 1967.

PHYTOCHEMISTRY

Phytochemistry is concerned with all the chemical activities of plant life. A few decades ago it was reasonable to divide plant chemistry into two general categories; the first was concerned with the plant as a photosynthetic organism, which fixes carbon dioxide from the air and in conjunction with a few mineral elements absorbed from the soil produces a variety of carbohydrates, proteins, fats, vitamins, etc., which are essential for the life of animals. Without green plants animal life would disappear from the earth.

In addition to the chemistry involved in photosynthetic activities, which is common to all plants containing chlorophyll, many thousands of other chemical processes take place in plants. Some of these chemical reactions are confined only to certain plants, thus they are not necessarily involved with processes essential for photosynthesis. However, as methods of detection become more sensitive, the distribution of such compounds is shown to be more widespread than was previously supposed. They have also been shown to undergo reactions with so-called essential constituents of green plants. Advances are being made very rapidly in our knowledge of these special compounds. Over one thousand new ones are being reported yearly. Books on various groups of these compounds are appearing in great profusion.

A discussion of the chemicals occurring in plants, including both those found in all green plants and those known, for want of a better name, as secondary products, involves the following groups: chlorophylls; carotenoids; mono- and oligo-saccharides; starch and inulin; hemicelluloses and gums; cellulose and chitin; pectic substances; flavanoides; glycosides; nonvolatile acids; amino acids; proteins; purines and pyrimidines; lipids; alkaloids; terpenes; miscellaneous volatile plant products; steroids; surface waxes; cutin and suberin; lignin; rubber, gutta percha, and chicle; vitamins; acetylenic compounds; sulfur and selenium compounds; plant growth regulants; and the gibberellins. Even this list is not complete. In the space available it is not possible to discuss all these classes; most of them are described in other articles in this volume.

As methods for the detection and identification of substances found in plants become increasingly sensitive one is truly amazed at the variety of substances occurring in different species, and even more so at the great number found in individual species.

With the use of activation analysis it is possible to detect the presence of every known inorganic element in any solution tested. Whereas methods for the detection of organic compounds are not this sensitive, there is some reason to suspect that the occurrence of organic compounds may be ubiquitous also. In any event, with the use of chromatographic methods, for the isolation and identification of organic compounds, the number of occurrences of reported various compounds has mushroomed over the past decade.

The widespread occurrence of certain compounds has become embarrassing to chemical plant taxonomists; so much so that one author has suggested that if a substance is reported to be present below a certain concentration it should not be taken too seriously.

An interesting aspect of plant chemistry is the number of instances in which certain chemicals, either essential or secondary products, occur in a great diversity as to concentration. Thus the dry matter of the Irish Potato consists almost entirely of starch, whereas certain other plants may contain almost none. Similarly the quantity of fats present in plants varies from a very high percentage to a very small amount. Also the number of compounds, again both essential and secondary, may be extremely high.

Even though the cardiac glycosides have very complicated structures, often a large number of different ones are found in the same tissue. Thus,

the leaves of *Strophanthus boivinii* Baill. contain 24 cardenolides and about 20 heart glycosides have been reported present in the seeds of *Acokanthera oppositofolia* Lam. A study of the fatty acids in limes and lemons has revealed the presence of 81 and 102 different acids in the two fruits, respectively.

A good example of the large number of compounds reported in a single species is the 1200 compounds found in tobacco and tobacco smoke. This does not include the individual components of complex substances such as the brown pigments and resins which have not been resolved. Admittedly the number of compounds in tobacco is influenced by the effects of curing; nevertheless these data give an indication of the tremendous complexity of the chemical systems in plants, the enormous number of enzymes which are present, and the extremely complicated set of controls which must be operating to determine when and how much of a given ingredient is to be formed.

Space does not permit inclusion of many important plant products treated in detail elsewhere in this volume. See articles on Cellulose, Starches, Sugars, Alkaloids, Fats, Gums, Vitamins, etc.

Amino Acids. The *amino acids* occur in plants as constituents of proteins and in the free state. An amino acid is defined as a compound which has both an amino and a carboxylic acid group. The general structure may be represented as $\underset{\text{NH}_2}{\text{RCHCOOH}}$ in which R represents an aliphatic, aromatic, or heterocyclic group. The more common exceptions to this generalization include amino acids with a second amino or carboxyl group, and those which have in addition an amide or guanidine group. The more common natural amino acids contain a free amino group alpha to the carboxyl; exceptions are proline and hydroxyproline, in which the amino group becomes an imino group forming part of a pyrrolidine ring. In general the natural amino acids contain one or more asymmetric centers. Most of the natural amino acids have the L-configuration. A number of D-amino acids have been found in bacterial cell wall hydrolyzates and in fungal antibiotics.

The amino acids commonly found as constituents of proteins include glycine, alanine, valine, leucine, isoleucine, asparatic acid, glutamic acid, asparagine, glutamine, proline, serine, threonine, lysine, arginine, histidine, cysteine, methionine, phenylalanine, tyrosine, and tryptophan. These 20 amino acids are incorporated into proteins following their initial activation by 20 specific aminoacyl-sRNA synthetase enzymes.

By chromatographic methods, most of these regular constituents of proteins can readily be identified as components of the free amino acid pool of all plants. The concentrations of the basic and sulfur-containing amino acids and of tryptophan and proline are usually considerably lower than those of the other amino acids. Even these amino acids are occasionally accumulated in very high amounts as for example, histidine in ripening banana fruits, and arginine in apple trees (up to 66% of the nitrogen in extracts of young twigs represented this amino acid). High concentrations of proline have also been observed in some species.

There is no relation between the concentration of amino acids in the soluble nitrogen pool and the quantities combined in the tissues's protein. There is a somewhat metabolically inert vacuolar pool, that seems to contain a large proportion of any amino acid that a plant accumulates in unusual amount. Quite often types are included which are not known to be incorporated into protein. Of these γ-aminobutyric acid is virtually ubiquitous, but is regarded as a secondary product with no clear physiological function.

There are about 200 plant constituents which have been strictly characterized as amino acids (these are nonprotein amino acids except for the relatively few protein amino acids) and there is a further large number of nonprotein amino acids which have been identified only by chromatographic parameters during surveys of plant material.

Plant Proteins. Very many different proteins occur in plants. There may be as many as a thousand in a single cell. Plant proteins consist of a specific sequence of amino acids residues, joined by peptide linkages. Proteins generally consist of 17 different amino acids, one imino acid and two amide residues, each of which is usually represented several to many times in a single protein.

In addition to the so-called protein amino acids, a considerable number of others have been cited as occurring in a few proteins, but unless critical chemical identification has been carried out, claims for new protein amino acids should not be accepted. However, there are several authentic cases, such as the occurrence of ϵ-trimethyllysine in several plant cytochromes c, hydroxyproline predominantly in the cell-wall protein, and iodotyrosines in marine algal protein. Molecules of an homologous enzyme with a specific activity, for example, ribonuclease, isolated from different species, will normally differ slightly in their amino acid sequences, according to the species from which they were isolated.

There is a contrast between the large number of specific functions which proteins perform and the simplicity of their basic design. This results because the function of proteins depends on the particular sequence of the amino acid residues in the protein. The simplicity of their basic design is even more impressive when one considers that there are about 200 different amino acids which occur free in plants.

Role of Minerals in Phytochemistry. Of the many elements found in nature only a relatively few (and not necessarily the most abundant) are essential for the growth of green plants. These include C, H, O, N, P, S, Ca, Mg, K, Fe, and the minor elements Mn, Cu, Cl, Zn, B, Mo, and in special instances, Co, Si, Se, and some others. Plants differ widely in the percentage composition of these elements and there often is no direct relation to the quantity of various elements present in plants and the composition of the media in which they are growing. In addition to the elements required for growth, plants absorb many nonessential elements found in their environment, often accumulating them to a marked degree. The formerly held concept that plants could control what they take in, whereas true to some degree, is essentially fallacious. Plants no doubt contain traces of all elements present in their environment. Plants absorb nonessential elements from the soil in which they are growing to such a degree that they serve as indicators of the mineralization present, and are, in fact, used for this purpose.

Plants growing in soil rich in selenium have become so adapted to this element that it has become essential for their growth. Organic compounds of selenium are important constituents of these species and take part in necessary metabolic processes.

A study of the mineral composition of food plants has shown, that in spite of the effects of the mineral content of the media in which they are grown, there is a certain consistency in content of certain elements. Thus, the sodium content, which is important for heart patients, may vary from a trace to one milligram per 100 grams of the edible portion in corn to 47, 60, and 126 milligrams in carrots, beets, and celery respectively.

The more or less consistent composition of food plants with regard to mineral elements does not support the natural-food faddists, who oppose the use of commercial fertilizers. Whether the growing of foods without commercial fertilizers alters food composition to an important degree has not been established as far as the writer is aware.

The role of the essential elements has been studied for many years and may be briefly summarized as follows. *Nitrogen* occurs in proteins and in other important cell constituents such as purines, pyrimidines, porphyrins, and coenzymes. *Phosphorus* is found in both the inorganic and organic form and is absorbed chiefly as the orthophosphate ion. Phosphate plays an important role in energy transfer, in respiration and photosynthesis, and in many other metabolic reactions. Phosphorus occurs in organic form in a large variety of important compounds. These include phytic acid, phospholipids, phosphorylated sugars and their intermediate breakdown products which are found in the glycolytic and alternate pathways, nucleoproteins and nucleic acids, purine and pyridine nucleotides, flavin nucleotides, and other enzymes. *Sulfur* participates in the formation of the sulfur-containing amino acids, cystine, cysteine, and methionine, and the sulfur bearing vitamins, biotin, thiamine, and coenzyme A. *Potassium* is essential as an activator for enzymes in both protein and carbohydrate metabolism. A deficiency affects respiration, photosynthesis, chlorophyll development, and the water content of leaves. It is necessary to maintain cell organization, permeability, and hydration. *Magnesium* activates a number of important enzymes involved in respiration, photosynthesis, and the synthesis of nucleic acids. It is the mineral constituent of the chlorophylls. *Calcium* is an important constituent of cell walls in the form of calcium pectate. Calcium is required for cell division and chromosome stability, cell expansion and middle lamella structure, mitochondrial production, cell hydration, and possibly also in determining ionic permeability. Although deficiency symptoms of *Boron* are striking, a specific role has not been established. There is some speculation that it may function in the translocation of sugar. *Chlorine* is required for the photosynthetic evolution of oxygen. *Copper* is a component of phenolases, laccase, and ascorbic oxidase. *Iron* is present in cytochromes and in nonheme iron proteins which are involved in photosynthesis, nitrogen fixation, and respiratory-linked dehydrogenases. *Manganese* function as an enzyme activator in respiration and nitrogen metabolism. In many reactions it can be replaced by other divalent cations such as those of magnesium, cobalt, zinc, and iron. *Molybdenum* functions in gaseous fixation and nitrate assimilation. *Zinc* serves as an activator of several enzymes such as alcohol dehydrogenases, carbonic anhydrases, alkaline phosphate, and others.

Halometabolites. The halogens, chlorine, fluorine, bromine, and iodine are found in all environments in which both land and sea plants are growing and often considerable quantities of these elements are taken up. They usually remain in ionic form, but there are many interesting exceptions. Chlorine has been reported in the fungal partner of the algal fungal combination. In general nonionic chlorine compounds have been found relatively rarely in higher plants. In 1959, it was reported in jaconine an alkaloid from *Senecio jacobaea* L. (Compositae). It was next reported in acetylenic compounds found in a number of species of the Compositae. Other nonionic chlorine compounds discovered in higher plants recently include a chlorinated amino acid, chloroindole compounds, chlorine containing alkaloids, and sesquiterpenes. Nonionic halogen compounds have been frequently found in the fungi (in addition to those associated with lichens). Many antibiotics contain nonionic halogen as part of the molecule.

A considerable number of species contain organically bound fluorine. Of these fluoroacetic acid is most commonly known.

LAWRENCE P. MILLER

Reference

Miller, Laurence P., Ed., "Phytochemistry," Vols. I, II, III, Van Nostrand Reinhold Co., New York, 1973.

PICKLING

Foods

Pickling, when applied to foods, is a process used to flavor and to protect perishable foods from undesirable action of enzymes and microorganisms. Acid, usually in the form of vinegar, with or without salt, sugar and spices is used, often in conjunction with heat.

Today's methods are outgrowths of the ancient processes of fermentation and of salting and drying. Since food practices in various parts of the world differed in so many respects, it is difficult to trace the sources of present technics. Many practices originated from methods of preservation by lactic acid bacterial fermentations.

Today's well-standardized methods of pickling produce many wholesome and appetizing foods. These may be grouped in three general classes: fruit products, vegetable products, and meat and fish products.

The preservation of fruits depends in part upon the natural presence or addition of organic acid to lower the pH enough so that they may be heat-processed at low temperature. In a few instances alcohol is used. In general, fruits are heated in a sugar-vinegar solution containing a suitable spice mixture, filled hot into containers, and covered with syrup at a minimum of 180°F. To avoid cracking or splitting of skins or undue shrivelling, some fruits require soaking in unheated syrup for several hours prior to final processing. Sufficient vinegar or lemon juice is added in preparation of some pickles to acidify the product to a degree required to insure safe processing at 180°F. In a broad sense jellies, jams and preserves may be considered pickled products.

The second group of products includes those that are eaten either in the fermented or salted state or after refreshing and conversion to edible pickled products. The former consist primarily of sauerkraut, vegetable blends, and salt and dill cucumber pickles. The latter include cucumber pickles, olives, pickled onions, cauliflower, beans and green tomatoes. Sauerkraut differs from salt and dill pickles in that it has been dry-salted with only 2 to 2½% salt, while the pickles are packed in a salt brine of sufficient concentration to contain finally 2½ to 5% salt.

Many vegetables, some fruits, and nuts are packed in higher concentrations of salt brines (5 to 15%) and may or may not undergo a typical lactic fermentation, depending on salt concentration. The vegetable is later removed from the brine, refreshed with water, and converted to one of several types of pickles by addition of acids, salt, sugar, and spices. As in the case of pickled fruits, prolonged presoaking in sugar solutions is often essential before final processing. In this process the sugar concentration in the pickle is raised slowly to avoid shrivelling. The concentrations of sugar, vinegar, and salt in some pickles are so high that pasteurization is not essential.

The flavor imparted and the keeping quality produced by pickling of hams, bacon, certain beef products, and other meat and fish products have no substitute.

Smoked fish and meat, often previously packed in a pickle brine, depend for preservation upon the substances added during pickling, the products developed in the food during pickling, the surface drying action during smoking, and the action of smoke products such as aldehydes, acids, and phenolic derivatives. Cold smoking is sometimes used in processing on farms. Preservation of some sausage meats or sausage depends partially upon a marked lactic acid fermentation similar to that occurring in vegetables.

The pickling of food products depends upon the combined action of a number of chemical substances classified as foods, sugars, organic acids, salt, spices, smoke derivatives, standard products of fermentation, and certain unidentified substances produced within the food as a result of its method of handling. Although certain additives of a non-food character, such as nitrates and nitrites, are used to effect desirable changes, the use of other chemical preservatives, such as benzoic acid, sulfites, boric acid, etc., is falling into disuse.

CARL S. PEDERSON

Cross-reference: *Foods.*

Metals

"Pickling" is a term applied to the process of removing oxides, scale and other impurities from metal surfaces. *Sulfuric acid* is the pickling agent most often used, largely because it is the lowest cost acid available. The operation is simple. The steel is immersed in tanks of acid in such a fashion that the acid can reach all surfaces. The acid first attacks the scale and exposes the clean metal surface. Attack on the metal then begins. When the scale is uniform and all surfaces are equally exposed to the acid, the work can be removed as soon as it is scale-free. This avoids any serious attack on the metal. When the descaling action is spotty because of nonuniform scale or other factors, overpickling often occurs in localized areas. This produces an excessively roughened surface and spoils the work of many applications.

An inhibitor is usually added to the pickling bath to reduce the rate of attack on the metal. Inhibitors have little effect on the rate of descaling. They vary in their effectiveness in protecting the metal. If used in large amounts, some will almost completely stop the attack on low carbon steel but are less effective for high carbon steel.

As the temperature or acid concentration of the pickling bath is increased, the pickling time is reduced. Benefits from increasing the acid concentration above 15% (by weight) are not significant at temperatures up to 212°F. However, the higher boiling point of more concentrated acid allows its use at even higher temperatures than 212°F, with an increase in pickling rate.

Electrolytic Pickling. Electrolytic action can be used to increase the speed of acid pickling. The electric current may be alternating or may be direct, with the work anodic or cathodic. Cathodic pickling is the most rapid method. However, the advantage decreases greatly with rise in bath temperature and acid concentration. For most installations, acid and heat are less costly than direct current, so electrolytic pickling is not economical. Cathodic pickling is sometimes considered objectionable because it charges the steel with hydrogen at a faster rate than chemical pickling.

Hydrochloric acid is a good pickling agent. It is used less than sulfuric acid because it is more costly and the fumes are a greater problem. It is receiving increased attention for large-scale operations that include regeneration of the acid and discarding of metal oxides. Interest stems from pickling benefits and simpler waste disposal needs. Also, small establishments may prefer hydrochloric acid because of its fast action at room temperatures and effectiveness for light acid dipping before electroplating.

The pickling rate for hydrochloric acid mixtures can be increased by raising the temperature. However, loss of HCl as vapor and corrosion of equipment become serious problems at temperatures above 130°F.

Phosphoric acid is also a good pickling agent and is intermediate in cost between sulfuric and hydrochloric acids. It leaves the pickled steel in a somewhat passivated state because of forming iron phosphates at the descaled surface. This is helpful in some cases where rust-resistance is needed but very unsatisfactory if the steel is to be electro-plated after pickling. The pickling rate is slower than with hydrochloric- or sulfuric-acid solutions. Heating is necessary for satisfactory pickling rates.

The relative economics of the acid pickling solutions and the operating details may change as ways are developed to treat the used solutions to avoid pollution by plant effluents.

CHARLES L. FAUST

Cross-references: *Foods, Metals, Electroplating.*

PIGMENTS

Pigments are small, discrete solid particles used for opacifying, decorative, or protective purposes and which are insoluble in the medium in which they are employed. They may be used either in a medium applied as a surface coating, e.g., paint or printing inks; or in the bulk of the material itself, as in the mass coloring of plastics and rubber. With the exception of extenders, many of which are in the 1–50 micron size range, pigment particles are typically 0.01–1 micron in diameter. Unlike light mixtures, pigment colors are subtractive so that a mixture of the primary pigment colors is black, not white.

The properties of a pigment (including hue, opacity, lightfastness, heat stability, solubility, ease of dispersion, and rheological characteristics in fluid media) are not solely a function of chemical structure, but are influenced significantly—and in some cases primarily—by such physical factors as particle size, size distribution, shape, crystal structure, and the nature of the pigment surface.

For brilliance, tinctorial strength, and variety of hues, the organic pigments far surpass the inorganic, and are therefore used in applications where these factors take precedence over economy, heat stability, lightfastness, and solvent bleed resistance —areas where the inorganic pigments generally excel. It is sometimes advantageous to use organic and inorganic pigments in combination. Lithol Rubine, for example, can be blended with Molybdate Orange to obtain enamels lower in cost and yet superior in bleed resistance to Toluidine Reds.

Inorganic Pigments

White Opaque Pigments. This class of pigments is used primarily to gain opacity and for tinting colored pigments, viz. adjusting the lightness and saturation of a color mixture. The most important member of this class is titanium dioxide, TiO_2. Its major attributes are brightness, excellent opacity, high tinting strength, and inertness.

The anatase form of TiO_2 tends to chalk and is used primarily as a self-cleaning exterior white. The rutile form has superior hiding power (opacity) and tinting strength and resists chalking as well.

The remaining pigments are used where their special properties are required, either alone or, more frequently, in conjunction with TiO_2.

Zinc oxide, ZnO, is used in coatings to control rheologic properties, impart hardness, promote adhesion, etc.

Basic carbonate white lead, $2PbCO_3 \cdot Pb(OH)_2$, is used primarily to enhance film durability and adhesion. Basic silicate white lead, approx. $2PbO \cdot SiO_2$, is essentially similar but is required in lesser amounts.

Zinc sulfide, ZnS, and lithopone, $ZnS \cdot BaSO_4$, have lower opacity than TiO_2 and are not widely used currently.

Antimony oxide, Sb_2O_3, is used primarily in the fire-retardant paints.

White Extender Pigments. White pigments in this class are used as film builders and reinforcing agents in coatings to lower costs or improve chemical or physical properties.

Several important pigments are: calcium carbonate, $CaCO_3$, calcium sulfate, $CaSO_4$, barytes or blanc fixe, $BaSO_4$, diatomaceous silica (remains of marine organisms), and china clays (hydrated aluminum silicates). These are used to change the rheology, add bulk, improve the durability, improve ease of dispersion, and control gloss.

Various magnesium and aluminum silicates (mica, talc, asbestine, etc.) are examples of products with fibrous or platelike particles which impart special properties such as resistance to cracking, reduction of permeability, etc.

Black Pigments. The most important black pigments are elemental carbon particles: carbon blacks, lamp blacks, and bone blacks. This group has the advantages of relative inertness, low cost, and excellent lightfastness.

A black iron oxide, Fe_3O_4, is finding increasing use in printing inks for man-machine readable characters.

Iron Oxides. Iron oxides are available as browns, maroons, reds, oranges and yellows. These products are obtained either as natural oxides or as synthesized colors. Their major attractions are low price and inertness. The natural oxides are cheaper but require a sacrifice of quality: brightness and strength. Their excellent solvent and chemical resistance, lightfastness, and good hiding power dictate their use when economy is important. Yellow and brown iron oxides are heat-sensitive at elevated temperatures.

Chrome Colors. Lead chromates are available as yellows, oranges, and reds, and are used in large volume because of their relatively low cost, moderately good lightfastness, brightness (versus iron oxides), resistance to solvents, and good hiding power. However, their disadvantages are their only fair chemical resistance and heat resistance.

The lead chromates contain several polymorphic forms and solid solutions: chrome orange (tetragonal, $PbCrO_4 \cdot PbO$); medium yellow (monoclinic, $PbCrO_4$); light yellow (monoclinic, $PbCrO_4 \cdot PbSO_4$); primrose yellow (orthorhombic, $PbCrO_4 \cdot PbSO_4$); and molybdate oranges and reds (monoclinic, $PbCrO_4 \cdot PbMoO_4 \cdot PbSO_4$). The molybdate oranges, 75–80% $PbCrO_4$, have an additional advantage in that they are comparable in brightness with some organics and have better hiding power.

Strontium chromate, $SrCrO_4$, and the zinc compounds, $K_2O \cdot 4ZnO \cdot 4CrO_3 \cdot 3H_2O$ and $ZnCrO_4 \cdot 4Zn(OH_2)$, are yellows used in making corrosion inhibiting primers.

Chromium oxide, Cr_2O_3, and hydrated chromium oxide, $Cr_2O_3 \cdot 2H_2O$ are green pigments. Chromium oxide is dull and weak in tinting strength; however, the color is essentially inert and low in cost. Hydrated chromium oxide is a very bright, transparent color with excellent lightfastness and solvent and chemical resistance, but is high in cost and has low tinting strength.

Chrome greens are low cost mixtures of a chrome yellow, usually a primrose, with prussian blue.

Cadmium Colors. The basic constituent of the solid solutions which make up the cadmium colors is the yellow cadmium sulfide, CdS. Solid solutions which contain increasing amounts of zinc sulfide, $CdS \cdot ZnS$ result in yellows which are increasingly greener in hue. If instead cadmium selenide, $CdS \cdot CdSe$ is used, oranges, reds and maroons are produced. The Mercadiums*, $CdS \cdot HgS$, are analogous to the cadmium sulfoselenides in hues but contain mercuric sulfide, rather than cadmium selenide, in solid solution with cadmium sulfide. The major differences between the two are that the Mercadiums* are more economical, and at
* Hercules Trademark.
quite elevated temperatures the sulfoselenides are more stable.

The cadmium colors are sold both pure and in large volume as the lithopone (a mixture containing approximately 60% $BaSO_4$). They are much brighter than iron oxides and have excellent lightfastness, solvent and chemical resistance, and good hiding power and heat resistance. However, they are high in cost, fade in some tint applications, and have some deficiencies in weathering.

Prussian Blue. The major inorganic blue is either ammonium or potassium ferric ferrocyanide. The color has a relatively low cost, good tinting strength acid resistance, bleed resistance, and has excellent lightfastness in full tone. Poor alkali and heat resistance and only fair lightfastness in weak tints limit the use of this pigment.

Organic Pigments

Precipitated dyestuff pigments are simply dyes which have been insolubilized by salt formation. Anionic dyestuffs such as Tartrazine Yellow and Orange II are precipitated as metal salts on an aluminum hydrate substrate. Cationic dyes, for example Methyl Violet and the Rhodamine Reds, are most commonly precipitated with a heteropoly acid such as a phosphomolybdic acid. These pigments excel in brilliance, cleanliness, and tinctorial strength; however, they are generally much inferior to the better azo pigments in heat and light stability and in alkali resistance. Use of precipitated dyestuff pigments is limited primarily to printing inks and metal decorating.

Fluorescent pigments are fluorescent dyes which have been immobilized in solution in a polymer matrix, a procedure necessitated by the fact that fluorescence is generally reduced drastically if the dyestuff is used in the solid state. The dyes employed, e.g., rhodamines, are those wherein the emitted light reinforces the normal reflected light of the compound, producing colors which are extremely bright, although relatively light fugitive.

Azo pigments are the coupling products of diazotized arylamines with, most commonly, acetoacetanilides, naphthols, or pyrazolones. They account for the greatest portion of organic reds, oranges, and yellows consumed. The azo pigments lacking salt-forming carboxy and/or sulfo groups have good alkali stability but, particularly in the case of the lower molecular weight products, are relatively poor in solvent bleed resistance. Conversely, metal carboxylate and sulfonate azo pigments are comparatively insoluble in lacquer and enamel solvents but are more sensitive to alkali.

High-performance pigments combine the high degree of brilliance, strength, and cleanliness associated with the classical organic pigments with the fastness properties generally found only among inorganic pigments. They are frequently employed in coatings, e.g., exterior automotive finishes, and in synthetic fibers where solvent bleed resistance, heat stability, and/or excellent lightfastness is required.

The most important of the high-performance pigments are the copper phthalocyanines. Phthalocyanine Blue, the copper chelate of unsubstituted phthalocyanine, is marketed primarily in two crystal modifications; a relatively red shade (α- form) and a greener shade (β- form). The α- form tends to grow in particle size and revert to the more stable β- polymorph in the presence of heat or certain solvents unless specially stabilized. Phthalocyanine Green is obtained by replacing essentially all of the 16 peripheral hydrogen atoms in Phthalocyanine Blue with chlorine, or if a yellower shade green is desired, with a mixture or chlorine and bromine. High-performance pigments of other hues have not yet approached the relatively low cost of the Phthalocyanine Blues and Green.

Other high-performance pigments include Azo Condensation, Flavanthrone, Anthrapyrimidine and Tetrachloroisoindolinone Yellows; Anthanthrone and Perinone Oranges; Dioxazine Violets; and Azo Condensation, Perylene, Thioindigo, and Quinacridone Reds, Maroons, and Violet. Like Phthalocyanine Blue, the Quinacridones also afford a practical example of polymorphism: β-Quinacridone is violet, whereas the γ- form is red.

CHARLES G. INMAN
N. WILLIAM WAGAR

References

Gaertner, H., "Modern Chemistry of Organic Pigments," *Journal of the Oil and Colour Chemists' Associaton,* **46,** 13 (1963).

Patton, T. C., Ed., "Pigment Handbook," Vol. I, John Wiley & Sons, New York, 1972.

Patterson, David, Ed., "Pigments—An Introduction to their Physical Chemistry," Elsevier Publishing Company, Great Yarmouth, England, 1967.

PLANETOLOGY, CHEMICAL

Planetology is the logical extension and generalization to the planets of the Earth Sciences such as geology, geochemistry and meteorology. However it is broader than any combination of these and embodies a new integration of many branches of science which have in the past been regarded as quite distinct. Because it pertains to the part of the energy spectrum in which the major planetary processes fall, chemistry occupies a central position in this scheme.

The terrestrial or inner planets in particular show how variations in planetary physical parameters are reflected in chemical behavior. Some of these parameters, such as the mean solar distances, planetary radii and densities, have been known reasonably well for some time. However, others of equal importance, such as the rotation rates and surface temperatures, have become well-known with advances in radio technology and planetary probes. Some of the pertinent physical data are shown in Table 1 which has been compiled with special regard for those factors which influence the chemical environment. Particularly important in this respect are the gravitational accelerations, solar distances and temperatures, which to a large degree govern the escape of the atmospheric constituents. These temperatures have been obtained chiefly by the measurement of microwaves emitted from the solid bodies of the planets. As far as is known all the temperatures fall rapidly with height in the atmospheres to some minimum or minima and then rise again to very high values ($> 1500°K$) at the top where they merge with that of the solar plasma. For the planet Venus, the surface temperatures have been revised upward by as much as $300°K$ in recent years, thus changing drastically the concept of chemical processes on this planet.

The great temperature variations among the planets should have a profound effect on the degree of interaction of the planetary surfaces with the lower atmospheres as well as on the stabilities of the compounds which comprise them. In the case of Earth, we know that equilibrium is attained only within small volumes of rock (of the order of a cm^3) and then only when crystallization occurs at temperatures usually exceeding $500°K$ deep within the crust during epochs of profound shearing and deformation of the layers. The rocks we observe on the surface are the exhumed products of this crystallization and are grossly out of equilibrium internally as well as with the atmosphere. This disequilibrium not only preserves these products of deep-seated processes for our inspection but also is necessary for life which depends on atmospheric oxygen.

Although Venus is the sister planet of Earth in size and density, the generally high temperatures which appear to prevail on this planet alter the chemical environment radically. Under these conditions the attainment of inorganic chemical equilibrium is greatly favored and both homogeneous reactions within the lower atmosphere and heterogeneous reactions between the atmospheric gases and the minerals of the lithosphere should proceed. One of the most abundant atmospheric constituents on Venus is CO_2, as has been known for a long time. The partial pressure of this constituent at the surface appears to be about 100 atmospheres to an order of magnitude. Given the high temperatures for the surface it seems very likely that this gas is produced from the lithosphere by reactions of the following type:

$$\underset{\text{calcite}}{CaCO_3} + \underset{\text{quartz}}{SiO_2} \rightleftharpoons \underset{\text{wollastonite}}{CaSiO_3} + \underset{\text{gas}}{CO_2} \quad (1)$$

$$\underset{\text{magnesite}}{MgCO_3} + \underset{\text{quartz}}{SiO_2} \rightleftharpoons \underset{\text{enstatite}}{MgSiO_3} + \underset{\text{gas}}{CO_2} \quad (2)$$

These are all common minerals in Earth's crust

TABLE 1. PHYSICAL DATA FOR THE TERRESTRIAL PLANETS.
BY PERMISSION OF THE JOURNAL OF CHEMICAL EDUCATION.

	Mean Solar Distance	Solar Radiation Intensity	Siderial Period of Revolution in days	Siderial Period of Rotation in days	Mean Radius	Mean Density in g/cm³	Surface Gravity	Visual Albedo	Velocity of Escape at Surface km/sec	Normal Range of Surface Temperature in °K
Mercury	0.387	6.68	88	88	0.38	5.46	0.38	0.058	4.3	250–625
Venus	0.723	1.91	225	~243*	0.961	5.06	0.88	0.76	10.4	500–900
Earth	1.0	1.0	365	1.0	1.000	5.52	1.00	0.39	11.3	225–315
Moon	1.0	1.0	365	27.3	2.273	3.33	0.16	0.072	2.4	75–400
Mars	1.524	0.43	687	1.0	0.523	4.12	0.39	0.148	5.1	175–315

*Retrograde.

under the proper conditions of pressure, temperature and composition. Thermochemical data indicate that reaction (1) is approximately in equilibrium at 100 atm CO_2 partial pressure and a temperature of 700°K, which is the approximate mean surface temperature. At 800°K this reaction should be displaced to the right with CO_2 liberation, and at 600°K it should be displaced to the left with CO_2 absorption at a CO_2 pressure of 100 atm. Thus it is possible that CO_2 is liberated in the hotter parts of the planet near the subsolar point and reacts again with the rocks at the poles and antisolar point. Various other volatile chemical compounds may also be transported by this type of mechanism.

We are accustomed to think of oxygen as an abundant atmospheric constituent, but at the high temperatures inferred for Venus this element would be expected to react rapidly with the mineral components such as Fe_2SiO_4 (olivine), $FeSiO_3$ (pyroxene) and Fe_3O_4 (ferrite) to form additional ferrite or Fe_2O_3 (hematite). These are major components of the basaltic lava flows which have been extruded from deep within Earth's crust or mantle and cover much of the continents as well as oceanic areas. If, on the basis of similarity of planetary scale, we are justified in assuming that Venus has followed a course of evolution parallel to Earth, we should expect these lavas to be present in the Venus crust also. However, there the reactions of oxygen with the iron compounds should prevent it from accumulating in the atmosphere beyond the value determined by the following equilibrium:

$$\underset{\text{ferrite}}{2Fe_3O_4} + \underset{\text{gas}}{\tfrac{1}{2}O_2} \rightleftharpoons \underset{\text{hematite}}{3\ Fe_2O_3}. \qquad (3)$$

At 700°K P_{O_2} for this reaction is only 10^{-20} atm, which is far too little to be detected by direct means. However, P_{O_2} may also be determined from the reaction

$$\underset{\text{gas}}{CO_2} \rightleftharpoons \underset{\text{gas}}{CO} + \underset{\text{gas}}{\tfrac{1}{2}O_2}, \qquad (4)$$

which relates the observable quantities P_{CO_2} and P_{CO} to the oxygen pressure. From observations the ration P_{CO_2}/P_{CO} is inferred to be approximately 10^6, which at 700°K corresponds to a value of P_{O_2} close to 10^{-20} atm. Consequently the Venus surface may be in equilibrium with respect to reaction (3).

Many other reactions involving such species as H_2, H_2O, NH_3 CH_4, etc., may be written. The observed quantity of water above the reflecting cloud layers is only 10^{-3} per cent of the total atmosphere. The concentration of H_2O below the clouds appears to be in the range of 0.1 to 1% as inferred from the Russian planetary probes. Consequently, at the prevailing temperatures the entire atmosphere appears to be very dry so that existence of liquid water on the surface seems impossible.

Although there may be only minute quantities of O_2 present in the atmosphere, the value $(P_{CO_2}/P_{CO}) = 10^6$ defines the atmosphere as quite oxidizing for many of the compounds referred to above. This and the apparent small quantity of water militate against the occurrence of any significant amounts of hydrogen and reduced carbon compounds such as NH_3 and the hydrocarbons. This aspect of the equilibrium considerations is particularly interesting, as it was once widely believed that hydrocarbons were abundant on the surface and in the clouds.

On Earth, and by inference on the other planets as well, the solar ultraviolet radiation which heats up the upper atmosphere also causes dissociation of molecules such as CO_2 and H_2O into lighter species such as CO, O, OH, and H and at the same time ionizes many of these particles. Atomic hydrogen along with He is light enough to escape from planets the size of Earth and Venus while even atomic oxygen and OH may escape from Mercury and Mars. On Earth and Venus CO and O (and by recombination also O_2) may accumulate in amounts exceeding those ordained by the thermochemical equilibrium and this must be taken into account in the analysis of the spectra and abundance calculations.

From the above considerations a zonal model of the Venus atmosphere emerges: Zone 1, near the surface constitutes the thermochemical reaction zone in which the molecular abundances are established. Because of the probable rapid decrease of temperature with height we must postulate above the reaction zone a zone of frozen thermochemical equilibria, Zone II, in which abundances established in zone I are preserved. It is chiefly this zone which is observed in the spectra. Above these is Zone III in which photochemical reactions dominate and which imposes some modification on the spectra of the lower zones.

Reactions analogous to those discussed for Venus may also account for the atmosphere of Mercury. The presence of this atmosphere is inferred from the observation that the dark side of this planet has a temperature considerably above absolute zero. Since Mercury always presents the same hemisphere to the sun these unexpectedly high temperatures can be accounted for only by assuming the heat is being transferred from the bright side by some fluid medium. Because of the great intensity of ultraviolet radiation in the Mercury orbit it might be expected that the common molecular species, CO_2, O_2 and H_2O would be dissociated. Also because of the small mass of Mercury the decomposition products, O, OH and H should escape. It is possible, however, that the more stable heavier species such as N_2, and CO might be able to exist in the lower atmosphere along with argon the common product of the radioactive decay of K^{40} in rocks. Because of the possible loss of oxygen, the atmosphere of Mercury may be highly reducing and the surface rocks may contain such substances as metallic iron and graphite. However it seems unlikely that the total surface pressure could rise above 10^{-3} atmosphere.

Conditions on the outer terrestrial planets Earth, Moon and Mars contrast sharply with those we have deduced for Venus and Mercury, chiefly because of the considerably lower surface temperatures of the former. In the case of the moon, the very small mass practically ensures the escape of all the common atoms and molecules. This process is self-reinforced by the permissive nature of the vanishingly thin atmosphere toward the ultraviolet radiation which dissociates and ionizes the atoms and molecules and thus furthers their escape. Specimen rocks obtained during lunar landings

of the Apollo series have confirmed the suspected low water content of the moon. Chemical analyses have also shown the lunar rocks to be high in titanium and iron and low in potassium and sodium as compared with Earth's crust. Examination of these rocks and the lunar surface features indicate that they are products of cataclysmic events involving the infall of large meteoroids and resultant volcanic activity. Radioactive dating of the rocks indicates that most of these events occurred about 4.5 billion years ago, although some also occurred much later.

Although Earth and Mars are cool and massive enough to retain their atmospheres, there is almost complete disequilibrium between the atmospheric gases and the surface rocks. This general disequilibrium on Earth has already been described. For Mars it can only be inferred from the kinetic properties of minerals and from the atmospheric compositions as given by the spectra.

Both Earth and Mars appear to show a rough correspondence to the ice-vapor equilibrium if temperature variations are taken into account. On Mars the partial pressure of water appears to correspond to the temperature of the polar regions which serve as sinks for water in the form of ice. Also on Earth there seems to be a fairly close approach to the equilibrium value of CO_2 for the reaction (2) since the observed partial pressure corresponds to this equilibrium at about 330°K, which is reasonably close to the surface temperature. This was first pointed out by Urey who attributed the attainment of equilibrium to the presence of liquid water, which should act as a catalyst.

The origin of our solar system is the most intriguing problem with which planetology has to deal. The chemical aspects of this problem have been investigated by a number of authors but in greatest detail by Urey, who treated the condensation of planetary matter from a gas of solar composition, the hypothetical solar nebula which forms the starting point for nearly all recent thought on this subject. The solar gas is quite reducing at the higher temperatures (> 2000°K) where the melting and vaporization of silicates and metals are important. Under these conditions iron can exist only as a metal and carbon is nearly all bound as CH_4 or CO. However as the temperature falls more oxidized compounds such as Fe_3O_4 become stable. It is obvious that the initial composition of planetary material depends on the temperature and pressure at which condensation occurred. Unfortunately the process is difficult to trace beyond this point since we can only guess at the mechanisms of accumulation, consolidation and differentiation of the planets which followed. This is especially true of the deep interiors where strange and unknown high pressure phases may exist. It therefore seems desirable to keep separate the hypotheses of the origins of planets from those about their contemporary states. However both aspects of planetary studies are expected to come under closer scrutiny as interplanetary investigations intensify.

ROBERT F. MUELLER

PLASTICIZERS

A plasticizer is a high-boiling solvent or softening agent, usually liquid, added to a polymer to facilitate processing or to increase flexibility or toughness. (Where these effects are achieved by chemical modification of the polymer molecule, e.g., through copolymerization, the resin is said to be "internally plasticized.")

Thermoplastic polymers are composed of long-chain molecules held together by secondary valence bonds. When incorporated in the polymer with the aid of heat or a volatile solvent, plasticizers replace some of these polymer-to-polymer bonds with plasticizer-to-polymer bonds, thereby facilitating movement of the polymer chain segments and producing the physical changes described. Thermosetting polymers, which consist of three-dimensional networks connected through primary valence bonds, are not usually amenable to such softening by external plasticizers. Not all the thermoplastics can be plasticized satisfactorily. Polyvinyl chloride polymers and copolymers, and cellulose esters respond particularly well to plasticizing and represent the major outlets for plasticizers.

The results obtained by addition of plasticizer vary with different polymers. In polyvinyl chloride, for example, plasticizer concentrations of 30–50% convert the hard, rigid resin to rubber-like products having remarkably high elastic recovery, while similar plasticizer concentrations in cellulose acetate produce tough but essentially rigid products.

The plasticizer field has grown tremendously since camphor was patented as plasticizer for nitrocellulose in 1870. The reported U.S. production of plasticizers in 1968 exceeded 1.3 billion pounds. Close to two-thirds of this production volume is used in vinyl chloride polymers and copolymers. Plasticized polyvinyl chloride is fabricated at elevated temperature into film and sheeting, and into molded and extruded articles; it is also processed in the form of dispersions, which may be applied as coatings and subsequently baked to form continuous films. These dispersions include plastisols, in which the plasticizer is the continuous phase; organosols, in which a volatile solvent is added; and water dirpersions. Plastisols are also used in molding.

Compatibility. Where the polymer-to-plasticizer attraction is strong, the plasticizer has high compatibility with the resin and is said to be of the "primary" or "solvent" type. With polyvinyl chloride this attraction is furnished particularly well by ester groups. Where the polymer-to-plasticizer attraction is low the plasticizer is of the "secondary" or "nonsolvent" type; it functions as a spacer between polymer chains but cannot be used alone because of limited compatibility. This is manifested by exudation of the plasticizer from the resin. Secondary plasticizers are often employed to take advantage of other desirable properties which they may impart. Where they are used merely to cheapen the formulation, they are usually referred to as "extenders."

Plasticizing Efficiency. It is customary when evaluating plasticizers in polyvinyl chloride to compare them at concentrations which produce a standard apparent modulus in tension, as measured at room temperature. Since the stress-strain relationship is generally nonlinear it is necessary to specify a given point on the stress-strain curve as well as the rate of loading or straining. The efficiency may be expressed as the concentration of a given plasticizer necessary to produce this standard modulus. Other properties, e.g., indentation hardness, may take the place of tensile modulus.

Such tests, while they constitute an adequate basis for routine evaluation of plasticizers, furnish only a rudimentary picture of the elastic properties of the plasticized resin. More complete studies supply valuable information. For example, tensile creep tests have shown that polyvinyl chloride resin plasticized with trioctyl phosphate will deform more in response to stresses of short duration than will resin plasticized with tricresyl phosphate; the reverse is true for stresses of long duration.

For many applications low-temperature flexibility of the plasticized composition is also important. Plasticizers of low viscosity and low viscosity-temperature gradient are usually effective at low temperature. There is also a close relationship between rate of oil extraction and low-temperature flexibility: plasticizers effective at low temperature are usually rather readily extracted from the resin. Plasticizers containing linear alkyl chains are generally more effective at low temperature than those containing rings. Low-temperature performance is evaluated by measurement of stiffness in flexure or torsion or by measurement of second-order transition point, brittle point or peak dielectric loss factor.

Permanence. Where a plasticizer is used in thin films it is important that it have low vapor pressure. For polyvinyl chloride a plasticizer vapor pressure of no more than 4 mm at 225°C has been suggested as a rough criterion, although much higher vapor pressures are tolerated in plasticizers for cellulose acetate. Volatile losses are determined, not only by the plasticizer vapor pressure, but also by the plasticizer-resin interaction; plasticizers of limited compatibility may exhibit unexpectedly high volatile loss. Comparisons are usually made by heating a plasticized sample in contact with activated carbon and measuring the weight loss of the sample. Resistance of the plasticizer to migration determines whether the plasticized composition will mar or soften varnished surfaces with which it comes in contact. Stability and water-and oil resistance are further factors in plasticizer permanence. Plasticizers which have poor oil resistance are usually also the worst offenders with respect to marring.

Commercial Plasticizers. *Phthalates.* These esters, prepared from *o*-phthalic anhydride, constitute the most important group of plasticizers from the stand-point of production and sales volume. Among these the dioctyl phthalates are the most widely used in vinyl chloride resins, where they are preferred because they offer a good compromise with respect to a wide range of properties: satisfactory volatility, good compatibility, fair low-temperature flexibility, and moderately low cost. Most popular of the group is the 2-ethylhexyl ester, known as DOP, but the isoctyl, n-octyl, and capryl esters also find use. Increased emphasis on low volatility and plastisol viscosity stability has in recent years led to use of octyl decyl and didecyl phthalates. The lower alkyl phthalates are generally employed in resins other than polyvinyl chloride. Dibutyl phthalate is used chiefly in nitrocellulose lacquers and polyvinyl acetate adhesives;

diethyl, dimethyl, and di(methoxyethyl) phthalates are used in cellulose acetate. For low toxicity, methyl and ethyl phthalyl ethyl glycolates are favored for cellulose acetate, and butyl phthalyl butyl glycolate for vinyl chloride polymers and nitrocellulose.

Phosphates. The phosphates, second only to phthalates in production volume, are favored for flame resistance and low volatility. Tricresyl phosphate (mixed meta and para isomers) is the most popular; it is used in polyvinyl chloride and in nitrocellulose lacquers. Resins plasticized with tricresyl phosphate are deficient in low-temperature flexibility. Diphenyl cresyl phosphate and triphenyl phosphate are other examples, the former for polyvinyl chloride, the latter for cellulose acetate. Diphenyl 2-ethylhexyl phosphate is preferred to tricresyl phosphate in polyvinyl chloride where its low toxicity and improved low-temperature flexibility are required. Tri(2-ethylhexyl) phosphate is outstanding among phosphates used in polyvinyl chloride with respect to low-temperature flexibility; in flame-and oil resistance, however, it is inferior to tricresyl phosphate. Tri(butoxyethyl) phosphate finds some use in synthetic rubber. (See **Phosphates**)

Esters of Aliphatic Dibasic Acids. This group consists of the adipates, sebacates, and azelates. These esters lead the field for low-temperature flexibility and efficiency in vinyls: they are useful in the preparation of low-viscosity stable plastisols. The principal disadvantages are high cost and poor solvent resistance. They are generally incompatible with cellulose acetate. Cheapest but most volatile are the adipates. Among the sebacates, the 2-ethylehexyl ester is outstanding in polyvinyl chlorides for low-temperature flexibility and low volatility; the butyl ester is used in polyvinyl butyral. The 2-ethylexyl ester of azelaic acid falls between the corresponding adipate and sebacate in price and properties.

Fatty Acid Esters. Monohydric alcohol esters of fatty acids usually have limited compatibility with vinyl chloride polymers. However, small amounts of these esters are useful for imparting low-temperature flexibility and softness and for their lubricant action during processing. Butyl oleate and stearate are leaders in this group. Methyl and butyl acetyl ricinoleates are used in nitrocellulose. Various glycol and glycerol esters of fatty acids have properties similar to the monohydric alcohol esters. Castor oil (glyceryl triricinoleate) and its derivatives are used in nitrocellulose. Triethylene glycol di(2-ethylbutyrate) is the leading polyvinyl butyral plasticizer.

Epoxidized esters of fatty acids are similar in their plasticizing properties to other fatty acid esters, with the added advantage that they improve heat and light stability of vinyl resins by acting as hydrogen chloride scavengers. Although generally used together with conventional stabilizers, they can also be used as the sole stabilizer.

Polymeric Plasticizers. Polyesters prepared from dicarboxylic acids and diols enjoy some use in polyvinyl chloride resins for specialized applications. Ranging in properties from somewhat viscous liquids to soft resins, these plasticizers have outstanding permanence: they are of very low volatility and generally show good resistance to extraction and migration. The acids used commercially in preparing them include adipic, sebacic and azelaic; among the diols employed are propylene glycol and 2-ethyl-1,3-hexanediol; monobasic acids may be included in the preparation to limit molecular weight. Reported molecular weights of commercial polyester plasticizers range from 850 to 8000. Those of highest molecular weight are the most permanent but are somewhat difficult to process because of high viscosity and slow solvation of the resin. Other disadvantages include poor low-temperature properties and high cost. Most polyesters are rated as secondary plasticizers and are blended with the more efficient monomeric types.

Butadiene-acrylonitrile copolymers may also be used in polyvinyl chlorides and are rated as primary plasticizers for these resins. Like polyesters, they are favored for permanence, particularly oil resistance, and because they permit high loading with filler without serious impairment of physical properties. They are difficult to incorporate in vinyl resin and are often used together with conventional plasticizers. GRS rubber and polyisobutylene are also blended with certain polymers for plasticizer-like action. Other addition polymers have been studied as plasticizers, among them a series of promising liquid polymers of dibutyl itaconate.

Other Plasticizers. Acetyl tributyl citrate is an outstanding nontoxic plasticizer for polyvinyl chloride in food packaging. Plastisol formulations containing this ester have exceptional viscosity stability. Acetyl triethyl citrate is a good plasticizer for the cellulosics and acetyl trioctyl citrate shows promise with vinylidene chloride polymers. Other compounds of diverse nature find application as plasticizers. These include tetra-*n*-butyl thiodisuccinate, camphor, *o*-nitrobiphenyl and partially hydrogenated isomeric terphenyls.

CHARLES J. KNUTH

Cross-references: *Solvents, Fatty Acids, Esters, Phosphates, Polyester Resin.*

PLATINUM AND ITS COMPOUNDS

Platinum is one of the group of six elements in Group VIII of the Periodic Table which because of numerous resemblances are collectively designated the platinum metals. They occur together in many places in the world, but the economically significant deposits comprise these elements disseminated at the parts-per-million level in ore bodies which are also significant sources of copper and nickel. These are at Sudbury, Ontario, Canada; the Bushveld Igneous Complex, South Africa; and near Norils'k, Siberia, U.S.S.R. and in the Kola Peninsula, U.S.S.R. The platinum metals rank between 71st and 80th among all elements in order of abundance in the earth's crust. Yet their cosmic abundance appears to be relatively much greater than this, and so it has been postulated that terrestrially these elements have concentrated preferentially in the earth's iron-nickel core. Such a distribution occurs during smelting of ores and concentrates of these metals under conditions which produce some fused base metals (iron-nickel-copper) and liquid slag. The precious metals are then found practically entirely in the metallic phase, and this is in fact the way these elements are concentrated from Canadian ore prior to refining them.

"Platina" or native platinum had been known as an unwanted adjunct of gold from Spanish America throughout the 18th century. At the beginning of the 19th century its composition was mainly established by the work of British and French scientists. Pure platinum was first produced in England by W. H. Wollaston in 1803. He and his collaborator S. Tennant are credited with identifying five of the six elements in this group.

The metal occurs mainly in the alloy native platinum in which it is the principal component. The other platinum metals, save perhaps osmium, are present at low concentrations, and there are a few per cent of base metals, iron and often copper, nickel or silver. Platinum also occurs in the other platinum-metal alloys such as iridosmine. There are a few compound mineral species, notably sperrylite, $PtAs_2$; cooperite, PtS; and braggite, $(Pt, Pd, Ni)S$. The over-all crustal abundance of platinum has been estimated as 0.005 ppm; although this is lower than the corresponding figure for palladium, platinum is regarded as having the greatest cosmic abundance among the elements in the group.

Some physical properties of platinum follow: atomic number 78; atomic weight 195.09; stable isotopes 192 (0.78%), 194 (32.8%), 195 (33.7%), 196 (25.4%), 198 (7.23%); density 21.45 g/cc (20°C); melting point 1772°C; hardness (annealed) 40–42 Vickers units; electrical resistivity 9.85 μ ohm-cm (0°C). The metal crystallizes with a cubic closest-packed lattice having $a = 3.9231$ Å and yielding a metallic radius (12-fold coordination) of 1.385 Å.

Alloys of platinum with other metals have been extensively studied from the points of view of improving mechanical or thermal properties, of reducing cost, or of broadening applications. There is extensive and in many cases complete miscibility on the part of platinum with the elements of Groups VIII and IB. The chief commercial alloys are with copper, gold, iridium, rhodium or ruthenium. Recently alloys with cobalt have received considerable attention because of their strong ferromagnetic properties. Improved hardness and strength are achieved by addition of iridium; alloys with up to 30% iridium remain workable. For high-temperature applications rhodium is the preferred hardening agent.

The most striking feature of the platinum metals is their tendency to resist oxidation, particularly into aqueous solution. From the energy relationships involved, it is evident that this nobility is due mainly to the very strong cohesion of the atoms in the solid state. The metals show much increased reactivity in finely divided form, and in highly dispersed condition such as the "blacks" they exhibit remarkable catalytic powers.

Platinum is not appreciably attacked by oxygen, though a superficial coating of PtO_2 probably forms and is volatilized above 1000°C. Individual acids have little effect on the metal, but aqua regia will dissolve it, forming chloroplatinic acid, H_2PtCl_6. Halogens react with hot metal, fluorine producing

PtF_4 and chlorine a mixture of $PtCl_2$, $PtCl_3$ and $PtCl_4$. Fused alkali or alkaline salts like cyanides or nitrites attack the metal strongly. Also numerous elements like carbon, phosphorus, silicon, arsenic, etc., combine or alloy with platinum at elevated temperature. Care must be taken to avoid heating compounds of these elements in platinum laboratory ware under reducing conditions. Even sooty burner flames or unburned gas may cause embrittlement of the metal through production of a carbide.

Despite the reluctance of the elements themselves to enter into direct chemical reactions, a vast number of their compounds have been characterized. The abundance of such compounds arises from several considerations. First, these elements may and do assume a wide range of oxidation states. Second, the metal atoms, generally with square-planar or octahedral coordination, form almost exclusively covalent attachments to the atoms or groups of atoms (ligands) with which they are combined. Third, the platinum group elements have electron configurations appropriate to forming strong bonds with a wide variety of donor atoms in the ligands. Accordingly, the number and variety of ligands that have been introduced into this structural framework in recent years has become enormous. The complex ions or molecules having this structure exhibit considerable kinetic as well as thermodynamic stability. Consequently the opportunity is also afforded for a good deal of isomerism. Only a few compounds can be mentioned here and under entries for the other platinum metals.

The common oxidation states of platinum in combination are +2 and +4. Higher fluorides, PtF_5 and PtF_6, are known and are extremely reactive; the latter in fact brought about formation of the first compound of an inert gas, $XePtF_6$. The dichloride is a brownish green solid, insoluble in water but dissolving in hydrochloric acid to form H_2PtCl_4. The tetrachloride is a red-brown solid, soluble in water. It may be obtained from H_2PtCl_6, chloroplatinic acid, which is formed when the metal dissolves in aqua regia. This is a strong acid, forming yellow or orange solutions; its potassium or ammonium salts are sparingly soluble in water. The latter is the compound in which this metal is separated and purified from the other metals of this group. When H_2PtCl_6 is treated with Na_2CO_3 a yellow or brown precipitate, $PtO_2 \cdot xH_2O$, platinic acid, is formed. This is amphoteric in character, and if separated and heated at 100°C it forms a brown or black solid PtO_2. This oxide reverts to the elements when heated above 200°C.

The solution chemistry of platinum is almost entirely based on complexes. Those of platinum (II) are square-planar, while those of platinum (IV) are octahedral. These form remarkably stable bonds and are relatively inert to ligand substitution. Because of their stability these complexes have played an important role in coordination chemistry, originally in Werner's working out of the nature of "secondary valency," and in more recent times in the study of geometric isomerism and the mechanism of substitution reactions.

Because of the character of the commercially exploited deposits of platinum metals, much of their production is linked to that of nickel and copper.

Residues from the electrolytic refining of copper and nickel, or of the carbonyl refining of the latter, yield precious metal concentrates. The richer South African deposits furnish considerable concentrate by gravity separation alone. The isolation of the individual platinum metals from the concentrates entails prolonged wet chemical operations for details of which more specialized references should be consulted. The total world production of platinum metals in 1968 was 3.4 million troy ounces, of which more than 90% was platinum and palladium. Owing to the nobility and high intrinsic value of these metals, considerable quantities are annually recovered from scrap, used appliances, spent catalyst, etc.

Most of the applications of platinum depend either on its catalytic powers or on its nobility. In the petroleum industry the metal catalyzes the isomerization of hydrocarbons in gasoline to improve its octane rating ("Platforming"), or the synthesis of particular hydrocarbons for petrochemical production. The chemical industry employs Pt-Rh gauze catalysts in the production of nitric acid from ammonia or of hydrogen cyanide. Finely dispersed platinum catalysts are used in production of pharmaceuticals and vitamins, most particularly as hydrogenation or dehydrogenation catalysts; also in the development of air pollution abatement devices such as mufflers for diesel engines.

Very pure platinum is used in resistance thermometers and in thermocouples. An alloy with 10% (sometimes 13%) rhodium forms the second element in the latter. Both these devices are used as means of interpolation in the International Practical Temperature Scale (revised 1968). The rhodium alloy is also used in windings for high-temperature electric furnaces, and for the spinnerets and bushings used in the production of rayon and glass fiber. Other applications in the chemical industry include the lining of small pressure vessels, or for the fabrication of larger reaction vessels from platinum-clad nickel or copper.

Alloyed usually with iridium the metal is used for jewelry, laboratory ware, electrodes (especially anodes for electrolytic oxidations), electrical contacts and electrodeposited printed circuits. An alloy with 4% tungsten is used in certain spark electrodes and for the grids of radar tubes. In recent years certain alloys with cobalt with strong magnetic properties have been developed, and are used in hearing aids, self-winding watches, and so forth.

W. A. E. McBryde

POISONS (CATALYST)

There are many cases in the nomenclature of science where a term which was used in early development lost its original suitability but was nevertheless retained more or less from force of habit. *Poison* as applied to the activity of a catalyst is a case in point. The literature is filled with terms designed to supplant the word *poison* in this sense, and some of them are real improvements. Currently such terms as inhibitors, selective adsorption, fouling, compound formation and many others are used to denote this concept. In one sense promoters serve as a negative catalyst (*poison*) to the reverse reaction. This, of course, aids the forward catalyst.

Just where the expression "catalyst poison" became scientific is not easy to establish, but the most obvious case is in the oxidation of SO_2 to SO_3 over platinum. Here poisoning with arsenic seems to have been a matter of coating the platinum with a thin layer of arsenic metal. It would be stretching the case to call this compound formation or an intermetallic compound. Hence poisoning in this case means a kind of solid adsorption.

In the various hydrogenation reactions over such catalysts as copper or chromium, sulfur is referred to as a poison, because either of these will form sulfides easily. In each case the catalyst would be poisoned, but for an entirely different reason chemically.

Almost any substance that deactivates a catalyst poisons it. For example, just as there is autocatalysis, there is autopoison. If NO_2 automatically catalyzes the action of nitric acid on copper, so will the selective adsorption of one or more reactants act as a poison.

In general, we seldom refer to poisoning of a gaseous catalyst. It is obvious that gaseous catalysts are less likely to be subject to the same type of contamination as solids. Hence the term poisoning is not used of nitrogen oxides in the chamber process, whereas it is applied to platinum used in contact catalyst for the same reaction.

The word "poison" is seldom used for liquid catalysts. In their case one refers to the formation of another compound. Intermediate compound formation at active points is a recognized phenomenon accounting for many catalytic mechanisms.

Enzymes are generally regarded as catalysts. The enzyme zymase is generally operative in solutions up to a certain percentage of alcohol; it might be proper to call this a case of autopoisoning because the product of the reaction kills the catalyst. Likewise there are many borderline cases of fermentation processes which, while not catalytic in the strictest sense, are nevertheless active only in media of a narrow pH range. They might be said to be inhibited, or their effect destroyed by a modest change in pH, since most of them act in a range of 4 to 9. Likewise most enzyme catalysts are poisoned by the heavy metal ions.

Along with the development of enzyme and acid-base catalysts some specific importance is attached to poisoning in such cases.

Since some enzymes must have coenzymes to function properly, and since these coenzymes have certain metal ions, it is evident that a system not meeting this requirement would be deactivated. Any chemical group that tends to react with a protein group with which the enzyme might be active would constitute a possible poison. Certain acyl and alkyl group may act in this manner. This is, of course, a special case and a special use of the term. Likewise in acid-base catalyzed reductions any change in the pH of the system would be a form of poisoning. This also is the matter of what words one uses to discuss the facts.

In addition to the usual interpretation of solid state catalyst poisons there have come into use lately two ideas. While in a sense they are not really poisons the result is similar and are sometimes so classed.

First there is the case of the area which adsorbed

material covers against the uncovered area. Since the area is not used, it is inert and is therefore non-catalytic. This too is a recent development in catalytic activity and is not specifically a poison. Second, a catalyst may be prone to adsorb one of the reaction products and hence render itself poisoned by its own refusal to leave the catalyst surface.

The problem of catalytic inhibition in one way or another is chiefly related to the solid catalysts. Since there are relatively more industrial processes using solid catalysts, and since economy as well as chemistry is involved, it is necessary that this area be given more attention.

Apparently there is no solid state catalyst which cannot be inhibited or poisoned in one way or another to at least a certain degree. This can be accounted for in several ways. Promoters are intentionally added materials to "fix" the catalyst, so that it will have the best effect on a certain reaction.

In this connection, mention should be made of the use of added matter to cut down excessive action in order to get the required amount of activity. For instance, several special metallic catalysts are used in the hydrogenation of organic compounds, the object being to remove double bonds, reduce the oxygen content, or possibly change the side chain composition in ring compounds. Such a reaction is desirable in the catalytic production of liquid or gaseous fuels, but not in organic synthesis.

It is possible to "poison" such a catalyst in such a way as to make it active for a special purpose. Thus reactions can be catalyzed over such a modified catalyst that would be impossible to catalyze over the pure metal. Nickel and cobalt are instances in which the activity can be modified by the use of small amounts of sulfur. This effect is probably caused by formation of traces of sulfide, or the sulfur may be adsorbed to some extent. The former is more likely.

The nature of any catalyst is highly specific for the reaction for which it is to be used. The literature of chemistry describes innumerable studies of this kind. Likewise the search seems to go on for the one material which meets all the requirements as the perfect catalyst for a given reaction. The best advice for the student, researcher or production man is to search the literature for references to poisons for whatever catalyst he plans to use.

F. A. Griffitts

Cross-references: *Catalysis, Promoters.*

Reference

Emmett, P. H., "Catalysis," Vol 1–8, Reinhold Pub. Corp., New York, 1954–58.

POLARIMETRY

Polarimetry in its broadest sense comprises all investigation of optical phenomena in which polarized light is involved. For the purposes of this discussion, the subject is limited to the measurement of the rotation of the plane of polarized light when such light passes through a layer of an optically active substance. One of the chief applications of polarimetry is in sugar analysis.

According to the modern wave theory, light can be regarded as an electromagnetic disturbance which travels as trains of waves oscillating transversely to the direction of propagation. (In this way it is different from sound, which consists of the wave transmission of longitudinal displacements.) Ordinary light contains elements in which all of the possible directions of transverse vibrations are involved. Light is said to be *plane polarized* when the transverse vibrations are in a single plane perpendicular to the direction of transmission.

Many different media transmit plane polarized light without any change in the plane which contains the direction of the vibration and the direction of the ray of light. However, when plane polarized light is passed through an optically active medium, the plane of polarization is rotated through a definite angle, the magnitude of which depends on the nature of the medium, the length of the layer traversed, and the wave length of the light. A substance is said to be dextrorotatory if the rotation is to the right when seen by an observer looking toward the source of light, and levorotatory if the left. Three types of substances exhibit optical activity: (1) certain crystals under certain circumstances, for example quartz crystals, when the polarized light passes parallel to the crystal axis; (2) many liquids such as the volatile oils; and (3) solutions of substances in which the molecules have an asymmetric structure. Ordinary sugar (sucrose) dissolved in water is an example of the third class.

The instrument employed to measure the amount of rotation of the plane of polarization is called a *polarimeter.* Its essential parts comprise the *polarizer,* a device for producing plane polarized light; the *analyzer,* a device for measuring the angle through which the plane of polarization is rotated; and a *tube* for containing the substance being measured. The observation tube usually rests in a trough, mounted between the polarizer and analyzer.

Polarizer. The polarizer almost universally used in the construction of polarimeters is a prism fabricated from a doubly refracting crystal such as calcite. In such crystals both the ordinary and extraordinary rays are plane polarized, in planes at right angles to one another, and the crystals are highly transparent for both beams over a very great range of wave lengths. This type of polarizer is constructed by cementing two sections of a calcite crystal in such a way that one of the beams undergoes total reflection. The nicol prism and its many modifications are based on this principle.

Polaroid polarizing filters consist of submicroscopic needles or threadlike pleochroic crystals of iodoquinine sulfate embedded in a suitable matrix such as cellulose nitrate or cellulose acetate, and all oriented in the same direction. These filters may be produced in almost any size desired.

Analyzer. A prism similar to the polarizer can also serve as the analyzer. In the simplest form of polarimeter, the analyzer is rotated until the observed illumination reaches a minimum. The optically active material is placed between the two prisms and the analyzer again rotated to give the darkest possible field. The difference in the angles observed by reading the circular scale of the instrument is the rotation of the substance. Under favorable conditions, the position of complete extinction can be read with an uncertainty as low as 3 minutes (0.05 degree).

Greater precision and greater convenience of working are provided by the use of the half-shade principle. In this optical arrangement the field of the instrument appears divided into two parts, and the setting is made by rotating the analyzing prism until the two halves of the field become equal in intensity. The simplest and one of the most effective half-shade systems is the Lippich. In front of the polarizing prism, there is mounted a second, small nicol prism which is turned toward the analyzer and covers one-half of the large nicol prism. If the large polarizing prism is rotated through an angle with respect to the half prism, a half-shade angle is formed equal to that angle of rotation.

Saccharimeters. The polarimeters described, employing monochromatic light sources, are capable of measuring the rotation of all optically active substances. Another type of polarimeter has been developed solely for use in the measurements of rotation of sugar solutions and closely related substances.

The leading factor in the development of the saccharimeter was the desire to utilize the advantages of white light illumination. This is accomplished by use of the quartz compensating system, which has now become the characteristic part of all saccharimeters. In quartz compensating instruments the polarizing and analyzing nicols remain stationary. The rotation of the sugar solution is measured by moving a wedge of quartz of the opposite rotation between the solution and the analyzer until the rotation of the wedge system exactly neutralizes, or compensates, the rotation of the sugar solution. A graduated scale attached to the movable wedge indicates the rotation of the solution.

A large percentage of the precision measurements in polarimetry has been made with sodium light.

Mercury Light. Several types of mercury lamps are available. These lamps take advantage of the fact that mercury vapor, heated to incandescence by an electric current in a vacuum, gives an intense line spectrum.

Units for Expressing Rotation (Specific Rotation). Modern polarimetry dates from the investigations of Biot extending from 1812 to 1840. It was he who worked out the fundamental physical laws upon which polarimetry is based. He recognized the difference between rotation produced by crystalline structure and that produced by liquids and dissolved solids.

The optical rotation of an optically active solid or liquid depends upon the layer thickness, the wave length, and the temperature. The optical rotation of a solid in solution, furthermore, depends upon the concentration of the optically active substance and the nature of the medium in which the substance is dissolved.

When the angular rotations of different optically active substances are compared under identical conditions of concentration, layer thickness, and wave length, each substance gives a characteristic value. When determined under standard conditions the rotation obtained is termed the *specific rotation* and designated by the symbol $[\alpha]$. Thus, $[\alpha_\lambda{}^t]$ represents specific rotation at temperature t and wave length λ.

Application of Polarimetry. By far the most general application of polarimetry is in the analysis of the products of the cane- and beet-sugar industry. The so-called direct polarization obtained by reading the solution of sugar on the saccharimeter is used as a measure of the sugar content for many of the steps of the refining process. This rotation is a true measure of the sucrose content only when the other constituents present have no effect on the optical rotation. The results obtained on impure products, although subject to error introduced by interfering materials in the solution, serve a useful purpose in factory control and in many trade transactions.

In addition to the use in the analysis of ordinary sugar (sucrose), polarimetry finds wide application in the analysis of other commercial sugars such as dextrose, lactose, and maltose, as well as for products containing these sugars. The analytical chemist uses the optical rotation as a means of identification and estimation of a wide variety of optically active organic materials, including such substances as volatile oils, alkaloids, and camphors. Most compounds of this class require the use of monochromatic light, since the rotatory dispersion of most of these materials differs markedly from that of quartz.

In investigations concerning the arrangement of atoms in space and the character of the atomic linkage, optical rotation methods are a valuable aid. Certain substances are optically active in the crystalline state only, in which case the activity is due to the crystalline structure. Other materials possess rotatory power in the crystalline, gaseous, liquid, or dissolved state. In the latter group, the optical activity is due to asymmetry of the molecule. Although optically active substances are known to be derived from many elements, those of carbon are the most numerous and the most important.

C. F. SNYDER

Cross-references: *Optical Rotation, Instrumentation, Asymmetry.*

POLAR MOLECULES

The term "polar" is applied to molecules in which there exists a permanent separation of positive and negative charge, or dipole moment. Such a moment was first postulated by P. Debye for organic molecules having structural asymmetry, in order to explain certain of the observed electrical properties. He chose for his model the simplified picture of an electric dipole contained in a spherical molecule, free to rotate into alignment with an applied electric field, subject, of course, to the viscous drag of the surrounding medium and to collisions with other molecules due to their thermal motion. Then the net dipole moment per mole (molar polarizability) is calculated statistically on the basis of the average fraction of the dipoles oriented in the direction of the field. At ordinary temperatures and field strengths the electrical energy involved in the orientation is much smaller than the thermal energy, kT; therefore, a very small fraction of the dipoles are actually aligned with the field. The molar polarizability is:

$$P = \frac{4\pi}{3} N \left(\alpha + \frac{\mu^2}{3kT} \right) \qquad (1)$$

where μ is the dipole moment, N is Avogadro's number and T is the absolute temperature. The polarizability α represents the induced moment per molecule (as opposed to the permanent moment) resulting from the distortion of electron orbits by the applied field. For nonpolar molecules it is the only contribution; it corresponds to the optical polarizability as measured by the refractive index. The above expression is an approximation in that it does not take into account the short range interaction of the dipoles with one another or saturation effects at high field strengths.

The molar polarizability may be related to the dielectric constant ϵ by the approximate equation of Clausius and Mosotti:

$$\frac{\epsilon - 1}{\epsilon + 2} \frac{M}{d} = P \qquad (2)$$

where M is the molecular weight and d is the density. At optical frequencies the polarizability is related to the refractive index n by replacing ϵ by n^2 in the above expression. The measurement of dipole moment may therefore be accomplished by a determination of dielectric constant in the vapor phase or in dilute solution. In the latter case the molar polarizability is extrapolated to infinite dilution to eliminate the effects of intermolecular interactions. The dipole moment is obtained from the temperature coefficient of the polarization, as may be seen from equation (1). Alternatively a companion measurement of the optical refractive index may be made in order to determine α; the difference between the total polarization and the optical contribution is then taken to represent the dipole contribution from which μ is calculated. Dipole moments are usually expressed in terms of Debye units (1 Debye = 10^{-18} e.s.u.).

Equation (1) is valid in the low frequency region where the dipoles are able to rotate in phase with the applied electric field. This rotation is subject to various restraints; in the simple case of a spherical molecule of radius a rotating in a fluid of viscosity η this leads to a relaxation time

$$\tau = \frac{4\pi a^3 \eta}{kT} \qquad (3)$$

corresponding to a frequency range of *anomalous dispersion* in which the molecules become unable to follow the oscillations of the applied field. This gives rise to an out-of-phase component of the dielectric constant representing a conductivity or *dielectric loss*, ϵ'', i.e., a dissipation of energy in the form of heat. Mathematically this is expressed as a complex dielectric constant

$$\epsilon = \epsilon' - j\epsilon'' \qquad (4)$$

$$\epsilon' - \epsilon'_\infty = \frac{\epsilon_0' - \epsilon'_\infty}{1 + \omega^2 \tau_e^2} \qquad (5)$$

$$\epsilon'' = \frac{(\epsilon_0' - \epsilon'_\infty) \omega \tau_e}{1 + \omega^2 \tau_e^2} \qquad (6)$$

$$\tau_e = \left(\frac{\epsilon_0' + 2}{\epsilon'_\infty + 2} \right) \tau \qquad (7)$$

ω is $2\pi \times$ the frequency, and the subscripts $_0$ and $_\infty$ refer to dielectric constant measured at very low and very high frequency, respectively. Cole and Cole showed for polar molecules having a unique relaxation time that a plot ϵ'' vs ϵ' is a semicircle centered as the ϵ'-axis. For more complicated molecules or high polymers the center becomes depressed below the ϵ'-axis and the curve may be further distorted. This behavior is generally characterized by a distribution of relaxation times as a result of the orientation of molecular segments of various shapes and sizes. Information regarding the freedom of orientation within the molecules may thus be gained.

These simple relationships between dipole moment and dielectric constant fail for concentrated solutions or pure polar liquids because they do not take into account the interaction of the dipoles, both permanent and induced, with one another. The calculations have been extended by Onsager, Kirkwood, and others, to include these effects. It becomes necessary to include in the theory a correlation factor which is a measure of the extent of nonrandom orientation of the dipoles, i.e., the tendency of the dipoles to aggregate parallel or antiparallel to one another. Experimental determination of this quantity yields further insight into the structure of polar liquids.

In the solid state most crystalline polar compounds exhibit low dielectric constants because the rotational freedom necessary for dipole orientation has been frozen out. A few compounds, however, do show a persistence of high dielectric constant to temperatures below the melting point, indicating rotational freedom in the solid. At some lower temperature the rotation ceases and the dielectric constant drops. This change in dielectric constant has frequently been used as a method of detecting second order transitions in polar compounds.

Some success in the correlation of dipole moment with molecular structure has been achieved. A series of moments assigned to individual bonds was developed empirically by Smyth, Pauling, and others from measurements on simple molecules. Some of these are presented below. They are intended for approximate calculations of dipole moments of complex molecules by the vectorial addition of the bond moments along the bond directions as determined by other means. It is assumed that there is no interaction between bonds; this generally results in the calculated values being higher than the measured moment because of inductive effects, i.e., electrons being shared in a bond between two atoms are not wholly available to a neighboring bond, so that neither bond attains its full moment. An approximate correction for this effect has been made by Eyring and co-workers. Despite these inadequacies, the calculated moments are often helpful in determining molecular structures, and have been often applied in the case of various substituted benzene-ring compounds. Of special importance is the ability to decide between a polar or nonpolar, i.e., symmetrical, structure.

More detailed quantum mechanical calculations of dipole moments have been less successful. Rather the experimentally determined moments have been used to assign varying degrees of ionicity and covalency to the bonds in establishing the

Bond Dipole Moments*

	$\mu \times 10^{18}$ (e.s.u.)
H—I	0.38
H—S	0.68
H—Br	0.78
H—N	1.31
H—Cl	1.08
H—O	1.51
H—F	1.94
C—H	0.4
C—I	1.19
C—S	0.9
C—Br	1.38
C—N	0.22
C—Cl	1.46
C—O	0.74
C—F	1.41
C≡C	0.0
C≡O	2.3
C≡S	2.6
C≡C	0.0
C≡N (Cyanide)	3.5
N≡C (Isocyanide)	3.0

*From C. P. Smyth, *J. Phys. Chem.*, **59**, 1121 (1955).

electronic hybridization structure. Quantum mechanics has also shown how it is possible to obtain extremely accurate measurements of the dipole moments from spectroscopic Stark splittings.

D. Edelson

Cross-references: *Bonding; Dipole Moment.*

POLAROGRAPHY

Polarography is that branch of electrochemistry which deals with the measurement of current-voltage behavior of microelectrodes. An electrolytic cell is arranged to consist of a large electrode, an electrolyte, and an electrode of small area (microelectrode). The most common microelectrode is the *dropping mercury electrode* which consists of a capillary tube of fine bore through which mercury flows slowly, forming a drop every few seconds. When a varying voltage is applied across the cell, the large electrode remains relatively unchanged in potential (unpolarized), while the microelectrode undergoes a change in potential (polarization).

A *polarograph* is an instrument for automatic registration of a current-voltage curve (*polarogram*). The original polarograph, invented by Professor Jaroslav Heyrovsky in Prague in 1925, consisted of (1) a revolving wire-wound potentiometer drum with a sliding contact which served to apply a continuously increasing voltage to the cell, (2) a galvanometer with shunts to measure the current, and (3) a cylindrical drum wrapped with photographic paper and mechanically connected to the potentiometer drum to revolve simultaneously with it. A light beam from the galvanometer recorded the current as a vertical deflection on the photographic paper, while the horizontal axis was proportional to the applied potential. Later instruments have utilized pen and ink recorders for registration of the current-voltage curves. Manually operated equipment of greater simplicity (although of lesser convenience) is also available.

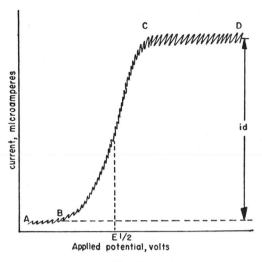

Fig. 1.

A typical polarogram is shown in Figure 1. Increasingly negative cathode potential is plotted to the right and increasing current upwards. The rising current region, followed by a horizontal plateau, is known as a *polarographic wave*. The oscillations represent the (partially damped) variations of current with the formation of a succession of mercury drops. Such a curve would be observed upon electrolysis of a solution containing a small concentration (10^{-5} to 10^{-2} molar) of a reducible material, such as cadmium ion, in the presence of a large concentration (0.1 to 1 molar or more) of *supporting electrolyte,* such as potassium chloride. The supporting electrolyte serves to decrease the electrolytic resistance of the cell, without undergoing electrode reactions in the region of applied potential involved. It also acts, because of its relatively high concentration, to carry a large part of the electric current by migration of its ions, so that the reducible material (which may be ionic) carries virtually none of the current. The effect is that the reducible material is transported almost entirely by diffusion rather than by electrical migration. This situation is essential to quantitative polarography.

Region AB (Figure 1) represents a region of *polarization* in which the potential of the dropping mercury electrode varies with the passage of only a very small current (the *residual current*). The residual current is largely due to the necessity of continually charging the growing mercury-solution interface as an electrical condenser. It may be determined separately by running a curve on the supporting electrolyte alone.

Region BC, the rising portion of the curve, represents a region of *depolarization* in which an increasing current flows due to an electrode reaction (e.g., the electrodeposition of heavy metal ions). If the concentration of reducible material is small, a region CD of *concentration polarization* occurs, when the cathode potential is made sufficiently negative. In this region, the concentration of reducible material at the electrode surface has reached a vanishingly small value. Only as much current can flow as is determined by the rate of diffusion of reducible material to the electrode surface, where it undergoes immediate reaction.

The current i_d, the total plateau current less the residual current, is known as the *diffusion current*. Its magnitude may be calculated with reasonable accuracy from an equation, based on diffusion theory, due to Ilkovic (1934):

$$i_d = 605nD^{1/2}Cm^{2/3}t^{1/6}$$

where i_d is the diffusion current (microamperes)

    n is the number of electrons involved per ion or molecule of reducible substance

    D is the diffusion coefficient of the reducible substance, (cm^2 per sec)

    m is the mass rate of flow of mercury (mg. per sec)

    t is the time for the formation of a mercury drop (sec)

    C is the concentration of reducible material (millimoles per liter).

The Ilkovic equation permits a quantitative comparison of diffusion currents with different capillary electrodes. The proportionality between concentration and diffusion current is the basis of quantitative analysis by the polarographic method.

The potential corresponding to a current equal to one-half the diffusion current value is known as the *half-wave potential*, $E_{1/2}$. For many electrode reactions its value in a given supporting electrolyte is independent of the concentration of reducible material, and the characteristics m and t of the capillary electrode. It is therefore characteristic of the electrode reaction and serves as the basis for qualitative analysis by the polarographic method.

All the considerations given above apply equally well to oxidation reactions as to reduction reactions. (The direction of the current and sign of the applied potential change are reversed.) In practice, however, the number of oxidation reactions which can be studied is rather limited.

Examples of reduction reactions suitable for polarographic determinations are: (1) deposition of metals as amalgams; (2) reduction of metal ions and their complexes to lower oxidation states; (3) reduction of anions (iodate, chromate, etc.); (4) reduction of molecular oxidants (nitric oxide, oxygen, hydrogen peroxide); (5) reduction of organic molecules (quinones, aldehydes, ketones, nitro compounds, peroxides, azo compounds, certain unsaturated hydrocarbons, certain halogenated compounds).

Examples of oxidation reactions are: (1) oxidation of strong ionic reducing agents such as titanium (III); (2) oxidation of hydroquinones.

Applications of the polarographic method include the quantitative determination of all classes of substances yielding polarographic waves. In particular, the method has been applied to the determination of traces of metals in alloys, ores, and organic materials, analysis of drugs and vitamins, determination of dissolved oxygen, etc. Polarographic measurements are useful in following reaction rates, determining mechanisms of electrode reactions, determining diffusion rates, and in determining formation constants of complexes.

HERBERT A. LAITINEN

Cross-reference: *Electrochemistry.*

POLONIUM

Polonium was discovered in 1898 by Pierre and Marie Curie and named after Mme. Curie's native land, Poland. It was the first element to be discovered as a direct result of its radioactivity, and at the time there was some question as to whether it was indeed a new element. In their first attempt the Curies separated polonium from about a ton of pitchblende and succeeded in concentrating it to such an extent that its specific radioactivity was some 400 times that of uranium, showing that it was not uranium or thorium. Marckwald (1902-03) noted the chemical similarity to tellurium and suggested the name "radiotellurium." Polonium was shown to be the last radioactive member of the radium series and Rutherford designated it radium F (RaF). Mme. Curie measured an atomic weight for radium of about 225 (now known to be 226.03), and from this value and the known decay sequence the approximate atomic weight of polonium could be deduced. Thus it was not until about 1905 that polonium was generally recognized as a new element and the homolog of selenium and tellurium predicted by Mendeleev in 1889.

It was the isotope ^{210}Po, a 138.4 day α-emitter, which led to the discovery of the element (number 84 in the Periodic Table). This isotope is a member of the uranium $(4n + 2)$ natural radioactive decay series. Some 27 isotopes of polonium are now known (all radioactive); seven of these are found in nature in the three naturally occurring $(4n, 4n + 2$ and $4n + 3)$ radioactive decay series. While ^{210}Po is the longest lived naturally occurring isotope two with longer lives have been produced artificially: the 2.9 year α-emitter ^{208}Po which may be produced by a (p,2n) reaction on ^{209}Bi and the 103 year α-emitter ^{209}Po which may be produced by a (d,2n) reaction on ^{209}Bi.

Chemical and Physical Properties. Two crystalline forms of metallic polonium have been recorded: α-Po, a simple cubic low temperature form, and β-Po, a simple rhombohedral high temperature form. It appears to change phase at about 36°C although both phases seem to co-exist between about 18 to 54°C, perhaps due to radiation damage effects. The metal is silvery-white in bulk, brownish in thin films, and black when precipitated from solution. It melts at 254°C and boils at 962°C to form an apparently colorless vapor containing Po_2 molecules.

Although the physical properties of the metal resemble those of elements in the same period (thallium, lead and bismuth) more than those of selenium and tellurium, its chemistry shows much similarity to that of tellurium. Polonium oxidizes slowly in air at room temperature forming mainly the basic dioxide. The metal dissolves in concentrated sulfuric or selenic acid. Both concentrated nitric acid and aqua regia dissolve it by oxidation to Po(IV). It also dissolves in dilute $(2N)$ hydrochloric acid to form the red or pink Po(II) which rapidly oxidizes to the yellow Po(IV) due to radiolysis reactions.

Polonium exhibits the oxidation states +6 (at least in solid compounds), +4, +2, 0, and −2 and there is evidence for the existence of +3.

As expected polonium salts are more basic than

the corresponding selenium and tellurium compounds.

Health Hazards and Radiolysis Problems. The high specific activity of ^{210}Po creates both a serious health hazard and a chemical problem due to radiation damage. Polonium when ingested tends to concentrate in certain organs, e.g., the kidneys, spleen and liver, where irreversible radiation damage is done by the α-particles. In addition polonium has a notorious reputation for migrating and getting spread all around a laboratory. It is not clear to what extent this behavior is due to its peculiar chemistry, to its radiolysis effects or to its high specific activity.

Because of the very high specific activity any work with weighable amounts of polonium must be carried out under strict control in a totally enclosed system. Because of the short range of α-particles and the relative freedom of ^{210}Po from β and γ-radiations, glove boxes provide adequate protection at least for moderate amounts (milligrams) if the glove material is impenetrable to polonium.

Because of radiation effects the concentration of ^{210}Po has a marked effect on the chemical procedures which can be used in its separation as well as the equipment which can be used. At tracer levels polonium may be separated by solvent extraction, ion-exchange or paper chromatography. However, aqueous solutions of even millimolar concentrations (about 1 curie/ml) exhibit a visible evolution of gas and the formation of oxidizing and reducing radiolysis products which render difficult any studies concerning the (II) or the (VI) valence states, the latter being questionable in solution. In addition the α-particles and their radiolysis products decompose other ionic or neutral polyatomic species present, particularly organic matter. Thus separations involving solvent extraction or ion exchangers are not feasible at higher concentrations of ^{210}Po.

Solid compounds are very difficult to study at the curie (220 μg) level as most of the α-particle energy is absorbed within the compound itself causing considerable radiation damage. Under these conditions salts of organic acids char within a short time, the iodate evolves iodine, and in any of its crystalline compounds every atom is displaced at least once a day.

Use of ^{210}Po for α-Particle, Neutron and Power Sources. During the Second World War large amounts of ^{210}Po were produced in nuclear reactors to be used as neutron sources by utilizing the (α,n) reaction on light elements such as Li, Be, B and F. After the war Po-Be neutron sources have been used largely because of the small amount of associated γ-ray activity compared to other sources.

^{210}Po α-energy standards may be prepared and safely used when covered with thin-rolled (as low as 0.0001 inch thick) stainless steel. ^{210}Po has also received attention for use as a power source for satellites and out-of-the-way places on earth. The usual method considered for this direct conversion, based on the thermocouple principle, involves the use of such semiconducting thermoelectric elements as lead telluride or cobalt silicide. The heat output of ^{210}Po is about 140 watts per gm. Here again the near lack of γ-rays is an advantage.

M. H. LIETZKE AND R. M. STOUGHTON

References

Bagnall, K. W., "The Chemistry of Selenium, Tellurium and Polonium," Amsterdam, London, and New York, Elsevier Publishing Company, 1966.

Figgins, P. E., "The Radiochemistry of Polonium," NAS-NS 3037, January 1961, Washington, Office of Technical Services, Dept. of Commerce.

Corliss, William R., and Harvey, Douglas G., "Radioisotopic Power Generation," Chapter 6, Englewood Cliffs, N.J., Prentice-Hall, Inc., 1964.

O'Kelley, G. D., Chapter 3.1, "Radioactive Sources" in *Methods of Experimental Physics,* Volume 5, "Nuclear Physics," Part B, New York, Academic Press, 1963.

POLYALLOMER RESINS

Polyallomers are block copolymers prepared by polymerizing monomers in the presence of anionic coordination catalysts. The polymer chains in polyallomers are composed of homopolymerized segments of each of the monomers employed. The structure of a typical polyallomer can be represented as:

$$(PPPPPPPP\cdots)_x(EEEE\cdots)_y$$

where P represents a propylene molecule and E an ethylene molecule. The number and length of the individual segments can be varied within wide limits depending on the process conditions, monomer concentrations, and catalyst systems employed. The copolymers exhibit the crystallinity normally associated only with the stereoregular homopolymers of these monomers.

The word polyallomer is derived from the Greek words *allos, meros,* and *poly. Allos* means "other" and denotes a differentiation from the normal. The word *meros* means "parts" and the prefix *poly* is added to show that these materials are polymeric. Since allomerism is defined as a constancy of crystalline form with a variation in chemical composition, the polyallomers are examples of allomerism in polymer chemistry.

Polyallomers can be synthesized in slurries by contacting the parent monomers with anionic coordination catalysts of the Ziegler-Natta type at temperaures of 70–80°C.[1,3,4,5] Polyallomers are also synhesized in solution at 140–200°C by contacting the parent monomers with hydrogen-reduced alpha-titanium trichloride and a lithium-containing cocatalyst.[4] In both the slurry and solution processes the polyallomers are formed by alternate polymerization of the monomers employed. In the synthesis of polyallomers from monomers which differ widely in polymerization rates it is sometimes desirable to alternately polymerize a single monomer and then a mixture of monomers to obtain the most desirable properties.[4]

Evidence for the crystalline nature of the polyallomers includes x-ray diffraction patterns, infrared spectra, and for olefin polyallomers, low solubility in hydrocarbon solvents.

The physical properties of polyallomers are generally intermediate between those of the homopolymers prepared from the same monomers, but frequently represent a better balance of properties than blends of the homopolymers. This is illustrated by comparing the properties of a propylene-ethylene polyallomer containing 2.5% ethylene with polypropylene, high-density polyethylene, and a blend of 5% high-density polyethylene and 95% polypropylene.

Compared to polypropylene, a propylene-ethylene polyallomer has a lower brittleness temperature, higher impact strength, and less notch sensitivity. Compared to high-density polyethylene, a propylene-ethylene polyallomer is harder, higher-melting, and of higher impact strength. Other advantages of this polyallomer over high-density polyethylene include its excellent resistance to environmental stress cracking and its low and uniform mold shrinkage, which minimizes sinks and voids in molded parts. A propylene-ethylene polyallomer thus overcomes the most serious property deficiencies of polypropylene (poor impact and low-temperature properties) and of high-density polyethylene (poor stress-crack resistance and excessive mold shrinkage).

Propylene-ethylene polyallomer is used in vacuum forming, blow molding, injection molding, film and sheeting, wire covering and pipe. Propylene-ethylene polyallomer is the easiest of all polyolefins to vacuum form. The excellent melt strength and broad processing range permit a deep draw. Reproduction of mold detail and surface finish is very good. In film and sheeting, propylene-ethylene polyallomer has optical properties equal to polypropylene but much higher impact strength, particularly at low temperatures. In wire covering, blow molding, and pipe extrusion operations, the combination of good processability and excellent environmental stress crack resistance is important.

H. J. Hagemeyer, Jr.
M. B. Edwards

Cross-reference: *Polymerization.*

References

1. Bier, G., *Angew. Chem.*, **73**, 194–196 (1961).
2. Hagemeyer, H. J., *Mod. Plastics,* **39**, 157 (1962).
3. Hagemeyer, H. J., and M. B. Edwards, *J. Polym. Sci.*, Part C, **4**, 731–742 (1963).
4. Hagemeyer, H. J. and M. B. Edwards, U.S. Patent 3,529,037 (1970).
5. Natta, G., *J. Polym. Sci.*, **34**, 542–543 (1959).

POLYAMIDE RESINS

Polyamide resins are synthetic polymers that contain an amide group, —CONH—, as a recurring part of the chain. Poly-alpha-aminoacids, i.e., proteins, whether natural or synthetic, are not normally included in this classification.

The polyamides trace their origin to the studies of W. H. Carothers, begun in 1928, on condensation polymerization, a process that involves the repetition many times of a reaction known to the organic chemist as a condensation reaction because it links two molecules together with the loss of a small molecule. Esterification and amidation are examples:

$$CH_3COOH + C_2H_5OH \rightleftharpoons CH_3COOC_2H_5 + H_2O$$

acetic acid · · · ethyl alcohol · · · ethyl acetate or ethyl ester of acetic acid

$$CH_3COOCH_3 + C_4H_9NH_2 \rightleftharpoons$$

methyl acetate · · · butyl amine

$$CH_3CONHC_4H_9 + CH_3OH$$

N-butylacetamide

Polymerization requires at least two reactive groups per molecule as in the following example:

$$n\ H_2N(CH_2)_6NH_2 + n\ HOOC(CH_2)_4COOH \rightleftharpoons$$

hexamethylenediamine · · · adipic acid

$$H\ [\ NH(CH_2)_6NHCO(CH_2)_4CO\ ]_nOH +$$

poly(hexamethylene adipamide)

$$(2n - 1)H_2O$$

The first truly high molecular weight polyamide was made this way in 1935 and lead to the development of the first wholly man-made fiber which became popularly known as nylon, a name coined by Du Pont for fiber-forming polyamides. Numerals representing the number of carbon atoms in first the diamine and then the diacid are used to identify the nylon. Thus, poly(hexamethylene adipamide) is nylon-66 ("six" "six", not "sixty six"), and poly(hexamethylene sebacamide) made from the 10-carbon sebacic acid is nylon-610 ("six" "ten").

Both reacting species may be present in the same molecule:

$$n\ H_2N(CH_2)_{10}COOH \rightleftharpoons$$

11-aminoundecanoic acid

$$H\ [\ NH(CH_2)_{10}CO\ ]_nOH + (n - 1)H_2O$$

poly(11-aminoundecanoic acid)

Here a single number is used to indicate the number of carbon atoms in the original monomer, i.e., nylon-11 ("eleven" not "one-one"). In some instances the cyclic analogue or lactam is more accessible than the amino acid and is polymerized by a ring-opening rather than condensation mechanism:

$$n\ CH_2 \begin{matrix} CH_2{-}CH_2{-}C{=}O \\ \\ CH_2{-}CH_2{-}NH \end{matrix} + H_2O \rightleftharpoons$$

epsilon-caprolactam

$$H\ [\ NH(CH_2)_5CO\ ]_nOH$$

polycaprolactam or polycaproamide
nylon-6

$$n\ (CH_2)_{11} \begin{matrix} {-}C{=}O \\ \\ {-}NH \end{matrix} + H_2O \rightleftharpoons$$

laurolactam

$$H\ [\ NH(CH_2)_{11}CO\ ]_nOH$$

nylon-12

Some of the monomers commonly used to prepare the nylon resins are tabulated below. Both petrochemical and vegetable products provide the source materials that are transformed into the reactive intermediates. The table correctly suggests that there is a wider choice in diacids than in diamines.

Monomer	Formula	Source(s)
hexamethylene diamine	$H_2N(CH_2)_6NH_2$	butadiene, furfural, or propylene
adipic acid	$HOOC(CH_2)_4COOH$	butadiene or cyclohexane
suberic acid	$HOOC(CH_2)_6COOH$	butadiene or acetylene
azelaic acid	$HOOC(CH_2)_7COOH$	oleic acid
sebacic acid	$HOOC(CH_2)_8COOH$	castor oil
dodecanedioic acid	$HOOC(CH_2)_{10}COOH$	butadiene
"dimer" acid	$HOOC\text{-}C_{34}H_{66}\text{-}COOH$	oleic and linoleic acids
caprolactam	$(CH_2)_5 \begin{smallmatrix} \text{---------CO} \\ \; \\ \text{---------NH} \end{smallmatrix}$	toluene, benzene, or cyclohexane
7-aminoheptanoic acid	$H_2N(CH_2)_6COOH$	cyclohexane or ethylene
capryllactam	$(CH_2)_7 \begin{smallmatrix} \text{---------CO} \\ \; \\ \text{---------NH} \end{smallmatrix}$	butadiene or acetylene
9-aminononanoic acid	$H_2N(CH_2)_8COOH$	soybean oil
11-aminoundecanoic acid	$H_2N(CH_2)_{10}COOH$	castor oil
laurolactam	$(CH_2)_{11} \begin{smallmatrix} \text{---------CO} \\ \; \\ \text{---------NH} \end{smallmatrix}$	butadiene

The most important commercial polyamide resins are nylons-66 and -6. Other commercial nylons include 610, 612, 11, and 12.

The above equations illustrate via the double arrows an important facet of polyamides—the equilibrium nature of the polymerization reactions. Achieving and maintaining useful molecular weights (about 10,000 or more) for plastics applications require low moisture contents in order to avoid the reverse reaction of hydrolysis. Most commercial nylons are processed at melt temperatures in excess of 200°C, and molecular weight stability requires a water content below 0.3 weight per cent. At a molecular weight of about 11,300, nylon-66 and nylon-6 have, on the average, 99 amide groups linking together 100 monomer units with one unreacted amine group and one unreacted acid group at the ends of the 700-atom long chain. This corresponds to 99% reaction. Because nylon plastics have average molecular weights in the 11,000 to 40,000 range, the need for pure materials and freedom from side reactions is seen to be essential for successful polymerization.

An average molecular weight (number average) is used because any one nylon comprises a broad range of molecular weights. Nylons typically have a "most probable distribution" in which the weight average is twice the number average.

As made, nylon-6 contains up to ten weight per cent of lactam monomer that has to be extracted for most applications. Nylons made from larger lactam rings contain only about one per cent monomer.

Not all polyamide resins are nylons, and not all nylons are polyamide resins. Some nylons such as nylon-4 or those with a high content of relatively inflexible rings are too unstable or have too high a melt viscosity to be melt processible and are not normally included in the "polyamide resin" category. However, spinning or casting from solution permits some such polymers to be converted into useful fibers or films. A class of polyamide resins distinct from the nylons is based upon polymerization of dimerized vegetable oil acids and poly-alkylene polyamines (e.g., ethylenediamine or diethylenetriamine). These are relatively low in molecular weight (2000–10,000), vary from liquids to low melting solids, and are more soluble and flexible than the higher molecular weight, more crystalline nylons.

Nylons are semicrystalline polymers with fairly sharp melting points varying from 180°C for nylon-12 to 270° for nylon-66. The melting point increases in zigzag fashion with increasing concentration of amide groups, being higher where there is an even number of chain atoms between the amide groups. For example, nylon-66 averages five CH_2-groups per CONH but has either four or six chain atoms between amide groups; nylon-6 also has five CH_2/CONH but always has five chain atoms between amides and melts 40°C lower than nylon-66. The degree of crystallinity and the morphology of nylons are more readily controlled by choice of processing conditions than other crystalline polymers such as polyethylene or polyacetal.

The nylons are typically tough and strong. Nylon-66 was the first thermoplastic to provide a combination of stiffness and toughness suitable for mechanical applications. Nylons were therefore the first members of the family of engineering thermoplastics that now include newer materials such as the polyacetals, polysulfones, and polycarbonates. Nylons are outstanding in withstanding repeated impact, in resistance to organic solvents, and in water resistance. A low coefficient of friction, an ASTM self-extinguishing rating, reasonable electrical properties, adequate creep resistance, good fatigue properties, and good barrier properties, particularly to oxygen, are also characteristic of the nylons. Properties change with temperature but are often acceptable in the interval of about −60 to 110°C. Properties may change also with relative humidity because nylons characteristically absorb moisture and water acts as a plasticizer, that is, stiffness decreases and toughness increases. Water absorption decreases as the amide group concentration in the nylon decreases; for example, nylon-6 at saturation contains 9.5% water and nylon-11, 1.9%. Nylons are somewhat notch sensitive, and parts are typically designed to avoid sharp corners. Nylons are attacked by strong acids, oxidizing agents, and a few specific salt solutions such as aqueous potassium thiocyanate or methanolic lithium chloride.

Control of crystallinity and morphology provides one tool for changing properties as desired with any given nylon. But there are many other tools for modification of nylons, and this viability has been an important factor in meeting specific market demands in the face of increased competition from newer materials. Copolymerization and plasticization are alternatives that provide lower melting and tougher compositions with a somewhat different balance of properties. Lubricants as processing aids, nucleating agents to accelerate crystallization and give a little stiffer product, molybdenum disulfide or graphite to improve lubricity in gears or bearings, colorants, fire retardants, carbon black for weather resistance, and antioxidants for better resistance to thermal oxidation are examples of additives employed to achieve specific effects. Combination with other polymers is another technique. Reacting nylon-66 with formaldehyde in alcohol solution has yielded a low melting derivative that can be cross-linked to provide a thermoset. Glass fiber reinforcement has proved to be particularly effective in nylons and has yielded products of exceptional strength, stiffness, and heat resistance.

The nylon resins are converted into useful shapes principally by injection molding or extrusion. These shapes are most often in finished form, but forming is sometimes employed to impart added strength via orientation. Machining of extruded stock shapes is often appropriate where the number of parts does not justify manufacture of a costly mold. Nylon-6 castings are made by the base catalyzed, anhydrous polymerization of monomer in molds that are relatively inexpensive because they do not have to sustain high pressures. Rota-

tional molding, fluidized bed coating, and electrostatic spray coating are also used, especially with low melting nylons such as nylon-11.

Football face guards, hammer handles gears, sprockets, journal bearings, bristles, filaments, refrigerant tubing, film as a cooking pouch, coil forms, casters, package strapping, loom parts, automobile dome lights, power tool housings, and virtually thousands of other diverse articles attest to the performance of nylon resins in electrical appliances, automotive parts, business equipment, consumer products, and other industrial and home applications.

The non-nylon polyamide resins include relatively low-melting solids that are used with or without modifiers in hot melt cements, heat seal and barrier coatings, inks for flexographic printing of plastic film, and other specialty adhesives and coatings such as varnishes to provide a glossy, transparent, protective layer over print. The large hydrocarbon side chains that are present in the dimerized vegetable oil acids used to make these polyamide resins contribute to their low crystallinity, low water absorption, and flexibility. The fluid, "reactive" polyamides are lowest in molecular weight and contain an excess of amine groups. Monomers with more than two amine groups per molecule such as diethylenetriamine ($H_2NCH_2CH_2$-$NHCH_2CH_2NH_2$) are used, and care must be taken during polymerization to avoid premature gelation. The excess amine groups permit interaction with epoxy, aldehyde, hydroxymethyl, anhydride, acrylic, and other groups in other resins to produce a variety of thermosetting formulations. Combinations with epoxy or phenol-formaldehyde polymers with or without added fillers are the most common. The epoxy compositions are cured at relatively low temperatures and are useful for bonding aluminum in aircraft, steel in automobiles, wood, leather, glass, ceramics, and other materials. The resin combination broadens the capacity of the adhesive to wet a variety of surfaces and enhances bond strengths. The phenolic-polyamide compositions require higher curing temperatures but withstand higher temperatures in use. They tend to form bonds less tough than the epoxy mixes but bond strongly to copper and other common components of printed circuits. The phenolic blends also produce acid and alkali resistant coatings useful as container linings or wire coatings.

M. I. KOHAN

Cross-reference: *Fibers, Synthetic.*

References

Floyd, D. E., "Polyamide Resins," Ed. 2, Reinhold, New York, 1966.

Kohan, M. I., "Nylon Plastics," publication in press.

Company and trade literature are often the best sources for the properties and processing behavior of specific polyamide resins.

"Encyclopedia of Polymer Science and Technology," Vol. 10, John Wiley and Sons, New York, 1969, pages 460 to 615.

POLYCARBONATE RESINS

Polycarbonates can be defined as linear polyesters of carbonic acid. The carbonate linkage joining the organic units in the polymer gives this thermoplastic resin its name. The carbonate radical is an integral part of the main polymer chain. Both the General Electric Co. and Farbenfabriken Bayer in Germany concurrently and independently conducted research that led to the definition of the properties of polycarbonates based on 4,4'-dihydroxyl-diphenyl alkanes. Procedures for producing polycarbonate thermoplastic resins commercially emerged rapidly from this research.

Processes

Three processes have gained importance in the production of polycarbonates from bisphenol A [2,2'-bis(4-hydroxyphenyl)propane]: (1) transesterification, (2) interfacial or emulsion polymerization, and (3) nonaqueous solution polymerization. In the first method, the reactants are bisphenol A and diphenyl carbonate. In the latter two, the reactants are bisphenol A and phosgene.

Transesterification. In the transesterification preparation of polycarbonates, the reactants are mixed together in a heated vessel under reduced pressure. This procedure is continued until the desired molecular weight has been attained; then the molten polymer is discharged and directly pelletized for use. Specifically, bisphenol A and a slight molar excess of diphenyl carbonate are heated under inert gas to a temperature of 200–200°C at 20–30 mm Hg. During this period, phenol is distilled and the phenyl carbonate of bisphenol is formed. As the rate of phenol distillation slows, the temperature is increased to 290–300°C and the pressure reduced to 0.2 mm Hg. These conditions are maintained for the elimination of diphenyl carbonate until the desired molecular weight has been reached.

The high temperatures required aceclerate potential decomposition and rearrangement reactions which primarily affect the color of the product. For this reason, certain conditions are required, e.g., the use of excess diphenyl carbonate, the use of high-purity raw materials and the addition of catalysts to reduce reaction time. Acidic type catalysts are not as effective as basic material. Alkali and alkaline earth metals, oxides, hydrides, and amides have been used; however, these must be neutralized in the final polymer.

The advantages of this process are (a) no solvent is needed, thus eliminating solvent recovery, and (b) no halogen is used directly, thus reducing the polymer purification steps found in the other processes. Disadvantages include (a) a limit on molecular weight due to the high melt viscosity of the resin, (b) color formation from prolonged high temperatures, and (c) mechanical problems associated with high-temperature, high-vacuum technology.

Interfacial or Emulsion Polymerization. The interfacial or emulsion polymerization technique employs the reaction of an aqueous alkaline solution of bisphenol with phosgene in the presence of an immiscible liquid which is a solvent for both phosgene and polymer. The resulting polymer solution is purified and then the polymer is recoverd from solution by any of several techniques.

The reaction is carried out by dissolving the bisphenol A in caustic solution, adding methylene chloride [or other chlorinated solvent], then bubbling phosgene into the agitated mixture. The pH of the solution is maintained above 10 by the addition of caustic as necessary. Temperatures below 40°C are desirable to decrease the rate of phosgene hydrolysis, which is an attendant side reaction. The polymerization proceeds in two steps. The first, which is the production of the chlorocarbonate ester of the bisphenol, occurs rapidly. The second step, the reaction of the chloro ester with additional sodium salt [of bisphenol A] is slow. It can be accelerated by the addition of tertiary amines or quarternary ammonium bases.

The resin solution formed is washed with water until the electrolyte [NaCl] is removed. Recovery of the resin can be accomplished in several ways. One method involves polymer precipitation by the addition of a nonsolvent. This yields a powder which is dried and then pelletized. In this method, the solvent and nonsolvent must be separated for reuse. Another method, which avoids this separation is the direct evaporation of the solvent and finally extrusion into pellets.

The advantages of this process are that very high molecular weight polymers can be made and that reaction is simple and at low temperature. The major disadvantage is the difficulty of washing the polymer solution free of electrolyte.

Nonaqueous Solution Polymerization. Polymerization in nonaqueous solution involves the reaction of bisphenol A and phosgene in the presence of an organic base which acts both as acid aceptor and solvent or organic base, plus an inert polymer solvent. The polymer solution is then purified by the removal of the organic base. Recovery of polymer can be effected by precipitation or solvent removal.

In this reaction the bisphenol A is dissolved in dry organic base or organic base plus an inert solvent, then phosgene is bubbled into the solution. The most commonly used base-solvent system is pyridine-methylene chloride. The base acts by removing hydrogen chloride as the salt.

At the end of the reaction, any excess base is converted to its salt and the resulting polymer solution water washed free of the salt. Finally, the polymer is recovered from solution and pelletized. The reaction temperature varies up to the boiling point of the solvent with no adverse effects.

High-molecular weight polymer can be prepared. Control of molecular weight in this process, as with the interfacial polymerization system, is accomplished by the addition of monofunctional materials.

The advantage of this system is that reaction is carried out in a homogeneous solution. The major disadvantage lies in the fact that the organic base must be recovered because of its cost.

Properties and Fabrication

The finished aromatic polycarbonate resins can be processed by virtually every method and technique used for plastics. They are, however, particularly suitable for injection molding and the majority of parts are produced in this manner. The extrusion processes can be used to produce film and sheet, tubing or rods. Film may be obtained from solution casting. Injection molding and extrusion of polycarbonate resins may be done at temperatures ranging from 450°F to 600°F. High pressures for injection molding are suggested and short mold-filling times are preferred. In general, pinpoint gates are not recommended with polycarbonate resins and lands should be as short as possible. Mold temperatures should be around 200°F. It is important that the resin be as free of moisture as possible to prevent degradation and poor appearance in molded parts.

Polycarbonates offer a combination of very useful properties unmatched by any other thermoplastic material. The outstanding properties are: (1) very high impact strength (16 ft-lb/in. notch) combined with good ductility, (2) excellent dimensional stability combined with low water absorption (0.35% immersed in water at room temperature), (3) high heat distortion temperature of 270°F, (4) superior heat resistance showing excellent resistance to thermal oxidative degradation, (5) good electrical resistance.

Its excellent toughness properties make it particularly important as a structural and engineering material able to replace hard-to-fabricate metal parts. Advantageous polycarbonate properties plus its ability to be injection molded also dictate its use in many applications where parts consist of several separate sub-assemblies. With polycarbonates, these sub-assemblies can be molded in one shot, thus increasing manufacturing efficiency as well as bettering final assembly properties.

The self-extinguishing properties of polycarbonate combined with its good dielectric and heat resistance make it suitable for use in electrical appliances and tools. Because of its low water absorption, polycarbonate is used as impellers in multistage water pumps.

Its toughness and its transparency enable polycarbonate to be utilized in such applications as lenses, safety shields, instrument windows, glazing applications, and outdoor lighting applications. Its freedom from toxicity has resulted in FDA approval of polycarbonates for food applications and in the medical field in such applications as heart valves, blood oxygenators, etc.

Polycarbonates have good resistance to inorganic and organic acids, solutions of neutral and acid salts, most alcohols, ethers, aliphatic hydrocarbons, oxidizing agents such as hydrogen peroxide, reducing agents and vegetable oils. Alkalies will slowly decompose polycarbonates although they can be used intermittently in dilute solutions of soaps and detergents, for cleaning purposes, for instance. Aromatic hydrocarbons, chlorinated hydrocarbons, ketones and esters will act on polycarbonates as solvents, partial solvents, plasticizers or crystallizing agents.

Although polycarbonates are produced in a great variety of colors, it is entirely possible to use first or second surface finishing by painting, printing, vacuum metallizing, electroplating and other well-known methods of the industry by choosing the recommended materials for this purpose. All normal solvent cementing and bonding methods are used for joining polycarbonate parts for final assembly. Because of their excellent machinability

polycarbonates lend themselves to final assembly machining and are used to produce prototype parts for engineering purposes.

J. TOME AND H. JACOBY

POLYELECTROLYTES

Polyelectrolytes are macromolecules with incorporated ionic constitutents. Polyelectrolytes may be cationic or anionic, depending on whether the fixed ionic constituents are positive or negative. Examples of cationic polyelectrolytes are polyvinyl-ammonium chloride and poly-4-vinyl-N-methylpyridinium bromide. Examples of anionic polyelectrolytes are potassium polyacrylate, polyvinylsulfonic acid, and sodium polyphosphate. If a polyelectrolyte contains both fixed positive and negative ionic groups, it is called a polyampholyte. Polyelectrolytes may be synthesized by polymerization of a monomer containing the ionic substituent, as for instance in the polymerization of acrylic acid to polyacrylic acid, or by attaching the ionic constituent by chemical means to an already existing macromolecule, as for instance in the quaternization of poly-4-vinyl-pyridine with methyl bromide, or in the preparation of sodium carboxymethylcellulose from natural cellulose. Many macromolecules occurring in nature are polyelectrolytes. Examples are gum arabic, which carries carboxylate groups; carrageenin, which contains sulfate groups; proteins, which carry both negative carboxylate and positive ammonium groups; and nucleic acids, which contain negative phosphate groups and basic purine and pyrimidine groups, which acquire positive charges at low pH. Inorganic long-chain polyphosphates have also been isolated from biological materials.

A solution of a polyelectrolyte in water or other suitable solvent conducts an electric current, indicating that the polyelectrolyte is ionized. Transference and electrophoresis measurements show that both the macroion and the counterions (gegenions) contribute to the conductance. Because the counterions are osmotically active, polyelectrolytes show much higher osmotic pressures and diffusion rates than do nonionogenic macromolecules. The osmotic pressure of a polyelectrolyte solution is greatly reduced by the addition of a simple electrolyte which distributes itself among the two sides of the membrane according to the thermodynamic theory of Donnan equilibrium. Polyelectrolytes are called weak if they carry weakly ionized groups such as —COOH, and strong if they carry strongly ionized groups such as —COONa. On titrating polyacrylic acid with sodium hydroxide, the pH increases much more slowly than it does in a corresponding titration of a monocarboxylic acid, thus indicating a pronounced buffering capacity. Even in the case of strong polyelectrolytes, the full osmotic activity of the counterions is not realized; as counterions leave the macroion, the electrostatic potential on the latter builds up making it increasingly difficult for additional counterions to escape. This binding effect becomes especially strong with multivalent counterions, whose effective concentration may be rendered several orders of magnitude smaller by a polyelectrolyte than their stoichiometric concentration.

The electric charge on the macroion has several important secondary effects. If the macroion is a flexible chain, intramolecular repulsion between charged segments will stretch out the macroion from a coiled to a more rod-like structure, resulting in much larger solution viscosities than are usually obtained with uncharged polymers under corresponding conditions. Intermolecular repulsion causes the macroions to arrange themselves so that they are as far from each other as is possible. With this ordering, the light scattering which is characteristic of solutions of ordinary macromolecules is greatly diminished, often to the vanishing point, as a result of destructive interference. These secondary effects of charge may be reduced by the addition of simple electrolytes which screen the charged elements from each other, and in some cases also lower the charge by specific counterion binding. At high enough concentrations of added salt, the light scattering of the polyelectrolyte may become sufficiently pronounced to allow its use for the determination of the molecular weight.

An interesting class of polyelectrolytes, denoted by polysoaps, is obtained by attaching soap-like molecules to the polymer chain. Such a polysoap is for instance produced by the quaternization of polyvinyl-pyridine with *n*-dodecyl bromide. The polysoap molecules differ from ordinary polyelectrolytes in that they may reach protein-like compactness in solution. They behave like prefabricated soap micelles and solubilize hydrocarbons and other compounds insoluble in water.

While the applications of polyelectrolytes for practical purposes depend on their general ionic properties, nevertheless large differences appear among individual members of the class in their applicability to a specific use. When polyelectrolytes are absorbed at interfaces, they affect the zeta-potential and a suspending action may result. Adsorption at growing crystal surfaces is also believed to be the reason for the high effectiveness of small amounts of certain polyelectrolytes in preventing or retarding the precipitation of calcium carbonate. The dispersion of clays by polyelectrolytes is applied in oil-well drilling. The ability of long-chain polyelectrolytes to bind together small particles has found uses in soil conditioning and in the flocculation of phosphate slimes. Because of their effect on the solution viscosity, certain polyelectrolytes are used as thickening agents. Because of their ability to bind di- and trivalent cations, some anionic polyelectrolytes are used in water softening and as enzyme inhibitors. When polyelectrolytes are adsorbed or otherwise incorporated into membranes, they make the latter permselective, hindering small ions of the same charge as the macroion from passing through the membrane while allowing free passage to small ions of opposite charge. The well-known ion-exchange resins are polyelectrolytes which have been cross-linked to prevent them from dissolving.

The most important and widespread use of polyelectrolytes is to aid in the removal of small suspended solids from waste water in the primary, secondary and dewatering stages of treatment. Increasing numbers of cities are now applying them for this purpose in municipal waste treatment plants with beneficial results.

ULRICH P. STRAUSS

Cross-references: *Ions; Polypeptides; Electrophoresis*

Reference

Chemical Week, pp. 55–56, March 3, 1971.

POLYESTER RESINS

Polyester resins are the polycondensation products of di-or poly carboxylic acids and di-or poly hydroxy alcohols or alkylene oxides. Saturated polyesters are thermoplastic and find their principal uses in fibers notably tirecords, and films. Unsaturated polyesters are curable or thermosetting and are used in coatings and molding operations. The unsaturated polyesters are a special type of alkyd resins, but unlike other types of alkyd resins are not usually modified with fatty acids or drying oils. The outstanding characteristic of these resins is their ability, when catalyzed, to cure or harden at room temperature under little or no pressure. They have therefore removed the size limitations placed on molded plastics by presses and other molding equipment. The polyester resins are widely used for low-pressure laminating, casting, coating applications (See Alkyd Resins). The unsaturated polyesters are usually crosslinked through their double bonds with a compatible monomer also containing ethylenic unsaturation. The cross-linked resins are thermosetting, and when fully cured are insoluble and infusible.

The principal unsaturated acids used in the manufacture of polyesters are maleic and fumaric acids. Saturated acids, usually ortho- or isophthalic acid and adipic acid may also be included in the formation. Fire resistance is imparted by using either acid or glycol type constituents of high halogen content, for example "HET-acid" (hexachloro endomethylene tetrahydrophthalic anhydride), tetrabromo phthalic anhydride, and others. The function of the saturated acids is to reduce the amount of unsaturation of the final resin, making it tougher and more flexible. The acid anhydrides are often applicable. The dihydroxy alcohols most generally used are ethylene, diethylene and dipropylene glycols.

Styrene diallyl phthalate and acrylate esters are the most common cross-linking agents for polyesters.

Production of saturated polyesters by direct esterification of terephthalic acid has recently been achieved by Teijin Ltd. and Toyobo Co. in Japan by *continuous* direct reaction of ethylene oxide with terephthalic acid to produce bis(metahydroxyethyl) terephthalate which yields polyester by a condensation reaction. By using ethylene oxide, instead of a glycol, the reaction speed is increased, and it becomes practical to work with relatively crude acid and glycol, applying the purification step for the final product only. This technique has been evolved with primary objectives in the fiber and textile fields, but seems likely to be expanded into the bulk esters.

Polyethylene terephthalate for textile fibers and

fabrics now is made by a continues process from ethylene glycol and terephthalic acid or its di-methyl ester. The raw materials are fed into a heated reaction zone where a very high vacuum causes rapid removal of the water of condensation, thus accelerating the reaction. The resultant material is extruded as fiber or film and stretched to orient the molecules.

Triallyl cyanurate is used to prepare resins with exceptional heat resistance.

The structure of an uncross-linked unsaturated polyester may be represented by the formula for ethylene glycol maleate:

$$—R—CH{=}R—CH{=}CH—R—$$

$$(R \text{ is } —OHOCH_2CH_2OCO—)$$

The manufacture of unsaturated polyesters at present is still a two-step batch process. In the first step, the acid and the glycol are fed to a stain-less steel or glass-lined cooking kettle. An inert gas, such as nitrogen or carbon dioxide, is blown through to keep out oxygen which would cause dis-coloration of the resin. The mixture is then grad-ually brought to the reaction temperature, which is generally about 200°C. Water produced in the reaction is removed, along with some of the glycol. In the second step, the product of the polycon-densation reaction is blended with the cross-linking agent in a separate kettle.

Inhibitors such as hydroquinone may be added at different points of the process to prevent premature cross-linking and to control the storage life of the uncatalyzed resin and the pot life of the catalyzed resin. The catalyst for the final cure of the resin is added shortly before the resin is used. Benzoyl peroxide is the usual catalyst for cures at tem-peratures above 50°C. Methyl ethyl ketone perox-ide is a popular catalyst for lower temperature cures, particularly in conjunction with cobalt naphthenate. Fibers are often incorporated in polyester resins by molding or premixing to im-prove physical properties. Fillers may also be applied to improve strength or to reduce cost. Re-inforcing fibers can be mixed with highly filled polyester resin as in sheet molding or bulk molding compositions, or they can be inserted in the mold as fiberglass preform or mat or in other woven or nonwoven forms.

Reinforcing fillers include such fibrous materials as glass, quartz, cotton, asbestos, silane treated sand, ramie and sisal. Bulk fillers such as clay, silicates and carbonates are used to improve the handling properties of the final products. Water up to 50%, has been used as a filler.

After World War II the peace uses were slow in getting started. Almost the entire production had been pre-empted by military applications, and thus necessitated reconsideration of and redevelop-ment of civilian applications. These however, have now come into widespread use, particularly in the fields of transportation, small boats, and building.

A large number of techniques are used in pro-ducing filled polyester products, larger than any other plastics. These include a continuous process for sheet, pipe, fish rod stock, match mold die-molding, (probably the largest volume) protrusion,

spraying, bay molding, hand layup, continuous lamination, centrifugal pipe molding, etc.

Cured polyester resins vary from soft to hard and rigid. Their electrical properties are excellent, they are dimensionally stable, and they can be ob-tained in many colors. The polyesters are adapt-able to most standard molding and laminating procedures. They are also used as casting resins, mainly for "potting" of electrical equipment, and as protective, decorative, and electrical insulating coatings.

From the standpoint of the electrical industry, the polyester resins have two outstanding prop-erties. They harden without any loss of solvent, resulting in insulating coatings which are non-porous and therefore very electrically resistant. In addition, they increase the mechanical strength of the equipment on which they are used. This, in turn, permits reduction of the amount of metal used in the equipment with savings in weight, space, and cost.

The electrical industry is also a large outlet for polyester film. When used in electrical equipment, the film permits the bulk of insulation to be re-duced and makes possible the use of higher work-ing temperatures. The film has a tensile strength of about 25,000 psi, which is several times that of any other plastic film.

Foamed resins can be produced from polyesters which range from rubbery foams to very rigid foams. They are produced by the reaction between polyesters of relatively high acid or hydroxyl number and di-isocyanate cross-linking agents, such as hexamethylene di-isocyanate, and tolyl di-isocyanate. However, this use of polyesters has declined in favor of polyethers.

The current trend is toward ever larger installa-tions, operated by financially strong organizations, and the use of high pressure to produce articles that need few finishing operations. Matched metal dies and hydraulic presses are used for large-volume production.

One of the largest volume application of rein-forced polyesters is in the building industry, where a high strength-to-weight ratio, translucency, low heat transmission, and weather resistance are im-portant advantages. They can be used in the form of laminated sheets or as sandwich panels with a core of foamed polyester resin. Another large use is for transportation at land and sea where great time savings can be effected with low-pressure molding technique. A third large application is for airplane parts, where light weight and high strength are particularly important. The reinforced poly-esters are also used for a wide range of industrial and consumer products, including tanks, fume hoods, pipe and ducting, furniture, luggage, and housings for mechanical parts and appliances.

As engineering data are accumulated, a wider ac-ceptance of the favorable properties of reinforced polyesters may be expected. Designs for buildings, based on their high strength, have already been de-veloped in which the number of supporting mem-bers is minimized. It even becomes possible to de-pend on only the walls to support a building.

However, price and finishing costs are still ob-stacles to the full scale penetration of the lower price range building fields such as "mobile homes" which in 1969 amounted to only 10,000,000 lbs. High-density foams have penetrated engineering profiles, and also the furniture industry.

The volume production of plastics for 1969 for marine use is 270,000,000 lbs, for land transportation the estimate for 1969 was 220,000,000 lbs for 1969 of which a major part was polyester. The building market was lower, but may soon surpass the others.

The U.S. production and sales of polyester resins have had an unbroken uptrend, as is shown by the following table.

U.S. PRODUCTION AND SALES OF POLYESTER RESINS
(IN THOUSANDS OF POUNDS AND DOLLARS)

Year	Production	Sales	
		Quantity	Value
1961	193,221	180,185	$ 62,174
1962	212,230	193,388	63,552
1963	254,858	228,592	77,211
1964	316,628	276,282	84,556
1965	398,884	343,605	99,331
1966	470,046	406,658	122,627
1967	513,492	449,183	125,139
1968	615,408	543,266	149,671
1969(p)	667,433	591,841	n.a.

(p) = preliminary. Latest information available.
n.a. = Not Available.

JOHAN BJORKSTEN

POLYETHYLENE

Dimers and other low molecular weight polymers of ethylene have been known for years but macromolecular products were not produced until 1933 when Faucett and Gibson obtained polyethylene by heating purified ethylene at about 200°C at extremely high pressures (1000–2000 atm.) in the presence of traces of oxygen. This type product was first produced commercially in the United Kingdom in 1939 and because of its excellent electrical properties was used extensively during World War II for insulation of coaxial cable.

Polyethylene is also characterized by low water absorption, low water vapor permeability of films, excellent resistance to impact and ease of processibility. After the cessation of hostilities the product was widely used as a thin film for packaging.

Ethylene ($CH_2\!=\!CH_2$) is produced by cracking propane. The purified monomer plus oxygen is passed through high-pressure towers or tubular reactors to yield 5–15 per cent of a highly branched macromolecule. A typical product contains 20 methyl or ethyl groups per one thousand carbon atoms. The unreacted ethylene is recycled.

The physical properties of polyethylene and many other polymers are a function of molecular weight, molecular weight distribution and branching. The short chain branching present in so-called high-pressure polyethylene decreases stiffness, tear strength, hardness, softening temperature, yield point and chemical resistance. In contrast, branching has a positive effect on permeability, toughness, and flex life.

Oxidation in air occurs readily at the side chains. These positions are also vulnerable to irradiation which may produce useful cross-linked products. The resistance to air oxidation or degradation in the presence of ultraviolet light is increased by the addition of stabilizers and carbon black. This polymer may be readily chlorinated or chlorosulfonated. The product obtained by chlorosulfonation is called "Hypalon."

The "low-density" polymer is characterized by excellent resistance to aqueous corrosives, and organic solvents at ordinary temperatures. However the amorphous polymers are adversely affected by surface-active agents. This type of polymer is produced at a rate of over 2 billion pounds annually in the United States and is sold under various trade names. It may be mixed with boron compounds or heavy metal salts to produce useful products for the atomic energy industry.

Typical properties of a low-density polymer of ethylene are: specific gravity 0.915, tensile strength 1500 psi, impact strength greater than 10 ft lbs/in notch, thermal expansion 17×10^{-5} in/in/°C, heat deflection temperature at 66 psi 105°F, dielectric strength 2.25, and water absorption 0.01 per cent.

Polyethylene is widely used as film for packaging, linings, barriers, or temporary shelters. This polymer may be injection molded or blow molded to form squeeze bottles, may be extruded as pipe or tubing and may be coated on metal or other surfaces. Finely divided polymer is available in sizes as small as 50 microns. This powder which behaves like a fluid may be applied as is or as an aqueous slurry to various surfaces which are subsequently heated to fuse the coating. The powdery product may also be applied by flame spraying. Polyethylene sheet may be cast or extruded. This sheet may be formed by heat and joined at higher temperatures. Partial oxidation of the surface facilitates printing.

Some of the disadvantages of this widely used inexpensive polymer are overcome by copolymerization with propylene, 2-butene, ethyl acrylate, or acrylic acid. Copolymers with small amounts of propylene or 2-butene have higher softening points. Useful vulcanizable elastomers are obtained when ethylene propylene and butadiene are copolymerized. Copolymers with small amounts of ethyl acrylate (EEA) or vinyl acetate (EVA) are more flexible and tolerate a large proportion of filler. These copolymers are compatible with many other thermoplastics and thermosetting resins. Low molecular weight polymers, "Epolene" and copolymers may be used in place of microcrystalline waxes in paraffin wax blends.

Copolymers of ethylene with monomers containing carboxyl groups may be further reacted with sodium, potassium, magnesium or zinc ions to produce transparent products with superior resistance to solvents and crazing. These "ionomers" such as "Suryln A" may be fabricated using techniques similar to those used for polyethylene.

In addition to amorphous copolymers, crystalline products called "polyallomers" having well-balanced physical properties are obtained from ethylene and propylene. These products are characterized by good flow during processing, and superior resistance to tearing and stress cracking.

Polyethylene is produced by metal alkyl catalysis at moderate pressure (2–75 atmospheres), and at moderate temperature (50–250°C) in an aliphatic hydrocarbon solvent. This crystalline polymer is available under the trade names "Hostalen" "Dylan," "Hi-fax" and "Marlex." These products will have less than 3 methyl and ethyl groups per thousand carbon atoms.

Properties of a typical high-density polyethylene molded specimen are: specific gravity 0.95, tensile strength 4000 psi, impact strength 8 ft lbs/in/notch, heat deflection temperature at 66 psi 160°F, and water absorption 0.01 per cent.

RAYMOND B. SEYMOUR

Cross-linked Polyethylene

Homopolymers and copolymers of ethylene may be readily cross-linked either by irradiation or by chemical reaction. Both techniques involve creation of free radicals by splitting off hydrogen from the polymer chain followed by linking at these free radical sites. With optimum cross-link density the polymer is no longer thermoplastic but rather has been converted essentially to one large, interconnected molecule which is insoluble, will not melt at elevated temperatures, and has excellent heat, chemical, and creep resistance, high impact strength, and is free from stress cracking.

Irradiation cross-linking is accomplished by exposing a preformed sheet or part to electron beam or gamma rays. The cross-link density is dependent on the time of exposure of the polymer chains and on ray penetration. For this reason, irradiation cross-linking is generally limited to products having thin wall sections. With thicker sections, cross-link density tends to be too high on the outer surfaces and too low in the interior of the part, and the cost per pound of irradiation would in addition be very high.

Chemical cross-linking is accomplished by intimately mixing a suitable cross-linking agent with polyethylene and exposing the mixture to heat. The cross-linkable composition is first formed into a desired product by conventional means, such as extrusion, molding, calendering, etc., and then heated. As the temperature of the resin mass increases, the cross-linking agent decomposes at an increasing rate, splitting off hydrogen from the polymer chains and thus creating free radical sites which immediately cross-link. The cross-link density is dependent on the amount of peroxide present and is as uniformly distributed as the degree of initial dispersion of the cross-linking agent. Thus there is uniformity throughout the resin mass with no areas having excessive or insufficient cross-links for optimum physical properties.

All grades of polyethylene may be chemically cross-linked, as well as such copolymers as ethylene/vinyl acetate and ethylene/ethyl acrylate. Those peroxides having the structure, R—C—O—O—C—R', where a tertiary carbon is adjacent to the oxygen are excellent cross-linking agents. Typical are dicumyl peroxide, 2,5-dimethyl-2,5-di(t-butyl peroxy)hexane, and 2,5-dimethyl-2,5-di(t-butyl peroxy)hexyne-3.

Several fillers may be used in cross-linked polyethylene in rather large quantities. Carbon black is the preferred filler, since it is believed to become part of the cross-linked matrix through chemical bonds to active sites on the carbon particle. While moderate loading of carbon black in polyethylene results in considerable embrittlement of the mixture, very high percentages of carbon black may be included in cross-linked polyethylene with no serious change in low-temperature brittle point. For example, a low-density cross-linked polyethylene composition, half of which is carbon black, will have a low-temperature brittle point below minus 55°C. A cross-linked mixture over 70% of which is carbon black, still has an Izod impact strength of 5 foot-pounds at room temperature. These compounds uncross-linked would be brittle and without commercial usefulness.

The degree of cross-linking may be determined in several ways. Most accurate but time consuming is a so-called swell and extract test. A sample is extracted in hot xylene or other solvent and the per cent swell and per cent extract determined. Low swell and low extract indicate a high degree of cross-linking. Another method involves measuring the resistance to permanent distortion, or set, at temperatures above the crystalline melting point of polyethylene. This test may be made in minutes, and while possibly not as accurate as solvent extraction, it is quite practical for product control work. Called the Cabot Corporation "XL-O-Meter" test, it first measures the incompressibility of the cross-linked sample above its crystalline melting point under a specific load. With removal of the load, the recovery of the sample is measured. High incompressibility and high recovery indicate a high degree of cross-linking.

Commercially, cross-linked polyethylene has found widest application as power cable insulation. These compounds based on low-density polyethylene are well-suited for most low-voltage applications such as line and building wire, service entrance cable, and similar constructions. Special cross-linkable compounds are also being introduced for higher voltage cables up to 37 KV service.

These compounds are applied to wire by conventional extrusion techniques and cross-linked in standard CV (continuous vulcanization) equipment in steam pressures up to 250 psi (205°C). At these temperatures, cross-linking can be completed theoretically in about 30 seconds, and therefore the speed of extrusion is dependent on the length of the CV tube, as well as the size of the wire and thickness of the insulation.

The cross-linkable polyethylene compounds used for wire insulation are essentially based on low-density resins. Cross-linkable linear polyethylene tends to make wire and cable too stiff. Crosslinkable compounds of the higher-density resins have found application in pipe and molded fittings. Such compounds can be fabricated by special techniques into products having heat and creep resistance at temperatures above 95°C, excellent low-temperature impact resistance, and outstanding resistance to a very wide range of chemicals. Pipe made of these materials is so chemical resistant that no solvent has been found that will permit so-called solvent-welded joints. Fortunately, the creep resistance of cross-linked linear polyethylene is so good that standard threaded joints will hold at pressures in excess of 100 psi and at water temperatures in excess of 95°C.

As well as compositions having excellent electrical insulating properties, cross-linked polyethylene compositions can be made having very low resistivity which may be classified as semiconductors. Resistivities less than 100 ohm-centimeters can be obtained without sacrificing the physical properties of high- and low-temperature resistance, creep and impact strength, and excellent chemical resistance.

Cross-linkable polyethylene compounds may be molded into useful products by compression, transfer, and injection-molding processes. It is important that the desired part be formed before cross-linking is initiated. Therefore, good temperature control of the process is necessary. Injection of the compound must be complete and the mass fully fluxed before exposure to cross-linking temperatures.

Cross-linked polyethylene displays strong plastic memory. In its elastic state above its crystalline melting point, cross-linked polyethylene may be formed or shaped into many configurations. If held in a shape while cooled below its crystalline melting point, the cross-linked resin will retain that shape indefinitely. On being reheated again above the crystalline melting point, the stressed portions will return very tenaciously to the original shape when cross-linked. This property has found practical use in such products as heat shrinkable tubing.

Cross-linked polyethylene should be considered as a type of material unlike other known thermoset and thermoplastic compositions. The variations in composition and wide range of properties possible with these compounds approach the ideal of a universal material more closely than most resins. The possibility of meeting a broad spectrum of conditions with economical compounds that can be processed in standard equipment ensures that cross-linked polyethylene will be widely used.

WENDELL J. POTTER

Crystallization of Polyethylene

The *chemical structure* of polyethylene is a long linear sequence of CH_2-units. As the name indicates, it is usually made by polymerization of the monomer ethylene, C_2H_4. The addition of each monomer unit to a growing chain increases the number of repeating units by two. On either end of the chain different atoms or groups of atoms, R_i and R_t, are found. The nature of R_i and R_t, the end groups, depends upon the *i*nitiation and *t*ermination mechanisms operating when polyethylene is synthesized. Both initiation and termination are random, so that the chain length is not uniform for all molecules in a sample, but a *wide variation of chain length* is found.

A useful polymer is so large that the presence of a different end group is of little importance for the overall composition of the molecule. It does, however, represent a defect in the molecule which ideally would consist of CH_2's only. Other defects are created by including occasionally a wrong repeating unit in the chain. These foreign repeating units may have been in the monomer as an impurity or they may be added deliberately to cause copolymerization. Branching and change by subsequent

oxidation are two other ways to introduce defects into the chemical structure of the polymer chain. Of particular importance is the spatial requirement of the defect in the crystalline state. Table 1 illustrates some of these defects.

TABLE 1. DEFECTS IN THE CHEMICAL STRUCTURE OF POLYETHYLENE

Chemical Symbol:	Description:
—CH₃	methyl end (or side group)
—CH=CH₂	vinyl end (or side group)
—CO—	CO foreign repeating unit
—CH₂CHOH—	vinyl alcohol copolymer unit
—CH(CH₂)ₓCH₃—	side branch of length X + 1
—CH₂CHOCOCH₃—	vinyl acetate copolymer unit
—CHCl—	chlorine side group
—CH— | (CH₂)ₓ | —CH—	cross link of length X between two chains

The polymerization represents a connection of repeating units which is permanent. Subsequent changes will not alter the neighbors along the chain. In a certain sense, one can look upon the polymer molecule as being already "crystallized" into a *one-dimensionally ordered array* of set configuration. The bonds which hold the molecule together are strong covalent bonds with about 50 to 100 times higher bond energy than the weak van der Waals type bonds which keep the chains together in the condensed phase.

The polymer molecules are not stretched out in an orderly way, but are free to rotate, although with some hindrance, around any C—C bond into two other positions which are called "gauche." Figure 1 indicates the three possible *conformations* around one C—C bond. Energetically, the *trans*-conformation is favored by about 800 calories per mole of repeating units over the gauche conformation. Change from *trans* to *gauche* is possible by rotation, but in order to effect the change, besides the 800 calories energy difference, another 3000 calories/mole must be provided to squeeze the groups bound to the two carbons past each other. The conformations between the *gauche* and *trans*-position are not favored energetically, reaching a maximum, the above-mentioned 3000 calories, for the position of alignment.

In a typical polyethylene molecule there are about ten thousand CH_2 units. The distance between two carbons is 1.54 Å. This leads to a molecule which has a stretched-out, zigzag chain of 12,600 Å when the *trans*-conformation is found around all bonds. The diameter of the zigzag chain is only about 3 Å. To visualize such a chain, it is useful to make a *model* by enlarging the chain dimensions ten million times. Then the contour length is 12,600 millimeters or about forty feet and the diameter is only one-eighth inch. These are the dimensions of a good-sized clothesline. If we

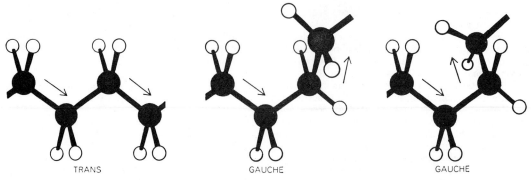

FIG. 1. Three stable conformations a polyethylene chain can assume around a single C-C-bond.

allow equal chance to any of the three conformations of Figure 1, the enormous total of 3^{10000} or 10^{4771} different conformations are possible. Only one of these is all *trans*- and thus fully stretched. Without any special ordering process, the molecule will assume a random coil conformation as is indicated in Figure 2. One can calculate for such a random coil only the mean of the end-to-end distance. For our "clothesline" this mean is less than one foot.

In the melt or in solution, the segments of the polymer chain random coils are rather mobile. In addition to many fast vibrations, it is possible for the molecule to change among the different conformations by rotating around the C—C bonds from one *gauche* to the *trans*-conformation or vice versa. The picture we get of a polymer molecule is that of a random coil which continually changes its conformation. The rate of change from one conformation into the other is determined by the amount of kinetic energy available to overcome the barriers to rotation or, in other words, the temperature.

If one *cools* a polymer to lower and lower temperature, the changes between different conformations become more and more sluggish. Outwardly the viscosity of the melt increases with decreasing temperature. Finally the conformations become fixed altogether and a solid results. The point within the time scale of the experiment at which the motion of the molecules from one conformation into another stops is called the *glass transition* temperature. For polyethylene this temperature for time scales of minutes to hours is at about $-35°C$. Below $-35°C$, polyethylene in the random coil conformation of the melt is glassy.

Before the transition temperature into the glassy state is reached, however, *crystallization* intervenes. In the crystalline state the chains of the polymer are also fixed, but in the extended all-*trans* conformation which allows them to be more closely packed and ordered. The closer packing (about 16%) allows stronger interaction (van der Waals bonds) between the chains. In addition, there is a reduction in energy of 800 calories for each mole of CH_2-units which have, on straightening, changed from the *gauche* to the *trans*-conformation. Over-all about 960 calories are evolved for each mole of CH_2-repeating units (14 grams) crystallizing.

From the description of the solidification of polyethylene, it becomes clear that there are some fundamental differences between the crystallization of flexible long-chain molecules and spherical atoms like, for example, gold. In the polymer crystal there is very strong bonding along the chain and only weak bonding between the chains. This anisotropy is reflected in the physical properties like high tensile strength along the chain direction and weakness at right angles to them. For polyethylene the C—C-chain bonds are as strong as the bonds in diamond, while the van der Waals bonds between the chains are as weak as those in paraffins and waxes. In gold, on the other hand, bonding is strong (but not as strong as in diamond) to all 12 neighbors of each atom.

A much farther-reaching difference is discovered on looking at the *melts* of the two substances. In

FIG. 2. Random coil of 125 CH_2 repeating units. The reference molecule described in the text would be 80 times as long. The two ends are marked by the larger atoms, representing end groups.

both cases the "repeating units" CH_2 and gold are largely disordered. They are still in the condensed state, which means that most of the bonds holding the atoms in close proximity are still intact; however, the "repeating units" no longer have a fixed equilibrium position, as they had in the crystal. In gold any one of the bonds of the now only approximately twelve neighbors can be broken and remade, after movement, with another gold atom. In polyethylene the same can take place, but only with the weak van der Waals bonds. The *strong covalent bonds remain* at all times. Disorder is present to about the same degree because of the many conformations polyethylene can take by rotation around the C—C bonds without bond breakage. In fact, the increase in disorder on melting as measured by the increase in entropy per mole of "repeating unit" is identical for gold and polyethylene. The difference in melting point (1063°C for gold and 142°C for polyethylene) is entirely due to the difference in energy between the respective melts and solids. Since the order along the chain remains perfect, we can look upon molten polyethylene as still being a flexible "crystal" in one dimension. Accordingly, we can also look for a *second "melting point"* which destroys this order. Usually this second "melting point" is called depolymerization or decomposition. In polyethylene this is a process which is not reversible. If one calculates the process for hypothetical reversal of the polymerization monomer gas results. The increase in disorder during this process is quite comparable to the evaporation of gold to the vapor phase; in fact, it is in line with any liquid being vaporized (Trouton's rule of constant entropy of evaporation). The two gases now are quite similar; little spherical or almost spherical molecules far apart with high translational mobility make up both the gold and polyethylene "vapor."

The "normal" way of producing crystalline polyethylene is to polymerize first and produce thus a disordered and highly entagled melt or solution. The second step is the crystallization. For a melt of our model polyethylene this would involve disentangling 30-foot long "clotheslines" and putting them in order. Obviously, this is much more difficult than putting a set of marbles into order one by one, a process similar to crystallization of gold.

The first successful investigations of the crystal structure of polyethylene were conducted in 1939 by Bunn in Britain. To help the crystallization he stretched extruded fibers of polyethylene. The macroscopic stretching helped the paralleling of the chains and, although far from perfect, sufficient order and orientation of the crystallites were achieved to elucidate the ideal structure.

The next big advance in understanding came when another possible mode of crystallization was tried, namely, crystallization from dilute solution. In dilute solution, single chains are separated from each other and one chain at a time can be added to a growing crystal. The big surprise was that polymer molecules behaved quite "logically." They crystallized in the same fashion as someone faced with the task of putting a large number of randomly coiled, separated clotheslines into order would proceed. He would wind up the lines into folded packets and stack them. Keller in England,

Fischer in Germany, and Till in the United States simultaneously in 1957 discovered that polyethylene crystallizes from the dilute solution in the form of single crystals containing folded polymer chains. Thin lamellae about 100 Å thick and 1000 Å wide with crystallographically regular surfaces were formed on slow cooling of the solution. With the above dimensions, 100,000 billion of these crystals weigh only one gram. The surface of these crystals is about one million square centimeters per gram. Although these crystals are exceedingly small, one can establish by electron diffraction that the chains are oriented at right angles to the lamellae; this means they must bend over every 100 Å, so that our folded standard molecule of 10,000 CH_2 would form 126 folds. The whole crystal would contain about 40,000 of these chains.

With the discovery of seemingly perfect but small single crystals, the field of solid-state physical chemistry of polymers had come of age in 1957. It was quite clear that for an understanding of the solid-state behavior the properties of the small crystallites must be studied. Out of this understanding one could then hope to go on and try to understand less perfect crystalline aggregates. In particular, it seemed possible to master the old problem of non-reproducibility of polymer properties. It became evident that the nonreproducibility stemmed from ignorance about the enormous variation possible in morphology. Only if the same morphological structure is present in different samples are they expected to have the same properties. No other material has as many variables as can be found in linear polymers.

As soon as one learned how to make these lamellae to millimeter size, optical microscopy could be used for their observation. The optical technique which allows quantitative observation is *interference microscopy*. A technique where the phase-lag suffered by light going through the sample is used to show up the crystal. Quantitative measurements of the phase-lag allow the determination of thickness with an accuracy of ±6 Å.

How does it come about then that always *thin folded-chain lamellae* seem to arise on slow crystallization of polymers? In recent years even from the melt such folded lamellae were grown. The thickness of all these lamellae is temperature dependent, ranging from about 100 to 1000 Å. The greater thicknesses were achieved from the melt, where higher crystallization temperatures are possible. The solution to this question is not quite given to date. However, it seems probable that what happens is that a long chain which is first absorbed on a growing crystal face finds it much faster to crystallize by folding together with a fixed fold-length than to straighten out completely, despite the fact that a completely straightened out chain in a crystal would be thermodynamically more stable. The sharp fold at the surface of the lamella causes a large surface free energy. The once laid down layer of folded chains is thus only a compromise solution and tries to "thicken" and straighten out. If the temperature is relatively low as in solution crystallization not much comes of this thickening before more and more layers of polymer chains are put on top of the original polymer chain. The crystal becomes metastable.

The thickening or growth of the fold length of chains already crystallized or absorbed on the growth face of the crystal has a certain temperature-dependent rate. The crystal growth, which adds more layers of folded chains to the polymer crystal, also has a rate which is temperature-dependent. At "normal" conditions of growth the latter rate is faster, so that folded crystal lamellae result, which can only be made to thicken to imperfect crystals afterwards at higher temperature.

All one has to do to grow thicker crystals according to this hypothesis is to alter the two rates relative to one another. Since the polymer crystal is strongly anisotropic, pressure from all sides suggests itself for such experiments. Hydrostatic pressure raises the crystallization temperature of polyethylene by 20°C for each 1000 atmospheres pressure. Crystallization under 5000 atmospheres from the melt gave crystals which can finally be called *true single crystals.* The pressure-crystallized polyethylene consists of lamellae up to 100 microns in size, and about 100,000 Å thick. The polymer chains are fully or almost fully extended. Chain folding is not an important feature in these samples.

Single crystals (folded or extended chain) arise only under very special conditions. From solution only a 5°C region in temperature and low concentration guarantee large-area lamellae. The extended chain crystals from the melt can be grown up to date only under pressure and in a temperature region of a few degrees. What grows under other conditions? From dilute solution on more rapid cooling *dendrites* arise—structures which are similar in appearance to snowflakes. The difference here is that the polymer dendrites occur much more readily, a result which one would expect, since it is so much easier to put little H_2O molecules into order in a compact ice block than to try to disentangle polymer chains.

Dendritic shapes also grow from the melt. Here, because of the high concentration, the dendrites are even more "feathery"; mostly the separate branches radiating out from the center are not resolvable by the optical microscope. All one can see is the regularity made visible by polarized light. These structures, called *spherulites,* make up most of crystallized polyethylene, their size and perfection varying widely depending upon crystallization conditions.

After this insight into polymer crystallization, it seems obvious that the properties of polyethylene must be determined to a large degree by the crystal defects. How much defects influence physical properties of solids is known from studies of metals. We can divide all properties into two categories, structure-sensitive and structure-insensitive properties. *Structure-insensitive properties* are, for example, the volume of a polymer sample, its heat content, specific heat, and its compressibility. Here the defect has only a minor influence. The volume requirement for a CH_2 which is concentrated in a defect is perhaps up to 20% larger than that of a crystallized one, a rather minor change, in particular if one remembers that most CH_2's are well crystallized. For the structure-insensitive properties it suffices to express the structure of the polymer by, for example, its "degree of crystallinity" which can be calculated from the per cent increase in volume of an ideal crystal. The volume of the latter can be calculated from the arrangement of the perfect crystal.

Much less is known about the structure-sensitive properties such as the ultimate strength of the polymer or the rate of diffusion of small molecules through the polymer. Here a small percentage of defects can alter the properties by many orders of magnitude. Very little exact information is available about this most important topic, but we can see now the direction which research must take to elucidate failure mechanisms and diffusion kinetics. To have any possibility of a description, the morphology of the actual sample must be known in all detail, in more detail than we are, in fact, able to provide at present for most samples. To have a chance of producing meaningful measurements on structure-sensitive properties samples must be used which have a reproducible morphology. Even though we may not be able at present to elucidate the structure in all detail, we are able to show that if two samples have the same structure they are reproducible.

BERNHARD WUNDERLICH

POLYIMIDES

Polyimides are heat-resistant polymers which have an imide group (—CONHCO—) in the polymer chain. Polyimides, such as Kaptan, Pyre-MI, Vespel, H-film, poly(amide-imides), such as Amoco AI-220, and poly(esterimides), such as Terebec FN and FH, are commercially available.

Poly(amide-imides) are prepared by the thermal degradation of a soluble poly[amide-(amic acid)]. The latter may be produced by the condensation of an aliphatic diamine with less than a molar equivalent of pyromellitic dianhydride or with a molar equivalent of a derivative of trimellitic anhydride, such as the acyl chloride in dimethylacetamide as shown in the following equation:

acyl chloride
of trimellitic
anhydride

aliphatic
diamine

poly[amide(amic acid)]

$$\xrightarrow[(-H_2O)]{\Delta} \text{poly(amide-imide)}$$

The poly(amide-imides) are soluble in dimethylacetamide but are insoluble in less polar solvents such as toluene and perchloroethylene. They are used for wire enamels, high-temperature adhesives, laminates and molded articles.

The poly(ester-imides) are produced by the thermal decomposition of the soluble poly(amic acids) which are obtained by the condensation of an aromatic diamine and the bis-(ester anhydride) of trimellitic anhydride as shown in the following equation:

It is customary to apply these polymers as the poly(amic -acids) and to dehydrate the film, coating, fiber or molded forms by heating to produce the polyimides. Polyimides are insoluble in most solvents but are attacked by alkalies, ammonia and amines. These heat resistant polymers are used

bis-(ester anhydride) of trimellitic anhydride $\longrightarrow$ poly(amic -acid)

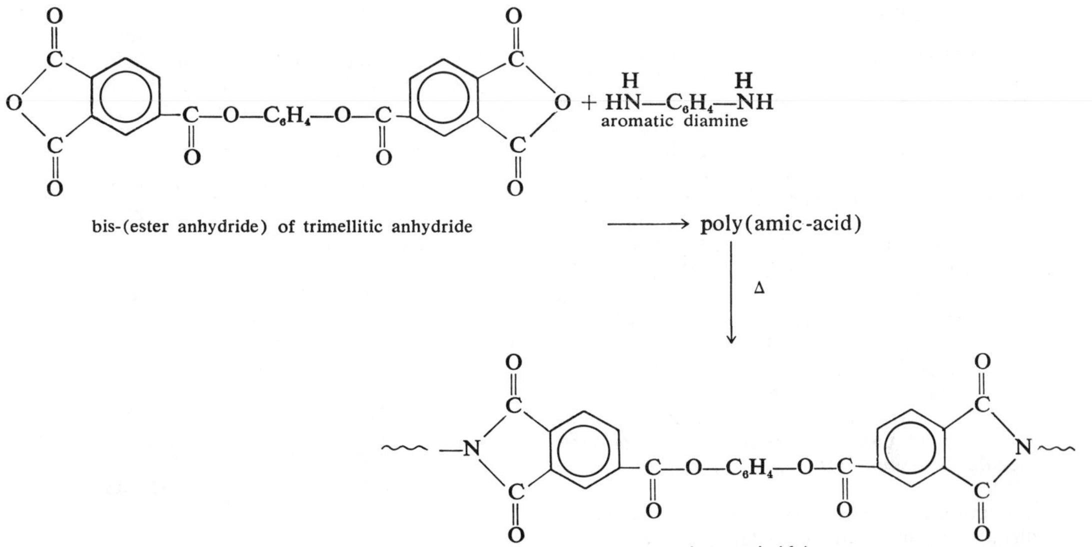

poly (ester-imide)

These poly(ester-imides) have good electrical properties. Their tensile modulus is about 400,000 psi at 25°C and approximately 50 per cent of this modulus is retained at 200°C. Poly(ester-imide) films fail when heated at 240°C for 1000 hrs.

Polyimides are produced by the thermal dehydration of the soluble poly(amic -acid) which is obtained by the condensation of a diamine, such as 4,4'-diaminophenyl ether and a dianhydride, such as pyromellitic dianhydride called PMDA as shown in the following equation:

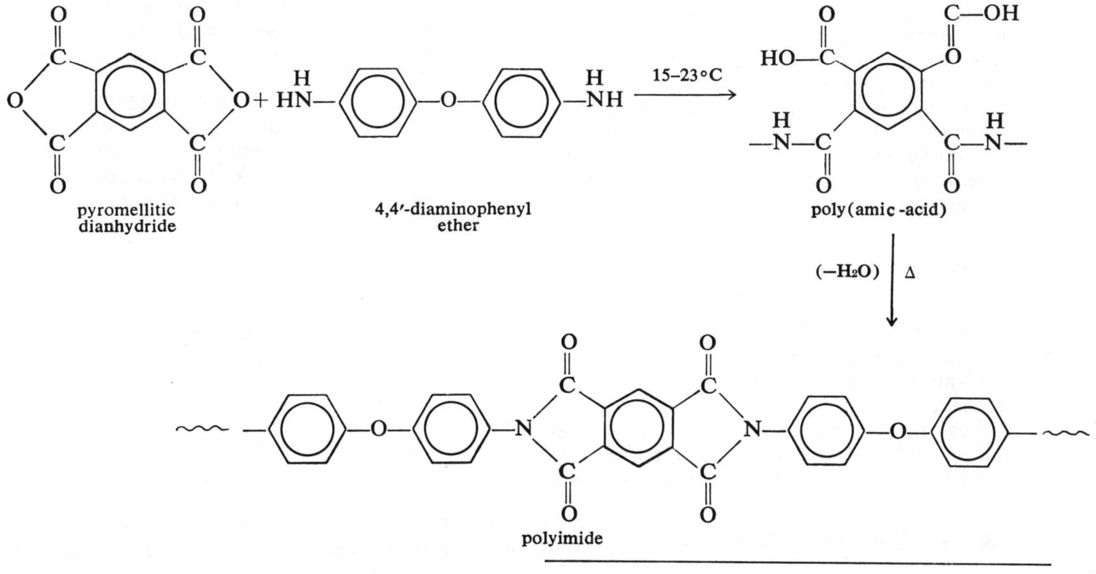

without fillers and with a graphite filler.

Polyimide films have excellent electrical properties and a tensile modulus of over 400,000 psi at 25°C. Over 60 per cent of this modulus is retained at 200°C. Polyimide wire enamels are stable for up to 100 thousand hours at 200°C. Polyimide fibers have a tenacity of 7 g/denier at 25°C and over 1000 hrs at 283°C is required to reduce the value to 1 g/denier.

The coefficient of linear expansion of polyimides is 4.0–5.0 × 10^{-5} in./in./°C. The heat deflection is

pyromellitic
dianhydride

4,4'-diaminophenyl
ether

poly(amic -acid)

$(-H_2O)$ | Δ

polyimide

680°F. Polyimides have been used as binders for abrasive wheels, high-temperature laminates, wire coatings, insulating varnishes and in aerospace applications. Polyimides are being produced at an annual rate in excess of 1 million pounds.

RAYMOND B. SEYMOUR

POLYMERIZATION

There exist chemical reactions, particularly in organic chemistry, which produce very large molecules by a process of repetitive addition. These are of great practical importance in the field of rubbers, plastics, coatings, adhesives and synthetic fibers. The initial materials which give rise to such reactions are called *monomers;* they have molecular weights between 50 and 250 and have certain reactive or functional groups which enable them to undergo *polymerization*. The large molecules, which are formed by a polymerization reaction are called *high polymers, macromolecules* or simply *polymers;* they usually consist of several hundred and in many cases even of several thousand monomeric units and, consequently, have molecular weights of many hundred thousands and even of several millions. The number of monomers contained in a polymer molecule determines its *degree of polymerization* (D.P.).

Polymerization processes never lead to macromolecules of uniform character but always to a more or less broad mixture of species with different molecular weights, which can be described by a *molecular weight distribution function*. The individual macromolecules of such a system belong to a *polymer-homologous series;* the molecular weight and the degree of polymerization of a given material have, therefore, always the character of *average* values.

There exist many ways to assemble small molecules to give large ones and, hence, there exist several types of polymerization reactions. The most important are the following:

(1) *Vinyl-type addition polymerization*. Many olefins and diolefins polymerize under the influence of heat and light or in the presence of catalysts, such as free radicals, carbonium ions or carbanions. Free radicals are particularly efficient in starting polymerization of such important monomers as styrene, vinylchloride, vinylacetate, methylacrylate or acrylonitrile. The first step of this process—the so-called *initiation* step—consists in the thermal or photochemical *dissociation* of the catalyst, and results in the formation of two *free radicals:*

$$\text{R—R} \xrightarrow[\text{or light}]{\text{heat}} \text{2R} \cdot \quad (1)$$

Catalyst molecule → two free radical type fragments

The most commonly used catalysts are peroxides, hydroperoxides and aliphatic azocompounds, which need activation energies between 25 and 30 kcal for decomposition.

The free radicals R·attack the monomer and react with its double bond by adding to it on one side and reproducing a new free electron on the other side:

$$\text{R} \cdot + \text{CH}_2\!\!=\!\!\text{CHX} \rightarrow \text{R—CH}_2\text{—CHX} \cdot \quad (2)$$

This step is called *propagation* reaction; it adds more and more monomer units to the growing chain and builds up the macromolecules while the free radical character of the chain end is maintained. Each single addition represents the reaction of a free radical with a monomer molecule—a process which requires an activation energy of 8–10 kcal.

Whenever two free radical chain ends collide with each other they can react in such a manner that the resulting products have lost their free radical character and are converted into normal stable molecules. One way is a process of *recombination*:

$$\text{R—(—CH}_2\text{—CHX—)—}_x\text{CH}_2\text{—CHX} \cdot +$$
$$\text{R—(—CH}_2\text{—CHX—)—}_y\text{CH}_2\text{—CHX} \cdot \rightarrow$$
$$\text{R—(—CH}_2\text{—CHX)—CH}_2\text{—CHX-}$$
$$\text{—CHX—CH}_2\text{—(CH}_2\text{—CHX—)—}_y\text{R}$$

where *one* macromolecule of the degree of polymerization $(X + y + 2)$ is formed. The other is a process of *disproportionation*:

$$\text{R—(—CH}_2\text{—CHX)}_x\text{—CH}_2\text{—CHX} \cdot +$$
$$\text{R—(—CH}_2\text{—CHX—)—}_y\text{CH}_2\text{—CHX} \cdot \rightarrow$$
$$\text{R—(—CH}_2\text{—CHX—)—}_x\text{—CH}_2\text{—CHX} \cdot +$$
$$\text{R—(—CH}_2\text{—CHX—)—}_y\text{CH}_2\!\!=\!\!\text{CH}_2\text{X}$$

$$(4)$$

where a hydrogen atom moves from one molecule to the other so that one of the two resulting molecules—the $(x + 1)$mer—has a double bond at its end, whereas the other one—the $(y + 1)$mer—has a saturated chain end. Reactions in the course of which free radicals are destroyed are called *termination* or *cessation* steps; they convert the transient reactive intermediates into stable polymer molecules.

Vinyl-type addition polymerization can also be carried out with *acidic catalysts* such as boron trifluoride or tin tetrachloride and with *basic catalysts* such as alkali metals or alkali alkyls. An example of the first case is the low-temperature polymerization of isobutene, which gives "Vistanex" and Butyl rubber; an example of the second type is the polymerization of butadiene with sodium, which leads to Buna rubber.

(2) Another important kind of addition polymerization is the formation of polyethers by the opening of epoxy ring compounds. Polyoxyethylene ("Carbowax") is produced by a sequence of additions of ethylene oxide to an alcohol or amine, as initiator:

$$\text{CH}_3\text{—CH}_2\text{OH} + \text{CH}_2\text{—CH}_2 \rightarrow$$
$$\diagdown \; \diagup$$
$$\text{O}$$
$$\text{CH}_3\text{—CH}_2\text{—O—CH}_2\text{—CH}_2\text{—OH} +$$
$$\text{CH}_2\text{—CH}_2 \rightarrow$$
$$\diagdown \; \diagup$$
$$\text{O} \qquad (5)$$
$$\text{CH}_3\text{—CH}_2\text{—O—CH}_2\text{—CH}_2\text{—O-}$$
$$\text{—CH}_2\text{—CH}_2\text{OH}, \quad \text{and so on}$$

No termination reaction occurs in this case; the reaction proceeds until all the monomer is used. This process is catalytically accelerated by the presence of alkali. A similar addition polymerization involving the opening of a ring compound is the conversion of caprolactam into polycaprolactam ("Perlon" or 6-nylon) under the influence of acidic or basic catalysts. All addition polymerizations are typical *chain reactions* with at least two or three different elementary steps cooperating in building up the resulting macromolecules.

(3) There exist other, different classes of reactions which form large molecules, namely processes in the course of which a *small fragment,* usually H_2O, is *split out* of two reacting monomers and where the monomers are chosen in such a manner that the removal of the fragment can be repeated many times. Multi step reactions of this type are called *polycondensations;* they involve the use of at least a pair of bifunctional monomers and proceed by a sequence of identical condensation steps. One important process of this type is the formation of polyesters from glycols and dicarboxylic acids. Thus the progressive removal of water from ethylene glycol and adipic acid leads to a soft, rubbery polyester ("Paracon")

$$HOCH_2—CH_2OH +$$
$$HOOC—(—CH_2—)—_4COOH →$$
$$HOCH_2CH_2—O—CO—(—CH_2—)—_4COOH +$$
$$HOCH_2—CH_2OH → HOCH_2—CH_2OCO =$$
$$—(—CH_2—)—_4COOCH_2—CH_2OH,$$ and so on.

As long as in processes of this type only *bifunctional* monomers are used, the resulting macromolecules are *linear* and, as a consequence, are of the soluble and fusible type. They can be used as fiber formers, rubbers or thermoplastic resins. If, however, some of the monomers are tri- or tetramethylolurea, the reaction leads to three-dimensional polymeric networks which are hard and brittle thermosetting resins, such as "Bakelite" or "Glyptal."

The preceeding classification of polymerization reactions concentrates essentially on the organic chemical character of the involved monomers and on the mechanism of their interaction. There exists, however, another classification which is concerned about the manner in which polymerization reactions are carried out in practice and which is of interest and importance whenever industrial application is contemplated. We shall, therefore, briefly enumerate here the most important *polymerization techniques.*

(1) Polymerization in the *gas phase* is usually carried out under pressure (several thousand psi) and at elevated temperatures (around 200°C); the most important example is the polymerization of ethylene to form polythene.

(2) Polymerization in *solution,* essentially under normal pressure and at temperatures from −70°C to 70°C; important examples are the production of Butyl rubber with boron trifluoride and the synthesis of the various "Vinylites" with benzoyl peroxide.

(3) Polymerization in *bulk* (or in block) under normal pressure in the temperature range from room temperature to about 150°C. The batch polymerization of methylmethacrylate to give "Lucite"

or "Plexiglass" and the continuous polymerization of styrene to give the various types of polystyrene can be quoted as examples.

(4) Polymerization in *suspension* (bead or pearl polymerization) under normal pressure in the range from 60 to 80°C operates with a suspension of globules of an oil-soluble monomer in water and uses a monomer soluble catalyst. Substantial quantities of polystyrene and polyvinyl acetate are made by this method.

(5) Polymerization in *emulsion* under normal pressure and in the temperature range from −20°C to 60°C uses a fine emulsion of oil-soluble monomers in water and initiates the reaction with a system of water-soluble catalysts. This method is probably the most important of all, because it is used in very large scale in the copolymerization of butadiene and styrene and in the polymerization of many other monomers, such as chloroprene and vinyl chloride, to produce latices of the various synthetic rubbers.

HERMAN F. MARK

Cross-references: *Free Radicals; Molecules; Radical.*

Polymerization by Oxidative Coupling

Polymerization by oxidative coupling is a new technique for the preparation of some new classes of high-molecular weight linear polymers. Schematically the reaction is represented below and involves the oxidative coupling of certain organic compounds containing two active hydrogen atoms to give a linear polymer. The hydrogens end up

$$nHRH \xrightarrow[\text{catalyst}]{\frac{n}{2} O_2} —(R)—_n + nH_2O$$

ultimately as water. High-molecular weight polymers have been prepared in this manner from phenols, diacetylenes and dithiols.

When 2,6-dimethylphenol is oxidized with oxygen in the presence of an amine complex of a copper salt as catalyst a high-molecular weight polyether (PPO®)* is formed.

The reaction is exothermic and proceeds rapidly at room temperature. The polymerization is generally performed by passing oxygen or air through a stirred solution of the catalyst and monomer in an appropriate solvent. When the desired molecular weight is attained, the polymer is isolated by dilution of the reaction mixture with a nonsolvent for the polymer. The precipitated polymer is then removed by filtration, washed thoroughly and dried.

* Registered Trade Mark.

The polymer is soluble in most aromatic hydrocarbons and chlorinated hydrocarbons and insoluble in alcohols, ketones and aliphatic hydrocarbons.

A large number of other 2,6-disubstituted phenols have been oxidatively coupled. A representative list of the results is presented below.

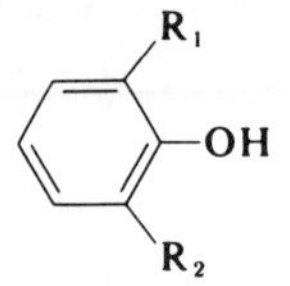

R_1	R_2	Principal Product
methyl	methyl	polymer
methyl	ethyl	polymer
methyl	i-propyl	polymer
methyl	t-butyl	diphenoquinone
methyl	phenyl	polymer
methyl	chloro	polymer
methyl	methoxy	polymer
ethyl	ethyl	polymer
i-propyl	i-propyl	diphenoquinone
t-butyl	t-butyl	diphenoquinone
methoxy	methoxy	diphenoquinone
nitro	nitro	no reaction

Polymer formation readily occurs if the substituent groups are relatively small and not too electronegative. When the substituents are bulky, the predominant product is the diphenoquinone formed by a tail-to-tail coupling. No appreciable reaction occurs when 2,6-dinitrophenol is oxidized even at 100°C.

A family of engineering thermoplastics based on the above technology includes PPO polyphenylene oxide, Noryl® thermoplastic resins (modified phenylene oxide) and glass reinforced varieties of each. The phenylene oxide resins are characterized by: (1) outstanding hydrolytic stability; (2) excellent dielectric properties over a wide range of temperatures and frequencies; and (3) outstanding dimensional stability at elevated temperatures. Because of these properties modified phenylene oxides are finding major application in the areas of business machine housings, appliances, automotive, TV and communications, electrical/electronic, and water distribution.

Oxidative polymerization of 2,6-diphenylphenol yields a crystallizable polymer that is characterized by a very high melting point (~480°C) and excellent electrical properties. It can be spun into a fiber with excellent thermal, oxidative and hydrolytic stability. It is marketed under the trademark Tenax®.

By performing the oxidation at elevated temperatures the phenols which would ordinarily yield polymers are converted instead to diphenoquinones. These quinones are readily reduced to the corresponding hydroquinones, compounds which promise to be useful as antioxidants and polymer intermediates.

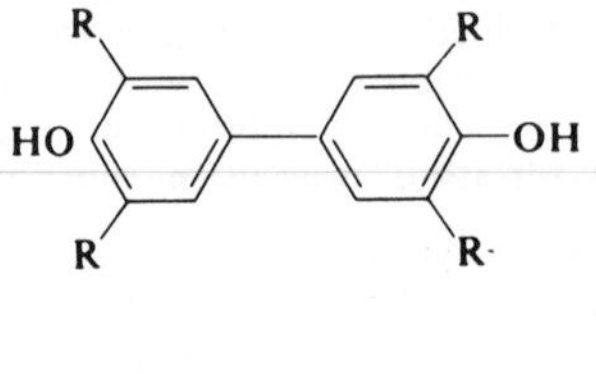

Oxidative coupling of diacetylenes yields another unusual class of polymers. From m-diethynylbenzene, for example, is obtained a high-molecular weight polymer that can be cast into a tough, flexible film.

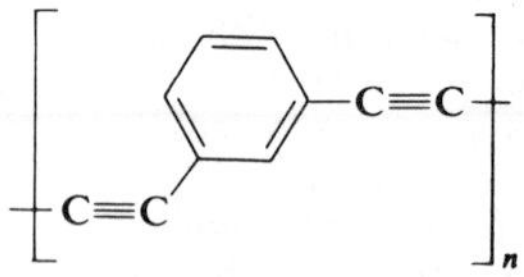

The polymer contains 96.75% carbon and on heating to about 350°F or above it spontaneously rearranges to an insoluble and infusible material. When ignited the hydrogen in the polymer burns leaving a carbon residue.

In the same manner dithiols can be converted to polydisulfides.

ALLAN S. HAY

Emulsion Polymerization

Since an aqueous system provides a medium for dissipation of the heat from exothermic addition polymerization processes, many commercial elastomers and vinyl polymers are produced by the emulsion process. This two-phase (water-hydrophobic monomer) system employs soap or other emulsifiers to reduce the interfacial tension and disperse the monomers in the water phase. Aliphatic alcohols may be used as surface tension regulators.

Formulas for emulsion polymerization also include buffers, free radical initiators, such as potassium persulfate ($K_2S_2O_8$), chain transfer agents, such as dodecyl mercaptan ($C_{12}H_{25}SH$). The system is agitated continuously at temperatures below 100°C until polymerization is essentially complete or is terminated by the addition of compounds such as dimethyl dithiocarbamate to prevent the formation of undesirable products such as cross-linked polymers. Stabilizers such as phenyl Beta-naphthylamine are added to latices of elastomers.

The final product in latex form may be used for water-type paints or coatings or the water may be removed from the finely divided high-molecular weight polymer. Separation may be brought about by the addition of electrolytes, freezing or spray drying.

It is believed that polymerization of hydrophobic monomers is initiated by free radicals in the aqueous phase and that the surface-active oligomers produced migrate to the interior of the emulsifier micelles where propagation continues. Monomer molecules dispersed in the water phase also solubilize by diffusing to the expanding lamellar micelles. These micelles disappear as the polymerization continues and the rate may be measured by noting the increase in surface tension of the system.

RAYMOND B. SEYMOUR

Polymerization, Radical

Polymerization is a process by which the small molecules of a substance, called the monomer, are joined together to form a much larger molecule, the polymer. In addition polymerization, polymer is the sole product of the reaction so that the monomer and polymer have essentially the same chemical composition; for example, monomeric styrene ($CH_2:CH \cdot C_6H_5$, i.e., C_8H_8) can be converted to polystyrene represented as $(C_8H_8)_n$ where n can be large, say 1000. In a polymerization of this type, polymer is formed by a stepwise reaction in which molecules of monomer are added one at a time to a reactive center; the center grows in size while retaining its reactivity.

In a radical polymerization, the reactive centers are free radicals and the process is a typical chain reaction. The monomers in radical polymerizations normally contain carbon-carbon double bonds in their molecules; styrene is typical. Usually radical polymerization is performed in the liquid phase. The chain reaction can be divided into the following steps:

(i) initiation —formation of a reactive free radical and its capture by monomer to form a center

(ii) propagation—reaction of a center with a molecule of monomer to form a larger center

(iii) termination—deactivation of a center so that it becomes incapable of further growth

(iv) transfer —reaction of a center with another molecule so that further growth of that particular center is prevented but a new center, capable of growth, is formed.

Commonly in radical polymerizations, initiation occurs continuously at a steady rate and is balanced by termination so that a steady concentration of growing centers (usually in the region of 10^{-8} mole/1) is established. The number of propagation reactions greatly exceeds the number of reactions of other types so that *macromolecules* are built up. The life-time of an active center is very much less than the duration of the whole process of polymerization and so the macromolecules are produced even in the earliest stages; there is not a continuous rise in the molecular weight of the polymeric product as found in polymerizations of certain other types. It is instructive to consider in some detail the component reactions in the overall process of radical polymerization.

Initiation. In principle, the simplest method for initiation is to add to the purified monomer a small amount of a substance which dissociates to fairly reactive free radicals. This initiator (or sensitizer) is chosen so that its decomposition ocurs at a suitable rate at the working temperature; thus azo*iso*butyronitrile is commonly used at about 60°C dissociating according to the equation:

$$(CH_3)_2C(CN) \cdot N:N \cdot C(CN)(CH_3)_2 \rightarrow 2(CH_3)_2C(CN) \cdot + N_2$$

The radical adds to monomer thus

$$(CH_3)_2C(CN) \cdot + CH_2:CHX \rightarrow (CH_3)_2C(CN) \cdot CH_2 \cdot CHX \cdot$$

forming the starting point of a polymer chain, i.e., an *end-group;* this reaction is the real initiation of

polymerization. Initiators of other types are also used, notably peroxides, both organic and inorganic. In some cases, the initiator is chosen to give free radicals under the influence of light; this process can be useful for initiating polymerizations at comparatively low temperatures. Two-component initiating systems are widely used in this connection, an example being

$$H_2O_2 + Fe^{2+} \rightarrow Fe^{3+} + OH^- + \cdot OH$$

which clearly would be selected for aqueous systems. At elevated temperatures or under the influence of external sources of energy (light, high-energy radiations, ultrasonics, mechanical work), many monomers polymerize apparently spontaneously without deliberate addition of sensitizer; the mechanisms of initiation under such circumstances are not completely understood.

Propagation. The propagation reaction in a radical polymerization can be represented by the general equation

$$P \cdot CH_2 \cdot CHX \cdot + CH_2:CHX \rightarrow P \cdot CH_2 \cdot CHX \cdot CH_2 \cdot CHX \cdot$$

which corresponds to the conversion of a carbon-carbon double bond to two carbon-carbon single bonds. The group $—CH_2 \cdot CHX—$ in the polymer chain is referred to as the *monomer* unit. If the growing center includes more than a few monomer units, the characteristics of the growth reaction are reasonably supposed to be independent of the size of the center.

The growth reaction is exothermic (in the region of 20 kcal/mole, i.e., about 80 kj/mole); under some circumstances, polymerizations may become self-heating and difficult to control. The growth reaction involves a decrease in entropy since a free molecule of monomer becomes organized in a polymer chain. The opposing effects of changes in enthalpy and entropy indicate that, for every polymerizing system, there is a *ceiling temperature* below which the growth reaction is favored thermodynamically but above which the reverse process is favored; the value of the ceiling temperature depends on the nature of the monomer and on its concentration in the system. Certain monomers, e.g., α-methyl styrene, were once thought not to polymerize by a radical mechanism but it is now clear that they will do so provided that the experiment is performed below the ceiling temperature.

The growth reaction shown above represents *head-to-tail* addition; the CHX groups occur at alternate sites along the main polymer chain and the unpaired electron is sited on a substituted carbon atom. Head-to-head addition, to give a polymer radical $P \cdot CH_2 \cdot CHX \cdot CHX \cdot CH_2 \cdot$, may occur occasionally but is likely to be followed by tail-to-tail addition to give $P \cdot CH_2 \cdot CHX \cdot CHX \cdot CH_2 \cdot CH_2 \cdot CHX \cdot$ which can be regarded as the normal growing radical. Head-to-head groupings may well be sites of instability in the polymer.

The substituted carbon atoms in the polymer chain are asymmetric. *Stereoregular polymers* are produced if all these carbon atoms have the same configuration (all *d* or all *l*) or if the *d* and *l* configurations occur alternately; pronounced stereoregularity is seldom achieved in radical polymerizations except perhaps at very low temperatures. When dienes are polymerized by a radical mechanism, the resulting polymers contain several distinct types of monomer unit, thus butadiene can give rise to $-CH_2 \cdot C(CH:CH_2)-$, $-CH_2 \cdot CH:CH \cdot CH_2-$ *cis,* and $-CH_2 \cdot CH:CH \cdot CH_2-$ *trans.*

Termination. In many radical polymerizations, termination occurs by interaction of pairs of growing radicals, either by combination to give $P \cdot CH_2 \cdot CHX \cdot CHX \cdot CH_2 \cdot P$ or by disproportionation to give $(P \cdot CH:CHX + P \cdot CH_2 \cdot CH_2X)$. The relative importances of these alternative processes depend upon the chemical nature of the monomer and, to a lesser extent, upon the temperature in the sense that the chance of disproportionation rises as the temperature is increased. Combination gives rise to a head-to-head grouping in the chain, and disproportionation to some unsaturated end-groups for molecules; both structural features may give rise to instability.

Termination can occur for polymer radicals of any size and so there is inevitably a wide *distribution of sizes* among the final molecules. The distribution can be predicted by application of kinetic principles and can be determined experimentally by fractionation of the whole polymer, e.g., by gel permeation chromatography. It is possible to quote only *average molecular weights* for polymers; they can be determined by several experimental methods, e.g., osmometry and viscometry.

The average *chain length* or *degree of polymerization* (DP) of the molecules in a sample of polymer is the average number of monomer units contained in them. The average *kinetic chain length* (ν) in a polymerization is the number of growth reactions which, on average, occur between an initiation step and the corresponding termination process. The relationship between degree of polymerization and kinetic chain length depends on the relative frequencies of combination and disproportionation (for 100% combination DP = 2ν; for 100% disproportionation DP = ν) but may also be affected by the occurrence of transfer reactions (see later).

Termination is commonly *diffusion-controlled,* i.e., it is governed by the rate at which the reactive sites in growing radicals can come together rather than by chemical factors. In viscous media, termination may be so seriously impeded that both the overall rate of polymerization and the degree of polymerization increase markedly. In systems where the polymer is insoluble in the reaction medium, polymer radicals may be trapped in the precipitated material and be able to grow but unable to participate in termination processes.

Transfer. The average molecular weight of a polymer produced in a particular system may be substantially reduced by occurrence of some types of transfer reactions. If the system contains certain substances, e.g., mercaptans, a growing polymer radical may abstract hydrogen thus

$$P \cdot + R \cdot SH \rightarrow P \cdot H + RS \cdot$$

giving a dead polymer molecular and a new radical which can react with monomer to reinitiate polymerization. If reinitiation is 100% efficient, the effect of transfer of this type is to reduce the average degree of polymerization without affecting the rate of polymerization or the kinetic chain length. In practice, transfer is commonly accompanied by retardation since some of the new radicals are consumed in side-reactions instead of reacting with monomer; this type of transfer is said to be degradative.

Other components of the polymerization mixture, including monomer and initiator, may engage in transfer reactions. They are particularly significant for allyl monomers for which *degradative transfer* to monomer is of such importance that rates and degrees of polymerization are very low. The radical produced in the reaction

$$P \cdot + CH_2:CH \cdot CH_2 \cdot O \cdot CO \cdot CH_3 \rightarrow$$
$$P \cdot H + CH_2:CH \cdot CH(O \cdot CH \cdot CH_3) \cdot$$

is so stabilized by resonance that it is not reactive enough to initiate efficiently.

Transfer to polymer, causing reactivation of a polymer molecule at some point along its length, leads to the growth of *branches.* The process can occur intermolecularly and also intramolecularly; the latter process is particularly important in the free radical polymerization of ethylene at high pressure where it leads to the production of numerous short branches which considerably affect the properties of the polymer.

Transfer to polymer, the subsequent growth of branches and termination of their growth by combination lead to *cross-linking* whereby the separate polymer molecules are united to form an insoluble three-dimensional network. Cross-linking is however much more likely to occur during the polymerization of those monomers which contain more than one carbon-carbon double bond per molecule. The monomer unit in the polymer first formed still possesses an unsaturated grouping which can participate in another polymerization chain. Certain monomers of this type however engage in a special type of reaction so that reaction of one double bond in a monomer is immediately followed by reaction of the second double bond; this type of growth is shown by, for example, methacrylic anhydride.

$$P \cdot + CH_2:C(CH_3) \cdot CO \cdot O \cdot CO \cdot C(CH_3):CH_2 \rightarrow$$

$$P \cdot CH_2 \cdot \overset{\cdot}{C}(CH_3) \cdot CO \cdot O \cdot CO \cdot C(CH_3):CH_2 \rightarrow$$

$$P \cdot CH_2 \cdot \underset{|}{C}(CH_3) \cdot CH_2 \cdot \overset{\cdot}{C}(CH_3) \cdot$$
$$\quad\quad\quad CO-O-CO$$

Inhibitors and Retarders. Various substances can reduce the rate at which a monomer is converted to polymer. Inhibitors completely suppress polymerizations whereas retarders only reduce the rate. The former deactivate very readily the primary radicals so that growth of polymer chains cannot begin; the latter deactivate growing polymer radicals so causing premature termination. Inhibitors are commonly used to stabilize monomers during storage. Many nitro compounds and quinones act as inhibitors and retarders.

Copolymerization. A process known as copolymerization can occur if reactive radicals are generated in a mixture of monomers; the resulting polymer molecules contain monomer units of more than one type. Copolymerization is of great significance academically, where it leads to information about the reactivities of monomers and radicals, and also industrially where it is used for the production of materials with special properties. Usually the composition of a copolymer is different from that of the mixture of monomers from which it is derived. For this reason, the average compositions of feed and copolymer drift during the course of a copolymerization. There are useful analogies between copolymerization and fractional distillation; special mixtures of monomers producing copolymers without change of composition are said to give azeotropic copolymerizations.

Extensive tables of so-called *monomer reactivity ratios* are available and make it possible to predict the compositions of copolymers formed from particular mixtures of monomers.

In many binary copolymerizations, there is a pronounced tendency for the two types of monomer unit to alternate along the copolymer chain. In extreme cases, there is almost perfect alteration, notably for pairs of monomers, e.g., maleic anhydride and stilbene, which do not polymerize on their own. Ternary copolymerizations are of practical importance; the kinetic treatments developed for binary copolymerizations can be extended to these systems.

J. C. BEVINGTON

References

Bevington, J. C., "Radical Polymerization," Academic Press, London, 1961.

Bamford, C. H., Barb, W. G., Jenkins, A. D., and Onyon, P. F., "The Kinetics of Vinyl Polymerization by Radical Mechanisms," Butterworths, London, 1958.

Ham, G. E., "Vinyl Polymerization," Marcel Dekker, New York, 1967.

North, A. M., "The Kinetics of Free Radical Polymerization," Pergamon, Oxford, 1966.

POLYMERS

Electroconductive Polymers

Most polymers are electrical insulators and have conductivities of $10^{-15}\Omega^{-1}cm^{-1}$ or less. However, there are several ways to arrive at compositions of polymeric nature that have higher conductivity. A simple way to obtain such a system is to use electrically conductive fillers such as metal powders or special types of carbon black. In these physical mixtures the polymer itself does not become conductive but acts only as an inert matrix to keep the conducting filler particles together. Conduction then occurs through chains of touching, conducting particles. Control of conductivity is limited in that it tends to be high as long as continuous chains of conducting particles are present. When fewer particles are present and an insufficient number of contacts between them can be established, the conductivity drops sharply. With metal fillers, conductivities of $10^2\Omega^{-1}cm^{-1}$ can be realized. The particle size and the effectiveness of dispersion are important. In polymer-filler systems the conducting particles may rearrange under the influence of thermal or mechanical cycles, and the bulk conductivity tends to change as a result of such cycles.

Ionic conductivity can be found in polyelectrolytes such as the salts of polyacrylic acid, sulfonated polystyrene or quaternized polyamines (ion-exchange resins). When dry, these materials have low conductivities. However, in the presence of small amounts of polar solvents or water—some of these polyelectrolytes are somewhat hygroscopic—electrical conductivity can be observed. The currents are carried by ions (protons, for instance). Such systems can only be used in cases where very small currents are expected. Large currents would result in observable electrochemical changes of the materials. In applications as antistatic electricity coatings, conductivities of $10^{-8}\Omega^{-1}cm^{-1}$ are sufficient.

Thermal decomposition of a large number of organic solids yields carbonaceous materials which are electrically conductive. It is believed that the conductive pyrolysis products are of polymeric nature, and that at high temperatures a carbon skeleton similar to graphite is formed. Since these products are insoluble, infusible mixtures, very little is known about their structure. Variation of the pyrolysis conditions leads to products with different conductivities. A well-studied example is polyacrylonitrile. Upon pyrolysis, an originally colorless piece of polyacrylonitrile (Orlon) fabric turns black with remarkable retention of its structure and becomes electrically conductive. Depending on the pyrolysis conditions, conductivities up to $10^{-1}\Omega^{-1}cm^{-1}$ have been obtained. It is believed that an aromatic system of condensed six-membered rings analogous to graphite is formed.

A number of polymers of more defined chemical structure exhibit *electronic* electrical conduction. According to one of the early concepts, long conjugated unsaturated chains would make good electronic conductors, assuming that resonance would render a fraction of the electrons in the molecules mobile, and thus give rise to electrical conductivity. Synthesis of long conjugated chains has been attempted by polymerization of acetylene derivatives (phenylacetylene), by dehydration or dehydrohalogenation of polyalcohols (polyvinylalcohol) or polyhalides (polyvinylchloride) and by polycondensations of suitable monomeric reaction partners, for instance, diamines with dialdehydes. In addition to the conjugated systems with only carbon in the chain and those with carbon and nitrogen, polymeric chelates have also been reported. Here the d-orbitals of the transition elements are supposed to form a part of the conjugated system.

Problems associated with the study and fabrication of these polymeric materials arise from the fact that many of them cannot be purified because crosslinking renders them infusible or insoluble, or both. Consequently the molecular weights and other structural details cannot be determined. Some noncrosslinked polymers of these types have been described with low molecular weight and low conductivities.

In another approach, the fact that crystalline monomeric charge transfer complexes exhibit electrical conductivity led to preparation of polymeric charge transfer complexes. These can be obtained from a polymeric electron donor and a monomeric electron acceptor or from a polymeric acceptor and a monomeric donor, the former type being the more common. These polymers are not cross-linked and some are soluble, but their conductivities are generally low.

Another example of an extension of the properties of monomeric compounds into the realm of polymers is the case of the 7,7,8,8-tetracyanoquinodemethan (TCNQ) compounds. Some monomeric, salt-like derivatives of TCNQ have conductivities of the order of $1\Omega^{-1}cm^{-1}$. Apparently stacks of TCNQ$^-$ ions and neutral TCNQ are responsible for these high conductivities. The polymeric TCNQ compounds consist of polycations, TCNQ$^-$ ions and neutral TCNQ, and have conductivities ranging from 10^{-10} to $10^{-3}\Omega^{-1}cm^{-1}$. These polymeric materials are soluble in organic solvents, can have high molecular weights (several million) and can be cast as films from solutions. Although the compounds are polyelectrolytes, they exhibit electronic conduction when dry. Among the many types of electrically conducting polymeric compositions the TCNQ derivatives seem to have an advantage because of an attractive combination of properties, namely controllable molecular weight, solubility, known chemical structure, fair chemical and thermal stability and electronic conduction controllable over several orders of magnitude.

The possibility of synthesizing polymeric superconductors has been proposed, but at the present time these ideas have not been confirmed by successful experiments.

JOHN H. LUPINSKI
KENNETH D. KOPPLE

References

1. "Physics and Chemistry of the Organic Solid State," Editors, Fox, Labes, Weissberger, Vol. II, Chapter 2, (1965); Vol. III, Chapter 3, (1967); Interscience, New York.
2. Gutman, F., and Lyons, L. E., "Organic Semiconductors," John Wiley & Sons, Inc., ·New York, 1967.

Inorganic Polymers

The boundaries of the class of materials which may be classified as inorganic polymers are hazy. Most inorganic materials can be considered polymeric since they are built up of a relatively simple atomic grouping repeated a very large number of times. Metals and simple ionic materials are easily excluded, but there still remains a large group of covalently bonded, regularly repeating materials. For example, many mineral silicates are based on the monomer $[SiO_4]^{4-}$ which is covalently bonded to form the large, two-dimensional sheets from which these materials are built. Still, we do not normally think of most of these inorganic, covalently bonded polymeric materials as polymers, because their behavior is so different from what we have come to expect of organic polymers. Such properties as high viscosity in the melt and in solution, rubbery elasticity, moldability, ability to form fibers, films, and so on, are not possessed by most of these materials. In a few cases, enough of them are present to suggest the underlying similarity in structure; for example, in the silicate minerals, crysotile asbestos forms fibers of excellent textile

quality. Such samples show the possibility of obtaining useful inorganic polymers.

In the light of this discussion inorganic polymers will be considered to be those materials in which the main polymiric chain contains no organic carbon and in which behavior similar to that of organic polymers can be developed.

The question "Why is there such a difference in behavior between the usual inorganic and organic polymeric materials?" is helpful in guiding such a development. The contrast must be due to differences in molecular structure. For example, in the case of quartz the $[SiO_4]^{4-}$ tetrahedra are covalently bonded together. The high regularity of the structure and the large number of cross-links per $[SiO_4]^{4-}$ unit lead to a material which is strong and dimensionally stable, but brittle. The same situation of over-crosslinking can be found with organic polymers. If the number of cross-links in quartz is reduced by substituting organic groups, such as methyl, for some of the oxygen-silicon linkages, the silicone polymers are produced. These polymers, the only commercial inorganic ones, show that inorganic materials which behave as organic polymers can be made. However, a number of obstacles are found which are not as troublesome with organic polymers. For example, six to eight membered rings are more stable than long-chains. In the case of organic materials, if chains can be formed initially, they have considerable stability. With inorganic materials, the bonds are much more labile (constantly forming and breaking) and the long chain may break down to a collection of smaller rings.

Other factors which influence the properties of polymers can be illustrated by examining the bond energies or bond strengths and the ionic character of bonds based on Si as contrasted to similar ones based on C.

Bond Energies

Si—Si	53 kcal	C—C	83 kcal
Si—O	106	C—O	86
Si—N	82	C—N	73
Si—C	78		

Ionic Character

Si—Si	0	C—C	0
Si—O	51%	C—O	22%
Si—N	30	C—N	7
Si—C	12		

From the bond energies we would expect the homo-atomic silane polymers with Si—Si bonds to be considerably less stable than the more familiar C—C chain polymers. This expectation fits the observed facts. On the other hand, one could expect little gain in stability in the carbon series by going to an ether linked chain (—C—O—C—), while in the silicon series a silicon-oxygen linkage is stronger than any of the others. This is reflected in the very good stability of the silicone polymers. From bond energies one might also expect that a chain of alternating Si and N atoms would have good stability.

In addition to pure thermal stability, if the polymer is to be heated in air, one must also consider oxidative stability. In the carbon series oxidation always leads to more stable species and tends to occur, but in the silicon series there is a much higher tendency towards reaction with oxygen. This is the principal reason for the low utility of the silane polymer. Finally, a third factor in polymer stability is the ease of attack by solvents, acids, bases, etc. This is largely determined by the ionic nature of the bonds involved. The silica based polymers should be more susceptible to such attack than carbon, since they have a higher per cent of ionic nature.

We do find that acidic or basic water solutions attack silicones when they are heated together under pressure. Their resistance is still high, however, because of other details of the way the polymer molecules are bound together.

The polymers which have been used to illustrate problems of inorganic polymer formation have been heteroatomic, that is, their chains are built from different atoms alternating with each other. The other structure mentioned has been homo-atomic—all the atoms in the chain are the same. There are only a few homoatomic polymers of any promise. Most elements will form only cyclic materials of low molecular weight if they polymerize at all. In addition to the silane polymers, black phosphorus, a high-pressure modification of the element, forms in polymeric sheets. (see **Black Phosphorus**) Boron has similar tendencies in its compounds. The outstanding member of this class is sulfur. A transition from S_8 rings to long sulfur chains takes place over a narrow temperature range around 159°C. An increase in the viscosity of the liquid by 2000 times or more, within a range of 25°, is the tangible evidence of polymerization. The material also forms rubbery, plastic and fibrous forms when chilled to room temperature. However, it has a strong tendency to revert to the cyclic form unless stable groups are placed at the end of the chain or copolymerization hinders the process. Attempts to improve the stability of polymeric sulfur have met with some success. This is the only homoatomic inorganic polymer which appears technically interesting at present.

The class of heteroatomic polymers, besides containing the silicones, offers more promise for useful materials. Most nonmetals and many of the less positive metals form heteroatomic compositions. In many cases they are high polymers. The silicones themselves behave quite the same as organic polymers and are used as oils, rubbers and resins. The rubber is vulcanized either by the reaction of organic peroxides with the methyl groups on the chain or by incorporating groups such as Si—OH or Si—OR which crosslink on exposure to moisture in the atmosphere. The properties in which they excel over organic polymers are high thermal stability, resistance to oxidation and inertness to organic reagents. These are usually the special properties one hopes to get from inorganic polymers. The polymers may be modified by substitution of other groups for the methyl groups on the side chain and by copolymerizations with other heteroatoms in which B—O—Si, Al—O—Si, Sn—O—Si, Ti—O—Al and other combinations are produced.

A similar class is the titanates. Three-dimensional Ti—O chains form pigments and pigment binders for paints and waterproofing compounds for use on cloth. Their properties can be modified by substituting monofunctional groups for some of the oxygen, for example, by forming esters to interupt the chains.

The polyphosphates have also been widely studied. Here the phosphate ion is found as a high polymer. The molecular weight of the polymer ranges from 250,000 to 2,000,000. The polyphosphates are water soluble and form fibers. No uses have been found for this class of materials. They hydrolyze slowly in atmospheric moisture and also embrittle on standing.

Other attempts to base a polymer on B—N heteratomic chains are being vigorously pursued although the B—O bond with an energy of 130 kilocalories is thermally more stable than the B—N bond at 100 kilocalories. The borates formed with B—O chain links, however, are too hydrolytically unstable and too thoroughly cyclized to be useful. B—N compounds are also plagued with the same weaknesses. However, the very high thermal stability of low molecular weight materials has encouraged the search for high polymers with the same basic structure. The combination of boron and nitrogen approximates that of carbon with carbon due to its location in the Periodic Table. (See **Boron and Compounds**).

One other area of materials deserves mention here. Coordination polymers are found when metal atoms are joined together by coordinating bonding involving some bridging group, e.g.

In view of the high thermal stability of monomeric chelation compounds, coordination polymers were expected to be promising for use at high temperatures. This has not proved to be the case. Thermal stabilities are usually lower than for low molecular weight materials. In addition, if the polymerization goes beyond a few monomer units, the materials tend to become insoluble and infusible so they cannot be fabricated into useful items. (See **Chelation**)

Many other polymeric inorganic materials are recognized but poorly understood because of the difficulty of characterizing cross-linked insoluble materials which decompose when attempts are made to dissolve them. Although a great deal of work has been done with inorganic polymers since 1940, the number of successful commercial materials in the field is small.

T. E. FERINGTON

Cross-references: *Polymerization; Glass; Silicones; Silicon and Compounds; Copolymers.*

Organic Polymers

Organic high polymers have a great number of different chemical structures, ranging from completely nonpolar to very polar and even ionic materials. They all clearly resemble each other, however. The basis for this resemblance is that many of their properties are governed by their high molecular weights, which range from 5,000 to tens of millions. For example, as the molecular weight increases in a given polymer family, the tensile strength of the polymer increases markedly. In some cases it approaches that of steel on a weight-for-weight basis, especially when oriented fibers are fabricated.

In a similar way, the viscosity of the molten material changes from a free flowing liquid at low molecular weights to the highly viscous polymeric liquid where flow may be observed only over a long period of time or under a considerable applied pressure. A property which shows up only in the case of high polymers is rubbery elasticity. Here again the development of a sufficiently long and flexible molecule is necessary before rubbery behavior develops.

Because of this striking dependence on molecular size, the measurement of molecular weight and dimensions is very important. Some of the most significant early work of Staudinger was the demonstration of the existence of large molecules joined by covalent bonds. (Others felt that such large molecules were not possible.) In order to determine these molecular properties the molecules must be dissolved. Thus each molecule can be separated from its neighbors and its effect measured independently. Solution properties such as osmotic pressure, light scattering and viscosity are used to measure the molecular weight of polymers. In the case of many natural polymers the ultracentrifuge has proved uniquely useful.

The osmotic pressure determination of molecular weights is based on the thermodynamic interaction of solvent and solute to lower the activity of the solvent. Experimentally, the solution is separated from the solvent by a semipermeable membrane. The solvent tends to pass through the membrane to dilute the solution and bring the activity of the solvent in both phases to equilibrium. The quantitative measurement of this tendency is obtained by allowing the liquid solution to rise in a vertical capillary connected to the solution compartment. The equilibrium height· it achieves or the rate at which it rises can be measured.

The measurements are converted to effective pressure (π) at zero polymer concentrations (c) and the average molecular weight ($\bar{M}_n$) gotten from the following relation:

$$\lim_{c \to 0} \frac{\pi}{c} = \frac{RT}{\bar{M}_n}$$

(R = gas constant; T = absolute temperature).

The light-scattering method is based on similar thermodynamic interactions. In any solution there are random variations in concentration and refractive index. These scatter some light out of a beam passing through the liquid. In a polymer solution the nature of the fluctuations and thus the amount of scattered light (τ) depend on the attractive forces between polymer and solvent molecules. This, in turn, depends on the polymer molecular weight ($\bar{M}_w$). The following equation describes the behavior:

$$\lim_{c \to 0} \frac{Hc}{\tau} = \frac{l}{\bar{M}_w}$$

$$(H = \text{a constant})$$

This method was developed by Peter J. W. Debye in 1944. Its evolution has been one of the most stimulating chapters of polymer physics.

In contrast to these thermodnamic methods, the viscosity molecular weight determination depends on the interference in the flow of the solvent caused by the dissolved molecules. In contrast to osmometry and light scattering, it has not been possible to develop the viscosity effect into an absolute measure of molecular weight. Rather, it must be calibrated, preferably by light scattering measurements.

The relationship between the measured limiting specific viscosity [η] and the molecular weight ($\bar{M}_v$) is as follows:

$$\lim_{c \to 0} \frac{\eta_{sp}}{c} = [\eta] = K\bar{M}_v{}^a$$

K and a are determined by calibration for a given polymer-solvent system.

$$\eta_{sp} = \frac{\eta_{\text{solution}}}{\eta_{\text{solvent}}} - 1$$

In each of the equations above a different symbol has been used for the molecular weight. Most polymers are heterodisperse, i.e., have many molecular weight species of the same chemical nature, thus the experiments yield an average molecular weight. Osmotic pressure gives a lower average ($\bar{M}_n$) or number average) because it emphasizes the effect of small molecules, while light scattering emphasizes the larger molecules and gives a higher average, ($\bar{M}_w$) or weight average). The viscosity average ($\bar{M}_v$) is between the two and closer to the weight.

Many natural polymers are monodisperse (all molecules have the same molecular weight). In this case the ultracentrifuge which separates materials according to their effective density in solutions is a most powerful tool for molecular weight determination. With poly-disperse materials, the interpretation of ultracentrifuge results becomes more complex and widespread application of this method to synthetic polymer molecular weight determination has not yet been achieved.

In addition to the primary effect of the great length of the molecule, the details of the distribution of functional groups along the polymer chain modify the behavior of these materials. This leads to differences in their applications. For example, natural rubber exists in the rubbery state at room temperature. If cooled below zero degrees Centigrade it becomes a hard, inflexible material, brittle and easily broken. On the other hand, if heated too far above room temperature, it begins to flow quite rapidly and behaves more like a fluid than a rubber. The same pattern is observed with other materials. For example, polystyrene, hard and brittle at room temperature, becomes rubbery when heated up sufficiently. The study of the mechanical behavior of polymers at various temperatures is called rheology. The temperature at which the rubbery material becomes glassy is called the glass transition temperature (T_g). This transition temperature depends on the nature of the backbone and the substituent groups on the polymer chain. Rubbery materials, e.g., polyisoprene, polychloroprene, polybutadiene, the copolymer of butadiene and styrene, etc., have molecular chains with considerable flexibility. Usually small side groupings and irregularities in the chain prevent them from coming together in a regular structure. Instead the molecules stay in an amorphous random packing much like a pile of cooked spaghetti. As with cooked spaghetti, there is a tendency for the whole mass to flow, by the movement of chains past one another. With natural rubber this flow at room temperature and above had to be inhibited by tying the chains together with chemical bonds before a useful product was obtained. This cross-linking is called vulcanization in the case of rubber. The process was discovered by Charles Goodyear in 1839. A similar cross-linking to inhibit the motion of the chains is necessary to make useful products from the newer synthetic rubbers also.

Other polymers are not rubbery at room temperature, instead they exist in the glassy state. They are amorphous, but because of more bulky substituents on the polymer chain the molecule is less flexible. There is less ease of molecular motion under applied stress at room temperature. Examples of such polymers are polystyrene and polymethylmethacrylate, which are transparent due to their amorphous, homogeneous nature. When heated they first become rubbery and then, at higher temperatures, show viscous flow so that they may easily be molded. When they are cooled to room temperature the rate of flow is vanishingly small due to the stiff chains; thus items made in this way can be used at ordinary temperatures if they are not required to bear too large a load. These common organic glasses are brittle and easily broken on impact. One of the interesting problems for polymer development is to obtain impact resistance without losing transparency and without increasing the cost by an excessive amount.

A related class of polymers is the crystalline, thermoplastic materials. These also are fabricated by heating to a high temperature so that they flow; but when they are cooled ordered regions develop within them, which makes them translucent. They have much tendency to flow because of these mechanical "cross-links" and have good dimensional stability. Polyethylene and polypropylene belong to this class. Here the chain is simple and regular so that different polymer molecules, or different parts of the same molecule, can pack next to each other. The same situation exists with "Teflon" (polytetrafluoroethylene).

As might be expected, there are polymers intermediate between the crystalline and glassy ones. For example, polyvinylchloride shows enough order to prevent its classification as glassy, but not enough to be considered crystalline.

Many crystalline polymers form part of another class of materials, the fiber-forming polymers. The formation of fibers of significant strength depends on the growth of ordered structures when the fiber is stretched. Thus crystalline materials, such as polypropylene, whose crystallites can grow on elongation, form strong fibers. In some cases fiber formation is aided by polar groups on the polymer chain. These interact with each other to give strong attractive forces which aid the molecular alignment that is needed. For example, polyacrylonitrile, the base polymer of "Orlon" and "Acrilan," has a —CN dipole in each monomer unit. The cooperative attraction of hundreds of these units along a chain gives a very strong cumulative effect. The fibers are strong even though they are not crystalline. The intermediate degree of order they develop is referred to as paracrystallinity.

In the case of other common fibers the polar groups are found in the polymer chain itself. In nylons the amide group,

$$\underset{\displaystyle -R-N-C-R-C-N-}{\overset{\displaystyle H \quad\;\; O \qquad\quad\; O \quad\; H}{}}$$

offers the possibility of both dipolar attraction and hydrogen bonding. The ester group in "Dacron" and similar polyesters contributes the —C=O dipole. (See **Fibers, Synthetic**)

The introduction of oxygen and nitrogen into the chain changes its flexibility, stability toward chemical reaction, resistance to solvents, strength and other properties. In this way quite extensive changes in behavior are obtained. Most of these heterochain polymers are prepared by condensation or ring opening reactions. The first important work in this field was the classic investigation of W. H. Carothers in the late 1920's on polyesters and polyamides. (See **Polyamides**)

The toughness, high melting points and high tensile strength of many of these polymers have since led to their widespread use in fibers, films and molded objects. In general these polymers are rather high cost materials, which are used, because of their unusual properties, in places where ordinary polymers are inadequate. For example, polysulfide rubbers show outstanding solvent resistance, while the silicones are a unique class of materials inert to many environmental conditions and unwettable by most liquids. Polycarbonates and acetals are so dimensionally stable that they can be used in place of metals in molded items. Many of the interesting developments in polymers over the last few years have involved new syntheses and new variations in structure of such heterochain materials.

The materials described above are thermoplastic resins, i.e., they all melt on being heated to sufficiently high temperatures and can be molded while molten. This characteristic is associated with molecular chains which are long and stringlike with few branches on them. If, however, the polymer chains have many covalent bonds linking them together into a network, a thermosetting resin develops which may flow at an early stage of its history, but is insoluble and infusible after the full crosslinking reaction has taken place. Any of the previous chain compositions can be used in making thermosetting resins if provision for crosslinking is made by using multifunctional monomers. Some are used more commonly, e.g., the epoxy resins, phenolformaldehyde, urea formaldehyde, melamine, etc. Separate articles describing the properties of many of these plastics are found in other parts of this encyclopedia. In general they are useful because of their inertness to solvents, resistance to dimensional change on heating, rigid dimensions, physical strength, chemical resistance and abrasion resistance.

Many of the varieties of polymers which have been discussed have analogs in polymers isolated from natural systems, for example, natural rubber is a polyisoprene with a purely carbon chain. Other natural polymers are based on the C—O—C bond. The cellulose and starch polymers, which are found in plants are composed of chains of six-membered carbon-oxygen rings joined through an oxygen linkage. Cellulose and starches differ from each other in the spatial orientation of the links joining the six membered rings.

Products from different sources in each class differ in degree of branching of the molecule and in amount of crosslinking.

In unmodified cellulose the hydroxyl groups give a large amount of hydrogen bonding which leads to insolubility in most solvents. On the other hand if these are changed by chemical reactions to ether or ester groups a much more tractable material results. Cellulose acetate, butyrate and nitrate; methyl and ethyl ether and carboxy methyl ether are widely used modified celluloses. Starches also are modified, but much less commercial success has been had with them.

The polymers described above have been chemically pure, although physically heterodisperse. It is often possible to combine two or more of these monomers in the same molecule to form a copolymer. This process produces still further modification of molecular properties and, in turn, modification of the physical properties of the product. Many commercial polymers are copolymers because of the blending of properties achieved in this way. For example, one of the important new polymers of the past ten years has been the family of copolymers of acrylonitrile, butadiene and styrene, commonly called ABS resins. The production of these materials has grown rapidly in a short period of time because of their combination of dimensional stability and high impact resistance. These properties are related to the impact resistance of acrylonitrile-butadiene rubber and the dimensional stability of polystyrene, which are joined in the same molecule.

Since they are organic materials most polymers are not water soluble, however, water solubility can be obtained by substituting the proper side groups on the polymer chain. Such polymers include the nonionic materials polyvinyl alcohol, polyethylene glycol, etc., where the strong dipolar and hydrogen bonding interactions cause the solubility; and the polyelectrolytes where ionizable groups such as the carboxylate, sulfonate, quaternary ammonium, etc., cause the solubility. Many of these water soluble polymers are used to increase the viscosity of water based systems. As little as 0.1% or 0.2% of the polymer is needed to produce a very viscous solution. They also are of wide biological interests.

T. E. FERRINGTON

Cross-references: *Crosslinking Polymers; Elastomers; Polymers, Inorganic; Block and Graft Polymers; Ladder Polymers; Polymers, Stereoregular.*

POLYPEPTIDES

Definition. Polypeptides are the class of compounds composed of amino acid units chemically bound one to the other through amide linkages, with the elimination of the elements of water. They may be represented by

$$-NH \cdot CHR \cdot CO_2H + NH_2 \cdot CHR' \cdot CO- \rightarrow$$

$$-NH \cdot CHR \cdot CO \cdots NH \cdot CHR' \cdot CO- + H_2O$$

where the characteristic amide linkage concerned with the union of the α-amino or α-imino group of one amino acid with the α-carboxyl of another is referred to as a *peptide bond* (dotted line). Compounds which include linkages of this type are commonly designated *peptides* or *polypeptides*.

Thus, peptides or polypeptides may be considered polymers of amino acids which are united *via* amide bonds to form chains which incorporate as few as two or as many as several thousand *amino acid residues*. A segment of a hypothetical peptide chain may be depicted

Like polymers which incorporate two, three, and four amino acid residues are known as di-, tri-, and tetrapeptides, respectively, etc.

Polypeptide Nature of Proteins. Although the realization that proteins are composed of amino acids had its inception as early as 1820 with the isolation of glycine from gelatin hydrolysates by Braconnot, it remained for Hofmeister and Fischer independently to postulate, some eighty years later, that the chemical union of these amino acids in the protein molecule was mediated through peptide linkages. As a consequence of such hypothesis, the protein molecule could, in its simplest form, be regarded as a linear polypeptide (I) of $x + 2$ residues

$$NH_2CHRCO-(NHCHRCO)_x-NHCHRCO_2H$$

I

wherein R represents a side chain of diverse composition, i.e., alkyl, aryl, heterocyclic, etc. A portion of the considerable body of evidence which has since accumulated in support of the polypeptide nature of proteins includes: (a) The characteristic purple color induced by the peptide bond in alkaline copper sulfate solution ("biuret reaction"), although very pronounced with proteins, becomes decreasingly so with decreasing size of the polypeptide chain and, consequently, with the lesser number of peptide bonds; (b) although the number of titratable amino and carboxyl groups in the intact protein is relatively small, a marked increase in these groups, which appear in nearly equal number, results upon complete hydrolysis; (c) proteolytic enzymes which catalyze the hydrolysis of proteins to amino acids induce the cleavage of synthetic peptides in like manner; (d) partial acid hydrolysis of proteins leads to the liberation of small peptides whose chemical structure may be ascertained; (e) infrared and x-ray analyses of proteins reveal the presence of the characteristic peptide bond found in smaller synthetic peptides of known composition; and (f) a parallel behavior is revealed by proteins and synthetic peptides upon their sequential degradation with various chemical agents.

Proteins are present in every living cell and are possessed of the most diverse biochemical activities. They may appear, for example, as enzymes, genes, virus, and hormones, and they compose a major portion of muscle, tendon, skin, and hair. Although little more than twenty different kinds of amino acids, all of the L-configuration, have thus far been demonstrated as protein constituents, a single protein may none the less contain several hundreds or even thousands of residues. Qualitative and quantitative differences in their amino acid composition, as well as differences in the arrangement of residues in the peptide chains, in the nature and number of lateral bonds between chains, and in the looping of chains, make theoretically possible countless numbers of distinct kinds of proteins which may exist in a multiplicity of structural conformations. X-ray analysis has, for example, represented one such conformation of the polypeptide chain of the protein molecule as a helical or spiral structure which contains 3.7 amino acid residues per turn of helix, with the pitch of the helix being about 5.4 Å. In any event, proteins differ from smaller open-chain polypeptides not only in that they embody a much larger variety of amino acid residues, but also because they possess a molecular weight of at least 10,000, an arbitrary figure which roughly corresponds to the lower limit of ability to be retained by cellophane on dialysis. Otherwise, differences in the general physical and chemical properties between peptides and proteins show a direct correlation to molecular size. (See **Proteins**).

Nomenclature. Since amino acid residues of the general type —NH·CHR·CO— are to all intents and purposes considered acyl substituents, their identification in any given peptide is denoted by the order in which they occur from the terminal amino end of the peptide chain. In addition, the configuration of each amino acid residue is designated by the small capital L or D prefix. As illustrative of this nomenclature scheme, the tetrapeptide which incorporates an N-terminal L-alanine residue followed by a glycine, an L-valine, and a D-alanine residue, as represented structurally by (II), may be denoted L-alanylglycyl-L-valyl-D-alanine.

$$NH_2CH(CH_3)CO—NHCH_2CO—NHCH·$$
$$(CH(CH_3)_2)CO—NHCH(CH_3)CO_2H$$
$$II$$

The number of stereoisomers theoretically possible for a peptide of known sequence may be given as 2^n, where n refers to the number of asymmetric centers. With the single exception of glycine, the α-carbon atom of all amino acids which have thus far been demonstrated to be components of naturally occurring peptides and proteins is asymmetric. Additional centers of asymmetry may be found in several other amino acids, such as the diasymmetric "isoleucines," "hydroxyprolines," and "threonines." In consequence of such asymmetry, a variety of configurational combinations becomes feasible in a peptide of known amino acid sequence. This type of stereoisomerism may be illustrated with the aforementioned tetrapeptide which, since it contains three centers of asymmetry (on the two alanine residues and one valine residue), may possess 2^3, or eight, theoretically possible optical isomers. These are depicted in the following:

A. L-L-L   L-D-D   L-L-D   L-D-L
B. D-D-D   D-L-L   D-D-L   D-L-D
    III   IV   V   VI

where the A and B forms of any given pair are optical antipodes and each pair (III to VI) of themselves represent one of four possible racemic modifications.

Abbreviated Names. It is apparent that the routine use of conventional peptide nomenclature could become quite cumbersome where more than only a few amino acid residues are involved. Thus, a system of abbreviated names has been adopted. Here, the first three letters of the name of a given amino acid are employed to denote the —NH·CHR·CO— portion of that amino acid, e.g., Gly and Ala for a glycine and an alanine residue, respectively. As illustrative of the use of such abbreviations, an empirical representation of $Gly_2AlaLeu$ could be made for the peptide designated glycylleucylglycylalanine. A more complete and precise depictation of this compound as H·Gly·Leu·Gly·Ala·OH can be made in those cases for which the sequence of amino acid residues is known. Here, the appropriate symbols are arranged in proper order and joined by a period. If residues derived from asparagine, glutamine, cysteine, or half-cystine are involved, these are generally represented as Asp-NH₂, Glu-NH₂, CySH, and CyS-, respectively. Functional derivatives of amino acid residues, such as γ-ethyl glutamate and S-benzylcysteine, could in analogous manner be depicted by Glu-OEt and CyS-Bz, while structural isomers, such as the leucine, isoleucine, and norleucine residues, are accordingly represented by Leu, Ileu and Nleu.

Classification of Peptides. On the basis of structural considerations, peptides may be classified according to whether they occur in *normal* or *abnormal* peptide linkage and possess *open chains* or are *cyclic*. Such classification is given in what follows:

(a) *Normal, open-chain* peptides contain x molecules of amino acid which are bound in peptide linkage via their α-amino or α-carboxyl groups, or both, with the elimination of $x - 1$ molecules of water, e.g.,

$$NH_2CHR^1CO_2H + NH_2CHR^2CO_2H +$$
$$NH_2CHR^3CO_2H + NH_2CHR^4CO_2H \rightarrow$$
$$NH_2CHR^1CO—NHCHR^2CO—NHCHR^3·$$
$$CO—NHCHR^4CO_2H + 3H_2O$$

Since the terminal residues of peptides of this type contain either a free α-amino or α-carboxyl group, they exist as dipoles. Such polypeptides may occur as straight chains, wherein only a single terminal α-amino and α-carboxyl group is present, e.g., glycylalanylglycine, or as branched chains, wherein the occurrence of several terminal α-amino and α-carboxyl groups result from the presence within the molecule of a diaminodicarboxylic acid, such as cystine, which may serve as a bridge between the two peptide chains. Although most proteins thus far examined appear to be normal, open-chain polypeptides of the type described above, the available evidence is not yet sufficiently adequate to exclude completely the existence of amide linkages which involve the side-chain acidic functions of aspartic or glutamic acids, or the basic functions of arginine lysine, or histidine.

(b) In *normal, cyclic polypeptides,* x molecules of the component amino acids are united in peptide linkage through their respective α-amino and α-carboxyl groups with the elimination of x molecules of water. In peptides of this type, the possession of acidic or basic properties is dependent upon the presence of the requisite functional groups in the side chains of the amino acid residues involved, inasmuch as free α-amino and α-carboxyl groups are lacking. The simplest members of this class of compounds are represented by the 2:5-diketopiperazines (IX), which may be regarded as condensation products of two amino acids (VII and VIII) with the loss of two molecules of water. A somewhat larger member of this class of compounds is exemplified by the cyclodecapeptide antibiotic, gramicidin S.

$$NH_2CHR^1CO_2H + NH_2CHR^2CO_2H \rightarrow$$

$$\text{VII} \qquad\qquad \text{VIII}$$

$$
\begin{array}{c}
\text{CO} - \text{NH} \\
R^1CH \qquad CHR^2 + 2H_2O \\
\text{NH} - \text{CO}
\end{array}
$$

$$\text{IX} \qquad\qquad \text{X}$$

(c) Polypeptides in *abnormal peptide linkage* include those compounds wherein the amide bond which unites the component amino acids does not involve both an α-amino and an α-carboxyl group. Such linkages may or may not implicate α-amino acids exclusively. Thus amide bonds associated with the side-chain carboxyl of glutamic or aspartic acid, or the ω-basic functions of lysine, arginine, or histidine, belong to this category, as do bonds concerned with the amino or carboxyl groups of such nonprotein amino acids as β-alanine. ϵ-Glycyllysine (XI) and γ-glutamylglycine (XII) are representative of the former type, whereas the latter is exemplified by carnosine, i.e., β-alanylhistidine (XIII). Frequently both normal and abnormal peptide linkages reside within the same molecule, as in glutathione (γ-glutamyl-cysteinylglycine).

$$
\begin{array}{c}
NH_2CH_2CO - NH(CH_2)_4CHNH_2 \\
| \\
CO_2H
\end{array}
$$

$$\text{XI}$$

$$
\begin{array}{c}
NH_2CH(CH_2)_2CO - NHCH_2CO_2H \\
| \\
CO_2H
\end{array}
$$

$$\text{XII}$$

$$
\begin{array}{c}
NH_2CH_2CH_2CO - NHCHCH_2C \text{———} N \\
| \qquad\qquad || \qquad || \\
CO_2H \quad CH \qquad CH \\
\diagdown NH \diagup
\end{array}
$$

$$\text{XIII}$$

Naturally Occurring Peptides. A large number of simple, open-chain polypeptides has been isolated *in vitro* through the partial acid hydrolysis and enzymic degradation of proteins, especially since the advent of paper and ion exchange chromatography. *In vivo* enzymic cleavages of proteins which arise from tissue autolysis, bacterial action, digestive processes, and the like undoubtedly result in the generation of a large variety of peptides of similar nature. Peptides of related structure have also been detected in blood plasma in relatively small amounts, and in red blood corpuscles, muscle, liver and other tissues, as well as in urine in somewhat larger amounts. The natural occurrence of peptides of this type may, in the main, be attributed to the fact that they represent either the end products of protein metabolism or some transitory stage in the interconversion of amino acids and proteins. Other naturally occurring peptides, however, appear to possess a more clearly defined physiological role as coenzymes, hormones, or antibiotics, and presumably serve to create a more optimal environment for the parent organism. The existence within these compounds of either abnormal peptide linkages or amino acid residues of unusual structure, constitution, or configuration conceivably imparts to the molecule an exceptionally high degree of resistance to the hydrolytic action of the proteolytic enzymes which it ordinarily may encounter.

Synthesis of Peptides. From a purely fundamental point of view, synthetic peptides may serve as substrates in the investigation of the specificity and mode of action of proteases and peptidases, and as models for studies on the constitution, physical properties, and chemical reactions of proteins. Convenient techniques of peptide synthesis therefore become a vital tool in the elucidation of knowledge of protein structure and function. Alternatively, when considered from a practical point of view, these same techniques may be utilized not only for the unequivocal demonstration of the constitution and structure of such naturally occurring substances as the polypeptide antibiotics, hormones, and coenzymes, but also may be employed for either the commercial production of these compounds *per se* or for the formulation of suitable analogs thereof. A host of related compounds, with varying degrees and shades of biological activity, thereby become potentially available.

In any consideration of methods that may be employed for the union of two amino acids via a peptide linkage, the direct condensation of the α-amino group of one amino acid and the α-carboxyl group of the other, with the elimination of a single molecule of water, is most readily and simply visualized. Achievement of peptide bond synthesis through direct condensation, is made difficult, however, not only by virtue of the dipolar nature of the amino acids involved, but by energy considerations as well. Since such condensation implicates the carboxyl function of one amino acid and the amino function of the other, it becomes essential to suppress the reactive character of the amino group of the first and of the carboxyl group of the second. In practice, therefore, appropriate substitution of these functional groups becomes a necessary prelude to condensation. The masking substituents must be of such nature as to permit their ready removal, subsequent to condensation, by procedures which will not induce breakdown of the labile peptide molecule. With the foregoing in mind, the sequence of events for the coupling of amino acids via peptide linkages may be enumerated as follows.

(a) *Masking of the amino function.* Implicit in the choice of a suitable blocking substituent for the amino acid are the conditions that not only should the derivative which incorporates such substituent be easily prepared, but also that subsequent removal of this substituent should proceed under conditions that would lead to no appreciable destruction of the remainder of the molecule. Various substituents have been found to satisfy these criteria. Selective methods employed for their ultimate removal include, among others, fission by acid in aqueous and nonaqueous media, cleavage with dilute alkali, hydrogenolysis, and hydrazinolysis.

(b) *Masking of the carboxyl function.* Appropriate substitution of the carboxyl group of an amino acid presents little difficulty and is conveniently achieved through the formation of the alkali salt, NH_2CHRCO_2Na, if the peptide synthesis is to be effected in aqueous media, or the ester derivative (methyl, ethyl, or benzyl), if nonaqueous media are to be employed. Regeneration of the carboxyl group is readily accomplished by neutralization with acid in the former instances, whereas saponification with alkali or selective cleavage with dilute acid is commonly utilized in the case of the methyl or ethyl ester derivatives. For the selective cleavage of benzyl esters, hydrogenolysis is generally employed.

(c) *Coupling.* Subsequent to adequate masking of the two reactant amino acids, as given in (a) and (b), there arises the problem of how condensation of the resulting derivatives should be achieved. This problem has been satisfactorily met by conversion of the unsubstituted carboxyl function of the acylated amino acid, e.g., $C_6H_5CH_2OCO$—$NHCHRCO_2H$, to the corresponding azide, acid chloride, mixed anhydride, activated ester, "carbodiimide" derivative, or other active intermediate, followed by reaction of the group so activated with the unprotected amino groups of the other reactant, e.g., NH_2CHRCO_2Et. Formation of a peptide bond may be alternatively attained through activation of the unprotected amino group of the other reactant, with such agents as phosgene, phosphorus trichloride, or tetraethyl pyrophosphite, succeeded by treatment of the "activated amine" with an acylamino acid.

(d) *Elongation of the peptide chain.* Additional amino acid residues may be affixed to an acyldipeptide or derivative thereof, in a manner comparable to that given in (c), after pertinent activation either of the C-terminal carbonyl function or of the amino moiety which results from selective removal of the acyl blocking group.

(e) *Removal of masking substituents.* After the desired acylated peptide derivative has been secured, final removal of the protecting groups on both ends of the peptide chain is accomplished with liberation of the free peptide.

The foregoing discussion presents a rather simplified picture of the procedures currently employed in the synthesis of peptides involving a variety of amino acids. In actuality, such picture becomes somewhat more complicated in synthesis wherein the possibility of racemization of optically active components must be considered, as well as in syntheses in which reactive basic or acidic radicals as occur in lysine, histidine, and arginne, or glutamic and aspartic acids, respectively, are involved. Still further complications are imposed by the presence of the hydroxyl groups of threonine, serine, hydroxyproline, and hydroxylysine, and by the sulfur atoms of cystine and methionine. The appropriate literature should be consulted for methods of peptide synthesis applicable to each of these amino acids, together with the limitations implicit in their use.

MILTON WINITZ

Cross-references: *Amino Acids; Proteins; Stereochemistry; Asymmetry.*

POLYPROPYLENE

This polymer is produced by the polymerization of purified propylene ($CH_3CH=CH_2$) using a stereospecific catalyst such as a slurry of aluminum alkyl in an aliphatic hydrocarbon solvent (Natta, 1954). Any atactic polymer may be extracted with hot heptane. Polypropylene is characterized by low density and superior resistance to heat, abrasion and stress cracking. Unmodified polymers of propylene are readily degraded by heat and light but stabilized formulations are suitable for use at elevated temperatures and in sunlight. Composites containing carbon black have excellent resistance to ultraviolet light radiation. Polymers containing flame retardants are available.

Molded or extruded polypropylene articles have good electrical resistivity and are inert to fungi, bacteria, and many aqueous solutions of alkalies and acids. Polypropylene is attacked by strong oxidizing agents such as chlorine and fuming nitric acid. It is insoluble in cold organic solvents but is softened and swollen by some hot solvents. Polypropylene films have low moisture vapor permeability.

This highly crystalline polymer has an identity period of 6.5Å which accounts for its helical structure. Copolymers of propylene and ethylene, butene-1 and styrene are available. One of the more important commercial products is a vulcanizable elastomeric terpolymer of ethylene, propylene and a diene such as butadiene.

The properties of a typical molded specimen of unfilled polypropylene are: specific gravity 0.90, tensile strength 5000 psi, flexural strength 7000 psi, impact strength 4 ft lbs/in. notch, thermal expansion 7×10^{-5} in./in./°C, heat-deflection temperature @ 264 psi 140°F and water absorption 0.01 per cent.

Over 250 million pounds of polypropylene are produced annually in the United States. Some of the trade marks used are "Moplen," "Olefane," "Pro-fax," "Escon" and "Alathon." The name "Olemer" is applied to for one of the commercial copolymers. Propylene may be injection-molded and blow-molded to produce bottles or cast as film. Oriented film resembles cellophane in clarity and accounts for approximately 25 per cent of all polypropylene used. It may be extruded as pipe, wire coating or filaments. The latter are used for rope, twine, brushes, and carpets.

RAYMOND B. SEYMOUR

POLYSACCHARIDES See CARBOHYDRATES, GUMS AND MUCILAGES.

POLYSTYRENE AND COPOLYMERS

Polystyrene is generally considered to be the thermoplastic whose commercial availability in the 1930's began the modern plastics industry. In today's industry, the term "polystyrene plastics" includes the simpler polymerized styrenes and also those largely styrene materials that are modified with rubbers, additives such as light-stabilizing agents, and with other monomers, to develop special properties which meet specific industrial needs. In this respect, polystyrenes are among the most versatile products known to the plastics industry.

Polystyrene plastics have generally excellent resistance to indoor environments. They can function continuously at temperatures of −30 to +160°F. Most of the available formulations have a load-bearing design strength of 800 to 1500 psi in continuous use.

Polystyrenes will burn at a slow rate, producing heavy, sooty, black smoke. Their self-ignition temperature is approximately 900°F. Exposure to outdoor environments discolors polystyrenes and induces brittleness. They are not recommended for long-term outdoor exposure. Most polystyrenes can be decorated by standard techniques. Hot stamping is a rapid method of marging parts. Silk screening, spray painting and standard printing may be used to decorate moldings and sheet.

General-Purpose Polystyrene. General-purpose polystyrene is the simple polymerized form of styrene monomer. It is characterized by hardness, rigidity, water-white clarity, heat and dimensional stability, excellent electrical resistance, ease of fabrication by all common thermoplastic molding techniques, excellent colorability in an almost unlimited range, and of course, low cost.

All general-purpose polystyrene is not identical, however. Homopolymers with different characteristics such as heat resistance, flow and setup are produced through changes in molecular make-up and the addition of flow additives (internal and external lubricants), fillers and stabilizers.

In general, the unmodified polystyrenes possess excellent resistance to most mineral and organic acids, alkalies, salts, lower alcohols, and aqueous solutions of these materials. They are softened or dissolved by many hydrocarbons, ketones, the higher aliphatic esters, and essential oils. In some instances susceptibility to chemical attack can be reduced by residual stress relief or annealing.

The choice of an unmodified polystyrene for an application usually involves a balance between heat distortion and processability. These formulations which exhibit high resistance to heat distortion in finished form require higher heats and pressures, and lower cyclic rates, for injection molding. In extrusion operations, a stiff-flowing polystyrene exhibits more uniformity and produces stronger parts than free-flowing formulations.

Most applications do not require a polystyrene with a high heat resistance. If this property can be compromised, a freer-flowing type can be chosen which will offer greater ease of molding, particularly when producing thin-walled parts.

General-purpose polystyrenes are commonly used in housewares, toys, novelties, packages and appliances where clarity and colorability are assets, and where impact and flex demands are not excessive.

Impact Polystyrene. Impact polystyrene, sometimes referred to as rubber-reinforced polystyrene, was developed for use in those applications where general purpose material did not have the impact strength or elongation characteristics necessary for proper performance.

As is true of general-purpose polystyrene, the impact grades are available in various combinations of flow, setup and heat resistance, and the relationships of these properties to each other remain the same. Chemical resistance of the impact grades is basically the same as for general-purpose. Impact polystyrenes are well suited to injection molding and extrusion, and can be handled on most standard equipment. They can be molded to yield parts of high gloss, fine detail and excellent dimensional stability. These moldings are well suited to automatic degating operations, cold punching and thermal or solvent welding.

Sheet stock is produced by the standard flat sheet extrusion method. High gloss can be obtained by passing the sheet through an infrared heating unit between the die and the cooling rolls. Another method of improving surface gloss is the lamination of biaxially oriented general-purpose polystyrene film to one surface at the cooling rolls.

Impact polystyrenes have found a tremendous market in the manufacture of refrigerator door liners and in packaging. They are also commonly used for appliance housings, radio and TV cabinets, closures, sporting goods, toys, cameras, furniture, luggage, pipe and fittings, automotive parts and women's shoe heels.

Heat-Resistant Polystyrene. Heat-resistant polystyrene was developed to impart dimensional stability to molded parts used in high-heat areas. It is available in both normal and high-impact resins both of which have good heat distortion, hardness, rigidity, ease of fabrication and low cost. However, moldability, as measured by viscosity, is sacrificed to achieve high-temperature performance.

The largest end use for these polymers is in small radio cabinets where the heat from vacuum tubes could cause warpage of general-purpose polystyrene. Other applications for heat-resistant impact polystyrenes exist in the television and housewares markets.

Light-Stable Polystyrene. The development of polystyrene formulations which exhibited good resistance to indoor light exposure carried the polystyrenes into the field of indoor light fixtures. Light-stable resins exhibit rigidity, hardness and dimensional stability, and generally sell at low cost.

The first of the light-stable polystyrenes met IES-NEMA-SPI standards and had a useful, nonyellowing indoor service life up to five years. Improvements in base polymers and stabilizers have almost doubled this useful life span in recent materials.

Flame Retardant Formulations. In several uses, the flammability of styrene polymers is reduced by the addition of flame retarding agents. Up to 25% loadings may be necessary, depending on the properties required and the effectiveness of the added compounds. Much art is available; however, that most widely practiced employs combinations of antimony oxide and halogen-containing organic compounds as additives. These additives usually dilute other polymer properties, such as tensile strength, impact strength, heat distortion, color, fabrication stability or aging characteristics. Any selection is a compromise of improved flammability characteristics, other required polymer properties and cost.

SAN and ABS Polystyrene. A copolymer of styrene and acrylonitrile (SAN) can accept greater stress than general-purpose polystyrene before breaking or crazing. A prime motive for its development was the achievement of increased strength in many different environments. Variation of the acrylonitrile in this copolymer gives diverse chemical and physical properties.

This copolymer is more rigid, has greater tensile strength and modulus, has greater hardness, and is more chemically resistant than general purpose polystyrene. It also offers clarity and chemical resistance not available in impact formulations.

The SAN copolymer has been modified into a terpolymer by the addition of rubber. Today, acrylonitrile-butadiene-styrene (ABS) forms the basis for a large family of materials whose characteristics are difficult to generalize. Variations in the content of the three monomers, and the methods of combining them, yield products whose characteristics are widely diverse, and whose fabrication economies are affected greatly.

The various plastics available in the ABS family can be pigmented to yield almost any hue. They are light in weight and are nonconductive. Although ABS plastics are not highly flammable, they will support combustion; this can be curtailed by the application of flame-retardant coatings. Prolonged outdoor exposure will result in property degradation. (See **ABS Plastics**).

Other Styrene Copolymers. Other vinyl monomers are copolymerized with styrene to give specific improved properties. Methyl methacrylate imparts its characteristic light and weathering stability advantages as it is added to styrene polymers. When properly fabricated, this copolymer will perform well in outdoor applications. Styrene-methyl methacrylate is finding use in the production of automotive parts, light diffusers, exterior signs, dial faces, crystal-clear brush backs, and similar items.

Improved heat distortion, special chemical properties and changed processing characteristics are available via combinations of chlorostyrene, t-butyl styrene and maleic anhydride with styrene. In general, properties are roughly additive and ideal combinations depend on the requirements of the user. When heat distortions are raised, the ease of fabrication usually decreases.

Heat distortions greater than 300°F are given by 25% maleic anhydride—75% styrene copolymers. These copolymers also give resistance to some of the hydrocarbon solvents which attack polystyrene. Compared to polystyrene they are more water sensitive and show lower impact values.

Chlorostyrene copolymers show greater resistance to burning and can be made flame retardant with greater ease. Materials containing t-butyl styrene are more sensitive to paraffinic solvents than

polystyrene but are more resistant to aromatic attack.

Expandable Polystyrenes. Technology developed in Germany opened a new market for polystyrene in the area of molding foamable beads. U.S. producers now manufacture expandable polystyrene that has low density, insulation value and cushioning properties. It is available in colors, and can be formulated with flame retardant ingredients incorporated.

This product can be extruded into planks and sheeting, and one manufacturer achieves decorative surface effects and variable density by injection molding.

Thermoforming or molding of expandable polystyrene suit it to applications in packaging, trays, sandwich structures and the like. Insulated containers, such as picnic jugs, are also made of molded expandable polystyrene. The material finds wide use in buoyancy and flotation applications.

Ecology-Disposal. Their economy and utility make polystyrene products widely employed for single-use or disposable items, such as convenience food packaging. These items are then found in waste streams which can be disposed of in land fill or through incineration. When these polymers are burned under appropriate conditions in well designed units the products of combustion are primarily water and carbon dioxide. They provide no special hazard. In many cases, the fuel value of the polymer is desirable to help dispose of other waste components which burn with greater difficulty.

Similarly, styrene polymers are relatively inert when placed in land fill disposal areas. No harmful materials are contributed to the soil or ground water due to the comparative biological and chemical stability of the products. Materials added to styrene polymers for special purposes (colorants, flow agents, plasticizers, etc.) must be considered separately in disposal.

D. J. GRIFFIN

Cross-references: *ABS Resins; Elastomers; Polymerization; Rubber.*

POLYURETHANES

Simple urethanes were prepared by Wurtz in the middle of the nineteenth century. These compounds were obtained by the reaction of alkanols, such as ethanol, and in isocyanate, such as phenyl isocyanate. Subsequently, it was found that this reaction was catalyzed by tertiary amines and salts of heavy metals. This reaction with primary alkanols was 30 times faster than with secondary alkanols or water and 200 times faster than with tertiary alkanols.

A reinvestigation of this reaction by O. Bayer in 1937 showed that macromolecules could be obtained in a stepwise reaction of dihydric alcohols and diisocyanates. The polyurethanes obtained from this reaction are now used widely as adhesives, coatings, fibers, elastomers, wire enamels, cellular products and molded articles.

The polyurethane obtained by the reaction of hexamethylene diisocyanate and 1,4-butandiol has better resistance to yellowing than the aromatic polyurethanes. These polymers are used as general purpose fibers and elastomers in Europe but their principal use in the USA is for the annual production of more than 500 million pounds of rigid and flexible polymeric foams. The principal disocyanates used are 2,4-tolylene diisocyanate and diphenylmethane diisocyanate. The principal glycols are ethylene glycol, glycerol, pentaerythritol, sorbitol, trimethylolpropane, sucrose, castor oil and polymers with terminal hydroxyl groups. The latter may be polyesters with low acid numbers, polyoxirane, poly(oxytetramethylene)glycol or block copolymers of oxirane and propylene oxide.

The general reaction for the production of polyurethanes is as follows:

$$\overset{\delta-}{C}-NR-\overset{\delta+}{N}-\overset{\delta+\delta-}{C} + HOROH \rightarrow$$
$$\underset{O}{\|} \qquad \underset{O}{\|}$$
diisocyanate \qquad dyhydric compound

$$\overset{H}{C}-NR-\overset{|}{N}-C-OROH$$
$$\underset{O}{\|} \qquad \underset{O}{\|}$$
urethane

The active terminal groups continue to react. It is customary to add an amine, such as diethylenetriamine and a heavy metal salt such as dibutyltin dilaurate as catalysts.

Diisocyanates have been added to alkyds to yield uralkyd coatings. Coatings with superior resistance to abrasion are also produced by moisture curing of prepolymers with residual isocyanate groups. The water present reacts with the residual isocyanate groups to produce an unstable carbamic acid which decomposes to carbon dioxide and an amine as shown in the following equation:

$$-\overset{\delta+}{N}-\overset{\delta-}{C} + \overset{\delta+\delta-}{HOH} \rightarrow \left[\overset{H}{\underset{|}{-N}}-C-OH \right]$$
$$\underset{O}{\|} \qquad \qquad \underset{O}{\|}$$
water

polymer with terminal isocyanate groups \qquad polymer with terminal carbamic acid groups

$$\rightarrow CO_2 + \overset{H}{\underset{|}{-N}H}$$
carbon dioxide

polymer with terminal amine groups

The residual isocyanate groups then react with the amine groups to yield a polyurea as shown by the following equation:

$$\overset{H}{\underset{|}{-N}H} + \overset{}{C}-N- \rightarrow \overset{H \qquad H}{\underset{|}{-N}-\overset{|}{C}-\overset{|}{N}-}$$
$$\underset{O}{\|} \qquad \qquad \underset{O}{\|}$$

polymer with terminal amine groups \qquad polymer with terminal isocyanate groups \qquad polyurea

Solutions of polyurethanes may also be used as coatings. Heat resistant coatings are prepared by heating a blocked isocyanate with a polymer containing free hydroxyl groups. The former which is obtained by the reaction of phenol and a polyisocyanate loses phenol and yields a polymer with free isocyanate groups when heated at 150°C. The

blocked isocyanates are used for the production of wire enamels.

Because of the versatility of the polyurethane reaction, it is possible to produce elastomers which range in hardness from a soft 5 shore A to a rigid 80 shore D hardness. These tough elastomers may have an elongation as high as 1000 per cent and a modulus as high as 4400 psi at 300 per cent modulus. These elastomers have been used as tires, sheets, fabric coatings, gaskets, seals, o-rings, and pump impellers.

Caulking materials, films and adhesives may be produced using modifications of the techniques used for the production of elastomers. Polyurethane films have been used for shoe uppers. Room temperature-cured caulking compositions have been used as binders for solid rocket propellants.

Cellular polymers with properties ranging from rigid to flexible products have been prepared using the previously discussed formulations. Products with densities ranging from 75 to 2 lbs/cu ft. have been obtained by controlling the amount of moisture present in the formulation or by adding gaseous propellants. Rigid products made with blowing agents have a thermal insulating K factor as low as 0.11 BTU/hr/ft²/°F/in as compared with cork which has a K factor of 0.26. It is customary to add silicones and mineral oil to regulate the cell size and to reduce shrinkage of these foams.

Since polyurethane foams develop their own surface films, it is possible to produce puncture-resistant light-weight sheets or intricately shaped articles. It is now possible to produce integral cushions in furniture or to fabricate light-weight fabric laminates. Thus the framework, the cushions and the covering of furniture may all be produced from appropriately selected polyurethanes.

Polyurethane foams may also be used to provide "instant shelters" and flotation devices. Volatile diisocyanates, such as tolylene diisocyanate may cause adverse respiratory effects if present in concentrations greater than 0.02 ppm. However, prepolymers with terminal isocyanate groups are less volatile and even the more volatile reactants may be used safely with adequate ventilation.

RAYMOND B. SEYMOUR

POLYVINYL RESINS

Polyvinyl Chloride

Since its introduction about 40 years ago, PVC has gained acceptance for a wide variety of applications, ranging from auto seat covers to siding for domestic housing to insulation for electrical wiring.

Many factors are responsible for the rapid growth of this material; these include:

(1) Compoundability into a wide variety of rigid or flexible forms
(2) Low cost
(3) Resistence to weathering
(4) Water and chemical resistance
(5) Ease of processing
(6) Self-extinguishing properties.

Chemistry

Vinyl chloride monomer is a gas which is stored and shipped under pressure sufficient to keep it in a liquid state. Boiling point is $-14°C$, freezing point $-160°C$ and density at $20°C$ is 0.91.

The monomer is produced by the reaction of hydrochloric acid with acetylene. This reaction can be carried out in either a liquid or gaseous state. In another technique ethylene is reacted with chlorine to produce ethylene dichloride. This is then catalytically dehydrohalogenated to produce vinyl chloride. The byproduct is hydrogen chloride. A recent innovation referred to as "oxychlorination" permits the regeneration of chlorine from HCl for recycle to the process. This process is now the most economical.

Polymerization may be carried out in any of the following manners:

(1) Suspension: a large particle size dispersion or suspension of vinyl chloride is made in water by addition of a small quantity of emulsifying agent. The product after polymerization and drying consists of granules.

(2) Emulsion: a larger quantity of emulsifier is employed, resulting in a fine particle size emulsion. The polymer after spray drying, is a finely divided powder suitable for use in organosols and plastisols.

(3) Solution: vinyl chloride is dissolved in a suitable solvent for polymerization. The resultant polymer may be sold in solution form, or dried and pelletized.

Emulsions may be polymerized by use of a water-soluble catalyst (initiator), such as potassium persulfate, or a monomer-soluble catalyst, such as benzoyl peroxide, lauroyl peroxide or azobisisobutyronitrile. Suspension and solution polymerizations employ the monomer soluble catalysts only. In addition to the above-mentioned initiators, diisopropyl peroxydi-carbonate may also be employed, where lower-temperature polymerization may be desired, e.g., to reduce branching and minimize degradation.

Because of the low level of emulsifiers and protective colloids, the suspension polymer types are most suitable for electrical applications and end uses requiring clarity. This form is also employed in the bulk of extrusion and molding applications. Cost is lower than for emulsion and solution forms. The emulsion or dispersion resins are employed mainly for organosol and plastisol applications where fast fusion with plasticizer at elevated temperature will occur as a result of the fine particle size of the resin.

Monomers such as vinyl acetate or vinylidene chloride may be copolymerized with vinyl chloride. Up to 15% of the comonomer may be employed. Vinyl acetate increases the solubility, film formation and adhesion. Processing or forming temperatures are generally lowered. Chemical resistance and tensile strength decrease with increasing amount of vinyl acetate.

Rigid Vinyls

These have been separated into two categories according to ASTM:

Type I is rigid PVC with excellent chemical resistance, physical properties and weathering resistance such as obtained from unplasticized high molecular weight PVC.

Type II has the added feature of high impact resistance but with slightly lower chemical and physical requirements.

Perhaps the most important applications for rigid PVC will be in building. This is a rapidly growing market. Fabrication is via extrusion. Examples of applications are pipe, siding, roofing shingles, panels, glazing, window and door frames, rain gutters and downspouts.

Blow-molded bottles, which exhibit excellent product resistance, and good clarity, are also expected to become an important outlet for rigid PVC.

Formulations for extrusion generally include light and heat stabilizers, lubricants, which facilitate molding, and colorants. These materials are generally purchased in a compounded ready to use cube form, in order to minimize irregularities in blending, etc.

The outstanding characteristics of these rigid vinyls are chemical, solvent and water resistance; resistance to weathering when properly stabilized, therefore permitting long-term outdoor exposure; and low cost. Abrasion and impact resistance are satisfactory.

A major deficiency is heat sensitivity. Here, degradation begins with the split-off of HCl. The resultant unsaturation leads to cross-linking and chain cission, causing a degradation of the physical properties. Maximum service temperature for continuous exposure should not exceed $150–175°C$. Cold flow or creep is another deficiency, which leads to dimensional changes in materials under constant load, e.g., water pipe under constant service pressures will tend to enlarge in diameter, resulting in decreased strength; long spans of pipe or siding may sag. Temperature accelerates this effect.

PVC has a high coefficient of expansion, one of the highest for all plastic materials, and substantially higher than metals and wood. Therefore, design allowances must be made to provide for movement in order to avoid buckling, breakage, etc.

Flexible PVC

An unusually wide variety of products and usages are possible with plasticized vinyls. Typical applications include floor and wall coverings, boots, rainwear, jackets, upholstery, garden hose, electrical insulation, film and sheeting, foams and many others.

The primary processing techniques are by means of extrusion, calendering and molding. Special techniques involve organosols and plastisols.

Plasticizers used to develop the desired flexibility and performance are selected on the basis of cost and application requirements, e.g., temperature; service life; exposure to solvents, chemicals, water, UV, food; tensile strength; abrasion resistance; flexibility; tear strength; etc.

Plasticizers must be classed as *primary,* where high compatibility exists, or *secondary,* where compatibility is limited, thus restricting the amount that can be tolerated. The addition of secondary plasticizers may impart special properties or simply reduce cost (extender plasticizer).

Primary plasticizers may be further subdivided.

The *phthalate* types are by far the most popular due to cost and ease of incorporation. Dioctyl phthalate and diisooctyl phthalate are typical of this class. They exhibit good general-purpose properties. *Phosphate* plasticizers are also important for general-purpose use. Typical of these are tritolyl phosphate and trixylenyl phosphate. These plasticizers also impart fire retardant properties. *Low-temperature* plasticizers, such as dibutyl sebacate, are used where good low-temperature flexibility is required. For maximum compatibility and minimum cost, a typical plasticizer combination would be a blend of 50% DOP and 50% dibutyl sebacate.

Polymeric plasticizers are generally polyesters with a relatively low molecular weight. They are used where resistance to high temperatures and freedom from migration and extraction are required. Polymerics are more difficult to incorporate, have poor low-temperature properties, and are expensive.

Epoxy plasticizers are epoxidized oils and esters. These are generally classed with the polymerics. However, molecular weight is lower. Therefore, resistance to extraction and heat are slightly inferior. Low-temperature properties are better and epoxies are more easily incorporated.

Extender plasticizers, which are used mainly to reduce cost, consist of chlorinated waxes, petroleum residues, etc. Incorporation of excessive amounts may result in exudation on aging. The chlorinated types decrease flammability.

Organosols and Plastisols

Plastisols are dispersions of powdered PVC resin in plasticizer. A typical composition would consist of 100 parts of PVC resin dispersed in 50 parts of DOP. The resultant paste when heated to 300°F fuses or "fluxes" into a solid plastic mass. Stability of this plastisol at room temperature may range from several weeks to several months depending on the plasticizers and resins employed.

An organosol is the same mixture as described above, with the addition of solvent to reduce viscosity. These find their major applications in coatings. The solvent is evaporated before fusion of the film. Various pigments, colorants, stabilizers and fillers may be added, depending on the desired properties. Emulsion polymerization resins are generally employed because of their fast fusion rates. Coarser particle sized PVC resins would require extended time at the elevated temperature.

Plastisols allow the use of inexpensive manufacturing techniques, such as slush and rotational molding, casting, dipping, etc. They are employed for the manufacture of a large variety of parts, e.g., toys, floor mats, handles and many others.

Foams are made by the addition of blowing agents to the plastisol. These may be continuously applied to a moving substrate which includes a pass at an elevated temperature where foaming occurs, followed by fusion of the plastisol.

Organosols find their major application in coatings, which may be applied by spray, dip, knife, roller, etc. Typical products are coated aluminum siding, fabrics, paper, industrial coatings, etc.

An important development about ten years ago was the use of plasticizers which crosslink upon application of heat and thus produce a more rigid end product. This extends the range of products obtainable by plastisol techniques into rigids. By varying the amount of crosslinking plasticizer incorporated, various levels of flexibility are obtained.

HAROLD A. SARVETNICK

Poly(vinylidene Chloride)

Poly(vinylidene chloride) is produced by the free radical chain polymerization of vinylidene chloride ($H_2C{=}CCl_2$) using suspension or emulsion techniques. The monomer which has a boiling point of 31.6°C was first synthesized in 1838 by Regnault who dehydrochlorinated 1,1,2-trichloroethane which he obtained by the chlorination of ethylene. As shown by the following equation, the commercial product which has been produced since the late 1930's is produced by a reaction similar to that used by Regnault.

$$H_2ClCCHCl_2 \;+\; Ca(OH)_2 \xrightarrow[-2H_2O]{90°C}$$

1,1,2-tri-   calcium
chloroethane   hydroxide

$$CaCl_2 \;+\; 2H_2C{=}CCl_2$$

calcium   vinylidene
chloride   chloride

Since this monomer readily forms an explosive peroxide (H_2C—CCl_2 with an O₂ across) it must be kept under a nitrogen atmosphere at −10°C in the absence of sunlight.

The commercial products are copolymers of vinylidene chloride and vinyl chloride called Saran, Pliovic, Velon, Cryovac and Rovana. These copolymers were patented by Wiley, Scott and Seymour in the early 1940's. A typical formulation for emulsion copolymerization contains vinylidene chloride (78 g), vinyl chloride (22 g), potassium persulfate (0.22 g), sodium bisulfite (0.11 g), Aerosol MA surfactant (3.58 g), nitric acid (0.07 g) and water (180 g). More than 95 per cent of these monomers are converted to copolymer when this aqueous system is agitated in an oxygen-free atmosphere for 7 hours at 30°C.

A typical formulation for suspension copolymerization contains vinylidene chloride (85 g), vinyl chloride (15 g), methylhydroxypropylcellulose (0.05 g), lauroyl peroxide (0.3 g) and water (200 g). More than 95 per cent of these monomers are converted to copolymer when this aqueous suspension is agitated in an oxygen-free atmosphere for 40 hrs at 60°C. The glass transition temperature of the homopolymer is −17°C. It has a specific gravity of 1.875 and a solubility parameter of 9.8.

Because of its high crystallinity, the homopolymer (PVDV) is insoluble in most solvents at room temperature. However, since the regularity of repeating units in the chain is decreased by copolymerization, Saran is soluble in cyclic ethers and aromatic ketones. This copolymer (100 g) is plasticized by the addition of α-methyl-benzyl ether (5 g), stabilized against ultraviolet light degradation by 5-chloro-2-hydroxybenzophenone (2.0 g) and heat stabilized by phenoxypropylene oxide (2.0 g).

The poly(vinylidene chloride-co-vinylchloride) may be injection molded and extruded. Extruded pipe and molded fittings which were produced in large quantity in the 1940's have been replaced to some extent by less expensive thermoplastics. A flat extruded filament is used for scouring pads and continuous extruded circular filament is used for the production of insect screening, filter clothes, fishing nets and automotive seat covers.

A large quantity of this copolymer is extruded as a thin tubing which is biaxally stretched by inflating with air at moderate temperatures before slitting. This product, called Saran Wrap, has a tensile strength of 15,000 psi. Since it has a high degree of transparency to light and a high coefficient of static friction (0.95) it is widely used for the protection of foods in the household. It has a low permeability value for gases such as oxygen and nitrogen.

Poly(vinylidene chloride-co-acrylonitrile) is widely used as a latex coating for cellophane, polyethylene and paper. Since this copolymer is soluble in organic solvents, it is also used as a solution coating. The resistance to vapor permeability and the ease of printing on polyethylene and cellophane is increased by coating with this vinylidene chloride copolymer.

The tensile strength of both film and fiber is increased tremendously by cold drawing 400–500 per cent. Thus, tensile strengths as high as 40,000 psi in the direction of draw have been obtained by cold drawing. The annual production of Saran is about 60 million pounds.

RAYMOND B. SEYMOUR

Poly(vinylidene Fluoride)

Poly(vinylidene fluoride) is produced by the free radical chain polymerization of vinylidene fluoride ($H_2C=CF_2$). This odorless gas which has a boiling point of −82°C is produced by the thermal dehydrochlorination of 1,1,1-chlorodifluoroethane or by the dechlorination of 1,2-dichloro-1,1-difluoroethane. As shown by the following equations, 1,1,1-chlorodifluoroethane may be obtained by the hydrofluorination and chlorination of acetylene and by the hydrofluorination of vinylidene chloride or of 1,1,1-trichloroethane.

$$HC\equiv CH \ + \ 2HF \longrightarrow H_3CCHF_2$$

acetylene   hydrogen   1,1,-difluoro-
               fluoride   ethane

$$\xrightarrow[-HCl]{Cl_2} H_3CCClF_2$$

1,1,1-choloro-
difluoroethane

$$H_2C=CCl_2 + \ 2HF \longrightarrow H_3CCClF_2$$

vinylidene   hydrogen −HCl
chloride   fluoride

$$H_3C-CCl_3 \ + \ 2HF \longrightarrow H_3CCClF_2$$

1,1,1-trichloro-   −2HCl
ethane

$$H_3CCClF_2 \xrightarrow[-HCl]{600°} H_2C=CF_2$$

1,1,1-chlorodi-   vinylidene
fluoroethane   fluoride

$$H_2ClCCClF_2 \xrightarrow[-Cl_2]{\Delta} H_2C=CF_2$$

1,2-dichloro-1,1-
difluoroethane

Poly(vinylidene fluoride) which is called Kynar is polymerized under pressure at 25–150°C in an emulsion using a fluorinated surfactant to minimize chain transfer with the emulsifying agent. Ammonium persulfate is used as the initiator. The homopolymer is highly crystalline and melts at 170°C. It can be injection molded to produce articles with a tensile strength of 7000 psi, a modulus of elasticity in tension of 1.2×10^5 psi, a notched izod impact resistance of 3.8 ft lbs/in. and a heat deflection of 300°F.

Poly(vinylidene flouride) is resistant to most acids and alkalies but it is attacked by fuming sulfuric acid. It is soluble in dimethylacetamide but is insoluble in less polar solvents. Copolymers have been produced with ethylene, tetrafluoroethylene, chlorotrifluoroethylene and hexafluoroethylene. The latter is an elastomer called Viton or Fluorel.

The homopolymer is used as a chemical resistant coating for steel, for tank linings, hose, and pump impellors. The elastomeric copolymer with hexafluoroethylene when cured with hexamethylenediamine is used as a seal, gasket, o-ring, tubing, coating and lining.

RAYMOND B. SEYMOUR

Poly(Vinyl Alkyl Ethers)

Poly(vinyl alkyl ethers) are products with properties which range from sticky resins to elastic solids. They are obtained by the low-temperature cationic polymerization of alkyl vinyl ethers having the general formula $ROCH=CH_2$. These monomers are prepared by the addition of the selected alkanol to acetylene in the presence of sodium alkoxide or mercury(II) catalyst. As shown by the following equations, the latter yields an acetal which must be thermally decomposed to produce the alkyl vinyl ether.

$$HC\equiv CH \ + \ ROH \xrightarrow[130-180°C]{Na^+, OR^-} H_2C=CHOR$$

acetylene   alkanol   alkyl vinyl ether

$$HC\equiv CH + 2ROH \xrightarrow{Hg^{++}}$$

acetylene   alkanol

$$H_3C-CH(OR)_2 \xrightarrow[(-ROH)]{cat.} H_2C=CHOR$$

acetal   200-300°C   alkyl vinyl
                               ether

These monomers are also produced by an oxidative process in which the alkanols are added directly to ethylene and the alkyl ethyl ethers are thermally decomposed to produce hydrogen and the alkyl vinyl ethers.

Commercial polymers have been produced from methyl, ethyl, isopropyl, n-butyl, isotubtyl, t-butyl, stearyl, benzyl and trimethylsilyl vinyl ethers. The poly(methyl vinyl ether) called PVM or Resyn is produced by the polymerization of the monomer by boron trifluoride in propane at −40°C in the presence of traces of an alkyl phenyl sulfide. The polymer may have isotactic, syndiotactic or stereoblock configurations depending on the solvent and catalyst used.

Nonpolar solvents favor the formation of ion pairs between the polymer cation and the counteranion and favor the production of isotactic polymers. Soluble catalysts, such as diethyl aluminum chloride and ethyl aluminum dichloride, also affect the stereoregularity of the polymer chains. The tendency for the formation of stereoregular polymers is decreased as the size of the alkyl group is increased. Typical structures of these polymers are shown below:

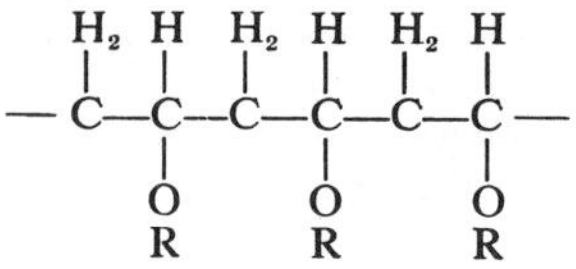

isotactic polymer

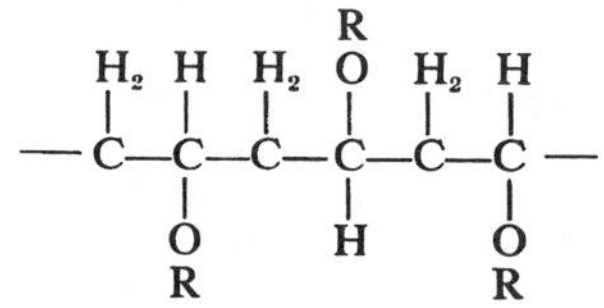

Syndiotactic polymer

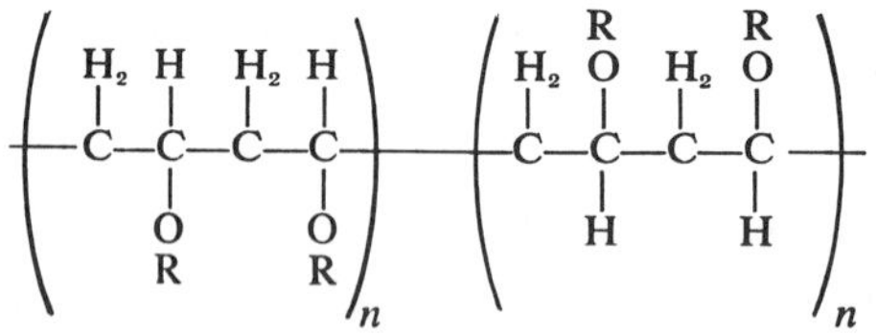

stereoblock polymer

Poly(methyl vinyl ether) is soluble in cold water but becomes insoluble in a reversible process when the temperature is raised to 35°C. This sticky polymer has a glass transition temperature of −20°C. It has been used as an adhesive and as a heat sensitizer for polymer latices.

Poly(vinyl ethyl ether) is soluble in ethanol, acetone and benzene. It is a rubbery product which may be cross-linked by heating with dicumyl peroxide. Poly(vinyl isobutyl ether), called Oppanol C, has a glass transition temperature of −5°C. It has been used as an adhesive for upholstery, cellophane and adhesive tape.

The processing properties of poly(vinyl chloride) have been improved by copolymerizing vinyl chloride with a small amount of vinyl alkyl ether. Copolymers of vinyl alkyl ethers and maleic anhydride, called Gantrez, are used as water soluble thickeners, paper additives, textile assistants and in cleaning formulations. Approximately 25 million pounds of poly(vinyl alkyl ethers) are produced annually in the U.S.A.

RAYMOND B. SEYMOUR

PORCELAIN ENAMEL

A porcelain enamel is a thin coating of glass fused to a metal. While porcelain enamels are usually applied to sheet iron, cast iron or aluminum, these vitreous coatings can be applied to a wide variety of metals including the heat resisting alloys. Coatings are available to give highly opaque whites or a wide range of colors. Porcelain enamels may use a ground coat, but direct-on enamels containing titania as an opacifier are in general use.

These special glasses used as porcelain enamels must melt at temperatures at which the metal retains its shape and must have thermal expansions which are compatible with the metal. The metal is usually subjected to special treatment so as to obtain good bonding between the glass and metal. Porcelain enamel coatings are available having a wide range of colors, varying whiteness or gloss, and good acid resistance or high-temperature thermal stability.

The enamel glasses are made by smelting at elevated temperatures various alkali, boron, silica, titania and fluorine-containing materials along with other special materials. These materials on smelting form a molten glass which is quenched and shattered (usually in water) to form small fragments called frit. The frit is ball milled with the necessary amount of water, using clay as a suspending agent to form a fine powder suspension. This prepared slip is applied to the metal by spraying or by dipping. The ware is dried and then heated in furnaces to 1350 to 1550°F where the glass particles melt to form a smooth coating. In some cases more than one coating may be applied. There has been extensive development of automatic and mechanized equipment for processing porcelain enamels.

Porcelain enamels are used on many products such as stoves, refrigerators, sinks, washing machines, bath tubs, kitchen utensils, architectural facings, chemical and food storage containers, and signs; special compositions and processing being required in each case to develop the best engineering properties.

R. L. COOK

References

"Porcelain Enamels," by A. I. Andrews.

POROMERIC MATERIALS

The term "poromeric" refers to a man-made material which has the outward appearance of leather and, more important, functions as leather, especially with regard to the ability to transmit water vapor. The multibillion dollar shoe industy makes the most extensive use of these materials. The garment and accessories industries are also major end-users. The pace of development has been so rapid that several types of materials have already evolved in the decade since first commercialization.

In general, a poromeric consists of three layers: (1) the substrate or supporting layer, (2) the coating or film layer, and (3) the topcoat or finish layer.

Substrate. The substrate performs two primary functions: (1) it dominates the physical properties of the structure, and (2) it contributes to the tactile aesthetics of the composite. Those familiar with consumer textiles will recognize such terms as hand, drape, roll and feel.

Cotton-based knits and wovens were among the early substrates, but these suffered from deficiencies in the balance between stress and strain, and between tear resistance and nonfraying edges. In addition, cotton deteriorates in wear from the humidity and the action of the chemicals in human perspiration.

The major technical breakthrough in substrates was the development of man-made fibers and of dry-laid nonwoven web technology. Man-made fibers have made available a reliable and economical source of controlled-property fibers. Fiber diameter, fiber length, surface finish, and other treatments, such as crimping and drawing, provide the opportunity for a variety of end-product properties. In addition, it is possible to produce fiber which shrinks when subjected to the proper combination of heat and moisture. Improved chemical and rot resistance are other improvements over cotton. Nylon and polyester fibers have found the greatest acceptance in poromerics, although acrylics and polyolefins have also been used.

Dry-laid nonwoven technology combined standard textile technology with the new fibers. Making a garnetted or carded web which may or may not be cross-lapped to provide anisotropy was one of the first advances. This was combined with the needle-punching process from felt-making to provide a three-dimensional structure with good internal strength. Another important development was the random air-laid web which provided anisotropy with more uniformity and control than a cross-lapped web.

A third factor which contributed to modern substrates was the development of a wide range of synthetic latices for binder materials. Acrylonitriles, acrylics, and urethanes are among the more commonly used materials, each imparting its own set of chemical and physical attributes to the substrate. Spray-bonding, dip-saturation, and padding are the more common ways of applying binder to the substrate. By varying the fiber, the binder, the web forming, and the binder application method, it is possible to produce substrates with a wide variety of physical and aesthetic properties. By varying the degree of needling, the depth of needle penetration and by needling from one or both sides, a density gradient is established through the thickness of the nonwoven layer, similar to that which occurs naturally in leather.

In some manufacturing schemes, a woven fabric is interposed between the substrate and the film layer to further modify the physical properties of the structure. In at least one commercial product, the nonwoven web is needle-punched around and through a scrim reinforcing web.

Coating. The coating or film layer, which may or may not be top-coated, contributes greatly to the aesthetics of the poromeric package. It has a major effect upon the appearance and also affects tactile properties. The coating layer takes the brunt of the abuse from scuffing, abrasion and snagging. In addition, by interacting with the substrate, the coating layer can position the neutral axis of the composite to minimize excessive strain and improve the product life under the conditions of constant flexing. The coating layer in some constructions is embossed to impart a leather-like grain to the material.

Early coatings were porous, plasticized polyvinyl chloride, which had a short life due to embrittlement and subsequent flex failure because of plasticizer loss through migration and extraction by perspiration and foot oils. Permanent-type plasticizers overcame the failings of these products.

Permeability in these products is obtained through several means. One involves forming the coating layer by sintering particles of a dry, pigmented polyvinyl chloride powder. The passageways formed from the packing of the particles provides a network of pores through which air and water vapor may pass. Another method used with PVC is to incorporate in the vinyl compound a material which can be leached or dissolved out, leaving a series of pores and passages. A third method is to incorporate a chemical blowing agent in the compound which, when heated, liberates quantities of gas to produce a cellular foamed product.

More recently developed materials are films of microporous polyurethane. Both polyester- and polyether-based resins are used. There are several methods of imparting microporosity. Blowing agents, mechanical foaming, and leaching processes have all been used. Another technique involves laying down a wet film of a solvent solution of the polymer. These materials have very little air permeability and the transfer of water vapor is controlled by the rate of diffusion through the thin polyurethane cell walls. Most of these materials require a topcoat to impart abrasion resistance, although a suede-like surface can be achieved by abrading the surface layer of the micropores.

A third generation of film materials is unique in that the entire poromeric structure is composed of a microporous polyurethane foam. There is no separate substrate involved. The manufacturing technique used provides a varying yet controlled cell structure from one surface to the other. What would usually be considered the substrate has relatively fewer but larger cells while the wear surface has a very fine and dense cellular structure.

Topcoating. This layer in the poromeric structure can serve as both a wear layer and a decorative layer. In the case of the plasticized vinyl materials, the top layer may be strictly a pattern which is overprinted with vinyl inks to alter the surface appearance. At times, a thin microporous layer of hard acrylic or wear-resistant urethane may be applied to enhance the scuff resistance or to lower the coefficient of friction. Various color shadings and pattern effects can be obtained by surface coating.

In the case of the microporous polyurethane films, the top coat of urethane or acrylic serves the main purpose of a wear-resistant layer. It may also function as an aesthetic component. There are some poromeric products which consist solely of a substrate and a thin urethane coating.

(*Note*: Two of the more important synthetic poromerics, "Corfam" and "Aztran," have been discontinued by their manufacturers. Editor)

JERROLD J. ABELL

POROSITY

Porosity of a solid, in its broadest sense, is defined as the volume of interstices measured in percentage of the over-all apparent volume of the solid. Too often it is confused with permeability, which is that property of a solid which determines the rate of flow of a fluid through it.

Porosity exists in two forms, closed and open.

Both the closed and open pores affect the apparent density of a solid. Only open pore structure is conductive to permeability. Porosity may also be characterized as natural or controlled. Natural porosity is found in animal and vegetable membranes and in minerals. Materials with controlled porosity of either closed or open type may be fabricated to meet the requirements of industry.

The most commonly used method of measuring open porosity is based on the increase in weight of a solid resulting from immersion in a liquid.

$$P = \frac{(W_2 - W_1) D_1}{W_1 \times D_2} \times 100 \qquad (1)$$

P = open porosity
W_1 = weight of sample in air
D_1 = apparent density of sample
W_2 = weight of sample after immersion
D_2 = density of liquid

For solids having very small pores, liquids with high penetration, such as carbon tetrachloride, are used. A more accurate method makes use of gas absorption. This method is, in fact, sensitive enough to detect the porosity of silica gel and similar materials which have capillaries so fine as not to permit penetration of liquids. Indeed, for the smallest capillaries (micropores) even gas procedures are inadequate. In closed pore structures the true porosity cannot be determined by these methods. In order to approach more closely a measurement of true porosity, the solid may be crushed and pulverized to expose all closed pores. Then the density of the pulverized material is compared with the bulk density.

The apparent volume of a substance consists of three parts: closed pores, open pores and solid material. Equation (1) is the expression for calculating the open porosity. Equation (2) is the expression for calculating the closed porosiy.

$$P_1 = \frac{V_t - (P + S)}{V_t} \times 100 \qquad (2)$$

where V_t = total volume (100%)
P_1 = percentage of closed porosity
P = percentage open porosity obtained from Equation (1)
S = percentage of solids obtained as follows:

$$S = \frac{W}{D.V_a} \times 100$$

where W = weight of crushed powder
and D = desnsity of crushed powder
V_a = apparent volume before crushing

Practical Applications. Porosity is found to be the controlling factor in many industrial applications. Porous ceramic plates and tubes are produced from refractory materials. These may be prepared by mixing and firing various grain sizes. They are used as filters, separators, diffusers, and dryers which find many applications in sanitary engineering and in the chemical and oil industries. (The filter bed, for example, illustrates the versatility in the application of the property of porosity in these branches of engineering.) Filter beds consist of porous media produced by mixing together sand or other materials of varying particle size and are so constructed as to allow selective filtration of a solid-liquid slurry. In like manner, these filter beds may be employed in the separation of liquids from gases, in which case the liquid is retarded by the bed.

The property of porosity is widely employed in the building industry for the purpose of fabricating light-weight and thermally insulating construction materials. Examples of such materials are insulating brick, expanded vermiculite, plaster, and foam glass. To illustrate, porosity in insulating brick is obtained by mixing the raw ingredients with organic material which is decomposed and oxidized in the firing operation, thus producing a porous structure.

Nature has provided woods of varying porosity. In the high-porosity category (soft woods) are found, for example, pines, red wood, and cedar. In the low-porosity class (hard woods) are birch, maple, and oak. Leather is typical of a class of materials known as poromerics.

The porosity of metals is an important variable. A casting of a metal such as iron is more porous than one of steel. Powdered metals may be formed into various shaps and then sintered to produce a porous structure which serve to filter, separate, and diffuse gases and liquids. Porosity is also an important property in the fabrication of some metal bearings which allow oil to penetrate and provide lasting lubrication.

The control of porosity in fibrous materials is one of the most important applications of this property. For example, the porosity of cloth. which varies depending upon the weave and warp. is controlled to provide lightness and warmth. Porous fabrics can be impregnated with hydrophobic impregnants, imparting water-repellant properties to the cloth without destroying the permeability. The treatment of clothes to provide waterproof garments is an illustration of this. Other important applications are related to the fibrous porous structure of paper. Included in this category are blotting paper, writing paper, and paper used in electrical capacitors, where for each use the requirements of porosity must differ to produce the effect desired.

Porosity in the rubber and plastics industries is controlled to produce sponge-like characteristics which differ radically from those of nonporous sheet rubber or hard resins. Such materials having a closed-pore structure are often called cellular plastics.

An aspect which cannot be gone into here is that which deals with the importance of porosity in the control and maintenance of life processes.

J. F. POTTER

Cross-references: *Adsorption; Carbon, Activated.*

PORPHYRINS

The porphyrins are intensely colored macrocyclic tetrapyrrole compounds widely distributed in animals, plants, and microorganisms. Iron porphyrins known as hemes (Fe^{2+}) and hemins (Fe^{3+}) occur in hemeproteins. Magnesium porphyrins are present as intermediates in the pathways of the biosynthesis of chlorophylls. The chlorophylls are magnesium complexes of chlorins (dihydroporphyrins). The occurrence of small quantities of copper, zinc, and manganese porphyrins in nature has been reported but, as biologically significant functions have not been found for these compounds, it has been suggested that such compounds may simply arise spontaneously (nonenzymatically) from the combination of the porphyrins with available metal ions. Metal-free porphyrins have no known biological function other than as biosynthetic intermediates and, under normal conditions, only occur in small amounts in nature. Somewhat larger amounts are found in animals with porphyria diseases, heavy metal poisoning, or certain anemias. Porphyrin accumulation has also been noted in certain microorganisms.

Protoporphyrin IX, the porphyrin found in several hemeproteins including hemoglobins, myoglobins, b-type cytochromes, catalases and peroxidases, was the first natural porphyrin to have its structure fully elucidated. Crystalline protoporphyrin IX iron (III) chloride ($C_{34}H_{32}N_4FeCl$) or hemin, as it is usually called, was found many years ago to be readily obtained when blood was treated with hot acetic acid and sodium chloride. After many years of research chiefly by Nencki, Kuster, Willstattre, and finally by Hans Fischer, the structure of hemin was elucidated and confirmed by synthesis in Fischer's laboratory in 1930 (see Fig. 1). Recent studies by infrared and nuclear magnetic resonance spectroscopy and by x-ray crystallography fully support Fischer's structure. Porphyrins were thus found to be distinguished by a macrocylic structure where four pyrrole-like rings are linked at alpha positions by carbon atoms. The four carbon bridge positions are known as *meso* positions and are designated as alpha, beta, gamma, and delta positions. The eight other peripheral positions available for substitution are called *beta* positions and are designated by the numbers 1, 2, . . . , 8.

Other hemeproteins have porphyrins structurally related to protoporphyrin IX. Thus in *c*-type cytochromes the porphyrin has been found to be identical with protoporphyrin IX except for ethyl groups at the 2 and 4 positions with thioether linkages to cysteine residues of the peptide chain. Porphyrin

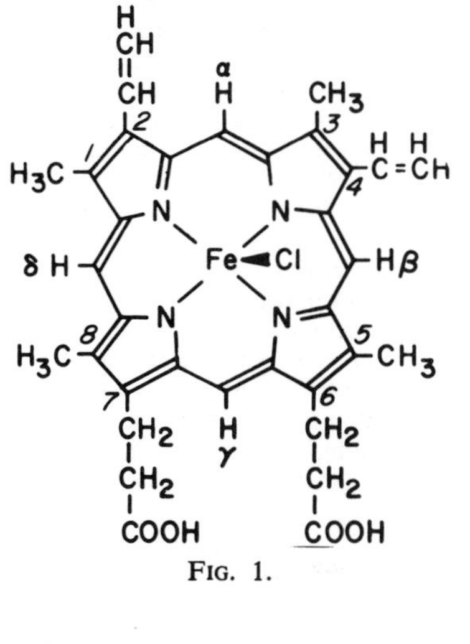

a of cytochrome *c* oxidase (cytochromes *a* and a_3) differs from protoporphyrin IX in having formyl and long chain alkyl (about 17 carbons) groups, presumably at the 8 and 2 positions respectively. Chlorocruoro or spirographis porphyrin has a 2-formyl group in place of the 2-vinyl group.

The structural similarities found among the naturally occurring porphyrins is expected in view of the fact that, as far as is now known, the same biosynthetic pathways are used by all forms of life. The pathways used for iron porphyrins and for chlorophylls are the same up to the stage of protoporphyrin IX. The ability to synthesize porphyrins from simple precursors is nearly universal among living forms; only a very few organisms require preformed tetrapyrrole compounds as growth factors.

Early studies in the use of isotopically labelled compounds in biological systems showed that acetate and glycine were incorporated into hemin in humans and rats. Glycine was able to provide all four nitrogen atoms, the *meso* carbons, and one carbon of each pyrrole and of each alpha position (next to the ring) for substituents in the 2, 4, 6, and 7 positions. These and further studies demonstrated that delta-amino-levulinic acid (ALA) and a monopyrrole, porphobilinogen (PBG), were early heme precursors. The pathway to form protoporphyrin IX appears to be as follows: ALA to PBG to uroporphyrinogen III to coproporphyrinogen III to protoporphyrinogen IX to protoporphyrin IX. Iron is inserted, presumably enzymatically, into protoporphyrin IX to form heme. Porphyrinogens are colorless reduced porphyrins where the four pyrrole rings are separated by saturated—CH_2— (methylene) bridges. Thus six reducing equivalents are required to convert porphyrins to porphyrinogens which are readily autoxidated back to porphyrins.

A large number of differently substituted porphyrins have been synthesized, in large measure as a result of the brilliant investigations carried out during the 1920's and 1930's in Hans Fischer's laboratory in Munich. In these syntheses dipyrryl intermediates (dipyrrylmethenes and dipyrrylmethanes) have been used extensively whereas monopyrroles and tetrapyrroles (bilanes) have been used less frequently. The great majority of condensation reactions which produce the porphyrin ring system proceed in very low yield. Yields of 1 to 3% are very common, and the desired product is frequently obtained in a yield of a few tenths of a per cent. Moreover, such condensations often produce several different porphyrins from which the desired product must be separated. When it is considered that the pyrrole intermediates required for porphyrin synthesis must themselves be prepared by several-step syntheses, the difficulty frequently encountered in preparing even small amounts of synthetic porphyrins becomes readily apparent.

By far the most widely used intermediates for porphyrin ring synthesis, the alpha-bromo-alpha-methyl-dipyrrylmethenes and alpha-bromo-alpha-(bromo-methyl) dipyrrylmethenes, have proved of great importance in establishing through synthesis the structure of naturally occurring porphyrins and their degradation products. Other important intermediates are alpha-unsubstituted dipyrrylmethanes

FIG. 1.

which often give porphyrins when condensed with formic acid. Reported syntheses where porphyrins have been obtained directly from monopyrryl inter-, mediates include are reaction of pyrrole itself with certain aldehydes to give alpha, beta, gamma, delta-tetrasubstituted porphyrins, frequently in excellent yield. The alpha-(hydroxymethyl) pyrroles and alpha (dialkylaminomethyl) pyrroles have also been convenient intermediates. Syntheses of porphyrins from monopyrryl intermediates which have two different beta-substituent groups suffer from the disadvantage that a mixture of isomeric porphyrins must be anticipated. In certain porphyrin syntheses both from monopyrryl, and dipyrryl intermediates the presence of metallic ions, such as copper, zinc, or magnesium, has been observed to increase the yield of porphyrin, and the metal may be subsequently removed by treatment with acid.

Porphyrins with free ring positions undergo electrophilic substitution reactions typical of an aromatic system. Acylation, halogenation, sulfonation and chloromethylation reactions can be carried at *beta* positions smoothly and in good yield. Nitration has been shown to take place at *meso* positions. In acid solution deuterium exchange for hydrogen has been shown to take place both at *meso* and at *beta* positions. Porphyrins undergo stepwise chemical or photochemical reduction to dihydroporphyrins (chlorins result when reductions take place at beta positions; phlorins are obtained when reduction takes place at *meso* positions), tetrahydroporphyrins, and hexahydroporphyrins. Chlorins are green and phlorins are blue in organic solvents. Tetrahydroporphyrins are purple. Drastic reduction with hydriodic acid cleaves a porphyrin into monopyrroles. Vigorous oxidation of a porphyrin yields substituted maleimides. Both the pyrroles from reduction and maleimides from oxidation have proved valuable in elucidating the structure of naturally occurring porphyrins.

Coordination compounds of porphyrins with a large number of metal ions have been prepared. With alkali metal ions (Li^+, Na^+, K^+) and with Cu^+ and Ag^+, two metal ions can be bound to one porphyrin molecule to give complexes which exhibit low stability under weakly acidic conditions. Mg^{2+}, Zn^{2+}, Cd^{2+} and transition metal ions form one-to-one complexes. With certain complexes (Pt^{2+}, Pd^{2+}, Ni^{2+} and Cu^{2+}) highly acidic conditions (e.g., concentrated sulfuric acid) may be required to displace the metal ion. With others such as those with Mg^{2+}, Zn^{2+}, and Cd^{2+}, the metal ion can be readily displaced under mildly acidic conditions. Additional (nonporphyrin) ligands may also bind with the metal ion. The ability of iron porphyrins to bind additional ligands such as oxygen, hydrogen peroxide, carbon monoxide, histidine, etc. and to assume different oxidation and spin states is of particular biochemical importance. Oxygen and carbon monoxide bind to low-spin (diamagnetic) Fe^{2+} complexes; here pi-bonding, with iron acting as the pi-donor, contributes importantly to the bonding. Fe^{2+} complexes are frequently very readily autoxidized. The explanation of those structural features of the hemoglobin and myoglobin molecules which permit these proteins to bind oxygen reversibly without undergoing oxidation remains an intriguing biochemical question.

The aromatic character of the porphyrin ring system as demonstrated by substitution reactions is also supported by the high ring current field effects observed in both nuclear magnetic resonance spectra and in diamagnetic susceptibility values. Also consistent with a high degree of electron delocalization are the quite extensive theoretical treatments of the highly characteristic (and most useful) electronic spectra of porphyrin derivatives. The porphyrin system conforms to the Huckel rule requirements for aromaticity with an "inner ring" of 18 pi-electrons and an "outer ring" of 18 pi-electrons.

Differences in peripheral substitution can markedly affect the properties of porphyrins. In general, it has been found that the more electron-withdrawing are the peripheral substituents the lower is the basicity of the central nitrogen atoms, the less strongly will pi-acceptor ligands such as oxygen and carbon monoxide be bound to low-spin Fe^{2+} complexes, the more strongly are (sigma) donor ligands such as amines bound to central metal ions, the less extensive is intermolecular association (presumably through pi-pi interactions between porphyrin molecules) in a solvent such as chloroform, the lower is the rate of autoxidation of Fe^{2+} porphyrins, and the more stable is the Fe^{2+} compared with Fe^{3+} oxidation state. Electronic spectra, characterized by very strong absorption near 400 mμ (the Soret band) and by weaker bands at both longer and shorter wave lengths, reflect differences in both type and the relative positions of substituent groups. Frequently, but by no means always, the more electron-withdrawing are the substituent groups the longer are the wave lengths of the absorption maxima. Different metal ions may result in characteristically different electronic spectra.

The increased insight into the effects of structure and medium on the properties and reactions of porphyrin compounds has stimulated, and promises to continue to stimulate, attempts to interpret structure-function relationships among hemeproteins.

W. S. CAUGHEY AND J. LYNDAL YORK

Cross-references: *Pyrroles; Iron, Biochemical and Biological Aspects.*

POTASSIUM

Potassium, one of the alkali metals, has an oxidation potential of 2.95 volts referred to hydrogen, and is extremely reactive. Its atomic number is 19 and atomic weight 39.098; it occurs in the form of its compounds quite widely, and is of particular significance in the metabolism of plants and animals. It ranks seventh of the elements in abundance, the lithosphere containing 2.59% of potassium. Water of the oceans contains 0.07% of potassium chloride and this has been the source of evaporite beds of commercial importance.

Potash, the term used in commerce for the salts of potassium, refers particularly to the oxide of potassium. Potassium chloride is called muriate of potash; potassium hydroxide is caustic potash. Potassium carbonate was prepared in ancient times by leaching wood ashes and was the lye used in soap-making.

Potassium salts found in the Stassfurt deposits of Germany in 1839 became the basis for much of the

German chemical industry. Soon after the start of the war in 1914 the rest of the world found itself cut off from the German source, and potash was produced in other countries from lake waters, from potash silicates and from plant residues at considerable cost. Potash had become a necessity for agriculture and for chemical industry.

In the United States, the Searles Lake deposit at Trona, California became the first major source of potash salts. Operation on this complex deposit started in 1918 and yielded several products, but the potash output was limited. Further exploration led to the discovery of potash in 1925 in oil well cores from New Mexico, chiefly in Eddy County. These deposits were developed, and provided adequate supplies of potash for the United States and some export. These deposits are all still in operation.

Major deposits were found in Saskatchewan, Canada, at a depth of about 3500 feet; the first attempt to develop these deposits in 1951 met with shaft sinking problems that prevented its completion. The first successful operation began in 1954 and production began in 1958. Now there are many operating mines and plants and Canadian plants have a capacity of over 10,000,000 tons of K_2O annually.

Exploration for potassium has shown that it is very widely distributed and deposits are being worked in every continent today. Evaporite beds usually show potassium chloride, but some beds show sulfates, always as double salts, generally with magnesium. The deposits are always associated with sodium chloride, the main solute in sea water. In fact, in evaporation of sea water 98% of the water must be evaporated before potassium salts start to crystallize.

While the major production comes from mining the salt deposits, some lake brines are worked, such as the Dead Sea in Israel and the Great Salt Lake in Utah. At Trona, California the complex deposit yields several different products on solution and separation by fractional crystallization.

Potassium metal was first prepared by Humphry Davy in 1807 by electrolysis of potassium hydroxide. Potassium is a soft, silvery-white metal and has a specific gravity of 0.85, a melting point of 63.2°C, and a boiling point of 760°C. Potassium consists chiefly of three isotopes of atomic weight 39, 40, and 41, which comprise 93.09, 0.0118 and 6.90%, respectively. Isotope 40 is radioactive and has a half-life of 1.4×10^9 years. About 89% of its activity is by beta emission to give an atom of calcium and the rest by electron capture and gamma-ray emission to give an atom of argon.

The radioactivity of potassium, though very small, accounts in large measure for the accumulation of heat in the earth. It also provides a possibility for dating geological specimens and a means for the analytical determination of potassium. Its radioactivity is about one-thousandth that of uranium and therefore offers no hazard. Analytical applications, however, require large samples, long counts, or heavy shielding from background radiation.

Elemental potassium shows a body-centered cubic lattice at room temperature and below, and the lattice constant of 5.31 Å has been reported.

The ionic radius of 1.33 Å is exhibited in its compounds.

The molar heats of formation of the chlorides of lithium, sodium, and potassium are 97.7, 98, and 104 kcal/g-mole. This order differs from that of the electrode potentials which are, respectively, −3.05, −2.71, and −2.92 volts.

Potassium reacts violently with water and with air, and is therefore stored in inert oils. It will form several oxides, potassium oxide (K_2O), peroxide (K_2O_2), superoxide (KO_2), and an ozonide (KO_3), showing a variety of oxides not formed by sodium. The superoxide is used in gas masks to produce oxygen.

Potassium will react with hydrogen at about 350°C to form its hydride; it will react with nitrogen under an electric charge to form azides. It will react with sulfur to form six different sulfides. Reaction with halogens and with several metal halides is explosive. Carbon monoxide and potassium will form an explosive carbonyl. Solid carbon dioxide and potassium will react with an explosion if subjected to shock. Even graphite will react to form a series of potassium carbides.

Because of its high reactivity, its compounds are very stable. However, also because of its high reactivity, it will undergo reactions to form unstable compounds as noted above.

Potassium will form alloys with many metals, and distinct compounds with many metals. Alloys of sodium and potassium can be prepared by heating potassium chloride with metallic sodium. Metallic potassium is now prepared by fractional distillation of the alloy.

The eutectic with sodium has about 77% potassium and has a melting point of −12.3°C. The alloys are more reactive than sodium. It is of interest to note that at room temperature oxygen reacts with the alloy to form sodium monoxide and potassium superoxide, while at higher temperatures sodium peroxide and the superoxide of potassium are formed.

Compounds of potassium are very important commercially. With its particular significance in metabolism of plants and animals, its use as fertilizer accounts for almost 95% of total production. In 1970 world production of potash (calculated to potassium oxide) was over 20 million tons. The United States, Canada, East Germany, West Germany, and the U.S.S.R. each produced well over 2 million tons. Most of this was potassium chloride, but a small amount was sulfate or nitrate. Agricultural consumption is heavy in the United States, Europe and Japan, but use is expanding throughout the world as heavier yields of food crops are required.

Potassium, an essential material for plant growth, is found in all soils. Plants use it either from soil water solution or from clay or other materials where it is subject to ion exchange. Potassium has not been found in any organic compounds in living tissue, but it is found in the cell fluid and absorbed on proteins. It enters into life reactions associated with nerve response, in heart muscle, in enzyme action, in respiration and in almost every life function.

In the plant, a deficiency of potassium can be detected readily by failure of function. Plants appear to maintain a level of potassium content in fruit or seed, so the size and yield is often determined by the available potassium.

Potassium compounds find a number of uses in industry. Potassium chloride is used in the manufacture of other compounds, chiefly through the hydroxide. Potassium chloride has a specific gravity of 1.98, is generally cubic in crystal form, and melts at 772°C. It is used as a substitute for salt in the diet when sodium chloride must be limited, as with some cardiac problems.

Potassium hydroxide is produced by the electrollysis of an aqueous solution of potassium chloride. It is marketed as a solid or as a 50% solution. It is used as the electrolyte in some alkaline storage batteries and in some fuel cells, chiefly because its electrical conductivity is higher than that of sodium hydroxide. It is also used in preparing many of the other compounds of potassium.

Potassium sulfate is used with calcium sulfate to control the rate of set and improve the strength of gypsum cements. However its chief use is as a fertilizer for crops such as citrus and tobacco where chloride is considered objectionable.

Potassium carbonate is produced largely by carbonation of potassium hydroxide, although many methods have been described for its production. Consequently, its high price limits its use in comparison with sodium carbonate which can be secured at lower cost. Potash is required in making glass and ceramics of quality such as fine tableware, television tubes and optical glass.

Potassium nitrate has been prepared from natural sources usually associated with guano depositions. It is now made synthetically from potassium chloride and nitric acid. Its use in gunpowder led to its early manufacture, but now its use as fertilizer in special applications is of greater importance. It is also used in curing meats, and was called nitre in old recipes.

A number of other compounds are used, where the potassium imparts some advantage, such as reduced solubility or faster reaction than the corresponding sodium salt. Potassium compounds are usually less hygroscopic than the corresponding sodium salts. Compounds used in considerable amounts are the chlorate, chromate, dichromate, permanganate, ferrocyanide, and ferricyanide. Historically, most laboratory chemicals were made as the potassium salts, but sodium salts have replaced this general use.

Analytically, potassium is determined gravimetrically by weighing as the chloroplatinate, cobaltinitrite, or tetraphenyl boron salt. However, volumetric procedures with tetraphenylboron are quick and accurate, and have come into general use. Flame photometric or atomic absorption methods are well adapted to the determination of potassium, particularly in low concentration. Radiometric methods measuring beta or gamma emission are also used.

E. A. SCHOELD

References

Mellor, J. W., "Comprehensive Treatise on Inorganic and Theoretical Chemistry," Volume 2, 1963 Supplement, John Wiley & Sons, New York.

Noyes, Robert, "Potash and Potassium Fertilizers," Noyes Development Corporation, 1966.

Schoeld, E. A., "Method for Flotation Concentration in Coarse Size Range," U.S. Patent 2,931,-502, July 2, 1956.

"World Survey of Potash," The British Sulfur Corporation, London, 1966.

"Hunger Signs in Crops," National Fertilizer Association, Washington, D.C., 1941.

Schoeld, E. A., "Potassium," pp. 552-561, "Encyclopedia of the Chemical Elements," Hampel, C. A., Editor, Van Nostrand Reinhold Co., New York, 1968.

POWDERED METALS

Metals in powdered or finely divided form are derived by two general means: (1) directly from the process producing the metal, or (2) by disintegration of massive metal by physical or chemical techniques.

In the first category are the metals which are not obtained in massive or consolidated form from the reduction process because the reduction conditions do not melt the metal product or do not cause it to consolidate (as by sintering). Examples are tungsten and molybdenum, made by the hydrogen reduction of their oxide powders at about 1000°C, very much below their respective melting points of 3410 and 2610°C. Finely divided metal powders result, which must then be converted to massive form by arc melting under vacuum or an inert atmosphere, or by sintering the compacted powder by resistance heating. Another example is the commercial production of high-purity nickel by the carbonyl (Mond) process. Impure nickel granules and carbon monoxide are reacted under pressure to form volatile nickel carbonyl, $Ni(CO)_4$, which when heated decomposes to give nickel powder and carbon monoxide.

Powdered metals are produced from massive metal or metal compounds for use in powder metallurgy or for pigment purposes by several physical or chemical methods. The physical methods include (a) grinding or ball-milling, sometimes at liquid nitrogen temperatures where most metals are brittle; (b) inert gas atomization whereby a molten metal stream is broken into particles by a cross-stream of inert gas and the particles solidified by cooling; (c) rotating a metal rod heated with an arc or plasma so that tiny droplets of metal are thrown from the rotating electrode; (d) pressurizing a molten metal with a soluble gas, like hydrogen, and then running the metal into a vacuum chamber where it disrupts into particles as the gas escapes.

The chemical methods include (a) the hydride process whereby a metal hydride is formed by exposure of the metal to hydrogen, and the hydride then decomposed at a higher temperature to yield metal particles; (b) the carbonyl process in which a metal carbonyl is formed by exposure of the metal or metal compound to carbon monoxide and the carbonyl then decomposed at a higher temperature to form metal particles; (c) formation of a binary alloy from which one component is leached chemically leaving the other component in particle form, as is done to produce Raney nickel from $NiAl_3$ by action of sodium hydroxide, which dissolves the aluminum; (d) hydrogen reduction of metal compounds used to produce metals like iron, copper, nickel, cobalt, tungsten and molybdenum; (e) reduction of metal compounds with sodium or other alkali metals or magnesium to produce, for example, tantalum, zirconium, hafnium, titanium and niobium; and (f) electrolytic reduction of compounds to form metals such as copper, iron, silver and tantalum. Electrodeposited manganese and chromium and the dendritic forms of many other metals deposited on cathodes are brittle and easily disintegrated to powder form.

Alloys such as bronze and stainless steel powders can be made by treating the massive alloys by one of the above methods or by mixing the powdered components in desired proportions. Refractory carbides, chiefly tungsten carbide, are made by mixing the powdered metal or metal oxide and carbon black in the correct ratio and heating the mixture. Cemented carbides are made by mixing cobalt powder with the powdered carbide or its metal and carbon components, compacting the mixture and sintering it by heating.

The uses of powdered metals, chiefly to make products by powder metallurgy techniques, require certain physical properties in the powder, e.g., particle size and size range, surface area, density and flow properties. Thus, the method of producing the powdered metal is chosen to yield particles of desired properties in the metals involved.

Another field of application for powdered metals is for pigments. Aluminum powder is the principal metal used for this purpose. A large portion of the several hundred million pounds of aluminum powder produced annually (280 million in 1969, 200 million in 1970) is consumed as a paint pigment. Other metals used for pigments include copper, zinc and bronze.

Because of the increased rate of surface reaction of finely divided metals with reactants like oxygen, nitrogen and water, special precautions must be taken during the preparation and handling of powdered metals to prevent such reactions. The means vary with the chemical properties of the metals concerned, but use of an inert atmosphere or a vacuum is most common. In addition to the hazard of dust explosions, a compound such as an oxide film on the surface of the metal particles interferes with the subsequent consolidation of the particles by powder metallurgy operations. Aluminum, magnesium, zirconium and magnesium-aluminum alloy powders are among the more hazardous powdered metals from the explosion standpoint, but proper inerting minimizes the hazard.

CLIFFORD A. HAMPEL

PRECIPITATION

The term "precipitation" in chemistry has several meanings. One meaning is the formation of an insoluble compound from a solution upon the addition of a properly selected reagent. The appearance of a precipitate under established conditions of acidity or basicity (pH), and observation of bulkiness, rate of settling, color, and behavior on further treatment fall in the field of qualitative chemical analysis. The formation of an insoluble or nearly insoluble compound so that there is complete or nearly complete precipitation of the salt or element sought is the sequence of events upon which gravimetric analysis rests.

Precipitation of an insoluble compound may be made the basis of volumetric analysis methods, called precipitation titration methods, either by visual observation of no further formation of a precipitate upon addition of more reagent solution, or by potentiometric readings. It may be favored by a reduction in the temperature of the solution. Of interest is the method of precipitation from a homogeneous solution, in which the precipitating agent is generated uniformly throughout the solution, by hydrolytic action, for example. Precipitation may also occur in a supersaturated solution; the precipitation equals the amount of solute by the amount which supersaturation exceeds saturation at equilibrium.

Precipitation of dust or of droplets suspended in a gas or in air, such as in the Cottrell precipitator, is termed electrostatic precipitation. In this process, the material to be precipitated is already present in a phase distinct from that of the medium; it is only necessary to provide it with an electric charge to bring about precipitation. Precipitation of lyophobic sols by electrolytes is a part of the study of the colloidal state.

In quantitative gravimetric analysis, the precipitation is performed so as to avoid contamination, which might be the result of occlusion (impurity within the newly formed crystal) or adsorption on the crystal surface. The general rule is to make a slow precipitation from a hot dilute solution. At the moment of its formation the precipitated compound might be slightly soluble; on gradually lowering the temperature, the separation increases, with a minimum of occlusion. It is also recommended that the precipitating solution be added drop by drop, with vigorous stirring. The best procedure is to filter the precipitate, to redissolve it, and then to reprecipitate it in the absence of contaminants. The amount of reagent added is in slight excess over the equivalent amount. Assuming that the precipitate is slightly soluble, the excess reagent insures more nearly complete precipitation by exceeding the value of the solubility product.

In the process of precipitation from homogeneous solution, the concentration gradients encountered in ordinary precipitation are avoided. Not only is the coprecipitation of interfering elements much less, but also the precipitate obtained is dense and compact, and readily filtered and washed. Urea is considered an almost ideal reagent for hydrolytic processes; the hydrolysis takes place evenly throughout the solution or reaction medium which is now homogeneous. The reaction for the hydrolysis is:

$$NH_2 \cdot CO \cdot NH_2 + H_2O \rightarrow 2NH_3 + CO_2$$

The method has been extended into a two-stage precipitation, in which the precipitation of the bulk of the element is precipitated but not all of it, by interrupting the hydrolysis by cooling. After filtration, the small residual balance is precipitated by warming again. Contamination by coprecipitation is almost completely avoided.

In a suspension, Stokes' law relates the size of the solid particle to its rate of fall under the influence of gravity in the liquid medium. The viscosity of the medium is also considered. Similarly, the velocity (constant) of fall of a spherical droplet (which must be larger than the mean free path of the molecules of the gas) through a gas with viscosity η is given by the expression

$$v = \frac{2}{9} \frac{ga^2}{\eta} \sigma$$

in which σ is the density of the droplet, and a its radius. With an observed steady rate of fall of small solid spheres of density d_1 of known dimensions, in a liquid of density d_2, the viscosity η of the latter may be computed by the relation:

$$V = \frac{2ga^2}{9\eta} (d_1 - d_2).$$

ELBERT C. WEAVER

PRINTING INKS

Printing ink is a colored fluid or paste composition capable of being deposited in thin films which reproduce a desired image by any one of several printing processes. A printing ink usually consists of one or more pigments dispersed in a vehicle which (except for newsinks) contains an organic polymer designated as the binder. The vehicle usually contains other ingredients such as solvents, oils and driers. Therefore, in some respects printing inks are similar to other disperse systems.

The composition of printing inks varies widely depending upon the printing method to be employed, the specific type of printing press and press speed, the surface to be printed, and conditions to which the printed matter is to be subjected during use. Virtually all printing inks, however, are dispersions of solids in a polymeric vehicle. The major functions of the pigments are to provide the desired opacity and color to the printed image. Other solids referred to as extenders may be used to increase the viscosity and pseudoplastic behavior of the ink and to sharpen the image definition during printing. The continuous phase or vehicle may contain several components including organic polymers to bind the colorant to the stock, solvents to carry both binder and pigments, organic metallic salts to catalyze oxidative drying where this occurs, and a variety of functional organic and inorganic compounds.

One classification for printing inks is based on the mechanism of drying the printed ink film. The choice of vehicle system is determined by the reproduction process, speed of printing, type of stock and end use requirements of the printed matter. One or a combination of the following mechanisms can be involved in forming the solid film: absorption of vehicle, oxidation, precipitation (by water or wax), absorption of radiation (infrared or ultraviolet) and volatization of solvent. Letterpress inks for newsprint dry by absorption of the vehicle in the porous stock and selective absorption of solvent rich phases from quick-set ink vehicles precipitate a resin-rich phase on the substrate surface. Oxidative drying inks are the oldest and depend on a broad spectrum of linseed, alkyd and phenolic vehicles (see Drying oils, Driers). Commercial printing and books are a few examples of printed matter produced with these systems. Inks which set or dry by precipitation are based on the insolubility of the resin binder after some new component is introduced into or picked up by the vehicle immediately after impression. Typical examples are steam-, moisture- and wax-set inks which frequently are used in food packaging. Gravure, flexographic and the classical form of heat-set inks dry rapidly as solvents are removed by evaporation under infrared and hot-air sources. Newly developed heatset inks with minimal or no solvent which in one case cross-link at elevated temperatures, or in another case, on exposure to ultraviolet radiation, provide drying systems useful for both sheet and web processes and should eliminate air pollution. Applications include publications, catalogs, brochures and record jackets.

Another classification usually applied to letterpress and lithographic inks includes scratch-proof and gloss inks in which the compositions of the vehicles are modified to provide the desired properties in the printed films.

Letterpress Inks. The most common and familiar process is letterpress wherein the inks are printed from surfaces raised above the plate level. The first requirement of a letterpress and most other printing inks is that they be capable of transfering from the ink fountain to distribution rolls. Here the ink must form a thin uniform film which on the final inking roll transfers to the printing portions of the plate. The ink film remains on the elevated surfaces rather than flowing away because of its rheological properties. Letterpress and lithographic inks are rheologically described as highly viscous and pseudoplastic; that is, an abnormally large resistance is developed to low shearing stresses and this resistance decreases as the shear rate is increased. The ink is then transferred directly from the plate to the surface of the stock. The type of press, the speed of printing, and the properties of the stock may require modification of the ink properties. In general, the higher the press speed, the lower the viscosity of the ink.

Flexographic Inks. Flexographic inks are distinguished by their low viscosity, volatility of much of the vehicle and the fact that they are printed from flexible rubber plates. Originally named "aniline inks" due to the initial use of dyes derived from aniline intermediates, the name has been changed in keeping with new formulations employing pigments and an expanded range of applications. Flexible plastic, or more usually, rubber printing plates make the process well adapted for rapid web printing on relatively simple presses of packaging materials including metallic foils, paper and plastic films. Economical rubber plates, rapid drying and absence of odors in the final printed product are the major advantages of the flexographic process.

Flexographic inks consist of a film-forming binder, a plasticizer, a solvent, and colorants of pigments, dyes or both. The choice of solvents is limited to those which will not affect the rubber printing plate; consequently alcohol, alcohol-hydrocarbon, glycol, glycol-ethers, and water are in widespread use. The film former used must be compatible with one or more of the above solvents, permit rapid solvent evaporation, and provide a tack-free, flexible film which adheres firmly to the substrate.

The expanding use of printed plastic film has caused original use of shellac or alcohol soluble nitrocellulose to broaden and now includes zein, ethyl cellulose, maleic resins, urea resins, polyamide resins, rosins, acrylic resins, water soluble resins, and proteins. Dyes are frequently used in flexographic inks when the original high sheen of metallic foils is to be retained in the printed areas. Pigments impart varying degrees of opacity to flexographic inks; consequently an opaque white ink is often used as a base upon which the other colors are printed.

Lithographic Inks. The planographic process derives its name from the fact that the ink is transferred from a plane surface, the printing areas of which are wet by the ink while the nonprinting areas are kept water wet and hence are ink-repellant. In the antiquated stone or direct lithography, the porous surface is made ink receptive with a greasy crayon or a suitable paper transfer. The nonprinting areas are made water-receptive and ink-repellant with gum arabic. Dampening rollers deposit just enough moisture on the nonprinting areas to insure deposition of ink from the following inking rollers only on the printing areas of the plate. The ink is then transferred directly from the stone to the paper.

Offset lithography, on the other hand, involves one transfer from a printing plate to a resilient rubber blanket, and a second transfer from the rubber blanket to the stock. The plate is curved to fit circumferentially on the first of three cylinders of identical diameters. The plate cylinder contacts the blanket cylinder, which in turn transfers the ink to the stock at the line of contact between blanket blanket and impression cylinders. Metal plates are prepared by a variety of processes, the majority of which depend on (1) surface preparation, (2) deposition of a photosensitive film, (3) exposure to a negative or positive version of the desired image, and (4) specific chemical or electrochemical treatments of the image and nonimage areas to impart ink receptivity and repellancy, respectively.

The lithographic process makes several distinctive demands on the ink. The thin ink films deposited require a higher than usual tinctorial strength in the ink. The pigment must not leave the ink to tint the slightly acidic fountain solution, and only a limited amount of fountain solution should emulsify in the ink. The ink should not show any tendency to wet the nonprinting areas of the plate. Lithographic printing inks dry by oxidation, absorption, penetration and heat-set processes. Lithographic ink vehicles vary in composition extensively depending on the drying processes, press speed and end use requirements. Alkyd and phenolic resin combinations provide the base for oxidative drying systems, resin-solvents combinations form quick- and heat-set vehicles, polyesters, epoxies or urethane resins constitute vehicles for thermally catalyzed heat-set vehicles and ethylenically unsaturated esters plus photoinitiators are representative reactive materials for ultraviolet or electron beam drying ink vehicles.

Printing of tin plate for containers is the largest single application of the letterpress offset process due to the ability of the rubber offset blanket to conform to the tin plate. This process imposes the additional requirements on the inks that they be capable of withstanding the baking temperatures of drying, the mechanical stresses of the forming, crimping and soldering without adhesion failure, chipping or discoloration and finally that they withstand pasteurization temperatures without change in color.

Gravure Inks. Intaglio or gravure printing in which the ink is transferred to the stock from etched or engraved recesses in a metal plate probably offers the broadest possible range of combinations of stocks and inks. Used in the printing of such diverse products as money, newspaper supplements, textiles and plastic films, cigarettes and food packaging, the process is most efficient for extended printing runs in which the cost of the relatively expensive engraved or etched plates is justified. The stock must be sufficiently flexible to be impressed into the plate surface recesses where the ink is contacted and removed. The ink film transferred can be significantly thicker than those from the previously described printing processes.

Inks for copper and steels plate engravings are formulated to enable the ink to be distributed on the plate mechanically or by hand, and then have the excess completely removed from the surface, again by manual operations. The ink must be non-abrasive, may not bleed in water in those operations where the paper stock is predampened, and must wipe clean. The inks dry by penetration and oxidation.

The much larger application of gravure inks is in the field of rotogravure in which the ink is of sufficiently low viscosity to flow into the multitude of microscopic ink cups on the plate surface as it revolves in the ink pan. The excess ink is removed from the surface by an oscillating steel doctor blade. Rotogravure inks dry by penetration and/or evaporation of the solvent. The broad range of resin-solvent combinations which can be employed in the process has led to the following classification system of the inks dependent on the major resin component or solvents:

Type A	hydrocarbon-soluble resins
C	nitrocellulose
E	alcohol soluble resin derivatives
T	chlorinated rubber
W	water-based
X	miscellaneous

During printing and in the end use of the printed film a variety of demands are made of the ink including rapid solvent release, adhesion to the stock, flexibility, rub resistance, and resistance to a variety of packaged materials.

Silk Screen. Silk screen printing is a comparatively simple process based on the direct reproduction of a stencil image by pressing ink by means of a rubber squeegee through a mesh screen directly onto the stock. The process is particularly well adapted to short runs or printing irregular shaped articles. The ink film thickness deposited depends upon the rheological properties of the ink and the size of the screen mesh: the larger the mesh the thicker the film. In general, these films are much thicker than obtained from any other printing process. The inks have to pass through the mesh readily without plugging, and release the screen easily and cleanly after printing. The inks have to be short and buttery for these purposes. A broad range of vehicles is available for use in the inks since metallic and plastic screens have largely supplanted the original silk screens.

Printing ink production is almost always a batch process, and the equipment employed usually depends upon the viscosity of the product. Letterpress, lithographic and silk screen inks are usually highly viscous and thixotropic. Pigments are purchased either predispersed in a vehicle or as dry powder. The latter are wet by the vehicle in a variable-speed impeller or change-can mixer which also disperse the pigments to a considerable degree. Final dispersion is accomplished on a three roll mill; the number of passes required is determined by the viscosity of the ink, the pigments used, and the desired degree of dispersion. The dispersion is usually evaluated on a fineness-of-grind-gage. An ink film is drawn down a calibrated, tapered path of diminishing depth. The method of reading is standardized as to the film thickness at which a predetermined arbitrary number of parallel scratches produced by oversized particles are observed.

Dispersions for lower viscosity inks, including flexographic and rotogravure inks, are usually prepared in a concentrated form in a ball, sand or shot mill. The mill charge, usually a premix, contains the pigment, the resin binder, the plasticizer, and some of the solvent. After completion of the dispersion process the balance of the solvent can be slowly added with less intensive agitation.

Colored Inks. The impact of color is nowhere more evident than in printed media. Color in printing inks is obtained from pigments which can be relatively opaque or transparent. Inks containing opaque pigments tend to hide the surfaces on which they are printed, and reflect those wave lengths which are not absorbed by the pigments. Transparent pigments, on the other hand, resemble optical filters in that the color of the original surface contributes to the color of the final printed image. Multicolor printing depends on the superimposition and juxtaposition of dots of two or more differently colored transparent ink films. In four-color process printing, for example, the yellow ink may be printed first, magenta ink second, cyan ink third, and the black ink fourth. Judicious choice of the colored inks that are printed one over the other and per cent of the area covered by each ink (determined by the printing plates) provides a wide range of hues and tones to reproduce the original subject.

In high-speed multicolor printing one ink is printed over a previously printed film without completely drying the ink first printed. The process depends on the ink film first printed being sufficiently cohesive that while the second film printed will adhere to the first, the second film will always split internally rather than in the first film. A similar relationship is required of the ink films printed in the third and fourth impressions. This phenomenon in multicolor priiting is referred to as "trapping."

WM. D. SCHAEFFER

Cross-references: *Rheology; Paints.*

PROMOTERS

A promoter is an important constituent of most commercial catalysts. It may be defined as a substance added during catalyst preparation which by itself has little catalytic effect but which when added to the catalyst in small amount imparts either better activity, stability or selectivity for the desired reaction than is realized without it. Since there are many ways in which a promoter can function, there are many types of promoters.

Structural Promoters. Structural promoters increase the surface area of the active component. Usually this also involves increasing catalyst stability by inhibiting loss of surface during usage. The classic case is the promotional action of alumina in iron synthetic ammonia catalyst. Structural promotion can be studied by means of surface area, x-ray diffraction, electron microscopy as well as by catalytic activity. The increase in structural stability can probably be related in most cases to an increase in the melting point of the active component as a result of the presence of the promoter. This can occur when the promoter forms a solid solution with the active component. In some cases a structural promoter may stabilize the structure of the carrier and only indirectly affect the surface area of the active component.

Selectivity Promoters. In some cases where more than one reaction is possible, it is desirable to have a selectivity promoter to guide the reaction along the proper path or prevent further reaction. More often than not, this probably involves poisoning the sites active for undesired side or secondary reactions. A probable example of this is the usage of potassium with chromia on alumina (dehydrocyclization) or iron (Fischer-Tropsch synthesis). In these cases the potassium possibly acts by poisoning the sites that are active in cracking (acid groups) and thereby decreases the gas yield. Thus, what is an inhibitor for one reaction may be a selectivity promoter for another.

In some cases where the desired reaction product can undergo further reaction to give poor yields (for example, naphthalene oxidation with a phthalic anhydride product), it is undesirable to have a very active fine-pore structure catalyst. Naphthalene molecules getting into such a pore system are very likely to be oxidized further than phthalic anhydride, resulting in poor yields. The promoter in such a case may simply fill up some of the fine-pore space leading to less opportunity for further reaction of the desired product.

Electronic Promoters. Reactions on metallic surfaces involving hydrogen have been shown to be related to the electronic character of the metallic system. Metals having many vacant orbitals or "holes" and having a high attraction for additional electrons (e.g., tantalum) strongly adsorb hydrogen, and the electrons from the hydrogen probably become part of the electronic system of the bulk metal. Metals without empty orbitals (copper and gold) have a low attraction for hydrogen and do not adsorb it strongly in the pure state. The first is a poor hydrogenation catalyst, probably because the adsorbed hydrogen is held too tightly to be available for reaction. The relatively poor catalytic power of unpromoted metals without empty orbitals can be attributed to lack of adsorbed hydrogen. Highest catalyst activity is realized with metals such as nickel and platinum, which have a few empty orbitals so that hydrogen is adsorbed but can be readily released to the other reactant. If a foreign substance is added, this will affect the number of empty orbitals and, therefore, the catalytic activity. If it improves the activity more than expected from averaging, it may be considered to be an electronic promoter.

Lattice Defect Promoters. The active centers of many oxide catalysts are thought to be located at lattice defects that occur near the surface. A small amount of impurity (promoter) can have a very large effect on the number of lattice defects, since each interstitial foreign atom may be the center of a lattice defect which extends for 10 atomic diameters or more. For interstitial substitution to occur, it is usually necessary that the foreign ion be about the same size as the one it replaces. Related to defect formation are the electrical conductivity and the valence state. Small amounts of promoters may affect the electrical conductivity of semiconductors many fold, and even change the stable valence state of neighboring atoms if the valence state differs from the atom it replaces. The dissociation pressure of the oxide may also be affected. This type of promotion can be studied by measurement of valence state and electrical conductivity, as well as by catalytic performance.

Adlineation Promoters. It is reasonable to expect that boundary zones between phases or crystals will have a catalytic activity different from the bulk phases. A promoter may act by creating or increasing the number of such interfaces. This type of promotion is discussed at length by Berkman *et al.*

Diffusion Promoters. In most commercial catalytic processes, the catalyst is employed in the form of granules, pellets or rings, since high pressure drops lead to channelling and high gas compression costs. Largely as a result of the use of large particle-size catalysts, diffusion into particles is often a rate determining factor particularly at high temperatures. The diffusion characteristics may be improved by incorporation of a diffusion promoter. Such a promoter must be capable of decreasing the resistance to diffusive flow without appreciably harming the physical strength or other catalyst properties. A system of interconnected macropores as well as micropores is preferable for obtaining both low diffusion resistance and high surface area. Examples include organic materials that burn out to leave a porous structure, diatomaceous or fullers earth and hydrous gels.

Phase Change Promoters. Catalyst surfaces must be considered as dynamic systems undergoing very rapid transitions. In many cases, such as oxidation catalysts, the catalytic action itself is probably dependent on the ability of local points on the surface to change back and forth between different oxidation states and crystal structures if other than surface atoms participate. Hence, a possible mechanism for promotion is by aiding oxidation state transitions. Transition rates might be utilized to study this type of promotion.

Dual Action Promoters. In some cases of catalysis more than one reaction must be catalyzed to achieve the over-all result. In such cases the promoter may act to catalyze one of the reactions. The best example of this dual action is the isomerization of paraffins with a nickel-silica-alumina gel catalyst. Neither the nickel on other carriers, such as silica gel, or silica-alumina gel by itself are active, but in combination high activity is achieved. Apparently, active centers for both hydrogen and proton transfer are needed for this reaction. The alumina in this case could be considered as a dual action type of promoter. It is probable that there are many cases of promotion of this type, although few have been established.

W. B. INNES

References

Moss, R. L., *Chem. Eng.* (*London*), No. 109; *Chem. abs.*, 114–141 (1966).

PROPELLANTS

Propellants are fuels which contain within themselves the necessary ingredient (i.e., oxidizer) for combustion or conversion of potential energy into useful or kinetic energy. When the fuel and oxidizer components exist in the same molecule, such as in nitromethane or nitrocellulose, or as an intimate mixture such as methyl alcohol and hydrogen peroxide, it is called a monopropellant. In bipropellant systems the fuel and oxidizer are stored separately until they are mixed in the combustion chamber.

On the basis of their physical state propellants are classified as liquid or solids. (Gaseous systems as such have found little practical use.) Application of liquids is restricted almost entirely to rockets while solid propellants are also used in guns and cannons. Liquids are sprayed into a combustion chamber where burning occurs; solids are formed into tubular or cylindrical shapes and burn on all exposed surfaces. Liquid propellants can exist as either mono- or bipropellants, while solid propellants are considered to be monopropellants. A third class of propellants, still in an experimental state of development, combines solid and liquid ingredients into a hybrid system. The solid is usually the fuel upon which the liquid oxidizer is sprayed.

The two conflicting requirements for monopropellants—that they be stable for storage yet readily combustible without added oxygen—have limited the number of liquids suitable for use. Examples of useful homogeneous liquid monopropellants include nitromethane, ethylene oxide and hydrazine. Mixtures of compounds include hydrogen peroxide and alcohol and ammonia and ammonium hydroxide. These systems are stable at ordinary temperatures, but react when heated under pressure or in the presence of a catalyst to give hot combustion gases.

The most common liquid systems are the bipropellants. To differentiate liquefied gases such as oxygen, hydrogen and ammonia from ordinary liquids like hydrazine and kerosene the terms "cryogenic" (i.e., refrigerated) and "storable," respectively, are often used. Systems which ignite spontaneously when mixed are called hypergolic. Nonhypergolic systems require an auxiliary ignition device.

The method by which the potential energy of propellants is converted to kinetic energy is different for rocket systems than for guns. The projectile in a gun is accelerated solely by the pressure of the combustion gases created by the burning propellant. In a rocket the gases are ejected at high velocities through a nozzle and the momentum change thus created provides thrust to the rocket. Performance of a propellant in either case is normally described in terms of the amount of energy released per unit weight of propellant. The gun ballistician uses the term "impetus" which he defines by means of data obtained in a closed chamber. Impetus has the units of foot-pounds/pound. In rockets, performance is normally described in terms of the amount of thrust developed per unit weight (or unit weight flow) of propellant. This characteristic, called specific impulse, is generally expressed as pound-seconds per pound or simply as seconds. If the combustion gases are considered ideal, specific impulse can be approximately defined by an equation derived in part from the ideal gas law. For comparative purposes, then, performance of both gun and rocket propellants is related to the identity RT/M and maximum performance from a propellant can thus be assumed to be achieved by maximizing the flame temperature and minimizing the molecular weight of the combustion gases.

Use of propellants possessing very high flame temperatures is restricted somewhat by the fact that dissociation in the combustion gases, which causes energy losses, increases with temperature as do also the problems associated with materials of construction (e.g., nozzle throat and gun bore erosion). The desire for low molecular weight accounts, in part, for the extensive use of chemical compounds containing elements of low atomic weight such as hydrogen, carbon and nitrogen and the attempt to maintain as much hydrogen as possible in the exhaust gases.

In most propellant systems the oxidizer is the most important ingredient. In liquids, oxidizers fall into two main classes—those based on oxygen and those based on the halogens (of which fluorine possesses the most useful properties). The more common oxygen-based oxidizers are liquid oxygen, nitric acid and hydrogen peroxide. Cryogenic liquid oxygen has an extremely low boiling point ($-297°F$) and consequently presents difficult storage and handling problems. Nitric acid is used as white fuming or red fuming, depending upon the amount of dissolved NO_2 present. A small amount of HF is frequently added to reduce corrosion problems. Hydrogen peroxide is used in concentrations to 100 per cent, although 90 per cent is more common. Since this oxidizer reacts with many materials, it also presents problems in storage and handling.

Of the fluorine-based oxidizers, cryogenic liquid fluorine provides highest performance. It has a low boiling point ($-306°F$) and is extremely reactive with most fuels. It is also very toxic and reacts with many common metals. Chlorine trifluoride and oxygen difluoride are other fluorine-based oxidizers of importance.

An almost limitless number of liquid compounds can serve as fuels. Included are almost all oxygen-deficient organic compounds as well as compounds containing inorganic hydrogen. As has already been noted hydrogen (i.e., liquid hydrogen) is the ideal chemical fuel. Its low density, however, poses a problem in storage and handling. Inorganic hydrogen combined with nitrogen results in ammonia or hydrazine, both possessing good performance characteristics (hydrazine is also storable). When combined with the light metals hydrogen gives hydrides with lithium, beryllium, boron and aluminum. Except for some of the borohydrides all are solids. The hydrides of boron have been tried (but without much success) in both rocket and turbojet (air-aspirated) systems.

Organic hydrogen when combined with carbon, oxygen and nitrogen leads to a multitudious array of organic chemicals, including saturated and unsaturated hydrocarbons, alcohols, amines, aromatics, ethers and nitroparaffins. Gasoline and petroleum fractions have found wide usage as rocket fuels, among which the special grade of kerosene identified as RP-1 is the most frequently used.

The early solid propellants exemplified by nitrocellulose-type compositions are called homogeneous propellants and are true monopropellants, since each molecule contains the necessary fuel and oxidizer components for combustion. About 1942 a second group called composite propellants was developed. They consist of mixtures of fuel and oxidizer in which the oxidizer is a finely divided crystalline material and the fuel is plastic in nature and serves to bind the mixture together into a solid structure. In recent years both solid oxidizers and metallic fuels have been incorporated into nitrocellulose or other energetic, homogeneous binder systems to produce so-called composite-modified propellants.

Homogeneous solid propellants have been and are still widely used in guns and cannons and in many small rockets. Early compositions contained only nitrocellulose and a stabilizer such as diphenylamine. Nitroglycerine was later added to increase energy level, leading to the "double-base" types of propellant. More recent formulations may contain as many as a dozen ingredients, each having a specific role in controlling performance. Unlike liquid propellants where rate of reaction can be controlled by a valve, mass consumption rate in solids can be changed only by variations in composition and in exposed burning surface. Performance is thus "built in."

Current composite solid propellant systems are identified primarily by their fuel binder. The first composite propellant was composed of potassium perchlorate in asphalt and found use in early jet-assist take-off rockets (JATO's) for aircraft. The low melting point of asphalt and its tendency to flow during storage, however, made its replacement desirable. The styrenes and methacrylates were used but were found to be too brittle. Synthetic rubbers—to remain even today as the major class of binder materials—soon replaced the plastics. A liquid polysulfide which could be cured with a catalyst was the first of the synthetics to be used. Later, the more versatile polyurethane rubbers, formed by the condensation polymerization of a diol and an isocyanate, came into use. However, to meet the severe structural requirements for military weapons, new synthetic rubbers were developed. Compounds based on polybutadiene were found to offer the best properties. The first one was a copolymer of polybutadiene and acrylic acid. Later acrylonitrile was added. Currently, a carboxy-terminated polybutadiene, cured with imine or epoxy curatives, appears to provide the best balance in overall performance.

Since propellants made with a synthetic rubber binder retain their rubbery characteristics over a relatively wide range of temperatures, most can be bonded directly to the rocket case. This is in contrast to the rigid binders or to the thermoplastic binders which must first be formed into grains, covered on their outer surface with a nonburning material and then loaded into the rocket case as "free-standing" grains. Most composite solid propellants that contain ammonium perchlorate oxidizer are manufactured by the casting process. The liquid prepolymer, plasticizers, curative, auxiliary fuel, ballistic catalyst and oxidizer are all mixed in a horizontal or vertical mixer. When well blended the mixture is poured directly into the rocket case. The cast motor is placed in an oven at a fairly high temperature (130–190°F) where the cure reaction takes place.

Most ammonium nitrate oxidized composite propellants are manufactured by either a molding or an extrusion process. This is necessary because useful compositions must contain about 80 weight per cent ammonium nitrate and such compositions are too viscous to be cast. The final propellant grains are generally used as free-standing grains, since they do not possess the mechanical properties required for a case-bonded grain.

Propellants have in the past been used principally by the military. Advent of the space age is shifting emphasis in that direction. Selection of one type of propellant over another depends in many cases on personal preference and on availability. A need for instant readiness has probably been the major factor in stimulating extensive use of solid propellants in military weapons. The controlled operation and stop-start capability of liquid rockets has certainly influenced their selection for space missions. At the present time both classes of propellants offer about the same overall performance. Liquid engines have complex pumping, metering and control systems but are controllable. Solid engines must have their performance built into their composition and structure. Once started they cannot be readily stopped and restarted. However, the engine is simple and possesses high reliability.

Industrial application of propellants has been small quantity-wise compared to the military and has been restricted almost entirely to solids. Industrial tools such as stud setting devices, oil-well perforating guns and industrial cannon for quarries are a few examples. Sporting ammunition uses a considerable quantity of solid propellants. However, an inherent, inborn fear of explosive materials by the general public will probably always keep civilian uses of propellants to the present low level.

FRANCIS A. WARREN

Cross-references: *Hydrazine; Oxygen, Peroxide; Fluorine; Hydrogen.*

PROSTAGLANDINS

The prostaglandins are a group of fatty acid derived molecules containing 20 carbon atoms. They have been detected in practically every part and organ in the human body with especially high concentrations in seminal fluid. They are very active biologically, causing smooth muscle preparations to contract at concentrations of 10^{-9} g/ml and having definite effects on human blood pressure at doses as low as 0.1 microgram per kilogram of body weight.

Chemically characterized in 1960, they are very new materials on the scientific time scale. In the short period of a decade, they have arrived at the forefront of pharmaceutical and biological research. More than 1500 references have appeared in the scientific literature, three-fourths of which have been published during the past three years.

The structures of the basic series of compounds are illustrated on top of page 922.

The systematic naming of these materials is based on the C_{20} prostanoic acid skeleton. The numbering system starts with the carboxyl carbon designated as number one and proceeds around the chain to the terminal methyl group. This is illustrated for structure *1*. The trivial names for these compounds begin with the letters PG. A third letter distinguishes the four basic series illustrated as *1–4* above and a subscript refers to the number of additional double bonds present. The PGE_2 series has another cis-double bonds at carbon 5 and the PGE_3 series has two additional cis- double bonds at positions 5 and 17.

A description of the biological activity of the prostaglandins dates back to 1930 when it was shown that human seminal plasma could cause either relaxation or excitation of the uterus. It was also noted that the effects observed could be correlated with a woman's past history of fertility. In 1933–34 two groups independently showed that these effects were caused by a new class of lipid materials different from anything known at the time. The topic lay more or less dormant until 1947 when research was again started in this area and rapid progress was made after 1956. In 1960, two prostaglandins, PGE_1 *1*, and PGF_1 *2*, were isolated in crystalline form and in 1962 their chemical structures were published.

The structure elucidation of the prostaglandins is a classic. With only a few milligrams of material, Swedish workers were able to determine the structure, with the exception of one stereochemical question. The relative configuration was unequivocally demonstrated by x-ray crystallography. The chemical work utilized sophisticated mass spectrometry and combination mass spectrometry-gas chromatography methods combined with careful isotopic labeling experiments. This was one of the first successful examples of the use of these combined techniques in organic chemistry.

The biosynthesis of the prostaglandins was elucidated simultaneously by two groups of research workers in 1964. The fact that certain C_{20} unsaturated fatty acids were essential body chemicals was known. It was postulated that these acids were the precursors of the prostaglandins. When dihomo-γ-linolenic or arachadonic acid was in-

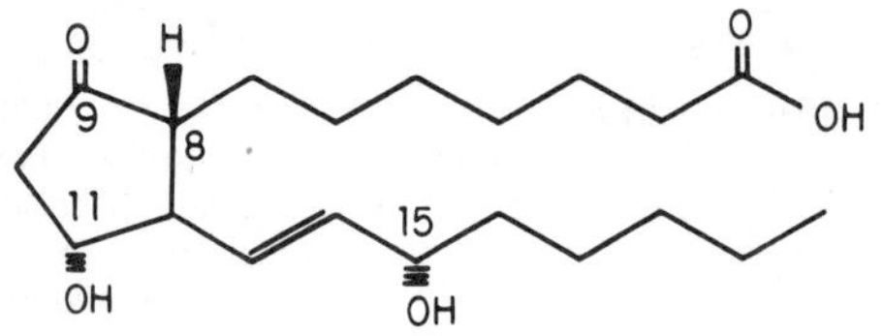

PGE₁ *1*

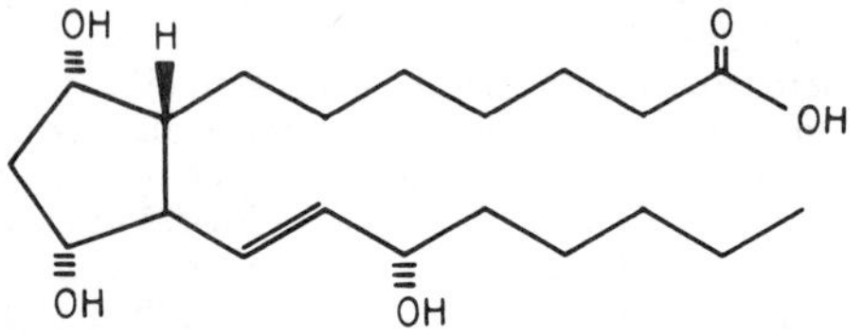

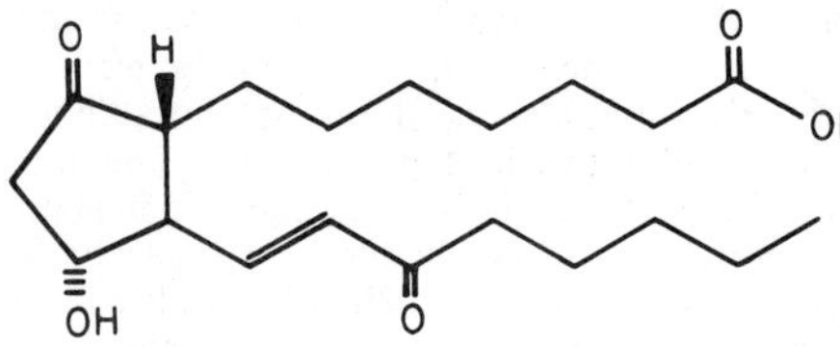

5

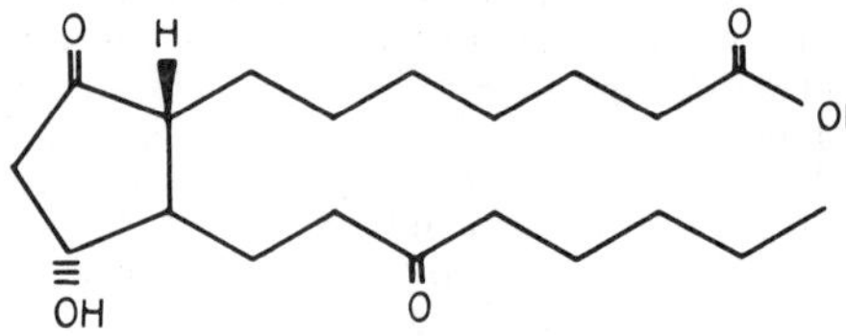

PGF₁ *2*

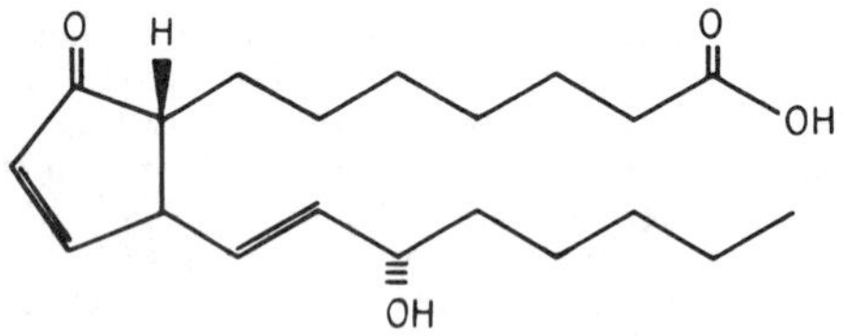

6

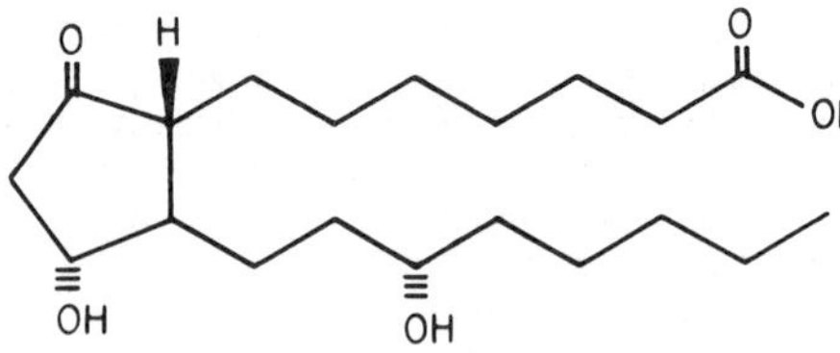

PGA₁ *3*

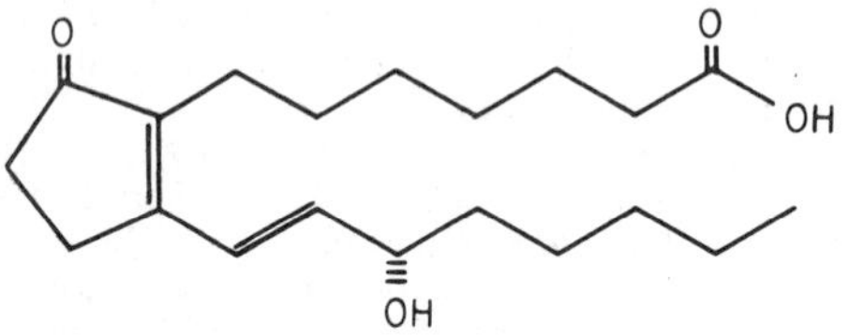

7

PGB₁ *4*

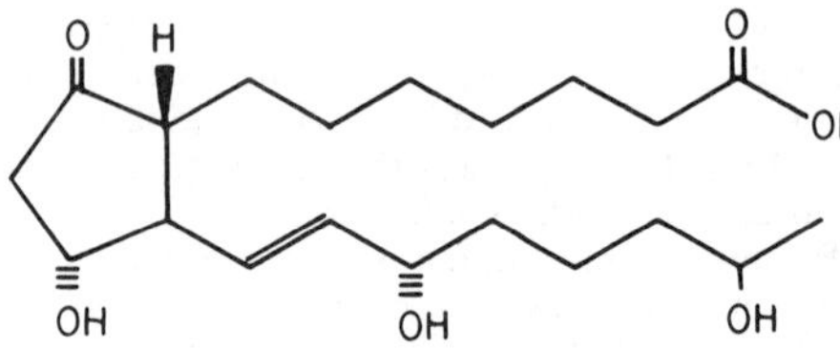

8

cubated with sheep seminal vesicles with the proper co-factors, E_1 and E_2 series prostaglandins, respectively, were produced. The mechanism of this reaction has been determined by clever use of pure $^{18}O_2$ or $^{16}O_2$ and known mixtures of these oxygen isotopes. Mass spectral data showed that the oxygens at C_9 and C_{11} in the five-membered ring were derived from the same molecule of gaseous oxygen. The postulated endo peroxide intermediate then reacts further to give either E or F series prostaglandins. There does not seem to be interconversion of the two series at the site of biosynthesis. A Dutch group has extended this finding and has synthesized unnatural prostaglandins using different, well chosen substrates.

The metabolism of the prostaglandins has been worked out in some detail especially for the guinea pig, but the work is far from complete. The prostaglandins rapidly disappear from the circulation. Approximately 90% of exogenous PGE₁ *1* is metabolized upon one pass through the lungs and/or liver. The primary metabolites formed in the largest quantities seem to be PGB₁ *4*, 15-keto PGE₁ *5*, 15-keto dihydro PGE₁ *6*, 13,14 dihydro PGE₁ *7*, and 19-hydroxy PGE₁ *8*. It is not known for certain whether the PGAs are metabolites or are products of biosynthesis. Recently

it has been shown that PGE₂ can be metabolized to PGF₂ in the liver. Compounds *7* and *8* are active biologically, however the rest are generally inactive. All of the prostaglandin type compounds can be further metabolized in a fashion which normally occurs with fatty acids.

The most recently discovered source of prostaglandin is the seawhip coral, *Plexaura homomalla,* which grows in abundance throughout the Caribbean area. Approximately 1.5% of this organism is PGA₂, 15-epi PGA₂, or their derivatives. The function of the prostaglandins in the gorgonian is not known.

Biosynthesis also supplies both natural materials and analogs. The precursor fatty acids are readily available, but sheep seminal vesicles, the best source of the necessary enzymes at present, are in short supply. Alternative sources of the enzyme and even more efficient chemical syntheses are certain to be announced soon, since research in this area is intense.

Several chemical syntheses of the prostaglandins and their metabolites or derivatives have been accomplished. In 1970, the synthesis of optically active PGE₂ was announced. Chemical modification and synthetic methods are being investigated in many pharmaceutical and university laboratories.

The hope is not only to find cheaper sources of starting materials, but to find prostaglandin derivatives or analogues that are more specific and more potent in their action.

Since the greatest natural concentration of these materials in humans is in seminal plasma, it would seem logical to assume that the prostaglandins play an important role in reproduction. It has been postulated, but not proved, that the prostaglandins are necessary for erection and ejaculation in the male. There are indications that one cause of male infertility is the lack of sufficient E_1 type prostaglandins.

The female reproductive organs are very sensitive to the prostaglandins. In vitro, the PGE compounds generally inhibit nonpregnant uterine contractions whereas F type materials have the opposite effect. Vaginal absorption into the blood stream of the prostaglandins has been demonstrated using radioactive labeled compounds. The muscle relaxant effects of these materials may be helpful for the transport of sperm to fertilize the ovum. There is strong evidence that these materials are involved in normal menstruation and parturition. High levels are present in the menstrual and amniotic fluids, respectively, during these events. It has been postulated that the symptoms of primary dysmenorrhea are due to an excess of prostaglandins.

The second dramatic effect of the prostaglandins is in the cardiovascular area. As mentioned previously, PGE_1 (1) can cause a significant drop in blood pressure when given intravenously at doses as low as 0.1 microgram per kilogram. Interestingly, the F series compounds cause the opposite effect, i.e., they elevate blood pressure. The only biological effects of the A series prostaglandins are blood pressure lowering and nateuresis. A considerable research effort has been expended to investigate these effects; however, the short duration of action as well as a rebound effect still prove to be limiting factors.

Another most interesting and biologically important facet of the prostaglandins is their relationship with cAMP (cyclic adenosine 3′, 5′ monophosphate) . cAMP mediates the activity of many hormones and has been termed the "third messenger" by some investigators. A partial list of these hormones includes adrenalin, vasopressin, luteinizing hormone, gastrin, glucagon, ACTH, thyroid stimulating hormone, serotonin, acetylcholine and histamine. These materials in turn influence fat mobilization, formation and metabolism of steroids, gastrin secretion, membrane permeability, kidney and central nervous system functions. The mechanism postulated for the interaction of the prostaglandins in this system is that they inhibit adenyl cyclase, the enzyme that converts ATP into cAMP. A further refinement of this mechanism is the negative feedback hypothesis, which means one of the hormones listed above initiates cAMP formation, the cAMP then initiates prostaglandin synthesis, which in turn inhibits the formation of more cAMP. While this theory has not been proved, research in this area is being aggresively pursued since it could provide an understanding of the cybernetic control of human biochemical phenomena at a basic level.

Recently a number of prostaglandin inhibitors have been announced. Several materials are now known that seem to interfere with the action of various prostaglandins at their target enzymes or receptor sites. Inhibitors of the synthesis of prostaglandins from its precursors are also available. Most of the latter materials are fatty acids, similar to the natural substrates provided by the body. A very important exception to this generalization are the salicylates, especially aspirin. Recent articles indicate that most of the effects, both positive and negative, of these salicylates and a number of other anti-inflammatory drugs, can be explained by invoking this blocking mechanism. This discovery could provide a long-sought biological rationale for the pharmacological activity of these useful chemicals and others of a similar nature.

The prostaglandins have biological effects in numerous other areas. Smooth muscle tissues other than the reproductive system are affected. Included in this list are blood vessels, arteries, pulmonary tissue, respiratory muscles, the eye, the spleen and gastrointestinal smooth muscles. They also affect the nervous system, both centrally and peripherally, are intimately connected with calcium metabolism and have some insulin-like properties.

The prostaglandins have been investigated clinically in humans in several areas. They have been used to induce abortion even in the second trimester of pregnancy by either intravenous or intravaginal administration. Induction of normal labor at term has also been accomplished with PGE_2 alone or after pretreatment with oxytocin. The prostaglandins have been tested in humans as nasal decongestants and in the treatment of asthma. There is also a strong indication that some types of ulcers can be treated using these materials.

The prostaglandins still have drawbacks in practice. The PGE series compounds are chemically unstable and in most cases must be administered to the patient intravenously. Side effects such as diarrhea can be a problem also. These difficulties are certain to be overcome soon since many university and pharmaceutical company laboratories are involved in research on related new compounds and on novel methods of administration. With all this in mind it is still safe to predict that the prostaglandins will assume an important position in medicine at least equivalent to that of the steroid hormones today.

RICHARD A. MUELLER

References

Bergström, S., and Samuelsson, B., Editors, "Prostaglandins," Proceedings of the 2nd Nobel Symposium, Interscience, New York, 1967.

Ramwell, P., and Shaw, J., Editors, "Prostaglandin Symposium of the Worcester Foundation for Experimental Biology," Interscience, New York, 1968.

Heinzeman, R. V., Editor, "Annual Reports in Medicinal Chemistry, Vol. 7," Academic Press, New York, 1972.

Chem. Eng. News, **50,** No. 42, 12-13 (Oct. 16, 1972).

Hinman, J. W., "Prostaglandins: A Report on Early Clinical Studies," *Postgraduate Medical Journal,* **46,** 562 (1970).

PROTACTINIUM

Protactinium, element 91, lies between thorium and uranium in the Periodic Table. It is one of the more intractable of the natural radioelements and only recently has a reasonable clear picture of its chemistry begun to emerge.

Of the fifteen isotopes known at present, ranging in mass from 225 to 237, only two, [234]Pa and [231]Pa, occur naturally and only [231]Pa and [233]Pa have sufficiently long half-lives, 32,480 years and 27.4 days, respectively, for use in chemical studies. In 1913 Fajans and Göhring discovered the first isotope of protactinium, UX_2([234]Pa), and named it brevium; the long-lived α-emitter, [231]Pa, was independently isolated from pitchblende in 1918 by Hahn and Meitner and by Soddy and Cranston. This isotope is the parent of actinium in the $(4n + 3)$ series,

$$^{235}U \xrightarrow{\alpha} {}^{231}Th \xrightarrow{\beta} {}^{231}Pa \xrightarrow{\alpha} {}^{227}Ac,$$

one ton of uranium at radioactive equilibrium containing approximately 340 mg. of [231]Pa. The various residues from uranium and radium refineries have always constituted the main source of protactinium-231 but since its distribution in any plant is markedly affected by the process used in the plant, many different residues have been used as sources. The intractable nature of many of these residues, coupled with the lack of knowledge of the chemistry of the element and the difficulties experienced in maintaining protactinium (V) in true solution, resulted in low recoveries; before 1960 the amount of pure protactinium available in the world was considerably less than 1 gram. However, research into the recovery of protactinium from the "ethereal sludge" which formed in the older uranium recovery processes culminated in the isolation, in 1960, of over 100 g. of essentially pure protactinium-231. In addition to the natural source, protactinium-231 can be obtained synthetically by neutron bombardment of ionium ([230]Th). Protactinium-233, produced by neutron bombardment of thorium-232, is the most important artificial isotope since it decays to the fissionable uranium-233.

$$^{232}Th(n,\gamma)^{233}Th \xrightarrow[\text{23.5 m}]{\beta} {}^{233}Pa \xrightarrow[\text{27.4 d}]{\beta} {}^{233}U.$$

In addition, owing to its convenient half-life and nuclear properties this isotope has frequently been employed for tracer investigations of the coprecipitation, solvent extraction and ion-exchange behavior of the element.

Protactinium exists naturally in the quinquevalent state, and although reduction to protactinium (IV) has been achieved both in solution and in the solid state, trivalent protactinium compounds are poorly characterised; it is unlikely that protactinium (III) could have more than a transient existence in aqueous solution. Protactinium (V) is reduced in solution only by powerful reagents

such as zinc amalgam and chromium (II) and the potential Pa (IV) = Pa (V) + e^- has been estimated at +0.1v; atmospheric oxygen and ultraviolet light oxidize protactinium (IV). Both valence states give colorless aqueous solutions. Solvent extraction and ion exchange data have provided evidence for the existence of both anionic and cationic protactinium (V) species in aqueous hydrofluoric, hydrochloric, hydrobromic, nitric, sulfuric and oxalic acid solutions. The nature of these ions is always very complex; it is unlikely that Pa^{5+} ions can exist in acid media and there is no evidence for the formation of the protactinyl ion, PaO_2^+. Further the complexes formed at other than tracer protactinium (V) concentrations in the above acids, apart from hydrofluoric and sulfuric acid, are thermodynamically unstable with respect to hydrolysis, which is followed by condensation to uncharacterized polynuclear species which generally have low solubility products. Such behavior is encouraged by low acidity and high protactinium (V) concentrations; the presence of fluoride or sulfate ions is essential for maintaining macro-amounts of protactinium (V) in true solution in aqueous media. Although little is known of the properties of protactinium (IV) in aqueous solution it appears to be less susceptible to hydrolysis than protactinium (V).

Protactinium (V) is extracted from aqueous acid solutions, apart from those containing fluoride ion, by a variety of organic solvents. The extraction by donor solvents such as the higher ketones and alcohols is particularly efficient and generally increases with the concentration of hydrogen and ligand ions in the aqueous phase but in the case of sulfuric and oxalic acid solutions the extraction coefficient decreases above a certain ligand concentration owing to the formation of multi-charged anionic complexes. All ligands so far investigated, including hydroxyl, are readily replaced by fluoride which induces protactinium (V) to assume a coordination number greater than 6 and consequently protactinium (V) can always be extracted from an organic phase into an aqueous phase containing free fluoride ion. Other media useful for recovering protactinium (V) from organic solvents include sulfate, oxalate and acidic hydrogen peroxide. Higher amines will extract anionic complexes from certain aqueous acid solutions and protactinium (V) solvates, and chelates, extract into inert solvents such as kerosene or benzene.

Protactinium (V) is sorbed on anion exchangers from aqueous hydrochloric, nitric and sulfuric acid, the distribution coefficient increasing with acidity in the first two solutions but decreasing in the last. The addition of traces of fluoride ion is effective in preventing sorbtion. Like thorium (IV), but unlike uranium (IV), protactinium (IV) is not sorbed on anion exchangers from solutions 6 to 12 *M* in HCl. Protactinium (IV) is extracted from this solvent as the thenoyltrifluoroacetonate but, unlike protactinium (V), does not extract into tributyl phosphate or methylisobutyl ketone.

Protactinium (IV) and (V) are precipitated from acid solutions on the addition of excess aqueous ammonia or potassium hydroxide and unidentified precipitates form on the addition of iodate, peroxide, phosphate, carbonate and phenylarsonate to solutions of either protactinium (IV) or (V) in dilute sulfuric acid. In contrast to the higher oxidation state protactinium (IV) is insoluble in aqueous media containing fluoride ion and the compound PaF_2SO_4 is precipitated on the addition of hydrofluoric acid to sulfuric acid solutions of protactinium (IV). Protactinium (V) is spontaneously deposited from aqueous acid solution on several metal surfaces but the mechanism of such deposition is not understood. Tracer protactinium (V) is quantitatively deposited on platinum or stainless steel cathodes from oxalic-nitric acid mixtures, ammonium fluoride solution and ammonium fluoride-ammonium chloride mixtures; anodic deposition on lead, nickel, platinum and gold takes place from sulfate or carbonate media. The most interesting feature of paper chromatography and ionography studies is the demonstration of the existence of a soluble protactinate ion, at tracer concentration, in very strong potassium hydroxide solution.

The preparative chemistry of protactinium has been much explored recently. The presently characterized solid compounds include the halides PaX_5 (X = F, Cl, Br and I), Pa_2F_9 and PaX_4 (X = F and Cl), the oxyhalides Pa_2OX_8 (X = F and Cl), $PaOX_3$ (X = Cl and Br), PaO_2X (X = Cl and Br) and $Pa_2O_3Cl_4$, some complex halides of the types M^IPaX_6 (X = F, Cl and Br), $M_2^IPaF_7$, $M_3^IPaF_8$, $(NH_4)_4PaF_8$ and $(NMe_4)_3PaCl_8$, the pentachloridephosphine oxide complexes $PaCl_5 \cdot R_3PO$ (R = Ph and Bz), the nitrates $PaO(NO_3)_3$ x H_2O (1 < x < 4) and $Pa_2O(NO_3)_4 \cdot 2CH_3CN$, some hexanitratocomplexes, M^IPa $(NO_3)_6$, and the isostructural sulfate, and selenate complexes $H_3PaO(SO_4)_3$ and $H_3PaO(SeO_4)_3$ respectively. The pentoxide, Pa_2O_5, and the dioxide, PaO_2, are known and several phases intermediate in composition between PaO_2 and Pa_2O_5 have been identified crystallographically. [Numerous quaternary and ternary perovskite-like oxides of the type $BaM_{0.5}^{III}Pa_{0.5}O_3, M^{II}PaO_3, M^IPaO_3$ and $PaO_2 \cdot 2M_2^VO_5$ are also known.] An oxysulfide PaOS and a nitride PaN_2 have also been characterized and metallic protactinium can be obtained by reduction of the tetrafluoride.

A. G. MADDOCK

PROTEINS

Proteins are large molecules found universally in the cells of living organisms, or in such biological fluids as blood plasma. They invariably contain carbon, hydrogen, oxygen and nitrogen, almost invariably contain sulfur, and sometimes contain phosphorus. They are specifically characterized by yielding a mixture of alpha amino acids when hydrolyzed by means of acids, alkalies or certain enzymes. (See **Amino Acids, Enzymes**).

Proteins are exceedingly diverse in properties and functions. Some are relatively inert fibers, such as the keratins of wool, hair or horn, or the collagens of tendon and connective tissue, which play an important structural role in animal organisms. Others are readily soluble in water or in dilute salt solutions, such as ovalbumin of egg white, serum albumin of blood plasma, or hemoglobin of red blood cells, and their molecules are not very far from spherical in shape. These are often called globular or corpuscular proteins in contradistinction to the fibrous proteins. Many of them can be obtained from water in crystalline form, and x-ray studies have shown that protein crystals are highly ordered systems—true crystals in every respect. All known enzymes, the essential catalysts of biological systems, are proteins, many being present in solution in cytoplasm or cellular secretions, others more or less firmly anchored to larger cellular structures. A number of hormones, such as insulin and several of the hormones of the pituitary gland, are also proteins, as are all the antibodies which are called forth in immunological reactions. Protein foods are essential to the nutrition of all animals. Proteins are thus of prime importance in the functioning of all living organisms.

Amino Acids Derived from Proteins. The amino acids which proteins yield on hydrolysis are all α amino acids. The simplest is glycine, with the formula $^+H_3N \cdot CH_2 \cdot COO^-$ or $H_2N \cdot CH_2 \cdot COOH$ (Either of the two formulas may be used to represent glycine; the latter is the more usual, but the former represents the structure more accurately).

All but two of the amino acids derived from proteins may be represented by a generalization of this formula, as $^+H_3N \cdot CHR \cdot COO^-$. There are more than twenty possible groups which can occupy the position denoted by R. All the amino acids found in proteins, except glycine, are optically active, because of the asymmetry of the α-carbon atom which is surrounded by H, R, —COO^- and —NH_3^+ groups. The relative configuration of all these groups is the same for all amino acids derived from proteins. These amino acids are all known as L-amino acids, their enantiomorphs as D-amino acids. A number of D-amino acids have been found in nature; they are present in certain antibiotics and in various other compounds. So far as is known, however, they are not found in proteins. Two amino acids found in proteins, proline and hydroxyproline, are five-membered ring structures, the amino (or imino) group being in the ring.

Many methods have been used in the past for determining the content of amino acid residues in proteins. The most widely used techniques today involve acid hydrolysis, followed by separation of the amino acids on an ion exchange resin, or by paper chromatography. In the procedure of W. H. Stein and S. Moore, the protein is generally first hydrolyzed by heating in a strongly acid solution—such as 6 N HCl at 100°C for 24 hours—and the amino acids in the hydrolyzate are then separated by passage of the solution over a suitable ion exchange resin, and collected in a series of fractions. The separated amino acids are then treated with ninhydrin, which produces colored derivatives which may then be determined colorimetrically. By this method a virtually complete amino acid analysis may be carried out on a few milligrams of protein. The process now requires only a few hours, and the analyses are made and recorded by automatic devices. Tryptophan is largely destroyed on acid hydrolysis, and must be determined separately on an alkaline or enzymatic hydrolyzate. Serine and threonine are also partially destroyed, and this destruction must be corrected for. (See **Amino Acids**).

Simple and Conjugated Proteins. Many proteins have been obtained which yield only alpha amino acids, and no other substances, on hydrolysis—for example, the hormone insulin and the enzyme pepsin. These are known as *simple* proteins. Others yield additional compounds beside the amino acids; these are known as *conjugated* proteins. There are numerous classes of conjugated proteins; they include the glyco-proteins and mucoproteins, which contain carbohydrate groups; lipoproteins, which contain fatty acids, cholesterol and phospholipids; heme proteins, such as hemoglobin and several oxidative enzymes, which contain iron-prophyrin (heme) groups; nucleoproteins, in which proteins are associated with nucleic acids; and many others. Many of the simpler viruses, such as tobacco mosaic or tomato bushy stunt virus, are nucleoproteins; and all viruses are constituted largely of proteins and nuleic acids.

Classification of Proteins. In the past proteins have commonly been classified according to their solubility in various solvents; for instance, *albumins,* which are readily soluble in pure water; *globulins,* which are insoluble in water but dissolve readily in aqueous salt solutions; *prolamines,* which are soluble in alcohol-water mixtures but insoluble in either pure alcohol or pure water; and various other classes. This terminology is largely arbitrary, although still used in the naming of many proteins.

Proteins which are known to exert a specific activity are often given names which distinguish this activity. This is particularly true of the enzyme proteins, which catalyze specific chemical reactions, and the protein hormones, which arouse specific physiological responses. (See **Hormones**)

Proteins as Acids and Bases; Isoelectric Points. Even the simplest of the amino acids, glycine, contains one free amino and one carboxyl group. Thus on addition of acid it acquires a positive charge:

$$^+H_3N \cdot CH_2 \cdot COO^- + H^+Cl^- \rightleftharpoons$$
$$^+H_3N \cdot CH_2 \cdot COOH + Cl^- + H_2O$$

On addition of alkali it acquires a negative charge:

$$^+H_3N \cdot CH_2 \cdot COO^- + Na^+OH^- \rightleftharpoons$$
$$H_2N \cdot CH_2 \cdot COO^- + Na^+ + H_2O$$

Therefore when placed in an electric field, glycine in acid solution migrates as a positively charged ion to the cathode; in alkaline solution it migrates as a negatively charged ion to the anode. At an intermediate pH value, near neutrality, the positive and negative charges on the molecule just balance one another, and the molecule does not move in either direction when placed in an electric field. The pH at which this occurs is known as the *isoelectric point* of the molecule. For glycine the isoelectric point is near pH 6, not far from neutrality; for glutamic acid, with two carboxyl groups and one amino group, it is more acid, near pH 3; for lysine, with two amino groups and one carboxyl, it lies at an alkaline pH value, near 10.

Proteins behave similarly to amino acids, except that one protein molecule may contain scores or hundreds of free carboxyl groups (from aspartic and glutamic acid residues), amino groups (from lysine residues), imidazole groups (from histidine residues) and guanidino groups (from arginine residues). According to the amino acid composition of the protein, and the pH of the medium, the state of charge on the protein molecule can be varied from a maximum positive value in strongly acid solution to a maximum negative value in strongly alkaline solution. When the solution is placed in an electric field (see **Electrophoresis**) the protein migrates as an anion or a cation, depending on the pH of the solution, the speed of migration increasing with the net charge of the protein. Each protein has a characteristic isoelectric point, depending on the relative number of acid and basic side chains in the molecule. Some proteins bind other ions than hydrogen ions—cations like calcium, for instance, or anions such as chloride—so that the net charge on the protein, and hence the isoelectric point, is somewhat dependent on the amount of these other ions present in the medium. Approximate isoelectric points of a few proteins are listed below.

Protein	Source	Isoelectric Point, pH
Bushy stunt virus	Tomato plant infection	4.1
β-Lactoglobulin	Cow's milk	5.2
Carboxypeptidase	Pancreas	6.0
Hemoglobin	Human red cells	6.7
Growth hormone	Anterior pituitary	6.8
Ribonuclease	Pancreas	9.5
Cytochrome C	Horse heart	10.7
Lysozyme	Egg white	10.7

Peptide Chains in Proteins; Their Structure and Biosynthesis. Amino acids in protein molecules are joined to one another by elimination of water to form peptide ($—CO \cdot NH—$) linkages. Thus the linkage of two amino acids, glycine and alanine, gives the dipeptide glycylalanine

$$^+H_3N \cdot CH_2COO^- + {}^+H_3N \cdot CH(CH_3)COO^- \rightleftharpoons$$
$$^+H_3N \cdot CH_2 \cdot CONH \cdot CH(CH_3)COO^- + H_2O.$$

If they are linked in the reverse order, the result is an isomeric dipeptide, alanylglycine. The polypeptide chains found in proteins are composed of dozens or hundreds of such amino acid residues linked together in an order that is precisely specified for any given protein. The information determining this order is contained in the sequence of bases in a portion of the deoxyribonucleic acid (DNA) that is found in the nucleus of the cell. (See **Nucleic Acids**). There are four kinds of bases in DNA—adenine (A), guanine (G), cytosine (C) and thymine (T). A sequence of three bases in DNA specifies a particular amino acid residue in the resulting protein. For instance the triplet sequence CAU, or CAC, in DNA, specifies the amino acid histidine. Any one of the four triplets ACU, ACC, ACG or ACA specifies the amino acid threonine. The entire set of specifications is known as the genetic code. These sequences in DNA do not serve directly to produce proteins. Instead the sequence of bases in DNA is transcribed to give a corresponding sequence in a particular form of ribonucleic acid (RNA), known as messenger RNA. This passes into the cytoplasm of the cell, and becomes attached to structures known as ribosomes; the attached messenger RNA then serves as a template, guiding the set of chemical reactions that involve the specific linking together of successive amino acid residues, with elimination of water as each link is formed, to yield the completed polypeptide chain. The details of this complex process cannot be discussed here, but we note that this synthesis does not proceed spontaneously; it requires large amount of free energy to form the polypeptide chain from the constituent amino acids. This energy is derived from "high energy" compounds in the cell, such as adenosine triphosphate (ATP). Just twenty different amino acids can be utilized to build polypeptide chains by this biosynthetic system. Some others can be formed by chemical reactions occurring after the chain has been laid down. Thus, in making collagen, some of the proline initially incorporated in the chain, according to the genetic code, is later converted into hydroxyproline, by a secondary chemical reaction, introducing an —OH group on one of the carbons in the proline ring. Also two residues of the amino acid cysteine, each containing a thiol (—SH) group, can later come together and form a disulfide (—S—S—) bond to give what is known as a cystine residue, joining two different peptide chains, or two portions of the same chain:

$$\cdots R_1CH-CO-HN \cdot CH \cdot CO-NH-CHR_2 \cdots$$
$$\begin{array}{c} | \\ CH_2 \\ | \\ S \\ | \\ S \\ | \\ CH_2 \\ | \end{array}$$
$$\cdots R_3CH \cdot CO-HN \cdot CH \cdot CO-NH-CHR_4 \cdots$$

Such cross links appear to be frequent in fibrous proteins of the keratin class, which contain large amounts of cystine residues. They have been definitely shown to exist, and their positions exactly located, in the hormone insulin from beef pancreas. In other cases, a single peptide chain may be folded back on itself, and fastened by a disulfide bridge so as to form a loop in the chain. One such loop has been shown to occur in the insulin molecule, and some proteins, such as serum albumin or ribonuclease, contain a considerable number of such folds and loops.

Some proteins, such as α- and β-casein of milk, contain phosphorus in the form of phosphate esters of hydroxyamino acids. These may in some cases be diesters, or pyrophosphate esters, which act as bridges between two peptide chains. Such esters may also serve to form a loop in a single peptide chain as with a disulfide bridge.

Determination of the order in which the amino acid residues are arranged in the peptide chains of a protein is a particularly difficult problem. The number of peptide chains present can be determined by labeling the chain ends which contain free α-amino groups, using such reagents as 2,4-dinitrofluorobenzene, which remain bound to the amino group even when the protein is subsequently hydrolyzed to amino acids. The number of such end groups, and the nature of the amino acid residues involved, may then be identified. Gentle partial hydrolysis of the protein, by acids or enzymes, which break it down into a series of smaller peptides, but not completely to amino acids, is often of great value. The resulting smaller peptides are then separated, for instance by their relative movement on a paper chromatogram (see **Chromatography**), by migration in an electric field, or in other ways. These peptides may then be labeled, hydrolyzed, and their structure worked out step by step. It may be possible, from the study of these smaller peptides, and the overlapping sequences found in them, to reconstruct the entire sequence of amino acid residues found in the original protein. The reasoning is not very unlike that involved in solving a crossword puzzle, but skill and judgement of a high order are required to obtain trustworthy results.

The first notable achievement in this field was the determination of the complete sequences of amino acid residues in the two peptide chains of the molecules of insulin from horse, sheep, pig, and one species of whale, by F. Sanger and his associates. Each of these insulins consists of an A (acidic) chain, 21 residues long, with four half-cystine residues; and a B (basic) chain, 30 residues long, containing two half-cystine residues.

Sanger showed that the two chains are linked together by two disulfide (S—S) bonds between the A and B chain, while a third disulfide bond links half-cystine residues numbers 6 and 11 of the A chain to form a loop within this chain. The whole structure is shown diagrammatically in Fig. 1. The structures of sheep and hog insulin are very similar; the B chain is identical in all three, and the A chains differ in only three places (residues 8, 9, and 10) from one species to another.

Similar studies have now been extended to a large number of other proteins. For instance, Fig. 2 shows the complete amino acid residue sequence of the enzyme ribonuclease from beef pancreas, which is composed of a single chain of 124 amino acid residues, cross-linked by 4 disulfide bonds as indicated in the figure. Lysozyme from egg white, an enzyme which digests material in bacterial cell walls, is composed of a single chain of 129 amino acid residues linked by four disulfide bonds; many other sequences in the peptide chains of proteins are now known. The longest peptide sequences that have been completely, or almost completely, determined are those of the subunits of the enzymes catalase (505 residues) and glutamic dehydrogenase (506 residues). Complete sequences are now known (1971) for well over a hundred proteins, and the number is increasing rapidly.

Association of Subunits in Proteins. Many proteins are composed of several polypeptide chains, held together by noncovalent bonds. The hemoglobins in the red blood cells of vertebrates are notable examples. These hemoglobins are composed of four polypeptide chains, each containing a heme group with a central iron atom that serves to bind oxygen reversibly. (See **Porphyrins**). Two of the four peptide chains are called α-chains, each containing 141 amino acid residues in human hemoglobin. The other two are called β-chains; each contains 146 residues. The complete sequence of amino acid

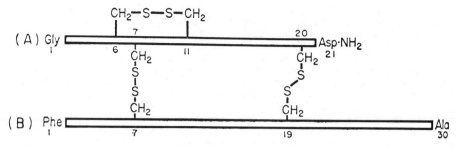

FIG. 1. A diagrammatic picture of the structure of the insulin molecule, according to Sanger, showing the two cross linkages by disulfide bonds between the A and B chains, and the disulfide loop between residues 6 and 11 in the A chain. The symbol Gly denotes a glycine residue with a free amino group at one end of the A chain; Asp. NH_2 denotes an asparagine residue, with a free carboxyl group, at the other end. In the B chain, a phenylalanine (Phe) residue and an alanine (Ala) residue occupy the corresponding positions. The amino acid residues are numbered in each chain, starting from the free amino terminal end.

residues in the α and β chains of human hemoglobin has been worked out and similar information is now available for many other species of hemoglobin. The complete hemoglobin molecule can thus be denoted briefly by the formula $\alpha_2\beta_2$. The total molecular weight of the four chains and the associated heme groups is close to 64,500. At high or low pH the $\alpha_2\beta_2$ structure dissociates reversibly according to the equation: $\alpha_2\beta_2 \rightleftharpoons 2\alpha\beta$; and under more drastic conditions the $\alpha\beta$ units dissociate into separate α and β subunits.

Spatial Conformations of Peptide Chains. The work of W. T. Astbury and others has shown that a large number of fibrous proteins, notably keratin, myosin, epidermin, and fibrinogen, exist normally in what

is known as the α conformation, which shows a characteristic diffraction pattern when the fibers are examined by x-rays. On stretching, sometimes accompanied by heat, they can be drawn out into what is known as the β conformation which is about twice as long as α, and shows a very different x-ray pattern. The evidence indicates that the peptide chains in the β structure are more or less fully extended, probably in a pattern that has been called a "pleated sheet" by L. Pauling and R. B. Corey, who have described such structures in detail. The more tightly coiled α conformation consists of more or less parallel peptide chains, each chain being wound into a helix, the coiling being due to the attractions between the C—O and the N—H groups of the dif-

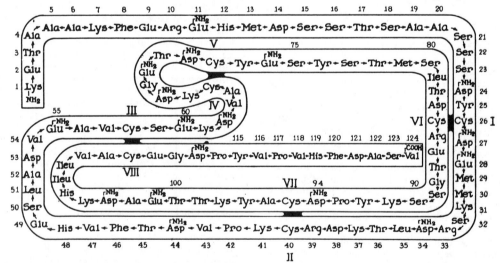

FIG. 2. The amino acid sequence of the enzyme ribonuclease from beef pancreas. This protein contains 124 amino acid residues, with lysine at the amino terminal end and valine at the carboxyl terminal end. The 3-letter symbols for the amino acid residues are formed by taking the first three letters of the name of each amino acid. Asparagine and glutamine are designated by AspNH, and GluNH₂, respectively. The positions of the four disulfide bonds are shown by the dark bands connecting pairs of Cys residues. This figure does not attempt to indicate the 3-dimensional structure, which is now known in considerable detail. The molecule contains several short stretches of helix, in the first 60 residues; there are regions of pleated sheet, from residue 70 on. The histidine residues at positions 12 and 119 are close together, and they play a major role in the catalytic action of the enzyme on ribonucleic acid and nucleotides. (This figure is from D. G. Smyth, W. H. Stein, and S. Moore, *Jr. Biol. Chem.*, **238**, 227 1963).

ferent peptide residues. Detailed calculations concerning the arrangements of the atoms in such helices—notably one structure commonly termed the α-helix—have been worked out by Pauling and Corey. Similar helical coils of amino acid residues exist within the framework of the structure of many globular proteins.

Collagen, the major fibrous protein of tendon and connective tissue, is built on quite a different pattern, the nature of which has been clarified by G. N. Ramachandran, F. H. C. Crick, A. Rich, W. Traub, and others. Essentially, it is a helical structure made up of three polypeptide chains closely associated and held together by hydrogen bonding and other noncovalent linkages. A complete turn of the helix involves 30 amino acid residues for each of the three chains, with an interval of 85.8Å along the fiber axis. Collagen is rich in the imino acids proline and hydroxproline; the five membered rings present in these give rise to special steric effects that have much to do with the collagen structure.

During the last few years x-ray diffraction studies of protein crystals have provided detailed knowledge of the 3-dimensional arrangement of certain globular proteins. Myoglobin from the heart of the sperm whale is a heme protein containing 153 amino acid residues (J. C. Kendrew *et al*). The peptide chain is arranged chiefly in successive stretches of α-helix, ranging from 6 to 24 amino acid residues in length. There are interruptions between successive helical segments involving one or more amino acid residues, so that the successive helices are often inclined in rather sharp angles to each other. The over-all shape is roughly that of an ellipsoid $45 \times 35 \times 25$Å. The whole structure is highly compact; it is stabilized by hydrogen bonding within the helical segments and also among some of the polar side chains of the amino acid residues; hydrophobic interactions between nonpolar side chains in the interior of the molecule are also of prime importance in stabilizing the structure. The lysozyme molecule from egg white (D. C. Phillips *et al*) is also a highly compact structure which however contains a sort of cavity into which it appears that substrate molecules can fit. The general shape of the lysozyme molecule is roughly ellipsoidal with dimensions about $45 \times 30 \times 30$Å. It contains only about 30% helix, and one region of pleated sheet.

Each of the four peptide chains in hemoglobin (2α and 2β chains) has a 3-dimensional arrangement very similar to that of myoglobin. The four chains pack together in an approximately tetrahedral arrangement, with a twofold axis of symmetry. When the hemoglobin molecule takes up oxygen, there are changes in the structure, both of the individual chains and of the arrangement of the chains relative to each other. M. F. Perutz has described this in detail. These changes provide the structural basis for the remarkable efficiency with which hemoglobin serves as an oxygen carrier in blood and also aids the transport of carbon dioxide in blood and its discharge in the lungs.

Molecular Weights of Proteins. Proteins are large molecules; even the smaller proteins generally have molecular weights over 10,000; while molecular weights of many millions are not uncommon. Molecular weights may be determined by such methods as osmotic pressure, light scattering, sedimentation equilibrium in the ultracentrifuge, sedimentation velocity and diffusion, and other procedures. Information regarding molecular size and shape is also obtained from study of viscosity, double refraction of flow, dispersion of the dielectric constant, and depolarization of fluorescent light from protein solutions. Some typical values for the molecular weights of a few proteins are given below:

Protein	M.W.
Insulin	5,733
Lysozyme	14,500
Ribonuclease	13,300
Chymotrypsinogen	21,000
Ovalbumin	43,000
Hemoglobin	64,500
Serum albumin	66,000
Gamma globulin	150,000
Edestin	310,000
Fibrinogen	340,000
Thyroglobulin	640,000
Myosin	500,000
Hemocyanin (octopus)	2,800,000
Hemocyanin (snail)	8,900,000
Tomato bushy stunt virus	7,600,000
Tobacco mosaic virus	40,000,000

Lysozyme, ribonuclease, chymotrypsinogen, ovalbumin and serum albumin are composed each of a single polypeptide chain; all the others listed in this table contain more than one chain. The two viruses are nucleoproteins, and contain many subunits. Tobacco mosaic virus, for instance, contains 5% RNA in the form of a long thin thread, about 300 nanometers (nm) long, which is the infectious part of the virus. The protein of the virus consists of more than 2100 subunits, each of the molecular weight about 18,000, and all identical. Each subunit is compact and globular. They are closely packed around the RNA in a helical arrangement, and serve as a protective coating. The resulting structure is a cylindrical rod 300 nm long and 15 nm in diameter.

Denaturation. Protein molecules are complex and delicate structures. Most globular proteins readily undergo a change known as denaturation. Denatured proteins are usually far less soluble, near their isoelectric points, than the native proteins from which they were derived. Proteins which crystallize readily in the native state become incapable of crystallization after denaturation; and other changes are observed. Denaturation can be produced by heating the protein solution briefly to temperatures near the boiling point, by exposure to acid and alkaline solutions, or to concentrated solutions of urea or guanidine hydrochloride, or to various detergents. Precipitation by reagents such as alcohol or acetone, at room temperature, may produce denaturation; although the same reagents at lower temperatures may be employed to precipitate the protein without denaturation. Denaturation often involves no change in the molecular weight of the protein, although sometimes the native protein may be split into two or more subunits on denaturation. True denaturation does not involve the splitting of peptide bonds; apparently it involves a breaking of many weak internal bonds within the native protein molecule, such as hydrogen bonds; the result is that the highly ordered structure of the native protein molecule is replaced by a much looser and more random structure. Denaturation is probably not a single unique process; different denaturing agents, applied in different ways and for different times, may give a whole series of different denatured products. There is now strong evidence that the denaturation of many proteins is reversible. For example, C. B. Anfinsen and his collaborators have treated ribonuclease (see Figure 2) with a reducing agent, thereby converting all the disulfide bonds to sulfhydryl groups. In concentrated urea solution this reduced ribonuclease is a typical denatured protein with no enzyme activity. On removing the urea, and slowly reforming the disulfide bonds by oxidation in air, the native protein is recovered, with full activity, and apparently identical in all respects with the original enzyme. Similar experiments have been carried out with lysozyme and other proteins. These findings suggest that the amino acid sequence of a polypeptide chain provides sufficient information to direct the formation of the correct 3-dimensional structure.

Fractionation. The complex mixtures of proteins found in various tissues may be fractionated in a great variety of ways. The most widely used in the past has been the "salting out" procedure, in which increasing amounts of such salts as ammonium sulfate, sodium sulfate, or phosphate buffers, were added to the system, and a series of protein fractions were thus precipitated. More recently, fractionation with organic precipitants such as ethanol or acetone—employed at low temperature so as to avoid denaturation—has been widely used. Many proteins (euglobulins) which are insoluble at the isoelectric point in the complete absence of salt, are readily dissolved by small amounts of salt. Specific ions, such as zinc, cadmium, or barium, sometimes have highly specific effects at low concentrations, precipitating some proteins and leaving others in solution. Most proteins are least soluble near their isoelectric points and more soluble in more acid or alkaline solution. The solubility of some proteins is greatly affected by change of temperature. All these facts, and others, have been utilized in working out systematic schemes of protein fractionation. Quite different techniques— electrophoresis, selective extraction by counter-current distribution, selective adsorption and elution, chromatographic procedures—have also been employed. The recently developed chromatographic methods permit separation of very closely related protein components in mixtures that were incapable of resolution by earlier procedures.

Proteins in Nutrition. In general, plants can synthesize their own proteins, given nitrogen in the form of ammonium salts, nitrates, or nitrites, and a few other simple compounds. Nitrogen-fixing bacteria can obtain the necessary nitrogen for protein synthesis from the atmosphere. Many organisms, however, must receive some of their nitrogen in the form of amino acids in order to synthesize protein. Gram negative bacteria, such as *Escherichia coli*, can often do very well with ammonium salts as the sole source of nitrogen. On the other hand, lactic acid bacteria require a large number of preformed amino acids for growth; this requirement has served as the basis for widely used methods of biological assay for the amino acid content of proteins and of various media.

Mammals can generally utilize ammonium salts in the synthesis of protein, provided certain essential amino acids are supplied in the diet. To maintain nitrogen balance in normal adult men, W. C. Rose has found that the following amino acids are essential; leucine, isoleucine, methionine, valine, threonine, phenylalanine, tryptophan and lysine. All the others can be synthesized if these are present. Rose found that growing rats require all of these and also histidine. Some arginine is also required to achieve optimum growth rates. In men, an arginine deficiency leads to an impairment of sperm production. All the other amino acids can be synthesized by men or rats, if these essential ones are present, together with an adequate supply of total nitrogen and other dietary constituents.

An adequate protein diet therefore involves two considerations: (1) adequacy of total protein supply for nitrogen metabolism, (2) adequacy of all the essential amino acids, to make the protein "complete" for synthetic purposes in the diet. Some plant proteins are deficient in one or two essential amino acids—for instance, lysine. If the protein of the diet is derived largely from such sources, addition of some of the missing amino acids to the diet may greatly improve its biological value.

Protein deficiency is a grave and widespread nutritional problem in many parts of the world, notably in Africa, Asia and South America, but to some extent in all the major continents. The increase of the world's supply of edible protein, and its proper distribution to the people, are therefore problems of major social import. (See **Nutrition**)

JOHN T. EDSALL

Cross-references: *Amino Acids, Nucleic Acids, Nutrition, Vitamins, Hormones, Molecules, Crystals, Polypeptides, Enzymes.*

References

R. E. Dickerson and I. Geis, "The Structure and Action of Proteins," (Harper and Row, New York, 1969. A short and lucid introduction, with many excellent drawings. A supplement with stereo drawings is also available.

F. Haurowitz, "The Chemistry and Function of Proteins," (Academic Press, New York 1962). A good general account of the proteins in one volume.

H. Neurath, Editor, "The Proteins," Second edition. A comprehensive treatise in five volumes, Volume VI in preparation. (Academic Press, New York, 1963–70).

PROTONS

The proton is an elementary particle with a mass of 1.6726×10^{-24} grams and a positive charge of 4.803×10^{-10} electrostatic units. Hence the charge is the same in magnitude as that of the electron but opposite in sign. In atomic mass units the proton has a mass of 1.007277.

The proton as a positively charged particle was

first observed as the lightest particle in the Kanal-strahlen or positive rays of a discharge tube. Although it had the same mass as the hydrogen atom, and was evidently the nucleus of the hydrogen atom, it was described by Rutherford in 1914 as the "long sought positive electron." Of course, the positive electron, or position, was discovered much later in 1932 by Anderson. By 1920 it was evident that the positively charged nucleus of the hydrogen atom represented an important unit in the structure of other atoms. The name proton (from the Greek *protos*, first) was ascribed to this particle by Rutherford.

The proton, like the electron, has an intrinsic angular momentum or spin and hence acts as an elementary magnet. The magnetic moment of the proton is 2.79278 nuclear magnetons [(15.2103 $\pm$ 0.0002) $\times$ 10^{-4} Bohr magnetons]. Since the proton mass is 1836 times the electronic mass, and since the magnetic moment is inversely proportional to mass, nuclear magnetic moments are less than electronic magnetic moments by a factor of about 1000. Individual protons, like electrons, can have spin quantum numbers of $+\frac{1}{2}$ and $-\frac{1}{2}$ only. The gyromagnetic ratio of the proton (ratio of magnetic moment to angular momentum) has been measured and found to have a value of (2.67520 $\pm$ 0.00006) $\times$ 10^{4} sec^{-1} $gauss^{-1}$. The presently accepted value of e/m for the proton is (9.57897 $\pm$ 0.00006) $\times$ 10^{3} emu/gram.

The proton exhibits a structure consisting of a positive core containing about one half of the total charge surrounded by a more diffuse region which is also positive. The antiparticle of the proton is the antiproton, discovered on the Berkeley bevatron. It is like the proton except that the charge, while equal in magnitude to that of the proton, is negative.

Every atomic nucleus contains a number of protons equal to its atomic number Z plus a number of neutrons N sufficient to make up its observed mass number A. Thus $A = N + Z$. A nuclear species characterized by its values of A and Z is called a *nuclide*. At all except extremely short distances (of the order of 10^{-13} cm) two protons will suffer an electrostatic repulsion, the force of which will vary inversely as the square of the distance between them. However, at separations of the order of 10^{-13} cm this longer range force it outweighed by a short-range (nuclear force) attraction. This attraction between two protons is about the same as that between two neutrons or between a proton and a neutron and is large compared to electromagnetic forces. For this reason and because of the Pauli Exclusion Principle the stable nuclides up to about $Z = 26$ (iron) have at most only a few more neutrons than protons. At about this atomic number the nuclear binding energy per nucleon (neutron or proton) is a maximum—some 9 MeV. Even as low as $_2^4$He the binding energy per nucleon is about 7 MeV. Thus, unlike electrostatic forces, the nuclear forces tend to saturate at a low number of nucleons. In larger atoms with large numbers of neutrons and protons the electrostatic repulsion of the protons increases since the total nucleon charge increases and the electrostatic repulsion begins to come into play. To compensate for this repulsion more neutrons are necessary to confer stability on the nucleus. However, protons and neutrons are particles that must obey the Pauli Exclusion Principle; hence as more neutrons and protons are added they must enter higher energy levels in the nucleus. Thus there is a limit to the number of extra neutrons that can cause a net increase in stability and consequently heavier nuclei become relatively less stable than lighter nuclei.

For use in nuclear reactions protons may be accelerated to high energies in a particle accelerator. The first nuclear reaction induced by accelerated particles involved striking a lithium nucleus with a high speed proton to produce two helium nuclei. In the last few decades radioactive isotopes of almost all elements have been made in nuclear reactions involving proton irradiations.

Protons enter importantly into stellar reactions. For example, the energy of stars like our sun appears to be generated by a cycle of nuclear reactions, the net result of which is the conversion of four protons to one helium nucleus with the liberation of 24.6 MeV plus the annihilation energy of two positrons.

M. H. LIETZKE AND R. W. STOUGHTON

References

Moore, W. J., "Physical Chemistry," 3rd Ed., Prentice-Hall, Inc., Englewood Cliffs, N.J., 1962.

Glasstone, Samuel, "Source Book on Atomic Energy," D. Van Nostrand Co., New York, 1950.

Halliday, David, "Introductory Nuclear Physics," John Wiley & Sons, New York, 1950.

PSYCHOTROPIC DRUGS

Psychotropic drugs are agents that are used therapeutically to alter the behavior, emotional states and/or mental functioning of psychologically disturbed humans. These drugs were formerly called tranquilizers, either "major" or "minor," and presented as promoting relaxation and inducing a sedating or calming effect. Evidence gathered over years of their use in a variety of clinical states has shown this term to be nonapplicable in many instances. Indeed, certain of these psychotropic drugs may actually induce a certain degree of motor stimulation, behavioral activation and/or muscle rigidity—yet at the same time altering those states or processes that had led the individual to seek —and be given—psychotherapeutic intervention. Further, the "major-minor" tranquilizer distinction is misleading because the inference is implied that the difference between them may be just a matter of degree. This does not seem to be the case; the "major" tranquilizers (phenothiazines, thioxanthene or butyrophenone derivatives) have pharmacologic and clinical actions quite different from the "minor" tranquilizers (benzodiazepine and glycerol derivatives).

As of the last few years, the convention has developed to more appropriately call these drugs in accordance with their main clinical value in psychiatry, e.g., "antipsychotics," "antidepressants" and "antianxiety" agents.

Antipsychotic Agents. Presently there are three major chemical categories of "antipsychotic" drugs marketed in the United States: phenothiazines, thioxanthenes and butyrophenones. Drugs derived from these chemical classes each have the capacity to do the following:

(1) Reduce psychotic manifestations (relieve emotional tension; ameliorate thought or behavior disorganization; decrease aggression, overactive or withdrawn behavior; and lessen hallucinatory or delusional phenomena) at doses that do not produce marked soporific effects.

(2) Directly affect the extrapyramidal motor system to produce a parkinson-like syndrome (tremors, rigidity, postural abnormalities and mask-like facies.)

(3) Produce homeostatic stabilization, yet cause little direct autonomic nervous system alteration.

Drugs of this group have been well documented as being useful in the treatment of both acute and chronic schizophrenia and of the manic phase of manic-depressive psychosis. Occasional reports exist as to their potential use in the treatment of anxiety states (in small doses), as well as in certain types of depressed states.

Phenothiazines. The prototype of this group of drugs is chlorpromazine, one of the most studied active medicinal substances of all time, a phenothiazine derivative. Subsequent to its introduction into psychiatry a number of variants were synthesized in an attempt to reduce toxicity and increase specificity. Structural changes consisted of substitutions of the halogen attached to the nucleus and/or replacement of the aliphatic with a piperidine or piperazine side-chain. These substitutions resulted in drugs of varying potency and that differed in the type of side effect—secondary pharmacologic activity—they produced. For example, generally speaking, one needs more milligrams of drugs with aliphatic (e.g., chlorpromazine, promazine, triflupromazine), or piperidine (e.g., thioridazine, mesoridazine, piperacetazine) side-chains than is needed with drugs with piperazine side-chains for the same degree of clinical effective action. The aliphatics have more marked sedative-hypnotic, motor-inhibiting, mood-dampening actions than do the piperazines, which are more likely to be activating and motor-stimulating. The aliphatics are more likely to produce hematologic, hepatotoxic, hypotensive, autonomic (ANS), or photosensitivity reactions than the piperazines. The piperazines, (e.g., perphenazine, trifluoperazine, fluphenazine), on the other hand, are more likely to produce extrapyramidal side effects and occasional hypertension. (A recent modification of pragmatic promise for clinical use involves a sustained release I.M. form of fluphenazine which requires an injection only once every two-four weeks.)

Phenothiazines exhibit a diversity of pharmacologic activity possessing antihistaminic, antiemetic, antispasmodic, local anesthetic, anticholinergic and adrenergic-blocking actions.

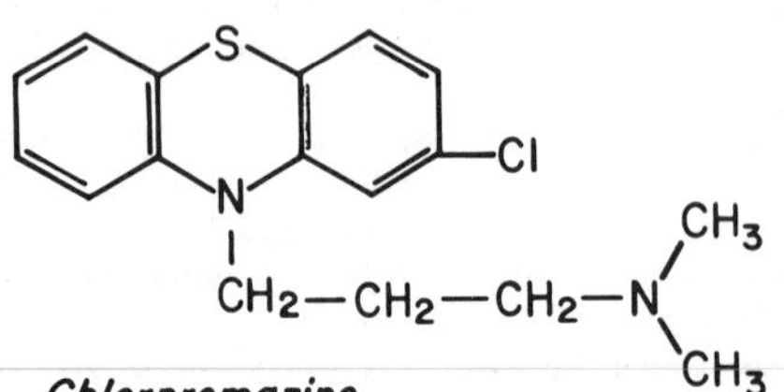

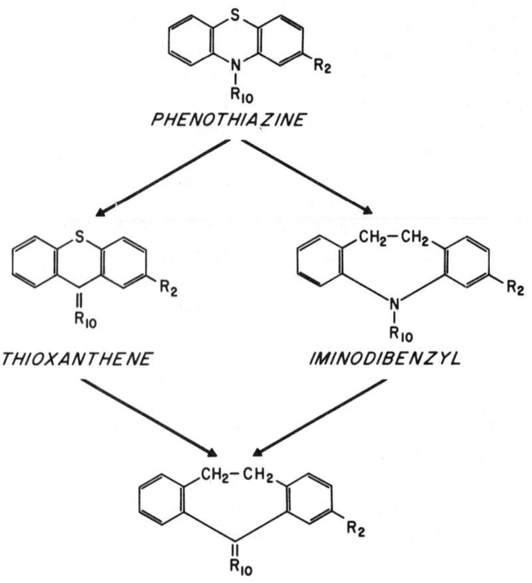

PHENOTHIAZINE

THIOXANTHENE

IMINODIBENZYL

DIBENZOCYCLOHEPTENE

FIG. 2.

Thioxanthenes. The search for drugs with greater clinical specificity and lesser toxicity led to the development of structurally similar agents in which the nucleus of the ring structure rather than the side-chain was altered. Focus was centered on a series of drugs which differed from the phenothiazine nucleus by having a double bond to a carbon in the 10 position of the central ring rather than a nitrogen. (See Figure 2). Drugs of this class are called thioxanthenes and also are effective antipsychotics. They also may possess either an aliphatic (e.g., chlorprothixene) or a piperazine (e.g., thiothixene) side-chain.

As a class, the available thioxanthenes (in comparison to phenothiazines) have been shown to produce fewer hepatotoxic, hematologic, and photosensitivity reactions but are somewhat more likely to produce extrapyramidal side effects. Within this class of drugs, the subclasses show the same differences and variations in dosage level and range, type of actions, and type of side effects associated with the phenothiazine subclasses.

Butyrophenones. A class of drugs that has also been found to be useful in the treatment of psychoses are the butyrophenones. They were arrived at by the synthesis of meperidine-related drugs during the search for drugs with morphine-like potency. The basic structure consists of a ketonic phenyl ring with a straight propylene chain with a piperidine nucleus. The ketonic phenyl ring has a fluorine substituent in the para-position and a tertiary alcohol group with a phenyl ring in the four position of the piperidine nucleus. Various halogen substituents on the para-position of this phenyl ring resulted in antipsychotic agents of varying potency and with a high capacity to produce extrapyramidal (EPS) side effects (e.g., haloperidol, trifluperidol). Further structural changes (replacement of the hydroxy group or the phenyl nucleus attached to the piperadine ring) resulted in a series of drugs (e.g.,

Chlorpromazine

FIG. 1.

spiroperidol, benzperidol, droperidol) with marked alteration in onset and duration of activity as well as alterations in neuroleptic potency. Droperidol has also been found to be one of the most specific antishock and antiemetic agents known.

Drugs of this class have a greater potential than phenothiazines and thioxanthenes to produce EPS but have far less of a tendency to cause ANS, hematologic or hepatotoxic side effects.

Antidepressant Agents. Antidepressants are a group of drugs that have as one common property the ability to elevate mood in depressed individuals. Originally, the term antidepressant had the connotations of possessing psychomotor stimulant and psychic energizing actions. Clinical experience has taught us that these two latter actions are not essential accompaniments to mood-elevation in depressed patients. In fact, certain of this group of drugs have potent hypnotic and motor-inhibiting effects. Two major types of antidepressants exist: MAO inhibitors and tricyclic compounds.

MAO Inhibitors. Drugs of this type have the ability to inhibit the enzyme monoamine oxidase leading to an increase in cerebral serotonin and norepinephrine levels. Two chemical classes of MAO inhibitors have been developed, hydrazide derivatives (e.g., iproniazid, isocarboxazid, nialamide, phenelazine) and nonhydrazide derivatives (e.g., trancylcypromine). These drugs came into psychiatry through an observation of the mood-elevating and slight motor-stimulating effects of the antitubercular drugs, isoniazid and iproniazid. The use of MAO inhibitors in psychiatry has decreased over the last five to 10 years because of the comparatively greater effectiveness of the tricyclic antidepressants as well as the marked toxic properties of the MAO inhibitors, particularly on the liver and the hematopoietic system.

Tricyclic Compounds. Agents of this type come from three structurally related chemical classes each with a heterocyclic nucleus (e.g., iminodibenzyl, dibenzocycloheptene, dibenzoxepins). They bear a likeness to phenothiazine nucleus differing mainly in that the sulfur atom in the five position of the central ring has been replaced by a two-methylene bridge (see Figure 2). Dibenzocycloheptene derivatives also possess the double bond to the carbon of the 10 position, characteristic of thioxanthenes.

The iminodibenzyls (e.g., imipramine, desipramine) were the first of these drugs introduced into psychiatry. The original drug of this class, imipramine, was initially proposed as a less toxic more effective chlorpromazine to which it structurally and pharmacologically in animals bears considerable likeness. It failed to demonstrate significant antipsychotic activity but was noted in passing to alter the mood state of depressed psychotic patients.

Further, chemical modifications led to the development of the dibenzocycloheptenes (e.g., amitriptyline, nortriptyline and protriptyline) and then to the dibenzoxepins (e.g., doxepin)—which has an oxygen atom in the central ring bridge.

These drugs all possess many of the phenothiazine pharmacologic properties but have considerably more anticholinergic effects. Unlike the phenothiazines, they facilitate central adrenergic transmission. The majority of these drugs can induce

hypnotic and motor-inhibiting effects that tend to disappear after a couple of weeks of continued administration—just when the antidepressant effects of the drugs are becoming clinically obvious.

Within these chemical classes of drugs, subclasses with side-chain similar to the phenothiazines have and are being synthesized and tested. The spectrum of differences in potency and other pharmacologic actions (i.e., sedation vs. stimulation) noted with the phenothiazines subclasses are applicable to the tricyclic subclasses.

Antianxiety Agents. Antianxiety agents have been developed from much more diverse chemical structures. They have as a common property that of reducing the anxieties experienced in emotional illnesses and from every day stresses. They, like the barbiturates (which were previously used as calmatives and sedatives), can induce muscle relaxation, are soporiphic, have anticonvulsant qualities, but do not have as great a potential for causing dependency. The two major chemical classes from which the most prescribed agents come are the glycerol and benzodiazepine derivatives.

Glycerol Derivatives. The earliest of the antianxiety agents of the modern psychotropic drug era was meprobamate, which was derived from mephenesin. Questions still persist as to the clinical effectiveness of this drug as an antianxiety agent although it has been one of the most widely prescribed drugs for this purpose.

Benzodiazepine Derivatives. Three drugs of this class are available (e.g., chlordiazepoxide, diazepam and oxazepam). In contrast to meprobamate, little conflict exists as to their efficacy in the treatment of anxiety states. Although equally effective in their antianxiety actions, they differ in other pharmacological properties with chlordiazepoxide having the most marked sedative, motor-inhibiting actions. Diazepam is reported to have some antidepressant, antispasmodic (especially in cases of status epilepticus), and occasionally stimulant-like activity. Oxazepam, supposedly the active metabolite of diazepam, has not been reported to have similar activity. Increased aggressive or assertive behavior has been noted with the first two drugs but not with the latter.

Marked adverse ANS, hematopoietic, or other physiologic reactions rarely occur with these drugs.

ALBERTO DiMascio

PURINES

Purine compounds are found in nature in plants, to a lesser degree in animals. They are derived from purine, which may be called 7-imidazo-(4,5d)-pyrimidine, imidazolo-4′,5′;4,5-pyrimidine, or 1,3,4,6-tetraazindene. Emil Fischer, who gave this ring system the name "purin" (from purum and uricum), introduced the following schematic representation and numbering,

$$N_1 = {}_6C - H$$
$$H - C^2 \quad {}_5C - {}^7N$$
$$\quad\quad\quad\quad\quad {}_8C - H$$
$$N^3 - {}^4C - {}_9N$$

and this symbol is still commonly used.

Purine ($C_5H_4N_4$, mol. wt. 120.06, m.p. 217°C) forms colorless crystals, soluble in water with neutral reaction to litmus or turmeric. Several syntheses are available, and many of its salts have been prepared. It is mainly used for organic syntheses, and in studies of metabolism.

The purine derivative of greatest biological importance is uric acid ($C_5H_4N_4O_3$), a white, tasteless and odorless crystalline compound. It is only slightly soluble in water, and decomposes above 250°C without melting. Uric acid is very widely distributed in animals, plants, and in bacteria, usually in form of its salts. It is the main nitrogenous end product of protein metabolism in reptilian and avian excrements (guano). Small amounts occur in normal human urine, blood, saliva, and cerebrospinal fluid; under pathological conditions the amount is increased, e.g., in gout, when uric acid crystals are deposited in the joints. Uric acid has been synthesized in a number of ways. Shaking with phosphorus oxychloride under pressure gives 2,6,8-trichloropurine (old name):

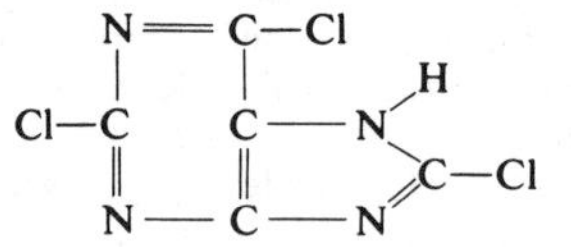

The 3 chlorine atoms may be catalytically removed, and replaced by hydrogen, to give purine. The reactivity of the chlorine atoms decreases from position 6, the most reactive, to 2, and finally 8; this behavior is most important for the synthesis of various purine derivatives.

Hypoxanthine (6-hydroxypurine, $C_5H_4N_4O$) is widely distributed in plants and animals, and is obtained by hydrolysis of nucleic acids.

Xanthine (2,6-dihydroxypurine, $C_5H_4N_4O_2$) occurs in tea leaves, in beet juice, and in the actively growing tissues of sprouting lupine seedlings; it is also found in urine, blood, liver tissue, and in urinary calculi.

Adenine (6-aminopurine, $C_5H_5N_5 \cdot 3H_2O$) is present in tea leaves and in beet juice. It is a constituent of the nucleic acids, from which it may be prepared by hydrolysis with mineral acids.

Guanine (2-amino-6-hydroxypurine, $C_5H_5N_5O$) is found in guano, fish scales, leguminous plants, sprouting lupine seedlings, and in various animal tissues. Its white crystals have a pearly luster, and are used in making artificial pearls. Guanine forms a white deposit in the tissues of swine affected with a kind of gout.

The guanine isomer, 2-hydroxy-6-aminopurine, (isoguanine) has been found in certain seeds and in insects.

The nucleic acids from yeast, and those from thymus tissue, contain the residues of the two aminopurines, adenine and guanine. 1-Methylxanthine, 7-methylxanthine, and 1,7-dimethylxanthine (paraxanthine) occur in urine. A number of alkaloids are purine derivatives.

Caffeine (theine coffeine) (1,3,7-trimethyl-2,6-diketopurine, $C_8H_{10}N_4O_2$)

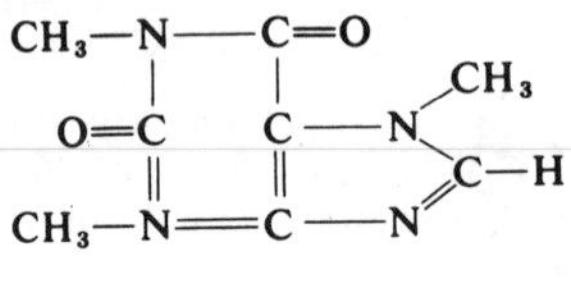

is the most widely used purine derivative. It is the main alkaloid in coffee beans (1 to 1.75%, Phillipine coffee 1.6 to 2.4%), tea leaves (1 to 4.8%), cola nuts (2.7 to 3.6%), Maté (Paraguay Tea), 1.25 to 2%), guarana paste (2.7 to 5.1%). Small amounts of it are present in cocoa. Caffeine is a stimulant for the central nervous systems, heart and muscle activity, and has diuretic effect.

Theobromine (3,7-dimethyl-2,6-diketopurine, $C_7H_8N_4O_2$) is the main alkaloid of the cocoa bean (unroasted 1.5 to 1.8%, roasted 0.6 to 1.4%); small amounts of it are found in tea leaves and in cola nuts. Its effect on the central nervous system is less than that of caffeine, but it is a stronger diuretic. Pharmaceutically its sodium compound is usually applied in form of double salts with sodium acetate, lactate, salicylate, or other salts.

Theophylline (1,3-dimethyl-2,6-diketopurine, $C_7H_8N_4O_2$) isomeric with theobromine, occurs in small quantities in tea leaves. Its physiological effects are similar to those of caffeine. Since it is an excellent diuretic, it is synthesized in relatively large quantity.

Finally, it should be mentioned that a number of purines, singly or in combination with pyrimidine derivatives, have been found to be important in the nutrition of bacteria. They act either as growth stimuli or are essential growth factors.

PAUL ROTHEMUND*

Cross-references: *Alkaloids, Nucleic Acids.*

*Deceased

PYROLYSIS

Pyrolysis is the name given to the transformation of a compound into another substance or substances by heat alone. Although the term frequently carries the implication of decomposition into smaller fragments, there are many examples of pyrolytic change which innvolve either isomerization with no change in molecular weight or building-up reactions giving rise to compounds of higher molecular weight. Hence, the term "pyrolysis" is broader in its scope than "thermal decomposition."

Hydrocarbons provide many good examples of pyrolysis both from theoretical and practical viewpoints. Methane resists thermal change unless very high temperatures are used, but at about 1300–1400°C it decomposes into hydrogen and a carbon black which is useful in rubber compounding. For this decomposition a tower full of checkerbrick is heated to the desired temperature by burning methane. Then the oxygen supply is cut off and methane alone is passed into the tower. The fluffy carbon produced is collected by blowing it through a filter, then the cycle of combustion and pyrolysis is resumed.

Other gaseous alkanes decompose at much lower temperatures, in the neighborhood of 400 to 600°C. Ethane gives rise primarily to ethylene and hydrogen, propane to both propylene plus hydrogen and to ethylene plus methane. The butanes decompose similarly into butylenes plus hydrogen but also into smaller olefins plus methane or ethane. These decompositions are of the free radical type. Thus, the initial effect of heat is to produce a small amount of some radical R, which will trigger the subsequent decomposition. This may be illustrated with butane:

$$R + CH_3CH_2CH_2CH_3 \rightarrow$$
$$RH + CH_2CH_2CH_2CH_2 (A) \text{ or}$$
$$CH_3CH_2CHCH_3 (B)$$
$$A \rightarrow CH_3CH_2(R') + CH_2{=}CH_2$$
$$B \rightarrow CH_3(R'') + CH_2{=}CHCH_3$$

To a much lesser extent, A and B may detach hydrogen to form C_4H_8. The R' and R'' radicals take the place of first R in promoting decomposition of more butane.

Liquid alkanes also follow a similar decomposition pattern and this is the basis of the thermal "cracking" of petroleum fractions in gasoline manufacture. The ethylene, propylene and butylenes of commerce also are made in reactions of this type.

Aromatic hydrocarbons pyrolyze somewhat differently. Benzene requires temperatures of 750–800°C and the product formed in biphenyl. This is the industrial synthesis of biphenyl. Toluene breaks down at lower temperatures (about 600–650°C), which suggests that the aliphatic position is involved rather than an aromatic position. In confirmation of this, 1,2-diphenylethane is the product formed. At a higher temperature (750°C) both toluene and 1,2-diphenylethane decompose profoundly, giving rise to benzene, naphthalene, anthracene, and phenanthrene in relative molar amounts of 11, 3, 1, 1, as well as other polycyclic hydrocarbons. At 750–825°C benzene arises not only from toluene but from a variety of organic compounds (as propylene, thiophene, methylthiophene, pyridine, picoline) by way of a 3-carbon fragment C_3H_3. This fragment, $\cdot CH{=}CH{-}CH{:}$, is both a methylene and a radical. It dimerizes to benzene, C_6H_6. Important also as an intermediate in these high-temperature processes is the C_4H_4 diradical, $\cdot CH{=}CH{-}CH{=}CH\cdot$, which is formed as a decomposition product of toluene, thiophene, pyridine, etc. It is a precursor of thermally-created naphthalene, phenanthrene, quinoline, and other products generally associated with coal tar.

Primary alcohols decompose in two ways on heating. They yield olefins and water, or they change into aldehydes and hydrogen. These changes occur at about 600°C. Tertiary alcohols decompose into olefins at somewhat lower temperatures, but if the high temperature is maintained there is considerable scission into ketones.

Ordinarily catalysts are used in any practical example of alcohol decomposition, since appropriate selection of catalyst promotes a unidirectional decomposition. Alumina is an efficient dehydration catalyst (alcohol to olefin), and brass is a good dehydrogenating catalyst (primary alcohol to aldehyde, or secondary alcohol to ketone). When an active catalyst is used, the transformation then is classified as catalytic, not pyrolytic.

Aliphatic amines resemble alcohols in their mode of pyrolysis. This is to be expected in view of one classification of amines as "ammono alcohols." Thus, ethylamine pyrolyzes bidirectionally at about 500°C into ethylene and ammonia, or into acetonitrile and hydrogen.

Many ethers have been subjected to pyrolysis, but one group in particular has been studied in great detail, namely, unsaturated ethers such as vinyl allyl ether, phenyl allyl ether, tolyl crotyl ether, etc. Such ethers isomerize at 180–250°C. Phenyl allyl ether changes into o-allylphenol, $CH_2{=}CH{-}CH_2{-}C_6H_4{-}OH$. Vinyl allyl ether pyrolyzes to 4-pentenal:

$$
\begin{array}{ccc}
CH_2{=}CH{-}O & & CH_2{-}CH{=}O \\
\quad\quad\quad | & \rightarrow & \quad\quad\quad | \\
CH_2{=}CH{-}CH_2 & & CH_2{-}CH{=}CH_2
\end{array}
$$

If the O in this ether is replaced by CH_2, as in a 1,5-hexadiene, a comparable isomerization occurs in the same temperature range:

$$
\begin{array}{ccc}
CH_2{=}CH{-}CHR & & CH_2{-}CH{=}CHR \\
\quad\quad\quad | & \rightarrow & \quad\quad\quad | \\
CH_2{=}CH{-}CH_2 & & CH_2{-}CH{=}CH_2
\end{array}
$$

The temperature for these changes is insufficient for cleavage of the compounds into radicals and the mechanism is regarded as ionic. The course of the change is intramolecular.

Among aldehydes and ketones, acetone may be singled out for treatment since its decomposition at 700°C is both the industrial and academic method for preparing ketene: $CH_3COCH_3 \rightarrow CH_2{=}C{=}O + CH_4$. For this reaction, an empty copper tube or glass tube serves well; or, a lamp with an electrically heated "Nichrome" or platinum filament is frequently used.

A wide variety of acids has been studied. Some acids decompose spontaneously at room temperature or lower. Others require intense heat to effect any change. For example, carbonic and carbamic acids are not isolable; dinitrocinnamic acid breaks down at 0°C; acetoacetic acid yields acetone at temperatures below 100°C; malonic acid changes into acetic acid at about 140°C; and 2-furoic acid decomposes into furan at 220°C. In contrast, acetic acid requires red heat, but it is interesting to note that ketene is formed from acetic acid in this process. This reaction competes with the pyrolysis of acetone for the industrial manufacture of ketene.

When acids contain other functional groups these groups may profoundly influence the manner of pyrolysis. α-Hydroxy acids yield inner esters, γ-hydroxy esters change into lactones, γ-amino esters form lactams, etc.

The pyrolysis of salts often is of value in chemical work. A well-known commercial synthesis of sodium oxalate depends on heating sodium formate to 350–450°C. Hydrogen is liberated. Dry distillation of the calcium or barium salts of acids at 430–490°C is a time-honored way of making ketones. The nature of the metallic part of the salt has a great effect on the yield of product. Thus, high yields of acetone are obtainable from lithium, barium, calcium or lead salts of acetic acid, whereas poor yields arise from the sodium or potassium salts. Silver and copper carboxylates behave quite differently, since they change on heating into the free acid (RCOOH) or its decarboxylation product, RH. If the salt is silver acetate, it pyrolyzes into acetic acid because of the thermostability of the latter, whereas if it is a heterocyclic salt such as silver quinolinecarboxylate it becomes decarboxylated to quinoline. This method is useful in decarboxylation reactions.

Methyl esters are much more thermostable than ethyl esters or higher homologs. Esters of the latter type decompose into acid olefin at about 500°C. This comparative ease of esters of the latter type is explained by assuming a quasi 6-membered ring involving the β-hydrogen with the carbonyl group in the transition state, followed by rupture. Heating phthalic anhydride at 700°C gives rise chiefly to biphenyl, presumably via benzyne (C_6H_4) since phenylpyridine is formed if a mixture of phthalic anhydride and pyridine is heated.

It will be recalled that alcohols also undergo pyrolysis into olefins. In this connection, therefore, it is interesting to contrast the pyrolytic behavior of methyl α-hydroxypropionate (methyl lactate) which may be regarded as an alcohol, and methyl α-acetoxypropionate, its acetic ester. Methyl lactate never pyrolyzes into methyl acrylate (in the manner of ethanol to ethylene), whereas methyl α-acetoxypropionate changes smoothly at 450–500°C into methyl acrylate:

$$CH_3—CHOAc—COOCH_2 \rightarrow$$

$$CH_2{=}CH—COOCH_3 + AcOH$$

Finally, it is appropriate to mention the fact that Wöhler's famous experiment was a pyrolytic one. He heated ammonium cyanate and obtained urea. Urea, in turn, is known to dissociate above its melting point (132°C) into ammonia and isocyanic (or cyanic) acid. This decomposition of carbamide, and the comparable decomposition of carbamic esters into isocyanates provides the clue to an understanding of their chemical behavior.

CHARLES D. HURD

Cross-references: *Aldehydes, Aliphatic Compounds, Aromatic Compounds, Coal Tar, Cracking, Free Radicals, Hydrocarbons, Ketenes, Ketones, Olefins.*

References

Hurd, C. D., "Pyrolysis of Carbon Compounds," Reinhold (N.Y.), 1929.

PYRROLES

The pyrrole ring is a pentagonal molecule containing one nitrogen and four carbon atoms. This ring is the fundamental building block of many pigments having physiological and commercial importance. Historically, this structure is of great significance because of the fact that it is present in indigo, an important article of ancient, medieval and modern commerce, as well as in the most precious ancient dye, Royal Purple, also called Tyrian Purple. The chemical synthesis of indigo, coupled with that of alizarin, was the turning point of the chemical phase of the industrial revolution and had far-reaching effects on agriculture and ultimately on political institutions. The chief pyrrole pigments, both in tonnage and in physiological activity, are the chlorophylls and the hemoglobins. The chlorophylls are the green leaf pigments which absorb the energy of the sun and initiate the process of photosynthesis. The hemoglobins are the oxygen-carrying pigments of higher animals. Hemoglobins in all species which have been investigated contain the same pyrrole pigment, heme; the globins, the proteins to which heme is bound, differ from species to species. Heme was produced synthetically in the laboratory of Hans Fischer. The most important commercial products derived from pyrrole are the phthalocyanins, a series of blue and green pigments possessing extraordinary stability towards light.

Pyrrole (C_4H_5N) is the parent substance of the series. It is a colorless oil, boiling at 129°C, density d_4^{20} 0.968. It becomes brown on continued exposure to air but may be stored without change in a vessel sealed under an atmosphere of purified nitrogen. It is also sensitive to strong acids. Alkylated pyrroles are even more sensitive to air than pyrrole itself. On the other hand, pyrroles with carbonyl groups attached to the ring are usually stable to the air and can be stored indefinitely without special precautions.

The classical structural formula for pyrrole, established through the researches of Adolph von Baeyer, is represented by:

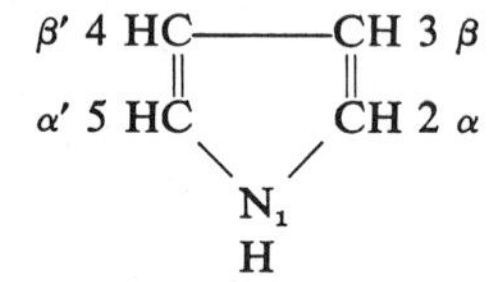

It is now recognized that the properties of pyrrole are best rationalized by its representation as a resonance hybrid. The major canonical forms contributing to its stabilization may be written as follows:

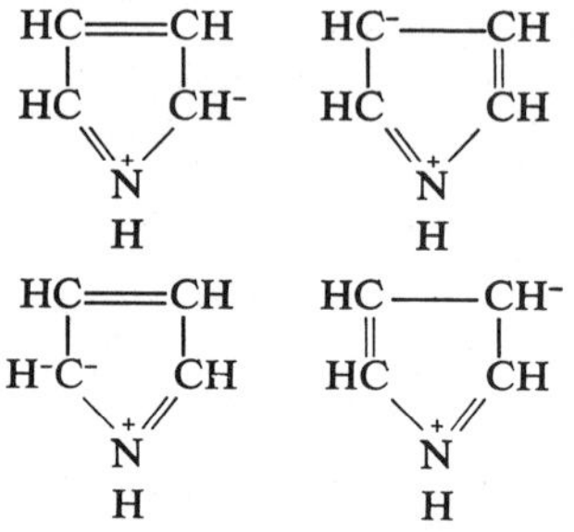

This formulation is consistent with the fact that pyrrole is no more basic in its reactions than the aliphatic amides and is a much stronger acid than ammonia, being roughly equivalent to diphenylamine in its acidity. Pyrroles with electron-attracting substituents, such as carbethoxy and cyano groups, may become as acidic as phenols and may be fractionated by pH control in aqueous solutions.

The formulation as a resonance hybrid is also consistent with the fact that electron-seeking reagents, such as the halogens or the alkyl halides, substitute on the carbons and not on the nitrogen, the α-position being favored over the β. Salts of pyrrole with acids are thus best represented by:

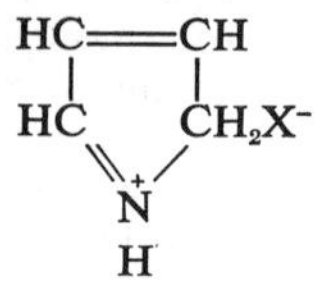

In the laboratory, pyrrole is usually made by the pyrolysis of ammonium mucate. Commercially, it has been separated from coal tar in small quantities but is now available synthetically from the reaction of ammonia with furane over an alumina catalyst. It can also be prepared by the action of ammonia on butynediol, made from acetylene and formaldehyde.

The most characteristic chemical property of the pyrroles is the ease with which they undergo nuclear substitution with electron-seeking reagents. In this respect they are analogous to the phenols. Pyrrole itself has nearly the same reactivity as resorcinol. Characteristic, for example, is the ease of iodination, even when tri-iodide ion is employed as the reagent, and the fact that acetylation in many cases does not require the addition of a catalyst such as aluminum chloride. One of the most useful reactions of the series is that of acid-catalyzed decarboxylation. This makes possible the protection of a position on the ring with an ester grouping and the hydrolysis and decarboxylation of this group when desired. Pyrroles react readily with aldehydes and ketones in the presence of acid. When formic acid or a pyrryl aldehyde is used, the product is a substitute dipyrrylmethene. Dipyrrylmethenes are yellowish orange pigments which are characterized by their high affinities for many metals, such as copper, nickel, iron or cobalt. They are used in the synthesis of porphyrins.

Mono-α-methylpyrroles may be condensed with bromine to form α-bromo-α^1-methyldipyrrylmethenes; methenes of this type are useful in the synthesis of porphyrins.

The biological significance of the porphyrines lies in their great affinity for some of the metals. This is because the four nitrogens of the porphyrin ring are presented to a metal at the correct positions for the formation of square planar bonds. Metals of the proper size and with a tendency to covalent bond formation in this configuration will form stronger complexes with porphyrins than with any other known type of organic compound. Metals capable of ionic bond formation and of fitting into the porphyrin ring will also form strong complexes, even though their normal structures are not square and planar.

Because of the shape of metalloporphyrins, copper, which is planar in the cupric state and tetrahedral in the cuprous state, is only capable of forming cupric porphyrin complexes. Thus the formation of a copper porphyrin derivative deprives the metal of its power to undergo a reversible oxidation-reduction reaction. Consequently, physiological copper catalysts are not porphyrin derivatives. In the case of iron, however, a porphyrin will occupy four of six vertices of an octahedron. Since both the ferrous and ferric derivatives have the same shape, oxidation-reduction reactions are not inhibited. The remaining positions of the octahedron may be occupied by many different types of substituents giving rise to a whole series of naturally occurring redox catalysts such as cytochromes, catalases, and peroxidases.

Investigations by Shemin and by Granick threw much light on the path of biological synthesis of the naturally occurring porphyrin pigments. Thus the use of radioactive tracers showed that the four nitrogens of the porphyrin are derived from glycine nitrogens and that the four bridges and four of the α carbons, come from the methylene group of glycine. All the remaining carbons arise from acetate ions. It is thus seen that the building blocks for the formation of porphyrins are always present in the diet. This confirms earlier nutritional findings that neither added chlorophyll nor added blood pigments are utilized in the formation of the organic skeletons of the physiological heme pigments.

A. H. CORWIN

Cross-references: *Porphyrins, Chlorophyll, Pigments, Phthalocyanines.*

Q

Quinoline compounds are characterized by the structure:

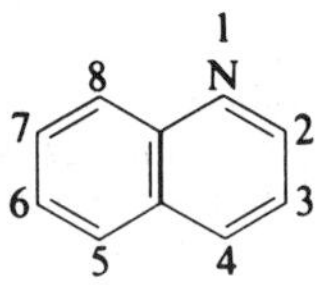

Quinoline, quinaldine (2-methylquinoline), and lepidine (4-methylquinoline), can be isolated from the tar-base fraction obtained from coal-tar distillates. While other quinoline compounds occur naturally (cinchona and angostura alkaloids), most derivatives are synthesized from monocyclic intermediates or, to a lesser extent, from quinoline, quinaldine and lepidine.

Quinoline compounds can be prepared by reaction of an aromatic amine with an appropriately substituted three-carbon fragment. In the Doebner-Miller and Skraup syntheses, this moiety is an α,β-unsaturated carbonyl compound formed in situ and produced by an aldol condensation in the former and by the dehydration of glycerol or a substituted glycerol in the latter. Thus, aniline and acetaldehyde yield quinaldine in a Doebner-Miller synthesis and p-toluidine and glycerol afford 6-methylquinoline in the presence of an oxidant in a Skraup synthesis. An α,β-unsaturated carbonyl compound, ethoxymethylenemalonic diethyl ester, is used directly in the Gould-Jacobs synthesis and on treatment with aniline affords 3-carbethoxy-4-hydroxyquinoline.

The three-carbon moiety is a β-dicarbonyl compound in the Combes, Knorr and Conrad-Limpach syntheses. The carbonyl groups can be keto and/or aldehydo groups in the Combes synthesis but β-ketoesters are employed in the latter two syntheses. Thus, aniline and acetylacetone yield 2,4-dimethylquinoline in a Combes synthesis and, depending on the reaction conditions, o-toluidine and acetoacetic ester produce either 2-hydroxy-4,8-dimethylquinoline (Knorr) or 4-hydroxy-2,8-dimethylquinoline (Conrad-Limpach).

Quinoline compounds can be prepared by the Friedlander synthesis, which consists of treating an o-aminoaryl carbonyl compound with an aliphatic carbonyl compound. Hence acetone and anthranilic acid afford 2-methyl-4-hydroxyquinoline. The Pfitzinger synthesis is an analogous reaction, but employs isatin as the nitrogenous reactant.

When isatin is treated with acetone, 2-methyl-4-carboxyquinoline is obtained.

Reactions: Quinoline is a weak base ($K_B = 3.2 \times 10^{-10}$) and undergoes many of the reactions of tertiary amines. In general quinoline compounds form quinolinium salts with Lewis acids or with reactive organohalides. When treated with base, N-alkylquinolinium salts carrying an alkyl substituent in the 2 or 4 position react with N-alkylquinolinium salts carrying no 2 or 4 substituent to form cyanine dyes. Cyanine blue is prepared in this manner.

Electrophilic substitution. Nitration of quinoline affords a mixture of the 8 and 5-nitroquinolines in which the former predominates. Similarly, quinoline-8-sulfonic acid is the main product of sulfonation. Bromination at 300°C yields 3-bromoquinoline but at 500°C 2-bromoquinoline is obtained.

Nucleophilic substitution. High yields of 2-amino or 2-hydroxyquinoline are obtained when quinoline is heated with alkali amides or hydroxides, respectively. When quinoline is heated with Grignard or organolithium reagents, the organic grouping enters the 2 position.

Oxidation and Reduction. The high-temperature oxidation of quinoline by sulfuric acid using a selenium catalyst represents a commercial synthesis of nicotinic acid. Depending on the nature and site of the substituents, substituted quinolines undergo oxidation to yield either pyridine or benzene derivatives or mixtures of both. Perbenzoic acid transforms quinoline into its N-oxide.

The catalytic hydrogenation of quinoline compounds yields a 1,2,3,4-tetrahydroquinoline as the main product. More strenuous conditions produce a decahydroquinoline. Accumulation of substituents in the 2,3 and 4 positions tends to favor the formation of a 5,6,7,8-tetrahydroquinoline.

Reactions of Substituted Quinolines. Except when substituted in the 2 and 4 positions of the quinoline ring, the reactions of substituents situated elsewhere parallel those of the correspondingly substituted naphthalene.

Alkyl derivatives: Quinoline compounds carrying a —C— R$_2$ in the 2 or 4 position undergo active methylene reactions, i.e., base-catalyzed condensations with aldehydes, ketones, epoxides, esters, acyl halides, reactive organohalides and carbon dioxide, as well as reactions such as the Michael and the

Mannich. Thus quinaldine and ethyl benzoate react in the presence of potassium ethoxide and form the ketone, 2(2-quinolyl) acetophenone.

Halogen derivatives. Halogen atoms substituted in the 2 or 4 positions undergo a facile replacement by hydroxyl, sulfhydryl, alkoxyl, ammono and amino groups as well as by anions like that derived from diethyl malonate. Thus, treatment of 2,7-dichloroquinoline with dilute acid demonstrates the differential reactivity of halogen atoms for 2-hydroxy-7-chloroquinoline is obtained.

Hydroxyl derivatives: The 2 and 4-hydroxyquinolines undergo reactions atypical of phenols in that treating them with phosphorus pentachloride effects conversion to the chloroquinoline, alkylation with alkyl halides or sulfates in an alkaline medium affords an N-alkyl derivative and their ethers can be cleaved with dilute acids. Spectroscopic evidence in the ultraviolet and infrared indicates that they exist in their tautomeric forms. Thus, 2-hydroxyquinoline exists as the quinolone, the latter structure appreciably stabilized through resonance with a dipolar structure. 8-Hydroxyquinoline forms insoluble chelate compounds with many metallic ions, e.g., aluminum, bismuth, cadmium, copper, magnesium, nickel and zinc among others. The use of this reagent as an analytical reagent has been investigated widely.

Amino derivatives. Anomolous results are obtained when a 2 or 4-aminoquinoline is treated with nitrous acid. Instead of the expected formation of the diazonium salt (which is formed from any of the five other aminoquinolines) the corresponding hydroxyquinoline is isolated. When concentrated hydrochloric or hydrobromic acid is used for diazotization, the corresponding halogen derivative is isolated.

Derivatives of 4-amino and 8-aminoquinoline, e.g., chloroquine and pamaquine are invaluable antimalerials.

Quinoline Compounds in Commerce. In addition to quinoline, lepidine, quinaldine, chloroquine, pamaquine, the cinchona alkaloids (antimalerials and cardiac depressants) and the cyanine dyes (photographic sensitizers), the following are marketed commercially: Nupercaine, a local anesthetic; Yatren 105 and Vioform, used as amebicides; chincophen, an analgesic and antipyretic; oxyquinoline sulfate, an antiseptic; and quinoline yellow, a paper dye.

The production of various quinoline compounds is given below:

	Production (pounds)	Sales (pounds)
Quinoline	782,000	781,000
Quinaldine	27,000	—
Quinoline yellow	68,000	55,000
Yatren 105	4,000	2,000

RALPH DANIELS

QUINONES

The quinones are unsaturated cyclic diketones with both the oxygen atoms attached to carbon atoms in simple, fused, or conjugated ring systems. They may be regarded as oxidation products of dihydroxy aromatic compounds with the substituent groups in positions corresponding to the *ortho* or *para* positions in the benzene ring. Thus *o*-benzoquinone can be prepared by the oxidation of catechol with silver oxide and *p*-benzoquinone by the oxidation of hydroquinone. The compound *m*-benzoquinone is unknown, and would not be feasible from structural considerations. In the naphthalene series six quinones are theoretically possible. The 1,2- and 1,4-quinones are similar structurally to the benzoquinones, but in 2,6-naphthoquinone the oxygen atoms are in adjacent rings.

A number of possibilities for quinone formation also exist in the anthracene series; the best known compound in this group is 9,10-anthraquinone, where both oxygens are substituted on the *meso* carbon atoms linking the two benzene rings. Unlike the benzoquinones, the rings are fully aromatic and additions to the double bonds do not take place readily. In the phenanthrene series compounds such as 3,4-phenanthrenequinone are known to exist, but the best known isomer is 9,10-phenanthrenequinone where the carbonyl groups are adjacent and two of the rings are aromatic in character.

A number of quinones are also known which can be regarded as derivatives of more highly condensed ring systems such as naphthacene, chrysene, and pyrene where the carbonyl groups may be adjacent, *para* to one another or contained in different rings. In 4,4′-diphenoquinone the two six-membered rings are not fused, but are linked by a double bond *para* to the carbonyl groups. In stilbenequinone and its derivatives the carbonyl groups are located in different rings which are conjugated through a two-carbon bridge. Acenaphthenequinone is commonly classified with the quinones, although this is not strictly correct since the carbonyl groups are contained in a five-membered ring which cannot be reduced to a fully aromatic ring system.

The most typical reaction of the quinones is their reversible reduction to aromatic dihydroxy compounds by chemical reagents or electrolysis. The oxidation potentials for the quinone-hydroquinone reactions depend upon the degree of conjugation of the reduced and oxidized ring systems. *p*-Benzoquinone has an oxidation potential of +0.715 volt while 1,4-naphthoquinone and 9,10-anthraquinone have potentials of 0.484 and 0.154 volt, respectively. Thus *p*-benzoquinone will readily oxidize 9,10-dihydroxyanthracene to the corresponding anthraquinone. The oxidation potentials of these systems are dependent on oxonium ion concentration; thus the quinhydrone and chloroanil (tetrachloro-*p*-benzoquinone) electrodes have been used for the measurement of pH in neutral or acid solutions. The carbonyl groups of these compounds will undergo typical reactions such as the combination of *p*-benzoquinone with one or two moles of hydroxylamine hydrochloride to yield *p*-benzoquinone monoxime or *p*-benzoquinone dioxime. *p*-Benzoquinone monoxime exists in tautomeric equilibrium with *p*-nitrosophenol.

The quinones will undergo substitution reactions in various ways depending upon the aromatic character of the ring system. *p*-Benzoquinone and 1,4-naphthoquinone can be chlorinated to yield tetrachloro-*p*-benzoquinone and 2,3-dichloro-1,4-naphthoquinone, respectively. The halogen atoms adjacent to the carbonyl group are labile and readily undergo nucleophilic substitution reactions with thiol or amino groups. The halogenation of anthraquinone is very difficult but a dibromoanthraquinone has been prepared. *p*-Benzoquinone is stable in cold concentrated chromic acid. Further oxidation with hydrogen peroxide opens the ring to yield diphenic acid. In the presence of alkaline permanganate one of the carbonyl groups is eliminated and the product is fluorenone. Phenanthrenequinone dissolves readily in sodium bisulfate solution to form an addition product. The double bonds in the ring system are aromatic in character.

H. P. Burchfield and George L. McNew

Cross-references: *Pigments, Dyeing, Hydroquinones, Vitamins.*

R

RACEMIZATION

The conversion of an optically active compound, i.e., one that rotates the plane of polarized light, into its racemic or optically inactive form is known as *racemization*. In this process half of the optically active compound is converted into its mirror image. The resultant mixture of equal quantities of the dextro- and levo-rotatory isomers is without effect on plane-polarized light due to external compensation (meso forms are internally compensated). Racemization of compounds possessing more than one asymmetric atom may yield products with residual optical activity due to the presence of unchanged centers of asymmetry. An example of this is found in the mutarotation of either α- or β-D-glucose in which only 1 of the 5 asymmetric centers is affected by the opening and closing of the hemiacetal ring.

Racemization may occur in molecules in which structural changes, such as those due to resonance, enolization, substitution or elimination of groups, temporarily destroy the asymmetry needed to maintain the optical activity. Also, Walden inversion of half of an optically active isomer can yield a racemate without the destruction of the center of asymmetry; this phenomenon is observed in the reaction of D-butanol-2 with $HClO_4$.

In the case of the optically active acids, racemization is postulated as the result of the enolization mechanism:

$$\begin{array}{ccc}
R & & R \qquad OH \\
\diagdown & & \diagdown \quad \diagup \\
C^*H-COOH & \rightleftharpoons & C=C \\
\diagup & & \diagup \quad \diagdown \\
R' & & R' \qquad OH
\end{array}$$

In this case it is seen that the hydrogen attached to the alpha carbon migrates to the oxygen of the carbonyl group forming a carbon-to-carbon double bond and thereby destroying the asymmetry previously found at the alpha carbon atom. Reversion of the enol to the acid reforms the center of asymmetry on a random basis (i.e., giving the same quantities of dextro and levo forms of the acid). Variations of pH, temperature, solvents and catalysts are likely to change the rates of racemization. The progress of the reaction is conveniently followed by examination of the reaction mixture with plane-polarized light in a polarimeter. A similar racemization of disodium L-cysteine in liquid ammonia in the presence of $NaNH_2$ is reported to occur through abstraction of the α-proton to form a carbanion followed by random recombination of the proton.

Although most racemization reactions have been observed to occur in solution, L-leucine on the surface of silicates has been completely racemized by heating to 200°C for 6 hours. Racemization has also been observed in optically active compounds which possess no asymmetric atom but owe their activity to hindered rotation around a single C—C bond. For example, 8'-methyl-1,1'binaphthyl-8-carboxylic acid has been resolved into its optically active forms and racemized by heating in dimethylformamide.

One of the earliest known examples of racemization was described by Pasteur in his studies of tartaric acid. By heating D-tartaric acid to 165°C in water he partially converted it into a mixture of D-, L- and *meso*-tartaric acids. The occurrence of some DL-tartaric acid along with the natural D-tartaric acid in the wine industry is explained on the basis of partial racemization.

The process of racemization has a number of practical application in the laboratory and in industry. Thus, in the synthesis of an optical isomer it is frequently possible to racemize the unwanted isomer and to separate additional quantities of the desired isomer. By repeating this process a number of times it is theoretically possible to approach a 100% yield of synthetic product consisting of only one optical isomer. An example of the utilization of such a process is found in the production of pantothenic acid and its salts. In this process the mixture of D- and L-2-hydroxy-3,3-butyrolactones are separated. The D-lactone is condensed with the salt of beta-alanine to give the biologically active salt of pantothenic acid. The remaining L-lactone is racemized and recycled.

The process of racemization is important in the survival and growth of living cells and is catalyzed by a group of enzymes called racemases. Alanine racemase, for example, is able to convert D-alanine to DL-alanine if a suitable alpha keto acid is also present. In this reaction the asymmetry of the alpha-carbon atom of alanine is lost as the amino acid is converted to the keto acid and back. This process is analogous to the well-known process of transamination (in which racemization seldom occurs).

Louis H. Goodson

Cross-references: *Asymmetry, Optical Rotation, Stereochemistry.*

RADIATION

Radiation in the oldest and strictest sense is the process in which energy is emitted by a body in the form of quanta or photons, each quantum having associated with it an electromagnetic wave possessing a frequency ν, and wave length λ. This in-

TABLE 1. RANGE OF ELECTROMAGNETIC WAVES

Type	Octaves	Wave length Range, A (1A = 10^{-8} cm)	Generation	Detection
γ-Rays	—	0.001–1.4 0.06–0.5 used in radiology	Emitted when atomic nuclei disintegrate (radioactivity)	As for x-rays but more penetrating
X-rays	14	0.006–1019 0.0001 and less in betatron	Emitted by sudden stoppage of fast moving electrons; emitted by stars and galaxies	a. Photography b. Phosphorescence c. Chemical action d. Ionization e. Photoelectric action f. Diffraction by crystals, etc.
Ultraviolet rays	5	136–3,900	Radiated from very hot bodies and emitted by ionized gases	Same as x-rays a–e: reflected, refracted by finely ruled gratings
Visible rays	1	3,900–7,700 Violet 3,900–4,220 Blue 4,220–4,920 Green 4,920–5,350 Yellow 5,350–5,860 Orange 5,860–6,470 Red 6,470–7,700	Radiated from hot bodies and emitted by ionized gases	Sensation of light; same as as ultraviolet rays
Infrared rays	9	7,700–4 × 10^6	Heat radiations	Heating effects on thermocouples, bolometers, etc. Rise in temperature of receiving body. Photography (special plates). Reflected, refracted, diffracted by coarse gratings
Solar radiation	—	Limiting wave lengths reaching earth; 2,960–53,000		

TABLE 1. CONTINUED

Type	Octaves	Wave length Range A (1 A = 10^{-8} cm)	Generation	Detection
Hertzian waves	28	1 × 10^6 to 3 × 10^{14}		
Short Hertzian	17	1 × 10^6 to 1 × 10^{11}	Spark-gap discharge oscillating triode valve, etc.	Coherer. Spark across minute gaps in resonant receiving circuit. Reflected, refracted, diffracted.
Radio	11	1 × 10^{11} to 3 × 10^{14}	Same	Coherer. Conversion to alternating current. Rectification with or without heterodyning and production of audible signals
Broadcasting band	—	2 × 10^{12} to 5.5 × 10^{12}		
Electric waves	—	3 × 10^{14} to 3.5 × 10^{16}	Coil rotating in magnetic field	Mechanical. Electrical. Magnetic. Thermal effects of alternating currents

cludes the radiant energy associated with the electromagnetic spectrum ranging from the highest energy, or shortest wave lengths, of secondary cosmic rays, the γ-rays from radioactive disintegration of atomic nuclei, x-rays, ultraviolet, visible light, infrared, radio or Hertzian and electric waves. All these waves, seemingly so different in properties and produced by vastly different methods, even though the origin in all cases may be termed the unrest of electric charges, are actually identical in every respect except length. They are not electrically charged and they are massless. All have the same velocity of propagation, namely, 30 billion cm/sec.; all may be refracted, reflected, diffracted and polarized; all display a dual nature, apparent in different properties, namely wave-like and corpuscular (the photons). Table 1 summarizes the ranges of the electromagnetic spectrum.

In a looser sense the term "radiation" also includes energy emitted in the form of particles which possess mass and may or may not be electrically charged. Beams of such particles may be considered as "rays." The charged particles may all be accelerated and high energy imported to "beams" in the particle accelerators such as cyclotrons, betatrons, synchotrons and linear accelerators. The article **Particles and Antiparticles,** discusses these, their relationships and complex significance.

GEORGE L. CLARK*

Cross-references: *Radioactivity, Atoms, Electrons, Mesons, Alpha Particles.*

*Deceased.

RADIATION CHEMISTRY

Radiation chemistry is concerned with the chemistry of processes produced by high-energy radiation. For the purpose of definition, two arbitrary distinctions between radiation chemistry and photochemistry have been suggested. In one, the existence of ionization (in addition to excitation) is designated as the distinctive feature of radiation chemistry; in the other, the absorption of a photon in a single elementary process (*e.g.,* by an atom, molecule, free radical, or ion) is designated as the distinctive feature of photochemistry. In practice there is seldom any ambiguity, however, because radiation energies in the range of a thousand to millions of electron-volts are generally used in radiation chemistry.

The emissions of nuclear reactions (neutrons, recoil particles, and alpha, beta, and gamma rays) and their artificially produced analogs (x-rays and highly accelerated electrons and atomic nuclei) are the radiations commonly used in radiation chemistry. The natural alpha emitters and x-ray machines were used as sources in most early work (*i.e.,* prior to about 1942). At the present time, the most generally used sources are cobalt-60 (1.17 and 1.33 MeV gamma rays, half-life of 5.3 years) and charged-particle accelerators such as the cyclotron, the linear accelerator (linac), and the Van de Graaff generator. Particle emissions of nuclear processes are used primarily for special purposes such as for studies of the effect of amount of energy deposition per unit length of particle track (linear energy transfer—LET) in the medium. For all the energetic-particle and electromagnetic-wave radiations employed in radiation chemistry, nearly all the energy directly responsible for chemical effects is deposited by interaction of a charged particle with matter. For example, most of the energy of energetic neutrons is dissipated by the energetic charged particles (primarily protons in hydrogenous systems) knocked out of molecules in momentum-transfer processes, and the energy of gamma or x-rays is dissipated by the electrons produced in Compton, photoelectric, or pair-production processes.

Radiation-chemical processes are characterized quantitatively by a yield. Early workers in the field observed a correlation between the amount of chemical change and the amount of ionization. By analogy with the expression of photochemical yields in terms of quantum yield (*i.e.,* molecules produced or converted per photon absorbed), radiation-chemical yields were expressed in terms of molecules produced or converted per ion pair. However, high-energy radiation produces excited molecules as well as normal and excited ions plus electrons. A measure of the relative amount of total excitation may be obtained by comparison of the ionization potentials of molecules, in the range 8–16 eV, with the average energy expenditure per ion pair formed, usually in the range 25–35 eV, as determined in ionization-chamber experiments with high-energy radiation. Thus, in general, over half the deposited energy is manifested as some form of excitation. Therefore, in order to avoid the implication of mechanism, radiation-chemical yields are generally expressed at present in terms of molecules produced or converted per 100 eV absorbed by the system. Such yields are denoted by the symbol G and for significant non-chain processes are in the range 0.1–20. Energy absorbed by a system is usually measured by use of a secondary standard such as a chemical dosimeter for which G has been determined accurately by methods such as calorimetry, ionization-chamber measurements, or measurement of the input of charged particles of accurately known energy from a Van de Graaff accelerator. The Fricke dosimeter (an aqueous solution of 0.4 M H_2SO_4, 10^{-3} M $FeSO_4$ and 10^{-3} M NaCl) is a universally accepted secondary standard with $G(Fe^{2+} \rightarrow Fe^{3+}) = 15.6$ for high-energy electrons, γ- and x-rays.

The primary objective of research in radiation chemistry is development of a complete picture of the sequence of events from deposition of radiation energy in a system to production of ultimate chemical effects. Between the initial act of energy deposition and the ultimate effect in systems exposed to high-energy radiation, a complicated array of elementary physical and chemical processes occurs; the processes involve a variety of transient species (ions, electrons, excited states, free radicals) of lifetimes that range from tens of milliseconds to hundredths of a picosecond (10^{-12} sec). Therefore, a major preoccupation of experimental and theoretical research in radiation chemistry is the development and application of methods for study of the formation and behavior of transient species resultant from the deposition of radiation energy in a system.

The transient species consequent on energy deposition in a medium have been studied by a variety of methods. One method for direct observation of a radiation-induced transient species is to use conditions that preclude those reactions by which the species would normally disappear; then its lifetime may be increased to such an extent that it becomes a persistent rather than a transient species, and it can be studied by some conventional technique. Thus, on irradiation of crystals and glasses at −196° or lower, positive and negative ions (the latter formed by attachment of electrons to certain additives), electrons, atoms and free radicals are immobilized and are identified by their characteristic absorption spectrum or by electron-spin-resonance (ESR) spectrometry. The behavior of mobile positive holes and electrons (*i.e.,* precursors of the immobilized positive and negative ions and electrons) can be inferred from the results of such experiments, and subsequent reactions of the immobilized species can be followed by the study of luminescence or changes in the ESR or absorption spectrum with increase in temperature or on exposure to light of selected wave lengths. Such studies have provided an insight into the nature and behavior of electrons in a variety of condensed media and provide a fundamental basis for interpretation of radiation effects in liquid and solid systems at ordinary temperatures.

The study of radiation-induced transient species in systems at ordinary temperatures has been accomplished by the application of ESR, luminescence, and conductivity techniques to systems under continuous irradiation and by the use of such techniques and absorption spectroscopy in pulse radiolysis. Pulse radiolysis has made possible observation of the time-dependent behavior of transient species produced by electron or x-ray pulses in the MeV energy range; pulse duration has been shortened from the much studied microsecond range into the nanosecond range and, recently, into the picosecond range. Application of the pulse radiolysis technique with absorption spectroscopy has demonstrated the existence of hydrated electrons in irradiated aqueous solutions (with a lifetime near 1 millisecond in very pure water) and has provided absolute specific rates for numerous reactions of this species (as well as of hydrogen atoms and hydroxyl radicals in aqueous solution). Behavior of the hydrated-electron precursor (the "dry" electron) has been inferred from the results of recent picosecond work. In numerous other systems, pulse radiolysis has permitted identification and study of the behavior of electrons and a variety of ions, excited states, and free radicals. In an application of nanosecond pulse radiolysis to ethanol glasses at $-196°$, dipole relaxation in the coulomb field of a trapped electron has been observed by the change from an absorption spectrum (immediately after a pulse) similar to that of an electron trapped in an alkane glass to the spectrum of the ethanol-solvated electron.

Luminescence (fluorescence or phosphorescence) from an irradiated system is a direct manifestation of transient excited species and, as such, has been used for elucidation of the elementary reactions of such species. The luminescence studied from an irradiated system may be that of the solvent or of a suitably chosen scintillator solute (present at a concentration that may be as low as 1 molecule in 100,000). Study of the spectrum and intensity of luminescence as a function of temperature, viscosity, and medium composition (concentrations of scintillator, quenchers, and other energy acceptors) has contributed to an understanding of such relaxation processes of excited molecules as internal conversion, intersystem crossing, fluorescence and phosphorescence, excited dimer formation, and excitation transfer. Radiation studies have been complemented by similar photoexcitation studies in which luminescence is studied as a result of direct excitation of molecules to specific states. The use of nanosecond pulses in both the photoexcitation and radiation-excitation work has provided an extensive accumulation of absolute specific rates for the various relaxation processes of excited molecules. Present photoexcitation studies employ lasers for work in the picosecond range.

Measurements of radiation-induced conductivity in liquids of low dielectric constant (ϵ less than 10) give yields of those sibling cation-electron pairs whose members escape from their mutual coulombic fields. Yields of such free ion pairs are in the range $G = 0.05$ to $G = 0.9$ for all (about 40) of the molecular dielectric liquids studied (G = 4.4 for such ion pairs in liquid argon). By the use of pulse techniques, values have been obtained for the electron mobility in a number of representative molecular dielectric liquids. In units of cm^2 sec^{-1} V^{-1}, the values are in the range 0.1–90; a characteristic anion mobility in such liquids is 0.001, and the mobility of an excess electron in liquid argon is near 500. Such results, particularly in combina-

tion with spectrophotometric observation of the electron and its decay in pulse experiments, provide a deep insight into the nature and behavior of electrons in dielectric liquids.

Much research in radiation chemistry has been concerned with the study of radiation effects that are persistent under ordinary conditions. From such ultimate effects (*e.g.*, over-all chemical changes), it is frequently possible to make reasonable inferences with respect to certain early effects and the transient species involved. A common technique for detection and study of the behavior of radiation-induced transient species is the addition to a system of a small concentration of some indicator substance (scavenger) that will intercept a particular transient species with production of a characteristic effect in the indicator or on the ultimate reaction products. In this manner, radioactive iodine and C^{14}—labeled ethylene have been used for the study of free radicals. The yields and behavior of solvent excited states have been determined by the use of scavengers that give characteristic products with known quantum yield from their excited triplet and singlet states; *e.g.*, the *cis-trans* isomerizations of diolefins, 2-olefins, and stilbenes have been used for study of triplet excited states. By study of the effect of pressure (up to 9000 atmospheres) on a product yield determined by a competition between two reactions (one being diffusion controlled) of the hydrated electron, the partial molal volume and cavity volume of an electron in water have been determined. The scavenger technique has been especially fruitful in the elucidation of elementary ionic processes in dielectric liquids. Scavengers specific for the solvent cation or the electron are used to probe the population of cation-electron pairs whose members do not escape from their mutual coulombic fields but undergo rapid recombination. Such "geminate" ion pairs do not contribute to the measured conductivity but have been observed in certain pulse experiments. Ion scavenging studies give total ion-pair yields in dielectric liquids of $G \approx 4$, in accord with gas-phase values.

For the interpretation of gas-phase radiation chemistry, much information is available with respect to energy-deposition processes and subsequent ionic processes from ion-chamber, ion-mobility, cloud-chamber, and mass-spectrometric studies. The mass spectrometer has become a powerful tool for the study of processes induced by collision of an isolated molecule with an electron of well-defined energy and for study of reactions of an ion with a molecule. Such studies provide a wealth of thermochemical data fundamental to all chemical science. Among such data are the energies associated with the following processes: formation of normal and excited ions plus electrons from molecules and free radicals, ion-pair formation (in this special context, the separation of a molecule into two oppositely charged ionic fragments), bond dissociation, attachment of a proton to one or more molecules, and attachment of an electron to a molecule or free radical, sometimes with dissociation of the anion. Since these are the dominant processes that occur at an early stage in the radiation chemistry of a system, a knowledge of such processes is fundamental for interpretation of radiation effects. Additional information (*e.g.*, electron reaction rates and thermalization times) is obtained from electron swarm experiments and by application of the microwave conductivity technique to pulse irradiated gases. With such information and information on the behavior of excited molecules derived from photochemical studies (which, in recent years, have been extended to the far ultra-violet, ~500 Å), much understanding of the radiation chemistry of gaseous systems has been acquired from continuous and pulse irradiation experiments that utilize applied electric fields, sensitizers, isotopic labeling, and scavengers for ions, electrons and free radicals.

For the most part, detailed statements regarding deposition of energy in condensed systems are hypotheses based on extrapolation from knowledge of gaseous systems. Experiments on the scattering of electrons by thin films (~100 Å) may reveal some of the details of energy deposition. Non-homogeneity is a characteristic feature of energy deposition in condensed systems. Because many of the subsequent processes are diffusion controlled, the chemical consequences of radiation absorption are dependent on the spatial distribution of primarily produced ionized and excited species. The LET or ionization density along a particle track varies directly with the square of particle charge and inversely with the square of particle velocity. A 1 MeV electron may transfer energy to about one molecule in every thousand along its track. In condensed systems, many such energy transfers result in several ionizations and excitations within a small volume called a spur so that the track of a 1 MeV electron resembles a string of widely separated beads (the spurs). However, the distance between successive energy transfers along the track of an alpha particle of similar energy may be only a few molecules; thus, the spurs overlap and the track is a cylinder of dense ionization and excitation. Such nonhomogeneity of energy deposition has been investigated by study of ESR relaxation times for electrons trapped in glasses at −196°. Theoretical treatments based on the Samuel-Magee diffusion model have been quite successful in quantitative interpretation of solute and LET effects in the radiation chemistry of aqueous solutions. For description of the nonhomogeneous kinetics of ionic processes in dielectric liquids, models have been presented which incorporate a solvent-dependent distribution of sibling cation-electron separation distances (at electron thermalization) and a corresponding distribution of cation-electron neutralization times. Such distributions determine the yield of free ion pairs and the kinetics of cation and electron scavenging.

The study of radiation-induced polymerization and of radiation effects in polymers has contributed to understanding of some of the elementary reactions of radiation chemistry. The radiation chemistry of inorganic solids and of solid-adsorbate (heterogeneous) systems, and the study of chemical reactions on radiation-induced catalytic or energy-storage sites in inorganic solids have been subjects of considerable interest. Initial processes of great importance in these systems are electron and positive-hole trapping at solid imperfections and the creation of such imperfections. Many of the same

methods (chemical, spectroscopic, ESR) are used for study of these systems, but the conceptual framework for elucidation of mechanism is derived largely from solid-state physics.

The concepts and information developed in basic radiation-chemistry research have been applied to development of industrial radiation-chemical processes, such as the treatment and synthesis of polymers and the manufacture of ethyl bromide, and to the synthesis of chemicals not readily synthesized by other means.

ROBERT R. HENTZ

References

Burton, M., and Magee, J. L., eds., "Advances in Radiation Chemistry," Wiley-Interscience, New York, 1969.
Ausloos, P., ed., "Fundamental Processes in Radiation Chemistry," Interscience, New York, 1968.
Haissinsky, M., ed., "Actions Chimiques et Biologiques des Radiations," Masson et C¹ᵉ, Paris, 1970.
Hart, E. J., and Anbar, M., "The Hydrated Electron," Wiley-Interscience, New York, 1970.
Phillips, G. O., and Cundall, R. B., eds., *Radiation Research Reviews*, Elsevier, Amsterdam, 1968.

RADICALS

In numerous chemical reactions, certain groups of atoms remain bound together. In reactions involving dissociation and ionization such groups are referred to as *radicals* or, more exactly, *radical ions,* for example, OH^-, $SO_4^=$, $CO_3^=$, NH_4^+, NO_3^-, $Al(H_2O)_6^{+++}$.

In reactions not involving dissociations where a group of atoms tenaciously stick together, for example, in the displacement of OH from an alcohol, R—OH, the OH is also referred to as a hydroxy radical, but more often as the hydroxy group. In the same compound R— may either be called an alkyl group or an alkyl radical. The latter is more often reserved for the case where a free alkyl radical, $R\cdot$, is meant.

A free radical is a substance carrying a magnetically noncompensated electron. This definition does not exclude atomic sodium or other atoms carrying unpaired electrons, but in common usage elements are referred to as free atoms rather than as free radicals. The term "free radical" is reserved for very short-lived alkyl free radicals or to the longer-lived but reactive large organic molecules of the triaryl methyl type. Nevertheless, a few simple molecules having an odd number of electrons, nitric oxide, chlorine dioxide, and nitrogen dioxide, have many of the properties associated with free radicals.

The properties of free radicals can best be considered by comparison with the properties of free atoms such as sodium or chlorine. Both of these atoms are extremely reactive due to the unpaired electron which is seeking to become paired.

When sodium vapor is passed into chlorine gas, it burns with a bright yellow flame, forming sodium chloride and liberating 35,000 calories of heat per mole:

$$Na\cdot + Cl_2 \rightarrow Na^+Cl^- + Cl\cdot + 35,000 \text{ cal.}$$

The combination of hydrogen and chlorine does not take place in the dark if both substances are dry. Sunlight, however, will initiate the reaction, which then proceeds by a chain mechanism in a violent, explosive manner:

$$Cl_2 + energy \xrightarrow{\text{sunlight}} 2Cl\cdot$$

$$Cl\cdot + H_2 \rightarrow HCl + H\cdot$$

$$H\cdot + Cl_2 \rightarrow HCl + Cl\cdot, \text{ etc.}$$

Only a few free chlorine atoms (theoretically, one) need be formed by the absorption of energy from the light to set off the mixture to a complete reaction. Actually, but rarely, the free $H\cdot$ or $Cl\cdot$ may combine with another free $H\cdot$ or $Cl\cdot$ which would stop the chain.

The chlorination of methane is another example of a free radical reaction propagated by a chain mechanism. If light is allowed to impinge on a mixture of chlorine and methane, a chain reaction is started in which methyl chloride (CH_3Cl), methylene chloride (CH_2Cl_2), chloroform ($CHCl_3$), and carbon tetrachloride (CCl_4) are all formed.

Paneth first prepared a free methyl radical in 1929 by the thermal decomposition of lead tetramethyl at low pressure in an inert atmosphere in a quartz tube. The existence of the free methyl was detected by allowing it to pick up a metal mirror of zinc, tellurium, antimony, or lead deposited in the quartz tube just beyond the point at which the tube was heated. The methyl free radical had a short half-life of about 0.006 second at 1 mm pressure. Free methyl radicals may also be obtained by the thermal decomposition of azomethane at 400° or from hydrocarbons or ethers at still higher temperatures (900°C).

$$CH_3\text{—}N{=}N\text{—}CH_3 \xrightarrow{400°} N_2 + 2CH_3\cdot$$

Gomberg's preparation of the first free radical in 1900 involving an unpaired electron on carbon was viewed with scepticism, since the idea that carbon always carried four bonds in all its compounds had become firmly grounded. He had treated triphenylchloromethane with metallic silver with the expectation of obtaining hexaphenylethane. Instead he obtained a substances which took up oxygen from the air and which he was finally forced to call a triphenylmethyl free radical. Actually, hexaphenylethane is first formed, but it dissociates in benzene to triphenylmethyl radicals to an extent of about 3%. Some substituted ethanes are known which are completely dissociated into triaryl methyls, e.g., hexa-p-biphenylethane.

Though stable in the absence of air, the triaryl methyl free radicals add a large number of reagents, e.g., iodine, oxygen, nitric oxide, hydrogen in the presence of platinum, water if a trace of iodine is present, aniline, phenol, hydrogen chloride.

The triaryl methyl radicals are colored; the more chance for resonance structures that are possible, the more highly colored it will be. Triphenylmethyl is yellow, for example, but tri-p-biphenylmethyl is a deep violet.

Numerous commercial polymerizations to produce plastics, synthetic fibers, molding compounds, and rubber are catalyzed by substances which have been shown to produce free radicals. One of the most widely used of these catalysts is benzoyl peroxide, which decomposes into free benzoxy radicals. The benzoxy free radical is capable of initiating a chain reaction which may continue indefinitely. It may also lose carbon dioxide to give a new free radical (phenyl), itself capable of initiating polymerization. The initiation step is the reaction of the free radical with a molecule of the monomer (vinyl chloride, for example) to be polymerized. Propagation goes without regeneration of any new fragment and in this way differs from the halogenation of methane. This type of chain reaction may be terminated by removal of hydrogen from a monomer, another chain, water, or in other ways. Other peroxides and hydroperoxides may decompose in a similar manner to yield free radicals which in turn cause polymerization in the same way.

LEALLYN B. CLAPP

RADIOACTIVITY

Radioactivity, discovered by Becquerel in 1896, is the spontaneous nuclear disintegration of atoms with emission of corpuscular or electromagnetic radiations. The principal types of radioactivity are α-disintegration, β-decay (electron emission, positron emission and electron capture) and isomeric transition. The radiations are ordinarily α-particles (doubly charged He nuclei), β-particles (electrons), and γ-rays. (See **Radiation, Particles and Antiparticles**). Radioactivity may be natural or artificial or induced. The first is exhibited by naturally occurring substances; the second is produced by nuclear reactions under controlled conditions. Natural radionuclides may be classified as follows: (1) *Primary,* which have lifetimes exceeding several hundred million years and which presumably have persisted from the time of nucleogenesis, to the present; they include the α-emitters U^{238}, U^{235}, Th^{232} and Sm^{147} and the β-active nuclides K^{40}, Rb^{187}, La^{138}, In^{115}, Lu^{176} and Re^{187}. (2) *Secondary,* which have geologically short lifetimes and are decay products of primary natural radionuclides. All presently known members of this class belong to the elements from Tl to U; those derived from U^{235} are members of the U, or Ra, series; those from U^{238} of the actinium series, and those from Th^{232} of the Th series. (3) *Induced,* which have geologically short lifetimes and are products of nuclear reactions occuring currently or recently in nature; examples are C^{14} (natural radiocarbon, by which dating may be done) produced by cosmic ray neutrons in the atmosphere, and Pu^{239} produced in uranium minerals by neutron capture. (4) *Extinct,* which have lifetimes too short for survival from the time of nucleogenesis to the present, but long enough for persistence into early geologic times with measurable effects; at present I^{129} is the only suspected member of this class.

The radioactive series mentioned above is a succession of nuclides each of which transforms by radioactive disintegration into the next until a stable nuclide results. The first member is called the *parent,* the intermediate members are called *daughters,* and the final stable member is called the *end product.* Three such series are encountered in natural radioactivity and many others in induced radioactivity, particularly among the heavy elements and fission products. In such a series the emission of an α-particle reduces the atomic number by 2(loss of 2+ charges from the nucleus) and the atomic weight by 4; the emission of a β-particle results in an increase of 1 in the atomic number and no change in atomic weight.

Artificial radioactivity, of course, refers to the unstable nuclear species which have been produced by bombardment of relatively stable nuclei by high-energy protons, electrons, neutrons, α-particles, etc., in atomic piles, cyclotrons, betatrons and other particle accelerators. All the transuranium elements, each with several isotopes, 93 to 105, elements such as 43, 61, 85, 87 and numerous isotopes of previously naturally occurring elements have been so produced. A disintegration scheme for neptunium 93, illustrates the numerous series:

$$[_{94}Pu^{241}] \xrightarrow[(10y)]{\beta-} [_{95}Am^{241}] \xrightarrow[(490y)]{\alpha}$$

$$_{93}Np^{237} \xrightarrow[(2.2 \times 10^6 y)]{\alpha} {}_{91}Pa^{233} \xrightarrow[(27.4d)]{\beta-}$$

$$_{92}U^{233} \xrightarrow[(1.62 \times 10^5 y)]{\alpha} {}_{90}Th^{229} \xrightarrow[(7 \times 10^3 y)]{\alpha}$$

$$_{88}Ra^{225} \xrightarrow[(14.8d)]{\beta-} {}_{89}Ac^{225} \xrightarrow[(10.0d)]{\alpha}$$

$$_{87}Fr^{221} \xrightarrow[(4.8m)]{\alpha} {}_{85}At^{217} \xrightarrow[(0.018s)]{\alpha}$$

$$_{83}Bi^{213} \begin{cases} \xrightarrow[47m]{\beta-(96\%)} {}_{84}Po^{213} \xrightarrow{\alpha} (3.2 \times 10^6 s) \\ \xrightarrow[\alpha(4\%)]{} {}_{81}Tl^{209} \xrightarrow[(2.2m)]{\beta-} \end{cases}$$

$$_{82}Pb^{209} \xrightarrow[(3.32h)]{\beta-} {}_{83}Bi^{209} \text{ (stable)}$$

Aside from the theoretical significance of radioactivity in terms of nuclear structures and stabilities, and of the synthesis of new atomic species, now up to hafnium, atomic number 105, there are some very practical aspects of the phenomenon, namely, chemical dating, the use of radioactive tracers in following chemical and biological reactions, and therapy with β- and γ-rays. As an example of use of a tracer may be cited the use of ThB, a natural source, an isotope of Pb, added in traces to storage battery plates to follow all the reactions and mechanism of charge and discharge. The pure β-emitting radio-isotopes useful in chemistry, biology and medicine are as in Table 1.

GEORGE L. CLARK*

Cross-references: *Chemical Dating, Isotopes, Iodine, Radium, Thorium, Transuranium Elements, Uranium.*

*Deceased.

RADIUM

Radium is the heaviest of the alkaline earth elements (Groups IIA of the Periodic Table). Its atomic number is 88 and the atomic weight of the principal isotope is 226.0254, with a half-life of 1620 years; 13 other isotopes are known which have half-lives of only a few weeks. The presence of radium in nature is due to its continual formation by the decay of uranium 238.

History. In 1898, Pierre and Marie Curie and G. Bémont found that the major part of the radioactivity in pitchblende was due to a radioactive substance. In the next few years, they managed to concentrate the substance in the barium fraction of their residues and gradually to separate it from barium by fractional crystallization. Examination of the spectrum of the fractions by Demarcay showed that new lines in the ultraviolet became more intense with respect to Ba as the purification continued; simultaneous measurements of the apparent atomic weight of the Ba fraction as the radioactivity was enriched showed a decided increase, indicating the presence of an element heavier than Ba. This was named radium. The final atomic weight, obtained by Marie Curie, of the purest fraction was 225.18, which was very close to the present value.

For the early work on radium and the studies of radioactivity that grew out of it, Marie Curie received several prizes from the French Academy of Science in 1902 and Pierre Curie was elected to membership in the Academy in 1905. Both later received the Nobel Prize for their work.

Occurrence. Radium occurs in all uranium ores, being supported by radioactive decay. Since it forms many water-soluble compounds, it is leached from the ores by ground water and is widespread over the earth's surface. In pitchblende radium is present to about 300 milligrams per 10^6 g of uranium. In the earth's surface there are about 1.8×10^{13} g of radium. In normal ocean water there is about 10^{-13} gram per liter and the earth's ground water contains from 10^{-12} to 10^{-11} gpl depending on the location.

Production. Radium is coprecipitated with silica, barium, and lead from uranium ores by an acid sulfate treatment. The precipitate (called white cake) is then treated with excess carbonate to convert the radium and barium salts to the carbonates. These are then treated with HCl to remove the lead and silica and the process is repeated

TABLE 1

Element	Isotope	Half life	Max. Energy, Mev.
Hydrogen	$_1H^3$	12.46 yrs	0.01795
Carbon	$_6C^{14}$	5568. yrs	0.155
Phosphorus	$_{15}P^{32}$	14.32 dys	1.701
Sulfur	$_{16}S^{35}$	87.1 dys	0.1670
Chlorine	$_{17}Cl^{36}$	4.4×10^5 yrs	0.714
Calcium	$_{20}Ca^{45}$	152 dys	0.254
Arsenic	$_{33}As^{77}$	38 hrs	0.700
Strontium	$_{38}Sr^{89}$	53 dys	1.463
Strontium	$_{38}Sr^{90}$	19.9 yrs	0.61
Yttrium from Sr^{90}	$_{39}Y^{90}$	61 hrs	2.18
Silver	$_{47}Ag^{111}$ (some γ)	7.6 dys	(1.04 (91%)) (0.80 (1%)) (0.70 (8%))
Bismuth	$_{83}Bi^{210}$ (some α)	5.02 dys	1.17

The most useful γ-ray emitters are $_{11}Na^{22}$, $_{11}Na^{24}$, $_{19}K^{42}$, $_{21}Sc^{46}$, $_{23}V^{48}$, $_{25}Mn^{52}$, $_{25}Mn^{54}$, $_{26}Fe^{59}$, $_{27}Co^{58}$, $_{27}Co^{60}$, $_{29}Cu^{64}$, $_{30}Zn^{65}$, $_{32}As^{76}$, $_{35}Br^{82}$, $_{51}Sb^{124}$, $_{53}I^{131}$, $_{55}Cs^{137}$, $_{79}Au^{198}$. Many of these isotopes have replaced x-ray generators, notably $_{27}Co^{60}$, which is used extensively in cancer therapy and in industrial radiography for penetrating thick metal sections *in situ*.

several times. The final product of pure mixed $BaCO_3$ and $RaCO_3$ is then converted to $BaBr_2$ and $RaBr_2$ which are separated from each other by fractional crystallization, giving a final product of up to 90% $RaBr_2$. For further purification ion exchange techniques are used.

Radium metal can be prepared from its salts by electrolytic reduction. This was first done by M. Curie and A. Debienes. A $RaSO_4$ can be precipitated by dilute H_2SO_4. Its solubility is 2.1×10^{-3} gpl at 20°C.

Commercial Uses. *Neutron Sources.* The alpha particles emitted by radium and its daughters are sufficiently energetic to cause nuclear reactions in light elements. The most important of these is the $_4Be^9 + _2He^4 = _6C^{12} + _0n^1$ which is the basic reaction of commercial Ra-Be neutron sources. In practice, the beryllium is in the form of finely divided powder (less than 300 mesh) which is intimately mixed and compressed with particles of $RaSO_4$ or a $RaSO_4$–$BaSO_4$ mixture of less than 30 micron diameter. This type of source can give about 1.7×10^7 neutrons/g of radium.

However, the neutron yield depends on the Be/Ra ratio; in general a ratio of 8/1 is used, which gives about 10^7 n/g of Ra. The yield also depends on the packing density and thus will vary from source to source.

A more constant, although weaker, source can be made by using the compound $RaBeF_4$ which may be prepared with a constant Be ratio. This will yield 1.84×10^6 n/sec/g of $RaBeF_4$ and is constant to 0.5%.

These Ra-Be sources emit neutrons having a broad spectrum with maximum energy of 13.0 MeV and the greatest intensity at about 5 MeV. If one uses the gamma rays emitted by the Ra (actually by its daughters) by enclosing the radium in a capsule in the Be so that none of the alpha particles will penetrate, then the neutrons are formed by the reaction $\gamma + _4Be^9 = _0n^1 + 2_2He^4$. The neutrons then will have an energy of less than 0.6 MeV and yield of about 2×10^5 n/sec/g of Ra.

Luminous Paints. An important commercial application of radium was in the manufacture of luminous paints. In these an alpha emitter such as radium was uniformly dispersed in an inorganic phosphor. Since the phosphorescence was excited by the alpha particles, no external source of energy was required. These paints were widely used for watch dials and instrument dials until the early 1950's. The availability of other, less hazardous alpha emitters has since then reduced the use of radium for this purpose.

Radon Sources and Radiotherapy. Although Ra^{226} itself does not emit very intense γ radiation, the radioactive daughters in equilibrium with it are intense gamma emitters. For this reason radium sources have long been used in medicine for the treatment of tumors by irradiation. Although the use of Ra^{226} for this purpose has been supplanted by the general availability of other radioactive isotopes, especially Co^{60}, it is still used in medicine, primarily for interstitial irradiations. In this method the Ra^{226} is sealed in small capsules or "needles" which are surgically implanted in the area to be irradiated and are removed after the treatment is ended.

A more usual technique is to fill the needles with the first daughter of Ra^{226}, *i.e.*, Rn^{222}. This has all the original gamma-emitting daughters in equilibrium with itself and has a half-life of 3.8 days. Since the total activity of the radon needles will decay to a safe level within a few weeks, there is no need to remove the Rn needles at the end of the treatment, and the surgical procedure is much less complex than in the case of treatment with Ra^{226} itself. The use of Ra^{226} for production of Rn^{222} is probably the most important medical use of radium at the present time.

Toxicology. The general problem of radium poisoning was brought to public attention by the disastrous effects of radium on a number of women who had been employed in a factory making luminous dials during the First World War. They had been in the habit of wetting the points of their brushes on their lips and had thus accumulated a large amount of radium internally. During a 10–15 year period about 24 of them died from osteogenic sarcoma or aplastic anemia. It was found that from 1.2 to 50/μg of radium remained in their bodies at the time of death. On the basis of physical and medical evidence of these and other cases the maximum permissible limit of fixed Ra^{226} in the body has been set at 0.1/μg at any time.

Due to its chemical similarity to calcium, radium tends to concentrate in the bone where the alpha radiation can cause breakdown of the red cell producing centers in the marrow, and also can lead to cancer of the bone. Fortunately, very sensitive tests, such as detection of exhaled radon can be used to detect amounts of radium that are even one tenth of the maximum permissible dosage. Most countries now require tests of this type to be made of all workers handling large amounts of radium.

A. M. FRIEDMAN

References

Curie, M., "Recherches sur les Substances Radioactives," Thesé, Faculté des Sciences de Paris, Paris, Gauthier-Villars, 1904.

Curie, P., Curie M., and Bermont, G., *Compt. Rend.* **127,** 1215 (1898).

Bagnall, K. W., "Chemistry of the Rare Radioelements, Polonium-Actinium," New York, Academic Press Inc., 1957.

Soddy, F., "The Chemistry of the Radio-elements," London, Longmans Green and Co., 1914.

Adams, J. O., and Lander, W. M., "The Natural Radiation Environment," Chicago, University of Chicago Press, 1964.

Gmelin, "Handbuch der Anorganischer Chemie," System Nr. 31, Berlin, G.m.b.H., 1928.

Rajewsky, "Review of Radium Poisoning," *Radiology,* **32,** 57 (1939).

National Bureau of Standards, "Radium Protection," National Bureau of Standards Handbook, H. 23.

"Thorpe's Dictionary of Applied Chemistry," Vol. X, 4th Edition, London, Longmans Green and Co., 1950.

Kirk, R. E., and Othmer, D. F., Eds., "Encyclopedia of Chemical Technology," Vol. 1, New York, Interscience, 1953.

RADON

Element 86, symbol Rn, is the heaviest of the noble gases; the atomic weight of isotope Rn^{222} = 222.02.

Physical Properties. Radon is a colorless monatomic gas at 25°C and one atmosphere pressure. When cooled or compressed, it condenses to a colorless liquid, which fluoresces in blue, blue-green, or lilac colors in different types of glass (effect of the intense α-bombardment of the glass). The solid fluoresces in blue, yellow, and orange-red colors (in succession) when cooled slowly to liquid air temperature in glass. Some physical properties of the element are

Melting point	−71°C
Boiling point	−61.8°C
Critical temperature	104.4°C
Critical pressure	62.4 atm
Density (liq., ~20°C)	5 g/cc
$\triangle H$ vaporization	3.94 kcal/mole
$\triangle H$ fusion	†0.65 kcal/mole
Heat capacity, C_P	†4.97 cal/degree/mole
Heat capacity, C_V	†2.98 cal/degree/mole
Viscosity, 0°C	†2.13 × 10⁻⁴ poise
Surface tension	†29 dyne/cm
Atomic diameter	†3.64 Å
Magnetic suscepti-	
bility, χ_A	†−57.4 × 10⁻⁶
Ionization potential	10.7 volt

†Estimated value.

The vapor pressure of liquid radon is given approximately by

$$\log P \text{ (mm)} = \frac{-862}{T} + 4.08,$$

where T is the temperature in degrees Kelvin. This equation is derived from vapor pressure data of Gray and Ramsay[1] from 202.6 to 377.5°K. Measurements of the vapor pressure of the solid by others appear to be much less reliable.

The emission spectrum of radon has been reported by many investigators. The strongest lines occur at 7055.42 and 7450.00 Å. The most complete spectrum, containing 172 lines from 3316 to 10161 Å, is that of Rasmussen.[2] In the ultraviolet region, Wolf[3] has observed 122 lines from 2378 to 3761 Å.

Radon is slightly soluble in water and appreciably soluble in organic liquids. Jennings and Russ[4] and Bagnall[5] give the coefficient of solubility (ratio of concentration in the liquid phase to that in the gas phase) for a number of liquids. The following are a few coefficients at 18° selected from their tables: water, 0.285; acetone, 6.30; benzene, 12.82; carbon disulfide, 23.14; ethyl alcohol, 6.17. The solubility coefficient of radon in water decreases regularly from 0.507 at 0°C to 0.106 at 100°C. The enthalpy of solution is 6.7 kcal/mole at 0°C and 4.7 kcal/mole at 35°C.

Isotopes. Eighteen isotopes of radon, all radioactive, are known at the present time. The half-lives, modes of decay, and emission energies are shown in Refs. 6 and 10. Atomic weights on the carbon-12 scale (calculated from known masses

and emission energies) are also included, where known.

The natural isotopes are found in very low concentrations in soil, the oceans, lake and well waters, and the atmosphere. Concentrations of Rn^{222} in the atmosphere have been measured most frequently and found to vary greatly with geographic location and time. Among the factors influencing the concentrations in air are: proximity to emanating ore bodies; porosity of soil; height of sampling station above ground; presence or absence of waste gases from the combustion of coal, natural gas, and other fuels; meteorological conditions. After rainfall, abnormally low levels of radon have been observed due to the low rate of diffusion of the gas through wet soil. The concentrations of Rn^{222} observed in many European and North American cities generally fall in the range of 2 to 80 × 10⁻¹⁴ curie per liter of air; occasionally, during temperature inversions and conditions of dense smoke or fog, higher concentrations have been observed. The lowest concentrations of radon, 0.3 to 2.6 × 10⁻¹⁵ curie per liter, have been noted in air over the oceans.

Chemical Properties. In their early experiments, Rutherford, Ramsay, and Soddy showed that radon (either Rn^{222} or Rn^{220}) remained uncombined when treated with the following substances: alkali metals; alkali hydroxides; sulfuric, hydrochloric, and nitric acids; air, hydrogen, and carbon dioxide at high temperatures; phosphorus pentoxide; burning phosphorus; platinum black, palladium black, zinc powder, magnesium powder, calcium oxide, and lead chromate, each at red heat. J. J. Thomson and F. Himstedt also found no evidence that radon reacted with common acids and bases or with copper and platinum heated to incandescence. More recently, H. Kading and N. Riehl have attempted to prepare radon compounds by photochemical and electrical discharge methods in mixtures of the element with water and air, bromine, and water, air, and iodine, but without success.

B. Nikitin and co-workers have found evidence that radon forms clathrate compounds similar to those known for xenon, krypton, and argon. In such compounds, the noble gas atoms are trapped in the lattice of the host substance during crystallization. There are no chemical bonds between the noble gas atoms and surrounding atoms in the usual sense but only weak Van der Waal forces, and the trapped atoms escape when the host crystal melts or dissolves. From phase studies with tracer amounts of radon and macro amounts of sulfur dioxide, hydrogen sulfide, phenol, p-chorophenol, and water, Nikitin has inferred the existence of such clathrates as $Rn \cdot 6H_2O$, $Rn \cdot 3C_6H_5OH$, and $Rn \cdot 3p\text{-}CiC_6H_4OH$.

In 1962, after the discovery of xenon fluorides, the reaction of radon and fluorine was examined, and evidence for the existence of a radon fluoride was obtained.[7] Tracer amounts of radon (5 microcuries to 2 millicuries) were shown to form a stable compound of very low volatility when heated to 400°C with fluorine in a nickel vessel. Although elemental radon distills in vacuum at −78°C and even lower temperatures, the compound prepared in this manner does not distill at room temperature and 2 × 10⁻⁶ mm Hg pressure; it moves from a heated vessel only at about 230 to 250°C and re-condenses in a cooler region. The compound is reduced by hydrogen at 500°C, with the liberation of elemental radon.

Solutions of oxidized radon have been prepared recently with fluoride solvents, such as bromine trifluoride, bromine pentafluoride, iodine penta-fluoride and hydrogen fluoride.[9] The radon has been shown to be present in these solutions as a positive ion. Since the behavior of radon corresponds to that of a metal in these solutions, radon can be classified as a "metalloid" element together with arsenic, antimony, tellurium, polonium and astatine.

The electron configuration of a neutral radon atom in the ground state is $5s^2 5p^6 5d^{10} 6s^2 6p^6$ (^1S_o). Previously this was considered to be too stable to be altered by chemical bond formation. The new chemistry of krypton, xenon, and radon indicates that charge transfer processes can occur, however, even with filled s and p orbitals, and that heavy noble gas atoms can behave as electron donors to highly electrophilic substances, such as fluorine.

Medical and Other Uses

Both radium and radon have been widely used as radiation sources for the treatment of cancer. The β-γ radiation is the same with either element, since it is produced by daughters in the decay chain, chiefly Bi^{214} and Pb^{214}. However, radon can be used with greater safety than radium since its activity decays to a small fraction of the initial value within several weeks. Measured amounts of radon are sealed into very small glass, gold, or platinum capillary tubes, called "seeds" or "needles," for therapeutic use. The "seeds" are implanted in patients and left in place for extended periods of time (sometimes permanently). In the treatment of skin diseases an ointment consisting of a solution of radon in petroleum jelly is also used by some practitioners. The medical use of radon and the methods of preparation of "seeds" and ointment are described at length by Jennings and Russ.[4]

Radon has been used as a gaseous tracer in leak detection and in the measurement of flow rates. Compressed into a small capsule, it has also been used as a "point source" of γ-rays in the radiography of metal welds and castings. Neutron sources have sometimes been prepared from radon-beryllium mixtures. Since α-particles emitted by radon and its daughters ionize gases effectively and promote chemical change, many kinetic studies have been made of radon-induced reactions, such as the synthesis and decomposition of ozone, ammonia, water, and hydrogen bromide; oxidation of methane, ethane, and cyanogen; polymerization of acetylene, ethylene, and cyanogen.[8]

LAWRENCE STEIN

References

1. Gray, R. W., and Ramsay, W., *J. Chem. Soc.,* 1073 (1909).
2. Rasmussen, E., *Z. für Physik,* **62,** 494 (1930); *ibid,* **80,** 726 (1933).
3. Wolf, S., *Z. für Physik,* **48,** 790 (1928).
4. Jennings, W. A., and Russ, S., "Radon: Its Technique and Use," London, John Murray, 1948.
5. Bagnall, K. W., "Chemistry of the Rare Radio-elements," pp. 105–113, London, Butterworths Scientific Publications, 1957.
6. "Chart of the Nuclides," Schenectady, New York, Knolls Atomic Power Laboratory, Eighth Edition, 1965.
7. Fields, P. R., Stein, L., and Zirin, M. H., "Noble-Gas Compounds," H. H. Hyman, Editor, pp. 113–119, Chicago, The University of Chicago Press, 1963; *J. Am. Chem. Soc.,* **84,** 4164 (1962).
8. Lind, S. C., Hochanadel, C. J., and Ghormley, J. A., "Radiation Chemistry of Gases," New York, Van Nostrand Reinhold Co. 1961.
9. Stein, L., *J. Am. Chem. Soc.,* **91,** 5396 (1969); *Science,* **168,** 362 (1970).
10. Stein, L., "Radon," pp. 589–594, in "Encyclopedia of the Chemical Elements," Hampel, C. A., Editor, Van Nostrand Reinhold, New York, 1968.

RARE EARTHS

The rare earth elements (also called "rare earths," "lanthanides," and "lanthanons") are a group of fifteen elements of similar properties placed in Groups III of the Periodic Table. Their properties are somewhat similar to those of scandium, yttrium, and actinium in the same group. The rare earths have atomic numbers 57 to 71 inclusive, and are, in serial order: lanthanum (La), cerium (Ce), praseodymium (Pr), neodymium (Nd), promethium (Pm), samarium (Sm), europium (Eu), gadolinium (Gd), terbium (Tb), dysprosium (Dy), holmium (Ho), erbium (Er), thulium (Tm), ytterbium (Yb), and lutetium (Lu). The rare earths from lanthanum to samarium are called the "cerium earths"; those from europium to dysprosium are "terbium earths." Those from holmium to lutetium are called "yttrium earths" because of their resemblance to yttrium which is similar to and always occurs with the rare earths.

The rare earths are inner transition elements characterized by progressive filling up of the $4f$ electrons without changing the outer $5s^2 5p^6$ $5d^{0 \text{ or } 1} 6s^2$ levels, resulting in a concurrent decrease in atomic size. This reduction in atomic radii with increasing atomic number is known as the *lanthanide contraction*. In this respect, and also in chemical behavior, the rare earths resemble the actinides, the second rare earth-like series beginning with actinium, which show a similar *actinide contraction*.

Occurrence. The only commercial minerals of importance are monazite and monazite sand, and bastnasite. Monazite is a rare earth-thorium phosphate which occurs as small crystals in many acid granites and is recovered from placers resulting from the weathering of such rocks. It also occurs in some pegmatites, but commercial pegmatite deposits are rare.

Bastnasite is a rare earth fluocarbonate which occurs in hydrothermal deposits associated with barite, fluorite, and calcite.

Important monazite deposits are placers in Idaho, beach sands in Florida and Georgia, Brazil, and

India, and massive pegmatic monazite in the Union of South Africa. Bastnasite deposits are found in California and New Mexico.

Other minerals, sometimes useful for the extraction of the heavier rare earths, are gadolinite, $Fe(RE)_2 \cdot Be_2(Si_2O_{10})$; allanite, $4(Ca, Fe)O \cdot 3(Al, R, Fe)_2O_3 \cdot 6SiO_2 \cdot H_2O$; and samarskite, a rare earth-iron-calcium-magnesium niobotantalate.

Process liquors from some uranium recovery operations contain recoverable thorium and yttrium earths, and these are important sources of yttrium.

The cerium earths predominate in monazite and bastnasite. Cerium is the most abundant rare earth, being comparable to boron in occurrence in the earth's crust. Promethium is found only in the products from uranium fission.

Properties. The elements are tervalent in most compounds. Cerium, praseodymium, and terbium exist also in the tetravalent state, and samarium, europium, and ytterbium form easily oxidized divalent compounds. Due to similarities in atomic structure, the properties of the rare earths are quite similar, and vary only slightly from one rare earth to the adjacent neighbor. Consequently, their separation is difficult unless use is made of oxidation states other than the tervalent. With increasing atomic number, the rare earths become less basic, the salts generally become more soluble, and the differences between adjacent members decrease.

With the exception of lanthanum and lutetium, rare earth compounds show characteristic sharp absorption bands in the ultraviolet and visible spectra. This absorption is responsible for the pastel colors of the colored rare earth salts (green Pr, pink Nd, yellow Sm and Ho, rose Er, and pale green Tm).

Rare Earth Metals. The metals have a silver-gray luster. Except for the more reactive ones (Eu and Yb), they do not tarnish in air if they are pure. Hardness increases generally with atomic number, but Eu and Yb can be cut with a knife. Densities (g/cc) vary from 6.2 for La to 9.8 for Lu. Melting points vary from 795°C (Ce) to 1652°C (Lu), and boiling points range from 1427 to 3470°C.

Being active reducing agents, the metals react slowly with water, and are soluble in dilute acids. They are pyrophoric, cerium igniting in air at 150–180°C. Above 200°C, they combine directly with halogens, and form nitrides with nitrogen above 1000°C. Interstitial hydrides approximating $RH_{2.8}$ are formed by absorption of hydrogen.

The mixture of rare earth metals made commercially without appreciable separation of rare earths is known as "misch" metal. It contains about 22% La, 50% Ce, 18% Nd, 5% Pr, 1% Sm, and 2% other rare earth metals. It is often sold as "cerium" metal.

Rare Earth Compounds. Common water-soluble salts are the acetates, chlorides, nitrates, and sulfates. The carbonates, oxalates, hydroxides, oxides, phosphates, and fluorides are insoluble. Cerous (Ce^{+3}) salts are similar to the other tervalent rare earth salts, while ceric (Ce^{+4}) salts are more like those of thorium. Tervalent acetates and sulfates show decreased solubility in hot solutions.

Acetates are made by treating the hydroxide, carbonate or oxide with acetic acid. Carbonates are precipitated by the addition of alkali carbonates to neutral rare earth solutions; they are only slightly soluble in excess of alkali carbonate. Fluorides are precipitated in hydrated form on adding hydrofluoric acid or soluble fluorides to rare earth solutions. The precipitates are insoluble in excess hydrofluoric acid and in mineral acids. Anhydrous fluorides are made by hydrofluorinating the oxides at elevated temperatures.

Chlorides, bromides and iodides are prepared by dissolution of the hydroxide, oxide, or carbonate in the halogen acid. The salts crystallize from water as hydrates. Anhydrous chlorides and bromides are made by heating oxides with the ammonium halide, followed by sublimation of the excess ammonium salts in vacuum. Nitrates are formed similarly to the other water-soluble salts. They form many double nitrates with alkali and alkaline earth nitrates. Oxalates are precipitated from slightly acid solution with oxalic acid or alkali oxalates. They are important in the analysis for rare earths, and in the separation of rare earths from other metals. Hydroxides are precipitated from solution by alkali and ammonium hydroxides. The sulfates are characterized by their formation of sparingly soluble double sulfates with alkali sulfates. The most important double sulfate is the sodium salt, $RE_2(SO_4)_3Na_2SO_4 \cdot 2H_2O$.

Extraction and separation. The extraction of rare earths from monazite is commercially important. The ore is opened by heating with sulfuric acid to form anhydrous rare earth and thorium sulfates and phosphoric acid, or by heating with sodium hydroxide solutions to form rare earth and thorium hydroxides and sodium phosphate. The reaction products are lixiviated in water, and if the alkaline method is used, the washed rare earth-thorium hydroxides are solubilized by treatment with acid.

Thorium is separated from the rare earths by fractional basicity precipitation or by precipitation of compounds such as thorium pyrophosphate. Rare earths are usually recovered from the thorium filtrates by precipitation of the double rare earth sodium sulfate. The double sulfate precipitate is converted to rare earth hydroxide which serves as the starting material for making commercial rare earth salts.

Cerium is separated from the rare earths by oxidation to the tetravalent ceric state, followed by basicity separations, crystallizations of insoluble ceric compounds, or by solvent extraction. Ceric salts are generally much less soluble than those of the tervalent rare earths. Crystallization of ammonium hexanitratocerate or ammonium trisulfatocerate, precipitation of basic ceric nitrates or sulfates, and fractional basicity separations with the hydrous oxides are commonly used procedures to separate and purify cerium.

Separation of the cerium-free rare earth mixture (often called "didymium") was formerly done by long series of fractional crystallizations. Modern separation processes use ion exchange or solvent extraction methods. All of the individual rare earth oxides are produced commercially in 99.9% purity, and in some cases to 99.9999% pure.

Uses. Contrary to general belief, rare earths are widely used. Annual world consumption is more than 25 million pounds rare earth oxide equivalent.

Most uses are based on mixtures extracted from ores without appreciable rare earth separation. The largest use is in gasoline cracking catalysts, followed by glass polishes, misch metal and ferrocerium alloys, glass and ceramics, and arc carbon cores.

Lighter flints are about 75% misch metal. Ferrocerium alloys and misch metal are used in making ductile iron, to overcome the deleterious effects of tramp elements in steel, to reduce chill in gray cast irons, and in the manufacture of special steels.

Specially prepared cerium oxide and mixed rare earth oxides are widely used in polishing spectacle and instrument lenses.

The red phosphor in color television video tubes is an yttrium or gadolinium oxide or oxysulfide activated with europium.

In glass, rare earths are used as decolorizers (Ce) and as coloring agents (Ce and Nd). Strong yellow ceramic stains are based on praseodymium oxide. Lanthanum oxide is used in silica-free optical glass and in fiber optics, and neodymium compounds are used in temperature-compensating ceramic capacitors and in laser glass.

The unusual nuclear and magnetic properties of some rare earths lead to their use in nuclear poisons (Gd, Eu) and permanent magnets (Sm). Yttrium-based garnet type materials are used in microwave devices, and as synthetic gems.

HOWARD E. KREMERS

RARE METALS

The term rare metals is an arbitrary but convenient one designating the many metals which are uncommon for several reasons. (1) The natural supply or abundance in the earth's crust may be small. (2) Even if fairly prevalent, the concentration may be so low as to require the handling and processing of huge amounts of undesirable material in order to extract even small quantities of the desired element in either combined or elemental form. (3) The chemical and physical properties of the element may be such that conversion to elemental form is difficult. (4) Even though available, the element may not have enough attractive properties or uses to create a demand for it in competition with other available materials on a cost basis; hence, it is not produced commercially.

Included under the category of rare metals is a majority of the known elements which are metals or semiconductors. The obvious exceptions among the metals are sodium, magnesium, chromium, iron, cobalt, nickel, copper, zinc, mercury, tin, aluminum, lead and antimony. To this can now be added titanium, molybdenum and manganese, whose annual United States production exceeds 10,000 tons of each metal.

In recent years many of the rare metals have become widely available and used in elemental form even though the quantity of each is small compared with that of the common metals. Among them are uranium, zirconium, hafnium, tantalum, lithium, indium, thorium, many of the rare earth elements, boron, gallium, beryllium, tellurium, and the semiconductors germanium and silicon.

The source, preparation, properties and uses of each of the rare metals are described in articles in this book, and in more detail in the references listed below.

CLIFFORD A. HAMPEL

References

Hampel, C. A., Editor, "Rare Metals Handbook," 2nd Ed., Van Nostrand Reinhold Co., New York, 1961.

Hampel, C. A., Editor, "The Encyclopedia of the Chemical Elements," Van Nostrand Reinhold Co., New York, 1968.

REACTION RATES

The velocity of a chemical reaction usually depends upon the concentration of reactants and upon their nature. This dependence is expressed in terms of a numerical constant (at a definite temperature and pressure) which has the units (concentration)$^{1-n}$ (time)$^{-1}$, where n is the order of the reaction. It is the specific reaction rate that we shall now examine.

For a long time it has been known that the velocities of almost all chemical reactions increase with a rise in temperature. Most of the rare exceptions to this rule, upon careful inspection, are found to result from a decrease in concentration of reactants with an increase in temperature, and even here the absolute rate constant does increase with the temperature rise. Crystal nucleation and denaturation of proteins seem to be bona fide exceptions. A rough rule is well known that many reactions in solution approximately double in velocity at ordinary temperatures for each increase in temperature of 10°C. Reactions are known which do not increase so rapidly in velocity, and others increase in velocity much more rapidly. A notable case of the latter is found in the rate of denaturation of ovalbumin. This is the process observed in boiling an egg. In this case a 10°C rise in termperature gives about 25 times the effect one would predict from the simple rule—a result familiar to those living in mountainous country.

The earliest attempts to discuss quantitatively the effect of temperature on specific reaction rate took the form of a power series in temperature. Such a series can of course fit observed data over a previously investigated range in temperature. However, results cannot be safely extrapolated, nor can one gain any insight into the nature of chemical reactions from such a treatment.

In 1889 Arrhenius considered the effect of temperature on the rate of conversion of sucrose. He proposed that an equilibrium existed between the ordinary and active sucrose molecules, and that only these active molecules (containing more than the average energy) underwent inversion. Then using the temperature dependence of this equilibrium constant, which had been previously examined by van't Hoff in 1884, he wrote:

$$k' = Ae^{-E/RT} \qquad (1)$$

In this expression k' is the specific reaction rate, E is the difference in energy between the active and the inert molecules, A is a factor which is only slightly temperature-dependent and generally can be assumed to be not dependent on temperature. The constant e is the basis of the natural logarithms. Since the earliest reactions considered were of the first order ($n = 1$) the name "frequency factor" was used for A, the units of which were then reciprocal time. This nomenclature has been continued and is unfortunately employed for reactions of other orders.

The Arrhenius equation is convenient and quite satisfactory in its description of the dependence of specific rate constant on temperature for most chemical reactions. However, it has its origin in thermodynamics and will not give information concerning the absolute velocity, as it fails to provide a theory for the frequency and entropy of activation.

The development of kinetic theory during the last half of the nineteenth century and the early years of the twentieth naturally led to attempts to apply these results to a discussion of reaction rates. In this case a reaction would be thought of as the result of a collision between reacting species. This may be expressed for the collision of A and B to give the compound AB as

$$k' = Ze^{-E/RT} \text{ cc/molecule sec} \qquad (2)$$

if we assume a standard state of one molecule per cc. The kinetic theory expression for Z is then

$$Z = \sigma_{A,B}{}^2 \left[8\pi kT \left(\frac{m_A + m_B}{m_A m_B} \right) \right]^{1/2} \qquad (3)$$

where $\sigma_{A,B}$ is the mean molecular diameter of A and B, k is Boltzmann's constant, T is the absolute temperature, and m_A and m_B are the masses of the molecules. An early application of this theory by Lewis to the data of Bodenstein for the reaction $H_2 + I_2 \rightleftharpoons 2HI$ gave results in surprisingly good agreement with the experiments. Further investigations, however, did not always produce results so satisfactory. In fact, reactions are known for which the calculated value of Z exceeds the measured value of A in the Arrhenius equation by a factor of 10^8. One also questions whether σ_{AB} calculated from viscosity data for gases is appropriate for system undergoing chemical reaction. In order to correct these difficulties a steric factor, P, is frequently introduced, giving

$$k' = PZe^{-E/RT} \qquad (4)$$

However, in any critical analysis such a factor must be viewed as an empirical relation, even though the calculated results are in excellent agreement with those measured experimentally.

We now arrive at a method for calculating A which is referred to as the absolute reaction rate theory. It rests on the idea that in a chemical reaction we may view the passage from the initial to the final state as a continuous change from the initial to the final configuration of the molecules. At some point an intermediate configuration called the activated complex or the transition state is reached. This activated complex is critical, for if it is once attained, the probability of proceeding to the final state is very large. From considerations of statistical mechanics one may show that the specific rate constant may be written for the process A + B + . . . → activated complex

$$k' = \frac{\kappa kT}{h} \frac{F\ddagger}{F_A F_B \ldots} e^{-E_0/RT} \qquad (5)$$

where κ is the probability that the activated state once achieved will go to final products and is called the transmission coefficient. For almost all reactions its value may be taken as unity; h is Planck's constant, and k as before is Boltzmann's constant. $F\ddagger$, F_A, F_B are the partition functions per unit volume for the activated complex and for the initial species which form it. The value of E_0 is the difference between the energy of the activated complex and the initial state at absolute zero; and R is the gas law constant. These partition functions may in principle be calculated and frequently relatively simple approximations give good results. However, calculation from first principles of sufficiently accurate values is not usually feasible.

From statistical mechanical considerations we may write equation (5) in the following form:

$$k' = \frac{\kappa kT}{h} K\ddagger \qquad (6)$$

$K\ddagger$ is the equilibrium constant between the activated and ground state. From the thermodynamic relation $\Delta F\ddagger = -RT \ln K\ddagger$ we may write:

$$K' = \frac{\kappa kT}{h} e^{-\Delta F\ddagger/RT} = \frac{\kappa kT}{h} e^{-\Delta H\ddagger/RT} e^{\Delta S\ddagger/R} \qquad (7)$$

where $\Delta F\ddagger$ is the free energy of activation, $\Delta H\ddagger$ the enthalpy of activation and $\Delta S\ddagger$ the entropy of activation. It should be clearly noted that the equation has been developed for an elementary reaction and that it should be applied only to such a reaction. Thus if one deals with a chain reaction it is necessary to evaluate the over-all rate constant in terms of the elementary steps. And it is to these individual steps that the absolute reaction rate theory can be applied. The names of Bodenstein, Polanyi, Christiansen and Herzfeld come immediately to mind when one speaks of chain reactions. The absolute reaction rate theory in the form of equations (5) and (7) has been applied with success to a wide range of solid, liquid and vapor phase reactions. It is equally useful in considering the rates of very rapid reactions which may occur in a flame, and the rates of those reactions which under ordinary conditions require geologic ages.

The activated complex may be viewed as an ordinary molecule with the exception that vibration in one direction—along the reaction coordinate—gives rise to decomposition. The lifetime of an activated complex is given by:

$$\tau = \frac{h}{2kT} \qquad (8)$$

which at ordinary temperatures is about 10^{-13} second. Thus we cannot observe them directly. However, it is possible to use experimental rate data to describe them accurately.

A detailed discussion of the activated state and the reaction coordinate leading from initial substances to products may be carried out in terms of a potential energy surface. If we consider a linear activated complex A-B-C it is evident that the knowledge of two distances A-B and B-C, for example, fixes the complex. A potential energy surface for the system would be a plot of the potential energy along a vertical axis orthogonal to the axes for the distances which describe the molecule. If the triatomic complex is not linear, three lengths will be necessary to describe it. In this case a four dimensional graph with the potential energy coordinate orthogonal to each of the three orthogonal space coordinates is required. For a general tetra-atomic activated complex six orthogonal space coordinates plus the energy coordinate are necessary. In theory one could obtain this surface from quantum mechanical considerations. Actually, for all but the simplest systems this is not feasible. There are also semi-empirical methods for the calculation of these potential energy surfaces, but these do not generally give sufficiently accurate surfaces for practical use in predicting kinetic data. In fact, for all but the very simplest reactions one examines the nature of the activated complex from experimental kinetic data.

The absolute reaction rate theory which we have discussed was formulated by H. Eyring (1935). Polanyi and Evans (1935) developed important aspects of the theory. Somewhat similar ideas were discussed by A. Marcelin, who considered molecules crossing a critical surface in phase space; by Rodebush (1923–35); by La Mer; and by O. K. Rice and H. Gershinowitz (1934–5), who stated that a reaction may take place if the system lies in a certain fraction of phase space. These latter investigators did not, however, use the properties of the potential energy surface in configuration space to define the activated complex. The development of potential energy surfaces by F. London (1928) and by Eyring and Polanyi provided the avenue for obtaining a clear conception of the activated complex. Eyring's formula generalized Pelzer and Wigner's treatment of the reaction $H + H_2$ (para) $= H_2$ (ortho) $+ H$.

The effect of pressure on the velocity of a chemical reaction may be stated as follows:

$$-\frac{\partial \ln k'}{\partial P} = \frac{\Delta V\ddagger}{RT} \qquad (9)$$

when $\Delta V\ddagger$ is the change in volume in going from the initial to the activates state. If the change in volume is positive (the activated state is more voluminous than the initial state) one observes a decrease in velocity of reactions with increasing pressure. If the value of $\Delta V\ddagger$ is negative, the specific rate constant will increase with increasing pressure. An interesting case is observed when one considers the synthesis of diamond from graphite. In this case an examination of the thermodynamic properties of the system indicates that a high pressure is necessary before diamond will become more stable than graphite. However, the value of $\Delta V\ddagger$ is positive for this reaction so that the high pressure necessary to stabilize diamond decreases the rate of its formation.

F. William Cagle, Jr., and Henry Eyring

Cross-references: *Chemical Engineering, Thermodynamics.*

REACTION TYPES

Chemical changes can be classified according to the sort of reacting substances or of the nature of the product or products.

(1) In a direct *combination*, one product is formed. Two substances combine and form a third: (a) $Cu + Cl_2 \rightarrow CuCl_2$ (combination of two elements); (b) $C_2H_2 + 2Br_2 \rightarrow C_2H_2Br_4$ (combination of an element and a compound); (c) $2NH_3 + H_2SO_4 \rightarrow (NH_4)_2SO_4$ (combination of two compounds).

(2) In a *decomposition*, one substance is changed into simpler parts: (a) $2H_2O \rightarrow 2H_2 + O_2$ (decomposition of a compound into elements); (b) $2KNO_3 \rightarrow 2KNO_2 + O_2$ (decomposition of a compound into another compound and an element); (3) $NH_4NO_3 \rightarrow 2H_2O + N_2O$ (decomposition of a compound into simpler compounds).

(3) In a *replacement* reaction, one element takes the place of another that is in a compound: (a) $Zn + 2HCl \rightarrow ZnCl_2 + H_2$ (zinc replaces hydrogen); (b) $2NaBr + Cl_2 \rightarrow 2NaCl + Br_2$ (chlorine replaces bromine); (c) $CH_3COOH + 3Cl_2 \rightarrow 3HCl + CCl_3COOH$ (chlorine replaces hydrogen).

(4) In a *double replacement* or metathesis, two compounds react and form two new compounds: (a) $2NaOH + H_2SO_4 \rightarrow Na_2SO_4 + 2H_2O$ (neutralization); (b) $2AgNO_3 + CaCl_2 \rightarrow Ca(NO_3)_2 + 2AgCl$ (s) (formation of a precipitate); (c) $Na_2SO_3 + 2HCl \rightarrow 2NaCl + H_2SO_3$ followed by $H_2SO_3 \rightarrow H_2O + SO_2$ (g) (escape of a gas). In this case the double replacement reaction is followed by the decomposition of the unstable sulfurous acid.

(5) *Polymerization* is a special case of combination. Sometimes molecules of the same kind (or different kinds) form huge clusters called polymers. In certain cases (called condensation) polymerization is accompanied by the loss of a simple compound, such as water or common salt. Ethylene (C_2H_4) a well-known hydrocarbon, polymerizes and forms polyethylene, and in a similar manner styrene ($C_6H_5CHCH_2$) forms polystyrene, and phenol (C_6H_5OH) and formaldehyde (HCHO) react and form Bakelite.

Chemical changes may also be classified according to energy changes that take place when the change goes on. *Exothermic* reactions are those in which energy is liberated: for example, $C + O_2 \rightarrow CO_2 + 94,300$ calories for 12 grams of carbon. An *endothermic* reaction, on the other hand, requires a continual supply of energy in order to proceed. In the process for making water gas, the coke at about 1400°C reacts with steam, but the reaction ceases to be practical at a temperature below 1000°C: $C + H_2O \rightarrow CO + H_2 - 31,100$ calories.

More restricted classifications may also be used. For example: The reaction of an acid with a base is called *neutralization:* $NaOH + HCl \rightarrow H_2O + NaCl$, or ionically $H^+ + OH^- \rightarrow H_2O$ or $H_3O^+ + OH^- \rightarrow 2H_2O$. The reaction of an acid with an alcohol is called *esterification;* $CH_3COOH + C_2H_5OH \rightarrow CH_3COOC_2H_5 + H_2O$. The reaction of a glyceryl ester of a fatty acid with a strong alkali is *saponification;* $(C_{17}H_{35}COO)_3C_3H_5 + 3NaOH \rightarrow C_3H_5(OH)_3 + 3C_{17}H_{35}COONa$. A reaction with water is *hydrolysis:* $CO_3^{2+} + H_2O \leftrightharpoons OH^- + HCO_3^-$.

Many reactions in which a transfer of electrons occurs are classed as *oxidation-reduction* (redox). For example, $6FeCl_2 + 14HCl + K_2Cr_2O_7 \rightarrow 6FeCl_3 + 2KCl + 2CrCl_3 + 7H_2O$. Cr goes from oxidation number 6+ to 3+, a gain of 3 electrons per atom (reduction); Fe goes from oxidation number 2+ to 3+, a loss of 1 electron per atom (oxidation).

Numerous types of reactions are named after chemists who have discovered or investigated that type, e.g., Fischer-Tropsch, Diels-Alder, Friedel-Crafts.

Induced nuclear changes, and radioactive disintegrations are *nuclear* reactions, and are not classed as chemical changes.

ELBERT C. WEAVER

Cross-references: *Molecules, Polymerization, Combining Weight, Reaction Rates.*

REAGENTS

In the past the term "reagent" has sometimes been loosely used to designate substances or mixtures added to others in the course of various kinds of chemical operations in the laboratory, but it is now usually restricted to those added for analytical purposes. Though reagents may be pure gases, liquids, or solids, they are, in practice, usually solutions of pure substances in water or other of the more common solvents. The concentration of such solutions may be expressed simply as weight of solute present in a given volume or weight of solvent, but it is more usual and convenient to express their concentration in terms of molarity or normality.

High purity is almost always a necessary characteristic of the chemicals used as reagents or for the preparation of solutions used as reagents. This is a matter of such importance that various groups or committees of analytical chemists have at different times concerned themselves with the problem of specifications and standards for reagent chemicals. The first committee of the American Chemical Society to deal wtih this problem was appointed in 1903 and continued until 1915. The second committee, which was much more active than the first, was appointed in 1917 and has continued with various changes in organization and personnel up to the present. The recommendations of this second committee were first published in the form of articles in *Industrial and Engineering Chemistry, Industrial and Engineering Chemistry (Analytical Edition),* and *Analytical Chemistry.* In 1941 this committee issued under the auspices of the society a pamphlet entitled "A. C. S. Analytical Reagents," which consists of a collection of reprints of the articles on specifications and standards published up to that date covering a total of 143 of the most important reagent chemicals. In 1950 the committee prepared and the society published "Reagent Chemicals," a book of officially approved specifications and standards for 177 chemicals.

Reagents may be classified in different ways; one, for example, is in accordance with the principal type of action that occurs when they are used. Thus they may be grouped as precipitating reagents, oxidizing reagents, selective solvents, and so on. Another important method of classification is in accordance with the extent to which they are selective in their action. A general reagent is one that under given experimental conditions reacts with a considerable number of substances. For example, hydrogen sulfide is a general reagent for metal ions. Such a reagent is also sometimes called a group reagent. A selective reagent is one that under a given set of conditions reacts in a characteristic way with only a few substances. For example, sulfuric acid is a selective precipitating reagent for metal ions, since lead, barium, strontium, and possibly calcium are the only ones that react. A specific reagent is one that under a given set of conditions yields a reaction with only a single substance. For example, dimethylglyoxime in acid solution is a specific precipitant for palladous ion. Though relatively few reagents are truly specific, this is a most desirable property in a reagent, and attempts are constantly being made to discover analytical reactions that are more specific.

Another very desirable property of a reagent is high sensitivity, which means that exceedingly small amounts of substance may be detected or determined. Feigl first proposed the term "identification limit" as a quantitative index of sensitivity. This is defined as the minimum quantity of a substance, expressed in micrograms, that can be clearly revealed by a given reagent under the optimum experimental conditions. Another such term which has been widely adopted is "concentration limit," proposed by Hahn. This is expressed either as the actual concentration of substance that may be detected in a solution, such as micrograms per milliliter, or as a ratio of weight of substance to volume of solution. Still another term is "dilution limit," which is simply the reciprocal of the concentration limit expressed as a ratio. However, the sensitivity of a given reagent may be much modified by the presence of a substance or substances other than the particular one that is to be detected or determined. Generally the presence of foreign substances causes a decrease in sensitivity, but they may also cause an increase. The term "limiting proportion" was introduced by Schoorl as an index of the effect of a foreign substance. This is defined as the ratio, under given experimental conditions, of the minimum quantity of a substance that can be detected to the quantity of some attendant foreign substance. The effect on sensitivity of the presence of foreign substances is of great practical importance.

In the search for better reagents the greatest emphasis in the past few decades has been on organic compounds as reagents for inorganic substances, and many such reagents with remarkable and useful properties have been discovered. Most useful are the organic reagents that form inner complex salts with metals, the classic example being dimethylglyoxime as a reagent for nickel. In general, these reagents are highly sensitive and are often very selective in their action. However, the formation of other types of complexes between organic molecules and inorganic ions has also been the basis of many new reagents. The effect of particular atomic groupings in organic compounds on their properties as reagents for metals has been investigated to such an extent that predictions can now be made as to whether a given compound will prove suitable as a reagent for a given inorganic substance. Many organic reagents that serve for the detection or determination of organic substances have also been introduced in recent years.

EARLE R. CALEY

Cross-reference: *Analytical Chemistry.*

References

Feigl, F., "Chemistry of the specific, selective, and sensitive reactions," New York, 1949.

Rosin, J., "Reagent chemicals and standards," 5th Ed., Princeton, 1967.

Welcher, F. J., "Chemical solutions; reagents useful to the chemist, biologist, and bacteriologist," Princeton, 1966.

Welcher, F. J., "Organic analytical reagents," 4 vols., New York, 1947–48.

REARRANGEMENT REACTIONS

Probably the most famous rearrangement reaction in history took place when Fredrich Wöhler (1800–1882) in 1828 evaporated an aqueous solution of ammonium cyanate. To his amazement, he discovered that the compound had changed in its character, but not in composition. That is, the atoms within the compound had rearranged into a new pattern, and in this case urea was formed.

$$NH_4OCN \rightarrow (NH_2)_2CO$$
ammonium cyanate   urea

Wöhler's experiment was significant because it was an example of changing an inorganic compound into an organic compound. Before this time, theorists had postulated a strict division between inorganic and organic compounds, and had thought that organic compounds must be made only by living organisms of some sort. Wöhler's rearrangement was one of the first significant reactions to disprove this artificial distinction.

Rearrangement reactions are encountered chiefly in organic chemistry. Rearrangements involve various attachments of hydrogen atoms. These reactions are distinct from resonance in which various distributions of electrons are involved. Many rearrangements are catalyzed by acids, especially by protons from mineral acids.

One example of rearrangement is the reaction that takes place when 1-butene changes to 2-butene in the presence of aluminum sulfate at 280°C.

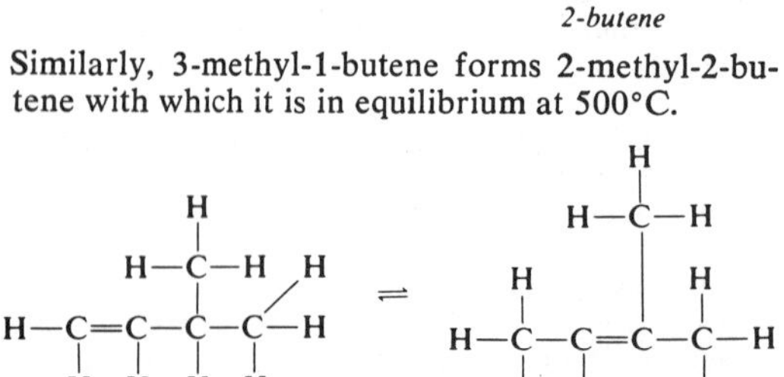

1-butene

2-butene

Similarly, 3-methyl-1-butene forms 2-methyl-2-butene with which it is in equilibrium at 500°C.

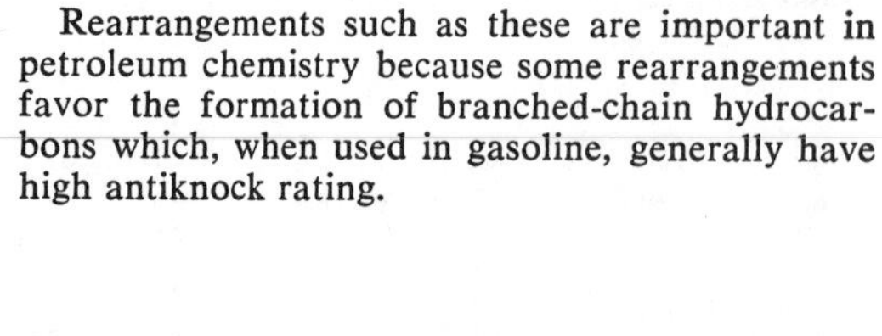

3-methyl-1-butene   *2-methyl-2-butene*

Rearrangements such as these are important in petroleum chemistry because some rearrangements favor the formation of branched-chain hydrocarbons which, when used in gasoline, generally have high antiknock rating.

Various theories have been proposed to explain rearrangements. A most fruitful theory that has tied together several apparently unrelated rearrangement reactions is that of Frank C. Whitmore (1887–1947). By its use, reaction mechanisms can be explained. Consider a compound in which a hydrogen atom is bonded covalently to a carbon atom which in turn is likewise bonded to another carbon atom. The first carbon atom also holds a strongly electronegative group or element (halogen, oxygen, etc.), and it is attached to a centrally located atom through a multiple bond. The electronegative group or atom causes the hydrogen atom to ionize. The resulting proton migrates from the electropositive part of the molecule. Sometimes further shifts occur to fill the place vacated by the hydrogen atom.

A few rearrangements are shown to illustrate typical cases. Many rearrangements are identified with the name of a chemist who investigated the sort of reaction that today bears his name.

The Fries reaction: This reaction is catalyzed by aluminum chloride. Esters of phenols rearrange to ortho- and para-hydroxy ketones. For example:

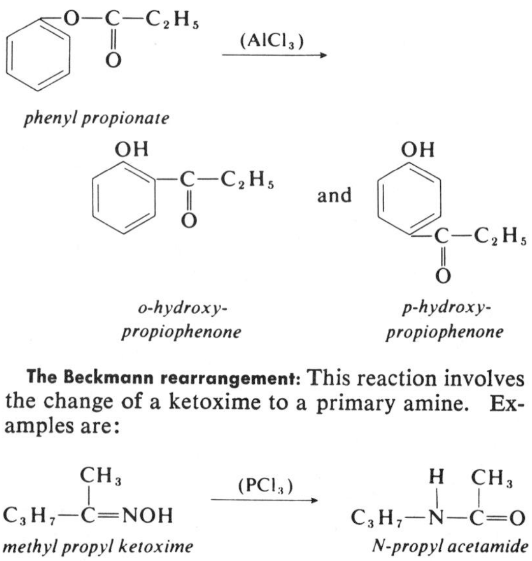

phenyl propionate

o-hydroxy-propiophenone   *p-hydroxy-propiophenone*

The Beckmann rearrangement: This reaction involves the change of a ketoxime to a primary amine. Examples are:

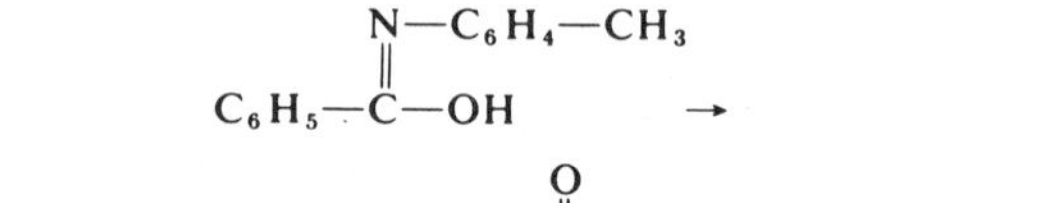

methyl propyl ketoxime   *N-propyl acetamide*

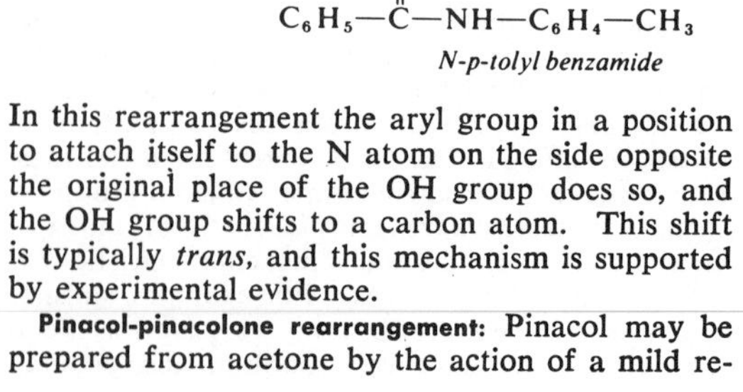

sym-phenyl p-tolyl ketoxime

N-p-tolyl benzamide

In this rearrangement the aryl group in a position to attach itself to the N atom on the side opposite the original place of the OH group does so, and the OH group shifts to a carbon atom. This shift is typically *trans*, and this mechanism is supported by experimental evidence.

Pinacol-pinacolone rearrangement: Pinacol may be prepared from acetone by the action of a mild reducing agent. The presence of a strong acid causes pinacol to rearrange and form pinacolone.

$$(CH_3)_2C(OH)-C(OH)(CH_3)_2 \xrightarrow{(H^+)}$$

pinacol

$$CH_3-\underset{O}{\overset{\parallel}{C}}-C(CH_3)_3 + H_2O$$

pinacolone

In the case of benzopinacol, for example, the Whitmore explanation assumes the formation of an intermediate carbonium ion. The pair of electrons originally covalently bonding carbon to oxygen of the hydroxyl group is kept by the OH group. The OH group thus becomes an ion (OH^-), and the carbon atom remains positively charged (carbonium ion). A hydrocarbon group comes from the adjacent carbon atom. Then the H of the OH group leaves the compound as a proton, depositing its original electron with the compound. Thus the original compound regains electronic balance, and the original proton (H^+) catalyst is restored for further action.

ELBERT C. WEAVER

Cross-references: *Claisen Condensation; Wurtz Reaction;* and other special reaction types.

RECYCLING

In the broadest sense, recycling refers to the operation of returning a material, recovered or separated at the end of a given processing step, back to the same or preceding processing step. In this sense recycling constitutes one of the most common features of chemical processing.

The material recycled may be in the same chemical and physical form as one of the materials initially charged to the process, or it may have been modified during the processing steps. All or only part of the desired material leaving a given processing step may be recycled; the quantity returned is, in general, at the discretion of the design engineer and the plant operator. The effect of recycling upon chemical and physical relationships during the processing step can often be predicted from thermodynamic and kinetic data, although in many complex systems experimental work in laboratory-scale or full-scale pilot plants is required to determine the effect of the nature and amount of material recycled.

Recycling operations may assume any of the following forms:

(1) *Return of Unconverted Reactants:* In many industrial processes conditions selected for the reaction result in an incomplete conversion of one or more reactants into products. If the attack percentage is small enough, economic considerations usually dictate separation of unconverted reactants from the product and recycling into the process along with fresh reactants.

Total recycling of a specified material is often limited by the accumulation of an undesirable impurity which remains in the recycle stream. In those ammonia processes utilizing recycling a purge on the recycle stream must be maintained to prevent accumulation of argon and other impurities within the reaction system.

(2) *Return of Part of an Intermediate or Final Product:* Recycling of part of a product stream back to the processing system frequently causes a modification of the processing conditions, the result being a more desirable product. This form of recycling is typified by the return of cycle stock back to the reaction system in the cracking of oil stocks. In the broadest sense, the return of reflux to the enriching section of distillation and extraction columns may be classified in this category.

(3) *Return of Auxiliary Process Materials:* Many operations and processes involve the use of auxiliary materials which leave the processing system essentially unchanged. If these materials are sufficiently valuable, they may be reused; in fact, economic considerations often dictate that total recycle be maintained if possible. Recycling of inert carriers and diluents, of solvent in extraction processes, and of catalyst in fluidized bed systems is quite common.

Effect on Thermal Cracking Operations. In general, at a fixed pressure and temperature, the yield of gasoline from a given charge stock increases with contact time in the tube still heater. However, a maximum yield is reached, and at longer contact times the yield actually decreases. When cracking begins, the primary decomposition reaction to gasoline predominates, and the gasoline yield increases with cracking time in a more or less linear fashion. As the maximum yield point is approached, the rate of polymerization of reactive species becomes increasingly dominant, and beyond the time corresponding to the maximum point, the polymerization rate is more rapid than the primary decomposition rate. The yield of gasoline correspondingly decreases, larger amounts of tar are formed, and coke is eventually produced. Molecular species present during the cracking process can be generally classified in increasing order of stability as follows:

Paraffins	Naphthenes
Olefins	Aromatics
Diolefins	

As paraffinic molecules are converted to some of the other molecular species, the cracking products tend to be more stable under conditions of cracking than the original charge. Thus the recycle stock is more "refractory" than fresh stock, and a greater contact time is required for the same yield.

The recycle stock and the fresh stock remain in the tube still heater the same length of time during any single cracking "pass." It follows that during any single pass the yield of gasoline from the *recycle stock* must be maintained well below the maximum point, for otherwise the maximum point of the *fresh stock* could be exceeded, resulting in excessive tar and coke production. In industrial practice the maximum crack per pass is limited to about one-half of the yield that can be obtained by a single cracking operation. Recycling thus has the effect of materially increasing the yield of gasoline without excessive formation of tars and coke.

The example of thermal cracking serves well to

illustrate the effect of recycling. Recycle stock is also blended with fresh stock in catalytic cracking and hydrocracking, operations which have substantially displaced thermal cracking in modern petroleum refinery practice.

J. Frank Valle-Riestra

REFORMATSKY REACTION

The Reformatsky reaction (1887) depends on interaction between a carbonyl compound, an α-halo ester, and activated zinc in the presence of anhydrous ether or ether-benzene, followed by hydrolysis. The halogen component, for example ethyl bromoacetate, combines with zinc to form an organozinc bromide that adds to the carbonyl group of the second component to give a complex readily hydrolyzed to a carbinol. The reaction

$$Zn + BrCH_2COOC_2H_5 \rightarrow$$
$$BrZnCH_2COOC_2H_5 \qquad \diagdown C=O \longrightarrow$$

is conducted by the usual Grignard technique except that the carbonyl component is added at the start. Magnesium has been used in a few reactions in place of zinc but with poor results, for the more reactive organometallic reagent tends to attack the ester group; with zinc this side reaction is not appreciable, and the reactivity is sufficient for addition to the carbonyl group of aldehydes and ketones of both the aliphatic and aromatic series. The product of the reaction is a β-hydroxy ester and can be dehydrated to the α,β-unsaturated ester; thus the product from benzaldehyde and ethyl bromoacetate yields ethyl cinnamate. α-Bromo esters of the types $RCHBrCO_2C_2H_5$ and $RR'CBrCO_2C_2H_5$ react satisfactorily, but

Ethyl β-phenyl-β-hydroxy-
propionate (b.p. 130°/6 mm.)

β- and γ-bromo derivatives of saturated esters do not have adequate reactivity. Methyl γ-bromocrotonate ($BrCH_2CH{=}CHCO_2CH_3$), however, has a reactive, allylic bromine atom and enters into the Reformatsky reaction.

L. F. Fieser
Mary Fieser

REFORMING

Reforming is a commercial process widely used for conversion of knocking or low-octane gasolines or naphthas to knock-resistant or high-octane fuels. The earliest reforming process was thermal reforming which is carried out noncatalytically at high temperatures (500–550°C) and high pressures (300–1000 psig). Thermal reforming is actually thermal cracking of heavy low-octane straight-run naphthas. (See Cracking). The octane number is increased through the formation of olefins, the production of additional aromatics and the conversion of higher molecular weight paraffins to more knock-resistant lower molecular weight paraffins.

Thermal reforming is a relatively inexpensive process which, however, is not good enough to produce the high-octane gasolines required by present car engines without prohibitive loss of liquid product. As a result thermal reforming has diminished in importance. For example, in 1965, thermal reforming capacity in the U.S.A. was about 75,000 barrels of charge per day; in 1970, it was 50,000 barrels of charge per day.

In the last 20 years another method of reforming using specific catalysts has been rapidly expanding. In the 1940's and early 1950's, processes using non-noble metal catalysts such as molybdena or chromia on alumina were introduced. However, at the present time, (1971) the vast majority of reforming capacity involves the use of platinum catalysts; the first commercial process using platinum was introduced in 1949.

The principal reactions involved in platinum reforming are: (1) dehydrogenation of six-membered naphthenes (cyclohexanes) into aromatic hydrocarbons of high-octane number; (2) dehydrocyclization of certain paraffins to aromatic hydrocarbons; (3) isomerization of five-membered naphthenes (cyclopentanes) into six-membered naphthenes which are subsequently dehydrogenated into aromatic hydrocarbons; (4) paraffin isomerization with a net increase in the amount of higher-branched paraffins, which are more knock-resistant; (5) cracking, which increases octane number by removing paraffins from the gasoline range. This latter reaction is minimized as much as possible since it causes significant loss of liquid product.

The catalyst actually used in platinum reforming is dual-functional: it consists of metallic platinum to provide hydrogenation-dehydrogenation activity and silica-alumina or a halogenated alumina to provide acid activity. A number of the reactions cited above utilize both types of activities or sites in going from reactant to final product. For example, in methylcyclopentane conversion to benzene, methylcyclopentane is first dehydrogenated to methylcyclopentene on a platinum site. The methylcyclopentene then migrates through the gas phase to an acid site where it is rearranged to cyclohexene. Finally, the cyclohexene is dehydrogenated to benzene on a platinum site.

Dehydrogenation of cyclohexanes and dehydrocyclization of paraffins are thermodynamically favored by both high temperature and low hydrogen pressures. However, under the conditions of platinum reforming, side reactions occur which lead to coke formation and deposition of the latter on the catalyst, causing rapid deactivation of the catalyst. To counteract this effect hydrogen is used with charge under pressures ranging from about 150 to 600 psig. The temperature of the process varies from about 480 to 540°C. Under these conditions coke formation and deposition on the catalyst are significantly reduced.

There are currently two major process variations of platinum reforming, both of which utilize a number (usually 3 to 5) of adiabatic fixed-bed reactors. In semiregenerative reforming the unit is operated until catalyst activity or selectivity (the relative balance between gas and liquid products) has deteriorated to the point where further operation is either impossible or uneconomic. The unit is then shut down for catalyst regeneration. In cyclic reforming appropriate piping and valving is provided to allow any single reactor to be regenerated while continuing over-all operation of the unit.

As octane demands on reforming have increased, there has been a trend toward lower pressure levels, since the loss of liquid product which must be incurred to achieve a given octane number is reduced. As the lead content of gasoline is reduced or eliminated, low-pressure reforming operation becomes particularly important.

A major improvement in reforming catalysts occurred in 1968 with the introduction of catalysts containing a mixture of platinum and rhenium as the hydrogenation-dehydrogenation component. These bimetallic catalysts are more stable in maintaining both high activity and selectivity over a given period than conventional platinum catalysts. Due to the increased stability, bimetallic catalysts are particularly useful in semiregenerative reforming where there is no capacity for frequent regeneration. This greater catalyst stability can be employed in a number of ways depending on the specific refinery situation. However, one particular advantage is that it increases the capabilities of semiregenerative units for lower-pressure reforming operation.

As stated above, cracking is generally an undesirable reaction in reforming since it results in significant loss of liquid product. In addition, the acid components used in platinum reforming catalysts tend to crack the more desirable highly branched paraffins more readily than the less desirable n-paraffins. However, in 1968, the Selectoforming process was introduced as an adjunct to platinum reforming. This process is designed to further upgrade the octane number of platinum reformates through the use of a shape-selective zeolite catalyst (see **Catalysts, Shape-Selective**) which preferentially cracks the n-paraffin components.

The total capacity of catalytic reforming units in the U.S.A. at the present time is about 3,000,000 barrels of charge per day.

A. J. SILVESTRI

REFRACTIVE INDEX

Isotropic substances. The refractive index n of a substance is commonly defined in terms of Snell's law, $n_1 \sin i_1 = n_2 \sin i_2$ where n's and i's are respectively the refractive index and the angle of incidence of light entering or leaving the substance. The refractive index can also be defined as the ratio of the velocity of light in vacuum, c, to that of the same light in the substance, thus: $n = c/v$. The angle i is measured between the *wave-normal* of the light and the line perpendicular to the surface where the light passes into or out of the substance. For gases,

nearly all liquids, most glasses, and crystalline substances with isometric (cubic) symmetry, this wave-normal coincides with the *ray*, or path along which the light energy actually travels. In these substances, the refractive index is independent of the direction, and the substances are therefore optically *isotropic*.

Anisotropic substances. Most crystals are not isometric, and in them the ray-direction differs from that of the wave-normal. The classic demonstration of this difference is the double refraction observed in Iceland spar, an optically clear grade of the mineral calcite, $CaCO_3$. Unpolarized light incident at right angles on a cleavage-plate of calcite is resolved into two components as it enters the plate. One of these components, the *Ordinary* ray ("O-ray"), passes straight through the plate, emerging on the far side in line with, and parallel to, the incident ray. It differs from the incident ray, however, in now being plane-polarized. The other component, the *Extraordinary* ray ("E-ray"), travels obliquely through the plate, so that it emerges at a point that is not in line with the incident ray. The E-ray nevertheless again becomes parallel to the direction of the incident ray; this can be used to demonstrate that the wave-normal direction within the calcite suffered no deviation by refraction where the light entered the plate. The E-ray also emerges as plane-polarized light, with vibration direction at right angles to that of the O-ray, and in a plane that contains (1) the E-ray within the crystal, (2) the wave-normal of the E-ray within the crystal, and (3) the *optic axis*, which in calcite is the direction of the crystallographic axis c, the triad or three-fold axis of symmetry.

Uniaxial crystals. Hexagonal, tetragonal, and trigonal crystals have exactly one unique direction, the axis of hexagonal, tetragonal, or trigonal symmetry, which is also the direction of the optic axis. Light passing through such crystals is resolved into two polarized components, of which the O-ray always vibrates at right angles to the direction of the optic axis; the E-ray vibrates in the *principal plane*, already defined as the plane containing the ray, its associated wave-normal, and the optic axis. The actual vibration direction (electric vector in electromagnetic theory) for the E-ray lies in the wavefront and makes an angle with the optic axis that is not 90°. Depending only upon this angle, the refractive index of the E-ray varies from that of the O-ray (obtained with vibrations at 90° to the optic axis) to a limit designated n_e when the vibration is parallel to the optic axis. The *principal refractive indices* are n_o and n_e; for a specified wave length of light, n_o and n_e are characteristic of the composition and the arrangement of atoms and interatomic bonds within the crystal; in most solids, they depend only slightly upon such other factors as temperature and pressure. An *optic axis* may be formally defined as a direction in an anisotropic substance, along which light waves are transmitted with the same velocity (and refractive index), without dependence upon the direction of their vibration. *Uniaxial* crystals have exactly one such direction. The other possible classes are "isoaxial," a rarely-used synonym for *isotropic*, in which every direction is the same as every other direction with respect to light; and *biaxial*, in which there are two directions of no double refraction.

Biaxial crystals. Crystals belonging to the orthorhombic, monoclinic, and triclinic systems of crystallization have no directions associated with rotational symmetry of higher order than that of a diad, or twofold axis of symmetry. Unpolarized light passing through such crystals is doubly refracted, forming two components or rays, either or both of which may behave rather like the E-ray in a uniaxial crystal. If the two components are designated X' or 1, and Z' or 2, with ray 1 having the smaller refractive index and ray 2 the larger, it is possible to find three *Principal Refractive Indices*, n_x, n_y, and n_z, such that

$$n_x \leqq n_1 \leqq n_y \leqq n_2 \leqq n_z.$$

An ellipsoidal *indicatrix* may be constructed with its semi-axes proportional to n_x, n_y, and n_z, in mutually perpendicular directions X, Y, and Z. This indicatrix must necessarily conform to any symmetry possessed by the crystal, so that in monoclinic crystals the symmetry diad (or the direction normal to the single plane of symmetry if there is no diad) must coincide in direction with one of these axes X, Y, or Z. In orthorhombic crystals, the three mutually perpendicular directions of symmetry (all diads or normal to planes of symmetry) determine the directions of X, Y, and Z, but their sequence (as *a, b, c* for the crystallographic axes) may take any relation to the sequence of X, Y, and Z (six possibilities include *a, b, c* respectively parallel to X, Y, Z; to Y, Z, X; to Z, X, Y; to Z, Y, X; etc.). In triclinic crystals, there is no necessary relation between the conventional crystallographic axes and the ellipsoid axes X, Y, and Z. In all crystals, it may be expected that Z will tend to be parallel to the direction, if any, of strongest interatomic bonds, and X conversely. Impressing higher symmetry, as that of a single triad or tetrad axis, requires the ellipsoid to become an ellipsoid of revolution about the direction of higher symmetry, and furthermore, that the ellipsoid axis X or Z becomes the optic axis. The presence of more than one triad or other higher axis requires the ellipsoid to have X = Y = Z, and it is a sphere.

In the general case with n_x, n_y, and n_z all different, the indicatrix is of course a triaxial ellipsoid.

Assuming a transparent plate cut with plane parallel faces from the interior of any crystal, the associated indicatrix may be used to predict the optical properties of the plate (or any other particular shape). For example, a beam of unpolarized light incident normally on the plate is resolved into two components which vibrate parallel to the surface of the plate in directions of the major and minor axes of the plane ellipse generated by intersection of the indicatrix with a plane through its center, parallel to the surface of the crystal plate. These two ellipsoid semi-axes are vectors representing the magnitudes of refractive indices n_1 and n_2 as well as the vibration directions of the respective components of the transmitted light. Snell's law applies to the corresponding wave-normals (not rays); the ray-directions are conjugate to the vibration directions in the planes containing wave-normal and vibration direction in each case.

Two central sections through a general ellipsoid are circular. Their respective normals are called the *optic axes* or binormals, and light passing through the crystal with wave-normals along these optic axes may vibrate in any direction in the circular sections. Such light passes through the crystal without change in its state of polarization, and because there are two such directions the crystal is said to be *biaxial*. The angle between the optic axes is conventionally designated 2V. If this angle is measured across the direction of Z, it may be designated $2V_z$; if across X, $2V_x$. The optic axes necessarily lie in the plane of X and Z. Historically, calcite was one of the first birefringent substances investigated. It was noticed that the E-ray is "repelled" by the optic axis, and then that in some other crystals it is "attracted." From these observations, the *optical character* of uniaxial crystals was designated respectively negative (−) and positive (+). By generalization to biaxial crystals, we find a continuum with uniaxial positive at one end, where the optic axis is Z, varying through conditions of biaxial with $2V_z$ very small, then increasing toward 90°; beyond that, as $2V_z$ becomes increasingly obtuse, $2V_x$ is increasingly acute, and in the limit $2V_x = 0°$ and the crystal is uniaxial negative.

Measurement of Refractive Index. The principal methods of refractometry are (1) prism method, in which the deviation of light by a prism is measured giving data from which the index is easily calculated; (2), method of total reflection, in which the critical angle is measured for total reflection of light passing from one substance into another of lower refractive index; (3) immersion method, where the index of refraction of a microscopic particle is compared with that of a surrounding medium such as a drop of oil of known refractive index, the comparison yielding the datum that the particle has a higher (or lower) index, and in some instances approximately how much higher (or lower). A set of calibrated liquids can thus be used to determine the index of the particle between arbitrarily close limits depending upon the operator's skill, patience, and equipment. The prism method and the method of total reflection can yield precision to 10^{-5} in refractive index, depending mainly on the size and perfection of the sample and its polished surface(s) and the precision with which an angle can be measured. The immersion method is more convenient in most cases, but its precision is usually no better than 2×10^{-3} to 5×10^{-3}; with special precautions this can be improved to perhaps 2×10^{-4}. Phase-contrast methods can be applied to improve the accuracy of the immersion method, using polarized light and a phase-contrast microscope.

HORACE WINCHELL

Cross-references: *Asymmetry; Crystals.*

REFRACTORIES

The American Society for Testing and Materials defines "refractory," the adjective, as "Resistant to high temperatures" and "refractories," the noun, as "Materials, usually nonmetallic, used to withstand high temperatures." Refractories are used in such diverse applications as the manufacture of metallurgical products, rocket nozzles and launch pads, in power generation, oil refining, home heating and in numerous other industries and processes where high temperatures exist.

Refractories are produced in various forms. Brick and other preformed shapes are usually fired but in some cases are chemically bonded. Brick can be formed by melting the refractory material and casting. Refractories are also available in granular form which can be mixed with water and cast like concrete or placed by pneumatic gunning. Premixed, plastic refractories are available to form monolithic units by ramming into place.

Most refractories are categorized as acid or basic and the selection of materials must be on the basis of the chemical environment as well as other factors such as maximum temperature, abrasion, temperature fluctuations, and atmosphere (oxidizing or reducing). Insulating refractories are designed with low thermal conductivity and are used for the purpose of saving heat.

Acid refractories range from those high in silica, SiO_2, and low in alumina, Al_2O_3, through those containing up to 45% Al_2O_3. High-alumina refractories which are considered amphoteric, or compatible with either acid or basic conditions, may range from 50% alumina to almost pure alumina. Mullite refractories are in the alumina-silica category. Basic refractories have as their major constituents magnesia, MgO, chromic oxide, Cr_2O_3, and/or lime, CaO. Fused cast, carbon, zircon, zirconia, and silicon carbide are types of refractories which have specialized applications.

Silica. Silica is the most prevalent mineral in the world and raw material of suitable purity can be found in many locations. As a refractory, it is unique because of its ability to withstand load to within a few degrees of its melting point. Several grades of silica brick have been developed. The purest grade, superduty, has upper limits of alumina as an impurity of 0.35% and maximum alkali content of 0.15%. For coke oven service, high-density silica brick with alumina contents up to 1.15% have been developed. This general category of refractories is used in glass tanks and open-hearth steelmaking furnaces but, wherever used, the furnaces must be carefully heated. Silica is subject to relatively large volume changes with small increases in temperature. Above 1200°F, however, the coefficient of thermal expansion is very low and large changes in temperature are not detrimental.

Another form of silica which has gained favor in some applications is fused silica. This product of a melting operation is carefully cooled, ground, rebonded into the desired shapes and fired. Fused silica is used for coke oven doors and jambs, nozzles for continuous casting, brazing fixtures, forms for hot-forming large airplane wing sections and other areas where rapid changes in temperature or large differential temperatures prevail. Prolonged heating above 2000°F will cause fused silica to devitrify and become temperature-sensitive.

Fireclay. Fireclay products are in the alumina-silica series of refractories. The alumina contents vary from 20 to 25% for low-duty and ladle brick to about 45% alumina for superduty fireclay brick. Generally the higher the alumina content, the more refractory is the product. Low-duty brick and ladle brick have softening points around 2600°F, while superduty brick melts at around 3200°F. Medium-duty and high-duty fireclay brick are the classes between these extremes.

The raw materials for fireclay products are found in many states including Georgia, Alabama, Maryland, Ohio, Pennsylvania, New Jersey, Kentucky, Missouri, Colorado, Texas and California.

Generally, these clays have varying amounts of impurities such as iron oxide and alkalies. When subjected to certain heat treatments a glassy bond forms. Various physical properties can be developed in each of the classes of fireclay brick, depending on degrees of compaction and firing and the chemical composition.

These refractories can also be obtained in the form of castables which have cement binders and their placement is similar to placing concrete. These castables can also be placed pneumatically. In either case, castables must be cured under moist conditions to develop the cold bond and then they are subsequently fired in place.

Another form of monolithic fireclay refractory is the plastic or ramming mix. This form requires tamping or ramming in place and drying out to obtain strength. As with castables, final firing is done in place.

Semisilica. Semisilica refractories fall between silica brick and fireclay brick and contain 70 to 80% silica with alkali contents generally being below 0.5%. At intermediate temperatures, they exhibit better load-bearing ability than fireclay brick and are generally applied where this property is needed. This material has been used in combustion chambers, domes and checkers of blast furnace stoves, checker chambers of open-hearth furnaces, certain areas of soaking pits and reheat furnaces, ceramic kilns, etc. Naturally occurring siliceous fireclay used for this type of refractory can be found primarily in New Jersey, but in some cases the composition has been synthesized.

High-Alumina. The alumina contents of the various classes of high-alumina refractories have been established by ASTM as 50, 60, 70, 80, 85, 90, and 99%. Raw materials for their production vary from fireclay enriched with bauxite or diaspore, through bauxite and diaspore to tabular, fused or Bayer process alumina. The lower three classes have increasingly higher melting points (refractoriness) but they are not noted for good load-bearing ability. In recent years, however, and because of the demands occasioned by service in blast furnace stoves and electric furnace roofs, manufacturers have developed 60 and 70% alumina brick with improved load-bearing ability. In addition to numerous metallurgical applications, the refractories in the lower range of alumina have been used in cement, lime and dolomite burning kilns, glass tanks, boilers, etc.

The 80% and higher alumina content refractories have good high-temperature load-bearing ability and are particularly adaptable to high-temperature kilns, certain areas of glass tanks and aluminum melting furnaces.

High-alumina brick usage has increased rapidly as the desire for producing steel and other commercial products at higher temperatures and faster rates became a reality.

Mullite. This class of refractory brick is treated separately because it is unique and has special properties. Theoretically, mullite is $3Al_2O_3 \cdot 2SiO_2$ or 71.8% Al_2O_3 and 28.2% SiO_2. Formerly, mullite brick was made using kyanite, sillimanite or andalusite, but these raw materials are so uncertain in quality and availability that mullite grain is produced by calcining at high temperatures mixtures of clay and alumina or clay and bauxite. Mullite brick with their good load-bearing ability, have recently been used in high temperature areas of blast furnace stoves and in glass tanks.

Magnesite. Refractories which consist essentially of the oxide magnesia in crystalline form are generally known as "magnesite" products. The grains used in magnesite refractories are formed by dead burning natural magnesites, $MgCO_3$, or magnesium hydroxides that are extracted from seawater or inland brines. High-purity grains containing 98% MgO are currently used in commercial refractory products. Magnesite grain is used alone and in combination with chromite or dolomite to form basic refractory brick and granular materials.

Magnesite brick produced for basic oxygen steelmaking furnaces uses coal tar pitch as an ingredient. The pitch residue (carbon) which is retained in the brick structure when heated in service, improves the slag resistance of the brick. Basic oxygen furnace bricks are either fired and impregnated with pitch or bonded with pitch. Pitch-bonded brick are commonly tempered before use by heating at low temperatures to remove the volatile components of the pitch.

Magnesite and chrome ore combinations are known as magnesite-chrome brick when magnesite grain is the predominant component and as chrome-magnesite brick when chrome ore predominates. Bricks of these combinations are produced as fired or unfired chemically bonded types. Products in which chrome ore predominates are used more in copper refining while magnesite-chrome products find greater use in the glass and steel industries. Magnesite-chrome brick that are fired to temperatures above 3000°F are used extensively in open-hearth and electric furnace construction. These high-fired brick are termed "direct-bonded" and are characterized by bonds that are formed between components by solid state diffusion.

Dolomite. Dolomite is calcined to form dead-burned grain, $CaO \cdot MgO$, for use in brick and granular materials. High-purity dolomite grain, containing less than 2% alumina, silica and iron oxide, is used in pitch-bonded and tempered brick produced for basic oxygen steelmaking furnace applications. Fired brick made from high-purity grain are being used in hot zones of cement kilns. Large quantities of loose granular dolomite are purchased annually by the steel industry for use in maintaining bottoms of open-hearth and electric furnaces. After calcining, dolomite is susceptible to hydration and will disintegrate during storage under atmospheric moisture conditions.

Fused-Cast Refractories. Although this class of material could be discussed under the headings of high-alumina or basic refractories, it has been decided to treat them as a separate class. They are formed by melting refractory materials in electric furnaces and casting in sand or graphite molds. Compositions can vary from 99% alumina, through mullite compositions to alumina-silica-zirconia. The alumina-silica series of cast refractories have found wide acceptance in the construction of glass tanks and as skid rails in steel reheating furnaces.

Basic compositions (approximately 60% MgO, 20% Cr_2O_3 with iron oxide and alumina) have been successfully used as open-hearth and electric furnace sidewall material and for open-hearth roofs. They have also been applied in the Kaldo process of steelmaking. Being low in porosity, the fused-cast refractories have good slag resistance but are subject to spalling if temperatures change suddenly.

Carbon. As a refractory, carbon has the capability to withstand temperatures in excess of 6300°F and is extremely resistant to spalling and slag attack. The main disadvantage of carbon is that it must be protected against oxidation above 600°F.

Raw materials involved in the manufacture of carbon refractories are calcined anthracite, lamp black, petroleum coke (which is low in ash content), and pitch. Large block stock is made by an extrusion process and subsequent baking at approximately 1800°F. One type of carbon brick is manufactured by a hot-pressing technique which fires the carbonaceous material electrically as forming pressure is applied.

Carbon or graphite may vary in composition from the extremely pure grades used as moderators and reflectors in nuclear reactor to materials containing 10 to 15% ash.

One of the most important uses for carbon as a refractory is the application in the hearth, tuyere breast and bosh of blast furnaces. The bosh application of carbon requires a material that has high thermal conductivity and good resistance to attack by alkalies. Carbon products are also utilized as blast furnace splasher plates and trough liners and are used extensively in electrolysis pots for the reduction of alumina to aluminum.

Graphite, a crystalline form of carbon available as a natural product or by subjecting amorphous carbon to high temperatures, has many diverse applications such as electrodes for electric arc furnaces, motor brushes, melting crucibles, pencil lead, and as an ingredient in clay-graphite refractory brick and stopper heads, the latter being used to control the flow of molten steel from ladles.

Zircon and Zirconia. Large deposits of high-purity zircon, 98–99% $ZrO_2 \cdot SiO_2$, have been found in Florida, Australia, India, Madagascar and Brazil. Zircon bricks have a relatively low coefficient of thermal expansion; consequently, they have good spalling resistance. They also are relatively resistant to siliceous fluxes and slags high in iron oxide. Nozzles for pouring steel have been successfully used, but because of high cost, the application of this refractory material has not been widespread.

Zirconia, ZrO_2, is even more expensive than zircon and has as its source baddeleyite, found principally in Brazil and Ceylon. Zirconia undergoes crystal structure changes with changes in temperature and, in the pure form, has poor spalling resistance. The addition of 4 to 5% CaO or MgO to zirconia will stabilize this refractory, making its use possible. Steel pouring nozzles of this material are available, as are crucibles for melting noble metals.

Silicon Carbide. This class of brick is noted for its abrasion resistance, high thermal conductivity and chemical inertness. The bricks generally contain 85–90% SiC, but for less severe service may contain as low as 60% SiC. Originally, most silicon carbide bricks were silicate-bonded, but in recent years silicon nitride-bonded silicon carbide or silicon oxynitride-bonded silicon carbide have been developed. The latter two have superior oxidation resistance over the silicate-bonded.

They have been used in the ceramic industry (as kiln furniture), in boilers and incinerators, and as muffles in kilns and furnaces. Recently, this type brick has been used in the throat of a blast furnace. In this application, the temperatures are not extremely high, but the lining must withstand the impact and abrasion of the burden as it is charged into the furnace.

A unique property of this refractory material is its electrical conductivity which permits its use as resistance heating elements in kilns and furnaces.

K. A. BAAB AND C. R. BEECHAN

REFRIGERANTS

Refrigeration may be generally considered as a process for producing cold in which a working substance known as a refrigerant is caused to undergo a physical change. Examples of refrigeration processes include the liquefaction of air by expansion through a nozzle, the cooling produced by passing an electric current through a bimetallic junction, and the cooling resulting from the melting of ice, the refrigerants being air, bimetallic junction, and ice, respectively. The most useful and widely used physical change, however, is that in which the refrigerant undergoes a change from the liquid to the vapor state. This discussion will be concerned only with working substances whose latent heat of vaporization produces the cooling effects.

The most commonly used process for making use of the change from liquid to vapor state of the refrigerant is the vapor compression cycle, which involves a recirculation of the refrigerant by means of a positive displacement compressor driven by a prime mover. In this process, high-pressure refrigerant vapor from the compressor enters a heat exchanger (condenser) in which it condenses, rejecting sensible heat from work done upon it during compression plus its latent heat of vaporization. The condensate flows through a restriction (expansion valve or other similar device) thus undergoing a drop in pressure, during which a sufficient portion of the refrigerant vaporizes to reduce the temperature of the remaining liquid to that consistent with the lower pressure. The liquid-vapor mixture then enters a heat exchanger (evaporator) and absorbs heat equivalent to the latent heat of vaporization of that portion of the refrigerant existing as liquid. The resulting low pressure vapor then passes to the compressor to complete the cycle. Thus it is evident that a refrigerant acts as an agent for transporting heat from a lower to a higher temperature level. The ratio of the heat rejected in the condenser or absorbed in the evaporator to the work necessary to effect the transport represents the most common way of expressing the efficiency of any given cycle, and is termed the coefficient of performance (COP). The COP for heating is greater than the value for cooling as a result of the work added to the refrigerant in the compressor.

Most early refrigeration units used ammonia as a refrigerant, but the first domestic refrigeration machines, requiring small units of light and cheap construction, used sulfur dioxide, ethyl chloride or methyl chloride. Ammonia is now mainly used in high tonnage applications involving attendance by trained personnel. The chemists' search for safe, noncorrosive, low-toxic refrigerants finally led to fluorinated and/or chlorinated hydrocarbons which offer a class of materials suitable for most applications, and which are used in practically all domestic refrigeration and air-conditioning machines. The continuing development of new applications presents a challenge to the chemist for the synthesis of refrigerants having still better physical, chemical and thermodynamic properties.

The choice of a refrigerant for a specific application involves careful consideration of (1) thermodynamic properties, (2) physical properties, and (3) chemical properties. The thermodynamic properties of a refrigerant determine its efficiency. When tabulated or graphically portrayed on a temperature-entropy or pressure-enthalpy diagram, these properties afford the designer a convenient means of studying the performance of a refrigerant in a given cycle. A high latent heat of vaporization is desirable, since it represents a measure of the refrigerating effect. However, it must be accompanied by a low specific heat of the liquid, to minimize the loss in heat of vaporization due to cooling of the warm liquid to evaporator temperature when passing through the expansion valve (throttling loss). Conversely, a high specific heat of the vapor is desirable to minimize super-heating of the vapor and attendant power loss during compression. The critical temperature of the refrigerant should be substantially higher than the maximum condensing temperature to minimize power consumption and throttling loss.

The physical and chemical properties of a refrigerant—more than its thermodynamic properties—determine its suitability for a given application. In general, the boiling point is the most important single property of a refrigerant, since it is related to latent heat, critical temperature, leakage, volume displacement, freezing point, oil solubility, safety, etc. There is a useful empirical correlation for comparing the capacity of refrigerants in a positive displacement machine as a function of their boiling points. This is derived from Trouton's Law (assumed value for Trouton's Law constant of 22) and the perfect gas law, and it can be applied to approximate a relative capacity index for a material when thermodynamic and physical data are not available.

The possibility of slight leaks permitting moisture-laden air to be drawn into the system makes vacuum operation undesirable (air causes high condensing pressures and consequent excessive power; moisture leads to freeze-up and corrosion). Therefore, a refrigerant which has an evaporating pressure slightly above atmospheric pressure at the lowest temperature expected under operating conditions would be preferred. Such a refrigerant would have the additional advantage of a lower condensing pressure and result in cheaper, lighter and safer equipment.

The ideal refrigerant would be completely inert; it would be harmless in itself and would not react with any substance within the system in which it is contained. As no known substance meets these requirements, all known refrigerants represent a compromise with respect to chemical properties to be settled by consideration of the application. Thermal stability of a refrigerant is important in order that undesirable decomposition or reaction products not be formed during compression. Hydrolysis can produce corrosive acids which damage equipment and evolve hydrogen, an undesirable inert. Even the relatively stable halogenated hydrocarbons are slowly hydrolyzed, forming HCl. High temperatures and open flames give toxic reaction products such as phosgene, hydrochloric acid, and carbon monoxide with many of the halogenated hydrocarbons which are stable at normal temperatures. Certain refrigerants react vigorously with specific materials desirable for system components, thus imposing engineering design limitations unless an alternate refrigerant can be selected. A system charged with ammonia requires the absence of copper or copper alloys; these are strongly attacked by aqueous ammonia, which may be present due to the unavoidable presence of small amounts of water. Refrigerant 22 (monochlorodifluoromethane) is a good solvent for hermetic system components such as gaskets and motor insulation materials. Refrigerant 12 (dichlorodifluoromethane) used in practically all domestic refrigeration systems requires highly refined "refrigeration grade" lubricants to insure long-service system stability.

The organic halides (e.g., methyl and ethyl chlorides) are good solvents for certain gasket materials; natural rubber and many neoprene compositions cannot be used with these refrigerants. Hermetically sealed refrigerating units, in which an electric motor is exposed to the refrigerant, present added complexity in the number of materials of construction which require refrigerant compatibility. Current application trends are to more numerous and larger units of the hermetic type, and the behavior of refrigerants with various insulating materials and other motor components must be considered.

Solubility and compatibility with lubricants is another important property of refrigerants. Ordinary oils react with sulfur dioxide, so special oils must be provided with this refrigerant. Highly refined oils containing a small amount of unsaturated hydrocarbons are necessary in units charged with methyl chloride and some halogenated hydrocarbons to prevent a condition known as "copper plating." This is caused by solution of copper in the oil followed by subsequent decomposition of the organic copper compound, especially at points of high temperature. The copper residue resulting from the decomposition can cause serious seizure of moving parts.

Safety of refrigerants must be considered as a relative term dependent upon the application. Hydrocarbons, ammonia, and some halogenated compounds are combustible, whereas the "Freons" are nonflammable. Mixtures of toxic vapors and air may attack the membranes of the eyes, or respiratory system, and/or be absorbed by the blood and carried to other parts of the body to which they may be injurious. The National Board of Fire Underwriters has tested the commonly used refrigerants and classified them into groups according to their toxicity.

The most toxic refrigerants are sulfur dioxide and ammonia which are listed in Underwriters' toxicity groups 1 and 2. Commonly used fluorocarbons, such as Refrigerants 11, 12, 22, 113, and 114 are listed in Underwriters' groups 5 and 6 and are generally considered to be harmless in the intended applications. These fluorocarbons are also listed in group 1 (least hazardous) of the ASA Standard B9 Safety Code.

The absorption cycle, although of less importance than the vapor compression cycle, is used in both domestic and large scale refrigeration units. A description of special qualifications for refrigerants used in the absorption cycle is beyond the scope of this article.

T. L. ETHERINGTON AND A. E. SCHUBERT

Cross-references: *Absorbents, Evaporation, Toxicology.*

REINFORCED PLASTICS

In reinforced plastics, the ancient Egyptian method of combining straw into wet earth to strengthen bricks, or the later practice of incorporating coarse bristles into mortar for stronger dwelling walls is abruptly brought into modern technology.

Although some component reinforced plastic materials have been known for over 100 years, the combined technical and commercial successes of reinforced plastics per se have come about only in the past 25 years; they resulted from two events: (1) the volume commercial production of continuous filament fiber glass for reinforcement (1935); and (2) the invention of combining styrene monomer with an alkyd-base resin to yield unsaturated polyester resins (1939). The product was the first low-pressure laminating resin which formed the matrix for the superior reinforcing materials to give end products with unique properties of strength, durability and adaptability to different products and markets.

The industry has expanded from modest beginnings to one with a total volume of over 1,000,000,000 pounds of product in 1970. All known forms of plastic molding have been utilized and many new ones devised to meet specific market needs.

Materials. The materials used are resins, reinforcing agents and fillers. The resins are chiefly polyester types (dibasic acids + dihydric alcohols) which cure by addition polymerization after incorporation of peroxide catalyst plus either natural promoters or application of heat. Cures are effected either at room temperature in open layup molds, or in heated press molds. An inhibitor such as hydroquinone is incorporated to prevent uncontrolled polymerization and to give longer shelf life.

Epoxy Resins are a reaction product of epichlorohydrin and bisphenol A; they cure either at room or elevated temperature after addition of selective amine hardening agents.

Phenol-formaldehyde Resins date from 1910; they cure by condensation reaction and hence require high-pressure molding. This limits their usage to powdered or liquid molding compounds, filled and reinforced.

Thermoplastic Resins are newcomers to reinforced plastics; all types are capable of being reinforced with fiber glass or other filamentary materials.

Glass Fibers form the major type of reinforcement and are produced by exuding molten glass from small-diameter orifices in a heated crucible. An organic film former is applied at fiber forming for lubrication, strand protection and eventual compatibility with the end-use plastic. The glass is converted into various forms or intermediates for use with the resins including rovings, chopped strand mats, chopped strands, woven roving, milled fibers, twisted and plied yarns, light-weight fabrics and many others, each intended to satisfy the requirements of a specific reinforced plastics molding method.

Asbestos is a natural fibrous mineral which is mined and processed to provide bulk filaments from minute lengths up to $\frac{1}{2}''$. The fibers may be mixed with various types of resin to form molding compounds.

Resin-to-Glass Bonding. An organic glass-sizing material is present on the glass in amounts less than 2% by weight. It contains a chemical cross-linker (silane or chrome complex) to assist in creating a reacted bond between glass and resin during the laminating process. The organic polymer collects in small nodules instead of a smooth film; hence, resin-to-glass bonding is not a continuous process. Electrical resistance and wet-strength retention of finished laminates are used to evaluate the effectiveness of the bond. Table 1 shows the

TABLE 1

	Rigid Polyester Casting	Cross-linking Agent-treated Glass Cloth: Polyester Laminate (Same Resin)
Original Electrical Resistance, megohms	50,000	40,000
Electrical Resistance after 24 hours' exposure, megohms	40,000	50
Electrical Resistance after 14 days, megohms	7,000	10
Tensile Wet-Strength Retention, % (2 hr boil in water)	—	80%*

*This figure compares to only 20% to 40% WSR for an identical glass cloth polyester laminate made without cross-linking agent on the glass fiber. The deterioration in insulation resistance of the laminate indicates existence of gaps in the bonding between the resin and the glass filaments. However, the high wet-strength retention of the treated glass cloth laminate assures the effectiveness of the chemical cross-linking agents at the actual bonding sites.

difference between treated and untreated cloth laminates vs. a rigid polyester casting in these two

respects. Mechanical fastening of glass by resin also contributes to strength of the glass-resin bond. The nature and appearance of the bonding elements between fiber glass and resins are excellently shown by the stereo-scanning electron microscope. In practice, laminates generally higher in glass content by weight exhibit higher strength and performance properties.

Molding Methods. Approximately 14 major methods of molding to produce reinforced plastics are in commercial usage. Some are adaptations of other fields of plastics molding and others are indigenous to reinforced plastics. The methods can be grouped into six major types of processes as follows:

A. *Low-temperature or Room-temperature Cure.*

Cure times are from 20 minutes to 2 hours from addition of peroxide catalyst plus accelerators to polyester resin at 3 to 5 poise viscosity in any one of the following methods:

Hand Layup involves application of a nonreinforced resinous skin (gel) coat to a prepared mold followed by laying in dry reinforcements and impregnating with the catalyzed, promoted resin. Cure occurs by action of promoter on the catalyst to attack the unsaturated resin bonds. Typical products are boats, custom or prototype automotive bodies, chemical ducts, tanks, and hoods and generally larger sized structures.

Sprayup accomplishes hand layup objectives by chopping fiber glass roving into a resin stream onto the work or mold where it is wet out or coalesced by rolling out and allowed to cure. Critical properties in both hand layup and sprayup are uniformity of distribution of the glass reinforcement and thoroughness of cure.

Pressure and Vacuum Bag Molding Plus Sandwich Construction. To eliminate voids and increase laminate density, a flexible membrane is placed over a wet hand layup in a suitable mold and vacuum applied until cure. Autoclaves and hydroclaves may also be used. Cloth is the main reinforcement and laminate glass contents of 25 to 55% are reached. It is also possible to produce high-modulus skin-honeycomb-skin structures for radomes, types of tooling, aircraft wing tips, tail sections and other modular units as dwellings or housing components.

Miscellaneous room temperature cure processes include specialized adaptations of hand layup, i.e., gun wetting, prewetting, confined flow pressure or vacuum impregnation, curing with ultraviolet-sensitized resin, and casting techniques.

B. *Intermediate—Temperature Cure Processes.*

These are based on resin systems wherein 1.5 to 3.0 poise viscosity polyesters are laid up at room temperature and require 150 to 180°F to initiate gel and cure, which completes itself with heat supplied from the exotherm generated.

Architectural and Industrial Paneling. The thinned, unfilled resin with light-stable and weather-resistant additives is impregnated into and cured with 20 to 25% glass reinforcement to provide the popular flat or corrugated translucent building panels. Variations include fabrication of attractive double-wall core panels from the cured sheet and also layup and cure of reinforced plastic faced plywood for rigidized containerization.

C. *High-temperature Curing Processes.*

Resins catalyzed with benzoyl peroxide, tertiary butyl hydroperoxide, tertiary butyl perbenzoate or others, are alloyed with glass reinforcement at temperature from 235 to 360°F utilizing well-designed and machined heated molds in compression, transfer or injection presses.

Premix and BMC Molding involves mixture of generally 30% of a 35 poise polyester resin with 50% clay, calcium carbonate and asbestos fillers and 5 to 25% short length ($\frac{1}{8}$" to $\frac{1}{4}$") glass-reinforcing fibers. New low-profile resins are used for improved molded surfaces, and superficial thickeners like $MgCO_3$ or MgO are incorporated (BMC) for better handling and molding properties (15 to 20 second mold cycles for small parts). Critical areas in premix molding are impact strength and avoidance of excessive glass fiber degradation during mixing.

Preform and SMC Matched Die Molding. In preform molding, chopped glass fiber is preshaped over a revolving screen and combined with filled and catalyzed resin in a matched die mold set at the compression-molding operation. In SMC molding, filled, superficially thickened, catalyzed low-profile resin is blended with chopped ($1\frac{1}{2}$" to 2") reinforcing fibers on a carrier film in a continuous throughput machine. This sheet molding compound is then charged to the die set in weighed amounts, thereby achieving faster, more economical molding cycles with less waste.

D. *Miscellaneous Reinforced Plastic Processes.*

Filament Winding. Dry or preimpregnated glass fiber roving is creeled, tensioned, impregnated with polyester or epoxy resin and then wound in a programmed pattern onto a rotating mandrel to produce reinforced plastic pipe and other symmetrical surfaces of revolution.

Rod Stock and Pultrusion. Continous-filament fiber glass roving when creeled, tensioned and drawn through a resin bath, limiting dies, and cured, is the fabrication technique used for fishing rods (solid and hollow), electronic components, guy or strain stand-off insulators, spring elements, and structural shapes such as T, U, I-beams, etc.

Centrifugal Casting. Fiber glass chopped strand mat reinforcement is placed against the inner wall of a rotatable, heated (usually), cylinder mold container (with bottom) or other configuration, and the mold spun or rotated. Resin fed in by spray or jet thoroughly impregnates the mat, forces the voids to the inner surface by centrifugal action, and cures to form such items as pipe, torpedo launching tubes, containers, water-softening tanks and other chemical or utility gear.

Casting, Potting or Encapsulation. Filled, usually nonreinforced, pourable mixes of resin plus clay, $CaCO_3$ or milled glass fiber are used to encapsulate electrical or electronic components.

Reinforced Structural Foam. Drum-wound, expandable reinforcing fiber glass mats may be impregnated with activated polyurethane or other thermosetting foams. The foam expands the glass mat during blowing and results in a greatly strengthened insulating, low-density cured product that is useful in refrigeration, furniture construction and other applications.

E. *Reinforced Thermoplastics.*

The major properties of all thermoplastics, such as per cent elongation, heat deflection and all physical strengths, are all substantially improved by incorporation of reinforcing glass fibers or asbestos. Three methods are described:

Injection Molding. Moldable glass-loaded pellets (of all thermoplastic resin types) or bulk glass fiber plus the clear resin pellets blended at the press hopper are injection-molded using standard screw equipment. Only minor press modifications are required to adapt to glass.

Glass Reinforced Rotationally Cast Thermoplastics. Fibers $\frac{1}{32}$", $\frac{1}{8}$" or $\frac{1}{4}$" are rotated and tumbled with polyethylene, butyrate, styrene, vinyl (dry or plastisol) resins in standard roto-casting equipment.

Strengthening Thermoformable Thermoplastic Sheet. Vacuum-formed acrylic or other thermoplastic sheet may take the place of a polyester gel coat, being coated with sprayup laminates as earlier described.

F. *Unique and Advanced Composites.*

Reinforcements include high-modulus glasses, high-silica and quartz batting and fibers, boron filaments, metal and metallic oxide "whiskers" plus carbon and graphite fibers, fabrics and yarns. Resins used are high-temperature, high-performance epoxy, phenolic, polyimide, and other matrices.

Conclusions

Table 2 presents range of physical properties possible in reinforced plastic molded parts. Table 3

TABLE 2. GENERAL PRODUCT PHYSICAL PROPERTIES

Parameter	Units	Range of Values Possible
Specific gravity	—	1.3 to 2.2
Reinforcement level	% by wt.	5 to 85
Tensile strength	psi	15,000 to 180,000
Compressive strength	psi	15,000 to 50,000
Flexural strength	psi	25,000 to 139,000
Impact strength	ft-lb/in. notch	2.5 to 30
Water absorption	R.T. 24 hr.-%	0.01 to 1.0
Wet-strength retention	%	60 to 95
Heat resistance, continuous	°F.	300 to 500
Burning rate	—	Slow to self-extinguishing
Volume resistivity 50% RH-73°F.	ohms/cc	10^{14}
Arc resistance	seconds	120 to 450

TABLE 3. MARKETS SUPPLIED AND
DISTRIBUTION

	Millions of Pounds		
	1968	1969	(Est) 1970
Aircraft & aerospace	32	42	26
Appliance parts, equipment and housings	27	34	57
Construction	108	124	152
Consumer products	53	63	95
Electrical	71	82	63
Marine craft & accessories	201	270	325
Pipes ducts, tanks (corrosion)	71	106	112
Transportation	167	220	266
Miscellaneous	50	53	83
TOTALS	780	994	1179

provides a 3-year record of market penetration for combined reinforced plastic products.

J. GILBERT MOHR

References

"Handbook of Reinforced Plastics of SPI," Oleesky-Mohr, Reinhold, 1964. See also revised edition, "Technology and Engineering of Reinforced Plastics/Composites of SPI," Mohr, Oleesky, Shook and Meyer, Van Nostrand Reinhold, 1973.
SPI RP/C Division Proceedings of Annual Technical and Management Conference, SPI, 250 Park Avenue, New York, New York 10017. See Also SPI Annual RP/C Market Survey.
"Modern Plastics Magazine," Annual January issue Market Survey plus annual "Modern Plastics Encyclopedia."

REINFORCING AGENTS

Reinforcing agents are solids, usually in a high state of subdivision, which, when dispersed in the substance to be reinforced, are capable of imparting improved mechanical properties. In its strictest sense this definition would include a number of instances of reinforcement in metallurgy, such as the enhancement of the tensile strength and hardness of iron by iron carbide in steel. It is customary, however, to apply the terms "reinforcement" and "reinforcing agent" mainly to systems in which the reinforced phase is an elastomer or a plastic. True reinforcing agents do not merely stiffen (i.e., raise Young's modulus) the polymer into which they are incorporated but produce increases in such properties as tensile strength, tear strength, resistance to abrasion, resistance to flexing and, in the case of plastics, impact resistance.

A reinforcing agent may function by any one or more of the following mechanisms: (1) mechanical, (2) physical adsorptive, and (3) chemisorptive. In the mechanical mechanism the reinforcing effect is furnished by the cohesive strength of the reinforcing agent. The reinforcing action of fibrous materials is in part due to mechanical interlocking

but in most of the important instances of polymer reinforcement this effect is relatively subordinate. Fabrics, which reinforce resins in laminates largely by this mechanism, are not reinforcing agents in the sense considered here.

The physical (Van der Waals) adsorption mechanism of reinforcement can be expected to operate whenever the reinforced phase wets the reinforcing agent. One would expect that finely divided fillers would be more effective reinforcing agents, in that they would contribute more loci of reinforcement and a larger interfacial free energy to the system, than an equal volume of large particles. This is generally found to be the case, the most effective reinforcing agents having particle diameters of the order of 100 to 1000Å. There exists in all instances an optimum loading of reinforcing agent, and in general, reinforcement will also pass through an optimum in particle size. The latter condition may, however, be affected by increasing difficulties in dispersing extremely fine fillers.

The chemisorptive mechanism of reinforcement, when present, is probably always superimposed on the Van der Waals' attractions between filler and polymer, for chemical reaction could hardly occur without intimate physical contact as prerequisite. The most important instance in which evidence for chemical reinforcement exists is the system carbon black-rubber (both natural and synthetic). Points of chemisorptive attachments of rubber molecules at the filler surface act in a manner similar to the crosslinkages between polymer molecules formed in vulcanization. Consequently, carbon black produces more stiffening than would be expected on the basis of its bulk alone. It is not known with certainty how much, aside from stiffening, the chemisorptive linkages contribute to reinforcement, nor is the exact nature of the linkage between rubber and carbon black completely understood. It appears that chemisorptive bonds are necessary for the development of superior abrasion resistance but not for high tensile strength.

Typical reinforcing agents for elastomers are given below. Of all reinforcing fillers for rubber, and of the carbon blacks in particular, the oil furnace blacks represent the most important category. Their use in automobile tires has become universal in the industry. The manufacture of channel blacks is rapidly being phased out and their application in off-the-road tires and wire carcasses are being taken over by oxidatively post-treated furnace blacks. Thermal blacks are used extensively in mechanical goods where low cost takes precedence over high performance.

Mineral fillers are used in non-tire applications whenever color is a factor. Among these the colloidal silicas are the most reinforcing. Clays and whiting are used in many low-cost, noncritical applications.

Some Reinforcing Agents for Elastomers

(1) Carbon Blacks	(2) Mineral Pigments
(a) Oil furnace blacks	(a) Silica
(b) Thermal decomposition blacks	(b) Silicates
	(c) Clays
	(d) Whiting

Organic reinforcing agents (lignin, proteins, phenolics, vinyl resins, etc.) have been investigated extensively but have not attained commercial significance.

The most useful loading of reinforcing agent is determined by the desired balance of physical properties. As the loading is increased from zero the modulus, tensile strength, tear strength, abrasion resistance and hardness increase while the ultimate elongation and resilience of the stock decrease. In the majority of cases the best balence of properties is obtained at 15 to 25 volume % of filler, depending largely on the particle size, with extremely fine particle fillers requiring smaller quantities for optimum reinforcement.

While the most important uses of reinforcing agents are in the field of elastomers there are also some significant instances of reinforcement of plastics by fillers. Polyethylene, when crosslinked, is reinforced by carbon black. One of the main applications of black-reinforced polyethylene is in plastic pipe. Asbestos fiber or shredded fabric is used to reinforce phenolics. Chopped glass fiber strand is employed in a number of plastics; it is usually given a vinyl silane finish to facilitate bonding to the resin.

The reinforcement of brittle plastics by rubber to impart impact resistance to the composition is practiced widely. The most notable examples are the high-impact polystyrenes and ABS (acrylonitrile-butadiene-styrene) resins. In these polymers the rubber is the discrete phase. Interfacial bonding is desirable and is provided by graft polymerization of some of the resin monomer (or monomers) onto the rubber.

GERARD KRAUS

Cross-references: *Rubber, Carbon Black, Elastomers, Adsorption, Clays.*

REPELLENTS

Animal

The need for animal repellents has long been recognized. The reduction of damage to communication cables, protection of packaged materials from rodent gnawing, avoidance of damage to trees, elimination of injury to shrubbery or furniture by household pets, and protection of germinating seeds and field crops are prime examples of the need for such repellents. Of special importance is the use of repellents in combination with insecticides or herbicides to prevent feeding by livestock until the level of the residues is reduced to safe limits. The search for new and better repellents is being carried on in commercial, university, and government laboratories both in the United States and abroad.

Probably the best multipurpose animal repellent is a technical preparation known as Faunatrol[R], which contains approximately 80% N,N,1,1-tetramethyl-2-butynylamine as the active ingredient. Prepared by a condensation of methylacetylene and dimethylamine, emulsions of the product varying in concentration from 0.1 to 6.0% may be sprayed on cattle- and sheep-grazing areas to prevent grazing, on communication cables to prevent attack by livestock, swine, and rodents, and around crop areas to prevent nocturnal attack by deer. Con-

centrations in the range mentioned have no deleterious effect on crops or rubber insulation and appear to be nontoxic orally to dogs, rats, mice, and rabbits.

For Rodents: Other amines which have been found to be highly efficacious in repelling rodents are the straight-chain, saturated amines possessing 10 to 18 carbon atoms, the trinitrophenylamines, guanidine, and N,N-diphenylguanidine. Likewise, the related amides serving the same purpose are N-isopropylisobutyramide, N-isopropylacrylamide, and the N,N-diethylamide of 4,6,6-trimethyl-2-hydroxy-3-cyclohexenecarboxylic acid. A number of nitriles are quite effective in repelling rodents from stored and packaged materials impregnated with these compounds; several prime examples are α-cyano-β-phenylacrylonitrile, dodecyl cyanide, and tridecyl cyanide. This is perhaps not surprising, since many nitriles and cyanoesters impregnated into cartons will keep away numerous species of cockroaches and other insects.

Other substances which have been patented for use as rodent repellents are N,N-dimethyl-S-tert-butylsulfenyl dithiocarbamate, arylnitroolefins, dodecanol, chlorinated camphenes, toxaphenes, the lactone of 2-hydroxybiphenyl-2'-carboxylic acid, and a number of tricyclohexyltin compounds (hydroxide, chloride, acetate, and bis(tricyclohexyltin) oxide). Tricyclohexyltin hydroxide is incorporated in the adhesive during corrugated paper manufacturing at 0.01 part by weight of the treated boxes, which are thus made repellent to Norway rats. Expanded polystyrene boards containing 1% of this compound in the polymer prior to its expansion for use in temporary greenhouse walls and floor are not attacked by rodents. Likewise, young apple trees sprayed in winter with an alkyd paint containing 1% tricyclohexyltin hydroxide show no rodent injury the following spring. Peanut bags fabricated by adding 0.1 part by weight of tricyclohexyltin acetate to the slurry are not destroyed by mice.

For Rabbits: Despite their size, rabbits can cause extensive damage to certain crops. Considerable repellency to rabbits can be imparted to pine seedlings by pressure-spraying them in the nursery with endrin or zinc incorporated into latex, wax, or asphalt, or nicotine in wax prior to planting. An emulsion very useful for spraying on outdoor surfaces as a weather-resistant covering repellent to rabbits consists of approximately 80% polybutenes having an average molecular weight of 1000 to 1900, 2.5% nonionic emulsifier, and 17.5% water. Water-dispersible thiram, which, like endrin, is very useful as an insecticide, appears to be promising as a repellent for cottontail rabbits.

For Deer: Thiram is the active ingredient for several preparations effective as a repellent for deer to prevent fraying and browsing of young trees in the spring. A mixture containing 10% of the wettable powder and 20% stable bitumen emulsion as an adhesive is nontoxic to foliage and shoots of pine, spruce, larch, Douglas fir, and dormant shoots of oak and beech. Also effective are 10% sprays of thiram, which should be applied annually, after growth has finished and before the deer arrive.

For Dogs: Dogs may pose a particular problem for mailmen and other delivery men approaching or attempting to enter a home having such pets. They often injure trees or other plants on which they urinate or defecate, besides offending the esthetic sense of the neighborhood community by so doing. Cleansing tissues or cloths moistened with several drops of ethyl butyl ketone, ethyl amyl ketone, methyl isoamyl ketone, methyl octyl ketone, or isobutyl heptyl ketone are effective repellents for approaching dogs. Geranyl and heptylidene-acetones also have this effect. These repellents may also be prepared in the form of aerosols or dusts. A patented dog repellent consists of a mixture of 8.5 parts of oil of citronella, 1 part of oil of anise, and 0.5 part of oil of eucalyptus. The odor of this mixture is intolerable to dogs. The weather persistence of the oil mixture is increased by the addition of poly(acrylic acid) or poly (methacrylic acid) dissolved in 3 parts of alcohol and 95 parts of acetone; the film left on evaporation of the solvents is said to be effective for 14 days. Another patented repulsive agent for dogs consists of 20 parts of 20% formic acid, 20 parts of 40% formaldehyde, 10 parts of technical butyric acid, 30 parts of 40% ammonium sulfide, and 20 parts of water.

For Fish: The United States Navy Department holds the patent on a repellent used for clearing fish from an area without polluting it to a level toxic to the fish. This is accomplished effectively by adding less than 1×10^{-7} milligram/liter of potassium phenylacetate to the water for a prolonged period of time.

MARTIN JACOBSON

Birds

For the most part, the methods used to control birds can be divided into two categories, those that are mechanical devices and those that are physiological and achieve this with chemicals. By these methods, either directly or indirectly, birds can be repelled by frightening, execution, birth control and environmental deterrency.

Mechanical Methods. *Landing Sites.* The most common method for repelling birds involves devices (mainly nonchemical) which interfere with desirable roosting, feeding, or nesting sites. These include wires, with or without electrical current, protective netting, and sharp projections. Chemical substances which act by physical methods are the sticky coating materials such as chicle, soybean lecithin, gun grease, and plastics. The substance is sprayed, painted, or spread on ledges, sills, trees or other landing sites. These substances probably act via the tactile receptors on the feet. The disadvantage of the mechanical methods as applied to landing sites is the cost of labor necessary for widespread usage.

General Application. There are a variety of methods used to repel birds which affect the sight and sound senses. They are not limited to landing sites but are useful to discourage birds from frequenting certain areas. Devices which project imitation sounds, loud explosive noises and light displays are examples. Since birds become accustomed with these physical disturbances, they often proceed unabated after familiarization.

Food Supply. Since most birds are dependent on volunteer food sources, regulation of food consumption of growing birds could be accomplished by chemical means. One approach would be the adding of indigestible bulk to the food source or in some way limiting the access of the birds to the feed. By controlling diet, it may be possible to introduce premature moulting in the females or interfere with their egg laying cycle.

Repellent chemicals used as treatments on crops or seeds and fruit buds could be a successful means of discouraging birds via the senses of taste, smell, or touch. While not a great deal is known about the effect of repellents which affect the olfactory organs of birds, it has been shown that smell is much less important than taste in the reactions of many avian species.

A great deal of research has been devoted to repellents which involve the gustatory receptors of the tongue. Because there are relatively very few "taste buds" in birds as compared with other animals, most of the common taste sensations classified as salt, sour, bitter, and sweet taste are quite nonuniformity discerned. Those which can be characterized by man may or may not be detected by birds. In addition, it is not readily known whether the aversion to certain chemicals is due to a taste sensation or a toxic effect. Examples of some of the chemicals used to affect taste include tetramethylthiuram disulfide and derivatives of anthraquinone.

Physiological Methods. *Poisons.* Certain birds have been discouraged from feeding areas by chemical frightening agents. These substances are divided into lethal and nonlethal types. In the latter more desirable case, success is largely dependent upon the development of temporary paralysis in birds through a reversible cholinesterase inhibition with a wide safety margin between the temporary immobilization and death. Useful chemicals include certain phenyl N-methyl carbamates and 4-aminopyridine.

There are a number of other formulations which have been applied to seeds to discourage the birds' eating habits. Certain colors, particularly green and yellow are effective for short periods before the unnaturally colored food items become familiar and the birds break through this deterrent barrier.

The protection afforded seeds and crops by a repellent on the food supply of the bird is dependent on the availability of other sources of food. If these are scarce, protection may be difficult and repellents may fail. Again, with the encroachment of civilization on the wildlife environment, the concern for the use of pesticides and their effects on both man and wildlife necessitates the development of chemicals of low hazard level.

Wetting Agents. Applications of certain chemical wetting agent solutions containing either linear alcohol ethoxylates (industrial synthetic detergents), certain soaps, or sucrose esters produce mortality in birds. While the exact mechanism of action is unknown, it is believed that the insulating property of the plumage is destroyed allowing cold temperature and evaporation to reduce body temperature to a lethal level. Recent experiments have shown this method to be somewhat limited but effective, inexpensive and environmentally safe. The low inherent toxicity of most surfactants when compared with other chemical avicides, makes their use fairly safe for decreasing the population of certain depredating birds. More work, however, is necessary to assure biodegradability of the useful detergents.

Chemosterilization. Certain chemical substances (side chain aza and diaza analogs of cholesterol) which drastically lower the cholesterol level of animals have been applied to certain female birds resulting in a severe decrease in their egg-laying capabilities. It has been demonstrated that when grains or feed sources common to the diet of a given bird species are impregnated with these cholesterol-lowering agents, the reproductive capabilities are drastically reduced. The exact mode of action has not been completely studied. It is known, however, that cholesterol is very vital for the ovulation process and since in fowl there are relatively high levels of circulating cholesterol, the lowering agents are effective. Because the chemicals used are stored in body tissue, they can be given for a period of up to 10 days with the interval of decreased ovulation lasting 6 months or longer. There have been no apparent serious deleterious effects noted in the birds that have taken the chemical.

It is apparent that the majority of methods used to control or repel bird populations are somewhat primitive. However, for the more sophisticated chemical and physiological methods to be more broadly used and successful, further studies on the physiology of the senses of birds will be necessary, particularly because of the variation which exists between species.

PAUL D. KLIMSTRA

References

Proceedings: Fourth Vertebate Pest Conference, West Sacramento, California, Mar. 1970.
Proceedings: Fifth Bird Control Seminar, Bowling Green, Ohio, Sept. 1970.

Insect

Insect repellents are substances which insects avoid approaching or contacting. These substances are generally mildly toxic, generally not insecticides, keep insects away from person, food, shelter, and prevent damage to plants, animals and manufactured products by making conditions unattractive and offensive to insects. An *effective* repellent is a substance which satisfies specific degrees of performance criteria and allows significant economic gain or physical comfort. Since repellency is relative, factors such as volatilization from treated surface, nature of surface, estimation of comfort or discomfort of person or animal, densitization of sensory receptors of insects, insect habits relative to climatic conditions and to methods and timing of application must be considered in evaluating effective repellency.[1, 7c, 15]

Repellents in one form or another have been used since early civilization to relieve man from insect bites, noises and transmitted diseases, e.g., malaria from mosquitoes, typhus from lice and diseases from other arthropod vectors. Smoke was early recognized as an effective repellent, but repellency towards man limited its usefulness.

Plants containing essential oils, e.g., geraniol citronellol, citronellal and other terpenes were rubbed over the body. Pitches, tars and various earth clays were also used. Although less restricting than insect nets, these early forms of repellents were not cosmetically satisfactory.

That many plants are considered resistant to insects may be due to a natural repellent substance(s). 5-Methoxy-2-benzoxazolinone in certain species of corn, for example, is one of the principal "resistant factors" of the corn borer. Catnip (nepetalactone) protects the plant *Nepeta cataria* from phytophagus insects. Ammonium nitrate in sweet clover species

5-Methoxy-2-benzoxazalinone   nepetalactone

M. infesta makes it 5–10 times more resistant to attack than susceptible clover *M. officinalis.*

Natural repellents are not restricted only to plant sources. Insect secretions produced *de novo* function principally to repel predacious enemies. Arthropods contain diverse chemicals produced in special exocrine glands and stored in sac-like reservoirs. When these chemicals are discharged they may be gaseous, fine sprays of liquids, oozing liquids which volatilize from the body, or foams. The best known classes of arthropods include Diplopoda (millepedes), Chilopoda (centipedes), Arachnida (spiders) and Insecta. Chemical classes isolated include terpenoids, benzoquinones, aromatic compounds and miscellaneous including hydrocarbons, acids, esters, carbonyl compounds, steroids, and nonexocrine substances. Examples are:

Terpenoids: citronellal, geraniol and related substances;
Benzoquinones: benzoquinone and homologs, methyl ether derivatives and homologs (best known and probably principal component of arthropod secretions);
Aromatics: cresols, benzaldehydes, benzoic acids;
Miscellaneous: formic, acetic, caprylic acids, hexylacetate, hexanols, heptanols, 2-heptanone.

The subject of arthropod defensive substances is far from exhausted and should provide stimulus for further research.[3a, 4, 6, 10, 10a, 11, 14, 16]

During the period of 1923–1940, essential oils were the substances of choice in oil and cream preparations. Nash's insect repellent mixture, for example, contained 40% citronella, 40% spirits of camphor and 20% oil of cedar. With the advent of World War II, it became necessary to find an effective repellent for the protection of troops against mosquitoes and other disease carrying vectors. The U. S. Department of Agriculture collected and synthesized 20,000 chemicals as repellents. More than 30% of these compounds were active, but not all were useful. Because of the stringent requirements for a good repellent, the success in finding a useful repellent averaged 2 per

1000 compounds. Although mosquitoes were of primary interest, biting flies (black, horse, deer, sand, horn), chigger mites, ticks and fleas were also of interest. Protection of livestock was concerned primarily with biting flies.[15]

The criteria for the ideal repellent are based on providing an individual protection. The important criteria are: a. broad spectrum in activity; b. durability; c. nonirritating; d. nonallergenic; e. not objectionable to user regarding odor, feel, staining of clothing; f. nontoxic; g. stable to heat and sunlight; h. compatible with a wide range of formulation additives.

For practical effectiveness, local application to skin or clothing is recommended. Use of repellents for large areas or large acreage crops appears uneconomical. Where practical, insecticides are employed for such purposes. Volatility and cost are problems. Nonvolatile repellents may serve as feeding deterrents, but are hampered in efficacy by growth dilution of plant tissue.

Compounds tested against mosquitoes and biting flies are generally applied to the skin or cloth. In addition to human forearms, canaries, mice, rabbits and guinea pigs are used for testing against biting flies. Testing against fleas, ticks, chigger mites and lice is done almost exclusively with applications to cloth.

The search for new repellents has been characterized by the lack of standardized procedures in the evaluation of repellents. Diverse criteria, different methods of testing, different host systems have made it very difficult to determine effectiveness of repellents. Methods used for skin repellents have included "time of first bite," "robin robin" and "probe time." The fact that some people are more attractive to mosquitoes than others complicates measurement of repellency.[2, 2a] For testing on cloth, generally impregnation may be made from acetone solution or preferably from an emulsion of repellent and suitable emulsifier.[7c, 15]

An attempt has been made to measure intrinsic repellency without the complication of variables inherent in host repellency via the dual-port olfactometer.[8, 13] Injection of chemical into a host stream on a standardized target area measures repellency in absolute units using calculations involving a molecular weight, injection and air flow rates. Although compounds may be active by this method, poor repellent properties may be encountered on skin due to an imbalance between potency and volatility.[8, 9, 13]

Many correlations between repellency and physical properties and structure of chemicals have been made. Such properties as boiling points within a homologous series, molecular weights, volatility, partition coefficients, absorption of compounds in the 449–471 cm⁻¹ region of the far infrared and studies on insect chemoreception have been used.[14, 17]

From the many compounds studied, the following generalizations can be made: (a) repellency does not seem to be restricted to any particular class of compounds, (b) there is not a sharp distinction in degrees of repellency between compounds, but rather a continuous graduation. General classes from which active compounds have been found include amides, aldehydes, imides, alcohols, esters, ethers, diols amide-esters, hydroxy esters and diesters. Recently acetals, carboxamide acetals, ortho esters or generally ketal types have been discovered to have repellency, e.g., dibenzylcyclopentanone ketal and N, N-dimethylforamide dibenzylacetal. Structure activity relationships of compounds have not been worked out effectively to allow prediction of activity with any degree of accuracy.

Repellents may be categorized as skin repellents, including livestock, clothes repellents, space repellents (inanimate surfaces) and "systemic repellents." Rarely, however, is a chemical repellent to all kinds of insects in these uses. Repellents best for human use do not work well on livestock (dynamics of absorption and evaporation from animal hairs). Further, skin repellents may also be used on clothing, but some of the best clothing treatments are not recommended for skin applications.

Several useful insect repellents have been developed since the U.S. Department of Agriculture's early efforts. Probably the best known skin repellent is N,N-diethyl-m-toluamide (DEET). It has been considered the best all-purpose insect repellent. Of the materials currently available, it approaches the criteria established for an ideal repellent. One drawback is its solvent effect on plastic and painted surfaces. Other substituted diethylbenzamides are also active, but differ in degrees of water leaching resistance. Other compounds recommended for general purpose use on skin include 2-ethyl-1,3-hexanediol (Rutgers 612), dimethyl carbate, dimethyl phthalate and Indalone.

Insect repellents for cattle are very important especially when combined with insecticides. Thus far, no good single acting repellent has been found. Pyrethrins, malathion, methoxychlor are used in combinations including the repellents butoxy propylene glycol, 2,3,4,5-bis (butyl-2-ene) tetrahydrofurfural, di-n-butyl succinate, di-n-propyl isocinchomeronate. A new structural type includes 3-chloropropyl-n-octyl sulfoxide (MKG 1207),[5] which has shown promise against biting flies and ticks. Generally, the combinations are used in sprays, dips, dusting powders, back and face rubs. Through the use of these agents, increases in milk and meat production can be realized.

Repellents remain effective longer when applied to clothing than when applied to skin. Although DEET can be used on clothing, it is not very persistent. One of the most effective single treatments of clothing is 2-n-butyl-2-ethyl propanediol. Although a single ingredient can be effective, added repellency for specific insects can be achieved through combination. One of the best developed is the so-called M-1960. It contains 30% 2-n-butyl-2-ethyl propanediol (mosquitoes), 30% N-n-butylacetanilide and 30% benzyl benzoate (mites and ticks) and 10% emulsifier and persists for several weeks. Another useful preparation is M-2086 containing 45% benzyl benzoate, 45% di-n-butyl phthalate (mites and ticks) and 10% emulsifier. In the evaluation of these repellents, the compounds must not only survive for long periods of time, but also must resist repeated washings.

Space repellents are easier to find because the stringent criteria for skin and clothing are not required. This type is characterized by long lasting effects over periods of months. These repellents are important in the treating of netting which, although restrictive to persons, still allows air flow through 0.25 inch openings. N,N-Dipropyl-2-[(p-methoxybenzyloxy)]-acetamide, for example, keeps yellow fever mosquitoes from passing netting openings up to 266 days.[7a] Other chemicals which can be classified as space repellents are those applied in dairy barns, to loading docks, patio and picnic areas, storage containers and areas in public buildings, etc. A newer candidate for such uses is the broad spectrum 2-hydroxyethyl-n-octyl sulfide (MGK-874).[12]

An ideal form of an insect repellent would be an orally acceptable systemic for human and animal use. None has been developed to date.

With added leisure time today and the need to increase flow of products to all parts of the country and throughout the world, insect repellents become more important. Possibilities open for the need to incorporate insect repellents into perfumes, cosmetic creams, suntan lotions and packaging and hauling containers. Aerosols, for example, are best suited to meet requirements for convenient travel size containers and for household and patio use. Protection of cornmeal, soy flour and dried milk for safe delivery to foreign parts is satisfied by impregnating multiwall paper bags. Treatment of bottle cartons and cases, and treatment of loading docks prevents invading insects from being transported from infested to noninfested areas.

Ecological factors indicate repellents as desirable; however, insecticides will continue to be the mainstay in controlling insects. The need for development and utilization of repellents will continue. Studies of intrinsic values of repellents, defensive secretions of arthropods, insect behavior-inducing chemicals may provide clues which will help in devising better topical and possibly systemic repellents. Through the use of better repellents, attractants, etc., significant reductions in insecticide quantities and thus residues may result.[3, 7, 8, 12]

A. J. Crovetti

Cross-reference: *Insecticides*

References

1. Abrams, J., and Dworkin, Z. Z., *Soap and Chem. Specialties,* **34,** 84 (1958).
2. Acree, F., et al, *Science,* **161,** 1346 (1968); *Chem. Eng. News,* **45,** (No. 23), 11 (1967).
3. *Aerosol Age,* **13,** 69 (June 1968);/Markarian, H., et al, *Soap and Chem. Specialties,* **41,** (No. 10), 94 (1965); (c) Michell, J. N., *Aerosol Age,* **13,** (No. 8), 44 (1968); (c) Reed, A. B., *Am. Perf & Cosmetics,* **84,** 115 (1969).
4. Akeson, W. R., et al, *Science,* **163,** 293 (1969).
5. Baker, G. J., *Soap and Chem. Specialties,* **38,** 111 (1962).
6. Beroza, M., in Beroza, M., Editor, "Chemicals Controlling Insect Behavior," pp. 155–161, Academic Press, New York, 1970.
7. Brett, R. L., *Soap and Chem. Specialties,* **43,** 100 (Dec. 1967); (a) *Chem. Week,* **103,** (No. 21), 25 (1968); (b) *J. Agr. & Food Chem.,* **6,** 88 (1958); (c) Kirk-Othmer, Editors, "Encyclopedia of Chemical Technology," **17,** 724, 2nd Ed. (1969).
8. Burton, D. J., *Am. Perf. & Cosmetics,* **84,** 41 (1969).
9. *Chem. Eng. News,* **47,** (No. 39), 52 (1969).
10. Crosby, D. G., in Gould, R. F., Editor, "Advances in Chemistry," p. 8, 1966; (a) Jacobson, M., *ibid,* p. 22.
11. Jacobson, M., *Ann. Rev. Entomology,* **11,** 403 (1966).
12. Moore, J. B., *Aerosol Age,* **11,** (No. 12), 63 (1966).
13. *Oil, Paint & Drug Reporter,* **193,** 5 (May 27, 1968).
14. Shembough, G. F., et al, in Metcalf, R. L., Editor, "Advances in Pest Control Research," Vol. 1, p. 277, Interscience Publishers Inc., New York, 1957.
15. Smith, C. N., *Soap and Chem. Specialties,* **34,** (No. 2), 105 (1958).
16. Weatherston, J., and Percy, J. E., in Beroza, M., Editor, "Chemicals Controlling Insect Behavior," p. 95, Academic Press, New York (1970).
17. Wright, R. H., *Nature,* **178,** 638 (1956); (a) Wright, R. H., and Kellogg, F. E., *Nature,* **195,** 404 (1962).

REPLACEMENT REACTIONS

When a piece of iron is placed in a solution of copper sulfate, the iron becomes coated with copper and the blue copper sulfate solution slowly changes to green ferrous sulfate

$$(Fe + CuSO_4 \rightarrow Cu + FeSO_4) \qquad (1)$$

In this reaction, iron replaces copper. Such a reaction is typical of *replacements*. In replacement reactions a free element reacts with a compound and takes the place of another element originally in the compound. Similarly, when zinc is placed in lead acetate solution, zinc replaces lead; and when tin is placed in silver nitrate solution, tin replaces silver. In the examples cited, sulfates, acetates, and nitrates have been specified. Any soluble compound could be used as well as the particular compounds mentioned.

A similar reaction takes place among nonmetals. While elementary fluorine replaces chlorine from chlorides, fluorine is so active that numerous side reactions take place simultaneously, and the replacement reaction may become obscured by other competing reactions at the time. Chlorine, a less active nonmetal than fluorine, replaces bromine, a still less active element, from bromides. For example,

$$2KCl + Cl_2 \rightarrow 2KCl + Br_2$$

The reactions described thus far take place in water solutions, and they are essentially ionic. In terms of ions they may be one of the following:

$$Cu^{2+} + SO_4^{2-} + Fe \rightarrow Cu + Fe^{2+} + SO_4^{2-} \qquad (2)$$

$$Cu^{2+} + Fe \rightarrow Cu + Fe^{2+} \qquad (3)$$

$$2K^+ + 2Br^- + Cl_2 \rightarrow 2K^+ + 2Cl^- + Br_2 \qquad (4)$$

$$2Br^- + Cl_2 \rightarrow 2Cl^- + Br_2. \qquad (5)$$

In even simpler terms, these reactions are electron-transfer reactions.

In the examples, iron atoms transfer two electrons each to copper ions, and in the second case two bromide ions transfer one electron each to a Cl_2 molecule. The readiness with which replacement reactions of this sort take place can be measured in terms of ease of electron transfer. The measurement, given in volts, is called the electrochemical ionization potential of the element. These potentials, arranged in sequence, make up a list of elements (or radicals) called the *electrochemical replacement series.*. In this list the voltage of a hydrogen electrode in a normal solution of H^+ ions at 25°C is given an arbitrary value of 0.00 volt. Metals that replace hydrogen from acids have a value in the scale greater than that for hydrogen. Metals replaced by hydrogen have values on the scale less than that for hydrogen. See **Electrode Potentials.**

While these values give the replacing value under the conditions described, other conditions, especially concentrations, may change the order in the list. For example, copper may be electroplated onto steel from a bath that contains copper ions, and no replacement plating takes place.

Organic replacements are common. Many reactions are known in which an element (or radical) takes the place of another. Such reactions may be called either replacement or substitution reactions. This discussion will be confined to reactions in which an element is a reactant. The reaction

$$C_2H_5I + NaOH \rightarrow C_2H_5OH + NaI \quad (6)$$

exhibits the replacement of iodine by the OH radical, but it does not qualify strictly as a replacement, because no free element is involved. In general discussions, however, the terms replacement and substitution are often used interchangeably. Chlorination reactions may be replacements. Benzene (C_6H_6), when chlorinated, becomes successively monochlorobenzene (C_6H_5Cl), dichlorobenzene ($C_6H_4Cl_2$, three compounds 1:2, 1:3, and 1:4 are known), trichlorobenzene ($C_6H_3Cl_3$), and so forth up to hexachlorobenzene, (C_6Cl_6), depending on the extent of chlorination. The by-product of each reaction is HCl. Similarly many other hydrocarbons are chlorinated. In general, the greater the extent to which hydrogen of a hydrocarbon becomes replaced by chlorine, the denser the compound becomes, and the less it tends to burn. Benzene, for example, burns readily, but hexachlorobenzene extinguishes fires.

Other halogens are vigorous replacing agents, especially fluorine. The reaction of fluorine or fluorinating agents on organic compounds produces such compounds as tetrafluorethylene (C_2F_4) and octafluorocyclobutane (C_4F_8) and dichlorodifluoro methane (CCl_2F_2).

Chlorination of glacial acetic acid at 100°C produces monochloroacetic acid,

$$CH_3COOH + Cl_2 \rightarrow CH_2ClCOOH + HCl. \quad (7)$$

The result of replacement of hydrogen on the carbon atom in the alpha position in respect to the carboxyl radical is to produce a stronger acid. Monochloroacetic acid, a colorless crystalline solid that melts at 61°, is used to make 2,4-D, a weed killer. Acid halides are more readily halogenated than their corresponding acids.

$$C_3H_7COBr + Br_2 \rightarrow C_3H_6BrCOBr + HBr. \quad (8)$$

Replacement of H by Br in cases of this sort forms an alpha-bromo derivative of the acid.

ELBERT C. WEAVER

RESINS, SYNTHETIC WATER-SOLUBLE

Synthetic water-soluble resins are materials used to thicken water solutions, suspend solids and retain moisture. Most important among them are polyvinyl alcohols, polyvinyl-pyrrolidones, polyvinylmethyl ethers, polyacrylic acid and its salts, polyacrylamides and ethylene oxide polymers. In many applications the synthetics have replaced natural gums (which see), such as starches, dextrins, alginates, etc. Favoring the synthetics are increased efficiency, so that lower dosages are needed for a given job; greater flexibility, so that polymers can be tailored to give specific properties; and, for control of water pollution, lower biological oxygen demand (BOD) when compared with the natural products.

More than 125 million pounds a year of synthetic water-soluble resins are consumed in the United States, entering into a broad range of products, such as: adhesives (remoistenable adhesive compounds, aqueous thickeners), beverages (clarification), ceramics (clay binders, glazing formulations), cosmetics (hair sprays, lotions, ointment formulations), detergents (anti-redeposition agents, builders), explosives (binders), fertilizers (binders, granulating agents), food products (thickeners, stabilizers, humectants, emulsifying aids), glass (binders for glass fibers), latex coatings (thickeners, pigment suspending agents), leathers (adhesives, leather drying), minerals processing (flocculants for settling fines), paper (coating binders, clay flocculants, surface sizes), petroleum production (water-loss agents, drilling-mud additives), pharmaceuticals (thickeners, suspending agents), plastics (protective colloids for emulsion polymerization), rubber (release agents for rubber products), and textiles (warp-size agents).

Physical Properties. Water-soluble polymers can be nonionic (do not ionize in water solutions), including ethyl hydroxyethylcellulose, hydroxyethylcellulose and polyethylene glycol; anionic (ionize in aqueous solution to produce polymeric species with negative charges), including sodium carboxymethylcellulose and polyacrylic acid salts; or cationic (polymeric species which upon ionization in water possess a positive charge), represented by the acrylamide copolymers.

There are also differences in viscosity behavior. Most resin solutions of moderate concentrations have pseudoplastic flow, the viscosity decreasing with increasing shear rate. These resins include ethyl hydroxyethylcellulose, hydroxyethylcellulose, sodium carboxymethylcellulose, polyethylene oxide and polyvinyl chloride. Other synthetic water-soluble resins are thixotropic. Flow of these materials results in a viscosity decrease. Thixotropic systems reestablish their structures upon standing undisturbed. Thixotropic flow is always combined with some other type of flow, the most common pairing being the thixotropic-pseudoplastic. Examples are solutions of sodium carboxymethylcellulose, cellulose ether, hydroxyethylcellulose and sodium carboxymethyl hydroxyethylcelluose, which can be thixotropic and pseudoplastic.

Patterns of Use. About 48% of all synthetic water-soluble resins are used to thicken and stabilize water-based products. Hydroxyethylcellulose and methylcellulose are examples of thickening and gelling agents that are incorporated into such products as latex paints and cosmetics. The resins most often used as stabilizing, suspending or dispersing agents for water-insoluble solids or liquids are hydroxyethylcellulose and polyvinyl alcohol in emulsion polymerization, and carboxymethylcellulose in drilling muds, paper manufacture, and in detergent formulations as soil-suspending aids.

About 17% of all synthetic water-soluble resins are used as film-forming agents. Typical examples are polyvinyl alcohol and polyethylene oxide for packaging films, or carboxymethylcellulose for textile warp sizing. Water-retention aids include another 12% of the resins, and are represented by polyethylene oxide and methylcellulose for cement-asbestos extrusion, or by hydroxyethylcellulose for oil-well cements. Typical coagulants (7%) for suspended solids or liquids are the polyacrylamides for water-treatment uses, while for binding agents (6%) carboxymethylcellulose and methylcellulose are representative—the former for welding rods and the latter for toothpaste.

On a product distribution analysis, detergents and laundry products account for about 16% of the total usage of synthetic water-soluble resins. Usage by other market areas includes: textiles (14%), with resins such as carboxymethylcellulose and polyvinyl alcohol; adhesives (12%), with polyvinyl alcohol being typical; paper (10%), using resins such as the polyacrylamides and polyethylenimine; paints (9%), in which hydroxyethylcellulose and methylcellulose are widely used; pharmaceuticals and cosmetics (7%), incorporating, among others, polyethylene oxide and polyvinylpyrrolidone.

ROBERT L. DAVIDSON

Cross-references: *Gums and Mucilages (natural); Cellulose*

RESONANCE

Since the middle of the 19th century it has been realized that some compounds, notably organic compounds customarily referred to as aromatic or conjugated, cannot be adequately described by any one single structural formula (valence-bond structure). Although great advances were made, particularly by F. Arndt and C. K. Ingold and their associates, no full understanding of the structure of such compounds was possible until the development of wave mechanics in the mid-1920's. The concept of quantum-mechanical resonance, developed by Heisenberg in the treatment of other phenomena, was applied to aromatic and conjugated molecule, notably by Pauling, to provide the first satisfactory representation of these compounds.

The term *resonance,* synonymous with *mesomerism,* as used today in chemistry, refers to the fact that many molecules cannot be represented by any single valence-bond structure, but must be considered as intermediate between two or more such structures. Such molecules are called *resonance*

hybrids. The term does *not* imply that the molecule resonates (in the physical sense) between the structures; the molecule is *not* well represented by one structure during part of its life time, by another during another part; rather, the structure is unique, but is most conveniently described by the mathematical expedient called resonance. Thus, benzene, e.g., cannot be represented by any one of the structures I–V alone, but is intermediate between all of them, and many others resembling III–V. Mathematically this is expressed by saying that the wave

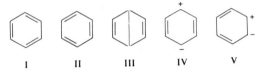

| I | II | III | IV | V |

function describing the molecule, ψ_G, is a linear combination of the wave function, $\psi_I \cdots$, describing the various structures:

$$\psi_G = a_I\psi_I + a_{II}\psi_{II} + \cdots.$$

Some of the properties of the molecule, particularly those depending on the charge distribution (bond lengths, dipole moments) are intermediate between the corresponding properties expected for the contributing structures. Other properties, particularly those depending on energy quantities (spectra, ionization potentials) are outside the limits suggested by the structures. The energy content of a resonance hybrid is *lower* than the corresponding value for *any* one of the contributing structures, and is found by solving a secular equation formed from certain integrals involving the various structures. The *resonance energy* is the difference in energy between the actual molecule and the energy predicted for the contributing structure of lowest energy.

All structures contributing to one resonance hybrid must fulfill a number of conditions: (1) they must represent identical, or nearly identical arrangements of atoms; (2) they must have the same number of unpaired electrons; and (3) their energy content must be of the same order of magnitude.

While resonance is most familiar in aromatic and conjugated compounds, the same concept enters into a satisfactory description of every electron pair bond: thus the Heitler-London treatment of H_2 involves resonance between the structures H—H, H^+H^- and H^-H^+. Further, in heteropolar bonds, resonance between covalent and ionic structures is important, thus HCl is a resonance hybrid of H—Cl and H^+Cl^-; a further contribution from H^-Cl^+ is so unimportant as to be negligible. This type of resonance, which is present in *every* chemical bond, is customarily included in our concept of the electron pair bond, and not treated explicitly.

Thus, resonance is a convenient mathematical tool which permits a description of molecules in terms of valence-bond structures particularly in those cases where one such structure is insufficient. Resonance, however, is only a tool, not a *real* phenomenon. If one is satisfied by a description of molecules not involving valence-bond structures, use of an alternate but equivalent tool, the *molec-*

ular orbital theory due primarily to Hückel, Hund and Mulliken, allows one to reach identical results without any use of the resonance concept. The term *resonance* was chosen because the mathematical treatment is analogous to the mathematical treatment of resonance phenomena in classical physics, and should *not* be understood to imply anything more than the analogy in the mathematical methods.

Finally, there are chemical applications of the physical phenomenon of resonance (e.g., nuclear magnetic resonance absorption) which have no relation to the resonance concept discussed in this article.

H. H. JAFFE

Cross-references: *Organic Chemistry, Valence Bond.*

RHENIUM AND COMPOUNDS

Many claims regarding the discovery of element number 75, a homolog of manganese, appear in the chemical literature. Credit for its actual discovery is now generally attributed to Walter Noddack and Ida Tacke, who announced in 1925 that they had detected it in platinum ores and columbite. Simultaneously O. Berg and I. Tackle published confirming evidence for the existence of element 75 in the above mentioned ores. This latter evidence was based on x-ray examination of concentrates obtained from the platinum and columbite ores. Rhenium, together with manganese and technetium constitute the VIIB family of elements in the Periodic System.

Rhenium does not occur in nature in the elementary or compound form as a distinct mineral species. It is however widely distributed in very minor amounts. Its average concentration in the earth's crust is of the order of 1×10^{-9}, or 0.001 part per million. Rhenium is reported to occur in a wide variety of minerals, notably molybdenite, rare earth minerals, columbite, tantalite, platinum ores, copper sulfides and oxides of manganese.

Contrary to expectation, rhenium occurs only in very minor amounts in certain manganese minerals, none of which serve as a source for the extraction of rhenium. Of the minerals investigated to date molybdenite, MoS_2, is the most promising commercial source. While no confirming evidence is at hand it seems quite probable that the rhenium in molybdenite occurs as the sulfide, either ReS_2 or Re_2S_7. It is of interest to note also that the molybdenite was originally associated with sulfide ores of copper.

Chemical Properties. The chemical activity of rhenium depends to a large extent on the manner in which it is prepared. If prepared in a fine state of subdivision by hydrogen reduction of one of its powdered salts it may actually be pyrophoric if steps are not taken to remove excess hydrogen. The finally divided metal apparently acts catalytically to oxidize the hydrogen, sufficient heat being liberated to ignite the metal. On the other hand a polished surface of the compact metal is not readily tarnished even on long exposure to air. Rhenium metal plated on strips of platinum, silver, tantalum, copper and brass retains a high luster as long as the specimens are preserved in a perfectly dry atmosphere. On exposure to moist air, however, the metal is gradually oxidized, the end product being perrhenic acid, $HReO_4$. Contrary to the behavior of rhenium plate on a metallic base, a bright plate of rhenium on a spectrographic graphite electrode has retained its initial appearance for several years, during which time no attempt was made to preserve it in a dry environment. The characteristic valence of rhenium is 7, but in addition to this most stable state it also exhibits valences of −1, 1, 2, 3, 4, 5 and 6.

The halogen elements, with the exception of iodine, react with rhenium with decreasing vigor in passing from fluorine to bromine. In no case, however, is the maximum valence state of rhenium realized. This is probably because even in the case of fluorine the rhenium atom is not capable of accommodating seven atoms of the halogen. All the halides of rhenium on contact with water or moist air undergo hydrolysis, and frequently disproportionation. The actual products formed will depend on the nature of the halide in question but generally include two or more of the following: complex halogen and oxyhalogen acids of the metal, rhenium (IV) oxide ReO_2, perrhenic acid, $HReO_4$, rhenium metal, hydrohalogen acids and free halogen.

Sulfides of the metal may be prepared by the action of sulfur vapor on the finely divided powder at elevated temperatures or by the action of hydrogen sulfide on acid solutions of the perrhenate. The best known sulfides of rhenium are ReS_2 and Re_2S_7.

Rhenium carbonyl, $Re(CO)_5$, has been prepared by treating Re_2O_7 with carbon monoxide at 250°C and 200 atm pressure.

Phosphides and arsenides of the metal have been prepared by direct union of the elements, but nitrogen is without effect on the metal even at elevated temperatures.

There appears to be some controversy in regard to the existence of rhenium carbide. There appears to be some evidence that the metal heated in an atmosphere of carbon monoxide or methane will decompose the latter two compounds with consequent incorporation into the lattice of rhenium a small amount of carbon.

Rhenium is not acted upon by hydrochloric acid. With strong oxidizing acids such as nitric and hot sulfuric acid the metal is vigorously attacked and converted to perrhenic acid. Hydrogen peroxide in ammoniacal solution converts the metal, its oxides and sulfides to ammonium perrhenate.

Uses. A number of possible applications of rhenium have been suggested but because of the high cost and present rarity of the element none has received any widespread commercial use. Perhaps the most promising use in the field of electronics, as suggested in the patent literature on the production of rhenium filaments. Other possible applications based on the high melting point of the metal are suggested in a patent issued to P. R. Mallory and Company. The alloy is described as one containing tungsten, molybdenum and rhenium for use in electrical make and break contact points. Alloys consisting of platinum and rhenium or platinum and rhenium together with iron, rhodium and iridium for use as thermocouple elements are the subject of several patents.

A refractory alloy consisting of more than 90%

tungsten, less than 10% rhenium and not more than 1% vanadium has been described. Addition of iridium, iron, nickel and cobalt to platinum-rhenium alloys improves the mechanical properties of the alloy and prevents grain growth.

The catalytic properties of rhenium, some of its compounds and a few of its alloys have been investigated. These catalysts may be in the form of the active metal powder, colloidal rhenium, colloidal rhenium adsorbed from suspension on activated charcoal, rhenium alloyed with platinum or silver, rhenium and tungsten alloys, and rhenium sulfide. Particular catalyst preparations of rhenium in one of the above forms has been used in the dehydrogenation of alcohols, synthesis of ammonia, hydrogenation of coal, coal tar and mineral oil, oxidation of ammonia and the oxidation of sulfur dioxide to sulfur trioxide.

Rhenium-platinum catalysts are used in an increasing number of petroleum refineries for the production of high-octane gasoline. The two metals in a ratio of about 1:1 are supported on an alumina base. Their use has increased throughput, has boosted reformate and aromatic yield, has decreased catalyst cost (rhenium sells for about one-third the price of platinum), and has increased catalyst life. This application accounts for a major portion of the use of rhenium.

A. D. Melavan

References

Hampel, C. A., "Encyclopedia of the Chemical Elements," pp. 594–598, Van Nostrand Reinhold Co., New York, 1968.

Melavan, A. D., "Rhenium," pp. 418–433 in "Rare Metals Handbook," Hampel, C. A., Editor, Reinhold Publishing Corp., New York, 1961.

Chem. Eng., **76,** pp. 80–82 (March 24, 1969) (includes detailed diagram of chemical reactions of rhenium).

RHEOLOGY

The science of rheology deals with the flow or deformation of matter. One is impressed by the wide spectrum of flow responses exhibited by different materials; the mobility of common liquid, the behavior of doughs and batters, the properties of such foods as syrups, jellies, butter and cheese, the extensibility of a rubber band, the flow of paint when brushed on a surface, and the moldability of glazier's putty are all manifestation of flow or deformation. The study of the flow characteristics of various materials and the development formulations to explain and predict such behavior of matter constitute the science of rheology.

Rheology as a separate and distinct science is comparatively young. Although some of the phenomena with which it is concerned had been investigated for many years, it was not until 1929 that the name "rheology" was adopted. At that time a Society of Rheology was organized, and has since been growing. The real importance of rheology in dealing with both scientific and technical problems assures its continued growth.

Some concept of the widespread applications of rheology may be conveyed by pointing out that its principles are applied in the study of the flow behavior of materials in such diverse fields as synthetic fibers, cellulose derivatives, synthetic resins, surface coatings, printing inks, rubber technology, lubrication, oil well drilling, adhesives, biology, foods, creep and failure of structural materials, plastics of all kinds, work hardening of metals, and many others.

In recent years, rheology has become very important because of its applications in the aircraft industry. The high speeds now realized result in extremely high stresses on structural members of the planes. Even more significant is the fact that the rate of application of these stresses is very high. It is through the application of rheological principles to problems in the design of structural materials, adhesives, etc. that equipment capable of withstanding these stresses is made possible. Indeed the demand for such information necessitated the development of new experimental equipment and techniques to investigate the rheological properties of materials at extremely high stress rates.

Newton is said to be the first to offer a model or theory of viscous flow. He assumed the rate of shear of two parallel plates separated by a thin film of liquid to be proportional to the tangential shearing force applied to the plates. The proportionality factor is the coefficient of viscosity of the liquid. The coefficient of viscosity is more specifically defined as the shearing force per unit area which results in a unit rate of shear.

Because of experimental difficulties, the Newton model or theory could not be tested in the laboratory. Many years later, Poiseulle (1846), in studying the flow of liquids through capillary tubes, discovered the rate of flow to be directly proportional to the pressure applied on the liquid. By combining Poiseuille's results with Newton's theory, an equation for the flow of liquids through capillaries was evolved. This equation is known as the Poiseuille Equation and has the following form:

$$\frac{V}{t} = \frac{\pi R^4 P}{8L\eta}$$

where $\frac{V}{t}$ = Volume of flow per second

R = Radius of tube
P = Pressure on the liquid
L = Length of tube
η = Coefficient of viscosity of the liquid

Poiseuille's equation adequately defines the flow behavior of a great many materials, among which are many liquids and solutions. Such systems are said to show Newtonian flow since they obey Newton's law.

The flow or deformation characteristics of many other materials under stress differ markedly. They differ particularly in their manner of response to varying rates of application of the deforming or stressing force. This fact gives rise to many different flow patterns and provides a basis for defining certain types of rheological behavior. Probably the most instructive manner for illustrating these types of flow behavior is by a plot of the deformation or strain rate as a function of the applied force or stress. Such a plot is known as a consistency curve. In actual practice one chooses suitable parameters

FIG. 1.   Newtonian Flow.

for plotting, such as those measured by the particular type of instrument or rheometer used. For example, if measurements are made on a rotational viscometer, the RPM of the driven cup or bob can be plotted as the ordinate and the torque on the suspended cup or bob as the abcissa. When measurements are made with capillary tubes the pressure on the liquid may be plotted against the rate of efflux of the fluid.

Broadly speaking, the consistency curves of various materials divide them into two groups. In the first group the consistency curve is a straight line beginning at the origin. These are the Newtonian materials mentioned above. Their rate of deformation is directly proportional to the stress. Figure 1 shows a consistency plot for a Newtonian liquid. The cotangent of angle α is proportional to the coefficient of viscosity of the material.

The second group comprises those systems whose rates of deformation are not directly proportional to the shear rate. They may deviate from direct proportionality in several ways. Inasmuch as these systems do not follow Newton's law, they are said to be non-Newtonian. This type of rheological behavior is characteristic of a great many materials of commercial importance. Several of the more important types of non-Newtonian flow will be described.

Plastic Flow. Some materials, such as common putty, show no deformation until the applied stress reaches a critical value. Thereafter, flow takes place. These are known as Bingham bodies, named after E. Bingham, who first described and characterized this type of flow. Such rheological behavior is called plastic flow. The critical stress value required to initiate flow is called the yield value. A typical consistency plot for a plastic system is shown in Figure 2. The stress OA, is the yield value. In systems showing plastic flow, the flow behavior beyond the yield point may be either Newtonian or non-Newtonian.

The use of the term "plastic" to describe a type of flow behavior should not be confused with the more common usage of the word which describes the group of products fabricated from synthetic resins and similar materials. In fact, the majority

of the so-called plastics do not exhibit plastic flow in the rheological sense. Furthermore, one should not assume a body to be plastic from a casual inspection of its deformation characteristics under stress. Some materials, notably glasses and bitumens, are not plastic although seemingly possessing a yield value. When submitted to careful examination on a suitable rheometer, bitumens are found to be pseudoplastic and glasses are Newtonian. It is because of their very high viscosity that a small deforming stress appears to produce no flow. If, however, the small stress is allowed to act for a sufficiently long time flow does in fact take place. In the instance of a truly plastic body, no flow or deformation would occur however long the small stress acted unless, of course, its magnitude exceeded the yield value of the body.

Pseudoplasticity. A great many systems, especially solutions of cellulosic esters and many of the synthetic resins, give consistency curves which are nonlinear, although they go to the origin. The consistency curve for this type of system is concave toward the strain axis. Figure 3 shows such a plot. It is to be noted that the rate of strain or flow increases at a greater rate than the applied stress. In other words, the apparent viscosity of such a system decreases as the shear rate or stress increases. Such a system possesses no true viscosity but gives a series of apparent viscosity values depending upon the instrumental and manipulative details used in making the measurements. A system exhibiting this type of flow is said to be pseudoplastic. When such a system is measured in a rotational viscosimeter wherein the rate of rotation of one of the coaxial cylinders can be increased continuously and then decreased continuously, the down curve and up curve coincide. It is obviously impossible to specify a viscosity for such a system since its viscosity is a function of the shear rate.

Thixotropy. Thixotropy describes the flow behavior of a system in which the "apparent" viscosity decreases as the system is disturbed by stirring, shaking, etc. Thixotropic breakdown is a reversible change. A thixotropic system differs from the pseudoplastic system described above in a very important manner. If one determines the consistency curve of a thixotropic system using the conventional rotational viscosimeter, he will observe that the up curve is concave toward the strain axis as in the case of the pseudoplastic body. However, the down curve is essentially a straight line and does not, of course, coincide with the up curve. Figure 4 illustrates the consistency plot for a thixotropic system. Certain metallic oxide hydrosols and bentonite suspensions in water are typical thixotropic systems. There are many others.

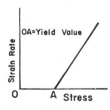

FIG. 2.

FIG. 3.   Pseudo-plastic flow.

FIG. 4.   Thixotropic flow.

Recently systems possessing antithixotropy have been reported. A dilute solution of polymethacrylic acid is said to increase in "apparent" viscosity after a short period of shear. Polyisobutylene in tetrahydronaphthalene shares this property. (See **Thioxotropy**).

Dilatancy. A system is said to be dilatant if its rate of increase of strain rate decreases with increased shear. The consistency curve is concave toward the shear axis. Among the better known systems exhibiting dilatant behavior are pastry doughs, highly pigmented paints and many other industrially important materials. In fact, the phenomenon of dilatancy is usually associated with suspensions, especially those containing high concentrations of suspended materials. One observes the phenomenon in stirring settled pigment from the bottom of a can of paint. A spatula or putty knife embedded in the pigment can be withdrawn with little tension if it is done slowly. However, if one attempts to withdraw the knife rapidly it requires a very considerable force to do so. In other words, at increased stress rates the flow rate decreases. (See Fig. 5)

In this short space only a few of the areas of concern to the rheologist have been briefly touched upon. No discussion of the very important studies on creep and stress relaxation of materials has been possible. Neither has the stress working of metals been considered. In spite of these limitations the very broad general area with which rheology deals has been presented.

GALE F. NADEAU

Cross-references: *Liquid State, Thixotropy.*

References

John D. Ferry, "Viscoelastic Properties of Polymers," New York, John Wiley and Sons, Inc., 1970.

F. R. Eirich, "Rheology, Theory and Applications," (5 volumes) New York, Academic Press, 1956–69.

H. Green, "Industrial Rheology and Rheological Structures," London, Chapman and Hall, 1949.

S. Middleman, "The Flow of High Polymers," New York, Interscience Publishers, 1968.

Scott Blair, G. W., "Elementary Rheology," New York, Academic Press, 1969.

RHODIUM AND COMPOUNDS

Rhodium is a member of the platinum group of metals, the six elements comprising the members of Group VIII in the second and third transition series of the Periodic Table. It was discovered and named in 1803 by W. H. Wollaston (England). He and S. Tennant working in partnership isolated pure platinum and revealed four additional elements as constituents of the native platinum from Spanish America. The name was given to the new element (*rhodos*, Greek = rose) because of the color of the chloride solutions in which it was first isolated.

The abundance of rhodium in the earth's crust has been estimated as 0.001 ppm. It invariably occurs associated with the other platinum metals, but makes up no more than 4% of the platinum-metal content of the presently worked deposits in Canada, Russia or South Africa. In these deposits rhodium is probably uncombined. One recently characterized rhodium mineral is hollingworthite, (Rh, Pd, Pt, Ir) AsS.

Selected physical properties include: atomic number 45; atomic weight 102.9055; anisotropic; density 12.41 g/cc (20°C); melting point 1963°C; hardness (annealed) 100–120 Vickers units; electrical resistivity 4.33 μ ohm-cm (0°C). The metal crystallizes with cubic closest-packed lattice, having $a = 3.8031$ Å, with metallic radius for 12-fold coordination = 1.34 Å. The element is harder than platinum or palladium and shows a high degree of work hardening. It also possesses a very high reflectivity which, taken with its freedom from tarnish, leads to uses for reflecting or ornamental surfaces.

Rhodium finds extensive use as an alloying element in platinum. It raises the melting temperature, hardness, and mechanical strength of platinum especially for the first 20% added. Rhodium-platinum alloys are also important in thermocouples.

This element is less subject to chemical attack than platinum or palladium. However, hot sulfuric or hydrobromic acid will react with the compact metal, and in finely divided form it is attacked by aqua regia. Fused alkalis, especially under oxidizing conditions, attack the metal, and so does fused sodium bisulfate. The halogens react with hot rhodium, fluorine forming RhF_3 (mainly) with RhF_4, and chlorine $RhCl_3$. Around 600°C in air the metal shows tarnish due to superficial formation of Rh_2O_3; this decomposes above 1100°C.

Although compounds have been identified with this element in each integral oxidation state from 0 to +6, rhodium shows a marked preference for the +3 state in its compounds. The yellow aqueous cation $Rh(H_2O)_6^{3+}$ has been well characterized in sulfate or perchlorate solutions, but in the presence of chloride ion some or all of the water molecules are replaced. Much of the solution chemistry of rhodium involves complexes. These have octahedral coordination and are rather inert to ligand exchange; in this respect there is some resemblance to Co(III). Two examples of these are $RhCl_6{}^{3-}$, the rose-colored species to which the element owes its name, and $Rh(NO_2)_6{}^{3-}$, colorless and utilized in one method of purifying rhodium by ion exchange.

Rhodium chloride, $RhCl_3$, as prepared from the elements at 300° is a red solid, insoluble in water. However, a water-soluble modification can be prepared by reacting the yellow hydrous oxide with hydrochloric acid, carefully crystallizing the tetrahydrate, and dehydrating this in gaseous HCl. This compound is a starting material for the synthesis of a wide variety of other rhodium compounds.

Most of the commercial applications of rhodium are to harden and strengthen platinum. It is the most satisfactory agent for this purpose when the platinum is to be used for high-temperature work. This is because these alloys show the least weight loss due to formation of volatile oxides. Thus, rhodium-platinum alloys are used for thermocouples, furnace windings, bushings for production of glass fiber, gauze catalysts for the oxidation of ammonia or the synthesis of hydrogen cyanide, and a number of other applications where high temperatures prevail.

Of all the platinum metals, rhodium seems to be the most satisfactory for electrodeposition. The deposited metal is very hard, and offers a highly reflecting tarnish-free surface. Plating solutions are usually based on sulfates or phosphates with various additives to improve the quality of the plate.

Rhodium has not found extensive use as a catalyst, but has been recommended for rather specific purposes, e.g., the hydrogenation of benzene to cyclohexane or the oxidation of primary alcohols to aldehydes.

W. A. E. McBryde

RUBBER, NATURAL AND SYNTHETIC

Natural and synthetic rubbers are polymeric materials possessing characteristic elastic properties. "Elastomer" is a generic term applied to any substance, including mixtures, that possesses rubberlike properties. Virtually all long-chain polymeric substances may be made to exhibit typical rubberlike behavior through the proper selection of temperature and/or plasticization. The commercially important rubbers are natural rubber and a considerable number of synthetics, e.g., SBR, "stereo" rubbers, nitrile rubber, butyl rubber, neoprene, Thiokol, polyurethanes, and others. Currently natural, SBR, and "stereo" rubbers comprise most of the consumption of new rubber in the U.S. and the discussion in this section will be devoted largely to these rubbers.

Production of Natural Rubber. Although a considerable variety of plants yield rubber-like material, the rubber tree, *Hevea brasiliensis,* cultivated on plantations within an equatorial belt approximately 1400 miles wide, accounts for over 95% of the natural rubber produced. Wild *Hevea brasiliensis* seeds, gathered in Brazil, once the chief source of natural rubber, were the progenitors of the trees now totaling over 11 million acres in plantations in southeast Asia, mainly Malaya and Indonesia, and other minor producing areas. On a broad average, 100 rubber trees per acre produce annually 400–

600 pounds of dry rubber, but improved stocks with yields in excess of 1800 pounds have been developed.

Dry natural rubber is obtained from the latex by a series of steps comprising coagulation, washing, sheeting-out, and drying. Coagulation is accomplished by adding a measured amount of acid, usually formic, to the latex diluted to a solids content of approximately 20%, and allowing the acidified mixture to stand until the coagulum floats to the surface of the serum. The two principal grades of dry natural rubber (crepe and smoked sheet) differ in sheeting-out and drying steps: the former is extensively macerated during the washing step and the final sheets are dried at temperatures below 95°F; the latter type after sheeting-out is partially dried in a smokehouse and finally in a drying chamber which may reach 140°F. Approximately 31 market grades of plantation rubber are recognized internationally.

Production of Synthetic Rubber. In 1941 the world production of synthetic rubber was approximately 80,000 tons; three years later it was 900,000 tons. The many fold increase was due chiefly to the creation of a synthetic rubber industry by the U.S. Government, which produced 760,000 tons in 1944 on a war emergency basis. The bulk of the synthetic rubbers currently consists of "cold" SBR (*S*tyrene *B*utadiene *R*ubber) copolymers of styrene and butadiene. Other synthetic rubbers are "stereo" rubber which consists predominantly of one stereo configuration and special-purpose rubbers: butyl from isobutylene; EPT a terpolymer of ethylene, propylene and a diene; neoprene from 2-chloro-1,3-butadiene; NBR types (*N*itrile-*B*utadiene *R*ubber); solution polymers which involve alkali metal or transition metal coordination catalysts; and many other elastomeric types.

As polymers, rubbers fall into one of two broad classes: condensation polymers such as Thiokols or polyurethanes and addition polymers such as SBR or butyl. The mechanism of formation is either a condensation type or an addition polymerization of a free radical (e.g., SBR) or an ionic (e.g., butyl rubber) type. Synthetic polymers can be produced by either bulk, solution, pearl, or emulsion processes. Thus SBR is an addition copolymer formed by a free radical mechanism in an emulsion.

A typical recipe in parts by weight which is used to prepare "cold" SBR at 41°F follows: butadiene, 72; styrene, 28; water, 180; emulsifier, 4.5; dodecyl mercaptan, 0.08–0.2; and a *p*-methane hydroperoxide-iron complex-sulfoxylate redox initiator; these rubbers together with high reinforcing furnace blacks (see **Carbon Black**) give vulcanizates that are superior to natural rubber in tire treads.

An endeavor of over a 75-year duration ended when *cis*-polyisoprene, the principal component of natural rubber, was synthesized. This synthetic natural rubber could be produced by means of lithium metal-based or transition metal coordination initiators. These initiators have been used to prepare high-*cis*, high-*trans*, and high-syndiotactic, as well as random mixtures of these configurations from diene monomers. Rubbery copolymers of diene monomers and monoolefins, such as styrene, have been prepared with random, block, or graft structures. Mixtures of *e*thylene, *p*ropylene, and a diene have been *t*erpolymerized (EPT) to form another potentially valuable class of vulcanizable elastomers.

Properties. The physical and chemical properties of natural rubber are fixed by biological factors, over which man has little control. In synthetic rubber a much greater variation of properties can be realized. However, in both natural and synthetic rubbers the properties are the resultant contributions of the fine structures and of the gross molecules of the polymer.

The chemical properties of rubber are of paramount importance. The building units of the synthetic polymer molecule account for its chemical behavior, such as breakdown or scorching during processing, cross-linking during vulcanization, chain scission during heating, deterioration while aging, and various reactions with different reagents. The high percentage of unsaturation and the isoprene unit, $-CH_2-CH=C(CH_3)-CH_2-$, are responsible for the unique chemical behavior of natural rubber. The marked ease of breakdown during processing, the tendency to scorch, the ease of vulcanization and tendency to overcure, and the poor aging resistance are characteristics attributable to the isoprene units. In contrast, butyl rubber is difficult to break down, requires special chemicals for vulcanization, and exhibits excellent aging resistance which are accounted for by the low unsaturation and the isobutene unit, $-(CH_3)C(CH_3)CH_2-$, in the polymer molecules. The chemical behavior of butadiene-styrene copolymers with their randomly arranged 1,4- and 1,2-butadiene and styrene units, $-CH_2-CH=CHCH_2-$, $-CH_2CH(CH=CH_2)-$, and $-CH(C_6H_5)-CH_2-$, is generally intermediate between those of natural and butyl rubbers. In "stereo" polybutadiene the 1,4-structure exists predominantly in the *cis* or *trans* configuration and the 1,2-structure in the isotactic or syndiotactic configuration. The structure variations do not affect the chemical properties greatly, but the physical properties of raw and vulcanized rubbers are significantly altered.

Resistance to solvents and the plasticization of rubbers can be interpreted in terms of the physicochemical interaction of building units of the polymer molecules. Interactions arising from the polarity of the building units, from induced dipoles, from dispersion forces, and from hydrogen bonding are dominant factors in determining the behavior of specialty rubbers such as neoprene, nitrile copolymers, Thiokols, polyurethanes, Koroseal, and others.

Another important characteristic of commercial rubber is processability, which indicates how the rubber behaves in various mixing, extruding, spreading, and conveying equipment used in manufacturing processes. The chemistry and configuration of building units, the adhesive and cohesive forces, and the plasticity are dominant factors here. Natural rubber masticates well, incorporates carbon black rapidly, adheres to metal rolls satisfactorily, and extrudes smoothly. Most synthetic rubbers process satisfactorily. Actually, large-scale commercial utilization predicates acceptable processing characteristics.

Properties of Vulcanizates. Raw rubbers have few end uses and the ultimate article is generally obtained by mixing the polymer with a variety of ingredients and vulcanizing this mixture at an elevated temperature (See **Vulcanization**). The various ingredients of a compounding formulation are mixed by mechanical means, such as on a rubber mill or in an internal (Banbury) mixer. Masterbatches are prepared by mixing the latex with a suspension or an emulsion of one or more ingredients and coprecipitating. Each ingredient of compounding formulations performs one or more functions. The types and amounts of ingredients are determined by the chemical and physical properties of the raw rubber and by the ultimate use of the vulcanized product. A generalized compounding formulation and the functions performed by the ingredients follow: (a) rubber (basic ingredient); (b) pigment (diluting, hardening, reinforcing); (c) softener (aiding processing, plasticizing, solvent-proofing); (d) vulcanization agent (cross-linking); (e) accelerator (accelerating cross-linking reaction); (f) activator (controlling vulcanization reaction); (g) and any of a number of ingredients performing special functions such as coloring; retarding deterioration caused by oxygen, ozone, light, or heat; lubricating; or tackifying.

The small differences in properties of natural rubber and synthetic *cis*-polyisoprene does not prevent the use of the latter as a replacement for natural rubber. Commercialization of synthetic polyisoprene ended the dominance of the rubber market by the natural product. *cis*-Polybutadiene processes poorly. However, this deficiency can be overcome in blends with other rubbers without loss of much of the dynamic and wear properties in tire treads.

In truck and passenger tires *cis*-polybutadiene is providing longer wearing tires, fewer tire failures, greater safety, and better over-all performance. The discovery of "stereo" rubbers is comparable to the discovery of "cold" rubber.

The manufacturing process for dry rubber, except for some special cases, involves basically mixing the various compounding ingredients with rubber, shaping the mixture to the desired form, and heating the mass until vulcanized. Many ingenious molds have been devised to hold the compounded rubber in the desired shape while heating to the final set. In certain articles, as in tires, a fabric is embedded in the body of the rubber for strength. In tire manufacturing the fabric is rubberized by spreading or calendering, the coated pieces are used to build up a premold shape, and the extruded tread is applied last before the composite is placed in the tire mold for vulcanization. Rubber articles can be vulcanized in molds, in air or steam-heated zones, in hydraulic presses, and on rotating drums. Latex is processed somewhat differently. (See **Latex**)

C. A. URANECK

Cross-references: *Elastomers, Polymerization, Latex, Carbon Black, Vulcanization.*

Cyclized Rubber

"Cyclized rubber" is the term which designates the hard, nonrubbery, resinous product obtained from natural rubber (*cis*-1,4 polyisoprene), balata (*trans*-1,4 polyisoprene), or synthetic polyisoprenes when these materials are made to undergo a special type of isomerization. In fact, prior to the discovery of *cis-trans* isomerization of diene polymers about 1957, cyclized rubber (or "cyclorubber") was often referred to as "isomerized rubber," and these terms have been used interchangeably for a long time. Although "cyclized rubber" is now the preferred name, it is only roughly descriptive of this resin, which does not have a unique molecular structure but rather consists of an assortment of mono-, bi-, tri-, tetra- and other polycyclic groups distributed at random throughout the polymer backbone, along with a small amount of isolated, unreacted isoprene units.

The cyclization process can be brought about in a variety of ways: by treating the polymers at somewhat elevated temperatures with strong acids such as H_2SO_4, with organic acids and their derivatives such as *p*-toluenesulfonic acid and its chloride, Lewis acids such as $SnCl_4$, $TiCl_4$, BF_3, $FeCl_3$, or with other catalysts of an acidic character. This process is considered to have a carbonium ion mechanism.

The detailed structure of the cyclized rubber molecule has been the subject of considerable controversy over the years. It has recently been elucidated, however, with the aid of infrared and nuclear magnetic resonance spectra of the cyclization products obtained from *cis* and *trans* polyisoprenes, squalene (a hexaisoprene model compound), and several deuterated forms of polyisoprene, as well as by a careful analysis of the kinetics of the cyclization process. Indeed, current scientific interest in cyclized rubber has been stimulated both by the desire to apply modern techniques to the resolution of the long-standing controversy regarding its microstructure, and by the increasing attention being paid to cyclization reactions generally in the chemistry of diene and other unsaturated polymers. The same techniques, for example, are being employed in the study of the possible formation of ladder, or double-chain, polymers in the cyclization of 3,4-polyisoprene, 1,2-polybutadiene, and comparable polymers (See **Ladder Polymers**).

The physical properties of cyclized rubber, and hence also the uses to which it has been put, have depended in large measure on the methods by which the resins have been prepared. The main applications for cyclized rubber have been as adhesives, thermoplastic molding materials, reinforcing agents for various natural and synthetic rubbers, and for use in coating compositions and printing inks.

Cyclized rubber as an isomer of natural rubber has been known since 1910 and an extensive technical, trade and patent literature has appeared dealing with its properties, uses and methods of preparation.

MORTON A. GOLUB

References

Mast, W. C., in "Kirk-Othmer Encyclopedia of Chemical Technology," 2nd Ed., Vol. 17, Interscience Publishers, New York, 1968, pp. 651–655.

Golub, M. A., in "Polymer Chemistry of Synthetic Elastomers," Part II, J. P. Kennedy and E. G. M. Tornquist, Editors, Interscience Publishers, New York, 1969, Chapt. 10.

RUBIDIUM AND COMPOUNDS

Rubidium, Rb, atomic number 37 and atomic weight 85.467, is the fourth member of the alkali metal group, Group IA of the Periodic Table. It was discovered in 1861 by Bunsen and Kirchhoff by the use of the spectroscope and is named for the prominent red lines of its spectrum. The element resembles the other alkali metals in its physical and chemical properties, and is monovalent in its compounds which are very stable to oxidation and reduction.

Occurrence. Rubidium is the 22nd most prevalent element in the earth's crust, almost as abundant as chlorine, but it is not found in any mineral as a principal constituent. Rather, it occurs widely dispersed in very low concentrations in potassium minerals and this accounts for the scarcity of the production and use of this rather prevalent element.

Evaporite deposits of salts such as carnallite, $KCl \cdot MgCl_2 \cdot 6H_2O$, contain 0.02 to 0.15% rubidium, and African lepidolite, a lithium mineral, contains as much as 1 to 1.5% Rb. The former offers a potential source of the element and the latter has been a major source when it was used as the raw material for lithium recovery, an operation no longer conducted in this country. Pollucite, a cesium aluminum silicate, also contains rubidium. All rubidium produced must come from the residues or concentrates which are by-products of the production of other alkali metals.

Derivation. The chief problem in extracting rubidium from natural sources is to separate it from much greater concentrations of other alkali metals and from magnesium, in the case of carnallite. Fractional crystallization of compounds such as alums, chlorostannates and ferrocyanides is used to separate rubidium compounds from those of other alkali metals, a laborious process requiring 25 or more recrystallizations. These compounds are then decomposed to yield RbCl from the chlorostannate and Rb_2CO_3 from the ferrocyanide. It may also be possible to use a liquid ion exchange process developed at Oak Ridge National Laboratory, now used to recover cesium compounds (see **Cesium**), for rubidium separation from solutions of alkali metal compounds. Details of these and other methods of preparing rubidium compounds from natural sources are described by Mosheim.

Rubidium Metal Preparation. There are several methods of making elemental rubidium, among them: (1) electrolysis of fused salts, such as the chloride or cyanide, under an inert atmosphere; (2) thermal decomposition of such compounds as the hydride and azide under vacuum or inert gas; and (3) chemical reduction of such compounds as the oxide, carbonate, hydroxide, halides, sulfate or chromates with metals like calcium, magnesium, barium, aluminum and zirconium at elevated temperatures. The vigorous reaction of rubidium with air and water requires extreme care in its preparation and handling.

Physical Properties. Among the most interesting physical properties of rubidium are its large ionic radius; low ionization potential, density, melting point and boiling point; and its high electrode potential. About 27% of natural rubidium is the beta-emitting rubidium-87 with a half-life of 6.3 ×

10^{10} years which decomposes to strontium and can be used to determine the age of rubidium-containing rocks.

Rubidium is a soft, silvery-white, body-centered cubic metal. Its melting point is 39°C; boiling point, 688°C; density, 1.532 g/cc (20°C); ionic radius, 1.48 Å; latent heat of fusion, 6.1 cal/g; latent heat of vaporization, 212 cal/g; specific heat, 0.080 cal/g/°C (0°C); thermal conductivity, 0.075 cal/sec/°C/cm (50°C); vapor pressure, 1 mm at 294°, 10 at 387°, 100 at 519° and 400 at 628°C; electron work function, 2.09 eV; electrical resistivity, 11.6 microhm-cm (0°C); coefficient of thermal expansion, 90×10^{-6} (20°C); viscosity, 0.6734 centipoise (38°C); electrode potential, -2.925 volts; and thermal neutron absorption cross section, 0.73 barns.

Chemical Properties. Rubidium is a very reactive element and is quite similar to potassium in its chemical reactions. It forms four oxides: yellow monoxide, Rb_2O; dark-brown dioxide or peroxide, Rb_2O_2; black trioxide, Rb_2O_3; and dark-orange tetroxide, Rb_2O_4. Upon exposure to air it becomes covered rapidly with a gray-blue film of a mixture of Rb_2O, Rb_2O_2 and Rb_2O_4. Rubidium hydroxide is a powerful base, and it, the fluoride and carbonate are very hygroscopic and corrosive.

Rubidium halides form double halide complexes with many other metals, such as Sb, Bi, Cd, Co, Cu, Fe, Pb, Mn, Hg, Ni, Th and Zn, which can be used in the separation and purification of rubidium from other alkali metals.

The soluble rubidium compounds are the carbonate, chloride, bromide, iodide, hydroxide, chlorate, nitrate, sulfate, sulfide and chromate. The relatively insoluble compounds include the perchlorate, permanganate, chloroplatinate, chlorostannate, fluosilicate and periodate.

Rubidium forms alloys with the alkali metals, the alkaline earth metals, mercury, antimony, bismuth and gold. Alloys with the last three metals have the property of releasing electrons under the influence of light, and can be employed in photoelectric tubes.

Applications. The uses of rubidium in photoelectric cells and as getters in vacuum tubes are the largest outlets for the metal. While rubidium can be applied as a fuel in the ion propulsion engine and in the thermionic mechanism and the magnetohydrodynamic principle of generating electricity, cesium is the more desirable element for these purposes.

Rubidium carbonate has been used in the preparation of special glasses, but the application is minor. Rubidium hydroxide can be used as an electrolyte in low-temperature alkaline storage batteries, but its high cost is a disadvantage.

A recent study by Carroll and Shapr indicates that rubidium (as RbCl) has promise as an antidepressant agent in the treatment of mania in man, an effect opposite to that of lithium (q.v.).

CLIFFORD A. HAMPEL

References

Mosheim, C. Edward, "Rubidium," pp. 604–610, "Encyclopedia of the Chemical Elements," Hampel, C. A., Editor, Van Nostrand Reinhold Co., New York, 1968.

Perel'man, F. M., "Rubidium and Cesium," Macmillan Co., New York, 1965.

Hampel, C. A., "Rare Metals Handbook," 2nd Ed., pp. 434–440, Reinhold Publishing Corp., New York, 1961.

Carroll, B. J., and Sharp, P. T., *Science,* **172,** 1355–57 (1971).

RUTHENIUM AND COMPOUNDS

Ruthenium is a member of the platinum group of metals. These occur in the second and third transition series (d-block) in Group VIII of the Periodic Table. The element was identified and named by K. Klaus (Russia) in 1844. The name is based on a latinized name for Russia, which for a century following the discovery of alluvial deposits of native platinum in the Urals mountains in 1824, was the world's major source of platinum metals.

It is interesting that, although ruthenium is estimated to comprise 0.001 ppm in the earth's crust, and by some authorities to be the least abundant of the platinum metals there, it ranks next only to platinum in estimates of abundance in meteorites and of cosmic abundance. Ruthenium probably occurs uncombined as a minor component in such alloys as native platinum, etc. There is also a rare mineral species laurite, RuS_2, often containing some osmium.

Selected physical properties include: atomic number 44; atomic weight 101.07; stable isotopes 96 (5.5%), 98 (1.9%), 99 (12.7%), 100 (12.6%), 101 (17.1%), 102 (31.5%), 104 (18.6%); density 12.45 g/cc (20°C); melting point 2314°C; hardness (annealed) 200–350 Vickers units; electrical resistivity 6.71 μ ohm-cm (0°C). The metal crystallizes in a hexagonal close-packed lattice in which $a = 2.7056$ Å and $c/a = 1.5820$; an average value for the metallic radius in this arrangement is 1.335 Å. Although certain thermal evidence has in the past been taken to imply the existence of allotropic modifications of the element, it is not now believed that such allotropy exists.

Ruthenium is used as a hardening agent for palladium and platinum for which purpose it is very effective. Although the element exhibits appreciable miscibility with a wide range of other metals, these alloys have not been extensively investigated.

The metal is not attacked by cold or hot acids including aqua regia. However it is rather easily attacked by alkaline oxidants such as hypochlorite solution or concentrated sodium hydroxide solution. Likewise fused alkaline substances like hydroxides, carbonates, cyanides, and peroxide attack the metal, especially if it is finely divided. Halogens attack the metal, and by reaction at elevated temperature, compounds like RuF_5 or $RuCl_3$ may be produced by direct combination. When $RuCl_3$ is prepared from the elements it is a black solid, insoluble in water and exists in two crystalline modifications. There is also a brown water-soluble form which is the point of departure for many synthetic reactions.

Ruthenium forms a volatile tetroxide, RuO_4, but differs from osmium in that the tetroxide does not apparently form directly from the elements. Heated in oxygen above 1000°C, ruthenium loses a little weight (*cf.* iridium, platinum), but this is explained by formation of a trioxide of some volatility at that temperature. There is also a black, solid dioxide, RuO_2, with the structure of rutile.

Compounds of ruthenium have been reported in all oxidation states from 0 to +8, but of these, +2, +3 and +4 are the most common. As is the case with all the platinum metals, there is a strong tendency to form complexes. Only in perchlorate solutions does the element show properties of an aquo cation, and even then it is extensively hydrolyzed, Ruthenium forms numerous polynuclear complexes; for instance, a species at one time formulated as $RuCl_5OH^{2-}$, produced when RuO_4 reacts with hydrochloric acid, has been shown to be $Ru_2Cl_{10}O^{4-}$ ($Cl_5Ru-O-RuCl_5^{4-}$). Ruthenium also has an unusually great tendency to form stable nitrosyl complexes containing the $RuNO^{3+}$ group. Many reactions of the element in aqueous solution can be "masked" in the presence of nitric acid or nitrite, and this fact is utilized in several chemical separations, e.g., of osmium from ruthenium.

The principal application of ruthenium is as a hardening agent for platinum or palladium. Small amounts of this element greatly increase the corrosion resistance of titanium. A limited number of alloys with high ruthenium content have been developed for special purposes to take advantage of their great hardness and freedom from corrosion. A number of catalytic applications have been proposed, but of these special mention is made of the selective reduction of carbonyl groups in organic compounds by hydrogen.

W. A. E. McBRYDE

S

SAFETY PRACTICE

General. Safety organization in any operation is simply a systematic method for protecting workmen from traumatic injury and from damage to their health as a result of the work which they are hired to carry on. Progressive managements have long recognized the necessity for such protection of workmen as a social obligation.

There is now also a legal obligation imposed by every State and by the Federal Government which requires every employer to conform to stated minimum standards of environment and practices to protect the safety and health of their employees. The Federal government and almost all state governments provide a staff of safety inspectors to enforce these minimum standards. Most chemical companies choose to provide conditions and practises which go beyond what is required by law or regulation.

One compelling reason for attention to safety matters is that it is good and profitable business activity. Partly, but not entirely, as a result of Workmen's Compensation laws which began to be enacted about 1911 and are now in force in every state, it costs far less to prevent accidents and injuries than to pay for them.

Accident prevention can be accomplished by organized and coordinated effort. It is a misapprehension that "accidents will happen." It is much more accurate to say that "accidents are caused," and since they are caused they can be prevented by removing the cause. This is pretty well demonstrated by the fact that the chemical industry, which is inherently somewhat hazardous, actually operates with injury rates at about $\frac{1}{4}$ the average for all industry.

Because safe practice must be an integral part of all phases of a chemical operation it has to be applied by the line organization even though it may be, and generally should be, directed and administered by a staff person with special knowledge and special training.

Fire Control. One feature of fires in chemical plants is likely to be gas or vapor explosions. It is possible to reduce the damage from such explosions by providing vents in the proper ratio. Venting is quite successful in reducing the peak pressures produced by dust explosions, where the rate of pressure rise is quite slow, but substantially less so in reducing the pressures of gas and vapor explosions where the rate of rise is apt to be very much faster. A more fundamental solution and one which is rapidly being adopted in the chemical industry is to eliminate the sources of explosive gases and vapors in buildings by putting such units as reactors, stills, screens, mills, and associated piping and conveyors out in the open and housing only the necessary controls and the operators within the buildings. In cold climates this may introduce something of a hardship during maintenance work but this can commonly be overcome by the use of temporary shelters.

Much of the equipment used in chemical plants is subjected to conditions leading to corrosion and erosion and is also apt to be under pressure. Under these conditions it is highly desirable that maintenance of the equipment be put on a scheduled rather than on an emergency basis. This scheduled maintenance, when it eliminates failures and makes the necessary repair at the convenience of the maintenance crew, is not only a much more efficient method of running the plant but eliminates one of the major sources of injury in plants.

Vessel Entry. Chemical plant maintenance very frequently involves entering vessels of various sorts. This is always a highly hazardous procedure and should be subject to a detailed written job procedure which should not be varied at all. The minimum requirements are that the vessel be clean and that the individual entering wear a lifeline and be attended by someone outside at all times. Unless the vessel has been cleaned and has been shown by specific test to have a safely respirable atmosphere and is provided with sufficient ventilation to maintain the atmosphere safely respirable, the individual going in should be provided with supplied atmosphere respiratory equipment of some type, either self-contained or, preferably, a hose mask with blower.

Hot Work. One of the maintenance operations which is a prolific source of fires and explosions in chemical plants is welding and cutting by either the gas or electric processes. To properly control welding, minimal rules should be set down for fire protection and for shielding for welding operations carried on anywhere in the plant. These should include the use of incombustible materials under welding which must be done above combustible surfaces, the use of screens to protect passers-by from arc flash and, by all means, the use of fire watchers who check back a half hour and an hour after the welding is finished to be certain that no smoldering fires have been left behind.

For control of welding in dangerous areas, certain plant areas where there may be flammable liquids or solids should be designated and any hot operations, including welding and grinding in such areas should require a permit signed by both the production supervisor of the area and the appropriate representative of the safety or fire protection department. Such permit should not be signed until the responsible supervisors have been convinced that the hot operation must be done in the area, in other words, that the material cannot be taken somewhere else and brought back after the work is finished, and also that the area where the work is to be done, has been effectively cleared of any flammable or explosive materials.

One of the necessities in chemical plant maintenance is welding on closed systems such as tanks, stills, pipe lines, reactors and similar vessels. A very substantial hazard both to the operator and to the plant can be produced if such equipment has not been freed of flammable and toxic materials before the welding is started. An acceptable procedure is to disconnect and blind all lines leading to the equipment which is to be entered or welded with the exception of one or two lines at the lowest point which are disconnected and left open for drainage. The tank should then be steamed if it has contained a volatile material that is not water soluble or washed with water if it has contained a water soluble volatile material. If steaming is to be depended upon it should be continued until the entire outside of the tank has been brought to a temperature of 140 to 160°F. After the tank has been cooled, scale and debris should be scraped down and removed and both the exterior and interior should be rinsed down with water. The men doing the cleaning should, of course, have protective clothing and respiratory protection unless continuing tests show that the atmosphere in the tank remains respirable. After the tank has been thoroughly cleaned and freed of volatile materials the welding can proceed in the usual way.

Storage and Handling. One major factor in the outstanding reduction in accidents and injuries in the chemical industry over the past several years has been the handling of materials in much larger quantities which has required mechanization of the handling and storage processes. It is always desirable, if it can be arranged, to handle chemicals within closed pipe lines, tanks, and conveyors so that people do not come in contact with them. Care should be taken in setting up storage areas to keep substances which will react if they come in contact physically separated so that they will not be mixed under emergency conditions such as might occur in a fire, explosion or flood. A good chemist is necessary to decide which materials should be separated and which may safely be stored together.

If liquids must be stored in carboys or drums they should be set on clean dunnage and the stores well separated so that it is possible to get at any type of material without moving other types. Carboys should preferably be in a single layer and should never be stacked more than 6 ft. high. Carboy handlers should always grip the carboy box by both the sides and bottom so that the carboy cannot drop out of the box and should keep the operator well away from the box.

Drums and barrels are best stored in racks. Drums which contain corrosive materials, in particular, should be stored with the bungs up and in such a manner that it is easy to get at each drum so that the bungs can be loosened once a week to vent off any gas which may have developed.

Storage tanks for chemicals should be preferably either above ground or in pits where they can be seen so as to reduce the possibility of undetected leakage. If they are above ground they should be surrounded by a diked area sufficient to contain the contents in case of failure of the tank. If the tanks contain volatile materials they should invariably be equipped with safety valves and vents which will vent off escaping material to a safe place. Low-pressure tanks should also have safety valves even if they contain nonvolatile materials because in case of fire the expansion of the air contained in an empty low-pressure tank may be sufficient to rupture it.

The filling and emptying connections are preferably made through the tops of tanks to minimize the possibility of loss of contents if a filling or emptying connection is broken. The tank roof should be substantial and should be provided with permanent walkways with hand rails and toe boards for access to the connections. Tank cars should also be loaded and unloaded through the dome rather than through the bottom connections.

The main consideration in the handling and storage of solid materials is to provide bins with sufficient slope to the bottoms that they can be emptied by gravity without the material arching over and requiring someone to enter the bin either from above or from below. There is also usually a dust problem in handling solids into and out of bins. It can generally be controlled by enclosing and exhausting the loading, transfer, and discharge points on conveyors and elevators.

Piping. Chemical pipe lines are preferably installed in trenches or tunnels. If they must be installed overhead, a serious effort should be made to isolate them so that they will not drip on people working or passing beneath.

There are several possibilities of injuries in working on chemical pipe lines. Failure of packing in valve barrels or of gaskets in flanges is one of the most prolific and is ordinarily protected against by tieing plastic shields over the valves around the stem and around the flanges to contain leaking material if possible but, in any event, to direct a stream of leaking material away from the person working on the fitting.

Because of the possibility of cross-connections, if the wrong valve is opened in most chemical plants it is very important that lines and valves be clearly and unmistakably identified and be put in places where they can be seen so that the identification can be recognized.

F. A. VAN ATTA

References

Fawcett, H. H., and W. S. Wood, "Safety and Accident Prevention in Chemical Operations," Interscience Publishers, New York, 1965.

Morgan, C. T. *et al.,* "Human Engineering, Guide to Equipment Design," McGraw-Hill, New York, 1963.

Steere, Norman V., Editor, "Handbook of Laboratory Safety," The Chemical Rubber Co., Cleveland, 1967.

Anonymous, "Fundamentals of Industrial Hygiene," The National Safety Council, Chicago, 1967.

Anonymous, "Accident Prevention Manual for Industrial Operations," Sixth Edition. The National Safety Council, Chicago, 1969.

SALTS

Common salt, or sodium chloride (NaCl), is an example of a class of compounds called *salts*. From one standpoint, salts are considered to be derived from acids. When the hydrogen of an acid is replaced by a metal [or its equivalent such as the ammonium (NH_4^+) ion] a salt is formed. Thus, sodium sulfate (Na_2SO_4) is formed from sulfuric acid (H_2SO_4). If the replacement of hydrogen is incomplete, the resulting compound is called an acid salt. Sodium hydrogen sulfate ($NaHSO_4$) is an example of an acid salt.

Also, salts may be considered to be related to hydroxides. When the hydroxyl ion (OH^-) is replaced by chloride (Cl^-), sulfate (So_4^{2+}), nitrate (NO_3^-), or other anion, a salt is formed. Thus, sodium sulfate or (Na_2SO_4) is formed from sodium hydroxide (NaOH). Basic salts retain one or more hydroxyl groups (or oxygen by loss of water). Bismuth subnitrate ($BiONO_3$) and lead hydroxy acetate (Pb $OHC_2H_3O_2$) are examples of basic salts.

From a more general standpoint, salts are compounds associated because of a group of common properties. Salts are compounds that have a relatively high melting point; they conduct electricity when melted; if soluble in water, their water solutions also conduct electricity. Sodium hydroxide (NaOH), a compound with strong basic properties, is classed as a salt according to this definition.

The explanation of these properties of salts is related to the ionic character of the compounds. Salts generally have electrovalent bonds in which one or more electrons are transferred from the cation (positive part, usually metallic) of the salt to the negative anion. Thus in the salt potassium nitrate (KNO_3, $K^+NO_3^-$), the potassium ion has one fewer electron than a potassium atom, and the nitrate ion one more. The crystal lattice that is potassium nitrate is composed of potassium ions and nitrate ions. These ions dissociate when the compound is melted or when it dissolves in a solvent. When ions are free to move in a liquid, they conduct electricity. (See **Ions**)

Many salts have water (or some other compound) loosely bonded to them in definite amounts. Examples of such hydrated salts are copper sulfate pentahydrate ($CuSO_4 \cdot 5H_2O$) and sodium carbonate decahydrate ($Na_2CO_3 \cdot 10H_2O$). (See **Hydration**)

When solutions that contain a mixture of ions crystallize, double salts may form. For example, from a solution of aluminum, potassium, and sulfate ions, common alum [$KAl(SO_4)_2 \cdot 12H_2O$] may crystallize. The alums, Mohr's salt [$(NH_4)_2SO_4 \cdot FeSO_4 \cdot 6H_2O$], and many others are examples of double salts. When in solution, double salts respond to tests for the presence of the several ions.

Complex salts, on the other hand, may not dissociate completely when they dissolve. For example the complex salt potassium chlorostannate (K_2SnCl_6) does not give the reactions of the stannic ion. Many other complex salts are known.

ELBERT C. WEAVER

Cross-references: *Ions, Electrochemistry, Acids, Bases, Saponification.*

SAMPLING

Sampling is the process of withdrawing a small portion of material to represent the entire lot in some test. It is one important step in the control of practically every manufacturing process. In the chemical field, samples are usually taken to determine the qualities of certain constituents, including moisture, and for certain physical properties, such as density, melting point, particle size, etc. The material may be either gas, liquid or solid. The size and location of the parent lot may range from a few pounds in a jar to the contents of a railroad car or to a naturally occurring deposit. Often, especially in the case of solids, several phases are present. In the sampling of such heterogeneous materials it is necessary that the various phases or classes of material be present in the sample in the same proportion as they occur in the parent lot. The sampling procedures are different in nearly every case, and are so numerous that it is far beyond the scope of this brief article to do more than give general principles and to bring attention to some sources of error.

Sampling operations may be divided into two parts: (1) obtaining the comparatively large gross sample from the parent lot, and (2) the reduction of this gross sample to the amount of material necessary for the particular test. In many cases, unfortunately, too little care is given to the first stage of the sampling. This can lead to serious error, because no matter how carefully the sample reduction and analytical work are done, the results applying only to the sample are always uncertain to the extent that the sample is not truly representative. It is because of this very reason that much valuable time and effort have been wasted, often justifying the criticism that "ten-dollar" analyses are made on "ten-cent" samples.

Sampling should always be done keeping in mind the test to be performed and the type of information desired. In cases where average values are sought, samples are composed of many well-mixed aliquots from every part of the original lot of material. In other cases, point or instantaneous values may be sought and the gross sample for this information must necessarily be unmixed and as small as possible. When "average" samples are used, information is lost concerning the variability existing in the material. Many tests require samples which must be taken and treated in special ways. For instance, samples of solids for a moisture determination must be taken quickly and immediately placed in vapor-tight containers. Guarding against loss of moisture must take precedence over the other niceties of sampling. Samples for physical properties (such as sieve analysis) must remain in an "as is" condition, while a sample for chemical properties must be ground to a powder to justify the small samples used in analyses and also to facilitate the dissolving.

In sampling material from a flowing stream, it is better to take all of the stream part of the time, than to take part of the stream all of the time. Several mechanical devices for performing this operation are on the market. If the measured variable in a flowing stream be subject to periodic fluctuations and an average value be desired, then care must be taken to have the sampling of different frequency, or better still to have the duration of one sampling period be longer than several cycles of the variable. On the other hand, if the instantaneous values of the variable are desired, the sampling period must be much shorter than the cycle of the variable.

The sampling of gases and liquids generally offers few problems because by their fluid nature they can usually be rendered quite homogeneous. However, poor samples are sometimes obtained by removing the sample from rapidly-flowing streams too close to a reaction zone and before complete mixing has occurred. Highly viscous fluids such as oils are sometimes poorly mixed. Liquid phases will, of course, separate into layers, and sometimes liquids will be found stratified into different density layers even though apparently only one phase is present. The sampling of dusts or muds is primarily a problem of removing very small solid particles from a fluid stream, filtration being the usual means.

The sampling of solids is a difficult problem because they are most likely to be heterogeneous, segregated by particle size and by particle density. Long experience has led to sampling specifications in many fields. In general, when sampling from piles, it is important to obtain aliquots distributed throughout the volume of the pile not just close to the surface. This can best be done as the pile is moved, or with a trier consisting of slotted concentric pipes that can be inserted into the pile and then opened. Transported solids are subject to vibration and are most likely to be segregated; most careful sampling is required in this case.

The size of the sample is governed by the fundamental principle that the inclusion or omission of the largest single fragment of impurity shall not affect the final results of the test within the desired accuracy. Application of this principle usually results in an original gross sample many times larger than is needed for the test. Samples of hundreds of pounds are not infrequent in engineering operations. In those cases where the material may be crushed to a smaller particle size, application of the above rule then allows smaller sample sizes. These subsamples may then be reground to even smaller particle sizes, which allows further reduction, and so on. In uranium ore operations, for example, repeated grinding coupled with intricate and precise sampling equipment are used on a large scale to realize a test sample, which is representative of many tons of ore. If, however, the material cannot be ground (for instance, a sample for sieve analysis), then the reduction of size of sample below a certain point can be done only with decreased accuracy.

Reducation of the gross sample is usually done by coning and quartering. This method requires simple equipment, but is very slow, laborious, and erratic in unskilled hands. Samples reduced on a riffle are generally excellent; however, erratic samples may be obtained when the riffle is fed too fast. Several mechanical sample reducers are commercially available and generally produce excellent results. Since sample reduction is essentially a particle-handling procedure, particle properties are those most sensitive to possible discrimination during the reduction operation. Any sample reduction must particularly guard against bias due to particle size, particle density, or particle shape.

When the particle size has been reduced to a fine powder, a few milligrams will contain thousands or even million of particles as is shown in the table.

Particle Size (assumed cubic)	Number of Particles per Gram (density $= 2$ g/ml)
1 mm	500
100 mesh (149 μ)	150,000
325 mesh (44 μ)	6,000,000
10 μ	500,000,000
1 μ	500,000,000,000

Since the grinding also acts as an excellent means of mixing, a grab sample of only a few milligrams may be taken from any part of the powdered sample and will constitute an excellent subsample. The grinding operation must not introduce impurities; in particular, the grinding implements should not contain the same elements as are being sought in the analysis.

The most common pitfalls in sampling may be briefly summarized as follows:

(1) The material taken for the sample is not representative of the main body.

(2) The material undergoes change of composition during preparation and/or storage of the sample, the chief offenders being oxidation, moisture loss or gain, absorption of carbon dioxide, etc. There may also be contamination with the materials of the grinding equipment.

(3) Segregation of particle sizes and particle densities may cause errors at any stage of operation from the very first sampling operation until the final portion used for the analysis or test. Many solid materials, apparently homogeneous, yield on crushing fine and coarse particles that differ widely in composition. The chances for error from this source are obvious.

The methods used to determine how good the sample is are inevitably statistical. There are two techniques of general usefulness which are known as the statistical control chart and acceptance sampling. The control chart consists of a record either of averages or ranges of the results plotted vertically and extended horizontally on a rough time scale so that trends are at once evident. Dotted lines are drawn across the chart to form the interval about the general average likely to contain the averages of the successive samples. The chart in effect makes the same comparison that an analysis of variance would make, i.e., the long term between samples variation is compared with the short term variation effective within the samples. The reader is referred to any one of several books on Quality Control, developed particularly for the control of specified qualities during a manufacturing process, such as: "Design and Analysis of Industrial Experiments," O. L. Davies, Editor, Hafner, N.Y., 1963; "Statistical Methods," G. W. Snedecor, Collegiate Press, Ames, Iowa, 1956; "Statistical Methods for Chemists," W. J. Youden, Wiley, 1951; "Statistical Methods for Research Workers," R. A. Fisher, Hafner, N. Y., 1950; "A.S.T.M. Manual on Quality Control," Philadelphia, 1951.

FRANK J. CARPENTER
VICTOR R. DEITZ

SAPONIFICATION

Saponification is a special case of hydrolysis in which an ester is converted into an alcohol and a salt of the appropriate acid by reaction with an alkali. Though the operation has numerous applications throughout the chemical industry, it is noteworthy because some 80% of standard soap is prepared by this method. The esters may be of mono- or polybasic acids and mono- or polyhydric alcohols, the physical conditions under which the reaction occurs being suitably varied to secure an adequate rate. The alkali most commonly used is sodium hydroxide, because of cost and water solubility, but other appropriate alkaline materials are suitable.

Since the preparation of soap is typical of a fairly complex reaction, chemically, and since it is common, it serves as a useful example of the saponification operation. The complication, of course, occurs because the usual esters used for soap are the glycerol esters of fatty acids, saturated and unsaturated. Thus the saponification of stearin (glycerol tristerate) is commonly shown as follows:

$$C_{17}H_{35}COO-CH_2$$
$$|$$
$$C_{17}H_{35}COO-CH + 3NaOH \rightarrow$$
$$|$$
$$C_{17}H_{35}COO-CH_2$$

$$3C_{17}H_{35}COONa + HO-CH_2$$
$$|$$
$$HO-CH$$
$$|$$
$$HO-CH_2$$

Actually, the saponification appears to progress stepwise, the first hydrolytic reaction taking place as follows:

$$C_{17}H_{35}COO-CH_2$$
$$|$$
$$C_{17}H_{35}COO-CH + NaOH \rightarrow$$
$$|$$
$$C_{17}H_{35}COO-CH_2$$

$$C_{17}H_{35}COO-CH_2$$
$$|$$
$$C_{17}H_{35}COO-CH + C_{17}H_{35}COONa$$
$$|$$
$$HO-CH_2$$

The diglyceride formed is subsequently split to the monoglyceride, which finally is converted to glycerol, if sufficient alkali is present. Thus, the reaction is a bimolecular one rather than quadrimolecular, as is commonly indicated. In actual practice, the fats used are complex glycerides of a number of saturated and unsaturated acids, rather than the stearin shown here.

Technologically, the saponification operation varies in degree of difficulty depending on the ester. The reaction rate differs for different esters, for one thing, but another determining factor is the contact area possible between the alkali and the ester. In the case mentioned above, the fat at the start is insoluble and immiscible in water, so that reaction in a nonagitated vessel would be very slow, occurring only at the limited interface. Actually, if the alkali is added slowly, there is enough soap formed to enable emulsification of the fat to occur. At this point the rate of reaction rises rapidly, since the contact area of a good emulsion is enormous.

As was indicated, the alkali is added slowly to avoid a "salting out" effect on the soap already formed, so as to obtain its optimum emulsifying effect. Further, the gradual alkali additions are usually made in three stages, approximating the formation of the diglyceride, monoglyceride and glycerol. Salting out of the soap-glyceride mass at intermediate stages, using sodium chloride, may also be used to promote glycerol removal and to enable the remaining material to be more completely saponified.

As is usual, the temperature at which saponification is carried out affects the reaction rate markedly, so it is desirable to operate at as high a temperature as the particular system will allow. For ordinary fats, operation at the boiling point of the soap-alkali solution usually gives an acceptable rate, but autoclave operation can be used to raise the temperature.

For more volatile esters, of course, it becomes still more important to use pressure vessels to increase the reaction rate, or at least reflux so as to control losses. Further, agitation must be exceptionally good with such materials, since the alkali salts of the shorter-chain acids frequently show little emulsifying activity and so require other means for developing high contact areas.

In a few specialized instances, where elevated temperatures must be avoided for special reasons, such as decomposition, techniques have been developed where "cold" saponification can be used. In such instances, a small quantity (ca. 1%) of an agent such as ethyl alcohol is added to the intimate mixture of fat and concentrated alkali to start the reaction, which then proceeds exothermically and rapidly, though usually not to completion.

The alkali used has an important bearing on the problems met during reaction. Though this is in part due to solubility problems encountered with the alkali metal hydroxides, more frequently it centers on the varying solubilities of the salts formed. Thus, though a large number of potassium salts of the higher fatty acids are fairly soluble in water, most of the comparable sodium salts have very limited solubility. Depending on what compounds are being sought by the saponification reaction—whether the alcohol, the salt itself or the acid which may be regnerated therefrom—the technical problems introduced by the limited solubility mentioned may be quite severe. This is especially true of the salts of the higher fatty acids since they form complex hydrates having a colloidal nature.

For special circumstances, the use of solvents other than water may be justified. The lower alcohols, such as ethanol, frequently offer advantages in completeness of reaction and rapid conversion rate, together with a minimum of problems due to colloidal association. Again, such high-boiling solvents as ethylene glycol monoethyl ether offer advantages in rapid reaction rate. If desired, of course, the alkali alcoholates, such as sodium methylate, can be used to completely exclude water from the reaction mass, for it is frequently the salt hydrates which introduce complications.

Saponification, of course, reaches an equilibrium,

the exact point depending on the ester and the quantity of reactant present. Ordinarily the most successful procedure is to use an excess of alkali to complete the reaction, the excess being removed by neutralization or similar means. By such methods very complete hydrolysis is possible.

Though saponification is the dominant reaction when the techniques described above are used, in many instances side reactions may occur which may profoundly modify the products. Oxidation, of course, is one of the more obvious things to guard against, since many times the esters being treated are unsaturated. Both isomerization and polymerization, however, may occur under the alkaline conditions obtaining, the unconjugated polyethenoid acids becoming conjugated during treatment. This especially appears to be true in the case of highly unsaturated compounds. However, by properly controlling reactants and conditions, saponification remains a very flexible and useful industrial and laboratory operation possible of wide application and at low cost.

S. B. ELLIOTT

Cross-references: *Esters, Metallic Soaps, Glycerides, Fats, Emulsion, Detergents, Soaps.*

SATURATION

The word "saturation" has several meanings. For example, a solution is said to be saturated in respect to a given solute when it contains the greatest quantity of that solute possible to be dissolved in a given quantity of solvent at a given temperature.

The atmosphere is saturated with water vapor and the relative humidity is 100% when the atmosphere contains the maximum quantity of moisture possible at that temperature.

A color is called saturated when it is free from admixture with white.

A magnetic field is said to be saturated when it contains the maximum flux possible under given conditions.

In organic chemistry, saturated compounds are those that contain the maximum quantity of hydrogen or equivalent elements. Such compounds lack double or triple bonds. In saturated organic compounds, carbon atoms are joined to adjacent carbon atoms by a covalent bond consisting of a shared electron pair, each carbon atom contributing one electron of the pair.

Ethane (CH_3—CH_3) is a saturated compound; ethylene (CH_2=CH_2) and acetylene ($CH \equiv CH$) are unsaturated. Thus, chemists often speak of the "degree of unsaturation," as this is indicative of the reactivity with other elements or compounds.

Ethane reacts with bromine by substitution, never by addition.

$$(C_2H_6 + Br_2 \rightarrow C_2H_5Br + HBr);$$

but both ethylene and acetylene react with bromine by addition:

$$C_2H_4 + Br_2 \rightarrow C_2H_4Br_2,$$
$$C_2H_2 + 2Br_2 \rightarrow C_2H_2Br_4$$

The general formula for the saturated hydrocarbons of the paraffin (or alkane) series is C_nH_{2n+2}.

Saturated alcohols, aldehydes, esters, ethers, amines, ketones, and carboxylic acids are characterized by lack of double bonds in their chains of carbon atoms. The following are examples of saturated compounds because they possess no double bonds, and each carbon atom is attached to its neighbor by a single bond (shared electron pair): C_2H_5OH, ethanol; CH_3CHO, acetaldehyde; $CH_3COOC_2H_5$, ethyl acetate; $(CH_3)_2O$, dimethyl ether; $C_2H_5NH_2$, ethyl amine; CH_3COOH, acetic acid; $(C_2H_5)_2CO$, diethyl ketone.

ELBERT C. WEAVER

Cross-references: *Paraffins; Bonding.*

SCANDIUM AND COMPOUNDS

Scandium, element number 21 in the Periodic Table, was discovered in 1876 by Lars Nilson in Sweden and was named after an area in the lower peninsula of that country. The existence of this element had been predicted earlier by Mendeleev in his classical studies of the Periodic Table, and the accuracy of his predictions still stand as a triumph of science.[1] Scandium occurs as the single isotope 45, which results in an atomic weight of 44.9559.

Scandium occurs in very low concentrations in silicate rocks in the earths crust and is about as abundant as beryllium and the rare earths (5 ppm), although one mineral, thortveitite, contains 30 to 40% scandium oxide. Thortveitite occurs in very small amounts as protruding or redisual prismatic crystals from the weathering of pegmatite dykes such as the black Norwegian uranite, and because of its light tan color and prismatic needle form, the thortveitite can be recognized and recovered. Because of the scarcity of this mineral and the lack of demand for scandium, it is estimated that less than 50 pounds of this mineral was obtained from quarries in Norway in the 50 years following its discovery in 1911; some thortveitite has been found in Madagascar. Large-scale uranium ore processing operations in the United States have included a solvent-extraction procedure, and the scandium present in the amount of only a few ppm follows the uranium into the organic layer, but does not strip out with the uranium into the hydrochloric acid wash; a hydrofluoric acid strip of the organic phase picks up the scandium and is the basis of this separation and recovery method that has made a considerable quantity of scandium oxide available on the market.[2] Interestingly, although scandium is in Group IIIB of the Periodic Table, it does not occur in ores with the other members of this group (yttrium and the rare-earth elements). This is due to the great difference in ion size between scandium on the one hand and yttrium and the rare earths on the other hand (see **Yttrium**), which finds scandium more associated with titanium and zirconium; Goldschmidt has discussed the interesting geochemistry of scandium in some detail.[3]

In the metallic state, scandium has three conduction electrons ($4s^2$, $3d^1$), and loses these when dissolved to give colorless aqueous solutions, with scandium in an oxidation state of +3. It forms an insoluble hydroxide, fluoride, phosphate, oxalate and other salts, and in most processing operations, the oxalate is ignited to give the oxide, Sc_2O_3, a white refractory material that is easily dissolved in mineral acids if it has not been heated to tempera-

TABLE 1. PHYSICAL PROPERTIES OF SCANDIUM

Atomic number	21
Atomic weight	44.9559 (100% isotope 45)
Melting point	1539°C (2802°F)
Boiling point	2727°C (4941°F)
Density	2.99 g/cc
Crystal structure, to 1335°C	Hexagonal close-packed, a = 3.308 ± 0.001 Å
	c = 5.267 ± 0.003 Å
	c/a = 1.59
above 1335°C	body-centered cubic
Atomic volume	15.0 cc/mole
Metallic radius	1.64 Å
Heat capacity 25°C	6.01 cal/mole/°C
Heat of fusion	3.85 kcal/mole
Heat of vaporization at 1630°C	87.6 ± 0.4
Linear coefficient of expansion 0–900°C	12×10^{-6}/deg
Electrical resistivity (room temperature)	66×10^{-6} ohm-cm
Thermal neutron absorption cross section	24.0 ± 1.0 barns/atom

tures above 1000°C. Scandium may be determined by gravimetric methods, although an EDTA titrimetric procedure is more convenient for many purposes.[4]

Metallic scandium may be prepared by first treating the oxide with anhydrous hydrogen fluoride to form ScF_3. This salt is reduced with calcium metal in a tantalum crucible at 1600°C in a purified argon atmosphere. The scandium is obtained as a solid ingot (m.p. 1539°C) which has a very slight yellowish color compared to other metals, and contains 3% or more tantalum present as pure tantalum dendrites. Careful sublimation in a high vacuum at 1500°C produces good quality metal. Although the metal is very reactive toward oxygen when heated, it tarnishes slowly in room air; it dissolves easily in acids with evolution of hydrogen.

The relatively low density of scandium metal (2.99 g/cc; Mg = 1.74; Al = 2.70; Ti = 4.5) coupled with its high melting point has aroused interest in it as a structural material for outer-space applications. However, the lack of reasonable quantities of this material at competitive prices has precluded development of this interest. The relative scarcity of scandium has also resulted in a comparative dearth of information on fabrication techniques for it, although it is known that it may be rolled, swaged and welded by slight modification of procedures that serve for other active metals.

There has been sufficient amounts of scandium metal available to make possible a number of studies of its alloying behavior with other metals. It is apparent that scandium is similar to yttrium and the rare earth metals in many alloying tendencies, with a number of isostructural intermetallic compounds existing in corresponding alloy systems. However, there are differences between scandium, and yttrium and the rare earth metals that provide interesting material for assisting in clarifying the role of atomic size, electronic structure and electronegativity as factors in determining alloying behavior. Although scandium forms a complete solid solution with titanium and an extensive solid solution with magnesium, no applications have developed from this information. Other chemical uses of this element have not been found so it has no current commercial use of any significance. Its interesting electronic structure (one "d" electron in

the metallic state) may find interest in this element developing in the future.

No extensive toxicity studies have been made of this element, but what information is on hand suggests that scandium is not a hazardous material. Properties of scandium are given in Table 1.

ADRIAN H. DAANE

References

1. Weeks, M. E., "The Discovery of The Elements," Easton, Pa., The Journal of Chemical Education, 1964.
2. Lash, L. D. and Ross, J. R., *J. Metal,* **13,** 555 (1961).
3. Goldschmidt, V. M., "Geochemistry," Oxford, Clarendon Press, 1954.
4. Fritz, J. S., and Pietrzyk, D. J., *Anal. Chem.,* **31,** 1157 (1959).
5. Wakefield, G. F., Spedding, F. H., and Daane, A. H., *Trans. AIME,* **218,** 608 (1960).
6. Daane, A. H., "Scandium," pp. 620–626, in "Encyclopedia of the Chemical Elements," C. A. Hampel, Editor, Reinhold Publishing Corp., New York, 1968.

SEEDS

The seed is the most advanced type of reproductive development in plants. Functionally it is a living plant whose development is arrested. The seed continues the species and increases its possibilities of multiplication and distribution. It characterizes seed plants (Spermatophytes) which include flowering plants (Angiosperms) and cone-bearing plants and relatives (Gymnosperms). The seed consists of an embryonic plant (embryo), a quantity of stored food either in the embryo or surrounding it, and the enclosing coats. The embryo usually consists of an axis on which a root (radicle) is developed at one extremity and a shoot (plumule) at the other. Plumule development ranges from a bud to a shoot bearing distinct leaves. Cotyledons (often designated "seed leaves") are also borne laterally on the embryo axis near its apex. They differ in number, nature, and function among species. In the flowering plants, monocotyledons (lilies, grasses) have one cotyledon, dicotyledons (broad-leafed plants) have two, while most gymnosperms have more than two, often in a whorl.

The role of the cotyledon is related to food supplies for the embryo. All seeds have a temporary or permanent food storage tissue, the endosperm, although the "endosperm" of gymnosperms differs in origin from that of angiosperms. In certain seeds much of the endosperm persists and constitutes the primary site of food storage in the mature seed. Such seeds, designated albuminous, include such diverse types as lily, asparagus, and cereals of the monocotyledons, and carrot, ash, and castorbean of the dicotyledons. In other seeds the food reserves in the endosperm are digested and translocated into the embryo prior to seed ripening, usually into the cotyledons, which may become enlarged and constitute the bulk of the seed. In this type, designated exalbuminous seed, the reserve foods are actually in the embryo, and the seed consists of embryo and seed coats. Examples are pea, lima bean, peanut, and squash. In certain cases the cotyledons are brought above the soil, develop chlorophyll, and carry on photosynthesis. The amount of food synthesized, however, is usually insignificant. In grasses the single cotyledon, the scutellum, is highly modified; it serves as a food-absorbing and translocating organ.

The conditions necessary for germination include: (1) favorable moisture supply, (2) favorable oxygen tension, (3) suitable temperature, and (4) for certain species, a specific temperature or light requirement. Water must be absorbed to soften the seed coats, increase their permeability to gases, hydrate the tissues, activate the enzymes, facilitate translocation of the digested (soluble) foods to points of growth in the embryo, and bring about a swelling of the seed which results in the rupture of the seed coats. The entrance of oxygen in larger quantities through the moist seed coats is accompanied by a marked increase in the respiratory rate. Respiration is the process by which energy of foods is made available to the organism. Most of the metabolic processes occurring in the germinating seed require an expenditure of energy. Seeds of most species will germinate in a less-than-atmospheric tension. Seeds require a suitable temperature for germination so that respiration and assimilation can proceed at an accelerated rate. Species vary in their temperature requirements but usually do not germinate above 50°C nor below 5°C and their optimum temperature is about 30°C.

Dormancy in seeds helps plants survive by preventing germination under conditions unfavorable for seedling survival. In certain cases it appears to involve a chemical inhibitor; with abscisic acid likely the most important. In other instances a temperature requirement exists, involving an exposure of 2 to 5°C (or lower) for 4 to 6 weeks—a process called stratification when artificially applied. Other seeds have a light requirement for germination. Studies of their spectral relations revealed the 'red/far-red' light effects. Red light (660 nm) converts the pigment involved, phytochrome, to its active form (P_{fr}), then such seeds germinate. When far-red light (730 nm) converts the phytochrome to its inactive (P_r) form, germination is inhibited. In sunlight more P_{fr} is present than P_r, hence light-requiring seeds germinate. Little evidence exists on how phytochrome is involved in seed germination although leads are developing. Light-sensitive seeds (many are weeds) buried in the soil remain dormant, even though imbibed, until a disturbance, like plowing, exposes them to light. That a smaller group of plants, less studied and understood, exists in which light inhibits seed germination, indicates the complexities. Both low temperature and light stimuli are probably related to a mechanism that either depletes the inhibitory substances or antagonizes them by producing stimulatory substances (promotive hormones?). Successful use of gibberellins to overcome dormancy in cold- or light-requiring seeds support such views. Other growth regulators (e.g., auxins, cytokinins) are probably concerned in seed germination; their roles are now being elucidated. The emerging concept is that in conjunction with gibberellins they act in a balanced interaction.

The nature of the food reserves stored in seeds varies with species: carbohydrates, fats, and proteins occurring in varying proportions. Most seeds can be divided into two groups, those whose storage material is carbohydrate (up to 80% carbohydrate) and those whose principal reserve is fat, i.e., triglycerides (up to 60% fat). The latter group is by far the larger, although among crop plants its predominance is less. Starch is an important reserve in certain economic plants (soybean is an important exception), the endosperm of cereals (wheat, corn) being especially rich in this carbohydrate. Other carbohydrates found in seeds are hemicelluloses (date palm, asparagus) and sugars, especially sucrose (fresh sweet corn). Fats occur in large quantities in the seeds of the castorbean, flax, cotton, soybean, and coconut plants. Proteins form a part of the food reserves of all seeds but are especially abundant in legumes.

More is known about the composition of seed fats than of the lipids (fat-like substances) of other plant parts. While there is greater diversity relative to the constituent fatty acids of seed fats than among the fatty substances of fruits, and particularly of leaves, nevertheless there is some conformity of composition along phylogenetic lines in that certain groupings of the principal fatty acids are found either in individual plant families or groups of families. The fats of the seed and fruit of the same species may be quite similar or they may differ widely. The fatty materials of leaves are primarily ionic lipids and glycolipids.

One of the characteristics of proteins in seeds is that while some are metabolically active, such as enzymes, others are largely metabolically inactive and constitute the storage proteins which vary with species. The nitrogen content of seeds usually comprises, in addition to proteins, a certain amount of free amino acids and amides. The amides are commonly glutamin and asparagin, while the free amino acids are largely the same as those forming part of the protein structure. Recent research has indicated that there are numerous free nonprotein amino acids in plants but the level of such amino acids in seeds is usually low. Dicotyledonous seeds usually contain globulins as their principal protein, very low prolamin, with glutelin varying from 0 to 50% of total proteins. Seed storage proteins usually have high nitrogen and proline contents, but low contents of lysine, tryptophan, and methionine. Alkaloids represent additional nitrogenous constituents of seed. Examples in seeds include piperine in *Piper nigrum,* ricinine in castorbeans, hyoscine in *Datura,* and lupinidine (sparteine) in lupine. The alkaloid content of the plant may not be a true indicator of its content in the seed. Often the content in the plant is the higher, but coffee is an exception where the caffeine content of the seed (bean) is much higher.

Proteins of the seeds of cereals have been extensively studied, particularly those of wheat because of its importance in baking. The protein gluten, which can be divided into a number of fractions (glutenin, gliadin, etc.), depending on the arbitrary conditions used for fractionation, gives bread wheat flour its unique ability to form a spongy, elastic, cohesive dough when mixed with water.

After digestion of the food reserves of the seed the soluble products are translocated toward the growing points of the embryo. The soluble carbohydrates are used in germination either for respiration or in the synthesis of cell constituents, particularly wall carbohydrates. In fatty seeds the fats are transformed into carbohydrates and utilized as such during germination. Protein reserves are not used in respiration but are used to synthesize organic nitrogenous compounds needed in the construction of new protoplasm.

JOHN C. FRAZIER

Cross-references: *Phytochemistry, Fats, Carbohydrates, Proteins, Soil Chemistry.*

References

Barton, L. V., Stokes, P., Evanari, M., Lang, A., Brown, R., and Wareing, P. F., in "Encyclopedia of Plant Physiology," **15**(2), 1965.

Galston, A. W., and Davies, P. J., "Control Mechanisms in Plant Development," Prentice-Hall, Englewood Cliffs, N.J., 1970.

Kansas Agricultural Experimental Station, Manhattan Kansas 66506

Koller, D., Mayer, A. M., Poljakoff-Mayber, A., and Klein, S., "Seed Germination," *Ann. Rev. Plant Physiol.,* **13,** 437 (1962).

Wareing, P. F., "Germination and Dormancy," in "Physiology of Plant Growth and Development" (Ed. Wilkins, M. B.), McGraw-Hill, London, 1969.

Wareing, P. F., and Phillips, I. D. J., "The Control of Growth and Differentiation in Plants," Pergamon Press, Oxford, 1970.

SELENIUM AND COMPOUNDS

Selenium (atomic number 34, atomic weight 78.96) was discovered, and given its name in 1817 by John Jacob Berzelius while studying a method for the production of sulfuric acid. Although discovered in 1817, selenium remained a laboratory curiosity for some fifty years. Finally in 1873 Willoughby Smith discovered, quite by accident, that the current resistance of selenium decreased as the intensity of illumination increased, and furthermore, that its resistance increased slightly as the temperature increased above 170°C (338°F). This led to the development of the photoelectric cell, and eventually brought selenium into the public eye.

Selenium rarely occurs in its native state. It is occasionally found in conjunction with native sulfur, and in the form of the metal selenides, such as clausthalite, PbSe; cucairite, $(AgCu)_2Se$; crookesite, $(CuAgTi)_2Se$; naumannite, Ag_2Se, and zorgite, $(ZnCu)_2Se$. Most frequently it is found as an accessory mineral in the ores of lead, copper and nickel, from which it is recovered as a by-product.

The principal sources of selenium in this country are the copper ores mined in Utah, Arizona and New Mexico. Other countries producing selenium as a by-product are Mexico, Canada, Australia, Sweden, Belgium, Japan and the Western zone of Germany.

The anode muds, or slimes, from electrolytic copper refineries provide the source of most of the world's selenium. Basically there are three main methods of recovery.

At present there are six producers of selenium in the United States and Canada with a total output of over two million pounds annually.

Physical Properties. Selenium is a member of the oxygen-sulfur family of elements (Group VIA of Periodic System) and resembles sulfur in presenting a number of allotropic forms. As many as six allotropic forms are sometimes claimed but only three allotropic forms of solid selenium are recognized, some of which are known in more than one condition (see **Allotropes**).

Amorphous selenium begins to soften at 40–50°C (104–122°F) but does not become completely fluid until the true melting point is reached, 217°C (423°F). It becomes elastic at 70°C (158°F).

The red powder, produced by reducing selenious acid, turns black on standing and yields the vitreous form on heating.

Monoclinic selenium, a metalloid resembling glass, is prepared by crystallization from carbon disulfide.

Hexagonal selenium is metallic, a good conductor of electricity, and is the most stable modification. It is slowly formed by heating amorphous selenium to 90–217°C (194–423°F), or monoclinic selenium at 120°C (248°F).

Gaseous selenium is prepared by heating any allotrope to the boiling point, 688°C (1270°F). There is evidence of some molecules of Se_8 being present below 550°C (1022°F). Between 550°C (1022°F) and 900°C (1652°F) the molecules Se_2 and Se_6 predominate, primarily Se_2 between 900°C (1652°F) and 1800°C (3272°F), while above 2000°C (3632°F) the molecules are monatomic.

Colloidal selenium is prepared by the reduction of dilute aqueous solutions of selenium dioxide, either by chemical or electrical means.

Exposure to light rays, and radium and Roentgen rays as well, even for extremely short periods of time, increases the conductivity of granular, crystalline selenium. The conductivity returns to normal in a short time when the illumination ceases. Various theories have been advanced to account for this phenomenon but none to date fully explain the effect.

The ordinary selenium cell, or photoelectric cell, consists primarily of an emitter, a metal surface coated with a thin layer of metallic selenium, and a collector, both enclosed in an evacuated container. It requires an external source of emf, and its response is proportional to the square root of the incident energy. Its resistance being low, its output cannot be readily amplified by vacuum tubes. It is not suited for precision measurements since its performance is variable and depends somewhat upon previous treatment.

The type of selenium photocell commonly used in photographic exposure meters is the photovoltaic type, also called the "barrier layer cell" or photo emf cell. It usually consists of a metallic surface covered with a thin film of selenium, on top of which is a transparent film of another metal, e.g., platinum. This cell generates its own emf. Properly constructed cells are very constant in operation, and when used with the proper external resistance, the response is nearly proportional to the incident energy.

Selenium also exhibits what may be considered unipolar conduction, thus permitting its use as a rectifier of alternating currents. The rectifying unit usually consists of a nickel-plated steel, or aluminum, disk, one face of which is coated with a thin layer of selenium. The selenium layer is in turn coated with a thin layer of a low-melting alloy, e.g., cadmium-bismuth alloy.

Selenium rectifiers have characteristics similar to copper oxide rectifiers, but, usually are smaller, more efficient and have a wider permissible temperature and voltage range.

Chemical Properties. The chemical behavior of selenium resembles that of sulfur and tellurium. Its normal valence is -2 in the hydride, $+4$ in the dioxide and its derivatives, and $+6$ in the selenates.

Selenium combines directly with oxygen, hydrogen, nitrogen, the halides and most of the more common metals.

Selenium and sulfur are miscible in all proportions, forming several series of mixed crystal.

Hydrochloric acid does not attack selenium. The element dissolves in concentrated sulfuric acid, and reacts readily with nitric acid, forming selenium dioxide.

Selenium is said to dissolve in cold, 66% solutions of sodium or potassium hydroxides. The reactions are complex.

Toxicity. Selenium and its compounds are referred to as highly toxic in the literature, the maximum allowable concentration in air being about 0.10 part per million. However, this toxicity is such that with normal commercial procedures large quantities of selenium are being handled in a great variety of ways with no toxic effects upon the workers. Contact with metallic or amorphous selenium has not been reported as a source of skin injury, although selenium salts have given rise to a contact dermatitis.

Selenium and the vapors of selenium, whether taken orally, inhaled as vapors, or adsorbed through the skin, produce gastrointestinal disturbances and have deleterious effects upon the lungs, liver, kidneys, and other organs. In acute stages there is a garlic odor to the breath.

Uses. The applicatons of selenium and its compounds are many and greatly diversified. It finds use in the glass industry as a decolorizer, and in the production of ruby glass. It is used in the manufacture of selenium rectifiers, photoelectric cells, and is added to copper and stainless steel to improve their machinability. Selenium is sometimes used as a vulcanizing agent for rubber. As an acid it is used to etch steel, the dioxide as an oxidizing agent, and the cadmium selenide as a red pigment. Selenium oxychloride is one of the most powerful solvents known. As the element or in compound form, selenium has also found applications in the lubricant, printing and photographic industries.

JOHN R. STONE

Cross-reference: *Tellurium.*

SEMICONDUCTORS

As the name implies, the electrical conductivity of semiconductors is intermediate between that of conductors and nonconductors, or insulators. The ability of any material to conduct electricity requires the presence of mobile charge carriers within the conducting medium. These carriers may be ions or electrons, or both, that are free to move when an electric field is applied. Ionic conduction usually produces gross changes in composition, either in the bulk of the medium or at the electrodes, and is a separate subject in itself. This discussion will be confined to electronic conduction in crystalline solids.

Most metals are good conductors. A one cm cube of a typical metal has an electrical resistance of the order of 5×10^{-6} ohm. The resistance of a similar cube of a typical insulator, like diamond, sulfur, quartz, or mica, is many orders of magnitude larger. At room temperature a one cm cube of sulfur has a resistance of about 10^{17} ohms. Between these two extremes are semiconductors like PbS, SiC, Cu_2O, Si, Ge and Se with resistivities in the range from about 10^{-2} to 10^9 ohm-cm. Many properties of semiconductors are in sharp contrast to those of metals. As the temperature is raised, the resistivity of a semiconductor decreases rapidly, while that of a metal increases relatively slowly. In general the thermoelectric power of a semiconductor is large compared to that of metals. The conductivity of semiconductors is usually quite sensitive to light, being higher when illuminated than when in the dark. Semiconductors exhibit a more pronounced magnetoresistive effect (change of resistivity in a magnetic field) than do metals. The Hall coefficient is usually much larger for semiconductors than metals.

Clearly semiconductors are basically different from conductors; in fact, at sufficiently low temperatures a perfect crystal of most semiconductors would behave as an insulator. Properties peculiar to semiconductors usually occur as a result of crystalline imperfections which include both lattice defects and chemical impurities. Explanation of differences in the properties of metals, semiconductors and insulators has been accomplished only in the past twenty years and is a major triumph of the quantum theory of solids.

In order to describe the properties of a molecule, it is necessary to examine the electronic binding structure between the atoms which comprise the molecule. The properties of a crystalline solid similarly depend on the nature of its electronic binding. One approach to the problem, which is perhaps most familiar to the chemist, involves a consideration of different solids in the light of their degree of covalent binding. The bond in the diatomic hydrogen molecule is a characteristic covalent bond. If a third H atom approaches this molecule, it is repelled because there is no close-lying energy level available for the third electron. In other words, from the point of view of valence, the H_2 molecule is saturated. This is not the case, however, with a typically metallic atom like Li because, though each atom contributes only one electron, there are eight available state of approximately the same energy arising from the $2s$ and $2p$ atomic levels. There are so many levels available that the Li atom can bind itself to almost as many atoms as can find space about it. The electrons do not belong to any one covalent bond but resonate among all of them. This does not result in a condition of saturated valence, because Li may be pictured as bound to its eight nearest neighbors by eight unsaturated covalent bonds. The bonds are unsaturated because each atom contributes one electron which must participate in eight electron-pair bonds. On a time average there is effectively one-fourth of an electron in each bond which could normally hold two electrons. Since the electrons participate in so many bonds, they are highly non-localized, and one would expect them to be mobile, and therefore, to contribute to electrical conductivity.

In contrast to this picture for metals is the case of a typical insulator like diamond. Here each atom contributes four electrons with eight close-lying energy states available for binding. In the diamond lattice each atom is surrounded by four nearest neighbors to form a tetrahedral structure. There are just sufficient electrons to saturate each of the four covalent bonds. As one would expect, the binding energy is much greater than in the case of unsaturated covalent bonds, and an electron is effectively localized in a given bond. The electrons are not mobile and cannot contribute to electrical conductivity.

Electronic binding in an intrinsic (pure) semiconductor is saturated, like that in an insulator, but the binding energy is not as great. An electron can be extracted from a covalent bond more easily. This can occur through the simple expedient of absorption of either heat energy from lattice vibration or incident light of suitable wave length. When an electron is ejected from a covalent bond by absorption of thermal or radiant energy, it is free to move throughout the crystal and contribute to conductivity until some other incomplete bond accepts it.

The bond from which the electron was ejected suffers a deficit or a state of incompletion, and this fact introduces a new concept in the theory of conductivity. The deficient bond readily accepts an electron from a nearby normal bond with the absorption of only a small amount of thermal energy from the lattice. This process can repeat itself, so that there is a random motion of the deficit. For each random jump of the deficit there is a reciprocal electron motion. Thus in semiconductors there are two quite distinct ways in which electrons can move: the electron of a thermally generated electron-deficit pair can move while the deficit remains fixed; another electron can move into the deficit leaving behind itself another deficit, or, what is equivalent, the deficit itself can move.

In an electric field a small uniform drift is superposed on the random motions of both electrons and deficits and conduction takes place by both processes simultaneously. The electron flow which takes place by virtue of deficits is equivalent to a flow of positively charged carriers in the opposite direction. To distinguish the two conduction processes it has proved convenient to regard the deficit flow as a current of positively charged holes. All these results apply equally well to an insulator, the difference being that considerably larger energies are required to produce the electron-hole pair.

In an intrinsic semiconductor there is a hole for each free electron. In an extrinsic semiconductor this is not the case, and the conductivity may be primarily by holes or by electrons. This behavior is brought about by the presence of impurity atoms. An impurity atom which supplies electrons in excess of those required to saturate the covalent binding scheme is called a donor impurity, and the resulting semiconductor is called n-type because the conductivity is primarily by negative electrons. Conversely, an impurity atom which furnishes fewer electrons than are required to complete the binding arrangement is called an acceptor impurity, and the resulting semiconductor is called p-type, because the conduction is due primarily to positive holes. The conductivity of an extrinsic semiconductor is proportional to the concentration of the donor or acceptor atoms. If both acceptor and donor impurities are simultaneously present and uniformly distributed, the conductivity is proportional to the difference of their concentrations. The reason for this is that the hole furnished by an acceptor atom is readily filled by the excess electron of a nearby donor. In so far as conductivity is concerned, this leaves the local binding scheme of the crystal negligibly different from that in a pure or intrinsic semiconductor. Semiconductor impurities are usually present in extremely minute concentrations. As little as 10^{-6} atomic per cent of arsenic will reduce the room temperature resistivity of germanium from 47 ohm-cm to 4 ohm-cm. Such low levels of concentration are beyond the range of ordinary methods of analysis, and the resistivity or Hall coefficient is generally used to measure the impurity concentration.

The energy band theory is an alternate approach to the description of solids. A molecular orbital type solution is determined for an electron in the periodic field of the lattice and in a self-consistent field due to all the other electrons. This point of view is more convenient for considering dynamic processes in the solid.

The differences in the properties of metals and semiconductors can be readily explained. The small increase in metallic resistivity with temperature is due to increased vibration of the atoms which reduces the electron mobility. This effect is also present in semiconductors but is overshadowed by a much increased generation of current carriers. From a dynamic consideration of the Hall effect, it develops that the Hall coefficient is inversely proportional to the current carrier density. Since the carrier density is much smaller in semiconductors, they exhibit a larger Hall coefficient than metals.

The thermoelectric power of a semiconductor is generally larger than that of metals because the carrier density in a semiconductor is a very sensitive function of the temperature. Similarly the photoconductivity of semiconductors arises from the generation of current carriers by incident radiant energy.

Prompted by the success of solid-state theory in elucidating their properties, widespread interest has developed in semiconductors and in a variety of electronic devices based on their properties. These devices include rectifiers, modulators, detectors, photocells, thermistors and transistors. Germanium and silicon are the two elements most suitable for use as semiconductors.

HENRY E. BRIDGERS

Cross-references: *Crystals, Impurities, Germanium, Silicon, Solid State.*

Organic Semiconductors

Semiconduction, in the sense of a low conducductivity which increases with increasing temperature, and photoconduction are common attributes of organic materials containing closed rings of alternating double and single bonds (conjugation, delocalized electrons) and of some other materials of less specific structure, such as proteins and technical polymers, as well. The phenomena are concerned with charge transfer in a bulk sample—preferably a single crystal—whether stimulated thermally, optically or electronically. For such transfer to occur, charge carriers must actually move from molecule to molecule, not just around a closed ring or down a long molecular chain. The explanation of semiconduction in organic materials, therefore, must look beyond the molecule itself. However, this is a problem for which as yet no adequate theoretical treatment exists, since the "slow step" in the conduction process has to be intermolecular carrier transport and the intermolecular distances are generally too large to allow application of the band theory of inorganic semiconduction, without extensive modification and sweeping assumptions. Only in materials for which a single molecular orbital wave function or group of functions can be written for a matter "domain" has the band model and the quantum mechanical approach achieved some success in explaining a few experimental transport properties. However, the intriguing possibility of "tailoring" the electrical properties of the substrate through organic chemistry has spurred considerable research activity in the field. Semiconducting organic solids fall into two major categories: well-defined substances (molecular crystals and crystalline complexes, isotactic and syndiotactic polymers, etc.) and disordered materials (atactic polymers and pyrolytic materials).

Anthracene, probably the most extensively studied organic semiconductor, is an example of a molecular crystal. Some of its transport properties are not unlike those familiar from studies of silicon and germanium. Hole and electron mobilities of the order of 1 cm^2/v-sec can be measured in zone-refined single crystals. These drift-mobility experiments employ photoconductivity, since the crystal is a poor conductor in the dark. There is also experimental evidence that carriers are produced only on the surfaces or at electrodes by excitation waves (or "excitons") generated at the point of photon absorption. Unfortunately, in order to observe the basic electronic properties of "small domain" organic crystals in general, and anthracene in particular, purification techniques as extensive as those applied to inorganic materials must be applied. Zone-refined anthracene single crystals are nearly free from space-charge effects due to carrier trapping and furthermore can show *electronic* (*n*-type) conductvity. Hole-conduction alone and many trapping effects, had been observed in less pure crystals and had contributed to confusion in experimental results and their theoretical interpretation in the past. A model, similar to the one-electron band model classic to inorganic semiconductors, is sometimes applied to the *perfect* aromatic molecular crystal. The free electron is said to spend most of its time in the neighborhood of one molecule or another, and very little time in the interstitial regions; its wave function would be a linear combination of excited molecular orbitals. Because these orbitals are almost completely isolated from one another, the effective mass of the charge carrier would have to be large. Free hole states can be treated in a similar fashion. But with a sufficiently large effective mass in the perfect crystal, a better picture is provided by the "hopping model," in which one assumes that the carrier spends a long time on some molecule; the perturbations arising from the presence of other molecules cause an occasional jump to a neighbor. Either the electron on the negative ion formed or the hole left behind may now be the means of conduction, by hopping to the adjacent molecule under applied field. The two charges are sufficiently far apart that they can act independently of one another.

In passing from anthracene to the charge-transfer complexes (intermolecular addition compounds composed of an electron donor like an aromatic hydrocarbon and an electron acceptor or Lewis acid, like a halogen molecule), we go from one extreme in which the optical evidence for delocalized intermolecular charge is limited or absent to the other in which it is so strong its effects overwhelm the optical properties associated with the private molecular energy levels. The metastable (Perylene)$_2$(I$_2$)$_3$ and (Pyrene)(I$_2$)$_3$ complexes and the unstable bromine-perylene complex show relatively high conductivities (0.01 to 1 mho/cm at room temperature), and they are optically opaque in thicknesses as low as 1000 Å in the near-infrared, visible and near-ultraviolet. The activation energies for conduction of these complexes are fairly low (0.14 to 0.01 ev). Conductivities up to 100 mho/cm are exhibited by other complexes where the electron exchange between donor and acceptor molecules in the complex has the effect of producing many nearly free carriers, even with very small carrier mobilities. The electrical properties of such complexes are often nearly like those of metals in many ways, showing degeneracy effects and small or even negligible activation energies for conductivity. With these substances, the band model appears to provide a fairly good theoretical framework for qualitative interpretations of their electrical properties: a successful analogy may be drawn between the donor-acceptor action of the chemical groups in the molecular complexes and the behavior of impurity atoms in germanium or silicon. The best examples of the high-conductivity complexes are the charge-transfer complex salts of the radical anion of tetra-cyanoquinodimethane,

which have the generalized structure $M^+(TCNQ^-)$ $(TCNQ)$, where TCNQ is tetracyanoquinodimethane and $TCNQ^-$ is the radical anion,

M^+ is a cation, usually organic, such as the triethylammonium ion. In a few cases, compounds lacking neutral TCNQ have been formed: 5,8-hydroxyquinoline with TCNQ gives the anion radical salt, $C_9H_7NO_2H^+TCNQ^-$, whose resistivity (powder) = 14 ohm-cm. However, quinoline with TCNQ gives $C_9H_7NH^+(TCNQ^-)(TCNQ)$, whose resistivity (single crystal) is 0.01 ohm-cm. These complexes have an electron density of one free electron per molecule and a resistivity of the order 100 to 0.01 ohm-cm or less, depending on direction (the electrical resistivity is highly anisotropic along the three principal crystal axes). For comparison, the resistivity of hyperpure silicon is about 100 ohm-cm, graphite about 0.001 ohm-cm, and most organic compounds 10^{10} to 10^{14} ohm-cm. The thermoelectric power of the TCNQ complexes is about -100 microvolts/°C—characteristic of a good metallic conductor. The triethylammonium salt has room temperature conductivity values of 4.0, 0.05, and 0.01 mho/cm in the three principal crystal directions, and these vary exponentially with temperature (semiconductor-, rather than conductor-like behavior) with a low activation energy of 0.14 eV. The thermoelectric power is in the direction of highest conductivity, indicating that electrons are the majority carriers in this material. No Hall voltage has been detected in this salt.

There is no evidence that a band model is applicable to semiconducting polymers and the intrinsic picture results in inconsistencies. Indeed, the electric properties of the polymers and similar organic compounds are not nearly so well established as those of the molecular crystals and charge-transfer complexes, partly because their structures are not so well characterized. Such materials are usually insulating, but appreciable conductivities can be developed by modification through chemical synthesis and/or pyrolysis. Such structural features as chain length or the number of conjugated carbon atoms per molecule, as expected, influence the electrical properties but no clear understanding of the relationshoips involved has yet been reported. Indeed, the actual chemical structures of the pyrolized polymers, in particular, are often not known.

Experimental measurements on organic semiconductors are hampered by the fact that many compounds are available only in powder-like form. It is common to study compressed pellets of these materials, which practice introduces serious intergranular effects. Also, unless these experiments are carried out in oxygen-free conditions, free radicals form; and there is some evidence for a biradical triplet-state overlap of molecular orbitals. In addition, electrode influences perturb the experimental results in unknown ways, particularly on high resistivity specimens. In a few specific cases it has been shown that good agreement between single crystals and compressed pellets can be realized, but in other cases there are violent discrepancies between the two. Loss tangent measurements or other AC resistance techniques oftentimes applied in such situations must be approached with extreme caution, as there is no *a priori* guarantee that the AC losses are related to the DC conductivity. The potential box model used to describe the semiconducting properties of organic substances indicates that under favorable conditions a solid free radical would behave as a semiconductor with very small or zero energy gap, and this has proved to be the case. The role of odd electrons in the conduction process in crystals of substances having only partial radical character is less well understood, but there appears to be a correlation between the semiconductivity and the concentration of free electrons in pyrolyzed polymers. The idea of conduction by electrons without activation has implications in biophysics. The conductivity of dry proteins is rather low to explain the rates of metabolic processes in terms of electron transfer in the classical sense and it has been suggested that a large permeability factor for the barrier to tunneling by the carriers exists where the concentration of polarizable electrons in the substrate is high and a strong field gradient is applied, allowing carriers of equivalent rest-state energies to progress by this means down the gradient. Such substances' semiconduction appears to be relatively insensitive to small amounts of impurities, to involve carriers of extremely low mobilities and to exhibit a proportional increase in thermopower with temperature up to a saturation value.

Among the organic dyes studied, the cationic dyes appear to be generally *n*-type semiconductors and the anionic dyes *p*-type, when such electrical behavior can be detected. It has been found possible to "dope" copper phthalocyanine polymers and the sign of the Hall coefficient is observed to change at an oxygen pressure of 490 mm. Hg, indicating conversion from *n*-type to *p*-type material; the change is reversible. At room temperature, water-vapor and other supplementary additives or dopants, whether purposely introduced either in the preparation of the compound (like an excess or deficiency of copper) or after the compound has been formed (e.g., H_2S, Br_2, or O_3), alter or modify the conductivity type and the thermoelectric power.

Knowledge of organic semiconduction has not yet progressed to the point where applications or devices using organic semiconductors can be fabricated, although rectifying contacts to organic semiconductors are well known. Since the properties of these compounds are likely to depart significantly from those familiar in inorganic semiconductor technology, the implication is that different applications and devices will result, possibly of industrial importance. It may well be, however, that the greatest contribution to human progress derived from studies of organic semiconductors will be in enlarging our understanding of certain important biological processes, such as photosynthesis, initiation of photosensitive carcinogenesis, and processes involved in vision, all of which are strongly concerned with activated electron transfer in organic polymeric media.

Industrial applications could include the use of polymeric semiconductors as anisotropic semiconductive coatings, as photo-voltage sources of photoelectric devices of other kinds, and as semiconductive layers or films to heat surfaces or remove charges from them. One promising application involves their use in electrically applied coatings. The semiconductive organics may also find employment as fuel-cell electrodes and as phosphor modifiers and coatings for color lighting. The possibilities could be almost as great in number as there are organic structures.

NORMAN J. JUSTER

SENSITIZING DYES

Sensitizing dyes are added to photographic silver halide emulsions when it is desired to extend the range of spectral sensitivity of the latter. When untreated with such dyes, photographic emulsions are ordinarily sensitive only to ultraviolet, violet and blue. Such emulsions are, therefore, unsuitable for the many purposes in which sensitivity to the longer wave length regions of the spectrum is required, i.e., green, yellow and red.

It was discovered accidentally by H. W. Vogel of Berlin in 1873 that sensitivity to these longer wave lengths could be imparted by adding certain dyes to the photographic emulsion by a plate-bathing procedure. From the outset, Vogel related the region of sensitivity imparted by one of these dyes to the light absorption of the dye. Thus, he saw that a dye that absorbed green strongly might act as a sensitizer for green. Subsequent work showed that, for dye sensitization to take place, the dye must be adsorbed on the surface of the silver halide particles, or "grains," (which are dispersed throughout the photographic emulsion), and that the band of sensitization due to the dye corresponds in wave length to the absorption of the dye-silver halide adsorption complex formed.

The essential function of a sensitizing dye is its ability to transfer to the silver halide particle on which it is adsorbed the energy which it absorbs from incident light. A photographic latent image is thus produced which can be developed in the usual way.

After Vogel's discovery, dye sensitizers gradually found their place in practical photography. Members of a number of dye classes were found to show some action as photographic sensitizers (though many dyes do not). The dyes that are the most effective for the purpose are now drawn almost entirely from the group of Polymethine Dyes of which the Cyanine Dyes are a sub-group. But even here, though many cyanines sensitize strongly, many others have little or no effect. For best results, sensitizing dyes should be extremely pure. As a general rule, also, they are most effective when used in relatively low concentrations.

By a curious coincidence, 1856, the year of W. H. Perkin's discovery of the first commercially produced aniline dye mauveine, also saw the isolation of the first cyanine dye. This discovery was made by C. H. Greville Williams, who had also been a student of A. W. Hofmann in London. An extremely careful worker, Williams was occupied in one series of papers with the characterization of quinoline obtained by the distillation with caustic alkali of alkaloids such as cinchonine. Quinoline obtained in this way, after quaternating with alkyl iodides, gave rise to beautiful reddish-blue dyes on treatment with silver oxide. Actually, the quinoline contained 4-methyl-quinoline (or lepidine) as an impurity, and it was later found that it was the condensation together of salts of quinoline and of lepidine that formed the dyes.

In any event, the yields of dye were low, and although the colors were beautiful the dyes were extremely acid-sensitive and far too fugitive for the dyeing of textiles. However, the dye from the amyl iodide salt was manufactured in small quantities by the Paris firm of Menier who gave it the name of "Cyanine" (from Greek, kyanos, blue-green) which has been retained as the name for the whole related family of dyes, whether blue or not. It was 17 years before Vogel discovered that Williams' Cyanine had marked photographic sensitizing properties (though the results would be considered poor by modern standards) and almost another quarter century before more effective relatives of Cyanine were used in photography in appreciable quantities.

In time it was established that in Williams' Cyanine two quinoline rings, each with N-amyl, were linked together through a —CH=group at the 4-position of each nucleus. The two rings shared a single positive (cationic) charge between them, the complex cation being electrically neutralized by an anion, usually iodide. The dye could thus be called a 4,4′-cyanine iodide.

Around 1900 attention became focused on quinoline dyes, isomeric with Williams' dye, but having one of the quinoline rings linked through the 2-position. These 2,4′- or isocyanines comprised the most important sensitizing dye class for the period from roughly 1900 to 1920; it was then found that better sensitizers could be obtained from heterocyclic nuclei other than quinoline, such as benzoxazole, benzothiazole, benzoselenazole and β-naphthothiozole.

Perhaps more important than any other was the development, initiated by Mills and Hamer, by which the single =CH—bridge connecting the rings in the earliest cyanines was expanded to =CH (—CH=CH)$_n$-, thus giving rise to "vinylogous" series of cyanines. With such chains, cyanines of the benzothiazole series are all useful photographic sensitizers for the values $n = 0, 1, 2, 3, 4$ and 5. The absorption maxima in methanol of these dyes lie at 423, 560, 650, 760, 870, and 980 mμ for dyes with $n = 0, 1, 2, 3, 4, 5$, respectively. In general, for symmetrical vinylogous series of dyes (i.e., cyanines with two like nuclei), the absorption maximum is shifted about 100 mμ to longer wave lengths for each additional vinylene group (—CH=CH) in the bridge. The regions of sensitivity imparted by these dyes correspond approximately to the absorption of the dyes in methanol, though shifted somewhat toward longer wave lengths. Thus, the symmetrical benzothiazole dye with $n = 1$ is a strong sensitizer for the green, with its maximum effect at about 600 mμ, while the dye with $n = 2$ is a strong red sensitizer with its maximum effect at about 700 mμ. The dyes with very long chain lengths ($n = 3, 4$ and 5) absorb in the infrared and sensitize in that region.

In the 1930's the dyes available as photographic sensitizers were augmented by a large group known as Merocyanines. Such dyes are un-ionized: in them a basic (electron-repelling) nucleus is linked to an acidic (electron-attracting) nucleus.

In aqueous solution, many sensitizing dyes tend to form "J-aggregates" in which the dye molecules are stacked in a regular arrangement reminiscent of a stack of playing cards. A new absorption band, often very sharp and located at longer wave lengths than the molecular absorption, is usually associated with the J-aggregate. Similar aggregates are also formed by such dyes on the surface of the silver halide particles of the photographic emulsion, and a strongly selective band of sensitization related to the J-aggregate absorption band may appear. Such sensitizations are often capable of being intensified by the addition of a "supersensitizer" which may be a colorless substance or a second dye.

It seems possible that the energy of light absorbed by a dye molecule forming part of a J-aggregate is often retained in the aggregate long enough to be dissipated as heat or as fluorescence. Such a dye layer will interact inefficiently with the silver halide, and the dye will not sensitize well. However, energy transfer to the silver halide may be favored at sites where the exciton motion is disturbed by the presence of a foreign molecule, and it is thought that the supersensitizing substance acts as such a perturbation after becoming part of a mixed aggregate. Energy transfer to silver halide is thus facilitated, with an increase in sensitizing power.

The molecules of most sensitizing dyes are planar: at least, that portion is planar which contains the essential chromophoric system. Even a slight distortion of this system out of planarity may have a marked depressant effect on the sensitizing action of the dye. However, a number of useful sensitizers contain cyclic substituents (e.g., phenyl, naphthyl) held in a plane normal to that occupied by the main chromophoric system of the dye.

Mention was made earlier that Williams' Cyanine was extremely fugitive to light. This also characterizes most of the cyanine and related dyes prepared in more recent times. However, the poor light-fastness of these dyes is not a bar to their use

in photography as sensitizers, for they are not exposed to light except for the (usually) brief period of exposure.

The term Orthochromatic is applied to photographic emulsions made sensitive to the green by means of dyes, while emulsions sensitized to the entire visible spectrum are called Panchromatic. Sensitization in the near infrared to 1200 mμ and beyond can be attained by the use of infrared sensitizers. Sensitivity to relatively narrow bands of the spectrum is a requirement in various processes of color photography. Here it is also necessary that sensitizers be nondiffusing from one layer to another in multicoated layers. These requirements have been met satisfactorily.

While the cyanine and related dyes have been developed primarily because of their role in photography, a number have been tested for possible chemotherapeutic action. As a result, one cyanine has been found to be effective against pinworm in human beings and is on the market under the name "Povan" or "Vanquin." Another, "Indocyanine Green" which absorbs in the infrared, is used for injection into the blood stream to give information regarding the circulation, such as may be necessary in the preoperative diagnosis of certain structural defects of the heart.

LESLIE G. S. BROOKER

Cross-reference: *Dyes.*

References

1. Hamer, F. M., "Cyanine Dyes and Related Compounds," Interscience, New York, 1964.
2. "Mees' Theory of the Photographic Process," 3rd Ed. Ed. by T. H. James, Macmillan, New York, 1966. Chapters 11 and 12.
3. Mees, C. E. K., "From Dry Plates to Ektachrome Film," Ziff-David Publishing Company, 1961.

SEQUESTERING AGENTS

Sequestering agents and metal-ion deactivators are chelating agents which form very stable, soluble complexes with metal ions. They are widely used in industry, chiefly to obviate undesirable properties of metal ions without the necessity of precipitating or removing these ions from solution. The terms "sequestering agent" and "metal-ion deactivator" are not strictly synonymous, but their meanings overlap to a considerable extent and they are often used interchangeably. Sequestering agents were first used to soften water by forming soluble complexes with calcium, magnesium, and ferrous ions, thus preventing the formation of the insoluble soaps and boiler scales which accompany the use of hard water. Other, similar uses were later developed. "Metal ion deactivation," on the other hand, implies a decrease in the catalytic activity of a metal ion. However, as many chelating agents can be used for both purposes, they are sold as "sequestering agents" as well as "metal-ion deactivators." For example, the polyphosphates and the ethylenediaminetetraacetates are used both in water softening and as "preservatives" for concentrated solutions of hydrogen peroxide. In the latter case, they destroy cupric and ferric ions, which catalyze the decomposition of peroxide.

The use of chelating materials as sequestering agents and metal-ion deactivators has increased rapidly in recent years. They have found application, for example, as masking agents in analytical chemistry and in electroplating, as decontaminating agents for radioactive deposits, as rust and scale removers, in tying up traces of metals which cause discoloration in dyes, in removing iron stains from fabrics, and in other applications where it is necessary to keep heavy metal ions in solution at high pH. They are employed as plant growth regulators, as antioxidants in drugs and in the prevention of allergies due to metal ions.

Since the oxidation and deterioration of many important organic materials are catalyzed by the ions of heavy metals, metal-ion deactivation is important in nonaqueous, as well as in aqueous, media. For this purpose, the chelating agent and the complexes which it forms must be soluble in the nonaqueous medium. Metal-ion deactivators find application as antioxidants in rubber, antigelling agents in fuel oil, antigumming agents in gasoline, and to retard the development of rancidity in edible fats and oils.

JOHN C. BAILAR, JR.

Cross-references: *Chelation; Coordination Compounds.*

SILICATES

Silica and its related products, the principal bases of our mineral building materials, make up about 60 percent of the earth's crust. The SiO_4 tetrahedron is the universal and primary structural unit. Because of the relative insolubility of most heavy metal silicates, the soluble silicates contain only traces of other elements, but the mineral silicates may contain any or all of about 42 metals and nonmetals besides the rare earths.

Silica

Silica has three important polymorphic crystalline modifications: high and low *quartz;* upper high, lower high, and low *tridymite;* high and low *cristobalite,* and also more or less hydrated amorphous forms.

Quartz, the most common form of silica and the most common mineral as well, is very generally available as *sand* and *quartzite.* Crystals of a size and purity satisfactory for optical and resonator plate use are very uncommon and are found primarily in Brazil. Large crystals for such uses are now grown from alkaline solutions at high temperatures and pressures.

Quartz is tough and resists the attack of all acids except H_2F_2 or NH_4HF_2 which form SiF_4. The rate of solution even in alkali is very small at room temperature, but the true solubility in caustic solutions is quite high. In water at room temperature, the solubility of quartz is probably about 0.001%. At 368°C the quartz solubility is 0.085% and in superheated steam it is as much as 0.77% at 600°C and 2000 bars.

The structures of *quartz, tridymite,* and *cristobalite* are so different that the transition from one to the other is slow compared to the transition from the higher to the lower forms. Tridymite has a more open hexagonal structure than quartz. The Si—O distance is 1.52 and the nearest O—O distance is 2.57Å. The high-low inversion temperature of cristobalite varies, apparently dependent on its history. Cristobalite and tridymite are found only in volcanic rock and have no industrial use.

Amorphous silica gels are manufactured commercially chiefly from soluble silicates. *Opal* is a natural gel. No definite silicic acids are known, but many salts of hypothetical metasilicic, orthosilicic, disilicic, and polysilicic acids exist. The water in silica gels may be replaced with organic liquids such as alcohol and ether to form organogels. Organoesters may also be formed.

Pyrogenic and *precipitated hydrated silica* pigments with an ultimate particle size below 25 mu are sold in tonnage quantities. A wide range of surface area and hydration is available.

Amorphous silica appears to reach an equilibrium solubility of about 0.012 between a pH of 6 and 8.5 and 0.010% at a pH between 2.1 and 2.7. Above 11.9 the total SiO_2, whether rapidly or slowly reactive, increases fairly rapidly and is about 0.5% at pH 11. The latter pH corresponds to a $SiO_2:Na_2O$ ratio of about 3. Concentrated solutions of such a ratio can be prepared. They are very viscous and the SiO_2 is probably more highly polymerized.

Vitreous silica or fused *silica glass* is important because of its low coefficient of expansion and resistance to other chemicals except H_2F_2. It is permeable to H_2 and He_2 at high temperatures.

Silicate minerals, about 800 in number, are classified according to the possible silicon-oxygen network as *Nesosilicates, Inosilicates, Phyllosilicates, Tectosilicates, Sorosilicates,* and *Cyclosilicates.*

Alkali Silicates. The soluble silicates are the alkali metal silicates, once widely called "waterglass" and the silicates of organic alkali cations. About 1.5 million tons (on a liquid basis) of the soluble silicates are sold annually in the form of glasses, liquids or crystals. Sodium trisilicate or "neutral" glass may refer to either solid or powdered glass with a ratio of $Na_2O:3SiO_2$. "Alkaline" glass has a ratio of about $Na_2O:2SiO_2$. Fifty or more uses take sodium silicates in carload quantities and a few take the more costly potassium silicates.

The crystalline soluble silicates are all synthetic. Their names refer only to the simplest analytical composition and a hypothetical series of silicic acids. Commercial anhydrous *sodium orthosilicate* and *sesquisilicate* (a sesquisodium silicate) are mixtures of crystals of NaOH and Na_2SiO_3. The hydrated orthosilicate is also a mixture of crystals, but the *sesquisilicate* and *metasilicate pentahydrate* are definite crystal forms.

The more siliceous compounds in all of these systems tend to dissolve only slowly in water. Sodium disilicate is very soluble but the rate of solution is very low even in hot water. The series of high ratio compounds from sodium tetrasilicate to a ratio of 1 Na_2O to 13 SiO_2 are nearly insoluble without added alkali in the solution. The lithium compounds appear to be only slightly soluble, but it is possible to mix neutral lithium salts with sodium silicate to some extent and to dissolve hydrated SiO_2 in cold lithium hydroxide solutions. Strong organic alkalies also dissolve reactive silica and in some cases crystalline products will precipitate. Double alkali salts have been reported in the mixed alkali systems.

The more alkaline crystalline sodium silicates are used industrially, primarily as detergents and builders in detergent compositions. They also protect against the corrosion of metals, alloys, enamels, and glazes. The glasses with weight ratios of about $Na_2O:3.2SiO_2$ and $Na_2O:2SiO_2$ are important articles of commerce. They are dissolved with steam and water in special pressure dissolvers, but if finely ground they may be partially dissolved in boiling water. Neither the glasses nor solutions crystallize readily under usual conditions. (See **Glass**)

The liquids are opalescent unless specially treated and filtered, but even the settled commercial opalescent liquids contain only a few tenths of a per cent of suspended solids since the glass is formed by melting purified glass sand with soda ash of high purity: $Na_2CO_3 + xSiO_2 = Na_2O \cdot xSiO_2 + CO_2$ and the heavy metal and alkaline earth silicates are nearly insoluble in alkaline silicate solutions. Solutions up to $Na_2O:2.7SiO_2$ may be formed by dissolving quartz in NaOH solution but for higher ratios, amorphous forms of silica must be used or part of the alkali removed—by base exchange, for instance.

A variety of aqueous solutions of sodium and potassium glasses ranging in ratio by weight from $1.5SiO_2$ to $4SiO_2$ per unit Na_2O and $2.1SiO_2$ to $2.5SiO_2$ (3.3–4.15 molecular ratio) per unit K_2O are available commercially. They are differentiated by weight ratio of silica to alkali and by density in degrees Baumé. The compositions are controlled primarily by determination of alkali and of viscosity, which is quite sensitive in the heavy solutions marketed at 0.5–600,000 poises at 20°C. After a certain critical concentration is reached, viscosity rises very rapidly. These viscous solutions, which lose water either to the air or to absorbent substrata, have a high concentration of colloidal silica and are widely used as adhesives for fibrous materials as well as for thin metal sheets and ceramics. The combination of alkali ions, silicate ions (crystalloidal silica) and colloidal silica accounts for the detergent qualities which permit them to build detergents. They readily provide and maintain a pH of about 10.5 or higher. The silica reacts rapidly with heavy metal oxides and ions to form a protective surface for most metals and alloys, preventing corrosion by the very efficient, modern cleaners or in other aqueous systems. A combination of setting by reaction and evaporation accounts for their use in cements and molded products, sealers, etc. for leaks, concrete and coatings.

Solutions of alkali metal-free quaternary ammonium silicates also contain both colloidal and crystalloidal silica which will remain after the organic cation is destroyed by heating. They are used commercially as binders for inorganic, incombustible coatings.

A continuous gel forms by condensation of the hydrated colloidal particles when the alkali is neutralized. In dilute solutions, "activated silica" is produced for aiding the coagulation of water. In more concentrated solutions, solid gels form which absorb water or gases or may be purified and treated with catalytic elements primarily for cracking and polymerizing petroleum products. By more complex procedures a fine pigment silica can be formed. Sols purified by removing the alkali salts by freezing, base exchange, etc. may be concentrated above 50% SiO_2 and are becoming of increasing commercial importance for fillers and film formation.

Reaction with metal salts similarly yields pigments and gelatinous precipitates. These are not crystalline but amorphous and are usually mixtures of metal and silicon oxide hydrates. A continuous gel of sodium aluminate with sodium silicate is the synthetic zeolite ($Na_2O:Al_2O_3:5SiO_2:xH_2O$) for base exchange at higher capacity than the natural zeolite such as "green-sand." Similar compositions crystallized by heat treatment form "molecular sieves." The familiar "silicate garden" is made up of fungoid growths from water-soluble crystals covered with a dilute solution of sodium silicate.

A synthetic calcium silicate sometimes having the formula $CaO:3.5SiO_2:1.8H_2O$ has been used as a reinforcing filler for rubber and similar compositions are used as absorbents and decolorizing agents. A magnesium silicate sometimes having the formula $2MgO:3SiO_2:2H_2O$ is a decolorizing agent and, with higher water content, has been prescribed for the treatment of stomach ulcers. Lead and zinc silicates are pigments and phosphors, respectively. The potassium silicates are used primarily as binders for phosphors in the manufacture of television tube screens and in coatings for welding rods or in special cements.

Strong solutions of neutral salts or ammonia cause the separation of a liquid layer of the colloidal silicate by coacervation. Water-miscible organic liquids such as alcohol or acetone have a similar effect. Certain reactive groups of proteins appear to add to or condense with the silicate molecule.

When the alkali is removed, addition or condensation reactions between the silica gel or sol colloid with organic esters, alcohols, amides, etc. become possible.

As the need for conservation of resources becomes more evident, the place of soluble silicates in a simple, complete, natural cycle from salt and silica makes imperative their use as replacements for phosphate detergent builders needed for fertilizers, for starch adhesives needed for food, and in the form of fine silica for carbon black rubber reinforcement agents from scarce petroleum supplies needed for energy.

Analysis. The analysis of silica and the silicates is quite simple, but requires careful attention to details. The solids other than the crystalline detergents are first brought into solution, if necessary, by fusion with sodium carbonate and leaching with steam and water. Silica is separated by double precipitation with acid and subsequent evaporation, ignition and final checking by evaporation with H_2F_2. Metallic cations are found in the filtrate. Alkalies and nonmetals, etc. are usually determined in separate samples, but in the soluble silicates alkali is determined by titration before evaporation for the precipitation of SiO_2.

JOHN H. WILLS

Cross-references: *Gels, Glass, Adhesives, Crystals, Silicone Resins, Clays.*

SILICON AND COMPOUNDS

Silicon (Si) is the second most abundant element, making up 25.7% of the earth's crust. (Atomic number, 14; atomic weight, 28.086; oxidation number ± 4; m.p., 1420°C; b.p. = 2355°C; density = 2.33 g per cc at 20°C; hardness = 7 (Mohs scale); outer electron arrangement = $3s^2 \, 3p^2$; covalent radius, Å = 1.173; electronegativity = 1.8.) It is a nonmetal, the second element in Group IVA of the Periodic Table. It is known only in one dark-colored crystalline form, namely, the octahedral form in which the silicon atoms have the diamond arrangement. The so-called amorphous form consists of minute crystals of this form. Silicon occurs in nature combined with oxygen in various forms of silicon dioxide and with oxygen and metals in the many types of silicates, but is never found uncombined.

Elementary silicon, first prepared by Berzelius in 1823, is prepared commercially by heating silicon dioxide with carbon (coke) in an electric furnace. A highly crystalline product is obtained by crystallization from aluminum or zinc. "Hyper-pure" silicon (99.97%) is prepared by reducing the tetrachloride with zinc. Further purification by zone refining removes phosphorus.

The Czochralski method is commonly employed to grow single crystals of silicon used in semiconductor devices. A single crystal "seed" is dipped into molten silicon held at the melting point and then slowly withdrawn.

Silicon is just below carbon in the Periodic Table. Since they both have two electrons in each of the outer s and p orbitals, silicon resembles carbon in a number of ways. For example, they both have tetrahedral arrangement in their molecules and optical isomerism occurs in silicon compounds. Silicon is less electronegative than carbon, resulting in its acidic properties being very weak. However, dilute solutions of orthosilicic acid, H_4SiO_4, can be made and salts of this acid occur in nature. The low electronegativity also results in its halides being somewhat ionized and reactive.

Silicon differs from carbon in that no silicon compounds containing double or triple bonds have been discovered; also it does not combine with itself to form structures containing more than six silicon atoms joined together. The Si—Si bond is weaker than the C—C bond and the Si—H bond weaker than the C—H bond, as pointed out below. Consequently the hydrides of silicon are few in number and much less stable than the hydrocarbons.

Bond Energies

Bond	Energy (kcal./mole)
Si—Si	42.5
C—C	58.6
Si—C	57.6
Si—O	89.3
C—O	70.0
Si—H	75.1
C—H	87.3

From the bond energy it is evident that the Si—C bond found in silicones is about as strong as the C—C bond. Since these bonds are much alike in character, the Si—C bond accounts in part for the high stability of silicones.

The most important chemical property of silicon is its tendency to combine with oxygen to form tetrahedral structures in which one silicon atom is surrounded by four oxygen atoms. This SiO_4 tetrahedron is the basic structure in silicon dioxide and in the silicates in which the tetrahedra are joined together into complex rings or chains. The Si—O bond distance in the tetrahedron is 1.62Å and the O—O distance is 2.7Å.

Silicon has a maximum coordination number of 6 in the fluosilicates which contain the $SiF_6^=$ ion.

Elementary silicon is rather inert at room temperatures, becoming active at higher temperatures, reacting directly with chlorine at 450°C to form silicon tetrachloride and at still higher temperatures reacts with many metals to form silicides. It also combines directly with many nonmetals as it does with carbon. (See **Carbides**.) Above 1710°C it reacts completely with oxygen to form silicon dioxide. With HF it reacts to form fluosilicic acid (H_2SiF_6) and H_2. Also it reacts with HCl to form silicon tetrachloride and H_2. With strong bases silicon reacts to form silicates and H_2.

Elementary silicon is used in the preparation of silicones, in various alloys of iron, aluminum, copper and manganese, as a deoxidizing agent in the manufacture of steel. Spring steel contains 1.80 to 2.20% Si which toughens and increases the elastic limit. Steels containing 3½ to 4% Si have much better magnetic properties. Tantiron and duriron containing 14 to 15% Si are very resistant to corrosion and are used in acid-resistant pipes and tanks. Silicon adds strength and corrosion resistance to aluminum alloys. Aluminum alloys containing up to 17% Si are used in making intricate castings having low shrinkage and good resistance to salt water corrosion.

Pure elementary silicon when doped with traces of elements such as boron and phosphorus is one of the best semiconductors. For such purposes single crystals of silicon must be used. It is widely used in power rectifiers, transistors, diodes and solar cells. The rectifiers are very important and have almost revolutionized the conversion of a-c to d-c in the electrolytic industries.

Silicon Compounds. *Fluosilicic Acid and the Fluosilicates.* Fluosilicic acid (H_2SiF_6) is an active acid. It is a colorless fuming liquid which is quite soluble in water. The commercial 30% solution is sold in plastic bottles as it attacks glass. Its salts are toxic and the sodium and potassium salts have been used as rat poisons and insecticides for chewing type insects such as crickets, cockroaches, etc. Sodium aluminum fluosilicate is a good mothproofing agent for woolen articles. Barium fluosilicate ($BaSiF_6$) is effective against Japanese beetles and Mexican bean beetles.

Hydrides. Of the silanes, a saturated series having the type formula Si_nH_{2n+2}, only the first six members are known. These compounds can be prepared by treating magnesium silicide (Mg_2Si) with 20% HCl in an atmosphere of H_2. One set of conditions produces the hydrides in the proportions 40% SiH_4, 30% Si_2H_6, 15% Si_3H_8, 10% Si_4H_{10}, and 5% $Si_5H_{12} + Si_6H_{14}$. Monosilane (SiH_4) and disilane (Si_2H_6) can be prepared in 70 to 80% yields by reaction of Mg_2Si with NH_4Br in liquid ammonia in a current of H_2. The silanes ignite spontaneously in air and become progressively less stable to heat with increase in chain length. SiH_4 is stable up to red heat while Si_6H_{14} (hexasilane) is unstable at room temperature. The silanes react violently with halogens to form silicon tetrahalides (SiX_4) and HX.

Silicon carbide (SiC), or "Carborundum," has a diamond-like structure and is nearly as hard as the diamond. It is prepared by heating fine SiO_2 and coke with a little salt and sawdust in an electric furnace. The crystals obtained are greenish blue to black. It is one of the most widely used abrasives for grinding and cutting metals. (See **Carbides**.)

Silicon dioxide or silica [$(SiO_2)_x$], melt. pt. 1710°C, is one of the most important compounds of silicon occurring in nature. It has a solubility (colloidal) in water of 0.0426 g/100 cc. H_2O at 90°C and a solubility of 0.0160 g/100 cc. H_2O at 25°C. This accounts for the formation of petrified wood, geyserite, etc.

All forms of silicon dioxide are composed of SiO_4 tetrahedra in which each oxygen is shared with another tetrahedron, with the result that each crystal is a giant molecule with the average stoichiometry SiO_2. Silica exists in three crystalline forms in nature, namely quartz, tridymite, and cristobalite. Quartz is most common, occurring in granite, chalcedony (agate), amethyst, flint, sand, etc. Several forms containing H_2O such as opal, geyserite and diatomaceous silica are found in nature. Silica gel used as an absorbent and catalyst carrier also contains some water. (See **Silicates**.)

Keatite, another crystalline form of silica, was produced by the first atomic bomb explosion in New Mexico.

Quartz is a piezoelectric substance and is used in stabilizing amplifier tube circuits. It is also used in apparatus for measuring electrical potentials up to several thousand volts and to measure instantaneous high pressures. Large crystals for such purposes are now grown from alkaline solutions at high temperatures and pressures.

Silica is unreactive toward most nonmetals, acids, and metal oxides at ordinary temperatures. It is attacked by fluorine and HF to form SiF_4, and by strong bases to form silicates. An example of the latter is the reaction of SiO_2 with NaOH to form water glass, a substance used as a glue, etc. At elevated temperatures it becomes active and can be reduced by carbon and several active metals; it reacts with metallic oxides and carbonates. An example of the latter is the formation of glass from SiO_2 and various metallic oxides or carbonates. Aluminum borosilicate glass makes good textile fibers.

Fused quartz transmits ultraviolet light and due to a low coefficient of expansion is resistant to breakage under temperature changes.

Diatomaceous silica, skeletal remains of water plants, known as diatoms, is widely used because of its high insulating value, high absorptive capacity, mildly abrasive character, and its finely divided character making it a good raw material for preparing other silicon compounds (see Diatomaceous Earth).

Silica occurs in considerable quantity in the tissues of many higher plants and to a much lesser degree in the tissues of animals. Some plants exclude silica and others such as rice absorb a great deal as it is beneficial for the normal growth of this plant. The silica in plants is usually opaline but quartz has been found in leaves of lantana, strawberries, black raspberries and other plants. Silica can deposit in all parts of a plant and the percentage is often very high. For example sorghum leaf sheath epidermis has been found to contain 11.137% SiO_2 on a dry weight basis.

Silicon Monoxide. SiO is the only divalent silicon compound for which there is any good evidence. If silica is heated with elementary silicon in vacuo at 1,450°C, practically the whole of the silicon sublimes as an oxide. The spectrum of the vapor shows it to be SiO. If the vapor is cooled rapidly, a light brown solid is obtained which gives no x-ray pattern and might be amorphous SiO. If the vapor is cooled slowly or the amorphous solid is heated, the x-ray pattern indicates that the product is silica plus elementary silicon. The amorphous powder burns to silica in air, and is oxidized by water at 400°C and by carbon dioxide at 500°C.

Silicon tetrachloride is a colorless liquid with a boiling point of 57.57°C and a density of 1.483 at 20°C. It is prepared commercially by heating silicon dioxide and carbon in a stream of chlorine. It is quite active chemically as the bonds are 30% ionic. It reacts readily with H_2O to form SiO_2 and HCl, making a dense smoke screen.

It reacts with alcohols to form esters of orthosilicic acid such as $Si(OC_2H_5)_4$ which is called tetraethyl orthosilicate or tetraethoxysilane. The most important reaction of $SiCl_4$ is the one with Grignard Reagents (RMgCl) to form compounds such as alkyl-trichlorosilanes ($RSiCl_3$), dialkyldichlorosilanes (R_2SiCl_2), trialkylchlorosilanes (R_3SiCl), and tetraalkylsilanes R_4Si. The hydrolysis of the chlorosilanes produces the various types of silicones.

Silicon tetrachloride is widely used in preparing organosilicon compounds such as silicones and makes excellent smoke screens for military purposes. This important compound was first prepared by Berzelius in 1823. Hexachlorodisilane Si_2Cl_6 and octachlorotrisilane Si_3Cl_8 are also known. Silicon tetrafluoride, silicon tetrabromide and silicon tetraiodide are known and are like silicon tetrachloride in properties.

Silicon disulfide, $(SiS_2)_x$, consist of infinite chains of SiS_4 tetrahedra with opposite edges in common.

This compound is prepared by heating the elements together and purifying by sublimation. It reacts with water to form hydrogen sulfide and silica. It is said to form thiosilicates when heated with alkaline sulfides.

Silicon nitride, Si_3N_4, is made directly by heating silicon in nitrogen at 1300 to 1380°C. It can also be made by heating tetraaminosilane, $Si(NH_2)_4$, which loses ammonia. This compound is very stable and probably exists as a giant molecule.

Aminosilanes are of interest as they can be used as intermediates for preparing silicones and silizane polymers. The NH_2 groups of dialkyl diaminosilanes react with water to form the corresponding silicols which condense to form silicones. Silizane polymers are formed by simply warming the alkylaminosilanes. Ammonia is lost and structures

of the type

$$-N\overset{H}{\underset{R}{\overset{|}{\underset{|}{Si}}}}-N\overset{R}{\underset{R}{\overset{|}{\underset{|}{Si}}}}-N\overset{H}{\underset{R}{\overset{|}{\underset{|}{Si}}}}-$$

are formed.

Even though $-N\overset{H}{\underset{R}{\overset{|}{\underset{|}{Si}}}}-$ groups react with mois-

ture, recent work indicates some useful polymers of this type may be developed.

Silicon can replace carbon atoms in many compounds and recent discoveries indicate that silicon may be used as a molecular variant in drugs. When a carbon atoms is replaced by a silicon atom in some biologically active compounds, little change in activity is observed. Meprobamate and silameprobamate, for example, show almost identical muscle-relaxing capacity. In some cases the addition of a silicon atom changes the type of biological activity. For example, 3,3-dimethyl-1-butanol carbamate is a slow-acting convulsant; whereas 3,3-dimethyl-3-sila-1-butanol carbamate is a fast-acting muscle relaxant.

F. C. LANNING

Cross-references: *Silicates, Silicone Resins.*

References

1. Wells, A. F., "Structural Inorganic Chemistry," 3rd Ed., pp. 765–816, Oxford, England, Clarendon Press, 1962.
2. Emeleus, H. J., and Sharpe, A. G., "Silicon Hydrides and Their Derivatives," Advances in Inorganic Chemistry, Vol. 11, pp. 249–307, New York and London, Academic Press, 1968.
3. Lanning, F. C., Ponnaiya, B. W. X., and Crumpton, C. F., "The Chemical Nature of Silicon in Plants," *Plant. Physiol.,* **33** (5): 339, 1958.
4. Lanning, F. C., "Silica and Calcium Deposition in the Tissues of Certain Plants," *Advancing Frontiers of Plant Sciences,* **13,** 55, 1965.

SILICONE RESINS

The chemistry of the silicones is based on the hydrides, or silanes, the halides, the esters, and the alkyls or aryls. The silicon oxides are composed of networks of alternate atoms of silicon and oxygen so arranged that each silicon atom is surrounded by four oxygen atoms and each oxygen atom is attached to two independent silicon atoms:

$$-\overset{\overset{\displaystyle |}{O}}{\underset{\underset{\displaystyle |}{O}}{Si}}-O-\overset{\overset{\displaystyle |}{O}}{\underset{\underset{\displaystyle |}{O}}{Si}}-O-\overset{\overset{\displaystyle |}{O}}{\underset{\underset{\displaystyle |}{O}}{Si}}-O-$$

Such a network can be described as a series of spiral silicon-oxygen chains cross-linked with each other by oxygen bonds. If some of the oxygen atoms are replaced with organic substituents, a linear polymer will result:

$$-\overset{\overset{\displaystyle R}{|}}{\underset{\underset{\displaystyle R}{|}}{Si}}-O-\overset{\overset{\displaystyle R}{|}}{\underset{\underset{\displaystyle R}{|}}{Si}}-O-\overset{\overset{\displaystyle R}{|}}{\underset{\underset{\displaystyle R}{|}}{Si}}-O-$$

$$R = CH_3$$

Taking into consideration the stability of structures involving C-Si bonds, it is evident that the basic chain itself must be comparable in its stability to that of silica and the silicate minerals, and if the R substituents contain no carbon-to-carbon bonds, as for example with methyl groups, the combination should have excellent thermal stability and chemical resistance.

Among the most efficient of silicone monomers in early use as building blocks in the preparation of silicone resins are the halogen alkyl or aryl silanes. Compounds of the type represented by the formula R_2SiCl_2 are capable of undergoing hydrolysis to form long-chain polymers of varying consistencies and viscosities with a predetermined number of molecules of R_3SiCl as chain stoppers. If cross links are desired, tri-functional compounds such as $RSiCl_3$ can be used.

Many silicone resins have been prepared through the use of silazine monomers, that is, compounds with amino groups attached directly to silicon. Di-2-pyridyldichlorosilane has also been described as an intermediate in the preparation of oils, emulsifying agents and resins. Fluorinated aromatic rings are found in many silicone resins. The esterification of dimethylbis-(*p*-carboxylatophenyl)-silane with glycol or glycerol yields thermoplastic materials. Trimethyl-*p*-hydroxyphenylsilane is acceptable as a monomer in resin formation when compounded with hexamethylenetetramine. Trichlorosilane is often a constituent of cohydrolysis monomeric mixtures.

Emulsifying agents have been prepared from quaternary ammonium salts with silicon in the cation. There is a large number of alkyd-silicon resins.

Water-soluble metal salts of alkyltrisilanols are efficient in the reduction of surface tension. A silicone putty is made by compounding a benzene-soluble silicone polymer with silica powder and an inorganic filler.

Chlorinated alkyl or aryl groups are often found in polysiloxane resins. One of the chief advantages of this type of halogenated product lies in its reduced tendency to burn. Whereas diphenyloxosilane polymer burns readily in a flame, the introduction of one, two or three chlorines in each benzene ring progressively reduces this tendency. Physically, these resins appear in many different forms, from horny through sticky-resinous to rubber-like, depending on the conditions of combination and composition.

Some of the most important types of silicone resins, useful as coating preparations, are analogous to the alkyds. Glycerol, for instance, is allowed to react with trialkylethoxysilanes, by which reaction one or more hydroxyls of the glycerol are replaced by R_3SiO. Polysiloxane resins with terminal ethoxyl groups can be used. Synthesis of silicone resins with terminal halogen or hydroxyl is also possible, though generally the halogen disappears in hydrolysis of further processing.

Ethyl silicate (tetraethoxysilane) is often used without modification as a water-repellent material for concrete and masonry in general. All, or nearly all, the ethoxyl groups are hydrolyzed by the moisture of the air to form cross-linked water-repellent polymers. The material is applied in desirable thickness, dissolved in some volatile solvent which soon evaporates. Silicone resins which are partially condensed before application, or even fully condensed, can also be used here. In the latter case, hardness is achieved on evaporation of the solvent. Certain silicone resins are useful as hydrophobic agents for the impregnation of paper and fabrics.

The simplest silicone resins are formed by the almost simultaneous hydrolysis and condensation (by dehydration) of various mixtures of methylchlorosilanes. Ice water often suffices for the first step, but advanced condensation to resinous materials of satisfactory thermosetting properties generally comes about on heating. As far as solvents are concerned, water alone has its disadvantages in that the organic materials are so slightly soluble therein. Mixed solvents are commonly used, generally water with such compounds as dioxane, one of the amyl alcohols, dibutyl ether or even an aromatic hydrocarbon. Warm water hydrolysis of di-*t*-butyldiaminosilane forms noncrystallizable liquids or resinous products, and this resinification can be controlled. Among catalysts for these condensations may be found ferric chloride, the hydroxyl ion, triethyl borate, stannic chloride and sulfuric acid.

A water-methylene dichloride mixture is satisfactory as a hydrolyzing agent on groups of compounds such as phenyltrichlorosilane, dimethyldichlorosilane and methyltrichlorosilane. The value of higher-boiling ethers lies in their ability to provide higher-boiling reaction systems.

Ferric chloride is sometimes an important constituent of the hydrolyzing mixture. The formation of gels during polymerization can be controlled. Patents are in existence covering the hydrolysis of chlorosilanes by pouring their solutions onto the surface of a swirling solution of the active electrolyte.

Alkaline hydrolysis of dialkyldialkoxysilanes can sometimes be used for the purpose of preparing silicone resins.

Triethyl orthoborate affects polymerization by dehydration, probably reacting with the water which it abstracts, to form boric acid and alcohol. This principle is commonly used. Antimony pentachloride is used on occasion and also sulfuric acid, but there is always danger that the latter will split off an alkyl or an aryl group. Sulfuric acid also sometimes induces equilibration. This *tendency* on the part of the acid *reacts* sometimes with the opposite effect.

Some polysiloxanes are curable with lead monoxide, with a consequent reduction in both curing time and temperature. High-frequency electrical energy vulcanizes in one case at least. Zirconium naphthenate imparts improved resistance to high temperatures. Barium salts are said to prevent "blooming." Sulfur dichloride is also used. Some resins are solidified by pressure vulcanization, using di-*t*-butyl peroxide. Improvements are to be found in lower condensation temperatures and shorter times of treatment.

Viscosity is often regulated by bubbling air through solutions of polysiloxanes or the liquid material itself. In this manner, alkyl side chains are oxidized and oxygen bridges set up between silicon atoms. Obviously, the greater the number of such cross links, other influences constant, the greater will be the viscosity.

Addition of glycerol, phthalic anhydride and "butylated melamine formaldehyde resins" is sometimes found to improve the thermosetting properties of silicone resins. Methylsilyl triacetate has the same effect in certain cases. Some silicone resins can be advantageously modified by the addition of polyvinyl acetal resins or nitroparaffins.

A solution of cellulose nitrate in butyl acetate, diluted with toluene, can be plasticized with dibutyl phthalate and tetraethoxysilane. After application to glass, the lacquer cannot be stripped from the surface even after soaking in boiling water. It has been stated, however, that tetraethoxysilane sometimes decreases the tensile strength of a lacquer in spite of its effect in increasing adhesive properties. A uniformly lustrous appearance is imparted to compounds containing plasticized ethylcellulose, cellulose fibers and pigment, by applying liquid polymerized alkyl polysiloxanes with an average radical/silicon ratio of between 1.85 and 2.20. Several resinous materials are known which contain sulfur connected to carbon but not directly to silicon.

Inorganic fillers include titanium dioxide, "Celite" and zinc oxide. Lithium or lead salts of acetic acid, stearic acid or phenol are sometimes used as fillers. Silica and alumina are also feasible. Trimethyl-β-hydroxyethylammonium bicarbonate has been used as a curing agent.

There are any number of review articles and patents covering the increase in serviceability of paints and varnishes which are admixed with silicone resins. Products suitable for use as plasticizers, paint vehicles, etc., are sometimes prepared from mixtures which include phthalic anhydride. Silicone paints, in general, show high adhesion and permit greater retention of color and tint. Of special importance here is the absence of color in the resin and the freedom from discoloration during baking and curing. At high temperatures, silicone varnishes show much higher electrical resistance than others. Paints admixed with silicone resins generally show increased resistance to alkalies and to the elevated temperatures involved in baking processes.

Treating spinnarets with silicon resins eliminates much of the plugging.

Aluminum alkoxides are successful as hardening agents.

Copolymers of adipic acid, glycerol, and 1,3-diacetoxymethyltetramethyldisiloxane, or similar compound, are sometimes used. The silicon adaptation of the alkyd resin possesses, in general, increased hardness and flexibility. In addition, there is greater stability at higher temperatures. Copolymers can also be prepared using amyldibutoxyboron. Antiknock properties are claimed for this type of product as well as increased heat resistance.

The most important development in the chemistry of silicone resins embodies the preparation of monomeric silanes with at least one alkenyl group attached to silicon. Hydrolyzable groups are also present, so that polymerization can take place in two ways—by the conventional hydrolytic processes followed by condensation and by addition polymerization on the double bond, usually through the catalytic activity of benzoyl peroxide or similar agent. Some of the products find use as textile finishes, lubricating oils, additives or molding compounds. Vinyl and allyl groups are most common, with an occasional methallyl.

Triethylsilyl acrylate can be induced to undergo hydrolysis of the ethoxyl radicals to a desired extent forming linear or cross linked polymers. Addition polymerization will also take place on the double bond of the acrylate radical. More stable monomers result from the use of allyl or vinyl groups instead of acrylates. The latter contain a silicon-oxygen-carbon linkage which is always more or less susceptible to hydrolysis.

Other copolymers of this type with vinyl acetate or vinyl butyral resins have been found satisfactory for use in the lamination of wood, glass and metals. Glass-resin adhesives are also known.

General resistance to external influences constitutes the most outstanding property of silicone resins. Ultimate failure and breakdown is probably attributable, more than do anything else, to eventual oxidation of the radicals.

Among the more recent developments in this field may be mentioned the use of certain silicone resins, particularly those containing vinyl groups, as adhesives. Others find value as wood-sealing products. A silicone-glycol copolymer has been reported with curing properties, and alkyd resins are now modified with silicones. Combination epoxide-silicone resins have been investigated. A harder type of silicone resins sometimes results from the processing of monomers containing H or olefinic group attached to silicon. Dental impression materials are coming more to the front from this source as well. Finally, more attention is being devoted to studies of the relation between chemical structure and thermodynamic properties.

Resistance to γ-radiation is especially valuable. "Mouth-tissue-simulating molding compositions" for dentures are on the market. Optical lenses prepared from silicone resins are not affected by hot climates. Mold release agents are still the subject of research, as are resins with long storage capabilities. Coating bananas with silicone resins reduces the possibility of bruising. Heat-transfer compositions must have high molecular weights. In oil wells, a new use has come to the fore in the prevention of sand flow. Transparent resins are still in demand, some of which are porous.

Compounding silicone resins with phenyl formaldehyde or melamine polymers seems to produce good results. Even asbestos can sometimes be used. Sulfur in the resin makes a product suitable for use as fuel hose, gaskets and gas tanks. Phosphorus is found in silicone resin coatings but the product is liable to be toxic. Titanium is found in waterproof coatings. Tin compositions have low toxicity. Some resins of high viscosity contain Si-O-N units. Cross linking, however, is not recommended. The tendency to become brittle is ever present although a few instances are known to be contrary.

HOWARD W. POST

Cross-reference: *Silicon.*

References

McGregor, R. R., "Silicones and their Uses," McGraw-Hill, New York, 1956.

Coates, G. E., "Organometallic Compounds," John Wiley and Sons, Inc., New York, 1956.

Hagihara, N., Kumada, M., and Okawara, R., "Handbook of Organometallic Compounds," W. A. Benjamin, Inc., New York, 1968. (trans.)

Fordham, S., "Silicones," Philosophical Library, New York 16, N.Y., 1961.

Eaborn, Colin, "Organosilicon Compounds," Academic Press, Inc., New York, 1960.

SILVER AND COMPOUNDS

The element silver lies between copper and gold in subgroup IB of the Periodic Table. Its atomic number is 47 and the electron configuration is 2: 8: 18: 18: 1. Silver is highly stable, as indicated by the 18 electrons in the fourth quantum level. The atomic weight of silver is 107.868 ± 0.001. Silver is near the noble end of the electropotential series of metals, having a standard potential of $+0.7978$ volt in normal ionic solution at 25°C. This is about half-way between the potentials of copper and gold.

Occurrence. Silver occurs in nature chiefly in the form of the sulfide, which may exist as the mineral argentite, Ag_2S, or may be dissolved or intimately associated with other sulfides, especially those of lead and copper. Other common silver minerals are cerargyrite (horn silver), $AgCl$; proustite, $3Ag_2S \cdot As_2S_3$; pyrargyrite, $3Ag_2S \cdot Sb_2S_3$; stephanite, $5Ag_2S \cdot Sb_2S_3$; and native silver. Most lead ores and most copper ores are argentiferous, though there are important exceptions. Recovery of silver and gold from those ores constitutes an important part of their metallurgical treatment.

Silver is also commonly associated in nature with gold. Not only does gold occur with silver in copper and lead ores, but native gold, which is the largest source of that metal, usually contains silver. Gold and silver are mutually soluble in each other in all proportions in the solid state.

Productions. The world's production of silver in 1970 was approximately 288,560,000 fine ounces. More than half is a by-product of lead and copper production; largely for this reason, as well as a growing industrial demand, silver production has been slowly increasing in recent years. The three leading countries in silver production are Canada with 17% in 1970; United States with 16%; and Mexico with 16%. Thus 49% of the world's silver came from North America, and an additional 17% from South America.

Chemistry. Under ordinary conditions silver does not oxidize in the atmosphere. Silver oxide (Ag_2O)

has the very low heat of formation of 434 g-cal per gram of oxygen at 18°C. The standard free energy of formation of silver oxide is zero at about 150°K, rising to positive values above that temperature. On heating to 300–350°C, the oxide decomposes, forming metallic silver.

Molten silver readily absorbs oxygen. It is frequently stated that molten silver will absorb about 20 times its own volume of oxygen near the melting point. The solubility for oxygen is a function of the temperature, and while there is no sharp increase in solubility in passing from solid to liquid, the solubility curve is rising steeply at this temperature. Thus molten silver during cooling and freezing releases a large volume of oxygen, giving rise to "sprouting" or "spitting," i.e., throwing out from the solidifying mass particles of silver which may either separate or remain attached, often in a stringer-like form, to the parent body of metal.

Silver combines readily with sulfur, both at room and at furnace temperatures. The heat of formation of the sulfide, Ag_2S, is 172 g-cal per gram of sulfur. In air containing traces of hydrogen sulfide or other forms of sulfur, silver tarnishes through formation of a thin film of silver sulfide. The surface film does not penetrate or result in continuing corrosion; it is readily removed by abrasion, chemical solution, or electrolytic reduction.

An interesting property of silver is its oligodynamic or germicidal effect. It effectively kills many lower organisms, although harmless to the higher animals. One of its applications is as a bactericide in the Katadyn process of sterilizing water.

Silver resists attack by most of the common acids, including acetic, benzoic, citric, formic, hydrochloric, lactic, oxalic, phosphoric, and sulfuric, under usual conditions but dissolves readily in nitric acid. It also resists attack by sodium and potassium hydroxide solution or the fused salts, and by sodium chloride solutions, glycerine, and phenol. Silver and silver sulfide dissolve in dilute solutions of sodium or potassium cyanide, the basis of the cyanidation process of extracting silver and gold from ore and making solutions for electroplating with silver.

From silver nitrate solution silver deposits electrolytically in coarse nonadherent crystals, while from double cyanide solutions with sodium or potassium it deposits in a fine-grained adherent coating which is easily buffed to a high luster, or an initially bright deposit can be obtained by means of certain additions to the plating bath.

Properties. Silver crystallizes in the face-centered cubic lattice, with $a = 4.0774$ kX at 20°C. Its density is 10.49 at 20°C; the density of liquid silver is 9.30 near the melting point. The melting point is 960.8°C and the normal boiling point approximately 2259°C. The electrical resistivity of silver is 1.59 (microhm–cm) at 20°C and the thermal conductivity 0.934 cal per sq cm per cm thickness per deg C per sec at 100°C. Its heat of fusion is 25 cal per gram and heat of vaporization approximately 565 cal per gram. The specific heat of the solid is given by the relation $Cp = 0.055401 + (0.14414 \times 10^{-4})T - (0.16216 \times 10^{-8})T^2$. The specific heat of the liquid is 0.086 cal per gram and of the vapor approximately 0.046 cal per gram. The magnetic susceptibility is -0.20×10^{-6} cgs

units per gram. Its optical reflectivity has the very high value of 95% in the visible wave length region when freshly polished. For pure annealed silver the ultimate tensile strength is 18,000 psi, elongation 60% in 2 inches. Cold work may increase the tensile strength to 40,000 psi with elongation of 7%. The modulus of elasticity in tension is 11,000,000 psi. Recrystallization of cold-worked silver begins at 150°C and the metal is completely softened at 300°C.

Silver has the highest electrical and thermal conductivities of any of the elements, and its ductility and malleability are rated second only to those of gold. Its electrical conductivity on the IACS scale is 106 (copper = 100).

Uses. The chief uses of silver center around its high electrical and thermal conductivity, high corrosion resistance, and the noncontamination of liquids with which it is in contact during their manufacture, transportation, or storage. The principal chemical uses are in construction of heat-exchanger coils and evaporating equipment, pipelines, in photographic manufacturing equipment, in apparatus for manufacture of acetate rayon, in pharmaceutical compounds, and as a hydrogenation or oxidation catalyst. Its largest uses are for sterling silverware, in photography, in silver brazing alloys, in silver plating, and in coinage. It is also used for electrical conductors and contact points, in certain bearing alloys, in jewelry, in dentistry and surgery, in optical goods as a backing for mirrors, in parts of rockets and missiles, and in electronic devices. A new use of large potential is in photochromic glass, which darkens when exposed to light and regains its clarity in darkness.

World consumption of silver has exceeded mine production every year since 1951. Nearly half has been used in coinage until 1966. To make more silver available for the rapidly growing industrial uses, legislation was passed in 1965 to reduce greatly the amount of silver in United States coins (formerly 90% silver, 10% copper), eliminating silver entirely from dimes and quarters and reducing the silver content of half-dollars to 40 per cent.

Alloys. The most important silver alloys are the following:

Silver-copper alloys. These include sterling silver (92.5 Ag, 7.5 Cu) and standard silver used for United States subsidiary coinage (90 Ag, 10 Cu).

Silver-copper-zinc alloys. These comprise most of the silver brazing alloys ("silver solders").

Silver-gold alloys. Silver is commonly used as an alloying agent in gold used in jewelry.

Silver-lead alloys. Silver in small percentages is sometimes added to lead used for insoluble anodes and is being tried as an addition to storage-battery grids. The effect is to reduce corrosion rates and thereby increase life.

Silver-cadmium alloys. The alloy containing 97 Cd, 2 Ag, and 1 Cu or Ni is used for automotive bearings.

Compounds. The most important industrial compound of silver is silver nitrate, $AgNO_3$. It is used in the preparation of photographic emulsions for coating films and plates. It solubility in water is 122 g per 100 g of H_2O at 0°C, 222 g at 20°C, and 952 g at 100°C.

Silver sulfide (Ag_2S) is malleable at room temperature, and becomes more plastic on heating. It is stable in air, but on heating to 250°C in hydrogen or 400°C in steam it decomposes, forming metallic silver.

Silver chloride (AgCl) when exposed to light first becomes pink, then violet, brown, and finally black.

An ammoniacal solution of silver oxide, Ag_2O, on standing for some time deposits black crystals of "fulminating silver," which explode violently when barely touched. The compound deposited is probably Ag_3N or $NHAg_2$.

Other silver compounds of some importance are silver bromide (AgBr); silver iodide (AgI) used for cloud seeding in artificial precipitation experiments; and silver thiosulfate, $(Ag_2S_2O_3)$. Of the many organic compounds, argyrol, a pharmaceutical, is perhaps the most important.

ALLISON BUTTS

Cross-references: *Photography, Metals, Electrochemistry.*

References

Addicks, L. (editor), "Silver in Industry," Reinhold Publishing Corp., New York, 1940.

Butts, A. (editor), with collaboration of C. D. Coxe, "Silver, Economics, Metallurgy, and Use," D. Van Nostrand Co., Princeton, N.J., 1967.

U.S. Bureau of the Mint, "Annual Report of the Director of the Mint," Washington, D.C.

SOAPS

Soaps are salts of long-chain, organic carboxylic acids ($R—COO^-X^+$). R is usually an unbranched saturated or unsaturated alkyl group, but it may possess branches or even ring groups. The cation (X) present is most usually sodium, although potassium and, to a lesser extent, various ammonium soaps are employed. These soaps are very much more water-soluble than the acids from which they are derived. The most characteristic properties of aqueous soap solutions, their ability to clean and to foam, have been known since ancient times. Soaps are also well recognized by their characteristic waxy feel.

Raw Materials. The most widely used raw materials for soaps are natural fats and oils which may be refined for soap manufacture. These are triglycerides, i.e., fatty acid esters of glycerin, the fatty acids of importance for soaps being essentially in the range 12-carbon to 18-carbon chain length. The chief sources of fatty acids for soap are tallow and coconut oil, the best known soaps being based on a mixture of 80% tallow fat and 20% coconut oil. Tallow, derived from the cattle and sheep meat industry, varies considerably in properties such as color and titer (solidification point of the fatty acids). The titer of tallow used for soap is generally 40°C or higher. The chain length makeup of its fatty acids is typically 2% C_{14}, 32.5% C_{16}, 14.5% C_{18}, 48.3% C_{18}^- (oleic), 2.7% $C_{18}^=$ (linoleic), and of coconut oil 8% C_8, 7% C_{10}, 48% C_{12}, 14.5% C_{14}, 8.8% C_{16}, 2% C_{18}, 6% C_{18}^-, 2.5% $C_{18}^=$. Partial replacements for tallow are lard (hog-fat), hardened marine oils, and, especially abroad, palm oil. Palm kernel oil, which has a fatty acid composition similar to that of coconut oil, is sometimes used to replace the latter.

In certain instances rosin acids (abietic acid, etc.) are used in soap making. Sodium hydroxide, as a concentrated aqueous solution, is the alkali most generally used for fat and oil saponification. Potassium or ammonium hydroxides are used to make more soluble or "soft" soaps; see below. Blends of alkalies are sometimes employed.

Soap Making. The essential process of soap making is the saponification of triglycerides in alkaline media. A substantial fraction of the total production of soap is still carried out by a batch process in large open kettles. Saponification is achieved by a combination of steam and added caustic soda. Thereafter the soap is salted out or "grained" to produce curdy kettle wax soap and lye which is separated for recovery of glycerin. Several washes ensue. The processes of "closing" of the soap, by adding steam and water, and graining are repeated several times and after the final wash-lye phase is removed, the appropriate amounts of water and heat are applied to produce neat soap overlying a lower quality "nigre" soap layer. The pumpable neat soap, containing about 30% water, is then processed. Several continuous processes for soap manufacture have been developed. Some involve preparation of fatty acids from the triglycerides as an intermediate step, whilst others involve direct saponification to the soap stage.[1].

Soap Processing. Neat soap, which contains about 30% water, is processed in several ways. For bar manufacture, chips are formed by cooling the pumpable mass on chill rolls; after drying, mixing and milling operations, screw extrusion of soap in the form of logs follows from a so-called plodder. For preparing spray-dried powders, the neat soap is pumped into large mixing vessels (crutchers), the appropriate additives blended, and the mixture then sprayed from nozzles into heated air for drying.

Properties. An understanding of the preparation and properties of soaps is facilitated by studying the phase diagrams of soap/water mixtures—work pioneered by J. W. McBain[2]. The properties of soap are dictated by its molecular structure, viz., a long hydrophobic hydrocarbon chain coupled to a hydrophilic ionic head group (COO^-X^+). In anhydrous, hydrate and concentrated solution form, soap molecules adlineate into regular, sandwich-like arrays of thickness approximately equal to two soap molecule lengths. This configuration allows hydrocarbon chains and polar groups each to pack side by side. The intermolecular distances in these double layer systems, tilt angle of the chains, and so on, are a sensitive compromise between the way the hydrocarbon chains and the head groups each wish to pack. This is reflected in the fact that, even in anhydrous systems, as many as ten different crystalline phases are encountered on heating the soap from room temperature to the melting point, $\sim$300°C. Hydrated soaps, such as those encountered in commercial bars, also exist in several different phases, the predominant phase(s) present having a marked influence on the properties of these soaps. In hot, concentrated soap/water systems above the Krafft temperature (see below) two important phases are encountered. The first is neat phase, already referred to: x-ray data indicate the basic bimolecular leaflet structure is still present but with several layers of water interposed between the adjacent ionic head group planes. At higher water contents and somewhat lower temperatures, soap exists in "middle phase," a phase avoided during soap manufacture because of its high viscosity. This high viscosity results from the structure of the basic unit aggregates of soap molecules, or "micelles," which have the form of long cylinders of diameter about two soap molecule lengths. At still higher water content, the soap/water system is isotropic and the soap micelles are more or less spherical, with sphere diameter again about two chain lengths. An important phenomenon related to the formation of micelles is the Krafft "point." Below this temperature, soap is rather insoluble above it, the solubility increases rapidly and soap exists in the micellar or liquid crystalline phases described above. For the sample soaps sodium stearate, palmitate and oleate and potassium stearate, the Krafft temperatures are roughly 70°, 60°, 25° and 48°C, respectively.

The dual hydrophobic/hydrophilic nature of soap explains its most characteristic property, viz., surface activity. This is a consequence of the spontaneous attempt of the hydrophobic chains of soap molecules in water to remove themselves from the aqueous environment by a process of adsorption at, for example, the water/air and water/oil interfaces. This results in a lowering of the surface and interfacial tension, and is responsible for the ability of soap to promote foam and emulsion stability. Micelle formation itself, in which the hydrocarbon chains point "in," i.e., away from the water, and the ionic groups "out," i.e., in contact with the water, is a manifestation of the same phenomenon. Below the critical micelle concentration (cmc) the dissolved soap exists as single ions and behaves as a typical electrolyte; above it there is a mixture of micelles and single ions and the solution surface tension becomes insensitive to concentration. The cmc's of typical soaps at 70°C are roughly $0.07, 0.7, 3, 20 \times 10^{-3}M$, for sodium stearate, palmitate, myristate, and laurate, respectively. In a homologous series of fatty acid soaps, characteristic "soap" properties, such as micelle formation, pronounced foaming, etc., first occur at a chain length of 7 carbons.

Because of its tendency to aggregate in solution, soap is known as an association colloid. An aqueous solution of soap micelles can in fact be considered a stabilized suspension of micro droplets of hydrocarbon in water. As such they have the ability to dissolve and suspend oily materials, one of the processes involved in detergency. By adsorption onto oily dirt and the substrate to be cleaned, soaps can lead to a reduction of the contact angle of water at the water/soil/substrate interface so facilitating removal of the dirt by mechanical agitation. By adsorbing onto particulate dirt, and also removing, by precipitation, calcium ions involved in dirt/substrate "bridges," soaps assist detergency of solid soils.

Types. Sodium soaps are by far the most widely used. To offset the limited solubility associated with longer chain-length saturated soaps (e.g., C_{16} and C_{18}), blending is necessary with shorter chain-length soaps, e.g., C_{12} and C_{14}, and unsaturated soaps, such as sodium oleate. Experience has shown the blend of 80 tallow fat and 20 coconut oil, referred to earlier, achieves the desired solubility characteristics. Sodium soaps, furthermore, tend to be "hard," while potassium and ammonium soaps are "soft" and more soluble; thus potassium soaps are often blended with sodium soaps to achieve improved solubility and foaming. For the same reason, potassium, ammonium and triethanolammonium soaps are used in liquid soaps and shampoos.

Soaps are sometimes "superfatted," i.e., contain a small amount of unneutralized free fatty acid. This contributes to the mildness of the soap towards skin, both by reducing the pH and probably also by deposition on the skin of "acid soaps" which have a relatively low solubility. Acid soaps, e.g., sodium acid stearate, are well-defined complexes of acid and soap with definite stoichiometry. They have been extensively studied.

Usage. The chief usage of soap today is in toilet bars. Recent trends, especially in Western Europe, are toward higher contents of coconut fatty acids, to achieve quicker and more copious foaming, but with a concomitant increase in superfatting for mildness. Emphasis is also placed on deodorant characteristics in soaps; the main antibacterial agents used for this purpose are the brominated salicylanilides, halogenated carbanilides and bishydroxyphenylalkanes. Floating soaps are soaps formulated to contain entrapped air; transparent soaps are produced by use of a number of devices, e.g., incorporation of ethanol, sugar and glycerin, or by processing within a critical phase boundary of the soap-electrolyte-water system.

Soaps are used widely in shaving creams. These are frequently formulated from a mixture of coconut oil and stearic acid with mixed sodium and potassium hydroxides. A small amount (4–8%) of stearic acid is often left unneutralized to give a pearly appearance, and glycerin may be added for body.

Until World War II soap was the major laundry surfactant. Its major drawback in this respect was its sensitivity to water hardness, i.e., formation of insoluble calcium and magnesium soaps which constitute undesirable lime soap scum. This disadvantage aside, soaps are known to be outstanding laundering materials. However, despite the use of various water softening and alkaline cleaning additives, such as phosphate and carbonates, soaps lost ground rapidly for this use to the very effective synthetic detergent/condensed phosphate mixtures.

There has been a steady increase in share by synthetic detergents of a steadily expanding market, with soaps tending to level off in sales and tonnage. Prediction as to the future, seemingly safe from tabulated data, is uncertain in the light of current concern about possible effects of detergent ingredients on the environment. Guidelines in this respect are by no means clear. However, if legislation against any of the currently used major detergent ingredients becomes widespread, there is a possibility that soap will assume an increasing share of the market. In this event synthetic fatty acids would achieve greatly increased importance. Packaged soap powders (or flakes) now constitute only a very small percentage of the total market and are used mainly as specialty products in the United States.

E. D. GODDARD

References

1. Ryer, F. V., "Soap" in Kirk-Othmer "Encyclopedia of Chemical Technology," 2nd Edition, Interscience, **18**, 415–432, New York, 1969.
2. McBain, J. W., and Lee, W. W., *Ind. Eng. Chem.,* **35**, 917 (1943); *Oil & Soap,* **20**, 17 (1943).

SODIUM AND COMPOUNDS

Some of chemistry's most famous names figure prominently in the discovery of metallic sodium and its early history. In studies of the electrolysis of potassium hydroxide and of sodium hydroxide, Sir Humphry Davy isolated first potassium and then sodium in the fall of 1807. The next year Gay-Lussac and Thenard obtained the metal by non-electrolytic means, reducing sodium hydroxide with iron at high temperature. The conflict between electrolytic and chemical reductions, launched at that early date, raged for over a century and was not resolved until the development of the Downs cell in 1921 made sodium the cheapest available nonferrous metal at a current price of $18\frac{3}{4}$ per pound.

Incentive for the development of a commercial method for sodium production was provided in 1824 by Oersted's discovery that sodium could be used to prepare pure aluminum by the reduction of aluminum chloride. This was developed into a workable process by Deville and Bunsen in 1854. Deville in the same year described a process for preparing sodium by reducing a mixture of sodium carbonate and lime with charcoal at a temperature above the boiling point of sodium, removing the sodium as a vapor. Although not too efficient, Deville's process remained in use for 30 years. It was displaced by an improved chemical reduction process patented in 1886 by Castner based on the reduction of sodium hydroxide by iron carbide. Castner's process was operated successfully in England for two years until Hall's discovery of the electrolytic route to aluminum abruptly eliminated the major market for sodium. Reluctant to abandon the sodium reduction process for aluminum manufacture, Castner set out to develop a still cheaper route to metallic sodium. By 1891 he had patented a process based on the electrolytic reduction of sodium hydroxide, thus completing the full circle which started in Davy's laboratory in 1807. While failing to challenge seriously Hall's electrolytic aluminum process, the Castner cell was operated commercially for thirty years. Annual production of sodium by this method had reached 30 million pounds by 1921 when the Downs cell was introduced.

Sources. Sodium is the sixth most abundant element, occurring naturally in many forms. Its major source is sodium chloride, the major inorganic constituent in sea water, but deposits of crystalline salt in underground salt domes and dry lake beds are a more convenient source. Evapora-tion of sea water is practiced commercially in some areas, particularly in hot, arid climates where solar heat is used to vaporize the water from large shallow pans. Sodium also occurs naturally in a wide variety of other minerals. Some of the most common are borax, $Na_2B_4O_7$; cryolite, Na_3AlF_6; Chile saltpeter, $NaNO_3$; and soda ash, Na_2CO_3.

Manufacture. Faraday in 1833 and a succession of later workers had attempted to obtain sodium by the electrolysis of sodium chloride. A number of these early efforts led to issued patents, but none was a commercial success until the Downs cell was introduced in du Pont's Niagara Falls plant about 1921 and patented in 1924. The high current efficiency which is the major advantage of the Downs cell is a result of good mechanical arrangements for the removal of sodium and chlorine and of the simplicity of the electrode reactions:

$$\text{Anode: } 2OH^- \rightarrow H_2O + \tfrac{1}{2}C_2 + e^-$$

$$\text{Cathode: } Na^+ + e^- \rightarrow Na$$

In contrast, the anode reaction in a Castner cell is more complex:

$$\text{Anode: } 2OH^- \rightarrow H_2O + \tfrac{1}{2}O_2 + 2e^-$$

The water formed at the anode is not easily removed from the molten NaOH electrolyte and diffuses to the cathode, where it reacts with half of the sodium produced:

$$Na + H_2O \rightarrow NaOH + \tfrac{1}{2}H_2$$

Thus, the overall reaction in a Castner cell is:

$$2NaOH \rightarrow Na + NaOH + \tfrac{1}{2}H_2 + \tfrac{1}{2}O_2$$

and an additional electron must be supplied to prepare the starting NaOH from NaCl. Thus, the maximum current efficiency theoretically possible for conversion of NaCl to Na via a Castner cell is $33\frac{1}{3}\%$

Production and Use. Total U.S. capacity in 1969 has been estimated at 378 million pounds per year. Usage in 1968 was 313 million pounds, with an estimated 1973 demand of 360 million pounds. Growth from 1958 to 1968 was 3.6% per year, and is estimated a 3% annually through 1973. Currently 83% of production is used in the synthesis of tetraethyllead and tetramethyllead, 6% in titanium reduction, and 11% in all other uses.

Physical Properties. Sodium is the most abundant of the alkali metals, Group IA of the Periodic Table, with an atomic number of 11 and atomic weight of 22.98977. It is a soft metal melting at 97.83°C and boiling at 882.9°C. Other physical properties are: heat of fusion, 622.2 cal/g-atom; heat of vaporization, 20,629.8 cal/g-atom at the b.p.; density 0.9674 g/cc (25°C) and 0.9270 for liquid at 97.83°C; thermal conductivity, 0.314 cal/cm²(sec)(°C/cm) at 25°C; electrical resistivity, 4.985 microhm-em at 25°C and 9.60 for liquid at 97.83°C; viscosity, 0.450 centipoise at 200°C, 0.293 at 500°C and 0.167 at 900°C; surface tension, 182.0 dyne/cm at 200°C and 152.0 at 500°C; and neutron absorption cross-section, 0.505 barns.

Principal Compounds. Sodium chloride is of of course the most important commercial compound, since most other sodium compounds are prepared from it either directly or indirectly. The other important compounds are listed below, with estimates of recent consumption.

Compound	Million of Pounds
Sodium hydroxide	17,000
Sodium carbonate	13,400
Sodium sulfate	3,000
Sodium-lead alloy	2,600
Sodium tripolyphosphate	2,300
Sodium silicate	1,370
Sodium tetraborate	1,120
Sodium linear dodecylbenzene-sulfonate	650
Sodium sulfite	460
Sodium bicarbonate	400
Sodium chlorate	366
Sodium bichromate	306
Sodium hydrosulfite	79
Sodium thiosulfate	60

Nonchemical Applications. Uses which depend upon sodium's physical rather than its chemical properties are based upon its excellent thermal and electrical conductivity. Thus sodium is an important high-temperature heat-exchange medium and is finding increasing usage as an electrical conductor. In addition to its high thermal conductivity, sodium has a number of other properties which make it an outstanding heat-exchange fluid. These include a large liquid range (98–883°C), low viscosity, excellent thermal stability, noncorrosivity toward steel, and a neutron absorption cross-section which is acceptably low for use in nuclear power plants.

As an electrical conductor, sodium is surpassed only by silver, copper, aluminum, and gold. Because of its lower density and lower cost, sodium is preferred to each of these conductors on a cost-effectiveness basis and would be used extensively except for the hazards associated with its high chemical reactivity. Recently an attempt has been made to circumvent these hazards by encapsulating sodium in polyethylene tubing. Such conductors are four times as flexible as copper, and maintain flexibility even down to −40°C. The sodium vapor lamp is another application of sodium as an electrical conductor. The yellow light from these lamps is not aesthetically pleasing, but is in a wave length range to which the eye is extremely sensitive, so that it is particularly effective for highway lighting.

Chemical Applications. Hydrogen is reduced by sodium at 200–300°C to form sodium hydride, which is used chiefly for descaling metal surfaces. Other mixed hydrides of some commercial significance include $NaBH_4$, $NaAlH_4$, Na_3AlH_6, $NaHAl(OCH_3)_3$, $NaH_2Al(C_2H_5)_2$, and $NaH_2Al(OCH_2CH_2OCH_3)_2$.

Sodium forms four distinct compounds with oxygen: the monoxide, Na_2O; the peroxide, Na_2O_2; the superoxide, NaO_2: and the ozonide, NaO_3. Sodium peroxide is the most important member of the series and is widely used as a strong oxidizing agent, particularly for bleaching textiles and wood pulp. It is prepared by heating sodium at 300–400°C with air or oxygen at elevated pressure.

In a clean system, sodium will dissolve in liquid ammonia to give a dark blue solution and, in more concentrated solutions, a bronze-colored phase as well. If metal catalysts such as iron, cobalt, or nickel are present, sodium reacts with ammonia to give sodamide:

$$Na + NH_3 \rightarrow NaNH_2 + \tfrac{1}{2} H_2$$

At 600°C in the presence of coke, sodium cyanamide is formed, and can be reacted with additional coke at 800°C to give sodium cyanide in a process developed by Castner:

$$2NaNH_2 + C \rightarrow Na_2CN_2 + 2H_2$$
$$Na_2CN_2 + C \rightarrow 2NaCN$$

Substitution of nitrous oxide for coke can lead to the synthesis of sodium azide:

$$2NaNH_2 + N_2O \rightarrow NaN_3 + NaOH + NH_3$$

Most metal chlorides are reduced by sodium to the free metal. The reduction of titanium tetrachloride by sodium to yield titanium is the most important of these reactions industrially and has been used in place of magnesium reduction in recent installations. The reduction of zirconium tetrachloride is also practiced commercially to produce zirconium.

Sodium is frequently used for deoxidizing most of the common structural metals, particularly lead, copper, iron, steel, chromium, and various alloys of these metals.

Of the common metals, those which do not form alloys with sodium include aluminum, boron, silicon, iron and chromium. The most important sodium alloys are those with lead and with mercury. Sodium-lead alloy is widely used in the synthesis of lead alkyls, while sodium amalgam is widely used for the production of sodium hydroxide via the electrolysis of sodium chloride in the amalgam cell.

Reactions with Organic Compounds. Since all reactions of sodium involve the transfer of an electron from the metal to the reacting species, it is surprising to realize that the dominant characteristic of every organic molecule which reacts with sodium is a localized center of high electron density. Typical examples of such nucleophilic centers are the unshared electron pairs of halogen, oxygen, or nitrogen atoms and the pi-electron clouds of olefinic and aromatic hydrocarbons. It is likely that the first step in all reactions is an attack by the nucleophilic center on the positively charged sodium nucleus, facilitated by the ready polarizability of the sodium electrons, which recede from the point of attack and permit formation of at least a partial ion-pair bond between the sodium cation and the nucleophilic center. Subsequent or perhaps simultaneous electron rearrangements then lead to the final products of the reaction. For heterogeneous reactions, these products must be removed from the sodium surface before further reaction can occur, and this removal frequently limits the overall reaction rate. Thus, choice of a solvent is frequently all-important in organosodium chemistry. The most commonly chosen solvents are paraffins, aromatics, ethers, and liquid ammonia, which dissolves sodium and thus permits a homogeneous reaction, so that product removal rate is not a key consideration. For heterogeneous reactions, the state of subdivision of the sodium frequently determines the reaction rate, which is often accelerated several thousand-fold by the use of sodium dispersions or solid carriers having a high surface area.

Sodium will displace hydrogen from a wide variety of organic compounds, as illustrated with an alcohol:

$$ROH + Na \rightarrow RO^-Na^+ + \tfrac{1}{2}H_2$$

Other compounds having labile hydrogens also undergo this reaction, such as nitroalkanes, nitriles, esters, aldehydes, and ketones. In these cases, it is the hydrogen in the position adjacent to the unsaturated group which is displaced, leading to a resonance-stabilized carbanion, as with acetophenone:

$$C_6H_5CCH_3 + Na \rightarrow [C_6H_5-C=CH_2 \leftrightarrow$$
$$\underset{O}{\|} \qquad \qquad \underset{O^-}{|}$$
$$C_6H_5-C-CH_2^-]\,Na^+ + \tfrac{1}{2}H_2$$
$$\underset{O}{\|}$$

In most cases, further reaction of the carbanion takes place. With esters, for instance, the carbanion can displace an alkoxide anion from a second ester molecule, as in the familiar Claisen condensation:

$$\underset{O}{\overset{O}{\|}} \qquad \underset{O}{\overset{O}{\|}}$$
$$Na^{+-}CH_2C-OEt + CH_3C-OEt \rightarrow$$

$$\underset{O}{\overset{O}{\|}} \qquad \underset{O}{\overset{O}{\|}}$$
$$CH_3C-CH_2-C-OEt + Na^{+-}OEt$$

$$\downarrow$$

$$Na^+O^- \qquad \underset{O}{\overset{O}{\|}}$$
$$CH_3C=CH-C-OEt + EtOH$$

Sodium will also displace hydrogen from acetylene, and from a few aromatic compounds such as fluorene, indene, cyclopentadiene, and triphenylmethane.

Displacement of a halogen by sodium is also a common reaction. One of the most familiar examples is the Wurtz reaction:

$$2RX + 2Na \rightarrow R-R + 2NaX$$

Another related reaction is the cleavage of diaryl ethers by sodium:

$$Ar-O-Ar + 2Na \rightarrow Ar^-Na^+ + Ar-ONa^+$$

Safe Handling Techniques. Sodium is available commercially in bricks, drums and tank cars from DuPont, Ethyl Corporation, and U.S. Industrial Chemicals Company. Full details on safe handling techniques are available in brochures supplied by these manufacturers. These should be consulted by anyone planning to work with sodium. Key precautions in working with sodium are maintaining an inert atmosphere and avoiding contact with water and other substances with which sodium reacts readily. Nitrogen, or occasionally helium or argon, is used as an inert blanket to exclude air from vessels and lines in sodium service. Dry powder fire extinguishers should be used on sodium fires; other common fire-fighting chemicals such as carbon dioxide, carbon tetrachloride, and water react with sodium. Contact with the skin and eyes can lead to both thermal burns and chemical burns resulting from the conversion of sodium into strong caustic by reaction with traces of moisture on the skin.

Sodium is usually transferred from one vessel to another on a commercial scale in the liquid form. This requires insulated lines to maintain the temperature above its melting point of 98°C. At temperatures below its melting point, sodium in brick form may introduced into a reaction vessel through the use of a nitrogen-purged charging port. At ambient temperatures sodium may also be transferred conveniently in the form of a dispersion in an inert hydrocarbon or ether solvent. These dispersions, with particles as small as one micron, are easily prepared by stirring molten sodium at high speed (10–20,000 rpm in a laboratory preparation; lower speeds in larger equipment) and then allowing the mixture to cool. A small amount of a surface-active agent is usually added, typically 0.5–1.0% of oleic acid. Such dispersions are easily handled and have the added advantage of increasing the reactivity of the sodium by several orders of magnitude because of the tremendous increase in surface area. Molten sodium may also be distributed over the surface of an inert solid carrier such as alumina, charcoal, sodium carbonate, or sodium chloride to give highly reactive, free-flowing powders of large surface area. These powders can be transferred at ambient temperature in inert-gas pneumatic systems; they are highly effective in fluidized-bed reactor systems, particularly those in which the solid carrier is one of the products of the sodium reaction, as in the reduction of metal chloride with sodium using sodium chloride as an inert carrier.

KENNETH L. LINDSAY

References

Sittig, M., "Sodium. Its Manufacture, Properties and Uses," Reinhold, New York, 1956.

Jackson, C. B., "Liquid Metals Handbook. Sodium-NaK Supplement," U.S. Govt. Printing Office, Washington D.C., 1955.

"Handling and Uses of the Alkali Metals," Advances in Chemistry Series No. 19, American Chemical Society, Washington, D.C., 1957.

Morton, A. A., "Solid Organoalkali Metal Reagents," Gordon and Breach, New York, 1964.

Watt, G. W., *Chem. Reviews,* **46,** 289–379 (1950).

Lindsay, Kenneth L., "Sodium," pp. 653–663, "Encyclopedia of the Chemical Elements," Hampel, C. A., Editor, Van Nostrand Reinhold Corp., New York, 1968.

SODIUM CHLORIDE

Of all the minerals known to man, salt by far is used in the greatest number and variety of applications. It has been estimated that there are more than fourteen thousand separate and distinct practical uses for salt and brine.

One of the basic uses is as a primary source of sodium and chlorine. Eleven basic chemicals are produced directly from salt as a raw material: soda ash, calcium chloride, chlorine, caustic soda, hydrogen, sodium, sodium sulfate, sodium bisulfate, hydrochloric acid, sodium cyanide and sodium hypochlorite. Whereas each of these compounds has important direct uses, they, too, are used as chemical building blocks in the manufacture of virtually every existing chemical product.

Rock Salt. Rock salt is produced by dry mining a deposit of salt underneath the earth's surface. This salt occurs as bedded or domed deposits in 28 states and several provinces of Canada. Michigan, New York, Texas, Ohio, Louisiana and Kansas account for the bulk of the production of rock salt in the United States. This, of course, excludes solar salt production at Great Salt Lake and on the West Coast.

These salt deposits were formed in past geologic ages by evaporation of impounded salt water under desert conditions. Two types of deposit are found. One occurs in essentially horizontal stratified beds, as originally laid down. The other type is a plug or dome formed by super-incumbent weight on deep-lying horizontal beds, which has caused upward flow toward the surface through zones of weakness.

All rock salt is the same basic substance, sodium chloride; however, due to the manner in which it was formed, variances in the physical characteristics appear. Some rock salts have higher impurity content—others have color or cleavage characteristics which serve to distinguish them. Commercial rock salt varies in purity from about 97% (Kansas) to about 99% (Louisiana), and averages about 98% for the Northern States. By far the most significant impurity in rock salt is the mineral anhydrite (calcium sulfate). Dolomite, quartz, calcite and iron oxides are found in minor quantities.

Salt is mined much like coal. A shaft is cut to the deposit; the working face is undercut, drilled and blasted. Rock salt is not purified in any way, but is crushed and screened to commercial sizes, and sold both bagged and in bulk.

Evaporated Salt. As its name implies, evaporated salt is produced by evaporating water from brine to form salt crystals. But not all evaporated salt is the same. There is evaporated granulated salt in which each crystal is a tiny cube and there is grainer or flake salt which is irregular in shape, frequently thin and flaky, and unusually soft.

The conventional process for producing evaporated salt starts with rock salt in its natural underground formation. Holes are drilled into these salt deposits, water is pumped into the deposits to dissolve the salt and the brine is brought to the surface for refining. In this operation all the insolubles are left behind in the well. The brine is then at least partially purified by the addition of chemicals to remove hardness and dissolved gases. The resulting semipure brine is evaporated in multiple effect vacuum pans where the salt crystallizes as perfect cubes of sodium chloride. The same kind of brine can be evaporated in open pans (called grainer pans) to produce grainer salt having crystals of a characteristic hopper structure. Purified salt is manufactured conventionally by taking the same brine and subjecting it to intensive chemical purification before it is fed to the vacuum pans for crystallization.

Granulated or vacuum evaporated salt is produced by boiling brine at less than atmospheric pressure in large tightly sealed evaporators commonly called "vacuum pans." These are installed either singly or in multiple units.

The underlying principle of the process is the lowering of the boiling point of the brine by decreasing the pressure of the vapor above the liquid. When evaporation has concentrated the brine to the point of saturation, salt begins to crystallize in perfectly cubical grains. The size of the crystals is controlled by the rapidity of evaporation—dependent in turn on degree of vacuum, temperature and agitation. When the crystals have grown to the proper size they drop to the bottom of the pans. Crystals drawn from the pans are washed, dried and cooled prior to being graded and stored in the proper bins ready for shipment in bulk or bag.

Grainer salt is produced by surface evaporation of brine in flat pans open to the atmosphere. Steam pipes a few inches above the tank bottom supply the heat. The salt crystals form on the surface of the brine and are held there by surface tension. They grow laterally to first form thin flakes, and then as they grow heavier, and tend to sink, they develop into hollow pyramids floating point down. These are known as hopper crystals. Eventually they sink and are scraped to one end of the pan, removed, dried and screened. The fragile hoppers break up, giving the name flake salt.

Solar Salt. Solar salt is formed from ocean water and salt lake brines, which are impounded in shallow lagoons and evaporated by the sun's heat. Evaporation takes place at the surface and produces flakes and hoppers in the same way as grainer salt, but much larger and coarser. The salt accumulates on the bottom of the lagoons, the mother liquor is drawn off and the salt harvested by scraping machinery. It is then washed, dried and screened.

Table Salt. A free-flowing agent is added to vacuum pan salt—often about 1% calcium silicate or magnesium carbonate. Also, to prevent the salt from caking, a few parts per million of sodium ferrocyanide are added. Iodized salt (usually table salt) has in addition 0.01% potassium iodide, and a stabilizer for the iodine, often consisting of 0.1% sodium carbonate and 0.1% sodium thiosulfate. (See **Iodine**.)

For livestock salting and certain refrigeration uses, both rock and evaporated salt are compressed into 50 pound blocks and 4 pound bricks. Evaporated salt is also pressed into small pellets, for regenerating zeolite water softeners (See **Water Conditioning**). All these products approach solid salt in density.

C. E. MacKinnon
Revised by Margaret L. Winter

SOIL CHEMISTRY

Soil chemistry includes all aspects of the study of soil as a chemical system. The eight chemical elements in soils which generally surpass 1% by weight are oxygen, silicon, aluminum, iron, calcium, magnesium, potassium, and sodium; the eleven elements making up 0.2 to 1% include titanium, hydrogen, phosphorus, manganese, fluorine, sulfur, strontium, barium, carbon, chlorine, and chromium. The most abundant minerals present in less-weathered soils are quartz, feldspars, micas and colloidal layer silicates including vermiculite, chlorite, and montmorillonite. Calcareous soils contain calcite and dolomite. More-weathered soils contain larger amounts of more resistant minerals such as kaolinite, halloysite, allophane, hematite, goethite, gibbsite, anatase, pyrolusite, tourmaline, and zircon. The organic matter, or humus, content of soils varies from less than 1% to over 80%. Generally, upland soils range from 1 to 8% organic matter, while less well-drained soils are frequently higher.

Soils developed under coniferous forests often accumulate acid organic matter at the surface; the resulting leaching through the soil of chelating organic acids bleaches (*podzolizes*) the mineral soil beneath. These soils are gray when plowed. Organic matter from hardwood trees and grasses which are high in bases, particularly calcium, accumulates in the soil and causes a dark color in the surface horizon. Poor drainage leads to the development of light-colored gray horizon within the soil column (*profile*), owing in part to the reduction of iron oxides to ferrous form. A bluish color is sometimes present, particularly when vivianite, $(Fe)_3(PO_4)_2 \cdot 8H_2O$, forms. Soluble soil salts, mainly chlorides and sulfates of sodium, calcium, and magnesium, when present in quantities over 0.1 to 0.7% cause a condition known as salinity or soil alkali. If much Na_2CO_3 is present, some organic matter is mobilized and together with FeS, colors the soil black, giving rise to the name black alkali.

The most reactive portion of the soil resides in colloidal organic matter, layer silicates, and hydrous oxides of iron, aluminum, and occasionally manganese and titanium. The colloids of soil have a negative electrostatic charge arising through carboxyls of organic compounds and through excess negative charge of oxygen in the silicate structure. The negative charge is neutralized by exchangeable cations, giving systems known as colloidal electrolytes. When these exchangeable ions, i.e., counterions, are hydrogen or aluminum, the colloids act as a moderately strong acid. Different colloids range in the strength of acidity as evidenced by the shapes of the titration curves, which are analogous to the shapes of those of weak and strong soluble acids. The colloids are hydrophilic and subject to flocculation in the presence of dilute salt solutions, owing to repression of the charge developed by dissociated cations. The flocculation is reversible.

Important chemical characteristics of the soil include the total exchange capacity for cations, expressed as total meq of cations per 100 gm of soil, and the base status, which is the percentage saturation of the negative charge with cations such as calcium, magnesium and sodium. The more productive soils are about 80% saturated with calcium and magnesium. Excessive hydrogen and aluminum saturation (much over 15%) is termed soil

acidity. Excess sodium saturation (12% or more) leads to dispersiveness of the soil and poor productivity.

There are also positive charges associated with aluminum and iron colloids of soils. These charges give rise to phenomenon known as anion exchange capacity, which is mainly concerned with phosphorus chemistry; the usual soluble anions such as nitrate, chloride, and sulfate are little held. Synthetic organic soil conditioners are long-chain organic molecules with carboxyl charges along the chain which react with the positive charges of the soil particles. These colloidal molecules can bind the soil particles into aggregates. Natural humus of soil acts in a similar way until oxidized by soil organisms.

Analytical methods employed in soil chemistry include the standard quantitative methods for the analysis of gases, solutions, and solids, including colorimetric, titrimetric, gravimetric, and instrumental methods. The flame emission spectrophotometric method is widely employed for potassium, sodium, calcium, and magnesium; barium, copper and other elements are determined in cation exchange studies. Occasionally arc and spark spectrographic methods are employed.

The most commonly made chemical determination is that of soil pH measurement, as an indicator of soil acidity. The glass electrode has proved the most satisfactory method for soil pH measurement because the moistened soil rapidly equilibrates in contact with the glass surface, no reagents are added to the soil, and the soil CO_2 tension is not disturbed by bubbling through of gases. Colorimetric indicators are also employed. Soils of pH 4.3 to 5 are highly acid, of pH of 5 to 6 are moderately acid, of pH 6.3 to 6.6 are very slightly acid, of pH 6.7 to 7.3 are considered neutral; soils of pH 8 to 9 are moderately alkaline, and of pH 9 to 11 are very alkaline. For acid soils, pH measurement serves as a guide to agricultural liming practices. For many crops, such as alfalfa, the soil is adjusted to pH 6.5 to 7 by the addition of ground limestone, the active ingredients of which are $CaCO_3$ in calcic limestone and $CaCO_3 \cdot MgCO_3$ in dolomitic limestone. A soil colloidal acid may be represented as HX. Then the liming reaction, by which the exchangeable calcium is increased, is

$$CaCO_3 + 2HX \rightarrow CaX_2 + CO_2 + H_2O$$

The proton donor, X, represents a variety of organic and inorganic donors, including the reaction $Al(OH_2)_6 \rightarrow Al(OH)(OH_2)_5 + H^+$. The reaction is hastened by fine grinding of the limestone, and liming materials are graded on the basis of fineness and $CaCO_3$ equivalence. Burned lime (CaO), marl ($CaCO_3$), and sugar refinery wastes [$Ca(OH)_2$] are also used in liming. For some crops, owing to disease susceptibility and preferences, soils are kept more acid, as low as pH 5.3. Calcium and magnesium, necessary to plant growth, are furnished to plants from exchangeable form.

Fine grains (finer than 20μ in diameter) of mica [$KAlSi_3Al_2O_{10}(OH)_2$], and potassium feldspar ($KAlSi_3O_8$) slowly undergo chemical weathering in soils with the release of potassium into exchangeable form. Other ions such as calcium, sodium, and iron are also released by mineral weathering.

In subhumid and arid regions the release of potassium is fast enough for crop production but must be supplemented by the addition of potash fertilizer salts in more leached soils of humid regions.

Soil chemists have rapid chemical tests for measurement of the amounts of plant-available K, P, and N, as well as other elements in soils which are essential to plants. When the quantity of an element is too low for efficient crop production, it is added as fertilizer, such as KCl, $Ca(H_2PO_4)_2$, or ammonium or nitrate salts. Large chemical fertilizer industries are required for mining, refining, and preparation of chemical salts for soil application as fertilizers.

Extraction of soils for analysis of the readily available nutrients include replacement of exchangeable cations by salt solutions, dilute acids, and dilute alkalies such as $NaHCO_3$. Fluoride solutions are employed to repress iron, aluminum, and calcium activity during the extraction of phosphorus. Extraction of the soil solution is effected by displacement in a soil column, often through the application of pressure across a pressure membrane. The soil solution is analyzed by conductance and elemental analysis methods. Also, the total elemental analysis of soils is made by Na_2CO_3 fusion of the soil followed by classical geochemical analysis methods.

Organic compounds of great variety have accumulated in soils as residues from plant and animal life of the soil. The more unstable compounds of these residues are rapidly oxidized to CO_2 and H_2O by biochemical processes, while the more stable fractions accumulate. Conjugated ring compounds containing the elements C, H, O, N, P, S, and several other elements in small quantities accumulate in relatively stable organic and organomineral colloidal complexes. Lignin-like, phytin-like, and nucleoprotein-like compounds are included. Sorption of the organic matter on mineral colloid surfaces, particularly on layer silicates, such as montmorillonite, helps to stabilize the organic matter against biochemical oxidation. In tropical soils, high stability of soil organic matter is imparted by coatings of aluminum hydroxide and red ferric oxide. Organic and iron oxide colloids, when fairly abundant, stabilize the soil into porous aggregates through which ample air and water can circulate. Decomposition of soil organic matter, especially when hastened by tillage, gradually releases HNO_3, H_2SO_4, and H_3PO_4 in amounts which are highly significant in nutrition of crops. Much of the nitrogen, sulfur, and phosphorus required by crops is furnished in this way.

The oxidation potential of well-aerated soils is low (−0.5v) and of reduced soils is high (+0.30v). These relationships are sometimes expressed by soil scientists as reduction potentials or redox potentials, in which case the algebraic signs are the opposite. The oxidation potential is advantageously measured with a platinum-blackened electrode in the soil in place in the field. Moderately good aeration is a requirement of a productive soil. The oxidation status may also be tested in the field by rapid spot tests for ferric and ferrous iron in soils. Most of the dilute acid-soluble iron is in ferric form in well drained soils. Localized spots of decomposing organic matter are important in reducing

small but important quantities of iron to ferrous form and manganese to divalent form so as to be available to plants. Moderately to highly alkaline soils sometimes have inadequate activity of the reduced forms of iron and manganese, particularly in the absence of sufficient organic matter. Small quantities of Cu, Zn, B, and Mo must be present in productive soils in forms which have enough activity to be available to growing plants.

M. L. JACKSON

Cross-references: *Colloid Chemistry, Geochemistry, Ions, Fertilizers, Agricultural Chemistry, Clays.*

SOLID SOLUTIONS, METALLIC

Metallic bonding is not completely understood, and as a consequence there is not unanimity in interpretation and classification of metallic structures. Rather than attempt to do justice to all schools of thought, the present discussion is based only on what appear to be the two dominant factors: electron-to-atom ratio and atomic size.

A metal is an array of positive ion cores, the atoms stripped of their valence electrons, imbedded in an electron gas made up of the valence electrons. This is essentially the free-electron model expounded about 50 years ago by H. A. Lorentz. This model was derived to account for the outstanding characteristics of metals, their high electrical and thermal conductivities. It, however, also provides an explanation for the generally observed crystal structures, simple, high-density, high-symmetry structures: BCC (body-centered cubic), FCC (face-centered cubic), and HCP (hexagonal-close-packed). The metallic bonding is due to the coulomb attraction between the electron gas and the ion cores. The three typically metallic structures all tend to minimize the average electron gas-ion distance and hence maximize the bonding energy.

The free electron gas model gives little reason for the preference for any particular one of the metallic structures, yet there is a remarkable regularity among the elemental structures. From this one might be led to believe that symmetry is an important aspect of the metallic bond. However, one can estimate the importance of symmetry by comparing the heat of transformation from one allotropic form to another, that is, from one symmetry to another, with the total bonding energy which is given by the heat of vaporization. In general, it is so found that the symmetry change affects the bonding energy by less than 1%.

From the electron gas model we might expect that we could regard a metallic solution as having an electron gas concentration given by the average valence of the components, the elements themselves being just points on a continuous bonding versus average valence curve. Thus, for example, if we would make up a series of, say, Mo-Rh alloys, we might expect and would find that the Mo-rich alloys are BCC, just like Mo; that the Rh-rich alloys are FCC, just like Rh; and that intermediate compositions are HCP, which is just the structure of Tc and Ru, the elements that lie between Mo and Rh.

Were average valence (or as metallurgists and physicists prefer to call it, the electron-to-atom ratio (e/a)) the only factor in metallic bonding, the many types of alloy structures that are known to occur presumably would not exist. We have already seen that the small symmetry dependence of the bond energy is capable of imparting a remarkable regularity among the crystal structures of the elements. Thus, also in alloy structures, we can expect to find that energetically relatively unimportant factors can determine the crystal structure. The most frequently invoked alloy structure determining factors are electronegativity, Brillouin zone geometry, and atomic size, but by far the most important of these is atomic size. We shall in fact limit the ensuing discussion to the role of this factor only.

Metallic solutions may be of two components (binary), three components (ternary), or more. However, multicomponent solutions do not introduce any fundamentally new structural considerations and in general we shall limit our discussion to binary solutions, A-B with A the solvent and B the solute. Various measures of atomic size have been employed and we shall adopt atomic volume, v_A and v_B, as the most appropriate measure.

Substitutional Solutions. When $v_B/v_A \cong 1$ a solution can be formed by simply replacing solvent by solute atoms. However, as v_B/v_A deviates from unity this replacement scheme introduces strain energy and the substitutional solution becomes less stable. The amount of strain energy for a given atomic size disparity depends on the elastic constants of the particular solution, but as a rough rule, substitutional solution extent is restricted if $\frac{1}{2} \lesssim v_B/v_A \gtrsim 2$. This is equivalent to Hume-Rothery's famous 15% radius difference rule.

Ordered Substitutional Solutions. A substitutional solution can to some extent relieve atomic size disparity strain energy by ordering, that is, by an alternation of large and small atoms. When there is just a tendency toward such an alternation the solution exhibits *short-range order.* However, at or in the neighborhood of certain stoichiometric ratios the alternation can be perfect or nearly perfect and then the solution exhibits *long-range order.* BCC substitutional solutions frequently exhibit long-range order based on AB stoichiometry with all cube corner sites A occupied and cube center sites B occupied. Examples of such solutions are CuZn, FeAl, and ZrRu. It should however be emphasized that while atomic size difference is a possible origin of ordering, other origins based on electronegativity or Brillouin zone effects are also possible, and it is probably fair to say that in no alloy has the origin of ordering been unambiguously identified.

FCC substitutional solutions frequently exhibit long-range order based on A_3B stoichiometry with all of the cube faces A occupied and all cube corners B occupied. Examples of so-ordered solutions are Cu_3Au, Au_3Cu, and Fe_3Pt.

Interstitial Solutions. When $v_B/v_A \ll 1$, the solute atoms may enter the solution in the interstices between solvent atoms. In general, only the light elements, hydrogen, boron, carbon, nitrogen, and oxygen, are sufficiently small to form interstitial solutions. As a rough rule one may take $v_B/v_A \lesssim \frac{1}{2}$ as the limit for interstitial solution formation. Just about all metals satisfy the atomic size requirement for interstitial solutions with the above-noted light elements, but in fact relatively few extensive interstitial solutions exist, illustrating the fact that the size factor is a necessary but not a sufficient condition for solution formation. Examples of extensive interstitial solutions are all of the Group IVB and VB metals with H, C, N, and O, Fe(FCC) −C but not Fe(BCC), Fe(FCC) −N but not Fe(BCC), and Pd-H.

Ordered interstitial solutions also occur: hydrides, carbides, and nitrides, of stoichiometry MX, M_2X, and M_4X. While oxygen does enter into interstitial solution, the oxides can generally not be considered as interstitial phases. By the time that the oxide composition is approached (MO, MO_2, MO_3) the electronegativity of the oxygen atom has resulted in the oxygen approaching an O^{--} ion and hence being in fact much larger than the metal M^{+-} ion. The interesting transition of the oxygen from a small interstitial solute, probably positive ion, to a large negative ion in the oxide does not appear to have received much attention.

Intermetallic (Size) Phases. For $v_B/v_A \gg 1$, the strain energy in a substitutional solution would be excessive so that a new packing, frequently resulting in very complex crystal structures, is assumed. The packing in the FCC and HCP lattices results in the maximum number of nearest-neighbor bonds, twelve, of equal-sized atoms. When, however, smaller atoms are packed around a larger atom it is clear that more than twelve nearest-neighbor bonds can be formed. Thus in the $v_B/v_A \gg 1$ intermetallic phases there may occur nearest-neighbor bonding (coordination) numbers of 14, 15, or 16. In general it would appear that not much chemical bonding significance should be placed in the coordination numbers or the symmetry of such structures since they are simply consequences of the problem of filling space by unequal-sized atoms.

The Laves phases, of which there are some 250 known, of stoichiometry A_2B are good examples of a size-determined intermetallic structure. For example, the $MgZn_2$ form of this structure can be looked at as formed from an HCP A lattice from which two out of every four atoms are removed and replaced by one B atom. This forms the stoichiometry A_2B with an ideal size ratio $v_B/v_A = 2$ and it is found that Laves phases tend to occur at about this size ratio.

Vacancy Defect Phases. When we form a substitutional solution with $v_B/v_A > 1$, the lattice can relax the strain by forming a lattice vacancy in the vicinity of the larger solute atom. The Al-rich part of the AlNi phase is an example of a defect solution. AlNi is a long-range order CuZu-type structure. Increased Al concentration is obtained not by replacing Ni atoms by Al atoms but rather by creating vacancies of the Ni lattice. The extent to which this is due to strain relief around the larger Al atom or to other effects is not known.

Many intermetallics, for example, Laves phases, exhibit wide deviations from stoichiometry. Whether this is accomplished by replacing atoms by a disparate-sized one or by vacancy formation is not yet known but it is probable that the latter mechanism is frequently employed.

P. S. RUDMAN

Cross-references: *Phase Rule, Solutions, Crystals and Crystallization, Allotropy.*

SOLID STATE

Solid state physics is that field which encompasses the experimental investigation and theoretical interpretation of the physical behavior of matter in the solid phase.

Structure. Most solids are crystalline because the energy of the ordered arrangement is less than for the disordered one. On the basis of the symmetry exhibited by the three-dimensional array of the atoms, crystals are categorized as belonging to one of seven crystal systems: triclinic, monoclinic, orthorhombic, tetragonal, hexagonal, cubic, and rhombohedral. Consideration of rotation-reflection axes leads to the further division of the seven systems into thirty-two crystal classes. Crystals may be classified according to the type of chemical binding present. Binding arises predominantly from electrostatic forces and quantum effects due to the motion of the atomic electrons. The five classes of binding with examples are: ionic (alkali halides), covalent (diamond), metallic (alkali metals), molecular (inert gases), and hydrogen-bonded (ice). Information about the arrangement of atoms in crystals is obtained by the diffraction of x-rays having wave lengths comparable with atomic spacings in solids. Extensive studies of amorphous solids and liquid-crystals accompany the development of uses for such substances.

Imperfections. Actual crystals are not perfect, and it is known that imperfections play an essential role in crystal growth, diffusion, absorption, luminescence, and other physical processes. Among the imperfections are lattice vacancies and interstitial atoms; these have a marked effect on the optical and electrical properties of the crystal. Such defects are produced in solids irradiated by nucleons and electrons. In alkali halides, a deviation from stoichiometry produces coloring (color centers) and accompanying absorption phenomena. A large part of the optical and electrical behavior of semiconductors may be attributed to free electrons (or holes) liberated by foreign atoms in the lattice. Certain types of crystal irregularity that may be associated with missing or extra planes of atoms in part of the crystal are termed dislocations; these are particularly important in determining mechanical properties and the mechanism of crystal growth. Impurity atoms, such as Cr in ruby, play an important role in solid state lasers and masers. These atoms are "pumped" to an energy level higher than the ground state and then are stimulated by electromagnetic radiation to emit the excess energy at the same frequency as the stimulating radiation and in phase, thus producing a coherent amplified beam.

Thermal Properties. Theories of the specific heat of solids start from the assumption that the vibrational energy of a system of N atoms is equal to the energy of a system of 3N harmonic oscillators. Thus, the main problem in the theory of specific heat is the determination of the frequency spectrum of the oscillators. The Debye theory, based upon the vibrational modes of a continuous medium with a frequency range extending up to a cut-off frequency, predicts that the specific heat of the lattice can be approximated at very low temperatures by a contribution directly proportional to the cube of

the absolute temperature T and as asymptotically approaching $3R$ (R = gas constant) at very high temperatures. The shape of the specific heat vs. T curve in the intervening range is determined by a parameter, the Debye temperature, which is characteristic of the lattice and which can be correlated with the lattice frequency spectrum. Experiment showed that in metals at low temperatures an additional specific heat directly proportional to T is to be added to the T^3 term predicted by Debye. This contribution is understood as arising from the conduction, or free, electrons in the solid, and thus this term is quite important in metals and detectable in some semiconductors. The thermal conductivity of a solid is made up of two contributions, one from the lattice and the other from the free electrons, with the latter term dominating in metals and the former in nonmetals. Actually the lattice contribution may be large, and thus the thermal conductivity of some semiconductors is comparable with that of metals.

Band Theory. The behavior of electrons in metals varies from one metal to another, but one can treat metals on the basis of the free electron theory, according to which an appreciable fraction of the electrons in a metal specimen (of the order of one per atom) is able to move freely within the sample subject only to the potential barriers at the surfaces. More information about the behavior of all solids, including non-metals, is obtained by considering that an electron passing through a crystal undergoes a periodic variation in potential energy which is correlated with the periodicity of the crystal lattice. Whereas the simple free electron model permits all values of the electron energy, the introduction of a periodic potential yields forbidden energy ranges for which solutions representing an electron moving through the crystal do not exist.

Near the top or bottom of an allowed energy band, the energy is approximately a quadratic function of the wave number (2π times the reciprocal of the electron wave length). This dependence permits the determination of an effective mass which is used in describing the motion of the electron, or its associated wave packet, in applied electric or magnetic fields. In metals either allowed energy bands are only half-filled in the ground state or there is overlapping between filled and empty allowed bands; in either case the electrons can readily make transitions to empty allowed states and thus be accelerated by applied fields. In insulators the electrons completely fill the states in an allowed band and a sizable (forbidden) energy gap exists between the top of this filled band and the bottom of the next band of allowed energy states which are empty. Thus, a field can accelerate electrons only if a sizable amount of energy is supplied by thermal activation, optical excitation, or very strong applied electric field.

Semiconductors, which have electrical resistivities falling between the ranges ascribed to metals and to insulators, are characterized by (1) narrower forbidden energy gaps than insulators, making it relatively easy to stimulate *intrinsic* conduction thermally, optically, or electrically, and (2) the introduction of states into the forbidden band by the presence of appropriate impurity atoms (choice of impurity depends on the semiconductor, e.g., Group

III or V elements in the Group IV semiconductors silicon and germanium). *Impurity* conduction occurs when electrons are excited from impurity levels just below the empty allowed band into that band, or when impurity levels just above the top of the filled band accept electrons from the filled band and leave in the filled band "holes," which act like free charge carriers of positive sign.

Effective mass values of electrons and holes have been experimentally determined for a number of solids by the technique of cyclotron resonance. This information, combined with other experimental results including magnetoresistance, optical absorption, and photoconductivity, yields a picture of the band structure as a function of the wave vector k (magnitude equal to the wave number and direction that of the electron momentum). Electron spin resonance gives values for the corrections due to spin-orbit interaction. Further data come from quantum phenomena (e.g., Shubnikov-de Haas effect) of the type in which magnetic field dependence of various properties such as magnetic susceptibility and electrical resistivity, measured at very low temperatures, shows an oscillatory character. All of these results are used to construct the surface of constant energy (Fermi surface) in k-space; this surface is characteristic of the solid and can be used to predict many of the physical properties of the material.

Electrical Properties. The electrical properties of a solid are primarily dependent upon the concentration of charge carriers (n) and the carrier mobility (μ), which is defined as the drift velocity acquired by a carrier in an electric field of unit intensity. An estimate of n is obtained by measuring the Hall coefficient, which is the ratio of the transverse electric field set up in the Y-direction divided by the product of the current density flowing in the X-direction and the magnetic field intensity applied in the Z-direction. In appropriate units the Hall coefficient of a metal equals $1/(ne)$, where e is the electronic charge. In semiconductors the same equation applies, if carriers of only one sign are present, except that the right side is to be multiplied by a statistical factor of the order of unity. The sense of the Hall electric field indicates the sign of the charge carrier. The electrical conductivity is given by $ne\mu$, and conductivities due to positive and negative carriers are arithmetically additive. The mobility is experimentally determined by combining conductivity and Hall coefficient measurements; theoretically the mobility is found by studying the collisions of charge carriers with lattice ions and impurities. In metals, in first approximation, n is temperature-independent and μ goes as $1/T$, so that the resistivity is directly proportional to T. As absolute zero is approached, the resistivity of a metal or alloy either approaches a constant value dependent on the disorder in the lattice, or the substance becomes a superconductor. In the latter case, the resistivity drops sharply, at a transition temperature, from the normal curve to a value immeasurably close to zero. In 1957 Bardeen, Cooper, and Schrieffer developed a fruitful theory according to which a quantum mechanical interaction of conduction electrons by pairs reduces their energy so that, in summation, there is a small energy gap between the normal and superconducting state that accounts for the phenomena characteristic of superconductivity. Niobium compounds in the superconducting state are used as the field windings of electromagnets to carry currents sufficient to generate fields of the order 100 kilogauss.

The resistivity of intrinsic semiconductors and alloys has the exponential temperature dependence given by proportion to exp $(E_g/2kT)$, where E_g is the width of the forbidden energy band and k is the Boltzmann constant. The resistivity of an impurity semiconductor is more complex in its temperature-dependence. Magnetoresistance refers to the increase observed in the resistance when a magnetic field is applied, usually transverse to the current flow. In a metal, in first approximation, the magnetoresistive ratio (resistance change divided by original resistance) is proportional to the square of the magnetic field strength with the proportionality factor containing μ^2. In semiconductors the behavior is more complicated, but the study of orientation effects has yielded essential information about the shapes of the energy bands.

Magnetic Properties. The magnetic characteristics of solids ultimately arise from the properties of orbital electronic motion and unpaired electron spins. A diamagnetic contribution is produced by an applied magnetic field in all atoms and ions, but this contribution may be more than balanced by a paramagnetic contribution. Diamagnetism may be thought of as an application of Lenz' Law to the orbital electronic motion. Diamagnetic susceptibilities are practically independent of temperature. Paramagnetism arises from the orientation of permanent magnetic dipoles with components parallel to the applied magnetic field. Since the orientation is influenced by thermal motion, paramagnetic susceptibilities are approximately inversely proportional to the absolute temperature. The permanent magnetic dipole moments arise from unpaired electron spins, especially in incompletely filled electron shells such as the $3d$ shell in elements 21 through 28 and the $4f$ shell in the rare earth elements. Ferromagnetism is characterized by a spontaneous magnetization such that the atomic magnetic moments throughout a small region (called a domain) are aligned to a high degree by a molecular field inherent in the material. This type of magnetization is a typical "cooperative" phenomenon, i.e., it arises from an interaction among atoms which can be attributed to exchange forces of a quantum mechanical nature. The application of an external magnetic field readily produces alignment of the domain moments with field and thus leads to the large magnetic induction characteristic of the ferromagnetic. If the sign of the exchange integral describing the interaction between atoms is negative instead of positive, ferromagnetism gives way to antiferromagnetism, a state in which neighboring spins are lined up antiparallel rather than parallel. The most readily observable result is that the susceptibility shows a maximum as a function of temperature. Ferrites (e.g., Fe_3O_4) demonstrate ferrimagnetism, which is attributed to antiparallel alignment of the two Fe^{3+} spins with the resultant molecular magnetic moment coming from the Fe^{2+} ion.

V. A. JOHNSON

Cross-references: *Crystals and Crystallization; Magnetochemistry; Semiconductors; Crystals, Color Centers in.*

References

Kittel, Charles, "Introduction to Solid State Physics," John Wiley and Sons, Inc., New York, 3rd edition, 1967.

Azaroff, Leonid V., and Brophy, James J., "Electronic Processes in Materials," McGraw-Hill Book Co., New York, 1963.

SOLUTIONS

The equilibrium of a saturated solution represents a balance between the potentials and entropies of the molecules present in the two phases. These depend upon pressure, temperature, and the kind and strength of the attractions between the molecules. The attractions may be classified as interactions between ions, dipoles, metallic atoms, and the "electron clouds" of nonpolar molecules, differing among themselves in kind, in range, and in strength. The potential energy of the molecules of a nonpolar liquid is measured appropriately for the purpose of solubility relations by its energy of vaporization per cc, called its "cohesive energy density." The square root of this quantity will be used below as a "solubility parameter" δ.

We consider, first, the mutual solubility of two nonpolar liquids, whose molecules have practically equal sizes, and equal attractive and repulsive forces. When they are brought into contact, thermal agitation will cause mutual diffusion until the two species are uniformly distributed. The mixing process has produced maximum molecular disorder, and therefore entropy, which is given by the expression, for 1 mole of solution,

$$\Delta S^M = -R(x \ln x_1 + x_2 \ln x_2), \qquad (1)$$

where R is the gas constant and x_1 and x_2 the respective mole fractions. The partial molal entropies of transfer of 1 mole from pure liquid to solution are

$$\bar{s}_1 - \bar{s}_1^0 = -R \ln x_1, \qquad (2)$$

for component 1, and with subscript 2, for the other component.

The partial molal free energies of transfer are related to the *fugacities* in pure liquid, f^0 (vapor pressure corrected for deviation from the perfect gas law), and in solution, by the equations,

$$\bar{F}_1 - \bar{F}_1^0 = -R \ln (f_1/f_1^0). \qquad (3)$$

and its counterpart.
Liquids such as are here postulated mix with no heat effect; therefore $F_1 - F_1^0 = -T(s_1 - s^0)$, etc.; therefore

$$f_1/f_1^0 = x_1 \qquad \text{and} \qquad f_2/f_2^0 = x_2 \qquad (4)$$

which is Raoult's law, and defines the *ideal* solution.

If one of the components of an ideal solution, e.g., component 2, is a solid, its fugacity, f_2^s, is less than the fugacity of the pure, supercooled liquid, and limits the amount that can dissolve to $x_2 = f_2^s/f_1^0$. The ratio f_2^s/f_2^0, can be calculated from its melting point and heat of fusion.

Most solutions deviate from Raoult's law. The curved lines in Fig. 1 represent positive deviations, with $f_1/f_1^0 > x$. The ratio f_1/f_1^0 is called *activity*, and $f_1/f_1^0 = a_1$ and $a_1/x_1 = \gamma_1$ (5) the *activity coefficient*.

Regular Solutions. The internal forces of a pair of liquids are seldom so nearly alike as to permit their mixture to obey Raoult's law very closely throughout the whole range of composition. In the absence of chemical interaction, the attraction between two different molecular species, provided their dipole moments are zero or small, is approximately the geometric mean of the attractions between the like molecules. Since a geometric mean is less than an arithmetic mean, the mixing is accompanied by expansion and absorption of heat. The partial molal heat of transfer per mole from pure liquid to solution is given with fair accuracy for many systems by the equation,

$$H_2 - H_2^0 = v_2\phi_1^2(\delta_2 - \delta_1)^2 \qquad (6)$$

and its cognate,
where $v \equiv$ molal volume, δ is a *solubility parameter,* the square root of the energy of vaporization per cc, and ϕ_2 is volume fraction.

Thermal agitation, except in the liquid-liquid critical region, suffices to give essentially maximum randomness of mixing, especially when one component is dilute, so that the entropy of mixing may be practically ideal, although the heat of mixing is not, and the partial molal free energy can be computed by combining the entropy and the heat terms, Eqs. (2), (5) and (6),

$$RT \ln a_2^s/x_2 = v_2\phi_1^2(\delta_1 - \delta_2)^2 \qquad (7)$$

This equation neglects the effects of expansion upon both the heat and the entropy, but the errors largely cancel when combined in Eq. (7).

A plot of a_2 vs. x_2 for symmetrical systems (i.e., $v_1 \approx v_2$) is shown in Fig. 1 for a series of values of the heat term. It shows how the partial vapor pressure of a component of a binary solution deviates positively from Raoult's law more and more as the components become more unlike in their molecular attractive forces. Second, the place of T in

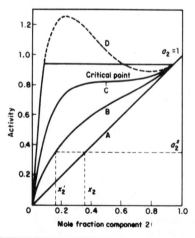

Fig. 1. Activity vs. Mole fraction for varying deviations from Raoult's Law.

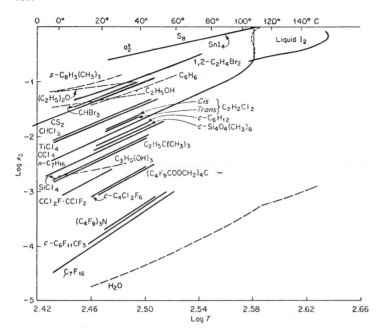

FIG. 2. Solubility of Iodine.

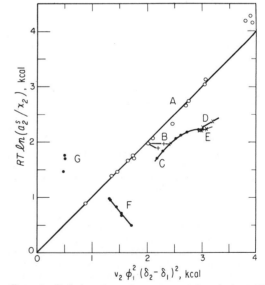

FIG. 3. Relation between energy of solution of iodine derived from measured solubility, x_2, and that calculated from solubility parameters. Line A, beginning at lower left: CS_2, $CHCl_3$, $TiCl_4$, *cis*-$C_{10}H_{18}$, *trans*-$C_{10}H_{18}$, CCl_4 *c*-C_6H_{12}, *c*-C_5H_{10}, $SiCl_4$, CCl_3CF_3, $CCl_2F\cdot CClF_2$, $C_4Cl_3F_7$, *c*-$C_4Cl_2F_6$, C_7F_{16}.
Line B, left to right: *c*-C_6H_{12} (on line A), *c*-$C_6H_{11}C_2H_5$ (below), *c*-$C_6H_{11}CH_3$, *c*-$C_6H_{10}(CH_3)_2$.
Line C, left to right, normal paraffins: $C_{16}H_{34}$, $C_{12}H_{26}$, C_8H_{18}, C_7H_{16}, C_6C_{14}, C_5H_{12}.
Line D, left to right: 2,3-$(CH_3)_2C_4H_8$, 2,2-$(CH_3)_2$ C_4H_8.
Line E, left to right: 2,2,3-$(CH_3)_3C_4H_7$, 2,2,4-$(CH_3)_3C_5H_9$.
Line F, top to bottom: C_6H_6, $C_6H_5CH_3$, *p*-C_6H_4 $(CH_3)_2$, *m*-$C_6H_4(CH_3)_2$, 1,3,5-$C_6H_3(CH_3)_3$.
Group G, from top: 1,2-$C_2H_4Cl_2$, CH_2Cl_2, 1,1-$C_2H_4Cl_2$.

the equation shows that the deviation is less the higher the temperature. Third, when the heat term becomes sufficiently large, there are three values of x_2 for the same value of a_2. This is like the three roots of the van der Waals equation, and corresponds to two liquid phases in equilibrium with each other. The criterion is that at the critical point the first and second partial differentials of a_2 and a_1 are all zero.

The presence of a dipole in one component adds a temperature-dependent component to its self-attraction and also induces a dipole in the other component. The effect can often be allowed for, for practical purposes, by an empirical adjustment of its solubility parameter.

If the dipole is hydrogen bonding, then this component is "associated," and it mixes less readily with a nonpolar second component.

If the components are, respectively, electron-donor and acceptor, or basic and acidic in the generalized sense of Gilbert Lewis, negative deviations from Raoult's law occur, with enhancement of solubility.

The effects of these various factors are well illustrated by solutions of iodine, I_2. In Fig. 2 are plotted the saturation values of log x_2 for iodine against log T. The slopes of the lines, when multiplied by R, give the entropy of transfer of iodine from solid to saturated solution. The solid lines are for violet solutions, from which chemical equilibria are absent. The positions of the lines are determined by the solubility parameters: how well is seen in Table 1, where δ-values are given for iodine in a spread of solvents calculated by means of Eq. 7 from the measured values of x_2. The broken lines indicate nonviolet solutions.

The factors that cause solutions of iodine to deviate from the behavior of regular solutions are illustrated in Fig. 3, in which values of the left hand member of Eq. 7 are plotted against those of the right for iodine solutions at 25°C. a_2^s is the activity of solid iodine; x_2 denotes measured solubility; v_2 is the extrapolated molal volume of liquid iodine, 59 cc.; ϕ_1 is the volume fraction of the solvent, ~1.0; $\delta_2 = 14.1$; δ_1 is the solubility parameter of the solvent. Illustrative values of x_2 and δ_1 are given in Table 1.

The points on line A are all for regular solutions, conforming to Eq. 7 over large ranges of x_2. Line B starts with a point for iodine in cycyohexane, next a point for methylcyclohexane, followed by one for dimethylcyclohexane. The point below is for ethylcyclohexane. Line C is for normal alkanes, from $C_{16}H_{34}$ to C_7H_{12}; groups D and E are for branched alkanes. Displacements from line A increase with increasing ratios of $-CH_3$ to $-CH_2$. The reason for this is not clear.

Line F contains points for aromatics, from benzene at the top to mesitylene at the bottom. All complex with iodine, altering its color. Group G consists of CH_2Cl_2 and 1,1- and 1,2-$C_2H_4Cl_2$, with strong dipoles, which enhance energy of vaporization without increasing solvent power for iodine.

TABLE 1.   δ-VALUES FOR I_2, 25°C

Solvent	Molal vol. cc.	δ_1	100 x_2
n-C_7F_{16}	227.0	5.7	.0185
$SiCl_4$	115.3	7.6	.499
Cyclo-C_6H_{12}	109.	8.2	.918
CCl_4	97.1	8.6	1.147
$TiCl_4$	110.5	9.0	2.15
CS_2	60.6	9.9	5.46
$CHBr_3$	87.8	10.5	6.16

Gases. Gas solubilities may be expressed as (1) volume of gas dissolved in unit volume of solvent, known as the *Ostwald coefficient,* designated by γ; (2) the volume of gas reduced to 0°C and 1 atmosphere dissolved in unit volume of solvent, known as the *Bunsen coefficient,* designated α; (3) the mole fraction, x; or (4) the moles per liter, c, dissolved at 1 atmosphere partial pressure. Henry's law, that the amount of gas dissolved is proportional to its partial pressure, holds rather well at moderate pressures in the absence of a chemical equilibrium. The fact that a substance is a gas at 1 atmosphere and ordinary temperatures indicates that its attractive forces are low and that consequently its solubility will be greater in solvents with low δ-values; also that solubility of different gases in the same solvent will be higher the higher the critical temperature of the gas.

The solubility of a number of gases at 1 atmosphere partial pressure and 25°C expressed as RT ln x_2 is plotted in Fig. 4 against the squares of the solubility parameters of a number of solvents. A high amount of regularity is evident for all except the gases SF_6 and CF_4, whose molecules attract molecules of the solvents very selectively. Similar irregularity is evident in the case of the solvent $(C_4F_9)_3N$. In all other cases the positions of missing points could be predicted with confidence.

Variations of solubility with temperature are illustrated in Fig. 4 for 10 gases in cyclohexane. The slopes of the lines times the gas constant R

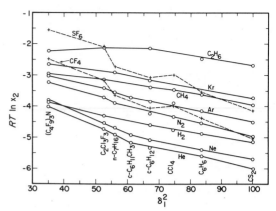

FIG. 4. Solubility of gases, log x_2, at 25°C and 1 atm vs square of solubility parameter of solvents.

give values for the entropy of solution. In decending from C_2H_6 to He the entropy increases from -8.7 cal/deg mole to $+8.1$ partly from increases in entropy of dilution, $-R$ ln x_2, but also because the successive gases attract the surrounding solvent molecules less and less strongly, but since they have the same kinetic energy they finally almost blow bubbles permitting more freedom of motion to adjacent molecules of solvent.

The foregoing interesting phenomena are treated at length in "Regular and Related Solutions," by J. H. Hildebrand, J. M. Prausnitz, and R. L. Scott, Van Nostrand Reinhold, New York, 1970.

Solid Solutions. The formation of a solid solution requires not only attractive forces which are not too different, but also identical crystal structures. The latter condition is found most frequently among solids whose molecules are rotating, giving highly symmetrical crystals. (See **Crystals**)

Metallic solutions, in the absence of compounds, follow the foregoing rules to a fair extent, but with added complications on account of the states of their electrons. The metals have a wide range of solubility parameters and exhibit many cases of incomplete miscibility in the liquid state.

Salt Solutions. The most obvious requirement necessary in a solvent for a salt is that it shall have a high dielectric constant, as is the case with water, liquid ammonia, hydrogen fluoride, and, in a smaller degree, methyl alcohol, in order to weaken the coulombic attraction of its ions for one another. It is possible to formulate the equilibrium between a solid salt and a solution of its ions by considering the changes in energy and entropy involved in vaporization of the solid to gaseous ions, and hydration of the ions. This would be relatively simple if the lattice energy of the solid and the hydration of the ions were solely electrostatic; but the process involves also van der Waals forces, polarization, covalent forces, hydrogen bonding, and entropy changes, which, in the case of water, are considerable, by reason of the ice structure persisting in water and the different structure of water of hydration. Consequently, such a breakdown of the problem, while it may serve to suggest comparisons, is better for explanation than for prediction.

The Periodic System offers the most useful guide

by virtue of the trends it reveals; e.g., the decreasing solubility in water of the sulfates and the increasing solubility of the hydroxides of the elements of Group II in descending the group.

Liquid ammonia, because of its lower dielectric constant, is in general a much poorer solvent for salts than water; but this is offset to some extent toward salts of electron-acceptor, Lewis acid cations by its greater basic, electron-donor character.

Insight into the nature of electrolytes in water solutions is afforded by their effects in varying concentration upon the freezing point of water, Δt, at varying concentrations, m moles per 1000 grams. In Fig. 5 $\Delta t/m$ is plotted against m on a logarithmic scale.

The molal lowering of nonelectrolytes is illustrated by sucrose and H_2O_2. These enter so easily into the hydrogen bonded structure of water that they give the theoretical lowering, 1.86° up to 0.1 M in the case of sucrose and to 10 M by H_2O_2.

Binary electrolytes, such as KCl, although completely ionized, even in the solid state, lower the freezing point less than $2 \times 1.86°$, even when as dilute as 10^{-3} M. This was at first attributed to incomplete ionization but is now explained by the long range of electrostatic forces. Note that Mg^{++} and SO_4^{--} are less independent than K^+ and Cl^-. $AgNO_3$, unlike KCl, etc., is a weak salt, and undissociated molecules increase rapidly with concentration. The ions nearer to an ion of one sign are those of opposite sign, therefore electric conductivity is less than the sum of ionic conductivities extrapolated to zero concentration.

This effect has been formalized in the concept of *ionic strength,* expressed as

$$I = \tfrac{1}{2}(m_1 z_1^2 + m_2 z_2^2 + m_3 z_3^2 + \cdots)$$

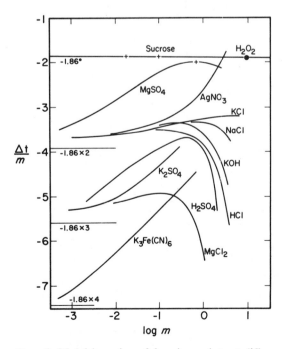

FIG. 5. Molal lowering of freezing points at different concentrations.

where z is the ionic charge. Applied to solutions of KCl, K_2SO_4 and $MgSO_4$ the values of I are respectively, 0.01, 0.03 and 0.04. Ionic strength is significant for dealing with the equilibrium and kinetic properties of an ion in mixtures of electrolytes.

Concentrated solutions are strongly affected by ionic hydration. Its strength depends upon ionic radius and charge, therefore it is in general stronger for cations than anions. K_2SO_4 and $MgSO_4$ both yield 3 ions, but the hydration is stronger for Mg^{++} than K^+, Na^+ than K^+, Na^+ than Ag^+. The line for K_2SO_4 ascends whereas the one for $MgSO_4$ plunges downward, (a) because the strong hydration of Mg^{++} diminishes the coulombic attraction of SO_4^{--} and (b) because it ties up molecules of water, decreasing the amount of solvent.

JOEL H. HILDEBRAND

SOLUBILIZATION

Solubilization, defined loosely, is the enhancement of the solubility of one substance, the solubilizate, by another substance, the solubilizing agent or solubilizer. More strictly, it is a process occurring in the presence of a solvent, whereby one species, the solubilizing agent, diminishes the activity coefficient of another species, the solubilizate, and both species are soluble thereafter, J. W. McBain, who coined this term, used it to denote the dissolution of an otherwise insoluble material brought about by interaction with micelles, a type of colloid, present in the solvent. The definition given here, however, is more inclusive than his original concept, and could be extended logically to systems whose characteristics are remote or completely apart from colloidal behavior. Practice nevertheless limits the term to usage in which there is either a close or a marginal relationship to micelles, and the literature of solubilization refers chiefly to systems in which the solubilizers are micelle formers. For example, potassium laurate solubilizes hydrocarbons in water, and calcium xenylstearate solubilizes water in hydrocarbons because of the micelle-forming nature of the respective solubilizing agents. However, the striking similarity among interactions between various agents and both soluble and insoluble species makes it undesirably arbitrary to restrict the term solubilization rigidly to its original usage.

Because absolute insolubility does not exist in nature, insolubility must be considered a matter of degree. Consequently, if an *apparently* insoluble species, in unlimited excess, is in contact with a solvent it must have a finite concentration and activity in the solvent at equilibrium. A solubilizing agent added to the system may interact with this species by coordination, hydrogen bonding, dipole interaction, complex formation, or in some other manner. In any case, the interaction results in a decrease of the effective concentration, or activity, of this species. Accordingly, more of the solubilizate progressively dissolves until its activity returns to the initial equilibrium value in the pure solvent, whereupon the activity coefficient is correspondingly less. If a species is freely soluble, or even infinitely miscible with a solvent, an interaction causes no *apparent* increase in the solubility of the species, but its activity, as evidenced by its osmotic behavior, nevertheless similarly decreases. The activity represents the tendency of the species to escape from the solution. Since solubility depends upon a balance between the opposing tendencies to enter and to leave the solution, the decreased activity is in effect equivalent to increased solubility. Solubilization is said to occur then, regardless of the independent solubility or insolubility in the pure solvent.

The salts of high-molecular weight organic acids are particularly important solubilizing agents. In nonpolar solvents such as hydrocarbons, they form colloidal aggregates known as association micelles. Most frequently such a micelle constitutes a limited number of salt monomers associated into a spheroidal cluster, with the polar ends of the salt monomers oriented toward the interior, and the nonpolar hydrocarbon ends at the periphery. Other polar species such as water, alcohol, acids, and dyes can be solubilized by these micelles in a variety of ways. In benzene solution, for example, zinc dinonylnaphthalene sulfonate can solubilize at least six moles of water for each equivalent weight of the salt present. The solution remains transparent, and no phase separation is observed. During the progressive addition of six moles of water per gram-equivalent of salt, the micelles expand to aggregations containing ten acid residues per unit, whereas the water-free micelles contain only seven. The water molecules are believed to be held in the polar core of the micelle where the environment is favorable to their retention.

Methanol, on the other hand, decreases the size of magnesium phenylstearate micelles. As methanol is solubilized by this salt in toluene solution, the micelle size decreases progressively from 23 salt monomers per aggregate to as little as 2 at a methanol concentration of 2% by weight. Each of these dimers is then associated with ten molecules of the alcohol. The partial pressure of methanol over the solution is demonstrably less than that over the salt-free methanol-toluene solution of equal methanol concentration. Rhodamine B dissolves very sparingly in pure benzene as the colorless and non-fluorescent base form. It is converted to the brilliantly fluorescent colored form by the addition of any of numerous micelle-forming solubilizers. No major changes in micelle size are believed to result from this solubilization, and it is postulated that a dye molecule replaces a monomer of the solubilizing agent in the matrix of the micelle.

In aqueous solutions, salts of high-molecular weight acids form micelles whose orientations are the reverse of those in nonpolar solvents. The hydrocarbon portions of the monomers, being insoluble in water, are oriented inward, whereas the ionic, or polar ends are oriented outward. Solubilization by these agents is complicated by dissociation of cations from surfaces of the aggregates and by the resulting surface charges developed. Both polar and nonpolar species such as hydrocarbons, dyes, alcohols, fats, organic acids, and a wide variety of soluble and insoluble species are solubilized in aqueous micellar solutions. Micelle enlargement is frequently said to follow from solubilization by these salts in aqueous solution, although the possibility of reduction of micelle size should not be excluded from consideration.

A nonpolar solubilizate such as hexane penetrates deeply into such a micelle, and is held in the nonpolar interior hydrocarbon environment, while a solubilizate such as an alcohol, which has both polar and nonpolar ends, usually penetrates less, with its polar end at or near the polar surface of the micelle. The vapor pressure of hexane in aqueous solution is diminished by the presence of sodium oleate in a manner analogous to that cited above for systems in nonpolar solvents. A 5% aqueous solution of potassium oleate dissolves more than twice the volume of propylene at a given pressure than does pure water. Dimethylaminoazobenzene, a water-insoluble dye, is solubilized to the extent of 125 mg per liter by a 0.05 M aqueous solution of potassium myristate. Bile salts solubilize fatty acids, and this fact is considered important physiologically. Cetyl pyridinium chloride, a cationic salt, is also a solubilizing agent, and 100 ml of its $N/10$ solution solubilizes about 1 g of methyl ethyl-butyl ether in aqueous solution.

Among other species that are good solubilizing agents are the nonionic compounds such as the polyethylene oxide-fatty acid condensates and the fatty esters of polyalcohols. A wide variety of nonionic solubilizing agents is possible, but most of those available are of variable composition. They can be effective in both aqueous and non-aqueous solutions.

The colloidal nature of some systems can disappear completely as solubilization proceeds. For instance, when methanol is solubilized by magnesium and sodium dinonylnaphthalene sulfonates, the aggregates decrease in size to a degree beyond which they can be considered micelles. In toluene solutions, micelles of these salts dissociate progressively on the addition of methanol increments until each of the particles in these solutions contains only one salt monomer when the methanol concentration reaches about 2% by weight. Probably the properties of some species which cause them to aggregate are those which make them good solubilizing agents, but it is evident that micelles are not a necessary condition for solubilization.

Accordingly, it is logical for solubilization to occur in systems which show no colloidal behavior, although frequently the effect in these cases is described by other proper terminology. Usually the term "solubilization" is applied in cases where the solubilizing agent is effective in small quantities, but arbitrary limitations of quantity might confuse the basic concept of solubilization. The terms cosolvency, hydrotropy, and "salting in" are used sometimes to describe effects which may be considered within the broad general scope of solubilization.

Applications of solubilization, although not always completely understood, range widely. A solubilizing agent can be used to bring an otherwise insoluble substance into solution where it is needed for a specific use, or it can be incorporated in a formulation to suppress the activity of an unwanted species which otherwise cannot be eliminated or prevented from occurring. In the pharmaceutical industry, drugs which are insoluble in pure water are solubilized by suitable agents to form homogeneous solutions. Dyes are solubilized for more efficient penetration and uniform coloring of fabrics. Soaps and detergents in aqueous solution are effective cleansing agents because they solubilize oily and greasy residues which may be flushed away from contaminated surfaces, although other effects may be equally important in the process. Removal of silver halides from photographic papers and films by aqueous fixing solutions may be considered solubilization by noncolloidal solubilizers. Certain oil-soluble salts dissolved in dry cleaning fluids can solubilize water. The water, which is solubilized in the micelles can in turn solubilize inorganic salts. The salts are then retained in the polar cores of the micelles where the water is held. This effect is referred to as secondary solubilization. In automotive fuels and lubricating oils, nonaqueous detergents are used to maintain engine cleanliness by solubilizing products of oxidation and combustion which tend to form sludges and gums, and to suppress the destructive effects of acids and other species generated in operation. Other solubilizing agents are used in these fluids to incorporate otherwise insoluble additives for oxidation and corrosion inhibition.

SAMUEL KAUFMAN

Cross-references: *Solubility; Solutions; Micelles; Colloid Chemistry.*

References

McBain, M. E. L., and Hutchinson, E., "Solubilization and Related Phenomena," Academic Press, Inc., New York (1955).

Shinoda, K., "Solvent Properties of Surfactant Solutions," Marcel Dekker, Inc., New York (1967).

Singleterry, C. R., "Micelle Formation and Solubilization in Nonaqueous Solvents," *J. American Oil Chemists Soc.,* **32,** 446 (1955).

Bascom, W. D., Kaufman S., and Singleterry, C. R., "Colloid Aspects of the Performance of Oil-Soluble Soaps as Lubricant Additives," Proceedings Fifth World Petroleum Congress (1959).

Kaufman, S., "Effect of the Cation on Solubilization by Oil-Soluble Sulfonates," *J. Colloid and Interface Science,* **25,** 401 (1967).

SOLVAY PROCESS

Ernest Solvay, chemist-industrialist, developed the modern process for the commercial manufacture of soda ash (commercial-grade Na_2CO_3) from common salt and limestone using ammonia as an intermediate in the reactions. The Solvay process is also sometimes called the ammonia-soda process. Soda ash produced by this process is relatively cheap; it first rivaled, then defeated the older Leblanc-process product.

Ernest Solvay was born at Rebecq, in the province of Brabant, Belgium, in 1838. At the age of sixteen, his father sent him to his uncle Semet, who was director of the gas works at Schaerbeek, a suburb of Brussels. Opportunities for an education were better there, although still limited. Young Solvay was able to obtain a certain amount of technical education, mainly in gas technology, together with plant practice at the gas works. He conceived the idea of condensing the residual ammoniacal liquors in order to make their handling more economical, and to facilitate their recovery by distillation. Among many experiments was one in which he observed that dissolving solid ammonium bicarbonate in sea water caused the precipitation of sodium bicarbonate—to him a discovery.

In 1861 the process using this observation was first installed in his small factory at Couillet, Belgium, and the commercial feasibility was demonstrated over the next few years. Actually the reactions were not new, but had been established by Fresnel in 1811. The essential reaction was the precipitation of sodium bicarbonate according to the reaction

$$NaCl + (NH_4)HCO_3 \rightleftarrows NaHCO_3 + NH_4Cl \quad (1)$$

The sodium bicarbonate was removed by filtration, washed and calcined to produce soda ash:

$$2NaHCO_3 \rightarrow Na_2CO_3 + H_2O + CO_2 \quad (2)$$

The carbon dioxide was returned to the process.

This seemingly simple process is not easy to carry out on a commercial scale. Five attempts had been made by different groups, and all had failed. The chief problem was the near-perfect recovery of the ammonia, since the circulating ammonia had a value greater than that of the product soda ash. However by efficient plant engineering, Solvay succeeded where others had failed. By 1874 he had constructed two large plants at Dombasle, near Nancy, France, and at Northwich, England. Soon after there were Solvay plants in every major country in the world, including the United States.

The commercial operation of the Solvay process is dependent on the availability of cheap salt and limestone. A solution of salt from a brine well is ammoniated and then passes downward through carbonation towers countercurrent to a rising stream of carbon dioxide to form ammonium bicarbonate. In the final "making" tower, sodium bicarbonate precipitates according to reaction (1). The carbon dioxide is produced from limestone in a conventional kiln according to the reaction

$$CaCO_3 \rightarrow CaO + CO_2 \quad (3)$$

The CaO is hydrated to form milk of lime, Ca$(OH)_2$, calcium hydroxide. Mother liquor from the filtration of the sodium bicarbonate goes to the ammonia recovery unit where it is treated with the calcium hydroxide and the ammonia distilled out:

$$2NH_4Cl + Ca(OH)_2 \rightarrow$$
$$CaCl_2 + 2NH_3 + 2H_2O \quad (4)$$

Thus the overall reaction, ignoring all recycling, is

$$CaCO_3 + 2NaCl \rightarrow CaCl_2 + Na_2CO_3 \quad (5)$$

The first Solvay process plant in the United States was constructed at Syracuse, New York in 1884. Today six manufacturers produce some five million tons annually of manufactured soda ash. In recent years the industry has fallen on hard times. Deposits of naturally-occurring soda ash in Wyoming can be exploited at about half the cost of constructing a Solvay process plant, and any growth of the industry will probably come from the natural product. By-product caustic from the manufacture of electrolytic chlorine is giving soda ash competition in some important areas. The existing plants are old and plagued by rising costs of labor, materials and maintenance. The most recent plant to be constructed in the United States was completed in 1935. Calcium chloride produced in the process is a waste which has little market, and disposal is increasingly difficult because of tightened environmental controls.

RICHARD STEPHENSON

SOLVENT EXTRACTION

Solvent extraction, particularly liquid-liquid extraction, is becoming one of the most widely applied processing techniques in the chemical and allied industries. It is one of the most important methods of producing and purifying such diverse things as vegetable oils, lubricating oils, various organic and fine chemicals, pharmaceuticals, and the heavy metals, uranium, thorium, and plutonium, for atomic energy use.

In the extractive metallurgy of the nonferrous metals increasing emphasis is being placed on solvent extraction studies both for primary recovery and for purification. Promising techniques are available for copper recovery and for the separation of copper, nickel, and cobalt. Two significant forces are combining to cause this trend. Ore grades for all metals are constantly falling so that many older metallurgical processes can not be economically applied. In this regard the recovery of nonferrous metals in general is working toward the area which that was first entered in uranium ore processing. Beyond this problem, plant operators and designers are facing continued pressure to minimize or avoid atmospheric pollution. Here, solvent extraction in combination with other hydrometallurgical techniques may show significant overall savings compared to older pyrometallurgical approaches.

In solvent extraction, a mixture of different substances is separated by the use of a liquid solvent. To effect a separation in this way, at least one of the components of the mixture must be insoluble or at least only partially soluble in the solvent. According to the nature of the mixture to be treated, the broad field is ordinarily broken down into two types of operation: (1) In liquid-liquid extraction the mixture to be treated is a liquid. (2) In leaching and washing, the mixture to be treated is either one of several solids or of solids and a liquid or liquids. In either event, the theoretical principles common to the various diffusional operations are applicable, and solvent extraction is in general analogous to the other "mass transfer operations," absorption, adsorption, and distillation. Consequently, the same major principles and types of data must be considered in the study of an extraction process as in any of the other separation processes. These are (1) equilibrium data which give the relationships between the concentrations of the various substances at equilibrium; (2) rate data which give the rates of diffusion and extraction in terms of the physical characteristics of the materials and "driving force" or departure from equilibrium existing in various parts of the system; and (3) material and, in special cases, energy balance data relating the quantities of the various materials involved and their energy content. Finally, the selection of equipment requires performance data from the various kinds of equipment that could be used.

There are a number of ways of carrying out an extraction operation; the simplest and least effective is the single-stage contacting of the entire quantities of feed material and associated solvent at one time. The first refinement of this simple operation is the division of the solvent into a number of portions, each of which in turn is contacted with the feed. This simple multiple contacting procedure allows a high degree of recovery of the material being dissolved by the solvent; however, the concentration of the combined extracts will be lowered proportionately each time a new portion of solvent is used and then added to the combined extracts. This difficulty of extract dilution can be largely overcome by resorting to either countercurrent multiple contacting or true continuous countercurrent operation, where in the ideal case the concentration of dissolved material in the discharged solvent can be very nearly that which would be in equilibrium with the concentration in the feed stock. As implied by the name, in either case the fresh solvent and feed enter at opposite ends of the system. Because of this, the nearly extracted feed is contacted with fresh solvent, allowing the maximum in recovery. Similarly, at the other end, the about-to-be-discharged solvent is contacted with fresh feed, allowing it to dissolve a maximum amount of the material being extracted.

Regardless of the method of contacting being used, two steps are characteristic of any extraction operation: (1) mixing of the solvent and the material to be extracted, and (2) settling or some other method of separating the resulting phases. The mixing and subsequent settling step taken together form an extraction stage or unit. The calculation of multistage extraction systems is ordinarily based upon a series of such stages or units of perfect performance, the "ideal" or "equilibrium" stage. A number of convenient graphical calculation methods have been developed. The most important of these is possibly the use of the triangular phase diagrams. The other commonly used graphical methods utilize either rectangular equilibrium-distribution diagrams or selectivity diagrams for the material being extracted.

Oil and Fat Recovery and Purification Processes. The two most important trends in oil and fat processing during recent years have both involved the use of solvent extraction. The first of these is the use of solvent extraction for recovering the oils from soybeans, cottonseed, and flaxseed. The solvent ordinarily used for seed oil extractions is hexane; isopropanol has also been used. Studies have been made using trichloroethylene, benzene, various alcohols, and petroleum ether for the extraction of oils from castor seed, wheat germ, milkweed seed, cottonseed, and soybeans. In general, such processes have been at a disadvantage economically due to the higher-priced solvent used. A more interesting recent trend in the technology of oils and fats has been the use of solvent extraction for the purification and separation of the various fats and oils. In the "Solexol" process, for example, liquid propane is used as a solvent at about 200°F. By lowering the temperature, a separation of the crude oil into edible oil, paint oil, a sterol concentrate, and finally a pigment lecithin can be effected. This same sort of technique is applied to a crude fish oil to separate oils of varying iodine-number, vitamin concentrates, stearin, and finally various fatty acids.

Glyceride oils may be extracted with furfural to yield fractions of different iodine-numbers or to make high vitamin concentrates. Furfural may also be employed in conjunction with propane for the solvent fractionating of fats and fat splitting.

Petroleum Hydrocarbon Processing. In the refining, extraction finds two important fields of application. Originally extraction was considered primarily as a method of purifying materials. The removal of color bodies and sludge formers from gasolines, naphthas, or distillates by extraction with aqueous ferric chloride-sodium chloride solutions and the caustic sweetening of distillates by contacting with aqueous sodium hydroxide are good examples of this sort of application.

More recently, extraction has become one of the more important ways of separating and purifying individual hydrocarbons or groups of hydrocarbons from various refinery streams. Liquid sulfur dioxide may be used to separate an oil or reformed naphtha into a fraction rich in aromatics and a paraffin-olefin fraction. Aqueous methanol is a selective solvent for removal of aromatic fractions from naphthas, and liquid ammonia may be used to recover the aromatic fraction from kerosenes and naphthas. In either case the aromatic fraction may be further separated into benzene and mixtures of toulene and xylenes.

Propane and propylene may be separated by extraction with furfural. Butylene and butadiene are being separated by a cuprous ammonium acetate solvent. Viscous materials such as monoethanolamine, diethanolamine, and the glycols may be used to extract the various light hydrocarbons, if the elevated temperatures are used.

The most important use of extraction in petroleum refining is in the production of lubricating oils. The most important processes volume-wise are propane deasphalting, furfural extraction, phenol extraction, cresylic acid-propane extraction, methyl ethyl ketone-benzol dewaxing, and propane dewaxing. About 95% of the lubricating oil manufactured in this country has been treated by one of these processes.

Pharmaceuticals. Extraction has become the most important means of recovering antibiotics and other fermentation products which are ordinarily produced in relatively dilute solutions. Penicillin is extracted from the filtered fermentation broth with amyl acetate and may be purified by a precipitation followed by reextraction with amyl or butyl acetate. Aureomycin and streptomycin may be recovered from their aqueous solutions by using isoamyl or amyl alcohol. In the initial production of streptomycin, the drug, after adsorption from the broth by activated carbon, was recovered as the hydrochloride by extraction with hydrochloric acid-alcohol solvent.

The various vitamin B_{12} factors may be extracted from fermentation broths by a double solvent such as benzene-phenol or carbon tetrachloride-cresol. Natural or synthetic vitamin A may be concentrated by extraction with acetonitrile-heptane.

Food Products. One of the economically most important extraction operations is the production of beet sugar, which currently accounts for about one-third of the world's sugar production. By the use of countercurrent extraction, a dark impure solution containing as high as 12% sucrose is produced. This syrup, by successive purifications and concentrations, may be processed to recover as high as 85 to 90% of the sucrose originally present in the beets as a chemically pure product.

An extraction process that has become of considerable importance in this country during the past few years is the manufacturing of soluble coffee products. By countercurrent extraction with hot water, practically all the solubles may be extracted from the roasted and ground coffee berries. This extract is ordinarily of such a concentration that it may be directly dried by drum or spray driers to produce the finished product.

Metals Purification. Solvent extraction has been found to be the most satisfactory method of purifying uranium, thorium, and plutonium for atomic energy purposes. Most of the extraction systems described in the literature have utilized nitric or hydrochloric acid solutions of the metals. The solvents utilized have been ethers or ether-like compounds, acetone or other ketones, and hydrocarbon solutions of organo-phosphates such as tributyl phosphate. The metals are usually recovered by re-extraction into water followed by a precipitation step of some sort. The use of solvent extraction to purify various other metals is possible, but it is seldom economically feasible.

The separation and recovery of several other rare or uncommon metals are feasible as technically and commercially successful processes by use of solvent extraction. Notable examples are the separate recovery of pure tantalum and niobium and of pure zirconium and hafnium from their respective naturally occurring mixtures which are otherwise most difficult to handle. This is very important in the case of producing hafnium-free zirconium for the cladding of uranium fuel elements in nuclear reactors, where its low neutron absorption cross section is a vital property. Hafnium has a high cross section and small amounts of it in zirconium renders the latter unsatitsfactory for reactor use. On the other hand, the high cross section of hafnium makes it a valuable control rod material and any zirconium in it reduces its effectiveness for this purpose.

DENNIS D. FOLEY

Cross-references: *Solvents, Solutions, Diffusion.*

SOLVENTS

A *solvent* is a substance capable of dissolving other substances to form a homogeneous system called a solution. Every concept of solvent is dependent on the solution concept. A *solution* is most explicitly defined as a system of two or more components in a single phase. A solution therefore may consist of gases, liquids or solids, or any combination thereof, provided they are in a single phase. The solvent usually is considered the dispersing medium and the component present in the largest amount. Another prevalent idea is that the solute is the component to separate first on cooling. The historical concept of a solvent as solely a medium to provide space for the solute to conform to the properties of a gas has been useful. It should be honorably retired as it may result in misleading conclusions and interpretations.

A solvent usually is thought of and used as a liquid. This idea is generally followed in recent books[1,2,3] and will be followed in this presentation. A solvent and a solute may consist of more than one component, but for practical purposes, they may be considered as a single entity. Some polycomponent solvents, frequently called mixed solvents, are liquids refined from natural sources which include ASTM refined solvent naphtha, turpentine, xylene, refined fusel oil, and petroleum ether and the blended solvents, or those mixes prepared to fulfill a specific need. Two-component systems will generally be used for discussion and illustration.

Solvent production in the United States is large. Following are some 1970 production figures:

	1970 Production lb $\times 10^9$
Benzene	8.6
1,2-Dichloroethane	6.5
Ethylbenzene	4.6
Ethylene glycol	2.5
Ethanol, synthetic	2.2
2-Propanol	2.1
Acetic acid	2.0
Cumene	1.8
Cyclohexane	1.8
Acetone	1.6

The principal uses of solvents are: media for chemical reaction, vehicles for coatings (paints, lacquers, printing inks, etc.), synthetic fibers, cleaners, separation, purification and physical processing in the chemical industry. When water is not considered, the use of inorganic solvents is small. If water is excluded, the coating industry is the largest user of solvents and chemical manufacture second.

Classification of Solvents. A solvent is not a property concept. The frequently used phrase "solvent property" was originally intended to mean "dissolving ability." The role of a solvent in many areas of present technology requires other attributes besides dissolving ability. Hence, the nature of a solvent as interpreted from its other attributes is a primary consideration for its classification and use. A solvent in itself is a complex system that may be characterized by a variety of chemical, physical, thermodynamic and use properties together with other attributes. The consideration of a substance as a solvent is always as a component, or a potential component, of a solution. The particular attribute to be considered depends on the system in which it will be used. The selection of a solvent to be used alone with a solute or which will be used to prepare a blended solvent is usually made from a class that has the most desirable characteristics. There are many ways that solvents may be classified but only a few involve multiple attributes.

The most common and best known classification is *chemical functionality.* This property is the first to be considered, as it is an index to the compatibility of the solvent and the solute. The several functionality classifications now in use substantially agree on the basic divisions but differ somewhat in arrangement and subdivisions. The following is a familiar basic arrangement:

1. Inorganic
   a. Liquefied gases, including ammonia, sulfur dioxide, hydrogen fluoride, carbonyl chloride.
   b. Miscellaneous, including antimony (III) chloride, selenium oxychloride.
   c. Water
2. Organic
   a. Hydrocarbons
   b. Hydroxy compounds.
   c. Ethers
   d. Carbonyls
   e. Acids
   f. Acid anhydrides
   g. Esters
   h. Halogenated hydrocarbons
   i. Nitro compounds
   j. Nitriles
   k. Amines
   l. Amides
   m. Sulfur compounds
   n. Polyfunctional compounds

There are several subdivisions for each of the basic classes.

Many of the *physical* and *thermodynamic properties,* generally referred to as physical properties, aid in the selection of a substance for a specific use. One or more properties may be considered in the selection, e.g., solubility parameter, dielectric constant, protolysis constant, boiling point, viscosity and dipole moment. Certain properties such as ebullioscopic and cryoscopic constants have limited application. There are many values derived from use characteristics, particularly in the coating field, that are very helpful in selecting solvents, such as the Kauri Butanol value, evaporation rate and flow characteristics imparted to the solution.

Some physical properties indicate the safe use of the solvent, such as flash point, minimum ignition temperature and flammable limits.

The physiological properties are often a major factor in the selection of a solvent. Odor is a primary consideration in perfumery and a major factor in the other areas of cosmetics, as well as in some pharmaceuticals and in spray bomb solvents. Taste must be a major consideration in all food products and in some pharmaceuticals. The toxicity to living organisms, particularly to humans, of solvent vapors is a dominant factor in the selection of solvents for many uses. The vapor toxicity is evaluated by the threshold limit value (the maximum amount of solvent vapor considered safe for the majority of humans eight hours a day, five days a week). Local air pollution regulations limit the quantity of organic solvents emitted as vapor into the atmosphere. Many solvent formulas have been changed to contain less noxious materials to meet ecological regulations.

Chemical Properties. Practically all use of solvents is dependent on their chemical nature although the dependency may not be readily apparent since it is evaluated by physical and thermodynamic information. One of the most useful classifications of solvents based on their chemical nature is *polar* and *nonpolar*. A polar molecule is one in which there exists a permanent separation of positive and negative charges, or one in which the centers of the positive and negative charges do not coincide. A nonpolar molecule is one that does not have a permanent separation of the charges or one in which the positive and negative centers coincide. The quantitative measure of polarity is the dipole moment, μ, expressed as

$$\mu = Er$$

where E is the magnitude of the total positive and negative charges and r is the distance between the centers of the charges. For nonpolar molecules $r = 0$, therefore $\mu = 0$. Nonpolar molecules are substantially chemically neutral and do not form coordinate covalent bonds readily. Polar molecules are chemically active and do form coordinate covalent bonds.

The properties of a solute may be important factors in the chemical nature of a solvent in a system. Boron trifluoride is a strong acid. Helium, one of the most nonpolar and inactive molecules, as a liquid solvent assumes a basic property in the presence of boron trifluoride. The solvent also plays an important role in the chemical nature of the solute. A substance may be acidic, basic or neutral depending on the solvent. Ammonium acetate, for instance, would be neutral in a nonionizing solvent. In an acidic solvent such as acetic acid, it is a strong base and in the basic solvent ammonia, it is a strong acid.

Some solvents act as either an acid or base depending on the nature of the solute. Such solvents are called *amphoteric*. Amphoprotic solvents ionize into acid, or positively charged particles, and base, or negatively charged particles. The best known of the amphoteric solvents is water.

$$H_2O + H_2O \rightleftharpoons H_3O^+ + OH^-$$

weak — strongest

acid base acid base

in water

Similarly ethanol can act as an acid or a base, but it is slightly less basic than water

$$C_2H_5OH + C_2H_5OH \rightleftharpoons C_2H_5OH_2^+ + C_2H_5O^-$$

Acetamide is a weakly acidic and weakly basic solvent and is a better acid-enhancing solvent than water

$$CH_3CONH_2 + CH_3CONH_2 \rightleftharpoons$$
$$CH_3CONH_3^+ + CH_3CONH^-$$

There are three types of amphoterism. One type is based on the formation of the *onium* ion; another based on *isomeric tautomeric* forms; the third is dependent on the solvent.

Hydrogen Bonding. Hydrogen bonding is the most common reaction with solvents. This is because hydrogen is present and available in more solvents than any other element or group of elements. Hydrogen bonding does not normally occur between nonpolar molecules containing hydrogen, e.g., hexane and decane do not show any indication of hy-

drogen bonding, whereas methanol and water exhibit two of the most common manifestations of hydrogen bonding: heat of solution and volume change. When 50 ml each of methanol and water are mixed, the resulting solution becomes quite warm and the volume shrinks to 96.36 ml at 25°. There are abnormal viscosity and refractive index changes also. The viscosities of water and methanol are 0.890 and 0.544 at 25°, respectively. A solution containing 40.3% methanol attains a viscosity of 1.672 cp. The refractive index-composition curve is a parabolic function. The molal heat of solution is an index to the extent of bonding, whether hydrogen or otherwise.

Molecules exist in many solvents in an associated state, the most commonly known of which are due to hydrogen bonding (see **Association**). Water is the best-known example. It is strongly bonded and is represented thus:

$$\left[\begin{array}{ccc} H & H & H^- \\ H-O \cdots H-O \cdots H-O \end{array} \right]_n$$

Hydrogen fluoride is one of the most strongly bonded substances with a bonding energy of about 10 kcal mole^{-1}:

$$[H-F \cdots H-F]_n$$

and the hydrogen-to-nitrogen bond is illustrated by ammonia which is less strongly bonded than water:

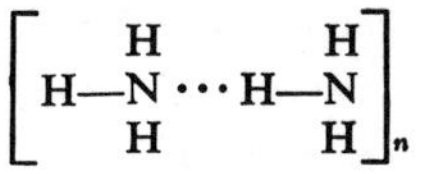

$$\left[\begin{array}{cc} H & H \\ H-N \cdots H-N \\ H & H \end{array} \right]_n$$

Carboxylic acids twice bonded by hydrogen to oxygen form dimers with a six-membered ring:

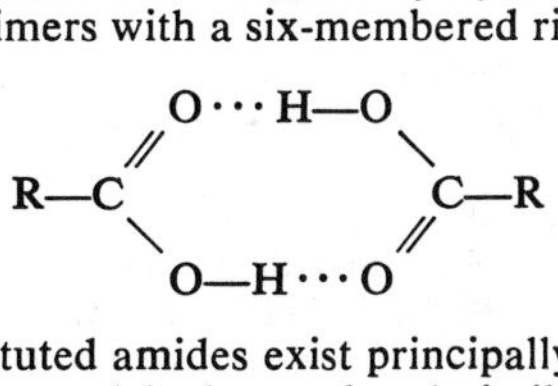

N-Unsubstituted amides exist principally in the *cis*-configuration and hydrogen bond similarly to carboxylic acids:

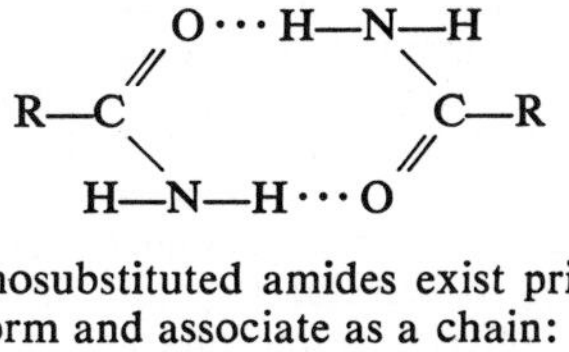

The N-monosubstituted amides exist principally in the *trans*-form and associate as a chain:

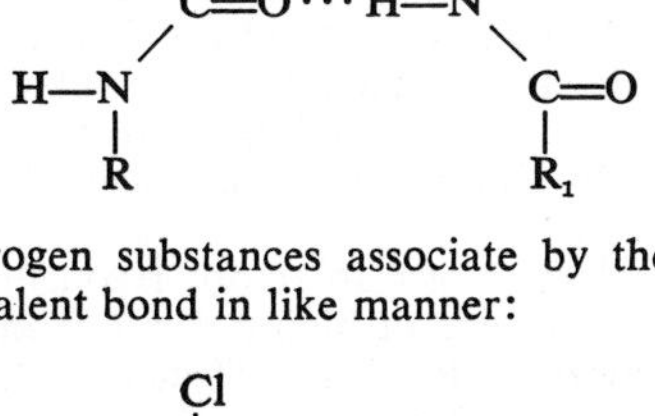

Nonhydrogen substances associate by the coordinate covalent bond in like manner:

It may be generalized that aprotic solvents are substantially noncoordinate covalent bonded and amphiprotic solvents are substantially coordinate covalent bonded.

A solvent molecule may exhibit different degrees of acidity or basicity under different circumstances, e.g., 2-nitropropane exists in two tautomeric forms:

$$H_3C-\underset{\underset{H}{|}}{\overset{\overset{CH_3}{|}}{C}}-NO_2 \rightleftharpoons H_3C-\overset{\overset{CH_3}{|}}{C}=N\overset{O^-}{\underset{OH}{\diagup}}$$

nitro form $\quad$ aci form
$K_1 = 2.12 \times 10^{-8}$ $\quad$ $K_1 = 7.73 \times 10^{-6}$

Carbonyl solvents exist in two tautomeric forms:

$$H-\underset{\underset{H}{|}}{\overset{\overset{R_1}{|}}{C}}-\overset{\overset{R_2}{|}}{C}=O \qquad H-\overset{\overset{R_1}{|}}{C}=\overset{\overset{R_2}{|}}{C}-OH$$

R_1 and R_2 are the same or different alkyl or similar groups or one or both may be hydrogen. The nitro and aci forms of the nitroalkanes are a variation of the keto and enol forms of the carbonyls.

Each solvent has its own extent of ionization which Kolthoff has termed *protolysis constant.* This is familiar for water as its ionization constant, but the term was nonspecific because it did not specify whether it was a selfionization or a solvent-induced phenomenon. The familiar protolysis constant of water, pK_s:

$$\underset{\text{acid}}{H_2O} + \underset{\text{base}}{H_2O} \rightleftharpoons \underset{\text{acid}}{H_3O^+} + \underset{\text{base}}{OH^-}$$
$$pK_s = H_3O^+ \times OH^- = 14.000 \text{ at } 25°$$

Similarly for methanol:

$pK_s = CH_3OH^+_2 \times CH_3O^- = 16.7$; pK_a in water is 15.5.

The ionic dissociation of a solvent increases with the dielectric constant. The ionization constants of solutes in a solvent are a function of the protolysis constant and the dielectric constant.

Solvents and Chemical Reactions. Solvents can change the extent of a chemical reaction. The preparation of many of the newer inorganic compounds has been made possible by some of the solvents that have become available in recent years such as N-methyl- and N,N-dimethylacetamide, dimethyl sulfoxide and hexamethylphosphoric triamide. Electrochemical reactions in nonaqueous systems are offering interesting and rewarding reaction possibilities. The yield of many chemical reactions may be increased by changing the solvent. For instance, Kornblum prepared primary and secondary nitroalkanes in decidedly improved yields by using dimethylformamide as the solvent instead of water. Hardy further improved the process by using dimethylsulfoxide as the solvent (see **Nitroparaffins**).

Chemical reactions are more frequent between solutes and solvents of high polarity than between those of low polarity. The molecules of the solvent water react with many substances, more especially with high polarity solutes, forming a variety of hydrates. As the polarity decreases the tendency to form *solvates* decreases. If the dielectric constant is used as the criterion of polarity the tendency to form solvates decreases, for example, in the order of water, ethanol and ammonia. The formation of solvates also depends on the difference in chemical quality of acid and base between the substance being solvated and the solvent.

Solubility. The rule of thumb *similia similibus solvuntur* was long the guide to solubility. It was a boon, but like practically all generalizing concepts it had its limitations. For instance, water and sugar are both rich in hydroxyl groups, therefore, water should be a good solvent for sugar, and it is. Chloroform and beryllium chloride contain about the same amount of chlorine. Chloroform should be a good solvent for the salt; the salt is insoluble. A thermodynamic method for estimating solubilities was developed some time ago. Only recently it has come into general use.

The "internal pressure" of a substance is a significant property in the theory of solutions and has been termed the *solubility parameter.* It may be defined as the energy of vaporization per cubic centimeter and expressed as

$$s = (\Delta H_v/V)^{\frac{1}{2}}$$

where ΔH_v is the molal heat of vaporization; V is the molal volume; s is the solubility parameter. The energy of vaporization usually is the easiest way to determine the solubility parameter of solvents. Determined heat of vaporization or that calculated from reliable vapor pressure measurements give the most reliable values.

The solubility parameter may be calculated from other properties. For example:
From surface tension (γ)

$$s = 4.1(\gamma/V^{\frac{1}{3}})^{0.43}$$

From the van der Waal constant for the gas (a)

$$s \cong a^{\frac{1}{2}}/V$$

From critical pressure (p_c)

$$s = 1.25p_c^{\frac{1}{2}}$$

and also from viscosity.

For nonpolar liquids this expression holds over small volume changes:

$$(\partial H/\partial V)_T = nH_v/V$$

but n is near unity for solvents adhering closely to Raoult's law. Also it can be shown that for nonpolar liquids

$$s = T(\partial P/\partial T)_V^{\frac{1}{2}}$$

A reliable method for the prediction of solubility is to compare the solubility parameters of the solvent and the solute. Solution occurs when the proposed components of the solution have about the same solubility parameters. For some solutes such as polymers, the extent of hydrogen bonding in the solvent and the solute should be about the same.

The solubility parameter may be estimated from solubility data provided the solubility parameter of one component is known. This is particularly useful for solid solutes in liquid solvents whose solubility parameter is known.

The solubility parameter is a function of temperature. Over a limited temperature range, the heat of vaporization approaches linearity and

$$H_v = H_v' - \Delta C_p (T - T')$$

and ΔC_p is the difference between the heat capacity of the liquid and the gas.

JOHN A. RIDDICK

Cross-references: *Association, Solubility, Solutions, Solubilization, Solvent Extraction, Toxicology.*

References

1. Riddick and Bunger, "Organic Solvents," Wiley-Interscience, New York, 1970.
2. Davis, "Acid-Base Behavior in Aprotic Organic Solvents," National Bureau of Standards Monograph 105, Washington, 1968.
3. Marsden and Mann, "Solvent Guide," 2nd Ed., Wiley-Interscience, New York, 1963.
4. Charlot and Tremillion, trans. by Harvey, "Chemical Reactions in Solvents and Melts," Pergamon, London, 1969.
5. Hildebrand and Scott, "Solubility of Nonelectrolytes," Reinhold, New York, 1950.
6. Wheland, "Resonance in Organic Chemistry," Wiley, New York, 1955.
7. Patty, "Industrial Hygiene and Toxicology," 2nd Rev. Ed., Vols. 1 and 2, Interscience, New York, 1963.

SPECIFIC GRAVITY

Specific gravity is defined as the ratio between the mass of the sample and the mass of an equal volume of a standard material. For liquids and solids, the standard material is usually water at 4°C (precisely 3.98°C), and for gases, air or sometimes hydrogen at the same temperature and pressure as the sample. In the laboratory, often density is measured, and the results expressed as specific gravity.

The density of water at 3.98°C is 0.999973 g/cm^3 or 1.000000 g/ml. Although precise measurements of density are made at a known temperature, gross measurements often disregard temperature. For example, the density of concentrated sulfuric acid is 1.834 g/ml. In such a case the acid was measured at 20°C. It is compared with water at 4°C, and its specific gravity expressed as 1.834_4^{20}. That is, sulfuric acid is 1.834 times as dense as an equal volume of water. Specific gravity, being a ratio, does not have measurement units as does density.

The specific gravity of aluminum, for example, is 2.7. The density of aluminum is 2.7 g/ml in the metric system, and 2.7×62.4 lb/ft^3 (the weight of 1 ft^3 of water) or 168.5 lb/ft^3 in the English system.

A rapid estimate of the specific gravity of a specimen can be made by using a set of liquids of known specific gravity and increasing in value. The sample is placed in each liquid successively until one is found in which it just floats. The specific gravity of the specimen and the liquid are then the same.

Specific gravity can be measured by using a weighed sample in a volumenometer. The sample is immersed in a liquid and the increase in volume is measured. The apparatus is held at constant temperature.

Solids. To find the specific gravity of a solid, a method that depends upon Archimedes' principle can be used. Weigh the solid. Immerse the solid in a liquid and reweigh while it is immersed. Divide the loss in weight by the density of the liquid (in the case of water, use 1). The weight divided by the loss of weight is the specific gravity.

A Jolly balance consists of a spring suspended from a fixed point and a wire basket immersed in a liquid to a fixed mark. The same steps as described in the previous paragraph are used, except that the basket is always immersed to a fixed mark.

Typical values for the specific gravity of solids range from gold 19.3 to cork 0.25.

Liquids. The most convenient method to measure the specific gravity of a liquid is to float a hydrometer in it. If the hydrometer has a narrow stem, readings may be made to one part in 1000. Hydrometers are used to measure the specific gravity of petroleum products, milk (lactometer), sirups in canneries, solutions in the chemical laboratory, etc. The pyknometer method can also be used for a liquid. The vessel is weighed three times: empty (W_1), full of liquid (W_2), full of water (W_3). ($W_2 - W_1$)/($W_3 - W_1$) is the specific gravity.

Another method is to use a solid of known weight. Reweigh the solid immersed in the liquid. Then weigh the solid immersed in water. The ratio of the loss of weights is the specific gravity because an equal volume of liquid was displaced in both cases.

The Westphal balance refines this method. A solid of known weight is used. It is suspended from the end of the arm of a balance. Then the beam is balanced while the solid is immersed in water. The solid is now immersed in the liquid under test, and the amount of weight needed to restore balance is recorded. Again, the specific gravity is the ratio of the buoyant forces.

A gravitometer is an instrument used for finding specific gravity of liquids. Consult the catalog of an apparatus dealer for details.

A column of liquid is balanced against a column of water connected to it in a U-shaped tube. The ratio of the heights of the columns gives the specific gravity of the liquid.

The specific gravity of typical liquids is: carbon tetrachloride 1.595, carbon disulfide 1.2628, glycerol 1.26, sea water 1.03, acetone 0.792, ethyl alcohol 0.7893, gasoline 0.68 to 0.72. Readings such as the above are changed to Baumé by using $B = (ks - k)/s$, where B is Baumé reading, s is specific gravity, and k usually 145 (possibly 146.78). This formula applies to values greater than one on the specific gravity scale.

Gases. To find the specific gravity of a gas, a light globe of large size is weighed (1) filled with dry air, (2) evacuated, (3) filled with the gas to be measured. The ratio of [(3)—(2)]/[(1)—(2)] is the specific gravity of the gas. The buoyant effect of the air on the globe must be considered.

Another method uses a delicate quartz fiber that has a quartz ball at one end and a hollow quartz sphere at the other. This apparatus is balanced while it is enclosed in dry air. When a gas other than air surrounds the apparatus, the buoyant effects on the two ends become different, and the displacement of the fiber is a measure of the specific gravity of the gas that surrounds it.

A third method for finding the specific gravity of a gas uses the time for it to pass through a given small opening under a given pressure and temperature. Then the time is measured for an equal volume of air at the same temperature and pressure to pass through the same opening. The times are proportional to the square roots of their specific gravities.

Typical values for specific gravities of gases, compared to air, are hydrogen 0.0656, ammonia 0.59, oxygen 1.11, carbon dioxide 1.53, chlorine 2.49.

ELBERT C. WEAVER

SPECTROSCOPY

Absorption

Absorption spectroscopy, in the most general sense, is the study of the absorption of radiant (electromagnetic) energy by a chemical species as a function of the energy incident upon that species. Within this broad definition, absorption processes are known throughout the electromagnetic spectrum ranging from the gamma region (nuclear resonance absorption or the Mössbauer Effect) to the radio region (nuclear magnetic resonance). In practice, however, absorption spectroscopy is limited to those absorption processes which are not followed by the emission of radiant energy of equal or less energy than that absorbed.

All radiant energy absorption processes involve the quantized excitation of a chemical species by the absorption of a photon. The energy transition may be translational, rotational, vibrational, electronic, or nuclear. If, after excitation, the chemical species gives off the excess energy by emitting a photon of less energy than that absorbed, fluorescence or phosphorescence is said to occur, depending upon the lifetime of the excited state. The emitted energy is what is normally studied. If the source of radiant energy and the absorbing species are in identical energy states, i.e., in resonance, the excess energy is often given up by the nondirectional emission of a photon of energy identical to that absorbed. Either absorption or emission may be studied, depending upon the chemical and instrumental circumstances. Examples of the employment of resonance processes include nuclear resonance absorption, atomic absorption spectroscopy, and nuclear magnetic resonance. If the emitted energy is studied, the term "resonant fluorescence" is often used. However, if the absorbing species releases the excess energy in small steps by the process of intermolecular collision or some other mode, it is commonly understood that this phenomenon falls within the realm of absorption spectroscopy. In this context, the terms absorption spectroscopy, spectrophotometry, and absorptimetry are often used synonomously.

Within the restrictions of the limited definition, most absorption spectrophotometry is done in the ultraviolet, visible, and infrared regions of the electromagnetic spectrum. These three regions are assigned the following approximate wavelength ranges respectively: 0.2–0.4 microns (μ); 0.4–0.7 μ;

0.7–25 μ, where $1 \mu \cong 10^{-6}$ m $= 10^{-4}$ cm $= 10^4$ Å $= 10^3$ nm. The ultraviolet and visible regions represent the excitation of outer electronic levels, while the infrared involves the excitation of vibrational levels. The infrared is sometimes broken down to include the near infrared (0.7–2 μ), but the distinction is more instrumental than molecular since instruments designed for the visible region can be used to penetrate the near infrared. More recent work has extended the 0.2–25 μ range, and commercial instruments are available for use in the vacuum (far) ultraviolet (0.01–0.2 μ) and the far infrared (25– 400 μ) regions. This article will be limited to a discussion of the 0.2 to 25 micron range.

Two important physical characteristics of an absorbing medium are important to the chemist. The first is the primary law of absorption (historically the combination of two independent laws), the Bouguer-Beer law, which expresses the absorptive capacity of the absorbing species at a specified wave length (or energy) as a function of the thickness of the absorbing medium and concentration of that species. The second is a plot of this absorptive capacity as a function of wave length (or energy) known as an absorption spectrum.

Bouguer's law, stated in 1729 and independently somewhat later by Lambert, relates the absorption of parallel monochromatic (monoenergetic) radiant energy by a homogeneous isotropic medium to the thickness of that medium at constant temperature and concentration. Beer's law (1852) relates the absorption of parallel monochromatic radiant energy by a homogeneous isotropic medium to the concentration of the absorbing species at constant temperature and thickness. The combined law is commonly called simply Beer's law. Expressed mathematically in the present American usage of terms:

$$\log_{10}(P_o/P) = \log_{10}(1/T) = A = abc$$

where P_o is the incident radiant power, P the emerging radiant power, T the transmittance, A the absorbance, a the absorptivity or absorption coefficient b the thickness of the absorbing medium, and c the concentration of the absorbing species. (In practice, P_o is not generally measured and more often is redefined as the radiant power transmitted by a standard.) The value of a depends, of course, on the units of b and c. For purposes of tabulation a is often expressed in the units 1/mole·cm, and this is given the special name molar absorptivity with the attendant special symbols ϵ or a_M. Many other names and symbols are more or less commonly used throughout the literature, and the reader is cautioned to make sure he understands the definition of any particular one he may encounter.

Just as a is ideally a constant characteristic of the absorbing species at a specific wave length, so a plot of a (or some mathematically related function) vs. wave length (or some related function) is also a characteristic of the absorbing species. Absorptive capacity is plotted along the ordinate expressed generally as T, $\%T$, A, $\log_{10} A$, a, or $\log_{10} a$; with wave length, energy, frequency, wave-number, or $\log_2$ wave length of the radiant energy plotted along the abscissa. Most popular usage is $\%T$ or A vs.

wave length in nanometers for the ultraviolet and visible regions, and %T or A vs. wave length in microns or wave-number in cm⁻¹ for the infrared. The geometrical shape of the spectrum is dependent upon the plotting system used, all other factors held equal, and an unambiguous labelling of the axes should accompany any such plot.

Points of maximum absorption are known as absorption peaks, and these are associated with a region of lesser absorption known as an absorption band. Often the two terms are used interchangeably. The location of these peaks along the abscissa is a good qualitative indication of the absorbing species and is independent of the plotting method used. However, chemical parameters such as pH, solvent, temperature, and pressure, along with instrumental parameters such as spectral band width, stray radiant energy, and speed of scan coupled with the response of the detetctor in automatic recording instruments may affect not only the value of a (hence the shape of a spectrum) but also the location of the absorption peak. Thus, the conditions under which a spectrum is obtained should be reported with the actual plot.

Applications. Since the absorption of radiant energy is a quantum mechanical phenomenon, important information can be obtained by means of absorption spectroscopy about such molecular characteristics as bond distances, bond strengths, and dipole moments. However, for the chemist the most common application of this technique is in the field of chemical analysis where qualitative and quantitative information about the absorbing species is desired.

For qualitative analysis, one major assumption is made; namely, that if the absorption spectrum of an unknown is identical with that of some known species, the unknown is identical with the known. Normally this assumption is quite valid; however, instances have been reported where two different species have given essentially identical absorption spectra. Moreover, since the absorption spectrum is dependent upon both the instrumental and chemical parameters employed, as noted above, it is important that the unknown be run under conditions identical to that of the standard. If log A is plotted on the ordinate, the geometrical shape of the spectrum is independent of concentration. If the unknown spectrum can then be superimposed upon the standard spectrum in all respects identification has been achieved. If two or more absorbing species are present, it is normally assumed their spectra are additive in the absence of any chemical reaction. Thus, if proper reference spectra are available, the qualitative analysis of a complex mixture is quite possible, just as it is in emission spectroscopy. In the infrared region particularly, qualitative information can be obtained about the sample even in the absence of a standard spectrum since the wave lengths and relative intensities of absorption bands can be correlated with structural characteristics. For example, there are assigned spectral positions for such functional groups as hydroxyl, carbonyl, and amide. Most of these assignments relate to organic compounds, although some work has been done with respect to inorganic structures, particularly in the far infrared region.

It is possible by means of infrared absorption spectrophotometry to perform a complete structural determination of an unknown, but such work has so far taken a large amount of time, requiring a specialist in molecular spectroscopy. Usually, when a structural determination of a new compound is required, in order to make the problem less difficult absorption spectrophotometric data are combined with nuclear magnetic resonance, mass spectrometric and chemical information, as well as any other technique which may apply to the compound at hand (such as nuclear resonance absorption for complexes of iron, or optical rotatory dispersion and circular dichroism spectroscopy for optically active compounds). However, because of the unique "fingerprint" aspect of an infrared spectrum, perhaps its greatest utility lies in identifying a previously known compound. Computer programs are now available for searching large catalogues of standard spectra, and some progress has been made in writing programs that "interpret" spectra. It is anticipated that with the advent of computer linked spectrophotometers having rapid scan capability, considerable progress in the rapid identification of a compound will be made in the near future.

Quantitative applications of absorption spectrophotometry depend upon the use of Beer's law, which relates absorptive capacity to concentration. Thus, if one knows a and b and measures A, c can be calculated. In principle a standard listing of values of a at specified wave lengths for various species is all that is necessary, with no calibration work needed. However, because a can vary depending upon experimental conditions, all analytical work of any accuracy in this area requires a spectrophotometric calibration for the instrument and chemical system under investigation. Ideally, a plot of A vs. c (a calibration curve) should give a straight line. However, such plots generally show a certain amount of curvature, over an extended concentration range, due primarily to the following effects: (1) failure of the instrument to comply with the requirements of Beer's law by having either a significant spectral band width (lack of monochromaticity) or some convergence or divergence in the incident beam; (2) the actual concentration of the absorbing species is different than the supposed or analytical concentration as a result of competing equilibria, degradation, polymerization, or some other chemical effect; (3) the absorbing medium fails to meet the requirements of homogeneity and isotropicity; and (4) instrumental aberrations are present, such as nonlinear or inaccurate photometry, stray radiant energy, and faulty absorption cells. In most situations all these effects enter in to some extent. Most workers prefer to plot A vs. c for calibration purposes, but there is no requirement that such be done. In most work, a is seldom computed, since analyses are normally done under conditions of constant thickness, and the calibration curve is all that is necessary.

Just as spectra are assumed to be additive, so for polycomponent systems in quantitative analysis are absorbances assumed to be additive. Hence, one may express the total absorbance at any wave length as the sum of several individual A's or abc products. To analyze for n constituents, n wave lengths must be chosen to set up n simultaneous equations of this type. For this purpose, it is necessary to calculate *a* for each constituent at each of the *n* wave lengths. The use of matrices greatly facilitates the actual computations. In the infrared region, examples exist of as many as twelve constituents having been simultaneously determined.

Prior to spectrophotometric measurement, often one or more chemical reactions, exclusive of any separation steps, must be undertaken to convert the desired constituent into some absorbing form. In the visible region this is commonly known as the "color reaction." Examples include the oxidation of manganous ion to the purple permanganate ion for the determination of manganese; the reduction of molybdosilicic acid to the corresponding heteropoly blue for the determination of silicon; and the reaction of ferrous ion with 1,10-phenanthroline to yield a red colored complex for the determination of iron. There are several thousand such systems known, relating to the determination of virtually every inorganic species and many organic species, particularly those of medical importance. With the exception of the technique of differential spectrophotometry, such methods are for the determination of trace rather than major constituents since the molar absorptivities of many species in the visible and ultraviolet regions are quite large (10^3–10^4 1/mole·cm). This capability to yield rapid yet accurate determinations of small amounts is perhaps the most important single aspect of absorption spectrophotometry. This has been most recently applied, with singular success, in the field of automated clinical analysis.

In the infrared region, work is normally limited to major constituents. The infrared equivalent of a color reaction is generally not necessary since most organic molecules absorb in that region of their own accord. However, infrared spectrophotometry is, for the most part, limited to nonaqueous situations, since the common cell materials are soluble in water. Although water insoluble infrared transmitting glasses are available, they are still quite expensive.

One application of spectrophotometry which is unique to the visible region of the electromagnetic spectrum is the specification of color or colorimetry. The term colorimetry is often improperly applied to the quantitative application of visible absorption spectroscopy. In fact, the term relates to the numerical specification of color, which, under the Commission Internationale de l'Éclairage (C.I.E.), has been standardized internationally. The process involves the calculation of three numbers, the tristimulus coefficients, by mathematically treating the spectrum (absorption or reflection) of a colored sample with respect to three arbitrary primaries, the source under which the sample is to be viewed, and the spectral sensitivity of the human eye. These three numbers, which are most useful for digital computing purposes, can be normalized to the corresponding trichromatic coefficients. These values can then be used to express the color of the sample in terms of dominant wave length (hue), excitation purity (saturation) and relative brightness. This application of visible spectrophotometry is of paramount importance in any industry which deals with a colored product.

G. CLARK DEHNE

Gamma Ray

Gamma rays, unlike beta particles, are emitted from radioactive nuclei in discrete energies. The spectrum of energies observed, and the relative intensities of these gamma rays, often uniquely characterize the radionuclide which emits them. Gamma ray spectroscopy is the technique whereby the energies and relative intensities of these gamma rays are determined. In the last several years a striking and fundamental change has taken place in the instrumentation used in gamma ray spectroscopy. Prior to 1965 essentially all gamma ray spectroscopy was done using sodium iodide [NaI (Tl)] scintillation detectors. Since that time, however, the technology of making germanium detectors by drifting lithium ions into the germanium matrix has developed to the point where large germanium ingots can be drifted with lithium to form highly efficient detectors with excellent resolution.

A sodium iodide scintillation detector consists of a polished crystal of NaI (thallium-activated) which is coupled optically to a photomultiplier tube. Gamma rays may interact with the crystal to produce light quanta (scintillations). The number of quanta produced is proportional to the energy absorbed in the detector. The light quanta are converted to electrical pulses by the photomultiplier tube. These pulses may then be processed electronically by a pulse height analyzer system to yield a gamma ray spectrum.

Not all the gamma rays are completely absorbed within the detector itself, but some expend only a part of their energy, resulting in the generation of a lower energy pulse. (This is true of both sodium iodide and germanium detectors.) Thus, in addition to the full energy peak produced for a gamma ray of a particular energy, the observed spectrum also contains a continuum of lower energies, called the Compton continuum. Because of this continuum the full energy peak of less intense gamma rays of lower energy may be obscured, making chemical separation or electronic manipulation necessary.

A lithium-drifted germanium [Ge(Li)] detector is a single crystal of ultrapure semiconductor germanium which has had lithium diffused into its bulk to produce a large-volume intrinsic diode (active volumes of 50 cm³ are common). Gamma rays interacting in the intrinsic region cause a charge separation to occur. These charges then are collected individually and amplified electronically. A disadvantage of these detectors is that they must be maintained under high vacuum at liquid nitrogen temperatures, making the system expensive, delicate and bulky. However, the single-step conversion process of the Ge(Li) diode results in a detector of vastly superior resolution. It is in fact the resolution of these detectors which has radically altered gamma ray spectroscopy during this time period. While a very good sodium iodide detector 3″ × 3″ cylinder may have a resolution of approximately 7%, roughly equivalent to 46 kilovolts full width at half-maximum (FWHM) for the 662 MeV gamma of Ba¹³⁷, present day Ge(Li) detectors frequently have resolutions of approximately 2 kilovolts FWHM at this energy. The relative efficiency of these detectors is still significantly poorer than that of sodium iodide, the better detectors being approximately 15% as efficient as the 3″ × 3″ NaI(Tl) cylinder, but the increase in resolution more than makes up for the decrease in efficiency.

Pulse height analyzers have developed apace with modern detector technology, and analog to digital converters (ADC) with digitizing rates of 100 MHz or more now permit analysis of fairly radioactive samples without the high dead times which occur with slower ADC's. Memory sizes for pulse height analyzers have increased from the 400 channel analyzer of the middle 1960s to the 4096 channel pulse height analyzers routinely available today. This increase in the number of channels is required to make efficient use of high-resolution detectors, since one would like to have at least 4 channels in a gamma ray peak. This may be illustrated by the case where a spectrum is being taken from zero to 3000 keV over 4096 channels (approximately 0.75 keV per channel) with the detector having a resolution of approximately 2.1 keV at the 1330 keV peak of cobalt-60. Under these conditions the peak will be in fact less than 4 channels wide (FWHM). Therefore as the resolution increases the number of channels in a peak decreases and a smaller energy spread must be taken or more channels used.

Although the data may be read out on a digital printer, the large amount of data to be handled has resulted in increasing use of magnetic tape units that are directly compatible with high-speed electronic computers. Also enjoying increasing popularity are small dedicated computers which are programmed to accept input from ADC's and use a portion of the computer memory for data storage and the remainder for the input data processing program. On essentially all analyzers the data are presented on an oscilloscope as they are accumulated, so that the operator may make decisions as to energy spread, counting time, resolution, etc. Many of the newer large memory analyzers are available at costs equal to or less than those of 400 channel analyzers 10 years ago.

The use of Ge(Li) gamma ray spectroscopy in decay scheme and fission product studies, and in activation analysis has revolutionized these fields. The tedious radiochemical separations once required to obtain pure nuclear species free from interfering gamma rays usually are no longer necessary. Now only a series of simple group separations is required, and the ultimate resolution between the nuclides is made from the gamma ray spectrum itself on the basis of the high resolution of these modern detectors. The standard qualitative analysis separation scheme has proved very satisfactory for making such group separations.

A number of very rapid chemical separations have been developed for use in special cases where specific radionuclides of very short half-life must be measured. It must be cautioned, however, there are occasionally gamma rays from two different nuclear species which are so close in energy that they cannot be resolved even with modern high resolution detectors. This is a hazard which some researchers have not avoided; however, since many radioactive species emit more than one gamma ray it is often possible to use a gamma ray other than the principal gamma ray for identification and analysis.

After the mixture has been fractioned either chemically or electronically, it is usually possible to identify quantitatively the elements present by their characteristic gamma ray spectra combined with measurement of radioactive decay rates. In principle it is possible to calculate absolute disintegration rates from gamma spectra knowing the detector efficiency, the geometry and the radioactive decay scheme of the isotope being measured. In practice, however, the desired accuracy can be obtained much more readily by comparison with a standard. This is particularly true in neutron activation analysis where irradiation of a standard together with the unknown eliminates the need for exact knowledge of neutron flux and absorption cross-sections in addition to the factors listed above.

Philip D. LaFleur and
Barbara A. Thompson

Optical

Spectroscopy is the branch of science dealing with measurement of radiant energy. Broadly, the field covers measurement of emitted and absorbed energy over the entire electromagnetic spectrum. Most measurements are made in the x-ray, ultraviolet, visible, and infrared spectral regions, and the term spectroscopy itself is commonly reserved for the field of ultraviolet, visible, and infrared emission spectroscopy. Spectroscopy is one of the most powerful tools available for the study of atomic and molecular structure and is used in the analysis of a wide range of samples.

Sir Isaac Newton laid the basis for spectroscopy when he demonstrated, in 1664, that white light could be resolved into component colors with a prism. In 1800 and 1801, Herschel and Ritter discovered the infrared and ultraviolet regions. Von Fraunhofer demonstrated, in 1814, the dark lines in the solar spectrum which bear his name. Kirchhoff and Bunsen produced emission spectra in 1861 and showed their relationship to Fraunhofer lines, thereby laying the basis for emission spectroscopy. Maxwell, in his theory of light propagation (1873), suggested the use of electrical and magnetic fields instead of the previously used elastic forces. From this, the idea of an electromagnetic spectrum of energy was derived. X-rays were discovered by Roentgen in 1891.

Spectral measurements in various wave length regions played an important role in the development of our understanding of atomic and molecular structure. X-ray emission spectra were used by Moseley to assign elements to their places in the Periodic Table. Ultraviolet, visible, and infrared emission spectra played a vital role in assigning the position of outer electrons in the atom. Ultraviolet and visible absorption spectra, due to shifts of electrons in molecules, have told much about this aspect of molecular structure. Infrared absorption spectra, arising from interatomic molecular vibrations, are used to establish structure of molecules. Most of these technics are important to the analytical chemist and provide him with a wide range of tools for analysis of inorganic and organic compounds. Indeed, there are few analyses which cannot be made by some phase of spectroscopy (although it is many times more feasible to use other methods).

Emission spectra arise when an excited atom or molecule returns to its normal state. Excited molecules give rise to bands of lines and are referred to as *band spectra*. Excited atoms give rise to discrete lines, the yellow lines of sodium being the most familiar example. The number of lines in an element's emission spectrum depends upon the number and position of the outermost electrons and upon the degree of excitation of the atom.

The type and complexity of equipment needed to produce and study emission spectra depend upon the work to be done. Several means of excitation are normally employed. Most common is the direct current arc; but flames, alternating current arcs, and high-voltage discharge sparks are also used. Between them, these sources are capable of exciting most lines of the various elements. The emitted radiation is dispersed with a prism or a diffraction grating spectrograph. A variety of optical arrangements with these two basic elements is possible, and each has its advantages. Spectra are commonly recorded on photographic plates or films but photoelectric records and direct feeds to computers find increasing use. The intensity of the lines and their position may be estimated visually or be measured with a densitometer.

Analysis of samples by emission spectroscopy is generally limited to metallic elements. Nonmetals give rise to emission spectra, but they are weak and hard to excite. Qualitative analysis depends on the fact that the spectrum of each element is different. Each possesses a few characteristic strong lines, one or more of which coincides with a line in the spectrum of almost any other element. Identification is generally done by comparison. The wave lengths of all except the very faintest lines of all elements have been tabulated. Comparison of the position of lines of an unknown with that of iron gives wave lengths of the unknown lines, whose source can then be checked by reference to tables. In practice, a library of reference spectra and familiarity with emission spectra are most useful, and a skilled spectrographer seldom needs to use an elaborate method for qualitative analysis.

The intensity of a line is proportional to the amount of the element producing the radiation in the sample, and measurements of intensity are used for quantitative analysis. In practice, only minor and trace elements in a sample can be determined in this manner. The usual method involves measuring the ratio of intensity of a strong line of the element sought to a weak line of the matrix. Quantitative analyses for trace and alloying elements in metals such as steel, aluminum, brasses, etc. are readily performed in this manner. Specialized instruments have been developed for making such analyses rapidly. Photocells placed at the focal points of the lines to be used in the analysis replace the photographic plate. The sample is burned, the amount of radiation is measured by the phototubes and the analysis recorded. A complete analysis requires but a few minutes. In less routine cases such speed is not generally possible, but most analyses can be completed in less than one hour. The precision of the method is about ±5 to 10%.

Since most spectrographic analyses are for trace elements, such precision is frequently better than can be achieved by other methods in the same amount of time.

R. E. KITSON

Microwave

The portion of the electromagnetic spectrum extending approximately from one millimeter (300,000 megacycles) to 30 centimeters (1,000 megacycles) is designated the microwave region. This region lies between the far infrared and the conventional radiofrequency regions. Spectroscopic applications of microwaves consist almost exclusively of absorption work, rather than of the emission type.

With some exceptions, absorptions of microwave energy represent changes of the absorbing molecules from one rotational energy level to another. The absorption spectrum is characteristic of the absorbing molecule as a whole, as contrasted for example to data characteristic of functional groupings in infrared absorption spectroscopy or of elemental composition in ultraviolet emission spectroscopy. The specimen in microwave spectroscopy ordinarily must be in the gaseous state and at low pressure. A few micrograms of the sample are sufficient for obtaining a spectrum. In general, any molecule possessing a permanent dipole or magnetic moment will absorb characteristically in the microwave region.

As in any other type of absorption spectroscopy, the apparatus required in microwave spectroscopy consists essentially of a source of radiation, a sample cell and a detector. The source is typically a klystron or other specially designed high-vacuum electron tube. The source output is monochromatic under any given set of conditions, so no separate monochromator is required in microwave spectroscopy. The frequency of the electromagnetic energy emitted by the klystron tube may be varied somewhat by mechanical and/or electrical adjustment, and different types are available to cover various portions of the microwave spectrum. The sample cell usually consists of wave guide, a long metal tube with rectangular cross section. The detector usually consists of a silicon crystal, although bolometers and other heat-type detectors are occasionally used.

In addition to these three basic components, a complete microwave spectrometer typically includes provision for modulation of the absorption spectrum, an a-c amplifier for the detector output, a final indicator consisting either of a cathode ray oscilloscope or a strip chart recorder, a sweep generator to vary synchronously the source frequency and one axis of the final indicator, a means of frequency measurement, a gas handling system, and the necessary electronic power supplies. The modulation, which is typically based upon the Stark effect, is employed in order to obtain sufficient sensitivity in the measurements and to provide additional data for the identification of the absorbing molecules and their structural characteristics. The a-c amplifier must be tuned to the frequency of the modulation. Methods of measuring frequencies in the microwave region can be accurate to within one part in several million.

A few fields of application of microwave spectroscopy may be listed to illustrate the scope of the method. Structural information concerning bond lengths, bond angles and nuclear masses may in many instances be calculated from suitable measurements of absorption frequencies. Internuclear distances, for example, can be calculated to within $\pm 0.002 \text{Å}$ from moments of inertia obtained from microwave spectra. Nuclear spin and quadrupole coupling infromation may be obtained from data on the fine structure of some absorption bands. Internal rotation, hindered or unhindered, of organic molecules may be studied by noting the fine structure which is imparted to certain rotational energy levels by internal rotation motions within the molecule.

Some useful analytical applications have been developed, particularly isotopic analyses and analyses of chemical isomers. Qualitative identifications are based upon the measured frequencies of characteristic absorption bands; the frequencies exhibited by a particular component are independent of what other components are present in the sample and of total sample pressure. The total absorption for a given line is proportional to the concentration of that substance in the sample. So quantitative determinations are based upon measurements of integrated line intensities, or of the product of peak height and width at half-maximum intensity, or simply of peak height under certain experimental conditions.

The microwave region of the electromagnetic spectrum has only relatively recently been exploited. For example, only one paper was published prior to 1946. Subsequent development has proceeded rather slowly, due in part to the lack of complete, commercially available microwave spectrometers. However, suitable equipment has been designed and constructed in several laboratories, and some complete units can now be purchased. The microwave region has been found to be a very fruitful one in molecular spectroscopy, and microwave spectroscopy is now taking its rightful place in the broad field of spectroscopy.

ROBERT B. FISCHER

STABILIZERS (Plastics)

To be formed, shaped or molded to the finished end product thermoplastic products have to be processed at elevated temperatures which can range from 200° to as high as 500°F and occasionally even higher for short periods of time. As these plastics basically consist of long-chain polymers, they are susceptible to degradation during the processing, especially at the high temperatures involved. Therefore, stabilizers are necessary to protect the polymer from degradation during its processing and especially to prevent changes in physical properties and color during its processing cycle, and after it has been manufactured to further protect it against weathering environment, i.e., heat and sunlight. These conditions in some cases can be just as drastic as the heat cycle. For instance, Arizona sunlight degrades plastics very readily unless they are suitably protected with UV absorbers.

At present, there is no one single stabilizer that will function to protect all the various thermoplastics now on the market so it is necessary to use in most cases a combination. They consist of a basic heat stabilizer, and antioxidant with perhaps an additive for light stability, specifically a UV light absorber.

As the various classes of plastics require different processing conditions and vary in their resistance to degradation, the basic function of the stabilizer must be adjusted to fit the polymer needs and whether to emphasize against heat or light or both.

For instance, some polymers are extremely susceptible to the heat cycle during processing, whereas others are vulnerable to outdoor exposure, particularly sunlight.

Polyvinyl Chloride Polymers. The PVC group of polymers is probably one of the largest outlets for stabilizers especially when one considers that over a billion pounds per year are produced of this thermoplastic. To stabilize them adequately against heat and light, a ratio of 2–3% of stabilizers of rather complex nature must be used.

PVC resins degrade very rapidly when heated, unless adequate protection or stabilization is employed, with a loss of hydrogen chloride which sets off a mechanism known as the "zipper reaction" (Fig. 1.).

This causes the further loss of hydrogen chloride and eventual total decomposition. A suitable stabilizer must first prevent the heat degradation; otherwise there will be a drastic change in color of the polymer from practically colorless to yellow, gold, brown and finally black. As these colors change there is also a change in mechanical properties which make the product useless for practical purposes.

As there is no satisfactory theoretical or practical solution to stabilization of this type of resin, the stabilization problem has been solved by purely empirical methods. An adequate stabilization system for PVC polymers needs a multiplicity of compounds usually based on a metallic soap plus organic phosphites and epoxy compounds. Some stabilizers give very satisfactory protection as far as processing (heat) is concerned, but fail in terms of light stability. It is necessary to stabilize the polymer for both; hence the common practice is to compound PVC with several different stabilizers, each performing its own particular function or exercising a "synergistic" action.

Polyolefin Plastics. Two general classes of polyolefin polymers are available, polyethylene and polypropylene. More recently the copolymers of these basic olefins plus modifying dienes are now being explored as rubber replacements; however, as these usually involve some form of vulcanization or cross-linking through the use of peroxides etc., the stabilization problems are more unique

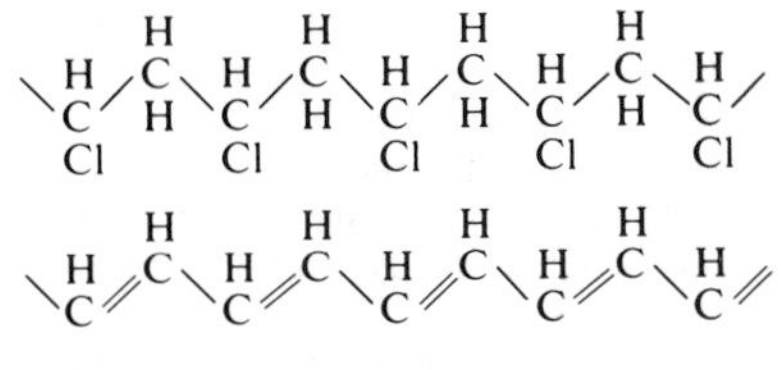

FIG. 1

than those of the purely thermoplastic olefin polymer. The polyethylene group does not give any particular problems in stabilization and antioxidants such as are generally used for rubber are adequate. The more recently developed blocked or hindered phenol type antioxidants perform quite satisfactorily.

Proper stabilization of polypropylene is very important for as the temperature is raised for processing rapid oxidation can occur, with chain scission and a loss in mechanical properties. Vulnerability of polypropylene can be seen when comparing the molecular structure of these polymers. In the latter, there is a methyl group attached to every other carbon in the polymer chain creating a greater number of tertiary hydrogens which are prone to oxidation, whereas polyethylene is more linear in structure and therefore has greater stability to oxidation at elevated temperatures.

The oxidation rate is readily demonstrated when these two polymers are exposed at 200°F to oxygen and the oxygen uptake measured. Unstabilized polypropylene will absorb oxygen at the rate of 0.12 cc/gm/min and have an induction period of eleven hours, whereas the linear polyethylene has an induction period almost fifteen times higher, and then only absorbs oxygen at 0.003 cc/gm/min. It is apparent that a strong stabilizer or antioxidant is needed for this polymer. Both polymers degrade rather rapidly when exposed to weathering conditions unless suitable ultraviolet light stabilizers are used.

Again, as in PVC polymers, no single system gives adequate protection to both heat and weathering conditions and it is necessary to use a combination of stabilizers.

Polyolefins which have been compounded with carbon black usually give good outdoor weathering.

Polystyrene. In the styrene family of plastics the simply polymerized styrenes, if made properly, are relatively stable to most of the processing conditions and resist oxidation. However, this polymer is extremely sensitive to light and unless stabilized properly will yellow rapidly outdoors as well as indoors. Stabilizers against light degradation are aliphatic and cyclic amines, amino alcohols and most recently the newer UV stabilizers such as the hydroxybenzophenones and the various substituted derivatives are employed. The hydroxybenzotriazoles are also used as well as some specific hindered phenols.

As polystyrene in its pure polymer form is rather brittle and does not have impact necessary for many end uses, the high-impact polystyrenes were developed based on rubber modified or styrene(diene) copolymers. As an unsaturated linkage is now introduced in such polymers they become more susceptible to oxidation. High-impact polystyrenes are also more difficult to extrude and require different type stabilizers to prevent yellowing during fabrication and minimize oxidation. Typical stabilizers are the hindered phenols such as the 2,6 ditertiary butyl para cresol, phosphite esters, alkylated phenols and for light stability the addition of suitable UV absorbers already mentioned. (See **Polystyrene**)

The group of thermoplastics discussed above are susceptible to environmental or processing conditions that require specific stabilization. Further, this group can be classified in that they are susceptible to oxidation during processing and light exposure. No one mechanism for the degradation of these plastics can be assigned. Theoretically, it can be postulated that the more unsaturated the backbone or structure of the polymer, the more susceptible to degradation and the greater need for stabilization, and vice versa. A good example is the group of acrylates, more specifically polymethylmethacrylate, which are quite stable even at elevated temperature. This group, therefore, needs only small amounts of stabilizers for particular environmental conditions such as outdoor exposure and protection against UV radiation. Other polymers which are not subject to rapid oxidation are the polyamides and polyesters. The latter group has good stability towards oxidative changes as during the polymerization the unsaturation is diminished; however, they do suffer from light aging and it is necessary to use large amounts of UV absorbers. As stabilizers for polyamides, various phenolic compounds as well as phosphite esters have been recommended.

Acetal Resins. Polymers of this type have a high melting point on the order of 350°F which means they need to be processed at elevated temperatures and will be susceptible to oxidation. Extensive patent literature exists on suggested stabilizers but the alkyldiene bis alkylated phenols are satisfactory for heat stabilization. Acetal polymer compositions for outdoor exposure have to be light stabilized and in general UV absorbers such as mentioned above are used.

Acetal copolymers are also on the market and it is claimed that by the use of copolymers stability is maintained at higher temperatures over a longer period of time. Again this is a molecular structure problem involving the copolymer unit which "ties knots" randomly along the polymer chain and these knots resist the "unzipping" reaction which causes degradation. Note the similarity of this degradation phenomena to that of PVC where the "zipper" reaction based on dehydrochlorination occurs. The same type of heat stabilizers as those for the pure polyacetals are used and again for outdoor exposure ultraviolet stabilizers are used, i.e., the hydroxybenzophenones, benzotriazoles, etc. (See **Acetal Resins**)

Fluorocarbon Plastics. The fluorocarbon polymers are thermoplastic and are analogous chemically to the polyolefins, but some or all of the hydrogen atoms have been replaced by fluorine atoms and this substitution gives excellent weather and oxidation resistance.

There are four species of fluorocarbon resins now commercially available: polytetrafluoroethylene, fluorinated ethylene propylene, chlorotrifluoroethylene and polycinylidene fluoride. For instance, the PTFE has thermostability up to 500°F and can be used continuously at this temperature for long periods of time.

The vinyl fluoride group of polymers exhibit typical properties of the family and are particularly resistant to weathering and oxidative conditions.

These polymers in general do not require any special additives except for unusual conditions where heat stabilizers may be added and/or UV absorbers to prolong the weathering ability of such compositions. (See **Fluorocarbon Resins**)

G. P. MACK

Cross-references: Antioxidants, Ultraviolet Absorbers.

STANDARDS

Standards are materials used as reference substances in analytical and testing work. They are either materials of definite composition and exceedingly high purity or typical specimens of some particular class of product, the composition of which with respect to one or more component having been carefully established. The very pure materials are generally referred to as *primary standards;* they are used to determine the concentration of solutions employed as standard solutions in volumetric analysis. Materials which have been carefully analyzed with respect to some component are referred to as *standard samples;* they are used to establish the validity of a new method of analysis or to check the functioning of methods and operators in routine work.

In addition to high purity and definite composition, a primary standard should possess certain other characteristics. It should react quickly and completely with the solution being standardized and the reaction should take place stoichiometrically, that is, according to some definite pattern for which there can be written a chemical equation involving only whole numbers. A satisfactory method of determining the equivalence-point in the reaction must exist. A primary standard should keep well and should be stable toward atmospheric water, carbon dioxide and oxygen. It should withstand a temperature of 110°C so that any superficial water can be removed before use by oven drying. Preferably also it should have a high equivalent weight so that the error in weighing will be minimized in the result.

A number of primary standard materials are known. Most of them are available commercially and a few can be obtained in tested and certified purity from the National Bureau of Standards, Washington, D.C. In the following list are included the primary standards in most common use. The equivalent weight given is for the reaction in which the primary standard is usually used; it should be emphasized, however, that the equivalent weight may vary from reaction to reaction. Thus, in the titration of sodium carbonate with hydrochloric acid, the reaction takes place in two stages: $Na_2CO_3 + HCl \rightarrow NaHCO_3 + NaCl$ (phenolphthalein as indicator) and $NaHCO_3 + HCl \rightarrow H_2CO_3 + NaCl$ (methyl orange as indicator); the equivalent weight in either reaction is the molecular weight of sodium carbonate divided by 1, or if the total hydrochloric acid for the two reactions is measured, as is customary, the equivalent weight is the molecular weight divided by 2.

Standard Samples. The program on standard materials at the National Bureau of Standards was initiated in the early years of this century. In its early phases it was concentrated heavily on iron and steel and the materials related to the ferrous metallurgical industry. The program was subsequently broadened to include a variety of other materials of significance in commercial analytical and testing practice. A complete description of the standards from the Bureau, their composition, cost, and the methods of ordering them will be found in NBS Miscellaneous Publication 260, issued October 1, 1965, available from the Superintendent of Documents, U.S. Government Printing Office, Washington 25, D.C., price $0.45. The following list is a summary of the types of materials available with the specific items in a few cases; the number given is the Bureau of Standards sample number.

Chemical Standards

84d Acid potassium phthalate	Acidimetric value
39g Benzoic acid	Acidimetric value
40e Sodium oxalate	Oxidimetric value
83a Arsenic trioxide	Oxidimetric value
136 Potassium dichromate	Oxidimetric value
17 Sucrose (cane sugar)	Calorimetric and saccharimetric values
41 Dextrose (glucose)	Reducing value

Microchemical Standards

140 Benzoic acid	For determination of carbon, hydrogen
141 Acetanilide	Nitrogen, carbon, hydrogen
142 Anisic acid	Methoxyl
143 Cystine	Sulfur, carbon, hydrogen, nitrogen
144 2-Chlorobenzoic acid	Chlorine
145 2-Iodobenzoic acid	Iodine
146 Tri-alpha-naphthyl phosphate	Phosphorus

pH Standards   pH (approx.)

185 Acid potassium phthalate	4.0
186Ib Potassium dihydrogen phosphate	
186IIb Disodium hydrogen phosphate	6.8
187a Borax	9.2
188 Potassium hydrogen tartrate	3.6

Melting-point Standards: Five metals, aluminum, copper, lead, tin and zinc, whose melting points are certified to 0.1°C or better.

Turbidimetric and Fineness Standards: Two samples of cement for checking sieves.

Thermoelectric Standards: Chromel and Alumel thermocouple wire

Spectrographic Standards

Steels. Some fifteen plain carbon and alloy steels for the spectrographic determination of the common alloying elements.

Aluminum Alloys. Four alloys, analyzed for copper, magnesium, silicon, manganese, iron, nickel, chromium, titanium, and zinc.

Tin Metal. Ten alloys analyzed for all the minor constituents.

Paint Pigment Standards: Some 25 pigments for color and tinting strength.

Rubber Compounding: Zinc oxide, sulfur, benzothiazyl-disulfide, etc.

Hydrocarbon and Organic Sulfur Compounds: An extensive list of the paraffins, alkyl cyclopentanes, alkyl cyclohexanes, monoolefins, diolefins, cyclomonoolefins, acetylenes, alkyl benzenes, naphthalenes, polycyclic aromatic hydrocarbons and organic sulfur compounds of interest to the petroleum industry.

Oils for Calibrating Viscosimeters

Radioactivity Standards: Samples of radon, radium, cobalt, beta-ray emitting materials, rocks and ores.

The standard samples issued by the Bureau of Standards are in general those materials which are typical of classes of products extensively used or manufactured in industry. These materials have been carefully prepared and mixed and then analyzed by a number of competent analysts both at the Bureau and in the industry. The names of the analysts, the methods used and the results are given on the certificates which accompany each lot of sample. The classes of materials covered are the following:

    Aluminum-base alloys

    Copper-base alloys

    Lead- and tin-base alloys

    Magnesium-base alloy

    Nickel-base alloy

    Zinc-base die-casting alloy

    Zinc spelter

    Iron and Steel. Comprising a large portion of the total number of samples and including plain carbon steels and alloy steels containing all of the common alloying elements.

    Steel-making alloys

    Bauxite and alumina refractories

    Iron ores, zinc ore, tin ore

    Phosphate rock

    Chrome refractory, fluorospar, clay, feldspar, glass sand, glass, limestone, silica brick, burned magnesite, titanium dioxide

In addition to the standard samples issued by the National Bureau of Standards, a large number of standards are available from certain large manufacturing companies, for example, an extensive list of aluminum alloys for spectrographic analysis from the Aluminum Company, magnesium standards from the Dow Chemical Company for spectrographic analysis. A list of such sources will be found in the booklet Report on Standard Samples for Spectrochemical Analysis, American Society for Testing Materials.

HARVEY DIEHL

STARCHES

Starches are widely distributed throughout the plant world as reserve polysaccharides, and are stored principally in seeds, fruits, tubers, roots, and stem pith. (See **Photosynthesis**) They usually occur as discrete particles or granules of 2–150 microns in diameter. The physical appearance and properties of granules vary widely from one plant to another, and may be used to classify starches as to origin. Some are round, some elliptical and some polygonal. Many have a spot, termed the hilum, which is the intersection of two or more lines or creases. The granules are anisotropic and show strong bire-fringence with two dark extinction lines extending from edge to edge of the granule with their intersection at the hilum. Some granules show a series of "striations" arranged concentrically around the hilum.

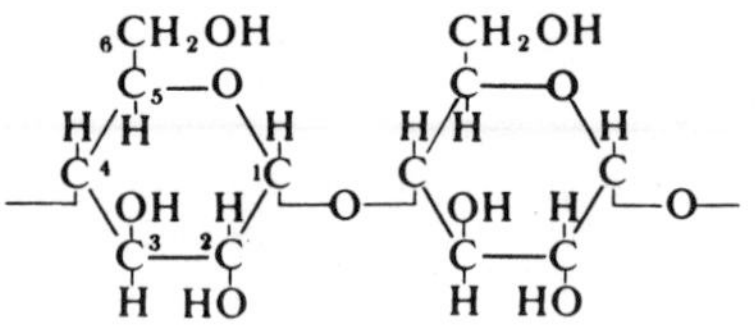

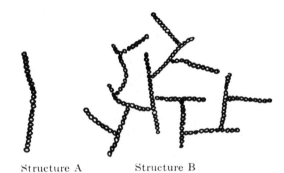

FIG. 1. Basic repeating unit of amylose and amylopectin.

Structure A   Structure B

FIG. 2.

Although starches are hydrolyzed only to D-glucose, they are not single substances but, except in very rare instances, are mixtures of two structurally different glucans. One, termed amylose, is a linear chain (structure A) of D-glucose units joined by α-1 → 4 links, and the other, known as amylopectin, is a bush-shaped structure (structure B) of 1 → 4 linked α-D-glucose units with α-1 → 6 links at the branch points which occur about every 25–27 sugar units. Most starches contain 22–26% amylose and 74–78% amylopectin. Amylose may be selectively precipitated from a hot starch dispersion by the addition of substances such as butanol, fatty acids, various phenols and nitrocompounds such as nitropropane and nitrobenzene. On cooling slowly, the amylose combines with the fractionating agent to form a complex which separates as microscopic crystals. Disruption of the complexes occurs when they are dissolved in hot water or extracted with ethanol.

Corn starch is the most important of the starches manufactured in the United States. Approximately 130 million bushels of corn are processed annually. Waxy corn starch is made from waxy corn. Most waxy starches are entirely free of amylose, and therefore consist solely of branched, amylopectin molecules.

Starch may be prepared from white potatoes by either batch or continuous processes. Commercial wheat starch is usually prepared by the Martin process. Starches from other sources (cassava, tapioca, etc.) are prepared in a manner similar to that for corn or potato starch.

Several different crystalline forms of starches may be obtained. Native cereal starches occur as the "A" form and tuber starches as the "B" form. Intermediates are designated as "C" types. In general, the B type is obtained by evaporation of pastes at room temperature or by precipitation by freezing or retrogradation, the C type at higher temperatures, and the A type at 80–90°. At higher temperatures amorphous starch is usually obtained. Precipitation of starch pastes with alcohols or some other precipitating agents gives a "V" type of starch with a more symmetrical arrangement of molecules.

One of the most important properties of starch granules is their behavior on heating with water. On heating, water is at first slowly and reversibly taken up and limited swelling occurs, although there are no perceptible changes in viscosity and birefringence. At a temperature characteristic for the type of starch, the granules undergo an irreversible rapid swelling and lose their birefringence, and the viscosity of the suspension increases rapidly. Finally, at higher temperatures starch diffuses from some granules and others are ruptured leaving formless sacs. Swelling may be induced at room temperature by numerous chemicals such as formamide, formic acid, chloral, strong bases and metallic salts. Magnesium sulfate impedes gelatinization.

When aqueous starch solutions are allowed to stand under aseptic conditions, they become opalescent and finally undergo precipitation, known as retrogradation, to give a starch with a "B" x-ray pattern. Amylose molecules retrograde more readily than amylopectin molecules. In certain cases, such as white potato starch, there is an initial preferential precipitation of amylose. Generally, retrogradation proceeds faster as the starch concentration is increased and as the temperature is decreased toward 0°.

Starch solutions have a high positive optical rotation because of the presence of α-D-glucosidic linkages. The values range from +180° to +220° for aqueous alkaline solutions. Osmotic pressure measurements on acetylated amyloses and amylopectins from several starches give values for which the molecular weights of the unacetylated components have been calculated to be 100,000–210,000 for amyloses and 1,000,000–6,000,000 for amylopectins. At present there is little information on the shape of the molecules although amylose molecules have been shown to be very flexible.

Starches behave as polyhydroxy alcohols and, in the presence of an impelling agent, are capable of ether formation with alkyl and acyl halides and alkyl sulfates, and of ester formation with both inorganic and organic acids. For the preparation of acyl derivatives, the impelling agent such as pyridine or an acid anhydride is necessary to absorb the water formed during acylation. Alkalies, especially sodium hydroxide, are normally used as impelling agents for etherification. The fully methylated ethers of amylose and amylopectin have been utilized extensively in structural determinations.

The enzymes which hydrolyze starch may be classified as: (1) liquefying (dextrogenic) or α-amylases, (2) saccharogenic or β-amylases, and (3) phosphorylases. α-Amylases decrease rapidly the solution viscosity and cause rapid and extensive degradation. The main products of degradation are malto-dextrins of about 6 D-glucose units. Malt α-amylase hydrolyzes starches to fermentable sugars in about 90% of the theoretical yield. The stage at which the solution is no longer colored by iodine is known as the *achroic point* and the reducing value calculated as maltose is termed the *achroic R-value*. In contrast to α-amylases, β-amylases cause only a slow change in the viscosity of starch solutions. With whole starches, the hydrolysis proceeds rapidly until about 50–53% of the theoretical amount of maltose is produced, and then very slowly until a limit of about 61–68% is reached. The unhydrolyzed residue is called a β-amylase limit dextrin and may be likened to a pruned amylopectin molecule. Phosphorylases such as P-enzyme bring about the reversible hydrolysis of starch to α-D-glucose 1-(dihydrogen phosphate).

For many industrial applications the properties of natural starches are changed by various treatments. These include the action of enzyme, acids or oxidizing agents on an aqueous suspension of the starch, or by heating essentially dry starch with or without the addition of small quantities of acids or alkalies. "Thin-boiling" starches are made by the treatment of starches with acids at temperatures below the gelatinization point, whereas "Lintner-Soluble Starch" is prepared by the treatment of raw starch with 7.5% hydrochloric acid for seven days at room temperature. Oxidized starches are prepared usually by the action of hypochlorite or peroxide. In the former case sodium or calcium hypochlorite is added to a slightly alkaline starch slurry and the reaction allowed to take place at 30–50°C until the desired degree of oxidation is reached. Excess oxidizing agent is neutralized by the addition of sodium bisulfite. The oxidized starch retains its granular structure, is colored by iodine, gives suspensions of greater clarity than the parent starch, and has a shorter cooking time, lower viscosity, increased adhesiveness, and a lower rate of congealing.

"British gums" or "Torrefection dextrins" are prepared by high-temperature treatment of starches. This treatment causes cross linking and an increase in end groups. Addition of small amounts of alkalies or acids to the starch prior to heating causes a more rapid change and yields products of a different nature.

Starches are hydrolyzed for the manufacture of sirups and D-glucose. In the manufacture of these products, a starch-water slurry is treated at 30–45 lbs. of steam pressure with hydrochloric acid at pH 1.5–2.0. For the preparation of sirups, the hydrolysis is stopped at 42% conversion of starch to D-glucose.

About one-third of the total corn starch and 95% of the total corn sirup is sold for food purposes. Other applications of starches and modified starches include adhesives, sizes for the textile, paper, and leather industries, binders in sand molding and manufacture of asbestos products, and as depressants in the flotation process of ore separation.

ROY L. WHISTLER

Cross-references: *Carbohydrates, Cellulose, Sugars, Agricultural Chemistry, Foods, Nutrition.*

STEEL

There are many hundreds of steels with widely differing properties, allowing a judicious selection for each particular application. The best choice is the steel that fulfills the requirements of the application and is most economical from a standpoint of material and processing costs.

In their simplest form, steels may be considered alloys of iron and up to 1.7% (by weight) carbon. Carbon is so potent in its effects, it must be specified to hundredths of a per cent. Alloys with higher amounts of carbon fall into various groups of cast and malleable irons with the upper commercial limit of about 4.5% carbon. The theoretical limit of carbon is 6.67%. Steel is produced in greater quantities than all other alloys due to its favorable mechanical properties, low cost and its versatility in heat treatment, i.e., it can be softened (annealed) for fabrication and then hardened for final application. This is principally due to two allotropic forms of iron: alpha iron (ferrite) and gamma iron (austenite), and a metastable phase (martensite).

At low temperatures, the iron atoms are in a body-centered cubic lattice structure (ferrite). At temperatures of 1333°F to 2065°F, depending on the carbon content, the energy level of the atoms is great enough to cause an allotropic change (transformation) in the crystalline structure to a face-centered cubic lattice (austenite). The two allotropic forms are similar in the respect that both are soft, ductile and low strength. Austenite is nonmagnetic and ferrite is magnetic.

The equilibrium structure at low temperatures is ferrite and iron carbide (Fe_3C or cementite). At 0.80% carbon, the phases exist as a eutectoid (pearlite) which consists of alternate plates or lamellae of ferrite and cementite. As a function of varying carbon, the pearlite decreases with corresponding increases in free ferrite (at lower carbon levels) or free cementite (at higher carbon levels). Cementite is very hard and brittle and thus limits the high-carbon steels. The bulk of the steels produced have low carbon contents (<0.80%) which yield relatively high toughness and ductility characteristics.

In austenite, the carbon atoms dissolve into the interstices between the iron atoms. Generally, most other elements form a solid solution which involves a substitution of atoms. The advantages of interstitial solutioning in steels is the fast diffusion rate of carbon at high temperatures which is important to rapid and economical heat treatments. In addition, the carbon solubility in austenite is about forty times greater than in ferrite. This solubility difference is largely responsible for the heat treatability of steels.

On heating, ferrite and cementite transform to austenite which on slow cooling decomposes back to ferrite and cementite*, all or part of in the form of lamellae (pearlite) or a feathery or acicular aggregate (banite). At the highest transformation temperatures, a coarse pearlite forms, then at successively lower temperatures, a finer pearlite, a

feathery banite and an acicular banite all of which achieve successively higher hardness. At a cooling rate fast enough to prevent the formation of the phases mentioned above, austenite transforms to the metastable phase, martensite. The reaction starts at the M_s temperature and is completed at a lower temperature (M_f temperature). The parameters of all the austenite decomposition reactions can be determined from TTT (Time-Temperature-Transformation) diagrams.

The metastable phase, martensite, has a body-centered tetragonal structure. This is almost the structure of the body-centered cubic ferrite except it contains considerable excess carbon in solution (supersaturation) that results in a distorted and highly strained lattice. This structure yields maximum hardness and strength. However, a steel in this condition is not useful since it is brittle. At a sacrifice in hardness and strength, heating to relatively low temperatures (tempering) is employed to increase toughness and ductility and to relieve the high internal stresses.

Tempering is an approach toward equilibrium conditions brought about by the diffusion of carbon atoms (from the martensite) to form spheroidal particles of cementite which changes the martensitic structure toward a body-centered cubic (ferrite) structure. The degree of tempering used depends on the balance of strength and ductility needed for each particular application. For a given hardness and strength, a tempered martensitic structure generally has the greatest toughness and ductility compared to other structures.

The ease in transforming to martensite, i.e., rate of decomposition of austenite, is known as hardenability. Steels of low hardenability have rapid rates of austenite decomposition to structures of pearlite, ferrite, cementite and bainite and, thus, are difficult to cool fast enough to form martensite. High-hardenability steels have slow austenite decomposition rates and, thereby can be cooled fast enough to form martensite before the other phases form.

Hardenability should not be confused with the mechanical property, hardness. The hardness of martensite is dependent on its carbon content and increases as a function of carbon to a maximum hardness of about R_c 65 to 68 at about an 0.80% carbon level.

The simplest form of steel has been defined as alloys of iron and carbon. Actually steels contain one or more additional elements, e.g., sulfur, manganese, chromium, nickel, vanadium, tungsten, molybdenum, columbium, titanium. The general types are carbon steels, low-alloy steels, tool steels, stainless steels, and heat-resistant steels.

Carbon Steels. These alloys of iron and varying amounts of carbon, are produced in the largest quantities and are the most economical. They are generally used in the as-wrought condition with varying structures of ferrite and cementite and are frequently used in low-strength structural applications.

The principal additional elements present in carbon steels (and all other steels) are phosphorus, sulfur, silicon and manganese. The sulfur and phosphorus are impurities. Silicon is added as a degasifier to remove oxygen and promote ingot soundness. Manganese neutralizes the detrimental

---

* The equilibrium phases and transformation temperatures can be determined from an iron carbon equilibrium diagram.

effects of FeS which causes the steel to be hot short, that is, brittle in hot working.

Low-Alloy Steels. These are a large group of steels alloyed with varying amounts and combinations of elements (up to 5%). They are the second greatest tonnage of steels produced and are often heat treated to tempered martensitic structures for higher strength structural applications such as bridges and ships, and for machine parts.

The principal advantage of alloy steels is the increased hardenability compared to carbon steels (all common alloying elements except cobalt increase hardenability, but differ in potency). Increased hardenability allows many diversified parts to be cooled at commercially feasible rates for the formation of martensite and to minimize distortion and cracking.

Certain alloying elements or combination of elements achieve higher properties such as ultimate strength, yield strength-to-tensile strength ratio, toughness and ductility, wear resistance and corrosion resistance. Many of these characteristics are due to solid-solution strengthening and to the formation of complex carbides during tempering.

The best known and most widely used classification system of carbon and low-alloy steels is the joint AISI and SAE. The steels are identified by a four-digit numerical system. The first number indicates the type steel, e.g., chromium, molybdenum, nickel-chromium or carbon steel. The second figure shows the percentage of the predominating alloying element and the last two or three digits represent the average carbon content in hundredths of a per cent.

Tool Steels. Tool steels heat-treated to tempered martensitic structures, are characterized by high hardness and are used in the working parts of tools and in dies. Special properties or combination of properties are obtained, e.g., high wear resistance, toughness and shock resistance, high-temperature hardness and low seizing and galling characteristics.

Depending on the tool steel, the alloying elements vary from 0.35 to 2.5% carbon and up to 3% manganese, 2% silicon, 4.25% nickel, 12% chromium, 5% vanadium, 18% tungsten, 9.5% molybdenum and 12% cobalt. The general types of tool steels are water-hardening, shock-resistant, cold-working, hot-working, high-speed and special-purpose.

Stainless Steels. Chromium additions of 12% or more attain a "passivity" condition which confers unusually good corrosion resistance characteristics to steels. (Nickel, molybdenum and copper also improve the corrosion behavior.) In addition, these steels have good resistance to oxidation (scaling) at moderately high temperatures.

Passivity is the formation of a very thin impervious chromium oxide film by means of oxidation. The film prevents further oxidation (corrosion). Passivity occurs only in oxidizing environments and some neutral environments. The impervious film cannot be formed or maintained in a reducing media, e.g., hydrochloric acid.

There are three types of stainless steels: martensitic, ferritic and austenitic.

Martensitic Stainless Steels. These steels are classified in the AISI 400 series, e.g., 403 and 420. The major alloying elements are 12% to 18% chromium and 0.10% to 1.2% carbon. Although corrosion resistance normally increases at the higher chromium levels, higher chromium additions are generally balanced with higher carbon additions which promotes the formation of chromium carbides, thereby rendering the extra chromium ineffective for corrosion protection.

The martensitics are usually heat treated to tempered martensitic structures and are characterized by high hardness and strength and moderate corrosion resistance. Typical applications are aircraft parts and cutlery.

Ferritic Stainless Steels. These are non-hardenable and non-heat treatable (cannot be transformed to austenite or martensite) alloys that have a ferrite structure, and are also included in the AISI 400 series, e.g., 405 and 446. Certain alloying elements are ferritizers (stabilizers of ferrite) e.g., chromium, aluminum and molybdenum. Enough of these elements prevents the allotropic change to austenite and a subsequent allotropic transformation to martensite. The ferrite is the equilibrium phase at all temperatures, resulting in relatively low strengths.

The major alloying elements of the ferritic stainless steels are 12 to 27% chromium, up to 4.5% aluminum and up to 0.35% carbon. In most of these alloys the carbon contents are very low, e.g., 0.08% maximum. Higher amounts of carbon in the lower chromium alloys would balance the compositions for allotropic changes and they would no longer be considered ferritics.

The ferritic stainless steels are used primarily for their corrosion and oxidation resistance. Typical applications are automobile trim and furnace parts.

Austenitic Stainless Steels. These alloys have a stable austenite structure at a wide range of temperatures, including room temperature, and are classified in AISI 300 series, e.g., 301, 316 and 321. Certain elements are austenitizers (stabilizers of austenite). Enough of these elements prevents the allotropic change to ferrite and suppresses the M_s temperature to such low temperatures that it is no longer practical or possible to transform to martensite. Thus, austenite is the equilibrium phase over a wide range of temperatures.

The major alloying elements in the austenitic alloys are 16% to 26% chromium plus 6 to 22% nickel. Other additions include carbon, molybdenum, titanium, columbium, copper and manganese. Manganese is used as a replacement for nickel in the AISI 200 type stainless steels.

This is the most widely used group of stainless steels. They have the best corrosion resistance for a wide range of applications and, in addition, moderate strengths at high temperatures. Typical applications are naval, chemical and aircraft parts.

Heat-Resistant Steels. These are alloys used for applications of about 700°F and higher. They include:

Alloy Steels. These are martensitic type steels containing up to 10% alloying elements and include a few of the low-alloy steels and low-carbon die steels.

In general, these steels are characterized by a resistance to tempering and/or a secondary hardening during tempering. Secondary hardening reactions involve the formation of complex carbides which maintain or even increase hardness with increasing tempering temperatures in a certain temperature range. In effect this achieves higher strengths and dimensional stability at high temperatures.

Ferritic and Martensitic Stainless Steels. This group includes most of the AISI 400 ferritic and martensitic stainless steels. In addition, there is a series of 403 modifications referred to by various names, such as, 422, Lapelloy and Greek Ascoloy. These alloys contain 12% to 14% chromium and up to 0.35% carbon with small additions of other alloying elements, e.g., molybdenum, columbium, tungsten, nickel and copper. Many of the minor alloying elements contribute to a resistance to tempering and a secondary hardening reaction.

Austenitic Stainless Steels. Most of the 300 type stainless steels are included in this group except for the lower alloy compositions such as 301 and 303.

Precipitation-Hardening Stainless Steels. This is a relatively new group of martensitic alloys that derive their high strength from the formation of martensite and the precipitation of intermetallic compounds during tempering. They have higher chromium contents and considerably better corrosion resistance than all other martensitic steels. The best known one is 17-4PH, which contains 17% chromium and 4% copper.

Semi-Austenitic Stainless Steels. This is also a relatively new group of martensitic alloys that are similar to the precipitation hardening stainless steels except they can be heat treated to an austenitic or martensitic structure stable at room temperature. The advanges are good formability characteristics and high strength. Parts in-process are heat treated to an austenitic structure for its excellent formability properties, then, the finished part is heat treated to a martensitic structure for its high strength. Typical alloys are 17-7PH, 15-7Mo, AM350 and AM355.

Maraging Steels. This is the newest group of martensitic steels which are heat-treatable to unusually high strength levels. Essentially, these steels are alloyed with about 18% nickel plus other elements such as molybdenum, aluminum and titanium. The latter two contribute to the strength by precipitation as intermetallic compounds.

PAUL A. BERGMAN

STEREOCHEMISTRY

As portrayed on the printed page of texts and journals, molecules have the appearance of being only two dimensional. During the last twenty years, the study of the structures and reactions of molecules in three dimensions has become one of the most important occupations of scientists in all areas of chemistry and biochemistry. The static stereochemistry of a molecule refers to its structure as depicted in three dimensions. The dynamic stereochemistry of a reaction concerns the changes in the spatial arrangements of the atoms in the reacting molecules, aside from any chemical changes that might occur (oxidation, reduction, etc.).

Two molecules are said to be stereoisomers if they possess identical chemical formulas with the same atoms bonded one to another, but differ in the manner these atoms are arranged in space.

Thus *sec*-butanol can exist in two forms, I and II, which cannnot be superimposed on each other.

I   II

This particular example represents one class of stereoisomers known as "enantiomers," which may be defined as two molecules that are mirror images but are nonetheless nonsuperimposable. Such molecules are said to possess opposite configuration. If these isomers are separated ("resolved"), the separate enantiomers have been found to rotate the plane of plane-polarized light. This phenomenon of "optical activity" has been known for well over a century. A 50-50 mixture of two enantiomers is optically inactive or "racemic," since the rotation of light by one enantiomer is precisely compensated by the rotation of light in the opposite direction by the other enantiomer.

Physiological activity is closely related to configuration. Thus the left-rotating, or *levo,* form of adrenalin is over ten times more active in raising the blood pressure than is the right-rotating, or *dextro* form. Many organic chemicals essential to plants and animals are optically active. Enzymes, which catalyze chemical reactions in the body, are frequently programmed to accept only one enantiomer. All the essential amino acids, generally of the formula III, are of the *levo* type, although

III

important exceptions exist. Recently, several amino acids were found by NASA in a meteorite that presumably originated from the asteroid belt between Mars and Jupiter. The proof that the amino acids were extraterrestrial came from the fact that they were racemic. Any terrestrial contaminants from laboratory handling would have been optically active and *levo.*

Many examples of optically active molecules contain an asymmetric carbon atom, that is, one with four different groups attached, as in I-III. A wide variety of other atoms may also be asymmetric (IV-VI). An asymmetric center is by no means a

IV   V   VI

necessary condition for enantiomerism. Well-known examples of nonsuperimposable mirror images without asymmetric atoms are allenes (VII), spiranes (VIII), biphenyls (IX), and various inor-

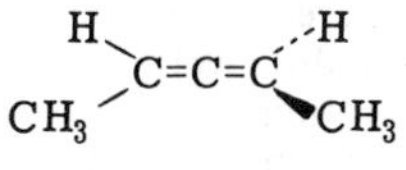

VII

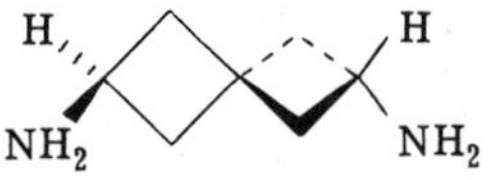

VIII

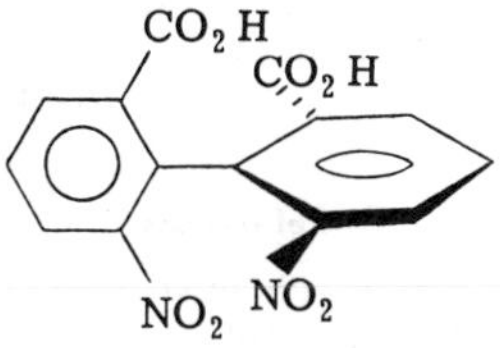

IX

ganic complexes. Molecules that can support optical activity are said to be "chiral," and to possess "chirality" (meaning "handedness," since the human hand is chiral).

The process of converting optically active materials into equal amounts of the enantiomers is called "racemization." Ordinarily this process requires breaking of bonds to form a symmetrical ("achiral") intermediate, and reforming the bonds to generate the racemic material. For special cases such as phosphines (IV) and sulfoxides (VI), in which one "substituent" is a nonbonding electron pair, configurational inversion may occur without breaking any bonds. For such molecules the tetrahedral enantiomers may interconvert through a metastable planar intermediate.

Stereoisomers that are not enantiomers are called "diastereoisomers." Three classes may be distinguished: configurational, geometrical, and conformational isomers. Configurational diastereomers include molecules with more than one chiral center. Thus 2,3-dichlorobutane can exist in three configurationally different forms, X-XII. Although

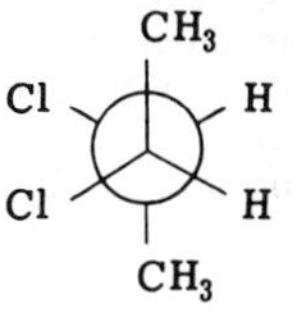

X 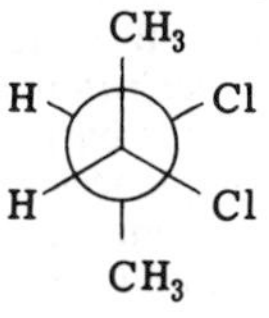   XI 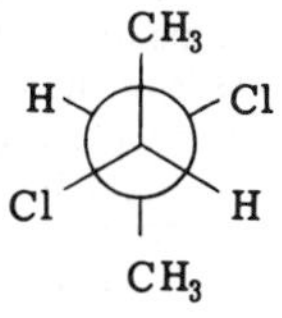   XII

forms X and XI are enantiomers, XII is a stereoisomer that is not a mirror image of X or XI. It is therefore termed a diastereomer of X and XI. Even though there are two asymmetric centers in XII, it is superimposable on its mirror image. The molecule is therefore achiral. The term *meso* is applied to molecules that contain chiral centers but are achiral as a whole. The molecules of nature frequently have many asymmetric centers. If a molecule has n centers, there can be 2^n stereoisomers, although this number may be reduced if some of the diastereoisomers are *meso* or if certain ring constraints are present. Glucose is one of the aldohexose sugars, which contain four chiral centers (disregarding the phenomenon known as anomerism). The naturally occurring *dextro*-glucose is enantiomeric to *levo*-glucose, and diastereomeric to the other fourteen isomers.

Geometrical isomers differ in the arrangement of

groups about certain bonds, rather than about a chiral center. 2-Butene may exist in *cis* (XIII) or *trans* (XIV) forms. In the former case the methyl

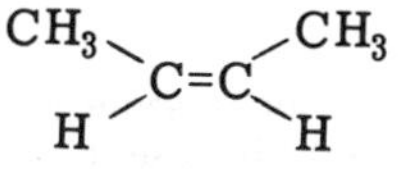

XIII 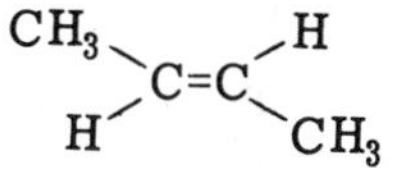   XIV

groups lie on the same side of the double bond, and in the latter on opposite sides. Chirality is not important in geometrical isomerism. The isomers may be interconverted if the double bond is broken to leave a residual single bond about which rotation may occur, followed by reformation of the double bond.

Geometrical isomerism may occur not only in alkenes (XIII, XIV), but also in oximes (XV), azo compounds (XVI), and many other doubly bonded

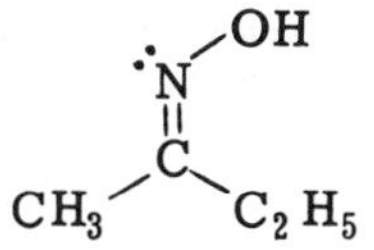

XV 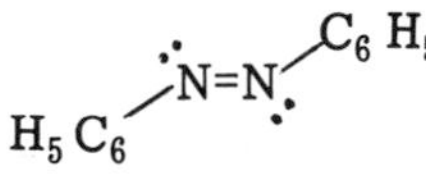   XVI

systems. More importantly, cyclic molecules exhibit this type of isomerism; the average plane of the ring serves as the reference. Molecule XVII is therefore named *cis*-1,2-dimethylcyclopropane, and XVIII is the *trans* isomer. Interconversion of XVII and XVIII would require breaking and reforming a ring bond.

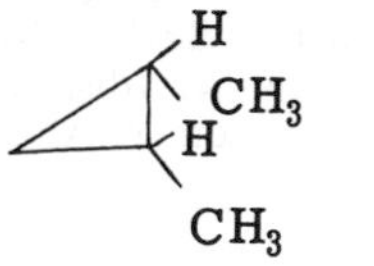

XVII 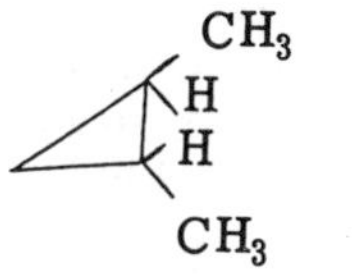   XVIII

Conformational isomers differ only in the arrangements of atoms obtainable by rotations about one or more single bonds. *meso*-2,3-Dichlorobutane can exist not only in the form XII, but also as the representation XIX. To take a simpler case, *n*-butane may exist in two conformational forms,

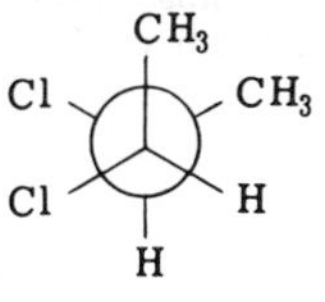

XIX 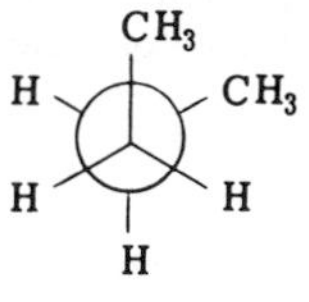   XX 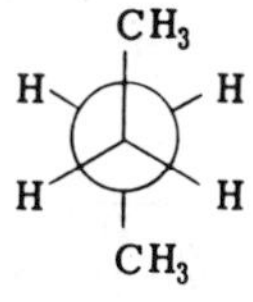   XXI

known as the *gauche* (XX) and the *anti* (XXI). These isomers may interconvert by rotation about the C-C single bond, a process that requires an energy of only 3-5 kilocalories/mole. No bonds are broken, in contrast to the manner by which geometrical isomers are interconverted. In a more complicated but very common case, substituents on the six-membered cyclohexane ring may assume either the equatorial (XXII) or axial (XXIII) positions. These isomers may interconvert by "ring re-

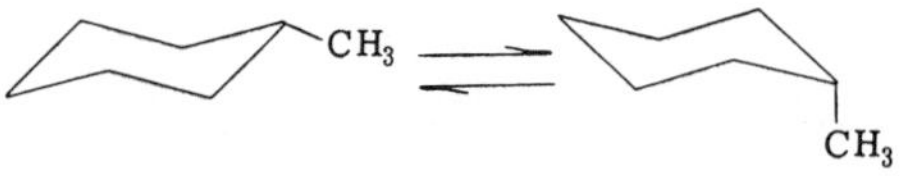

$$\text{(1)}$$

XXII   XXIII

versal" (Eq. 1) which consists of a sequence of single-bond rotations. β-D-(+)-Glucose has the specific conformation of XXIV, in which all the substituents are equatorial.

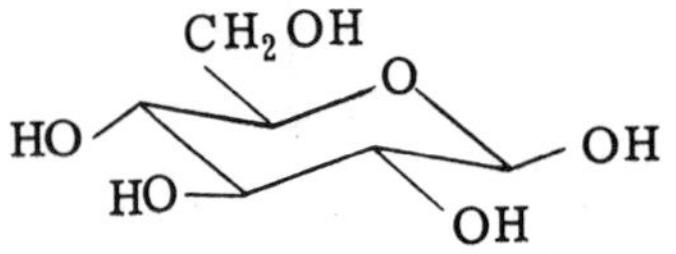

XXIV

Stereochemistry is one of the most important characteristics of a reaction mechanism. In the nucleophilic displacement of iodide ion on *sec*-butyl bromide (Eq. 2), the reaction is known to occur with inversion. If one disregards the identity of the

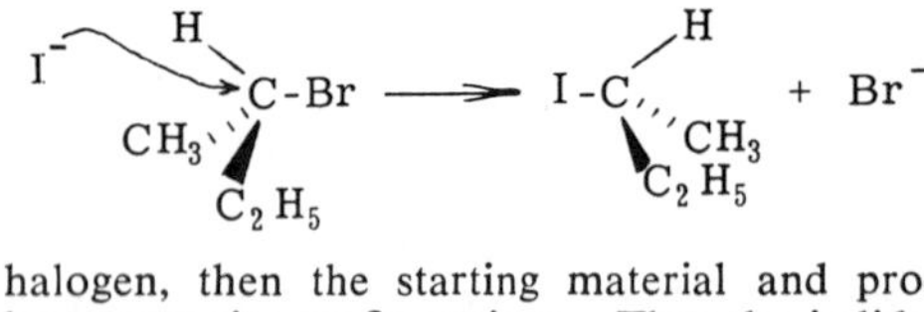

$$\text{(2)}$$

halogen, then the starting material and product have opposite configurations. Thus the iodide ion must attack the C-Br bond from the backside, thereby effecting an inversion of the chiral center.

Important mechanistic consequences may also derive from geometrical isomerism. *cis*-3,4-Dimethylcyclobutene may ring-open to form either *cis,trans*- or *trans,trans*-2,4-hexadiene. The methyl groups may rotate away from each other (disrotation, Eq. 3) to form the *trans, trans* isomer. Alter-

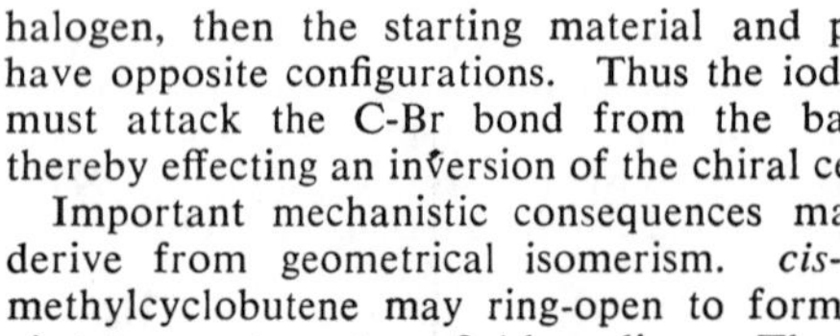

$$\text{(3)}$$

natively, they may rotate in the same direction (conrotation, Eq. 4) to form the *cis,trans* isomer. An alternative disrotatory mode to form a *cis,cis*

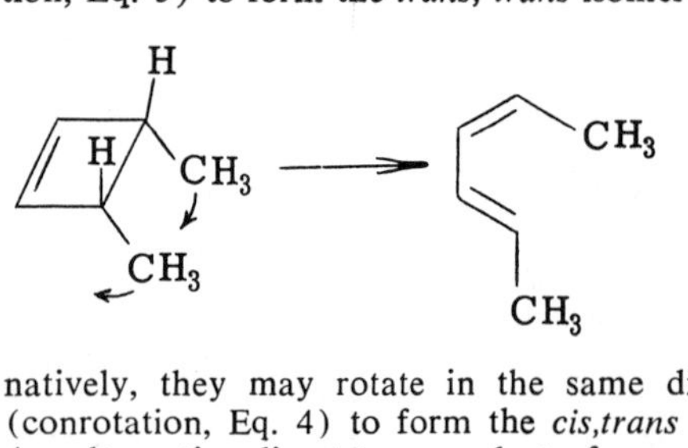

$$\text{(4)}$$

isomer by rotation of the methyl groups toward each other need not be considered because the methyl groups cannot pass by each other. Recent experimental and theoretical consideration of this problem has demonstrated that only the conrotatory mode is permitted for this ring opening. Thus the *cis*-dimethylcyclobutene always forms only the *cis,trans*-hexene.

The fundamentals of static stereochemistry, as outlined above, are now reasonably well understood. Applications of configurational, geometrical, and conformational isomerism to the study of organic, inorganic, and biochemical reaction mechanisms are steadily contributing to a deeper understanding of the ways that Nature chooses to carry out her chemical changes.

JOSEPH B. LAMBERT

References

Eliel, E. L., "Stereochemistry of Carbon Compounds," McGraw-Hill Book Co., Inc., New York, N.Y., 1962.

Lambert, J. B., "The Shapes of Organic Molecules," *Scientific American*, Vol. **222**, No. 1, 1970, pp. 58–70.

Mislow, K., "Introduction to Stereochemistry," W. A. Benjamin, Inc., New York, N.Y., 1966.

STEREOISOMERISM

If one considers a molecular unit consisting of three unlike atoms A, B, and C which are connected ("bonded") to each other to form a linear system, three arrangements are possible: A—B—C, A—C—B, and B—A—C. By employing the simple test of superimposability it is seen that these arrangements are nonidentical and are termed *structural isomers* (Gr. *isos*, equal). Structural isomers are thus chemical species that have the same molecular formula (the same number and types of atoms) but differ in the sequence in which the atoms are bonded. Of great significance is the fact that these isomers are separated by an energy barrier: in order to convert, for example, A—B—C into A—C—B an input of energy would be necessary to break the existing A—B and B—C bonds, followed by a rearrangement of the sequence of the atoms to yield A—C—B. The rearrangement process is termed an *isomerization* and the magnitude of the energy required (the barrier) for the conversion has important consequences regarding the number of arrangements that may exist for structural isomers. This will be discussed shortly.

Turning to a four-atom arrangement consisting of A_2B_2, four structural isomers are possible if linearity of the system is again assumed: A—A—B—B; A—B—A—B; A—B—B—A; B—A—A—B. Each arrangement can be characterized in terms of interatomic distances (A to B, B to B, and A to A) and such distances would be distinctive for each isomer. It is not mandatory, however, that any one of these A_2B_2 combinations exists in a linear form and the consequences of nonlinearity will be viewed. If (for the ABBA case) planarity of the unit still exists but the internuclear angles A—B—B are set at, say, 120°, two arrangements are apparent:

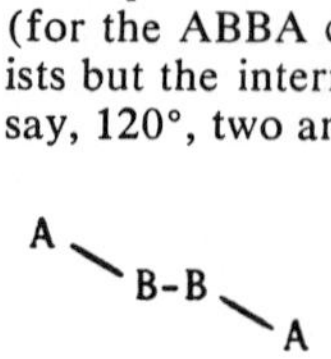

1 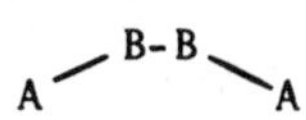   **2**

Arrangements **1** and **2** are clearly different forms of the same structural isomer and are called *stereo-*

isomers (Gr. *stereo*, solid or space). The relationship between **1** and **2** may be expressed in terms of the geometry of the molecule and hence these forms have also been called *geometrical isomers*. (Recalling the provisos set down for this system—planarity and angles—the difference between **1** and **2** is merely the location of one A with respect to the other in the unit.)

The statement made earlier about the energy requirement for the conversion of one structural isomer into another will now be examined. Conversion of **2** into **1** may be viewed in the simplest manner as involving a rotation of 180° about the B-B internuclear axis:

$$A \overset{B-C-B}{\diagdown} \diagdown A \quad \xrightarrow{180°} \quad 1$$

2

This operation involves no reshuffling of the atomic arrangement (or constitution) within the unit but does represent an isomerization. How readily such a conversion occurs then depends upon the size of the energy barrier to rotation. Looking at specific examples, the compound CH_3—N=N—CH_3, exists in two different and stable stereochemical arrangements (**3**, *cis* and **4**, *trans*) corresponding to **1** and **2** above:

$$CH_3 \diagdown_{N=N} \diagup CH_3 \qquad \diagup_{N=N} \diagdown CH_3$$
$$\qquad\qquad\qquad CH_3$$

3   **4**

The C—N—N—C atoms are coplanar in each form and the interconversion of **3** and **4** requires a relatively high amount of energy. On the other hand, hydrogen peroxide (H_2O_2), which may be thought of as existing in similar *cis* and *trans* stereochemical arrangements (**5** and **6**), does *not* exhibit

$$H \diagdown_{O-O} \diagup H \qquad \diagup_{O-O} \diagdown H$$
$$\qquad H$$

5   **6**

stereoisomerism and only one isolable H_2O_2 is known. The difference in the two systems lies in the fact that rotation about the O—O single bond in H_2O_2 is relatively "free" (i.e., requires little energy) whereas the interconversion of **3** and **4** is a higher energy process that necessitates breaking a π bond before rotation may occur.

We now consider the case of four atoms (called *ligands*) that are attached to a center atom. If the general case consists of grouping A_{abcc}, square planar and tetrahedral structures, among others, may result:

$$a \diagdown_{A} \diagup b \qquad a \diagdown_{A} \diagup c \qquad \overset{a}{\underset{c}{\diagup|\diagdown}} c$$
$$c \diagup \diagdown c \qquad c \diagup \diagdown b \qquad b$$

7   **8**   **9**

The square planar forms **7** and **8** are stereoisomers which are nonequivalent to the single nonplanar tetrahedral arrangement **9**. Now if grouping A_{abcd} is examined one predicts three square planar stereoisomers and the following *two* tetrahedral stereoisomers (**10** and **11**):

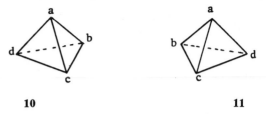

10   **11**

The relationship of **10** to **11** is that of the right hand to the left hand and these nonsuperimposable stereoisomers are mirror images or *enantiomers* (Gr. *enantio-*, opposite). Thus, *chiral* (Gr. *cheir*, hand) molecules are those which possess mirror images and arise when appropriate conditions of geometry and number and types of ligands are present in a system.

A molecule that has a mirror image is also said to be dissymmetric while one that does not (an achiral molecule) have an enantiomer is nondissymmetric. The classification of a given structure as dissymmetric or nondissymmetric is based upon the presence (or lack) of symmetry elements (axes, planes) in the structure. The reader is referred to the citations given at the end of this article for detailed discussions on this topic.

It is important to note that in either **10** or **11** the magnitude of any internuclear angle (e.g., a—A—c), or any bond length (e.g., A—b), or the distance between any two ligands is exactly the same. This is not true in the case of the achiral square planar isomers that may be written for the A_{abcd} system.

The single most important physical property that differentiates enantiomers is their ability to rotate the plane of plane polarized light. This property is called *optical activity* and is displayed only by chiral molecules. Thus, stereoisomers which are also chiral are known as *optical isomers*. Chiral molecules that rotate polarized light in a clockwise fashion are termed *dextrorotatory* (*d*) while those that rotate the beam counterclockwise are *levorotatory* (*l*). Enantiomers have optical rotations of the same magnitude but of different signs (*d* or *l*).

The structures **10** and **11** denoted above contain a single chiral center A, the atom to which the ligands are attached. If two different such centers, A and B, are in a molecule the number of optical isomers is increased to four:

A±   A±
                                                B±

one chiral center   two different chiral centers

where + and − refer to the handedness or *configuration* at the chiral center. In the AB system, each optical isomer will have a mirror image whose configurations are opposite at the chiral centers.

A+   A−   A+   A−
B+   B−   B−   B+

mirror images   mirror images

The relationship of A+B+ or A−B− to A+B− (or A−B+) cannot be an enantiomeric one for the obvious reason that a given optical isomer may have only one mirror image. Instead, the relationship is said to be *diastereomeric* (Gr. *dia,* apart). Any given AB optical isomer will therefore have one enantiomer and two diastereomers.

We again examine a specific case. In 3-chloro-2-butanol, CH_3CH—$CHCH_3$, and A±B± situation—

exists whose four isomers (**12-15**) are shown in three-dimension so that the mirror image relation-

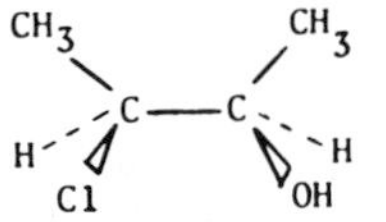

12

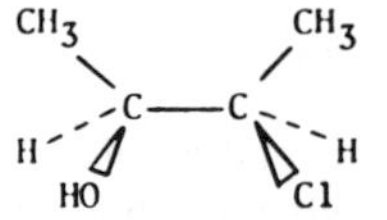

13

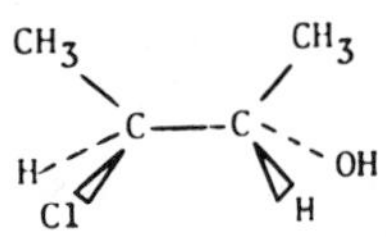

14

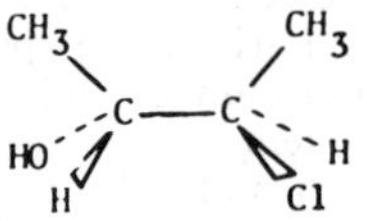

15

ship of **12** to **13** and **14** to **15** is readily apparent. (The carbon atoms in the middle of the carbon chain are the chiral centers A and B.) Taking **12** (which is assumed to be the A+B+ combination) if one views down the bond axis of the chiral centers from the right a projection of the molecule results which shows the orientation of ligands to

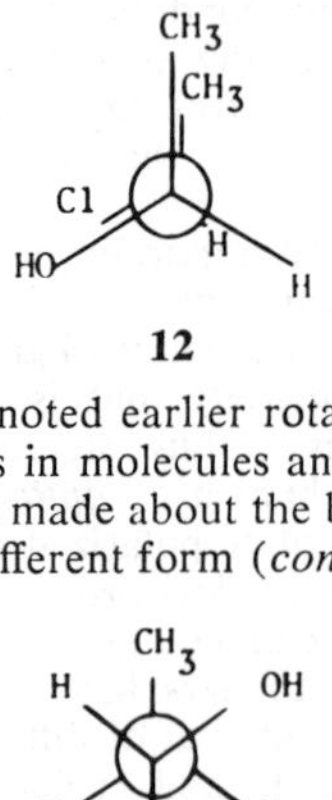

12

one another. As noted earlier rotation may occur about single bonds in molecules and if a clockwise rotation of 180° is made about the bond axis of the chiral centers, a different form (*conformation*), **16**, results.

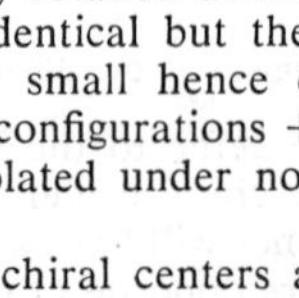

16

Conformations **16** and the *infinite number* of others that are obtained by rotation about the single bond in **12** are all nonidentical but the energy barrier separating them is small hence only one chiral compound having configurations ++ at the chiral centers may be isolated under normal conditions for **12**.

If two *identical* chiral centers are present in a molecular unit the number of stereoisomers is reduced to three: A+A+ and A−A− represent a pair of enantiomers but A+A− and A−A+ are identical arrangements. The +− form is said to be a *meso*

or optically inactive diastereomer of the active forms A+A+ and A−A−.

The three tartaric acids may be used to illustrate the method employed in depicting three-dimensional molecules in two-dimension projections.

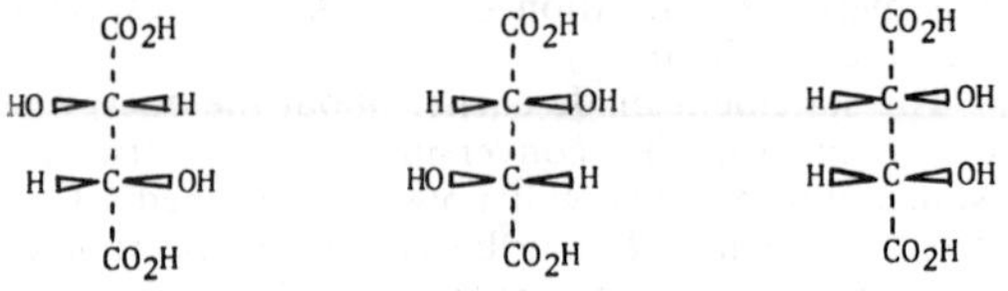

enantiomers

meso

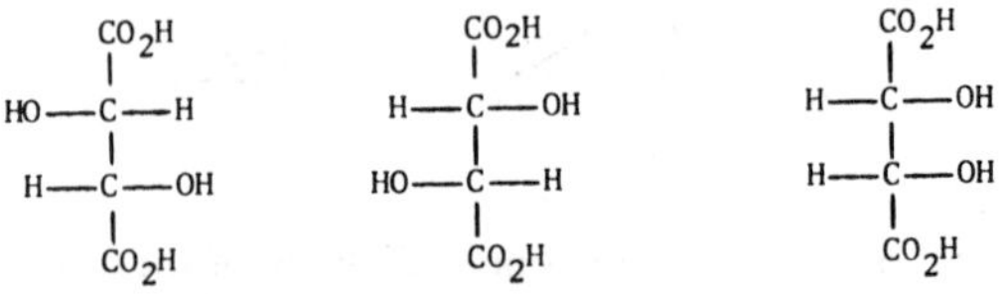

D   enantiomers   L   meso

The top formulas show the three-dimensional relation of the groups along the main carbon chain (dashed groups lie below the plane of the paper, bold above) while the bottom formulas correspond to projections in two dimensions obtained by lifting the dashed substituents into the plane of reference and pushing the bold groups down into the plane. In this process a unique projection is obtained for each three-dimensional molecule. The symbols D and L under the formulas for the enantiomeric tartaric acids are notations used to relate the configuration at the bottom chiral center to that of a standard compound, glyceraldehyde. The reader may consult the references for detailed discussions of the D,L and other systems of configuration notation.

The presence of a chiral center is a sufficient, but not a necessary condition, for the existence of chirality in a molecule. For example, numerous biphenyl derivatives may exist in chiral pairs if the

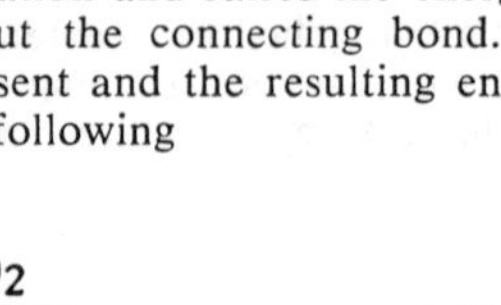

size of the R groups is large enough to restrict rotation about the single bond connecting the two rings. The restriction causes the rings to adopt a nonplanar orientation and raises the energy barrier to rotation about the connecting bond. No chiral center is present and the resulting enantiomers in the example following

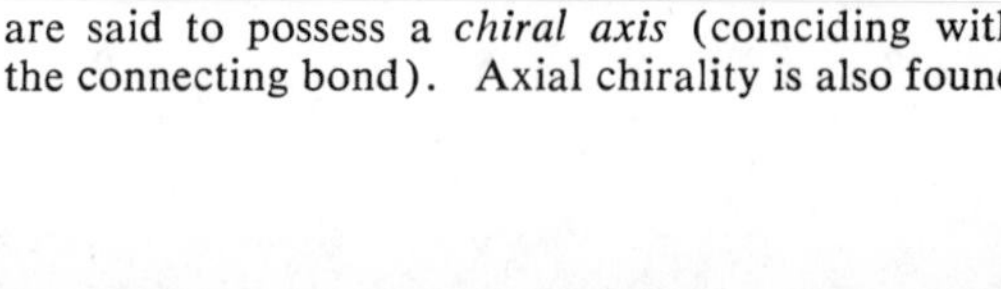

are said to possess a *chiral axis* (coinciding with the connecting bond). Axial chirality is also found in the compounds known as allenes

e.g.

$$H, Cl, Cl, H \quad C=C=C$$

and in spiranes

e.g.

$$CH_3 \quad CH_3 \quad H \quad H$$

while chiral structures such as *trans*-cyclooctene

$$H \qquad H$$

are said to possess a *chiral plane*. Each of the last three structures has a mirror image.

It is clear, then, that the number of chiral isomers that may exist for a given structural isomer is 2^n, where n = the number of different chiral elements (centers, axes, or planes). When identical chiral elements are present, the 2^n formula does not hold.

The examples used above to illustrate centers of chirality included carbon atoms which were attached to four unlike ligands, resulting in localized tetrahedral geometry. Numerous other atoms may serve as chiral centers, however, and these include the Group IVA elements silicon, germanium, and tin; the Group VA elements nitrogen, phosphorus, antimony, and arsenic; and the Group VIA elements sulfur, selenium, and tellurium. Under conditions of bonding to three or four dissimilar ligands, chiral molecules containing these atoms as chiral centers may be isolated. Also, the geometric form about the chiral center need not be tetrahedral, for octahedral complexes of the transition metals or their ions (Co^{3+}, Cr^{3+}, etc.) may be chiral when substituted by the proper number and type of ligands.

Conformations in Six-Membered Rings. The nonplanar ring compound, cyclohexane, which contains six contiguous —CH_2— units, may exist in chair or boat forms (hydrogens are not shown):

"boat"

"chair"   "chair"

These forms are conformations of the C_6H_{12} structure and are separated by energy barriers which

restrict, but do not prohibit, interconversion of the conformations. Substitution of a group on the ring yields a single nonchiral structure that exists in two main conformations, one with the substituent R *axial* and the other *equatorial* to the main plane of the ring.

$$R \quad H \qquad R \quad H$$

R *axial* (**17**)   R *equatorial* (**18**)

Conformation **18** is in a lower energy state and predominates in the equilibrium mixture. The introduction of a second substituent into the ring gives rise to three structural isomers (shown here in projection):

$$R \quad R \qquad R \qquad R$$

19   **20**   **21**

In **19** and **20**, both carbon atoms in the ring to which the R groups are bonded are identical chiral centers and hence one pair of enantiomers and one meso diastereomer exist for each structure. In **21** no chiral center is present but the isomers that are possible in this case (shown below in the preferred conformations)

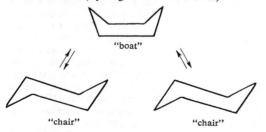

trans   *cis*

may be viewed as having a diastereomeric relationship.

Consequences of Molecular Chirality. A mixture containing an equal number of molecules of enantiomers is known as a *racemic modification*. The preparation and reactions of these modifications (as well as the individual enantiomers themselves) represent important aspects of the study of stereochemistry.

If a molecule CH_3—C—CH_2CH_3 is converted
$\qquad\qquad\qquad\quad \parallel$
$\qquad\qquad\qquad\quad O$

into CH_3—CH—CH_2CH_3 by some achiral reagent
$\qquad\qquad\quad |$
$\qquad\qquad\quad OH$

(e.g., hydrogen gas and a catalyst) the chiral center in the product molecule may be considered as being generated by approach of a hydrogen from the top *or* bottom "face" of the planar C—C—O grouping
$\qquad\qquad\qquad\qquad\qquad\qquad\qquad\quad |$
$\qquad\qquad\qquad\qquad\qquad\qquad\qquad\quad C$

in reactant molecule

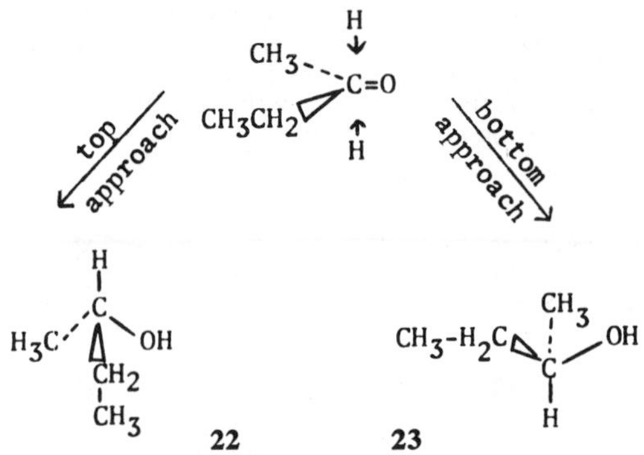

22   **23**

The molecule (**22**) produced by "top" approach is the enantiomer of that (**23**) resulting from "bottom" approach. In fact, the pathways leading to each are enantiomeric, hence are of equal energy. The overall result is thus the production of a racemic modification, since one approach is as probable as another.

Now if a similar reaction were conducted with a chiral substrate that has a preexisting chiral center (A+), the combinations of configurations at the centers in the product molecules would be A+B+ and A+B−. These are diastereomers, the pathways involving their information are diastereomeric (unequal energy!), and hence they are produced in unequal amounts. Such a case is illustrated as follows:

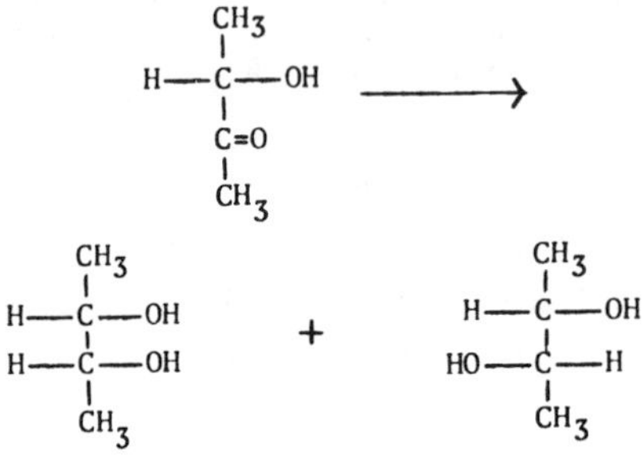

diastereomers

Diastereomeric reaction pathways may be obtained in numerous other ways. An interesting case is represented by the biological reduction of acetaldehyde-1-d, CH_3—C=O. The product is the chiral
|
D

structure CH_3—CH—OH and a racemic modifica-
|
D

tion might be expected from the reduction. In fact, the transfer of a hydrogen during the enzymatic reduction (an enzyme is a large chiral molecule) to one face of the acetaldehyde is diastereomeric with the transfer to the opposite face, hence (very) unequal amounts of the enantiomers are formed. These interactions may be described as E+ .... A+ and E+ .... A−, where E+ represents the chirality of the enzyme and A+ or A− represent the incipient chirality in the reduced acetaldehyde molecules.

The last example reflects in a modest way the importance of the study of stereoisomerism. Bio-

logical conversions represent a glorious array of diastereomeric reactions and interactions: a given chiral amino acid is metabolized but its enantiomer is not; a certain complex drug (often a chiral molecule) alleviates pain but its enantiomer is inactive; and subtle changes in structure alter a given chiral compound's action completely in the human body. The rationales for these observations are being sought by many investigators and a detailed knowledge of the subject of stereoisomerism is essential in finding the answers to the many questions involved.

ALEX T. ROWLAND

References

1. Eliel, E. L., "Stereochemistry of Carbon Compounds," McGraw-Hill Book Company, Inc., New York, 1962.
2. Mislow, K., "Introduction to Stereochemistry," W. A. Benjamin, Inc., New York, 1965.
3. Belloli, R., "Resolution and Stereochemistry of Asymmetric Silicon, Germanium, Tin, and Lead Compounds," *J. Chem. Ed., **46**,* 640 (1969).
4. Eliel, E. L., "Recent Advances In Stereochemical Nomenclature," *J. Chem. Ed., **48**,* 163 (1971).

STEREOREGULAR POLYMERS

The properties of natural and synthetic high polymers and their applications in plastics, fibers, elastomers, adhesives and coatings are determined in large part by (a) their average molecular weights, (b) the forces between the long chain molecules and (c) geometrical considerations, especially the degree of regularity of repeating units. Molecular weight characteristics and forces have received attention for many years. In contrast, the importance of stereoregularity vs. irregularity of chemical substituents became fully appreciated only around 1950[1,2]. In general, strong interchain forces and regularity promote normal crystallinity, along with high strength, high softening temperatures, hardness and insolubility in common solvents.

Some polymers have regular structure free of diastereoisomerism because of the symmetry of the monomers, such as vinylidene chloride, CH_2=CCl_2 and isobutene CH_2=$C(CH_3)_2$. However, the term stereoregular polymer is generally reserved for stereoregular polymer structures derived from unsymmetrical monomers which can be obtained by special ionic methods of polymerization (usually from heterogeneous systems). These polymerization processes often using complex catalysts such as Ziegler-Natta activated transition metal catalysts, e.g., from AlR_3 and polymeric $TiCl_3$, have been called stereoregulated, stereospecific or oriented polymerizations. In 1964 the writer suggested the word stereopolymerization to describe those special ionic polymerization systems for treating unsymmetrical ethylenic monomers to obtain either normally crystalline polymers with DDDDD regularity, permanently amorphous polymers with irregular DLDDL sequences or intermediate structures according to conditions chosen. Such control of polymer stereoisomerism was achieved first with vinyl alkyl ethers, but the first stereoregular polymers to become the basis of a major new industry were the stereoregular or isotactic propylene polymers (heat-resistant molding plastics and fibers)[2].

Crystallinity has been one of the principal effects by which stereoregularity or tacticity has been studied in polymers. However, as expected, not all stereoregular polymers are equally crystallizable. Differences in chemical reactivity, nuclear magnetic resonance and infrared have given useful information about stereoregularity. However, for comparing polymers from a given monomer x-ray diffraction and solubility data are most reliable for estimating tacticity.

Interest in controlling steric configurations and stereoisomerism in polymers by polymerization conditions developed only slowly. Staudinger and Schwalbach in 1931 suggested that invariable low crystallinity in polyvinyl acetate might be caused by diastereoisomerism, that is randomness in D and L positions of the acetate groups along the chain molecules[1]. Branching and deviations from head to tail addition were studied meanwhile as types of isomerism. However not until 1948 were examples of stereoregulated polymerizations of a vinyl-type monomer disclosed by Schildknecht and coworkers. In both early types of stereopolymerizations vinyl isobutyl ether diluted by liquid propane could be treated at low temperatures[1]. Addition of gaseous boron fluoride gave very rapid polymerizations to rubberlike substantially amorphous high polymers. Careful addition of cold boron fluoride etherate, immiscible with liquid propane at $-78°C$ or above produced a slow growth or proliferous polymerization to form normally crystalline polymers.

This suggested that it might be possible to prepare normally crystalline polymers from other unsymmetrical monomers of the type $CH_2{=}CHY$ in ionic heterogeneous systems.

The discovery of stereopolymerizations of 1-alkenes by use of Ziegler-Natta catalysts in heterogeneous systems in 1954, and subsequent studies of polymer structure, attracted worldwide attention to this field. Propylene and 1-butene, which are monoallylic compounds, had not been homopolymerized by conventional ionic or free radical conditions to give linear high polymers suitable as plastics or other synthetic materials. Short branches in polyethylenes had been shown to reduce crystallinity and hardness. Natta and coworkers demonstrated the ability of catalysts such as those from reaction of aluminum alkyls with titanium halides to form normally crystalline, surprisingly high softening polymers from propylene[2,3]. A helix of three monomer units explained the regularity required for crystallization and the identity period observed from x-ray diffraction. Stereoregular isotactic polymers of crystal melting ranges shown in the table were prepared by slow heterogeneous ionic polymerizations using special catalysts at moderate pressures and temperatures.

The polymers as prepared to this time have stereoregularities which in most cases are much below 100%

Stereoregular propylene polymer plastics were first supplied by Esso, Farbwerke Hoechst, Hercules and Montecatini. These polymer products have outstanding utility, for example, heat resistance superior to that of polyethylenes in sterilizable hospital devices. By copolymerization and control of the degree of stereoregularity brittleness at low temperatures can be avoided. Stereoregular 1-butene and isobutylethylene polymers are also manufactured, but the isotactic polymers from styrene and from methyl methacrylate are too brittle for much use. Crystallizable polystyrenes also have been prepared by heterogeneous anionic polymerizations (Lewis basic catalysts) and crystalline methyl methacrylate polymers can be prepared using Grignard catalysts.

Soluble catalysts derived from organoaluminum compounds and vanadium halides promote formation of atactic elastomeric propylene polymers and copolymers such as ethylene-propylene-diene terpolymer rubbers (EPDM). These have become commercially important recently. Amorphous adhesive propylene homopolymers have some commercial use. Syndiotactic or DLDL propylene polymers have been reported but their structures

CRYSTAL MELTING RANGES OF SOME STEREOREGULAR POLYMERS

Isotactic Stereopolymers	Melting Range	Stereopolymers	Approximate Maximum Melting Point
Propylene	165–176°C	Isotactic 1,2-butadiene	120°C
1-Butylene	120–136	Syndiotactic 1,2-butadiene	154
1-Amylene	60–70	*Cis*-1,4-butadiene	+1
Isopropylethylene	300–310	*Trans*-1,4-butadiene	148
Isobutylethylene	235–250	*Cis*-1,4-isoprene	22
Isoamylethylene	about 110	*Trans*-1,4-isoprene	65
4,4-Dimethyl-1-pentene	>380	*Cis*-1,4-(2,3-dimethyl butadiene)	190
Styrene	230–250	*Trans*-1,4-(2,3-dimethyl butadiene)	260
Isobutyl vinyl ether	100–130		
Methyl methacrylate	160		

and properties have not been completely established.

Isomeric isoprene polymers are formed biologically as natural rubber (cis-1,4) and balata or gutta-percha (largely trans-1,4). Modified Ziegler-Natta type catalysts, colloidal lithium or lithium alkyls (in absence of ethers) were found to give predominantly cis-1,4 polymer rubbers from isoprene. Cis-1,4-polybutadiene, so-called synthetic natural rubbers, have become important in tires and in graft copolymerization with styrene for high-impact plastics. Different conditions of polymerization give rigid trans-1,4 diene polymers resembling balata.

Although precise mechanisms of the stereopolymerizations are yet uncertain, several characteristics become evident. The reactions are predominantly heterogeneous, ionic reactions at low or moderate temperatures. The more stereoregular polymers grow as a separate phase upon the solid or immiscible liquid catalyst. However, some monomers such as vinyl isobutyl ether can form somewhat isotactic polymer fractions even from homogeneous solutions of Lewis acid and monomer. In contrast, the polymers from vinyl isopropyl ether, which apparently are stereoregular when obtained by slow growth polymerization using boron fluoride etherate catalysts, nevertheless do not crystallize readily. BF_3 can be used to form isotactic vinyl isobutyl ether polymers if it is applied in a separate phase of methylene chloride immiscible with liquid propane. Relatively polar solvents which favor separation of gegen ions from their growing macroions generally impair stereospecificity.

Although the Ziegler-Natta catalyst systems for stereopolymerization of 1-olefins were regarded by Natta[2,3], Mark[4] and others as examples of anionic polymerizations, the writer considers them as a special type of cationic polymerization. A consistent system relating monomer structure to response to catalyst types is only possible if propylene containing an electron repelling methyl group attached to the ethylene nucleus polymerizes with Lewis acid catalysts (cationic polymerization). Propylene as an allyl compound lacks sufficient electron withdrawal from the ethylene group to homopolymerize by free radical initiation (peroxide, azo catalysts or ultraviolet light) and it also lacks sufficient electron donation (as in isobutene) for homopolymerization by conventional cationic systems[5]. Ziegler-Natta catalysts have been observed to homopolymerize some other monoallyl compounds. An intensive study of the literature by the writer and Mabel D. Reiner showed no well-characterized homopolymers of high molecular weight obtained from monoallyl compounds by free radical or conventional ionic catalyst systems.

Transition metal catalysts for polymerization of 1-alkenes similar to those of Ziegler were developed in DuPont laboratories and have been called coordination catalysts[6].

Outside of vinyl addition polymerizations some crystallizable stereoregular polymers also have been prepared. An example is isotactic polymer from

$$CH_3$$
$$|$$
propylene oxide —CH CH$_2$O— made by using

ferric chloride complex catalysts for proliferous type reactions[4].

After the demonstrations of preparation of stereoregular polymers having novel properties by means of special ionic methods, the possibilities of free radical methods were examined extensively. It must be concluded that in free radical systems the structures of homopolymers and copolymers can be little influenced by specific catalysts and other reaction conditions, but are determined largely by monomer structure. This is consistent with the relative uniformity of comonomer reactivity ratios in radical copolymerizations. However, it has been found possible to obtain somewhat more syndiotactic structure, DLDL, than normally obtained by radical reactions, at low temperatures and by selecting solvents. Examples are polyvinyl chlorides of higher than usual crystallinity from polymerizations at low temperature e.g., −50°C under ultraviolet light.

Although they do not crystallize, polyvinyl acetates prepared at low temperatures apparently are more syndiotactic since they yield more than usually crystalline polyvinyl alcohols by saponification. Monomers of high polarity such as vinyl trifluoracetate by radical polymerization can form relatively syndiotactic polymers from which more crystalline polyvinyl alcohols can be prepared by saponification.

C. E. SCHILDKNECHT

Cross-references: *Polymerization; Catalysis*

References

1. Schildknecht, C. E., Zoss, A. O., Gross, S. T., Davidson, H. R., and Lambert, J. M., *Ind. Eng. Chem.*, **40**, 2104 (1948); Schildknecht, C. E., "Vinyl and Related Polymers," Wiley, New York, 1952.
2. Natta, G., Pino, P., Corradini, P., Danuso, F., Mantica, E., Mazzanti, G., Moraglio, G., *J. Am. Chem. Soc.*, **77**, 1708 (1955).
3. Natta, G., *Science*, **147**, 261 (January 15, 1965).
4. Gaylord, N. G., and Mark, H. F., "Linear and Stereoregular Addition Polymers," Interscience, New York, 1959; cf., C. E. Schildknecht, *ACS Polymer Preprints*, **13**, 253 (1972).
5. Schildknecht, C. E., *Polym. Eng. and Sci.*, **6**, 240 (July 1966); "Allyl Compounds and Their Polymers (Including Polyolefins)," Wiley, New York, 1973.
6. Billmeyer, F. W., Jr., "Textbook of Polymer Science," 2nd Ed., p. 319, Interscience-Wiley, New York, 1971.

STERIC HINDRANCE

The interference that results when two or more atoms or groups tend to occupy the same point in space is called *steric hindrance*. This type of interference can have profound effects on the physical and chemical properties of molecules. Thus, the chemical reactivity of a specific functional group in a compound can be markedly reduced by the introduction of various bulky groups in certain positions in the molecule (usually adjacent to the functional group). Molecular spectra, stereochemical factors, bond distances, acidity and basicity are just a few of the many properties that can also be affected by the presence of steric hindrance in a molecule.

The basic concepts were put forward in the latter half of the nineteenth century to explain certain peculiarities of organic reactions. Kehrmann, in an investigation of the formation of oximes from quinones and hydroxylamine, found that a quinone of the general type I, where R represented a halogen atom or alkyl group, always tended to form an oxime of type II in preference to the isomeric oxime III (although the latter could also be formed to a smaller extent at the same time). A disubstituted quinone of type IV, on the other hand, was only capable of forming a single oxime of type V, whereas a tetrasubstituted quinone, VI, formed no oxime at all.

Kehrmann suggested in 1888 that the "space-filling" properties of the groups R were responsible for this phenomenon.

In 1894 Victor Meyer drew attention to the great difficulty of preparing certain ortho-substituted aromatic esters. Thus, while 3,5-dimethyl-benzoic acid, VII, could readily be esterified by reaction with an alcohol and acid catalyst, 2, 4, 6-trimethyl-benzoic acid, VIII, could not be (such esters can be made,

however, by the action of alkyl halides on their silver salts). Many other groups in the ortho-position, such as Cl, Br, NO_2, etc., were also found to make esterification more difficult.

Meyer, like Kehrmann, postulated that the space-filling properties of the bulky ortho-groups were responsible for this state of affairs. He also employed the term "steric hindrance" to describe the phenomenon, a term which is still used.

Notwithstanding the great importance of steric factors it should be pointed out that electronic and catalytic effects also play a predominant role in esterification under different circumstances.

A very interesting class of compounds which are capable of exhibiting steric hindrance are the biphenyls. In 1922 Christie and Kenner showed that the freedom of rotation of the two phenyl groups about the central carbon-carbon bond in such compounds could be completely restricted by the presence of certain bulky substituents. In that year they resolved 6,6'-dinitro-diphenic acid, IX, into two optically active (enantiomorphic) forms (see **Stereochemistry**). Stereoisomerism is possible because the benzene rings are incapable of becoming

coplanar with each other, thus preventing the molecule from possessing a plane of symmetry at any time.

Later it was shown that in some cases only two ortho groups can provide enough steric hindrance to prevent coplanarity, as in diphenyl-2,2'disulfonic acid, X. With this class of compounds the effect of size of the interfering groups is the predominant factor. If such sterically hindered molecules are sufficiently activated by heat, for example, the hindrance to complete rotation can often be overcome and a formerly optically active compound will become racemized, i.e., optically inactive.

It should be emphasized that the so-called freedom of rotation about a single bond does not mean *complete* freedom. Even in a simple molecule such as ethane there is still some interference between the hydrogen atoms of the two methyl groups which is theoretically capable of preventing rotation. However, the interference is small enough so that at ordinary temperatures the rotation of the groups in ethane can be considered to occur in a continuous manner. (More accurately, this hindrance to rotation in ethane is now regarded as being largely due to the repulsion of the electron clouds in the carbon-hydrogen bonds). The term *conformation* is used frequently today to denote any one of an infinite number of possible spatial arrangements in a molecule (which normally cannot be isolated) that result from the rotation of atoms or groups about a single bond.

The interference between hydrogen atoms has some interesting consequences in the cycloalkanes. Until fairly recently it was believed that cyclopropane, cyclobutane and cyclopentane, the first three members of this series, were planar molecules. In cyclopropane the carbon-carbon bond angle is only 60°, as against the normal tetrahedral value of approximately 109° 28'; this molecule should therefore exhibit the most strain ("angle" or "Baeyer" strain) and be the least stable; cyclobutane having a carbon-carbon bond angle of 90° should exhibit less strain and cyclopentane (108°) the least.

It is now known that this particular type of strain is only one of the many factors that are involved and that these molecules gain in stability if the hydrogen atoms are "staggered" rather than "eclipsed" causing the rings to become nonplanar or "puckered." In effect, therefore, some of the strain caused by the repulsions of the hydrogen atoms ("steric" or "eclipsing" strain) is reduced at the expense of "angle" strain. In cyclopentane it has been shown that the puckering is not fixed at any one place but "travels" around the ring in a dynamic fashion. In such a manner the molecule is rendered more stable than it would otherwise be if the hydrogen atoms were allowed to exert their maximum repulsive forces.

In cyclohexane, which exists in two relatively strain-free forms in dynamic equilibrium with one another, it appears that the tetrahedral angle is not distorted at all in the so-called "chair" form, XI, but that slight distortions occur in the "boat" form, XII. In recent years it has been shown that the majority of molecules are in the chair form at room

XI   XII

temperature, as the carbon-hydrogen bonds are all nicely "staggered" in that form causing the repulsions between the hydrogen atoms to be at a minimum (this is clearly brought out in molecular models). In the cycloalkanes immediately above cyclohexane, however, there is a considerable amount of interference between the hydrogen atoms as it is not possible for all of them to assume positions sufficiently far apart from each other.

Steric Inhibition of Resonance. Examples of the effect of steric hindrance on resonance are very numerous in organic chemistry. For example, the coupling of a simple diazonium compound such as benzene diazonium chloride occurs readily with tertiary amines such as N,N-dimethylaniline, to give an azo dye. However, coupling fails to occur with the 2,6-dimethyl derivative.

Resonance in N,N-dimethylaniline creates a center of high electron density in the paraposition, which is necessary for coupling to occur. In form XIIIb the presence of the double bond between the nitrogen and the benzene ring causes the methyl groups and the nitrogen atom to be in the same plane as the benzene ring. In the 2,6-dimethyl derivative the resonance cannot occur as the steric hindrance between the methyl groups on the benzene ring and on the tertiary nitrogen prevents the dimethylamino group from becoming coplanar.

XIII-a   XIII-b

Steric inhibition of resonance has profound ef-

fects on other properties of molecules such as their spectra and dipole moments.

RICHARD L. BENT

Cross-references: *Alicyclic and Cyclic Compounds, Conformational Analysis, Resonance, Stereochemistry, Stereoisomerism.*

References

Newman, M. S., Editor, "Steric Effects in Organic Chemistry," John Wiley and Sons, Inc., 1956.
Gray, G. W., Editor, "Steric Effects in Conjugated Systems," Butterworths Scientific Publications, London, 1958.
Eliel, E. L., "Stereochemistry of Carbon Compounds," McGraw-Hill Book Company, Inc., 1962.
Pauling L., and Hayward, R., "The Architecture of Molecules," W. H. Freeman and Company, 1964.

STERILIZATION (Industrial)

Sterilization refers to the complete elimination of microbial life. This is accomplished commercially by subjecting the material to dry heat, heat in the presence of steam, filtration, the action of chemicals, or irradiation. These cover the important means of accomplishing sterilization, the remaining approaches being used only to a very limited degree.

The material to be sterilized must be packaged in such a manner that once sterility is achieved, recontamination by microorganisms, both airborne and by contact, is prevented. Liquids are placed in vials which are hermetically sealed or sealed by plugs of cotton which are tight enough to filter microorganisms out of any air which may be drawn into the container by changes in pressure, but which allow the movement of steam or air to equalize pressures. Surgical dressings, gloves, glassware, and similar materials may be packaged in paper or fabric either of which is permeable to steam and air but not to microorganisms. The wrapping is so overlapped or sealed that there are no open channels through which airborne organisms may enter. In all packaging it must be remembered that recontamination is prevented by hermetic sealing, providing a filter through which air must pass during the "breathing" of the package, or by making the channel of entry so tortuous that microorganisms are deposited on the walls of the channel and not carried into the product. The last method is the least reliable.

The achievement of sterility is tested by transferring portions of the material aseptically from the container into sterile tubes on flasks containing a quantity of sterile Fluid Thioglycollate Medium which is highly favorable to bacterial growth. Portions of the material are also transferred to tubes containing Sabouraud Liquid Medium for the detection of yeasts and filamentous fungi. Replicate samples may also be implanted into several other kinds of media which may encourage specific types of microorganisms whose nutrient requirements are unusual. The tubes are closed with sterile cotton plugs or metal caps, then divided and stored at several temperatures which encourage the growth of different kinds of microorganisms. At the end of

7 to 14 days the growth of the microorganisms in the contaminated material will appear as a pellicle, a sediment or uniform turbidity. The medium in tubes containing sterile material will remain clear.

It has been amply demonstrated that a combination of moisture and heat is far more effective in killing bacteria than dry heat. This is believed due to coagulation of the cell proteins which occurs at a lower temperature than their destruction by oxidation. Sterilization of materials with steam is, therefore, the most common method employed. This takes place in an autoclave which is a pressure vessel having a jacket surrounding all but the door. Material, properly packaged, is loaded into it and the door closed and sealed. The material is then subjected to steam at temperature of 240°F for 30 minutes. This is sufficient to produce sterility if all air in the package is replaced by steam. This can be done by the gradual displacement of air downward through the load by the entering steam, while bleeding it out of the sterilizer with an outlet placed at its lowest point. When steam issues from this outlet, the timing of the operation begins.

The air may also be removed by drawing a vacuum on the chamber before the steam is introduced. In the commercial sterilization of large quantities of dressings, elaborate procedures have been developed in which repeated applications of vacuum and steam are used to exhaust all the air. In such installations, thermocouples connected to suitable recording devices are used throughout the chamber to insure the attainment of minimum sterilizing temperatures. Liquids in ampules are sterilized in the same type of autoclave, except that there is no need to adopt precautions about penetration, since these already contain the necessary water. Those drugs which can withstand these conditions are profitably sterilized in this manner, but unfortunately many are inactivated or modified and must be sterilized by methods described later.

Some drugs, particularly biologicals, cannot be heated at all without undergoing decomposition or deactivation. These are sterilized in solution by filtration through sterile bacteria-retaining filters of various kinds. They are then filled aseptically into sterile containers and closed with sterile closures.

Commercial chemical sterilization is practically confined to the use of formaldehyde and ethylene oxide. The formaldehyde is used in an autoclave similar to a steam autoclave but equipped with a formaldehyde generator. This is merely a small attached chamber in which 37% formaldehyde aqueous solution can be heated and vaporized into the chamber, the solution supplying the required water also. The advantage of formaldehyde lies in the fact that much lower temperatures are required with its use than with steam alone. Ethylene oxide may be used as the pure gas but more commonly is employed as a mixture in various proportions with carbon dioxide and with the fluorinated hydrocarbons in order to reduce its flammability and explosiveness. The material to be sterilized is placed in an autoclave and a high vacuum is drawn. The gas is then injected and pressure built up to correspond to the required concentration. The time of exposure, concentration of gas, and temperature of sterilization all vary with the nature of the material and must be determined by test runs. The use of ethylene oxide gives the advantage of moderate processing temperature, together with low moisture requirement making it possible to process materials adversely affected by the previous methods described.

Irradiation of materials to produce sterility has taken many forms, but only three types are commercially significant; irradiation by ultraviolet, irradiation by an electron beam and irradiation by a radioactive source.

The use of ultraviolet radiation is confined to thermolabile materials which cannot be filtered, such as plasma, milk, and similar biologic materials. It requires an intense source of ultraviolet and the use of an extremely thin layer of the fluid being sterilized. Due to the difficulties involved, this method is generally used in the absence of other, more readily applied techniques. The ability of high-energy radiation to destroy bacteria in a material without significant rise in temperature has made the electron beam technique of great interest in the processing of extremely heat sensitive materials. It has been found that to sterilize a spore-bearing culture a dose of 2.5 megarads is necessary. To give such doses in a reasonable time interval by electron radiation requires massive and expensive equipment. The electron beam consists of negatively charged particles with relatively poor penetration and limits the use of the Van de Graaff generator and linear accelerator to materials of small cross-section and low density. The type of radiation emitted by a radioactive source is gamma radiation, or electromagnetic radiation. It is neither particulate nor charged and, therefore, has great penetrating power. A radiation installation employs a cobalt-60 source and an irradiation vault of considerable size and complexity, in which the volume of material is slowly moved around the source. A dose of 2.5 megarads accumulated over a period of hours is required to assure sterility. A gamma-radiation plant can be increased in capacity by the simple expedient of increasing the cobalt-60 charge, thus reducing the time a product must move through the irradiation vault to accumulate a dosage of 2.5 megarads. The effect of radiation on the properties of a product cannot be predicted and each product must be assessed individually. Gamma radiation has particular application in the sterilization of large and bulky items, and may have its greatest potential in the sterilization of disposable items made of rubber, plastics and metals.

J. N. MASCI

Cross-reference: *Disinfectants.*

STEROIDS

Steroids are an important class of chemical compounds found in virtually all forms of plant and animal life. They are usually crystalline, and contain one or more hydroxyl groups; hence, the name sterol (Gr. stereos, solid) was originally given to this group of materials. Sterols isolated from animal sources are called *zoosterols,* those from plants *phytosterols,* and those from yeast and fungi *mycosterols.* Generally, sterols have been used in medicine to treat a variety of conditions ranging from endocrine hormonal alterations to coronary insufficiency. They have been prescribed in many physical forms ranging from crude extract elixirs to more refined and sophisticated single-entity tablets.

Structure. The nucleus of the steroid is a tetracyclic ring system known as perhydro-1,2-cyclopentanophenanthrene. It is composed of one five- and three six-membered rings fused in an angular fashion. The term "stereochemistry" refers to the spatial orientation of the atoms within the molecule. While much of steroid chemistry depends on the stereochemistry of the molecules involved, the relationship of the term "steroid" is only fortuitous. There are six centers of asymmetry (C-5, 8, 9, 10, 13, 14) in the steroid nucleus leading to a possibility of 64 different stereoisomers. The rings of the steroid template are often referred to by letters (A, B, etc.). The positions represented by R, R_1, and R_2 are varied in some of the natural and synthetic steroids.

The stereochemistry of steroids is due to the ring juncture and to the substituents located at various positions on the rings. Groups which are above the plane of the ring system are called β-oriented (solid line) while those below the plane are α-oriented (dotted line). The absolute configuration of the naturally occurring steroids was established as a result of the correlation with D-glyceraldehyde. X-ray analysis of crystal structure has been a useful technique to prove the stereochemistry of some of the basic steroids. By this means, ring juncture and group position placement were confirmed.

The nomenclature of steroids is most often based on rules approved and published by the IUPAC. Briefly, the names of steroids are based on certain fundamental structures, usually hydrocarbons. The prefixes and suffixes which are attached to the names indicate the nature of the substituents. The positions of these substituents are indicated by the number of the carbon atom to which they are attached. Groups of unknown configuration are given the notation δ and their bonds indicated by wavy lines. The stereochemistry at C-5 is always designated in the nomenclature by the 5α or 5β prefix unless there is a double bond present. The stereochemistry of other ring junctions is assumed to be as described in the general formula of the parent compound. When a ring is cleaved, the prefix *seco* is added and the original steroid numbering is retained. The prefix *des* or *de* is used to denote removal of a functional group or a whole ring from a normal parent steroid nucleus.

Specific Sterols. *Cholestanes.* For the most part, sterols are monohydric alcohols containing a hydroxyl group at the C-3 position which is usually β. The best known representatives belong to the cholesterol family. Cholesterol usually is the principal sterol of the animal organism. It is present in all tissue but primarily in the spinal cord, the brain, and gallstones. Commercially it is obtained from the alkali saponification of cattle spinal cords. Much of the early steroid structure elucidation and

characterization was done with the readily available cholesterol. Analogs of cholesterol are numerous (examples are stigmasterol, ergosterol, and lanosterol) and include the 7-dehydro structures, which through ultraviolet irradiation lead to lumisterol and tachysterol, and culminate in the vitamin-D family of compounds.

cholesterol

cholic acid

Bile Acids. Another large class of naturally occurring steroids are the bile acids. They are generally hydroxylated cholanic acid derivatives an example of which is cholic acid. These substances are produced by the liver upon the degradation of cholesterol. In bile, they occur largely as water soluble sodium salts of peptide conjugates. Their primary mammalian function is to promote the resorption of fats and other water insoluble substances. The carbon skeleton of the bile acids is similar to cholesterol, but lacks the C-5,6 double bond and instead possesses a C-5β juncture. In addition, the side chain is terminated by a C-24 carboxyl group.

Sex Hormones. The sex hormones are steroidal substances secreted by the gonads when stimulated by proteinic hormones of the anterior lobe of the pituitary gland. Experiments have definitely proved that the testes and ovaries normally secrete chemical agents of hormonal character that control sexual processes and secondary sex characteristics. In the broad sense, they are growth substances of general importance to health and well-being. Naturally occurring representatives of the sex hormones include members belonging to the estrogen, androgen, and progestogen families of compounds. Historically, estrogens were the first steroid sex hormones isolated. They are characterized structurally by an aromatic ring A; therefore, the substituent (H or CH_3) present at C-10 in other natural steroids is lacking. While there are numerous natural estrogens, the most common and abundant are estrone, estradiol, and estriol. These substances are ovarian hormones which with progesterone, control the human female sexual cycle. In animals, estrogens possess in varying degree, the abil-

estrone   estradiol

estriol

ity to produce characteristic changes of estrus. The commercial source of these substances is either chemical degradation of diosgenin (from the dioscorea plant), extraction from pregnant mare's urine, or by total synthesis from nonsteroidal substances. Many synthetic variations of the estrogen-type molecule have been made. The commercially important ones are the 17α-substituted estradiol derivatives which when combined with a synthetic progestogen can alleviate menstrual disorders or control fertility.

Androgens are compounds which are responsible for the development of the male sex organs and secondary sex characteristics. The most prominent members of this series of compounds, which are naturally occurring, are testosterone, androsterone, etiocholanolone and dehydroepiandrosterone. All four of these substances resemble the natural estrogens in that there is an oxygen function at C-3 and C-17; however, they lack the aromatic ring A and possess a C-10 angular methyl group.

testosterone   dehydroisoandrosterone

androsterone   etiocholanolone

Testosterone is the most active of this group and is the functioning hormone found in the male testes, ovary (female) and adrenal cortex. The other androgens mentioned are found in the urine as excretory metabolites of which androsterone is the principal one. Androgens are either isolated from natural sources such as testes (synthesized by the Leydig cells) and urine, produced by degration of diosgenin or cholesterol, or prepared by total synthesis from nonsteroidal molecules. Numerous chemical transformations have been applied to the androstane molecule. Commercially, such derivatives are of use as growth-promoting substances

(anabolic property) and as replacement therapy for hypogonadal males (androgenic property). Some of the important structurally modified androgens are those related to the estrane molecule where the C-10 methyl group has been replaced by a hydrogen. These substances are also referred to as 19-nor steroid derivatives (e.g., 19-nortestosterone).

Another important naturally occurring mammalian hormone is progesterone. This substance has been termed "the hormone of pregnancy" because of its importance just prior to and during the gestation period. It is secreted by the corpus luteum (material which surrounds the egg). The uterine mucosa which grows under the influence of estradiol, is proliferated by progesterone for potential reception of a fertilized ovum. If there is no fertilization, the corpus luteum regresses and the uterine mucosa is flushed away during menstruation. Commercially, progesterone has been obtained from cow ovaries, the degradation of cholesterol and diosgenin, and by total synthesis from nonsteroidal intermediates. A variety of derivatives of progesterone have been prepared as well as other substances having the estrane and androstane nucleus. Because of the profile of physiological activity which mimics that of progesterone, these substances as a group are often referred to as progestogens or progestins. The principal commercial utility for progestogins is for treating menstrual disorders, for maintenance of pregnancy, and when used with estrogens for control of fertility. From this latter concept originated the first commercially successful means for the physiological control of conception.

progesterone   cortisol

Adrenal Cortical Hormones. The adrenal cortex is a gland which produces a number of vitally important steroid hormones essential for maintenance of life. Deficiencies of these hormones produce a serious imbalance in electrolytes and improper carbohydrate and protein metabolism. About 30 compounds have been isolated from adrenal cortical extracts. Included is a series of very important steroids which have been identified. These are called corticoids. They are, in general, related structurally to progesterone but possess additional oxygen functions, notably at positions C-11, C-17, and C-21. These additional groups impart marked pharmacological properties. The major and most important hormone secreted by the human adrenal cortex is cortisol. The administration of cortisone, which causes a marked rise in plasma levels of cortisol, dramatically relieves the symptoms of patients suffering from rheumatoid arthritis. The early source of these corticoids was the hog adrenal. With the increased demand for these substances other sources and processes were discovered.

These include manufacture from desoxycholic acid and preparation from other sterols. A vast number of chemical modifications of the corticoid molecule have been performed in order to enhance activity and decrease some of the undesirable biological properties. Some of the analogs of cortisone are presently used to treat adrenal insufficiency, rheumatoid arthritis, and various types of inflammatory conditions (systemic and topical).

Another important steroid originally isolated from the adrenal cortex is aldosterone. In general, this substance differs from the coritcoids by an aldehyde group in place of the methyl group at C-13 (ketalized with the 11β-hydroxyl group). Aldosterone is known as a mineralocorticoid because of its ability to retain sodium in the body. Its presence is essential in the right amount to regulate the composition and volume of body fluid by promoting reabsorption of sodium along the distal tubule of the nephron (secretory unit in the kidney which forms urine). Aldosterone which has been synthesized with difficulty, is generally not used therapeutically. However, due to its occasional abnormally high presence in the body, substances which block the action of aldosterone are of therapeutic value. Commercially, spironolactone is used to competitively block aldosterone and thereby normalize electrolyte balance.

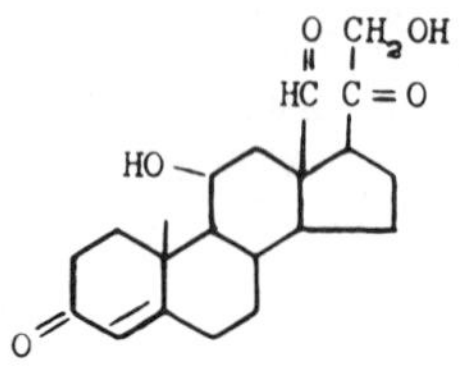

aldosterone

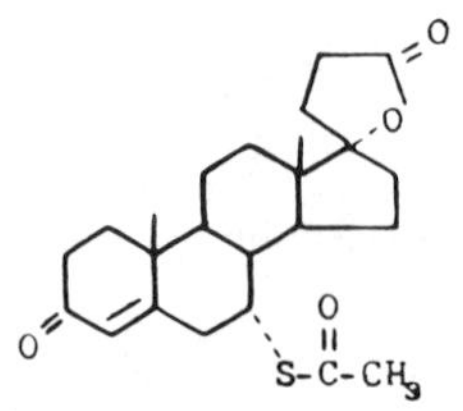

spironolactone

Other Steroids. Other steroids from natural sources (primarily plant) include the cardiac glycosides, the sapogenins, and the steroid alkaloids. Cardenolides (strophanthidin) and bufadienolides (bufotalin) which are related to digitalis are examples of cardic glycosides. These steroids have in common a ring A/B cis juncture and an unsaturated lactone ring attached to the C-17 position. In addition, there may be two or more hydroxyl groups present as glycoside derivatives in the natural environment.

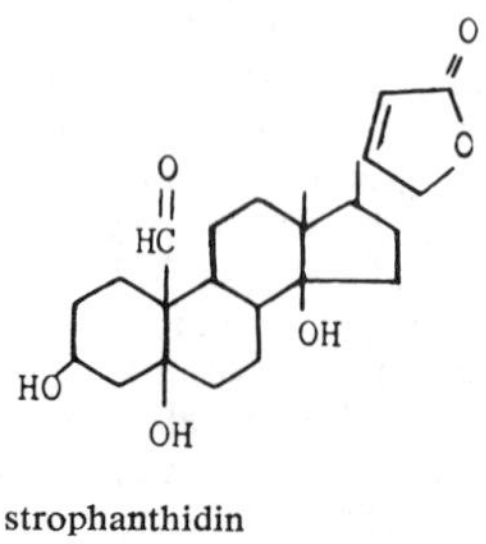

strophanthidin

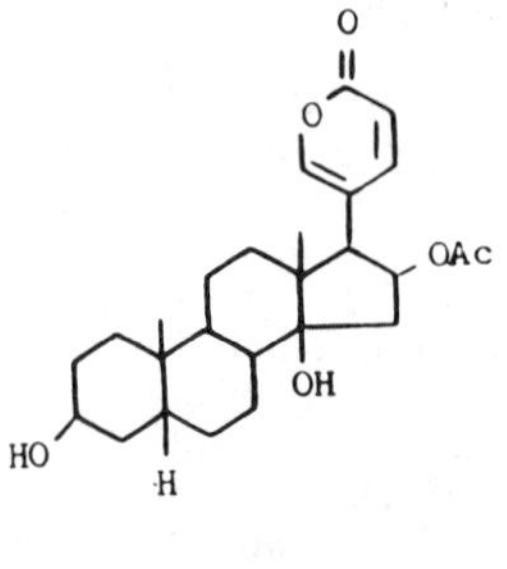

bufotalin

Many of these glycoside derivatives possess powerful cardiotonic activity. Digitalis is a pharmaceutical preparation made from the leaves of purple foxglove. It is a mixture of glycosides and of inestimable value for the treatment of heart disease. The therapeutic index, however, is very small since an excessive dose causes death by severe heart contraction.

The term sapogenin is used to describe a group of plant steroid saponin glycosides that form colloidal, soapy solutions in water. The various sapogenins are closely related with variations of the number and position of hydroxyl groups, the stereochemistry of the ring A/B junction, and unsaturation at C-5 being the most prominent. At present, sapogenins are of importance because of their facile conversion to useful steroid intermediates. One of the most important sapogenins commercially is diosgenin. It can be readily degraded to very useful steroid intermediates of the androstane and pregnane series. The primary source for many of these sapogenins is plants located in tropical regions. The highest concentration is often found in the roots of yam-like substances.

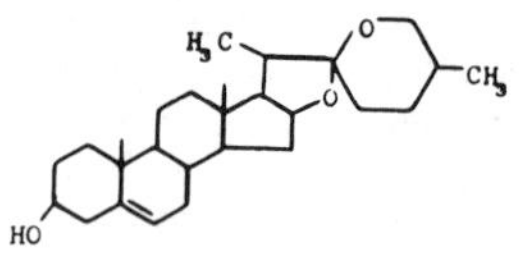

diosgenin

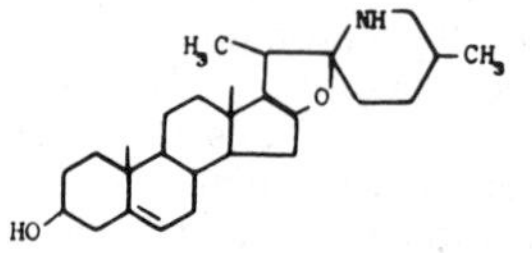

solasodine

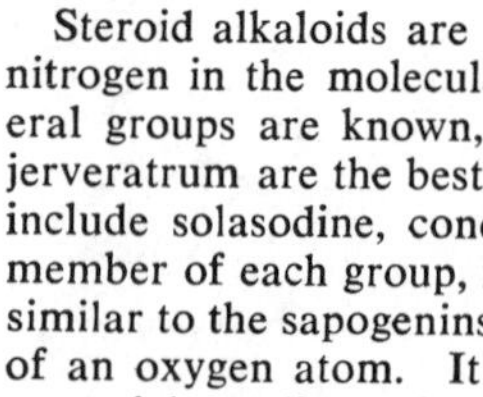

conessine

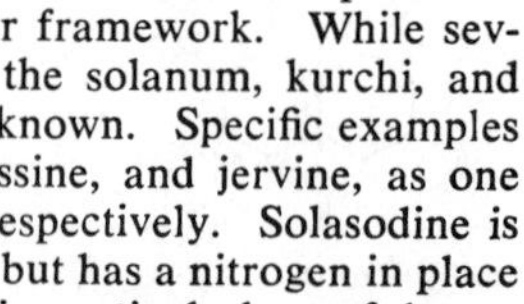

jervine

Steroid alkaloids are substances which possess a nitrogen in the molecular framework. While several groups are known, the solanum, kurchi, and jerveratrum are the best known. Specific examples include solasodine, conessine, and jervine, as one member of each group, respectively. Solasodine is similar to the sapogenins but has a nitrogen in place of an oxygen atom. It is particularly useful as a potential candidate for degradation to useful steroid intermediates. Conessine is a pentacyclic alkaloid containing two nitrogen atoms. It can also be degraded to potentially useful intermediates. The jervine alkaloids represent a departure from the normal tetracyclic ring arrangement. These compounds possess the C-nor-D-homo ring system and because of this have been used for preparing other interesting C-nor-D-homo derivatives.

Insects and other arthropods do not synthesize the steroid nucleus, but sterols are essential for vital functions including moulting. In addition, fungi require sterols for sexual reproduction. One steroid, ecdysone, is produced by the insect protho-racic gland upon activation by the brain. This sub-stance is known as a moulting hormone which with the juvenile hormone, a nonsteroid, regulate insect development. There are numerous other steroids related to ecdysone which play a role in the insect replication process.

Biogenesis. The vast family of steroids are pre-sumably derived from the same parent compound, isoprene, C_5H_8. From C^{14} tracer studies, it has been demonstrated by a variety of organisms and systems that similar biosynthetic reactions occur wherever steroids are produced. In the presence of ATP, acetate is activated by combination with coenzyme A (CoA). Acetyl CoA condenses with acetoacetyl CoA to give 3-hydroxy-3-methylglutaryl CoA. The thioester grouping is reduced by NADPH to af-ford 3,5-dihydroxy-3-methylvaleric acid (mevalonic acid). Phosphorylation by ATP produces meva-lonic acid pyrophosphate. Decarboxylation and dehydration by ATP gives isopentenyl pyrophos-phate. Six of these isoprene analogs then condense to a hydrocarbon polyisoprene chain called squa-lene which occurs in nature. This substance is probably the starting material from which all ster-oids are made. This enzymatic transformation in-volves a specific folding of the chain to form a series of potential chair or boat fused rings. Such a procedure involves a nonstop sequence which in-cludes molecular rearrangement and eliminations to form lanosterol which in turn, by loss of methyl groups from C-4 to C-14, yields zymosterol. This substance affords cholesterol, which is common in both plants and animals. The conversion of choles-terol to bile acids and to various steroid hormones in animals is accomplished by a gradual oxidative degradation of the side chain.

Analytical Techniques. A wide variety of analytical methods have been utilized in the steroid field to identify compounds, prove structure, assess homo-geneity, and establish stereochemistry. Among the instrumental techniques are colorimetry, rotatory dispersion, ultraviolet, and infrared spectroscopy, nuclear magnetic resonance, and mass spectros-copy. Of the variety of color reactions used as "wet" analysis methods, the widely used Lieber-mann-Burchard reaction is the most important. The separation of free sterols from their esters has been accomplished by forming an insoluble com-plex with digitonin. Various chromatographic methods are particularly popular not only for the separation, but the identification of sterols. These include paper, column, thick and thin layer, vapor phase, and electrophoresis chromatography.

PAUL D. KLIMSTRA

Cross-references: *Cholesterol; Hormones, Steroid; Bile Acids.*

Reference

Fieser, L. F., and Fieser, M., "Steroids," Reinhold Publishing Corp., New York, 1959.

STOICHIOMETRY

Stoichiometry may be defined as the mathematics of chemical reactions and processes. As such it re-lates to all the quantitative aspects of chemical changes, both mass and energy. Stoichiometry is based on the absolute laws of conversion of mass and of energy and on the chemical law of combin-ing weights. This basis makes stoichiometry as ex-act as any other branch of mathematics.

The law of conservation of mass dictates that, re-gardless of the nature of the changes undergone in a physical or chemical process, the total mass of all the materials in the system remains the same, even though the physical states and chemical composi-tions of the materials may change. Likewise, the law of conservation of energy is based upon the fact that the total energy in a reacting system re-mains constant even though the level or form of the energy may change. In radioactive transforma-tions, however, a slight correction must be applied to the law of the conservation of mass. Mass and energy have been found to be interconvertible, so that in general the total energy of the system re-mains constant even though there may be small mass changes.

The above concepts form the basis for weight and heat balance calculations. Such calculations are of great significance in engineering practice for the purpose of evaluating performance of existing operations or designing new manufacturing facili-ties and equipment.

The basic laws of conservation specifically state that matter or energy in a given system cannot be created or destroyed, and accordingly this requires that the following equality holds true:

$$\text{Input} = \text{output} + \text{accumulation}$$

For continuous, steady flow systems the change in in-process inventory is zero during any interval of time. In this case, therefore, the above expression reduces to the simplified form of input = output.

In making material weight balances the above relation may be applied to a single unit of the oper-ation, or to the over-all operation with reference to the separate elements and/or the total mass enter-ing and leaving the system. This method of analy-sis can best be exemplified by means of a synthetic problem: Let it be assumed that consideration is being given to a continuous, steady flow system, to which X pounds of material is fed per minute and from which Y pounds of useful product are ana-lyzed to contain $a\%$ and $b\%$ by weight of a certain constituent, respectively. It is desired to determine the extent of unmeasured loss, Z pounds per min-ute, incurred from the system and the average con-centration, c, of the said constituent in the waste stream.

First, it may be written that input = output. By dividing each side of this equality by an element of time, this relationship can then be transformed into the following expression:

$$\text{Rate of input} = \text{rate of output}$$

Then, a total weight balance can be written to ex-press this statement of equality in terms of the quantities specified in the problem:

$$X = Y + Z \qquad (1)$$

A similar balance may also be written in terms of the constituent in question:

$$\frac{a}{100} X = \frac{b}{100} Y + \frac{c}{100} Z \qquad (2)$$

Finally, by algebraic solution of the two simultaneous equations it follows that

$$Z = X - Y \qquad (3)$$

and

$$c = \frac{aX - bY}{X - Y} \qquad (4)$$

The above example is only a simple illustration of a weight balance. Similarly, the reaction between elements and compounds may be symbolically expressed to portray the principle of conservation of matter. For example, if hydrogen is completely burned to water, the reaction between it and oxygen can be represented as follows:

(Hydrogen) + (Oxygen) → (Water)

$$H_2 + \tfrac{1}{2}O_2 \rightarrow H_2O$$

(2.02 Wgt. units) + (16.00 Wgt. units) →

(18.02 Wgt. units)

It would be found that these materials would always react in the same relative proportions to form water in an amount equal to the total weight of reactants. The relative weights indicated are equal to the molecular weights of the materials in question. Even if a reaction does not go to completion, the quantities which did react would be proportional to the combining weights expressed in the balanced chemical equation.

Since the element of time is usually involved as the basis of a stoichiometric calculation, proper quantitative deductions often depend on adequate knowledge of other laws or principles, such as those governing rates of reaction and those pertaining to chemical equilibria. When materials in the gaseous state are involved, the general gas laws are of great utility.

Another independent relation for a system is obtainable by applying the law of conservation of energy, which requires that energy input equals energy output. A valid equality of this type must include all forms of energy such as potential energy, kinetic energy, internal energy, flow work, electrical energy, etc. This type of equality results in the so-called "total energy balance." Another very useful but similar expression is the Bernoulli mechanical energy balance for steady mass flow of fluids. However, heat energy is very frequently the only primary effect in a process so that, in such cases, the total energy balance can be simplified to the very advantageous expression of heat input equals heat output. This constitutes the basis for heat balances which, together with weight balances, are the most useful tools in any stoichiometric calculations.

Since chemical reactions involve combination of atoms or molecules to form new compounds or decomposition of compounds to form simpler ones, it is most convenient in stoichiometric calculations to employ molecular units rather than weight units. This particular kind of unit is called a "mole" and represents the quantity of substance numerically equal to its molecular weight. This weight quantity may be based on any system of weight units de-

sired, and it is thus necessary to designate this basis by referring to pound moles, gram moles, etc.

A particular chemical reaction may be written to embody both laws of conservation of mass and energy as demonstrated below:

$$FeS + \tfrac{7}{4}O_2 \rightarrow \tfrac{1}{2} Fe_2O_3 + SO_2 + 268{,}000 \text{ Btu}$$

This equation states that 1 pound mole of ferrous sulfide reacts with 7/4 moles of oxygen to form ½ mole of ferric oxide and 1 mole of sulfur dioxide, accompanied by a release of heat amounting to 268,000 Btu. However, to assign a specific meaning to the numerical value for this heat release, it is customary to specify a reference temperature and pressure for the reaction, these being 25°C and 1 atmosphere in the example cited. In making heat balance calculations, it is then convenient to choose these conditions as the datum level and then calculate the heat input and heat output quantities above or below the reference state.

WALTER C. LAPPLE

STORAGE OF CHEMICALS

A storage facility depends upon the physical and chemical nature of the material being stored.

Solid chemicals generally handled in open top conveyances, stored in outdoor piles and transferred by drag-scrapers, conveyor belts and crane loaders, etc. are normally common raw materials such as coal, sulfur, stone and various metallic ores. Processed or finished chemicals normally are stored in covered piles or bins, bunkers and silos. As an example prilled ammonium nitrate is produced via the normal neutralization reaction and prilled in a conventional tower. The resulting production is stored in indoor pile configurations and handled with conventional conveyor facilities. The low-order explosive characteristic of this material dictates the exercise of precautions which limit the possibility of fire involvement. Safer water solutions of ammoniated ammonium nitrate (32% nitrogen) are a commercial means of distribution for this material for agricultural application.

Bins, bunkers, hoppers and silos have application where containment is desirable and where some in-storage conditioning is required. Agricultural grains, soya beans, wheat and corn normally require moisture control as well as heat build-up attenuation. Roof structure slope is normally dictated by the angle of repose of the stored material and the shape of the discharge cone is controlled by the internal friction of the material as well as the friction of the stored material against the material of construction. Bridging characteristics also play an important part in the design configuration detailed for storage containers.

Gases. Elements and compounds normally in gaseous form at ambient temperature and atmospheric pressure are stored in several ways: (1) as gases under slight pressure in gasholders; (2) as gases at ambient temperature under high pressure in pressurized tanks or cylinders; (3) as liquefied gases in tanks or cylinders under pressure and at ambient temperature; or (4) as liquefied gases at atmospheric pressure in tanks at a temperature which will maintain their liquid state.

Storage of chemicals as gases is generally restricted to those materials which are above their critical temperature at normal ambients. Methane, hydrogen, helium and argon are commercial gases which generally fit into this category, and various gasholder designs are normally detailed for their storage. Wet-seal as well as dry Wiggins-type gasholders have been employed in such storage facilities.

However, storage of gases in the liquid state, the last two methods listed above, is more commonly and widely used, due chiefly to the greater capacity of a given storage facility for a liquid than for a gas. For example, methane and natural gas in liquid form at $-258°F$ occupy 1/600th the volume that they do in gaseous form. For the same reason, other materials like chlorine, ammonia, carbon dioxide, nitrogen, oxygen and argon are stored in the liquid state.

Huge volumes of natural gas are stored as gas in underground porous formations of depleted natural gas fields which represent probably the largest storage capacity for a single chemical commodity. Over 300 such reservoirs in the United States have a capacity of over 5,000 billion cubic feet of gas, equal to about 111 million tons. Helium is also stored in similar formations. An increasing number of storage plants for liquefied natural gas, LNG, have been built whose capacity is now over 27 billion cubic feet or 600,000 tons of gas. These are held at atmospheric pressure and a temperature of $-260°F$.

Liquids. Tanks and other vessels for the storage of normally liquid chemicals comprise the greatest number of units for the largest variety of chemicals in the chemical process industries. They are utilized to store incoming liquid raw materials, to store in-plant liquids from step to step in the manufacturing process, to store the plant products prior to shipment, and by customers to store chemical products as they are received. While the design and materials of construction (including linings) depend on the specific chemical involved, there are several code and engineering requirements that are common to the construction of storage facilities for liquid chemicals.

The specification of storage vessels or tanks for the containment of liquid chemicals or commodities is initially dependent upon the quantity to be stored. In general, the stored volume is dictated by the production rate and it is normal to establish storage requirements at 20 times daily production. It is obvious that shipping schedules or turnaround periods must be capable of preventing production shutdowns due to lack of storage capacity.

Required storage volumes on the order of 30,000 gallons can be readily specified in horizontal non-fired pressure vessels per ASME Section VIII Div. 1 code requirements. These vessels are also readily available as shop fabricated tanks and can be ordered in integral saddles or skid mounted configurations. Storage volumes which involve quantities up to ten 30,000 gal blimps have been considered economical; however, quantities beyond this capacity are normally stored in spheres designed to ASME Section VIII Div. 1. It should be noted that the ASME code applies to vessels designed for liquid or gas pressures exceeding 15 psi internal or external.

In general low-pressure (7″ water column maximum gas pressure) flat-bottom storage tank specifications for petroleum products or conventional petrochemical commodities can be properly established as API 650 "Welded Steel Tanks for Oil Storage" latest edition published by the American Petroleum Institute's Division of Refining. The basic API 650 code is applicable to steel storage vessels designed for liquids whose specific gravity is one or greater. The design can accommodate various roof designs; column supported cones, self supporting (slopes greater than 2″ in 12″ to maximum slopes 9″ in 12″ or 37°) cones, and self supporting umbrellas or domes. The roofs must be capable of withstanding a uniformly distributed live load (snow) of 25 lb/sq ft as well as the dead load of the roof plate and supporting rafters and girders. The supporting colums are restricted to a maximum l/r (unsupported length to radius of gyration ratio) of 180.

The API 650 code is generally applied to flat-bottom storage tanks for use at atmospheric pressure and normal ambient temperatures. Because of the phenomenon of notch ductility associated with conventional steels (A283C) it is generally agreed that the minimum use temperature for A283C is $+20°F$ for thicknesses up to and including ½″. Minimum use temperatures are established from U.S. Weather Bureau isotherm plots showing the lowest one-day mean for the tank's location and adding 15°F to this value.

Conventional A283C plate can be used for all thicknesses to the maximum allowed by code 1½″, for use temperatures equal to or greater than 50°F.

The design stress used in API 650 for the shell courses is 21,000 psi. The thickness of each course is calculated using the conventional hoop stress correlation and establishing the maximum point of pressure for the course being designed as one foot above its lower edge. The above considerations give for any course thickness the following relationship:

$$t = \frac{2.6\,D(H-1)\text{SpGr}}{21,000 \times \text{joint efficiency}} \tag{1}$$

where t is in inches and D and H are in feet. For a joint efficiency of 85% dictated by the code, Eq. (1) reduces to:

$$t = .0001456\,D(H-1)\text{SpGr} \tag{2}$$

The tank diameter D is generally taken as the internal diameter (ID) with shell courses having the same ID although the code calls for the center-line diameter of the shell plates unless otherwise specified by the customer. Flush ID's make for a better design in the event lining is required or a floating roof is to be installed. No thinner course can be placed below a thicker course and minimum plate thicknesses are established based on tank diameter primarily dictated by stability considerations during erection of these upper courses.

Tank diameter is normally established by volume requirements and allowable soil bearing values. An allowable soil-bearing value of 3000 lb/ft would limit the liquid height to 48 ft for a specific gravity of 1.0. This liquid height in an API 650 vessel would infer that the contained volume would extend to the top of the top angle or to the bottom of any overflow which limits the tank's filling height. The tank diameter would then result from the following relationship:

$$V = 0.14D^2H \qquad (3)$$

where V is volume in barrels (42 gallon barrels), D is shell diameter in feet, and H is liquid height in feet.

The allowable soil bearing value is generally the responsibility of the owner and any soils consultant he may employ. Settlement expected, as predicted by such soils experts is normally accounted for by establishing a uniform slope for the grade from the tank's center to the shell. A slope of 1″ in 10 ft is normally specified. Settlement of supporting roof structure columns somewhat greater than this value may result in standing water on the roof itself. Normal roof slopes are usually specified as ¾″ in 12″ and for conventional roof plate thicknesses of 3/16″ and top angles 3 × 3 × ¼ to 60 ft diameter and 3 × 3 × ⅜ for diameters greater than 60 ft. Tank pressures can be specified to 1 oz/sq in. (1.732″ water column) although a more conventional setting for the relief value is ¾ oz/sq in. or 1.3″ W.C. Maximum vacuum settings are established by minimum plate thicknesses for the shell at 1 oz/sq in., however, it is normal to specify a ½ oz setting in detailing the characteristics of the pressure-vacuum vapor conservation relief valve on the vessel, if one is required to attenuate emissions or to reduce product losses.

The size of the relief valve is taken from design considerations given in API RP 2000 "Venting Atmospheric and Low Pressure Storage Tanks" latest edition. In general this correlation establishes relief capacity based on thermal breathing losses of approximately 2 ft³/hr for each square foot of roof and shell area for stock having a flash point below 100°F. For stock having a flash point above 100°F, the breathing capacity is taken as 0.6 of the indicated rate. In addition to breathing or thermal flow rates (vacuum requirements are also established by this correlation) pump-in (100 bbl/hr) displacements are established as 600 ft³/hr for 100°F flash point material or higher and 1200 ft³/hr for flash points below 100°F. The theoretical displacement for any fluid is 560 ft³/hr of displaced vapor for 100 bbl/hr pumped in or drawn out. Vent sizing based on the above considerations can either be dictated by the breathing volume and pump-in rate or by the breathing volume and pump-out rate. In general, for reasonably low pumping rates (1000 bbl/hr or 700 gal/min) the vacuum breathing rate controls primarily because of the lower set point of ½ oz/sq. in. Emergency relief capacity is generally afforded by the roof to top angle joint which is considered frangible and is weaker than the shell or shell to bottom joints.

H. D. KERFMAN

STRONTIUM AND COMPOUNDS

Strontium, symbol Sr, atomic number 38, atomic weight 87.62 is one of the alkaline earth elements located in Group IIA of the Periodic Table between calcium and barium. Its physical and chemical properties are intermediate between those of calcium and barium.

Strontium was first detected by Crawford of Edinburgh in 1790, although some sources credit its discovery to William Cruikshank, also a Scotsman, in 1787. Both men worked with strontianite, $SrCO_3$, found at Strontian, Scotland, from which the element's name is derived.

Strontium exists in nature as a mixture of four stable isotopes: Sr^{88} (82.74%), Sr^{86} (9.75%), Sr^{87} (6.96%) and Sr^{84} (0.55%). Several artificial radioactive isotopes have been prepared, of which Sr^{90} has received the most attention because of its presence in radioactive fallout.

Occurrence. Strontium comprises about 0.0002% of igneous rocks and about 0.02 to 0.03% of the earth's crust.

There are two principal ores: celestite, $SrSO_4$, and strontianite, $SrCO_3$, the former the more plentiful. Strontium minerals are found in Arkansas, Arizona, California (e.g., Strontium Hills), New York and West Virginia in the United States. The United Kingdom is the world's largest producer. An important new source of celestite is Nova Scotia.

Derivation. Celesite, the $SrSO_4$ mineral, is the chief source of strontium compounds made in the United States. It is usually first converted to the carbonate by treatment with sodium carbonate or to the sulfide by reduction with coke at high temperatures. The carbonate and the sulfide are used to produce other strontium compounds.

Metallic strontium is derived by several methods. The electrolysis of fused KCl and $SrCl_2$ yields strontium metal. The oxide may be reduced with aluminum by heating a mixture of SrO and Al in a vacuum at high temperature so that the strontium distills out of the reaction zone. High-purity metal can also be made by heating strontium hybride in a vacuum at 1000°C. and by distilling the mercury from a strontium amalgam.

Elemental strontium is not produced commercially in more than small quantities because calcium and barium are more abundant and serve all the purposes for which metallic strontium might be applied.

Physical Properties. Strontium is a hard silver-white metal somewhat softer than calcium, and is malleable, ductile, machinable and capable of being drawn into wire. Its melting point, 770°C, is intermediate between those of calcium, 851°C, and barium, 710°C. Similarly, its density of 2.6 g/cc falls between those of calcium, 1.54, and barium, 3.5.

Chemical Properties. The chemistry of strontium closely parallels that of calcium. Strontium is a reactive metal and in the air quickly forms an oxide coating which is somewhat protective at room temperature. It forms only divalent compounds. Its base-forming characteristics are less pronounced than those of barium and greater than those of calcium. An active reducing agent, strontium reacts vigorously with water to liberate hydrogen and form $Sr(OH)_2$ and with acids to form hydrogen and the strontium salt of the acid. At elevated temperatures it reduces the halides and oxides of many metals to produce the corresponding metal.

Noted for the brilliant crimson color which its volatile compounds impart to flames, the element burns brightly when heated in air, oxygen, chlorine,

TABLE 1. PHYSICAL PROPERTIES OF STRONTIUM

Symbol	Sr
Atomic number	38
Atomic weight	87.62
Atomic volume, cc/g-atom	34.5
Atomic radius, Å	2.13
Ionic radius, Å	1.13
Electron configuration	2–8–18–8–2
Electron distribution	$1s^2, 2s^2, 2p^6, 3s^2,$
	$3p^6, 3d^{10}, 4s^2, 4p^6, 5s^2$
Density, g/cc, 20°C	2.6
Melting point, °C	770
Boiling point, °C	1380
Latent heat of fusion, cal/g	25
Latent heat of vaporization, cal/g, 1380°C	447
Specific heat, cal/g/°C, 20°C	0.176
Vapor pressure, atm, 877°C	0.01
1081°C	0.1
1279°C	0.5
1380°C	1.0
Electrical resistivity, microhm-cm, 20°C	23
Temperature coefficient of resistivity	5.0×10^{-3}
Ionization potential (gaseous element), volts	
1st electron	5.69
2nd electron	10.98
Surface tension, dyne/cm	165
Thermal neutron absorption cross section, barns	1.21

bromine gas and sulfur. When it reacts with oxygen both the oxide, SrO, and the peroxide, SrO_2, may be formed.

Chemical Compounds and Their Uses. Strontium forms compounds that are the counterparts of the corresponding calcium compounds. For example, the sulfide, chloride, bromide, iodide, nitrate, etc. are soluble, while the carbonate, fluoride, sulfate, oxalate and phosphate are insoluble. One major difference is that $Sr(OH)_2$ is quite soluble in hot water, 22 g/100 g H_2O at 100°C, while $Ca(OH)_2$ is not.

Two oxides exist: SrO, which resembles CaO and is made by thermal decomposition of the hydroxide, carbonate or nitrate, and the peroxide, SrO_2, formed by the addition of an alkali to an aqueous solution of a strontium salt containing hydrogen peroxide. The latter is a bleaching agent.

Strontium hydroxide, $Sr(OH)_2$, can be prepared by the action of water on SrO, and by heating SrS or $SrCO_3$ in steam at 500–600°C. Strontium hydroxide reacts with organic acids to yield lubricant soaps and greases that are structurally stable and resist oxidation and thermal breakdown over wide ranges of temperature.

Strontium nitrate, $Sr(NO_3)_2$, is used in flares, pyrotechnic devices and tracer bullets because of the intense red color it imparts to compositions used for such purposes. It is made by the action of nitric acid on $SrCO_3$. Strontium chlorate, $Sr(ClO_3)_2$, is similarly used. The preparation of fireworks, flares, etc. represents a large application of strontium compounds in this country.

Strontium sulfide, SrS, is prepared by the carbon reduction of the sulfate in a furnace or by heating

$SrCO_3$ with hydrogen sulfide. It has luminescent properties and is used in some luminous paints; it also is a depilatory.

Strontium is one of the small group of elements (others are the alkali metals, calcium and barium) which form an ionic hydride that is a stable crystalline compound. SrH_2 has a density of 3.27 g/cc (higher than that of the metal) and melts at 650°C. The hydride is a strong reducing agent and reacts readily with water and acids to liberate hydrogen.

Other compounds are made by methods used to make the corresponding calcium compounds.

Probably the major consumption of strontium chemicals is in the manufacture of strontium ferrite ceramic permanent magnets which are used in small electric motors. This relatively new market is growing at a rapid rate. The magnet is made by molding under pressure a wet mixture of strontium carbonate and ferric oxide and then sintering the shape to yield strontium ferrite. Another new outlet for strontium carbonate is in the glass of the front plate of color television tubes to decrease x-ray emission.

Biological and Toxicological Aspects. Strontium resembles calcium in its metabolism and behavior in the body. It is similar to calcium also in its low toxicity level, as contrasted with the poisonous nature of soluble barium compounds.

However, the artificial isotopes Sr^{89} and Sr^{90} are extremely hazardous. Sr^{89}, half-life 50 days, emits beta radiation of 1.5 MeV. Sr^{90}, the more dangerous because of its half-life of 28 years, emits beta particles of 0.61 MeV. These isotopes are among the most hazardous handled in laboratory and plant operations and extreme care must be used to prevent exposure to them.

The chief problem with Sr^{89} and more especially with Sr^{90} is that they seek out and are deposited in bones where, like natural strontium (which does not contain either Sr^{89} or Sr^{90}), they replace calcium in the normal bone structure. In this location they act as a source of internal radiation that damages bone marrow and blood-forming organs and induces cancer.

Much work has been conducted on the presence of and hazard caused by fallout of Sr^{90} from nuclear explosions, mainly those occurring in the atmosphere. The Sr^{90} deposited on grass and fodder eaten by cows enters the human body through the milk from cows and is deposited in the bones and teeth, particularly those of children. It appears possible to remove most of this damaging isotope from milk by treatment with a vermiculite adsorbent.

The bromide and chloride of strontium are used in medicine for certain nervous disorders, and a few other strontium compounds are used for the treatment of rheumatism and gout.

There appears to be no specific value of strontium in plant and animal life such as is associated with other trace elements.

CLIFFORD A. HAMPEL

References

"Strontium Chemicals," *Chem. Met. Eng.*, **53**, No. 1, 152–155 (Jan. 1946).

Mellor, J. W., "A Comprehensive Treatise on Inorganic and Theoretical Chemistry." Vol. 3, New York, Longmans, Green & Co., Inc. 1946.

Hampel, C. A., "Encyclopedia of the Chemical Elements," pp. 663–665, Van Nostrand Reinhold Corp., New York, 1968.

Ephraim, F., "Inorganic Chemistry," 3rd Ed., P. C. Thorne and A. M. Ward, Editors, London, Gurney and Jackson, 1939.

"Minerals Yearbook," Washington, D.C., U.S. Bureau of Mines, published annually.

Mantell, C. L., "Rare Metals Handbook," 2nd Edition, C. A. Hampel, Editor, p. 28, New York, Reinhold Publishing Corp., 1961.

STRUCTURAL ANTAGONISM

Compounds structurally related to a particular biologically active compound have been observed to antagonize the biological action of the latter. Enzymology probably gave the first example of such action, in that the products of certain enzyme systems were reported to inhibit their formation to an extent greater than that attributable to mass action effects. In these enzyme reactions, the products were structurally similar to the starting material (substrate). Many such antagonistic effects were demonstrated in *pharmacology,* such as the effects of numerous structural analogues of epinephrine (e.g., propadrine) in antagonizing the action of epinephrine on smooth muscle. In *nutrition,*

OH

—OH

$CHOH—CH_2—NH—CH_3$

Epinephrine

$CHOH—CH(NH_2)—CH$

Propadrine

certain amino acids were found to inhibit growth, and a structurally related amino acid would prevent (reverse) this growth inhibition. Following the earlier postulation of many similar concepts in *chemotherapy,* the discovery of the biological antagonism between the sulfonamide drugs (e.g.

SO_2NH_2 COOH

NH_2 NH_2

Sulfanilamide   *p*-Aminobenzoic Acid

sulfanilamide) and *p*-aminobenzoic acid was a major advance in the establishment of fundamental concepts of structural antagonisms and a rational approach to research in chemotherapy. The basic concepts of structural antagonism that have evolved from several areas of scientific research are the following.

(1) Each of the enzymes (biological catalysts for conversion of nutrients to other, often essential, products of cell metabolism) specifically catalyzes one type of reaction, with a high degree of specificity for the conversion of one particular substrate to a product. The enzyme, a large protein molecule, forms a complex with the substrate which is attached at a specific site on the protein molecule, and the complex then reacts further under normal conditions to form the product. The specificity of the enzyme for a particular substrate is believed to result from the existence on the enzyme of a structural pattern of intramolecular groups, into which the substrate molecule fits and is linked by a specific pattern of electrostatic and other interatomic forces.

(2) Compounds structurally related to the substrate can similarly interact with the enzyme, but the resulting complex either cannot undergo the normal reaction, or can yield only a modified product, different in structure from the normal product. If such compounds structurally related to the substrate cannot perform the function of the normal substrate in a biological system, such structural analogues are called *inhibitors* (or inhibitory analogues, or antagonists of the substrate).

(3) The inhibitor may combine with the enzyme either by reacting at the same site on the molecule as the normal substrate, competing with the latter and preventing its attachment at this site (competitive inhibition), or the mechanism of inhibition may involve attachment of the inhibitor to the enzyme at some other site on the molecule independent of the normal substrate attachment in such a manner that the inhibitor is still capable of forming a complex with the enzyme-substrate complex as well as the free enzyme (*noncompetitive* inhibition).

(4) Coenzymes (cofactors of enzymes) which must combine with the enzyme protein molecule (apoenzyme) to form the active complete catalyst (or holoenzyme) can be antagonized in a similar manner by structural analogues of the coenzymes, but in many such cases the equilibria of analogue-enzyme combination are not rapidly attained.

The fundamental concepts of structural antagonisms have been applied to a wide variety of biologically active compounds, such as vitamins, amino acids (the components of proteins), purines and pyrimidines (components of nucleic acids), hormones, etc.

The liberation of histamine is thought to be an important factor in anaphylactic shock and in many allergic conditions. While some substances counteract the effects of histamine by other mechanisms, certain structural analogues appear to be direct and specific histamine antagonists. Many of the antihistamines structurally resemble histamine.

Vitamin K or its probable precursor, 2-methyl-1,4-naphthoquinone, appears to function in the formation of prothrombin which affects the rate of blood coagulation, and appears to reverse the action of dicumarol, the hemorrhagic substance in spoiled sweet clover hay.

Most structural analogues of a biologically active compound are inert, many are antagonists, but some actually replace the natural active compound in performing the same biological function. For example, oxybiotin (*o*-heterobiotin) replaces the vitamin biotin for many organisms, and is utilized without conversion to biotin. The analogue apparently is converted to an analogue of the coenzyme form of biotin, and this coenzyme analogue actually performs the normal function of biotin.

In other cases, an analogue replaces its corresponding substrate or growth factor for some organisms but inhibits growth of others. *p*-Aminosalicyclic acid inhibits growth of *Mycobacterium tuberculosis* H37Rv, and *p*-aminobenzoic acid reverses the toxicity; however, for a certain mutant strain of *Escherichia coli, p*-aminosalicylic acid promotes growth in lieu of *p*-aminobenzoic acid. Cases are known in which an analogue replaces some of the biological functions of a vitamin but inhibits other functions.

Structural antagonisms, particularly in cases of competitive inhibition, offer a basis for the study of biochemistry in living organisms. Effects exerted upon systems which are inhibited by antagonists of a particular cell metabolite or nutrient include not only those exerted by the metabolite or nutrient itself, but also those exerted by the limiting precursors of the metabolite, by the products of the inhibited enzyme reaction, by substances which increase the effective enzyme concentration, and by compounds which influence the rates of destruction of either the metabolite or the inhibitory analogue. Testing techniques which elucidate the mechanisms of effects exerted by such secondary reversing agents of the inhibited biological system, have been termed *inhibition analysis,* and offer a new method for the study of biochemistry.

WILLIAM SHIVE

Cross-references: *Pharmacology, Nutrition, Inhibitors, Proteins.*

SUBLIMATION

In the strict sense of the term, sublimation refers to the change of a solid to a gaseous phase without an intermediate liquid phase. The term is used loosely, however, and other changes that are similar are commonly called sublimation also.

When vapor-pressure curves are plotted on a set of pressure-temperature axes, and a vapor-pressure line separates the gaseous from the solid phase in the diagram, then the substance for which the curve is plotted can be sublimed. Sublimation can be accomplished by a change in temperature (usually raising), by a change in pressure (usually lowering), or by changing both temperature and pressure.

Ice and snow have an appreciable vapor pressure, and both sublime. Even on a very cold day, ice disappears slowly, and no liquid water forms. Camphor, naphthalene, paradichlorobenzene, solid carbon dioxide, arsenic trioxide, and many other substances sublime readily. In some cases, sublimation is the method used to purify the substance.

The extensive use of solid carbon dioxide (Dry Ice) as a refrigerant is based on the fact that it has no liquid phase at ordinary room pressure, but the solid sublimes directly from the solid to colorless, odorless gas. Carbon dioxide, however, does have a liquid phase at ordinary room temperature, and

the liquid is sold compressed in steel tanks as "liquid carbonic gas." In this form, among other uses, it is used for fire protection.

Iodine sublimes, although when iodine is heated in an open vessel the liquid phase of the element can be noticed. The vapor pressure of iodine is low, 0.2 mm at room temperature (20°C), and it increases with an increase of temperature. If iodine is heated, but kept below its melting point (114°C), sublimation takes place. If iodine is heated to its melting point at ordinary room pressure, then a solid-liquid-vapor transition takes place. (See **Iodine**).

When ammonium chloride is heated, the compound dissociates completely into ammonia and hydrogen chloride: $NH_4Cl \rightleftharpoons NH_3 + HCl$. If this experiment is conducted in an open vessel, some of the ammonia escapes because it has a higher rate of diffusion than the denser hydrogen chloride gas. If, however, the ammonia does not escape, then ammonium chloride (sal ammoniac) re-forms when the gases cool. While this over-all operation with ammonium chloride is equivalent to sublimation, dissociation and recombination are not generally associated with or implied by the term sublimation.

ELBERT C. WEAVER

SUGARS

Some 400 billion tons of sugars are produced each year by photosynthetic reactions. Water plants produce about 390 of these. Of the ten which appear on land, 2.45 billion tons are produced under cultivation. These figures demonstrate the significance of the sugars, which are a group of organic compounds related by molecular structure that comprise the simpler members of the general class of *carbohydrates.* Each consists of a chain of two to seven carbon atoms (usually five or six). One of the carbons carries aldehydic or ketonic oxygen which may be combined in acetal or ketal forms. The remaining carbon atoms usually bear hydrogen atoms and hydroxyl groups. In general, sugars are more or less sweet, water-soluble, colorless, odorless, optically active substances which lose water, caramelize and char when heated.

Some sugars exist as discrete units called *monoses* or *monosaccharides.* Others are coupled into di-, tri-, and higher *saccharides.* Polymers of less than ten units usually are called *oligosaccharides*; higher polymers are called *polysaccharides.* Sugars combined by acetal linkages with nonsugar alcohols form a great variety of *glycosides.* Of the few sugars which are found in the free state, only sucrose, glucose, maltose, fructose, and lactose have commercial significance as pure compounds. The remainder are found in large numbers of *polysaccharides* and *glycosides.*

Sugars are detected by tests for the carbonyl groups. The violet color which appears when a solution with alpha naphthol is underlayered with concentrated sulfuric acid constitutes the Molisch test. The silver mirror which is deposited on a clean glass surface when aqueous ammoniacal silver is reduced constitutes the Tollens test. The brick-red cuprous oxide which precipitates from hot tartrate stabilized alkaline cupric solution is a positive Fehling test.

The sucrose or trehalose type of sugars which have no free carbonyl sugars are not oxidized by such reagents. Hence, they are named *nonreducing sugars* in contrast to the remainder which are *reducing sugars.*

Ketose sugars are distinguished from *aldose* sugars by the fiery-red color developing in the presence of hot hydrochloric acid and resorcinol (the Seliwanoff test).

Mixtures of sugars can be separated by chromatography. The individual sugars can be identified by the physical properties of their derivatives. Among such useful derivatives are the phenylhydrazones, osazones, semicarbazones, and oximes; anilides of the onic acids; acetate, benzoate and para-toluenesulfonate esters; the methyl and triphenylmethyl ethers; and the isopropylidine and benzylidine derivatives.

D-*Glucose* (dextrose) ($C_6H_{12}O_6$) is known commercially as corn sugar. It occurs in the free state in fruit juices and together with fructose in honey. Crystallized honey has deposited dextrose hydrate crystals. Much larger quantities occur polymerized as starch or as cellulose or combined with fructose in sucrose or with galactose in lactose. Glucose is a normal component of the blood and as such it is called *blood sugar*. In the disease, diabetes mellitus, glucose appears in the urine.

The commercial source is acid- or enzyme-hydrolyzed corn starch. After vigorous hydrolysis the material is neutralized, concentrated, passed over bone char, crystallized, centrifuged and dried. When crystallized from water at temperatures above 115°C, an anhydrous beta form is produced, (m.p. 148°C). At temperatures between 50° and 115°C an anhydrous alpha form is produced (m.p. 146°C). The most common form, however, is the monohydrate of the alpha form, called dextrose hydrate (m.p. 80–85°C) $[\alpha]_D^{20} + 52.74°$ (H_2O, $c10$). The beta form dissolves more rapidly than the alpha and hence commands some price premium.

More than 450,000 tons of the material are produced each year in the United States. Glucose finds use in the confectionery, baking and canning industries, in the preparation of caramel color and in intravenous alimentation.

The sirup remaining after glucose production is called *hydrol* or *corn molasses* and it is consumed in cattle feeding and fermentations.

Glucose is absorbed directly through the walls of the intestine and appears in the blood stream both in the free state and as phosphorylated derivatives. Glucose and its derivatives are delivered to the cells where they are degraded to provide both energy and structural materials for the synthesis of body components. In periods of glucose surplus the liver and muscles convert the excess to *glycogen* which is the animal carbohydrate-storage form. Supplies in excess of the limited glycogen storage capacity may be converted to adipose tissue.

Fructose (levulose) ($C_6H_{12}O_6$) occurs with dextrose in honey, but more commonly they are combined in sucrose. As a fructose polymer, it is the storage carbohydrate in the tubers of the dahlia and the Jerusalem artichoke, as well as the Hawaiian ti plant. Fructose has been produced to a limited extent from the Jerusalem artichoke, as a by-product of the dextran industry, and from invert sugar.

The price for some years has hovered in the range of $1.25 per pound and few markets were found. However, beginning in 1969, a syrup of fructose was offered for sale at about a cent a pound less than going sucrose prices. The solids in the product were about 95% fructose produced by enzymic isomerization of glucose. A number of organisms elaborate the intercellular enzyme, glucose isomerase. Several strains of *Streptomyces* appear to give higher yields of fructose and they can be activated by xylose, eliminating the need for the arsenic activators employed in early studies.

Problems of economics and taste purity tend to limit the market for fructose in this form. One economic factor is the relatively high cost of the active enzyme. There are indications that whoever solves the problems of economical production of a high quality product will fall heir to a lucrative business. The market should be significant in products designed for diabetics for absorption and utilization occurs with only limited dependence on available insulin levels.

D-Fructose is the most soluble of the common sugars. It crystallizes in an anhydrous beta form, m.p. 102–104°C $[\alpha]_D^{20}$ $-123°$ $\rightarrow$ $92.4°$ (H_2O, $c4$), which is hygroscopic. In addition, fructose is the sweetest of the sugars. The relative sweetness of fructose, however, depends on both concentration and temperature. When diluted to taste extinction at room temperature it is about 70% sweeter than sucrose. In 10% solution it is only 20% sweeter. At 5°C fructose is 43% sweeter than sucrose. At 40°C they are equal and at 60°C fructose is only 79% as sweet as sucrose.

Fructose is absorbed by the walls of the small intestine, but from a dilute solution, before it appears in the blood-stream, it is converted into dextrose. From a concentrated solution much of the fructose appears unchanged in the blood-stream.

Sucrose ($C_{12}H_{22}O_{11}$) is known to the English-speaking peoples as *sugar* and by equivalent terms in other languages. It has been found in the juice of every land plant examined for it. There is some evidence that sucrose is one of the earlier products of photosynthesis and, as such, may be considered as a starting product for all living things.

Production. The world production of sucrose in 1971 was 90,574,000 tons (raw value), of which 80,849,000 was produced freed from molasses (centrifugal). Of this amount, about 63% was cane sugar and the remainder beet. In the same year, U.S. consumption was 10,512,820 tons, or about 102 pounds per capita. Of this, 7,299,950 tons were from cane and 3,212,873 tons from beets. Comparatively minor amounts of sucrose are consumed as maple sugar and sorghum sirup.

Cane sugar is produced by shredding and crushing the cane. The dark, cloudy juice is neutralized with lime, filtered, concentrated, crystallized, and centrifuged to produce *raw sugar,* which averages more than 97% sucrose. This raw sugar is refined by being washed, dissolved, passed over bone char, filtered, concentrated, crystallized, centrifuged and dried to produce *refined white sugar,* which contains 99.96% sucrose. Raw sugar for the American refiners comes principally from Hawaii, Philippines, Louisiana, Florida, Dominican Republic, Brazil, Peru, Mexico, and Puerto Rico.

The mother liquors from both raw and refined sugar processes are reworked to produce additional yields of the respective sugars. When the concentration of impurities rises so high that it is uneconomical to produce more sugar, the residue is a dark, viscous sirup, called molasses. The raw process produces *blackstrap* or *cane final molasses*. Refineries produce *refiner's sirup, sugar house molasses* or *treacle*. These products are fed to animals, fermented to alcohol, or to pharmaceutical preparations (dextran, citric acid, penicillin).

Intermediate stages of recovery produce *soft sugars* or *brown sugars* in which relatively small crystals of pure sucrose are coated with layers of colored molasses. Of the thirteen color grades produced in earlier years four (Nos. 6, 8, 10 & 13) now serve the market. Soft sugars have characteristic flavors and aromas and contain 2 to 4% water.

Beet sugar is produced by extraction from beet slices called *cossettes*. The extract is limed, filtered, concentrated, passed over bone char, crystallized, centrifuged and dried to produce a refined sugar. Raw beet sugar is not produced in the United States. Beet molasses contains little invert sugar compared to cane molasses; hence, recovery of additional amounts of sugar is possible by precipitation with alkaline earth hydroxides. In the Steffen-process, lime is added to molasses to precipitate sucrose as a calcium sucrate from which the sugar is released by carbon dioxide. In the United States, sugar beets are produced in 20 of the 27 states west of the Mississippi and north of the Ohio rivers.

Raffinose, monosodium glutamate and betaine are by-products of beet molasses.

Physical Properties. Sucrose crystallizes from water in a characteristic monoclinic form; m.p. 184°C $[\alpha]_D^{20}$ + 66.53° (H_2O, $_c26$). It is readily soluble in water, dilute ethanol, N,N-dimethyl formamide and anhydrous ammonia; it is practically insoluble in anhydrous ethanol, ether, chloroform and anhydrous glycerol. The commercially practical limit of solubility in water is about 67%.

Sugar's most outstanding physical property is sweetness. In establishing relative sweetness scales sucrose usually is asigned a value of 100. On this scale 10% solutions have relative sweetnesses of: fructose 120, glycerol 77, gluscose 69 and lactose 39.

Chemical Properties. Sucrose is a disaccharide in which an α-D-glucosyl residue in the 6-membered ring form is combined with a β-D-fructoside residue in the 5-membered ring form.

One of the three outstanding chemical properties of sucrose is its instability to acid. In acid solution it splits to *invert sugar* (containing equal parts of *dextrose* and *levulose*) at a thousand times the rate of the hydrolysis of *maltose*. Inversion of sugar is practised quite widely to decrease the tendency of sugar to crystallize in candy, ice cream, baked goods and soft drinks, as well as to increase the sweetness.

A second important chemical property is its stability to alkali. A new but burgeoning field of sucrochemistry depends to a large degree on this property. Strong alkaline catalysts facilitate ether formation and ester interchanges which are the pathways by which sugar is being converted to surfactants, plastics, plasticizers, and surface-coating materials.

A third unusual property of sugar is that both carbonyl groups of the two monosaccharides are involved in the central bond; hence sugar is not oxidized in the Fehling test. The aldehyde group is detectable neither by the Tollens test nor the Schiff test. Some of sugar's widespread use in food depends on the fact that its reducing groups are not available to react with food proteins in the Maillard browning reactions.

Metabolic Properties. Sucrose does not normally appear in the blood of animals. If administered intravenously it is excreted unchanged by the kidneys. The reaction is so rapid that intravenous sucrose has been proposed as a diuretic. In the walls of the intestines, sucrose is hydrolyzed to glucose and fructose, which appear in the blood stream in the free state as well as in the phosphate derivatives. The reaction is so fast that orally administered sucrose affects the blood-sugar level faster than any other food including even invert sugar.

Maltose, or malt sugar does not often appear in the free state in nature. It occurs in sprouted grain, and in the malting stage of the brewing process. Maltose is produced commercially by degrading starch with a beta-amylase.

Malt sirup, an extract of malt treated to remove the malt flavor and concentrated is called *maltose sirup*. Under war conditions when sugar supplies were limited, (e.g., in 1945) 100,000 tons of maltose sirup were consumed. Maltose finds little commercial use except in infant carbohydrate supplements. Malt sirup is used principally as a flavor.

Maltose is a reducing disaccharide in which an α-glucosyl residue is attached to the hydroxyl oxygen of carbon 4 of another gluscose unit. Both glucoses are in their 6 membered ring forms. The sugar crystallizes from water as the monohydrate of the beta-form m.p. 102–103°C, $[\alpha]_D^{20}$ +111.7° → 130.4° (H_2O, $_c4$). It is about a third as sweet as sucrose. Maltose is hydrolyzed in the digestive tract by the pancreatic alpha-amylases to glucose.

Lactose, or milk sugar, has been found to comprise from 2 to 6% of the milks of all mammals studied, including that of the whale. It has been estimated that the total U.S. milk supply contains more than 273,000 tons of lactose; less than 10% is recovered and marketed as the sugar. Lactose is produced by precipitating the protein, filtering, concentrating, crystallizing, centrifuging, dissolving, decolorizing, concentrating, crystallizing, centrifuging and drying. The commercial product is the monohydrate of the alpha-form (m.p. 201.6°C). If crystallized at temperatures above 93.3°C, the anhydrous beta form is produced (m.p. 252.2°C).

Lactose is much less soluble in water but in other solvents its solubility is similar to that of sucrose.

In lactose a beta-galactosyl residue is combined to the hydroxyl oxygen on the fourth carbon of a glucose unit, and both moieties are in six membered ring form. Hence, lactose is a reducing sugar. It is readily fermented to lactic acid by milk souring organisms such as *Lactobacilli*. The Torula yeasts ferment it to alcohol. In solution it is readily oxidized to lactobionic acid which on further reaction yields galactaric and saccharic acids, subsequently tartaric, oxalic and carbonic acids. Alkali and heat degrade the sugar to various saccharinic acids and on down to lactic acid.

Digestion takes place in the small intestine where a pancreatic beta-galactosidase splits lactose into galactose and glucose. Galactose is absorbed readily from the intestine. However, it appears that the blood galactose is converted by the liver into glucose. The galactose in cerebrosides and in milk are produced *in situ* by conversion of blood glucose.

JOHN L. HICKSON

SULFONATION

Sulfonation in its broadest sense includes all methods of converting organic compounds to sulfonic acids or sulfonates, containing the structural group $C—SO_2—O$ or, in some cases, $N—SO_2—O$. The term is applied mainly to use of the common sulfonating agents, namely, concentrated sulfuric acid, oleum and other reagents containing sulfur trioxide in labile form, and sulfur trioxide itself. The only other widely used method of making sulfonates is by the action of alkali metal sulfites on alkyl halides, known as the Strecker reaction.

Reaction of vegetable oils with concentrated sulfuric acid has also been called sulfonation, and the resulting products are known in industry as sulfonated oils, although they are predominantly sulfate esters characterized by the structural group $C—O—SO_3$. To chemists these are sulfated oils, and the formation of sulfuric acid esters from either alcorols or olefins is more properly termed sulfation.

Benzene reacts with concentrated sulfuric acid and with oleum, respectively, as follows:

$$C_6H_6 + H_2SO_4 \rightarrow C_6H_5SO_2OH + H_2O$$

$$C_6H_6 + SO_3 \rightarrow C_6H_5SO_2OH$$

The acid reaction product is benzenesulfonic acid, and its salts are called benzenesulfonates. Similar reactions of other aromatic compounds proceed readily over a wide range of conditions. The relatively great ease of sulfonation of aromatic compounds is a traditionally important property for distinguishing them from aliphatic compounds. Also side reactions are less common and yields are usually greater than in the sufonation of aliphatics.

Reaction time, temperature, $\%SO_3$ in the sulfonating reagent, and proportion of reagent to substrate are important variables in the sulfonation process. Agitation is also important, because of limited mutual solubility of the reacting materials. Special solvents and catalysts may be used, but are not often essential. Sulfonation of benzene beyond the monosulfonic stage yields mainly benzene-1, 3-disulfonic acid. The second stage of reaction requires a higher temperature or a stronger acid than the first. If the disulfonic acid is made with use of concentrated sulfuric acid, temperatures as high as 260°C are used, but it is more satisfactory to use a strong oleum at temperatures up to 100°C.

In sulfonation of more complex aromatic compounds, temperature may affect not only reaction rate but also the nature of the reaction product. For example, change of temperature in sulfonation of naphthalene can change the composition of the resulting monosulfonic acid from about 95% alpha isomer at room temperature to practically 100% beta isomer at 200°C.

Purification of sulfonated products, which is often difficult, is largely avoided by adjusting conditions of sulfonation so as to obtain a product suitable for practical use without purification or after mere drying. Inorganic sulfate, the chief impurity, can be held to a minimum by the judicious use of oleum. Adding water to the acid mix at the end of sulfonation often gives an upper layer containing nearly all of the sulfonic acid and a lower layer containing most of the excess of sulfuric acid. Conversion of the acid mix to calcium salts favors a more complete separation, since calcium sulfate is much less soluble in water than many calcium sulfonates.

At ordinary temperature, the gaseous paraffins are inert to concentrated sulfuric acid, but are slowly absorbed and sulfonated by oleum. Their susceptibility to sulfonation increases with temperature and with molecular weight, but direct sulfonation of aliphatic hydrocarbons is in general an unsatisfactory way of obtaining sulfonates. The preferred method is the Strecker reaction, as in the reaction:

$$CH_3I + Na_2SO_3 \rightarrow NaI + CH_3SO_3Na$$

A good yield of mixed sulfonic acids can be obtained by passing chlorine and sulfur dioxide through paraffin-base petroleum fractions, the main reaction being:

$$C_nH_{2n+2} + Cl_2 + SO_2 \rightarrow C_nH_{2n+1}SO_2Cl + HCl$$

The sulfonylchloride is easily hydrolyzed, e.g., during neutralization.

Olefins are more easily sulfonated than paraffins, but with complications. With concentrated sulfuric acid the main reaction is sulfation, as in formation of ethylsulfuric acid from ethylene. Both sulfation and sulfonation occur when oleum or sulfur trioxide act on olefins. For example, the vapor phase reaction of ethylene with sulfur trioxide gives mainly carbylsulfate.

$$C_2H_4 + 2SO_3 \rightarrow \begin{array}{c} CH_2SO_2O— \\ | \quad \quad \backslash \\ CH_2—O—SO_2 \end{array}$$

Complete hydrolysis of carbylsulfate yields isethionic acid ($HOCH_2CH_2SO_2OH$), a sulfonic acid of some importance in manufacture of wetting and cleansing agents.

Polymerization is a common side reaction in sulfonation of olefins, also oxidation of the hydrocarbon with formation of sulfur dioxide. Alicyclic hydrocarbons are somewhat more susceptible to sulfonation than their open chain analogues, with similar side reactions.

The action of sulfuric acid and related sulfonating agents on aliphatic compounds other than hydrocarbons is varying and complex. However, simple sulfation with negligible side reaction is possible in the case of most saturated primary alcohols, from which the alkyl sulfuric acids are easily formed and converted to neutral salts. The most important products made in this way are mixed sodium alkyl sulfates from higher alcohols, principally lauryl.

Sulfonated Products in Industry. About 700 of the dyes which have been offered for sale are sulfonates. They constitute a majority of all commercial dyes and an overwhelming majority of all commercial sulfonates. Dye intermediates are prominent among the few sulfonates commonly found in published market price lists; these include three isomeric naphtholdisulfonic acids and five isomeric naphthylaminosulfonic acids.

Of the synthetic colors approved for use in foods, drugs or cosmetics in the U.S., more than half are sulfonates. The sulfa-drugs, so-called, are sulfonated products, and so is chloramin-T, which has been extensively used as an antiseptic.

Stearonapththalenesulfonic acid, better known as Twitchell reagent, was once the principal catalyst for hydrolysis of fats, and has been largely replaced by sulfonic acids recovered from the acid treatment of mineral oils.

Alkylbenzenesulfonates, used as cleaning agents, now account for the main tonnage of sulfonated products.

Some sulfonates lose their SO_3 group when they are used as intermediates. The important example is the production of phenols by fusion of sulfonates with caustic alkali, as shown in the reaction:

$$C_6H_5SO_3Na + 2NaOH \rightarrow$$
$$C_6H_5ONa + Na_2SO_3 + H_2O$$

The phenolate is converted to free phenol by acidulation. Alpha- and betanaphthol are made in this way from the corresponding sulfonates, and there are other examples, including a part of our supply of phenol itself. Arylamines, e.g., 2-amino-anthraquinone, are similarly made by reaction of ammonia and aryl sulfonates.

However, most sulfonates are produced to serve as end products, and the one important property which most of them have in common is solubility in water. In fact, the main purpose for which sulfonation is practiced is to increase the solubility of organic compounds in water.

A. S. RICHARDSON

Cross-references: *Detergents, Dyes and Dyeing, Sulfur and Compounds.*

SULFUR AND COMPOUNDS

Occurrence. Sulfur is widely distributed in nature. It has been detected in certain stars, novae, cosmic clouds, and the sun. It is present in the meteorites that come to earth from cosmic space, and is found in the oceans, the atmosphere, the earth's crust, the lunar crust, and in practically all plant and animal life. One finds large masses of the sulfides in the earth's crust which on the average contains 0.06% sulfur. It is found concentrated in deposits as the element and as sulfides and sulfates. Sulfur is important in the building of plant and animal tissue and significant quantities of free and combined sulfur are found in coal and petroleum. It is a constituent of foods and of practically all animal fluids, secretions, and excretions.

Sulfur occurs in combination with many elements and is a major constituent of many minerals. Sulfides, such as pyrite, FeS_2; marcasite, FeS_2; pyrrhotite, Fe_nS_{n+1}; chalcopyrite, $CuFeS_2$; chalcocite, CuS; galena, PbS; and sphalerite, ZnS, are important sources of sulfur, as by-products of the recovery of their metal values.

In the United States, Mexico, Italy, Japan, South America, Russia, and Poland, commercial sources of elemental sulfur are found in quantity and purity sufficient to justify commercial recovery. There are in general three types of deposits in which free or native sulfur is found.

Solfataras are sulfur deposits associated with volcanic vents, and are formed by the oxidation of hydrogen sulfide. Mineral springs or fumerarole deposits are seen encrusting hot springs such as those in Yellowstone Park. But the gypsum type is by far the most important group of deposits. Large deposits occur in sedimentary rock and may be the result of the reduction of gypsum (probably biological), although other theories regarding deposition have been advanced.

Much of the world's production of sulfur today comes from sedimentary deposits, especially from the calcareous horizons of the cap rock of shallow salt domes in the coastal regions of Texas and Louisiana, and the Isthmus of Tehuantepec, Mexico. Other sedimentary deposits of commercial interest are in Western Texas, in Italy, in Iraq, in the Vistula River area in Poland, and in the U.S.S.R. Explorations in the late 1960's in the Sverdup Basin in Canada's far northern Arctic archipelago revealed deposits of elemental sulfur.

Recovery. In the United States, Mexico, and Poland, sulfur is mined by the Frasch process. Sulfur melts at about 116°C. The Frasch process takes advantage of this fact and melts the sulfur underground by injecting water, heated above this temperature, into the deposit. The melted sulfur is then pumped to the surface.

Various schemes have been devised and applied for the recovery of elemental sulfur from pyrites, coke oven gas, smelter fumes, the sulfur dioxide in stack gases, and the hydrogen sulfide evolved in the refining of petroleum and the producing of natural gas. Sour natural gas has become the most important source of recovered sulfur. The quantity of recovered sulfur produced in the world probably exceeds that produced by the Frasch process.

Physical Properties.

The element sulfur, atomic number 16, and atomic weight 32.06, is the second element of Group VIA of the Periodic Table. There are ten isotopes. The approximate distribution of the four stable isotopes is 95.1% S^{32}, 0.74% S^{33}, 4.2% S^{34}, and 0.016% S^{36}. The six radioactive isotopes are S^{29}, S^{30}, S^{31}, S^{35}, S^{37}, and S^{38}.

Pure solid sulfur has a pale yellow color. Commercial sulfur may be sulfur-yellow, straw or honey-yellow, yellowish-brown, or reddish to yellowish-gray. Sulfur is tasteless and odorless. It is very soluble in carbon disulfide, less soluble in aromatic solvents, indifferently so in aliphatic, and insoluble in water. Pure sulfur melts at 115.2°C and boils at 444.6°C. Sulfur is a poor conductor of heat and electricity. The ignition temperature of sulfur in air is 261°C. On Mohs scale its hardness varies from 1.5 to 2.5.

On melting, sulfur is a straw-yellow, transparent liquid, and at temperatures not far above the melting point the liquid is supposed to consist of octatomic molecules. Between 120 and 160°C, it becomes more fluid with rising temperature and turns dark brown. At about 250°C it turns brownish-black. Apparently the structure of the liquid undergoes an abrupt change at about 160°C and this transformation is accompanied by the absorption of 2.751 cal per g. The sudden and enormous increase in viscosity which occurs at this temperature is but one indication of a structural alteration. Other properties show marked discontinuities. Beyond 230°C the viscosity decreases but the color remains dark up to the boiling point, 444.6°C. If sulfur at the boiling point is cooled slowly, it passes through the changes described in the reverse order. Elastic sulfur is made by rapidly chilling liquid sulfur which has been heated to elevated temperatures.

All diverse allotropic forms of liquid and solid sulfur are composed of polyatomic molecules, either simple rings or unbranched helical chains. At temperatures below 160°C, depending on pressure, the stable molecule is the eight-membered ring, which constitutes the lambda component of the liquid, the alpha and beta are crystalline forms stable below 1.5 kilobars, five crystalline are forms possibly stable at higher pressures, and there are many metastable crystalline forms.

At 1000°C its vapor consists chiefly of S_2 molecules. At lower temperatures the vapor is a mixture of all molecules, presumably rings containing three to ten or more sulfur atoms.

Chemical Properties. Sulfur units directly with almost all elements except gold, iodine, platinum, and the inert gases. The valence of sulfur in the sulfides is two; in the sulfites, sulfur dioxide, and many organic molecules four; and in the sulfates and sulfur trioxide six.

Hydrogen sulfide, H_2S, is formed by the reaction of hydrogen with molten sulfur, or by the attack of acids on metal sulfides. It is a colorless gas having the odor of rotten eggs. It is poisonous when inhaled. It burns in air with a pale blue flame to give sulfur dioxide and water. It is stable at room temperature and may be stored in liquid form in steel cylinders. It is used as a reducing agent. Hydrogen sulfide and acetylene condense to form thiophene, an important intermediate in the production of antihistamines.

Hydrogen disulfide, H_2S_2, and hydrogen trisulfide, H_2S_3, are known, as are the pentasulfide, H_2S_5, and the hexasulfide, H_2S_6.

The hydrogen atoms in hydrogen sulfide can be replaced by metals to form the normal sulfide, or the acid sulfide. Sulfides of the alkali metals dissolve readily in alcohol and in water and are used in the paper and leather industries. Sulfates can be reduced to the sulfides by heating with carbon. Heavy metal sulfides, insoluble in water, can be made by passing hydrogen sulfide through a solution of the metal salt. This reaction with hydrogen sulfide is used in quantitative analysis.

Sulfur combines with simple sulfides like Na_2S or NaSH to form polysulfides having the general formula Na_2S_x, in which x varies from 2 to 5. The polysulfides, oxidized to form thiosulfates, are used in the tanning industry.

Sulfur unites with calcium metal to form calcium sulfide. The sulfides of strontium and barium are made by the reduction of the corresponding sulfate. Calcium hydrosulfide, $Ca(SH)_2 \cdot 6H_2O$, is prepared by passing hydrogen sulfide into a suspension of calcium hydroxide.

Lime-sulfur solution, consisting of calcium sulfide and polysulfides, is made by boiling sulfur with a suspension of lime. It is used as an insecticide and a fungicide. Oxysulfides of the alkaline earths also have been produced.

Molybdenum forms a number of sulfides such as molybdenum trisulfide, MoS_3. Tungsten disulfide, WS_2, is made by the reaction of sulfur with the metal. The trisulfide, WS_3, has been formed by melting together wolframite, carbon, sulfur, and sodium carbonate. Uranium burns in sulfur vapor to form the disulfide, US_2. Uranium sesquisulfide, U_2S_3, and uranium monosulfide, US, have been described.

The sulfides of carbon, silicon, and boron are made by heating sulfur with these elements at high temperatures. Carbon disulfide, CS_2, boils at 46.25°C, while SiS_2 and B_2S_3 are solid compounds difficult to volatilize. CS_2 is important commercially, being used in the preparation of rayon, carbon tetrachloride, and as a solvent. It is a dangerous fire and explosion hazard due to its low ignition temperature and extremely low flash point. The greatest caution must be used in handling and transporting it.

The phosphorus sulfides, P_4S_3, P_4S_5, P_4S_7, and P_4S_{10} are formed by heating mixtures of the elements. (See **Phosphorus Sulfides**.)

Tetraarsenic tetrasulfide, As_4S_4, is made by heating sulfur and arsenic, as is tetraarsenic hexasulfide, As_4S_6.

Stibnite, the principal ore of antimony, is antimony trisulfide, Sb_2S_3. It and antimony pentasulfide, Sb_2S_5 are made by heating sulfur with antimony.

Selenium and sulfur are miscible in all proportions in the solid and in the liquid. Some evidence indicates the existence of eight-membered rings, Se_4S_4, Se_2S_6, and SeS_7. Tellurium and sulfur are miscible in all proportions in the liquid, but not in the solid. There seem to be no confirmed tellurium sulfides.

Sulfur burns in air to form sulfur dioxide and a small amount of sulfur trioxide. The sulfur dioxide produced is used as such or processed to produce a variety of chemical compounds, the most important of which is sulfuric acid. Sulfur dioxide also is made by burning pyrites, pyrrhotite, or chalcopyrite, and by decomposition of sulfates and sulfites. Sulfuric acid, often called the work horse of chemical industry, is used in making fertilizers, refining petroleum, pickling steel, and in making a host of products including detergents, synthetic resins, synthetic fibers, explosives, paints, and plastics. Approximately 80% of all sulfur consumed is converted to sulfuric acid.

Sulfur dioxide at room temperature and pressure is a colorless, irritating gas. In liquid form, boiling point −10.02°C, melting point −75.46°C, it is available at low cost in steel cylinders. It is an excellent reducing agent. Because of its ease of liquefaction, it is employed in refrigeration processes. It is used in petroleum refining. Water solutions of the gas are applied as a fungicide and as a preservative of beverages and foods. Both sulfurous acid and sulfites find extensive use as bleaching agents for textiles, hats, feathers, and dried fruits.

Sulfur trioxide, SO_3, is formed in the first stage of the contact process for the manufacture of sulfuric acid. In this process sulfur dioxide and oxygen are heated together in the presence of a catalyst, such as platinum or vanadium pentoxide, to form sulfur trioxide which is combined with water to form H_2SO_4.

The alkali metal sulfates are formed by the action of sulfuric acid on the metal oxide or hydroxide. Potassium sulfate, K_2SO_4, sodium sulfate, Na_2SO_4, and Epsom salt, $MgSO_4 \cdot 7H_2O$, occur naturally. Glauber's salt, $Na_2SO_4 \cdot 10H_2O$, is made by the double decomposition of sodium chloride and magnesium sulfate. Cupric sulfate, $CuSO_4$, the most important salt of copper, is made by treating scrap copper with sulfuric acid. Calcium sulfate occurs naturally as anhydrite, $CaSO_4$, and as gypsum, $CaSO_4 \cdot 2H_2O$. The sulfates of barium and strontium occur as the minerals, barites and celestite.

The following sulfur halides have been identified: symmetrical disulfur dihalides and polysulfur dihalides of F, Cl and Br; thiothionyl fluoride (unsymmetrical S_2F_2), SCl_2, and probably SF_2, SF_4 and SCl_4, S_2F_{10}, SF_6, $SClF_5$, and $SBrF_5$. Of these, S_2Cl_2, SCl_2, and SF_6 are commercially important. The sulfur iodides exist only in solution and are very unstable. The most important oxyhalides are the thionyl halides, SOF_2, $SOFCl$, $SOCl_2$, and $SOBr_2$. Sulfuryl fluoride, SO_2F_2, and chloride, SO_2Cl_2, and a great many oxyfluoride derivatives also exist.

Commercial Forms of Sulfur. Elemental sulfur is sold on the basis of the long ton (2240 pounds). A number of grades or terms used to describe the commercial products are set forth below.

Native sulfur, mined by the hot-water process, is termed *Frasch sulfur. Recovered sulfur* is elemental sulfur produced from hydrogen sulfide obtained from sour natural gas, or petroleum refinery gas, and from gases manufactured from coal. *Crude domestic dark sulfur* on the U.S. market analyzes more than 99% sulfur. It contains small quantities of hydrocarbon derivatives of petroleum and is commercially free of arsenic, selenium, and tellurium. *Crude domestic bright sulfur* on the U.S. market is 99.5%–99.9% pure and commercially free from arsenic, selenium, and tellurium. *Virgin block sulfur,* or broken-rock brimstone, is sublimed or refined sulfur. *Roll sulfur,* or stick or cannon sulfur, is refined sulfur cast into convenient shapes. *Flowers of sulfur* is crude sulfur which has been refined by sublimation. *Amorphous sulfur* is the insoluble residue produced by extracting flowers of sulfur with carbon disulfide. *Commercial flour sulfur* is produced by grinding sulfur. *Colloidal sulfur* is a suspension of fine particles of sulfur in water. Crude sulfur shipped in solid form contains about 50% fines with lumps up to 8 inches in size. Liquid shipments of crude sulfur constitute about 90% of the sulfur transported in the United States.

Uses of Sulfur. The uses of sulfur are many and varied. Often it is considered along with salt, coal, and limestone to be one of the four basic raw materials of chemical industry. Eighty per cent of the brimstone consumed in the U.S. is burned to form sulfur dioxide for conversion to sulfuric acid, or for use in producing wood pulp, for bleaching, etc. The following classification, illustrates the economic importance of sulfur: (1) acid and chemicals, (2) phosphate fertilizers, (3) ammonium sulfate, (4) pulp and paper, (5) textiles, (6) rubber, (7) agriculture (other than fertilizers).

Sulfur and its compounds act both as direct and indirect fertilizers. They correct alkali in soils. They react with soil and release its nutrient elements. They act as a soil ameliorant. Sulfur, usually in the form of organic matter and sulfates, improves soil structure, increases its waterholding capacity, modifies soil reaction, and stimulates growth of soil microorganisms.

Sulfuric acid is used to treat phosphate rock to make phosphate fertilizer materials. The products of this treatment, often with further processing, are triple superphosphate, diammonium phosphate, single superphosphate, and phosphoric acid.

Protein is the essential constituent of all living cells and is required for building new tissue in the body or replacing old tissue. Protein, as far as human nutrition is concerned, is synthesized from ten basic amino acids. The most important of the sulfur-containing amino acids is methionine, which is vital to growth. Another is cystine, the major constituent of skin, hair, and nails. Other important sulfur-containing substances are enzymes and hormones, which regulate complex biological reactions, and vitamin B_1, or thiamine.

All sulfur requirements for human and animal nutrition must be supplied by foods. The ultimate source of this sulfur is the soil. Sulfur is absorbed from the soil by plants as the sulfate ion, and is transformed within the plant to complex organic sulfur compounds. Sulfur in food plants is transferred in one or more steps of the food chain to the bodies of animals and human beings.

Compounds of sulfur have many commercial uses. Sodium thiosulfate, commonly called "hypo," is used in photography. It is employed as an "antichlor" in the textile and paper industries to remove the excess chlorine used in bleaching. Calcium polysulfide is used in the manufacture of paper pulp, in tanning leather, and in the manufacture of chemicals. Matches contain sodium sesquisulfide. Ammonium sulfate is used as a fertilizer in agriculture. Sulfur hexafluoride serves as an excellent gaseous dielectric. Sulfur monochloride is used as a vulcanizing agent, and sulfuryl chloride as an oxidizing and chlorinating agent. Sulfamic acid is used in the tanning of leather and in the production of dyes and cleaning compounds. Salts of sulfamic acid are used in flameproofing paper and textiles.

Industrially, organic sulfur compounds find important application in the formulation of perfumes, in the manufacture of dyes (i.e., the sulfur blacks), in the manufacture of rubber goods, in the stabilization of refined lubricants as well as in the manufacture of special lubricants. A relatively new large-scale use is the application of sulfonated fats and oils as detergents. Sulfur compounds are acid-resistant cements and polysulfide rubbers.

Many synthetic detergents depend on the use of sulfonic or sulfuric acid in their preparation. Thiourea is used in making resins and plastics. Organic polysulfides are employed as linings for aircraft fuel tanks and protective coatings. Compounds containing sulfur such as the sulfa drugs and penicillin are used in medicine. Mercaptans are employed as warning agents in fuel gas systems.

WERNER W. DUECKER
JAMES R. WEST

SULFURIC ACID

Sulfuric acid is the most important of the inorganic acids and has by far the widest industrial use, both in tonnage and in variety of application. The production of 100% H_2SO_4 was 28.7 million tons in 1969, with estimates of 48.8 million for 1975 and 69.3 million for 1980. One reason for this is the fact that it is relatively cheap; another is that other acids can be made by adding sulfuric acid to their salts. On the other hand, sulfuric acid is not liberated from its salts by other acids; on heating, other acids volatilize, leaving the sulfates unaffected. The sulfates decompose at higher temperature.

Near 100% concentration, sulfuric acid is an oily liquid and is strongly corrosive. It is miscible with water in all proportions, with evolution of so much heat that the process is hazardous. In slight dilution it freezes easily, well above the ice point. At greater dilutions it can be cooled to −40°C or below without freezing. It also has high electrical conductivity, and has considerable use as an electrolyte. It is a polar compound and readily ionizes in aqueous solution.

Distinct from 100% acid are the fuming sulfuric acids (sometimes called *oleums*); these contain an excess of sulfur trioxide, and are made in a number of concentrations. Their freezing point curves have a number of peaks and valleys, but in the main they freeze more easily than the acid itself.

Sulfuric acid is made from sulfur or pyrite (FeS_2) by combustion, as a result of which sulfur dioxide is formed, more or less diluted with air. The acid can be made from any mixture of sulfur dioxide and air, for example, from copper blast-furnace gas. A method known as the chamber process was formerly used, featuring a series of lead-lined reaction chambers; it was a slow and expensive operation which is now virtually obsolete. Far more efficient is the so-called contact process, in which the sulfur dioxide-air mixture is passed over and through a solid catalyst at high temperature. This first yields sulfur trioxide, which reacts with water and forms sulfuric acid. As this reaction may proceed with explosive violence, the sulfur trioxide is usually dissolved in concentrated sulfuric acid, with which it is completely miscible. The catalysts used in the contact process can be either platinum powder on an asbestos carrier, or vanadium pentoxide. The latter is preferable, as it is less sensitive to impurities and is not so readily poisoned.

Sulfuric acid has innumerable industrial uses; the manufacture of fertilizers is an important volume application, as is also its general use in chemicals. It is also used as a sulfonating agent, in the

making of dyes and pigments, in petroleum refining, and in rayon manufacture. Its electrical properties are exploited by its use in storage batteries and electroplating baths. It is perhaps the most versatile inorganic chemical known.

Sulfuric acid should be handled and shipped with the greatest care and should not be stored near organic materials, which it decomposes. It is violently corrosive to the skin and requires a white label when shipped. Tank cars and barges are used for tonnage quantities. Its safety tolerance when admixed with air is 1 milligram per cubic meter.

ELBERT C. WEAVER

SUPEROXIDES

Superoxide compounds are characterized by the presence in their structure of the O_2^- ion. The O_2^- ion has an odd number of electrons (13) and, as a result, all superoxide compounds are paramagnetic. At room temperature all superoxides have a yellowish color. At low temperature many of them undergo reversible phase transitions which are accompanied by a color change to white. Superoxide compounds known to be stable at room temperature are:

Sodium superoxide	NaO_2
Potassium superoxide	KO_2
Rubidium superoxide	RbO_2
Cesium superoxide	CsO_2
Calcium superoxide	$Ca(O_2)_2$
Strontium superoxide	$Sr(O_2)_2$
Barium superoxide	$Ba(O_2)_2$
Tetramethylammonium superoxide	$(CH_3)_4NO_2$

The superoxides are generally prepared by one of three methods:

(1) Direct oxidation of the metal, metal oxide, or metal peroxide with pure oxygen or air. All alkali metal superoxides, with the exception of lithium have been prepared in this manner. The superoxides of potassium, rubidium, and cesium form quite readily upon direct oxidation of the molten metal in air or oxygen at atmospheric pressure. Attempts to prepare sodium superoxide under the same conditions result in the formation of sodium peroxide, Na_2O_2. As a result, it was generally felt, prior to 1949, that sodium superoxide was not stable enough to be synthesized. However, in 1949 this superoxide was prepared for the first time, in good yield and purity, by the direct oxidation of sodium peroxide at 490°C under an oxygen pressure of 298 atm. Sodium superoxide is now commercially available and is prepared by a high-temperature, high-pressure, direct oxidation of the peroxide. It is now known that the pale yellow color common in commercial grade sodium peroxide is due to the presence of 5 to 10% sodium superoxide.

(2) Oxidation of an alkali metal dissolved in liquid ammonia with oxygen. All the alkali metal superoxides have been prepared by this method. Although lithium superoxide (LiO_2) has not been isolated in a room temperature-stable form, it has been demonstrated that when lithium is oxidized in liquid ammonia at −78°C the superoxide does form and is stable at that temperature.

(3) Reaction of hydrogen peroxide with strong bases. Hydrogen peroxide can be caused to react with strong inorganic bases to form intermediate peroxide compounds which disproportionate to yield superoxides. The alkaline earth metal superoxides, and sodium, potassium, rubidium, cesium, and tetramethylammonium superoxide have been obtained via this process. Claims have also been made for the synthesis of lithium superoxide via this method; however, such claims have not been adequately substantiated.

Using the formation of potassium superoxide as an example, the reactions involved in this process are:

$$2KOH + 3H_2O_2 \rightarrow K_2O_2 \cdot 2H_2O_2 + 2H_2O$$

followed by

$$K_2O_2 \cdot 2H_2O_2 \rightarrow 2KO_2 + 2H_2O.$$

From the commercial point of view the most important of the superoxides is KO_2. This compound has been in large scale commercial production for many years. It is manufactured in very good yield and purity by air oxidation of the molten metal. This compound is utilized in self-contained breathing devices which are widely used in fire fighting operations and in mine rescue work. The function of the superoxide is to provide oxygen and to remove exhaled carbon dioxide. This unique capability of superoxides is explained by the following chemical reactions:

$$2KO_2(s) + HOH(v,l) \rightarrow 2KOH(s,soln) + 3/2O_2(g)$$

and

$$2KOH(s,soln) + CO_2(g) \rightarrow K_2CO_3(s,soln) + H_2O$$

where s = solid, v = vapor, l = liquid, g = gas, and soln = solution.

Up to 34% of the weight of potassium superoxide is available as breathing oxygen. The lower molecular weight NaO_2 is capable of supplying up to 43% of its weight as oxygen. Thus, sodium superoxide is a better oxygen storage compound. However, it has not been widely used due to its relatively high cost of approximately $20/lb. The cost of KO_2 is approximately $4/lb. One spectacular use of sodium superoxide has been for revitalization of the air in the cabins of the Soviet manned space vehicles. The use of superoxides for maintaining proper oxygen and carbon dioxide levels in the atmospheres of space vehicles, space stations, and submarines is of great interest to workers involved with the development of these systems.

The handling and storage of superoxides requires care and caution. Chemically they are powerful oxidizing agents and strong bases and as a result, they react vigorously with acids and organic materials. All superoxides are extremely hydroscopic, thus their safe storage requires the use of tightly sealed, clean, dry containers.

The chemical bond between the superoxide ion, O_2^-, and the metal ion is ionic in nature. Melting points of potassium, rubidium and cesium superoxide have been determined, and in keeping with the ionic nature of the compounds, the melting temperatures are high, in the order of 400°C.

The most reliable technique for the analysis of superoxides is that developed by Seyb and Kleinberg. In this method the superoxide sample is treated with a mixture of glacial acetic acid and diethyl or dibutyl phthalate. The superoxide reacts with the acetic acid to yield oxygen, hydrogen peroxide, and potassium acetate. The amount of superoxide in the sample is related to the amount of oxygen evolved which is measured with a gas buret. The stoichiometry of the analytical reaction is:

$$2KO_2 + 2HC_2H_3O_2 \rightarrow 2KC_2H_3O_2 + H_2O_2 + O_2$$

It is important that a sufficiently dilute glacial acetic acid-diethyl phthalate mixture be used. Contact of undiluted glacial acetic acid with the superoxide will result in a violent and uncontrollable reaction.

As a result of the paramagnetic nature of superoxides, it is possible to determine their purity by means of paramagnetic susceptibility measurements. The use of this method is limited by its poor accuracy.

A. W. PETROCELLI

SURFACE CHEMISTRY

Surface chemistry deals with the behavior of matter, where such behavior is determined largely by forces acting at surfaces. Since only condensed phases, i.e., liquids and solids, have surfaces, studies in surface chemistry require that at least one condensed phase be present in the system under consideration. The condensed phase may be of any size ranging from colloidal dimensions to a mass as large as an ocean. Interactions between solids, immiscible liquids, liquids and solids, gases and liquids, gases and solids, and different gases on a surface fall within the province of surface chemistry.

Surface forces determine whether one material will wet and spread on a substrate, e.g., whether a liquid will wet a solid and spread into crevices and pores to displace air. This seemingly simple phenomenon is of cardinal importance in determining the strength of adhesive joints and of reinforced plastics; it establishes the printing and writing qualities of inks; lubricants will wet and spread over entire surfaces or be confined to limited working areas depending upon built-in wetting or nonwetting properties; ores are floated if the surrounding liquid is readily displaced by air bubbles; the dispersion of pigments in paints depends upon wetting of the individual particles by the liquid; the action of a foam breaker frequently depends upon its ability to spread on the foam; secondary oil recovery often involves displacement of oil from sand by water; wetting is also a factor in detergency; water and soil repellancy depend upon nonwetting.

Wetting or nonwetting often depends upon the adsorption of a solute at a surface or interface. The bulk liquid phase either advances or recedes, depending upon the nature of the solute and the condensed phase. However, there are many phenomena where adsorption is essential to the process but wetting is not a factor. For example, toxic gases and cigarette tars are removed by adsorption on suitable substrates; color bodies are removed from vegetable oils by adsorption on activated clays; heterogeneous catalysis requires the adsorption of reactants on the catalytic surface; dyeing of fabrics is an adsorption process; dispersions and emulsions are stabilized by the adsorption of suitable solutes and flocculated by the adsorption of other solutes; foaming depends upon adsorption; chromatography is a preferential adsorption process; the action of many corrosion inhibitors depends upon their adsorption on metal surfaces.

The spreading of an insoluble monolayer is a process analogous to adsorption with a number of specialized applications. Thus, cetyl alcohol is spread as a monolayer on reservoirs to retard the evaporation of water. Some antifoaming agents act by spreading as monolayers.

Because of the widespread applications of surface chemistry, practically all industries, knowingly or otherwise, make use of the principles of surface chemistry. Countless cosmetic and pharmaceutical products are emulsions—lotions, creams, ointments, suppositories, etc. Food emulsions include milk, margarine, salad dressings and sauces. Adhesive emulsions, emulsion paints, self-polishing waxes, waterless hand cleaners and emulsifiable insecticide concentrates are commonplace examples of emulsions, which fall within the province of surface chemistry. Other products which function in accordance with the principles of surface chemistry include detergents of every variety, fabric softeners, antistatic agents, mold releases, dispersants and flocculants.

Surface forces are merely an extension of the forces acting within the body of a material. A molecule in the center of a liquid drop is attracted equally from all sides, while at the surface the attractive forces acting between adjacent molecules results in a net attraction into the bulk phase in a direction normal to the surface. Because of unbalanced attraction at the surface, the tendency is for these molecules to be pulled from the surface into the interior, and for the surface to shrink to the smallest area that can enclose the liquid. The work required to expand a surface by 1 sq cm in opposition to these attractive forces is called the surface tension.

This concept applies equally well to solids. Molecules in a solid surface are also in an unbalanced attractive field and possess a surface tension or surface free energy. While the surface tension of a liquid is easily measured, this is much more difficult to do for a solid, since to increase the surface extraneous work must be done to deform the solid.

In the case of solid or liquid solutions it is frequently observed that one component of the solution is present at a greater concentration in the surface region than in the bulk of the solution. Thus, for an ethanol-water system, the surface region will contain an excess of ethanol. The concentration of water will be higher at the surface than in the bulk, if the solute is sulfuric acid. Molybdenum oxide dissolved in glass will concentrate at the surface of the glass. The concentrating of solute molecules at a surface is called adsorption.

If a clean solid is exposed to the atmosphere, molecules of one or more species present in the atmosphere will deposit on the surface. If the clean solid is immersed in a solution, molecules of one or another species present in the solution will be apt to concentrate at the solid-liquid interface. These phenomena are also referred to as adsorption.

All adsorption processes result from the attraction between like and unlike molecules. For the ethanol-water example given above, the attraction between water molecules is greater than between molecules of water and ethanol. As a consequence, there is a tendency for the ethanol molecules to be expelled from the bulk of the solution and to concentrate at the surface. This tendency increases with the hydrocarbon chain-length of the alcohol. Gas molecules adsorb on a solid surface because of the attraction between unlike molecules. The attraction between like and unlike molecules arises from a variety of intermolecular forces. London dispersion forces exist in all types of matter and always act as an attractive force between adjacent atoms and molecules, no matter how dissimilar they are. Many other attractive forces depend upon the specific chemical nature of the neighboring molecules. These include dipole interactions, the hydrogen bond and the metallic bond.

There is an additional explanation for the tendency of a solute such as ethanol to concentrate at the surface of a liquid, which originates with Langmuir. According to his "principle of independent surface action" each portion of a molecule behaves independently of other portions of the molecule in its attraction to other molecules or functional groups on a molecule. The attraction between the CH_3CH_2 portion of the ethanol molecule and water arises from relatively weak London dispersion forces, as compared with the additional attraction of strong hydrogen-bonding forces acting between the hydroxyl group of the alcohol and water. Hydrogen bonding is also responsible for the strong attraction between water molecules. As a consequence, not only is the alcohol concentrated at the surface, it is also oriented with the hydroxyl group toward the water and the hydrocarbon chain directed outward. Since the attraction between adjacent hydrocarbon molecules is less than that between adjacent water molecules, hydrocarbon liquids have lower surface tensions than water, and the surface tension of an aqueous alcohol solution is intermediate between that of liquid hydrocarbons and water.

As noted earlier, the phenomenon of adsorption is encountered in diverse applications. Medical applications are often the most complex and the least understood. For example, replacement hearts and kidney machines require plastics that can be kept in contact with human blood for long periods of time. However, foreign material in contact with blood results in clotting. The material first becomes coated with adsorbed protein. Some time later the clotting process begins, apparently due to activation of the Hageman factor, one of the proteins in blood, at the blood-material interface. The activation initiates a chain reaction that results in the conversion of fibrinogen to fibrin. It has been suggested that the Hageman factor is helical in form and that adsorption results in an unfolding of the protein helix with exposure of certain active sites which then initiate the clotting of blood. Other proteins are also adsorbed and their biological function may be altered, but little is known about this.

The ideal surface for contact with human blood

is the surface of blood vessels, and the immediate surface contains heparinoid complexes. Heparin, a negatively charged polysaccharide, has been bonded to silicon rubber and other polymers. In one procedure, a quaternary ammonium compound is first adsorbed on the polymer substrate and heparin is in turn adsorbed on the positively charged surface. Chemical bonding of heparin has also been achieved. Such surfaces do not cause clotting of contacted blood.

As noted earlier, the phenomenon of adsorption is encountered in such diverse applications as the separation of components in chromatography, the removal of toxic gases by activated charcoal, heterogeneous catalytic reactions and the dyeing of fabrics. The surface area of solids is most commonly determined by the adsorption of nitrogen on the surface of the solid at −195°C. Nitrogen is assigned an area of 16.2 Å² per molecule. The method is due to Brunauer, Emmett and Teller and is referred to as the BET method for determining surface area.

The fundamental adsorption equation is due to Gibbs. In the case of a solution containing a single solute,

$$\Gamma = \frac{1}{RT} \left(\frac{\overset{\bullet}{\partial \gamma}}{\partial \ln a} \right)_T$$

where Γ is the excess of solute at the surface as compared with the concentration of solute in the bulk liquid expressed in moles per sq cm, R is the gas constant, T is the absolute temperature γ is the surface tension of the solution, and a is the activity of the solute. Where the solution is sufficiently dilute, the concentration of the solute may be substituted for its activity.

The equation also applies to the adsorption of a gas on a solid. At low gas pressures, p, the equilibrium pressure of the gas can be substituted for a, the activity of the solute. The amount of gas adsorbed v/V is equivalent to the surface excess Γ, where v is equal to the volume of gas adsorbed per gram of solid and V is the molar volume of the gas. The total free energy change at constant pressure is $\Sigma \delta \gamma$, where Σ is the area per gram of solid.

When a drop of liquid is placed on the surface of a solid, it may spread to cover the entire surface, or it may remain as a stable drop on the solid. There is a solid-liquid interface between the two phases. In the case of liquids that do not spread on the solid, the bare surface of the solid adsorbs the vapor of the liquid until the fugacity of the adsorbed material is equal to that of the vapor and the liquid.

The equation relating contact angle to surface tension, generally ascribed to Young or Dupre, is

$$\gamma_{Se} = \gamma_{SL} + \gamma_L \cos \theta$$

where γ_{Se} is the surface tension of the solid covered with adsorbed vapor, γ_{SL} is the solid-liquid interfacial tension, γ_L is the surface tension of the liquid and θ is the contact angle.

As a general rule, organic liquids and aqueous solutions will spread on high-energy surfaces, such as the clean surfaces of metals and oxides. The rule has a number of exceptions. For one, certain organic liquids will deposit a low-energy film by adsorption on higher-energy surfaces over which the bulk liquid will not spread.

Zisman discovered that there is a critical surface tension characteristic of low-energy solids, such as plastics and waxes. Liquids that have a lower surface tension than the solid will spread on that solid, while liquids with a higher surface tension will not spread. Examples of critical surface tension values for plastic solids in dynes per cm are: "Teflon," 18; polyethylene, 31; polyethylene terephthalate, 43; and nylon, 42–46. As one indication of the way this information can be used in practical applications, one can consider the bonding of nylon to polyethylene. If nylon were applied as a melt to polyethylene, it would not wet the lower-energy polyethylene surface and adhesion would be poor However, molten polyethylene would spread readily over solid nylon to provide a strong bond.

There are a large number of materials that exhibit a pronounced tendency to concentrate at surfaces and interfaces and thus alter the surface properties of matter. These materials are called surface-active agents or surfactants. Depending upon the manner in which they are used or the purpose they serve in specific applications, they may be referred to as detergents, emulsifying agents, foaming agents or foam stabilizers, antibacterial agents, fabric softeners, flotation reagents, antistatic agents, corrosion inhibitors, or by other names. There are two general ways of classifying surfactants. According to solubility, they are classified as water or oil soluble. The other classification is according to change type. Those that do not ionize are called nonionic surfactants. If they ionize and the surface-active ion is anionic, the material is an anionic surfactant. If the surface-active ion carries a positive charge, it is called a cationic surfactant.

Only molecules with certain specific types of configurations exhibit surface activity. In general, these molecules are composed of two segregated portions, one of which has low affinity for the solvent and tends to be rejected by the solvent. The other portion has sufficient affinity for the solvent to bring the entire molecule into solution. Water-soluble soaps are probably the oldest surfactants. The long hydrocarbon chain has a low affinity for water and is referred to as the hydrophobic or nonpolar portion of the molecule. The carboxylate group has a high affinity for water and is called the hydrophilic or polar portion.

LLOYD OSIPOW

Cross-references: *Adsorption, Colloid Chemistry, Interfaces, Surface Tension, BET Theory, Monomolecular Films.*

References

Davies, J. T., and Rideal, E. K., "Interfacial Phenomena," Academic Press, New York, 1963.

Gould, R. F., Ed., "Interaction of Liquids at Solid Substrates," American Chemical Society, Washington, D.C., 1968.

Osipow, L. I., "Surface Chemistry: Theory and Industrial Applications," Reinhold Publishing Corp., New York, 1962.

Schwartz, A. M., Perry, J. W., and Berch, J., "Surface Active Agents and Detergents," Vol. II, Interscience Publishers, New York, 1958.

SURFACE TENSION

The interface between any two phases (liquid/gas, liquid/liquid, solid/solid, solid/gas, solid/liquid) is characterized by an interfacial free energy, γ; for the interface between a liquid and a gas γ is called the surface tension. In minimizing its surface free energy, a liquid behaves as if an elastic membrane were stretched over its surface. Thus liquids form spherical drops, rise in capillaries whose walls they wet well and are depressed in capillaries whose walls they do not wet well. Surface tension (more generally, interfacial free energy) plays a dominant role in the formation and stabilization of emulsions, the behavior of bubbles and foams, the sintering of powders, the separation of powdered solids by flotation, the cleaning of fabrics by detergent solutions, in the use of adhesives and in all adsorption phenomena.

The surface tension of a liquid is the work required to increase its surface area in a reversible process by one unit. In Figure 1, if a film of liquid is confined within the area MBCN by a freely moveable wire MN on the wire frame ABCD, it will be found that a force of F dynes is required to keep the film from contracting. Since the film has two sides, if the frame is 0.5 cm wide the force acts on a length 1.0 cm of surface, and γ is equal to F/cm. An infinitesimal increase of F will move the wire MN toward AD. When it has moved 1 cm to M'N', the work done on the film is F ergs for 1 cm² increase in surface area. If F is decreased infinitesimally, this work can be recovered as the film contracts. Thus, in the CGS system, the units of γ are ergs/cm² or, equivalently, dynes/cm, and these are the units commonly used.

Since γ is a measure of reversible work, it is in thermodynamic terms a free energy* and may be expressed as $\gamma = h - Ts$, where h is specific surface energy (more precisely, for measurements at constant pressure, it is an enthalpy), s is specific surface entropy, and T is absolute temperature. The h term arises because a molecule in the surface has a smaller number of neighbors to attract it than does a molecule in the interior of the liquid; it

therefore has a higher (less stable) potential energy. The s term arises because the surface layer of the liquid is less dense than the interior (this also contributes to h) and thus has a more random molecular arrangement and correspondingly a higher entropy. If h and s were independent of T, γ would decrease linearly with increasing T; for most liquids over short ranges of temperature this is approximately true but a more accurate empirical description is $\gamma = \gamma_0 (1 - T/T_c)^{1.23}$, where T_c is the critical temperature of the liquid. Since γ depends on intermolecular forces, one expcts it to be related to other thermodynamic properties; a number of approximate empirical relations are known. Molecular theories of surface tension have been developed, but the good ones are complicated; the simple ones are rather approximate. From fairly simple reasoning it may be shown that the specific surface enthalpy $h = \alpha(L/V)(V/N)^{1/3}$ where V/N is the average volume per molecule in the liquid and (L/V) is the energy of vaporization per unit volume; α is a factor ranging between $\frac{1}{8}$ and $\frac{1}{2}$ which depends on the details of molecular packing and bonding in the surface layer. For many organic liquids (excluding those that form strong hydrogen bonds) γ itself is given approximately by a similar relation: at room temperature $\gamma \simeq 0.040 (L/V)^{1.15} V^{1/3}$, where L/V is in calories/cc. Note that $h = \gamma - T d\gamma/dT$; since $d\gamma/dT$ is negative, h is greater than γ by a factor that depends on temperature. At room temperature, h/γ is 1.14 for mercury, 1.64 for water, and 2.3 for n-octane.

The magnitude of γ ranges from a fraction of a dyne/cm (liquid helium) to several thousand dynes/cm. Molten metals span the range from a few hundred to 1800 (platinum) and above; some molten glasses around 1000°C lie in the range 200–500. Values for some common liquids at 20°C are water 72.8, mercury 475, benzene 28.9, ethanol 22.3, n-octane 21.8 dyne/cm.

By thermodynamic reasoning, it may be shown that the equilibrium pressure difference across a general curved liquid interface is $\gamma(1/r_1 + 1/r_2)$ where r_1 and r_2 are the principal radii of curvature. This equation is basic to several methods for measuring surface tension (bubble pressure, shape of pendant drops or of sessile drops). Either from this equation or from the notion of surface tension, one obtains $z = 2\gamma \cos \theta / \rho \, gr$, where z is the height to which a liquid of density ρ in a capillary of radius r will rise above the hydrostatic level; g is the acceleration due to gravity, and θ is the contact angle between the liquid surface and the wall of the capillary. Other important methods for measuring surface tension include the force to lift a wire loop or frame from an interface, the force to raise a thin plate through an interface, and measurement of the oscillating cross-sectional shape of a thin noncircular jet of liquid.

For a sphere, each of the principal radii of curvature is equal to the radius of the sphere; thus the pressure inside a spherical liquid drop of radius r is greater than that on the outside by $2\gamma/r$. In consequence, the vapor pressure p is greater than that of the bulk liquid, p_0, in approximate accordance with $\ln_e p/p_0 = 2\gamma M/\rho rRT$, where M is the

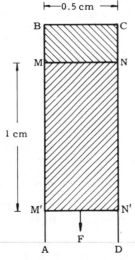

—0.5 cm—

1 cm

F

FIG. 1.

* Per unit area.

molecular weight of the vapor, ρ is the density of the liquid, and R is the gas constant.

Similarly, very small particles of a solid are more soluble than large ones. This relative instability of small particles becomes important when the particles are very small. It is responsible for super-cooling in the condensation of clean vapors or the freezing of clean liquids; in the absence of surfaces or foreign particles on which a new phase can condense, the very small particles which represent the first stage in the growth of a new phase are unstable if the degree of supercooling is small. The effect of curvature on vapor pressure is often used to measure the size of very small pores in porous solids such as silica gel; volatile liquids will condense in such pores at partial pressures below their normal vapor pressure.

The surface tension of a solution may be either higher or lower than that of the pure solvent. "Surface-active" solutes (in a given solvent) are those which decrease γ; the concentration of such solutes in the surface region is higher than in the bulk phase in accordance with the Gibbs adsorption isotherm. So-called surfactants show this behavior in extreme form; at a concentration of only 0.01 wt. per cent, some of these materials will lower the γ of water to 25 dynes/cm. A freshly formed surface of a solution of a surfactant takes some time to develop its equilibrium surface tension since the surfactant must diffuse to the surface if γ is to be decreased. Another way to lower the surface tension of a liquid is to spread on its surface a "monolayer" of nearly insoluble surfactant (e.g., oleic acid on water). Such monolayers can be studied by measuring the force required to decrease their area with a moveable barrier. If the area per molecule is sufficiently reduced, the force required to decrease the area further rises steeply. In this condition the monolayer is "close-packed" and the area per molecule is the average molecular cross-sectional area parallel to the surface. Some close-packed monolayers greatly decrease the rate of evaporation of the underlying liquid.

The surface free energy, γ_s, of a solid is difficult to measure because a solid does not change shape freely in response to surface forces. For some glasses and for silver, γ_s has been determined by measuring the least tensile force required to keep a very thin filament from contracting at temperatures high enough for creep to occur at an appreciable rate. In general, any solid whose melt at high temperature has a high γ can be expected to have an even higher value of γ_s in the solid form at room temperature. The "high-energy" surfaces of such solids (e.g., metals) have a strong tendency to adsorb many gases or vapors, either chemically or physically, because such adsorption usually lowers γ_s. For example, γ_s for the oxidized surface of a metal is much lower than for the clean metal. For similar reasons, low-surface-energy liquids spread spontaneously on high-energy surfaces. In recent years much success has attended the study of "low-energy" solid surfaces in terms of the contact angles made on them by pure liquids covering a range of surface tensions. On crystalline solids, nonequivalent exposed crystal faces will in general have different values of γ_s.

The surface free energy per gram of a finely divided substance can be quite substantial; for particles of silver 100 Å in diameter dispersed on a catalyst support, it is over 100 cal/gm. For water in droplets of one micron diameter, on the other hand, it amounts to only about 0.1 cal/gm. Surface tension and other surface effects are very important in biological systems, in which matter is finely subdivided by a variety of interfaces.

J. N. WILSON

Cross-references: *Interfaces, Colloid Chemistry, Emulsions, Foams, Adsorption, Surface Chemistry, Monomolecular Films.*

SYNERESIS

Many gels exhibit the phenomenon of exuding small quantities of liquid. This process was termed "syneresis" by Thomas Graham and was first noted in 1861. Syneresis may be defined as the spontaneous separation of an initially homogeneous colloid system into two phases—a coherent gel and a liquid. The liquid is actually a dilute solution whose composition depends on the original gel. When the liquid appears, the gel contracts, but there is no net volume change. Syneresis is reversible if the colloid particles do not become too coagulated immediately after their formation.

Heller, in 1937, classified three types of syneresis as to cause: (1) syneresis of desorption, caused by the particle becoming less hydrophilic with time; (2) syneresis of aggregation whereby discrete gel particles may unite into a denser gel portion; and (3) syneresis of contraction, where a gel with fibrillar structure contracts and squeezes out the intermicellar liquid. Most commonly, syneresis is the visible manifestation of further slow coagulation which follows the initial setting of the gel, the gel forming process itself being an enmeshing of the hydrous particles into a network. It may be further explained as the exudation of liquid held by capillary forces between the heavily hydrated particles constituting the framework of the gel. The phenomenon was termed by Ostwald one of the most characteristic properties of gels.

A common example of syneresis is found when a mold of gelatin remains in the icebox for some time. A general shrinkage of the body of the gel occurs and a liquid collects around the edge of the mold. The liquid is a dilute solution of the original composition. The total volume of the system remains constant and thus syneresis should not be considered as merely the opposite of imbibition. Agar gels are valuable as bacteriological culture media, depending on their degree of syneresis. A gel which synerizes too much may interfere with the counting of colonies found on plates and with the isolation of pure cultures. Large amounts of exudate may cause the gel to slip in its container and induce the spread of colonies. The agar substitutes now coming into use must be examined carefully for this tendency.

Intravenous injections of gelatin or gums have become a routine procedure in cases of excessive hemorrhage. The edema of hemorrhage seems to be related to colloid osmotic pressure. The loss of proteins from the blood stream reduces the pressure in the capillaries so that the tissue colloids are able to withdraw liquid from the blood. These hydrated tissue colloids then undergo syneresis, so that large volumes of liquid collect within the tissues instead of being excreted normally through the kidneys. The injection of hydrophilic colloids to replace blood protein raises the colloid osmotic pressure.

Protoplasm is a true colloid and therefore secretes when it contracts. The secretion is a form of syneresis. If the surface properties of colloids change, their water-holding capacity is affected. The secretions of the ductless glands are thought to be the result of syneresis and in this case, a nerve impulse is sufficient to alter the surface properties and bring about secretion of the hormones.

The separation of serum from a blood clot, or of whey in milk souring or cheese making; the sweating of butter; the separation of liquid from lean meat on heating; and the serum exuding from a wound or blister are all examples of syneresis.

R. A. CARPENTER

Cross-references: *Gels, Colloid Chemistry.*

SYNERGISM

Synergism is a term commonly used to denote positive interaction between components in a mixture. As such it means the opposite of antagonism. Actually, the words refer to a departure from the expected value of a property of a mixture when the actual value is compared with that calculated from knowledge of the property in the individual components. Synergism and antagonism represent the degree to which a model of a situation fails to match reality. This failure is due to an interaction of the components on each other or on a third entity.

Interaction is a test of the knowledge of a system. Continued inspection and rearrangement of the model used can eliminate or isolate interactions due to specific components and permit quantitative evaluation of the departure from ideality. Physical laws concerning collegative properties are always accompanied by limits which define the ideal situation. This is when all the components contribute to a property of a mixture in proportion to their concentration in the mixture. Nature is never "ideal" and, therefore, predictions of the behavior of a system which are based on these laws are predestined to be inaccurate.

Interaction, whether it be synergistic or antagonistic, is detected and determined quantitatively by comparing the observed value with the norm or ideal value as represented by:

$$\frac{P_1C_1 + P_2C_2}{C_1 + C_2}$$

where C_1 and C_2 represent the concentration of each component and P_1 and P_2 are the values of the property of each pure component. The validity of this measure is entirely dependent on the use of significant and additive quantities. The choice of a norm, which, if exceeded, indicates the occurrence of interaction, is not difficult from the practical point of view where evidence of improved performance, lower cost and time saving is sufficient. The difficulty arises when a measure is sought that deviations from it will be quantitatively significant.

Synergism has often carried the connotation of mystery or inexplicable phenomena. The modern science of statistical design and analysis of experiments and data has done much to resolve problems arising from a multitude of interacting variables. Today, the appearance of synergism is merely a reminder that insufficient knowledge of the system was available for accurate prediction.

Examples of synergism abound in industrial chemistry. Styrene-methylmethacrylate copolymers are stronger than either polymer alone. Definite compressive strength maxima occur at 30 and 50% styrene, due probably to the dissymmetry introduced into the chain. The modern synthetic detergents such as alkyl aryl sulfonates are produced together with large amounts of sodium sulfate formed during neutralization. This so-called inert material greatly increases the surface activity of the sulfonate. A household fly spray of pyrethrum in kerosene is 10 times as effective if a small amount of sesame oil is added. Sesame oil alone, either pure or diluted with kerosene exhibits no toxicity toward houseflies.

Macaluso classifies synergism as to three manners of occurrence: "(1) As a result of combined stepwise action such that one component acts on the result of the other component. (2) As a result of direct mutual interaction of the components such that a new entity or product is formed with a structure and properties of its own. (3) As a result of complimentary functioning or indirect interaction of the components, each acting in its own way toward the same effect."

An awareness of synergism and antagonism is an essential part of the "know-how" of the modern scientist. The application of synergism need not depend on chance discovery but can be expected, sought after, and planned for.

Synergism can also be applied to creative thinking. Stuber has pointed out that two or more heads working on a properly organized basis can be many times more effective than the same number of heads working individually or on a loosely organized basis. This process of mutual mental stimulation can be utilized at all levels in chemistry and chemical engineering—from top management to lab assistant. Management can harness this force by organizing discussion sessions at which participants will seek to generate new ideas by thinking of and writing down every possible idea that might help solve a problem.

R. A. CARPENTER

T

TANK CARS FOR CHEMICALS

The Department of Transportation (DOT) and various Federal Administrations (Railroad, Highway) are charged with the administration of shipping regulations for interstate commerce. Tank car configurations and classification of regulated materials are incorporated in DOT's CFR Title 49 available from the Government Printing Office. Identical data is published in Agent R. M. Graziano's Tariff latest edition.

The classifications and specifications for shipping containers are listed for the following categories: corrosive liquid, flammable compressed gas, flammable liquid, flammable solid, nonflammable compressed gas, oxidizing material, poisonous gas or liquid, Class A, and poisonous liquid or solid, Class B. Materials which do not come under any of the definitions listed are considered nonregulatory (N.R.); however, combustible liquids listed as N.R. must be carried in a container fitted with a safety valve if the flash point is below 150°F Tag's open cup.

Approved ancillary items (relief safety valves, protective housing, gauging and sampling devices, thermometer well etc.) must be established by the Committee on Tank Cars, the American Association of Railroads. Tank Car classification designations are assigned a prefix either ARA, AAR, ICC or DOT. The prefix ARA (American Railroad Association) became obsolete in 1927, when the designations AAR and ICC were adopted. Tank cars classified under both ICC and ARA can be used to transport certain regulatory commodities as authorized in Agent R. M. Graziano's latest Tariff in addition to nonregulatory commodities. Tank cars under AAR may only be used for nonregulatory materials. The ICC prefix was superceded by DOT in 1967 when the responsibility for interstate regulation was assigned to the Department of Transportation. The term "Pressure Tank Car" is generally applied to those cars having safety valves set for 75 psig and higher.

As an example a tank car designated DOT 111A100W would be described as a steel fusion-welded tank without an expansion dome and could be insulated or noninsulated. The 2% outage (ullage space or vapor space) required must be provided for in the tank shell. Typical commodities for which this tank car is generally supplied would be kerosene, gasoline, fuel oil, vegetable oils and phosphorus. The car tank's design thicknesses would be based on the well-known hoop stress relationship for thin shell cylinders, that is:

$$t = \frac{Pd}{2se}$$

where t = thickness in inches thinnest plate, P = calculated bursting pressure psi, d = inside diameter in inches, s = minimum ultimate tensile strength in psi, and e = efficiency of longitudinal welded joint taken as 90%.

The circumferential joint stresses due to internal pressure (on ends) are generally one-half of the hoop stress thus these latter stresses do not control the design. The end heads are designated ellipsoidal wherein the major axis equals the inside shell diameter and the minor axis is one-half of the major axis. A head of this configuration would give plate thicknesses identical to those generated by hoop stress in the shell. A one-piece head without a welded seam would allow the use of 100% joint efficiency. A minimum thickness of $7/16''$ is designated for the shell assembly and the calculated burst pressure is to be not less than 500 psi. The establishment of a maximum overpressure to 85 psi for the relief valve set at 75 psi gives a safety factor of 5.9, that is, the actual stress level is 1/5.9 of the minimum tensile strength. Bottom outlets are allowed in this classification with the stipulation that such outlets if they are extended 6'' or more below the shell must be "V" grooved so that the root wall thickness would not exceed $3/8''$.

In contrast, a DOT 112A400W tank car is used for liquefied petroleum gas (LPG) having a vapor pressure at 115°F not exceeding 300 psi. This same tank car can be used to transport anhydrous ammonia and vinyl chloride since the vapor pressure of these materials at 115°F will not exceed the relief setting of 300 psi established for this class. This multiple-purpose tank car must have valve components which are compatible with the commodities, thus all copper-bearing materials are prohibited primarily due to the ammonia service. In lieu of insulation, the upper 2/3 of the tank shell must be painted with a light-reflective paint. Bottom outlets and washout taps are prohibited, and the liquid and vapor openings must be fitted with excess flow valves to protect against sudden loading or unloading line depressurization to atmosphere. Shell thickness is established as previously outlined for the DOT 111A100W tank car however, the burst pressure is taken as 1000 psi with the maximum overpressure established as 330 psi. Minimum shell thickness is $11/16''$.

In general, the tank car manufacturer does not establish the suitability of a particular protective lining for a specific commodity. This responsibility is normally left to the shipper/producer of the commodity whose manufacturing experience has developed extensive compatibility data for a series of linings, surface preparation required and proper application procedures to be used. These data would also indicate that a preferred choice might be an aluminum alloy, stainless clad or rubber-lined steel.

Typical tank car orders might involve the following:

33,500 gal. liquefied petroleum gas (LPG)
14,300 gal. 54% phosphoric acid, 90 ton chlorine
13,250 gal. molten sulfur and 19,650 gal. carbon dioxide.

The actual quantity of material to be carried depends on the required car design, capacity of the wheel trucks supplied and the specific gravity of the commodity involved.

Specific recommendations are readily available from tank car manufacturers or carriers for specific materials listed in previously noted tariffs. For new materials not previously classified, application for DOT and AAR approval must be obtained and application forms are available from the American Association of Railroads.

There are many tank car configurations which are available and which have been specifically designed for a particular application. An example would be an insulated DOT 105A600 car designed to carry anhydrous hydrogen chloride. This car is detailed to have a water capacity of 12,600 gallons and is supplied with two individually fitted carbon dioxide tanks which served to maintain the anhydrous hydrogen chloride liquid at a temperature below the relief valve set point for the (14-day design) tank car trip. The carbon dioxide equilibrium temperature is 0°F at 300 psi, the set point for the CO_2 vessels. For reasonably short rail trips with low winter ambients and initial low-temperature HCl charging, the specific heat value of the anhydrous hydrogen chloride is great enough to insure equilibrium pressures substantially below the relief valve set point during the trip thus auxiliary refrigeration (CO_2 charge) is not required.

Another specialty car insulated 18,000 water gallons DOT 111A100-W-1 is designed to carry coal tar pitch charged at 325°F to the vessel. The load is heated in transit to 450°F $\pm$ 5°F for subsequent discharge at the terminal point. This car was fitted with an internal combustion engine, pump, combustion chamber, heat exchange fluid loop and combustion safety controls. The heat exchange loop consisted of a series of half oval strips welded to the exterior surface of the car tank itself. This configuration is normally preferred over internal coils since it obviates product contamination possibilities. Net heat absorption is detailed as 300,000 Btu/hr.

Multicompartmented car tanks are available as well as heavily insulated specialty cars designed for the transport of liquid oxygen, liquid argon, liquid nitrogen and liquefied hydrogen transported at −425°F.

H. D. KERFMAN

TANNING

The leather industry is divided into two broad categories—light and heavy leathers. Light leather includes shoe upper, lining, glove, garment, handbag, and novelty leathers and is made predominantly from light to medium weight cattlehides and from the skins of calf, goat, and sheep. Heavy leather includes sole, upholstery, bag, case, strap, and harness and is made from heavy cattlehides.

Light leathers are usually tanned with basic chromium sulfate referred to as "chrome" or "mineral" tanning. Some shoe lining and novelty leathers are tanned with "vegetable" tannins. Combinations of these tannages are used in some instances. Heavy leather is universally tanned with "vegetable" tannins which are complex mixtures of naturally occurring phenolic compounds.

Chrome Tanning. Basic chromium sulfate is used for chrome tanning. It is manufactured from chrome ore imported from Rhodesia, New Caledonia, and Turkey. It is applied in aqueous solution to skins "pickled" with sulfuric acid and sodium chloride to a pH of 2.0–2.5. In this pH range the chrome penetrates the skin or hide rapidly in a revolving drum. It reacts with the collagen when the skins are neutralized slowly to a pH of 3.5–4.0. The hydrothermal stability, shrink temperature, or "melting point" of the skin is raised to 90–100°C, and the protein is no longer susceptible to attack by bacteria.

The main mechanism of chrome tanning is usually considered to be the coordination of chromium to the carboxyl groups of collagen followed by the formation of secondary valences with the amino groups.

Vegetable Tanning. While there are many vegetable tannins, the most important ones are quebracho, chestnut, and wattle. Quebracho is extracted from the very hard and heavy wood of the quebracho tree which grows in Paraguay and Argentina. Chestnut is extracted from chestnut wood growing mainly in France and Italy. Wattle is extracted from the bark of black acacia trees, grown as a 7–10 year crop, mainly in south and east Africa.

Blends of vegetable tannins are always used in the manufacture of heavy leather in order to make use of the complementary properties of these tannins. Unlike chrome tanning, the initial pH of the liquor is in the range of 4.0–5.0 to permit penetration into hides that still contain calcium hydroxide from the unhairing. The pH is lowered gradually to 3.0–3.5 by countercurrent movement of stronger and more acidic liquors in a series of vats. Additional tanning is completed in drums with strong liquors.

The mechanism of vegetable tanning is the formation of numerous hydrogen bonds between the many phenolic hydroxyl groups of the tannins and the oxygen and nitrogen atoms in collagen. Carboxyl groups occur in some vegetable tannins and form salt links. The relatively low hydrothermal stability of such leather is due to the weakness of the hydrogen bonds.

Alum, Formaldehyde, and Oil Tanning. These old tanning agents are used alone or in combination with chrome to produce specialty leathers such as white leather, glove leather and chamois. Alum is also used in the "tawing" of furs. They are relatively weak tanning agents and do not produce the high hydrothermal stability of chrome-tanned leather.

Zirconium Tanning. Basic zirconium sulfate is the only new mineral tanning agent. Its tanning action was discovered by I. C. Somerville in 1931.

It is more expensive than chrome and is used for special purposes such as the manufacture of white grain and suede leather and for filling the flanks and shoulders of cattle "side" leather tanned with chrome for shoe upper leather.

While the mechanism of tanning is related to that of basic chromium sulfate, there is one striking difference, namely, that basic zirconium sulfate reacts with collagen at a much lower pH. For this reason it is applied to skins pickled to a pH of 1.0–1.5 at which appreciable tanning takes place even while the zirconium salt is penetrating the skin. The acidity is then neutralized gradually. At a higher pH the tanning action is so rapid that the grain is drawn and the center may be under-tanned. However, zirconium can be applied to chrome-tanned leather at a pH of 2.5–3.5 without deleterious effect. Used alone, a higher amount of zirconium than chrome is required, partly due to the high molecular weight of the zirconium salt. Basic zirconium sulfate is a good filling and plumping agent. Unlike chrome, increasing amounts of zirconium increase the firmness of the leather.

Syntans. Synthetic tanning agents or syntans were developed by Edmund Stiasny beginning in 1911. Originally, both phenolsulfonic acid-formaldehyde, and naphthalenesulfonic acid-formaldehyde condensation products were made. For many years only naphthalene syntans were available partly because of the cost of phenol but also due to the shortcomings of the original phenolic syntans. Today a great many syntans are available for different purposes. They are usually referred to as auxiliary or replacement tans, depending upon their efficiency when used alone.

The naphthalene syntans are not tanning agents in the sense that they do not produce satisfactory leather when used alone. The "leather" is non-putrescible but is flat, tinny and unsalable. However, these syntans serve many useful purposes. Their main use, for bleaching chrome-tanned skins for utility white leather, was developed by Thomas Blackadder. The leather yellows readily, but this is reduced by a pigment finish. The sodium salts used on acidic chrome leather act as a dye-resist for acid dyes and permit greater penetration and uniformity of color, more particularly for pastel shades.

High quality phenolic syntans were eventually produced. They are based on phenolsulfone condensation products. The sulfone linkage, unlike the methylene bridge, confers substantial light resistance. Because of the large number of phenolic hydroxyl groups, these syntans are technically adequate to produce white leather when used as the only tanning agent. They are used with other tanning agents to produce such effects as textured grain leather with a deep, fine pattern.

World War II accelerated the search for an adequate replacement for vegetable tannins. This was developed by H. G. Turley and his associates. It is made by condensing phenol with formaldehyde and then sulfonating. It was "tailor-made" to produce sole leather with the same degree of tannage and yield of leather as vegetable tannins. In view of the complexity and high molecular weight of natural tannins, it is too much to expect sole leather made from a synthetic tanning material alone to be identical with that made from a blend of vegetable tans. The leather lacks the solidity and firmness desired in sole leather and imparted, for example, by chestnut extract. Its resistance to light is as low as that of quebracho. This substitute can be blended with vegetable tannins and, with processing adjustments, could be used as the only tanning agent in an emergency. Paradoxically, it is now used in larger volume for retanning chrome-tanned light leathers. The cost limits its use on sole leather generally to the extract wheel, following the yard tannage, to confer a desirable surface effect.

Melamine-Formaldehyde Tanning. Tanning with trimethylolmelamine was discovered before 1943 by W. O. Dawson. It is particularly interesting from a technical standpoint because it was the first organic tannage capable of producing leather that does not shrink in boiling water. In this respect it resembles an inorganic tannage. This is presumably due to the formation of numerous covalent crosslinks by a Mannich-type reaction. When too much is used the leather is weakened due in part to the properties of the resulting three-dimensional polymer and in part to too many crosslinks. When used alone, the leather is very white and light-fast. The tannage is somewhat astringent and expensive so that it is customarily used as a retannage of chrome-tanned leather for whites or for filling flanks.

Later it was found by W. Windus that it was not necessary to start with the preformed methylol-melamine. Melamine and formaldehyde can be added directly to depickled skins, and the tanning agent is formed *in situ*. The low solubility of the melamine is an automatic regulator of the rate of tanning which is convenient for uniform penetration of a skin.

Resin Impregnations. The first use of a resin for impregnation and tanning was developed by George D. Graves and J. S. Kirk in the late 1930's using a polymer of maleic anhydride and styrene. The skins are depickled to a pH of 4.0–5.0 and drummed with a water solution of the sodium salt of the polymer. The skins are then acidified to precipitate the polymer. At this stage the shrink temperature is low, and the polymer can be washed out. It is permanently insolubilized by converting it to an aluminum, chrome, or zirconium salt. The resin does cause some complications in dyeing. However, when used alone or in combination with basic zirconium sulfate it produces a white suede with an excellent nap and a silky feel.

A more recent development is based on impregnation with polymers of urea or dicyandiamide and formaldehyde. These are dispersed with such agents as the sodium salt of a naphthalene syntan and then precipitated in the leather by acidification. The impregnant is white and is used to produce white leather or to improve the cutting value of the flanks of chrome-tanned side leather.

Resorcinol-Formaldehyde Tanning. It was discovered by W. Windus in 1946 that a synthetic tanning agent could be produced in the presence of the hide or skin in the customary dilute aqueous solution at room temperature, by using a very reactive phenol, such as resorcinol or pyrogallol, with an aldehyde, such as formaldehyde or furfural. These compounds are added to pickled skins, and the pH is lowered to approximately 1.0 to increase the rate of reaction and tanning. A Mannich-type reaction between the low-molecular weight resorcinol-formaldehyde condensation product and the epsilon-amino group of the lysine residues of collagen was postulated.

In most respects the properties of the leather resemble those of a vegetable-tanned leather with one striking exception. The hydrothermal stability of resorcinol-formaldehyde-tanned leather is high. On skins it is easy to produce leather that does not shrink in boiling water.

Dialdehyde Tanning. The tanning action of glutaraldehyde was announced simultaneously and independently in 1957 by E. M. Filachione and his associates and by L. Seligsberger. This five-carbon dialdehyde is markedly superior to formaldehyde in the stability of its combination with collagen and the properties it confers to leather. It is now widely used in basic research for the fixation of proteins.

Aldehydes are the only tanning agents that react most rapidly in slightly alkaline solution. When glutaraldehyde is used alone the pH is raised gradually from 4.0 to 8.5. The maximum hydrothermal stability is 85°C and the color of the leather is a light brown. Since essentially all light leather is chrome-tanned, tanners prefer to modify the leather by retanning with other tanning agents. This is the common method of using glutaraldehyde. Tanning is rapid at an elevated temperature and the normal pH of 3.5.

Glutaraldehyde enhances the perspiration resistance, alkaline stability and washability of leather. This has led to its application where these properties are in need of improvement. Uses include upper leather for work and nurses' shoes and hunting boots, lining leather for shoes and hat sweat bands, insole leather, and leather for golf, hockey and fliers' gloves and for garments. William F. Happich and his associates developed, by the use of a combination of glutaraldehyde and chrome, a sheepskin with the wool on (shearling) which is washable, thus making it practical for use as a medical pad by hospitals. A shearling is ideal for the prevention and cure of bed sores.

The mechanism of tanning is presumably the same as that for formaldehyde, namely, cross-linking of polypeptide chains through the epsilon-amino groups of the lysine residues. Spatial relationships and bifunctionality which permit greater crosslinking appear to account in large part for its superiority to formaldehyde.

WALLACE WINDUS

References

Gustavson, K. H., "Chemistry of Tanning Processes," Academic Press, Inc., New York, 1956.
O'Flaherty, F., Roddy, W. T., and Lollar, R. M., "Chemistry and Technology of Leather," Vol. 2, Reinhold Publishing Corp., New York, 1958.
J. Am. Leather Chemists Assoc., **52**, 2 and 17 (1957); **59**, 448 (1964).

TANTALUM AND COMPOUNDS

Tantalum, Ta, atomic number 73, is a metallic element located in Group V of the Periodic Table directly below columbium (niobium), which it closely resembles. Its atomic weight is 180.9479; its specific gravity, 16.6; and its melting point, 2996°C. It has a characteristic bluish-gray color. Only Ta^{181} occurs in nature but several radioactive isotopes have been prepared artificially. Tantalum has a valence of +5 in its important compounds.

In 1802, Ekeberg announced the discovery of a new element which he named tantalum after the mythological Tantalus. The year before, Hatchett had reported his discovery of columbium in a mineral of American origin. It is believed that both men were working with mixtures of the oxides of tantalum and columbium since neither reported the presence of two elements. In 1844, Rose reported that two elements were present in the oxide extracted from columbite. One, he said, was tantalum and the other he named niobium (after Niobe, daughter of Tantalus). Marignac, in 1866, made his important contribution to the chemistry of tantalum and columbium when he separated the two elements by means of the difference in solubilities of their complex fluorides of potassium. Ductile tantalum was first produced in Germany by W. von Bolton in 1903, and became the first practical metallic filament for electric lamps. Tantalum was first produced in the United States in the laboratory of Fansteel Inc. in 1922 by C. W. Balke. It has been in continuous commercial production since that time.

In nature, tantalum is almost always found associated with columbium and its commercially important mineral source is tantalite or columbite. It does not occur in the free state. Tantalite and columbite are variations of the same mineral, $(Fe, Mn)(Ta, Cb)_2O_6$, the choice of name being determined by the predominance of Ta or Cb. Tantalite-columbite is metallic black and produces a streak which varies from light brown to black. The crystal structure is orthorhombic; the hardness is 6 to 6.5; specific gravity varies from 5.2 to 7.8.

Because of its refractory nature, tantalum metal is produced by powder metallurgy rather than by smelting. Tantalum is obtained from the pulverized ore by liquid extraction with hydrofluoric acid and methyl isobutyl-ketone. The metal powder which for many years was produced by the electrolytic reduction of K_2TaF_7 is now usually produced by the sodium reduction of K_2TaF_7 or by the carbothermic reduction of Ta_2O_5. Consolidation is accomplished by pressing and sintering in a vacuum, by electron-beam melting, or by a combination of the two methods. The ingots are forged, rolled or drawn at room temperature.

Shaped articles such as pump or valve parts can be given a corrosion-resistant coating by vapor deposition of tantalum from a mixture of a tantalum halide and hydrogen at about 1000°C.

Reactions. Massive tantalum is completely inert to most corrosive gases and liquids below 200°C, the important exceptions being fluorine, hydrofluoric acid, strong alkali solutions, fuming sulfuric acid (oleum), and concentrated sulfuric acid (above 150°C). It is most easily dissolved in a mixture of nitric and hydrofluoric acids. It reacts with all but the noble gases above 300°C and can absorb more than 700 volumes of hydrogen at red heat. Tantalum powder at high temperatures reacts with carbon to form TaC, with boron to form TaB$_2$ and with oxygen to form Ta$_2$O$_5$. Commercial, fine tantalum powder is not pyrophoric. Tantalum oxide (Ta$_2$O$_5$) is a refractory compound which is formed by igniting in air Ta, TaH, TaC or the hydrated oxide. Tantalic acid (Ta$_2$O$_5 \cdot n$H$_2$O) is formed when tantalum halides or other tantalum salts are hydrolyzed. It is acidic in its reactions and forms orthohexatantalates (4R$_2$O$\cdot$3Ta$_2$O$_5 \cdot n$H$_2$O), and pertantalates (RTaO$_8 \cdot n$H$_2$O). All except the potassium salt (4K$_2$O$\cdot$3Ta$_2$O$_5 \cdot$16H$_2$O) are quite insoluble in water.

The halides of tantalum are formed by direct reaction with the halogens, fluorine at room temperature, chlorine at 200°C, bromine at 300°C and iodine beginning at about 300°C. TaF$_5$ is soluble in water but all the others hydrolyze on contact with water or moist air to form oxyhalides and finally tantalic acid. Tantalum pentafluoride forms a series of double fluorides with alkali metal fluorides, the most important being potassium fluorotantalate, K$_2$TaF$_7$. These complex fluorides are soluble in diluted hydrofluoric acid but hydrolyze in hot water.

Tantalum Compounds. Of the many compounds of tantalum which have been prepared, only three can be considered important items of commerce. Tantalum carbide (TaC) is used in combination with cobalt and tungsten carbide in the manufacture of hard carbide tools and dies. Tantalum oxide, Ta$_2$O$_5$, and potassium fluorotantalate are used in the manufacture of Ta metal.

Alloys. Tantalum forms alloys with columbium, tungsten, zirconium, iron, nickel, cobalt and many other metals. However, it does not alloy with mercury, copper, zinc, calcium and the alkali metals. All known alloys of tantalum except Ta-W alloys containing 10% or less of W are less resistant to corrosion than the unalloyed metal. The important tantalum alloys (other than ferrocolumbium, which is used to inhibit intergranular corrosion in austenitic steels) are T-111 which contains 7–9% W, 2.4% Hf and <40 ppm C, and T-222 with 9.6–11.2% W, 2.2–2.8 Hf and 80–170 ppm C. These alloys are of interest for use in nuclear power systems because of their good creep characteristics and their corrosion resistance to hot liquid metals.

Uses. The uses of tantalum are determined by its unique properties and its relatively high cost. Currently, the principal uses are electronic applications, 70%; chemical equipment, etc., 20%; and aerospace, nuclear, and other applications, 10%.

The tantalum capacitor has had its most important development since the late 1940's and has become the standard for reliable performance for these devices. It answers the demand for miniature units which have high capacitance-to-size ratio, wide useful temperature range, and unlimited standby life. The thin, stable, oxide film has a high dielectric constant and excellent resistance to corrosion by acid electrolytes. Three types of construction are in use: the conventional foil, the porous anode in an aqueous salt or acid electrolyte and the porous anode with solid MnO$_2$ electrolyte. Each type has its own application advantages.

Tantalum sheet and wire are used in the construction of components for special-purpose vacuum tubes. It has good high-temperature strength, low vapor pressure and a scavenging effect which helps maintain low tube pressures during use. The metal can be formed easily into shapes and resists strong acid cleaning solutions.

Tantalum has been used in the manufacture of corrosion-resistant chemical equipment since about 1930. Its high strength, ductility, weldability and remarkable corrosion resistance to acid chemicals make it well suited for this use. The high cost of the metal is compensated by its inertness to acid chemicals, its high heat-transfer rate and its tolerance to extreme temperature changes. Heat-transfer units capable of transmitting 4 million BTU per hour using 150-pound steam are in daily use.

Tantalum is completely inert to body fluids and tissues. In surgery it has been used as suture wire, cranial repair plates and in the form of gauze for abdominal muscle support in hernia repairs. According to Jay A. Nadel, M.D. (*Investigative Radiology*, **3**, 1968) Ta powder has been used in x-ray studies of respiratory systems. Fine Ta powder, insufflated into the airways of the lungs of living dogs, makes possible excellent roentgenograms because of retention of the powder by mucous surfaces and its opacity to x-rays. The powder was mostly eliminated by coughing and ciliary activity within four days without damage to the lungs.

DONALD F. TAYLOR

References

1. Hampel, C. A., Ed., "Rare Metals Handbook," 2nd ed., Van Nostrand Reinhold Co., New York, 1967.
2. Miller, G. L., "Tantalum and Niobium," Academic Press Inc., New York, 1959.
3. Taylor, D. F., "Tantalum and Compounds," Kirk-Othmer, Eds., "Encyclopedia of Chemical Technology," Vol. 19, 630–652, John Wiley & Sons Inc., New York, 1969.
4. Sisco, F. T., and Epremian, Eds., "Columbium and Tantalum," John Wiley & Sons Inc., New York, 1963.

TAUTOMERISM

Certain substances exist under ordinary laboratory conditions as equilibrium mixtures between two or more species. In their chemical reactivity, therefore, they show the behavior to be expected of either or both of the two structures involved. Such substances are said to be tautomeric and the phenomenon is described as tautomerism. Classic examples are the ketone and enol form of acetoacetic ester, quinone monoxime and its tautomer, *p*-nitrosophenol, or the aldehyde and hemiacetal forms of the saccharides. Synonyms for "tautomerism" which have been employed are, among others, "dynamic isomerism," "pseudomerism" and "kryptomerism."

A number of subdivisions of the general phenomena of tautomerism have been made and will be discussed briefly.

Tautomerism involving only changes in the positions of hydrogen atoms and multiple links (the most common kind) is also called "prototropy," i.e., "keto-enol tautomerism" of acetoacetic ester and its relatives. A second example is the "3-carbon-tautomerism" studies by Thorpe, Ingold, Linstead and others such as the tautomerism between α,β- and β,γ-unsaturated acids and related compounds.

"Anionotropy" is used to refer to tautomerism between allylic isomers and related substances when a double bond and an atom or group which can form a stable anion are interchanged in proceeding from one tautomer to the other. Thus, crotyl and methylvinylcarbinyl chloride may sometimes be said to be in tautomeric equilibrium.

Ring-chain tautomerism is used to describe an equilibrium between a cyclic and an acyclic substance.

A fourth subdivision has been made with the use of the term "valency tautomerism" to describe interconversions in which only the bonding (and not the order of attachment) of atoms takes place. Examples are found in the behavior of cyclooctatriene and cycloheptatriene which are found in equilibrium under conditions with their bicyclic "valency tautomers."

It is obvious that the distinction of tautomerism from other types of isomerism is often a subtle one. The word "tautomerism," however, has been retained specifically to refer to isomerism when the isomers are in equilibrium with each other under ordinary laboratory conditions.

D. Y. CURTIN

Cross-reference: *Isomers.*

TELLURIUM AND COMPOUNDS

In 1782, F. J. Muller von Reichenstein extracted from a gold ore a substance which he thought might be an unknown element. He sent a sample to the famous Swedish chemist, Torbern Bergman, for analysis. However, because of the small size of the specimen, Bergman was unable to perform adequate tests. In 1789, Kitaibel, a Hungarian chemist, independently discovered the same element. The element remained unnamed until 1798 when M. H. Klaproth suggested the name tellurium, from the Latin *tellus,* the earth.

For a long time, tellurium was thought to be an exception to the law of periodicity of the elements, because its atomic weight exceeds that of iodine. It was not until 1913 that this discrepancy was explained, when the classic x-ray experiments of Moseley showed that the elements are periodic, not in their atomic weights, but in their "Moseley numbers" or, as they are known today, their atomic numbers.

Tellurium is widely distributed in small amounts in the earth's crust with the average concentrations in crystal rocks of tellurium and its sister elements selenium and sulfur being reported as 0.002, 0.09, and 520 ppm, respectively. Tellurium is most commonly found in small amounts in sulfide deposits in the form of independent minerals such as altaite, $PbTe$; calaverite, $AuTe_2$; coloradoite, $HgTe$; rickardite, Cu_4Te_3; petzite, Ag_3AuTe_2; sylvanite, $(Ag,Au)Te_2$; tetradymite, Bi_2Te_2S. There are also oxidized minerals, including TeO_2 as well as tellurium in native forms.

An important source of tellurium is the sulfide ores of the Canadian Shield, where tellurium occurs as a dispersed element in discrete minerals in forms of chalcopyrite and pyrrhotite. Ores containing

tellurium are also mined in the United States in Montana, Utah, Arizona and New Mexico. In South America the major source of tellurium is the lead and copper ores of Peru.

Most of the Free World's production of tellurium is centered in North America, with the current annual production on this continent being around 350,000 lb. The estimated annual total Free World production is about 500,000 lb.

The principal commercial source of tellurium (and selenium) is the anode slime formed in electrolytic copper refining. Following the removal of copper and selenium, tellurium can be extracted from the slimes by caustic leaching and neutralization of the resulting leach liquor. The precipitate so obtained is an impure tellurium dioxide. This material is further refined by re-solution and re-precipitation, and metallic tellurium produced by a suitable reduction process. A very satisfactory process is one in which tellurium is electrodeposited on stainless steel cathodes from a strongly caustic solution of sodium tellurite. The resulting electrodeposit is removed from the cathodes and cast in suitable form or pulverized for sale as refined or commercial-grade tellurium.

Refined tellurium contains impurities such as lead, copper and iron, as well as selenium and oxygen. This product can be upgraded to high-purity tellurium, successfully by vacuum distillation or zone melting. Although neither vacuum distillation nor zone melting removes selenium, this element can be removed by treatment of molten tellurium with hydrogen, which converts selenium to hydrogen selenide.

Tellurium is the fourth member of the Group VIA of the Periodic System (i.e., the oxygen group). The atomic weight and atomic number are 127.60 and 52, respectively. Eight stable isotopes of tellurium are known having mass numbers of 120, 122, 123, 124, 125, 126, 128 and 130 and abundances of 0.08, 2.46, 0.87, 4.61, 6.99, 18.71, 31.79, and 34.49%, respectively. As would be expected from its position in the periodic chart, tellurium is somewhat similar, chemically, to selenium (the third member of the oxygen group) and, to a lesser degree to sulfur (the second member). The most common oxidation states of tellurium are -2, $+4$, and $+6$. Tellurium melts at 450°C and boils at 990°C.

Tellurium exists in stable form as hexagonal crystalline material having trigonal symmetry. Hexagonal crystalline tellurium is characterized by a marked anisotropy in many of its physical properties including the following: electrical conductivity, linear compressibility, linear coefficient of thermal expansion, optical absorption and index of refraction with respect to polarized light, and galvanomagnetic properties.

Tellurium is soluble in oxidizing acids and caustic. It is insoluble in nonoxidizing acids and in carbon disulfide. Tellurium forms binary tellurides with most of the elements in the Periodic Table. The most common method of synthesis involves heating stoichiometric amounts of the two elements in an evacuated ampoule. Hydrogen telluride, although highly unstable much above 0°C, will precipitate some metal ions from solution as the tellurides. Metal chlorides can yield the corresponding telluride when heated with tellurium in a stream of hydrogen. Ternary tellurides are also formed by heating the elements in question in an evacuated ampoule.

The halides of tellurium range from a colorless gas through a pale colored liquid, and white crystals to deeply colored crystals having a melting point above 360°C. Although these compounds are considered generally to be covalent, there is evidence that $TeCl_4$ exists in an ionized form, $TeCl_3^+Cl^-$, in the solid state. Not all of the halides are formed in which tellurium has a valency of 2, 4 and 6. Thus, tellurium hexafluoride, TeF_6, and ditellurium decafluoride, Te_2F_{10}, are the only hexagonal fluorides formed, while tellurium diiodide, TeI_2, is the only dihalide which has not yet been prepared.

Like sulfur and selenium, tellurium readily forms oxides which can be hydrolyzed to oxyacids. The chemistry of these oxides and oxyacids of tellurium reflects the more metallic nature of the element and its amphoteric behavior.

When tellurium is treated with a strong oxidizing agent such as nitric acid, TeO_2 is formed. A better method of preparation is the burning of tellurium in a mixture of NO_2 and air or oxygen. TeO_2 is a solid at room temperature. It is practically insoluble in water but dissolves in both acids and bases. Salts (tellurites) may be formed by reacting TeO_2 with bases or carbonates. It is difficult to oxidize TeO_2 to TeO_3. The trioxide may be prepared by heating telluric acid which, in turn, is obtained by treating tellurium with strong oxidizing agents, e.g., chloric acid or a mixture of HNO_3 and PbO_2. Tellurium monoxide, TeO, has not yet been formed in the solid state, although there is considerable evidence for its existence in the vapor. The pentoxide, Te_2O_5, has been reported as having been formed by heating orthotelluric acid, or a-TeO_3, at 406° for 25 hours, and is said to be stable at room temperature.

Of the oxyacids, H_2TeO_3, H_6TeO_6, and H_2TeO_5, orthotelluric acid, H_nTeO_n, is of particular interest, in view of the marked difference in its structural and ionization properties from those of the corresponding sulfur and selenium acids and of its tendency to polymerize to form metatelluric acid, $(H_2TeO_4)_n$. In contrast to sulfuric and selenic acids, telluric acid is very weak. The primary and secondary dissociation constants for H_6TeO_6 are 2×10^{-8} and 1×10^{-11}, respectively.

The simplest sulfur compound of tellurium which has been prepared is tellurium sulfide. This compound is formed by reaction between tellurium and the sulfide of zinc, cadmium or mercury. The higher sulfides, TeS_2 and TeS_3, which are analogous to the oxides, are precipitated from tellurite solution by treatment with sodium sulfide or hydrogen sulfide. During formation of the disulfide in basic solution, the ions $TeS_3^=$ and $TeS_2O^=$ are formed. Of the oxysulfides, the red compound $TeSO_3$ is of particular interest since it can be formed by reaction between tellurium and concentrated sulfuric acid and can be used as a qualitative test for tellurium.

The organic compounds of tellurium include tellurols or telluromercaptans, RTeH; tellurides and ditellurides, R_2Te and R_2Te_2; tellurium dihalides, R_2TeX_2; telluroxides, R_2TeO; tellurones, R_2TeO_2; telluronium compounds, $(R_3Te)X$ and $(RTe_2$-

R
|
CHCOOH)X; tellurinic acid and derivatives, RTeOOH; and telluroketones, RC_2Te. Although alkyl tellurols can be synthesized, aromatic tellurols have not yet been isolated.

Considerably less is known about the toxicology of tellurium and its compounds as compared with selenium. In general it appears that tellurium is less toxic than selenium. The principal observation with tellurium is the pronounced garlic breath which may last for a considerable period after exposure. Other symptoms of tellurium intoxication are giddiness, nausea, transient headaches, somnolence, metallic taste and dryness in the mouth. Care should be exercised in handling tellurium, particularly in cases where the formation of fumes of tellurium or tellurium dioxide is possible.

Additions of tellurium in the order of 0.04% by weight to leaded, resulfurized, low-carbon open hearth steels and to high-strength alloy steels at the 150,000 psi tensile strength level produce a very marked improvement in machinability with no deleterious effects on mechanical properties.

Tellurium is a powerful chilling agent in iron castings. This property has been employed to control chill depth and provide a tough abrasion resistant surface. In grey cast iron, tellurium also increases the soundness and reduces subsurface pinholes.

In malleable cast iron the addition of tellurium permits the use of a higher silicon content which results in a much shorter annealing time. In addition, tellurium promotes the formation of spheroidal graphite during annealing thereby improving the ductility.

Tellurium copper containing 0.5% tellurium has been employed for many years in applications, e.g., welding and soldering tips, requiring an alloy having good machinability and high electrical and thermal conductivity. Its relative machinability is 90 compared with 100 for free-cutting brass and both the electrical and thermal conductivity are about 90% that of copper.

Lead alloys containing about 0.05% tellurium and 0.06% copper have a higher fatigue strength and recrystallization temperature than ordinary lead. These properties have led to the application of such alloys in cable sheathing and in chemical equipment.

Copper-, lead-, and tin-base alloys have been developed as bearing alloys for automotive use. Tellurium at a concentration of around 0.1–0.2% appears to promote a more uniform dendritic structure within the alloy.

Tellurium and certain tellurium compounds such as tellurium diethyldithiocarbamate are used in natural and synthetic rubbers as curing agents or accelerators. Tellurium can improve the aging and mechanical properties of the rubber, notably the resistance to heat and abrasion.

Tellurium has been investigated for its catalytic activity particularly in processes where a catalyst of moderate activity but high selectivity is required, e.g., in the controlled oxidation of propylene to acrolein. Traces of tellurium added to platinum catalysts can serve to direct the hydrogenation of nitric oxide to hydroxylamine and preventing the reaction from yielding ammonia.

The most promising applications of semiconducting tellurium compounds lie in the area of thermoelectricity either for power generation as with lead telluride or for refrigeration as with bismuth telluride. These applications are based, respectively, on the Seebeck effect which gives rise to a current upon heating the junction of two dissimilar conductors, e.g., p- and n-type PbTe, and the Peltier effect which results in a heat transfer when a current is passed through the junction of two dissimilar conductors, e.g., p- and n-type Bi_2Te_3. The properties of the tellurium-containing materials which have been developed to date appear to be adequate for a number of applications, e.g., power generation at remote locations for signaling devices and relay stations and thermoelectric cooling for electronic equipment, specialized small refrigerators, and air conditioning.

Many tellurides possess semiconducting properties which have been examined extensively together with thermal, crystallographic and magnetic properties. The optical properties of many of these materials have been investigated.

Of special interest is cadmium telluride which has numerous potential applications as an infrared detector, γ-ray detector and counter and solar cell. Considerable attention is focused on chalcogenide glasses of tellurium such as Te-As-Ge-Si which as amorphous semiconductors have possible uses in switching devices on account of their peculiar current-voltage characteristics. There is little doubt that semiconducting tellurium-bearing materials will find increasing utilization in numerous solid state devices.

RAY E. HEIKS (deceased)
Revised by W. CHARLES COOPER

References

Chizhikov, D. M., and Schastlivyi, F. P., "Tellurium and Tellurides," Collet's Wellingborough, England, 1970.

Cooper, W. C., Editor, "Tellurium," Van Nostrand Reinhold, New York, 1971.

Cooper, W. C., "Tellurium," pp. 694–701, "Encyclopedia of the Chemical Elements," C. A. Hampel, Editor, Van Nostrand Reinhold, New York, 1968.

"Selenium and Tellurium Abstracts," published by the Chemical Abstracts Service of the American Chemical Society for the Selenium-Tellurium Development Association, Inc., 475 Steamboat Road, Greenwich, Connecticut 06830.

Sindeeva, N. D., "Mineralogy and Types of Deposits of Selenium and Tellurium," Interscience, New York, 1964.

TEMPERATURE SCALES

The 1954 Kelvin Scale. In 1954, the International Committee of Weights and Measures adopted as the primary standard of temperature the Kelvin scale as measured by a gas thermometer with a single fixed thermostatic point, the *triple point of water* (obtained experimentally by freezing a portion of the H_2O contained in a sealed evacuated chamber filled only with H_2O liquid and vapor). To this single point, the value 273.16 kelvins (273.16 K) is assigned. (Since 1967, the Committee has adopted the *kelvin,* abbr. K, as the unit of absolute temperature, and of temperature interval, both; the former notations *degree Kelvin* and °K are no longer proper form). The other common temperature scales are now defined in terms of the Kelvin scale by the relations: Celsius, $t(°C) = T(K) - 273.15$; Rankine, $T(°R) = (9/5)T(K)$; Fahrenheit, $t(°F) = T(°R) - 459.67$, all numbers are to be taken as exact.

Historical. The first gas-filled thermoscope was made by Galileo in 1597. Alcohol-in-glass thermometers were in use in the Academy of Florence by the mid-1600's. Huygens in 1665 and Newton in 1680 proposed the ice and steam points as thermostatic fixed points. In 1701 Newton utilized a linseed oil thermometer on which he noted the ice point as 0°, body heat as 12° and observed the steam point at 33–34°. Daniel Fahrenheit of Danzig, an instrument maker active in England and Holland, constructed a permanently sealed mercury-in-glass thermometer in 1714, and lowered his 0° mark to the coldest ice-salt mixture he could obtain, retaining Newton's 12° for body heat under the armpit. Later he divided these graduations by 2, then once again by 4, obtaining a scale with body heat as 96°, on which he observed the ice point at 32° and the steam point at 212°. The belief that Fahrenheit's 180° interval between ice and steam points derived from the 180 degrees of the semicircle has been refuted; his 32° and 212° values were observed. Fahrenheit also observed the supercooling of water, with return to the ice point upon crystallization, and the barometric variations of the steam point. His thermometers became popular because the small size of his degree and the low null point obviated fractional and negative readings in many cases. His scale has persisted in popular and engineering usage in the English-speaking world, and somewhat in Holland, to the present day.

In 1730, René de Réaumur of France introduced a thermometer based on ice point 0°, steam point 80°. This scale persisted in popular, albeit illegal, usage in Germany and France almost to the present day. In 1742 Anders Celsius of Uppsala introduced a thermometer with ice point 100° and steam point 0°, later this was inverted to 0°C and 100°C, respectively, and has become known colloquially as the "centigrade" scale, the preferred scale in general and scientific usage throughout the world. Mercury thermometers were extended to −40°C and 360°C, the approximate melting and boiling points of mercury.

Unavoidable variations in the bore of glass capillaries and the thermal hysteresis of glass made it impossible for any two thermometers to agree exactly, and thus to define an unambiguous scale, except at the fixed points. This difficulty was overcome through the discovery of the gas laws. In gas thermometers, the thermometric quantity is the limiting value at low pressures of the pressure-volume product $(pV)_0$ of a given mass of gas. Experiment has shown that this defines a temperature scale independent of the nature of the gas. William Thomson (Lord Kelvin) showed that the gas thermometric scale coincides with a thermodynamic temperature scale definable from the Carnot efficiency of heat engines, independent of the nature of the working substance of the engine. In 1854, Kelvin suggested that this scale should be based on only a single thermostatic point. The idea was revived in 1939 by W. F. Giauque of Berkeley, and finally adopted internationally in 1954. In the interim, the "old" Kelvin scale was developed with a degree equal in size to the Celsius degree. To establish the "old" Kelvin temperature of any particular thermostatic bath, e.g., boiling benzene (80°C), it became necessary to carry out pV measurements at three temperatures and to solve the equation

$$\frac{T(\text{Benzene B.P.}) \text{ in } °K}{100 \text{ degrees}} =$$

$$\frac{(pV)_0 \text{ at Benzene B.P.}}{(pV)_0 \text{ at steam point} - (pV)_0 \text{ at ice point}}$$

There were numerous difficulties: (1) a change of 1 torr in pressure alters the B.P. of water 0.04°C; (2) the F.P. of air-saturated water (Celsius zero standard) is 0.0024°C lower than the F.P. of air-free water; however, melting ice generates air free water; (3) the relative precision of the denominator must of necessity always be poorer than that of its less precise term (the steam term) by a ratio of about 4 or 5-to-1, regardless of further refinements in technique, i.e., the steam point must be held to better than 0.01°C to obtain a precision of only 0.04 or 0.05 °K in T(ice point) on the "old" Kelvin scale. The result was T(°K) = t(°C) + a constant equal to 267(Gay-Lussac), 273(Regnault), 273.16 (Birge 1941).

The best "old" Kelvin values of the ice point were reviewed in 1954, they ranged from 273.13 to 273.18°K with Dutch and British physicists reporting the lower values, Americans and Germans the higher. A compromise was then adopted: the two fixed point definition was abandoned, and the triple point of water (4.58 torr, 0.0100°C, of which 0.0024° comes from air removal and 0.0076° from the Clapeyron pressure slope for ice/water equilibrium) was substituted for the ice point and assigned the value 273.16 K. The Kelvin temperature of, e.g., boiling benzene, can now be obtained from only two sets of pV measurements via the equation: [T(Benzene B.P.) in kelvins/273.16 K] = [(pV)₀ at Benzene B.P./(pV)₀ at water triple point], with greater precision and less experimental labor.

Fixed Reference Points. For the calibration of thermometers and thermocouples, the following basic reference points are internationally recognized (under a standard atmosphere of 101325 newtons per square meter, except the triple points), together with their Kelvin values: NBP O_2 90.18 K; ice/air saturated water 273.15; TP H_2O 273.16; NBP H_2O 373.15; NBP S 717.75; NMP Sb 903.65; NMP Ag 1233.95; NMP Au 1336.15 K. Secondary reference points: NBP He 4.22 K; NBP H_2 20.37; NBP Ne 27.06; NBP N_2 77.34; NSubP CO_2 194.65; NMP Hg 234.29; transition point $Na_2SO_4 \cdot 10H_2O$ 305.53; TP Benzoic Acid 395.51; NBP Naphthalene 491.11; NMP Sn 505.00; NBP Benzophenone 579.05; NMP Cd 594.05; NMP Pb 600.45; NBP Hg 629.73; NMP Zn 692.65; NMP Al 933.25 K. Interpolation formulas are also provided for use with Pt-resistance

thermometers between the O_2 and Sb points, with Pt/Pt-Rh thermocouples between the Sb and Au points; and with optical pyrometers above the Au point. The precision of these assigned values is approximately as follows: about 0.05 K between the O_2 and ice points, a few thousandths K between the ice and steam points, around 0.1 K at the S point, and around 1 K at the Au point. At higher temperatures, the following Celsius values have also been assigned (1948 Int. C scale): freezing Cu in a reducing atmosphere 1083°C; freezing Ni 1453; freezing Co 1492; freezing Pd 1552; freezing Pt 1769; freezing Rh 1960; freezing Ir 2443; melting W 3380°C.

ANDRE J. DE BETHUNE

References

Encyclopaedia Britannica (1966), articles on "Heat" and "Thermometry."

Partington, J. R., "Advanced Treatise on Physical Chemistry," Vol. I, Longmans Green, London (1949).

Zemansky, Mark W., "Heat and Thermodynamics," 4th Ed., McGraw-Hill, New York (1957).

Preston, T., "Theory of Heat," 4th Ed., Macmillan, London (1929).

American Institute of Physics Handbook, McGraw-Hill, New York (1957).

National Bureau of Standards, Technical News Bulletin, pp. 121–124 (June, 1968).

Gamgee, A., Proc. Camb. Phil. Soc., **7**, Part 3, 95 (1890).

TERPENES-TERPENOIDS

The class of organic compounds known as terpenes is characterized by the presence of the repeating carbon skeleton of isoprene:

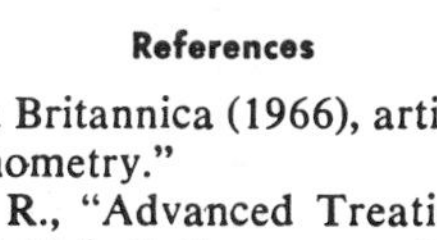

These compounds are widely distributed in nature. The name "terpene" used properly refers to the hydrocarbons which are exact multiples of the skeletal isoprene unit. However the name "terpene," sometimes used loosely, includes not only hydrocarbons but also other functional types of naturally occurring organic compounds which contain the reoccurring isoprene skeleton. In the strictest sense, the names "terpenoid" or "isoprenoid" should be used instead of the more loosely applied usage of terpene.

Terpenoids are divided into subclasses as follows:

Subclass Name	No. of Carbon Atoms	No. of Isoprene Skeletal Units
Hemiterpenoids	C_5	1
Monoterpenoids	C_{10}	2
Sesquiterpenoids	C_{15}	3
Diterpenoids	C_{20}	4
Sesterterpenoids	C_{25}	5
Triterpenoids	C_{30}	6
Tetraterpenoids	C_{40}	8
Polyterpenoids	$(C_5)_n$	n

Together they possess a wide variety of functional groups and structures. Nearly every common functional group is represented. Acyclic, monocyclic and polycyclic structures are observed. The greatest structural variation within a single subclass is to be found among sesquiterpenoids.

The combination of skeletal isoprene units in a regular fashion was exemplified early in the study of terpenoids by nearly all of these compounds. On the basis of this regularity the *regular isoprene rule* was formulated and was taken to mean that terpenoids would possess structures built from a regular "head-to-tail" arrangement of isoprene units. However, as structures of more terpenoids were elucidated, departures from the regular rule were observed. In time "irregular" structures were also accommodated through postulated rearrangement of a regular isoprenoid chain to the "irregular" isoprene skeleton. Rearrangements could occur subsequent to or concommitant with natural cyclization. Thus, the adaptation of the *regular isoprene rule* now finds expression in the *biogenetic isoprene rule*, a rule which is supported experimentally. Particularly significant examples of naturally occurring compounds conforming to the biogenetic isoprene rule are lanosterol, a triterpenoid alcohol associated with cholesterol in wool fat, and gibberellic acid, an important plant-growth regulating substance which is the product of a diterpenoid precursor.

The various terpenoid subclasses are not equally distributed in nature. Representatives of the low-molecular weight end of the terpenoid spectrum are seldom encountered as stable isolable natural products. Isoprene itself has not been detected in plants or animals, but the existence of two highly reactive hemiterpenoid substances in living cells is well established. These are the isomeric γ,γ-dimethylallyl pyrophosphate and isopentenyl pyrophosphate, the two being ubiquitous in living organisms and represent the fundamental isoprene building block in terpenoid biogenesis. One source of evidence for the existence of these hemiterpenoids is the presence of the γ,γ-dimethylallyl unit as a substituent of other classes of natural products, often phenols. The origin of these truly vital hemiterpenoids has been found to be mevalonic acid (3,5-dihydroxy-3-methylpentonoic acid), a six-carbon acid produced by the coenzyme A-assisted condensation of three moles of acetic acid. Decarboxylation and dehydration of mevalonic acid pyrophosphate are known to give the five-carbon unit of isopentenyl pyrophosphate. Isopentenyl pyrophosphate and γ,γ-dimethylallyl pyrophosphate are not only the links between the various subclasses of terpenoids but also biogenetically connect seemingly unrelated plant constituents such as steroids and some types of phenolics and alkaloids.

Monoterpenoids and to a lesser extent sesquiterpenoids are the chief components of the volatile oils readily obtained by the distillation of leaves, wood, and blossoms of a broad array of plants. Sesquiterpenoids are among the most universally distributed natural products. Iridolactone, a monoterpenoid, occurs as a defensive secretion of an ant species belonging to the genus *Iridomyrmex*. The iridoids, which is the general name given to the structural type exemplified by iridolactone, is an important

group of monoterpenoids. Another representative of this group is loganin which along with the amino acid tryptophan provides the carbon atoms of a group of indole alkaloids. Medically important quinine and reserpine are members of this group of alkaloids. The "resin acids" are a group of diterpenoid carboxylic acids which form the major non-volatile part of natural resins often obtained from conifers. Examples are abietic, pimaric and isopimaric acids. Sesterterpenoids, the most recently discovered terpenoid subclass, are produced by insects and fungi. The tricyclic cereoplasteric acid has been isolated from the waxy coating secreted by the insect *Cereoplastes albolineatus*. The fungus responsible for the leaf spot disease in corn produces the acyclic geranylnerolidol and the tricyclic ophiobola-7,18-dien-3α-ol. Triterpenoids are found mainly in plants where they occur in resins and plant sap as free triterpenoids, esters or glycosides. A few are observed in animal sources, for example, the acyclic squalene and the tricyclic lanosterol. The connection between squalene and lanosterol is a vitally important one since these two triterpenoids, along with the hemiterpenoid isopentenyl pyrophosphate, are intermediates in the mevalonic acid based biogenesis of steroids. Tetraterpenoids are frequently referred to as "carotenoids." These constitute a group of natural pigments containing long systems of conjugated carbon-carbon double bonds which are responsible for their color. β-Carotene is the principal pigment of the carrot but this pigment has been isolated from other plant sources also. Finally the presence of polyisoprenoids (natural rubber and gutta-percha) in nature shows that the isoprene unit, like the simple sugar and amino acids, has been used to form linear macromolecules.

The use of terpenoids, usually as mixtures prepared from plants, dates from antiquity. The several "essential oils" produced by distillation of plant parts contained the plant "essences." These oils have been employed in the preparation of perfumes, flavorings, and medicinals. Examples are: oils of clove (local anesthetic in toothache), lemon (flavoring), lavender (perfume), and juniper (diuretic). Usually essential oil production depends on a simple technology which often involves steam distillation of plant material. The perfume industry of Southern France uses somewhat more sophisticated procedures in the isolation of natural flower oils since these oils are heat sensitive. The separation of oils from citrus fruit residues in California and Florida is done by machine.

The oleoresinous exudate or "pitch" of many conifers, but mainly pines, is the raw material for the major products of the naval stores industry. The oleoresin is produced in the epithelial cells which surround the resin canals. When the tree is wounded the resin canals are cut. The pressure of the epithelial cells forces the oleoresin to the surface of the wound where it is collected. The oleoresin is separated into two fractions by steam distillation. The volatile fraction is called "gum turpentine" and contains chiefly a mixture of monoterpenes but a smaller amount of sesquiterpenes is present also. The nonvolatile "gum rosin" consists mainly of the diterpenoid resin acids and smaller amounts of esters, alcohols and steroids. "Wood turpentine," "wood rosin" and a fraction of intermediate volatility, "pine oil" are obtained together by gasoline extraction of the chipped wood of old pine stumps. "Pine oil" 'is largely a mixture of the monoterpenoids terpineol, borneol and fenchyl alcohol. "Sulfate turpentine" and its nonvolatile counterpart, "tall oil," are isolated as by-products of the kraft pulping process. "Tall oil" consists of nearly equal amounts of saponified fatty acid esters and resin acids.

Turpentine is used in syntheses by the chemical and pharmaceutical industries. It also is used as a paint thinner and as a component of polishes and cleaning compounds. Pine oil finds application as a penetrant, wetting agent and preservative, especially by the textile and paper industries, and as an inexpensive deodorant and disinfectant in specialty products. The resin acids are used in the production of ester gum, "Glyptal" resins and are indispensable in paper sizing.

A few individual terpenoids, as well as less expensive mixtures of these compounds, find practical applications. Some examples are: the diterpenoid Vitamin A, the sesquiterpenoid santonin (as an anthelmintic), and the pyrethrins, pyretholone esters of the monoterpenoid chrysanthemic acid (used as an insecticide). A number of sesquiterpenoid lactones of the germacranolide, guaianolide and elemanolide types have shown promise as tumor inhibitors.

The different terpenoid content of plants has served as a finger printing method helpful in botanical identification, especially in cases where differentiation by morphological characteristics has failed. The striking difference in the chemotaxonomy of Jeffery and ponderosa pines serves as an example. The turpentine from the former species consists almost entirely of the paraffinic hydrocarbon n-heptane. Turpentine from ponderosa pine consists largely of the monoterpenes β-pinene and Δ^3-carene.

Besides being of considerable biochemical and botanical interest and of importance in the industrial arts, the food, perfumary, and pharmaceutical trades, the terpenoids have been also a continuing challenge to the organic chemist. The earliest work on the volatile oils was very difficult since the oils were usually complex mixtures and as a consequence individual compounds were isolated as liquids of uncertain purity. Physical constants such as molecular refraction and melting points of solid derivatives, especially those from which the compound could be regenerated were of importance in structural investigations. Organic chemistry relied heavily on oxidative degradation techniques. Dehydrogenation of cyclic terpenoids to aromatic systems and the synthesis of these aromatics played an important role in structure determination. In recent times much of the previous difficulty of obtaining pure samples of terpenoids has been overcome through the use of various chromatographic techniques. Gas-liquid phase chromatography has been used to good advantage in separating both microgram quantities and much larger amounts of the more volatile monoterpenoids and sesquiterpenoids. Much of the type of information formerly obtained only by degradative procedures and dehydrogenations can now be obtained through mass spectrometry. Through nuclear magnetic resonance, infra-

red and ultraviolet spectrometry, structural information can be obtained in a small fraction of the time that was formerly needed to gain the same amount of information.

Acid-promoted cyclization of acylic terpenoids are common. Geraniol, or more readily its trans isomer nerol, can be cyclized with acids to p-menthane derivatives. More importantly, citral, 2,6-dimethyl-2-octen-8-al, when condensed with acetone gives pseudoionone. The latter when cyclized with acid gives a mixture of α- and β-isomers which in turn are used in the preparation of perfumes, as an intermediate in a number of industrial syntheses of Vitamin A, and also in a commercial synthesis of the plant hormone abscisic acid. Polyclic terpenoids are prone to rearrangement of the carbon skeleton. The acid catalyzed rearrangement of the monoterpene camphene to derivatives of isobornyl alcohol are well known and have been the subject of extensive theoretical studies. Diterpenoids and triterpenoids undergo "backbone" rearrangement through the migration of hydride and methyl groups. Frequently these migrations are stereospecific and are acid promoted.

Studies of terpenoid chemistry have also involved syntheses. Several of the complex sesquiterpenoid structures have been confirmed or, in some cases, correctly established through synthesis. Many elegant new general synthetic methods have been developed as a result of attempts to synthesize terpenoids. The chemistry of terpenoids present a continually growing area of chemical research, perhaps the equal of any in complexity, subtlety, and variety.

ROBERT T. LALONDE

References

Ruzicka, L., "History of the Isoprene Rule," *Proceedings of the Chemical Society,* 341 (1959).

Corforth, J. W., "Terpenoid Biosynthesis," *Chemistry in Britain,* **4,** 102 (1968).

Simonsen, J. L., "The Terpenes," Vol. I (1947) and Vol. II (1949), Second Edition, Vol. III (1952) Vol. IV and V (1957).

TEXTILE PROCESSING

Textile chemistry in its broadest meaning includes chemical processing, properties, compositions, and changes therein, of fibers, yarns, fabrics, and the lubricants, dyestuffs, sizes, finishes, detergents, and other materials deliberately, naturally or accidentally applied thereto.

Fibers can be classed according to their chemical composition as organic or inorganic. They also are classified by source as natural and man-made. (See **Fibers, Natural and Synthetic**)

Man-made fibers as delivered to textile processing plants usually need little or no scouring. Natural fibers usually must be freed from contaminants before they can be converted into yarns and fabrics. Historically, the cleaning of natural fibers is more probably the origin of modern chemistry than is the frequently cited alchemy of the Middle Ages. Textile chemistry is a direct descendant of the textile art of prehistoric man.

Scouring, or cleaning, of the raw material fibers is a process, the origin of which is lost in prehistoric time. Linen is an example of this. Somewhere, sometime, in the development of civilization someone observed that when a stalk of the flax plant soaked in water for a long time it lost its woody material and became a bundle of soft pliable fibers. To basket weavers the plaiting of these fibers into long strands was an easy step. By the time man learned how to twist fibers into yarn he also knew how to clean flax by retting (rotting).

The exact stage in man's progress at which he learned about alkalies is unknown. However, all primitive dyeing techniques had a preliminary step in which the raw fiber was steeped in water made alkaline with the leachings of wood ash. This solution was used by the Chinese as early as 2000 B.C. to remove gum from silk. To the primitive degumming bath was added powdered dried sea worms. This last provided some proteolytic enzymes while furnishing protective protein. This art is duplicated, in its components, in modern silk degumming baths based on extended chemical research and development.

The scouring of wool, according to Pliny, depended on soapwort. Other sources indicate that dung and putrid urine were also widely used. In fact the use of urine persisted into the late 19th century. Fullers earth, gypsum and absorbent clays also were used in historic times to remove the natural grease, wax and soil from wool.

Freezing the wool grease and fracturing it off the wool by beating has been used as have conventional dry-cleaning solvents to scour wool.

Soap used as a pomade by the Gauls was probably not used as a scouring assistant for textile until about 200 A.D. At this time it began to replace soapwort in wool scouring. Its use was limited to fine wools for some centuries. By the 15th Century soap was used in degumming silk. In the 16th Century it was known to have been used to scour cotton and as a bleaching assistant. From that time to the mid-thirties of this century soap has been the major cleansing agent for all fiber.

The new synthetic detergents have replaced soap because of their resistance to pH changes, greater efficiency, stability in presence of metallic ions and lower processing costs (See **Detergents**).

Water is the medium of textile wet processing and its supply and purification is an important part of textile chemistry and processing. (See **Water Conditioning**).

Fibers which have been retted, degummed, scoured or otherwise cleaned are ready to be twisted into yarns and woven or knitted into fabrics. Those fibers such as silk which are of extremely long lengths are called filaments and require only twisting together to produce textile yarns.

The short fibers such as cotton or wool are known as staple fibers. They are characterized by short, easily measurable lengths usually less than eight inches. Staple fibers can be made by cutting filaments into short lengths. Regardless of origin, staple fibers must be processed in a complex series of mechanical operations to produce textile yarns. They are first separated, straightened in the initial steps of opening and carding. To facilitate this process oils and lubricants (carding assistants) are added to reduce fiber surface friction. Some of these ASSISTANTS or AUXILIARIES are subsequently

removed while others stay in place into the final consumer item.

Textile auxiliaries and assistants include natural and synthetic:

1. Fats, oils, and waxes used for lubricating fibers, yarns, and fabrics.
2. Film-forming and adhesive materials such as starches, gums, resins, classed as "slashing" compounds and used in temporary protective coatings ("sizings," or "dressings"), as well as thermosetting resins used to produce "wash and wear" and "permanent crease" finishes.
3. Surfactants, such as soaps, detergents, wetting and penetrating agents, softeners, used in scouring or improving the "feel" or appearance of textiles. (See **Aminoplast Resins, Detergents, Gums, Starch**)

The carded straightened fibers can be aligned in more parallel arrangement by combing or they can be twisted into yarn directly from the carding machine.

The twisted yarn, whether from staple fibers or filaments, is usually "sized" or "slashed" with a protective coating before being woven or knitted into a fabric.

There is a group of fabrics which do not require yarns for their production, i.e., the so-called nonwoven fabrics. The oldest form of these is felt. The practice of working fibers together, without spinning, into a dense uniform mass is called felting, and a very old art. It is known to most of the wandering herdsmen of the world. It has a chemical background, particularly in the fur felt industry. Here, nitric acid was originally used to improve the felting of furs. In the 19th Century mercuric chloride and arsenic were added to improve the process. Today textile chemistry has eliminated these poisonous materials and replaced them with hydrogen peroxide. Recently, nonwovens have been bonded without felting by use of thermosensitive resins and fibers. The fabrics can be bleached, dyed, printed and finished in the same manner as woven fabrics.

Once the fabric is completed, the sizing or slashing compound is usually removed in the first or scouring step of the fabric finishing process. Soaps or detergents in water are used to remove the oils, waxes, sizes and stains. Water-soluble materials are no problem. Starch and glue sizes can be removed by enzyme conversion to water-soluble materials. (See **Enzymes**). Solvent-soluble materials require special scouring baths containing dispersions of, or consisting entirely of organic solvents.

The washed fabrics can be dried and sold as is or, more likely, be mercerized, bleached, dyed, printed, coated and otherwise treated to develop desirable physical appearance and properties.

Cotton and other cellulosic yarns and fabrics can be "mercerized," or treated with 18–20% sodium hydroxide at cool or room temperatures. This shrinks and increases the strength of cotton yarns and fabrics. In 1890 Lowe showed that mercerizing while yarn or fabric was kept from shrinking developed a very lustrous fabric. This was "mercerizing" until very recently when cotton "stretch" fabrics have been made by slack mercerization. These fabrics are characterized by "controlled elasticity" which is developed by the degree and direction of shrinking permitted in the caustic step.

Stretch fabrics are also produced by the use of small amounts of truly elastic yarns such as natural and synthetic rubbers or polyurethanes, in combination with considerably larger amounts of non-elastic yarns. Stretch fabrics are also made with textured yarns. These are continuous filament thermoplastic man-made yarns which have been twisted, folded, strained or knit into compacted shapes, and thermally set in those shapes. The set yarns are then opened up by untwisting, deknitting or stretching. Like coiled or flat springs they tend to return to the set, compacted shape and show elastic recovery. Staple fiber yarns impregnated with thermosetting resins and similarly treated develop similar elastic recovery. To develop "stretch" fabrics the yarns are woven or knitted under tension into open fabrics which are permitted to contract in finishing. The amount of contraction or shrinkage developed in finishing is the reverse of "stretch" in the finished fabric.

When fabrics have been washed, bleached, mercerized, dyed or printed it is the usual practice to improve their appearance and feel by "finishing."

At one time finishing included stiffening with starches and gums, weighting with clays and metal salts, drying to width and pressing or calendering the fabrics to produce a smooth, glossy surface. These procedures are still important, but advances in textile chemistry have produced an entirely new group of textile finishes and materials. This group includes the thermosetting resins which can be formed or cured in place in the fabric while the fabric is held extended, carefully folded, or pressed firmly between smooth or engraved surfaces. On extended fabrics resins can produce stability, crease resistance, water repellency and similar effects. When the fabric is folded, calendered, pressed or polished, the resins result in durable pleats, permanent embossed designs or long-lasting glazes.

On light-weight fabrics the resins provide slip and fray resistance. The resins used are primarily aminoplasts. "Wash and wear" fabrics are the products.

A recent development in this treatment has been the application of the resin to the fabric in the finishing plant with curing of the resin only after the fabric has been tailored into a garment. The curing is completed after the garment has been pressed or pleated in final form. "Permanent Press" is the result.

Mixtures of natural and man-made-thermoplastic fibers also can be used to produce wash and wear fabrics.

Textile finishing can *dull* or *deluster* the fabric by formation in place of barium or calcium sulfate, by inclusion of finely divided pigments in the film of the finishing agent. Clays and chalk also dull but are used primarily as "fillers" or "weighters" to reduce the cost per pound of the finished fabric.

An important continuing development in textile processing is the application of resins and dyestuffs from organic solvents. The use of such solvents, and the proper recovery, offers a probable way to reduce air and water pollution. In addition it permits the use of water sensitive agents in treating textiles.

Most textile materials are subject to attack by insects, fungi and microbials. To stop or retard such attacks fungicides (mildew-proofing), moth-proofing agents and bacteriacidal agents can be included in finishing compounds. (See **Fungicides, Pesticides**)

"Waterproofing" of fabrics is a very old finishing process. Originally natural waxes and tars were used. Today, waxes, metallic soaps, natural and man-made rubbers, fatty pyridinium salts are used to coat, saturate, or only surface fabrics to develop water proofing or water repellency. (See **Metallic Soaps, Rubber**)

Where human comfort is a factor, waterproofed fabrics should be air-permeable. The wax-metallic soap and the pyridinium compounds do this very well. Recently microporous coatings of various elastomers have become important because of their almost complete water resistance and satisfactory air permeability.

                                                    MARTIN H. GURLEY, JR.

THALLIUM AND COMPOUNDS

In the middle of the Nineteenth century thallium was discovered by W. Crooke through the use of spectroscopy. He named the element from the bright green line on the spectroscope, since *thallus* in Latin refers to the bright green tint of new vegetation. The abundance of thallium in the earth's crust is estimated to be the same order of magnitude as mercury. It is one of the less known metals because of its limited application.

The first important application of thallium was discovered in 1920 when a German company introduced a proprietary rat poison, whose active ingredient was thallium sulfate. The consumption of metallic thallium is not published; however, its availability is sufficient and should satisfy any new use which does not exceed several thousand pounds annually.

Although thallium is present in greater amounts in the earth's crust in potash minerals, it is almost exclusively recovered from the sulfide minerals during the smelting operations of lead and zinc. The principal domestic producer is the American Smelting and Refining Company which recovers thallium from the cadmium-containing flue dusts treated at the Globe Plant in Denver, Colorado. As previously stated, the thallium compounds are concentrated in the flue dusts of lead and zinc smelting operations. The extraction of thallium from flue dusts is dependent upon the solubility of thallium and its compounds (oxides and sulfates) in hot water. The purification of thallium is also accomplished by taking advantage of the difference in solubility in water of thallium compounds and compounds of the impurities.

The metal may be produced by the electrolysis of compounds such as carbornate, sulfate and perchlorate or by precipitation of metallic thallium by zinc.

Thallium (Tl) is in Group IIIA of the Periodic Table along with boron, aluminum, gallium and indium. The element usually has a bluish-grey tinge, but a freshly cut surface will have a metallic luster which rapidly dulls on exposure to air. It is somewhat softer than lead and harder than indium which has a Brinell hardness of 1; the relationship of the hardness of these elements is approximately $H_{Pb}:H_{Tl}:H_{In}$ 4:2:1. The metal has high malleability but low tensile properties.

The crystal structure is hexagonal close-pack (lattice constants are a=3.457Å, c=5.525Å. There is an allotropic transformation from the hexagonal close-pack structure to body center cubic structure which is stable above 230°C.

Thallium is a superconductor at temperatures in the region of liquid helium similar to its close neighbors Pb and Hg. Below 2.32°K all measurable resistance disappears.

Measuring the change of volume in metals during solidification, Endo found a 3.23% change in volume for thallium during solidification.

The cubic compressibility as determined by Richards and White at 20°C over a pressure range of 100–500 megabars is 2.83×10^{-6} for thallium as compared to 2.40×10^{-6} for lead.

As may be seen in the following table, thallium resembles lead in a number of physical properties.

	Thallium	Lead
Atomic no.	81	82
Atomic weight	204.37	207.2
Density—20°C g/cm³	11.85	11.34
Density—20°C lb/in³	0.4281	0.4097
Atomic volume cm³/g-atom	17.24	18.27
Melting point	303°C	327.4°C
Boiling point	1457°C	1737°C
Specific heat— 20°C cal/g/°C	0.031	0.031
Heat of fusion—cal/g	5.04	5.89

On exposure to air thallium will readily oxidize to Tl_2O. In water containing air it is oxidized to thallium hydroxide, but in oxygen-free water it does not react at ordinary temperatures. Two groups of compounds are formed—thallous, with a valence of 1 and thallic with a valence of 3. The more numerous and stable salts are of the thallous group. Thallium is dissolved by nitric acid, dilute sulfuric and very slowly by hydrochoric acid. The acetate, nitrite, nitrate, perchlorate, hydroxide, carbonate, sulfate and ferricyanide are water-soluble compounds.

Thallous chloride is only slightly soluble in cold water, whereas the chlorides of the common impurities, zinc, iron, copper and cadmium, are soluble. Thallium sulfate may readily be prepared from the chlorides and other salts or by dissolving the metal in concentrated sulfuric acid. Hydrogen sulfide will precipitate Tl_2S from acetic acid, alkaline or neutral solutions but not from solutions containing mineral acids. Thallium compounds, particularly thallium sulfate, have been used for the destruction of rodents and ants. Water solutions of thallous formate-malonate mixtures are used for high-density sink-float separations.

Certain thallium compounds, such as thallium oxysulfide and mixed thallium bromide iodide crystals have found application in detecting radiation in the infrared range. A thallium sulfide cell has been used as a semiconductor device, which depends on the change of electrical resistance with the amount of radiation falling on it. Thallium oxide resistive glazes have been patented for use in printed circuit electronics. Thallium in combination with arsenic and sulfur forms low melting point glasses which have been proposed for some semiconductor applications.

Although the alloy systems of thallium have been widely investigated, there are at present no important commercial alloys. Thallium readily alloys with most elements and forms eutectic systems with a number. An unusually low (minus 60°C) melting point results from the alloying of mercury and thallium (8½% thallium). A ternary eutectic alloy of thallium (14.3%) mercury (69.7%) and indium (16.0%) has a freezing point of minus 63°C. This alloy has been proposed for use in thermometers and thermostats. It also has been recommended for use in applications where low resistance electrical and thermal contacts are difficult to make.

H. E. Howe, revised by S. C. Carapella, Jr.

Reference

Howe, H. E., "Thallium," pp. 706–711, in "Encyclopedia of the Chemical Elements," C. A. Hampel, Editor, Van Nostrand Reinhold Co., New York, 1968.

THERMOCHEMISTRY

Thermochemistry is concerned with the measurement and interpretation of heat changes which accompany changes of state and chemical reactions. According to the first law of thermodynamics (the law of the conservation of energy) the energy of a system is a single-valued function of its state. Therefore the change in energy associated with the conversion of the system from one state to another is independent of the path followed by the system during the conversion.

It is impossible to define the total absolute energy of a system; we are restricted to considerations of changes in energy content. It is, nevertheless, convenient to speak of the energy content of a system in a certain state, meaning thereby the energy necessary to bring the system to that state from some other state which is arbitrarily selected as a reference state and in which the energy of the system is set equal to zero.

The energy changes associated with chemical processes in most cases involve energy in the form of heat and mechanical or electrical work. If work is performed during a change from an initial to a final state, the heat effect accompanying the change is not equal to the total energy change. If the heat *absorbed* by the system is Q and the work *performed* by the system is W, the *increase* in energy is

$$\Delta E = E_2 - E_1 = Q - W \qquad (1)$$

where E_1 and E_2 are, respectively, the energy contents of the system in the initial and final states. In most cases W is the work of expansion or contraction of the system under a finite pressure, and for a reversible process then takes the form

$$W = \int_{V_1}^{V_2} P dV \qquad (2)$$

where P and V are pressure and the volume of the system. In the important special case of a reversible process taking place at constant pressure this becomes

$$W = PV_2 - PV_1 = P\Delta V \qquad (3)$$

Thus, from equation (1), the heat absorbed by the system during such a process is

$$Q = (E_2 + PV_2) - (E_1 + PV_1) \qquad (4)$$

The property $E + PV$ is represented by H, the *enthalpy*. Since H_1 and H_2 are dependent only on the initial and final states, it is evidently permissible to speak of the change in enthalpy (ΔH) accompanying *any* process, regardless of whether the initial and final pressures are the same or the process is carried out reversibly. The enthalpy change in a process for which the initial and final pressures are equal is called the heat absorbed "at constant pressure,"

$$Q_p = H_2 - H_1 = \Delta H = \Delta E + P\Delta V \qquad (5)$$

without in any way implying that the pressure remains constant throughout the process. The heat absorbed in a process for which the initial and final volumes are equal,

$$Q_v = E_2 - E_1 = \Delta E \qquad (6)$$

is similarly called the heat absorbed "at constant volume." In processes involving only condensed phases at moderate pressures the work term $P\Delta V$ is usually negligible. ΔH and ΔE are quantities of primary interest in thermochemistry.

Thermochemical data are expressed either in absolute joules (one joule is the energy dissipation per second in one absolute ohm when one absolute ampere is flowing), calories (one calorie equals, by definition, 4.1840 absolute joules), or kilocalories (1000 calories). The unit of mass is usually the mole, except in certain macromolecular systems where the mole is indeterminate. In such systems, the unit of mass is the gram, or a mole of some arbitrarily specified submolecular unit.

The enthalpy change in a chemical reaction is equal to the sum of the enthalpies of the products less that of the reactants, each substance being in an appropriately specified state. This quantity can in some cases be evaluated by direct observation of the heat absorbed when the reaction is carried out in a calorimeter. More frequently the reaction is not susceptible to direct calorimetric observation, and indirect methods have to be employed. The most important of these employ (a) heats of formation, and (b) equilibrium data.

The *standard enthalpy of formation* of a substance is defined as the heat absorbed when it is formed from its elements in their standard states, usually in an isothermal process. Enthalpies of formation are in most cases derived from heats of combustion, including the heats of combustion of graphite and hydrogen. The enthalpy change in a reaction is then given by

$$\Delta H = \sum_{\text{products}} \Delta H_f - \sum_{\text{reactants}} \Delta H_f \qquad (7)$$

In the case of a reversible reaction van't Hoff's law states that

$$\Delta H° = RT^2 \left(\frac{\partial \ln K}{\partial T}\right)_p \qquad (8)$$

where K is the equilibrium constant for the reaction. The superscript zero indicates that the enthalpy change refers to the reaction with reactants and products all in their standard states. The selection of standard states is based primarily on convenience; for example, for reactions in solution the standard state of each participating solute is usually taken as the hypothetical state of unit activity at a concentration of one molal or one molar. Two points may be noted: (1) ΔH values are generally insensitive to activity variations, so that in practice $\Delta H°$ will usually be indistinguishable from ΔH at reasonable concentrations of reactants and products; and (2) since equilibrium measurements *per se* give no information concerning the size of the mole, nonthermodynamic information is required for establishing the amount of reaction to which the value of $\Delta H°$ refers. This latter consideration is particularly important in connection with reactions involving macromolecules.

The quantities

$$C_v = (\partial E / \partial T)_v; \quad C_p = (\partial H / \partial T)_p \qquad (9)$$

are called respectively the heat capacity at constant volume and the heat capacity at constant pressure. If in heating a substance in a calorimeter from temperature T_1 to temperature T_2 at constant pressure, Q calories are absorbed per mole, the mean heat capacity over this temperature range is

$$\widetilde{C}_p = Q/(T_2 - T_1) = (H_2 - H_1)/(T_2 - T_1) \qquad (10)$$

If the temperature interval does not exceed a few degrees, $\widetilde{C}_p$ may be safely identified with C_p. The observed heat absorption must be corrected for the heat capacity of the calorimeter, and for any change in the distribution of the substance between phases.

The enthalpy change in a reaction varies with temperature according to the equation

$$(\partial(\Delta H)/\partial T); = \Delta C_p \qquad (11)$$

where

$$\Delta C_p = \sum_{\text{products}} C_p - \sum_{\text{reactants}} C_p \qquad (12)$$

ΔC_p may be expressed, over more or less restricted temperature ranges, by an empirical equation of the form

$$\Delta C_p = a + bT + cT^2 \qquad (13)$$

Equation (11) then gives on integration

$$\Delta H - \Delta H_0 = aT + \tfrac{1}{2}bT^2 + \tfrac{1}{3}cT^3 \qquad (14)$$

where ΔH_0 is an integration constant.

For any infinitesimal process carried out reversibly at constant pressure at temperature T, the entropy change dS is

$$dS = dH/T \qquad (15)$$

From equation (9) it follows, after integration,

that

$$S - S_o = \int_0^T (C_p/T)\,dT$$

where C_p is the heat capacity at constant pressure. If the system under consideration is composed of a pure substance which forms perfect crystals at $T = 0$, the third law of thermodynamics states that $S_o = 0$. If the substance undergoes phase transitions in the temperature range of interest, equation (16) must be suitably modified. Entropies are additive, so that the entropy change in a reaction is given by

$$\Delta S = \sum_{\text{products}} S - \sum_{\text{reactants}} S \qquad (17)$$

For an isothermal process

$$\Delta G = \Delta H - T\Delta S \qquad (18)$$

where ΔG is the free energy change. It is thus possible to evaluate free energy changes by purely thermochemical measurements. For a reversible reaction

$$\Delta G^\circ = -RT \ln K \qquad (19)$$

where ΔG° is the free energy change when reactants and products are all in their standard states.

For reactions involving substances in solutions and other mixtures, the thermodynamic properties are functions of the composition. Consideration of these complications is beyond the scope of this discussion.

The bulk of thermochemical data is derived from calorimetric measurements. Because of the difficulties inherent in the control and measurement of heat, the design of calorimetric equipment is extremely varied and illustrates many ingenious modifications.

JULIAN M. STURTEVANT

Cross-references: *Calorimetry; Thermodynamics.*

References

Lewis, G. N., and Randall, M., "Thermodynamics," revised by K. S. Pitzer and L. Brewer, McGraw-Hill Book Co., New York, 1961.

McCullough, J. P., and Scott, D. W., eds., "Experimental Thermodynamics," vol I, Butterworths, London, 1968.

Rossini, F. D., ed., "Experimental Thermochemistry," vol I, Interscience Publishers, New York, 1956.

Skinner, H. A., ed., "Experimental Thermochemistry," Vol. II, Interscience Publishers, New York, 1962.

Sturtevant, J. M., "Calorimetry," in "Physical Methods of Chemistry," A. Weissberger and B. W. Rossiter, eds., Interscience Publishers, New York, 1971.

THERMODYNAMICS

Thermodynamics treats systems whose states are determined by thermal parameters, such as temperature, in addition to mechanical and electromagnetic parameters. By *system* we mean a geo-metric section of the universe whose boundaries may be fixed or varied, and which may contain matter or energy or both. The *state* of a system is a reproducible condition, defined by assigning fixed numerical values to the measurable attributes of the system. These attributes may be wholly reproduced as soon as a fraction of them have been reproduced. In this case the fractional number of attributes determines the state, and is referred to as the *number of variables of state* or the *number of degrees of freedom* of the system.

The concept of *temperature* can be evolved as soon as a means is available for determining when a body is "hotter" or "colder." Such means might involve the measurement of a physical parameter such as the volume of a given mass of the body. When a "hotter" body, A, is placed in contact with a "colder" body, B, it is observed that A becomes "colder" and B "hotter." When no further changes occur, and the joint system involving the two bodies has come to equilibrium, the two bodies are said to have the same temperature. It is a fact of experience that two bodies which have been shown to be individually in equilibrium with a third, will be in equilibrium when placed in contact with each other, i.e., will have the same temperature. This statement is sometimes called the *zeroth law* of thermodynamics. The physical parameter used to specify the "hotness" of the third body might be adopted as a quantitative measure of temperature, in which case, that body becomes the thermometer.

From what has been said above it is apparent that temperature can only be measured at equilibrium. Therefore thermodynamics is a science of equilibrium, and a thermodynamic state is necessarily an equilibrium state. Furthermore, it is a macroscopic discipline, dealing only with the properties of matter and energy in bulk, and does not recognize atomic and molecular structure. Although severely limited in this respect, it has the advantage of being completely insensitive to any change in our ideas concerning molecular phenomena, so that its laws have broad and permanent generality. Its chief service is to provide mathematical relations between the measurable parameters of a system in equilibrium so that, for example, a variable like the pressure may be computed when the temperature is known, and *vice versa*.

It provides these relations with the aid of three postulates or *laws* (in addition to the zeroth law) which we shall now proceed to explain.

The *first law* of thermodynamics may be expressed in the following form,

$$dE = Dq - Dw \qquad (1)$$

where dE represents a differential change in the *internal energy* of a system during some change defined by the passage of the system from one thermodyamic state to another. Dq and Dw are, respectively, the *heat* absorbed by the system and the *work* done by the system on its environment during the differential change. The small d preceding E in (1) implies that dE is an exact differential, or that E depends only on the state of the system, and is independent of the path of the change. The large D in Dq and Dw implies that in general this is not true for the *heat* and *work*.

To clearly understand what is meant by *heat* and *work* and to gain further insight into the meaning of (1), consider a system set up in such a manner that it can exchange energy only with its surroundings in the form of well defined mechanical or electrical work. Such an arrangement might involve surrounding the system by an *adiabatic* wall. Then it is observed that the system always undergoes the same change when a given amount of work is given to it or extracted from it. This prompts us to define a quantity, E, the internal energy, to be associated with each state of the system, whose change (and therefore whose value to within a constant) can be measured by the amount of work passing between system and environment when the change of state is performed under adiabatic conditions.

If, now, the same change of state (defined by the thermodynamic states between which the system passes) is conducted under nonadiabatic conditions, it is found that the work involved does not equal the previously measured change in E. Conservation of energy then demands that an exchange take place between the system and its environment which is not recognizable as work. This quantity of energy is defined as *heat,* and is nothing more than a discrepancy term designed to transform (1) into the equality which it is. Heat, therefore, has no measurable existence of its own. It is incorrect to say that a body contains a particular amount of heat, even though the quantity of heat passing in or out of a system is measurable because the work done by the system and its internal energy can both be measured.

The *second law* of thermodynamics may be expressed as follows,

$$T\,dS - Dq \geq 0 \tag{2}$$

Here T stands for the temperature and S denotes the *entropy* of the system. Dq has the same meaning as in (1). To fully understand the entropy as well as the inequality in (2) it is necessary to understand *reversible* and *irreversible* processes. A *reversible* process is one in which an infinitesimal change in driving force of the process completely reverses the process in all of its detailed aspects. For example, consider a gas enclosed in a cylinder by a weighted frictionless piston. Let the pressure exerted by the weighted piston be just insufficient (by an infinitesimal amount) to contain the gas in equilibrium. A slow expansion occurs which can be completely reversed by an infinitesimal change in driving force, i.e., by an infinitesimal increase in the weight of the piston.

Another example of a reversible process, more chemical in nature, is the slow discharge of an electrochemical cell, maintained slow by the action of an impressed electromotive force. A slight increase in this electromotive force not only reverses the flow of current, but also the chemical reaction of discharge, and causes a slow electrolysis to occur.

A distinguishing feature of a reversible process is its physical impossibility. It occurs infinitely slowly and may be aptly described as a sequence of equilibrium states. Nevertheless, calculations of phenomena as they would occur, were reversibility truly possible, are feasible (for example, it is possible to calculate the volume work performed by a gas as it passes through a sequence of equilibrium states or the electrical work performed by a cell as it discharges through a sequence of equilibrium states), and it is this aspect which lends importance to the concept of reversibility.

By contrast, an *irreversible* process is any real process, occurring at a finite rate, and with any degree of violence. In (2) the inequality refers to an irreversible, and the equality to a reversible change of state. As indicated by the use of d in front of S, the differential of S is exact, so that S is a function only of the thermodynamic state of the system. This is part of the postulate. Equation (2) also specifies how S is to be measured, for if the change of state is conducted reversibly,

$$dS = \frac{Dq}{T} \tag{3}$$

Measurement of the heat absorbed by a system in a change conducted reversibly then yields dS directly by division by T.

An immediate consequence of (2) is that in a system isolated from its surroundings, so that Dq is zero, any irreversible change is necessarily accompanied by an increase in entropy, while for a reversible change no variation in entropy occurs. Since irreversible changes are to be associated with real changes this is tantamount to the assertion that the entropy of an isolated system increases for real spontaneous changes so that it is maximized when the system achieves equilibrium, i.e., when no further real spontaneous change is possible. In line with this assertion, (2) shows that $dS = 0$, for a reversible change, i.e., one involving a system in equilibrium.

The quantity S seems somewhat more abstract than E. Logically, however, the two are on the same basis. E is generally more familiar than S since the latter is a thermal quantity, associated specifically with thermodynamics. A nonthermodynamic, molecular interpretation of S can be given. This asserts that S measures the logarithm of the relative probability of a given state. It is not surprising, then, that S increases in a spontaneous process.

In its early development thermodynamics was an engineering subject. A good deal of its inquiries were directed toward the efficiency of machines. The inequality, (2), tells immediately how to get the most work out of a system. Thus, by substituting the value of Dq, obtained through solution of (1), the following result is obtained,

$$DW \leq TdS - dE \tag{4}$$

the inequality still referring to an irreversible process. Since the right member contains exact differentials only, its value is determined by the states between which the change occurs rather than by the path of the change. Therefore it is apparent that the left side (the work performed by the system) is maximized when the equality holds, i.e., when the system operates reversibly or when the change is conducted over such a path that it occurs reversibly.

A machine which converts heat into work generally operates in cycles, abstracting an amount of heat, Q_2, from a reservoir at a higher temperature, T_2, performing a certain amount of work, W, and returning an amount of heat, $-Q_1$, to a reservoir at the temperature, T_1. The efficiency, η, of the machine is defined as the fraction of the amount of heat absorbed at the higher temperature which is converted into work during one cycle. Thus

$$\eta = W/Q_2 \qquad (5)$$

After one cycle the machine returns to its original state, and, therefore the total changes in its internal energy, E, and entropy, S, are zero. From what has been said above, the maximum efficiency is achieved when the machine operates reversibly. But, then, since dS is zero and the total change of entropy is Q_2/T_2 plus Q_1/T_1, we have

$$\frac{Q_2}{T_2} + \frac{Q_1}{T_1} = 0 \qquad (6)$$

Furthermore, since dE is also zero, equation (1) demands that

$$Q_1 + Q_2 = W \qquad (7)$$

The simultaneous solution of (6) and (7) yields

$$\eta = \frac{W}{Q_2} = \frac{T_2 - T_1}{T_2} \qquad (8)$$

This shows that no machine operating in cycles can be 100% efficient, i.e., that it is impossible to convert heat entirely into work without effecting permanent changes. The latter statement is often taken as an alternative form of the second law.

The universal efficiency η, depending only, as it does, on the temperatures of the two reservoirs, can be used to define the thermodynamic scale of temperature. Thus, two reservoirs have the ratio of thermodynamic temperatures

$$\frac{K_2}{K_1} = \frac{1}{1-\eta} \qquad (9)$$

where η is the efficiency of a machine (any machine) operating reversibly between them. If to (9) we add the requirement

$$K_2 - K_1 = 100 \qquad (10)$$

when the high-temperature reservoir is boiling water, and the low temperature is a mixture of ice and water, then K proves to be identical with T, the Kelvin temperature. The thermodynamic temperature has the advantage of being independent of the properties of any special substance.

The *third law* of thermodynamics asserts that *the entropy of a system at the absolute zero of temperature (T or $K = 0$) is zero, provided that the system is in its lowest energy state.* In reality this cannot be said to be strictly a law of thermodynamics, since it presumes an acquaintance with the detailed structure of the system, especially in regard to its spectrum of energy states. Despite these non-thermodynamic overtones, the third law is extremely useful in many applications. For example, if a system is observed to undergo a change at absolute zero involving a loss of entropy, one may conclude that originally it was not in its lowest energy state. Another application involves the calculation of the equilibrium constant of a chemical system from purely thermal measurements. In

this brief exposition it is impossible to discuss the third law in greater detail.

Some of the most important applications of thermodynamics are achieved by substituting Dq from (1) into (2), i.e., by combination of the first and second laws. The result is

$$D\phi = dE + DW - T\,dS \leq 0 \qquad (11)$$

where the inequality still stands for an irreversible spontaneous process. $D\phi$ is used as a shorthand for the three terms on its right. $D\phi$ assumes a special form depending upon the work which the system is capable of doing. For example, if the system can only do volume work,

$$DW = p\,dV \qquad (12)$$

where p is the pressure and V the volume of the system. Then

$$D\phi = dE + p\,dV - T\,ds \leq 0 \qquad (13)$$

It is convenient to find some function ψ, dependent only the thermodynamic state of the system which imitates $D\phi$ when certain restricted changes of state are carried out. For example, let $\psi = E$. E initiates $D\phi$ for changes of state in which V and S are maintained constant, since then dV and dS are zero. Suppose the initial state of the system is one of equilibrium. Then no spontaneous change can occur and all possible changes are the reverse of spontaneous ones. Then

$$D\phi = (dE)_{s,v} > 0 \qquad (14)$$

for all possible changes, where the subscripts S and V indicate the maintained constancy of those variables. Equation (14) implies that if the initial state is one of equilibrium E is a minimum for all displacements along paths of constant entropy and volume.

The displacement involved is *not* one leading *out of equilibrium*, for then functions like S could not be given experimental meaning along its path. Rather, additional forces are added to the system against which the system can perform other than volume work, so that a new state of equilibrium is reached subject to these new forces. This will be elaborated below.

Another choice of ψ might be

$$\psi = F = E + pV - TS \qquad (15)$$

the so-called Gibbs' free energy. This imitates $D\phi$ along paths of constant temperature and pressure, and can be shown (in the same manner employed in the case of E) to be minimized along these paths in a state of equilibrium. This minimization may be expressed by

$$(dF)_{T,p} = 0 \qquad (16)$$

Here again a displacement from equilibrium is not implied. Rather an additional force is added leading to a new equilibrium. For example, suppose the system represents a solution involving a chemical reaction. If the reaction is set up in a chemical cell, no electromotive force will persist when chemical equilibrium has been achieved. The state of the system is determined by variables of state including temperature, pressure and composition.

If, now, a new force is added in the form of an impressed electromotive force electrolysis occurs and a new state of equilibrium is achieved subject to the magnitude of the impressed electromotive force. The variables of state are greater by one (the E.M.F.) than in the original equilibrium state. It is to this kind of displacement that (16) refers.

It can be shown that the change in F, ΔF, can be measured by the electrical work performed in the displacement. Equation (16) can be used as a differential equation connecting the variables of state in the initial equilibrium state. So can (14), for that matter, by writing dF more explicitly in terms of the variations of variables of state.

In this manner it can be shown that when a solute is distributed between the two phases 1 and 2, at equilibrium, and F_1 and F_2 are the free energies of the two phases, then

$$\left(\frac{\partial F_1}{\partial n_1^{\,1}}\right)_{T,p,n^1} = \left(\frac{\partial F_2}{\partial n_1^{\,2}}\right)_{T,p,n^2} \tag{17}$$

where $n_1^{\,1}$ and $n_1^{\,2}$ are the numbers of moles of the solute in phases 1 and 2, respectively, and n^1 and n^2 represent the mole numbers of other components. For many solutes it is observed that

$$\left(\frac{\partial F}{\partial n}\right)_{T,p} = K + RT \ln c \tag{18}$$

where K and R are constants at any one temperature and pressure and c is the concentration of the solute. Substitution of (18) into (17) with appropriate attachment of subscripts yields

$$\frac{c_1^{\,2}}{c_1^{\,1}} = \exp\left[-\left(\frac{K_2 - K_1}{RT}\right)\right] = K \tag{19}$$

and defines the partition coefficient k.

In closing, we note that since F is minimized at constant temperature and pressure in the equilibrium state, any reaction system which is in a state in which F is not minimized at the temperature and pressure in question drifts toward the state in which it is. A knowledge of F as a function of composition thus permits the prediction of the direction in which a particular reaction will proceed.

Many other applications are possible, for example, other thermodynamic potentials like F and E can be invented with consequent utility.

HOWARD REISS

Cross-references: *Thermochemistry, Stoichiometry.*

THIOLS

A thiol (mercaptan) can be regarded as a compound formed by the replacement of a hydrogen atom of hydrogen sulfide by an aliphatic, aromatic or heterocyclic radical. Thiols are represented by the general formula, R—S—H. The —S—H group is known as a sulfhydryl group as well as a mercapto group. Simple thiols are named as follows: CH_3—S—H, methanethiol; H—S—CH_2—CH_2—CH_2—S—H, 1,2-ethanedithiol; the name mercapto may be used as a prefix for polyfunctional compounds such as H—S—CH_2—CO_2H, mercaptoacetic acid.

Thiols are less soluble in water and possess lower boiling points than the corresponding alcohols (R—O—H). This can be attributed to the inability of sulfur to form strong hydrogen bonds as compared to the oxygen atom in alcohols. Thiols show a weak absorption band in the infrared at 2600–2550 cm^{-1}. The low molecular weight thiols have an extremely disagreeable odor and are toxic as is hydrogen sulfide. The nose can detect one volume of ethanethiol in 50 billion volumes of air.

A variety of thiols occur in crude petroleum fractions. Besides having a disagreeable odor these substances may poison metal catalysts when the petroleum is refined. Butanethiol is found as a component of skunk secretion. The aroma of thiols such as propanethiol in onions or 2-propene-1-thiol in garlic may result from enzymatic action during processing. Thiols contribute to the aroma of various foods such as beer, cooked vegetables, bakery products, cheese and coffee.

The preparation of alkanethiols is accomplished: (1) by reaction of alkyl halides or sulfates (RX) with sodium hydrogen sulfide (NaSH) in refluxing alcohol solution; (2) by reduction of disulfides with finely divided zinc by boiling with 50% sulfuric acid-50% water; (3) by hydrolysis of alkyl dithiocarbamates, which are made by reaction of carbon disulfide and ammonia (to form the ammonium salt of dithiocarbamic acid) followed by reaction with alkyl halides; (4) basic hydrolysis of S-alkylthioureas, which are made by reaction of thiourea and alkyl halides; and (5) by the reaction of Grignard reagents with sulfur.

(1) R—X + NaSH → R—SH + NaX

(2) R—S—S—R + Zn/H_2SO_4 →
$$2R\text{—SH} + ZnSO_4$$

(3) CS_2 + 2NH_3 → NH_2—$\overset{\displaystyle S}{\overset{\|}{C}}$—S$^-$ NH_4^+

NH_2—$\overset{\displaystyle S}{\overset{\|}{C}}$—S$^-NH_4^+$ + R—X →
$$NH_2\text{—}\overset{\displaystyle S}{\overset{\|}{C}}\text{—S—R} + NH_4X$$

NH_2—$\overset{\displaystyle S}{\overset{\|}{C}}$—S—R + H_2O → R—SH

(4) NH_2—$\overset{\displaystyle S}{\overset{\|}{C}}$—$NH_2$ + R—X → NH_2—$\overset{\displaystyle S}{\overset{\|}{C}}$—S—R $\;\overset{\displaystyle NH_2^+\ X^-}{}$

NH_2—$\overset{\displaystyle S}{\overset{\|}{C}}$—S—R $\overset{\displaystyle NH_2^+\ X^-}{}$ + NaOH/H_2O →
$$R\text{—SH} + NH_2\text{—}\overset{\displaystyle O}{\overset{\|}{C}}\text{—}NH_2$$

(5) R—MgX + S → RSMgX

R—S—MgX + H_2O → R—SH

Two important processes have been developed for the commercial manufacture of thiols. These involve the reaction of hydrogen sulfide with olefins or alcohols. Depending upon the type of catalyst employed, the type of thiol formed can be controlled:

$$R—CH{=}CH_2 + H_2S \xrightarrow[\text{catalysis}]{\text{acid}} R—\underset{\underset{SH}{|}}{CH}—CH_3$$

$$R—CH{=}CH_2 + H_2S \xrightarrow[\text{catalysis}]{\text{free-radical}}$$
$$R—CH_2—CH_2—SH$$

Potassium tungstate on alumina is an effective catalyst in the vapor phase reaction of an alcohol with hydrogen sulfide to yield thiols.

Thiols are weak acids and form salts (mercaptides) when treated with aqueous solutions of strong bases. The mercaptides formed from heavy metals (mercury, lead, silver) are insoluble in water. The ready formation of insoluble mercury salts gave rise to the name mercaptan (L. *mercurium captans,* seizing mercury). Standardized aqueous silver nitrate is one method that can be used for the determination of thiols. The mercaptide ion like the sulfide ion reacts with aqueous sodium nitroprusside to give a purple color which is used as a qualitative test for the presence of a sulfhydryl group.

The sulfur-hydrogen bond is much weaker than the oxygen-hydrogen bond in alcohols, therefore, it is not surprising that thiols are readily oxidized to disulfides:

$$\underset{\text{thiol}}{2RS—H} \xrightarrow[\text{agent}]{\text{oxidizing}} \underset{\text{disulfide}}{R—S—SR}$$

Mild oxidizing agents (iodine, dimethyl sulfoxide, oxygen in the presence of amines) are used since strong oxidizing agents (nitric acid, potassium permanganate, chromic acid) oxidize disulfides with cleavage of the S—S bond to give sulfonic acids ($R—SO_3H$). The conversion of a thiol to a disulfide is frequently carried out by adding iodine to an alcoholic or acetic acid solution of the thiol. This procedure can be used as a quantitative method for the determination of thiols. Other reagents using the oxidation of thiols to disulfides have recently been developed and are frequently used in protein chemistry. A common reagent used is Ellman's Reagent 5,5'-dithio-bis-(2-nitrobenzoic acid) which reacts in aqueous media at pH 8 with sulfhydryl groups to yield a highly colored anion which can be analyzed spectrophotometrically. The sweetening process for gasoline (doctor process) utilizes the oxidation of thiols to the less odorous disulfides by means of sodium plumbite solution containing a small amount of free sulfur.

Thioesters are formed by the reaction of acid anhydrides or acyl halides with thiols. This is the usual procedure because of the unfavorable position of the equilibrium when thiols are treated with carboxylic acids.

$$R—SH + (R'CO)_2O \rightarrow R—S—\overset{O}{\overset{||}{C}}—R' + R'CO_2H$$
$$\text{thioester}$$

$$R—SH + R'COCl \rightarrow R—S—\overset{O}{\overset{||}{C}}—R' + HCl$$

The acid-catalyzed reaction of thiols with aldehydes and ketones leads to the formation of thioacetals and thioketals. These are often useful intermediates in organic synthesis involving aldehydes and ketones.

$$2R—SH + R'CHO \rightarrow \underset{\text{thioacetal}}{R'—CH{-}(SR)_2} + HOH$$

$$2R—SH + R'_2CO \rightarrow \underset{\text{thioketal}}{R'_2—C{-}(SR)_2} + HOH$$

The reaction of thiols with olefins leads to the formation of thioethers. Depending upon the reaction conditions (ionic or free-radical) the type of thioether formed can be controlled.

$$R—S—H + R'—CH{=}CH_2 \xrightarrow[\text{catalysis}]{\text{acid}}$$
$$R—\underset{\underset{CH_3}{|}}{S}—CH—R'$$

$$R—S—H + R'—CH{=}CH_2 \xrightarrow[\text{conditions}]{\text{free-radical}}$$
$$R—S—CH_2—CH_2—R'$$

The sulfhydryl group plays an important role in the chemistry of living systems. The activity of various enzymes, the structural aspects of proteins, and reactivity of biochemical substrates are dependent on the presence of sulfhydryl groups. Some important biochemical compounds containing the —SH group are coenzyme A, pantheine, dimercaptolipoic acid, glutathione, and cysteine.

Acyl derivatives (thioesters) of coenzyme A, especially acetyl coenzyme A, are important products of lipid (fat) and carbohydrate metabolism. For the cell to use fatty acids as a source of energy, the fatty acid must be converted to acetyl coenzyme A which is further metabolized to carbon dioxide and water via the citric acid cycle (Krebs cycle).

The amino acid cysteine is responsible for a large fraction of the total sulfur found in proteins. Cysteine is readily oxidized to cystine with the formation of a disulfide bond.

$$2HS—CH_2—\underset{\underset{\underset{\text{cysteine}}{NH_2}}{|}}{CH}—CO_2H \rightarrow$$

$$\underset{\underset{\underset{\text{cystine}}{NH_2}}{|}}{(S—CH_2—CH}—CO_2H)_2$$

This facile interconversion is responsible for the "permanent wave" of hair. Hair is composed of the protein keratin which is 15–20% cystine. The disulfide bonds present in keratin are responsible for its structural properties. The "permanent wave" requires the reduction of the disulfide bonds of the keratin, shaping the hair (keratin) and then oxidation of the sulfhydryl groups to form new disulfide bonds.

All cells contain glutathione, a tripeptide containing cysteine. It can exist as reduced glutathione which contains a free —SH group or as oxidized glutathione which has a disulfide bond. Glutathione may function as a reducing agent in cells to keep the many sulfhydryl-containing enzymes in the reduced state.

Uses. Some important industrial uses for thiols include emulsion-polymerization modifiers, polymer stabilizers, intermediates in the manufacture of agricultural chemicals (insecticides, herbicides and fungicides), odorants in the gas industry and rubber vulcanization accelerators. Pharmaceutically important thiols include 2,3-dimercaptopropanol which is valuable in the treatment of arsenical and mercurial poisoning; 3-mercaptovaline which is used in the treatment of Wilson's disease; and 6-mercaptopurine which is used in the treatment of leukemia.

MARVIN R. BOOTS AND SHARON G. BOOTS

References

Noller, C. R., "Chemistry of Organic Compounds," 3rd., W. B. Saunders Company, Philadelphia, 1965.

Kharasch, N., Ed., "Organic Sulfur Compounds," Vol. 1, Pergamon Press, Inc., New York, 1961.

Kharasch, N., and Meyers, C. Y., Eds., "Organic Sulfur Compounds," Vol. 2, Pergamon Press, Inc., New York, 1966.

White, A., Handler, P., and Smith, E. L., "Principles of Biochemistry," 4th ed., McGraw-Hill, New York, 1968.

Turk, S. D., in "Kirk-Othmer, Encyclopedia of Chemical Technology," Vol. 20, H. F. Mark, J. J. McKetta, Jr., D. F. Othmer and A. Standen, Eds., 2nd ed., pp. 205–218, Interscience, New York, 1969.

Maricin R. Boots, Ph.D., Associate Professor, Department of Chemistry and Pharmaceutical Chemistry, Virginia Commonwealth University, Richmond, Va. 23219

Sharon G. Boots, Ph.D., Research Fellow, Oittre.

THIXOTROPY

Substances that coagulate to a gel when left standing undisturbed but which liquefy again when agitated by stirring or vibration are said to be thixotropic. In its broadest sense the term includes all cases of isothermal, reversible work-softening; it was first applied, however, to the isothermal, reversible sol-gel-sol transformation. The term "thixotropy" was derived by Peterfi and Freundlich from the Greek words *thixis*, meaning the touch or the striking, and *trepo*, to turn or to change. Familiar substances exhibiting thixotropy are colloidal suspensions of montmorillonite clay, certain paints and printer's inks, household shortening and some shaving creams; many other thixotropic systems are known, however, and it is thought that in certain circumstances protoplasm, blood and other biological fluids may be thixotropic.

Thixotropy is only one of several rheological phenomena characterized by a time dependency; moreover, as exemplified by the tendency of ferric- and aluminum-oxide gels to crystallize on long standing, in many thixotropic systems still other time-dependent phenomena may be in progress concurrently. There is, therefore, some confusion surrounding the use of the term. The reversibility and essentially isothermal nature of the sol-gel-sol transformation serve to differentiate thixotropic gels from those that can be reliquefied by heating. These criteria also distinguish thixotropy from ordinary strain-softening. For the kind of "imperfect thixotropy" characterized by an undisturbed state that is not a true gel and a disturbed state that is not a truly fluid sol, the term "false body" has been proposed. This distinction has most often been attempted in the literature of paint technology. In some cases thixotropy and "false body" may result from phenomena sufficiently different to warrant differentiation; usually, however, a clear distinction is not possible and in its broadest meaning thixotropy includes all such cases of isothermal, reversible work-softening.

At this point it is appropriate to differentiate thixotropy from its near opposite: dilatancy. In this case the contrast is sharp. A dilatant substance is one that develops increasing resistance to deformation or flow with increasing rate of shear. Reversible work-hardening is a term often used synonymously for dilatancy, but strict definitions of these two terms have yet to be agreed upon by rheologists. The classic example of dilatancy is the familiar phenomenon of wet sand becoming firm and appearing to dry suddenly when walked upon. This effect is caused by a shear-produced increase in volume of the initially close-packed, water-saturated sand. Many quicksands, on the other hand, are water-saturated mixtures of clay and sand that are rendered thixotropic by the clay present. Some clay suspensions have been found to be dilatant, however. A common dilatant substance that may be found in most homes is a thick dispersion of ordinary starch.

The diversity of thixotropic systems is now known to be so great that it is doubtful that a single descriptive theory can be devised to encompass them all. In general, however, the phenomenon is limited to aerosols and colloidal suspensions. Particle shape is important and thixotropy generally is enhanced if the suspended particles are anisometric. Particle shapes of common thixotropic clays include flat lamella, discs, plates, thin sheets rolled into tubes, and lath-like forms. In gels of high polymers the particles commonly are fibrillar and may have many branches. Other important factors are the pH of the system and the types and concentrations of electrolytes present. Many nonthixotropic suspensions may be rendered thixotropic by properly adjusting the pH of the system or by adding a suitable electrolyte in the proper amount. For a given system it will generally be found that the gel strength and the time of gelation can be optimized by adjusting the concentration of a suitable electrolyte. For clay suspensions with negatively charged micelles, the hydroxides and halides of lithium and sodium offer the most possibilities; in this case the nature of the cation is the most important factor. For positively charged micelles it is the nature of the anion in the electrolyte that is most important.

Although there are undoubtedly exceptions, in general the sol state may be visualized on the molecular level as a suspension of colloidal particles oscillating in Brownian motion. Translatory and rotary motions produce interparticle collisions of which some result in more or less temporary interparticle bonds. Gelation proceeds as more and more such interparticle linkages are formed. As the suspended particles coagulate into a more or less continuous network, Brownian motion is observed to be greatly diminished or to cease altogether. Subsequent liquefaction is accomplished by any mechanical disturbance of an amplitude sufficient to rupture the interparticle bonds.

Whether or not the particles in the gel actually contact each other is a question very early propounded by Freundlich. Except for a few gels of colloidal graphite in which interparticle contacts seem to have been proved, this question for most systems remains unresolved. The nature of the interparticle bonds also is still imperfectly understood, although it is generally agreed that electrostatic and van der Waals attractive forces and the osmotic effects of the diffuse electric double layer of the colloidal micelles certainly are all involved. In some cases, particularly in aqueous clay suspensions, there is evidence that the suspending fluid may participate in the interparticle bonding and contribute to the gel strength. In thixotropic systems of high polymers steric interaction undoubtedly is a factor important in determining gel strength.

Thixotropy was first studied and evaluated by determining the time of gelation. This is still considered to be an important property of thixotropic gels, but thixotropy is now most often studied in rotational or falling ball viscometers that permit evaluation of additional properties of the system. In either of these instruments the same sample may be retained throughout the study but because of greater facility in applying theory to measurement, the rotational viscometer usually is preferred. In this instrument the rate of shear is determined as a function of shear stress and time under specified conditions of temperature and pressure. Thixotropy is revealed by hysterisis loops in plots of the resulting data.

Some thixotropic suspensions gel more rapidly if they are rolled or tapped gently—a phenomenon called rheopexy. Thixotropic suspensions almost invariably gel faster as the temperature is raised. This can be attributed to the increase in frequency of interparticle collisions caused by the effect of temperature on Brownian motion. The effect of gently rolling or tapping rheopectic suspensions may be similar but it is also possible that the "activation energy" probably associated with gelation may in part be supplied by such treatment.

D. M. ANDERSON

Cross-references: *Clay, Colloid Chemistry, Rheology.*

THORIUM AND COMPOUNDS

Thorium is the 90th element in the Periodic System and is classed as the first in the actinide series, which is characterized by the filling of the 5+ electron shell. It is the second heaviest of the naturally occurring elements with an atomic weight of 232.0381 and with essentially one isotope, Th^{232}. In chemical combination it is normally a quadrivalent cation. A very electropositive element, it lies between beryllium and magnesium in reducing strength. For its potential ($Th^0 = Th^{4+}$) a value of $-1.899v$ has been calculated. Thorium was discovered in 1828 by J. J. Berzelius in a mineral (later named thorite) from Brevig, Norway.

Its abundance in the earth's crust is estimated to be 12 ppm, widely scattered throughout the world. Usually it is found in igneous rocks in concentrations too low to be of interest economically. The most important mineral commercially is monazite sand, a mixture of rare earth phosphates containing

thoria. It is found in India (seacoast of Travancore), Brazil, and the Carolinas; domestic deposits are known in Florida, California, and Idaho. Typical thorium contents of the different monazite concentrates are: India 9–10%, Brazil 6–7%, and domestic 3–4%. Other thorium minerals are: thorite, a silicate similar to zircon; thorianite, mixed oxides containing uranium (Ceylon); Aeschynite, columbates and titanates of cerium and thorium; and the element is a minor constituent of many rare earth minerals.

Uses. The first large use of thorium was in the Wellsbach mantle, based on the brilliant white luminescence produced by heating a mixture of cerium and thorium oxides. A woven fabric (long fiber cotton, ramie, or artificial silk) is treated with a solution of thorium and cerium nitrates to give an ignited mixture of 99% ThO_2 and 1% CeO_2 by weight. The cerium concentration is quite critical to attain the peak luminosity. Phosphorus and silica contents must be low in fabric and oxides or brittleness occurs. Additions of magnesium and beryllium nitrates are made to improve ash strength. A more voluminous ash is produced by addition of 0.5–1% thorium sulfate. The use of thoria in the Wellsbach mantle is rapidly declining.

Increasing use is being made of thorium in electronic work. The metal is an excellent "getter" for the removal of oxygen and nitrogen in discharge tubes, and as it has a low work function, consequently, the starting potential of a thorium electrode is low.

Thoriated tungsten cathodes also show improved thermionic emission, owing, it is believed, to the formation of a monatomic thorium layer. Additionally, the thorium deoxidizes the tungsten and by forming a layer of thoria at the tungsten grain boundaries inhibits grain growth at high temperatures. This is particularly important in preventing the embrittlement of tungsten incandescent lamp filaments.

The addition of small amounts of ThO_2 to various metals such as nickel has an excellent dispersion-strengthening effect. The beneficial increase in mechanical properties is especially evident in the creep and stress-rupture values for these mixtures.

Thorium has been found to increase the high-temperature mechanical properties of magnesium alloys. As little as 2% thorium increases the creep resistance properties of magnesium alloys.

Perhaps the largest potential use for thorium is as a fertile material in breeder nuclear reactors. Its use in breeder reactors would increase potential fuel reserves tremendously since fissionable material (U^{233}) can be obtained from nonfissionable Th^{232} by the following reaction:

$$_{90}Th^{232} + _0n^1 \rightarrow$$

$$_{90}Th^{232} \xrightarrow{-\beta^-} _{91}Pa^{233} \xrightarrow{-\beta^-} _{92}U^{233}$$

Other uses for thorium are: (1) in high-temperature (above 1900°C) refractories, either as thorium oxide (ThO_2) or thorium sulfide (ThS_2); (2) as catalyst in Fischer-Tropsch process (thoria mixed with cobalt, magnesia and kieselguhr); (3) as catalyst (thoria) for organic synthesis as well as reactions such as $NH_3 \rightarrow HNO_3$, $CO_2 \rightarrow CO$, and $SO_2 \rightarrow SO_3$; (4) for sun lamp electrodes as source of ultraviolet radiations thorium carbide is mixed with iron carbide; (5) as metal in photoelectric cells sensitive to certain ultraviolet regions; (6) as metal target material in x-ray tubes for a monochromatic x-ray source.

Properties. The appearance of massive metal is silvery-white, turning gray on air exposure. Densities have been reported from 11.2–11.7 g/cc (11.72 g/cc calculated from x-ray data). Thorium has a melting point of 1755 ± 10°C. As a metal it has good working qualities and can be treated as mild steel. The massive metal is stable in air but as a powder it is pyrophoric. It is quite reactive combining with hydrogen (300–400°C), nitrogen (650°C), halogens (450°C), sulfur (450°C), carbon, and phosphorus. Hydrochloric acid attacks the metal vigorously leaving a rather large residue (12–25%). With nitric acid the metal shows an initial vigorous reaction but soon becomes passive. There is a slow attack of the metal with hydrogen evolution by dilute hydrofluoric, nitric, and sulfuric acids, and by concentrated perchloric and phosphoric acids. Complete and rapid dissolution of the metal in hydrochloric or nitric acid is obtained by use of catalytic amounts of fluoride or fluosilicate. Alkali hydroxides do not attack it.

Metallic thorium exists in two allotropic forms. At room temperature and up to 1380°C it has a face-centered cubic structure; above 1380°C it is body-centered cubic. The boiling point is estimated to lie in the region 3500–4200°C.

Compounds. Thorium compounds are colorless except for a colored anion. Thorium compounds and those of quadrivalent cerium and zirconium are similar in many respects. Crystalline thorium compounds are highly hydrated, containing 4 to 12 moles of water per mole. In solution, thorium ion has a coordination number of 8 and is heavily hydrated (15 to 20 moles of water per mole are reported for the nitrate). There are many double salts of thorium, especially with the alkali metals and ammonia. Those of the halides, sulfates, and nitrates have been actively studied. Complexes of thorium in aqueous solution form readily with ions such as: fluoride, iodate, sulfate, carbonate, oxalate, tartrate, citrate, pyrophosphate, molybdate, and many others. Hydrolysis of thorium salts occurs, but is less than for other quadrivalent cations.

The oxide, ThO_2, (m.p. 3220 ± 10°C) is an extremely refractory material, quite resistant to acids and alkalies.

Thorium nitrate, $Th(NO_3)_4$, has many hydrates, $12H_2O$, $6H_2O$, $4H_2O$, and $2H_2O$. The anhydrous salt has been prepared. It is an extremely soluble salt, not only in water but also in such organic compounds as alcohols, ethers, ketones, and esters. This latter property has been utilized in purification methods.

The halides, ThX_4, and the oxyhalides are known. Except for the fluoride they are soluble salts.

Thorium sulfate, $Th(SO_4)_2$, has several hydrates, 8, 6, 4, and 2 moles of water, and an anhydrous form. A basic salt, $ThO(SO_4)$, has been studied.

The existence of thorium carbonate has been questioned. The hydrated basic carbonate, $ThOCO_3$, is known and is quite insoluble in water. However, with excess carbonate a complex is formed which is water-soluble.

Water-insoluble compounds important in analytical chemistry are: thorium fluoride, ThF_4, thorium peroxide, Th_2O_7, thorium oxalate, $Th(C_2O_4)_2$, thorium molybdate, $Th(MoO_4)_2$, and thorium hydroxide, $Th(OH)_4$.

Other compounds are: hydrides, ThH and ThH_4; sulfides, ThS, ThS_2, Th_2S_3, and Th_7S_{12}; oxysulfide, $ThOS$; carbides, ThC, and ThC_2; phosphides, ThP and Th_3P_4; nitrides, Th_2N_3 and Th_3N_4; sulfite, ortho-, meta-, and pyrophosphates; chromate; molybdate; boride, selinide; silicide; and telluride.

At the persent time thorium and its compounds are of interest to the United States Government through the Atomic Energy Commission. Most thorium research in the United States is therefore under its supervision as well as the dissemination of all information thereby gained.

ROBERT L. MARTIN

Cross-references: *Fission, Radioactivity.*

References

Cuthbert, F. L., "Thorium Production Technology," Addison-Wesley, 1958.
"The Metal Thorium," Proceedings of the Conference on Thorium, American Society for Metals, 1958.
Wilhelm, H. A., "Thorium," pp. 711–718, in "Encyclopedia of the Chemical Elements," C. A. Hampel, Editor, Van Nostrand Reinhold Co., New York, 1968.

TIN AND COMPOUNDS

Tin, Sn, atomic number 50, atomic weight 118.69 is in Group IVA of the Periodic Table. There are ten naturally occurring isotopes. The four valence electrons of tin occupy the $5s^2$ and $5p^2$ orbitals. Accordingly, tin forms two series of compounds, the stannous tin series in which tin has an oxidation number of +2 after losing only the $5p$ electrons, and the stannic tin series with an oxidation number of +4 corresponding to the loss or sharing of all four outer electrons.

Sources. The major free world tin-producing countries in order of their production are: Malaysia (41%), Bolivia (16%), Thailand (11%), Indonesia (9%), Nigeria (5%), Congo (5%), and Australia (5%). Production in China and Russia is substantial. No workable deposits have been found in the United States.

Cassiterite, SnO_2, is the main source of tin; 80% of the world's tin ore occurs in low-grade alluvial deposits. The principal method of mining is bucket-line dredging in a placer but gravel pump, hydraulic mining and various open-pit methods are also used.

Underground lode deposits in Bolivia and Cornwall are recovered by blasting, drilling, crushing, and grinding followed by gravity separation and smelting. Refined tin from the more than 25 world smelters has a minimum purity of 99.8% tin. Mining, benefication, smelting and refining processes are discussed in *The Encyclopedia of the Chemical Elements*(3).

Physical Properties. Tin has a density of 7.29 g/cc in the familiar beta form (white tin), and the alpha allotrope which forms at well below 0°C has a density of 5.77 g/cc. The melting point is 231.9°C, boiling point 2270°C, electrical resistivity 11 microhm-cm at 0°C, thermal conductivity 0.145 cal/cm/cm²/sec/°C at 100°C, specific heat 0.053 cal/g at 25°C, latent heat of fusion 14.2 cal/g, and latent heat of vaporization 520 cal/g.

The technological value of tin is dependent in large part on a blending of many characteristic properties. For example, the low melting point, 232°C, of tin allows it to be melted in iron pots with simple heating. Casting qualities are good in either metal or rubber molds. Cast bars or slabs are readily rolled into sheet, extruded into tubes or pipes, or drawn into wire because it is soft and pliable. In the molten state it wets and adheres to most common metals. Liquid tin is highly fluid, drains rapidly leaving a thin even coating on fabricated articles which is silvery-white in color, lustrous and nontoxic. A tin coating has both lubricating properties and conformability which aid in forming and drawing operations. A tin coating makes possible the joining of metals which are difficult to solder. Tin in the liquid state has a low surface tension and surface smoothness and these characteristics combined with its high boiling point, 2270°C, are fully utilized in the production of "float glass." Tin coatings can be applied to most metals by electrodeposition. Tin readily forms alloys with other metals imparting hardness and strength to the alloys. Tin as a minor constituent in alloy systems (cast iron, titanium, zirconium) can have specific and unusual effects such as improving the wear resistance, high-temperature strength and corrosion resistance. For additional data on physical, mechanical, and thermal properties see Faulkner(1) and Hedges(2).

Chemical Properties. Oxygen or dry air combine with tin to form an invisible and protective oxide film on the surface. The oxide film thickens and has a yellowish tinge as the temperature is raised. In the absence of oxygen (as for example in a vacuum-packed food container) a reversal of potential of the tin-iron couple occurs making the tin anodic to the steel. If a small amount of tin dissolves, it is nontoxic and tasteless whereas a corresponding amount of dissolved iron would affect the appearance and flavor of the food. Tin does not react directly with nitrogen, hydrogen, carbon dioxide, hydrogen sulfide or gaseous ammonia. This is important in melting and alloying practices and in exposure to polluted air. Sulfur dioxide, when moist, attacks tin as do the halogens. Tin reacts with both strong acids and strong bases, but it is relatively resistant to weak acids and to neutral salt solutions. Nonaqueous solutions of lubricating oils, organic solvents and gasoline have little effect on tin. Tin forms an unusually wide range of both inorganic and organic compounds. These are discussed below.

Major Industrial Applications. Tinplate provides an outlet for half the primary tin used in the United States (25,000 long tons). Tinplate, which is low-carbon steel strip rolled to 0.01–0.006 inch and thinly coated with tin by electrodeposition, is the ideal metal container material(2,4).

About two-thirds of the primary and secondary tin is used in the United States in the form of tin alloys(2). Tin when alloyed with lead forms a group of alloys known as *soft solders*. Moving the percentage of tin up or down alters the melting characteristics and uses of the alloys. The tin-base solders have no serious competitors in the field of low-temperature joining.

Copper-tin alloys, with or without modifying elements, are classed under the general name of tin-bronzes. *Phosphor-bronzes* (5–10% tin) are extra-strength alloys used for ship propellers, gears, bushings, springs, and pressure-tight castings. They may be modified with lead and/or zinc to improve machinability.

The high tin-antimony-copper alloys, by varying the percentage of the three constituents, find applications as babbit bearings, modern pewter and die-casting alloys. Tin-base babbitt contains 4 to 8% antimony and 4–8% copper, balance tin. Modern pewter usually contains 91% tin- 7% antimony-2% copper, but for softer types, 95% tin is used. Lead-based alloys containing tin and antimony are also used as bearing alloys and for type metal. About 30% of the tin used as alloys is recycled every year.

Tin and tin alloy coated steel, cast iron, copper, copper alloys and aluminum articles have functional and decorative uses in electronics, food processing equipment and automotive parts.

Uses for tin in the form of chemical compounds surged upwards in 1942 with the invention of the electrolytic tinplate processes. Stannous sulfate, chloride, fluoride and fluoborate salts and the potassium or sodium stannates are used for the electroplating of tin and tin alloys. Stannous fluoride is well-known as a toothpaste additive. Stannous oxide, SnO, and stannic chloride, $SnCl_4$, are primarily used as starting materials for the preparation of other tin compounds. Organotin stabilizers are used to prevent the breakdown and darkening of the polyvinyl chloride plastics when subjected to heat and light. The most important stabilizers are dibutyltin maleate, dilaurate, mercaptides and dioctyltin maleate. During the 1950's, organotin compounds with biocidal and fungicidal properties were discovered. The biocidal properties are associated only with compounds where the tin atom is combined directly with three carbon atoms (trialkyl and triaryl compounds) and where the total number of carbon atoms in the molecule is about 12. Hundreds of tons of tin are now consumed in the form of organotin fungicides in wood preservation, paper making and textile manufacture; as disinfectants in hospitals; as antifoulants for paints; as agricultural sprays for root crops, fruit and nut trees. Leaders in the field of biocides are tributyltin oxide, chloride and fluoride, triphenyltin acetate and hydroxide, and tricyclohexyl tin chloride(5). The trialkyl and triaryl tin compounds have toxic properties and their use as biocides must be under strict supervision; they must be adequately labeled to indicate hazards in handling.

Consumption. Free world consumption of virgin tin is around 177,000 long tons. Six nations account for about 70% of the total. The United States consumes 30%; Japan 13%; United Kingdom 10%; West Germany 7%; and France 6%. World per capita consumption is about 0.16 lb. The United States uses an average of 58,000 long tons of virgin tin and another 23,000 long tons of secondary tin mostly recovered from tin alloy scrap and scrap tinplate. Uses of virgin tin are divided as follows: tinplate 50%, solder 25%, bronze 7%, bearings 4%, and chemicals 4%, miscellaneous uses-the remainder. Secondary tin is used mainly for bronze, solder, white metal and chemicals. The average price of tin over the last five years was $1.58/pound.

ROBERT M. MACINTOSH

References

1. Faulkner, C. J., "The Properties of Tin," Tin Research Institute, Inc., Columbus, Ohio, Publication No. 218, 1965.
2. Hedges, E. S., Editor, "Tin and Its Alloys," St. Martins' Press, Inc., New York, 1960.
3. MacIntosh, R. M., "Tin," pp. 722–732, in "Encyclopedia of the Chemical Elements," C. A. Hampel, Editor, Van Nostrand Reinhold Co., New York, 1968.
4. Hoare, W. E., et al, "The Technology of Tinplate," St. Martins' Press, Inc., New York, 1965.
5. van der Kerk, G. J. M., "Present Trends and Perspectives in Organotin Chemistry," Tin Research Institute, Inc., Columbus, Ohio, Publication No. 413, 1970.

TITANIUM AND COMPOUNDS

Discovery. The element titanium was discovered in 1790 by William Gregor; his analysis of the black sand occurring along the beach in the County of Cornwall showed 45.25% of a white metallic oxide up to that time unrecognized. A few years later Kloprath noticed the close agreement between Gregor's findings and his own investigation of the oxide recovered from "red schorl" (rutile) found in Hungary. Acknowledging Gregor's priority, Klaproth applied to the new element the name titanium from the mythological Titans.

Occurrence. Titanium is widely distributed in nature. It is ninth in abundance of the elements making up the crust of the earth and accounts for 0.63% of the total. Of the metals suitable for structural use, it is exceeded in amount only by aluminum, iron and magnesium. All the rocks brought back from the moon by Apollo 11 contain an unusually high percentage of titanium—up to 12%. However, the Apollo 12 stones contain about 2% which is more in line with earth rocks. Two new titanium minerals, ferropseudobrookite and a chromium titanium spinel, occur in the moon rocks.

The more common titanium minerals are ilmenite (iron titanate), and rutile (titanium dioxide), which are mined as ores of titanium.

Deposits of ilmenite and rutile of commercial importance occur in Florida, New Jersey, New York, North Carolina and Virginia in the United States, in Argentina, Australia, Brazil, Canada, Egypt, India, Malaya, Norway, Republic of South Africa, Senegal, Sierro Leone, The Congo, and U.S.S.R. The largest deposits of rutile are in Australia.

Derivation. The production of elemental titanium is fraught with many difficulties since the hot metal combines actively with oxygen, nitrogen and moisture of the air as well as with carbon and most refractories. It decomposes the present refractory

materials, absorbing the oxygen. These contaminants render the metal so hard and brittle as to be useless. Once picked up, there is no practical method of removing these impurities. This difficulty is overcome by carrying out the reduction of $TiCl_4$ with magnesium or with sodium in a mild steel vessel under an atmosphere of argon or helium at around 800°C.

Although the actual reaction mechanism is probably more complex, the net reduction is represented by the equation: $TiCl_4 + 2Mg \rightarrow Ti + 2MgCl_2 +$ heat. The magnesium chloride produced is tapped and drained from the pot in the molten form.

After removal from the furnace, the vessel is cooled and then opened in a dry room to prevent moisture pickup by the hygroscopic magnesium chloride. The pot with its reaction product of titanium along with some magnesium chloride and magnesium is chucked in a lathe, and the mixed products are turned out in chips up to 0.5 in size. The turnings are later distilled in a vacuum to remove $MgCl_2$ and excess Mg and yield titanium sponge.

The sponge form titanium must be melted into an ingot before it can be rolled, forged or extruded. This is accomplished by first pressing it into bars which serve as consumable electrodes in the arc melting of the titanium in a water-cooled copper crucible the size and shape of the desired ingot. An inert atmosphere is maintained.

A very similar procedure employs sodium as the reducing agent.

Titanium of very high purity is prepared by the thermal decomposition of titanium tetraiodide; however, the method does not seem to be amenable to large scale operation and the cost is high.

Physical Properties. The silvery white metal has a density of 4.507 g/cc at 20°C, a melting point of 1668°C and a boiling point of 3260°C. Its electrical conductivity is only 3.6% that of copper; specific heat at 20°C is 0.13 cal/g/°C; linear coefficient of expansion from 20 to 300°C is 8.2 microinches per inch per °C; thermal conductivity is 0.041 cal/cm/sec/°C; modulus of elasticity is 15.5×10^6 psi; tensile strength at 25°C is 20,000 psi yield and 34,000 psi ultimate; latent heat of fusion is 5 kcal/mole; and its hardness is 80 to 100 Vickers. Superconductivity occurs below 1.73°K. The thermal neutron absorption cross section is 5.8 barns and the work function is 4.17 eV. Titanium is 45% lighter than steel but just as strong, and although 60% heavier than aluminum it is more than twice as strong.

It has an atomic number of 22 and an atomic mass of 47.90. Proportions of principal isotopes are 46, 10.82; 47, 10.56; 48, 100; 49, 7.50; and 50, 7.27. Titanium occurs in two modifications; alpha which crystallizes in the hexagonal system is stable up to 882°C, and beta having a body-centered cubic lattice is stable above this transition temperature. Titanium can be machined and worked with the same equipment and in the same way as stainless steel. However, fusion welding requires an inert atmosphere of argon or helium to prevent contamination by O, N and H with consequent brittleness at the weld.

Chemical Properties. Titanium is a member of the first transition series of elements and has an electron configuration of 2, 8, 10, 2. Consequently it shows variable valence, has both metallic and nonmetallic characteristics, and yields colored ions. However, the titanic ions carrying a positive charge of 4 are colorless. The metallic role of titanium is exhibited in such compounds as the chloride, nitrate, phosphate and sulfate, while the nonmetallic characteristic appears in a long series of titanates, as exemplified by calcium, iron and sodium titanates. It has an atomic radius of 1.54 Å and a covalent radius of 1.32 Å. Ionization energy is 6.83 volts and electronegativity is 1.6. Titanium metal is very resistant to corrosion and is unaffected by atmospheric conditions and by sea water. Nitric acid has no appreciable effect, but it is attacked rapidly by concentrated sulfuric and hydrochloric acids, and slowly by dilute sulfuric acid. It is not affected by strong alkalies, sulfur or sulfur compounds, chlorinated solvents, chlorides and wet chlorine. The metal ignites in air at 1200°C and burns with incandescence. Heat of combustion to the dioxide is 4700 cal/g. In contact with liquid oxygen titanium can be ignited by impact. Titanium dust is very explosive. It is one of the few elements that will burn in an atmosphere of nitrogen, that is, it reacts to form the nitride with the liberation of heat and light. At 700°C it decomposes steam to form the oxide with the liberation of hydrogen. It combines directly with the halogens to form the corresponding tetrahalides.

Uses. Unfortunately, all the known processes for winning titanium from its ores in commercial quantities are difficult, slow and expensive. At the present high price most of the output goes into military hardware, largely for airplane parts. Smaller amounts go into guided-missile parts, airborne equipment and space vehicles. The largest industrial uses are in the construction of parts of commercial airplanes and of chemical plant equipment. Of the total consumption of 23,687 tons of titanium ingots in 1970, jet engines accounted for 53%, aircraft frames took 25%, and space vehicles and missiles consumed 9%. Industrial uses accounted for only 13%.

Although only a small proportion of the titanium metal consumed goes into chemical plant equipment, its use is of increasing importance to the chemical industry. More and more chlorine plants are installing titanium towers, pumps, heat exchangers and other units, because of the metal's excellent resistance to wet chlorine. Other applications of titanium equipment take advantage of its corrosion resistance to chloride solutions, including sea water, to nitric acid, to chlorine dioxide and other bleaching agents, and to sulfur dioxide.

As the price of producing the metal is reduced, it has great potential uses where the properties of light weight, high strength and resistance to corrosion are important factors.

Alloys. Titanium used as a construction material is usually in the form of alloys, most of which have higher strength than pure titanium without being unduly affected with respect to corrosion resistance. The alloy of Ti-6Al-4V, for example, used in chemical equipment, has a room-temperature tensile strength at the yield point of about 120,000 psi, while that of commercial (99.2%) titanium is about 63,000 psi. The alloying agents

are seldom present in more than 10–15% concentrations.

Alloying elements in titanium fall into two groups: those that strengthen and stabilize the alpha or room-temperature modification and those that strengthen the high-temperature, beta, modification. Agents of the first group are oxygen, nitrogen, carbon and aluminum, and those falling into the second group are iron, manganese, chromium, molybdenum and vanadium.

Compounds. *Oxides.* Titanium forms four well-defined oxides: the monoxide, TiO, which shows weakly basic properties; dititanium trioxide Ti_2O_3, which is decidedly basic; the dioxide, TiO_2, which is amphoteric; and the trioxide or pertitanic acid, TiO_3, which exhibits acid properties only.

Industrially, the dioxide is by far the most important compound of titanium. Titanium dioxide for pigment use is produced by hydrolytic precipitation from titanyl sulfate solution on heating, and by the vapor phase oxidation of anhydrous titanium tetrachloride with oxygen or air at high temperatures. Titanium dioxide exists in three crystal modifications: anatase, brookite and rutile.

As a result of its extreme whiteness and brightness and its high index of refraction, titanium dioxide is widely used as a white pigment in paints, lacquers and enamels, paper, floor coverings, rubber, coated fabrics and textiles, printing ink, plastics, roofing granules, welding rods, synthetic fabrics, ceramics and cosmetics. A synthetic rutile may be cut and polished to form gem stones that outsparkle diamonds. Rutile has an index of refraction of 2.61 to 2.90 as compared with 2.4117 for diamond, although its hardness is only 6 to 6.5.

In recent years eight companies in the United States produced about 655,000 tons/year of titanium dioxide pigment, almost equally divided between the rutile and anatase types.

Other Inorganic Compounds. Titanic sulfate is produced in large quantities as an intermediate step in the manufacture of titanium dioxide pigments by the action of concentrated sulfuric acid on finely ground ilmenite. Such solutions contain the titanyl sulfate, $TiOSO_4$, and the experimental evidence indicates that the titanyl salt is the only sulfate of tetravalent titanium stable enough to persist under ordinary conditions.

Titanous sulfate, $Ti_2(SO_4)_3$, may be prepared readily by the electrolytic reduction of the tetravalent salt in water solution. It has a blue color, as do the other titanium (III) salts, and is not hydrolyzed in solution but is easily oxidized to the tetravalent state. It is a strong reducing agent.

Anhydrous titanium tetrachloride is produced by the action of chlorine on titanium dioxide and coke in a fluidized bed. It is a colorless liquid with a boiling point of 136°C and a density of 1.74 g/cc at 10°C; it is soluble in organic solvents as well as in water. The anhydrous tetrachloride is reduced to titanium metal by magnesium or sodium at elevated temperatures in an inert atmosphere of argon or helium, and it is oxidized with air or oxygen to titanium dioxide and chlorine at high temperatures. Both these reactions are employed commercially. It fumes strongly in moist air and dissolves in water to form solutions which are hydrolyzed readily to yield titanium dioxide and hydrochloric acid. The anhydrous liquid is employed in producing smoke screens and as a catalyst in many organic reactions.

Titanous chloride may be prepared by reducing titanic chloride with aluminum, magnesium, zinc, tin or hydrogen, or by electrolytic reduction of the solution. Titanium dichloride may be prepared by the reduction of the tetrachloride with sodium amalgam, by the extreme reduction of a stream of the tetrachloride in a current of hydrogen at 700°C, and by dissolving the metal in hydrochloric acid. Solutions of the dichloride are unstable and readily convert to the trichloride.

Titanium tetrafluoride may be obtained by the action of fluorine or of anhydrous hydrofluoric acid on metallic titanium at elevated temperatures. The product obtained by dissolving titanium dioxide in aqueous hydrofluoric acid is probably fluorotitanic acid, H_2TiF_6.

The tetrabromide and tetraiodide are produced by the action of bromine or iodine vapor, respectively, on heated titanium metal, or by the action of the corresponding halide vapor on a mixture of titanium dioxide and carbon at elevated temperatures.

Titanic phosphate precipitated from aqueous solutions on addition of a soluble phosphate to salts of tetravalent titanium is more or less basic and consequently of variable composition, so that in general more uniform products may be obtained by heating the dioxide with an alkali metal phosphate or phosphoric acid.

Titanium forms a number of sulfides under nonaqueous conditions: TiS_2, dititanium trisulfide, Ti_2S_3, and the monosulfide, TiS.

The tetranitrate is formed by dissolving hydrous titanic oxide or titanates, sodium titanate or barium titanate, in nitric acid.

Titanium carbide formed by heating titanium dioxide with carbon at a high temperature is a hard crystalline solid used in making cutting tools and grinding stones.

The element forms an extensive series of titanates with the alkali, alkaline earth and heavy base metals.

Organic Compounds. Many organic compounds of titanium, such as esters, alcoholates, phenylates and complexes, have been prepared. Examples are titanium phthalate, titanium oxalate, titanium gluconate, titanium tetraethylate, titanium stearate, triphenylcatechol titanate and butyl titanate. Titanium is a constituent of a number of permanent metallized azo dyes. Compounds have been made in which there is a titanium-to-carbon bond. An example is $C_6H_5Ti(OC_3H_7)_3$. A very large number of complex compounds of titanium have been prepared.

JELKS BARKSDALE

References

Barksdale, Jelks, "Titanium, Its Occurrence, Chemistry and Technology," Second Edition, Ronald Press Co., New York, 1966.

Barksdale, Jelks, Section on Titanium in "Economic Geography of Industrial Minerals," Reinhold Publishing Corp., New York, 1956.

Miller, J. A., "Titanium, A Materials Survey," U.S. Bureau of Mines, Information Circular 7791, Washington, D.C., 1957.

Ogden, H. R., Chapter on "Titanium" in "Rare Metals Handbook," C. A. Hampel, Editor, Second Edition, Reinhold Publishing Corp., New York, 1961.

"Minerals Yearbook," U.S. Bureau of Mines, Washington, D.C., Yearly Volumes.

TITRATION AND VOLUMETRIC ANALYSIS

Titration is the process by which an unknown quantity of a particular substance is determined by adding to it a standard reagent which it reacts with in a definite and known proportion. The addition of the standard reagent (a reagent of known concentration and often referred to as the titrant) is controlled and measured by some device and some method of indication is required to signal when the quantity of standard reagent added is just sufficient to quantitatively react with the substance to be determined. Therefore, knowing the proportion in which the substances react, and having determined the quantity of the one substance (the titrant) necessary to react in this proportion, the unknown quantity of the substance in the reaction vessel can be readily calculated.

Suppose, for example, that the quantity of hydrochloric acid in a reaction flask is to be determined by titration using a standard $0.1000M$ sodium hydroxide solution. It is known that one mole of sodium hydroxide will neutralize one mole of hydrochloric acid according to the reaction:

$$HCl + NaOH \rightarrow NaCl + H_2O$$

Therefore, when the hydrochloric acid in the sample is completely neutralized, the number of moles of standard sodium hydroxide added equals the moles of hydrochloric acid present at the start of the titration. If some type of *indicator* signals that the titrated solution is neutral after the delivery of 20.00 ml of $0.1000M$ sodium hydroxide, the moles of sodium hydroxide added equals

$$20.00 \text{ ml} \times 0.1000 \, \frac{\text{moles}}{\text{liter}} \times \frac{1 \text{ liter}}{1000 \text{ ml}}$$

$$= 0.002000 \text{ mole}$$

and, as stated above, this equals the number of moles of hydrochloric acid in the titrated sample. For this example, the weight of hydrochloric acid in the sample is:

$$0.002000 \text{ mole} \times 36.47 \, \frac{\text{grams}}{\text{mole}}$$

$$= 0.07294 \text{ gram HCl}$$

The points in a titration where the amount of titrant added is just sufficient to combine in a stoichiometric or empirically reproducible proportion with the substance to be determined is referred to as the *equivalence point*. The experimentally observed signal for termination of a titration is referred to as determination of the *end point*. The end point should coincide or be very close to the equivalence point. Any difference between equivalence and end points is referred to as the *indicator blank*.

Titrations are almost always performed with solutions, but it is also conceivable to perform them with gaseous, solid and molten state substances if proper equipment is available. The most common equipment for solution titrations consists of a reaction flask or beaker containing the substance to be quantitatively determined and a burette for controlling and measuring the titrant delivered into the reaction flask. It is also necessary to provide some method of mixing the reactants, either by shaking the reaction vessel or use of a mechanical stirrer.

There are various methods of classification of titrations. One method is according to the type of reaction which occurs between the titrant and the substance to be determined. The types according to this classification are acid-base, oxidation-reduction (redox), precipitation and complexation titrations.

Titrations are also classified according to the method of determining the end point of a titration. The classical method utilizes visual color indicators which undergo a sharp change of color to indicate the end point. The mechanism of indicator color change depends on the type of reaction which occurs between the titrant and substance to be determined. For an acid-base titration, the color indicator is itself a weak acid or base which on the acid side of the equivalence point has one color and on the base side another color. An easily detectable intermediate color should correspond closely to the equivalence point. For the above example of titration of hydrochloric acid with sodium hydroxide the color indicator should undergo a perceptible color change when the solution is neutral, pH = 7. For an oxidation-reduction titration the color indicators are usually substances which have one color in the reduced form and another color in the oxidized form. Indicators which change color by various mechanisms exist for precipitation and complexation titrations.

Instrumental methods of detecting the end point are based on the measurement of many different physical properties of the solution. The most generally used electrochemical methods of end point detection utilize various types of electrodes in the titration vessel across which changes of either voltage, current or resistance are measured, and are classified as potentiometric, amperometric and conductometric titrations, respectively. Spectrophotometric or photometric titrations utilize the changes in absorbed radiation to indicate the end-point. Less commonly used instrumental titration methods utilize changes of such physical properties as solution temperature, dielectric constant, refractive index, viscosity, velocity of sound, radioactivity, scattered light, etc. and are described as thermometric, high-frequency, refractometric, viscometric, sonic, radiometric, turbidimetric, and nephelometric titrations, or by other usually self-descriptive terms. Coulometric titrations are so named for the mode of preparation of titrant. Titrant is prepared either in the reaction flask in the presence of the substance to be determined or externally by electrolysis of a suitable solution, and from the measured number of coulombs required to reach the end point, the quantity of the sought-for substance is calculated. One of the above mentioned indicator devices is usually suitable for detecting the end point.

Automatic titrations can be performed whereby either the titration curve is automatically recorded or the delivery of titrant is automatically terminated at the end-point. Such systems release the operator for sample preparation and can greatly decrease the expense per sample determination.

Nonaqueous titrations are those which use solvents other than water for both the titrant and substance to be determined. These titrations are of great importance in titration of organic materials. The important determination of water by titration is named after the man who developed the procedure and is called the Karl Fischer titration.

HOWARD V. MALMSTADT

Amperometric Titrations

An amperometric titration is a regular titration in which the equivalence point is determined by measuring the diffusion current which flows through the system as the result of imposing a constant voltage across two electrodes in the solution. The location of the equivalence point is determined from a graph of current vs. milliliters of titrant. Usually it occurs as the intersection of two straight lines. Therefore, it is not necessary to give any special concern to the area around the equivalence point during the titration. The intersection of the two straight lines is readily accomplished by dependable extrapolation.

The diffusion current measured is the result of an electrode reaction taking place under the driving force of the applied voltage, and is proportional to the concentration of the electroactive species. The electrode reaction may involve the species being titrated, the titrant, the product of the reaction of either, or of an "indicator." Each situation gives a different shaped plot of current vs. volume of titrant. The titrations may be classified according to the shape of curve which they give.

Classification of Titrations. One type of curve is that obtained when the species being titrated is being oxidized or reduced at the electrode (most of the electrode reactions used are reductions). The current will decrease due to the removal of the species by the titrant. When the titration is complete, the current flow becomes essentially zero and remains zero upon the addition of excess titrant. As an example, consider the titration of lead ion with sulfate. At an applied voltage of -0.65 volt vs. S.C.E. (saturated calomel electrode as a reference) the lead ion is reducible at the dropping mercury electrode, and a diffusion-controlled current is obtained, exactly as in polarographic analysis. Now, when the concentration of lead ion is reduced by reaction with the sulfate to form insoluble lead sulfate, the diffusion-controlled current is decreased accordingly, because it is proportional to the lead ion concentration. This continues until no lead ion remains. No species reducible at the applied voltage of -0.65 volt vs. S.C.E. is then present, and the current approaches zero and remains so upon addition of excess sulfate. By measuring the current at four or five points both before and after the equivalence point, straight lines may be constructed, and their intersection denotes the equivalence point. Current readings are generally corrected for the volume change caused by the titrant before plotting.

A second type of titration curve involves the titration of a nonactive species with an active (oxidizable or reducible at an electrode) titrant. An example is the titration of barium ion with chromate. Here no current is obtained until all the barium has been precipitated by the added chromate. Then the current will increase as the excess chromate is reduced at the indicator electrode at the applied potential. A third classification is that in which both the species being titrated and the titrant provide a diffusion current by reduction at the indicator electrode. A typical example would be the titration of cupric ion with α-nitroso-β-naphthol.

Choice of Applied Voltage. The voltage applied across the electrodes must be sufficient to cause the desired oxidation or reduction reaction to take place. It must not be so high that it causes undesired reactions to occur at the electrodes. In general the potential is conveniently found by examining a polarogram of a solution containing the active ion. A voltage corresponding to the flat, diffusion-current portion of the curve is the most desirable. Thus, in the first example, it is desired to have lead ion reduced as long as it is present. A polarogram of lead nitrate indicates a diffusion current occurring due to the reduction of lead between -0.05 and -1.0 volt vs. S.C.E. A polarogram of sulfate does not show any reduction wave in this region. Thus, any value in this range would be satisfactory, provided the actual titration sample contains no other reducible ions.

Concentration Range. Polarographic measurement is most commonly made with solutions $10^{-2} - 10^{-6}$ M. Since the currents obtained in amperometric titrations are the same, solutions of ions in this general range will be the most satisfactory. The wide range is possible by using a series of shunts which makes possible the measurement of definite fractions of the total diffusion current.

Relation to Other Methods. As indicated, amperometric titrations are directly related to polarographic measurement. In effect, it is a polarographic measurement of concentration of the active species after increments of the titration. The method is also related to the "dead-stop" end point. In the latter, two indicator electrodes are used instead of one indicator and one reference as in amperometric. A constant potential applied will cause the electrode reactions to occur. If one of the species necessary for one of the electrode reactions is removed by titration, the current drops to zero at the end point as with the amperometric titration. This cessation of current flow is noted on a sensitive galvanometer.

WARREN W. BRANDT

Automatic Titration

In the determination of a substance by a titration procedure there are many manipulations, all of which can be removed from the hands of the chemist and automatically performed. The automation of some or all of the titration steps has been referred to very generally as *automatic titration,* although the degree of automation varies considerably with specific methods.

An ideal automatic titrator might be considered as an instrument into which the chemist places a prepared sample, presses a button, and a short time later reads the answer from a report sheet. The ideal titrator would automatically perform all manipulations for a wide variety of titrations with a minimum or preferably no pre-titration considerations or adjustments. An ideal automatic volumetric analyzer would consist of an ideal automatic titrator combined with other instrumentation to select automatically a representative sample, eliminate interferences, insert the prepared sample into the titrator and clean and set-up equipment for the next determination. Although instruments having many of these characteristics are available for a few specific titrations, no one apparatus is, at present, generally applicable to all titrations, and it is necessary for the analyst to decide what components, instruments, and techniques are best suited for his specific requirements.

The classical or manual titration procedure is familiar to all chemists, and it illustrates the steps in performing a titration. The titrant is prepared by weighing a reagent grade chemical and diluting to volume in a volumetric flask. A buret is filled with the standard titrant and the stopcock manually operated to deliver increments of titrant into the reaction flask, and the solution mixed by shaking or stirring with a rod. During these manipulations the person performing the titration must continuously and closely observe the end-point detection system for a sudden change marking the end-point, or he must observe and record several readings from which a titration curve may be plotted.

In considering the automation of the titration procedure the components are conveniently divided into two groups; the stirring unit, end point detection system, and the control unit, for starting and stopping the delivery of titrant and stirring, in one group; and the titrant preparation and delivery system and the data presentation system in the other group.

There are at present three basic types of automatic titrators which decrease or eliminate technician time in buret manipulations and end-point detection. These methods have been applied mainly with the potentiometric end-point system, but they are also applicable with certain modifications for other end-point detection techniques. One type of automatic titrator stops the delivery of titrant at a pre-set end point value. A succesful instrument of this type automatically anticipates and seeks the equivalence-point potential and turns the buret off when it is reached. It is especially good for slow reactions. This type of automatic titrator requires knowledge of an absolute equivalence point value which must be pre-set on the instrument prior to titration. Any solution phenomena which cause changes of the absolute equivalence point value can, of course, lead to errors.

Another type of automatic titrator records the titration curve, and the end-point is read from the plot after the titration is completed. This method has been applied for potentiometric, thermometric, conductometric, high-frequency, and spectrophotometric titrations.

A third type of automatic titrator automatically determines the inflection point (end-point) of an ordinary titration curve and turns off the titrant delivery when it is reached. This is accomplished by electronically producing a voltage which is proportional to either the second or third derivative of the ordinary titration curve, and this derivative voltage is ideally suited for triggering a relay system which turns the buret off at the inflection point. This instrument is called the automatic derivative titrator and is applicable with most methods of end point detection, including potentiometric spectrophotometric, turbidimetric, amperometric, and chemiluminescent.

There are several advantages which characterize the derivative automatic titrator. The equipment is simple, compact, and relatively inexpensive. The end point value does not have to be known and usually no instrument adjustments are required— it is only necessary to push the start button. The end point detection systems and control unit can be simplified because absolute values are not important, only the relative changes in the measured quantities.

There has also been much progress on the elimination of technician time for preparation and storing of standard titrant and filling and reading burets. In certain cases it is possible to generate the titrant electrolytically at a suitable generator anode or cathode, and the milliequivalents of titrant delivered can be determined readily by integrating the current from the start of the titration to the end point and applying Faraday's law. The data presentation system can be made to show the titration result directly in numbers on a dial or printed tape.

Various automatic burets have been developed whereby the titrant is forced through a tube with a precision plunger driven by a motor. The movement of the plunger and volume of titrant delivered is determined by the number of revolutions of the motor. The plunger and gear system can be made so that a revolution counter will read directly in milliliters, and a very high accuracy is possible.

One novel type of autotitrator has been developed for continuous titration in a process stream. The solution to be titrated is pumped into a reaction vessel at a constant rate and the titrant is pumped in at a variable rate automatically regulated to keep the combined solutions at a predetermined equivalence potential. This potential is continuously compared with a fixed potential and the difference potential is used to regulate the flow rate by the titrant pump. The flow rate of the titrant is proportional to the concentration of the substance being determined..

All titrations that have been performed manually can be performed automatically by one or more of the commercially available titrators, and these instruments are now widely used for nearly all types of aqueous and nonaqueous titrations.

H. V. MALMSTADT

Conductometric Titrations

Conductometric titrations are titrations in which the end-point determination depends upon a change in conductivity of the solution being titrated. The method depends upon the involvement of ions in the titration reaction such that rate of change of the conductivity changes at the end point. The individual ion equivalent conductivities at infinite dilution for a series of ions are:

H^+	315.2	OH^-	173.8
Li^+	33.0	Cl^-	65.5
Na^+	43.2	Br^-	67.3
K^+	64.2	I^-	66.3
Ca^{++}	51.9	$SO_4^=$	68.5
Ba^{++}	55.4	NO_3^-	61.6

The conductivities of the salt or acid solutions formed from these ions are roughly proportional to the sum of the individual ion conductivities.

For example, if an acid is titrated with a base, the conductivity of the acid is first measured; then as the base is added acid and base are converted to salt and undissociated water having a lower conductivity. Thus, the conductivity is decreased as the acid is neutralized until the end point is reached. With the addition of more base, the conductivity rises again, due to the presence of the good conductor, hydroxyl ion, in the base. The original decrease in conductivity as the base is added is approximately linear, as is the subsequent (to the end point) increase in conductivity. Thus, if a plot is made of conductivity vs ml base added, the lines before and after the end point may be extrapolated until they cross. The crossing point is the end point (Fig. 1).

A large number of reactions involving ions which are rapid and complete enough to be used quantitatively can be done conductometrically. Generally, rather concentrated solutions are used to get a good rate of change of conductivity as the titrant is added. The method is most accurate in titrations giving the largest change in rate of change of conductivity at the end point. Generally speaking, the accuracy is comparable to that of other methods of end-point indications and has the advantages of requiring simple apparatus, working in highly colored solutions and being very generally applicable. The disadvantages are the necessity for the reaction to be ionic, the failure of the method in cases where a high concentration of supporting electrolyte is present (e.g., sulfuric acid in a di-

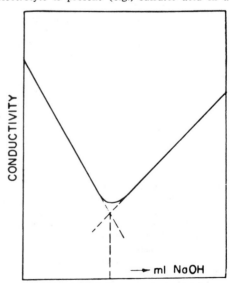

FIG. 1. Typical plot of titration of strong univalent acid with NaOH.

chromate oxidation) and the requirement of plotting data as the reaction proceeds in order to find the end point.

F. R. DUKE

Coulometric Titrations

A coulometric titration is a process involving the electrolytic generation of a reagent which reacts with the substance to be analyzed with an equivalency between the coulombs used and the amount of substance involved, as expressed by Faraday's Laws in the form:

$$\frac{W_x}{E_x} = \frac{Q}{96,487}$$

where W_x is the weight of the substance x, of equivalent weight E_x, Q is the number of coulombs involved, and 96,487 is the number of coulombs per Faraday. Coulometric titration procedures were introduced in 1938 by Szebelledy and Somogyi, but extensive exploitation of the technique did not occur until about ten years later.

If uncorrected current-time data are to be employed for the evaluation of Q, the essential apparatus is indicated in Fig. 1, A. If a constant current source is neither available nor desirable, the essential apparatus is indicated in Fig. 1, B. In both cases, suitable end-point detection is necessary.

A 45-volt "B" battery, connected in series with a large variable resistor serves as a simple, essentially constant-current source, the large resistor ensuring that the current is little affected by small changes in cell resistance. Other sources using vacuum tubes or transistors, mechanical designs, or combinations of the two are available; some of these are produced commercially. One electromechanical device claims a constancy of $\pm 0.01\%$, but a wide range exists, both in sophistication and cost.

In cases where precise potentials must be maintained at an electrode in order to cause a desired reaction, or where a separation of certain species is desired, a constant potential source is required. Controlled potentials at mercury electrodes are especially useful and various potentiostats have been designed to maintain such constant potentials.

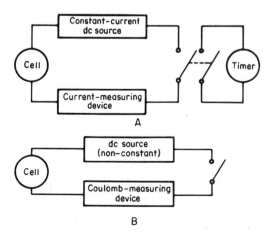

FIG. 1. Schematic diagrams for coulometric titration apparatus.

The current may be measured with a suitable galvanometer, milliammeter, or ammeter with an accuracy compatible with the rest of the titration. Timing may be done with a stop-watch; but an electric, or electronic, timer properly switched to start and stop with the electrolysis current start and stop is more convenient.

Coulometric titrations may be accomplished by the internal method, i.e., placing the sample solution in contact with the electrode at which the titrant is generated. The external method may be used in case it is necessary or desirable to separate the sample solution from the generator electrode. The kinds of cells uesd depend upon whether the internal or external method is to be used but, even within the two categories, the type seems limited only by the ingenuity of the designer.

If the internal method is to be used, the cell consists of a generator electrode and an auxiliary electrode to complete the electrical circuit, both assembled in a container provided with a stirring mechanism. Usually, the two electrode compartments must be separated by some device which prevents significant diffusion of the liquids between the two compartments. The kind of generator electrode employed depends upon the nature of the electrode process desired.

When an undesirable reaction may occur between the sample and the electrode, or some other complication arises, the external method may be necessary or desirable. In such an event, the design of the apparatus is modified to permit the titrant to be generated at the electrode in a cell disconnected from the sample solution. The reagent precursor solution is allowed to flow past the electrode, the titrant is generated and carried by excess solution to the sample held in a separate container. Provision must be made to overcome the time lag existing between the generation of the titrant and its contact with the sample.

In all coulometric titrations the end point must be detected. Since very small numbers of coulombs can be measured, it is apparent that the accuracy of a coulometric titration may depend ultimately upon the accuracy of the end-point detection. Most of the usual visual and other methods for end-point detection have been used, but amperometric, electrometric, and spectrophotometric methods are generally preferred because they may be incorporated into automatic equipment. One method involves the polarization of an electrode and the detection of a change in current when the end-point is reached. Claims of success at concentrations as low as 0.01 microgram per milliliter have been made for some of these methods. Another method involves end-point detection by amperometric means as follows: The generator current is stopped as soon as the indicator current begins to change, and the time is noted. The titrant is then generated in three or four short, measured periods of time and the corresponding indicator currents noted. The linear plot of indicator current vs. time is extrapolated back to zero indicator current to give the time required for the titration.

In theory, a coulometric titration may be adapted to the determination of any species capable of accepting or donating electrons. Neutralization, traditional oxidation-reduction, and Lewis-type acid-base reactions are thus included; theoretically, any reagent in such a reaction could be determined quantitatively by a coulometric titration technique. Certain practical difficulties arise in some cases, but often they may be overcome. For example, some chlorination and bromination reactions, normally too slow for coulometric titrations, may be speeded up by a suitable catalyst.

Traditional acids may be titrated in a cathode compartment, while bases may be titrated in an anode compartment (at the same time, if desirable); suitable end-point detection and precautions concerning carbon dioxide are assumed.

Coulometric titrations involving most of the common, and many of the less common, oxidizing and reducing agents have been reported. Electro-generated halogens (including F^-), $Ce(IV)$, $Fe(II)$, $Sn(II)$, $Ti(III)$, and $Cr(VI)$ ions have been used. In the less-common class should be mentioned $Ag(II)$ and $Mn(III$ and $IV)$ ions, $Cu(I)$ as the chloride complex, $Mn(I)$ and $Mo(V)$ in cyano complexes, and the transuranium elements, such as Np and Pu. The determination of water with electro-generated Karl Fischer reagent, the generation of free radicals to determine anthracene and similar compounds, the determination or organic reducing agents, double bonds, pyridine compounds, and the generation of various species in solid electrolytes are also of interest. Increased applications to analyses in nonaqueous solvents for H^+, traces of chlorine, etc., as well as methods for the determination of rarer elements such as W, Pa, Y, Ru, and Am have been described.

Of particular recent interest are the several methods for monitoring SO_2, HCN, NH_3 and other noxious gases in the atmosphere via continuous coulometric titrations, as are the applications to the measurement of pesticides, insecticides, drugs, and fertilizers. Controlled-potential coulometric titration, applicable to substances which undergo oxidation or reduction in multiple stages such as U(VI) to U(IV) through U(III), has also been reported.

Recent instrumental developments have emphasized simplification of apparatus for student applications, development of solid-state constant-current sources, automatic coulometric titrators for continuous analysis of gas and liquid streams, and the incorporation of computers for calculations.

The advantages of the coulometric titration techniques are: The successful handling of the submicrogram and microgram, as well as macrogram, amounts of sample. No standard solutions are required because the coulomb (or the electron) is the primary standard. Numberous recent high-precision determinations lend support to Tutundzic's proposal that the Faraday be made the primary chemical standard.

The method is limited by the following: An equivalency must exist between the coulombs involved and the measured species. Fairly rapid reactions are essential. Such cumbersome equipment may be required for suitable end-point detection in solutions of low concentration, that the equipment constitutes a disadvantage. Also at low concentrations, diffusion, kinetic, and adsorption problems may become acute.

K. A. VAN LENTE AND R. E. VAN ATTA

References

Bard, A. J., "Electrolysis and Coulometric Analysis, *Anal. Chem.*, **40,** 64R (1968).

Bard, A. J., "Electrolysis and Coulometric Analysis," *Anal. Chem.*, **42** 22R (1970).

Lagowski, J. J., "Titrations in Nonaqueous Solvents," *Anal Chem.*, **42**, 305R (1970).

Milner, G. W. C., and Phillips, G., "Coulometry in Analytical Chemistry" Analytical Sciences Division, A. E. R. E., Harwell, Pergamon Press, 1967.

Nonaqueous Titrations

Many organic acids and bases are insoluble in water and must therefore be dissolved in some nonaqueous solvent prior to their determination by titration. Alcohols and alcohol-water mixtures have been used for many years as solvents for organic acids and bases. The scope of acid-base titrations in aqueous or alcoholic solution is limited, however. Aliphatic amines can be titrated as bases in water or alcohol, but aromatic amines and many nitrogen heterocyclics are too weakly basic to be determined. All types of amines can be titrated if the proper nonaqueous medium is chosen. In general, amines which have a pK_b of 0 to 6.5 can be titrated accurately in water; in a suitable nonaqueous solvent amines which have a pK_b (in water) of about 0 to 13 can be titrated. The range of acidic compounds which can be determined by alkalimetric titration is also much greater in nonaqueous solvents. The speed and accuracy of acid-base titrations are as great in nonaqueous solvents as in water.

Titration of Bases. The solvent used for any titration should have the following characteristics:

(1) It should readily dissolve the substance to be determined. The products of titration should either be soluble or should by crystalline precipitates. Gelatinous precipitates are to be avoided.

(2) It should be available commercially at moderate price. Preferably the solvent should not require preliminary purification.

(3) The solvent should not enter into disturbing side reactions with the sample to be titrated or with the titrant.

(4) It should be essentially neutral or slightly acidic, but should not be basic. This requirement is essential if weak bases are to be titrated.

Consider the general, net reaction for titration of a weak base with a strong acid:

$$B \quad + SH^+ \rightarrow BH^+ + \quad S \qquad (1)$$

(base)　(acid)　(acid)　(base)

Here B is the base titrated and SH^+ is the titrant, represented as a solvated proton. If the solvent (S) has basic properties, as does water, the equilibrium will be unfavorable near the soichoimetric point and the end point will not be sharp. The effect of a basic solvent on the equilibrium is particularly great because of the mass effect due to the large amount of solvent present as compared with the other reactants. Alcohol is a somewhat better solvent than water because it is a little less basic. For best results, however, a solvent such as glacial acetic acid is much more satisfactory because it has extremely weak basic properties.

From reaction (1) it will be seen that the acid titrant used should be as strong an acid as possible. In aqueous solution all of the strong mineral acids react completely with water to form the same acidic species (H_3O^+), hence it makes little difference which one is chosen as the titrant.

The end point in acidimetric titrations can be detected either potentiometrically or by means of visual indicators. In acetic acid, either crystal violet or methyl violet serves very well as a visual indicator for the titration of most amines. Very weak organic bases should be titrated potentiometrically.

In general the following classes of compounds can be titrated as bases in nonaqueous solution.

(1) Aliphatic amines and amino acids.

(2) Aromatic amines. Exceptions are those which are extremely weakly basic because of electron-with-drawing groups such as *o*-nitro, N-phenyl, 2,4-dichloro, etc.

(3) Basic nitrogen heterocyclics such as pyridine, quinoline, thiazole, etc. Pyrrole type compounds are too weakly basic to be titrated.

(4) Ammonium, amine and alkali metal salts of carboxylic acids, nitric, sulfuric and hydrochloric acid. Sulfates are titrated to bisulfate in mixed acetic acid and dioxane.

Titrimetric methods are available which make it possible to distinguish quantitatively between various types of amines in mixtures. One such method is based on differences in basic strength of various amines. Other methods make possible the separate determination of primary, secondary and tertiary amines.

Titration of Acids. In selecting conditions for the titration of weak acids in nonaqueous solvents, essentially the same principles apply as for the titration of bases. Thus, for the titration of weakly acidic compounds, the solvent chosen should not have acidic porperties and the titrant should be as strong a base as possible. Anhydrous ethylenediamine, dimethylformamide, butylamine, pyridine and acetone have found wide use as solvents. Sodium, potassium and lithium alcoholates in benzene-methanol or in ethylenediamine-ethanolamine are suitable titrants. Recently tetraalkylammonium hydroxides in benzene-methanol have been shown to be very useful for titratiing weak acids. One advantage of the latter is that the titration products are almost always soluble in the commonly used organic solvents.

The end point in the titration of acids may be determined potentiometrically using a glass-calomel,, antimony-glass, antimony-calomel or other electrode system. The titration curves obtained are not always reproducible, but it is still possible to locate the end point accurately in most cases. Thymol blue, azo violet, *o*-nitroaniline and a number of other visual indicators are available for nonaqueous titrations of acids.

The following classes of compounds can be titrated as acids if the proper conditions are used.

(1) Carboxylic acids, acid halides and anhyhydrides. In anhydrous media carboxylic acid anhydrides are titrated as monobasic acids: $(RCO)_2O + CH_3ONa \rightarrow RCO_2Na + RCO_2CH_3$.

(2) Phenols of all types.

(3) Enols and imides. In general all 1,3-dicar-

bonyl compounds can be titrated if there is a hydrogen atom on the 2-carbon or nitrogen. A cyano group has electron withdrawing properties similar to the carbonyl group. Thus, ethylcyanoacetate ($N\equiv C—CH_2COOEt$) or even malononitrile ($N\equiv C—CH_2—C\equiv N$) can be titrated as a monobasic acid.

(4) Aryl sulfonamides of primary amines.

(5) Aryl and diaryl thioureas.

(6) Ammonium and amine salts of most acids.

(7) Miscellaneous acidic compounds such as dithiooxamide, nitromethane, carbon dioxide, etc.

JAMES S. FRITZ

Spectrophotometric Titration

The detection of the end point in a titration by measuring the changes of absorbed narrow bandpass radiation is referred to as *spectrophotometric or photometric titration*. In the general manual spectrophotometric titration procedure, the titrant is added to the sought-for constituent contained in a suitable titration vessel; the solution is mixed; and the absorbance values are read from the meter of a spectrophotometer or filter photometer after the addition of each increment of titrant. A plot of absorbance versus milliliters of titrant is then made, and the shape of the titration curve depends on the combination of reactant, reaction product, titrant, and/or indicator that undergoes changes of absorbance at the preset wavelength during the titration.

There are various general shapes of titration curves which are probable with the spectrophotometric end point detection method, and these are sketched in Figure 1. Types 1 through 7 are without and types 8 through 10 with added indicators. The general shape of the titration curve can change for the same chemical reaction by merely changing the preset wavelength. The following conditions

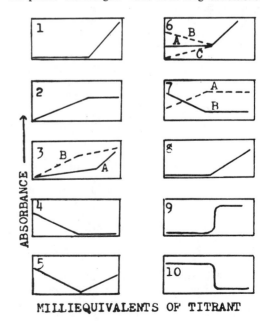

MILLIEQUIVALENTS OF TITRANT

FIGS. 1.1–1.10.

for all examples are true at a specific preset wavelength, and dilution of the sample by the titrant is assumed to be negligible.

Figure 1.1 shows a typical titration curve for the case where the reactant (sought-for constituent) and reaction product do not absorb, but the titrant does absorb. An example of this is the titration of nonabsorbing ferrous iron with a titrant such as permanganate which absorbs strongly at the preset wave length.

If a solution containing a small amount of copper is buffered and titrated with a standard EDTA titrant, a spectrophotometric titration curve similar to Figure 1.2 can be obtained by presetting the wave length to 260 millimicrons. In this case the reactant and titrant do not absorb significantly at the present wave length, but the reaction product, the copper-EDTA complex, has a large absorptivity.

In Figure 1.3 the reactant does not absorb and the reaction product and titrant do absorb, but in Fig. 1.3A the absorptivity of the titrant is greater than that of the reaction product, and the reverse is true in Figure 1.3B. The type of curve illustrated in Figure 1.4 is quite common and results from absorption by the reactant and essentially no absorption by titrant and reaction product. A V-shaped curve results from the reactant and titrant absorbing and reaction product not absorbing as illustrated in Figure 1.5. Several shapes result from the reactant, reaction product and titrant all absorbing; these are illustrated in Figures 1.6A, B and C. In 6A the absorptivity of the reaction product is equal, in 6B less than and in 6C greater than the absorptivity of the reactant. Both reactant and reaction product adsorb in Figure 1.7 and titrant does not absorb, but the absorptivity of reaction product is greater than that of the reactant in 7A and less in 7B.

In Figure 1.8, neither the reactant, titrant nor reaction product absorbs in a useful wave length range, and an indicator is necessary. An interesting example of this type is the titration of a small quantity of thorium with EDTA, with a copper salt as indicator. The thorium is quantitatively complexed with EDTA before the copper commences to complex. As the EDTA is added, the absorbance remains constant until essentially all the thorium is complexed; then the titrant combines with copper to form the copper-EDTA complex which has a high absorptivity at the preselected wave length, and the absorbance increases continuously as the concentration of indicator-EDTA complex increases.

In all the above titration curves the end points are found by the intersection of two straight lines and are similar in this respect to conductometric and amperometric titration curves. The curves in Figures 1.9 and 1.10 are quite similar in shape to potentiometric titration curves, and are most typical of titrations with added indicators. Nearly all visual color indicators for aqueous and nonaqueous acid-base, redox, and complexation titrations give curves of these general types. A rapid change of either pH, redox potential or pM at the equivalence point causes a large sudden concentration change for one form of the indicator. In Figure 1.9 the form of the indicator that absorbs at the preset wave length suddenly increases in concentration

giving a rapid rise of absorbance, and in Figure 1.10 the concentration of the absorbing form of indicator suddenly decreases and the absorbance decreases. These rapid changes of absorbance are readily detected by the eye if the absorption bands are in the visible, as with the classical visual indicators; but various photocells are more sensitive, selective and also suitable at wave lengths outside the sensitivity range of the eye. The output electrical signals from the light detector are readily plotted, recorded or used to terminate the titration automatically at the equivalence point.

Although conventional spectrophotometers or filter photometers can be adapted for spectrophotometric titrations, new instruments designed specifically for easy insertion and removal of various-sized titration beakers, effective stirring and sensitive and selective detection are now available. These new instruments can provide for automatic recording of the entire titration curve or termination at the end point in many cases, which greatly simplifies the method. Titration to a preset absorbance value or the derivative method which is independent of absolute absorbance can be used for automatic end point detection.

All titrations presently using visual color indicators and many other titrations that can give curves similar to those outlined and illustrated here can be performed by spectrophotometric titrations. Therefore, the spectrophotometric method should rank with potentiometric and amperometric as one of the most useful instrumental methods of end point detection.

The accuracy of spectrophotometric titrations depends primarily on the reaction between sought-for constituent and titrant in the most suitable solvent, and accuracies of 0.1% are common. The accuracy of determinations by conventional absolute spectrophotometric measurements are seldom more accurate than 1% and can be much worse because of various solution and instrumental phenomena. For example, quantitative results for samples of low absorbance can have considerable errors because of imperfections or dust and dirt on the sample cell windows; quantitative results for samples of high absorbance can have significant errors because of stray light in the instrument. Unsuspected impurities absorbing at the same wave lengths as the substance to be determined or solution turbidity can cause large percentage errors. Whereas these phenomena cause changes of absolute absorbance they merely cause a shifting of the titration curve on the absorbance axis and do not affect the end point, unless, of course, an absorbing impurity should react with the titrant.

H. V. MALMSTADT

TOXICOLOGY

The manufacturer who seeks the assistance of a toxicologist or pharmacologist is primarily interested in answers to three questions:

(a) What is the degree of toxicity, and what are the general effects produced by a certain compound?

(b) How may the substance enter the body and what precautions must be taken during the manufacture, handling, transport, etc., of the compound in order to avoid injury to the industrial population?

(c) What are the possibilities of injurious effect to the consumers of (a) the compound, or (b) to some preparation of which it may become a part?

In order to plan a sound investigation of these questions, it is necessary for the toxicologist to become acquainted with the purpose, manner, and general, specific or exceptional conditions which pertain to the use of the material to be investigated. He needs to know: (a) the chemical and physical properties of the substance under consideration (and if possible, he should be supplied with an analytical method for its determination); (b) the commercial method of preparation of the compound or the types and quantities of impurities present in the "purified" and in the "commercial" product; (c) the preparations or mixtures of which the compound is to become a part, and the concentration of the compound in them; (d) the stability of the compound and possible changes that it may undergo (as caused by light, age, oxidation, heat, evaporation, etc.) before or after it becomes part of a final product; (e) in addition, he has to determine by what route (skin, respiratory tract, or gastroenteric tract) absorption of the material is most likely to occur by the population engaged in its manufacture, handling and distribution; and (f) by what route, absorption of the compound or of a product containing the compound (in its original or in a changed form) is likely to occur among the population at large.

Acute (Immediate) Toxicity. Nearly every toxicological investigation is initiated by a preliminary study or by screening tests. These serve two purposes: (1) to detect potentially hazardous compounds, and to warn the manufacturer that extensive investigation may be needed should the commercial possibilities justify the development of the substance; and (2) to inform the manufacturer of the potential hazards that may be associated with the manufacture, handling, etc., of the product. In screening a compound, it is generally administered by the three methods by which substances most commonly gain entrance to the tissues: oral, cutaneous, and inhalation.

On the basis of his experience with compounds of similar chemical structure, the toxicologist chooses the range of dosage to be administered in order to determine the lethal dose in animals. In screening tests, which they call *range-finding tests* (RF), Smyth and Carpenter and Smyth, Carpenter, Weil and Pozzani use the following doses when closely spaced dosages are required because of the steepness of the dosage response curve: 1.00, 1.26, 1.58, 2.00, 2.52, 3.16, 3.98, 5.00, 6.30, 7.95, 10.00, 12.60, etc. This is a repeating series found by taking the antilogarithms of the numbers 1, 2, 3, etc., For most of their RF tests, they use an integral g/kg and 2, 4, and 8 times that dosage; e.g., 1, 2, 4, 8 or 3, 6, 12, 24 g/kg.

In 1943 Deichmann and LeBlanc introduced the determination of the *Approximate Lethal Dose* (ALD) (*J. Ind. Hyg. and Tox.*, **25**, 415, 1943). The doses are 50% progressions of 0.001. Any unit of measure (grams, milligrams, milliliters, etc.) may be employed. The doses were determined ex-

perimentally and found to be spaced sufficiently to practically preclude the possibility of killing an animal with one dose, while failing to kill with the next higher dose. The intervals between the doses are small enough, on the other hand, that this method is useful in range-finding. Only a few (not previously starved) animals need be used. In all likelihood, results will be decisive: i.e., all animals treated with doses up to a certain level will survive, and all those treated with higher doses will die. The *Approximate Lethal Dose* (ALD) is the lowest concentration that kills. For all practical purposes an ALD (determined with six to eight animals) is an approximate LD_{50} (*J. Ind. Hyg. and Tox.*, **30**, 373, 1948).

Oral Toxicity. The albino rat is the animal of choice for oral screening tests; more data have been collected for this than for any other species. Whenever possible, liquid compounds are administered undiluted by stomach tube. Water is the choice when a solvent is required. For water-insoluble materials, propylene glycol or a vegetable oil like corn oil may be employed. At times aqueous suspensions are preferred, utilizing methyl cellulose, carboxymethyl cellulose, or acacia, as suspending agent.

Inhalation Toxicity. Smyth and Carpenter recommend a simple procedure which determines the hazard (not the concentration) associated with the exposure to certain vapors or dusts. They place six male albino rats in a 9-liter desiccator through which contaminated air is passed at a rate of 2.5 liters per minute. Fifty ml of the solution to be tested is placed in a 250-ml gas washing bottle fitted with a fritted disk bubbler. The exposure is carried out at room temperature and continued until half the animals have died or until the exposure period has reached eight hours. The test may be repeated and attempts made to determine the period of time of exposure which will produce, in addition, zero and 100% mortality. To test the effect of dust, 50 gm of a material is placed, in layers, above layers of glass beads in a 100 ml burette having a diameter of $\frac{1}{2}''$. The air, after passing through the burette and having become contaminated with dust, is passed as above, at a rate of 2.5 liters per minute, into the desiccator. Exposure is again permitted until the desired degree of mortality is produced. In drawing conclusions from an inhalation experiment, it is well to remember that an animal, if placed in a gas chamber, suffers exposure not only by inhalation, but most likely also by ingestion and through the skin.

Skin Toxicity. In testing for the ability of the material to pass through the skin, guinea pigs or rabbits are usually employed. Draize tests each substance upon rabbits with abraded skin as well as upon others with intact skin. His dosages are 3.9, 6.0, and 9.4 ml/kgm. The material to be tested, which may be a dry powder, a liquid, or a solution of the substance, is placed under a rubber sleeve which is slipped onto the trunk of an animal. Exposure is permitted for 24 hours. In our laboratory, the dosages applied to the abdominal skin of rabbits are selected from those listed in Table 1 in an effort to determine the *Approximate Lethal Dose* in mg of compound per kg body weight. Generally six rabbits are treated. Exposure time is limited to

TABLE 1. DETERMINATION OF THE APPROXIMATE LETHAL DOSE (ALD) AND TABULATION
OF TOXICITY CLASSES

(Data in this table refer only to effects resulting from ingestion of a single dose of a compound)

Toxicity Rating	Single Oral Dose, as suggested by Hodge and Sterner (1949) per kgm	Doses Suggested For ALD by Deichmann and LeBlanc (1943) per kgm	Commonly Used Term	Probable Lethal Oral Dose for Man
1	1 mg or less	1.0 mg	Extremely toxic	A taste or one grain
2	1 to 50 mg	1.5 mg 2.2 mg 3.3 mg 5.0 mg 7.0 mg 10.0 mg 16.0 mg 24.0 mg 37.0 mg	Highly toxic	One teaspoon (4 ml)
3	0.05 to 0.5 gm	0.055 gm 0.08 gm 0.12 gm 0.18 gm 0.28 gm 0.42 gm	Moderately toxic	One ounce (30 gm)
4	0.5 to 5.0 gm	0.62 gm 0.94 gm 1.4 gm 2.1 gm 3.2 gm 4.7 gm	Slightly toxic	One pint (250 gm)
5	5.0 to 15.0 gm	7.1 gm 10.7 gm	Practically nontoxic	One quart
6	15.0 gm and more	16.0 gm 24.0 gm 36.0 gm	Relatively harmless	More than one quart

Doses suggested for Range-Finding and R.F. LD$_{50}$ by Smyth and Carpenter (1944) (1954) are the following: 1.00, 1.26, 1.58, 2.00, 2.52, 3.16, 3.98, 5.00, 6.30, 7.95, 10.00, 12.60 etc. This repeating series was found by taking the antilogarithms of 1, 2, 3 etc.

six hours. After an exposure, the unabsorbed portion of a material is removed by washing an animal under the tap with a mild soap and running water. The results present the ALD. Before treatment, the hair is clipped very short over the abdomen.

Hodge and Sterner summarized the discussions of a group of toxicologists who met at the April 1947 AIHA meeting in Buffalo, N.Y. to propose meaningful terms for degrees or classes of toxicity. Table 1, which includes an estimate of the probable lethal dose for man, presents the recommendations of this group for the characterization of the toxicity of a compound. In Table 1, there have been inserted the dosages of Deichmann and LeBlanc for the determination of the *Approximate Lethal Dose*. A footnote gives the dosages employed by Smyth, Carpenter, Weil and Pozzani. According to Heyroth: "Such data are usually adequate to safeguard the scientists and workmen engaged in the development of processes for the manufacture of a new compound. When it appears likely that the commercial promise of a new material will lead to its manufacture on a large scale and its widespread distribution, then more precise quantitative data become advisable."

Frequently the *Approximate Lethal Dose* needs to be supplemented by a determination of the LD$_{50}$. This requires the use of a larger number of animals. It was Trevan who proposed the term "LD$_{50}$," meaning the dose which is lethal to 50% of animals. When a suitable range of doses is administered, a certain dose will be found that, even though it may have an effect on the animal, will not kill (LD$_0$), while a higher dose may kill every animal (LD$_{100}$). Between LD$_0$ and LD$_{100}$ we have a dosage range which indicates the percentage of animals killed, increasing in direct proportion to increased dosage.

Eye Toxicity. As recently described by M. Keplinger, damage to the eye can have severe consequences. Therefore, a determination of potential effects of chemicals in the eye is made. The two areas of concern are local effects such as irritation, anesthesia, mydriasis or miosis, and systemic effects due to absorption from the lacrimal sac.

Very small quantities of some chemicals (certain organophosphates, for example) can cause marked systemic effects, including death, through contact with the eye. With such compounds, goggles are necessary to prevent accidental splashing into the eyes.

Local effects in the eye, especially irritation, are usually determined first by introducing 50 mg or 0.1 ml of the chemical into the eye of the rabbit. The irritation is scored at 1 minute, 1, 24 and 72 hours and at 7 days following the initial instillation. The criteria used for the descriptive rating are the frequency, the extent and persistence of irritation or damage which occur in the three ocular tissues—cornea, iris and conjunctiva. Each tissue is rated with regard to density of damage and area of damage, according to a standard scoring system. This test is usually conducted in four to six rabbits. The totals of the irritation scores are divided by the number of animals to give a mean eye-irritation score. The maximum possible score at any one examination and scoring period is 110 points, which indicates maximal irritation and damage to all three ocular tissues. A score of zero obviously indicates no irritation whatsoever.

In this system, special emphasis is placed upon irritation or damage to the cornea. It is felt that this indicates a more serious hazard. It is good practice to observe the animals until the scores in the eyes return to zero (even if it is longer than seven days) or until it is concluded that permanent damage to the eye has occurred.

The descriptive ratings are evaluated and given a verbal evaluation as follows:

Nonirritating, practically nonirritating, minimally irritating, mildly irritating, moderately irritating, severely irritating, extremely irritating, maximally irritating and permanent damage.

The eye of the rabbit is rather sensitive and may show much more irritation than the eye of man. Shampoos, soaps and detergents are examples of materials which cause very marked effects in the eye of the rabbit but do not cause as marked effects in man. The monkey has been used as an experimental animal for eye irritation studies in an attempt to find an animal model which is more similar to man. While the monkey is used, particularly with those compounds which cause rather severe damage to the eye of the rabbit, it is not used nearly as commonly as the rabbit for these tests. (*J. Occ. Med.*, **13**, 2, 1971).

Reproduction Studies provide valuable information on fertility, gestation, viability, lactation and survival of the offspring. These tests are usually conducted on mice or rats for no less than three generations. Usually the compound to be tested is incorporated into the diet, and feeding of the parent generation is continued through weaning of the second litter. These offspring are selected for breeding the second generation. The offspring are exposed to the test compound via the uterus, the mother's milk, and the subsequent diet. (For details see "Pesticides Symposia," edited by Wm. B. Deichmann, published by Halos and Associates, Inc., 9703 S. Dixie Hwy, Miami, Florida 33156; p. 125–138; authors: Keplinger, Deichmann and Sala.)

Teratogenicity. Some compounds, administered to a pregnant female, cause deformities in the offspring. If this occurs and the young are delivered, the deformed young are cannibalized. For this reason the protocol for a teratogenic study differs from that of a reproduction study. In a teratogenic study the compound is administered to the pregnant animal, usually during the critical period of organogenesis but sometimes throughout the entire gestation period. Prior to delivery, the fetuses are taken by Caesarian section and examined for abnormalities. Rabbits, rats and mice most commonly are used for such studies; however, other species, particularly primates, have been used. (M. L. Keplinger, *J. Occ. Med.*, **13**, 6, 1971).

Mutagenicity. These studies are conducted to discover if a compound is capable of producing genetic damage. Some ancillary methods for mutagenic studies using microorganisms, plants and insects are available. These provide certain information, but it has been stated that they do not provide data directly applicable to potential effects on mammalian cells. Some methods utilizing mammalian cells also are available. One of these, the dominant lethal study in mice, is a study for detecting certain mutagenic changes in male germinal cells. Male mice are treated and then are mated with untreated females. The effects are then measured in the female. The host mediated assay is directly or indirectly related to the mammal. In this study mutagenic changes induced in microorganisms or cultured mammalian cells, resident in the peritoneal cavity of treated host animals, are determined. Mutations may include changes in pigmentation and nutritional requirements. Another type of mutagenic study utilizing mammalian cells is the cytogenic study. These studies involve karyotyping of cells (leukocytes, germinal cells, etc.) of treated animals. (M. L. Keplinger, *J. Occ. Med.*, **13**, 6, 1971).

Chronic Toxicity. The primary purpose of a chronic toxicity test is to administer (by ingestion, inhalation, etc.) a certain compound for a period of time sufficiently prolonged to bring to light effects that cannot be produced by tests of a few weeks to months. Chronic inhalation tests are needed by industrial hygienists for setting the maximum allowable concentration of toxic vapors and dusts in the air to which workers may be exposed for eight hours daily for the working lifetime.

Generally, rats are fed an experimental diet for two years, and dogs are fed for a period of two to three years. (These experiments are not designed to test for carcinogenic action.) In most laboratories it is customary to feed diets which are contaminated with three or four different dosage levels of a certain compound. An inherent difficulty in a chronic rat feeding experiment lies in the fact that half, or more, of even the control (unexposed) rats do not survive a two-year period. In such an experiment it becomes the responsibility of the pathologist to distinguish between the micropathological changes of old age and those induced by the compound.

Carcinogenicity. Lifetime rat and mouse feeding studies are usually employed to determine possible tumorigenic or carcinogenic properties of a compound. Recently the hamster has also been used for such studies. The points to consider in testing for carcinogenicity include the following:

1) Feeding of the compound for as long a period as possible—preferably the lifespan of the animal.
2) Starting the feeding of the compound to weanling males and females.
3) Administration of two or three dosages (the highest dose being the highest tolerable dose) in addition to the feeding of the uncontaminated diet.
4) Use of a large enough group of animals to provide an adequate number of survivors to live out a full lifespan.
5) Sacrificing of old animals before they die to obtain material suitable for histopathologic study.
6) Conduction of reproduction studies.
7) Careful consideration of feeding a well-balanced basic diet, one that will neither inhibit nor support the formation of tumors.
8) Careful selection of the strain of animals, keeping in mind the interpretation of positive data with animals that are hypo- or hypersensitive to the compound.

For the testing of bladder carcinogens, the beagle dog is the animal of choice. (*J. Nat. Can. Inst.,* **43**, 263, 1969.)

Federal Legislation: According to Alan T. Spiher, Jr., three federal laws, in addition to numerous state and local regulations, are of concern to the toxicologist in planning and interpreting his experiments. The individual state and local regulations are too diverse to be considered here; but for any compound which may find its way into interstate commerce, (1) the Federal Food, Drug, and Cosmetic Act, as amended, (2) the Federal Insecticide, Fungicide, and Rodenticide Act, and (3) the Federal Hazardous Substances Labeling Act, have the most important bearing. Chemicals on which toxicological studies are conducted may be divided into several groups:

(1) *Functional food additives:* those additives added to food deliberately to accomplish some technical effect. Such additives would include preservatives, antioxidants, flavoring agents, emulsifiers, etc.

(2) *Incidental food additives:* present in food only because of the use of an additive somewhere in the growth, development, production, or handling of the food. Such additives would include migrants from packaging materials, tissue residues of therapeutic agents administered to animals, such as insecticides, fungicides, herbicides, etc.

(3) *Accidental food contaminants:* only rarely found in food because of some unintentional occurrence. These chemicals might include anything not included in the two previous categories.

(4) Economic poisons used on raw agricultural products and in manufactured food.

(5) Economic poisons which do not come in contact with food. The term "economic poison" is defined under the Federal Insecticide, Fungicide, and Rodenticide Act, and includes insecticides, fungicides, rodenticides, and herbicides. It is any substance or mixture of substances intended for preventing, destroying, repelling or mitigating any insects, rodents, nematodes, fungi, weeds, and other forms of plant or animal life or viruses, except viruses in or on living man or other animals, which the Secretary shall declare to be a pest, and any substance or mixture of substances intended for use as a plant regulator, defoliant, or desiccant. (See **Pesticides**).

(6) Industrial chemicals and household chemicals which have toxic potential to those who handle them.

(7) Drugs.

(8) Cosmetics.

All substances which are, or may become, components of food, drugs, and cosmetics, are regulated by the Federal Food, Drug, and Cosmetic Act. Economic poisons which may be in or on food are regulated by both the Food, Drug, and Cosmetic Act, and the Insecticide, Fungicide, and Rodenticide Act. Economic poisons which do not become a component of food are subject only to the latter act.

Chemicals which do not fall under either of the two acts mentioned above, but which are hazardous substances intended or suitable for household use, are regulated by the Federal Hazardous Substances Labeling Act. This act defines a hazardous substance as "any substance or mixture of substances which (i) is toxic, (ii) is corrosive, (iii) is an irritant, (iv) is a strong sensitizer, (v) is flammable, or (vi) generates pressure due to decomposition, heat, or other means if such substance or mixture of substances may cause substantial personal injury or substantial illness during, or as approximate result of, any customary or reasonably foreseeable handling or use, including reasonably foreseeable ingestion by children."

The Federal Food, Drug, and Cosmetic Act, as amended, provides, among other things, that a food shall be deemed to be adulterated "(A) if it bears or contains any added poisonous or added deleterious substance (other than one which is (i) a pesticide chemical in or on a raw agricultural commodity; (ii) a food additive; or (iii) a color additive) which is unsafe within the meaning of section 406, or (B) if it is a raw agricultural commodity and it bears or contains a pesticide chemical which is unsafe within the meaning of paragraph 408 (a); or (C) if it is or it bears or contains any food additive which is unsafe within the meaning of section 409."

Sections 408 and 409 provide for the promulgation of regulations establishing the conditions under which a pesticide or a food additive may be safely used. Such regulations are promulgated only upon the conclusion by FDA based on adequate data that the proposed uses are safe. In reaching the conclusion that the proposed use of any additive in food is safe, the proponent must supply adequate toxicological information to establish the safety of the additive under the proposed conditions of use. This decision may also take into account other similar or related substances in the food supply found in the total diet of the consumer. Additionally, the law requires that the tolerance limitation for a food additive shall be established at a level no higher than that which the Secretary finds to be reasonably required to accomplish the physical or other technical effect for which such additive is intended. Procedural regulations promulgated by the Food and Drug Administration and found in Title 21 of the Code of Federal Regulations spell out in greater detail the toxicological information to be supplied as a part of pesticide or food additive petitions.

The Federal Food, Drug, and Cosmetic Act, as amended, is also concerned with chemicals which may be used as drugs and cosmetics. The law provides in section 505, that before a new drug may be shipped in interstate commerce, a new drug application must have been submitted and approved. One of the requisites for such approval is that sufficient toxicological evidence be presented to show that the drug is safe for use as directed in the labeling. With respect to cosmetics, the Act provides in section 601 (a) that a cosmetic may not contain any poisonous or deleterious substance which may render it injurious to health when used according to directions. No provision in the law requires pretesting of cosmetics before they enter interstate commerce, a requirement which is in effect for pesticides, food additives, and drugs.

The Federal Hazardous Substances Labeling Act defines the term "hazardous substance," as, among other things, any substance or mixture of substances which is toxic. The term "toxic" is further defined as applying to any substance which has the capacity to produce personal injury or illness to man through ingestion, inhalation, or absorption through any body surface, and relates the degree of toxicity and other hazardous terms to LD_{50} figures for animals or other appropriate tests prescribed by regulations and requiring the evaluation of the toxicological significance of animal experiments.

The Federal Insecticide, Fungicide, and Rodenticide Act requires that, before any economic poison may be shipped in interstate commerce, it must be registered by the manufacturer with the Pesticide Regulation Division of the Agricultural Research Service of the Department of Agriculture. The requirements for approval of the registration include the submission of satisfactory labeling and data to establish that the appropriate claims are supportable. The law states that the labeling must include, among other things, an adequate caution statement, and adequate directions for use. In the case of a chemical highly toxic to man, an antidote statement is also required. It is in regard to the requirement of adequate directions *for use* that this law interacts with the Federal Food, Drug, and Cosmetic Act, for if the label contains directions for applying, for instance, an insecticide on crops at a given rate which will result in contamination of the food, when marketed, in excess of the tolerance set by the Food and Drug Administration, then the directions for use will not be considered adequate and registration will not be approved.

The provisions of these federal laws make it plain that in these days of enlightened legislation for the protection of the public, a great deal of toxicological information is necessary before a chemical, falling into one of the categories under discussion, can be marketed. Furthermore, with the passage of new laws and an increasingly strict interpretation of old ones, the amount of legally required toxicological information is constantly increasing.

WM. B. DEICHMANN
W. E. MACDONALD

TRACE ELEMENTS ANALYSIS

Trace elements in chemical systems are often important either in their own right or as indicators of chemical processes occurring in the system as a whole. However, analytical chemistry is frequently the greatest limitation on our ability to study trace element behavior, and new analytical methods seek to overcome these limits in specific applications. Some of the factors important in trace elements analysis are the following:

Sensitivity. Trace elements may occur either thinly dispersed, such as in biological or geological samples, or at relatively high concentrations but in small amounts, such as in dust particles or on crystal surfaces. In either case an analytical method must be sensitive, but the former also requires a method which works well with samples at high dilution. Methods in which wet chemical steps are included permit the analysis of samples as large as necessary to overcome sensitivity limitations inherent in the method of determination, and these are suitable for dispersed elements. Sensitive color reactions in solution, which detect submicrogram amounts of many elements, and pile neutron activation analysis, which detects most elements at nanogram levels and below, are examples. The analysis of trace elements at higher concentrations in very small samples may be accomplished nondestructively by x-ray fluorescence analysis or electron probe microanalysis.

Nondestructive Analysis. Analysis of very valuable samples or samples where speed or other considerations are important is often best performed by a method where emission from the irradiated sample is a measure of the amount of the trace element present. Irradiation with x-rays or an electron beam is nondestructive for many sample materials and gives rise to x-rays characteristic of the elements present, thus permitting selective analysis of complex mixtures even for elements at relatively low concentrations. Neutron activation analysis can be nondestructive under certain circumstances, especially where the matrix of the sample does not become very radioactive or where the trace element radioactivity decays by emitting characteristic gamma rays which can be counted selectively.

Selectivity. In the analysis of mixtures of trace elements, determination of an element without interference from other elements present may save time and improve accuracy. In colorimetry the most desired color-producing reagents are those which are selective as well as sensitive. Electrochemical methods of analysis of aqueous solutions may be highly selective by means of careful control of current and voltage relationships. Spectroscopy in general affords selectivity by focusing on particular wave lengths of emitted radiation, such as in x-ray fluorescence, infrared spectroscopy, fluorimetry, emission spectroscopy, and flame photometry, or of absorbed radiation, as in atomic absorption.

Accuracy. For trace element determination accuracy is often difficult to achieve, either because of insensitivity of the analytical method or because of contamination of the sample during the analysis in the laboratory. Nondestructive analysis avoids many contamination difficulties by avoiding chemical reagents, and neutron activation analysis has greater contamination control than other wet chemical methods by inducing radioactivity by neutron capture in the pile before chemical handling. In some analytical methods the matrix of the sample may affect the apparent absolute values of trace element concentrations, and a good method should be free of this complication. Thus, in emission spectrographic analysis samples and standards of the same matrix should be compared, in x-ray fluorescence analysis interference in x-ray intensities between different elements should be avoided, and in activation analysis appreciable absorption of the neutron flux during irradiation should not occur.

Speed. If all conditions of sensitivity and accuracy are satisfied, speed of analysis may be the determining factor in choosing an analytical method. Nondestructive methods are frequently rapid, and in any method the calculations should be performed as quickly as possible, such as by a digital computer. In general, ease of analysis as well as the reliability of the results obtained will continue to make the analytical chemist search for new and better ways of trace elements analysis for many specialized applications.

JOHN W. WINCHESTER

Cross-references: *Absorption Spectroscopy, X-rays.*

TRANSFERENCE NUMBERS

The total current passing through an electrolyte in the fluid state, whether in solution or in the fused state, is carried by all the ions present in the fluid state. That portion of the total current carried by any species of ion is called the *transference number* of the ion. The symbol for transference number used here will be t (t^+ for the transference number of a positive ion and t^- for the transference number of a negative ion).

The simplest case to consider is that of an uniunivalent electrolyte in solution. The equivalent conductance Λ of an electrolyte at any concentration, c, may be defined as the conductance of the electrolyte in a solution of the concentration in question when placed between electrodes 1 cm apart and of sufficient area to encompass the volume of solution containing one gram equivalent weight of the electrolyte. If the conductance of a 1-cm cube of the solution, called the specific conductance, is L, then

$$\Lambda = \frac{1000\,L}{c} \qquad (1)$$

where c is the normality of the electrolyte in solution.

If $\mathbf{F}$ is the faraday of electricity (96,487 coulombs), α the degree of dissociation of the electrolyte and u^+ and u^- the respective mobilities of the positive and negative ions in centimeters per second, then from conductance theory

$$\Lambda = \mathbf{F}\alpha(u^+ + u^-) \qquad (2)$$

Now the current flowing through a solution of electrolyte enclosed in a tube of length l and cross section A under an impressed electromotive force E across the tube is given by the equation

$$I = \frac{ELA}{l} = 0.001\,EAc\alpha\mathbf{F}(u^+ + u^-)/l \qquad (3)$$

Of this total current, the parts I^+ and I^- carried by the positive and negative ions are

$$I^+ = 0.001EAc\alpha\mathbf{F}u^+/l \qquad (4)$$

and

$$I^- = 0.001^-EAc\alpha\mathbf{F}u^-/l \qquad (5)$$

respectively.

The fractions of the total current carried by the two ions are, respectively:

$$t^+ = \frac{I^+}{I} = \frac{0.001EAc\alpha\mathbf{F}u^-/l}{0.001EAc\alpha\mathbf{F}(u^+ + u^-)/l} \qquad (6)$$

$$= \frac{u^+}{u^+ + u^-}$$

and

$$t^- = \frac{I^-}{I} = \frac{u^-}{u^+ + u^-} \qquad (7)$$

Thus

$$t^+ + t^- = 1 \qquad (8)$$

The ion-forming portion of an electrolyte is called the ion constituent. If α is the degree of dissociation of an electrolyte, we can write the positive and negative ion constituent mobilities respectively, as

$$U^+ = \alpha u^+; \quad U^- = \alpha u^- \qquad (9)$$

and Λ becomes from Eqs. (2) and (9)

$$\Lambda = \mathbf{F}(U^+ + U^-) \qquad (10)$$

Also Eqs. (6) and (7) become

$$t^+ = \frac{U^+}{U^+ + U^-}; \quad t^- = \frac{U^-}{U^+ + U^-} \qquad (11)$$

and from Eqs. (10) and (11)

$$U^+ = t^+ \Lambda/F; \quad U^- = t^- \Lambda/\mathbf{F} \qquad (12)$$

Thus, by measuring the transference number, t, of an ion and the equivalent conductance, Λ, of the solution of the electrolyte in which the ion occurs, the ion's constituent mobility can be calculated. If the degree of dissociation, α, of the electrolyte is also ascertained, the ion mobility can also be calculated by means of Eq. (9).

Now Λ of the electrolyte is equal to the sum of the equivalent conductance λ^+ and λ^- of the respective positive and negative ion constituents of the electrolyte, that is,

$$\Lambda = \lambda^+ + \lambda^- \qquad (13)$$

and from Eqs. (10) and (13)

$$\lambda^+ = FU^+; \quad \lambda^- = FU^- \qquad (14)$$

and from Eqs. (12) and (14)

$$\lambda^+ = t^+\Lambda; \quad \lambda^- = t^-\Lambda \qquad (15)$$

Thus, the ion constituent equivalent conductances can be calculated from measured values of transference numbers and equivalent conductance of the pertinent electrolyte.

Transference numbers can be measured using either the Hittorf, moving boundary, or analytical method. Transference numbers can also be found by measuring the potentials of the proper cells with and without transference and taking the ratios of these respective potentials.

The Hittorf method for measuring transference numbers is based on the fact that different ions move with different velocities. The positive cation moves toward the negatively charged cathode and the negative anion moves toward the positive anode during a transference number measurement. The ion that moves the faster will carry a larger fraction of the current, ion charge and other factors being equal, and will tend to increase in the region around the electrode to which it is attracted and to decrease in concentration in the region around the electrode from which it is migrating.

At the same time the electrode reaction taking place during the transference experiment may increase the ion concentration in the region surrounding the electrode from which the ion is migrating, and may decrease the concentration in the region around the electrode toward which the ion is moving. By analyzing a solution initially and finally for the number of equivalents of an ion in a given weight of solvent around an electrode and then calculating the equivalents of the ion added to the electrode portion due to the electrode reaction by a definite measured current flowing through the solution for a measured time, the transference number of an ion can be found.

Transference numbers can be measured more accurately by the moving boundary method. In this method two solutions with different colors or different refractive indices are placed in a tube graduated in fractions of milliliters. Electrodes are inserted in the bottom and top of the tube and a difference in potential is applied to the electrodes. The movement of the boundary is observed and the volume swept out by the moving ion is recorded.

Hittorf transference numbers can be obtained from transference numbers measured by the method of moving boundaries by proper correction of the observed volume through which a boundary sweeps. The volume correction Δv is found by taking into account the change in volume in an electrode portion due to the change in molal and partial molal volumes arising from the change in amounts of the elements, ions and compounds in the electrode region.

True transference numbers are those uninfluenced by movements of solvent by migration of solvated ions. In obtaining true transference numbers, a method similar to that of Hittorf is used. However, in water solvent, for example, a second nondissociated solute, such as sugar, is added and the changes of the amounts of both electrolyte and water in an electrode portion is referred to this second solute. The ratio of reference substance to water changes due to the water carried along by the migrating ions as water of solvation. In the case of the solute sugar, the change in concentration can be found using a polarimeter.

Transference numbers vary slightly with concentration, and their dependences on concentration in the case of different valence types of electrolytes have been subjected to theoretical treatment.

Transference numbers are used in mobility, electromotive force, and conductivity theories. Trans-ference measurements are very important in the determination of the extent of solvation of ions.

EDWARD S. AMIS

Cross-references: *Electrochemistry, Ions, Solutions.*

TRANSITION ELEMENTS

In the normal building-up of the chemical elements by adding electrons one by one, simultaneously increasing the nuclear charge one by one, the completion of an outermost shell octet is always followed by the formation of a new outermost principal quantum level having a principal quantum number one unit higher. Thus, no atom of any element ever possesses, in its isolated condition, more than 8 electrons in its outermost shell. But when the s orbital of the new shell has been filled, then before the p orbitals of that shell become occupied, the filling of that outer shell is interrupted long enough to fill in the underlying empty d orbitals if there are any. Such d orbitals are designated by the orbital quantum number 2, which requires that the principal quantum level in which they appear must be number 3 or higher, since the orbital quantum number has a maximum value of $n - 1$. From the third principal quantum shell on, then, each shell includes five d orbitals. The sequence of filling in building up the elements successively is that electrons begin to occupy these d orbitals only after the s orbital of the next higher shell has been filled.

Once the filling of the underlying d orbitals has been begun, then these d orbitals become available for participation in chemical bonding, along with the s and p orbitals of the outermost shell. Elements which by virtue of this partial occupancy of their underlying d orbitals can employ them together with outer shell orbitals for bonding are called "transition elements." This ability to use both the outermost shell and the underlying d orbitals clearly distinguishes an element as a transition element, for major group elements show no evidence of using other than their outermost shell electrons and orbitals in the formation of bonds. This difference results in distinct differences in the chemistry of the two principal classes of elements, major group and transition, making it convenient to discuss these classes separately.

From the fourth period of the periodic system on, the filling of the outermost shell s orbital is followed by a succession of transition elements, nine in number, which correspond to the filling of the five d orbitals through the ninth electron. The tenth d electron, when the outermost s orbital contains two electrons, produces a major group element because with a $d^{10}s^2$ structure, the underlying d orbitals can no longer participate in bonding and only the outermost shell orbitals and electrons are available. Thus, although zinc, cadmium, and mercury are commonly included as the last transition element in their respective periods, they do not show the characteristic and unique properties of transition elements that are based on underlying d orbital participation. The first series of transition elements therefore begins with element number 21, scandium, and continues through copper, 29. The second series of transition elements begins with yttrium number 39, and ends with silver 47. A third series of transition elements begins with lanthanum number 57 but does not end until gold 79, and a fourth series is started at actinium number 89 and to date progresses through the highest atomic number yet discovered, 105.

The reason for the extended third and fourth transition series is that the filling of the $(n-1)$ d orbitals is interrupted after lanthanum and actinium by the filling of the $(n-2)$ f orbitals, thus adding fourteen more elements to each series. These interruptions within interruptions are called "inner transition elements" and are discussed in a separate article.

By virtue of the fact that none of the transition elements has more than two electrons in the outermost principal quantum level of its atoms, all transition elements have a surplus of outer shell vacancies over electrons and therefore exhibit the typical properties of the metallic state. Their metallic bonding is much stronger than that of any major group metal and on the average considerably stronger than that of the inner transition metals. Carbon has by far the greatest atomization energy of any of the major group elements. Its bonding energy is nevertheless exceeded by four of the transition metals, niobium, tantalum, tungsten, and rhenium, and nearly equalled by five more of these metals. This relatively high strength of bonding imparts great mechanical strength, as well as high melting points and very low volatility. It also contributes to the valuable resistance which most of these metals in bulk exhibit toward corrosion. Therefore, taken as a class, the transition elements have their greatest importance in their practical applications as free metals and their alloys. Included, of course, are all forms of iron and steel, titanium, copper, chromium, tungsten, and many other valuable metals of construction.

Among the transition elements are the densest of ordinary substances. A dozen transition metals are denser than lead, and at 22.5 grams per milliliter, osmium is about twice as dense.

With the probable exception of Group IIIB, scandium, yttrium, and lanthanum, all the transition elements exhibit variable oxidation states in their chemistry. For example, titanium, with d^2s^2 valence electrons, can lose two, three, or four electrons partially, thus exhibiting oxidation states of +2, +3, and +4. It is very interesting that within each transition group, again excepting IIIB, the lower oxidation states appear more stable for the first member of the group whereas the second two members of the group are much more similar to one another than to the first member and are more stable in the higher oxidation states than in the lower. No simple explanation seems yet available. A good illustration is afforded by Group VIIB, in which Mn_2O_7 is strongly oxidizing and explosively unstable whereas rhenium metal in moist air is oxidized spontaneously to Re_2O_7, which is thermally stable and relatively nonoxidizing. Technetium is much more similar to rhenium than to manganese.

Although the d orbitals that impart special interest to the transition elements are of equal energy in an isolated atom, under the influence of surrounding atoms of different elements this equality of energy is removed. The basic reason for this is that the distribution of d orbitals in space around the nucleus gives them different orientation with respect to a given neighboring atom. The energy of a d orbital that is directed toward an electron pair of a neighboring atom is raised above that of a d orbital not so directed. The effects of these energy differences are discussed elsewhere (See **Ligand Field Theory**). It may be of interest here to mention, however, that the difference created between d orbital energies frequently corresponds to a frequency of visible light, so that absorption in the visible region is possible through promotion of a d electron from a lower to a higher energy d orbital. Consequently, compounds of the transition elements commonly show color, a variety of colors being exhibited to correspond to the different oxidation states.

Reliable values of electronegativity for the transition elements are not yet available, for their chemistry is complicated by variation of oxidation states and ligand field effects. However, qualitatively these elements are probably mostly quite similar in electronegativity, having values ranging between those of calcium and magnesium. Some of the platinum metals are probably considerably higher in electronegativity, and gold may be the most electronegative of all the metallic elements, at least of the transition series.

Most commonly, the simpler compounds of the transition elements are those with the nonmetals, especially oxygen, sulfur, and the halogens. In compounds of a particular transition element with a nonmetal, in which the transition element exhibits a variety of oxidation states, the variation in properties follows that expected from consideration of electronegativity and partial charge. The partial negative charge acquired by each nonmetal atom will depend on the competition. The higher the ratio of nonmetal atoms to metal atoms, the greater the competition among nonmetal atoms for the electrons of the metal, and therefore the lower the partial negative charge on each nonmetal atom. The compound will correspondingly tend to be lower melting, more volatile, chemically more oxidizing, and more susceptible to hydrolysis. For example, the (II) halides of the transition elements are all highly condensed solids of relatively high melting point and very low volatility, in keeping with their relatively polar bonding and high negative charge on halogen. They are not halogenating agents and they are not appreciably hydrolyzed. The halides of these same transition elements in higher oxidation states tend to be much lower melting, volatile compounds with appreciable halogenating ability and easily hydrolyzed.

The oxides of any given transition element in various oxidation states similarly exhibit the same kind of relationship to one another. The oxide of the lowest oxidation state is always the most stable, highest melting, least volatile, most basic, and least oxidizing compound, in keeping with the condition of its oxygen atoms which bear highest partial negative charge. Oxides of higher oxidation states of this transition element invariably are less stable, lower melting, more volatile, less basic or more acidic, and more oxidizing (more easily reduced), for therein the oxygen atoms have smaller partial negative charge because of the competition among oxygen atoms for the electrons from the metallic atom.

The elements at the end of each transition series,

copper, silver, and gold, have ground state electronic configurations that denote a penultimate shell of 18 electrons below the valence shell of one. The implication that these elements are therefore univalent major group elements is proved erroneous by the existence of many compounds of copper (II), silver (II), and gold (III), demonstrating the use of underlying *d* orbitals and electrons. Nevertheless, these elements are quite unusual among transition elements. In physical properties, such as density, melting point, and boiling point, they are intermediate between the nickel and zinc groups. In electrical and thermal conductivity they are in a class by themselves, far excelling all other elements.

That phase of the chemistry of transition elements which has attracted greatest attention and is currently of greatest interest generally is the chemistry of their coordination compounds. Thousands of such compounds have been made and no upper limit to the number that might be made could be set.

R. T. SANDERSON

Cross-references: *Electronic Configurations; Periodic Law and Periodic Table; Coordination Compounds; Ligand Field Theory.*

TRANSPORTATION AND DISTRIBUTION OF CHEMICALS

Dimensions

Freight Commodity Statistics are compiled annually by the Bureau of Accounts of the Interstate Commerce Commission. For Class I Rail Carriers, that is, those with annual operating revenues of at least $5,000,000, the latest figures are for the year 1970, some of which have been extracted as follows:

Statistics for water transportation are referred to later in this article.

Regulation

Transportation in the United States is regulated in the public interest. The Interstate Commerce Commission and the Department of Transportation share regulatory functions.

Interstate Commerce Commission. The 19th Century saw the birth and phenomenal growth of railroads. In the United States the new type of economical transportation developed along with the growth of the young country. Monopolistic rights were granted. Fortunes were made. The railroad was universally hailed and welcomed. But as the century wore on free competition began to have a ruinous effect on rail transportation as prices (freight rates) were cut to drive out competition. The Supreme Court ruled in 1886 that Congress had jurisdiction over interstate commerce. In 1887 the Act to Regulate Commerce, later to become the Interstate Commerce Act, was passed. The Interstate Commerce Commission is charged with administering the Act. An organization chart of its Sections and Bureaus can be obtained from the Commission's office in Washington, D.C. Administration of the Act in the public interest encompasses, broadly put, the financial health of the carriers subject to the Act, the granting and denying of operating rights, and the authority to establish freight rates and practices.

Freight rates are not fixed by the Interstate Commerce Commission in the way that telephone rates are made. Transportation companies—that is, rail-

	All Traffic	Chemicals & Allied Products	%
Carloads	26,478,726	1,608,681	6.1
Revenue	$11,351,054,869	$1,136,631,290	10.0

For Class I Motor Carriers (annual operating revenues of at least $1,000,000) the latest figures are for 1969, as follows:

	All Traffic	Chemicals & Allied Products	%
Truckloads	18,352,612	2,092,816	11.4
Revenue	$3,885,893,657	$537,279,899	13.8

Statistics were assembled annually for a number of years by the Bureau of Economics of the Interstate Commerce Commission based on a one per cent waybill sample of audited revenue carload waybills terminated by Class I railroads, including state to state distribution of traffic and revenue. Unfortunately this fountain of hard data was dried up for reasons of economy and the last terminating year was 1966. The data for subsequent years have been sorely missed and as a consequence the one per cent waybill sample may be reactivated. In the meantime, the 1966 statistics are still useful for making broad comparisons.

roads, pipelines, motor carriers, water carriers and freight forwarders—make their rates by the negotiating process, publish them and file them with the I.C.C. Unless protests are filed with the Commission, they generally go into effect. The Commission may review rates on its own motion, and occasionally does so. It will entertain protests filed before the rates take effect or complaints after that date.

The Commission has attempted to simplify the freight pricing structure by creating a national uniform classification of commodities and a rate for each class based on mileage. Sound as it might appear in theory, such a rigid system of class rates is not responsive to commercial reality. A class-mileage scale would arbitrarily determine markets, influence unduly the establishment of production locations without relation to real commercial considerations, and force the carriers into an economic straight jacket that would erode commercial competition.

When regulation of transportation was first enacted the railroads were, beyond doubt, monopolists. This situation has changed. Motor carriers and barges create competitive forces on the freight prices of some commodities. To many, therefore, regulation no longer appears to be needed. Advocates of de-regulation maintain, further, that government has usurped management prerogatives and does a poor job of it. They also attack the justice of the actions and decisions of the Commission. Nevertheless, those who favor continued regulation look upon the Interstate Commerce Commission as the bulwark of a national system of transportation. In their view, legislation to abolish the Commission would have to provide for some other means of restraining carriers from mutual self-destruction.

Department of Transportation. The Department of Transportation is a new Cabinet-level department. Created in 1966 by means of Public Law 89–670, it began to function on April 1, 1967. About a dozen agencies, already operating, were placed under its aegis and several new organizations were formed. Transferred from the I.C.C. were the Bureau of Railroad Safety and the Bureau of Motor Carrier Safety. The Coast Guard was transferred from the Treasury Department.

The chemical industry encounters the department most often through its Office of Hazardous Materials and its implementing Board, the Hazardous Materials Regulations Board. The Board has ultimate responsibility to regulate the transportation by rail, highway, water, pipeline and air of explosives and other dangerous articles, including radioactive materials, etiologic agents, flammable liquids, flammable solids, oxidizing materials, corrosive liquids, compressed gases, and poisonous substances.

Regulations for the transportation of hazardous materials, also procedures to implement them, were inherited by the new department from those agencies that were formerly charged with those responsibilities. Also, the Board began to formulate its own rules, inviting shippers, carriers and the public to consult and advise. Evidence of the regulations is visible on the highways. For example, a truck-trailer carrying a corrosive liquid carries signs reading CORROSIVES; one carrying a flammable liquid, FLAMMABLE. There are also extremely precise regulations with regard to packaging and labeling.

The U.S. Coast Guard requires cargo signs and cards to be displayed in the pilot house and on the tank barge for all tows of commodities that are designated as Dangerous Cargoes in Sub-Chapter O of the Coast Guard regulations, 46 CFR 150.

In addition to D.O.T. regulations there are regulations prescribed by local communities and regions. Transportation in tunnels and on bridges is regulated. The Port of New York Authority, for example, prescribes rules that are based upon laws of the States of New Jersey and New York.

Safety

In-plant safety is a common industry practice. In the chemical industry the concern for safety reaches out beyond the plant into transportation.

The Manufacturing Chemists Association, representing 90% of the production capacity of basic industrial chemicals in the United States and Canada, created the Chem-Card system. Each Chem-Card covers a specific chemical possessing hazardous properties that is moved in tank trucks. The card describes the nature of the hazard and what to do in case of an emergency. The driver carries the card in his vehicle. A manual of all such cards is provided for the ready reference of those local officials who make judgments as to actions to be taken at the scene for the public safety.

The Manufacturing Chemists Association has also designed a Chemical Transportation Emergency Center (CHEMTREC) to alert public safety and fire fighting personnel as to the hazards in connection with a specific accident, with 24 hours a day telephone response to accident calls. Individual corporations have similar systems.

Additional information on the subject of the safe transportation of chemicals can be obtained from the Manufacturing Chemists Association, Washington, D.C., the National Fire Protection Association, Boston, the Bureau of Explosives of the Association of American Railroads and other sources referred to in this article.

Kinds of Transportation

Common and Contract Carriage. Transportation in the United States is performed, in the main, by common carriers. These are carriers that are operated as profit-making private businesses in the service of the public. Our railroads are common carriers. Many of our truck lines are common carriers, holding themselves out to be fit, willing and able to transport commodities, and possessing certificates to do so in interstate and intrastate commerce issued to them by the Interstate Commerce Commission and by State Commissions. Other truck lines are contract carriers to whom have been issued permits to perform certain defined contracts to carry a described commodity or commodities between specified places for a named shipper or consignee.

Barge transportation on the inland waterways of the United States is performed by common and contract carriers. The common carriers hold themselves out to serve the public. The contract carriers operate under contract, very much as the motor carriers do.

A kind of contract carriage by rail exists in the form of guaranteed annual volume arrangements. In such arrangements the tonnage and freight rates are agreed upon by the two parties and published in a tariff. Legally, the provisions of the tariff are available to all. In fact, however, the arrangements are usually tailored to fit but one situation. Nevertheless, the rail carrier is technically a common carrier and not a contract carrier.

Rail Transportation. The national rail transportation system is privately owned. It is made up of about 70 Class I railroads, and approximately 420 Class II railroads. These privately owned corporations are operated for profit. Each owns its own locomotives, its own freight cars (with some exceptions) and its own tracks, stations, yards, signal equipment, and rights of way, etc.

A freight car of one railroad may move on the tracks of any railroad in the system. Revenue received for the joint haul is divided on a prearranged basis. Cars must be constructed to conform with requirements for physical interchanges, and with weight and size limitations of bridges, tunnels and rails. The Interstate Commerce Act requires railroads to provide transportation, including railroad cars. By interpretation, the railroads may fulfill this obligation, in the case of special types of cars, like tank cars, by paying an allowance to private car owners for providing them.

Railroad car service is a substantial problem for the chemical shipper. A nation-wide box car shortage, particularly at grain loading times, has become chronic. Interstate Commerce Commission orders for the redistribution of cars in the direction of heavy seasonal loading, such as the mid-West during harvest time, have been stop-gap and ineffectual. Other remedies include a proposal to create a quasi-governmental nonprofit corporation to purchase cars and create a pool.

Private car service is a different kind of problem, but a serious one, particularly for the chemical industry. Because tank cars are almost exclusively furnished by shippers, an allowance for their use is paid by the carrier to the owner. The unprecedented growth of the chemical industry after 1945 was accompanied by improvements and innovations in the design of highly specialized tank cars and by an increase in their production. Some of these cost little more than a box car. But the great majority are very sophisticated vehicles whose cost runs from $20,000 to $90,000. The allowances paid by the railroads seldom cover the full cost. The Interstate Commerce Commission's decisions in I&S 8135 and I&S 8406 provide a comprehensive record of this subject for the student willing to dig.

Motor Transportation. In the section dealing with statistics it was indicated that 11% of all U.S. truck traffic consists of chemicals and allied products. A great proportions of this is carried in tank trucks. Of 70,000 tank trailers in the U.S. approximately 25% are used for the transportation of chemicals.

The design of tank trailers is severely limited by several requirements. Formulas established by the Federal Government and by the several states dictate weight, length and width, etc. These particulars vary from state to state. Full information on the subject is published by American Trucking Associations, Inc., Washington, D.C.

The National Tank Truck Carriers, Inc., Washington, D.C. publishes Commodity and Equipment Data Sheets and a Hazardous Commodity Handbook. These provide data gathered from authoritative sources, comprised of commodity names and characteristics, recommendations and requirements as to equipment and fittings; also methods, precautions and problems in handling.

In addition to common and contract carriage, a shipper has available to him the right to transport a commodity owned by him anywhere in the United States. This is private carriage. It may be done in motor vehicles owned and operated by the owner of the goods; or by a lease arrangement. The feature that distinguishes private carriage from carriage-for-hire is that ownership of the goods and control of the vehicle and drivers are exclusively in the same hands.

Private carriage is appropriate for chemicals, subject as they are to a multiplicity of government regulations that concern the type of vehicle, including special fittings, linings and hoses, hours and routes of travel, safety procedures and clothing, and the growing demands of ecology. The direct enforceable control that the chemical shipper can exert over his own vehicle and driver tends to provide equipment and service precisely tailored and operated for a particular product and a specific customer. The Private Carrier Conference, Inc., Washington, D.C. can provide information on this subject.

Water Transportation. The growth of the petrochemical industry after World War II was most pronounced where natural gas and large quantities of water occurred.

The great and sudden increase in waterside chemical plants was encouraged by voter acceptance of the proposition that public money spent on waterway improvement would produce benefits to society worth many times their cost. Deeper, straighter and safer channels, bank stabilization and flood controls were designed, authorized, budgeted and built. These improvements made it possible to employ larger towboats with larger tows just at a time when technology was capable of building them.

In 1938 sternwheelers were still being built. The largest modern towboat on the rivers was approximately 750 horsepower. By 1960 there were some towboats of 9000 horsepower and many in the 3000–4000 h.p. range. Transportation costs dropped from about four mills to about two mills per ton-mile.

Barge transportation of chemicals bred its own inventiveness. To take advantage of powerful, high-speed towboats, integrated sets of barges were designed. The bow barge was raked for speed and the dimensions of the several barges were designed to allow a single locking through locks 110 feet wide and 1200 feet long that were being built to replace the old 600-foot locks. Barges were equipped with tanks and fittings of special steels. Coatings, linings and gasket materials were applied to protect against corrosion by the chemical being carried, or to protect the chemical. Automated control systems were installed, both in barges and in towboats. High-speed pumps became standard equipment.

Statistics on water transportation are published by the Corps of Engineers. In 1970 over 61 million tons of chemicals were carried on the lakes and rivers and along the coasts of the United States. Waterways are maintained by the Corps of Engineers. There are 15,675 miles of commercially navigable waterways 9 feet deep or deeper and 9,868 miles under nine feet.

Coastwise traffic uses the seas and oceans and enjoys much deeper drafts, and the revolution that has taken place during the last quarter century in the transportation of chemicals has been particularly dramatic in its effect upon the carriage of chemicals by sea. The virtual death of the American merchant marine in the coastwise trades has removed from the oceans general cargo ships that, prior to 1942, plied between Atlantic Coast ports and between Atlantic and Gulf ports. They succumbed to increasing costs of vessel construction, higher operating costs, higher stevedoring costs and antiquated port facilities that were unable to generate revenues for modernization.

While prewar and wartime ships were fading from the scene, a new breed was being conceived and built to accommodate the expanding needs of the chemical industry. They were the specially designed chemical tankers. There are approximately 20 of these ships, most of them very sophisticated in terms of construction and fittings. They are half again as fast as the 10-knot Liberties. Additional information can be obtained at the Maritime Administration, Office of Ship Operations. For barge information, consult American Waterways Operators, Inc., Washington, D.C. and Water Transport Association, New York.

Air Transportation. A complete set of regulations covering restricted articles, their classifications, and required labelling is available from the International Air Transport Association in Montreal. While the tonnages of chemicals shipped by air are not great compared with other methods, air carriage of small quantities is often necessary; these IATA regulations should be in the hands of all chemical manufacturers. Their validity is recognized by the U.S. Department of Transportation.

Pipelines. The pipeline is here to stay as a transporter of chemicals. In the United States chemical pipelines, regulated and private, carry helium, ethylene, liquid ethane, butylene, butadiene, acetone, propylene, hydrogen, carbon monoxide, nitrogen, oxygen, ammonium nitrates, urea mixtures, ammonium phosphate mixtures, anhydrous ammonia, liquid sulfur and clear 95% saturated solution of rock salt.

Chemical pipelines are planned, under construction or operating in Mexico, Canada, European countries and Australia.

Williams Brothers Engineering Company, Tulsa, Oklahoma is well established in this field and can furnish additional information.

Freight Transportation Pricing

The little monster that is known as a freight rate has repelled the efforts of the average person to comprehend it. This is understandable. Other publicly regulated services have rates that reveal a pattern of uniformity. The telephone charge from New York to Chicago is the same as from Chicago to New York. Postal rates are made on the rigid zone and mileage basis. What is it about freight rates that makes them different?

The first step is to abandon the word "rate" with its connotation of scale, or mileage measure. Substitute instead the word "price." The second step is to clear from your mind the false notion that the Interstate Commerce Commission, or any Government body, fixes that price. In the great majority of transactions, the price is freely negotiated, a concept that is common to our economic system. Two free entrepreneurs—one a carrier, the other a purchaser of the service—negotiate a price for the service that is to be sold by the carrier to the ship-

per. In this negotiation the laws of the market place shape the end result. Competition from motor carriers, barge lines, or private carriage, or from a combination of them, will affect the outcome of the negotiations between a railroad and a shipper. The personalities of the individuals, ancillary considerations in connection with past or future transactions, the prospect of the negotiated price producing new business or an increase in future profits— these factors and others weigh more or less in the determination of the price, just as they would if the negotiations did not involve a regulated industry.

Regulation, however, does introduce some additional considerations. The carrier, by law, must more than cover his out-of-pocket costs. He may not set a price that is unreasonably high, or unduly discriminatory, preferential or prejudicial. As a consequence, the price negotiations must take into account the cost of the transportation service and the location of the user in relation to other users who might be unduly preferred or prejudiced or discriminated against. To weigh the merits and values of such regulatory concepts, the parties are guided by the Interstate Commerce Act and related laws and a body of case law. The carrier is free to publish the negotiated rate without further consultation. He also is permitted to fix prices after conference with and in conjunction with competing carriers. This exemption from the several antitrust acts responds to a need for economic orderliness in a national transportation system that is composed of mutual competitors who are required to serve the public interest. Economic regulation is the price that regulated carriers pay for the freedom to agree upon rates among themselves.

Disputes concerning the lawfulness of freight rates are adjudicated by the Interstate Commerce Commission and may be appealed up through the U.S. Supreme Court.

Trends

There are trends visible that may forecast future developments in transportation and distribution. Electronic data processing and computers are increasingly being utilized to record and report inventories, select replacement sources, schedule deliveries, keep track of cars, and produce rate tariffs. Greater use is being made of "intermodal" transportation—barge-storage-truck, for example—and of unit-trains and annual volume rates. Containers, trailers-on-flat-cars, and lighters-aboard-ships offer economies for certain commodities.

Economic regulation of transportation will almost certainly be continued but the trend seems to be toward somewhat freer competition.

The growing pressure of social forces will influence tomorrow's transportation and distribution. Concern for the quality of life has already resulted in Federal and State regulations which have required the expenditure of substantial sums for plant pollution control. The future may find requirements in many areas for closed systems in the transfer, transportation, and consumption of commodities which have pollutant characteristics.

Safe transportation of hazardous materials, already the subject of voluminous government regulations and of intensive industry research will probably receive more attention. The man who directs a chemical transportation system will have to look beyond engineering design and beyond calculated return on investment. He will be influenced more and more by the social impact of his decisions, and by the political climate in which his ideas and plans will be nurtured.

D. L. CAMPBELL KERR

TRANSURANIUM ELEMENTS

The elements of the periodic system of atomic numbers 89 (actinium) through 103 are so chemically similar to another group of elements, the rare earths or lanthanide series (atomic numbers 57–71) that their electronic structures must also be similar, and they have been given the name actinides. An inner electron shell, consisting of fourteen $5f$ electrons, is filled in progressing across the series. The elements immediately beyond the actinide series are sometimes called transactinide elements and differ in properties from the actinides. All known elements with atomic number greater than 92, uranium, have half-lives too short to have existed since their original creation. They were discovered and produced by nuclear reactions and synthesis and are called the transuranium elements. The presently known (1972) transuranium elements have the following names and symbols:

93, neptunium (Np); 94, plutonium (Pu);
95, americium (Am); 96, curium (Cm);
97, berkelium (Bk); 98, californium (Cf);
99, einsteinium (Es); 100, fermium (Fm);
101, mendelevium (Md); 102, nobelium (No);
103, lawrencium (Lr); 104, rutherfordium (Rf) or kurchatovium (Ku); 105, hahnium (Ha) or nielsbohrium (Ns). The names and symbols for elements 104 and 105 have been proposed but are not yet official.

Chemically the actinide elements are very similar, although the observed differences are those expected and anticipated from their unique position in the periodic system as part of a second rare-earth series. All have trivalent ions, which form inorganic complex ions and organic chelates. Also in common are acid-insoluble trifluorides and oxalates, soluble sulfates, nitrates, chlorides, and perchlorates. Neptunium, plutonium, and americium have higher oxidation states in aqueous solution (similar to uranium), but the relative stability of these states to the common trivalent ion becomes progressively less as one proceeds to the higher atomic numbers. This is a direct consequence, indeed an identifying feature, of the actinide role as a second rare earth type transition series.

Elements 104 and 105 fall in Groups IVB and VB of the Periodic Table and are similar in chemical properties to hafnium and tantalum, respectively.

The electronic configurations of the gaseous atoms of the transuranium elements (beyond the electronic structure of radon) as determined by spectroscopic means and atomic beam experiments are as follows (with predicted values given in parentheses):

neptunium, $5f^46d7s^2$; plutonium, $5f^67s^2$;
americium, $5f^77s^2$; curium, $5f^76d7s^2$;
berkelium, $5f^97s^2$;

californium, $5f^{10}7s^2$; einsteinium, $5f^{11}7s^2$;
fermium, $5f^{12}7s^2$; mendelevium, $5f^{13}7s^2$;
nobelium, $5f^{14}7s^2$; lawrencium, $5f^{14}6d7s^2$;
104, $5f^{14}6d^27s^2$; 105, $5f^{14}6d^37s^2$.

One of the most important methods for study and elucidation of chemical behavior of the actinide elements has been ion-exchange chromatography. Adsorption on and elution from ion-exchange columns has made possible the identification and separation of trace quantities of most of the actinides and in particular the transuranium elements. The behavior of each actinide element in this respect is very similar to its analogous rare earth element. This has made it possible to detect as little as one or two atoms when this small a number has been made in some of the transmutation experiments.

Neptunium (Np, atomic number 93, after the planet Neptune). Neptunium was the first of the synthetic transuranium elements to be discovered; the isotope Np^{239} was produced by McMillan and Abelson in 1940 at Berkeley, California, as the result of the bombardment of uranium with cyclotron-produced neutrons. The isotope Np^{237} (half-life of 2.2×10^6 years) is currently obtained in kilogram quantities as a by-product from nuclear reactors. Trace quantities of the element are actually found in nature due to transmutation reactions in uranium ores produced by neutrons which are present from natural sources.

Neptunium metal has a silvery appearance, is chemically reactive, melts at 637°C and exists in at least three structural modifications: α-neptunium, orthorhombic, density = 20.45 g/cm³ (25°C); β-neptunium (above 280°C), tetragonal, density (313°C) = 19.36 g/cm³; γ-neptunium (above 577°C), cubic, density (600°C) = 18.0 g/cm³.

Neptunium gives rise to five ionic oxidation states in solution: Np^{+3} (pale purple), analogous to the rare earth ion Pm^{+3}, Np^{+4} (yellow green), NpO_2^+ (green blue), NpO_2^{++} (pale pink), and NpO_6^{-5} or NpO_5^{-3} (green). These latter oxygenated species are in contrast to the rare earths which exhibit only simple ions of the (II), (III), and (IV) oxidation states in aqueous solution. The element forms tri- and tetrahalides such as NpF_3, NpF_4, $NpCl_3$, $NpCl_4$, $NpBr_3$, NpI_3, and oxides of various compositions such as are found in the uranium-oxygen system, including Np_3O_8 and NpO_2.

Plutonium (Pu, atomic number 94, after the planet Pluto). Plutonium was the second transuranium element to be discovered; the isotope Pu^{238} was produced in 1940 by Seaborg, McMillan, Kennedy and Wahl at Berkeley, California, by deuteron bombardment of uranium in the 60-inch cyclotron. Of by far the greatest importance is the isotope Pu^{239} (half-life of 24,360 years), produced in extensive quantities in nuclear reactors from natural uranium:

$$U^{238}(n,\gamma)\ U^{239} \xrightarrow{\ \beta-\ } Np^{239} \xrightarrow{\ \beta-\ } Pu^{239}$$

Plutonium has assumed the position of dominant importance among the transuranium elements because of its successful use as an explosive ingredient in nuclear weapons and the place which it holds as a key material in the development of industrial utilization of nuclear energy, one pound being equivalent to about 10,000,000 kilowatt hours of heat energy. Its importance depends on the nuclear property of being readily fissionable with neutrons and its availability in quantity.

Plutonium also exists in trace quantities in naturally occurring uranium ores. It is formed in much the same manner as neptunium, by irradiation of natural uranium with the neutrons which are present.

Plutonium metal can be prepared, in common with neptunium and uranium, by reduction of the trifluoride with alkaline earth metals. The metal has a silvery appearance, is chemically reactive, and melts at 639.5°C. It exhibits at least six crystalline modifications: α-plutonium (below 115°C), monoclinic, density = 19.737 g/cm^3 (25°C); β-plutonium (below 185°C) BC monoclinic, density = 17.77 g/cm^3 (150°C); γ-plutonium (below 310°C), orthorhombic, density = 17.19 g/cm^3 (210°C); δ-plutonium (below 453°C) FC cubic, density = 15.92 g/cm^3 (320°C); δ'-plutonium (below 480°C) tetragonal, density = 15.99 g/cm^3 (465°C); and ϵ-plutonium (above 480°C) BC cubic, density = 16.48 g/cm^3 (500°C).

Plutonium also exhibits five ionic valence states in aqueous solutions: Pu^{+3} (blue lavender), Pu^{+4} (yellow brown), PuO$_2^+$ (presumably pink), PuO$_2^{+2}$ (pink orange), and PuO$_6^{-5}$ or PuO$_5^{-3}$ (green). The ion PuO$_2^+$ is unstable in aqueous solutions, disproportionating into Pu^{+4} and PuO$_2^{+2}$; the Pu^{+4} thus formed, however, oxidizes the PuO$_2^+$ into PuO$_2^{+2}$, itself being reduced to Pu^{+3}, giving finally Pu^{+3} and PuO$_2^{+2}$.

Plutonium forms binary compounds with oxygen: PuO, PuO$_2$, and intermediate oxides of variable composition; with the halides: PuF$_3$, PuF$_4$, PuCl$_3$, PuBr$_3$, PuI$_3$; with carbon, nitrogen and silicon; PuC, PuN, PuSi$_2$; in addition, oxyhalides are well known: PuOCl, PuOBr, PuOI.

Because of the high rate of emission of alpha particles, and the physiological fact that the element is specifically absorbed by bone marrow, plutonium, as well as all of the other transuranium elements are radiological poisons and must be handled with special equipment and precautions.

Americium (Am, atomic number 95, after the Americas). Americium was the fourth transuranium element to be discovered; the isotope Am^{241} was identified by Seaborg, James, Morgan and Ghiorso late in 1944 at the wartime Metallurgical Laboratory (now the Argonne National Laboratory) of the University of Chicago as the result of successive neutron capture reactions by plutonium isotopes in a nuclear reactor:

$$Pu^{239}(n,\gamma)\ Pu^{240}(n,\gamma)\ Pu^{241} \xrightarrow{\beta^-} Am^{241}$$

Americium is produced in kilogram quantities. Since the isotope Am^{241} can be prepared in relatively pure form by extraction as a decay product over a period of years from intensely neutron-irradiated plutonium containing Pu^{241}, this isotope is used for much of the chemical investigation of this element. Better suited is the isotope Am^{243} due to its longer half-life (7370 years as compared to 433 years for Am^{241}). A mixture of the isotopes Am^{241}, Am^{242}, and Am^{243} in which the Am^{243} predominates, can be prepared by intense neutron irradiation of Am^{241} according to the reactions Am^{241}(n,γ) Am^{242}(n,γ) Am^{243}. The isotope may also be prepared by direct,

prolonged neutron irradiation of Pu^{239}, followed by chemical separation from plutonium and curium isotopes:

$$Pu^{239}(n,\gamma)Pu^{240}(n,\gamma)Pu^{241}(n,\gamma)$$
$$Pu^{242}(n,\gamma)Pu^{243} \xrightarrow{\beta^-} Am^{243}$$

Americium can be obtained as a silvery white reactive metal by reduction of americium trifluoride with barium vapor at 1000°–1200°C. It appears to be more malleable than uranium or neptunium and tarnishes slowly in dry air at room temperature. The density is 13.67 g/cm^3 (20°C) with melting point of 1176°C. It has a double hexagonal, close-packed crystalline structure at temperatures up to 1079°C (α-americium) and a face-centered cubic structure above 1079°C (β-americium).

The element exists in four oxidation states in aqueous solution: Am^{+3} (light salmon), Am^{+4} (pink-red), AmO$_2^+$ (yellow), and AmO$_2^{+2}$ (light tan). The trivalent state is highly stable and difficult to oxidize. AmO$_2^+$, like plutonium, is unstable with respect to disproportionation into Am^{+3} and AmO$_2^{+2}$. The ion Am^{+4} is unstable in solution unless in the form of a fluoride complex, although tetravalent solid compounds are well known. Divalent americium exists in solid compounds.

Americium dioxide, AmO$_2$, is the important oxide; Am$_2$O$_3$ and, as with previous actinide elements, oxides of variable composition between AmO$_{1.5}$ and AmO$_2$ are known. The halides AmF$_2$ (in CaF$_2$), AmF$_3$, AmF$_4$, AmCl$_2$ AmCl$_2$, AmI$_2$, and AmI$_3$ have also been prepared.

Curium (Cm, atomic number 96, after Pierre and Marie Curie). Although curium comes after americium in the periodic system, it was actually known before americium and was the third transuranium element to be discovered. It was identified by Seaborg, James, and Ghiorso in the summer of 1944 at the wartime Metallurgical Laboratory in Chicago as a result of helium-ion bombardment of Pu^{239} in the Berkeley, California, 60-inch cyclotron. It is of special interest because it is in this element that the first half of the transition series of actinide elements is completed.

The isotope Cm^{242} (half-life 163 days) produced from Am^{241} by the reactions Am^{241}(n,γ) Am^{242} $\xrightarrow{\beta^-}$ Cm^{242} was used for much of the early work with macroscopic quantities, although this was difficult due to the extremely high specific alpha activity. A better, but still far from ideal, isotope for the investigation of curium is Cm^{244}. Its somewhat longer half-life of 18 years still presents a problem of relatively high specific alpha activity but it has been used extensively because of its availability as the result of production by the reactions Am^{243}(n,γ) Am^{244} $\xrightarrow{\beta^-}$ Cm^{244}. Much better suited for such investigations are the longer-lived isotopes Cm^{247} (half-life 16,000,000) years and Cm^{248} (half-life 350,000 years) which are becoming available in increasing quantities as the result of their production in high-neutron-flux reactors through the successive capture of neutrons in the reactions Cm^{244}(n,γ Cm^{245}(n,γ) Cm^{246}(n,γ) Cm^{247}(n,γ) Cm^{248}; the difficulties of this long production chain are compounded by the need to separate the Cm^{247} and Cm^{248} from the remaining lower-mass number isotopes by mass spectrometric methods in order to obtain final products of the desired low specific alpha activity. Cm^{248} is also produced in relatively high isotopic purity as the alpha-decay daughter of Cf^{252}.

Curium metal, which melts at 1340°C, resembles the other actinide metals quite closely. Also, in common with other metals, it can be prepared by heating curium trifluoride with barium vapor at 1275°C.

Curium exists in aqueous solution predominantly as Cm^{+3}, although Cm^{+4} exists as complex fluoride ion. Solid CmO_2 and CmF_4 have been prepared as well as solid Cm_2O_3, CmF_3, $CmCl_3$, $CmBr_3$ and CmI_3. It has been demonstrated that curium(III) cannot be reduced in aqueous solutions.

Berkelium (Bk, atomic member 97, after Berkeley, California). Berkelium, the eighth member of the actinide transition series, was discovered in December 1949 by Thompson, Ghiorso and Seaborg and was the fifth transuranium element synthesized. It was produced by cyclotron bombardment of Am^{241} with helium ions at Berkeley, California.

Berkelium metal has been prepared on a microgram scale by alkali metal reduction of the trifluoride. It exists in a double hexagonal, close-packed phase (presumably the low-temperature phase) and a face-centered cubic phase, and melts at 986°C.

Berkelium exhibits two ionic oxidation states in aqueous solution, Bk^{3+} (yellow-green), and somewhat unstable Bk^{+4} (yellow), as might be expected by analogy with its rare earth homolog terbium. Solid compounds include Bk_2O_3, BkO_2 (and presumably oxides of intermediate composition), BkF_3, BkF_4, $BkCl_3$, $BkBr_3$, and BkI_3.

The existence of Bk^{249} with a half-life of 314 days makes it feasible to work with berkelium in weighable amounts so that its properties can be investigated with macroscopic quantities. This isotope can be prepared by the intense neutron bombardment of Cm^{244} as the result of the capture of successive neutrons by the reactions

$$Cm^{244}(n,\gamma) \quad Cm^{245}(n,\gamma) \quad Cm^{246}(n,\gamma) \quad Cm^{247}(n,\gamma)$$

$$Cm^{248}(n,\gamma) \; Cm^{249} \xrightarrow{\beta^-} Bk^{249}$$

Californium (Cf, atomic number 98, after the state and University of California). Californium, the sixth transuranium element to be discovered, was produced by Thompson, Street, Ghiorso and Seaborg in January 1950 by helium-ion bombardment of microgram quantities of Cm^{242} in the Berkeley 60-inch cyclotron.

The best isotope for the investigation of the chemical and physical properties of californium is Cf^{249} (half-life 352 years), produced in pure form as the beta-particle decay product of Bk^{249} which is available in only limited quantity because its production from lighter isotopes requires multiple neutron-capture reactions over long periods of time in high-neutron-flux reactors. Mixtures of californium isotopes produced by reactions such as

$$Bk^{249}(n,\gamma) \; Bk^{250} \xrightarrow{\beta^-} Cf^{250}(n,\gamma) \; Cf^{251}(n,\gamma) \; Cf^{252}$$

are also used but these have the disadvantage of high specific radioactivity, especially the spontaneous fission decay of Cf^{252}.

Californium metal can be prepared by the reduction of CfF_3 with lithium. It is quite volatile and can be distilled at temperatures of the order of 1100–1200°C. It appears to exist in two different cubic crystalline modifications.

Californium exists mainly as Cf^{+3} in aqueous solution (emerald green), but it is the first of the actinide elements in the second half of the series to exhibit the (II)-oxidation state, which becomes progressively more stable on proceeding through the heavier members of the series. It also exhibits the (IV) state in CfF_4 and CfO_2, which can be prepared under somewhat intensive oxidizing conditions. Solid compounds also include Cf_2O_3 (and higher intermediate oxides), CfF_3, $CfCl_3$, $CfBr_2$, $CfBr_3$, and CfI_3.

Einsteinium (Es, atomic number 99, after Albert Einstein). Einsteinium, the seventh transuranium element to be discovered, was identified by Ghiorso *et al,* in December 1952 in the debris from a thermonuclear explosion in work involving the University of California Radiation Laboratory, the Argonne National Laboratory, and the Los Alamos Scientific Laboratory. The isotope produced was the 20-day Es^{253}, originating from beta decay of U^{253} and daughters.

Einsteinium can be investigated with macroscopic quantities using the isotopes Es^{253} (half-life 20.5 days), Es^{254} (half-life 276 days) and Es^{255} (half-life 38.3 days), whose production by the irradiation of lighter elements is severely limited because of the required long sequence of neutron-capture reactions over long periods of time in high-neutron-flux reactors. Most of the investigations have used the short-lived Es^{253} because of its greater availability, but the use of Es^{254} will increase as it becomes more available. In any case the investigation of this element is very difficult due to the high specific radioactivity and small available quantities of the isotopes.

Einsteinium metal has been prepared, and it has been determined that it is even more volatile than californium metal.

Einsteinium exists in normal aqueous solution essentially as Es^{+3} (green), although Es^{+2} can be produced under strong reducing conditions. Solid compounds such as Es_2O_3, $EsCl_3$, $EsOCl$, $EsBr_2$, $EsBr_3$, EsI_2, and EsI_3 have been made.

Fermium (Fm, atomic number 100, after Enrico Fermi). Fermium, the eighth transuranium element to be discovered, was identified by Ghiorso, *et al,* early in 1953 in the debris from a thermonuclear explosion in work involving the University of California Radiation Laboratory, the Argonne National Laboratory, and the Los Alamos Scientific Laboratory. The isotope produced was the 22-hour Fm^{255}, originating from the beta decay of U^{255} and daughters

No isotope of fermium has yet been isolated in weighable amounts; thus all the investigations of this element have been done with tracer quantities. The longest-lived isotope is Fm^{257} (half-life about 80 days) whose production in high-neutron-flux reactors is extremely limited because of the very long sequence of neutron-capture reactions that is required.

Despite its very limited availability, fermium, in the form of the 3.24-hour-Fm^{254} isotope, has been identified in the "metallic" zero-valent state in an atomic-beam magnetic resonance experiment. This established the electron structure of elemental fermium in the ground state as $5f^{12}7s^2$ beyond the radon structure).

Fermium exists in normal aqueous solution almost exclusively as Fm^{+3}, but strong reducing conditions can produce Fm^{+2} which has greater stability than Es^{+2} and less stability than Md^{+2}.

Mendelevium (Md, atomic number 101, after Dmitri Mendeleev). Mendelevium, the ninth transuranium element to be discovered, was first identified by Ghiorso, Harvey, Choppin, Thompson and Seaborg in early 1955 as a result of the bombardment of the isotope Es^{253} with helium ions in the Berkeley 60-inch cyclotron. The isotope produced was Md^{256} which decays by electron capture to Fm^{256}, which in turn decays predominantly by spontaneous fission with a half-life of about 1.5 hours. The first identification was notable in that only of the order of one to three atoms per experiment were produced. The extreme sensitivity for detection depended on the fact that its chemical properties could be accurately predicted as eka-thulium and there was a high sensitivity for detection because of the spontaneous fission decay. The chemical properties have been investigated solely by the tracer technique and seem to indicate that the predominant oxidation state in aqueous solution is the (III) state, although the (II) state is quite stable and a (I) state seems to exist. There seem to be no isotopes of sufficiently long half-life to make it possible to isolate this element in weighable quantity.

Nobelium (No, atomic number 102, after Alfred Nobel). Nobelium, the tenth transuranium element to be discovered, was identified by Ghiorso, Sikkeland, Walton and Seaborg in 1958 as a result of the bombardment of curium with C^{12} ions in the heavy ion linear accelerator (HILAC) at Berkeley, California. The isotope produced was No^{254}, now known to decay with a half-life of 55 seconds. An earlier claim to discovery of element 102, based on work performed at the Nobel Institute for Physics in Stockholm, was shown to be erroneous, but the name nobelium proposed by the Stockholm workers has been accepted by the discoverers at Berkeley. By use of the 3-minute No^{255}, some of the chemical properties of nobelium have been determined. The (II) oxidation state appears to be most stable and the (III) state can be achieved only with difficulty by use of strong oxidizing agents.

Lawrencium (Lr, atomic number 103, after Ernest O. Lawrence). Lawrencium, the eleventh transuranium element to be discovered, was identified in 1961 by Ghiorso, Sikkeland, Larsh and Latimer as a result of the bombardment of californium with boron ions in the HILAC at Berkeley, California. The isotope produced was probably Lr^{258} which decays with a half-life of 5 seconds. Solvent extraction experiments with a few atoms of the longer lived Lr^{256} (half-life 35 seconds) have shown that lawrencium ions appear to exist in the (III) state in aqueous solution. This is probably the only stable oxidation state that will be found for lawren-

cium. The element Lr is the last member of the 14-member actinide series.

-Element 104 is the first transactinide element; it falls in Group IVB of the Periodic Table under hafnium. The first claim to the discovery of element 104 was made by G. N. Flerov and associates at Dubna, USSR, in 1964, who bombarded Pu^{242} with Ne^{22} ions to produce a radioactivity which decayed by spontaneous fission with a half-life of 0.1 to 0.3 second. Flerov proposed the name kurchatovium (Ku) for element 104, after I. V. Kurchatov. In 1969 Ghiorso and co-workers at Berkeley questioned the existence of the isotope claimed by Flerov and associates. Ghiorso and associates bombarded Cf^{249} with C^{12} ions to produce 104^{257} and 104^{259}, and Cm^{248} with O^{18} ions to form 104^{261}. All three isotopes were characterized by their half-lives (4, 3, and 70 seconds, respectively), emission of characteristic alpha particle radiation during decay, and the identification of their known nobelium daughters. On the basis of this work, the Berkeley group claimed the discovery of element 104 and has proposed the name rutherfordium (Rf), after Lord Ernest Rutherford.

Element 105 is the second transactinide element; it is in Group VB of the Periodic Table under tantalum. In 1967, Flerov *et al.* made a preliminary announcement of the discovery of two isotopes of element 105 which decayed by alpha particle emission, resulting from the reaction of Ne^{22} ions on Am^{243}. The radioactivity properties of the two isotopes were poorly defined. Early in 1970 the same group made a separate claim to the discovery of element 105, citing observation of an isotope which decayed by spontaneous fission with a half-life of 2 seconds. Ghiorso and his group, rejecting observation of spontaneous fission as a definitive identification of a new isotope, reported the preparation of 105^{260} by the reaction of N^{14} ions on Cf^{249}. The isotope decayed by characteristic alpha particle emission to Lr^{256}, which was identified also. Following announcement of the Berkeley results, the Dubna laboratory then reported that it had identified an alpha particle emitter with decay properties similar to those observed at Berkeley. Ghiorso *et al.* have proposed the name hahnium (Ha) for element 105, after Otto Hahn. The Soviet group has proposed that this element be named nielsbohrium (Ns), after Niels Bohr.

GLENN T. SEABORG

Cross-references: *Isotopes, Nucleonics, Periodic Law, Elements.*

TRITIUM

Tritium is a radioactive isotope of hydrogen, the nucleus consisting of one proton and two neutrons. The atom of tritium has three mass units and an isotope weight of 3.01703. Tritium has a half-life of 12.4 years and decays by β-emission to give He^3. The energy of the β-particle is 0.018 mev. No γ radiation is emitted in this process and it is known as a pure β-emitter. The name "tritium" comes from the Greek *tritos* meaning third. A tritium atom is sometimes represented by the symbol T and its nucleus, called a "triton," is indicated by t. In 1934 Rutherford, with Oliphant and Harteck, bombarded a stationary deuterium with deuterons. The result was the emission of a proton, leaving a third isotope of hydrogen of mass three as the residual nucleus. In this way, tritium was produced by a deuteron-proton reaction. The equation is as follows:

$$_1H^2 + _1H^2 = _1H^3 + _1H^1$$

In this equation, $_1H^3$ is the new isotope which has been called tritium. Tritium is unstable and does not occur naturally. It can be obtained only in nuclear transmutation processes. Tritium can also be produced by the bombardment of Be^9 with deuterons. This is given in the following equation:

$$_4Be^9 + _1H^2 = _4Be^8 + _1H^3$$

In the future it is believed that *tritons,* nuclei of tritium, will be useful as projectiles in the transmutation of other elements. In 1948, Pool and Kundu reported that tritons were used as projectiles in the bombardment of both Rh^{103} and Co^{59} undergoing reactions of the triton-proton type. Reactions of this kind have features in common with the Oppenheimer-Phillips reactions. The triton consists of two neutrons and a proton. It is believed that, under the influence of the electrostatic field of the target nucleus, the proton is repelled, leaving a temporary combination of two neutrons called a dineutron. In 1950, Hemmendinger performed an experiment in which tritium was bombarded by tritons and it is believed that the dineutron was produced. The following nuclear equation for this process was proposed:

$$_1H^3 + _1H^3 = _2He^4 + _0n^2$$

Tritium may be produced by the bombardment of Li^6 with low energy neutrons, giving the following equation:

$$_3Li^6 + _0n^1 = _1H^3 + _2He_4$$

For hydrogen tracer studies the stable isotope, deuterium, or the radioactive isotope, tritium, may be used. Tritium emits a β-particle of very low energy, so that special techniques of measurement must be employed. The main difficulty in using the isotopes of hydrogen as tracers is that the masses are so different from that of ordinary hydrogen that noticeable differences are observed in reaction rates. This would be especially true of tritium, since its mass is three times that of ordinary hydrogen. For example, large proportions of heavy water are harmful to certain living organisms. This is presumably due to the low reaction rates or low mobilities with deuterium as compared with ordinary hydrogen.

When tritium is used as a label for an organic compound and the mobilities of hydrogen or tritium is not involved, then in this case, the tracer technique is good, and the difference in reactivity from that of ordinary hydrogen is probably of no significance. Among its other tracer applications, tritium has found use as a tracer in studying photosynthesis, although the results so far obtained are not too conclusive. The nuclei of triton has a relative charge of one, a rest mass of 3.01703 mass units, a spin of $\frac{1}{2}$, obeying Fermi statistics. It has a magnetic moment of 2.97968 nuclear magnetons. The above data on tritium should be compared with that of deuteron. The nucleus of deuterium has a relative charge of one, a relative mass of two,

just twice that of normal hydrogen, a rest mass of 2.01472 mass units, and a spin of one. It obeys Bose statistics and it has a magnetic moment of 0.857648 nuclear magnetons. This isotope of hydrogen was discovered by H. Urey in 1932.

The binding energy per nucleon, in the lighter as well as the heavier nuclei, is less than for nuclei of intermediate mass number. In other words, by a process of fusion, when the combination of two or more lighter nuclei is fused, a huge liberation of energy results. Tritium, being one of the lighter nuclei, is used for purposes of this kind, as in the hydrogen bomb.

JEROME BREWER

Cross-references: *Hydrogen, Isotopes, Nucleonics, Lithium, Fusion.*

TUNGSTEN AND COMPOUNDS

Tungsten (Wolfram), W, atomic number 74, atomic weight 183.85, is a Group VI metallic element. It is heavy, specific gravity 19.3, and has the highest melting pont, 3400°C, of any metal and of any element except carbon. In massive form, it is light gray in color, has a low vapor pressure, and is a good electrical conductor. Five isotopes, 180, 182, 183, 184 and 186, occur naturally. Tungsten exhibits valences from 6 to 2, being acidic in the higher and basic in the lower.

The name "tungsten" was first used by A. Cronstedt in 1755 and is derived from the Swedish "tung," meaning heavy, and "sten," meaning stone. The name "wolfram" has an earlier origin which is somewhat obscure. Tungstic acid was first extracted from a tungsten mineral (scheelite) by C. W. Scheele in 1781. Two years later, J. J. and F. d'Elhujar extracted tungstic acid from wolframite and reduced it to "wolfram" powder by heating with charcoal. The first patents for producing steels containing tungsten were applied for in 1857, but it was not until 1908 that W. D. Coolidge obtained a patent for producing ductile tungsten wire. High-speed steels using tungsten were developed early in the first world war. Hard carbide cutting tools, made by liquid phase sintering mixtures of WC and cobalt powders, were developed by Krupp in Germany in 1927.

Tungsten does not occur in the free state. It is stated to be 26th in order of abundance of the elements of the earth's crust and to be more plentiful than cadmium, zirconium, mercury and silver and less plentiful than antimony, nickel, tin and chromium.

Its important minerals are classified in two groups, scheelite and wolframite. Scheelite, which is native calcium tungstate ($CaWO_4$), varies in color from white through yellow to brown and in specific gravity from 5.4 to 6.1. Its hardness is 4.5 to 5; its streak is white. Ultraviolet light causes the mineral to fluoresce with a bright blue color. The wolframite group consists of hübnerite, wolframite and ferberite, which are chemical variations of $(Fe,Mn)WO_4$. The mineral is called hübnerite when it contains 80% or more $MnWo_4$ and ferberite when it contains 80% or more $FeWo_4$. The color varies from brown to black, the hardness 5 to 5.5, and the specific gravity from 7.1 to 7.5.

Tungsten minerals are found in some pegmatites,

batholiths and sills. These minerals are normally found associated with cassiterite, sulfides, arsenides, quartz, feldspar and many other minerals. Tungsten ores rarely contain more than 2% of tungsten minerals and are usually concentrated to above 60% WO_3 by standard methods of ore benefication before shipment. The important deposits of tungsten ores are found in China, United States, Korea, Bolivia, Brazil, Peru, Russia, Portugal, Australia, Indochina, The Democratic Republic of the Congo, and Malaya. In the United States, producing deposits are found in California, Nevada, Idaho, Colorado and South Carolina. Political and economic situations determine the production rate of the deposits.

Tungsten is not produced by conventional smelting techniques because of its chemical reactivity and its high melting point, 3400°C. The manufacture of pure ductile tungsten consists of (1) chemical extraction and purification of tungstic acid or ammonium paratungstate from the ore; (2) reduction to metal powder; and (3) conversion of this powder into massive metal.

(1) Wolframite is pulverized and may be decomposed to a soluble alkali tungstate either by fusing with an alkali hydroxide and leaching in water or by boiling for a longer time in a strong alkali solution. The alkali tungstate is purified by recrystallization and then converted to tungstic acid, H_2WO_4, by neutralization of the solution with HCl. Scheelite is pulverized and decomposed with hydrochloric acid to form insoluble tungstic acid. This crude acid may be purified by dissolving it in ammonium hydroxide and crystallizing the filtered solution to produce ammonium paratungstate which, when heated, results in tungstic oxide, WO_3. Techniques for removing specific impurities are proprietary to the manufacturers.

(2) Tungstic oxide or tungstic acid is reduced to the metal powder by heating at about 850°C in pure dry hydrogen. The properties of the oxide or acid, the rate of hydrogen flow and the temperature and time of reduction have significant effects on the characteristics of the metal powder produced. The average particle size of hydrogen-reduced tungsten powder varies from 0.5 to about 10 microns. Metal powder for less critical uses is made by reducing the oxide with carbon or with hydrocarbons.

(3) Pure tungsten powder is pressed in a mechanical or a hydrostatic press to form green compacts which are sintered in hydrogen to produce dense crystalline ingots. Sintering is accomplished by heating the compacts by induction or by passing a heavy electric current through them. The ingots can then be rolled or swaged hot to form plates, rods, etc. A new method which employs the plasma reduction of a tungsten halide is used to produce tungsten vessels, tubes and forging blanks.

Reactions. Tungsten is a transition element. It forms compounds in valences of 6 to 2, being acidic in the higher and basic in the lower valences. It is metallic in its elemental form. Massive tungsten resists attack by solutions of the alkalies, ammonia and single acids, but it dissolves rapidly in a mixture of hydrofluoric and nitric acids. It reacts with oxygen above 400°C, but is not affected by hydrogen or ammonia at high temperatures.

Tungsten forms the binary compounds W_2C, WC, WB, WB_2 and WSi_2 by direct synthesis above 1400°C. By direct reaction with the proper halide, it forms WF_6 at room temperature, WCl_6 at about 250°C and WBr_5 and WI_4 at red heats. The halides hydrolyze on contact with water or moist air to form oxyhalides and finally tungstic acid.

Tungsten trioxide, anhydrous tungstic acid, is formed by heating the metal, the lower oxides, the carbides, or tungstic acid in air. The lower oxide (WO_2) can be prepared by the restricted oxidation of W or by the partial reduction of WO_3.

Tungstic acid (H_2WO_4) is the canary yellow precipitate which forms when a hot solution of an alkali tungstate is acidulated with HCl. Tungsten forms normal tungstates ($R_2O \cdot WO_3 \cdot nH_2O$), paratungstates in which the R_2O to WO_3 ratio may be 3 to 7 or 5 to 12, and metatungstates ($R_2O \cdot 4WO_3 \cdot nH_2O$).

Uses. The first important industrial use of tungsten began in 1908 when the newly developed ductile tungsten wire was used as filaments for electric lamps. Because of its high melting point and low vapor pressure, tungsten has never been replaced in this application.

Currently, the principal uses of tungsten by weight are tungsten carbide as hard carbide cutting and wear-resistant tool inserts, 48%; high temperature-resistant alloys, 19%; pure tungsten as wire, electrical contact points and aerospace shapes, 16%; and high-speed tool steels, 17%.

Interesting uses for pure tungsten include "make and break" contact points for automotive ignition systems, x-ray targets, nonconsumable welding rods, metal-to-glass seals, and as heating elements for vaporizing aluminum and other metals for coating mirrors, automobile head lamps, etc.

Sintered (not alloyed) combinations of tungsten and a lower melting metal such as silver or copper are used as welding electrodes and to make contact points for circuit breakers, motor starters and such devices by which heavy currents are controlled. The silver or copper contribute low resistances to the heavy currents and the tungsten inhibits mechanical wear and damage caused by arcing.

Tungsten alloys containing about 90% W and 10% Ni and Cu are machinable and, because of their high density, are used in the construction of vibration-damping devices and gyroscope rotors. Alloys having moderate tungsten contents are used where high-temperature strength is required. High-speed steels contain lesser amounts of tungsten.

Tungsten carbide, WC, is made commercially by direct synthesis or by reacting WO_3 and lampblack at temperatures above 1400°C. Fine WC powder, 1 to 10 mircons, is blended with cobalt (and other carbides), pressed into compacts and liquid phase sintered to make inserts for cutting and wear resistant tools and dies.

DONALD F. TAYLOR

References

(1) Hampel, C. A., Ed., "Rare Metals Handbook," 2nd ed., Reinhold Publishing Corp., New York, 1967.

(2) Smithells, C. J., Ed., "Tungsten," Chapman & Hall Ltd., London, 1926.

(3) Li, K. C., and Wang, Cy, "Tungsten," Reinhold Publishing Corp., New York, 1955.

U

ULTRAVIOLET STABILIZERS

It has been recognized for many years that organic materials are degraded by exposure to ultraviolet radiation. Thus, the ultraviolet radiation in sunlight causes yellowing and weakening of natural fibers, chalking of paint films, burning of the skin, and embrittlement and discoloration of plastics. While light degradation is widely observed, it has been the rapid growth of synthetic polymers and the necessity for stabilizing them for use in outdoor applications that has provided the impetus for development of ultraviolet stabilizers and their widespread use.

At this point, a brief discussion of the energy relationships involved in ultraviolet degradation of plastics will aid in understanding the processes taking place. Ultraviolet radiation is defined as electromagnetic radiation in the region between 4 and 400 nm in wave length. While radiation in this region can be obtained in the laboratory from artificial light sources, it is only the region from 290–400 nm that is important in the degradation of plastics used outdoors. This is because the ozone in the upper atmosphere effectively absorbs the shorter wave lengths (4–290 nm). Light in the 290–400 nm region has an energy content ranging from 95 to 71 kcal per mole which is sufficient to break the bonds of organic molecules. It is this bond cleavage that initiates the reactions leading to the degradation observed in organic substances exposed to ultraviolet light.

The visible and infrared light contained in the sun's radiation above 400 nm is of relatively low energy and is not appreciably involved in degradation processes.

When a polymer absorbs light, it may reradiate the absorbed energy at much longer wave lengths (heat), at slightly longer wave lengths (luminescence), or the energy may be transferred to another molecule. When none of these processes is operative, the absorbed light energy may cause bond breaking leading to degradation. By incorporating an ultraviolet absorbing compound into the plastic, it is possible to essentially eliminate all of the above processes, since the absorber even at concentrations as low as 0.5% can effectively compete for the incident ultraviolet radiation, thus protecting the plastic from degradation. A second approach to stabilization is to incorporate an additive which, though not an absorber in itself, can accept energy from the polymer substrate, and thus, leave the polymer intact. Since protection of plastics against light degradation can be achieved by these two mechanisms, the broader term "ultraviolet stabili-zer" or "light stabilizer" is used to refer to such additives.

Ultraviolet absorbers continue to be the most widely used stabilizers. Such products must have long-term stability to ultraviolet light, be relatively nontoxic, heat stable, have little color, must not sensitize the substrate, and must be priced at levels which the plastics processor can tolerate. The principal classes of chemicals meeting these requirements at present are the 2-hydroxybenzophenones, the 2-(2'-hydroxyphenyl)benzotriazoles, substituted acrylates, and aryl esters. Typical compounds representative of these classes are 2-hydroxy-4-octoxy-benzophenone, 2-(2'-hydroxy-5'-methylphenyl)ben-zotriazole, ethyl-2-cyano-3,3-diphenyl acrylate, dimethyl p-methoxybenzylidene malonate, and p-tert-octylphenyl salicylate.

The particular absorber to be used in a given application depends on several factors. One important criterion is whether the absorber will strongly absorb that portion of the ultraviolet spectrum responsible for degradation of the plastic under consideration. Compatibility, volatility, thermal stability, and interactions with other additives and fillers are other items that must be considered. When used in food wrappings, Food and Drug Administration approval must be obtained. While one or more of these considerations may rule out a given stabilizer or influence the choice of one class over another, the final selection must await the results of extensive accelerated and long-term tests.

At this point, it should be indicated that much effort has gone into the development of accelerated testing procedures. Many of the devices and techniques employed are based on knowledge gained in the evaluation of dyes, textiles, and rubber. For example, the carbon-arc Fade-O-Meter and the Xenon-arc Weather-O-Meter have been adapted from the dye field for use in plastics evaluation. Extensive use is also made of the fluorescent sunlamp, fluorescent blacklight, S-1 sunlamp, Hanovia lamp, and others. Such instruments are very useful for comparison of one stabilizer with others and for evaluating total stabilizing formulations in particular polymers. Nevertheless, no accelerated weathering device has yet been found which can accurately predict the outdoor weatherability of a broad range of polymers. Accelerated outdoor weathering is carried out in Phoenix, Arizona, where high levels of ultraviolet radiation occur and the temperature is high. To determine the lifetime under more humid conditions, tests are often conducted in the vicinity of Miami, Florida.

For extended outdoor applications, most polymers require some degree of light stabilization.

There are wide variations in the inherent stability of different polymers ranging from less stable ones, such as polypropylene, to the highly light-stable poly(methyl methacrylate). Because of the dramatic growth of polyolefins, and particularly polypropylene, over the past several years, there has been an upsurge in requirements for ultraviolet absorbers. The hydroxybenzophenones, such as 2-hydroxy-4-octoxy benzophenone, have been widely used for stabilization of polypropylene. The benzotriazoles have also achieved commercial importance in this application. End uses of polypropylene requiring ultraviolet absorbers include upholstery fabrics, indoor-outdoor carpeting, lawn furniture, ropes, and various crates and boxes. Polyethylene is also stabilized with the hydroxy benzophenone absorbers. Applications include baskets, beverage cases, bags for fertilizer, and films for greenhouses.

Polystyrene light stabilization has been achieved with a variety of ultraviolet absorbers including the benzophenones, benzotriazoles, and salicylates. While yellowing of polystyrene occurs in many applications, it is particularly noticeable in diffusers used with fluorescent lights. This problem has been effectively solved by using ultraviolet light absorbers. In this instance, superior stabilization is achieved when the ultraviolet absorber is used in conjunction with specific antioxidants.

The hydroxybenzophenones, hydroxyphenylbenzotriazoles, and substituted acrylates are all used for stabilization of polyvinyl chloride. This polymer is growing at a substantial rate, and increasing uses are developing for light-stabilized grades. Among current uses, may be mentioned auto seat covers, floor tiles, light diffusers, vinyl-coated fabrics, siding, and exterior trim. Since the processing of polyvinyl chloride requires the use of a heat stabilizer, care must be exercised to avoid undesirable interactions between the heat and light stabilizers.

The stabilization of polyesters is generally achieved with the hydroxybenzophenone and hydroxyphenylbenzotriazole absorbers. The choice of absorber depends on the curing catalysts and promoters used. The stabilization of fire retardant grades of polyesters offers a greater problem than the standard grades because the halogenated monomer acids used are appreciably more sensitive to ultraviolet light than the unhalogenated acids (phthalic, isophthalic). Applications for light-stabilized polyesters include sheets for roofs and skylights and various surface coatings.

Cellulosic plastics are used in a number of outdoor applications with signs being one of the principal areas of use. This plastic can be stabilized reasonably well with the aryl esters of salicyclic acid. It is of interest to note that these esters undergo a photochemical rearrangement in the plastic to derivatives of hydroxy benzophenone. The hydroxy benzophenones may be added initially to effect stabilization.

As noted earlier, poly(methyl methacrylate) plastic has excellent resistance to ultraviolet radiation. Nevertheless, in long-term outdoor applications or in lighting fixtures, small amounts of ultraviolet absorbers are employed to retard the yellowing and degradation in physical properties which would otherwise occur.

The previous discussion illustrates how widely ultraviolet absorbers are used for stabilization of plastics against degradation by ultraviolet light. The second principal method for light stabilization is the use of energy transfer agents. Important stabilizers currently in use which function by this mechanism are nickel complexes of 2,2'-thiobis(4-tert-octylphenol). For example, the butyl amine adduct of this complex is widely used. The nickel salts of mono alkyl esters of 3,5-di-tert-butyl-4-hydroxy-benzyl phosphonic acid are also useful stabilizers. When color is not a consideration, nickel dialkyl dithiocarbamates, and nickel acetophenone oximes may be used. Thus far, the nickel stabilizers have been used primarily in polypropylene and to some extent in polyethylene. They are especially useful in polypropylene fibers since stabilization by energy transfer is less dependent on sample thickness than is stabilization by ultraviolet absorption. Some of the nickel stabilizers have the further advantage that they act as dye acceptors and thus aid printing and dyeing of fibers and other items made from polyolefins.

Thus far, the discussion on ultraviolet stabilizers has been concerned only with their use for stabilization of plastics. While this is the principal use, the stabilizers which function by absorption are also widely used to prevent ultraviolet light from damaging furniture, clothing, and other articles. For example, a thin plastic film (6–10 mil) containing a high concentration of absorber is useful for covering a store window exposed to the sun. Such a film absorbs the ultraviolet radiation and thus prevents damage to the articles behind the film. The clarity and lack of color of the film permits customers to see the articles readily. Surface coatings containing ultraviolet absorbers are used in the same way to protect items such as flooring and furniture. Ultraviolet absorbing coatings may also be used to protect plastics, but generally, it is more practical to incorporate the absorber in the plastic itself. The incorporation of ultraviolet absorbers into plastic sunglasses is important for protecting the eye from ultraviolet radiation damage. Suntan lotions contain compounds such as monoglyceryl p-aminobenzoate that permit the longer wave lengths (330–400 nm) in the ultraviolet which cause tanning to pass through to the skin. At the same time, the more highly energetic short wave lengths (290–330 nm) which cause burning are strongly absorbed. When no tanning is desired, creams containing hydroxy benzophenones may be used, since these products remove a high percentage of the 290–400 nm radiation.

Highly satisfactory formulations have been developed for light stabilization of a wide range of polymers. Studies are continuing not only toward empirical development of superior stabilizing formulations, but also toward understanding the mechanisms of the degradation and inhibition process involved. This dual approach can be expected to yield products which will meet the increasingly severe demands that will result as plastics find their way into new outdoor uses.

W. B. HARDY

References

"Stabilization of Polymers and Stabilization Processes," Advances in Chemistry Series, No. 85, 1968.

Daniels, T. C., and Kumler, W. D., *J. Am. Pharm. Assoc.*, **37**, p. 474, 1948.

Hirt, R. C., and Searle, N. Z., "Bibliography on Ultraviolet Degradation and Stabilization of Plastics," SPE Transactions, Vol. 2, p. 32, January, 1962.

Voight, J., "Die Stabilisierung der Kunststoffe gegen Licht und Wärme," New York, Springer-Verlag, 1966.

URANIUM AND COMPOUNDS

The discovery of uranium goes back to 1789, when Klaproth demonstrated the existence of an unusual new substance in pitchblende ores. He named the new element uranium, in honor of the discovery of the planet Uranus. It was the element in which the property of radioactivity was first discovered by Becquerel in 1896. It has the atomic number 92 and is now known to be the third member of the rare earth-like transition series (the actinide series), in which the inner $5f$ electron shell is being filled. The gaseous atom has the electronic structure of radon (atomic number 86) plus three $5f$, one $6d$ and two $7s$ electrons.

Uranium is found to some extent in the majority of rocks and sediments making up the earth's crust. The concentration appears to vary from about 0.2 to 10 ppm in most igneous rocks, running somewhat higher in rocks enriched in sodium, potassium and silicon. At an average abundance of 4 ppm there would seem to be about 10^{13} to 10^{14} tons of uranium in the upper 20 kilometers of the earth's crust.

Uranium is considerably less abundant in concentrated deposits. The black unoxidized deposits containing pitchblende (uraninite), an oxide intermediate between UO_2 and U_3O_8 in composition, are by far the most important and common of the different types of uranium deposits known today. The bulk of these ores occurs in sandstones and conglomerates; however, significant reserves have also been found in vein deposits. Coffinite, a hydrous uranium silicate, is also usually associated with these deposits. Brannerite, a uranium titanate, is an important component of some black ore deposits. The yellow oxidized uranium deposits are of less importance. The most common of these are the deposits of carnotite, a potassium uranium vanadate, which are abundant in the Colorado Plateau region of the United States. Many other types of uranium ores have been found, but they are relatively unimportant.

Since the advent of the discovery of the release of nuclear energy from uranium, many countries have carried on extensive exploration for this valuable substance. Until the early 1950's the largest known sources of uranium were in the pitchblende vein deposits of the Belgian Congo and the Great Bear Lake area in Canada and the carnotite-type deposits of the Colorado Plateau. These sources, however, were overshadowed by the development in the mid-1950's of much larger reserves of pitchblende and coffinite ores in the United States, Canada and South Africa. Important deposits are also known to exist in France, Australia, USSR, Czechoslovakia and other African countries. Less important reserves have been found in other countries in Europe, South America and Asia.

Deposits of phosphate rock and carbonaceous shales, containing low concentrations (50–150 ppm) but overall very large quantities of uranium, may become important sources of uranium as high-grade resources dwindle. Although limited quantities of uranium can be obtained economically as a by-product of processing phosphate rock to make chemical fertilizers, recovery of most of the uranium can be achieved only at several times present costs.

Natural uranium consists of three isotopes: U^{238} (99.28%), U^{235} (0.71%), and U^{234} (0.006%). The chemical atomic weight is thus largely determined by the U^{238} and is 238.029. All of these isotopes have unstable nuclei, and U^{238} and U^{235} are the parents of rather complicated decay chains which give rise to natural alpha and beta radio activities. Both chains eventually decay into stable lead isotopes, but the U^{238} chain passes through Ra^{226}, the normal radium of commerce which was and still is very valuable for certain therapeutic uses. Since the half-life relationships of all of the members of the chains are known, and further since the half-lives of the U^{238} and U^{235} parents (4.51 × 10^9 and 7.13 × 10^8 years, respectively), are known to be comparable to the age of the earth, the ratios of various uranium and lead isotopes have been used to date the age of formation of the oldest minerals and, in turn, the age of the earth itself.

Three isotopes of uranium are of prime importance to the development of nuclear fuels and explosives: U^{238} and U^{235}, in natural uranium, and U^{233} whose short half-life (1.62 × 10^5 years) precludes the possibility of its natural occurrence as it is not being formed in any of the natural decay series.

U^{235} is the key to the utilization of uranium. Although its natural abundance is only 0.71%, it undergoes fission so readily with slow neutrons that a self-sustaining fission chain reaction can be brought about in a nuclear reactor. Such a nuclear reactor can be constructed from natural uranium and a suitable moderator (graphite, heavy water, etc.) alone. (See **Fission, Nuclear**)

While the abundant isotope, U^{238}, will not in itself undergo fission with slow neutrons in a nuclear reactor, it does capture neutrons to eventually produce Pu^{239}.

$$U^{238}(n,\gamma)U^{239} \xrightarrow{\beta^-} Np^{239} \xrightarrow{\beta^-} Pu^{239}$$

Pu^{239}, like U^{235}, is also very fissionable and is used both as a concentrated form of nuclear fuel and as a nuclear explosive. It is therefore possible to convert most of the natural uranium into fissionable nuclear fuels instead of realizing the energy from the rare isotope U^{235} alone and thereby multiply some 100-fold the amount of energy which can be realized from uranium. This nuclear conversion of U^{238} into P^{239} can be performed in "breeder" reactors where it is possible to produce more new fissionable material than the fissionable material consumed in sustaining the chain reaction.

Instead of being used to fuel natural uranium nuclear reactors, the rare isotope U^{235} can be concentrated by physical methods (usually gaseous diffusion) and used directly as a more concentrated fuel or explosive. Uranium slightly enriched in U^{235} (a few per cent) is often used to fuel nuclear power reactors for the generation of electricity. The third isotope of importance, U^{233}, can only be produced in quantity by irradiating natural thorium (Th^{232}) with neutrons from a nuclear chain reaction:

$$Th^{232}(n,\gamma)Th^{233} \xrightarrow{\beta^-} Pa^{233} \xrightarrow{\beta^-} U^{233}$$

Since thorium itself is not readily fissionable, U^{233} must be produced initially at the expense of some other fissionable material. However, in "breeder" reactors, thorium can be converted into a usable nuclear fuel and thus increase the world's total nuclear power resources.

Metallic uranium initially prepared is a lustrous, silvery metal which melts at 1132°C and boils at about 3818°C. It can be prepared by reduction of the halides by alkali and alkaline earth metals or by reduction of various oxides (UO_3, U_3O_8, UO_2) by carbon, aluminum, or calcium at high temperature. Electrolysis of uranium compounds in molten salt baths of alkali halides also gives good results.

The metal crystallizes in an orthorhombic form stable at room temperature but undergoes two transitions at 667°C and 772°C, respectively, on heating. The calculated density of alpha-uranium (the room-temperature form) from x-ray data is 18.97. It is a malleable, ductile metal, slightly paramagnetic, and highly reactive chemically; initially lustrous metallic pieces eventually become coated with a dull black oxide coating. Pyrophoric metal can be prepared, and even massive metal will combine directly with oxygen, the halides, and gaseous hydrogen, nitrogen, carbon monoxide, and carbon dioxide. It is not attacked by alkalies but reacts directly with water and dissolves readily in acids to give compounds of uranium(III) and uranium(IV).

The commercial uses of uranium before the discovery of fission were somewhat limited. It had been used in filaments of electric lamps and as an additive to special steels. Compounds of uranium had found some use in photography and dyeing and as ceramic and glass coloring agents. Uranium carbide is a good catalyst in making synthetic ammonia. At present, however, large quantities of very pure uranium are used in connection with the nuclear energy program in many countries to fuel reactors and produce plutonium, and for concentrating and extracting the light and rare isotope, U^{235}. Many governments have exercised controls over the mining, manufacture, and uses of this valuable natural resource.

Uranium ions representing four different oxidation states have been prepared in aqueous solutions: U^{+3} (red), U^{+4} (green), UO_2^+ (unstable), and UO_2^{++} (yellow). The U^{+3} ion is highly unstable, since it will liberate hydrogen from water, and UO_2^+ representing uranium(V) disproportionates with the production of U^{+4} and UO_2^{++}. Although definitely unstable, millimolar solutions of uranium(V) do appear capable of preparation. All of these ions appear to form complexes with the inorganic ions chloride, sulfate, carbonate, etc., and with numerous organic chelating agents.

Since the (IV) and (VI) compounds represent the most important oxidation states, the chemistry of uranium may be related to the two corresponding oxides, UO_2 and UO_3. The former dissolves in acid solutions to form U^{+4} solutions and also gives rise to solid salts of the type UCl_4 and $U(SO_4)_2 \cdot 9H_2O$. The UO_2^{++} ion in acid solution leads to uranyl derivatives of the type $UO_2(NO_3)_2 \cdot 6H_2O$ and UO_2Cl_2. Closely related are the basic derivatives of the trioxide, the uranates, and the polyuranates. Alkali metal diuranates, $M_2U_2O_7$, are well known, and the simple uranates such as Na_2UO_4 have also been prepared.

No trivalent uranium compounds can be precipitated from aqueous uranium(III) solutions because of the very rapid oxidation to uranium(IV). The common water-soluble salts of uranium(IV) are the chloride, bromide, sulfate, and perchlorate; all of these solutions are hydrolyzed. Chloride and bromide ions form weak complexes in dilute solutions; sulfate ion complexes U^{+4} more strongly. Uranium(IV) oxalate, phosphate, fluoride, molybdate, arsenate, ferricyanide, and hydroxide are insoluble in nearly neutral solutions.

The interesting oxygenated cations MO_2^{++} and MO_2^+ are almost unique to the actinide elements. The uranyl ion, UO_2^{++}, behaves in many respects as a simple doubly charged ion and in many respects resembles Ca^{++}. There also appears to be evidence that more complicated polymeric species of this ion, such as $U_2O_5^{++}$, $U_3O_8^{++}$, etc., are formed in concentrated uranyl solutions as a result of hydrolytic phenomena.

In addition to UO_2 and UO_3, UO, U_2O_5, and U_3O_8 are also known. There is also some evidence for the existence of U_4O_7 and U_6O_{17}, and several crystal modifications of U_3O_8 are known. In fact, the phase relationships for the uranium-oxygen system are so complex because of solid solution formation that it is possible to obtain almost any oxide ranging in composition between UO and UO_3. The same is also true to a somewhat lesser extent with other transuranium elements.

Three uranium carbides have been prepared: a monocarbide, UC; the dicarbide, UC_2; and the sesquicarbide, U_2C_3. The first two are formed by either direct reaction of carbon with molten uranium or by the action of carbon monoxide on the metal at high temperatures. The sesquicarbide, stable below 1800°C, can be formed by heating mixtures of UC and UC_2; the sesquicarbide will form only if the reaction mixture is subjected to mechanical stress.

Uranium metal also reacts directly with nitrogen to give nitrides such as UN, U_2N_3, and UN_2. Both nitrides and carbides are semimetallic compounds; they can be converted to U_3O_8 by ignition in air and also react with water.

At temperatures of 250°–300°C uranium metal will react with hydrogen to form a hydride, UH_3. At higher temperatures the substance loses hydrogen reversibly. The compound is conveniently used as a starting material for the preparation of reactive uranium powder and is itself reactive enough to be used directly to prepare halides, carbides and nitrides; it is thus a very important compound from which a great many uranium compounds can be conveniently prepared. An interesting observation on UH_3 is that it exhibits ferromagnetism at low temperatures.

The halides of uranium also constitute a numerous and important group of compounds. The most important compound of this class is the hexafluoride, UF_6. This substance very fortunately has a high vapor pressure at room temperature (120 mm) and is thus well suited for use in the gaseous-diffusion process for the separation of the isotopes U^{238} and U^{235}. It can be prepared from uranium dioxide as follows:

$$UO_2 + 4HF \xrightarrow{500°C} UF_4 + 2H_2O$$

$$UF_4 + F_2 \xrightarrow{350°C} UF_6$$

A series of oxyhalides is also known: UO_2F_2, $UOCl_2$, UO_2Cl_2, UO_2Br_2, etc. They are all soluble in water and become increasingly less stable in going from fluoride to iodide.

Many other physical and chemical properties of uranium and its compounds are now known, and it is interesting to note that this element which remained rather obscure and unknown for 150 years has been the object of more scientific investigation in recent times than almost any of the other more "common" elements.

GLENN T. SEABORG

Cross-references: *Radioactivity, Fission, Transuranium Elements, Isotopes.*

V

The term "valence" was introduced into chemistry to denote the relative combining power of an atom of a given element. By assigning integral valence numbers to the elements concerned and "balancing" these in each compound, the relative proportions of the elements in many different compounds could be accounted for. Thus, giving hydrogen and chlorine each a valence of 1, oxygen a valence of 2, and nitrogen a valence of 3, the valence-balancing principle leads to the formulas HCl, H_2O, NH_3, Cl_2O, NCl_3, and N_2O_3, designating the relative numbers of atoms of these elements in compounds which they form with each other. From these and the atomic weights, the relative masses of the elements in each of these compounds can be deduced.

It was soon found that a given element does not always exhibit the same valence in different compounds; also that, in dealing with inorganic compounds, it is necessary to assign a positive or negative sign to each valence number. Valence-balancing then consists of algebraic addition, yielding a zero sum for each compound. In such substances as $FeCl_2$, $FeCl_3$, FeO and Fe_2O_3, chlorine and oxygen are assigned the "polar valence numbers" -1 and -2, respectively, and iron is considered as having a valence of $+2$ in some compounds and $+3$ in others. In the compounds HCl, HClO, $HClO_2$, $HClO_3$ and $HClO_4$, the valences of hydrogen and oxygen are taken as $+1$ and -2. To give overall neutraliity, the valence of chlorine must then be -1, $+1$, $+3$, $+5$, and $+7$, respectively, in the compounds listed.

In organic chemistry, "nonpolar valence numbers," without plus and minus signs, are used. Balancing is frequently done with the aid of structural formulas, in which a bond line connecting the symbols for two atoms represents a unit valence for each. Conforming to the valence numbers of 4, 2 and 1 for carbon, oxygen and hydrogen, respectively, the structural formulas of some simple organic compounds are:

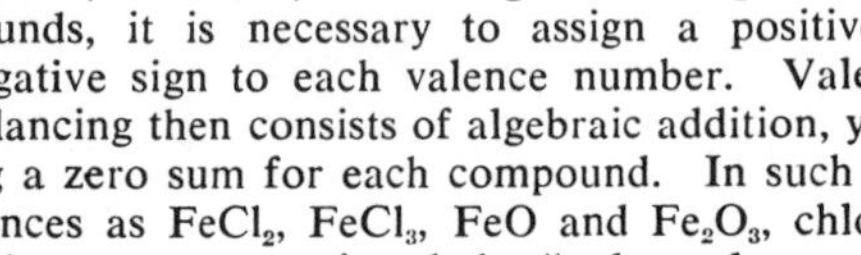

In compounds such as C_2H_4, C_2H_2 and H_2CO, the same valence numbers are used for the elements as before, but two or three bond lines are drawn between adjacent symbols in the structural formulas, to represent "double bonds" and "triple bonds":

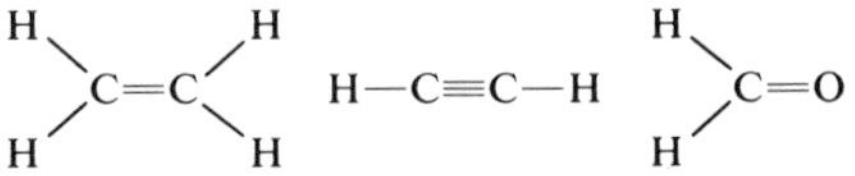

Valences and their regularities are related to the structures of the atoms and to the ways in which the atoms interact to form compounds. These relationships are now well understood. The ability of an atom to combine with other atoms depends primarily on the number of electrons in its outermost or "valence" shell. In an atom of an electropositive element, this number is small. Moreover, these electrons are not tightly held and, in a suitable environment, they may be transferred to other atoms. This atom then exhibits a positive valence equal to the number of electrons so transferred. On the other hand, an atom of an electronegative element, with a larger number of electrons in its valence shell, not only holds them tightly, but, on occasion, adds enough electrons from other atoms to complete a "stable" shell (usually containing eight electrons). Such an element then exhibits a negative valence, equal to the number of added electrons.

Valence electrons can also be *shared,* usually in pairs, between two atoms. A shared electron pair is part of the valence shell of each of the two sharing atoms. The sharing is usually unequal, one of the two bonded atoms holding both electrons more tightly than does the other atom. Positive and negative valence numbers can be assigned as if electron transfer had occurred, i.e., as if the shared electrons both belonged completely to one of the two atoms.

In organic compounds, chemical combination normally occurs by electron-pair sharing, rather than by electron transfer. The two atoms joined by an electron-pair bond (a "covalent bond") both hold the shared electrons tightly. The nonpolar valence number of each element in these compounds is usually just the number of electron-pairs which each atom of that element shares with other atoms.

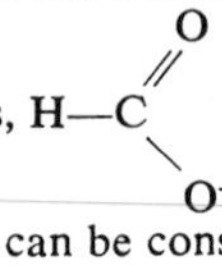

Organic compounds of nitrogen, in which that element is considered to have a valence of 5, are exceptions to this last statement. Each nitrogen atom shares only four electron pairs with other atoms. The fifth valence represents an electron transferred from the nitrogen atom to another atom. A compound such as sodium formate is likewise composed of formate ions, $H-C$, and sodium ions, Na^+. Each oxygen can be considered to have its normal valence of 2, but one of the valences of one of the oxygens represents an electron transferred from a sodium atom.

There is evidence to show that the two oxygen atoms in a formate ion are equivalent. In present-day terminology, there is resonance between two equivalent structures:

$$H-C\overset{\displaystyle O}{\underset{\displaystyle O^-}{\Big<}} \;\;\leftrightarrow\;\; H-C\overset{\displaystyle O^-}{\underset{\displaystyle O}{\Big<}}$$

Whereas, in most organic compounds each electron pair is shared between two atoms, in this ion at least one electron pair is shared between three atoms (O, C, O): the orbital determining the paths of the electrons of the electron pair encloses all three nuclei. Each oxygen atom is still considered to have a valence of 2, although it cannot be specified which oxygen atom has the extra electron. It actually belongs to the resonating system, not to a particular oxygen atom.

Many inorganic and mixed organic-inorganic compounds do not conform to the valence system as just outlined. They are often designated as "secondary valence compounds." It was long considered that in these there are interatomic forces of a different type, weaker than those responsible for the binding between atoms in "primary valence compounds." Now it is known that the forces, in most cases, are of the same type, involving the sharing of electron pairs between adjacent atoms. In primary valence compounds (to which the valence rules apply), one can imagine the complete structure as being formed from uncharged atoms entirely by *primary valence reactions,* in which one electron of each shared pair comes from each of the two atoms sharing the pair:

$$A\cdot + \cdot B \rightarrow A:B$$

In the formation of a secondary valence compound there is, in addition, at least one *secondary valence reaction,* in which a bond is formed by the addition of a lone pair in the valence shell of one atom to the valence shell of another atom:

$$A + :B \rightarrow A:B$$

As an example, the two primary valence compounds, boron trifluoride and ammonia, react to give a stable compound, $BF_3 \cdot NH_3$. The valence electron distribution in this compound is as represented by the following "dot formula":

$$\begin{array}{ccc} :\!\ddot{F}\!: & H & \\ :\!\ddot{F}\!:\!B & :\!\ddot{N}\!:\!H & \\ :\!\ddot{F}\!: & H & \end{array}$$

The B—N bond has been formed by the addition of a lone pair, in the nitrogen valence shell in the ammonia molecule, to the boron valence shell. The numbers of electron pair bonds around the boron and nitrogen atoms do not equal their usual valences.

Another class of compounds not conforming to valence limitations is that sometimes designated as "inclusion compounds' or "clathrate compounds."

These are solid compounds, which can be considered as addition products (without electron-sharing or electron transfer) of two or more component compounds. In most cases, one component forms (often by hydrogen-bonding) a rigid framework, with holes into which molecules or ions of the other component fit. For example, water molecules can form a hydrogen-bonded structure much like that of ice, but with holes regularly distributed therein; into these holes any of various small organic molecules can fit. Sizes and shapes, rather than valence considerations, determine which ones will do this. (See **Clathrate Compounds**).

The concept of valence is thus a useful one, especially for purposes of classification and correlation, but it should be realized that it involves an oversimplification. The valence numbers express certain properties of the elements, as regards their capacities for combination, but they do not distinguish between the processes of electron-sharing and electron transfer, nor do they make allowance for differences in the strengths and polarities of bonds. Moreover, they are not applicable to chemical combinations in which both electrons of a bond are supplied by the same atom nor to inclusion compounds.

MAURICE L. HUGGINS

Cross-references: *Bonding, Molecules, Combining Number, Oxidation.*

VANADIUM AND COMPOUNDS

Vanadium is a member of Group VB of the periodic system and of the first transition series. Its atomic number is 23, atomic weight 50.9414, melting point 1890°C, and specific gravity 6.11 at 20°C.

The pure metal has a bright silvery luster, is very ductile, and can be cold-worked readily. Small amounts of nitrogen and oxygen decrease its ductility, and commercial grades of vanadium may be quite brittle. It is stable in air or oxygen at ordinary temperatures, but combines rapidly with oxygen, nitrogen, and carbon at elevated temperatures, forming V_2O_5, VN, or VC, respectively. Lower oxides such as VO_2, V_2O_3, and VO are also formed. The metal is resistant to water, sea water, HCl, and H_2SO_4, but is attacked by HNO_3.

History. N. G. Sefstrom in 1830 discovered a new element in a Swedish iron ore which he named vanadium after Vanadis, the Norse goddess of beauty. Earlier, in 1801, Andres Manuel del Rio had described an element which he found in a lead ore at Zimapar, Mexico as erythronium, but later decided the element was chromium. Wohler subsequently showed that erythronium and vanadium were identical.

Commercial use of vanadium started in 1906 with production of ferrovanadium by aluminothermic reduction to meet demands of the automobile industry. Marden and Rich prepared the pure metal in 1927 using calcium reduction.

Occurrence. Vanadium is widely distributed and ranks twenty-second among elements in the earth's crust, but few deposits contain over 1%. It occurs in small amounts in iron ores, particularly titaniferous magnetites, petroleum crudes, asphaltites, and phosphate rock. The important minerals containing larger amounts of vanadium are:

Patronite, a complex sulfide, $V_2S_5 + nS$

Carnotite, $K_2O \cdot 2UO_3 \cdot V_2O_5 \cdot 3H_2O$

Roscoelite, $2K_2O \cdot 2Al_2O_3 \cdot (Mg,Fe)O \cdot 3V_2O_5 \cdot$ $10SiO_2 \cdot 4H_2O$

Descloizite, $4(Pb,Zn)O \cdot V_2O_5 \cdot H_2O$

Cupro-descloizite, $4(Cu,Pb)O \cdot (V,As)_2O_5 \cdot H_2O$

Vanadinite, $Pb_5(VO_4)_3Cl$

The high-grade patronite ores found in Peru, coning up to 38% vanadium, were the first to be mined extensively. Other important sources were the carnotite and roscoelite ores of the western United States, vanadinite from Mexico and South America, descloizite and cupro-descloizite from Southwest Africa and Zambia, and European iron ores. With the advent of the atomic age, the main source became vanadium recovered as a coproduct with uranium from carnotite and related ores in this country. In recent years, the titaniferous magnetites of South Africa, U.S.S.R., Finland, and Norway have become the major source.

Extraction Procedures. Ores, slags, and boiler residues containing vanadium in a lower state of oxidation must be roasted with salt, soda ash, or sodium sulfate to convert the vanadium to water-soluble sodium vanadate, $NaVO_3$. The roasted product is then leached in water or soda ash solution, and then sometimes leached with dilute acid. Oxidized ores such as carnotite can be leached with sulfuric acid without roasting. Vanadium is precipitated from water-leach or soda-leach solutions by adding sulfuric acid to a pH of about 2.5. The precipitate is sodium polyvanadate, approximately $(H,Na)_4V_6O_{17}$ or $(H,Na)_6V_{10}O_{28}$. Acid leach liquors containing vanadium are treated by solvent extraction or ion exchange to separate uranium and any impurities present.

The roasting and leaching procedure is sometimes used to extract vanadium from titaniferous magnetite, but a more common procedure, used in the U.S.S.R. and South Africa, is to smelt the magnetite and produce vanadium-bearing pig iron. The pig iron is blown with oxygen and the vanadium is recovered in a slag containing 15 to 25% V_2O_5. The slag is roasted and leached to extract vanadium or is used to produce ferrovanadium by a direct reduction process.

Production of Metal. Reduction of fused oxide with aluminum produces a commercial grade of metal containing 90% vanadium. The reaction is exothermic and is carried out without external heat. Metal containing over 99% vanadium is produced by calcium reduction of V_2O_5 in a steel bomb or by reduction of VCl_3 with magnesium or sodium in an argon atmosphere. Carbon reduction of V_2O_3 at 1700°C in a vacuum has also been used to produce high-purity vanadium. The purest metal, 99.95%, is produced by electrolytic refining of impure grades in a fused salt bath.

Ferrovanadium containing 50–55% or 70–80% vanadium is produced by aluminum reduction of a mixture of fused oxide and either iron scale or punchings. Although external heat is not needed if iron scale is used, better vanadium recovery is realized by using an electric furnace. Silicon metal or ferrosilicon can be used instead of aluminum to reduce the fused oxide, but this requires a two-step

process for high vanadium recovery. Ferrosilicon is also used for producing ferrovanadium from vanadium-bearing slags; some of the iron in the slag can be removed by a preliminary ferrosilicon reduction step. Carbon reduction in an electric furnace to produce ferrovanadium has been discontinued. Commercial V_2C for use as an additive to steel is made by carbon reduction of V_2O_4 in a vacuum. Aluminum-vanadium alloys for the titanium industry are made by aluminum reduction of fused oxide.

Chemistry. Vanadium forms compounds in four valence states. Pentavalent vanadium is nominally acidic with well-defined amphoteric properties. The oxide, V_2O_5, dissolves in alkalies to form a series of meta-, ortho-, pyro-, and polyvanadates, depending upon the pH. V_2O_5 reacts with halides to form $VOCl_3$, VOF_3, and $VOBr_3$. Reaction of metallic vanadium with fluorine yields VF_5.

The tetravalent oxide, VO_2, is formed by reduction of V_2O_5 with ammonia or other reductants. It is amphoteric and dissolves in acids to give $(VO)^{2+}$ salts such as $VOSO_4$ and $VOCl_2$. These salts form deep blue solutions. Chlorination of vanadium at 300°C produces VCl_4.

The basic trivalent oxide, V_2O_3, is formed by reduction of V_2O_5 with hydrogen or ammonia at 650°C. It is insoluble in water or alkalies and does not react readily with acids. The chloride, VCl_3, is made by reduction of VCl_4.

The divalent oxide, VO, is best prepared by reduction of V_2O_5 with hydrogen at 1700°C. It has a metallic luster and good electrical conductivity, and possibly was mistaken for metal by early experimenters. It is insoluble in water or alkalies, but dissolves in acids to form divalent salts.

Analysis. Oxidation-reduction procedures are commonly used for the quantitative determination of vanadium. A very satisfactory method involves reduction of the pentavalent ion to tetravalent ion with excess ferrous ammonium sulfate, destruction of the excess with ammonium persulfate and then titration with permanganate to the pink end point. This method is widely used in industrial laboratories, tungsten being the only interfering element.

Uses. Vanadium consumption in the United States has grown steadily, increasing from 2016 net tons contained vanadium in 1960 to 4802 net tons in 1971. Nearly 80% is used in steel, primarily for low-alloy high-strength steels containing less than 0.10% vanadium. These small percentages refine the grain size and improve the mechanical properties significantly. Another important use is for high-speed tool steels. Nonferrous alloys consume about 10% of the total vanadium, largely for production of titanium alloy containing 6% aluminum and 4% vanadium. The major chemical uses of vanadium are for catalytic purposes and ceramic stains and glazes. V_2O_5 is an excellent oxidation catalyst and is used for the manufacture of sulfuric acid, phthalic anhydride, maleic anhydride, and nylon. $VOCl_3$ is used as a catalyst in making ethylene-propylene rubber. Vanadium and vanadium alloys may have nuclear applications because of the low neutron cross-section of the element.

Toxicology and Biological Effects. Vanadium compounds are toxic to humans and animals, and should be considered industrial hazards. Dust from vanadium and its compounds is irritating to the respiratory tract, but the effects are not long lasting nor cumulative.

In small amounts vanadium reduces the synthesis of cholesterol by the human body. This may have significance in the treatment of cardiovascular disease.

H. W. RATHMANN

References

Rostoker, W., "The Metallurgy of Vanadium," John Wiley and Sons, Inc., New York, 1958.

Kirk-Othmer, "Encyclopedia of Chemical Technology," Second Edition, Vol. 21, John Wiley and Sons, Inc., New York, 1970.

Hampel, C. A., Editor, "Rare Metals Handbook," Second Edition, Reinhold Publishing Corp., New York, 1961.

"Encyclopedia of the Chemical Elements," Hampel, C. A., Editor, Reinhold Publishing Corp., New York, 1968.

VAN DER WAALS FORCES

Johannes Diderik van der Waals was born at Leiden in 1837 and studied at the university in the same town. He is best known for his contributions to the theory of gaseous and liquid states. He assigned quantitative values to the volume of molecules and to the attraction between molecules (van der Waals forces) in his equation of state. This enabled him to formulate an equation which expresses the relation between pressure and volume of a substance in the gaseous as well as in the liquid state and which gives far more accurate values of pressure-volume relationships than Boyle's Law. van der Waals also demonstrated that the gaseous and liquid states can merge continuously. These theories were published in 1873 in his doctoral dissertation entitled "On the continuity of the gaseous and liquid state."

Van der Waals forces, in the most general sense, are forces between atoms and/or molecules other than those leading to chemical bonding. While the distinction between these and chemical forces is not always sharp (e.g., there are loosely bound, so-called van der Waals molecules like O_4, HgA), the binding strength for the two types is ordinarily quite different: it is of the order of electron-volts in the case of chemical forces, millivolts in the case of van der Waals forces.

As a function of distance, van der Waals forces between *unexcited* structures (i.e., atoms or molecules in their normal states) usually show the behavior illustrated in Figure 1, where the potential energy (negative integral over force times displacement) is plotted against the distance of separation between the molecules. Beyond the point R_0 the energy curve rises, hence the forces are attractive. For smaller R the forces are repulsive.

The repulsive forces arise from the interpenetration of the electronic clouds surrounding the molecular nuclei. Because their calculation involves certain terms called exchange integrals, they are sometimes called *exchange forces*. Qualitatively, they are responsible for the mechanical rigidity, the "impenetrability" of the molecules and are sometimes approximated in computation by a straight vertical line at some characteristic distance R'. This

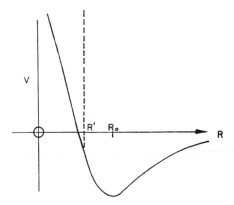

FIG. 1. Graph of a typical intermolecular potential.

distance is known as the "gas kinetic diameter" if the interaction takes place between similar molecules.

In the restricted sense, van der Waals forces are the forces which correspond to the attractive region beyond R_0 in Figure 1. If repulsive forces were absent, the constant b in van der Waals' equation would be zero; the constant a owes its existence to the attractive forces. Hence, it may be said by way of definition that van der Waals forces in the general sense are the interactions which cause deviations from the ideal gas law; in the specific sense they are the interactions giving rise to van der Waals' a. Henceforth, this article is devoted to the latter. Aside from the departures just mentioned, they are likewise responsible for surface tension, friction, changes of phase, cohesion of most solids, viscosity, the Joule-Thomson effect and many others.

The physical origin of these forces was once sought in gravitational interactions which, however, are now known to be very much too small to produce such effects. The first modern conjecture was made by Debye (1920), who supposed that permanent multipoles, primarily quadruples, located in one of the structures, induced by electrostatic polarization small multipoles in the other. This *induction effect* does take place but is in general too small to account for the observations. Keesom, in 1921, presented a theory explaining van der Waals forces in terms of the alignment of permanent multipoles, which are shown to arrange themselves preferably in a manner leading to attraction. While this *alignment effect* is an important component of the van der Waals forces among polar molecules, it does not tell the whole story. In particular, neither induction nor alignment provides a way of explaining the van der Waals forces between symmetric atoms, such as the rare gases.

The principal missing factor in this account was provided by F. London in 1930. In developing the theory of what he called *dispersion forces,* he showed that quantum physics admits a mechanism crudely characterized as follows. The electrons in one atom (e.g., helium) revolve, as it were, in ignorance of those in the other so long as the atoms are very far apart. When they are brought together, they revolve more or less in phase in the two structures, the more so the closer they are together, and the result of this phase agreement is an attractive force. Dispersion forces are universally present in all interactions; they are not limited to molecules having permanent polarity.

The potential energy, V_D, arising from this effect has in general a much shorter range than the mutual energies of multipoles; for the simplest molecules (rare gases, O_2, N_2, etc.) they cease to be of importance at a distance of separation of 10^{-7} cm. They depend on distance in accordance with the formula

$$V_D = -\frac{A}{R^6} - \frac{B}{R^8} - \frac{C}{R^{10}} - \cdots$$

hence they disappear in a manner proportional to R^{-6} at large distances. In this expression, A/R^6 is said to represent the dipole-dipole, B/R^8 the dipole-quadrupole, and C/R^{10} the quadrupole-quadrupole effects. The latter were first calculated by the writer in 1931. Formally, the series has an indefinite number of terms, but its convergence is asymptotic, and it is generally inadvisable to retain terms beyond those written above.

Only for sufficiently symmetric molecules are A, B, and C independent of the orientation of the molecules in space. In general these parameters are functions of angles; hence the forces are not central forces. When more than two molecules interact, the dispersion effect is very nearly additive, i.e., the total energy is approximately a sum of pairwise components V_D. Additivity does not hold for the repulsive forces.

A general rule for the relative importance of van der Waals forces, for their dependence on angles or indeed distance, cannot be given. In the simple example of H_2O, at a distance of separation equal to the kinetic diameter, induction, dispersion, and alignment effects are all of comparable magnitude and function approximately in the ratio of $1:3:8$. In larger molecules the situation is very complex.

The forces between multipole molecules are not properly given by Figure 1, for they can be repulsive as well as attractive. But the average over-all orientations, weighted by the Boltzmann factor, is an attractive force. The dispersion effect is always attractive if the molecules are in their lowest energy states. When one or both of the molecules are excited, there are usually several modes of interaction, some attractive and others repulsive even at large distances of separation.

Information regarding the numerical values of van der Waals forces is mostly semiempirical, derived with the aid of theory from an analysis of chemical or physical data. Attempts to calculate the forces from first principles have had a measure of success only for the simplest systems, such as H—H, He—He and a few others. When judging the difficulties of such calculations, one must bear in mind that the energies sought are of the same order of magnitude as the *errors* in the best atomic energy calculations.

HENRY MARGENAU

VAPOR PRESSURE

Vapor pressure is the term applied to the driving force behind the apparently universal tendency for liquids and solids to disperse into the gaseous phase. All known liquids and solids possess this fundamental property, although in some cases it is too minute to be measurable. A typical liquid will exert a vapor pressure which is constant and reproducible. This pressure is dependent only upon the temperature of the system, and increases with increasing temperatures.

The molecular theory explains the phenomenon of vapor pressure through molecular activity. The molecules of a liquid are in rapid motion, even though they are in contact with each other. This motion or activity increases with temperature. At the vapor-liquid interface, this motion results in diffusion of some molecules from the liquid into the vapor. The attraction between molecules is strong, and some of the molecules dispersed into the vapor return to the liquid. The net number of molecules escaping produces the vapor pressure. For all practical purposes, this vapor pressure can be assumed constant whether the system is at equilibrium or not, due to the extremely high rate of molecular diffusion at the interface of the two phases.

In solids, the attractive forces of the molecules are so dominant that each is more or less frozen in place. Some diffusion does occur, however, as evidenced by the evaporation of ice, the odor of moth balls, and the slow diffusion or alloy formation of some metals kept in intimate contact. This vapor pressure increases with temperature, but is also a function of the molecular arrangement of the solid. As some solids such as sulfur are heated and molecular rearrangements take place, forming another allotrope of the same element, the vapor pressure changes sharply as this rearrangement occurs. The diffusion of a solid into the vapor state is called sublimation. The energy required to accomplish this transfer is high, since it is the sum of the heat normally required to melt the substance as well as vaporize it.

Vapor pressure can only be exhibited when the molecular activity is at a low enough level to permit continuous contact of the molecules and thus formation of a liquid. The maximum temperature at which this is possible is a fundamental property, and is called the critical temperature. Above this temperature the material cannot be compressed to form a liquid, and only one phase results. This temperature is 374.0°C for water, and −240.0°C for hydrogen.

The fundamental relationship between temperature and vapor pressure can be derived from thermodynamic laws. With certain limiting assumptions, the Clausius-Clapeyron equation is most often applied:

$$\frac{dp}{dT} = \frac{qp}{RT^2}$$

where q is the heat of vaporization; in calories per gram mole; $\frac{dp}{dt}$ is the slope of vapor pressure versus temperature curve at the point in question in cm Hg per °C; p is the pressure in cm Hg; R is the gas constant in calories per °C per gram mole; and T is the temperature in °K. Because of these limiting assumptions, the integrated form of this equation is used in practice primarily as a guide to develop methods of correlating and plotting vapor pressure data.

An excellent and versatile empirical vapor pressure correlation was developed by F. D. Othmer, P. W. Maurer, C. J. Molinary, and R. C. Kowalski (*Ind. Eng. Chem.,* **49** (1957). The reference equation is:

$$\log P = \left(\frac{L}{L'}\right) \log P' + C$$

where $\frac{L}{L'}$ is the ratio of molar latent heats of the substance in question to a reference substance. The ratio is the key to the accuracy of this method; it very neatly minimizes effects of deviations from ideality.

The vapor pressure of a solution containing a nonvolatile substance (e.g., salt in water) is lower than that of the pure liquid. This phenomenon can again be explained by interference of the liquid molecular activity by the dissolved substances. The relationship between this vapor pressure depression and the concentration of the dissolved substance is valid for most substances at low concentrations. It was found to be dependent on the relative numbers of molecules of the solute and the solvent, and allowed accurate determinations of molecular weights of unknown solutes. If the Clausius-Clapeyron equation given above is combined with the above concentration relationship, it can be shown that:

$$\Delta T = \frac{RT^2}{q} \cdot C$$

where ΔT is the elevation of the boiling point, and C is the mole ratio of solute to solvent. This defines the effect of any solute on the vapor pressure exhibited by any solvent of latent heat q.

In the same manner, the vapor pressure of one component of a solution of two liquids has a different relationship with temperature than if it were pure. For many liquid mixtures, such as most hydrocarbon mixtures, the vapor pressures of the components vary directly from that exhibited in the pure form as their molar concentration in the solution. This relationship is known as Raoult's law:

Partial Pressure $= P_o x$

where x is the molar concentration of the component in the liquid, the P_o is the vapor pressure of the pure component of the same temperature as the mixture. A mixture following this rule is called an "ideal" solution, and its total volume is the sum of its components' volumes.

When most mixtures are reduced by boiling, logically the more volatile component is removed faster and heavier material concentrates. There are important exceptions to this rule, and these are called azeotropic or constant boiling mixtures. In essence, due to association or some other reason for deviation from Raoult's law, there are certain mixtures where the boiling temperature is either above that of the high boiling component or below that of the more volatile component. As the composition changes, at some point the boiling temperature change has to reverse itself to approach the remaining major component. At this point the composition of the liquid and vapor are identical and the mixture behaves as a single component, boiling away with no change in temperature or composition, the "CBM," (constant boiling mixture). The composition of CBM's is frequently pressure sensitive.

If the gas phase above the liquid is also "ideal," the partial pressure of a component in this phase is equal to the total system pressure times the mole fraction of the component in the gas phase. This is called Dalton's law:

Partial Pressure $= P_t y$

Combining these two formulae, it can be seen that:

$$\frac{y}{x} = \frac{P_o}{P_t} = K$$

for any particular temperature. This relationship can largely define many very complex liquid mixtures if the pressures used in the correlation are corrected by experimental data for deviation from the ideal.

For work of any precision, it is essential that the correlative methods described above be only used to help fit experimental data, particularly if the system is at elevated pressure.

A different relationship results if two liquids are relatively immiscible in each other. Molecular interference is minimal, and the total pressure exerted is equal to the sum of those of the individual pure components. The fundamental property of vapor pressure is thus dependent on temperature and composition of the material considered. These known and reproducible relationships have great technical application.

Vapor pressure relations can be used to determine heats of solution, heats of sublimation, and heats of fusion. Problems dealing with the solution of gases of liquids and adsorption of gases by solids are best handled by vapor pressure concepts. In dealing with solutions of miscible liquids, the most simple and useful relationship involves plotting (for the most volatile component) the mole fraction y in the vapor against x, the mole fraction in the liquid. The ratio of y/x is called the phase equilibrium constant K, and is used for definition of bubble points and dewpoints of simple and complex hydrocarbon mixtures over temperature and pressure ranges to near the critical.

Some equations of state also define vapor pressure relationships. One of the most recognized is that developed by Benedict, Webb and Rubin due to its ability to predict P-V-T properties in the two-phase region plus describing behavior of the superheated vapor. (Ref., *J. Chem. Phys.,* **8,** 334 (1940); ibid., **10,** 747 (1942); *Chem. Eng. Progr.,* **47,** (1951).

DOUGLAS L. ALLEN

Cross-references: *Distillation, Gas Laws, Solvents.*

VENOMS

Animal venoms are the most complex poisons known to man. Most venoms and animal poisons are mixtures of many different substances. Some venom components may be proteins, including enzymes, polypeptides or glycopeptides, while others are carbohydrates, aminopolysaccharides, lipids, free amino acids, amines, steroids, alkaloids, furans and formic acid. A single venom may contain 20 to 30 different components, even though the toxic moiety may only be composed of several, or even a single substance, such as the puffer fish toxin tetrodotoxin, an amino perhydroquinazoline. It is obvious that the various components of venoms have evolved and adapted in a remarkable way, in most instances as part of the animal's offensive or defensive armament; and although there are some fractions of venoms for which we have not yet found a specific pharmacological property that seems related to the design of the venom, there are some synergistic actions of certain whole venoms that are not present when one separates the various fractions and studies their individual toxicological properties. Also, it is quite possible that the biological properties of some venom fractions played an important role in the animal's posture in eons past, and that these characteristics are no longer essential to the function of the venom in its present ecological niche.

Not only are the venoms and animal poisons complex in their chemical structure, but they are remarkably diversified and complicated in their modes of action. Some venoms have an effect, either directly or indirectly, on almost every organ system, and few are tissue-specific, although some may have a marked effect on one organ or tissue and little on another. No venom should ever be considered, as is so often done, as a "neurotoxin" or a "cardiotoxin" or a 'hemotoxin" and thus permit its other biological properties to be minimized or dismissed. Many errors in understanding the properties of venoms have occurred because of this oversimplified classification. It is best to consider all animal venoms as complex mixtures capable of producing one or more deleterious changes in several organ systems or tissues, and of inducing these changes concurrently.

Finally, it is evident that, quite aside from the specific and combined activities of the various fractions of a venom, there can be produced in the envenomated organism certain autopharmacological substances which may not only complicate the poisoning but which may, in themselves, produce more serious consequences than the venom. It takes on the order of 150 simultaneous bee stings to produce death in a nonsensitized human but the sting of a single bee can be fatal to the sensitized individual. The release of such autopharmacological substances as histamine, bradykinin and adenosine, for example, can be so overwhelming to the individual that their effects may be far more deleterious than those of the venom.

The study of the chemistry of animal poisons is further complicated by the fact that qualitative as well as quantitative differences in the composition of these toxins may exist, not only from species to species within the same animal genus, but also from individual to individual within the same species. A poison may vary with the individual animal at different times of the year or under different environmental conditions, or even with the age and sex of the animal.

Venomous animals have sometimes been divided into two groups on the basis of the use to which the animal puts its venom. In general, most venoms delivered from the oral pole, usually in association with salivary or maxillary glands, are used by the animals in offense, as in the gaining of food. The venoms of such animals tend to have a higher enzyme content and lethal index than those delivered from the aboral pole. Venoms delivered from the aboral pole tend to be associated with the defensive posture of the animal and, in general, arise from dermal tissues and contain few or no enzymatic constituents. As a whole the aboral venoms are less lethal, but they appear to contain greater amounts of substances which are pain-producing. *Venomous* animals should be differentiated from *poisonous* animals. The former term is applied to those creatures having a gland or group of highly specialized secretory cells, a venom duct (although this is not always found), and a structure for delivering the venom. Poisonous animals have no such apparatus; poisoning is a result of ingestion.

Snake Venoms. Although there are fewer than 4000 species of snakes, approximately 10% of the species are venomous. Venomous snakes are usually divided into four families: the sea snakes (*Hydrophidae*); the elapids (*Elapidae*), composed of the cobras, mambas, kraits and coral snakes; the vipers (*Viperidae*), which include the gaboon viper, puff adder, Russell's viper, the horned vipers of the Sahara, saw-scaled vipers and the European viper; and the pit vipers (*Crotalidae*), the rattlesnakes, copperheads, and water moccasins of the United States, the bushmaster of South America and certain primitive pit vipers of Asia.

The venoms of all these snakes are complex mixtures, chiefly proteins, many of which exhibit enzymatic activity. The snake venoms studied to date appear to contain between 11 to 25 different components. The enzymes of snake venoms have been studied in some detail. The more important of these are:

proteinases	desoxyribonuclease
L-arginine-ester	
hydrolases	phosphomonoesterase
transaminase	phosphodiesterase
hyaluronidase	5'-nucleotidase
L-amino acid oxidase	ATPase
cholinesterase	alkaline phosphatase
phospholipase A	acid phosphatase
phospholipase B and C	DPNase
ribonuclease	endonucleases

Proteinases have been found in most snake venoms. Almost all *Crotalidae* venoms so far examined appear to be rich in proteolytic enzyme activity. The *Viperidae* venoms have lesser amounts of proteinase, while the *Elapidae* and *Hydrophidae* venoms have little or no proteolytic enzyme activity. This enzyme is responsible for the digestion of tissue proteins and peptides. Venoms rich in proteinase produce marked tissue changes and destruction.

Hyaluronidase has been found in every snake venom examined to date. It hydrolyzes the hyaluronic acid gel of the spaces between cells and fibers, particularly in connective tissue, and thus reduces the viscosity of these tissues. This breakdown in the hyaluronic acid barrier probably allows other fractions of the venom to penetrate the tissues. The amount of enzyme is probably related to the extent of the edema and swelling caused by the venom.

L-Amino acid oxidase has been found in over 70 snake venoms, chiefly in viper and crotalid venoms. It is absent from sea snake venoms, and either absent or present in low concentrations in the elapid venoms. It catalyzes the oxidation of L-α-amino acids and of α-hydroxy acids. It is the most active of the known amino acid oxidases, and ophio-L-amino acid oxidase is probably a group of homologous enzymes. It is probably not a toxic component of snake venom, but its action is integrated with the digestive function of the poison.

Cholinesterase has been identified in the venoms of at least 50 snakes. In general, elapid venoms are rich in the enzyme, while viperid and crotalid venoms either do not contain the enzyme or possess it in only small amounts. The enzyme catalyzes the hydrolysis of acetylcholine to choline and acetic acid. Some studies have shown the enzyme is acetylcholinesterase. In normal tissues the enzyme prevents the excessive accumulation of acetylcholine at cholinergic synapses and at the neuromuscular junction. Acetylcholine is found in large amounts in some elapid venoms, in limited amounts in other elapid and certain of the viperid and crotalid poisons, and not at all in many other snake venoms, regardless of family.

Phospholipase A, B, C, and D are catalysts involved in the hydrolysis of lipids. Snake venoms phospholipase A catalyzes the hydrolysis of one of the fatty ester linkages in diacyl phosphatides, forming lysophosphatidase and releasing both saturated and unsaturated fatty acids. Some snake venoms may contain several enzymes with phospholipase A activity.

Phosphomonoesterase has properties of an orthophosphoric monoester phosphohydrolase. Its optimum activity is at pH 9.5, and in some venoms it has an isoelectric point above pH 8.6. Phosphodiesterase has been found in almost all snake venoms tested. It is an orthophosphoric diester phosphohydrolase which also releases 5'-nucleotides from polynucleotides, thus acting as an exonucleotidase.

5'-Nucleotidase is a common constituent of all snake venoms. In most instances it is the most active phosphatase in the venom. It is a 5'-ribonucleotide phosphohydrolase which catalyzes the hydrolysis of 5'-mononucleotides, yielding the ribonucleoside and orthophosphate.

There was a time when all the biological effects of snake venoms were attributed to the enzymatic activities contained in these poisons. It is now generally recognized that the lethal and perhaps some of the more deleterious properties are related to the nonenzymatic components. Although some of these proteins were being studied three decades ago, it was not until the mid-sixties that definitive chemical experiments were carried out.

By means of chromatography, crystallization, electrophoresis, ultracentrifugation, amino acid analysis, and sequence of amino acids a number of toxins have been isolated and characterized from the venoms of the sea snake, the cobra, and certain vipers and rattlensakes. In the sea snake *Laticauda semifasciata* the lethal property is associated with two peptides having molecular weights of approximately 7,000. These comprise 30% of the crude venom. In the cobra *Naja nigricollis* a peptide has been isolated which is approximately nine times more lethal than the crude venom of which it comprises about 3%. This peptide has been purified to homogenicity and the amino acid sequence-characterized. A lethal toxin having a molecular weight of about 12,000 has been isolated from the venom of *Vipera palestinae*. In the venom of the rattlesnake *Crotalus viridis helleri* a protein having a molecular weight of approximately 30,000, and two peptides having molecular weights of about 6,000 each have been isolated and characterized on the basis of their pharmacological effects.

The recent studies on these nonenzymatic proteins and peptides indicate their importance as the lethal fractions of snake venoms, and in some cases their role as neurotoxins. However, the enzymes are no doubt responsible for other specific pharmacological effects and certainly play a significant role in the over-all design of the venom.

Marine Poisons. For the most part, the 2,000 or so species of venomous and poisonous marine animals are widely distributed throughout the marine fauna, from the unicellular protistan *Gonyaulax* to certain of the chordates. They are found in almost all the seas and oceans of the world. It is generally believed that most of the noxious marine animals have been identified, although a number of forms have not yet been adequately described.

The toxins of the marine animals are far more varied in structure than those of the reptiles. The brittle star *Ophiocomina nigra* discharges a viscous substance on stimulation that is characterized as a highly sulfated mucopolysaccharide which is acidic and has a pH of 1.0, while puffer fish poison, tetrodotoxin, is an amino perhydroquinazoline. The toxins of the venomous fishes appear to be proteins having molecular weights approaching 800,000, while still other marine toxins are amines and lipids.

Paralytic shellfish poison, saxitoxin, mussel poison and *Gonyaulax* poison are generally employed synonymously. The toxin is stored and concentrated for the most part in the hepatopancreas of certain mussels or the siphon of some clams, which have been exposed to certain *Gonyaulax* species. The poison is one of the most lethal substances known. Its chemical properties are summarized in Table 1.

Coelenterate or cnidarian venoms, those of the jellyfishes, Portuguese man-of-war, sea wasps, sea anemones and certain corals vary considerably in their structure. The stinging unit is the nematocyst, which within its capsule contains proteins, sulfur-containing amino acids, hydroxyproline, glycine, tyrosine, arginine, proline, alanine, glutamic acid, aspartic acid, a succinoxidase inhibitor, hexosamine, uronic acid, orthodiphenols, mineral salts, alkaline and acid phosphatases, 5'-nucleotidase, cholinesterase, and 5-hydroxytryptamine. A number of recent studies, with partially purified toxins, indicate that the lethal and certain nerve-muscle effects of cnidarian toxins are polypeptides, but their exact nature has not yet been determined.

In the phylum Echinodermata, the starfishes, brittle stars, sea urchins and sea cucumbers, there are both venomous and poisonous species. Perhaps

TABLE 1. SOME CHEMICAL AND BIOLOGICAL PROPERTIES OF PARALYTIC SHELLFISH POISON

Property	Mussel poison	Saxitoxin	G. catenella poison
Molecular formula	$C_{10}H_{17}N_7O_4 \cdot 2HCl$	$C_{10}H_{17}N_7O_4 \cdot 2HCl$	$C_{10}H_{17}N_7O_4 \cdot 2HCl$
Molecular weight	372	372	372
N-content (Kjehdahl)	26.1	26.8	26.3
Diffusion coefficient	4.9×10^{-6}	4.9×10^{-6}	4.8×10^{-6}
Specific optical rotation	130°	128°	128°
pKa	8.3, 11.5	8.3, 11.5	8.2, 11.5
Molecular extinction of oxidation product	6000	—	—
Sakaguchi test	Negative	Negative	Negative
Benedict-Behre test	Positive	Positive	Positive
Jaffe test	Positive	Positive	Positive
Aromatic structures	Present	Present	Present
Carbonyl groups	None	None	None
Reduction with H_2	Nontoxic dihydro derivative	Nontoxic dihydro derivative	Non toxic dihydro derivative
Lethality	5300 mouse units/mg.	5200 mouse units/mg.	5100 mouse units/mg.
Proposed structure	Tetrahydro purine derivative	Tetrahydro purine derivative	Tetrahydro purine derivative

Saxitoxin

the most important venomous members of the phylum are certain genera of sea urchins, which have small pincer-like organs, the pedicellariae, distributed over their entire body surfaces between their spines. Extracts from these structures have been shown to contain 8 immunologically distinct proteins, all of which are lethal to mice, and against which specific neutralizing antibodies have been made. Highly lethal materials have been separated by conventional fractionation procedures. The most lethal fraction ($LD_{50} = 2 \times 10^4$ mg protein N) behaves like an enzyme. Further purification has resulted in the separation of 5 substances, 3 of which exhibit kinin-like activity. The material from one has been found to correspond to synthetic bradykinin.

The most active toxin in extracts of certain sea cucumbers is called holothurin A. The formula is $C_{50-52}H_{81-85}O_{25-26}$, and the molecular weight has been calculated to be 1,155. The provisional structure is:

COMPOUND	R
HOLOTHURIN	$-OSO_3^- Na^+$
DeH	$-H$

SUGAR	SYMBOL
D-GLUCOSE	G
D-XYLOSE	X
D-QUINOVOSE	Q
3-O-METHYLGLUCOSE	G-OMe

HOLOTHURIN A

Fish toxins can be divided into two major groups: those from poisonous fishes and those from venomous fishes. Fish poisoning, or icthyotoxism, can be further subdivided into: (a) icthyosarcotocism (fish whose toxin is within their musculature, viscera or skin), (b) icthyootoxism (fish whose gonads or roe are toxic, or where there is a relationship between gonadal activity and the production of the toxin); (c) icthyohemotoxism (fish which have a toxin in their blood); and (d) icthyocrinotoxism (fish which produce a toxin by glanular secretion from their skin but otherwise lack a true venom apparatus). The toxins of these poisonous fishes vary remarkably and there does not appear to be any common structural basis on which they might be compared. Tetrodotoxin, from the puffer fish, has the formula $C_{11}H_{17}N_3O_8$ and the structure:

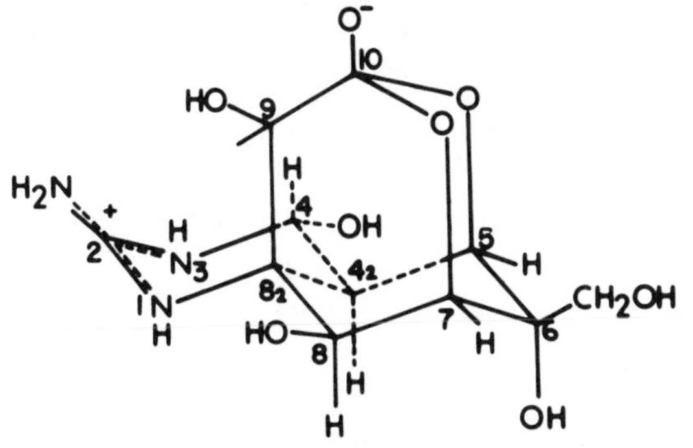

while the toxic substances released from the skin of the boxfish *Ostracion lentiginosus* has the formula $C_{23}H_{46}O_4NCl$ and the structure:

Both these toxins have interesting pharmacological properties which have already been put to use as tools in biology and medicine.

The venoms of the stingrays, scorpionfishes, lionfishes and stonefishes appear to be somewhat related chemically and pharmacologically; and to a lesser extent so do the venoms of the weeverfishes and most of the other venomous piscines. A common property of almost all fish venoms is their relative instability. Most of the lethal property, as well as certain of the other deleterious activities, is lost on heating, or even when venom-containing extracts are allowed to stand at room temperature for a short time. Until very recently studies on these toxins have been greatly hampered by difficulties in obtaining a stabilized product. Recent preliminary studies indicate that the toxic fractions of fish venoms are proteins having molecular weights between 50,000 and 800,000, that they are relatively free of enzymes, and that the pain-producing fraction is much more stable than the more deleterious components and may be a kinin-like material.

Arthropod Venoms. Of the 800,000 or so species of arthropods, less than one-tenth are venomous, and only a small number of these are of a potential danger to man. Almost all spiders, approximately 20,000 species, are venomous. Fortunately for man, only a relatively small number of the spiders have fangs long and strong enough to penetrate the human skin. There are some 500 species of scorpions, some of which must be considered lethal. In the Hymenoptera, the bees, wasps, yellow jackets and ants there are numerous genera and species of potential danger to man. Among the ticks, caterpillars, assassin bugs, certain moths, butterflies, grasshoppers and other arthropods there are a number of poisonous species.

The arthropods are the most highly developed, specialized and versatile of the invertebrates. It seems reasonable, therefore, to assume that their venoms should be the most complex and diversified.

Some of these animals, such as the spiders, use their venom in the gaining of food, while others, like the scorpions, use it chiefly in defense. Still others produce and release a substance from their body surface, or special glands, which because of its odor, irritant or certain other properties repels other arthropods, and even reptiles and mammals.

The venom of the bee is a good example of the complexity of an arthropod toxin. This poison contains lipids, small peptides, apamin, melittin and other proteins and enzymes, apic acid, sugars, free bases and amino acids, and a number of unidentified components. The pharmacological properties of this venom, as well as most of its components, have been studied. The danger of envenomation by honey bees, in man, is not usually a serious one, unless the individual is sensitive to the venom. There is still considerable question as to which components of bee venom causes sensitization. They may not necessarily be the more toxic ones.

Scorpion venoms contain 10–15 proteins and at least six nonproteins. The lethal and more deleterious components appear to be proteins of low molecular weight. Two proteins having neurotoxic activity have been separated from certain African scorpions. The fractions were homogeneous in the ultracentrifuge and in the chromatographic homogeneity test on Amberlite. Amino acid composition and sedimentation equilibrium gave molecular weights of 7,249 and 6,822. End group analysis and alkylation studies showed that both toxins consist of a single peptide chain ending in lysine at the N-terminal and threonine or glycine at the C-terminal. The American scorpions *Centruroides* and *Vejovis* also appear to have similar toxic low molecular weight proteins, although in *Vejovis* the effect of the toxins on the nervous system is not nearly as severe.

The venoms of the spiders have not been studied in detail, chiefly because of the small yield obtained from each spider. It may take the milkings from 1,000 spiders to accumulate a milligram of venom. The venom of the black widow spider *Latrodectus* sp. contains as many as 5 proteins and at least 5 nonproteins, but the nature of these components has not yet been determined. The venom of the brown spider, *Loxosceles* sp., is also a complex mixture of unknown proteins.

In the ant venoms we have an example of remarkable chemical diversification. Some ants sting, some bite, and some spray their venom into the wound produced by their mandibles. Formic acid is a component of many species of ants. It is particularly concentrated in certain spraying formicine ants, where it may account for 20% of their total body weight. Another important toxin in formicine ants is dendrolasin, the first furan to have been demonstrated in animals. Its formula is $C_{15}H_{22}O$. In the dolichoderines, another group of defense substances, cyclopentanoid monoterpenes, are found: they are iridomyrmecin ($C_{10}H_{16}O_2$), isoiridomyrmecin and isodihydronepetalactone. The venom of *Pseudomyrmex pallidus* is a basic protein without free amino acids, while in the myrmicine ant *Solenopsis* the venom appears to be a protein. In another myrmicine ant the toxin is *d*-limonene. In addition, acetic, propionic, isobutyric and isovalerianic acids have been isolated from this ant.

Some butterflies are toxic because their larvae feed on plants of the family Aristolochiaceae. Aristolochic acid-I is apparently carried through the pupal stage to the adult butterfly, where it is thought to play a significant role in the adult's protection. Some butterflies have a high concentration of an acetylcholine-like compound, while others, which feed on plants containing cardiac glucosides, are also rejected by birds. HCN is released by some Zygaenidae, when traumatized.

Amphibian Toxins. The toxins of such amphibians as toads, frogs, salamanders and newts are, for the most part, produced in the dermal tissues or by glands in the animal's skin. Almost all of the amphibian toxins play some role in defense. It is thus remarkable that they have evolved in such a diversified way, for some are biogenic amines, peptides, proteins, steroids and even steroidal alkaloids.

One of the most poisonous substances known, bactrachotoxin:

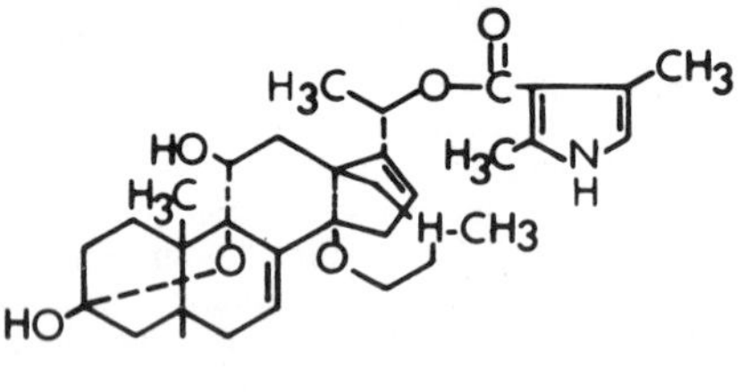

is four times more lethal than tetrodotoxin, or 250 times more lethal than either curare or strychnine. The toxin is found in the skin of a very small Colombian frog, *Phyllobates aurotaenia.* It is used by the Indians in western Colombia as arrow or blowgun poison. A pair of potent frog toxins, also steroidal alkaloids, are found in the small Neotropical frog, *Dendrobates pumilio.* Their formulas are $C_{19}H_{33}O_2N$ and $C_{19}H_{33}O_3N$. Another potent toad toxin is bufotalin:

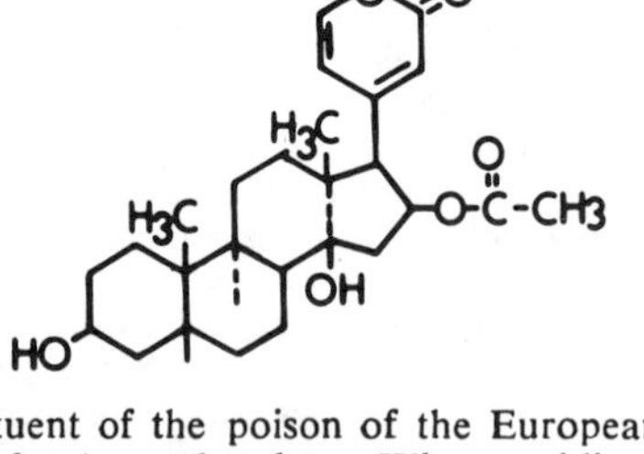

a constituent of the poison of the European toad, *Bufo vulgaris,* and others. When oxidized with chromic anhydride it yields a ketone, bufotalone.

Samandarine is the main alkaloid of the skin of *Salamandra maculosa taeniata,* a saturated secondary amine, $C_{19}H_{31}O_2N$:

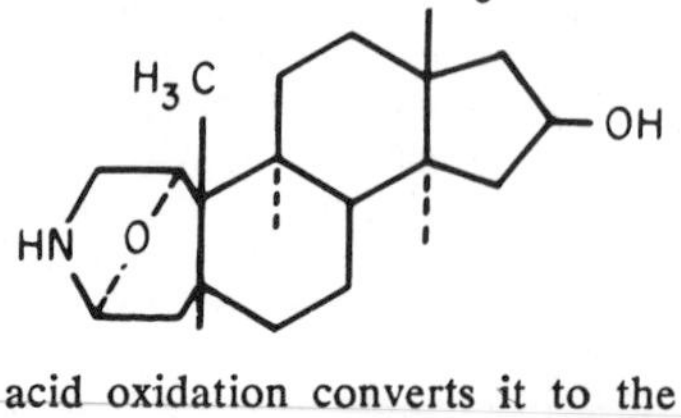

Chromic acid oxidation converts it to the corresponding ketone, samandarone. Samandarone is the main alkaloid of *S. maculosa maculosa.* An-

other important salamander toxin is the alkaloid samandaridine, $C_{21}H_{31}O_3N$, also a secondary amine.

Tarichatoxin (tetrodotoxin) appears to be limited in the amphibia to the family Salamandridae; primarily to certain members of the genus *Taricha,* in lesser amounts in *Notophthalmus, Cynops,* and trace amounts in *Triturus.* In the adult *Taricha* the toxin is concentrated in the skin, ovaries, muscle and blood.

FINDLAY E. RUSSELL AND ARNOLD F. BRODIE

References

Eisner, T., "Chemical defense against predation in arthropods," In: "Chemical Ecology," edited by Sondheimer, E., and Simeone, J. B., pp. 157–219, Academic Press: New York, 1970.

Habermann, E., "Bee and Wasp Venoms," *Science,* **177,** 314–322 (28 July 1972).

Kaiser, E., and Michl, H., "Die Biochemie der tierischen Gifte," Franz Deuticke: Wien, 1958.

Russell, F. E., "Marine toxins and venomous and poisonous marine animals," In: "Advances in Marine Biology," Vol. III, edited by Russell, F. S., pp. 255–384, Academic Press: London, 1965.

Russell, F. E., "Pharmacology of animal venoms," *Clin. Pharmacol. Therap.,* **8,** 849–873 (1967).

Russell, F. E., and Saunders, P. R., Editors, "Animal Toxins," Oxford: Pergamon Press, 1967.

VINYL RESINS

The term *vinyl resin* includes all polymers produced by chain reaction polymerization, such as poly(vinyl acetate), poly(vinyl chloride), poly(vinyl alkyl ethers), poly(vinyl fluoride), polyvinylpyrrolidone, polystyrene and acrylic resins. Products such as poly(vinyl acetals) and poly(vinyl alcohol) are also classified as vinyl resins. However, only vinyl chloride and its polymers will be discussed in this section.

Poly(vinyl chloride) was first obtained by Liebig and reported by Regnault in 1835. The original product was obtained by exposing a solution of ethanolic potassium hydroxide and 1,2-dichloroethane to sunlight and removing the insoluble poly(vinyl chloride). This polymer was investigated extensively by Ostromislensky and Klatte prior to World War I, but a commercial product was not available until 1927.

There was little use for this polymer prior to the discovery of its plasticization and heat stabilization by Semon and Susich and Figentscher, respectively, in the 1930's. It is of interest to note that the worldwide annual production of poly(vinyl chloride) which was less than 100 million pounds in 1940 increased to over 12 billion pounds in 1970. The annual production in the U.S.A. is greater than 3 billion pounds.

The pioneer technique for the production of vinyl chloride was by the thermal dehydrochlorination of 1,2-dichloroethane in the presence of alkalies. As shown by the following equation, this precursor was produced by the catalytic chlorination of ethylene.

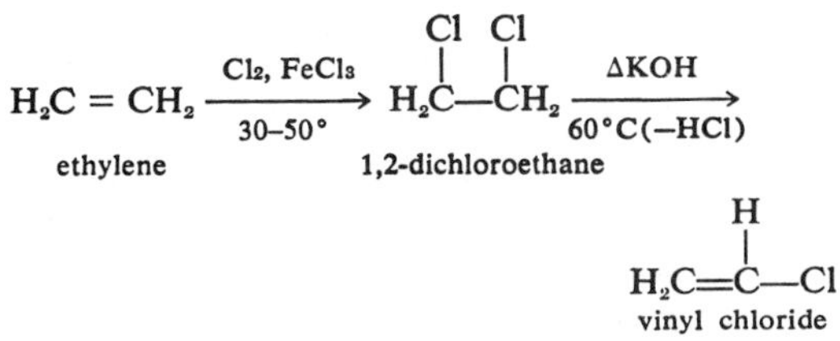

$$H_2C = CH_2 \xrightarrow[30-50°]{Cl_2,\ FeCl_3} H_2C\overset{\overset{Cl}{|}}{-}CH_2\overset{\overset{Cl}{|}}{} \xrightarrow[60°C(-HCl)]{\Delta KOH}$$

ethylene   1,2-dichloroethane

$$H_2C{=}C\overset{\overset{H}{|}}{-}Cl$$

vinyl chloride

In the 1940's, most of the vinyl chloride monomer was produced by the more economical thermal dehydrochlorination of 1,2-dichlorethane over a barium chloride catalyst at 500°C. Since the byproduct from this process was hydrochloric acid, it was essential that there be a use or a market for this acid near the monomer production facility.

Much of the vinyl chloride monomer in the 1950's and early 1960's was produced by the catalytic hydrochlorination of acetylene using carbon or activated alumina impregnated with mercury(II) chloride at 150°C. The conversion was about 25 per cent per pass and the hydrogen chloride and acetylene mixture was recycled. As shown in the following equation, the precursor was obtained by the thermal cracking, partial oxidation, or the electrical discharge dehydrogenation of alkanes such as ethane in the Wulff, Sachse and Schoch processes, respectively.

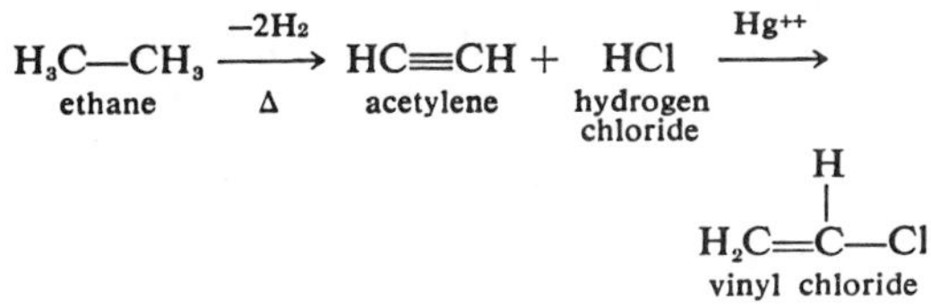

$$H_3C{-}CH_3 \xrightarrow[\Delta]{-2H_2} HC{\equiv}CH + HCl \xrightarrow{Hg^{++}}$$

ethane   acetylene   hydrogen chloride

$$H_2C{=}C\overset{\overset{H}{|}}{-}Cl$$

vinyl chloride

Much of the vinyl chloride monomer is now produced by the oxychlorination process in which the chlorine is recovered from the byproduct hydrogen chloride produced by the thermal dehydrochlorination process. This economical process is an adaptation of the classical Deacon process formerly used for the production of chlorine as shown by the following equation:

$$4HCl + O_2 \xrightarrow{\Delta,\ pressure} 2Cl_2 + 2H_2O$$

hydrogen   oxygen   chlorine   water
chloride

Vinyl chloride is a colorless gas which has a boiling point of 14°C. Poly(vinyl chloride), which is commonly called PVC, has a solubility parameter of 9.7 but because of its high degree of crystallinity, it is not readily dissolved by solvents having similar solubility parameter values. PVC is soluble in cyclohexanone and tetrahydrofuran which have solubility parameter values of 9.9 and 9.5, respectively.

Some of the early commercial polymer was produced by the bulk and solution polymerization of vinyl chloride, but most of the present polymer is obtained by free radical polymerization in the suspension and emulsion processes. A spray dried product from the latter process is preferred for plastisol manufacture since the resultant particles are coated with the surface active agent which prevents the attack by liquid plasticizer until the dispersion of PVC in the plasticizer is heated at 300°F or higher.

Benzoyl peroxide was used as an initiator for the free radical bulk polymerization of vinyl chloride as early as 1914. The polymer is insoluble in its monomer and the product precipitates as formed. Since this initial product is made up of macroradicals, propagation continues in the swollen precipitated product to produce a high-molecular weight polydisperse product.

The major production of PVC is by the free radical polymerization of vinyl chloride monomer suspended in an agitated aqueous solution containing small amounts of water soluble polymers, such as poly(vinyl alcohol). As shown by the following equations, free radicals may be produced by the decomposition of benzoyl peroxide or azobisisobutyronitrile.

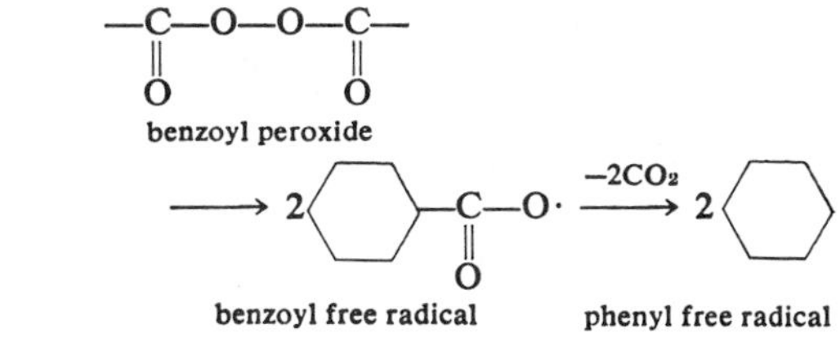

benzoyl peroxide

benzoyl free radical   phenyl free radical

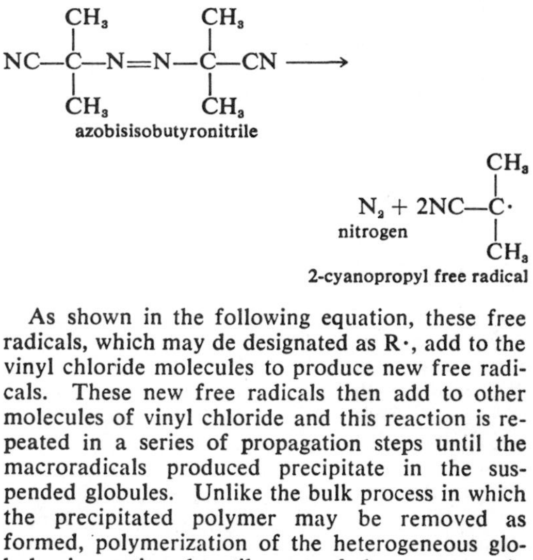

azobisisobutyronitrile

$$N_2 + 2NC\overset{\overset{CH_3}{|}}{\underset{\underset{CH_3}{|}}{C}}\cdot$$

nitrogen

2-cyanopropyl free radical

As shown in the following equation, these free radicals, which may de designated as R·, add to the vinyl chloride molecules to produce new free radicals. These new free radicals then add to other molecules of vinyl chloride and this reaction is repeated in a series of propagation steps until the macroradicals produced precipitate in the suspended globules. Unlike the bulk process in which the precipitated polymer may be removed as formed, polymerization of the heterogeneous globules is continued until most of the monomer in these globules has been converted to a polymer.

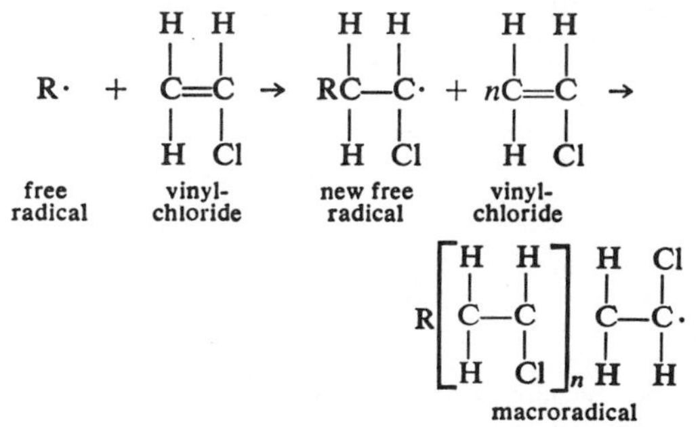

free radical + vinyl chloride → new free radical + n vinyl chloride →

macroradical

A typical formulation for suspension polymerization in an inert atmosphere is as follows: vinyl chloride, 30–50 g, poly(vinyl alcohol), 0.001g, trichloroethane, 0.1g, lauroyl peroxide, 0.001g, and water 90 g. As shown in the following equation, the trichloroethane serves as a modifier or transfer agent which reduces the chain length by producing dead polymer and new free radicals.

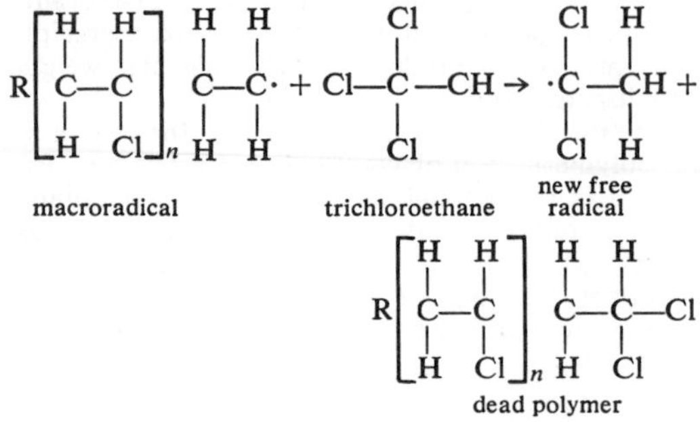

$$R\begin{bmatrix} \overset{H}{\underset{|}{C}}-\overset{H}{\underset{|}{C}} \\ \overset{|}{H} \quad \overset{|}{Cl} \end{bmatrix}_n \overset{H}{\underset{|}{C}}-\overset{H}{\underset{|}{C}}\cdot + Cl-\overset{Cl}{\underset{|}{C}}-CH \rightarrow \cdot\overset{Cl}{\underset{|}{C}}-CH +$$

macroradical   trichloroethane   new free radical

$$R\begin{bmatrix} \overset{H}{\underset{|}{C}}-\overset{H}{\underset{|}{C}} \\ \overset{|}{H} \quad \overset{|}{Cl} \end{bmatrix}_n \overset{H}{\underset{|}{C}}-\overset{H}{\underset{|}{C}}-Cl$$

dead polymer

The product obtained from the suspension process is centrifuged and dried in drum dryers.

Potassium persulfate and hydrogen peroxide have been used as initiators for the polymerization of vinyl chloride in emulsion systems. Since percarbonates may be prepared in situ, they may also be used advantageously as initiators. Fatty acid soaps, alkyl sulfates or aryl sulfonates may be used as emulsifiers. Polymerization in emulsion is much faster than bulk or suspension polymerization. Excellent yields of polymer are obtained in a few hours when the emulsion technique is used. The polymer is obtained by coagulation or by spray drying of the emulsion.

The dried PVC is blended with fillers, pigments, stabilizers, processing aids and plasticizers. This blended product may be sheeted on a two-roll mill or extruded to obtain a compounded PVC resin. The latter may be converted to a film or sheet by passing the stock through a four-roll calender or extruding it through a circular or slit die. Pipe filaments or other profiles, such as building sidings, gutters and drain spouts, are obtained by extruding PVC through appropriately designed dies. Coated wire is also produced by extruding poly(vinyl chloride) while wire is being drawn through a circular die. The wide use of plasticized PVC sheet is indicated by the use of over three million pounds of 20-gauge sheet for lining a 500-acre pond at Moab, Utah.

Both plasticized and rigid poly(vinyl chloride) may be injection molded to produce pipe fittings, phonograph record blanks, and other articles. The term rigid PVC applies to formulations containing less than 5 per cent plasticizer.

In addition to modification by the addition of plasticizers and fillers, PVC may be modified by blending with polypropylene, acrylonitrile, butadiene styrene copolymer (ABS), or acrylic resins. The heat deflection temperature may be increased by chlorination at low temperature. Copolymerization with vinyl acetate yields a copolymer called Vinylite which is more soluble and has greater impact resistance than PVC. Plastic foams are produced by adding blowing agents, such as *p,p′-oxybis*-(benzenesulfonyl hydrazide), which decomposes to produce nitrogen when heated.

A typical formulation for PVC sheet is as follows: PVC 100 g, acrylic resin impact modifier 13.8 g, acrylic resin processing modifier 5.0 g, dibasic lead phosphite stabilizer 50 g, epoxidized soybean oil stabilizer 5.0 g, stearic acid 0.25 g, and plasticizer 0–50g. Typical rigid and plasticized PVC sheets have the following properties:

Property	Rigid PVC	Plasticized PVC
Specific gravity	1.35	1.25
Tensile strength, psi	7500	2500
elongation, per cent	20	300
compressive strength, psi	10,000	1200
flexural yield strength, psi	12,000	—
notched izod impact strength, ft lbs/in	1	15
deflection temperature, °F at 264 psi	150	—
index of refraction	1.53	—
softening point, °F	175	—

RAYMOND B. SEYMOUR

VIRUSES

Viruses are infectious agents mainly characterized by their small size and ability to reproduce only within living cells. Viral diseases appear to have existed as far back as history records, even though knowledge of the physical and chemical nature of viruses is very recent. Among numerous examples of human viral diseases may be listed mumps, measles, smallpox, yellow fever, poliomyelitis, influenza, herpes, certain warts, dengue, and probably the common cold. A few viral diseases of domestic animals are hoof-and-mouth disease of cattle, hog cholera, swine influenza, rabies, equine encephalomyelitis, blue tongue of sheep, Newcastle disease of chickens, and canine distemper.

There are numerous viral diseases of plants and they have often been given picturesque names somewhat descriptive of the symptoms of the disease. Some examples are tobacco mosaic, sugar beet curly top, little peach, aster yellows, buckskin disease of cherry, tulip break, quick decline of citrus, tomato spotted wilt, chlorotic streak of sugar cane, potato yellow dwarf, and sudden death of cloves.

Viral diseases are not confined to higher forms of life, for viruses have been shown to infect fish, frogs, insects, algae, fungi, and even bacteria.

Among different viruses, or even among different strains of the same virus, great differences may occur in the amount of virus produced in a given infected host. In terms of virus particles per infected cell, the yield may range from one per cell to several million. Or in terms of milligrams of virus per liter of infective sap (in the case of plant viruses), the yields may range from 2–2000 mg per liter. In the extreme case of tobacco mosaic virus, the virus may account for as much as 10% of the total dry matter and to one-third of the nitrogen of the diseased tissue.

The small size of viruses distinguishes them from other infectious agents, for they are smaller than most of these and can be passed through filters which, in general, will retain bacteria and other microorganisms. It was in fact from filtration experiments by Iwanowski, and more especially by

Beijerinck, that viruses came to be recognized at the dawn of the twentieth century as a distinct class of disease agents.

Each virus is represented by particles of a characteristic shape and size which may or may not differ from those of another virus. Thus, some viruses are essentially spherical, others are rod-shaped, sperm-like, fibrous, or irregular bodies. The diameter of the particles of one of the smallest viruses, that of hoof-and-mouth disease of cattle, is about 20 nm (8×10^{-7} inch) while the particles of a pox virus are approximately 200 nm wide and 300 nm long. Between these two extremes are found the majority of viruses.

Because of this minute size, only a few of the very largest viruses can be observed with a conventional microscope employing visible light; however, the individual particles of all viruses can be seen in the electron microscope, where their images are magnified several thousand times. Many techniques have been developed with the electron microscope in order to make visible not only the virus particles which have been isolated from infected tissues, but also those in contact with cells or even within tissue sections.

Much can also be learned about the size and shape of virus particles by indirect methods, some of which were employed many years before the electron microscope was generally available as a laboratory tool. Viruses were discovered by filtration experiments and the sizes of viruses can be estimated by filtering through membranes of graded porosities and testing the filtrates for infectivity. It is also possible to determine the size and shape of virus particles from sedimentation, diffusion, and density data, and such an approach affords an excellent means of studying the properties of virus particles in solution, both in an untreated condition and after a variety of chemical and physical manipulations. The sizes of viruses can also be estimated from light-scattering data and from target diameters calculated from x-ray inactivation studies. The diameters of virus particles have also been obtained from low-angle x-ray scattering diagrams.

Despite the variation in size and shape of different viruses, all possess the common property of multiplying mainly within living cells. These cells may be in an organism or in a culture medium. The RNA of the QB bacterial virus has been duplicated in a cell-free medium provided with the proper nucleotides, enzymes and some QB-RNA as template. The product is infectious.

The ability to mutate, a property possessed by most living things, is also a characteristic of viruses. Thus, spontaneous mutations, occurring with low but significant frequency, result in new strains of a virus possessing properties similar to but distinguishable from those of the parent strain.

While the capacities to reproduce and to mutate are commonly associated with living things, some viruses, notably certain plant viruses, have been obtained in crystalline form, a property usually linked with inanimate objects. Hence these viruses, possessing properties of both living and lifeless things, may be considered intermediate forms. Tobacco mosaic virus was the first virus to be obtained in crystalline form and to be identified as a definite chemical entity. This discovery, reported by Stan-

ley in 1935, inaugurated a whole new field of biochemical investigation on the nature of viruses and virus diseases.

Tobacco mosaic virus proved to be a nucleoprotein containing 95% protein and 5% ribonucleic acid. As other plant viruses were isolated, purified and analyzed, they also were found to be nucleoproteins although most proved to have proportions of nucleic acid and protein which differed from tobacco mosaic virus. It now appears that nucleic acid and protein are common constituents of all viruses, and hence deserve emphasis in considering the fundamental properties of these disease agents. The nucleic acid is especially important, for, even when freed of all protein, it remains infectious and causes production of whole virus. Thus, the genetic function of the virus appears to reside in the nucleic acid and changes made in this material are critical because they can cause either inactivation or mutation (heritable changes in propertites) of the virus. Some other constituents occasionally found in purified viruses include lipid (neutral fat, phospholipid, fatty acids and aldehydes, and cholesterol) and polysaccharide. A number of insect viruses in the mature state occur in curious forms in which infectious, rod-like particles are embedded in protein matrixes called inclusion bodies or capsules.

Some novel, infectious agents causing symptoms characteristic of plant viruses, but having particles which are quite different, have recently been shown to cause potato spindle tuber, tomato bunchy top, citrus exocortis and chrysanthemum stunt diseases. These agents appear to be single-stranded ribonucleic acids with molecular weights of only 100,000 or less and they have no protein associated with them. They are called viroids.

Quite commonly the proteins of intact viruses are not attacked by proteolytic enzymes; consequently, treatment with proteolytic enzymes can often be used as a step in purifying viruses, since normal tissue proteins are usually hydrolyzed by such enzymes. However, the enzyme carboxypeptidase, which acts upon the carboxyl-terminal residues of peptides and proteins, has been found to hydrolyze the C-terminal residues of a number of undenatured viruses. Using this enzyme as a structural chemical tool, evidence has been obtained that the protein coat of the huge tobacco mosaic virus molecule (particle weight, 40×10^6) is made up of about 2130 identical protein subunits of molecular weight 17,531. The complete amino acid sequence has been determined for the protein subunits, each of which was found to consist of 158 acid residues. Similar analyses have been made on dozens of spontaneous and chemically induced mutants of tobacco mosaic virus. The proteins have been found to be identical in some cases and to differ in proportions of 1 to 17 amino acids in other cases.

The protein of tobacco mosaic virus can be obtained by degrading the virus in cold 67% acetic acid or in dilute alkali at about pH 10.5. If the resulting protein, separated from nucleic acid, is caused to aggregate by proper adjustment of the pH, protein rods are obtained that look much like virus particles but which are noninfectious. If such aggregation is done in the presence of the virus nucleic acid (a single strand of ribonucleic acid with a molecular weight of about 2×10^6 which can be obtained by extracting the virus with 80% aqueous phenol), rod-like particles are obtained that not only look like virus but are infectious and very much the same as natural virus in all ways tested. The coaggregation of virus protein and nucleic acid to yield infectious particles is called reconstitution.

The chemically reactive groups of virus proteins are the same as those of common proteins, namely, amino, indole, phenolic, guanidino, sulfhydryl, amide, carboxyl, and perhaps aliphatic hydroxyl. Various reagents can react with such groupings but no change in virus activity results unless the nucleic acid is also affected. However, it may be possible to alter the virus protein coat, such as by cross-linking reactions with formaldehyde or by heat, so that the nucleic acid cannot subsequently be released within the cell. Such a virus is effectively inactivated. Carboxyl terminal residues of virus protein can be removed without inactivating the virus.

The amount of nucleic acid in viruses ranges from about 1% for Newcastle disease virus to about 50% for T2 bacterial virus. Both ribonucleic acid (RNA) and deoxyribonucleic acid (DNA) are found in all major types of viruses but never both in the same virus. The purine and pyrimidine bases of viral nucleic acids correspond to those commonly observed in nature, with few exceptions. For example, certain bacterial viruses appear to have 5-hydroxymethylcytosine in place of cytosine.

The nucleic acid components of viruses have been shown to react with a variety of reagents such as substituted B-chloroethyl sulfides (mustards), formaldehyde and other aldehydes, nitrous acid, hydroxylamine, N-bromosuccinimide, and dimethyl sulfate. Nitrous acid has proved most effective in the production of mutants. These mutants are thought to result largely from the oxidative deamination of cytosine to uracil and of adenine to hypoxanthine (which then acts like guanine) in the nucleic acid chain. Such changes presumably alter the genetic code contained in the nucleic acid and result in the synthesis of a slightly different viral protein or production of an altered disease pattern in the next cycle of replication of the virus.

Although most highly purified virus preparations are singularly lacking in enzymatic activities, such properties have been reported for some viruses. Among the enzymes found closely associated with viruses are a lysozyme found in certain bacterial viruses, influenza neuraminidase, and the RNA polymerases of reovirus, cytoplasmic polyhedrosis virus, and wound tumor virus.

Viruses, owing mainly to the substantial amount of protein in their makeup, are antigenic substances. As such they elicit the formation of antibodies in animals and these antibodies account, at least in part, for the immunity acquired after infection with a virus. Also this phenomenon provides a basis for the use of vaccines in preventing or moderating viral infections. (See **Antibodies and Antigens**). The serological reactions commonly employed with viruses include the precipitin, complement fixation, and neutralization tests. One or more of these may be applied to demonstrate the presence or absence of antigenic relationships among viruses, to detect impurities in virus preparations, for quantitative measure of virus, and rather extensively for diagnosis of virus disease.

In 1911, Rous reported the transmission of a malignant growth, a chicken sarcoma, by means of a cell-free filtrate which contained a sarcoma virus. This provoked a lively interest in the possibility that viruses cause cancer and led subsequently to the discovery that some animal and plant viruses do induce tumors. Work with these tumor viruses revealed several singular phenomena. For example, in some cells the Bryan strain of the Rous virus does not form a complete, infectious particle unless the cells are also infected with a helper virus. Further, Rous virus contains RNA, but unlike many other RNA viruses, the RNA appears to be transcribed through DNA as part of the replicative process. In another case, that of the Shope papilloma, most of the viral nucleic acid formed in the infected tissue does not get a protein coat if the host is a domestic rabbit whereas in infected, wild, cottontail rabbits a high proportion of the virus appears in complete particles. The stability of the viral infectivity depends heavily on the protein coat and hence it is difficult to demonstrate the presence of infectious virus in tumors from the domestic rabbit. In the cases of such viruses as the SV40 virus of monkeys or the polyoma virus of mice, the viral DNA is sometimes integrated into host DNA where it is not demonstrable in the usual ways as infectious virus. The relationship between these phenomena and oncogenicity of the viruses has not been clearly established, but much has been learned about the transformation of normal into malignant cells by viruses. Viral particles similar to those of the Rous sarcoma virus (C-type particles) seem to be related in some way to several human leukemias. Strong evidence has been obtained that the Burkitt lymphoma of children is caused by a DNA-containing, herpes-type of virus although definitive proof is lacking.

The complete details of reproduction are not known for any virus but the following outline summarizes some of the steps. First a virus particle adsorbs to a more or less specific receptor site on the surface of a susceptible cell. There may or may not be interaction between virus and receptor, such as that which causes a partial opening of the poliovirus particle, but in any case the next step is penetration. Either whole virus penetrates by an engulfment process or, as in the case of some bacterial viruses, the nucleic acid of the virus is injected. If whole virus enters the cell, the protein coat of the virus is stripped off enzymatically to release the nucleic acid. Virus nucleic acid makes its way to specific sites, either in the nucleus or cytoplasm depending on the virus, and there initiates essential steps in the self-duplication process. These steps, involving the host genetic material in some cases and quite independent of it in others, include the synthesis of enzymes needed to catalyze the formation of new virus nucleic acid and protein. When a supply of viral parts has accumulated, assembly of new virus particles occurs. In some cases, such as that of tobacco mosaic virus, assembly is almost spontaneous, resembling a crystallization of protein subunits around the viral nucleic acid. In other cases, the process is more complicated. Eventually, new virus progeny are released from the infected cell either by a burst or lysis mechanism which destroys the infected cell or by a gradual process in which virus particles are completed at the cell membrane and then ooze out individually ready to start a new cycle of infection. Current methods for the control of virus diseases are in some cases quite unsatisfactory and improvement may await the further elucidation of the nature of virus multiplication. Several human and animal viral diseases are sucessfully controlled by the use of vaccines containing active or inactive virus. The control of smallpox by means of a mild, active-virus vaccine is a well-known case of this sort. In other instances in which vaccines have been developed and tested, the results have frequently been less satisfactory owing to such difficulties as inability to obtain sufficiently concentrated and purified virus, instability of the virus, and transient immunity elicited by some viruses even under optimum conditions. In addition, a number of viruses have shown a pronounced tendency to mutate during epidemic cycles, eventually culminating in strains of virus against which a given vaccine is no longer effective. Some animal and plant viral diseases are substantially checked by control of the insects which transmit the disease from one individual to the next. Also, with infected plants, a measure of control has sometimes been achieved by breeding and introduction of resistant varieties or by removing and destroying infected individuals as soon as they appear.

Finally, it should be noted that the destructive effects of viruses are occasionally harnessed for the benefit of man. Thus, the deliberate introduction of myxoma virus has been a means of limiting the rabbit population in certain areas where it had become a problem. Likewise, the dissemination of an insect virus has been used to control an alfalfa caterpillar and hence to increase the alfalfa crop. However, it seems likely that the greatest benefit to man to be derived from viruses is the knowledge to which the study of viruses may lead concerning the nature of life itself and the fundamental process of life. For viruses exert a profound effect on the vital functions of the cells which they invade and they, themselves, possess some of the attributes of both living and lifeless matter.

C. A. KNIGHT AND W. M. STANLEY*

* Deceased.

Cross-references: *Nucleic Acids, Proteins, Polypeptides.*

Interferon

Interferon is the name given to an antiviral system which is activated in response to virus infection and other inducing substances. It has been demonstrated in animals, and to a lesser extent in man, that interferon is an important body defense against virus infections, which may in part explain the extremely low death rate associated with most of these infections.

Interferon was discovered in 1957 by Drs. Isaacs and Lindenmann, who observed that virus-infected cells of the chicken produced a soluble protein which, when transferred to uninfected cells, rendered them resistant to viruses. It was subsequently demonstrated that the infected cells produced the interferon protein which was capable of stimulating other body cells to produce the antiviral protein—a second protein of the interferon system which has been demonstrated only indirectly.

The interferon protein (the initially synthesized protein of the interferon system) has been reasonably well characterized. In comparison with antibody, which is the better known defensive protein produced by infected animals, the interferon protein has several distinctive properties. Unlike antibody, which is produced by specialized cells of the body, interferon can be produced by virtually any cell following suitable stimulation. Antibody combines directly with virus, causing loss of virus infectivity, whereas interferon does not inactivate virus directly but instead reacts with susceptible cells, which then become resistant to subsequent virus multiplication.

The antiviral activity of antibody is specifically directed against only the same antigenic type of virus which stimulated antibody formation. In contrast, there is no antiviral specificity of interferon, in that interferon which is stimulated by one virus will inhibit the growth of virtually all animal viruses. Different amounts of interferon are required to inhibit different viruses. Unlike antibody, the antiviral action of interferon is relatively specific for the animal species, that is, interferon produced by the cells of one species generally will protect cells of only the same or closely related species. Within animals, the antiviral action of interferon is predominantly close to the site of its formation where it is present in highest concentration. In contrast, antibody circulates throughout the body. Interferon preparations which are induced by different viruses in one animal species are usually indistinguishable; interferons from cells of different animal species may be distinguished not only by the species specificity of their antiviral action but also by differences in sedimentation velocity, antigenicity, and heat stability. Other properties of interferons include stability over pH range 1–10, weak antigenicity of presently available amounts of interferon, and molecular weight range of 12,000 to 200,000 daltons.

Available biological and biochemical evidence indicates that the information for interferon production is normally present in the cellular DNA in unexpressed (repressed) form. Virus infection appears to result in the expression of the DNA information (derepression) resulting in the formation of interferon messenger RNA which then directs the synthesis of the interferon protein molecules.

The time required for initiation of interferon production varies from as little as 1 hour to as long as 30 hours, depending on the inducing virus and the cell type. Production of interferon by a cell occurs over a period of several hours. In addition to newly produced interferon, there is evidence that an inactive form of interferon may be present in certain body cells and be released in its active form following stimuli such as endotoxin and cycloheximide. The total amount of interferon produced by many infected animals has been estimated to exceed 100,000 units and perhaps may be as great as several million units. A unit of interferon is that amount required to minimally inhibit virus replication in a tissue culture assay system.

The interferon systems within different cell types are not always activated by the same stimuli. As a group, viruses, as well as nucleic acid inducers of interferon, may stimulate almost any cell type within the body. In comparison, there are a number of inducers which seem only to stimulate a narrow range of cells among which are lymphocytes and macrophages. These inducers include bacteria, their endotoxins and lipids, phytohemagglutinin and other mitogens, cycloheximide, diethylaminoethoxyfluorenone (tilorone), polysaccharides, chlamydiae, mycoplasmas, protozoa, synthetic polyanionic polymers and antigens which react with sensitized lymphocytes.

Interferon is released into extracellular fluid almost immediately after production within the cell. Released interferon reacts with surrounding cells to induce the virus resistance within a few hours. Studies in tissue culture systems indicate cellular uptake of only a small fraction of the applied interferon during the period of development of resistance.

Establishment of the virus-resistant state by reaction of interferon with a cell requires cellular protein and RNA synthesis. This suggested that interferon induces the formation of a new antiviral protein. The information for this second protein, like that of the interferon protein, is thought to reside in the cellular DNA in repressed form. Interferon may derepress the DNA and thus permit the formation of messenger RNA which in turn directs synthesis of the antiviral protein. Available evidence indicates that the proposed antiviral protein modifies the cell's ribosomes so that viral messenger RNA cannot bind to them in order to be translated into essential virus-specified proteins.

Interferon holds great promise of becoming an important agent for prevention and therapy of virus infections. Part of its potential utility comes from its ability to inhibit the multiplication of virtually all viruses and its lack of detectable toxicty. Theoretically, interferon could be injected into an infected tissue in advance of or supplementary to normally produced interferon. The problems to be overcome before such application include: (1) production of large quantities of interferon; (2) the species barrier which requires that interferon, for use in man, must be prepared in human cells, or cells of closely related species; and (3) the tendency for protection to be localized at the site of injection and to spread poorly to infected sites. The problems may be largely overcome by current attempts to develop drugs like nucleic acids and tilorone which can stimulate the production of large amounts of endogenous interferon. It has already been possible to prevent a wide variety of virus infections by prior injection of interferon into animals and in at least one study in man. Early studies with a nucleic acid inducer also indicate some effectiveness in prevention of experimental human respiratory infections.

SAMUEL BARON, M.D.

Cross-references: *Antibodies and Antigens; Proteins.*

References

1. Isaacs, Alick, "Interferon," in "Advances in Virus Research," Kenneth M. Smith and Max A. Lauffer, (Eds.), Academic Press, New York, pp. 1–38, 1963.
2. Finter, Norman B., (Ed.), "The Interferons," North Holland, Amsterdam, 1966.
3. Vilcek, Jan, "Interferon," Little Brown and Co., Boston, 1970.
4. Merigan, T. C., (Ed.), Symposium on Interferon and Host Response to Virus Infection, *Archives of Internal Medicine,* **126,** 49–157 (1970).

VISCOSE

The manufacture of regenerated cellulose fibers and films is largely based on the use of "viscose," the industrial name given to an orange-red, viscous solution of sodium cellulose xanthate in aqueous sodium hydroxide. The composition of viscose varies widely, depending upon its intended use and may contain 5 to 10% cellulose, 5 to 9% sodium hydroxide, various sulfur by-products, and regeneration retardants of complex chemical composition.

The basic raw materials used for viscose are cellulose, sodium hydroxide, carbon disulfide and water. Virtually all viscose is prepared from wood pulp obtained from either hard or soft woods purified by either a sulfite or sulfate pulping process. These pulping processes produce a chemical grade of highly purified cellulose with an alpha-cellulose content of 88 to 99%. The remaining content of the purified pulp consists of beta- and gamma-celluloses and very low concentrations of ligneous resin impurities.

Although viscose can be manufactured by either continuous or batch processes, the batch process accounts for most of the viscose production since it affords greater process flexibility. The viscose process consists of several steps carried out in the following order: (1) steeping; (2) pressing; (3) shredding; (4) aging; (5) xanthation and (6) ripening.

In the steeping or mercerizing steps, cellulose is immersed in 17 to 20% sodium hydroxide solution at a temperature between 18 and 25°C. The primary function of steeping is to remove hemicellulose impurities from the pulp. Simultaneously a chemical reaction occurs with the hydroxyl groups of the cellulose reacting with sodium hydroxide to form alkali cellulose:

$$R_{cell}OH + NaOH \rightleftharpoons R_{cell}ONa + H_2O$$

After steeping, the alkali cellulose is pressed to a wet weight equivalent to 2.5 to 3.0 times the original pulp weight. Pressing removes the low molecular weight cellulose fractions which are soluble in the sodium hydroxide solution and reduces excess sodium hydroxide. The pressed alkali cellulose is mechanically worked in shredders, revolving blade mixers, to yield finely divided, fluffy particles known as "crumb." The shredding process increases the surface area of the alkali cellulose to increase its reactivity in the subsequent process steps.

The alkali cellulose crumbs are aged at a constant temperature between 18 and 30°C to depolymerize the cellulose under controlled conditions. The depolymerization, which is an oxidative reaction, reduces the average molecular weight of the original pulp by a 2- to 3-fold factor. Reduction in the degree of polymerization of the cellulose is necessary to obtain a viscose solution of suitable viscosity for spinning.

The aged alkali cellulose crumbs are placed in rotating churns and reacted with carbon disulfide under controlled temperatures, usually in the range of 20 to 35°C. The amount of carbon disulfide varies with the end use intended and may range from 20 to 50% based on the dry cellulose. The usual concentration range for carbon disulfide is 28 to 40%. The conversion of alkali cellulose to cellulose xanthate is represented by the reaction:

$$R_{cell}ONa + CS_2 \rightleftharpoons R_{cell}OCSSNa$$

This heterogeneous reaction is not carried to completion and the average degree of xanthate substitution is of the order of one xanthate group per anhydroglucose unit. Side reactions accompany the cellulose xanthation reaction and by-products such as sodium trithiocarbonate are formed which are responsible for the orange color of the xanthate crumb and the resulting viscose solution.

The viscose solution is prepared by dissolving the cellulose xanthate crumbs in aqueous sodium hydroxide at 15–20°C under high-shear mixing conditions. The resulting solutions contain 5 to 10% cellulose and 5 to 9% sodium hydroxide. The viscose solution is then filtered to remove insoluble fiber or foreign matter and is deaerated (degassed) and stored to achieve required chemical and colloidal changes. The storage or ripening of the viscose at a controlled temperature in the range of 15 to 30°C results in a partial decomposition and redistribution of the xanthate groups on the cellulose molecules. These chemical reactions cause changes in the state of solution of the cellulose and a consequent increase in the ease of coagulation of the cellulose.

Viscose additives or modifiers are used in the production of high-tenacity rayon fibers since they affect the xanthate decomposition rate during spinning. Additives such as aliphatic and aromatic amines, polyoxyethylene glycol, ethoxylated organic amines and amides, and quaternary salts of ethoxylated primary amines are added to the viscose during the mixing step or just prior to spinning at concentrations normally less than 5% based on the weight of cellulose.

In the spinning of rayon, the viscose solution is metered through a spinneret containing large numbers of very fine holes into a coagulation-regeneration bath composed of sulfuric acid, zinc sulfate and sodium sulfate. The viscose is coagulated very rapidly and the cellulose xanthate decomposes to form continuous filaments of regenerated cellulose. Similarly, in the production of films, the viscose solution is metered through a long slot orifice into a coagulation bath composed of sulfuric acid and sodium sulfate. The physical and structural properties of the fiber or film produced by these processes is influenced by many factors such as viscose composition and viscosity, the coagulation-regeneration conditions, and purification processes.

Decomposition of the cellulose xanthate is accompanied by release of carbon disulfide which may be recovered from the coagulation-regeneration baths by volatization and absorption processes. By-product viscose compounds such as sodium trithiocarbonate and sodium sulfide are likewise decomposed in the acid coagulation-regeneration baths with the formation of carbon disulfide and hydrogen sulfide. Currently, the rayon industry is studying processes to permit recovery of hydrogen sulfide.

In 1970 about 800 million pounds of rayon fiber and about 340 million pounds of cellophane film were produced by the viscose process in the United States.

JOSEPH W. SCHAPPEL

VITAMINS

Historically vitamins are those organic substances other than fats, proteins and carbohydrates which are essential in small amounts in the diets of higher animals. The existence of such nutrients became generally recognized during the first two decades of the 20th century, and were commonly associated with certain deficiency diseases which resulted from their lack—e.g., night blindness (xerophthalmia), beriberi, scurvy, rickets and pellagra.

Since the substances themselves were unknown, they were originally given letter names A, B, C, etc. These names in some cases are still retained. The recognized vitamins are now known chemically and constitute a heterogeneous group of substances which belong in a group only because of having nutritional functions.

The interest in vitamins has shifted in the minds of modern biochemists away from the diseases they prevent to a recognition that they are fundamental working parts of the machinery of living things in the healthy state. For many of the vitamins, we now know how they fit in; for a number of others, there is vast ignorance. No one can deny that without a knowledge of vitamins there is no possibility of grasping the elementary facts of the chemistry of living things.

"Vitamin A" activity is possessed by several related substances: carotenes, vitamin A alcohol, vitamin A_2 alcohol and esters, and other derivatives of these alcohols. These substances are most frequently associated with the visual pigments for the building of which they are essential raw materials. 'Vitamin A' is essential for reproduction, for example, and has far broader meaning. Vitamin A acid, which does not serve as a raw material for visual pigments can, however, perform other essential functions, but its action is largely unknown.

The complex situation with respect to vitamin A has some parallel in the cases of vitamins D, E and K. In the case of vitamin D, the two more important forms are calciferol and vitamin D_3. Several other naturally occurring or artificially produced substances have "vitamin D activity"; that is, they are able under suitable conditions to prevent or cure the disease rickets. Various forms of vitamin D are not equally effective, and their relative effectiveness varies for different species of animals. Calciferol, for example, (derived from irradiated ergosterol) is relatively ineffective for chickens. Because vitamin D is a physiological entity rather than a chemical entity, it is measured in physiological units rather than in milligrams or grams.

Vitamin E activity is likewise possessed by three related naturally occurring substances α-, β-, and γ-tocopherols. Any one of the three will prevent sterility in rats under specified conditions. α-Tocopherol is the most active.

Vitamin K activity is possessed by two naturally occurring substances "vitamin K_1" and 'vitamin K_2." Either one of these two, as well as any one of several synthetic substances, has under specified conditions the effect of decreasing the clotting time of the blood (originally observed in chickens).

All the aforementioned vitamins are soluble in fats and fat solvents and may be associated in nature with fatty substances. They are sometimes still referred to as "fat soluble vitamins." It is interesting that in every case more than one naturally occurring substance by itself possesses the characteristic physiological activity of the group to which it belongs.

Among the so-called "water-soluble vitamins" a different situation exists. The characteristic physiological activity is often possessed by a single compound, and when more than one active form is known, they are quite closely related chemically and are readily interconvertible.

"Vitamin C" (antiscorbutic) activity is possessed almost exclusively by ascorbic acid and its (reversible) oxidation product dehydroascorbic acid. While some similar synthetic substances have slight activity, they are not important. Aside from being essential in an obscure way for the incorporation of hydroxyproline into collagen, the mechanism of its action is largely unknown.

The name "vitamin B" has a most interesting history. Originally, of course, it was given to what was supposed to be a single substance, present in rice, bran, yeast, wheat germ, "protein-free milk," etc. Subsequent investigations, in which the use of yeasts and bacteria as test organisms has played an important role, have revealed that the physiological activity originally observed was due to the *additive effect* of a considerable number of substances, *each one* of which is of itself essential. If and when the designations B_1, B_2, B_3, etc., are used, they have an entirely different meaning from the parallel use of D_1, D_2, D_3, or K_1 and K_2, because in the case of the D and K vitamins one form can replace another. In the case of the B vitamins each form is a distinctly different substance with different functions, and each member of the family is separately indispensable. No one B vitamin can replace any other. The use of letter names for the B vitamins should be discontinued. The expression "vitamin B complex" is also objectionable at this time.

There are specific cases in which a particular species is able to produce a specific vitamin for itself and does not require it in the diet. Rats, for example, do not require dietary ascorbic acid, but it is nonetheless indispensable as a constituent of their tissues.

The fact that vitamins are needed in relatively small amounts suggests that they may play catalytic roles in body chemistry. The catalytic roles have been elucidated in the case of several of the B vitamins, but in general such roles have not been established for the other vitamins, and it would be going beyond present knowledge to say that all vitamins function catalytically.

All of the so-called "B vitamins" seem to have this in common: they appear to be universal constituents of every living cell. Thus, they appear to be present in every type of tissue and in every type of organism. This does not appear to be true of the other vitamins. In the light of present knowledge, certain catalytic substances, notably lipoic acid and coenzyme Q, should be regarded as vita-

mins, even though they may not be routinely required by the standard man or standard mammal. There is evidence that they are sometimes required. If a substance needs *always* to be required in the diet in order to be a vitamin, then niacinamide would have to be excluded, because in the presence of an abundance of tryptophane, niacinamide can be produced endogenously.

The problem of the functions of the vitamins is a large one and cannot be discussed here in detail. Elementary students of nutrition are sometimes given serious misconceptions by learning such sayings as the following: "calcium is for bone," "iron is for blood," "phosphorus is for brain," "vitamin A is for the eyes," "vitamin B, is for nerves," "vitamin B_{12} is for blood formation," "vitamin K is for blood coagulation," etc. Such sayings contain a small part of the truth, but they cover up the fact that the elements calcium, iron, phosphorus and a host of B vitamins are probably absolutely essential for the life and well being of every cell in the body, whether muscle, nerve, liver, kidney, lung or what not.

Knowledge with respect to vitamin functions is farthest advanced in the area related to the B vitamins. These are often indispensable parts of recognized enzyme systems which control fundamental chemical operations that every kind of cell has to perform in order to get its living energy.

In order for cells to get energy out of fuel, oxidations have to take place. *Nicotinamide* and *riboflavin* (two of the B vitamins) function in catalysts for different stages of cellular oxidations. Without these complex catalysts no oxidation can take place and the cells simply cannot live. Pyruvic acid is a key intermediate in the energy-yielding processes of cells; *thiamine* and *lipoic acid* are parts of catalytic systems which carry the energy metabolism on from this point. In building up as well as tearing down processes, acetyl (C_2) groups are intimately involved. *Pantothenic acid* is a part of enzymes which are essential to all the transformations involving this group. In a somewhat parallel manner *folic acid* and cobalamine (vitamin B_{12}) are thought to be involved in the transformations involving single carbon (C_1) units. Amino acids from proteins must be metabolized, decarboxylated, and interacted; *pyridoxal, pyridoxamine* and *pyridoxin,* three interconvertible forms of what was originally called vitamin B_6, are parts of essential enzymes, necessary for several of these fundamental processes.

It is evident, therefore, on the basis of these facts that a deficiency of any one of these key substances could not only impair but actually stop the vital functioning of any and every cell in the body. On the basis of this generalized need it is not incomprehensible that a very large number of disease symptoms involving every part of the body and practically every special organ have been ascribed to vitamin deficiency and allegedly have been benefited by vitamin administration.

In specific cases there may be localized areas that are first to reveal vitamin deficiencies, and these symptoms may be used for diagnostic purposes. In riboflavin deficiency in humans, for example, the region around the corners of the mouth become inflamed (cheilosis). This does not mean that riboflavin is needed merely for this region; it means that for reasons not fully known, this particular area is peculiarly susceptible to riboflavin lack and acts as an indicator. Probably by the time one demonstrates riboflavin deficiency in this manner, many cells and tissues throughout the body have their metabolism somewhat impaired by the deficiency, and the administration of riboflavin would benefit a multitude of tissues in addition to those near the corners of the mouth.

In thiamine deficiency there is nerve involvement, but the lack of other B vitamins is known to affect nerves adversely too; and the total lack of any one of these essential catalytic elements would cause nerves to cease functioning entirely. The first sign of thiamine deficiency is generally regarded as loss of appetite. We are not well acquainted with the machinery which involves the production of appetite, but it would appear that the cells which are involved in this machinery are peculiarly susceptible to thiamine deficiency, and that this symptom can be used as a diagnostic sign. The lack of other essential nutrients may also destroy appetite, and it would be but a minute fraction of the truth to say "thiamine is for appetite production."

Cobalamine (vitamin B_{12}) deficiency is associated in the popular mind with anemia. This is based upon the fact that it is an essential link in the process of building red blood cells, and the machinery involved in this process is peculiarly susceptible to cobalamine lack. This same catalytic substance is essential for other cells too, nerve cells for example, and these are impaired by its lack. Other B vitamins and other nutrients, iron for example, are also essential for building red blood cells and breaking any link in the chain is enough to disrupt the whole process.

Pantothenic acid deficiency is of a type which does not show itself by any one convenient outward sign. When it becomes severe, many different tissues and organs are seriously affected, among which the adrenal glands are prominent. In chickens where pantothenic acid deficiency was first noted, it produced what was called "chick dermatitis." Actually, however, limiting the pantothenic acid intake not only affects the skin adversely, the chickens grow at a fraction of the normal rate, all the internal organs are impaired as to size and are deficient in pantothenic acid and in the essential enzymes necessary to deal with the C_2 fragment both in energy-yielding and building-up processes. This is an example of generalized deficiency.

With all this as a background, we are prepared to discuss briefly vitamin requirements—a discussion which involves a crucial but sometimes overlooked point. While vitamins are notoriously effective in small amounts, the *quantities* present in diets are of the utmost importance. Sometimes the question is asked "How many vitamins are present in white bread (or some other food)?" This is approximately equivalent to asking how many vitamins exist, because bread, especially if it contains even a trace of milk, probably contains traces of every known vitamin. The important questions are, "How *much* of the various vitamins does bread contain?" or "How many vitamins are present in bread in abundant amounts?"

The latter question is difficult to answer, partly because of uncertainty as to what is meant by "abundant." It is not always appreciated that "vitamin deficiencies" as observed in experimental animals or humans always involve an inadequate supply—never a total lack. Polished rice as an exclusive article of diet contains just enough thiamine (about ¼ the minimum for health) to keep people alive for a time, but suffering from beri-beri. If people were deprived of thiamine entirely by eating exclusively a thiamine-free rice, they could not reproduce and death would promptly claim the whole population.

A detailed discussion of vitamin requirements cannot be given here. The subject is a large one and is complicated by many factors, among which is the fact that intestinal microorganisms may produce or destroy vitamins. Not the least of the complications is the individual variation in needs. According to the findings of biochemical genetics, relating genes to potentialities for enzyme building, each individual must because of his distinctive set of genes, have distinctive nutritional needs quantitatively speaking. How large the differences in needs may be for two individuals is not known, but in some cases there are indications that the differences are large. The possibility presents itself that many reasonably well individuals may have mild nutritional deficiencies, including vitamin deficiencies, even though they may eat wisely according to prevailing standards.

The most acceptable maxims for insuring, as far as possible, against nutritional deficiency, including vitamin deficiency, are: Eat diversified foods. Consume substantial amounts of milk and dairy products. Include in the diet some raw foods, a variety of vegetables including yellow and green ones, and a variety of fruits. Good protein, such as that in meat or cheese, is essential. Avoid quantities of refined foods, starches, sugar and alcohol.

The importance of diversity can hardly be overemphasized. No vitamin or other nutritional essential need be obtained from a single source, and individuals, in order to supply their needs, need not consume foods that are distasteful to them. Unfortunately, however, it is not easy to eat wisely and at the same time cheaply.

In concluding the discussion of vitamins, it seems well to point out that while most of the vitamins have probably been discovered to date, there are still others to be recognized in the future. Only when they have all been discovered will we have a complete picture of nutritional needs. A nutritional chain is as strong as its weakest link and health cannot be maintained when even one link is weak.

ROGER J. WILLIAMS

Cross-references: *Nutrition, Foods, Biochemistry.*

Reference

Kutsky, R. J., "Handbook of Vitamins and Hormones," Van Nostrand Reinhold Co., New York, 1973.

VULCANIZATION

The term "vulcanization" is generally considered specific to rubber technology, although it has been loosely applied to other cases as well ("vulcanized oils," "vulcanized fibers," etc.). It was presumably coined to describe the process, discovered by Charles Goodyear, a Connecticut inventor, in 1839 (and independently by Thomas Hancock in England), whereby the application of sulfur and heat to natural rubber convert it to a stronger, less plastic material, more resistant to temperature changes. The name itself (based on Vulcan) strongly emphasizes the use of heat, but is really intended to cover the process as a whole, whereby raw rubber becomes a tougher, more durable material.

It is the vulcanization process which gave rise to the whole technology of rubber Thus, it was obviously necessary to develop means whereby the sulfur and other ingredients could be mixed and dispersed in the rubber. This in turn required the "breaking-down" and softening of the rubber to permit mixing. The processing of rubber prior to vulcanization is thus referred to as "compounding." It consists of a preliminary mastication of the raw rubber on a differential-speed roll mill or in an internal mixer (Banbury), followed by the addition and mixing of the compounding ingredients. In the case of sulfur vulcanization, the "compound" contains, besides elemental sulfur, a vulcanization accelerator, zinc oxide and stearic acid, which are necessary for rapid and efficient vulcanization, as well as other ingredients such as an antioxidant, softeners and carbon black or other fillers. The compound is then "cured" in a mold for 20 to 60 minutes at temperatures of 130–140°C to ensure vulcanization.

The technology of vulcanization has developed greatly since the early days, especially with the advent of synthetic elastomers. Since it is now understood that vulcanization is a process of crosslinking the linear chain molecules of rubber, it is possible to select suitable vulcanizing agents for the different synthetic rubbers, depending on the chemical structure of the polymeric chains. Thus, sulfur vulcanization is only suitable for elastomers containing a greater or lesser amount of unsaturation, other methods being used for other elastomers. For example, Neoprene (polychloroprene), although an unsaturated elastomer, is most efficiently vulcanized by the reaction between some of its reactive chlorine atoms with metal oxides (ZnO, MgO). Similarly, silicone elastomers require peroxides, urethane elastomers can be vulcanized with diisocyanates, while elastomers containing carboxyl group substituents (acrylic acid copolymers) may be vulcanized either by metal oxides or by diepoxides.

It should be noted, however, that sulfur vulcanization systems are still highly preferred, largely because of their ease of handling and control, as well as their low cost. Furthermore, it is undoubtedly true that a wealth of experience is available about sulfur vulcanization. Hence, even in the development of new elastomers, it was often found more convenient to incorporate a small proportion of unsaturation (if none was originally present) in the polymer chain, so that sulfur vulcanization became feasible. This has actually been done in the case of urethane and silicone elastomers, butyl rubber and the ethylene-propylene terpolymers.

It should also be mentioned, however, that other

agents than sulfur can be, and have been, used even with unsaturated elastomers. Thus, peroxides like dicumyl peroxide (Dicup) crosslink natural rubber, presumably by attacking the more reactive allylic hydrogen atom lying α to the double bond and permitting two adjacent chains subsequently to combine. Other reactive compounds which can also remove this α-hydrogen, e.g., quinone dioxime, dinitrosobenzene, etc., can therefore also cause crosslinking of the polymer.

Despite the widespread use of sulfur vulcanization, the mechanism of the chemical reaction involved has remained obscure. This is due to the fact that this process involves a complex series of reactions, which have been especially difficult to elucidate, since the extent of reaction is relatively small and the final product (vulcanized rubber) insoluble and not tractable to analysis.

It has been shown that the sulfur forms crosslinks either by direct attack on the double bond in the polymer or by substituting for the hydrogen on the α-carbon atom (allylic). Furthermore, the sulfur crosslinks between chains may range from monosulfides to polysulfides. Also, some of the sulfur is used up in nonvulcanization reactions to form cyclic sulfide structures along the polymer chain. These cyclic sulfides add nothing to the crosslinking reaction and, instead, act as active sites for oxidative attack and poor aging. It is also known that the proportion of these side reactions increases with increasing time and temperature of vulcanization, and decreases with efficient accelerators.

The crosslinking action during vulcanization becomes apparent in the changes in physical behavior of the vulcanizate. Thus, the "modulus," or force required for a given elongation, rises continuously as vulcanization proceeds, as does the tensile strength. However, both of these properties reach a maximum, after which they may show a downturn (known as "reversion"). This downturn may be due to more than one cause. Thus, it is known that the reactions occurring during vulcanization can lead to chain scission as well as to crosslinking. Hence a downturn in modulus values with increasing cure is a sign of such a degradation process becoming predominant. However, a downturn in tensile strength need not necessarily reflect chain scission, but may be due to continued crosslinking to too high a density of crosslinks. This can be corroborated by a continued *rise* in modulus values, indicating that crosslinking is still predominating. In other words, the tensile strength seems to depend on an *optimum* crosslink density while the modulus shows a monotonic rise with increasing crosslinking.

In the same way, swelling measurements with suitable solvents indicate the progress of vulcanization. Thus, as curing proceeds, the sol fraction (soluble portion) of the rubber decreases down to values of less than 1%, while the swelling of the insoluble fraction (gel) also shows a gradual decrease. Thus, both modulus and swelling measurements give a direct measure of the degree of crosslinking present.

Vulcanization can be effected with ionizing radiation—a phenomenon undergoing extensive experimentation.

MAURICE MORTON

W

WASTES TREATMENT

The demands for water by public water supplies and for industrial uses have been increasing at such prodigious rates that there is grave concern in regard to the quantity and quality of our surface and ground water resources. In some industrial centers, there is likewise concern regarding the quality of the air we breathe. Both of these natural resources are being contaminated by wastes from human population and from industrial processes. This discussion is confined to the pollution due to water-borne wastes from industry, which has grown to be greater than the pollution from human wastes. Methods for the treatment of human wastes (sewage) are well established, and this type of pollution is rapidly coming under control.

Water pollution by the waste products of industry and man is due to suspended solids which cause unsightly and often odorous sludge blanks, or which discolor the waters; to organic matter which in itself may be inoffensive, but which will decompose in the stream or lake and use up all the dissolved oxygen, thus killing aquatic life by suffocation; to acids, alkalies or toxic chemicals which render the water poisonous to aquatic life or unsatisfactory for water supply purposes because of tastes and odors; and to pathogenic bacteria which cause disease.

Classification of Wastes. A classification of industrial wastes is developed below with two ideas in mind, i.e., nature of the pollutant and possible method of removing it.

(1) *Solids:* (a) Dry solids which may be disposed of on dumps or waste land, and (b) solids suspended in water which will settle as a sludge or float to produce objectional scum.

(2) *Oils and Greases:* Lubricating oils, greases, cutting oil emulsions, and vegetable oils, which produce sleeks or scum on receiving bodies of water.

(3) *Soluble Materials:* Included in this classification are (a) inorganics as brines, acids, alkalies, salts; and (b) toxic ions which kill fish and other aquatic life (ammonia phosphates, cyanides, copper, chromium, cadmium, lead, mercury, zinc); (c) organic materials from food processing industries and other organic materials which are susceptible to decomposition by bacteria; and (d) synthetic organic compounds which are inert or toxic and which are not amenable to biological decomposition.

(4) *Volatile Compounds,* such as gasoline, alcohols, aldehydes and ethers, which may cause explosions in sewers, or sulfides and related compounds which are toxic or may cause odor nuisances.

Methods of Reduction. The first step is the development of a control department to study each process within the industry with the point of view of stopping carelessness in operation which causes losses of water, processed material, and waste end products. Large reductions in unnecessary losses have been accomplished in many industries by this "good housekeeping" program. Another preliminary step is the separation of relatively pure cooling and condenser waters from those waste waters which are concentrated and which must be treated.

Obviously the first step in treatment is the removal of all possible dry solids from equipment and floors before flushing with water, and the screening of all drains so that such materials do not enter the sewers. Fine screens are often used to remove coarse solids from wash waters. Solids which pass through screens may be removed in a mechanically cleaned sedimentation tank. These settled solids are then filter-pressed, reclaimed, burned, digested or pumped as a slurry to lagoons. If the solids are colloidal in nature or will not settle readily, sedimentation is preceded by pH adjustment and addition of flocculating agents such as polyelectrolytes, lime and alum, sodium aluminate, ferrous sulfate, ferric chloride or demulsifiers. (See **Water Conditioning**). Acid or alkaline wastes are neutralized; often mixing one waste with another will cause the necessary coagulation. In petroleum, vegetable oil and mining industries, flotation with or without chemical treatment is as important as sedimentation. Chemical oxidation or reduction of the wastes is occasionally necessary as a part of the chemical treatment preceding flocculation and sedimentation. Settled solids resulting from chemical coagulation may be valuable and recoverable, but often they are so diluted with the precipitated chemical floc that they are lagooned, dried on sand beds or vacuum-filtered and disposed on a dump.

However, if these solids are of a putrescible nature, they should be digested anaerobically, following the practices of sewage sludge digestion. The digestion consists largely of a methane fermentation; it produces a gas which may be used to generate power in gas engines and to heat the digestion tank. The anaerobic digestion process has also been applied to distillery slops, packing house wastes and other food-processing wastes to destroy organic matter in solution and suspension without previous sedimentation. This process is very efficient in the destruction of organic matter in concentrated wastes and has efficiencies of 80 to 90% removal. Because of the very high concentration of the influent wastes, the effluents are still quite strong and often must be treated on trickling filters.

If the waste contains soluble organic matter, chemical flocculation and sedimentation will not be effective in its removal. If the soluble material is easily destroyed by chemical oxidation, chlorination is often effective. However, if the organic matter is complex and subject to decomposition by aerobic bacteria, the standard procedures of sewage treatment may be applied. These procedures are known as (1) standard or low-rate trickling filters, (2) high-rate trickling filters with recirculation of part of the effluent through the filters, and (3) the activated sludge process. In trickling filters a growth of organisms builds up as a slime on the stone in the filter, which stone may vary in depth from 3 to 15 feet. The waste is sprayed onto the surface of the stone filter and trickles down over the growth-covered stone. The aerobic biological growths utilize the organic matter in the waste as food for energy and growth, thus converting it to cell tissue, CO_2, H_2O and nitrates.

In the activated sludge process, similar growths in the form of a flocculant sludge are suspended in aeration tanks into which air is introduced through porous diffusers, or by surface agitation. The suspended growths, called "activated sludge," are then removed by sedimentation and the clarified effluent is discharged to the recovery body of water. (See **Activated Sludge**).

Recycling of wastes within a process in place of fresh make-up water is used in many industries. For example, (1) the rinse water in metal plating operations may be returned to metal plating baths as make-up water; (2) water from the wet milling of corn is recycled in the process and eventually used as make-up water for the corn steeping process; and (3) white water in the paper making industry is often completely recycled within the mill. Tremendous savings have been made in these industries by recycling or bottling-up of the wastes.

Evaporation of wastes is often possible if the resulting solids are of value. The waste must be sufficiently concentrated for the evaporation to be economical. Recycling in the wet process corn milling industry allows the build-up of corn solubles to a point where evaporation is economical.

Lagooning is often used for the disposal of the solids in slurries. Lagoons or oxidation ponds may be effective in the destruction of organic wastes, particularly in warm and dry climates. The mechanism of purification in oxidation ponds is similar to that in trickling filters and activated sludge, except that the biological flora are augmented by the blue-green algae which increase the dissolved oxygen in the water by photosynthesis.

Intermittent spraying of screened and settled wastes from canneries and packing houses on grass lands and wooded areas has been quite satisfactory. If the soil is of a porous nature, large volumes of waste waters can be absorbed.

A new type of waste is causing concern due to the rapid expansiton of the atomic energy program. The wastes from atomic fission are radioactive, the activity varying from a few minutes to thousands or millions of years, depending on the radioisotopes present. The activity may be reduced to safe levels by adsorption on chemical flocs, on such biological flocs as activated sludge, by ion-exchange resins, and by dilution with inert materials; but the more concentrated wastes are buried in waste areas or encased in lead and concrete and buried at sea.

Although it would be impossible to list all wastes, the following grouping will give the reader some general idea of the extent and types of some of the most troublesome and strongest wastes. *Food processing wastes* result from (1) slaughter-house and meat packing, (2) canneries and frozen foods, (3) creameries and milk products, (4) fermentation wastes from the production of beer, wine, distilled liquor, yeasts and antibiotics, (5) corn, soybean and edible vegetable products, and (6) beet sugar, cane sugar and molasses. *Apparel wastes* may result from (1) laundry and dry cleaning, (2) tanneries, (3) textile manufacture, (4) wool scouring and other pretreatment of material occurring fibers. *Chemical wastes* result from the (1) synthetic production of chemical compounds such as rubber, plastics, rayon, nylon, paints, fungicides and heavy chemicals, (2) pulp and paper manufacturing, (3) oil production and distillation, (4) metal finishing and plating, (5) acid mine drainage and (6) radioactive wastes.

Evaluation. In general, laboratory tests are preformed according to "Standard Methods of Water and Sewage Analysis, 13th Ed." (APHA). *Suspended matter* is determined by filtration through a Gooch crucible and the *dissolved solids* are determined by evaporation of the filtrate from the suspended matter determination. The results are expressed as milligrams or micrograms per liter and may be converted into pounds or tons per day or per unit of product.

The most important test is the *biochemical oxygen demand* (BOD) test, because it is a measure of the amount of oxygen which will be used in the stream by the biological decomposition of the organic matter which may be present. This is done by incubating a sample of the waste with standard dilution water for 5 days (or more) and determining the decrease in dissolved oxygen. These results are then expressed in mg/l of oxygen demand and often are converted into the *population equivalent* of the waste, i.e., it is said that the waste is equivalent to the sewage from a given population. In making this calculation the per capita contribution of oxygen demand is 0.17 lb per day. This test is applicable to most of the food processing and apparel wastes, but of course falls down as a measure of organic matter which is not susceptible to biological decomposition.

The standard method for determining the chemical oxygen demand (COD) consists in vigorously oxidizing the organic matter with potassium dichromate in concentrated sulfuric solution at boiling temperature. The COD often bears a relationship with the BOD of a given waste, but this relationship is different for most wastes and for different effluents in the steps of purification of a particular waste. It cannot be applied directly to stream purification studies or predictions.

The determination of pH, acidity and alkalinity are applicable to wastes containing acid and alkalies. Metal ions are determined in the metal finishing wastes. These latter determinations are difficult because of interferences and require rather tedious extraction procedures. Phenols and cyanides are of particular interest because only very small quantities are allowable in water supplies. These tests are very delicate and are expressed as micrograms per liter or parts per billion. Chlorine gas is used both to destroy certain chemical wastes and/or to sterilize bactetria. The amount of chlorine required to meet particular specifications is determined and recorded as the "Chlorine Requirement." The test specifies that the pH, temperature and time of reaction should be controlled, as near as possible, to the actual plant process conditions.

W. D. HATFIELD

Cross-references: *Water Conditioning, Activated Sludge, Chlorine, Radioactive Waste Management.*

Radioactive Waste Management

The production of nuclear power as well as many operations in factories, laboratories, and hospitals produce quantities of radioactive material that are no longer suitable for use. The management of this radioactive material becomes more critical as the use of nuclear materials increases. Only very small fractions of the radioactivity can be disposed of to the environment and the majority must be stored under conditions which will prevent any accidental release.

Major sources of radioactive materials are the mining and milling of uranium and thorium, the processing of spent fuel from nuclear reactors and the radioisotopes used in hospitals, laboratories and industrial work. The wastes can be in the form of solids, liquids or gases and usually can be classified by their level of radioactivity as well as their physical state. A further generalization may be made that wastes fall into two groups: (a) large volume-low activity waste, and (b) low volume-high activity waste.

It should be noted that nuclear reactors do not produce large quantities of wastes during normal operation. Their waste is generated during the reprocessing of the nuclear fuel when it is dissolved and the uranium or plutonium separated for reuse as fuel. The fission products released in this process are minimized by allowing the fuel to decay for about three months before dissolution. The amount produced is still large; for example, it is estimated that in 1980 the United States may be generating 50,000 megawatts of electricity with nuclear reactors. This would produce about 18,000 megacuries of fission products per year (after 100 days decay) or equilibrium inventory of almost 300,000 megacuries.

Relatively large volumes of low activity radioactive wastes are disposed of by discharge to the normal sewage system. This includes excreta from hospital patients treated with radioisotopes and laboratory and industrial solutions where the activity is so dilute that it is no longer useful. In the United States, federal or state regulations require that the concentration of activity be such that the wastes leaving the premises should not exceed specified levels, set to minimize human exposure.

With still higher concentrations it is not possible to dispose of the radioactivity to the environment and it is necessary that it be stored. Liquids are difficult to retain and most modern practices attempt to convert liquid wastes to solids which are less bulky and easier to handle. Solidification can be accomplished by evaporation, by precipitation or coagulation and by ion exchange on resins, zeolites or clays. The solid material may then be stored with less possibility of leakage or other transfer to the environment. The clays have a particular advantage in that they may be fired to form an impervious mass. The low volume-high activity liquid wastes generated in fuel reprocessing have usually been stored in tanks underground.

The major radionuclide that will be released in normal practice is tritium. This will be produced in the form of tritiated water and it will not be possible to separate this from normal water before discarding to the environment.

Most solid wastes fit the large volume-low activity category and it is customary to reduce their volume by incineration or compaction. Care is necessary in incineration to prevent the loss of volatile radioactive materials.

Gases and airborne solids are generally treated by filtration and scrubbing. Fortunately most of the gaseous radioactive materials have very short half-lives and can be released to the environment without hazard. Such releases are also subject to federal or state regulations. Longer-lived radionuclides such as krypton-85 may be collected and stored either in solution or as a compressed gas. If the krypton is not contained it would contribute about 20,000 megacuries to the world inventory.

The storage of solid material by shallow burial at selected sites licensed or controlled by the government is the most general solution to waste management. The storage areas are fenced off and guarded so that accidental exposure is minimized. Most of the present sites are designed for surface burial but investigation of storage is abandoned mines and salt formations is expected to lead to safer storage since the possibility of leaching with ground water is removed.

Certain types of disposal such as pumping radioactive wastes into deep wells or ocean disposals of solids and liquids are no longer carried out in the United States on a large scale. Emphasis is on complete containment and on conversion of the nuclear by-products into useful forms.

JOHN H. HARLEY

References

"Regulations regarding waste disposal," Code of Federal Regulations, Title 10, Chapter I, Part 20.
"Management of Radioactive Wastes," Safety Series Booklets 12 and 19, International Atomic Energy Agency, Vienna, Austria (1965 and 1966).
"Ionizing Radiation: Levels and Effects," A report of the United Nations Scientific Committee on the Effects of Atomic Radiation to the General Assembly, with annexes Vol. 1: Levels, United Nations, New York, 1972.

Spent Sulfite Liquor

In the manufacture of cellulose pulp by the sulfite process, wood chips are heated under pressure in an aqueous bisulfite solution containing an excess of sulfur dioxide. During this digestion the lignin of the wood is made soluble by sulfonation, and the hemicelluloses of the wood are hydrolyzed to pentose and hexose sugars. The soluble lignosulfonates and sugars together with some extraneous materials, comprising roughly one-half of the wood, dissolve in the cooking liquor which is then separated from the cellulosic pulp as "spent sulfite liquor." In the United States alone about three million tons of sulfite pulp are manufactured annually, and this production results in over three million tons of spent sulfite liquor solids.

Spent sulfite liquor, as it is discharged from the pulp mill, is a clear, amber-colored liquid having a pH of from 2 to 4 and containing from 8 to 15 per cent solids, depending upon the pulping conditions and the efficiency of liquor collection. Spent sulfite liquors vary considerably from mill to mill depending upon the wood species employed, base used for preparation of cooking liquor, and type of pulp produced. Traditionally, calcium-base liquors have been used by the industry, although in recent years other bases such as magnesium, sodium, and ammonium have been introduced for purposes of chemical recovery, faster digestion, and specialty pulp production. For calcium-base liquors, an approximate percentage composition for softwood spent sulfite liquor solids is 55 lignosulfonates, 14 hexose sugars, 6 pentose sugars, 12 sugar acids and residues, 3 resins and extractives, and 10 ash; and for hardwood liquor solids is 42 lignosulfonates, 5 hexose sugars, 20 pentose sugars, 20 sugar acids and residues, 3 extractives, and 10 ash.

Because calcium-base liquors are not amenable to practical chemical recovery, they have always been sewered into adjacent waterways, and as such, have constituted a public nuisance. Since the last part of the nineteenth century, pressures for the abatement of these pollutional effluents have motivated research directed toward the utilization of spent sulfite liquor and its two important components—lignosulfonates and sugars. Recent public interest in water pollution has added renewed incentive for finding practical uses. A variety of utilization schemes for spent sulfite liquor and its individual components have been proposed, and a limited number of these have proved to be commercially and economically practical and revenue producing.

Whole spent sulfite liquor found a variety of uses based on the adhesive properties of its carbohydrate content and upon the dispersing and surface-active properties of its lignosulfonate content. In addition, in the composite liquor, the properties of one component are influenced by those of the others, giving a new set of properties characteristic of spent sulfite liquor. The relatively low-grade utilization scheme of roadbinding is the largest user of crude spent sulfite liquor, accounting for an estimated 125 million pounds of spent liquor solids per year. Unfortunately, the use is limited to warm-weather months and to secondary roads not too distant from the mill.

Another large-scale use for crude spent sulfite liquor, amounting to approximately 65 million pounds of solids per year, is that of binder for animal feed pellets. Concentrated liquors and liquor solids from the calcium, magnesium, ammonium, and sodium-based processes have recently found exceptional acceptance as pelletizing agents for the production of animal and poultry feeds and are used in amounts up to 4% of the finished pellets. In addition to the nutritional values of the carbohydrate and inorganic constituents, the lignosulfonates in the liquors are said to lubricate the pellet die and to reduce power consumption during pellet formation. Stronger, denser pellets with less tendency to break are produced.

The hexose sugars in spent sulfite liquor can be fermented to ethyl alcohol, and the commercial production of alcohol from coniferous spent liquors is common practice in European pulp mills. Where ethyl alcohol is available from petroleum and from black strap molasses, the sugars of spent sulfite liquor ordinarily cannot compete economically, but one large mill in the United States and two in Canada are producing alcohol commercially from coniferous spent sulfite liquors. The hexose and pentose sugar contents of both coniferous and hardwood spent sulfite liquors are fermented commercially to *Torula* yeast by one mill in the United States. *Torula* yeast is used as a protein concentrate and as a good source of vitamin B in animal feeds, dehydrated soups, crackers, prepared baby cereals, specialty diet products, and other foods.

The recent concern over synthetic sweeteners has generated a great interest in sweeteners such as xylose and its hydrogenation product, xylitol, and it is expected that sulfite pulp producers will initiate xylose production from their spent liquors.

The lignosulfonate content of spent sulfite liquors has found a number of tonnage uses in industry based upon both their physical and chemical properties. The dispersant properties of the lignosulfonates, combined with their ability to sequester many metallic ions, have contributed to their success as additives for the preparation of oil well drilling muds, a market which currently amounts to more than 90 million pounds of lignosulfonate solids annually. Approximately 30 million pounds per year are utilized as concrete additives and about 15 million pounds per year are employed for water treatment. A host of miscellaneous uses based upon these important physical and chemical properties account for about another 100 million pounds of purified or processed lignosulfonates. Some of the more important uses are in the tanning industry where they are employed along with vegetable and chrome tanning agents, in agriculture for metal complexing and for dispersing properties, and in mining for stabilizing soil formations and for fluidized transportation of crushed ores.

The 4-hydroxy-3-methoxyphenylpropane structure of coniferous lignosulfonates suggested their use as a raw material for the production of organic chemicals. The profitable conversion of the lignosulfonate content of spent sulfite liquor into pure organic chemicals has been pursued by the pulp and paper industry for many years. As early as 1937 it was demonstrated that vanillin could be produced commercially from lignosulfonates and, within 15 years, substantially all of the flavoring vanillin in the United States and Canada was produced by alkaline oxidation of lignosulfonates of coniferous woods. However, flavoring vanillin requirements utilized only a negligible percentage of the millions of tons of available lignosulfonates, and therefore, the use of vanillin as a raw material for chemical production was investigated.

Processes were evolved for the simple conversion of vanillin to its derived acid, vanillic acid. Many new esters, amides, ethers, and other derivatives of vanillic acid were prepared and evaluated for a variety of uses. For example, ethyl vanillate was found to be less toxic to humans than sodium benzoate, but very toxic to specific microorganisms. As such, it found usefulness in the treatment of human diseases and as a preservative in foodstuffs. At present, it is manufactured commercially for the specific treatment of the progressive disseminated form of the two mycotic diseases, histoplasmosis and coccidioidomycosis (valley fever). Similarly, in Europe the diethylamide of vanillic acid has found widespread use as an analeptic agent for the control of respiration and blood pressure. This product has been introduced recently into the United States, and its production should require considerable amounts of vanillin. In addition, vanillin is the raw material employed for the commercial manufacture of *l*-Dopa (*l*-dihydroxyphenylalanine), a compound recently found to be a specific treatment for Parkinson's syndrome in essentially all of its manifestations. The demand for vanillin for the production of these drugs may increase several fold over today's market requirements, which exceed 2.5 million pounds in the United States alone. In addition to vanillin, other pure organic chemicals such as acetovanillone, vanillil, oxalic acid, and conidendrin are available from the commercial processing of coniferous spent sulfite liquors.

In recent years many of our sulfite mills for several reasons ceased pulping coniferous woods, and in their stead, now pulp aspens, cottonwoods, alder, birch, and other readily available deciduous woods. The spent sulfite liquor from these pulping operations contains lignosulfonates derived from 3,5-dimethoxy-4-hydroxyphenylpropane structural elements as well as the 4-hydroxy-3-methoxyphenylpropane elements, and aromatic compounds with 3,5-dimethoxy-4-hydroxy-substitution are produced in degradation reactions along with the 4-hydroxy-3-methoxy-substituted compounds produced alone from coniferous spent liquors. In addition, woods of the family Salicaceae such as aspens, cottonwoods, and willows give spent sulfite liquors containing considerable amounts of *p*-hydroxybenzoic acid. Current studies on syringaldehyde and other 3,5-dimethoxy-4-hydroxyphenyl derivatives indicate greater chemical utilization of deciduous-wood spent sulfite liquors in the future.

In the processing of spent sulfite liquor it has been found that specific fractions of the lignosulfonate content might be much more effective for a particular end use than is the entire product. Accordingly, the lignosulfonates have been fractionated on the basis of molecular weights and degree of sulfonation by a variety of methods including solvent precipitation, ion-exchange, ion-exclusion, electrodialysis, reverse osmosis, gel-filtration, and reactive solvent fractionation. By use of these procedures combined with chemical processing, a host of lignosulfonate products essentially tailormade for specific end uses are now available from commercial producers.

Due to public concern and pressures from regulatory agencies sulfite pulp mills of recent construction obviate possible stream pollution and concern over disposal and utilization by pulping with magnesium-base cooking liquors. Magnesium-base spent liquors can be evaporated and burned to yield magnesium oxide and sulfur dioxide which are reused in preparing fresh cooking liquors. Under these conditions, the organic constituents of the spent liquor are employed only for their fuel value in a manner similar to that employed in the alkaline pulping industry. In other instances, to obviate pollution, older mills have initiated evaporation and burning of spent sulfite liquor without chemical recovery. In such cases, these mills have almost invariably changed to soluble bases such as sodium or ammonium so that cooking cycles could be shortened and more concentrated cooking liquors could be employed resulting in higher pulp productions and much less water to be evaporated. Other mills have changed to the production of higher yield pulps using sulfite cooking liquors of less acidity. The spent liquors from these pulping operations are relatively new and have not been investigated to any extent.

                                                                                          I. A. PEARL

References

Pearl, I. A., "Chemistry of Lignin," pp. 225 ff, pp. 292 ff, Marcel Dekker, New York, 1967.

Pearl, I. A., "Utilization of By-Products of the Pulp and Paper Industry," *Tappi,* **52,** No. 7: 1253 (1969).

Solid Waste

The waste problem in the United States today involves the disposal or treatment of between 165 and 170 million tons per year which is over 4½ lbs. per day for every man, woman and child in the nation. Methods currently being used to treat and dispose of solid waste are to say the least, archaic, although perhaps more research is being done in this field than any other waste disposal area.

Solid waste treatment can be divided roughly into five categories: (1) preparation, (2) separation, (3) disposal, (4) chemical processing, and (5) recovery or recycle.

Preparation. Preparation of solid waste is used to make the waste material more easily handled, and to condition it for subsequent disposal. Major preparation techniques now in use are compaction, size reduction and pulping.

Compaction of solid waste involves compressing the waste into a smaller volume. Average reduction ratios are between 2½ and 3 to 1. The advantages of the compaction system are reduction in the number of trips to a disposal area and a product which is more readily acceptable for landfill. It also extends the length of life of the landfill because of the higher bulk density. Compaction is usually achieved by pneumatic or hydraulic machines and normally is not recommended as a method of pretreatment before incineration.

The second preparation method is size reduction or shredding. Shredders, hammer mills or grinders reduce the solid waste to a consistent particle size, facilitating handling and simplifying the use of the material in landfill and incineration. High horsepower requirements for such equipment offset some of its advantages.

A more recent method which has been used with some degree of success, especially on shipboard installations, is pulping. This is essentially a wet grinding technique which masticates the waste with water and then squeezes out most of the residual water to form a uniform pulp, which then can be fed to a landfill or possibly incinerated. Pulped waste with a controlled amount of water can be pumped over long distances, dewatered and discarded.

Separation. Separation of solid waste is usually performed in the dry state. This involves separation by density in various types of sizing or jigging machines such as vibrating screens or separators, or it can take place in the wet state by the use of flotation, where the heavier materials will sink to the bottom of the flotation tank and the light materials will rise to the top. Magnetic separation is used for the removal of ferrous metals from mixed waste. Glass and metallic materials are effectively removed from bulk waste by various gravity and magnetic separation techniques. Unless there is a significant difference in density there is doubt that separation techniques will be effective.

Disposal. There are two major types of disposal methods in use in the United States. The first is the sanitary landfill, which involves filling a large natural void, such as an abandoned strip mine or quarry, with solid waste and periodically covering it with clean soil. The application of the solid waste and the soil is similar to a multilayer cake. In order to keep rodent infestation to a minimum, each day's sanitary fill must be covered with clean soil. Unfortunately, most sanitary landfills are not sanitary. Many of the materials put into it will decompose naturally in the soil, but others will not. Plastics and metal will remain forever in their original state. Leaching of chemicals by rainwater can often be a problem by the pollution of watersheds. Newest techniques in sanitary landfill utilize a plastic membrane covering the bottom of the landfill. Ground water is pumped out of the bottom of the plastic container and treated before it is run off to the watershed.

Incineration, the burning of solid waste with air, is a positive means of destruction but it has not been acceptable in the last few years because of air pollution. Good incinerators, which will not cause air pollution, are entirely possible with our present technology, and when used in combination with a sanitary landfill, can extend the life of the sanitary landfill by at least 900%. Incineration, naturally, will only handle combustible materials; noncombustibles must first be separated from the waste to increase capacity.

Heat recovery in conjunction with incineration is a possibility and it has been tried only in a few cases but offers significant advantages for future systems.

Chemical Processing. There are a number of chemical processing possibilities but only several are worth mentioning. One which is of considerable interest today is the pyrolysis of solid waste. Pyrolysis is accomplished by an indirect heating of the waste to extract fuel gas, oxygenated aliphatic and cyclic compounds, other combustible liquids and carbonaceous residue. The pyrolysis of waste is expensive and therefore has been tried only on an experimental basis, but it offers hope for the future as a means of realizing a financial return from

waste. The destructive distillation of automobile tires has proved to be an economic and a feasible method for the recovery of chemicals and the disposal of a difficult waste product.

Various specific wastes can be treated by hydrolysis or extraction for the recycle of particular materials, but this is not true of every type.

Some materials can be sintered for the recovery of metals. Calcium compounds can be calcined in high-temperature retorts and metals can be melted to form scrap.

Recovery or Recycle. Several types of waste can be recycled directly back into the manufacturing process. The most obvious of these is scrap automobiles. By shredding and compacting steel from the bodies of old automobiles, scrap can be baled and fed back into the steel mills to produce more new steel. The cast iron from the engine blocks can be recycled similarly. The aluminum industry is currently attempting to recycle aluminum containers, such as beer and soda cans, in a similar manner. Although such recycle methods are expensive they will ultimately become commonplace.

Composting, the natural degradation of waste by bacteria, has been quite successful in Europe and in Japan but not in the United States because of the lack of sufficient economic incentive. Large composting plants of 300 to 500 tons per day cause solid waste to degrade under relatively low temperatures at accelerated rates to produce a dry, pathogenically pure material which can be recycled to the land. The problem with composting is that the reduction in volume is only about 30% and unless the product is used for landfill there is not a sufficient agricultural market for its ultimate use. It is a good soil conditioner but not a replacement for chemical fertilizer. Nevertheless, it does produce a safe, usable product and therefore is better than many methods of disposal now in use.

R. D. Ross

WATER

Water is essential to all life, both animal and vegetable, and comprises about 70 to 90% of the weight of living organisms. It is the dispersion medium for all biochemical reactions which constitute the living process, and takes part in many of these reactions.

About 71% of the earth's surface is seawater, containing about 3.5% salt, mostly NaCl. About 98% of the volume of the hydrosphere is seawater. The remaining 2% is water vapor and fresh water in lakes and streams and in the ground. Water is continuously evaporated into the atmosphere from all exposed water surfaces and from growing plants, is condensed into clouds and precipitated back to the earth's surface as rain, snow, sleet and hail. Part of the land precipitation runs off over the surface to brooks, rivers and lakes and to the sea, and the remainder soaks into the ground to replenish the groundwater and furnish water for plant growth and transpiration, with the residual flowing slowly through the ground to join the surface water flow. About 75% of the rainfall on continental United States is lost to the atmosphere through evapotranspiration from growing plants. Nevertheless, most of the habitable land masses on earth have adequate fresh water resources for man's needs, provided the supplies are developed and the quality of the water is maintained.

Man uses water not only for drinking and culinary purposes but also for bathing, laundering, heating and air-conditioning, for agriculture, stock-raising and gardens, for industrial processes and cooling, for water, steam and atomic power, for fire protection, for disposal of waste waters, for fishing, swimming, boating and other recreational purposes, for fish and wildlife propagation, and for navigation. For many of these uses, the water must be withdrawn from a watercourse, but for others, it is not withdrawn. Most of the water withdrawn, except for agricultural use, is returned to the watercourse with pollutants, a temperature rise, or both, to be reused by downstream communities with further impairment of quality each time.

The drinking water supplies for most urban communities must be treated to destroy pathogenic germs and viruses, and to reduce the concentration of one or more of the following impurities: tastes and odors, turbidity, color, hardness, salinity, iron and manganese. The used water becomes municipal sewage or wastewater, which is returned to the watercourse through the sewage system. Wastewaters from industry may discharge to the sewage system or directly to the watercourse after suitable treatment or cooling. In order to protect the quality of the receiving waters, most wastewaters must receive some treatment. All wastewaters containing human sewage should be heavily chlorinated for disinfection. Additional treatment is primarily for the purpose of maintaining an adequate supply of dissolved oxygen in the receiving waters (3 ppm or more) during low flows and warm weather for the protection of aquatic life, and consists of removing organic matter (biochemical oxygen demand) and ammonia at points of heavy pollution, but may also include removal of floating solids, suspended matter and grease. Persistent pesticides and toxic substances, such as Ag, As, Ba, Cd, Cr, CN, Cu, F, Hg, Pb, and Se should be kept out of the drinking water supply and should not be discharged in wastewater or stormwater runoff in sufficient amounts to harm aquatic life or to impair the use of the receiving waters for drinking. Such substances are best controlled by stopping their use.

Most municipal sewage systems are combined, receiving both sanitary sewage continuously and stormwater runoff during rainstorms. Since the peak rates of discharge of storm water are from 50 to 200 times the average dry-weather flow of sanitary sewage, most intercepting sewers and treatment plants are designed for only 2 to 5 times the dry-weather flow, and overflows of mixed sewage and storm water occur nearly every time it rains. Only about 3% of the year's production of sanitary sewage, but about 10 to 35% of pollutants thus reach the receiving waters without treatment. This defect can be remedied by the construction of complete separate systems of sanitary sewers, or by providing storage for the overflows and pumping therefrom during and between storms at low rates for treatment. The capacity of the treatment plants must be increased about 40% to care for the additional storm water.

Most of the marine fisheries available to man are contained on the continental shelf out to depths of about 10 meters. It is thus evident that harmful wastewater should not be discharged into tidal estuaries, but that disinfected municipal sewage without toxic substances may be discharged in seawater at greater depths, provided it is sufficiently mixed with seawater to avoid anaerobic deposits of sludge. Such disposal should improve the propagation of marine life if carefully managed.

Pure water does not exist in nature. Rainwater dissolves the gases of the atmosphere and entrains particulates in dusts and smokes. When it runs off over ground, it scours silts, clays and leaves into streams, impairing turbidity, color, and tastes and odors. When it seeps through the topsoil, it dissolves CO_2 produced by bacterial decomposition of organic matter in the topsoil, thereby lowering the pH of the water, and it dissolves some decayed organics. When it seeps through the pores or crevices in underlying rocks, it dissolves some of the rocks, such as calcium and magnesium carbonates, and iron and manganese compounds, to increase the hardness, alkalinity and salinity. The suspended and colloidal impurities may be removed by coagulation with alum or ferric sulfate; settling and filtration, and some of the hardness (Ca, Mg, Fe and Mn) may be removed similarly by precipitating the metals with lime and soda during coagulation. Water for some industrial uses and for boilers requires almost complete removal of dissolved salts by demineralization or by distillation. Seawater may be made suitable for drinking, at considerable cost, by distillation or demineralization. Intelligent and economic solutions of water treatment problems require a considerable knowledge of the physics and chemistry of water, and of the equilibrium constants of the reactions involved in coagulation, softening, corrosion control and disinfection. Much of this information is available for a 25°C water temperature, but very little is available for other temperatures from 0°C to 100°C.

A molecule of H_2O vapor has a volume of about 18.85 Å³. A sphere of equal volume has a diameter of 3.3 Å. Assuming that liquid water is composed of such spheres, free to rotate and move about, but in contact with one another, the porosity between them would be 0.476 if they were packed rectilinearly (at the corners of cubes) and 0.258 if they were in a closest-packing arrangement. The corresponding densities at 1 atm would be 0.825 and 1.17, the arithmetic average of which is 0.9975, the density of water at 23°C. This suggests that the packing of molecules of pure water might follow a probability curve of densities between 0.825 and 1.17.

According to Pauling, the volume per H_2O molecule in ice is about 32.3 Å³; the density 0.92, and the crystal structure is held together by directional hydrogen bonds with a bond energy of about 5 kcal/mole (169 psi). Some German experiments in 1885 gave a tensile strength for ice of 142 to 223 psi, in confirmation of the strength of the hydrogen bond. The heat of melting of ice at 0°C is about 1.436 kcal/mole, indicating that about 71% of the bonds are still intact in the water after melting, although the liquid is without tensile strength. This may be explained if it is assumed that each 0-atom in water is bonded to 3 others or less instead of 4 as in ice, thus replacing the tetrahedral ice crystal with groups of 4, 3 and 2 and with single H_2O molecules. The kinetic energy of water molecules (Brownian motion) at 0°C is sufficiently great to assure fluidity even with transient local densities approaching 1.17, and it increases with increase in temperature to 100°C by 1.8 kcal/mole, sufficient for a further reduction of about 50% in intact hydrogen bonds which would leave a residual of about 35% still intact, mostly in groups of 2 and 3 H_2O molecules. The variation in strength of ice and in residual hydrogen bonds in water suggests a statistical variation in the strength of hydrogen bonds. The heat of vaporization of water is about 10.5 kcal/mole at 25°C, and about 9.72 at 100°C, possibly indicating a slight increase in the number of single molecules with increase in temperature.

The mass density of pure water at 1 atm is a maximum of about 1.0 at 4°C, and decreases to about 0.958 at 100°C. The viscosity is about 1.79 centipoises at 0°C and reduces to about 0.284 at 100°C. The surface tension in contact with air at 1 atm is 75.6 dynes/cm at 0°C and 58.9 at 100°C. The dielectric constant of water which affects the solubility of electrolytes is 88.0 at 0°C and 55.33 at 100°C. All of these properties appear to be related to the residual intact hydrogen bonds in liquid water.

Water ionizes into H^+ and OH^-, the ion product constant $K_w = \alpha_{H^+} \, \alpha_{OH^-}$ varying with the temperature as follows:

Temp. °C	0	10	20	25	30	40	50	60	80	100
$K_w \times 10^{14}$	0.11	0.29	0.68	1.00	1.47	2.91	5.48	9.65	23	52

In pure water $\alpha_{H^+} = \alpha_{OH^-} = \sqrt{K_w}$. At 25°C, $\sqrt{K_w} = 10^{-7}$, and only one water molecule out of 555 million is ionized. Any aqueous solution in which $\alpha_{H^+} = \alpha_{OH^-}$ is *neutral*. At 10°C, $\alpha_{H^+} = \sqrt{K_w} = 10^{-7.27}$. If α_{H^+} exceeds $10^{-7.27}$, at 10°C, the solution is *acid*, and if α_{H^+} is less than $10^{-7.27}$, the solution is *alkaline*. The pH of a solution is log $\frac{1}{\alpha_{H^+}}$, the exponent of 10 without sign in the numerical value of α_{H^+}. Any electrolyte which ionizes in water to yield H^+ or combines with OH^- is an *acid*, to yield OH^- or combine with H^+ is an *alkali*, to yield or react with neither is a *neutral salt*.

Among the water-soluble substances of importance to water quality for which the ionization constants are available at 25°C, but not at other temperatures, are H_2S, H_2SO_3, NH_3, SiO_2, H_3BO_3, H_3PO_4, $CaCO_3$, $MgCO_3$, $Mg(OH)_2$, $Ca(SiO_3)_2$, $Mg(SiO_3)_2$, $Ca(OH)_2$, $Ca_3(PO_4)_2$, $Al(OH)_3$, $Fe(OH)_2$, $FeCO_3$, $Fe(OH)_3$, $Al(PO_4)$, $AgCl$, $BaSO_4$, CaF_2, $Cu(OH)_2$, $CaHPO_4$, $CaSO_4$, FeS, $FePO_4$, MgF_2, $MnCO_3$, $Mn(OH)_2$, MnO_2 and $Zn(OH)_2$.

The alkalinity of water is given by the following:

$$2[Alk] = 2[CO_3^{--}] + [HCO_3^-] + [HSiO_3^-] +$$
$$[H_2BO_3^-] + 2[HPO_4^{--}] + [H_2PO_4] + [HS]^-$$
$$[OH^-] + [NH_3]$$

where $[Alk] = Alk \times 10^{-5}$ and Alk is in mg/l of equivalent $CaCO_3$. The most common constituent of alkalinity is HCO_3^-. Aqueous CO_2 ionizes to HCO_3^-, which ionizes to CO_3^{--} at higher pH. The equilibrium constants are: $K_{CO_2} = \dfrac{\alpha_{H^+} \, \alpha_{HCO_3^-}}{\alpha_{CO_2}}$ and

$K_{HCO_3} = \dfrac{\alpha_{H^+} \alpha_{CO_3^-}}{\alpha_{HCO_3^-}}$, and vary with the temperature as follows:

Temp. °C	0	10	20	25	30	40	50
$K_{CO_2} \times 10^7$	2.61	3.34	4.05	4.31	4.52		
$K_{HCO_3^-} \times 10^{11}$	2.36	3.24	4.20	4.69	5.13	6.03	6.73

A mineral analysis of a water may be checked by comparing the concentrations of positive and negative ionic charges which should be equal.

THOMAS R. CAMP*

* Deceased.

WATER CONDITIONING

The term "water conditioning" is used broadly to cover all the processes used in treating water to remove or reduce undesirable impurities to specified tolerances.

Water impurities and the treatments required may be grouped under three headings: (1) suspended matter, color and organic matter, (2) dissolved mineral matter, and (3) dissolved gases; (1) may include sediment, turbidity, color, microorganisms, tastes, odors and other organic matter. (2) consists chiefly of the bicarbonates, sulfates and chlorides of calcium, magnesium and sodium. Small amounts of silica and alumina are commonly present. Other constituents which may be present are iron, manganese, fluorides, nitrates, potassium, and some mine water and surface waters (contaminated with mine drainage or trade wastes) may be acid, usually with sulfuric acid. Dissolved gases which may be present are oxygen, nitrogen, carbon dioxide, hydrogen sulfide and methane.

Sedimentation. With waters containing large amounts of coarse, easily settled, suspended matter, sedimentation (plain sedimentation) is often of value in reducing the load on the filters and effecting economies in amounts of chemicals used for coagulation. Sedimentation may be carried out in sedimentation tanks or basins or in reservoirs. Detention periods vary over a wide range—from a few hours up to one or more months.

Coagulation. Coagulation is employed to form, by cataphoresis and entanglement, larger aggregates with the turbidity, color, microorganisms, and other organic matter present in the water. These larger particles, known as the "floc," may then be removed by filtration through a sand or Anthrafilt filter or by settling and filtration. The coagulant employed is either an aluminum or iron salt, usually the sulfate. Aluminum sulfate is the most widely used coagulant. Others are ferric sulfate, ferrous sulfate (must be oxidized by air or chlorine) and sodium aluminate. Most favorable pH values for aluminum coagulants usually range from 5.5 to 6.8 and for iron from 3.5 to 5.5 and above 9.0 but there are exceptions. Coagulation aids are ground clay (not too finely pulverized) and activated silica.

Filtration. Filtration is effected by flowing the coagulated or coagulated and settled water downward through a bed of fine filter sand or Anthrafilt in either a pressure type or gravity type filter. Flow rates in industrial practice range up to 3 gpm per sq. ft. of filter bed area while in municipal practice maximum flow rate is usually 2 gpm per sq. ft.

Chlorination. Chlorination is the most widely used disinfecting or sterilizing process. Where daily water requirements are not large, it is common practice to use a hypochlorite, but for large plants liquefied chlorine gas is used. Chlorination may be practiced before filtration (prechlorination), after filtration (post-chlorination), or both before and after.

Taste and Odor Removal. Except for sulfur waters, most tastes and odors are organic in nature. Activated carbon is widely used for their removal. In powdered form, it may be added to the water being treated in coagulation and settling equipment. In such installations, aeration is frequently used as preliminary treatment. In granular form, it is used in filters (activated carbon filters or purifiers). As substances producing tastes and odors are usually extremely small in amount, activated carbon filters are frequently operated for 6 months to one or more years before replacement of bed is necessary.

Hardness Removal (Water Softening). *Sodium Cation Exchanger (Zeolite) process.* This is the most widely used water-softening process in industrial, commercial, institutional and household applications. Hard water is softened by flowing it, usually downward, through a bed (2' to over 8' in thickness) of a granular or bead type sodium cation exchanger in either a pressure-type (most widely used) or gravity-type water softener. As water comes in contact with sodium cation exchanger, hardness (calcium and magnesium ions) is taken up and held by the exchanger which gives up to the water an equivalent amount of sodium ions. At end of softening run (4 to over 24 hours in industrial practice and 1 to over 2 weeks in household use), softener is cut out of service, regenerated and returned to service ($\frac{1}{2}$ to $1\frac{1}{2}$ hours). Regeneration is effected in 3 steps: (1) backwashing to cleanse and hydraulically regrade the bed (2) salting with specified amount of common salt (sodium chloride) solution, usually 10 to 15% in strength, which removes calcium and magnesium from the exchanger and restores sodium to it and (3) rinsing to remove calcium and magnesium chlorides and excess salt.

Hydrogen Cation Exchanger Process. Calcium, magnesium, sodium and other cations are removed by flowing water (usually downward) through a bed (2' to over 8' in thickness) in an acid-proof pressure-type (most widely used) or gravity-type shell. As water comes in contact with hydrogen cation exchanger, calcium, magnesium, sodium and other cations are taken up by the exchanger which gives up to the water an equivalent amount of hydrogen ions. At the end of operating run (4 to over 24 hours), unit is cut out of service, regenerated and returned to service ($1\frac{1}{4}$ to 2 hours). Regeneration is effected in 3 steps (1) backwashing to cleanse and hydraulically regrade the bed (2) acid treatment with sulfuric or hydrochloric acid which removes metallic cations from the bed and restores hydrogen to it and (3) rinsing to remove salts (sulfates or chlorides) and excess acid. The carbon dioxide formed from the bicarbonates may be reduced to below 5 to 10 ppm by aeration and the sulfuric and hydrochloric acids formed from chlorides and sulfates may be (1) neutralized with an alkali (usually caustic soda) (2) neutralized by sodium bicarbonate content of a sodium cation exchanger

softened water (in which case aeration follows neutralization) or (3) removed by an anion exchanger.

Cold Lime (or Lime Soda) Process. Chemicals used may be (1) lime plus a coagulant or (2) lime plus soda ash plus a coagulant. Dosages vary according to composition of raw water and result desired such as (a) calcium alkalinity reduction, (b) calcium and magnesium alkalinity reduction, (c) reduction of total hardness without excess chemicals and (d) excess chemical treatment. Precipitates produced are calcium carbonate and magnesium hydroxide. Rated residuals without excess chemicals are 35 ppm (2 gpg) for calcium and 33 ppm (2 gpg) for magnesium, both expressed as calcium carbonate. Operating results will range between these and theoretical solubilities. With excess chemicals, total hardness may be lowered to 16 ppm (1 gpg). The process is best carried out in the sludge blanket-type of equipment in which the treated water is filtered upward through a suspended blanket of previously formed sludge. Detention periods range from one to two hours. Usually, treated water is filtered before going to service but where small amounts of turbidity are unobjectionable, filters may be omitted.

Hot Lime Soda Process. In this process, treatment with lime and soda ash is carried out at temperatures around the boiling point in closed, steel pressure tanks. Heating is usually accomplished with exhaust steam and pressures most widely used range from 5 to 10 psig, but higher pressures up to but seldom above 20 psig are also used. At these temperatures, the reactions proceed swiftly and precipitates formed are larger than those in cold lime soda process so no coagulant is needed. Detention period is one hour and deaeration effected in primary heater is sufficient to lower dissolved oxygen content to 0.3 milliliter per liter which is sufficient for low-pressure boilers. For high-pressure boilers, either an integral or separate deaerator is used and this will bring the dissolved oxygen down to less than 0.005 ml/l. With 20 to 30 ppm excess soda ash, the hardness will be reduced to 25 ppm ($1\frac{1}{4}$ gpg). Softening to practically zero hardness may be effected by either (1) two-stage hot lime soda phosphate treatment in which effluent from hot lime soda softener is treated with sodium phosphate or (2) the filtered effluent is passed through a sodium cation exchanger. Anthrafilt filters are usually employed with hot process softeners.

Demineralization (Deionization). Metallic cations are removed by a hydrogen cation exchanger. Anions are removed by an anion exchanger. Depending on the hookup used, the carbon dioxide formed from the bicarbonates may be removed mechanically by an aerator, degasifier or vacuum deaerator, or chemically by a strongly basic anion exchanger. Strongly basic anion exchangers will remove both strongly ionized acids, such as sulfuric and hydrochloric, and weakly ionized acids, such as silicic and carbonic. Weakly basic anion exchangers will remove only strongly ionized acids.

Iron and Manganese Removal. In clear, deep ground waters, iron and/or manganese may occur as soluble, colorless, divalent bicarbonates. These may be removed (1) by oxidation plus settling (if necessary) plus filtration (2) by cation exchange with sodium or hydrogen cation exchangers or (3) filtration through an oxidizing (manganese zeolite) filter. In (1) addition of an alkali or lime may be needed to build up the pH value so as to speed up the oxidation. Iron and/or manganese in organic (chelated) form may usually be removed by coagulation, settling and filtration. In acid waters, these metals may be removed by neutralization (plus increase of pH), aeration, settling and filtration.

Fluoride Removal. Fluorides may be reduced to below 1 ppm by filtration through a bed of a specially prepared, granular bone char (bone black). Regeneration is effected with caustic soda solution followed by treatment with dilute phosphoric acid.

Dissolved Gases. *Oxygen and Nitrogen* may be removed (1) hot in a deaerating heater (deaerator) or (2) cold in a vacuum deaerator. *Carbon dioxide* may be removed in (1) an aerator, (2) a deaerating heater or (3) a vacuum deaerator or it may be neutralized with lime or an alkali or by filtration through a bed of granular calcite. *Hydrogen sulfide* may be removed by (1) aeration followed by chlorination, (2) treatment with flue gas plus aeration followed by chlorination or (3) filtration through an oxidizing manganese zeolite filter (household use). If sulfur content and pH values are high, (1) may effect but little removal but, in some cases, with fairly long detention periods, sulfur bacteria may effect notable reductions.

ESKEL NORDELL

Cross-references: *Ion Exchange, Ion Exchangers, Water.*

WAXES

The English term "wax" is derived from the Anglo-Saxon *weax,* which was the name applied to the natural material gleaned from the honeycomb of the bee. In modern times the term wax has taken on a broader significance, as it is generally applied to all wax-like solids, natural or synthetic, and to liquids when they are composed of monohydric alcohol esters. Unlike the ordinary oils of animal and vegetable origin and the animal tallows, the waxes, with a few exceptions, are free from glycerides, which are common constituents of oils and fats. Bayberry wax is a vegetable tallow which happens to have all the physical characteristics of a wax, and has always been classed as such.

Animal and Vegetable Waxes. The most important insect wax from an economic viewpoint is *beeswax,* secreted by the hive-bee. Wax scales are secreted by eight wax glands on the underside of the abdomen of the worker bee. These wax wafers are used by the bee in building its honeycomb. From $1\frac{1}{2}$ to 3 pounds of wax can be obtained from the combs when they are scraped. The crude wax must be rendered and refined before it can be sold as "yellow beeswax." When this is bleached, it is known as "white beeswax."

The chemical components of beeswax are alkyl esters of monocarboxylic acids (71–72%), cholesteryl esters (0.6–0.8%), coloring matter (0.3%), lactone (0.6%), free alcohols (1–$1\frac{1}{2}$%), free wax acids (13.5–14.5%), hydrocarbons (10.5–11.5%), moisture and mineral impurities (0.9–2%). Myricyl palmitate ($C_{46}H_{92}O_2$) is the principal constituent of the simple alkyl esters (49–53%); the simple esters include alkyl esters of unsaturated fatty acids.

The complex esters include hydroxylated esters the chief component of which is believed to be ceryl hydroxypalmitate, $C_{42}H_{84}O_3$. The principal free wax acid component is cerotic acid ($C_{26}H_{52}O_2$). The principal hydrocarbon is hentriacontane ($C_{31}H_{64}$).

The uses of beeswax are many, including church candles, electrotypers and pattern makers wax, cosmetic creams, adhesive tape, munition shells, modelling of flowers, shoe paste constituent, etc. The United States consumes about 8 million pounds of beeswax annually, more than half of which it imports from foreign countries.

Although there are many other kinds of insect waxes, only two are of economic importance namely, shellac wax and Chinese insect wax. Shellac wax is derived from the lac insect, a parasite that feeds on the sap of the lac tree indigenous to India. The commercial wax is not ordinarily the native Indian lac wax, but is a by-product recovered from the dewaxing of shellac spar varnishes. Lac wax melts at 72–80°C, whereas commercial shellac wax melts at 80–84.5°C. Its high melting point and dielectric properties favor its use in the electrical industry for insulation. Chinese insect wax is the product of the scale insect.

The land animal waxes are either solid or liquid. *Woolwax,* derived from the wool of the sheep, is of great economic value. It is better known as anhydrous lanolin, and is of a stiff, soft, solid consistency. The only representative of liquid animal wax is "mutton bird oil" obtainable from the stomach of the mutton bird.

The unsaponifiables of woolwax, known as "woolwax alcohols," are in considerable demand by cosmetic and pharmaceutical industries. Woolwax has a great affinity for water, of which it will absorb 25 to 30%. Refined woolwax is kneaded with water to produce a water-white, colorless ointment, known as hydrous lanolin or "lanolin USP." Anhydrous lanolin is widely used in cosmetic creams, since it is readily absorbed by the skin. It is also used in leather dressings and shoe pastes, as a superfatting agent for toilet soap, as a protective coating for metals, etc. United States consumption of wool wax is about 1.5 million lb/year.

The marine animal waxes are both solid and liquid. The solid marine animal waxes are represented by a wax of considerable economic importance, namely *spermaceti,* derived from a concrete obtained from the head of the sperm whale. The liquid waxes of marine animals are represented by sperm oil obtained from the blubber and cavities in the head of the sperm whale. Spermaceti is the wax used in the candle which defines our unit of candle power; it is used chiefly as a base for ointments, cerates, etc. Sperm oil contains a considerable amount of esters made up of unsaturated alcohols and acids, both of which are susceptible to hydrogenation. Hydrogenated sperm oil is the equivalent of spermaceti wax and harder than the commercial pressed spermaceti. Both yield cetyl alcohol as the unsaponifiable. There is a fairly large demand for cetyl alcohol in the manufacture of lipstick, shampoo, and other cosmetics. Sperm oil itself is an excellent lubricant for lubricating spindles of cotton and woolen mills, or wherever there is need for a very light, limpid, nongumming lubricant.

The waxes obtained from plants occur in the leaves, stems, barks, fruit, flowers, and roots. The leaves of palm trees furnish wax of great economic importance. Particularly is this true of the product furnished by harvesting the leaves of the carnauba palm. The wax is removed from the leaves by sun-drying, trenching, threshing and beating; the powdered wax is melted in a clay or iron pot over a fire, strained, cast into blocks, and broken into chunks for shipment from Brazil. *Carnauba wax* dissolves well in hot turpentine and/or naphtha, from which solvents it gels on cooling; it has a good solvent retention power. Its hardness, luster, and favorable behavior with solvents make it a highly valued ingredient in shoe pastes, floor polishes, carbon paper, etc. A small amount of carnauba, such as 2.5 per cent, when added to paraffin will raise the melting point of the latter enormously (e.g., from 130 to 170°F), making it a very useful ingredient in the production of inexpensive high-melting blended waxes. United States consumption is provided by imports from Brazil which amount to over 11 million lb/year.

The chemical composition of carnauba wax comprises 84–85% of alkyl esters of higher fatty acids. Of these esters only 8–9% (wax basis) are simple esters of normal acids. The other esters are acid esters 8–9%, diesters 19–21%, and esters of hydroxylated acids 50–53% (was basis) of which about one-third are unsaturated. It is the hydroxylated saturated esters that give carnauba its extreme hardness, whereas the esters of the hydroxylated unsaturated fatty acids produce the outstanding luster to polishes.

Ouricury, carandá, and raffia are commercial palm leaf waxes of lesser importance. Ouricury wax has a very high content of esters of hydroxylated carboxylic acids and is used as a substitute for carnauba in carbon papers, etc. Carandá and raffia waxes have very low contents of these acids and make unsatisfactory substitutes for carnauba.

The most important wax obtainable from the stems of plants is *candelilla,* obtained in Mexico and the southwestern United States. To recover the wax the plant stalks are pulled up by the roots and boiled in acidulated water. On cooling, the congealed wax is removed from the surface of the water in the tank. The crude wax is given an additional refinement before it is placed on the market. Candelilla wax is brownish in color, and melts at 66–78°C. Most vegetable waxes are essentially alkyl esters of aliphatic acids; candelilla, on the other hand, contains 51 to 59% of hydrocarbons and less than 30% of esters. The chief hydrocarbon is hentriacontane ($C_{31}H_{64}$), common to other vegetable waxes. The hydrocarbons melt at 68°C, and the esters at 88–90°C. Candelilla is often used in conjunction with carnauba in leather dressings, floor waxes, etc. It is also used in sound records, electrical insulators, candle compositions, etc. Imports average about 1600 tons per year.

Because of the enormous tonnage of sugar cane processed in Cuba and elsewhere, it is possible to recover an appreciable tonnage of *sugarcane wax* as a by-porduct. The crude wax contains about one-third each of wax, resin, and oil, and hence needs considerable refinement by selective solvents before it can become of value for industrial use.

The refined sugarcane wax is dull yellow in color, melts at 79–81°C and is hard and brittle. It has a durometer hardness of 85–96. It is chemically composed of 78–82% of wax esters, 14% free wax acids, 6–7% free alcohols, and 3–5% hydrocarbons. A proportion of the esters are sterols—sitosterol and stigmasterol—combined with palmitic acid, which are responsible for the good emulsification properties of the wax itself in the preparation of polishes and the like. The proportions of sterols in the refined wax is far less than in the crude wax.

Of waxes obtained from fruits, *japanwax* is the only one of great economic importance, particularly to the Asiatic countries. The wax occurs as a greenish coating on the kernels of the fruit of a small sumac-like tree. Japanwax is actually a vegetable tallow, since it is comprised of 90–91% of glycerides. Peculiarly the glycerides include 3–6.5% (wax basis) of alkyl esters of dicarboxylic acids as well as monocarboxylic acids. The chief dicarboxylic acid is known as japanic acid [$(CH_2)_{19}(COOH)_2$] which is present with lower as well as higher homologs. The dicarboxylic acids have 19 to 23 carbon atoms, whereas the monocarboxylic acids of the simple glycerides present have 16 to 20 carbons.

The textile industries in the past have been large users of japanwax since it is a source of emulsifying softening agents. Other industries using japanwax include those engaged in the manufacture of rubber, soap, polishes, pomades, leather dressings, cordage, etc. Japanwax is a relatively soft but firm wax, which melts at 48.5–54.5°C. About 3000 tons are normally produced per year in China, and twice that amount in Japan.

Other fruit waxes include *bayberry* wax, used in making Christmas candles since the days of the Pilgrims. The wax of rice bran is coming into commercial use, but waxes of the cranberry, apple, grapefruit, etc. are only of academic interest.

Waxes from grasses include bamboo leaf wax, esparto wax, and hemp fiber wax. Esparto wax is a hard, tough wax with a melting point of 73–78°C, and is the most important grass wax. Most of the esparto wax produced is consumed in the British Isles. It is chiefly useful as a substitute for carnauba. Waxes obtained from roots of various species of plants are minute in quantity and of no economic importance.

Mineral Waxes. The fossil waxes are associated with fossil remains which have not been bituminized, that is, converted to hydrocarbons by geological change. A fossil wax, chemically speaking, is composed largely of saponifiables, such as wax acids and esters. Fossil waxes of nearly pure ester composition are occasionally found in fragments of prehistoric plant life, still in a state of preservation as to the original wax constituents. Not far removed from fossil wax of the pure ester composition is *montan wax,* a natural mineral wax which is essentially an ester wax that has undergone partial bituminization. Montan wax is commercially extracted from the nonasphaltic insoluble pyrobitumen with which it is associated, by means of selective solvents such as alcohol and benzene, or by means of benzene alone. Crude montan wax is black and contains about 30% of resins, which is reduced to 10% or less upon refinement. The chemical components of montan wax (deresinified) are alkyl esters of fatty acids (40%), alkyl esters of hydroxy fatty acids (18%), free wax acids (18%), free monohydric alcohols (3%), resins (<12%), and ketones (<10%). There are a number of industrial uses for montan wax: electrical insulation, leather finishes, polishes, carbon papers, shoe pastes, brewer's pitch, etc. The crude wax is also used as a basic material in the manufacture of many synthetic waxes where its montanic acid content is utilized in making derivatives.

The principal source of the montan wax consumed in the United States is imports in annual amounts of some 3.4 million lb from Germany and Czechoslovakia, where it is derived from lignites and brown coals. Also, one California company extracts it from lignite.

Peat wax has somewhat the same composition as montan wax. It has only been produced on a limited scale in Ireland, where it is processed from the native peat. It has asphaltic constituents that tend to make it incompletely miscible with paraffin waxes.

The earth waxes are naturally occurring mineral waxes consisting of hydrocarbons with some oxygenated resinous bodies which can be eliminated by ordinary refining procedures. The earth wax of great economic importance is *ozocerite,* which originally was called ceresin wax. Important sources are the Carpathian mountains in Europe and to a far lesser extent Utah. Chemically speaking, ozocerite has hydrocarbons of a type different from those found in paraffin wax, giving it unique physical properties. The melting point of pure Galician ozocerite is 73°C. It is less soluble in organic solvents than paraffin. When added to paraffin wax in amounts of 15% or thereabouts, it will reduce the paraffin crystals to micro size and improve the tensile strength. Crude imported ozocerite has a dielectric constant of 2.37–2.43, the refined 2.03, and domestic (Utahwax) 2.63. Ozocerite is used in the electrical industry, paste polishes, cosmetics, wax flowers, crayons, etc. Of all the waxes it has the greatest affinity for oil.

Petroleum Waxes. Petroleum is the largest single source of hydrocarbon waxes. The production of petroleum waxes in the United States has risen from 386,181,000 lb in 1916 to 1,347,920,000 lb in 1951 and 1.6 billion lb annually in recent years. About one-third of the petroleum waxes are exported. The annual consumption in recent years has been about 1 billion lb and represents some 95% of the total waxes of all kinds (domestic and imported) consumed. The largest single use of petroleum waxes is in paper coatings which require about 53% of the total. The second largest use is in candles, which consume 140 million lb per year in the United States. The third greatest use in electrical equipment. In contrast, in the early 1950's the largest single use for petroleum waxes was in the manufacture of paper containers for dairy products, and the second was in waxed wrappers for bread. Both outlets are now dominated by plastics.

Crude petroleums differ greatly in both the nature of their hydrocarbons (paraffinic, aromatic, naphthenic, etc.), as well as in their available content of wax. Wax distillates are obtainable with the batch-type, continuous-type, and pipe-still processes, but not from the cracking process. There are, broadly speaking, three principal types of wax encountered in crude oil, namely *paraffin wax, slop wax,* and *petrolatum.* The ordinary procedure in producing slack wax is to pump the paraffin distillate at a temperature of 80–100°F to the paraffin sheds (wax plant), where it is allowed to repose in tanks to promote settling at a temperature between 0 and 32°F. It is then pumped through a bank of cooling units (wax chillers) to hydraulic presses, which squeeze out the wax from the chilled distillate. The product is a soft solid known as *slack wax.* Slack wax finds uses in the industries, but most of it is sweated, pressed, and further refined to produce the various grades of fully refined paraffin wax of commerce. Some of the slack may be "pudged" to the extent that it still contains several per cent of oil; it is then called *scale wax.* Most of the scale waxes produced have a melting point (drop) of 126–130°F (ASTM) and are used in waterproofing thread in the fabrication of cotton duck and canvas, waxing kraft papers, builders' papers, cement bag stock, roofers' felt, car liners, and match splints. Scale wax of very low oil content and higher melting point is used in the manufacture of crayons.

Paraffin wax contains 14 hydrocarbons ranging from $C_{18}H_{38}$ to $C_{32}H_{66}$, solidifying between 27.0 and 68.9°C (80.5 and 156.0°F). *Petrolatum wax,* which is a microcrystalline wax, has hydrocarbons ranging from $C_{34}H_{70}$ to $C_{43}H_{88}$, inclusive. Its solidifying range is 71.0–83.8°C (159.7–182.7°F). *Slop wax* (by-product from the heavy distillate in the coking process) has 13 hydrocarbons ranging from C_{26} to C_{43}, solidifying between 55.7 and 83.3°C (132.2 and 182.0°F). *Rod wax* (collected from the sucker rods in the field) has 8 hydrocarbons ranging from C_{35} to C_{41}, solidifying at 73.9 to 82.5°C (165.0–180.5°F).

Fully refined paraffin wax as regularly offered in the market is graded according to its melting point. There are also special refined grades offered by some refining companies, such as the so-called hard block fully refined paraffins with melting points of 138–140°F, and 143–145°F. The tensile strength of a paraffin wax is greatly influenced by the oil content. Ordinarily a well-refined paraffin of 130°F melting point will have a tensile strength of about 250 psi. The addition of 1 or 2% of an oil-absorbent wax of microcrystalline structure will increase the tensile strength to 350 psi.

Microcrystalline Waxes. Microcrystalline petroleum waxes are characterized not only by microcrystalline structure but by very high average molecular weight, manifested by a much higher viscosity than that of paraffin wax. The chlorophyll present in plants is considered to be a microcrystalline wax.

Microcrystalline waxes are obtained as by-products from (a) the dewaxing of "lube oil raffinates," (b) the deoiling of petrolatum produced from deasphaltic residual oil, or (c) the deasphalting and deoiling of settlings of tanks holding crude oil in the oil field. These types of microcrystalline waxes are sometimes referred to as "motor-oil wax," "residual oil microcrystalline wax," and "tank-bottom microcrystalline wax," respectively. They have also been referred to as "micro wax," "petrolatum wax," and "petroleum ceresin," respectively.

The production of microcrystalline wax in the United States is in excess of 400,000,000 lb per annum. Much of the wax is exported, principally to Europe. The domestic consumption is estimated at 250,000,000 lb per year. The *micro waxes* are graded with 145–150°F, and 160–165°F ASTM melting points, and are refined by selective solvent extraction from the crude wax, a "mobile slurry." A yield of 25–27% of refined wax of the lower melting point is claimed from S.A.E. 20 motor-oil distillate. These waxes are paraffin-like. *Petrolatum waxes* are of a 145–175°F melting point range. The *petroleum ceresins* which are refined from deposits taken from tanks near the wells, called lease tanks, or in the refinery storage, have melting points which range between 165 and 195°F. In the solvent dewaxing processes the solvents for effectively separating the microcrystalline waxes vary with the refinery methods and the character of the feed stock.

A microcrystalline wax derived from petroleum may be defined as a solid hydrocarbon mixture, of average molecular weight range of 490 to 800, considerably higher than that of paraffin wax, which is 350 to 420. The viscosity (SUS at 210°F) of a microcrystalline wax is within the range of 45 to 120 seconds. The lower limit corresponds to 5.75 and the upper limit to 25.1 centistokes at 210°F. The penetration value (ASTM) is of wide variation, namely 3 to 33, although sticky oily laminating waxes are encountered with as high a penetration as 60. Microcrystalline waxes have an occluded oil content which is not easily set free as it is in paraffin waxes. Therefore, a microcrystalline wax which has an oil content of 1 to 4% is virtually a dry wax. A microcrystalline wax which shows a penetration 20 to 30, which is desirable for many needs, will have an oil content of 5.5 to 10.5%.

When a microcrystalline wax is added to melted paraffin it acts like a solute with paraffin as the solvent; the melting point of the blend is greatly elevated, and the crystallization of the paraffin is depressed. The behavior is that of a two-phase system until about 15% of the microcrystalline wax has been added. With the addition of 15 per cent of the petrolatum wax (M.P. 188°F) the melting point of the paraffin is elevated from 130 to 160°F. For many industrial uses microcrystalline wax is admixed with paraffin wax.

The uses of microcrystalline waxes include adhesives, barrel lining, beater size for paper stocks, beer can lining, carbon papers, cheese coatings, cosmetic creams, drinking cups, electrical insulation, floor wax, fruit coating, glass fabric impregnation, heat sealing compounds, laminants for paper, ordnance packing, paper milk bottles, shoe and leather treatments, vegetable coatings, wax emulsions, wax figures and toys, and other miscellaneous purposes.

Synthetic Waxes. These include the following types:

(1) Long-chain polymers of ethylene with OH or other stop-length groupings at end of chain. An example is polyethylene wax of about 2000 molecular weight.

(2) Long-chain polymers of ethylene oxide combined with a dihydric alcohol, namely polyoxyethylene glycol, ("Carbowax").

(3) Chlorinated naphthalenes, ("Halowaxes").

(4) Waxy polyol ether-esters, as for example, polyoxyethylene sorbitol.

(5) Synthetic hydrocarbon waxes prepared by the water-gas synthesis in which carbon monoxide (CO) is reduced by hydrogen (H₂) under pressure, at a predescribed temperature, by means of a catalytic agent. (Fischer-Tropsch waxes "F-T 200" and "F-T 300").

(6) Wax-like ketones, straight-chain and cyclic: (a) Symmetrical ketones produced by the catalytic treatment of the higher fatty acids. (b) Unsymmetrical ketones produced by the Friedel-Crafts' condensation of fatty acids and the like with cyclic hydrocarbons. Examples of the straight-chain ketones are laurone, palmitone, and stearone, and of the cyclic ketones are phenoxyphenyl heptadecyl ketone.

(7) Amide derivatives of fatty acids. The length of the chain may be increased by heating the fatty acid, e.g., stearic acid, with an amino alcohol.

(8) Imide (*N*) condensation products that are wax-like are those of the condensation reaction of one mole of phthalic anhydride with one mole of a primary aliphatic amine to produce a phthalimide. Phthalimide waxes are used in polishes and carbon paper.

(9) Polyoxyethylene fatty acid esters are produced by the reaction of polyethylene glycols with fatty acids. The commercial products are waxy solids which include "Carbowax 4000 (Mono) Stearate." Some of the products act as plasticizers and lubricants for plastics. The polyethylene glycols are soluble in water.

(10) Miscellaneous synthetic waxes (unclassified).

In addition to the above waxes there is a group of synthetic wax-like emulsifiable materials extensively employed in the industries. They are the polyhydric alcohol fatty acid esters, such as ethylene glycol monostearate, glyceryl monostearate, glycerol distearate, and a number of others.

ALBIN H. WARTH

Cross-references: *Hydrocarbons, Petroleum.*

WILLGERODT REACTION

This reaction, discovered in 1887, is conducted by heating a ketone, for example ArCOCH₃, with an aqueous solution of yellow ammonium sulfide (sulfur dissolved in ammonium sulfide), and results in formation of an amide derivative of an arylacetic acid and in some reduction of the ketone. The dark reaction mixture usually is refluxed with

$$ArCOCH_3 + (NH_4)Sx \rightarrow$$

$$ArCH_2CONH_2 + ArCH_2CH_3$$

alkali to effect hydrolysis of the amide, and the arylacetic acid is recovered from the alkaline solution. Although the yields are not high, the process sometimes offers the most satisfactory route to an arylacetic acid, as in the preparation of 1-acenaphthylacetic acid from 1-acetoacenaphthene, a starting material made in 45% yield by acylation of the hydrocarbon with acetic acid and liquid hydrogen fluoride. The product is obtained in better yield and is more easily purified than that from an alternate process consisting in hypochlorite oxidation, conversion to the acid chloride, and Arndt-Eistert reaction.

$$C_{12}H_9COCH_3 \xrightarrow[78\%]{\begin{array}{c}1.\ KOCl\\2.\ SOCl_2\end{array}}$$

1-Acetoacenaphthene

$$C_{12}H_9COCl \xrightarrow{CH_2N_2}$$

$$[C_{12}H_9COCHN_2] \xrightarrow[64\%,\ \text{from acid chloride}]{Ag_2O,\ Na_2S_2O_3}$$

Diazo ketone

$$C_{12}H_9CH_2^\cdot COOH$$

1-Acenaphthylacetic acid

A modification of the Willgerodt reaction that simplifies the procedure by obviating the necessity of a sealed tube or autoclave consists in refluxing the ketone with a high-boiling amine and sulfur (Schwenk, 1942). Morpholine, so named because of a relationship to an early erroneous partial formula suggested for morphine, is suitable and is made technically by dehydration of diethanolamine. The reaction is conducted in the absence of water, and the reaction product is not the amide but the thioamide; this, however, undergoes hydrolysis in the same manner to the arylacetic acid.

L. F. FIESER
MARY FIESER

WITTIG REACTION

This reaction provides an excellent method for the conversion of a carbonyl compound to an olefin, as follows:

$$(C_6H_5)_3 \xrightarrow{CH_3Br}$$

Triphenylphosphine

$$(C_6H_5)_3\overset{+}{P}CH_3(Br^-) \xrightarrow[-C_6H_6-LiBr]{C_6H_5Li}$$

Methyl triphenyl-
phosphonium bromide

$$(C_6H_5)_3P{=}CH_2 \leftrightarrow (C_6H_5)_3\overset{+}{P}{-}\overset{-}{CH_2}$$

Wittig Reagent

$$\begin{array}{c}(C_6H_5)_2C{=}O\\ \downarrow\\ (C_6H_5)_3P\cdots CH_2\\ \vdots\quad\ \ \vdots\\ O\cdots C(C_6H_5)_2\end{array} \longrightarrow$$

$$\begin{array}{cc}(C_6H_5)_3P & + & CH_2\\ \parallel & & \parallel\\ O & & C(C_6H_5)_3\end{array}$$

The reagent is unstable and so is generated in the presence of the carbonyl compound by dehydrohalogenation of the alkyltriphenylphosphonium bromide with phenyllithium in dry ether in a nitrogen atmosphere. There are various modifications such as the phosphonate, in which diethyl benzylphosphonate, cinnamaldehyde and sodium methoxide yield 1,4 diphenylbutadiene.

G. L. CLARK*

* Deceased.

WOLFF-KISHNER REACTION

The Wolff-Kishner method of reduction was discovered independently in Germany (Wolff, 1912) and in Russia (Kishner, 1911). A ketone (or aldehyde) is converted into the hydrazone, and this derivative is heated in a sealed tube or an autoclave with sodium ethoxide in absolute ethanol.

$$\underset{}{\Large>}C{=}O \xrightarrow{H_2NNH_2} \underset{}{\Large>}C{=}NNH_2 \xrightarrow[200°]{NaOC_2H_5,}$$

$$\underset{}{\Large>}CH_2 + N_2$$

After preliminary technical improvements, Huang Minlon (1946) introduced a modified procedure by which the reduction is conducted on a large scale at atmospheric pressure with efficiency and economy. The ketone is refluxed in a high-boiling water-miscible solvent (usually di- or triethylene glycol) with the aqueous hydrazine and sodium hydroxide to form the hydrazone; water is then allowed to distil from the mixture till the temperature rises to a point favorable for decomposition of the hydrazone (200°); and the mixture is refluxed for three or four hours to complete the reduction.

L. F. FIESER
MARY FIESER

Cross-reference: *Clemmensen Reaction.*

WOOD

Wood is a vascular tissue which occurs in all higher plants. The most important commercial sources of wood are the gymnosperms, or softwood trees and the dicotyledonous angiosperms, or hardwood trees. Botanically, wood serves the plant as supporting and conducting tissue, and it also contains certain cells which serve in the storage of food. The trunks and branches of trees and shrubs are composed of wood, except for the very narrow cylinder of pith in the center and the bark which covers the outside. Botanists refer to wood by its Greek name, *xylem*.

A new cylinder of wood is laid down each year around the previously formed wood in the tree. This new growth originates in the cambium, a very narrow growing layer, which elaborates both the wood and the bark, and which separates these two tissues from each other. Each year's growth of wood forms a new concentric ring in the woodystem, as viewed in cross-section. These are termed the annual rings. Each of these has an inner part (toward the pith) which is laid down in the early growing season and is termed the spring wood and an outer layer (towards the bark) which is laid down later and is known as summer wood. The cell walls of the latter are often thicker, forming a denser structure than the spring wood.

Most of the cells of wood are long, narrow hollow fibers and tubular-shaped cells arranged with their long axes parallel to the axis of the tree trunk. Certain food storage cells lie in radial bands, termed wood rays, which are perpendicular to the tree axis. The walls of this complex system of plant cells form the basic framework and material of all wood substance. All wood substance is composed of two basic chemical materials, *lignin*, and a polysaccharidic system, which is termed *holocellulose*. The latter embraces *cellulose* and the *hemicelluloses,* a mixture of pentosans, hexosans and polyuronides, and in some instances small amounts of pectic materials. Wood cell wall tissue also always retains small amounts of mineral matter (ash).

The outer portion of the cell wall, known as the primary wall, is heavily lignified. The intercellular substance, termed the middle lamella, is mainly lignin. The lignin of the middle lamella and primary walls thus serves as a matrix in which the cells are imbedded. Dissolution and removal of the lignin results in separation of the wood fibers. This is the underlying principle in the manufacture of chemical pulps from wood for paper or other cellulose products. (See **Paper.**)

About one-fourth or more of the lignin is in the middle lamella-primary wall complex. The remainder is within the holocellulose system of the cell walls.

Besides the cell wall tissue, which is the basic material of all wood substance, wood contains a variety of materials, many of which may be extracted by selected solvents. These extraneous components lie mainly within the cavities (lumen of the cells and on the surfaces of the cell walls. These "extraparietal substances" include a wide range of chemically different materials, such as essential oils, aliphatic hydrocarbons, fixed oils, resin acids, resinols, tannins, phytosterols, alkaloids, dyes, proteins, water-soluble carbohydrates, cyclitols, and salts of organic acids. The amount and composition of these extraneous substances vary greatly. The occurrence of certain of these substances is often very specific as to genera or species. Generally, however, the total amount of the extraneous components is only a few per cent of the total weight of the wood.

The chemical composition of the extractive-free wood, i.e., the cell wall substance varies less than do the extractives. However, it is by no means constant, there being major differences between hardwoods and softwoods, and often between different genera and even between species. There is even some variability in chemical composition within the same log. For example, there is more lignin in the thinner-walled spring wood than in the summer wood, and more lignin in wood ray cells than in the tracheids or fibers. The heart-wood tissue often contains greater deposits of extraneous (extractive) components than the sapwood. There are major differences between most softwoods (conifers) and most temperate zone hardwoods (broad-leaf trees). Usually the softwoods have greater amounts of extraneous components extractable by organic solvents. Generally the lignin content of softwoods is higher than that of hardwoods, *viz.,* in the order of 25–30% compared to 17–24%. Also there is a major difference in the pentosan content between these two groups of woods, the hardwoods containing usually about 17–22% and the softwoods about 8–14%. There are exceptions to these generalizations, however.

The components of the cell wall substance of wood are exceedingly difficult to separate. Separations are rarely complete and generally bring about drastic chemical changes, especially in the lignin and molecular size degradation of the polysaccharides. To a considerable extent the components are apparently interpenetrating polymer systems. The long-chain linear polysaccharides tend to be parallel to the fiber axis and to form areas of varying degrees of crystallinity, as shown by x-rays and other physical and chemical properties. The lignin appears to be an amorphous tridimensional polymer.

Wood forms one of the world's most important chemical raw materials. It is the primary source of cellulose for the pulp and paper and cellulose industries. These industries are well up in the group of 10 major industries of the United States. For paper, rayon, films, lacquers, explosives and plastics, which comprise the greatest chemical uses of wood, it is the cellulose component (plus certain amounts of hemicellulose) of wood that is of value. The lignin forms a major industrial waste as a by-product of the paper and cellulose industries. Its major use is in its heat value in the recovery of alkaline pulping chemicals. A variety of minor uses for lignin have been developed, such as for the manufacture of vanillin, adhesives, plastics, oil-well drilling compounds and fillers for rubber.

Wood wastes from the lumber and woodworking industries form a great potential source of sugars and alcohol by acid hydrolysis of its polysaccharides followed by fermentation. Wood hydrolysis processes, however, are not yet economically competitive with other sources of sugars and alcohol in this country and many other areas of the world.

Wood is also an industrial source of charcoal, tannin, rosin, turpentine, and various other essential oils and pharmaceutical products.

Wood is an abundant, renewable and relatively cheap raw material. Its place in chemical industry is most likely to continue to grow.

EDWIN C. JAHN

Cross-references: *Cellulose, Hemicellulose, Lignin.*

Wood Preservatives

Wood preservative is the generic name given to a material applied to wood to prevent its destruction by fungi, wood-boring insects, marine borers and fire. A common characteristic of these materials is toxicity to those organisms that attack wood, or in the case of fire retardants the ability to control combustion in terms defined by the Underwriters Laboratory. In addition, a satisfactory wood preservative must also (a) be capable of penetrating wood, (b) remain in the wood for extended periods without losing its effectiveness due to chemical breakdown, (c) be harmless to humans and animals, (d) be noncorrosive and (e) be available in quantity at a reasonable cost. For certain uses, the preservative may be required to be colorless, odorless, nonswelling and paintable.

The principal wood preservatives in use today are classified as (a) preservative oils, (b) toxic chemicals in organic solvents, and (c) water-soluble salts.

The most important of the preservative oils is coal-tar creosote and its solutions in the form of creosote-coal tar and creosote petroleum. Coal-tar creosote is defined by the American Wood Preservers Association as—"A distillate of coal tar produced by high-temperature carbonization of bituminous coal; it consists principally of liquid and solid aromatic hydrocarbons and contains appreciable quantities of tar acids and tar bases; it is heavier than water and has a continuous boiling range of at least 125°C, beginning at about 200°C." This material is one of the oldest wood preservatives and is regarded by many as the best substance known for protection against all forms of wood-destroying organisms. For normal service conditions, minimum retentions of 8–10 lbs. per cubic foot of wood penetrated are sufficient. For extreme conditions, retentions as high as 35 lbs. per cubic foot are desirable. Currently, coal-tar creosote and creosote solutions are used as preservatives for about 60% of all wood products treated.

Of the large number of toxic chemicals that are oil-soluble, only three are recognized by the American Wood Preservers Association and only one of these is a commercially important wood preservative, pentachlorophenol (C_6Cl_5OH). Although "penta" is effective against fungi and insects, it will not protect against marine borers, and hence cannot be used as a wood preservative for salt water installations. Penta was used to treat approximately 24% of the wood treated in 1969.

The two most common solvents for carrying penta into the wood are a relatively high-boiling No. 2 fuel oil and a very low-boiling liquified petroleum gas, butane. When fuel oil is used it remains in the wood and although the end product is brighter and cleaner than creosote-treated wood, it has a somewhat oily character and cannot readily be painted. When LP gas is used as the solvent, the butane is recovered and the penta is deposited in the wood as a dry crystalline material, hence the wood retains its color and is readily paintable. The latter treatment is patented and licensed by Koppers Company, Inc.

Water-borne preservatives are divided into two categories. One group which includes acid copper chromate, chromated zinc chloride, copperized chromated zinc arsenate and fluor-chrome-arsenate-phenol is used where the wood is not subjected to excessive leaching. The second group, ammoniacal copper arsenite and three types of chromated copper arsenate which react to become practically water insoluble, are used at about 0.6 pound per cubic foot when wood is placed in ground contact under severe service conditions.

Water-borne preservatives penetrate wood easily and the solution presents no problem in flammability or health hazards. Disadvantages of water soluble preservatives include the swelling and shrinking of the treated material, reduction in bending strength and stiffness as a result of failure to redry following treatment, and less protection against weathering and mechanical wear than provided by either preservative oils or oil soluble chemicals in which the solvent remains in the wood.

In addition to the general preservative categories discussed there is a fourth group known as "proprietary preservatives" which are composed of various combinations of toxic materials and solvents. These are sold under trade names and in some cases are protected by patents. An example of a widely used proprietary is Wolman salts (Tanalith) which is a combination of sodium fluoride, disodium arsenate, sodium chromate and dinitrophenol. It is not practical to prepare a comprehensive list of "proprietary preservatives" indicating their composition. Many of these materials have proved effectiveness; however, before large-scale purchases are undertaken The American Wood Preservers Institute, 2600 Virginia Avenue, Washington, D.C. 20037, should be consulted for advice as to the effectiveness of the chosen preservatives.

While wood preservatives can be applied by simple means such as brushing, spraying and cold soaking, well over 90% of all commercial wood treatment is by one of the "pressure processes." Although the details of the individual processes vary, the type of equipment and general procedure used are similar. Treatment takes place in closed cylinders 6 to 9 feet in diameter and up to 180 feet in length. The wood to be treated is placed in the retort, submerged in the preservative and subjected to pressures in the order of 200 psi. The preservative is often at an elevated temperature and in some cases the wood is given a preliminary pressure or vacuum period prior to admitting the preservative into the retort.

The most satisfactory nonpressure process is known as "thermal" treatment. The wood to be treated is immersed in a preservative at an elevated temperature. This causes the air in the wood cells to expand so that when the wood is transferred into a preservative bath of lower temperature, the air contracts forming a partial vacuum and atmospheric pressure forces the liquid into the wood.

Although the fire retardant treatment of wood is essentially the injection of water soluble salts into the wood by pressure treating methods, it deserves separate menton if for no other reason than the higher salt retention (4 lbs/cu. ft. vs. 0.6 lbs/cu. ft.) required. Most fire retardant treatments are proprietary and are specified by the Underwriters Laboratory on the basis of flame spread ratings. The latter's list of building materials will give details of the ratings including those of a recently introduced leach-resistant fire retardant suitable for outdoor service. Common fire-retardant chemicals include diammonium phosphate, ammonium sulfate, sodium tetraborate and boric acid which are used in various combinations. Two widely used proprietary fire retardants are trade-named Pyresote and Non Com.

Over 250 million cubic feet of wood have been treated in each year 1965–1969. This application of preservatives and fire retardants has greatly extended the usefulness of wood since it adds the assurance of a long service life to the other inherent advantage which wood possesses as an engineering material.

WILLIAM T. NEARN

WURTZ REACTION

A method of synthesizing hydrocarbons discovered by Wurtz (1855) consists in treatment of an alkyl halide with metallic sodium, which has a strong affinity for bound halogen and acts on methyl iodide in such a way as to strip iodine from the molecule and produce sodium iodide. The reaction involves two molecules of methyl iodide and two atoms of sodium:

$$H-\underset{\underset{H}{|}}{\overset{\overset{H}{|}}{C}}\!+\!I + 2Na + I\!+\!\underset{\underset{H}{|}}{\overset{\overset{H}{|}}{C}}-H \rightarrow$$

$$H-\underset{\underset{H}{|}}{\overset{\overset{H}{|}}{C}}-\underset{\underset{H}{|}}{\overset{\overset{H}{|}}{C}}-H + 2NaI$$

Actually, the reaction probably proceeds through the formation of methylsodium, which interacts with methyl iodide:

$$CH_3I \xrightarrow{\ Na\ } CH_3Na \xrightarrow{\ CH_3I\ } CH_3CH_3$$

The Wurtz reaction can be applied generally to synthesis of hydrocarbons by the joining together of hydrocarbon residues of two molecules of an alkyl halide (usually the bromide or iodide). With halides of high molecular weight the yields are often good, and the reaction has been serviceable in the synthesis of higher hydrocarbons starting with alcohols found in nature, for example:

$$2C_{20}H_{41}Br \xrightarrow[\text{31\% yield}]{Na} C_{40}H_{82}$$
Dihydrophytyl bromide 31% yield Perhydrolycopene

$$2n\text{-}C_{16}H_{33}I \xrightarrow[\text{70–80\% yield}]{Mg\ (ether)} C_{32}H_{66}$$
Cetyl iodide 70–80% yield n-Dotriacontane

Cetyl iodide (from the alcohol of spermaceti wax) has been converted into the C_{32}-hydrocarbon both by the action of sodium amalgam in alcohol-ether and, as shown in the equation, with use of magnesium in place of sodium.

A general expression for the Wurtz synthesis is:

$$2RX + 2Na \rightarrow R \cdot R + 2NaX$$
Alkyl halide Alkane

It might appear that the synthesis could be varied by use of two different alkyl halides, with the linking together of the two hydrocarbon fragments, for example:

$$CH_3CH_2CH_2CH_2I + ICH_2CH_2CH_3 \xrightarrow{2Na}$$
n-Butyl iodide n-Propyl iodide

n-Heptane
$$CH_3CH_2CH_2CH_2CH_2CH_2CH_3$$

The reaction mixture, however, contains many millions of molecules of each halide, and there is nearly as much opportunity for interaction of like as of unlike molecules. Some butyl iodide molecules will react with molecules of propyl iodide and yield heptane as pictured, but some will combine with other molecules of the same kind and produce octane. The total result can be represented as follows:

$$CH_3CH_2CH_2CH_2CH_2I +$$

$$CH_3CH_2CH_2I \xrightarrow{2Na} \begin{cases} n\text{-}C_6H_{14}, \text{ b.p. } 69° \\ n\text{-}C_7H_{16}, \text{ b.p. } 98° \\ n\text{-}C_8H_{18}, \text{ b.p. } 126° \end{cases}$$

The reaction affords a mixture of which the unsymmetrical product n-heptane can be expected to constitute no more than one-half, and since the three hydrocarbon components are similar and do not differ greatly in boiling pont, isolation of even a small amount of n-heptane in a moderately homogeneous condition would obviously be difficult. It is therefore impracticable to utilize an unsymmetrical Wurtz reaction in synthesis, for the inevitable result is:

$$RX + R'X \rightarrow RR' + RR + R'R'$$

L. F. FIESER
MARY FIESER

X

XENON

The element xenon (Xe), atomic number 54 and atomic weight 131.3, is the next to heaviest of the monatomic noble gases, helium, neon, argon, krypton, xenon, and radon, which comprise Group 0 of the Periodic Table. Nine stable isotopes of mass numbers 124, 126, 128, 129, 130, 131, 132, 134, and 136 exist and fifteen radioactive species with mass numbers from 121 to 144 have been produced artificially by the fission of uranium and by other nuclear reactions.

This extremely rare element was first discovered by Sir William Ramsey and M. W. Travers in 1898 as one of the constituents of air. The xenon content of dry air is about 0.000008% by volume and slight traces of the gas are found in natural gases in the earth. Small quantities of the gas are produced commercially by fractional distillation of air.

Xenon has a normal boiling point at 1 atmosphere of $-108.12°C$; melting point at 1 atmosphere $-111.8°C$; critical temperature 16.6°C; critical pressure 57.64 atm; critical density 1.100 gm/cc; latent heat of vaporization at the normal boiling point 23.0 cal/gm; first electron ionization potential 12.1 volts. It is most easily identified by its characteristic spectrum. The crystal structure of the solid is face-centered cubic with the length of a cube edge = 6.24×10^{-8} cm at 88°C.

The 54 electrons of the xenon atom form a stable configuration of closed shells with 2, 8, 18, 18 and 8 electrons, respectively. As a consequence the xenon atom attracts other atoms with comparatively weak interatomic forces and like other members of the noble gas family was long classified as chemically inert. A few special crystalline compounds held together by weak van der Waal's attractive forces are found to exist under a limited range of conditions—for example $Xe \cdot nH_2O$ (n about 5 or 6) and $Xe \cdot 2C_6H_5OH$ (see **Clathrate Compounds**). For a wide range of new compounds see **Noble Gas Compounds**.

Because of its expense xenon has much more limited use than the other inert gases. It is used to a limited extent in gaseous discharge tubes for lighting purposes, in negative glow lamps, and in thyrotrons and rectifiers in place of mercury in applications where low-temperature operation is required. It has proved useful in gas discharge lamps where extremely short flashes of light are required for high speed photography.

HENRY A. FAIRBANK

References

Selig, Henry, "Xenon," pp. 796–804 in "Encyclopedia of the Chemical Elements," C. A. Hampel, Editor, Van Nostrand Reinhold Co., New York, 1968.

Hyman, H. H., Editor, "Noble Gas Compounds," University of Chicago Press, Chicago, 1963.

Malm, J. G., Selig, H., Jortner, J., and Rice, S. A., "The Chemistry of Xenon," *Chem. Rev.,* **65,** 199 (1965).

XEROGRAPHY

Xerography is the name given to a relatively new process of photography which is based on photoconductivity and electrostatic phenomena. The mechanism of image formation in xerography is photoelectrical, whereas in conventional photography it is photochemical. The xerographic plate consists of a thin photoconductive insulating layer on a conductive base. The plate is sensitized by electrical charging, and optical images are transformed into electrical images, during exposure, by photoconduction within the sensitized medium. Image development and fixing are essentially dry, since no aqueous solutions are used.

Xerography, formerly known as "electrophotography," was invented by Chester F. Carlson.* The process was developed and adapted to commercial applications through the joint research efforts of the Haloid Company** of Rochester, New York, and Battelle Memorial Institute of Columbus, Ohio. It is now used extensively in photocopying, duplicating, and the enlarging and printing of microfilm. Other applications are rapidly coming into use. Xerography is inherently fast. Processing, from exposure to finished print, can be accomplished in less than one second. Xerographic prints can be made on almost any kind of paper, and on other surfaces such as wood, metal, glass, plastics, cloth, ceramics, etc.

The sequence of process steps in xerography is as follows:

(1) The surface of the xerographic plate is sensitized by electrical charging.

(2) In the sensitized condition electrical charges are uniformly distributed over the plate surface.

(3) The plate is exposed by projecting an image onto the plate. Electrical charges are conducted away from the areas exposed to light, leaving an electrostatic image.

(4) The image is developed by applying an electrically-charged powder to the plate surface. This powder must be charged to a polarity opposite to that of the electrostatic image. The powder adheres only to the image areas.

---

* Deceased.
** Now the Xerox Corp.

(5) A sheet of paper is placed over the developed image, and an electrical charge of the same polarity as that of the electrostatic image is applied to the paper.

(6) The charged paper attracts the powder image from plate to paper.

(7) The paper is removed from the plate and the print is heated to fix the powder image.

The Photosensitive Surface. Xerographic plates are prepared by depositing a smooth layer of a photoconductive insulating material on a conductive base by vacuum evaporation, or by applying a lacquer containing the photoconductive material. The film is very thin (about 0.001″ to .002″). In the dark the film must be a very good electrical insulator so that, after sensitizing by electrical charging, it will retain electrical charges on its surface. When the surface of the photoconductive insulator is illuminated, the film becomes conductive, and electrical charges are transported through the coating to the base plate. Thus, the xerographic plate becomes discharged in proportion to the amount of light impinging on the plate surface. When properly handled, the xerographic plate is not damaged by repeated sensitization and exposure.

The mechanism of electrical image formation in xerographic plates is reversible, whereas image formation in silver emulsions is essentially irreversible. Thus, the xerographic plate can be re-used many times.

Amorphous selenium and zinc oxide are the most commonly used photoconductive materials for xerographic plates, although sensitized organic polymers have been introduced recently, and selenium alloyed with such elements as tellurium, arsenic and antimony have been proposed for increased photosensitivity. In a modified form of the process, paper coated with zinc oxide dispersed in a resin binder is used. The electrical image in this case is produced, developed, and fixed directly on the zinc oxide coating, and the coated paper sheet becomes the finished print.

Sensitizing the Plate. Making a print by xerography starts with sensitizing the xerographic plate. This is usually done by placing the plate in a corona charging device where corona emission from parallel wires sprays electrical charges uniformly over the photoconductive film. This operation must be done in the dark, and from this point on through development of the image on the plate, the plate must be shielded from light. Other methods of sensitizing involve induction charging, α-particle ionization, and conductive roller charging.

Xerographic plates can be sensitized with either positive or negative charges. The polarity of charging used depends on the particular characteristics of the photoconductive coating; e.g., amorphous selenium performs better when sensitized with a positive polarity, whereas zinc oxide-resin coatings require a negative polarity. The magnitude of the electrical charge on the plate after sensitization is measured by the potential to which the surface of the plate is raised. Usually this potential is in the range of 200 to 600 volts.

Xerographic Developers. A variety of materials can be used to develop xerographic images. However, such materials must be formulated to provide specific properties depending on the developing and fixing techniques employed and the requirements established for the xerographic prints. In general, a developer material must be pulverizable or dispersable into fine particles, must be capable of accepting and retaining electrical charges, and should have no adverse effects on the xerographic plate. The average particle size of developing powders for general copying work should be about 10 microns. For high-resolution imaging, particle size should be 0.1 micron or less.

Developing powders, or toners, for cascade, and magnetic brush development are usually made of resins or resin blends in which a pigment, such as carbon black, has been dispersed. The resins are selected to provide the proper triboelectric relationship with the brush fibers or carrier beads, a melting point within the proper range for heat fixing (105–150°C), or solubility in the solvents used for vapor fixing.

Liquid electrophoretic toners are now used frequently to develop electrostatic images on zinc oxide-coated paper. These developers consist of fine pigment particles dispersed in dielectric liquids with control agents added to charge the particles to the required polarity and to fix them to the paper surface.

A wide variety of colored pigments and dyes can be used in xerographic developers, and the making of color prints using xerographic techniques has been demonstrated.

Applications to Copying and Duplicating. Numerous copying machines utilizing xerographic principles are now in commercial use. These are of two types: (1) Machines employing reusable photoconductive films, such as amorphous selenium and sensitized organic polymers, in which the developed image is transferred to plain paper, and (2) Machines using sensitized zinc oxide coated paper in which the image is developed and fixed on the coated paper.

Automation of process steps has resulted in machines that produce copies at high rates of speed. Machines capable of speeds up to 60 copies per minute are currently in use.

Applications to X-rays. Some insulating materials are photoconductive in the x-ray and gamma ray regions. Amorphous selenium xerographic plates exhibit this property. This is the basis of a process known as "Xeroradiography." The steps involved are analogous to those of xerography, although the techniques are somewhat different.

The xeroradiographic plate is sensitized by corona charging, exposed to x-rays by conventional radiographic techniques, and developed in a special type of powder-cloud chamber. The x-ray image can be viewed directly on the plate or it can be transferred to adhesive-coated paper to provide a permanent record.

R. M. Schaffert

Cross-references: *Photography, Developers.*

References

"*Electrophotography,*" R. M. Schaffert, Focal Press (London, New York) 1965.

"*Xerography and Related Processes,*" Ed. J. H. Dessauer, H. E. Clark, Focal Press (London, New York) 1965.

X-RAYS

X-rays are electromagnetic radiation of very short wave length, generally of the order of 10^{-8} cm. They were discovered in late 1895 by the German physicist, W. C. Röntgen, and hence are frequently referred to as Röntgen rays. In the course of investigating the luminescence produced by cathode rays Röntgen enclosed a discharge tube in a box made of thin black cardboard. When the tube was operated in a darkened room Röntgen observed that a nearby sheet of paper, coated on one side with barium platinocyanide, fluoresced brilliantly. After proving that whatever was responsible for this phenomenon originated in the discharge tube, he concluded that it was some form of penetrating rays, which he called x-rays.

In the early part of 1896 investigators in several countries observed that when x-rays were passed through gases, ions were produced and the gases acquired the ability to conduct electricity. However, for several years after their discovery there was no clear understanding of the nature of x-rays, and it was not until 1912 that definite evidence was obtained that the rays were an electromagnetic radiation analogous to light but of shorter wave length. In that year, following a suggestion of von Laue, it was demonstrated by W. Friedrich that on passing a beam of x-rays through a crystal of copper sulfate a definite diffraction pattern was obtained. Soon thereafter it was shown by the Braggs, English physicists, that the scattering of x-rays could be represented as a "reflection" by successive planes of atoms in the crystal. They derived an equation to describe the conditions for obtaining reflection maxima. The Bragg equation may be written $n \lambda = 2 d \sin \theta$, where λ is the wave length of the x-rays; d is the distance between successive reflecting planes in the crystal; θ is the glancing angle; and n is the order of the reflection, specified by the values 1, 2, 3. . . .

In Röntgen's experiments the x-rays were produced by cathode rays striking the walls of the discharge tube. However, better results were obtained by allowing the cathode rays to fall on a piece of metal, called an anticathode, placed in their path; the x-rays were then emitted from the anticathode. In 1911, in England, C. G. Barkla observed that when different metals were used as anticathodes there was a broad continuous emission (so-called white x-radiation) and superimposed upon this were several series of characteristic x-ray lines peculiar to each element.

In 1913 H. G. J. Moseley made systematic measurements of the frequencies ν of the characteristic x-ray lines. He then calculated a quantity, to which he gave the symbol Q, that was related to the square root of the frequency of the characteristic x-rays for each element. Upon examining the results Moseley noticed that Q increased by a constant amount in passing from one element to the next in the Periodic Table. He concluded that there is in the atom a fundamental quantity which increases by regular steps in passing from one element to the next and that this quantity was the charge on the atomic nucleus, now called the atomic number Z. Hence he was able to express the relationship between the frequencies of the characteristic x-rays and the atomic number by $\nu^{1/2} = a (Z - b)$, where for each series of x-ray lines a and b are constant for all the elements. When this relationship was plotted for the x-ray lines of the elements, discontinuities appeared in the plot wherever there were missing elements in the Periodic Table. These vacant spaces have since been filled. This work provided the final convincing evidence that the atomic number and not the atomic weight governs the observed periodicity of the chemical properties of the elements.

In the early 1920's Arthur Compton found the necessary conditions for reflecting x-rays from mirrors, diffracting from ruled optical gratings, and refracting in prisms. Hence x-rays, like light, are transverse electromagnetic vibrations. In the broadest sense, including the photoradiation resulting from the stopping of high-energy charged particles in matter, they occupy a broad range in the electromagnetic spectrum, from 10^3 to 10^{-4} Å or shorter.

The fact that x-rays may be diffracted finds wide use in the determination of the structures of crystals. In the first application of this technique Bragg, using single crystals, determined the structures of sodium and potassium chlorides. However, the simplest technique for obtaining x-ray diffraction data is the powder method devised independently by A. W. Hull and by P. Debye and P. Sherrer. Instead of a single crystal with a definite orientation to the x-ray beam, a mass of finely divided crystals with random orientations is used. Out of these random orientations there will be some at the proper angle for x-ray reflection from each set of planes in the crystals. The reflected x-ray beams are then recorded photographically. On a flat film plate the observed pattern consists of a series of concentric circles. In order to index the crystal the observed spacings between lines are compared with those predicted on the basis of theoretical calculations.

For precise structure investigations the rotating-single-crystal method, developed by E. Schiebold, is widely used. A crystal, exposed to an x-ray beam, is rotated slowly. During the course of the exposure successive planes of the crystal pass through the orientation necessary for Bragg reflection, each producing a spot on film or a record on a suitable electronic radiation detector. The spots or records are indexed and their intensities measured. The reconstruction of the crystal structure from the intensities of the various x-ray diffraction maxima is accomplished by Fourier synthesis. In this process the central problem is the deduction of the phases of terms. For this purpose there are various methods available.

X-rays enter prominently into nuclear processes. One example is the x-ray emission that follows the capture of an orbital electron, usually from the lowest or K-level, by the nucleus of an atom. When such capture occurs an electron from one of the higher energy levels will immediately move in to fill the vacant position, and the excess energy will be emitted as the corresponding characteristic x-ray. Since the orbital-electron capture must precede the electronic transition and the emission of x-rays, the x-rays emitted will be characteristic of the product nucleus with an atomic number one unit less than the original nucleus. For example when ^{49}V decays by K-capture the x-rays emitted are characteristic of titanium.

Nuclear levels in atoms may also be excited by x-rays. This may result in a nucleus being excited to a higher, nonmetastable level, from which it returns, not to the ground level, but to a metastable one. Usually the nucleus cannot move up directly very readily to the metastable level since the transition is forbidden.

With the development of solid state detectors it is now possible to use x-rays in such advanced analytical applications as the electron microprobe and the scanning electron microscope. In the electron microprobe a thin collimated beam of electrons is allowed to strike the specimen. Identification of the target material is made by observing the characteristic x-rays produced. The advantage of this technique is that extremely small specimens may be examined. In the scanning electron microscope the image is produced by very low energy secondary electrons which are emitted when an electron beam strikes the specimen. The characteristic x-rays produced may be used for identification purposes. In industry advantage is taken of the great penetrating power of high-energy x-rays in such applications as the examination of castings for defects. In addition to analytical applications x-rays are used today in radiolysis and in medicine. In the latter field x-rays are used in radiographic examinations and in x-ray therapy, such as in the treatment of cancer and other neoplasms.

M. H. LIETZKE AND R. W. STOUGHTON

References

Glasstone, Samuel, "Source Book on Atomic Energy," D. Van Nostrand Co., New York, 1950.

Moore, W. J., "Physical Chemistry," 3rd Ed., Prentice-Hall, Inc., Englewood Cliffs, N.J., 1962.

Y

YTTRIUM

Yttrium, element number 39 in the Periodic Table, was discovered in 1794 by the Finnish chemist Gadolin, and is named after the small town in Sweden, Ytterby; the elements terbium, erbium and ytterbium are also named after this town.[1] Because of the "lanthanide contraction" in the rare earth group, yttrium always occurs as a constituent of rare earth ores, and is usually included with these elements in any consideration of elements in the Periodic Table. Yttrium is thus chemically much closer to the rare earths than to scandium, another member of Group IIIB of the Periodic Table.[2] Recent interest in yttrium as a component of moderators for nuclear reactors (a consequence of the low neutron cross section of yttrium) occasioned a large increase in studies dealing with the chemistry of this element, and as a consequence the scientific literature shows much more information on yttrium than would normally have been the case.[3]

Yttrium is present in the earth's crust to the extent of 40 ppm, compared to 46 ppm for cerium, 18 ppm for lanthanum, and 10 ppm or less for most of the other rare earths; copper is present in the amount of 45 ppm, but this is quite misleading, as the copper ores are very concentrated, while the yttrium ores are generally quite dilute. In these ores, yttrium is present as a phosphate (xenotime, monazite), silicate (gadolinite) or other forms, and is found in Norway, although more recent sources have been found in India, Brazil, Canada and the states of Idaho, Nevada, California and South Carolina in the U.S. In processing the ores, an electromagnetic concentration is followed by a high-temperature sulfuric acid digestion and aqueous leach. The resulting solution is loaded on ion-exchange resin beds which are eluted with an EDTA (ethylenediaminetetraacetic acid) solution that is buffered with ammonia and also contains copper ion which serves to prevent precipitation of the insoluble EDTA in the resin bed.[4] The rare earths come off in inverse order of their atomic numbers; yttrium comes between holmium and dysprosium, and would appear to have an atomic number of 66.5 on a plot of elution order versus atomic number. On most plots of a particular physical or chemical property of the rare earth elements versus atomic number, yttrium falls somewhere between gadolinium and erbium, testifying to the rare earth-like nature of yttrium.

In aqueous solution, yttrium loses the three conduction electrons of the metallic form ($5s^2, 4d^1$) to give the colorless ion. Insoluble salts of yttrium include the oxalate, hydroxide, fluoride and phosphate, although the oxalate forms the most filterable precipitate and is used in most gravimetric analyses in which the oxalate is ignited to form the oxide. Titrimetric and spectrophotometric methods of analysis have been used for yttrium, but gravimetric procedures have found widest usage.[5]

Yttrium oxide is a white refractory powder that dissolves in mineral acids. Yttrium fluoride is a white powder (colorless crystal) that is relatively inert in room air, but the other yttrium halides are extremely hygroscopic and hydrolyze in room air to form the oxyhalides or the solutions of the ions. Yttrium salts with the common anions are colorless.

Yttrium metal may be prepared by first converting the oxide to the fluoride with anhydrous hydrogen fluoride. Calcium or magnesium metal is used to reduce the fluoride to the metal, and because of the large amount of yttrium metal needed for nuclear reactor moderator applications, several alternate procedures for scaling up processes for preparing the metal have been devised.[3] These processes were utilized to prepare over 14 tons of yttrium metal of greater than 99.6% purity as part of the U.S. Atomic Energy Commission nuclear reactor program.

Yttrium is finding much usage in the color television industry where Y_2O_3-Eu and YVO_4-Eu phosphors provide the vivid red color that allows intense colors that have become popular in the late 1960's. Prior to this time, the weak red phosphors that were available made it necessary to tone down the other phosphors available to give a balanced color picture, and the weaker picture was not accepted by the public. The YVO_4-Eu phosphor is prepared by adding alkali to a solution of YCl_3, $EuCl_3$ and NH_4VO_3 which produces a mixture of the hydrous oxides in which the Y/Eu ratio is 19/1. This precipitate is washed and then ignited to 1200°C to give the desired phosphor, which has the approximate composition of $Y_3Fe_5O_{12}$.[6]

Yttrium oxide is also used much in the preparation of yttrium-iron garnets that are used as microwave filters. The superior performance of this material permits the pickup of lower power radiation and the use of smaller radar collectors due to their noise elimination action.

Yttrium metal has been found to form some useful alloys. Small amounts of yttrium improve the corrosion resistance of chromium, niobium, tantalum and molybdenum, probably by the formation of an adherent coating of a mixed oxide whose thermal expansion matches that of the substrate metal well enough to prevent flaking-off of the coating.[7] Yttrium added to iron and iron-base alloys spheroidizes the graphite and improves the workability, high-temperature recrystallization resistance and grain refinement qualities of the al-

TABLE 1. PHYSICAL PROPERTIES OF YTTRIUM

Atomic number	39
Atomic weight	88.905 (100% isotope 89)
Melting point, °C	1509
Boiling point, °C	3200
Density, g/cc	4.472
Crystal structure	
Room temperature to 1490°C	Hexagonal close-packed
	a = 3.6457 Å, c = 5.7305 Å
1490–1509°C	Body-centered cubic, a = 3.90 Å
Atomic volume, cc/mole	19.86
Metallic radius, Å	1.802
Heat capacity, 25°C, cal/mole/°C	6.50
Heat of fusion, kcal/mole	4.1
Heat of vaporization, 25°C, kcal/mole	93
Thermal conductivity, cal/sec/cm/°C	0.0240
Linear coefficient of expansion, per °C	10.8×10^{-6}
Thermal neutron absorption cross section, barns/atom	1.31 ± 0.08
Electrical resistivity, microhm-cm 25°C	65

loys. Yttrium dissolves in magnesium to make an alloy of some potential interest, although it does not seem to be singularly improved over the cheaper alloys of misch-metal with magnesium.[8] It is interesting that yttrium forms a complete solid solution with thorium, but is immiscible with uranium —completely opposite behavior with two metals of direct interest in nuclear technology.[9]

A significant amount of study of the toxicity of yttrium has revealed no hazards associated with the normal handling of yttrium-containing materials.

Table 1 above gives some of the physical properties of yttrium.

A. H. DAANE

References

1. Weeks, M. E., "The Discovery of The Elements," Easton, Pa., The Journal of Chemical Education, 1960.
2. Goldschmidt, V. M., "Geochemistry," Oxford, Clarendon Press, 1954.
3. Spedding, F. H., and Daane, A. H., "The Rare Earths," New York, John Wiley and Sons, Inc., 1961.
4. Powell, J. E., Ch. 5, *ibid.*
5. Banks, C. V., and Klingman, D. W., Ch. 23, *ibid.*
6. Woyski, M. W., and Silvernail, W. J., Ch. 21, *ibid.*
7. Collins, J. F., Calkins, V. P., and McGurty, J. A., Ch. 20, *ibid.*
8. Leontis, T. E., Ch. 19, *ibid.*
9. Gschneidner, K. A., and Waber, J. T., Ch. 17, *ibid.*
10. Daane, A. H., "Yttrium," pp. 810–821, in "Encyclopedia of the Chemical Elements," C. A. Hampel, Editor, Reinhold Publishing Corp., New York, 1968.

Z

ZINC AND COMPOUNDS

Zinc has been used as a constituent of brass and bronze for more than 2,000 years. It is probable that zinc was first recognized as a distinct metal by metal workers in India and China prior to 1000 A.D. but it received no attention as a metal in Europe until about 1500 A.D. It should be noted that in our highly industrialized civilization here in the United States, zinc ranks fourth in tonnage of metal produced, following steel, aluminum and copper in that order.

Occurrence. Zinc occurs to the extent of about 120 g per ton of the earth's crust. It is usually found in nature as the sulfide, ZnS, known as sphalerite or zinc blende; it is often associated with sulfides of other metals, especially lead, cadmium, copper and iron. This association is particularly intimate in many of the deposits and explains why their development was delayed. Most other zinc minerals have probably been formed as oxidation products of the sulfides; however, they represent only minor sources of zinc.

Production. About five and one-half million tons of recoverable zinc were mined throughout the world in 1970. Annual United States mine production has remained at about 550,000 to 600,000 tons for the past decade (547,000 tons in 1970). Canada has the largest mine production, 1,211,000 tons recoverable zinc in 1970, with Russia's production estimated at about 615,000 tons. Other large producers are Australia, Peru, Japan, and Mexico.

Production of slab zinc does not follow mine production due to movements of ores and concentrates from one country to another. As a result, in this area of production the United States ranks first (1970 figures) with 887,000 tons, Japan second with 745,000 tons, Russia third with 615,000 tons, followed in order by Canada, West Germany, Belgium, France and Poland. It should be noted that considerable zinc metal is imported into the United States, 271,000 tons in 1970, chiefly from Canada.

Interestingly in the United States, Tennessee has the largest mine production of recoverable zinc (120,000 tons), followed by New York (58,000 tons), Colorado (52,000 tons), and Missouri (also 52,000 tons). It should also be remarked that the United States in 1970 produced 74,000 tons of secondary distilled slab zinc for a grand total of 961,000 tons of slab metal for the year.

Consumption. The United States consumed 1,-165,000 tons of slab zinc in 1970, a drop from the peak consumption of 1,410,000 tons in 1966. This decrease reflects the decline in general industrial activity, particularly in die castings (zinc's major use) for automobiles, in the manufacture of appliances and in construction work. The average price of zinc (St. Louis) is about 15¢/lb.

Concentration. In concentrating the zinc ore, it is first crushed to appropriate size and the sulfide mineral is separated from the gangue rock by gravity concentration, differential density (sink-float), differential flotation, or combinations of these methods depending on the character of the ore.

Reduction to Metal. The zinc sulfide concentrates thus obtained are roasted to a crude oxide or calcine; frequently the sulfur dioxide gases from the roasting are collected and processed for the manufacture of sulfuric acid. The reduction of the roasted concentrates or calcine to metal is accomplished by one of two methods; (1) distillation in retorts or furnace or, (2) leaching with dilute sulfuric acid followed by electrolytic deposition of the zinc from a purified solution. Currently the electrolytic zinc process in the United States is producing somewhat less than 45% the primary zinc.

In the first procedure the retort may be a batch horizontal type or a continuous vertical type heated externally either by fuel or by electrothermic means. Coal or coke is used as the reducing agent, about 0.5 to 0.8 ton being required per ton of slab zinc produced. The effluent gas from the retort containing zinc vapor and carbon monoxide passes into condensers of various types where the zinc is collected in the liquid state and then cast into slabs. If the calcine contains sufficient recoverable impurities (cadmium, germanium, lead, arsenic), a volatilization step may be introduced prior to retorting, the vapors condensed and collected for recovery.

In the second or electrolytic process, the crude zinc oxide or calcine is leached with sulfuric acid (usually spent electrolyte) and the resultant zinc sulfate solution is carefully purified. The final purified solution which is slightly acid is electrolyzed using lead anodes and aluminum cathodes. The electrolysis regenerates sulfuric acid which is recycled to the leaching tanks. The impurities in the original concentrates, such as lead, gold and silver, appear in the leaching and electrolytic tank residues which are usually shipped to a lead plant for recovery.

In the early 1950's the Imperial Smelting Corporation, Ltd., of England developed a blast furnace or vertical type smelter which has the unique advantage of permitting the treatment of a mixed zinc-lead concentrate and recovering both metals together with any silver or gold present. The zinc metal thus produced conforms to Prime Western.

Furnaces of this type have been and are being constructed at various places throughout the world.

Grades. The zinc obtained by any of the above recovery processes is usually melted and recast into slabs weighing 55 pounds. There are five standard grades of slab zinc, the specifications for which have been established by the American Society for Testing Materials (B6-67). They are designated Special High grade, High grade, Intermediate, Brass Special and Prime Western, the first named being the purest and having a minimum zinc content (by difference) of 99.990%. Prime Western is the least pure having a zinc content (by difference) of 98.0%. Electrolytic zinc is either Special High grade or High grade. Slightly more than 50% of the zinc metal consumed is in the form of Special High grade followed by Prime Western (about 28%) and Brass Special (10%). Each of the five grades has certain specific applications.

Properties. Zinc is in Group IIB, Period 4 of the Periodic Table. It has fifteen isotopes ranging in mass from 60 to 72; the important naturally occurring ones in decreasing order of abundance being 64 (48.89%), 66 (27.81%), 68 (18.5%), 67 (4.11%), and 70 (0.62%). Other physical constants are presented below:

and serves as the starting point for all zinc industrial compounds. It is generally manufactured by variations in the processes used for the metallurgical recovery of zinc, the essential difference being that the zinc vapor, instead of being condensed as a metal, is oxidized in a separate chamber, and the resultant zinc oxide collected in bagrooms. Some zinc oxide, usually where high purity is desired, is made by the direct distillation of high-purity zinc metal followed by oxidation of the zinc while in the vapor form. Over 210,000 tons of zinc oxide were used in 1968, the rubber industry being the largest single (112,000 tons) user, followed by paints (26,000 tons), chemicals (23,000 tons), photocopying (22,000 tons) and ceramics (10,000 tons). Leaded zinc oxide has been decreasing steadily in importance, only about 8,000 tons having been produced in 1968.

On the basis of zinc content the manufacture of zinc sulfate ranks next to zinc oxide in importance. Its two chief uses are as a constituent of the precipitating bath in the production of viscose rayon fiber and in agriculture as a trace element in animal nutrition and also as an addition agent in trace quantities to the soil as a fertilizer and in sprays for control of certain plant diseases especially in the

Atomic weight	65.38
Atomic number	30
Color	Bluish-white
Crystal structure	Hexagonal
Hardness, Mohs Scale	2.5
Ductility	Fair
Density (cast)	7.133 g/cc
Melting point	419.5°C
Boiling point	907°C
Heat of fusion (cal/mole at 419.5°C)	1595
Heat of vaporization (cal/mole at 907°C)	27560
Specific heat (25°C)	0.0925 cal/g
Electrochemical equivalent	0.3388 mg/coulomb
Electrode potential Zn^{++} ($H_2O = 0.0$ volt)	−.76

Zinc is divalent and amphoteric. The metal is slowly oxidized in air, moisture accelerating the rate of attack. It reacts with mineral acids and at elevated temperatures with the halogens. It also reacts with potassium hydroxide or sodium hydroxide forming zincates together with evolution of hydrogen. Zinc forms a complex ion with excess ammonia.

Zinc as an element is not inherently toxic; however, many of the compounds have a very unpleasant effect when injected or inhaled. Rapid elimination of zinc from the body is probably responsible for the low toxicity. Biological studies have quite well established the physiological importance of zinc as a trace element in animal nutrition.

Compounds. Zinc compounds and pigments are widely used for a large number of purposes. Their U.S. output in 1968 (the last year for which firm figures are currently available) excluding lithopone, reached a total of more than 335,000 tons having an equivalent zinc content of about 220,000 tons.

Zinc oxide is the most important zinc chemical

citrus industry. Other uses for zinc sulfate are in the froth flotation of minerals, the clarification of glue and in electrogalvanizing. Lithopone is a coprecipitated zinc sulfide-barium sulfate pigment and finds considerable use in paints, floor coverings, coated fabrics and in other applications of a white pigment.

The next most important compound of zinc in point of tonnage is zinc chloride. It is used as a deodorant and in disinfecting and embalming fluids. It is used alone or with phenol or chromated for preserving railway ties and for fireproofing lumber, with ammonium chloride as a flux for soldering and etching metals, and galvanizing; in the manufacture of parchment paper, vulcanized fiber and dry batteries; as a mordant in printing and dyeing textiles; in mercerizing cotton, sizing and weighting of fabrics.

Other compounds of zinc find a variety of industrial uses such as zinc acetate in the preservation and fireproofing of wood, and as a mordant in dyeing, zinc chromate as a pigment, zinc silicate as a fluorescent material, zinc sulfide as a white pigment, and as a phosphor where it exhibits unique luminescent properties.

Uses. Zinc is chemically active and alloys readily with other metals, these two factors having considerable bearing on many of its uses. Its relatively high position in the electromotive series makes it extremely valuable in protecting iron and steel against corrosion. (See **Galvanizing**).

As for the major applications of the metal, its use in zinc-base alloys (die-casting) has increased rapidly in the last thirty years and beginning with 1961 it has been the most important single use, passing the 500,000 tons mark in 1964. However, the preliminary figures for 1970 indicate a consumption of only about 450,000 tons for this purpose, a drop of about 130,000 tons from the previous year. This decrease is attributed to the drop in activity in the auto and appliance industries.

Next use in importance in 1970 as it has been since 1961 when it was displaced from the top position is the galvanizing industry with 427,000 tons. This was only about 50,000 tons less than in the previous years.

Third in rank is the use in brass with about 130,000 tons, about 50,000 lower than the previous year, thus providing the single largest percentage loss—almost 28%.

For the first time, the estimated use of the metal in 1970 of about 44,000 tons for the manufacture of zinc oxide exceeded the use in rolled zinc (40,000 tons) in the forms of sheet, strip, rod and wire which have varied industrial uses.

Another zinc product which exhibits a slow upward trend in its use is zinc dust with a production of about 51,000 tons in 1970. It commands a price premium of about 3 or 4¢/lb over slab zinc.

A. PAUL THOMPSON

References

1. Mathewson, C. H., "Zinc: The Science and Technology of Metal, Its Alloys and Compounds," Reinhold Publishing Company, New York, 1959.
2. Sections on "Zinc and Zinc Alloys" and "Zinc Compounds" by A. W. Schlecten and A. Paul Thompson, "Encyclopedia of Chemical Technology," 2nd ed. pp. 555–614, John Wiley & Sons, Inc., New York, 1970.
3. Chapter on "Zinc," "Mineral Facts and Problems," U.S. Bureau of Mines Bulletin 650 (1970).
4. "Minerals Yearbooks," U.S. Bureau of Mines, Washington, D.C.
5. "Year Book of the American Bureau of Metal Statistics," 50th Annual Issue (1970) New York.

ZIRCONIUM AND COMPOUNDS

Zirconium, atomic number 40, atomic weight 91.22, is located in Group IVB in the Periodic Table below titanium. Its naturally occurring isotopes are 90, 91, 92, 94 and 96. The density of zirconium is 6.506 g/cc; melting point, 1856°C; boiling point, 4377°C; heat of fusion, 5.5 kcal/mole; heat of vaporization, 124 kcal/mole; heat capacity, C_p, 6.06 cal/g-atom/°C; thermal conductivity, 0.0505 cal/cm/°C/sec (20°C); electrical resistivity, 0.56 microhm-cm; and thermal neutron absorption cross section, 0.18 barns/atom. It is a strong ductile

metal, amenable to fabrication, with good heat transfer properties, excellent corrosion resistance and good bonding characteristics. This makes zirconium especially suitable for use in thermal nuclear reactors.

In 1789, while studying some semiprecious stones from Ceylon, N. H. Klaproth discovered the element zirconium. However, the investigation of the chemical properties of the element proceeded very slowly because of the quite refractory nature of the mineral zircon, which was the main source of supply until about the turn of the century. Zircon ($ZrSiO_4$) is found in the sands of Florida and Australia. Other sources are Brazil and India. Gem zircon has a hardness of 7–7.5 and a specific gravity of 4–4.8, which is the highest specific gravity of any commercial gem.

Zirconium comprises about 0.028% of the earth's crust and its known deposits are greater than the combined amounts of nickel, copper, zinc, lead, tin, and mercury. It may be extracted from baddeleyite or zircon by fusion with $NaHSO_4$ or $Na_2S_2O_7$. The problem of production of the metal from the ore is one of extreme interest to the scientist at the present time because of the extremely valuable properties which zirconium and its oxides show at high temperatures.

The Kroll process employs the sublimation of $ZrCl_4$ into molten magnesium metal resulting in zirconium metal by displacement. Following $MgCl_2$ removal, metallic zirconium is melted in the absence of oxygen or nitrogen.

A second method, a very costly one, involves the dissociation of zirconium tetraiodide and the deposition of zirconium on a metal filament at 1200°C. In practice, the crude metal and a limited amount of iodine are placed in a glass container fitted with a tungsten filament. The system is evacuated and the impure metal is heated somewhat below the temperature of the filament. Pure zirconium bars produced by the iodide process are very costly and of low tensile strength, being nearly as weak as copper. Less then 0.5% of magnesium, carbon, or nitrogen renders zirconium metal nonductile. Investigations of the reduction of quadrivalent zirconium in aqueous solution by electrolysis and by the polarographic method indicate that quadrivalent zirconium cannot be reduced to an intermediate valence state by electrolysis of solutions of zirconium salts.

Process for the electrowinning of zirconium depend on the formation of sodium or potassium in a fused salt. The zirconium compound then reacts chemically with the alkali metal produced by electrolysis.

Hafnium and zirconium always occur together in nature and because of their similarity of properties separation is very difficult. Zirconium contains from 1.5 to 2% of hafnium, or the same proportion present in the ore from which the metal was produced. For most chemical purposes hafnium can be regarded as a heavy isotope of zirconium.

One of the better ways of separating zirconium from hafnium is by the preparation of the addition compounds $3ZrCl_4 \cdot 2POCl_3$ and $3HfCl_4 \cdot 2POCl_3$, followed by distillation, the latter boiling 5° below the zirconium compound. The most convenient method for separating hafnium from zirconium is by the use of thenoyltrifluoroacetone by solvent extraction.

Zircon heated with soda, lime, potash is converted into the zirconate in the presence of acids. $ZrSiO_4$ may be heated with carbon in the electric arc furnace, the main reaction being $ZrSiO_4 + 4C \rightarrow ZrC + SiO + 3CO$. Zirconium carbide is extremely hard and refractory, melting above 3500°C (face-centered cubic symmetry) and resists chemical attacks at room temperature. When removed from the arc furnace it will burn in air, leaving ZrO_2, this being the most economical source of reasonably pure ZrO_2 now available. Hot zirconium carbide burns in chlorine or bromine to form $ZrCl_4$ or $ZrBr_4$ and at 1000°C combines with iodine to form ZrI_4.

Zirconium oxychloride ($ZrOCl_2 \cdot 8H_2O$) is a soluble salt resulting in a strongly acid solution when dissolved in water and is the starting material for a number of industrial processes. It is used in the preparation of acid dyes (primarily sulfonic or carboxylic acid groups) which are extensively used in pigment manufacture. Zirconium oxychloride precipitates these acid dyes from aqueous solution forming compounds of intensively colored solids. Such pigments are used for printing inks. The tetrahalides of zirconium are solids, subliming at 300°C and above. They are readily hydrolyzed and must be kept moisture-free. Normal oxy-acid salts such as $Zr(SO_4)_2$ and $Zr(NO_3)_4$ undergo extremely easy hydrolysis forming the species ZrO^{++}. The more complex $[Zr_4(OH)_8]^{8+}$ and possibly $[Zr_2O_3]_n^{2+}$ exist in aqueous solution as well as low polymers of hydrated $ZrOOH^+$. X-ray diffraction studies have shown that the zirconyl ion in the zirconyl halides is a tetramer $[Zr(OH)_2 \cdot 4H_2O]_4^{8+}$ and that the species also exists in aqueous solution.

Zirconium tetrachloride ($ZrCl_4$) acts as a catalyst in certain organic reactions as the Friedel-Crafts reaction. It is used as an additive to molten magnesium metal adding strength to magnesium castings and forms addition products with compounds containing oxygen or nitrogen. Zirconium tetrafluoride, unlike the other halides, is stable in the presence of water. Fluorozirconic acid is reacted upon by alkalies to form complex salts of the composition $M_2^1ZrF_6$, $M_3^1ZrF_7$ and $M_5^1ZrF_9$.

The zirconium atom seems unable to form a true hydroxide. This is probably due to the instability of two or more hydroxyl groups in the presence of each other on the same metal atom. When alkali is added to zirconium oxychloride a precipitate of $ZrO_2 \cdot xH_2O$ is formed called hydrous zirconia. High-purity zirconia (more than 99% pure ZrO_2) may be produced from zircon ($ZrSiO_4$) by the use of plasma-arc technology. This material has considerable absorptive properties, particularly for oxygen-containing anions. Hydrous zirconia will adsorb sulfate ions from a dilute solution of sodium sulfate to such an extent that no precipitate will form by the addition of barium chloride to the filtrate.

Urushiol (the active irritant of poison ivy), a dihydroxybenzene compounds, is precipitated quantitatively by hydrous zirconia, making the latter useful for the treatment of poison ivy dermatitis. It is also effective as a body deodorant.

Although zirconium forms no known compound containing a zirconium carbon bond, many compounds exist where oxygen or nitrogen form the link between the two. The coordination number of zirconium toward oxygen is 6 or 8, and at times 7. The compound, zirconium tetracetylacetonate, resulting from the chelation of acetylacetone with Zr^{+4} is a very good example of a coordination number of 8. Zirconium will also bond to ring structures such as cyclopentadiene through delocalized electrons to form the "sandwich" compound zirocene.

Disulfatozirconylic acid, $Zr(SO_4)_2 \cdot 4H_2O$, is the most common sulfate. It is used in tanning leather, in precipitating amino acids and in deodorizing fish oils. Zirconium is known to form electron deficient polymers such as zirconium borohydride, $Zr(BH_4)_4$, as well as polymeric metal alkoxides.

Alphahydroxycarboxylic acids react with zirconium to form tri- and tetra-carboxylatozirconylic acids. These find use in treating dermatitis and in deodorants.

The stable carboxylate, H_2ZrO_2Ac (zirconyl acetate) is formed by the action of basic zirconyl hydroxide on acetic acid. This compound finds extensive use in making textiles water repellent. Basic zirconium acetate finds use as a weighting filler for silk. Other salts find use in the fireproofing, mordanting and water repelling of fabrics.

Although zirconium is essentially quadrivalent, lower valency states of zirconium do exist. They are much less stable than those of titanium and our knowledge of them is confined to the chlorides, bromides, iodides and fluorides such as $ZrCl_3$ and $ZrCl_2$. There is evidence for the existence of ZrH_2, Zr_2O_3, Zr_2O_5 and ZrO_3 although some are not too well established as definite compounds.

Coordinate bond formation resulting in a loose association between zirconium and the coordinating substance helps to explain most examples of zirconium catalysis. The incompleteness of the two outermost shells contribute to the catalytic properties of this element. Examples of zirconium catalysis include ammonia synthesis, conversion of SO_2 to SO_3, organic oxidations, organic syntheses, esterifications, dehydrations, cracking, polymerizations and alkylations.

Powered zirconium metal is almost ideal for ammunition primers. It finds use in smokeless flashlight powders, blasting caps for electric ignition, pyrotechnics and in the manufacture of vacuum radio and photoelectric tubes.

Metallic zirconium is of outstanding interest in the field of atomic energy because of its very low tendency to absorb slow neutrons combined with its resistance to corrosion and relatively high melting point (1860°C). It finds important uses in the fabrication of nuclear fuels and other reactor components because of this property of low-neutron-capture cross section and its favorable alloying characteristics.

Commercial zircon is used in the ceramic industries, particularly as an addition to porcelain (spark plugs) to reduce the coefficient of expansion and increase the dielectric strength. Zirconium dioxide is an excellent opacifier in ceramics, makes a good paint, and is used under the name of "Konstrastin" for the x-ray observation of the digestive processes.

Zirconium's extremely high resistance to corrosion will probably make it more prominent in surgical instruments, pins, screws for bone repairs, etc. Zirconium oxide will survive extreme temperatures over long periods of time. This property has made it a promising material for the outer surfaces of manned spaceships.

WINSTON R. DEMONSABERT

Reference

McClain, J. H., "Zirconium," pp. 830–839 in "Encyclopedia of the Chemical Elements," C. A. Hampel, Editor, Van Nostrand Reinhold Co., New York, 1968.

Index